## Springer Collected Works in Mathematics

More information about this series at http://www.springer.com/series/11104

Moscow, 1973

Israel M. Gelfand

# Collected Papers II

*Editors*
Semen G. Gindikin
Victor W. Guillemin
Alexandr A. Kirillov
Bertram Kostant
Shlomo Sternberg

Reprint of the 1988 Edition

Springer

*Author*
Israel M. Gelfand (1913 – 2009)
Department of Mathematics
Rutgers State University of New Jersey
New Brunswick, NJ
USA

*Editors*
Semen G. Gindikin
Department of Mathematics
Rutgers State University of New Jersey
New Brunswick, NJ
USA

Victor W. Guillemin
Massachusetts Institute of Technology
Cambridge, MA
USA

Alexandr A. Kirillov
University of Pennsylvania
Philadelphia, PA
USA

Bertram Kostant
Massachusetts Institute of Technology
Cambridge, MA
USA

Shlomo Sternberg
Harvard University
Cambridge, MA
USA

ISSN 2194-9875
Springer Collected Works in Mathematics
ISBN 978-3-662-48720-4 (Softcover)
978-3-540-19035-6 (Hardcover)

Library of Congress Control Number: 2012954381

Mathematics Subject Classification (2010): 20Gxx, 22xx, 32xx

Springer Heidelberg New York Dordrecht London


Printed on acid-free paper

Springer-Verlag GmbH Berlin Heidelberg is part of Springer Science+Business Media
(www.springer.com)

# IZRAIL M. GELFAND

# COLLECTED PAPERS

## VOLUME II

Edited by

S.G. Gindikin V.W. Guillemin A.A. Kirillov
B. Kostant S. Sternberg

SPRINGER-VERLAG
BERLIN HEIDELBERG NEW YORK
LONDON PARIS TOKYO

Professor Izrail M. Gelfand
Member of the Academy of Sciences of the USSR
A. N. Belozersky Laboratory of Molecular Biology and
Bioorganic Chemistry, Building "A", Moscow State University
Moscow, GSP-234, 119899, USSR

*Editors:*

Professor Semen G. Gindikin
A. N. Belozersky Laboratory of Molecular Biology and
Bioorganic Chemistry, Building "A", Moscow State University
Moscow, GSP-234, 119899, USSR

Professor Victor W. Guillemin
Department of Mathematics, Massachusetts Institute of Technology
Cambridge, MA 02139, USA

Professor Aleksandr A. Kirillov
Moskovskii Universitet, Mehmat, Moscow 117234, USSR

Professor Bertram Kostant
Department of Mathematics, Massachusetts Institute of Technology
Cambridge, MA 02139, USA

Professor Shlomo Sternberg
Department of Mathematics, Harvard University, Science Center
One Oxford Street, Cambridge, MA 02138, USA

Mathematics Subject Classification (1980):
20G 05, 20G 15, 20G 20, 20G 25, 20G 35, 20J 05, 22D 10, 22D 20, 22D 30, 22E, 32A 07, 32A 35, 32M 05, 32M 10, 32M 15, 33A 35, 33A 45, 33A 75, 43A 80, 43A 85, 43A 90, 06B 15, 06D 05, 10D 05, 10D 40, 12A 67, 12A 64, 17B 10, 17B 15, 17B 35, 17B 56, 20C 05

ISBN 978-3-540-19035-6 Springer-Verlag Berlin Heidelberg New York
ISBN 978-3-540-19035-6 Springer-Verlag New York Berlin Heidelberg

Library of Congress Cataloging-in-Publication Data
(Revised for vols. 2–3)
Gel'fand, I. M. (Izrail' Moiseevich)
Collected papers.
In English, French, and German. Includes bibliographies;
1. Mathematics. 2. Gel'fand, I. M. (Izrail' Moiseevich). 3. Mathematicians–Soviet Union–Biography.
I. Gindikin, S. G. (Semen Grigor'evich)
II. Title. QA3.G38 1987 510 87-32254



2141/3140-543210 Printed on acid-free paper

# Preface

I am very grateful to Springer-Verlag for publishing my three volumes of Collected Papers. I was asked to write a survey of the papers included: I do not think that the author has this right. It seems to me that whatever a man achieves in science comes from above. For this reason, he connot be the judge of his own work. But I hope that at the end of the third volume I will be able to write what I have tried to do in these papers and did not succeed or did not fully succeed to do. It may be that some of these problems remain interesting today. From the bibliography you can see that a lot of papers are written together with some of my colleagues and friends. I want to express my deep gratitude to them. I very much enjoyed working with them and I learned a lot from my contact with them.

I am very grateful to all my friends who did a lot of work, especially in translating, correcting and generally editing these three volumes. Without them it would not have been possible to get this edition ready.

Thanks also very much to Professors S. G. Gindikin, V. W. Guillemin, A. A. Kirillov, B. Konstant and S. Sternberg.

Moscow, September 1987

I. M. Gelfand

# Foreword

The second volume of I. M. Gelfand's Collected Papers consists of articles dealing with group representation theory. This is one of the main topics in I. M. Gelfand's work and embraces more than forty years of his creative activity.

Representation theory clearly shows how manifold this activity was. The articles brought together in the second volume are assembled into nine sections, each of which essentially develops a new, fruitful direction, usually linking representation theory with other branches of mathematics. Many well known mathematicians, beginning with D. A. Rajkov, M. A. Najmark, and M. I. Graev, and going on to D. A. Kazhdan, I. N. Bernstein, and A. V. Zelevinskij, learned representation theory in the "Gelfand school" and wrote with him their best papers in this domain.

Representation theory has been "very lucky": such perspective and powerful mathematicians as E. Cartan, H. Weyl, Harish-Chandra, A. Selberg, A. Weil, R. Langlands, and P. Deligne worked in the field. But even against this background, I. M. Gelfand's contribution knows no equal in range, depth of approach, and beauty of results.

First of all, let us point out I. M. Gelfand's main idea: infinite-dimensional unitary representations of classical groups admit description as explicit and beautiful as finite-dimensional representations; moreover, in a certain sense, infinite-dimensional representations are simpler: finite-dimensional representations are "singular points" in the set of all representations. How unexpected this idea was can be seen from the fact that it contradicts E. Wigner's assumption that infinite-dimensional representations of the Lorentz group are extremely complicated and probably related to von Neumann factors and measure theory.

A second – more constructive – idea is that the basis of harmonic analysis on semisimple groups is a certain commutative ring. The role of this ring may be played either by the centre of the enveloping algebra (Casimir-Laplace operators), or the Banach algebra of spherical functions with respect to a maximal compact subgroup, or other, more complicated, and at present unsufficiently studied, rings.

An important discovery of I. M. Gelfand was the role of "integral geometry" in representation theory. Papers grouped together in the third section give an idea of the diversity of applications of this approach, known as the horysphere method.

The next idea, originating in algebraic geometry, consists in viewing the underlying field as a parameter. In this way representations of complex, real, $p$-adic, and finite groups become part of a more general scheme.

The category approach to representation theory led I. M. Gelfand to the deep theory of Verma modules that explains the classical results of H. Weyl, E. Cartan, and B. Kostant, as well as recent achievements in the theory of so-called Harish-Chandra modules. The notion of the B-G-G resolution is also successfully used in the latest work of representation theory of infinite-dimensional Lie algebras related to string theory.

The so-called Gelfand-Tsetlin basis for finite-dimensional representations of classical groups is very popular, especially among theoretical physicists. This is an excellent example of the "inverse influence" of infinite-dimensional representation theory on its prototype – the theory of finite-dimensional representations.

Another aspect of the algebraic geometry approach to representation theory was developed in our joint paper with I. M. Gelfand on the "birational classification" of Lie algebras. I. M. Gelfand's original idea was that enveloping algebras of Lie algebras are true objects of study in non-commutative algebraic geometry. The fecundity of this approach has now been substantiated by numerous papers. It should also be noted that the presently very popular theory of $D$-modules and its application to representations of semi-simple groups appeared under the influence of these papers and discussion about them at Gelfand's seminars.

Finally, I. M. Gelfand's cycle of papers with various coauthors on representation theory of infinite-dimensional Lie groups considerably clarified results obtained by physicists in this field and brought to the attention of mathematicians this very important branch of representation theory.

I feel that any mathematician will find a great deal of interesting and instructive material in I. M. Gelfand's papers on representation theory. Unfortunately, published articles convey only a small part of the extraordinary creative atmosphere which surrounds I. M. Gelfand – as anyone who has been at his seminars (which have been taking place at the Mechanics and Mathematics Department of Moscow University for 40 years now) at least once will readily agree.

Moscow, June 1988 A. Kirillov

# Table of contents

## Part I. General problems of representation theory



## Part II. Infinite-dimensional representations of semisimple Lie groups

**Part III. Geometry of homogeneous spaces; spherical functions; automorphic functions**



**Part IV. Models of representations; representations of groups over various fields**

## Part V. Verma modules; resolutions of finite-dimensional representations



## Part VI. Enveloping algebras and their quotient skew-fields



## Part VII. Finite-dimensional representations



## Part VIII. Indecomposable representations of semisimple Lie groups and of finite-dimensional algebras; problems of linear algebra

## Part IX. Representations of infinite-dimensional groups



## Appendix

# Part I

# General problems of representation theory

# 1.

## (with D. A. Rajkov)

# Irreducible unitary representations of locally bicompact groups

Mat. Sb. **13** (55), 301–316 (1942)
[Transl., II. Ser., Am. Math. Soc. **36** (1964) 1–15]. Zbl. **166**:401

A linear (finite-dimensional) representation of a topological group $G$ is defined, as is well known, to be a homomorphic mapping from a group to the topological group of matrices over the field of complex numbers. A representation is called unitary if all the matrices occurring in the representation are unitary. A representation by $n$th order matrices is called irreducible if there is no proper subspace of $n$-dimensional complex space left invariant under all the matrices occurring in the representation. We say that the linear representations of the group $G$ form a complete system if for every element of the group other than the identity there is a representation sending it to a matrix different from the identity.

For compact Lie groups, the existence of a complete system of finite-dimensional irreducible unitary representations was proved by Peter and Weyl [9]. All that is used in their approach, besides the compactness of the groups in question, is the existence on these groups of an invariant measure. Having constructed an invariant measure on any locally compact group satisfying the second axiom of countability, Haar [3] automatically extended the Peter-Weyl result to arbitrary compact groups satisfying the second axiom of countability. For commutative locally compact groups with the second axiom of countability, the existence of a complete system of irreducible unitary representations (in fact, all one-dimensional) was proved by J. von Neumann [7]. [1] But the existence of a complete system of finite-dimensional irreducible unitary representations for a general locally compact group with the second axiom of countability could not be proved, since there are such groups admitting no such representation other than the trivial representation as the identity matrix. There arises the question of infinite-dimensional representations of locally bicompact groups.

The natural generalisation of a representation as unitary matrices is a representation as unitary operators in a Hilbert space. We will mean by a *unitary representation* of the topological group $G$ a homomorphic mapping from the group to the topological group of unitary operators in a complex Hilbert space $\mathfrak{H}$; more precisely, a function $U_g$ mapping the group $G$ into the totality of unitary operators acting on $\mathfrak{H}$ and having the following two properties: 1) $U_{gh} = U_g U_h$ for all $g, h \in G$, and 2) $U_g$ is a strongly continuous function of $g$, i.e.,

1) Cf. also the authors' article [2], where this result is obtained by means of commutative normed rings.

$\| U_{g'}\eta - U_g\eta \| \to 0$ whenever $g' \to g$ for any $\eta \in \mathfrak{H}$.

*In the present article we demonstrate the existence of a complete system of irreducible unitary representations for an arbitrary locally bicompact group.*

The proof will depend upon the close connection between unitary representations and positive definite functions.

1. A function $\phi(g)$ defined on a group $G$ is called *positive definite* if $\sum_{k=1}^{n} \sum_{l=1}^{n} \phi(g_l^{-1} g_k) \lambda_k \overline{\lambda_l} \geq 0$ holds for all finite systems of elements $g_1, \cdots, g_n \in G$ and complex numbers $\lambda_1, \cdots, \lambda_n$.

*Every continuous positive definite function $\phi(g)$ defined on a topological group $G$ gives rise to a representation of the group as unitary operators on some Hilbert space.*

The space is constructed as follows. We define a scalar product in the totality of all complex functions $\lambda(h)$ $(h \in G)$ that differ from zero at only a finite number of points:

$$(\lambda, \mu) = \sum_h \sum_{h'} \phi(h'^{-1} h) \lambda(h) \overline{\mu(h')}, \tag{1}$$

where the sum extends formally over all the elements of the group $G$ (but only a finite number of summands are different from zero). Identifying those functions $\lambda$ and $\mu$ for which $(\lambda - \mu, \lambda - \mu) = 0$ and taking the completion of the resulting linear system under the norm $|\lambda| = \sqrt{(\lambda, \lambda)}$, we obtain a Hilbert space. Let us denote it by $\mathfrak{L}_2(\phi)$.

Every element $g \in G$ gives rise to a left translation operator $T_g$, whose value at the function $\lambda(h)$ is given by the formula $T_g\lambda(h) = \lambda(g^{-1}h)$. The operator $T_g$ is unitary. In fact,

$$(T_g\lambda, T_g\mu) = \sum_h \sum_{h'} \phi(h'^{-1} h) \lambda(g^{-1} h) \overline{\mu(g^{-1} h')},$$

whence, using the substitution $h \to gh$, $h' \to gh'$, we obtain

$$(T_g\lambda, T_g\mu) = \sum_h \sum_{h'} \phi(h'^{-1} g^{-1} gh) \lambda(h) \overline{\mu(h')} = (\lambda, \mu).$$

Since the functions $\lambda(h)$ are everywhere dense in $\mathfrak{L}_2(\phi)$, $T_g$ has a unique extension to a unitary operator on all of $\mathfrak{L}_2(\phi)$. $T_g$ is a strongly continuous function of $g$: $|T_{g'}\eta - T_g\eta| \to 0$ as $g' \to g$ for every $\eta \in \mathfrak{L}_2(\phi)$. In fact, since

$$|T_{g'}\eta - T_g\eta|^2 = |T_{g^{-1}g'}\eta - \eta|^2 = 2[(\eta, \eta) - \Re(T_{g^{-1}g'}\eta, \eta)],$$

it suffices to show that $(T_g\eta, \eta) \to (\eta, \eta)$ as $g \to e$, where $e$ is the identity of the group $G$. But when $\eta = \lambda(h)$, we have, by (1) and the continuity of $\phi(g)$,

$$(T_g \lambda, \lambda) = \sum_h \sum_{h'} \phi(h'^{-1} gh) \lambda(h) \overline{\lambda(h')}$$

$$\to \sum_h \sum_{h'} \phi(h'^{-1} h) \lambda(h) \overline{\lambda(h')} = (\lambda, \lambda) \text{ as } g \to e.$$

Now, whatever $\eta \in \mathfrak{L}_2(\phi)$ and $\epsilon > 0$, there is a $\lambda(h)$ such that $|\eta - \lambda| < \frac{\epsilon}{3}$. Choosing a neighborhood of the identity $V$ such that $|T_g \lambda - \lambda| < \frac{\epsilon}{3}$ when $g \in V$, we obtain for those $g$

$$|T_g \eta - \eta| \le |T_g \eta - T_g \lambda| + |T_g \lambda - \lambda| + |\lambda - \eta| = 2|\lambda - \eta| + |T_g \lambda - \lambda| < \epsilon.$$

Obviously $T_{gh} = T_g T_h$. Thus the operators $T_g$ make up a unitary representation of the group $G$. The function

$$\xi_0(h) = \begin{cases} 1 \text{ for } h = e, \\ 0 \text{ for } h \neq e \end{cases}$$

is a cyclic vector for $\mathfrak{L}_2(\phi)$, in the sense that the vectors $T_h \xi_0$ generate the space $\mathfrak{L}_2(\phi)$. In fact, the functions $\lambda(h)$ are everywhere dense in $\mathfrak{L}_2(\phi)$, and each of them is a linear combination of the vectors $T_h \xi_0$: $\lambda = \sum_h \lambda(h) T_h \xi_0$. In view of (1),

$$\phi(g) = (T_g \xi_0, \xi_0). \tag{2}$$

*Conversely, every unitary representation of a topological group G gives rise to continuous positive definite functions.* Indeed, if $U_g$ are unitary operators in a Hilbert space $\mathfrak{H}$ making up a representation of the group $G$, then $\phi(g) = (U_g \xi, \xi)$ will be a continuous positive definite function for any vector $\xi \in \mathfrak{H}$; moreover, $\phi(g) \not\equiv 0$ whenever $\xi \neq 0$. If $\mathfrak{H}$ has a cyclic vector $\xi_0$, i.e., the vectors $U_g \xi_0$ generate the space $\mathfrak{H}$, then $\mathfrak{H}$ is isomorphic to the space $\mathfrak{L}_2(\phi_0)$ arising from the function $\phi_0(g) = (U_g \xi_0, \xi_0)$. The isomorphism is given by the correspondence $\sum_h \lambda(h) U_h \xi_0 \leftrightarrow \lambda(h)$. The operators $U_g$ in $\mathfrak{H}$ correspond to the translation operators $T_g$ in $\mathfrak{L}_2(\phi_0)$.

2. Let $\phi(g)$ and $\psi(g)$ be positive definite functions. We shall write $\psi(g) \ll \phi(g)$ if $\phi(g) - \psi(g)$ is itself a positive definite function. If $\phi(g)$ is continuous and $\psi(g) \ll \phi(g)$, then $\psi(g)$ is also continuous. Indeed, as M. G. Kreĭn showed [5], every positive definite function $\phi(g)$ satisfies the inequality

$$|\phi(g) - \phi(h)|^2 \le 2\phi(e)[\phi(e) - \Re\phi(h^{-1}g)] \text{ for all } g, h \in G. \tag{3}$$

But if $\psi(g) \ll \phi(g)$, then $\phi(e) - \psi(e) \ge 0$ and $\phi(e) - \psi(e) - \Re[\phi(h^{-1}g) - \psi(h^{-1}g)] \ge 0$, whence

$$|\psi(g) - \psi(h)|^2 \le 2\psi(e)[\psi(e) - \Re\psi(h^{-1}g)] \le 2\phi(e)[\phi(e) - \Re\phi(h^{-1}g)].$$

Thus, as a consequence of the continuity of $\phi(g)$ at the point $e$, we obtain the continuity of $\psi(g)$ for all $g$.

We shall call a continuous positive definite function $\phi(g)$ *elementary* if, whenever $\psi(g) \ll \phi(g)$, it follows that $\psi(g) = \alpha\phi(g)$. Trivial examples of elementary positive definite functions are afforded by the functions $\phi(g) \equiv c$, where $c$ is a non-negative constant.

**Theorem 1.** *A unitary representation of the group $G$, arising from an elementary continuous positive definite function $\phi(g)$, is irreducible.*

**Proof.** Let $P$ be a projection operator in $\mathfrak{L}_2(\phi)$ commuting with all the $T_g$. Then $\psi(g) = (T_g P\xi_0, \xi_0)$ will be a positive definite function satisfying $\psi(g) \ll \phi(g)$. Indeed, for any function $\lambda(g)$ differing from zero only at a finite number of points $g_1, \cdots, g_n \in G$,

$$\sum_{k=1}^{n}\sum_{l=1}^{n} \psi(g_l^{-1}g_k)\lambda(g_k)\overline{\lambda(g_l)} = \sum_{k=1}^{n}\sum_{l=1}^{n} (\lambda(g_k)\, T_{g_k} P\xi_0, \lambda(g_l)\, T_{g_l}\xi_0)$$

$$= (P\sum_{k=1}^{n} \lambda(g_k)\, T_{g_k}\xi_0, \sum_{l=1}^{n} \lambda(g_l)\, T_{g_l}\xi_0) = (P\lambda, \lambda)$$

and

$$\sum_{k=1}^{n}\sum_{l=1}^{n} \phi(g_l^{-1}g_k)\lambda(g_k)\overline{\lambda(g_l)} = (\lambda, \lambda).$$

But $0 \le (P\lambda, \lambda) \le (\lambda, \lambda)$. These inequalities thus indicate the positive definiteness respectively of the functions $\psi(g)$ and $\phi(g) - \psi(g)$. Since $\phi(g)$ is assumed elementary, $\psi(g) = \alpha\phi(g)$. Consequently, $(P\lambda, \lambda) = \alpha(\lambda, \lambda)$ for all $\lambda(g)$ differing from zero only at a finite number of points. Since such $\lambda(g)$ are everywhere dense in $\mathfrak{L}_2(\phi)$, we have $P = \alpha E$. Thus we have shown that the only projection operators in $\mathfrak{L}_2(\phi)$ commuting with all the $T_g$ are $P = 0$ and $P = E$. But this implies the irreducibility of the representation $T_g$, since, if the operators $T_g$ reduced a subspace of $\mathfrak{L}_2(\phi)$, they would commute with the operator of projection onto that subspace.

**Theorem 2.** *Positive definite functions giving rise to irreducible unitary representations of a group $G$ are all elementary.*

**Proof.** Let $U_g$ be the unitary operators in a Hilbert space $\mathfrak{H}$ making up an irreducible representation of the group $G$. Then each nonzero vector $\xi \in \mathfrak{H}$ will be cyclic. Suppose $\phi(g) = (U_g\xi, \xi)$, $\psi(g)$ is a positive definite function, and $\psi(g) \ll \phi(g)$. For any two vectors $\lambda = \Sigma\, \lambda(h)\, U_h\xi$ and $\mu = \Sigma\, \mu(h)\, U_h\xi$, where the functions $\lambda(h)$ and $\mu(h)$ are each different from zero only at a finite number of points, we put

$$(B\lambda, \mu) = \sum_h \sum_{h'} \psi(h'^{-1}h)\lambda(h)\overline{\mu(h')},$$

$B$ is a selfadjoint operator and

$$0 \leq (B\lambda, \lambda) \leq (\lambda, \lambda) = \sum_h \sum_{h'} \phi(h'^{-1}h)\lambda(h)\overline{\lambda(h')}.$$

Since the vectors $\lambda$, once again, are everywhere dense in $\mathfrak{H}$, $B$ can be uniquely extended to a selfadjoint operator on all of $\mathfrak{H}$. $B$ commutes with all the $U_g$. Indeed, for any $g, h \in G$, we have

$$(U_g BU_h \xi, U_h \xi) = (BU_h \xi, U_{g^{-1}h}\xi) = \psi(h^{-1}gh) = (BU_g U_h \xi, U_h \xi),$$

and, since the vectors $U_h \xi$ form a basis for $\mathfrak{H}$,

$$(U_g B\eta, \eta) = (BU_g \eta, \eta) \text{ for all } \eta \in \mathfrak{H};$$

consequently $U_g B = BU_g$. Now the selfadjoint operator $B$ can only commute with the operators $U_g$ constituting an irreducible system if $B = \alpha E$. For if the operators $U_g$ commuted with $B$, they would commute with its spectral functions $E(\Delta)$. But then, as we have seen, the projection operator $E(\Delta)$ must be $0$ or $E$. Thus, the entire spectrum of the operator $B$ must be concentrated at a point $\alpha$ and $B = \alpha E$. From this it follows that

$$(BU_g \xi, \xi) = \alpha(U_g \xi, \xi), \text{ i.e., } \psi(g) = \alpha\phi(g),$$

Q.E.D.

A topological group $G$ admits a complete system of irreducible unitary representations if for each element $g_0 \neq e$ there is an irreducible representation $U_g$ in some Hilbert space $\mathfrak{H}_0$ such that $U_{g_0} \neq E$. We shall say that there is a *complete system of elementary continuous positive definite functions* on $G$ if for each element $g_0 \neq e$ there is an elementary continuous positive definite function $\phi_0(g)$ such that $\phi_0(g_0) \neq \phi_0(e)$. From the previous results, we immediately obtain

**Theorem 3.** *In order for a topological group $G$ to admit a complete system of irreducible unitary representations, it is necessary and sufficient that there be a complete system of elementary continuous positive definite functions on it.*

Indeed, if $\phi_0(g_0) \neq \phi_0(e)$, then the representation $T_g$ arising from the function $\phi_0(g)$ satisfies $T_{g_0}\xi_0 \neq \xi_0$ by virtue of (2), and so $T_{g_0} \neq E$. Conversely, if $U_{g_0} \neq E$, there is a vector $\xi_0$ such that $(U_{g_0}\xi_0, \xi_0) \neq (\xi_0, \xi_0)$; hence $\phi_0(g) = (U_g \xi_0, \xi_0)$ satisfies $\phi_0(g_0) \neq \phi_0(e)$.

We shall show that this necessary and sufficient condition for the existence of a complete system of irreducible unitary representations is fulfilled on any locally bicompact group.

3. As is known, A. Haar [3] demonstrated that every locally compact group satisfying the second axiom of countability admits a left-invariant measure $m$, i.e., a non-negative set function (which, generally speaking, is allowed to assume the value infinity), countably additive on the algebra $(B)$ of all Borel sets and invariant under "left translation":

$$m(gE) = m(E) \text{ for all } E \in (B) \text{ and all } g \in G.$$

As was subsequently noticed by several authors, Haar's construction, in the form given by Banach [1], is applicable to any locally bicompact group. In this case, just as in the situation examined by Haar, every open set with compact closure has finite positive measure, and the measure of any set is the greatest lower bound of the measures of those sets containing it that are countable unions of open sets with compact closures.

In what follows, we shall always assume that $G$ is a locally bicompact group. Let $m$ be a *left-invariant Haar measure* 1) defined on it.

For any fixed $h \in G$, $m(Eh)$ is also a left-invariant Haar measure. By the uniqueness theorem [8, 4, 11], $m(Eh) = l_h m(E)$, where $l_h$ is a number independent of $E$. From this it follows immediately that $l_{gh} = l_g l_h$ [8]. It can be shown that $l_h$ is a continuous function of $h$.

Let $\mathfrak{L}_1$ denote the space of all measurable absolutely integrable complex functions $x(g)$ with the norm 2)

$$\| x \| = \int | x(h) | \, dh.$$

To every $g \in G$, there correspond in $\mathfrak{L}_1$ a left translation operator $T_g$: $T_g x(h) = x(g^{-1}h)$ and a right translation operator $T'_g$: $T'_g x(h) = x(hg^{-1})$. The operators $T_g$ are unitary. The fundamental properties of Haar measure assure that, for any function $x(h) \in \mathfrak{L}_1$, as $g \to g_0$, the relations

$$\int | x(g^{-1}h) - x(g_0^{-1}h) | \, dh \to 0 \text{ and } \int | x(hg^{-1}) - x(hg_0^{-1}) | \, dh \to 0 \qquad (4)$$

---

1) We nowhere make explicit use of the local compactness of the group $G$. All our conclusions depend solely upon the properties of Haar measure. Hence it would be possible, without imposing any special topological restrictions on the group, to require simply that it admit a measure sharing the fundamental properties of a left-invariant Haar measure, described in a purely axiomatic manner (for this it would be necessary to determine some domain of definition for the measure) [10]. This would not, however, yield any generalisation of our results, since, as A. Weil [12] proved, the completion of any group with a Haar measure is a locally bicompact group, and then the completeness of the system of all irreducible unitary representations of the group is an immediate consequence of the completeness of the system of irreducible unitary representations of the completed, locally bicompact group.

2) Wherever the domain of integration is not indicated, it is intended that the integral extend over the entire group.

hold. Thus the operators $T_g$ and $T'_g$ are strongly continuous functions of $g$.

Let $x, y \in \mathfrak{L}_1$. Then

$$x * y = \int x(h^{-1}g)\, y(h)\, dh$$

exists for almost all $g$ and is $\in \mathfrak{L}_1$, since $\| x*y \| \leq \| x \| \; \| y \|$. Let us introduce the notation

$$x^*(g) = l_g^{-1}\, \overline{x(g^{-1})}.$$

Since [8]

$$\int x(g)\, dg = \int x(g^{-1})\, l_g^{-1}\, dg,$$

then whenever $x \in \mathfrak{L}_1$, $x^* \in \mathfrak{L}_1$ also, for $\| x^* \| = \| x \|$. It is easy to see that $x^{**} = x$ and $(x*y)^* = y^* * x^*$.

We shall call a linear functional $L$ defined on $\mathfrak{L}_1$ *positive* if $L(x*x^*) \geq 0$ for all $x \in \mathfrak{L}_1$. Let $\phi(g)$ be an essentially bounded, integrally positive definite function, i.e.,

$$\operatorname{sup\,ess} |\phi(g)| < \infty, \int\int \phi(h^{-1}g)\, x(g)\, \overline{x(h)}\, dg\, dh \geq 0 \text{ for all } x \in \mathfrak{L}_1.$$

Then, as is easy to check, the functional $L_\phi(x) = \int \phi(g)\, x(g)\, dg$ is positive.

Theorem 4. *Every positive linear functional $L$ arises from a continuous positive definite function $\phi(g)$:*

$$L(x) = L_\phi(x) = \int \phi(g)\, x(g)\, dg \text{ for all } x \in \mathfrak{L}_1.$$

Proof. The bilinear functional

$$(x, y) = L(x*y^*)$$

enjoys all the properties of a scalar product. Indeed, the properties $(x_1 + x_2, y) = (x_1, y) + (x_2, y)$ and $(\lambda x, y) = \lambda(x, y)$ are obvious, while the property $(y, x) = \overline{(x, y)}$ follows from the reality of the left-hand side and of the two extreme terms on the right-hand side of the equation

$$L((x+\lambda y)*(x+\lambda y)^*) = L(x*x^*) + \lambda L(y*x^*) + \overline{\lambda} L(x*y^*) + \overline{\lambda}\lambda L(y*y^*).$$

Let $\mathfrak{L}_2(L)$ denote the Hilbert space obtained from $\mathfrak{L}_1$ by identifying those $x$ and $y$ for which $(x-y, x-y) = 0$ and completing under the norm

$$|x| = \sqrt{(x, x)}.$$

If $x \in \mathfrak{L}_1$, we have: $|x|^2 = L(x*x^*) \leq |L| \; \|x\| \; \|x^*\| = |L| \; \|x\|^2$, whence

$$|x| \leq \sqrt{|L|}\; \|x\|.$$

The idea behind the proof of the theorem is now that the space $\mathfrak{L}_2(L)$, constructed via the positive linear functional $L$, is just one of the previously defined spaces $\mathfrak{L}_2(\phi)$, namely the one arising from a certain continuous positive definite function $\phi(g)$ obtained as follows. The closure of $\mathfrak{L}_1$ in $\mathfrak{L}_2(L)$ contains a "function" $\xi_0 = \xi_0(h)$ all of whose "mass" is concentrated at the point $h = e$. The "functions" $T_g \xi_0$ generate the space $\mathfrak{L}_2(L)$, and $\mathfrak{L}_2(L)$ is the $\mathfrak{L}_2(\phi)$ arising from the positive definite function $\phi(g) = (T_g \xi_0, \xi_0)$. This function also gives rise to the functional $L$. The vector $\xi_0$ is naturally constructed as a limit of functions $x \in \mathfrak{L}_1$ whose masses shrink to the point $e$.

Let $\{V\}$ be the system of all neighborhoods of the identity $e$ of the group $G$, in the natural ordering. A *unit*, denoted $e_V(g)$, is any function from $\mathfrak{L}_1$ satisfying the conditions

$$e_V(g) \geq 0, \; e_V(g) = 0 \text{ outside } V, \; e_V(g^{-1}) = e_V(g), \; \| e_V \| = 1.$$

For any $x \in \mathfrak{L}_1$ we have

$$\begin{aligned} \| x * e_V^* - x \| = \| x * e_V - x \| &= \int | \int [T_h x(g) - x(g)] \, e_V(h) \, dh \, | \, dg \\ &\leq \int ( \int | T_h x(g) - x(g) | \, dg ) \, e_V(h) \, dh \leq \sup_{h \in V} \| T_h x - x \| \end{aligned}$$

and this, because of the first relation in (4), implies

$$\| x * e_V^* - x \| \longrightarrow 0 \text{ as } V \longrightarrow e.$$

Hence it follows that

$$(x, e_V) = L(x * e_V^*) \longrightarrow L(x) \text{ as } V \longrightarrow e \text{ for all } x \in \mathfrak{L}_1.$$

Since $e_V$ is uniformly bounded in $\mathfrak{L}_2(L)$: $| e_V | \leq \sqrt{|L|}$, and since the $x \in \mathfrak{L}_1$ are everywhere dense in $\mathfrak{L}_2(L)$, it follows from the relation just obtained that $\lim_{V \to e} (\eta, e_V)$ exists for any $\eta \in \mathfrak{L}_2(L)$. Indeed, given $\epsilon > 0$, choose $x \in \mathfrak{L}_1$ such that $|\eta - x| < \dfrac{\epsilon}{3\sqrt{|L|}}$, and for this $x$ choose a neighborhood $V$ of the identity such that $|(x, e_{V_1}) - (x, e_{V_2})| < \frac{\epsilon}{3}$ for any $V_1, V_2 \subset V$. Then for any such $V_1$ and $V_2$, we will have

$$\begin{aligned} |(\eta, e_{V_1}) - (\eta, e_{V_2})| &\leq |(\eta, e_{V_1}) - (x, e_{V_1})| + |(x, e_{V_1}) - (x, e_{V_2})| \\ &+ |(x, e_{V_2}) - (\eta, e_{V_2})| \leq |\eta - x| \, |e_{V_1}| + \frac{\epsilon}{3} + |\eta - x| \, |e_{V_2}| < \epsilon. \end{aligned}$$

Thus, as $V \longrightarrow e$, $e_V$ converges weakly to some $\xi_0 \in \mathfrak{L}_2(L)$ for which

$$(x, \xi_0) = L(x).$$

Now, choosing $x = e_V$ and passing to the limit as $V \longrightarrow e$, we obtain $(\xi_0, \xi_0) = \lim_{V \to e} L(e_V)$. But $e_V * e_V^* = e_V * e_V$, in turn, is the function $e_{V^{-1}V}$. Hence $(e_V, e_V) = L(e_{V^{-1}V}) \longrightarrow (\xi_0, \xi_0)$. Consequently,

$$|e_V - \xi_0|^2 = (e_V, e_V) - (\xi_0, e_V) - (e_V, \xi_0) + (\xi_0, \xi_0) \longrightarrow 0,$$

i.e., $e_V$ converges strongly to $\xi_0$ as $V \longrightarrow e$.

It is easily seen that

$$T_g x * (T_g y)^* = x * y^*.$$

It follows that the operators $T_g$ are unitary on $\mathfrak{L}_2(L)$: $(T_g x, T_g x) = (x, x)$ for any $x \in \mathfrak{L}_1$, and since $\mathfrak{L}_1$ is dense in $\mathfrak{L}_2(L)$, $T_g$ can be uniquely extended to a unitary operator on $\mathfrak{L}_2(L)$. Since, for any $x \in \mathfrak{L}_1$,

$$|T_{g'} x - T_g x| \leq \sqrt{|L|}\ \|T_{g'} x - T_g x\| \longrightarrow 0 \text{ as } g' \longrightarrow g,$$

$T_g$ on $\mathfrak{L}_2(L)$ is a strongly continuous function of $g$. Indeed, let $\eta \in \mathfrak{L}_2(L)$. Choose $x \in \mathfrak{L}_1$ such that $|x - \eta| < \frac{\epsilon}{3}$, and then pick a neighborhood $V$ of the element $g$ such that $|T_{g'} x - T_g x| < \frac{\epsilon}{3}$ for $g' \in V$. Then

$$|T_{g'}\eta - T_g \eta| \leq |T_{g'}\eta - T_{g'}x| + |T_{g'}x - T_g x| + |T_g x - T_g \eta|$$
$$= 2|\eta - x| + |T_{g'}x - T_g x| < \epsilon.$$

Finally, it is clear that $T_{gh} = T_g T_h$. Thus $T_g$ is a unitary representation of the group $G$. Hence it follows that

$$\phi(g) = (T_g \xi_0, \xi_0)$$

is a continuous positive definite function.

Let us show that the functional $L$ arises from this function $\phi(g)$; more precisely, that

$$L(x) = \int (T_g \xi_0, \xi_0)\, x(g)\, dg \text{ for all } x \in \mathfrak{L}_1.$$

Indeed, since $e_V$ converges strongly to $\xi_0$, $L(T_g e_V) = (T_g e_V, \xi_0)$ converges uniformly to $(T_g \xi_0, \xi_0)$, and so

$$\int (T_g \xi_0, \xi_0)\, x(g)\, dg = \lim_{V \to e} \int L(T_g e_V)\, x(g)\, dg. \tag{5}$$

But

$$\int L(T_g e_V)\, x(g)\, dg = L\left(\int T_g e_V\, x(g)\, dg\right). \tag{6}$$

In fact, let $E$ be any measurable set. Put $m_L(E) = L(f_E)$, where $f_E(g)$ is the characteristic function of the set $E$. It is easy to check that $m_L(E)$ is a countably additive measure (complex, in general) and that

$$L(y) = \int y(h)\, dm_L(h)$$

for any function $y \in \mathfrak{L}_1$. Applying the Fubini theorem, we obtain

$$\int L(T_g e_V)\, x(g)\, dg = \int \left( \int e_V(g^{-1}h)\, dm_L(h) \right) x(g)\, dg$$
$$= \int \left( \int e_V(g^{-1}h)\, x(g)\, dg \right) dm_L(h) = L\left( \int T_g e_V x(g)\, dg \right),$$

i.e., equation (6). From equations (5) and (6) it follows that

$$\int (T_g \xi_0, \xi_0)\, x(g)\, dg = \lim_{V \to e} L\left( \int T_g e_V x(g)\, dg \right) = \lim_{V \to e} L\left( \int e_V(g^{-1}h)\, x(g)\, dg \right)$$
$$= \lim_{V \to e} L\left( \int e_V(g^{-1})\, x(hg)\, dg \right) = \lim_{V \to e} L\left( \int e_V(g)\, x(hg)\, dg \right).$$

But, according to the second relation in (4),

$$\left\| \int e_V(g)\, x(hg)\, dg - x(h) \right\|$$
$$= \int \left| \int [x(hg) - x(h)]\, e_V(g)\, dg \right| dh \le \int \left( \int |x(hg) - x(h)|\, dh \right) e_V(g)\, dg$$
$$\le \sup_{g \in V} \int |x(hg) - x(h)|\, dh \longrightarrow 0 \text{ as } V \longrightarrow e.$$

Hence

$$\lim_{V \to e} L\left( \int e_V(g)\, x(hg)\, dg \right) = L(x),$$

and the theorem is proved.

It follows from Theorem 4 and the preceding remarks that *every essentially bounded, integrally positive definite function coincides almost everywhere with a continuous positive definite function.* Indeed, if $\psi(g)$ is an essentially bounded integrally positive definite function, and $\phi(g)$ is the continuous positive definite function giving rise to the positive functional $L_\psi(x) = \int \psi(g) x(g)\, dg$, then

$$\int \psi(g)\, x(g)\, dg = \int \phi(g)\, x(g)\, dg \text{ for all } x \in \mathfrak{L}_1,$$

and this is possible only if $\psi(g)$ and $\phi(g)$ are equal almost everywhere.

4. We shall call a positive definite function $\phi(g)$ *normalised* if $\phi(e) = 1$. The totality $\mathfrak{P}$ of all continuous positive definite functions $\phi(g)$ for which $\phi(e) \le 1$ is obviously convex, in other words, whenever it contains two functions $\phi(g)$ and $\psi(g)$, it contains the "line segment" joining them, i.e., contains each function $\lambda\phi(g) + (1-\lambda)\psi(g)$, $0 \le \lambda \le 1$. An *extreme point* of a convex set is a point of the set which is not an interior point of any line segment contained in the set. Obviously, the function $\phi(g) \equiv 0$ is an extreme point of the set $\mathfrak{P}$. *Any other extreme point of the set $\mathfrak{P}$ is a normalised elementary positive definite function.* Indeed, let $\phi(g) \not\equiv 0$ be an extreme point of the set $\mathfrak{P}$, and suppose $\psi(g) \ll \phi(g)$, $\psi(g) \not\equiv 0$, $\psi(g) \not\equiv \phi(g)$, i.e., $\phi(g) - \psi(g)$ is a positive definite function different both from zero and from $\phi(g)$. Since a positive definite function achieves its maximum in absolute value at $g = e$, we have

$\phi(e) > \psi(e) > 0$. Moreover, it is clear that $\phi(e) = 1$, since otherwise $\phi(g)$ would be an interior point of the line segment $\left[0, \frac{\phi(g)}{\phi(e)}\right]$. Hence

$$\phi(g) = \psi(e)\,\frac{\psi(g)}{\psi(e)} + [1 - \psi(e)]\,\frac{\phi(g) - \psi(g)}{1 - \psi(e)},$$

whence, by the extremality of the function $\phi(g)$, it follows that $\frac{\psi(g)}{\psi(e)} = \frac{\phi(g) - \psi(g)}{1 - \psi(e)}$ or $\psi(g) = \psi(e)\,\phi(g)$; thus $\phi(g)$ is an elementary function. Conversely, *every normalised elementary continuous positive definite function is an extreme point of the set* $\mathfrak{P}$. Indeed, suppose $\phi(g)$ is an elementary positive definite function, $\phi(e) = 1$, and $\phi(g) = \lambda\psi(g) + (1 - \lambda)\,\chi(g)$, where $\psi, \chi \in \mathfrak{P}$. Obviously then, $\psi(e) = \chi(e) = 1$, and, since $\lambda\psi(g) \ll \phi(g)$, $\lambda\psi(g) = \alpha\phi(g)$. Setting $g = e$, we obtain $\alpha = \lambda$, i.e., $\psi(g) = \phi(g)$, hence also $\chi(g) = \phi(g)$; thus $\phi(g)$ is an extreme point of the set $\mathfrak{P}$.

So *normalised elementary continuous positive definite functions can be interpreted as the nonzero extreme points of the convex set* $\mathfrak{P}$ *of all continuous positive definite functions* $\phi(g)$ *with* $\phi(e) < 1$.

Making use of the correspondence established by Theorem 4 between continuous positive definite functions and positive functionals, we shall show that there are enough such extreme points.

We equip the totality of all continuous positive definite functions with the "weak topology", identifying them with the linear functionals on $\mathfrak{L}_1$ to which they give rise. Precisely, a neighborhood of a continuous positive definite function $\phi_0(g)$ is defined as the set, given any finite number of functions $x_1, \cdots, x_n \in \mathfrak{L}_1$ and any $\epsilon > 0$, of all continuous positive definite functions $\phi(g)$ satisfying

$$\left| \int \phi_0(g)\, x_k(g)\, dg - \int \phi(g)\, x_k(g)\, dg \right| < \epsilon \quad (k = 1, \cdots, n). \tag{7}$$

Theorem 5. *The set of all continuous positive definite functions* $\phi(g)$ *satisfying* $\phi(e) \leq 1$ *is the smallest weakly closed convex set containing all the normalised elementary continuous positive definite functions and the function* $\phi(g) \equiv 0$.

Proof. Let $\mathfrak{R}\mathfrak{L}_1$ denote the closed linear subspace spanned over the reals by the set of all functions $x \in \mathfrak{L}_1$ satisfying $x^* = x$. If $L$ is a positive linear functional on $\mathfrak{L}_1$, then $L(x^*) = \overline{L(x)}$ for all $x \in \mathfrak{L}_1$. Indeed, as we have seen earlier,

$$L(y * x^*) = \overline{L(x * y^*)}.$$

Taking $y = e_V$ here, and letting $V \rightarrow e$, we obtain the desired relation

$L(x^*) = \overline{L(x)}$. Therefore, a positive linear functional given on $\mathfrak{L}_1$ can be viewed as a real linear functional defined on $\Re\mathfrak{L}_1$, which is positive in the previous sense. Moreover, every real linear functional defined on $\Re\mathfrak{L}_1$ obviously has a unique extension to a linear functional on $\mathfrak{L}_1$, given by the formula

$$L(x) = L(\Re x) + iL(\Im x), \text{ where } \Re x = \frac{x + x^*}{2}, \ \Im x = \frac{x - x^*}{2i}.$$

Hence positive linear functionals that are different on $\mathfrak{L}_1$ are already different on $\Re\mathfrak{L}_1$. It is easy to see that both the norm and the weak topologies on the space of positive linear functionals do not depend upon the interpretation of the functionals as being defined on $\mathfrak{L}_1$ or only on $\Re\mathfrak{L}_1$.

For any positive linear functional $L = L_\phi$, we have

$$|L_\phi| = \sup\operatorname{ess} |\phi(g)| \phi(e). \tag{8}$$

Thus the convex set $\mathfrak{P}$ of all continuous positive definite functions $\phi(g)$ satisfying $\phi(e) \leq 1$ can be interpreted, using Theorem 4 and taking into consideration equation (8), as the convex set $\mathfrak{P}'$ of all positive linear functionals $L$ on $\Re\mathfrak{L}_1$ lying in the ball $|L| \leq 1$. But it is easy to see that $\mathfrak{P}'$ is weakly closed in the space of all real linear functionals defined on $\Re\mathfrak{L}_1$. Consequently, by the Kreĭn-Mil'man theorem [6], $\mathfrak{P}'$, and hence $\mathfrak{P}$, is the weakly closed convex hull of its extreme points. Finally, the previously established correspondence between extreme points of the set $\mathfrak{P}$ and normalised elementary continuous positive definite functions completes the proof of Theorem 5.

It can be proved that, *if the group $G$ is not discrete, the function $\phi(g) \equiv 0$ lies in the weak closure of the set of all normalised continuous elementary positive definite functions.*

5. For a locally bicompact group, because it admits a Haar measure, there is a simple well-known way of constructing continuous positive definite functions. Let $\mathfrak{L}_2$ denote the Hilbert space consisting of all square integrable measurable functions with the scalar product

$$(x, y) = \int x(h)\overline{y(h)}\, dh.$$

The operators $U_g x(h) = x(g^{-1}h)$ obviously form a unitary representation of the group $G$ in $\mathfrak{L}_2$.

From this it follows that for any function $x(h) \in \mathfrak{L}_2$,

$$(U_g x, x) = \int x(g^{-1}h)\overline{x(h)}\, dh$$

is a continuous positive definite function.

The availability of this construction shows at once that *on a locally bicompact group $G$ there exists, for each element $g_0 \neq e$, a continuous positive* *definite function $\phi_0(g)$ such that $\phi_0(g_0) \neq \phi_0(e)$; furthermore, for each neighborhood $V$ of the identity $e$, there is a normalised continuous positive definite function vanishing outside $V$.*

Indeed, choose any neighborhood $W$ satisfying the condition $WW^{-1} \subset V$, and then pick some function $x_W(h) \in \mathfrak{L}_2$ satisfying $(x_W, x_W) = 1$ and $x_W(h) = 0$ outside $W$. The function $\phi(g) = (U_g x_W, x_W)$ will have the required properties.

Theorem 6. *On a locally bicompact group $G$ there is a complete system of elementary continuous positive definite functions.*

Proof. Suppose, arguing by contradiction, that there is an element $g_0 \neq e$ such that $\zeta(g) = 1$ for every normalised elementary continuous positive definite function $\zeta(g)$. Choose some real normalised continuous positive definite function $\phi_0(g)$ satisfying $\phi_0(g_0) \neq 1$; as was shown above, this is always possible so long as $g_0 \neq e$. Let $V$ be a neighborhood of the identity such that

$$1 - \phi_0(g) < \epsilon \text{ and } |\phi_0(g_0) - \phi_0(g_0 g)| < \epsilon \text{ for all } g \in V, \tag{9}$$

where $\epsilon$ is a fixed positive number. Let $f_V(g)$ be the characteristic function of the set $V$. Consider a neighborhood (7) of the function $\phi_0(g)$ defined by the functions $x_1(g) = \frac{1}{m(V)} f_V(g)$ and $x_2(g) = \frac{1}{m(V)} f_V(g_0^{-1} g)$, and by the fixed number $\epsilon$. Theorem 5 assures that this neighborhood contains at least one function of the form

$$\phi(g) = \lambda_1 \zeta_1(g) + \cdots + \lambda_k \zeta_k(g),$$

where $\zeta_1, \cdots, \zeta_k$ are normalised elementary continuous positive definite functions and $\lambda_1 \geq 0, \cdots, \lambda_k \geq 0$, $\lambda_1 + \cdots + \lambda_k \leq 1$. Then this same neighborhood will also contain the real part $\psi(g) = \Re\phi(g)$, which is as much a positive definite function as $\phi(g)$ is. By assumption, $\psi(g_0) = \psi(e) \leq 1$. Since, by all our choices,

$$\left| \int [\phi_0(g) - \psi(g)]\, x_1(g)\, dg \right| = \left| \frac{1}{m(V)} \int_V [\phi_0(g) - \psi(g)]\, dg \right| < \epsilon,$$

we find, taking into consideration the first inequality in (9):

$$1 - \psi(e) = \frac{1}{m(V)} \int_V [1 - \psi(e)]\, dg \leq \frac{1}{m(V)} \int_V [1 - \psi(g)]\, dg$$

$$\leq \frac{1}{m(V)} \int_V [1 - \phi_0(g)]\, dg + \left| \frac{1}{m(V)} \int_V [\phi_0(g) - \psi(g)]\, dg \right| < 2\epsilon. \tag{10}$$

Furthermore, since $\phi_0(g_0 g) - \psi(g_0 g) = [\phi_0(g_0) - \psi(g_0)] - [\phi_0(g_0) - \phi_0(g_0 g)] + [\psi(g_0) - \psi(g_0 g)]$, we have

$$\left|\frac{1}{m(V)}\int_V [\phi_0(g_0 g) - \psi(g_0 g)]\, dg\right| \geq |\phi_0(g_0) - \psi(g_0)|$$

$$- \left|\frac{1}{m(V)}\int_V [\phi_0(g_0) - \phi_0(g_0 g)]\, dg\right| - \left|\frac{1}{m(V)}\int_V [\psi(g_0) - \psi(g_0 g)]\, dg\right|.$$

But by successive application of the Schwarz inequality, M. G. Kreĭn's inequality (3), and inequality (10), we obtain

$$\left|\frac{1}{m(V)}\int_V [\psi(g_0) - \psi(g_0 g)]\, dg\right|^2 \leq \frac{1}{m(V)}\int_V |\psi(g_0) - \psi(g_0 g)|^2\, dg$$

$$\leq \frac{2}{m(V)}\int_V [\psi(e) - \psi(g)]\, dg \leq \frac{2}{m(V)}\int [1 - \psi(g)]\, dg < 4\epsilon.$$

In addition, the second inequality in (9) assures that

$$\left|\frac{1}{m(V)}\int_V [\phi_0(g_0) - \phi_0(g_0 g)]\, dg\right| < \epsilon.$$

Thus, taking into consideration the inequality $1 - \psi(e) < 2\epsilon$, which is contained in (10), we find

$$\left|\frac{1}{m(V)}\int_V [\phi_0(g_0 g) - \psi(g_0 g)]\, dg\right|$$

$$> |\phi_0(g_0) - \psi(g_0)| - \epsilon - 2\sqrt{\epsilon} = |\phi_0(g_0) - \psi(e)| - \epsilon - 2\sqrt{\epsilon}$$

$$> [1 - \phi_0(g_0)] - [1 - \psi(e)] - \epsilon - 2\sqrt{\epsilon} > 1 - \phi_0(g_0) - 3\epsilon - 2\sqrt{\epsilon}. \quad (11)$$

On the other hand, by our choices we have

$$\left|\int [\phi_0(g) - \psi(g)]\, x_2(g)\, dg\right| = \left|\frac{1}{m(V)}\int_V [\phi_0(g_0 g) - \psi(g_0 g)]\, dg\right| < \epsilon. \quad (12)$$

Combining (11) and (12) we obtain

$$1 - \phi_0(g_0) < 4\epsilon + 2\sqrt{\epsilon}.$$

But this is impossible for arbitrarily small $\epsilon$, since $1 - \phi_0(g_0)$ is by assumption a (constant) positive number. This contradiction proves the theorem.

Our main result follows immediately from Theorems 3 and 6:

**Theorem 7.** *Every locally bicompact group admits a complete system of irreducible unitary representations.*

## BIBLIOGRAPHY

[1] S. Banach, *Sur la mesure de Haar* (S. Saks, *Théorie de l'intégrale*, Appendix 1, Warsaw, 1933).

[2] I. M. Gel'fand and D. A. Raĭkov, *On the theory of characters of commutative topological groups*, Dokl. Akad. Nauk SSSR 28 (1940), 195–198. (Russian) MR 2, 217.

[3] A. Haar, *Der Massbegriff in der Theorie der kontinuierlichen Gruppen*, Ann. of Math. (2) 34 (1933), 147–169.

[4] S. Kakutani, *On the uniqueness of Haar's measure*, Proc. Imperial Acad. Tokyo 14 (1938), 27–31.

[5] M. G. Kreĭn, *On a certain ring of functions on a topological group*, Dokl. Akad. Nauk SSSR 23 (1939), 749–752. (Russian) MR 1, 337.

[6] M. Kreĭn and D. Mil'man, *On extreme points of regular convex sets*, Studia Math. 9 (1940), 133–138. MR 3, 90.

[7] J. von Neumann, *Almost periodic functions in a group*, Trans. Amer. Math. Soc. 36 (1934), 445–492.

[8] ———, *On the uniqueness of Haar's measure*, Mat. Sb. (N.S.) 1 (43) (1936), 721–734. (Russian)

[9] F. Peter and H. Weyl, *Die Vollständigkeit der primitiven Darstellungen einer geschlossenen kontinuierlichen Gruppe*, Math. Ann. 97 (1927), 737–755.

[10] D. A. Raĭkov, *Harmonic analysis on commutative groups with the Haar measure and the theory of characters*, Trudy Mat. Inst. Steklov. 14 (1945). (Russian) MR 8, 133.

[11] ———, *A new proof of the uniqueness of Haar's measure*, Dokl. Akad. Nauk SSSR 34 (1942), 211–212. (Russian) MR 4, 219.

[12] A. Weil, *Sur les groupes topologiques et les groupes mesurés*, C. R. Acad. Sci. Paris 202 (1936), 1147–1149.

Translated by:
F. E. J. Linton

# 2.

## (with M. A. Najmark)

# Unitary representations of the group of linear transformations of the straight line

Dokl. Akad. Nauk SSSR **55** (1947) 567–570. Zbl. **29**:5

*(Communicated by A. N. Kolmogoroff, Member of the Academy, 20. VIII. 1946)*

Let $R$ be the group of linear transformations $y=\alpha x+\beta$ of the real axis, where $\alpha$ runs over all positive numbers, and $\beta$ runs over all real numbers. In his survey of the theory of probability Kolmogoroff has put the following problem: To find the general form of all positive-definite functions on $R$ and their decomposition into elementary positive-definite functions.

It follows from (1) that this problem is equivalent to the following one: To find the general form of all unitary representations of $R$ in the separable Hilbert space and their decomposition into irreducible unitary representations.

The present note gives a solution of this problem, which the authors happened to find while engaged on a more comprehensive study of unitary representations. The methods used are due in their algebraical part to Frobenius. As developed here they can also be applied with some modifications to any solvable Lie group. But that question will be discussed elsewhere. In particular, the general form of the unitary representations of the group $K$ of all transformations $w=\alpha z+\beta$, where $\alpha$, $\beta$ run over all complex numbers with the exception of $\alpha=0$ can be obtained in an analogous way.

The group $R$ contains the following commutative subgroups: 1) The group $\mathfrak{T}$ of translations $x\to x+\beta$; we denote its elements by $t_\beta$; 2) the group $\mathfrak{S}$ of tensions* $x\to\alpha x$; we denote its elements by $s_\alpha$.

Let there be given a unitary representation of $R$ in a separable space $\mathfrak{H}$ (the case of non-separable space can be easily reduced to the considered one by decomposing the space into a direct sum of cyclic subspaces). Denote by $T_\beta$ and $S_\alpha$ the operators corresponding to $t_\beta$ and $s_\alpha$. Then

$$T_{\beta_1}T_{\beta_2}=T_{\beta_1+\beta_2};\; S_{\alpha_1}S_{\alpha_2}=S_{\alpha_1\alpha_2};\; T_\beta S_\alpha=S_\alpha T_{\alpha\beta} \tag{1}$$

Denote by $\mathfrak{N}$ the set of all $f\in\mathfrak{H}$ satisfying $T_\beta f=f$ for all $\beta$. If $f\in\mathfrak{N}$ then $T_\beta S_\alpha f=S_\alpha T_{\alpha\beta}f=S_\alpha f$, hence $S_\alpha f\in\mathfrak{N}$. Thus, $\mathfrak{N}$ and therefore $\mathfrak{M}=\mathfrak{H}-\mathfrak{N}$ reduce our representation so that $T_\beta\equiv 1$ in $\mathfrak{N}$ and $S_\alpha$ is an arbitrary representation of $\mathfrak{S}$, that is $S_\alpha$ is a representation of the multiplicative group of all positive numbers.

Consider the representation in $\mathfrak{M}$; $T_\beta f=f$ can be fulfilled in $\mathfrak{M}$ for all $\beta$ if only $f=0$. By a well-known theorem of Stone (2) the equality (1) implies: $T_\beta f=\int\limits_{-\infty}^{+\infty}e^{i\lambda\beta}dE(\lambda)f$, where $E(\lambda)$ is a resolution of the identity. Further by the Hellinger-Hahn theory (2) the space $\mathfrak{M}$ can be realized as a finite or infinite direct sum of spaces $\mathfrak{M}_k$ $(k=1,2,3,\dots)$ of functions

* This group is isomorphic to the factor-group $R/\mathfrak{T}$.

$f_k(\lambda)$ such that $\|f_k\|^2 = \int_{-\infty}^{+\infty} |f_k(\lambda)|^2 d\sigma_k(\lambda) < +\infty$, where $\sigma_k(\Delta)$ is a positive completely-additive function on the Borel sets of the real axis and $\sigma_k(-\infty, +\infty) = 1$. Furthermore, every function $\sigma_k(\Delta)$ is absolutely continuous with respect to $\vartheta_{k-1}(\Delta), \ldots, \sigma_1(\Delta)$. By the definition of $\mathfrak{M}$ none of $\sigma_k(\Delta)$ is concentrated at zero. We shall write $f \sim \{f_k(\lambda)\}$, then $T_\beta f \sim \{e^{i\lambda\beta} f_k(\lambda)\}$.

The operator $S_\alpha$ can be given by a matrix $\|S_{kj}(\alpha)\|$ where $S_{kj}(\alpha)$ is a bounded operator from $\mathfrak{M}_j$ into $\mathfrak{M}_k$, so that

$$S_\alpha f \sim \left\{ \sum_j S_{kj}(\alpha) f_j(\lambda) \right\}$$

and, since $S_\alpha$ is unitary,

$$\sum_{k} \|f_k\|^2 = \sum_k \left\| \sum_j S_{kj}(\alpha) f_j \right\|^2$$

Put $S_{kj}(\alpha)1 = \pi_{kj}(\alpha; \lambda)$; it follows from (1) that

$$S_{kj}(\alpha) f_j(\lambda) = \pi_{kj}(\alpha; \lambda) f_j(\lambda/\alpha) \tag{2}$$

hence the unitarity condition can be put in the form

$$\sum_k \int_{-\infty}^{+\infty} |f_k(\lambda)|^2 d\sigma_k(\lambda) = \sum_k \int_{-\infty}^{+\infty} \left| \sum_j \pi_{kj}(\alpha; \lambda) f_j(\lambda/\alpha) \right|^2 d\sigma_k(\lambda) \tag{3}$$

Putting in (9) $f_k(\lambda) = 0$ for $k \neq 1$ and $f_1(\lambda) =$ the characteristic function of a set $\Delta$, we obtain $\sigma_1(\Delta) = \sum_k \int_{\Delta\alpha} |\pi_{k1}(\alpha; \lambda)|^2 d\sigma_k(\lambda)$, hence

$$\sigma_1(\Delta\alpha) = \sum_k \int_\Delta \left| \pi_{k1}\left(\frac{1}{\alpha}, \lambda\right) \right|^2 d\sigma_k(\lambda).$$

As all $\sigma_k(\Delta)$, $k = 2, 3, \ldots$, are absolutely continuous with respect to $\sigma_1(\Delta)$, the last equality means that $\sigma_1(\Delta\alpha)$ is absolutely continuous with respect to $\sigma_1(\Delta)$. We call generally a non-negative Borel set function $\sigma(\Delta)$ on a topological group $\mathfrak{G}$ a differentiable measure if for any «translated» set $\Delta g$ the function $\sigma(\Delta g)$ is absolutely continuous with respect to $\sigma(\Delta)$. Thus, $\sigma_1(\Delta)$ is a differentiable measure on the multiplicative group of additive numbers.

Put $d\sigma_1(\lambda\alpha) / d\sigma_1(\lambda) = \omega(\lambda, \alpha)$. Then $\omega(\lambda, \alpha_1)\ \omega(\lambda\alpha_1, \alpha_2) = \omega(\lambda, \alpha_1, \alpha_2)$. Hence, repeating the argument of p. 573, we see that

$$\sigma_1(\Delta) = \int_\Delta \omega_1^2(\lambda) |\lambda|^{-1} d\lambda$$

As all $\sigma_k(\Delta)$ are absolutely continuous with respect to $\sigma_1(\Delta)$, we have also: $\sigma_k(\Delta) = \int_\Delta \omega_k^2(\lambda) |\lambda|^{-1} d\lambda$.

Denote by $E_k$ the set of $\lambda$ values, satisfying $\omega_k(\lambda) = 0$. Evidently $E_1 \subset E_2 \subset E_3 \subset \ldots$ and $E_1$ is one of the sets $(0)$, $(0, +\infty)$, $(-\infty, 0)$. Put $\varphi_k(\lambda) = \omega_k(\lambda) f_k(\lambda)$ and define the norm of $\varphi_k$ by

$$\|\varphi_k\|^2 = \int_{-\infty}^{+\infty} |\varphi_k(\lambda)|^2 |\lambda|^{-1} d\lambda$$

then $\|\varphi_k\| = \|f_k\|^2$, hence $f_k \to \varphi_k$ is an isometric mapping of $\mathfrak{M}_k$ on to the space $\mathfrak{P}_k$ of all $\varphi_k$ with finite $\|\varphi_k\|^2$ satisfying the condition $\varphi_k(\lambda)=0$ on $E_k$; evidently $\mathfrak{P}_1 \supset \mathfrak{P}_2 \supset \mathfrak{P}_3 \supset \ldots$ So, we can realize $\mathfrak{M}$ as a direct sum of $\mathfrak{P}_k$. The mapping $f_k \to \varphi_k$ carries $S_{kj}(\alpha)$ into operators on $\varphi_k(\lambda)$. We denote them also by $S_{kj}(\alpha)$. In virtue of (2), $S_{kj}(\alpha)\varphi_j(\lambda) = \pi'_{kj}(\alpha; \lambda)\varphi_j(\lambda/\alpha)$, where $\pi'_{kj}(\alpha, \lambda) = \omega_k(\lambda)\pi_{kj}(\alpha; \lambda)\omega_j^{-1}(\lambda/\alpha)$ for $\lambda/\alpha \bar{\in} E_j$ and is not defined at all for $\lambda/\alpha \in E_j$. The unitarity condition (3) will now become

$$\sum_k \int_{-\infty}^{+\infty} |\varphi_k(\lambda)|^2 |\lambda|^{-1} d\lambda = \sum_k \int_{-\infty}^{+\infty} \Big|\sum_j \pi'_{kj}(\alpha, \alpha\lambda)\varphi_j(\lambda)\Big|^2 |\lambda|^{-1} d\lambda \tag{4}$$

Denote by $n_0 (\leqslant +\infty)$ the number of the $\mathfrak{M}_k$ and by $\mathfrak{H}_0$ the unitary space of all sequences $\{\xi_k\}$ such that $\|\xi\|^2 = \sum_{k=1}^{n_0} |\xi_k|^2 < +\infty$ (if $n_0=\infty$) and by $\mathfrak{H}_k$, $k=1, 2, 3, \ldots$, its finite-dimensional subspace defined by the conditions: $\xi_k = \xi_{k+1} = \ldots = 0$; $f \sim \{\varphi_k(\lambda)\}$ means that the elements of $\mathfrak{M}$ can be realized as vector-function $\varphi(\lambda)$, such that $\varphi(\lambda) \in \mathfrak{H}_0$ for $\lambda \bar{\in} \sum E_k$ and $\varphi(\lambda) \in \mathfrak{H}_k$ for $\lambda \in E_k$. If we now write $f \sim \varphi(\lambda)$, then

$$\|f\|^2 = \int_{-\infty}^{+\infty} \|\varphi(\lambda)\|^2 |\lambda|^{-1} d\lambda; \quad T_\beta f \sim e^{i\lambda\beta}\varphi(\lambda); \quad S_\alpha f \sim A(\alpha; \lambda)\varphi(\lambda/\alpha) \tag{5}$$

where $A(\alpha; \lambda)$ is the operator defined by the matrix $\pi'_{kj}(\alpha; \lambda)$.

(4) can now be written in the form

$$\int_{-\infty}^{+\infty} \|\varphi(\lambda/\alpha)\|^2 |\lambda|^{-1} d\lambda = \int_{-\infty}^{+\infty} \|A(\alpha; \lambda)\varphi(\lambda/\alpha)\|^2 |\lambda|^{-1} d\lambda$$

This implies that for a. a. (abbreviated for *almost all*) $\lambda \bar{\in} \alpha \sum E_k$, $A(\alpha; \lambda)$ is an isometric operator with the domain $\mathfrak{H}_0$, and for a. a. $\lambda \in \alpha E_k$ with the domain $\mathfrak{H}_k$. Further, by the definition of $\pi'_{kj}(\alpha; \lambda)$ the range of $A(\alpha; \lambda)$ is $\mathfrak{H}_0$ for a. a. $\lambda \bar{\in} \alpha \sum E_k$ and $\mathfrak{H}_k$ for a. a. $\lambda \in E_k$. Then for a. a. $\lambda \bar{\in} \alpha \sum E_k$ and $\lambda \in E_k$ the operator $A(\alpha; \lambda)$ maps $\mathfrak{H}_0$ onto $\mathfrak{H}_k$ isometrically, which is possible only in the case where $\lambda \bar{\in} \sum E_k$ for a. a. $\lambda \bar{\in} \alpha \sum E_k$. But this takes place if and only if, with the exception of a set of measure zero, $\sum E_k$ is one of the sets $(0)$, $(-\infty, 0)$, $(0, +\infty)$. A similar argument shows that the same is also true for each $E_k$.

We now use (1). By (5) it is equivalent to $A(\alpha_1; \lambda)\ A(\alpha_2; \lambda/\alpha_1) = (A\,\alpha_1\alpha_2; \lambda)$, which holds for a. a. triplets $(\alpha_1, \alpha_2, \lambda)$. Put $B(\alpha, \beta) = A(\alpha/\beta; \alpha)$; this operator is isometric and defined for a. a. pairs $(\alpha/\beta, \alpha)$ and hence by Fubini's theorem also for a. a. pairs $(\alpha, \beta)$. Now the last equality can be put in the form

$$B(\alpha, \beta)B(\beta, \gamma) = B(\alpha, \gamma) \tag{6}$$

this holds for a. a. triplets $(\alpha, \beta, \gamma)$. But then a $\beta$-value $\beta=\beta_0$ is in existence, such that $B(\alpha, \beta_0)B(\beta_0, \gamma) = B(\alpha, \gamma)$ for a. a. pairs $(\alpha, \gamma)$. Put $B(\alpha, \beta_0) = B(\alpha)$, $B(\beta_0, \gamma) = C(\gamma)$; then $B(\alpha, \gamma) = B(\alpha)\ C(\gamma)$ and therefore $A(\alpha; \lambda) = B(\lambda)\,C(\lambda/\alpha)$. Putting this in (6) we get: $B(\lambda)\,C(\lambda/\alpha_1)\,B(\lambda/\alpha_1)\,C(\lambda/\alpha_1\alpha_2) = B(\lambda)\,C(\lambda/\alpha_1\alpha_2)$. Hence $B(\lambda)\,C(\lambda) = 1$ for a. a. $\lambda$ so that

$$A(\alpha; \lambda) = B(\lambda)\,B^{-1}(\lambda/\alpha) \tag{7}$$

Now put $\psi(\lambda) = B^{-1}(\lambda)\varphi(\lambda)$; then $\|\psi(\lambda)\| = \|\varphi(\lambda)\|$. Write $f \sim \psi(\lambda)$; then $\|f\|^2 = \int_{\infty}^{+\infty} \|\psi(\lambda)\|^2 \frac{d\lambda}{|\lambda|}$, $T_\beta f \sim e^{i\lambda\beta}\psi(\lambda)$, $S_\alpha f \sim \psi(\lambda/\alpha)$.

By the definition of $\mathfrak{H}_0$, $\psi(\lambda)=\{\psi_1(\lambda), \psi_2(\lambda), \ldots\}$, where $\psi_k(\lambda)$ are scalar functions and $\|\psi(\lambda)\|^2=\sum_{k=1}^{n_0}|\psi_k(\lambda)|^2$, whence $\|f\|^2=\sum_{k=1}^{n_0}\int_{-\infty}^{+\infty}\frac{|\psi_k(\lambda)|^2}{|\lambda|}d\lambda$.

Write $f\sim\{\psi_k(\lambda)\}$, then

$$T_\beta f\sim\{e^{i\lambda\xi}\psi_k(\lambda)\},\quad S_\alpha f\sim\{\psi_k(\lambda/\alpha)\} \tag{8}$$

Let now the representation be irreducible; then either $\mathfrak{M}$ or $\mathfrak{N}$ is (0). In the first case $\mathfrak{M}$ is one-dimensional and $S_\alpha$ is the character of the multiplicative group of positive numbers. In the second case $\mathfrak{H}_0$ is one-dimensional so that $\psi(\lambda)$ can be considered a scalar complex-valued function. In this case the representation is irreducible if and only if $\mathfrak{H}$ consists of functions $\psi(\lambda)$ different from zero only for positive or only for negative $\lambda$-values. These types of irreducible representations evidently correspond to the decomposition of the real axis into transitive manifolds with respect to the transformation $\lambda\longrightarrow\alpha\lambda$.

If we pass in this case to the Fourier-transform $f(x)=\frac{1}{\sqrt{\pi}}\int_{-\infty}^{+\infty}\frac{\psi(\lambda)}{\sqrt{|\lambda|}}e^{i\lambda x}d\lambda$, we can easily obtain the following theorem:

Theorem. *Every irreducible unitary representation of $R$ is equivalent to one of the following types:*

I. *$\mathfrak{H}$ is one-dimensional; $T_\beta\equiv 1$, $S_\alpha$ is the character of the multiplicative group of positive numbers.*

II. *$\mathfrak{H}$ consists of all square sommable functions $f(x)$, $-\infty<x<+\infty$, which are boundary values on the real axis of functions analytic in the upper half-plane; $T_\beta f(x)=f(x+\beta)$; $S_\alpha f(x)=\sqrt{\alpha}f(\alpha x)$.*

III. *$\mathfrak{H}$ consists of all square sommable functions $f(x)$, $-\infty<x<+\infty$, which are the boundary values on the real axis of functions analytic in the lower half-plane; $T_\beta f(x)=f(x+\beta)$; $S_\alpha f(x)=\sqrt{\alpha}f(\alpha x)$.*

Received
20.VIII.1946.

## REFERENCES

[1] I. Gelfand and D. Raikov, Rec. Math., 13 (55), No. 2—3, 301 (1943).
[2] M. H. Stone, Linear Transformations in Hilbert Space, N. Y., 1932.

# 3.

# Center of the infinitesimal group ring

Mat. Sb., Nov. Ser. **26** (68) (1950) 103–112. Zbl. **35**:300

1. Let $R$ be a Lie algebra (also called an infinitesimal Lie group). Choose a basic $e_1,\ldots,e_n$ in $R$. Then the commutator in $R$ is given by the formulas

$$[e_i, e_k] = \sum_j c_{ik}^j e_j.$$

Consider the ring $K$ whose elements are formal polynomials in $e_1,\ldots,e_n$. We shall consider two such polynomials equal if one can be transformed into the other in a finite number of steps, each step consisting in replacing $e_i e_k - e_k e_i$ by $[e_i, e_k] = \sum_j c_{ik}^j e_j$. The ring $K$ will be called the *infinitesimal group ring*.

Our goal is to find the center of the ring $K$, i.e. the set of such of its elements which commute with all the elements in $K$. This is an important problem in representation theory. Indeed, for an irreducible representation or its multiple a characteristic property is that the elements of the center are represented by the operators that are scalar multiples of the identity. In the case of a semisimple Lie group one element of the center—a polynomial of second degree—has been constructed in [1].

The problem is also of great interest for the investigation of transformation groups on manifolds. Consider a manifold $M$ with transformation group $\mathfrak{G}$ acting on it. The transformation group is a Lie group. The set of all functions on $M$ constitutes a linear space, where the transformation group is represented by shift operators. The elements $e_1,\ldots,e_n$ of the infinitesimal Lie group correspond to the differential operators $X_1,\ldots,X_n$ (infinitesimal shift operators, also known as Lie operators). Each element of $K$ is a polynomial in $e_1,\ldots,e_n$ and it is represented by the corresponding polynomial of the Lie operators. Among the Lie operators there are differential operators which correspond to the center of $K$. We shall call them the Laplace operators of the transformation group. Such a name is warranted by the fact that for transformation groups of constant curvature spaces (e.g., of a Euclidean space) these operators are polynomials in Laplace operators of the respective constant curvature space.

One can show without difficulty that for a given transformation group the set of common eigenfunctions for all Laplace operators constitutes a representation space for that group. This is a generalization of a well known way of obtaining representations of the rotation group with the help of spherical functions.

In this paper we establish a simple relation between the center of the ring $K$ and the invariants of the adjoint group (Theorems 1 and 2). Then, for simple groups (e.g. the group of unimodular matrices, the group of orthogonal matrices, etc.) we find the center of that ring. The results are then applied to splitting tensor representations of these groups.

2. It is well known that a Lie group $\mathfrak{G}$ can be considered as a group of linear transformations of the Lie algebra $R$ (the adjoint group). Denote by $\bar{R}$ the linear space dual to the algebra $R$. Since for each linear transformation $A$ acting in $R$ there is a corresponding dual linear transformation $\bar{A}$ which acts in $\bar{R}$, one can consider our group also as a transformation group acting in the space $\bar{R}$.

Consider a multilinear function $l(\xi, \eta, \zeta)$ where $\xi, \eta, \zeta$ are vectors in the space $\bar{R}$, i.e. a function of the form

$$\sum_{i,j,k} a^{ikj} \xi_i \eta_k \zeta_j$$

(we consider a function depending on three vectors only to simplify the notation). If such a function does not change under the transformations which correspond to the elements of the group $\mathfrak{G}$ we call it an *invariant function* or an *invariant form.* Let us now write out the invariance condition explicitly in terms of the structural constants $c^j_{ik}$ of the Lie algebra $R$.

To each element $a$ of the algebra $R$ there corresponds a linear transformation $A$ of the whole algebra defined by the formula

$$Ae_k = [a, e_k].$$

Any transformation of the adjoint group is a product of transformations $e^{tA}$. Therefore, a multilinear function is invariant if and only if it is invariant under the transformations of the form $e^{t\bar{A}}$, i.e. iff

$$l(e^{t\bar{A}}\xi, e^{t\bar{A}}\eta, e^{t\bar{A}}\zeta) = l(\xi, \eta, \zeta).$$

Taking the derivative with respect to $t$ and letting $t = 0$ one has

$$l(\bar{A}\xi, \eta, \zeta) + l(\xi, \bar{A}\eta, \zeta) + l(\xi, \eta, \bar{A}\zeta) = 0.$$

Consider the linear transformation $E_p$ which corresponds to the element $e_p$ of the algebra $R$. It takes $e_i$ into $\sum_j c^j_{ip} e_j$. Now, for the vectors of the dual space $R$, the coordinates transform (under the dual transformation) in the same way as the vectors of the basis in the space $R$, i.e. a vector with the coordinates $\xi_1, \ldots, \xi_n$ is taken by the dual transformation into the vector with coordinates $\xi'_i = \sum_j c^j_{ip} \xi_j$. Thus, invariance condition takes the form:

$$\sum_{i,k,l} (a^{ikl} c^{i_1}_{ip} \xi_{i_1} \eta_k \zeta_l + a^{ikl} \xi_i c^{k_1}_{kp} \eta_k \zeta_l + a^{ikl} \xi_i \eta_k c^{l_1}_{lp} \zeta_{l_1}) = 0. \qquad (1)$$

Let us now formulate the conditions under which an element $P = \sum_{i,k,l} a^{ikl} e_i e_k e_l$ of the ring $K$ commute with all the elements $e_p$ $(p = 1, \ldots, n)$. Since the elements $e_p$ generate the whole of $K$ these conditions are necessary and sufficient for $P$ to be in the center of $K$. In order to find them, we shall use the fact that each of the elements in $K$ can be uniquely presented in the form

$$aI + \sum_i a^i e_i + \sum_{i,k} a^{ik} e_i e_k + \cdots,$$

where the coefficients $a^{ik}$, $a^{ikl}, \ldots$ are symmetric under permutations of their indices. Denoting $PQ - QP$ by $[P, Q]$ one can easily check that

$$[PQ \ldots R, S] = [P, S]Q \ldots R + P[Q, S] \ldots R + \ldots + PQ \ldots [R, S].$$

Therefore one has:

$$\begin{aligned}[P, e_p] &= \sum_{i,k,l} (a^{ikl}[e_i, e_p]e_k e_l + a^{ikl} e_i [e_k, e_p] e_l + a^{ikl} e_i e_k [e_l, e_p]) \\ &= \sum_{i,k,l} (a^{ikl} c_{ip}^{i_1} e_{i_1} e_k e_l + a^{ikl} c_{kp}^{k_1} e_i e_{k_1} e_l + a^{ikl} c_{l_p}^{l_1} e_i e_k e_{l_1}). \qquad (2)\end{aligned}$$

Thus it follows from (1) and (2) that if a multilinear function is invariant under the transformations of the adjoint group then the corresponding element of the ring $K$ commutes with all the elements $e_p$ and therefore lies in the center of the ring $K$. Conversely, let $P$ be in the center of the ring $K$. One can write it in such a way that the corresponding multilinear form is symmetric. One can easily see from the formula (2) that the multilinear form corresponding to the element $[P, e_p]$ is also symmetric. It follows then that the element $P$ is in the center of the ring $K$ if and only if the elements $[P, e_p]$ $(p = 1, \ldots, n)$ correspond to multilinear forms which vanish identically. Comparing the formulas (1) and (2) one has the following theorem.

**Theorem 1.** *For an element P*

$$P = aI + \sum_i a^i e_i + \sum_{i,k} a^{ik} e_i e_k + \sum_{i,k,l} a^{ikl} e_i e_k e_l + \ldots$$

*to lie in the center of the ring K it is sufficient that the forms*

$$\sum_i a^i \xi_i, \sum_{i,k} a^{ik} \xi_i \eta_k, \sum_{i,k,l} a^{ikl} \xi_i \eta_k \zeta_l, \ldots$$

*be invariant under the adjoint group action. If, in addition, the element P is written in such a form that the coefficients $a^{ik}$, $a^{ikl}, \ldots$ are symmetric then the above condition is also necessary.*

Consider now some examples.

(1) The group $\mathfrak{G}$ of $n \times n$ matrices with non-vanishing determinant.

The Lie algebra $R$ in this case consists of all $n \times n$ matrices with the commutator given by the formula

$$[X, Y] = XY - YX.$$

For each element $A$ of the group there is a corresponding linear transformation of $R$ which takes any element $X$ of the Lie algebra into $AXA^{-1}$. The invariants of the adjoint group are most easily computed if one takes the basis in $R$ consisting of matrices $E_{ik}$. The matrix $E_{ik}$ has 1 at the intersection of the $i$-th row and the $k$-th column and zeros in all other places. The element $A$ transforms these matrices in the following manner:

$$E_{ik} \to AE_{ik}A^{-1}.$$

Thus, an element $\xi$ of $R$ with the coordinates $\xi_{ik}$ is transformed by the formula $\xi \to A'\xi A'^{-1}$ where $A'$ is the transpose of the matrix $A$.

The invariants of such a matrix $\xi$ are:

$$\sum_i \xi_{ii}, \sum_{i_1,i_2} \xi_{i_1i_2}\xi_{i_2i_1},\ldots, \sum_{i_1,\ldots,i_n} \xi_{i_1i_2}\xi_{i_2i_3}\ldots\xi_{i_ni_1} \tag{3}$$

(trace $\xi$, trace $\xi^2,\ldots,$trace $\xi^n$).

Any other entire rational invariant is an entire rational function of the invariants (3). The polynomials (3) correspond to the following elements of the center:

$$\sum_i E_{ii}, \sum_{i_1,i_2} E_{i_1i_2}E_{i_2i_1}\ldots \sum_{i_1,\ldots,i_n} E_{i_1i_2}E_{i_2i_3}\ldots E_{i_ni_1}.$$

Theorem 2 below implies that they generate the center of the group ring.

(2) The group $\mathfrak{G}$ of $n \times n$ orthogonal matrices.

The algebra $R$ consists of all $n \times n$ skew-symmetric matrices with the commutator $[A,B]=AB-BA$. Take for the basis in $R$ the set of matrices $F_{ik}$ $(i<k)$ which have 1 at the intersection of the $i$-th row and $k$-th column $(i<k)$, $-1$ at the intersection of the $k$-th row and $i$-th column and zeros in all other places. A transformation $A$ of the adjoint group takes $F_{ik}$ into the matrix $AF_{ik}A^{-1}$. Let $\xi_{ik}$ $(i<k)$ be coordinates of a vector in $R$ and denote by $\xi$ a skew-symmetric matrix with the entries $\xi_{ik}$ for $i<k$. A transformation of the adjoint group takes $\xi$ into $A\xi A^{-1}$.

The invariants of the matrix $\xi$ under the adjoint group transformations are

$$\sum_{i_1,i_2} \xi_{i_1i_2}\xi_{i_2i_1}, \sum_{i_1,i_2,i_3,i_4} \xi_{i_1i_2}\xi_{i_2i_3}\xi_{i_3i_4}\xi_{i_4i_1},\ldots, \sum_{i_1,\ldots,i_k} \xi_{i_1i_2}\ldots\xi_{i_{k-1}i_k}\xi_{i_ki_1}, \quad k=\left[\frac{n}{2}\right],$$

and they correspond to the following elements of the center of the ring $K$:

$$\sum_{i_1,i_2} F_{i_1i_2}F_{i_2i_1}, \sum_{i_1,i_2,i_3,i_4} F_{i_1i_2}F_{i_2i_3}F_{i_3i_4}F_{i_4i_1},\ldots, \sum_{i_1,i_2,\ldots,i_k} F_{i_1i_2},\ldots,F_{i_{k-1}i_k}F_{i_k}F_{i_1}.$$

Here we assume that $F_{ki}=-F_{ik}$ $(i<k)$. In this case it is more convenient to use another set of invariants, that is the sum $\Delta_2$ of the principal minors of order 2, the sum $\Delta_4$ of the principal minors of order 4, etc. No minors of odd order are involved since for a skew-symmetric matrix all the minors of odd order vanish. For $n$ even the last of those invariants, i.e. the determinant of the matrix, is the full square of some polynomial in the entries of the matrix:

$$\Delta_n = S^2.$$

The polynomial $S$ is also invariant but only under the proper orthogonal transformations, i.e. under the transformations of determinant $+1$. These transformations constitute a connected component of the orthogonal transformation group and for that reason $S$ is entitled to a place in the system of invariants. Then the complete system of the adjoint group invariants is $\Delta_2, \Delta_4,\ldots,\Delta_{n-2}, S$ for $n$ even and $\Delta_2, \Delta_4,\ldots,\Delta_{n-1}$ for $n$ odd. Once again, Theorem 2, which is to be proved shortly, implies that the corresponding elements of the ring $K$ generate the whole of the ring $K$.

3. Consider the ring of all polynomials in $\xi_1, \ldots, \xi_n$ (where $\xi_i$ are coordinates of some vector in $\bar{R}$) invariant under the adjoint group transformations. If the coefficients of such a polynomial satisfy the symmetry condition then the corresponding element of the ring $K$ lies in the center of $K$. Conversely, for each element $P$ of the center there is a unique invariant polynomial which is obtained by writing $P$ in such a form that its coefficients are symmetric.

Let us now show that the generators of the ring of invariant polynomials correspond to the elements of the ring $K$ which generate its center. Indeed, let

$$P = \sum_{i_1, i_2, \ldots, i_k} a^{i_1 i_2 \ldots i_k} e_{i_1} \ldots e_{i_k}$$

and

$$Q = \sum_{i_1, i_2, \ldots, i_l} b^{i_1 \ldots i_l} e_{i_1} \ldots e_{i_l}$$

be two elements of the ring $K$. Their direct multiplication yields an element of the ring $K$, the coefficients of which are not symmetric. To make them symmetric we use the relations $e_i e_k = e_k e_i + \sum_j c^j_{ik} e_j$. But these do not affect the coefficients of the leading terms Thus the product of two polynomials corresponds to the product of the respective elements of the ring $K$ modulo terms of lesser degree. Simple inductive reasoning now shows that the generators of the ring of invariant polynomials correspond to the generators of the center of the ring $K$. Therefore, we have prove the following theorem.

**Theorem 2.** *Denote by $Z$ the ring of polynomials in the coordinates $\xi_i$ of a vector from $\bar{R}$ invariant under the adjoint group transformations. Write each of the polynomials in such a form that its coefficients are symmetric. Now, for each polynomial*

$$\sum_{i_1, i_2, \ldots, i_k} a^{i_1 i_2 \ldots i_k} \xi_{i_1} \ldots \xi_{i_k}$$

*consider the element*

$$\sum_{i_1, i_2, \ldots, i_k} a^{i_1 i_2 \ldots i_k} e_{i_1} \ldots e_{i_k}$$

*of the ring $K$. The set of all such elements constitutes the center of the ring $K$. Under this correspondence the generators of the ring $Z$ correspond to the generators of the ring $K$.*

The theorem implies that the elements of the center constructed in Examples (1) and (2) generate the center.

4. Consider now some applications of the above results.

Let $\mathfrak{G}$ be a Lie group and $g \to U_g$ its finite-dimensional representation which sends each element $g$ into a linear transformation $U_g$ in a finite-dimensional linear space $M$.

Each element $a$ of the Lie algebra $R$ is then uniquely represented by some linear transformation $A_a$ and under this correspondence the element $[a, b]$ goes into the linear transformation $A_a A_b - A_b A_a$ and the element $\lambda a + \mu b$ goes into the transformation $\lambda A_a + \mu A_b$.

One can extend any representation of the Lie algebra to the representation of the ring $K$ by representing each polynomial $P(e_1,\dots,e_n)$ by the transformation $P(A_{e_1},\dots,A_{e_n})$. Conversely, each representation of the ring $K$ is also a representation of the algebra $R$ and, consequently, it corresponds to some representation of the group $\mathfrak{G}$. We note here that the ring $K$ could naturally be called "infinitesimal group ring of the group $\mathfrak{G}$". One can, therefore, instead of considering the representations of the group, study representations of the ring $K$, i.e. of the "infinitesimal group ring of the group $\mathfrak{G}$".

Consider a representation of the ring $K$ which is a multiple of an irreducible representation (i.e. a direct sum of equivalent irreducible representations). The Schur lemma then implies that elements of the center of the ring $K$ are represented by the operators which are scalar multiples of the identity. When $\mathfrak{G}$ is a semisimple group the reverse statement is also true, that is, if in a finite-dimensional representation the elements of the center of the ring $K$ go into scalar multiples of the identity then the representation is a multiple of an irreducible representation. Indeed, each finite-dimensional representation of a semisimple group is completely reducible; on the other hand one can show that for a non-equivalent irreducible representation of a semisimple group the corresponding homomorphisms of the center are different. Therefore the generators of the center go into scalar multiples of the identity only if the representation is a multiple of an irreducible representation.

As an example, consider splitting the tensor representation of the linear group into representations which are multiples of irreducible representations.

Let $M$ be a linear space of covariant tensors $x_{i_1 i_2 \dots i_f}$ of rank $f$. Each matrix $a_i^k$ of the linear group transforms these tensors by the formula

$$x'_{i_1\dots i_f} = a_{i_1}^{k_1} a_{i_2}^{k_2} \dots a_{i_f}^{k_f} x_{k_1} \dots {}_{k_f}.$$

If $c_i^k$ is a matrix which defines the element $c$ of the infinitesimal group then it corresponds to the following tensor transformation:

$$x'_{i_1\dots i_f} = x_{i_1\dots i_{f-1}k_f} c_{i_f}^{k_f} + x_{i_1\dots i_{f-2}k_{f-1}i_f} c_{i_{f-1}}^{k_{f-1}} + \dots .$$

Thus, an element $c_i^k$ of the Lie algebra corresponds to the transformation in the space of tensors which has the matrix

$$c(i_1,\dots,i_f;k_1,\dots,k_f) = \delta_{i_1}^{k_1}\delta_{i_2}^{k_2}\dots\delta_{i_{f-1}}^{k_{f-1}} c_{i_f}^{k_f} + \delta_{i_1}^{k_1}\delta_{i_2}^{k_2}\dots c_{i_{f-1}}^{k_{f-1}}\delta_{i_f}^{k_f} + \dots .$$

Let us now compute the transformations which correspond to the generators $\sum_i E_{ii}$, $\sum_{i,k} E_{ik}E_{ki}$, $\sum_{i,k,l} E_{ik}E_{kl}E_{li}, \dots$ of the center of the group ring. One has

$$\sum_i E_{ii}(i_1,\dots,i_f;k_1,\dots,k_f) = \sum_i \delta_{i_1}^{k_1}\delta_{i_2}^{k_2}\dots\delta_{i_{f-1}}^{k_{f-1}} c_{ii;i_f}^{k_f} + \sum_i \delta_{i_1}^{k_1}\delta_{i_2}^{k_2}\dots c_{ii;i_{f-1}}^{k_{f-1}}\delta_{i_f}^{k_f} + \dots$$

(here $c_{ik;i_p}^{k_p}$ is a matrix of the element $E_{ik}$, i.e. $c_{ik;i_p}^{k_p} = 1$ for $i_p = i$, $k_p = k$ and $c_{ik;i_p}^{k_p} = 0$ otherwise).

Thus

$$\sum_i E_{ii}(i_1,\dots,i_f;k_1,\dots,k_f) = n\delta_{i_1 i_2\dots i_f}^{k_1 k_2\dots k_f}.$$

Now

$$\sum_{i,k,l} E_{ik}(i_1,\ldots,i_f;l_1,\ldots,l_f)E_{ki}(l_1,\ldots,l_f;k_1,\ldots,k_f)$$
$$= \sum_{i,k,p,q,l} {}^{l_1\ldots l_{p-1}\; l_{p+1}\ldots l_f}_{i_1\ldots i_{p-1}\; i_{p+1}\ldots i_f}\, {}^{k_1\ldots k_{q-1}k_{q+1}\ldots k_f}_{l_1\ldots l_{q-1}l_{q+1}\ldots l_f}\, c^{l_p}_{ik;i_p}\, c^{k_q}_{ki;i_q}$$

and since

$$\sum_{i,k} c^{l_p}_{ik;i_p} c^{k_p}_{ki;l_p} = \delta^{k_q}_{l_p}\delta^{l_p}_{l_q},$$

one has

$$\sum_{i,k} E_{ik}E_{ki} = \sum_{p,q,l} \delta^{l_1\ldots l_{p-1}l_{p+1}\ldots l_f}_{i_1\ldots i_{p-1}i_{p+1}\ldots i_f}\,\delta^{k_1\ldots k_{q-1}k_{q+1}\ldots k_f}_{l_1\ldots l_{q-1}l_{q+1}\ldots l_f}\,\delta^{l_p}_{l_q}\delta^{k_q}_{i_p}$$
$$= \sum_{p,q} \delta^{k_1\ldots k_{p-1}k_{p+1}\ldots k_{q-1}k_{q+1}\ldots k_f}_{i_1\ldots i_{p-1}i_{p+1}\ldots i_{q-1}i_{q+1}\ldots i_f}\,\delta^{k_q}_{i_p}\delta^{k_p}_{i_q}$$

because

$$\sum_{l} \delta^{l_q}_{i_q}\delta^{k_p}_{l_p}\delta^{l_p}_{l_q} = \delta^{k_p}_{i_q}.$$

Thus the operator $\sum_{i,k} E_{ik}E_{ki}$ acts on the tensor $x_{i_1\ldots i_f}$ in the following way: it permutes the indices in those places whose numbers are $p$ and $q$ (i.e. $i_p$ and $i_q$) for all $p$ and $q$ (including $p=q$) and then sums up the results.

The operator $\Delta_3 = \sum_{i,k,j} E_{ik}E_{kj}E_{ji}$ is computed in the same way:

$$\Delta_3 = \sum_{i,k,j,l,m} E_{ik}(i_1,\ldots,i_f;l_1,\ldots,l_f)E_{kj}(l_1,\ldots,l_f;m_1,\ldots,m_f)\,E_{ji}(m_1,\ldots,m_f;k_1,\ldots,k_f)$$
$$= \sum_{i,k,j,l,m,p,q,r} c^{l_p}_{ik;i_p} c^{m_q}_{kj;l_q} c^{k_r}_{ji;m_r}\, \underset{p}{\delta'^{l_1\ldots}_{i_1\ldots}}\, \underset{q}{\delta'^{m_1\ldots}_{l_1\ldots}}\, \underset{r}{\delta'^{k_1\ldots}_{m_1\ldots}}$$

where $\underset{p}{\delta'^{l_1\ldots}_{i_1\ldots}}$ means that the indices $i_p$ and $l_p$ are dropped and, similarly, $\underset{q}{\delta'^{m_1\ldots}_{l_1\ldots}}$ means that the dropped indices are $l_q$ and $m_q$.

Now

$$\sum_{i,k,j} c^{l_p}_{ik;i_p} c^{m_q}_{kj;l_q} c^{k_r}_{ji;m_r} = \sum_{i,k,j} \delta^{i}_{i_p}\delta^{l_p}_{k}\delta^{k}_{l_q}\delta^{m_q}_{j}\delta^{k_r}_{i}\delta^{j}_{m_r} = \delta^{k_r}_{i_p}\delta^{l_p}_{l_q}\delta^{m_q}_{m_r}$$

and we finally get

$$\Delta_3 = \sum_{p,q,r} \underset{p,q,r}{\delta'}{}^{k_1\ldots k_f}_{i_1\ldots i_f}\,\delta^{k_r}_{i_p}\delta^{k_p}_{i_q}\delta^{k_q}_{i_r}$$

where $\underset{p,q,r}{\delta'}{}^{k_1\ldots}_{i_1\ldots}$ means that the indices $i_p, i_q, i_r$ and $k_p, k_q, k_r$ are dropped.

Thus, the operator $\Delta_3$ acts on the tensor $x_{i_1\ldots i_f}$ in the following way: one takes all possible triples $p, q, r$ including those which comprise the same indices but in a different order and those for which some of the integers $p, q, r$ coincide; for each triple $p, q, r$ we consider a new tensor obtained by cyclic transmutation of the indices $i_p, i_q, i_r$ and then sum up the results for all possible triples $p, q, r$.

The operators $\Delta_4, \ldots, \Delta_n$ are computed in the same manner.

It follows from the above that to split a tensor representation into multiples of irreducible representations one has to equate each of the operators $\Delta_k$ to $\lambda_k E$

where $E$ is an identity operator. The resulting decomposition of the representation space provides the desired splitting of the representation. It should be noted that if the rank of the tensors is $f$ then to obtain the decomposition it is enough to use the operators only up to $\Delta_f$ inclusive.

The resulting decomposition of the tensor into components which are multiples of irreducible representations is unique and in that respect it differs from the decomposition of tensors into irreducible components. The latter is not unique because, starting with tensors of rank three, some irreducible representations they are decomposed into appear more than once.

Finally, we consider the decomposition of tensors which include both covariant and contravariant indices.

The linear transformation $a_i^k$ sends the tensor $x_{i_1\ldots i_f}^{i_{f+1}\ldots i_g}$ into the tensor

$$x'^{i_{f+1}\ldots i_g}_{i_1\ldots i_f} = x^{k_{f+1}\ldots k_g}_{k_1\ldots k_f} a^{k_1}_{i_1}\ldots a^{k_f}_{i_f} b^{i_{f+1}}_{k_{f+1}}\ldots b^{i_g}_{k_g}$$

where $a_i^l b_l^k = c_i^k$. Therefore, the matrix $c_i^k$ of the infinitesimal group corresponds to the following linear transformation of the tensors:

$$x'^{i_{f+1}\ldots i_g}_{i_1\ldots i_f} = c^{k_1}_{i_1} x^{i_{f+1}\ldots i_g}_{k_1 i_2\ldots i_f} + \ldots + c^{k_f}_{i_f} x^{i_{f+1}\ldots i_g}_{i_1 i_2\ldots k_f} - c^{i_{f+1}}_{k_{f+1}} x^{k_{f+1}\ldots i_g}_{i_1\ldots i_f} - \ldots .$$

Let us compute the operator $\Delta_2 = \sum_{i,k} E_{ik}E_{ki}$. One has

$$\begin{aligned}\sum_{i,k} E_{ik}E_{ki} = {} & \sum c^{l_p}_{ik i_p} c^{k_q}_{ki;l_q} \underset{p}{\delta'}{}^{l_1\ldots l_f}_{i_1\ldots i_f} \underset{q}{\delta'}{}^{k_1\ldots k_f}_{l_1\ldots l_f} \delta^{i_{f+1}\ldots}_{k_{f+1}\ldots} \\ & + \sum c^{i_p}_{ik;l_q} c^{l_q}_{ki;k_q} \delta^{k_1\ldots k_f}_{i_1\ldots i_f} \underset{p}{\delta'}{}^{i_{f+1}\ldots i_g}_{l_{f+1}\ldots l_g} \underset{q}{\delta'}{}^{l_{f+1}\ldots l_g}_{k_{f+1}\ldots k_g} \\ & - \sum c^{l_p}_{ik;i_p} c^{l_q}_{ki;k_q} \underset{p}{\delta'}{}^{l_1\ldots l_f}_{i_1\ldots i_f} \delta^{k_1\ldots k_f}_{l_1\ldots l_f} \delta^{i_{f+1}\ldots i_g}_{l_{f+1}\ldots l_g} \underset{q}{\delta'}{}^{l_{f+1}\ldots l_g}_{k_{f+1}\ldots k_g} - \sum \ldots\end{aligned}$$

where the notation is the same as above. The last sum is quite similar to the third and is therefore omitted.

The computations similar to those in the previous case give the following rule for computing the action of the operatot $\Delta_2$ on the tensor $x^{i_{f+1}\ldots i_g}_{i_1\ldots i_f}$: take all possible pairs of integers $p, q$ (including those which differ only in the order of $p$ and $q$ and those for which $p$ and $q$ coincide); if the indices $i_p$ and $i_q$ are both of the same type, they are simply permuted; otherwise we replace the tensor by another one:

$$-x^{i_{f+1}\ldots i_{q-1}\alpha i_{q+1}\ldots i_g}_{i_1\ldots i_{p-1}\alpha i_{p+1}\ldots i_f} \delta^{i_q}_{i_p}$$

(the tensor is contracted over $\alpha$); finally, the results are summed up for all possible $p$ and $q$.

One can show that general rule for computing the tensors ${}_k x^{i_{f+1}\ldots i_g}_{i_1\ldots i_f}$ can be formulated as follows (we give it here for $\Delta_5$ for the sake of clarity).

Suppose we are given a tensor $x^{i_{f+1}\ldots i_g}_{i_1\ldots i_f}$. To compute the tensor

$$\Delta_5 x^{i_{f+1}\ldots i_g}_{i_1\ldots i_f}$$

take all possible combinations of five numbers $p, q, r, s, t$ (including those which differ only in the order of the numbers and those which include numbers occurring more than once). Then we take all cyclic permutations in the tensor $x^{i_{f+1}\ldots i_g}_{i_1\ldots i_f}$ replacing $i_p$ by $i_q$, $i_q$ by $i_r$, $i_r$ by $i_s$, and so on, provided all the permuted indices are of the same type. If there are indices of different type, for example, if $i_p$ is a lower index and $i_q$ is an upper index, then we contract the tensor over them and then multiply it by $\delta_{i_p}^{i_q}$. The resulting tensor is multiplied by $(-1)^v$ where $v$ is the number of upper indices among the indices $i_p, i_q, i_r, i_s, i_t$. Summing the results over all possible combinations of five indices yields the result, i.e.

$$\Delta_5 x_{i_1 \ldots i_f}^{i_{f+1} \ldots i_g}.$$

Decomposition of the tensor representation of the orthogonal group can be done in a similar way.

Received July 3, 1948

**References**

1. Casimir, H., van der Waerden, B.L.: Algebraischer Beweis der vollständigen Reducibilität der Darstellungen halbeinfacher Liescher Gruppen. Math. Ann. **111** (1935) 1–12.

# 4.

# Spherical functions on symmetric Riemannian spaces

Dokl. Akad. Nauk SSSR **70**, (1950) 5–8
[Transl., II. Ser., Am. Math. Soc. **37** (1964) 39–43]. Zbl. **38**:274

1. Let $\mathfrak{G}$ be a Lie group, and $\mathfrak{H}$ a compact subgroup of $\mathfrak{G}$. We consider the manifold whose elements are the left cosets of $\mathfrak{H}$ in $\mathfrak{G}$. We may regard $\mathfrak{G}$ as a group of transformations acting on $X$. Namely, if $x \in X$ is a coset $\mathfrak{H}g_0$, then by $y = xg$ we shall mean the coset $\mathfrak{H}g_0 g$.

Let $R$ be the set of all summable functions $f(g)$ satisfying the condition $f(h_1 g h_2) = f(g)$ for almost all $g$ and for all $h_1, h_2 \in \mathfrak{H}$ (i.e., functions which are constant on double cosets of $\mathfrak{H}$). We define the product of such functions by the formula $f = f_1 \times f_2$, where $f(g) = \int f_1(gg_1^{-1}) f_2(g_1)\, dg_1$. If we define the addition of such functions as usual and put $\|f\| = \int |f(g)|\, dg$, then $R$ is a normed ring. If we introduce in $R$ the involution $f^*(g) = \overline{f(g^{-1})}$, then $R$ becomes a normed ring with involution [1]. Also, as in [2], it can be shown that $R$ is semisimple. A linear functional $L(f)$ on $R$ is called positive if $L(f^* f) \geq 0$ for all $f \in R$. A positive functional is given by $L(f) = \int f(g) \phi(g)\, dg$, where $\phi(g)$ is a positive definite function satisfying the condition

$$\phi(h_1 g h_2 \equiv \phi(g) \text{ for arbitrary } h_1, h_2 \in \mathfrak{H}, \tag{1}$$

i.e., it is constant on double cosets of $\mathfrak{H}$.

*If a positive definite function $\phi(g)$, satisfying* (1), *cannot be expressed in the form $\phi(g) = \phi_1(g) + \phi_2(g)$, where the $\phi_i(g)$ are positive definite, satisfy* (1), *and are linearly independent, then such a decomposition is impossible even if the $\phi_i(g)$ do not satisfy* (1), *i.e., $\phi(g)$ is an elementary positive definite function.*

Let us consider the unitary representations corresponding to an elementary positive definite function [3, 1]. *The representations corresponding to an elementary function $\phi(g)$ satisfying* (1) *are characterized by the existence of a vector which is invariant with respect to the transformations corresponding to elements $h \in \mathfrak{H}$.*

The functions $\phi(g)$ may be transferred at will to the manifold $X$. Indeed, let $x_1, x_2$ be the cosets $\mathfrak{H}g_1, \mathfrak{H}g_2$, respectively. Put $K(x_1, x_2) = \phi(g_2^{-1} g_1)$. From the equation $\phi(h_1 g h_2) = \phi(g)$ it follows that $K(x_1, x_2)$ does not depend on the choice of representatives from the coset.

The kernel $K(x_1, x_2)$, corresponding to the function $\phi(g)$, satisfies the following conditions: (1°) $K(x_1 g, x_2 g) = K(x_1, x_2)$, and (2°) $K(x_1, x_2)$ is a positive definite kernel. The converse assertion is also true.

If we fix some point $x_0$ and put $K(x_0, x) = K(x)$, then the function $K(x)$

satisfies the condition $K(xg_1) = K(x)$ for all transformations leaving $x_0$ fixed, i.e., $K(x)$ is constant on "spheres" with center $x_0$.

The functions $K(x)$ (or $K(x_1, x_2)$) which correspond to elementary positive definite functions will be called *zonal spherical functions*.

**2. Positive definite functions associated with a symmetric Riemannian space.** In this section we shall assume that there exists an involutory automorphism of the group $\mathfrak{G}$ under which the elements of $\mathfrak{H}$ ($\mathfrak{H}$ compact) are left fixed (Condition A). Then $X$ may be considered as a symmetric Riemannian space with the transformations $\mathfrak{G}$ acting as its motions.

**Theorem 1.** *If Condition A is satisfied, then the ring $R$ is commutative. In general, commutativity of the ring $R$ is equivalent to the condition that for each continuous unitary representation of the group $\mathfrak{G}$, the subspace of vectors invariant under all transformations from $\mathfrak{H}$ is at most 1-dimensional.*

Let $\mathfrak{M}$ be the set of all two-sided maximal ideals of the ring $R$.[1)] The homomorphism corresponding to the two-sided maximal ideal $M$ has the form $f \longrightarrow \int_{\mathfrak{G}} f(g)\, \phi_M(g)\, dg$, where $\phi_M(g)$ is an elementary positive definite function satisfying (1). Applying a theorem on the decomposition of positive functionals into elementary ones, [1], we obtain:

**Theorem 2.** *If Condition A is satisfied, then every positive definite function, satisfying $\phi(h_1 g h_1) = \phi(g)$ for $h_1, h_2 \in \mathfrak{H}$, can be represented in the form*

$$\phi(g) = \int_{\mathfrak{M}} \phi_M(g)\, d\sigma(M),$$

*where $\sigma$ is some nonnegative set function on $\mathfrak{M}$.*

This theorem can be reformulated in the following way:

**Theorem 2′.** *Let $K(x_1, x_2)$ be a positive definite form defined on a symmetric Riemannian space and invariant with respect to the motions, i.e., $K(x_1 g, x_2 g) = K(x_1, x_2)$. Then $K$ has a representation of the form*

$$K(x_1, x_2) = \int_{\mathfrak{M}} K_M(x_1, x_2)\, d\sigma(M),$$

*where $K_M(x_1, x_2)$ is a zonal spherical function and $\sigma$ is a nonnegative set function on $\mathfrak{M}$.*

**3. Products of spherical functions. Mean-value theorem.** Denote by $S$ the collection of double cosets of $\mathfrak{H}$ or, otherwise described, the collection of all "spheres," where "spheres" are considered equivalent if they can be interchanged

---

1) The ring $R$ may possess maximal ideals which are not two-sided. See [1], p. 476, where the corresponding ring is examined in the case of Lobačevskian space.

by some motion. It is clear that the dimension of $S$ is equal to the number of independent invariants of pairs of points in $X$.[2)] We transfer to $S$ the existing measure on $\mathfrak{G}$.

We may consider each function $f(g) \in R$ as a function $f(s)$ defined on $S$. Thus the multiplication in $R$ is given by the formula $f = f_1 \times f_2$, where

$$f(s) = \int_S f_1(s_1) f_2(s_2)\, a(s_1, s_2, s)\, ds_1\, ds_2,$$

the integration being carried out with respect to the measure existing on $S$.

The function $a(s_1, s_2, s)$ has the following geometric interpretation. Let us consider points $x_1, x_2 \in X$, each of which lies on a sphere $s_1$ with center at the point $x_1$, and denote by $P_1$ the set of all points belonging to spheres with center $x_1$ and sufficiently close to $s_1$. Analogously, let $P_2$ be the set of all points on spheres with center $x_2$ and sufficiently close to $s_2$. Then $a(s_1, s_2, s) = \lim [\mu(P_1 \cap P_2)/\mu(P_1)\mu(P_2)]$, where $\mu$ is the measure on $X$.

Since the zonal functions are constant on "spheres," let us denote them by $\phi_M(s)$. From the rule of multiplication in $R$ we obtain the following formula:

$$\phi_M(s_1)\phi_M(s_2) = \int_S a(s_1, s_2, s)\phi_M(s)\, ds. \tag{2}$$

Thus we obtain

Theorem 3. *For the zonal spherical functions, multiplication is given by formula* (2).

A mean-value theorem is valid for the spherical functions; namely, if we have a sphere $s$ with center at $x_1$, then the integral over $s$ of the function $K_M(x)$ is equal to $\phi_M(s) K_M(x_1)$. Actually, this theorem is true for the nonzonal spherical functions, but for lack of space we do not give the proof. For zonal functions the mean-value theorem coincides with the law of multiplication and gives another definition of the function $a(s_1, s_2, s)$.

4. Differential equations of spherical functions. Let $e_1, e_2, \cdots, e_n$ be a basis of the infinitesimal Lie group (Lie algebra), and let $E_1, E_2, \cdots, E_n$ be the corresponding differential operators (infinitesimal transformations) on $X$. As is well known, we may consider the group $\mathfrak{G}$ as a group of linear transformations of the space (the adjoint group). Let $K$ be the ring whose elements are formal polynomials in $e_1, \cdots, e_n$. There we regard two polynomials as equal if one can be obtained from the other by a finite sequence of substitutions $e_i e_k - e_k e_i$ for

---

[2)] Translator's note: "invariants of pairs of points" means functions of pairs of points which do not change under the motions of the group. Cf. pp. 195–196 of Amer. Math. Soc. Transl. (2) 21 (1962).

$[e_i, e_k] = \Sigma c_{ik}^j e_j$. We shall now determine the center of the ring $K$, i.e., the polynomials $P(e_1, \cdots, e_n)$ which commute with every $e_i$.

Theorem 4. *In order that a polynomial* $P = aI + \Sigma a^i e_i + \Sigma a^{ik} e_i e_j + \cdots$, *belonging to* $K$, *commute with every* $e_i$, *it is sufficient that the forms* $\Sigma a^i \xi_i$, $\Sigma a^{ik} \xi_i \eta_k, \cdots$ *be invariants of the adjoint group. If* $P$ *can be written in such a way that the coefficients* $a^{ik}$, $a^{ikl}, \cdots$ *are symmetric, then this condition is also necessary.*

From this we obtain the following: if in each $P(e_1, \cdots, e_n)$ we substitute for $e_1, \cdots, e_n$ the differential Lie operators $E_1, \cdots, E_n$, then we obtain differential operators commuting with all transformations $g \in \mathfrak{G}$ of the manifold $X$ (the motions on $X$).

Consider some examples. Let $\mathfrak{G}_2$ be the group of all real nonsingular matrices. The $e_i$ become the matrices $E_{ik}$ which have 1 as the $ik$th entry and 0 elsewhere. Applying Theorem 4, we have

Theorem 5. *In the group of all real nonsingular matrices, the following polynomials in the Lie operators* $E_{ik}$ *commute with all* $E_{ik}$: $\Delta_1 = \Sigma E_{ii}$, $\Delta_2 = \sum_{i_1, i_2} E_{i_1 i_2} E_{i_2 i_1}$, $\Delta_3 = \sum_{i_1, i_2, i_3} E_{i_1 i_2} E_{i_2 i_3}, \cdots$, $\Delta_n = \Sigma E_{i_1 i_2} \cdots E_{i_n i_1}$. *Every other polynomial commuting with all the* $E_{ik}$ *is a polynomial in* $\Delta_1, \cdots, \Delta_n$.

In the case of the group of matrices with determinant 1, $\Delta_1$ is missing. If $\mathfrak{G}$ is the group of $2 \times 2$ matrices of determinant 1, we may interpret it as a group of motions of the Lobačevskian plane. In general, in the case of a symmetric Riemannian space, among our operators is a second-order operator, the Laplacian (the second differential parameter of Beltrami) of the space. In the same way as for the unimodular group, one can easily obtain the associated operators for other semisimple groups. We remark that the number of independent operators on a symmetric Riemannian space is equal to the number of independent invariants of pairs of points. If the group $\mathfrak{G}$ is semisimple, then among the $P(E_1, \cdots, E_n)$ which commute with all motions there is a finite set such that all remaining ones are polynomials in this finite set (the number of independent ones is equal to the number of independent invariants of pairs of points in $X$). We shall denote these independent elements by $\Delta^{(1)}, \cdots, \Delta^{(n)}$, and obtain the Laplace operators of the symmetric space $X$.

Theorem 6. *The zonal spherical functions satisfy the equation* $\Delta^{(i)} \phi_M(x) = \lambda_i \phi_M(x)$, *where* $\lambda_i$ *is a scalar.*

Thus the zonal spherical functions may be characterized as eigenfunctions of the Laplace operator which are constant on spheres with center at some point.

Theorem 6 is obtained as a special case of the following theorem:

**Theorem** 7. *Suppose we have a continuous unitary representation* $g \longrightarrow Tg$ *of the group* $\mathfrak{G}$ *into the Hilbert space* $\mathfrak{H}$. *With each element* $e_i$ *of the Lie algebra there is associated an operator* $A_i$ *(generally unbounded) in* $\mathfrak{H}$ [4]. *If* $P(e_1, \cdots, e_n) \in K$ *commutes with each* $e_i$, *then* $P(A_1, \cdots, A_n)$ *is a multiple of the identity.*

In the case where $\mathfrak{G}$ is a complex semisimple Lie group and $\mathfrak{H}$ is a maximal compact subgroup, the spherical functions are explicitly calculated in [5].

The questions treated in this paper were considered also by M. G. Kreĭn, whose results partly overlap with the results presented here.

## BIBLIOGRAPHY

[1] I. M. Gel'fand and M. A. Naĭmark, *Normed rings with involutions and their representations*, Izv. Akad. Nauk SSSR Ser. Mat. 12 (1948), 445–480. (Russian) MR 10, 199.

[2] D. A. Raĭkov, *To the theory of normed rings with involution*, Dokl. Akad. Nauk SSSR 54 (1946), 387–390. (Russian) MR 8, 469.

[3] I. M. Gel'fand and D. A. Raĭkov, *Irreducible unitary representations of locally bicompact groups*, Mat. Sb. (N. S.) 13 (55) (1943), 301–316; English transl., Amer. Math. Soc. Transl. (2) 36 (1964), 1–15. MR 6, 147.

[4] L. Gårding, *Note on continuous representations of Lie groups*, Proc. Nat. Acad. Sci. U. S. A. 33 (1947), 331–332. MR 9, 133.

[5] I. M. Gel'fand and M. A. Naĭmark, *Supplementary and degenerate series of representations of the complex unimodular group*, Dokl. Akad. Nauk SSSR 58 (1947), 1577–1580. (Russian) MR 9, 329.

Translated by:
Stephen Puckette

# Part II

# Infinite-dimensional representations of semisimple Lie groups

# 1.

## (with M. A. Najmark)

# Unitary representations of the Lorentz group

J. Phys., Acad. Sci. USSR **10** (1946) 93–94. Zbl. **61**:253

**The usual (finite-dimensional) representations of the Lorentz group do not leave invariant any positive definite form, *i. e.*, are not unitary. In this note there are indicated all unitary (hence infinite-dimensional) irreducible representations of the Lorentz group.**

In connection with general investigations of the authors on the theory of group representations they have considered as an example such representations of the Lorentz group, which leave invariant a positive definite quadratic form.

As the question seems to be of interest for physics and has been discussed already*, we shall present here our results on this object. The connection of these results with those of Dirac will be considered in another note.

In the usual representations of the Lorentz group the quantities to be transformed (*e. g.* vectors, tensor) depend on subscripts assuming each a finite number of values, *i. e.* these representations are finite-dimensional. It can be proved, that none of finite-dimensional representations leaves any positive definite quadratic form invariant.

Hence the representations leaving a positive quadratic form invariant must be infinite-dimensional, *i. e.* the quantities in these representations depend on subscripts assuming an infinite number of values.

We shall indicate here all irreducible unitary representations of the Lorentz group. They are of special interest because every unitary representation of the Lorentz group is in a certain sense decomposable in a sum of irreducible representations.

These irreducible representations will be given here in such a form, that the quantities to be transformed depend on two subscripts $x$, $y$ assuming each all real values. Then these quantities $f_{x,y}$ can be considered as functions $f(x, y)$ of two real variables or putting $z = x + iy$ as functions $f(z)$ of a complex variable $z$.

It is known that every Lorentz transformation can be given by a matrix $\begin{pmatrix} \alpha & \beta \\ \gamma & \delta \end{pmatrix}$, where $\alpha, \beta, \gamma, \delta$ are complex numbers satisfying the condition $\alpha\delta - \beta\gamma = 1$. The matrix of the corresponding Lorentz transformation can be written in terms of $\alpha, \beta, \gamma, \delta$ in the following form:

$$\begin{pmatrix} R(\alpha\bar{\delta} + \gamma\bar{\beta}), & -I(\alpha\bar{\delta} - \gamma\bar{\beta}), & R(\alpha\bar{\beta} - \gamma\bar{\delta}), & R(\alpha\bar{\beta} + \gamma\bar{\delta}) \\ I(\alpha\bar{\delta} + \gamma\bar{\beta}), & R(\alpha\bar{\delta} - \gamma\bar{\beta}), & I(\alpha\bar{\beta} - \gamma\bar{\delta}), & I(\alpha\bar{\beta} + \gamma\bar{\delta}) \\ R(\alpha\bar{\gamma} - \beta\bar{\delta}), & -I(\alpha\bar{\gamma} - \beta\bar{\delta}), & \frac{\alpha\bar{\alpha} - \beta\bar{\beta} - \gamma\bar{\gamma} + \delta\bar{\delta}}{2}, & \frac{\alpha\bar{\alpha} - \beta\bar{\beta} + \gamma\bar{\gamma} - \delta\bar{\delta}}{2} \\ R(\alpha\bar{\gamma} + \beta\bar{\delta}), & -I(\alpha\bar{\gamma} + \beta\bar{\delta}), & \frac{\alpha\bar{\alpha} + \beta\bar{\beta} - \gamma\bar{\gamma} - \delta\bar{\delta}}{2}, & \frac{\alpha\bar{\alpha} + \beta\bar{\beta} + \gamma\bar{\gamma} + \delta\bar{\delta}}{2} \end{pmatrix},$$

* P. A. M. Dirac, Proc. Roy. Soc., A, **183**, 284 (1945).

The matrices $\begin{pmatrix}\alpha & \beta\\ \gamma & \delta\end{pmatrix}$ and $\begin{pmatrix}-\alpha & -\beta\\ -\gamma & -\delta\end{pmatrix}$ define the same Lorentz transformation.

All irreducible representations of the Lorentz group can be divided into two classes.

The representations of the first class are described as follows:

*If a Lorentz transformation is given by a matrix* $\begin{pmatrix}\alpha & \beta\\ \gamma & \delta\end{pmatrix}$, *then in a representation of the first class the quantity* $f(z)$ *transforms into*

$$|\beta z+\delta|^{-2+n+i\rho}(\beta z+\delta)^{-n} f\left(\frac{\alpha z+\gamma}{\beta z+\delta}\right).$$

*The positive form which is invariant under these transformations is* $\int\limits_{-\infty}^{\infty}\int\limits_{-\infty}^{\infty}|f(x+iy)|^2dxdy$ *so that the quantities are functions* $f(z)$ *satisfying the condition*

$$\int\limits_{-\infty}^{\infty}\int\limits_{-\infty}^{\infty}|f(x+iy)|^2\,dx\,dy<+\infty.$$

Here $n$ is a fixed integer and $\rho$ — a fixed real number. Every number pair $(n, \rho)$ defines a representation. The representations corresponding to $(n, \rho)$ and $(-n, -\rho)$ are equivalent. All other representations of this class are non-equivalent. For even $n$ these representations are single-valued and for odd $n$ — two-valued.

The representations of the second class are described as follows:

*If a Lorentz transformation is given by a matrix* $\begin{pmatrix}\alpha & \beta\\ \gamma & \delta\end{pmatrix}$, *then in the representation of the second class the quantity* $f(z)$ *transforms into* $|\beta z+\delta|^{-2-\rho} f\left(\frac{\alpha z+\gamma}{\beta z+\delta}\right)$. *The positive form which is invariant under these transformations is*

$$\iiiint\limits_{-\infty}^{\infty}|z_1-z_2|^{-2+\rho} f(z_1)\overline{f(z_2)}\,dx_1dy_1dx_2dy_2 \quad (z_1=x_1+iy_1,\ z_2=x_2+iy_2),$$

*so that the quantities are functions* $f(z)$ *satisfying the condition*

$$\iiiint\limits_{-\infty}^{\infty}|z_1-z_2|^{-2+\rho} f(z_1)\overline{f(z_2)}\,dx_1dy_1dx_2dy_2<+\infty \quad (z_1=x_1+iy_1,\ z_2=x_2+iy_2).$$

Here $\rho$ is a fixed real number satisfying the condition $0<\rho<2$. Every such number $\rho$ defines a representation of the second class. All such representations are single-valued and for distinct $\rho$ they all are non-equivalent. The representations of the second class join in a natural manner to the representations of the first class for $n=0$.

The representations here described can also be given in another form which seems to be useful for applications. This form can be obtained in the following manner. Let $\xi_1, \xi_2, \xi_3$ be the coordinates of the three-dimensional space; put further $\xi_0=ct$ so that $\xi_0, \xi_1, \xi_2, \xi_3$ can be considered as the coordinates of the space-time. Let further $\xi_0, \xi_1, \xi_2, \xi_3$ satisfy the condition $\xi_0^2-\xi_1^2-\xi_2^2-\xi_3^2=0$, which is the equation of the light cone. If we put now $z=x+iy=\frac{\xi_1+i\xi_2}{\xi_0-\xi_3}$, we obtain a correspondence between the numbers $z$ and the straight lines of the light cone. Hence for the subscript of the quantities $f(z)$ can also be taken the variable straight line of the light cone, so that these quantities can be treated as functions of this straight line. All our formulae for the representations can then be rewritten in these new variables. A more detailed exposition of this question will be given in another note.

The representations here considered but in an infinitesimal form were obtained before by the first of the authors (*I. G.*) and reported in spring of 1944 in the Institute for Physical Problems of the Academy of Sciences of the USSR. The representations in the infinitesimal form and other allied topics will be considered in another paper.

Detailed proofs of the results here stated will be published in one of the mathematical journals.

# 2.

(with M. A. Najmark)

# Unitary representations of the Lorentz group

Izv. Akad. Nauk SSSR, Ser. Mat. **11** (1947) 411–504. Zbl. **37**:153

In this paper all unitary irreducible representations of the unimodular complex group of second order are determined; This group is locally isomorphic to the Lorentz group.

It is proved also that an arbitrary unitary representation can be decomposed into these irreducible representations.

As far as possible, the theorems are proved in such a way that similar proofs can be carried out for all complex semisimple groups.

The finite-dimensional representations of the Lorentz group are well known and these representations are used in different questions of geometry and physics. The set of all finite-dimensional representations consists of tensor and spinor representations.

The problem of determining all unitary representations of the Lorentz group has remained, up to now, unsolved.

In this paper we describe all infinite-dimensional irreducible unitary representations of the Lorentz group[1]. The final results are given by simple formulas: §5, formulas (56), (65) and §8, formulas (187), (188).

The applications of our results are no more difficult and, in some respects, even simpler than the applications of finite-dimensional spinor representations.

It is proved also that any unitary representation of the Lorentz group can be decomposed into these irreducible representations; the knowledge of the irreducible representations thus gives the complete description of all unitary representations of the Lorentz group[2].

Further, we determine the decomposition of the regular representation into irreducible representations. It turns out that the decomposition of the regular representation does not include all irreducible representations (as in the case of commutative and compact groups) but only the representations of the so-called "principal series". This fact is not due to complications related to the set theory but is quite "classic". We obtain also an analog of the Plancherel formula.

---

[1] For a summary of the main results see [3].

[2] In a recently published paper by Dirac some unitary representations of the Lorentz group are given. As the authors have shown, these representations are reducible and can be decomposed into irreducible ones with $n = 0$ or $\rho = 0$ (in the notation of this paper; see §5 and §8).

In our representations, to each element $g$ of the Lorentz group we associate a unitary operator $U_g$ in an infinite-dimensional space; this operator has no trace in the usual sense. However, an operator $\int_Q U_g d\mu(g)$ for compact $Q$ has a trace. Hence, we can define the trace of the operator $U_g$ and the character of the representation [←6, formula (94)].

Formulas of representation theory are usually quite elegant. Therefore, the authors have tried not only to give the existence proofs but to write down each result in the final form.

The methods used here can also be applied to an arbitrary complex semisimple Lie group.

Instead of the Lorentz group we study the group $G$ of complex matrices of second order with determinant equal to 1. The Lorentz group is locally isomorphic to this group and both groups have the same representations.

## §1. Some subgroups of the group $G$

Several subgroups of the group $G$ are very important for our construction, namely, operators $U_g$ of a representation act in a function space; elements of this space are functions on the cosets of the group $G$ by some subgroups.

1. *The subgroup $K$*. Denote by $K$ the set of all matrices $k$ such that

$$k = \begin{pmatrix} k_{11} & k_{12} \\ 0 & k_{22} \end{pmatrix}. \tag{1}$$

Evidently $K$ is a subgroup of $G$. Since the determinant of $k$ is equal to $k_{11}k_{22}$ we have

$$k_{11}k_{22} = 1. \tag{2}$$

Thus, a matrix $k$ is determined by two independent complex parameters, for example, by $k_{12}$ and $k_{22}$. A left action $k \to k'k$ is the linear transformation of these parameters:

$$\begin{aligned} k_{12} &\to k'_{11}k_{12} + k'_{12}k_{22}, \\ k_{22} &\to k'_{22}k_{22}. \end{aligned} \tag{3}$$

The determinant of this transformation is $k'_{11}k'_{22} = 1$. Let us put

$$k_{12} = u_{12} + iv_{12}, \quad k_{22} = u_{22} + iv_{22}.$$

Then (3) can be rewritten as the linear transformation of the variables $u_{ik}$, $v_{ik}$ whose Jacobian is the square of the determinant of the transformation (3), i.e. also equal to 1.

Hence the left-invariant measure in $K$ is given by

$$d\mu_e(k) = du_{12}dv_{12}du_{22}dv_{22}. \tag{4}$$

Similarly, the right action $k \to kk'$ is a transformation

$$\begin{aligned} k_{12} &\to k_{11}k'_{12} + k_{12}k'_{22}, \\ k_{22} &\to k'_{22}k_{22} \end{aligned} \tag{5}$$

of the variables $k_{12}, k_{22}$; the Jacobian of this transformation is $\Delta = k_{22}^2$. Hence the right-invariant measure on $K$ is

$$d\mu_r(k) = |k_{22}|^{-4} du_{12} dv_{12} du_{22} dv_{22}. \tag{6}$$

Let us write

$$\beta(k) = \frac{d\mu_l(k)}{d\mu_r(k)}. \tag{7}$$

Then, from (4) and (6),

$$\beta(k) = |k_{22}|^4. \tag{8}$$

This formula together with (3) and (5) implies

$$\beta(k_1 k_2) = \beta(k_1)\beta(k_2). \tag{9}$$

2. *The subgroup H*. Denote by $H$ the set of all matrices $h$ such that

$$h = \begin{pmatrix} h_{11} & 0 \\ h_{21} & h_{22} \end{pmatrix}. \tag{10}$$

Evidently $H$ is a subgroup of $G$. As before,

$$h_{11} h_{22} = 1. \tag{11}$$

The subgroup $H$ is related to the subgroup $K$ as follows. Let us take the matrix $s = \begin{pmatrix} 0 & -1 \\ 1 & 0 \end{pmatrix}$. Then the automorphism $g \to s^{-1} g s$ transforms $H$ into $K$ and $K$ into $H$. With the notation

$$h_{21} = u_{21} + i v_{21}, \qquad h_{22} = u_{22} + i v_{22}$$

we therefore get the following expressions for differentials of the left- and right-invariant measures in $H$:

$$d\mu_l(h) = |h_{22}|^{-4} du_{12} dv_{12} du_{22} dv_{22}, \tag{12}$$

$$d\mu_r(h) = du_{12} dv_{12} du_{22} dv_{22}. \tag{13}$$

Thus

$$\frac{d\mu_l(k)}{d\mu_r(h)} = |h_{22}|^{-4}. \tag{14}$$

3. *The subgroup Z*. Denote by $Z$ the set of all matrices $z$ such that[3]

$$z = \begin{pmatrix} 1 & 0 \\ z & 0 \end{pmatrix}. \tag{15}$$

Evidently $Z$ is a subgroup of $H$; the multiplication of matrices $z_1, z_2$ reduces to

[3] Below we will denote by the same letter the element of the subgroup $Z$ and the complex parameter determining this element. This will never lead to misunderstanding. The same should be said about the next subgroup, Z.

the addition of the parameters $z_1$ and $z_2$. So the subgroup $Z$ is commutative and it is isomorphic to the additive group of complex numbers. It follows that the invariant measure in $Z$ is

$$d\mu(z) = dx\,dy, \tag{16}$$

where $z = x + iy$

4. *The subgroup* $\mathrm{Z}$. Denote by $\mathrm{Z}$ the set of all matrices $\zeta$ such that

$$\zeta = \begin{pmatrix} 1 & \zeta \\ 0 & 1 \end{pmatrix}. \tag{17}$$

Evidently $\zeta$ is a subgroup of $K$; this subgroup also is isomorphic to the additive group of complex numbers. Hence, the invariant measure in $\mathrm{Z}$ is defined by

$$d\mu(\zeta) = d\xi\,d\eta, \tag{18}$$

where $\zeta = \xi + i\eta$.

5. *The subgroup* $D$. Denote by $D$ the set of all diagonal matrices of the group $G$, i.e. of all matrices $\delta$ of the form

$$\delta = \begin{pmatrix} 1/\lambda & 0 \\ 0 & \lambda \end{pmatrix}. \tag{19}$$

Evidently $D = K \cap H$. The group $D$ is isomorphic to the multiplicative group of the complex numbers.

Put $\lambda = \sigma + i\tau$. Because of the above isomorphism, the invariant measure in $D$ is given by

$$d\mu(\delta) = \frac{d\sigma\,d\tau}{|\lambda|^2}. \tag{20}$$

## §2. Some relations between the group $G$ and the subgroups $H, K, Z, \mathrm{Z}, D$

We will need some theorems about the presentation of an element in $G$ as a product of the elements of the subgroups defined above. These theorems will be useful in the study of coset spaces of $G$ by these subgroups.

I. *Each element* $k \in K$ *can be uniquely presented in the form*

$$k = \delta\zeta \tag{21}$$

*and in the form*

$$k = \zeta\delta. \tag{22}$$

Indeed, by (1), (17) and (19), the equality (21) is equivalent to the equalities

$$k_{12} = \lambda^{-1}\zeta, \quad k_{22} = \lambda \tag{23}$$

which determine $\lambda$ and $\zeta$.

The equality (22) is equivalent to the equalities

$$k_{12}=\lambda\zeta, \qquad k_{22}=\lambda \tag{24}$$

which also uniquely determine $\lambda$ and $\zeta$.

Similarly,

II. *Each element $h\in H$ can be uniquely presented in the form*

$$h=\delta z \tag{25}$$

*and in the form*

$$h=z\delta. \tag{26}$$

III. *Each element $g\in G$ with $g_{22}\neq 0$ can be uniquely presented in the form*

$$g=kz \tag{27}$$

*and in the form*

$$g=zk. \tag{28}$$

Indeed, because of (1) and (15), the equality (27) is equivalent to the equalities

$$g_{12}=k_{12}, \qquad g_{21}=k_{22}z, \qquad g_{22}=k_{22} \tag{29}$$

from which $k_{12}$, $k_{22}$ and $z$ are uniquely determined. From the last of the equalities (29) and the condition $k_{11}k_{22}=1$ it follows that the condition $g_{22}\neq 0$ is necessary for the representation (27).

The proof of (28) is quite similar.

The set of elements $g\in G$ satisfying $g_{22}=0$ (i.e. which cannot be presented in the forms (27) and (28) is the submanifold of lower dimension in the group $G$. Hence this submanifold has the invariant measure zero.

IV. *Each element $g\in G$ with different eigenvalues $\lambda_1$ and $\lambda_2$ (where $\lambda_1\lambda_2=1$) and with $g_{12}\neq 0$ can be represented in the form*

$$g=z^{-1}kz \tag{30}$$

*where $k_{11}=\lambda_1$ and $k_{22}=\lambda_2$. If we take eigenvalues in this order then $k_{12}$ and $z$ in* (30) *are determined uniquely.*

Let $x=\|z_{pq}\|$ be a nonsingular matrix which satisfies the condition

$$xg=\delta x, \qquad \delta=\begin{pmatrix}\lambda_1 & 0\\ 0 & \lambda_2\end{pmatrix}. \tag{31}$$

This condition is equivalent to equalities

$$x_{p1}g_{1q}+x_{p2}g_{2q}=\lambda_p x_{pq}, \qquad (p,q=1,2), \tag{32}$$

which for fixed $p$ give the values of $x_{p1}$ and $x_{p2}$ up to a multiplicative constant. When $p=q=2$ we have

$$z_{21}g_{12}+x_{22}(g_{22}-\lambda_2)=0.$$

If $x_{22}=0$ then, in view of the condition $g_{12}\neq 0$, we have $x_{21}=0$ and the solution is trivial. So, for the existence of a nontrivial solution, we must suppose $x_{22}\neq 0$.

The statement III implies that a matrix $x$ can be presented in the form $x=k_1z$. Then by (31)

$$g=x^{-1}\delta x=z^{-1}k_1^{-1}\delta k, z=z^{-1}kz$$

and obviously

$$k_{11}=\lambda_1, \quad k_{22}=\lambda_2.$$

From equalities (30), (1) and (15) it follows that

$$k_{12}=g_{12}, \quad -zk_{12}+k_{22}=g_{22}.$$

These equations determine $k_{12}$ and $z$ uniquely.

So, there are two different representations (30) which correspond to two permutations of the eigenvalues $\lambda_1$ and $\lambda_2$.

## §3. Some relations between integrals over the group $G$ and over the subgroups $K, H, Z, \mathrm{Z}, D$

The results of §1 and §2 enable us to express the integration over the group $G$ through iterated integrations over some subgroups.

1. Let $f(k)$ be an integrable function on $K$. In view of (21) we can write $f(k)=f(\delta\zeta)$ and we can consider $f$ as a function in $\delta$ and $\zeta$.

Further, because of (4),

$$\int f(k)d\mu_l(k)=\int f(k)du_{12}dv_{12}du_{22}dv_{22}. \tag{33}$$

On the other hand, the equality $k=\delta z$ is equivalent to

$$k_{12}=\zeta/\lambda, \quad k_{22}=\lambda. \tag{34}$$

The complex Jacobian of the transformation (34) from the variables $k_{12}, k_{22}$ to the variables $\zeta$ and $\lambda$ is equal to $1/\lambda$. The transformation (34) defines in the integral (33) the transformation from the real variables $u_{ik}, v_{ik}$ to the real variables $\xi, \eta, \sigma$ and $\tau$. The Jacobian of this transformation is equal to the square of the absolute value of the complex Jacobian. Thus, because of formulas (18) and (20), the equality (33) can be written in the form

$$\int f(k)d\mu_l(k)=\int\int f(\delta\zeta)d\mu(\zeta)d\mu(\delta). \tag{35}$$

In complete analogy with (35) we obtain

$$\int f(k)d\mu_r(k)=\int\int f(\zeta\delta)d\mu(\zeta)d\mu(\delta), \tag{36}$$

$$\int f(h)d\mu_l(h)=\int\int f(\delta z)d\mu(z)d\mu(\delta), \tag{37}$$

$$\int f(h)d\mu_r(h)=\int\int f(\delta z)d\mu(z)d\mu(\delta). \tag{38}$$

2. Now let $f(g)$ be an integrable function on $G$. Because of formula (27) $f(g)=f(kz)$ for almost all $g$ and we can consider $f(g)$ as a function in two variables, $k$ and $z$.

It is easy to see that

$$\int f(g)d\mu(g)=\int f(g)\frac{d\alpha_{12}d\beta_{12}d\alpha_{21}d\beta_{21}d\alpha_{22}d\beta_{22}}{|g_{22}|^2} \tag{39}$$

The transformation (27) from $g$ to $k$ and $z$ is equivalent to the transformation (29) from the variables $g_{12}, g_{21}, g_{22}$ to the variables $k_{12}, k_{22}, z$. The complex Jacobian of this transformation is equal to $k_{22}=g_{22}$. Thus we can rewrite (39) in the form

$$\int f(g)d\mu(g)=\int\int f(kz)du_{12}dv_{12}du_{22}dv_{22}dxdy,$$

i.e., in view of (4) and (16),

$$\int f(g)d\mu(g)=\int f(kz)d\mu_l(k)d\mu(z). \tag{40}$$

In complete analogy with this formula we obtain

$$\int f(g)d\mu(g)=\int f(\zeta h)d\mu_r(h)d\mu(\zeta). \tag{41}$$

3. Let $f(g)$ be a function on $G$ and $\varphi(k)$ be a function on $K$; denote by $\lambda_g$ and $1/\lambda_g$ the eigenvalues of $g$; let $k_g$ be the element of $K$ defined by the condition $g=z^{-1}k_g z$. With the function $\beta(k)$ defined in §1.1 we have

$$\int\int f(z^{-1}kz)\varphi(k)d\mu_l(k)=\int f(g)\frac{\sum\varphi(k_g)\beta^{\frac{1}{2}}(kg)}{|\lambda_g-1/\lambda_g|^2}d\mu(g)=\int f(g)\frac{\sum\varphi(k_g)\beta^{\frac{1}{2}}(k_g)}{|(g_{11}+g_{22})^2-4|}d\mu(g) \tag{42}$$

where the sum consists of two terms corresponding to the two representations $g$ in the form $z^{-1}kz$.

To prove this formula we remove from $K$ the elements with $|k_{11}|=|k_{12}|=1$. This operation cuts the group $K$ in two connected components, $K_1$ and $K_2$. Neither of these components contains two matrices which differ only by permutation of the diagonal elements. Since the set of removed elements has the measure zero, the second and the third integrals in (42) can be written as the sum of the integrals over $K_1$ and $K_2$. But because of the statement IV in §2, if $z$ runs over the subgroup $Z$ and $k$ runs over one of the components $K_1$ or $K_2$ of the group $K$, then $g=z^{-1}kz$ runs over the group $G$ (and only once).

The equality $g=z^{-1}kz$ is equivalent to the equalities

$$g_{12}=k_{12},\quad g_{21}=-k_{12}z^2+(\lambda-1/\lambda)z,\quad g_{22}=-k_{12}z+\lambda,\quad (\lambda=k_{22}).$$

The complex Jacobian of this transformation from $g_{12}, g_{21}, g_{22}$ to $k_{12}, z, \lambda$ is

$$\frac{\partial(g_{12},g_{21},g_{22})}{\partial(k_{12},z,\lambda)}=\begin{vmatrix}1 & 0 & 0\\ -z^2 & -2k_{12}z+\left(\lambda-\dfrac{1}{\lambda}\right) & \left(1+\dfrac{1}{\lambda^2}\right)\\ -z & -k_{12} & 1\end{vmatrix}$$

$$=\frac{1}{\lambda}\left(\lambda-\frac{1}{\lambda}\right)(-k_{12}z+\lambda)=\frac{1}{\lambda}\left(\lambda-\frac{1}{\lambda}\right)g_{22}.$$

Therefore, the Jacobian of the corresponding real transformation from the variables

$\alpha_{pq}, \beta_{pq}$ to the variables $u_{12}, v_{12}, \sigma, \tau, x, y$ is equal to

$$|\lambda|^{-2}|\lambda - 1/\lambda|^2, |g_{22}|^2 = \beta^{-\frac{1}{2}}(k)|\lambda - 1/\lambda|^2 \cdot |g_{22}|^2.$$

Denote by $k'_g$ the element of $K_1$ for which $g = z^{-1}k'_g z$. Then

$$\int_{K_1} [\int f(z^{-1}kz)\varphi(k)d\mu(z)]d\mu_l(k) = \int_{K_1} [\int f(z^{-1}kz)\varphi(k)dxdy]du_{12}dv_{12}d\sigma d\tau$$
$$= \int f(g)\varphi(k'_g)\beta^{\frac{1}{2}}(k'_g)\cdot|\lambda_g - 1/\lambda_g|^{-2}|g_{22}|^{-2}d\alpha_{12}d\beta_{12}d\alpha_{21}d\beta_{21}d\alpha_{22}d\beta_{22}.$$

This means

$$\int_{K_1} [\int f(z^{-1}kz)\varphi(k)d\mu(z)]d\mu(k) = \int f(g)\frac{\varphi(k'_g)\beta^{\frac{1}{2}}(k'_g)}{|\lambda_g - 1/\lambda_g|^2}d\mu(g).$$

To obtain the formula (42) we add to this equality the similar equality for $K_2$.

## §4. Cosets of $G$ by Z and by $K$

1. *Cosets by* Z. We denote by $\tilde{h}$ right cosets of $G$ by Z, i.e. sets of elements $g \in G$ of the form $\zeta g_0$ for a fixed $g_0$. The set of all these cosets, i.e. the corresponding homogeneous space, will be denoted by $\tilde{H}$. The right multiplication by an element $g \in G$ sends coset $\tilde{h}$ into another coset $\tilde{h}'$ and therefore yields the transformation of the space $\tilde{H}$.

So, to each element $g \in G$ corresponds the transformation of the space $\tilde{H}$. We write it as follows: $\tilde{h}' = \tilde{h}\bar{g}$.

If $g_{22} \neq 0$, then by the statement III in §2, we have $g = \zeta h$. This equality means that $g$ and $h$ belong to the same coset $\tilde{h}$. In view of the uniqueness of the representation $g = \zeta h$ the coset $\tilde{h}$ contains only one element of $H$. Identifying $\tilde{h}$ and $h$ we identify $H$ with the subset of $\tilde{H}$ consisting of elements whose representatives satisfy $g_{22} \neq 0$. The complement in $\tilde{H}$ to this subset is a submanifold of the lower dimension; it plays the role of the infinity in $H$.

So, up to the manifold of lower dimension, the space $H$ coincides with $\tilde{H}$.

The transformation $\tilde{h}' = \tilde{h}\bar{g}$ can be considered as the transformation of $H$ and we can write $h' = h\bar{g}$. It is easy to write down this transformation in the parameters which define matrices $h, h'$ and $g$.

The equality $h' = h\bar{g}$ means that matrices $h'$ and $hg$ belong to the same coset, i.e. we have $hg = \zeta h'$ or, in parameters,

$$\begin{pmatrix} h_{11} & 0 \\ h_{21} & h_{22} \end{pmatrix}\begin{pmatrix} g_{11} & g_{12} \\ g_{21} & g_{22} \end{pmatrix} = \begin{pmatrix} 1 & \zeta \\ 0 & 1 \end{pmatrix}\begin{pmatrix} h'_{11} & 0 \\ h'_{21} & h'_{22} \end{pmatrix}.$$

From this,

$$h'_{21} = g_{11}h_{21} + g_{21}h_{22}, \qquad h'_{12} = g_{12}h_{21} + g_{22}h_{22}. \tag{43}$$

In other words, the transformation $h' = h\bar{g}$ implies the linear transformation (43) of the variables $h_{21}, h_{22}$. The complex Jacobian of this transformation equals the determinant of the matrix $g$, i.e. it is equal to 1. Therefore, the real determinant of the corresponding real transformation of the variables $u_{21}, v_{21}, u_{22}, v_{22}$ ($h_{21} =$

$u_{21}+iv_{21}, h_{22}=u_{22}+iv_{22}$) is also 1. Hence, it follows that the right-invariant measure in $H$ is invariant under the transformation $h'=h\bar{g}$ (see (13)).

2. *Cosets by K.* We denote by $\tilde{z}$ right cosets of $G$ by $K$, i.e. sets of elements $g \in G$ such that $g=kg_0$ for a fixed $g_0$. The space of these cosets we denote by $\tilde{z}$. The right multiplication by $g$ yields the transformation $\tilde{z} \to \tilde{z}'$ of the space $\tilde{Z}$. We denote this transformation by $\bar{g}$ and write $\tilde{z}'=\tilde{z}g$.

If $g_{22} \neq 0$, then by the statement III in §2 we have $g=kz$. This equality means that the matrices $g$ and $z$ lie in the same class $\tilde{z}$. The decomposition $g=kz$ is unique and therefore the class $\tilde{z}$ contains only one element $z \in Z$. Identifying $\tilde{z}$ and $z$ we identify $Z$ with the complement to one point in $\tilde{Z}$. In fact, if $g=s_0$ (see §1.2), then

$$kg_0=ks=\begin{pmatrix} k_{11} & k_{12} \\ 0 & k_{22} \end{pmatrix}\begin{pmatrix} 0 & -1 \\ 1 & 0 \end{pmatrix}=\begin{pmatrix} k_{12} & -k_{11} \\ k_{22} & 0 \end{pmatrix}.$$

Since $k_{12}$ and $k_{22}$ are arbitrary, the matrix on the right side of this equality can be an arbitrary matrix from $G$ with $g_{22}=0$. All such matrices belong to the one coset $\tilde{z}_0=\{ks\}$ which plays the role of the point at infinity in $\tilde{Z}$.

The transformation $\tilde{z}'=\tilde{z}\bar{g}$ can be considered as the transformation of $Z$ and we write $z'=z\bar{g}$; the element $z\bar{g}$ is defined for all but one element $z \in Z$. It is not defined for the element $z$ which $\bar{g}$ maps into the element at infinity.

The transformation $z'=z\bar{g}$ is the transformation of the parameter $z$ which defines the matrix $z$. It can be easily written down in explicit form. Since $z'$ and $zg$ belong to the same coset by $K$, i.e. $zg=kz'$, we have

$$\begin{pmatrix} 1 & 0 \\ z & 1 \end{pmatrix}\begin{pmatrix} g_{11} & g_{12} \\ g_{21} & g_{22} \end{pmatrix}=\begin{pmatrix} k_{11} & k_{12} \\ 0 & k_{22} \end{pmatrix}\cdot\begin{pmatrix} 1 & 0 \\ z' & 1 \end{pmatrix}.$$

Thus

$$g_{11}z+g_{21}=k_{22}z', \qquad g_{12}z+g_{22}=k_{22}$$

and finally

$$z'=\frac{g_{11}z+g_{21}}{g_{12}z+g_{22}}. \tag{44}$$

So *the transformation $z'=z\bar{g}$ in the subgroup $Z$ is reduced to the transformation* (44) *of the complex parameter $z$.*

The Jacobian of this transformation is equal to the square of the absolute value of the derivative $dz'/dz$, i.e. it is equal to

$$|g_{12}z+g_{22}|^{-4} \tag{45}$$

This expression can be written in group theory notation. Let $\beta(g)=\beta(k)$ for $g=kz$. Thus we extend the function $\beta(k)$ to the almost all group $G$ and it is easy to check that

$$\beta(g)=|g_{22}|^4 \tag{46}$$

and

$$\beta(kg)=\beta(k)\cdot\beta(g). \tag{47}$$

So the expression (45) coincides with $\beta^{-1}(z\bar{g})$ and

$$\frac{d\mu(z\bar{g})}{d\mu(z)} = \beta^{-1}(z\bar{g}). \tag{48}$$

## §5. The principal series of irreducible representations of the group $G$

Now we turn to the construction of the irreducible representations of the group $G$. To do this, we first consider some reducible representations of this group.

1. *The regular representation of the group G.* We denote by $\mathfrak{H}_G$ the Hilbert space of square integrable functions on the group $G$ with inner product defined by

$$(f_1, f_2) = \int f_1(g)\overline{f_2(g)}\,d\mu(g). \tag{49}$$

Let us put

$$V_{g_0} f(g) = f(gg_0). \tag{50}$$

From the invariance of the measure $\mu(g)$ it follows that the formula (50) defines a unitary operator in $\mathfrak{H}_G$. Obviously, $V_g$ is the unitary representation of the group $G$. We call it the regular representation of this group.

For a finite or a compact group the decomposition of the regular representation into irreducible representations contains all irreducible representations of this group. Later we will see that for the group $G$ this is not so.

2. *Quasi-regular representation of the group G.* Let us denote by $\mathfrak{H}_H$ the Hilbert space of square integrable functions on $H$ (with respect to the right-invariant measure). The inner product is defined by the formula

$$(f_1, f_2) = \int f_1(h)\overline{f_2(h)}\,d\mu_r(h). \tag{51}$$

Let us put

$$U_g f(h) = f(h\bar{g}). \tag{52}$$

In view of §1.4, the measure $\mu_r(h)$ is invariant under the transformation $h' = h\bar{g}$ and therefore $U_g$ is a unitary operator in $\mathfrak{H}_H$. Obviously, $U_g$ is a unitary representation of the group $G$. We call it the quasi-regular representation of $G$. Later we will see that the quasi-regular representation can be decomposed into the same irreducible representations as the regular representation but in the decomposition of the quasi-regular representation each irreducible representation is contained no more than twice.

3. *Decomposition of the quasi-regular representation of the group G into irreducible representations.* Let $f = f(h)$ be an arbitrary element of $\mathfrak{H}_H$. From (12), (13) and (37) we have

$$\|f\|^2 = \int |f(h)|^2 d\mu_r(h) = \int |f(h)|^2 \beta(h)\,d\mu_l(h) = \iint |f(\delta z)\beta^{\frac{1}{2}}(\delta)|^2 d\mu(\delta)\,d\mu(z). \tag{53}$$

So, for almost every $z$, the function $f(\delta z)\beta^{\frac{1}{2}}(\delta)$ is square integrable on the group $D$.

Let $X$ be the group of characters of the group $D$, i.e. the group of characters of the multiplicative group of complex numbers. Consider the Fourier transform of the function $f(\delta z)\beta^{\frac{1}{2}}(\delta)$:

$$f_\chi(z) = \int f(\delta z)\beta^{\frac{1}{2}}(\delta)\chi(\delta)d\mu(\delta). \tag{54}$$

By the Plancherel theorem $f_\chi(z)$, as a function in $\chi$, is a square integrable function on $X$. With the proper normalization of the measure on $X$ we have

$$\int |f_\chi(z)|^2 d\mu(\chi) = \int |f(\delta z)\beta^{\frac{1}{2}}(\delta)|^2 d\mu(\delta).$$

So the equality (53) can be written in the form

$$\| f \|^2 = \int\int |f_\chi(z)|^2 d\mu(z) d\mu(\chi). \tag{55}$$

Let $\mathfrak{H}_Z$ be the Hilbert space of square integrable functions on $Z$ with inner product given by

$$(f_1, f_2) = \int f_1(z)\overline{f_2(z)}d\mu(z). \tag{56}$$

The formula (55) means that the space $\mathfrak{H}_H$ is the continuous direct sum of Hilbert spaces $\mathfrak{H}^\chi = \mathfrak{H}_z$.

Let us now consider the transformation of $f_\chi(z)$ when we come from $f(h)$ to $f(h\bar{g})$. We put

$$f'(h) = f(hg); \qquad h' = hg; \qquad h = \delta z; \qquad h' = \delta' z'. \tag{57}$$

Since $h'$ and $hg$ belong to the same right coset by Z, we have $hg = \zeta h'$ and therefore, in view of (57), $\delta z g = \zeta\delta' z'$. Hence

$$zg = \delta^{-1}\zeta\delta' z' = \delta^{-1}\zeta\delta\delta^{-1}\delta' z'. \tag{58}$$

It is easy to check that $\delta^{-1}\zeta\delta$ is an element of Z; we denote this element by $\zeta'$. Further we denote $\delta^{-1}\delta' = \delta_1$ so that we have

$$\delta' = \delta\delta_1. \tag{59}$$

Then the equality (58) can be written

$$zg = \zeta'\delta_1 z'. \tag{60}$$

The equality (60) means that $zg$ and $z'$ belong to the same coset by $K$, i.e. $z' = z\bar{g}$; this equality implies also that $\delta_1$ depends only on $z$ and $g$ and does not depend on $\delta$. So we have

$$h\bar{g} = h' = \delta' z' = \delta\delta_1 z\bar{g}.$$

But then we can write

$$\begin{aligned} f'_\chi(z) &= \int f(\delta\delta_1 z\bar{g})\beta^{\frac{1}{2}}(\delta)\chi(\delta)d\mu(\delta) = \int f(\delta z\bar{g})\beta^{\frac{1}{2}}(\delta\delta_1^{-1})\chi(\delta\delta_1^{-1})d\mu(\delta) \\ &= \beta^{-\frac{1}{2}}(\delta_1)X(\delta_1)\int f(\delta z\bar{g})\beta^{\frac{1}{2}}(\delta)\overline{\chi(\delta)}d\mu(\delta) \end{aligned}$$

and we obtain

$$f'_\chi(z) = \beta^{-\frac{1}{2}}(\delta_1)\chi(\delta_1)f_\chi(z\bar{g}). \tag{61}$$

We see that $f'_\chi(z)$ depends only on $f_\chi(z)$ and does not depend on $f_{\chi_1}(z)$ for $\chi_1 \neq \chi$.

This means that the representation $U_g$ is decomposed into a continuous direct sum of the representations $U_{\chi;g}$ (in the spaces $\mathfrak{H}_z$) defined by

$$U_{\chi;g}f(z) = \beta^{-\frac{1}{2}}(\delta_1)X(\delta_1)f(z\bar{g}). \tag{62}$$

The formula (62) can be written in a more convenient form: let us put $\chi(g) = \chi(\delta)$ for $g = \zeta\delta z$. Then the equality (60) implies that $\chi(\delta_1) = \chi(zg)$ and $\beta(\delta_1) = \beta(zg)$. The definition (62) can be rewritten as follows:

$$U_{\chi;g}f(z) = \beta^{-\frac{1}{2}}(zg)\chi(zg)f(z\bar{g}). \tag{63}$$

Let us check the unitarity of the representation $U_{\chi;g}$. Since $|\chi(\delta)| = 1$ we have

$$\| U_{\chi;g}f \|^2 = \int \beta^{-1}(zg)|f(z\bar{g})|^2 d\mu(z).$$

In view of the formula (48) this expression coincides with

$$\int |f(z)|^2 d\mu(z) = \| f \|^2.$$

Thus, to each character $\chi$ of the group $D$ corresponds the unitary representation $U_{\chi;g}$. We call the set of these representations *the principal series of representations of the group G.*

Put, as above, $\delta = \begin{pmatrix} \lambda^{-1} & 0 \\ 0 & \lambda \end{pmatrix}$. Then $\chi(\delta)$ can be considered as a character of the multiplicative group of complex numbers $\lambda$. If $\lambda = re^{i\varphi}$, then $\chi(\lambda) = \chi(r)\cdot\chi(e^{i\varphi})$. $\chi(r)$ is a character of the multiplicative group of positive numbers and, therefore, $\chi(r) = r^{i\rho}$, where $\rho$ is a real number.

Further $\chi(e^{i\varphi})$ is a character of the additive group of numbers $\varphi$ and therefore

$$\chi(e^{i\varphi}) = e^{-in\varphi}$$

where $n$ is an integer. So we have

$$\chi(\lambda) = r^{i\rho}e^{-in\varphi} = |\lambda|^{i\rho}\left(\frac{\lambda}{|\lambda|}\right)^{-n} = |\lambda|^{n+i\rho}\lambda^{-n}. \tag{64}$$

Using the expression (64) for $\chi(\lambda)$ we can give a more detailed (compared to (63)) definition of the operators $U_{\chi;g}$. First, let us remark that if $zg = \zeta\delta z$, where $z = \begin{pmatrix} 1 & 0 \\ z & 1 \end{pmatrix}$ and $\delta = \begin{pmatrix} \lambda^{-1} & 0 \\ 0 & \lambda \end{pmatrix}$, then $\lambda = g_{12}z + g_{22}$.

We will consider $f(z)$ as a function of the parameter $z$; then the equalities (44) and (64) and the above formula for $\lambda$ imply that (63) can be written in the form

$$U_{\chi;g}f(z) = |g_{12}z + g_{22}|^{n+i\rho-2}(g_{12}z + g_{22})^{-n} f\left(\frac{g_{11}z + g_{21}}{g_{12}z + g_{22}}\right) \tag{65}$$

Thus a representation $U_{\chi;g}$ is determined by two parameters $\rho$ and $n$, where $\rho$ is an arbitrary real number and $n$ is an arbitrary integer.

4. *The irreducibility of the representations of the principal series.*

**Theorem 1.** *All representations of the principal series are irreducible.*

*Proof*. Let us put in formula (65) $g = z_0 = \begin{pmatrix} 1 & 0 \\ z_0 & 1 \end{pmatrix}$. Then this equality takes the form

$$U_{\chi;z_0} f(z) = f(z + z_0); \tag{66}$$

let us further put $g = \delta = \begin{pmatrix} \lambda^{-1} & 0 \\ 0 & \lambda \end{pmatrix}$; then (65) implies

$$U_{\chi;\delta} f(z) = |\lambda|^{n+i\rho-2} \lambda^{-n} f(z\lambda^{-2}). \tag{67}$$

We must prove that each bounded operator $A$ which commutes with all the operators $U_{\chi;g}$ is a multiple of the identity. In fact, moreover, we will prove that each bounded operator which commutes with all the operators $U_{\chi;z_0}$ and $U_{\chi;\delta}$ is a multiple of the identity, i.e. that the operators $U_{\chi;z_0}$ and $U_{\chi;\delta}$ form an irreducible system of operators.

In other words the representation $U_{\chi;g}$ for $g = h$ is an irreducible representation of the group $H$ (and for $g = k$ it is an irreducible representation of the group $K$).

Let us consider instead of function $f(z)$ their Fourier transforms

$$\varphi(p) = \varphi(p_1 + ip_2) = \frac{1}{2\pi} \iint f(x + iy) e^{-i(xp_1 + yp_2)} dx dy = \frac{1}{2\pi} \iint f(z) e^{-i\operatorname{Re}(z\bar{p})} dx dy. \tag{68}$$

The transformation from $f(z)$ to $\varphi(p)$ is a unitary operator from the space $\mathfrak{H}_z$ to the space $\mathfrak{H}_\varphi$ of square integrable function in $p_1$ and $p_2$; the inner product in this space is defined by the formula

$$(\varphi_1, \varphi_2) = \int \varphi_1(p) \overline{\varphi_2(p)} \, dp_1 dp_2. \tag{69}$$

The operators $U_{\chi;z_0}$, $U_{\chi;\delta}$ and $A$ correspond to some operators in the space $\mathfrak{H}_\varphi$. We denote these operators again by $U_{\chi;z_0}$, $U_{\chi;\delta}$ and $A$. Then we have

$$U_{\chi;z_0} \varphi(p) = \frac{1}{2\pi} \iint f(z + z_0) e^{i\operatorname{Re}(z\bar{p})} dx dy = \frac{1}{2\pi} \iint f(z) e^{i\operatorname{Re}[(z - z_0)\bar{p}]} dx dy$$

$$= e^{i\operatorname{Re}(z_0\bar{p})} \frac{1}{2\pi} \iint f(z) e^{-i\operatorname{Re}(z\bar{p})} dx dy.$$

This means

$$U_{\chi;z_0} \varphi(p) = e^{i\operatorname{Re}(z_0\bar{p})} \varphi(p). \tag{70}$$

Further

$$U_{\chi;\delta} \varphi(p) = |\lambda|^{n+i\rho-2} \lambda^{-n} \frac{1}{2\pi} \iint f(z\lambda^{-2}) e^{-i\operatorname{Re}(zp)} dx dy.$$

The change of variables $z = \lambda^2 z_1$ gives

$$U_{\chi;\delta} \varphi(p) = |\lambda|^{n+i\rho-2} \lambda^{-n} |\lambda|^4 \frac{1}{2\pi} \iint f(z_1) e^{-i\operatorname{Re}(z_1 p\bar{\lambda}^2)} dx dy,$$

i.e.

$$U_{\chi;\delta} \varphi(p) = |\lambda|^{n+i\rho+2} \lambda^{-n} \varphi(p\bar{\lambda}^2). \tag{71}$$

In view of formula (70) the operator $A$ commutes with all operators of multiplication by $e^{i\mathrm{Re}(z_0\bar{p})}$ and therefore with limits of linear combinations of such functions; hence it commutes with all operators of multiplication by an essentially bounded function in $p$ and therefore $A$ is also an operator of multiplication on the essentially bounded function

$$A\varphi(p) = \omega(p)\varphi(p). \tag{72}$$

From (71) and (72) we have

$$AU_{\chi;\delta}\varphi(p) = \omega(p)|\lambda|^{n+i\rho+2}\lambda^{-n}\varphi(p\bar{\lambda}^2) \tag{73}$$

and

$$U_{\chi;\delta}A\varphi(p) = |\lambda|^{n+i\rho+2}\lambda^{-n}\omega(p\bar{\lambda}^2)\varphi(p\bar{\lambda}^2). \tag{74}$$

As the operator $A$ commutes with all operators $U_{\chi;\delta}$, comparing (73) and (74) we see that $\omega(p) = \omega(p\bar{\lambda}^2)$ for arbitrary $\lambda$ and therefore $\omega(p) = \text{Const.}$ for almost all $p$. Because of (72) this means that the operator $A$ is an multiple of the identity. The theorem is proved.

5. *Elementary spherical harmonics.* A representation of the principal series with $n = 0$ has the following property:

*In the representation space there exists a vector $f_0$ invariant under the operators of the representation which correspond to unitary matrices from the group G.*

Let us find an explicit expression for this vector. A unitary matrix in $G$ has the form

$$g = \begin{pmatrix} \alpha & \beta \\ -\bar{\beta} & \bar{\alpha} \end{pmatrix}$$

where

$$|\alpha|^2 + |\beta|^2 = 1. \tag{75}$$

The operator of a representation $U_{0,\rho;g}$ transforms $f_0(z)$ into

$$|\beta z + \bar{\alpha}|^{i\rho-2} f_0\left(\frac{\alpha z - \bar{\beta}}{\beta z + \bar{\alpha}}\right).$$

Therefore the function $f_0(z)$ should satisfy the condition

$$|\beta z + \bar{\alpha}|^{i\rho-2} f_0\left(\frac{\alpha z - \bar{\beta}}{\beta z + \bar{\alpha}}\right) = f_0(z) \tag{76}$$

for all $\alpha$, $\beta$ with $|\alpha|^2 + |\beta|^2 = 1$.

In the formula (76) we first take $z = 0$. Then

$$|\alpha|^{i\rho-2} f_0\left(-\frac{\bar{\beta}}{\bar{\alpha}}\right) = C \tag{77}$$

where $C = f_0(0)$. Next we take $z = -\dfrac{\bar{\beta}}{\bar{\alpha}}$; by (75) we have

$$1 + |z|^2 = |\alpha|^{-2}.$$

From this expression for $\alpha$ and from (77) we obtain

$$f_0(z) = C(1 + |z|^2)^{-1+i\rho/2}. \tag{78}$$

It is easy to check that conversely the function $f_0(z)$ satisfies the condition (76), i.e. this function is invariant under all operators corresponding to unitary matrices from the group $G$.

Therefore *for each of the representations* $U_{0,\rho;g}$ *there exists, up to a scalar multiple, only one such vector* $f_0$.

Let us put $C = 1/\sqrt{\pi}$ in equality (78); then, as it is easy to verify,

$$\|f_0\|^2 = \int |f_0(z)|^2 d\mu(z) = 1.$$

We call the function

$$\psi_\rho(g) = (U_{0,\rho;g} f_0, f_0) \tag{79}$$

*the elementary spherical function* corresponding to the representation $U_{0,\rho;g}$.

From the definition of $\psi_\rho(g)$ it follows that this function is positive definite on the group $G$ and is invariant under the right and left action ($g' = gg_0$ and $g' = g_0 g$) of a unitary matrix $g_0$.

Let us find an explicit formula for the spherical function $\psi_\rho(g)$. Each matrix $g$ can be presented in the form $g = ua$ where $u$ is a unitary matrix and $a$ is a hermitian matrix. Further, each hermitian matrix $a$ can be presented in the form $a = u_1^{-1}\delta u_1$ where $u_1$ is a unitary matrix and $\delta = \begin{pmatrix} \lambda^{-1} & 0 \\ 0 & \lambda \end{pmatrix}$, $0 < \lambda \leqq 1$. Therefore, it is sufficient to calculate the function $\psi_\rho(g)$ only for $g = \delta$.

Since

$$U_{0,\rho;\delta} f_0(z) = \lambda^{i\rho-2} f_0(z\lambda^{-2}),$$

we have

$$\begin{aligned}\psi_\rho(\delta) &= \int \lambda^{i\rho-2} f_0(z\lambda^{-2})\overline{f_0(z)}\, d\mu(z) \\ &= \lambda^{i\rho-2}/\pi \int (1+|z|^2\lambda^{-4})^{-1+i\rho/2}(1+|z|^2)^{-1-i\rho/2} d\mu(z) \\ &= 2\lambda^{i\rho-1}\int_0^\infty (1+r^2\lambda^{-4})^{-1+i\rho/2}(1+r^2)^{-1-i\rho/2} r\,dr.\end{aligned}$$

Let us make the change of variables

$$x = \frac{1-\lambda^4}{1+r^2}$$

in the last integral. Then

$$\begin{aligned}\psi_\rho(\delta) &= \lambda^{i\rho-2}(1-\lambda^4)^{-1}\int_0^{1-\lambda^4} (1-x)^{-1-i\rho/2} dx \\ &= \frac{2\lambda^{-i\rho+2}(1-\lambda^4)^{-1}}{i\rho}(\lambda^{2i\rho}-1) = \frac{2}{i\rho}\frac{\lambda^{i\rho}-\lambda^{-i\rho}}{\lambda^2-\lambda^{-2}}.\end{aligned}$$

Finally, we get the following expression for the elementary spherical function which corresponds to the representation $U_{0,\rho;g}$ (where $\lambda = e^t$):

$$\psi_\rho(\delta) = \frac{2\sin\rho t}{\rho\,\mathrm{sh}\,2t}. \tag{80}$$

## §6. The trace of the representation of the principal series

1. *The group ring.* Operators $U_{\chi;g}$ of the principal series, being unitary operators in a Hilbert space, do not have a trace in the usual sense. Nevertheless, we will show that it is possible to define the trace of an operator $U_{\chi;g}$ if, instead of the representation of the group, we consider a representation of the group ring. Let us recall necessary definitions. Let $R$ be the set of functions $x(g)$ that are integrable over $G$; define addition and multiplication by a complex number in the trivial way. We define the multiplication of two such functions as convolution:

$$(x_1 * x_2)(g) = \int x_1(g y_2^{-1}) x_2(g_1) d\mu(g_1). \tag{81}$$

The set $\tilde{R}$ of all elements $\lambda e + x(g)$ where $e$ is a formal unity, with the operations defined above, is called the group ring of the group $G$. In this ring we define the norm by

$$\| \lambda e + x(g) \| = |\lambda| + \int |x(g)| d\mu(g) \tag{82}$$

and the involution by

$$(\lambda e + x(g))^* = \bar{\lambda} e + \overline{x(g^{-1})}. \tag{83}$$

In this way, $\tilde{R}$ becomes a complete normed ring with involution.

Let $W_g$ be a unitary representation of the group $G$. To each element of the ring $a = \lambda e + x(g)$ we associate the operator

$$W_a = \lambda E + \int x(g) W_g d\mu(g). \tag{84}$$

The integral in formula (84) converges in the operator norm.

It is easy to check that the mapping $a \to W_a$ is a representation of the ring $\tilde{R}$. From (84) it follows that to an element $a^*$ corresponds the adjoint operator $W_a^*$, i.e. $W_{a^*} = (W_a)^*$.

Conversely, to each representation $W_x$ of the ring $\tilde{R}$ satisfying the condition $W_{x^*} = (W_x)^*$ corresponds the representation $W_g$ of the group $G$ from which $W_x$ can be obtained by formula (84) with $\lambda = 0$.

2. *Formulas for the trace.* Now let $W_g$ coincide with one of the representations $U_{\chi;g}$ of the principal series. To each element $x(g) \in R$ we associate an operator

$$U_{\chi;x} f(z) = \int x(g) U_{\chi;g} f(z) d\mu(g) = \int x(g) \beta^{-\frac{1}{2}}(zg) \chi(zg) f(z\bar{g}) d\mu(g). \tag{85}$$

The convergence of the integral is the mean convergence in the space $\mathfrak{H}_z$.

Let $x(g)$ be a function such that the integral (85) converges in the usual sense. Let us define $f(g)$ as $f(z)$ for $g = kz$. Then $f(g)$ is defined for almost all $g \in G$ and in view of the invariance of the $d\mu(g)$ we have

$$\begin{aligned} U_{\chi;x} f(z) &= \int x(g) \beta^{-\frac{1}{2}}(zg) \chi(zg) f(zg) d\mu(g) \\ &= \int x(z^{-1} g) \beta^{-\frac{1}{2}}(g) \chi(g) f(g) d\mu(g). \end{aligned}$$

From this and from the equality (40) it follows that

$$\begin{aligned} U_{\chi;x} f(z) &= \iint x(z^{-1} g z_1) \beta^{-\frac{1}{2}}(k z_1) \chi(k z_1) f(k z_1) d\mu_e(k) d\mu(z_1) \\ &= \iint x(z^{-1} k z_1) \beta^{-\frac{1}{2}}(k) \chi(k) f(z_1) d\mu_e(k) d\mu(z_1). \end{aligned}$$

This equality means that the operator $U_{\chi;x}$ has the kernel

$$K(z, z_1; \chi) = \int x(z^{-1}kz_1)\beta^{-\frac{1}{2}}(k)\chi(k)d\mu_e(k) \tag{86}$$

so that

$$U_{\chi;x} f(z) = \int K(z, z_1; \chi) f(z_1) d\mu(z_1). \tag{87}$$

Assume that for a given function $x(g)$ the integral (86) converges absolutely and the kernel $K(z, z_1; \chi)$ is a Hilbert-Schmidt kernel, i.e.

$$\int |K(z, z_1; \chi)|^2 d\mu(z) d\mu(z_1) < +\infty. \tag{88}$$

If this is true not only for $x(g)$ but for $|x(g)|$ as well, then passage from (85) to (87) is justified. On the other hand, the integral in (88) is the trace of the operator $U^*_{\chi;x}U_{\chi;x} = U_{\chi;x^*x}$, the kernel of this operator being

$$K_1(z, z_1; \chi) = \int K(z, z_2; \chi)\overline{K(z_2 z_1; \chi)} d\mu(z_2). \tag{89}$$

Let us take $x_1 = x^*x$; then for the trace of the operator $U_{\chi;x^*x} = U_{\chi;x_1}$ we have the formula

$$S(U_{\chi;x_1}) = \int K_1(z, z; \chi) d\mu(z) \tag{90}$$

and the condition (88) can be rewritten as follows:

$$S(U_{\chi;x_1}) = \int K_1(z, z_1; \chi) d\mu(z) < +\infty. \tag{91}$$

Thus, *if the integral* (86) *converges absolutely and the condition* (91) *for the function* $|x(g)|$ *is satisfied then the operator* $U_{\chi;x_1} = U_{\chi;x^*x}$ *has the trace which is given by* (90).

Using the expression (86) for the kernel we can obtain formulas for the trace, namely, from (91) it follows that

$$S(U_{\chi;x_1}) = \int [x_1(z^{-1}kz)\beta^{-\frac{1}{2}}(k)\chi(k)d\mu_l(k)] d\mu(z). \tag{92}$$

From this and from the equalities (35) and (42) we have

$$S(U_{\chi;x_1}) = \iiint x_1(z^{-1}\delta\zeta z)\beta^{-\frac{1}{2}}(\delta)X(\delta)d\mu(\delta)d\mu(\zeta)d\mu(z), \tag{93}$$

$$S(U_{\chi;x_1}) = \int x_1(g)\frac{\chi(\lambda_g) + \overline{\chi}(\lambda_g)}{|\lambda_g - \lambda_g^{-1}|^2} d\mu(g) = \int x_1(g)\frac{\chi(\lambda_g) + \overline{\chi(\lambda_g)}}{|(g_{11} + g_{22})^2 - Y|} d\mu(g), \tag{94}$$

where, as in (42), $\lambda_g$ is an eigenvalue of the matrix $g$.

3. *The existence of the trace.* Let us denote by $Q_c$ the compact set of the elements of the group $G$ that satisfy the condition

$$|g_{pq}| \leqslant C \qquad (p, q = 1, 2). \tag{95}$$

Since

$$g^{-1} = \begin{pmatrix} g_{22} & -g_{12} \\ -g_{21} & g_{11} \end{pmatrix}$$

we see that $g \in Q_c$ implies $g^{-1} \in Q_c$. Let us also denote by $R'$ the set of functions $x(g) \in R$ that are continuous on the group $G$ and vanish outside some set $Q_c$. If $x_1(g), x_2(g) \in R'$ then obviously

$$x_1(g) + x_2(g) \in R', \qquad \lambda x_1(g) \in R' \qquad \text{and} \qquad (x_1(g))^* \in R'.$$

If, further, $x_1(g), x_2(g) \in R'$ then the convolution

$$x(g) = (x_1 * x_2)(g) \in R'.$$

To see this, we denote by $Q_{c_1}Q_{c_2}$ the set of all products $g_1 g_2$, $g_1 \in Q_{c_1}$, $g_2 \in Q_{c_2}$; we have $Q_{c_1}Q_{c_2} \subset Q_{2c_1c_2}$. Now, if $x_1(g)$ and $x_2(g)$ vanish outside $Q_{c_1}$ and $Q_{c_2}$ respectively, then the integral in (81) vanishes except when $g_1^{-1} \in Q_{c_1}$; on the other hand, in this integral $g_1$ should belong to $Q_{c_2}$ because outside $Q_{c_2}$ we have $x(g_2) = 0$, So the function $(x_1 * x_2)(g)$ can be nonzero only when $g \in Q_{c_1}Q_{c_2} \subset Q_{2c_1c_2}$, i.e. $(x_1 * x_2)(g) \in R'$.

I. *For any function* $x(g) \in R'$ *the integral in* (86) *converges absolutely for all* $z$ *and* $z_1$.

To check this let us put

$$z^{-1} = \begin{pmatrix} 1 & 0 \\ -z & 1 \end{pmatrix}, \quad z_1 = \begin{pmatrix} 1 & 0 \\ z_1 & 1 \end{pmatrix}, \quad k = \begin{pmatrix} k_{11} & k_{12} \\ 0 & k_{22} \end{pmatrix} \, (k_{11} = k_{22}^{-1})$$

and write $x(g)$ as a function in the parameters $g_{pq}$:

$$x(g) = x(g_{11}, g_{12}, g_{21}, g_{22}).$$

From (4) and (8) we get

$$\begin{aligned} \int |x(z^{-1}kz_1)| \beta^{-\frac{1}{2}}(k) d\mu_e(k) = \int |x(k_{22}^{-1} + k_{12}z_1, k_{12}, -k_{22}^{-1}z + z_1(-k_{12}z + k_{22}),& \\ -k_{12}z + k_{22})| \cdot |k_{22}|^{-2} du_{12} dv_{12} du_{22} dv_{22}.& \qquad (96) \end{aligned}$$

Let a function $x(g)$ vanish outside $Q_c$. Then the integral (96) is taken over $k_{pq}$ such that

$$|k_{12}| < C, \quad |-k_{12}z_1 + k_{22}| < C, \quad |k_{22}^{-1} + k_{12}z| < C.$$

These inequalities imply

$$|k_{22}| < C + |k_{12}| \cdot |z_1| < C(1 + |z_1|), \quad |k_{22}^{-1}| < C \quad (1 + |z|),$$

i.e. for fixed $z$ and $z_1$, the integral in the variables $k_{12}, k_{22}$ is taken over a bounded region which does not contain a vicinity of $k_{22} = 0$. So the integral (96) exists and the statement I is proved. Therefore we have proved *the existence of the kernel* $K(z, z_1; \chi)$ *for any function* $x(g) \in R$ that belong to $R'$.

Now let

$$T_\delta = \beta^{-\frac{1}{2}}(\delta) \int\int x(z^{-1}\delta\zeta z) d\mu(\zeta) d\mu(z) \qquad (97)$$

and let us prove that:

II. *For any function* $x(g) \in R'$ *the integral* (97) *converges absolutely for arbitrary* $\delta$.

To prove this we put

$$z = \begin{pmatrix} 1 & 0 \\ z & 0 \end{pmatrix}, \quad \delta \begin{pmatrix} \lambda^{-1} & 0 \\ 0 & \lambda \end{pmatrix}, \quad \zeta = \begin{pmatrix} 1 & \zeta \\ 0 & 1 \end{pmatrix}$$

and, as above, we write $x(g)$ as a function in the parameters $g_{ij}$:

$$x(g) = x(g_{11}, g_{12}, g_{21}, g_{22}).$$

Then the equality (97) can be rewritten in the form

$$T_\delta = |\lambda|^{-2} \iiiint x\left(\frac{1+\zeta z}{\lambda}, \frac{\zeta}{\lambda}, -\frac{1+\zeta z}{\lambda} + \lambda z, -\frac{\zeta z}{\lambda} + \lambda\right) dx\,dy\,d\xi\,d\eta. \tag{98}$$

We substitute in this integral the variables $g_{12}, g_{21}$ for $\zeta, z$:

$$\frac{\zeta}{\lambda} = g_{12}, \qquad -\frac{1+\zeta z}{\lambda} z + \lambda z = g_{21}. \tag{99}$$

From the second of the formulas (99) we get the equation

$$g_{12} z^2 - \left(\lambda - \frac{1}{\lambda}\right) z + g_{21} = 0. \tag{100}$$

Its solution is

$$z = \frac{\lambda - \lambda^{-1} + \sqrt{(\lambda - \lambda^{-1})^2 - Y g_{12} g_{21}}}{2 g_{12}}$$

Further,

$$\frac{1+\zeta z}{\lambda} = \lambda^{-1} + g_{12} z = \tfrac{1}{2}(\lambda + \lambda^{-1}) \pm \tfrac{1}{2}\sqrt{(\lambda - \lambda^{-1})^{-} - Y g_{12} g_{21}},$$

$$-\frac{z\zeta}{\lambda} + \lambda = -g_{12} z + \lambda = \tfrac{1}{2}(\lambda + \lambda^{-1}) \mp \tfrac{1}{2}\sqrt{(\lambda - \lambda^{-1})^2 - Y g_{12} g_{21}}.$$

The complex Jacobian of the transformation (99) from $\zeta, z$ to $g_{12}, g_{21}$ is

$$\frac{\partial(\zeta, z)}{\partial(g_{12}, g_{21})} = \frac{\partial \zeta}{\partial g_{12}} \cdot \frac{\partial z}{\partial g_{21}} = \pm \lambda \frac{1}{\sqrt{(\lambda - \lambda^{-1})^2 - Y g_{12} g_{21}}}.$$

We write $\mu = \lambda + \lambda^{-1}$ and obtain

$$\begin{aligned} T_\delta = \iiiint \Big[ & x\left(\frac{\mu}{2} + \tfrac{1}{2}\sqrt{\mu^2 - Y - Y g_{12} g_{21}}, g_{12}, g_{21}, \frac{\mu}{2} - \tfrac{1}{2}\sqrt{\mu^2 - Y - Y g_{12}, g_{21}}\right) \\ & + x\left(\frac{\mu}{2} - \tfrac{1}{2}\sqrt{\mu^2 - Y - Y g_{12} g_{21}}, g_{12}, g_{21}, \frac{\mu}{2} + \tfrac{1}{2}\sqrt{\mu^2 - Y - Y g_{12} g_{21}}\right) \Big] \\ & \cdot \frac{d\alpha_{12}\, d\beta_{12}\, d\alpha_{21}\, d\beta_{21}}{|\mu^2 - Y - Y g_{12} g_{21}|} (g_{12} = \alpha_{12} + i\beta_{12}, g_{21} = \alpha_{21} + i\beta_{21}). \end{aligned} \tag{101}$$

This formula shows that $T_\delta$ in fact depends only on $\mu = \lambda + \lambda^{-1}$, i.e. the trace of the matrix $\delta$.

The function $x(g)$ belonging to $R'$ is bounded, i.e. there exists $C_1$ such that $|x(g)| < C_1$; besides, $x(g)$ vanishes outside some $Q_c$. From (101) we have

$$|T_\delta| \leqq 2C_1 \iiiint_{Q_c} \frac{d\alpha_{12}\, d\beta_{12}\, d\alpha_{21}\, d\beta_{21}}{|\mu^2 - Y - Y g_{12} g_{21}|}. \tag{102}$$

Introduce polar coordinates in the integral (102) by

$$g_{12} = r, e^{i\varphi_1}, \qquad g_{21} = r_2 e^{i\varphi_2}$$

and write

$$\mu^2 - Y = Yre^{i\theta}. \tag{103}$$

Then, from inequality (102), we get

$$T_\delta \leqq \pi C_1 \int_0^c \int_0^c \left[ \int_0^{2\pi} \frac{d\theta}{|r - r_1 r_2 e^{i\theta}|} \right] r_1 r_2 dr_1 dr_2. \tag{104}$$

That means

$$T_\delta \leqq 2\pi^2 C_1 \int_0^c \int_0^c F\left(\frac{r}{r_1 r_2}\right) dr_1 dr_2 \tag{105}$$

where

$$F(u) = \frac{1}{2\pi} \int_0^{2\pi} \frac{d\theta}{|u - e^{i\theta}|}. \tag{106}$$

Let us consider some properties of the function $F(u)$. When $|u|$ is sufficiently small, then

$$F(u) = \frac{1}{2\pi} \int_0^{2\pi} (1 - 2u\cos\theta + u^2)^{-\frac{1}{2}} d\theta = \frac{1}{2\pi} \int_0^{2\pi} \left[ \sum_{n=0}^{\infty} P_n(\cos\theta) u^n \right] d\theta \tag{107}$$

where $P_n(z)$ is the $n$-th Legendre polynomial. Write

$$C_n = \frac{1}{2\pi} \int_0^{2\pi} P_n(\cos\theta)\, d\theta. \tag{108}$$

Obviously

$$C_0 = 1, \qquad C_1 = 0, \qquad |C_n| \leqq 1 \tag{109}$$

and therefore for $0 < u < 1$

$$|F(u) - 1| < \frac{u^2}{1-u}. \tag{110}$$

Furthermore, when $u > 0$,

$$F\left(\frac{1}{u}\right) = \frac{1}{2\pi} \int_0^{2\pi} \frac{d\theta}{|u^{-1} - e^{i\theta}|} = \frac{1}{2\pi} \int_0^{2\pi} \frac{u\,d\theta}{|u - e^{i\theta}|};$$

that means

$$F\left(\frac{1}{u}\right) = uF(u). \tag{111}$$

Finally, for $u > 1$,

$$F(u) < \frac{1}{2\pi} \int_0^{2\pi} \frac{d\theta}{u-1} = \frac{1}{u-1}. \tag{112}$$

The function $F(u)$ has the unique singularity in $u = 1$. Since $F(u)$ can be expressed via the elliptic integral

$$\int_0^{\pi/2} \frac{d\theta}{\sqrt{1 - k^2 \sin^2\theta}}$$

with $k^2 = Yu/(1+u)^2$, this singularity is of the logarithmic type. So $\int_a^b F(u)du$ converges for an arbitrary finite interval $(a, b)$.

We return now to the integral in (105); since the function $F(r/r_1r_2)$ is symmetric in the variables $r_1$ and $r_2$, the integral over the region $0 \leqq r_1 \leqq C$, $0 \leqq r_2 \leqq C$ is twice the integral over the region $0 \leqq r_1 \leqq C$; $0 \leqq r_2 \leqq r_1$.

Let us make the change of variables

$$r_1 = r_1, \qquad r_2 = \frac{u}{r_1}.$$

in the last integral. In these variables the integral is equal to

$$\int_0^{c^2}\left[\int_{\sqrt{u}}^{c} F\left(\frac{r}{u}\right)\frac{dr}{r_1}\right]du = \frac{1}{2}\int_0^{c^2} F\left(\frac{r}{u}\right)\ln\frac{C^2}{u}\,du.$$

We put $u = vr$ and the integral on the right side takes the form

$$\frac{r}{2}\int_0^{C^2/r} F\left(\frac{1}{v}\right)\ln\frac{C^2}{rv}\,dv. \tag{113}$$

The convergence of the integral (113) implies the absolute convergence of the integral in formula (97).

III. *For any function $x(g) \in R'$ the integral* (98) *converges uniformly in $\lambda$ in any bounded region of the $\lambda$-plane that does not contain a vicinity of the point $\lambda = 0$.*

The uniform convergence of the integral (98) in $\lambda$ is equivalent to the uniform convergence of the integral (101) in $v = \frac{1}{4}\mu^2 - Y$ in any bounded region of the $v$-plane. In view of the estimate (104) it is sufficient to prove the uniform convergence of the integral (104) in $r$ over an arbitrary finite interval $(a, b)$ in the positive part of the $r$-axis. To do this it is sufficient to prove the uniform convergence in $r$ of the integral (105) which in its turn follows from the uniform convergence of the integral (113).

In other words, we must show that for sufficiently small $\varepsilon_1, \varepsilon_2 > 0$

$$\frac{r}{2}\int_{1-\varepsilon_1}^{1+\varepsilon_2} F\left(\frac{1}{v}\right)\ln\frac{C^2}{rv}\,dv < \varepsilon$$

for any $r \in (a, b)$. But this is clear because the last integral can be written in the form

$$\frac{r}{2}\int_{1-\varepsilon_1}^{1+\varepsilon_2} F\left(\frac{1}{v}\right)\ln\frac{C^2}{v}\,dv - \frac{r}{2}\ln r\int_{1-\varepsilon_1}^{1+\varepsilon_2} F\left(\frac{1}{v}\right)dv$$

and our statement follows from the boundedness $r/2$ and $r/2\ln r$ and from the convergence of the integrals

$$\int_0^{1+\varepsilon_2} F\left(\frac{1}{v}\right)\ln\frac{C^2}{v}\,dv \quad \text{and} \quad \int F\left(\frac{1}{v}\right)dv.$$

As the integrand in (98) is a continuous function in $\mu$ and therefore in $\lambda$, the uniform convergence of the integral implies that $T\delta$ is a continuous function in $\lambda$. So it is proved that:

IV. *If $x(g) \in R'$ then $T_g$ is a continuous function in $\lambda$.*

As the function $x(g)$ vanishes outside some compact $Q_c$ and the element $\lambda$ and $\lambda^{-1}$ of the matrix $\delta$ are the eigenvalues of the matrix $g = z^{-1}\delta\zeta z$, we conclude that $T_\delta = 0$ for all sufficiently small and sufficiently large $|\lambda|$. Hence:

V. *If $x(g) \in R'$ the integral* (93) *converges absolutely.* In fact the formula (93) can be written as follows

$$S(U_{\chi;x_1}) = \int T_g \chi(\delta) d\mu(\delta). \tag{114}$$

This integral is the Fourier transform of the integrable function $T\delta$ and therefore it converges.

If we change the function $x(g) \in R'$ for the function $|x(g)|$, which also belongs to $R'$, we obtain the absolute convergence of the integral (93).

**Theorem 2.** *For any function $x(g) \in R'$ the operator $U_{\chi;x_1}$ which corresponds to the function $x_1(g) = (x^* * x)(g)$ has the trace which is given by formulas* (93) *and* (94). *The trace is a continuous function in $X$.*

*Proof.* In view of II, the existence of the trace follows from the convergence of the integral (86) for $x_2(g) = |x(g)|$ and the convergence of the integral (93) for $(x_2^* * x_2)(g)$. The first condition is satisfied by statement I. Since $(x_2^* * x_2)(g)$ also belongs to $R'$ the second condition is satisfied by $V$.

The continuity of the trace follows from the statement that by (114), the trace is the Fourier transform of an integrable function. The theorem is proved.

Comparing the formula (94) with the expression (85) for $U_{X;x}$ we see that it is natural to attach to the operator $U_{\chi;g}$ the trace

$$S(U_{\chi;g}) = \frac{\chi(\lambda_g) + \bar{\chi}(\lambda_g)}{|\lambda_g - \lambda_g^{-1}|^2}. \tag{115}$$

This trace exists for all elements $g \in G$ with the exception of those with eigenvalues $\pm 1$.

**Theorem 3.** *If $\chi_1$ and $\chi_2$ are two different characters of the group $D$ and $\chi_1(\delta) = \chi_2(\delta)$ then the representations $U_{\chi_1;g}$ and $U_{\chi_2;g}$ are not equivalent.*

*Proof.* If these representations are equivalent then the corresponding representations $U_{\chi_1;x}$ and $U_{\chi_2;x}$ of the group ring are also equivalent. But then the operators $U_{\chi_1;x}$ and $U_{\chi_2;x}$ have the same trace. Let us apply this result to the functions

$$x = x_1 \pm x_2, \qquad x = x_1 \pm ix_2$$

where $x_1, x_2 \in R'$. We have

$$S(U_{\chi_1;x_1^* x_2}) = S(U_{\chi_1;x_1^* x_2})$$

Let $x_1(g)$ be a function vanishing in some neighbourhood of the unity of the group, $g = e$, and let $x_2(g)$ be a function equal to zero in some neighbourhood of the element $g = \delta = \begin{pmatrix} \lambda^{-1} & 0 \\ 0 & \lambda \end{pmatrix}$. We normalize these functions and consider the limit of the both parts of the equality

$$S(U_{X_1;x_1^* x_2}) = S(U_{X_2;x_1^* x_2})$$

when the chosen neighbourhoods shrink to the corresponding points. In view of the formula (94) for the trace we get

$$X_1(\lambda) + \overline{X_1(\lambda)} = X_2(\lambda) + \overline{X_2(\lambda)}$$

for arbitrary $\lambda$. This is possible only if either $X_1(\lambda) = X_2(\lambda)$ or $X_1(\lambda) = X_2(\lambda)$.

**Theorem 4.** *For any integrable function $x(g)$ the operator $U_{X;x}$ is compact.*

*Proof.* If $x_2(g) = (x_1^* \cdot x_1)(g)$ and $x_2(g) \in R'$ then the operator

$$U_{X;x_2} = U_{X;x_1}^* \cdot U_{X;x_1}$$

has the trace and hence this operator is compact.

On the other hand for arbitrary $\varepsilon > 0$ there exists a function $x_1(g) \in R'$ such that

$$\int |x(g) - x_1(g)| \, d\mu(g) < \varepsilon.$$

But then

$$\| U_{X;x} - U_{X;x_1} \| = \| \int [x(g) - x_1(g)] U_{X;g} d\mu(g) \| \leqq \int |x(g) - x_1(g)| \, d\mu(g) < \varepsilon,$$

i.e. the operator $U_{X;x}$ is the limit of the operators $U_{X;x_1}$ with $x_1 \in R'$. So the operator $U_{X;x}$ is also compact.

## §7. Decomposition of the regular representation of the group $G$ into irreducible representations; analog of the Plancherel theorem

We will show that the regular representation of the group $G$ can be decomposed into a continuous direct sum of the representations of the principal series. The main role in the proof of this fact is played by a formula similar to the Plancherel formula for commutative groups. First we deduce this formula.

Given a function $x(g)$, we construct some auxiliary functions related to it.

1. *Functions $X$, $\varphi$ and $\Phi$.* Let $x(g)$ be an integrable function on $G$. We introduce the following parameters in $G$:

$$p_1 = \frac{g_{11}}{g_{12}}; \quad p_2 = \frac{1}{g_{12}}; \quad p_3 = \frac{g_{22}}{g_{12}}. \tag{116}$$

The function $x(g)$ written in the parameters $p_1, p_2, p_3$ will be denoted by $\tilde{x}(p_1, p_2, p_3)$.

The Jacobian of the transformation (116) is $(g_{12}^3 g_{22})^{-2}$ and

$$\int |x(g)| \, d\mu(g) = \int |\tilde{x}(p_1, p_2, p_3)| \cdot |p_2|^{-6} d\mu(p_1) d\mu(p_2) d\mu(p_3) \tag{117}$$

where $d\mu(p_k)$ is the product of the real and imaginary parts of $p_k$, $k = 1, 2, 3$.

Formula (117) means that the function $\tilde{x}(p_1, p_2, p_3) \cdot |p_2|^{-6}$ is integrable (in $p_1, p_2, p_3$). Let us suppose that the function $\tilde{x}(p_1, p_2, p_3) \cdot |p_2|^{-4}$ is also integrable. Then the Fourier transform of this function, which we denote by $X(\zeta_1, \zeta_2, \zeta_3)$, exists

i.e.

$$X(\zeta_1,\zeta_2,\zeta_3)=\frac{1}{(2\pi)^3}\int \tilde{x}(p_1,p_2,p_3)\cdot|p_2|^{-4}e^{-i\operatorname{Re}(\bar{p}_1\zeta_1-\bar{p}_2\zeta_2+\bar{p}_3\zeta_3)}d\mu(p_1)d\mu(p_2)d\mu(p_3). \tag{118}$$

Consider now the function

$$\varphi(z_1,z_2,\lambda)=|\lambda|^{-2}\int x(z_1^{-1}\delta\zeta z_2)d\mu(\zeta) \tag{119}$$

and its Fourier transform in $z_1$ and $z_2$:

$$\Phi(w_1,w_2,\lambda)=\frac{1}{(2\pi)^2}\int \varphi(z_1,z_2,\lambda)e^{i\operatorname{Re}(\bar{w}_1z_1-\bar{w}_2z_2)}d\mu(z_1)d\mu(z_2). \tag{120}$$

If

$$g=z_1^{-1}\delta\zeta z_2,\quad z_j=\begin{pmatrix}1 & 0\\ z_j & 1\end{pmatrix},\quad j=1,2;\quad \delta=\begin{pmatrix}\lambda^{-1} & 0\\ 0 & \lambda\end{pmatrix};\quad \zeta=\begin{pmatrix}1 & \zeta\\ 0 & 1\end{pmatrix},$$

then

$$g_{11}=\frac{1+\zeta z_2}{\lambda},\quad g_{12}=\frac{\zeta}{\lambda},\quad g_{22}=-\frac{z_1\zeta}{\lambda}+\lambda$$

and we have

$$p_1=z_2+\zeta^{-1},\quad p_2=\lambda\zeta^{-1},\quad p_3=-z_1+\lambda^2\zeta^{-1}. \tag{121}$$

The equality (119) can be written in these parameters as follows:

$$\varphi(z_1,z_2,\lambda)=|\lambda|^{-2}\int \tilde{x}(z_2+\zeta^{-1},\lambda\zeta^{-1},-z_1+\lambda^2\zeta^{-1})d\mu(\zeta). \tag{122}$$

We put this expression for $\varphi(z_1,z_2,\lambda)$ in the formula (120) and obtain for $\Phi(w_1,w_2,\lambda)$ the formula

$$\Phi(w_1,w_2,\lambda)$$
$$=\frac{1}{(2\pi)^2}\int \tilde{x}(z_2+\zeta^{-1},\lambda\zeta^{-1},-z_1+\lambda^2\zeta^{-1})\cdot|\lambda|^{-2}e^{i\operatorname{Re}(\bar{w}_1z_1-\bar{w}_2z_2)}d\mu(z_1)d\mu(z_2)d\mu(\zeta).$$

In this integral we change variables by formula (121); the complex Jacobian of this transformation is $\lambda\zeta^{-2}=p_2\lambda^{-1}$, so the function $\Phi$ can be now expressed as follows:

$$\Phi(w_1,w_2,\lambda)$$
$$=\frac{1}{(2\pi)^2}\int \tilde{x}(p_1,p_2,p_3)\cdot|p_2|^{-4}e^{i\operatorname{Re}[\bar{w}_1(\lambda p_2-p_3)-\bar{w}_2(p_1-p_2\lambda^{-1})]}\cdot d\mu(p_1)d\mu(p_2)\cdot d\mu(p_3)$$
$$=\frac{1}{(2\pi)^2}\int \tilde{x}(p_1,p_2,p_3)\cdot|p_2|^{-4}e^{-i\operatorname{Re}[\bar{w}_2p_1-(\bar{w}_1\lambda+\bar{w}_2\lambda^{-1})p_2+\bar{w}_1p_3]}\cdot d\mu(p_1)d\mu(p_2)d\mu(p_3).$$

Since we assumed the existence of the integral

$$\int|\tilde{x}(p_1,p_2,p_3)|\cdot|p_2|^{-4}d\mu(p_1)d\mu(p_2)d\mu(p_3),$$

the integrals (119) and (120) also exist. Moreover, comparison with (118) shows that

$$\frac{1}{2\pi}\Phi(w_1,w_2,\lambda)=X\left(w_2,w_1\bar{\lambda}+\frac{w_2}{\bar{\lambda}},w_1\right). \tag{123}$$

The sum $w_1\bar{\lambda}+w_2/\bar{\lambda}$ does not change if we replace $\lambda$ by $\bar{w}_2/\lambda\bar{w}_1$. Therefore, the equality (123) implies that the function $\Phi(w_1, w_2, \lambda)$ satisfies the condition

$$\Phi\left(w_1, w_2, \frac{\bar{w}_2}{\lambda\bar{w}_1}\right)=\Phi(w_1, w_2, \lambda). \tag{124}$$

Conversely, if a function $\Phi$ satisfies the condition (124), then the formula (123) determines the function $X$ uniquely. We can put

$$X(\zeta_1, \zeta_2, \zeta_3)=\frac{1}{2\pi}\Phi\left(\zeta_3, \zeta_1, \frac{\bar{\zeta}_2+\sqrt{\bar{\zeta}_2^2-Y\bar{\zeta}_1\bar{\zeta}_3}}{2\bar{\zeta}_3}\right). \tag{124'}$$

In view of formula (124) the choice of sign for the function $\sqrt{\bar{\zeta}_2^2-Y\bar{\zeta}_1\bar{\zeta}_3}$ is irrelevant.

The functions $\varphi$ and $\Phi$ will play an essential role in the subsequent considerations. This can already be seen from the fact that, in view of (86), their Fourier transforms over the multiplicative group of complex numbers $\lambda$

$$\int\varphi(z_1, z_2, \lambda)|\lambda|^{n+i\rho}\lambda^{-n}\frac{d\mu(\lambda)}{|\lambda|^2}=K(z_1, z_2, X)\equiv K(z_1, z_2, n, \rho)$$

$$(\lambda=\sigma+i\tau, d\mu(\lambda)=d\sigma\, d\tau) \tag{125}$$

is the kernel of the operator $U_{X;x}=U_{n,\rho;x}$.

Similarly,

$$\int\Phi(w_1, w_2, \lambda)|\lambda|^{n+i\rho}\lambda^{-n}\frac{d\mu(\lambda)}{|\lambda|^2}=L(w_1, w_2, n, \rho), \tag{126}$$

where $L(w_1, w_2, n, \rho)$ is the Fourier transform in $z_1$ and $z_2$ of the kernel $K(z_1, z_2, n, \rho)$.

From the formula (124) a relation between functions $L(w_1, w_2, n, \rho)$ and $L(w_1, w_2, -n, -\rho)$ follows, namely, it is easy to see that

$$L(w_1, w_2, -n, -\rho)=\left|\frac{w_1}{w_2}\right|^{n+i\rho}\left(\frac{\bar{w}_1}{\bar{w}_2}\right)^{-n}L(w_1, w_2, n, \rho). \tag{127}$$

2. *An analog of the Plancherel formula.* We now go on to the proof of the formula which is an analog of the Plancherel formula for commutative groups.

Let us assume first that a function $x(g)=\tilde{x}(p_1, p_2, p_3)$ is measurable, bounded and that it vanishes outside a region

$$|p_1|\leqq C_1, \quad |p_3|\leqq C_3, \quad 0<\varepsilon_2\leqq|p_2|\leqq C_2. \tag{128}$$

We denote the set of all such functions by $\mathfrak{H}'_G$. Then from the formulas

$$X(\zeta_1, \zeta_2, \zeta_3)=\frac{1}{(2\pi)^3}\int\tilde{x}(p_1, p_2, p_3)\cdot|p_2|^{-4}e^{-i\operatorname{Re}(\bar{p}_1\zeta_1-\bar{p}_2\zeta_2+\bar{p}_3\zeta_3)}\cdot d\mu(p_1)d\mu(p_2)d\mu(p_3);$$

$$\cdot X''_{\zeta_2\bar{\zeta}_2}(\zeta_1, \zeta_2, \zeta_3)=-\frac{1}{4(2\pi)^3}\int\tilde{x}(p_1, p_2, p_3)\cdot|p_2|^{-2}e^{-i\operatorname{Re}(\bar{p}_1\zeta_1-\bar{p}_2\zeta_2+\bar{p}_3\zeta_3)}$$

$$\cdot d\mu(p_1)d\mu(p_2)d\mu(p_3),$$

it follows by the usual Plancherel theorem that

$$\begin{aligned} &-\int X(\zeta_1,\zeta_2,\zeta_3)\overline{X''_{\zeta_2\bar{\zeta}_2}(\zeta_1,\zeta_2,\zeta_3)}\,d\mu(\zeta_1)d\mu(\zeta_2)d\mu(\zeta_3)\\ &=\tfrac{1}{4}\int|\tilde{x}(p_1,p_2,p_3)|^2\cdot|p_2|^{-6}d\mu(p_1)d\mu(p_2)d\mu(p_3)=\tfrac{1}{4}\int|x(g)|^2d\mu(g). \end{aligned} \tag{129}$$

If we replace the function $X(\zeta_1,\zeta_2,\zeta_3)$ by $\Phi(w_1,w_2,\lambda)$ then we get

$$\tfrac{1}{4}\int|x(g)|^2d\mu(g)=-\frac{1}{(2\pi)^2}\int\Phi(w_1,w_2,\lambda)\overline{\Phi''_{\lambda\bar{\lambda}}(w_1,w_2,\lambda)}\,d\mu(w_1)d\mu(w_2)d\mu(\lambda). \tag{130}$$

Now let us assume that a function $\tilde{x}(p_1,p_2,p_3)$ has continuous second derivatives in all variables and vanishes outside a region of the type (128). We denote the set of all such functions by $\mathfrak{H}''_G$. Obviously $\mathfrak{H}''_G\subset\mathfrak{H}'_G\subset\mathfrak{H}_G$.

From the formula (122) for $\varphi(z_1,z_2,\lambda)$ it follows that if $\tilde{x}\in\mathfrak{H}''_G$ then there exists a continuous derivative $\varphi''_{\lambda\bar{\lambda}}(z_1,z_2,\lambda)$. Moreover, the function $\Phi''_{\lambda\bar{\lambda}}(w_1,w_2,\lambda)$ is the Fourier transform of the function $\varphi''_{\lambda\bar{\lambda}}(z_1,z_2,\lambda)$ in $z_1$ and $z_2$. Therefore, for a fixed $\lambda$, we have

$$\int\Phi(w_1,w_2,\lambda)\overline{\Phi''_{\lambda\bar{\lambda}}(w_1,w_2,\lambda)}\,d\mu(w_1)d\mu(w_2)=\int\varphi(z_1,z_2,\lambda)\overline{\varphi''_{\lambda\bar{\lambda}}(z_1,z_2,\lambda)}\,d\mu(z_1)d\mu(z_2)$$

and formula (130) can be written in the form

$$\frac{1}{4}\int|x(g)|\,d\mu(g)=-\frac{1}{(2\pi)^2}\int\varphi(z_1,z_2,\lambda)\overline{\varphi''_{\lambda\bar{\lambda}}(z_1,z_2,\lambda)}\,d\mu(z_1)d\mu(z_2). \tag{131}$$

By assumption, $x(g)\in\mathfrak{H}''_G$. Therefore the integral is taken over a region

$$|z_2+\zeta^{-1}|\leqq C_1,\quad 0<\varepsilon_2\leqq|\lambda\zeta^{-1}|\leqq C_2,\ |-z_1+\lambda^2\zeta^{-1}|\leqq C_3.$$

Therefore, for fixed $z_1$ and $z_2$, the function $\varphi(z_1,z_2,\lambda)$ vanishes outside a region

$$0<\varepsilon\leqq|\lambda|\leqq C.$$

So the integral

$$K(z_1,z_2,n,\rho)=\int\varphi(z_1,z_2,\lambda)\cdot|\lambda|^{n+i\rho}\lambda^{-n}\frac{d\mu(\lambda)}{|\lambda|^2} \tag{132}$$

Converges. Let us put

$$\lambda=e^te^{i\theta},\ -\pi\leqq\theta\leqq\pi,\ \varphi(z_1,z_2,\lambda)=\psi(z_1,z_2,t,\theta).$$

Then formula (131) can be written as follows:

$$\int|x(g)|^2d\mu(g)=-\frac{1}{(2\pi)^2}\int_{-\pi}^{\pi}\int_{-\infty}^{\infty}\left[\int\psi(z_1,z_2,t,\theta)\overline{\Delta\psi(z_1,z_2,t,\theta)}\,d\mu(z_1)d\mu(z_2)\right]dtd\theta \tag{133}$$

where $\Delta\psi=(\partial^2\psi/\partial t^2)+(\partial^2\psi/\partial\theta^2)$. The function $\psi$ has continuous derivatives in $t$ and $\theta$ and vanishes outside the set $|t|\leqq C$.

Moreover, the formula can be written in the form

$$K(z_1,z_2,n,\rho)=\int_{-\pi}^{\pi}\int_{-\infty}^{\infty}\psi(z_1,z_2,t,\theta)e^{it\rho}e^{-in\theta}dtd\theta \tag{134}$$

where integration in $t$ is, in fact, taken over a region of the form $|t| \leqq c$. The formula (134) then implies that

$$-(n^2+\rho^2)K(z_1, z_2, n, \rho) = \int_{-\pi}^{\pi}\int_{-\infty}^{\infty} \Delta\psi(z_1, z_2, t, \theta)e^{it\rho}e^{-in\theta}dtd\theta.$$

By the usual Plancherel theorem we have

$$\frac{1}{(2\pi)^2}\sum_{n=-\infty}^{\infty}\int_{-\infty}^{\infty}|K(z_1, z_2, n, \rho)|^2(n^2+\rho^2)d\rho = -\int_{-\infty}^{\infty}\int_{-\pi}^{\pi}\psi(z_1, z_2, t, \theta)\Delta\psi(z_1, z_2, n, \theta)dtd\theta.$$

So from the formula (133) it follows that

$$\int|x(g)|^2d\mu(g) = \frac{1}{(2\pi)^2}\sum_{n=-\infty}^{\infty}\int_{-\infty}^{\infty}[\iint|K(z_1, z_2, n, \rho)|^2d\mu(z_1)d\mu(z_2)](n^2+\rho^2)d\rho. \quad (135)$$

Furthermore, the formula (127) shows that

$$|L(w_1, w_2, -\eta, -\rho)| = |L(w_1, w_2, n, \rho)|$$

and therefore

$$\begin{aligned}\int|K(z_1, z_2, n, \rho)|^2d\mu(z_1)d\mu(z_2) &= \int|L(w_1, w_2, n, \rho)|^2d\mu(w_1)d\mu(w_2)\\ &= \int|L(w_1, w_2, -n, -\rho)|^2d\mu(w_1)d\mu(w_2) = \int|K(z_1, z_2, -n, -\rho)|^2d\mu(z_1)d\mu(z_2).\end{aligned} \quad (136)$$

Now we can rewrite the formula (135) in the form

$$\int|x(g)|d\mu(g) = \frac{1}{8\pi^2}\sum_{n=-\infty}^{n=\infty}\int_{0}^{\infty}[\iint|K(z_1, z_2, n, \rho)|^2d\mu(z_1)d\mu(z_2)](n^2+\rho^2)d\rho. \quad (137)$$

We call formula (137) the analog of the Plancherel formula.

Let us note that the integral

$$\iint|K(z_1, z_2, n, \rho)|^2d\mu(z_1)d\mu(z_2)$$

is the trace of the operator $U_{n,\rho;x^*x}$. Therefore, formula (137) can be rewritten in the form

$$\int|x(g)|^2d\mu(g) = \frac{1}{8\pi^2}\sum_{n=-\infty}^{\infty}\int_{0}^{\infty}S(U_{n,\rho;x^*x})(n^2+\rho^2)d\rho. \quad (138)$$

3. *Inversion formulas.* The formula (137) was obtained under the assumption that $\tilde{x}\in\mathfrak{H}_G''$. It is easy to generalize this formula to all functions belonging to $\mathfrak{H}_G$. In doing so we will obtain analogs to the inversion formulas in the Plancherel theorem.

Denote by $\mathfrak{H}$ the set of all measureable functions $K(z_1, z_2, n, \rho)$ $(n=0, \pm 1, \pm 2, \ldots;$ $0 \leqq \rho < +\infty)$ such that

$$\sum_{n=-\infty}^{\infty}\int_{0}^{\infty}[\iint|K(z_1, z_2, n, \rho)|^2d\mu(z_1)d\mu(z_2)](n^2+\rho^2)d\rho < +\infty. \quad (139)$$

We will consider $\mathfrak{H}$ as the Hilbert space with the usual addition and multiplication by scalars and with inner product given by

$$(K, K') = \frac{1}{8\pi^2} \sum_{n=-\infty}^{\infty} \int_0^{\infty} [\iint K(z_1, z_2, n, \rho) K'(z_1, z_2, n, \rho) d\mu(z_1) d\mu(z_2)](n^2 + \rho^2) d\rho. \tag{140}$$

The formula (137) shows that the mapping $x(g) \to K(z_1, z_2, n, \rho)$ is an isometric transformation from $\mathfrak{H}_G''$ to $\mathfrak{H}$. Since $\mathfrak{H}_G''$ is dense in $\mathfrak{H}_G$ this isometry can be uniquely extended to the isometric transformation of the whole space $\mathfrak{H}_G$.

This transformation is easy to describe. First, let a function $\tilde{x} \in \mathfrak{H}_G$ vanish outside a set of the type (128). Then, in $\mathfrak{H}_G''$ exists a sequence of functions $x_m(g) = \tilde{x}(p_1, p_2, p_3)$ which vanishes outside the same set and which are such that

$$\int |x(g) - x_m(g)|^2 d\mu(g) \to 0 \qquad \text{for } m \to \infty;$$

in the parameters $p_1, p_2, p_3$ we get

$$\int |\tilde{x}(p_1, p_2, p_3) - \tilde{x}_m(p_1, p_2, p_3|^2 \cdot |p_2|^{-6} d\mu(p_1) d\mu(p_2) d\mu(p_3) \tag{141}$$

We associate the function $K_m(z_1, z_2, n, \rho)$ to the function $\tilde{x}_m$; since the sequence $\tilde{x}_m$ converges in $\mathfrak{H}_G$, the sequence $K_m$ converges in $\mathfrak{H}$ (by isometry of the transformation). Denote its limit by $K(z_1, z_2, n, \rho)$. Then formula (137) is true for the function $x(g)$ and $K(z_1, z_2, n, \rho)$. From the definition of the norm in $\mathfrak{H}$ it follows also that we can choose a subsequence (denote it again by $K_m(x_m)$) such that

$$K_m(z_1, z_2, n, \rho) \to K(z_1, z_2, n, \rho) \qquad \text{for } m \to \infty \tag{142}$$

for all $n$ and almost all $z_1, z_2, \rho$.

We change the variables in the integral (141) from $p_1, p_2, p_3$ to $z_1, \zeta, \lambda$ by means of the formulas

$$p_1 = z_2 + \zeta^{-1}, \qquad p_2 = \lambda\zeta^{-1}, \qquad p_3 - z_1 + \lambda^2\zeta^{-1}.$$

In the variables $z, \zeta, \lambda$, (141) takes the form

$$\int |\tilde{x}(z_2 + \zeta^{-1}, \lambda\zeta^{-1}, -z, +\lambda^2\zeta^{-1}) - \tilde{x}_m(z_2 + \zeta^{-1}, \lambda\zeta^{-1}, -z_1 + \lambda^2\zeta^{-1})$$
$$\cdot |\lambda|^{-6} d\mu(\zeta) d\mu(\lambda) d\mu(z_1) \to 0 \qquad \text{for } m \to \infty.$$

Hence, by (122), we have

$$\int |\varphi(z_1, z_2\lambda) - \varphi_m(z_1, z_2, \lambda)| \cdot |\lambda|^{-4} d\mu(\lambda) d\mu(z_1) \to 0 \qquad \text{for } m \to \infty.$$

We again choose a subsequence (and again denote it by $\tilde{x}_m$) such that

$$\int |\varphi(z_1, z_2, \lambda) - \varphi_m(z_1, z_2, \lambda)| \cdot |\lambda|^{-4} d\mu(\lambda) \to 0 \qquad \text{for } m \to \infty \tag{143}$$

for almost all $z_1, z_2$. But for fixed $z_1$ and $z_2$ the integration in $\lambda$ is taken over the region $0 < \varepsilon \leqq |z| \leqq C$. Therefore, from (143), it follows that

$$\int |\varphi(z_1, z_2, \lambda) - \varphi_m(z_1, z_2\lambda)| \frac{d\mu(\lambda)}{|\lambda|^2} \to 0 \qquad \text{for } m \to \infty;$$

hence

$$\int \varphi_m(z_1, z_2, \lambda) |\lambda|^{n+i\rho} \lambda^{-n} \frac{d\mu(\lambda)}{|\lambda|^2} \to \int \varphi(z_1, z_2, \lambda) |\lambda|^{n+i\rho} \lambda^{-n} \frac{d\mu(\lambda)}{|\lambda|^2} \qquad \text{for } m \to \infty.$$

This means that

$$K_m(z_1, z_2, n, \rho) \to \int \varphi(z_1, z_2, \lambda) \cdot |\lambda|^{n+i\rho} \lambda^{-n} \frac{d\mu(\lambda)}{|\lambda|} \quad \text{for } m \to \infty.$$

Comparing this result with formula (142) we see that

$$K(z_1, z_2, n, \rho) = \int \varphi(z_1, z_2, \lambda) \cdot |\lambda|^{n+i\rho} \lambda^{-n} \frac{d\mu(\lambda)}{|\lambda|^2}$$

$$= \int x(z_1^{-1} \delta \zeta z_2) |\lambda|^{n+i\rho-2} \lambda^{-n} \frac{d\mu(\lambda)}{|\lambda|^2} d\mu(\zeta), \tag{144}$$

i.e. $K(z_1, z_2, n, \rho)$ is the kernel of the operator $U_{n,\rho ix}$.

So we have proved that *if a function* $x(g) \in \mathfrak{H}_G$ *vanishes outside a set of the type* (128) *then the passage to the kernel* $K(z_1, z_2, n, \rho)$ *is an isometric transformation* $\mathfrak{H}_G \to \mathfrak{H}$; *in other words, for such* $x(g)$ *and* $K(z_1, z_2, n, \rho)$ *formula* (137) *is true.*

Now let $x(g)$ be an arbitrary function. Denote by $\hat{x}(g)$ a function which is equal to $x(g)$ on the set (128) and which is zero outside this set. The above statement is true for $\hat{x}(g)$. Hence *for an arbitrary function* $x(g)$ *the last integral in formula* (144) *converges by norm in* $\mathfrak{H}$ *and for the function* $K(z_1, z_2, n, \rho)$ *the formula* (137) *is true.*

Now we show that each function $K \in \mathfrak{H}$ is the image of some $x(g)$ under this transformation. Moreover we find the formula for the transformation.

Denote by $\mathfrak{H}^1_\Phi$ the set of all measurable functions $\Phi(w_1, w_2, \lambda)$ that satisfy (124) and

$$\|\Phi\|_1^2$$

$$= \int |\Phi(w_1, w_2, \lambda)|^2 d\mu(w_1) d\mu(w_2) \frac{d\mu(\lambda)}{|\lambda|^2} + \int |\Delta_\lambda \Phi|^2 \cdot |\lambda|^2 d\mu(w_1) d\mu(w_2) d\mu(\lambda) < +\infty. \tag{145}$$

Going from $\Phi(w_1, w_2, \lambda)$ to $K(z_1, z_2, n, \rho)$ we map $\mathfrak{H}^1_\Phi$ isometrically into the space $\mathfrak{H}^1$ of all functions $K(z_1, z_2, n, \rho)$ that satisfy the condition

$$\|K\|_1^2 = \sum_{n=-\infty}^{\infty} \int_0^{\infty} \left[\iint |K(z_1, z_2, n, \rho)|^2 d\mu(z_1) d\mu(z_2)\right] \cdot [(n^2 + \rho^2)^2 + 1] d\rho < +\infty.$$

Obviously, all functions $K(z_1, z_2, n, \rho)$ that vanish when $|n| > n_0$ or $\rho > \rho_0$ belong to $\mathfrak{H}^1$. So $\mathfrak{H}^1$ is dense in $\mathfrak{H}$ (in the norm of $\mathfrak{H}$). For $\Phi \in \mathfrak{H}^1_\Phi$, the integral

$$\frac{1}{(2\pi)^2} \int |\Phi_\lambda(w_1, w_2, \lambda)|^2 d\mu(w_1) d\mu(w_2) d\mu(\lambda) = \int |X_{\zeta_2}(\zeta_1, \zeta_2, \zeta_3)|^2 d\mu(\zeta_1) d\mu(\zeta_2) d\mu(\zeta_3).$$

exist. We put

$$\tilde{x}(p_1, p_2, p_3) \cdot |p_2|^{-4} \frac{p_2}{2} = \frac{1}{(2\pi)^3} \int X_{\zeta_2}(\zeta_1, \zeta_2, \zeta_3) e^{i \operatorname{Re}(\bar{p}_1 \zeta_1 - \bar{p}_2 \zeta_2 + \bar{p}_3 \zeta_3)} d\mu(\zeta_1) d\mu(\zeta_2) d\mu(\zeta_3).$$

This integral converges in the mean and by the usual Plancherel theorem we have

$$\tfrac{1}{4} \int |\tilde{x}(p_1, p_2, p_3)| \cdot |p_2|^{-6} d\mu(p_1) d\mu(p_2) d\mu(p_3) = \int |X_{\zeta_2}|\zeta_1, \zeta_2, \zeta_3)|^2 d\mu(\zeta_1) d\mu(\zeta_2) d\mu(\zeta_3)$$

$$= \frac{1}{(2\pi)^2} \int |\Phi_\lambda(w_1, w_2, \lambda)|^2 d\mu(w_1) d\mu(w_2) d\mu(\lambda).$$

Hence, it follows that $x(g)=\tilde{x}(p_1, p_2, p_3)$ belongs to $\mathfrak{H}_G$. So the constructed isometry maps $\mathfrak{H}_G$ onto some closed subspace of the space $\mathfrak{H}$ and this subspace contains $\mathfrak{H}^1$. But $\mathfrak{H}^1$ is dense in $\mathfrak{H}$, so this subspace coincides with $\mathfrak{H}$.

In other words, the transformation $x(g)\to K(z_1, z_2, n, \rho)$ is an isometry $\mathfrak{H}_G$ on $\mathfrak{H}$.

Now we construct the converse transformation $\mathfrak{H}$ on $\mathfrak{H}_G$. First let $K(z_1, z_2, n, \rho)\in\mathfrak{H}$ be a function vanishing outside a set

$$|z_1|\leqq C_1, \quad |z_2|\leqq C_2, \quad |n|\leqq n_0, \quad 0\leqq \rho\leqq \rho_0.$$

To this function corresponds the function $x(g)\in\mathfrak{H}_G$. We put

$$y(g)=\frac{1}{8\pi^4}\sum_{n=-\infty}^{\infty}\int_0^{\infty}[\int K(z_1, z\bar{g}, n, \rho)\beta^{-\frac{1}{2}}(zg)\overline{\chi_{n,\rho}(zg)}d\mu(z)](n^2+\rho^2)d\rho, \tag{146}$$

and we will prove that $y(g)=x(g)$. Let $x(g)$ be a function from $R'$; we consider the integral

$$\int\overline{x'(g)}\,y(g)d\mu(g)$$
$$=\frac{1}{8\pi^4}\sum_{n=-\infty}^{\infty}\int_0^{\infty}[\int\int\overline{x'(g)}K(z, z\bar{g}, n, \rho)\beta^{-\frac{1}{2}}(zg)\overline{\chi_{n,\rho}(zg)}d\mu(g)d\mu(z)](n^2+\rho^2)d\rho. \tag{147}$$

The inner integral is the result of the substitution $z_2=z_1=z$ in the image of the function $K(z_1, z_2, n, \rho)$ considered as a function in $z_2$ under the operator $U_{-n,-\rho;\bar{x}'}$. Let $\overline{K'(z_1, z_2, n, \rho)}$ be the kernel of this operator; then the kernel of the operator $U_{n,\rho;x'}$ is $K'(z_1, z_2, n, \rho)$. Now the formula (147) can be written in the form

$$\int\overline{x'(g)}y(g)d\mu(g)=\frac{1}{8\pi^4}\sum_{n=-\infty}^{\infty}\int_0^{\infty}[\int\int\overline{K'(z, z_1, n, \rho)}\cdot K(z, z_1, n, \rho)d\mu(z)d\mu(z_1)](n^2+\rho^2)d\rho.$$

In other words we get

$$\int y(g)\overline{x'(g)}d\mu(g)=(K, K'). \tag{148}$$

We can approximate an arbitrary function $x(g)\in\mathfrak{H}_G$ by functions $x'(g)$. The corresponding $K'$ then tends to some element from $\mathfrak{H}$. Since $K\in\mathfrak{H}$, the right side of (148) has a limit and therefore the left side has a limit. It follows that $y(g)\in\mathfrak{H}_G$ and that the equality (148) is true for an arbitrary $x'\in\mathfrak{H}_G$ and for an arbitrary $K'\in\mathfrak{H}$.

On the other hand we have the similar equality

$$\int x(g)\overline{x'(g)}d\mu(g)=(K, K')$$

for the function $x(g)$ whose image by isometric mapping to $\mathfrak{H}_G$ to $\mathfrak{H}$ is $K(z_1, z_2, n, \rho)$.

Therefore, $x(g)=y(g)$ so that

$$x(g)=\frac{1}{8\pi^4}\sum_{n=-\infty}^{\infty}\int_0^{\infty}[\int K(z, z\bar{g}, n, \rho)\beta^{-\frac{1}{2}}(zg)\overline{\chi_{n,\rho}(zg)}d\mu(z)](n^2+\rho^2)d\rho \tag{149}$$

or, more explicitly,

$$x(g)=-\frac{1}{8\pi^4}\sum_{n=-\infty}^{\infty}\int_0^{\infty}\left[\int K\left(z, \frac{g_{11}z+g_{21}}{g_{12}z+g_{22}}, n, \rho\right)\right.$$

$$\left.\cdot|g_{12}z+g_{22}|^{n+i\rho-2}\cdot(g_{12}z+g_{22})^{-n}d\mu(z)\right](n^2+\rho^2)d\rho \tag{150}$$

and formula (137) is true for the functions $x(g)$ and $K(z_1,z_2,n,\rho)$. Now let $K(z_1,z_2,n,\rho)$ be an arbitrary function in $\mathfrak{H}$. Let

$$\hat{K}(z_1,z_2,n,\rho)=\begin{cases}K(z_1,z_2,n,\rho) & \text{when } |z_1|\leqq C_1, |z_2|\leqq C_2, |n|\leqq n_0, 0\leqq\rho\leqq\rho_0,\\ 0 & \text{outside this region.}\end{cases}$$

For the function $\hat{K}(z_1,z_2,n,\rho)$ integral (150) converges by norm in $\mathfrak{H}_G$ to a function $x(g)\in\mathfrak{H}_G$ and the mapping $\hat{K}(z_1,z_2,n,\rho)\to x(g)$ is the inverse transformation of $\mathfrak{H}$ to $\mathfrak{H}_G$. So we have proved:

**Theorem 5.** *For an arbitrary function $x(g)\in\mathfrak{H}_G$ the integral*

$$K(z_1,z_2,n,\rho)=\int x(z_1^{-1}\delta\zeta z_2)\beta^{-\frac{1}{2}}(\delta)\chi_{n,\rho}(\delta)d\mu(\delta)d\mu(\zeta) \tag{151}$$

*converges by norm in $\mathfrak{H}$ and belongs to $\mathfrak{H}$. Conversely, for arbitrary $K(z_1,z_2,n,\rho)\in\mathfrak{H}$, the integral*

$$x(g)=\frac{1}{8\pi^4}\sum_{n=-\infty}^{\infty}\int_0^{\infty}\left[\int\beta^{-\frac{1}{2}}(zg)\overline{\chi_{n,\rho}(zg)}K(z,z\bar{g},n,\rho)d\mu(z)\right](n^2+\rho^2)d\rho \tag{152}$$

*converges by norm in $\mathfrak{H}_G$ and belongs to $\mathfrak{H}_G$. These formulas give the mutually inverse isometric transformations of $\mathfrak{H}_G$ to $\mathfrak{H}$ and of $\mathfrak{H}$ to $\mathfrak{H}_G$ and for corresponding $x(g)$ and $K(z,z_2,n,\rho)$ formula (137) is true.*

The formula (150) can be written in another form by introducing the variable $v=g_{12}z+g_{22}$. Then we obtain

$$x(g)=\frac{1}{8\pi^4}|g_{12}|^{-2}\sum_{n=-\infty}^{\infty}\int_0^{\infty}\left[\int K\left(\frac{v-g_{22}}{g_{12}},\frac{g_{11}v-1}{g_{12}v},n,\rho\right)\right.$$

$$\left.\cdot|v|^{-n-i\rho-2}v^n d\mu(v)\right](n^2+\rho^2)d\rho$$

or, if we also introduce the parameters $p_1,p_2,p_3$,

$$\tilde{x}(p_1,p_2,p_3)=\frac{1}{8\pi^3}|p_2|^2\sum_{n=-\infty}^{\infty}\int_0^{\infty}\left[\int K\left(p_2v-p_3,p_1-\frac{p_2}{v},n,\rho\right)\right.$$

$$\left.\cdot|v|^{-n-i\rho-2}v^n d\mu(v)\right](n^2+\rho^2)d\rho. \tag{153}$$

4. *The decomposition of the regular representation into irreducible representations.* Theorem 5 enables us to decompose the regular representations into irreducible components. To each function $x(g)\in\mathfrak{H}_G$ we associate the kernel (see (151))

$$K(z_1,z_2,X)=\int x(z_1^{-1}kz_2)\beta^{-\frac{1}{2}}(k)\chi(k)d\mu_e(k),$$

that is a Hilbert–Schmidt kernel. In particular, the function $f(z_2)=K(z_1,z_2,X)$

lies in $\mathfrak{H}$ for almost all $z_1, X$. Therefore, to each function $x(g)\in\mathfrak{H}_G$ we associate an element of the space $\mathfrak{H}_z$ which depends on $z_1$ and $X$. For any $g_0\in G$ the function $x(gg_0)$ also belongs to $\mathfrak{H}_G$. We consider the transformation of the $K(z_1, z_2, X)$ when we repplace the function $x(g)$ by $x(gg_0)$. Assume first that $x(g)\in R'$ and therefore that $x(gg_0)\in R'$. Denote the kernel corresponding to $x(gg_0)$ by $K_1(z_1, z_2, X)$. We have

$$K_1(z_1, z_2, X) = \int x(z_1^{-1}kz_2g_0)\beta^{-\frac{1}{2}}(k)X(k)d\mu_e(k). \tag{154}$$

Let $z_2g_0 = k_1z_2'$ so that $z_2' = z_2\overline{g_0}$. Then the equality (154) can be written as follows:

$$\begin{aligned}K_1(z_1, z_2, X) &= \int x(z_1^{-1}kk_1z_2')\beta^{-\frac{1}{2}}(k)X(k)d\mu_e(k)\\ &= \int x(z_1^{-1}kk_1z_2')\beta^{\frac{1}{2}}(k)X(k)d\mu_r(k) = \int x(z_1^{-1}kz_2')\beta^{\frac{1}{2}}(kk_1^{-1})X(kk_1^{-1})d\mu_r(k)\\ &= \beta^{-\frac{1}{2}}(k_1)\overline{X(k_1)}\int x(z_1^{-1}kz_2')\beta^{\frac{1}{2}}(k)X(k)d\mu_r(k)\end{aligned}$$

or

$$K_1(z_1, z_2, X) = \beta^{-\frac{1}{2}}(k_1)\bar{X}(k_1)K(z_1, z_2', X) = \beta^{-\frac{1}{2}}(z_2g_0)\bar{X}(z_2g_0)K(z_1, z_2\bar{g}_0, X). \tag{155}$$

This means that the mapping of $x(g)$ to $x(gg_0)$ corresponds to the transformation of the function $f(z_2) = K(z_1, z_2, X)$ by the operator $U_{\bar{X};g_0}$.

The operator $U_{X;\bar{g}_0}$ being continuous, this statement is true for an arbitrary function $x(g)\in\mathfrak{H}_G$. We can consider the space $\mathfrak{H}$ as the direct sum of the spaces $\mathfrak{H}_{z_1,n,\rho} = \mathfrak{H}_z$; this sum is defined as follows: $\mathfrak{H}$ consists of all functions $f = f_{z_1;n,\rho}$ with values in $\mathfrak{H}_z$ such that

$$\sum_{n=-\infty}^{\infty}\int_0^{\infty}\int \|f_{z_1,n,\rho}\|^2(n^2+\rho^2)d\rho\, d\mu(z_1) < +\infty. \tag{156}$$

The inner product of two such functions $f = f_{z_1,n,\rho}$ and $f' = f'_{z_1,n,\rho}$ is defined by

$$(f, f') = \frac{1}{8\pi^4}\sum_{n=-\infty}^{\infty}\int_0^{\infty}\int (f_{z_1,n,\rho} f'_{z_1,n,\rho})(n^2+\rho^2)d\mu(z_1)d\rho.$$

Hence the mapping of $\mathfrak{H}_G$ to $\mathfrak{H}$ that was constructed on p. 70 is an isometric mapping of $\mathfrak{H}_G$ to the direct sum of the spaces $\mathfrak{H}_{z_1,n,\rho} = \mathfrak{H}_z$. To the operator $V_{g_0}x(g) = x(gg_0)$ in the regular representation corresponds, by (155), the operator $U_{\bar{X},g_0}$ of the principal series acting in the space $\mathfrak{H}_{z_1,n,\rho}$. This means that the regular representation is decomposed into the direct sum of the irreducible representations $U_{\bar{X},g}$ of the principal series.

**Theorem 6.** *The regular representation of the group G can be decomposed into the direct sum of the irreducible representations of the principal series.*

5. *The unitary equivalence of the representations* $U_{X;g}$ *and* $U_{\bar{X};g}$. We now use our analog of the Plancherel theorem to prove:

**Theorem 7.** *The representations* $U_{X;g}$ *and* $U_{\bar{X};g}$ *are unitary equivalent.*

*Proof.* To each function $x(g)\in R'$ we associated two kernels:

$$K(z_1, z_2, \bar{X}) = \int x(z_1^{-1}\delta\zeta z_2)\bar{X}(\delta)\beta^{-\frac{1}{2}}(\delta)d\mu(\zeta) \tag{157}$$

and

$$K(z_1, z_2, X) = \int x(z_1^{-1}\delta\zeta z_2)X(\delta)\beta^{-\frac{1}{2}}(\delta)d\mu(\zeta) \tag{158}$$

of the operators $U_{X;g}$ and $U_{\bar{X};g}$. Then, by formula (94) for the trace, we have

$$S(U_{X;x^*x}) = S(U_{\bar{X};x^*x})$$

or

$$\int\int |K(z_1, z_2, X)|^2 d\mu(z_1)d\mu(z_2) = \int\int |K(z_1, z_2, \bar{X})|^2 d\mu(z_1)d\mu(z_2). \tag{159}$$

Let $\mathfrak{H}_K$ be the Hilbert space of functions $K(z_1, z_2)$ satisfying

$$\int\int |K(z_1, z_2)|^2 d\mu(z_1)d\mu(z_2) < +\infty. \tag{160}$$

When $x(g)$ runs over all $R'$, then in analogy to the Plancherel formula, the functions $K(z_1, z_2, X)$ and $K(z_1, z_2, \bar{X})$ runs over the dense sets $\mathfrak{S}_1$ and $\mathfrak{S}_2$ in $\mathfrak{H}$. On the other hand, by (159) the correspondence

$$K(z_1, z_2, X) \leftrightarrow K(z_1, z_2, \bar{X})$$

is an isometric transformation of $\mathfrak{S}_1$ to $\mathfrak{S}_2$ and hence it can be uniquely prolonged as a unitary operator $W$ in $\mathfrak{H}_K$.

Now let $\{\varphi_1(z), \varphi_2(z), \ldots\}$ be a complete orthonormal system of functions in $\mathfrak{H}_Z$. Then an arbitrary $K(z_1, z_2) \in \mathfrak{H}$ can be represented as the sum

$$K(z_1, z_2) = \sum_{k=1}^{\infty} \varphi_k(z_1) f_k(z_2) \tag{161}$$

with

$$\| K \|^2 = \int\int |K(z_1, z_2)|^2 d\mu(z_1)d\mu(z_2) = \sum_{k=1}^{\infty} \int |f(z_2)|^2 d\mu(z_2). \tag{162}$$

This means that $\mathfrak{H}_K$ can be considered as a direct countable sum of spaces $\mathfrak{H}_Z$.

On the other hand, when we map $x(g)$ to $x(gg_0)$, the kernel $K(z_1, z_2, \bar{X})$, considered as a function in $z_2$, is transformed by the representation $U_{\bar{X};g}$. This transformation corresponds to the transformation of each component $f_k(z_2)$ by the representation $U_{\bar{X};g}$.

In other words, the transformation of the kernel $K(z_1, z_2, X)$ that corresponds to the right action $x(g) \to x(gg_0)$ is the representation of the group $G$ that is a multiple of the irreducible representation $U_{\bar{X};g}$. Denote this representation by $U'_{g_0}$.

Similarly, the transformation of the kernel $K(z_1, z_2, X)$ that corresponds to the action $x(g) \to x(gg_0)$ is a representation of the group $G$ which is a multiple of the representation $U_{X;g_0}$; denote this representation by $U''_{g_0}$.

So we associate the kernels $U'_{g_0}K(z_1, z_2, X)$ and $U''_{g_0}K(z_1, z_2, X)$ to the function $x(gg_0)$. By definition, the operator $W$ transforms the first kernel into the second, i.e. $U'_{g_0}$ into $U''_{g_0}$. Therefore, the representations $U'_{g_0}$ and $U''_{g_0}$, which are mutliples of the representations $U_{\bar{X};g_0}$ and $U_{X;g_0}$, respectively, are equivalent. But then, as we will show in the next paper, the representations $U_{\bar{X};g_0}$and $U_{X;g_0}$ are also equivalent.

The operator $V$ that transforms $U_{X;g_0}$ into $U_{\bar{X};g_0}$ can be written explicitly if, instead of considering the space $\mathfrak{H}_Z$ of square integrable functions $f(z)$, we consider their Fourier transforms

$$\varphi(w) = \frac{1}{2\pi}\int f(z)e^{-i\,\mathrm{Re}(z\bar{w})}d\mu(z). \tag{163}$$

By the Plancherel theorem,

$$\int |f(z)|^2 d\mu(z) = \int |\varphi(w)|^2 d\mu(w). \tag{164}$$

Instead of $U_{X;g}$ we will write $U_{n,\rho;g}$; then, by definition of the operator $V$, we have

$$VU_{-n,-\rho;g} = U_{n,\rho;g} V. \tag{165}$$

For $g = z_0 = \begin{pmatrix} 1 & 0 \\ z_0 & 1 \end{pmatrix}$ we have $U_{n,\rho;z_0}\varphi(w) = e^{i\mathrm{Re}(\bar{w}z_0)}\varphi(w)$ and equality (165) takes the form

$$Ve^{i\mathrm{Re}(\bar{w}z_0)}\cdot\varphi(w) = e^{i\mathrm{Re}(\bar{w}z_0)}V\varphi(w). \tag{166}$$

So the operator $V$ commutes with the operator of multiplication by $e^{i\mathrm{Re}(\bar{w}z_0)}$ for arbitrary $z_0$; it follows that $V$ is an operator of multiplication by a bounded function

$$V\varphi(w) = \omega(w)\varphi(w). \tag{167}$$

Now let us put $g = \delta = \begin{pmatrix} \lambda^{-1} & 0 \\ 0 & \lambda \end{pmatrix}$ in (165). Then, by formula (65),

$$U_{n,\rho;\delta}\varphi(w) = |\lambda|^{n+i\rho+2}\lambda^{-n}\omega(w\bar{\lambda}^2)\varphi(w\bar{\lambda}^2) \tag{168}$$

and for $g = \delta$ the equality (165) takes the form

$$\omega(w)|\lambda|^{-n-i\rho+2}\lambda^{n}\varphi(w\bar{\lambda}^2) = |\lambda|^{n+i\rho+2}\lambda^{-n}\omega(w\bar{\lambda}^2)\varphi(w\bar{\lambda}^2).$$

Hence,

$$\omega(w\bar{\lambda}^2) = |\lambda^2|^{-n-i\rho}(\lambda^2)^n\omega(w).$$

Substituting 1 for $w$ and $w$ for $\bar{\lambda}^2$, we obtain

$$\omega(w) = C|w|^{-n-i\rho}\bar{w}^n \tag{169}$$

where $C = \omega(1)$. Obviously we can assume that $C = 1$. Then (167) takes the form

$$V\varphi(w) = |w|^{-n-i\rho}\bar{w}^n\varphi(w). \tag{170}$$

This is an explicit expression for the operator $V$.

## 8. The complementary series of irreducible representations of the group $G$

1. *The definition of the representations of the complementary series.* In §7 we have shown that the regular representation can be decomposed into irreducible representations of the principal series. Now we show that the group $G$ has other irreducible representations.

The representations of the principal series have the form

$$U_g f(z) = \alpha(zg) f(z\bar{g}), \tag{171}$$

where

$$\alpha(\delta) = \beta^{-\frac{1}{2}}(\delta)\chi(\delta) \tag{172}$$

and

$$\alpha(\delta\zeta z) = \alpha(\delta). \tag{173}$$

Here $\chi(\delta)$ is a character of the group $D$ such that

$$|\chi(\delta)|=1. \tag{174}$$

The inner product in $\mathfrak{H}_z$ is defined by

$$(f_1,f_2)=\int f_1(z)\overline{f_2(z)}\,d\mu(z). \tag{175}$$

We will omit the condition (174) and try to choose a metric in the set of functions $f(z)$ in such a way that formula (171) would still give a unitary representation of the group $G$.

It is easy to see that with the choice (175) of the metric the condition (174) on the character is necessary. More generally, let a metric be given by the formula

$$(f_1,f_2)=\int f_1(z)\overline{f_2(z)}\,d\sigma(P_z) \tag{176}$$

where $\sigma(P)$ is a positive completely additive function of the compact set $P\subset Z$. Put $g=z_0$ in (171). We have

$$U_{z_0}f(z)=f(zz_0).$$

Hence, the unitarity condition $\|U_{z_0}f\|^2=\|f\|^2$ shows that

$$\int|f(zz_0)|^2\,d\sigma(P_z)=\int|f(z)|^2\,d\sigma(P_z).$$

So, up to a multiplicative constant, $\sigma(P)$ is an invariant measure in $Z$ and a metric (176), up to a multiplicative constant, coincides with (175). But then the unitarity condition on an operator $U_g$ implies that

$$\int\beta^{-1}(zg)\cdot|\chi(zg)|^2|f(z\bar{g})|^2\,d\mu(z)=\int|f(z_1)|^2\,d\mu(z_1).$$

Substituting $z_1=zg$ in the second integral and using the formula (48), we obtain

$$\int\beta^{-1}(zg)\cdot|\chi(z\bar{g})|^2\cdot|f(z\bar{g})|^2\,d\mu(z)=\int|f(zg)|^2\beta^{-1}(zg)\,d\mu(z).$$

Hence $|\chi(zg)|=1$.

Therefore, with a metric of the form (176), and therefore of the form (175) as well, we obtain representations of the principal series only.

Now we try a metric of the form

$$(f_1,f_2)=\iint K(z_1,z_2)f_1(z_1)\overline{f_2(z_2)}\,d\mu(z_1)\,d\mu(z_2) \tag{177}$$

where $f_1,f_2$ are bounded measurable functions such that the integral (177) converges absolutely.

The unitarity condition for the operator (171) means that

$$\begin{aligned}\iint K(z_1,z_2)\beta^{-\frac{1}{2}}(z_1g)\beta^{-\frac{1}{2}}(z_2g)\chi(z_1g)\overline{\chi(z_2g)}\,f_1(z_2\bar{g})\overline{f_2(z_2\bar{g})}\,d\mu(z_1)\,d\mu(z_2)\\ =\iint K(z_1',z_2')f_1(z_1')\overline{f_2(z_2')}\,d\mu(z_1')\,d\mu(z_2').\end{aligned} \tag{178}$$

We change the variables in the second integral: $z_1'=z_1\bar{g},z_2'=z_2\bar{g}$. In these new variables this integral takes the form

$$\iint K(z_1\bar{g},z_2\bar{g})\beta^{-1}(z_1g)\beta^{-1}(z_2g)f_1(z_1\bar{g})\overline{f_2(z_2\bar{g})}\,d\mu(z_1)\,d\mu(z_2)$$

and (178) shows that

$$K(z_1\bar{g}, z_2\bar{g}) = \beta^{\frac{1}{2}}(z_1 g)\beta^{\frac{1}{2}}(z_2 g)\chi(z_1 g)\overline{\chi(z_2 g)} \tag{179}$$

or

$$K\left(\frac{g_{11}z_1 + g_{21}}{g_{12}z_1 + g_{22}}, \frac{g_{11}z_2 + g_{21}}{g_{12}z_2 + g_{22}}\right)$$
$$= K(z_1, z_2)\cdot|g_{12}z_1 + g_{22}|^2\cdot|g_{12}z_2 + g_{22}|^2\chi(g_{12}z_1 + g_{22})\overline{\chi(g_{12}z_2 + g_{22})}. \tag{180}$$

For $g = z_0$ we get

$$K(z_1 + z_0, z_2 + z_0) = K(z_1, z_2);$$

hence $K(z_1, z_2) = K_1(z_1 - z_2)$.

For the function $K_1$ the condition (180) is

$$K_1\left(\frac{z_1 - z_2}{(g_{12}z_1 + g_{22})(g_{12}z_2 + g_{22})}\right)$$
$$= K_1(z_1 - z_2)\cdot|g_{12}z_1 + g_{22}|^2\cdot|g_{12}z_2 + g_{22}|^2\chi(g_{12}z_1 + g_{22})\overline{\chi(g_{12}z_2 + g_{22})}. \tag{181}$$

Putting $z_2 = 0$ here and choosing $g_{12}$ so that $g_{12}z_1 + g_{22} = 1$, we obtain

$$K_1\left(\frac{z_1}{g_{22}}\right) = K_1(z_1)\cdot|g_{22}|^2\bar{\chi}(g_{22}). \tag{182}$$

On the other hand, putting $z_1 = 0$ and choosing $g_{12}$ so that $g_{12}z_2 + g_{22} = 1$ we obtain

$$K_1\left(-\frac{z_2}{g_{22}}\right) = K_1(-z_2)\cdot|g_{22}|^2\chi(g_{22}). \tag{183}$$

Comparing (182) and (183) we see that $X(g_{22})$ is real. From (182) it also follows that

$$K_1(z) = C\cdot|z|^{-2}\chi^{-1}(z) \tag{184}$$

where $C$ is a constant. Because of the measurability of the representation, $\chi(z)$ is a real measurable function. From the condition $\chi(z_1 z_2) = \chi(z_1)\cdot\chi(z_2)$ if follows that

$$\chi(z) = |z|^{-\rho} \tag{185}$$

where $\rho$ is a real number. By (184) we then have

$$K(z_1, z_2) = C|z_1 - z_2|^{-2+\rho} \tag{186}$$

and formula (177) for the inner product can be written

$$(f_1, f_2) = C\int\int|z_1 - z_2|^{-2+\rho}f_1(z_1)\overline{f_2(z_2)}d\mu(z_1)d\mu(z_2). \tag{187}$$

The representation itself is given by

$$U_g f(z) = |g_{12}z + g_{22}|^{-2-\rho}f\left(\frac{g_{11}z + g_{21}}{g_{12}z + g_{22}}\right). \tag{188}$$

It is clear that the integral (187) converges if and only if $\rho > 0$. Besides, a metric (187) must be positive definite, that is we should have

$$(f,f) = C\int\int |z_1 - z_2|^{-2+\rho} f(z_1) f(z_2) d\mu(z_1) d\mu(z_2) > 0. \tag{189}$$

Let us find $\rho$ satisfying this condition. First, let $0 < \rho < \frac{1}{2}$ and let $f(z)$ be a function vanishing outside some circle $|z| \leq C$ and so smooth that its Fourier transform

$$\varphi(w) = \frac{1}{2\pi}\int f(z) e^{-i\mathrm{Re}(\bar{z}w)} d\mu(z) \tag{190}$$

is integrable. Denote by $\mathfrak{H}'$ the set of such functions $f(z)$ and by $H'$ the set of their Fourier transforms.

Clearly, when $w = 0$, we have

$$\varphi(0) = \frac{1}{2\pi}\int f(z) d\mu(z). \tag{191}$$

It is known that

$$f(z) = \frac{1}{2\pi}\int \varphi(w) e^{i\mathrm{Re}(\bar{z}w)} d\mu(w). \tag{192}$$

For arbitrary $z_2$ the integral

$$\int |z_1 - z_2|^{-2+\rho} f(z_1) d\mu(z_1) = \int |z|^{-2+\rho} f(z + z_2) d\mu(z)$$

$$= \frac{1}{2\pi}\int |z|^{-2+\rho} [\int \varphi(w) e^{i\mathrm{Re}[(z+z_2)\bar{w}]} d\mu(w)] d\mu(z)$$

converges. In the last integral we change the variable $z = re^{i\theta}$ and write $w = r_1 e^{i\theta_1}$. Then

$$\int |z_1 - z_2|^{-2+\rho} f(z_1) d\mu(z_1) = \frac{1}{2\pi}\int_0^{2\pi}\int_0^{\infty} r^{-1+\rho} [\int \varphi(w) e^{i\mathrm{Re}\, r r_1 e^{i(\theta-\theta_1)}} e^{i\mathrm{Re}(z_2\bar{w})} d\mu(w)] dr\, d\theta.$$

The absolute value of the integral in square brackets does not exceed $\int |\varphi(w)| d\mu(w)$ and hence this integral converges uniformly in $\theta$. Therefore, the last integral is equal to

$$\int_0^{\infty} r^{-1+\rho} [\int \varphi(w) e^{i\mathrm{Re}(z_2\bar{w})} I_0(rr_1) d\mu(w)] dr$$

where $I_0$ is the Bessel function of order zero.

It is easy to see that this integral converges absolutely in both variables. So it is equal to

$$\int \varphi(w) e^{i\mathrm{Re}(z_2\bar{w})} \left[\int_0^{\infty} r^{-1+\rho} I_0(rr_1) dr\right] d\mu(w)$$

$$= 2^{-1+\rho} \frac{\Gamma(\rho/2)}{\Gamma(1-\rho/2)} \int \varphi(w) e^{i\mathrm{Re}(z_2\bar{w})} r_1^{-\rho} d\mu(w)$$

$$= 2^{-1+\rho} \frac{\Gamma(\rho/2)}{\Gamma(1-\rho/2)} \int \varphi(w) e^{i\mathrm{Re}(z_2\bar{w})} \cdot |w|^{-\rho} d\mu(w)$$

where the last integral converges absolutely.

We multiply both parts of the equality by $\overline{f(z_2)}$ and integrate in $z_2$. Then, in view of (190), we obtain

$$\iint |z_1 - z_2|^{-2+\rho} f(z_1)\overline{f(z_2)}\, d\mu(z_1)\, d\mu(z_2) = 2^\rho \pi \frac{\Gamma(\rho/2)}{\Gamma(1-\rho/2)} \int |w|^{-\rho} \cdot |\varphi(w)|^2\, d\mu(w). \tag{193}$$

The equality (193) is proved for $0 < \rho < \frac{1}{2}$. But when $f(z)$ (and hence $\varphi(w)$ also) is fixed, both sides of this equality are analytic functions in $\rho$ in the strip region $0 < \operatorname{Re}\rho < 2$ and therefore (193) is true in the whole of this strip. In particular, it is true for $0 < \rho < 2$.

Formula (193) shows that *the metric* (193) *is positive definite for* $0 < \rho < 2$ *and* $C > 0$.

For $\rho = 2$

$$(f, f) = C \left| \int f(z)\, d\mu(z) \right|^2 \tag{194}$$

so in this case the metric is also positive definite.

It is easy to see that when $\rho > 2$ the metric stops being positive definite. Indeed, let, for example, $2 < \rho < 4$. Then (193) is still true for functions $f(z)$ which satisfy the condition

$$\varphi(0) = \frac{1}{2\pi} \int f(z)\, d\mu(z) = 0. \tag{195}$$

As the constant in front of the integral on the right side of (193) is negative, the metric will be positive for functions $f(z)$ which satisfy (195) if $C < 0$. On the other hand, if we take a nonnegative $f(z)$ we obtain

$$(f, f) = C \int |z_1 - z_2|^{-2+\rho} f(z_1) f(z_2)\, d\mu(z_1)\, d\mu(z_2) < 0.$$

Let $0 < \rho < 2$; define the inner product in $\mathfrak{H}'$ by

$$(f_1, f_2) = \int |z_2 - z_1|^{-2+\rho} f_1(z_1)\overline{f_2(z_2)}\, d\mu(z_1)\, d\mu(z_2) \tag{196}$$

and the inner product in $H'$ by

$$(\varphi_1, \varphi_2) = 2^\rho \pi \frac{\Gamma(\rho/2)}{\Gamma(1-\rho/2)} \int |w|^{-\rho} \varphi_1(w)\overline{\varphi_2(w_2)}\, d\mu(w_1)\, d\mu(w_2). \tag{197}$$

The equality (193) shows that the formula (190) gives an isometric correspondence between $\mathfrak{H}'$ and $H'$. This correspondence can be uniquely prolonged to the isometric correspondence between the complements of $\mathfrak{H}'$ and $H'$ with respect to inner products (195) and (196). Denote these complements by $\mathfrak{H}_\rho$ and $H_\rho$. Denote also by $\tilde{H}_\rho$ the set of all functions $\varphi(w)$ satisfying

$$\int |w|^{-\rho} \cdot |\varphi(w)|^2\, d\mu(w) < +\infty. \tag{198}$$

Let us prove that $H_\rho = \tilde{H}_\rho$. The set $H'$ is dense in the space of functions $\varphi(w)$ with the norm

$$\int |\varphi(w)|^2\, d\mu(w) < +\infty.$$

Let $\Delta$ be a compact set in the $w$-plane that does not contain the point $w=0$ and let $\varphi_\Delta(w)$ be a measurable function vanishing outside $\Delta$. Then, for arbitrary $\varepsilon_1>0$, there exists a function $\varphi'(w)\in H'$ such that

$$\int|\varphi_\Delta(w)-\varphi'(w)|^2 d\mu(w)<\varepsilon_1. \tag{199}$$

Let $Q_1$ be the set $|w|\leqq\varepsilon_2$ and $Q_2$ be the set $|w|>\varepsilon_2$. Denoting by $C$ a number such that $|\varphi_\Delta(w)-\varphi'(w)|<C$ for all $w$ and introducing the variables $r$ and $\theta$ by $w=re^{i\theta}$, we have

$$\int|w|^{-\rho}|\varphi_\Delta(w)-\varphi'(w)|^2 d\mu(w)$$

$$=\int_{Q_1}|w|^{-\rho}\cdot|\varphi_\Delta(w)-\varphi'(w)|^2 d\mu(w)+\int_{Q_2}|w|^{-\rho}\cdot|\varphi_\Delta(w)-\varphi'(w)|^2 d\mu(w)$$

$$\leqq C^2\int_0^{2\pi}\int_0^{\varepsilon_2} r^{-\rho} r\,dr\,d\theta+\varepsilon_2^{-\rho}\int_{Q_2}|\varphi_\Delta(w)-\varphi^1(w)|^2 d\mu(w)<\frac{2\pi e^2}{2-\rho}\varepsilon_2^{2-\rho}+\varepsilon_2^{-\rho}\varepsilon_1.$$

We have shown that it is possible to approximate a function $\varphi_\Delta(w)$ by functions from $H'$ in the metric of the space $\tilde{H}_\rho$. On the other hand the set of functions $\varphi_\Delta(w)$ is dense in $\tilde{H}_\rho$. Therefore $H'$ is dense in $\tilde{H}_\rho$ and $H_\rho$ as complement of $H'$ coincides with $\tilde{H}_\rho$.

Hence, $H_\rho$ is the set of functions $\varphi(w)$ satisfying (198).

Denote by $\mathfrak{H}'_\rho$ the set of measurable bounded functions such that

$$\iint|z_1-z_2|^{-2+\rho}|f(z_1)|\cdot|f(z_2)|d\mu(z_1)d\mu(z_2)<+\infty. \tag{200}$$

**Lemma.** *If* $0<\rho<2$ *then for an arbitrary function* $f(z)\in\mathfrak{H}'_\rho$ *the integral*

$$\varphi(w)=\frac{1}{2\pi}\int f(z)e^{-i\operatorname{Re}(z\bar{w})}d\mu(z) \tag{201}$$

*converges in the norm of* $H_\rho$ *and*

$$\|f\|^2=\|\varphi\|^2, \tag{202}$$

*where*

$$\|f\|^2=\int|z_1-z_2|^{-2+\rho}f(z_1)f(z_2)d\mu(z_1)d\mu(z_2). \tag{203}$$

*Proof.* First let $f(z)$ be a measurable bounded function vanishing outside a set $|z|\leqq C$. Then the integral (201) converges in the usual sense.

Let us prove that for these $f(z)$'s and corresponding $\varphi(w)$'s the equality (202) is satisfied.

To do this we choose from $\mathfrak{H}'$ a uniformly bounded sequence of functions $f_m(z)$ vanishing outside the set $|z|\leqq C$ such that

$$\int|f(z)-f_m(z)|^2 d\mu(z)\to 0 \quad \text{for } m\to\infty. \tag{204}$$

Then

$$\int|f(z)-f_m(z)|d\mu(z)\to 0 \quad \text{for } m\to\infty. \tag{205}$$

Decompose the set of all pairs $(z_1, z_2)$ into sets $\varepsilon_1, \varepsilon_2$ defined by

$$\varepsilon_1: |z_1 - z_2| \leqslant \varepsilon,$$
$$\varepsilon_2: |z_1 - z_2| > \varepsilon.$$

Denoting by $C'$ the constant such that $|f(z)| < C'$, $|f_m(z)| < C'$, we have

$$\left| \int_{\varepsilon_1} |z_1 - z_2|^{-2+\rho} [f(z_1) - f_m(z_1)] \overline{f(z_2)} d\mu(z_1) d\mu(z_2) \right|$$

$$\leqq 2C' \int_{\varepsilon_1} |z_1 - z_2|^{-2+\rho} |f(z_2)| d\mu(z_1) d\mu(z_2)$$

$$\leqq 2C_\pi'^2 C^2 \int_0^\varepsilon \int_0^{2\pi} r^{-2+\rho} r dr d\theta = \frac{4\pi^2 C'^2 C^2 \varepsilon^\rho}{\rho}$$

and

$$\left| \int_{\varepsilon_2} |z_1 - z_2|^{-2+\rho} [f(z_1) - f_m(z_1)] \overline{f(z_2)} d\mu(z_1) d\mu(z_2) \right|$$

$$\leqq \varepsilon^{-2+\rho} C_\pi' C^2 \int |f(z_1) - f_n(z_1)| d\mu(z_1).$$

These estimates show that

$$\int |z_1 - z_2|^{-2+\rho} [f(z_1) - f_n(z_1)] \overline{f_n(z_2)} d\mu(z_1) d\mu(z_2) \to 0 \quad \text{for } m \to \infty.$$

In a similar way we can prove that

$$\int |z_1 - z_2|^{-2+\rho} f_m(z_1) \overline{[f(z_2) - f_m(z_2)]} d\mu(z_1) d\mu(z_2) \to 0 \quad \text{for } m \to \infty.$$

Summing up these formulas we obtain

$$\| f_m \|^2 \to \| f \|^2 \quad \text{for } m \to \infty. \tag{206}$$

To each function $f_m(z)$ we associate by (201) the function $\varphi_m(w)$; it has been proved that

$$\| f_m \|^2 = \| \varphi_m \|^2. \tag{207}$$

Let us prove that $\| \varphi_m - \varphi \| \to 0$ for $m \to \infty$. All functions $\varphi_m$, $\varphi$ are bounded by the same constant $C'' = \pi C^2 C'$. Moreover, by (204)

$$\int |\varphi_m(w) - \varphi(w)|^2 d\mu(w) = \int |f_m(z) - f(z)|^2 d\mu(z) \to 0 \quad \text{for } m \to \infty.$$

Let $F_1$ and $F_2$ be the regions in the $w$-plane define by

$$F_1: |w| \leqq \varepsilon,$$
$$F_2: |w| > \varepsilon.$$

Then

$$\int_{F_1} |\varphi(w) - \varphi_m(w)|^2 \cdot |w|^{-\rho} d\mu(w) \leqq 4C''^2 \int_{F_1} |w|^{-\rho} d\mu(w) = 8\pi C''^2 \int_0^\varepsilon r^{-\rho+1} dr = \frac{8\pi C''^2 \varepsilon^{2-\rho}}{2-\rho}$$

and

$$\int_{F_2} |\varphi(w) - \varphi_m(w)|^2 \cdot |w|^{-\rho} d\mu(w) \leqq \varepsilon^{-\rho} \int |\varphi_m(w) - \varphi(w)|^2 d\mu(w).$$

These estimates imply that

$$\|\varphi-\varphi_m\|^2=2^\rho\pi\frac{\Gamma(\rho/2)}{\Gamma(1-\rho/2)}\int|\varphi(w)-\varphi_m(w)|^2\cdot|w|^{-\rho}d\mu(w)\to 0 \quad \text{for } m\to\infty$$

and hence $\|\varphi_m\|\to\|\varphi\|$ for $m\to\infty$.

Taking the limit of both sides of (207) and using (206) we obtain that $\|f\|^2=\|\varphi\|^2$.

Now let $f(z)$ be an arbitrary function from $\mathfrak{H}'_\rho$. Denote by $f_\varepsilon(z)$ the function defined by

$$f_C(z)=\begin{cases} f(z) & \text{if } |z|\leqq C,\\ 0 & \text{if } |z|>C.\end{cases}$$

Denote by $\varphi_c(w)$ the corresponding function $\varphi(w)$. We have just proved that $\|f_C\|^2=\|\varphi_C\|^2$. Similarly, if $C_1>C$,

$$\|f_{C_1}-f_C\|^2=\|\varphi_{C_1}-\varphi_C\|^2.$$

Because of the assumed convergence of the integral (200), the left side of this equality tends to zero for $C_1$, $C_2\to\infty$. Hence, $\lim_{C\to\infty}\varphi_C=\varphi$ exists in the sense of convergence in the norm in $H_\rho$. In other words, the integral (201) converges in the norm in $H_\rho$. Taking the limit of both sides of the equality $\|f_C\|=\|\varphi_C\|$ as $C\to\infty$, we obtain $\|f\|=\|\varphi\|$. The lemma is proved.

This lemma shows that $H'_\rho$ can be considered as a part of $\mathfrak{H}_\rho$.

As we have seen above, the operator $U_g$, defined by (188), is a unitary operator in the space $\mathfrak{H}'_\rho$ with respect to the metric (196). This operator can be uniquely prolonged to a continuous operator $U_g$ in $\mathfrak{H}_\rho$ and to a unitary operator $U_g$ in $H_\rho$.

So we have proved:

**Theorem 8.** *For* $0<\rho<2$ *the formula* (188) *defines a unitary representation of the group G in the space* $\mathfrak{H}_\rho$.

For $\rho=2$ the equality (194) shows that $\int f(z)d\mu(z)=0$ implies $f=0$. Therefore, for $\rho=2$, the space $\mathfrak{H}_\rho$ is one-dimensional and the corresponding representation is the identity representation.[4] For all other $\rho$, $0<\rho<2$, the representation is infinite-dimensional. For $\rho=0$, the representation coincides with the representation of the principal series that corresponds to the character $X\equiv 1$.

Let us note also that the formula (188) can be formally obtained from the formula (65) of §5 for the principal series, with $n=0$ and $\rho$ replaced by $\rho^i$.

We call the set of representations obtained above *the complementary series of representations* of the group $G$.

[4] It is possible, however, to define a metric on the set of functions $f(z)$ satisfying $\int f(z)d\mu(z)=0$ in such a way that for $\rho=2$ (188) would determine the representation of the principal series with $n=2$, $\rho=0$ (see p. 182). So for $\rho=2$ we have, in fact, two representations: the identity representation and the representation of the principal series with $n=2$, $\rho=0$.

**Theorem 9.** *All representations of the complementary series are irreducible.*

The proof of this theorem is similar to the proof of the corresponding theorem for the principal series.

2. *The trace of representations of the complementary series.* Let $U_{\rho_1;g}$ be a representation of the complementary series. We consider the corresponding representation of the group ring $U_{\rho_1;x}$ and prove that for $x(g)\in R'$ the operator $U_{\rho_1;x^*\times x}$ has a trace.

Let the operator $U_{\rho_1;x}$ act on a function $f(z)\in\mathfrak{H}'$. Repeating the reasoning of §6.6 we see that on such functions the operator has a kernel

$$K(z_1,z_2,\rho_1)=\int x(z_1^{-1}\delta\zeta z_2)|\lambda|^{-2+\rho_1}\frac{d\sigma d\tau}{|\lambda|^2}d\mu(\zeta) \tag{208}$$

where

$$\delta=\begin{pmatrix}\lambda^{-1} & 0\\ 0 & \lambda\end{pmatrix},\quad \lambda=\sigma+i\tau.$$

This kernel can be expressed in terms of the function $\varphi(z_1,z_2,\lambda)$ defined by formula (119), namely, by this formula,

$$K(z_1,z_2,\rho_1)=\int\varphi(z_1,z_2,\lambda)\cdot|\lambda|^{-\rho_1}\frac{d\sigma d\tau}{|\lambda|^2}. \tag{209}$$

Comparing the expression for the function $\varphi(z_1,z_2,\lambda)$ with the definition of the function $T\delta$ by the formula (97) we see that

$$T_\delta=\int\varphi(z,z,\lambda)d\mu(z). \tag{210}$$

If $x(g)\in R'$ then, as was shown in §6.3, the function $T_\delta$ is continuous and vanishes outside some compact set of the group $D$. So the integral

$$\int T_\delta\cdot|\lambda|^{-\rho_1}\frac{d\sigma d\tau}{|\lambda|^2}=\int\varphi(z,z,\lambda)|\lambda|^{-\rho_1}\frac{d\sigma d\tau}{|\lambda|^2}d\mu(z)=\int K(z,z,\rho_1)d\mu(z) \tag{211}$$

converges. The convergence is absolute because when $x(g)\in R'$ then $|x(g)|$ also belongs to $R'$.

Now we prove that if $x(g)=(x_1\times x_1^*)(g)$ and $x_2(g)\in R'$ then the integral (211) gives the trace of the operator $U_{\rho_1;x}$.

By assumption, $x(g)=(x_1\times x_1^*)(g)$. Let $\varphi_1(z_1,z_2,\lambda)$ be the function $\varphi$ that corresponds to $x_1(g)$. We will find the relationship between $\varphi$ and $\varphi_1$.

More generally, let $x_1(g)$ and $x_2(g)$ be two functions in $R'$ and $x(g)=(x_1\times x_2)(g)$. We will find the expression for the function $\varphi$ corresponding to $x=x_1\times x_2$ through the functions $\varphi_1$ and $\varphi_2$ corresponding to $x_1$ and $x_2$.

By the definition of the convolution

$$x(g)=\int x_1(gg_1)x_2(g_1^{-1})d\mu(g_1).$$

Hence,

$$\begin{aligned}\varphi(z_1, z_2, \lambda) &= |\lambda|^{-2} \int x_1(z_1^{-1}\delta\zeta z_2 g_1) x_2(g_1^{-1}) d\mu(g_1) d\mu(\zeta)\\ &= |\lambda|^{-2} \int x_1(z_1^{-1}\delta\zeta g_1) x_2(g_1^{-1} z_2) d\mu(g_1) d\mu(\zeta)\\ &= |\lambda|^{-2} \int x_1(z_1^{-1}\delta\zeta\delta_1\zeta_1 z) x_2(z^{-1}\zeta_1^{-1}\delta_1^{-1} z_2) d\mu(\delta_1) d\mu(z) d\mu(\zeta) d\mu(\zeta_1).\end{aligned}$$

Let

$$\zeta\delta_1 = \delta_1\zeta', \qquad \zeta_1^{-1}\delta_1^{-1} = \delta_1^{-1}\zeta_1'.$$

Then it is easy to check that

$$\varphi(z_1, z_2, \lambda) = |\lambda|^{-2} \int x_1(z_1^{-1}\delta\delta_1\zeta' z) x_2(z^{-1}\delta_1^{-1}\zeta_1' z_2) d\mu(\delta_1) d\mu(z) d\mu(\zeta') d\mu(\zeta_1')$$

so that

$$\varphi(z_1, z_2, \lambda) = \int \varphi_1(z_1, z, \lambda\lambda_1) \varphi_2(z, z_2, \lambda_1^{-1}) \frac{d\sigma_1 d\tau_1}{|\lambda_1|} d\mu(z). \tag{212}$$

Now let $\varphi^*(z_1, z_2, \lambda)$ be the function $\varphi$ that corresponds to the function $x^*(g) = \overline{x(g^{-1})}$. We will find the connection between $\varphi$ and $\varphi^*$. We have

$$\varphi^*(z_1, z_2, \lambda) = |\lambda|^{-2} \int \overline{x(z_2^{-1}\zeta^{-1}\delta^{-1} z_1)} d\mu(\zeta).$$

Let $\zeta^{-1}\delta^{-1} = \delta^{-1}\zeta_1$. Then $d\mu(\zeta)/d\mu(\zeta_1) = |\lambda|^4$ and

$$\varphi^*(z_1, z_2 \lambda) = |\lambda|^2 \int x(z_2^{-1}\delta^{-1}\zeta_1 z_1) d\mu(\zeta_1)$$

or

$$\varphi^*(z_1, z_2, \lambda) = \varphi\left(z_2, z_1, \frac{1}{\lambda}\right). \tag{213}$$

From (212) and (213) we see that if the function $\varphi_1(z_1, z_2, \lambda)$ corresponds to a function $x_1(g)$, then

$$\varphi(z_1, z_2, \lambda) = \iint \varphi_1(z_1, z, \lambda\lambda_1) \overline{\varphi_1(z_2, z, \lambda_1)} \frac{d\sigma_1 d\tau_1}{|\lambda_1|^2} d\mu(z) \tag{214}$$

corresponds to the function $x(g) = (x_1 \times x_1^*)(g)$. So we have

$$\int \varphi(z_1, z_1, \lambda) d\mu(z_1) = \iiint \varphi_1(z_1, z, \lambda\lambda_1) \overline{\varphi_1(z_1, z, \lambda)} \frac{d\sigma_1 d\tau_1}{|\lambda_1|^2} d\mu(z) d\mu(z_1), \tag{215}$$

and when $\lambda = 1$,

$$\int \varphi(z_1, z_1, 1) d\mu(z_1) = \iiint |\varphi_1(z_1, z_1, \lambda_1)|^2 \frac{d\sigma_1 d\tau_1}{|\lambda_1|^2} d\mu(z) d\mu(z_1). \tag{216}$$

Now we come from the function $\varphi_1$ to its Fourier transform $\Phi_1$ by means of formula (120). The formula (216) can be written in the form

$$\int \varphi(z_1, z_1, 1) d\mu(z_1) = \iiint |\Phi_1(w_1, w_2, \lambda_1)|^2 \frac{d\sigma_1 d\tau_1}{|\lambda_1|^2} d\mu(w_1) d\mu(w_2), \tag{217}$$

so the integral on the right side of (217) is finite. But then we can replace $\varphi_1$ by $\Phi_1$ in the equality (215). We get

$$\int \varphi(z_1, z_1, \lambda) d\mu(z_1) = \iiint \Phi_1(w_1, w, \lambda\lambda_1) \overline{\Phi_1(w_1, w, \lambda_1)} \frac{d\sigma_1 d\tau_1}{|\lambda_1|^2} d\mu(w_1) d\mu(w) \quad (218)$$

where the integral converges absolutely.

Let

$$L_1(w_1, w_2, \rho_1) = \int \Phi_1(w_1, w_2, \lambda) |\lambda|^{-\rho_1} \frac{d\sigma d\tau}{|\lambda|^2}. \quad (219)$$

Multiplying both sides of the equality (218) by $|\lambda|^{-\rho_1}$ and integrating with respect to $d\sigma d\tau/|\lambda|^2$, we obtain

$$\int \varphi(z_1, z_1, \lambda) \cdot |\lambda|^{-\rho_1} \frac{d\sigma dx}{|\lambda|^2} d\mu(z_1) = \iint L_1(w_1, w, \rho_1) \overline{L_1(w_1, w, -\rho_1)} d\mu(w_1) d\mu(w). \quad (220)$$

Let us find the relationship between $L_1(w_1, w, \rho)$ and $L_1(w_1, w, -\rho)$. To do this we multiply both sides of (124) by $|\lambda|^{-\rho_1}$ and integrate with respect to $d\sigma d\tau/|\lambda|^2$; we get

$$L_1(w_1, w_2, -\rho_1) = |w_1|^{-\rho_1} \cdot |w_2|^{\rho_1} L_1(w_1, w_2, \rho_1). \quad (221)$$

Because of formulas (211) and (221) the equality (220) can be written in the form

$$\int K(z, z, \rho) d\mu(z) = \iint |L_1(w_1, w_2, \rho_1)|^2 \cdot |w_1|^{-\rho_1} |w_2|^{\rho_1} d\mu(w_1) d\mu(w_2). \quad (222)$$

From the definitions of the functions $K$, $L_1$ and $\Phi$ it follows that

$$L_1(w_1, w_2, \rho_1) = \frac{1}{(2\pi)^2} \int K(z_1, z_2, \rho_1) e^{i\mathrm{Re}(z_1\bar{w}_1 - z_2\bar{w}_2)} d\mu(z_1) d\mu(z_2). \quad (223)$$

So the function $L_1(w_1, w_2, \rho_1)$ is the kernel of the operator $U_{\rho_1;x_1}$ in the space $H'_{\rho_1}$.

Let

$$\psi(w) = |w|^{-(\rho_1/2)} \varphi(w).$$

This formula gives an isometric mapping of the space $H_\rho$ on the space $H$ of functions $\psi(w)$ such that

$$\int |\psi(w)|^2 d\mu(w) < +\infty.$$

In this space the operator $U_{\rho_1;x}$ has the kernel

$$\tilde{L}(w_1, w_2, \rho_1) = L(w_1, w_2, \rho_1) \cdot |w_1|^{-(\rho_1/2)} |w_2|^{(\rho_1/2)} \quad (224)$$

and the equality (222) takes the form

$$\int K(z, z, \rho_1) d\mu(z) = \iint |\tilde{L}(w_1, w_2, \rho_1)|^2 d\mu(w_1) d\mu(w_2). \quad (225)$$

This formula shows that $\tilde{L}(w_1, w_2, \rho_1)$ is a Hilbert–Schmidt kernel in the space $H$. Both sides of this equality are equal to the trace of the operator $U_{\rho_1;x^*x}$. Our statement is proved.

Repeating the calculations of §6.2 we can prove that the expression for the trace can be written in the form

$$S(U_{\rho_1;x}) = \int x(g) \frac{|\lambda_g|^{\rho_1} + |\lambda_g|^{-\rho_1}}{|\lambda_g - \lambda_g^{-1}|^2} d\mu(g). \quad (226)$$

**Theorem 10.** *If* $x(g) = (x_1^* \times x_1)(g)$ *and* $x_1(g) \in R'$ *then the operator* $U_{\rho_1;x}$ *has a trace and this trace is given by formulas* (225) *and* (226).

This theorem, as in §6, immediately implies:

**Theorem 11.** *For an arbitrary integrable function* $x(g)$ *an operator* $U_{\rho_1;x}$ *of the irreducible representation of the complementary series is compact.*

**Theorem 12.** *Representatios of the complementary series that correspond to different* $\rho_1$ *from the interval* $0 < \rho_1 < 2$ *are not equivalent to each other and they are not equivalent to representations of the principal series.*

It is easy to see that for a function $x(g) \in R'$ the trace $S(U_{\rho_1;x})$ is a continuous function in $\rho_1$ in the interval $0 < \rho_1 < 2$ but at the point $\rho_1 = 2$ the trace is discontinuous because the limit of the expression (226) as $\rho_1 \to 2$ is equal to

$$\int x(g) \frac{|\lambda_g|_+^{-2} |\lambda_g|^2}{|\lambda_g - \lambda_g^{-1}|^2} d\mu(g). \tag{227}$$

and the trace of the identity representation is equal to

$$\int x(g) d\mu(g).$$

The expression (227) can be written as

$$\int x(g) d\mu(g) + \int x(g) \frac{|\lambda_g|^2 \bar{\lambda}_g^{-2} + |\lambda_g|^{-2} \bar{\lambda}^2}{|\lambda_g - \lambda_g^{-1}|^2} d\mu(g). \tag{228}$$

This shows that for $\rho_1 \to 2$ the representation of the complementary series, in some sense, tends to a reducible representation, namely, to the direct sum of the identity representation and the representation of the principal series that corresponds to the character $X(\lambda) = |\lambda|^2 \bar{\lambda}^{-2}$, i.e. with $n = 2$, $\rho = 0$ (see the note on p. 81).

## §9. Decomposition of an arbitrary unitary representation of the group $G$ into representations of the principal and complementary series

1. *An estimate of the norm in the ring* $R$. We prove now that the representations described are, up to equivalence, all unitary irreducible representations of the group $G$. Moreover we prove that an arbitrary representation can be decomposed into a direct sum of the representations of the principal and complementary series.

First we get an estimate for the norm $\|x\| = \int |x(g)| d\mu(g)$ of a function $x(g)$. To obtain this estimate let us make some notes.

I. *Let* $f(z_1, z_2)$ *be a continuous function with integrable partial derivatives in* $z_1, \bar{z}_1, z_2, \bar{z}_2$ *of first order. Then*

$$\iint |f(z_1 + \zeta_1, z_2 + \zeta_2) - f(z_1, z_2)| d\mu(z_1) d\mu(z_2) \leqq A_1 |\zeta_1| + A_2 |\zeta_2| \tag{229}$$

*where*

$$A_1 = \iint [|f'_{z_1}(z_1, z_2)| + |f'_{\bar{z}_1}(z_1, z_2)|] d\mu(z_1) d\mu(z_2), \tag{230}$$

$$A_2 = \iint [|f'_{z_2}(z_1, z_2)| + |f'_{\bar{z}_2}(z_1, z_2)|] d\mu(z_1) d\mu(z_2). \tag{231}$$

In fact let

$$\varphi(t)=f(z_1+t\zeta_1, z_2+t\zeta_2).$$

Then the left side of (229) is equal to

$$\iint_1 |\varphi(1)-\varphi(0)|\,d\mu(z_1)d\mu(z_2)=\iint\left|\int_0^1 \varphi'(t)dt\right|d\mu(z_1)d\mu(z_2)$$

$$\leqq\int_0^1\left[\iint|\varphi'(t)|\,d\mu(z_1)d\mu(z_2)\cdot dt\leqq\int_0^1\{\iint[|\zeta_1|\cdot|f'_{z_1}(z_1+t\zeta_1, z_2+t\zeta_2)|\right.$$

$$+|\zeta_1|\cdot|f'_{\bar z_1}(z_1+t\zeta_1, z_2+t\zeta_2)|+|\zeta_2|\cdot|f'_{z_2}(z_1+t\zeta_1, z_2+t\zeta_2)|$$

$$\left.+|\zeta_2|\cdot|f'_{\bar z_2}(z_1+t\zeta_1, z_2+t\zeta_2)|\right]d\mu(z_1)d\mu(z_2)\}dt.$$

Changing the variables $z_1'=z_1+t\zeta_1$, $z_2'=z_2+t\zeta_2$, we see that the last expression is

$$[A_1|\zeta_1|+A_2|\zeta_2|]\int_0^1 dt=A_1|\zeta_1|+A_2|\zeta_2|.$$

II. *If, moreover, the function $f(z_1, z_2)$ is itself also integrable, then for any $\varepsilon$ from the interval $0\leqq\varepsilon\leqq 1$ we have*

$$\iint|f(z_1+\zeta_1, z_2+\zeta_2)-f(z_1, z_2)|\,d\mu(z_1)d\mu(z_2)\leqq(A_1+B)|\zeta_1|^\varepsilon+(A_2+B)|\zeta_2|^\varepsilon \quad (232)$$

*where*

$$B=2\iint|f(z_1, z_2)|\,d\mu(z_1)d\mu(z_2). \quad (233)$$

In fact, in addition to the inequality (229) we have the obvious inequality

$$\iint|f(z_1+\zeta_1)|z_2+\zeta_2)-f(z_1, z_2)|\,d\mu(z_1)d\mu(z_2)\leqq B.$$

Hence,

$$\iint|f(z_1+\zeta_1, z_2+\zeta_2)-f(z_1, z_2)|\,d\mu(z_1)d\mu(z_2)\leqq(A_1|\zeta_1|+A_2|\zeta_2|)^\varepsilon B^{1-\varepsilon}$$

$$\leqq(A_1^\varepsilon|\zeta_1|^\varepsilon+A_2^\varepsilon|\zeta_2|^\varepsilon)B^{1-\varepsilon}\leqq(A_1+B)|\zeta_1^\varepsilon)+(A_2+B)|\zeta_2|^\varepsilon,$$

because if $0\leqq\varepsilon\leqq 1$ and $A>0$, $B>0$ then $(A+B)^\varepsilon\leqq A^\varepsilon+B^\varepsilon$, $A^\varepsilon B^{1-\varepsilon}\leqq A+B$.

Now, to obtain an estimate for the norm

$$\|x\|=\int|x(g)|\,d\mu(g)$$

we assume first that $x(g)\in\mathfrak{H}_G$ and that to $x(g)$ corresponds the function $\varphi(z_1, z_2, \lambda)$ with the continuous derivative $\varphi_{\lambda\bar\lambda}(z_1, z_2, \lambda)$ for all $z_1, z_2, \lambda$.

Let

$$\psi(z_1, z_2, \lambda)=\varphi_{\lambda\bar\lambda}(z_1, z_2, \lambda) \quad (234)$$

and let us assume that the function $\psi(z_1, z_2, \lambda)$ has continuous partial derivatives in $z_1, \bar z_1, z_2, \bar z_2$ up to third order; assume also that the integral

$$\int|\psi(z_1, z_2, \lambda)|\cdot|\lambda|^k d\mu(z_1)d\mu(z_2)d\mu(\lambda)$$

converges and that the similar integrals for these derivatives of $\psi$ are also finite for $K=0, \pm 1, \pm 2, \pm 2\pm\varepsilon$ with some $\varepsilon$, $0<\varepsilon<1$.

From formulas in §7.1 it follows that

$$\tilde{x}(p_1, p_2, p_3) = -a\cdot|p_2|^2 \int \psi(p_2 v - p_3, p_1 - p_2/v, v)\,d\mu(v); \qquad a = \frac{4}{(2\pi)^3}. \tag{235}$$

In fact, the right side of (235) is equal to

$$-\frac{4}{(2\pi)^5}\cdot|p_2|^2 \int \Phi''_{\lambda\bar{\lambda}}(w_1, w_2, v)e^{-i\,\mathrm{Re}[\bar{w}_1(p_2 v - p_3) - \bar{w}_2(p_1 - p_2/v)]}\cdot d\mu(w_1)d\mu(w_2)d\mu(v)$$

$$= -\frac{4}{(2\pi)^3}|\rho_2|^2 \int X''_{\zeta_2\bar{\zeta}_2}(\zeta_1, \zeta_2, \zeta_3)e^{i\,\mathrm{Re}(\bar{p}_1\zeta_1 - \bar{p}_2\zeta_2 + \bar{p}_3\zeta_3)}\cdot d\mu(\zeta_1)d\mu(\zeta_2)d\mu(\zeta_3) = \tilde{x}(p_1, p_2, p_3).$$

Hence it follows that

$$\int |x(g)|\,d\mu(g) = \int |\tilde{x}(p_1, p_2, p_3)|\,|p_2|^{-6} d\mu(p_1)d\mu(p_2)d\mu(p_3)$$
$$= a \int |\int \psi\left(p_2 v - p_3, p_1 - \frac{p_2}{v}, v\right) d_\mu(v).$$

Let $G_1$ and $G_2$ be two subsets of $G$ defined as follows:

$$G_1\colon |p_2| \leqq 1,$$
$$G_2\colon |p_2| > 1.$$

Then

$$\int_{G_2} |x(g)|\,d\mu(g) \leqq a \iint_{G_2} \left|\psi\left(p_2 v - p_3, p_1 - \frac{p_2}{v}, v\right)\right|\cdot|p_2|^{-4} d\mu(p_1)d\mu(p_2)d\mu(p_3)d\mu(v)$$

$$= a \int_{|p_2|>1} [\int |\psi(z_1, z_2, v)|\,d\mu(z_1)d\mu(z_2)d\mu(v)]\,|p_2|^{-4} d\mu(p_2)$$

$$= \pi a \int |\psi(z_1, z_2, v)|\,d\mu(z_1)d\mu(z_2)d\mu(v).$$

So we have

$$\int_{G_2} |x(g)|\,d\mu(g) \leqq \pi a \int |\psi(z_1, z_2, v)|\,d\mu(z_1)d\mu(z_2)d\mu(v). \tag{236}$$

Let us now estimate the integral over $G_1$. To do this we put

$$\varphi(t) = \psi\left(p_1 v t - p_3, p_1 - \frac{p_2 t}{v}, v\right)$$

and apply to the function $\varphi(t)$ the identity

$$\varphi(1) = \varphi(0) + \varphi^1(0) + \int_0^1 (1-t)\varphi''(t)dt.$$

We get

$$\psi\left(p_2 v t - p_3, p_1 - \frac{p_2}{v}t, v\right) = \psi(-p_3, p_1, v) + \psi'_{z_1}(-p_3, p_1, v)p_2 v + \psi'_{\bar{z}_1}(-p_3, p_1, v)\bar{p}_2\bar{v}$$
$$- \psi'_{z_2}(-p_3, p_1, v)p_2 v^{-1} - \psi'_{\bar{z}_2}(-p_3, p_1, v)\bar{p}_2\bar{v}^{-1} + \Psi \tag{237}$$

where

$$\Psi = \int_0^1 \varphi''(t)(1-t)dt.$$

By definition (234) of the function $\psi$,

$$\int \psi(z_1, z_2, v)v^k d\mu(v) = 0 \quad \text{for } k = 0, \pm 1.$$

So, integrating the equality (237), we obtain

$$\int \psi\left(p_2 v - p_3, p_1 - \frac{p_2}{v}, v\right) d\mu(v) = \int \Psi d\mu(v) = |p_2|^2 \int_0^1 (1-t)\left\{\int \left[\frac{p_2^2}{|p_2|^2} v^2 \psi''_{z_1 z_1} + \frac{\bar{p}_2^2}{|p_2|^2}\bar{v}^2 \psi''_{\bar{z}_1\bar{z}_1} + \frac{p_2^2}{|p_2|^2} v^{-2}\psi''_{z_2 z_2} + \frac{\bar{p}_2^2}{|p_2|^2}\bar{v}^{-2}\psi''_{z_2\bar{z}_2} - 2v\bar{v}^{-1}\psi''_{z_1\bar{z}_2} - 2\bar{v}v^{-1}\psi''_{\bar{z}_1 z_2} - 2\frac{p_2^2}{|p_2|^2}\psi''_{z_1 z_2} - 2\frac{\bar{p}_2^2}{|p_2|^2}\psi''_{\bar{z}_1 z_2} + 2|v|^2\psi''_{z_1\bar{z}_1} + 2|v|^{-2}\psi''_{z_2\bar{z}_2}\right] d\mu(v)\right\} dt \tag{238}$$

where all derivatives of $\psi$ are taken at the point $(p_2 vt - p_3, p_1 - (p_2 t/v), v)$. When $p_2 = 0$ the inner integeral is equal to

$$2\int [\psi''_{z_1\bar{z}_1}(-p_3, p_1, v)\cdot|v|^2 - \psi''_{z_1\bar{z}_2}(-p_3, p_1, v)v\bar{v}^{-1} - \psi''_{\bar{z}_1 z_2}(-p_3, p_1, v)\bar{v}v^{-1} + \psi''_{z_2\bar{z}_2}(-p_2, p_4, v)|v|^{-2}]d\mu(v).$$

We will assume below that the function $\psi(z_1, z_2, v)$ satisfies the condition[5]

$$\int [\psi''_{z_1\bar{z}_1}(z_1, z_2, v)\cdot|v|^2 - \psi''_{z_1\bar{z}_2}(z_1, z_2 v)v\bar{v}^{-1} - \psi''_{\bar{z}_1 z_2}(z_1, z_2, v)\bar{v}v^{-1} + \psi''_{z_2\bar{z}_2}(z_1, z_2, v)|v|^{-2}]d\mu(v) = 0 \tag{239}$$

for all $z_1, z_2$. Then the inner integral in (238) is equal to zero when $p_2 = 0$ and the value of this integral does not change if we subtract from each derivative of $\psi$ in the point $(p_2 vt - p_3, p_1 - (p_2/v), v)$ its value in the point $(-p_3, p_1, v)$.

Consider, for example, the integral

$$\int \left[\psi''_{z_1 z_1}\left(p_2 vt - p_3, p_1 - \frac{p_2}{v}t, v\right) - \psi''_{z_1, z_1}(-p_3, p_1, v)\right] d\mu(p_1)d\mu(p_3).$$

Applying to this integral the inequality (232) we get that it does not exceed

$$(A_1 + B)|p_2 vt|^\varepsilon + (A_2 + B)\left|\frac{p_2}{v}t\right|^\varepsilon \leq t[(A_1 + B)|v|^\varepsilon + (A_2 + B)|v|^{-\varepsilon}]|p_2|^\varepsilon$$

where

$$A_1 = \int \{|\psi'''_{z_1 z_1 z_1}(z_1, z_2, v)| + |\psi'''_{z_1 z_1 \bar{z}_1}(z_1, z_2, v)|\} d\mu(z_1)d\mu(z_2),$$
$$A_2 = \int \{|\psi'''_{z_1 z_1 z_2}(z_1, z_2, v)| + \psi'''_{z_1 z_1 \bar{z}_2}(z_1, z_2, v)|\} d\mu(z_1)d\mu(z_2),$$
$$B = 2\int |\psi''_{z_1 z_1}(z_1, z_2, v)| d\mu(z_1)d\mu(z_2).$$

---

[5] From further considerations, it follows that if all other conditions on the function $\psi(z_1, z_2, v)$ are satisfied, then the condition (239) is necessary for the integrability of $x(g)$.

By applying such an estimate to each function in the integral (238) we get that

$$|p_2|^{-4}\int\left|\int\psi\left(p_2 v - p_3, p_1 - \frac{p_2}{v}, v\right)d\mu(v)\right|d\mu(p_1)d\mu(p_3)$$

$$\leqq |p_2|^{-2+\varepsilon}C'\int_0^1 t(1-t) = C|p_2|^{-2+\varepsilon}$$

where $C=\frac{1}{6}C'$ and $C'$ is the sum of the integrals of the form

$$\int|\hat{\psi}(z_1,z_2,v)|\cdot|v|^k d\mu(z_1)d\mu(z_2)d\mu(v). \tag{240}$$

Here $K$ is equal to one of the numbers $\pm\varepsilon$, $\pm 2\pm\varepsilon$ and $\hat{\psi}$ is a second or third order derivative of the function $\psi$ in $z_1, \bar{z}_1, z_2, \bar{z}_2$. Integrating this equality in $p_2$ we get

$$\int_{G_1}|x(g)|d\mu(g) \leqq aC \int_{|p_2|\leqq 1}|p_2|^{-2+\varepsilon}d\mu(p_2) = 2\pi aC\int_0^1 r^{-1+\varepsilon}dr = \frac{2\pi aC}{\varepsilon}.$$

We add this inequality to the inequality (236) and obtain

$$\int|x(g)|d\mu(g) \leqq \pi a\left(C_1 + \frac{2C}{\varepsilon}\right) \tag{241}$$

where

$$C_1 = \int|\psi(z_1,z_2,v)|d\mu(z_1)d\mu(z_2,d\mu(v). \tag{242}$$

The constant $C$ does not exceed some constant $C_2$ which in its turn is equal to the sum of all integrals (240) with $k=0$, $k=\pm 2\pm\varepsilon$. Therefore, finally,

$$\int|x(g)|d\mu(g) \leqq \pi a\left(C_1 + \frac{2C_2}{\varepsilon}\right). \tag{243}$$

Later we will need an estimate for $\int|x(g)|d\mu(g)$, not in terms of the function $\varphi(z_1,z_2,\lambda)$, but in terms of its Fourier transform in $z_1$ and $z_2$, i.e. the function $\Phi(w_1,w_2,\lambda)$.

Let

$$I = \int(1+|z_1|^4\cdot|z_2|^4\cdot|v|^{2+\varepsilon})\cdot|\psi(z_1,z_2,v|^2 d\mu(z_1)d\mu(z_2)d\mu(v) < +\infty \tag{244}$$

and

$$I' = \int(1+|z_1|^4\cdot|z_2|^4\cdot|v|^{2+k+\varepsilon})\cdot|\hat{\psi}(z_1,z_2,v)|^2\cdot|v|^k d\mu(z_1)d\mu(z_2)d\mu(v) < +\infty \tag{245}$$

where the function $\hat{\psi}$ is defined as above and $k=0$, $2+\varepsilon$, $-2-\varepsilon$.

By the Schwartz inequality,

$$\left[\int|\psi(z_1,z_2,v)|d\mu(z_1)d\mu(z_2)d\mu(v)\right]^2 \leqq C_0 I$$

where

$$C_0 = \int\frac{d\mu(z_1)d\mu(z_2)d\mu(v)}{1+|z_1|^4\cdot|z_2|^4\cdot|v|^{2+\varepsilon}}$$

We obtain a similar estimate for integrals of the type (240). Therefore

$$\int|x(g)|d\mu(g) \leqq A(I^{\frac{1}{2}} + \sum I'^{\frac{1}{2}})\varepsilon^{-2} \tag{246}$$

where $A$ is some constant.

In terms of the function $\Phi(w_1, w_2, \lambda)$ we get

$$I = \int [|\Phi''_{\lambda\bar{\lambda}}(w_1, w_2, \lambda)|^2 + |\lambda|^{2+\varepsilon} \cdot |\Delta w_1 \Delta w_2 \Phi''_{\lambda\bar{\lambda}}(w_1, w_2, \lambda)|^2] d\mu(w_1) d\mu(w_2) d\mu(\lambda) \tag{247}$$

and

$$I' = \int [|\Phi''_{\lambda\bar{\lambda}}(w_1, w_2, \lambda) w_1^{\alpha_1} \bar{w}_2^{\beta_1} w_2^{\alpha_2} \bar{w}_2^{\beta_2}|^2 \cdot |\lambda|^k + |\Delta w_1 \Delta w_2 \Phi''_{\lambda\bar{\lambda}} \cdot w_1^{\alpha_1} \bar{w}_4^{\beta_1} w_2^{\alpha_2} \bar{w}^{\beta_2}|^2 \cdot |\lambda|^{2k+2+\varepsilon}] d\mu(w_1) d\mu(w_2) d\mu(\lambda) \tag{248}$$

where $\Delta w_1$ and $\Delta w_2$ are Laplace operators in $w_1$ and $w_2$ and $\alpha_1, \beta_1, \alpha_2, \beta_2$ are nonnegative integers such that

$$\alpha_1 + \alpha_1 + \alpha_2 + \beta_2 = 2 \quad \text{or} \quad 3.$$

2. *Functions $\tilde{\chi}$ and $\tilde{\Phi}$ in the case of an even function $\Phi$.* Assume that $\Phi$ (see §7.1) is an even function in $\lambda$, i.e.

$$\Phi(w_1, w_2 - \lambda) = \Phi(w_1, w_2, \lambda).$$

Obviously this condition is equivalent to

$$X(\zeta_1, \zeta_2, \zeta_3) = X(\zeta_1, -\zeta_2, \zeta_3)$$

i.e. to condition that $X$ is an even function in $\zeta_2$. Let

$$\tilde{\Phi}(w_1, w_2, \lambda) = \Phi\left(w_1, w_2, \lambda \sqrt{\frac{\bar{w}_2}{\bar{w}_1}}\right).$$

Since $\Phi$ is an even function in $\lambda$, the choice of sign of the $\sqrt{(\bar{w}_2/\bar{w}_1)}$ is unimportant.

Obviously the function $\tilde{\Phi}$ is also an even function in $\lambda$. In addition, from the condition (124) it follows that

$$\tilde{\Phi}\left(w_1, w_2, \frac{1}{\lambda}\right) = \Phi\left(w_1, w_2, \frac{1}{\lambda}\sqrt{\frac{\bar{w}_2}{\bar{w}_1}}\right) = \Phi\left(w_1, w_2, \frac{\bar{w}_2}{w_1}\lambda \sqrt{\frac{\bar{w}_1}{w_2}}\right) = \Phi\left(w_1, w_2, \lambda \sqrt{\frac{\bar{w}_2}{\bar{w}_1}}\right),$$

that is

$$\tilde{\Phi}\left(w_1, w_2, \frac{1}{\lambda}\right) = \tilde{\Phi}(w_1, w_2, \lambda). \tag{249}$$

Conversely, let $\tilde{\Phi}$ be an even function in $\lambda$ which satisfies the condition (249). If we put

$$\Phi(w_1, w_2, \lambda) = \tilde{\Phi}\left(w_1, w_2, \lambda \sqrt{\frac{\bar{w}_1}{\bar{w}_2}}\right), \tag{250}$$

then we get the function $\Phi$, which is even in $\lambda$ and satisfies (124).

Replacing $\Phi$ by $\tilde{\Phi}$ we replace the kernel $L(w_1, w_2, n_1, \rho)$ by the kernel

$$\tilde{L}(w_1, w_2, n, \rho) = \int \tilde{\Phi}(w_1, w_2, \lambda) |\lambda|^{n+i\rho} \lambda^{-n} \frac{d\sigma d\tau}{|\lambda|^2}. \tag{251}$$

From equality (249) it follows that

$$\tilde{L}(w_1, w_2, -n, -\rho) = \tilde{L}(w_1, w_2, n, \rho). \tag{252}$$

It is easy to establish the connection between $L$ and $\tilde{L}$. To do this we note first that $L$ and $\tilde{L}$ are zero for odd $n$. This follows from the fact that $\Phi$ and $\Phi$ are even functions in $\lambda$.

Substituting $\lambda\sqrt{(\bar{w}_2/\bar{w}_1)}$ in formula (126) for $\lambda$ we get

$$L(w_1, w_2, n_1, \rho) = \int \Phi\left(w_1, w_2, \lambda\sqrt{\frac{\bar{w}_2}{\bar{w}_1}}\right)|\lambda|^{n+i\rho}\lambda^{-n}\left|\frac{\bar{w}_2}{\bar{w}_1}\right|^{(n+i\rho)/2}\left(\frac{\bar{w}_2}{\bar{w}_1}\right)^{-n/2}\frac{d\sigma d\tau}{|\lambda|^2}$$
$$= \left|\frac{w_2}{w_1}\right|^{(n+i\rho)/2}\left(\frac{\bar{w}_2}{\bar{w}_1}\right)^{-n/2}\int \tilde{\Phi}(w_1, w_2, \lambda)\cdot|\lambda|^{n+i\rho}\lambda^{-n}\frac{d\sigma d\tau}{|\lambda|^2},$$

that is

$$L(w_1, w_2, n, \rho) = \left|\frac{w_2}{w_1}\right|^{\frac{(n+i\rho)}{2}}\left(\frac{\bar{w}_2}{\bar{w}_1}\right)^{-n/2}\tilde{L}(w_1, w_2, n_1, \rho). \tag{253}$$

Now let

$$\tilde{\chi}(\zeta_1, v, \zeta_3) = \chi(\zeta_1, \sqrt{\zeta_1\zeta_3}\bar{v}_1\zeta_3). \tag{254}$$

Since $\chi$ is an even function in $\zeta_2$, the choice of sign in front of $\sqrt{\zeta_1\zeta_3}$ is unimportant. Besides, it is clear that $\tilde{\chi}$ is an even function in the variables $v$.

Conversely, if $\tilde{\chi}$ is a function even in $v$, we can define a function $\chi$ which is even in $\zeta_2$ by

$$\chi(\zeta_1, \zeta_2, \zeta_3) = \tilde{\chi}\left(\zeta_1, \frac{\zeta_2^2}{\sqrt{\zeta_1\zeta_3}}, \zeta_3\right). \tag{255}$$

The functions $\tilde{\Phi}$ and $\tilde{\chi}$ are connected by a simple formula, which can be easily obtained from (123). Let us substitute in this formula $\lambda\sqrt{(\bar{w}_2/\bar{w}_1)}$ instead of $\lambda$. We have

$$\frac{1}{2\pi}\Phi\left(w_1, w_2, \lambda\sqrt{\frac{\bar{w}_2}{\bar{w}_1}}\right) = \chi\left(w_2\sqrt{w_1 w_2}\left(\bar{\lambda} + \frac{1}{\bar{\lambda}}\right).w_1\right),$$

that is

$$\frac{1}{2\pi}\tilde{\Phi}(w_1, w_2, \lambda) = \tilde{\chi}\left(w_2, \lambda + \frac{1}{\lambda}, w_1\right). \tag{256}$$

3. *The functions $\tilde{\chi}$ and $\tilde{\Phi}$ in the case when $\Phi$ is odd.* Now let $\Phi$ be an odd function in $\lambda$ i.e.

$$\Phi(w_1, w_2 - \lambda) = -\Phi(w_1, w_2\lambda).$$

Obviously, it is equivalent to the fact that $\chi$ is odd function in $\tau_2$, and hence $\tilde{x}$ is an odd function in $p_2$.

Let

$$\tilde{\Phi}(w_1, w_2, \lambda) = \left|\frac{w_1}{w_2}\right|^{\frac{1}{2}} \left(\frac{\bar{w}_2}{\bar{w}_1}\right)^{\frac{1}{2}} \Phi\left(w_1, w_2, \lambda\sqrt{\frac{\bar{w}_2}{\bar{w}_1}}\right). \tag{257}$$

Since $\Phi$ is an odd function in $\lambda$ the choice of sign of the function $\sqrt{(\bar{w}_2/\bar{w}_1)}$ is unimportant.

Obviously the function $\tilde{\Phi}$ is also an odd function in $\lambda$. From (124) it follows that

$$\tilde{\Phi}\left(w_1, w_2, \frac{1}{\lambda}\right) = \left|\frac{w_1}{w_2}\right|^{\frac{1}{2}} \left(\frac{\bar{w}_2}{\bar{w}_1}\right)^{\frac{1}{2}} \Phi\left(w_1, w_2, \lambda^{-1}\sqrt{\frac{\bar{w}_2}{\bar{w}_2}}\right)$$

$$= \left|\frac{w_1}{w_2}\right|^{\frac{1}{2}} \left(\frac{\bar{w}_2}{\bar{w}_1}\right)^{\frac{1}{2}} \Phi\left(w_1, w_2, \frac{\bar{w}_2}{\bar{w}_1}\lambda\sqrt{\frac{\bar{w}_1}{\bar{w}_2}}\right),$$

that is

$$\Phi\left(w_1, w_2, \frac{1}{\lambda}\right) = \tilde{\Phi}(w_1, w_2, \lambda). \tag{258}$$

Conversely, let $\tilde{\Phi}(w_1, w_2, \lambda)$ be an odd function in $\lambda$ which satisfies (258). If we put

$$\Phi(w_1, w_2, \lambda) = \left|\frac{\bar{w}_2}{\bar{w}_1}\right|^{\frac{1}{2}} \left(\frac{\bar{w}_1}{\bar{w}_2}\right)^{\frac{1}{2}} \tilde{\Phi}\left(w_1, w_2, \lambda\sqrt{\frac{\bar{w}_1}{\bar{w}_2}}\right) \tag{259}$$

we get the function $\Phi$ that is odd in $\lambda$ and satisfies the condition (124).

Replacing $\Phi$ by $\tilde{\Phi}$ we replace the kernel $L(w_1, w_2, n, \rho)$ by the kernel $\tilde{L}(w_1, w_2, n, \rho)$ which is defined by (251) and, as in p. 91, the equality (252) remains true. Repeating the corresponding arguments from §9.2, we obtain that

$$L(w_1, w_2, n, \rho) = \tilde{L}(w_1, w_2, n, \rho)\left|\frac{w_2}{w_1}\right|^{(n-1+i\rho)/2} \left(\frac{\bar{w}_2}{\bar{w}_1}\right)^{-(n-1)/2}. \tag{260}$$

Let us note that in this case the functions $L$ and $\tilde{L}$ are zero for even $n$.

Now let

$$\tilde{\chi}(\zeta_1, \mu, \zeta_3) = \left|\frac{\zeta_1}{\zeta_3}\right|^{\frac{1}{2}} \sqrt{\frac{\zeta_3}{\zeta_1}}\, \chi\left(\zeta_1, \zeta_1\sqrt{\frac{\zeta_3}{\zeta_1}}\bar{\mu}, \zeta_3\right). \tag{261}$$

Then from (123) it follows that

$$\frac{1}{2\pi}\tilde{\Phi}(w_1, w_2, \lambda) = \tilde{\chi}\left(w_2, \lambda + \frac{1}{\lambda}, w_1\right). \tag{262}$$

Finally, if $\chi$, and hence $\Phi$, are arbitrary functions, we can decompose them in to the sum of an even and an odd functions in $\lambda$ and we define the functions $\tilde{\chi}$ and $\tilde{\Phi}$ as the sum of the functions that correspond to these summands. Obviously in this case the formulas (258) and (262) are also true.

4. *Degenerate functions $\tilde{\chi}$ and $\tilde{\Phi}$*. Let us consider functions $\tilde{\chi}$ and $\tilde{\Phi}$ of the form

$$\tilde{\chi}(\zeta_1, v, \zeta_3) = \chi_1(\zeta_1)\chi_2(v)\chi_3(\zeta_3), \tag{263}$$

$$\tilde{\Phi}(w_1, w_2, \lambda) = \Phi(w_1)\Psi(w_2)\omega(\lambda). \tag{264}$$

By (258) and (262) we have

$$\tilde{\chi}(w_1) = \chi_3(w_1), \qquad \Psi(w_2) = \chi_1(w_2), \qquad \frac{1}{2\pi}\omega(\lambda) = \chi\left(\lambda + \frac{1}{\lambda}\right)$$

and

$$\omega\left(\frac{1}{\lambda}\right) = \omega(\lambda). \tag{265}$$

The functions of the form (263) and (264) will be called degenerate. To a degenerate function $\tilde{\Phi}$ corresponds the function

$$\Phi(w_1, w_2, \lambda) = \Phi(w_1)\Psi(w_2)\omega\left(\lambda\sqrt{\frac{\bar{w}_1}{\bar{w}_2}}\right) \tag{266}$$

in the case of even $\omega(\lambda)$ and the function

$$\Phi(w_1, w_2, \lambda) = \left|\frac{w_2}{w_1}\right|^{\frac{1}{2}} \left(\frac{\bar{w}_1}{\bar{w}_2}\right)^{\frac{1}{2}} \Phi(w_1)\Psi(w_2)\omega\left(\lambda\sqrt{\frac{\bar{w}_1}{\bar{w}_2}}\right) \tag{2.67}$$

in the case of odd $\omega(\lambda)$.

Denote by $\|\Phi\|_1^2$ the right side of the equality (145) and introduce the norm $\|\Phi\|_1^2$ on functions $\tilde{\Phi}$. We denote this norm by $\|\tilde{\Phi}\|_1$.

Let $F$ be the set of all degenerate functions $\tilde{\Phi}(w_1, w_2, \lambda)$ that satisfy the following conditions:

1°. The function $\Phi(w_1)$ vanishes outside a region $0 < \varepsilon_1 \leqq |w_1| \leqq C_1$ and has continuous derivatives up to fourth order.

2°. The function $\Psi(w_2)$ vanishes outside a region $0 < \varepsilon_2 \leqq |w_2| \leqq C_2$ and has continuous derivatives up to fourth order.

3°. The function $\omega(\lambda)$ vanishes outside a region $0 < \varepsilon \leqq |\lambda| \leqq C$ and has continuous derivatives up to eighth order.

4°.
$$\int \omega(\lambda) \cdot |\lambda - \lambda^{-1}|^2 \frac{d\mu(\lambda)}{|\lambda|^2} = 0.$$

To each function $\tilde{\Phi} \in F$ corresponds the function $x(g) \in \mathfrak{H}_G$. This function $x(g)$ is integrable. Indeed from 1°, 2°, 3° it follows that the norm $\|\tilde{\Phi}\|_1$ is finite; in accordance with p. 000 it is sufficient to show that the corresponding function $\psi(z_1, z_2, \lambda) = \varphi''_{\lambda\bar{\lambda}}(z_1, z_2, \lambda)$ satisfies the condition (239) of p. 88.

From the definition of the function $\Phi(w_1, w_2, \lambda)$ we have

$$\varphi(z_1, z_2, \lambda) = \frac{1}{(2\pi)^2} \int \Phi(w_1, w_2, \lambda) e^{-i\mathrm{Re}(z_1\bar{w}_1 - z_2\bar{w}_2)} d\mu(w_1) d\mu(w_2)$$

so the condition (239) means that

$$\int \Phi''_{\lambda\bar{\lambda}}(w_1, w_2, \lambda) |w_1\bar{\lambda} + w_2\bar{\lambda}^{-1}|^2 d\mu(\lambda) = 0. \tag{268}$$

Integrating by parts we get

$$\int \Phi(w_1, w_2, \lambda)|w_1\bar{\lambda} - w_2\bar{\lambda}^{-1}|^2 \frac{d\mu(\lambda)}{|\lambda|^2} = 0. \tag{269}$$

If $\tilde{\Phi} \in F$ and $\omega(\lambda)$ is an even function, then the left side of (269) can be written as

$$\int \Phi(w_1)\Psi(w_2)\omega\left(\lambda\sqrt{\frac{\bar{w}_1}{\bar{w}_2}}\right)\cdot|w_1\bar{\lambda} - w_2\bar{\lambda}^{-1}|^2 \frac{d\mu(\lambda)}{|\lambda|^2}$$

$$= \Phi(w_1)\Psi(w_2)|w_1 w_2| \int \omega(\lambda)|\lambda - \lambda^{-1}|^2 \frac{d\mu(\lambda)}{|\lambda|^2}.$$

The last expression is zero because of 4°.

The proof of (239) for an odd function $\omega(\lambda)$ is similar.

So, to each function $\tilde{\Phi} \in F$ corresponds an integrable function $x(g)$. Denote by $\mathfrak{A}$ the set of all these function.

**Lemma 1.** *Let $\mathcal{E}$ be a complete normed Banach space and let $f_1(x), f_2(x), \ldots, f_n(x)$ be a finite number of additive and distributive functions defined on some linear manifold $L \subset \mathcal{E}$ that is dense in $\mathcal{E}$. If no nonzero linear combination of these functionals is a bounded functional in $\mathcal{E}$ then the set $L'$ of the elements from L satisfying conditions*

$$f_1(x) = f_2(x) = \ldots = f_n(x) = 0 \tag{270}$$

*is also dense in L.*

*Proof.* Obviously we can assume that any nonzero linear combination of these functionals does not vanish identically on $L$. Otherwise we can decrease the number of functionals in such a way that it would be true. With this assumption it is possible to choose elements $x_1, x_2, \ldots, x_n$ in $L$ such that

$$f_i(x_k) = \begin{cases} 0, & i \neq k \\ 1, & i = k \end{cases} \qquad (i, k, = 1, 2, \ldots, n).$$

An arbitrary element from $L$ can be represented in the form

$$x = \lambda_1 x_1 + \ldots . + \lambda_n(x_n) + y$$

where $y \in L'$ and $\lambda_i = f_i(x)$.

Let $L'$ be non dense in $\mathcal{E}$; then there exists a bounded functional $f_0(x) \not\equiv 0$ such that $f_0(y) = 0$ for all $y \in L'$. At the same time $f_0$ is not identically zero on $L$ since $L$ is dense in $\mathcal{E}$.

Therefore, the equation $f_0(x) = 0$ determines an $n-1$-dimensional subspace in the $n$-dimensional space of all linear combinations $\lambda_1 x_1 + \ldots + \lambda_n x_n$. Replacing $x_1, \ldots, x_n$ and $f_1, \ldots, f_n$ by some linear combinations, we can assume that this subspace consists of all elements of the form $\lambda_2 x_2 + \ldots + \lambda_n x_n$. Then $f_0(x_1 \neq 0$ and we can normalize $f_0(x)$ so that $f_0(x_1) = 1$. Therefore

$$f_0(\lambda_1 x_1 + \ldots + \lambda_n x_n + y) = \lambda_1,$$

$$f_1(\lambda_1 x_1 + \ldots + \lambda_n x_n + y) = \lambda_1,$$

i.e. $f_1(x)=f_0(x)$ on $L$. But this is impossible because $f_0(x)$ is a bounded functional and $f_1(x)$ is, by assumption, unbounded.

**Lemma 2.** *Let $U_g$ be a unitary representation of the group G and let*

$$\int x(g)U_g d\mu(g)=0 \tag{271}$$

*for all functions $x(g)\in\mathfrak{A}$. Then $U_g$ is the identity representation, i.e. $U_g\equiv E$.*

*Proof.* Consider any function $\Phi_0$ satisfying the conditions:

1*. $\tilde{\Phi}_0(w_1,w_2,\lambda)$ vanishes outside a set

$$0<\varepsilon_1<|w_1|<C_1,\qquad 0<\varepsilon_2<|w_2|<C_2,\qquad 0<\varepsilon<|\lambda|<C \tag{272}$$

and has continuous partial derivatives in $w_1,\bar{w}_1,w_2,\bar{w}_2$ up to second order and in $\lambda,\bar{\lambda}$ up to sixth order.

2*. $$\int \Phi_0(w_1,w_2,\lambda)\cdot|\lambda-\lambda^{-1}|^2\frac{d\mu(\lambda)}{|\lambda|^2}=0$$

can be approximated in the norm $\|\tilde{\Phi}\|_1$ by linear combinations of functions $\tilde{\Phi}\in F$.

In fact it is possible to choose a sequence

$$\tilde{\Phi}^{(m)}=\sum_k \Phi_k^{(m)}(w_1)\Psi_k^{(m)}(w_2)\omega_k^{(m)}(\lambda)$$

of linear combinations of degenerate functions $\tilde{\Phi}$ satisfying conditions 1°, 2°, 3° (see p. 93) so that this sequence converges uniformly to $\tilde{\Phi}_0$ in the region (272).

Now let $\omega_0(\lambda)$ be a fixed function, satisfying 3° and such that

$$\int\omega_0(\lambda)|\lambda-\lambda^{-1}|^2\frac{d\mu(\lambda)}{|\lambda|^2}=1.$$

Let

$$\omega_{1k}^{(m)}(\lambda)=\omega_k^{(m)}(\lambda)-\omega_0(\lambda)\int\omega_k^{(m)}(\lambda)|\lambda-\lambda^{-1}|^2\frac{d\mu(\lambda)}{|\lambda|^2}$$

and

$$\tilde{\Phi}_1^{(m)}=\sum_k \Phi_k^{(m)}(w_1)\Psi_k^{(m)}(w_2)\omega_{1k}^{(m)}(\lambda). \tag{273}$$

The function $\Phi_1^{(m)}$ is already a linear combination of functions from $F$ and the sequence $\Phi_1^{(m)}$ also converges uniformly together with required derivatives to $\tilde{\Phi}_0$ in the region (272).

Indeed, the integral

$$\int\tilde{\Phi}^{(m)}(w_1,w_2,\lambda).|\lambda-\lambda^{-1}|^2\frac{d\mu(\lambda)}{|\lambda|^2}$$

and its derivatives of the proper order tend to zero uniformly in $w_2,w_2$. Hence the expression

$$\omega_0(\lambda)\int\tilde{\Phi}^{(m)}(w_1,w_2,\lambda)\cdot|\lambda-\lambda^{-1}|^2\frac{d\mu(\lambda)}{|\lambda|^2}$$

and its derivatives of proper order tend to zero uniformly in $w_1, w_2, \lambda$. Therefore, the sequence $\Phi_1^{(m)}$ also converges to $\tilde{\Phi}_0$ and the sense of the norm $\|\tilde{\Phi}\|_1$.

An integrable function $x_0(g)$ corresponds to the function $\tilde{\Phi}_0$. Denote by $\mathfrak{A}_0$ the set of these functions $x_0(g)$.

From the statement we have just proved and from the estimate (246) it follows that a function $x_0(g)$ can be approximated in the norm $\|x\|$ by linear combinations of functions from $\mathfrak{A}$. So the equality (271) is also true for functions $x_0(g) \in \mathfrak{A}_0$.

Denote by $X$ the set of functions $\chi$ satisfying the conditions:

1**. $\chi$ vanishes outside a set of the form

$$0 < \varepsilon_1 \leqq |\zeta_1| \leqq C_1, \quad |\zeta_2| \leqq C_2, \quad 0 < \varepsilon_3 \leqq |\zeta_3| \leqq C_3$$

and has continuous derivatives up to second order in $\zeta_1, \bar{\zeta}_1, \bar{\zeta}_3, \zeta_3$ and up to sixth order in $\zeta_2, \bar{\zeta}_2$.

$$2^{**}. \quad \int \chi(\zeta_1, \zeta_2, \zeta_3) d\mu(\zeta_2) = 0.$$

To each function $\chi$ corresponds the function $\tilde{\Phi}$ satisfying the conditions 1*, 2* and therefore the function $x(g) \in \mathfrak{A}_0$. Indeed it is easy to see that the conditions 2** and 2* are equivalent.

Let

$$\chi(\zeta_1, \zeta_2, \zeta_3) = \chi_1(\zeta_1)\chi_2(\zeta_2)\chi_3(\zeta_3) \tag{274}$$

where $\chi_1(\zeta_1)$ and $\chi_2(\zeta_2)$ vanishes outside sets

$$0 < \varepsilon_1 \leqq |\zeta_1| \leqq |C_1|, \quad 0 < \varepsilon_3 \leqq |\zeta_3| \leqq C_3$$

and have continuous derivatives up to second order and $\chi_2(\zeta_2)$ vanishes outside a set $|\zeta_2| \leqq C_2$, satisfies the condition

$$\int \chi_2(\zeta_2) d\mu(\zeta_2) = 0 \tag{275}$$

and has continuous derivatives up to sixth order.

We denote the sets of such functions by $X_1, X_2, X_3$, respectively.

When $\chi_1, \chi_2, \chi_3$ satisfy these conditions, the function $\chi$, defined by (274), belongs to $X$ and for the corresponding function $x(g) = \tilde{x}(p_1, p_2, p_3)$ the equality (271) is true.

Let

$$y(p_1, p_2, p_3) = \tilde{x}(p_1, p_2, p_3)|p_2|^{-4}.$$

The function $y(p_1, p_2, p_3)$ is the Fourier transform of the function $\chi$, so it is a product

$$y(p_1, p_2, p_3) = y_1(p_1) y_2(p_2) y_3(p_3) \tag{276}$$

where

$$y_1(p_1) = \frac{1}{2\pi} \int \chi_1(\zeta_1) e^{i\mathrm{Re}(\bar{p}_1\zeta_1)} d\mu(\zeta_1), \tag{277}$$

$$y_2(p_2) = \frac{1}{2\pi} \int \chi_2(\zeta_2) e^{-i\mathrm{Re}(\bar{p}_2\zeta_2)} d\mu(\zeta_2), \tag{278}$$

$$y_3(p_3) = \frac{1}{2\pi} \int \chi_3(\zeta_3) e^{i\mathrm{Re}(\bar{p}_3\zeta_3)} d\mu(\zeta_3). \tag{279}$$

Moreover

$$\|x\| = \int |x(g)|d\mu(g) = \int |\tilde{x}(p_1,p_2,p_3)|\cdot|p_2|^{-6}d\mu(p_1)d\mu(p_2)d\mu(p_3)$$
$$= \int |y_1(p_1)|d\mu(p_1)\cdot\int |p_2|^{-2}|y_2(p_2)|d\mu(p_2) - \int |y_3(p_3)|d\mu(p_3). \tag{280}$$

Denote by $Y, Y_1, Y_2, Y_3$ the sets of functions defined by formulas (276), (277), (278), (279) where $\chi_1, \chi_2, \chi_3$ belong to $X_1, X_2, X_3$. Denote also by $\tilde{Y}_1, \tilde{Y}_2, \tilde{Y}_3$ the closures of the sets $Y_1, Y_2, Y_3$ in the norms:

$$\|y_1\|_1 = \int |y_1(p_1)|d\mu(p_1), \quad \|y_3\|_1 = \int |y_3(p_3)|d\mu(p_3) \; \|y_2\|_2 = \int |y_2(p_2)|\cdot|p_2|^{-2}d\mu(p_2).$$

Let $\tilde{Y}$ be the set of functions of the form (276) where $y_1 \in \tilde{Y}_1$, $y_2 \in \tilde{Y}_2$, $y_3 \in \tilde{Y}_3$. Each function from $\tilde{Y}$ is the limit in the norm $\|x\|$ of a sequence of functions from $Y$. So the equality (271) is true for all functions

$$x(g) = \tilde{x}(p_1,p_2,p_3) = y(p_1,p_1,p_3)|p_2|^4$$

with $y \in \tilde{Y}$.

Let us find the closures of $\tilde{Y}_1, \tilde{Y}_2, \tilde{Y}_3$. For $\tilde{Y}_1$ we use an obvious estimate:

$$\|y_1\|_1^2 \leq C\int (1+|p_1|^4)|y_1(p_1)|^2 d\mu(p_1) = C\|X_1\|_1^2 \tag{281}$$

where

$$C = \int \frac{d\mu(p)}{1+|p_1|^4}, \quad \|\chi_1\|_1^2 = \int [|\chi_1(\zeta_1)|^2 + |\Delta_{\zeta_1}\chi_1(\zeta_1)|^2]d\mu(\zeta_1). \tag{282}$$

Let a function $\chi_1(\zeta_1) \in X$ satisfy the conditions:

(a) $\chi_1(\zeta_1)$ vanishes outside a set $|\zeta_1| < C$ and has continuous derivatives up to second order.

(b) A function $\chi_1(\zeta_1)$ and its derivatives up to second order vanish when $\zeta_1 = 0$.

Then the function can be approximated by the functions $\chi_1 \in X$ in the sense of uniform convergence of functions and their derivatives up to second order, i.e. in the sense of norm $\|X_1\|_1$.

Therefore, $\tilde{Y}_1$ contains all functions that are Fourier transforms of the functions $\chi_1$ satisfying (a) and (b).

Denote by $L_1$ the set of functions $y_1(p_1)$ with the finite norm $\|y_1\|_1$.

The set $L_1'$ of functions $y_1(p_1)$ for which

$$\int (1+|p_1|^4)|y_1(p_1)|^2 d\mu(p_1) < +\infty$$

is dense in the norm $\|y_1\|_1$.

Denote by $X_1^{(a)}$ the set of functions $\chi_1$, satisfying (a) only. This set is dense in the set of functions $\chi_1$ with the finite norm $\|\chi_1\|_1'$.

The estimate (281) implies that the set of functions $Y_1^{(a)}$ which correspond to $\chi_1 \in X^{(a)}$ is dense in $L_1'$ and hence in $L$ with respect to the norm $\|y_1\|_1$.

The condition (b) means that

$$\int y_1(p_1)p_1^\alpha \bar{p}_1^\beta d\mu(p_1) = 0 \tag{283}$$

for all nonnegative $\alpha$ and $\beta$ such that $\alpha+\beta \leq 2$. But when $\alpha+\beta > 0$ the integral (283) is a nonbounded functional in $L_1$ and no linear combination of these functionals except zero is a bounded functional in $L_1$.

By lemma 1 the set $Y_1^{(b)}$ of functions from $Y_1^{(a)}$ satisfying (283) for $0<\alpha+\beta<2$ is also dense in $L_1$. In the case $\alpha=\beta=0$, condition (283) takes the form

$$\int y_1(p_1)d\mu(p_1)=0. \tag{284}$$

Therefore, $\tilde{Y}_1$ contains all functions from $Y_1^{(b)}$ that satisfy the condition (284).

Since $Y_1^{(b)}$ is dense in $L_1$ its closure in norm $\|y_1\|_1$ consists of all functions $y_1$ with a finite norm satisfying (284).

Indeed, let $y_1$ be an arbitrary function from $L_1$ that satisfies (284). Then there exists a sequence $y_1^{(n)}\in Y_1^{(b)}$ such that

$$\|y_1-y_1^{(n)}\|_1\to 0 \quad \text{for} \quad n\to\infty.$$

Hence,

$$\int y_1^{(n)}(p_1)d\mu(p_1)\to\int y_1(p_1)d\mu(p_1). \tag{285}$$

Let $y_1^0(p_1)$ be a fixed function from $Y_1^{(b)}$ such that

$$\int y_1^0(p_1)d\mu(p_1)=1.$$

Let

$$\tilde{y}_1^{(n)}(p_1)=y_1^{(n)}(p_1)-y_1^0(p_1)\int y_1^{(n)}(p_1)d\mu(p_1).$$

By (285) the sequence $\tilde{y}_1^{(n)}(p_1)$ also converges to $y_1(p_1)$ in the norm $\|y_1\|_1$. The functions $\tilde{y}_1(p_1)$ belong to $\tilde{Y}_1$ and therefore $y_1(p_1)$ belongs to $\tilde{Y}_1$.

So *the set $\tilde{Y}_1$ consists of all functions $y_1(p_1)$ that satisfy the conditions*

$$\int|y_1(p_1)|d\mu(p_1)<+\infty, \quad \int y_1(p_1)d\mu(p_1)=0. \tag{286}$$

A similar result is true for $\tilde{Y}_3$.

Now let us consider $\tilde{Y}_2$. First we estimate the norm

$$\|y_2\|_2=\int|y_2(p_2)|\cdot|p_2|^{-2}d\mu(p_2).$$

Obviously

$$\|y_2\|_2^2\leqq C\int|y_2(p_2)|^2\cdot|p_2|^{-4}(1+|p_2|^4)d\mu(p_2) \tag{287}$$

where

$$C=\int\frac{d\mu(p_2)}{1+|p_2|^4}.$$

Further,

$$\int\limits_{|p_2|>1}|y_2(p_2)|^2\cdot|p_2|^{-4}d\mu(p_2)\leqq\int|y_2(p_2)|^2d\mu(p_2). \tag{288}$$

Let us estimate the integral

$$\int\limits_{|p_2|\leqq 1}|y_2(p_2)|^2\cdot|p_2|^{-4}d\mu(p_2).$$

We assume that in addition to the condition $y_2(0)=0, y_2$ also satisfies the conditions:

$$y'_{2p_2}(0)=y'_{2\bar{p}_2}(0)=0.$$

Let

$$\varphi(t)=y_2(tp_2).$$

Then the identity

$$\varphi(1) = \varphi(0) + \varphi'(0) + \int_0^1 (1-t)\varphi''(t)dt$$

can be written in the form

$$y_2(p_2) = \int_0^1 (1-t)\left[\sum_{\alpha+\beta=2} p_2^\alpha \bar{p}_2^\beta y''_{2p_2^\alpha \bar{p}_2^\beta}(tp_2)\right]dt.$$

Hence

$$|y_2(p_2)| \leqq \tfrac{1}{2}|p_2|^2 \sum_{\alpha+\beta=2} \max_{p_2} |y_{2p_2^\alpha \bar{p}_2^\beta}(p_2)|$$

and therefore

$$\int_{|p_2|\leqq 1} |y_2(p_2)|^2 \cdot |p_2|^{-4} d\mu(p_2) \leqq \frac{\pi}{4}\left(\sum_{\alpha+\beta=2} \max_{p_2} |y''_{2p_2^\alpha \bar{p}_2^\beta}(p_2)|^2\right).$$

So we have

$$\| y_2 \|_2^2 \leqq \| y_2 \|_3 = \| \dot{\chi}_2 \|_3 \tag{289}$$

where

$$\| y_2 \|_3^2 = C_1 \int |y_2(p_2)|^2 (1 + |p_2|^4) d\mu(p_2) + C_2 \left(\sum_{\alpha+\beta=2} \max_{p_2} |y''_{2,p_2^\alpha \bar{p}_2^\beta}(p_2)|\right)^2. \tag{290}$$

Therefore

$$\| \chi_2 \|_3^2 = C_1 \int [|\chi_2(\zeta_2)|^2 + |\Delta_{\zeta_2}\chi_2|^2] d\mu(\zeta_2) + C_2' \left(\sum_{\alpha+\beta=2} \int |\zeta_2|^{\alpha+\beta} |\chi_2(\zeta_2)| d\mu(\zeta_2)\right)^2 \tag{291}$$

with some constants $C_1, C_2, C_2'$.

Denote by $L_2$ the set of functions $y_2(p_2)$ with the finite norm $\| y_2 \|_2$ and by $L_3$ the set of functions $y_2(p_2)$ with the finite norm $\| y_2 \|_3$ that vanish outside some neighbourhood of the point $p_2 = 0$.

Obviously the set $L_3$ is dense in $L_2$ in the norm $\| y_2 \|_2$.

Further, it is obvious that $X_2$ is dense in norm $\| \chi_2 \|_3$ in the set of functions $\chi_2$ with a finite norm $\| \chi_2 \|_2$ satisfying (275). From the estimate (289) it follows that $Y_2$ is dense in $L_3$ and therefore in $L_2$ in the norm $\| y_2 \|_2$.

Hence $\tilde{Y}_2$ consists of the functions $y_2(p_2)$ such that

$$\int |y_2(p_2)| \, |p_2|^{-2} d\mu(p_2) < +\infty.$$

Let

$$\varphi(g) = (U_g f_1, f_2).$$

All functions $\varphi(g)$ are continuous and bounded.

Moreover, if $\varphi(g) = (U_g f_1, f_2)$, then all functions $\varphi(gg_0)$) and $\varphi(g_0 g)$ are the functions of the same form. Indeed,

$$\varphi(gg_0)=(U_{gg_0}f_1,f_2)=(U_gU_{g_0}f_1,f_2)$$

and

$$\varphi(g_0g)=(U_{g_0g}f_1,f_2)=(U_gf_1,U_{g_0}^{-1}f_2).$$

By (271), $\varphi(g)$ satisfy the condition

$$\int x(g)\varphi(g)d\mu(g)=0 \tag{292}$$

for all

$$x(g)=\tilde{x}(p_1,p_2,p_3)=|p_2|^4 y(p_1,p_2,p_3)$$

such that $y\in\tilde{Y}$.

Let us note that for dense sets of elements $f_1$ and $f_2$ the function $\varphi(g)$ and hence $\varphi(g_0g)$ and $\varphi(gg_0)$ have the required number of derivatives in $g$. In fact let a function $a(g)$ be differentiable and vanish outside a compact. If in (292) we replace $f_1$ by $\int a(g)U_gf_1\,d\mu(g)$ then $\varphi(g)$ becomes differentiable. The set of such $f_1$'s is dense (cf. similar arguments in 000).

Let us write (292) in parameters $p_1,p_2,p_3$ (changing $\varphi(g)$ to $\varphi(p_1,p_2,p_3)$). We have

$$\int y(p_1,p_2,p_3)\varphi(p_1,p_2,p_3)|p_2|^{-2}d\mu(p_1)d\mu(p_2)d\mu(p_3)=0$$

so that

$$\int y_1(p_1)y_2(p_2)|p_2|^{-2}y_3(p_3)\varphi(p_1,p_2,p_3)d\mu(p_1)d\mu(p_2)d\mu(p_3)=0 \tag{293}$$

for all $y_1,y_2,y_3$ from $\tilde{Y}_1$, $\tilde{Y}_2$, $\tilde{Y}_3$. Let us rewrite (293) in the form

$$\int |y_2(p_2)|p_2|^{-2}[\int\varphi(p_1,p_2,p_3)y_1(p_1)y_3(p_3)d\mu(p_1)d\mu(p_3)]d\mu(p_2)=0.$$

This equality holds for an arbitrary integrable function $y_2(p_2)|p_2|^{-2}$. Hence,

$$\int\varphi(p_1,p_2,p_3)y_1(p_1)y_3(p_3)d\mu(p_1)d\mu(p_3)=0 \tag{294}$$

for almost all $p_2$. But the integral in (294) is a continuous function in $p_2$ and therefore the condition (294) holds for all $p_2$.

Now let $y_0(p)$ be a fixed function such that

$$\int y_0(p)d\mu(p)=1$$

and $y_1(p_1)$, $y_3(p_3)$ be arbitrary integrable functions. Let

$$\tilde{y}_1(p_1)=y_1(p_1)-y_0(p_1)\int y_1(p_1)d\mu(p_1),$$
$$\tilde{y}_3(p_3)=y_3(p_3)-y_0(p_3)\int y_3(p_3)d\mu(p_3).$$

Then

$$\int\tilde{y}_1(p_1)d\mu(p_1)=\int\tilde{y}_3(p_3)d\mu(p_3)=0$$

and hence $\tilde{y}_1$ and $\tilde{y}_3$ belong to $\tilde{Y}_1$ and $\tilde{Y}_3$, respectively. Therefore, the condition (294) is true for $\tilde{y}_1$ and $\tilde{y}_3$. Further, let

$$\varphi_1(p_2,p_3)=\int\varphi(p_1,p_2,p_3)y_0(p_1)d\mu(p_1),$$
$$\varphi_2(p_1,p_2)=\int\varphi(p_1,p_2,p_3)y_0(p_3)d\mu(p_3)-\int\varphi(p_1,p_2,p_3)y_0(p_1)y_0(p_3)d\mu(p_1)d\mu(p_3).$$

Substituting $\tilde{y}_1$ and $\tilde{y}_2$ for $y_1$ and $y_3$ in (294) we get

$$\begin{aligned}&\int\varphi(p_1,p_2,p_3)y_1(p_1)y_3(p_3)d\mu(p_1)d\mu(p_3)\\&\quad=\int[\varphi_1(p_2,p_3)+\varphi_2(p_1,p_2)]y_1(p_1)y_3(p_3)d\mu(p_1)d\mu(p_3)\end{aligned}$$

for all integrable functions $y_1(p_1)$ and $y_3(p_3)$. So $\varphi(p_1, p_2, p_3)$ is a sum

$$\varphi(p_1, p_2, p_3) = \varphi_1(p_2, p_3) + \varphi_2(p_1, p_2). \tag{295}$$

Since with $\varphi(g)$ the functions $\varphi(gg_0)$ and $\varphi(g_0 g)$ also satisfy (292), we obtain that these functions are also of the form (295).

Let $g = \begin{pmatrix} 0 & 1 \\ -1 & 0 \end{pmatrix}$; the shift $g \to gg_0$ changes the parametrs as follows:

$$p_1 \to -\frac{1}{p_1}, \qquad p_2 \to \frac{p_2}{p_1}, \qquad p_3 \to p_2 - \frac{p_2^2}{p_1}.$$

So the function $\varphi$ which is of the form (295) can be expressed, on the other hand, in the form

$$\varphi(p_1, p_2, p_3) = \varphi_3\left(\frac{p_2}{p_1}, p_3 - \frac{p_2^2}{p_1}\right) + \varphi_4\left(-\frac{1}{p_1}, \frac{p_2}{p_1}\right). \tag{296}$$

We prove that from (295) and (296) it follows that $\varphi$ does not depend on $p_3$. Because of our assumptions we see that $\varphi$ has continuous derivatives. From (295) it follows that $\partial\varphi/\partial p_3 = \partial\varphi_1/\partial p_3$, i.e. $\partial\varphi/\partial p_3$ does not depend on $p_1$. Differentiating (296) we get that

$$\frac{\partial\varphi}{\partial p_3} = \frac{\partial\varphi_3\left(\dfrac{p_2}{p_1}, p_3 - \dfrac{p_2^2}{p_1}\right)}{\partial p_3}$$

does not depend on $p_1$ for arbitrary $p_2$ and $p_3$.

Write

$$\frac{\partial\varphi_3(u, v)}{\partial v} = \psi(u, v).$$

We see that along all curves $u = p_2/p_1$, $v = p_3 - p_2^2/p_1$ ($p_1$ and $p_3$ being fixed) the function $\psi(u, v)$ is a constant. The equation of these curves can be written as

$$v = p_3 - p_2 u \tag{297}$$

with arbitrary $p_2$ and $p_3$. Since any pair of points $(u_1, v_1)$ and $(u_2, v_2)$ lie on some curves (297) the function $\psi(uv) = \partial\varphi_3/\partial v$ is a constant. In a similar way one can obtain that $\partial\varphi_3/\partial\bar{v}$ is a constant and we conclude that

$$\varphi_3(u, v) = C_1(u)v + C_2(u)\bar{v} + C(u).$$

Since $\varphi_3$ is a bounded function, we have $C_1(u) = C(u) = 0$ and therefore

$$\varphi(p_1, p_2, p_3) = C\left(\frac{p_2}{p_1}\right) + \varphi_4\left(-\frac{1}{p_1}, \frac{p_2}{p_1}\right)$$

so that $\varphi$ does not depend on $p_3$.

Quite similarly, using the transformation $g \to g_0 g$ with $g_0 = \begin{pmatrix} 0 & 1 \\ -1 & 0 \end{pmatrix}$ we can show that $\varphi(p_1, p_2, p_3)$ does not depend on $p_1$. Therefore, $\varphi(p_1, p_2, p_3) = \varphi(p_2)$.

Using the same transformation, $g \to gg_0$, $g_0 = \begin{pmatrix} 0 & 1 \\ -1 & 0 \end{pmatrix}$, we obtain that this is possible only if $\varphi(p_1, p_2, p_3)$ is a constant.

We have proved that

$$\varphi(g) = (U_g f_1, f_2)$$

does not depend on $g$ for a dense set of pairs $f_1, f_2$ and therefore for all pairs $f_1, f_2$. So

$$(U_g f_1, f_2) = (U_e f_1, f_2) = (f_1, f_2).$$

That means $U_g = E$. Lemma 2 is proved.

5. *Positive functionals on R*. It is known that to each unitary representation $U_g$ of the group $G$ corresponds a functional on the ring $R$, namely

$$F(x) = (U_x f_0, f_0) \tag{298}$$

where $f_0$ is a fixed element in the space of representation. This functional is positive, that is $F(x^* x) \geqq 0$ for all $x \in R$.

Conversely, starting from a positive functional, we can construct a cyclic representation of the group $G$, taking $R$ as a representation space and defining the inner product in $R$ by $(x, y) = F(y^* x)$. Here an element $x$ is assumed to be equivalent to zero if $(x, x) = 0$. The operator $U_{g_0}$ of the representation is then defined by

$$U_{g_0} x(g) = x(g_0^{-1} g).$$

Let us take a nonzero positive functional $F(x)$ in $R$. This functional can be considered as a functional on degenerate functions of the form (263). If the representation is not the identity one then, by Lemma 2, we can choose an element $f_0$ such that the functional $F$ is nonzero on the set of all these functions.

To a degenerate function $\tilde{\chi}$ corresponds a degenerate function $\tilde{\Phi}$ given by

$$\tilde{\Phi}(w_1, w_2, \lambda) = \Phi(w_1) \Psi(w_2) \omega(\lambda) \tag{299}$$

where

$$\Phi(w_1) = \chi_3(w_1), \qquad \Psi(w_2) = \chi_1(w_2), \frac{1}{2\pi} \omega(\lambda) = \chi_2\left(\lambda + \frac{1}{\lambda}\right)$$

so that $\omega(\lambda)$ satisfies the condition

$$\omega\left(\frac{1}{\lambda}\right) = \omega(\lambda). \tag{300}$$

Let $\Phi(w_1)$, $\Psi(w_2)$ be arbitrary functions with continuous derivatives up to fourth order in $w_1, \bar{w}_1$, $w_2, \bar{w}_2$ and that vanish outside some region of the form

$$0 < \varepsilon_1 \leqq |w_1| \leqq C_1, \qquad 0 < \varepsilon_2 \leqq |w_2| \leqq C_2.$$

Also, let $\omega(\lambda)$ be an arbitrary function which satisfies (300) and the conditions[6]:

[6] Let us note that the condition (3), in fact, follows from other conditions (see the footnote on p. 000)

(1) $\int\limits_{|\lambda|>1} |\omega(\lambda)| d\mu(\lambda) < +\infty$;

(2) to a function $\tilde{\Phi} = \Phi(w_1)\Psi(w_2)\omega(\lambda)$ corresponds an integrable function $x(g)$;

(3) $$\int \omega(\lambda)|\lambda - \lambda^{-1}|^2 \frac{d\mu(\lambda)}{|\lambda|^2} = 0.$$

Let us estimate the norm of the function $x(g)$ in (2) from below. By the formula (42) in §3 we have

$$\int |T_\delta| |\lambda - \lambda^{-1}|^2 \frac{d\mu(\lambda)}{|\lambda|^2} \leqq 2\int |x(g)| d\mu(g) \tag{301}$$

where $T_\delta$ is the function introduced in §6.3.

On the other hand,

$$T_\delta = \int \varphi(z, z, \lambda) d\mu(z),$$

and therefore

$$\int \left| \int \varphi(z, z, \lambda) d\mu(z) \right| \cdot |\lambda - \lambda^{-1}|^2 \frac{d\mu(\lambda)}{|\lambda|^2} \leqq 2\int |x(g)| d\mu(g). \tag{302}$$

The degenerate function $\Phi$, by assumption, has continuous partial derivatives in $w_1, \bar{w}_1, w_2, \bar{w}_2$ up to fourth order and vanishes outside some set of the form

$$0 < \varepsilon_1 \leqq |w_1| \leqq C_1, \quad 0 < \varepsilon_2 \leqq |w_2| \leqq C_2.$$

Therefore we can present $\varphi(z, z, \lambda)$ as the Fourier integral:

$$\begin{aligned}
\varphi(z, z, \lambda) &= \frac{1}{(2\pi)^2} \int \left[ \int \Phi(w_1, w_2, \lambda) e^{-i\mathrm{Re}[\bar{z}(w_1 - w_2)]} d\mu(w_1) d\mu(w_2) \right] d\mu(z) \\
&= \frac{1}{(2\pi)^2} \int\int \left[ \int \Phi(w_2 + w, w_2, \lambda) d\mu(w_2) \right] e^{-i\mathrm{Re}(\bar{z}w)} d\mu(w) d\mu(z) \\
&= \frac{1}{2\pi} \int \Phi(w_2, w_2, \lambda) d\mu(w_2) = \frac{1}{2\pi} \int \tilde{\Phi}(w, w, \lambda) d\mu(w) = C_0 \omega(\lambda)
\end{aligned}$$

where

$$C_0 = \frac{1}{2\pi} \int |\Phi_0(w_1)|^2 d\mu(w_1)$$

and a function $\Phi_0(w)$ will be defined on p. 105.

Now (302) can be written as

$$C_0 \int |\omega(\lambda)| \cdot |\lambda - \lambda^{-1}|^2 \frac{d\mu(\lambda)}{|\lambda|^2} \leqq 2\int |x(g)| d\mu(g). \tag{303}$$

This is the required estimate from below for $\|x\|$.

The functional $F(x)$ can be considered as a functional on functions of the form (299). Because of the arguments above it cannot be zero on all such functions.

Now we consider the transformation of functions $\Phi$ and $\tilde{\Phi}$ caused by replacement of $x(g)$ by $x^*(g)$.

Denote by $\Phi^*$ and $\tilde{\Phi}^*$ the functions $\Phi$ and $\Phi$ corresponding to $x^*(g)$. Then the formula (213) in §8 implies

$$\Phi^*(w_1, w_2, \lambda) = \overline{\tilde{\Phi}\left(w_2, w_1, \frac{1}{\lambda}\right)}. \tag{304}$$

Let, first, $\Phi$ be an even function in $\lambda$. Then

$$\Phi^*(w_1, w_2, \lambda) = \Phi^*\left(w_2, w_1, \lambda\sqrt{\frac{\bar{w}_1}{\bar{w}_2}}\right) = \overline{\tilde{\Phi}\left(w_2, w_1, \frac{1}{\lambda}\sqrt{\frac{\bar{w}_1}{\bar{w}_2}}\right)}$$

so that

$$\tilde{\Phi}^*(w_1, w_2, \lambda) = \tilde{\Phi}\left(w_2, w_1, \frac{1}{\lambda}\sqrt{\frac{\bar{w}_1}{\bar{w}_2}}\right) = \overline{\Phi(w_2, w_1, \lambda)}. \tag{305}$$

Now let $\Phi$ be an odd function in $\lambda$. Then

$$\tilde{\Phi}^*(w_1, w_2, \lambda) = \left|\frac{w_1}{w_2}\right|^{\frac{1}{2}} \left(\frac{\bar{w}_2}{\bar{w}_1}\right)^{\frac{1}{2}} \Phi^*\left(w_1, w_2, \lambda\sqrt{\frac{\bar{w}_2}{\bar{w}_1}}\right)$$

$$= \left|\frac{w_2}{w_1}\right|^{\frac{1}{2}} \left|\frac{\bar{w}_1}{\bar{w}_2}\right|^{\frac{1}{2}} \Phi\left(w_2, w_1, \frac{1}{\lambda}\sqrt{\frac{\bar{w}_1}{\bar{w}_2}}\right) = \tilde{\Phi}\left(w_2, w_1, \frac{1}{\lambda}\right)$$

so that formula (305) is true in this case also. So this formula is true for a general function $\Phi$.

Therefore, when we replace $x$ by $x^*$, the function $\tilde{\Phi}$ of the form (299) must be replaced by the function

$$\tilde{\Phi}^*(w_1, w_2, \lambda) = \overline{\Phi(w_2)}\,\overline{\Psi(w_1)}\,\overline{\omega(\lambda)} \tag{306}$$

of the same form.

Let us now find the transformations of functions $\Phi$ and $\tilde{\Phi}$ caused by the convolution of the functions $x(g)$.

Let $x(g) = (x_1 \times x_2)(g)$. Denote by $\Phi, \Phi_1, \Phi_2$ and $\tilde{\Phi}, \tilde{\Phi}_1, \tilde{\Phi}_2$ the functions corresponding to $x(g)$, $x_1(g)$, $x_2(g)$.

By (214) from §8,

$$\Phi(w_1, w_2, \lambda) = \int\int \Phi_2(w_1, w, \lambda\lambda_1)\Phi_2(w, w_2, \lambda_1^{-1})\frac{d\sigma_1\, d\tau_1}{|\lambda_1|^2}\, d\mu(w). \tag{307}$$

It is easy to check that the same formula is true for functions $\tilde{\Phi}$, namely, if $\tilde{\Phi}_1$ and $\tilde{\Phi}_2$ corresponding to $x_1$ and $x_2$ then the function

$$\tilde{\Phi}(w_1, w_2, \lambda) = \int \tilde{\Phi}_1(w_1, w, \lambda\lambda_1)\tilde{\Phi}_2(w. w_2, \lambda_1^{-1})\frac{d\sigma_1 d\tau_1}{|\lambda_1|^2}\, d\mu(w) \tag{308}$$

corresponds to the convolution $x = x_1 \times x_2$. In particular, if $\tilde{\Phi}_1$ and $\tilde{\Phi}_2$ are of the form

$$\tilde{\Phi}_1(w_1, w_2, \lambda) = \Phi_1(w_1)\Psi_1(w_2)\omega_1(\lambda),$$
$$\tilde{\Phi}_2(w_1, w_2, \lambda) = \Phi_2(w_1)\Psi_2(w_2)\omega_2(\lambda),$$

then to the function $x_1 \times x_2$ corresponds the function

$$\tilde{\Phi}(w_1, w_2, \lambda) = \Phi_1(w_1)\Psi_2(w_a)\cdot \int \Psi_1(w)\Phi_2(w)d\mu(w)\cdot \int \omega_1(\lambda\lambda_1)\omega_2(\lambda_1^{-1})\frac{d\sigma_1\, d\tau_1}{|\lambda_1|^2} \quad (309)$$

which has a similar form with

$$\omega(\lambda) = \int \omega_1(\lambda\lambda_1)\omega_2(\lambda_1^{-1})\frac{d\sigma_1\, d\tau_1}{|\lambda_1|^2}.$$

This function $\omega(\lambda)$ also satisfies the conditions (1), (2), (3). The validity of (1) and (2) is obvious and the validity of (3) follows from the footnote on page 102. It is possible also to verify (3) directly: since $\omega(1/\lambda) = \omega(\lambda)$, the condition (3) can be rewritten as

$$\int \omega(\lambda)(1 - \lambda^{-2})d\mu(\lambda) = 0, \qquad d\mu(\lambda) = d\sigma\, d\tau, \qquad \lambda = \sigma + i\tau.$$

But by the assumption,

$$\int \omega_1(\lambda)(1 - \lambda_1^{-2})d\mu(\lambda_1) = 0, \qquad \int \omega_2(\lambda_2)(1 - \lambda_2^{-1})d\mu(\lambda_2) = 0$$

and we have

$$\begin{aligned} &\int \omega(\lambda)(1 - \lambda^{-2})d\mu(\lambda) \\ &= \int \omega_1(\lambda\lambda_2^{-1})\omega_2(\lambda_2)(1 - \lambda^{-2})d\mu(\lambda)\frac{d\mu(\lambda_2)}{|\lambda_2|^2} \\ &= \int \omega_1(\lambda_1)\omega_2(\lambda_2)(1 - \lambda_1^{-2}\lambda_2^{-2})d\mu(\lambda_1)d\mu(\lambda_2) \\ &= \int \omega_1(\lambda_1)(1 - \lambda_1^{-2})d\mu(\lambda_1)\int \omega(\lambda_2)d\mu(\lambda_2) \\ &\quad + \int \omega_1(\lambda_1)\lambda_1^{-2}d\mu(\lambda_1)\int \omega_2(\lambda_2)(1 - \lambda_2^{-2})d\mu(\lambda_2) \\ &= 0. \end{aligned}$$

6. *The commutative rings $\Omega$ and $\Gamma$.* We return now to the positive functional $F(x)$. Let us choose a function $\Phi_0(w)$ such that

$$\int |\Phi_0(w)|^2 d\mu(w) = 1.$$

Let us assume also that $\Phi_0$ has continuous derivatives in $w, \bar{w}$ up to fourth order and that it vanishes outside a set $0 < \varepsilon \leqq |w| \leqq C$.

Let

$$\Phi(w_1, w_2, \lambda) = \Phi_0(w_1)\overline{\Phi_0(w_2)}\omega(\lambda) \quad (310)$$

where $\omega(\lambda)$ satisfies the conditions (1) and (2) for this $\Phi_0(w)$. Then

$$\tilde{\Phi}(w_1, w_2, \lambda) = \Phi_0(w_1)\overline{\Phi_0(w_2)}\,\overline{\omega(\lambda)}$$

and because of (309) the convolution of two functions of the form (310) gives a function of the same form with

$$\omega(\lambda) = \int \omega_1(\lambda\lambda_1)\omega_2(\lambda_1^{-1})\frac{d\sigma_1 d\tau_1}{|\lambda_1|^2}. \quad (311)$$

In other words, the set of all functions $\tilde{\Phi}$ of the form (310) is a commutative ring which is isomorphic to the ring $\Omega$ of functions $\omega(\lambda)$ satisfying the conditions (1), (2) and (3). The multiplication in $\Omega$ is defined by formula (311) and involution $*$ by the formula

$$\omega^*(\lambda) = \overline{\omega(\bar{\lambda})} = \overline{\omega\left(\frac{1}{\lambda}\right)}. \tag{312}$$

If the representation is not the identity then by the results of p. 102 we can choose an element $f_0$ in the representation space and function $\Phi_0$ in such a way that the corresponding functional $F$ is not zero for some function $\omega(\lambda)\in\Omega$ which has continuous derivatives up to eighth order and vanishes outside a set $0 < \varepsilon \leqq |\lambda| \leqq C$.

When the function $\Phi_0(w)$ is fixed, then to a function $\omega(\lambda)$ corresponds an integrable function $x(g)$. Define the norm of $\omega(\lambda)$ (denote it by $\|\omega\|$) as the norm of this function $x(g)$: $\|x\| = \int |x(g)| d\mu(g)$. We estimate the norm $\|\omega\|$ using the estimate (246) for the norm $\|x\|$. In our case

$$I \leqq \sum_{\alpha+\beta\leqq 6} C_{\alpha\beta} \int |\lambda|^{2\alpha+2\beta-2+\varepsilon} \cdot [\omega_{\lambda^\alpha \bar{\lambda}^\beta}(\lambda)|^2 d\mu(\lambda) + C\int |w_{\lambda\bar{\lambda}}(\lambda)|^2 d\mu(\lambda) \tag{313}$$

where $C_{\alpha\beta}, C$ are integrals in $w_1$, $w_2$ of the products

$$\Phi_0(w_1)\overline{\Phi_0(w_2)}|w_1|^{\alpha_1} \cdot |w_2|^{\alpha_2},$$

$\alpha$ and $\beta$ are positive integers, and $\alpha_1, \alpha_2$ are integers.

Furthermore,

$$I' \leqq \sum_{\alpha+\beta\leqq 6} b_{\alpha\beta} \int |\lambda|^{2\alpha+2\beta-2+\varepsilon+2k} |\omega_{\lambda^\alpha \bar{\lambda}^\beta}(\lambda)|^2 d\mu(\lambda) + b\int |\omega_{\lambda\bar{\lambda}}(\lambda)|^2 \lambda^k d\mu(\lambda) \tag{314}$$

where $b_{\alpha\beta}, b$ are similar to $C_{\alpha\beta}, C$.

Let

$$\lambda = e^t e^{i\theta}, \qquad -\pi \leqq \theta \leqq \pi, \qquad u = t + i\theta, \qquad \omega(\lambda) = \gamma(t, \theta).$$

This mapping transforms the ring $\Omega$ into the ring $\Gamma$ of functions $\gamma(t, \theta)$ satisfying the conditions:

1′ $\gamma(-t, -\theta) = \gamma(t, \theta)$;

2′ $\int_{-\pi}^{\pi} \int_{-0}^{\infty} |\gamma(t, \theta)| e^{2t} dt d\theta < +\infty$;

3′ to the function $\tilde{\Phi} = \Phi(w_1)\psi(w_2)\omega(\lambda)$ where $\omega(\lambda) = \gamma(t, \theta)$ corresponds an integrable function $x(g)$;

4′ $\int_{-\pi}^{\pi} \int_{0}^{\infty} \gamma(t, \theta)(e^{2t} - e^{2i\theta}) dt d\theta = 0.$

The multiplication in $\Gamma$ is defined as the convolution of $\gamma_1(t, \theta)$ and $\gamma_2(t, \theta)$, i.e. by the formula

$$\gamma(t, \tau) = \int_{-\pi}^{\pi} \int_{0}^{\infty} \gamma_1(t - t_1, \theta - \theta_1)\gamma_2(t_1, \theta_1) dt, d\theta, \tag{315}$$

and involution is given by the formula

$$\gamma^*(t, \theta) = \overline{\gamma(-t, -\theta)} = \overline{\gamma(t, \theta)}.$$

Furthermore,

$$\lambda \frac{\partial}{\partial \lambda} = \frac{\partial}{\partial u}, \quad \bar{\lambda} \frac{\partial}{\partial \bar{\lambda}} = \frac{\partial}{\partial \bar{u}}$$

so the estimates for $I$ and $I'$ can be written in the form

$$I \leqq \sum_{\alpha+\beta \leqq 6} C_{\alpha\beta} \int_{-\pi}^{\pi} \int_{-\infty}^{\infty} |\gamma_{u^\alpha \bar{u}^\beta}|^2 e^{\varepsilon t} dt d\theta + C \int_{-\pi}^{\infty} \int_{-\infty}^{\infty} |\gamma_{u\bar{u}}|^2 e^{2t} dt d\theta, \tag{316}$$

$$I' \leqq \sum_{\alpha+\beta \leqq 6} b_{\alpha\beta} \int_{-\pi}^{\pi} \int_{-\infty}^{\infty} |\gamma_{u^\alpha \bar{u}^\beta}|^2 e^{(2k+\varepsilon)t} dt d\theta + b \int_{-\pi}^{\pi} \int_{-\infty}^{\infty} |\gamma_{u\bar{u}}|^2 e^{(k-2)t} d\theta dt. \tag{317}$$

Putting together these formulas and the estimate (246) and using 1′ we get

$$\int |x(g)| d\mu(g) \leqq \frac{1}{\varepsilon^2} C_0 \left[ \sum_{\alpha+\beta \leqq 6} \int_{-\pi}^{\pi} \int_{0}^{\infty} |\gamma_{u^\alpha \bar{u}^\beta}|^2 e^{(4+\varepsilon)t} dt d\theta \right]^{\frac{1}{2}} \tag{318}$$

with some constant $C_0$.

Denote the right side of (318) by $\|\gamma\|_1$ and denote by $\Gamma'$ the set of functions satisfying 1′, 2′, 4′ and such that the norm $\|\gamma\|_1$ is finite. From (318) it follows that the condition 3′ is also satisfied; therefore $\Gamma' \subset \Gamma$.

If $\gamma$ and $x$ correspond to one another we define $\|\gamma\| = \|x\| = \int |x(g)| d\mu(g)$. Thus we obtain a norm in $\Gamma$.

We denote the completion of the ring $\Gamma$ by the norm $\|\gamma\|$ by $\tilde{\Gamma}$.

We can consider the positive functional $F$ on $R$ which we have chosen as a positive functional on $\tilde{\Gamma}$; we will prove in another paper that this positive functional has the form

$$F(\gamma) = \int_{\mathfrak{M}} M(\gamma) dC(\Delta_M), \tag{319}$$

where $M(\gamma)$ is the value of the element $\gamma$ of the ring $\tilde{\Gamma}$ on the symmetric[7] maximal ideal $M$ of this ring, $C(\Delta)$ is completely additive nonegative function of a Borel set $\Delta$ in the space $\mathfrak{M}$ of all maximal ideals of the ring $\tilde{\Gamma}$ and the integral is taken over all this space.

Now we consider these maximal ideals of the ring $\tilde{\Gamma}$ in more detail.

Denote by $\Gamma''$ the set of functions $\gamma(t, \theta)$ that satisfy the condition 1′ and 4′ (p. 106), have continuous derivatives in $t$ and $\theta$ up to eighth order and vanish outside a set $|t| \leqq C$.

Obviously $\Gamma'' \subset \Gamma'$. Let some maximal ideal $M$ of the ring $\tilde{\Gamma}$ be given. Assume that there exists a function $\gamma_0(t, \theta) \in \Gamma''$ which is not zero on $M$ and such that

$$\int_{-\pi}^{\pi} \int_{-\infty}^{\infty} \gamma_0(t, \theta) e^{2t} dt d\theta = 0. \tag{320}$$

Normalize this function $\gamma_0$ by the condition $M(\gamma_0) = 1$.

---

[7] A maximal ideal $M$ is said to be symmetric if $M^* = M$, i.e. if together with any $\gamma$ it also contains the conjugate element

$$\gamma^*(t, \theta) = \overline{\gamma(t, \theta)}.$$

Consider then the function

$$\gamma_1(t,\theta)=\tfrac{1}{2}[\gamma_0(t+t_0,\theta+\theta_0)+\gamma_0(t-t_0,\theta-\theta_0)];$$

this function also belongs to $\Gamma''$. Besides, since derivatives of $\gamma_0$ are uniformly continuous, it follows that $\gamma_1$ is a continuous function in $t_0$ and $\theta_0$ by the norm $\|\gamma\|_1$. Therefore, $\gamma_1$ is a continuous function in $t_0$ and $\theta_0$ by the norm $\|\gamma\|$. But then $M(\gamma_1)$ is also a continuous function in $t_0$ an $\theta_0$.

We put

$$X(t_0,\theta_0)=M(\gamma_1)=\tfrac{1}{2}M\{\gamma_0(t+t_0,\theta+\theta_0)+\gamma_0(t-t_0,\theta-\theta_0)\}. \qquad (321)$$

It follows from the above that $X(t_0,\theta_0)$ is a continuous function in $t_0$ and $\theta_0$. Moreover, this function satisfies a functional equation which can be obtained as follows.

Let $\gamma(t,\theta)$ be an arbitrary function from $\Gamma$. Then

$$\int_{-\pi}^{\pi}\int_{-\infty}^{\infty}\gamma(t-t'+t_0,\theta-\theta'+\theta_0)\gamma_0(t',\theta')dt'd\theta'$$

$$=\int_{-\pi}^{\pi}\int_{-\infty}^{\infty}\gamma_0(t-t'+t_0,\theta-\theta'+\theta_0)\gamma(t',\theta')dt'd\theta'$$

so that

$$\gamma(t+t_0,\theta+\theta_0)\times\gamma_0(t,\theta)=\gamma_0(t+t_0,\theta+\theta_0)\times\gamma(t,\theta).$$

If we substitute here $-t_0,-\theta_0$ for $t_0,\theta_0$ and add the result of the substitution to the above equality, we obtain

$$\tfrac{1}{2}[\gamma(t+t_0,\theta+\theta_0)+\gamma(t-t_0,\theta-\theta_0)]\times\gamma_0(t,\theta)=\gamma_1(t,\theta)\times\gamma(t,\theta).$$

Hence $\gamma\to M(\gamma)$ is a homomorphism; we have

$$\tfrac{1}{2}M\{\gamma(t+t_0,\theta+\theta_0)+\gamma(t-t_0,\theta-\theta_0)\}\cdot M(\gamma_0)=M(\gamma_1)M(\gamma). \qquad (322)$$

We put in particular

$$\gamma(t,\theta)=\gamma_0(t+t_1,\theta+\theta_1)+\gamma_0(t-t_1,\theta-\theta_1).$$

Using the definition (321) of the function $X|t_0,\theta_0|$ and the equality $M(\gamma_0)=1$ we obtain from (322)

$$X(t_0+t_1,\theta_0+\theta_1)+X(t_0-t_1,\theta_0-\theta_1)=2X(t_0,\theta_0)X(t_1,\theta_1). \qquad (323)$$

This is the required functional equation for $X(t,\theta)$.

It is known that the function

$$X(t,\theta)=\cos(\rho t-n\theta)=\tfrac{1}{2}(e^{i\rho t}e^{-in\theta}+e^{-i\rho t}e^{in\theta}) \qquad (324)$$

is a continuous solution of this equation.

We prove that $n$ should be an integer. In fact, the function $\gamma_0(t,\theta)$ satisfies the condition

$$\gamma_0(t,-\pi)=\gamma_0(t,\pi)$$

so it can be continued to a periodic function in $\theta$ with the period $2\pi$. From the identity

$$\gamma_0(t+t_0,\theta+\theta_0+2\pi)+\gamma_0(t-t_0,\theta-\theta_0+2\pi)$$
$$=\gamma_0(t+t_0,\theta+\theta_0)+\gamma_0(t-t_0,\theta-\theta_0)$$

follows that

$$X(t_0,\theta_0+2\pi)=X(t_0,\theta_0).$$

This is possible only when $n$ is an integer.

Now let $\gamma(t,\theta)$ be an arbitrary function from $\Gamma$ that vanishes for $|t|>C$. Let us consider the expression

$$\tilde{\gamma}(t,\theta)=(\gamma_0\times\gamma)(t,\theta)=\int_{-\pi}^{\pi}\int_{-\infty}^{\infty}\gamma_0(t-t',\theta-\theta')\gamma(t',\theta')dt'd\theta'. \tag{325}$$

Here the integration is, in fact, taken over the region $|t'|\leqq C$.

As $\gamma\to M(\gamma)$ is a homomorphism, we get

$$M(\tilde{\gamma})=M(\gamma_0)M(\gamma)=M(\gamma). \tag{326}$$

On the other hand, if we change the variables $t'\to -t',\theta'\to-\theta'$ in the integral (325) and use the fact that $\gamma(t,\theta)$ is an even function, then we can write the expression for $\tilde{\gamma}(t,\theta)$ as

$$\tilde{\gamma}(t,\theta)=\int_{-\pi}^{\pi}\int_{-\infty}^{\infty}\gamma_0(t+t',\theta+\theta')\gamma(t',\theta')dt'd\theta'. \tag{327}$$

Taking the halfsum of the left and right sides of (325) and (327), we get

$$\tilde{\gamma}(t,\theta)=\int_{-\pi}^{\pi}\int_{-\infty}^{\infty}\frac{\gamma_0(t+t',\theta+\theta')+\gamma_0(t-t',\theta-\theta')}{2}\gamma(t,\theta')dt'd\theta'. \tag{328}$$

We approximate the integral (328) by a sum

$$\gamma_a(t,\theta)=\sum_{k,j}\frac{\gamma_0(t+t_k,\theta+\theta_j)+\gamma_0(t-t_k,\theta-\theta_j)}{2}\int_{\Delta k}\int_{\Delta'_j}\gamma(t',\theta')dt'd\theta'. \tag{329}$$

Then from the definition (321) of the function $X(t,\theta)$ it follows that

$$M(\gamma_a)=\sum_{k,j}X(t_k,\theta_j)\int_{\Delta k}\int_{\Delta'_j}\gamma(t',\theta')dt'd\theta'.$$

When $\Delta t_k\to 0$, $\Delta'\theta_j\to 0$ this expression tends to

$$\int_{-\pi}^{\pi}\int_{-\infty}^{\infty}X(t,\theta)\gamma(t,\theta)dt\,d\theta. \tag{330}$$

Let us prove that this limit is equal to $M(\tilde{\gamma})$. Because of the continuity of $M(\gamma)$ in the norm $\|\gamma\|$ it is enough to prove that $\gamma_a\to\tilde{\gamma}$ in the norm $\|\gamma\|$ when $\Delta t_k\to 0,\Delta'\theta_j\to 0$. By (318) it is enough to prove that $\gamma_a\to\tilde{\gamma}$ in the norm $\|\gamma\|_1$.

The functions $\tilde{\gamma}(t,\theta)$ and $\gamma_a(t,\theta)$ vanish outside a set $|t|\leqq C$ so we must prove that on this set the function $\gamma_a(t,\theta)$ and its derivatives up to sixth order tend uniformly to $\tilde{\gamma}(t,\theta)$ and its derivatives. But this is obvious because the function $\gamma_0(t,\theta)$ and its derivatives up to sixth order are uniformly continuous. So we have

$$M(\tilde{\gamma})=\int_{-\pi}^{\pi}\int_{-\infty}^{\infty}\gamma(t,\theta)X(t,\theta)dt\,d\theta.$$

Comparing this with (326), we get

$$M(\gamma)=\int_{-\pi}^{\pi}\int_{-\infty}^{\infty}\gamma(t,\theta)X(t,\theta)dt\,d\theta=\int_{-\pi}^{\pi}\int_{-\infty}^{\infty}\gamma(t,\theta)\cos(\rho t-n\theta)dt\,d\theta. \quad (331)$$

From the equality $\gamma(t,\theta)=\gamma(-t,-\theta)$ it follows that

$$M(\gamma)=\int_{-\pi}^{\pi}\int_{0}^{\infty}\gamma(t,\theta)\cos(\rho t-n\theta)dt\,d\theta. \quad (332)$$

Let $\gamma$ satisfy the condition $\gamma^*=\gamma$, i.e.

$$\overline{\gamma(t,\theta)}=\gamma(t,\theta).$$

If the ideal $M$ is symmetric, then it takes real values on such function. So we see that if the function $\gamma(t,\theta)$ is real, then the expression (332) is real. This is possible only when function $\cos(\rho t-n\theta)$ takes only real values. Therefore we have only two cases:

(a) $\rho$ is real,

(b) $n=0$ and $\rho$ is imaginary: $\rho=i\rho_1,\rho_1>0$.

In the latter case the expression for $M(\gamma)$ takes the form

$$M(\gamma)=\int_{-\pi}^{\pi}\int_{0}^{\infty}\gamma(t,\theta)ch\rho_1 t\,dt\,d\theta. \quad (333)$$

In the case (a) the right side of (332) is continuous in the norm $\|\gamma\|_1$ on a set $\Gamma'_C$ of functions $\gamma\in\Gamma'$ which are bounded together with all derivatives up to sixth order by the same constant $C$. To show this we put

$$\|\gamma\|_0=\int_{-\pi}^{\pi}\int_{0}^{\infty}|\gamma(t,\theta)|sh^2 t\,dt. \quad (334)$$

Then (303) implies that

$$C_0\|\gamma\|_0\leqq\|\gamma\|\leqq\|\gamma\|_1, \quad (335)$$

so it is enough to prove that the right side of (332) is continuous on $\Gamma'_C$ in the norm $\|\gamma_0\|$. This follows from the estimate

$$\int_{-\pi}^{\pi}\int_{0}^{\infty}|\gamma(t,\theta)|dt\,d\theta\leqq 2\pi\varepsilon C+\frac{1}{sh^2\varepsilon}\int_{-\pi}^{\pi}\int_{\varepsilon}^{\infty}|\gamma(t,\theta)|sh^2 t\,dt\,d\theta$$

$$\leqq 2\pi\varepsilon C+\frac{1}{sh^2\varepsilon}\int_{-\pi}^{\pi}\int_{0}^{\infty}|\gamma(t,\theta)|sh^2 t\,dt\,d\theta.$$

Denote by $\Gamma''_C$ the set of functions from $\Gamma''$ that are bounded together with their derivatives up to sixth order by he same constant $C$.

Each function $\gamma\in\Gamma'_C$ is the limit (in the norm $\|\gamma\|$) of some sequence of elements from $\Gamma''_C$. Indeed let $\gamma^{(N,\varepsilon_1)}_{(t,\theta)}$ be a function from $\Gamma''_C$ defined by

$$\gamma^{(N,\varepsilon_1)}(t,\theta)=\begin{cases}\gamma(t_1\theta) & \text{for} \quad |t|<N-\varepsilon,\\ 0 & \text{for} \quad |t|>N.\end{cases}$$

Then

$$\int_{-\pi}^{\pi}\int_{0}^{\infty}|\gamma_{u^\alpha\bar{u}^\beta}-\gamma^{(N,\varepsilon_1)}_{u^\alpha\bar{u}^\beta}|^2e^{(Y+\varepsilon)t}dtd\theta=\int_{-\pi}^{\pi}\int_{N-\varepsilon_1}^{N}|\gamma_{u^\alpha\bar{u}^\beta}-\gamma^{(N,\varepsilon_1)}_{u^\alpha\bar{u}^\beta}|^2e^{(Y+\varepsilon)t}dtd\theta$$
$$+\int_{-\pi}^{\pi}\int_{N}^{\infty}|\gamma_{u^\alpha\bar{u}^\beta}|^2e^{(Y+\varepsilon)t}dtd\theta$$
$$\leqq 8\pi C^2\int_{N-\varepsilon_1}^{N}e^{(Y+\varepsilon)t}dt+\int_{-\pi}^{\pi}\int_{N}^{\infty}|\gamma_{u_\alpha\bar{u}_\beta}|^2e^{(Y+\varepsilon)t}dtd\theta.$$

Take $N$ so large that the second summand is less than $\varepsilon_2$ and after that take $\varepsilon_1$ so small that the second summand is less than $\varepsilon_2$. Then we have

$$\|\gamma-\gamma^{(N,\varepsilon_1)}\|<2\varepsilon_2.$$

The formula (332) is true for all functions from $\Gamma''_C$. As the right side of this formula is continuous in the norm $\|\gamma\|_1$ it is true also for all functions from $\Gamma'_C$ for arbitrary $C>0$.

In the case (b), the right side of (333) is continuous by norm

$$\|\gamma\|_{\rho_1}=\int_{-\pi}^{\pi}\int_{0}^{\infty}|\gamma(t,\theta)|ch\rho_1 tdtd\theta.$$

On the other hand, we have seen that with an appropriate choice of $N$ and $\varepsilon_1$ we have

$$\|\gamma-\gamma^{(N_1\varepsilon_1)}\|_{\rho_1}<\varepsilon_2.$$

In a similar way, it can be shown that if $\gamma\in\Gamma'_C$ and $\|\gamma\|_{\rho_1}$ is finite then

$$\|\gamma-\gamma^{(N,\varepsilon_1)}\|_{\rho_1}<\varepsilon_2.$$

The equality (333) is true for $\gamma^{(N,\varepsilon_1)}$. Taking the limit, we obtain that this equality is true for all functions $\gamma\in\Gamma'_C$ with a finite norm $\|\gamma\|_{\rho_1}$.

We have assumed that our ideal is not zero on at least one function $\gamma_0$ from $\Gamma''$ satisfying the condition

$$\int_{-\pi}^{\pi}\int_{0}^{\infty}\gamma_0(t,\theta)e^{2t}dtd\theta=0. \tag{336}$$

If this is not so, then the above arguments imply that the ideal is zero on all functions $\gamma$ from $!'_C$ satisfying (336).

Let us choose a function $\gamma_0(t,\theta)$ from $\Gamma'_C$ such that

$$\int_{-\pi}^{\pi}\int_{0}^{\infty}\gamma_0(t,\theta)e^{2t}dtd\theta=1.$$

For arbitrary function $\gamma(t,\theta)\in\Gamma'_C$ let

$$\gamma_1(t,\theta)=\gamma(t,\theta)-\gamma_0(t,\theta)\int_{-\pi}^{\pi}\int_{0}^{\infty}\gamma(t,\theta)e^{2t}dt\,d\theta.$$

Then $\gamma_1(t,\theta)$ also belongs to $\Gamma'_C$; moreover it satisfies the condition (336). Hence

$M(\gamma_1)=0$, that is

$$M(\gamma)=C\int_0^{\infty}\int_{-\pi}^{\pi}\gamma(t,\theta)e^{2t}dtd\theta$$

where $C=M(\gamma_0)$.

Applying this formula to the convolution of two functions, we get that $C^2=C$, so that $C=1$ or $C=0$. In the second case the ideal is zero on all functions from $\Gamma'_C$.

Let $\rho_1>2$; we put

$$\gamma_{\varepsilon_1}(t,\theta)=t^6e^{-t(\rho_1+\varepsilon_1)} \quad for \quad t\geqq 0,$$
$$\gamma_{\varepsilon_1}(t,\theta)=t^6e^{t(\rho_1+\varepsilon_1)} \quad for \quad t<0.$$

This function is bounded together with its derivatives up to sixth order. It is clear also that the norm $\|\gamma\|_1$ is finite if we choose the number $\varepsilon$ in (318) in such a way that $0<\varepsilon<\frac{1}{2}(\rho_1-2)$. Let

$$\gamma'_{\varepsilon_1}(t,\theta)=\gamma_{\varepsilon_1}(t,\theta)-\gamma_0(t,\theta)\int_{-\pi}^{\pi}\int_0^{\infty}\gamma_{\varepsilon_1}(t,\theta)(e^{2t}-e^{2i\theta})dtd\theta$$

where $\gamma_0(t,\theta)$ is a fixed function from $\Gamma''$. Then $\gamma\in\Gamma'_C$ with some $C$. The norm $\|\gamma'_{\varepsilon_1}\|_{\rho_1}$ also being finite, we have

$$M(\gamma'_{\varepsilon_1})=2\pi\int_0^{\infty}t^6(e^{i\rho_1}+e^{-i\rho_1})e^{-t(\rho_1+\varepsilon_1)}dt-M(\gamma_0)\int_{-\pi}^{\pi}\int_0^{\infty}\gamma_{\varepsilon_1}(t,\theta)(e^{2t}-e^{2i\theta})dtd\theta.$$

It is easy to see that this expression is unbounded when $\varepsilon_1\to 0$.

On the other hand, if the functional $M(\gamma)$ is continuous then

$$|M(\gamma'_{\varepsilon_1})|<C\|\gamma'_{\varepsilon_1}\|_1.$$

For $\varepsilon_1\to 0$ the right side of this inequality tends to $C\|\gamma'_0\|_1$ where

$$\gamma'_0(t,\theta)=t^6e^{-i\rho_1}-\gamma_0(t,\theta)\int_{-\pi}^{\pi}\int_0^{\infty}t^6e^{-i\rho_1}(e^{2t}-e^{2i\theta})dtd\theta \quad for \quad t\geqq 0.$$

There $M(\gamma'_{\varepsilon_1})$ must be bounded when $\varepsilon_1\to 0$.

This contradiction shows that in the case (b) only values $0<\rho\leqq 2$ are admissible.

We have assumed that the functional $M(\gamma)$ is not zero on at least one function $\gamma\in\Gamma''$. If $M(\gamma)=0$ on all functions from $\Gamma''$ then the above arguments imply that $M(\gamma)=0$ on all functions $\gamma$ from arbitrary $\Gamma'_C$. Combining this remark with (319), (332) and (333) we obtain the following result.

For an arbitrary function $\gamma$ from $\Gamma'_C$ the positive functional $F(\gamma)$ on $\Gamma$ is of the form

$$F(\gamma)=\sum_{n=-\infty}^{\infty}\int_0^{\infty}\left[\int_{-\pi}^{\pi}\int_0^{\infty}\gamma(t,\theta)\cos(\rho t-n\theta)dtd\theta\right]dC_n(\rho)$$
$$+\int_0^{2}\left[\int_{-\pi}^{\pi}\int_0^{\infty}\gamma(t,\theta)ch\rho_1 tdtd\theta\right]dC(\rho_1) \tag{337}$$

where $C_n(\rho)$ and $C(\rho_1)$ are nonnegative nondecreasing functions in $\rho$ and $\rho_1$.

In the ring $\tilde{\Omega}$ we obtain for the functional $\Gamma(\omega)(\omega(\lambda)=\gamma(t,\theta)\in\Gamma'_C)$ the following expression:

$$F(\omega)=\sum_{n=-\infty}^{\infty}\int_0^{\infty}\left[\int\omega(\lambda)\frac{|\lambda|^{n+i\rho}\lambda^{-n}+|\lambda|^{-n-i\rho}\lambda^{n}}{2}\frac{d\mu(\lambda)}{|\lambda|^2}\right]dC_n(\rho)$$
$$+\int_0^2\left[\int\omega(\lambda)\frac{|\lambda|^{\rho_1}+|\lambda|^{-\rho_1}}{2}\frac{d\mu(\lambda)}{|\lambda|^2}\right]dC(\rho_1). \tag{338}$$

Since $\omega(\lambda)=\omega(\lambda^{-1})$ this formula can also be written in the form

$$F(\omega)=\sum_{n=-\infty}^{\infty}\int_0^{\infty}\left[\int\omega(\lambda)|\lambda|^{n+i\rho}\lambda^{-n}\frac{d\mu(\lambda)}{|\lambda|^2}\right]dC_n(\rho)$$
$$+\int_0^2\left[\int\omega(\lambda)|\lambda|^{-\rho_1}\frac{d\mu(\lambda)}{|\lambda|^2}\right]dC(\rho) \tag{339}$$

7. *The decomposition of a given unitary representation of the group G into irreducible representations.* Now we apply the above results to the solution of the main problem, that is, to the decomposition of a given unitary representation into irreducible representations.

Let a function $\Phi_0$ be chosen as on p. 105. We put

$$\tilde{\Phi}(w_1,w_2,\lambda)=\psi(w_1)\Phi_0(w_2)\omega(\lambda)$$

where a function $\psi(w_1)$ has continuous derivatives up to fourth order and vanishes outside a region $0<\varepsilon_1\leqq|w_1|\leqq C_1$ and where $\omega(\lambda)=\gamma(t,\theta)\in\Gamma'_C$.

Functions $x(g)$ which correspond to these functions $\tilde{\Phi}$ are integrable. Denote the set of these functions by $S_{\Phi_0}$.

Denote also by $\mathfrak{M}_{\Phi_0}$ the set of all finite sums

$$x(g)=\sum_k x_k(g_k^{-1}g)$$

where $x_k(g)\in S_{\Phi_0}$. This set consists of integrable functions and is invariant under the transformation $x(g)\to x(g_0^{-1}g)$. Additionally, this set is linear. The linear set of elements

$$f=\int x(g)U_g f_0\,d\mu(g)$$

which is invariant under action of all operators of the representation $U_g$ corresponds to this set in $\mathfrak{H}$. We may again denote this by $\mathfrak{M}_{\Phi_0}$ and we consider the functions $x(g)$ as elements of the space $\mathfrak{H}$.

In accordance with p. 102, the inner product of two such functions $x(g)$ and $y(g)$ is defined by the formula

$$(x,y)=F(y^*x).$$

Hence it is the finite sum of the expressions $F(x_1^*x_{g_0})$ where $x_1,x\in S_{\Phi_0}$ and $x_{g_0}(g)=x(g_0^{-1}g)$. So it is enough to find $F(x_1^*x_{g_0})$.

Let the functions

$$\tilde{\Phi}(w_1,w_2,\lambda)=\Psi(w_1)\Phi_0(w_1)\omega(\lambda),$$
$$\tilde{\Phi}_1(w_1,w_2,\lambda)=\Psi_1(w_1)\Phi_0(w_2)\omega_1(\lambda)$$

correspond to functions $x(g)$, $x_1(g)$ from $S_{\Phi_0}$. Let us find the function $\tilde{\Phi}$ corresponding to $x_1^* x_{g_0}$. Note first that for appropriate functions $\tilde{L}$ we have

$$\tilde{L}_1(w_1, w_2, n, \rho) = \Psi(w_1)\Phi_0(w_2)a_{n,\rho},$$
$$\tilde{L}_1(w_1, w_2, n, \rho) = \Psi_1(w_1)\Phi_0(w_2)a'_{n,\rho},$$

where

$$a_{n,\rho} = \int \omega(\lambda)|\lambda|^{n+i\rho}\lambda^{-n}\frac{d\mu(\lambda)}{|\lambda|^2},$$
$$a'_{n,\rho} = \int \omega_1(\lambda)|\lambda|^{n+i\rho}\lambda^{-n}\frac{d\mu(\lambda)}{|\lambda|^2}.$$

Hence

$$L_1(w_1, w_2, n, \rho) = l(w_1, n, \rho)l_0(w_2, n, \rho),$$
$$L_1(w_1, w_2, n, \rho) = l_1(w_1, n, \rho)l_0(w_2, n, \rho),$$

where for even $n$

$$l_0(w_2, n, \rho) = |w_2|^{(n+i\rho)/2}\bar{w}_2^{-n/2}\Phi_0(w_2), \tag{340}$$
$$l(w_1, n, \rho) = |w_1|^{-(n+i\rho)/2}\bar{w}_1^{n/2}\Psi(w_1)a_{n,\rho}, \tag{341}$$
$$l_1(w_1, n, \rho) = |w_1|^{-(n+i\rho)/2}\bar{w}_1^{n/2}\Psi_1(w_1)a'_{n\rho}, \tag{342}$$

and for odd $n$

$$l_0(w_2, n, \rho) = |w_2|^{(n-1+i\rho)/2}\bar{w}_2^{-(n-1)/2}\Phi_0(w_2), \tag{343}$$
$$l(w_1 n, \rho) = |w_1|^{-(n-1+i\rho)/2}\bar{w}_1^{(n-1)/2}\Psi(w_1)a_{n,\rho}, \tag{344}$$
$$l_1(w_1, n, \rho) = |w_1|^{-(n-1+i\rho)/2}\bar{w}_1^{(n-1)/2}\Psi_1(w_1)a'_{n,\rho}. \tag{345}$$

For the corresponding kernels $K(z_1, z_2, n, \rho)$ we have

$$K(z_1, z_2, n, \rho) = f(z_1, n, \rho)f_0(z_2, n, \rho), \tag{346}$$
$$K_1(z_1, z_2, n, \rho) = f_1(z_1, n, \rho)f_0(z_2, n, \rho), \tag{347}$$

where

$$f_0(z_2, n, \rho) = \frac{1}{2\pi}\int l_0(w_2, n, \rho)e^{i\mathrm{Re}(z_2\bar{w}_2)}d\mu(w_2), \tag{348}$$

$$f(z_1, n, \rho) = \frac{1}{2\pi}\int l(w_1, n, \rho)e^{-i\mathrm{Re}(z_1\bar{w}_1)}d\mu(w_1), \tag{349}$$

$$f_1(z_1, n, \rho) = \frac{1}{2\pi}\int l_1(w_1, n, \rho)e^{-i\mathrm{Re}(z_1\bar{w}_1)}d\mu(w_1). \tag{350}$$

As was shown in §7.4, the kernel

$$K_{g_0}(z_1, z_2, n, \rho) = f_{g_0}(z_1, n, \rho)f_0(z_2, n, \rho) \tag{351}$$

corresponds to a function $x_{g_0}(g) = x(g_0^{-1}g)$ where

$$f_{g_0}(z_1, n, \rho) = |g_{12}^0 z_1 + g_{22}^0|^{n+i\rho-2}(g_{12}z_1 + g_{22}^0)^{-n} f\left(\frac{g_{11}^0 z_1 + g_{21}^0}{g_{12}^0 z_1 + g_{22}^0}, n, \rho\right) \quad (352)$$

and

$$g_0 = \begin{pmatrix} g_{11}^0 & g_{12}^0 \\ g_{21}^0 & g_{22}^0 \end{pmatrix}.$$

Hence, to an arbitrary function $x(g) \in \mathfrak{M}_{\Phi_0}$ corresponds a kernel of the form

$$K(z_1, z_2, n, \rho) = f(z_1, n, \rho) f_0(z_1, n, \rho)$$

where the function $f_0(z_2, n, \rho)$ is fixed. In other words, to each function $x(g) \in \mathfrak{M}_{\Phi_0}$ we can associate the function $f(z_1, n, \rho)$. Because of (352) the shift $x(g) \to x(g_0^{-1}g)$ corresponds to the transformation of this function considered as a function in $z_1$ by a representation $U_{n_1, \rho; g_0}$ of the principal series.

We now return to initial functions $x$ and $x_1$ from $S_{\Phi_0}$. In view of (347) and (351) it follows that to the functions $x_1^* x_{g_0}$, the kernel

$$K_0(z_1, z_2, n, \rho) = \int \overline{K_1(z, z_1, n, \rho)} K_{g_0}(z, z_2, n, \rho) d\mu(z)$$
$$= \overline{f_0(z_1, n, \rho)} f_0(z_2, n, \rho) a_{n,\rho}^{(0)}, \quad (353)$$

corresponds, where

$$a_{n,\rho}^0 = \int \overline{f_1(z, n, \rho)} f_{g_0}(z, n, \rho) d\mu(z). \quad (354)$$

Now (353) implies that the corresponding kernel $L_0$ is of the form

$$L_0(w_1, w_2, n, \rho) = l_0(w_1, n, \rho) l_0(w_2, n, \rho) a_{n,\rho}^{(0)}. \quad (355)$$

Denote by $\tilde{L}_0$, $\tilde{\Phi}_0$ the corresponding functions $\tilde{L}$ and $\tilde{\Phi}$. By (340) and (343) we have

$$\tilde{L}_0(w_1, w_2, n, \rho) = \overline{\Phi_0(w_1)} \Phi_0(w_2) a_{n,\rho}^{(0)}.$$

Hence

$$\tilde{\Phi}_0(w_1, w_2, \lambda) = \overline{\Phi_0(w_1)} \Phi_0(w_2) \omega_0(\lambda), \quad (356)$$

where

$$\omega_0(\lambda) = \frac{1}{(2\pi)^2} \sum_{n=-\infty}^{\infty} \int_{-\infty}^{\infty} a_{n,\rho}^{(0)} |\lambda|^{-n-i\rho} \lambda^n d\rho \quad (357)$$

Let

$$\varphi(z, \lambda) = \frac{1}{(2\pi)^2} \sum_{n=-\infty}^{\infty} \int_{-\infty}^{\infty} f(z, n, \rho) |\lambda|^{-n-i\rho} \lambda^n d\rho, \quad (358)$$

$$\varphi_1(z, \lambda) = \frac{1}{(2\pi)^2} \sum_{n=-\infty}^{\infty} \int_{-\infty}^{\infty} f_1(z, n, \rho) |\lambda|^{-n-i\rho} \lambda^n d\rho. \quad (359)$$

The formula (358) defines $\varphi(z, \lambda)$ not only for $x(g) \in S_{\Phi_0}$ but for any function $x(g) \in \mathfrak{M}_{\Phi_0}$, because to each function $x(g) \in \mathfrak{M}_{\Phi_0}$ corresponds a function $f(z, n, \rho)$. In the case when $x(g)$, $x_1(g) \in S_{\Phi_0}$ it is easy to express these functions in terms of functions $\omega(\lambda)$, $\omega_1(\lambda)$, $\Psi(w_1)$, $\Psi_1(w_1)$. If, for example, $\omega(\lambda)$ is an even function in $\lambda$, then by formulas (341) and (349) we have

$$\varphi(z,\lambda)$$
$$=\frac{1}{(2\pi)^3}\sum_{n=-\infty}^{\infty}\int_{-\infty}^{\infty}\left[\int a_{n,\rho}|w_1|^{-(n+i\rho)/2}\bar{w}_1^{n/2}\Psi(w_1)e^{-i\mathrm{Re}(z_1\bar{w}_1)}d\mu(w_1)\right]|\lambda|^{-n-i\rho}\lambda^n d\rho$$
$$=\frac{1}{(2\pi)^3}\int\left[\sum_{n=-\infty}^{\infty}\int_{-\infty}^{\infty}a_{n,\rho}|w_1|^{-(n+i\rho)/2}\bar{w}_1^{n/2}|\lambda|^{-n-i\rho}\lambda^n d\rho\right]\Psi(w_1)e^{-i\mathrm{Re}(z_1\bar{w}_1)}d\mu(w_1),$$

that is

$$\varphi(z,\lambda)=\frac{1}{2\pi}\int\omega(\lambda\sqrt{\bar{w}_1})\Psi(w_1)e^{-i\mathrm{Re}(z_1\bar{w}_1)}d\mu(w_1). \tag{360}$$

If $\omega(\lambda)$ is an odd function in $\lambda$, then similarly

$$\varphi(z,\lambda)=\frac{1}{2\pi}\int|w_1|^{\frac{1}{2}}\cdot\bar{w}_1^{-\frac{1}{2}}\omega(\lambda\sqrt{\bar{w}_1})\Psi(w_1)e^{-i\mathrm{Re}(z_1\bar{w}_1)}d\mu(w_1). \tag{361}$$

Now let us substitute in formula (357) for $\omega_0(\lambda)$ the expression (354) for $a_{n,\rho}^{(0)}$. Taking into account (352), (358), and (359), we get

$$\omega_0(\lambda)=\int|g_{12}^0z+g_{22}^0|^{-2}\overline{\varphi_1[z,\lambda(g_{12}^0z+g_{22}^0)]}\,\varphi\left[\frac{g_{11}^0z+g_{21}^0}{g_{12}^0z+g_{22}},\lambda_1\lambda\right]\frac{d\mu(\lambda)}{|\lambda|^2}d\mu(z). \tag{362}$$

Let us prove that the function $\omega_0(\lambda)=\gamma_0(t,\theta)$ belongs to $\Gamma_0$. In the formula (360) for $\varphi(z,\lambda)$ the integration in $w_1$ is, in fact, taken over the region $0<\varepsilon_1\leqq|w_1|\leqq C_1$. Therefore, all derivatives $\varphi_{\lambda^\alpha\bar{\lambda}^\beta}(z,\lambda)$ exist for $\alpha+\beta\leqq 6$. Moreover, the function $\varphi(z,\lambda)$ vanishes for all $z$ outside a region $0<\varepsilon\leqq|\lambda|\leqq C$. If we formally differentiate both parts of (362) in $\lambda$ we get

$$\frac{\partial^{\alpha+\beta}\omega_0(\lambda)}{\partial\lambda^\alpha\partial\bar{\lambda}^\beta}=\int|g_{12}^0z+g_{22}^0|^{-2}\varphi_1[z,\lambda_1(g_{12}^0z+g_{22}^0)]\varphi_{\lambda^\alpha\bar{\lambda}^\beta}$$
$$\cdot\left[\frac{g_{11}^0z+g_{21}^0}{g_{12}^0z+g_{22}^0},\lambda_1\lambda\right]\lambda_1^\alpha\bar{\lambda}_1^\beta\frac{d\mu(\lambda_1)}{|\lambda_1|^2}d\mu(z). \tag{363}$$

Hence, by the Schwartz inequality,

$$\left|\frac{\partial^{\alpha+\beta}\varphi_0(\lambda)}{\partial\lambda^\alpha\partial\bar{\lambda}^\beta}\right|^2\leqq\int|\varphi_1[z,\lambda_1(g_{12}^0z+g_{22}^0)]|^2\frac{d\mu(\lambda_1)}{|\lambda_1|^2}d\mu(z)\int|g_{12}^0z+g_2^0|^{-4}\left|\varphi_{\lambda^\alpha\bar{\lambda}^\beta}\right.$$
$$\left.\cdot\left[\frac{g_{11}^0z+g_{21}^0}{g_{12}^0z+g_{22}^0},\lambda_1\lambda\right]\right|\cdot|\lambda_1|^{2(\alpha+\beta)}\frac{d\mu(\lambda_1)}{|\lambda_1|^2}d\mu(z)$$
$$=|\lambda|^{-2(\alpha+\beta)}\int|\varphi_1(z,\lambda_1|^2\frac{d\mu(\lambda_1)}{|\lambda_1|^2}d\mu(z)\int|\varphi_{\lambda^\alpha\bar{\lambda}^\beta}(z,\lambda_1|^2|\lambda_1|^{2(\alpha+\beta)}\frac{d\mu(\lambda_1)}{|\lambda_1|^2}d\mu(z)$$
$$=|\lambda|^{-2(\alpha+\beta)}\int|\omega_1(\lambda_1\sqrt{\bar{w}_1})|^2|\Psi_1(w_1)|^2\frac{d\mu(\lambda_1)}{|\lambda_1|^2}d\mu(w_1)\cdot\int|\omega_{\lambda^\alpha\bar{\lambda}^\beta}$$

$$\cdot(\lambda_1\sqrt{\bar{w}_1})|^2|w_1|^{2(\alpha+\beta)}|\lambda_1|^{2(\alpha+\beta)}\frac{d\mu(\lambda_1)}{|\lambda_1|^2}d\mu(w_1)$$

$$=|\lambda|^{-2(\alpha+\beta)}\int|\omega_1(\lambda_1)|\frac{d\mu(\lambda_1)}{|\lambda_1|^2}\cdot\int|\Psi_1(w_1)|^2d\mu(w_1)\cdot\int|\omega_{\lambda^\alpha\bar{\lambda}^\beta}(\lambda_1)|^2$$

$$\cdot|\lambda_1|^{2(\alpha+\beta)}\frac{d\mu(\lambda_1)}{|\lambda_1|^2}\cdot\int|\Psi(w_1)|^2d\mu(w_1).$$

This estimate shows that the integral (363) converges uniformly in $\lambda$ in an arbitrary region of the form $0<\varepsilon\leqq|\lambda|\leqq C$ and hence differentiation in $\lambda$, $\bar{\lambda}$ in the integral is permissible. This estimate implies also that the function

$$|\lambda|^{\alpha+\beta}\left|\frac{\partial^{\alpha+\beta}\omega_0(\lambda)}{\partial\lambda^\alpha\partial\bar{\lambda}^\beta}\right|$$

is bounded. Let $\gamma_0(t,\theta)=\omega_0(\lambda)$; then

$$|\lambda|^{\alpha+\beta}\left|\frac{\partial^{\alpha+\beta}\omega_0(\lambda)}{\partial\lambda^\alpha\partial\bar{\lambda}^\beta}\right|=\left|\frac{\partial^{\alpha+\beta}\gamma}{\partial u^\alpha\partial\bar{u}^\beta}\right|.$$

Therefore $\left|\frac{\partial^{\alpha+\beta}\gamma}{\partial u^\alpha\partial\bar{u}^\beta}\right|$ is bounded when $0\leqq\alpha+\beta\leqq 6$. Assume that

$$\left|\frac{\partial^{\alpha+\beta}\gamma}{\partial u^\alpha\partial\bar{u}^\beta}\right|\leqq C. \tag{364}$$

Let us prove the finiteness of the norm $\|\gamma_0\|_1$. It is enough to show that

$$\int_{-\pi}^{\pi}\int_0^\infty\left|\frac{\partial^{\alpha+\beta}\gamma}{\partial u^\alpha\partial\bar{u}^\beta}\right|e^{(y+\varepsilon)t}dtd\theta<+\infty. \tag{365}$$

In fact, in this case, the inequality (364) implies

$$\int_{-\pi}^{\pi}\int_0^\infty\left|\frac{\partial^{\alpha+\beta}\gamma}{\partial u^\alpha\partial u^\beta}\right|^2e^{(y+\varepsilon)t}dtd\theta\leqq C\int_{-\pi}^{\pi}\int_0^\infty\left|\frac{\partial^{\alpha+\beta}\gamma}{\partial u^\alpha\partial\bar{u}^\beta}\right|e^{(y+\varepsilon)t}dtd\theta<+\infty.$$

Introducing the function $\omega_0(\lambda)$ we see that we have to show the finiteness of the integral

$$I=\int|\lambda|^m\left|\frac{\partial^{\alpha+\beta}\omega_0}{\partial\lambda^\alpha\partial\bar{\lambda}^\beta}\right|\frac{d\mu(\lambda)}{|\lambda|^2} \tag{366}$$

where $m=\alpha+\beta+y+\varepsilon$. This integral depends on $g_{11}^0, g_{12}^0, g_{21}^0, g_{22}^0$. Thus, it is enough to prove, for example, the existence of the integral $\int|I|d\mu(g_{21}^0)$ for fixed $g_{12}^0$, $g_{22}^0$. But from formula (363) for $\partial^{\alpha+\beta}\omega/\partial\lambda^\alpha\partial\bar{\lambda}^\beta$ it follows that

$$\int|I|d\mu(g_{21}^0)<\int|g_{12}^0z+g_{22}^0|^{-2}\varphi_1[z,\lambda_1(g_{12}^0z+g_{22}^0)]\left|\varphi_{\lambda^\alpha\bar{\lambda}^\beta}\left(\frac{g_{11}^0z+g_{21}^0}{g_{12}^0z+g_{22}^0},\lambda_1\lambda\right)\right|$$

$$\cdot|\lambda_1|^{\alpha+\beta}|\lambda|\frac{d\mu(\lambda_1)}{|\lambda_1|^2}\frac{d\mu(\lambda)}{|\lambda|^2}d\mu(z)d\mu(g_{21}^0). \tag{367}$$

Introducing the new variables

$$v_1 = \lambda_1(g_{12}^0 z + g_{22}^0); \qquad v = \lambda_1 \lambda; \qquad u = \frac{g_{11}^0 z + g_{21}^0}{g_{12}^0 z + g_{22}^0} = \frac{\dfrac{1 + g_{12}^0 g_{21}^0}{g_{22}^0} z + g_{21}^0}{g_{12}^0 z + g_{22}^0}; \qquad z = z,$$

we get

$$\int I \, | d\mu(g_{21}^0) < |g_{22}^0|^2 \int |g_{12}^0 z + g_{22}^0|^{2+\varepsilon} \cdot |\varphi_1(z, v_1)| \cdot |\varphi_{\lambda^\alpha \bar{\lambda}^\beta}(u, v) \cdot |v|^m \cdot |v_1|^{-y-\varepsilon}$$

$$\cdot \frac{d\mu(v_1)}{|v_1|^2} \frac{d\mu(v)}{|v|^2} d\mu(z) d\mu(u) = |g_{22}^0|^2 \int |g_{12}^0 z + g_{22}^0|^{2+\varepsilon} \cdot |\varphi_1(z, v_1)|$$

$$\cdot |v_1|^{-y-\varepsilon} \frac{d\mu(v_1)}{|v_1|^2} d\mu(z) \cdot \int |\varphi_{\lambda^\alpha \bar{\lambda}^\beta}(\varphi, v)| \cdot |v|^m \frac{d\mu(v)}{|v|^2} d\mu(u).$$

So it is enough to prove that

$$\int |\varphi_{\lambda^\alpha \bar{\lambda}^\beta}(z, \lambda)| \cdot |\lambda|^m \frac{d\mu(\lambda)}{|\lambda|^2} d\mu(z) < +\infty \tag{368}$$

and

$$\int |g_{12}^0 z + g_{22}^0|^{2+\varepsilon} \cdot |\varphi_1(z, \lambda) \cdot |\lambda|^{-y-\varepsilon} \frac{d\mu(\lambda)}{|\lambda|^2} d\mu(z) < +\infty. \tag{369}$$

Since the functions $\varphi(z, \lambda)$ and $\varphi_1(z, \lambda)$ for all $z$ vanish outside sets of the form $0 < \varepsilon \leqq |\lambda| \leqq C$, the integrals

$$I_1 = \int \varphi_{\lambda^\alpha \bar{\lambda}^\beta}(z, \lambda) d\mu(z),$$
$$I_2 = \int |g_{12}^0 z + g_{22}^0|^{2+\varepsilon} \cdot |\varphi_1(z, \lambda)| \, d\mu(z),$$

satisfy the same condition. Therefore it is enough to prove that the integrals $I_1$ and $I_2$ are finite. But this follows from (360) and (361): the functions $\varphi_{\lambda^\alpha \bar{\lambda}^\beta}(z, \lambda)$ and $\varphi_1(z, \lambda)$ are the Fourier transforms with respect to $w_1$ of functions that have continuous derivatives up to second and fourth order in $w_1, \bar{w}_1$.

So we have proved that $\omega_0(\lambda) \in \Gamma'_C$ with some $C > 0$.

Therefore, corresponding to the function $x_1^* x_{g_0}$ is the function $\tilde{\Phi}$ that is equal to $\Phi_0(w_1) \Phi_0(w_2) \omega_0(\lambda)$ with $\omega_0(\lambda) \in \Gamma'_C$. The inner product $(x_{g_0}, x_1) = F(x_1^* x_{g_0})$ is the value of the functional $F(\omega)$ at the function $\omega_0 = \gamma_0$ from $\Gamma'_C$ that is defined by the formula (339) from p. 113. So

$$(x_{g_0}, x_1) = \sum_{n=-\infty}^{\infty} \int_0^{\infty} \left[ \int \omega_0(\lambda) |\lambda|^{n+i\rho} \lambda^{-n} \frac{d\mu(\lambda)}{|\lambda|^2} dC_n(\rho) + \int_0^2 \left[ \int \omega_0(\lambda) |\lambda|^{-\rho_1} \frac{d\mu(\lambda)}{|\lambda|^2} \right] dC(\rho_1). \right. \tag{370}$$

From formulas (354) and (357) it follows that for a real $\rho$

$$\int \omega_0(\lambda) |\lambda|^{n+i\rho} \lambda^{-n} d\mu(\lambda) = \int \overline{f_1(z, n, \rho)} f_{g_0}(z, n, \rho) d\mu(z). \tag{371}$$

In other words, this integral is the inner product in the space $\mathfrak{H}_z$. Let us now calculate the integral

$$\int \omega_0(\lambda) |\lambda|^{-\rho_1} \frac{d\mu(\lambda)}{|\lambda|^2}, \qquad 0 < \rho_1 \leqq 2.$$

We put

$$f(z,\rho_1)=\int\varphi(z,\lambda)\cdot|\lambda|^{\rho_1}\frac{d\mu(\lambda)}{|\lambda|^2},\tag{372}$$

$$f_1(z,\rho_1)=\int\varphi_1(z,\lambda)\cdot|\lambda|^{\rho_1}\frac{d\mu(\lambda)}{|\lambda|^2}.\tag{373}$$

The function $f(z,\rho_1)$ is defined for an arbitrary $x(g)\in\mathfrak{M}_{\Phi_0}$. Under the transformation $x(g)\to x_{g_0}(g)=x(g_0^{-1}g)$ the function $f(z,n,\rho)$ in the kernel

$$K(z_1,z_2,n,\rho)=f(z,n,\rho)f_0(z_2,n,\rho)$$

transforms into

$$f_{g_0}(z,n,\rho)=|g_{12}^0z+g_{22}^0|^{n+i\rho-2}(g_{12}^0z_1+g_{23}^0)^{-n}f\left(\frac{g_{11}^0z+g_{21}^0}{g_{12}^0z+g_{22}^0},n,\rho\right)$$

so that the function $\varphi(z,\lambda)$ transforms into

$$\varphi_{g_0}(z,\lambda)=|g_{12}^0z+g_{22}^0|^{-2}\varphi\left[\frac{g_{11}^0z+g_{21}^0}{g_{12}^0z+g_{12}^0},\lambda(g_{12}^0z+g_{22}^0)\right].\tag{374}$$

Hence $f(z,\rho_1)$ transforms into

$$f_{g_0}(z,\rho_1)=|g_{12}^0z+g_{22}^0|^{-2-\rho_1}f\left(\frac{g_{11}^0z+g_{21}^0}{g_{12}^0z+g_{22}^0},\rho_1\right).\tag{375}$$

In other words, under the transformation $g\to g_0^{-1}g$, the function $f(z,\rho_1)$ considered as a function in $z$ transforms by a representation $U_{\rho_1;g_0}$ of the complementary series.

From the definition (372) of the function $f(z,\rho_1)$ we see that if $\omega(\lambda)$ is even then

$$\begin{aligned}f(z,\rho_1)&=\frac{1}{2\pi}\int\omega(\lambda\sqrt{\bar w_1})\Psi(w_1)e^{-i\mathrm{Re}(z,\bar w_1)}|\lambda|^{-\rho_1}\frac{d\mu(\lambda)}{|\lambda|^2}d\mu(w_1)\\&=\frac{1}{2\pi}\int a_{\rho_1}\Psi(w_1)|w_1|^{(\rho_1/2)}e^{-i\mathrm{Re}(z_1\bar w_1)}d\mu(w_1)\end{aligned}$$

where

$$a_{\rho_1}=\int\omega(\lambda)|\lambda|^{-\rho_1}\frac{d\mu(\lambda)}{|\lambda|^2}.$$

This means that the function $f(z,\rho_1)$ is the Fourier transform in $w_1$ of the function

$$\psi(w_1,\rho_1)=a_{\rho_1}\Psi(w_1)|w_1|^{-\rho_{\frac{1}{2}}}.$$

It is easy to check that this statement also holds for the odd function $\omega(\lambda)$.

The function $\psi(w_1,\rho_1)$ considered as a function in $w_1$ belongs to the space $H_{\rho_1}$ (see §8.1). By the above, under the transformation $x(g)\to x(g_0^{-1}g)$ it transforms into

$$\psi_{g_0}(w_1,\rho_1)=U_{\rho_1;g_0}\psi(w_1,\rho_1)$$

where $U_{\rho_1;g_0}$ is an operator of the complementary series in the space $H_{\rho_1}$. Let us also note that $\omega(\lambda^{-1})=\omega(\lambda)$ and hence $a_{-\rho_1}=a_{\rho_1}$. Therefore

$$\psi(w_1,-\sigma_1)=a_{-\rho_1}\Psi(w_1)|w_1|^{\rho\frac{1}{2}}=|w_1|^{-\rho_1}\psi(w_1\rho_1). \qquad (376)$$

Let us now return to the integral $\int\omega_0(\lambda)|\lambda|^{-\rho_1}(d\mu(\lambda)/|\lambda|^2)$. From the definition (362) of the function $\omega_0(\lambda)$ it follows that this integral is equal to

$$\int\left[\int\overline{\varphi_1(z,\lambda)}\,\varphi_{g_0}(z,\lambda\lambda_1)\frac{d\mu(\lambda_1)}{|\lambda_1|^2}d\mu(z)\right]\cdot|\lambda|^{-\rho_1}\frac{d\mu(\lambda)}{|\lambda|^2},$$

and from the estimates (367), (368) and (369) it follows that this integral converges absolutely. So we can change the order integration:

$$\int\omega_0(\lambda)|\lambda|^{-\rho_1}\frac{d\mu(\lambda)}{|\lambda|^2}=\int\overline{f_1(z,-\rho_1)}f_{g_0}(z,\rho_1)d\mu(z)$$

$$=\int\overline{\psi_1(w,-\rho_1)}\psi_{g_0}(w,\rho_1)d\mu(w_1).$$

Because of (376) the last inequality can be written in the form

$$\int\omega_0(\lambda)|\lambda|^{-\rho_1}\frac{d\mu(\lambda)}{|\lambda|^2}=\int|w|^{-\rho_1}\overline{\psi_1(w,\rho_1)}\psi_{g_0}(w,\rho_1)d\mu(w) \qquad (377)$$

so that this integral is the inner product of the functions $\psi_1(w,\rho_1)$ and $\psi_{g_0}(w,\rho_1)$ in the space $H_{\rho_1}$.

When $\rho_1=2$ all functions $f_1(z,2)$ satisfy the condition

$$\int f_1(z,2)d\mu(z)=0. \qquad (378)$$

This is clear for those functions $f_1(z,2)$ that corespond to functions $x(g)\in S_{\Phi_0}$ because in this case the Fourier transform of the function $f_1(z,2)$ is equal to $a_2\Psi(w_1)\cdot|w_1|^{\rho\frac{1}{2}}$ and this function vanishes at $w_1=0$.

On the other hand, the validity of the equality (378) does not cease if we apply the transformation $x(g)\to x(g_0^{-1}g)$; in fact, this transformation maps $f_1(z,2)$ into

$$|g_{11}^0z+g_{21}^0|^{-4}f_1\left(\frac{g_{11}^0z+g_{21}^0}{g_{12}^0z+g_{22}^0},2\right).$$

So the equality (378) is true for all $f_1(z,2)$ that correspond to functions $x(g)$ from $\mathfrak{M}_{\Phi_0}$. Therefore, the Fourier transform of the function $f_1(z,2)$, i.e. the function $\varphi_1(z,2)$, vanishes at $w=0$ so that the integral

$$\int\left|\frac{\varphi_1(w,2)}{w}\right|^2d\mu(w)$$

exists. Hence for $\rho_1=2$ we have

$$\int\omega_0(\lambda)|\lambda|^2\frac{d\mu(\lambda)}{|\lambda|^2}\int|w|^{-2}\overline{\psi_1(w,2)}\psi_{g_0}(w,2)d\mu(w). \qquad (379)$$

To each function $f_1(z,2)$ we associate the function $f(z)$ that is square integrable and satisfies $f'_{\bar z}(z)=f_1(z,2)$. We define, in the set of these functions $f(z)$, the inner product by

$$(f,f)=\int|f(z)|^2 d\mu(z).$$

The equality (379) shows that the correspondence $f(z)\sim f_1(z,2)$ is isometric. If $f(z)$ is transformed by an operator $U_{2,0;g_0}$ of the principal series into the function

$$(g_{12}^0 z+g_{22}^0)^{-2} f\left(\frac{g_{11}^0 z+g_{21}^0}{g_{12}^0 z+g_{22}^0}\right),$$

then $f_{\bar{z}}'(z)$ is transformed into the function

$$|g_{12}^0 z+g_{22}^0|^{-4} f\left(\frac{g_{11}^0 z+g_{21}^0}{g_{12}^0 z+g_{12}^0}\right),$$

similarly to the function $f_1(z,2)$. Therefore, when $\rho_1=2$, the representation is equivalent to the representation of the principal series with $n=2$ and $\rho=0$.

Substituting the expressions (371) and (372) into the formula (370) for inner product, we have

$$(x_{g_0},x_1)=\sum_{n=-\infty}^{\infty}\int_0^{\infty}[\int f_1(z,n,\rho)f_{g_0}(z,n,\rho)d\mu(z)]dC_n(\rho)$$

$$+\int_0^2[\int|w|^{-\rho_1}\overline{\psi_1(w,\rho_1)}\psi_{g_0}(w,\rho_1)d\mu(w)]dC(\rho_1). \tag{380}$$

This formula is deduced under the assumption that $x,x_1\in S_{\Phi_0}$. But both parts of the equality (380) are invariant under the transformation $x(g)\to x(g_0^{-1}g)$, $x_1(g)\to x(g_0^{-1})$. Therefore, this equality is true for all functions $x,x_1\in\mathfrak{M}_{\Phi_0}$.

Let us consider the set $\mathfrak{H}_0$ of the pairs of functions $f=\{f(z,n,\rho),\psi(w,\rho_1)\}$ such that

$$\sum_{n=-\infty}^{\infty}\int_0^{\infty}[\int|f(z,n,\rho)|^2 d\mu(z)]dC_n(\rho)+\int_0^2[\int|w|^{-\rho_1}|\psi(w,\rho_1)|^2 d\mu(w)]dC(\rho_1)<+\infty. \tag{381}$$

Here the measurability by $\rho$ and $\rho_1$ must be understood in respect to $C_n(\rho)$ and $C(\rho_1)$.

Define in the usual way addition and multiplication by a scalar in $\mathfrak{H}_0$. Define also the inner product by

$$(f,f_2)=\sum_{n=-\infty}^{\infty}\int_0^{\infty}[\int f(z_1,n,\rho)\overline{f_1(z,n,\rho)}d\mu(z)]dC_n(\rho)$$

$$+\int_0^2[\int|w|^{-\rho_1}\psi(w,\rho_1)\overline{\psi_1(w,\rho_1)}d\mu(w)]dC(\rho_1).$$

The formula (380) for $g_0=e$ implies that the correspondence

$$x(g)\sim\{f(z,n,\rho),\psi(w,\rho)\}$$

is an isometric mapping of $\mathfrak{M}_{\Phi_0}$ into $\mathfrak{H}_0$. Denote by $\tilde{\mathfrak{M}}_{\Phi_0}$ the closure of the manifold $\mathfrak{M}_{\Phi_0}$ in the space $\mathfrak{H}$. The above correspondence can be uniquely extended to an isometric mapping of $\tilde{\mathfrak{M}}_{\Phi_0}$ into $\mathfrak{H}_0$.

Let us prove that the image of $\mathfrak{M}_{\Phi_0}$ under this mapping coincides with $\mathfrak{H}_0$. To do this, we apply the Fourier transformation in $z$ to the functions $f(z, n, \rho)$:

$$\psi(w, n, \rho) = \frac{1}{2\pi} \int f(z, n, \rho) e^{i\mathrm{Re}(z\bar{w})} d\mu(z).$$

If $x(g) \in S_{\Phi_0}$ then from (341) and (349) it follows that for even $n$

$$\psi(w, n, \rho) = |w|^{-(n+i\rho)/2} \bar{w}^{n/2} \Psi(w) a_{n,\rho} \tag{382}$$

and for odd $n$

$$\psi(w, n, \rho) = |w|^{-(n-1+i\rho)/2} \bar{w}^{(n-1)/2} \Psi(w) a_{n,\rho} \tag{383}$$

where

$$a_{b,\rho} = \int \omega(\lambda) |\lambda|^{n+i\rho} \lambda^{-n} \frac{d\mu(\lambda)}{|\lambda|^2}.$$

Here $\Psi(w)$ is an arbitrary function that has continuous derivatives up to fourth order and that vanishes outside a set $0 < \varepsilon, < |w < C_2$, and $\omega(\lambda)$ is an arbitrary function that has continuous derivatives up to eighth order and that vanishes outside a set of the form $0 < \varepsilon \leqq |\lambda| \leqq C$. Therefore, linear combinations of the functions of the type (382), (383) are dense in $\mathfrak{H}_0$ and hence the image of the space $\mathfrak{M}_{\Phi_0}$ coincides with $\mathfrak{H}_0$.

So we have an isometric mapping

$$x(g) \sim \{f(z, n, \rho), \psi(w_1, \rho_1)\}$$

of the space $\mathfrak{M}_{\Phi_0}$ onto $\mathfrak{H}_0$. The formula (380) shows that the action of the operator $U_g$ of the representation in the space $\mathfrak{M}_{\Phi_0}$ reduces to the actions of operators $U_{n,\rho;g_0}, U_{\rho_1;g_0}$ of the representations of the principal and the complementary series on functions $f(z, n, \rho)$, $\psi(w, \rho)$. This means that our representations (in the invariant space $\mathfrak{M}_{\Phi_0}$ is equivalent to the direct sum of representations of principal and complementary series. In particular, if this representation is irreducible then it is equivalent to one of the representations of these series.

Let us consider now the given representation in the orthogonal complement $\mathfrak{H} \backslash \mathfrak{M}_{\Phi_0}$. Repeating the same procedure, choose in this space an invariant subspace in which our representation is equivalent to a direct sum of the representations of the principal and the complementary series. After a finite number of such steps, we obtain an invariant subspace $\mathfrak{N}$ for which there are only two possibilities:

(a) $\mathfrak{N} = \{0\}$.

(b) for an arbitrary vector $f \in \mathfrak{N}$ and for arbitrary function $x(g) \in \mathfrak{A}$ (see p. 94) we have

$$\int x(g)(U_g f, f) d\mu(g) = 0.$$

By lemma 2 from p. 95 in this case $U_g \equiv E$ in $\mathfrak{N}$.

So we have proved the following theorems:

**Theorem 13.** *An arbitrary representation of the group G can be decomposed into a direct sum of the representations of principal and complementary series.*

**Theorem 14.** *Each irreducible unitary representation of the group G is equivalent to one of the representations of principal or complementary series.*

Let us note also that for any complex $\rho$ from the strip $I$ and an arbitrary integer $n$ our formulas also give a representation of the group $G$; however this representation is defined not in Hilbert space but in the space $L_p$ of functions $f(z)$ such that

$$\|f\|^p = \int |f(z)|^p d\mu(z) < +\infty,$$

where

$$p = \frac{4}{\rho + 2}.$$

It is easy to check that the operator $U_{n,\rho;g}$ preserves the norm $\|f\|$ in the space $L_p$.

## References

1. Dirac, P.A.M.: Proc. Roy. Soc. A **183** (1945) 284
2. Gelfand, I.: About one-parametric groups of operators in a normed space. Dokl. Acad. Nauk SSSR **XVV** (1939) 711–716.
3. Gelfand, I., Najmark, M.: Unitary representations of the Lorentz group. J. Phys **X** (1946) 93–94.

# 3.

## (with M. A. Najmark)

# On unitary representations of the complex unimodular group

Dokl. Akad. Nauk SSSR **54** (1946) 195–198. Zbl. **29**:5

*(Communicated by A. N. Kolmogoroff, Member of the Academy, 20. VIII. 1946)*

The complex unimodular group of $n$-th order is generally understood to be the group of all matrices $g = \| g_{pq} \|_{p,q=1,\ldots,n}$ of $n$-th order with complex $g_{pq}$ and with the determinant $=1$.

Finite-dimenisonal irreducible representations of $\mathfrak{G}$ have been completely investigated in the classical works of Cartan and Weyl. On the other hand, finite dimensional representations of $\mathfrak{G}$ are known not to be unitary; an exception is only the unit representation, *i.e.*, the representation mapping all group elements into the unit operator. The finding out of all irreducible unitary (hence, infinite-dimensional) representations of $\mathfrak{G}$ remained an open question up to the present.

Briefly the solution of this problem, as obtained by the authors, is dealt with in the present paper. It will be discussed in detail elsewhere. The results here reported can be extended to arbitrary semi-simple Lie groups; this question will be the subject of a subsequent communication. To begin with, some preliminary notions are given, which are necessary for the exposition of the principal results obtained. Later the irreducible representations of $\mathfrak{G}$ will be described and some other allied results set forth.

### § 1. Some Subgroups of $\mathfrak{G}$

The representations of $\mathfrak{G}$ will be further realized as operators in the space of functions of cosets of $\mathfrak{G}$ with respect to some of its subgroups. Let us consider these subgroups.

1. Subgroup $K$. Denote by $K$ the set of all $k = \| k_{pq} \| \in \mathfrak{G}$ such that $k_{pq} = 0$ for $p > q$. This implies: $k_{11} k_{22} \ldots k_{nn} = 1$. Put $k_{pq} = u_{pq} + i v_{pq}$. Then the left- and right-invariant measures $\mu$ and $\mu_r$ in $K$ are given by $d\mu_r(k) = |k_{22}|^{-4} |k_{33}|^{-6} \ldots |k_{nn}|^{-2n} du\, dv$, $d\mu_c(k) = |k_{33}|^2 |k_{44}|^4 \ldots |k_{nn}|^{2n-4} du\, dv$, where $du$, $dv$ denote the product of all differentials of $u_{pq}$ and $v_{pq}$ respectively, with the exception of $u_{11}$ and $v_{11}$. Hence, if we denote $\beta(k) = d\mu_e(k)/d\mu_r(k)$, then

$$\beta(k) = |k_{22}|^4 |k_{33}|^8 \ldots |k_{nn}|^{4n-4}.$$

2. Subgroup $H$. We denote so the set of all $h = \| h_{pq} \| \in \mathfrak{G}$, satisfying the condition $h_{pq} = 0$ for $p < q$. It is easy to write for $H$ the results similar to those indicated in No. 1 for $K$.

3. Subgroup $Z$. Denote by $Z$ the set of all $z = \| z_{pq} \|$ such that $z_{pq} = 0$ for $p < q$ and $z_{pp} = 1$. Evidently, $Z \subset H$. Put $z_{pq} = x_{pq} + i y_{pq}$, $p > q$. There exists in $Z$ a two-sided invariant measure $\mu$, which is

given by $d\mu(z) = dx\,dy$, where $dx$, $dy$ denote the products of differentials of all $x_{pq}$, $y_{pq}$ for $p > q$.

4. Subgroup Z. Thus we denote the set of all $\zeta = \|\zeta_{pq}\| \in \mathfrak{G}$ such that $\zeta_{pq} = 0$ for $p > q$ and $\zeta_{pp} = 1$. Evidently, $\mathrm{Z} \subset K$. Further, there exists in Z a two-sided invariant measure, which is given by an equality similar to the given in No. 3 for $Z$.

5. Subgroup $D$. Thus will be denoted the set of all diagonal matrices $\delta$ of $\mathfrak{G}$. The diagonal elements of $\delta$ will be denoted by $\delta_1, \delta_2, \ldots, \delta_n$. Evidently, $\delta_1\delta_2 \ldots \delta_n = 1$; $D$ is commutative and $D = K \cap H$. Put $\delta_p = \sigma_p + i\tau_p$; then $d\mu(\delta) = |\delta_2|^{-2} |\delta_3|^{-2} \ldots |\delta_n|^{-2} d\sigma\, d\tau$, where $d\sigma$, $d\tau$ are the products of the differentials of $\sigma_p$ and $\tau_p$, $p = 2, 3, \ldots, n$.

## § 2. Some Relations between $\mathfrak{G}$ and its Subgroups $H$, $K$, $Z$, Z and $D$

In studying cosets of $\mathfrak{G}$ with respect to its subgroups we will make use of the important fact that its elements are representable, with some exceptions, as products of elements of its subgroups.

I. Every element $k \in k$ can be represented, and in a unique manner too, in the form $k = \delta\zeta$, and also in the form $k = \zeta_1\delta_1$, where $\delta_1\delta_1 \in D$ and $\zeta_1\zeta_1 \in \mathrm{Z}$.

A similar circumstance takes place for $H$, $Z$ and $D$.

II. Denote by $\begin{pmatrix} p_1, p_2, \ldots, p_k \\ q_1, q_2, \ldots, q_k \end{pmatrix}$ the minor of $g \in \mathfrak{G}$ which is formed of its rows and columns with numbers $p_1, p_2, \ldots, p_k$ and $q_1, q_2, \ldots, q_k$, respectively. Put $g_m = \begin{pmatrix} m, m+1, \ldots, n \\ m, m+1, \ldots, n \end{pmatrix}$. If all $g_m$ are different from zero, then $g$ can be represented, and in a unique manner too, in the form $g = kz$ and also in the form $g = \zeta h$.

So, the exceptional elements $g$ form in $\mathfrak{G}$ a manifold of a lower dimension. The elements of $k$, $z$, $\zeta$ and $h$ in these representations can be expressed in terms of $g$ as follows:

$$k_{pq} = \frac{1}{g_{q+1}} \begin{pmatrix} p, q+1, \ldots, n \\ q, q+1, \ldots, n \end{pmatrix}, \quad p < q; \quad k_{pp} = \frac{g_p}{g_{p+1}}$$

$$z_{pq} = \frac{1}{g_p} \begin{pmatrix} p, p+1, \ldots, n \\ q, p+1, \ldots, n \end{pmatrix}, \quad p > q$$

$$\zeta_{pq} = \frac{1}{g_q} \begin{pmatrix} p, q+1, \ldots, n \\ q, q+1, \ldots, n \end{pmatrix}, \quad p < q$$

$$h_{pq} = \frac{1}{g_{p+1}} \begin{pmatrix} p, p+1, \ldots, n \\ q, p+1, \ldots, n \end{pmatrix}, \quad p > q; \quad h_{pp} = \frac{g_p}{g_{p+1}}, \quad g_{n+1} = 1$$

III. Every element $k$ with different diagonal elements can be represented, and in a unique manner too, in the form $k = \zeta^{-1}\delta\zeta$, and then $\delta_p = k_{pp}$.

A similar proposition holds for $H$.

IV. Every element $g$ with different eigenvalues $\lambda_1, \lambda_2, \ldots, \lambda_n$ can be written in the form $g = x^{-1}\delta x$, $x \in \mathfrak{G}$. The diagonal elements of $\delta$ are $\lambda_1, \lambda_2, \ldots, \lambda_n$, which may be written in any order and the elements of the $p$-th row of $x$ are the components of the eigenvector of $g$ corresponding to $\lambda_p$, hence are defined up to an arbitrary factor ($p = 1, 2, \ldots, n$).

Applying II to $x$, we have in virtue of III: Let any arrangement of the eigenvalues $\lambda_1, \lambda_2, \ldots, \lambda_n$ of $g$ on the diagonal of $\delta$ be given. If then all minors $x_m$ are different from zero, then $g$ can be represented in the form $g = z^{-1}\zeta^{-1}\delta\zeta z$, and also in the form $g = z^{-1}kz$. For a given arrangement of the diagonal elements of $\delta$ and $k$ the $z$ and $\zeta$ are defined uniquely. Hence for all $g$ with the exception of a manifold of lower dimension in $\mathfrak{G}$ there exist $n!$ such representations corresponding to all different permutations of the diagonal elements of $\delta$ and $k$.

## § 3. Relations between the Integrals over $\mathfrak{G}$ and over the Subgroups $K, H, Z, \mathrm{Z}, D$

The results of § 2 may be used in order to reduce the integration over $\mathfrak{G}$ to a repeated integration over its subgroups. All the formulae which follow are understood to mean that the existence of the integrals on the left-hand side implies their existence on the right-hand side, and conversely.

I. For $K$, $H$, $Z$, $\mathrm{Z}$ and $D$:

$$\left.\begin{aligned}\int_K f(k)\,d\mu_e(k) &= \int_D\int_{\mathrm{Z}} f(\delta\zeta)\,d\mu(\delta)\,d\mu(\zeta);\\ \int_K f(k)\,d\mu_r(k) &= \int_D\int_{\mathrm{Z}} f(\zeta\delta)\,d\mu(\delta)\,d\mu(\zeta)\end{aligned}\right\} \tag{1}$$

$$\left.\begin{aligned}\int_H f(h)\,d\mu_e(h) &= \int_D\int_Z f(\delta z)\,d\mu(\delta)\,d\mu(z);\\ \int_H f(h)\,d\mu_r(h) &= \int_D\int_Z f(z\delta)\,d\mu(\delta)\,d\mu(z)\end{aligned}\right\} \tag{2}$$

$$\int_K f(k)\,d\mu_e(k) = \int_{\mathrm{Z}}\int_D f(\zeta^{-1}\delta\zeta)\,\beta^{1/2}(\delta)\prod_{p<q}|\delta_p-\delta_q|^2\,d\mu(\delta)\,d\mu(\zeta) \tag{3}$$

II. For some normalization of the invariant mesure in $\mathfrak{G}$:

$$\int_{\mathfrak{G}} f(g)\,d\mu(g) = \int_K\int_Z f(kz)\,d\mu(z)\,d\mu_e(k) = \int_H\int_{\mathrm{Z}} f(\zeta h)\,d\mu(\zeta)\,d\mu_r(h) \tag{4}$$

and

$$\int_Z\int_K f(z^{-1}kz)\,\varphi(k)\,d\mu_e(k)\,d\mu(z) =$$

$$= \int_{\mathfrak{G}} f(g)\prod_{p<q}|\lambda_g^{(p)}-\lambda_g^{(q)}|^{-2}\sum\varphi(k_g)\,\beta^{1/2}(k_g)\,d\mu(g) \tag{5}$$

where $\lambda_g^{(1)},\ldots,\lambda_g^{(n)}$ are the eigenvalues of $g$; $k_g$ is defined by $g=z^{-1}k_gz$; and the sum in $\sum\varphi(k_g)\beta^{1/2}(k_g)$ is taken over all $k_g$ which are obtained for all possible permutations on its principal diagonal of the eigenvalues of $g$.

## § 4. The Cosets with Respect to $Z$ and $K$

1. The right cosets with respect to $K$. Denote by $\tilde{Z}$ the homogeneous space of the right cosets of $\mathfrak{G}$ with respect to $K$. The elements of $\tilde{Z}$ will be denoted by $\tilde{z}$. The right multiplication of $\tilde{z}$ by $g$ carries it in a new coset, which will be denoted by $\tilde{z}\bar{g}$. Then the group $\mathfrak{G}$ can be considered as the group of transformations $\tilde{z}\to\tilde{z}\bar{g}$ of $\tilde{Z}$. If $g=kz$, then in the coset $\tilde{z}$ containing $g$ there is precisely one element $z\in Z$. We then identify $\tilde{z}$ with $z$. By II § 2, $Z$ fills up the whole $\tilde{Z}$ with the exception of a manifold of lower dimension. Hence we can transfer into $\tilde{Z}$ the invariant measure in $Z$ and re-write the first formula (1) in a corresponding manner.

2. Two-sided cosets with respect to $K$. Denote by $\mathfrak{z}$ the sets of all elements $k_1gk_2$ for fixed $g$. These sets will be called two-sided cosets of $\mathfrak{G}$ with respect to $K$. Denote by $(\sigma_1,\sigma_2,\ldots,\sigma_n)$ a permutation of $1, 2, \ldots, n$. In each two-sided coset $\mathfrak{z}$ there is one and only one element $s=\|s_{pq}\|$ defined by the conditions: $s_{pq}=0$ for $q\neq\sigma_p$; $s_{p\sigma_p}=1$, $p=2,\ldots,n$, $s_{1\sigma_1}=\pm1$, where the sign $+$ or $-$ is used

according as $\sigma_1, \sigma_2, \ldots, \sigma_n$ is an even or an odd permutation of $1, 2, \ldots, n$. Hence there is precisely $n!$ two-sided cosets of $\mathfrak{z}$ with respect to $K$.

3. The cosets with respect to Z. Denote by $\tilde{H}$ the homogeneous space of the right cosets of $\mathfrak{G}$ with respect to Z; the elements of $\tilde{H}$ will be denoted by $\tilde{h}$. To the right multiplication of the elements of $\tilde{h}$ with $g$ there corresponds the transformation $\tilde{h} \to \tilde{h}\bar{g}$ in $\tilde{H}$. If $g = \zeta h$, then in the coset $\tilde{h}$ containing $g$ there is precisely one element $h \in H$. We then identify $h$ with $\tilde{h}$. In virtue of II § 2, $H$ fills up the the whole $\tilde{H}$ with the exception of a manifold of lower dimension. Hence we can transfer into $\tilde{H}$ the right-invariant mesure of $H$. In virtue of (4) this mesure is an invariant mesure with respect to the transformations $\tilde{h} \to \tilde{h}\bar{g}$.

The transformations $\bar{g}$ can also be considered as transformations in $H$; then $h\bar{g}$ is not defined for all, but for nearly all $h \in H$.

Received
20. VIII. 1946.

# 4.

## (with M. A. Najmark)

# The principal series of irreducible representations of the complex unimodular group

Dokl. Akad. Nauk SSSR **56** (1947) 3–4. Zbl. **29**:5

*(Communicated by A. N. Kolmogoroff, Member of the Academy, 14. X. 1946)*

In this note, which is a continuation of (1), a set of irreducible representations of $\mathfrak{G}$ is considered, which will be called the principal series of irreducible representations.

We shall use the results and notations of (1).

§ 1. The Regular Representation of $\mathfrak{G}$. Let $\mathfrak{H}$ be the Hilbert space of square-summable functions $f(g)$ on $\mathfrak{G}$ with the scalar product $(f_1, f_2) = \int_{\mathfrak{G}} f_1(g)\overline{f_2(g)}\,d\mu(g)$. Introduce in $\mathfrak{H}$ an operator $U_{g_0}$ by $U_{g_0}f(g) = f(gg_0)$. In virtue of the invariance of $d\mu(g)$ $U_{g_0}$ is unitary. Evidently $U_{g_0}$ is a representation of $\mathfrak{G}$; it is said to be the regular representation of $\mathfrak{G}$.

In the case of a finite or compact group the regular representation has the remarkable property that its decomposition into irreducible representations contains all the irreducible representations of the given group. It will further be seen that this property does not hold in the case of the complex unimodular group.

§ 2. Quasi-regular Representation of $\mathfrak{G}$. Let $\mathfrak{H}$ be the Hilbert space of square-summable functions on $H$ (cf. (1), 3, § 4) with the scalar product $(f_1, f_2) = \int_H f_1(h)\overline{f_2(h)}\,d\mu_r(h)$.

The operator $V_g f(h) = f(\overline{hg})$ is a unitary representation of $\mathfrak{G}$. We shall describe it as a quasi-regular representation of $\mathfrak{G}$. It will be shown to be decomposable into the same irreducible representations as the regular representation, but contain each of $U_{\chi;g}$ only once.

§ 3. The Principal Series of Irreducible Representations. Let us decompose $V_g$ into irreducible representations. By (1), 3, § 4

$$\int_H |f(h)|^2\,d\mu_r(h) = \int_Z d\mu(z)\int_D |\beta^{1/2}(\delta)f(\delta z)|^2\,d\mu(\delta) \tag{1}$$

Let $\chi(\delta)$ be the character of the group $D$. Put

$$f_\chi(z) = \int_D f(\delta z)\,\beta^{1/2}\overline{\delta_\chi(\delta)}\,d\mu(\delta)$$

Then by the known generalization of Plancherel's theorem for commutative groups (cf. (2)) (1) can be rewritten in the form

$$\int_H |f(h)|^2\,d\mu_r(h) = \int_X d\mu(\chi)\int_Z |f_\chi(z)|^2\,d\mu(z) \tag{2}$$

where $d\mu(\chi)$ is the differential of the invariant mesure on the character group $X$ of $D$. Equality (2) means that $\mathfrak{H}$ of § 2 is the continual direct sum of Hilbert spaces $\mathfrak{H}_z$ of square summable functions $f(z)$ on $Z$.

Put now $\beta(g)=\beta(\delta)$; $\chi(g)=\chi(\delta)$ for $g=\delta\zeta z$, and $\alpha_\chi(g)=\beta^{-1/2}(g)\chi(g)$. Then $\beta(g)$, $\chi(g)$ and $\alpha_\chi(g)$ are defined for almost all $g\in\mathfrak{G}$.

If we pass from $f(h)$ to $f(h\bar{g})$, then $f_\chi(z)$ goes into $\alpha_\chi(zg)f(z\bar{g})$, where $z\to z\bar{g}$ is the transformation of [1], 3, § 4. This means that $V_g$ is decomposed into a continual direct sum of representations in $\mathfrak{H}_z$ defined by

$$U_{\chi;\,g}f(z)=\alpha_\chi(zg)f(z\bar{g}) \tag{3}$$

So, to every character $\chi$ of $D$ there corresponds a representation $U_{\chi;\,g}$. All these representations are unitary.

Theorem 1. *All representations $U_{\chi;\,g}$ are irreducible.*

Let $\delta^s$ be the matrix which we get from $\delta$ by a permutation $s$ of its diagonal elements. Then for fixed $\chi$ and $s$ equality $\chi^s(\delta)=\chi(\delta^s)$ defines a character $\chi^s$ of $D$.

Theorem 2. *For any permutation $s$ the $U_{\chi;\,g}$ corresponding to $\chi$ and $\chi^s$ are equivalent; conversely, if $U_{\chi;\,g}$ and $U_{\chi';\,g}$ are equivalent, then $\chi'=\chi^s$ for some $s$.*

We shall call $U_{\chi;\,g}$, $\chi\in X$, the principal series of irreducible representations of $\mathfrak{G}$.

Equality (3) can be written in a more detailed form if we consider $f(z)$ as a function of all $z_{pq}$, $p>q$ defining the matrix $z$. It is sufficient to use the equality $zg=k\cdot z\bar{g}$ defining $zg$ and the formulae of [1], II, § 2.

If we put $z\bar{g}=z'=\|z'_{pq}\|$, then in virtue of these formulae we have

$$U_{\chi;\,g}f(z_{pq})=|g'_2|^{m_1+i\rho_1-2}|g'_3|^{m_2-m_1+i(\rho_2-\rho_1)-2}\dots$$
$$\dots|g'_n|^{m_{n-1}-m_{n-2}+i(\rho_{n-1}-\rho_{n-2})-2}g_2'^{\,-m_1}g_3'^{\,-(m_2-m_3)}\dots g_n'^{\,-(m_{n-1}-m_{n-2})}f(z'_{pq})$$

where

$$z'_{pq}=\frac{1}{g'_p}\begin{vmatrix} g'_{p,q} & g'_{p,p+1} & \dots & g'_{p,n}\\ g'_{p+1,q} & g'_{p+1,p+1} & \dots & g'_{p+1,n}\\ \dots & \dots & \dots & \dots\\ g'_{n,q} & g'_{n,p+1} & \dots & g'_{n,n}\end{vmatrix},\qquad p>q$$

$$g'_p=\begin{vmatrix} g'_{p,p} & g'_{p,p+1} & \dots & g'_{p,n}\\ g'_{p+1,p} & g'_{p+1,p+2} & \dots & g'_{p+1,n}\\ \dots & \dots & \dots & \dots\\ g'_{n,p} & g'_{n,p+1} & \dots & g'_{n,n}\end{vmatrix}$$

$$g'_{pq}=\sum_{s=1}^{p-1}z_{ps}g_{sq}+g_{pq}$$

and where $m_1, m_2, \dots, m_{n-1}$ are the integers, and $\rho_1, \rho_2, \dots, \rho_{n-1}$ are the real numbers defining the character $\chi$ by

$$\chi(\delta)=|\delta_2|^{m_1+i\rho_1}\delta_2^{-m_1}|\delta_3|^{m_2+i\rho_2}\delta_3^{-m_2}\dots|\delta_n|^{m_{n-1}+i\rho_{n-1}}\delta_n^{-m_{n-1}}$$

In case $n=2$ $U_{\chi;\,g}$ goes into the principal series of representations of the Lorentz group, described in [3].

Received
14.X.1946.

REFERENCES

[1] I. Gelfand and M. Neumark, C. R. Acad. Sci. URSS, 54, No. 3 (1946). [2] M. Krein, ibid., 30, No. 6 (1941). [3] I. Gelfand and M. Neumark, J. of Physics, 10, No. 2 (1946).

# 5.

(with M. A. Najmark)

# Complementary and degenerate series of representations of the complex unimodular group

Dokl. Akad. Nauk SSSR **58** (1946) 1577–1580. Zbl. **37**:304

In previous notes [1, 2] we have described the principal series of irreducible unitary representations of the complex unimodular group $\mathfrak{G}$. Here we describe the so-called complementary and degenerate series of representations of this group. We will keep the notations of [1, 2].

First we have to generalize some subgroups, considered in [1]. These subgroups are defined quite similarly, but numbers are replaced by matrices.

## 1. Some subgroups of the group

Let $n_1, \ldots, n_r$ be positive integers satisfying $n_1 + \ldots + n_r = n$ where $n$ is the order of the group $\mathfrak{G}$. Denote by $g_{pq}$ a matrix with $n_p$ rows and $n_q$ columns. Any matrix $g \in \mathfrak{G}$ can be represented as a matrix $\| g_{pq} \|_{p,q=1,\ldots,r}$.

1. *The subgroup K.* Denote by $K$ the set of all $k = \| k_{pq} \| \in \mathfrak{G}$ such that $k_{pq} = 0$ for $p > q$. Thus,

$$\mathrm{Det}(k_{11})\,\mathrm{Det}(k_{22})\ldots\mathrm{Det}(k_{rr}) = 1.$$

Let $d\mu_l(k)$ and $d\mu_r(k)$ be differentials of the left- and right-invariant measure on the subgroup $K$. Writing $\beta(k) = d\mu_l(k)/d\mu_r(k)$ we have

$$\beta(k) = |\Lambda_1|^{-2n_2 - 2n_3 - \ldots - 2n_r} |\Lambda_2|^{2n_1 - 2n_3 - \ldots - 2n_r} \ldots |\Lambda_r|^{2n_1 + \ldots + 2n_{r-1}} \tag{1}$$

where

$$\Lambda_j = \mathrm{Det}(k_{jj}), \quad j = 1, \ldots, r.$$

2. *The subgroup Z.* Denote by $Z$ the set of matrices $\| z_{pq} \|_{p,q=1,\ldots,r}$, such that $z_{pq} = 0$ for $p < q$ and $z_{pp} = 1$ where $1$ is the identity matrix of the order $n_p$. The two-side invariant measure on $Z$ is given by $d\mu(z) = \prod_{p>q} dx_{pq} dy_{pq}$, where $dx_{pq}$ (respectively, $dy_{pq}$) is the product of differentials of real (respectively, imaginary) parts of elements of the matrix $z_{pq}$.

3. *The subgroup D.* Denote by $D$ the set of matrices $\delta \in \mathfrak{G}$ such that $\delta_{pq} = 0$ for $p \neq q$. We denote the matrix $\delta_{pp}$ by $\delta_p$ and use the notation $\delta = [\delta_1, \ldots, \delta_r]$. In a similar way one can also define the subgroups $H$ and $Z$.

I. Any element $k$ can be uniquely represented in the form $k=\delta\zeta$ and in the form $k=\zeta\delta$.

A similar decomposition exists for elements $h$.

II. If $g_{n_1+\ldots+n_{p+1}}=0, p=1,2,\ldots,r-1$, then $g$ can be uniquely represented in the form $g=kz$ and in the form $g=\zeta h$. The matrix elements $k_{pq}^{(\lambda,\mu)}, \zeta_{pq}^{(\lambda,\mu)}, h_{pq}^{(\lambda,\mu)}$ of matrices $k,\zeta,h$ are determined by elements of the matrix $g$ as follows:

$$h_{pq}^{(\lambda,\mu)}=\frac{1}{g_{n_1+\ldots+n_{p+1}+1}} \cdot\begin{pmatrix} n_1+\ldots+n_p+\lambda, n_1+\ldots+n_{q+1}+1, n_1+\ldots+n_{q+1}+2,\ldots,n \\ n_1+\ldots+n_q+\mu, n_1+\ldots+n_{q+1}+1, n_1+\ldots+n_{q+1}+2,\ldots,n \end{pmatrix}, \tag{2}$$

$$p\geqq q, \quad 0\leqq\lambda\leqq n_p, \quad 0\leqq\mu\leqq n_q.$$

$$k_{pq}^{(\lambda,\mu)}=\frac{1}{g_{n_1+\ldots+n_{q+1}+1}} \cdot\begin{pmatrix} n_1+\ldots+n_p+\lambda, n_1+\ldots+n_{p+1}+1, n_1+\ldots+n_{p+1}+2,\ldots,n \\ n_1+\ldots+n_q+\mu, n_1+\ldots+n_{p+1}+1, n_1+\ldots+n_{p+1}+2,\ldots,n \end{pmatrix}$$

$$p\leqq q, \quad 0\leqq\lambda\leqq n_p, \quad 0\leqq\mu\leqq n_q \tag{3}$$

$$z_{pq}=h_{pp}^{-1}h_{pq}, \qquad \zeta_{pq}=k_{pq}k_{qq}^{-1}. \tag{4}$$

Moreover,

$$\det k_{pp}=\det h_{pp}=\frac{g_{n_1+\ldots+n_p+1}}{g_{n_1+\ldots+n_{p+1}+1}}. \tag{5}$$

If $r=n$, $n_1=\ldots=n_r=1$, these formulas transform into the corresponding formulas from [1].

As we did in [1], we will consider right cosets by these generalized subgroups Z and $K$, and $K$, and transformations $\tilde{h}-\tilde{h}\bar{g}$ and $\tilde{z}\rightarrow\tilde{z}\bar{g}$ of these cosets. According to [1], classes $\tilde{h}$ and $\tilde{z}$ can be represented (up to a submanifold of lower dimension) by elements of subgroups $H$ and $Z$, respectively. We will also need two-sided cosets $\mathfrak{z}$ by the subgroup $K$.

The group $K$ considered in n.2, §4 of [1] is a subgroup of the generalized group $K$. So the results of [1] imply that any two-sided coset by the generalized group $K$ is also determined by a permutation of numbers $(1,\ldots,n)$; however, now different permutations can determine the same coset.

## 2. Relations among the integrals over $\mathfrak{G}$ and over the subgroups $K$, $H$, $Z$, Z and $D$

As in [I], there are the following relations between these integrals:

$$\int_K f(k)d\mu_l(k)=\int_D\int_Z f(\delta\zeta)d\mu(\delta)d\mu(\zeta) \tag{6}$$

$$\int_K f(k)d\mu_r(k)=\int_D\int_Z f(\zeta\delta)d\mu(\delta)d\mu(\zeta) \tag{7}$$

$$\int_H f(h)d\mu_l(h) = \int_D \int_Z f(\delta z)d\mu(\delta)d\mu(z) \tag{8}$$

$$\int_H f(h)d\mu_r(h) = \int_D \int_Z f(z\delta)d\mu(\delta)d\mu(z) \tag{9}$$

$$\int_{\mathfrak{G}} f(g)d\mu(g) = \int_K \int_Z f(kz)d\mu(z)d\mu_l(k) = \int_H \int_Z f(\zeta h)d\mu(\zeta)d\mu_r(h). \tag{10}$$

## 3. Principal series of the unitary representations of the group

In what follows, we consider all the generalized subgroups $K$, $Z$, $Z$, $D$ for some fixed values of $n_1, \ldots, n_r$.

Let

$$\beta(kz) = \beta(k), \quad \chi(\delta\zeta z) = \chi(\delta) = \chi_2(\Lambda_2)\ldots\chi_r(\Lambda_r),$$

where $\chi_2, \ldots, \chi_r$ are arbitrary characters of the multiplicative group of all complex numbers, so that $\chi_j(\Lambda_j) = |\Lambda_j|^{m_j + i\rho_j} \Lambda_j^{-m_j}$, $j = 2, \ldots, r$, with integral $m_j$ and real $\rho_j$.

Denote by $\mathfrak{H}_z$ the space of all square integrable functions $f(z)$ on $Z$ and define the operator $U_g$ in the space $\mathfrak{H}_z$ by the formula

$$U_g f(z) = \beta^{-1/2}(zg)\chi(zg)f(z\bar{g}). \tag{11}$$

Then the mapping $g \rightarrow U_g$ determines a unitary representation of the group $\mathfrak{G}$.

We will call the set of all these representations the principal series of representations of the group $\mathfrak{G}$. For different choices of $n_1, \ldots, n_r$ we get, in general, different principal series.

For $r = n$, $n_1 = \ldots = n_r = 1$ we get the principal series from [1]; it will be called nondegenerate. All other principal series will be called degenerate. For the nondegenerate series a function $f(z) \in \mathfrak{H}_z$ depends on $n(n-1)/2$ complex parameters; in other cases $f(z)$ depends on fewer complex parameters.

**Theorem 1.** *All representations of (degenerate or nondegenerate) principal series are irreducible.*

## 4. Complementary series of unitary representations of the group

Assume now that $n_1 = n_2 = \cdots = n_{2\tau} = 1$ for some $\tau \geqq 1$. Denote by $Z'$ the set of all $z' \in Z$ such that $z'_{2j,2j-1} \neq 0$ for $0 < j \leqq \tau$ and $z'_{pq} = 0$ otherwise. Consider the set $\Sigma$ of all pairs $\{z, z'z\}$ where $z$ runs over $Z$, and $z'$ runs over $Z'$. Define the action of $g$ on $\Sigma$ by $\{z, z'z\}\ \bar{g} = \{z\bar{g}, (z'z)\bar{g}\}$. The group $\mathfrak{G}$ can be considered, therefore, as the group of transformations of $\Sigma$, and $\Sigma$ is a transitive space under this action of $\mathfrak{G}$. If $zg = k \cdot z\bar{g}$, then $\{z, z'z\}\bar{g} = \{z\bar{g}, z'\bar{k} \cdot z\bar{g}\}$.

Let

$$\chi(\delta) = \prod_{j=1}^{\tau} |\Lambda_{2j-1}|^{m_j + \sigma_j + i\rho_j} \Lambda_{2j-1}^{-m_j} |\Lambda_{2j}|^{m_j - \sigma_j + i\rho_j} \Lambda_{2j}^{+m_j} \prod_{j=2\tau+1}^{r} |\Lambda_j|^{m_j + i\rho_j} \Lambda_j^{-m_j} \tag{12}$$

where $m_1 = 0, m_2, \ldots, m_\tau, m_{2\tau+1}, \ldots, m_r$ are integers, $\sigma_j$, $\rho_j$ are real, and $0 < \sigma_j < 1$ for $j = 1, \ldots, \tau$. The numbers $m_2, \ldots, m_r$, $\sigma_j$, $\rho_j$, $j = 1, \ldots, \tau$ will be the parameters determining the given representation. Let also

$$a(z') = \prod_{j=1}^{\tau} |z'_{2j,2j-1}|^{-2+2\sigma_j}$$

and denote by $\mathfrak{H}'$ the space of all functions $f(z)$ such that the integral

$$(f,f) = \int a(z')f(z)\overline{f(z'z)}d\mu(z)d\mu(z')$$

converges absolutely. Here $d\mu(z') = \prod_{j=1}^{\tau} d\mu(z_{2j,2j-1})$. The condition $0 < \sigma_j < 1$ implies that $(f,f) \geqq 0$ for any function $f \in \mathfrak{H}'$. Therefore one can define the inner product in $\mathfrak{H}'$ by the formula

$$(f_1,f_2) = \int a(z')f_1(z)\overline{f_2(z'z)}d\mu(z)d\mu(z').$$

Denote by $\mathfrak{H}$ the completion of $\mathfrak{H}'$ in the norm $\sqrt{(f,f)}$; $\mathfrak{H}$ is a Hilbert space. Define now $\chi(\delta\zeta z) = \chi(\delta)$, $\beta(\delta\zeta z) = \beta(\delta)$; then $\chi(g)$ and $\beta(g)$ are defined for almost all $g \in \mathfrak{G}$. For $f \in \mathfrak{H}'$ define

$$U_g f(z) = \beta^{-1/2}(zg)\chi(zg)f(z\bar{g}). \tag{13}$$

The set $\mathfrak{H}'$ is invariant under the action of the operator $U_g$, and $(U_g f, U_g f') = (f,f')$, so that $U_g$ is a unitary operator in $\mathfrak{H}'$. Therefore it has a unique extension to a unitary operator $U_g$ in $\mathfrak{H}$. The mapping $g \to U_g$ is a unitary representation of the group $\mathfrak{G}$ in the space $\mathfrak{H}$. The set of all representations $g \to U_g$ for all possible values of parameters $m_2, \ldots, m_\tau, m_{2\tau+1}, \ldots, m_r$, $\rho_j$, $\sigma_j$, $0 < \sigma_j < 1$, $j = 1, \ldots, \tau$ is called a *complementary series of representations of the group* $\mathfrak{G}$. For different choices of $\tau$, $n_1 = \ldots = n_{2\tau} = 1$, $n_{2\tau+1}, \ldots, n_r$ we obtain, in general, different complementary series of representations. If $n_{2\tau+1} = \ldots = n_r = 1$, so that $r = n$, we obtain the so-called *nondegenerate* complementary series. All other complementary series are called *degenerate*.

For $\sigma_j > 1$ the form $(f,f)$ is indefinite. For $\sigma_j = 1$ we might have $(f,f) = 0$ for some $f(z) \neq 0$; factorizing by the subspace of all such $f$'s, we obtain some representations of degenerate principal series.

**Theorem 2.** *All representation of (degenerate or nondegenerate) complementary series are irreducible.*

Formulas (11) and (13) for the operators $U_g$ of the principal and complementary series representations can be made more precise by using parameters for the matrices $z$. To do this one has to use formulas (1), (2), (3) and (4). The principal and complementary series for the unimodular group of second order were introduced by the authors in [3] and studied in detail in [4].

Submitted July 8, 1947

## References

1. Gelfand, I.M., Najmark, M.A.: Dokl. Akad. Nauk SSSR (3) **54** (1946)
2. Gelfand, I.M., Najmark, M.A.: Dokl. Akad. Nauk SSSR (1) **56** (1947)
3. Gelfand, I., Najmark, M.: J. Phys. (2) **10**, (1947)
4. Gelfand, I.M., Najmark, M.A.: Izv. Akad. Nauk SSSR, Ser. Mat. (5) **11** (1947).

# 6.

## (with M. A. Najmark)

# The trace in principal and complementary series representations of the complex unimodular group

Dokl. Akad. Nauk SSSR **61** (1984) 9–11. Zbl. **38**:299

In the notes [1–3] principal and complementary series of unitary irreducible representations of the complex unimodular group were constructed. The operators of these representations, being unitary representations in an infinite-dimensional space, do not have, in general, the trace in the usual sense. To associate a trace to these operators, we have to pass from representations of the group to representations of its group ring.

We recall that the group ring $R$ consists of expressions of the form $a = \lambda e + x(g)$ where $e$ is the identity element of the ring $R$, $\lambda$ is a complex number and $x(g)$ is an arbitrary integrable function of $\mathfrak{G}$. Addition and multiplication by a constant are defined in the usual way, and multiplication in $R$ is given by the formula

$$(\lambda_1 e + x_1)(\lambda_2 e + x_2) = \lambda_1\lambda_2 e + \lambda_1 x_2(g) + \lambda_2 x_1(g) + \int x_1(gg_1^{-1})x_2(g_1)d\mu(g_1).$$

Define involution in $R$ by the formula

$$(\lambda e + x(g))^* = \bar{\lambda}e + \overline{x(g^{-1})}.$$

If $g \to U_g$ is a representation of the group $\mathfrak{G}$, then

$$\lambda e + x \to U_{\lambda e + x} = \lambda E + \int x(g)U_g d\mu(g) \tag{1}$$

is a representation of the group ring $R$. Formula (1) establishes a one-to-one correspondence between representations of the ring $R$ and unitary representations of the group $\mathfrak{G}$.

1. *Trace in nondegenerate series.* Let $g \to U_g$ be a representation of a principal or complementary nondegenerate series. Then from formula (4) in [1] one has

$$\begin{aligned} U_x f(z) &= \int x(g)U_g f(z)d\mu(g) \\ &= \int x(g)\beta^{-1/2}(zg)\chi(zg)f(z\bar{g})d\mu(g) = \int K(z,z')f(z')d\mu(z') \end{aligned}$$

where

$$K(z,z') = \int x(z^{-1}kz')\beta^{-1/2}(k)\chi(k)d\mu_e(k).$$

Therefore, the trace $S(U_x)$ of the operator $U_x$ is given by

$$\int K(z,z)d\mu(z) = \int x(z^{-1}kz)\beta^{-1/2}(k)\chi(k)d\mu_e(k)d\mu(z)$$

and formula (5) in [1] implies that

$$S(U_x) = \int x(g) \prod_{p \leqq q} |\lambda_g^{(p)} - \lambda_g^{(q)}|^{-2} \sum \chi(k_g) d\mu(g) = \int x(g) \frac{\sum \chi(k_g)}{D(g)} d\mu(g), \tag{2}$$

the summation in (2) being taken over all $k_g$ obtained by all possible arrangements of the eigenvalues $\lambda_g^{(p)}$ of the matrix $g$ in the main diagonal, and $D(g)$ being the discriminant of the characteristic polynomial of $g$. This formula shows that under a suitable choice of $x(g)$ the operator $U_x$ has a trace.

Let $s$ be an arbitrary permutation of indices $1, \ldots, n$; denote by $\delta_s$ the matrix obtained from $\delta$ by applying the permutation $s$ to its diagonal elements. Let $\chi_s(\delta) = \chi(\delta_s)$.

A permutation $s$ is said to be *admissible* if it transforms a function $\chi(\delta)$ of the form

$$\chi(\delta) = \prod_{j=1}^{\tau} |\delta_{2j-1}|^{m_j + \sigma_j - i\rho_j} \delta_{2j-1}^{-m_j} |\delta_{2j}|^{m_j - \sigma_j + i\rho_j} \delta_{2j}^{-m_j}$$

$$\cdot \prod_{j=2\tau+1}^{m} |\delta_j|^{m_j + i\rho_j} \delta_j^{-m_j}, \quad m_1 = 0, \quad 0 < \sigma_j < 1, \tag{3}$$

into the function $\chi_s(\delta)$ of the same form.

Using formula (2) for the trace, we obtain the following theorem.

**Theorem.** *Two representations $g \to U'_g$, $g \to U''_g$ of the principal nondegenerate series, corresponding to characters $\chi'$* and *$\chi''$ are equivalent if and only if $\chi'_s = \chi''$ for some permutation $s$. Two representations $g \to U'_g$, $g \to U''_g$ of the complementary nondegenerate series, corresponding to functions $\chi'$ and $\chi''$ of the form (3), are equivalent if and only if $\chi'_s = \chi''$ for some admissible permutation $s$.*

Representations of the nondegenerate complementary series are nonequivalent to representations of the nondegenerate principal series.

For $n = 2$ the results of this section are contained in [4].

2. *Trace in degenerate series.* Now let $g \to U_g$ be a representation of either the degenerate principal or complementary series, corresponding to a partition $n = n_1 + n_2 + \ldots + n_r$ (see [3]).

In a similar fashion to §1, we obtain that the trace $S(U_x)$ of an operator $U_x$ is given by

$$S(U_x) = \int x(z^{-1}kz) \beta^{-1/2}(k) \chi(k) d\mu_e(k) d\mu(z) = \int x(zkz^{-1}) \beta^{-1/2}(k) \chi(k) d\mu_e(k) d\mu(z) \tag{4}$$

where $K$ and $Z$ are now corresponding generalized subgroups.

Let us make the substitution $g = zkz^{-1}$ in the last integral and find the Jacobian of the corresponding transformation. We have

$$gz = zk, \qquad dg \cdot z + g \cdot dz = dz \cdot k + z \cdot dk$$

so that $z^{-1}g^{-1}dgz = -z^{-1}dz + k^{-1}z^{-1}dzk + k^{-1}dk$, or

$$z^{-1}g^{-1}dg \cdot z = -z^{-1}dz + \zeta^{-1}\delta^{-1}z^{-1}dz\delta\zeta + k^{-1}dk. \tag{5}$$

The determinants of the linear transformations $u \to z^{-1}uz$ and $u \to \zeta^{-1}u\zeta$ of the space of matrices are both equal to 1. Therefore one has to compute only the determinant of the transformation $w'_{pq} = w_{pq} + \delta_p^{-1} w_{pq}\delta_q$, $p > q$, where $(w_{pq}) = z^{-1}dz$. For fixed $p$ and $q$ this is a linear transformation of the space of matrices $w_{pq}$. As the transformation $w_{pq} \to \delta_p^{-1} w_{pq} d_q$ is the Kronecker product of linear transformations $\delta_p^{-1}$ and $\delta_q$, the required determinant is equal to $\prod |\lambda_{p,i}^{-1}\lambda_{q,j} - 1|^2$ where $\lambda_{p,i}$, $i = 1, \ldots, n_p$ are eigenvalues of $\delta_p$ and the product is taken over all $i$ between 1 and $n_p$, all $j$ between 1 and $n_q$, and then over all $p$ and $q$ satisfying $p > q$.

Therefore formula (5) implies that

$$d\mu_e(k)d\mu(z) = \prod |\lambda_{p,i}^{-1}\lambda_{q,j} - 1|^2 d\mu(g)$$
$$= \frac{|D(\delta_1)||D(\delta_2)|\ldots|D(\delta_r)|}{|D(\delta)|} |\Lambda_1|^{2n_1} |\Lambda_2|^{2n_1 + 2n_2} \ldots |\Lambda_r|^{2n_1 + \ldots + 2n_r} d\mu(g) \quad (6)$$

where $k = \delta\zeta$.

If a permutation of eigenvalues of a matrix $[\delta_1, \ldots, \delta_r]$ permutes only eigenvalues of one of matrices $\delta_j$, it does not change the right-hand side of (6). The set of all such permutations forms a subgroup $\sigma$ of order $n_1! n_2! \ldots n_r!$ of the symmetric group $\mathfrak{S}$ of all permutations of $n$ eigenvalues. Two permutations from the same left coset of the group $\mathfrak{S}$ by the subgroup $\sigma$ give equal expressions (6), and those from different cosets give, in general, nonequal expressions.

Substituting (6) into (4) and using formula (1) from [3] for $\beta(k)$ we obtain the following final formula for the trace of a representation of a degenerate (principal or complementary) series.

$$S(U_x) = \int x(g) \left[ \sum \chi(\delta) \frac{|D(\delta_1)|\ldots|D(\delta_r)|}{|D(\delta)|} |\Lambda_1|^{-n_1} \ldots |\Lambda_r|^{-n_r} \right] d\mu(g) \quad (7)$$

where $\delta = [\delta_1, \delta_2, \ldots, \delta_r]$ is the diagonal matrix formed by eigenvalues of the matrix $g$, and the sum has $n!/n_1! \ldots n_r!$ terms corresponding to cosets of the group $\mathfrak{S}$ by the subgroup $\sigma$.

Two representations of degenerate series are equivalent if and only if their traces coincide. Using formula (7) for the trace one can get a theorem similar to the one form Section 1.

Submitted April 29, 1948

## References


1. Gelfand, I.M., Najmark, M.A.: Dokl. Akad. Nauk SSSR (3)**54** (1946)
2. Gelfand, I.M., Najmark, M.A.: Dokl Akad. Nauk SSSR (1)**56** (1947)
3. Gelfand, I.M., Najmark, M.A.: Dokl Akad. Nauk SSSR (8)**58** (1947)
4. Gelfand, I.M., Najmark, M.A.: Izv Akad. Nauk SSSR, Ser. Mat. (5)**11** (1947)

# 7.

(with M. A. Najmark)

# On the connection between representations of a complex semisimple Lie group and its maximal compact subgroup

Dokl. Akad. Nauk SSSR **63** (1948) 225–228. Zbl. **35**:15
(Submitted by academician I.G. Petrovsky on September 28, 1948)

Any irreducible unitary representation of a complex semisimple Lie group $\mathfrak{G}$ can be considered as a representation of its maximal compact subgroup $\mathfrak{U}$. Then it can be decomposed into finite-dimensional representations of the group $\mathfrak{U}$. In this paper we study the relation between the representations of the group $\mathfrak{G}$ and the subgroup $\mathfrak{U}$. The reasoning will be carried out for the case when $\mathfrak{G}$ is a complex unimodular group and $\mathfrak{U}$ is its subgroup of unitary matrices indicating, however, how the results are formulated in the general case. We use the notation and results of [2].

1. *Unitary subgroup of the group* $\mathfrak{G}$. Denote by $\mathfrak{U}$ the set of all unitary matrices $u$ in the complex unimodular group $\mathfrak{G}$ and by $\Gamma$ the set of all diagonal matrices in $\mathfrak{U}$; $\mathfrak{U}$ is a maximal compact subgroup of the group $\mathfrak{G}$.

I. Each matrix $g \in \mathfrak{G}$ can be presented in the form $g = ku$ where $k \in K$ and $u \in \mathfrak{U}$.

If $g = k_1 u_1$ and $g = k_2 u_2$ then $u_2 = \gamma u_1$ for some $\gamma \in \Gamma$ and therefore the relation $g = ku$ defines $u$ up to an arbitrary left factor $\gamma \in \Gamma$.

This means that:

II. Each right coset $\tilde{z}$ of the group $\mathfrak{G}$ with respect to the subgroup $K$ contains one and only one right coset of the group $\mathfrak{U}$ with respect to the group $\Gamma$.

Denote by $\tilde{u}$ right cosets of the group $\mathfrak{U}$ with respect to the subgroup $\Gamma$ and by $\tilde{\mathfrak{U}}$ the set of all such cosets. By the proposition II we may identify each coset $\tilde{z}$ with the coset $\tilde{u}$ contained in $\tilde{z}$ and thereby identify the spaces $\tilde{Z}$ and $\tilde{\mathfrak{U}}$. The transformation $\tilde{z}' = \tilde{z}\bar{g}$ can then be considered as a transformation $\tilde{u}' = \tilde{u}\bar{g}$ of the corresponding cosets $\tilde{u}$.

2. *Integral relations.* In the space $\tilde{\mathfrak{U}}$ there is a measure $d\mu(\tilde{u})$ invariant under the transformation $\tilde{u} \to \tilde{u}\bar{u}_0$ and after an appropriate normalisation of the invariant measures $d\mu(u)$, $d\mu(\tilde{u})$, $d\mu(\gamma)$, $d\mu(g)$, $d\mu_l(k)$ one has:

$$\int f(u)d\mu(u) = \int d\mu(\tilde{u}) \int f(\gamma u) d\mu(\gamma),$$

$$\int x(g)d\mu(g) = \int d\mu(\tilde{u}) \int x(ku) d\mu_l(k), \tag{1}$$

$$\int f(\tilde{u})d\mu(\tilde{u}) = \int f(\tilde{u}\bar{g}) \frac{\beta(\tilde{u}(\bar{g})}{\beta(ug)} d\mu(\tilde{u}), \qquad \text{i.e. } \frac{d\mu(\tilde{u}\bar{g})}{d\mu(\tilde{u})} = \frac{\beta(\tilde{u}\bar{g})}{\beta(ug)}. \tag{2}$$

If $f(\gamma u) = f(u)$ one can set $f(u) = f(\tilde{u})$. (In the following we make no distinction between functions $f(u)$ and $f(\tilde{u})$.) Relations (1) then imply that, provided $\int d\mu(\gamma) = 1$, one has

$$\int f(u)d\mu(u) = \int f(\tilde{u})d\mu(\tilde{u})$$

and, consequently,

$$\int x(g)d\mu(g) = \int d\mu(u) \int x(ku)d\mu_l(k). \tag{3}$$

The above results are also valid for an arbitrary complex semisimple group $\mathfrak{G}$. The role of the subgroup $K$ is now played by the subgroup generated by the positive root vectors of the infinitesimal group of the group $\mathfrak{G}$ and the role of the group $\mathfrak{U}$ by the maximal compact subgroup and the role of the group $D$ by the maximal commutative Lie subgroup of the group $\mathfrak{G}$ containing a regular element. Finally, $\Gamma = \cap D$ and $\beta(k) = d\mu_l(k)/d\mu_r(k)$.

3. *Principal non-degenerate series of irreducible representations of the group* $\mathfrak{G}$. The principal non-degenerate series of irreducible representations of the group $\mathfrak{G}$ has been described by the authors in [2]. Now we give another description of that series.

Let $\mathfrak{H}_{\tilde{\mathfrak{u}}}$ be the set of all functions $f(\tilde{u})$ such that $\|f\|^2 = \int |f(\tilde{u})|^2 d\mu(\tilde{u}) = \int |f(u)|^2 d\mu(u)$ where $f(\gamma u) = f(u) = f(\tilde{u})$ (see section 2). Then $\mathfrak{H}_{\tilde{\mathfrak{u}}}$ is a Hilbert space.

Let us look for a representation $g \to U_g$ of the group $\mathfrak{G}$ in the form

$$U_g f(u) = \alpha(u, g) f(u\bar{g}), \tag{4}$$

where the function $\alpha(u, g)$ satisfies the relation $\alpha(\gamma u, g) = \alpha(u, g)$ and $u\bar{g}$ is any element of the coset $\tilde{u}\bar{g}$ (it does not matter which one since $f(\gamma u) = f(u)$). The condition $U_{g_1}U_{g_2} = U_{g_1 g_2}$ implies that

$$\alpha(u, g_1)\alpha(u\bar{g}_1, g_2) = \alpha(u, g_1 g_2). \tag{5}$$

It follows from here as well as from the unitarity condition on the operator $U_g$ and from relation (2) that

$$U_g f(u) = \frac{\alpha(ug)}{\alpha(u)} f(u\bar{g}), \tag{6}$$

where

$$\alpha(\zeta\delta u) = \alpha(\delta)\alpha(u), \qquad \alpha(\delta) = \beta^{-1/2}(\delta)\chi(\delta), \tag{7}$$

and $\chi(\delta)$ is a character of the group $D$. This can be written in the form

$$\chi(\delta) = |\lambda_2|^{m_2 + i\rho_2}\lambda_2^{-m_2}|\lambda_3|^{m_3 + i\rho_3}\lambda_3^{-m_3}\ldots|\lambda_n|^{m_n + i\rho_n}\lambda_n^{-m_n}. \tag{8}$$

Here $m_p$ are integers and $\rho_p$ are arbitrary real numbers.

Note that multiplication of a function $f(\tilde{u})$ by a unimodular function $\omega(\tilde{u})$ does not affect the norm $\|f\|$ and takes the function $\alpha(u)$ into $\alpha(u)/\omega(\tilde{u})$.

Consequently, the function $\alpha(u)$ is defined uniquely by the given representation up to a factor $\omega(\tilde{u})$ such that $|\omega(\tilde{u})| = 1$.

If $g = \zeta\delta u$ then $gg^* = \zeta\delta\delta^*\zeta^*$; since $\zeta^* \in Z$, the formula (2, 13) in [2] implies that

$$|\lambda_p|^2 = \Delta_p/\Delta_{p+1}, \qquad p = 1, 2, \ldots, m \tag{9}$$

where $\Delta_p$ is the minor of the matrix $gg^*$ composed of its last $n-p+1$ rows and columns and $\Delta_{n+1}=1$. There is no sense in trying to define arguments of $\lambda_p$ because $g=\zeta\delta u=\zeta\delta\gamma\cdot\gamma^{-1}u$.

For an arbitrary semisimple group $\mathfrak{G}$ the principal non-degenerate series is also given by formulas (6) and (7). The respective subgroups and functions $\beta(k)$ are those defined at the end of section 2. In formulas (6) and (7) $\chi(\delta)$ denotes the character of the group D.

The complementary non-degenerate series of representations is also described by formulas (6) and (7) if, instead of imposing the condition $|\chi(\delta)|=1$, one appropriately defines the function $\chi(\delta)$ and a metric in the space of functions $f(\tilde{u})$.

4. *Spherical functions.*

**Theorem 1.** *Let $U_g$ be a representation of the non-degenerate series of the group $\mathfrak{G}$ corresponding to the character $\chi(\delta)$. Then the necessary and sufficient condition for the existence of a vector in the representation space invariant with respect to all representations $U_u$, $u\in\mathfrak{U}$, is that $\chi(\gamma)=1$, i.e. that $m_2=m_3=\ldots=m_n=0$. If that condition is satisfied there is only one such vector up to a constant factor.*

Indeed, in view of (6) such a vector satisfies the relation

$$\frac{\alpha(uu_0)}{\alpha(u)}f_0(uu_0)=f_0(u),$$

i.e. $\alpha(uu_0)f_0(uu_0)=\alpha(u)f_0(u)$. Therefore, $\alpha(u)f_0(u)=C$ where $C$ is a constant; hence $f_0(u)=C\alpha^{-1}(u)$. Since $f_0(\gamma u)=f_0(u)$ this implies that $\alpha(\gamma u)=\alpha(u)$, i.e. $\alpha(\gamma)\alpha(u)=\alpha(u)$. Thus $\chi(\gamma)=\alpha(\gamma)=1$; $f_0(u)=C\alpha^{-1}(u)$ and the theorem is proved.

If $\alpha(\gamma)=1$ one has $\alpha(u)=\alpha(\tilde{u})$.

Since $|\alpha(\tilde{u})|=1$ then, letting $\omega(\tilde{u})=\alpha(\tilde{u})$ (cf. the end of section 3), one can assume that $\alpha(u)=1$ and $f_0(u)\equiv 1$. With the appropriate normalisation of $d\mu(\tilde{u})$ the function $f_0(u)\equiv 1$ is a normalised vector invariant with respect to $U_u$.

Let

$$\varphi(g)=(U_g f_0, f_0). \tag{10}$$

The irreducibility of the representation $U_g$ implies that $\varphi(g)$ is an elementary positive-definite function (see [3]) and it follows from the definition of $f_0$ that $\varphi(g)$ is constant on two-side consets by $\mathfrak{U}$: $\varphi(ug)=\varphi(gu)=\varphi(g)$.

The function $\varphi(g)$ is called a spherical function of the representation $U_g$.

Each matrix $g$ can be represented in the form $g=u_1\delta u_2$, where $\delta$ is a diagonal matrix with positive diagonal entries. Therefore it is sufficient to find $\varphi(\delta)$. Formulas (10), (1,7) in [2], (6), (7), (8) and (9) and equalities $f_0=1$, $m_2=m_3=\ldots=m_n=0$ imply that for any representation of the principal non-degenerate series one has

$$\varphi(\delta)=\int\alpha(u\delta)d\mu(u)=\int\Delta_2^{i(\rho_2/2)-1}\Delta_3^{i(\rho_3-\rho_2/2)-1}\ldots\Delta_n^{i(\rho_n-\rho_{n-1}/2)-1}d\mu(u), \tag{11}$$

where $\Delta_p$ is the minor of the matrix $u\delta^2u^*$ composed of its last $n-p+1$ rows and columns.

Computing the integral in (11) one has the following theorem.

**Theorem 2.** *Let $U_g$ be a representation of the principal non-degenerate series of the group $\mathfrak{G}$ corresponding to the character $\chi(\delta) = |\lambda_2|^{i\rho_2}|\lambda_3|^{i\rho_3}\dots|\lambda_n|^{i\rho_n}$. Then the corresponding spherical function is defined by the formula*

$$\varphi(\delta) = \left(\frac{2}{i}\right)^{n(n-1)/2} \frac{1}{\prod\limits_{1 \leq p < q \leq n} (\rho_q - \rho_p)(\lambda_q^2 - \lambda_p^2)} \begin{vmatrix} 1 & 1 & \dots & 1 \\ \lambda_1^{i\rho_2} & \lambda_2^{i\rho_2} & \dots & \lambda_n^{i\rho_2} \\ \lambda_1^{i\rho_n} & \lambda_2^{i\rho_n} & \dots & \lambda_n^{i\rho_n} \end{vmatrix}. \qquad (12)$$

Theorem 1 holds without any changes for representations of the principal degenerate series of an arbitrary complex semisimple group. In that case the formula for spherical functions is

$$\varphi(\delta) = c\frac{\sum\limits_s \pm \chi(\delta_s)}{\sum\limits_s \pm \beta^{-1/2}(\delta_s)},$$

where $\delta_s$ denotes the element into which the element $\delta$ goes under the automorphism $s$ of the group $S$. Then sign $\pm$ is determined by the parity of the element $s$, and the scalar $c$ is defined from the condition $\varphi(e) = 1$.

5. *Decomposition into representations of unitary subgroup.* The next theorem is a generalisation of Theorem 1.

**Theorem 3.** *Let $U_g$ be a representation of the non-degenerate series of the group $\mathfrak{G}$ corresponding to the character $\chi(\delta)$. The representation $U_u$ (corresponding to the same character) of the unitary subgroup $\mathfrak{U}$ contains a given irreducible representation $c(u)$ of that subgroup if and only if the representation space of $c(u)$ contains a weight vector of that representation of the weight $\chi(\gamma)$. The number of times the representation $c(u)$ is contained in the representation $U_u$ coincides with the number of linearly independent weight vectors of the representation $c(u)$ of the weight $\chi(\gamma)$.*

Theorem 3 also holds for representations of the non-degenerate series of an arbitrary complex semisimple group.

It follows from Theorem 3 that in the decomposition of the representation $U_u$ the representation $c(u)$ with the lowest highest weight occurs only once, and that weight is equal to $\chi(\gamma)$.

Received September 28, 1948

**References**

1. Gelfand, I.M., Najmark, M.A.: Izv. Akad. Nauk SSSR, Ser. Mat. **11** (1947) 411
2. Gelfand, I.M., Najmark, M.A.: Mat. Sb. (63) **21** (1957) 3, 405
3. Gelfand, I.M., Raikov, D.A.: Mat. Sb. (55) **13** (1943) 301
4. Weil, A.: *L'* integration dans les groupes topologiques et ses applications. Paris 1940

# 8.

(with M. A. Najmark)

# An analogue of the Plancherel formula for the complex unimodular group

Dokl. Akad. Nauk SSSR **63** (1948) 609–612. Zbl. **38**:18

The aim of this note is to prove an analogue of the Plancherel formula for the complex unimodular group of arbitrary order $n$. For $n = 2$ this formula was proved by the authors [1, 2]. However, for $n > 2$ some new difficulties appear that were absent in the case $n = 2$; these difficulties lead to some new additional constructions. As these new difficulties already appear for $n = 3$ and because of lack of space, we consider the case $n = 3$; only the final result will be formulated for arbitrary $n$. Everywhere in this note we freely use notations and results from chapter 1 of [2].

## §1. Some auxiliary functions

Let $x(g)$ be a function on the group $\mathfrak{G}$ satisfying the conditions

$$\int |x(g)|\, d\mu(g) < +\infty. \quad \int |x(g)|^2\, d\mu(g) < +\infty$$

(we will see later that the first condition is superfluous), and let $U_{\chi;g}$ be a principal nondegenerate series representation of the group $\mathfrak{G}$. To $x(g)$ we associate the operator $U_x = \int x(g) U_{\chi;g} d\mu(g)$ with the kernel

$$K(z', z'', \chi) = \int x(z'^{-1}\delta\zeta z'')\beta^{-1/2}(\delta)\chi(\delta) d\mu(\delta) d\mu(\zeta);$$

this kernel is $(2\pi)^{n-1}$ times the Fourier transform (with respect to $\delta$) of the function[1]

$$\varphi(z', z'', \delta) = \beta^{-1/2}(\delta) \int x(z'^{-1}\delta\zeta z'') d\mu(\zeta)$$

Our aim is to express $\int |x(g)|^2 d\mu(g)$ in terms of the function $K(z', z'', \chi)$, or, equivalently, in terms of the function $\varphi(z', z'', \delta)$. To do this we have to construct some auxiliary functions.

Denote by $Z_1, Z_2, Z_1, Z_2$ sets of matrices from $Z$ and $Z$ satisfying the conditions $z_{32} = 0$, $z_{31} = z_{21} = 0$, $\zeta_{23} = 0$, $\zeta_{12} = \zeta_{13} = 0$ respectively. These sets are subgroups of the groups $Z$ and $Z$ respectively.

Each element has a unique decomposition of the form $z = z_1 \cdot z_2$ and a unique decomposition of the form $z = z_2 \cdot z_1$.

A similar statement for $\zeta, \zeta_1$ and $\zeta_2$ is also true.

---

[1] We do not consider here the convergence of integrals. However, this can be done. We can also explicitly indicate which classes of functions correspond to each other under these integrals transformations.

Now let

$$\varphi(z', z'', \delta) = \varphi(z'_2 z'_1, z''_2 z''_1, \delta) = \varphi(z'_1, z''_1, z'_2, z''_2, \delta)$$

and denote by $\Phi(w'_1, w''_1, z'_2, z''_2, \delta)$ the Fourier transform of the function $\varphi$ in $z'_1, z''_1$; here $w_1 = (w_{12}, w_{13})$. Next, let $\omega \in Z_2, \omega_{23} = w_{12}:w_{13}$; define

$$\tilde{\Phi}_1(w'_1, w''_1, z'_2, z''_2, \delta) = \beta^{-1/2}(\delta'\delta'')\Phi_1(w'_1, w''_1, z'''_2, z^{\mathrm{IV}}_2, \delta'^{-1}\delta\delta'')$$

where $z'_2\omega' = \delta'\zeta'_2 z'''_2$, $z''_2\omega'' = \delta''\zeta''_2 z^{\mathrm{IV}}_2$, and denote by $\Phi_2(w', w'', \delta) = \Phi_2(w'_1, w''_1, w'_2, w''_2, \delta)$ the Fourier transform of the function $\tilde{\Phi}_1$ in $z'_2, z''_2$, so that $w = (w_1, w_2)$.

Introduce parameters in the group $\mathfrak{G}$ by the formula $g = z'^{-1}\delta s z''$, $x(g) = x(z'^{-1}\delta s z'') = x(z', z'', \delta) = x(z'_1, z''_1, z'_2, z''_2, \delta)$, where $s$ is the matrix defined by $s_{11} = s_{12} = s_{21} = s_{23} = s_{32} = s_{33} = 0$, $s_{13} = s_{31} = 1$, $s_{22} = -\bar{1}$. Denote by $X_1(w'_1, w''_1, z'_2, z''_2, \delta)$ the Fourier transform of the function $x$ in $z'_1, z''_1$, and let

$$\begin{aligned} &\tilde{X}_1(w'_1, w''_1, z'_2, z''_2, \delta) \\ &\quad = (2\pi)^{-4}\int x(z'^{-1}\omega'^{-1}z'^{-1}_2\delta s z''_2\omega'' z''_1)\exp iR(\bar{w}'_1 z_1 - \bar{w}''_1 z''_1)d\mu(z'_1)d\mu(z''_1). \end{aligned}$$

Substituting into this formula the decomposition $\omega' z'^{-1}_2\delta s z''_2\omega'' = \dot{z}'^{-1}_1\dot{z}'^{-1}_2\dot{\delta} s\dot{z}''_2\dot{z}''_1$ we get

$$\tilde{X}_1(w'_1, \omega''_1, z'_2, z''_2, \delta) = X_1(w'_1, w''_1, \dot{z}'_2, \dot{z}''_2, \delta)\exp iR(\bar{w}''_1\dot{z}''_1 - \bar{w}'_1\dot{z}'_1).$$

Finally, let $X_2(w', w'', \delta) = X_2(w'_1, w''_1, w'_{23}, w''_{23}, \delta)$ be the Fourier transform of the function $\tilde{X}_1$ in $z'_2, z''_2$. Then[2]

$$\begin{aligned} \Phi_2(w', w'', \delta) &= \beta^{-1/2}(\delta)\int X_2(w', w'', \delta^{(5)}) \\ &\quad\times \exp iR(\bar{w}''_{12}z^{(6)}_{21} - \bar{w}'_{12}z^{(5)}_{21} + \bar{w}''_{23}z^{(6)}_{32} - \bar{w}_{23}z^{(5)}_{32})d\mu(\zeta_1)\cdot d\mu(\zeta_2) \end{aligned}$$

where $\delta\zeta = z^{(5)-1}\delta s z^{(6)}$.

The transformation of $X_2$ to $\Phi_2$ defined by this formula can be decomposed into the product of pointwise transformations and Fourier transformations as follows. Let $s^{-1}\zeta = \zeta^0\delta^0 z^0$; then $\delta\zeta = \delta\cdot s\zeta^0 s^{-1}\cdot\delta^{-1}\cdot\delta\cdot s\delta^0 s^{-1}\cdot s z^0$ so that $z^{(5)} = \delta\cdot s\delta^0 s^{-1}\cdot\delta^{-1}$, $z^{(6)} = z^0$, $\delta^{(5)} = \delta\cdot s\delta^0 s^{-1}$.

Therefore, denoting by $\lambda_1, \lambda_2, \lambda_3$ the diagonal entries of the matrix $\delta$, we have (see (2.13), (2.14b) and (2.15) in [2]):

$$\begin{aligned} &z^{(5)}_{21} = -\lambda_2\lambda_1^{-1}\zeta_{23}\zeta_{13}^{-1},\ z^{(5)}_{23} = -\lambda_3\lambda_2^{-1}\zeta_{12}\Delta^{-1},\ z^{(6)}_{21} = \zeta_{23}\Delta^{-1}, \\ &z^{(6)}_{32} = \zeta_{12}\zeta_{13}^{-1},\ \lambda^{(5)}_1 = \lambda_1\zeta_{13},\ \lambda^{(5)}_2 = \lambda_2\Delta\zeta_{13}^{-1},\ \lambda^{(5)}_3 = \lambda_3\Delta^{-1}, \end{aligned}$$

where $\Delta = \begin{vmatrix} \zeta_{12} & \zeta_{13} \\ 1 & \zeta_{23} \end{vmatrix}$.

Choose $\lambda_1$ and $\lambda_2$ as parameters in $\delta$, so that $X_2(w', w'', \delta) = X_2(w', w'', \lambda_1, \lambda_2)$ and change variables in the formula for the relation between $\Phi_2$ and $X_2$ from $\zeta_{12}$, $\zeta_{13}$ to $\xi_1, \xi_2$ where $\lambda_1\zeta_{13} = \xi_1^{-1}, \lambda_2\Delta\zeta_{13}^{-1} = \xi_2^{-1}$. Then we have

$$\begin{aligned} \Phi(w', w'', \delta) = \int X_2(w', w'', \xi_1^{-1}, \xi_2^{-1})\exp iR\{&\bar{w}'_{12}\lambda_2\xi_1\zeta_{23} + \bar{w}''_{12}\lambda_1\lambda_2\xi_1\xi_2\zeta_{23} \\ &+ \bar{w}'_{23}\lambda_3\xi_2\zeta_{23}^{-1}(\lambda_2^{-1}\xi_2^{-1} + 1) + \bar{w}''_{23}(\lambda_2\xi_2^{-1} + 1)\zeta_{23}^{-1}\}|\xi_1|^{-6}|\xi_2|^{-4} \\ &\cdot d\mu(\xi_1)d\mu(\xi_2)d\mu(\zeta_{23}). \end{aligned}$$

[2] The notation $d\mu(\zeta_1)\cdot d\mu(\zeta_2)$ means that we have first to integrate in $\zeta_1$ and then in $\zeta_2$.

Let[3]

$$(2\pi)^{-1}\int X_2(w', w'', \xi_1^{-1}, \xi_2^{-1})|\xi_1|^{-6}\exp iR(\xi_1 u)d\mu(\xi_1) = X_3(w', w'', u, \xi_2).$$

Then

$$\Phi_2(w', w'', \delta) = 2\pi \int X_3(w', \omega'', (\bar{w}'_{12}\lambda_2 + \bar{w}''_{12}\lambda_1\lambda_2\xi_2)\zeta_{23}, \xi_2)$$
$$\cdot \exp iR\{\zeta_{23}^{-1}(\bar{w}'_{23}\lambda_3\xi_2 + \bar{w}''_{23})(1 + \lambda_2^{-1}\xi_2^{-1})\}|\xi_2|^{-4}|\zeta_{23}|^{-2}d\mu(\xi_2)d\mu(\zeta_{23}).$$

After the change of variables $\zeta'_{23} = \zeta_{23}(\lambda_2\bar{w}'_{12} + \lambda_1\lambda_2\xi_2\bar{w}''_{12})$ and then $\zeta'^{-1}_{23} = \alpha$, $\xi_2\zeta'^{-1}_{23} = \beta$ we have

$$\Phi_2(w', w'', \delta) = (2\pi)^{-2}\int X_4(w', \omega'', \alpha, \beta)\exp iR(\bar{\sigma}_1\alpha + \bar{\sigma}_2\beta)d\mu(\alpha)d\mu(\beta) \qquad (1)$$

where

$$X_4(w', w'', \alpha, \beta) = (2\pi)^3 X_3(w', w'', \alpha, \beta)\exp iR(\bar{w}''_{12}\bar{w}'_{23}\beta^2\alpha^{-1} + \bar{w}'_{12}w''_{23}\alpha^2\beta^{-1})|\beta|^{-4},$$
$$\sigma_1 = \bar{\lambda}_1 w'_{12}w''_{23} + \bar{\lambda}_2 w'_{12}w''_{23} + \bar{\lambda}_3 w'_{13}w''_{23},$$
$$\sigma_2 = \bar{\lambda}_1\bar{\lambda}_2 w''_{12}w''_{23} + \bar{\lambda}_1\bar{\lambda}_3 w''_{12}w'_{23} + \bar{\lambda}_2\bar{\lambda}_3 w'_{13}w'_{23}.$$

Formula (1) shows that $\Phi_3$ depends only on $w', w'', \sigma_1$ and $\sigma_2$, so that we can set $\Phi_2 = \Phi_3(w', w'', \sigma_1, \sigma_2)$. Formula (1) is the analogue of the main formula (123) in [1].

## §2. The proof of the analogue of the Plancherel formula

According to formulas (3.5) and (3.20) in [2], we have

$$\int |x(g)|^2 d\mu(g) = \int |x(s^{-1}g)|^2 d\mu(g) = \int |x(s^{-1}\zeta\delta z)|^2 \beta(\delta)d\mu(\zeta)d\mu(z)d\mu(\delta).$$

Writing $s^{-1}\zeta s = z'^{-1}, s^{-1}\delta s = \delta'$, i.e. $\lambda'_1 = \lambda_3$, $\lambda'_2 = \lambda_2$, and $x(z'^{-1}\delta' sz) = x(z'_1, z_1, z'_2, z_2, \lambda'_1, \lambda'_2)$ for $z = z_2 z_1, z' = z'_2 z'_1$ we get

$$\int |x(g)|^2 d\mu(g)| = \int x(z'_1, z_1, z'_2, z_2, \lambda_1, \lambda_2)|^2|\lambda_1|^6$$
$$\cdot|\lambda_2|^2 d\mu(z'_1)d\mu(z_1)d\mu(z'_2)d\mu(z_2)d\mu(\lambda_1)d\mu(\lambda_2)$$
$$= \int |X_1(w'_1 w''_1, z'_2, z''_2, \lambda_1, \lambda_2)|^2|\lambda_1|^6|\lambda_2|^2$$
$$\cdot d\mu(w'_1 d\mu(w''_1)d\mu(z'_2)d\mu(z''_2)d\mu(\lambda_1)d\mu(\lambda_2)$$

where $d\mu(w_1) = d\mu(w_{12})d\mu(w_{23})$. Passing in the last integral successively to functions $\tilde{X}_1, X_2, X_3, X_4$ and to $\Phi_2, \tilde{\Phi}_1, \Phi_1, \varphi$ by means of formulas from §1, we get

$$\int |x(g)|^2 d\mu(g) = \frac{1}{6\pi^6}\int L\varphi\cdot\overline{\varphi(z', z'', \delta)}d\mu(\lambda_2)d\mu(\lambda_3)d\mu(z')d\mu(z'')$$

where

$$L = \frac{\partial}{\partial\lambda_2}\frac{\partial}{\partial\lambda_3}\left(\lambda_2\frac{\partial}{\partial\lambda_2} - \lambda_3\frac{\partial}{\partial\lambda_3}\right)\frac{\partial}{\partial\bar{\lambda}_2}\frac{\partial}{\partial\bar{\lambda}_3}\left(\bar{\lambda}_2\frac{\partial}{\partial\bar{\lambda}_2} - \bar{\lambda}_3\frac{\partial}{\partial\bar{\lambda}_3}\right).$$

Passing now from the function $\varphi$ to its multiplicative Fourier transform

---

[3] $d\mu(w)$ for $w = u + iv$ always means $du \cdot dv$.

$K(z', z'', \chi)$ in $\delta$ we obtain the formula that represents an analogue of the Plancherel formula for the group $\mathfrak{G}$ in the case $n = 3$:

$$\int |x(g)|^2 d\mu(g) = \frac{1}{6(2\pi)^{10}} \sum_{m_1=-\infty}^{\infty} \sum_{m_2=-\infty}^{\infty} \int_{-\infty}^{\infty} \int_{-\infty}^{\infty} (m_2^2 + \rho_2^2)(m_3^2 + \rho_3^2)$$
$$\cdot [(m_2 - m_3)^2 + (\rho_2 - \rho_3)^2] d\rho_2 \, d\rho_3 \int |K(z', z'', m_2, \rho_2, m_3, \rho_3)|^2$$
$$\cdot d\mu(z') d\mu(z'') \tag{2}$$

The above proof can be extended to the unimodular group $\mathfrak{G}$ of arbitrary order $n$. As a result, we get the following theorem.

**Theorem.** *Let* $x(g)$ *be an arbitrary function satisfying the condition* $\int |x(g)|^2 d\mu(g) < \infty$. *Then*

$$\int |x(g)|^2 d\mu(g) = \frac{1}{n!(2\pi)^{(n-1)(n+2)}}$$
$$\sum_{m_1, \ldots, m_n = -\infty}^{\infty} \int_{-\infty}^{+\infty} \cdots \int \left[\int |K(z', z'', \chi)|^2 d\mu(z') d\mu(z'')\right] a(\chi) d\rho_2 \ldots d\rho_n \tag{3}$$

*where*

$$K(z', z'', \chi) = \int x(z'^{-1} \delta \zeta z'') \beta^{-1/2}(\delta) \chi(\delta) d\mu(\delta) d\mu(\zeta)$$
$$a(\chi) = \prod_{1 \leq p < q \leq n} [(n_p - n_q)^2 + (\rho_p - \rho_q)^2], \qquad n_1 = \rho_1 = 0 \tag{4}$$

*and the integral in (4) is convergent in the metric determined by the right-hand side of (3).*

This theorem enables us to invert formula (4) and to decompose the regular representation of the group $\mathfrak{G}$ into the representations of the nondegenerate principal series, in a similar fashion to [1], §7 in the case $n = 2$.

If $\int |x(g)| d\mu(g) < +\infty$ then $K(z', z'', \chi)$ is the kernel of the operator $U_{\chi;x}$ of the group ring representation, and the integral $\int |K(z', z'', \chi)|^2 d\mu(z') d\mu(z'')$ in (3) is the trace of the operator $U_{\chi;x^*x}$ (see [3]).

Submitted October 2, 1948

**References**


1. Gelfand, I.M., Najmark, M.A.: Izv. Akad. Naŭk SSSR, Ser. Mat. **11** (1947) 411–504
2. Gelfand, I.M., Najmark, M.A.: Mat. Sb. (63) **21** (1946) 2 405–434
3. Gelfand, I.M., Najmark, M.A.: Dokl. Akad. Naŭk SSSR (1) **61** (1948) 9–11.

# 9.

(with A. M. Yaglom)

# General relativistically invariant equations and infinite-dimensional representations of the Lorentz group

Zh. Ehksp. Teor. Fiz. **18** (1948) 703–733

*Abstract.* The paper presents formulas describing the general form of relativistically invariant equations of order 1 with respect to the wave function $\psi$ which has either finitely or infinitely many components. It also reveals when such an equation can be obtained from an invariant Lagrange function. Possible spin values of particles described by the equations are considered, as well as the corresponding mass spectrum. The paper also contains examples of finite-dimensional and infinite-dimensional invariant equations including those infinite dimensional equations that correspond to positive values of charge and energy[1].

## 1. Introduction

1. In this paper we study general relativistically invariant equations of the form[2]:

$$L^k(\partial\psi/\partial x^k) + i\kappa\psi = 0 \tag{1.1}$$

where the wave function $\psi(x^0, x^1, x^2, x^3) = \psi(ct, x, y, z)$ has either finitely or infinitely many components (in other words, the function $\psi$ takes values in either finite-dimensional or infinite-dimensional space $R$), $L^0, L^1, L^2, L^3$ are matrices defining a linear transformation of $\psi$ and $\kappa$ is a real scalar different from zero[3]. We assume here that under the Lorentz transformation

$$x^i \to x'^i = l^i_k x^k \tag{1.2}$$

the components of $\psi$ also undergo a linear transformation

$$\psi \to \psi' = S\psi \tag{1.3}$$

and the set of the operators $S$ forms a representation $D_R$ of the Lorentz group acting in $R$. Equation (1.1) is called relativistically invariant if it does not change

---

[1] A short account of the results of this paper has been published in Dokl. Akad. Naūk SSSR **59** (1948) 655.

[2] As usual a repeated index implies summation over all values of that index (i.e. 0, 1, 2, 3).

[3] One can easily see that systems of equations of higher order can also be reduced to the form (1.1) provided that the number of equations coincides with the number of unknown functions; a simple example of such a reduction can be found in [1]. It is also clear that $\kappa$ in (1.1) can be considered as a non-singular matrix (i.e. a matrix that has an inverse).

under the Lorentz transformation (1.2) of the coordinates $x^i$ and the simultaneous transformation (1.3) of $\psi$.

The simplest finite-dimensional relativistically invariant equation of the form (1.1) (the adjective "finite-dimensional" here means that the corresponding function $\psi$ has a finite number of components) is the famous Dirac equation. Other examples are proved by the equations describing particles with spin 0 and 1 and written in that form by Duffin in [1], and the Pauli–Fierz equations [2] for particles with spin greater than one[4]. Many other relativistically invariant equations studied by different authors in recent years (see, e.g. [3–7]) are also of the same form. General finite-dimensional relativistically invariant equations of the form (1.1) have been considered in [8–11]. Recently, some examples of relativistically invariant equations with respect to the function $\psi$ having infinitely many components have also been suggested (see [12–15])[5].

In this paper we consider equation (1.1) in the general form which includes both finite-dimensional and infinite-dimensional cases. Besides the question of the general form of equations invariant with respect to the proper Lorentz group, we also consider here conditions for the equation (1.1) to be invariant with respect to the full Lorentz group, which includes reflections and conditions under which such an equation can be obtained from an invariant Lagrange function. In the last sections of the paper, the problem of mass and spin spectrum corresponding to an invariant equation is considered and there are examples of equations with positive charge and energy density. We make no attempt here to discuss any possibility of applying individual equations for the description of some real observed particles. We believe that the experimental data available on elementary particles present a good opportunity for investigating all possible ways of describing such particles (see, especially, [16]) but can in no way be considered sufficient for associating to individual particles (with the exception of the electron) concrete equations (see also our comments on the Bhabha equation for the proton following formulas (7.9) and (7.9′) of this paper).

2. Substituting the transformed values of $\psi$ and $x^i$ into equation (1.1) one obtains the following condition which is both necessary and sufficient for the equation to be relativistically invariant:

$$L^i = l^i_j S L^j S^{-1}. \tag{1.4}$$

If this condition is satisfied for infinitesimal Lorentz transformations then condition (1.4) is satisfied for all transformations of the proper Lorentz group. Suppose now that to each infinitesimal Lorentz transformation

$$x^i \to x'^i = x^i + \varepsilon^i_k x^k = x^i + g^{ij}\varepsilon_{jk} x^k \tag{1.5}$$

[4] The statement in [8, 10] that Pauli–Fierz equations cannot be represented in the form (1.1) is based on a misunderstanding (see also footnote[26] to the present paper).

[5] Some of those papers consider $\psi$ as a vector having infinitely many components, others treat it as a function depending on some auxiliary variables and satisfying some integrability conditions. Evidently, the second approach can always be reduced to the first one; components of a vector in that case are coefficients in the decomposition of our function with respect to some complete system of functions.

where $g^{00} = -g^{11} = -g^{22} = -g^{33} = 1$,

$$g^{ij} = 0 \quad (i \neq j), \qquad \varepsilon_{jk} = -\varepsilon_{kj}$$

there corresponds the following transformation of $\psi$:

$$\psi \to \psi' = \psi' + (\tfrac{1}{2})\varepsilon^i_k I^k_i \psi = \psi + (\tfrac{1}{2})\varepsilon_{ik} I^{ki} \psi. \tag{1.6}$$

Six linear operators $I^{ki} = -I^{ik}$ $(i, k = 0, 1, 2, 3)$ (infinitesimal operators of the representation) satisfying the commutation relations

$$[I^{ik}, I^{jl}] = I^{ik} I^{jl} - I^{jl} I^{ik} = -g^{il} I^{kl} + g^{il} I^{kj} + g^{kj} I^{il} - g^{k} I^{ij} \tag{1.7}$$

completely define the representation $D_R$. Substituting formulas (1.5) and (1.6) into (1.4) one has

$$[L^i, I^{jk}] = g^{ij} L^k - g^{ik} L^j \qquad (i, j, k = 0, 1, 2, 3). \tag{1.8}$$

Thus the problem of finding all equations invariant with respect to the proper Lorentz group is reduced to finding all representations of the Lorentz group $I^{jk}$ and of all operators $L^i$ satisfying conditions (1.7) and (1.8).

3. In the following we assume that the representation $D_R$ of the Lorentz group is decomposed into irreducible representations. It is known that all finite-dimensional irreducible representations of the proper Lorentz group are either tensorial or spinorial representations; infinite-dimensional irreducible representations of that group have recently been found by M.A. Najmark together with one of the authors of the present paper [17, 18][6]. Instead of giving here formulas from [17, 18] which describe the representations we immediately present in Section 2 formulas for the corresponding infinitesimal operators $I^{jk}$ (see formulas (2.2) and (2.3) below). In that way, a unified description of both infinite-dimensional and finite-dimensional irreducible representations is achieved. For that reason we do not use here the usual spinorial notation since for our purposes the notation used here is much more convenient than the spinorial one.

In Section 3 formulas for irreducible representations given in Section 2 are used to find all equations invariant under transformations of the proper Lorentz group. The main result of this section is given by formulas (3.13) and (3.14) which define the matrix $L^0$ for general relativistically invariant equations of the form (1.1) (relation (3.1) shows that the operators $L^k$, $k = 1, 2, 3$, can be found from $L^0$). In particular, these formulas can be used in the finite-dimensional case, replacing formulas (66), (67), (47) and (48) in [8], which are more complicated and less convenient for the investigation of corresponding equations.

Section 4 deals with equations invariant with respect to the full Lorentz group. Firstly, all finite-dimensional and infinite-dimensional representations of the Lorentz group are found in Section 4.1. The results obtained there (see formulas

[6] The papers [17, 18] list all irreducible unitary representations of the Lorentz group. By allowing $\rho$ in the formulas there to assume complex values one obtains non-unitary infinite-dimensional irreducible representations as well (see [18], p. 504 and also footnote[11] to the present paper). Irreducible infinite-dimensional representations of the Lorentz group have also been considered in [19] and [15].

(4.15) and (4.21)) are quite similar to those known for the finite-dimensional case (cf. [20], sections 88–89 and 162). Then, in Section 4.2 we establish conditions under which an equation invariant with respect to the proper Lorentz group is also invariant under reflections (see formulas (4.24)).

Section 5 studies equations that can be obtained from invariant Lagrange functions. Equations of this kind are especially interesting since there is a general procedure for them which associates to each equation physical quantities in an invariant way (charge, energy, momentum, etc.) satisfying conservation laws and thus providing a physical interpretation for the whole theory[7].

Conditions under which there exists an invariant Lagrange function are of two types: (a) conditions on the transformation law for $\psi$ (i.e. on the representation $D_R$), and (b) conditions that should be satisfied by the matrices $L^k$. These two types are as follows:

(a) There should exist an invariant hermitian bilinear form[8] $(\psi_1, \psi_2)$ and that form is required to be non-singular. Only under that condition can one construct scalars, vectors, etc. that are invariantly related to $\psi$. For such a form to exist the representation $D_R$ should satisfy some conditions formulated in Section 5.1. It turns out that these conditions are automatically satisfied for all finite-dimensional representations of the full Lorentz group (though not for all infinite-dimensional representations).

(b) Equation (1.1) can be obtained from an invariant Lagrange function if and only if an invariant bilinear form $(\psi_1, \psi_2)$ satisfies the condition

$$(L^0\psi_1, \psi_2) = (\psi_1, L^0\psi_2) \tag{1.9}$$

for all $\psi_1, \psi_2$, i.e. $(L^0\psi, \psi)$ is required to be real (see Section 5, formula (5.30) and below). The Lagrange function itself is then of the form

$$\mathscr{L} = (1/2i)\{(L^k\partial\psi/\partial x^k, \psi) - (\psi, L^k\partial\psi/\partial x^k) + \kappa(\psi, \psi). \tag{1.10}$$

It follows hence in the usual manner that to each solution $\psi$ of the equation (1.1) there corresponds the current vector:

$$s^k = (\varepsilon/\hbar)(L^k\psi, \psi) \tag{1.11}$$

and the energy-momentum tensor:

$$T_j^k = (1/2i)\{(L^k\partial\psi/\partial x^j, \psi) - (\psi, L^k\partial\psi/\partial x^j). \tag{1.12}$$

---

[7] In the general case no physical quantities can be associated to a relativistically invariant equation of the form (1.1). For this reason any investigation of invariant equations that does not analyze whether it is possible to obtain the equation from an invariant Lagrange function seems to be of little physical interest.

[8] This invariant form is often denoted in articles on invariant equations by $\psi_2^+\psi_1$. Such a notation, in our view, has the defect of introducing a new symbol $\psi^+$ with the sole function of appearing as a left factor before $\psi$. Another widely used notation for this form is $\psi_2^* A\psi_1$ where $\psi_2^*$ is a "row" which is a Hermitian conjugate to the "column" $\psi_2$ and $A$ is a square Hermitian matrix. However, using this notation one has to remember all the time that under the Lorentz transformation the matrix $A$ transforms differently from the matrices $L^i$, $I^{jk}$, etc. We have chosen here to denote an invariant bilinear form by $(\psi_1, \psi_2)$ in order to stress its analogy with the scalar product of vectors.

At the end of Section 5 we establish which operators $L^0$ satisfy condition (1.9) (see (5.32), (5.33)).

In Section 6 we consider the problem of finding rest masses and spins of particles described by equation (1.1). It turns out that if $D_R$ splits into a finite number of irreducible representations, some of which are infinite-dimensional, then equation (1.1) usually corresponds to an infinite sequence of increasing spin values and the corresponding mass spectrum tends to zero with increasing spin. In exceptional cases, all sufficiently large values of spin may correspond to one and the same value of mass; what is impossible is an unlimited increase in mass with the increase in spin. We also show there that for equations obtained from Lagrange functions complex values of mass always correspond to solutions $\psi$ for which charge, energy and momentum vanish.

Finally, in Section 7 we consider some examples of relativistically invariant equations with the special emphasis on equations having positive charge or positive energy values. Considering finite-dimensional equations satisfying this condition we come to the following very simple equations: the Dirac equation describing a particle of spin 1/2, the Duffin equation for particles with spin 0 and 1, and the Pauli–Fierz equation which describe a particle of spin 3/2. Among our examples of infinite-dimensional equations there are two very simple equations which deserve special mention: one corresponds to integer spin values and another to half-integer spin values and both have positive densities of both charge and energy. They show that the general Pauli theorem [21], which plays an important role in the theory of finite-dimensional equations, is no longer valid in the infinite-dimensional case. They also show that the infinite-dimensional representation provides a possibility for describing in the existing framework particles which have charges of one and the same sign.

Other examples include equations that describe particles with half-integer values of spin, definite energy and indefinite charge, as well as particles with integer spin, definite charge and indefinite energy, i.e. with properties opposite to those peculiar to finite-dimensional equations. More examples of the latter have been given in the recent work by Harish-Chandra [15].

## 2. Irreducible representations of the proper Lorentz group

In this section we give formulas describing irreducible representations (both finite-dimensional and infinite-dimensional) of the proper Lorentz group.

Each irreducible representation of that group is defined by a pair of numbers $(k_0, k_1)$ where $k_0$ is either integer or half-integer[9] and $k_1$ is an arbitrary complex number. Denote by $R_{(k_0,k_1)}$ the space of that representation. Coordinate vectors in $R_{(k_0,k_1)}$ will be denoted by $\xi_p^k$ where both indices $k$ and $p$ are simultaneously either integers or half-integers.

For each $k$, index $p$ can assume the following $(2k+1)$ values:

$$p = k, k-1, k-2, \ldots, -k+1, -k.$$

---

[9] As usual, by half-integer we mean 1/2 of an odd integer.

Index $k$ runs over the following infinite number of values:

$$k = |k_0|, |k_0| + 1, |k_0| + 2, \ldots$$

so that the space $R_{(k_0,k_1)}$ is, in general, infinite-dimensional. The exception is the case when $2k_1$ is an integer of the same parity as $2k_0$ and $|k_1| > |k_0|$; then $k$ assumes only a finite number of values:

$$k = |k_0|, |k_0| + 1, \ldots, |k_1| - 1$$

and the space $R_{(k_0,k_1)}$ is finite-dimensional.

It is more convenient to consider instead of $I^{jk}$ the following six operators:

$$\begin{aligned} &H^0 = I^{32}, \quad H^+ = I^{21} + iI^{13}, \quad H^- = I^{21} - iI^{13}, \\ &F^0 = I^{01}, \quad F^+ = I^{03} + iI^{02}, \quad F^- = I^{03} - iI^{02}. \end{aligned} \tag{2.1}$$

The irreducible representation defined by the pair of numbers $(k_0, k_1)$ is given by the formulas:

$$\left.\begin{aligned} &H^0 \xi_p^k = ip\xi_p^k, \\ &H^+ \xi_p^k = \sqrt{(k+p)(k-p+1)}\, \xi_{p-1}^k, \\ &H^- \xi_p^k = -\sqrt{(k+p+1)(k-p)}\, \xi_{p+1}^k; \end{aligned}\right\} \tag{2.2'}$$

$$\left.\begin{aligned} 2F^0 \xi_p^k &= -\sqrt{(k+p)(k-p)}\, B_k \xi_p^{k-1} + pA_k \xi_p^k + \sqrt{(k+p+1)(k-p+1)}\, B_{k+1} \xi_p^{k+1}, \\ 2iF^+ \xi_p^k &= \sqrt{(k+p)(k+p-1)}\, B_k \xi_{p-1}^{k-1} + \sqrt{(k+p)(k-p+1)}\, A_k \xi_{p-1}^k \\ &\quad + \sqrt{(k-p+1)(k-p+2)}\, B_{k+1} \xi_{p-1}^{k+1}, \\ 2iF^- \xi_p^k &= \sqrt{(k-p)(k-p-1)}\, B_k \xi_{p+1}^{k-1} - \sqrt{(k+p+1)(k-p)}\, A_k \xi_{p+1}^k \\ &\quad + \sqrt{(k+p+1)(k+p+2)}\, B_{k+1} \xi_{p+1}^{k+1}; \end{aligned}\right\} \tag{2.2''}$$

where[10]

$$A_k = 2k_0 k_1 / k(k+1), \tag{2.3'}$$

$$B_k = \sqrt{(k^2 - k_0^2)(k^2 - k_1^2)/k^2(k^2 - 1/4)}. \tag{2.3''}$$

[10] The sign of the square root in (2.3″) can be chosen in any way. Changing the signs of some of the coefficients $B_k$ is equivalent to changing the signs of some of the coordinate vectors $\xi_p^k$. Considering a more general situation, one can choose another set of coordinate vectors $\xi_p'^k = \alpha(k)\xi_p^k$ where $\alpha(k)$ is a scalar coefficient depending on $k$. Then one can retain relations (2.2′) while replacing (2.2″) by

$$\begin{aligned} 2F^0 \xi_p^k &= -\sqrt{(k+p)(k-p)}\, C_k \xi_p^{k-1} + pA_k \xi_p^k + \sqrt{(k+p+1)(k-p+1)}\, D_{k+1} \xi_p^{k+1} \\ 2iF^+ \xi_p^k &= \sqrt{(k+p)(k+p-1)}\, C_k \xi_{p-1}^{k-1} + \sqrt{(k+p)(k-p+1)}\, A_k \xi_{p-1}^k \\ &\quad + \sqrt{(k-p+1)(k-p+2)}\, D_{k+1} \xi_{p-1}^{k+1}, \\ 2iF^- \xi_p^k &= \sqrt{(k-p)(k-p-1)}\, C_k \xi_{p+1}^{k-1} - \sqrt{(k+p+1)(k-p)}\, A_k \xi_{p+1}^k \\ &\quad + \sqrt{(k+p+1)(k+p+2)}\, D_{k+1} \xi_{p+1}^{k+1} \end{aligned} \tag{2.2'''}$$

where $A_k$ is defined from (2.3′) and $C_k$ and $D_k$ can take any values such that

$$C_k D_k = (k^2 - k_0^2)(k^2 - k_1^2)/k^2(k^2 - 1/4). \tag{2.3'''}$$

Formulas (2.3′) and (2.3″) involve either products or squares of $k_0$ and $k_1$. Therefore, the pairs $(k_0, k_1)$ and $(-k_0, -k_1)$ define the same representation. In all other cases the representations defined by $(k_0, k_1)$ and $(k_0', k_1')$ are not equivalent. One gets all unitary irreducible representations of the proper Lorentz group by either taking $k_1$ to be purely imaginary or by taking $k_0 = 0$, $k_1$ real and $0 < |k_1| \leqq 1$ (see footnote[21] to this paper). Then for $k_1$ purely imaginary one has representations of the principal series (of the first class) of unitary irreducible representations (see [17, 18]) and for $k_0 = 0$, $k_1$ real and $0 < |k_1| \leqq 1$ one has representations of the complementary series (of the second class); parameters $k_0$ and $k_1$ in this paper are related to the parameters $n$, $\rho$, defining irreducible unitary representations in [17, 18] in the following way[11]:

$$n = 2k_0, \quad \rho = 2ik_1 \text{ for } k_1 \text{ purely imaginary,}$$

$n = 2k_0$, $\rho = 2k_1$ for $k_0 = 0$, $k_1$ real, $0 < |k_1| \leqq 1$.

All irreducible finite-dimensional representations of the Lorentz group are also covered by formulas (2.2) and (2.3): one obtains these if $2k_1$ is an integer of the same parity as $2k_0$ and $|k_1| > |k_0|$. The parameters $k_0, k_1$ are related to the parameters $p$ (the number of unpointed indices of a symmetric spinor) and $q$ (the number of pointed indices) which are generally used to describe finite-dimensional irreducible representations of the Lorentz group in the following way:

$$p = |k_1 + k_0| - 1, \qquad q = |k_1 - k_0| - 1. \tag{2.5}$$

Note that formulas (2.3′) and (2.3″) are symmetric with respect to $k_0$ and $k_1$. Thus for any integer or half-integer $k_1$, representations corresponding to the pairs $(k_0, k_1)$ and $(k_1, k_0)$ are given by the same formulas and differ only in the supply of vectors $\xi_p^k$ ($k = |k_0|, |k_0| + 1, \ldots$ in the first case and $k = |k_1|, |k_1| + 1, \ldots$ in the second case). In particular, if $2k_0$ and $2k_1$ are of the same parity and $|k_1| > |k_0|$ then the pair $(k_0, k_1)$ defines a finite-dimensional representation and $(k_1, k_0)$ defines

---

[11] In other words for $k_1$ purely imaginary, $k_0$ real and $0 < |k_1| \leqq 1$ our representation is equivalent to the representation acting in the space of complex functions $f(z)$ in one complex variable $z$ satisfying some integrability conditions. Under that representation, each Lorentz transformation defined by the complex matrix $\begin{pmatrix} \alpha & \beta \\ \gamma & \delta \end{pmatrix}$ of order 2, where $\alpha\delta - \beta\gamma = 1$, goes into the following operator acting in the space of functions $f(z)$:

$$f(z) \to f\left(\frac{\alpha z + \gamma}{\beta z + \delta}\right) |\beta z + \delta|^{22(k_0 - k_1 - 1)} (\beta z + \delta)^{-2k_0}$$

$$= f\left(\frac{\alpha z + \gamma}{\beta z + \delta}\right) (\beta z + \delta)^{-(k_1 + k_0 + 1)} \overline{(\beta z + \delta)}^{-(k_1 - k_0 + 1)}. \tag{2.4}$$

The same formula defines a representation equivalent to the representation corresponding to the pair $(k_0, k_1)$ for all other values of the parameters $k_0$, $k_1$ as well. The only thing that is different is the supply of functions composing the space $R_{(k_0, k_1)}$. It is different for different pairs $(k_0, k_1)$ (cf. [18], p. 504). In particular, if $2k_1$ is an integer of the same parity as $2k_0$ and $|k_1| > |k_0|$ then the space $R_{(k_0, k_1)}$ is that of homogeneous polynomials of degree $|k_1 + k_0| - 1$ with respect to $z$ and of degree $|k_1 - k_0| - 1$ with respect to $\bar{z}$; formula (2.4) then gives a convenient way of writing down finite-dimensional representations.

an infinite-dimensional one given by the same formulas and acting in the space spanned by the vectors $\xi_p^k$ where $k = |k_1|, |k_1|+1, \ldots$ (the "tail" of the finite-dimensional representation).

Infinitesimal space rotations are evidently given by operators $I^{32}$, $I^{21}$ and $I^{13}$, or, alternatively, by $H^0$, $H^+$ and $H^-$. Formulas (2.2′) show that under our representations space rotations correspond to operators taking each coordinate vector $\xi_p^k$ into a linear combination of coordinate vectors with the same index $k$. Thus each irreducible representation of the Lorentz group, considered as a representation of the group of space rotations, splits into a direct sum of finite-dimensional irreducible representations that act in the subspaces consisting of linear combinations of coordinate vectors $\xi_p^k$ having one and the same value of the index $k$. Formulas (2.2′) then coincide with usual formulas for the $(2k+1)$-dimensional irreducible representation of the rotation group[12].

We remind the reader that irreducible representations of the rotation group are single-valued for integer $k$ and two-valued for half-integer $k$. This implies that our irreducible representations of the Lorentz group are single-valued for integer $k_0$ and two-valued for half-integer $k_0$. This follows immediately from the formula (2.4) defining the representation operator corresponding to the matrix $\begin{pmatrix} \alpha & \beta \\ \gamma & \delta \end{pmatrix}$.

## 3. Equations invariant with respect to the proper Lorentz group

Let us now proceed to find equations of the form (1.1) invariant under transformations of the proper Lorentz group. We have already seen (see formula (1.8) and below) that the problem is reduced to finding four matrices $L^i$ $(i = 0, 1, 2, 3)$ satisfying relations (1.8). Those relation imply, in particular, that

$$L^k = [L^0, I^{0k}] \quad (k = 1, 2, 3) \tag{3.1}$$

and, consequently, the operator $L^0$ (together with the infinitesimal operators $I^{jk}$, i.e. the transformation law for $\psi$) uniquely define all other operators $L^k$. For that reason, we shall be looking in the following only for the matrix $L^0$. Note that in order to find the most important physical properties of the particle described by equation (1.1) (i.e. possible values of mass and spin, charge, energy and momentum) it is sufficient to know only the matrix $L^0$, and in that respect our approach is the most expedient one for the subsequent analysis of the equation.

Substituting (3.1) into (1.8) one obtains the following necessary and sufficient conditions which describe when a relativistically invariant equation can be

[12] It can be shown that all irreducible representations of the group of space rotations are finite-dimensional, that each of them is given by formulas (2.2′) for $k = 0, 1/2, 1, 3/2, 2, \ldots$ and that any (either finite-dimensional or infinite-dimensional) representation of that group can be decomposed into finite-dimensional irreducible representations. The difference of the rotation group from the Lorentz group is due to the fact that the first of them is a compact group while the second is a non-compact group (see L.S. Pontryagin, "Topological Groups", Moscow, 1938, in Russian).

constructed from the matrix $L^0$:

$$[L^0, I^{jk}] = 0 \quad (j, k = 1, 2, 3); \tag{3.2}$$

$$[[L^0, I^{0j}], I^{0j}] = L^0 \quad (j = 1, 2, 3); \tag{3.3}$$

$$[[L^0, I^{0j}], I^{0k}] = 0 \quad (j \neq k;\ k = 1, 2, 3); \tag{3.4}$$

$$[[L^0, I^{0i}], I^{jk}] = g^{ij}[L^0, I^{0k}] - g^{ik}[L^0, I^{0j}] \quad (i, j, k = 1, 2, 3). \tag{3.5}$$

However one can easily verify that relation (3.2) and relations (3.3) for $j = 1$ together with commutation rules (1.7) for $I^{jk}$ imply relations (3.5), (3.4) and relation (3.3) for $j = 2, 3$. Therefore, one has to find only those operators $L^0$ that satisfy equalities (3.2) and relation (3.3) for $j = 1$. Substituting now for the operators $I^{jk}$ by the operators (2.1) we finally arrive at the following set of necessary and sufficient conditions for $L^0$ to be a coefficient before $\partial\psi/\partial x^0$ in a relativistically invariant equation:

$$[L^0, H^0] = 0, \qquad [L^0, H^+] = 0, \qquad [L^0, H^-] = 0, \tag{3.6}$$

$$[[L^0, F^0], F^0] = L^0. \tag{3.7}$$

We now assume that the representation $D_R$ is decomposed into a sum (not necessarily finite) of our irreducible representations. The irreducible components of the representation $D_R$ will be numbered by index $\tau$, each $\tau$ corresponding to a pair of numbers $(k_0, k_1)$ (this correspondence will be denoted by $\tau \sim (k_0, k_1)$) defining the representation numbered by $\tau$ (of course, there can be several $\tau$ corresponding to one and the same pair $(k_0, k_1)$). The representation space $R$ is then decomposed into a direct sum of subspaces $R_\tau = R_{(k_0, k_1)}$ which are representation spaces for irreducible representations of the Lorentz group with the corresponding values of $\tau$. Vectors $\xi_p^k$ of the subspace $R_\tau = R_{(k_0, k_1)}$ introduced in the preceding section will be denoted by $\xi_{p\tau}^k$. The set of all $\xi_{p\tau}^k$ forms the set of coordinate vectors for the space $R$. The elements of the matrix $L^0$ in that coordinate system will be denoted by $c_{p\tau;p'\tau'}^{k;k'}$:

$$L^0 \xi_{p\tau}^k = \sum_{\tau' k' p} c_{p\tau;p'\tau'}^{k;k'} \xi_{p'\tau'}^{k'}. \tag{3.8}$$

Equalities (3.6) show that $L^0$ commutes with all operators representing transformations of the rotation group defined by the representation $D_R$ of the Lorentz group. It easily follows from this that

$$c_{p\tau;p'\tau'}^{k;k'} = c_{\tau\tau'}^k \delta_{kk'} \delta_{pp'}, \tag{3.9'}$$

i.e.

$$L^0 \xi_{p\tau}^k = \sum_{\tau'} c_{\tau\tau'}^k \xi_{p\tau'}^k. \tag{3.9}$$

Indeed, the first equality (3.6) implies that

$$L^0 H^0 \xi_{p\tau}^k - H^0 L^0 \xi_{p\tau}^k = 0$$

for all $\xi_p^k$. Substituting in that relation the formulas (2.2′) and (3.8), one has

$$\sum_{\tau', k', p'} (p - p') c_{p\tau;p'\tau'}^{k;k'} \xi_{p'\tau'}^{k'} = 0$$

which immediately implies that $c^{k;k'}_{p\tau;p'\tau'}=0$ for $p'\neq p$, i.e.

$$c^{k;k'}_{p\tau;p'\tau'}=c^{k;k'}_{p\tau\tau'}\delta_{pp'}. \tag{3.10}$$

The last two equalities of (3.6) now yield, in view of (2.2′), (3.8) and (3.10),

$$\sqrt{(k+p)(k-p+1)}\,c^{k;k'}_{p-1,\tau\tau'}-\sqrt{(k'+p)(k'-p+1)}\,c^{k;k'}_{p\tau\tau'}=0,$$
$$\sqrt{(k'+p)(k'-p+1)}\,c^{k;k'}_{p-1,\tau\tau'}-\sqrt{(k+p)(k-p+1)}\,c^{k;k'}_{p\tau\tau'}=0,$$

whence

$$c^{k;k'}_{p\tau\tau'}=0 \quad \text{for } k'\neq k$$

and

$$c^{k;k}_{p\tau\tau'}=c^{k;k}_{p-1,\tau\tau'}.$$

The last equality shows that $c^{k;k}_{p}$, does not depend on $p$ and thereby completes the proof of relations (3.9′) and (3.9)[13].

To complete the computation of $L^0$ it remains to find those $c^{k}_{\tau\tau'}$ that satisfy relation (3.7), i.e. for which

$$[[L^0,F^0],F^0]\xi^k_{p\tau}=L^0\xi^k_{p\tau} \tag{3.11}$$

for all vectors $\xi^k_{p\tau}$. Substituting (2.2′) and (3.9) into (3.11) one comes to the following system of linear equations with respect to the unknowns $c^k_{\tau\tau'}$:

$$\left.\begin{aligned}
&B_k(\tau')B_{k+1}(\tau')c^{k+1}_{\tau\tau'}-2B_k(\tau')B_{k+1}(\tau)c^k_{\tau\tau'}+B_k(\tau)B_{k+1}(\tau)c^{k-1}_{\tau\tau'}=0,\\
&B_k(\tau)B_{k+1}(\tau)c^{k+1}_{\tau\tau'}-2B_k(\tau)B_{k+1}(\tau')c^k_{\tau\tau'}+B_k(\tau')B_{k+1}(\tau')c^{k-1}_{\tau\tau'}=0,\\
&B_k(\tau')[A_{k-1}(\tau')+A_k(\tau')-2A_k(\tau)]c^k_{\tau\tau'}=B_k(\tau)[2A_{k-1}(\tau')-A_{k-1}(\tau)-A_k(\tau)]c^{k-1}_{\tau\tau'},\\
&B_k(\tau)[A_{k-1}(\tau)+A_k(\tau)-2A_k(\tau')]c^k_{\tau\tau'}=B_k(\tau')[2A_{k-1}(\tau)-A_{k-1}(\tau')-A_k(\tau')]c^{k-1}_{\tau\tau'},\\
&2B_{k+1}(\tau')B_{k+1}(\tau)c^{k+1}_{\tau\tau'}-\{[B_{k+1}(\tau')]^2+[B_{k+1}(\tau)]^2+[B_k(\tau')]^2+[B_k(\tau)]^2\\
&\quad+[A_k(\tau')-A_k(\tau)]^2\}c^k_{\tau\tau'}+2B_k(\tau')B_k(\tau)c^{k-1}_{\tau\tau'}=0,\\
&2(k+1)^2B_{k+1}(\tau')B_{k+1}(\tau)c^{k+1}_{\tau\tau'}-\{(k+1)^2[B_{k+1}(\tau')]^2+(k+1)^2[B_{k+1}(\tau)]^2\\
&\quad+k^2[B_k(\tau')]^2+k^2[B_k(\tau)]^2\}c^k_{\tau\tau'}+2k^2B_k(\tau')B_k(\tau)c^{k-1}_{\tau\tau'}=4c^k_{\tau\tau'}.
\end{aligned}\right\} \tag{3.12}$$

Here $A_k(\tau)$ and $B_k(\tau)$ denote the values of (2.3′) and (2.3″) corresponding to the pair $(k_0,k_1)$ defining the representation marked by the index $\tau$.

Thus our task is now reduced to that of investigating and solving the system of linear equations. Solving any three of the equations (3.12) with respect to $c^{k-1}_{\tau\tau'}$, $c^k_{\tau\tau'}$ and $c^{k+1}_{\tau\tau'}$, and then substituting those values into other equations, one can easily verify that $c^k_{\tau\tau'}$ is different from zero only if one of the numbers of the pair $(k_0',k_1')$ defining the representation $\tau'$ is equal to the corresponding number of the pair $(k_0,k_1)$ defining the representation $\tau$ with another number of the pair $(k_0',k_1')$

[13] The general form (3.9) of an operator commuting with a completely reducible representation (i.e. one represented as a sum of irreducible representations) of some group can also be easily obtained with the help of Schur's lemma (see, e.g., [22], section 13). We have chosen, however, not to invoke that lemma here, using instead an approach which will be repeatedly used in the following.

differing from the corresponding number of $(k_0, k_1)$ by $\pm 1$, i.e. if $(k_0', k_1')$ is equal to $(k_0+1, k_1)$, $(k_0-1, k_1)$, $(k_0, k_1+1)$ or $(k_0, k_1-1)$[14], and then

$$c^k_{\tau\tau'} = c_{\tau\tau'}\sqrt{(k+k_0+1)(k-k_0)},$$
$$c^k_{\tau'\tau} = c_{\tau'\tau}\sqrt{(k+k_0+1)(k-k_0)}, \tag{3.13}$$

for $(k_0', k_1') = (k_0+1, k_1)$ and

$$c^k_{\tau\tau'} = c_{\tau\tau'}\sqrt{(k+k_1+1)(k-k_1)},$$
$$c^k_{\tau'\tau} = c_{\tau'\tau}\sqrt{(k+k_1+1)(k-k_1)}, \tag{3.14}$$

for $(k_0', k_1') = (k_0, k_1+1)$.

Here $c_{\tau\tau'}$ and $c_{\tau'\tau}$ are arbitrary complex numbers.

Formulas (3.9), (3.13) and (3.14) provide a complete solution for the problem of finding all equations invariant under transformations of the proper Lorentz group. We see that each such equation is completely defined by a set of complex numbers $c_{\tau\tau'}$ corresponding to all possible pairs $\tau, \tau'$ of "adjacent"[15] representations. In particular, in the finite-dimensional case, these formulas are equivalent to the more complicated formulas found in [8, 9].

For a relativistically invariant equation to be non-splitting (which means that it cannot be represented as a set of unrelated invariant equations, each with respect to its own separate group of components of the wave function $\psi$) it is evidently necessary that each pair of irreducible components of $D_R$ can be connected by a chain of "adjacent" representations. Note that in the infinite-dimensional case, the transformation law of $\psi$ may correspond to a representation that is decomposed into a continuous sum (integral) of irreducible representations (see, e.g., [18], section 9). Our result implies, however, that for a non-splitting equation the representation $D_R$ includes at most a countable number of irreducible non-equivalent components. Thus, for a non-splitting equation, decomposition of the representation giving the transformation law of $\psi$ cannot include integral terms unless it includes a continuous set of equivalent irreducible components.

Note that in another coordinate system

$$\xi'^k_{p\tau} = \alpha(\tau)\xi^k_{p\tau} \tag{3.15}$$

where $\alpha(\tau)$ are complex numbers depending on $\tau$, formulas (2.2′) and (2.2″) which define operators $I^{jk}$ are of the same form while the numbers $c_{\tau\tau'}$ defining the matrix $L^0$ go into

$$c'_{\tau\tau'} = c_{\tau\tau'}\frac{\alpha(\tau)}{\alpha(\tau')}. \tag{3.16}$$

---

[14] This result shows (cf. [20], pp. 70, 71) that a Kronecker product of a vector and a quantity transformed according to the irreducible representation defined by the pair $(k_0, k_1)$ is decomposed into four irreducible representations defined by the pairs $(k_0+1, k_1)$, $(k_0-1, k_1)$, $(k_0, k_1+1)$ and $(k_0, k_1-1)$.

[15] We say that irreducible components $\tau$ and $\tau'$ of the representation $D_R$ are "adjacent" if there exists $k$ such that $c^k_{\tau\tau'}$ is different from zero. Thus, for the representation numbered by $\tau \sim (k_0, k_1)$, "adjacent" representations are defined by the pairs $(k_0+1, k_1)$, $(k_0-1, k_1)$, $(k_0, k_1+1)$ and $(k_0, k_1-1)$.

Therefore, any two equations defined by the numbers $c_{\tau\tau'}$ and, respectively, $c'_{\tau\tau'}$ and related by (3.16) describe one and the same wave field. Thus the parameters of the equation (1.1) that are really essential are not the numbers $c_{\tau\tau'}$ but their products of the form

$$c_{\tau_1\tau_2}c_{\tau_2\tau_3}\cdots c_{\tau_{n-1}\tau_n}c_{\tau_n\tau_1}$$

which do not change under (3.16).

For example, for the representation $D_R$ which splits into four irreducible representations of the proper Lorentz group (so that $\tau$ runs over 1, 2, 3, 4) with the "adjacency" scheme

$$\begin{array}{ccc} 1 & \rightleftarrows & 2 \\ \upharpoonleft\downharpoonright & & \upharpoonleft\downharpoonright \\ 3 & \rightleftarrows & 4 \end{array}$$

the essential parameters are not given by the set of all eight numbers $c_{\tau\tau'}$ corresponding to such a scheme but by the five products

$$c_{12}c_{21},\ c_{24}c_{42},\ c_{43}c_{34},\ c_{31}c_{13}\ \text{and}\ c_{12}c_{24}c_{43}c_{31}.$$

All equations for which these five numbers coincide describe one and the same field but referred to different coordinate systems[16]. If, among irreducible representations into which the representation $D_R$ is decomposed, there are equivalent components $\tau_1, \tau_2, \ldots$ then the coordinate transformation

$$\xi'^k_{p\tau_i} = \sum_j \alpha(\tau_i, \tau_j)\xi^k_{p\tau_j}, \tag{3.17}$$

where $\|\alpha(\tau_i, \tau_j)\|$ is an arbitrary complex matrix, does not change operators $I^{jk}$ either. The essential parameters for equation (1.1) are then those combinations of the numbers $c_{\tau\tau'}$ which do not change under (3.17) and their number is diminished accordingly.

## 4. Equations invariant with respect to the full Lorentz group

1. *Irreducible representations of the full Lorentz group.* Using the results of section 2 one can now find irreducible representations of the full Lorentz group, i.e. of the group that includes both the proper Lorentz transformations and reflections (improper transformations).

Any representation of that group is completely defined if one specifies the operators $I^{jk}$ (or, equivalently, operators $H^0, H^+, H^-, F^0, F^+, F^-$) and the operator $T$ corresponding to the reflection

$$x^0 \to x^0, \quad x^1 \to -x^1, \quad x^2 \to -x^2, \quad x^3 \to -x^3. \tag{4.1}$$

[16] The statement in the paper [8] by Bhabha equivalent to the statement that only products of the pairs of numbers $c_{\tau\tau'}c_{\tau'\tau}$ are essential (and, consequently, that one can always assume that $c_{\tau\tau'} = c_{\tau'\tau}$) is a mistake.

These operators should satisfy commutation relations (1.7) for $I^{jk}$, the relation

$$T^2 = E \quad (E \text{ is an identical operator})^{17}, \tag{4.2}$$

and relations

$$[T, H^0] = 0, \quad [T, H^+] = 0, \quad [T, H^-] = 0, \tag{4.3}$$

$$[T, F^0]_+ = TF^0 + F^0 T = 0, \tag{4.4}$$

$$[T, F^+]_+ = 0, \quad [T, F^-]_+ = 0, \tag{4.5}$$

showing how reflections commute with the infinitesimal transformations of the Lorentz group. Conversely, any set of operators $I^{jk}$ and $T$ satisfying these relations defines a representation of the full Lorentz group. Note that formula (4.5) follows from (4.3), (4.4) and (1.7) and can, therefore, be omitted.

Operators $I^{jk}$ define a representation of the proper Lorentz group. One can show that for an irreducible representation of the full group, the representation of the proper group is either irreducible or splits into two non-equivalent irreducible representations (cf. [20], sections 88–89; it also easily follows from the following considerations). Considering both cases together, we denote the coordinate vectors of the representation space by $\xi^k_{p\tau}$ where index $\tau$, as in section 3, numbers irreducible representations of the proper Lorentz group and, therefore, takes either one or two values. The operators $I^{jk}$ act on $\xi^k_{p\tau}$ according to formulas (2.2′), (2.2″) and the task of finding all representations of the full group is reduced to finding all operators $T$ satisfying (4.2), (4.3) and (4.4).

Proceeding now in the same manner as in section 3 (where we had to find $L^0$; see text around formulas (3.8)–(3.12)) we have from (4.3) that

$$T\xi^k_{p\tau} = \sum_{\tau'} t^k_{\tau\tau'} \xi^k_{p\tau'}. \tag{4.6}$$

Also, by (4.4) one has

$$TF^0 \xi^k_{p\tau} + F^0 T\xi^k_{p\tau} = 0.$$

Substituting now $F^0 \xi^k_{p\tau}$ from (2.2″) and $T\xi^k_{p\tau}$ from (4.6) one can verify that equation (4.4) is equivalent to the following three relations:

$$A_k(\tau) t^k_{\tau\tau'} + A_k(\tau') t^k_{\tau\tau'} = 0, \tag{4.7}$$

$$B_k(\tau) t^k_{\tau\tau'} + B_k(\tau') t^{k-1}_{\tau\tau'} = 0, \tag{4.8′}$$

$$B_k(\tau') t^k_{\tau\tau'} + B_k(\tau) t^{k-1}_{\tau\tau'} = 0. \tag{4.8″}$$

Formula (4.7) now implies that $t^k_{\tau\tau'}$ is different from zero only if

$$A_k(\tau) + A_k(\tau') = 0,$$

i.e.

$$k_0 k_1 + k_0' k_1' = 0. \tag{4.9}$$

---

[17] For a two-valued representation one has to require only that $T^2 = \pm E$. Therefore, one can always replace $T$ by $iT$. This fact, however, is not essential for our purposes.

According to formulas (4.8′) and (4.8″), a necessary condition for $t^k_{\tau\tau'}$ to be different from zero is

$$[B_k(\tau)]^2 = [B_k(\tau')]^2,$$

i.e.

$$k^2(k_0^2 + k_1^2) - k_0^2 k_1^2 = k^2(k_0'^2 + k_1'^2) - k_0'^2 k_1'^2. \tag{4.10}$$

Another condition requires that $T$ be non-singular (since $T^2 = E$). This implies that if one has two irreducible representations $\tau$ and $\tau' \neq \tau$ then both representations split into the same set of representations of the rotation group, i.e. $k_0' = \pm k_0$. Since the simultaneous change of sign of $k_0$ and $k_1$ does not change the representation, one can replace relations (4.9) and (4.10) by

$$\begin{aligned} &k_0' = k_0, \quad k_1' = -k_1 \quad \text{for} \quad k_0 \neq 0, \quad k_1 \neq 0; \\ &k_0' = k_0, \quad k_1' = k_1 \quad \text{for} \quad k_0 = 0 \quad \text{or} \quad k_1 = 0. \end{aligned} \tag{4.11}$$

Thus each irreducible representation of the full Lorentz group either (a) is an irreducible representation of the proper group defined by the pair of numbers $(k_0, k_1)$, one of which is zero, or (b) can be decomposed into a direct sum of two irreducible representations of the proper group defined by the pairs $(k_0, k_1)$ and $(k_0, -k_1)$ where $k_0 \neq 0$ and $k_1 \neq 0$. Consider now both cases separately.

(a) In this case a full system of coordinate vectors is given by the vectors $\xi^k_p$ (the index $\tau$ here takes only one value and can therefore be omitted). By formula (4.6),

$$T\xi^k_p = t^k \xi^k_p. \tag{4.12}$$

It follows from (4.8) and (4.2) that

$$t^k = -t^{k-1} \tag{4.13}$$

and that

$$t^k = \pm 1 \tag{4.14}$$

for all $k$.

The operator $T$ given by formulas (4.12), (4.13) and (4.14) together with the operators (2.2′), (2.2″) defines the irreducible representation of the full Lorentz group. According to (4.12), (4.13) and (4.14) one gets two such representations by setting either

$$T\xi^k_p = (-1)^{[k]} \xi^k_p \tag{4.15'}$$

where $k$ is the integer part of $k$, or, correspondingly,

$$T\xi^k_p = (-1)^{[k]+1} \xi^k_p. \tag{4.15''}$$

It is easy to verify that these representations are not equivalent. Thus one obtains the following result: each irreducible representation of the proper Lorentz group defined by the pair $(0, k_1)$ or by the pair $(k_0, 0)$ can be continued to an irreducible representation of the full Lorentz group in two ways; the operator corresponding to the reflection (4.1) is either of the form (4.15′) or (4.15″).

(b) Denote the representations of the proper group into which our representation of the full group is decomposed by indices 1 and $\dot{1}$. Then the coordinate vectors will be denoted by $\xi^k_{p1}$ and $\xi^k_{p\dot{1}}$ and

$$T\xi^k_{p1} = t^k_{1\dot{1}}\xi^k_{p\dot{1}}, \quad T\xi^k_{p\dot{1}} = t^k_{\dot{1}1}\xi^k_{p1}. \tag{4.16}$$

By (4.11) one has $[B_k(1)]^2 = [B_k(\dot{1})]^2$ and, consequently, one can assume that $B_k(1) = B_k(\dot{1})$ (see footnote[10]). Then relation (4.8) implies that

$$t^k_{1\dot{1}} = -t^{k-1}_{1\dot{1}}, \quad t^k_{\dot{1}1} = -t^{k-1}_{\dot{1}1} \tag{4.17}$$

and also, according to (4.2), that

$$t^k_{\dot{1}1}t^k_{1\dot{1}} = 1. \tag{4.18}$$

The operator $T$ can be written in a more simple form in another system of coordinate vectors. Indeed, let $\xi'^k_{p1} = \alpha_1\xi^k_{p1}$ and $\xi'^k_{p\dot{1}} = \alpha_{\dot{1}}\xi^k_{p\dot{1}}$ where $\alpha_1, \alpha_{\dot{1}}$ are arbitrary complex numbers. Then instead of $t^k_{1\dot{1}}$ and $t^k_{\dot{1}1}$ one has

$$t'^k_{1\dot{1}} = t^k_{1\dot{1}}(\alpha_1/\alpha_{\dot{1}}), \quad t'^k_{\dot{1}1} = t^k_{\dot{1}1}(\alpha_{\dot{1}}/\alpha_1). \tag{4.19}$$

In particular, for $\alpha_{\dot{1}}/\alpha_1 = (-1)^{[k]}t^k_{1\dot{1}}$ one has

$$t'^k_{1\dot{1}} = (-1)^{[k]}, \quad t'^k_{\dot{1}1} = (-1)^{[k]}. \tag{4.20}$$

Thus, each pair of irreducible representations of the proper Lorentz group defined by the numbers $(k_0, k_1)$ and $(k_0, -k_1)$ where $k_0 \neq 0$, $k_1 \neq 0$ defines the unique representation of the full group. The operator corresponding to the reflection (4.1) under this representation can be written in the form

$$T\xi^k_{p1} = (-1)^{[k]}\xi^k_{p\dot{1}}, \quad T\xi^k_{k\dot{1}} = (-1)^{[k]}\xi^k_{p1}. \tag{4.21}$$

This formula, together with formulas (2.2), (2.3) for infinitesimal operators, defines the representation of the full group.

2. *Equations invariant with respect to the full Lorentz group.* For an equation to be invariant with respect to the full Lorentz group it is necessary, first of all, that the representation $D_R$ can be continued to the representation of the full group. The results of section 4.1 show that this is possible if the decomposition of $D_R$ into irreducible representations of the proper group for each representation $\tau \sim (k_0, k_1)$ also includes the representation $\dot{\tau} \sim (k_0, -k_1)$ (for $k_0 = 0$ or $k_1 = 0$ index $\dot{\tau}$ coincides with $\tau$). Reflection (4.1) then corresponds to the transformation of $\psi$ defined by formulas (4.21) and (4.15).

Let

$$L^k \partial\psi/\partial x^k + i\kappa\psi = 0$$

be an equation invariant under transformations of the proper Lorentz group. Substituting reflection (4.1) into (1.4) for the Lorentz transformation $\|l^i_j\|$ one gets an additional condition for the invariance of equation (1.1) under transformations of the full group:

$$L^0 T = TL^0. \tag{4.22}$$

With the use of formulas (4.16) and (4.12) for $T$ and formula (3.9) for $L^0$ equation (4.22) can be written in the form:

$$c^k_{\tau\tau'}t^k_{\tau'\dot{\tau}'} = t^k_{\tau\dot{\tau}}c^k_{\dot{\tau}\dot{\tau}'}. \tag{4.23}$$

Substituting now for $c^k_{\tau\tau'}$ and $t^k_{\tau\tau'}$ their values from (3.13), (3.14), (4.15) and (4.20), one has

$$c_{\tau\tau'} = c_{\dot{\tau}\dot{\tau}'} \quad \text{for} \quad \tau \neq \dot{\tau}, \quad \tau' \neq \dot{\tau}' \tag{4.24'}$$

(clearly, if $\tau$ and $\dot{\tau}$ are a pair of "adjacent" representations then $\dot{\tau}$ and $\dot{\tau}'$ are also a pair of "adjacent" representations) and

$$\begin{aligned} c_{\tau\tau'} &= \pm c_{\tau\dot{\tau}'} \quad \text{for} \quad \tau = \dot{\tau}, \quad \tau' \neq \dot{\tau}', \\ c_{\tau\tau'} &= \pm c_{\dot{\tau}\tau'} \quad \text{for} \quad \tau \neq \dot{\tau}, \quad \tau' \neq \dot{\tau}'. \end{aligned} \tag{4.24''}$$

The choice of the sign $(+)$ or $(-)$ in (4.24″) depends on whether the operator $T$ is given by formula (4.15′) or by formula (4.15″). The equation is invariant with respect to the full group under any choice of sign provided that once the sign is chosen for $\tau = \dot{\tau}$ it has the same value for all $\tau'$ "adjacent" to $\tau$. Condition (4.23) imposes no restrictions on $c_{\tau\tau'}$ for $\tau = \dot{\tau}$ and $\tau' = \dot{\tau}'$ [18].

Note that equations invariant with respect to the full Lorentz group should be considered equivalent if they can be transformed into one another by a change of basis that affects neither the form of operators $I^{jk}$ nor the operator $T$. Therefore one can allow only such coordinate transformations (3.15) for which

$$\alpha(\tau) = (\tau) \tag{4.25}$$

(see examples in Section 7). One should also change the definition of a non-splitting equation. An equation invariant with respect to the full Lorentz group will be called non-splitting if it cannot be represented as a set of several equations invariant with respect to the same group such that each of them involves only part of the components of the wave function $\psi$. It is quite possible that the matrices $L^0, L^1, L^2$ and $L^3$ and $I^{jk}$ may split so that equation (1.1) considered as an equation invariant with respect to the proper Lorentz group will be a splitting equation; however, as an equation invariant with respect to the full group it will be non-splitting because individual equations it splits into will be connected by the transformation $T$.

## 5. Equations obtained from an invariant Lagrange function

1. *Invariant bilinear Hermitian form* $(\psi_1, \psi_2)$. Now we establish conditions under which one can construct a non-degenerate[19] invariant bilinear Hermitian form out of $\psi$. First we consider the case of the wave function $\psi$ that transforms according to some representation of the proper Lorentz group, and we shall be looking for

---

[18] Note that if there are "adjacent" representations one can choose several ways of defining the operator $T$ which do not affect the invariance of the equation. Since an invariant equation is defined not only by the matrices $L^k$ but also by the transformation law for $\psi$ (i.e. the representation $D_R$) one actually has in such cases several different equations. Examples of such equations differing only in the form of operator $T$ are provided by scalar and pseudo-scalar equations and by vector and pseudo-vector equations (see Section 7).

[19] A bilinear form $(\psi_1, \psi_2)$ is called non-degenerate if there is no vector $\psi_1$ such that $(\psi_1, \psi) = 0$ for all vectors $\psi$. Non-degeneracy of a form indicates that there is no coordinate system in which it can be written as depending only on some part of the components of the vectors $\psi_1$ and $\psi_2$.

a form invariant under transformations of that group. As in section 3 we assume that the representation $D_R$ is decomposed into a direct sum of irreducible representations corresponding to different values of the index $\tau$. Vectors $\psi_1$ and $\psi_2$ can be represented as linear combinations of the coordinate vectors $\xi^k_{p\tau}$:

$$\psi_1 = \sum_{\tau,k,p} x^k_{p\tau} \xi^k_{p\tau}, \qquad \psi_2 = \sum_{\tau,k,p} y^k_{p\tau} \xi^k_{p\tau}.$$

Then any linear Hermitian form can be written as[20]

$$(\psi_1, \psi_2) = \sum_{\tau,k,p,\tau',k',p'} a^{k;k'}_{p\tau;p'\tau'} x^k_{p\tau} \overline{y^{k'}_{p'\tau'}} \tag{5.1}$$

where the numbers

$$a^{k;k'}_{p\tau;p'\tau'} = (\xi^k_{p\tau}, \xi^{k'}_{p'\tau'}) \tag{5.2}$$

satisfy the relation

$$a^{k;k'}_{p\tau;p'\tau'} = \overline{a^{k';k}_{p'\tau';p\tau}}. \tag{5.3}$$

We have to find out for which $a^{k;k'}_{p\tau;p'\tau'}$ the form (5.1) is invariant, i.e. for which $a^{k;k'}_{p\tau;p'\tau'}$ the relation

$$(S\psi_1, S\psi_2) = (\psi_1, \psi_2) \tag{5.4}$$

holds for all operators $S$ corresponding to Lorentz transformations (1.2) and for all vectors $\psi_1$ and $\psi_2$. Condition (5.4) can evidently be replaced by the following set of relations:

$$(S\xi^k_{p\tau}, S\xi^{k'}_{p'\tau'}) = (\xi^k_{p\tau}, \xi^{k'}_{p'\tau'}) \tag{5.5}$$

where $\xi^k_{p\tau}, \xi^k_{p'\tau'}$ are arbitrary coordinate vectors. Going now from the operators $S$ to the infinitesimal transformation $I^{ij}$ one has

$$(I^{ij} \xi^k_{p\tau}, \xi^{k'}_{p'\tau'}) + (\xi^k_{p}, I^{ij} \xi^{k'}_{p'\tau'}) = 0 \tag{5.6}$$

or, replacing $I^{ij}$ by the operators (2.1),

$$(H^0 \xi^k_{p\tau}, \xi^{k'}_{p'\tau'}) + (\xi^k_{p\tau}, H^0 \xi^{k'}_{p'\tau'}) = 0, \tag{5.7}$$

$$(H^+ \xi^k_{p\tau}, \xi^{k'}_{p'\tau'}) + (\xi^k_{p\tau}, H^- \xi^{k'}_{p'\tau'}) = 0, \tag{5.8}$$

$$(F^0 \xi^k_{p\tau}, \xi^{k'}_{p'\tau'}) + (\xi^k_{p\tau}, F^0 \xi^{k'}_{p'\tau'}) = 0, \tag{5.9}$$

$$(F^+ \xi^k_{p\tau}, \xi^{k'}_{p'\tau'}) + (\xi^k_{p\tau}, F^- \xi^{k'}_{p'\tau'}) = 0. \tag{5.10}$$

The last of these relations is a consequence of the three preceding ones in view of the commutation relations $[H^+, F^0] = iF^+$, $[H^-, F^0] = -iF^-$. Therefore, one has to satisfy only the relations (5.7), (5.8) and (5.9).

In view of (2.2), relation (5.7) yields

$$ip(\xi^k_{p\tau}, \xi^{k'}_{p'\tau'}) - ip'(\xi^k_{p\tau}, \xi^{k'}_{p'\tau'}) = 0,$$

[20] The bar always denotes a complex conjugate number.

i.e.

$$(p-p')a^{k;k'}_{p\tau;p'\tau'}=0, \tag{5.11}$$

and, consequently, $a^{k;k'}_{p\tau;p'\tau'}=0$ for $p \neq p'$:

$$a^{k;k'}_{p\tau;p'\tau'}=a^{k;k'}_{p\tau\tau'}\delta_{pp'} \tag{5.12}$$

Letting now $p'=p-1$ in (5.8) one easily obtains the following relation:

$$\sqrt{(k+p)(k-p+1)}\,a^{k;k'}_{p-1,\tau\tau'}-\sqrt{(k'+p)(k'-p+1)}\,a^{k;k'}_{p\tau\tau'}=0.$$

Taking now a complex conjugate relation and exchanging $k,\tau$ with $k',\tau'$ we have, in view of (5.3),

$$\sqrt{(k'+p)(k'-p+1)}\,a^{k;k'}_{p-1,\tau\tau'}-\sqrt{(k+p)(k-p+1)}\,a^{k;k'}_{p\tau\tau'}=0.$$

Those two relations imply that

$$a^{k;k'}_{p\tau;p'\tau'}=a^{k}_{\tau\tau'}\delta_{kk'}\delta_{pp'} \tag{5.13}$$

(cf. a similar approach used in section 3 to derive formula (3.9′)).

Letting now $p'=p$ and $k'=k, k+1$ and $k-1$ in (5.9), one has, respectively,

$$A_k(\tau)a^{k}_{\tau\tau'}+\overline{A_k(\tau')}a^{k}_{\tau\tau'}=0, \tag{5.14}$$

$$B_k(\tau)a^{k}_{\tau\tau'}-\overline{B_k(\tau')}a^{k-1}_{\tau\tau'}=0, \tag{5.15}$$

$$\overline{B_k(\tau')}a^{k}_{\tau\tau'}-B_k(\tau)a^{k-1}_{\tau\tau'}=0. \tag{5.16}$$

Substituting here (2.3′) and (2.3″) one has that $a^{k}_{\tau\tau'}$ can be different from zero only if

$$k_0k_1+k'_0\overline{k'_1}=0 \tag{5.17}$$

(we recall that $k_0$ is real) and

$$k_0^2+k_1^2=k_0'^2+k_1'^2. \tag{5.18}$$

But a non-degenerate invariant bilinear form exists only if for each irreducible representation $\tau$ there exists $\tau'$ such that $a^{k}_{\tau\tau'}\neq 0$ for all $k$. By (5.17) and (5.18) it follows from this that a non-degenerate invariant Hermitian bilinear form with respect to $\psi$ may exist only if in the representation $D_R$ together with each irreducible representation $\tau \sim (k_0, k_1)$ there is also the irreducible representation $\tau^* \sim (k_0, \overline{-k_1})$ (for $k_0=0$ or $k_1=0$ one can evidently assume that $\tau^* \sim (k_0, \overline{k_1})$). In particular, one can construct an invariant bilinear form from the wave function transformed according to an irreducible representation of the proper Lorentz group only if that representation satisfies the condition $\tau^*=\tau$, i.e. if either $k_0=0$ and $k_1$ is a real number or if $k_1$ is purely imaginary.

For finite-dimensional representations $k_1$ is real and, consequently, by definition, $\tau^*$ coincides with $\dot{\tau}$ (see section 4). Therefore, for each finite-dimensional representation of the full Lorentz group, there is always a non-degenerate invariant Hermitian form. The same is also true for infinite-dimensional representations splitting into irreducible representations with real $k_1$.

By (5.15)

$$a^k_{\tau\tau^*} = \frac{\overline{B_k(\tau^*)}}{B_k(\tau)} a^{k-1}_{\tau\tau^*}. \tag{5.19}$$

Now by definition of $B_k$ (formula (2.3″)) $\overline{B_k(\tau^*)}/B_k(\tau) = \pm 1$, and, therefore, relation (5.19) can be written as

$$a^k_{\tau\tau^*} = \pm a^{k-1}_{\tau\tau^*}, \tag{5.20}$$

and, consequently,

$$a^k_{\tau\tau^*} = \varepsilon^k a_{\tau\tau^*}, \quad \text{where} \quad \varepsilon^k = \pm 1, \quad \varepsilon^{k_0} = 1. \tag{5.21}$$

For $\tau = \tau^*$ one has the $(+)$ sign in (5.20) if $B_k(\tau)$ is real and $(-)$ if $B_k(\tau)$ is purely imaginary[21].

The numbers

$$a_{\tau\tau^*} = \overline{a_{\tau^*\tau}} \tag{5.22}$$

in (5.21) can evidently take arbitrary values; different values of $a_{\tau\tau^*} = \overline{a_{\tau^*\tau}}$ correspond to different invariant bilinear forms. However, one can always transform any invariant form to a simple canonical form by going to another system of coordinate vectors.

Indeed, denote by $\tau_1, \tau_2, \ldots$ equivalent irreducible representations in $D_R$ corresponding to the pair of numbers $(k_0, k_1)$ and by $\tau_1^*, \tau_2^*, \ldots$ those corresponding to the pair $(k_0, -\bar{k}_1)$. One can easily see that a non-degenerate invariant form $(\psi_1, \psi_2)$ may exist only if there is an equal number of both kinds of representations in $D_R$ (that number may be infinite). Now let

$$\xi'^k_{p\tau_i} = \sum_j \alpha(\tau_i, \tau_j)\xi^k_{p\tau_j}, \quad \xi'^k_{p\tau_i^*} = \sum_j \alpha(\tau_i^*, \tau_j^*)\xi^k_{p\tau_j^*} \tag{5.23'}$$

be a new system of coordinate vectors. They can always be chosen in such a way that

$$a_{\tau_i\tau_j^*} = \begin{matrix} 1 & \text{for} & i = j, \\ 0 & \text{for} & i \neq j. \end{matrix} \tag{5.24'}$$

Similarly, denote by $\tau_1, \tau_2, \ldots$ the set of equivalent representations in $D_R$ for which $\tau^*$ is equivalent to $\tau$ (i.e. either $k_1$ is purely imaginary, or $k_0 = 0$ and $k_1$ is real). By introducing a new system of coordinate vectors

$$\xi'^k_{p\tau_i} = \sum_j \alpha(\tau_i, \tau_j)\xi^k_{p\tau_j}, \tag{5.23''}$$

one can always achieve that

$$a_{\tau_i} = \begin{matrix} \pm 1 & \text{for} & i = j, \\ 0 & \text{for} & i \neq j. \end{matrix} \tag{5.24''}$$

21 This implies, in particular, that an irreducible representation is unitary (i.e. its operators preserve a positive definite bilinear form) only if either $k_1$ is purely imaginary or if $k_0 = 0$ and $|k_1|$ is real, $0 < |k_1| \leqq 1$.

The reduction of $a_{\tau\tau^*}$ to the form (5.24′) and (5.24″) is quite similar to the process of reducing any quadratic form to a sum of squares.

If a Hermitian bilinear form invariant with respect to the proper Lorentz group is also invariant with respect to reflections (i.e. the full Lorentz group) then the numbers $a_{\tau\tau^*}$ have to satisfy the following relations:

$$a_{\tau\tau^*} = a_{\dot{\tau}\dot{\tau}^*}. \tag{5.25}$$

In this case one can always choose a coordinate system in such a way that for each irreducible representation $\tau \sim (k_0, k_1)$ of the proper Lorentz group there exists only one representation $\tau^* \sim (k_0, -\overline{k_1})$ such that $a_{\tau\tau^*} \neq 0$. Moreover, one can always choose it in such a way that

$$\begin{aligned} a_{\tau\tau^*} &= +1 \text{ or } -1 \text{ for } k_1 \text{ real or purely imaginary,} \\ a_{\tau\tau^*} &= +1 \text{ for other values of } k_1. \end{aligned} \tag{5.24‴}$$

For $\tau \sim (0, k_0), \tau^* \sim (0, k_0)$ both $\tau$ and $\tau^*$ considered as representations of the full Lorentz group are either both described by formula (4.15′) or by formula (4.15″).

Note that if there are irreducible components of $D_R$ corresponding to $k_1$ real or purely imaginary then, according to (5.24‴), one has several essentially different invariant forms according to the choice of sign of $a_{\tau\tau^*}$. Different invariant forms lead to different types of invariant equations (1.1) obtained from a Lagrange function (see the example in section 7.2).

2. *Equations obtained from an invariant Lagrange function.* Any real invariant Lagrange function for an equation of the form (1.1) can always be represented in the form

$$\mathscr{L} = (1/2i)\{(\Lambda^k \partial\psi/\partial x^k, \psi) - (\psi, \Lambda^k \partial\psi/\partial x^k)\} + \kappa(\psi, \psi), \tag{5.26}$$

where $\Lambda^k (k = 0, 1, 2, 3)$ are four matrices satisfying relation (1.4) (i.e. all conditions of sections 3 and 4.2) and the brackets (e.g. $(\Lambda^k \partial\psi/\partial x^k, \psi)$) denote some bilinear invariant form defined by the formulas of the first part of this section (it is the general form of a real scalar bilinear in $\psi$ and linear in $\partial\psi/\partial x^k$).

Denote by $\Lambda^{k*}$ the matrix of a linear operator satisfying the condition

$$(\Lambda^k \psi_1, \psi_2) = (\psi_1, \Lambda^{k*} \psi_2) \tag{5.27}$$

for all $\psi_1, \psi_2$.

Let us now find equations obtained by the variation of the integral

$$\int\!\!\int\!\!\int\!\!\int \mathscr{L}\, dx^0 dx^1 dx^2 dx^3.$$

Since

$$\begin{aligned} \delta\mathscr{L} &= \frac{1}{2i}\left\{\left(\Lambda^k \frac{\partial\psi}{\partial x^k}, \delta\psi\right) + \left(\Lambda^k \frac{\partial\delta\psi}{\partial x^k}, \psi\right) - \left(\delta\psi, \Lambda^k \frac{\partial\psi}{\partial x^k}\right) - \left(\psi, \Lambda^k \frac{\partial\delta\psi}{\partial x^k}\right)\right\} \\ &\quad + \kappa\{(\psi, \delta\psi) + (\delta\psi, \psi)\} \\ &= \frac{1}{i}\left\{\left(\tfrac{1}{2}(\Lambda^k + \Lambda^{k*})\frac{\partial\psi}{\partial x^k} + i\kappa\psi, \delta\psi\right) - \left(\delta\psi, \tfrac{1}{2}(\Lambda^k + \Lambda^{k*})\frac{\partial\psi}{\partial x^k} + i\kappa\psi\right)\right\} \\ &\quad + \frac{\partial}{\partial x^k}\frac{1}{2i}\{(\Lambda^k \delta\psi, \psi) - (\psi, \Lambda^k \delta\psi)\} \end{aligned}$$

and dropping the divergence (the last term) one arrives at the following Euler–Lagrange equations:

$$(1/2)(\Lambda^k + \Lambda^{k*})\partial\psi/\partial x^k + i\kappa\psi = 0. \tag{5.28}$$

Thus, for an equation obtained from a Lagrange function (5.26), the coefficient infront of $\partial\psi/\partial x^k$ is given by the matrix

$$L^k = (1/2)(\Lambda^k + \Lambda^{k*}). \tag{5.29}$$

Evidently, such an operator $L^k$ satisfies the equality

$$(L^k\psi_1, \psi_2) = (\psi_1, L^k\psi_2) \quad (k = 0, 1, 2, 3) \tag{5.30}$$

for all $\psi_1, \psi_2$ (in other words, $L^{k*} = L^k$). Thus, one cannot obtain all relativistically invariant equations of the form (1.1) from real invariant Lagrange functions: a necessary condition is that relation (5.30) should be satisfied for some invariant bilinear form $(\psi_1, \psi_2)$, i.e. the quadratic form $(L^k\psi, \psi)$ $(k = 0, 1, 2, 3)$ should be real. Provided condition (5.30) is satisfied, equation (1.1) can always be obtained from a real[22] invariant Lagrange function; the latter can be defined by formula (1.10).

Note that condition (5.6) and relations (3.1) imply that (5.30) holds provided condition (1.9) is satisfied for all $\psi_1, \psi_2$, i.e. if the quadratic form $(L^0\psi, \psi)$ is real for all $\psi$.

Let us now see for which operators $L^0$ condition (1.9) is satisfied. Instead of (1.9) we can require that

$$(L^0\xi^k_{p\tau}, \xi^{k'}_{p'\tau'}) = (\xi^k_{p\tau}, L^0\xi^{k'}_{p'\tau'}) \tag{5.31}$$

for all $\xi^k_{p\tau}$ and $\xi^{k'}_{p'\tau'}$. In view of (3.9), (3.13), (3.14), (5.13) and (5.21), equality (5.31) can be written in the form

$$c_{\tau\tau'}a_{\tau'\tau'*} = \overline{c_{\tau'*\tau*}}a_{\tau\tau*}. \tag{5.32}$$

This is the final form of the condition which has to be satisfied by the matrix $L^0$ (i.e. the numbers $c_{\tau\tau'}$) if equation (1.1) is to be obtained from an invariant Lagrange function. In particular, if one chooses the basis for which the relations (5.24‴) are satisfied, one has

$$\begin{aligned} c_{\tau\tau'} &= \pm\overline{c_{\tau'*\tau'}} \text{ for } k_1 \text{ purely imaginary,} \\ c_{\tau\tau'} &= \overline{c_{\tau'*\tau*}} \text{ otherwise.} \end{aligned} \tag{5.33}$$

---

22 Note that starting with a complex Lagrange function

$$\mathscr{L} = (\Lambda^k\partial\psi/\partial x^k + i\kappa\psi, \psi) \tag{5.26'}$$

one obtains as the Euler–Lagrange equations the following two simultaneous systems of equations:

$$\Lambda^k(\partial\psi/\partial x^k) + i\kappa\psi = 0 \tag{5.28'}$$

and

$$\Lambda^{k*}(\partial\psi/\partial x^k) + i\kappa\psi = 0 \tag{5.28''}$$

in the notation of footnote[8]. They are obtained by taking the variation of the Lagrange function with respect to $\psi$ and, respectively, ($\psi^+$ or $\psi^*$). Thus, a complex Lagrange function generally yields an overdetermined system of equations. It is not overdetermined only if condition (5.30) is satisfied (equation (5.28″) then coincides with (5.28′)) and in that case one can use either a real or a complex Lagrange function.

Looking for essential parameters of equation (1.1) obtained from a Lagrange function, one has to take into account the fact that only those coordinate transformations can be allowed that do not affect the invariant bilinear form. In other words, one has to assume that in formula (3.15)

$$\alpha(\tau)\alpha(\tau^*) = 1 \tag{5.34}$$

(see the example of Section 7.2).

We also note that an equation obtained from an invariant Lagrange function should only be considered as splitting provided it splits into several equations, each of which can also be obtained from an invariant Lagrange function. However, an equation splitting into several equations with respect to separate groups of components of the wave function has to be considered as non-splitting if its Lagrange function (and, consequently, expressions for current, energy, momentum, etc.) involves all components of $\psi$.

## 6. Rest mass and spin values

In general, each relativistically invariant equation of the form (1.1) describes particles that may have different states corresponding to different rest mass and spin values. To find all mass and spin values possible for a given particle it is sufficient to consider flat waves in the rest mass system, i.e. solutions of equation (1.1) of the form

$$\psi(x^0, x^1, x^2, x^3) = \psi^{(0)} e^{-ip_0 x^0} \tag{6.1}$$

where $\psi^{(0)}$ does not depend on coordinates $x^i$. Substituting (6.1) into (1.1) one has

$$p_0 L^0 \psi^{(0)} - \kappa\psi^{(0)} = 0,$$

and, consequently, $p_0 = \kappa/\lambda$ where $\lambda$ is some eigenvalue of the matrix $L^0$. It follows this that all values of the rest mass of the particle described by equation (1.1) can be found from the formula

$$m_i = (\kappa c/\hbar)(1/\lambda_i) \tag{6.2}$$

where $\lambda_i$ runs over all non-zero eigenvalues of the matrix $L^0$.

If the matrix $L^0$ can be reduced to the diagonal form and if all its eigenvalues are different from zero, then the number of linearly independent solutions of equation (1.1) of the form (6.1) equals the number of components of the wave function $\psi$. Otherwise the number of components of the wave function is greater than the number of linearly independent flat waves in the rest mass system; in that case it is usually said that the equation has "redundant components" or "additional conditions"[23].

Formula (3.9) shows that matrix $L^0$ splits into "boxes" $\| c^k_{\tau\tau'} \|$ corresponding to different values of $k$. Each eigenvalue $\lambda_i^{(k)}$ of the matrix $\| c^k_{\tau\tau'} \|$ is also an eigenvalue

[23] An equation of the form (1.1) is usually called an equation without additional conditions if the matrix $L^0$ can be reduced to the diagonal form, and it is called an equation without redundant components if $L^0$ has no zero eigenvalues (cf. [11]).

of the matrix $L^0$ of multiplicity $(2k+1)$ (since the index $p$ in $\xi^k_{p\tau}$ takes on $2k+1$ values). The respective $2k+1$ eigenvectors $\psi^{(0)}_p$ of the matrix $L^0$ transform between themselves under space rotations corresponding to an irreducible representation of the rotation group. The solutions of equation (1.1) of the form (6.1) corresponding to these eigenvectors describe different rest states of the particle with the spin $k$ differing by the values of the projections of the spin on some axis. Therefore the possible spin values for a given particle are those values of $k$ for which the matrix $\| c^k_{\tau\tau'} \|$ has non-zero eigenvalues and the multiplicities of such eigenvalues define the numbers of different states with the spin $k$.

For infinite-dimensional irreducible representations of the Lorentz group, formulas (2.2) (or, more precisely, the set of $k$ values within them) show that equations involving wave functions $\psi$ that transform according to an infinite-dimensional representation generally describe particles that can have arbitrary integer or, respectively, half-integer spin states, with the spin values greater than some fixed values $k_0$ (in some cases a part of the "boxes" $\| c^k_{\tau\tau'} \|$ may correspond only to "redundant components" or "additional condition"). The rest mass values for the states with a given spin $k$ are defined by formula (6.2) from the non-zero eigenvalues $\lambda^{(k)}_i$ of the matrix $\| c^k_{\tau\tau'} \|$.

Suppose now that the representation defining the transformation law of the wave function $\psi$ (representation $D_R$) splits into a final number of irreducible representations, some of which are infinite-dimensional. In that case all the matrices $\| c^k_{\tau\tau'} \|$ are of finite order. In fact, the orders of the matrices $\| c^k_{\tau\tau'} \|$ will, evidently, coincide for all $k$ with the possible exception of several very low values. It follows from the basic formulas (3.13) and (3.14) that coefficients of the characteristic equations for the matrices $\| c^k_{\tau\tau'} \|$ are square roots of polynomials in $k$[24]. Hence, in general, the roots of such an equation increase infinitely in absolute value as $k \to \infty$[25]; only in exceptional cases, may coefficients of the characteristic equation all turn out to be independent of $k$, i.e. all eigenvalues of the matrices $\| c^k_{\tau\tau'} \|$ coincide for all $k$ with the possible exception of several very low values. Thus, any mass spectrum corresponding to this type of equation either tends to zero as $k \to \infty$ (the general case) or its values coincide for all $k$ sufficiently large (the exceptional case).

This result explains the failure of all attempts to construct a relativistically invariant equation with increasing mass spectrum. Of course, one can obtain equations with increasing mass spectrum by using an infinite number of irreducible representations. Adding for each spin value $k$ a sufficient number of new irreducible representations with $k_0 = k$, one apparently can obtain any mass value corresponding to that spin value without affecting the states of the particle with spin less than $k$. Such a construction, however, seems rather complicated since, as a rule, equations for which $D_R$ splits into an infinite number of irreducible representations also have a decreasing mass spectrum.

Note also that in this section we assume that the matrix $L^0$ has only real eigenvalues. Indeed, according to formula (6.2) a complex eigenvalue would

---

[24] In fact they are always polynomials in $k$.

[25] In general, eigenvalues of matrices $\| c^k_{\tau\tau'} \|$ increase as $k$ as $k \to \infty$.

correspond to a complex value of the rest mass; solution (6.1) then exponentially increases as $t \to \infty$ (or $t \to -\infty$). For an equation obtained from an invariant Lagrange function, formula (1.9) implies that an eigenvector $\psi^{(0)}$ of the matrix $L^0$ corresponding to the eigenvalue $\lambda$ satisfies the condition

$$(\lambda - \bar{\lambda})(\psi^{(0)}, \psi^{(0)}) = 0.$$

Therefore, an eigenvalue $\lambda$ may be complex only if $(\psi^{(0)}, \psi^{(0)}) = 0$. Thus, there can be no complex eigenvalues for the matrix $L^0$ if we agree to consider only those equations for which $L^0$ has no eigenvectors on the 'null cone" (i.e. when $(\psi^{(0)}, \psi^{(0)}) = 0$). The last restriction makes sense too because formulas (1.11) and (1.12) imply that the states of the wave function of the form (6.1) for which $(\psi^{(0)}, \psi^{(0)}) = 0$ would correspond to zero charge and energy densities.

## 7. Equations with positive values of charge or energy. Examples of invariant equations

1. *Equations with positive values of charge or energy.* In this section we consider only those equations that satisfy all conditions of sections 3–5. First of all, we shall be interested in finding equations that describe particles with positive definite full energy or positive (or negative) definite full charge.

According to formula (1.11) equation (1.1) yields the following charge density:

$$\rho = s^0/c = (\varepsilon/\hbar c)(L^0\psi, \psi). \tag{7.1}$$

It is sufficient to check that the charge density is positive for a flat wave satisfying condition (1.1); then the full charge is non-negative for all sums of such flat waves. For each flat wave there exists a corresponding "rest mass" coordinate system in which it takes the form (6.1). Therefore, for any invariant equation, the full charge is positive definite if for all eigenvectors $\psi^{(0)}$ of the matrix $L^0$ corresponding to non-zero eigenvalues

$$(L^0\psi^{(0)}, \psi^{(0)}) \geqq 0. \tag{7.2}$$

Similarly, the full energy is positive definite if the energy density

$$W = -T^0_0 = -(1/2i)\{(L^0\partial\psi/\partial x^0, \psi) - (\psi, L^0\partial\psi/\partial x^0)\} \tag{7.3}$$

is positive for all flat waves of the form (6.1), i.e. if for all eigenvectors $\psi^{(0)}$ of the matrix $L^0$ corresponding to non-zero eigenvalues one has

$$(\psi^{(0)}, \psi^{(0)}) \geqq 0. \tag{7.4}$$

In particular, it follows from (7.2) and (7.4) that for finite-dimensional equations of the form (1.1) with a diagonizable matrix $L^0$ the Dirac equation is the only one to have positive definite full charge and Duffin's equations for particles with spins 0 and 1 are the only ones to have positive definite energy, a result recently obtained in [11] in a somewhat different form.

Since in [11] this result was obtained with the use of cumbersome computations due to the use of the spinor notation of representations, we show here how it can be obtained from our formulas. If $L^0$ can be reduced to a diagonal form, condition (7.2), evidently, is equivalent to the requirement that

$$(L^0\psi,\psi)\geqq 0 \tag{7.2'}$$

for all vectors $\psi$, i.e. that the Hermitian quadratic form $(L^0\psi,\psi)$ should be non-negatively defined. Now if the representation $D_R$ contains irreducible components $\tau\sim(k_0,k_1)$ and $\dot{\tau}\sim(-k_0,k_1)$ for which $c^k_{\tau\dot{\tau}}=0$ (which is, in particular, always true provided there exists at least one $\tau$ such that $k_0\neq 1/2$) then the form $(L^0\psi,\psi)$ includes no terms containing $x^k_{p\tau}\bar{x}^k_{p\tau}=|x^k_{p\tau}|^2$ and $|x^k_{p\dot{\tau}}|^2$ but includes terms with $x^k_{p\tau}$ and $x^k_{p\dot{\tau}}$ of order one, which immediately implies that it is not non-negatively defined. Now, if that representation contains an irreducible representation $\tau\sim(1/2,k_1)$ for which $k_1>3/2$, then, taking into account the fact that for finite-dimensional representations one always has $a^k_{\tau\dot{\tau}}=-a^{k-1}_{\tau\dot{\tau}}$ (see (5.19) and (2.3″)), one concludes that the coefficients of $|x^k_{p\tau}|^2$ and $|x^{k-1}_{p\tau}|^2$ in the form $(L^0\psi,\psi)$ have opposite signs, making it impossible for the form to be always of the same sign. Thus, condition (7.2′) may be satisfied only if $D_R$ splits into irreducible representations corresponding to the pairs $\tau\sim(1/2,3/2)$ and $\dot{\tau}\sim(-1/2,3/2)$. It is easy to verify, however, that if each of those representations occurs more that once, equation (1.1) can be split into independent subequations. Thus, the only finite-dimensional equation with positive charge and matrix $L^0$ that can be reduced to the diagonal form is the Dirac equation for which $\tau=1,\dot{1}$, $1\sim(1/2,3/2)$, $\dot{1}\sim(-1/2,3/2)$ and $L^0$ (i.e. the matrix $\|c^{1/2}_{\tau\tau'}\|$) is of the form

$$\begin{Vmatrix}0 & 1\\ 1 & 0\end{Vmatrix}$$

(a real scalar $c_{11}$ in $L^0$ can, evidently, be included in $\kappa$).

Let us now proceed to finding finite-dimensional equations having positive full energy and a matrix $L^0$ that can be reduced to the diagonal form. In that case condition (7.4) is equivalent to the requirement that

$$(L^0\psi,L^0\psi)\geqq 0 \tag{7.4'}$$

for all vectors $\psi$, i.e. that the Hermitian form $(L^0\psi,L^0\psi)=((L^0)^2\psi,\psi)$ be non-negative. Now let $D_R$ include irreducible components $\tau\sim(k_0,k_1)$ and $\dot{\tau}=(-k_0,k_1)$ where $k_0\neq 0$. One can easily see that if there is no irreducible representation $\tau'$ in $D_R$ such that $c^k_{\tau\tau'}\neq 0$ and, simultaneously $c^k_{\dot{\tau}\tau'}\neq 0$, then the coefficient of $|x^k_{p\dot{\tau}}|^2$ in the form $(L^0\psi,L^0\psi)$ vanishes; if such a representation $\tau'$ does exist then that coefficient is proportional to $a^k_{\tau\dot{\tau}}\,|c^k_{\tau\tau'}|^2$. But a representation $\tau'$ for which $c^k_{\tau\tau'}\neq 0$ and $c^k_{\dot{\tau}\tau'}\neq 0$ may exist only if $\tau\sim(1,k_1)$, $\dot{\tau}\sim(-1,k_1)$ and $\tau'\sim(0,k_1)$. If $k_1>2$ then the coefficients of $|x^k_{p\dot{\tau}}|^2$ and $|x^{k-1}_{p\dot{\tau}}|^2$ in the form $(L^0\psi,L^0\psi)$ have the opposite signs, and, consequently, condition (7.4′) is not satisfied. Similarly, if equation (1.1) includes "adjacent" components of the form $\tau\sim(0,k_1)\rightleftarrows(0,k_1+1)\sim\tau'$ where $k_1>1$ then the coefficients of $|x^k_{p\tau}|^2$ and $|x^{k-1}_{p\tau}|^2$ in $(L^0\psi,L^0\psi)$ have different signs. Therefore, for the equations of the type we are interested in, only the following "adjacency schemes" are possible: $(1,2)\rightleftarrows(0,2)\rightleftarrows(-1,2)$ and $(0,1)\rightleftarrows(0,2)$. Therefore, $D_R$ may be decomposed into representations defined by the pairs $(1,2)$, $(-1,2)$, $(0,2)$ and $(0,1)$. One can now check without difficulty that the representations $\tau_0\sim(0,1)$,

$\tau_1 \sim (0,2)$, $\tau_2 \sim (1,2)$ and $\dot{\tau}_2 \sim (-1,2)$ cannot all be present in the decomposition of $D_R$ into irreducible representations (otherwise, the cóefficients of $|x^0_{0\tau_1}|^2$ and $|x^1_{p\tau_2}|^2$ in $(L^0\psi, L^0\psi)$ have opposite signs) and that each of the four representations may be present only once (otherwise the equation splits). Finally, there are only two possible "adjacency schemes" for equations with matrix $L^0$ that can be reduced to the diagonal form and with positive energy:

$$(1)\ (0,1) \rightleftarrows (0,2) \text{ and } (2)\ (1,2) \rightleftarrows (0,2) \rightleftarrows (-1,2).$$

For these schemes the energy is in fact positive for the following choice of the matrix $L^0$ (and only for that $L^0$ up to some transformation of the coordinate system):

$$(1)\ \|c^0_{\tau\tau'}\| = \left\|\begin{matrix} 0 & 1 \\ 1 & 0 \end{matrix}\right\|, \qquad \|c^1_{\tau\tau'}\| = \|0\|,$$

i.e.

$$L^0 = \left\|\begin{matrix} 0 & 1 & 0 & 0 & 0 \\ 1 & 0 & 0 & 0 & 0 \\ 0 & 0 & 0 & 0 & 0 \\ 0 & 0 & 0 & 0 & 0 \\ 0 & 0 & 0 & 0 & 0 \end{matrix}\right\|. \tag{7.5'}$$

$$(2)\ \|c^0_{\tau\tau'}\| = \|0\|,\ \|c^1_{\tau\tau'}\| = \left\|\begin{matrix} 0 & 1 & 1 \\ 1 & 0 & 0 \\ 1 & 0 & 0 \end{matrix}\right\|,$$

$$\text{i.e. } L^0 = \left\|\begin{matrix} 0 & 0\,0\,0 & 0\,0\,0 & 0\,0\,0 \\ 0 & 0\,0\,0 & 1\,0\,0 & 1\,0\,0 \\ 0 & 0\,0\,0 & 0\,1\,0 & 0\,1\,0 \\ 0 & 0\,0\,0 & 0\,0\,1 & 0\,0\,1 \\ 0 & 1\,0\,0 & 0\,0\,0 & 0\,0\,0 \\ 0 & 0\,1\,0 & 0\,0\,0 & 0\,0\,0 \\ 0 & 0\,0\,1 & 0\,0\,0 & 0\,0\,0 \\ 0 & 1\,0\,0 & 0\,0\,0 & 0\,0\,0 \\ 0 & 0\,1\,0 & 0\,0\,0 & 0\,0\,0 \\ 0 & 0\,0\,1 & 0\,0\,0 & 0\,0\,0 \end{matrix}\right\|. \tag{7.5''}$$

Equation (7.5′) is for the particle of spin 0 and (7.5″) is for the particle of spin 1. Depending on how one defines the operator $T$ one has in the case (1) a scalar (for $t^k = (-1)^{[k]}$ and $\tau \sim (0,1)$) or a pseudoscalar (for $t^k = (-1)^{[k]+1}$) equation and in the case (2) a vector (for $t^k = (-1)^{[k]}$ and $\tau \sim (0,2)$) or a pseudovector (for $t^k = (-1)^{[k]+1}$) equation.

2. *Equation for a particle of spin* 3/2. There also exist equations with positive energy or charge for which the matrix $L^0$ cannot be reduced to the diagonal form. For example, such are the equations studied by Pauli and Fierz in [2] and by

Ginzburg in [4]. The simplest of these equations can be found by considering invariant equations with the wave function $\psi$ transforming according to the representation that splits into the following four finite-dimensional irreducible representations: $1 \sim (1/2, 3/2)$, $\dot{1} \sim (-1/2, 3/2)$, $3 \sim (1/2, 5/2)$ and $\dot{3} \sim (-1/2, 5/2)$. Here the "adjacency scheme" is of the form

$$\begin{array}{ccc} (1/2,3/2) & \rightleftarrows & (1/2,5/2) \\ \uparrow\downarrow & & \uparrow\downarrow \\ (-1/2,3/2) & \rightleftarrows & (-1/2,5/2) \end{array}$$

Here $k$ can take values 1/2 and 3/2 and the matrix $L^0$ (i.e. the "boxes" $\| c_{\tau\tau'}^{1/2} \|$, $\| c_{\tau\tau'}^{3/2} \|$) are given by the formula

$$\begin{array}{cc} k=1/2 & k=3/2 \\ \begin{pmatrix} 0 & c_{1\dot{1}} & c_{13} & 0 \\ c_{\dot{1}1} & 0 & 0 & c_{\dot{1}\dot{3}} \\ c_{31} & 0 & 0 & c_{3\dot{3}} \\ 0 & c_{\dot{3}\dot{1}} & c_{\dot{3}3} & 0 \end{pmatrix}, & \begin{pmatrix} 0 & 2c_{3\dot{3}} \\ 2c_{\dot{3}3} & 0 \end{pmatrix}. \end{array}$$

In order for the equation to be invariant under reflections (section 4, formula (4.24)) the following relations have to be satisfied:

$$c_{1\dot{1}} = c_{\dot{1}1}, \quad c_{13} = c_{\dot{1}\dot{3}}, \quad c_{31} = c_{\dot{3}\dot{1}}, \quad c_{3\dot{3}} = c_{\dot{3}3}. \tag{7.6}$$

For the numbers $a_{\tau\dot{\tau}}$ defining an invariant bilinear form there are the following two alternatives that cannot be reduced to one another:

$$\text{(a)}\ a_{1\dot{1}} = +1, \quad a_{3\dot{3}} = +1$$

and

$$\text{(b)}\ a_{1\dot{1}} = +1, \quad a_{3\dot{3}} = -1.$$

These two possibilities correspond to the following conditions that have to be satisfied if our equation is to be obtained from an invariant Lagrange function (section 5, formula (5.32)):

$$\text{(a)}\ c_{1\dot{1}} = \overline{c_{\dot{1}1}}, \quad c_{13} = \overline{c_{3\dot{1}}}, \quad c_{\dot{1}\dot{3}} = \overline{c_{31}}, \quad c_{3\dot{3}} = \overline{c_{33}} \tag{7.7'}$$

and

$$\text{(b)}\ c_{1\dot{1}} = \overline{c_{\dot{1}1}}, \quad c_{13} = -\overline{c_{3\dot{1}}}, \quad c_{\dot{1}\dot{3}} = -\overline{c_{31}}, \quad c_{3\dot{3}} = \overline{c_{\dot{3}3}}. \tag{7.7''}$$

Thus, for an equation satisfying all conditions of sections 3–5, the matrix $L^0$ has to be of the form:

$$\text{(a)} \quad \begin{array}{cc} k=1/2 & k=3/2 \\ \begin{pmatrix} 0 & c_{11} & c_{13} & 0 \\ c_{11} & 0 & 0 & c_{13} \\ \overline{c_{13}} & 0 & 0 & c_{33} \\ 0 & \overline{c_{13}} & c_{33} & 0 \end{pmatrix}, & \begin{pmatrix} 0 & 2c_{33} \\ 2c_{33} & 0 \end{pmatrix}; \end{array}$$

(b) $k=1/2$ $k=3/2$

$$\begin{pmatrix} 0 & c_{11} & c_{13} & 0 \\ c_{11} & 0 & 0 & c_{13} \\ -\overline{c_{13}} & 0 & 0 & c_{33} \\ 0 & -\overline{c_{13}} & c_{33} & 0 \end{pmatrix}, \quad \begin{pmatrix} 0 & 2c_{33} \\ 2c_{33} & 0 \end{pmatrix},$$

where $c_{11}$, $c_{33}$ are arbitrary real numbers and $c_{13}$ is a complex number. One can simplify that expression further by going to a new coordinate system. Any coordinate transformations preserving all infinitesimal operators $I^{jk}$, the operator $T$ and the invariant bilinear form are, in view of (3.15), (4.25) and (5.34), given by the formulas:

$$\begin{aligned} \xi'^k_{p1} &= e^{i\theta_1}\xi^k_{p1}, & \xi'^k_{\dot{p}\dot{1}} &= e^{i\theta_1}\xi^k_{\dot{p}\dot{1}}, \\ \xi'^k_{p3} &= e^{i\theta_0}\xi^k_{p3}, & \xi'^k_{\dot{p}\dot{3}} &= e^{i\theta_3}\xi^k_{\dot{p}\dot{3}}. \end{aligned} \tag{7.8}$$

Under such a transformation the elements $c_{11}$ and $c_{33}$ of the matrix $L^0$ evidently do not change while $c_{13}$ goes into

$$c'_{13} = c_{13}e^{i(\theta_1-\theta_3)}.$$

Thus the essential parameters of the equation (1.1) in that case are three real numbers: $c_{11}$, $c_{33}$ and $|c_{13}|$.

Choosing $\theta_1$ and $\theta_3$ in the relations (7.8) accordingly and dividing the whole equation by $2c_{33}$ (which only changes the scalar $\kappa$) we can reduce our matrix $L^0$ to the following form

(a) $k=1/2$ $k=3/2$

$$\begin{pmatrix} 0 & \alpha & \beta & 0 \\ \alpha & 0 & 0 & \beta \\ \beta & 0 & 0 & 1/2 \\ 0 & \beta & 1/2 & 0 \end{pmatrix}, \quad \begin{pmatrix} 0 & 1 \\ 1 & 0 \end{pmatrix};$$

(b) $k=1/2$ $k=3/2$

$$\begin{pmatrix} 0 & \alpha & \beta & 0 \\ \alpha & 0 & 0 & \beta \\ -\beta & 0 & 0 & 1/2 \\ 0 & -\beta & 1/2 & 0 \end{pmatrix}, \quad \begin{pmatrix} 0 & 1 \\ 1 & 0 \end{pmatrix},$$

where $\alpha$ is a real number and $\beta$ is a real positive number. The eigenvalue of $L^0$ corresponding to the spin 3/2 are here equal to $\pm 1$ and the eigenvalues corresponding to the spin 1/2 are defined from the characteristic

$$\text{(a)}\ \lambda^4 - (\alpha^2 + 1/4 + 2\beta^2)\lambda^2 + (\alpha/2 + \beta^2)^2 = 0 \tag{7.9'}$$

or, respectively,

$$\text{(b)}\ \lambda^4 - (\alpha^2 + 1/4 - 2\beta^2)\lambda^2 + (\alpha/2 + \beta^2)^2 = 0. \tag{7.9''}$$

Clearly, in case (a) the eigenvalues $\pm\lambda_1$ and $\pm\lambda_2$ of the matrix $\|c^{1/2}_{\tau\tau'}\|$ are always real, and in case (b) they are real provided

$$\alpha^2 + 1/4 \geqq 2\beta^2$$

and

$$|\alpha - 1/2| \geqq 2\beta.$$

The resulting equations describe the particle that can be in the state of spin 3/2 and mass $m^{3/2} = \kappa c/\hbar$ or in the two states of spin 1/2 and mass equal to $m_1^{1/2} = (\kappa c/\hbar)(1/\lambda_1)$ or, respectively, $m_2^{1/2} = (\kappa c/\hbar)(1/\lambda_2)$. The ratios $m_1^{1/2}/m^{3/2}$ and $m_2^{1/2}/m^{3/2}$ of the "lower" masses to the "upper" mass can, of course, be made equal to any given values by appropriate choice of the parameters $\alpha$ and $\beta$. However, if we place no conditions for the charge and the energy to be positive, none of our equations will have any precedence over the others. For that reason there is no foundation in Bhabha's statement in [5] that the proton is described by the equation of type (a) with $\alpha = 3/2$ and $\beta = \sqrt{15}/2$ which gives the charge and the energy of both signs since that particular equation has no precedence over an innumerable number of other equations.

Since the equation we consider now is a finite-dimensional equation for the particle with half-integer spin, the energy corresponding to that equation cannot be positive definite (see [21]). In case (a) the matrix $L^0$ is symmetric and, consequently, can be reduced to the diagonal form. Hence, in view of section 7.1, none of the equations of type (a) has a positive definite full charge. The results of section 7.1 also imply that in case (b) the full charge may be positive only for those equations for which the matrix $L^0$ cannot be reduced to the diagonal form. That implies, in particular, that the charge cannot be positive unless the characteristic equation (7.9″) has multiple roots. Therefore, in order to find all equations with positive charge one has only to check those equations of type (b) for which (7.9″) has multiple roots, i.e. one has to consider the following cases:

(1) $\alpha = -1/2$ (roots (7.9″): $\lambda_{1,2} = \sqrt{1/4 - \beta^2}$, $\lambda_{3,4} = -\sqrt{1/4 - \beta^2}$);
(2) $\alpha = 1/2 \pm 2\beta$ (roots (7.9″): $\lambda_{1,2} = 1/2 \pm \beta$, $\lambda_{3,4} = -(1/2 \pm \beta)$) and
(3) $\alpha = -2\beta^2$ (roots (7.9″): $\lambda_{1,2} = 0$, $\lambda_3 = 1/2 - 2\beta^2$, $\lambda_4 = -1/2 + 2\beta^2$).

It is easy to see that cases (1) and (3) here yield for $\beta \neq 1/2$ the matrices $L^0$ that can be reduced to the diagonal form and, therefore, the corresponding equations are not those with positive charge. In case (2) the matrix $L^0$ cannot be reduced to the diagonal form and for all eigenvectors $\psi^{(0)}$ of that matrix corresponding to $k = 1/2$ one has $(\psi^{(0)}, \psi^{(0)}) = 0$. Thus, all corresponding equations are equations with positive charge. For $\beta \neq 1/2$ the resulting equation describes the particle with spin states 1/2 and 3/2 such that the spin 1/2 corresponds to the wave function for which $(\psi, \psi) = (L^0\psi, \psi) = 0$ (i.e. both the energy and the charge are zero). Considering only those equations that have no such "zero" states, one has a single equation with positive charge when $L^0$ is defined by the formula:

$$\overset{k=1/2}{\begin{pmatrix} 0 & -1/2 & 1/2 & 0 \\ -1/2 & 0 & 0 & 1/2 \\ -1/2 & 0 & 0 & 1/2 \\ 0 & -1/2 & 1/2 & 0 \end{pmatrix}}, \quad \overset{k=3/2}{\begin{pmatrix} 0 & 1 \\ 0 & 1 \end{pmatrix}}. \tag{7.10}$$

Here all eigenvalues of $L^0$ for $k=1/2$ vanish and equation (1.1) becomes the Pauli–Fierz equation 2 for the particle of spin 3/2[26].

3. *Examples of infinite-dimensional invariant equations.* The simplest infinite-dimensional equation satisfying all conditions of sections 3–5 is the equation with the wave function transforming according to the irreducible representation of the proper (as well as the full) Lorentz group defined by the pair of numbers $(0,1/2)$ or $(1/2,0)$ (according to the results of section 3 it is the only case when $D_R$ can be an irreducible representation of the proper group); recall that the pair $(0,1/2)$ is equivalent to the pair $(0,-1/2)$. Here $k=0,1,2,\ldots$ or, respectively, $k=1/2,\ 3/2,\ 5/2,\ldots$; the matrices $\|c^k_{\tau\tau'}\|$ are actually all scalars and by (3.13) and (3.14) one has

$$\|c^k_{\tau\tau'}\| = \|k+1/2\|. \tag{7.11}$$

The corresponding equation is, in that case, particularly simple and for that reason we write it out in full. Here $\psi$ is a vector having components $\psi^k_p$ ($p=k,k-1,\ldots,-k$ and $k=0,1,2,\ldots$ or, respectively, $k=1/2,\ 3/2,\ 5/2,\ldots$) and the equation is of the form:

$$\begin{aligned}&\frac{\partial}{\partial x^0}(k+1/2)\psi^k_p + 1/2\frac{\partial}{\partial x^1}\{\sqrt{(k+p+1)(k-p+1)}\,\psi^{k+1}_p + \sqrt{(k+p)(k-p)}\psi^{k-1}_p\}\\ &\quad+\frac{1}{4}\left(\frac{\partial}{\partial x^2}+i\frac{\partial}{\partial x^3}\right)\{\sqrt{(k+p+1)(k+p+2)}\,\psi^{k+1}_{p+1} - \sqrt{(k-p-1)(k-p)}\psi^{k-1}_{p+1}\}\\ &\quad-1/4(\partial/\partial x^2 - i\partial/\partial x^3)\{\sqrt{(k-p+1)(k-p+2)}\psi^{k+1}_{p-1}\\ &\quad-\sqrt{(k+p-1)(k+p)}\psi^{k-1}_{p-1} + i\kappa\psi^k_p = 0.\end{aligned} \tag{7.12}$$

Since the representations defined by the pairs $(0,1/2)$ and $(1/2,0)$ are unitary, it is clear that for all eigenvectors $\psi^{(0)}$ of the matrix $L^0$ one has $(\psi^{(0)},\psi^{(0)})\geqq 0$ and, consequently, the value of energy corresponding to our equations is positive. The charge corresponding to these equations is also positive: here all eigenvalues of $L^0$ are positive and, consequently, in all cases one has $(L^0\psi^{(0)},\psi^{(0)})\geqq 0$. Thus we have constructed to equations; one describes particles with integer spin and another describes those particles with half-integer spin for which both charge and energy are positive, a fact impossible for finite-dimensional equations where the charge

[26] It is easy to see that in this case the matrix $c^{1/2}$ has only second-order elementary factors and, consequently, satisfies the relation $(c^{1/2})^2=0$. It follows from this that $L^0$ here satisfies the relation $(L^0)^4=(L^0)^2$. Thus, the statement of Harish-Chandra [10] that one knows no equation satisfying that relation is not correct and is closely connected with the statement there that Pauli–Fierz equations cannot be represented in the form (1.1).

*Note added in proof.* Recently Harish-Chandra has published a work (Proc. Roy. Soc., A 192, 195, 1948) specially devoted to finding equations of the form (1.1) for which $(L^0)^4=(L^0)^2$.

Two such equations he has found there are much more complicated than the Pauli–Fierz equations; they describe particles that can have several different states (with spins 1/2 and 3/2) corresponding to one and the same mass value. The statement that those equations, unlike Pauli–Fierz equations, do not involve any "additional conditions" is again based on a misunderstanding (cf. footnotes [4] and [26] to the present paper).

cannot be positive for integer spin and the energy cannot be positive for half-integer spin.

The mass spectrum corresponding to the equation (7.12) is evidently decreasing: the mass value corresponding to the spin $k$ is $m^k = m_0/(k+1/2)$ where $m_0 = \kappa c/\hbar$.

For the next example, consider equations for which the representation $D_R$ splits into two irreducible representations of the proper Lorentz group. Clearly, an equation may satisfy all conditions of sections 3–5 only if those irreducible representations are defined by the pairs $(1/2, k_1)$ and $(-1/2, k_1)$ where $k_1$ is either purely imaginary or real (in the last case for $k_1$ integer or half-integer the pairs of representations defined by the pairs $(1/2, k_1)$ and $(-1/2, k_1)$ are also possible). For $k_1$ purely imaginary there are two more ways to define an invariant bilinear form $(\psi_1, \psi_2)$; let us assume that it is chosen in such a way that it is positive. In that case $L^0$ (i.e. the matrices $\| c^k_{\tau\tau'} \|$) for all our equations may be reduced to the form:

$$\| c^k_{\tau\tau'} \| = \begin{Vmatrix} 0 & k+1/2 \\ k+1/2 & 0 \end{Vmatrix}. \tag{7.13}$$

Thus, to all those equations there corresponds the same mass spectrum $m^{(k)} = m_0/(k+1/2)$.

For $k_1$ purely imaginary one has the series of equations for particles with half-integer spin, positive energy and charge of both signs. For $k_1$ real and the representations $(1/2, k_1), (-1/2, k_1)$ one has the equations with energy of both signs and charge that is either positive or not dependent on whether $0 \leqq |k_1| \leqq 3/2$ or $|k_1| > 3/2$. If $k_1$ is half-integer and greater than 1/2 one has the equation that splits into a finite-dimensional equation and its "tail", i.e. an infinite-dimensional equation corresponding to the representation $D_R$ splitting into $\tau \sim (k_1, 1/2)$ and $\dot{\tau} \sim (k_1, -1/2)$ (in particular, for $k_1 = 3/2$ the finite-dimensional equation is the Dirac equation). For $k_1$ integer or half-integer and $D_R$ splitting into $\tau \sim (k_1, 1/2)$ and $\dot{\tau} \sim (k_1, -1/2)$ (in particular, for the "tails" of finite-dimensional equations) one has positive charge and energy of both signs; spin values corresponding to those equations are integer for $k_1$ integer and half-integer for $k_1$ half-integer.

## References

1. Duffin, R.J.: Phys. Rev. **54** (1938) 1114; Kemmer; N. Proc. Roy. Soc. A **173** (1939) 91
2. Fierz, M. Pauli, W.: Proc. Roy. Soc. A **173** (1939) 211
3. Kramers, H.A., Belinfante, F.J., Lubanski, J.K.: Physica **8** (1941) 597; Lubanski J.K.: Physica **9** (1942) 310
4. Ginzburg, V.L.: ZhETF **13** (1943) 33, (Russian)
5. Bhabha, H.J.: Proc. Ind. Acad. Sci. A **21** (1945) 241
6. Petiau, G.: J. Phys. Rad. **8**, (1946) 7 124, 181
7. Potier, R., Compt. Rend. **224** (1947) 1332; **226** (1948) 63, 314
8. Bhabha, H.J.: Rev. Mod. Phys. **17** (1945) 200
9. Potier, R.: Compt. Rend. **222** (1946) 638, 855, 1076
10. Harish-Chandra: Phys. Rev. **71** (1947) 793
11. Wild, E.: Proc. Roy. Soc. A **191** (1947) 253
12. Dirac, P.A.M.: Proc. Roy. Soc. A **183** (1945) 284
13. Ginzburg, V. L., Tamm, I.E.: ZhETF **17** (1947) 227 (Russian)
14. Izmailov, S.: ZhETF **17** (1947) 629 (Russian)

15. Harish-Chandra: Proc. Roy. Soc. A. **189** (1947) 372
16. Alihanian, A., Alihanov, A., Morozov, V., Mushelishvili, G., Hrimian, A.: Dokl. Acad. Naŭk SSSR, **58** (1947) 1321 (Russian)
17. Gelfand, I., Najmark M.: J. Phys. **10** (1946) 93
18. Gelfand, I.M. Najmark, M.A.: Izv. Dokl. Acad. Naŭk SSSR, Ser. Mat. **11** (1947) 411
19. Bargmann, V.: Ann. Math. **48** (1947) 568
20. Cartain, E.: The Theory of Spinors. Hermann, Paris 1966. (Russian translation: Moscow, IL, 1947)
21. Pauli, W.: Phys. Rev. **58** (1940) 716; Pauli, W.: Rev. Mod. Phys. **13** (1941) 203
22. van der Waerden, B.L.: Die gruppentheoretische Methode in der Quantenmechanik. J.W. Edwards, Ann Arbor 1944

# 10.

## (with M. I. Graev)

# On the structure of the ring of rapidly decreasing functions on a Lie group

Dokl. Akad. Nauk SSSR **124** (1959) 19–21. Zbl. **103**:336

A careful scrutiny of representations of semisimple Lie groups makes it evident that a key question in representation theory concerns the structure of the ring of infinitely differentiable functions on that group. After the structure of that ring is determined the treatment of almost all the problems of representation theory, e.g., finding all irreducible unitary representations, finding all irreducible representations, spherical functions with respect to an arbitrary (even non-compact or discrete) subgroup, etc., are based on firm ground. In this paper we study the structure of that ring for the group of complex matrices of order 2 with determinant 1. The case of an arbitrary complex, and then of an arbitrary real, semisimple Lie group will be considered in the following papers.[1]

Consider the group $G$ of complex matrices of order 2 $g = \begin{Vmatrix} \alpha & \beta \\ \gamma & \delta \end{Vmatrix}$, $\alpha\delta - \beta\gamma = 1$. Irreducible representations of that group are given by the well known formula

$$T_g f(z) = f\left(\frac{\alpha z + \gamma}{\beta z + \delta}\right)(\beta z + \delta)^{n_1 - 1}\overline{(\beta z + \delta)^{n_2 - 1}}, \tag{1}$$

where $n_1, n_2$ are complex numbers such that $n_1 - n_2 = n$ is an integer[2].

We call a function $x(g)$ rapidly decreasing if $|x(g)| = o(\|g\|^{-n})$ for all $n$. The derivative of the function $x(g)$ is an expression of the form $P(D)x(g)$ where $P(D)$ is an arbitrary polynomial in infinitesimal right and left shift operators $D_i$ on the group (Lie operators). Denote by $\Gamma$ the set of infinitely differentiable functions $x(g)$ such that for any $P(D)$ the derivative $P(D)x(g)$ is a rapidly decreasing function. Taking convolution for the multiplication law in $\Gamma$ and introducing a topology in a natural way one gets a ring which we call the basic group ring of the group $G$.

---

[1] For the group of real matrices of order 2 with determinant 1 such an investigation has been carried out in the interesting papers [1, 2] by Ehrenpreis and Mautner who, as their work evidently shows, realised very well the importance of that problem. However, the method they used to solve it for the group of real matrices of order 2 (decomposition by spherical functions, i.e., the use of matrix elements) cannot be extended to other groups without giving rise to great and unjustified difficulties. The problem is that their investigation is carried out with the help of matrix elements and these are known only for the group of real matrices of order 2. Also, for the group of real matrices of order 2 there are no relations which arise in the general case and ultimately define connections between representations at different integral points (cf. condition 5). For the case of the complex group of matrices of order 2 see also [4].

[2] For integer $n_1, n_2$ the representation is semireducible.

Let $x(g) \in \Gamma$. For each $n_1, n_2$ the operator $x(g) T_g dg$, where $T_g$ is defined by formula (1), is an operator with a kernel $K(z_1, z_2; n_1, n_2)$. We call the kernel $K(z_1, z_2; n_1, n_2)$ the Fourier transform of the function $x(g)$. Then the convolution of functions $x_1(g)$ and $x_2(g)$ corresponds to the convolution of the corresponding kernels:

$$K(z_1, z_2; n_1, n_2) = \int K_1(z_1, z_; n_1, n_2) K_2(z, z_2; n_1, n_2) dz\, d\bar{z}.$$

Thus the ring $\Gamma$ of functions on the group becomes the ring of kernels with the usual multiplication law for kernels. The goal of the present paper is to describe the set of kernels $K(z_1, z_2; n_1, n_2)$.

Now we formulate necessary and sufficient conditions for a function $K(z_1, z_2; n_1, n_2)$ to be a Fourier transform of a function $x(g)$ from $\Gamma$.

1. $K(z_1, z_2; n_1, n_2)$ is an infinitely differentiable function in $z_1$ and $z_2$.

2. $K(z_1, z_2; n_1, n_2)$ has the following asymptotics with respect to $z_1$ and $z_2$. The function

$$K(z_1, z_2; n_1, n_2) z_1^{-n_1+1} \bar{z}_1^{-n_1+1} z_2^{n_1+1} \bar{z}_2^{n_2+1}$$

after the substitution $z_1' = z_1^{-1} z_2' = z_2^{-1}$ becomes an infinitely differentiable function in $z_1', z_2'$ in some neighbourhood of each point where $z_1' = 0$, or $z_2' = 0$, or $z_1' = z_2' = 0$. It follows then that for large $z_1, z_2$

$$K(z_1, z_2; n_1, n_2) \sim C z_1^{n_1-1} \bar{z}_1^{n_1-1} z_2^{-n_1-1} \bar{z}_2^{-n_2-1}.$$

3. $K(z_1, z_2; n_1, n_2)$ and also all its derivatives with respect to $z_1, \bar{z}_1, z_2, \bar{z}_2$ are entire functions in $n_1 + n_2$ of first order. The function $K$ and its derivative satisfy some uniformity conditions with respect to $z_1, z_2$. It is more convenient to formulate these conditions not for kernels $K(z_1, z_2; n_1, n_2)$ but for their "Mellin transforms" $\varphi(z_1, z_2; \lambda)$, namely, there exists a function $\varphi(z_1, z_2; \lambda)$ such that

$$K(z_1, z_2; n_1, n_2) = \int \varphi(z_1, z_2; \lambda) \lambda^{n_1-1} \bar{\lambda}^{n_2-1} d\lambda\, d\bar{\lambda}.$$

The function satisfies the following conditions: for any $n > 0$

$$|\varphi(z_1, z_2; \lambda)| = o(|\lambda|^{-n}) \quad \text{for } \lambda \to \infty;$$
$$|\varphi(z_1, z_2; \lambda)| = o(|\lambda|^{n}) \quad \text{for } \lambda \to 0$$

uniformly in any finite domain in $z_1, z_2$.

The function $\varphi(z_1, z_2; \lambda)$ satisfies some additional conditions in the neighbourhood of any point where $z_1 = \infty$ or $z_2 = \infty$ or $z_1 = z_2 = \infty$. Now we formulate these conditions.

First of all note that if a function $x(g)$ corresponds to the function $\varphi(z_1, z_2; \lambda)$ then the function $x(g_0^{-1} g)$ corresponds to the function

$$|\alpha_0 - \beta_0 z_1|^{-2} \varphi\left(\frac{\gamma_0 - \delta_0 z_1}{\alpha_0 - \beta_0 z_1}, z_2, \frac{\lambda}{\alpha_0 - \beta_0 z_1}\right).$$

Letting, in particular, $\gamma_0 = -\beta_0 = 1$, $\alpha_0 = \delta_0 = 0$, one concludes that the conditions satisfied by $\varphi(z_1, z_2; \lambda)$ should also be satisfied in a finite domain by the function

$$\varphi_1(z_1, z_2; \lambda) = |z_1|^{-2} \varphi\left(\frac{1}{z_1}, z_2; \frac{\lambda}{z_1}\right).$$

This gives conditions on the function $\varphi$ in the neighbourhood of any point where $z_1 = \infty$. Similar conditions on $\varphi(z_1, z_2; \lambda)$ in the neighbourhood of any point where $z_2 = \infty$ are obtained if one applies a right shift operator to $x(g)$ instead of a left shift operator. Finally, taking both right and left shifts together one obtains the condition on $\varphi(z_1, z_2; \lambda)$ in the neighbourhood of the point $z_1 = z_2 = \infty$.

4. The functions $K(z_1, z_2; n_1, n_2)$ and $K(z_1, z_2; -n_1, -n_2)$ are related, namely, there exists an operator $B$ such that

$$BK(n_1, n_2) = K(-n_1, -n_2)B.$$

Explicitly,

$$\begin{aligned}&\int K(z_1, z_2 - z; n_1, n_2) z^{-n_1-1} \bar{z}^{-n_2-1} dz\, d\bar{z}\\ &\quad = K(z_1 - z, z_2; -n_1, -n_2) z^{-n_1-1} \bar{z}^{-n_2-1} dz\, d\bar{z}\end{aligned}$$

(the convolution should be understood as a regularised value of the integral [3]).

5. For any integer $n_1$ and $n_2$ the following relations hold:

$$\frac{\partial^{n_1}}{\partial z_1^{n_1}} K(z_1, z_2; n_1, n_2) = (-1)^{n_1} \frac{\partial^{n_1}}{\partial z_2^{n_1}} K(z_1, z_2; -n_1, n_2)$$

for $n_1 = 1, 2, \ldots$ and, likewise,

$$\frac{\partial^{n_2}}{\partial \bar{z}_1^{n_2}} K(z_1, z_2; n_1, n_2) = (-1)^{n_2} \frac{\partial^{n_2}}{\partial \bar{z}_2^{n_2}} K(z_1, z_2; n_1, -n_2)$$

for $n_2 = 1, 2, \ldots$.

There is no more relation which should be noted separately, notwithstanding the fact that it is a consequence of those above.

6. For any positive integers $n_1, n_2$ the integral $\int K(z_1, z_2; n_1, n_2) z_2^{k_1} \bar{z}_2^{k_2} dz_2 d\bar{z}_2$ is a polynomial in $z_1 \bar{z}_1$ of order $n_1 - 1$ when $k_1 = 0, \ldots, n_1 - 1$; $k_2 = 0, \ldots, n_2 - 1$. The integral $\int K(z_1, z_2; -n_1, -n_2) z_1^{k_1} \bar{z}_1^{k_2} dz_1 d\bar{z}_1$ for any positive integers $n_1, n_2$ and $k_1 = 0, \ldots, n_1 + 1$, $k_2 = 0, \ldots, n_2 + 1$ is a polynomial in $z_2, z_2$ of order $n_2 - 1$.

The momentum of the function $x(g)$ is, by definition, an integral $\int x(g) a(g) dg$ where $a(g)$ is a matrix element of a finite-dimensional representation of the group $G$. Note that for each kernel $K(z_1, z_2;\, n_1, n_2)$ one can compute the momenta by means of the following formulas:

$$\begin{aligned}&\int x(g)(\delta + \beta z_1)^{n_1 - k_1 - 1} \overline{(\delta + \beta z_1)}^{n_2 - k_2 - 1} (\gamma + \alpha z_1)^{k_1} \overline{(\gamma + \alpha z_1)}^{k_2} dg\\ &\quad = \int K(z_1, z_2; n_1, n_2) z_2^{k_1} \bar{z}_2^{k_2} dz_2 d\bar{z}_2, \qquad k_1 = 0, \ldots, n_1 - 1, \qquad k_2 = 0, \ldots, n_2 - 1;\\ &\int x(g)(\alpha - \beta z_2)^{n_1 - k_1 + 1} \overline{(\alpha - \beta z_2)}^{n_2 - k_2 + 1} (-\gamma - \delta z_2)^{k_1} \overline{(-\gamma - \delta z_2)}^{k_2} dg\\ &\quad = \int K(z_1, z_2; -n_1, -n_2) z_1^{k_1} \bar{z}_1^{k_2} dz_1 d\bar{z}_1, \qquad k_1 = 0, \ldots, n + 1, \qquad k_2 = 0, \ldots, n_2 + 1.\end{aligned}$$

In particular, letting $n_1 = n_2 = 1$, one has

$$\int x(g)dg = \int K(z_1, z_2; 1, 1)dz_2 d\bar{z}_2.$$

Due to lack of space we cannot give here an important interpretation of conditions 4, 5, and 6 in representation terms.

Received September 24, 1988

## References

1. Ehrenpreis, Z., Mautner, F.J.: Ann. Math. **61** (1955) 406
2. Ehrenpreis, L., Mautner, F.J.: Trans. Am. Math. Soc. **84** (1957) 1
3. Gelfand, I.M., Shilov, G.E.: Generalised functions. Moscow 1958. (Russian)
4. Zhelobenko, D.P.: Dokl. Akad. Naūk **124** (1958) 4

# 11.

## (with M. A. Najmark)

# Unitary representations of classical groups*

Tr. Mat. Inst. Steklova **36** (1950) 1–288 (in Russian). Zbl. **41**:362

Transl. of the Introduction, Chap. 9 'Spherical functions' and
Chap. 18, 'Transitivity classes for the set of pairs. Another way of describing representations of the complementary series'.

*Dedicated to the memory of Vyacheslav Vasilyevich Stepanov*

### Introduction

1. Let $\mathfrak{G}$ be a group. We say that a representation of $\mathfrak{G}$ is given if a rule is specified associating to each element $g \in \mathfrak{G}$ a linear transformation $T_g$ acting in a linear space $\mathfrak{H}$ such that the product of elements of the group corresponds to the composition of linear transformations:

$$T_{g_1 g_2} = T_{g_1} T_{g_2} \tag{1}$$

with the unit element of the group corresponding to the identical transformation. We assume below that $H$ is a Hilbert space with scalar product satisfying the following conditions:

$$(\xi, \eta) = \overline{(\eta, \xi)},$$
$$(\alpha\xi_1 + \beta\xi_2, \eta) = \alpha(\xi_1, \eta) + \beta(\xi_2, \eta),$$
$$(\xi, \xi) \geqq 0$$

where $(\xi, \xi) = 0$ if and only if $\xi = 0$.

We assume the operators $T_g$ to be unitary[1], and if the group $G$ is a continuous one the representation is also assumed to be continuous in the sense that $(T_g \xi, \eta)$ is a continuous function of $g$ for any fixed $\xi$ and $\eta$. If those conditions are satisfied a representation $g \to T_g$ is called a *unitary representation of the group* $\mathfrak{G}$.

A representation is called *reducible* if there is a subspace $\mathfrak{H}_1$ in $\mathfrak{H}$ (different from $\mathfrak{H}$ and (0)) invariant under the action of all operators $T_g$. Otherwise a representation is called *irreducible*.

An important role played by irreducible unitary representations stems from

---

* The Russian original consists of nine parts divided into 36 chapters subdivided into sections. Only the Introduction, Chapter 9 from Part II and Chapter 18 from Part IV have been translated into English for publication in this volume. The reader will therefore find a few references to, for example, formulas which appear in those parts of the original which have not been translated into English.

[1] An operator $u$ is called *unitary* if it is invertible and does not change the length of vectors, i.e. if $(u\xi, u\xi) = (\xi, \xi)$ for all $\xi$.

the fact that the study of any unitary representation of the group $G$ is usually reduced to irreducible ones.

After the existence of irreducible unitary representations for locally compact groups had been proved in [12] the question arose as to how can one actually find those representations for the case of the most interesting groups. The principal role among Lie groups is played by simple Lie groups which (with the exception of five separate groups that will not be considered here) are reduced to the following classes of groups:

1°. group $A_n$ of all complex matrices of order $n+1$ with determinant 1;

2°. the group of all complex orthogonal matrices (it is usually convenient to consider separately the group of orthogonal matrices of even order $2n$ (group $B_n$) and separately the group of orthogonal matrices of odd order $(2n+1)$ (group $D_n$);

3°. the group of all complex matrices that leave invariant a non-degenerate skew-symmetric bilinear form[2] (such a form exists only in space with even dimension $2n$)—group $C_n$ ($n$ is called the rank of the group).

Those classes of simple groups are called *classical groups*.

The classical methods of representation theory relating representations of a finite group to the representations of its normal subgroup and the representations of its quotient group can apparently be extended to arbitrary Lie groups (see Gelfand and Najmark [5] and the subsequent elaboration in Mackey [29]).

Thus, in the problem of describing representations of an arbitrary Lie group, the central role is played by the study of representations of semi-simple Lie groups.

Since any semi-simple Lie group is a direct product of simple Lie groups, the study of representations of a semi-simple Lie group is therefore reduced to the study of representations of simple Lie groups.

The problem of finding all finite-dimensional representations of complex simple Lie groups[3] was solved by Cartan [26, 27] in 1914. None of those representations, with the exception of the identical one, is unitary.

The aim of the present paper is to give a list of all irreducible unitary representations of the classical groups listed above. The result is to a certain degree paradoxical: infinite-dimensional representations turn out to be in many respects simpler than finite-dimensional ones. In any case our results provide a unified scheme which includes as a speical case both finite-dimensional and unitary representations.

Moreover, it turns out as a result of our investigation that the unitarity condition for a representation is, to a certain extent, an unnatural one. Essentially, the

---

[2] The group of orthogonal matrices is the group of matrices that leave invariant the sum of the squares of coordinates, i.e. a symmetric bilinear form. Groups of types 2° and 3° are, therefore, the groups leaving invariant a bilinear form, symmetric and skew-symmetric, respectively.

[3] More precisely, Cartan's works show that each finite-dimensional representation of such a group is defined by a system of integers $m_1 \geqq \ldots \geqq m_n$. Conversely, for each $n$-tuple $(m_1, \ldots, m_n)$ a construction was given leading to the irreducible representation defined by the integers $m_1, \ldots, m_n$. However, using that construction, it would be difficult to write the representations explicitly, i.e. to write explicitly the matrices corresponding to the group elements. For the corresponding explicit formulas for infinitesimal representations see [15, 16].

formulas provided here cover, in a sense, all (and not only unitary) irreducible representations of classical groups. However, at the present stage in the development of functional analysis, the authors do not know yet how one can give a precise formulation for the problem of finding "all" (and not only unitary) irreducible representations.

Indeed, in the following the representation operators are realised as operators acting in a space of functions. It turns out that one can introduce into that space an infinite number of non-equivalent norms for which the representation operators are bounded. Now the definition of equivalence of two representations requires the isomorphism of representation spaces in which the representation operators act. Thus there appears to be a vast variety of different representations created by an extremely fine difference in the structure of functional spaces.

It turns out that infinite-dimensional representations of any of those groups are defined by $n$ integers $m_1,\ldots,m_n$ and $n$ complex numbers $\rho_1,\ldots,\rho_n$. When the numbers $\rho_1,\ldots,\rho_n$ are purely imaginary the representation is unitary. For integer $\rho_1,\ldots,\rho_n$ the representation includes a finite dimensional component.

We now proceed to the actual description of representations of classical groups giving thus a brief account of the whole paper.

2. For simplicity's sake we first consider the case of the group of matrices of order 2 with determinant 1, so that the elements of $\mathfrak{G}$ are matrices $\begin{pmatrix}\alpha & \beta\\ \gamma & \delta\end{pmatrix}$ satisfying the condition $\alpha\delta-\beta\gamma=1$. The representation space $\mathfrak{H}$ in which the operators $T_g$ are acting is the space of functions $f(z)$ depending on the complex variable $z$. Each representation is defined by an integer $m$ and a number $\rho$. The operator $T_g$ corresponding to the numbers $m$ and $\rho$ associates to each function $f(z)$ the function[4]

$$T_g f(z)=f\left(\frac{\alpha z+\gamma}{\beta z+\delta}\right)|\beta z+\delta|^{m+\rho-2}(\beta z+\delta)^{-m}. \tag{2}$$

If one introduces a scalar product in $\mathfrak{H}$ by the formula

$$(f_1,f_2)=\iint f_1(z)\overline{f_2(z)}dxdy \tag{3}$$

then for purely imaginary $\rho=i\rho'$ one can easily check that the operators $T_g$ are unitary. Thus, for $\rho$ purely imaginary formulas (2) and (3) define a unitary representation of the group $\mathfrak{G}$. It turns out that the pairs $m$, $\rho$ and $-m$, $-\rho$ correspond to equivalent representations. There is, however, one more possible way of defining a unitary representation of the group $\mathfrak{G}$ using formula (2), namely, defining the scalar product by the formula[5]

$$(f,f)=\int|z_1-z_2|^{-2+\rho}f(z_1)\overline{f(z_2)}dx_1dy_1dx_2dy_2,\quad 0\leqq\rho\leqq 2, \tag{4}$$

and taking operators $T_g$ defined by the same formula (2) for $m=0$ and $\rho$ real, one

[4] It is somewhat more convenient to write $\rho-2$ instead of $\rho$.

[5] For $\rho>2$ formula (2) defines a scalar product with a finite number of negative squares (see [6], p. 453 of the Russian version).

also obtains a unitary representation of the group $\mathfrak{G}$. Besides those given by the above formulas, there is only one more irreducible unitary representation of the group $\mathfrak{G}$, i.e. the identical one which associates to each element $g \in \mathfrak{G}$ the identical linear transformation.

Thus, all irreducible unitary representations of the group of matrices of order 2 are those given by the formula (2): (i) for $n$ integer and $\rho$ purely imaginary, the scalar product being given by formula (3); (ii) $n = 0, 0 < \rho < 2$ and the scalar product given by formula (4); and (iii) the identical representation of the group $G$.

It is interesting to note that formula (2) covers finite-dimensional representations as well, namely, suppose that $\rho$ is an integer such that $m + \rho - 2$ is even. Then, writing $|\beta z + \delta|^{m+\rho-2}$ as $(\beta z + \delta)^{(m+\rho-2)/2}(\overline{\beta z + \delta})^{(m+\rho-2)/2}$ one can reduce formula (2) to the form

$$T_g f(z) = f\left(\frac{\alpha z + \gamma}{\beta z + \delta}\right)(\beta z + \delta)^{m_1}(\overline{\beta z + \delta})^{m_2} \tag{5}$$

where $m_1 = (-m + \rho - 2)/2$ and $m_2 = (m + \rho - 2)/2$. For the space $\mathfrak{H}$ we take the set of only those functions that are polynomials in $z$ and $\bar{z}$ of degree at most $m_1$ in $z$ and $m_2$ in $\bar{z}$, i.e.

$$f(z) = a_1 + a_2 z + a_1 \bar{z} + \dots + a z^{n_1} \bar{z}^{n_2}. \tag{6}$$

The set of such polynomials forms a finite-dimensional space. One can easily check that the operator $T_g$ transforms any polynomial $f(z)$ given by formula (6) into the polynomial of the same form. Thereby one obtains a representation, and one can prove that all irreducible finite-dimensional representations of the group $G$ of matrices of order 2 with determinant 1 can be obtained in such a way.

Actually, the whole picture of irreducible representations of the group $\mathfrak{G}$ should be given by formula (2) for all integer values of $m$ and complex $\rho$. For integer $m$ and $\rho$ purely imaginary or for $m = 0$ and $0 < \rho < 2$[6] the representations are unitary. They are irreducible for all $\rho$ with the exception of those values of $\rho$ that are real and integer of the same parity as $m$. In that case, the representation splits into a finite-dimensional one and an infinite-dimensional one (see, e.g. above, where the finite-dimensional representations are described). However, our considerations in the present paper are limited to unitary representations for which the problem can be formulated in precise terms.

3. Consider now irreducible unitary representations of the group of matrices of order $n$ with determinant 1. As in the case of these groups of matrices of order 2, each representation operator acts in the space of functions defined on the points of some space (the points $z$ of the complex plane $Z$ in the above case). Thus, in order to define representations of the group $\mathfrak{G}$, we first describe the space $Z$ which replaces the complex plane $Z$ in the case of the group of matrices of order 2.

**Definition 1.** *Denote by $Z$ the set of complex matrices $z = \|z_{pq}\|$ of order $n$ for*

[6] For $\rho > 2$ quadratic form (4) is indefinite.

*which* $z_{pp} = 1$ $(p = 1, \ldots, n)$ *and* $z_{pq} = 0$ *for* $p < q$. *(For* $n = 2$ *one has* $z = \begin{pmatrix} 1 & 0 \\ z_{21} & 1 \end{pmatrix}$, *i.e. each element in* $Z$ *is defined by a point* $z_{21}$ *of the complex plane.)*

In the following we use the following simple proposition.

Almost all matrices $g$ can be uniquely represented in the form

$$g = kz, \tag{7}$$

where $z \in Z$ and $k$ is a matrix such that $k_{pq} = 0$ for $p > q$. The exceptions are those matrices $g$ for which at least one of the minors

$$\begin{vmatrix} g_{pp} & g_{p,p+1} & \cdots & g_{pn} \\ \cdots & \cdots & \cdots & \cdots \\ g_{np} & g_{n,p+1} & \cdots & g_{nn} \end{vmatrix}, \qquad p = 2, 3, \ldots, n$$

vanishes.

Let $z \in Z$ be an arbitrary matrix in $G$. Representing $zg$ in the form

$$zg = kz_1 \tag{8}$$

denote the element $z_1$ by $z\bar{g}$, i.e. $z_1 = z\bar{g}$ (the element $z$ goes into $z_1$ under the transformation $g$). For the matrices of order 2 one can compute without difficulty that the transformation from $z$ to $z_1$ means that if

$$z = \begin{pmatrix} 1 & 0 \\ z' & 1 \end{pmatrix}, \quad z_1 = \begin{pmatrix} 1 & 0 \\ z_1' & 1 \end{pmatrix}, \quad g = \begin{pmatrix} \alpha & \beta \\ \gamma & \delta \end{pmatrix} \quad \text{and} \quad z_1 = z\bar{g},$$

then

$$z_1' = \frac{\alpha z' + \gamma}{\beta z' + \delta}.$$

In the general case, representations turn out to be defined in the same way as for the group of matrices of order 2.

The representation space $\mathfrak{H}$ consists of functions $f(z)$ where $z \in Z$ (i.e. $z$ is defined by $n(n-1)/2$ parameters $z_{pq}$, $p < q$). Representation operators $T_g$ are defined in the following way:

$$T_g f(z) = f(z\bar{g}) \alpha(z, g), \tag{9}$$

where $\alpha(z, g)$ is a fixed function of $z$ and $g$ characterising the representation.

For the group of matrices of order 2 the function $\alpha(z, g)$, as shown by formula (2), is of the form:

$$\alpha(g, z) = |g_{12} z + g_{22}|^{m+\rho-2} |g_{12} z + g_{22}|^{m}$$

(here we change our notation and write $\begin{pmatrix} g_{11} & g_{12} \\ g_{21} & g_{22} \end{pmatrix}$ instead of $\begin{pmatrix} \alpha & \beta \\ \gamma & \delta \end{pmatrix}$ which we used in formulae (2)).

In the general case of arbitrary $n$ the function $\alpha(z, g)$ is constructed in the following way.

In the decomposition $zg = k_1 z_1$ we denote by $\delta_1, \delta_2, \ldots, \delta_n$ the diagonal elements of the triangular matrix $k_1$. Then the function $\alpha(z, g)$ is defined by the numbers

$m_1, \ldots, m_n$ and $\rho_1, \ldots, \rho_n$, where $m_1, \ldots, m_n$ are integers, and is given by the formula

$$\alpha(z, g) = |\delta_1|^{m_1 + \rho_2} \delta_1^{-m_1} |\delta_2|^{m_2 + \rho_2} \delta_2^{-m_2} \ldots |\delta_n|^{m_n + \rho_n} \delta_n^{-m_n}. \tag{10}$$

The explicit expression for $\alpha(z, g)$ in terms of the elements of matrices $z$ and $g$ can be easily given.

Since the group $G$ consists of matrices with determinant 1, one has $\delta_1 \ldots \delta_n = 1$. Therefore, integers $m_1, \ldots, m_n$ and, similarly, the numbers $\rho_1, \ldots, \rho_n$ in formula (10) are defined up to a common summand, and one can assume that

$$m_1 + \ldots + m_n = 0$$

and

$$\rho_1 + \ldots + \rho_n = 0.$$

Occasionally, another normalisation is used: $\rho_1 = m_1 = 0$.

In order to have unitary representations one has to define scalar products in the space $\mathfrak{H}$ of functions $f(z)$. This will be done at a later stage.

4. In order to define unitary representations of the unimodular group, one has to introduce a scalar product into the set of functions $\mathfrak{H}$ defined in the preceding sections, and to find those $m_k$ and $\rho_k$ for which the representation operators $T_g$ are unitary. There are two different possibilities. The first gives the so-called principal series of representations and the second gives complementary series to be described in Sect. 7 of this introduction.

The principal series of unitary representations of the unimodular group $\mathfrak{G}$ is defined in the following way. Consider the space $H$ of functions $f(z)$, $z \in Z$, for which there exists the norm

$$\| f \|^2 = \int | f(z)^2 \prod_{p>q} dx_{pp} dy_{pq} \tag{11}$$

$$(z_{pq} = x_{pq} + i y_{pq} \text{ for } p > q).$$

To each element $g \in \mathfrak{G}$ there corresponds a unitary operator $T_g$ given by the formula

$$T_g f(z) = f(z\bar{g}) \alpha(z, g) \tag{12}$$

where $\alpha(z, g)$ is a fixed function defining the representation and given by formulas (10) and (12). Numbers $m_k$ in formula (10) are integers, and the $\rho_k$ are purely imaginary. If the norm is defined by formula (11), and $T_g$ is defined by formula (12) then the representation is unitary if and only if the numbers $\rho_k$ appearing in the definition of $\alpha(z, g)$ are purely imaginary. Those representations are called represetations of the *principal series.*

By induction on $n$ one can prove that those representations are irreducible. Thus, the representations of the principal series are described by numbers $m_i$ and $\rho_i$. Consider now the question: when are two different representations of the principal series equivalent? The following theorem holds.

Two representations of the principal series given by the numbers $m_1, \ldots, m_n$; $\rho_1, \ldots, \rho_n$ and $m'_1, \ldots, m'_n$; $\rho'_1, \ldots, \rho'_n$, respectively, are equivalent if and only if there exists a permutation $s$ taking the numbers $m'_i$ into $m_i$ and numbers $\rho'_i$ into $\rho_i$.

The theorem is proved in Part V by extending the theory of characters of non-commutative finite groups to classical groups.

5. The principal series of unitary representations of the unimodular group, which we have just introduced, is sometimes more conveniently defined in a different way. Let us first consider the case of the group of matrices of order 2. The representation operators are then given by the formula

$$T_g f(z) = f\left(\frac{\alpha z + \gamma}{\beta z + \delta}\right) |\beta z + \delta|^{-n+\rho-2} (\beta z + \delta)^n.$$

In some cases that formula is inadequate because the formula $z' = \alpha z + \gamma / \beta z + \delta$ is not defined everywhere (namely, there is no $z'$ corresponding to the point $z = -\delta/\beta$) and because under the above definition of the integral,

$$\int |f(z)|^2 dx dy,$$

the measure of the whole complex plane is infinite. Therefore, it is very convenient to transform the complex plane by the stereographic projection to the surface of the sphere and consider each representation in terms of functions $f(\tilde{u})$ of the point $\tilde{u}$ of the surface of the sphere.

We now examine the second way of defining a representation based on the generalisation of the notion of stereographic projection to the case of arbitrary $n$. Denote by $\mathfrak{U}$ the set of all unitary matrices of order $n$ with determinant 1. There is the following easy proposition:

Each matrix $g \in \mathfrak{G}$ can be represented in the form

$$g = ku \tag{13}$$

where $k = \| k_{pq} \|$ is (as before) a triangular matrix such that $k_{pq} = 0$ for $p > q$ and $u \in \mathfrak{U}$.

The matrix $u \in \mathfrak{G}$ here is determined uniquely up to a diagonal unitary matrix $\gamma$. Denote by $\tilde{u}$ the class of matrices $\{\gamma u\}$, where $u$ is fixed and $\gamma$ runs over diagonal unitary matrices. Then the decomposition $g = ku$ defines the class $\tilde{u}$ uniquely.

The set of all $\tilde{u}$ we denote by $\tilde{\mathfrak{U}}$.

Consider the class $\tilde{u}$ for the group of matrices of order 2. Any unitary matrix $u$ of order 2 with determinant 1 is of the form

$$u = \begin{pmatrix} \alpha & \beta \\ -\bar{\beta} & \bar{\alpha} \end{pmatrix}$$

where $|\alpha|^2 + |\beta|^2 = 1$. Let the matrix $\gamma$ be of the form $\begin{pmatrix} \gamma & 0 \\ 0 & \gamma^{-1} \end{pmatrix}$ where $|\gamma| = 1$. Then the class $\{\gamma u\}$ consists of matrices

$$\begin{pmatrix} \gamma\alpha & \gamma\beta \\ -\gamma^{-1} & \beta\gamma^{-1}\alpha \end{pmatrix}$$

where $\alpha$ and $\beta$ are fixed and $\gamma$ runs over those complex numbers for which $|\gamma| = 1$.

Thus, each class $\tilde{u}$ is defined by a pair of numbers $(\alpha, \beta)$, for which $|\alpha|^2 + |\beta|^2 = 1$, determined up to a factor $\gamma$ such that $|\gamma| = 1$. It is easy to see that the set of $\tilde{u}$ is homeomorphic to the surface of the sphere.

For an arbitrary $n$ we now consider the following transformation.

Let $z$ be matrix in $Z$. Representing it in the form

$$z = ku, \tag{14}$$

denote by $\{\gamma u\}$ the class containing $u$. Then for each $z \in Z$ there is a unique $\tilde{u}$. Conversely, to each $\tilde{u}$, with the exception of those lying in some subspace of lesser dimension, there corresponds a unique $z$. For $n = 2$ the set of exceptional $u$ consists of a single point corresponding to the infinite point on the complex plane $Z$. For $n = 2$ the correspondence between $z$ and $\tilde{u}$ given by formula (14) is the stereographic projection.

The transformation $g$ defined on $Z$ by the formula $z \to z\bar{g}$ can be extended to $\mathfrak{U}$.

Indeed, let $\tilde{u} \in \tilde{\mathfrak{U}}$. The element $ug$ can be represented in the form

$$ug = ku_1.$$

Thus $g$ acts on $\tilde{u}$ by associating to $\tilde{u}$ the class $\tilde{u}_1$. We shall write it as $\tilde{u}_1 = \tilde{u}\bar{g}$.

Consider now the space $\mathfrak{H}$ of functions $f(\tilde{u})$ and define for each element $g \in \mathfrak{G}$ a linear operator $T_g$ acting in $\mathfrak{H}$ by the formula

$$T_g f(\tilde{u}) = f(\tilde{u}\bar{g})\frac{\alpha_1(ug)}{\alpha_1(u)} \tag{15}$$

where the function $\alpha_1(g)$ satisfies the relation

$$\alpha_1(ku) = \alpha_1(k)\alpha_1(u)$$

for all $k \in K$ and $u \in \mathfrak{U}$ and its computed in the same way as the function $\alpha(zg)$ of the preceding section. (For the computation of $\alpha_1(g)$ see §8.)

The integral of $f$ over $\tilde{\mathfrak{U}}$ is uniquely determined by the condition

$$\int_{\tilde{\mathfrak{U}}} f(\tilde{u}u_0)\,d\tilde{\mu}(\tilde{u}) = \int_{\tilde{\mathfrak{U}}} f(\tilde{u})\,d\tilde{\mu}(\tilde{u}). \tag{16}$$

Under our normalisation the integnal of unity, i.e. the measure of the whole of $\tilde{\mathfrak{U}}$ is equal to

$$\frac{\pi^{n(n-1)/2}}{1!2!\ldots(n-1)!}.$$

The Hilbert space $\mathfrak{H}$ consists of all functions $f(\tilde{u})$ for which the integral

$$\int |f(u)|^2\,d\mu(\tilde{u})$$

exists.

The second way of defining the principal series of representations of unimodular group is very convenient for the study of relations between representations of the unimodular group and those of the subgroup of unitary matrices[7], namely, for

[7] The subgroup of unitary matrices is a maximal compact subgroup of the group of unimodular matrices.

each irreducible representation $g \to T_g$ of the group $\mathfrak{G}$ one can consider its restriction to the set of elements $g \in \mathfrak{U} \subset \mathfrak{G}$ which gives a representation of the unitary subgroup $G$. Using the second way of describing principal series representations, one can find irreducible components of the representation $g \to T_g$ for $g \in \mathfrak{U}$ and the number of times each of them occurs in the representation.

In order to formulate the result we have to recall the usual way of describing representations of the group $\mathfrak{U}$. Let $u \to A_u$ be a unitary irreducible (but not necessarily finite-dimensional) representation of the group $\mathfrak{U}$ acting in the space $\mathfrak{H}_1$. Consider the subgroup $\Gamma$ of diagonal matrices

$$\gamma = \begin{pmatrix} \gamma_1 & 0 & \dots & 0 \\ 0 & \gamma_2 & \dots & 0 \\ \dots & \dots & \dots & \dots \\ 0 & 0 & \dots & \gamma_n \end{pmatrix}$$

in $\mathfrak{U}$ such that $\gamma_k = 1$ so that each $\gamma_k$ can be written as $\gamma_k = e^{i\varphi_k}$. Since the matrices $\gamma \in \Gamma$ commute, the corresponding operators $A_\gamma$ also commute. Therefore, operators $A_\gamma$ form a system of unitary commuting operators in a finite-dimensional space $\mathfrak{H}$. Hence, there exists an orthogonal basis consisting of eigenvectors $\xi_1, \dots$ of operators $A_\gamma$, i.e.

$$A_\gamma \xi = \lambda \xi$$

where the eigenvalue $\lambda$ depends on $\gamma: \lambda = \chi(\gamma)$. Since $A_{\gamma'\gamma''} = A_{\gamma'\gamma''}$ one has $\chi(\gamma'\gamma'') = \chi(\gamma')\chi(\gamma'')$. This functional equation is easily solved giving

$$\chi(\gamma) = \gamma_1^{k_1} \gamma_2^{k_2} \dots \gamma_n^{k_n} \tag{17}$$

where $k_1, \dots, k_n$ are integers.

The vector $\xi$ is called *the weight* vector and the set of integers $(k_1, k_2, \dots, k_n)$ is called its *weight*.

For the case of the unitary group with determinant 1 the only difference is that $\gamma_1, \dots, \gamma_n = 1$, so that the numbers $k_1, k_2, \dots, k_n$ are defined up to common summand. It is therefore convenient to normalise them by taking, for example, $k_1 = 0$.

Let us now turn to the problem of decomposing the representation $g \to T_g$ $(g \in \mathfrak{U})$ into irreducible representations of the group $\mathfrak{U}$. Let the representation $g \to T_g$ be defined by the numbers $(\rho_1, \dots, \rho_n)$ and $(m_1, \dots, m_n)$. Then the following theorem holds.

Let $T_g$ be a representation of the principal series of the group $\mathfrak{G}$ defined by the numbers $(\rho_1, \dots, \rho_n)$ and $(m_1, \dots, m_n)$. The corresponding representation $k \to T_u$ of the unitary subgroup $\mathfrak{U}$ contains a given irreducible representation $u \to A_u$ of the subgroup $\mathfrak{U}$ if and only if the representation $u \to A_u$ involves the weight $(m_2, \dots, m_n)$. One can also find the number of times a given representation $u \to A_u$ occurs in the decomposition of the representation $T_g$ into irreducible ones. It turns out that this number is equal to the number of linearly independent vectors of weight $(m_1, \dots, m_2)$ in the representation space $\mathfrak{H}$ in which the representation $u \to A_u$ acts.

In particular, when in the representation space $\mathfrak{H}$, where the operators $T_g$ act, one can show there is a vector $\xi_0$ invariant under the action of the operators $T_u$

for $u \in \mathfrak{U}$. It turns out that if the representation $T_g$ is defined by the numbers $(m_1, \ldots, m_n)$ and $(\rho_1, \ldots, \rho_n)$ then the necessary and sufficient condition for that is that $m_2 = \ldots = m_n = 0$. The vector $\xi_0$ invariant under all $T_u$, $u \in \mathfrak{U}$, is then defined uniquely up to a scalar factor.

One can associate so-called *spherical functions*, which are defined below, to each group and its given subgroup. The name is suggested by the case of the rotation group in three-dimensional space and its subgroup of rotations around a given axis when those spherical functions become the well-known spherical functions of the usual calculus. We have no possibility here of considering in more detail the interesting general theory of spherical functions, so we give a somewhat more formal but at the same time more operative definition.

Let $g \to T_g$ be a unitary representation of the group $\mathfrak{G}$ and $\mathfrak{U}$ its subgroup. Suppose that in the space $\mathfrak{H}$, where the operators $T_g$ act, there is a vector $\xi_0$ invariant under the action of the operators $T_u$, $u \in \mathfrak{U}$, i.e. $T_u \xi_0 = \xi_0$. The zonal, spherical function corresponding to that representation is the function

$$\varphi(g) = (T_g \xi_0, \xi_0). \tag{18}$$

One can easily see that $\varphi(gu) = \varphi(ug) = \varphi(g)$. Indeed, as example,

$$\varphi(gu) = (T_{gu} \xi_0, \xi_0) = (T_g T_u \xi_0, \xi_0) = (T_g \xi_0, \xi_0) = \varphi(g). \tag{19}$$

In our case $\mathfrak{G}$ is the group of complex unimodular matrices, and $\mathfrak{U}$ is the subgroup of unitary matrices. We have already mentioned that if for the representation $g \to T_g$ one has $m_2 = \ldots = m_n = 0$, then there exists a unique (up to a scalar factor) vector invariant with respect to $T_u$. Then the zonal spherical function corresponding to that representation satisfies the relations $\varphi(gu) = \varphi(ug) = \varphi(g)$. Each matrix can be represented in the form $g = hu$, where $h$ is a Hermitian positive definite matrix and $u$ is a unitary matrix, so that $\varphi(g) = \varphi(hu) = \varphi(h)$. Now any Hermitian positive definite matrix $h$ can be represented in the form $h = u_1^{-1} \varepsilon u_1$ where $u_1$ is a unitary and $\varepsilon$ is a diagonal matrix with positive scalars $\lambda_1, \ldots, \lambda_n$ on the diagonal. Hence, $\varphi(h) = \varphi(u_1^{-1} \varepsilon u_1) = \varphi(\varepsilon)$. Thus, it is sufficient to compute the spherical function on diagonal matrices $\varepsilon$. The computations yield

$$\varphi(\varepsilon) = \left(\frac{2}{i}\right)^{n(n-1)/2} \frac{1!2!\ldots(n-1)!}{\prod\limits_{1 \leqq p < q \leqq n} (\rho_q - \rho_p) \prod\limits_{1 \leqq p < q \leqq n} (\lambda_q^2 - \lambda_p)} \cdot \begin{vmatrix} 1 & 1 & \ldots & 1 \\ \lambda_1^{i\rho_2} & \lambda_2^{i\rho_2} & \ldots & \lambda_n^{i\rho_2} \\ \ldots & \ldots & \ldots & \ldots \\ \lambda_1^{i\rho_n} & \lambda_2^{i\rho_n} & \ldots & \lambda_n^{i\rho_n} \end{vmatrix} \tag{20}$$

where one has to set $\rho_1 = 0$ in (20).

Formula (20) is obtained by direct computation of the expression $(T_g \xi_0, \xi_0)$. Since vectors $\xi$ of the space $\mathfrak{H}$ are functions and the scalar product is an integral, the computation of $(T_g \xi_0, \xi_0)$ is reduced to an integration. The technique of computing that integral is very instructive and is used in the following once more (Part VI) for a different reason and in a more complicated setting.

6. In Sects. 4 and 5 we considered the so-called principal series of irreducible representations of the unimodular group $\mathfrak{G}$. Those representations were realised by operators acting in the space of functions of the matrix $x$, i.e. the function depending on $n(n-1)/2$ complex variables. However, those representations do not cover all representations of the group $\mathfrak{G}$. In this and the following sections we give a short account of the so-called degenerate and complementary series of irreducible unitary representations of the group $\mathfrak{G}$. Roughly speaking, degenerate series contain representations realised in the space of functions that depend not on $n(n-1)/2$ but on a smaller number of variables. The most "degenerate" of all is the identical representation of the group. Let us now describe the representations of the degenerate series.

Consider a decomposition of $n$ into a sum of $r$ ($r \leqq n$) integers:

$$n = n_1 + \ldots n_r. \tag{21}$$

Denote by $g_{pq}$ a matrix consisting of $n_p$ rows and $n_q$ columns. Any matrix $g \in \mathfrak{G}$ of order $n$ can be considered as being composed of matrices $g_{pq}$.

Now we introduce the subgroup $Z$ (which should actually be denoted $Z_{n_1,\ldots,n_r}$) consisting of all matrices $z = \| z_{pq} \|$ for which

$$z_{pq} = 0 \quad \text{for } p < q \quad \text{and} \quad z_{pp} = 1, \quad p, q = 1, \ldots, r. \tag{22}$$

Here 0 denotes the zero matrix containing $n_p$ rows and $n_q$ columns, and 1 is a unit matrix of order $n_p$. Thus, any matrix $z \in Z_{n_1,\ldots,n_r}$ depends not on $n(n-1)/2$ but on a lesser number of parameters. There is the following simple proposition:

Almost every element $g \in \mathfrak{G}$ can be uniquely represented in the form

$$g = kz, \tag{23}$$

where $z \in Z_{n_1,\ldots,n_r}$, and $k$ is a matrix composed of matrices $k_{pq}$, $k = \| k_{pq} \|$ such that $k_{pq} = 0$ for $p > q$. The exceptions are those $g$ for which at least one of the right-bottom-corner minors of order $n_p + \ldots + n_r$, $p = 2, 3, \ldots, r$, vanishes.

Let us now define the action of $g \in \mathfrak{G}$ on the points $z \in Z$. Representing the element $zg$ in the form

$$zg = k_1 z_1, \tag{24}$$

denote $z_1$ by $z\bar{g}$ ($z_1$ is obtained from $z$ by the action of $g$). Consider now the space $H$ of functions $f(z)$ for which the integral

$$\| f \|^2 = \int | f(z) |^2 \prod_{p>q} d\mu(z_{pq}) \tag{25}$$

exists (here $d\mu(z_{pq})$ denotes the product of all differentials of the real and imaginary parts of entries of the matrix $z_{pq}$).

We now define the representation $g \to T_g$ in the space $\mathfrak{H}$ in the following way. Let

$$T_g f(z) = f(z\bar{g}) \alpha(z, g). \tag{26}$$

To complete the definition one has to fix a function $\alpha(z, g)$. Different representations are obtained for different choices of the function $\alpha(z, g)$. We now write the formula for $\alpha(z, g)$. Denote by $\Lambda_p$ the determinants of the matrices $k_{pq}$ which compose the matrix $k_1$ in the decomposition

$$zg = k_1 z_1.$$

Then

$$\alpha(z,g) = |\Lambda_2|^{m_2+\rho_2-(n_1+n_2)}\Lambda_2^{-m_2} \cdot |\Lambda_3|^{m_3+\rho_3-(n_1+2n_2+n_3)}\Lambda_3^{-m_3} \dots |\Lambda_r|^{m_r+\rho_r-(n_1+2n_2+\dots+2n_{r-1}+n_r)}\Lambda_r^{-m_r},$$

where $m_2, \dots, m_r$ are integers and $\rho_2, \dots, \rho_r$ purely imaginary. Thus, the series defined by the decomposition $n = n_1 + \dots + n_r$ consists of unitary representations defined by the integers $(m_2, \dots, m_r)$ and the purely imaginary numbers $(\rho_2, \dots, \rho_r)$.

Note that for $r = n$ and $n_1 = n_2 = \dots = n_r = 1$ the representation turns into a principal non-degenerate series of unitary representations as described in Sect. 4. For all other decompositions one obtains a degenerate series. The most degenerate of all is the identical representation. Formally it is contained in our scheme for $r = 1$ and $n = n_1$. The next most degenerate is the series corresponding to $r = 2$ and $n = (n-1) + 1$. Representations of this series are realised by operators in the space of square-integrable functions $f(z_1, \dots, z_{n-1})$ in $(n-1)$ complex variables $z_1, \dots, z_{n-1}$. The operators $T_g$ are defined by the formula

$$T_g f(z_i) = f\left(\frac{\sum_{j=1}^{n-1} z_j g_{jk} + g_{nk}}{\sum_{j=1}^{n-1} z_j g_{jn} + g_{nn}}\right)\left|\sum_{j=1}^{n-1} z_j g_{jn} + g_{nn}\right|^{m+\rho-n}\left(\sum_{j=1}^{n-1} z_j g_{jn} + g_{nn}\right)^{-m}. \tag{27}$$

Therefore, in this case $z$ is defined by $n-1$ complex numbers, and the operator $z \to z\bar{g}$ is a projective transformation in the $(n-1)$-dimensional projective space.

For degenerate series the following theorem holds:

All representations of degenerate series are irreducible.

Let us also note that the representations corresponding to a decomposition $n = n_1 + \dots + n_r$ are equivalent to representations given by $n = n'_1 + \dots + n'_r$ where $n'_1, \dots, n'_r$ is a permutation of integers $n_1, \dots, n_r$.

The representations introduced above are called *principal series representations*; the representations acting in the space of functions depending on $n(n-1)/2$ variables (Sects. 4 of this Introduction) are called representations of the *principal non-degenerate series* while those introduced in this section are called representations of the *principal degenerate series.*

7. There is one more type of irreducible unitary representations of the group $\mathfrak{G}$, the so-called *complementary series* of representations.

This new possibility for constructing unitary representations stems from the fact that one can give a different definition of scalar product in the space of functions $f(z)$. (For the representations of the group of matrices of order 2, that scalar product is given by the formula $\int |z_1 - z_2|^{-2+\rho} f(z_2)\overline{f(z_2)}\, dx_1\, dy_1\, dx_2\, dy_2$, see formula (4).)

We now formulate the final result. Each representation is constructed in the Hilbert space $\mathfrak{H}_r$ consisting of functions $f(z)$ where $z \in Z$ and $Z$ is a set of matrices

$z = \|z_{pq}\|$ such that $z_{pp} = 1$ and $z_{pq} = 0$ for $p < q$. In order to define the scalar product in the space of functions $f(z)$ we need a new notation.

Let $\tau$ be a fixed integer. Denote by $\dot{z}$ the matrices $z$ of the form

$$\dot{z} = \begin{pmatrix} I_m & 0 & 0 & \dots & 0 \\ 0 & u_1 & 0 & \dots & 0 \\ 0 & 0 & u_2 & \dots & 0 \\ \dots & \dots & \dots & \dots & \dots \\ 0 & 0 & 0 & \dots & u_\tau \end{pmatrix} \tag{28}$$

where $I_m$ is a unit matrix of order $m = n - 2\tau$ and $u_i$ are matrices of order 2 of the form

$$u_i = \begin{pmatrix} 1 & 0 \\ z_i & 1 \end{pmatrix}, \quad z_i \neq 0, \quad i = 1, \dots, \tau.$$

Denote now by $\hat{z}$ those matrices in $Z$ which have zeroes in the places where in the matrix $\dot{z}$ there are the elements $z_i$, $i = 1, \dots, \tau$. One can easily show that each matrix $z$ can be uniquely represented in the form

$$z = \dot{z}\hat{z}. \tag{29}$$

Thus, each function $f(z)$ can be represented as a function of $\dot{z}$ and $\hat{z}$, i.e. $f(z) = f(\hat{z}, z) = f(\hat{z}, z_1, \dots, z)$.

The scalar product of the functions $f_1(z)$ and $f_2(z)$ is defined in the following way:

$$(f_1, f_2) = \int \prod_{p=1}^{\tau} |z'_p - z''_p|^{-2+2\sigma_p} f_1(\hat{z}, z'_1, \dots, z') \overline{f_2(\hat{z}, z''_1, \dots, z'')} d\mu(z_1) \dots d\mu(z_\tau) d\mu(\hat{z}) \tag{30}$$

where $d\mu(\hat{z})$ is the product of the differentials of the real and imaginary parts of the elements of matrix $\hat{z}$, the other differentials having a similar meaning. Also, $\sigma_p$ are fixed numbers satisfying inequalities $0 < \sigma_p < 1$, $p = 1, \dots, \tau$.

The operators $T_g$ are defined by the formula

$$T_g f(z) = f(z\bar{g})\alpha(zg) \tag{31}$$

where the function $\alpha(zg)$ is constructed in the following way.

Let $zg = kz_1$ and let $k_{pp}$ $(p = 1, \dots, n)$ be diagonal elements of the matrix $k$. Denoting them in the following way, $\delta_1, \dots, \delta_{n-2\tau}, v_1, \lambda_1, v_2, \lambda_2, \dots, v_\tau, \lambda_\tau$, we set $\alpha(zg) = \chi(zg)\beta_1(zg)$[8], where

$$\begin{aligned} \chi(zg) &= \prod_{p=2}^{n-2\tau} |\delta_p|^{m_p + i\rho_p} \delta_p^{-m_p} \prod_{p=1}^{\tau} |v_p|^{m_p + i\sigma'_p + \sigma_p} v^{-m_p} |\lambda_p|^{m_p + i\sigma'_p - \sigma_p} \lambda_p^{-m_p}, \\ \beta_1(zg) &= |\delta_2|^{-2} |\delta_3|^{-4} \dots |\delta_n|^{-2n+2}. \end{aligned} \tag{32}$$

Here $m_p$ are integers defining the representations, and $\rho_p$ and $\sigma'_p$, $\sigma_p$ are real numbers.

Formally, the complementary series can be defined also for $\sigma_p > 1$. Formulas

[8] In the following we write $\beta^{-\frac{1}{2}}(zg)$ instead of $\beta_1(zg)$.

for $T_g$ are in that case of the same form. However the scalar product is no longer positive definite. Thus, a representation is unitary if

$$0 < \sigma_p < 1. \tag{33}$$

Thus, formulas for operators $T_g$ of the representations of complementary series differ from those of the principal series by the definition of the function $\alpha(g)$, namely, some of the purely imaginary parameters $\rho$ of the principal series are replaced by complex parameters $i\sigma'_p \pm \sigma_p$.

Therefore, the formula $T_g f(z) = f(z\bar{g})\ (zg)$ with complex $\rho_n$ provides a unified algebraic scheme for defining representation operators. However, a positive definite invariant quadratic form (which we take as a scalar product) in the representation space does not exist for any values of complex parameters. It exists only in the following cases:

(i) If all the integers $m_i$ are different then the parameters $\rho$ have to be purely imaginary.

(ii) If there is a pair of integers $m$ (for example, $m_k$ and $m_{k+1}$) that coincide, then $i\rho_k$ has to be complex conjugate to $i\rho_{k+1}$.

The first possibility corresponds to the principal series and the second to complementary series. Degenerate complementary series are introduced in a fashion quite similar to that in which non-degenerate complementary series are introduced.

8. A major tool in the theory of finite-dimensional representations is the theory of characters. Characters of a finite-dimensional representation are defined in the following way.

Consider a group $G$ and let $g \to A_g$ be its finite-dimensional irreducible representation. Consider the trace (the sum of diagonal elements) of the matrix of the operator $A_g$ and denote it by $S(A_g)$. Then the function of $g: \pi(g) = S(A_g)$ is called the *character of the representation.* Since $S(BA_gB^{-1}) = S(A_g)$ for all $B$ equivalent representations give the same character. Therefore, the characters may serve to distinguish different (non-equivalent) representations. Note also that since $S(A_{g_0}A_gA_{g_0}^{-1}) = S(A_g)$ the character $\pi(g)$ has the following property:

$$\pi(g_0 g g_0^{-1}) = \pi(g) \tag{34}$$

for all $g$. We now try to generalise the notion of character to the case of finite-dimensional unitary representations of the unimodular group $\mathfrak{G}$.

Note that, in general, a unitary operator of the representation $g \to T_g$ is defined in an infinite-dimensional space and has, therefore, no trace. For example, for $g = e$ one has $T_g = E$ and the sum of diagonal elements of the identical operator is $+\infty$. To make sense of the expression $S(T_g)$ we use a peculiar summation procedure of the following type. Consider a continuous function $x(g)$ vanishing outside some compact set in $\mathfrak{G}$. Consider now the integral $\int x(g) T_g d\mu(g)$ where $\int \ldots d\mu(g)$ means that the integration is over all the parameters of the unimodular group $\mathfrak{G}$. It turns out that the following remarkable property is true.

The operator

$$K = \int x(g) T_g d\mu(g) \tag{35}$$

is completely continuous. Moreover, the operator $K$ is defined by a kernel $K(z_1, z_2)$ which is of the Hilbert-Schmidt type, i.e. there exists the integral $\int |K(z_1, z_2)|^2 d\mu(z_1) d\mu(z_2)$. Therefore, the trace of the operator $K$ may be defined in the following way:

$$S(K) = \int K(z, z) d\mu(z).$$

Using formula (35) one can prove that $S(K)$ is of the form

$$S(K) = \int x(g) \pi(g) d\mu(g). \tag{36}$$

The function $\pi(g)$ is called the *character of the representation* $x \to T_g$. It has all the usual properties of characters, e.g.,

(i) The character functions $\pi(g)$ for two representations coincide if and only if those representations are equivalent;

(ii) $\pi(g_0^{-1} g g_0) = \pi(g)$, etc.

The function $\pi(g)$ can be computed. Before writing out the result, note that the relation $\pi(g_0^{-1} g g_0) = \pi(g)$ implies that for any matrix $g$ with different eigenvalues the function $\pi(g)$ depends only on the eigenvalues of the matrix $g \in \mathfrak{G}$ up to their permutation. Denote the set of those eigenvalues by $\delta_1, \ldots, \delta_n$, let $\delta = (\delta_1, \ldots, \delta_n)$ and consider the function

$$\chi(\delta) = |\delta_1|^{-m_1 + \rho_1} \delta_1^{m_1} |\delta_2|^{-m_2 + \rho_2} \delta_2^{m_2} \ldots |\delta_n|^{-m_n + \rho_n} \delta_n^{-m_n} \tag{37}$$

where $m_1, \ldots, m_n$, $\rho_1, \ldots, \rho_n$ are the numbers defining the representation $g \to T_g$ of the principal series (as we have already said, the numbers $\rho_1, \ldots, \rho_n$ are then purely imaginary). Then the following formula holds:

$$\pi(g) = \frac{\sum \chi_s(\delta)}{\prod\limits_{i<k} |\delta_i - \delta_k|^2} \tag{38}$$

where the sum is taken over all $n!$ permutations $s$ of the numbers $\delta_1, \ldots, \delta_n$. Note that if there is a pair of coinciding eigenvalues of the matrix $g$ then the function $\pi(g)$ is not defined. (In particular, $\pi(e) = +\infty$ which is quite natural since the trace of the identical operator $E$ equals $+\infty$.)

An exceptionally simple formula (38) leads to the following important result.

Two representations of the principal series are equivalent if and only if the numbers $(m_1, \ldots, m_n)$, $(\rho_1, \ldots, \rho_n)$ of the first representation can be obtained by a permutation of the numbers corresponding to the second representation.

The trace $(g)$ for representations of the complementary series is defined by the same formula (38), the only difference being that the numbers $m_i$ and $\rho_i$ in the definition of the function $\chi(\delta)$ are now those defining representations of the complementary series. This shows once again that it is natural to consider non-unitary representations as well, i.e. those corresponding to complex values of the parameters. Those representations should also have a trace defined by the same formula (38).

The trace formulas for degenerate series can be obtained in the same way as for the non-degenerate case.

The trace formulas also throw some light on the relation between non-degenerate and degenerate representations. We will not describe that relation in the general case and shall consider only the group of matrices of order 2.

For the unimodular group of matrices of order 2 there is only one degenerate representation, namely, the identical one, the trace of which is the constant function $\pi_0(g)$ equal to 1.

The trace of the principal series of representations of the group of matrices of order 2 is given by the formula

$$\pi(g)=\frac{|\lambda|^{n+i\rho}\lambda^{-n}+|\lambda|^{-n-i\rho}\lambda^{n}}{|\lambda-\lambda^{-1}|^2} \tag{39}$$

where $\lambda$ and $\lambda^{-1}$ are eigenvalues of the matrix $g$ ($\operatorname{Det} g=1$). Here $n$ and $\rho$ are the numbers defining the representation. For a representation of the complementary series which is defined in that notation for $n=0$, $0<\rho_1<2$, the trace is given by the formula

$$\pi(g)=\frac{|\lambda|^{\rho_1}+|\lambda|^{-\rho_1}}{|\lambda-\lambda^{-1}|^2} \tag{40}$$

which is obtained from formula (39) for $m=0$ by analytical continuation in $\rho$.

The only degenerate representation for the group of matrices of order 2 is the identical representation. Its trace is, therefore, given by the formula

$$\pi(g)\equiv 1.$$

One can show that for $\rho\to 2$ the representation of the complementary series tends to a reducible representation, namely, to the direct sum of the identical representation and the representation of the principal series corresponding to $m=2$, $\rho=0$. That can also be deduced from the trace formula. Indeed, for $\rho_1=2$ the trace (40) of the complementary series tends to

$$\frac{|\lambda|^2+|\lambda|^{-2}}{|\lambda-\lambda^{-1}|^2}.$$

Consider the following identity:

$$\frac{|\lambda|^2+|\lambda|^{-2}}{|\lambda-\lambda^{-1}|^2}=1+\frac{|\lambda|^2\lambda^{-2}+|\lambda|^{-2}\lambda^2}{|\lambda-\lambda^{-1}|^2}. \tag{41}$$

The first term on the right-hand side of (41) is the trace of the identical representation while the second is the trace of the representation of the principal series with $\rho=2$, $m=2$. It turns out that for representations of the group of any order the situation is the same, namely, as we have already said, complementary series arise in those cases when the parameters (see formula (32)) satisfy inequalities (33). If some of the parameters $\sigma_k$ approach a limit, so that those inequalities turn into equalities, then the representation approaches a reducible representation which is a direct sum of representations of both non-degenerate and degenerate series.

One can also show that for some values of complex parameters in the function $\chi(\delta)$ the representation (which is no longer unitary) may include finite-dimensional representations of that group.

9. We now proceed to consider an analogue of the Fourier integral related to representations of the unimodular group.

Consider a function $f(g)$ on the group $\mathfrak{G}$ vanishing outside some compact set and having a sufficient number of derivatives. Let $g \to T_g$ be a unitary representation of the group $\mathfrak{G}$. Consider the operator

$$A = \int f(g) T_g d\mu(g), \tag{42}$$

where integration is over the invariant measure on the group $G$. The problem is to reconstruct the function $f(g)$ knowing the operators $A$ for all irreducible representations[9]. It turns out that the following interesting fact exists: it is sufficient to know $A$ only for the representations $g \to T_g$ of the principal series.

It has been mentioned in the proceding section that the operator $A$ can be defined by a kernel $k(z_1, z_2)$ where $z_1, z_2 \in Z$. Clearly, that kernel also depends on the representation $g \to T_g$, i.e. on the numbers $m_1, \ldots, m_n$ and $\rho_1, \ldots, \rho_n$. Thus, the problem is to invert formula (42), i.e. to express $f(g)$ through $K(z_1, z_2, m_k, \rho_k)$. First we give the formula expressing $f(e)$ through $K(z_1, z_2, m_k, \rho_k)$ where $e$ is the unit element in the group. The formula for arbitrary $f(g)$ is then easily obtained by a simple shift. Denote

$$\int K(z, z, m_k, \rho_k) d\mu(z) = \varphi(\chi), \tag{44}$$

where $\chi$ denotes the set of numbers $m_k$ and $\rho_k$. Then the following formula holds:

$$f(e) = \sum_{m_2, \ldots, m_n} \int_{-\infty}^{\infty} \cdots \int_{-\infty}^{\infty} \varphi(\chi) \omega(\chi) d\rho_2 \ldots d\rho_n \tag{45}$$

where

$$\omega(\chi) = \frac{2^{\frac{1}{2}n(n-1)}}{n!(2\pi)^{(n-1)(n+2)}} \prod_{p<q} [(m_p - m_q)^2 + (\rho_p - \rho_q)^2]. \tag{46}$$

These formulas yield an analogue of the Plancherel theorem. The formula is of the form

$$\int |f(g)|^2 d\mu(g) = \sum_{m_2, \ldots, m_n} \int_{-\infty}^{\infty} \cdots \int_{-\infty}^{\infty} S(A^* A) \omega(\chi) d\rho_2 \ldots d\rho_n \tag{47}$$

where $A_\chi$ is the operator defined by formula (42), i.e.[10]

$$A_\chi = \int f(g) T_g d\mu(g)$$

and $S(A_\chi^* A_\chi)$ denotes the trace of the operator $A_\chi^* A_\chi$.

---

[9] If the group $\mathfrak{G}$ is the additive group $X$ of real numbers, one has the usual Fourier integral. In that case irreducible representations are one-dimensional and are defined by the formula $x \to e^{itx}$ where $x \in X$ and $t$ is a fixed number defining the representation. The problem now is how to find $f(x)$ starting from

$$A(t) = \int_{-\infty}^{+\infty} f(x) e^{itx} dx. \tag{43}$$

The problem is solved by the well-known Fourier inversion formula.

[10] Operators $A_\chi$ depending on $\chi$, i.e. on the set of numbers $m_i$, $\rho_i$, can be naturally called the Fourier coefficients of the function $f(g)$; cf. formula (43) of footnote 9.

This result shows that a regular representation of the unimodular group is decomposed only into representations of the principal non-degenerate series. A similar situation also takes place for other groups, e.g. for solvable ones. However, for solvable Lie groups (see, e.g. [5]) representations that are not components of the regular one can, in a sense, be regarded as "limits" of those into which the regualar one is decomposed, while for the unimodular groups those "extra" representations are analytic continuations of them.

The proofs of the results formulated in this section are rather instructive. The important role here is played by a special choice of coordinates which, to a certain extent, recall the usual elliptical coordinates[11].

10. The second part of the paper is devoted to the study of unitary representations of other series of classical groups, i.e. the representations of the group of orthogonal and of the group of symplectic matrices. The corresponding investigation carried out in the first part for the group of unimodular matrices is useful not only because of its similarity, but also because the second part essentially relies on the results of the first part.

Consider, for example, the group of orthogonal matrices of even order. We consider those matrices as linear operators in the $2n$-dimensional complex space leaving invariant the form

$$(\xi,\eta)=\sum_{i=0}^{2n}\xi_i\eta_i. \tag{48}$$

Let us find $2n$ vectors $e_n$, $e_{n-1},\dots,e_1$, $e_{-1},\dots,e_{-n}$ satisfying the conditions

$$\begin{aligned}(e_i,e_k)&=0, \quad i\neq -k\\ (e_i,e_{-i})&=1.\end{aligned}$$

If we denote the coordinates of the vector $\xi$ in the basis $e_i$ $(i=-n,\dots,n)$ by $\xi_i'$, then $(\xi,\eta)$ takes the form

$$(\xi,\eta)=\sum\xi_i'\eta_{-i}'. \tag{49}$$

In the new basis $e_i$ $(i=-n,\dots,n)$ our group is the group of matrices that preserve the form (49). It turns out that, as in Sect. 3, in the new basis one can introduce $Z$ as the set of matrices $z=\|z_{pq}\|$ satisfying the condition

$$z_{pp}=1, \quad z_{pq}=0 \text{ for } p<q \tag{50}$$

and belonging to our group. All the formulas and theorems, similar to those of the first part, can be proved in the same way.

In conclusion we formulate the following problem. Let $\mathfrak{G}$ be a Lie group, $g\to T_g$ be a unitary representation of the group $\mathfrak{G}$. Consider an integrable function $x(g)$ on $\mathfrak{G}$ and the operator

$$A=\int x(g)T_g d\mu(g)$$

---

[11] A completely different way of proving the results of this section is given in [6] for $n=2$ and outlined for the general case in [11].

where integration is made with respect to the invariant measure. Prove or disprove that the operator $A$ is completely continuous.

We know that for classical groups the statement is true. It can be checked without difficulty that it is also true for the group of linear transformations of the straight line [5].

The fact that the statement is true for semisimple Lie groups ensures that for them (and probably for all Lie groups) there are no set-theoretical complications in the study of irreducible unitary representations, the problem of decomposing any representation into irreducible ones, etc.

## 9. Spherical functions

1. *Condition for the existence of the vector invariant with respect to all operators* $T_u$. Consider now the following problem.

Under what conditions does there exist a vector $f_0(\tilde{u})$ in the representation space $\mathfrak{H}_1$ invariant with respect to all operators $T_u$ corresponding to the elements $u$ of the unitary subgroup $\mathfrak{U}$?

According to formula (8.18*) the desired vector $f_0(u)$ should satisfy the condition

$$\frac{\alpha_1(uu_0)}{\alpha_1(u)} f_0(uu_0) = f_0(u) \tag{9.1}$$

for each $u_0 \in \mathfrak{U}$ for almost all $u \in \mathfrak{U}$. Hence

$$\alpha_1(uu_0) f_0(uu_0) = \alpha_1(u) f_0(u)$$

for almost all pairs $(u, u_0)$ and, consequently, for almost all pairs $(uu_0, u_0)$. Therefore $\alpha_1(u) f_0(u) = \text{const.}$ for almost all $u \in G$. Since $f_0(\gamma u) = f_0(u)$, it is possible if and only if $\alpha_{11}(\gamma u) = \alpha_1(u)$, i.e. if $\alpha_1(\gamma) = 1$. The explicit formula (for $\alpha_1$) implies that condition $\alpha_1(\gamma) = 1$ is equivalent to the equalities $m_2 = m_3 = \ldots = m_n = 0$. Under that condition, $f_0(u) = c\alpha_1^{-1}(u)$, i.e. up to a constant factor there exists only one such vector.

Thus, the following theorem is proved.

**Theorem 3.**** *A vector in the representation space invariant with respect to all representation operators, corresponding to the unitary subgroup, exists if and only if* $\chi(\gamma) = 1$, *i.e. if* $m_2 = m_3 = \ldots = m_n = 0$.

*If that condition is satisfied there exists up to a scalar factor only one such vector.*

2. *Definition of spherical function and its integral formula.* Since for $\alpha_1(\gamma) = 1$ the function $\alpha_1(u)$ is constant on each class $\tilde{u}$, one can set $\alpha_1(u) = \alpha_1(\tilde{u})$. On the other hand, the function $\alpha_1(u)$ is defined by a given representation up to a factor $\kappa(\tilde{u})$ of

* See formula (15) in the Introduction.

** Theorems 1 and 2 (dealing with Gauss' decomposition from and with the irreducibility of principal series) are not included here.

absolute value 1 (see §8). Letting $\kappa(u) = \alpha_1(\tilde{u})$ one can achieve $\alpha_1(u) = 1$. Then $f_0(\tilde{u}) = c$. Normalising the measure $d\mu(\tilde{u})$ in such a way that $\int d\mu(\tilde{u}) = 1$ and letting $f_0(\tilde{u}) = 1$ one obtains the normalised vector $f_0$. For that vector we define

$$\varphi(g) = (T_g f_0, f_0); \tag{9.2}$$

the function $\varphi(g)$ is called the spherical function corresponding to the given representation $T_g$. It is an elementary positive definite function satisfying the condition

$$\varphi(ug) = \varphi(gu) = \varphi(g). \tag{9.3}$$

The first statement follows from the fact that $T_g$ is an irreducible unitary representation of the group $G$ (see [12]) while the second is a consequence of the definition of the vector $f_0$, $T_u f_0 = f_0$. Therefore

$$\varphi(gu) = (T_g T_u f_0, f_0) = (T_g f_0, f_0) = \varphi(g),$$
$$\varphi(ug) = (T_u T_g f_0, f_0) = (T_g f_0, T_u^{-1} f_0) = (T_g f_0, f_0) = \varphi(g).$$

Formula (9.3) implies that it is sufficient to compute $\varphi(g)$ for diagonal matrices $\delta$ with positive elements. Indeed, any matrix $g \in \mathfrak{G}$ can be represented in the form $g = au$ where $a$ is a positive definite Hermitian matrix and $u$ is a unitary matrix. The matrix $a$ can then be represented in the form $a = u_1 \delta u_1^{-1}$ where $u_1$ is unitary and $\delta$ is a diagonal matrix with positive diagonal elements. Thus, $g = u_1 \delta u_2^{-1}$. Hence $\varphi(g) = \varphi(\delta)$ and it is sufficient to compute only $\varphi(\delta)$. Note that $\varphi(\delta)$ is a symmetric function of the diagonal elements $\lambda_p$, $p = 1, \ldots, n$, of the matrix $\delta$ since any permutation of these elements can be obtained by a unitary representation of the form $\delta' = u^{-1} \delta u$.

Since $f_0(u) \equiv 1$ and $\alpha_1(u) \equiv 1$, one has, using formula (8.18), $u_\delta f_0 = \alpha_1(u\delta)$, whence

$$\varphi(\delta) = (u_\delta f_0, f_0) = \int \alpha_1(u\delta) d\mu(u). \tag{9.4}$$

On the other hand formulas (8.14) for $m_2 = m_3 = \ldots = m_n = 0$ and (8.15) yield the following expression for $\alpha_1(u\delta)$:

$$\begin{aligned} \alpha_1(u\delta) &= \left(\frac{\Delta_2}{\Delta_3}\right)^{i\frac{1}{2}\rho_2 - 1} \left(\frac{\Delta_3}{\Delta_4}\right)^{i\frac{1}{2}\rho_3 - 2} \cdots \left(\frac{\Delta_{n-1}}{\Delta_n}\right)^{i\frac{1}{2}(\rho_n - 1) - n + 1} \Delta_n^{i\frac{1}{2}\rho_n - n + 1} \\ &= \Delta_2^{i\frac{1}{2}\rho_2 - 1} \Delta_3^{i\frac{1}{2}(3 - \rho_2) - 1} \ldots \Delta_n^{i\frac{1}{2}(\rho_n - \rho_{n-1}) - 1}, \end{aligned} \tag{9.5}$$

where $\Delta_p$ $(p = 2, \ldots, n)$ are the minors of the matrix

$$b = (u\delta)(u\delta)^* = u\delta^2 u^*,$$

composed of the last $n - p + 1$ rows and columns. The matrices $b, u, \delta^2$ can be considered as matrices of operators $b, u, \delta^2$ acting in the $n$-dimensional space $R_n$ with respect to a fixed orthonormal basis $f_1, f_2, \ldots, f_n$ and can, therefore, be written in the following way:

$$b_{pq} = (bf_q, f_p) = (u\delta^2 u^* f_q, f_p) = (\delta^2 u^* f_q, u^* f_p). \tag{9.6}$$

Let

$$u^* f_n = e_1, \quad u^* f_{n-1} = e_2, \ldots, u^* f_1 = e_n \tag{9.7}$$

$$i\tfrac{1}{2}(\rho_n - \rho_{n-1}) - 1 = \sigma_1,\ i\tfrac{1}{2}(\rho_{n-1} - \rho_{n-2}) - 1 = \sigma_2, \ldots, i\tfrac{1}{2}\rho_2 - 1 = \sigma_{n-1};\ \delta^2 = a. \tag{9.8}$$

Then vectors $e_1, e_2, \ldots, e_n$ form an orthonormal basis in $R_n$. If $u$ runs over all elements of the group $\mathfrak{U}$, then $e_1, e_2, \ldots, e_n$ runs over all such bases up to a factor in $e_1$ of absolute value 1. One has

$$\begin{aligned} \Delta_n &= (ae_1, e), \\ \Delta_{n-1} &= \begin{vmatrix} (ae_1, e_1) & (ae_1, e_2) \\ (ae_2, e_1) & (ae_2, e_2) \end{vmatrix}, \\ &\cdots\cdots\cdots\cdots \\ \Delta_2 &= \begin{vmatrix} (ae_1, e_1) & (ae_1, e_2) \ldots (ae_1, e_{n-1}) \\ (ae_2, e_1) & (ae_2, e_2) \ldots (ae_2, e_{n-1}) \\ \cdots & \cdots \\ (ae_{n-1}, e_1) & (ae_{n-1}, e_2) \ldots (ae_{n-1}, e_{n-1}) \end{vmatrix}, \end{aligned} \tag{9.9}$$

so that, in view of the explicit form of the inner product in the representation space

The spherical function $\varphi(\delta)$ of the principal series representation $T_g$ is

$$\varphi(\delta) = \int (ae_1, e_1)^{\sigma_1} \begin{vmatrix} (ae_1, e_1) & (ae_1, e_2) \\ (ae_2, e_1) & (ae_2, e_2) \end{vmatrix}^{\sigma_2} \cdots \begin{vmatrix} (ae_1, e_1) & (ae_1, e_2) \ldots (ae_1, e_{n-1}) \\ (ae_2, e_1) & (ae_2, e_2) \ldots (ae_2, e_{n-1}) \\ \cdots & \cdots \\ (ae_{n-1}, e_1) & (ae_{n-1}, e_2) \ldots (ae_{n-1}, e_{n-1}) \end{vmatrix}^{\sigma_{n-1}} d\mu(e), \tag{9.10}$$

where $d\mu(e)$ is the measure on the set of all orthonormal bases $\{e_1, \ldots, e_n\}$ of the $n$-dimensional space invariant under their unitary transformations, $a = \delta^2$, and the numbers $\sigma_p$ are defined by formula (9.8).

3. *Computation of the spherical function.* The integral (9.10) is computed by induction. First, let $n = 2$. Then

$$(ae_1, e_1) = b_{22} = \tilde{\lambda}_1 |u_{21}|^2 + \tilde{\lambda}_2 |u_{22}|^2, \quad |u_{21}|^2 + |u_{22}|^2 = 1,$$

where $\tilde{\lambda}_1 = \lambda_1^2$, $\tilde{\lambda}_2 = \lambda_2^2$ are eigenvalues of the operator $a$. Take $t = |u_{22}|^2$ and $\theta_1 = \arg u_{21}, \theta_2 = \arg u_{22}$ for parameters in the group $U$. Then the invariant measure in $U$, i.e. in the group of unitary matrices of order 2, is given by the formula:

$$d\mu(u) = \frac{1}{(2\pi)^2} dt d\theta_1 d\theta_2$$

and formula (9.10) takes the form[12]:

$$\varphi(\delta) = \frac{1}{(2\pi)^2} \int_0^1 [\tilde{\lambda}_1(1-t) + \tilde{\lambda}_2 t]^{\sigma_1} dt \int_0^{2\pi} \int_0^{2\pi} d\theta_1 d\theta_2 = \frac{1}{(\sigma_1 + 1)(\tilde{\lambda}_2 - \tilde{\lambda}_1)} (\tilde{\lambda}_2^{\sigma_1 + 1} - \tilde{\lambda}_1^{\sigma_1 + 1}),$$

[12] Formula (9.11) for spherical functions has been obtained previously by the authors in [6].

i.e.

$$\varphi(\delta)=\frac{\tilde{\lambda}_1^{\sigma_1+1}-\tilde{\lambda}_2^{\sigma_1+1}}{(\sigma_1+1)(\mu_2-\mu_1)}. \tag{9.11}$$

Suppose now that for the group of order $n-1$ the following formula holds:

$$(\delta)=c\,\frac{\begin{vmatrix} 1 & 1 & \dots & 1 \\ \tilde{\lambda}_1^{\sigma_{n-2}+1} & \tilde{\lambda}_2^{\sigma_{n-2}+1} & \dots & \tilde{\lambda}_{n-1}^{\sigma_{n-2}+1} \\ \dots & \dots & \dots & \dots \\ \tilde{\lambda}_1^{\sigma_2+\dots+\sigma_{n-2}+n-3} & \tilde{\lambda}_2^{\sigma_2+\dots+\sigma_{n-2}+n-3} & \dots & \tilde{\lambda}_{n-1}^{\sigma_2+\dots+\sigma_{n-2}+n-3} \\ \tilde{\lambda}_1^{\sigma_1+\sigma_2+\dots+\sigma_{n-2}+n-2} & \tilde{\lambda}_2^{\sigma_1+\sigma_2+\dots+\sigma_{n-2}+n-2} & \dots & \tilde{\lambda}_{n-1}^{\sigma_1+\sigma_2+\dots+\sigma_{n-2}+n-2} \end{vmatrix}}{\prod_{l=1}^{n-2}\prod_{k=1}^{l}(\sigma_k+\dots+\sigma_l+l-k+1)\prod_{n-1\geqq p\geqq q\geqq 1}(\tilde{\lambda}_p-\tilde{\lambda}_q)}$$

$$c=1!\,2!\dots(n-2)!, \tag{9.12}$$

where $\tilde{\lambda}_1,\tilde{\lambda}_2,\dots,\tilde{\lambda}_{n-1}$ are eigenvalue of the operator $a$. We have to prove that it then also holds for the group of order $n$. Since it has already been proved to hold for $n=2$, it is then true for all $n$.

The proof is made by taking the integral in (9.10) first over all systems $e_1,e_2,\dots,e_{n-1}$ with $e_n$ fixed and then over all $e_n$. But, for $e_n$ fixed, different systems $e_1,\dots,e_{n-1}$ are obtained from one another by a unitary transformation in $R_{n-1}$ that does not change the last determinant in (9.10), which may therefore be rewritten in the following way:

$$\varphi(\delta)=\int\begin{vmatrix}(ae_1,e_1) & (ae_1,e_2)\dots(ae_1,e_{n-1}) \\ \dots & \dots \\ (ae_{n-1},e_1) & (ae_{n-1},e_2)\dots(ae_{n-1},e_{n-1})\end{vmatrix}^{\sigma_{n-1}} d\mu(e_n)$$

$$\cdot\int(ae_1,e_1)^{\sigma_1}\begin{vmatrix}(ae_1,e_1) & \dots(ae_1,e_{n-2}) \\ \dots & \dots \\ (ae_{n-2},e_1) & \cdots(ae_{n-2},e_{n-2})\end{vmatrix}^{\sigma_{n-2}} d\mu(e_1,\dots,e_{n-1}). \tag{9.13}$$

Let $\mathfrak{M}$ be the subspace of all vectors in $R_n$ orthogonal to $e_n$ and $P$ the projection operator on that space. The inner integral in (9.13) is the function $\varphi$ corresponding to the operator $PaP$ in the $(n-1)$-dimensional space[13] $\mathfrak{M}$. By the induction hypothesis, this integral is given by formula (9.12) where one has to substitute the eigenvalues of the operators $PaP$ for $\tilde{\lambda}_1,\dots,\tilde{\lambda}_{n-1}$.

[13] We consider the operator $PaP$ instead of $a$ because the subspace spanned by the vectors $e_1,\dots,e_{n-1}$ is not invariant with respect to $a$.

The determinant

$$\begin{vmatrix} (ae_1, e_1) & (ae_1, e_2) & \dots (ae_1, e_{n-1}) \\ \dots & \dots & \dots \\ (ae_{n-1}, e_1) & (ae_{n-1}, e_2) & \dots (ae_{n-1}, e_{n-1}) \end{vmatrix}$$

can also be expressed through the eigenvalues $\mu_1, \mu_2, \dots, \mu_{n-1}$ of the operator $PaP$. Indeed, it is the determinant of the matrix $PaP$ and, consequently, equals $\mu_1\mu_2\dots\mu_{n-1}$. Thus, the integral (9.13) takes the form

$$\varphi(\delta) = \int (\mu_1\mu_2\dots\mu_{n-1})^{\sigma_{n-1}} \frac{\begin{vmatrix} 1 & 1 & \dots & 1 \\ \mu_1^{\sigma_{n-2}+1} & \mu_2^{\sigma_{n-2}+1} & \dots & \mu_{n-1}^{\sigma_{n-2}+1} \\ \dots & \dots & \dots & \dots \\ \mu_1^{\sigma_2+\dots+\sigma_{n-2}+n-3} & \mu_2^{\sigma_2+\dots+\sigma_{n-2}+n-3} & \dots & \mu_{n-1}^{\sigma_2+\dots+\sigma_{n-2}+n-3} \\ \mu_1^{\sigma_1+\sigma_2+\dots+\sigma_{n-2}+n-2} & \mu_2^{\sigma_1+\sigma_2+\dots+\sigma_{n-2}+n-2} & \dots & \mu_{n-1}^{\sigma_1+\sigma_2+\dots+\sigma_{n-2}+n-2} \end{vmatrix}}{\prod_{l=1}^{n-2}\prod_{k=1}^{l} (\sigma_k + \dots + \sigma_l + l - k + 1) \prod_{n-1 \geqq p \geqq q \geqq 1} (\mu_p - \mu_q)} d\mu(e_n) \tag{9.13'}$$

where $\mu_1, \mu_2, \dots, \mu_{n-1}$ are eigenvalues of the operator $PaP$ and, therefore, depend on $e_n$.

Now we shall write the equation satisfied by these eigenvalues.

If $\mu$ is one of the eigenvalues and $f$ is the corresponding eigenvector, then $Paf = \mu f$, i.e.

$$af = \mu f + \xi e_n. \tag{9.14}$$

Let $f_1, f_2, \dots, f_n$ be a basis in the whole space $R_n$ in which operator $a$ is diagonal, and let $af_k = \tilde{\lambda}_k f_k$, $k = 1, 2, \dots, n$. Denote by $\xi_1, \xi_2, \dots, \xi_n$ and $u_1, u_2, \dots, u_n$ the projections of $f$ and $e_n$ in that basis. Then, projecting equality (9.14) on $f$, one gets: $\tilde{\lambda}_k\xi_k = \mu\xi_k + \xi u_k$. Hence $\xi_k = \xi(u_k/(\lambda_k - \mu))$. On the other hand $(f, e_n) = 0$, i.e. $\sum_{k=1}^{n} \xi_k u_k = 0$. Substituting $\xi_k$ into this equality, one has the desired equation on $\mu_1, \dots, \mu_{n-1}$:

$$\sum_{k=1}^{n} \frac{|u_k|^2}{\tilde{\lambda}_k - \mu} = 0. \tag{9.15}$$

Let $t_k = |u_k|^2$, $\theta_k = \arg u_k$. Then one has in the integral (9.13):

$$d\mu(e_n) = \frac{1}{(2\pi)^n} dt_1 dt_2 \dots dt_{n-1} d\theta_1 d\theta_2 \dots d\theta_n. \tag{9.16}$$

Let us now pass from the integration parameters $t_1, t_2, \dots, t_{n-1}$ to the parameters $\mu_1, \mu_2, \dots, \mu_{n-1}$. To achieve this one has to find the domain of values the latter can take and compute the Jacobian matrix of the transformation.

Let us arrange the numbers $\tilde{\lambda}_1, \tilde{\lambda}_2, \ldots, \tilde{\lambda}_n$ in such a way that

$$\tilde{\lambda}_1 < \tilde{\lambda}_2 < \ldots < \tilde{\lambda}_n. \tag{9.17}$$

In view of (9.15) the numbers $\mu_1, \mu_2, \ldots, \mu_{n-1}$ are the roots of the equation

$$\sum_{k=1}^{n} \frac{t_k}{\tilde{\lambda}_k - \mu} = 0 \text{ where } \sum_{k=1}^{n} t_k = 1, \quad t_k > 0. \tag{9.18}$$

Hence

$$\tilde{\lambda}_1 < \mu_1 < \tilde{\lambda}_2 < \mu_2 < \tilde{\lambda}_3 < \mu_3 < \ldots < \tilde{\lambda}_{n-1} < \mu_{n-1} < \tilde{\lambda}_n, \tag{9.19}$$

and these inequalities provide integration limits for parameters $\mu_k$.

Let us prove that, conversely, for any given numbers $\mu_k$, satisfying inequalities (9.19) one can uniquely find the numbers $t_k$ satisfying (9.17). Thus it will be proved that each of the parameters $\mu_k$ can take any value from the interval $(\lambda_k, \lambda_{k+1})$. The numbers $t_k$ are defined from the equations

$$\sum_{k=1}^{n} \frac{t_k}{\tilde{\lambda}_k - \mu_i} = 0, \quad \sum_{k=1}^{n} t_k = 1, \quad i = 1, 2, \ldots, n-1. \tag{9.20}$$

Let $v_i = 1/\mu_i$; $i = 1, \ldots, n-1$; $v_n = 0$. Then equations (9.20) can be written in the form

$$\sum_{k=1}^{n} \frac{t_k}{1 - v_i \tilde{\lambda}_k} = \begin{cases} 0 & \text{for } i = 1, 2, \ldots, n-1, \\ 1 & \text{for } i = n. \end{cases} \tag{9.21}$$

By the Cauchy Lemma (see, e.g. Ref. [32], p. 276) for the determinant of this system one has

$$\det \left\| \frac{1}{1 - v_i \tilde{\lambda}_k} \right\|_{i,k=1,2,\ldots,n} = \frac{\prod\limits_{1 \leq k < i \leq n} (v_i - v_k) \prod\limits_{1 \leq k < i \leq n} (\tilde{\lambda}_i - \tilde{\lambda}_k)}{\prod\limits_{i,k=1} (1 - v_i \tilde{\lambda}_k)}. \tag{9.22}$$

Hence

$$t_p = (-1)^{n+p} \det \left\| \frac{1}{1 - v_1 \tilde{\lambda}_k} \right\|_{\substack{i=1,2,\ldots,n-1 \\ k=1,\ldots,p-1,p+1,\ldots,n}} : \det \left\| \frac{1}{1 - v_i \tilde{\lambda}_k} \right\|_{i,k=1,\ldots,r}$$

$$= (-1)^{n+p} \frac{\prod\limits_{1 \leq i_2 < i_1 \leq n-1} (v_{i_1} - v_{i_2}) \prod\limits_{1 \leq k_2 < k_1 \leq n} (\tilde{\lambda}_{k_1} - \tilde{\lambda}_{k_2})}{\prod\limits_{i=1}^{n-1} \prod\limits_{\substack{k=1 \\ k \neq p}}^{n} (1 - v_i \tilde{\lambda}_k)} :$$

$$: \frac{\prod\limits_{1 \leq i_2 < i_1 \leq n} (v_{i_1} - v_{i_2}) \prod\limits_{1 \leq k_2 < k_1 \leq n} (\tilde{\lambda}_{k_1} - \tilde{\lambda}_{k_2})}{\prod\limits_{i=k=1}^{n} (1 - v_i \tilde{\lambda}_k)}$$

$$= (-1)^{n+p} \frac{\prod\limits_{k=1}^{n} (1 - v_n \tilde{\lambda}_k) \prod\limits_{i=1}^{n} (1 - v_i \tilde{\lambda}_p)}{\prod\limits_{k=1}^{n-1} (v_n - v_k) \prod\limits_{k=1}^{p-1} (\tilde{\lambda}_p - \tilde{\lambda}_k) \prod\limits_{k=p+1}^{p} (\tilde{\lambda}_k - \tilde{\lambda}_p)}. \tag{9.23}$$

Since $\nu_n = 0$, the last equality takes the form

$$t_p = (-1)^{p+1} \frac{\prod\limits_{i=1}^{n-1} (1-\nu_i \tilde{\lambda}_p)}{\prod\limits_{k=1}^{n-1} \nu_k \prod\limits_{k=1}^{p-1} (\tilde{\lambda}_p - \tilde{\lambda}_1) \prod\limits_{k=p+1}^{n} (\tilde{\lambda}_k - \tilde{\lambda}_p)} = \frac{\prod\limits_{i=1}^{n-1} (1-\nu_i \tilde{\lambda}_p)}{\prod\limits_{k=1}^{n-1} \nu_k \prod\limits_{k \neq p} (\tilde{\nu}_k - \tilde{\lambda}_p)}$$

or, since $\nu_k = 1/\mu_k$, $k = 1,\ldots,n-1$, one has

$$t_p = \frac{\prod\limits_{i=1}^{n-1} (\mu_i - \tilde{\lambda}_p)}{\prod\limits_{i \neq p} (\tilde{\lambda}_i - \tilde{\lambda}_p)}. \tag{9.24}$$

Inequalities (9.19) imply that the signs of both the numerator and denominator are equal to $(-1)^{p-1}$ and, consequently, the whole expression is positive. Thus, the numbers $t_p$ satisfy all the required conditions and the statement is proved.

Let us now find the Jacobian $D(t_1,\ldots,t_{n-1})/D(\;_1,\ldots,\;_{n-1})$. Formula (9.24) implies

$$\frac{\partial t_p}{\partial \mu_i} = \frac{\prod\limits_{j=1, j\neq i}^{n-1} (\mu_j - \tilde{\lambda}_p)}{\prod\limits_{j \neq p} (\tilde{\lambda}_i - \tilde{\lambda}_p)} = \frac{\prod\limits_{j=1}^{n-1} (\mu_j - \tilde{\lambda}_p)}{\prod\limits_{j \neq p} (\tilde{\lambda}_i - \tilde{\lambda}_p)} \frac{1}{\mu_i - \tilde{\lambda}_p};$$

and, consequently,

$$\begin{aligned} \frac{D(t_1,\ldots,t_{n-1})}{D(\;_1,\ldots,\;_{n-1})} &= \frac{\prod\limits_{j,p=1}^{n-1} (\mu_j - \tilde{\lambda}_p)}{\prod\limits_{\substack{i,p=1 \\ j \neq p}}^{n-1} (\tilde{\lambda}_i - \tilde{\lambda}_p)} \operatorname{Det} \left\| \frac{1}{\mu_i - \tilde{\lambda}_p} \right\|_{i,p=1,\ldots,n-1} \\ &= \prod_{1 \leq k < i \leq n-1} (\mu_i - \mu_k) : \prod_{1 \leq i < p \leq n} (\tilde{\lambda}_i - \lambda_p). \end{aligned}$$

Inequalities (9.19) imply that the absolute value of the determinant is

$$\prod_{1 \leq k < i \leq n-1} (\mu_i - \mu_k) : \prod_{1 \leq p < i \leq n} (\tilde{\lambda}_i - \lambda_p). \tag{9.25}$$

Substituting this into (9.13′) we have:

$$\varphi(\delta) = \frac{1}{\prod\limits_{1 \leq i < p \leq n} (\tilde{\lambda}_p - \tilde{\lambda}_i) \prod\limits_{l=1}^{n-2} \prod\limits_{k=1}^{l} (\sigma_k + \ldots + \sigma_l + l - k - 1)}$$

$$\cdot \int_{\tilde{\lambda}_{n-1}}^{\tilde{\lambda}_n} \int_{\tilde{\lambda}_{n-2}}^{\tilde{\lambda}_{n-1}} \cdots \int_{\tilde{\lambda}_1}^{\tilde{\lambda}_2} \begin{vmatrix} 1 & 1 & \ldots & 1 \\ \mu_1^{\sigma_{n-2}+1} & \mu_2^{\sigma_{n-2}+1} & \ldots & \mu_{n-1}^{\sigma_{n-2}+1} \\ \mu_1^{\sigma_{n-3}+\sigma_{n-2}+2} & \mu_2^{\sigma_{n-3}+\sigma_{n-2}+2} & \ldots & \mu_{n-1}^{\sigma_{n-3}+\sigma_{n-2}+2} \\ \ldots & \ldots & \ldots & \ldots \\ \mu_1^{\sigma_1+\ldots+\sigma_{n-2}+n-2} & \mu_2^{\sigma_1+\ldots+\sigma_{n-2}+n-2} & \ldots & \mu_{n-1}^{\sigma_1+\ldots+\sigma_{n-2}+n-2} \end{vmatrix}$$

$$\cdot \mu_1^{\sigma_{n-1}} \mu_2^{\sigma_{n-2}} \ldots \mu_{n-1}^{\sigma_{n-1}} d\mu_1 d\mu_2 \ldots d\mu_{n-1} \tag{9.26}$$

The factors $\mu_1^{\sigma_{n-1}}, \mu_2^{\sigma_{n-1}}, \ldots$, can be transferred to the respective columns of the determinant. Integrating, one has for (9.26)

$$\left|\begin{matrix} \dfrac{1}{\prod\limits_{k=1}^{n-1}(\sigma_k+\ldots+\sigma_{n-1}+n-k)} & & \\ \tilde{\lambda}_2^{\sigma_{n-1}+1}-\tilde{\lambda}_1^{\sigma_{n-1}+1} & \ldots & \tilde{\lambda}_n^{\sigma_{n-1}+1}-\tilde{\lambda}_{n-1}^{\sigma_{n-1}+1} \\ \tilde{\lambda}_1^{\sigma_{n-2}+\sigma_{n-1}+2}-\tilde{\lambda}_1^{\sigma_{n-2}+\sigma_{n-1}+2} & \ldots & \tilde{\lambda}_n^{\sigma_{n-2}+\sigma_{n-1}+2}-\tilde{\lambda}_{n-1}^{\sigma_{n-2}+\sigma_{n-1}+2} \\ \ldots\ldots & \ldots & \ldots\ldots \\ \tilde{\lambda}_2^{\sigma_1+\ldots+\sigma_{n-1}+n-1}-\lambda_1^{\sigma_1+\ldots+\sigma_n+n-1} & \ldots & \tilde{\lambda}_n^{\sigma_1+\ldots+\sigma_{n-1}+n-1}-\lambda_{n-1}^{\sigma_1+\ldots+\sigma_{n-1}+n-1} \end{matrix}\right|$$

$$= \frac{1}{\prod\limits_{k=1}^{n-1}(\sigma_k+\ldots+\sigma_{n-1}+n-k)} \cdot \left|\begin{matrix} 1 & 1 & \ldots & 1 \\ \tilde{\lambda}_1^{\sigma_{n-1}+1} & \tilde{\lambda}_2^{\sigma_{n-1}+1} & \ldots & \tilde{\lambda}_n^{\sigma_{n-1}+1} \\ \tilde{\lambda}_1^{\sigma_{n-2}+\sigma_{n-1}+2} & \tilde{\lambda}_2^{\sigma_{n-2}+\sigma_{n-1}+2} & \ldots & \tilde{\lambda}_n^{\sigma_{n-2}+\sigma_{n-1}+2} \\ \\ \tilde{\lambda}_1^{\sigma_1+\ldots+\sigma_{n-1}+n-1} & \tilde{\lambda}_2^{\sigma_1+\ldots+\sigma_{n-1}+n-1} & \ldots & \tilde{\lambda}_n^{\sigma_1+\ldots+\sigma_{n-1}+n-1} \end{matrix}\right|.$$

Substituting this into (9.26) one obtains

$$\varphi(\delta) = \frac{1}{\prod\limits_{1\leq i<p\leq n}(\tilde{\lambda}_p-\tilde{\lambda}_i)\prod\limits_{l=1}^{n-1}\prod\limits_{k=1}^{l}(\sigma_k+\ldots+\sigma_l+l-k+1)} \cdot \left|\begin{matrix} 1 & 1 & \ldots & 1 \\ \tilde{\lambda}_1^{\sigma_{n-1}+1} & \tilde{\lambda}_2^{\sigma_{n-1}+1} & \ldots & \tilde{\lambda}_n^{\sigma_{n-1}+1} \\ \tilde{\lambda}_1^{\sigma_{n-2}+\sigma_{n-1}+1} & \tilde{\lambda}_2^{\sigma_{n-2}+\sigma_{n-1}+2} & \ldots & \tilde{\lambda}_n^{\sigma_{n-2}+\sigma_{n-1}+2} \\ \\ \tilde{\lambda}_1^{\sigma_1+\ldots+\sigma_{n-1}+n-1} & \tilde{\lambda}_2^{\sigma_1+\ldots+\sigma_{n-1}+n-1} & \ldots & \tilde{\lambda}_n^{\sigma_1+\ldots+\sigma_{n-1}+n-1} \end{matrix}\right|. \tag{9.27}$$

The same expression is obtained from formula (9.12) by substituting $n$ instead of $n(n-1)$. Thus, formula (9.27) is proved for all $n$.

It remains to substitute into this formula expressions for $\sigma_1, \sigma_2, \ldots, \sigma_{n-2}$ given by formula (9.8) and expressions $\tilde{\lambda}_i = \lambda_i^2$ where $\tilde{\lambda}_i$ is expressed through the diagonal elements of the matrix $\delta$. Then we come to the following theorem:

**Theorem 4.** *If $T_g$ is an irreducible representation of the principal series of the group $\mathfrak{G}$ corresponding to the character $\chi(\delta) = |\lambda_2|^{i\rho_2}|\lambda_3|^{i\rho_3}\ldots|\lambda_n|^{i\rho_n}$, then the corresponding spherical function is defined by the formula*

$$\varphi(\delta)=\left(\frac{2}{i}\right)^{n(n-1)/2}\frac{1}{\prod\limits_{1\leqq p<q\leqq n}(\rho_q-\rho_p)\prod\limits_{1\leqq p<q\leqq n}(\lambda_q^2-\lambda_p^2)}\cdot\begin{vmatrix}1 & 1 & \dots & 1\\ \lambda_1^{i\rho_2} & \lambda_2^{i\rho_2} & \dots & \lambda_n^{i\rho_2}\\ \lambda_1^{i\rho_3} & \lambda_2^{i\rho_3} & \dots & \lambda_n^{i\rho_3}\\ \\ \lambda_1^{i\rho_n} & \lambda_2^{i\rho_n} & \dots & \lambda_n^{i\rho_n}\end{vmatrix}.$$

## 18. Transitivity classes for the set of pairs. Another way of describing representations of the complementary series

This section is devoted to the study of some questions related to structure of the space of pairs $(z'_1, z'_2)$ which, as we have already seen in the Introduction to this chapter, is related, in its turn, to the study of complementary series.

1. *Two-sided cosets of G with respect to K.* Let $Z'$ denote, as before, the space of left cosets of $G$ with respect to the subgroup $K$ of triangular matrices.

Consider the set of all pairs $(z'_1, z'_2)$ where $z'_1, z'_2 \in Z'$. We have to find the set $M$ of pairs $(z'_1, z'_2)$ satisfying the following conditions:

(i) If $(z'_1, z'_2)\in M$ then $(z'_1 g, z'_2 g)\in M$.

(ii) The space $M$ is transitive, i.e. for any two pairs $(z'_1, z'_2)\in M$ and $(z''_1, z''_2)\in M$ there exists an element $g\in G$ such that $(z''_1, z''_2)=(z'_1\bar{g}, z'_2\bar{g})$.

We now show that there exist $n!$ such spaces $M$ and give their description.

Clearly, in each class $M$ there are pairs of the form $(e', z'_2)$ where $e'$ is the unit coset of $G$ with respect to $K$. Indeed, if $(z'_1, z'_2)$ is an arbitrary pair in $M$ and $g_1$ is an element of the coset $z'_1$ then $(e', z'_2\bar{g}_1^{-1})\in M$.

To describe $M$ it is sufficient to describe the set of pairs of the form $(e', z')\in M$.

If $(e', z')\in M$ then for each $k\in K$ we have $(e', z'\bar{k})\in M$. Indeed, $(e'k, z'\bar{k})=(e', z'\bar{k})$ and, consequently, $(e', z'\bar{k})\in M$. Conversely, if $(e', z')$ is any pair in $M$ then any other pair $(e', z'')$ in $M$ is also of the form $(e', z'\bar{k})$. Indeed, the transitivity of $M$ implies that there exists $g$ such that $(e'\bar{g}, z'\bar{g})=(e', z'')$. But $e'\bar{g}=e'$ implies that $g\in K$.

Thus the problem of describing $M$ is reduced to the study of the set of elements $z'\bar{k}$ where $z'$ is fixed. Since $z'$ is a coset of $G$ with respect to $K$, i.e. a set of elements of the form $k_1 g_0$ where $g_0$ is fixed, the problem is ultimately formulated in the following way:

Describe the set of elements in $\mathfrak{G}$ of the form $k_1 g_0 k_2$ where $g_0$ is fixed and $k_1, k_2$ are arbitrary elements from $K$. Such a set in $G$ is called a two-sided coset of $\mathfrak{G}$ with respect to $K$.

The following proposition holds:

Any element $g\in\mathfrak{G}$ can be represented in the form

$$g=k_1 s k_2 \tag{18.1}$$

where $k_1, k_2 \in K$ and $s$ is a matrix having only one non-zero element in each row and in each column. All these elements except one can be chosen to be equal to $+1$ and that one is determined from the condition $\det(s) = 1$.

Thus, there are $n!$ essentially different $s$.

Now we prove this statement: we show that for each element $g \in \mathfrak{G}$ one can choose $k_1$ and $k_2$ in such a way that $k_1 g k_2$ is an element of $s$.

Multiplication of matrix $g$ by $k_1$ from the left is reduced to multiplying the $n$-th row by a scalar, replacing the $(n-1)$-st row by a linear combination of the $n$-th and $(n-1)$-st rows, etc. Multiplication by $k_2$ on the right is reduced to multiplying the first column in $g$ by a scalar, replacing the second column by a linear combination of the first two columns, etc. Using these operations, we have to reduce $g$ to a form such that it has only one non-zero element in each row and in each column. Consider the first non-zero element in the last row. Denote it by $g_{nk_1}$. Then, multiplying the $k_1$-th column by a scalar, replacing the $(k_1+1)$-st column by a linear combination of the $k_1$-th and $(k_1+1)$-st columns, then replacing the $(k_1+2)$-nd by a linear combination of the $k_1$-th, $(k_1+1)$-st and $(k_1+2)$-nd columns and so on, we can show that $g_{n,k_1+1} = g_{n,k_1+2} = \ldots = g_{nn} = 0$ and $g_{nk_1} = 1$. Then, multiplying the last row by the numbers $-g_{n-1,k_1}, -g_{n-2,k_1}, \ldots, -g_{1k_1}$ and adding it to the $(n-1)$-st, $(n-2)$-nd, ..., 1-st row, respectively, we can make all the elements of the $k_1$-th column vanish except $g_{nk_1}$. Eliminating the last row and the $k_1$-th column from the matrix, repeat the same operation with the remainder, etc. Finally, one comes to the matrix $s$ having only one non-zero element in each row and in each column. All these elements are equal to 1 with the exception of the last one, which is either $+1$ or $-1$ (which is determined by the condition that the determinant of the matrix be equal to 1).

Thus, we have proved that there exist $k_1$ and $k_2$ such that $k_1 g k_2 = s$ or

$$g = k_1^{-1} s k_2^{-1}, \tag{18.2}$$

as required.

Let us now prove that $s$ in the representation of $g$ in the form

$$g = k_1 s k_2$$

is unique.

Let $g = k_1' s_1 k_2'$ and $g = k_1'' s_2 k_2''$, i.e. $k_1' s_1 k_2' = k_1'' s_2 k_2''$, or $k_1 = s_1^{-1} k_2 s_2$ where $k_1 = k_1''^{-1} k_1'$ and $k_2 = k_2'' k_2'^{-1}$. Suppose that the permutation $s_1$ takes $i$ into $n_i$ and the permutation $s_2$ takes $i$ into $m_i$. If $k_{ii}$ are diagonal elements of the matrix $k_2$ and $a_{pq}$ are elements of the matrix $k_1 = s_1^{-1} k_2 s_2$ then $a_{n_i m_i} = \pm k_{ii}$. Since $a_{pq} = 0$ for $p > q$ and $k_{ii} \neq 0$ one has $n_i \leqq m_i$ for all $i$. But any two permutations

$$\begin{pmatrix} 1 \ldots n \\ m_1 \ldots m_n \end{pmatrix} \quad \text{and} \quad \begin{pmatrix} 1 \ldots n \\ m_1' \ldots m_n' \end{pmatrix}$$

for which $m_i' \leqq m_i$, for $i = 1, \ldots, n$, must coincide, i.e. $s_1 = s_2$.

In the representation (18.1) matrices $k_1$ and $k_2$ are not defined uniquely. Indeed, let $k_1' s k_2' = k_1'' s k_2''$, i.e. $s k_2 s^{-1} = k_1$ where $k_1 = k_1''^{-1} k_1'$, $k_2 = k_2'' k_2'^{-1}$. Denoted by $K_s$ the subgroup of the group $K$ consisting of those elements for which $k \in K$ and

$s^{-1}ks \in K$. Thus $k_2 \in K_s$; put $k_2 = k_s$. Finally, we have $k_2 k_2'^{-1} = k_s$, i.e. $k_2'' = ksk_2'$. Therefore, in the decomposition $g = k_1' s k_2'$ the element $k_2'$ is determined up to a left factor belonging to $K_s$. For given $k_2'$ the matrix $k_1'$ is defined uniquely.

Note that for $s = e$ matrix $k_s$ coincides with $k$. For

$$s_0 = \begin{pmatrix} 0 & 0 & \ldots & 0 & 1 \\ 0 & 0 & \ldots & 1 & 0 \\ & & \ldots\ldots\ldots & & \\ 0 & 1 & \ldots & 0 & 0 \\ 1 & 0 & \ldots & 0 & 0 \end{pmatrix} \tag{18.3}$$

we have $K_{s_0} = D$ (i.e. it consists of diagonal matrices)[14].

Thus we have proved that a two-sided coset of $\mathfrak{G}$ with respect to the subgroup $K$ is uniquely defined by the permutation $s$ and consists of all elements of the form $g = k_1 s k_2$. Denote this coset by $\mathfrak{G}_s$.

What is the dimension of $\mathfrak{G}_s$? To answer this question let us first find the dimension of $K_s$. Let $k = \|k_{pq}\| \in K$ and let $s$ be a permutation $\begin{pmatrix} 1 \ldots n \\ m_1 \ldots m_n \end{pmatrix}$. Then the matrix $sks^{-1} = k'$ has the elements $k'_{pq} = k_{m_p m_q}$. The requirement $k'_{pq} = 0$ for $p > q$ implies $k_{m_p m_q} = 0$ for $p > q$. We have, therefore, to impose the condition: $k_{m_p m_q} = 0$ if $p > q$ and $m_p > m_q$.

Denote by $\tau$ the number of transpositions in the permutation $s$. Therefore, the number of conditions imposed on $k$ equals $\tau$. Since the dimension of $K$ is $\frac{1}{2}n(n+1)$ then the dimension of $K$ is $\frac{1}{2}n(n+1) - \tau$.

Let us now find the dimension of $\mathfrak{G}_s$. Each element $g \in \mathfrak{G}_s$ can be represented in the form $g = k_1 s k_2$ where $k_2$ is defined up to a factor belonging to $K_s$, and $k_1$ is defined by $k_2$ uniquely. Therefore, the dimension of $\mathfrak{G}_s$ is

$$2\tfrac{1}{2}n(n+1) - (\tfrac{1}{2}n(n+1) - \tau) = \tfrac{1}{2}n(n+1) + \tau.$$

Thus, the dimension of a two-sided coset $\mathfrak{G}_s$ of the group $\mathfrak{G}$ with respect to the subgroup $K$, defined by a permutation $s$, equals $\frac{1}{2}n(n+1) + \tau$ where $\tau$ is the number of transpositions in the permutation $s$.

We have described two-sided cosets of $\mathfrak{G}$ with respect to $K$ and, consequently, transitive subspaces $\mathfrak{M}$ of pairs $(z_1', z_2')$. Each such subspace $\mathfrak{M}$ is given by some permutation $s$ and for that reason we denote it by $\mathfrak{M}_s$. Let us find the dimension of $\mathfrak{M}_s$. For pairs of the form $(e', z')$ we have already seen that $z'$ is a left coset belonging to $\mathfrak{G}_s$. Hence the dimension of the set of such $z'$ is $\frac{1}{2}n(n+1) + \tau - \frac{1}{2}n(n+1) = \tau$. Since the dimension of $Z'$ is $\frac{1}{2}n(n-1)$ then the dimension of $\mathfrak{M}_s$ is $\frac{1}{2}n(n-1) + \tau$.

---

[14] This result can also be formulated in another way: each element $g$ can be represented in the form

$$g = ksz. \tag{18.4}$$

Indeed, the matrix $gs_0$ where the matrix $s_0$ is defined by formula (18.3) can be represented in the form $gs_0 = k_1 s_1 k_2$. Hence $g = k_1 s_1 s_0^{-1} s_0 k_2 s_0^{-1}$. Since $s_0 k_2 s_0^{-1} \in H$ and the diagonal elements of $k_2$ may be considered equal to 1, then, denoting $s_1 s_0^{-1}$ by $s$ and $s_0 k_2 s^{-1}$ by $z$, one has the required statement.

In particular, for

$$s_0 = \begin{pmatrix} 0 & 0 & \dots & 0 & 1 \\ 0 & 0 & \dots & 1 & 0 \\ & & \dots\dots\dots & & \\ 0 & 1 & \dots & 0 & 0 \\ 1 & 0 & \dots & 0 & 0 \end{pmatrix}$$

the dimension of $\mathfrak{M}_{s_0}$ is $n(n-1)$.

**References**

5. Gelfand, I.M., Najmark, M.A.: Unitary representations of the group of linear transformations of the straight line. Dokl. Akad Naūk SSSR (7) **XV** (1947) 571–574 (Russian).
6. Gelfand, I.M., Najmark, M.A.: Unitary representations of the Lorentz group. Izv. Akad. Naūk SSSR, Ser. Mat. **11** (1947) 411–504 (Russian).
11. Gelfand, I.M., Najmark, M.A.: Plancherel formula for complex unimodular group. Dokl. Akad. Naūk SSSR **63** (1948) 609–612 (Russian).
12. Gelfand, I.M., Rajkov, D.A.: Irreducible unitary representations of locally bi-compact groups. Mat. Sb. (55) **13** (1943) 301–316 (Russian).
15. Gelfand, I.M., Tsetlin, M.L.: Finite-dimensional representations of the group of unimodular matrices. Dokl. Akad. Naūk SSSR (5) **71** (1950) 825–828 (Russian).
16. Gelfand, I.M., Tsetlin, M.L.: Finite-dimensional representations of the group of orthogonal matrices. Dokl. Akad. Naūk SSSR (6) **71** (1950) 1017–1020 (Russian).
26. Cartan, E.: Bull. Soc. Math. France **41** (1913).
27. Cartan, E.: J. Math. (6) **10** (1914).
29. Mackey, G.W.: Proc. Nat. Acad. Sci. (9) **35** (1949).
32. Weyl, H.: Classical groups, their invariants and representations. Princeton University Press 1939, Princeton, N.J. (Russian translation, Moscow 1947).

# 12.

## (with M. I. Graev)

# Unitary representations of the real simple Lie groups

Dokl. Akad. Nauk SSSR **86** (1952) 461–463. Zbl. **49**:358

1. All irreducible unitary representations of the complex classic groups were described in [1]. Unitary representations of the real simple groups were constructed for the compact groups and the group of real nonsingular matrices of second order [2, 3]. Here we study the unitary representations of the real simple Lie groups. These representations have some specific features compared to the complex case.

2. A representation of a complex Lie group is constructed, roughly speaking, as follows [1]. We choose the subgroup of a group $G$ generated by positive root vectors (the subgroup $Z$ in the notation of [1]) and consider the space of right cosets (the space $H$ in the notation of [1]). Further, let us denote by $D$ the commutative subgroup generated by zero root vectors and by $\chi(\delta)$ a character (not necessarily with modulus one) of this subgroups. A representation of a nondegenerate series acts in the space of functions on $H$ which satisfy the functional equation $f(\delta h) = \chi(\delta) f(h)$. Operators of representation are given by motions in the space of cosets and the scalar product is defined in an appropriate way. The construction of representations of degenerate series is quite similar.

To apply this construction to the real Lie groups we must change it. In fact, if $G$ is a real form of some complex simple Lie group $G_{\text{compl}}$ then the subgroup $Z$ generated by positive root vectors of the group $G_{\text{compl}}$ and its conjugate groups $Z^g = g^{-1}Zg$, where $g \in G_{\text{compl}}$, can have quite different intersections with $G$. It might also happen that all the groups $Z^g$ intersect $G$ only in the unity element $e$ (for example when $G$ is compact). Nevertheless, it turns out that there exists a generalization of the space of functions on the space of cosets $G/Z$.

Let us denote by $\chi_e$ Lie operators of a group $Z^{g_1}$ where $g_1 \in G_{\text{compl}}$. The set of functions on $G_{\text{compl}}$ which are constant on the cosets $G_{\text{compl}}/Z^{g_1}$ coincides with the solution space of the system of differential equations $\chi_e f = 0$. This system is hyperbolic and the above cosets are its characteristics. In the case of a real group $G$ we can formally construct the same Lie operators $\chi_e$ (as usual

$$\frac{\partial}{\partial z} = \frac{1}{2}\left(\frac{\partial}{\partial x} - i\frac{\partial}{\partial y}\right), \quad \frac{\partial}{\partial \bar{z}} = \frac{1}{2}\left(\frac{\partial}{\partial x} + i\frac{\partial}{\partial y}\right)$$

where $z = x + iy$) and we can consider the space of solutions of the system $\chi_e f(g) = 0$ where functions $f(g)$ are now defined on $G$. But if the intersection $G \cap Z^{g_1}$ has a lower dimension, then the system $\chi_e f(g) = 0$ is not hyperbolic. The solutions of this system are in this case defined on the space of cosets $G/G \cap Z^{g_1}$ and usually these solutions are analytic in some of the parameters of this space. When two subgroups $G \cap Z^{g_1}$ and $G \cap Z^{g_2}$ ($g_1$ and $g_2$ belong to $G_{\text{compl}}$) are not conjugate, the function spaces on corresponding cosets of $G$ are actually different. Operators of representations act on homogeneous functions in these functions spaces, i.e. on the functions of which $f(\delta h) = \chi(\delta) f(h)$ where $h \in G/G \cap Z^{g_1}$ and $\delta \in D$. Here, as above, $\chi(\delta)$ is a character of the group $D$. Among all such spaces we must choose those for which the constructed representations are unitary. Let us mention that the classification of the manifolds on which different function spaces are defined is equivalent to the decomposition of all cosets $G_{\text{compl}}/Z$ (the space $H$ in the notation of [1]) into transitive classes with respect to right multiplication on elements $g \in G$. We also note that for a compact group this construction is closely related to the theory of finite-dimensional representations.

A detailed description of the different series of irreducible representations, calculation of their characters, and a proof that these representations are all irreducible unitary representations will be given later. In this paper we describe the principal nondegenerate series of irreducible unitary representations of the real unimodular $n$-th order group.

3. *The principal nondegenerate series of irreducible unitary representations of the real unimodular n-th order group*. There are $[n/2]+1$ such series, $D_0, \ldots, D_{[n/2]}$. We will describe representations of $D_m$ for an arbitrary $m$[1]. Let us take $r_1 = \ldots = r_m = 2$ and $r_{m+1} = \ldots = r_{m+\tau} = 1$ $(2m+\tau = n)$. Then a matrix $g \in G_{\text{compl}}$ is, up to the conjugation, a block matrix $g = \|g_{pq}\|$ $(p, q = 1, \ldots, m+\tau)$ where $g_{pq}$ is a matrix with $r_p$ rows and $r_q$ columns. Further, we denote by $\dot{z}$ a block matrix $\|Z_{pq}\|$ of order $n = 2m+\tau$ where if $p \leqq m$, $z_{pp} = \begin{pmatrix} 1 & 0 \\ z_p & 1 \end{pmatrix}$ and $\operatorname{Im} z_p \neq 0$, if $p > m$, $z_{pp} = 1$ and if $p \neq q$, $z_{pq} = 0$. We denote $\hat{k}$ any real $n \times n$ matrix $\|k_{pq}\|$ where $k_{pq} = 0$ when $p > q$ and by $\hat{z}$ we denote any real matrix $\|x_{pq}\|$ where $x_{pp} = 1$ and $x_{pq} = 0$ when $p < q$. Any real $n \times n$ matrix $g$ (with the exception of those forming a manifold of a lower dimension) can be uniquely represented in the form $\hat{k}\hat{z}$.

A representation of the series $D_m$ is defined by $m$ positive integers $K_1, \ldots, K_m$, by $m+\tau-1$ real numbers $\rho_1 \ldots \rho_{m+\tau-1}$, and by an integer $p (0 \leqq p < \tau)$. If $m \neq [n/2]$ then operators of the representation act on the functions $f(\dot{z}, \hat{z}) = f(z_1, \ldots, z_m, \hat{z})$, where $\dot{z}$ and $\hat{z}$ are the matrices described above and $f(z_1, \ldots, z_m, \hat{z})$ is an analytic function in each variable $z_s$ in the domains $\operatorname{Im} z_s > 0$ and $\operatorname{Im} z_s < 0$. The scalar product in this space is given by the formula

$$(f_1, f_2) = \int f_1(z_1, \ldots, z_m, \hat{z}) \overline{f_2(z_1, \ldots, z_m, \hat{z})} \prod_{s=1}^{m} |\operatorname{Im} z_s|^{k_s - 2} d\mu(\hat{z}) d\dot{z} \tag{1}$$

where $d\dot{z} = \prod_{s=1}^{m} dx_s dy_s$.

When $m = [n/2]$ the series $D_m$ consists of two parts, $D_m^+$ and $D_m^-$. For $D_m^+$ ($D_m^-$) the operators act on functions $f(x_1, \ldots, z_m, \hat{z})$ which are analytic in each $z_s$ when $\prod_{s=1}^{m} \operatorname{Im} z_s > 0$ ($\prod_{s=1}^{m} \operatorname{Im} z_s < 0$).

[1] $m$ is the number of nonreal eigenvalues which have matrices belonging to the subgroup $D_m$.

Operators $T_g$ of the representation are given by the formula

$$T_g f(z_1, z_2, \ldots, z_m, \hat{z}$$
$$= f(z'_1, \ldots, z'_m, \hat{z}') \cdot \prod_{s=1}^{m} (\beta_s z_s + \delta_s)^{-k_s} |d_s|^{k_s/2} \cdot \prod_{s=1}^{m+\tau-1} |d_s|^{i\rho_s} \cdot \prod_{s=m+1}^{m+p} \operatorname{sign} d_s \beta^{-\frac{1}{2}}(\hat{k}). \tag{2}$$

Here $\hat{z}'$ and $\hat{k}$ are determined from the condition $\hat{z}g = \hat{k}z'$, $z'_s = (\alpha_s z_s + \gamma_s / \beta_s z_s + \delta_s)$, and $\alpha_s, \beta_s, \gamma_s \delta_s$ are the elements of the matrix $k_{ss} = \begin{pmatrix} \alpha_s & \beta_s \\ \gamma_s & \delta_s \end{pmatrix}$ where $s = 1, \ldots, m$. Finally $d_s = \det K_{ss}$ and

$$\beta(\hat{k}) = |d_2|^{r_1 + r_2} |d_3|^{r_1 + 2r_2 + r_3} |d_{m+\tau}|^{r_1 + 2r_2 + \ldots + 2r_{m+\tau-1} + r_{m+\tau}}.$$

As a special case we get the principal series representations of the unimodular group of the second order obtained by Bargmann [2].

**Theorem 1.** *Representations of the principal series $D_0, D_1, \ldots, D_{[n/2]}$ are irreducible.*

### Reference


1. Gelfand, I.M., Najmark, M.A.: Tr. Mat. Inst. Akad. Naūk SSSR **36** (1950).
2. Bargmann, V.: Ann Math. (3) **48** (1947).
3. Harish-Chandra: Proc. Nat. Acad. Sci. **3** (1951) 6, 10.

# 13.

(with M. I. Graev)

# Unitary representations of real unimodular groups (Principal non-degenerate series)

Izv. Akad. Naūk SSSR, Ser. Mat. **17** (1952) 189–249. Zbl. **52**:341
(Submitted by Academician A.N. Kolmogorov)

The paper describes principal non-degenerate series of unitary representations of the group of real unimodular matrices of arbitrary order. The representations are proved to be irreducible.

## Introduction

1. All irreducible unitary representations of classical complex groups have been considered in [1]. The representations were given there by operators acting in the spaces of functions defined one some other spaces in which the groups acts.

A study of representations of real semisimple groups (as compared with one of complex groups) reveals some new features. This becomes already clearly visible if one compares representations of the group of complex unimodular matrices of order 2 which have been considered in [2] with representations of the group of real unimodular matrices of order 2 considered in [5]. Let us explain the difference in the case of principal series of representations for groups of both kinds.

Representations of the principal series of the group of complex matrices of order 2 are constructed in the following way. Consider the space of square integrable functions $f(z)$ in the complex variable $z = x + iy$:

$$\int_{-\infty}^{+\infty}\int_{-\infty}^{+\infty} |f(z)|^2 dx dy < +\infty.$$

Each representation of the principal series is defined by an integer $m$ and a number $\rho$. The representation associates to each matrix $g = \begin{pmatrix} \alpha & \beta \\ \gamma & \delta \end{pmatrix}$ the operator taking each function $f(z)$ into the function

$$T_g f(z) = f\left(\frac{\alpha z + \gamma}{\beta z + \delta}\right) |\beta z + \delta|^{m+\rho-2} (\beta z + \delta)^{-m}.$$

The group of real matrices has two principal series of representations. In the first of them one considers sequare integrable functions $f(x)$ defined on the real axis:

$$\int_{-\infty}^{+\infty} |f(x)|^2 dx < +\infty.$$

Each representation is defined by a real number $\rho$ and associates to each real matrix $g=\begin{pmatrix}\alpha & \beta\\ \gamma & \delta\end{pmatrix}$ the operator

$$T_g f(z)=f\left(\frac{\alpha x+\gamma}{\beta x+\delta}\right)|\beta x+\delta|^{i\rho-1}$$

or the operator

$$T_g f(x)=f\left(\frac{\alpha x+\gamma}{\beta x+\delta}\right)|\beta x+\delta|^{i\rho-1}\operatorname{sign}(\beta x+\delta).$$

This series is the most natural counterpart of the above series of representations of the complex group.

There is, however, one more principal series of representations of the real group. Representations of this series are defined by a positive integer $m$. Each of them is realised in the space of functions $f(z)$ in the complex variable $z=x+iy$ that are analytic in the half-plane $y>0$ and for which the integral

$$\int_{-\infty}^{+\infty}\int_{0}^{\infty}|f(z)|^2 y^{m-2}dydx$$

converges. The representation associates to each real matrix $g=\begin{pmatrix}\alpha & \beta\\ \gamma & \delta\end{pmatrix}$ the operator

$$T_g f(z)=f\left(\frac{\alpha z+\gamma}{\beta z+\delta}\right)\quad(\beta z+\delta)^{-m}.$$

Another part of this series may be constructed in the space of functions analytic in the half-plane $y<0$[1].

Some natural questions now arise concerning representations of real simple groups. In particular, what is the origin of the series defined on analytic functions, why are there several series instead of one, etc.

In this paper we consider representations of real unimodular groups of arbitrary order. Here one already has to deal with all the features peculiar to representations of real groups. In the sequel we consider only principal series of representations of the real unimodular group. Other series of representations as well as characters

---

[1] We note that this series may also be defined as a representation on the real axis because each function $f(z)$ analytic in the half-plane is completely defined by its boundary values $f(x)$. Then the scalar product is not given by the formula

$$\int_{-\infty}^{+\infty}\int_{0}^{\infty}f_1(z)\overline{f_2(z)}y^{m-2}dy\,dx,$$

but by the formula

$$\int_{-\infty}^{+\infty}\int_{-\infty}^{+\infty}|x_1-x_2|^{m-2}f_1(x_1)\overline{f_2(x_2)}dx_1\,dx_2.$$

This interpretation, however, does not consider all functions but only those that are boundary values of complex functions analytic in the upper (respectively, lower) half-plane.

of representations, the Plancherel theorem and other aspects will be considered elsewhere[2].

2. Let us see how one constructs representations of all principal non-degenerate series of the real unimodular group.

First of all we recall that representations of the group $\mathfrak{G}$ of complex unimodular matrices of order $n$ have been constructed in [1] in the space of functions $f(z)$ defined on the right coset space of the group $\mathfrak{G}$ with respect to the subgroup $K$ of complex triangular matrices. Of course, any subgroup $g_0^{-1}Kg_0$ conjugate to the subgroup $K$ defines the same coset space and, therefore, the same representations.

Let us now turn to the group $G$ of real unimodular matrices. Here one has to distinguish between the subgroup $K$ of triangular matrices and the subgroups $K^{g_0}=g_0^{-1}Kg_0$ of the complex group $\mathfrak{G}$ conjugate to it because these subgroups may have different intersections with the real group $G$. Subgroups $K^{g_0}$ corresponding to different intersections with $G$[3] may result in different series of representations of the group $G$.

Representations corresponding to a given subgroup $K^{g_0}=g_0^{-1}Kg_0$ are constructed in the space of functions defined on the set of right cosets of the real group $G$ with respect to the intersection $G\cap K^{g_0}$. However, the representation space does not include all functions but only those that are analytic in some of the variables.

It is very interesting to trace the origin of the analyticity condition. The reasons leading to it can be explained in the following way.

First $\mathfrak{G}$ be a complex group and $K$ its subgroup. The space of functions constant on right cosets of the group $\mathfrak{G}$ by the subgroup $K$ may be described by the condition that they remain constant under infinitesimal shifts corresponding to left multiplication of elements of the group $\mathfrak{G}$ by infinitesimal elements from $K$. Since infinitesimal shifts are described by Lie operators, those functions satisfy the system of equations $X_i f=0$ where $X_i$ are Lie operators corresponding to infinitesimal elements from $K$.

Let us now turn to the real group. Consider the space of functions $f(g)$ defined on the group $G$ of real matrices. Let $K_B$ be the subgroup of real triangular matrices and $K^{g_0}=g_0^{-1}K_B g_0$ be a subgroup in the group $\mathfrak{G}$ of complex matrices conjugate to $K_B$. Suppose that the matrix $g_0$ which defines the subgroup $K^{g_0}$ is real. Then $K^{g_0}$ is a subgroup of the real group $G$. Similarly to the case of the complex group let us require that the functions $f(g)$ should remain constant under infinitesimal shifts corresponding to left multiplication of elements from the group $G$ by infinitesimal elements from $K^{g_0}$. Once again, this requirement may be written in the form of a system of linear differential equations $X_i f=0$ which have to be satisfied by the function $f(g)$. Coefficients of these equations are analytic functions in the matrix elements of $g_0$. Therefore one can formally write Lie operators $X_i$ also in the case when the matrix $g_0$ is not real and then require that the functions $f(g)$ should satisfy the corresponding system of differential equations $X_i f=0$.

---

[2] A way of constructing representations in the general case of a semisimple real Lie group has been outlined by the authors in [3].

[3] That is, those intersections are not subgroups conjugate in $G$.

If the matrix $g_0$ is real the resulting system of differential equations is hyperbolic (as for the complex unimodular group). Its characteristics are precisely right cosets of the real group $G$ by the subgroup $K^{g_0}$. Equations $X_i f = 0$ then mean that the function $f(g)$ is constant on right cosets of $G$ by $K^{g_0}$. If the matrix $g_0$ is complex, the corresponding system of differential equations on $f(g)$ is not hyperbolic[4]. Then its solutions are functions constant on right cosets of the group $G$ by the intersection $G \cap g_0^{-1} K g_0$, where $K$ is the group of complex triangular matrices that are analytical with respect to some of the parameters of the group $G$. Thus, the requirement that a function be constant with respect to some of the parameters of the group is replaced by the analyticity condition if one goes from the case of a complex group to the case of a real one.

Representations of the group $G$ of real matrices are, as stated above, constructed in spaces of functions defined on the coset space of the group $G$ with respect to its intersection with some of the subgroups $g_0^{-1} K g_0$ conjugate to the group $K$ of complex triangular matrices. The resulting coset spaces may also be described in a different way. First, we remind the reader how this was done in [1] for the case of the complex unimodular group $\mathfrak{G}$. In that paper the subgroup $Z$ of triangular matrices that have zeros over the main diagonal and 1 on this diagonal is considered, i.e. the group of matrices $z = \|z_{pq}\|$ such that $z_{pp} = 1$ and $z_{pq} = 0$ for all $p < q$ $(p, q = 1, \ldots, n)$. It is proved that almost every matrix $g$ in the complex group $\mathfrak{G}$ can be uniquely represented in the form $g = kz$ where $k$ is a triangular matrix in $K$ and $z$ is a matrix in $Z$. Accordingly, the right coset space of the group $\mathfrak{G}$ by its subgroup $K$ can be identified (up to a manifold of lower dimension) with the group $Z$. Each matrix $g$ of the complex group $\mathfrak{G}$ acts on $Z$ taking each element $z$ in $Z$ into an element $z_1$ for which $zg = kz_1$ where $k$ is a matrix in the subgroup $K$.

The space $Z$ is not transitive under transformations corresponding to the elements of the group $G$ of real matrices and, therefore, it breaks down into several transitive subspaces. Consider such a transitive subspace in $Z$ and let $z$ be an element in it. The stationary subgroup for $z$ is the intersection $G \cap z^{-1} K z$. The space under consideration can then be identified with the coset space of the group $G$ by the subgroup $G \cap z^{-1} K z$. Conversely, almost all the elements of the complex group $\mathfrak{G}$ can be represented in the form $g = kz$ for some $k \in K$ and $z \in Z$ and, consequently, almost all of the subgroups conjugate to the subgroup $K$ of triangular matrices are of the form $z^{-1} K z$. Therefore, every coset space of the group $G$ of real matrices by its intersections with a subgroup conjugate to the group $K$ may be identified with some subspace in $Z$ transitive under transformations corresponding to the elements of the group $G$. Thus, representations of the group $G$ of real matrices can be realised in the spaces of functions defined on some subspace in $Z$ transitive under transformations by the elements of the group $G$.

---

[4] Indeed, in that case the characteristics of the system should be the cosets of the group $G$ with respect to its intersection with the subgroup $g_0^{-1} K g_0$. Since that intersection is, in part, imaginary, some of the characteristics are also imaginary. One notes that the analyticity of functions with respect to some of the parameters plays the role of their being constant on cosets corresponding to an "imaginary subgroup" of the group $G$.

In the present paper we give only those transitive spaces in $Z$ that are necessary for the description of principal non-degenerate series of representations of the real group $G$.

3. Let us describe representations of the group $G$ of real unimodular matrices found in this paper.

We begin by describing those subspaces in $Z$ transitive with respect the real group $G$ that yield the desired series.

Consider the set of such matrices $z$ in $Z$ that can be represented as a product $z = \dot{z}x$ of two matrices $\dot{z} = \| \dot{z}_{pq} \|$ and $x = x_{pq}$ in $Z$ of the following form. The entries $z_{12}, z_{34}, \ldots, z_{2m-1,2m}$ of the matrix $z$ are complex numbers with non-vanishing imaginary parts ($m$ is a fixed integer); other entries of the matrix $z$ that are under the diagonal vanish. The matrix $x$ is real such that $x_{12} = x_{34} = \ldots = x_{2m-1,2m} = 0$. It turns out that if 2m is less that the order $n$ of matrices in $Z$ then the set of such elements $z$ forms the subspace $Z_m$ transitive under transformations by the elements of the real group $G$. For $2m = n$ that set breaks into two transitive spaces $Z_m^+$ and $Z_m^-$. The first consists of such matrices $z = \dot{z}x$ for which the product $\prod_{p=1}^{m} \operatorname{Im} \dot{z}_{2p-1,2p}$ of imaginary parts of the elements $z_{2p-1,2p}$ of the matrix $z$ is positive; the second consists of such matrices for which the same product is negative.

For parameters which define each element $z = \dot{z}x$ of the transitive space $Z_m$ we take the entries $z_1 = \dot{z}_{12}, z_2 = \dot{z}_{34}, \ldots, z_m = \dot{z}_{2m-1,2m}$ of the matrix $z$ and real entries $x_{pq}$ of the matrix $x$.

For $m < n/2$ the representation is realised in the space of functions $f(z) = f(z_1, \ldots, z_m, x)$ defined on $Z_m$ and depending on $m$ complex variables $z_1 = x_1 + iy_1$, $z_2 = x_2 + iy_2, \ldots, z_m = x_m + iy_m$ and $(n(n-1)/2) - m$ real variables $x_{pq}$. The functions are required to be analytic with respect to each complex variable $z_p$ separately on the upper and lower half-planes. (One does not require that they can be continued analytically through the real axis.) We also require that the integral

$$\int |f(z)|^2 \prod_{p=1}^{m} |y_p|^{n_p - 2} \prod_{p=1}^{m} dx_p dy_p d\mu(x)$$

converge for all positive integers $n_1, n_2, \ldots, n_m$. (Here $d\mu(x)$ stands for the product of all differentials $dx_{pq}$ of the parameters corresponding to the matrix $x$.)

The representation associates to each real matrix $g$ of order $n$ the operator which takes the function $f(z) = f(z_1, \ldots, z_m, x)$ into the function

$$T_g f(z) = f(z\bar{g})\alpha(z, g). \tag{1}$$

Here $z\bar{g}$ denotes the transformation of $z$ by the matrix $g$. The function $\alpha(z, g)$ is defined by $m$ positive integers $n_1, n_2, \ldots, n_m$; $m + \tau - 1$ real numbers $\rho_1, \rho_2, \ldots, \rho_{m+\tau-1}$ $(\tau = n - 2m)$; and $\tau - 1$ parameters $\varepsilon_1, \varepsilon_2, \ldots, \varepsilon_{\tau-1}$ taking values either 0 or 1.

Now we describe the function $\alpha(z, g)$ corresponding to a given set of numbers $n_p, \rho_q$ and $\varepsilon_r$ $(p = 1, \ldots, m;\ q = 1, \ldots, m + \tau - 1;\ r = 1, \ldots, \tau - 1)$. With that in mind consider the real matrix $x$ in the decomposition $z = \dot{z}x$ of the matrix $z$. It can be shown that the matrix $xg$ can be uniquely represented in the form $xg = kx_1$ where

$x_1$ is a matrix of the same structure as the matrix $x$ and the matrix $k$ is of the form

$$k = \begin{pmatrix} \alpha_1 & \beta_1 & 0 & 0 & \dots & 0 & 0 & 0 & \dots & 0 \\ \gamma_1 & \delta_1 & 0 & 0 & \dots & 0 & 0 & 0 & \dots & 0 \\ k_{31} & k_{32} & 2 & 2 & \dots & 0 & 0 & 0 & \dots & 0 \\ k_{41} & k_{42} & 2 & 2 & \dots & 0 & 0 & 0 & \dots & 0 \\ \hdotsfor{10} \\ k_{2m-1,1} & k_{2m-1,2} & k_{2m-1,3} & k_{2m-1,4} & \dots & \alpha_m & \beta_m & 0 & \dots & 0 \\ k_{2m,1} & k_{2m,2} & k_{2m,3} & k_{2m,4} & \dots & \gamma_m & \delta_m & 0 & \dots & 0 \\ k_{2m+1,1} & k_{2m+1,2} & k_{m+1,3} & k_{2m+1,4} & \dots & k_{2m+1,2m-1} & k_{2m+1,2m} & \lambda_1 & \dots & 0 \\ \hdotsfor{10} \\ k_{n1} & k_{n2} & k_{n3} & k_{n4} & \dots & k_{n,2m-1} & k_{n,2m} & k_{n,2m+1} & \dots & \lambda_\tau \end{pmatrix}.$$

Let

$$\Lambda_p = \begin{cases} \begin{vmatrix} \alpha_p & \beta_p \\ \gamma_p & \delta_p \end{vmatrix} & \text{for } p \leqq m, \\ \lambda_{p-m} & \text{for } p = m+1, m+2, \dots, m+\tau \end{cases}$$

($\tau = n - 2m$). The function $\alpha(z, g)$ can then be given by formula

$$\alpha(z,g) = \prod_{p=1}^{m} \left( \frac{\beta_p z_p + \delta_p}{\sqrt{|\alpha_p \delta_p - \beta_p \gamma_p|}} \right)^{-n_p} \prod_{p=1}^{m+\tau-1} |\Lambda_p|^{i\rho_p} \prod_{p=m+1}^{m+\tau-1} (\operatorname{sign} \Lambda_p)^{\varepsilon_p}$$
$$\{|\Lambda_2|^{r_1+r_2} |\Lambda_3|^{r_1+2r_2+r_3} \dots |\Lambda_{m+\tau}|^{r_1+2r_2+\dots+2r_{m+\tau-1}+r_{m+\tau}}\}^{-1/2}. \tag{2}$$

Here $r_1 = \dots = r_m = 2$, $r_{m+1} = \dots = r_{m+\tau} = 1$.)

Thus, for $m < n/2$ the representation is realised in the above space of functions. It is defined by $m$ positive integers $n_1, n_2, \dots, n_m$, $m + \tau - 1$ real numbers $\rho_1, \rho_2, \dots, \rho_{m+\tau-1}$, and $\tau - 1$ parameters $\varepsilon_1, \varepsilon_2, \dots, \varepsilon_{\tau-1}$ which takes values of either 0 or 1. The operator $T_g$ of the representation is given by formulas (1) and (2).

For $m = n/2$ the representation is realised in the space of functions of the same from that are not defined on the whole space $Z_m$ but on the one of its two transitive parts $Z_m^+$ and $Z_m^-$. Each representation is defined by $m$ integers $n_1, n_2, \dots, n_m$ and $n = 1$ real numbers $\rho_1, \rho_2, \dots, \rho_{m-1}$. Operators of the representation are again defined by formulas (1) and (2).

Taking $m = 0, 1, \dots, [n/2]$ one obtains $[n/2] + 1$ different principal non-degenerate series of representations of the real group $G$[5].

[5] We note that the integer $m$ equals the number of pairs of conjugate eigenvalues for matrices of the stationary subgroup corresponding to a fixed element of the space $Z_m$.

The equivalence problem is solved with the use of character formulas which will be given later. The formulas also throw some light on the origin of different series. Roughly speaking, $[n/2] + 1$ different series exist because there are $[n/2] + 1$ different types of cosets in the group of real matrices. For the series defined by an integer $m$ the character vanishes identically on those matrices that have more than $m$ pairs of complex eigenvalues.

4. One should not think that the series described here are the only ones that yield unitary irreducible representations of the group $G$ of real matrices. A description of all the remaining series will be given later. Here we only give an example of one more series of unitary representations which exists for the group of matrices of even order. Representations of that series are distinguished by the fact that they are defined in the space of purely analytic functions (i.e. functions that are analytic with respect to all of their parameters).

For simplicity's sake consider the group $G$ of real matrices of order 4.

It turns out that the space $Z$ of complex matrices on which the group $G$ acts splits into two transitive subspaces $Z^+$ and $Z^-$ which are of the same dimension as $Z$ and the boundary space of lower dimension. Our representations are realised in the space of functions defined on one of the transitive spaces $Z^+$ or $Z^-$.

With the appropriate choice of parameters in $Z^+$ and $Z^-$ each of them can be considered as a space of pairs $(z', z'')$ of points in the complex space $\mathfrak{M}$ of dimension 3, $z'(z'_1, z'_2\, z'_3)$ and $z''(z''_1, z''_2, z''_3)$, that satisfy the condition

$$\Delta(z', z'') = \begin{vmatrix} z'_1 & z'_2 & z'_3 & 1 \\ z''_1 & z''_2 & z''_3 & 1 \\ \bar{z}'_1 & \bar{z}'_2 & \bar{z}'_3 & 1 \\ \bar{z}''_1 & \bar{z}''_2 & \bar{z}''_3 & 1 \end{vmatrix} > 0$$

or, respectively, the condition $\Delta(z', z'') < 0$.

The group $G$ acts in the space $\mathfrak{M}$ in such a way that each matrix $g = \|g_{pq}\|$ takes a point $z(z_1, z_2, z_3)$ of that space into the point $z\bar{g}$ which has the coordinates

$$\hat{z}_i = \frac{z_i g_{1i} + z_2 g_{2i} + z_3 g_{3i} + g_{4i}}{z_1 g_{14} + z_2 g_{24} + z_3 g_{34} + g_{44}} \qquad (i = 1, 2, 3).$$

The representation is constructed in the space of functions $f(z', z'')$ depending on six complex variables $z'_i$, $z''_j$ $(i, j = 1, 2, 3)$ which are defined on one of the two spaces above. It is required that the functions should be analytic with respect to all variables $z'_i$ and $z''_j$ on the respective space. We also require that the integral

$$\int |ff(z', z'')|^2 |\Delta(z', z'')|^{m-4} dz'_1 dz'_2 dz'_3 dz''_1 dz''_2 dz''_3$$

converge for the given value of $m$. (Here $dz$ stands for the product $dxdy$ where $z = x + iy$.)

Each representation is defined by a positive integer $m$ and associated to each matrix $g = \|g_{pq}\|$ of order 4 is the operator taking the function $f(z', z'')$ into the function

$$T_g f(z', z'') = f(z'\bar{g}, z''\bar{g})[(z'_1 g_{14} + z'_2 g_{24} + z'_3 g_{34} + g_{44})(z''_1 g_{14} + z''_2 g_{24} + z''_3 g_{34} + g_{44})]^{-m}.$$

The series just described is especially interesting in the theory of automorphic forms in several variables.

5. We now give a brief account of the contents of individual sections of the present paper.

Section 1–3 are of a preliminary nature. Here one considers some of the subgroups of the group $\mathfrak{G}$ of complex matrices and of the group $G$ of real matrices. Their study is necessary for describing the representations of principal non-degenerate series of the group $G$.

Section 4 describes some subspaces in the subgroup $Z$ of complex triangular matrices, transitive with respect to the action of the group $G$.

In Section 5 we prove that the spaces of functions in which representations of the principal series are realised consist of functions analytic with respect to some of the parameters of the group $G$.

Finally, in Section 6, we construct all principal non-degenerate series of unitary representations of the real group $G$.

In Section 7 we prove that the representations found in the preceding section are irreducible. The proof is based on the same idea as that used in the case of the group of complex matrices (see [1]) although it does not go through as easily because of some additional difficulties peculiar to the group of real matrices.

## 1. Some subgroups of the groups $\mathfrak{G}$ and $G$

In order to give an analytical description of representations of the principal series one has to consider some subgroups of the complex unimodular group $\mathfrak{G}$ and of the real unimodular group $G$.

In the following all matrices will be written in block form. Fixing an integer $m \leqq n/2$ let $r_1 = \ldots = r_m = 2$ and $r_{m+1} = \ldots = r_{m+\tau} = 1$ $(\tau = \mathrm{n} - 2m)$. Each matrix $g$ in $\mathfrak{G}$ or in $G$ will be written in the form

$$g = \| g_{pq} \| \quad (p, q = 1, \ldots, m + \tau), \tag{1.1}$$

where $g_{pq}$ denotes a matrix consisting of $r_p$ rows and $r_q$ columns. Using this notation consider some subgroups of the groups $\mathfrak{G}$ and $G$.

1. Subgroups $\hat{K}'_{m,\tau}$ and $\hat{K}_{m,\tau}$. The subgroup $\hat{K}'_{m,\tau}$ consists of matrices $k' = \| k'_{pq} \|$ of the complex group $\mathfrak{G}$ satisfying the condition

$$k'_{pq} = 0 \qquad \text{for } p > q. \tag{1.2}$$

The subgroup $\hat{K}_{m,\tau}$ consists of real matrices $k = \| k_{pq} \|$ satisfying condition (1.2)

2. Subgroups $Z'_{m,\tau}$ and $Z_{m,\tau}$. The subgroup $Z'_{m,\tau}$ consists of matrices $\zeta' = \| \zeta'_{pq} \|$ of the complex group $\mathfrak{G}$ satisfying the conditions

$$\zeta'_{pq} = 0 \quad \text{for } p > q \text{ and } \zeta'_{pq} \text{ are unit matrices} \tag{1.3}$$

The subgroup $Z_{m,\tau}$ consists of real matrices $\zeta = \| \zeta_{pq} \|$ satisfying conditions (1.3).

3. Subgroups $D'_{m,\tau}$ and $D_{m,\tau}$. The subgroup $D'_{m,\tau}$ consists of matrices $d' = \| d'_{pq} \|$ of the complex group $\mathfrak{G}$ satisfying the condition

$$d'_{pq} = 0 \qquad \text{for} \qquad p \neq q. \tag{1.4}$$

The subgroup $D_{m,\tau}$ consists of real matrices $d = \| d_{pq} \|$ satisfying the condition (1.4).

Each matrix $d$ of the subgroup $D'_{m,\tau}$ or, respectively, of the subgroup $D_{m,\tau}$ will be written as a row,

$$d = [d_1, d_2, \ldots, d_{m+\tau}], \tag{1.5}$$

where $d_p = d_{pq}$ $(p = 1, \ldots, m + \tau)$.

4. Subgroups $\hat{D}'_{m,\tau}$ and $\hat{D}_{m,\tau}$. The subgroup $\hat{D}'_{m,\tau}$ consists of matrices

$$\hat{d}' = [\hat{d}'_1, \hat{d}'_2, \ldots, \hat{d}'_{m,\tau}]$$

of the group $\mathfrak{G}$ satisfying the condition

$$|\det \hat{d}'_p| = 1 \qquad (p = 1, 2, \ldots, m + \tau). \tag{1.6}$$

The subgroup $\hat{D}_{m,\tau}$ consists of real matrices satisfying condition (1.6).

5. Subgroup $\dot{D}_{m,\tau}$. The subgroup $\dot{D}_{m,\tau}$ consists of real matrices

$$\dot{d} = [\dot{d}_1, \dot{d}_2, \ldots, \dot{d}_{m+\tau}]$$

satisfying the conditions

$$d_{m+1} = \ldots = d_{m+\tau-1} = 1, \qquad |d_{m+\tau}| = 1,$$

while the matrices $\dot{d}_p$ $(p \leqq m)$ are of the form

$$\dot{d}_p = \begin{pmatrix} h_{11}^{(p)} & 0 \\ h_{21}^{(p)} & h_{22}^{(p)} \end{pmatrix}, \qquad |\det \dot{d}_p| = 1 \qquad \text{and} \qquad h_{22}^{(p)} > 0. \tag{1.7}$$

It is easy to see that the right invariant measure on the subgroup $\dot{D}_{m,\tau}$ is defined by the formula

$$d\mu_r(\dot{d}) = d\mu_r(\dot{d}_1)d\mu_r(\dot{d}_2)\ldots d\mu_r(\dot{d}_m) \tag{1.8}$$

where $d\mu_r(\dot{d}_p)$ $(p = 1, \ldots, m)$ are right invariant measures on the groups of matrices $\dot{d}_p$ of order 2. These measures, in their turn, are expressed by the formulas

$$d\mu_r(d_p) = dh_{21}^{(p)} dh_{22}^{(p)} \tag{1.9}$$

(see [2]).

Consider the element

$$\dot{z}_0 = [v_1^0, v_2^0, \ldots, v_m^0, 1, \ldots, 1] \tag{1.10}$$

of the group of triangular matrices $Z$ where

$$v_1^0 = v_2^0 = \ldots = v_m^0 = v^0 = \begin{pmatrix} 1 & 0 \\ i & 1 \end{pmatrix}. \tag{1.11}$$

An element $\dot{d}$ of the subgroup $\dot{D}_{m,\tau}$ acts on $z_0$ taking it into

$$\dot{z} = \dot{z}_0 \bar{\dot{d}}. \tag{1.12}$$

Evidently, one has

$$\dot{z} = [v_1, v_2, \ldots, v_m, 1, \ldots, 1] \tag{1.13}$$

where

$$v_p = \begin{pmatrix} 1 & 0 \\ z_p & 1 \end{pmatrix}, \qquad z_p = x_p + iy_p \quad (p = 1, \ldots, m). \tag{1.14}$$

It follows from (1.12) that

$$v_0 \bar{d}_p = v_p,$$

and, consequently,

$$z_p = \frac{h_{11}^{(p)} i + h_{21}^{(p)}}{h_{22}^{(p)}} = \pm \frac{1}{(h_{22}^{(p)})^2} i + \frac{h_{21}^{(p)}}{h_{22}^{(p)}},$$

i.e.

$$x_p = \frac{h_{21}^{(p)}}{h_{22}^{(p)}}, \qquad y_p = \pm \frac{1}{(h_{22}^{(p)})^2}.$$

Hence

$$dx_p = \frac{1}{h_{22}^{(p)}} dh_{21}^{(p)} - \frac{h_{21}^{(p)}}{(h_{22}^{(p)})^2} dh_{22}^{(p)},$$

$$dy_p = \mp \frac{2}{(h_{22}^{(p)})^3} dh_{22}^{(p)}.$$

Therefore

$$dx_p\, dy_p = 2 \frac{1}{(h_{22}^{(p)})^4} dh_{21}^{(p)}\, dh_{22}^{(p)} = 2 |\operatorname{Im} z_p|^2 dh_{21}^{(p)}\, dh_{22}^{(p)}.$$

Using (1.9) we obtain the formula

$$d\mu_r(d_p) = \frac{1}{2(\operatorname{Im} z_p)^2} dx_p\, dy_p. \tag{1.15}$$

Thus, the right invariant measure on the group $D_{m,\tau}$ is expressed in the parameters $x_p, y_p$ $(p = 1, \ldots, m)$ in the following way:

$$d\mu_r(\dot{d}) = \frac{1}{2^m} \prod_{p=1}^{\infty} (\operatorname{Im} z_p)^{-2} dx_1\, dy_1 \ldots dx_p\, dy_p. \tag{1.16}$$

6. Subgroup $U_{m,\tau}$. The subgroup $U_{m,\tau}$ consists of real matrices

$$u = [u_1, u_2, \ldots, u_m, 1, \ldots, 1]$$

where $u_p$ is an orthogonal matrix

$$u_p = \begin{pmatrix} u_1^{(p)} & -u_2^{(p)} \\ u_2^{(p)} & u_1^{(p)} \end{pmatrix}, \qquad (u_1^{(p)})^2 + (u_2^{(p)})^2 = 1. \tag{1.17}$$

7. Subgroup $C_{m,\tau}$. The subgroup $C_{m,\tau}$ consists of real matrices

$$c = [c_1, c_2, \ldots, c_{m+\tau}]$$

satisfying the following condition: for each $p \leq m$ the matrix $c_p$ is a diagonal matrix of the form

$$c_p = \begin{pmatrix} \sqrt{\lambda_p} & 0 \\ 0 & \sqrt{\lambda_p} \end{pmatrix}, \quad \lambda_p > 0. \tag{1.18}$$

8. Subgroups $X_{m,\tau}$ and $Z_{m,\tau}$. The subgroup $X_{m,\tau}$ consists of real matrices $x = \| x_{pq} \|$ satisfying the condition

$$x_{pq} = 0 \quad \text{for } p < q \quad \text{and} \quad x_{pp} \text{ are unit matrices.} \tag{1.19}$$

It is easy to see that group $X_{m,\tau}$ admits the invariant measure

$$d\mu(x) = \prod_{p>q} d\mu(x_{pq}), \tag{1.20}$$

where $d\mu(x_{pq})$ stands for the product of differentials of all entries of the matrix $x_{pq}$ (cf. [1], formula (11.18)).

The subgroup $\hat{Z}_{m,\tau}$ consists of complex matrices $\hat{z} = \| z_{pq} \|$ satisfying condition (1.19).

In the following we usually omit the index $\tau$ in the notation of the subgroups.

## 2. Canonical decompositions

1. Decomposition of the groups $\mathfrak{G}$ and $G$. According to [1] almost every matrix of the complex group $\mathfrak{G}$ can be uniquely represented in the form

$$g = \hat{k}'\hat{z}, \tag{2.1}$$

where $\hat{k}' \in \hat{K}'_m$ and $\hat{z} \in \hat{Z}_m$. The only exceptions are the matrices for which at least one of the minors $g_{r_1+r_2+\ldots+r_p+1}$ vanishes.

We remind the reader that $g_p$ denotes the minor

$$g_p = \begin{vmatrix} g_{pp} & \cdots & g_{pn} \\ \cdots & \cdots & \cdots \\ g_{np} & \cdots & g_{nn} \end{vmatrix}.$$

The same statement holds for real matrices $g \in G$, namely, almost every matrix of the real group $G$ can be uniquely represented in the form

$$g = \hat{k}x, \tag{2.1'}$$

where $\hat{k} \in \hat{K}_m$ and $x \in X_m$.

Now we give analytic expressions for the entries of the matrices $k$ and $x$ in formulas (2.1) and (2.1') in terms of the entries of the matrix $g$ (the formulas were obtained in [1]).

Denote by $\begin{pmatrix} p_1 p_2 \ldots p_m \\ q_1 q_2 \ldots q_m \end{pmatrix}$ the minor of the matrix $g$ consisting of the entries from the rows $p_1, p_2, \ldots, p_m$ and the columns $q_1, q_2, \ldots, q_m$. By $g_m$ we denote, as above, the minor

$$\begin{pmatrix} m & m+1 & \dots & n \\ m & m+1 & \dots & n \end{pmatrix}.$$

Let $\hat{k} = \| k_{pq} \|$ and $x = \| x_{pq} \|$. Then the entries $k_{pq}^{(\lambda,\mu)}$ of the matrices $k_{pq}$ can be found from the formula

$$k_{pq}^{(\lambda,\mu)} = \frac{1}{g_{r_1+\dots+r_q+1}} \begin{pmatrix} r_1+\dots+r_{p-1}+\lambda, r_1+\dots+r_q+1, r_1+\dots+r_q+2,\dots,n \\ r_1+\dots+r_{q-1}+\mu, r_1+\dots+r_q+1, r_1+\dots+r_q+2,\dots,n \end{pmatrix}. \tag{2.2}$$

The determinant $\Lambda_p$ of the matrix $k_{pp}$ is

$$\Lambda_p = \frac{g_{r_1+\dots+r_{p-1}+1}}{g_{r_1+\dots+r_p+1}} \tag{2.3}$$

and matrices $x_{pq}$ are found from the relations

$$x_{pq} = h_{pp}^{-1} h_{pq}, \tag{2.4}$$

where the entries $h_{pq}^{(\lambda,\mu)}$ of the matrix $h_{pq}$ are defined by the formula

$$h_{pq}^{(\lambda,\mu)} = \frac{1}{g_{r_1+\dots+r_p+1}} \begin{pmatrix} r_1+\dots+r_{p-1}+\lambda, r_1+\dots+r_p+1, r_1+\dots+r_p+2,\dots,n \\ r_1+\dots+r_{q-1}+\mu, r_1+\dots+r_p+1, r_1+\dots+r_p+2,\dots,n \end{pmatrix}. \tag{2.5}$$

(See also [1], formulas (12.16), (12.13), (12.14) and (12.15).)

2. Decomposition of the groups $\hat{K}'_m$ and $\hat{K}_m$. Each element $k' = \| k'_{pq} \|$ of the group $\hat{K}'_m$ can be uniquely represented in the form

$$k' = \zeta' d' \tag{2.6}$$

where $\zeta' \in Z'_m$ and $d' \in D'_m$. The elements of the matrix $d'$

$$d' = [d'_1, d'_2, \dots, d'_{m+\tau}]$$

are defined by the formula

$$d'_p = k'_{pp} \quad (p = 1, \dots, m+\tau). \tag{2.7}$$

Each element $k$ of the real group $\hat{K}_m$ can be uniquely represented in the form

$$k = \zeta d \tag{2.8}$$

where $\zeta \in Z_m$ and $d \in D_m$.

In its turn, each element $d'$ of the group $D'_m$ can be represented in the form

$$d' = c\hat{d}' \tag{2.9}$$

where $c \in C_m$ and $\hat{d} \in \hat{D}'_m$. That presentation is unique if one requires that the matrix

$$c = [c_1, c_2, \dots, c_{m+\tau}]$$

satisfy the condition

$$c_p > 0 \quad \text{for} \quad p > m. \tag{2.10}$$

Similarly, each element $d$ of the real group $D_m$ can be represented in the form

$$d = c\hat{d} \tag{2.11}$$

where $c \in C_m$ and $\hat{d} \in \hat{D}_m$. One easily checks that such a representation is unique if the matrix

$$\hat{d} = [\hat{d}_1, \hat{d}_2, \ldots, \hat{d}_{m+\tau}]$$

satisfies the condition

$$\hat{d}_{m+1} = \ldots = \hat{d}_{m+\tau-1} = 1. \tag{2.12}$$

According to [2] every real matrix $g_2$ of order 2 with determinant $\pm 1$ can be represented in the form

$$g_2 = u_2 h_2 \tag{2.13}$$

where $u_2$ is an orthogonal matrix and $h_2$ is a triangular matrix of the form

$$h_2 = \begin{pmatrix} h_{11} & 0 \\ h_{21} & h_{22} \end{pmatrix}.$$

Such a representation is unique if we require that $h_{22} > 0$.

It follows then that the elements $\hat{d}$ of the group $\hat{D}_m$ satisfying condition (2.12) can be uniquely represented in the form

$$\hat{d} = u\dot{d} \tag{2.14}$$

where $u \in U_m$, $\dot{d} \in \dot{D}_m$.

Thus, each element $k$ of the group $\hat{K}_m$ can be uniquely represented in the form

$$k = \zeta c u \dot{d} \tag{2.15}$$

where $\zeta \in Z_m$, $c \in C_m$, $u \in U_m$ and $\dot{d} \in \dot{D}_m$.

In view of (2.1′) and (2.15) almost every element $g$ of the group $G$ can be uniquely represented in the form

$$g = \zeta c u \dot{d} x \tag{2.16}$$

where $\zeta \in Z_m$, $c \in C_m$, $u \in U_m$, $\dot{d} \in D_m$ and $x \in X_m$.

## 3. Subgroups $S_{m,\tau}$ and $H_{m,\tau}$. Some integral relations

In the following a special role is played by the subgroups $S_{m,\tau}$ and $H_{m,\tau}$ (or, for short, $S_m$ and $H_m$) of the real group $G$.

1. *Subgroup $S_m$*. Consider the set of all matrices the group $G$ that are of the form

$$s = \zeta c u \quad (\zeta \in Z_m, c \in C_m, u \in U_m). \tag{3.1}$$

It is immediately clear that such matrices constitute a subgroup of the group $G$ which we denote by $S_m$.

Formula (3.1) implies that elements $s = \|s_{pq}\|$ of the group $S_m$ satisfy the following conditions

$$\begin{aligned}&s_{pq} = 0 \quad \text{for } p > q;\\ &\text{for } p \leq m \text{ the matrix } s_{pp} \text{ is a scalar multiple of an orthogonal matrix;}\\ &\text{for } p > m \quad s_{pp} \text{ is an arbitrary number.}\end{aligned} \tag{3.2}$$

Now we find a formula for the measure on the group $S_m$. Let

$$\det s_{pp} = \Lambda_p, \qquad p = 1, \ldots, m + \tau. \tag{3.3}$$

Then

$$\Lambda_1 \Lambda_2 \ldots \Lambda_{m+\tau} = 1. \tag{3.4}$$

For $m > 0$ we set

$$\tilde{s}_{11} = \frac{1}{\sqrt{\Lambda_1}} s_{11}. \tag{3.5}$$

The matrix $\tilde{s}_{11}$ is orthogonal (for $m = 0$ we set $\tilde{s}_{11} = 1$). The parameters defining the matrices $\tilde{s}_{11}$ and $s_{pq}$ $(p < q, q > 1)$ will now be taken for the set of parameters in $s$.

Consider a left shift $s' = s^0 s$ defined by the matrix $s^0 = \|s^0_{pq}\|$. It acts in the following way:

$$\begin{aligned}\tilde{s}'_{11} &= \tilde{s}^0_{11} \tilde{s}_{11}, \qquad s'_{12} = s^0_{11} s_{12} + s^0_{12} s_{22}, \ldots,\\ s'_{1,m+\tau} &= s^0_{11} s_{1,m+\tau} + \ldots + s^0_{1,m+\tau} s_{m+\tau,m+\tau},\\ s'_{22} &= s^0_{22} s_{22}, \qquad s'_{23} = s^0_{22} s_{23} + s^0_{23} s_{33}, \ldots,\\ s'_{2,m+\tau} &= s^0_{22} s_{2,m+\tau} + \ldots + s^0_{2,m+\tau} s_{m+\tau,m+\tau}\\ &\ldots\ldots\ldots\ldots\ldots\ldots\ldots\ldots\ldots\ldots\\ s'_{m+1,m+\tau} &= s^0_{m+\tau,m+\tau} s_{m+\tau,m+\tau}.\end{aligned} \tag{3.6}$$

The determinant of this transformation is

$$\begin{aligned}\Delta^0_l &= (\Lambda^0_1)^{r_2+r_3+\ldots+r_{m+\tau}} (\Lambda^0_2)^{r_2+r_3+\ldots+r_{m+\tau}} (\Lambda^0_3)^{r_3+\ldots+r_{m+\tau}} (\Lambda^0_{m+\tau})^{r_{m+\tau}}\\ &= (\Lambda^0_3)^{-r_2} (\Lambda^0_4)^{-r_2-r_3} \ldots (\Lambda^0_{m+\tau})^{-r_2-r_3-\ldots-r_{m+\tau-1}}.\end{aligned}$$

Therefore, the left invariant measure on $S_m$ is defined by the formula

$$d\mu_l(s) = \frac{1}{|\Delta_l|} d\mu(\tilde{s}_{11}) \prod_{p=2}^{m+\tau} d\mu(s_{pp}) \prod_{p<q} d\mu(s_{pq}). \tag{3.7}$$

Here $d\mu(\tilde{s}_{11})$ is a left invariant measure on the group of orthogonal matrices $\tilde{s}_{11}$ (for $m = 0$ it drops out), the $d\mu(s_{pp})$ are invariant measures on the groups of matrices $s_{pp}$ and $d\mu(s_{pq})$ is the product of differentials of all entries in the matrices $s_{pq}$. In exactly the same way, one can show that the right invariant measure on $S_m$ is given by the formula

$$d\mu_r(s) = \frac{1}{|\Delta_r|} d\mu(\tilde{s}_{11}) \prod_{p=2}^{m+\tau} d\mu(s_{pp}) \prod_{p<q} d\mu(s_{pq}), \tag{3.8}$$

where

$$\Delta_r = (\Lambda_2)^{r_1+r_1}(\Lambda_3)^{r_1+r_2+r_3}\dots(\Lambda_{m+\tau})^{r_1+r_2+\dots r_{m+\tau}}. \tag{3.9}$$

Letting

$$\beta(s) = \frac{d\mu_l(s)}{d\mu_r(s)}$$

one has

$$\beta(s) = |\Lambda_2|^{r_1+r_2}|\Lambda_3|^{r_1+2r_2+r_3}\dots|\Lambda_{m+\tau}|^{r_1+2r_2+\dots+2r_{m+\tau-1}+r_{m+\tau}}. \tag{3.10}$$

2. *Subgroup $H_m$.* Consider the set of matrices $h$ in the group $G$ of the form

$$h = \dot{d}x \qquad (\dot{d}\in\dot{D}_m, x\in X_m). \tag{3.11}$$

It is evident that they constitute a subgroup of the group $G$ which we shall denote $H_m$.

In view of (3.11) elements $h = \|h_{pq}\|$ of the group $H_m$ satisfy the following conditions:

$h_{pq} = 0 \quad$ for $p < q$;

for $p \leq m$ matrices $h_{pp}$ are matrices of order 2 of the form

$$h_{pp} = \begin{pmatrix} h_{11}^{(p)} & 0 \\ h_{21}^{(p)} & h_{22}^{(p)} \end{pmatrix} \tag{3.12}$$

where $h_{22}^{(p)} > 0$ and $\det h_{pp} = \pm 1$
for $\quad p = m+1,\dots,m+\tau-1 \quad h_{pp} = 1$
and $h_{m+\tau,m+\tau} = \pm 1$.

Proceeding now in the same way as for the group $S_m$ but taking into account the fact that $|\det h_{pp}| = 1$ $(p = 1,\dots m+\tau)$ one has the following result.

The right invariant measure on the group $H_m$ is defined by the formula

$$d\mu_r(h) = \prod_{p=1}^{m} d\mu_r(h_{pp}) \prod_{p<q} d\mu(h_{pq}). \tag{3.13}$$

Here the $d\mu_r(h_{pp})$ are right inveriant measures on the groups of matrices $h_{pp}$ and $d\mu(h_{pq})$ is the product of differentials of all elements of the matrix $h_{pq}$.

3. *Decomposition of the group G.* In view of (2.16) one has the following.

Almost every element of the group $G$ (with the exception of a subspace of a lower dimension) can be uniquely represented in the form

$$g = sh \tag{3.14}$$

where $s\in S_m$, $h\in H_m$.

4. *Integral relation on the group $H_m$.* Consider the function $f(h)$ which is integrable on $H_m$. Since each element $h\in H_m$ can be uniquely represented in the form $h = \dot{d}x$ then any function $f(h) = f(\dot{d}x)$ can be considered as depending on $\dot{d}\in\dot{D}_m$ and $x\in X_m$.

The elements of the matrices $h = \|h_{pq}\|$ are expressed through the elements of the matrices $x = \|x_{pq}\|$ and $d[d_1, d_2, \ldots, d_{m+\tau}]$ by the formulas

$$h_{pp} = \dot{d}_p, \qquad h_{pq} = \dot{d}_p x_{pq} \quad \text{for } p > q.$$

Taking into account the fact that $|det\, \dot{d}_p| = 1$ one has

$$d\mu_r(h_{pp}) = d\mu_r(\dot{d}_p), \qquad d\mu(h_{pq}) = d\mu(x_{pq})$$

(for a fixed $\dot{d}_p$).

Formulas (1.8), (1.20) and (3.13) for measures on the groups considered now yield

$$\int f(h) d\mu_r(h) = \int d\mu(x) \int f(\dot{d}x) d\mu_r(\dot{d}). \tag{3.15}$$

5. *Integral relation on the group G.* On the group $G$ there exists a measure $d\mu(g)$ that is both right and left invariant. Consider a function $f(g)$ integrable on $G$. Since almost every element $g \in G$ can be uniquely represented in the form $g = sh$ the function $f(g) = f(sh)$ can be considered as depending on $s \in S_m$ and $h \in H_m$.

Now we obtain a formula reducing integration on $G$ to iterated integrations on $S_m$ and $H_m$.

The equality $g = sh$ establishes analytic relations between the parameters through which the measure on the group $G$ is expressed and the corresponding parameters in $S_m$ and $H_m$. Thus, there is a formula

$$\int f(g) d\mu(g) = \int d\mu_r(h) \int f(sh) \omega(s, h) d\mu_l(s) \tag{3.16}$$

where $\omega(s, h)$ is a positive function depending on $s$ and $h$.

One can easily see, however, that $\omega(s, h)$ is in fact a constant $c$ defined by the choice of normalisation for the measures $d\mu(g)$, $d\mu_l(s)$ and $d\mu_r(h)$.

Indeed, the invariance of the measure $d\mu(s)$ yields

$$\int f(g) d\mu(g) = \int f(s_0 g) d\mu(g).$$

Hence, in view of (3.16) and the left invariance of the measure $d\mu_l(s)$ one has:

$$\begin{aligned} \int d\mu_r(h) f(sh) \omega(s, h) d\mu_l(s) &= \int d\mu_r(h) \int f(s_0 sh) \omega(s, h) d\mu_l(s) \\ &= \int d\mu_r(h) \int f(sh) \omega(s_0^{-1} s, h) d\mu_l(s). \end{aligned}$$

Since $f(sh)$ is an arbitrary function one has

$$\omega(s, h) = \omega(s_0^{-1} s, h)$$

which means that $\omega(s, h)$ does not depend on $s$. Similarly, the invariance of the measure $d\mu(g)$ taken together with the right invariance of the measure $d\mu_r(h)$ and formula (3.16) yields

$$\omega(s, h) = \omega(s, h h_0^{-1})$$

and, consequently, the function $\omega(s, h)$ does not depend on $h$. This, $\omega(s, h)$ is a constant $c$ and therefore

$$\int f(g) d\mu(g) = c \int d\mu_r(h) \int f(sh) d\mu_l(s). \tag{3.17}$$

6. *Transformation of the measure in $H_m$.* Using formula (3.17) one can see how the measure $d\mu_r(h)$ is changed under the transformation $h_1 = h\bar{g}$.

First we show that on the group $S_m$ there is a formula

$$\int f(ss_1)d\mu_l(s) = \beta^{-1}(s_1)\int f(s)d\mu_l(s). \tag{3.18}$$

Indeed, going from the left invariant measure $d\mu_l(s)$ to the right invariant measure $d\mu_r(s)$ according to formula (3.9) one has

$$\int f(ss_1)d\mu_l(s) = \int f(ss_1)\beta(s)d\mu_r(s) = \int f(s)\beta(ss_1^{-1})d\mu_r(s).$$

Now the function $\beta(s)$ defined by the formula (3.10) has the following properties:

$$\beta(s_1 s_2) = \beta(s_1)\beta(s_2), \quad \beta(s^{-1}) = \beta^{-1}(s). \tag{3.19}$$

Thus

$$\int f(s)\beta(ss_1^{-1})d\mu_r(s) = \int f(s)\beta^{-1}(s_1)\beta(s)d\mu_r(s) = \beta^{-1}(s_1)\int f(s_1)d\mu_l(s)$$

and formula (3.18) is proved.

Substituting a function of the form $x(g) = f(h)\varphi(s)$ into (3.17) one has

$$\int x(g)d\mu(g) = c\int f(h)d\mu_r(h)\int \varphi(s)d\mu_r(s). \tag{3.20}$$

Let $g_0$ be an arbitrary element in $G$. Let

$$hg_0 = s_1 h_1, \quad \text{and, consequently, } h\bar{g}_0 = h_1. \tag{3.21}$$

Then $gg_0 = shg_0 = ss_1 h_1$ and $x(g\bar{g}_0) = f(hg_0)\varphi(ss_1)$. The invariance of the measure $d(g)$ yields

$$\int x(g)d\mu(g) = \int x(gg_0)d\mu(g)$$

and thereby, in view of (3.17) and (3.18),

$$\begin{aligned}\int f(h)d\mu_r(h)\int \varphi(s)d\mu_l(s) &= \int f(h\bar{g}_0)d\mu_r(h)\int \varphi(ss_1)d\mu_l(s)\\ &= \int f(h\bar{g}_0)\beta^{-1}(s_1)d\mu_r(h)\int (s)d\mu_l(s).\end{aligned}$$

Hence

$$\int f(h)d\mu_r(h) = \int f(h\bar{g}_0)\beta^{-1}(s_1)d\mu_r(h) \tag{3.22}$$

where $s_1$ is determined from the equality

$$hg_0 = s_1 h_1. \tag{3.23}$$

Writing (3.22) in the form

$$\int f(h\bar{g}_0)d\mu_r(h\bar{g}_0) = \int f(h\bar{g}_0)\beta^{-1}(s_1)d\mu_r(h)$$

we obtain the following formula for the measure transformation in $H_m$:

$$\frac{d\mu_r(h\bar{g}_0)}{d\mu_r(h)} = \beta^{-1}(s_1). \tag{3.24}$$

## 4. Some transitive subspaces in $Z$

Consider the subgroup $Z$ of complex triangular matrices $z = \|z_{ij}\|$ where $z_{ii} = 1$ and $z_{ij} = 0$ for $i < j$ $(i, j = 1, \ldots, n)$. According to [1] almost every element $g$ of the complex group $\mathfrak{G}$ can be uniquely represented in the form $g = kz$ where $z \in Z$ and the matrix $k$ vanishes below the diagonal. Therefore each element $g$ of the group $\mathfrak{G}$ defines the transformation $z \to z_1 = z\bar{g}$ of the group $Z$ where $z_1$ is defined by the relation $zg = kz_1$.

In this section we describe some subspaces in the subgroup $Z$ transitive with respect to the transformations of $Z$ by elements of the real group $G$.

1. *Subspaces* $\dot{Z}_{m,\tau}$, $Z^+_{m,\tau}$ *and* $Z^-_{m,\tau}$. Denote by $\dot{Z}_{m,\tau}$ the set of matrices $\dot{z}$ in $D_{m,\tau}$ of the form

$$\dot{z} = [v_1, v_2, \dots, v_m, 1, \dots, 1] \tag{4.1}$$

where $v_p$ are matrices of order 2 of the form

$$v_p = \begin{pmatrix} 1 & 0 \\ z_p & 1 \end{pmatrix}, \qquad \operatorname{Im} z_p \neq 0 \quad (p = 1, 2, \dots, m). \tag{4.2}$$

For $m = n/2$ we also introduce the sets $\dot{Z}^+_{m,0}$ and $\dot{Z}^-_{m,0}$. The set $\dot{Z}^+_{m,0}$ consists of matrices $\dot{z}^+$ in $\dot{Z}_{m,0}$ of the form

$$\dot{z}^+ = [v_1, v_2, \dots, v_m] \tag{4.1a}$$

where the $v_p$ are the matrices of the form (4.2) satisfying the condition

$$\operatorname{Im} z_1 \operatorname{Im} z_2 \dots \operatorname{Im} z_m > 0. \tag{4.3}$$

Similarly the set $\dot{Z}^-_{m,0}$ consists of the matrices $\dot{z}^-$ in $\dot{Z}_{m,0}$ of the form (4.1a) where the $v_p$ are the matrices of order 2 of the form (4.2) satisfying the condition

$$\operatorname{Im} z_1 \operatorname{Im} z_2 \dots \operatorname{Im} z_m < 0. \tag{4.3a}$$

For $m \neq n/2$ any two elements of the set $\dot{Z}_{m,\tau}$ can be transformed into one another by a transformation of the group $Z$ corresponding to some $g \in G$. For $m = n/2$ the statement is also true for the sets $\dot{Z}^+_{m,0}$ and $\dot{Z}^-_{m,0}$.

*Proof.* Let $v_p$ and $v'_p$ be two matrices of order 2 of the form (4.2). Then, as is well known, there exists a non-singular real matrix $g_p$ of order 2 such that

$$v_p g_p = k_p v'_p \tag{4.4}$$

where $k_p$ is a matrix of the form

$$k_p = \begin{pmatrix} k^{(p)}_{11} & k^{(p)}_{12} \\ 0 & k^{(p)}_{22} \end{pmatrix}, \tag{4.5}$$

i.e. $v_p \bar{g}_p = v'_p$. If complex numbers $z_p$ and $z'_p$ belong to one and the same half-plane $\operatorname{Im} z > 0$ or $\operatorname{Im} z < 0$ then $\det g_p > 0$. Otherwise, $\det g_p < 0$. We can always assume $|\det g_p| = 1$. It is now easy to see that for $m \neq n/2$ matrix (4.1) goes into the matrix

$$\dot{z}' = [v'_1, v'_2, \dots, v'_m, 1, \dots, 1]$$

under the transformation corresponding to the real unimodular matrix $g \in D_{m,\tau}$ of the form

$$g = [g_1, g_2, \dots, g_m, c_{m+1}, \dots, c_{m+\tau}] \tag{4.6}$$

where $c_{m+1},\dots,c_{m+\tau}$ are arbitrary numbers satisfying the condition

$$c_{m+1}\dots c_{m+\tau}\det g_1 \det g_2\dots\det g_m = 1.$$

For $m=n/2$ the corresponding matrix $g$

$$g=[g_1,g_2,\dots,g_m] \tag{4.6a}$$

is unimodular if and only if the sign of the expression

$$\operatorname{Im} z_1 \operatorname{Im} z_2\dots\operatorname{Im} z_m$$

is the same for the matrices $\dot{z}$ and $\dot{z}'$, i.e. if and only if the matrices $\dot{z}$ and $\dot{z}'$ both belong either to $\dot{Z}^+_{m,0}$ or to $\dot{Z}^-_{m,0}$.

**Definition.** *For $m \neq n/2$ denote by $Z_{m,\tau}$ the subspace in $Z$ transitive under transformations by elements $g \in G$ and containing the subspace $\dot{Z}_{m,\tau}$. For even $n$ denote by $Z^+_{n/2,0}$ and $Z^-_{n/2,0}$ the transitive subspaces in $Z$ containing the subspaces $\dot{Z}^+_{n/2,0}$ and $\dot{Z}^-_{n/2,0}$ respectively.*

2. *Analytic description of the subspaces $Z_{m,\tau}$, $Z^+_{n/2,0}$ and $Z^-_{n/2,0}$.* Every matrix $z$ in $Z_{m,\tau}$ ($m \neq n/2$) can be uniquely represented in the form

$$z=\dot{z}^x \quad (\dot{z}\in\dot{Z}_{m,\tau},\ x\in X_{m,\tau}). \tag{4.7}$$

For $m=n/2$ every matrix $z^+$ in $Z^+_{m,0}$($z^-$ in $Z^-_{m,0}$) can be uniquely represented in the form

$$z^+=\dot{z}^+x \qquad (z^-=\dot{z}^-x) \tag{4.7a}$$

where $\dot{z}^+\in\dot{Z}^+_{m,0}$ ($\dot{z}^-\in\dot{Z}^-_{m,0}$) and $x\in X_{m,0}$. Conversely, any matrix that can be represented in one of the above ways belongs to the respective transitive subspace[6].

*Proof.* Consider the case $m \neq n/2$. First of all, the matrices of the form $\dot{z}x$ for $\dot{z}\in\dot{Z}_m$ and $x\in X_m$ belong to the subspace $Z_m$. This immediately follows from the transitivity of $Z_m$ and the simple fact that the elements $\dot{z}$ and $\dot{z}x$ belong to the same transitive subspace.

Conversely, take in $Z_m$ the matrix

$$z_0=[v_1^0,\dots,v_m^0,1,\dots,1] \tag{4.8}$$

where $v_p^0$ are the matrices of the form

$$v_p^0=v_1^0=\begin{pmatrix}1 & 0\\ i & 1\end{pmatrix},\qquad p=1,\dots,m. \tag{4.9}$$

Let us now show that every matrix $z_0\bar{g}$, where $g\in G$ (if it is defined in $Z$), can be represented in the form $z_0\bar{g}=\dot{z}x$. To begin with, almost every matrix $g\in G$ can be represented in the form $g=\hat{k}x$ where $\hat{k}\in\hat{K}_m$, $x\in X_m$. The exceptions are those matrices $g$ for which at least one of the minors $g_{r_1+\dots+r_p+1}$ (see Section 2)

[6] In the following the index $\tau$ is usually omitted.

vanishes. Since we have assumed that the matrix $z_0\bar{g}$ is defined the minors $(z_0 g)_s$ of the matrix $z_0 g$ $(s=2,\dots,n)$ cannot vanish. Note that the matrix $z_0 g$ is obtained from the matrix $g$ by adding to each row the number $2p$ $(p=1,\dots,m)$ of the preceding row multiplied by $i$. It follows from this that

$$g_{r_1+\dots+r_p+1}=(z_0 g)_{r_1+\dots+r_p+1}$$

and hence none of the minors $g_{r_1+\dots+r_p+1}$ vanishes. Thus, the matrix $g$ can be represented in the form $g=\hat{k}x$.

Consider a matrix $\hat{k}=\|k_{pq}\|$. The matrices $k_{pp}$ $(p=1,\dots,m)$ are non-singular and therefore the matrices $v_p^0 k_{pp}$ can be represented in the form

$$v_p^0 k_{pp}=k_p v_p$$

where $k_p$ and $v_p$ are matrices of the form

$$k_p=\begin{pmatrix} k_{11}^{(p)} & k_{12}^{(p)} \\ 0 & k_{22}^{(p)} \end{pmatrix},\qquad v_p=\begin{pmatrix} 1 & 0 \\ z_p & 1 \end{pmatrix},\qquad \operatorname{Im} z_p\neq 0.$$

It follows immediately that the matrix $z_0\hat{k}$ can be represented in the form $z_0\hat{k}=k\dot{z}$ where $k\in K$, $\dot{z}\in\dot{Z}_m$. Thus, $z_0 g=k\dot{z}x$, i.e.

$$z_0\bar{g}=\dot{z}x.$$

Representation of the form $z=\dot{z}x$ is unique because the group $D'_m$ containing the whole space $\dot{Z}_m$ has only one element in common with $X_m$ and that is the unit matrix.

The above reasoning holds for the subspaces $Z_m^+$ and $Z_m^-$ as well. For the subspace $Z_m^{-1}$ one should start with the element $z_0$ of the form (4.8) where

$$v_1^0=\begin{pmatrix} 1 & 0 \\ -i & 1 \end{pmatrix}\quad\text{and}\quad v_p^0=\begin{pmatrix} 1 & 0 \\ i & 1 \end{pmatrix}\ (p=2,\dots,m). \tag{4.9a}$$

3. *Description of the space $Z_m$ by means of the subgroup $H_m$.* According to section 3 every element $g\in G$ that can be represented in the form

$$g=\hat{k}x\qquad (k\in\hat{K}_m,\quad x\in X_m)$$

can also be uniquely represented in the form $g=sh$ where $s\in S_m$ and $h\in H_m$. In its turn every element $h\in H_m$ can be uniquely represented in the form $h=\dot{d}x$.

Consider an element $z_0$ in the space $Z_m$ defined by the formulas (4.8) and (4.9). To each element $h$ in $H_m$ we associate the element $z_0\bar{h}$ in $Z_m$ in the following way:

$$h\to z_0\bar{h}. \tag{4.10}$$

It is easy to see that formula (4.10) defines a one-to-one mapping from $H_m\bar{\dot{d}}$ onto $Z_m$. Indeed, if $h=\dot{d}x$ then, evidently,

$$\dot{d}x\to(z_0\bar{\dot{d}})x \tag{4.10a}$$

and the correspondence $\dot{d}\to z_0\bar{\dot{d}}$ is a one-to-one mapping of the subgroup $\dot{D}_m$ onto $\dot{Z}$ (cf. section 1.5).

Let us now show that the transformation $h\to h\bar{g}$ of the group $H_m$ by elements

$g \in G$ defined in the following way:

$$h\bar{g} \text{ is an element } h_1 \text{ in } H_m \text{ such that } hg = sh_1, \tag{4.11}$$

coincides with the transformation in $Z_m$. In other words, the elements in $H_m$ and $Z_m$ which correspond to each other according to (4.10) are again taken by any $g \in G$ into the corresponding elements under (4.10).

Note that the unit element $e$ in $H_m$ corresponds to the element $z_0$ in $Z_m$. Our statement will be proved if we show that the element $e\bar{g}$ in $H_m$ corresponds to the element $z_0\bar{g}$ in $Z_m$ for any $g \in G$. Let $g = sh$. Evidently, $d\bar{g} = h$. On the other hand

$$\bar{z}_0 g = (z_0\bar{s})\bar{h}.$$

Now the matrix $s$ can be represented in the form $s = \zeta c u$ for some $\zeta \in Z_m, c \in C_m, u \in U_m$. It is easy to see that $z_0\bar{\zeta} = z_0$, $z_0\bar{c} = z_0$ and $z_0\bar{u} = z_0$. Hereby $z_0\bar{s} = z_0$, and, consequently, $z_0\bar{g} = z_0\bar{h}$ and the statement is proved.

Note that $S_m$ is the stationary subgroup for the element $z_0$.

For $m = n/2$ each of the subspaces $Z_m^+$ and $Z_m^-$ can be described in a similar fashion using the respective subgroups in $H_m$.

Our representations will be realised by operators acting in the space of functions $f(\dot{z}x) = f(z_1, \ldots, z_m, x)$ defined on the above subspaces in $Z_m$.

## 5. Analyticity of the functions $f(z_1, \ldots, z_m, x)$

According to [1] representations of the complex unimodular group $\mathfrak{G}$ are constructed in the space of functions defined on the space $Z'$ of right cosets of the group $\mathfrak{G}$ with respect to the subgroup $K$ of complex triangular matrices. Alternatively, they can be defined in the coset space of $\mathfrak{G}$ by any subgroup $K_1$ conjugate to $K$.

For the real unimodular group $G$ one has to distinguish between the subgroup $K$ of triangular matrices and the subgroups $K' = g_1^{-1} K g_1$ of the complex group $\mathfrak{G}$ conjugate to it. Indeed, intersections of those subgroups with the group $G$ of real matrices may be different and, consequently, one obtains different representations of the group $G$ of real matrices.

Representations corresponding to a given subgroup $K' = g_1^{-1} K g_1$ are constructed in the space of functions defined on the right coset space of the group $G$ with respect to the intersection $G \cap K'$. That space, however, does not, in general, have to contain all functions but only those that are analytic with respect to some of the variables.

The reasoning leading to the analyticity condition may be explained in the following way. Let us first go back to the complex group $\mathfrak{G}$. The space of functions constant on right cosets of the group $\mathfrak{G}$ by the subgroup $K$ may be described by the condition that they remain constant under infinitesimal shifts corresponding to left multiplication of elements of the group $\mathfrak{G}$ by infinitesimal elements from $K$. Since infinitesimal shifts are given by operators $X_i$ those functions should satisfy the system of equations

$$X_i f = 0 \tag{5.1}$$

where the $X_i$ are Lie operators corresponding to infinitesimal elements from $K$.

Let us now turn to the real group. Consider the space of functions $f(g)$ defined on the group $G$ of real matrices. Let $K_B$ be the subgroup of real triangular matrices and $K' = g_1^{-1} K_B g_1$ be a subgroup in the group $\mathfrak{G}$ of complex matrices conjugate to $K_B$. Suppose that the matrix $g_1$ which defines the subgroup $K'$ is a real one. Then $K'$ is a subgroup of the real group $G$. Let us require (in analogy with the complex group) that the functions $f(g)$ remain constant under infinitesimal shifts corresponding to left multiplication of elements from the group $G$ by infinitesimal elements from $K'$. This requirement can again be written as a system of linear differential equations $X_i f = 0$ which have to be satisfied by the function $f(g)$. Coefficients of these equations are analytic functions in the matrix elements of $g_1$. Therefore one can also formally write Lie operators $X_i$ in the case when the matrix $g_1$ is not real and then require that functions $f(g)$ should satisfy the corresponding system of differential equations $X_i f = 0$.

If the matrix $g_1$ is real the resulting system of differential equations is hyperbolic (as for the complex unimodular group) Its characteristics are precisely right cosets of the real group $G$ by the subgroup $K'$. If the matrix $g_1$ is complex the corresponding system of differential equations on $f(g)$ is not hyperbolic. Its solutions are functions constant on right cosets of the group $G$ with respect to the corresponding subgroup (namely, the subgroup $G \cap g_1^{-1} K g_1$ where $K$ is the subgroup of complex triangular matrices) that are analytic with respect to some of the parameters of the group. Thus, if one goes from the case of the complex group to the case of the real group, the requirement that a function be constant with respect to some of the parameters of the group is replaced by the analyticity condition.

In the following we consider the subgroup

$$K' = z_0^{-1} K_B z_0$$

where $z_0$ is an element of the subspace $Z_m$ defined by formulas (4.8) and (4.9). In this section we show that solutions of the corresponding system of differential equations are functions that are constant on right cosets of the group $G$ with respect to the subgroup $z_0^{-1} K z_0 \in G$ where $K$ is the group of complex triangular matrices and analytic with respect to some of the parameters. In order specify those parameters, we note that the subgroup $z_0^{-1} K z_0 \in G$ is a stationary subgroup for the element $z_0$ of the space $Z_m$. Therefore, the right coset space with respect to that subgroup can be identified with the space $Z_m$ and functions can be defined on $Z_m$. We now prove that those functions should be analytic with respect to each of the complex parameters $z_1, z_2, \ldots, z_m$ of the space $Z_m$ (see section 4).

First we describe the elements of the group $K_B$ of real triangular matrices. Since it is a subgroup of the group $\hat{K}_m$ of block trinangular matrices its elements (according to section 2) can be uniquely represented in the form

$$k = \zeta c \hat{d} \tag{5.2}$$

where $\zeta \in Z_m$, $c = [c_1, c_2, \ldots, c_{m+\tau}] \in C_m$ and the matrix $\hat{d} = [\hat{d}_1, \hat{d}_2, \ldots, \hat{d}_{m+\tau}] \in \hat{D}_m$ satisfies the condition (2.12):

$$\hat{d}_{m+1} = \ldots = \hat{d}_{m+\tau-1} = 1.$$

The matrices $\hat{d}_p (p = 1, \ldots, m)$ are of the form

$$d_p = \begin{pmatrix} k_{11}^{(p)} & k_{12}^{(p)} \\ 0 & k_{22}^{(p)} \end{pmatrix}, \qquad |\det \hat{d}_p| = 1. \tag{5.3}$$

Let us find the form of the matrix $z_0^{-1} k z_0$. Note, first of all, that $z_0^{-1} \zeta z_0$ is a matrix belonging to the group $Z'_m$ of complex block triangular matrices and that $z_0^{-1} c z_0 = c$. A direct computation shows that

$$z_0^{-1} d z_0 = \bar{d} = [\bar{d}_1, \bar{d}_2, \ldots, \bar{d}_{m+\tau}] \tag{5.4}$$

where

$$\bar{d}_p = \begin{pmatrix} k_{11}^{(p)} + i k_{12}^{(p)} & k_{12}^{(p)} \\ k_{12}^{(p)} + i(k_{22}^{(p)} - k_{11}^{(p)}) & k_{22}^{(p)} - i k_{12}^{(p)} \end{pmatrix} \quad (p \leqq m). \tag{5.5}$$

Hereby we conclude that each element $k'$ of the subgroup $K'$ can be represented in the form

$$k' = \zeta' c d \tag{5.6}$$

where $\zeta'$ is an element of the group $z_0^{-1} Z_m z_0$ which is a subgroup of the group $Z'_m$, $c \in C_m$ and $\bar{d} \in D'_m$ is defined by formulas (5.4) and (5.5).

According to section 2 almost every element $g$ of the real group $G$ can be uniquely represented in the form

$$g = \zeta c \hat{d} x \tag{5.7}$$

where $\zeta \in Z_m$, $c \in C_m$, $x \in X_m$ and $\hat{d} = [\hat{d}_1, \hat{d}_2, \ldots, \hat{d}_{m+\tau}] \in \hat{D}_m$ satisfies condition (2.12):

$$\hat{d}_{m+1} = \ldots = \hat{d}_{m+\tau-1} = 1.$$

Therefore almost every function defined on the group $G$ can be considered as a function in $\zeta \in Z_m$, $c \in C_m$, $\hat{d} \in D_m$, $x \in X_m$ and written in the form

$$f(g) = f(\zeta c \hat{d} x). \tag{5.8}$$

In order to write out the differential equations which have to be satisfied by $f(g)$ we have to consider infinitesimal elements of the subgroups $z_0^{-1} Z_m z_0$, $C_m$ and the subgroup $\bar{D}$ containing element $\bar{d}$ defined by formulas (5.4) and (5.5). It is easy to see, however, that the system of differential equations $X_i^{(Z)} f = 0$ corresponding to infinitesimal elements from $z_0^{-1} Z_m z_0$ is equivalent to the system of differential equations corresponding to infinitesimal elements of the group $Z_m$ itself. Therefore, solutions of that system of equations are functions constant on cosets of the group $G$ with respect to $Z_m$. Thus one can write

$$f(\zeta c \hat{d} x) = f(c \hat{d} x). \tag{5.9}$$

Similar considerations show that the differential equations $X_i^{(C)} f = 0$ corresponding to infinitesimal elements from $C_m$ imply that the functions $f(c \hat{d} x)$ actually do not depend on $c \in C_m$ either and hence

$$f(c \hat{d} x) = f(\hat{d} x). \tag{5.10}$$

Consider now infinitesimal elements from $D$ and the corresponding system of differential equations on $f$.

In (5.5) let

$$k_{11}^{(p)} = 1 + dt_1^{(p)}, \qquad k_{12}^{(p)} = dt_2^{(p)}.$$

Then one has the infinitesimal element

$$\delta\bar{d} = [\delta\bar{d}_1, \delta\bar{d}_2, \ldots, \delta\bar{d}_m, 0, \ldots, 0] \tag{5.11}$$

where the matrices $\delta\bar{d}_p$ $(p = 1, \ldots, m)$ are of the form

$$\delta\bar{d}_p = \begin{pmatrix} dt_1^{(p)} + i dt_2^{(p)} & dt_2^{(p)} \\ dt_2^{(p)} - 2i dt_1^{(p)} & -dt_1^{(p)} - i dt_2^{(p)} \end{pmatrix}. \tag{5.12}$$

Let

$$\hat{d} = [\hat{d}_1, \hat{d}_2, \ldots, \hat{d}_{m+\tau}] \tag{5.13}$$

where

$$\hat{d}_p = \begin{pmatrix} \alpha_p & \beta_p \\ \gamma_p & \delta_p \end{pmatrix} \quad (p \leq m). \tag{5.14}$$

Then

$$\delta\bar{d}\cdot\hat{d} = [\delta\bar{d}_1\cdot\hat{d}_1, \delta\bar{d}_2\cdot\hat{d}_2, \ldots, \delta\bar{d}_m\cdot\hat{d}_m, 0, \ldots, 0] \tag{5.15}$$

where

$$\delta\bar{d}_p\cdot\hat{d}_p = \begin{pmatrix} \alpha_p dt_1^{(p)} + (\gamma_p + i\alpha_p) dt_2^{(p)} & \beta_p dt_1^{(p)} + (\delta_p + i\beta_p) dt_2^{(p)} \\ -(\gamma_p + 2i\alpha_p) dt_1^{(p)} - i(\gamma_p + i\alpha_p) dt_2^{(p)} & -(\delta_p + 2i\beta_p) dt_1^{(p)} - i(\delta_p + i\beta_p) dt_2^{(p)} \end{pmatrix} \tag{5.16}$$

Therefore the Lie operators corresponding to the infinitesimal shifts $dt_1^{(p)}$ and $dt_2^{(p)}$ are of the form

$$X_1^{(p)} f = \frac{\partial f}{\partial \alpha_p}\alpha_p + \frac{\partial f}{\partial \beta_p}\beta_p + \frac{\partial f}{\partial \gamma_p}(-\gamma_p - 2i\alpha_p) + \frac{\partial f}{\partial \delta_p}(-\delta_p - 2i\beta_p),$$

$$X_2^{(p)} f = \frac{\partial f}{\partial \alpha_p}(\gamma_p + i\alpha_p) + \frac{\partial f}{\partial \beta_p}(\delta_p + i\beta_p) - i\frac{\partial f}{\partial \gamma_p}(\gamma_p + i\alpha_p) - i\frac{\partial f}{\partial \delta_p}(\delta_p + i\beta_p).$$

Our requirement is that the functions $f(\hat{d}x)$ satisfy the equations $X_1^{(p)} f = 0$ and $X_1^{(p)} f = 0$, i.e.

$$\frac{\partial f}{\partial \alpha_p}\alpha_p + \frac{\partial f}{\partial \beta_p}\beta_p + \frac{\partial f}{\partial \gamma_p}(-\gamma_p - 2i\alpha_p) + \frac{\partial f}{\partial \delta_p}(-\delta_p - 2i\beta_p) = 0, \tag{5.17}$$

$$\frac{\partial f}{\partial \alpha_p}(\gamma_p + i\alpha_p) + \frac{\partial f}{\partial \beta_p}(\delta_p + i\beta_p) - i\frac{\partial f}{\partial \gamma_p}(\gamma_p + i\alpha_p) - i\frac{\partial f}{\partial \delta_p}(\delta_p + i\beta_p) = 0, \tag{5.18}$$

where $p = 1, 2, \ldots, m$. Substituting equation (5.17) multiplied by $i$ from (5.18) one has

$$\frac{\partial f}{\partial \alpha_p}\gamma_p + \frac{\partial f}{\partial \beta_p}\delta_p - \frac{\partial f}{\partial \gamma_p}\alpha_p - \frac{\partial f}{\partial \delta_p}\beta_p = 0. \tag{5.19}$$

The equations defining the characteristics for (5.19) are of the form

$$\frac{d\alpha_p}{\gamma_p} = \frac{d\beta_p}{\delta_p} = \frac{d\gamma_p}{-\alpha_p} = \frac{d\delta_p}{-\beta_p}. \tag{5.20}$$

They are easily integrated and yield

$$\alpha_p\delta_p - \beta_p\gamma_p = C_1, \quad \alpha_p\beta_p + \gamma_p\delta_p = C_2 \quad \text{and} \quad \beta_p^2 + \delta_p^2 = C_3. \tag{5.21}$$

Since elements $\alpha_p, \beta_p, \gamma_p, \delta_p$ of the matrix $\hat{d}_p$ satisfy the condition

$$\alpha_p\delta_p - \beta_p\gamma_p = \pm 1$$

and in view of equations (5.19), the functions $f(\hat{d}x)$ depend on the variables

$$x_p = \frac{\alpha_p\beta_p + \gamma_p\delta_p}{\beta_p^2 + \delta_p^2}, \qquad y_p = \frac{\alpha_p\delta_p - \beta_p\gamma_p}{\beta_p^2 + \delta_p^2} \tag{5.22}$$

and on the elements of the matrix $x$.

Consider the parameters $x_p$ and $y_p$. Let

$$z = z_0\bar{d}. \tag{5.23}$$

It is easy to see that

$$z = [v_1, v_2, \ldots, v_m, 1, \ldots, 1] \in \dot{Z}_m \tag{5.24}$$

where

$$v_p = \begin{pmatrix} 1 & 0 \\ z_p & 1 \end{pmatrix} \tag{5.25}$$

and

$$z_p = \frac{i\alpha_p + \gamma_p}{i\beta_p + \delta_p} = \frac{\alpha_p\beta_p + \gamma_p\delta_p}{\beta_p^2 + \delta_p^2} + i\frac{\alpha_p\delta_p - \beta_p\gamma_p}{\beta_p^2 + \delta_p^2} = x_p + iy_p. \tag{5.26}$$

Thus the numbers $z_p = x_p + iy_p$ are the parameters corresponding to the element $\hat{d}x$ in the space $Z_m$.

According to (2.14) the element $\hat{d}$ can be represented in the form $\hat{d} = u\dot{d}$ where $u \in U_m, \dot{d} \in \dot{D}_m$. Since $z_0\bar{u} = z_0$ and, consequently, $z_0\hat{d} = z_0\dot{d}$, the function $f(\hat{d}x) = f(u\dot{d}x)$ does not depend on $u$.

Thus functions $f(g)$ defined on $G$ and satisfying differential equations $X_i^{(Z)}f = 0$, $X_i^{(C)}f = 0$ and (5.19) are constant on right cosets of the group $G$ with respect to the group $S_m$. According to section 4 they can be considered as functions

$$f(\dot{z}x) = f(z_1, z_2, \ldots, z_m, x)$$

defined on the space $Z_m$.

Consider now equations (5.18). We now show that they imply that the functions $f(z_1, z_2, \ldots, z_m, x)$ are analytic in each of the complex variables $z_p (p = 1, \ldots, m)$ with the fixed values of other variables.

In order to do this consider operators $\partial/\partial z$ and $\partial/\partial\bar{z}$. If $z$ is a complex variable $z = x + iy$, then, by definition

$$\partial/\partial z = 1/2(\partial/\partial x - i\partial/\partial y), \qquad \partial/\partial\bar{z} = 1/2(\partial/\partial x + i\partial/\partial y). \tag{5.27}$$

If $f(z)$ is a function in the complex variable $z$ then the condition that it is analytic can be written as

$$\frac{\partial f}{\partial\bar{z}} = 0. \tag{5.28}$$

We have to show, therefore, that equations (5.18) imply

$$\frac{\partial f}{\partial z_p} = 0 \qquad (p = 1, \ldots, m). \tag{5.29}$$

Multiplying (5.18) by $i$ one can rewrite it in the form

$$\left(i\frac{\partial f}{\partial\alpha_p} + \frac{\partial f}{\partial\gamma_p}\right)(i\alpha_p + \gamma_p) + \left(i\frac{\partial f}{\partial\beta_p} + \frac{\partial f}{\partial\delta_p}\right)(i\beta_p + \delta_p) = 0. \tag{5.30}$$

Letting

$$i\alpha_p + \gamma_p = z'_p, \qquad i\beta_p + \delta_p = z''_p, \tag{5.31}$$

equation (5.30) can be written as

$$\frac{\partial f}{\partial z'_p} z'_p + \frac{\partial f}{\partial z''_p} z''_p = 0. \tag{5.32}$$

It remains to show that the left-hand side of that equation is a scalar multiple of $\partial f/\partial\bar{z}_p$. This follows directly from the following statement:

If $f(z)$ is a function in the complex variable $z = x + iy$ and $z = z_1/z_2$, $z_1 = x_1 + iy_1, z_2 = x_2 + iy_2$ then

$$\frac{\partial f}{\partial\bar{z}_1} z_1 + \frac{\partial f}{\partial\bar{z}_2} z_2 = 2i\frac{z_2}{\bar{z}_2}\operatorname{Im} z\frac{\partial f}{\partial\bar{z}}. \tag{5.33}$$

All the statements of this section also hold for $m = n/2$. One has only to start with the element $z_0$ defined by formulas (4.8) and (4.9a) and work to the space $Z^-_{n/2}$.

We note that for $m \neq n/2$ the functions $f(z_1, z_2, \ldots, z_m, x)$ are defined for each $z_p$ (with the fixed values of other variables) both in the upper half-plane $\operatorname{Im} z_p > 0$ and in the lower half-plane $\operatorname{Im} z_p < 0$. In view of the statement proved above these functions are analytic in either half-plane.

For $m = n/2$ the functions $f(z_1, z_2, \ldots, z_m, x)$ are defined for each $z_p$ (with fixed values of other variables) only in one of the half-planes $\operatorname{Im} z_p > 0$ or $\operatorname{Im} z_p < 0$ chosen according to the sign of the expression

$$\operatorname{Im} z_1 \ldots \operatorname{Im} z_{p-1} \operatorname{Im} z_{p+1} \ldots \operatorname{Im} z_m.$$

In that half-plane the functions are analytic in $z_p$.

## 6. Description of representations of the principal non-degenerate series

Representations of the principal non-degenerate series $d_{m,\tau}$ will be constructed in the space $\mathfrak{H}_m$ of functions $f(z_1, z_2, \ldots, z_m, x)$ defined on the space $Z_m$ and analytic (according to section 5) with respect to the parameters $z_1, z_2, \ldots, z_m$.

According to section 4 the space $Z_m$ can be identified with the group $H_m$ and one can write

$$f(z_1, z_2, \ldots, z_m, x) = f(h) \quad (h \in H_m).$$

We define a scalar product in the space $\mathfrak{H}_m$ by the formula

$$(f_1, f_2) = \int f_1(h)\overline{f_2(h)}\,\omega(h)\,d\mu_r(h) \tag{6.1}$$

where $\omega(h)$ is a positive function which has yet to be defined.

The representation associates to each element $g$ of the group $G$ the unitary operator $T_g$ acting in the space $\mathfrak{H}_m$ defined by the formula

$$T_g f(h) = f(hg)\alpha(h, g). \tag{6.2}$$

Here $\alpha(h, g)$ is a fixed function defining the representation which has to be specified. By $hg$ we denote the element $h_1$ in $H_m$ into which the element $h$ is taken by $g$, i.e. $hg = s_1 h_1$ where $s_1 \in S_m$.

The condition $T_{g_1} T_{g_2} = T_{g_1 g_2}$ implies (as in [1]) the following functional equation on $\alpha(h, g)$:

$$\alpha(h, g_1 g_2) = \alpha(h\bar{g}_1, g_2)\alpha(h, g_1). \tag{6.3}$$

Let

$$\alpha(e, g) = \alpha(g) \tag{6.4}$$

where $e$ is a unit element in $G$. Letting $h = e$, $g_1 = h$, $g_2 = g$ in (6.3) one has

$$\alpha(hg) = \alpha(h, g)\alpha(h),$$

i.e.

$$\alpha(h, g) = \frac{\alpha(hg)}{\alpha(h)}. \tag{6.5}$$

Letting $h = e$, $g_1 = s \in S_m$, $g_2 = g$ in (6.3) and taking into account the fact that $e\bar{s} = e$ one has

$$\alpha(sg) = \alpha(s)\alpha(g). \tag{6.6}$$

Let us find the condition on the function $\alpha(h, g)$ which has to be satisfied in order for the representation operators $T_g$ to be unitary, i.e.

$$(T_g f, T_g f) = (f, f) \tag{6.7}$$

for all $f(h)$ in $\mathfrak{H}_m$.

According to (6.1) and (6.2) this equality can be written in the form

$$\int |(h, g)|^2 |f(h\bar{g})|^2 \omega(h)\,d\mu_r(h) = \int |f(h_1)|^2 \omega(h_1)\,d\mu_r(h_1).$$

Substituting $h_1 = h\bar{g}$ into the second integral and using formula (3.22) one has

$$\int |f(h_1)|^2 \omega(h_1)\,d\mu_r(h_1) = \int |f(h\bar{g})|^2 \omega(h\bar{g})\beta^{-1}(s_1)\,d\mu_r(h)$$

where $s_1 \in S_m$ is defined by the condition

$$hg = s_1 h_1 \tag{6.8}$$

and the function $\beta(s)$ is determined by formulas (3.10) and (3.3).

Thus

$$\int |f(h\bar{g})|^2 |\alpha(h,g)|^2 \omega(h) d\mu_r(h) = \int |f(h\bar{g})|^2 \omega(hg) \beta^{-1}(s_1) d\mu_r(h).$$

Since $f(h)$ is an arbitrary function we have

$$|\alpha(h,g)|^2 \omega(h) = \omega(h\bar{g}) \beta^{-1}(s_1)$$

or

$$|\alpha(h,g)| = \beta^{-\frac{1}{2}}(s_1) \sqrt{\frac{\omega(h\bar{g})}{\omega(h)}}. \tag{6.9}$$

Condition (6.9) is the unitarity condition of the operators $T_g$.

Letting $h = e$ in (6.9) and taking

$$g = sh \tag{6.10}$$

we have

$$|\alpha(g)| = |\alpha(sh)| = \beta^{-\frac{1}{2}}(s) \sqrt{\frac{\omega(h)}{\omega(e)}}. \tag{6.11}$$

In view of (6.6) this can be written as

$$|\alpha(s)||\alpha(h)| = \beta^{-\frac{1}{2}}(s) \sqrt{\frac{\omega(h)}{\omega(e)}}. \tag{6.12}$$

Hence

$$|\alpha(h)| = \sqrt{\frac{\omega(h)}{\omega(e)}} \tag{6.13}$$

and

$$|\alpha(s)| = \beta^{-\frac{1}{2}}(s). \tag{6.14}$$

Let

$$\alpha(s) = \beta^{-\frac{1}{2}}(s) \chi(s). \tag{6.15}$$

In view of (6.14) we have

$$|\chi(s)| = 1. \tag{6.16}$$

Taking into account the fact that

$$\alpha(s_1 s_2) = \alpha(s_1)\alpha(s_2) \quad \text{and} \quad \beta(s_1 s_2) = \beta(s_1)\beta(s_2),$$

for all $s_1, s_2 \in S_m$ one has

$$\chi(s_1 s_2) = \chi(s_1)\chi(s_2) \quad (s_1, s_2 \in S_m). \tag{6.17}$$

Formulas (6.16) and (6.17) imply that $\chi(s)$ is a character of the group $S_m$. Any element $s$ of the group $S_m$ can be written in the form $s = \zeta c u$ where $\zeta \in Z_m, c \in C_m$, $u \in U_m$ (see Section 3). Then

$$\chi(s) = \chi(\zeta)\chi(cu).$$

Now, for all elements $\zeta$ in $Z_m$, one has $\chi(\zeta)=1$. Thus

$$\chi(\zeta cu)=\chi(cu)=\chi(c)\chi(u). \tag{6.18}$$

Let us now find the formulas for the characters $\chi(c)$ and $\chi(u)$. Any element $c$ of the group $C_m$ is of the form

$$c=[c_1,c_2,\ldots,c_{m+\tau}] \tag{6.19}$$

where $c_p$ is a matix of order 2 of the form

$$c_p=\begin{pmatrix}\sqrt{\lambda_p} & 0\\ 0 & \sqrt{\lambda_p}\end{pmatrix},\quad \lambda_p>0 \tag{6.20}$$

for $p\leqq m$ and $c_p=\lambda_p$ for $p=m+1,\ldots,m+\tau$. It follows from this that for $m\neq n/2$ the group $C_m$ is isomorphic to the direct product of $m$ copies of the multiplicative group of positive real numbers and of $\tau-1$ copies of the multiplicative group of all real numbers. For $m=n/2$ the group $C_m$ is isomorphic to the direct product of multiplicative groups of positive real numbers. Therefore, the characters of the group $C_m$ are of the form

$$\chi(c)=|\lambda_1|^{i\rho_1}|\lambda_2|^{i\rho_2}\ldots|\lambda_{m+\tau-1}|^{i\rho_{m+\tau-1}}\left(\frac{\lambda_{m+1}}{|\lambda_{m+1}|}\right)^{\varepsilon_{m+1}}\cdots\left(\frac{\lambda_{m+\tau-1}}{|\lambda_{m+\tau-1}|}\right)^{\varepsilon_{m+\tau-1}}. \tag{6.21}$$

Here $\rho_1,\rho_2,\ldots,\rho_{m+\tau-1}$ are arbitrary real numbers, the parameters $\varepsilon_p(p=m+1,\ldots,m+\tau-1)$ take values of either 0 and 1 and, accordingly, the factor $(\lambda_p/|\lambda_p|)^{\varepsilon_p}$ is either 1 or sign $\lambda_p$.

Any element $u$ of the group $U_m$ is of the form

$$u=[u_1,u_2,\ldots,u_m,1,\ldots,1], \tag{6.22}$$

where $u_p$ is an orthogonal matrix of order 2:

$$u_p=\begin{pmatrix}u_1^{(p)} & -u_2^{(p)}\\ u_2^{(p)} & u_1^{(p)}\end{pmatrix},\quad (u_1^{(p)})^2+(u_2^{(p)})^2=1. \tag{6.23}$$

It follows hence that the group $U_m$ is isomorphic to the direct product of $m$ copies of the group orthogonal matrices of order 2. Therefore, the characters of the group $U_m$ are of the form

$$\chi(u)=(u_1^{(1)}+iu_2^{(1)})^{n_1}(u_1^{(2)}+iu_2^{(2)})^{n_2}\ldots(u_1^{(m)}+iu_2^{(m)})^{n_m} \tag{6.24}$$

where $n_1,n_2,\ldots,n_m$ are integers.

Let us now go back to the function $\alpha(h,g)$. Let

$$hg=s_1h_1. \tag{6.8}$$

Then, using (6.5) and (6.6), one has

$$\alpha(h,g)=\frac{\alpha(s_1h_1)}{\alpha(h)}=\frac{\alpha(h_1)}{\alpha(h)}\alpha(s_1)$$

whence, in view of (6.15),

$$\alpha(h,g)=\frac{\alpha(h_1)}{\alpha(h)}\beta^{-\frac{1}{2}}(s_1)\chi(s_1). \tag{6.25}$$

Now we find an analytic expression of the function $\alpha(h,g)$ in the parameters of the space $Z_m$.

Write an element $h$ in the form

$$h=\dot{d}x \tag{6.26}$$

where $\dot{d}\in\dot{D}_m$ and $x\in X_m$. Almost every element $xg$ can be uniquely represented in the form

$$xg=\hat{k}x_1 \tag{6.27}$$

where $\hat{k}\in\hat{K}_m$, $x_1\in X_m$.

The entries of the matrices $\hat{k}$ and $x_1$ in formula (6.27) are rational functions in the entries of the matrix $x$. They can be found using the formulas of the section 2.1 if one takes the matrix $g_1=xg$ instead of the matrix $g$.

Let

$$\hat{k}=\|k_{pq}\|, \tag{6.28}$$

and

$$k_{pp}=\begin{pmatrix}\alpha_p & \beta_p\\ \gamma_p & \delta_p\end{pmatrix}\quad (p=1,\dots,m), \tag{6.29}$$

$$\det k_{pp}=\Lambda_p\quad (p=1,\dots,m+\tau). \tag{6.30}$$

Here $\alpha_p,\beta_p,\gamma_p,\delta_p$ and $\Lambda_p$ are expressed through the elements of the matrix $g_1=xg$ using the formulas of section 2.1.

Denote by $d$ the matrix

$$d=[d_1,d_2,\dots,d_{m+\tau}]\quad \text{where } d_p=k_{pp}. \tag{6.31}$$

Then $\hat{k}=\zeta_1 d$ where $\zeta_1=\hat{k}d^{-1}\in Z_m$.

The matrix $s_1$ in (6.8) can be represented in the form

$$s_1=\zeta cu\quad (\zeta\in Z_m, c\in C_m, u\in U_m) \tag{6.32}$$

where $c$ is of the form (6.19)–(6.20) and $u$ is of the form (6.22)–(6.23).

Finally, the element $h_1=h\bar{g}$ can be represented in the form

$$h_1=\dot{d}'x'\quad (\dot{d}'\in\dot{D}_m, x'\in X_m). \tag{6.33}$$

It follows from (6.8), (6.26) and (6.27) that

$$\dot{d}\hat{k}x_1=s_1h_1$$

whence

$$\dot{d}\zeta_1 dx_1=\zeta cu\dot{d}'x'$$

or

$$\zeta_2\dot{d}\,dx_1=\zeta cu\dot{d}'x' \tag{6.34}$$

where $\zeta_2=\dot{d}\zeta_1\dot{d}^{-1}\in Z_m$. Since the decomposition $g=\zeta dx$, where $\zeta\in Z_m$, $d\in D_m$,

$x \in X_m$, the relation (6.34) implies that

$$x' = x_1 \tag{6.35}$$

and

$$\dot{d}d = cu\dot{d}'. \tag{6.36}$$

Equality (6.36) yields

$$|\det c_p| = |\det d_p| = |\Lambda_p| \quad \text{for } p = 1, \ldots, m \tag{6.37}$$

and

$$c_p = d_p = \Lambda_p \quad \text{for } p = m+1, \ldots, m+\tau-1. \tag{6.37a}$$

Therefore, in view of (6.21) one has

$$\chi(c) = |\Lambda_1|^{i\rho_1}|\Lambda_2|^{i\rho_2}\ldots|\Lambda_{m+\tau-1}|^{i\rho_{m+\tau-1}}\left(\frac{\Lambda_{m+1}}{|\Lambda_{m+1}|}\right)^{\varepsilon_{m+1}}\left(\frac{\Lambda_{m+\tau-1}}{|\Lambda_{m+\tau-1}|}\right)^{\varepsilon_{m+\tau-1}}. \tag{6.38}$$

Consider an element $z_0$ in the space $Z_m$ (see formulas (4.8) and (4.9)). To each element $h = \dot{d}x$ there corresponds an element $\dot{z}x$ in the space $Z_m$ and to each element $h = \dot{d}'x_1$ an element $\dot{z}'x_1$ where

$$\dot{z} = z_0\bar{\dot{d}}, \quad \dot{z}' = z_0\bar{\dot{d}}' \text{ are elements in } \dot{Z}_m. \tag{6.39}$$

Let

$$\dot{z} = [v_1, v_2, \ldots, v_m, 1, \ldots, 1], \quad \dot{z}' = [v_1', v_2', \ldots, v_m', 1, \ldots, 1] \tag{6.40}$$

where

$$v_p = \begin{pmatrix} 1 & 0 \\ z_p & 1 \end{pmatrix}, \quad v_p' = \begin{pmatrix} 1 & 0 \\ z_p' & 1 \end{pmatrix}, \quad p = 1, \ldots, m \tag{6.41}$$

and let

$$\dot{d} = [\dot{d}_1, \dot{d}_2, \ldots, \dot{d}_{m+\tau}], \quad \dot{d}' = [\dot{d}_1', \dot{d}_2', \ldots, \dot{d}_{m+\tau}']. \tag{6.42}$$

Formulas (6.39) now mean that

$$v_p = v_0\bar{\dot{d}}_p, \quad v_p' = v_0\bar{\dot{d}}_p',$$

i.e.

$$v_0\dot{d}_p = k_p v_p, \quad v_0\dot{d}_p' = k_p' v_p' \quad (p = 1, \ldots, m) \tag{6.43}$$

where $v_0 = \begin{pmatrix} 1 & 0 \\ i & 1 \end{pmatrix}$ and $k_p, k_p'$ are triangular matrices of the form

$$k_p = \begin{pmatrix} k_{11}^{(p)} & k_{12}^{(p)} \\ 0 & k_{22}^{(p)} \end{pmatrix}, \quad k_p' = \begin{pmatrix} k_{11}'^{(p)} & k_{12}'^{(p)} \\ 0 & k_{22}'^{(p)} \end{pmatrix}. \tag{6.44}$$

A direct computation yields

$$k_{22}^{(p)} = |\operatorname{Im} z_p|^{-\frac{1}{2}}, \quad k_{22}'^{(p)} = |\operatorname{Im} z_p'|^{-\frac{1}{2}}. \tag{6.45}$$

Formula (6.36) implies

$$\dot{d}_p d_p = c_p u_p \dot{d}_p' \quad (p = 1, \ldots, m). \tag{6.46}$$

It follows from this that

$$v_0\overline{\dot{d}_p d_p} = v_0\overline{c_p u_p \dot{d}_p'}.$$

Using the fact that $v_0\overline{c_p u_p} = v_0$ and formulas (6.43) one has

$$v_p \bar{d}_p = v'_p. \tag{6.47}$$

Formula (6.47) now gives the expression for $z'_p$ in terms of $z_p$ and the elements $\alpha_p, \beta_p, \gamma_p, \delta_p$ of the matrix $d_p$:

$$z'_p = \frac{\alpha_p z_p + \gamma_p}{\beta_p z_p + \delta_p}. \tag{6.48}$$

On the other hand the formula (6.47) means that

$$v_p d_p = \tilde{k}_p v'_p \tag{6.49}$$

where $\tilde{k}_p$ is a triangular matrix of the form

$$\tilde{k}_p = \begin{pmatrix} \tilde{k}_{11}^{(p)} & \tilde{k}_{12}^{(p)} \\ 0 & \tilde{k}_{22}^{(p)} \end{pmatrix}. \tag{6.50}$$

A direct computation shows that

$$\tilde{k}_{22}^{(p)} = \beta_p z_p + \delta_p. \tag{6.51}$$

Multiplying (6.46) on the left by $v_0$ and using commutativity between $v_0$ and $c$ we have

$$v_0 \dot{d}_p d_p = c_p v_0 u_p \dot{d}'_p$$

or

$$k_p v_p d_p = c_p (v_0 u_p v_0^{-1}) k'_p v'_p. \tag{6.52}$$

The matrix $v_0 u_p v_0^{-1}$ is a triangular matrix of the same form as matrices $k_p, k'_p$ and $\tilde{k}_p$. One can easily check by a simple computation that

$$(v_0 u_p v_0^{-1})_{22} = u_1^{(p)} - i u_2^{(p)}.$$

Equality (6.52) yields

$$k_{22}^{(p)} \tilde{k}_{22}^{(p)} = \sqrt{\lambda_p} (v_0 u_p v_0^{-1})_{22} k_{22}^{\prime(p)},$$

i.e. using the values of the elements computed above,

$$|\operatorname{Im} z_p|^{-\frac{1}{2}} (\beta_p z_p + \delta_p) = \sqrt{|\Lambda_p|} (u_1^{(p)} - i u_2^{(p)}) |\operatorname{Im} z'_p|^{-\frac{1}{2}}.$$

Hence

$$u_1^{(p)} - i u_2^{(p)} = \left| \frac{\operatorname{Im} z_p}{\operatorname{Im} z'_p} \right|^{-\frac{1}{2}} |\Lambda_p|^{-\frac{1}{2}} (\beta_p z_p + \delta_p). \tag{6.53}$$

Formula (6.24) for the character $\chi(u)$ now takes the form

$$\chi(u) = \prod_{p=1}^{m} \left| \frac{\operatorname{Im} z_p}{\operatorname{Im} z'_p} \right|^{n_p/2} \prod_{p=1}^{m} (\beta_p z_p + \delta_p)^{-n_p} |\Lambda_p|^{n_p/2}. \tag{6.54}$$

Using formula (6.25) ana taking into account the fact that

$$\chi(s_1) = \chi(c)\chi(u)$$

where $\chi(c)$ and $\chi(u)$ are given by formulas (6.38) and (6.54), one gets the following formula for $\alpha(h,g)=\alpha(\dot{z}x,g)$:

$$\alpha(\dot{z}x,g)=\frac{\alpha(\dot{z}'x_1)}{\alpha(zx)}\prod_{p=1}^{m}\left|\frac{\operatorname{Im} z_p}{\operatorname{Im} z'_p}\right|^{n_p/2}\prod_{p=1}^{m}(\beta_p z_p+\delta_p)^{-n_p}|\Lambda_p|^{n_p/2}$$
$$\cdot\prod_{p=1}^{m+\tau-1}|\Lambda_p|^{i\rho_p}\prod_{p=m+1}^{m+\tau-1}\left(\frac{\Lambda_p}{|\Lambda_p|}\right)^{\varepsilon_p}\beta^{-\frac{1}{2}}(s_1) \tag{6.55}$$

where, according to (3.10),

$$\beta(s_1)=|\Lambda_2|^{r_1+r_2}|\Lambda_3|^{r_1+2r_2+r_3}\dots|\Lambda_{m+\tau}|^{r_1+2r_2+\dots+2r_{m+\tau-1}+r_{m+\tau}}. \tag{6.56}$$

According to section 5 the function $\alpha(\dot{z}x,g)$ is analytic with respect to $z_1,z_2,\dots,z_m$ and, consequently, the formula for $\alpha(\dot{z}x,g)$ cannot include

$$\prod_{p=1}^{m}\left|\frac{\operatorname{Im} z_p}{\operatorname{Im} z'_p}\right|^{n_p/2}.$$

Therefore we set

$$\alpha(\dot{z}x)=\prod_{p=1}^{m}|\operatorname{Im} z_p|^{n_p/2}. \tag{6.57}$$

The formula for $\alpha(\dot{z}x,g)$ now takes the form

$$\alpha(\dot{z}x,g)=\prod_{p=1}^{m}(\beta_p z_p+\delta_p)^{-n_p}|\Lambda_p|^{n_p/2}\prod_{p=1}^{m+\tau-1}|\Lambda_p|^{i\rho_p}\prod_{p=m+1}^{m+\tau-1}\left(\frac{\Lambda_p}{|\Lambda_p|}\right)^{\varepsilon_p}\beta^{-\frac{1}{2}}(s_1). \tag{6.58}$$

It remains to find the scalar product in the space $\mathfrak{H}_m$. By formulas (6.13) and (6.57) one has:

$$\frac{\omega(h)}{\omega(\varepsilon)}=\prod_{p=1}^{m}|\operatorname{Im} z_p|^{n_p}$$

or

$$\omega(h)=c\prod_{p=1}^{m}|\operatorname{Im} z_p|^{n_p} \tag{6.59}$$

and the scalar product is of the form

$$(f_1,f_2)=c\int f_1(h)\overline{f_2(h)}\prod_{p=1}^{m}|\operatorname{Im} z_p|^{n_p}d\mu_r(h). \tag{6.60}$$

According to formulas (3.15) and (1.16) the right invariant measure expressed through the parameters of the space $Z_m$ is of the form

$$d\mu_r(h)=\frac{1}{2^m}\prod_{p=1}^{m}(\operatorname{Im} z_p)^{-2}d\dot{z}\,d\mu(x)$$

where $dz=dx_1\,dy_1\dots dx_m\,dy_m$ $(z_p=x_p+iy_p)$. Therefore, letting

$$f(h)=f(z_1,z_2,\dots,z_m,x)$$

and choosing an appropriate scalar factor one obtains the following formula for the scalar product in the space $\mathfrak{H}_m$:

$$(f_1, f_2) = c \int f_1(z_1, \ldots, z_m, x) \overline{f_2(z_1, \ldots, z_m, x)} \prod_{p=1}^{m} |\operatorname{Im} z_p|^{n_p - 2} d\dot{z}\, d\mu(x) \tag{6.61}$$

where

$$\begin{aligned} z_p &= x_p + iy_p \quad (p = 1, \ldots, m) \\ d\dot{z} &= dx_1\, dy_1 \ldots dx_m\, dy_m. \end{aligned} \tag{6.62}$$

Instead of considering the power function $x^\lambda$ $(x \geqq 0)$ it is more convenient to consider the function $x^\lambda/\Gamma(\lambda + 1)$. For $\lambda = -1$ it turns into the delta-function $\delta(x)$. Accordingly, we take

$$\frac{1}{c} = \prod_{p=1}^{m} \Gamma(n_p - 1)$$

in the formula (6.61).

The formula for the scalar product in the space $H$ now takes the form

$$(f_1, f_2) = \int f_1(z_1, \ldots, z_m, x) \overline{f_2(z_1, \ldots, z_m, x)} \prod_{p=1}^{m} \frac{|\operatorname{Im} z_p|^{n_p - 2}}{\Gamma(n_p - 1)} d\dot{z}\, d\mu(x). \tag{6.61a}$$

For the integral to converge it is necessary that the integers $n_p$ be positive. In the case when one or more of the numbers $n_p$ are equal to 1 the corresponding factors $|\operatorname{Im} z_p|^{n_p - 2}/\Gamma(n_p - 1)$ in (6.61a) should be considered as delta-functions or, equivalently, one has to interpret the integral in (6.61) as a limit with $n_p \to 1$.

All the above considerations are also true for $m = n/2$ and lead to the series $d^+_{n/2}$ and $d^-_{n/2}$ realised in the spaces of functions defined on the spaces $Z^+_{n/2}$ and $Z^-_{n/2}$.

Now we formulate our main result:

The principal non-degenerate series of unitary representations of the real unimodular group $G$ of order $n$ are decomposed into $(n + 1)/2$ series $d_0, d_1, d_2, \ldots, d_{(n-1)/2}$ for $n$ odd and into $(n/2) + 2$ series $d_0, d_1, \ldots, d_{n/2-1}, d^+_{n/2}, d^-_{n/2}$ for $n$ even.

Representations of each series $d_m$ for $m \neq n/2$ are defined by $m$ positive integers $n_1, n_2, \ldots, n_m$; $m + \tau - 1$ real numbers $\rho_1, \rho_2, \ldots, \rho_{m+\tau-1}$ $(2m + \tau = n)$; and $\tau - 1$ parameters $\varepsilon_{m+1}, \varepsilon_{m+2}, \ldots, \varepsilon_{m+\tau-1}$ taking the values either 0 or 1.

The representation corresponding to a given set of numbers $n_p, \rho_q$ and $\varepsilon_s$ $(p = 1, \ldots, m;\ q = 1, \ldots, m + \tau - 1;\ s = m + 1, \ldots, m + \tau - 1)$ is constructed in the space of functions $f(\dot{z}x) = f(z_1, z_2, \ldots, z_m, x)$ defined on the transitive space $Z_m$ that for each variable $z_p$ (other variables being fixed) are analytic separately in the upper half-plane $\operatorname{Im} z_p > 0$ and in the lower half-plane $\operatorname{Im} z_p < 0$.

The scalar product in that space is given by formula (6.61):

$$(f_1, f_2) = \int f_1(z_1, \ldots, z_m, x) \overline{f_2(z_1, \ldots, z_m, x)} \prod_{p=1}^{m} \frac{|\operatorname{Im} z_p|^{n_p - 2}}{\Gamma(n_p - 1)} d\mu(x) dx_1\, dy_1 \ldots dx_m\, dy_m$$

where

$$z_p = x_p + iy_p \quad (p = 1, \ldots, m).$$

Operators of the representation $T_g$ are given by the formula

$$T_g f(z_1, z_2, \ldots, z_m, x) = f(z'_1, z'_2, \ldots, z'_m, x_1)\alpha(\dot{z}x, g) \tag{6.63}$$

where

$$z'_p = \frac{\alpha_p z_p + \gamma_p}{\beta_p z_p + \delta_p} \quad (p = 1, \ldots, m),$$

$$\alpha(\dot{z}x, g) = \prod_{p=1}^{m} (\beta_p z_p + \delta_p)^{-n_p} |\Lambda_p|^{n_p/2} \prod_{p=1}^{m+\tau-1} |\Lambda_p|^{i\rho_p} \prod_{p=m+1}^{m+\tau-1} \left(\frac{\Lambda_p}{|\Lambda_p|}\right)^{\varepsilon_p}$$

$$\cdot \{|\Lambda_2|^{r_1+r_2}|\Lambda_3|^{r_1+2r_2+r_3}\ldots|\Lambda_{m+\tau}|^{r_1+2r_2+\ldots+2r_{m+\tau-1}+r_{m+\tau}}\}^{-\frac{1}{2}} \tag{6.64}$$

$$(r_p = 2 \quad \text{for} \quad p \leq m \quad \text{and} \quad 1 \quad \text{for} \quad p > m)$$

or, equivalently,

$$\alpha(\dot{z}x, g) = \prod_{p=1}^{m} \left[\frac{\beta_p z_p + \delta_p}{\sqrt{|\alpha_p\delta_p - \beta_p\gamma_p|}}\right]^{-n_p} \prod_{p=1}^{m+\tau-1} |\Lambda_p|^{i\rho_p} \prod_{p=m+1}^{m+\tau-1} \left(\frac{\Lambda_p}{|\Lambda_p|}\right)^{\varepsilon_p}$$

$$\cdot |\Lambda_1|^{n-\frac{3}{2}}|\Lambda_2|^{n-2-\frac{3}{2}}\ldots|\Lambda_m|^{\tau+\frac{1}{2}}|\Lambda_{m+1}|^{\tau-1}|\Lambda_{m+2}|^{\tau-2}\ldots|\Lambda_{m+\tau-1}|.$$

Here the elements of the matrix $x_1$ and $\alpha_p, \beta_p, \gamma_p, \delta_p$ $(p = 1, \ldots, m)$ as well as $\Lambda_p$ $(p = 1, \ldots, m + \tau)$ are rational functions in the elements of the matrix $x$ and can be found by the formulas given in section 2.1.

For $m = n/2$, representations of the series $d_m^+$ and $d_m^-$ are defined by $m$ positive integers $n_1, n_2, \ldots, n_m$ and $m - 1$ real numbers $\rho_1, \rho_2, \ldots, \rho_{m+\tau-1}$.

Each representation of those series corresponding to the system of numbers $n_p$ and $\rho_q$ $(p = 1, \ldots, m;\ q = 1, \ldots, m - 1)$ is constructed in the space of functions

$$f(\dot{z}x) = f(z_1, z_2, \ldots, z_m, x)$$

defined on the transitive space $Z_m^+$ or, respectively, on $Z_m^-$. With respect to each of the variables $z_p$ (all other variables being fixed) these functions are defined and analytic either in the upper or in the lower half-plane, the choice of which being determined according to the sign of the expression

$$\operatorname{Im} z_1 \ldots \operatorname{Im} z_{p-1} \operatorname{Im} z_{p+1} \ldots \operatorname{Im} z_m.$$

The scalar product is of the same form as for $m \neq n/2$.

Operators of the representation $T_g$ are given as in the case $m \neq n/2$ by the formula

$$T_g f(z_1, z_2, \ldots, z_m, x) = f(z'_1, z'_2, \ldots, z'_m, x_1)\alpha(\dot{z}x, g) \tag{6.65}$$

where

$$z'_p = \frac{\alpha_p z_p + \gamma_p}{\beta_p z_p + \delta_p} \quad (p = 1, \ldots, m),$$

$$\alpha(\dot{z}x, g) = \prod_{p=1}^{m} (\beta_p z_p + \delta_p)^{-n_p} |\Lambda_p|^{n_p/2} \prod_{p=1}^{m+\tau-1} |\Lambda_p|^{i\rho_p} |\Lambda_2|^{-2}|\Lambda_3|^{-4}\ldots|\Lambda_m|^{-(n-2)}, \tag{6.66}$$

or, equivalently,

$$\alpha(\dot{z}x, g) = \prod_{p=1}^{m} \left[\frac{\beta_p z_p + \delta_p}{\sqrt{|\alpha_p\delta_p - \beta_p\gamma_p|}}\right]^{-n_p} \prod_{p=1}^{m-1} |\Lambda_p|^{i\rho_p} |\Lambda_1|^{n-2}|\Lambda_2|^{n-4}\ldots|\Lambda_{m-1}|^2.$$

## 7. Irreducibility of representations of principal non-degenerate series

1. In this section we consider, together with the group $G_n$, the group $\tilde{G}_n$ of real matrices of order $n$ with determinant $\pm 1$.

One can easily see that for $\tau > 0$ formulas (6.63) and (6.64) define, when $g$ runs over the group $\tilde{G}_n$, a unitary representation of that group realised in the same space of functions as is the corresponding representation of the group $G_n$.

For $\tau = 0$ spaces $Z^+_{n/2}$ and $Z^-_{n/2}$ are not transitive under transformations by elements of the group $\tilde{G}_n$. But their sum

$$Z_{n/2} = Z^+_{n/2} + Z^-_{n/2}$$

is a transitive space. Consider a representation of the group $\tilde{G}_n$ realised in the space of functions

$$f(\dot{z}x) = f(z_1, \ldots, z_m, x)$$

defined on the transitive space $Z_{n/2}$ that are analytic in each of the variables $z_p$ (with fixed values of other variables) separately in the upper and lower half-planes. The scalar product in this space is given by formula (6.61):

$$(f_1, f_2) = \int f_1(z_1, \ldots, z_m, x)\overline{f_2(z_1, \ldots, z_m, x)}$$

$$\cdot \prod_{p=1}^{m} \frac{|\operatorname{Im} z_p|^{n_p - 2}}{\Gamma(n_p - 1)} d\mu(x) dx_1\, dy_1 \ldots dx_m\, dy_m$$

where the integral is taken over the whole space $Z_{n/2}$. Operators of the representation $T_g$ are defined by formulas (6.65) and (6.66) precisely as for the group $G_n$. It is easy to see that the representation is a unitary one.

Thus one obtains principal non-degenerate series of unitary representations of the group $\tilde{G}_n$ which, as for $G_n$, will be denoted by $d_m$ $(m = 0, 1, \ldots, [n/2])$.

The main result of that section is the following theorem.

**Theorem 1.** *All representations of principal non-degenerate series of the group* $G_n$ *of real matrices with determinant* $\pm 1$ *are irreducible.*

For $m = n/2$ the representation of the group $G_n$ of real matrices with determinant $+1$ defined by the representation of the group $\tilde{G}_n$ of the series $d_{n/2}$ is decomposed into representations of series $d^+_{n/2}$ and $d^-_{n/2}$ corresponding to the same numbers $n_p$ and $\rho_q$ as the original representation of the group $G_n$. For that reason Theorem 1 immediately implies irreducibility of representations of the group $G_n$ of the series $d^+_{n/2}$ and $d^-_{n/2}$ (for $n$ even). Since for $n$ odd the group $\tilde{G}_n$ is a direct product of the group $G_n$ by the cyclic group of order 2, Theorem 1 implies irreducibility of representations of principal non-degenerate series of the group $G_n$ for $n$ odd.

2. Instead of the group $\tilde{G}_n$ we now introduce a subgroup of $\tilde{G}_n$ and prove that its representations are irreducible. It may seem at first sight that the proof of the irreducibility of representations of a subgroup of the group $\tilde{G}_n$ is no simpler than the proof of the irreducibility of representations of the whole group $\tilde{G}_n$. However,

considering instead of $\tilde{G}_n$ an appropriate subgroup of $\tilde{G}_n$, it is possible to carry through the proof of irreducibility by induction in $n$.

For such a subgroup of the group $G_n$, we take, for $n$ even, the subgroup $\tilde{\mathfrak{A}}_n$ of all matrices $g$ satisfying the condition

$$g_{pn}=0 \qquad \text{for } p=1,2,\ldots,n-1 \tag{$*$}$$

and the condition

$$g_{nn}>0. \tag{$**$}$$

For $n$ odd we take the subgroup $\mathfrak{A}_n$ of all matrices with determinant $+1$ satisfying just the condition $(*)$.

There is the following lemma.

**Lemma 1.** *If $T_g$ is any of the representations of principal non-degenerate series of the group $\tilde{G}_n$ then for $n$ even $T_g$ is an irreducible representation of the subgroup $\tilde{\mathfrak{A}}_n$ and for $n$ odd $T_g$ is an irreducible representation of the subgroup $\mathfrak{A}_n$.*

Evidently, Theorem 1 is an immediate consequence of that Lemma.

3. Lemma 1 will be proved by induction in $n$. Assuming that the lemma is proved for the group of matrices of order $(n-1)$ let us prove it for the group of matrices of order $n$.

Consider several cases according to the parity of $n$ and the type of the series of representations[7].

*Case I.* $n$ is odd. Each representation of the group $G_n$ of the series $d_{m,\tau}$ $(2m+\tau=n)$ is defined by $m$ positive integers $n_1,n_2,\ldots,n_m$, $m+\tau-1$ real numbers $\rho_1,\rho_2,\ldots,\rho_{m+\tau-1}$. In this case the representation is realised in the space $\mathfrak{H}$ of functions

$$f(z)=f(\dot{z}x)=f(z_1,\ldots,z_m,x)$$

defined on the space $Z_{m,\tau}$ and analytic in each of the variables $z_1,z_2,\ldots,z_m$. The norm in the space $\mathfrak{H}$ is given by the formula

$$\|f\|^2=\int|f(z_1,\ldots,z_m,x)|^2\prod_{p=1}^{m}\frac{|\operatorname{Im} z_p|^{n_p-2}}{\Gamma(n_p-1)}\cdot$$

$$\cdot d\mu(x)dx_1\,dy_1\ldots dx_m\,dy_m \qquad (z_p=x_p+iy_p). \tag{7.1}$$

Operators of the representation $T_g$ are given by the formula

$$T_gf(z)=f(z\bar{g})\left[\prod_{p=1}^{m}\frac{\beta_pz_p+\delta_p}{\sqrt{|\alpha_p\delta_p-\beta_p\gamma_p|}}\right]^{-n_p}\prod_{p=1}^{m+\tau-1}|\Lambda_p|^{i\rho_p}$$

$$\cdot\prod_{p=m+1}^{m+\tau-1}\left(\frac{\Lambda_p}{|\Lambda_p|}\right)^{\varepsilon_p}|\Lambda_p|^{n-\frac{3}{2}}|\Lambda_2|^{n-\frac{1}{2}}$$

$$\ldots|\Lambda_m|^{\tau+\frac{1}{2}}|\Lambda_{m+1}|^{\tau-1}|\Lambda_{m+2}|^{\tau-2}\ldots|\Lambda_{m+\tau-1}|. \tag{7.2}$$

---

[7] The proof of irreducibility in cases I and II in fact coincides with the proof of irreducibility of representations of principal non-degenerate series of complex unimodular group (see [1]).

Denote by $z'$ the matrix of order $(n-1)$, the entries of which coincide with the entries $z_{pq}$ of the matrix $z \in Z_m$, for $p<n$, $q<n$. Note that the elements $z_{nq}$ of the bottom row of the matrix $z \in Z_{m,\tau}$ coincide with the elements $x_{nq}$ of the matrix $x \in X_{m,\tau}$ in the representation $z = \dot{z}x$ of the matrix $z$ where $\dot{z} \in \dot{Z}_{m,\tau}$, $x \in X_{m,\tau}$. For this reason the function $f(z) \in \mathfrak{H}$ can be represented in the form

$$f(z) = f(z', x_{n1}, x_{n2}, \ldots, x_{n,n-1}) \tag{7.3}$$

where $z' \in Z_{m,-1}$, and for the fixed value of $z'$ the function $f$ is square integrable with respect to $x_{n1}, x_{n2}, \ldots, x_{n,n-1}$.

Consider the Fourier transform

$$(z', w_1, \ldots, w_{n-1}) = \frac{1}{(2\pi)^{(n-1)/2}} \int f(z', x_{n1}, \ldots, x_{n,n-1}) \cdot e^{-i(x_n w_1 + \ldots + x_{n,n-1} w_{n-1})} dx_{n1} \ldots dx_{n,n-1}. \tag{7.4}$$

This representation is a unitary mapping of the space $\mathfrak{H}$ on the space $H$ of functions $\varphi(z', w_1, \ldots, w_{n-1})$ with the norm

$$\| \varphi \|^2 = \int | \varphi(z', w_1, \ldots, w_{n-1})| \cdot \prod_{p=1}^{m} |\operatorname{Im} z_p|^{n_p - 2} dz'\, dw_1 \ldots dw_{n-1}.^8 \tag{7.5}$$

Therefore all operators of the representation can be considered as being operators in $H$.

Let $A$ be a bounded operator in $H$ commuting with all operators $T_g$, $g \in \mathfrak{A}_n$. The statement of the lemma in the case we are now considering will be proved if we show that any such operator $A$ is a scalar multiple of the identity.

The group $\mathfrak{A}_n$ contains the subgroup $X_{m,\tau}$ and, in particular, it contains all the elements $x^0 \in X_{m,\tau}$ defined by the condition

$$x^0_{pq} = 0 \quad \text{for} \quad p < n. \tag{7.6}$$

In the space $\mathfrak{H}$ there is a shift operator $T_{x^0}$:

$$T_{x^0} f(z) = f(zx^0). \tag{7.7}$$

On the other hand, condition (7.6) implies that under the shift $z \to zx^0$ the elements $z'$ do not change and $x_{nq} \to x_{nq} + x^0_{nq}$, $q = 1, 2, \ldots, n-1$. Therefore formula (7.7) can be written in the form

$$T_{x^0} f(z', x_{n1}, \ldots, x_{n,n-1}) = f(z', x_{n1} + x^0_{n1}, \ldots, x_{n,n-1} + x^0_{n,n-1}). \tag{7.8}$$

Hence, using (7.4), one has that the operator $T_{x^0}$ in the space $H$ is of the form

$$T_x{}^0 \varphi(z', w_1, \ldots, w_{n-1}) = e^{i(x^0_{n1} w_1 + \ldots + x^0_{n,n-1} w_{n-1})} \varphi(z', w_1, \ldots, w_{n-1}). \tag{7.9}$$

Under our assumption, the operator $A$ commutes with will operators $T_{x_0}$ for any $x^0_{n1}, \ldots, x^0_{n,n-1}$ and, consequently, with all operators of multiplication by bounded functions $\omega(w_1, \ldots, w_{n-1})$. Therefore the operator $A$ is of the form

$$A\varphi(z', w_1, \ldots, w_{n-1}) = a(w_1, \ldots, w_{n-1}) \varphi(z', w_1, \ldots, w_{n-1}) \tag{7.10}$$

---

[8] To simplify the typography the expression $d\mu(x) dx_1\, dy_1 \ldots dx_m\, dy_m$ in (7.1) will be written simply as $dz$ with $dz'$ denoting the corresponding expression for $z'$.

here $a(w_1,\dots,w_{n-1})$ is an operator in the space of functions $f(z')$ with the norm

$$\|f\|^2 = \int |f(z')|^2 \prod_{p=1}^{m} \frac{|\operatorname{Im} z_p|^{n_p-2}}{\Gamma(n_p-1)}\, dz' \tag{7.11}$$

defined for almost all $w_1,\dots,w_{n-1}$ and uniformly bounded with respect to those variables.

Consider the subgroup of matrices in $\mathfrak{A}_n$ satisfying the additional condition

$$g_{nq} = 0 \qquad \text{for } q = 1,2,\dots,n-1. \tag{7.12}$$

Denote this subgroup by $\mathfrak{A}^0$. The operator $A$ has to commute with all operators $T_g$, $g \in \mathfrak{A}^0$. Let us find these operators $T_g$.

By formulas (5.44) and (5.45) in [1] we have (see also formulas given in section 2.1 of the present paper) the following formulas for the elements of the matrix $\hat{z} = \bar{z}g$:

$$\hat{z}_{pq} = \frac{\begin{vmatrix} g'_{pq} & g'_{p,p+1} & \cdots & g'_{p,n-1} \\ \cdots & \cdots & \cdots & \cdots \\ g'_{n-1,q} & g'_{n-1,p+1} & \cdots & g'_{n-1,n-1} \end{vmatrix}}{\begin{vmatrix} g'_{pp} & g'_{p,p+1} & \cdots & g'_{p,n-1} \\ \cdots & \cdots & \cdots & \cdots \\ g'_{n-1,p} & g'_{n-1,p+1} & \cdots & g'_{n-1,n-1} \end{vmatrix}} \qquad \text{for } g < p < n \tag{7.13}$$

where $g'_{pq}$ are elements of the matrix $g' = zg$ and

$$\hat{x}_{nq} = \hat{z}_{nq} = \sum_{s=1}^{n-1} x_{ns} \frac{g_{sq}}{g_{nn}} \qquad \text{for } q = 1,\dots,n-1. \tag{7.14}$$

Denote by $g^0$ the matrix of order $(n-1)$ for which

$$g^0_{pq} = g_{pq} \cdot c, \qquad q,p < n \tag{7.15}$$

where the real constant $c$ is chosen to make the determinant of $g^0$ equal to 1 or $-1$. Evidently, the sign of $\det g^0$ coincides with the sign of the element $g_{nn}$ in $g$. We also assume that the sign of $c$, which can be chosen at our description, also coincides with the sign of $g_{nn}$. Now let

$$b = g_{nn} c^9. \tag{7.16}$$

Then $b > 0$. Formulas (7.13) and (7.14) imply that under the transformation $z \to z\bar{g}$ the matrix $z'$ goes into $z'\bar{g}^0$ and $x_{nq}$ goes into $\hat{x}_{nq}$ where

$$b\hat{x}_{nq} = \sum_{s=1}^{n-1} g^0_{sq} x_{ns}. \tag{7.17}$$

Let us compute $\alpha(z,g)$. According to the above results (see (6.30), (2.3)) the expressions $\Lambda_p$ in the formula for $\alpha(z,g)$ are computed in the following way:

$$\Lambda_p = \frac{\tilde{g}_{r_1+\dots+r_{p-1}+1}}{\tilde{g}_{r_1+\dots+r_p+1}} \tag{7.18}$$

---

9 Since $\det g = 1$ equality (7.15) implies that $g_{nn} = \pm c^{n-1}$ and equality (7.16) implies that $b = \pm c^n$.

where $\tilde{g}_k$ denotes, as usual, the principal minor of the matrix $\tilde{g} = xg$:

$$\tilde{g}_k = \begin{vmatrix} \tilde{g}_{kk} & \dots & \tilde{g}_{kn} \\ \dots & \dots & \dots \\ \tilde{g}_{nk} & \dots & \tilde{g}_{nn} \end{vmatrix}.$$

Where also recall that $r_1 = r_2 = \ldots = r_m = 2$ and $r_{m+1} = \ldots = r_{m+n} = 1$.

The matrix $z' \in Z_{m,\tau-1}$ can be represented in the form

$$z' = \dot{z}'x', \quad \dot{z}' \in \dot{Z}_{m,\tau-1}, \quad z' \in X_{m,\tau-1}.$$

The matrices $\dot{z}'$ and $x'$ are obtained from the corresponding matrices $\dot{z}$ and $x$ composing the representation $z = \dot{z}x$ by dropping out the bottom row and the last column. Let $\tilde{g}' = x'g^0$. Then the principal minors $\tilde{g}_p$, $p < n$ of the matrix $\tilde{g} = xg$ are equal to the principal minors $\tilde{g}_p$ of the matrix $\tilde{g}'$ of order $(n-1)$ multiplied by a scalar factor equal to the product of $g_{nn}$ by some power of $c$. Therefore the expressions

$$\Lambda'_p = \frac{\tilde{g}'_{r_1} + \ldots + r_{p-1} + 1}{\tilde{g}'_{r_1} + \ldots + r_p + 1}, \qquad p = 1, 2, \ldots, m + \tau - 1$$

are equal to the corresponding expressions $\Lambda_n$ multiplied by a constant factor which is a power of $c$. Therefore, taking into account the fact that

$$g_{nn} = c^{1-n} \operatorname{sign} c \quad \text{and} \quad \Lambda'_{m+\tau-1} = \frac{1}{\Lambda'_1 \ldots \Lambda'_{m+\tau-1}}$$

one has

$$\alpha(z, g) = \alpha_0(g_{nn})\alpha'(z', g^0) \tag{7.19}$$

where $\alpha_0(g_{nn})$ is a function of the form

$$\alpha_0(g_{nn}) = g^s_{nn}|g_{nn}|^{t+i\rho} \tag{7.20}$$

and

$$\alpha'(z', g^0) = \prod_{p=1}^{m} \left[\frac{\beta_p z_p + \delta_p}{\sqrt{\alpha_p\delta_p - \beta_p\gamma_p}}\right]^{-n_p} \prod_{p=2}^{m+\tau-2} |\Lambda'_p|^{i(\rho_p - \rho_{m+\tau-1})}$$
$$\cdot \prod_{p=1}^{m+\tau-2} \left(\frac{\Lambda'_p}{|\Lambda'_p|}\right)^{\varepsilon_p} |\Lambda'_p|^{(n-1)-\frac{3}{2}} \ldots |\Lambda'_m|^{(\tau-1)+\frac{1}{2}}$$
$$\cdot |\Lambda'_{m+1}|^{(\tau-1)-1} \ldots |\Lambda'_{m+\tau-2}| \left(\frac{\Lambda'_{m+\tau-1}}{|\Lambda'_{m+\tau-1}|}\right)^{\varepsilon_{m+\tau-1}}. \tag{7.21}$$

The function $\alpha'(z', g^0)$ differs for $\tau > 1$ from the function defining the unitary representation of the group $\tilde{G}_{n-1}$ of the series $d_{m,\tau-1}$ only by a constant factor

$$(\Lambda'_{m+\tau-1}/|\Lambda'_{m+\tau-1}|)^{\varepsilon_{m+\tau-1}}.$$

For $\tau = 1$ the function coincides with the function defining the unitary representation of the series $d_{m,0}$ of the group $G_{n-1}$.

Thus for $g \in \mathfrak{A}^0$

$$T_g f(z', x_{n1}, \ldots, x_{n,n-1}) = \alpha_0(g_{nn})\alpha'(z', \overline{g^0}) f(z'\overline{g^0}, \hat{x}_{n1}, \ldots, \hat{x}_{n,n-1}) \tag{7.22}$$

or

$$T_g f(z', x_{n1}, \ldots, x_{n,n-1}) = \alpha_0(g_{nn}) T_{g^0} f(z', \hat{x}_{n1}, \ldots, \hat{x}_{n,n-1}) \tag{7.23}$$

where $\hat{x}_{n1}, \ldots, \hat{x}_{n,n-1}$ are defined by formula (7.17) and $T_{g^0}$ is a representation of the group $\tilde{G}_{n-1}$ defined by the equality

$$T_{g^0} f(z') = \alpha'(z', g^0) f(z'\overline{g^0}). \tag{7.24}$$

The operator $T_g$ acts on the function $\varphi(z', w_1, \ldots, w_{n-1})$ in the following way:

$$T_g \varphi(z', w_1, \ldots, w_{n-1}) = \alpha_0(g_{nn}) \alpha'(z', g^0) \frac{1}{(2\pi)^{(n-1)/2}} \int f(z'\overline{g^0}, \hat{x}_{n1}, \ldots, \hat{x}_{n,n-1})$$
$$\cdot e^{-i(x_{n1}w_1 + \ldots + x_{n,n-1}w_{n-1})} dx_{n1} \ldots dx_{n,n-1}.$$

In the variables $\hat{x}_{n1}, \ldots, \hat{x}_{n,n-1}$ one has

$$T_g \varphi(z', w_1, \ldots, w_{n-1}) = \alpha_0(g_{nn}) \alpha'(z', g^0) \Delta \cdot \frac{1}{(2\pi)^{(n-1)/2}} \int f(z'\overline{g^0}, \hat{x}_{n1}, \ldots, \hat{x}_{n,n-1})$$
$$\cdot e^{-i(\hat{x}_{n1}\hat{w}_1 + \ldots + \hat{x}_{n,n-1}\hat{w}_{n-1})} d\hat{x}_{n1} \ldots d\hat{x}_{n,n-1}, \tag{7.25}$$

or, in other words,

$$T_g \varphi(z', w_1, \ldots, w_{n-1}) = \alpha_1(g_{nn}) \Delta \cdot T_{g^0}(z', \hat{w}_1, \ldots, \hat{w}_{n-1}). \tag{7.26}$$

Here $\Delta$ is the Jacobian of the transformation (7.17) of the variables $x_{nq}$ and $\hat{w}_q$ defined by the formula

$$\frac{1}{b} \hat{w}_q = \sum_{s=1}^{n-1} \hat{g}_{sq} w_s \tag{7.27}$$

where $\hat{g} = g_{sq}$ is the inverse to the matrix which is the transpose of $g^0$ (i.e. $\hat{g} = (g^0)^{*-1}$).

Operator $A$ has to commute with the operator $T_g$. Taking into account the above formulas one has

$$a(w_1, \ldots, w_{n-1}) \cdot \Delta \cdot \alpha_0(g_{nn}) T_{g^0} \varphi(z', \hat{w}_1, \ldots, \hat{w}_{n-1})$$
$$= \Delta \cdot \alpha_0(g_{nn}) T_{g^0} a(\hat{w}_1, \ldots, \hat{w}_{n-1}) \varphi(z', \hat{w}_1, \ldots, \hat{w}_{n-1}).$$

Therefore

$$a(w_1, \ldots, w_{n-1}) T_{g^0} = T_{g^0} a(\hat{w}_1, \ldots, \hat{w}_{n-1}).$$

Hence

$$a(\hat{w}_1, \ldots, \hat{w}_{n-1}) = T_{g^0}^{-1} a(w_1, \ldots, w_{n-1}) T_{g^0} \tag{7.28}$$

for almost all $w_1, \ldots, w_{n-1}$.

In (7.15) let $g^0 = e$ ($e$ is a unit matrix). Then formula (7.27) takes the form

$$\frac{1}{b} \hat{w}_q = w_q, \qquad b > 0$$

and $T_{g^0} = 1$. Therefore relation (7.28) takes the form

$$a(bw_1, \ldots, bw_{n-1}) = a(w_1, \ldots, w_{n-1}) \tag{7.29}$$

where $b>0$. Therefore the function $a(w_1,\dots,w_{n-1})$ is constant on the half-lines going out from the point $(0,0,\dots,0)$. It can thus be considered as a function depending on $w\in\mathfrak{W}$ where $\mathfrak{W}$ is the $(n-2)$-dimensional sphere $w_1^2+\dots+w_{n-1}^2=1$.

Since the mapping $g^0\to\hat{g}=(g^0)^{*-1}$ is an isomorphism, relation (7.27) can be considered as a transformation of $\mathfrak{W}$ by the group $\tilde{G}_{n-1}$ of matrices of order $(n-1)$ with determinant $\pm 1$.

Evidently, $\mathfrak{W}$ is a transitive space and it can therefore be considered as a right coset space with respect to some subgroup of the group $\tilde{G}_{n-1}$.

We can consider the function $a(w)$ to be defined almost everywhere on the group $\tilde{G}_{n-1}$ assuming it to be constant on each such coset. Then relation (7.28) can be written as

$$a(gg^0)=T_{g^0}^{-1}a(g)T_{g^0}, \tag{7.30}$$

and this equality holds for almost all $g\in\tilde{G}_{n-1}$ for each fixed value of $g^0$.

Let $g_1=gg^0$; then (7.30) can be written as

$$a(g_1)=T_{g^{-1}g_1}^{-1}a(g)T_{g^{-1}g_1}=T_{g_1}^{-1}T_g a(g)T_g^{-1}T_{g_1}$$

whence

$$T_{g_1}a(g_1)T_{g_1}^{-1}=T_g a(g)T_g^{-1}. \tag{7.31}$$

Relation (7.31) holds for almost all pairs $(g^0,g), g^0,\ g\in\tilde{G}_{n-1}$ and therefore, by Fubini's theorem, for almost all pairs $(gg^0,g)=(g_1,g), g_1,g\in\tilde{G}_{n-1}$. Thus for almost all $g\in\tilde{G}_{n-1}$ the operator $T_g a(g)T_g^{-1}$ does not depend on $g$. Denote it by $a$. Then one has

$$a(g)=T_g^{-1}aT_g \tag{7.32}$$

for almost all $g\in\tilde{G}_{n-1}$.

The function $a(g)$ is, by definition, constant on right cosets of $\tilde{G}_{n-1}$ by the stationary subgroup of $\mathfrak{W}$. Let us start with the point $(0,0,\dots,1)$. Its stationary subgroup is the set $\mathfrak{A}'$ of matrices $l=\|l_{pq}\|\in\tilde{G}_{n-1}$ such that the matrix $l^{*-1}=\|\hat{l}_{pq}\|$ satisfies the condition

$$\hat{l}_{n-1,q}=0 \quad \text{for } q=1,2,\dots,n-2, \quad \hat{l}_{n-1,n+1}>0,$$

and therefore also the condition

$$l_{q,n-1}=0 \quad \text{for } q=1,2,\dots,n-2, \quad l_{n-1,n-1}>0.$$

Thus $\mathfrak{A}'$ is a subgroup of $\tilde{\mathfrak{A}}_{n-1}$.

On the right-hand side of relation (7.32) one has a continuous function of $g$ in the norm in $\mathfrak{H}$. For almost all $g$ it coincides with the function $a(g)$ which is constant on cosets of the group $\tilde{G}_{n-1}$ with respect to the subgroup $\tilde{\mathfrak{A}}_{n-1}$. Thus, by Fubini's theorem, it is constant on almost all cosets up to its values on a set of measure zero. Therefore, its continuity implies that $a(g)$ is constant on all cosets without exception.

Taking $g=l$ on the right-hand side of (7.32) one has

$$T_l^{-1}aT_l=a \tag{7.33}$$

and, consequently, the operator $a$ commutes with each operator $T_l, l\in\tilde{\mathfrak{A}}_{n-1}$.

We have already noted that for $\tau = 1$ the operators $T_{g^0}$ are operators of the unitary representation of the group $\tilde{G}_{n-1}$ from the series $d_{m0}$. For $\tau > 1$ formulas for $T_{g^0}$ compared to those for the series $d_{m,\tau-1}$ include the factor

$$(\Lambda_{m+\tau-1}/|\Lambda_{m+\tau-1}|)^{m+\tau-1}.$$

But for $\tau > 1$ one has

$$\Lambda'_{m+\tau-1} = \tilde{g}'_{n-1} = x_{n-1,1}g^0_{1,n-1} + \ldots + x_{n-1,n-2}g^0_{n-2,n-1} + g^0_{n-1,n-1}. \qquad (7.34)$$

If $g^0 = l \in \tilde{\mathfrak{A}}_{n-1}$ then $g^0_{1,n-1} = \ldots = g^0_{n-2,n-1} = 0$ and $\Lambda'_{m+\tau-1}$ is a constant. Thus operators $T_l$ coincide with operators $T'_l$ of the unitary representation up to a constant factor. Therefore, the operator $a$ commutes with each operator $T'_l$, $l \in \tilde{\mathfrak{A}}_{n-1}$. By the inductive assumption about the irreducibility of representations of the group $\tilde{\mathfrak{A}}_{n-1}$ of order $(n-1)$ the operator $a$ is a scalar multiplication: $a = \alpha \cdot 1$. Then (7.32) implies that $a(g) = \alpha \cdot 1$ for almost all $g \in \tilde{G}_{n-1}$. By (7.10) operator $A$ coincides with $\alpha \cdot 1$.

Thus each bounded operator $A$ in $H$ commuting with all operators $T_g, g \in \mathfrak{A}_n$ is a scalar multiple of the identity, and, consequently, $T_g$ is an irreducible representation of the group $\mathfrak{A}_n$.

*Case II.* $n$ is even, $\tau > 0$. The space $\mathfrak{H}$ in which the representation $T_g$ of the group $\tilde{G}_n$ is realised and operators of the representation are defined in exactly the same way as in case I.

We have to prove that $T_g$ is the representation of the group $\tilde{\mathfrak{A}}_n$ in that case.

The proof is carried out almost exactly as in the case I. The only point at which different considerations are required is the following.

As in case I we consider the subgroup $\mathfrak{A}^0$ of matrices in the group $\tilde{\mathfrak{A}}_n$ satisfying condition (7.12):

$$g_{nq} = 0 \qquad \text{for } q = 1, 2, \ldots, n-1.$$

Since $n-1$ is odd a real scalar $c$ in (7.15),

$$g^0_{pq} = q_{pq} \cdot c, \qquad pq < n,$$

can always be chosen in such a way that the determinant of the corresponding matrix $g^0$ of order $(n-1)$ is equal to 1. Of course, that condition does not define $c$ in a unique fashion. The sign of $c$ coincides with that of the determinant of the matrix $g$.

Now, as in case I, we set $b = g_{nn} \cdot c$. The number $b$ here can, however, be either positive or negative. The resulting relation (7.29) therefore means that for almost all $w_1, \ldots, w_{n-1}$ the function $a(w_1, \ldots, w_{n-1})$ depends only on the ratio $w_1 : w_2 : \ldots : w_{n-1}$. Hence it can be considered to be defined on the points of the real projective space $\mathfrak{W}$, in which each point is defined by some ratio $w = (w_1 : w_2 : \ldots : w_{n-1})$.

The space $\mathfrak{W}$ can be considered as a right coset space of the group $G_{n-1}$ with respect to some of its subgroups. Starting with the point $(0:0:\ldots:1)$ we have the stationary subgroup $\mathfrak{A}_{n-1}$. Then the proof of the lemma in case II is accomplished as in case I.

*Case III.* $n$ is even, $\tau = 0$. According to section 7.1 each representation of the group $\tilde{G}_n$ is defined by $m = n/2$ positive integers $n_1, n_2, \ldots, n_m$ and $m-1$ real numbers $\rho_1, \rho_2, \ldots, \rho_{m-1}$. It is realised in the space of functions

$$f(z) = f(\dot{z}x) = f(z_1, \ldots, z_m, x)$$

defined on the space $Z_{m,0}$ and is analytic with respect to each of the variables $z_1, \ldots, z_m$. The norm in the space $\mathfrak{H}$ is defined by the formula

$$\|f\|^2 = \int |f(z_1, \ldots, z_m, x)|^2 \prod_{p=1}^{m} \frac{|\operatorname{Im} z_p|^{n_p - 2}}{\Gamma(n_p - 1)} d\mu(x)$$

$$\cdot dx_1\, dy_1 \ldots dx_m\, dy_m \qquad (z_p = x_p + iy_p) \tag{7.35}$$

and representation operators $T_g$ are given by the formula

$$T_g f(z) = f(z\bar{g}) \prod_{p=1}^{m} \left[ \frac{\beta_p z_p + \delta_p}{\sqrt{|\alpha_p \delta_p - \beta_p \gamma_p|}} \right]^{-n_p} \cdot \prod_{p=1}^{m-1} |\Lambda_p|^{i\rho_p} |\Lambda_1|^{n-2} |\Lambda_2|^{n-4} \ldots |\Lambda_{m-1}|^2. \tag{7.36}$$

Denote by $z'$ the matrix of order $(n-1)$ composed of the elements $z_{pq}$ of the matrix $z \in Z_m$, for $p < n$ and $q < n$. Then $z' \in Z_{m-1,1}$ and the function $f(z)$ can be represented in the form

$$f(z) = f(z', x_{n1}, \ldots, x_{n,n-2}, z_m). \tag{7.37}$$

This function is square integrable with respect to the variables $x_{n1}, \ldots, x_{n,n-2}$. On the other hand, considered only as a function in $z_m$ with the values of other variables fixed, it is analytic with respect to $z_m$ separately in the upper and lower half-planes. Therefore it is equivalent to a pair of functions such that one of them is analytic in the upper and the other in the lower half-plane.

For such a function the integral

$$\int |f|^2 \frac{|\operatorname{Im} z_m|^{n_m - 2}}{\Gamma(n_m - 1)} dx_m\, dy_m$$

has to converge.

Now consider a Fourier transform of $f(z', x_{n1}, \ldots, x_{n,n-2}, z_m)$ with respect to the variables $x_{n1}, \ldots, x_{n,n-2}, z_m$. First we make the following observations.

If a function $f_1(z)$ of the complex variable $z = x + iy$ is analytic in the upper half-plane $y > 0$ and for some non-negative integer $k$ one has

$$\int |f_1(z)|^2 \frac{y^{k-1}}{\Gamma(k)} dx\, dy < +\infty, \tag{7.38}$$

then the integral

$$\frac{1}{2\pi} \int_{-}^{+} f_1(z) e^{-itz} dx \tag{7.39}$$

converges and does not depend on $y$. Denoting this integral by $\varphi_1(t)$ we have that the function $\varphi_1(t)$ vanishes identically for $t < 0$. The function $f_1(z)$ is expressed through the function $\varphi_1(t)$ by the formula

$$f_1(z) = \frac{1}{\sqrt{2\pi}} \int_0^\infty \varphi_1(t) e^{itz} dx \tag{7.40}$$

where

$$\int |f_1(z)|^2 \frac{y^{k-1}}{\Gamma(k)} dx\, dy = \frac{1}{2^k} \int_0^\infty |\varphi_1(t)|^2 t^{-k} dt. \tag{7.41}$$

Therefore there is an isometric correspondence between the functions $f_1(z)$ which are analytic in the upper half-plane with the norm

$$\| f_1 \|^2 = \int_{-\infty}^{+\infty} \int_0^\infty |f_1(z)|^2 \frac{y^{k-1}}{\Gamma(k)} dy\, dx$$

and the functions $\varphi_1(t)$ defined on the half-line $0 \leqq t < +\infty$ with the norm

$$\| \varphi_1 \|^2 = \frac{1}{2^k} \int_0^\infty |\varphi_1(t)|^2 t^{-k} dt.$$

Now we give an outline of the proof of this statement. For $k > 0$ the convergence of (7.38) implies the convergence of the integral

$$\int_{-\infty}^{+\infty} |f_1(z)|^2 dx = \int_{-\infty}^{+\infty} |f_1(x, y)|^2 dx$$

for almost all $y$.

Therefore one can define the function

$$\varphi_1(t, y) = \frac{1}{\sqrt{2\pi}} \int_{-\infty}^{+\infty} f_1(x, y) e^{-ixt} dx. \tag{a}$$

By the Plancherel formula one has

$$\int_{-\infty}^{+\infty} |\varphi_1(t, y)|^2 dt = \int_{-\infty}^{+\infty} |f_1(x, y)|^2 dx$$

and hence

$$\int_0^\infty dy \int_{-\infty}^{+\infty} |\varphi_1(t, y)|^2 \frac{y^{k-1}}{\Gamma(k)} dt = \int_0^\infty dy \int_{-\infty}^{+\infty} |f_1(x, y)|^2 \frac{y^{k-1}}{\Gamma(k)} dx < +\infty. \tag{b}$$

On the other hand the function $f_1(z) = f_1(x, y)$ is analytic and therefore satisfies the differential equation

$$\frac{\partial f_1}{\partial x} + i \frac{\partial f_1}{\partial y} = 0.$$

Relation (a) now yields the equation on the function $\varphi_1(t, y)$:

$$it\varphi_1 + i \frac{\partial \varphi}{\partial y} = 0.$$

Thus

$$\varphi_1(t, y) = e^{-ty} \varphi_1(t) \tag{c}$$

where $\varphi_1(t)$ is a function depending only on $t$.

Substituting this in (b) we have

$$\int_0^\infty dy \int_{-\infty}^{+\infty} |\varphi_1(t)|^2 \frac{y^{k-1}}{\Gamma(k)} e^{-2ty} dt < +\infty.$$

Convergence of this integral immediately implies that $\varphi_1(t)=0$ for $t<0$. Multiplying both sides of (a) by $e^{ty}$ and using relation (c) we have

$$\varphi_1(t) = \frac{1}{\sqrt{2\pi}} \int_{-\infty}^{+\infty} f_1(z) e^{-itz} dx \qquad (z = x+iy).$$

The function $f_1(z)$ can be expressed through $\varphi_1(t)$ by the formula

$$f_1(z) = \frac{1}{\sqrt{2\pi}} \int_0^\infty \varphi_1(t) e^{itz} dt.$$

Relations (b) and (a) also yield

$$\int_0^\infty dy \int_{-\infty}^{+\infty} |f_1(x,y)|^2 \frac{y^{k-1}}{\Gamma(k)} dx = \int_0^\infty |\varphi_1(t)|^2 dt \int_0^{+\infty} e^{-2ty} \frac{y^{k-1}}{\Gamma(k)} dy.$$

Thus

$$\int_{y>0} |f_1(z)|^2 \frac{y^{k-1}}{\Gamma(k)} dx\, dy = \frac{1}{2^k} \int_0^\infty |\varphi_1(t)|^2 t^{-k} dt.$$

Conversely, if the function $\varphi_1(t)$ defined for $0 \leqq t < \infty$ is such that the integral

$$\int_0^\infty |\varphi_1(t)|^2 t^{-k} dt$$

converges then the function $f_1(z) = f_1(x,y)$ defined by formula (7.40) is analytic in the half-plane $\operatorname{Im} z > 0$.

Finally, if $k=0$ then $y^{k-1}/\Gamma(k)$ should be interpreted as a delta function. Integral (7.38) is then of the form:

$$\int_{-\infty}^{+\infty} |f_1(x)|^2 dx$$

where $f_1(x)$ is the boundary value of $f_1(z)$ on the real axis. Therefore, one can define the function

$$\varphi_1(t) = \frac{1}{\sqrt{2\pi}} \int_{-\infty}^{+\infty} f_1(x) e^{-itx} dx.$$

As we know (see, e.g. [4]), $\varphi_1(t)=0$ for $t<0$ and

$$\int_{-\infty}^{+\infty} |f_1(x)|^2 dx = \int_0^\infty |\varphi_1(t)|^2 dt.$$

We also have that

$$f_1(z) = \frac{1}{\sqrt{2\pi}} \int_0^\infty \varphi_1(t) e^{itz} dt.$$

Hence the function $\varphi_1(t)$ can be expressed as an integral (7.39).

A similar statement is true for each function $f_2(z)$ analytic in the lower half-plane $y<0$. The corresponding function $\varphi_2(t)$ identically vanishes for $t>0$.

Thus one can associate to each pair of functions $f_1(z)$ and $f_2(z)$ the function $\varphi(t)=\varphi_1(t)+\varphi_2(t)$ defined on the real axis. It is given by the formula

$$\varphi(t)=\frac{1}{\sqrt{2\pi}}\int_{-\infty}^{+\infty} f_1(z)e^{-itz}dx+\frac{1}{\sqrt{2\pi}}\int_{-\infty}^{+\infty} f_2(z)e^{-itz}dx$$

where the first integral is taken along the straight line $y=c>0$ and the second integral is taken along the straight line $y=-c<0$.

Denote the pair of functions $f_1(z)$ and $f_2(z)$ by $f(z)$. Then the last formula can be written in the form

$$\varphi(t)=\frac{1}{\sqrt{2\pi}}\int f(z)e^{-itz}dx \tag{7.42}$$

where the integral is taken along the lines $y=c$ and $y=-c$.

Transformation (7.42) is an isometric mapping of the space of functions $f(z)$ which are analytic separately in the upper and lower half-planes with the norm

$$\|f\|^2=\int_{-\infty}^{+\infty}\int_{-\infty}^{+\infty}|f(z)|^2\frac{|y|^{k-1}}{\Gamma(k)}dy\,dx$$

on the space of functions $\varphi(t)$ defined on the real axis with the norm

$$\|\varphi\|^2=\frac{1}{2^k}\int_{-\infty}^{+\infty}|\varphi(t)|^2|t|^{-k}dt.$$

Let us now go back to functions $f(\dot{z}x)$ from the space $\mathfrak{H}$. The above considerations show that such functions admit a Fourier transform:

$$\varphi(z',w_1,\dots,w_{n-2},\xi)=\frac{1}{(2\pi)^{(n-1)/2}}\int f(z',x_{n1},\dots,x_{n,n-2},z_m)$$
$$\cdot e^{-i(x_{n1}w_1+\dots+x_{n,n-2}w_{n-2}+z_m\xi)}dx_{n1}\dots dx_{n,n-2}dx_m. \tag{7.43}$$

That transformation is an isometric mapping of the space $\mathfrak{H}$ onto the space $H$ of functions $\varphi(z',w_1,\dots,w_{n-2},\xi)$ defined for $z'\in Z_{m-1,1}$, $-\infty<w_k<+\infty$, $-\infty<\xi<+\infty$ with the norm

$$\|\varphi\|^2=\frac{1}{2^{n_m-1}}\int|\varphi(z',w_1,\dots,w_{n-2},\xi)|^2$$
$$\cdot\prod_{p=1}^{m-1}\frac{|\operatorname{Im} z_p|^{n_p-2}}{\Gamma(n_p-1)}|\xi|^{1-n_m}dz'\,dw_1\dots dw_{n-2}d\xi. \tag{7.44}$$

Introduce instead of $\xi$ a new variable $w_{n-1}$ defined by the formula

$$\xi=w_{n-1}+x_{n-1,1}w_1+\dots+x_{n-1,n-2}w_{n-2}. \tag{7.45}$$

The space $H$ can then be considered as a space of functions $\varphi(z',w_1,\dots,w_{n-1})$ with the norm

$$\|\varphi\|^2 = \frac{(n_m-1)!}{2^{n_m-1}} \int |\varphi(z', w_1, \dots, w_{n-1})|^2 \prod_{p=1}^{m-1} \frac{|\operatorname{Im} z_p|^{n_p-2}}{\Gamma(n_p-1)} \cdot$$

$$\cdot |w_{n-1} + x_{n-1,1} w_1 + \dots + x_{n-1,n-2} w_{n-2}|^{1-n_m} dz' \, dw_1 \dots dw_{n-1}. \tag{7.46}$$

The formula (7.43) then takes the form

$$\varphi(z', w_1, \dots, w_{n-1}) = \frac{1}{(2\pi)^{(n-1)/2}} \int f(z', x_{n1}, \dots, x_{n,n-2}, z_m) \cdot e^{-i\left(\sum_{s=1}^{n-2} x_{ns} w_s + z_m\left(w_{n-1} + \sum_{s=1}^{n-2} x_{n-1,s} w_s\right)\right)} dx_m \dots dx_{n,n-2} dx_m$$

or

$$\varphi(z', w_1, \dots, w_{n-1}) = \frac{1}{(2\pi)^{(n-1)/2}} \int f(z', x_{n1}, \dots, x_{n,n-2}, z_m) \cdot e^{-i\left(\sum_{s=1}^{n-2} (x_{ns} + z_m x_{n-1,s}) w_s + z_m w_{n-1}\right)} dx_{n1} \dots dx_{n,n-2} dx_m. \tag{7.47}$$

This formula can be represented in a somewhat different fashion. With that in mind we note that the elements of the bottom row of the matrix $z \in Z_{m,0}$ can be expressed through the elements of the matrices $\dot{z} \in \dot{Z}_{m,0}$ and $x \in X_{m,0}$ in the representation $z = \dot{z}x$ by the formulas

$$z_{nq} = x_{nq} + z_m x_{n-1,q}, \quad q = 1, \dots, n-2, \quad z_{n,n-1} = z_m. \tag{7.48}$$

Let

$$u_q = x_{nq} + x_m x_{n-1,q}, \quad q = 1, \dots, n-2, \quad u_{n-1} = x_m. \tag{7.49}$$

Each function $f(z) \in \mathfrak{H}$ can be represented in the form

$$f(z) = f(z', u_1, \dots, u_{n-2}, z_m).$$

The variables $x_{n1}, \dots, x_{n,n-2}, x_m$ in (7.47) can be expressed through the new variables $u_1, u_2, \dots, u_{n-1}$. One has

$$(z', w_1, \dots, w_{n-1}) = \frac{1}{(2\pi)^{(n-1)/2}} \int f(z', u_1, \dots, u_{n-2}, z_m) \cdot e^{-i(z_{n1} w_1 + \dots + z_{n,n-2} w_{n-2} + z_{n,n-1} w_{n-1})} du_1 \dots du_{n-1}. \tag{7.50}$$

Consider the transformation

$$\psi(z', w_1, \dots, w_{n-1}) = \left| w_{n-1} + \sum_{s=1}^{n-2} x_{n-1,s} w_s \right|^{-(n_m-1)/2} \varphi(z', w_1, \dots, w_{n-1}). \tag{7.51}$$

It is an isometric mapping of the space $H$ onto the space $H_1$ of functions $\psi(z', w_1, \dots, w_{n-1})$ with the norm

$$\|\psi\|^2 = \frac{1}{2^{n_m-1}} \int |\psi(z', w_1, \dots, w_{n-1})|^2 \prod_{p=1}^{m-1} \frac{|\operatorname{Im} z_p|^{n_p-2}}{\Gamma(n_p-1)} dz' \, dw_1 \dots dw_{n-1}. \tag{7.52}$$

Note that if under the mapping (7.51) $\varphi \to \psi$, then the action of each operator $T_g$ on the function $\psi$ in the space $H_1$ is given by the formula:

$$T_g\psi = \left| w_{n-1} + \sum_{s=1}^{n-2} x_{n-1,s} W_c \right|^{-(n_m-1)/2} T_g\varphi. \tag{7.53}$$

Let $A$ be a bounded operator in the space $H_1$ commuting with all operators $T_g, g \in \tilde{\mathfrak{A}}_n$. We have to show that it is a scalar multiple of the identity.

Consider matrices $x^0 \in Z_{m,0}$ in the group $G_n$ satisfying condition (7.6): $x^0_{pq} = 0$ for $p < n$.

The operator $T_{x^0}$ in $\mathfrak{H}$ can be considered as a shift operator

$$T_{x^0} f(z) = f(zx^0). \tag{7.54}$$

On the other hand, condition (7.6) implies that under the shift $z \to zx^0$ elements of the matrix $z'$ do not change and $z_{nq} \to z_{nq} + x^0_{nq}$. Therefore, by definition of $u_q, u_q \to u_q + x^0_{nq}$ and $x_m \to x_m + x^0_{n,n-1}$. Hence formula (7.54) can be written in the form

$$T_{x^0} f(z', u_1, \ldots, u_{n-2}, z_m) = f(z', u_1 + x^0_{n1}, \ldots, u_{n-2} + x^0_{n,n-2}, z_m + x^0_{n,n-1}). \tag{7.55}$$

Now, in view of (7.50), operator $T_{x^0}$ in the space $H$ is of the form

$$T_{x^0}\varphi(z', w_1, \ldots, w_{n-1}) = e^{i(x^0_{n1}w_1 + \ldots + x^0_{n,n-1}w_{n-1})}\varphi(z', w_1, \ldots, w_{n-1}). \tag{7.56}$$

By (7.53) this operator has the same form in the space $H_1$. As in case I we conclude from this that the operator $A$ in the space $H_1$ is of the form

$$A\psi(z', w_1, \ldots, w_{n-1}) = a(w_1, \ldots, w_{n-1})\psi(z', w_1, \ldots, w_{n-1})$$

where $a(w_1, \ldots, w_{n-1})$ is an operator in the space of functions $f(z')$ with the norm

$$\|f\|^2 = \int |f(z')|^2 \prod_{p=1}^{m-1} \frac{|\operatorname{Im} z_p|^{n_p - 2}}{\Gamma(n_p - 1)} dz'$$

defined for almost all $w_1, \ldots, w_{n-1}$ and uniformly bounded with respect to those variables.

Consider the subgroup $\mathfrak{A}^0$ of matrices in the group $\tilde{\mathfrak{A}}_n$ satisfying the additional condition (7.12),

$$g_{nq} = 0 \quad \text{for} \quad q = 1, 2, \ldots, n-1,$$

and compute operators $T_g, g \in \mathfrak{A}^0$.

As in case II, let

$$g^0_{pq} = g_{pq} \cdot c, \quad p, q < n$$

where $c$ is chosen in such a way that the matrix $g^0$ of order $(n-1)$ is unimodular. Now let $b = g_{nn} \cdot c$. Formulas (7.13) and (7.14) imply that under the transformation $z \to z\bar{g}$ the matrix $z'$ goes into $z'\overline{g^0}$ and $z_{nq}$ goes into $\hat{z}_{nq}$ where

$$b\hat{z}_{nq} = \sum_{s=1}^{n-1} g^0_{sq} z_{ns}. \tag{7.57}$$

Letting $\hat{u}_q = \operatorname{Re} \hat{z}_{nq}$ we also have

$$b\hat{u}_q = \sum_{s=1}^{n-1} g^0_{sq} u_s. \tag{7.58}$$

Let us compute $\alpha(z,g)$. The formula for $\alpha(z,g)$ includes $\Lambda_p$ which is computed by the following formula

$$\Lambda_p=\frac{\tilde{g}_{2(p-1)+1}}{\tilde{g}_{2p+1}}, \qquad p=1,\dots,m-1 \tag{7.59}$$

where $\tilde{g}_m$ is a principal minor of the matrix $\tilde{g}=xg$.

Each matrix $z'\in Z_{m-1,1}$ can be represented in the form

$$z'=\dot{z}'x', \qquad \dot{z}'\in\dot{Z}_{m-1,1}, \qquad x'\in X_{m-1,1}.$$

Let

$$\Lambda'_p=\frac{\tilde{g}'_{2(p-1)+1}}{\tilde{g}'_{2p+1}}$$

where $\tilde{g}'_m$ denotes the principal minors of the matrix $\tilde{g}'=x'g^0$. As in case I it is easy to see that the $\Lambda'_p$ differ from the corresponding values of $\Lambda_p$ by a constant factor which is equal to some power of $c$.

Let us now compute the factor

$$\frac{\beta_m z_m+\delta_m}{\sqrt{|\alpha_m\delta_m-\beta_m\gamma_m|}}$$

in $\alpha(z,g)$. With this in mind we represent the matrix $xg$ in the form $xg=\hat{k}x_1$ where $\hat{k}=\|\hat{k}_{pq}\|\in\hat{K}_m$ and $x_1\in X_m$. Then the formulas of section 6 yield

$$\alpha_m=k_{n-1,n-1}, \qquad \beta_m=k_{n-1,n}, \qquad \gamma_m=k_{n,n-1}, \qquad \delta_m=k_{nn}.$$

But a direct computation gives

$$k_{n-1,n-1}=\sum_{s=1}^{n-2}x_{n-1,s}g_{s,n-1}+g_{n-1,n-1}, \qquad k_{n-1,n}=0,$$

$$k_{n,n-1}=\sum_{s=1}^{n-2}x_{ns}g_{s,n-1}+g_{n,n-1}, \qquad k_{nn}=g_{nn}.$$

Thus

$$\frac{\beta_m z_m+\delta_m}{\sqrt{|\alpha_m\delta_m-\beta_m\gamma_m|}}=\sqrt{g_{nn}}\left|\sum_{s=1}^{n-2}x_{n-1,s}g_{s,n-1}+g_{n-1,n-1}\right|^{-\frac{1}{2}}$$

$$=\sqrt{g_{nn}|c|}\left|\sum_{s=1}^{n-2}x_{n-1,s}g^0_{s,n-1}+g^0_{n-1,n-1}\right|^{-\frac{1}{2}}. \tag{7.60}$$

Now

$$|\Lambda_m|=|\alpha_m\delta_m-\beta_m\gamma_m|=g_{nn}|c|^{-1}\left|\sum_{s=1}^{n-2}x_{n-1,s}g^0_{s,n-1}+g^0_{n-1,n-1}\right|.$$

The formula for $\alpha(z,g)$ now takes the following form:

$$\alpha(z,g)=\alpha_0(g_{nn})\alpha'(z',g^0), \tag{7.61}$$

where $\alpha_0(g_{nn})$ is a function of the form (7.20) and $\alpha'(z',g^0)$ is a function of the form

$$\alpha'(z', g^0) = \prod_{p=1}^{m-1}\left[\frac{\beta_p z_p + \delta_p}{\sqrt{|\alpha_p\delta_p - \beta_p\gamma_p|}}\right]^{-n_p} \prod_{p=1}^{m-1} |\Lambda'_p|^{i\rho_p}$$

$$\cdot |\Lambda'_1|^{n-2}|\Lambda'_2|^{n-4}\ldots|\Lambda'_{m-1}|^2 \left|\sum_{s=1}^{n-2} x_{n-1,s}g^0_{s,n-1} + g^0_{n-1,n-1}\right|^{n_m/2}. \quad (7.62)$$

Thus one obtains the following formula for operators $T_g, g\in\mathfrak{A}^0$ in the space $\mathfrak{H}$:

$$T_g f(z', u_1, \ldots, u_{n-2}, z_m) = \alpha_0(g_{nn})\alpha'(z', g^0) f(z'\overline{g^0}, \hat{u}_1, \ldots, \hat{u}_{n-2}, \hat{z}_m). \quad (7.63)$$

Let us now see how an operator $T_g$ acts on the function $\varphi(z', w_1, \ldots, w_{n-1})$ in $H$. We have

$$T_g\varphi(z', w_1, \ldots, w_{n-1}) = \frac{1}{(2\pi)^{(n-1)/2}}\alpha_0(g_{nn})\alpha'(z', g^0)$$

$$\cdot\int f(z'\overline{g^0}, \hat{u}_1, \ldots, \hat{u}_{n-2}, \hat{z}_m) e^{-i(z_{n1}w_1 + \ldots + z_{n,n-1}w_{n-1})} du_1\ldots du_{n-1}.$$

But by (7.57)

$$z_{n1}w_1 + \ldots + z_{n,n-1}w_{n-1} = \hat{z}_{n1}\hat{w}_1 + \ldots + \hat{z}_{n,n-1}\hat{w}_{n-1}$$

where $\hat{w}_q$ is given by the formula

$$\frac{1}{b}\hat{w}_q = \sum_{s=1}^{n-1} \hat{g}_{sq} w_s \quad (7.64)$$

where the matrix $\hat{g} = \|\hat{g}_{sq}\|$ is an inverse to the transpose of $g^0$, i.e. $\hat{g} = (g^0)^{*-1}$.

Therefore in the new variables $u_1, \ldots, u_{n-1}$ one has

$$T_g\varphi(z', w_1, \ldots, w_{n-1}) = \frac{\Delta}{(2\pi)^{(n-1)/2}}\alpha_0(g_{nn})\alpha'(z', g^0)$$

$$\cdot\int f(z'\overline{g^0}, \hat{u}_1, \ldots, \hat{u}_{n-2}, \hat{z}_m) e^{-i(\hat{z}_{n1}\hat{w}_1 + \ldots + \hat{z}_{n,n-1}\hat{w}_{n-1})} du_1\ldots du_{n-1}$$

where $\Delta$ denotes the Jacobian of the transformation (7.58). In other words

$$T_g\varphi(z', w_1, \ldots, w_{n-1}) = \frac{\Delta}{(2\pi)^{(n-1)/2}}\alpha_0(g_{nn})\alpha'(z', g^0)\varphi(z'\overline{g^0}, \hat{w}_1, \ldots, \hat{w}_{n-1}). \quad (7.65)$$

Finally, let us find the formula for the operator $T_g$ acting in the space $H_1$. Suppose that under the transformation $z' \to z'\overline{g^0}$ the elements $x_{n-1,q}$ of the bottom row of matrix $z'$ go into the elements $\hat{x}_{n-1,q}$. Then, using (7.65), (7.51) and (7.53), we obtain the following formula for the operator $T_g$ in the space $H_1$:

$$T_g\psi(z', w_1, \ldots, w_{n-1}) = \frac{\Delta}{(2\pi)^{(n-1)/2}}\alpha_0(g_{nn})\alpha'(z', g^0).$$

$$\cdot\left|\frac{\sum_{s=1}^{n-2}\hat{x}_{n-1,s}\hat{w}_s + \hat{w}_{n-1}}{\sum_{s=1}^{n-2}x_{n-1,s}w_s + w_{n-1}}\right|^{n_{m_2}-1} \psi(z'\overline{g^0}, \hat{w}_1, \ldots, \hat{w}_{n-1}). \quad (7.66)$$

Consider the expression

$$\frac{\sum_{s=1}^{n-2} \hat{x}_{n-1,s}\hat{w}_s + \hat{w}_{n-1}}{\sum_{s=1}^{n-2} x_{n-1,s}w_s + w_{n-1}}.$$

It is easy to see that $\hat{x}_{n-1,s}$ can be expressed in terms of the elements of the bottom row of the matrix $g' = z'g^0$ by the formula

$$\hat{x}_{n-1,s} = \frac{g'_{n-1,s}}{g'_{n-1,n-1}}. \tag{7.67}$$

Here

$$g'_{n-1,n-1} = \sum_{s=1}^{n-2} x_{n-1,s}g^0_{s,n-1} + g^0_{n-1,n-1}. \tag{7.68}$$

Therefore, using (7.64) and (7.67) we have

$$\begin{aligned} \sum_{s=1}^{n-2} \hat{x}_{n-1,s}\hat{w}_s + \hat{w}_{n-1} &= \frac{1}{g'_{n-1,n-1}} \sum_{s=1}^{n-1} g'_{n-1,s}\hat{w}_s \\ &= \frac{b}{g'_{n-1,n-1}} \sum_{s=1}^{n-2} \left( \sum_{p=1}^{n-1} x_{n-1,p}g^0_{ps} \right)\left( \sum_{q=1}^{n-1} \hat{g}_{sq}w_s \right). \end{aligned} \tag{7.69}$$

But $g = (\hat{g}^0)^{*-1}$ and, consequently, the elements of the matrices $g^0$ and $\hat{g}$ are related by the formula

$$\sum_{s=1}^{n-1} g^0_{ps}\hat{g}_{qs} = \delta_{pq}$$

where $\delta_{pq}$ is the Kronecker delta. Thus formula (7.69) can be written in the form

$$\sum_{s=1}^{n-2} \hat{x}_{n-1,s}\hat{w}_s + \hat{w}_{n-1} = \frac{b\left( \sum_{s=1}^{n-1} x_{n-1,s}w_s + w_{n-1} \right)}{\sum_{s=1}^{n-2} x_{n-1,s}g^0_{s,n-1} + g^0_{n-1,n-1}}$$

and therefore

$$\frac{\sum_{s=1}^{n-2} \hat{x}_{n-1,s}\hat{w}_s + \hat{w}_{n-1}}{\sum_{s=1}^{n-2} x_{n-1,s}w_s + w_{n-1}} = \frac{b}{\sum_{s=1}^{n-2} x_{n-1,s}g^0_{s,n-1} + g^0_{n-1,n-1}}. \tag{7.70}$$

Formulas (7.62) and (7.70) now imply:

$$\left| \frac{\sum_{s=1}^{n-2} \hat{x}_{n-1,s}\hat{w}_s + \hat{w}_{n-1}}{\sum_{s=1}^{n-2} x_{n-1,s}w_s + w_{n-1}} \right|^{n_{m_2}-1} \alpha'(z', g^0) = \prod_{p=1}^{m-1} \left[ \frac{\beta_p z_p + \delta_p}{\sqrt{|\alpha_p\delta_p - \beta_p\gamma_p|}} \right]^{-n_p}$$

$$\cdot \prod_{p=1}^{m-1} |\Lambda'_p|^{i\rho_p} |\Lambda'_1|^{n-2} |\Lambda'_2|^{n-4} \ldots |\Lambda'_{m-1}|^2$$

$$\cdot \left| \sum_{s=1}^{n-2} x_{n-1,s} g^0_{s,n-1} + g^0_{n-1,n-1} \right|^{\frac{1}{2}} b^{(n_m-1)/2}. \tag{7.71}$$

As has been shown above,

$$\sum_{s=1}^{n-2} x_{n-1,s} g^0_{s,n-1} + g^0_{n-1,n-1} = \frac{|c|}{g_{nn}} |\Lambda_m|.$$

On the other hand, $|\Lambda_1 \Lambda_2 \ldots \Lambda_m| = 1$ and, consequently,

$$\left| \sum_{s=1}^{n-2} x_{n-1,s} g^0_{s,n-1} + g^0_{n-1,n-1} \right|^{\frac{1}{2}} = k |\Lambda'_1|^{-\frac{1}{2}} |\Lambda'_2|^{-\frac{1}{2}} \ldots |\Lambda'_{m-1}|^{-\frac{1}{2}} \tag{7.72}$$

where $k$ is a constant equal to some power of $c$.

Thus formula (7.66) for the operator $T_g$ in the space $H_1$ takes the form

$$T_g \psi(z', w_1, \ldots, w_{n-1}) = \frac{1}{(2\pi)^{(n-1)/2}} \alpha_1(g_{nn}) \tilde{\alpha}(z', g^0) \psi(z' \overline{g^0}, \hat{w}_1, \ldots, \hat{w}_{n-1}). \tag{7.73}$$

Here $\alpha_1(g_{nn})$ is again a function of the form (7.20) and

$$\tilde{\alpha}(z', g^0) = \prod_{p=1}^{m-1} \left[ \frac{\beta_p z_p + \delta_p}{\sqrt{|\alpha_p \delta_p - \beta_p \gamma_p|}} \right]^{-n_p} \prod_{p=1}^{m-1} |\Lambda'_p|^{i\rho_p}$$

$$\cdot |\Lambda'_1|^{(n-1)-\frac{3}{2}} |\Lambda'_2|^{(n-1)-\frac{7}{2}} \ldots |\Lambda'_{m-1}|^{\frac{3}{2}}. \tag{7.74}$$

It is easy to see that $\tilde{\alpha}(z', g^0)$ coincides with the function defining the unitary representation of the group $G_{n-1}$ from the series $d_{m-1,1}$.

Thus, for all $g \in \mathfrak{A}^0$,

$$T_g \psi(z', w_1, \ldots, w_{n-1}) = \frac{\alpha_1(g_{nn})}{(2\pi)^{(n-1)/2}} T_{g^0} \psi(z', \hat{w}_1, \ldots, \hat{w}_{n-1}) \tag{7.75}$$

where $T_{g^0}$ is a unitary representation of the unimodular group $G_{n-1}$ of order $(n-1)$ from the series $d_{m-1,1}$ defined by the relation

$$T_{g^0} f(z') = \tilde{\alpha}(z', g^0) f(z', g^0) \tag{7.76}$$

and $\hat{w}_1, \ldots, \hat{w}_{n-1}$ are given by formula (7.64).

The proof that an operator $A$ commuting with all operators $T_g, g \in \tilde{\mathfrak{A}}_n$, is a scalar multiple of the identity is now completed exactly in the same way as in case II.

To complete the proof of Lemma 1 it remains to consider the case $n = 2$.

For $n = 2$ we have two series of representations of the group $\tilde{G}_2$: $d_{0,2}$ and $d_{1,0}$.

There is no $z'$ in this case. Therefore, for both the series $d_{0,2}$ and $d_{1,0}$ the formula for the operator $A$ in the corresponding functional space is of the form

$$Af(w) = a(w) f(w) \tag{7.77}$$

where $a(w)$ is a scalar function defined and bounded for almost all $w$. In that case relation (7.29) means that $a(bw) = a(w)$. Thus, $a(w)$ is a constant. But then $A$ is an operator of multiplication by a scalar. Therefore, irreducibility is proved in this case as well.

Received November 12, 1952

## References

1. Gelfand, I.M., Najmark, M.A.: Unitary representations of classical groups. Tr. Mat. in-ta im. Steklova **36** (1950). (Russian).
2. Gelfand, I.M., Najmark, M.A.: Unitary representations of the Lorentz group. Izv. Akad. Nauk SSSR, Ser. Mat. **12** (1948) 411–504. (Russian).
3. Gelfand, I.M., Graev, M.I.: Unitary representations of real simple Lie groups. Dokl. Akad. Nauk SSSR **XXXVI** (1952) 461–463. (Russian).
4. Titchmarsh, E.C.: Vvedenie v teoriju integralov Furje. M.-L. 1948. (English transl.: Introduction to the theory of Foūrier integrals, 2nd edn. Oxford: Clarendon Press 1948.)
5. Bargmann, V.: Irreducible unitary representations of the Lorentz group. Ann. Math. **48** (1947) 568–640.

# 14.

## (with M. I. Graev)

# On a general method for decomposition of the regular representation of a Lie group into irreducible representations

Dokl. Akad. Nauk SSSR **92** (1952) 221–224. Zbl. **53**:15

1. The problem of decomposition of the regular representation of a group into irreducible representations is an analog of the expansion of a function into the Fourier integral (an analog of the Plancherel formula). This problem is rather important in the general theory of representations. Here we discuss only Lie groups for which this problem is, in fact, equivalent to the following problems:

*The first problem.* Assume that for any given function $x(g)$ on a Lie group $G$ all integrals $\int x(g)\chi_\alpha(g)dg$ are given. Here $\chi_\alpha(g)$ are the characters of those irreducible representations which are involved in the decomposition of the regular representation. We have to find an integral of $x(g)$ over an arbitrary generic conjugate class. As a rule, the solution of this problem presents no difficulties. For a semisimple Lie group it reduces essentially to the following problem: how to express a function on the maximal commutative subgroup in terms of its Fourier coefficients. For the complex unimodular group this subgroup consists of all diagonal matrices and for other semisimple groups it can be determined similarly.

*The second problem.* Assume that for any function $x(g)$ on a Lie group $G$ its integrals over generic conjugate classes are given. How are we to find $x(e)$ where $e$ is the identity of the group? We study the second problem because it presents the main difficulties for Lie groups. In the case of finite groups the solution of the problem is trivial since the unity element is itself one of generic conjugate classes. In a Lie group the unity element forms a "singular" conjugate class whereas a generic class is usually a manifold. However, for a compact Lie group the unity element is the limit of generic classes and therefore the problem can be easily solved. In this case, $g(e)$ is the limit of the integrals over conjugate classes divided by the volume of the corresponding classes[1]. For noncompact Lie groups this simple method fails because in this case the unity element is not a limit of generic classes. Take, for example, the group $GL(n)$. If all eigenvalues tend to 1 then the corresponding conjugate classes converge not to the identity matrix but to the conjugate class consisting of elements whose Jordan representation has only one block. The identity matrix is a singular point in this class[2].

[1] A simple analog of this problem for compact groups is presented by the problem of finding $f(0,0)$ if the integrals $f(x,y)$ over circles $x^2+y^2=c^2$ are known.

[2] A simple analog of this problem for noncompact groups is presented by the problem of finding $f(0,0)$ if the integrals $f(x,y)$ over hyperbolae $x^2-y^2=c^2$ are known.

This is why, up to now, the solution of the second problem for noncompact Lie groups has encountered grave difficulties. At the same time, in all particular cases where the solution has been found the final formulas for compact and noncompact groups are very similar. This similarly is not casual. Here we suggest a solution method which applies to both compact and noncompact Lie groups. Its application to concrete groups is quite simple.

2. We use the following generalization of some results of M. Riesz [3]. Consider an integral

$$R(\lambda)=\int_{-\infty}^{\infty}\cdots\int_{-\infty}^{\infty} f(x_1,\dots,x_n)\left|\sum_{i,j=1}^{n}\alpha_{i_j}x_ix_j\right|^{\lambda/2}dx_1,\dots,dx_n \tag{1}$$

where $\sum\alpha_{i_j}x_ix_j$ is a nonsingular quadric in real variables $x_1,\dots,x_n$[3], $\lambda$ is a complex number and $f(x_1,\dots,x_n)$ is a smooth function with compact support. This integral converges for $\operatorname{Re}\lambda>-n$ and is an analytic function in this domain. For $\operatorname{Re}\lambda<-n$ the integral (1) is understood to be the analytic continuation of $R(\lambda)$. We assume that $n$ is odd. Then $R(\lambda)$ is an analytic function whose only singularities are simple poles at $\lambda=-n,-(n+2),\dots,-(n+2k),\dots$. A simple calculation shows that the residues of $R(\lambda)$ are given by the formulas

$$\operatorname*{Res}_{\lambda=-n} R(\lambda)=c_0 f(0,\dots,0), \tag{2}$$

$$\dots\dots\dots\dots\dots\dots\dots\dots$$

$$\operatorname*{Res}_{\lambda=-(n+2k)} R(\lambda)=c_k\Delta^k f|_{x_1=\dots=x_n=0}, \tag{3}$$

$$\dots\dots\dots\dots\dots\dots\dots\dots$$

Here $\Delta$ is the differential operator $\sum_{i,j=1}^{n}\beta_{i_j}x_ix_j$ where $\|\beta_{i_j}\|=\|\alpha_{i_j}\|^{-1}$. The number $c_0,\dots,c_k,\dots$ are independent on $f(x_1,\dots,x_n)$.

3. We illustrate the idea of the method in the case of a complex semisimple Lie group. Real semisimple groups will be discussed in a separate paper. Since the solution is the same for all complex semisimple groups, we present here the simplest case, namely that of the unimodular group. Later we give a hint on how to generalize it to an arbitrary complex semisimple group. Instead of the unimodular group it is more convenient to discuss the group $G$ of all complex nonsingular matrices with a real determinant since the manifold of these matrices has an odd dimension, namely $2^{n2}-1$.

In the case of an arbitrary matrix group a generic conjugate class is a set of matrices which are conjugate to a given diagonal matrix $\delta$ with different eigenvalues. We denote the integral of a function $x(g)$ over such a class by $I\delta$ and the problem consists in finding $x(e)$ from a given $I\delta$. Now we give a more rigorous definition of $I\delta$. Let $g$ be a matrix from $G$ and let $\delta$ be the diagonal form of $g$. We describe a matrix $g$ by a set of parameters, some of these parameters being the elements of $\delta$.

[3] M. Riesz considered the case when a quadric is positive definite or has only one minus. The result for an arbitrary nonsingular quadric can be obtained in a similar way.

The set of all other parameters is denoted by $\bar{g}$. Then any function on $G$ is a function in $\delta$ and $\bar{g}$ so that

$$\int x(g)dg = \int x(\bar{g}, \delta)\omega(\bar{g}, \delta)d\bar{g}d\delta \tag{4}$$

where $\omega(\bar{g}, \delta)$ is the Jacobian of the transformation to the parameters $\bar{g}$ and $\gamma$. We define

$$I_\delta = \frac{1}{|D(\delta)|}\int x(\bar{g}, \delta)\omega(\bar{g}, \delta)d\bar{g} \tag{5}$$

where $D(\delta)$ is the discriminant of the characteristic polynomial of the matrix $\delta$. $I_\delta$ is called an integral of the function $x(g)$ over the conjugate class determined by $\delta$. The function $I_\delta$ is a smooth symmetric function in the elements of $\delta$. We can now rewrite formula (4) as follows:

$$\int x(g)dg = \int I_\delta \cdot |D(\delta)|d\delta. \tag{6}$$

Let $\varphi(g) = \varphi(\delta_g)$ be a function of the eigenvalues of $g$. Then in view of (6) we have

$$\int x(g)\cdot|\varphi(g)|^{\lambda/2}dg = \int I_\delta \cdot |D(\delta)|\,|\varphi(\delta)|^{\lambda/2}d\delta = R(\lambda). \tag{7}$$

We will take as the function $\varphi$ some function which is a nonsingular quadric of parameters $\delta$ and therefore near the identity matrix $e$ it is a nonsingular quadric of all $2n^2 - 1$ parameters of $g$ up to the terms of highest order. For example, if the eigenvalues of a matrix $g$ are $\lambda_k = e^{\tau_k + i\varphi_k}$ then we can take $\varphi(g) = \sum_{k=1}^n (\tau_k^2 - \varphi_k^2)$ or $\varphi(g) = \sum_{k=1}^n \tau_k\varphi_k$. In this case both integrals in formula (7) will be essentially integrals of the type (1). Let us calculate the residues of these integrals for $\lambda = -(2n^2 - 1)$. The first integral in (7) is taken over a manifold of dimension $2n^2 - 1$ and according to formula (2), its residue in $\lambda = -(2n^2 - 1)$ is equal to $c_0 x(e)$. The second integral in (7) taken over a manifold of dimension $2n - 1$ and when $\lambda = -(2n^2 - 1)$ by (3) the residue of this integrals is

$$\Delta^{n(n-1)}\{I_\delta \cdot |D(\delta)|\}_{\delta = e}.$$

Here $\Delta$ is the homogeneous differential operator of second order in $\tau_k$ and $\varphi_k$ which corresponds to the quadric $\varphi(\delta)$. If we put $(\partial/\partial\bar{\mu}_k) = \frac{1}{2}((\partial/\partial\tau_k) - i(\partial/\partial\varphi_k)$ and $(\partial/\partial\mu_k) = \frac{1}{2}(\partial/\partial\tau_k) + i(\partial/\partial\varphi_k))$ and use the properties of the function $|D(\delta)|$ we get

$$\Delta^{n(n-1)}\{I_\delta \cdot |D(\delta)|\}_{\delta = e} = \tilde{c}\prod_{i<j}\left(\frac{\partial}{\partial\mu_i} - \frac{\partial}{\partial\mu_j}\right)\left(\frac{\partial}{\partial\bar{\mu}_i} - \frac{\partial}{\partial\bar{\mu}_j}\right)I_\delta|_{\delta = e} \tag{8}$$

where each difference $(\partial/\partial\varphi_i) - (\partial/\partial\varphi_n)$ must be formally replaced by $(\partial/\partial\varphi_i)$. So we have

$$x(e) = c\prod_{i<j}\left(\frac{\partial}{\partial\mu_i} - \frac{\partial}{\partial\mu_j}\right)\left(\frac{\partial}{\partial\bar{\mu}_i} - \frac{\partial}{\partial\bar{\mu}_j}\right)I_\delta|_{\delta = e} \tag{9}$$

with the substitution $(\partial/\partial\varphi_i)$ for $(\partial/\partial\varphi_i) - (\partial/\partial\varphi_n)$. Now to return to the unimodular group, we must replace $(\partial/\partial\tau_i) - (\partial/\partial\tau_n)$ by $(\partial/\partial\tau_i)$ in formula (9). The constant $c$ can be easily calculated from formulas (2), (3), and (8). Formula (9) was originally obtained in [1] and [2].

4. This proof can be carried out with no substantial changes for any complex semisimple group.

In this case we must take as $\delta$ the elements of the corresponding commutative subgroup, we must take as the function $D(\delta)$ the product of second degrees of all the roots of the group, and, for a function $\varphi(\delta)$, we can choose, for example, the real or the imaginary part of the second Cartan invariant of the group.

Another proof of the analog of the Plancherel formula for complex semisimple Lie groups was given by Harish-Chandra in [4].

## References

1. Gelfand I.M., Najmark, M.A.: Dokl. Akad. Naūk SSSR (6) **63** (1948).
2. Gelfand I.M., Najmark, M.A.: Tr. Mat. Inst. Akad. Naūk SSSR **36** (1950).
3. Riesz, M.: Acta Math. **81** (1949) 1–2.
4. Harish-Chandra: Proc. Nat. Acad. Sci. (10) **37** (1951).

# Part III

# Geometry of homogeneous spaces; spherical functions; automorphic functions

# 1.

## (with F. A. Berezin)

# Some remarks on the theory of spherical functions on symmetric Riemannian manifolds[1]

Tr. Mosk. Mat. O.-va. **5** (1956) 311–351
[Transl., II. Ser., Am. Math. Soc. **21** (1962) 193–238]. Zbl. **72**:17

CONTENTS



1) The principal results of the article were presented at the meeting of the Moscow Mathematical Society of 6 December 1955.

The theory of spherical functions as functions on the ordinary sphere arose in the end of the 18th and the beginning of the 19th centuries in the works of Legendre, Laplace, and Jacobi. Consideration of these functions was dictated by the demands of mathematical physics and analysis. The next important step was made only in the 20th century by É. Cartan and H. Weyl, who gave the contemporary definition of spherical functions and connected their theory with the theory of Lie groups and their linear representations.

## §1. Introduction

In nos. 1, 2, and 3 of this section, we give the basic known facts of the theory of spherical functions on symmetric spaces. In no. 4, we describe briefly the content of the present work.[1)]

**no. 1. Definition of spherical functions.** Spherical functions are usually defined on a manifold $M$, in which a group $G$ of transformations acts transitively. (Such manifolds are called homogeneous.) We shall suppose that the group $G$ is a Lie group. Let $G_0$ denote the subgroup of $G$ that leaves the point $x_0$ of the manifold fixed (such a subgroup is called stationary). The set of elements of $G$ such that $gx_0 = x$ ($gx$ denotes the point $x \in M$ transformed by $g$) is obviously a right coset of $G$ with respect to $G_0$. Thus there exists a one-to-one correspondence between points of $M$ and right cosets of $G$ with respect to $G_0$. Hence $M$ is often called the manifold of right cosets and is denoted by $G/G_0$.

We consider an irreducible representation $\phi$ of the group $G$, which associates with every element $g$ a unitary operator $T_g^{(\phi)}$ in a Hilbert space $H_\phi$. Suppose furthermore that $H_\phi$ contains a vector $\xi$ of length $1$ that is invariant under $G_0$, that is, such that $T_g\xi = \xi$ if $g \in G_0$. Then with every vector $\eta \in H_\phi$ we can associate a function $f_{\phi,\eta}(g) = (\eta, T_g^{(\phi)}\xi)$; obviously the function $f_{\phi,\eta}(g)$ is constant on the right cosets of $G$ with respect to $G_0$ and is hence a function on the manifold $M = G/G_0$. These functions are called spherical.

We infer a number of consequences from the definition.

1. *The spherical functions belonging to one and the same representation are a linear family.* In fact

$$\lambda f_{\phi,\eta}(g) + \mu f_{\phi,\eta'}(g) = \lambda(\eta, T_g^{(\phi)}\xi) + \mu(\eta', T_g^{(\phi)}\xi) = (\lambda\eta + \mu\eta', T_g^{(\phi)}\xi) = f_{\phi,\lambda\eta+\mu\eta'}(g).$$

2. *Different functions correspond to different vectors.* If this were not the case, then there would be a vector $\eta$ such that $(\eta, T_g^{(\phi)}\xi) = 0$ for all $g$. But, since the representation $\phi$ is irreducible, every vector in $H_\phi$ can be approximated by linear combinations of vectors $T_g^{(\phi)}\xi$. Hence $\eta$ is orthogonal to $H_\phi$ and there-

1) The present work is in large measure a development of the work [5].

fore we have $\eta = 0$.

3. *If* $f_{\phi,\eta}(g)$ *is a spherical function, then its translates* $f_1(g) = f_{\phi,\eta}(g_1^{-1}g)$ *are also spherical functions.* We consider the vector $\eta_1 = T_{g_1}^{(\phi)}\eta$:

$$f_{\phi,\eta_1}(g) = (\eta_1, T_g^{(\phi)}\xi) = (T_{g_1}^{\phi}\eta, T_g^{(\phi)}\xi) = (\eta, T_{g_1^{-1}g}^{\phi}\xi) = f_{\phi,\eta}(g_1^{-1}g).$$

Since the function $f_{\phi,\eta}(g)$ depends only on the coset $x$ to which $g$ belongs, we can phrase the preceding fact in slightly different terms: *if* $f(x)$ *is a spherical function, then its translate* $f(g^{-1}x)$ *is also a spherical function.* Taken together, the above three statements mean that *if we carry over to the family of spherical functions belonging to the representation* $\phi$ *the scalar product from* $H_\phi$, *that is, if we set* $(f_{\phi,\eta_1}, f_{\phi,\eta_2}) = (\eta_1, \eta_2)$, *then the linear transformation* $T_g f(x) = f(g^{-1}x)$ *induce a representation in this family that is equivalent to the representation* $\phi$.

Among the spherical functions belonging to a representation $\phi$, a special role is played by functions $f_{\phi,\xi'} = (\xi, T_g\xi')$, where $\xi'$ like $\xi$ has the property that $T_g\xi' = \xi'$ for $g \in G_0$. These are the so-called zonal spherical functions. Evidently a zonal function has the property that $f(g) = f(g'gg'')$ if $g'$ and $g''$ are elements of the subgroup $G_0$. It follows from this that if we regard a zonal function $f$ as a function on the manifold $G/G_0$, then $f(x) = f(gx)$ when $g \in G_0$. The set of elements of the group $G$ of the form $g'gg''$, where $g'$ and $g''$ run independently through $G_0$, is called a two-sided coset of the group $G$ with respect to $G_0$. There corresponds to it a submanifold on the manifold $G/G_0$, transitive with respect to the stationary subgroup. We call this submanifold a sphere with center in the invariant point.

In the present work, we consider functions not on arbitrary homogeneous manifolds, but only on symmetric Riemannian spaces.[1)]

**no. 2. Some facts about symmetric spaces.**[2)] Definition. A homogeneous space is called symmetric, if its group of rigid motions $G$ admits an involutory automorphism $*$ which describes the stationary subgroup $G_0$, that is, such that the equation $g^* = g$ is satisfied by elements of $G_0$ and by no other elements. A homogeneous space is Riemannian (that is, has an invariant positive definite metric) if and only if its stationary subgroup is compact.

Besides the metric, a symmetric space can have still other invariants of pairs of points, that is, functions of pairs of points that do not change under rigid motions. All of these invariants are symmetric, since for every pair of points on a symmetric space, there is a rigid motion which represents them.

1) Spherical functions on symmetric Riemannian spaces were first considered by É. Cartan [11].

2) In this paragraph, we describe the results of É. Cartan relating to symmetric spaces. See [1], [2], [3].

The maximal number of independent invariants of pairs of points is called the rank of a symmetric space. The complete collection of invariants of two points is called the complex distance between these points.

In order for two points to be transformable one into the other with transformations from the stationary subgroup, it is necessary and sufficient that their complex distances to the invariant point should be equal.

We have agreed to call a manifold transitive with respect to the stationary subgroup a sphere with center at the invariant point. The complex distance of an arbitrary point of this sphere from the center is naturally enough called the complex radius of the sphere. In the theory of symmetric spaces it is proved that two spheres are congruent if their complex radii coincide.

One of the most important properties possessed by symmetric spaces is the fact that they can be imbedded in a natural way into the group of rigid motions. This is carried out in the following way. We consider in the group $G$ the manifold $\Theta$, consisting of elements $\theta$, characterized by the equality $\theta^* = \theta^{-1}$. The formula

$$\theta \to g\theta g^{*-1} \tag{1.1}$$

describes rigid motions in $\Theta$. The connected component of the manifold $\Theta$ that contains the identity of the group is a model of the symmetric space, imbedded in the group. (In fact, one can show that the transformations (1.1) act transitively in this component. Furthermore, the stationary subgroup $G_0$ of the point $e \in \Theta$ is described by the automorphism $*$ and hence, in agreement with definition, is the stationary subgroup of the symmetric manifold $G/G_0$.)

We list a few typical examples of symmetric spaces.

1. The manifold $H_n$ of positive definite matrices of the $n$-th order with determinant $1$. Rigid motions in it are given by the formula

$$h \to gh\overline{g'}, \tag{1.2}$$

where $g$ is an arbitrary complex unimodular matrix, $g'$ is the transposed matrix, and $\bar{g}$ is the complex-conjugate matrix. Thus the group of rigid motions of the manifold $H_n$ is the group of all complex unimodular matrices of order $n$, $SL(n)$. It is obvious from formula (1.2) that the stationary subgroups for the point $E \in H_n$ is the group of unitary unimodular matrices of order $n$, customarily denoted by $SU(n)$. The automorphism $*$ that describes this subgroup operates as follows on elements of the entire group: $g^* = \bar{g}'^{-1}$. It is also evident that the elements of $H_n$ satisfy the condition $h^* = h^{-1}$. Thus $H_n$ is simultaneously a model of the space $SL(n)/SU(n)$, imbedded in the group of rigid motions.

Since the stationary group acts on elements of $H_n$ according to the formula $h \to uhu^{-1}$, and since each Hermitian matrix can be brought into diagonal form by transformations of this type, the sphere with center at $E$ is the class of Hermitian matrices with given eigenvalues. The complex radius of this sphere can evidently be taken to be the collection of coefficients of the characteristic polynomial of the matrix $h$ or the collection of its eigenvalues, considered up to permutations. The rank of the space $H_n$ is $n - 1$.

2. The manifold of unitary matrices of order $n$ with determinant $1$. Rigid motions here are given by the formula

$$u \to u_1^{-1} u u_2, \tag{1.3}$$

where $u_1$, $u_2$ are unitary matrices.

The group of rigid motions is consequently the direct product $G = SU(n) \times SU(n)$. The elements of the group $G$ are pairs $g = (u_1, u_2)$. The automorphism describing the stationary subgroup acts as follows: $g^* = (u_1, u_2)^* = (u_2, u_1)$. Thus the stationary subgroup is the group $G_0$ consisting of the pairs $g = (u, u)$. For a model of the space imbedded in $G$, we have the manifold $\Theta$, consisting of the pairs $\theta = (u, u^{-1})$. It is easy to verify that the general formula (1.1) for rigid motions in $\Theta (\theta \to g \theta g^{*-1})$ is equivalent to (1.3) in this case. It follows from (1.3) that the stationary subgroup of the point $E$ acts on elements of the manifold according to the formula $u \to u_1 u u_1^{-1}$. Hence (similar to the first example), a sphere with center $E$ is the class of adjoint unitary matrices, and the complex radius of this sphere is the collection of eigenvalues of a matrix, considered up to permutations. The rank of the space is equal to $n - 1$.

3. The $n$-dimensional sphere. The group of rigid motions is the group of all proper orthogonal transformations of $n + 1$-dimensional space, $O(n + 1)$. The stationary subgroup is the subgroup leaving invariant a point of the sphere and, evidently, the vector connecting the center of the sphere with this point. This is consequently the group of all proper orthogonal transformations of $n$-dimensional space, $O(n)$. The automorphism describing the stationary subgroup is given by the matrix

$$I = \begin{pmatrix} -1 & & & & 0 \\ & 1 & & & \\ & & \cdot & & \\ & & & \cdot & \\ & & & & \cdot \\ 0 & & & & 1 \end{pmatrix}$$

and acts on elements of the group according to the formula $g^* = IgI^{-1}$. A sphere with center at the invariant point we have called a surface of transitivity for the stationary subgroup. In the given case this is evidently an ordinary $(n - 1)$-dimensional sphere. There is only one invariant of pairs of points on the sphere – the distance between them in the sense of spherical geometry. Hence the rank of the sphere is equal to $1$.

4. $n$-dimensional Lobačevskian space. The group of rigid motions is the group of all rigid motions of Lobačevskian space, which is isomorphic to the group of all proper transformations of $(n+1)$-dimensional space preserving the quadratic form $-x_1^2+x_2^2+\cdots+x_{n+1}^2$. The stationary subgroup is the subgroup leaving invariant a point of the surface $-x_1^2+x_2^2+\cdots+x_{n+1}^2=r^2$: this is again $O(n)$.[1] The involutory automorphism is given by the same matrix and the same formula as in the preceding case.

As in the preceding case, a "sphere" with center at the invariant point is an ordinary $(n-1)$-dimensional sphere (only this time lying on the surface $-x_1^2+x_2^2+\cdots+x_{n+1}^2=r^2$), and the complex distance becomes the ordinary distance in Lobačevskian space. The rank of the space is equal to $1$.

5. Euclidean space with the usual group of rigid motions in it, consisting of parallel translations and rotations around a fixed point. A sphere is an ordinary sphere, and the involutory automorphism describing a stationary subgroup is reflection in a point. The distance is the ordinary distance, and the rank of the space is equal to $1$.

We now make a remark that is important for the sequel. As is known, there exists a theorem stating that a Hermitian matrix can be reduced to diagonal form by the help of a unitary matrix, that is, for every Hermitian matrix $h$, there exists a unitary matrix $u$, such that $uhu^{-1}=\epsilon$, where $\epsilon$ is a diagonal matrix, defined uniquely except for the order of its eigenvalues. On the other hand, it is known that every unimodular matrix can be uniquely represented in the form $g=u\cdot h$, where $u$, $h$ are respectively a unitary and a Hermitian positive definite matrix. Thus we find that every unimodular matrix can be represented in the form

$$g=u_1\epsilon u_2,$$

where $u_1$, $u_2$ are unitary matrices and $\epsilon$ is a diagonal matrix with positive eigenvalues, defined uniquely except for permutations. It is of great significance that $\int_G f(g)\,dg$, taken with respect to the invariant measure, can be rewritten as follows:

$$\int_G f(g)\,dg=\int f(u_1\epsilon u_2)\,du_1\,du_2\,d\mu\epsilon, \tag{1.4}$$

where $du_1$, $du_2$ are the differentials of the invariant measure on the group of unitary matrices and $d\mu\epsilon$ is the differential of a certain measure on the group of diagonal matrices with positive eigenvalues.

Analogous facts hold in the general case. We consider a maximal commutative subgroup $E$ contained in $\Theta$. (In the example just given, this is the group of all diagonal matrices with positive eigenvalues; in the case of the manifold of unitary

1) The geometry arising on the surface $-x_1^2+x_2^2+\cdots+x_{n+1}^2=r^2$ as a result of the rigid motions cited in the text is the well-known model of $n$-dimensional Lobačevskian geometry.

matrices, this is the group of diagonal unitary matrices.) It turns out that every element of the group $G$ can be written in the form

$$g = u_1 \epsilon u_2, \tag{1.5}$$

where $u_1$, $u_2$ are elements of the stationary subgroup $G_0$ and $\epsilon \in E$. Here the element $\epsilon$ is defined by formula (1.5) uniquely, up to automorphisms of the group $E$, which can be given by the formula $\epsilon \to u \epsilon u^{-1}$. Automorphisms of this form comprise a finite group, the group $S$. In the case of Hermitian positive definite matrices, and also in the case of unitary matrices, the group $S$ is the group of all possible permutations of the roots of the matrix $\epsilon$.

In correspondence with the decomposition (1.5), the integral over the group of rigid motions can be written in the form of formula (1.4), also in the general case.

If $\alpha$ designates a root of the subalgebra corresponding to the group $E$, then the measure $d\mu\epsilon = \prod_{\alpha>0} \sin \frac{(\alpha, t)}{2} dt$ in the case of a compact symmetric space, and $d\mu\epsilon = \prod_{\alpha>0} \operatorname{sh} \frac{(\alpha, t)}{2} dt$ 1) in the case of a non-compact symmetric space. $dt$ is the differential of the invariant measure of the group $E$. (If $(t_1, \cdots, t_n)$ are canonical parameters in $E$, then $dt = c\,dt_1 \cdots dt_n$, where $c$ is a normalizing constant.)

In the case of the space of Hermitian positive definite matrices, we have

$$d\mu\epsilon = \prod_{p<q} (e^{t_p} - e^{t_q})^2 dt_1 \cdots dt_{n-1}, \tag{1.6}$$

where $\epsilon_p = e^{t_p}$ is a diagonal element of the matrix $\epsilon$. In the case of the manifold of unitary matrices, we have

$$d\mu\epsilon = \prod_{p<q} (e^{it_p} - e^{it_q})^2 dt_1 \cdots dt_{n-1}, \tag{1.6'}$$

where $\epsilon_p = e^{it_p}$ is a diagonal element of the matrix $\epsilon$.

Formulas (1.6) and (1.6') are very similar to each other. This circumstance is not accidental; it is explained by the fact that the symmetric spaces under consideration are dual to each other. Another example of a pair of dual symmetric spaces are the $n$-dimensional sphere and $n$-dimensional Lobačevskian space (examples 3 and 4).

In general, if we consider a symmetric space $M = G/G_0$ with a semisimple group of rigid motions and such that it does not decompose into the topological product of other symmetric spaces, then it turns out that there also exists exactly one manifold $M = G'/G_0$ of the same type as $M$, and with stationary subgroup isomorphic to the stationary subgroup of the manifold $M$ (see for example, [1]). Such manifolds will be called dual in the sense of Cartan. Besides isomorphic stationary subgroups, dual manifolds have the same dimension and equal rank. Further-

1) We suppose that each root in these formulas occurs a number of times equal to its multiplicity. We suppose that our symmetric spaces cannot be decomposed into the direct product of symmetric spaces.

more, one of them is always compact and its group of rigid motions is compact, while the other is homeomorphic to Euclidean space. Not only formulas (1.6), but in general all formulas relating to spherical functions on dual spaces, are extremely similar to each other.

The group of rigid motions of Euclidean space (example 5) is not semisimple, and it has no dual.

The following important concept is the concept of a group ring.

**no. 3. The group ring.** We consider the set $\widetilde{R}$ of summable functions on the group $G$ and introduce in $\widetilde{R}$ the operation

$$f_1 f_2(g) = \int f_1(g g_1^{-1}) f_2(g_1)\, dg_1. \tag{1.7}$$

The integral on the right side is taken with respect to the invariant measure in the group (it always exists in the cases that interest us) and is extended over the entire group manifold.

The set of functions constant on two-sided cosets with respect to a certain subgroup $G_0$ forms, as one can easily see, a subring $R$ of this ring.

The rings $\widetilde{R}$ and $R$ play a fundamental rôle in the theory of representations.

If the group $G_0$ is described by an involutory automorphism, that is, if $G/G_0$ is a symmetric space, the ring $R$ and the family of spherical functions belonging to an irreducible representation possess closely connected properties [4], [5], [7].

1) *The ring $R$ is commutative.*

2) *The family of spherical functions contains only one (up to coefficients) zonal function.*

In this case, the following important theorem [5] also holds: *the mapping of the ring $R$ into the field of complex numbers, defined by the formula*

$$f(g) \to \int f(g)\phi(g)\, dg,$$

*where $\phi(g)$ is a zonal spherical function belonging to a certain irreducible representation, is a homomorphism, and all homomorphisms of this ring into the field of complex numbers can be obtained in this way.*

**no. 4. The content of the present work.** In §2, we consider symmetric spaces that are group manifolds of compact semisimple Lie groups or are dual to such spaces (they can always be represented as manifolds of cosets $G/G_0$, where $G$ is the corresponding complex group and $G_0$ is a maximal compact subgroup of $G$).

For these manifolds we study the law of multiplication in the ring $R$ of functions that are constant on two-sided cosets with respect to the stationary subgroup.

It was shown in the work [5] that if the manifold $G/G_0$ is symmetric, then multiplication in the ring $R$ can be given not only by the general formula $\int f_1(g g_1^{-1}) f_2(g_1)\, dg_1$, but also by the more convenient formula

$$f_1 \cdot f_2(t) = \int a(t;\, t^{(1)},\, t^{(2)})\, f_1(t^{(1)})\, f_2(t^{(2)})\, dt^{(1)}\, dt^{(2)}, \tag{1.8}$$

where the integration is carried out over the manifold of two-sided cosets ($t$ is a point of this manifold).

In §2 we give the differential equation of the function $a$ for the case of the symmetric spaces described above.

As an inference from the considerations of this section, we obtain a theorem concerning eigenvalues of the product of Hermitian positive definite matrices (this theorem was proved earlier, using a suggestion of I. M. Gel'fand, by Lidskiĭ [6], using a different method).

In §3, we introduce the functional equation for zonal spherical functions, which has the form

$$\Phi_\rho(t^{(1)}) \cdot \Phi_\rho(t^{(2)}) = \frac{1}{\mu(S_{t^{(1)},\,t^{(2)}})} \int_{x \in S_{t^{(1)},\,t^{(2)}}} \Phi_\rho(x)\, d\mu x, \tag{1.9}$$

where the integral standing on the right side is taken over the "sphere" $S_{t^{(1)},\,t^{(2)}}$ with complex radius $t^{(1)}$ and center at complex distance $t^{(2)}$ from the initial point. $\mu(S_{t^{(1)},\,t^{(2)}})$ is the volume of this sphere.

It is shown in the work [5] that zonal spherical functions satisfy the relation

$$\Phi_\rho(t^{(1)}) \cdot \Phi_\rho(t^{(2)}) = \int a(t;\, t^{(1)},\, t^{(2)})\, \Phi_\rho(t)\, dt, \tag{1.10}$$

where the function $a(t;\, t^{(1)},\, t^{(2)})$ is the same as in formula (1.8).

Combination of (1.9) and (1.10) leads to the following result:

$$\int a(t;\, t^{(1)},\, t^{(2)})\, f(t)\, dt$$

is the mean value of the function $f(t)$ over the sphere with complex radius $t^{(1)}$ and center at complex distance $t^{(2)}$ from the initial point.

Using this property of the function $a$, we compute it for the case where the space is the $n$-dimensional sphere.

The functional equation (1.9) is found first in the work [7] of Godement, although its geometric meaning is not pointed out there.

In §4, we prove the "theorem about the mean" from the functional equation (1.9), which is a generalization of the known theorem of Darboux. On a symmetric space of rank $n$, there are $n$ independent "operators of Laplace" $\Delta_x^{(k)}$, remarkable for the fact that a zonal spherical function is an eigenfunction for each of them (the index $x$ shows that the operator is applied to functions depending on the variable $x$). On the space of functions that depend only on the complex distance $t$ from a variable point to the initial point, the operator $\Delta_x^{(k)}$ turns into the operator $\Delta_t^{(k)}$, which it is natural to call the "radial part" of the operator $\Delta^{(k)}$.

If we denote by $\psi(t, x)$ the mean value of the function $f$ over the sphere of radius $t$ with center at the point $x$, then we have the formula ("theorem of the mean")

$$\Delta_x^{(k)} \psi(t, x) = \Delta_t^{(k)} \psi(t, x). \tag{1.11}$$

One can show that under certain natural additional assumptions, a zonal spherical function is the only eigenfunction for the Laplace operator. Hence it is a function of its eigenvalues: $\Phi_\rho(t) = \Phi(t, \lambda^{(1)}, \ldots, \lambda^{(n)})$ ($\lambda^{(k)}$ is an eigenvalue of the operator $\Delta^{(k)}$). As is shown in the works [5] and [13], the operators $\Delta^{(k)}$ are permutable among themselves. Furthermore, they can be chosen to be self-adjoint. Hence the operator $\Phi(t, \Delta_x^{(1)}, \ldots, \Delta_x^{(n)})$ has a meaning. In §4 we prove the following theorem. The operator $\Phi(t, \Delta_x^{(1)}, \ldots, \Delta_x^{(n)})$ is the averaging operator over the sphere with center at the point $x$ and complex radius $t$. In the case where the symmetric space is the Euclidean plane with the usual rigid motions in it, the equations (1.11) become a single one:

$$\left(\frac{\partial^2}{\partial x_1^2} + \frac{\partial^2}{\partial x_2^2}\right)\psi(r; x_1, x_2) = \frac{1}{r}\frac{d}{dr}\left(r\frac{d}{dr}\right)\psi(r; x_1, x_2),$$

which is Darboux's equation [18]. The second theorem has the following form in this case: $J_0(r\sqrt{-\Delta})$, where $J_0(x)$ is the Bessel function of order zero, is the averaging operator on the circumference of radius $r$. For the case of spaces with constant curvature, a result analogous to ours was obtained by Olevskiĭ [12].

In §5, we consider a ring closely connected with the algebra of representations of a compact semisimple Lie group. The elements of the algebra of representations are different representations of the group $G$, the sum of representations is their direct sum,1) and the product of representations is the Kronecker product.2)

The elements of the ring $\Xi$ under study are additive functions $\xi$ of representations, in which the law of multiplication is defined as follows:

$$\xi_1 \cdot \xi_2(\psi) = \sum_{\psi_1} \xi_1(\hat{\psi}_1 \times \psi)\,\xi_2(\psi_1). \tag{1.12}$$

The sum is extended over all irreducible representations, and $\hat{\psi}$ is the representation contragredient to $\psi$.3)

The ring $\Xi$ is constructed in fairly close analogy to the ring $R$ considered

---

1) The sum of representations $\phi$ and $\psi$, acting in spaces $R_\phi$ and $R_\psi$ respectively, is the representation $\phi + \psi$, which acts in the space $R_{\phi+\psi} = R_\phi \dot{+} R_\psi$ according to the formula $T_g^{(\phi+\psi)}(\xi_\phi + \xi_\psi) = T_g^{(\phi)}\xi_\phi + T_g^{(\psi)}\xi_\psi$.

2) The Kronecker product of representations $\phi$ and $\psi$ is the representation $\phi \times \psi$ acting in the space $R_{\phi\times\psi} = R_\phi \times R_\psi$ of formal bilinear forms $\eta = \sum a_{\xi_\phi \xi_\psi} \xi_\phi \times \xi_\psi$, where $\xi_\phi \in R_\phi$ and $\xi_\psi \in R_\psi$, according to the formula

$$T_\eta^{(\phi\times\psi)} = \sum a_{\xi_\phi \xi_\psi} T_g^\phi \xi_\phi \times T_g^\psi \xi_\psi.$$

3) The representation $\psi_1$ is called contragredient to $\psi$ (and is written $\hat{\psi}$) if the operators of this representation are written in a certain basis by matrices $A'^{-1}(g)$, where $A(g)$ are the matrices of operators of the representation $\psi$ ($'$ denotes the transpose). Evidently $\hat{\hat{\psi}} = \psi$.

in §2. In particular, the law of multiplication in $\Xi$ can be given not only by formula (1.12) but also by a formula analogous to (1.8):

$$\xi_1 \cdot \xi_2(\psi) = \sum_{\psi_1, \psi_2} \xi_1(\psi_1)\, \xi_2(\psi_2)\, G(\psi_1, \psi_2; \psi),$$

where the sum is extended over all pairs of irreducible representations.

The function $G(\psi_1, \psi_2; \psi)$ for irreducible $\psi_1, \psi_2, \psi$ is equal to the multiplicity with which the representation $\psi$ enters in the decomposition of the Kronecker product $\psi_1 \times \psi_2$. For the function $G(\psi_1, \psi_2; \psi)$ we establish a finite difference equation analogous to the differential equation satisfied by the function $a(t; t^{(1)}, t^{(2)})$, defining the law of multiplication in $R$. As is shown in §5, there exists a deep duality between the function $a$ giving the law of multiplication in the center of the group ring and the function $G$ giving multiplication of representations.

As will be shown in another place, an analogous duality exists between matrix elements of a matrix of an irreducible representation of the group $SU(2)$ (if the matrix is written in the weight basis) and the so-called "coefficients of Clebsch-Gordan". (The coefficients of Clebsch-Gordan give the decomposition of the Kronecker product of the spaces $R_\phi$ and $R_\psi$, in which the irreducible representations $\phi$ and $\psi$ act, into irreducible subspaces: if $\xi_\phi^{(k)} \in R_\phi$ is a vector of weight $k$ and $\eta_\psi^{(l)} \in R_\psi$ is a vector of weight $l$, then $\zeta_\pi^{(m)} = \sum_{k+l=m} C(\phi, \psi, \pi; k, l, m)\, \xi_\phi^{(k)} \times \eta_\psi^{(l)} \in R_\pi$ is a vector of weight $m$ belonging to the irreducible subspace $R_\pi$ of the space $R_\phi \times R_\psi$. The coefficients $C(\phi, \psi, \pi; k, l, m)$ are called the Clebsch-Gordan coefficients.)

Another example of such a duality are the formulas of Gel'fand and Cetlin for matrix elements of irreducible representations of the algebra of complex matrices with trace $0$ [14] and the formulas for co-ordinates in the group of unitary matrices [9], pp. 179-181. In all of these cases the duality consists in the fact that functions of discrete arguments satisfy finite difference equations analogous to differential equations satisfied by functions of real variables that correspond to them.

## §2. The law of multiplication in the center of the group ring of a compact semisimple group and in rings connected with certain other manifolds

In this section we consider the case where the symmetric space is the group manifold of a compact semisimple Lie group or the manifold $G/U$ of cosets of a complex semisimple group $G$ with respect to a maximal compact subgroup $U$. At first we consider the case where the group is the group $SU(n)$ of unitary matrices of order $n$ with determinant $1$, and then we take up the general case.

**no. 5. Characters as spherical functions.** We consider the manifold of the compact group $G$ (its elements will be denoted by the letter $g$), and define rigid motions in it with the aid of the formula

$$g \to g_1^{-1} g g_2.$$

Then the manifold of the group $G$ turns into a symmetric space (see no. 2) with the group of rigid motions $\widetilde{G} = G \times G$, the elements of which are the pairs $\widetilde{g} = (g_1, g_2)$. The stationary subgroup of the point $e$ is obviously the group $G_0$ acting on points of the manifold according to the formula

$$g \to g_1^{-1} g g_1.$$

Elements of the group $G_0$ are thus the pairs $(g, g)$.

In agreement with definition (see no. 1), spherical functions on the manifold $G = \widetilde{G}/G_0$ are functions $(\eta, T_{\widetilde{g}}^{(\phi)} \xi)$, where $\phi$ is an irreducible unitary representation of the group $\widetilde{G}$ in a Hilbert space $H_\phi$, where the vectors $\xi, \eta \in H_\phi$, and $\xi$ is invariant with respect to the stationary subgroup, that is, $T_{\widetilde{g}}^{(\phi)} \xi = \xi$ if $\widetilde{g} \in G_0$.

In order to compute the spherical functions, it is therefore necessary to find representations of the group $\widetilde{G}$ in the space of which there exists a vector invariant with respect to $G_0$.

This is done in the following fashion. We consider two irreducible representations $\phi$ and $\psi$ of the group $G$. Let their dimensions be $m$ and $n$ respectively. We consider further the space of $m \times n$ rectangular matrices and define in it a representation of the group $\widetilde{G}$ by the formula

$$T_{\widetilde{g}} C = T_{(g_1, g_2)} C = A_{g_1^{-1}}^{(\phi)} C A_{g_2}^{(\psi)}, \tag{2.1}$$

where $A_g^{(\phi)}$ and $A_g^{(\psi)}$ are the matrices of the operators $T_g^{(\phi)}$ and $T_g^{(\psi)}$, written in some basis. Since the group $G$ is compact, bases can be chosen so that these matrices are unitary. Then the formula

$$(C, C) = \operatorname{sp} C \cdot \overline{C'}$$

defines a scalar product in the space of matrices $C$ which is invariant with respect to the operators $T_{(g_1, g_2)}$.

It is not hard to show that the representation just constructed for the group $\widetilde{G}$ is irreducible. In the theory of representations it is proved (see for example, [8]) that every irreducible representation of the group $\widetilde{G} = G \times G$ can be constructed in this way. It follows from (2.1) that the operators from $G_0$ act on vectors of the space according to the formula

$$T_{(g, g)} C = A_{g^{-1}}^{(\phi)} C A_g^{(\psi)}.$$

Hence an invariant vector is a matrix $C$ such that

$$A_{g^{-1}}^{(\phi)} C A_g^{(\psi)} = C. \tag{2.2}$$

In agreement with Schur's lemma, the last equality can be satisfied only when the representations $\phi$ and $\psi$ are equivalent. In this case, we can take the matrices $A_g^{(\phi)}$ and $A_g^{(\psi)}$ as being equal, and then it follows from (2.2) that $C = \lambda E$. Thus, in the case under consideration, there exists only one zonal spherical function (see no. 3) $(\xi, T_{\widetilde{g}}\xi)$, where $\xi = \lambda E$. Demanding that $\|\xi\| = 1$, we obtain finally that the zonal function is equal to $\frac{1}{\dim \psi} \operatorname{sp} T^{(\psi)}_{g_1^{-1} g_2}$. In order to transfer this function from the group $\widetilde{G} = G \times G$ to the manifold $G$, which is the manifold of cosets $\widetilde{G}/G_0$, we choose a representative $(g, e)$ in each coset. Thus we obtain the following assertion.

*On the manifold of a compact group $G$, regarded as a homogeneous space under the motions $g \to g_1^{-1} g g_2$, the zonal spherical functions are functions of the form*

$$\Phi_\psi(g) = \frac{1}{\dim \psi} \operatorname{sp} T_g^{(\psi)},$$

*where $T_g^{(\psi)}$ is an operator of the irreducible representation $\psi$, and* $\dim \psi = \operatorname{sp} T_e^{(\psi)}$ *is the dimension of the representation $\psi$.*

The function $\chi_\psi(g) = \operatorname{sp} T_g^{(\psi)}$ is called the character of the representation $\psi$ and for semisimple groups has been computed by H. Weyl [8]. Using Weyl's formulas for $\chi_\psi(g)$ and for $\dim \psi$, we find that in the case of the unitary group the zonal spherical function is equal to

$$\Phi_\psi(g) = \frac{1!\cdots(n-1)! \begin{vmatrix} \epsilon_1^{\rho_1} & \cdots & \epsilon_n^{\rho_1} \\ \cdots & \cdots & \cdots \\ \epsilon_1^{\rho_n} & \cdots & \epsilon_n^{\rho_n} \end{vmatrix}}{\prod_{i<k} (\rho_i - \rho_k) \prod_{i<k} (\epsilon_i - \epsilon_k)}, \tag{2.3}$$

where $\epsilon_1 = e^{it_1}, \ldots, \epsilon_n = e^{it_n}$ are the eigenvalues of the matrix $g$, $\rho_1 > \rho_2 > \cdots > \rho_n$ are integers determining the representation $\psi$.

We remark that the function so obtained is closely connected with the spherical function computed in [9] for the case where the group $G$ is the group of all complex matrices with determinant $1$, and $G_0$ is the subgroup of unitary matrices. (The symmetric space in this case can be interpreted as the set of Hermitian positive definite matrices.) In fact, the function $\phi_\rho(t)$ computed there is equal to $\Phi_{-i\rho}(it)$. This relation is connected with the circumstance that the symmetric spaces mentioned are dual to each other.[1)]

**no. 6. The ring $R$ of functions that are constant on two-sided cosets of the group $\widetilde{G} = G \times G$ with respect to the subgroup $G_0$.**[2)] A two-sided coset with re-

1) In order for the formula $\phi_\rho(t) = \Phi_{-i\rho}(it)$ to be valid, we must replace the function $\phi_\rho(t_1, \ldots, t_n)$ computed in [9] by $\phi_\rho(t) = \phi_\rho(\frac{t_1}{2}, \ldots, \frac{t_n}{2})$.

2) As is shown a few lines further on, this ring is isomorphic to the ring of functions on the group $G$ having the property that $f(g_1^{-1} g g_1) = f(g)$. It is easy to show that in view of this property, the equality $f \cdot f_1(g) = f_1 \cdot f(g)$ holds for an arbitrary function $f_1(g)$ on the group $G$, where $\cdot$ denotes multiplication in the group ring, i.e., the convolution (1.7). Hence the ring $R$ is the center of the group ring of the group $G$.

spect to the subgroup $G_0$ consists of all elements of the group $\tilde{G}$ of the form $\tilde{g}_1 \tilde{\tilde{g}} \tilde{g}_2$, where $\tilde{g}_1$, $\tilde{g}_2$ run independently through $G_0$. Hence, on the manifold of cosets $\tilde{G}/G_0$, there corresponds to a two-sided coset a submanifold on which the group $G_0$ acts transitively. Functions that are constant on two-sided cosets are thus functions on the manifold $\tilde{G}/G_0$ that are constant on these submanifolds.

As already remarked, elements of the subgroup $G_0$ act on the manifold $\tilde{G} = G/G_0$ in accordance with the formula $g \to g_1^{-1} g g_1$. Hence functions that are constant on manifolds transitive with respect to $G_0$ are functions characterized by the fact that

$$f(g) = f(g_1^{-1} g g_1). \tag{2.4}$$

In the case where $G$ is the unitary unimodular group, an arbitrary matrix $g$ can be transformed into a diagonal matrix $\epsilon$ by means of transformations of the form $g \to g_1 g g_1^{-1}$, where the eigenvalues of $\epsilon$ are determined up to a permutation. Hence condition (2.4) means that $f(g)$ is a symmetric function of the roots $\epsilon_p = e^{it_p}$ of the matrix $g$ or of their logarithms $t_p$. In the last case, it is evidently necessary also to demand that the function be periodic (with period $2\pi$) in each argument. Furthermore, the condition of unimodularity implies that the variables $t_p$ are connected by the relation

$$t_1 + \cdots + t_n = 0. \tag{2.5}$$

We thus obtain that *the ring $R$ is the ring of symmetric functions, periodic in each argument, defined for real values $t_1, \cdots, t_n$ that satisfy condition* (2.5).

As has already been recalled (no. 3), the homomorphisms of $R$ into the field of complex numbers are given by the formula

$$f \to \int_G f(g) \Phi_\rho(g)\, dg, \tag{2.6$'$}$$

where $\Phi_\rho(g)$ is a zonal spherical function. We require its precise form in order to study the law of multiplication in $R$.

In view of the fact that the integrated function in (2.6′) depends only on the eigenvalues $\epsilon_p = e^{it_p}$ of the matrix $g$, this formula can be rewritten in the form[1)]

$$f \to \int_D f(t) \Phi_\rho(t)\, \omega(t)\, dt, \tag{2.6}$$

where $D$ is the region $-\pi \le t_k \le \pi$ $(t_1 + \cdots + t_n = 0)$ $\omega(t) = \prod_{p<q} (e^{it_p} - e^{it_q})^2$, $dt = c\,dt_1 \cdots dt_{n-1}$ ($c$ is an appropriately chosen constant).

Along with the ring $R$, we consider the two following auxiliary rings: the ring $R'$, consisting of the skew symmetric functions of $t_1, \cdots, t_n$ that satisfy the other properties of elements of the ring $R$, and the ring $R''$, consisting of all symmetric functions satisfying the same conditions.

---

1) See [8].

We give the homomorphisms of $R'$ into the field of complex numbers by the formula

$$x(t) \to \int_D x(t)\, \psi_\rho(t)\, d\mu t, \tag{2.7}$$

where

$$\psi_\rho(t) = \Phi_\rho(t) \prod_{p<q} (e^{it_p} - e^{it_q}) = \frac{1! \cdots (n-1)!}{\Pi(\rho_p - \rho_q)} \begin{vmatrix} e^{it_1\rho_1} \dots e^{it_n\rho_1} \\ \dots\dots\dots\dots \\ e^{it_1\rho_n} \dots e^{it_n\rho_n} \end{vmatrix}. \tag{2.8}$$

The rings $R$ and $R'$ are isomorphic to each other, the isomorphism being established with the aid of the operator of multiplication by the function [1] $\Delta = \Pi(e^{it_p} - e^{it_q})$: if $f(t) \in R$, then the corresponding element of $R'$ is equal to

$$x(t) = \Delta(t)\, f(t).$$

We introduce multiplication in the ring $R''$ in the following way: for $z_1, z_2 \in R''$,

$$z_1 \cdot z_2(t) = \int_D z_1(t_1 - \tau_1, \dots, t_n - \tau_n)\, z_r(\tau_1 \dots \tau_n)\, d\tau. \tag{2.9}$$

Homomorphisms of this ring into the field of complex numbers are given by the formula

$$z \to \frac{1}{n!} \int_D z(t_1, \dots, t_n) \sum_s e^{i(t_1\rho_{k_1} + \dots + t_n\rho_{k_n})}, \tag{2.10}$$

where the sum is extended over all possible permutations of the indices $s = \begin{pmatrix} 1 & \dots & n \\ k_1 & \dots & k_n \end{pmatrix}$ and $\rho_1, \dots, \rho_n$ are integers.

The rings $R'$ and $R''$ are isomorphic to each other, the isomorphism being realized with the help of the operator [2]

$$L = L_t = \frac{1}{1! \cdots (n-1)!} \prod_{i<j} \left(\frac{\partial}{\partial t_i} - \frac{\partial}{\partial t_j}\right),$$

which maps the ring $R''$ onto the ring $R'$.[3]

The operator $L$ has the two following important properties.

1. *It carries a symmetric function into a skew symmetric function, and conversely.*

2. *If a function $x$ is defined and skew symmetric in a symmetric convex region $D$, and satisfies the equation $Lx = 0$ in this region, then $x \equiv 0$.* The first of these properties is evident, and the second is proved in the following way. The general solution of the equation $Lf = 0$ has the form [4]

1) A proof of this fact is found in [4].

2) As has been said, the variables $t_1, \dots, t_n$ are not independent. However, it will be more convenient for us right now to suppose them independent, and all functions extended in any fashion with preservation of periodicity and the condition of symmetry from the place $t_1 + \dots + t_n = 0$ over the entire space.

3) A proof is found in [4].

4) The operator $L$ is hyperbolic. The derivation of its general solution differs in no

$$f = f_{12}(t_1 + t_2, t_3, \cdots, t_n) + f_{13}(t_1 + t_3, t_2, t_4, \cdots, t_n) + \cdots + f_{n-1n}(t_{n-1} + t_n, t_1, \cdots, t_{n-2}).$$

If the function $f$ is skew symmetric, then carrying out alternations on it, we obtain the function once again. Carrying out alternations on the function $f_{ij}(t_i + t_j, \cdots)$, we obtain $0$, since it is constant on each connected component of the intersection of the line $t_i + t_j = c$, $t_1 = c_1, \cdots$ with the region $D$, and in view of the convexity of $D$, this intersection is connected.

**no. 7. The law of multiplication in** $R$. We consider the equation

$$L_\tau \tilde{G}(t, \tau) = \frac{1}{n!} \sum_s \delta(t_1 - \tau_{i_1}, \cdots, t_n - \tau_{i_n}). \tag{2.11}$$

[The sum is extended over all permutations $s = \begin{pmatrix} 1 & \cdots & n \\ i_1 & \cdots & i_n \end{pmatrix}$, $\delta(x_1, \cdots, x_n)$ is the $n$-dimensional "$\delta$-function" on the torus $(e^{it_1} \ldots e^{it_n})$.[1])]

As was shown at the end of the preceding number, if the function $x$ is skew symmetric in a convex region and satisfies there the equation $Lx = 0$, then $x \equiv 0$. Hence the equation has a unique solution in the class of functions skew symmetric with respect to $\tau_1, \cdots, \tau_n$. Since the right part of the equation is symmetric in $t_1, \cdots, t_n$ and the operator $L_\tau$ does not affect these arguments, the function $\tilde{G}(t, \tau)$ is symmetric in $t_1, \cdots, t_n$.

We also note that in view of the periodicity of the right side of the equation (2.11), the function $\tilde{G}(t, \tau)$ is also periodic in each of the arguments $(t_i, \tau_j)$.

In the sequel we shall show that the function $\tilde{G}(t, \tau)$, in addition to the equation (2.11), also satisfies the equation

$$L_t \tilde{G}(t, \tau) = \frac{1}{n!} \sum_s \pm \delta(t_1 - \tau_{i_1}, \cdots, t_n - \tau_{i_n}) \tag{2.11'}$$

(+ or − is written as the permutation $s$ is even or odd, and $\delta(x_1, \cdots, x_n)$ is the "$\delta$-function" on the torus $e^{it_1}, \cdots, e^{it_n}$) and that it can be written in the form

$$\tilde{G} = \delta(t_1 + \cdots + t_n - \tau_1 - \cdots - \tau_n) G(t, \tau).$$

Therefore the operator inverse to $L$, if this last is considered only on the symmetric and skew symmetric functions defined for $t_1 + \cdots + t_n = 0$, exists and is given by the Green's function $G(t, \tau)$.

We return to the law of multiplication in the rings $R'$ and $S$. We apply $L$ to both sides of the equality (2.9):

$$Lz = \int_D Lz_1(t - \tau) z_2(\tau) d\tau.$$

---

essential way from the derivation of the general solution of the equation of the vibrating string

$$0 = \frac{\partial^2 f}{\partial t^2} - \frac{\partial^2 f}{\partial x^2} = \left(\frac{\partial}{\partial t} - \frac{\partial}{\partial x}\right)\left(\frac{\partial}{\delta t} + \frac{\partial}{\partial x}\right) f.$$

1) In other words,

$$L\tilde{G} = \sum_{k_1 \cdots k_n} \sum_s \delta(t_1 - \tau_{i_1} + 2k_1\pi, \cdots, t_n - \tau_{i_n} + 2k_n\pi).$$

$Lz$ and $Lz_1$ belong to $R'$. Recalling that the operator $L$ maps the ring $R''$ isomorphically onto the ring $R'$ and that the inverse mapping $L^{-1}$ exists, we can write the last equality in the form

$$x(t) = \int_D x_1(t-\tau) L^{-1} x_2(\tau)\, d\tau.$$

Using the fact that the operator $L^{-1}$ is given by the Green's function $G(t, \tau)$, we obtain:

$$x_1 \cdot x_2(t) = \int_D \int_D G(\tau, \tau_1) x_1(t-\tau_1) x_2(\tau_1)\, d\tau\, d\tau_1.$$

Using the periodicity of all of the functions appearing in the integrand, we can write the last equality in the form

$$x_1 \cdot x_2(t) = \int_D \int_D G(t-\tau_1, \tau_2) x_1(\tau_1) x_2(\tau_2)\, d\tau_1\, d\tau_2. \tag{2.12}$$

Thus we obtain the final law of multiplication in the ring $R'$.

We alternate the function $G(t-\tau_1, \tau_2)$ in each argument; the function obtained in this way, we denote by $\widetilde{a}(t; \tau_1, \tau_2)$. Evidently the integral standing on the right side of the equality (2.12) does not change if we replace the function $G(t-\tau_1, \tau_2)$ in it by $\widetilde{a}(t; \tau_1, \tau_2)$.

Using the fact that the operator of multiplication by $\Delta(t) = \prod_{p<q} (e^{it_p} - e^{it_q})$ maps the ring $R$ isomorphically onto $R'$, we obtain the following assertion.

*The operation of multiplication in $R$ is given with the aid of the function*

$$a(t; \tau_1, \tau_2) = \frac{\Delta(\tau_1)\Delta(\tau_2)}{\Delta(t)} \widetilde{a}(t, \tau_1, \tau_2)$$

*according to the formula*

$$f_1 \cdot f_2(t) = \int_D \int_D a(t; \tau_1, \tau_2) f_1(\tau_1) f_2(\tau_2)\, d\tau_1\, d\tau_2.$$

**no. 8. The connection between the functions $\psi_\rho(t)$[1] and $G(t, \tau)$.** We have seen above that the function $\widetilde{G}$ is the unique solution of the equation

$$L_\tau \widetilde{G} = \frac{1}{n!} \sum_k \sum_s \delta(t_1 - \tau_{i_1} + 2k_1\pi, \ldots, t_n - \tau_{i_n} + 2k_n\pi), \tag{2.11''}$$

that is symmetric in the first and skew symmetric in the second argument.

In just the same way, it is easy to see that the function $\psi_\rho(t)$ is the unique solution of the equation

$$L_t \psi_\rho(t) = \frac{1}{n!} \sum_s e^{i(t_1\rho_{k_1} + \cdots + t_n\rho_{k_n})}, \tag{2.13}$$

that is symmetric in the argument $t$ and skew symmetric in $\rho$.

It follows from this that the decomposition of the function $\widetilde{G}$ in a Fourier series has the following form:

---

1) $\psi_\rho(t) = \dfrac{1!\,2!\cdots(n-1)!}{\prod_{i<k}(\rho_i - \rho_k)} \begin{vmatrix} \epsilon_1^{\rho_1} & \ldots & \epsilon_n^{\rho_1} \\ \epsilon_1^{\rho_n} & \ldots & \epsilon_n^{\rho_n} \end{vmatrix}$. We recall that this function gives homomorphisms of the ring $R'$ into the field of complex numbers.

$$\widetilde{G}(t, \tau) = \sum_{\rho} e^{-i(\rho_1 t_1 + \cdots + \rho_n t_n)} \psi_\rho(\tau). \tag{2.14}$$

In fact, one shows that the function defined by the last equality satisfies the equation (2.11) simply by applying the operator $L_\tau$ to both sides of the equality (2.14). Furthermore, the antisymmetry of the function $\psi_\rho(\tau)$ with respect to $\tau$ implies that $G(t, \tau)$ is antisymmetric with respect to $\tau$. The last condition, as we saw above, uniquely defines the solution of the equation (2.11).

We are now in a position to prove the formulas recalled earlier

$$L_t \widetilde{G}(t, \tau) = \frac{1}{n!} \sum_k \sum_s \pm \delta(t_1 - \tau_{i_1} + 2k_1\pi, \cdots, t_n - \tau_{i_n} + 2k_n\pi) \tag{2.11'''}$$

and

$$\widetilde{G}(t, \tau) = \delta(t_1 + \cdots + t_n - \tau_1 - \cdots - \tau_n) G(t, \tau).$$

The first of these formulas is proved by applying the operator $L_t$ to both sides of the equality (2.14):

$$L_t e^{i(t_1\rho_1 + \cdots + t_n\rho_n)} = \frac{\prod_{k<l} (\rho_k - \rho_l)}{1! \cdots (n-1)!} e^{i(\rho_1 t_1 + \cdots + \rho_n t_n)},$$

$$\psi_\rho(\tau) = \frac{1! \cdots (n-1)!}{\Pi(\rho_k - \rho_l)} \begin{vmatrix} e^{i\tau_1\rho_1} \cdots e^{i\tau_1\rho_n} \\ \cdots\cdots\cdots\cdots \\ e^{i\tau_n\rho_1} \cdots e^{i\tau_n\rho_n} \end{vmatrix}. \tag{2.8'}$$

Expanding the determinant in (2.8′) and summing, we obtain the required result.

To obtain the second formula, we write the equality (2.14) in somewhat altered form:

$$\widetilde{G}(t, \tau) = \sum_{\rho} e^{-i(\rho_1 t_1' + \cdots + \rho_n t_n')} \psi_\rho(\tau') e^{-i(\rho_1 + \cdots + \rho_n)(\tau'' - t'')}.$$

Here

$$t_1' + \cdots + t_n' = \tau_1' + \cdots + \tau_n' = 0, \quad t'' = t_1 + \cdots + t_n, \quad \tau'' = \tau_1 + \cdots + \tau_n.$$

(It is easy to see that the expression for $\widetilde{G}(t, \tau)$ is valid, starting from the explicit expression for $\psi_\rho(\tau)$.)

In the last equality, we first sum over $\rho_1 + \cdots + \rho_n$ and then over the sets of $\rho_1, \cdots, \rho_n$ no two of which differ by a common summand for $\rho_k$, and obtain in this way the needed formula.

We now express $\psi_\rho(t)$ in terms of $\widetilde{G}$. The formula (2.14) yields the formula

$$\psi_\rho(t) = \frac{1}{(2\pi)^{n-1}} \int_{-\frac{\pi}{2} < x_k < \frac{\pi}{2}} e^{i(x_1\rho_1 + \cdots + x_n\rho_n)} \widetilde{G}(x, t)\, dx_1 \cdots dx_n. \tag{2.14'}$$

Formulas (2.14) and (2.14′) give the connection between the function $\widetilde{G}(t, \tau)$ and the function $\psi_\rho(t)$, defined for all values of $\tau$.

Going from the functions $\widetilde{G}$ and $\psi_\rho(t)$, defined for all values of $t$, to the functions $G$ and $\psi_\rho(t)$, defined for $t_1 + \cdots + t_n = 0$, that is, to functions defined on $SU(n)$, we obtain finally:

*The functions* $\psi_\rho(t)$ *and* $G(t, \tau)$ *are connected with each other by the relations*

$$G(t, \tau) = \sum_\rho e^{-i(\rho_1 t_1 + \cdots + \rho_n t_n)} \psi_\rho(\tau), \tag{2.15}$$

$$\psi_\rho(t) = \frac{1}{(2\pi)^{n-1}} \int_D e^{i(\rho_1 x_1 + \cdots + \rho_n x_n)} G(x, t)\, dx. \tag{2.15'}$$

The sum in formula (2.15) is extended over all possible collections $\rho = (\rho_1, \cdots, \rho_n)$ no two of which differ by a common summand for $\rho_k$, the integral in formula (2.15') is taken over the region $D$, which is defined by the conditions $-\pi \leq x_k \leq \pi$, $x_1 + \cdots + x_n = 0$; $dx = c\,dx_1 \cdots dx_{n-1}$ ($c$ is a certain constant).

Thanks to the symmetry of the function $\psi_\rho(t)$ in $\rho$ and of the function $G(x, t)$ in $x$, the formula (2.15) and (2.15') can be re-written in an equivalent form:

$$G(t, \tau) = \sum_{\rho_1 > \rho_2 > \cdots > \rho_n} \left(\sum_s e^{i(t_1 \rho_{k_1} + \cdots + t_n \rho_{k_n})}\right) \psi_\rho(\tau), \tag{2.16}$$

$$\psi_\rho(t) = \frac{1}{(2\pi)^{n-1}} \int_{\substack{x_1 > x_2 > \cdots > x_n \\ x_1 + x_2 + \cdots + x_n = 0 \\ -\pi < x_k < \pi}} \left(\sum_s e^{i(x_1 \rho_{k_1} + \cdots + x_n \rho_{k_n})}\right) G(x, t)\, dx. \tag{2.16'}$$

The sum over $s$ in these equalities denotes, as always, the sum taken over all permutations of the indices $k_1, \cdots, k_n$.

We recall that a collection of integers $\rho_1 > \cdots > \rho_n$, defined up to a common summand for $\rho_k$, gives an irreducible representation of the group $G$ and at the same time a representation of the group $\tilde{G} = G \times G$, in which the "diagonal subgroup" $G_0$ has an invariant vector, that is, a representation realized in spherical functions on the manifold $G = \tilde{G}/G_0$.

Thus, *in the formula* (2.16), *the outer sum is extended over all irreducible representations of the group* $\tilde{G} = G \times G$ *realized in spherical functions, or, equivalently, over all irreducible representations of the group* $G$.

**no. 9. Generalization of the results of nos. 7 and 8 to the case of an arbitrary semisimple group.** The reasoning of nos. 7 and 8 can be carried over with no change whatsoever to the case of an arbitrary compact semisimple group. We shall formulate the principal concepts and results of these nos. in general terms.

The ring $R$ is isomorphic to the ring of functions on the Cartan subalgebra $H$ of the algebra $G$, symmetric with respect to the Weyl group $s$ and periodic with respect to the lattice $\Gamma$, consisting of the inverse images of the identity under the canonical mapping of the subalgebra $H$ onto the corresponding subgroup.[1] The ring $R'$ is the ring of functions on the algebra $H$ that are skew symmetric with respect to $S$ and periodic with respect to $\Gamma$.

---

1) Concerning the canonical mapping, see for example [10].

Finally, the ring $R''$ consists of the same functions as $R$, and the law of multiplication in $R''$ is given by the formula

$$z_1 \cdot z_2(t) = \int_D z_1(t - t_1) z_2(t_1) \, dt_1,$$

where the integral is taken over the fundamental parallelepiped of the lattice $\Gamma$.

*The operator $L$ is the product of the operators of differentiation in the direction of all positive roots $\alpha$*; the functions $G$, $\tilde{a}$ and $a$ are defined just as in no. 7. The final result of no. 7 is stated in the following way:

*The law of multiplication in the ring $R$ can be given by the formula*

$$f_1 \cdot f_2(t) = \int_D \int_D a(t; \tau_1, \tau_2) f_1(\tau_1) f_2(\tau_2) \, d\tau_1 \, d\tau_2. \tag{2.17}$$

*The function*

$$a(t; \tau_1, \tau_2) = \frac{\Delta(\tau_1)\Delta(\tau_2)}{\Delta(t)} \tilde{a}(t; \tau_1, \tau_2),$$

*where* $\Delta(t) = \prod_{\alpha<0} \sin \frac{(\alpha, t)}{2}$ *and the function $\tilde{a}(t; \tau_1, \tau_2)$ is obtained from the function $G(t - \tau_1, \tau_2)$ by alternation of the function $G(t - \tau_1, \tau_2)$ with respect to $t$, $\tau_1$, and $\tau_2$. The function $G(t, \tau)$ is a solution of the equation*

$$L_\tau G = \frac{1}{n(S)} \sum_s \delta(t - s\tau),$$

*skew symmetric in $\tau$ (in the last equality, $n(S)$ denotes the number of elements of the group $S$, $s\tau$ is the vector $\tau$, transformed with $s \in S$).*

The computations of no. 8 are also carried over almost without change to the general case, where by the function $\psi_\rho(t)$ (which gives homomorphisms of the ring $R'$ into the field of complex numbers), we must mean

$$\psi_\rho(t) = \frac{\xi(\rho, t)}{d_\Lambda},$$

where $\xi(\rho, t)$ is the "$\xi$-function" of H. Weyl, $\rho = \Lambda + g$ ($\Lambda$ is the highest weight of a representation of the group $G$, and $g$ is the partial sum of positive roots), $d_\Lambda$ is the dimension of the representation with highest weight $\Lambda$.

*The connection between the functions $\psi_\rho(t)$ and $G(t, \tau)$ is given in the general case by the formulas*

$$G(t, \tau) = \sum_\rho \sum_s e^{i(s\rho, t)} \psi_\rho(\tau),$$

$$\psi_\rho(t) = \frac{1}{(2\pi)^n} \int_{D'} \sum_s e^{i(s\rho, x)} G(x, t) \, dx.$$

The outer sum in the first formula is extended over all irreducible representations of the group $G$, and the integral in the second formula is taken over the fundamental region for the group $\tilde{S}$, generated by the Weyl group $S$ and translations by vectors of the lattice $\Gamma$.

**no. 10. The manifold of cosets of a complex semisimple group with respect to a maximal compact subgroup.** For these manifolds, which, as we have recalled, are dual to the manifolds of the corresponding compact semisimple groups, almost all of the results of nos. 7 and 8 remain in force, and similar arguments can be used. The definitions of the rings $R$, $R'$, and $R''$ are carried over with no change. The homomorphisms of the ring $R$ into the field of complex numbers are given with the aid of the spherical functions $\phi_\rho(t)$, computed by Gel'fand and Naĭmark [9]:

$$\phi_\rho(t) = \frac{\sum_s \pm e^{i(\rho, st)}}{d_\rho \prod_{\alpha>0} \operatorname{sh} \frac{(\alpha, t)}{2}}.$$

For integers $\rho$, these functions are connected with the spherical functions $\Phi$, considered in the preceding nos., that is, with characters of the corresponding compact groups, divided by the dimension, by the formula

$$\phi_\rho(t) = \Phi_{-i\rho}(it).$$

Homomorphisms of the ring $R'$ into the field of complex numbers are given by the aid of the function

$$\psi_\rho(t) = \prod_{\alpha>0} \operatorname{sh} \frac{(\alpha, t)}{2} \phi_\rho(t) = \frac{\sum_s \pm e^{i(\rho, st)}}{d_\rho}.$$

Just as in the compact case, the principal rôle for finding the law of multiplication in $R$ is played by the solution of the differential equation

$$L_\tau G(t, \tau) = \frac{1}{n(S)} \sum_s \delta(t - s\tau),$$

where $L$ is the same operator as in the preceding case, and $\delta$ this time is the ordinary $\delta$-function of Dirac in $n$-dimensional space, and not the function on the torus, as previously. In correspondence with this, the function $G$ is also a function in space and not on the torus; this is an essential simplification.

The function $\tilde{a}$, giving the law of multiplication in the ring $R'$, is expressed just as in the preceding case, by $G$: $\tilde{a}(t; \tau_1, \tau_2)$ is the function $G(t - \tau_1, \tau_2)$ alternated with respect to $t, \tau_1, \tau_2$. The function $a$, giving the law of multiplication in $R$, is expressed in terms of the function $\tilde{a}$ as follows:

$$a = \frac{\Delta(\tau_1) \cdot \Delta(\tau_2)}{\Delta(t)}, \; \tilde{a}(t; \tau_1, \tau_2),$$

where

$$\Delta(t) = \prod_{\alpha>0} \operatorname{sh} \frac{(\alpha, t)}{2}.$$

The law of multiplication in the ring $R$ is given by a formula analogous to the corresponding formula for the case of a compact group:

$$f_1 \cdot f_2(t) = \int a(t; \tau_1, \tau_2) f_1(\tau_1) f_2(\tau_2)\, d\tau_1 d\tau_2.$$

The difference from the preceding case is that the integral is taken over the entire

$(\tau_1, \tau_2)$ space.

*There exist relations between the functions* $G(t, \tau)$ *and* $\psi_\rho(t)$ *analogous to the corresponding relations of no. 8:*

$$G(t, \tau) = \frac{1}{(2\pi)^{\frac{n}{2}}} \int \sum_s e^{i(\rho, st)} \psi_\rho(t)\, d\rho,$$

$$\psi_\rho(t) = \frac{1}{(2\pi)^{\frac{n}{2}}} \int \sum_s e^{i(\rho, sx)} G(x, t)\, dx.$$

The first integral is taken over all irreducible representations of the group, which can be realized with the aid of spherical functions. The second integral is taken over the fundamental region for the group $S$.

We turn to the function $G(t, \tau)$, which plays a principal rôle in the computation of the law of multiplication for the ring $R$. As has been said, this function is now not periodic. It has another remarkable property. We consider in the space of the variables $t$ the region $\Delta(\tau)$, which is the convex hull of the points $s\tau$ ($s\tau$ denotes as always the element $s$ of the group $S$, applied to the point $\tau$).

1. *The function* $G(t, \tau)$, *considered as a function of* $t$ *for fixed* $\tau$, *is equal to zero outside of the region* $\Delta(\tau)$.

We consider further in the space of variables $\tau$ the analogous region $\Delta(t)$.

2. *The function* $G(t, \tau)$, *considered as a function of* $\tau$ *for fixed* $t$, *is equal to zero within the region* $\Delta(t)$.

We shall carry out the proof, having in view as the manifold the manifold of positive definite unimodular Hermitian matrices. (Its group of rigid motions is the group of complex unimodular matrices, and its stationary subgroup is the group of unitary unimodular matrices. The manifold itself is dual to the manifold of the last group.)

In this case, the region $\Delta(\tau)$ is the convex hull of the points

$$t = (\tau_{i_1}, \cdots, \tau_{i_n}), \quad \tau_1 + \cdots + \tau_n = 0.$$

**Proof.** The second property follows from the fact that the function $G(t, \tau)$ for fixed $t$ is a skew symmetric function of $\tau$ and satisfies the equation $L_\tau G(t, \tau) = 0$ within the region $\Delta(t)$, and the region $\Delta(t)$ is convex.

To prove the first property, we shall show that it is equivalent to the second. For this, we consider the function $G$ as a function of both variables $t$ and $\tau$. The collection of regions $\Delta(\tau)$ for variable $\tau$ fills out in the space of the variables $t$ and $\tau$ the interior of a cone $\Delta_\tau$. We consider the plane manifold $t_1 = c_1, \cdots, t_n = c_n$. Its intersection with the exterior of the cone $\Delta_\tau$ will be denoted by $\tilde{\Delta}(t)$. The region $\tilde{\Delta}(t)$ is obviously a polyhedron, the faces of which are given by the equations $r_i = t_j$ (the same equations that give the faces of the polyhedron $\Delta(r)$ and the cone $\Delta_r$). Its vertices are intersections of faces – the points $st = (t_{i_1}, \cdots, t_{i_n})$, where $s = \begin{pmatrix} 1 & \cdots & n \\ i_1 & \cdots & i_n \end{pmatrix}$ runs through all possible permutations.

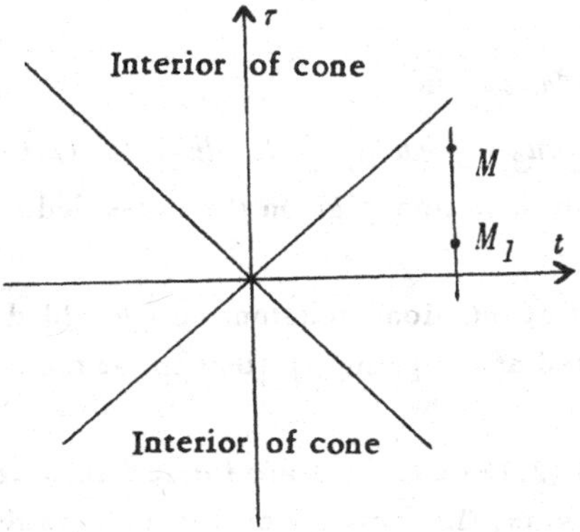


Figure 1.

We shall show that the region $\Delta(t)$ is convex (and hence coincides with the region $\Delta(t)$ introduced earlier) and that every point of the exterior of the cone $\Delta_r$ belongs to some region $\widetilde{\Delta}(t)$.

Both of the last assertions are easiest to verify in the following way. Let $M$ and $M_1$ be two arbitrary points from the exterior of $\Delta_r$, lying on the manifold $t_1 = c_1, \cdots, t_n = c_n$. Draw a straight line through them, and consider any straight line parallel to this line, lying on the manifold $t_1 = 0, \cdots, t_n = 0$. Pass a plane $Q$ through these two straight lines. The intersection of $Q$ with $\Delta_r$ is obviously a cone on the plane $Q$. Since $\Delta_r$ has the property that its intersection with the plane manifold $r_1 = c_1, \cdots, r_n = c_n$ is convex, the same is true of the cone on the plane $Q$. But such a cone on a plane can only be a pair of intersecting straight lines (Figure 1). The rest is evident.

The function $G(t, r)$ is skew symmetric in $r$ and satisfies the equation $L_r G(t, r) = 0$ within the symmetric region $\Delta(t)$, which has just been proved to be convex. We saw above that functions satisfying both of these conditions are identically equal to zero. Consequently $G(t, r) = 0$ outside the cone $\Delta_r$. This proves the theorem.

Going from the function $G(t, r)$ to the functions $\widetilde{a}(t; t^{(1)}, t^{(2)})$ and $a(t; t^{(1)}, t^{(2)})$, which are closely connected with it, we find also for them analogous results. In particular, for the function $a(t; t^{(1)}, t^{(2)})$, this result is formulated as follows. *The function $a(t; t^{(1)}, t^{(2)})$, considered as a function of $t$ for fixed $t^{(1)}$ and $t^{(2)}$, is equal to zero outside the convex hull of the points*

$$t^{(1)} + st^{(2)} = (t_1^{(1)} + t_{k_1}^{(2)}, \cdots, t_n^{(1)} + t_{k_n}^{(2)}).$$

**no. 11. Eigenvalues of the product of Hermitian matrices.** The last result of no. 10 can be formulated in the following way. *The logarithms of eigenvalues of the product of two Hermitian positive definite matrices* (the collection of logarithms of roots of a matrix forms a vector in $n$-dimensional space) *cannot lie outside the convex hull of vectors of the form $t^{(1)} + st^{(2)}$, where $t^{(1)}$ and $t^{(2)}$ are the sets of logarithms of the roots of the factors, and $s$ is a permutation of the roots.*

We consider a "generalized function", constant on two-sided cosets,

$$\widetilde{\delta}_{g_1}(g) = \int \delta(u_1 g_1^{-1} u_2 g u_3)\, du_1\, du_2\, du_3,$$

where $\delta(g)$ is Dirac's $\delta$-function.

Using the invariance of integration and the normalization $\int du = 1$, we obtain the equality

$$\int f(g)\widetilde{\delta}_{g_1}(g)\, dg = \int f(g)\,\delta(u_1 g_1^{-1} u_2 g u_3)\, dg\, du_1\, du_2\, du_3 =$$
$$= \int f(u_2^{-1} g_1 u_1^{-1} u_3^{-1})\, du_1 du_2 du_3 = \int f(u_1 g_1 u_2)\, du_1\, du_2, \qquad (2.18)$$

that is, $\int f(g)\,\delta_{g_1}(g)\, dg$ is the mean value of the function $f(g)$ on the two-sided coset that contains $g_1$.

Formula (2.18) also shows that in the set of functions constant on two-sided cosets, $\widetilde{\delta}_{g_1}(g)$ is the usual $\delta$-function centered at the point $g_1$ (that is, at the two-sided coset containing $g_1$).

We note that in this case the integral (2.18) can be transformed into an integral over the manifold of two-sided cosets. In fact, since the integrand $\widetilde{\delta}_{g_1}(g)\cdot f(g)$ is constant on every two-sided coset, one can first carry out the integration on each coset, and then over the manifold of such cosets. As a result of this, we obtain

$$\int f(g)\widetilde{\delta}_{g_1}(g)\, dg = \int f(t)\widetilde{\delta}_{t_1}(t)\,\Delta^2(t)\, dt = f(t_1). \qquad (2.18')$$

Here $t$, $t_1$ are the collections of logarithms of eigenvalues of the matrices $g$ and $g_1$, respectively, and $\Delta^2(t)$ is a density equal to $\prod_{p<q} (e^{t_p} - e^{t_q})^2$ (see no. 2).

We shall find the convolution of a function $f$, constant on two-sided cosets, with the function $\widetilde{\delta}_{g_1}, g)$:

$$f\cdot\widetilde{\delta}_{g_1}(g) = \int f(g_2^{-1} g)\widetilde{\delta}_{g_1}(g_2)\, dg_2 = \int f(u_1 g_1^{-1} u_2 g)\, du_1\, du_2.$$

In particular, if $f = \widetilde{\delta}_{g_2}(g)$, we obtain

$$\widetilde{\delta}_{g_2}\cdot\widetilde{\delta}_{g_1}(g) = \int \widetilde{\delta}_{g_2}(u_1 g_1^{-1} u_2 g)\, du_1 du_2 = \int \delta(u_1 g_2^{-1} u_2 g_1^{-1} u_3 g u_4)\, du_1 \cdots du_4 =$$
$$= \int \delta(u_1 g_2^{-1} u_2 u_3 g_1^{-1} u_4 u_5 g u_6)\, du_1 \cdots du_6.$$

The last integral can be different from zero only for those $g$ for which there exist $u_1, \cdots, u_6$ such that

$$(u_1 g_2^{-1} u_2)\cdot(u_3 g_1^{-1} u_4)\cdot(u_5 g u_6) = e,$$

or, equivalently, such that $u_5 g u_6 = (u_4^{-1} g_1 u_3^{-1})\cdot(u_2^{-1} g_2 u_1^{-1})$. The last equality means that the two-sided cosets containing $g_1$ and $g_2$ also contain representatives $\widetilde{g}_1$ and $\widetilde{g}_2$ such that $g$ belongs to the same two-sided coset as the product $\widetilde{g}_1\cdot\widetilde{g}_2$. Or, in other words, the element $g$ belongs to the product of the two-sided cosets containing $g_1$ and $g_2$ respectively.

Recalling on the other hand, that multiplication in the ring $R$ is given by the formula

$$f_1\cdot f_2(t) = \int a(t;\, t^{(1)},\, t^{(2)})\, f(t^{(1)})\, f(t^{(2)})\, dt^{(1)}\, dt^{(2)},$$

where $t$ is the collection of logarithms of eigenvalues of the matrix $g$, and applying it to the functions $\tilde{\delta}_{g_1}(g)$ and $\tilde{\delta}_{g_2}(g)$, we obtain, using (2.18′), that

$$\tilde{\delta}_{g_1} \cdot \tilde{\delta}_{g_2}(g) = \frac{a(t;\, t^{(1)},\, t^{(2)})}{\Delta^2(t^{(1)})\,\Delta^2(t^{(2)})},$$

or, in agreement with the property of the function $a$ established above: *the element $g = u_1 \epsilon u_2$ can belong to the product of the two-sided cosets of the elements $g_1 = u_3 \epsilon_1 u_4$ and $g_2 = u_5 \epsilon_2 u_6$ only when the logarithms of the eigenvalues of the matrix $\epsilon$ form a vector lying within the convex hull of the vectors $t^{(1)} = s t^{(2)}$, where $t^{(1)}$ and $t^{(2)}$ are the collections of logarithms of eigenvalues of the matrices $\epsilon_1$ and $\epsilon_2$, and $s$ is an arbitrary permutation of eigenvalues.*

In order to obtain our theorem from this, we consider as elements $g_1$ and $g_2$ Hermitian positive definite matrices $h_1$ and $h_2$. It follows from the equality

$$h_1 \cdot h_2 = h_2^{-\frac{1}{2}} \left( h_2^{\frac{1}{2}} h_1 h_2^{\frac{1}{2}} \right) h_2^{\frac{1}{2}}$$

that the roots of the matrix $h_1 \cdot h_2$ coincide with the roots of the matrix $h^2 = h_2^{\frac{1}{2}} h_1 h_2^{\frac{1}{2}}$.

We now show that the matrices $h = \left( h_2^{\frac{1}{2}} h_1 h_2^{\frac{1}{2}} \right)^{\frac{1}{2}}$ and $h_1^{\frac{1}{2}} h_2^{\frac{1}{2}}$ belong to one and the same two-sided coset. In fact, the matrix $h_1^{\frac{1}{2}} h_2^{\frac{1}{2}}$ belongs to the same class as the matrix $\tilde{h}$, defined by the relation $u\tilde{h} = h_1^{\frac{1}{2}} h_2^{\frac{1}{2}}$. From this relation, it follows that $\tilde{h}^2 = h_2^{\frac{1}{2}} h_1 h_2^{\frac{1}{2}}$, that is, $\tilde{h} = h$.

Since the matrix $h$ belongs to the same two-sided coset as $h_1^{\frac{1}{2}} h_2^{\frac{1}{2}}$, it follows from what was proved above that the collection of logarithms of its roots is a point of $n$-dimensional space which cannot lie outside of the convex hull of the points $\frac{1}{2}t^{(1)} + s\frac{1}{2}t^{(2)}$, where $\frac{1}{2}t^{(1)}$ and $\frac{1}{2}t^{(2)}$ are points the co-ordinates of which are the collections of logarithms of the roots of the matrices $h_1^{\frac{1}{2}}$ and $h_2^{\frac{1}{2}}$. Consequently, the set of doubled logarithms of the roots of the matrix $h^2$, which coincide with the roots of the matrix $h_1 \cdot h_2$, cannot lie outside of the convex hull of the points $t^{(1)} + st^{(2)}$, where $t^{(1)}$ and $t^{(2)}$ are the collections of logarithms of the matrices $h_1$ and $h_2$. The theorem is proved. The proof just given is not elementary. On the suggestion of one of the authors, an elementary proof of this theorem has been obtained by V. B. Lidskiĭ [6].

The theorem just proved can be given an elegant probabilistic interpretation. We consider an $n$-dimensional ellipsoid with fixed lengths of semi-axes. Its position is thus determined by the directions of the semi-axes, which form an orthonormal repère $X$. If we fix a certain initial repère $X_0$, then the position of the repère $X$ with respect to $X_0$ is given uniquely by a unitary transformation carrying

$X_0$ into $X$. We shall suppose that the position of $X$ depends upon chance in such a way that the density of probability is equal to the invariant measure on the unitary group.

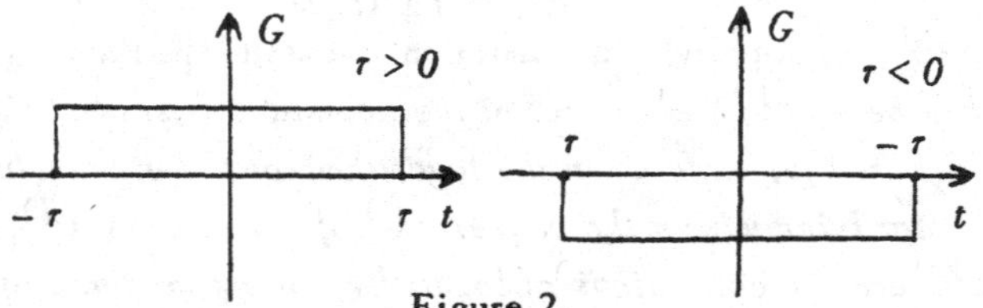


Figure 2.

We obtain a random ellipsoid with fixed lengths of semi-axes, in which, roughly speaking, all positions in space are equally likely. We now consider random compression of space along the $n$ mutually perpendicular axes. Suppose here that the coefficients of compression are given, and that the position of the axes depends upon chance in such a way that the density of probability again coincides with the invariant measure on the unitary group (that is, the positions of the axes in space are equally likely). As a result, the random ellipsoid goes into

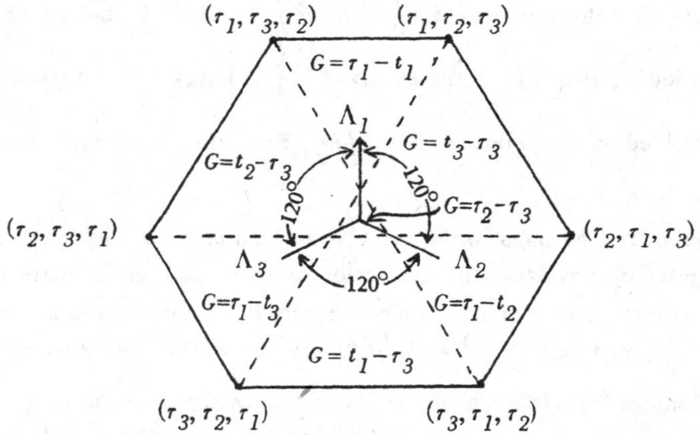


Figure 3.

another random ellipsoid, the lengths of whose axes will depend upon chance. The function

$$\frac{a(t; t^{(1)}, t^{(2)})}{\Delta^2(t^{(1)})\,\Delta^2(t^{(2)})}$$

is the density of the distribution of the logarithms $t_1, \cdots, t_n$ of the semi-axes of the new ellipsoid. $t^{(1)} = (t_1^{(1)}, \cdots, t_n^{(1)})$ are the logarithms of the semi-axes of the original ellipsoid, $t^{(2)} = (t_1^{(2)}, \cdots, t_n^{(2)})$ is the collection of logarithms of the eigenvalues of the affine transformation. One finds in particular that the logarithms of the lengths of the semi-axes of the new ellipsoid cannot lie outside of the convex hull of the points $t^{(1)} + st^{(2)}$.

In concluding this section, we give the explicit form of the functions $G(t, r)$ for the manifolds $SL(2)/SU(2)$ and $SL(3)/SU(3)$. For the manifold $SL(2)/SU(2)$, the graph of the function $G(t, r)$ as a function of $t$ for fixed $r$ is given in Figure 2. We remark that for $|r| < \pi$, the function $G(t, r)$ for $SU(2)$ is given by the same graph, if we consider the manifold of this group as a homogeneous space with rigid motions $u \to u_1^{-1} u u_2$. For the manifold $SL(3)/SU(3)$, the function $G(t, r)$ as a function of $t$ for fixed $r$ $(r_1 > r_2 > r_3)$ is defined by Figure 3. For odd permutations of the $r_k$, the function changes its sign. In Figure 3, co-ordinates $t_1, t_2, t_3$ of the point $t$ are the scalar products of the radius vector of this point with the vectors $\Lambda_1, \Lambda_2, \Lambda_3$.

## §3. The functional equation for zonal spherical functions

In this section, as in the preceding one, we do not use the explicit form of a spherical function, and hence the theorems proved are true in the most general case.

**no. 12. The functional equation.** As already recalled, the zonal functions $\Phi_\rho(g)$ give homomorphisms of the ring $R$ of functions constant on two-sided cosets of $G$ with respect to $G_0$ into the field of complex numbers, by the formula

$$M_\rho f = \int f(g)\Phi_\rho(g)\, dg.$$

Applying the homomorphism $M_\rho$ to the convolution of the functions $f_1 \cdot f_2$, after a short computation, we obtain

$$M_\rho(f_1 \cdot f_2) = \int f_1(g_1) f_2(g_2) \Phi_\rho(g_1 g_2)\, dg_1 dg_2. \tag{3.1}$$

Since $M_\rho$ is a homomorphism, we have

$$M_\rho(f_1 \cdot f_2) = M_\rho(f_1) M_\rho(f_2).$$

Hence, on the other hand,

$$M_\rho(f_1 \cdot f_2) = \int f_1(g_1) f_2(g_2) \Phi_\rho(g_1) \Phi_\rho(g_2)\, dg_1\, dg_2. \tag{3.2}$$

We now use the following facts from the theory of symmetric spaces (see no. 2). Every element $g$ can be represented in the form $g = u_1 \epsilon u_2$, where $u_1, u_2$ are elements of the stationary subgroup $G_0$, and $\epsilon$ is a representative of a two-sided coset with respect to $G_0$ ($\epsilon$ can be thought of as running through a certain commutative subgroup). Here the integral with respect to $dg$ can be replaced by an integral with respect to $du_1 du_2 d\mu\epsilon$ (see formula (1.4)). Since all of the functions in the integrand of the integral (3.1) are constant on two-sided cosets, that is, depend only upon $\epsilon$, the integral can be transformed in the following way:

$$M_\rho(f_1 \cdot f_2) = \int f_1(\epsilon_1) f_2(\epsilon_2) \Phi_\rho(u_1 \epsilon_1 u_2 u_3 \epsilon_2 u_4)\, d\mu\epsilon_1\, d\mu\epsilon_2\, du_1\, du_2\, du_3\, du_4.$$

Carrying out the integration with respect to $du_1$ and $du_4$, and making use of the invariance to replace the integral with respect to $du_2 du_3$ by the single integral with respect to $du$, we obtain:

$$M_\rho(f_1 \cdot f_2) = \int f_1(\epsilon_1) f_2(\epsilon_2)\, d\mu\epsilon_1 d\mu\epsilon_2 \int \Phi_\rho(\epsilon_1 u \epsilon_2)\, du.$$

(We suppose, as is ordinarily done, that the measure $du$ is normed so that $\int_{G_0} du = 1.$)

The integral (3.2) is equal to

$$M_\rho(f_1 \cdot f_2) = \int f_1(\epsilon_1) f_2(\epsilon_2) \Phi_\rho(\epsilon_1) \Phi_\rho(\epsilon_2)\, d\mu\epsilon_1 d\mu\epsilon_2.$$

Comparing the last two expressions with each other, and using the arbitrariness in the choice of the functions $f_1$, $f_2$, we obtain finally:

$$\Phi_\rho(\epsilon_1)\Phi_\rho(\epsilon_2) = \int \Phi_\rho(\epsilon_1 u \epsilon_2)\, du. \tag{3.3}$$

This is the sought-for functional equation, since, repeating all of the computations in the reversed order, it is easy to see that its fulfillment is not only necessary, but also sufficient, in order for the function $\Phi_\rho$ to give a homomorphism of the ring $R$ into the field of complex numbers.

An elegant geometric interpretation can be given for the equation just obtained.

For this, we remark that every function on a symmetric space, when extended over the group as a function constant on cosets with respect to the stationary subgroup $G_0$, becomes a function of the form $f(g) = \widetilde{f}(gg^{*-1})$. (Evidently a function of this form is constant on right cosets with respect to $G_0$.) This assertion is proved in the following way. Let $M$ be an arbitrary homogeneous space and $\widetilde{f}(x)$ a function on $M$. Then the function $f(g) = \widetilde{f}(gx_0)$ ($x_0$ is a point stationary under motions of the stationary subgroup $G_0$) is an extension of the function $\widetilde{f}(x)$ over the group, as a function constant on right cosets with respect to $G_0$. We apply this construction for a symmetric space. As is known (see no. 2) there exists a manifold $\Theta$ in the group, the elements of which are characterized by the fact that $\theta^* = \theta^{-1}$, and which serves as a model of the symmetric space. Also, rigid motions in $\Theta$ are described by the formula $\theta \to g\theta g^{*-1}$. The elements $e \in \Theta$ serves as the point $\theta_0$ left invariant by the stationary subgroup $G_0$. Hence the function $f(g) = \widetilde{f}(g\theta_0 g^{*-1}) = \widetilde{f}(gg^{*-1})$ serves as the extension of the function $\widetilde{f}(\theta)$ over the group $G$. In particular, its value on the manifold $\Theta$ is equal to $f(\theta) = \widetilde{f}(\theta^2)$. Hence, in order to obtain the original function $\widetilde{f}(\theta)$, we must replace $\theta$ by $\theta^{\frac{1}{2}}$ in the function $f$: $\widetilde{f}(\theta) = f(\theta^{\frac{1}{2}})$.

Using this remark, we rewrite (3.3)

$$\widetilde{\Phi}_\rho(\epsilon_1^2)\, \widetilde{\Phi}_\rho(\epsilon_2^2) = \int \widetilde{\Phi}_\rho(\epsilon_1 u \epsilon_2^2 u^{-1} \epsilon_1)\, du,$$

where $\epsilon_1 g_0 \epsilon_2^2 g_0^{-1} \epsilon_1$ is an element of the manifold $\Theta$, lying on the sphere of radius $\epsilon_2$ with center at the point $\epsilon_1^2$. The last integral is obviously the mean value over this sphere.

Replacing $\epsilon_1$ by $\epsilon_1^{\frac{1}{2}}$ and $\epsilon_2$ by $\epsilon_2^{\frac{1}{2}}$, we obtain the following necessary result. *A function $f$ on the manifold is zonal for a certain representation if and only* *if the product* $f(\epsilon_1)\cdot f(\epsilon_2)$ *of the values of this function at the points* $\epsilon_1$ *and* $\epsilon_2$ *is equal to the mean value of the function over the "sphere" with complex radius* $\epsilon_1$*, the center of which lies at complex distance* $\epsilon_2$ *from the origin of co-ordinates.*

**no. 13. Example.** As as example, we consider the $n$-dimensional sphere $S_n$. It is a symmetric space; the group acting transitively on it is the rotation group.

We introduce the usual co-ordinates $\theta_1, \cdots, \theta_n$ on the sphere with the aid of the usual formulas

$$x_1 = \sin\theta_1 \cdots \sin\theta_n,$$
$$x_2 = \sin\theta_1 \cdots \cos\theta_n,$$
$$\cdots\cdots\cdots\cdots\cdots\cdots.$$

As stationary subgroup, we choose the subgroup $G_0$ leaving fixed the vector $(0, \cdots, 0, 1)$, or, equivalently, leaving the last co-ordinate of every vector fixed. Its trajectories are obviously $(n-1)$-dimensional spheres lying in the $n$-dimensional space $x_{n+1} = \text{const}$. Hence $x_{n+1}$ or $\theta_1$ will serve as a parameter defining a sphere with center at the invariant point.

The last parameter is also the "complex radius" of the sphere. (In our case it is a one-parameter function.)

The invariant volume on the sphere is given by the density

$$d\mu = \frac{\Gamma(\frac{n+1}{2})}{2\left[\Gamma(\frac{1}{2})\right]^{n+1}} \sin^{n-1}\theta_1 \sin^{n-2}\theta_2 \cdots \sin\theta_{n-1}\, d\theta_1 \cdots d\theta_n.$$

(This density is normalized so that $\int_{S_n} d\mu = 1$.) Hence the integral over our space of a function constant on surfaces of transitivity for $G_0$, that is, of a function depending only on $\theta_1$, is equal to

$$\frac{\Gamma(\frac{n+1}{2})}{\Gamma(\frac{1}{2})\Gamma(\frac{n}{2})} \int_0^{\pi} f(\theta) \sin^{n-1}\theta d\theta.$$

The density denoted in the general case by $d\mu\epsilon$ is thus in our case equal to $\dfrac{\Gamma(\frac{n+1}{2})}{\Gamma(\frac{1}{2})\Gamma(\frac{n}{2})} \sin^{n-1}\theta d\theta$. We now consider the point $M$ (Figure 4) lying on a surface of transitivity, that is, on the $(n-1)$-dimensional sphere $S$ with center $O$ at distance $\theta_1$ from the invariant point $M_0$ and radius $\theta_2$. If the distance from $M$ to $M_0$ is equal to $\theta$, then the zonal function at the point $M$ depends only upon $\theta$. We consider the geodesic triangle $M_0MO$. If the angle $MOM_0$ is equal to $\phi$, then

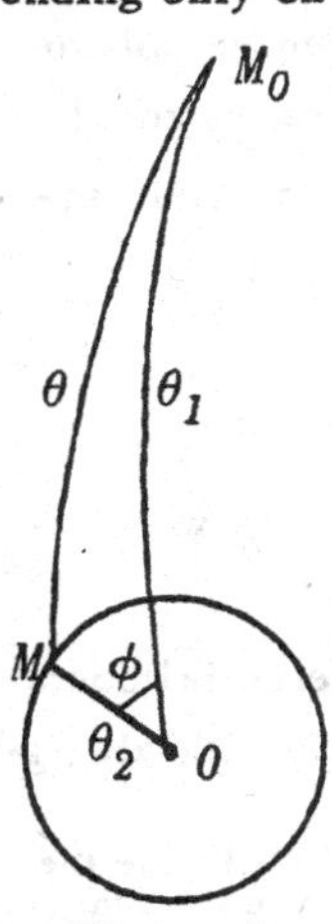

Figure 4.

$$\cos\theta = \sin\theta_1 \sin\theta_2 \cos\phi + \cos\theta_1 \cos\theta_2.$$

Before integrating our function over $S$, we cut $S$ by the surfaces $\theta_1 = \text{const}$. These surfaces are also $(n-1)$-dimensional spheres, and hence their intersections with $S$ are $(n-2)$-dimensional spheres. Our function must therefore be integrated over the set of these spheres, the measure in which is defined in accordance with the preceding density by

$$d\mu = \frac{\Gamma(\frac{n}{2})}{\Gamma(\frac{1}{2})\Gamma(\frac{n-1}{2})} \sin^{n-2}\theta d\theta.$$

The angle $\phi$ between the radii $OM$ and $OM_0$ will now obviously play the rôle of the parameter $\theta$.

In this way, we now obtain (taking $\cos\theta$ instead of $\theta$ for the argument of the zonal function):

$$\Phi_\rho(\cos\theta_1)\Phi_\rho(\cos\theta_2) =$$

$$= \frac{\Gamma(\frac{n}{2})}{\Gamma(\frac{1}{2})\Gamma(\frac{n-1}{2})} \int_0^\pi \Phi_\rho(\sin\theta_1 \sin\theta_2 \cos\phi + \cos\theta_1 \cos\theta_2) \sin^{n-2}\phi d\phi.$$

Making the substitution

$$\sin\theta_1 \sin\theta_2 \cos\phi + \cos\theta_1 \cos\theta_2 = \cos\theta,$$
$$\sin\theta_1 \sin\theta_2 \sin\phi\, d\phi = \sin\theta\, d\theta$$

in the integral, we obtain

$$\Phi_\rho(\cos\theta_1)\Phi_\rho(\cos\theta_2) = \frac{\Gamma(\frac{n}{2})}{\Gamma(\frac{1}{2})\Gamma(\frac{n-1}{2})} \int_{\theta_1-\theta_2}^{\theta_1+\theta_2} \Phi_\rho(\cos\theta) \frac{\sin\theta}{\sin\theta_1 \sin\theta_2} \sin^{n-3}\phi d\theta. \tag{3.4}$$

For $n = 2$, formula (3.4) gives the law of multiplication for Legendre polynomials. The formula has a particularly simple form for $n = 3$. In this case, zonal functions have the form $\Phi_\rho(\cos\theta) = \frac{\sin\rho\theta}{\rho\sin\theta}$, and formula (3.4) can be verified immediately.[1)]

## §4. Theorems concerning the mean

**no. 14. Differential equations for mean values.** In the works [5], [14], the "Laplace operators" $\Delta_x^{(1)}, \ldots, \Delta_x^{(n)}$ for the symmetric space $M = G/G_0$ were introduced.

The operators $\Delta_x^{(1)}, \ldots, \Delta_x^{(n)}$ are generators in the ring of differential operators on $M$, commuting with all Lie operators on $M$ (and consequently, with all rig-

1) The three-dimensional sphere is the group manifold of the group $SU(2)$. Hence the function $a$ obtained here, giving the law of multiplication can be obtained from the function $G(t, r)$ for $SU(2)$, computed in §2.

id motions). In the case where the function $f$ depends only upon the complex distance from the point $x \in M$ to the origin, that is, to the point $x_0$ that is invariant under rigid motions from $G_0$, the function $\Delta_x^{(k)} f(x)$ obviously has the same property. The operator induced by $\Delta_x^{(k)}$ in this set of functions will be called the radial part of the operator $\Delta_x^{(k)}$ and will be denoted by $\Delta_t^{(k)}$.

We consider the operator $V(t, x)$ which gives the mean value of the function over the sphere of radius $t$ and center at the point $x$. Then we have

**Theorem.** *The function* $\psi(t, x) = V(t, x) f$ *satisfies the system of equations*

$$\Delta_x^{(k)} \psi(t, x) = \Delta_t^{(k)} \psi(t, x). \tag{4.1}$$

**Proof.** We remark first of all that the relations (4.1) are satisfied if the function $f$ is a zonal spherical function. In fact, the functional equation for a zonal function is: $\Phi_\rho(t_1)\Phi_\rho(t_2) = V(t_1, x)\Phi_\rho$, where the distance from the point $x$ to the origin is equal to $t_2$.

The function $\Phi_\rho(t)$ is an eigenfunction for the Laplace operator (see [5]). Hence

$$\Delta_x^{(k)} V(t_1, x)\Phi_\rho = \Delta_x^{(k)} \Phi_\rho(t_1)\Phi_\rho(t_2) = \lambda_\rho^{(k)}\Phi_\rho(t_1)\Phi_\rho(t_2)$$

and in just the same way

$$\Delta_{t_1}^{(k)} V(t_1, x)\Phi_\rho = \Delta_{t_1}^{(k)}\Phi_\rho(t_1)\Phi_\rho(t_2) = \lambda_\rho^{(k)}\Phi_\rho(t_1)\Phi_\rho(t_2).$$

Thus relation (4.1) is correct for zonal spherical functions.

To extend the relations (4.1) to arbitrary functions, we use the so-called "Plancherel theorem," which states that linear combinations of zonal functions and their translates will arbitrarily approximate any function on $M$ from a sufficiently wide class (for example, any function with summable square). Plancherel's theorem was proved by É. Cartan [11] under the hypothesis that the space $M$ is compact. However, it remains true in the general case.

We remark that the relation (4.1) is linear with respect to the function appearing in it. Consequently, once it is proved for zonal functions $\Phi_\rho(g)$, it is proved also for all possible linear combinations of these functions.

On the other hand, Laplace operators, as already noted, commute with translations. The mean value operator also obviously has this property.

Thus the relation (4.1) is carried over to linear combinations of translates of zonal functions, that is, it is true for all functions.

As a special case of the theorem just proved, when the space $M$ is Euclidean $n$-dimensional space and the group $G$ is the group of all rigid motions in it, we obtain the known theorem of Darboux. In this case, there exists only one Laplace operator $\Delta = \frac{\partial^2}{\partial x_1^2} + \cdots + \frac{\partial^2}{\partial x_n^2}$ and the equations (4.1) go over to Darboux's equation (see for example [18], vol. 2).

**no. 15. Expression of the averaging operator by means of Laplace's operator.** As has already been recalled, zonal functions are eigenfunctions for Laplace's operators. Since they are constant on spheres with centers at the invariant point, they are furthermore eigenfunctions for the radial parts of Laplace's operators, that is, they satisfy the system of equations

$$\Delta_t^{(k)} \Phi_\rho(t) = \lambda^{(k)} \Phi_\rho(t). \tag{4.2}$$

Besides the system (4.2), zonal spherical functions satisfy the complementary condition

$$\Phi_\rho(0) = 1. \tag{4.3}$$

In fact, in agreement with definition (no. 1), a zonal spherical function as a function on the group is equal to $\Phi_\rho(g) = (\xi, T_g \xi)$, where the vector $\xi$ has length $1$. Hence $\Phi_\rho(e) = 1$, and from this (4.3) follows obviously.

It will be proved elsewhere that: 1) the system (4.2) with the additional condition (4.3) has a unique solution that is a function on the manifold $M$, and 2) the spectra of the operators $\Delta_x^{(k)}$ and $\Delta_t^{(k)}$ coincide. Hence the equality $\Phi_\rho(t) = \Phi(t, \lambda^{(1)}, \dots, \lambda^{(n)})$ has a meaning, where the zonal spherical function $\Phi(t, \lambda^{(1)}, \dots, \lambda^{(n)})$ as a function of the $\lambda^{(k)}$ is defined on the spectrum of the operators $\Delta_x^{(k)}$. On the other hand, it follows from the works [5] and [13] that the operators $\Delta_x^{(k)}$ commute with each other. Furthermore, they can be chosen self-adjoint (in the space of functions on $M$ with summable square for the invariant measure). Hence the operator $\Phi(t, \Delta_x^{(1)}, \dots, \Delta_x^{(n)})$ has a meaning.

**Theorem.** *Let $\Phi(t, \lambda^{(1)}, \dots, \lambda^{(n)})$ be a zonal spherical function. Then the operator $\Phi(t, \Delta_x^{(1)}, \dots, \Delta_x^{(n)})$ is the averaging operator over the sphere with center at the point $x$ and complex radius $t$.*

**Proof.** Let $t$ be the complex distance from the point $x$ to the invariant point $x$. We apply the operator $\Phi(t_1, \Delta_x^{(1)}, \dots, \Delta_x^{(n)})$ to the function $\Phi(t, \lambda^{(1)}, \dots, \lambda^{(n)})$: $\Phi(t_1, \Delta_x^{(1)}, \dots, \Delta_x^{(n)}) \Phi(t, \lambda^{(1)}, \dots, \lambda^{(n)}) = \Phi(t_1, \lambda^{(1)}, \dots, \lambda^{(n)}) \Phi(t, \lambda^{(1)}, \dots, \lambda^{(n)})$. The last expression, in view of the functional equation, is the mean value of the function $\Phi_\rho(t) = \Phi(t, \lambda^{(1)}, \dots, \lambda^{(n)})$ over the sphere with center at the point $x$ and complex radius $t_1$. As before, we denote this operator by the letter $V(t_1, x)$, and find (replacing $t_1$ by $t$)

$$\Phi(t, \Delta_x^{(1)}, \dots, \Delta_x^{(n)}) \Phi_\rho = V(t, x) \Phi_\rho. \tag{4.4}$$

Since the operators $\Phi(t, \Delta^{(1)}, \dots, \Delta^{(n)})$ and $V(t, x)$ are both linear and commute with translations, the identity (4.4) holds for arbitrary linear combinations of the functions $\Phi_\rho$ and their translations, that is, for all functions. The theorem is proved.

As an example, we consider $n$-dimensional Euclidean space. The equations (4.2) here assume the following form:

$$\frac{1}{r^{n-1}} \frac{d}{dr} \left(r^{n-1} \frac{d}{dr}\right) f = \left(\frac{d^2}{dr^2} + \frac{n-1}{2} \frac{d}{dr}\right) f = \lambda f.$$

This equation has, as is known, a unique solution defined by the condition $f(0, \lambda) = 1$. It is not hard to see that $f(r, \lambda) = f(r\sqrt{-\lambda})$. Consequently, in correspondence with the preceding considerations, the operator $f(r\sqrt{-\Delta})$, where $\Delta = \frac{\partial^2}{\partial x_1^2} + \cdots + \frac{\partial^2}{\partial x_n^2}$, is the averaging operator on the sphere of radius $r$ and center at the point $x = (x_1, \cdots, x_n)$.

## §5. The algebra of representations and functions on it

In this section, we consider the algebra $\Psi$ of representations of a compact semisimple Lie group $G$ and a ring of functions on this algebra. Just as in §2, all computations are carried out for the case in which the group is the group of unitary unimodular matrices. Then can, however, be easily carried over to the general case.

**no. 16. Simplest properties of the algebra of representations.** We first recall the basic definitions.

Elements of the algebra $\Psi$ are representations $\psi$ of the group $G$, the sum $\psi_1 + \psi_2$ is the direct sum of these representations (see no. 4), and the product $\psi_1 \times \psi_2$ is their Kronecker product (see no. 4).

It follows easily from the definition that both of these operations are commutative and that the distributive law

$$\psi_3 \times (\psi_1 + \psi_2) = \psi_3 \times \psi_1 + \psi_3 \times \psi_2$$

holds. The algebra $\Psi$ has a unit: this is the one-dimensional representation $\epsilon$ in which the number $1$ is put into correspondence with each element of the group $G$. Clearly $\epsilon \times \psi = \psi$.

For every representation $\psi$ the contragredient representation $\hat{\psi}$ is defined (see no. 4). From its definition, the following properties easily follow:

a) $\hat{\hat{\psi}} = \psi$, $\widehat{\psi_1 \times \psi_2} = \hat{\psi}_1 \hat{\psi}_2$, $\widehat{\psi_1 + \psi_2} = \hat{\psi}_1 + \hat{\psi}_2$.

b) The identity representation $\epsilon$ appears in the decomposition of $\psi \times \hat{\psi}$ into irreducible representations.

It is proved in the theory of linear representations that if $\psi$ is an irreducible representation, then $\psi \times \hat{\psi}$ contains $\epsilon$ only once, and furthermore, $\hat{\psi}$ is the only irreducible representation such that its product with $\psi$ contains $\epsilon$.

As a corollary of this, one finds that if $\psi$ is irreducible, then $\hat{\psi} \times \psi_1$ contains $\epsilon$ as many times as $\psi_1$ contains $\psi$.

**no. 17. The ring $\Xi$ of functions on the algebra of representations.** The elements of $\Xi$ are additive functions $\xi(\psi)$: if $\psi = \psi_1 + \psi_2$, then $\xi(\psi) = \xi(\psi_1) + \xi(\psi_2)$. Since every element of $\Psi$ is the sum of irreducible representations, the function

$\xi$ is completely determined by its values on irreducible representations.

From this point on, we shall always understand by the letter $\psi$ an irreducible representation.

Multiplication in the ring $\Xi$ is defined by the formula

$$\xi_1 \cdot \xi_2(\psi) = \sum_{\psi_1} \xi_1(\hat{\psi}_1 \times \psi)\, \xi_2(\psi_1). \tag{5.1}$$

The sum is extended over all irreducible representations. A norm can be introduced in $\Xi$, setting

$$\| \xi \| = \Sigma \,|\, \xi(\psi) \,|\, \dim \psi, \tag{5.2}$$

$\dim \psi$ denotes the dimension of the space $R_\psi$ in which the representation $\psi$ acts.

We shall show that the norm just introduced has the usual properties:

1. $\| \xi \| \geq 0$.
2. If $\| \xi \| = 0$, then $\xi = 0$.
3. $\| \xi_1 + \xi_2 \| \leq \| \xi_1 \| + \| \xi_2 \|$.
4. $\| \xi_1 \cdot \xi_2 \| \leq \| \xi_1 \| \cdot \| \xi_2 \|$.

The first three properties are obvious, since $\dim \psi > 0$. To prove the fourth property, we note that $\xi(\psi)$ can be represented in the form

$$\xi(\psi) = \sum_{\psi_1} \xi(\psi_1) \delta(\hat{\psi}_1 \times \psi), \quad \text{where} \quad \delta(\psi) = \begin{cases} 1, & \text{if } \psi = \epsilon, \\ 0, & \text{if } \psi \neq \epsilon. \end{cases}$$

We first prove property 4 for functions $\xi_{\psi_1}(\psi) = \delta(\hat{\psi}_1 \times \psi)$:

$$\delta(\psi_1 \times \psi) \cdot \delta(\hat{\psi}_2 \times \psi) = \sum_{\psi_3} \delta(\hat{\psi}_1 \times \hat{\psi}_3 \times \psi) \cdot \delta(\hat{\psi}_2 \times \psi_3) = G(\psi_1, \psi_2; \psi). \tag{5.3}$$

The function $G(\psi_1, \psi_2; \psi)$ has the following meaning: it shows how many times the representation $\psi$ goes into the Kronecker product $\psi_1 \times \psi_2$. In fact,

$$\delta(\hat{\psi}_2 \times \psi_3) = \begin{cases} 1, & \text{if } \psi_2 = \psi_3, \\ 0, & \text{if } \psi_2 \neq \psi_3. \end{cases}$$

Thus the only term of the sum that is different from zero is equal to $\delta(\hat{\psi}_1 \times \hat{\psi}_2 \times \psi) = \delta(\widehat{\psi_1 \times \psi_2} \times \psi)$. The representation $\widehat{\psi_1 \times \psi_2} \times \psi$ contains $\epsilon$, as already shown, as many times as the representation $\psi$ appears in the decomposition of the representation $\psi_1 \times \psi_2$ into irreducible representations. Thus

$$\| \delta(\hat{\psi}_1 \times \psi) \cdot \delta(\hat{\psi}_2 \times \psi) \| = \Sigma \,|\, G(\psi_1, \psi_2; \psi) \,|\, \dim \psi = \Sigma\, G(\psi_1, \psi_2; \psi) \dim \psi =$$
$$= \dim \psi_1 \times \psi_2 = \dim \psi_1 \times \dim \psi_2 = \| \delta(\hat{\psi}_1 \times \psi) \| \cdot \| \delta(\hat{\psi}_2 \times \psi) \|.$$

Consequently property 4 holds for these functions.

We proceed to the general case:

$$\xi_1 \cdot \xi_2(\psi) = \sum_{\psi_0} \xi_1(\hat{\psi}_0 \times \psi) \cdot \xi_2(\psi_0) = \sum_{\psi_0} \sum_{\psi_1, \psi_2} \xi_1(\psi_1)\, \xi_2(\psi_2) \delta(\hat{\psi}_1 \times \hat{\psi}_0 \times \psi) \delta(\hat{\psi}_2 \times \psi_0).$$

Changing the order of summation, we obtain:

$$\xi_1 \cdot \xi_2(\psi) = \sum_{\psi_1, \psi_2} \xi_1(\psi_1)\xi_2(\psi_2) \sum_{\psi_0} \delta(\hat{\psi}_1 \times \hat{\psi}_0 \times \psi)\delta(\hat{\psi}_2 \times \psi_0) =$$
$$= \sum \xi_1(\psi_1)\xi_2(\psi_2) G(\psi_1, \psi_2; \psi).$$

Thus

$$\| \xi_1 \cdot \xi_2 \| = \| \sum_{\psi_1, \psi_2} \xi_1(\psi_1)\xi_2(\psi_2) G(\psi_1, \psi_2; \psi) \| =$$
$$= \sum_{\psi} | \sum_{\psi_1, \psi_2} \xi_1(\psi_1)\xi_2(\psi_2) G(\psi_1, \psi_2; \psi) | \dim \psi \leq$$
$$\leq \sum_{\psi} \sum_{\psi_1, \psi_2} | \xi_1(\psi_1)\xi_2(\psi_2) G(\psi_1, \psi_2; \psi) | \dim \psi.$$

Changing the order of summation and noting that $G(\psi_1, \psi_2; \psi) \geq 0$, we obtain:

$$\| \xi_1 \cdot \xi_2 \| \leq \sum_{\psi_1, \psi_2} | \xi_1(\psi_1) | \cdot | \xi_2(\psi_2) | \sum_{\psi} G(\psi_1, \psi_2; \psi) \dim \psi =$$
$$= \sum_{\psi_1, \psi_2} | \xi_1(\psi_1) | \cdot | \xi_2(\psi_2) | \dim \psi_1 \times \psi_2 =$$
$$= \sum | \xi_1(\psi_1) | \cdot | \xi_2(\psi_2) | \dim \psi_1 \cdot \dim \psi_2 = \| \xi_1 \| \cdot \| \xi_2 \|.$$

Thus the norm in the ring $\Xi$ satisfies all of the necessary properties.

On the way, we have obtained the following result. *The law of multiplication in $\Xi$ is given by the formula*

$$\xi_1 \cdot \xi_2(\psi) = \sum_{\psi_1, \psi_2} \xi_1(\psi_1)\xi_2(\psi_2) G(\psi_1, \psi_2; \psi), \tag{5.4}$$

*where the function $G$ has a simple group-theoretic meaning: it shows how many times the representation $\psi$ appears in the decomposition of the Kronecker product of representations $\psi_1 \times \psi_2$ into irreducible representations.*

We consider the family of functions $\chi_t(\psi)$, which are the characters of the irreducible representations. (The character of a representation depends upon the representation $\psi$ and on the class of conjugate elements of the group. This last is defined by a parameter $t = (t_1, \cdots, t_n)$.)

We shall show that all homomorphisms of the ring $\Xi$ into the field of complex numbers are given by the formula

$$\xi \to M_t(\xi) = \sum_{\psi} \xi(\psi) \chi_t(\psi).$$

For this, we transform the expression $M_t(\xi_1 \cdot \xi_2)$:

$$M_t(\xi_1 \times \xi_2) = \sum_{\psi} \xi_1 \cdot \xi_2(\psi)\chi_t(\psi) = \sum_{\psi} \chi_t(\psi) \sum_{\psi_1, \psi_2} \xi_1(\psi_1)\xi_2(\psi_2) G(\psi_1, \psi_2; \psi) =$$
$$= \sum_{\psi_1, \psi_2} \xi_1(\psi_1)\xi_2(\psi_2) \sum_{\psi} \chi_t(\psi) G(\psi_1, \psi_2; \psi).$$

Since $G(\psi_1, \psi_2; \psi)$ is the multiplicity with which the representation $\psi$ enters in the Kronecker product $\psi_1 \times \psi_2$, we have

$$\sum_{\psi} \chi_t(\psi) G(\psi_1, \psi_2; \psi) = \chi_t(\psi_1 \times \psi_2) = \chi_t(\psi_1) \cdot \chi_t(\psi_2).$$

We thus finally obtain

$$M_t(\xi_1 \cdot \xi_2) = \sum_{\psi_1, \psi_2} \xi_1(\psi_1)\xi_2(\psi_2)\chi_t(\psi_1)\chi_t(\psi_2) = M_t(\xi_1)\cdot M_2(\xi_2). \qquad (5.5)$$

It is thus proved that the correspondence $\xi \to M_t(\xi)$ is a homomorphism of the ring $\Xi$ into the field of complex numbers.

We shall now show that all homomorphisms of this ring into the field of complex numbers can be obtained in this way.

In fact, if this were not the case, then there would exist a function $\xi(\psi) \not\equiv 0$ such that

$$M_t(\xi) = f(t) = \sum_{\psi} \xi(\psi)\chi_t(\psi) = 0$$

for all $t$. But since the $\chi_t(\psi)$ form a complete orthogonal system in the space of functions $f(t)$, we have

$$\|f\|^2 = \sum_{\psi} |\xi(\psi)|^2 \quad \text{and} \quad \xi(\psi) = 0.$$

**no. 18. The law of multiplication in $\Xi$.** This no. is devoted to a more detailed study of the ring $\Xi$. Its content is in many ways analogous to that of nos. 6-8.

The character of the representation $\psi$ is given by the formula

$$\chi_t(\psi) = \chi_t(\rho) = \frac{\begin{vmatrix} e^{it_1\rho_1} & \dots & e^{it_1\rho_n} \\ \dots & \dots & \dots \\ e^{it_n\rho_1} & \dots & e^{it_n\rho_n} \end{vmatrix}}{\prod\limits_{p<q} (e^{it_p} - e^{it_q})}, \qquad (5.6)$$

and the dimension of the representation by

$$\dim \psi = \chi_0(\psi) = 1! \cdots (n-1)! \prod_{p<q} (\rho_p - \rho_q) = \mu(\rho_1, \cdots, \rho_n).$$

The numbers $\rho_1, \cdots, \rho_n$, which define the representation, are integers satisfying the relation $\rho_1 > \rho_2 > \cdots > \rho_n$.

Two collections $\rho = (\rho_1, \cdots, \rho_n)$ and $\rho' = (\rho_1', \cdots, \rho_n')$ give equivalent representations if and only if the numbers $\rho_k$ and $\rho_k'$ differ by a common summand for all $k$. (In this case and only in this case, $\chi_t(\rho) = \chi_t(\rho')$.) This fact follows easily from (5.6) if we take into account the fact that, in the group of unitary matrices with determinant $1$, $t_1 + \cdots + t_n = 2k\pi$. In this way, elements of the ring $\Xi$ are functions $\xi(\psi) = \xi(\rho_1, \cdots, \rho_n)$, defined for integers $\rho_1, \cdots, \rho_n$ such that $\rho_1 > \cdots > \rho_n$ and coinciding in case the arguments differ by a common summand.

We consider a ring $\Xi'$, isomorphic to the ring $\Xi$. The elements of $\Xi'$ are functions $\xi'(\rho_1, \cdots, \rho_n)$, defined for all integer values of the arguments, skew symmetric in $\rho_1, \cdots, \rho_n$ and coinciding if the arguments differ by a common summand. Each function $\xi'(\rho_1, \cdots, \rho_n) \in \Xi'$ is completely defined by its values for $\rho_1 > \cdots > \rho_n$. This circumstance permits us to establish an isomorphism between the rings $\Xi$ and $\Xi'$. The norm in the ring $\Xi'$ is given by the formula

$$\|\xi'\| = \frac{1}{n!} \sum_{\rho} |\xi'(\rho_1, \cdots, \rho_n)| \, |\mu(\rho_1, \cdots, \rho_n)|,$$

and the law of multiplication by the formula

$$\xi_1' \cdot \xi_2' = \frac{1}{(n!)^2} \sum_{\rho^{(1)}\rho^{(2)}} \xi_1'(\rho^{(1)}) \xi_2'(\rho^{(2)}) G'(\rho^{(1)}, \rho^{(2)}; \rho),$$

where the function $G'$, skew symmetric in $\rho^{(1)}$, $\rho^{(2)}$, and $\rho$, is defined by the rule that for $\rho_1^{(1)} > \cdots > \rho_n^{(1)}$, $\rho_1^{(2)} > \cdots > \rho_n^{(2)}$, $\rho_1 > \cdots > \rho_n$, $G'(\rho^{(1)}, \rho^{(2)}; \rho) = G(\rho^{(1)}, \rho^{(2)}; \rho) = G(\psi_1, \psi_2; \psi)$. Finally, homomorphisms of the ring $\Xi'$ into the field of complex numbers are given by the formula $\xi' \to \frac{1}{n!} \sum_\rho \xi'(\rho) \chi(t, \rho)$. The function $\chi(t, \rho)$ is given for all integer $\rho_1, \cdots, \rho_n$ by formula (5.6) and for $\rho_1 > \cdots > \rho_n$, we have $\chi(t, \rho) = \chi_t(\rho)$.

The ring $\Xi'$ is constructed in considerable analogy with the ring $R'$ (§2).

For the study of the law of multiplication in $\Xi'$ [that is, for the study of the function $G'(\rho^{(1)}, \rho^{(2)}; \rho)$], we introduce, by analogy with what was done in §2, an operator $\Lambda$, but now not a differential operator, but a finite difference operator,

$$\Lambda = \Lambda_\rho = \frac{1}{n!} \sum_{i<j} \left(\frac{\partial}{\partial \rho_i} - \frac{\partial}{\partial \rho_j}\right),$$

where

$$\frac{\partial \xi(\rho_1, \cdots, \rho_n)}{\partial \rho_k} = \xi(\rho_1, \cdots, \rho_{k-1}, \rho_k + 1, \rho_{k+1}, \cdots, \rho_n) - \xi(\rho_1, \cdots, \rho_k, \cdots, \rho_n).$$

Just as for the operator $L$, the following assertions hold for the operator $\Lambda$: $\Lambda$ carries a function symmetric in $\rho_1, \cdots, \rho_n$ into a skew symmetric function, and conversely; if $\Lambda\xi = 0$ and $\xi$ is a skew symmetric function, then $\xi = 0$. The first of these assertions is obvious, and the second, in complete analogy with the corresponding assertion from no. 6, follows from the fact that the general solution of the equation $\Lambda_\eta = 0$ has the form

$$\eta = \eta_{12}(\rho_1 + \rho_2, \rho_3, \cdots, \rho_n) + \eta_{13}(\rho_1 + \rho_3, \cdots, \rho_n) + \cdots + \eta_{n-1, n}(\rho_{n-1} + \rho_n, \rho_1, \cdots, \rho_{n-2}).$$

We consider the equation

$$\Lambda_r G = \frac{1}{n!} \sum_s \widetilde{\delta}(\rho - sr), \tag{5.7}$$

where $s$ is a certain permutation of the indices $1, \cdots, n$ $(sr = (r_{i_1}, \cdots, r_{i_n}))$, and

$$\widetilde{\delta}(r) = \begin{cases} 1, & \text{if } r_1 = r_2 = \cdots = r_n, \\ 0, & \text{if at least one of the differences } r_i - r_j \text{ is different from } 0. \end{cases}$$

We shall later express the function $G'(\rho^{(1)}, \rho^{(2)}; \rho)$ in terms of the function $G(\rho, r)$, which is a solution of this equation.

In agreement with the properties just established for the operator $\Lambda$, the equation (5.7) has a unique solution in the class of functions skew symmetric in $r_1, \cdots, r_n$. Since the right side of the equation (5.7) is symmetric in $\rho_1, \cdots, \rho_n$, and the operator $\Lambda_r$ does not affect these variables, the function $G(\rho, r)$ is symmetric in the first argument.

As in no. 8, it is easy to verify the identity

$$\Lambda e^{i(t_1\rho_1+\cdots+t_n\rho_n)} = \frac{1}{n!}\prod_{p<q}(e^{it_p}-e^{it_q})e^{i(t_1\rho_1+\cdots+t_n\rho_n)}.$$

Hence the function $\chi(\rho, t)$, given by formula (5.6), is a solution of the equation

$$\Lambda\chi(\rho, t) = \frac{1}{n!}\sum_s e^{i(t_1\rho_{k_1}+\cdots+t_n\rho_{k_n})}. \tag{5.8}$$

Since the function $\chi(\rho, t)$ belongs to the class of functions skew symmetric in $\rho_1, \cdots, \rho_n$, it is the only solution of the equation (5.8) in this class.

As in no. 8, it follows from the foregoing remarks that

$$G(\rho, r) = \frac{1}{(2\pi)^{n-1}} \int_{\substack{-\pi<t_k<\pi \\ t_1+\cdots+t_n=0}} e^{-i(\rho_1 t_1+\cdots+\rho_n t_n)}\chi(r, t)\,dt, \tag{5.9}$$

$$\chi(r, t) = \sum G(\rho, r)e^{i(\rho_1 t_1+\cdots+\rho_n t_n)}. \tag{5.9'}$$

*The formulas* (5.9) *show that the function* $G(\rho, r)$ *has the following interpretation: it shows the multiplicity with which the weights* $\rho_1, \cdots, \rho_n$ *enter in the representation defined by the numbers* $r_1, \cdots, r_n$.

It follows from (5.9) that $\Lambda_\rho G(\rho, r) = \frac{1}{n!}\sum \pm\delta(\rho - sr)$. Hence the operator $\Lambda$ has an inverse in the class of symmetric and skew symmetric functions. This inverse operator is given by the "Green's function" $G(\rho, r)$:

$$(\Lambda^{-1}x)(\rho) = \sum_r G(\rho, r)x(r),$$

if the function $x(r)$ is skew symmetric, and

$$(\Lambda^{-1}x)(r) = \sum_\rho G(\rho, r)x(\rho),$$

if $x(\rho)$ is a symmetric function.

Finally, we consider the ring $\Sigma$ of functions symmetric in $\rho_1, \cdots, \rho_n$, coinciding for arguments that differ by a common summand.

The law of multiplication in $\Sigma$ is given by the formula

$$\zeta_1\cdot\zeta_2(r) = \sum_\rho \zeta_1(r_1-\rho_1, \cdots, r_n-\rho_n)\zeta_2(\rho_1, \cdots, \rho_n).$$

Homomorphisms of this ring into the field of complex numbers are given by the formula

$$\zeta \to \frac{1}{n!}\sum_\rho \zeta(\rho)\sum_s e^{i(\rho_1 t_{k_1}+\cdots+\rho_n t_{k_n})}. \tag{5.10}$$

The operator $\Lambda$ maps the ring $\Sigma$ isomorphically onto the ring $\Xi'$. The proof differs in no essential way from the arguments found in [5], and hence we omit it.

Repeating verbatim the arguments of no. 7, we obtain:

$$\xi_1'\cdot\xi_2'(\rho) = \sum_{\rho^{(1)},\rho^{(2)}} \xi_1'(\rho^{(1)})\xi_2'(\rho^{(2)})G(\rho^{(1)}-\rho^{(2)}, \rho).$$

We recall on the other hand

$$\xi_1' \cdot \xi_2'(\rho) = \frac{1}{n!n!} \sum_{\rho^{(1)}, \rho^{(2)}} \xi_1'(\rho^{(1)}) \xi_2'(\rho^{(2)}) G'(\rho^{(1)}, \rho^{(2)}; \rho),$$

where $G'(\rho^{(1)}, \rho^{(2)}; \rho)$ is a function skew symmetric in each argument, and that the function $G'(\rho^{(1)}, \rho^{(2)}; \rho)$ is completely determined by the last equality and the conditions of symmetry.

*Hence, between the functions* $G(\rho, r)$ *and* $G'(\rho^{(1)}, \rho^{(2)}; \rho)$, *the relation*

$$G'(\rho^{(1)}, \rho^{(2)}; \rho) = \sum_{s_1, s_2} \pm G(s_1\rho_1 - s_2\rho_2; \rho) \tag{5.11}$$

*exists.*

The plus sign appears if the permutations $s_1$ and $s_2$ are both even or both odd. The minus sign appears in the contrary case.

We recall that the function $G(\rho, r)$ is the multiplicity with which the weight $\rho = (\rho_1, \cdots, \rho_n)$ appears in the representation defined by the numbers $r = (r_1, \cdots, r_n)$, while the function $G(\rho^{(1)}, \rho^{(2)}; \rho) = G(\psi_1, \psi_2; \psi)$ shows how many times $\psi$ appears in the decomposition of the representation $\psi_1 \times \psi_2$ into irreducible representations. Hence *formula* (5.11) *shows the connection between two problems of the theory of representations: the decomposition of the Kronecker product of two irreducible representations and the determination of all weights of an irreducible representation.*

Another connection between these problems was pointed out by H. Weyl [8].

We now give the explicit form of the function $G(\rho, r)$ for the groups $SU(2)$ and $SU(3)$.

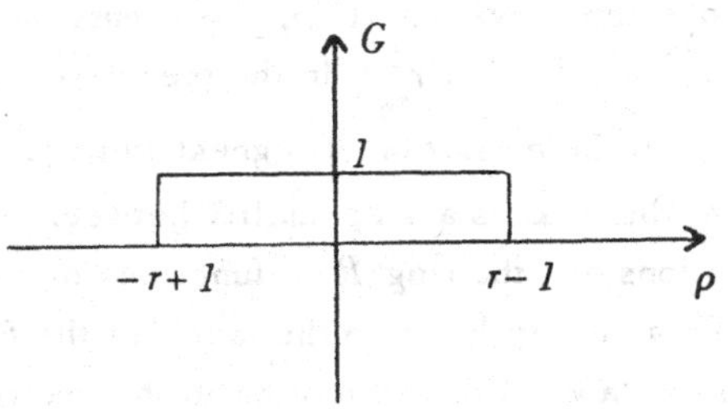


Figure 5.

For $SU(2)$, the function is given by its graph (Figure 5). Here we suppose that $G(\rho, r)$ is different from zero only at integer points of the same parity as $r-1$.

For $SU(3)$, its form is clear from Figure 6.

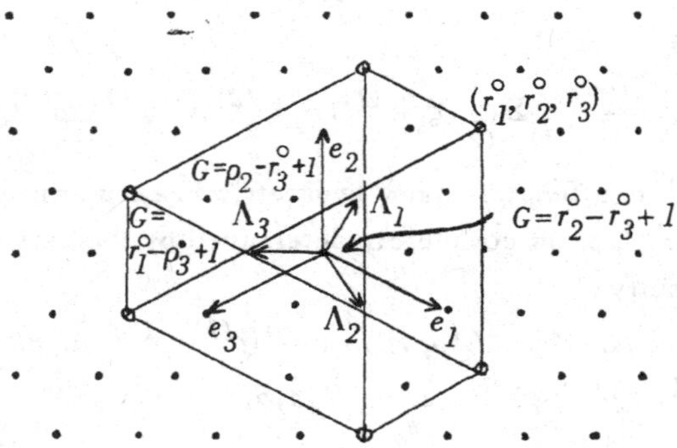


Figure 6.

In this figure, the numbers $\rho_k = \dfrac{(\xi, \Lambda_k)}{(\Lambda_k, \Lambda_k)}$ are the co-ordinates of the vector $\xi$. The vectors $\Lambda_1, \Lambda_2, \Lambda_3$ are chosen in such a way that

$$(\Lambda_i, \Lambda_j) = \begin{cases} \frac{2}{3}, & \text{if } i = j, \\ -\frac{1}{3}, & \text{if } i \neq j. \end{cases}$$

Hence $\rho_1 + \rho_2 + \rho_3 = 0$.

The lattice appearing in Figure 6 is formed by integer linear combinations of $\Lambda_1, \Lambda_2, \Lambda_3$: this is the lattice of the weights of the irreducible representations of $SU(3)$. $(\overset{0}{r}_1, \overset{0}{r}_2, \overset{0}{r}_3)$ is the highest weight of a representation, defined by the numbers $r_1 = \overset{0}{r}_1 + 2$, $r_2 = \overset{0}{r}_2 + 1$, $r_3 = \overset{0}{r}_3$.[1)] The function $G(\rho, r)$ is equal to zero outside of the region outlined, and also inside it at the points not congruent with $(\overset{0}{r}_1, \overset{0}{r}_2, \overset{0}{r}_3)$ modulo the sublattice generated by the vectors

$$e_1, e_2, e_3 \; (e_1 = \Lambda_3 - \Lambda_2,\; e_2 = \Lambda_1 - \Lambda_3,\; e_3 = \Lambda_2 - \Lambda_1).$$

Just as in no. 9, one can prove that $G(\rho, r) = 0$ outside the convex hull of the points $s(\overset{0}{r}_1, \ldots, \overset{0}{r}_n) = (\overset{0}{r}_{i_1}, \ldots, \overset{0}{r}_{i_n})$. In the present case, however, this fact is a trivial consequence of the definition of highest weight.

As we have shown, there exists a deep duality between the ring $\Xi$ of additive functions of representations and the ring $R$ of functions on the group $G$, constant on two-sided cosets. This duality leads to the fact that the functions $G(t, \tau)$ and $G(\rho, r)$, through which the laws of multiplication in the two rings are defined, satisfy similar equations: the function $G(t, \tau)$, depending on the continuous arguments $t$ and $\tau$, has a differential equation, and the function $G(\rho, r)$, depending on discrete arguments, has an analogous difference equation. As we have seen from the examples (Figures 2, 3, 5, and 6), these functions are very similar to each other. The same similarity persists also in the general case: the space of the variables $t, \tau$ can be divided into regions, in each of which the function $G(t, \tau)$

1) Since the numbers $r_1, r_2, r_3$ are defined up to a common summand, they can be chosen so that $\overset{0}{r}_1 + \overset{0}{r}_2 + \overset{0}{r}_3 = 0$.

is a polynomial in $(t^1, \cdots, t^n; r^1, \cdots, r^n)$. The space of the variables $(\rho_1, \cdots, \rho_n; r_1, \cdots, r_n)$ can likewise be divided up into regions, and the function $G(\rho, r)$ in these regions is equal to the very same polynomials in the variables $(\rho_1, \cdots, \rho_n; r_1, \cdots, r_n)$. The difference lies only in the fact that in the latter case, the power $x^n$ must be interpreted as a "symbolic power": $x^n = x(x-1)\cdots(x-n+1)$.

**no. 19. Generalization to the case of an arbitrary semisimple group.** Like the results of §2, the results of this section can be generalized to the case of an arbitrary semisimple group. Here, if $\frac{\partial}{\partial \rho_\alpha}$ denotes the operator of differentiation in direction of the root $\alpha$, then the finite difference operator $\Lambda$ can be written in the form

$$\Lambda = \frac{2^N}{n(S)} \prod_{\alpha > 0} \operatorname{sh}\left(\frac{1}{2}\frac{\partial}{\partial \rho_\alpha}\right),$$

where $N$ is the number of positive roots and $n(S)$ is the order of the Weyl group $S$. (The operator $e^{h\frac{\partial}{\partial x}}$, applied to the function $f(x)$, is equal to $e^{h\frac{\partial}{\partial x}} f(x) = f(x+h)$.)

With the above definition of the operator $\Lambda$, we have:

$$\Lambda \chi(\rho, t) = \frac{1}{n(S)} \sum_{s \in S} e^{i(\rho, st)}.$$

The remaining generalizations are obvious.

The definition of the rings $\Xi$ and $\Xi'$, and also of the functions $G(\rho, r)$ and $G(\rho^{(1)}, \rho^{(2)}; \rho^{(3)})$ are carried over with no change at all to the case of an arbitrary semisimple group. Here $G(\rho, r)$ is the multiplicity of the weight $\rho = (\rho_1, \cdots, \rho_n)$ in the representation given by the vector $r = (r_1, \cdots, r_n)$, $r = \Lambda + g$, where $\Lambda$ is the highest weight of the representation and $g$ is the partial sum of positive roots; the function $G(\rho^{(1)}, \rho^{(2)}; \rho^{(3)}) = G(\psi_1, \psi_2; \psi_3)$ shows how many times the representation $\psi_3$ appears in the decomposition of the Kronecker product $\psi_1 \times \psi_2$ into irreducible representations ($\rho^{(i)} = \Lambda_i + g_i$, where $\Lambda_i$ is the highest weight of the representation $\psi_i$ and $g$ is the partial sum of positive roots).

**no. 20. Generalization of the results of nos. 10 and 11 to arbitrary semisimple and some non-semisimple groups.** Let $G$ be a semisimple Lie group, $G_0$ a maximal compact subgroup. As is known, $G$ contains a regular commutative subgroup $H$ having the property that for every $g \in G$, there exist $g_1 \in G_0$, $g_2 \in G_0$, such that $g_1 g g_2 \in H$, where, if $h = g_1 g g_2 \in H$ and $h' = g_1' g g_2' \in H$, then the elements $h$ and $h'$ go into each other with the aid of the " H. Weyl group" $S$ – a finite group of transformations of $H$. In other words, every two-sided coset of $G$ with respect to $G_0$ has a finite number of representatives in $H$, going into each other with the aid of the group $S$. The group $H$ admits a one-to-one mapping onto its Lie algebra $\mathfrak{H}$ (the so-called "canonical mapping"), with the aid of which canonical co-ordinates are introduced in $H$. The operators of the group $S$, carried over to the algebra $\mathfrak{H}$ by this mapping, are linear transformations on the algebra $\mathfrak{H}$. Thus, *every two-*

*sided coset of $G$ with respect to $G_0$ is given by a vector $t \in \mathfrak{H}$ of Euclidean space, where the same two-sided coset corresponds to two vectors if and only if they go into each other with the aid of a transformation from the group $S$.* (In the case where $G = SL(n)$ is the group of complex unimodular matrices of order $n$ and $G_0 = SU(n)$ is the group of unitary unimodular matrices of order $n$, $H$ is the group of diagonal matrices with positive eigenvalues, and canonical co-ordinates of an element of $H$ are the logarithms of its eigenvalues, then the group $S$ is the group of permutations of the eigenvalues.)

The following theorem was proved in no. 10.

**Theorem 1.** *Let $G$ be a complex semisimple group, and $G_0$ a maximal compact subgroup of $G$. Let $g_1 \in G$ belong to a two-sided coset of $G$ with respect to $G_0$ given by the vector $t^{(1)} \in H$, and let $g_2 \in G$ belong to the two-sided coset given by the vector $t^{(2)} \in H$. Then the vector giving the two-sided coset containing $g_1 g_2$ lies within the convex hull of the vectors $t^{(1)} + st^{(2)}$, where $st^{(2)}$ is the vector $t^{(2)}$ acted on by the transformation $s$ of the group $S$.*

In the present paragraph, this theorem is generalized to the case of real semisimple groups and certain non-semisimple groups.

**A.** Let $G$ be a real semisimple group, and $G_0$ a maximal compact subgroup of $G$. The manifold of cosets $G/G_0$ is, as is known, a symmetric space, and there exists a submanifold $\Theta$ in $G$ serving as a model for it. (Rigid motions in $\Theta$ are given by the formula $\theta \to g\theta g^{*-1}$, where $*$ is the involutory automorphism that describes the subgroup $G_0$.) The subgroup $H \subset G$, the elements of which are representatives of the two-sided cosets of $G$ with respect to $G_0$, is a maximal commutative subgroup contained in the manifold $\Theta$. All maximal commutative subgroups contained in $\Theta$ are conjugates of each other.

We consider the complex semisimple group $[G]$ for which $G$ serves as a real form. Let $G_{[0]}$ be a maximal compact subgroup of $[G]$, and $\Theta_{[\ ]} \subset [G]$ a model of the symmetric space $[G]/G_{[0]}$. Let $H_{[\ ]} \subset \Theta_{[\ ]}$ be a maximal commutative subgroup contained in $\Theta_{[\ ]}$, the elements of which serve as representatives of the two-sided cosets of $[G]$ with respect to $G_{[0]}$, and let $S_{[\ ]}$ be the Weyl group for $H_{[\ ]}$. In agreement with [2] and [15], we may suppose that the following inclusions hold:

$$\left.\begin{aligned} G_0 &\subset G_{[0]}, \\ \Theta &\subset \Theta_{[\ ]} \\ H &\subset H_{[\ ]}. \end{aligned}\right\} \tag{5.12}$$

Besides this, one can show that

$$S \subset S_{[\ ]}.$$

The group $S_{[\ ]}$, as a group of linear transformations in $\mathfrak{H}_{[\ ]}$, preserves the invariant positive definite metric in $\mathfrak{H}_{[\ ]}$ (the Cartan metric). Hence the ends of the vectors of the form $t^{(1)} + st^{(2)}$ lie on the sphere with center at the point $t^{(1)}$ and consequently are extreme points of the convex hull of the vectors $t^{(1)} + st^{(2)}$.

Now let $g_1 \in G$, $g_2 \in G$, $t^{(1)} \in \mathfrak{H}$ the vector giving the coset containing $g_1$, and $t^{(2)} \in \mathfrak{H}$ the vector giving the coset containing $g_2$. It follows from (5.12) that the algebra $\mathfrak{H}$ is a subspace in the algebra $\mathfrak{H}_{[\ ]}$. In agreement with Theorem 1, the vector $t$ giving the coset containing $g_1g_2$ lies within the convex hull of the vectors $t^{(1)}+ st^{(2)}$, where $s$ is an arbitrary element of $S_{[\ ]}$. On the other hand, it is obvious that $t \in \mathfrak{H}$. Hence $t$ lies in the intersection of the convex hull of the vectors $t^{(1)} + st^{(2)}$, $s \in S_{[\ ]}$, and $\mathfrak{H}$. Since the ends of the vectors $t^{(1)} + st^{(2)}$ are extreme points for this convex hull, its intersection with $\mathfrak{H}$ is the convex hull of the vectors $t^{(1)} + st^{(2)}$ when $s \in S_{[\ ]}$ and $st^{(2)} \in \mathfrak{H}$. As will be shown in another place, these vectors can be obtained using only elements $s \in S \subset S_{[\ ]}$. Thus we have proved

**Theorem 2.** *Let $G$ be a semisimple Lie group, and $G_0$ a maximal compact subgroup. Let $g_1, g_2 \in G$ belong to two-sided cosets of $G$ with respect to $G_0$ that are given, respectively, by vectors $t^{(1)}, t^{(2)} \in \mathfrak{H}$. Then the vector giving the coset containing $g_1g_2$ lies within the convex hull of the vectors $t^{(1)} + st^{(2)}$.*

B. Let $G$ be a semisimple Lie group, and $G_0$ a maximal compact subgroup. The Lie algebra $\mathfrak{G}$ of the group $G$ is decomposed into the direct sum of subspaces $\mathfrak{G} = \mathfrak{G}_0 \dot{+} \mathfrak{R}$, where $\mathfrak{G}_0$ is the subalgebra corresponding to the subgroup $G_0$ and $\mathfrak{R}$ is the complementary subspace, invariant under the operators of the representation of the group $G_0$, induced by the regular representation of $G$. (The space $\mathfrak{R}$ is the tangent space to the manifold $\theta$, the model of the space $G/G_0$.) We denote by the letter $\Gamma$ the representation arising in this way in $\mathfrak{R}$. With the aid of the group $G_0$ and its representation $\Gamma$, one can construct the group $\tilde{G} = G_0 \times \mathfrak{R}$ (see [16]) having commutative radical. Elements of $\tilde{G}$ are pairs: $\tilde{g} = (g_0, \mathfrak{r})$, $g_0 \in G_0$, $\mathfrak{r} \in \mathfrak{R}$, and multiplication is given by the formula $\tilde{g}_1\tilde{g}_2 = (g_{01}g_{02}, \Gamma(g_{02}^{-1})\mathfrak{r}_1 + \mathfrak{r}_2)$. We denote the group so obtained by $\tilde{G}$, in token of the fact that it is constructed with the help of the semisimple group $G$.

Let $H \subset G$ be a commutative subgroup, consisting of representatives of two-sided cosets of $G$ with respect to $G_0$. The subalgebra $\mathfrak{H}$ of the algebra $\mathfrak{G}$ corresponding to it lies in the subspace $\mathfrak{R}$. It is easy to verify that every two-sided coset of $\tilde{G}$ with respect to $G_0$ has a finite number of representatives in $\mathfrak{H}$, and that they are all carried into each other with the help of the group $S$, the same group with the aid of which the vectors of $\mathfrak{H}$ (giving the two-sided cosets of $G$ with respect to $G_0$) are carried into each other. The following theorem holds.

**Theorem 3.** *Let $\tilde{G}$ be a Lie group with commutative radical, constructed with the help of a semisimple group $G$. Let $G_0 \subset \tilde{G}$ be a maximal compact subgroup of*

*$\tilde{G}$, and let $g_1, g_2 \in \tilde{G}$ belong to two-sided cosets with respect to $G_0$, representatives of which in $\mathfrak{H}$ are $t^{(1)}$ and $t^{(2)}$ respectively. Then $g_1 g_2 \in \tilde{G}$ belongs to a two-sided coset with respect to $G_0$ whose representative in $\mathfrak{H}$ lies within the convex hull of the vectors $t^{(1)} + st^{(2)}$.*

The proof of Theorem 3 is carried out in two steps: one first considers the case in which the original group $G$ is a complex semisimple group, and then the general case.

In the case where the original group $G$ is complex and semisimple, the zonal spherical functions on the manifold $\mathfrak{N} = \tilde{G}/G_0$ can be explicitly computed. These functions are obviously functions on $\mathfrak{H}$, and they can be given by the formula

$$\Phi_\rho(t) = c \frac{\sum_s \pm e^{i(\rho, st)}}{\prod_{\alpha>0} (\alpha, \rho) \cdot \prod_{\alpha>0} (\alpha, t)}, \tag{5.13}$$

where $\alpha > 0$ is a root of the algebra $\mathfrak{H} \subset \mathfrak{G}$, positive in the sense of a certain lexicographic ordering, and $c$ is a factor independent of $\rho$ and $t$, defined by the condition $\phi_\rho(0) = 1$. (Computation of the function $\Phi_\rho(t)$ will be carried out in another place.)

The proof of Theorem 3 for this case uses the explicit form of the function $\Phi_\rho(t)$. It differs in no way from the proof of Theorem 1, found in nos. 7-10. (It is necessary merely to replace the function

$$\Delta(t) = \prod_{\alpha>0} \operatorname{sh} \frac{(\alpha, t)}{2},$$

by $\Delta(t) = \prod_{\alpha>0} (\alpha, t)$ everywhere where the former function occurs in nos. 7-10.)

Now let $G$ be an arbitrary real semisimple group, $[G]$ the complex group for which $G$ is the real form, $\tilde{G} = G_0 \times \mathfrak{N}$ the group with commutative radical constructed with the aid of $G$, $\mathfrak{H} \subset \mathfrak{N}$ the subspace of $\mathfrak{N}$ consisting of representatives of two-sided cosets of $\tilde{G}$ with respect to $G_0$, and $S$ the Weyl group for $\mathfrak{H}$. Let $G_{[0]}$, $\mathfrak{N}_{[\ ]}$, $\mathfrak{H}_{[\ ]}$, and $S_{[\ ]}$ be the analogous groups constructed from $[G]$. The inclusions (5.12) imply, obviously, the following inclusions:

$$G_0 \subset G_{[0]},$$
$$\mathfrak{N} \subset \mathfrak{N}_{[\ ]},$$
$$\mathfrak{H} \subset \mathfrak{H}_{[\ ]},$$
$$S \subset S_{[\ ]}.$$

The rest of the proof of Theorem 3 repeats the arguments of paragraph A of the present no. and can hence be omitted.

In no. 11, Theorem 1 is used to infer properties of the eigenvalues of the product of Hermitian matrices. Similar properties can be inferred from Theorems 2 and 3.

For example, let $\mathfrak{N}$ be the space of symmetric matrices with trace $0$ and group of rigid motions acting according to the formula: $\mathfrak{n} \to g_0 \mathfrak{n} g_0^{-1} + \mathfrak{n}_1$ ($g_0$ is an orthogonal matrix, and the group $\widetilde{G}$ is formed from the group $G$ of real unimodular matrices). The indicated inference from Theorem 3 is: the collection $(t_1, \cdots, t_n) = t$ of eigenvalues of the sum $\mathfrak{n}_1 + \mathfrak{n}_2$ *of symmetric matrices with trace* $0$ *is a vector of* $(n-1)$*-dimensional space, lying within the convex hull of the vectors* $t^{(1)} + st^{(2)}$, *where* $t^{(i)} = (t_1^{(i)}, \cdots, t_n^{(i)})$ *is the collection of eigenvalues of the matrix* $\mathfrak{n}_i$ *and* $s$ *is an arbitrary permutation of the eigenvalues.* This theorem was obtained earlier by Lidskiĭ [6].

## BIBLIOGRAPHY


[1] É. Cartan, *Geometry of Lie groups and symmetric spaces*, Foreign Literature Press, Moscow, 1949.

[2] É. Cartan, *Sur certaines formes riemanniennes remarquables des géométries à groupe fondamental simple*, Ann. Sci. École Norm. Sup. (3) **44** (1927), 345-467.

[3] É. Cartan, *Sur une classe remarquable d'espace de Riemann*, Bull. Soc. Math. France **54** (1926), 214-264; **55** (1927), 118-134.

[4] I. M. Gel'fand and M. A. Naĭmark, *Unitary representations of a unimodular group containing an identity representation of the unitary subgroup*, Trudy Moskov. Mat. Obšč. 1 (1952), 423-475. (Russian)

[5] I. M. Gel'fand, *Spherical functions in symmetric Riemannian spaces*, Dokl. Akad. Nauk SSSR **70** (1950), 5-8. (Russian)

[6] V. B. Lidskiĭ, *On characteristic numbers of the sum and product of symmetric matrices*, Dokl. Akad. Nauk SSSR 75 (1950), 769-772. (Russian)

[7] R. Godement, *A theory of spherical functions*, I, Trans. Amer. Math. Soc. **73** (1952), 496-556.

[8] H. Weyl, *The classical groups, their invariants and representations*, Princeton Univ. Press, Princeton, N. J., 1939.

[9] I. M. Gel'fand and M. A. Naĭmark, *Unitary representations of the classical groups*, Trudy Mat. Inst. Steklov., vol. 36, Izdat. Akad. Nauk SSSR, Moscow-Leningrad, 1950. (Russian)

[10] E. B. Dynkin and A. L. Oniščik, *Compact global Lie groups*, Uspehi Mat. Nauk (N.S.) 10 (1955), 3-74. (Russian)

[11] É. Cartan, *Sur la détermination d'un système orthogonal complet dans un espace de Riemann symétrique clos*, Rend. Circ. Mat. Palermo **53** (1929), 217-252.

[12] M. N. Olevskiĭ, *Some mean value theorems in spaces of constant curvature*,

Dokl. Akad. Nauk SSSR 45 (1944), 95-98. (Russian)

[13] I. M. Gel'fand, *The center of an infinitesimal group ring*, Mat. Sb. N.S. 26 (68), (1950), 103-112. (Russian)

[14] I. M. Gel'fand and M. L. Cetlin, *Finite-dimensional representations of the group of unimodular matrices*, Dokl. Akad. Nauk SSSR 71 (1950), 825-828. (Russian)

[15] F. I. Karpelevič, *Surfaces of transitivity of a semisimple subgroup of the group of motions of a symmetric space*, Dokl. Akad. Nauk SSSR 93 (1953), 401-404. (Russian)

[16] F. A. Berezin, *Linear finite-dimensional representations of Lie groups with a commutative radical*, Dokl. Akad. Nauk SSSR 93 (1953), 759-761. (Russian)

[17] Harish-Chandra, *On some applications of the universal enveloping algebra of a semisimple Lie algebra*, Trans. Amer. Math. Soc. 70 (1951), 28-96.

[18] R. Courant and D. Hilbert, *Methoden der mathematischen Physik*, vol. I, Interscience, New York, 1943.

Translated by:
Edwin Hewitt

# 2.
# (with S.V. Fomin)

# Geodesic flows on manifolds of constant negative curvature

Usp. Mat. Nauk 7 (1) (1952) 118–137
[Transl., II. Ser., Am. Math. Soc. 1 (1955) 49–65]. Zbl. 66:361

Dynamical systems with continuous spectrum have been little studied, and the known examples of such systems are few.* Among the first such systems to be studied are the very interesting cases called the geodesic flows on manifolds of constant negative curvature. Indeed, the latter represent considerably more than simply one of the examples of dynamical systems with continuous spectrum. In fact, it follows from the variational principles of mechanics that the change in the course of time of the states of any mechanical system having $n$ degrees of freedom may be expressed as the geodesic motion in an $n$-dimensional manifold under a suitable metric.** The assumption that the geodesics are considered on a space of *negative* curvature has to do with the fact that in the case of positive or zero curvature we may obtain, as is known, dynamical systems decomposable into separate trajectories.

The geometry of manifolds of negative curvature is characterized by the instability of its geodesics. More precisely, if the curvature at every point lies between fixed negative bounds, then the distance between two infinitely near geodesics increases exponentially (at least in one direction). This circumstance implies that if the geodesics cannot go to infinity (for example, if the given manifold is compact), then the points on each geodesic are distributed uniformly on the whole surface (in contradistinction, for example, to the geodesics on a sphere).

Geodesic flows on manifolds of negative curvature have been considered in the last two decades by several authors (E. Hopf, G. A. Hedlund, et al.). The basic results obtained here are presented, e.g., in the paper of E. Hopf [3], where the literature on this question is also cited. These results essentially consist of the proof, under the usual conditions, of the ergodicity of such systems and of the existence of strong mixing.

The present paper is dedicated to the investigation of geodesic flows on manifolds of constant negative curvature. The group-theoretic method exploited here is not immediately applicable to the case of variable curvature. On the other hand, for manifolds of constant negative curvature it affords the possibility of finding the spectrum of the corresponding dynamical system; namely, of showing that it is Lebesgue— a result which up to this time has been obtainable by no other method. Knowledge of the spectrum permits the immediate establishing for the case of constant curvature the theorems previously proved by Hopf and Hedlund on ergodicity and strong mixing.

The determination of the spectral multiplicity of the geodesic flow is a very

*Despite the fact that the continuous spectrum is in a sense the *probable* case and the point spectrum the *singular* (cf. [1], p. 103).

**Cf., for example, [2], p. 184 et seq.

interesting question. Here connections are revealed between unitary representations of the group of motions of the Lobačevskiĭ plane and the theory of automorphic forms. It is possible that corresponding generalizations of the relationships found here will come to light in the general theory of automorphic forms (automorphic forms, connections with quadratic forms, matrices, etc.).

To begin with we consider the two-dimensional case, i.e., the geodesic flow on a surface (§1–3). Next the $n$-dimensional case is taken up (§4). The following paragraph (§5) contains the generalization of the method used in the paper to the case of arbitrary dynamical systems in homogeneous spaces. Here it is made clear that the geodesic flow on a surface of constant negative curvature on the one hand, and the well-known construction of dynamical systems with pure point spectrum on compact abelian groups on the other hand, are each special cases of one and the same construction–dynamical systems on the cosets of some locally compact Lie group.

The basic tool in the following paper is the theory of unitary representations of locally compact Lie groups.

**§1. Definition of the geodesic flow on a surface of constant negative curvature.**

Let $F$ be some surface of constant negative curvature. We shall assume this surface complete in the sense that every geodesic on it may be continued indefinitely. By a *line element* of such a surface is meant some point of it together with a specific direction at that point, or, what amounts to the same thing, with a fixed geodesic passing through the point. Let $H$ be the set of all such line elements. We set up a dynamical system on the manifold $H$ by requiring that each element of $H$ move with unit velocity along the geodesic corresponding to its line element. An invariant measure $\mu$ on $H$ is determined in the following manner: if $d\sigma$ is the area differential on $F$ and $d\theta$ the differential of angle defined by the direction of the geodesic, then

$$d\mu = d\sigma\, d\theta. \tag{1}$$

This dynamical system is called *the geodesic flow on the surface $F$.*

We now determine what the geodesic flow on $F$ is in group-theoretic terms, for this will be needed in what follows.

For any surface $F$ of constant negative curvature the Lobačevskiĭ plane (which we represent as the upper half of the complex plane) is the universal covering surface. Therefore every such surface $F$ may be obtained from the complete Lobačevskiĭ plane by identifying points of this plane which are congruent under some discrete group $D$ of non-euclidean motions.* Consequently the manifold $H$ may be obtained from the manifold of line elements of the Lobačevskiĭ plane by similar identifications.

It may be established that there is a one-to-one correspondence between the line elements of the Lobačevskiĭ plane and its motions. In fact, if one of the line elements is fixed, then for any other there exists one and only one motion taking the first element into the second. From this follows the correspondence between all the motions of the Lobačevskiĭ plane and its line elements. Exactly as is shown above, two motions of the hyperbolic plane correspond to the same line

*It is easily seen that $D$ must be isomorphic to the fundamental group of the surface $F$.

element of the surface $F$ in case they belong to the same coset of the subgroup $D$. In this manner the manifold $H$ may be considered as the set of all cosets of some discrete subgroup in the group of all motions of the Lobačevskiĭ plane.

As is well known, the group of all hyperbolic motions is isomorphic to the group $G$ of all second order real matrices of determinant one. (In fact, if the Lobačevskiĭ plane is realized as the upper half plane, then its motions consist of linear-fractional transformations $w=\frac{\alpha z+\beta}{\gamma z+\delta}$, with $\alpha$, $\beta$, $\gamma$, $\delta$, real and $\alpha\delta-\beta\gamma=1$.) In this manner we finally *represent the manifold $H$ as the totality of cosets $Dg$ in the group $G$ of real matrices of the second order with unit determinant, where $D$ is some discrete subgroup.* (A typical example for such a subgroup $D$ is the modular group, consisting of all integral matrices belonging to $G$.)

It is easy to verify that the uniform motion of line elements along geodesics may be expressed in terms of the group in the following manner. *Let $\{g_t\}$ be the one-parameter subgroup of $G$ consisting of elements of the form*

$$g_t=\begin{Vmatrix} e^{-t} & 0 \\ 0 & e^{t} \end{Vmatrix}.$$

*Then the point (i.e., the coset) $Dg$ moves in time $t$ to the point $Dgg_t$.*

We now consider the whole Lobačevskiĭ plane and take as the *initial* line element on it the point $i$ and the direction the imaginary semi-axis. This axis itself will determine a geodesic along which the line element will move. Under the defined mapping of line elements of the group $G$, the *initial* element corresponds to the identity $e$ of $G$. Then the element $g_t \in G$ corresponds to the line element consisting of the point $e^{-2t}i$ and the direction of the axis. Thus the mapping

$$e\longrightarrow g_t$$

actually establishes the motion of the *initial* element on the geodesic corresponding to it. Making use of the expression for distance in non-euclidean geometry

$$\rho(a,\, b)=-\frac{1}{2}\ln\left(\frac{b-\alpha}{b-\beta}:\frac{a-\alpha}{a-\beta}\right),$$

where $\alpha$ and $\beta$ are the ends of the geodesic passing through the points $a$ and $b$, we find that the velocity of the motion is one.

Consider now an arbitrary line element of the Lobačevskiĭ plane. It is obtained from the initial element with the aid of some motion $g\in G$. This motion transforms the geodesic along which the initial element moves (i.e., the imaginary half-axis) into the geodesic on which the given line element moves. The *initial* geodesic may be identified with the subgroup $\{g_t\}\subset G$. Consequently the second of the geodesics we are considering may be identified with the totality of elements of the form $gg_t$. From this it follows that the mapping

$$g\longrightarrow gg_t$$

determines the required motion of the line element.

Going from the complete Lobačevskiĭ plane to some surface $F$ involves replacing the elements $g$ by the cosets $Dg$.

By direct substitution of the variables it is easy to show that the invariant measure $\mu$ in the space of line elements, defined by $d\mu=d\sigma\, d\theta$, coincides with that measure which is determined on $H$ in a natural fashion as a measure on the cosets of $G$ mod $D$. We shall go through the analogous calculation in §3, where we must go from a dynamical system defined on the coset space $G$ mod $D$ to the corresponding surface $F$ (realized as a fundamental region of the group $D$ in the upper half plane.)

For the case when $D$ is the modular group we have the following interpretation of the dynamical system $H$. Consider the set of all possible lattices in the plane. A one-to-one correspondence may be established between the elements of $H$ and the lattices in the euclidean plane with cell area equal to one. In fact, the elements of the group $G$ (matrices of the second order) may be considered as affine mappings of the euclidean plane. Of these, the elements of the modular group $D$ and only those have the property that they take the integral lattice into itself. Each element of $G$ not belonging to $D$ takes the lattice of integers into some other lattice; two elements $g_1$ and $g_2$ take the integer lattice into the same lattice if and only if $g_1$ and $g_2$ belong to the same coset mod $D$. Thus, in this manner $H$ may be realized as the totality of possible lattices in the euclidean plane. The motions $g_t$ stretch the cells $e^t$ times in one direction and shrink them in the other (the hyperbolic deformation).

## §2. The Dynamical System $H$ and Unitary Representations of the Group $G$. The Spectrum.

Let $\mathfrak{H}$ designate the space of all measurable and square-summable (with respect to the invariant measure $\mu$) functions $\phi(h)$ on $H$. Each element $g \in G$ defines a unitary operator $U_g$ in $\mathfrak{H}$:

$$U_{g_0} f(h) = f(hg_0) \qquad (h = Dg \in H). \tag{1}$$

These operators form a unitary representation of the group $G$. Since every representation of any group is simultaneously a representation of each of its subgroups, we obtain a unitary representation of the one-parameter subgroup $\{g_t\}$ which defines the motion in the dynamical system $H$.

Let us now begin our principal task–the determination of the spectrum of the dynamical system $H$.This spectrum is that of the one-parameter group of unitary opertators $U_t$,*corresponding to the elements of the subgroup $\{g_t\}$. In order to determine this spectrum we use the following fact: every unitary representation of a locally compact group may be decomposed, in a sense which we shall make precise below, into irreducible representations.** Therefore we shall first find the spectrum of the operators $U_t$ in each of the irreducible representations of the group $G$. From this, with the help of the above mentioned decomposition of an arbitrary representation into irreducible parts, it is easy to obtain the spectrum of the operators $U_t$ in the representation (1), i.e., the spectrum of the dynamical system.

*More precisely, the group of operators $U_t$, as any continuous one-parameter group of unitary operators, may be represented, according to the theorem of Stone (cf., for example, [5], pg. 100) in the form

$$U_t = e^{itA},$$

where $A$ is some hypermaximal operator. The spectrum of $A$ is called the spectrum of the dynamical system.

**This result was first obtained by A. N. Kolmogorov and announced February 2, 1944 in a session of the Moscow Mathematical Society. Later this theorem was obtained independently by Mautner [6].

The irreducible unitary representations of the real unimodular group of the second order were first obtained in Bargmann's paper [4]. Not counting the identity representation, they reduce to the following types.

1) The principal series.

Let $g=\left\|\begin{smallmatrix}\alpha & \beta\\ \gamma & \delta\end{smallmatrix}\right\|$ be a real second order matrix with determinant 1. We consider the space of functions square-summable on the line. The scalar product is defined in the ordinary manner, i.e., as

$$\int_{-\infty}^{\infty} f_1(x)\overline{f_2(x)}\,dx.$$

To every element $g \in G$ corresponds the operator

$$U_g f(x) = f\left(\frac{\alpha x+\gamma}{\beta x+\delta}\right)|\beta x+\delta|^{-1+i\rho}, \tag{3}$$

where $\rho$ is an arbitrary real number.

The second principal series is determined in the same space by the formula

$$U_g f(x) = f\left(\frac{\alpha x+\gamma}{\beta x+\delta}\right)|\beta x+\delta|^{-1+i\rho}\operatorname{sign}(\beta x+\delta). \tag{3'}$$

2) The complementary series.

Every representation of this series consists of the operators defined for $g \in G$ by the formula

$$U_g f(x) = f\left(\frac{\alpha x+\gamma}{\beta x+\delta}\right)|\beta x+\delta|^{-1-\rho} \tag{4}$$

acting in the space of measurable functions of the line for which the scalar product is determined as follows:

$$(f_1, f_2) = \int_{-\infty}^{\infty}\int_{-\infty}^{\infty} f_1(x_1)\overline{f_2(x_2)}\,|x_1-x_2|^{-1-\rho}\,dx_1\,dx_2.$$

The parameter $\rho$ takes on all real values between $0$ and $1$.

3) The discrete series.

In the space of functions $f(z)$ of the complex variable $z$, analytic in the half plane $\operatorname{Im} z>0$, let a scalar product be defined by the formula

$$(f_1, f_2) = \int f_1(z)\overline{f_2(z)}\,|y|^{n-1}\,dx\,dy \qquad (z=x+iy),$$

where the integral is taken over the whole upper half plane and $n$ is a natural number.

To each element $g=\left\|\begin{smallmatrix}\alpha & \beta\\ \gamma & \delta\end{smallmatrix}\right\|$ we correspond the operator

$$U_g f(z) = f\left(\frac{\alpha z+\gamma}{\beta z+\delta}\right)(\beta z+\delta)^{-n} \qquad (n=1, 2, \ldots). \tag{5}$$

The number $n$ is called the weight of the given representation.

In addition, there exists a second discrete series, operating on those functions which are conjugates of analytic functions, with the corresponding weights being negative.

Let us find the spectrum of the operators $U_t$, corresponding to the elements

$$g = g_t = \left\|\begin{matrix} e^{-t} & 0\\ 0 & e^{t}\end{matrix}\right\| \tag{6}$$

in each of these representations. Substituting in (3) the terms $\alpha$, $\beta$, $\gamma$, and $\delta$ from (6), we obtain for the principal series

$$U_t f(x) = f(e^{-2t}x)\, e^{-t+i\rho t}. \tag{7}$$

Set $U_t = e^{iAt}$. Differentiating equation (7) with respect to $t$ at $t=0$ (which is legitimate) we obtain the following expression for the operator $A$:

$$iAf(x) = -2xf'(x) + f(x)(\rho i - 1).$$

This is a differential operator of the first order. Its spectrum is doubly-multiple Lebesgue.* In fact, for $x>0$ set $x=e^t$, $f(x)=e^{-t/2}\phi(t)$. Then in the space of functions $f(x)$ defined for $x>0$, the operator $iA$ takes the form

$$iA\varphi(t) = -2\frac{d\varphi}{dt} + i\rho\varphi,$$

i.e., $A-\rho E = 2i\,d/dt$ and, as is known, the spectrum of $A$ in the subspace of functions defined for $x>0$ is simply Lebesgue. Similarly we get the same result for the space of functions defined for $x<0$.

In exactly the same way it can be proved that the operators $U_t$ have Lebesgue spectrum in the cases of the discrete or complementary series. Since there are no other irreducible representations of the group $G$, this proves that in every irreducible representation of $G$ the one parameter subgroup of operators $U_t$ (corresponding to the elements (6)) has Lebesgue spectrum.

In order to find the spectrum of the operators $U_t$ in the space $\mathfrak{H}$ some auxiliary considerations are necessary.

Definition 1. Let $\sigma(a)$ be some non-decreasing, continuous from the right, real function of the parameter $a$, $-\infty<a<\infty$, and let $\mathfrak{H}_a$ be some (separable) Hilbert space for each $a$. Each space $\mathfrak{H}_{a_0}$ will be realized as a space of sequences $\{x_1(a_0), x_2(a_0), \ldots, x_n(a_0), \ldots\}$. Then the Hilbert space $\mathfrak{H}$, the elements of which are all the sequences of $\sigma$-measurable functions $\{x_n(a)\}$, $n=1, 2, \ldots$, such that

$$\int_{-\infty}^{\infty}\sum_{n=1}^{\infty} |x_n(a)|^2\, d\sigma(a) < \infty,$$

is called the generalized direct sum of the spaces $\mathfrak{H}_a$ with respect to the measure $\sigma(a)$ and is written

$$\mathfrak{H} = \int \oplus\, \mathfrak{H}_a.$$

To each $x\in\mathfrak{H}$ for fixed $a=a_0$ there corresponds a sequence of numbers $x_n(a_0)$ such that $\sum |x_n(a_0)|^2<\infty$, that is, some element of $\mathfrak{H}_{a_0}$. In this manner each element $x\in\mathfrak{H}$ may be considered as a function of $a$, taking values in the corresponding $\mathfrak{H}_a$: $x \leftrightarrow x(a)$. We set

$$(x, y) = \int_{-\infty}^{\infty} (x(a), y(a))\, d\sigma(a),$$

where $(x(a), y(a))$ designates the scalar product in $\mathfrak{H}_a$.

The above mentioned theorem of Kolmogorov-Mautner on the decomposition of

* We call the spectrum of an operator *absolutely continuous*, if all its spectral measures are absolutely continuous. An absolutely continuous spectrum is called *Lebesgue*, if each of its corresponding spectral measures is equivalent to ordinary Lebesgue measure.

an arbitrary unitary representation of a locally compact group into irreducible parts may be formulated in the following way. Let $\{U_g\}$ be some unitary representation of $G$ acting in the Hilbert space $\mathfrak{H}$. Then $\mathfrak{H}$ may be split into a generalized direct sum of spaces $\mathfrak{H}_\alpha$, in each of which there is defined a unitary representation $U_{g\alpha}$ of $G$ such that for almost all $a$ the corresponding representations $U_{g\alpha}$ are irreducible, and the operators $U_g$ may be expressed in the form:

$$U_g(x) = \{U_{g\alpha}(x(\alpha))\}.$$

In order to apply this theorem to the determination of the spectrum of the dynamical system $H$ we need the following:

**Lemma.** *Let $U$ be a unitary operator defined in the space $\mathfrak{H}=\int\oplus\mathfrak{H}_\alpha$ such that almost all of the operators $U_\alpha$ have Lebesgue spectrum. Then the operator $U$ in $\mathfrak{H}$: $Ux=\{U_\alpha(x(a))\}$ itself has Lebesgue spectrum.*

**Proof.** Let $E_\alpha(\Delta)$ be the resolution of the identity for the operator $U_\alpha$ in $\mathfrak{H}_\alpha$. It is possible to extend the function $E_\alpha(\Delta)$ from the intervals to every Borel set. Then the fact that $U_\alpha$ has Lebesgue spectrum means that $(E_\alpha(M)x(a), x(a))$ equals zero for every set $M$ of Lebesgue measure zero, and is non-zero for a suitable choice of $x(a)$ for each $M$ of positive Lebesgue measure. Clearly, if $E(M)$ is the resolution of the identity for $U$, then

$$(E(M)x, x) = \int_{-\infty}^{\infty} (E_\alpha(M)x(\alpha),\ x(\alpha))\, d\sigma(\alpha). \tag{8}$$

The fact that $(E(M)x, x)=0$ for every $x$ if the Lebesgue measure of $M$ equals zero follows immediately from (8). On the other hand, if $(E(M)x, x)=0$ for every $x$, then $(E_\alpha(M)x(a), x(a))=0$ for almost all $a$ (in the sense of the measure $\sigma(a)$) and, since almost all $U_\alpha$ have Lebesgue spectrum, the Lebesgue measure of the set $M$ is zero.

Now we can quickly find the spectrum of the dynamical system $H$. In fact, since every unitary representation of a locally compact group may be decomposed (in the sense of definition 1) into irreducible unitary representations, from the lemma we obtain:

**Theorem 1.** *The geodesic flow on an arbitrary surface of constant negative curvature has Lebesgue spectrum.*

Let us assume now that the discrete subgroup $D$ of $G$ is such that the space $H$ has finite measure (in other words, we consider the geodesic flow on a surface of finite area). We shall show that this dynamical system is ergodic.

In this case, the representation in $\mathfrak{H}$ is decomposed into a sum of irreducible representations of the above type and the identity representation. The identity representation is contained in these irreducible parts with multiplicity one; this follows immediately from the fact that the manifold $H$ is acted on transitively by the group of transformations $G$ in the sense that any point of $H$ may be carried into any other point by some element of $G$, and consequently there are no functions on $H$ other than constants which are invariant with respect to all $g\in G$.

Let us now show that every bounded measurable function on $H$, invariant with respect to all the $U_t$, is constant. In fact, if $x\in\mathfrak{H}$ is invariant relative to the $U_t$, then under the decomposition of $\mathfrak{H}$ into $\mathfrak{H}_\alpha$ (cf. the lemma) $x(a)\in\mathfrak{H}_\alpha$ will be invariant with respect to $U_{ta}$ for almost all $a$. But because the spectrum of $U_t$ was proved continuous in the infinite-dimensional representations, $x(a)=0$ for the $a$ corresponding to these. Since the identity representation is contained with multiplicity one, the invariant function must be constant. Thus we have obtained: *the geodesic flow on a surface of constant negative curvature having finite area is ergodic.* This result was first proved by Hopf by another method.

As is known (cf., for example, [1], pg. 101), every ergodic dynamical system with Lebesgue spectrum is strongly mixing. Thus the occurrence of strong mixing in $H$ follows from its ergodicity and Theorem 1.

## §3. The Connection with Automorphic Forms. The Multiplicity of the Spectrum.

In Theorem 1 the question of the multiplicity of the spectrum of the dynamical system $H$ remained open. Let us try to fill in this gap. Clearly the spectrum of the dynamical system $H$ will be of countable multiplicity if $\mathfrak{H}$ decomposed into infinitely many irreducible components. A complete clarification of the question of which irreducible representations occur in the decomposition of the representation in $\mathfrak{H}$ has been successful only for the discrete series.

Every function on $H$ may be considered as a function on all of $G$ which is constant on cosets of $D$. An element $g \in G$, i.e., a matrix of the second order, may be written as a pair of complex numbers:

$$g = \begin{Vmatrix} g_{11} & g_{12} \\ g_{21} & g_{22} \end{Vmatrix} \to (\omega_1, \omega_2),$$

where

$$\omega_1 = g_{11} + i g_{12}, \qquad \omega_2 = g_{21} + i g_{22}.$$

The numbers $\omega_1$ and $\omega_2$ are not independent; in order that the determinant of each matrix $g \in G$ be equal to $1$, we must have

$$\operatorname{Im} \overline{\omega}_1 \omega_2 = 1. \tag{9}$$

Now let

$$d = \begin{Vmatrix} m & n \\ p & q \end{Vmatrix} \tag{10}$$

be some element of the discrete subgroup $D$. Then

$$dg = \begin{Vmatrix} m & n \\ p & q \end{Vmatrix} \begin{Vmatrix} g_{11} & g_{12} \\ g_{21} & g_{22} \end{Vmatrix} = \begin{Vmatrix} mg_{11} + ng_{21} & mg_{12} + ng_{22} \\ pg_{11} + qg_{21} & pg_{12} + qg_{22} \end{Vmatrix}$$

Thus $dg$ corresponds to the pair of complex numbers

$$\omega_1' = m\omega_1 + n\omega_2, \quad \omega_2' = p\omega_1 + q\omega_2,$$

i.e., the pair $(\omega_1', \omega_2')$ is obtained from the pair $(\omega_1, \omega_2)$ by means of a linear transformation defined by the matrix (10). Hence each coset of $G$ mod $D$ may be realized as the totality of pairs of complex numbers (satisfying condition (9)), equivalent with respect to the linear transformations corresponding to elements of the subgroup $D$. Set

$$\frac{\omega_1}{\omega_2} = \tau, \quad \frac{\omega_1'}{\omega_2'} = \tau'.$$

Then

$$\tau' = \frac{\omega_1'}{\omega_2'} = \frac{m\omega_1' + n\omega_2'}{p\omega_1' + q\omega_2'} = \frac{m\tau + n}{p\tau + q}$$

(a linear fractional substitution).

Instead of the pair $(\omega_1, \omega_2)$ we may consider the pair $(\omega_2, \tau)$. The restriction corresponding to (9) is

$$(\bar{\tau} - \tau)\,|\,\omega_2\,|^2 = 2i. \tag{9'}$$

In this manner, if $\tau$ is given, $|\omega_2|$ is determined uniquely.

As is known, for every discrete group of linear fractional transformations of the complex plane a fundamental domain, $\mathfrak{T}$, may be taken, containing one and only one element from each class of points $\tau$ which are equivalent with respect to the mappings of $D$. Then two pairs of the form $(\omega_2, \tau)$ and $(\omega_2', \tau)$, where $\tau \in \mathfrak{T}$, can be equivalent either when $\omega_2 = \omega_2'$ (then they simply coincide) or when $\tau$ is a fixed point of some mapping $\tau' = \frac{m\tau + n}{p\tau + q}$, belonging to the discrete group $D$. But the set of fixed points can only be countable (since $D$ is countable). Therefore the correspondence between the points of the manifold $H$ and the pairs

$$(\omega_2, \tau), \quad \tau \in \mathfrak{T}, \quad \omega_2 = \sqrt{\frac{2i}{\bar{\tau} - \tau}}\, e^{i\theta}, \quad 0 \leqslant \theta \leqslant 2\pi \tag{11}$$

is one-to-one everywhere except on a countable set, which has measure zero and which we may neglect.* Consequently, every function $f(h)$ of $\mathfrak{H}$ may be considered as a function of a pair of complex numbers satisfying (11).

The invariant measure on $G$, and consequently on the coset space $H$, is expressed in the following form:

$$d\mu = \frac{dg_{12}\, dg_{21}\, dg_{22}}{|\,g_{22}\,|}$$

(the reader may easily check the invariance of this expression by direct calculation). Let us find the expression for this invariant measure in terms of the variables $\tau$ and $\omega_2$. Set $\tau = x + iy$ and take $x$ and $y$ as new variables. For the third parameter we shall take $\theta$, the argument of the complex parameter $\omega_2$. The parameters $x$, $y$, and $\theta$ are expressed in terms of $g_{12}$, $g_{21}$, $g_{22}$ in the following way:

$$x = \frac{g_{21}}{g_{22}\,(g_{21}^2 + g_{22}^2)} + \frac{g_{12}}{g_{22}}, \quad y = -\frac{1}{g_{21}^2 + g_{22}^2}, \quad \theta = \operatorname{arctg} \frac{g_{22}}{g_{21}}.$$

The Jacobian of the transformation from $g_{12}$, $g_{21}$, $g_{22}$ to $x$, $y$, $\theta$ equals

$$\frac{2}{g_{22}\,(g_{21}^2 + g_{22}^2)^2} = \frac{2y^2}{g_{22}}.$$

Consequently, in the variables $x$, $y$, and $\theta$ we have

$$d\mu = \frac{dx\, dy\, d\theta}{2y^2}.$$

Thus $\mathfrak{H}$ may be considered as the space of functions

$$f(\omega_2, \tau) = f(\theta, \tau),$$

measurable and square-summable, where the scalar product is defined by

* This passage from the coset $h$ to the pair $(\omega_2, \tau)$ of complex numbers satisfying conditions (11), is essentially nothing else than the returning from the group-theoretic realization of the dynamical system to the totality of line elements of some surface of constant negative curvature.

$$\int_0^{2\pi} d\theta \int_{\mathfrak{T}} f_1(\theta, \tau) \overline{f_2(\theta, \tau)} \frac{dx\, dy}{y^2} \qquad (\tau = x + iy).$$

Instead of considering functions $f(\omega_2, \tau)$ defined only in the fundamental region $\tau \in \mathfrak{T}$, $0 \leq \theta < 2\pi$, we may consider functions defined for all $\tau$, but then, in order for such a function really to define a function on $H$, it must satisfy the following invariance condition: for any element

$$d = \left\| \begin{matrix} m & n \\ p & q \end{matrix} \right\|, \quad d \in D$$

we have the equality

$$f\left[(p\tau + q)\,\omega_2, \frac{m\tau + n}{p\tau + q}\right] = f(\omega_2, \tau).$$

We shall now clarify in what way representations of the discrete series may be part of a given representation.

Let $\mathfrak{H}$ be a Hilbert space in which there is a representation of the group $G$ and let representations of the discrete series actually occur in this representation. We consider the compact subgroup in $G$ consisting of matrices of the form

$$\left\| \begin{matrix} \cos\varphi & \sin\varphi \\ -\sin\varphi & \cos\varphi \end{matrix} \right\|. \tag{12}$$

Since this subgroup is commutative, $\mathfrak{H}$ may be composed into a direct sum of one-dimensional subspaces, invariant with respect to the operators corresponding to the elements of the given subgroup. Let $\xi$ be a unit vector in such a one-dimensional subspace; then

$$U_\varphi \xi = e^{im\varphi} \xi, \tag{13}$$

where $U_\phi$ is the operator corresponding to (12). Since $U_{2\pi} = E$, the number $m$ must be an integer; it is called the weight of the vector $\xi$. In each irreducible representation of the discrete series corresponding to a given integer $n\,(n = 1, 2, \ldots)$, there is one and, within a proportionality factor, only one vector of weight $m$ for $m = n$, $n+1, \ldots$ (or $n, n-1, \ldots$ for negative $n$). In fact, in the above realization of the representations of the discrete series, the equation (13) is transformed into the functional equation:

$$f\left(\frac{z\cos\varphi - \sin\varphi}{z\sin\varphi + \cos\varphi}\right)(z\sin\varphi + \cos\varphi)^{-n} = e^{im\varphi} f(z),$$

an identity whose solution, within a constant multiple, is the function

$$f(z) = \frac{(z+i)^{\frac{m-n}{2}}}{(z-i)^{\frac{m+n}{2}}}.$$

For such a function $\int f(z)\overline{f(z)}\,|y|^{n-1}\,dx\,dy < \infty$ for $m \geq n$.

Now consider the one-parameter subgroups in $G$ consisting, respectively, of the matrices

$$\left\| \begin{matrix} e^{-t} & 0 \\ 0 & e^{t} \end{matrix} \right\| \quad \text{and} \quad \left\| \begin{matrix} \operatorname{ch} t & \operatorname{sh} t \\ \operatorname{sh} t & \operatorname{ch} t \end{matrix} \right\|.$$

The corresponding one-parameter subgroups of unitary operators, according to Stone's theorem, may be expressed as

$$e^{iA_1t} \text{ and } e^{iA_2t},$$

where $A_1$ and $A_2$ are hypermaximal operators. Let us consider the operators

$$A_+ = A_1 + iA_2 \text{ and } A_- = A_1 - iA_2.$$

It is possible to prove the following. The operator $A_+$ transforms every vector of weight $m$ into a vector of weight $m+1$, and $A_-$ takes vectors of weight $m$ into those of weight $m-1$.* In particular, the operator $A_-$ takes the vector $\xi_n$ (the vector of minimal weight belonging to the given representation of the discrete series) into zero: $A_-\xi_n = 0$. We remark that in the representations of the principal or complementary series the operators $A_-$ and $A_+$ do not annihilate any vector. Thus, given an arbitrary representation of the group $G$, in order to find which of the representations of the discrete series occur in its decomposition into irreducible parts, it is nesessary to find all solutions of the equation $A_-\xi_n = 0$, occurring among weight vectors of weight $n$. The number of linearly independent solutions of this equation will equal the number of times the irreducible representation corresponding to the given number $n$ occurs in the decomposition of the given representation into irreducible parts. Similarly, to find the representations for negative $n$, it is necessary to consider the solutions of $A_+\xi_n = 0$.

Let us now apply this to representations in the space of square-summable functions on the cosets of $G$ mod $D$. First we find the form that the condition for a vector to be a weight vector takes. Matrices (12) take the pair $(\omega_1, \omega_2)$ into the pair $(\omega_1 e^{i\phi}, \omega_2 e^{i\phi})$. Thus

$$U_\varphi f(\omega_1, \omega_2) = f(\omega_1 e^{i\varphi}, \omega_2 e^{i\varphi}).$$

Consequently, condition (13) becomes

$$f(\omega_1 e^{i\varphi}, \omega_2 e^{i\varphi}) = e^{im\varphi} f(\omega_1, \omega_2).$$

Clearly this condition is satisfied by a function of $\omega_2, \tau$ having the form $\omega_2^m \psi(\tau)$.

Let us find the operators $A_+$ and $A_-$. For the subgroup of matrices $\begin{pmatrix} e^{-t} & 0 \\ 0 & e^t \end{pmatrix}$ the operators $U_t$ are given by

$$U_t f(\omega_2, \tau) = f\left(\omega_2 \operatorname{ch} t - \bar{\omega}_2 \operatorname{sh} t, \frac{\tau \operatorname{ch} t - \frac{\bar{\omega}_2}{\omega_2}\bar{\tau} \operatorname{sh} t}{\operatorname{ch} t - \frac{\bar{\omega}_2}{\omega_2} \operatorname{sh} t}\right).$$

Since $U_t = e^{itA_1}$, it follows that $A_1 f = -i\frac{d}{dt} U_t f|_{t=0}$. Substituting the expression for $U_t f$ and differentiating, we obtain

$$A_1 f = -i\left(\bar{\omega}_2 \frac{\partial f}{\partial \omega_2} + \frac{\bar{\omega}_2}{\omega_2}(2\tau - \bar{\tau})\frac{\partial f}{\partial \tau}\right), \text{where } \frac{\partial}{\partial \tau} = \frac{1}{2}\left(\frac{\partial}{\partial \tau_1} - i\frac{\partial}{\partial \tau_2}\right), \quad \tau = \tau_1 + i\tau_2.$$

Similarly,

$$A_2 f = \bar{\omega}_2 \frac{\partial f}{\partial \omega_2} + \frac{\bar{\omega}_2}{\omega_2}\bar{\tau}\frac{\partial f}{\partial \tau}.$$

*The proof of this assertion is completely analogous to the proof of the lemma in §2 of the paper of Gel'fand and Shapiro, published in this number of Uspehi, page 3 (there the corresponding operators are called $H_+$ and $H_-$).

Hence,

$$A_{-}f=(A_1-iA_2)f=-2i\left(\bar{\omega}_2\frac{\partial f}{\partial\omega_2}+\frac{\bar{\omega}_2}{\omega_2}(\tau-\bar{\tau})\frac{\partial f}{\partial\bar{\tau}}\right).$$

Thus the problem in this case of finding solutions of $A_{-}f=0$ among weight vectors is that of solving the equation

$$n\omega_2^{n-1}\bar{\omega}_2\psi(\tau)+\omega_2^{n-1}\bar{\omega}_2(\bar{\tau}-\tau)\frac{\partial}{\partial\bar{\tau}}\psi(\tau)=\bar{\omega}_2\omega_2^{n-1}\left[n\psi(\tau)+(\bar{\tau}-\tau)\frac{\partial\psi}{\partial\bar{\tau}}\right]=0.$$

Setting $\psi(\tau)=(\bar{\tau}-\tau)^n\,\Phi(\bar{\tau})$, we find that $\frac{\partial\Phi}{\partial\tau}=0$ and, consequently, $\Phi(\bar{\tau})$ is an arbitrary analytic function of $\bar{\tau}$. Because of relation (9′) we have

$$\omega_2^n\psi(\tau)=\omega_2^n(\bar{\tau}-\tau)^n\Phi(\bar{\tau})=(2i)^n\frac{1}{\bar{\omega}_2^n}\Phi(\bar{\tau}).$$

The condition that the function be constant on cosets gives

$$\frac{1}{(p\bar{\omega}_1+q\bar{\omega}_2)^n}\Phi\left(\frac{m\bar{\tau}+n}{p\bar{\tau}+q}\right)=\frac{1}{\bar{\omega}_2^n}\Phi(\bar{\tau});$$

i.e.,

$$\Phi\left(\frac{m\bar{\tau}+n}{p\bar{\tau}+q}\right)=\Phi(\bar{\tau})(p\bar{\tau}+q)^n.$$

Thus the function $\Phi(\bar{\tau})$ must be an automorphic form* of weight $n$, corresponding to the discrete group $D$.

For functions $f(\omega_2,\tau)$ belonging to $\mathfrak{H}$ the scalar product $(f_1,f_2)$ is defined as the integral of $f_1\bar{f}_2$, where the integral is taken with respect to $\tau$ over the fundamental region $\mathfrak{T}$ in its non-euclidean metric. Integration with respect to $\omega_2$ need not be considered, since the product $f_1\bar{f}_2$ of functions of the desired type depends only on $\tau$ and $\bar{\tau}$. In fact, from (9′) we find (including, for simplicity, the factor $(2i)^n$ in the function $\Phi(\tau)$)

$$f_1\bar{f}_2=\frac{1}{\bar{\omega}_2^n}\Phi_1(\bar{\tau})\cdot\overline{\frac{1}{\bar{\omega}_2^n}\Phi_2(\bar{\tau})}=\frac{1}{|\omega_2|^{2n}}\Phi_1(\bar{\tau})\overline{\Phi_2(\bar{\tau})}=$$

$$=\left(\frac{\bar{\tau}-\tau}{2i}\right)^n\Phi_1(\bar{\tau})\overline{\Phi_2(\bar{\tau})}=\Phi_1(\bar{\tau})\overline{\Phi_2(\bar{\tau})}\,|y^n|.$$

Hence,

$$(f_1,f_2)=2\pi\int_{\mathfrak{T}}\Phi_1(\bar{\tau})\overline{\Phi_2(\bar{\tau})}|y|^{n-2}\,dx\,dy. \tag{14}$$

Thus the following theorem is verified:

**Theorem 2.** *If $H$ is the homogenous space of $G$ mod $D$, where $G$ is the group of real unimodular matrices of the second order and $D$ is a discrete subgroup, then in the decomposition of the unitary representation in $\mathfrak{H}$ into irreducible parts each representation of the discrete series with given weight $n$ occurs as many times as there exist automorphic forms $\Phi(\tau)$ of weight $n$ for which the integral* (14) *taken over the fundamental domain is finite.*

If $D$ is such that the fundamental domain $\mathfrak{T}$ is compact, then automorphic forms

*Concerning automorphic forms cf., for example, [7].

of weight $n$ which are square-integrable exist for infinitely many $n$. In fact, we have the following

**Theorem 3.** *For any discrete subgroup $D$ of the group of linear-fractional transformations of the upper half plane into itself such that the fundamental domain $\mathfrak{T}$ of $D$ is compact, and for any integer $n \geq 2$, there exist automorphic forms $\Phi(\tau)$ of weight $2n$ satisfying*

$$\int_{\mathfrak{T}} |\Phi(\tau)|^2 |y|^{n-2} dx\, dy < \infty. \tag{15}$$

**Proof.** Following the usual terminology, we shall call vertices and sides of the fundamental domain $\mathfrak{T}$ which lie on the real axis vertices and sides, respectively, of the second kind. If $\mathfrak{T}$ is compact there are neither vertices nor sides of the second kind. Therefore, in the case of a compact $\mathfrak{T}$, in order that a function $\Phi(\tau)$ satisfy condition (15) it is necessary and sufficient that it satisfy

$$\int_{\mathfrak{T}} |\Phi(\tau)|^2 dx\, dy < \infty.$$

It is easy to show that such automorphic forms actually exist. For this it is sufficient to consider the $\theta$-series: $\theta(z) = \Sigma (p_i z + q_i)^{-2m}$, where the sum is taken over all elements $d_i = \left\| \begin{smallmatrix} m_i & n_i \\ p_i & q_i \end{smallmatrix} \right\|$ of the discrete subgroup $D$. This series converges for $m \geq 2$ (cf.[7], pg. 113 et seq.) and represents an automorphic form satisfying (15).

Clearly, compactness of the fundamental region $\mathfrak{T}$ implies the compactness of the surface of constant negative curvature whose geodesic flow we are considering.

The existence, for infinitely many $n$, of automorphic forms of weight $n$ satisfying condition (15) may be proved for other discrete subgroups, in particular, for the modular group. It is probably true for any discrete subgroup $D$ for which $H$ has finite measure.

On the basis of Theorems 1, 2, and 3 we may formulate the following result: *the geodesic flow on every compact surface of constant negative curvature is ergodic, strongly mixing, and has Lebesgue spectrum of countable multiplicity.*

## § 4. The Geodesic Flow on $n$-dimensional Manifolds of Constant Negative Curvature.

Up to this point we have considered the geodesic flow on surfaces of constant negative curvature. In order to extend the above group-theoretic constructions to the geodesic flow in $n$-dimensional manifolds $(n \geq 3)$ of constant negative curvature it is necessary to introduce some modifications.

Let $G$ be the group of motions of the $n$-dimensional Lobačevskiĭ space. We consider the totality of possible orthogonal $n$-frames in this space. Fixing one of them, we may establish a one-to-one correspondence between the orthogonal $n$-frames and elements of the group $G$, since there obviously exists one and only one motion taking a given frame, considered as the initial one, into another fixed $n$-frame. Consider further the totality of $n$-frames having the first vector in common with the initial frame. It is clear that we may identify this collection with the line element determined by the given point and vector. The frames belonging to this set, represented as elements of the group $G$, form a compact subgroup $K$. Moreover, every other set of $n$-frames with common origin and first vector may be identified with its corresponding line element. Thus those motions taking the *initial* set of $n$-frames into some other such collection form a coset $gK$ of the subgroup $K$. In this manner we may establish a one-to-one correspondence between the line elements of $n$-dimensional Lobačevskiĭ space and the cosets $gK$ of a compact subgroup $K$ in its group of motions, $G$. If a manifold of constant negative curvature is being considered, then its line elements are obtained, as in the two dimensional case, by identifying line elements of hyperbolic space which are congruent with respect to some discrete subgroup of motions (the fundamental group of the manifold). Thus we finally obtain: the manifold $H$ of line elements of any space of constant negative curvature may be realized as the totality of double cosets $h = DgK$ of the group $G$ of motions of Lobačevskiĭ space by a discrete subgroup $D$ and a compact subgroup $K$.

The geodesic flow is the motion in the manifold of line elements in which each element moves along its geodesic with unit velocity. Let us formulate this motion in group-theoretic terms.

Consider the *initial* $n$-frame in Lobačevskiĭ space and assume that it moves with velocity $1$ along the geodesic determined by the *initial* line element. Since a motion of an $n$-frame uniquely determines a motion of the whole space, this process determines some one-parameter subgroup $\{g_t\}$ of $G$. Then the motion in the manifold of line elements is realized in the following manner:

$$gK \to gg_tK.$$

In fact, the geodesic passing through the initial element may be identified with the totality of cosets $\{g_tK\}$. The geodesic passing through any other element is obtained from the given one with the help of some motion $g$, i.e., it is represented by the totality of cosets of the form $\{gg_tK\}$. In order that two elements $g'$ and $g''$ translate the geodesic $\{g_tK\}$ into one and the same geodesic, it is necessary and sufficient that $g'$ and $g''$ belong to the same coset of $K$.* The *initial* element $K$ in time $t$ goes into the element $g_tK$. Consequently the element $gK$ in time $t$ goes into the element $gg_tK$. From this it follows that the motion on the manifold $H$ has the form

$$D_gK \to Dgg_tK.$$

Consider the case $n=3$. The investigation of the general case would be carried out in the same way if only the facts about all irreducible unitary representations of the corresponding group of motions were known.

The group of motions of three-dimensional Lobačevskiĭ space is isomorphic to the group of all complex matrices of the second order with determinant $1$. The motion in $H$ may be realized in the following way, just as in the two-dimensional case: in time $t$ the point $DgK$ goes into $Dgg_tK$, where

$$g_t = \begin{Vmatrix} e^{-t} & 0 \\ 0 & e^{t} \end{Vmatrix}.$$

*The condition that the element $g_t$ takes a coset $gK$ into a coset mod $K$ is expressed in the following manner: if $k \in K$, then $k' = g_{-t}kg_t \in K$ for all $t$.

In order to determine the spectrum of this dynamical system we first consider the dynamical system which may be defined in the set $H^*$ of one-sided cosets of $D$ in $G$, namely, where $Dg$ goes into $Dgg_t$. For the system $H^*$ the spectrum may be found just as for two dimensions, that is, the problem reduces to that of determining the spectrum in the irreducible representations of $G$.

The irreducible unitary representations of the complex unimodular group of second order matrices (the Lorentz group) have been found by I. M. Gel'fand and M. A. Naimark [8]. They fall into the following two types:

1) The principal series: if $g=\left\|\begin{matrix}\alpha & \beta\\ \gamma & \delta\end{matrix}\right\|$, then

$$U_g f(z) = |\beta z+\delta|^{n+i\rho-2}(\beta z+\delta)^{-n} f\left(\frac{\alpha z+\gamma}{\beta z+\delta}\right), \tag{16}$$

where $\rho$ takes on all real values and $n$ all integral values. $f(z)$ is an arbitrary function square-summable on the complex plane. The scalar product is taken in the usual way.

2) The complementary series:

$$T_g f(z) = (\beta z+\delta)^{-2-\rho} f\left(\frac{\alpha z+\gamma}{\beta z+\delta}\right), \tag{17}$$

where $0<\rho<2$ and the scalar product is defined as:

$$(f_1, f_2) = \int |z_1 - z_2|^{-2+\rho} f_1(z_1) f_2(z_2)\, d\mu(z_1)\, d\mu(z_2)$$

$$(\text{if}\quad z = x+iy,\ \text{then}\quad d\mu(z) = dx\,dy).$$

As in the case of the real group, it is easy to show that the operators $U_t$ corresponding to the subgroup $g_t \in G$ have simple Lebesgue spectrum in every irreducible representation. Consequently, the dynamical system $H^*$ has Lebesgue spectrum.

Let $\mathfrak{H}^*$ be the space of functions square-summable on $H^*$ and $\mathfrak{H}$ the subspace of those functions constant on the two-sided cosets $DgK$. This subspace is invariant relative to the operators $U_t$ and coincides with the functions square-summable on $H$. Since a unitary operator which has absolutely continuous spectrum on the whole space has absolutely continuous spectrum on every invariant subspace, it follows that the geodesic flow on a three-dimensional manifold of constant negative curvature has absolutely continuous spectrum.

We shall show that this spectrum must of necessity be Lebesgue.

In fact, the whole space $\mathfrak{H}^*$ decomposes into a generalized direct sum of spaces in which irreducible unitary representations of $G$ operate, having the form described above. The functions belonging to $\mathfrak{H}$, i.e., constant on cosets of $K$, split into components in the irreducible spaces which are invariant relative to the operators $U_k$ (i.e., operators constructed from elements $k \in K$):

$$U_k f(z) = f(z). \tag{18}$$

In the three-dimensional case the subgroup $K$ is realized as the set of matrices of the form

$$\left\|\begin{matrix} e^{-it} & 0 \\ 0 & e^{it}\end{matrix}\right\|.$$

Thus condition (18), in the complementary series (17), for instance, has the form

$$f(e^{-2it} z) = f(z),$$

i. e., it implies that $f(z)$ depends only on $|z| = r$. Functions satisfying condition (18) generate in the space corresponding to the irreducible representation a subspace invariant relative to the operators $U_t$ (corresponding to the elements $g_t = \left\| \begin{matrix} e^{-t} & 0 \\ 0 & e^t \end{matrix} \right\|$). The space $\mathfrak{H}$ is obviously made up of the generalized direct sum of these subspaces. We shall find the spectrum of $U_t$ in each of them. For the complementary series

$$U_t f(r) = e^{-t(2+\rho)} f(e^{-2t} r).$$

Setting $U_t = e^{iAt}$, we obtain

$$iAf(r) = -(2+\rho) f(r) - 2f'(r).$$

This is a differential operator of the first order with simple Lebesgue spectrum. Analogous considerations work for the principal series. Thus $\mathfrak{H}$ is the generalized direct sum of subspaces in each of which $U_t$ has Lebesgue spectrum. With the help of the lemma in §2 we obtain

**Theorem 4.** *The geodesic flow in any three-dimensional manifold of constant negative curvature has Lebesgue spectrum.*

In case the manifold has finite volume we immediately conclude that the system is ergodic and strongly mixing.

## §5. Dynamical Systems in Homogeneous Spaces.

It is not difficult to perceive the general group-theoretic pattern which lies at the base of the whole exposition in the preceding paragraphs.

Let $G$ be an arbitrary locally compact Lie group, $D$ a discrete and $K$ a compact subgroup of $G$, and $H$ the set of double cosets $h = DgK$. Further, let $\{g_t\}$ be an arbitrary one-parameter subgroup of $G$, all elements of which commute with $K$, i. e., $g_{-t} K g_t = K$. In $H$ define the one-parameter group of motions $s_t$ by setting

$$s_t(DgK) = Dgg_t K.$$

From the commutativity of $\{g_t\}$ with $K$ it follows that every coset is indeed carried into another coset by $s_t$. In $H$, as in ordinary coset spaces, a measure invariant under $s_t$ is defined in a natural manner. Thus $H$ becomes a dynamical system with invariant measure.

As in the cases analyzed above, the determination of the spectrum of the dynamical system $H$ reduces to the study of the irreducible unitary representations of $G$. If we restrict ourselves to semi-simple Lie groups, then the study of their representations reduces to the study of the representations of simple Lie groups. The irreducible unitary representations of the simple Lie groups have been found by Gel'fand and Naimark [9].

As is known (cf., e.g. [1], pg. 94), for every countable group of real numbers $\Lambda$ there exists a compact strictly ergodic dynamical system with pure point spectrum which coincides with $\Lambda$. The usual construction of this system on the group of characters of $\Lambda$ is a special case of the construction described above and is obtained from it when $G$ is commutative. In fact, to consider a simple example, let

$G$ be the euclidean plane, $K$ the identity subgroup, $D$ the subgroup of points with integer coordinates, and $\{g_t\}$ some line $y = ax$, where $a$ is irrational. Then the corresponding dynamical system is the ordinary torus with everywhere dense winding. The manifold $H = G/D$ in this case is itself a group, and the determination of the spectrum of the system immediately reduces to the consideration of the characters of $H$, i. e., of the irreducible unitary representations, which are all one-dimensional since the group is commutative.

## BIBLIOGRAPHY


[1] V. A. Rohlin, *Selected topics in the metric theory of dynamical systems*, Uspehi Matem. Nauk (N.S.) 4, no. 2(30), 57–128 (1949). (Russian)

[2] N. S. Krylov, *Contributions to the foundations of statistical physics*, Izdat. Akad. Nauk SSSR, Moscow-Leningrad, 1950. (Russian)

[3] E. Hopf, *Statistik der geodätischen Linien in Mannigfaltigkeiten negativer Krümmung*, Ber. Verh. Sächs. Akad. Wiss. Leipzig 91, 261–304 (1939).

[4] A. I. Plesner and V. A. Rohlin, *Spectral theory of linear operators*, II, Uspehi Matem. Nauk (N.S.) 1, no. 1(11), 71–191 (1946). (Russian)

[5] V. Bargmann, *Irreducible unitary representations of the Lorentz group*, Ann. of Math. (2) 48, 568–640 (1947).

[6] F. I. Mautner, *Unitary representations of locally compact groups*, Ann. of Math. (2) 51, 1–25 (1950).

[7] L. R. Ford, *Automorphic functions*, McGraw-Hill, New York, 1929.

[8] I. M. Gel'fand and M. A. Naĭmark, *Unitary representations of the Lorentz group*, Izvestiya Akad. Nauk SSSR. Ser. Mat. 11, 411–504 (1947). (Russian)

[9] I. M. Gel'fand and M. A. Naĭmark, *Unitary representations of the classical groups*, Trudy Mat. Inst. Steklov. 36 (1951). (Russian)

# 3.
## (with M. I. Graev)

# Geometry of homogeneous spaces, representations of groups in homogeneous spaces and related questions of integral geometry

Tr. Mosk. Mat. O.-va **88** (1959) 321–390
[Transl., II. Ser., Am. Math. Soc. **37** (1964) 351–429]. Zbl. **136**:434

**Contents**



## §1. Introduction

1. Let $X$ be a manifold on which a transformation group $G$ operates

$$x \longrightarrow xg.$$

The group $G$ is assumed to be transitive, i.e., two arbitrary points of $X$ can be transformed into each other by a transformation of the group $G$.

We consider the space of scalar or vector functions $f(x)$ on $X$. In this space the representation of the group $G$, $g \longrightarrow T_g$, is given by

$$T_g f(x) = f(xg)\, a(x, g). \tag{1.1}$$

The problem is to decompose this representation into irreducible representations.

The difficulties depend on the class of functional spaces which one considers. It is very interesting, for example, to bring in the space of generalized functions on $X$. In the present paper we confine ourselves to the simpler case of a space of functions with integrable square; here the operators of the representation $T_g$ are assumed to be unitary.

In such a general setting one cannot say much about the solution of the problem. However, it is remarkable that for complex semisimple Lie groups $G$ the decomposition problem for the representation (1.1) can be effectively solved in a general form.

Another space $\Omega$ associated with the initial homogeneous space $X$ is the geometrical object playing a fundamental role for the solution of the decomposition problem for the representation. The structure of this space $\Omega$ gives detailed information on the representation: irreducible components, the multiplicity of each component and effective formulas giving the decomposition of the representation in the space $X$ into irreducible ones. Certain submanifolds in the space $X$ form the points of $\Omega$; we call them horospheres.

If $X$ is a Lobačevskiĭ space and $G$ a group of motions in $X$, these manifolds are the classical horospheres. As is known, a horosphere in a Lobačevskiĭ space is a limit sphere, i.e., a cell whose closure is a sphere with one point at infinity. They are remarkable because they are homogeneous submanifolds in a Lobačevskiĭ space in which the euclidean geometry is realized.

The following device is used for the decomposition of the representation in the space $X$ into irreducible representations.

Each bounded function $f(x)$ (belonging to some appropriate class of functions, $x \in X$) is associated with a function $\phi(\omega)$ on the horosphere space $\Omega$. The function $\phi(\omega)$ is obtained by integration of $f(x)$ over the horosphere $\omega$ by means of a suitably selected measure.

For the function $\phi(\omega)$ the decomposition of the representation into irreducible representations is carried out quite simply, as will be shown in this paper. The decomposition problem of the representation in $X$ turns out to be related to the following problem of integral geometry: given $\phi(\omega)$, find $f(x)$. Thus the close relationship of the three following subjects is established:

1. Decomposition of the representation in $X$ into irreducible representations and its connection with the general theory of spherical functions.

2. Certain general notions in geometry of homogeneous spaces.

3. Certain basic questions in integral geometry.

The work as planned consists of several chapters. In the first and second chapters we will study the case of a complex semisimple Lie group $G$. Later on the real semisimple Lie groups will be considered.

In the present article (which contains the first chapter) homogeneous spaces $X$ are studied with stability subgroups of the simplest kind. In a realization by matrices these subgroups can be characterized as groups not containing Jordan matrices. All compact subgroups of the group $G$ are such subgroups and also the commutative subgroups of the group $G$ which are contained in Cartan subgroups.

As a special case we obtain the decomposition of the Kronecker product of two irreducible representations of the group $G$ into irreducible representations. In this context it turns out that infinite-dimensional representations of semisimple Lie groups are easier to deal with than finite-dimensional ones. The representation of the Kronecker product of two infinite-dimensional representations is given by a formula considerably simpler than the classical one for the finite-dimensional case.

2. We describe now in greater detail certain notions in the geometry of homogeneous spaces and also the problem itself.

We first consider the group $G$ of all complex matrices of order $n$ with determinant 1. The group $G$ can be realized as a transformation group of an $n$-dimensional affine space or of an $(n-1)$-dimensional projective space. However, the usual affine space $R_n$ and the usual projective space $P_{n-1}$ are very special cases and do not represent (for our purposes) adequate examples of homogeneous spaces on which the group $G$ acts.

Let us introduce two most natural spaces related with the group $G$ of complex matrices. In the usual affine space let us consider the following sequence: the point $x$, the straight line $l_1$ passing through the point $x$, the plane $l_2$ containing $l_1$ and so on up to the $(n-1)$-dimensional hyperplane $l_{n-1}$. We will call this sequence a generalized line element of the affine space.[1)] Analogously we define generalized line elements of the projective space.

*We call the set of all generalized line elements of the usual affine space the fundamental* (*or universal*) *affine space related to the group of complex matrices. Analogously we define the fundamental* (*or universal*) *projective space related to the group* $G$.

We consider the subgroup $Z_h$ of $G$ consisting of the elements which leave fixed a certain point $h$ of the affine fundamental space $H$. As is known [1], the subgroup $Z_h$ is conjugate to the group $Z$ of matrices of the form

$$\zeta = \begin{Vmatrix} 1 & \zeta_{12} & \dots & \zeta_{1n} \\ 0 & 1 & \dots & \zeta_{2n} \\ \cdot & \cdot & \cdot & \cdot \\ 0 & 0 & \dots & 1 \end{Vmatrix}.$$

Now let $h'$ be some other point of $H$. Displacing $h'$ by the elements of the subgroup $Z_h$, we then obtain a certain subspace $\omega$ in $H$ (the $Z_h$-orbit of $h'$).

The intrinsic geometry of $\omega$ coincides with the geometry of the homogeneous space $Z_h/Z_{h,h'}$ of the right cosets of the group $Z_h$ with respect to the subgroup $Z_{h,h'}$ (of the elements in $Z_h$ leaving fixed $h'$). It is known that the subgroups

1) Translator's note. Usually called a flag.

$Z_h$ are integral groups, typical for the euclidean geometry.

Consequently, we will call *euclidean subgroups* of the group $G$ the subgroups $Z_h$ of the group $G$ of complex matrices leaving in place one of the points of the fundamental affine space $H$.

Now let $X$ be an arbitrary homogeneous space whose group of motions is the group $G$ of all complex matrices of order $n$ with determinant 1. We ask which geometrical images in $X$ are the most natural analogues to the hyperplanes.

The three-dimensional Lobačevskiĭ space can serve as a typical example of a homogeneous space. It seems to be the easiest choice to consider the general planes as hyperplanes in a Lobačevskiĭ space. However, this does not work. The natural analogues to the hyperplanes are in a Lobačevskiĭ space the horospheres (i.e., spheres of infinite radius), because a Lobačevskiĭ geometry is realized also on the planes of a Lobačevskiĭ space. We want to pick out an object on which an euclidean geometry exists. The horospheres possess this property.

We realize the Lobačevskiĭ space as the space of all positive definite Hermitian matrices

$$p = \begin{Vmatrix} a & b \\ \bar{b} & c \end{Vmatrix}$$

with determinant $ac - b\bar{b} = 1$.

The motions in the space are defined by the formula

$$p \to g^* pg,$$

where $g = \begin{Vmatrix} \alpha & \beta \\ \gamma & \delta \end{Vmatrix}$ runs through the group $G$ of the complex matrices of second order with determinant $\alpha\delta - \beta\gamma = 1$. The spheres in the Lobačevskiĭ space with center in the point $p_0$ are formed by the motions of the group $G$ leaving in place the point $p_0$. We assume now that $p_0$ does not belong to the Lobačevskiĭ space and that it is a degenerate Hermitian matrix. Then the motions in the Lobačevskiĭ space leaving the point $p_0$ fixed define a horosphere. Let us suppose, for example, that $p_0 = \begin{Vmatrix} 0 & 0 \\ 0 & 1 \end{Vmatrix}$. The stability subgroup $C$ of the point $p_0$ consists of the matrices $k = \begin{Vmatrix} e^{i\phi} & \zeta \\ 0 & e^{-i\phi} \end{Vmatrix}$. The horospheres obtained through the motions of the subgroup $C$ may be given by the equations

$$a = \text{const},$$

where $a$ denotes an element of the matrix $p$: $p = \begin{Vmatrix} a & b \\ \bar{b} & c \end{Vmatrix}$.

It is not difficult to see that the same surfaces are defined by the motions $C$

of the euclidean group $Z$ of matrices

$$\zeta = \begin{Vmatrix} 1 & \zeta \\ 0 & 1 \end{Vmatrix}.$$

Thus the horospheres in a Lobačevskiĭ space are generated by the motions belonging to a fixed euclidean subgroup of the group of all motions.

In an arbitrary homogeneous space $X$ whose group of motions is the group $G$ of complex matrices with determinant 1, we can single out subspaces whose geometry is close to the euclidean geometry. We will consider subspaces generated by the motions of a certain euclidean subgroup of the group $G$. We will call such a subspace a horosphere of the space $X$.

3. We now give the algebraic definition of the euclidean subgroups and the basic homogeneous spaces connected with the group of complex unimodular matrices. The group $G$ of complex unimodular matrices is a simple Lie group. The euclidean subgroup $Z$ of matrices of the form

$$\zeta = \begin{Vmatrix} 1 & \zeta_{12} \dots \zeta_{1n} \\ 0 & 1 \quad \dots \zeta_{2n} \\ \cdot & \cdot \quad \cdot \quad \cdot \quad \cdot \\ 0 & 0 \quad \dots 1 \end{Vmatrix}$$

is the subgroup generated by all vectors belonging to positive roots (with respect to the Cartan subgroup $D$[1] which consists of the diagonal matrices in accordingly chosen lexographical order). Consequently, the euclidean subgroup of the group $G$ of complex matrices can be defined as the subgroup generated by all vectors belonging to positive roots according to a certain given order.

The fundamental affine space $H$ is by itself the space $G/Z$ of right cosets of the group $G$ of complex matrices with respect to the subgroup $Z$, generated by the vectors belonging to positive roots.

The fundamental projective space connected with the group $G$ of complex matrices is the space $G/K$ of right cosets of the group $G$ with respect to the subgroup $K$ of all triangle matrices

$$k = \begin{Vmatrix} k_{11} & k_{12} \dots k_{1n} \\ 0 & k_{22} \dots k_{2n} \\ \cdot & \cdot \quad \cdot \quad \cdot \quad \cdot \\ 0 & 0 \quad \dots k_{nn} \end{Vmatrix}.$$

1) We recall that a Cartan subgroup of a semisimple Lie group is an arbitrary maximal commutative subgroup containing elements in general position.

The subgroup $K$ is generated by all vectors belonging to non-negative roots.

In such a form these notions can be transferred without any change to the case of an arbitrary complex semisimple Lie group. The subgroup $Z$ of a complex semisimple Lie group $G$ generated by all vectors belonging to positive roots is called the euclidean subgroup of $G$, and the one generated by all vectors belonging to non-negative roots is the affine subgroup of the group $G$.

With the complex semisimple group $G$ we associate two homogeneous spaces: the space $G/Z$, namely, the fundamental affine space, and the space $G/K$, the fundamental projective space.

Now let $X = G/C$ be an arbitrary homogeneous subspace with the motion group $G$. Analogously to the euclidean subspaces we single out in this space surfaces generated by the motions of a certain euclidean subgroup of the group $G$. These surfaces will be called horospheres in the space $X$.

4. With each homogeneous space $X = G/C$ certain representation classes are associated. These representations are constructed in the following way: We consider the Hilbert space of the functions $f(x)$ in the space $X$ for which

$$\|f\|^2 = \int |f(x)|^2 \, dx < \infty. \tag{1.2}$$

Here $dx$ designates the measure on $X$ (invariant with respect to the transformations of the group $G$, if such a measure exists). In this Hilbert space an operator $T_g$ is associated with each element $g$ of the group $G$:

$$T_g f(x) = f(xg)\, a(x, g). \tag{1.3}$$

The operators $T_g$ have to be unitary and they should form the representation of the group $G$, i.e., $T_{g_1} T_{g_2} = T_{g_1 g_2}$ for two arbitrary elements $g_1$ and $g_2$ of the group $G$. These two requirements can be written in the form of functional equations for the factor $a(x, g)$:

$$d(xg) = |a(x, g)|^2 \, dx,$$

$$a(xg_1, g_2)\, a(x, g_1) = a(x, g_1 g_2).$$

These representations will be called *representations of the group $G$ in the space $X$.*

It is possible to associate also a vectorial representation of the group $G$ with the space $X$. This representation belongs to the space of vector functions

$$f(x) = (f_1(x), \cdots, f_n(x)),$$

for which

$$\|f\|^2 = \int f(x) f^*(x) \, dx < \infty.$$

The operators of the representation are again given by the formula (1.3), where

$a(x, g)$ is already a quadratic matrix of order $n$. In the present article we discuss only scalar representations.

The representations in the fundamental projective space $G/K$ associated with the complex semisimple Lie group $G$ were studied in [1]. All these representations are irreducible and form a so-called basic nondegenerate series of irreducible representations. On the other hand, let $K'$ be some extension of the affine subgroup $K$. The representation in the "degenerate projective space" $G/K'$ is also irreducible and forms one of the so-called degenerate basic series of irreducible representations of the group $G$.

In the general case the representations of the complex semisimple Lie group $G$ in the homogeneous space $X$ are reducible. Thus it is natural to raise the problem of decomposing these representations into irreducible ones.

5. We first decompose the quasiregular representations of a complex semisimple Lie group into irreducible ones. A representation of the group $G$ given by the formula

$$T_g f(h) = f(hg) \tag{1.4}$$

in the fundamental affine space $H = G/Z$ is called quasiregular.

The decomposition of a quasiregular representation of the group $G$ into irreducible representations is realized in the following way. Let $D$ be a Cartan subgroup of the group $G$. It will turn out that this group can be realized as the group of the continuous transformations of "left displacements"

$$h \longrightarrow \delta h$$

in the space $H$, permutable with the transformations of the group $G$. Let $\chi(\delta)$ be an arbitrary character for the subgroup $D$. We get

$$\phi(h, \chi) = \int \chi^{-1}(\delta) f(\delta h)\, d\delta, \tag{1.5}$$

where the integration is taken over $D$. We find that a quasiregular representation of the group $G$ induces in the space of functions $\phi(h, \chi)$ an irreducible representation belonging to a basic nondegenerate series and that the quasiregular representation is decomposed into a direct sum of representations thus obtained.

We note that each of the irreducible representations is contained in the decomposition of the quasiregular representation with finite multiplicity. This multiplicity does not exceed the number of all automorphisms $s$ of the Cartan subgroup $D$ induced by the interior automorphisms of the group $G$.

Analogously the "degenerate quasiregular representation" is decomposed into irreducible representations, i.e., the representation in the space $G/Z'$ where $Z'$ is a certain extension of the euclidean subgroup $Z$.

6. We come to the decomposition problem of a representation in the homogeneous space $X$ into irreducible representations.

We assume that:

1. The horospheres in $X$ are closed manifolds,

2. On each horosphere $\omega$ there exists a measure $d_\omega x$ invariant with respect to its sliding group, i.e., to the subgroup of the group $G$ preserving the horosphere $\omega$.

These assumptions are satisfied for the space $X = G/C$, where $C$ is a compact or Cartan subgroup of the group $G$. Such spaces will be considered in §§5 and 6.

Under the above mentioned assumptions the measures $d_\omega x$ on separate horospheres can always be normalized in such a way that they are preserved under the transformations of the group $G$ translating one horosphere into an other. In other words, the following equality holds:

$$d_{\omega g}(xg) = d_\omega x.$$

On the other hand, by the assumption 1, an arbitrary function which is finitary on the space $X$ is finitary on each horosphere $\omega$.

We now consider a representation in the space $X$. For simplicity we assume that the operator of the representation has the form

$$T_g f(x) = f(xg). \tag{1.6}$$

Let $f(x)$ be a finitary, infinitely often differentiable function on the space $X$. We integrate $f(x)$ over each horosphere $\omega$ in the space $X$:

$$\phi(\omega) = \int_\omega f(x)\, d_\omega x. \tag{1.7}$$

It is not difficult to see that if the function $f(x)$ is transformed by the initial representation (1.6), then the corresponding function $\phi(\omega)$ is transformed in the following way:

$$T_g \phi(\omega) = \phi(\omega g). \tag{1.8}$$

Thus we can consider the representation in the horosphere space of $X$ instead of the representation in the space $X$ itself.

We will find out how to decompose the representation (1.8) in the horosphere space into irreducible representations.

First we consider the case where the horosphere space is transitive. Let $L_{\omega_0}$ be the sliding group of a certain horosphere $\omega_0$, i.e., the subgroup of the group $G$ leaving $\omega_0$ in place. The space of all horospheres is the homogeneous space $G/L_{\omega_0}$. The sliding group $L_{\omega_0}$ either coincides with a certain euclidean subgroup $Z$ of $G$ or it is a certain extension of it. Let $L_{\omega_0} = Z$; then the representation (1.8) in the horosphere space is part of a quasiregular representation. But if $L_{\omega_0} \supset Z$, then this representation is part of a degenerate quasiregular representation of the group $G$. In both cases the decomposition of the representation is carried out in an elementary way, as was indicated in part 5.

Let us determine the number of parameters on which the irreducible representations appearing in the decomposition of the representation (1.8) in the horosphere space depend. We will call this number rank of the homogeneous space. In order to find the rank of the space $X$ we consider the group $D'$ of "left displacements" in the horosphere space, i.e., the group of the continuous transformations of this space permutable with the transformations of the group $G$. The (real) dimension of the group $D'$ is the rank of the space $X$.

We remark that in the case of a symmetric space this notion of rank of a space coincides with the notion of rank of a symmetric space in the sense of Killing-Cartan.

We have considered the case where the horosphere space of $X$ is transitive. In the general case, however, the space is not transitive. If it is stratified into transitive subspaces then the dimension of the base of this stratification will be called the *class* of the space $X$. Together with the rank the class is an important geometric property of a homogeneous space.

A quasiregular or degenerate quasiregular representation of the group $G$ belongs to each transitive subspace of horospheres. From this it follows that the representation of the group in the space of all horospheres is decomposed into a (continuous) direct sum of quasiregular and degenerate quasiregular representations (or of parts of them).

We recall now that the irreducible representations of the group are contained in the decomposition of the quasiregular and degenerate quasiregular representations with finite multiplicity. Consequently, the class of the space $X$ characterizes the "continuous" multiplicity with which the irreducible representations of the group $G$ are contained in the representation (1.8) of this group in the horosphere space of $X$.

We now return to the representation in the space $X$. Knowing the decomposition of the representation in the horosphere space into irreducible representations, we can define the irreducible components of the initial representation in the space $X$. For this it is sufficient to make use of the relation (1.7).

It remains to show that all irreducible components of the representation in the space $X$ are obtained. In other words, it is necessary to show that the function

$f(x)$ can be decomposed into the irreducible components thus found and that we can also get in explicit form the decomposition itself.

It is evident that the problem will be solved if we can find the inversion of the formula (1.7)

$$\phi(\omega) = \int_{\omega} f(x)\, d_{\omega}x,$$

i.e., if we express the value of the function $f(x)$ at an arbitrary point of $X$ by its integral over all possible horospheres. Thus our problem is reduced to a problem of integral geometry: we have to determine the function $f(x)$ on $X$ from the values of its integral over all possible horospheres in the space $X$.

The solution of a problem of similar type for the case where $X$ is a general euclidean space is well known (see, e.g., [5, 11]).

In [1, 3, 4] this problem was solved for the case where $X$ is a group space of a complex semisimple Lie group in which the motions are defined as right displacements:

$$g \longrightarrow gg_0.$$

In the present article the problem is solved for a more general class of homogeneous spaces on which the complex semisimple Lie group $G$ acts.

§§2 and 3 are of a preliminary character. In §4 general notions in the geometry of arbitrary homogeneous spaces are discussed. The relationship between these notions, the representation theory and the integral geometry is established. These notions will be introduced in connection with the example of the most interesting class of homogeneous spaces.

In §5 representations in the spaces with compact stability subgroup will be studied and in §6 representations in the spaces whose stability groups are subgroups of Cartan groups. The decomposition problem in these spaces will be completely solved.

The decomposition of the product of two irreducible representations (belonging to a general nondegenerate series of a complex semisimple Lie group) into irreducible representations is added at the end of this paper. This decomposition was obtained earlier for the Lorentz group by an other method, by M. A. Naĭmark (see [10]).

## §2. Representations in homogeneous spaces

**1. Definition of the measure in a homogeneous space.** Let $X$ be a homogeneous space with the group of motions $G$:

$$x \longrightarrow xg.$$

Throughout this article we suppose that $G$ is a complex connected semisimple Lie group. As is well known, the space $X$ can be realized as space $G/C$ of right cosets of the group $G$ with respect to the stability group $C$ of one of the points of the space $X$.

In the space $X$ we define the measure $dx$.

In what follows $dx$ is assumed to be a relatively invariant measure on $X$, if such a measure exists.

We note that a measure is called relatively invariant if

$$d(xg) = \rho(g)\, dx, \tag{2.1}$$

where $\rho(g)$ is a single-valued continuous function on the group $G$. From relation (2.1) follows $\rho(g_1 g_2) = \rho(g_1)\rho(g_2)$ for arbitrary elements $g_1$ and $g_2$ of the group $G$. In the special case where $G$ and the subgroup $C$ are unimodular groups we have $\rho(g) = 1$, i.e., $dx$ is an invariant measure.

But if the space $X$ does not possess a relatively invariant measure, then $dx$ will be understood to be a quasi-invariant measure, i.e., a measure such that $d(xg)$ and $dx$ are equivalent measures for arbitrary elements $g$ of the group $G$.

As is well known, such a measure always exists and is unique to within equivalence. Here the following integral relation holds:

$$\int f(g)\rho(g)\, d_r g = \int_{G/C} dx \int f(cg)\, d_r c, \tag{2.2}$$

where $d_r g$ and $d_r c$ denote right invariant measures of $G$ and $C$, respectively; the function $\rho(g)$ is strictly positive, is bounded from above and from below in each compact manifold, and satisfies the equation [1)]

$$\frac{\rho(cg)}{\rho(g)} = \frac{\Delta(c)}{\delta(c)}, \tag{2.3}$$

where

$$\delta(c) = \frac{d_r(c^{-1})}{d_r c}, \qquad \Delta(g) = \frac{d_r(g^{-1})}{d_r g}.$$

The measure $dx$ possesses the following property instead of the invariance property. Let the element $g_x$ translate the base point $x_0$ of the space $X$ into the point $x$. Then

$$d(xg) = \frac{\rho(g_x g)}{\rho(g_x)}\, dx.$$

The function $\rho(g)$ can always be assumed to be continuous, infinitely often differentiable, and normed by the equation $\rho(e) = 1$. Then from the relation (2.3)

1) Conversely, a quasi-invariant measure $dx$ on $X$ is defined by (2.2) for each such function $\rho(g)$.

we get

$$\rho(c) = \frac{\Delta(c)}{\delta(c)},$$

$$\rho(cg) = \rho(c)\rho(g).$$

**2. Representation of a group in a homogeneous space.** We consider the set of functions $f(x)$, given on $X$, for which

$$\|f\|^2 = \int |f(x)|^2\,dx < \infty. \tag{2.4}$$

We associate to each element $g$ of the group $G$ the operator $T_g$ of the form

$$T_g f(x) = f(xg)\,a(x, g). \tag{2.5}$$

We impose the following conditions on the factor $a(x, g)$:

1°. For any two elements $g_1$ and $g_2$ of $G$ and for almost all $x \in X$ the following equality holds:

$$a(x, g_1 g_2) = a(xg_1, g_2)\,a(x, g_1). \tag{2.6}$$

This condition means that

$$T_{g_1} T_{g_2} = T_{g_1 g_2},$$

i.e., that the operator $T_g$ forms a representation of the group $G$.

2°.

$$|a(x, g)|^2 = \frac{d(xg)}{dx}\ ^{1)} \tag{2.7}$$

for almost all $x$ with arbitrary fixed $g$. This condition means that the operator $T_g$ is unitary.

The representation of the group $G$ defined in this way will be called the *representation of $G$ in the homogeneous space $X$*.

The factor $a(x, g)$ can be obtained from the functional equation (2.6) and the relation (2.7):

$$a(x, g) = \left[\frac{d(xg)}{dx}\right]^{\frac{1}{2}} \chi_x(g), \tag{2.8}$$

where

$$|\chi_x(g)| = 1. \tag{2.9}$$

From relation (2.6) follows also

$$\chi_x(cg) = \chi_x(c)\,\chi_x(g), \tag{2.10}$$

where $c$ is an arbitrary element of the stability group $C_x$ of the point $x$. Hence

1) That is, $d(xg) = |a(x, g)|^2\,dx$, where $dx$ is the measure in the space $X$.

we get, in particular, that the function $\chi_x(c)$, if it is defined and continuous on the stability group $C_x$, is a character on this subgroup.

Besides, if $x_0$ is a fixed point of $X$ and $g_x$ an element of $G$ transforming $x_0$ into $x$, then there follows from relation (2.6)

$$a(x, g) = \frac{a(x_0, g_x g)}{a(x_0, g_x)}. \tag{2.11}$$

Conversely, let $\chi_{x_0}(g)$ be an arbitrary function satisfying the condition $|\chi_{x_0}(g)| = 1$ and $\chi_{x_0}(cg) = \chi_{x_0}(c)\chi_{x_0}(g)$, where $c$ runs through the stability group of the point $x_0$. Then, corresponding to (2.8) we get

$$a(x_0, g) = \left[\frac{d(xg)}{dx}\right]^{\frac{1}{2}}_{x=x_0} \chi_{x_0}(g)$$

and we define $a(x, g)$ by means of the formula (2.11).

It is easy to see that this function satisfies the conditions (2.6) and (2.7). We will indicate a condition which is sufficient for the equivalence of two representations in $X$.

*Let the two representations be given by*

$$T_g f(x) = f(xg)\frac{a(x_0, g_x g)}{a(x_0, g_x)}, \tag{2.12}$$

$$T'_g f(x) = f(xg)\frac{a'(x_0, g_x g)}{a'(x_0, g_x)}. \tag{2.12'}$$

*Then if* $a(x_0, c) = a'(x_0, c)$, *where* $c$ *runs through the stability group of* $x_0$, *the representations* (2.12) *and* (2.12′) *are equivalent.*

For in fact, in the case the ratio $\dfrac{a(x_0, g_x)}{a'(x_0, g_x)}$ does not depend on the choice of the representative $g_x$, and therefore it is a function of $x$.

We consider the mapping

$$f(x) \to \varphi(x) = f(x)\frac{a(x_0, g_x)}{a'(x_0, g_x)}.$$

This mapping is isometric since $\left|\dfrac{a(x_0, g_x)}{a'(x_0, g_x)}\right| = 1$. On the other hand, it transforms the representation (2.12) into the representation $\widetilde{T}_g$:

$$\widetilde{T}_g\varphi(x) = (T_g f(x))\frac{a(x_0, g_x)}{a'(x_0, g_x)} = f(xg)\frac{a(x_0, g_x g)\,a(x_0, g_x)}{a(x_0, g_x)\,a'(x_0, g_x)}$$
$$= \varphi(xg)\frac{a'(x_0, g_x g)}{a(x_0, g_x g)}\,\frac{a(x_0, g_x g)\,a(x_0, g_x)}{a(x_0, g_x)\,a'(x_0, g_x)},$$

i.e., $\widetilde{T}_g\phi(x) = \phi(xg)\,\dfrac{a'(x_0, g_xg)}{a'(x_0, g_x)}$, and therefore

$$\widetilde{T}_g = T'_g,$$

Q.E.D.

3. **Examples.** Let us consider some very important examples of representations in homogeneous spaces.

1°. *Regular representation.* Let $X$ be the space $G$ itself and let the motions in this space be the right translations:

$$g \longrightarrow gg_0.$$

The representation is constructed in the space of the functions $f(g)$ for which

$$\int |f(g)|^2\, dg < \infty;$$

the operator of the representation is given by

$$T_{g_0} f(g) = f(gg_0). \tag{2.13}$$

(It is not difficult to show that in this case we have always $a(x, g) \equiv 1$.) This representation is called a *regular representation of the group* $G$.

2°. *Quasiregular representation.* Let $G$ be a complex semisimple Lie group, let $D$ be a Cartan subgroup of $G$, and let $Z$ be the subgroup of $G$ generated by the vectors belonging to positive roots (in a certain lexicographical order). We form the space $H = G/Z$ of right cosets of $G$ with respect to $Z$. It is known that $Z$ is unimodular and consequently an invariant measure $dh$ exists in $H$.

We consider a set of functions $f(h)$ on $H$ for which

$$\int |f(h)|^2\, dh < \infty.$$

We give the operator $T_g$ of the representation by the formula

$$T_g f(h) = f(hg). \tag{2.14}$$

This representation is called a *quasiregular representation of* $G$.

For example, let $G$ be a complex unimodular group of second order, i.e., the group of complex matrices

$$g = \begin{Vmatrix} \alpha & \beta \\ \gamma & \delta \end{Vmatrix}$$

with determinant $\alpha\delta - \beta\gamma = 1$. In this case $Z$ is the subgroup of matrices of the form $\zeta = \begin{Vmatrix} 1 & \zeta \\ 0 & 1 \end{Vmatrix}$, and $H = G/Z$ is a two-dimensional complex affine space. The quasiregular representation is thus constructed in the space of functions $f(z_1, z_2)$

of the complex variables $z_1$ and $z_2$ for which

$$\int |f(z_1, z_2)|^2 \, dz_1 \, dz_2 < \infty.$$

The operator of the representation is given by the formula

$$T_g f(z_1, z_2) = f(z_1\alpha + z_2\gamma, z_1\beta + z_2\delta).$$

We will consider the quasiregular representations in detail in §3.

3°. *Representation on a Lobačevskiĭ plane.* Let $X$ be the upper halfplane of the complex plane ($\operatorname{Im} z > 0$) and $G$ be a group of real unimodular matrices of second order

$$g = \begin{Vmatrix} \alpha & \beta \\ \gamma & \delta \end{Vmatrix}, \quad \alpha\delta - \beta\gamma = 1.$$

We associate to each matrix $g$ a linear transformation of the halfplane $X$

$$z \to \frac{\alpha z + \gamma}{\beta z + \delta}.$$

As is well known, the halfplane $X$ with the group of motions $G$ represents one of the models of a Lobačevskiĭ plane. The invariant measure on $X$ is given by

$$y^{-2} \, dx \, dy \quad (z = x + iy).$$

The representation of $G$ is realized in the space of the functions $f(z) = f(x, y)$ defined in the upper halfplane for which

$$\iint |f(x, y)|^2 y^{-2} \, dx \, dy < \infty.$$

We define the operator of the representation $T_g$ by the formula

$$T_g f(z) = f\left(\frac{\alpha z + \gamma}{\beta z + \delta}\right) |\beta z + \delta|^m (\beta z + \delta)^{-m}, \tag{2.15}$$

where $m$ is an arbitrary fixed integer. It is not difficult to show, on the basis of part 2 of this section, that any other representation of $G$ in $X$ is equivalent to a representation of the form (2.15).

4°. *Representation in the space of Hermitian matrices.* Let $G$ be the group of *complex* unimodular matrices of second order

$$g = \begin{Vmatrix} \alpha & \beta \\ \gamma & \delta \end{Vmatrix}, \quad \alpha\delta - \beta\gamma = 1.$$

We consider the three-dimensional space $X$ of Hermitian matrices

$$p = \begin{Vmatrix} a & b \\ \bar{b} & c \end{Vmatrix}$$

of given signature and with given determinant. We define the motions in this space by the formula

$$p \to g^* pg.$$

There are three types of such spaces: the space of positive definite (or negative definite) Hermitian matrices, the space of nondefinite Hermitian matrices, and finally the space of degenerate Hermitian matrices $p$, satisfying the condition $p \geq 0$ (or $p \leq 0$).

It is well known that the space of positive definite Hermitian matrices is a Lobačevskiĭ space. It can be realized on one of the sheets of the two-sheeted hyperboloid $x_1^2 + x_2^2 + x_3^2 - x_4^2 = -1$ in the four-dimensional real space. The space of the nondefinite Hermitian matrices is realized on the surface of the one-sheeted hyperboloid $x_1^2 + x_2^2 + x_3^2 - x_4^2 = 1$. Finally the space of degenerate Hermitian matrices is realized on one nappe of a cone $x_1^2 + x_2^2 + x_3^2 - x_4^2 = 0$.

All these spaces are homogeneous spaces of the form $X = G/C$. In the case of a Lobačevskiĭ space $C$ is a unitary subgroup of the group $G$ (or a subgroup conjugate to it). In the case of the space of indefinite Hermitian matrices $C$ is the subgroup of all real matrices. In the case of the space of degenerate Hermitian matrices the subgroup $C$ consists of the matrices of the form

$$\zeta = \begin{Vmatrix} e^{i\varphi} & \zeta \\ 0 & e^{-i\varphi} \end{Vmatrix}.$$

We will find an invariant measure on each of these spaces. We choose $a$ and the real and imaginary parts of $b$ as coordinates of the point $p$. If

$$g^* pg = \begin{Vmatrix} a' & b' \\ \bar{b}' & c' \end{Vmatrix},$$

then $da'\,db' = \left| \frac{a'}{a} \right| da\,db$; thus the invariant measure on $X$ is given by the formula

$$dp = \frac{da\,db}{|a|},$$

where $db$ denotes the product of the differentials of the real and imaginary parts of the complex variable $b$. The representation of the group is constructed in the space of the functions $f(p) = f(a, b)$, defined on $X$, for which

$$\int |f(a, b)|^2 \frac{da\,db}{|a|} < \infty.$$

The operator of the representation is given by the formula

$$T_g f(p) = f(g^* pg)\, a(p, g). \tag{2.16}$$

The factor $a(p, g)$ can be easily calculated from the basic results in part 2 of this section.

In the case where the space $X$ consists of nondegenerate Hermitian matrices we get

$$a(p, g) \equiv 1,$$

which is correct up to transition to an equivalent representation.

Indeed, each function $\chi(c)$ on the stability group $C$ of $X$ satisfying the condition $\chi(c_1c_2) = \chi(c_1)\chi(c_2)$ is identically equal to unity. We now consider the space of degenerate matrices $X = G/C$, where $C$ consists of the matrices $\zeta = \begin{Vmatrix} e^{i\phi} & \zeta \\ 0 & e^{-i\phi} \end{Vmatrix}$. Any character on the subgroup $C$ is given by the formula

$$\chi(\zeta) = e^{-in\phi},$$

where $n = 0, \pm 1, \pm 2 \cdots$. Thus there exists a discrete series of representations in the space $X$. We will find explicit formulas for the operator of the representations. As is well known, almost every matrix $g$ of $G$ can be written uniquely in the form $g = \zeta h$, where $\zeta = \begin{Vmatrix} e^{i\phi} & \zeta \\ 0 & e^{-i\phi} \end{Vmatrix}$, $h = \begin{Vmatrix} \epsilon & 0 \\ z & \epsilon^{-1} \end{Vmatrix}$, $\epsilon > 0$. In agreement with this result we extend the function $\chi(\zeta)$ over the entire group $G$, assuming that

$$\chi(\zeta h) = \chi(\zeta).$$

Therefore, if $g = \begin{Vmatrix} \alpha & \beta \\ \gamma & \delta \end{Vmatrix}$, then $\chi(g) = \delta^n |\delta|^{-n}$.

We consider now in $X$ the point $p_0 = \begin{Vmatrix} 0 & 0 \\ 0 & 1 \end{Vmatrix}$. The matrix $g = \begin{Vmatrix} \alpha & \beta \\ \gamma & \delta \end{Vmatrix}$ transforms $p_0$ into $g^* p_0 g = \begin{Vmatrix} \gamma\bar{\gamma} & \gamma\delta \\ \gamma\bar{\delta} & \delta\bar{\delta} \end{Vmatrix}$.

Thus the stability group of $p_0$ is exactly the subgroup $C$ of the matrices of the form $\zeta = \begin{Vmatrix} e^{i\phi} & \zeta \\ 0 & e^{-i\phi} \end{Vmatrix}$. Let $p = \begin{Vmatrix} a & b \\ \bar{b} & c \end{Vmatrix}$ be an arbitrary matrix of $X$. Let $g_p$ be a matrix transforming $p_0$ into the point $p$. It is easily seen that $g_p$ can be chosen as a matrix of the form

$$g_p = \begin{Vmatrix} 0 & -\frac{1}{\sqrt{a}} \\ \sqrt{a} & \frac{b}{\sqrt{a}} \end{Vmatrix}.$$

We then have

$$\chi(g_p) = b^n |b|^{-n}, \quad \chi(g_p g) = \frac{\left(a^{\frac{1}{2}}\beta + ba^{-\frac{1}{2}}\delta\right)^n}{\left|a^{\frac{1}{2}}\beta + ba^{-\frac{1}{2}}\delta\right|^n}.$$

Consequently,

$$a(p, g) = \frac{\chi(g_p g)}{\chi(g_p)} = \frac{(a\beta + b\delta)^n b^{-n}}{|a\beta + b\delta|^n |b|^{-n}}.$$

Therefore, the representations of the group $G$ in the space $X$ of degenerate Hermitian matrices are given by the formula

$$T_g f(p) = f(g^* pg)(a\bar{b}\beta + b\bar{b}\delta)^n |a\bar{b}\beta + b\bar{b}\delta|^{-n},$$

where

$$p = \begin{Vmatrix} a & b \\ \bar{b} & c \end{Vmatrix}, \quad ac - b\bar{b} = 0, \ a > 0 \text{ and } g = \begin{Vmatrix} \alpha & \beta \\ \gamma & \delta \end{Vmatrix},$$
$$\alpha\delta - \beta\gamma = 1, \ n = 0, \pm 1, \pm 2, \ldots.$$

5°. The above-mentioned construction can be extended to the case of an arbitrary $n$. We consider the group $G$ of complex, unimodular matrices of order $n$.

Let $X_{n_1, n_2}$ be the space of all Hermitian matrices $p$ of order $n$ with given determinant, having $n_1$ positive and $n_2$ negative eigenvalues ($n_1 + n_2 \leq n$).

The motions in $X_{n_1, n_2}$ are given by the formula

$$p \to g^* pg.$$

As is well known, $X_{n_1, n_2}$ is transitive with respect to $G$.

The representation of $G$ in the space $X_{n_1, n_2}$ is again given by the formula

$$T_g f(p) = f(g^* pg)\, a(p, g).$$

If the matrices of $X_{n_1, n_2}$ are not degenerate, then the stability group of the point $p$ is a real form of $G$. Hence it follows that, up to transition to an equivalent representation,

$$a(p, g) \equiv 1.$$

The representations in the space $X_{n, 0}$ of positive definite matrices will be considered in this article (§5). The general case will be examined in the second part.

6°. *Representation in a symmetric space.* Let $G$ be a complex semisimple Lie group, and let $\theta$ be an involutive automorphism in $G$. We denote by $G_0$ the set of elements $g$ remaining fixed under the automorphism $\theta$. Then $G_0$ is a real form of the group $G$. We consider the representation in the symmetric space $X = G/G_0$. It is not difficult to show that up to equivalence it is given by the formula

$$T_g f(x) = f(xg).$$

In 4° and 5° we indicated special cases of such representations.

7°. *Still another representation in a group space.* Let $G$ be a complex semisimple Lie group and $X$ its group representation. We define the motions in $X$ as left and right displacements

$$g \to g_1^{-1} g g_2, \quad g_1, g_2 \in G.$$

Then the representation of $G \times G$ in the space $X$ can by defined by

$$T_{g_1}, g_{g_2} f(g) = f(g_1^{-1} g g_2),$$

where

$$\| f \|^2 = \int | f(g) |^2 dg < \infty.$$

8°. Let $G$ be a complex semisimple Lie group, and let $X_1$ and $X_2$ be two homogeneous spaces on which the group $G$ acts transitively. We consider two representations $T'_g$ and $T''_g$ of the group $G$ in the spaces $X_1$ and $X_2$, respectively.

$$T'_g f(x_1) = f(x_1 g) a_1(x_1, g),$$

where

$$\| f \|^2 = \int | f(x_1) |^2 dx_1 < \infty,$$

and

$$T''_g \varphi(x_2) = \varphi(x_2 g) a_2(x_2, g),$$

where

$$\| \varphi \|^2 = \int | \varphi(x_2) |^2 dx_2 < \infty.$$

We consider now the set of the functions $F(x_1, x_2)$, defined on the direct product of the spaces $X_1$ and $X_2$, for which

$$\| F \|^2 = \iint | F(x_1, x_2) |^2 dx_1 dx_2 < \infty.$$

On this set we determine the representation $T_g$ by the formula

$$T_g F(x_1, x_2) = T'_g T''_g F(x_1, x_2),$$

where the operator $T'_g$ acts on $F$ as on a function of $x_1$ and $T''_g$ as on a function of $x_2$. The representation so obtained is called the *Kronecker representation* or *tensor product* of the initial representations $T'_g$ and $T''_g$.

In the general case the space $X$ is not transitive, but splits into transitive components $X^{(\alpha)}$. Thus the Kronecker product of the initial representations splits into a direct sum of the representations of the group $G$ on the spaces $X^{(\alpha)}$.

The Kronecker product of the representations of the fundamental nondegenerate series of the group $G$ is an important example. As is well known (see [1]), in this case we have $X_1 = X_2 = G/K$, where $K$ is an affine subgroup of $G$. The space

$X_1 \times X_2 = X$ is not transitive; however, as was shown in [1], we get a transitive space $X'$ if we remove a manifold of lower dimension from $X$. The stability subgroups of the points of $X'$ are conjugate to a Cartan subgroup $D$ of the group $G$. Thus $X' = G/D$. Consequently: *the Kronecker product of two representations of a fundamental series of* $G$ *is a representation in the space* $G/D$, *where* $D$ *is a Cartan subgroup of* $G$. We will consider this representation in §7.

## §3. Quasiregular representation and its decomposition into irreducible representations

For the sake of completeness we go into more details of a quasiregular representation of a complex semisimple Lie group $G$. This representation occupies a special place in the representation theory of groups in homogeneous spaces. In §4 it will be shown that the decomposition problem of a representation of $G$ in a homogeneous space $X$ into irreducible representations can be reduced to the analogous problem for a quasiregular representation.

1. **Definition of a quasiregular representation.** We recall that a quasiregular representation of a group $G$ is a representation in the space $H = G/Z$, where $Z$ is a euclidean subgroup of $G$, i.e., a subgroup generated by all vectors belonging to positive roots.

In the case where $G$ is the group of complex matrices with determinant 1, the subgroup $Z$ consists of the matrices of the form

$$\zeta = \begin{Vmatrix} 1 & \zeta_{12} & \dots & \zeta_{1n} \\ 0 & 1 & \dots & \zeta_{2n} \\ \cdot & \cdot & \cdot & \cdot \\ 0 & 0 & \dots & 1 \end{Vmatrix}.$$

In this case the space $H$ can be treated as the space of generalized affine line elements of a usual $n$-dimensional affine space.[1)]

The space $H$ will be called the *fundamental affine space connected with the group* $G$.

In the space of functions $\phi(h)$ for which

$$\|\phi\|^2 = \int |\phi(h)|^2 \, dh < \infty$$

($dh$ denotes the invariant measure on $H$) the operator of the quasiregular

1) Let us recall that a "generalized affine line element" is a sequence of line elements of a usual affine space: the point $x$, the straight line $l_1$ passing through $x$, the surface $l_2$ containing $l_1$ and so on. (Translator's note: that is, a flag).

representation is given by the formula

$$T_g \phi(h) = \phi(hg)$$

where $hg$ denotes the point of $H$ into which the point $h$ was transferred by the element $g$ of $G$.

**2. Geometry of the fundamental affine space $H$.** Let $h$ be a point of $H$, and let $Z_h$ be its stability subgroup. We stratify the space $H$. We consider two points $h_1$ and $h_2$ as belonging to one and the same layer $z$ if their stability groups coincide: $Z_{h_1} = Z_{h_2}$. We note that the point $hg$ belongs to the same layer as $h$ if and only if $g^{-1} Z_h g = Z_h$, i.e., if $g$ is an element of the normalizer $K_h$ of the subgroup $Z_h$ in $G$.

We will call the normalizers $K$ of the euclidean subgroups $Z$ *affine subgroups of the group* $G$. These normalizers $K$ are subgroups of $G$ generated by all vectors belonging to non-negative roots. The space $Z = G/K$ of right cosets of $G$ with respect to its affine subgroup $K$ will be called the *fundamental projective space connected with* $G$.

We thus see that the space of the layers $z$ of the stratified space $H$ is the fundamental projective space $Z$. Each of the layers of the space $H$ is homeomorphic to the factor space $K/Z$. But as is known [1], we have $K = ZD$ where $D$ is a Cartan subgroup of the group $G$, i.e., a maximal commutative subgroup of $G$ containing elements in general position in $G$. Consequently, *each of the layers of the space $H$ is homeomorphic (as a space) to a Cartan subgroup $D$ of $G$.*

We consider now continuous transformations of the space $H$ permutable with the motions of the group $G$. We call them "left displacement" transformations in the space $H$. Let the left displacement $\delta$ transform the point $h \in H$ into the point $\delta h$. Then the permutability condition ($\delta$ permutable with the motions of $G$) is written in the form $(\delta h) g = \delta(hg)$.

Let for example $G$ be a group of complex matrices $g = \begin{Vmatrix} \alpha & \beta \\ \gamma & \delta \end{Vmatrix}$ of second order with determinant 1. $H$ is in this case the space of the pairs $(z_1, z_2)$ of complex numbers; the transformations of the group $G$ in $H$ are

$$(z_1, z_2) \to (z_1\alpha + z_2\gamma, \ z_1\beta + z_2\delta).$$

The similitude transformations are left displacement transformations

$$(z_1, z_2) \to (\delta z_1, \delta z_2),$$

where $\delta$ is an arbitrary complex number.

We establish the properties of left displacement transformations in the general case.

First of all, we remark that the left displacement transformations preserve the layer $z$. In fact, let the transformation $\delta$, $h \longrightarrow \delta h$, be given, permutable with all transformations of the group $G$:

$$(\delta h)\, g = \delta(hg).$$

Then we have

$$(\delta h)\, Z_h = \delta(hZ_h) = \delta h,$$

i.e., the elements $h$ and $\delta h$ have one and the same stability group. Consequently, the elements $h$ and $\delta h$ belong to one and the same layer.

**Lemma 3.1.** *Let $h_1$ and $h_2$ be elements of $H$ belonging to the same layer $z$. Then there exists one and only one left displacement $h \longrightarrow \delta h$ transforming the point $h_1$ into the point $h_2$.*

**Proof.** Let $h$ be an arbitrary point of $H$, and let the transformation $g_h$ of the group transfer $h_1$ into $h$. Then we have the following relation for $\delta$:

$$\delta h = \delta(h_1 g_h) = (\delta h_1)\, g_h = h_2 g_h. \tag{3.1}$$

Consequently, the transformation $\delta$ is unique if it exists. We will see that the formula (3.1) really defines a transformation in $H$, i.e., the expression $h_2 g_h$ does not depend on the choice of the transformation $g_h$ of the group $G$ transferring $h_1$ into $h$. This is evident because the equations $h_1 g = h_1 g'$ and $h_2 g = h_2 g'$ are equivalent in so far as both separately are equivalent to the condition

$$g' g^{-1} \in Z_{h_1} = Z_{h_2}.$$

We will finally see that the transformation $\delta$ obtained by the formula (3.1) is permutable with the transformations of the group $G$. In fact, we have

$$\delta(hg) = \delta(h_1 g_h g) = h_2 g_h g = (\delta h)\, g.$$

Hence the lemma is proved.

To know how the transformation $\delta$ acts on the space $H$ it is sufficient, by the lemma, to know how it operates on one of the layers $z$ of the space $H$.

We fix in the space $H$ a layer $z_0$ with stability group $K$. The transitive group of transformations $D = K/Z$ acts on this layer. Since this group is commutative, the transformations of the group are permutable with the transformations of the group $K$ on the layer $z_0$.

Thus we conclude

**Theorem 3.1.** *The group of transformations in $H$ permutable with the transformations of the group $G$ is isomorphic to the Cartan group $D$. This isomorphism can be established in the following way. We consider in the space $H$ the layer $z_0$ with the stability group $K \supset D$. Each element $\delta \in D$ is associated to a left displacement transformation coinciding on $z_0$ with the transformation $\delta$.*

**3. The function $\beta(\delta)$ and integral relations in the space $H$.** Let $dh$ be a measure on the fundamental affine space $H$ which is invariant with respect to the group $G$. We consider another measure $d_\delta h$ in the space $H$ obtained by the formula

$$d_\delta h = d(\delta h)$$

where $\delta$ is a fixed element of the Cartan subgroup $D$. This measure is also invariant with respect to the group $G$. In fact, for an arbitrary element $g \in G$ it is

$$d_\delta(hg) = d[(\delta h)g] = d(\delta h) = d_\delta h.$$

Consequently, the measure $d_\delta h$ differs from the measure $dh$ only by a constant factor. We get

$$\beta(\delta) = \frac{d(\delta h)}{dh}. \tag{3.2}$$

It is easy to see that the function $\beta(\delta)$ on $D$ satisfies the following relation:

$$\beta(\delta_1\delta_2) = \beta(\delta_1)\beta(\delta_2)$$

for arbitrary $\delta_1$ and $\delta_2$ of $D$.

Now let $dz$ be the measure on a fundamental projective space $Z$. Then the following integral relation holds for the function $f(h)$ on $H$:

$$\int f(h)\rho^{-1}(h)\,dh = \int dz \int f(\delta h)\,d\delta, \tag{3.3}$$

where $d\delta$ is an invariant measure on the Cartan group $D$ and $\rho(h)$ a weight factor depending on the choice of the measure $dz$.

In (3.3) we replace $f(h)$ by $f(\delta_0^{-1}h)$. Then the right-hand side of the equation does not change but the left-hand side is transformed into

$$\int f(\delta_0^{-1}h)\rho^{-1}(h)\,dh = \int f(h)\rho^{-1}(\delta_0 h)\beta(\delta_0)\,dh.$$

We conclude: *The weight factor $\rho(h)$ in the formula (3.3) satisfies the relation*

$$\rho(\delta h) = \beta(\delta)\rho(h), \tag{3.4}$$

*where* $\beta(\delta) = \dfrac{d(\delta h)}{dh}$.

It is easy to see that conversely the formula (3.3) defines a certain measure $dz$ on the projective space $Z$ if the function $\rho(h)$ satisfies the condition (3.4).

**4. Decomposition of a quasiregular representation of the group $G$ into irreducible representations.** On the space $H$ we define the function $\rho(h)$ satisfying the condition (3.4):

$$\rho(\delta h) = \beta(\delta)\rho(h);$$

on the fundamental projective space $Z$ let $dz$ be the measure defined by the integral relation

$$\int f(h)\rho^{-1}(h)\,dh = \int dz \int f(\delta h)\,d\delta. \tag{3.5}$$

We now consider the Fourier transform of the function $\phi(\delta h)\rho^{1/2}(\delta h)$

$$\psi(h, \chi) = \int \varphi(\delta h)\rho^{1/2}(\delta h)\chi^{-1}(\delta)\,d\delta, \tag{3.6}$$

where $\chi$ is a character on the Cartan group $D$. The function $\psi(h, \chi)$ satisfies the following functional relation, which is immediately obtained from (3.6):

$$\psi(\delta h, \chi) = \chi(\delta)\psi(h, \chi). \tag{3.7}$$

It is natural to call such functions (for $\chi$ fixed) *homogeneous functions of degree of homogeneity* $\chi$.

The function $\phi(h)$ can be decomposed into the homogeneous functions $\psi(h, \chi)$: By the formula for the inverse Fourier transform we have

$$\varphi(\delta h)\rho^{1/2}(\delta h) = \frac{1}{(2\pi)^n}\int \psi(h, \chi)\chi(\delta)\,d\chi. \tag{3.8}$$

The right-hand side of the equation (3.8) is to be understood as the integral over the real and the sum over all integral indices defining the character $\chi(\delta)$, and $n$ is the (real) dimension of the group $D$.

By the classical Plancherel formula we have

$$\int |\varphi(\delta h)|^2 \rho(\delta h)\,d\delta = \frac{1}{(2\pi)^{2n}}\int |\psi(h, \chi)|^2\,d\chi.$$

Here both parts of the equation depend only on the layer $z$. Hence, integrating with respect to $z$ and using the integral relation (3.5), we get

$$\int |\varphi(h)|^2\,dh = \frac{1}{(2\pi)^{2n}}\iint |\psi(h, \chi)|^2\,dz\,d\chi. \tag{3.9}$$

Thus *the transformation* (3.6) *gives an isometric mapping of the space of the functions* $\phi(h)$, *on which a quasiregular representation is given, on the space of functions with the norm*

$$\|\psi\|^2 = \frac{1}{(2\pi)^{2n}}\iint |\psi(h, \chi)|^2\,dz\,d\chi. \tag{3.10}$$

We examine how the operators of the representation act in the space of the functions $\psi(h, \chi)$. We have by definition

$$T_g\psi(h, \chi) = \int [T_g\varphi(\delta h)]\rho^{1/2}(\delta h)\chi^{-1}(\delta)\,d\delta,$$

i.e.,

$$T_g\psi(h, \chi) = \int \varphi(\delta hg)\rho^{1/2}(\delta h)\chi^{-1}(\delta)\,d\delta. \tag{3.11}$$

By condition (3.4) we have

$$\rho(\delta h) = \rho(\delta h g)\frac{\rho(h)}{\rho(hg)}.$$

Therefore, formula (3.11) gives

$$T_g\psi(h, \chi) = \psi(hg, \chi)\left[\frac{\rho(hg)}{\rho(h)}\right]^{-1/2}. \tag{3.12}$$

It is not difficult to see that the operator $T_g$ preserves the norm defined by the formula [1])

$$\|\psi\|^2 = \int |\psi(h, \chi)|^2\, dz.$$

Therefore *the functions* $\psi(h, \chi)$ *with* $\chi$ *fixed are transformed by a certain unitary representation of the group* $G$. *In* [1] *it was proved that this representation is irreducible.* [2]) *Consequently the formulas* (3.8) *and* (3.9) *give the decomposition of a quasiregular representation of the group* $G$ *into irreducible representations.*

We note [1] that two irreducible representations realized in the spaces of the functions $\psi(h, \chi_1)$ and $\psi(h, \chi_2)$ are equivalent if and only if

$$\chi_2(\delta) = \chi_1(\delta^s).$$

Here $s$ denotes an automorphism of the Cartan group $D$ induced by a certain interior automorphism of the group $G$. The group of all such automorphisms is called Weyl group. This group is always finite. Consequently, the multiplicity with which each irreducible representation "in general position" is contained in the decomposition of the quasiregular representation is always finite and equal to the number of elements in the Weyl group $S$.

Let $Z'$ be the subgroup of $G$ generated by the vectors belonging to negative roots. Then, as is well known, any element of $G$, with the exception of a set of elements of smaller dimension, can be uniquely represented in the form

$$g = \zeta\delta z \tag{3.13}$$

where $\zeta \in Z$, $\zeta \in D$ and $z \in Z'$.

Therefore, the points of a fundamental affine space $H$ can be identified with the elements of the subgroup $H' = DZ'$ and the points of a fundamental projective space with the elements of the subgroup $Z'$.

By (3.7) we have

$$\psi(\delta z, \chi) = \psi(\delta)\psi(z, \chi). \tag{3.14}$$

---

1) By (3.7) $|\psi(h, \chi)|$ depends only on $z$.

2) Such a set of irreducible representations of the group $G$ is in [1] called a basic nondegenerate series of irreducible unitary representations.

Consequently, the irreducible representation of a basic nondegenerate series of the group $G$ can also be realized in the space of the functions $\psi(z)$ on the subgroup $Z'$ for which

$$\|\psi\|^2 = \int |\psi(z)|^2 \, dz < \infty. \tag{3.15}$$

Here $dz$ can be thought of as an invariant measure on $Z'$. The operator of the representation $T_g$ is given by the formula

$$T_g \psi(z) = \psi(z\overline{g}) \, \rho^{-1/2}(zg) \, \chi(zg), \tag{3.16}$$

where

$$\rho^{-1}(\zeta\delta z) = \frac{d\,(z\overline{\delta})}{dz}, \quad \chi(\zeta\delta z) = \chi(\delta).$$

In an analogous way we can obtain the decomposition of a representation in the space $G/Z_1$, where $Z_1$ is a certain extension of the euclidean group $Z$. It turns out that if $Z_1$ is a subgroup of the affine group $K$, the representation in the space $G/Z_1$ is decomposed into the irreducible representations of the basic nondegenerate series (however, not every representation of this series is contained in the decomposition). If $Z_1$ is not contained in the affine subgroup, then the representation in the space $G/Z_1$ is decomposed into irreducible representations of one of the so-called degenerate series (see [1]).

## §4. Horospheres in homogeneous spaces and the problem of decomposition of representations into irreducible ones

In what follows (with the exception of the example of the Lobačevskiĭ plane) we consider spaces with a complex semisimple group of motions $G$.

**1. Horospheres in the space $X$.** We will first examine the fundamental affine space of the group $G$: $H = G/Z$. Let $Z_h$ be a subgroup of the group $G$, leaving fixed a certain point $h$ of $H$. We have already in §3 called $Z_h$ the euclidean subgroup of the group $G$. We note that the subgroup $Z_h$ is completely determined by the projective layer $z$ [1)] to which $h$ belongs. So from now on we write $Z_z$ instead of $Z_h$.

Let $X$ be an arbitrary homogeneous space with the group of motions $G$.

Let the elements of $Z_z$ act on the point $x \in X$. The subspace so obtained can be considered as a natural analogue to the hyperplane. By analogy with the Lobačevskiĭ geometry we call every such subspace a *horosphere in* $X$.

Thus: **Definition 4.1.** By a *horosphere in the homogeneous space* $X = G/C$ we mean the set $\omega = xZ_z$ of points of the form $x\zeta$, where $x$ is a fixed point of

1) We recall that a layer $z$ is a set of elements of $H$ possessing one and the same stability group.

the space $X$ and $\zeta$ runs through the elements of the euclidean subgroup $Z_z$ (i.e., the stability group of a certain point $h$ of the fundamental affine space $H$ belonging to the layer $z$).

We call $z$ the *direction of the horosphere* $\omega$.

**2. Examples.** 1°. We first consider the Lobačevskiĭ plane.

We realize it as upper halfplane of the complex plane ($\operatorname{Im} z > 0$) with the group of motions $G$,

$$z \to \frac{\alpha z + \gamma}{\beta z + \delta},$$

where $g = \begin{Vmatrix} \alpha & \beta \\ \gamma & \delta \end{Vmatrix}$ is a real matrix with determinant 1.

The euclidean subgroup $Z$ of the group $G$ is formed by the matrices $\zeta = \begin{Vmatrix} 1 & t \\ 0 & 1 \end{Vmatrix}$. Let $z_0$ be a point of the halfplane. The $Z$-orbit of $z_0$ is given by the parametric equation

$$z = \frac{z_0}{z_0 t + 1}.$$

It is not difficult to see that the orbit is tangent to the real axis at the point $z = 0$.

In a Lobačevskiĭ geometry these orbits are called horocycles. A horocycle is the limit of a sequence of circles with one point moving to infinity.

If we subject the points of the halfplane to the motions of a subgroup conjugate to $Z$, we get again a family of horocycles tangent to the real axis, but at a different point. Therefore, the direction of the horocycle is characterized by the point of contact with the real axis.

2°. We consider the three-dimensional Lobačevskiĭ space. We realize it as the space of all positive definite Hermitian matrices

$$p = \begin{Vmatrix} a & b \\ \bar{b} & c \end{Vmatrix}$$

with $ac - b\bar{b} = 1$. The motions in this space are

$$p \to g^* p g,$$

where $g$ runs through the group $G$ of complex matrices with determinant 1. The euclidean subgroup $Z$ of the group $G$ is formed by the matrices $\zeta = \begin{Vmatrix} 1 & \zeta \\ 0 & 1 \end{Vmatrix}$. We consider the surfaces described by the point $p_0$ subject to the motions of $Z$. Each such surface is given by the parametric equation

$$p = \zeta^* p_0 \zeta, \tag{4.1}$$

where $p_0$ is a point on the surface and $\zeta$ runs through the euclidean subgroup $Z$.

It is not difficult to see that these surfaces are formed by all matrices $p = \begin{Vmatrix} a & b \\ \bar{b} & c \end{Vmatrix}$ for which $a = \text{const}$. We will show that these surfaces are horospheres in the Lobačevskiĭ space.

We note that the horospheres in the Lobačevskiĭ space (in the realization chosen here) are generated by the motions leaving fixed a certain degenerate Hermitian matrix.

We consider the degenerate matrix $p_1 = \begin{Vmatrix} 0 & 0 \\ 0 & 1 \end{Vmatrix}$. Its stability group consists of the matrices $k = \begin{Vmatrix} e^{i\phi} & \zeta \\ 0 & e^{-i\phi} \end{Vmatrix}$. The corresponding horospheres consist of the matrices of the form $p = k^* p_0 k$. If $p_0 = \begin{Vmatrix} a_0 & b_0 \\ \bar{b}_0 & c_0 \end{Vmatrix}$, then the equation of the horosphere can be given in the form $a = a_0$. But our surfaces (4.1) are given by the same equation. Consequently, they are horospheres.

3°. Let $X$ be the space of all positive definite Hermitian matrices $p$ of order $n$ with determinant 1. The motions in $X$ have the form

$$p \longrightarrow g^* pg,$$

where $g$ runs through the group $G$ of all complex matrices with determinant 1. The space $X$ can also be realized as the space of all positive definite Hermitian forms $p(x, x)$ with discriminant 1 in the $n$-dimensional complex linear space $R_n$. (Here $x$ is a point in $R_n$.)

The elements of $G$ define a linear transformation $x \longrightarrow xg$ in the space $R_n$. Then the motions in the space $X$ are

$$p(x, x) \longrightarrow p'(x, x) = p(xg, xg). \tag{4.2}$$

We establish the equations of the horospheres in the space $X$. Let $Z_z$ be some euclidean subgroup of $G$. The realization of this subgroup by matrices depends on the choice of a base $e_1, \cdots, e_n$ in which the elements of the subgroup $Z_z$ are written in the form[1)]

$$\zeta = \begin{Vmatrix} 1 & \zeta_{12} & \dots & \zeta_{1n} \\ 0 & 1 & \dots & \zeta_{2n} \\ \cdot & \cdot & \cdot & \cdot \\ 0 & 0 & \dots & 1 \end{Vmatrix}. \tag{4.3}$$

1) We note that we consider only the bases in $R_n$ which can be transfomed into one another by transformations of the group $G$.

It is not difficult to see that the transformations of $Z_z$ preserve the determinants

$$\Delta_1 = p(e_1, e_1), \quad \Delta_2 = \begin{vmatrix} p(e_1, e_1) & p(e_1, e_2) \\ p(e_2, e_1) & p(e_2, e_2) \end{vmatrix}, \ldots$$

$$\ldots, \Delta_{n-1} = \begin{vmatrix} p(e_1, e_1) & p(e_1, e_2) & \ldots & p(e_1, e_{n-1}) \\ p(e_2, e_1) & p(e_2, e_2) & \ldots & p(e_2, e_{n-1}) \\ \cdot & \cdot & \cdot & \cdot \\ p(e_{n-1}, e_1) & p(e_{n-1}, e_2) & \ldots & p(e_{n-1}, e_{n-1}) \end{vmatrix}. \tag{4.4}$$

Conversely, as is well known, it is possible to reduce every positive definite Hermitian form $p(x, x)$ by a triangle transformation $\zeta$ to the form

$$p(x, x) = \Delta_1 |x_1|^2 + \frac{\Delta_2}{\Delta_1} |x_2|^2 + \ldots + \frac{\Delta_n}{\Delta_{n-1}} |x_n|^2. \tag{4.5}$$

Consequently, if the determinants $\Delta_1, \cdots, \Delta_{n-1}$ corresponding to two Hermitian forms $p(x, x)$ and $p'(x, x)$ with discriminant $\Delta_n = 1$ are pairwise equal, then these forms are obtained one from the other by a transformation $\zeta$, belonging to the subgroup $Z_z$.

Thus it is proved that *the horospheres in the space $X$ of all positive definite Hermitian forms $p(x, x)$ with discriminant 1 are determined in the following way. We choose in the space $R_n$ a base $e_1, \cdots, e_n$ and consider the set of all Hermitian forms $p(x, x)$ satisfying the conditions*

$$p(e_1, e_1) = c_1, \quad \begin{vmatrix} p(e_1, e_1) & p(e_1, e_2) \\ p(e_2, e_1) & p(e_2, e_2) \end{vmatrix} = c_2, \ldots$$

$$\ldots, \begin{vmatrix} p(e_1, e_1) & p(e_1, e_2) & \ldots & p(e_1, e_{n-1}) \\ p(e_2, e_1) & p(e_2, e_2) & \ldots & p(e_2, e_{n-1}) \\ \cdot & \cdot & \cdot & \cdot \\ p(e_{n-1}, e_1) & p(e_{n-1}, e_2) & \ldots & p(e_{n-1}, e_{n-1}) \end{vmatrix} = c_{n-1}, \tag{4.6}$$

*where $c_1, c_2, \cdots, c_{n-1}$ are positive constants. This set of Hermitian forms is a horosphere in the space $X$. Changing $c_1, \cdots, c_{n-1}$, we get a family of parallel horospheres. Different bases $e_1, \cdots, e_n$ in $R_n$ lead therefore to different families of horospheres.* [1)]

**3. Class of a homogeneous space.** The transformations of the group $G$ in the space $X$ transform each horosphere into a horosphere.

In fact, let the horosphere $\omega$ pass through the point $x$ and have the direction

---

1) If some of the constants $c_1, \cdots, c_{n-1}$, are negative, the equations (4.6) give horospheres in one of the spaces of nondefinitive Hermitian forms of given signature.

$z$, i.e., $\omega = xZ_z$. Then $\omega g = xg(g^{-1}Z_z g) = xgZ_z g$, which means that $\omega g$ is again a horosphere.

If $X$ is a Lobačevskiĭ plane or a three-dimensional Lobačevskiĭ space, then the set of horospheres in $X$ is transitive with respect to the group of transformations $G$: any two horospheres can be transformed into each other by a transformation from $G$. In the general case, however, the space of the horospheres is not transitive but splits up into transitive components.

**Definition 4.2.** We will say that the space $X$ is a space of class 0 if the set of horospheres in this space is transitive or if it splits into a finite number of transitive subsets.

We now assume that the set $\Omega$ of horospheres in $X$ is a stratified space in which the layers are transitive subsets of horospheres. Then by the *class of the space* $X$ we will mean the real dimension of the base of the stratified space $\Omega$.

In Part II it will be shown that the notion of class has an important group-theoretical meaning: it determines the number of parameters involved in the irreducible representations (equivalent among themselves) of the group $G$ that appear in the representation of this group in the homogeneous space $X$.

We will give a condition for $X$ to be a space of class 0:

**Lemma 4.1.** *Let $K$ be an affine subgroup of the complex semisimple Lie group $G$, and let $X = G/C$ be a homogeneous space on which the group $G$ operates. If there exists in $G$ a finite subset $S$ such that $CSK = G$, then $X$ belongs to the class 0. In particular, if $CK = G$, i.e., any element of $G$ can be given by $g = ck$, $c \in C$, $k \in K$, then the space of horospheres in the group $G$ is transitive.*

**Proof.** Let $x_0$ be a fixed point in $X$ whose stability group is $C$. An arbitrary horosphere in $X$ has the form

$$\omega = x_0 g_1 Z g_2,$$

where $Z$ is a fixed euclidean subgroup of $G$ and $g_1$ and $g_2$ are certain elements of $G$. Let $D$ be a Cartan subgroup of $G$. Then $K = DZ$. By our assumptions we can represent $g_1$ in the form

$$g_1 = cs\delta\zeta,$$

where $c \in C$, $s \in S$, $\delta \in D$ and $\zeta \in Z$. Consequently,

$$\omega = x_0 s Z g_2,$$

i.e., the horosphere $\omega$ can be obtained by displacing the horosphere $\omega_s = x_0 s Z$.

In particular, if the set $S$ contains only a single element of $G$, then any horosphere is obtained by translating the horosphere $x_0 Z$, i.e., the set of horospheres is transitive. Q.E.D.

The symmetric space $X = G/C$, where $C$ is a real form of the group $G$, is an example of a space of class 0. In fact, the real form $C$ of the complex semisimple Lie group $G$ satisfies always the conditions of the lemma.

If $C$ is a compact form of $G$, then we represent any element of the group $G$ in the form $g = ck$, where $c \in C$, $k \in K$. Consequently, the horosphere space of $X$ is transitive in this case.

We consider now the group space of the group $G$ in which the motions are defined as right displacements

$$g \longrightarrow gg_0.$$

An arbitrary horosphere in this space is given by

$$\omega = g_1^{-1} Z g_2,$$

where $g_1$ and $g_2$ are fixed elements of $G$ and $Z$ is a fixed euclidean subgroup. The two pairs $(g_1, g_2)$ and $(g_1', g_2')$ define the same horosphere if and only if

$$g_1' = \delta\zeta_1 g_1, \quad g_2' = \delta\zeta_2 g_2,$$

where $\zeta_1$ and $\zeta_2$ are elements of $Z$ and $\delta$ an element of the Cartan subgroup $D$. Hence it follows that *the horospheres in the group space form a manifold. This manifold is isomorphic to the manifold of the pairs* $(h_1, h_2)$ *of points of the fundamental affine space in which the pairs* $(h_1, h_2)$ *and* $(\delta h_1, \delta h_2)$ *are identified.*

If the pair $(h_1, h_2)$ corresponds to the horosphere $\omega$, then the pair $(h_1, h_2 g)$ corresponds to the horosphere $\omega g$. Therefore *the manifold of the horospheres of the group space is stratified by transitive submanifolds, each of which is identical to a fundamental affine space* $H$. *The projective space* $Z$ *is the base of this stratification. Thus the class of the group space of* $G$ *is equal to the real dimension of the projective space* $Z$.

**4. Regular subgroups of the group** $G$. We describe the space of all horospheres of the homogeneous space $X = G/C$. We fix some point $x_0$ in $X$. Any horosphere in $X$ can then be represented in the form

$$\omega = x_0 g_1 Z g_2, \tag{4.7}$$

where $g_1$ and $g_2$ are elements of $G$. Conversely formula (4.7) associates a certain horosphere to each pair $g_1$, $g_2$ of elements of $G$.

We identify in the space $G \times G$ of the pairs $(g_1, g_2)$ those pairs of elements from $G$ which define one and the same horosphere. We will call the resulting topological space $\Omega$ the *horosphere space* of $X$.

It is not difficult to see that the topology introduced in the horosphere space does not depend on the choice of $x_0 \in X$ and that therefore it is completely defined by $X$.

We remark that $\Omega$ does not have to be a Hausdorff space. In the present article we will consider spaces $X$ for which the horosphere space is a manifold. The general case will be treated in the second part.

Definition 4.3. We will call the subgroup $C$ of $G$ *regular in* $G$ if in the homogeneous space $X = G/C$ the horosphere space forms a manifold.

By virtue of the above remark "regularity" is a property invariant under transition to a conjugate subgroup.

At the end of the preceding section we already showed that the unit subgroup of $G$ is a regular subgroup. Namely, the horosphere space of the group space is the manifold of the pairs $(h_1, h_2)$ of points of the fundamental affine space where the pairs $(h_1, h_2)$ and $(\delta h_1, \delta h_2)$ are identified. We give another example of a regular subgroup.

First of all, we note that if the horosphere set in the space $X = G/C$ is transitive, then this set is a manifold and therefore $C$ is regular.

In fact, let $\omega_0$ be a horosphere in the space $X$, and let $L_{\omega_0}$ be the sliding group of $\omega_0$, i.e., the subgroup of transformations preserving $\omega_0$. Then the set of all horospheres forms a homogeneous space $\Omega = G/L_{\omega_0}$, and therefore it is a manifold.

We also show that *any compact subgroup* $C$ *of* $G$ *is a regular subgroup.*

Let at first $C = \mathfrak{U}$ be a maximal compact subgroup of $G$. As was shown above, in this case the set of horospheres is transitive and therefore a manifold. Let us describe this manifold: We consider the point $x_0$ with stability group $\mathfrak{U}$, and let

$$\omega_0 = x_0 Z.$$

The sliding group $L_{\omega_0}$ of $\omega_0$ consists of the matrices $g$ satisfying the equation

$$x_0 Z g = x_0 Z$$

or the equivalent equation

$$\mathfrak{U} Z g = \mathfrak{U} Z. \tag{4.8}$$

We decompose the Cartan group $D$ into a direct product

$$D = \Gamma \times E$$

of the torus subgroup $\Gamma$ and the vector subgroup $E$. As is known, there exists a maximal compact subgroup $\mathfrak{U}_1$ of $G$ such that $\mathfrak{U}_1 \supset \Gamma$ and $ZE\mathfrak{U}_1 = G$. The decomposition of the elements of $G$ according to $g = \zeta \epsilon u$, $\zeta \in Z$, $\epsilon \in E$, and $u \in \mathfrak{U}_1$ is unique. Since all maximal compact subgroups of $G$ are conjugate among themselves, it is possible to assume $\mathfrak{U}_1 = \mathfrak{U}$, without loss of generality.

It is easy to see then that $L_{\omega_0} \supset Z\Gamma$. On the other hand, if $L_{\omega_0} \neq Z\Gamma$, then the subgroup $L_{\omega_0}$ has to have a nonunique intersection with $E$. If $\epsilon \neq 1$ is then an element of this intersection, by (4.8) it can be written in the form

$$\epsilon = u\zeta,$$

where $u \in \mathfrak{U}$, $\zeta \in Z$. But by the uniqueness of the decomposition of the form $g = u\zeta\epsilon$ it is impossible.

Thus *the sliding group of the horosphere* $\omega_0 = x_0 Z$ *has the form* $L_{\omega_0} = Z\Gamma$, *where* $\Gamma$ *is a compact subgroup of the Cartan group* $D$.

Hence it follows that two horospheres $x_0 Z g_1$ and $x_0 Z g_2$ coincide if and only if $g_2 = \gamma\zeta g_1$, where $\zeta \in Z$ and $\gamma$ belongs to the compact subgroup $\Gamma$ of the Cartan group $D$. Thus we proved the following result.

*Let* $\mathfrak{U}$ *be a maximal compact subgroup of* $G$. *Then the horospheres in the space* $X = G/\mathfrak{U}$ *form a transitive manifold. This manifold coincides with the fundamental affine space in which the points* $h$ *and* $\gamma h$ *are identified and where* $\gamma$ *runs through the compact subgroup* $\Gamma$ *of* $D$.

Now let $C$ be an arbitrary compact subgroup of $G$. Then $C$ is a subgroup of a certain maximal compact subgroup of $G$. Without loss of generality we can assume that $C$ is a subgroup of the above group $\mathfrak{U}$.

Let $x_0$ be a point of the space $X = G/C$ whose stability group is $C$. It is obvious that any horosphere in $X$ can be represented in the form

$$\omega = x_0 \tilde{u} Z h,$$

where $\tilde{u}$ is an element of the space $\tilde{\mathfrak{U}} = \mathfrak{U}/C$ of right cosets of $\mathfrak{U}$ with respect to the subgroup $C$ and $h$ is an element of the fundamental affine space $H = G/Z$. Thus each horosphere in the space $X = G/C$ is given by the pair $(\tilde{u}, h)$.

When do two pairs $(\tilde{u}_1, h_1)$ and $(\tilde{u}_2, h_2)$ define the same horosphere? If

$$x_0\tilde{u}_1 Z h_1 = x_0 \tilde{u}_2 Z h_2, \tag{4.9}$$

then all the more

$$\mathfrak{U} Z h_1 = \mathfrak{U} Z h_2. \tag{4.10}$$

Consequently,

$$h_2 = \gamma h_1,$$

where $\gamma \in \Gamma$, $\delta \in D$. But then we get from (4.9)

$$\tilde{u}_1 Z = \tilde{u}_2 \gamma Z,$$

whence

$$\tilde{u}_1 = \tilde{u}_2 \gamma.$$

Therefore: *Let* $C$ *be a subgroup of the compact group* $\mathfrak{U}$. *The space of horospheres in* $X = G/C$ *coincides in this case with the space of the pairs* $(\tilde{u}, h)$, *where* $\tilde{u}$ *is a point of the space* $\tilde{\mathfrak{U}} = \mathfrak{U}/C$ *of right cosets of the group* $\mathfrak{U}$ *with respect to the subgroup* $C$, *and* $h$ *is a point of the fundamental affine space* $H$. *The pairs* $(\tilde{u}, h)$ *and* $(\tilde{u}\gamma^{-1}, \gamma h)$ *are identified;* $\gamma$ *is an element of the compact subgroup* $\Gamma$ *of the Cartan group* $D$.

Now it follows immediately that *the horospheres in the space* $X = G/C$, *where* $C$ *is a compact subgroup, form a manifold and that* $C$ *is therefore a regular subgroup.*

We decompose the Cartan subgroup $D$ into a direct product of a compact subgroup $\Gamma$ and a vector group $E$. Then, as is known, we can represent an arbitrary element $g \in G$ unambiguously in the form $g = \zeta\epsilon u$, $\zeta \in Z$, $\epsilon \in E$, $u \in \mathfrak{U}$.

In view of this the horosphere space of $X = G/C$ can also be described as the space of the triples $(\tilde{u}, \epsilon, u)$, where $\tilde{u} \in \tilde{\mathfrak{U}} = \mathfrak{U}/C$, $\epsilon \in E$, $u \in \mathfrak{U}$. The triples $(\tilde{u}, \epsilon, u)$ and $(\tilde{u}\gamma^{-1}, \epsilon, \gamma u)$, $\gamma \in \Gamma$, are identified.

We establish at the same time how the horosphere space is stratified by transitive subspaces.

Each transitive subspace of $\Omega$ is formed by the horospheres $x_0\tilde{u}Zh_1$, where $\tilde{u}$ is fixed. It is not difficult to check that this subspace $\Omega_{\tilde{u}}$ is a manifold. It coincides with the fundamental affine space in which the points $h$ and $\gamma h$ are identified, where $\gamma$ runs through the subgroup of elements from $\Gamma$ for which $\tilde{u}\gamma = \tilde{u}$.

In fact, two pairs $(\tilde{u}, h)$ and $(\tilde{u}, \gamma h)$ are identified if and only if $\tilde{u} = \tilde{u}_1\gamma^{-1}$. On the other hand, two manifolds $\Omega_{\tilde{u}_1}$ and $\Omega_{\tilde{u}_2}$ coincide if and only if $\tilde{u}_2 = \tilde{u}_1\gamma$, $\gamma \in \Gamma$.

Consequently, *the base of the stratified horosphere space* $\Omega$ *is the space* $\tilde{\mathfrak{U}} = \mathfrak{U}/C$ *of right cosets of the maximal compact subgroup* $\mathfrak{U}$ *with respect to the subgroup* $C$ *in which the points* $\tilde{u}$ *and* $\tilde{u}\gamma$ *are identified, where* $\gamma$ *runs through the compact subgroup* $\Gamma$ *of the Cartan group* $D$.

**5. Parallel horospheres. Rank of a homogeneous space.** We consider the homogeneous space $X = G/C$. We first assume that the horospheres in the space $X$ are translated into one another by transformations of $G$. Thus $G$ appears as a transitive group of transformations in the horosphere space $\Omega$. We define in $\Omega$ left displacement transformations. We consider continuous transformations in the transitive horosphere space $\Omega$ permutable with the transformations of $G$. We call them left displacement transformations. We will denote by $\delta\omega$ the horosphere into which the left displacement transforms the horosphere $\omega$. The condition of permutability of $\delta$ with the motions of $G$ is given by

$$\delta(\omega g) = (\delta\omega)\, g.$$

**Definition 4.4.** We will say that two horospheres $\omega_1$ and $\omega_2$ of a transitive family are *parallel* if they can be transformed into each other by a left displacement $\delta$: $\omega_2 = \delta\omega_1$. The element $\delta$ will be called the *compound distance* of $\omega_1$ to $\omega_2$.

From the definition it follows immediately that parallel horospheres have the same sliding group $L_\omega = L_{\delta\omega}$.

Conversely, *if two horospheres* $\omega_1$ *and* $\omega_2$ *have one and the same sliding group, then there exists one and only one left displacement transforming* $\omega_1$ *into* $\omega_2$.

**Proof.** Let there exist such a transformation $\delta$. If $\omega = \omega_1 g$ is an arbitrary horosphere, then we have

$$\delta\omega = \delta(\omega_1 g) = (\delta\omega_1) g = \omega_2 g. \tag{4.11}$$

The transformation is therefore unique. On the other hand, formula (4.11) actually defines a left displacement.

Indeed, the right-hand side of equation (4.11) does not depend on the choice of the element $g$ translating the horosphere $\omega_1$ into the given horosphere $\omega_2$: if $\omega_1 g = \omega_1 g'$, then also $\omega_2 g = \omega_2 g'$, since the sliding groups of $\omega_1$ and $\omega_2$ coincide.

The permutability of the transformation $\delta$, defined by (4.11), with the transformations of $G$ is established immediately;

$$\delta(\omega g_0) = \delta(\omega_1 g g_0) = \omega_2 g g_0 = (\delta\omega) g_0.$$

Now let $\omega$ be an arbitrary horosphere in $X$, and let $L_\omega$ be its sliding group. Two horospheres $\omega$ and $\omega g$ have the same sliding group $L_\omega$ if and only if

$$g^{-1} L_\omega g = L_\omega,$$

i.e., $g$ belongs to the normalizer $N(L_\omega)$ of the group $L_\omega$ in the group $G$. From the above proposition we conclude:

*The group of all left displacements in the transitive horosphere space is isomorphic to the factor group* $N(L_\omega)/L_\omega$, *where* $L_\omega$ *is the sliding group of the fixed horosphere* $\omega$ *and* $N(L_\omega)$ *is the normalizer of this group in* $G$.

**Definition 4.5.** Let $X$ be a homogeneous space and $\Omega$ the set of horospheres in $X$. If the horosphere space $\Omega$ is transitive with respect to the transformations of $G$, then the real dimension of the group of all left displacements in the horosphere space will be called the *rank of the space* $X$. If $\Omega$ is not transitive, then we decompose it into transitive subspaces $\Omega_a$. The highest of the dimensions of the left displacement groups in the subspaces $\Omega_a$ will then be called the *rank of the space* $X$.

In what follows it will be proved that the rank of a space has an important group-theoretic meaning: it defines the number of parameters on which the non-equivalent irreducible representations depend, which are contained in the decomposition of the representation of $G$ in $X$.

If $C$ is a real form of a complex semisimple Lie group, then there exists a close connection between the above notion of the rank of a symmetric space $X = G/C$ and the rank of a subgroup $C$ in the sense of Killing-Cartan.

We recall that the maximal number of independent parameters of a Lie group with respect to its inner automorphisms is called the rank of this Lie group in the sense of Killing-Cartan. If $C$ is a real form and $D$ is a Cartan subgroup of the complex semisimple Lie group $G$, then the rank of the subgroup $C$ coincides with the complex dimension of $D$.

We first assume that $C$ is a maximal compact subgroup of $G$. In this case, as was proved in part 4 of this section, the set of horospheres in the space $X = G/C$ is transitive; the sliding group of $\omega_0 = x_0 Z$ is $L_{\omega_0} = Z\Gamma$, where $\Gamma$ is a compact subgroup of $D$. The normalizer $N(L_{\omega_0})$ of the subgroup $L_{\omega_0}$ coincides with the normalizer of the group $Z$

$$N(L_{\omega_0}) = ZD.$$

Because the real dimension of $\Gamma$ is equal to the complex dimension of $D$, the rank of the space $X = G/C$ is equal to the complex dimension of $D$.

Therefore: *The rank of a symmetric space $X = G/C$, where $C$ is a maximal compact subgroup of the group $G$, is equal to the rank of the subgroup $C$ in the sense of Killing-Cartan.*

It is not difficult to verify that this assertion is true also for the case where $C$ is an arbitrary real form of the group $G$.

**6. Measure on a horosphere.** We consider an arbitrary transitive family of horospheres $\Omega_\alpha$. Let

$$\omega = xZ_z$$

be an arbitrary horosphere of this family, with the given direction $z$. Let $L_\omega$ be the sliding group of $\omega$, i.e., the set of elements of $G$ transforming the horosphere into itself. Then $L_\omega \supset Z_z$.

The space $\Omega_\alpha$ can be identified with the space $G/L_\omega$ of right cosets of the group $G$ with respect to the group $L_\omega$. On the other hand, the points of the horosphere $\omega$ form a homogeneous space with respect to the subgroup $L_\omega$:

$$\omega = L_\omega/(C_x \cap L_\omega),$$

where $C_x$ is the stability group of the point $x \in \omega$.

It is possible to give on the horosphere $\omega$ a measure quasi-invariant with respect to the transformations of $L_\omega$. We denote this measure by $d_\omega x$. In the special case where $L_\omega$ is a unimodular group and the subgroup $C_x$ is compact or commutative (and consequently $C_x \cap L_\omega$ is a unimodular group), $d_\omega x$ can be considered as invariant measure on $\omega$.

We assume that an invariant measure $d_\omega x$ is given on the horosphere and let $\omega' = \omega g$ be another horosphere of this transitive family. We will introduce a measure on this horosphere by

$$d_{\omega'}(xg) = d_\omega x.$$

It is evident that this measure is invariant on the horosphere $\omega'$ and that it does not depend on the choice of the element $g$ transforming $\omega$ into $\omega'$.

Thus, *if it is possible to define a measure* $d_\omega x$, *invariant with respect to the sliding group* $L_\omega$ *of* $\omega$, *on one of the horospheres* $\omega$ *of a transitive family, then the measure can be defined in the same way on every horosphere of this family. Also, the measures can always be normalized in such a way that they are preserved by the transformations of* $G$ *that transform one horosphere into another*

$$d_{\omega g}(xg) = d_\omega x.$$

**7. Representations on the horosphere space and their relation to quasiregular representations.** We suppose here that the horosphere space is stratified by the transitive subspaces $\Omega_a$. Let the representation in $\Omega$ be given. In order to decompose this representation into irreducible representations it is evidently sufficient to decompose the corresponding representations of $G$ in the subspaces $\Omega_a$.

We examine therefore some transitive subset $\Omega_a$ in the space of all horospheres. Let $\omega_0$ be a certain horosphere of $\Omega_a$, and let $L_{\omega_0}$ be its sliding group. Then

$$\Omega_a = G/L_{\omega_0}.$$

On the other hand, the subgroup $L_{\omega_0}$ is an extension of a certain euclidean subgroup $Z$ of the group $G$. Consequently, there is a quasiregular or degenerate quasiregular representation of $G$ realized in the horosphere space $\Omega_a$.

The problem is easier if the sliding group $L_{\omega_0}$ coincides with the euclidean subgroup $Z$. In this case the representation in $\Omega_a$ is given by

$$T_g\phi(\omega) = \phi(\omega g),$$

where

$$\|\phi\|^2 = \int |\phi(\omega)|^2\, d\omega < \infty,$$

and is a quasiregular representation of $G$.

We now consider the more general case where $L_{\omega_0} \neq Z$, but

$$L_{\omega_0} \subset K, \tag{4.12}$$

where $K = ZD$ is an affine subgroup of $G$. This is the case for the spaces $X = G/C$, where $C$ is a compact or Cartan subgroup of $G$.

We describe the representations in the horosphere space $\Omega_\alpha$ for this case.

Let $H = G/Z$ be a fundamental affine space.

Since $Z \subset L_\omega$, the horosphere space $\Omega_\alpha = G/L_\omega$ is a contraction of the space $H$. Namely, two points $h_1$ and $h_2$ of the space $H$ correspond to one and the same horosphere $\omega$ if and only if

$$h_2 = \delta h_1,$$

where $\delta$ is an element of the factor group $\hat{D} = L_\omega/Z$.

It is now possible to consider the function $\phi(h)$ in the fundamental affine space satisfying the additional condition

$$\phi(\delta h) = \chi_0(\delta)\phi(h)$$

(where $\delta$ runs through the subgroup $\hat{D} = L_\omega/Z$ and $\chi_0$ is a given character on this subgroup) instead of the function $\phi(\omega)$ in the horosphere space $\Omega_\alpha$.

These functions $\phi(h)$ form a subspace of the space of the quasiregular representations.[1] The decomposition of this subspace into irreducible representations is performed in the same way as the decomposition of the space of all quasiregular representations (see §3). This will be done in detail for certain cases in §§5 and 6.

We remark that only irreducible representations of the nondegenerate series appear in this decomposition. The number of parameters on which these irreducible representations depend is equal to the dimension of the factor group $D/\hat{D}$. Since $D = K/Z$ and $\hat{D} = L_\omega/Z$, we get $D/\hat{D} = K/L_\omega$. Thus the irreducible representations into which the representation of the horosphere space $\Omega_\alpha$ is decomposed depend on $r$ real parameters, where $r$ is the rank of the initial space $X$.

We studied the representations of the group $G$ on the horosphere space for the case where the sliding group $L_\omega$ of the horospheres $\omega$ is contained in the affine subgroup $K$. These representations are always decomposed into a direct sum of irreducible representations of the basic nondegenerate series. The decomposition is carried out by the same elementary method as in the case of the quasiregular representation of the group $G$ (see §3).

---

1) The term "subspace" is not completely adequate. In the case where the subgroup $\hat{D}$ is not compact, the space of these functions is an irreducible direct summand in the space of the quasiregular representations.

The situation is different if the sliding group $L_\omega$ is not contained in an affine subgroup. Then the representation of the group $G$ in the horosphere space $\Omega_a$ is decomposed into irreducible representations of one of the degenerate series. Thus if the number of transformations translating the horospheres into themselves is increased, degenerate series will appear in the decomposition of the representations. This case will be examined in another paper.

**8. Decomposition problem of a representation in a homogeneous space $X$ into irreducible representations (outline of a general method for the solution).** Let the representation in the homogeneous space $X$ be given by

$$T_g f(x) = f(xg)\, a(x, g). \tag{4.13}$$

This representation is in general not irreducible. Therefore *the decomposition problem for the representation* (4.13) *in the space $X$ exists.* Below we will state a general method for the solution of this problem.

We make the following assumptions about the space $X$:

1°. The horospheres in the space $X$ are connected subspaces; on each horosphere $\omega$ there exists a measure $d_\omega x$ invariant with respect to the subgroup of transformations of the horosphere $\omega$ into itself.

As was shown in part 6 of this section the measure on a horosphere can be normalized in such a way that it is preserved by the group $G$ transforming one horosphere into another:

$$d_{\omega g}(xg) = d_\omega x. \tag{4.14}$$

2°. The horosphere space of $X$ is stratified into subspaces transitive with respect to $G$.

These conditions are satisfied for the space $X = G/C$, where $C$ is a compact subgroup or a subgroup of the Cartan group $D$ (or finally, is the product of a compact subgroup and a subgroup $D_1 \subseteq D$). The representations in these spaces are studied in the present paper. Spaces of more general form will be considered in the second part of this work.

Let $f(x)$ be a function on the representation space. We assume that it is continuous, infinitely often differentiable and different from zero only in a sufficiently small neighbourhood of a given point in $X$.

We compare the function $f(x)$ to its integrals over all possible horospheres $\omega$ in the space $X$. We get

$$\phi(\omega) = \int_\omega K(\omega, x) f(x)\, d_\omega x. \tag{4.15}$$

We choose the kernel $K(\omega, x)$ so that $\phi(\omega)$ is transformed by a certain representation in this way:

$$T_g \phi(\omega) = \phi(\omega g)\, b(\omega, g), \tag{4.16}$$

while the initial function $f(x)$ is transformed by (4.13). We have by hypothesis

$$\phi(\omega g)\, b(\omega, g) = \int_\omega K(\omega, x)\, a(x, g)\, f(xg)\, d_\omega x,$$

i.e.,

$$\int_{\omega g} K(\omega g, x)\, b(\omega, g)\, f(x)\, d_{\omega g} x = \int_\omega K(\omega, x)\, a(x, g)\, f(xg)\, d_\omega x. \tag{4.17}$$

Replacing $x$ by $xg$ in the first integral of (4.17) and taking into account that $d_{\omega g}(xg) = d_\omega x$, we get

$$\int_\omega K(\omega g, xg)\, b(\omega, g)\, f(xg)\, d_\omega x = \int_\omega K(\omega, x)\, a(x, g)\, f(xg)\, d_\omega x.$$

Hence

$$K(\omega g, xg)\, b(\omega, g) = K(\omega, x)\, a(x, g). \tag{4.18}$$

Let $x$ be an arbitrary point of the space and $\omega$ be a horosphere passing through $x$. Then we get from (4.18) the condition for $b(\omega, g)$

$$b(\omega, g) = a(x, g) \tag{4.19}$$

for any element $g$ of $G$ leaving fixed the point $x$ and the horosphere $\omega$. If this condition is satisfied, the equation (4.18) defines the corresponding kernel $K(\omega, x)$. Evidently the equation (4.18) determines the kernel uniquely up to a constant factor, if $b(\omega, g)$ is given, on each transitive set of horospheres.

We treat the simplest case where $a(x, g) \equiv 1$. In this case we can choose $b(\omega, g) \equiv 1$ and $K(\omega, x) \equiv 1$. Formula (4.15) then becomes

$$\phi(\omega) = \int_\omega f(x)\, d_\omega x. \tag{4.20}$$

The functions $\phi(\omega)$ are now transformed by the representation

$$T_g \phi(\omega) = \phi(\omega g).$$

In part 8 of this section we proved that the decomposition of the space of the functions $\phi(\omega)$ into irreducible subspaces is reduced to the decomposition of quasiregular and degenerate quasiregular representations and can therefore be carried out in an elementary way. On the other hand, having obtained this decomposition, we find the irreducible components of the initial representation in the space $X$ by means of formula (4.15) (direct theorem).

Furthermore, we have to show that the representation in $X$ is really decomposed into the above irreducible representations, and we have to obtain the explicit formulas for this decomposition (inverse theorem). This problem is more difficult to solve, since the formula (4.15) has to be inverted. It is necessary to describe $f(x)$ by its integrals over all possible horospheres in $X$.

The problem can now be stated in this way: The generalized function (Delta-"function") concentrated in a point of the homogeneous space $X$ has to be decomposed into generalized functions concentrated on the horospheres in the space $X$. Thus the decomposition problem is reduced to a problem of integral geometry.

In §§5 and 6 this problem is solved for the space $X = G/C$, where $C$ is a compact or a Cartan subgroup of $G$.[1)]

The inversion of formula (4.20) is carried out in the following way. Let $\omega_z$ be a horosphere of direction $z$ passing through the given point $x$, and let $\delta\omega_z$ be a horosphere parallel to $\omega_z$ at the "compound distance" $\delta$. We form the integral

$$I(x, \delta) = \int \phi(\delta\omega_z)\, dz,$$

where $dz$ denotes a certain measure on the fundamental projective space $Z$ depending in general on the choice of the point $x$. Then it turns out that

$$f(x) = L\{\beta^{\frac{1}{2}}(\delta)\, I(x, \delta)\}_{\delta=e},$$

where $L$ is a certain differential operator on the Cartan group $D$, which will be described in detail in §5.

We note that the functions $\phi(\omega)$ obtained by integration of $f(x)$ in the space $X$ over all possible horospheres $\omega$ satisfy certain additional conditions of symmetry. This will become clear in detail in §§5 and 6, where concrete results are given; these conditions will be described in general form in the second part of this work.

## §5. Representations in spaces with compact stability subgroup

We first consider a regular representation of a complex semisimple Lie group $G$, i.e., a representation in the space of functions given on the entire group $G$. The decomposition of a regular representation of $G$ into irreducible representations is well known (see [1, 3, 4, 11]). Here we will give only the geometrical treatment of the decomposition problem of a regular representation. This permits us to single out the geometrical concepts and facts necessary in the study of representations in more general homogeneous spaces.

At the end of §5 we consider the representations in a homogeneous space $X = G/C$, where $C$ is an arbitrary compact subgroup of the group $G$. Here also the solution of the decomposition problem in the space $X$ will be given from the geometrical point of view.

**1. Horospheres in a group space.** We showed in §4 that the horospheres in a

1) In fact, the results of §§5 and 6 are found without special difficulty on spaces of a somewhat more general kind, which were mentioned at the beginning of this article.

group space form a manifold. Each horosphere in this manifold is given by a pair $h_1$, $h_2$ of points of the fundamental affine space:

$$\omega = (h_1, h_2).$$

Two pairs $(h_1, h_2)$ and $(\delta h_1, \delta h_2)$ represent one and the same horosphere.

The element $g$ of $G$ transforms each horosphere $\omega = (h_1, h_2)$ into the horosphere $\omega g = (h_1, h_2 g)$. Thus the manifold of all horospheres in the space $G$ is stratified into submanifolds that are transitive with respect to $G$. If $\omega = (h_1, h_2)$, then $h_1$ gives the transitive manifold of horospheres to which $\omega$ belongs and $h_2$ denotes the horosphere $\omega$ in this manifold.

We will introduce a measure on each of the horospheres in the group space. The sliding group of each horosphere coincides with the euclidean subgroup of $G$. Consequently, it is possible to introduce a measure $d_\omega g$ on each horosphere invariant with respect to its sliding group. The measures on different horospheres can be normalized in such a way that they are preserved by the transformations of the group $G$ translating one horosphere into another (condition of consistency):

$$d_{\omega g_0}(g g_0) = d_\omega g. \tag{5.1}$$

These measures can be introduced in the following way, for example: Let $\omega = (h_1, h_2)$. This means that each point of the horosphere $\omega$ is represented in the form

$$g = g_{h_1}^{-1} \zeta g_{h_2}, \tag{5.2}$$

where $g_{h_1}$ and $g_{h_2}$ are fixed representatives of the classes $h_1$ and $h_2$ and $\zeta$ is an element of the euclidean subgroup $Z$. The formula (5.2) establishes the correspondence between the points of the horosphere and the elements of the subgroup $Z$ in a one-to-one manner. Let $d\zeta$ be an invariant measure in the subgroup $Z$. Further, let $\rho(h)$ be a positive function in the fundamental affine space $H$, satisfying the relation

$$\rho(\delta h) = \beta(\delta)\,\rho(h), \tag{5.3}$$

where $\beta(\delta) = \dfrac{d(\delta h)}{dh}$ (see part 3 of §3). We then get

$$d_\omega g = \rho(h_1)\, d\zeta. \tag{5.4}$$

Evidently the measure so defined does not depend on the choice of the representatives $g_{h_1}$ and $g_{h_2}$. We will show that this measure does not depend either on the coordinates by which the horosphere $\omega$ was given. In fact, if we describe the same horosphere by the coordinates $\delta h_1$ and $\delta h_2$, then the corresponding measure on $\omega$ is defined by the formula

$$d'_\omega g = \rho(\delta h_1)\, d\zeta',$$

where $\zeta'$ and $\zeta$ are connected by the relation

$$\delta^{-1}\zeta'\delta = \zeta.$$

But then, as is well known [1], $d\zeta' = \beta^{-1}(\delta)\,d\zeta$. But from §3, we have $\rho(\delta h_1) = \beta(\delta)\rho(h_1)$, so that we get $d'_\omega g = \rho(h_1)\,d\zeta$, i.e., $d'_\omega g = d_\omega g$.

It is easy to see that the measure (5.4) on $\omega$ is invariant with respect to the sliding group and that it satisfies the consistency condition.

**2. Symmetry in the group space.** Let $S$ be the Weyl group of the group $G$, i.e., the group of automorphisms $s$ of the Cartan group $D$ induced by the inner automorphisms of the group $G$ itself

$$\delta \to \delta^s.$$

The elements $\delta \in D$ for which

$$\delta = \delta^s$$

only if $s = e$ will be called *elements in general position in the Cartan group* $D$. We now consider classes of conjugate elements of the group $G$ containing elements in general position in $D$. The elements of $G$ belonging to this class will be called *elements in general position in* $G$. It is known that all elements of $G$ except a manifold of lower dimension are elements in general position.

We show that the group space has the following symmetry.

**Theorem 5.1.** *Let $g_1$ and $g_2$ be two points of the group space such that $g_1^{-1}g_2$ is an element in general position in $G$. In other words, the point $g_1$ is transformed into the point $g_2$ by an element in general position in $G$. If it is possible to construct through these points two parallel horospheres $\omega$ and $\delta\omega$ at the compound distance $\delta$ from one another, then it is also possible to construct through $g_1$ and $g_2$ parallel horospheres $\omega'$ and $\delta^s\omega'$, where $s$ is an arbitrary element of the Weyl group.*

**Proof.** We describe at first the set of horospheres passing through a fixed point $g$ of the group space. The horosphere

$$\omega = g_1^{-1}Zg_2$$

contains the point $g$ if and only if for a certain $\zeta \in Z$ the equality $g = g_1^{-1}\zeta g_2$ holds. But then $g_2 = \zeta^{-1}g_1g$. Consequently, if $h$ denotes the residue class with respect to $Z$ to which $g_1$ belongs, then $\omega = (h, hg)$. Thus *an arbitrary horosphere passing through the point $g$ of the group space has the form*

$$\omega = (h, hg).$$

Since $(\delta h, \delta hg) = (h, hg)$, *each such horosphere is defined by its direction* $z$.

Now let two parallel horospheres $\omega$ and $\delta\omega$ pass through the points $g_1$ and $g_2$ at the compound distance $\delta$. It follows from the remark above that

$$\omega = (h, hg_1)$$

where $h$ is a certain element of the fundamental affine space $H$. But then

$$\delta\omega = (h, \delta hg_1).$$

This means that

$$\delta\omega = g_h^{-1} Z \delta g_h g_1,$$

where $g_h$ is an element of the class $h$. Thus we have

$$g_2 = g_h^{-1} \zeta \delta g_h g_1,$$

where $\zeta$ is a certain element of $Z$. Hence

$$g_2 g_h^{-1} = g_h^{-1} \zeta \delta g_h. \tag{5.5}$$

It is obvious that, conversely, it is possible to draw parallel horospheres at the compound distance $\delta$ from each other if the relation (5.5) is satisfied. However, as is known, the factor $\delta$ in the relation (5.5) is determined by $g_2 g_1^{-1}$ only up to an automorphism of $S$. Hence the proof is complete.

The factor $\delta$ in relation (5.5) is defined by $g_2 g_1^{-1}$ uniquely up to an automorphism of $S$. On the other hand, the class $h$ to which $g_h$ belongs is determined uniquely by the decomposition (5.5) if $\delta$ is fixed. Hence it follows that the number of pairs of parallel planes which can be traced through the points $g_1$ and $g_2$ is in general finite and equal to the number of elements of the Weyl group $S$.

**3. Irreducible representations of the group $G$ appearing in the decomposition of a regular representation.** We consider a regular representation of the group $G$, i.e., a representation of the form

$$T_{g_0} f(g) = f(gg_0), \tag{5.6}$$

realized in the space of functions with integrable square on the group $G$. Let $f(g)$ be a finitary infinitely often differentiable function on the group space $G$. We associate with this function the integrals over the horospheres in the space $G$:

$$\phi(\omega) = \int_\omega f(g)\, d_\omega g. \tag{5.7}$$

The representation (5.6) in $G$ induces in the space of the functions $\phi(\omega)$ a representation of the form

$$T_g \phi(\omega) = \phi(\omega g). \tag{5.8}$$

As has already been noted, each horosphere $\omega$ in the group space $G$ is given by a pair of points $(h_1, h_2)$ of the fundamental affine space and consists of the group elements of the form

$$g = g_{h_1}^{-1} \zeta g_{h_2},$$

where $g_{h_1}$ and $g_{h_2}$ are elements of the displacement classes $h_1$ and $h_2$, $\zeta \in Z$. The measure $d_\omega g$ on the horosphere $\omega$ can be obtained by the formula

$$d_\omega g = \rho(h_1)\, d\zeta.$$

Using this formula, (5.7) for $\phi(\omega)$ becomes

$$\phi(\omega) = \phi(h_1, h_2) = \rho(h_1) \int f(g_{h_1}^{-1} \zeta g_{h_2})\, d\zeta. \tag{5.9}$$

Formula (5.8) for the operator $T_g$ assumes the form

$$T_g \phi(h_1, h_2) = \phi(h_1, h_2 g). \tag{5.10}$$

Thus we obtain the representations in the horosphere space from the regular representations by means of formulas (5.7) and (5.8). By the general theory, discussed in §4, there arises a quasiregular representation in each transitive manifold of horospheres. In the given case this also follows directly from formula (5.10).

To obtain the irreducible components of the representation in the horosphere space we have to exclude the functions which satisfy the homogeneity condition

$$\psi(\delta\omega, \chi) = \chi(\delta)\psi(\omega, \chi).$$

These functions $\psi(\omega, \chi)$ form an invariant subspace. If $\omega$ runs through a transitive set, then the representation of the group $G$ in the space of the homogeneous functions $\psi(\omega, \chi)$ is irreducible.

In analogy with the function $\rho^{1/2}(h)$ (see part 3 of §3) we define a positive function $r(\omega)$ satisfying the condition

$$r(\delta\omega) = \beta^{1/2}(\delta)\, r(\omega). \tag{5.11}$$

We now define the homogeneous functions $\psi(\omega, \chi)$ by the formula

$$\psi(\omega, \chi) = \int \phi(\delta\omega)\, r(\delta\omega)\, \chi^{-1}(\delta)\, d\delta \tag{5.12}$$

(see §3, (3.6)).

If the horosphere $\omega$ is defined by a pair of points $(h_1, h_2)$ of the fundamental affine space, $\omega = (h_1, h_2)$, then we can write

$$r(\omega) = \rho^{-1/2}(h_1)\, \rho^{1/2}(h_2). \tag{5.13}$$

In fact, formula (5.12) can be written in coordinates:

$$\psi(\omega, \chi) = \psi(h_1, h_2, \chi) = \rho^{-1/2}(h_1) \int \phi(h_1, \delta h_2)\, \rho^{1/2}(\delta h_2)\, \chi^{-1}(\delta)\, d\delta. \tag{5.14}$$

Replacing in (5.12) $\phi(\omega)$ by the initial function $f(g)$ we get the following theorem.

**Theorem 5.2.** *The irreducible components of the representation*

$$T_{g_0} f(g) = f(g g_0)$$

*in the group space are given by the formula*

$$\psi(\omega, \chi) = r(\omega) \int \beta^{1/2}(\delta) \chi^{-1}(\delta)\, d\delta \int_{\delta\omega} f(g)\, d_{\delta\omega} g, \tag{5.15}$$

*where* $r(\omega)$ *is a weight factor, satisfying the condition* (5.11). *The integral* (5.15) *converges if* $f(g)$ *is a finitary function on the group* $G$. *The functions* $\psi(\omega, \chi)$ *are homogeneous functions in the horosphere space, their homogeneity degree is* $\chi$, *i.e.,*

$$\psi(\delta\omega, \chi) = \chi(\delta)\psi(\omega, \chi). \tag{5.16}$$

Writing (5.15) in coordinates, we get the same formula as in [1]:

$$\psi(h_1, h_2, \chi) = \rho^{1/2}(h_1) \int f(g_{h_1}^{-1} \zeta \delta g_{h_2}) \rho^{1/2}(\delta h_2) \chi^{-1}(\delta)\, d\zeta\, d\delta. \tag{5.17}$$

**4. Symmetry relations for the functions $\phi(\omega)$ and $\psi(\omega, \chi)$.** In part 2 of this section we described the symmetry in the group space $G$. This symmetry of a group space induces an additional relation for the functions $\phi(\omega)$ and $\psi(\omega, \chi)$, which we will now establish.

**Theorem 5.3.** *Let* $g$ *be an arbitrary point of* $G$. *We draw through this point all possible horospheres. Each of these horospheres is completely defined by its direction* $z$; *so for brevity we will denote these horospheres by* $\omega_z$. *Next we translate each horosphere through* $g$ *parallel to itself through the compound distance* $\delta$. *Then we integrate* $\phi(\omega)$ *over all these horospheres. Let*

$$I(g, \delta) = \beta^{1/2}(\delta) \int \phi(\delta\omega_z)\, dz \tag{5.18}$$

*or in coordinates*

$$I(g, \delta) = \beta^{1/2}(\delta) \int \phi(h, \delta h g)\, dz. \tag{5.18'}$$

*Here* $z$ *is a point of the fundamental projective space* $Z$ *associated with the point* $h$ *of the fundamental affine space* $H$, *and* $dz$ *denotes a measure in* $Z$ *connected with the invariant measure* $dh$ *in* $H$ *by the integral relation*

$$\int f(h) \rho^{-1}(h)\, dh = \int dz \int f(\delta h)\, d\delta.$$

*Then, if* $\delta$ *is an element in general position in the Cartan group* $D$, *the integral* (5.18) *converges and is preserved by the replacement of* $\delta$ *by* $\delta^s$

$$I(g, \delta) = I(g, \delta^s) \tag{5.19}$$

*for any element* $s$ *of the Weyl group* $S$.

**Proof.** We consider the set of points $M_{g,\delta}$ of the space $G$ situated on the horospheres $\delta\omega_z$. By Theorem 5.1 this point set coincides with the point set on the horospheres $\delta^s\omega_z$. Thus the expressions $I(g, \delta)$ and $I(g, \delta^s)$ represent the integral of the initial function $f(g)$ over one and the same subspace in $G$. To prove relation (5.19) it remains to show that the measures used in the integrations coincide too. From (5.9) we have

$$I(g, \delta) = \beta^{1/2}(\delta) \int \rho(h) f(g_h^{-1} \zeta \delta g_h g)\, d\zeta\, dz. \tag{5.20}$$

Here the expression $\beta^{1/2}(\delta)\rho(h)\, d\zeta\, dz$ defines the measure on the manifold $M_{g,\delta}$ over which the integral is taken. By [1], the integral (5.20) converges if $\delta$ is an element in general position.

On the other hand, let the pairs $(\zeta, h)$ and $(\zeta', h')$ be connected by the relation

$$g_h^{-1} \zeta \delta g_h = g_{h'}^{-1} \zeta' \delta^s g_{h'},$$

where $\delta$ and $s$ are fixed and $g_h$ and $g_{h'}$ are representatives of the classes $h$ and $h'$. Then [1] the integral relation holds

$$\beta^{1/2}(\delta) \int \rho(h) f(g_h^{-1} \zeta \delta g_h)\, d\zeta\, dz = \beta^{1/2}(\delta^s) \int \rho(h') f(g_{h'}^{-1} \zeta' \delta^s g_{h'})\, d\zeta'\, dz'.$$

By this relation the measure $\beta^{1/2}(\delta)\rho(h)\, d\zeta\, dz$ on the manifold $M_{g,\delta}$ is preserved if $\delta$ is replaced by $\delta^s$. Consequently, $I(g, \delta) = I(g, \delta^s)$. Q.E.D.

Using the expression (5.12) for the function $\psi(\omega, \chi)$ we get the

**Corollary.** *The functions $\psi(\omega, \chi)$ into which the regular representation is decomposed satisfy the following symmetry condition:*

$$\int r^{-1}(\omega_z) \psi(\omega_z, \chi)\, dz = \int r^{-1}(\omega_z) \psi(\omega_z, \chi_s)\, dz, \tag{5.21}$$

*where $\omega_z$ is a horosphere passing through a fixed point of the space $G$, while $r(\omega)$ is a weight factor satisfying the relation $r(\delta\omega) = \beta^{1/2}(\delta) r(\omega)$ and the character $\chi_s$ is defined by the formula*

$$\chi_s(\delta) = \chi(\delta^s). \tag{5.22}$$

*In another notation the symmetry relation has the form*

$$\int \frac{\rho^{1/2}(h)}{\rho^{1/2}(hg)} \psi(h, hg, \chi)\, dz = \int \frac{\rho^{1/2}(h)}{\rho^{1/2}(hg)} \psi(h, hg, \chi_s)\, dz. \tag{5.23}$$

**Remark.** We note that the function $I(g, \delta)$ defined by (5.20) does not depend on the choice of the measure $dz$ in the fundamental projective space $Z$. In fact, we get

$$\beta^{1/2}(\delta) \int f(g_h^{-1} \zeta \delta g_h g)\, d\zeta = F(h).$$

Then formula (5.20) becomes

$$I(g, \delta) = \int \rho(h) F(h)\, dz,$$

where

$$\rho(\delta h) F(\delta h) = \rho(h) F(h)$$

for any $\delta \in D$. Consequently, we have for any finitary function $u(h)$ on the space $H$

$$\int \rho(h) F(h)\, dz \int u(\delta h)\, d\delta = \int dz \int \rho(\delta h) F(\delta h)\, u(\delta h)\, d\delta$$
$$= \int F(h)\, u(h)\, dh.$$

Therefore, the expression $\rho(h) F(h)\, dz$ does not depend on the choice of the measure $dz$ and of the corresponding factor $\rho(h)$.

**5. Decomposition of a regular representation into irreducible representations.** In order to decompose a regular representation into irreducible representations it is necessary to find the inversion of the formula

$$\phi(\omega) = \int_{\omega} f(g)\, d_{\omega} g. \tag{5.24}$$

In other words, the finitary, infinitely often differentiable function $f(g)$ on the group $G$ has to be described by its integrals over the horospheres in the group space. This problem was essentially solved in [1, 3, 4, 11]. We formulate here the results in geometrical terms.

In what follows the differential operator $L$ on the Cartan subgroup $D$ plays an important role. This operator $L$ is defined in the following way.

Let $\mathfrak{g}$ be the Lie algebra of the group $G$, and let $\mathfrak{h}$ be a regular subalgebra of the algebra $\mathfrak{g}$. To each element of $\mathfrak{g}$ corresponds a certain Lie operator, i.e., an operator of infinitesimal displacement on the group. We examine a set of roots of the algebra $\mathfrak{g}$. To each root $\alpha$ corresponds a certain element $h_\alpha$ of the regular subgroup $\mathfrak{h}$ corresponds to each root $\alpha$ and consequently also a differential Lie operator $D_\alpha$ on the Cartan group $D$. We also consider the operator $\bar{D}_\alpha$ obtained from $D_\alpha$ by replacing all coefficients by their conjugates.

Then the operator $L$ is defined as

$$L = \prod_{\alpha > 0} D_\alpha \bar{D}_\alpha \tag{5.25}$$

(where the product is taken over all positive roots $\alpha$).

It will be shown that the following inversion of formula (5.24) is valid:

$$f(g) = cL\{\beta^{1/2}(\delta) \int \phi(\delta\omega_z)\, dz\}|_{\delta = e}, \tag{5.26}$$

where $\omega_z$ is a horosphere passing through the given point $g$, and $\delta\omega_z$ is the horosphere parallel to it at the compound distance $\delta$, with $c$ a numerical factor.

Therefore: *In order to describe the function $f(g)$ by the function $\phi(\omega)$ it is necessary to draw all possible horospheres through the given point $g$ and their parallel horospheres at the compound distance $\delta$. The integral of $\phi(\omega)$ over all* *these horospheres is a certain function of* $\delta$. *(We denoted it in part 4 of this section by* $I(g, \delta)$.*) Applying* $L$ *to this function and then putting* $\delta = e$, *we get the value of the initial function at the point* $g$ *up to a constant factor.*

If the horosphere space is provided with a coordinate system, then we can write (5.26) in the form

$$f(g) = cL\{\beta^{1/2}(\delta)\int \varphi(h, \delta hg)\, dz\}_{\delta=e} \tag{5.26'}$$

(see part 4). We will not derive this formula independently but we will rewrite it in the form in which it was obtained in [1]. Replacing in (5.26') $\phi(h, \delta hg)$ by its definition, we get

$$f(g) = cL\left\{\beta^{1/2}(\delta)\int \rho(h)\, f(g_h^{-1}\zeta\delta g_h g)\, d\zeta\, dz\right\}_{\delta=e}.$$

The last relation was established in [1] (see also [3, 4, 11]).

An analogous inversion formula holds in general homogeneous spaces related to the group $G$. In part 7 of §5 and in §6 the inversion formula will be obtained for a space of the form $X = G/C$, where $C$ is a compact subgroup or a subgroup of a Cartan group. The general case will be treated in the second part.

Formula (5.26) leads to the decomposition of the regular representation into irreducible representations in the following way.

The irreducible components of the regular representation are given, according to part 3 of this section, by

$$\psi(\omega, \chi) = \int \phi(\delta\omega)\, r(\delta\omega)\, \chi^{-1}(\delta)\, d\delta, \tag{5.27}$$

where $r(\omega)$ is a certain positive function satisfying (5.11).

In its turn each function $\phi(\omega)$ can be decomposed into the homogeneous functions $\psi(\omega, \chi)$. By the formula for the inverse transform we have

$$\varphi(\delta\omega)\, r(\delta\omega) = \frac{1}{(2\pi)^n}\int \psi(\omega, \chi)\, \chi(\delta)\, d\chi, \tag{5.28}$$

where $n$ is the real dimension of the Cartan group $D$. Hence

$$\beta^{1/2}(\delta)\, \varphi(\delta\omega) = \frac{1}{(2\pi)^n}\, r^{-1}(\omega)\int \psi(\omega, \chi)\, \chi(\delta)\, d\chi.$$

Replacing this expression in the inversion formula (5.26), we get

$$f(g) = \frac{c}{(2\pi)^n}\int P(\chi)\, d\chi \int r^{-1}(\omega_z)\, \psi(\omega_z, \chi)\, dz, \tag{5.29}$$

where the integration is taken over the set of horospheres $\omega_z$ passing through the point $g$ and $P(\chi)$ denotes the polynomial defined by

$$P(\chi) = L\chi(\delta)\,|_{\delta=e}. \tag{5.30}$$

In part 4 of this section a symmetry relation for $\psi(\omega, \chi)$ was established, namely,

$$\int r^{-1}(\omega_z)\psi(\omega_z, \chi)\,dz = \int r^{-1}(\omega_z)\psi(\omega_z, \chi_s)\,dz,$$

where $\chi_s(\delta) = \chi(\delta^s)$. Making use of this relation and of the obvious equality $P(\chi_s) = P(\chi)$, we get

**Theorem 5.4.** *Let $\mathfrak{X}$ be the set of all characters of the Cartan subgroup D. We select in $\mathfrak{X}$ the maximal open subset $\mathfrak{X}_0$ such that if $\chi \in \mathfrak{X}_0$ then $\chi_s \notin \mathfrak{X}_0$ with $s \neq e$. The decomposition of the regular representation of G into irreducible representations is then given by the formula*

$$f(g) = \frac{cN}{(2\pi)^n}\int_{\mathfrak{X}_0} P(\chi)\,d\chi \int r^{-1}(\omega_z)\psi(\omega_z, \chi)\,dz, \tag{5.31}$$

*where the inner integral is taken over the set of all horospheres passing through g.*

*Here*

$$\psi(\omega, \chi) = r(\omega)\int \beta^{1/2}(\delta)\chi^{-1}(\delta)\,d\delta \int_{\delta\omega} f(g)\,d_{\delta\omega}\,g. \tag{5.32}$$

*The polynomial $P(\chi)$ in (5.31) is obtained from (5.25) and (5.30); $r(\omega)$ is a weight factor satisfying (5.11), and N and n designate the order of the Weyl group S and the dimension of the Cartan group, respectively. The integrals (5.31) and (5.32) converge if $f(g)$ is a finitary and sufficiently often differentiable function.*

Using formula (5.11) for $r(\omega)$ and the expression for the invariant measure on the horospheres (see part 2 of this section), we can represent (5.31) and (5.32) in the following form:

$$\begin{aligned} f(g) &= \frac{cN}{(2\pi)^n}\int_{\mathfrak{X}_0} P(\chi)\,d\chi \int \frac{\rho^{1/2}(h)}{\rho^{1/2}(hg)}\psi(h, hg; \chi)\,dz \\ &= \frac{cN}{(2\pi)^n}\int_{\mathfrak{X}_0} P(\chi)\,d\chi \int \frac{\rho^{1/2}(h)}{\rho^{1/2}(hg^{-1})}\psi(hg^{-1}, h, \chi)\,dz, \end{aligned} \tag{5.31'}$$

where

$$\psi(h_1, h_2, \chi) = \rho^{1/2}(h_1)\int f(g_{h_1}^{-1}\zeta\delta g_{h_2})\rho^{1/2}(\delta h_2)\chi^{-1}(\delta)\,d\zeta\,d\delta. \tag{5.32'}$$

The integral in (5.31′) is taken over the fundamental projective space $Z$ and not over $H$, since

$$\psi(\delta h, \delta hg, \chi) = \psi(h, hg, \chi).$$

From [1] and [11] it follows that there exists a formula analogous to the Plancherel formula for the functions $f(g)$ and $\psi(\omega, \chi)$. Namely, we consider the space $\widetilde{\Omega}$ obtained from the horosphere space $\Omega$ by identifying all parallel horospheres. Then

$$\int |f(g)|^2\,dg = \frac{c}{(2\pi)^{2n}}\int_{\mathfrak{X}} P(\chi)\,d\chi \int |\psi(\omega, \chi)|^2\,d\widetilde{\omega}, \tag{5.33}$$

where $d\tilde{\omega}$ is a certain measure in the space $\tilde{\Omega}$. In the above-mentioned articles this formula was written in the form

$$\int |f(g)|^2 dg = \frac{c}{(2\pi)^{2n}} \int_{\mathfrak{X}} P(\chi)\, d\chi \int |\psi(h_1, h_2, \chi)|^2 dz_1\, dz_2, \tag{5.33'}$$

where $z_1$ and $z_2$ are points of the fundamental projective space $Z$ associated with the points $h_1$ and $h_2$ of the fundamental affine space.

The relation established by (5.31) and (5.32) can be continued to a one-to-one isometric correspondence between the group space of the functions $f(g)$ on $G$ for which

$$\|f\|^2 = \int |f(g)|^2 dg < \infty, \tag{5.34}$$

and the space of functions satisfying the homogeneity condition

$$\psi(\delta\omega, \chi) = \chi(\delta)\psi(\omega, \chi) \tag{5.35}$$

and the symmetry condition (5.21)

$$\int r^{-1}(\omega_z)\psi(\omega_z, \chi)\, dz = \int r^{-1}(\omega_z)\psi(\omega_z, \chi_s)\, dz, \tag{5.36}$$

for which

$$\|\psi\|^2 = \frac{c}{(2\pi)^{2n}} \int_{\mathfrak{X}} P(\chi)\, d\chi \int |\psi(\omega, \chi)|^2 d\tilde{\omega} < \infty. \tag{5.37}$$

Therefore, *the symmetry condition* (5.36) *imposed on the homogeneous function* $\psi(\omega, \chi)$ *turns out to be sufficient in order that this formula permit the inversion*

$$f(g) = \frac{c}{(2\pi)^n} \int P(\chi)\, d\chi \int r^{-1}(\omega_z)\psi(\omega_z, \chi)\, dz.$$

**6. Representation in the space $X = G/C$, where $C$ is an arbitrary compact subgroup.** We consider in this section the representation of $G$ in the space $X = G/C$, where $C$ is an arbitrary compact subgroup. We define the operator of the representation by

$$T_g f(x) = f(xg). \tag{5.38}$$

It is possible to consider the more general case where the operator of the representation is defined by

$$T_g f(x) = f(xg)\, a(x, g). \tag{5.39}$$

However, this would not change anything essentially. Moreover, the following assertion holds.

*Let $C$ be a subgroup of a simply connected group $\tilde{C} \subset G$. (This is automatically the case if $G$ is a classical Lie group, since the maximal compact subgroup of a classical group is simply connected.) If for a certain point $x \in X$ the factor $a(x, g)$ in* (5.39) *is a continuous function on $G$, then the representation* (5.39) *is equivalent to the representation* (5.38).

**Proof.** Without loss of generality it is possible to assume that the stability group of $x$ is $C$. We then have for any element $c \in C$

$$a(x, cg) = a(x, c)\, a(x, g).$$

Besides $|a(x, g)| = 1$.

To prove the assertion it is sufficient to show that

$$a(x, c) \equiv 1.$$

We assume that $a(x, c) \not\equiv 1$. The group $C$, being a compact Lie group, possesses an everywhere dense subset of elements of finite order. Consequently, there exists in $C$ an element $c_0$ of finite order $k$ for which $a(x, c_0) \neq 1$. Let $l$ be an arbitrary path in $\widetilde{C}$ connecting the point $e$ with the point $c_0$. By joining the paths $c_0 e, c_0^2 l, \cdots, c_0^{k-1} l$ successively to $l$ we get a closed path $L$ whose beginning and end are in $e$. The function $a(x, g)$ defines a continuous mapping of the closed path $L$ into a circle. The index of this mapping is different from zero (as is easily seen, it is equal to $k/k_1$, where $k_1$ denotes the smallest number for which $a(x, c_0^{k_1}) = 1$). But this is in contradiction to the simple connectedness of the subgroup $\widetilde{C}$. Q.E.D.

**7. Decomposition of a representation in the space $X = G/C$ into irreducible representations.** We consider the natural mapping of the group space $G$ into the space $X = G/C$ of right cosets of the group $G$ with respect to the compact subgroup $C$. By this mapping each function $f(x)$ in the space $X$ is associated with a function $f(g)$ on the group $G$, $f(g)$ constant on the cosets of $G$ with respect to $C$. If $f(x)$ is bounded on $X$, then the corresponding function $f(g)$ is bounded on $G$.

On the other hand, the horospheres on $G$ become horospheres in the space $X$ by a natural mapping of the group $G$ into the space $X$. Different horospheres may be mapped into the same horosphere: Two horospheres $\omega_1$ and $\omega_2$ in $G$ are mapped into one and the same horosphere in $X$ if and only if the elements of the horosphere $\omega_2$ are obtained from the elements of $\omega_1$ by left multiplication with a certain element $c \in C$:

$$\omega_2 = c\omega_1.$$

The class of the space $X$ and also its rank may therefore decrease in comparison with group space $G$.

As a consequence of the above mentioned facts we can consider the function $f(g)$ on the group space $G$ instead of the function $f(x)$ on the space $X$; $f(g)$ is constant on the cosets of the group $G$ with respect to the subgroup $C$. Instead of the integrals of the functions $f(x)$ over the horospheres in the space $X$ we can consider the integrals

$$\phi(\omega) = \int_{\omega} f(g)\, d_{\omega} g \tag{5.40}$$

of the corresponding functions $f(g)$ over the horospheres in the group space. The values of the function $\phi(\omega)$ on the horospheres $\omega$ and $c\omega$ are connected by an additional relation. Namely, since $f(cg) = f(g)$ for any element $c \in C$, we have

$$\varphi(c\omega) = \int_{c\omega} f(cg)\, d_{c\omega}(cg) = \int_{\omega} f(g) \frac{d_{c\omega}(cg)}{d_{\omega} g}\, d_{\omega} g.$$

We get

$$\frac{d_{c\omega}(cg)}{d_{\omega} g} = \sigma(c, \omega). \tag{5.41}$$

It is easy to see that $\sigma(c, \omega)$ in fact depends only on $c$ and on the transitive set of horospheres to which the horosphere $\omega$ belongs.

Thus *we have the additional relation for the function* $\phi(\omega)$:

$$\phi(c\omega) = \sigma(c, \omega)\, \phi(\omega), \tag{5.42}$$

*where* $c$ *is an arbitrary element of the compact group* $C$, *and the function* $\sigma(c, \omega)$ *is defined by the formula* (5.41).

In the assumed realization the factor $\sigma(c, \omega)$ has the following form:

$$\sigma(c, \omega) = \sigma(c, h_1, h_2) = \frac{\rho(h_1 c)}{\rho(h_1)}, \tag{5.43}$$

where $\rho(h)$ is a weight factor in the integral relation defining the measure $dz$ on the fundamental projective space $Z$:

$$\int f(h)\, \rho^{-1}(h)\, dh = \int dz \int f(\delta h)\, d\delta.$$

We recall that $\rho(h)$ satisfies the relation $\rho(\delta h) = \beta(\delta)\rho(h)$.

We will show that the factor $\rho(h)$ can always be chosen in such a way that

$$\rho(hc) = \rho(h) \tag{5.44}$$

for any element $c \in C$.

In fact, we decompose the Cartan group into a direct product

$$D = \Gamma \times E$$

of the compact subgroup $\Gamma$ and the vector subgroup $E$. As is well known, any element $g$ of the group $G$ can be written uniquely in the form

$$g = \zeta \epsilon u,$$

where $\zeta \in Z$, $\epsilon \in E$ and $u$ is an element of the compact maximal subgroup $\mathfrak{U}$. Thus every point of the fundamental affine space is given by a pair of elements $\epsilon \in E$, $u \in \mathfrak{U}$:

$$h = (\epsilon, u).$$

We can assume that $C$ is a subgroup of the group $\mathfrak{U}$. Then for $h = (\epsilon, u)$ we have $hc = (\epsilon, uc)$ for an arbitrary $c \in C$. Thus we obtain

$$\rho(h) = \rho(\epsilon, u) = \beta(\epsilon).$$

Obviously,

$$\rho(\delta h) = \beta(\delta)\rho(h) \text{ and } \rho(hc) = \rho(h)$$

for any $\delta \in D$ and $c \in C$.

If the factor $\rho(h)$ satisfies the condition

$$\rho(hc) = \rho(h),$$

we have from (5.42) and (5.43) that

$$\phi(c\omega) = \phi(\omega).$$

This means that if two horospheres of the group space $G$ are mapped into one and the same horosphere in the space $X = G/C$, then the values of the function $\phi(\omega)$ on these horospheres coincide. Thus $\phi(\omega)$ can be considered as an integral over the horospheres in the space $X$.

By Theorem 5.4., we therefore have

**Theorem 5.5.** *The decomposition of the representation*

$$T_g f(x) = f(xg)$$

*in the space $X = G/C$ into irreducible representations is given by the formula*

$$f(x) = \frac{cN}{(2\pi)^n} \int_{\mathfrak{X}_0} P(\chi)\, d\chi \int r^{-1}(\omega_z)\, \psi(\omega_z, \chi)\, dz, \tag{5.45}$$

*where $\omega_z$ denotes the horosphere in $X$ passing through $x$ with the direction $z$; the notation is the same as in Theorem 5.4. on page 400. The irreducible components $\psi(\omega, \chi)$ are given by the formula*

$$\psi(\omega, \chi) = r(\omega) \int \beta^{1/2}(\delta)\, \chi^{-1}(\delta)\, d\delta \int_{\delta\omega} f(x)\, d_{\delta\omega} x. \tag{5.46}$$

*The integrals in (5.45) and (5.46) converge if $f(x)$ is a bounded and sufficiently often differentiable function on the space $X$.*

The functions $\psi(\omega, \chi)$ defined by (5.46) satisfy the homogeneity condition

$$\psi(\delta\omega, \chi) = \chi(\delta)\psi(\omega, \chi) \tag{5.47}$$

and the symmetry relation

$$\int r^{-1}(\omega_z)\, \psi(\omega_z, \chi)\, dz = \int r^{-1}(\omega_z)\, \psi(\omega_z, \chi_s)\, dz. \tag{5.48}$$

The formulas (5.45) and (5.46) can be written in the following way. Using the natural mapping of the group $G$ on $X = G/C$ we can define the horospheres in $X$ as horospheres in $G$ in coordinate form $\omega = (h_1, h_2)$. Here the pairs $(h_1, h_2)$

and $(h_1 c, h_2)$, $c \in C$, define the same horosphere.

Let $x_0$ be a fixed point in $X$, and let $C$ be its stability group. We denote by $g_x$ an arbitrary element of $G$ transforming the point $x_0$ into the point $x$. We introduce a positive function on the fundamental affine space $H$ satisfying the conditions

$$\rho(\delta h) = \beta(\delta)\rho(h), \quad \rho(hc) = \rho(h)$$

for any $\delta \in D$, $c \in C$. Let $dz$ be the measure in the fundamental projective space $Z$ defined by the integral relation

$$\int f(h)\rho^{-1}(h)\,dh = \int dz \int f(\delta h)\,d\delta,$$

where $dh$ is a measure on $H$ invariant with respect to the group $G$.

Then

$$f(x) = \frac{cN}{(2\pi)^n} \int_{\mathfrak{X}_0} P(\chi)\,d\chi \int \frac{\rho^{1/2}(h)}{\rho^{1/2}(hg_x^{-1})} \psi(hg_x^{-1}, h, \chi)\,dz, \tag{5.49}$$

where

$$\psi(h_1, h_2, \chi) = \rho^{1/2}(h_1)\rho^{1/2}(h_2)\int f(x_0 g_{h_1}^{-1}\zeta\delta g_{h_2})\beta^{1/2}(\delta)\chi^{-1}(\delta)\,d\zeta d\delta \tag{5.50}$$

(see formulas (5.31′) and (5.32′)).

We notice that the value of the function $\psi(h_1, h_2, \chi)$ does not depend on the choice of the "coordinates" $h_1$ and $h_2$ by which the horosphere $\omega = (h_1, h_2)$ is defined in the space $X$. Namely, we have for any $c \in C$

$$\psi(h_1 c, h_2, \chi) = \psi(h_1, h_2, \chi).$$

Formulas (5.49) and (5.50) are meaningful also in the case where the function $\rho(h)$ does not satisfy the condition $\rho(hc) = \rho(h)$. In this case, however, the functions $\psi(h_1, h_2, \chi)$ are not always, in the strict sense, functions of the horospheres alone, since they also depend on the choice of the coordinates $h_1$ and $h_2$ by which these horospheres are defined. The following relation holds:

$$\psi(h_1 c, h_2, \chi) = \frac{\rho^{1/2}(h_1 c)}{\rho^{1/2}(h_1)} \psi(h_1, h_2, \chi) \tag{5.51}$$

for any $c$ of the stability group $C$.

A formula analogous to the Plancherel formula is valid for the functions $f(x)$ and $\psi(\omega, \chi)$, since we have only to let $dc$ be an invariant measure on $C$ such that

$$\int dc = 1.$$

We normalize the measure $dx$ in the space $X$ by the condition

$$\int_G f(g)\,dg = \int_{G/C} dx \int f(cg)\,dc.$$

Then for the function $f(g)$, which is constant on every coset of $G$ with respect to the subgroup $C$, we have

$$\int_G |f(g)|^2 dg = \int_X |f(x)|^2 dx.$$

Consequently, formula (5.33) becomes

$$\int |f(x)|^2 dx = \frac{c}{(2\pi)^{2n}} \int_{\mathfrak{X}} P(\chi)\, d\chi \int |\psi(\omega, \chi)|^2 d\tilde{\omega}.$$

This is analogous to the Plancherel formula for a representation in the space $X$. As in the case of a regular representation, the relation established by the formulas (5.45) and (5.46) can be continued to a one-to-one isometric correspondence between the space of the functions $f(x)$ on $X$ for which

$$\|f\|^2 = \int |f(x)|^2 dx < \infty$$

and the space of the functions $\psi(\omega, \chi)$ satisfying the homogeneity condition (5.47) and the symmetry condition (5.48) for which

$$\|\psi\|^2 = \frac{c}{(2\pi)^{2n}} \int_{\mathfrak{X}} P(\chi)\, d\chi \int |\psi(\omega, \chi)|^2 d\tilde{\omega} < \infty.$$

Thus the conditions imposed on the functions $\psi(\omega, \chi)$ are sufficient in order that the formula

$$f(x) = \frac{c}{(2\pi)^{2n}} \int_{\mathfrak{X}} P(\chi)\, d\chi \int r^{-1}(\omega_z)\, \psi(\omega_z, \chi)\, dz$$

permit inversion.

**8. Condition on the characters $\chi$ of the irreducible representations appearing in the decomposition of the representation in the space $X = G/C$.** Not all irreducible representations of the basic nondegenerate series are necessarily contained in the decomposition of the representation in the space $X$.

We first recall that we associate each element $\delta$ of $D$ with a left displacement transformation in the horosphere space of $X$

$$\omega \longrightarrow \delta\omega.$$

Let $\hat{D}$ be the subgroup of these elements $\delta$ of $D$ which lead to identity transformations, i.e.,

$$\delta\omega = \omega$$

for any horosphere $\omega$. Then the following statement is correct.

*Only those irreducible representations appear in the decomposition of a representation in the space $X$ whose corresponding character $\chi(\delta)$ satisfies the condition*

$$\chi(\delta) = 1$$

*for any* $\delta \in \hat{D}$.

In fact, if $\delta\omega = \omega$, then

$$\psi(\delta\omega, \chi) = \psi(\omega, \chi).$$

On the other hand, we have by the homogeneity condition for $\psi(\omega, \chi)$

$$\psi(\delta\omega, \chi) = \chi(\delta)\psi(\omega, \chi).$$

Consequently, if $\chi(\delta) \neq 1$, then the function $\psi(\omega, \chi)$ is identically zero.

We now describe the subgroup $\hat{D}$. If the horospheres $\omega = (h_1, h_2)$ and $\delta\omega = (\delta^{-1}h_1, h_2)$ coincide, we have

$$\delta^{-1}h_1 = h_1 c$$

for a certain element $c \in C$.

We decompose the Cartan group into a direct product

$$D = \Gamma \times E$$

of the compact subgroup $\Gamma$ and the vector subgroup $E$. As was already noted, each point of the fundamental affine space $H$ can be given by the pair $(\epsilon, u)$, where $\epsilon \in E$ and $u$ is an element of the maximal compact subgroup $\mathfrak{U}$ of $G$:

$$h = (\epsilon, u).$$

If $\delta = \gamma_1\epsilon_1$, $\gamma_1 \in \Gamma$, $\epsilon_1 \in E$, then

$$\delta h = (\epsilon\epsilon_1, \gamma_1 u).$$

Without loss of generality, we can assume that $C \subset \mathfrak{U}$. If now $\delta h = hc$, with $\delta = \gamma_1\epsilon_1$ and $h = (\epsilon, u)$, then

$$(\epsilon\epsilon_1, \gamma_1 u) = (\epsilon, uc).$$

Hence we have $\epsilon_1 = e$ and therefore $\hat{D} \subset \Gamma$. On the other hand, we have $u^{-1}\gamma_1 u = c$.

*Thus the element* $\delta$ *of* $D$ *belongs to the subgroup* $\hat{D}$ *if and only if the class of elements conjugate with* $\delta$ *in the maximal compact subgroup* $\mathfrak{U}$ *lies completely in* $C$.

Thus it follows in particular that if $C = \mathfrak{U}$, then $\hat{D} = \Gamma$. *The decomposition of the representation in the space* $X = G/\mathfrak{U}$ *contains only those irreducible representations whose characters satisfy the condition*

$$\chi(\Gamma) = 1.$$

These irreducible representations will be called representations of class 1.

Now let $C$ be a proper subgroup of the maximal compact group $\mathfrak{U}$. From the above we have $\delta \in \hat{D}$ if and only if all elements in $\mathfrak{U}$ conjugate to $\delta$ belong to the subgroup $C$. But only the elements of the center of the group $\mathfrak{U}$ possess this property. Since this center coincides with the center of the entire group $G$, we conclude:

*If $C$ is a proper subgroup of the maximal compact group $\mathfrak{U} \subset G$, then*

$$\hat{D} = C \cap \mathfrak{Z},$$

*where $\mathfrak{Z}$ is the center of the group $G$.*

**9. Representation in a space of positive definite hermitian matrices.** We consider the following example. Let $G$ be the group of complex matrices of order $n$ with determinant 1, and let $X$ be the space of all Hermitian positive definite matrices of order $n$ with determinant 1. The elements $g \in G$ induce in $X$ a transformation of the form

$$x \longrightarrow g^* xg.$$

We have to decompose the representation

$$T_g f(x) = f(g^* xg) \tag{5.52}$$

of the group $G$ in the space $X$ into irreducible representations.

It will be sufficient to make use of the formulas (5.49) and (5.50). We fix in $X$ a unit matrix $x_0 = e$. Its stability group is the subgroup of all unitary matrices. Therefore $X$ is a homogeneous space of the form $X = G/\mathfrak{U}$, where $\mathfrak{U}$ is the maximal compact subgroup of $G$.

We recall that the horospheres in $X$ form a manifold. Therefore, every horosphere in $X$ can be represented in the form $\omega = (h_0, h)$, where $h_0$ is a fixed point of the fundamental affine space, i.e., the space $H = G/Z$ of cosets of $G$ with respect to the euclidean subgroup $Z$. Thus we can consider the function

$$\psi(h_0, h, \chi) = \psi(h, \chi)$$

($h_0$ fixed) instead of the function $\psi(h_1, h_2, \chi)$. We will take the coset $Z$ as $h_0$. For definiteness, let $Z$ be that euclidean subgroup of $G$ which consists of matrices of the form

$$\zeta = \begin{Vmatrix} 1 & \zeta_{12} & \ldots & \zeta_{1n} \\ 0 & 1 & \ldots & \zeta_{2n} \\ \cdot & \cdot & \cdot & \cdot \\ 0 & 0 & \ldots & 1 \end{Vmatrix}. \tag{5.53}$$

Further, let $D$ be the Cartan group of the group $G$ consisting of the diagonal matrices

$$\delta = \begin{Vmatrix} \delta_1 & & \\ & \ddots & \\ & & \delta_n \end{Vmatrix} = \begin{Vmatrix} e^{\tau_1 + i\varphi_1} & & \\ & \ddots & \\ & & e^{\tau_n + i\varphi_n} \end{Vmatrix}, \tag{5.54}$$

where $\tau_1 + \tau_2 + \cdots + \tau_n = 0$, $\phi_1 + \phi_2 + \cdots + \phi_n = 0$. We choose as parameters of the matrix $\delta$ the numbers $\tau_1, \tau_2, \cdots, \tau_n$ and $\phi_1, \phi_2, \cdots, \phi_n$. An arbitrary character on the subgroup $D$ is given by

$$\chi(\delta) = e^{i(\tau_2\rho_2+\ldots+\tau_n\rho_n+\varphi_2 m_2+\ldots+\varphi_n m_n)} .$$

In part 8 of this section it was shown that the decomposition of the representation in the space $X$ contains only the irreducible representations of class 1, i.e., the representations corresponding to the characters

$$\chi(\delta) = e^{i(\tau_2\rho_2+\ldots+\tau_n\rho_n)} . \tag{5.55}$$

Formula (5.50) gives:

*An irreducible component of the representation* (5.52) *in the space* $X$ *is given by the formula*

$$\psi(h; \rho_2, \ldots, \rho_n) = \rho^{1/2}(h)\int f(g_h^* \delta^* \zeta^* \zeta \delta g_h)\, \beta^{1/2}(\delta)\, e^{-i(\tau_2\rho_2+\ldots+\tau_n\rho_n)}\, d\zeta\, d\tau_2 \ldots d\tau_n, \tag{5.56}$$

*where* $\delta$ *runs through the subgroup* $D$ *of the diagonal matrices,* $\zeta$ *runs through the subgroup* $Z$ *of the matrices of the form* (5.53); $h$ *is an element of the space of the cosets* $H = G/Z$ *and* $g_h$ *is an arbitrary representative of the coset* $h$. *The invariant measure* $d\zeta$ *on the group* $Z$ *is defined as the product of the differentials of the real and imaginary parts of the matrix elements of* $\zeta$.

By [1] the function $\beta(\delta)$ on the subgroup $D$ of the diagonal matrices $\delta$ is given by

$$\beta(\delta) = |\delta_2|^4 |\delta_3|^8 \cdots |\delta_n|^{4n-4}. \tag{5.57}$$

On the other hand, the factor $\rho(h)$ can be defined in the following way. We consider the subgroup $Z$ of the matrices of the form

$$z = \begin{Vmatrix} 1 & 0 \ldots 0 \\ z_{21} & 1 \ldots 0 \\ \cdot\ \cdot & \cdot\ \cdot\ \cdot\ \cdot \\ z_{n1} & z_{n2} \ldots 1 \end{Vmatrix} . \tag{5.58}$$

As is well known [1], any matrix of $G$, except the elements of a manifold of matrices of smaller dimension, can be uniquely represented in the form

$$g = \zeta \delta z, \tag{5.59}$$

where $\zeta \in Z$, $\delta \in D$, $z \in Z$. Thus it is possible to associate an element of the group $G$ of the form $\delta z$, $\delta \in D$, $z \in Z$ to almost every point $h$ of the fundamental affine space in a one-to-one manner. Then we get

$$\rho(h) = \rho(\delta z) = \beta(\delta). \tag{5.60}$$

We note that the functions $\psi(h; \rho_2, \cdots, \rho_n)$ satisfy the homogeneity condition: if

$$\delta_0 = \left\| \begin{matrix} e^{\tau_1^0 + i\varphi_1^0} & & \\ & \ddots & \\ & & e^{\tau_n^0 + i\varphi_n^0} \end{matrix} \right\|,$$

then

$$\psi(\delta_0 h; \rho_2, \ldots, \rho_n) = e^{i(\tau_2^0\rho_2 + \ldots + \tau_n^0\rho_n)} \psi(h; \rho_2, \ldots, \rho_n). \tag{5.61}$$

We will find the decomposition of the function $f(x)$ in the space $X$ into the functions $\psi(h, \chi)$. For this we use formula (5.49). We first express $\psi(hg_x^{-1}, h, \chi)$ by $\psi(h, \chi) = \psi(h_0, h, \chi)$.

Let $E$ be the subgroup of the diagonal matrices

$$\varepsilon = \left\| \begin{matrix} \varepsilon_1 & & \\ & \ddots & \\ & & \varepsilon_n \end{matrix} \right\|, \tag{5.62}$$

where $\epsilon_1, \cdots, \epsilon_n$ are positive and $\mathfrak{U}$ is a subgroup of the unitary matrices. Then any element of the group $G$ can be represented uniquely by

$$g = \zeta \epsilon u, \tag{5.63}$$

where $\zeta \in Z$, $\epsilon \in E$, $u \in \mathfrak{U}$. Let $g_h$ be an arbitrary element from the coset $h$. We decompose $g_h g_x^{-1}$ into a product

$$g_h g_x^{-1} = \zeta \epsilon u. \tag{5.64}$$

We recall that the function $\psi(h_1, h_2, \chi)$ satisfies the homogeneity condition (5.47) and the relation (5.51). From these relations we obtain

$$\psi(hg_x^{-1}, h, \chi) = \beta^{1/2}(hg_x^{-1}) \beta^{-1/2}(\varepsilon) \chi^{-1}(\varepsilon) \psi(h_0, h, \chi). \tag{5.65}$$

The elements $\epsilon_1, \cdots, \epsilon_n$ of the matrix can be simply expressed in terms of $h$ and $x$. In fact, from (5.61) we have $g_h^{*-1} g_x^* g_x g_h^{-1} = \zeta^{*-1} \epsilon^{-2} \zeta^{-1}$, and from $g_x^* g_x = x$ we therefore get

$$g_h^{*-1} x g_h^{-1} = \zeta^{*-1} \epsilon^{-2} \zeta^{-1}. \tag{5.66}$$

Hence it follows immediately that *the elements* $\epsilon_1, \cdots, \epsilon_n$ *of the matrix* $\epsilon$ *are defined by the formula*

$$\varepsilon_k^2 = \frac{\Delta_{k-1}(g_h^{*-1} x g_h^{-1})}{\Delta_k(g_h^{*-1} x g_h^{-1})}, \tag{5.67}$$

*where* $\Delta_k(g)$ *denotes the* $k$*th principal minor of the matrix* $g$.

In the basic formula (5.49) we have:

*The decomposition of the function* $f(x)$ *in the space* $X$ *into the functions*

$\psi(h;\ \rho_2, \cdots, \rho_n)$ *is given by the formula*

$$f(x) = c\int P(\rho_2, \ldots, \rho_n)\, d\rho_2 \ldots d\rho_n \int \beta^{1/2}(h)\, \varepsilon_2^{-2-i\rho_2}\varepsilon_3^{-4-i\rho_3} \ldots \ldots \varepsilon_n^{-2n+2-i\rho_n}\psi(h;\rho_2, \ldots, \rho_n)\, dz, \tag{5.68}$$

*where the elements* $\epsilon_2, \cdots, \epsilon_n$ *are defined by the relation* (5.67). *Integration over* $\rho_2, \cdots, \rho_n$ *means in this case integration over the manifold*

$$-(\rho_2 + \cdots + \rho_n) < \rho_2 < \rho_3 < \cdots < \rho_n.$$

We can identify the fundamental projective space $Z$ over which we integrated with the subgroup of the triangle matrices of the form (5.58). In this identification $dz$ denotes the product of the differentials of the real and the imaginary parts of the elements of the matrix $z$.

As in [1] the polynomial $P(\rho_2, \cdots, \rho_n)$ and the constant $c$ in (5.68) are determined by

$$P(\rho_2, \ldots, \rho_n) = \prod_{p<q} (\rho_p - \rho_q)^2 \prod_{p=2}^{n} \rho_p^2; \tag{5.69}$$

$$c = \frac{(-1)^{\frac{n(n-1)}{2}}}{(2\pi)^{n^2-1}}. \tag{5.70}$$

It is now convenient to pass from the homogeneous functions $\psi(h;\ \rho_2, \cdots, \rho_n)$ to the functions $\psi(z;\ \rho_2, \cdots, \rho_n)$ given on the subgroup $Z$ of the triangle matrices of the form (5.58). Then we get the following final result:

*The decomposition of the representation* $T_g f(x) = f(g^* xg)$ *of the group of complex Hermitian matrices of order* $n$ *with determinant* 1 *in the space* $X$ *of positive definite matrices is given by the formula*

$$f(x) = \frac{(-1)^{\frac{n(n-1)}{2}}}{(2\pi)^{n^2-1}} \int \prod_{p<q} (\rho_p - \rho_q)^2 \prod_{p=2}^{n} \rho_p^2\, d\rho_2 \ldots d\rho_n \times \int \Delta_1^{-2-i\rho_2}(z^{*-1}xz^{-1})\, \Delta_2^{-2+i(\rho_2-\rho_3)}(z^{*-1}xz^{-1}) \ldots \ldots \Delta_{n-1}^{-2+i(\rho_{n-1}-\rho_n)}(z^{*-1}xz^{-1})\, \psi(z;\rho_2, \ldots, \rho_n)\, dz, \tag{5.71}$$

*where*

$$\psi(z;\rho_2, \ldots, \rho_n) = \int \varepsilon_2^{1-i\rho_2}\varepsilon_3^{3-i\rho_3} \ldots \varepsilon_n^{2n-3-i\rho_n}\, d\varepsilon_2\, d\varepsilon_3 \ldots d\varepsilon_n \int f(z^*\epsilon^*\zeta^*\zeta\epsilon z)\, d\zeta. \tag{5.72}$$

Here $\Delta_k(g)$ is the $k$ th principal minor of the matrix $g$;

$$z = \left\| \begin{matrix} 1 & 0 \dots 0 \\ z_{21} & 1 \dots 0 \\ \dots & \dots \\ z_{n1} & z_{n2} \dots 1 \end{matrix} \right\|, \quad \zeta = \left\| \begin{matrix} 1 & \zeta_{12} \dots \zeta_{1n} \\ 0 & 1 \dots \zeta_{2n} \\ \dots & \dots \\ 0 & 0 \dots 1 \end{matrix} \right\|, \quad \varepsilon = \left\| \begin{matrix} \varepsilon_1 & 0 \dots 0 \\ 0 & \varepsilon_2 \dots 0 \\ \dots & \dots \\ 0 & 0 \dots \varepsilon_n \end{matrix} \right\|;$$

and $dz$ and $d\zeta$ denote the products of the differentials of the real and imaginary parts of the elements of the corresponding matrices $z$ and $\zeta$. Integration over $\rho_2, \cdots, \rho_n$ in (5.71) here means integration over

$$-(\rho_2 + \cdots + \rho_n) < \rho_2 < \rho_3 < \cdots < \rho_n$$

(or over any manifold obtained by an automorphism of the Weyl group). In particular, we obtain the decomposition of the representation of the group of complex matrices of second order in the space of the positive definite Hermitian matrices

$$x = \left\| \begin{matrix} a & b \\ \bar{b} & c \end{matrix} \right\|, \quad ac - b\bar{b} = 1,$$

by the formulas

$$f(a, b) = -\frac{1}{8\pi^3} \int_0^\infty \rho^2 d\rho \int (a - \bar{b}z - bz + cz\bar{z})^{-1-i\frac{\rho}{2}} \psi(z, \rho)\, dz,$$

where

$$\psi(z, \rho) = \int_0^\infty \varepsilon^{1-i\rho} d\varepsilon \int f\left[\frac{1}{\varepsilon^2} + \bar{z}\zeta + z\zeta + \varepsilon^2 |z|^2 (1 + |\zeta|^2)\zeta + \varepsilon^2 \bar{z} (1 + |\zeta|^2)\right] d\zeta$$

Here $z = x + iy$ and $\zeta = x' + iy'$ are complex numbers

$$dz = dxdy, \quad d\zeta = dx'\, dy'.$$

## §6. Representations in spaces of the form $X = G/C$, where $C$ is a subgroup of the Cartan group $D$

We consider here the representations in the space $X = G/C$, where $G$ is a complex semisimple Lie group and $C$ is an arbitrary subgroup of its Cartan group $D$. These representations are realized in the space of the functions $f(x)$ on $X$ for which

$$\|f\|^2 = \int |f(x)|^2\, dx < \infty.$$

In the case where $C$ is a unimodular group it is possible to consider the measure $dx$ on $X$ as invariant with respect to the group $G$. In what follows, however, we will not impose such an assumption on the measure $dx$.

The operators of the representation have the form

$$T_g f(x) = f(xg)\, a(x, g),$$

where the factor $a(x, g)$ satisfies the conditions

$$a(x, g_1g_2) = a(xg_1, g_2)\, a(x, g_1),$$
$$d(xg) = |a(x, g)|^2\, dx.$$

**1. Horospheres in the space $X$.** We first describe the horosphere space of $X$. Let $x_0$ be a point in the space $X$ whose stability group is $C$. We can represent an arbitrary horosphere in the form

$$\omega = x_0 g_1^{-1} Z g_2.$$

It is evident that if the elements $g_1'$ and $g_2'$ of the group $G$ are connected with the elements $g_1$ and $g_2$ by the relation

$$g_1' = \delta\zeta_1 g_1 c, \quad g_2' = \delta\zeta_2 g_2, \tag{6.1}$$

where $\zeta_1, \zeta_2 \in Z$, $\delta \in D$ and $c \in C$, then the horospheres $\omega = x_0 g_1^{-1} Z g_2$ and $\omega' = x_0 g_1'^{-1} Z g_2'$ are the same. It is not difficult to verify that the relation (6.1) is not only sufficient but also necessary for the coincidence of the horospheres $\omega$ and $\omega'$.

Thus *the horosphere space of $X = G/C$ coincides with the space of the pairs $(h_1, h_2)$ of points of the fundamental affine space $H = G/Z$; here the pairs $(h_1, h_2)$ and $(\delta h_1 c, \delta h_2)$, where $\delta \in D$ and $c \in C$, have to be identified.*

In what follows we will consider only the horospheres in general position in $X$. We define these horospheres in the following way: We first consider in the fundamental affine space $H$ all possible transformations of the form

$$h \longrightarrow \delta h \delta_1^{-1}, \tag{6.2}$$

where $\delta$ and $\delta_1$ run through the Cartan group $D$. These transformations are identity transformations if and only if $\delta$ and $\delta_1$ coincide and lie in the center of the group $G$.

We select in $H$ the set of those points $h$ which are preserved only by the identity transformations of the form (6.2). In other words, if

$$h = \delta h \delta_1^{-1},$$

then the elements $\delta$ and $\delta_1$ coincide and lie in the center of the group $G$. We say that these points $h$ are *in general position in the space $H$*. It is easy to see that the set of points in general position is obtained from $H$ by removing a manifold of smaller dimension.

Incidentally let us mention one property of points in general position in $H$. *If $h_0$ is a point in general position in $H$ and if the pair $(\delta_1, \delta_2)$ runs through a nonbounded set in the space $D \times D$, then the point $h = \delta_1 h_0 \delta_2^{-1}$ runs through a nonbounded set in $H$.*

*Horospheres of the form $\omega = (h_1, h_2)$, where $h_1$ is a point in general position in $H$, will be called horospheres in general position in $X$.*

We now describe the transitive manifolds of horospheres. The element $g$ of $G$ transforms the horosphere $\omega = (h_1, h_2)$ into the horosphere $\omega g = (h_1, h_2 g)$. Therefore, if $\omega = (h_1, h_2)$, then $h_1$ denotes the transitive manifold of horospheres to which $\omega$ belongs and $h_2$ designates the horosphere $\omega$ in this manifold.

Thus each horosphere manifold is formed by the horospheres

$$\omega = (h_1, h),$$

where $h_1$ is fixed and $h$ runs through the entire fundamental affine space $H$. Two horospheres coincide, $\omega = (h_1, h) = (h_1, h') = \omega'$, if and only if there exist elements $\delta \in D$ and $c \in C$ such that

$$h' = \delta h, \quad h_1 = \delta h_1 c^{-1}.$$

Consequently, if $h_1$ is a point in general position in the space $H$, then $\delta = c$ and $\delta$ lies in the center of the group $G$.

Therefore, *each transitive manifold of horospheres in general position in $X$ is obtained from the fundamental affine space $H$ by identifying the points $h$ and $\mathfrak{z}h$, where $\mathfrak{z}$ runs through the center of the group $G$ belonging to the subgroup $C$.*

We will introduce a measure on each horosphere in general position in $X$. From the above theorem the sliding group of a horosphere in general position is the group $Z\mathfrak{Z}_c$, where $Z$ is an euclidean subgroup of $G$ and $\mathfrak{Z}_c$ is a subgroup of the center $\mathfrak{Z}$ of the group $G$ contained in $C$. Consequently, this sliding group is unimodular and therefore there exists a measure $d_\omega x$ on the horosphere $\omega$ in general position, with $d_\omega x$ invariant with respect to the sliding group. It is possible to norm the measure $d_\omega x$ in such a way that it is consistent with the measure on other horospheres, i.e., that it is preserved by the transformations of the group $G$ transforming one horosphere into another.

We can introduce $d_\omega x$ in the same way as in §5; namely, by letting $\omega = (h_1, h_2)$. This means that each point $x$ of the horosphere $\omega$ is given by

$$x = x_0 g_{h_1}^{-1} \zeta g_{h_2}, \tag{6.3}$$

where $x_0$ is a point of $X$ whose stability group is $C$, $\zeta \in Z$ and $g_{h_1}$, $g_{h_2}$ are fixed representatives of the cosets $h_1$ and $h_2$. Formula (6.3) establishes a one-to-one relation between the points of the horosphere $\omega$ and the elements of the subgroup $Z$. Let $d\zeta$ be an invariant measure in the subgroup $Z$. Then, as in part 1 of §5, we get

$$d_\omega x = \rho(h_1)\, d\zeta, \tag{6.4}$$

where $\rho(h)$ is a positive function on $H$ satisfying the condition

$$\rho(\delta h) = \beta(\delta)\rho(h). \tag{6.5}$$

It is not difficult to see that this measure does not depend on the choice of the representatives $g_{h_1}$ and $g_{h_2}$ of the cosets $h_1$ and $h_2$ and therefore it does not change if the coordinates $h_1$ and $h_2$ are replaced by $\delta h_1$ and $\delta h_2$. We will show that this measure does not change either by transition from the coordinates $h_1, h_2$ to the coordinates $h_1, \mathfrak{z}h_2$, where $\mathfrak{z}$ is an element of the center of $G$ belonging to the subgroup $C$. In fact, if we define the horosphere $\omega$ by the coordinates $h_1, \mathfrak{z}h_2$, then the corresponding measure is expressed by the formula

$$d'_\omega x = \rho(h_1)\, d\zeta',$$

where $\zeta' = \mathfrak{z}\zeta\mathfrak{z}^{-1} = \zeta$ and consequently $d'_\omega x = d_\omega x$.

It is easy to see that the measure thus introduced is invariant with respect to the sliding group of the horosphere and that it satisfies the compatibility condition $d_{\omega g}(xg) = d_\omega x$.

We note that this measure depends on the choice of the coordinates describing the transitive manifold of horospheres. If we replace $h_1$ by $h_1 c$, $c \in C$, the measure is multiplied by $\dfrac{\rho(h_1 c)}{\rho(h_1)}$. The function $\rho(h)$ can be chosen so that the condition

$$\rho(hc) = \rho(h) \tag{6.6}$$

is satisfied for any $c \in C$. In this case, the measure will not depend on the special coordinate $h_1$ which we attach to the transitive manifold of horospheres. However, the condition (6.6) does not produce any essential simplification in the arguments or in the calculations, and in what follows we will not impose this additional condition on the function $\rho(h)$.

**2. Irreducible representations appearing in the decomposition of a representation in the space $X = G/C$.** Let $f(x)$ be a finitary, infinitely often differentiable function on the space $X$. To it we associate its integrals over the horospheres in $X$

$$\varphi(\omega) = \int_\omega K(\omega, x) f(x)\, d_\omega x. \tag{6.7}$$

In coordinates this integral is

$$\varphi(\omega) = \varphi(h_1, h_2) = \rho(h_1) \int K(g_{h_1}, g_{h_2}, \zeta) f(x_0 g_{h_1}^{-1} \zeta g_{h_2})\, d\zeta, \tag{6.8}$$

where $g_{h_1}$ and $g_{h_2}$ are representatives of the cosets $h_1$ and $h_2$.

It is assumed that the kernel $K(g_{h_1}, g_{h_2}, \zeta)$ does not depend on the choice of the representatives $g_{h_1}$ and $g_{h_2}$. This means that for any element $\zeta_1, \zeta_2 \in Z$

$$K(\zeta_1 g_{h_1}, \zeta_2 g_{h_2}, \zeta_1 \zeta \zeta_2^{-1}) = K(g_{h_1}, g_{h_2}, \zeta). \tag{6.9}$$

We require that the mapping given by (6.8) transform the function

$$T_g f(x) = f(xg)\, a(x, g)$$

into the function

$$T_g \varphi(h_1, h_2) = \varphi(h_1, h_2 g).$$

This means that

$$\int K(g_{h_1}, g_{h_2} g, \zeta)\, f(x_0 g_{h_1}^{-1} \zeta g_{h_2} g)\, d\zeta$$
$$= \int K(g_{h_1}, g_{h_2}, \zeta)\, a(x_0 g_{h_1}^{-1} \zeta g_{h_2}, g)\, f(x_0 g_{h_1}^{-1} \zeta g_{h_2} g)\, d\zeta.$$

From this equality we obtain the functional relation for the kernel $K(g_{h_1}, g_{h_2}, \zeta)$:

$$K(g_{h_1}, g_{h_2} g, \zeta) = K(g_{h_1}, g_{h_2}, \zeta)\, a(x_0 g_{h_1}^{-1} \zeta g_{h_2}, g). \tag{6.10}$$

We define the kernel by means of relations (6.9) and (6.10).

By (6.9) we have

$$K(g_{h_1}, g_{h_2}, \zeta) = K(g_{h_1}, \zeta g_{h_2}, e).$$

On the other hand, from (6.10) we get

$$K(g_{h_1}, \zeta g_{h_2}, e) = K(g_{h_1}, e, e)\, a(x_0 g_{h_1}^{-1}, \zeta g_{h_2}).$$

But

$$a(x_0 g_{h_1}^{-1}, \zeta g_{h_2}) = a(x_0, g_{h_1}^{-1} \zeta g_{h_2})\, a^{-1}(x_0, g_{h_1}^{-1}).$$

Therefore,

$$K(g_{h_1}, g_{h_2}, \zeta) = a(x_0, g_{h_1}^{-1} \zeta g_{h_2})\, a^{-1}(x_0, g_{h_1}^{-1})\, K(g_{h_1}, e, e) \tag{6.11}$$

We remark that the kernel is not uniquely defined. We can take any function as $K(g_{h_1}, e, e)$ for which

$$a^{-1}(x_0, g_{h_1}^{-1})\, K(g_{h_1}, e, e)$$

does not depend on the representative $g_{h_1}$ of the coset $h_1$.

It is natural to put

$$K(g_{h_1}, e, e) = a(x_0, g_{h_1}^{-1}). \tag{6.12}$$

Then we have

$$K(g_{h_1}, g_{h_2}, \zeta) = a(x_0, g_{h_1}^{-1} \zeta g_{h_2}). \tag{6.13}$$

*Thus, if we associate with the finitary function $f(x)$ on $X$ the function*

$\phi(h_1, h_2)$ *on the horosphere space of* $X$ *defined by*

$$\varphi(h_1, h_2) = \rho(h_1) \int a(x_0, g_{h_1}^{-1}\zeta g_{h_2}) f(x_0 g_{h_1}^{-1}\zeta g_{h_2})\, d\zeta, \tag{6.14}$$

*then to the transform of* $f(x)$

$$T_g f(x) = f(xg)\, a(x, g) \tag{6.15}$$

*corresponds the function*

$$T_g \varphi(h_1, h_2) = \varphi(h_1, h_2 g). \tag{6.16}$$

Therefore, in each transitive subspace of horospheres the functions $\phi(h_1, h_2)$ are transformed in accordance with a quasiregular representation.

We remark that the integral (6.14) depends not only on the horosphere over which it was taken but also on the choice of the coordinates by which the horosphere is given. The horosphere can be described by $(h_1, h_2)$, but also by $(h_1 c, h_2)$, where $c \in C$. On the other hand, we have

$$\varphi(h_1 c, h_2) = \rho(h_1 c) \int a(x_0, c^{-1} g_{h_1}^{-1}\zeta g_{h_2}) f(x_0 g_{h_1}^{-1}\zeta g_{h_2})\, d\zeta.$$

Hence $a(x_0, c^{-1} g_{h_1}^{-1} \zeta g_{h_2}) = a(x_0, c^{-1})\, a(x_0, g_{h_1}^{-1} \zeta g_{h_2})$ implies

$$\varphi(h_1 c, h_2) = a(x_0, c^{-1}) \frac{\rho(h_1 c)}{\rho(h_1)} \varphi(h_1, h_2). \tag{6.17}$$

We will also prove the following proposition:

*Let* $f(x)$ *be a finitary function on the space* $X$, *and let* $\phi(h_1, h_2)$ *be the function defined by* (6.14). *Let* $h_1$ *and* $h_2$ *be fixed points, where* $h_1$ *is a point in general position and* $\delta$ *runs through the Cartan group* $D$. *Then* $\phi(h_1, h_2)$ *is a bounded function of* $\delta$.

**Proof.** Let the function $f(x)$ be different from zero only on a certain compact region $U \subset X$. If $\phi(h_1, \delta h_2) \neq 0$ for a certain $\delta \in D$, then the horosphere $\omega = (h_1, \delta h_2) = (\delta^{-1} h_1, h_2)$ has a nonempty intersection with the set $U$. Without loss of generality it can be assumed that the coset $h_2$ in $H = G/Z$ coincides with the subgroup $Z$. Then $\omega = x_0 g_{h_1}^{-1} \delta Z$, where $g_{h_1}$ is a representative of the coset $h_1$. We may also assume that the complete preimage of the set $U$ in the group $G$ under the natural mapping of $G$ onto $X = G/C$ has the form $CV$, where $V$ is a compact set in $G$. However, in this case the set $C g_{h_1}^{-1} \delta Z$ in the group $G$ has a nonempty intersection with the compact set $V$. Consequently, the set $Z\delta^{-1} g_{h_1} C$ has a nonempty intersection with the compact set $V^{-1}$. Hence it follows that for a certain $c \in C$ the point $\delta^{-1} h_1 c$ of the space $H$ belongs to a fixed compact set in $H$. Since $h_1$ is in general position, it follows from the remark on p. 413, the element $\delta \in D$, for which $\phi(h_1, \delta h_2) \neq 0$, runs through a bounded set in $D$. Q.E.D.

For to find the irreducible components of the representation (6.16) we can now as in §5 make use of the formulas for a quasiregular representation. These irreducible components are defined in accordance with §3 (see formula (3.6)) by the following formula:

$$\psi(h_1, h_2, \chi) = \rho^{-1/2}(h_1) \int \varphi(h_1, \delta h_2) \rho^{1/2}(\delta h_2) \chi^{-1}(\delta)\, d\delta. \tag{6.18}$$

Substituting in (6.18) the expression (6.14) for the function $\phi(h_1, h_2)$ we get

**Theorem 6.1.** *The irreducible components of the representation*

$$T_g f(x) = f(xg)\, a(x, g)$$

*of the complex semisimple Lie group $G$ in the space $X = G/C$, where $C$ is a subgroup of the Cartan group $D$, are given by the formula*

$$\psi(h_1, h_2, \chi) = \rho^{1/2}(h_1) \rho^{1/2}(h_2) \times \int a(x_0, g_{h_1}^{-1}\zeta\delta g_{h_2}) f(x_0 g_{h_1}^{-1}\zeta\delta g_{h_2}) \beta^{1/2}(\delta) \chi^{-1}(\delta)\, d\zeta\, d\delta. \tag{6.19}$$

*Here $x_0$ is a fixed point in $X$, $h_1$ and $h_2$ are elements of the fundamental affine space $H = G/Z$, i.e., cosets of the group $G$ with respect to the euclidean subgroup $Z$; and $g_{h_1}$, $g_{h_2}$ are representatives of these cosets. The integral (6.19) converges if $f(x)$ is a bounded function on the space $X$, and $h_1$ is a point in general position in $H$. The function $\psi(h_1, h_2, \chi)$ satisfies the homogeneity condition*

$$\psi(h_1, \delta h_2, \chi) = \psi(\delta^{-1} h_1, h_2, \chi) = \chi(\delta) \psi(h_1, h_2, \chi), \tag{6.20}$$

*where $\delta \in D$, and also the relation*

$$\psi(h_1 c, h_2, \chi) = a(x_0, c^{-1}) \frac{\rho^{1/2}(h_1 c)}{\rho^{1/2}(h_1)} \psi(h_1, h_2, \chi) \tag{6.21}$$

*for any $c \in C$.*

**3. Condition on the character $\chi$ of an irreducible representation appearing in the decomposition of a representation in the space $X = G/C$.** We look for the irreducible representations actually contained in the decomposition of a representation in the space $X$.

Let $\mathfrak{Z}_c$ be the set of elements of the center $\mathfrak{Z}$ of $G$ belonging to the subgroup $C$

$$\mathfrak{Z}_c = \mathfrak{Z} \cap C.$$

From the functional relation (6.10) for the kernel $K(g_{h_1}, g_{h_2}, \zeta)$ we get

$$K(g_{h_1}, g_{h_2}\mathfrak{z}, \zeta) = K(g_{h_1}, g_{h_2}, \zeta)\, a(x_0 g_{h_1}^{-1}\zeta g_{h_2}, \mathfrak{z}).$$

*We show that the function $a(x, \mathfrak{z})$ is a character on the subgroup $\mathfrak{Z}_c$ and it does not depend on the choice of the point $x$.*

In fact, from the functional relation for $a(x, g)$ we have for $\mathfrak{z}_1, \mathfrak{z}_2 \in \mathfrak{Z}_c$

$$a(x, \mathfrak{z}_1\mathfrak{z}_2) = a(x, \mathfrak{z}_1) a(x\mathfrak{z}_1, \mathfrak{z}_2) = a(x, \mathfrak{z}_1) a(x, \mathfrak{z}_2).$$

Since $\mathfrak{Z}_c$ is compact, it also follows from this that

$$|a(x, \mathfrak{z})| = 1.$$

Since the group $\mathfrak{Z}_c$ is finite, the function $a(x, \mathfrak{z})$ can take on only a finite number of values. But being a continuous function it cannot depend on $x$ (where it is defined and continuous with respect to $x$).

We put

$$a(x, \mathfrak{z}) = \chi_0(\mathfrak{z}), \quad \mathfrak{z} \in \mathfrak{Z}_c. \tag{6.22}$$

Thus

$$K(g_{h_1}, g_{h_2}\mathfrak{z}, \zeta) = K(g_{h_1}, g_{h_2}, \zeta) \chi_0(\mathfrak{z}).$$

From this follows

$$\varphi(h_1, \mathfrak{z}h_2) = \varphi(h_1, h_2\mathfrak{z}) = \varphi(h_1, h_2) \chi_0(\mathfrak{z}),$$

and therefore

$$\psi(h_1, \mathfrak{z}h_2, \chi) = \psi(h_1, h_2, \chi) \chi_0(\mathfrak{z});$$

on the other hand, by the homogeneity condition we have

$$\psi(h_1, \mathfrak{z}h_2, \chi) = \psi(h_1, h_2, \chi) \chi(\mathfrak{z}).$$

Consequently, *the functions* $\psi(h_1, h_2, \chi)$ *defined by Theorem* 6.1 *differ from zero only under the condition*

$$\chi(\mathfrak{z}) = a(x, \mathfrak{z}), \tag{6.23}$$

*where* $\mathfrak{z}$ *runs through the subgroup* $\mathfrak{Z}_c = \mathfrak{Z} \cap C$ *of the center* $\mathfrak{Z} \subset G$.

**4. Symmetry relations for the functions $\phi(h_1, h_2)$ and $\psi(h_1, h_2, \chi)$.** The functions $\phi(h_1, h_2)$ and $\psi(h_1, h_2, \chi)$ satisfy an additional symmetry relation besides the homogeneity condition and the conditions (6.17) and (6.21). This symmetry condition results from the symmetry in the space $X$. It can be obtained in almost exactly the same way as in the case of a space with compact stability group (see §5).

Let $x = x_0 g_x$ be a fixed point of the space $X$. Through this point we draw all possible horospheres and construct their parallel horospheres at the compound distance $\delta$. As a result we get a set of horospheres of the form $\omega = (h, \delta h g_x)$. Each of these horospheres is given by the point $z$ of the fundamental projective space $Z = G/K$ associated with the point $h$ of the fundamental affine space $H$. We integrate the function $\phi(h_1, h_2)$ over these horospheres. Let

$$I(g_x, \delta) = \beta^{1/2}(\delta) \int \varphi(h, \delta h g_x) \, dz. \tag{6.24}$$

Replacing $\phi(h_1, h_2)$ by its definition in terms of the original function $f(x)$ on the space $X$ we get

$$I(g_x, \delta) = \beta^{1/2}(\delta) \int \rho(h)\, a(x_0, g_h^{-1}\zeta\delta g_h g_x)\, f(x_0 g_h^{-1}\zeta\delta g_h g_x)\, d\zeta\, dz. \tag{6.25}$$

If this integral converges, we obtain in exactly the same way as in §5 (Theorem 5.3) the following theorem.

**Theorem 6.2.** *The functions $\phi(h_1, h_2)$ in the horosphere space defined by the formula (6.14) satisfy the following relation: if*

$$I(g_x, \delta) = \beta^{1/2}(\delta) \int \varphi(h, \delta h g_x)\, dz, \tag{6.24}$$

*then for any element $\delta$ in general position in the Cartan group $D$ and for any element $s$ of the Weyl group $S$ we have*

$$I(g_x, \delta) = I(g_x, \delta^s). \tag{6.26}$$

**Corollary.** *The functions $\psi(h_1, h_2, \chi)$ defined by Theorem 6.1 satisfy the symmetry relation*

$$\int \frac{\rho^{1/2}(h)}{\rho^{1/2}(hg_x)} \psi(h, hg_x, \chi)\, dz = \int \frac{\rho^{1/2}(h)}{\rho^{1/2}(hg_x)} \psi(h, hg_x, \chi_s)\, dz, \tag{6.27}$$

*where the character $\chi_s$ is defined by the formula*

$$\chi_s(\delta) = \chi(\delta^s). \tag{6.28}$$

We will show that the integral (6.25) converges if the function $f(x)$ is different from zero only in a sufficiently small neighborhood of the point $x$. To prove this we first prove the following lemma.

**Lemma 6.1.** *We suppose that the function $f$ on the space $X$ differs from zero only in a sufficiently small neighborhood $U_x$ of the point $x = x_0 g_x$. We put*

$$F(g) = f(x_0 g g_x).$$

*Further let $u(g)$ be a continuous function on the group $G$, constant on the classes of conjugate elements and bounded on the Cartan group $D$. Then the function*

$$\hat{F}(g) = u(g) F(g)$$

*is bounded on the group $G$.*

**Proof.** We consider the neighborhood $U_0$ of the point $x_0$ in the space $X = G/C$. The preimage of this neighborhood in the group $G$ (at least if $U_0$ is sufficiently small) can be given in the form $CV_0$, where $V_0$ is a compact set in $G$ containing the unit element $e$. We suppose that the function $f(x)$ is different from zero only in the neighborhood $U_0 g_x$ of the point $x$. If $\hat{F}(g) \neq 0$, then $x_0 g g_x \in U_0 g_x$ or equivalently

$$g = \delta v,$$

where $\delta \in D$ and $v \in V_0$.

Without loss of generality we can assume that the group $G$ permits an exact matrix representation and that in this representation the diagonal matrices correspond to the element $\delta$ of the Cartan group $D$. We construct the characteristic polynomial of the matrix $g = \delta v$. Its coefficients have the form

$$a_k = (-1)^k \sum v_{i_1,\dots,i_k} \delta_{i_1} \dots \delta_{i_k},$$

where $\delta_{i_1}, \dots, \delta_{i_k}$ are the diagonal elements of the matrix $\delta$ and $v_{i_1,\dots,i_k}$ are the principal minors of the matrix $v$. If the original neighborhood $U_0$ of the point $x_0 \in X$ and also the set $V_0$ in the group $G$ are sufficiently small, then the matrix $v$ of $V_0$ can be as close to the unit matrix as desired. Consequently, the coefficients $v_{i_1,\dots,i_k}$ can be considered as bounded in modulus from above and from below by certain positive numbers.

Since $u(g) \neq 0$, the eigenvalues of the matrix $g$ are bounded. Consequently, the coefficients of the characteristic polynomial of the matrix $g$ are bounded. But then by the above remark the eigenvalues of the matrix $\delta$ are bounded too.[1]) Hence one sees immediately that the matrices $g$ for which $\hat{F}(g) \neq 0$ form a bounded set in the group $G$. Thus the lemma is proved.

We show now the convergence of the integral (6.25). As is well known [1], this integral can be represented in the form

$$I(g_x, \delta) = \beta^{1/2}(\delta) \int \rho(u)\, a(x_0, u^{-1}\zeta\delta u g_x)\, f(x_0 u^{-1}\zeta\delta u g_x)\, d\zeta du, \qquad (6.29)$$

where $u$ runs through the maximal compact subgroup $\mathfrak{U} \in G$ and $du$ is an invariant measure in $\mathfrak{U}$. By the lemma, with $\delta$ fixed, the function under the integral differs from zero only if the element $u^{-1}\zeta\delta u$ belongs to a certain bounded set in $G$. But then $\zeta$ belongs to a bounded set in the subgroup $Z$. Consequently, the integral in (6.29) is actually taken over a certain compact set and therefore this integral exists.

**5. Decomposition of a representation in the space $X = G/C$ into irreducible representations.** We will now find the decomposition of a function $f(x)$ on the space $X$ into the functions $\psi(h_1, h_2, \chi)$ by which the irreducible representations of the group $G$ are realized.

For this we have to obtain the inversion of the formula

$$\varphi(h_1, h_2) = \rho(h_1) \int a(x_0, g_{h_1}^{-1}\zeta g_{h_2})\, f(x_0\, g_{h_1}^{-1}\zeta g_{h_2})\, d\zeta, \qquad (6.30)$$

---

1) In fact, if $k$ of the eigenvalues $\delta_1, \dots, \delta_k$, of the matrix $\delta$, increase unboundedly and the others stay bounded, then by the statement about the coefficients $v_{i_1,\dots,i_k}$, the coefficient $a_k$ also increases beyond all bounds.

i.e., to express the finitary function $f(x)$ on $X$ by its integrals over the horospheres.

We consider on $G$ the function

$$F(g) = u(g)\, T_g f(x) = u(g)\, a(x, g)\, f(xg), \tag{6.31}$$

where $u(g)$ is a continuous function constant on the classes of conjugate elements and bounded on the Cartan group $D$. We suppose that the function $f(x)$ is different from zero only in a sufficiently small neighborhood of the point $x = x_0 g_x$. Then by Lemma 6.1 the function $F(g)$ is bounded on the group $G$. On the other hand, we suppose that the function $f(x)$ is continuous and sufficiently often differentiable and that the function $a(x, g)$ is continuous and differentiable with respect to $g$ in the neighborhood of $g = e$. Then the function $F(g)$ is also continuous and sufficiently often differentiable.

But then the following inversion formula is valid for the function $F(g)$ (see §5, formula (5.26')):

$$F(e) = cL\left\{\beta^{\frac{1}{2}}(\delta)\int \Phi(h, \delta h)\, dz\right\}_{\delta=e}, \tag{6.32}$$

where

$$\Phi(h_1, h_2) = \rho(h_1)\int F(g_{h_1}^{-1}\zeta g_{h_2})\, d\zeta, \tag{6.33}$$

and $L$ is the differential operator described in part 5 of §6.

The function $\Phi(h, \delta h)$ is immediately expressed by the function $\phi(h_1, h_2)$. In fact, we have

$$\Phi(h, \delta h) = u(\delta)\, \rho(h) \int a(x, g_h^{-1}\zeta\delta g_h)\, f(x g_h^{-1}\zeta\delta g_h)\, d\zeta.$$

Comparing this expression with (6.30) for the function $\phi(h_1, h_2)$, we get

$$\Phi(h, \delta h) = u(\delta)\, \frac{\rho(h)}{\rho(hg_x^{-1})}\, a^{-1}(x_0, g_x)\, \varphi(hg_x^{-1}, \delta h),$$

where $g_x$ is an element of $G$ transforming the point $x_0$ into the point $x$.

On the other hand, it is always possible to assume $u(g)$ equal to unity in the neighborhood of a unit element of the group $G$. Then the formula (6.32) gives:

**Theorem 6.3.** *Let $f(x)$ be a continuous infinitely often differentiable function on $X$, different from zero only in a sufficiently small neighborhood of the point $x = x_0 g_x$. We suppose also that the function $a(x, g)$ is continuous and sufficiently often differentiable in the neighborhood of the point $g = e$. If*

$$\varphi(h_1, h_2) = \rho(h_1)\int a(x_0, g_{h_1}^{-1}\zeta g_{h_2})\, f(x_0 g_{h_1}^{-1}\zeta g_{h_2})\, d\zeta, \tag{6.34}$$

*then*

$$f(x) = ca^{-1}(x_0, g_x)\, L \left\{ \beta^{1/2}(\delta) \int \frac{\rho(h)}{\rho(hg_x^{-1})} \varphi(hg_x^{-1}, \delta h)\, dz \right\}_{\delta = e} \tag{6.35}$$

*where L is the differential operator on the Cartan group D (described in part 5 of §6) and c a constant factor.*

**Replacing $\phi(h_1, h_2)$ in formula (6.35) by the function $\psi(h_1, h_2, \chi)$ and using the symmetry relation established in part 4 of this section, we get the final result:**

**Theorem 6.4.** *Let $\mathfrak{X}$ be the set of characters $\chi$ on the Cartan group D satisfying the condition*

$$\chi(\mathfrak{z}) = a(x, \mathfrak{z}), \tag{6.36}$$

*where $\mathfrak{z}$ runs through the elements of the center of the group G belonging to the stability group $C \subset D$. We pick out in $\mathfrak{X}$ the maximal open subset $\mathfrak{X}_0$ such that if $\chi \in \mathfrak{X}_0$, then $\chi_s \notin \mathfrak{X}_0$ for $s \neq e$. Then the decomposition of the representation of the form $T_g f(x) = f(xg)\, a(x, g)$ in the space $X = G/C$, where C is a subgroup of D, into irreducible representations is given by the formula*

$$f(x) = \frac{cN}{(2\pi)^n} a^{-1}(x_0, g_x) \int_{\mathfrak{X}_0} P(\chi)\, d\chi\, \frac{\rho^{1/2}(h)}{\rho^{1/2}(hg_x^{-1})} \psi(hg_x^{-1}, h, \chi)\, dz, \tag{6.37}$$

*where*

$$\begin{aligned} \psi(h_1, h_2, \chi) = \rho^{1/2}(h_1)\, \rho^{1/2}(h_2) \int a(x_0, g_{h_1}^{-1}\zeta\delta g_{h_2}) \\ \times f(x_0\, g_{h_1}^{-1}\zeta\delta g_{h_2})\, \beta^{1/2}(\delta)\, \chi^{-1}(\delta)\, d\zeta\, d\delta. \end{aligned} \tag{6.38}$$

*Here the following notations are used: $x_0$ is a fixed point in X; $g_x \in G$ transforms the point $x_0$ into x; $h_1$, $h_2$ are elements of the space of cosets $H = G/Z$ of the group G with respect to the euclidean subgroup Z; $g_{h_1}$, $g_{h_2}$ are representatives of the cosets $h_1$ and $h_2$; $d\zeta$ is an invariant measure on Z, dz is the measure in the fundamental projective space connected with the invariant measure on H by the relation*

$$\int f(h)\rho^{-1}(h)\, dh = \int dz \int f(\delta h)\, d\delta;$$

*c, n, N are constants; $P(\chi)$ is a polynomial of the indices defining the character* (see its definition on p. 399). *The integrals (6.37) and (6.38) converge if $f(x)$ is a continuous infinitely often differentiable function different from zero only in a sufficiently small neighborhood of a certain point in X. The functions $\psi(h_1, h_2, \chi)$ satisfy the homogeneity condition*

$$\psi(h_1, \delta h_2, \chi) = \psi(\delta^{-1} h_1, h_2, \chi) = \chi(\delta)\, \psi(h_1, h_2, \chi), \tag{6.39}$$

*and the relation*

$$\psi(h_1 c, h_2, \chi) = a(x_0, c^{-1}) \frac{\rho(h_1 c)}{\rho(h_1)} \psi(h_1, h_2, \chi) \tag{6.40}$$

*for any $c \in C$.*

It turns out that there exist formulas analogous to the Plancherel formulas for the functions $f(x)$ and $\psi(h_1, h_2, \chi)$:

$$\int |f(x)|^2 dx = \frac{cN}{(2\pi)^{2n}} \int_{\mathfrak{X}_0} P(\chi)\, d\chi \int |\psi(h_1, h_2, \chi)|^2 dz_1 dz_2.$$

Moreover, the formulas (6.37) and (6.38) completely define the single-valued isometric relation between the space of the functions for which

$$\|f\|^2 = \int |f(x)|^2\, dx < \infty,$$

and the space of the functions $\psi(h_1, h_2, \chi)$, $\chi \in \mathfrak{X}_0$, satisfying the conditions (6.39) and (6.40) for which

$$\|\psi\|^2 = \frac{cN}{(2\pi)^{2n}} \int_{\mathfrak{X}_0} P(\chi)\, d\chi \int |\psi(h_1, h_2, \chi)|^2\, dz_1 dz_2 < \infty.$$

The proof of this proposition will be given in another paper.

## §7. Decomposition of the Kronecker product of two irreducible representations of the group $G$ into irreducible representations

1. We will make use of the results of §6 for the solution of the following problem.

Let $G$ be a complex semisimple Lie group, and let there be given two irreducible unitary representations of this group belonging to the basic nondegenerate series (see §3)

$$T_g^1 f(x) = f(zg)\, a_1(z, g) \tag{7.1}$$

and

$$T_g^2 f(z) = f(zg)\, a_2(z, g), \tag{7.1'}$$

where $f(z)$ is a function with integrable square on the fundamental projective space $Z = G/K$.

We form the Kronecker (tensor) product of the representations (7.1) and (7.1′). This representation is realized in the space of the functions $f(z_1, z_2)$ for which

$$\|f\|^2 = \int |f(z_1, z_2)|^2\, dz_1\, dz_2 < \infty.$$

The operator of the representation is given by the formula

$$T_g f(z_1, z_2) = f(z_1 g, z_2 g)\, a_1(z_1, g)\, a_2(z_2, g). \tag{7.2}$$

The representation (7.2) has to be decomposed into irreducible representations. The manifold of the pairs $(z_1, z_2)$ is not transitive with respect to the group $G$. However, as was proved in [1], after removal from this manifold of a manifold of lower dimension we get a transitive manifold of the form $X = G/D$, where $D$ is the Cartan group of $G$.

For example, if $G$ is a complex unimodular group of matrices of order $n$, then almost any pair $(z_1, z_2)$ is obtained by a group displacement of the pair

$$x_0 = (z_1^0, z_2^0), \tag{7.3}$$

where $z_1^0$ is a coset in the space $Z = G/C$ containing the unit matrix and $z_2^0$ is a coset containing the matrix

$$s_0 = \begin{Vmatrix} 0 \ldots 0 & 1 \\ 0 \ldots 1 & 0 \\ & \\ \pm 1 \ldots 0 & 0 \end{Vmatrix}$$

The stability group of the point $x_0$ is the group of the diagonal matrices. Thus we can consider the representation (7.2) as a representation in the space $X = G/D$. Consequently, we can apply the formulas obtained in §6 for the decomposition of this representation into irreducible representations.

*The decomposition of* (7.2) *is given, according to Theorem* 6.4, *by the formula*

$$\begin{aligned} f(z_1, z_2) = \frac{cN}{(2\pi)^n} a_1^{-1}(z_1^0, g_{z_1, z_2}) a_2^{-1}(z_2^0, g_{z_1, z_2}) \\ \times \int_{\mathfrak{X}} P(\chi)\, d\chi \int \frac{\rho^{1/2}(h)}{\rho^{1/2}(hg_{z_1, z_2}^{-1})} \psi(hg_{z_1, z_2}^{-1}, h, \chi)\, dz, \end{aligned} \tag{7.4}$$

*where*

$$\begin{aligned} \psi(h_1, h_2, \chi) = \rho^{1/2}(h_1)\rho^{1/2}(h_2) \int a_1(z_1^0, g_{h_1}^{-1}\zeta\delta g_{h_2}) \\ \times a_2(z_2^0, g_{h_1}^{-1}\zeta\delta g_{h_2}) f(z_1^0 g_{h_1}^{-1}\zeta\delta g_{h_2}, z_2^0 g_{h_1}^{-1}\zeta\delta g_{h_2}) \beta^{1/2}(\delta)\chi^{-1}(\delta)\, d\zeta\, d\delta. \end{aligned} \tag{7.5}$$

*Here* $g_{z_1, z_2}$ *denotes an element of the group* $G$ *which transforms the pair* $x_0 = (z_1^0, z_2^0)$ *into the pair* $(z_1, z_2)$, *i.e.*,

$$z_1^0 g_{z_1, z_2} = z_1, \quad z_2^0 g_{z_1, z_2} = z_2. \tag{7.6}$$

*The remaining notations are the same as in Theorem* 6.4.

Both integrals converge if the function $f(z_1, z_2)$ is continuous and infinitely often differentiable, and different from zero only in a sufficiently small domain of the space $X = G/D$. The characters $\chi$ of the irreducible representations appearing in the decomposition satisfy the condition

$$\chi(\mathfrak{z}) = a_1(z_1, \mathfrak{z})\, a_2(z_2, \mathfrak{z}), \tag{7.7}$$

where $\mathfrak{z}$ runs through the center of the group $G$.

Since the function $a_1(z_1, g)\, a_2(z_2, g)$ can become infinite in the point $(z_1^0, z_2^0)$, we will give a somewhat different form to the formulas (7.4) and (7.5). Let

$(z_1', z_2')$ be an arbitrary fixed point of the manifold, and let $g_0$ be an element of the group $G$ transforming the point $(z_1^0, z_2^0)$ into the point $(z_1', z_2')$. Substituting in (7.4) and (7.5) the function $\psi(h_1, h_2, \chi)$ in the expression

$$a_1(z_1^0, g_0)\, a_2(z_2^0, g_0)\, \psi(h_1, h_2, \chi),$$

we get the following formula for the decomposition of the Kronecker product of two irreducible representations:

$$f(z_1, z_2) = \frac{cN}{(2\pi)^n} a_1^{-1}(z_1', g_0^{-1} g_{z_1, z_2})\, a_2^{-1}(z_2', g_0^{-1} g_{z_1, z_2}) \times \int_{\mathfrak{X}_0} P(\chi)\, d\chi \int \frac{\rho^{1/2}(h)}{\rho^{1/2}(h g_{z_1, z_2}^{-1})} \psi(h g_{z_1, z_2}^{-1}, h, \chi)\, dz, \tag{7.4'}$$

where

$$\psi(h_1, h_2, \chi) = \rho^{1/2}(h_1)\, \rho^{1/2}(h_2) \int a_1(z_1', g_0^{-1} g_{h_1}^{-1} \zeta \delta g_{h_2}) \times a_2(z_2', g_0^{-1}\; g_{h_1}^{-1} \zeta \delta g_{h_2})\, f(z_1^0 g_{h_1}^{-1} \zeta \delta g_{h_2},\; z_2^0 g_{h_1}^{-1} \zeta \delta g_{h_2})\, \beta^{1/2}(\delta)\, \chi^{-1}(\delta)\, d\zeta\, d\delta. \tag{7.5'}$$

2. We consider as an example a proper Lorentz group $G$. It is possible to realize this group as the group of all complex matrices of second order

$$g = \begin{Vmatrix} \alpha & \beta \\ \gamma & \delta \end{Vmatrix}$$

with determinant $\alpha\delta - \beta\gamma = 1$.

Let there be given two irreducible unitary representations (of the basic series) of the group $G$. By [1] these representations are realized in the space of functions $f(z)$ with integrable square, where $z$ is a complex variable. The operators of the representation are given by the formula

$$T_g^1 f(z) = f\left(\frac{\alpha z + \gamma}{\beta z + \delta}\right) |\beta z + \delta|^{i\rho_1 + m_1 - 2} (\beta z + \delta)^{-m_1} \tag{7.8}$$

and

$$T_g^2 f(z) = f\left(\frac{\alpha z + \gamma}{\beta z + \delta}\right) |\beta z + \delta|^{i\rho_2 + m_2 - 2} (\beta z + \delta)^{-m_2}. \tag{7.8'}$$

The Kronecker product of these representations is realized in the space of the functions $f(z_1, z_2)$ of two complex variables and given by the formula

$$T'_g f(z_1, z_2) = f\left(\frac{\alpha z_1 + \gamma}{\beta z_1 + \delta}, \frac{\alpha z_2 + \gamma}{\beta z_2 + \delta}\right) \times |\beta z_1 + \delta|^{i\rho_1 + m_1 - 2} (\beta z_1 + \delta)^{-m_1} |\beta z_2 + \delta|^{i\rho_2 + m_2 - 2} (\beta z_2 + \delta)^{-m_2}. \tag{7.9}$$

We notice that the fundamental affine space related to the group $G$ is here the space of the pairs $(z_1, z_2)$ of complex numbers $(z_1, z_2) \neq 0$.

On the other hand, it is possible to define almost every horosphere in the space $X = C/D$ by the coordinates $\omega = (h_0, h)$, where $h_0$ is a fixed point in general position in $H$. For definiteness we put $h_0 = (1, 1)$.

We introduce a new notation:

$$\psi(h_0, h, \chi) = \psi(z_1, z_2; \rho, m). \tag{7.10}$$

Let $g_0 = \begin{Vmatrix} 1 & 1 \\ 0 & 1 \end{Vmatrix}$. Then in the same notation as in the formulas (7.4′) and (7.5′) we have $z_1' = 0$, $z_2' = 1$. Formula (7.5′) defining the irreducible components of the representation becomes

$$\begin{aligned} \psi(z_1, z_2; \rho, m) = |z_2|^2 \int |(1-\zeta)\delta z_2|^{m_1+i\rho_1-2} [(1-\zeta)\delta z_2]^{-m_1} \\ \times |\zeta\delta z_2|^{m_2+i\rho_2-2} (\zeta\delta z_2)^{-m_2} |\delta|^{-i\rho+m} \delta^{-m} \\ \times f\left(\frac{z_1}{z_2} - \frac{1}{z_2^2\delta^2(1-\zeta)}, \frac{z_1}{z_2} + \frac{1}{z_2^2\delta^2\zeta}\right) d\zeta\, d\delta. \end{aligned} \tag{7.11}$$

Here $d\zeta$ and $d\delta$ denote the products of the differentials of the real and imaginary parts of the complex variables $\zeta$ and $\delta$.

Passing in (7.11) to the new variables of integration we obtain, after elementary transformations,

$$\begin{aligned} \psi(z_1, z_2; \rho, m) = |z_2|^{i\rho-m} z_2^m (-1)^{m_1} \int |t_1|^{\frac{m_2-m_1-m}{2} + i\frac{\rho_2-\rho_1+\rho}{2} - 1} t_1^{\frac{m+m_1-m_2}{2}} \\ \times |t_2|^{\frac{m_1-m_2-m}{2} + i\frac{\rho_1-\rho_2+\rho}{2} - 1} t_2^{\frac{m+m_2-m_1}{2}} |t_1 - t_2|^{\frac{m-m_1-m_2}{2} - i\frac{\rho_1+\rho_2+\rho}{2} - 1} \\ \times (t_1 - t_2)^{\frac{m_1+m_2-m}{2}} f\left(t_1 + \frac{z_1}{z_2}, t_2 + \frac{z_1}{z_2}\right) dt_1\, dt_2, \end{aligned} \tag{7.12}$$

where $t_i$ $(i = 1, 2)$ is a complex variable and $dt_i$ denotes the product of the differentials of the real and imaginary parts of $t_i$.

We will find an additional condition on the indices $\rho$ and $m$. The center of the group $G$ contains, besides the unit matrix, also the matrix

$$\mathfrak{z} = \begin{Vmatrix} -1 & 0 \\ 0 & -1 \end{Vmatrix}.$$

In view of this relation (7.7) gives $(-1)^m = (-1)^{m_1}(-1)^{m_2}$, *i.e., the indices* $m$ *of the irreducible representations contained in the decomposition have the same parity as* $m_1 + m_2$.

The decomposition of the function $f(z_1, z_2)$ into the functions $\psi(z_1, z_2; \rho, m)$ is given from (7.5) by the formula (we omit here the elementary transformation where the additional relations for the functions $\psi(h_1, h_2; \rho, m)$ are used)

$$f(z_1, z_2) = \frac{-1}{8\pi^3} \sum_m \int_0^\infty (\rho^2 + m^2) K(z_1, z_2; z; \rho, m)\, \psi(z, 1; \rho, m)\, dz\, d\rho, \tag{7.13}$$

where

$$K(z_1, z_2; z; \rho, m) = |z - z_1|^{-\frac{m_2-m_1-m}{2} - i\frac{\rho_2-\rho_1+\rho}{2} - 1} \times (z - z_1)^{-\frac{m+m_1-m_2}{2}} |z_2 - z|^{-\frac{m_1-m_2-m}{2} - i\frac{\rho_1-\rho_2+\rho}{2} - 1} \times (z_2 - z)^{-\frac{m+m_1-m_2}{2}} |z_2 - z_1|^{-\frac{m-m_1-m_2}{2} + i\frac{\rho_1+\rho_2+\rho}{2} - 1} (z_2 - z_1)^{-\frac{m_1+m_2-m}{2}}. \tag{7.14}$$

The summation in (7.13) is taken over the non-negative values of $m$ having the same parity as $m_1 + m_2$.

The decomposition of the representation (7.9) into irreducible representations was obtained earlier by another method, which is described in the paper of M. A. Naĭmark [10].

## BIBLIOGRAPHY


[1] I. M. Gel'fand and M. A. Naĭmark, *Unitary representations of the classical groups*, Trudy Mat. Inst. Steklov. 36 (1950). (Russian) MR 13, 722.

[2] ——, *Unitary representations of the Lorentz group*, Izv. Akad. Nauk SSSR Ser. Mat. 11 (1947), 411–504. (Russian) MR 9, 495.

[3] I. M. Gel'fand and M. I. Graev, *On a general method of decomposition of the regular representation of a Lie group into irreducible representations*, Dokl. Akad. Nauk SSSR 92 (1953), 221–224. (Russian) MR 15, 601.

[4] ——, *Analogue of the Plancherel formula for the classical groups*, Trudy Moskov. Mat. Obšč. 4 (1955), 375–404. (Russian) MR 17, 173.

[5] A. A. Hačaturov, *Determination of the value of the measure for a region of n-dimensional Euclidean space from its values for all half-spaces*, Uspehi Mat. Nauk 9 (1954), no. 3 (61), 205–212. (Russian) MR 16, 229.

[6] É. Cartan, *Collected works. Geometry of Lie groups and symmetric spaces*, IL, Moscow, 1949. (Russian)

[7] F. Kleĭn, *Non-euclidean geometry*, ONTI, Moscow, 1935. (Russian)

[8] K. Iwasawa, *On some types of topological groups*, Ann. of Math. (2) 50 (1949), 507–558. MR 10, 679.

[9] J. Tits, *Sur certaines classes d'espaces homogènes de groupes de Lie*, Acad. Roy. Belg. Cl. Sci. Mém. Coll. in 8° 29 (1955), no. 3. MR 17, 874.

[10] M. A. Naĭmark, *On irreducible linear representations of a proper Lorentz group*, Dokl. Akad. Nauk SSSR 97 (1954), 969–972 (Russian); see also *Decomposition of the tensor product of irreducible representations of a proper Lorentz group into irreducible representations*, Trudy Moskov. Mat. Obšč. 8 (1959), 121–154. (Russian) MR 16, 218; MR 22 # 4966.

[11] Harish-Chandra, *The Plancherel formula for complex semisimple Lie groups*, Trans. Amer. Math. Soc. 76 (1954), 485–528. MR 16, 111.

[12] F. Bruhat, *Sur les représentations induites des groupes de Lie* (Dissertation), pp. 98–205, Gauthier-Villars, Paris, 1956.

[13] I. M. Gel'fand and Z. Ja. Šapiro, *Homogeneous functions and their extensions*, Uspehi Mat. Nauk 10 (1955), no. 3(65), 3–70. (Russian) MR 17, 371.

[14] I. M. Gel'fand and G. E. Šilov, *Generalized functions and operations on them*, Fizmatgiz, Moscow, 1958. (Russian) MR 20 # 4182.

Translated by:
Dorothee Aeppli

# 4.

## (with I. I. Piatetski-Shapiro)

# Unitary representations in homogeneous spaces with discrete stationary groups

Dokl. Akad. Nauk SSSR **147** (1962) 17–20
[Sov. Math., Dokl. **3** (1962) 1528–1531]. Zbl. **119**:270

Let $G$ be a semisimple Lie group, $\Gamma$ a discrete subgroup of $G$. We denote by $X$ the space of left sets of $\Gamma$ in $G$. Every element $g \in G$ corresponds, obviously, with a motion in $X$, carrying $x$ into $xg$. denote by $L_2(X)$ the totality of all square-summable functions $f(x)$ on $X$. As is easily seen, the erators $T_g$: $T_g f(x) = f(xg)$ are unitary on $L_2(X)$. It is well known [2], that if $X$ is compact, a reprentation in $L_2(X)$ decomposes into the sum of a countable number of unitary representations. In the se in which $X$ is not compact, this is already false. In this case, as a rule, the "spectrum" of an educible representation is mixed, i.e., continuous and discrete.

In the present article there is indicated an effective process for isolating the discrete spectrum m the continuous spectrum. This process is a development of the method of orispheres [7]. In subquent work [8] we shall study the structure of continuous "spectra" with the help of the method of ispheres.

We denote by $L_2^0(X)$ the totality of all functions on $L_2(X)$ whose integrals over any compact orihere are equal to zero (the definition of orisphere is given below). The following theorem holds:

**Theorem.** *If $\Gamma$ is a tame subgroup, then $L_2^0(X)$ decomposes into a countable number of irreducible itary representations of the group $G$.**

In subsequent work we show that there exists a generalization of this theorem, namely, that every

*For the definition of a tame subgroup, see below.

irreducible unitary representation of the group $G$ lying discretely in $L_2(X)$ either is contained in $L_2^0(X)$ or is the "analytic continuation" of a representation of continuous spectrum belonging to $L_2^0(X$

The present work consists of three parts: 1) the definition of certain concepts relating to an arbitrary real semisimple Lie group and to a homogeneous space; 2) the definition of a tame discrete subgroup; 3) an outline of proof of the theorem on the decomposition of $L_2^0(X)$ into the sum of a countable number of irreducible unitary representations of the group $G$.

**1.** Let $G$ be a real semisimple Lie group; $g(t)$, an arbitrary one-parameter subgroup of $G$. The set $Z \subset G$ consisting of all $z \in G$ for which

$$\lim_{t \to +\infty} g(-t)\, z g(t) = 1 \qquad (1)$$

is called the *orispherical subgroup connected with the group* $g(t)$.

Let $X$ be a homogeneous space of the group $G$. The orbits of the orispherical groups are called the *orispheres* in $X$. An orisphere is called *compact* if the set of points of which it consists is compact.

Let $A$ be a commutative subgroup of $G$, consisting of semisimple transformations with real eigenvalues. $A$ is called the *moving subgroup* of the group $Z$ if $A$ is generated by one-parameter subgroups $h_1(t), \cdots, h_\nu(t)$, for which (1) holds, and if every $g \in G$, for which (1) takes place for all the subgroups $h_1(t), \cdots, h_\nu(t)$, belongs to $Z$.

**2.** At this point there is given the definition of a tame discrete subgroup of an arbitrary semisimple group. Let us agree to call a set $Y \subset X$ *cylindrical* if it is possible to stratify it into compact orispheres which are nonintersecting among themselves and which are sent into one another by the transformations of $G$.

A discrete subgroup $\Gamma$ of the semisimple group $G$ is called *tame* if the factor space $X = G/\Gamma$ possesses a finite covering of *tame restricted cylindrical sets*, and moreover the intersection of any two of these is compact. The definition of the concept of cylindrical sets which are tame and restricted is given below.

It is easily verified that for every tame discrete subgroup $\Gamma$ of a semisimple group $G$ the factor space $G/\Gamma$ has finite volume. Although there are grounds for believing that the reverse assertion is always true, at the present time this has been proved only for groups of real matrices of order two. This result is contained in the work of K. Siegel.

We pass now to the definition of tame restricted cylindrical sets. It is not difficult to see that for an arbitrary cylindrical set $Y$ there exists an orispherical subgroup $Z$ and a set $S \in G$ whose image in $G/\Gamma$ is $Y$, with the following properties: 1) for arbitrary $g \in S$ and $z \in Z$ there exists $\delta \in \Delta = \Gamma \cap Z$ such that $\delta z g \in S$; 2) if $g_1, g_2 \in S$ and $g_1 g_2^{-1} \subset \Gamma$, then $g_1 g_2^{-1} \subset \Delta = \Gamma \cap Z$. Condition 1) asserts that the image of $S$ consists of orispheres, and condition 2), that these orispheres do not intersect.

If the set $Y$ is compact, then we can choose $S$ in such a way that it even possesses the following properties: 3) there exists a neighborhood of unity $V$ in $G$ such that $g_1^{-1} \gamma g_2 \in V$ implies $\gamma \in \Delta$, where $g_1, g_2 \in S$, $\gamma \in \Gamma$; 4) for arbitrary $z \in Z$ there exists a compact neighborhood of unity $V_z$ such that $g^{-1} z g \in V_z$ for arbitrary $g \in S$. Condition 3) is a somewhat strengthened form of condition 2). Conditions 3) and 4) play a central role in the work.

Let us agree that a cylindrical set $Y$ for which there exists $S \subset Z$ possessing the properties 1)–4) listed above is called *restricted*. Generally speaking, restricted cylindrical sets are not compact, and in this connection, as is apparent, there is a geometrical reason for the existence of discrete subgroup semisimple groups for which the factor space $G/\Gamma$ is of finite volume and at the same time not compact.

Let $Y$ be a restricted cylindrical set. It is possible to show that all elements $g \in S$ are represented in the form $g = zat$, where $a$ is contained in a certain moving subgroup $A$ of the group $Z$, and $t$ contained in a certain compact set $T$ in $G$. It is well known that $A$ is contained in the normalizer the group $Z$. Let us agree further that the normal subgroup $\tilde{Z}$ of the group $Z$ is called *A-admissible* simply *admissible*, if the Lie algebra of the group $Z$ is a sum of root spaces.* The restricted cylindrical subset is called *tame* if: 5) for arbitrary admissible subgroups $\tilde{Z}$ of $Z$ (among them also for $Z$ self) the factor space $\tilde{Z}/(\tilde{Z} \cap \Gamma)$ is compact.

Thus, finally, the tame restricted cylindrical sets are characterized by the existence for them of $S$, and $A$, for which 1)–5) hold.

In a certain special case 5) coincides with a condition existing in the work of Harish-Chandra [6].

We remark, finally, that at the present time all known examples of irreducible** discrete subgroups semisimple Lie groups with finite volume of the factor space, with the exception of subgroups of the group of real matrices of second order, are arithmetic groups in the sense of Borel and Harish-Chandra [5].

A. Borel and Harish-Chandra [5] proved that the volume of the factor-space $G/\Gamma$ for such groups finite. As is apparent, it is possible to show by their methods that an arbitrary arithmetic group is tame discrete subgroup of the corresponding Lie group. In any case, for the classical discrete subgroups, such as, for example, the group of integral matrices, it can be proved without difficulty that they are tame.

3. We denote by $L_2^0(X)$ the totality of functions $f(x) \in L_2(X)$ for which the integrals over any compact orisphere are equal to zero. At this point we show that the operator $T_\phi = \int_G \phi(g)\, T_g\, dg$ on the space $L_2^0(X)$ is fully continuous for an arbitrary continuous finite function $\phi(g)$. From this it follows once that $L_2^0(X)$ decomposes into the sum of a countable number of irreducible representations of the group $G$.

In fact, it will be proved that the operator $T_\phi$ is a nuclear operator in $L_2^0(X)$ if $\phi$ is a finite, infinitely differentiable function of the form $\int_G \psi(gg_1)\, \overline{\psi(g_1)}\, dg_1$, where $\psi(g)$ is a certain continuous finite function.

Let $Y$ be a certain tame "restricted" cylindrical set. According to the definition, $Y$ is stratified a canonical way on a compact orisphere. Each compact orisphere, generally speaking, is stratified also on compact orispheres. We shall call the fibering obtained thus the *subordinate fibering*.

We now denote by $L_2^0(Y)$ the totality of functions of $L_2(X)$ which are zero outside $Y$ and whose integrals over an orisphere of arbitrary fibering subordinate to the canonical fibering of $Y$ are equal to zero. It is sufficient to prove the nuclearity of the operator $PT_\phi P$, where $P$ is the projection operator $L_2(X)$ on $L_2^0(Y)$.

As is well known [2], the operator $T_\phi$ is an integral operator on $X$ with the kernel

*The totality of all $\xi$ in the Lie algebra of the group $Z$ for which $[a, \xi] = \beta(a)\xi$ for arbitrary $a$ in the Lie algebra of the group $A$ is called the *root space* corresponding to the root $\beta(a)$.

**The discrete subgroup $\Gamma$ of the group $G$ is called *irreducible* if it is impossible to represent it in the form a product of two subgroups $G_1$ and $G_2$ such that the group $\Gamma_1 \times \Gamma_2$, $\Gamma_k \subset G_k$, is a subgroup of $\Gamma$ of finite index.

$$K(x_1, x_2) = \Sigma \phi(g_1^{-1} \gamma g_2), \tag{2}$$

where $g_1$, $g_2$ are representatives of the cosets $x_1$, $x_2$. It is easy to show that if the function $\phi(g)$ is distinct from zero only on a sufficiently small neighborhood of unity of the group $G$, then for $x_1, x_2 \in Y$

$$K(x, x_2) = \Sigma \phi(g_1^{-1} \delta g_2), \tag{3}$$

where $g_1$, $g_2$ are representatives of cosets belonging to $S$.

The function $K$ is not bounded on $X$, therefore the trace of the operator $T_\phi$ in $L_2(X)$ does not exist. However, we succeed in reducing the estimate of the trace of the operator $PT_\phi P$ to the estimate of $x$ derivatives of the function $K$ in directions lying in the orispheres. With the help of (4) the necessary estimate of the derivatives may be deduced simply enough.

**4.** It is not known whether the asymptotic formulas for the distribution of the "indices" of the representations entering into $L_2(G/\Gamma)$ formulas which, in case $G/\Gamma$ is compact, have been established for the distribution of the indices of the representations entering into $L_2^0(G/\Gamma)$ remain true when $\Gamma$ is a tame discrete subgroup of the semisimple group $G$. In particular, for the case when $G/\Gamma$ is compact, the following asymptotic relation is true. An irreducible unitary representation, on which $L_2(G/U)$ decomposes ($U$ is a maximal compact subgroup of $G$), may be naturally assigned indices $\rho$, running through the following region.

We denote by $\mathfrak{A}$ the Cartan subalgebra of the symmetric space $G/U$. Let, further, $\mathfrak{A}^+$ denote the totality of vectors of $\mathfrak{A}$ for which $(a, \alpha) \geq 0$ for all positive roots $\alpha$. $\mathfrak{A}^+$ represents itself as the natural range of a "index" $\rho$. Let, further, $B_n$ be the expanding sequence of subregions $\mathfrak{A}^+$. We denote by $N(B_n)$ the number of irreducible representations entering into $L_2(X)$ with indices contained in $B_n$. Then

$$N(B_n) \sim C_G C_\Gamma \int_{B_n} \prod_{\alpha > 0} (\rho, \alpha)^{\nu_\alpha} \, d\rho,$$

where $C_\Gamma$ is the volume of $G/\Gamma$; $C_G$ is a constant, depending only on $G$; $\nu_\alpha$ is the multiplicity of the root $\alpha$.

This formula is proved with the use of certain results of F. I. Karpelevič and S. G. Gindikin [9].

Received 11/AUG/62

## BIBLIOGRAPHY

[1] I. M. Gel'fand and S. V. Fomin, Uspehi Mat. Nauk 7 (1952), no. 1 (47), 118.

[2] I. M. Gel'fand and I. I. Pjateckiĭ-Šapiro, ibid. 14 (1959), no. 2 (86), 171.

[3] A. Selberg, Sbornik per. Mat. 1 (1957), 4.

[4] ———, ibid. 6 (1962), 3.

[5] A. Borel and Harish-Chandra, Bull. Amer. Math. Soc. 67 (1961), 579.

[6] Harish-Chandra, Proc. Nat. Acad. Sci. U.S.A. 45 (1959), 570.

[7] I. M. Gel'fand and M. I. Graev, Trudy Moskov. Mat. Obšč. 8 (1959), 321.

[8] I. M. Gel'fand and I. I. Pjateckiĭ-Šapiro, Dokl. Akad. Nauk SSSR 147 (1962) 275 = Soviet Math. Dokl. 3 (1962), 1574.

[9] S. G. Gindikin and F. I. Karpelevič, ibid. 145 (1962), 252 = Soviet Math. Dokl. 3 (1962), 962.

Translated by: W. D. Maurer

# 5.

## (with I. I. Piatetski-Shapiro)

# Unitary representations in a space $G/\Gamma$, where $G$ is a group of $n$-by-$n$ real matrices and $\Gamma$ is a subgroup of integer matrices

Dokl. Akad. Nauk SSSR **147** (1962) 275–278
[Sov. Math., Dokl. **3** (1962) 1574–1577]. Zbl. **119**:271

This work is concerned with the investigation of unitary representations in a space $X = G/\Gamma$, ere $G$ is a group of $n$-by-$n$ real matrices and $\Gamma$ is a subgroup of integer matrices. The fundamental ol used will be the method of orispheres. We recall the definitions of orispheres and of orisphere suboups.

Let $G$ be a real semi-simple Lie group, and let $g(t)$ be some one-parameter subgroup of the group The collection $Z \subset G$ consisting of all the $z$ for which $\lim\limits_{t\to\infty} g(-t)\, z\, g(t) = 1$ is said to be the *isphere subgroup associated with the group* $g(t)$. The orbits of the orisphere groups are called *oriheres in* $X$. An orisphere is said to be *compact* if the set of points of which it consists is compact.

If $G$ is a group of $n$-by-$n$ real matrices, then there exist as many nonconjugate orisphere subgroups there are representations of $n$ in the form of a sum of positive summands $n = k_1 + k_2 + \cdots + k_s$ epresentations differing in order being considered as distinct.) The corresponding orisphere subgroups ve the form

$$\begin{pmatrix} E_{k_1} & * & \dots & * \\ 0 & E_{k_2} & \dots & * \\ \cdot & \cdot & \dots & \cdot \\ 0 & 0 & \dots & E_{k_s} \end{pmatrix}, \tag{1}$$

ere there are zeros below the diagonal, arbitrary numbers above it, and $E_{k_i}$ denotes an identity max of order $k_i$.

We denote the subgroup (1) by $Z_{k_1,\dots,k_s}$. The following theorem describes the structure of all the mpact orispheres in the space $X = G/\Gamma$.

**Theorem.** *Each compact orisphere in $X$ is the image of a set $Z_{k_1,\dots,k_s}\, g_0$ under a natural mapng of $G$ onto $X$, where $g_0$ designates an arbitrary fixed element of the group $G$. The orispheres ${}_{1},\dots,{}_{k_s}\, g_0$ and $Z_{k'_1,\dots,k'_{s'}}\, g'_0$ will be taken into one another by motions from $G$ if and only if $= k'_1, \dots, k_s = k'_s$ and $s = s'$.*

We give the proof of this theorem briefly. It follows from the compactness of the factor-space ${}_{1},\dots,{}_{k_s}/\Gamma \cap Z_{k_1,\dots,k_s}$ that the image of the set $Z_{k_1,\dots,k_s}\, g_0$ is a compact orisphere in $X$. It is so clear that the images of the sets $Z_{k_1,\dots,k_s}\, g_0$ and $Z_{k'_1,\dots,k'_{s'}}\, g'_0$ in $X$ are taken into each other motions from $G$ if and only if the groups $Z_{k_1,\dots,k_s}$ and $Z_{k'_1,\dots,k'_{s'}}$ are conjugate in $G$, and thus and only if $s = s'$, $k_1 = k'_1, \dots, k_s = k'_s$. The proof that each compact orisphere in $X$ is the image a set $Z_{k_1,\dots,k_s}\, g_0$ is more complicated. For that proof it is first necessary to show that if $Z$ is an isphere subgroup in $G$ such that the factor-space $Z/\Gamma \cap Z$ is compact, then $Z$ has the form

$h^{-1} Z_{k_1,\dots,k_s} h$, where $h$ is some matrix with rational elements. Further, using the unique factorization in the ring of intergers, it is shown that $Z$ has the form $\gamma^{-1} Z_{k_1,\dots,k_s} \gamma$, where $\gamma \in \Gamma$.

We turn now to the consideration of functions on $X$. We denote by $H^0_{k_1,\dots,k_s}$ the collection of all the functions of $L_2(X)$, the integrals of which are zero on all the compact orispheres of the form $Z_{k_1,\dots,k_s} g_0$. It is easy to see that the space $H^0_{k_1,\dots,k_s}$ is invariant with respect to the translation operators of $T_g$. Further, we denote by $H'_{k_1,\dots,k_s}$ the intersection of all the spaces $H^0_{k'_1,\dots,k'_s}$, corresponding to the groups $Z_{k'_1,\dots,k'_s}$, that contain $Z_{k_1,\dots,k_s}$ as a proper subgroup. It is easy to verify that $H^0_{k_1,\dots,k_s} \subset H'_{k_1,\dots,k_s}$. We next denote by $H_{k_1,\dots,k_s}$ the factor-space $H'_{k_1,\dots,k_s}/H^0_{k_1,\dots,k_s}$. We define in $H_{k_1,\dots,k_s}$ in a natural way the operation of the operators of $T_g$. Further, we denote by $H^0$ the collection of all the functions $f(x)$ from $L_2(X)$, the integrals of which are equal to zero on all the compact orispheres.

**Theorem 2.** *The space $L_2(X)$ is isomorphic to the sum of the spaces $H_{k_1,\dots,k_s}$ and $H^0$:*

$$L_2(X) \cong \sum_{k_1+\dots+k_s=n} H_{k_1,\dots,k_s} + H^0. \tag{2}$$

We have shown previously [1] that $H^0$ decomposes into the sum of an even number of irreducible unitary representations of the group $G$. In the present article we are concerned with the decomposition into irreducible representations of the space $H_{k_1,\dots,k_s}$.

Along with its direct interest, the significance of this problem also consists in the fact that it leads to a series of remarkable analytic functions closely related to functions such as Riemann's classical zeta-function and its generalizations. We now present the definitions of these functions for $H_{k_1,\dots,k_s}$. The more general definition for arbitrary regular groups will be given later.

We consider the set $\Omega_1$ of all the compact orispheres of greatest dimension. Since any motion takes a compact orisphere again into a compact orisphere, the action of the group $G$ may be naturally defined in $\Omega_1$.

It follows from Theorem 1 that the group $G$ acts transitively in $\Omega_1$. We associate with each function $f(x) \in L_2(X)$ its integral over the compact orispheres of $\Omega_1$, which we denote by $\check{f}(\omega)$. Let $\mathfrak{L}$ designate the collection of functions $\check{f}(\omega)$ on $\Omega_1$ that are integrals of functions $f(x) \in L_2(X)$. Then $\mathfrak{L}$ is clearly isomorphic to $H_{1,1,\dots,1}$.

We denote by $L_2(\Omega)$ the collection of all the functions on $\Omega_1$ with integrable squares. The decomposition of $L_2(\Omega)$ into irreducible representations may be realized, as is well known, in the following manner. We denote by $A$ the connected component of the group of left translations of the space $\Omega$, i.e., the group of the transformations $\omega \to a\omega$ that commute with motions.

The group $A$ is identified in a natural way with the group of diagonal matrices with positive elements on their diagonals. By $a^{\kappa}$ we denote $\exp(\kappa \ln a)$, where $\ln a$ denotes the canonical transformation of $A$ into its Lie algebra. We set

$$\Phi(\kappa, \omega) = \int_A \phi(a\omega)\, a^{-\kappa} j^{1/2}(a)\, da, \tag{3}$$

where $j(a)$ is the Jacobian of the transformation $\omega \to a\omega$ which depends only on $a$.

Formula (3), where $\kappa$ is pure imaginary, gives a decomposition of $L_2(\Omega)$ into irreducible repre-ntations, where two representations with "norms" $\kappa$ and $\tilde{\kappa}$ are equivalent if and only if $\tilde{\kappa}(a) \equiv$ $a^\sigma$) for all $a$, where $a^\sigma$ denotes $a$ with permuted eigenvalues and $\sigma$ is the permutation.

It is possible to show that $\mathfrak{L}$ decomposes into the sum of two spaces, $\mathfrak{L}'$ and $\mathfrak{L}_1$, where $\mathfrak{L}'$ con-sts of only a constant and $\mathfrak{L}_1 \subset L_2(\Omega)$. It turns out that, in contrast to $L_2(\Omega)$, each irreducible repre-ntation occurs one time in $\mathfrak{L}_1$. Thus $\Phi(\kappa, \omega)$ for functions of $\mathfrak{L}_1$ in points corresponding to equiva-nt irreducible representations turn out to be dependent. We assume

$$\Phi(\sigma\kappa, \omega) = \zeta_\sigma(\kappa)\, \Phi(\kappa, \omega). \tag{4}$$

The functions $\zeta_\sigma(\kappa)$ are of fundamental importance. They are meromorphic over all of complex ace and satisfy the functional equation

$$\zeta_{\sigma_1\sigma_2}(\kappa) = \zeta_{\sigma_1}(\sigma_2\kappa)\, \zeta_{\sigma_2}(\kappa). \tag{5}$$

e latter is equivalent to the fact that each representation in $\mathfrak{L}_1$ occurs one time.

Using the functional equation (4), it is possible to show that

$$\zeta_\sigma(\kappa) = \prod_{\alpha} \theta((\kappa, \alpha)),$$

ere $\theta(s) = B\left(\frac{1}{2}, \frac{s}{2}\right) \zeta(s)\, \zeta^{-1}(s+1)$, $\zeta(s)$ is Riemann's classical zeta-function, and $z$ runs through me set of roots which for each $\sigma$ consists of all the negative roots $\alpha$ for which $\sigma\alpha$ is a positive ot; the parantheses denote Cartan scalar multiplication; $\sigma\alpha$ is a root obtained from $\alpha$ by the permu-tion $\sigma$.

The spaces $H_{k_1,\dots,k_s}$ may be analyzed similarly. It turns out that each of them is the sum of two aces $\hat{\mathfrak{L}}_{k_1,\dots,k_s}$ and $\mathfrak{L}'_{k_1,\dots,k_s}$, the first of which is imbedded in $L_2(\Omega_{k_1,\dots,k_s})$ and decomposes o the same irreducible representations as $L_2(\Omega_{k_1,\dots,k_s})$ but contains them in a lower multiplicity.* e latter property leads to functional equations for the functions which arise here as the analogues of $(\kappa)$. $\mathfrak{L}'_{k_1,\dots,k_s}$ consists of representations corresponding to the singular points of the functions $(\kappa)$ and of their analogues.

We present the proof for $H_{1,\dots,1}$ in brief. (We denote the scalar product of two functions $\check{f}_1$, $\in \mathfrak{L}$ by $[\check{f}_1, \check{f}_2]$.) We also agree, for any two functions $\phi_1(\omega)$ and $\phi_2(\omega)$ on $\Omega$, for which the inte-al $\int \phi_1 \bar{\phi}_2\, d\omega$ converges absolutely, to denote the magnitude of that integral by $(\phi_1, \phi_2)$.

It is easy to show that for every finite continuous function $\phi(\omega)$ on $\Omega$ the expression $(\check{f}, \phi)$ has aning for all $\check{f} \in \mathfrak{L}$ and is a linear functional on $\mathfrak{L}$. According to a theorem of Rees, there exists $\mathfrak{L}$ a vector $M_\phi$ such that $(f, \phi) = [\check{f}, M_\phi]$ for all $\check{f} \in \mathfrak{L}$. The operator $M$ plays an important role further investigations. There exists for $M$ a comparatively simple explicit expression, which we ll introduce now. We first agree on the following notation: $\Delta = \Gamma \cap Z$; $\Delta'$ is the normalizer of $\Delta$ $\Gamma$; $Z = Z \cap g^{-1} Zg$, where $g \in G$; $D_g$ is the collection of left co-sets of the group relative to the bgroup $Z_g$. We now set

* $\Omega_{k_1,\dots,k_s}$ designates the space of the orispheres associated with the group $z_{k_1,\dots,k_s}$.

$$M_{g_0}\ (\varphi\ (\omega)) = \int\limits_{Z_{g_0}} \varphi^*\ (g_0 z g)\ dz, \tag{6}$$

where $\phi^*(g)$ designates the function on $G$ corresponding to the function on $\Omega$.

The following fomula is valid:

$$M_\phi = \Sigma M_\gamma, \tag{7}$$

here $\gamma$ runs over the set of all the representatives of the two-sided co-sets $\Delta' \gamma \Delta$ of the group $\Gamma$ with respect to the subgroups $\Delta'$ and $\Delta$.

Further arguments are based on the consideration of the form $(M_{\phi_1, \phi_2})$ and on the study of the Mellin transformation of functions of the form $M_\phi$, where $\phi$ designates a continuous finite function.

Received 11/AUG/62

## BIBLIOGRAPHY

[1] I. M. Gel'fand and I. I. Pjateckiĭ-Šapiro, Dokl. Akad. Nauk SSSR 147 (1962), 17 = Soviet Math. Dokl. 3 (1962), 1528.

Translated by:
Sue Ann Walker

# Part IV

# Models of representations; representations of groups over various fields

# 1.

## (with M. I. Graev)

# Categories of group representations and the problem of classifying irreducible representations

Dokl. Akad. Nauk SSSR **146** (1962) 757–760
[Sov. Math., Dokl. **3** (1962) 1378–1381]

1. Suppose $G$ is an arbitrary (discrete or continuous) group and $N$ is one of its subgroups. Con-ider the set $X$ of representations $T_\chi (\chi \in X)$ of the group $G$, induced by the irreducible unitary repre-entations of the subgroup $N$.* To each representation $T_\chi$ we associate the ring $R_\chi$ of all linear perators which commute with the operators of the representation. To each pair of representations $\chi_1$, $T_{\chi_2}$ operating on $H_1$ and $H_2$ respectively, we associate the linear family $R_{\chi_2\chi_1}$ of all linear appings of $H_1$ into $H_2$ which commute with the operators of the representations. The family of epresentations $T_\chi$ together with the rings $R_\chi$ and the linear spaces $R_{\chi_2\chi_1}$ will be called the *cate-ory of representations of the group G relative to the subgroup N*. There then arises the problem of escribing this category. This problem is interesting for a number of reasons: its solution leads to a onvenient classification of irreducible unitary representations of the group $G$ (occurring in a regular epresentation) and gives the key to the construction of special functions, connected with the group $G$ nd so on.

In the present note we study the group $G$ of unimodular matrices of second order $g = \begin{pmatrix} \alpha & \beta \\ \gamma & \delta \end{pmatrix}$, $\delta - \beta\gamma = 1$, with elements from a finite field $K$ of order $k$ (as is well known, $k$ is a power of a prime ; we will assume $p \neq 2$). We will describe the categories of representations, relative to the subgroup of matrices of the form $\zeta_t = \begin{pmatrix} 1 & t \\ 0 & 1 \end{pmatrix}$. On the basis of this description we will give a convenient clas-ification of all irreducible group representations of $G$ and thus effectively give its construction (the atter problem concerns the definition of a Bessel function over a finite field). Finally we will formu-ate similar results for the group of matrices over the field of real numbers.

2. To each irreducible representation of the subgroup $Z$ we assign the additive character $\chi(t)$, $\in K (\chi(t_1 + t_2) = \chi(t_1)\chi(t_2))$. We construct the representation $T_\chi$ of $G$, induced by this character, the space $H_\chi$ of functions $f(g)$ on $G$ (with values in the field of complex numbers), which satisfy e condition $f(\zeta_t g) = \chi(t) f(g)$ for any $g \in G$ and $t \in K$. The operator of the representation has the orm $T_\chi(g_0) f(g) = f(gg_0)$. In such a manner, we obtain for all $k$ ($k$ being the order of the field $K$) the nduced representations $T_\chi$.

The representation $T_\chi$ may be also realized in the space of functions $f(x) \equiv f(x_1, x_2)$ on the 'affine plane'' $(x_1, x_2 \in K, (x_1, x_2) \neq (0, 0))$; the set of such pairs $(x_1, x_2)$ is identical with the set of osets $G/Z$). In this realization the operator of the representation has the form

---

*Recall the definition of an induced representation for the case of a *unimodular* subgroup $N$ (for the general efinition see, e.g., [1]). Let $c(n)$ be a unitary representation of the subgroup $N$ which operates in the space $H$ ith norm $\|h\|$. The representation $T(g)$ of the group $G$, induced by the representation $c(n)$, is constructed in the ilbert space of functions $f(g)$ on $G$ with values in $H$ such that $f(ng) = c(n) f(g)$ for any $n \in N$ and $\|f(\tilde{g})\|^2 \, d\tilde{g} < \infty$ where $\tilde{G} = G/N$, and $d\tilde{g}$ is an invariant measure on $\tilde{G}$. The operator of the representation $T(g)$ s a shift operator: $T(g_0) f(g) = f(gg_0)$.

We note that in the case of a Lie group it is useful, except for a unitary one, to consider a nonunitary repre-entation operating in certain naturally defined kernel spaces.

$T_\chi(g)\, f(x) = f(xg)\, a(x, g)$, where $xg = (\alpha x_1 + \gamma x_2, \beta x_1 + \delta x_2)$ and $a(x, g)$ is a fixed function satisfying the relations $a((0, 1), \zeta_t) = \chi(t)$, $a(x, g_1 g_2) = a(x, g_1)\, a(xg_1, g_2)$. Thus we obtain essentially the category of representation of the group $G$, associated with the affine plane.

3. Description of the ring $R_\chi$ and the space $R_{\chi_2\chi_1}$. The operators $A \in R_{\chi_2\chi_1}$ have the form $Af(g) = (1/k) \sum \phi(gg_1^{-1})\, f(g_1)$ (the sum runs over all $g_1 \in G$), where $\phi(g)$ is an arbitrary function on $G$ satisfying for any $g \in G$, $t_1, t_2 \in K$ the relations

$$\phi(\zeta_{t_2} g \zeta_{t_1}) = \chi_2(t_2)\, \phi(g)\, \chi_1(t_1). \tag{1}$$

From this we obtain when $\chi_1 = \chi_2 = \chi$ the result that *$R_\chi$ is isomorphic to the ring of functions $\phi(g)$ which satisfy relation* (1) *with multiplication defined by* $\phi_1(g) * \phi_2(g) = (1/k) \sum_{g_1} \phi_1(gg_1^{-1})\, \phi_2(g_1)$.*

In particular, the ring $R_1$, corresponding to the character $\chi \equiv 1$, is isomorphic to the ring of functions, constant on two-sided cosets of the subgroup $Z$. We first describe this ring $R_1$.

Suppose $A_\lambda$ is the characteristic function of the two-sided cosets of $Z$ with a representation $\begin{pmatrix} \lambda^{-1} & 0 \\ 0 & \lambda \end{pmatrix}$, and $B_\lambda$ is the characteristic function with representation $\begin{pmatrix} 0 & -\lambda^{-1} \\ \lambda & 0 \end{pmatrix}$. Then *$A_\lambda$, $B_\lambda$ form a basis in the linear space $R_1$* (which means that the dimension of $R_1$ is equal to $2k - 2$). *The operation of multiplication in $R_1$ is given by the formulas* $A_{\lambda_1} A_{\lambda_2} = A_{\lambda_1\lambda_2}$, $A_{\lambda_1} B_{\lambda_2} = B_{\lambda_1\lambda_2}$, $B_{\lambda_1} A_{\lambda_2} = B_{\lambda_1\lambda_2^{-1}}$, $B_{\lambda_1} B_{\lambda_2} = \sum_\lambda B_\lambda + kA_{-\lambda_1\lambda_2^{-1}}$. From these formulas it is easily shown that the center of the ring $R_1$ consists of elements of the form $A = \alpha_1 A_1 + \alpha_{-1} A_{-1} + \sum \beta_\lambda (A_\lambda + A_\lambda^{-1}) + \gamma' \sum_{\lambda'} B_{\lambda'} + \gamma'' \sum_{\lambda''} B_{\lambda''}$, where $\lambda'$ runs through the set of all squares, and $\lambda''$ is the complementary set (this means that the dimension of the center is $\frac{1}{2}(k + 5)$). Moreover, the elements $A$ of a ring satisfying the condition that $A'A = c_{A'} A$ for any element $A'$ of the ring are, to within a multiple, of the form $A = \pm\sqrt{k\pi(-1)} \sum \pi(\lambda) A_\lambda + \sum \pi(\lambda) B_\lambda$, where $\pi(\lambda') = 1$, $\pi(\lambda'') = -1$ (the symbols $\lambda'$, $\lambda''$ are as above) or of the form $A = \sum_\lambda (A_\lambda + B_\lambda)$, $A = \sum_\lambda (-kA_\lambda + B_\lambda)$. From these results it immediately follows that: *the ring $R_1$ is the direct sum of $\frac{1}{2}(k + 5)$ total matrix rings, of which 4 are composed of matrices of order* 1, *and the remaining $\frac{1}{2}(k - 3)$ are composed of matrices of order* 2.

Now suppose $\chi \not\equiv 1$. Then the elements of the ring $R_\chi$, namely the functions $\phi(g)$ are equal to zero on the two-sided cosets of $Z$ with representatives $\begin{pmatrix} \lambda^{-1} & 0 \\ 0 & \lambda \end{pmatrix}$, $\lambda \neq \pm 1$. Now we form in the space $R_\chi$ a basis $A_1$, $A_{-1}$, $B_\lambda$. We define each of the functions $A_1$, $A_{-1}$, $B_\lambda$ by the following conditions: they equal 1 respectively on the matrices $\begin{pmatrix} 1 & 0 \\ 0 & 1 \end{pmatrix}$, $\begin{pmatrix} -1 & 0 \\ 0 & -1 \end{pmatrix}$ or $\begin{pmatrix} 0 & -\lambda^{-1} \\ \lambda & 0 \end{pmatrix}$ and are different from zero only on the two-sided cosets of $Z$, containing these matrices. *The operation of multiplication in $R_\chi$, $\chi \not\equiv 1$, is given by the formulas* $A_{-1}A_{-1} = A_1$, $B_\lambda A_{-1} = A_{-1} B_\lambda = B_{-\lambda}$, $B_{\lambda_1} B_{\lambda_2} = \sum_\lambda \chi\left(\frac{\lambda}{\lambda_1\lambda_2} + \frac{\lambda_1}{\lambda\lambda_2} + \frac{\lambda_2}{\lambda\lambda_1}\right) B_\lambda + k\delta_{\lambda_1+\lambda_2} A_1 + k\delta_{\lambda_1-\lambda_2} A_{-1}$, *and $A_1$ is the unit of the ring.* (Here $\delta_t = 1$ when $t = 0$ and $\delta_t = 0$ when $t \neq 0$.) From this we conclude that for $\chi \not\equiv 1$ *the ring $R_\chi$ is commutative and has dimension $k + 1$.*

Taking into account relation (1), it is easy to determine the dimension of the spaces $R_{\chi_2\chi_1}$, $\chi_1 \neq \chi_2$. *If the characters $\chi_1(t)$ and $\chi_2(t)$ are connected by the relation $\chi_2(t) = \chi_1(\lambda^2 t)$ for some $\lambda \neq 0$, then the dimension of $R_{\chi_2\chi_1}$ is equal to $k + 1$; otherwise the dimension of $R_{\chi_2\chi_1}$ is equal to*

*Note that the ring $R_\chi$ is semisimple, i.e., can be decomposed into a direct sum of total matrix rings (this follows from the complete reducibility of the representations of a finite group).

– 1.

4. **The classification of irreducible representations of the group $G$.** We will divide the representations $T_\chi$, $\chi \not\equiv 1$ into two classes: $T_{\chi_1}$ and $T_{\chi_2}$ belong to the same class if $\chi_2(t) = \chi_1(\lambda^2 t)$ for some $\lambda \neq 0$. On the basis of § 3 we obtain: *the representation $T_\chi$, $\chi \not\equiv 1$ is a direct sum of $k+1$ pairwise inequivalent representations. If the representations $T_{\chi_1}$ and $T_{\chi_2}$ belong to the same class, then they are equivalent; if, however, $T_{\chi_1}$ and $T_{\chi_2}$ belong to different classes, then they have $k-1$ common irreducible summands. The representation $T_1$ corresponding to the character $\chi \equiv 1$ contains 4 irreducible representations, each with multiplicity* 1 *and* $\frac{1}{2}(k-3)$ *irreducible representations, each with multiplicity* 2. *The representations $T_\chi$, $\chi \not\equiv 1$ and $T_1$ have $\frac{1}{2}(k+1)$ general irreducible components; two of them appear in $T_1$ with multiplicity* 1, *and the remaining ones with multiplicity* 2. We can now classify all irreducible representations of the group $G$ according to the multiplicity with which they occur in the representations $T_\chi$. (Note that a regular representation of the group $G$ is a direct sum of representations $T_\chi$; which means that each irreducible representation of the group $G$ is contained in at least one of the representations $T_\chi$.)

We obtain the following types of irreducible representations: 1) A representation of the "basic" series; which occurs in all $T_\chi$, but in $T_1$ it occurs with multiplicity 2; the number of such representations is $\frac{1}{2}(k-3)$, the dimension of each being $k+1$. 2) A representation of the "analytic" series; occurs in all $T_\chi$, except $T_1$; the number of such representations is $\frac{1}{2}(k-1)$, the dimension of each being $k-1$. 3) One representation of dimension $k$; occurs in all $T_\chi$, but in $T_1$ it appears with multiplicity 1. 4) Two representations of dimension $\frac{1}{2}(k+1)$; each of them occurs in a representation $T_\chi$, $\chi \not\equiv 1$, which belongs only to one of the two classes; in $T_1$ it occurs with multiplicity 1. 5) Two representations of dimension $\frac{1}{2}(k-1)$; each of them occurs in representation $T_\chi$, $\chi \not\equiv 1$, which belongs to only one of the two classes and does not occur in $T_1$. 6) The identity representation, which is contained only in $T_1$.

5. All irreducible representations of the group $G$ may be obtained by decomposing the space $H_\chi$ into irreducible subspaces. In the case $\chi \equiv 1$, the irreducible subspace is the subspace of homogeneous functions of a given degree of homogeneity $\pi$ ($\pi(t)$ is a multiplicative character): $f(tx_1, tx_2) = \pi(t) f(x_1, x_2)$ for any $t \neq 0$.* Note that in these subspaces are realized only half of all irreducible representations. Here we will give the description of all irreducible subspaces of the space $H_\chi$, $\chi \not\equiv 1$. It is interesting that the analogues for homogeneous functions when $\chi \not\equiv 1$ are the Bessel functions.

Let us consider the functions $\phi(g) \in R_\chi$ which satisfy the condition

$$\psi(g) * \phi(g) = c_\psi \, \phi(g) \tag{2}$$

for any $\psi \in R_\chi$; $c_\psi$ is a complex number (the mapping $\psi \to c_\psi$ is a homomorphism of the ring $R_\chi$). The function $\phi(g)$ will be normed by the condition $\phi(e) = 1$, where $e$ is the group identity. Between such functions $\phi(g)$ and the irreducible subspaces of the spaces $H_\chi$, $\chi \not\equiv 1$, there exists a one-to-one mapping. An irreducible subspace corresponding to the function $\phi(g)$ consists of all functions representable in the form $\phi(g) * f(g)$, where $f(g)$ runs through $H_\chi$. The problem of describing all functions $\phi(g)$ which satisfy condition (2) is worked out below.

In view of (1) it is sufficient to know the values of the functions $\phi(g)$ only on the matrices $\pm e$ and $\begin{pmatrix} 0 & -\lambda^{-1} \\ \lambda & 0 \end{pmatrix}$.

*This does not take into account the cases when $\pi(t) \equiv 1$ or $\pi(t) = \pm 1$. In each case the space of homogeneous functions is the sum of two irreducible subspaces.

We denote by $\phi(\pm e)$ and $\phi(\lambda)$ the values of function $\phi(g)$ on these matrices. Then for $\phi(\lambda)$ we obtain the following functional equation:

$$k\varphi(\lambda_1)\varphi(\lambda_2) = \varphi(-e)\sum_{\lambda}\chi\left(\frac{\lambda}{\lambda_1\lambda_2} + \frac{\lambda_1}{\lambda\lambda_2} + \frac{\lambda_2}{\lambda\lambda_1}\right)\varphi(\lambda) + $$
$$+ \delta_{\lambda_1-\lambda_2}\varphi(-e) + \delta_{\lambda_1+\lambda_2}\varphi(e) \tag{3}$$

for any $\lambda_1, \lambda_2 \neq 0$; $\phi(-e)\,\phi(\lambda) = \phi(-\lambda)$; $\phi(-e) = \pm\,\phi(e)$, where $\phi(e) = 1$. There exist two classes of solutions to equation (3). The functions of the first class are given by the formulas

$$\phi(\lambda) = J_\pi(\lambda; \chi) \equiv \frac{1}{k}\sum_t \chi(-\lambda^{-1}(t + t^{-1}))\,\pi(t); \qquad \phi(-e) = \pi(-1),$$

where $\pi(t)$ is a multiplicative character on $K$; the summation runs over all $t \neq 0$ from $K$. For the description of the second class of solutions of the equation (3), we extend the field $K$, by adjoining all of its quadratic roots. In this extension we consider the set $K_1$ of elements "equal mod the identity" (i.e., such that the product of the element by its conjugate is equal to 1). This set is a group with multiplication of order $k + 1$. The functions of the second class are given by the formula

$$\phi(\lambda) = K_\pi(\lambda; \chi) \equiv -\frac{1}{k}\sum_t \chi(-\lambda^{-1}(t + t^{-1}))\,\pi(t); \qquad \phi(-e) = \pi(-1),$$

where $\pi(t)$ is a multiplicative character on $K_1$, $\pi \neq 1$; the summation runs over all $t \neq 0$ from $K_1$.* The functions $J_\pi(\lambda; \chi)$ and $K_\pi(\lambda; \chi)$ are naturally called Bessel functions over the finite field $K$. (The formulas for $J_\pi(\lambda; \chi)$ and $K_\pi(\lambda; \chi)$ are analogous to the formulas of the integral representation of the ordinary Bessel functions.)

6. We formulate analogous results for the group $G_B$ of unimodular matrices of the second order over the field of real numbers. It can be shown that for this group, the ring $R_\chi$, $\chi(t) = e^{i\sigma t}$ is commutative where $\chi \neq 1$. Moreover in the decomposition of the representation $T_\chi$ into irreducible representations of the basic continuous series and half of the irreducible representations of the discrete series. (The other half of the representations of the discrete series occur in the decomposition of the representation $T_{\bar\chi}$, $\bar\chi(t) = e^{-i\sigma t}$.) A realization of irreducible unitary representations (of the basic series) may also be found by the methods in §5 for the case of a finite field. The search for the functions $\phi(g)$ which satisfy condition (2) leads to the solution of a differential equation of the form $\lambda^2\phi''(\lambda) + 3\lambda\phi'(\lambda) + (4\sigma^2/\lambda^2 - \nu)\,\phi(\lambda) = 0$. The solution of this equation is the function $\lambda^{-1}J_a(2\sigma/\lambda)$, $a = \pm\sqrt{1+\nu}$, where $J_a(x)$ is the Bessel function.

Received 3/JULY/62

BIBLIOGRAPHY

[1] G. W. Mackey, Ann. of Math. (2) 55 (1952), 101.

[2] Frobenius, *The theory of characters and group representations*, Kharkov, 1937. (Russian)

[3] E. Hecke, Abh. Math. Seminar Univ. Hamburg 235 (1928), 6.

Translated by:
J. G. Ceder

*We note that functions of the first class correspond to irreducible representations contained in $T_1$, and functions of the second class to those not contained in $T_1$.

# 2.
## (with M. I. Graev)
# Construction of irreducible representations of simple algebraic groups over a finite field

Dokl. Akad. Nauk SSSR **147** (1962) 529–532
[Sov. Math., Dokl. **3** (1962) 1646–1649]. Zbl. **119**:269

1. As is known from [5, 6, 1], the irreducible unitary representations of any complex simple Lie roup $G$ may be obtained by the following construction. We take the uniform space $X = G/Z$ of right osets of the group $G$ by its maximal nilpotent subgroup $Z$. In the space of functions $f(x)$ on $X$ we onsider the representation $T(G)\, f(x) = f(xg)$. Decomposing $T(g)$ into irreducible representations, we et all irreducible representations of a "fundamental series" (i.e. contained in a regular representa- ion of $G$). Going on to simple algebraic groups over other fields, our construction gives this time nly a small portion of the irreducible representations.

In this note we give a construction and classification of irreducible representations of simple lgebraic Dickson-Chevalley groups over a *finite* field [3, 4]. By virtue of the Chevalley construction ach of these groups is determined by specifying a simple Lie algebra over the complex number field nd also a finite field $K$. In particular, the following infinite series of simple groups correspond to he classical simple Lie algebras: a) the factor group of the group of unimodular matrices of order $n$ with elements from $K$) with respect to its center, b) the commutator-group of the orthogonal projec- ive group (with elements from $K$), c) the factor group with respect to its center of a symplectic group with elements from $K$).

2. The construction and classification of all irreducible representations of Dickson-Chevalley roups is based on the following two fundamental theorems.

**Theorem 1.** *Let $G$ be a Dickson-Chevalley group, and $Z$ be its maximal nilpotent subgroup. We onsider the various characters $\chi(\zeta)$, i.e. the univariate representatives of the subgroup $Z$. To each haracter we associate the representation of $G$ induced by it. (This representation is realized in the pace of functions $f(g)$ on $G$ such that $f(\zeta g) = \chi(\zeta)\, f(g)$ for any $\zeta \in Z$. An operator of the repre- entation has the form $T_\chi(g_0)\, f(g) = f(gg_0)$.) We may establish that any irreducible representation f the group $G$ is contained in at least one of the representations $T_\chi(g)$.*

**Equivalent formulation of Theorem 1.** *In any representation $T(g)$ of $G$ there is a vector which s characteristic for every operator $T(\zeta)$, $\zeta \in Z$.*

We now introduce on the set of characters $\chi(\zeta)$ a partial ordering: we shall say that $\chi_1 < \chi_2$, f on any root subgroup of $Z$, if $\chi_2(\zeta) \equiv 1$ then also $\chi_1(\zeta) \equiv 1$. The maximal characters $\chi$ we call haracters *of general aspect.*

**Theorem 2.** *Let $\chi$ be a character of general aspect. Then each irreducible representation of $G$ s contained in a representation $T_\chi(g)$ with multiplicity not greater than 1.*

**Equivalent formulation of Theorem 2.** *If $\chi$ is a character of general aspect, then in any irreduci- le representation $T(g)$ of $G$ there is no more than one characteristic vector with respect to the op- rators $T(\zeta)$, $\zeta \in Z$, with characteristic value $\chi(\zeta)$.*

Theorems 1 and 2 replace Cartan's theory of highest weights which is not applicable to the groups we are considering.*

For simplicity we give the proofs of Theorems 1 and 2 only for the group $G_n$ of unimodular matrices of order $n$ over the field $K$. The maximal nilpotent subgroup $Z$ of $G_n$ consists of the triangular matrices $\|\zeta_{ij}\|$, where $\zeta_{ij} = 1$, and $\zeta_{ij} = 0$ for $i > j$.

Proof of Theorem 1. It is sufficient to prove the theorem for representations of the subgroup $G_n' \subset G_n$ of matrices $\|g_{ij}\|$, in which $g_{11} = 1$, and $g_{i1} = 0$ for $i > 1$. We give a proof by induction on $n$, assuming the theorem holding for $n-1$. (For the case $n = 2$ the theorem is evident.) Let $T(g)$ be a representation of $G_n'$ operating on some space $H$. In $G_n'$ we look at the subgroup $Z_n'$ of matrices $\|\zeta_{ij}\|$ in which $\zeta_{ij} = 0$ for $i > 1$, $i \neq j$, $\zeta_{ii} = 1$. This subgroup is commutative, and thus $H$ is decomposed according to the vectors characteristic with respect to the operators $T(\zeta')$, $\zeta' \in Z_n'$, with characteristic values $\chi(\zeta') = \chi_1(\zeta_{12}) \cdots \chi_{n-1}(\zeta_{1n})$ (the $\chi_i$ are additive characters on $K$). The characters $\chi$ may be considered as points of an $(n-1)$-dimensional linear space over $K$. The subgroup $Z_n'$ is a normal divisor in $G_n'$, and thus the operators $T(g)$, $g \in G_n'$, send these characteristic vectors again into characteristic vectors; mappings into the set of characters $\chi(\zeta')$ are determined by the same elements $g \in G_n'$. It is easy to see that any $\chi(\zeta')$ can be sent by these maps into some $\chi_0(\zeta')$ not depending on $\zeta_{13}, \cdots, \zeta_{1n}$. Let $H_{\chi_0}$ be the subspace of vectors with the characteristic value $\chi_0(\zeta')$. This subspace is invariant with respect to the operators $T(g')$, where $g'$ runs through the subgroup $G_n'' \cong G_{n-1}'$ of matrices from $G_n'$ in which $g_{1i} = 0$ for $i > 1$, $g_{22} = 1$, $g_{j2}$ 0 for $j > 2$. Accordingly by the induction hypothesis, there exists in $H_{\chi_0}$ a vector characteristic for all $T(\zeta)$, $\zeta \in Z \cap G_n''$. Obviously this vector will be characteristic for every operator $T(\zeta)$, $\zeta \in Z$. Theorem 1 is proved. For the other classes of Dickson-Chevalley groups the proof goes differently.

Proof of Theorem 2. We must show that *the ring $R_\chi$ of operators which permute with the operators $T_\chi(g)$, where $\chi$ is a character of general aspect, is commutative* (this being the equivalent formulation of Theorem 2). The ring $R_\chi$ is isomorphic to the ring of all functions $f(g)$ on the group which satisfy for any $\zeta_1, \zeta_2 \in Z$ the relation $f(\zeta_1 g \zeta_2) = \chi(\zeta_1) f(g) \chi(\zeta_2)$, with the convolution operation in the role of a product. Any matrix from $G_n$ may take on the form $g = \zeta_1 \delta s \zeta_2$, where $\zeta_1, \zeta_2 \in Z$, $\delta$ is a diagonal matrix (i.e. a matrix from a Cartan subgroup), and $s$ is the matrix of a substitution (i.e. an element of a Weyl group). Therefore the functions $f \in R_\chi$ need only be determined on the matrices $\delta s$. With no loss of generality we may assume that a character of general aspect $\chi$ is in the form $\chi(\zeta) = \chi_1(\zeta_{12} + \cdots + \zeta_{n-1,n})$, where $\chi_1$ is an additive character on $K$. Let $\mathfrak{M}$ be a set of matrices $\delta s$ on which the functions $f \in R\chi$ may take on values other than 0. It is easily seen that the matrices $\delta s \in \mathfrak{M}$ satisfy the condition $(\delta s)^\theta = \delta s$, where $\theta$ is an involutory antiautomorphism of $G$ preserving $Z$.** We take as a basis in $R_\chi$ the functions $f_{\delta s}$ (where $\delta s \in \mathfrak{M}$) given by the following conditions: $f_{\delta s}(g) = 1$ for $g = \delta s$, $f_{\delta s}(\delta_1 s_1) = 0$ for $\delta_1 s_1 \neq \delta s$. It is possible to show that the convolution $f(g) = f_{\delta_1 s_1} \circ f_{\delta_2 s_2}$ of two such functions can be expressed by the following formula:

$$f(\delta s) = \frac{N}{N_{s_1} N_{s_2}} \sum \chi(\zeta)\, \chi(\zeta_1)\, \chi(\zeta_2),$$

*In the case of classical complex Lie groups the integral form of Cartan's theory of highest weights is Gel'fand and Naĭmark's construction [1]. This construction is the special case of the construction given here for $\chi \equiv 1$.

**This antiautomorphism has the form $g^\theta = s_0 g' s_0$, where the prime denotes transposition and $s_0$ is a substitution matrix with units along the secondary diagonal.

here the sum is taken over all possible triples of matrices $\zeta$, $\zeta_1$, $\zeta_2$ from $Z$ bound by the relation $s = \zeta_1 \delta_1 s_1 \zeta \delta_2 s_2 \zeta_2$; $N$ is the order of the subgroup $Z$, $N_{s_i}$ is the order of the subgroup $s_i Z s_i^{-1} \cap Z$. ne commutativity of $R_\chi$, i.e. the fact that $f_{\delta_1 s_1} \circ f_{\delta_2 s_2} = f_{\delta_2 s_2} \circ f_{\delta_1 s_1}$ follows immediately from this rmula. For this it is sufficient to note that the condition on the matrices $\delta s \in \mathfrak{M}$ ensures that the quality $\delta s = \zeta_1 \delta_1 s_1 \zeta \delta_2 s_2 \zeta_2$ is equivalent to the equality $\delta s = \zeta_2^\theta \delta_2 s_2 \zeta^\theta \delta_1 s_1 \zeta_1^\theta$; and by the as- umption we made about the character $\chi$ we have $\chi(\zeta^\theta) = \chi(\zeta)$ for any $\zeta \in Z$. For other Dickson- hevalley groups the proof of Theorem 2 goes analogously.

3. We shall say that an irreducible representation of a Dickson-Chevalley group $G$ is *fundamental* we can find in it a characteristic vector of general aspect with respect to the subgroup $Z$; we say e representation is *degenerate* if there is in it no such vector. We may show that the number of ndamental representations is expressed in the form of a polynomial in the order $k$ of the underlying eld of degree $l$ ($l$ being the dimension of the Cartan subgroup); the number of degenerate representa- ons is expressed as a polynomial of lower degree. In this sense the fundamental representations omprise almost all of the irreducible representations of $G$.* To construct the fundamental represen- tions of $G$ we must decompose the representation $T_\chi(g)$ corresponding to the character of general spect into irreducible representations. Since the ring $R_\chi$ of those operators which permute with $\chi(g)$ is commutative, the task is reduced to that of finding those functions $\phi$ in $R_\chi$ such that $\circ\psi = c_\psi \phi$ for any $\psi \in R_\chi$. We may show that these functions are given by the functional relation

$$\varphi(\delta_1 s_1)\,\varphi(\delta_2 s_2) = \lambda \sum \chi(\zeta^{-1})\,\chi(\zeta_1^{-1})\,\chi(\zeta_2^{-1})\,\varphi(\delta s),$$

here the sum is taken over all matrices $\delta s \in \mathfrak{M}$ and $\zeta, \zeta_1, \zeta_2 \in Z$ which satisfy the relation $s = \zeta_1 \delta_1 s_1 \zeta \delta_2 s_2 \zeta_2$. The solutions of this equation, namely the functions $\phi(\delta s)$, we shall call the essel functions associated with the group $G$. For the group of unimodular matrices of second order ese functions are in the explicit form found in [2].

4. The irreducible representations of $G$ fall into a series. We relate two representations to one d the same series if to each $\chi(\zeta)$, $\zeta \in Z$ in these representations these corresponds the same mber of independent characteristic vectors (or in other words, if these representations are contained one and the same $T_\chi(g)$ and moreover with the same multiplicity). As an example we shall find e series of irreducible representations of the group $G_3$ of unimodular third order matrices (in the se when the number $k-1$ is not divisible by 3).

$G_3$ admits three series of fundamental representations "of general aspect": 1) $(k-2)(k-3)/6$ the representations have degree $(k+1)(k^2+k+1)$; they appear in every $T_\chi(g)$, moreover in $\chi(g)$ for $\chi \equiv 1$ they appear with multiplicity 6, while in the $T_\chi(g)$ for $\chi \not\equiv 1$ and $\chi_0(\chi_0$ is a aracter of general aspect) they appear with multiplicity 3; 2) $k(k-1)/2$ of the representations have gree $k^3-1$; they appear only in the $T_\chi(g)$ for $\chi \not\equiv 1$, in each with multiplicity 1; 3) $k(k+1)/3$ the representations have degree $(k-1)(k^2-1)$; they appear only in $T_{\chi_0}(g)$.

There are still two "singular" series of fundamental representations: 4) $k-2$ representations of gree $k(k^2+k+1)$ and 5) one representation of degree $k^3$. Finally there are three series of de- nerate representations: 6) $k-2$ representations of degree $k^2+k+1$; 7) one representation of de- ee $k^2+k$ and 8) one identity representation (cf. [7] where the characters of the irreducible repre- ntations of $G_3$ are computed).

* We note that the degree of a fundamental representation is always a polynomial in $k$ of degree $N$ ($N$ be- g the order of $Z$); the degree of any degenerate representation is expressed as a polynomial in $k$ of lower degree.

In the case of the group of unimodular matrices of order $n$ each series of fundamental representations "of general aspect" is given by the partition $n = n_1 + \cdots + n_r$ of $n$ into a sum of positive integers. It can be shown that the degree of these representations of these series is

$$\frac{(k-1)(k^2-1)\dots(k^n-1)}{(k^{n_1}-1)\dots(k^{n_r}-1)} \quad \text{(cf. [8])}.$$

## BIBLIOGRAPHY


[1] I.M. Gel'fand and M.A. Naĭmark, Trudy Mat. Inst. Steklov. 36 (1950).

[2] I.M. Gel'fand and M.I. Graev, Dokl. Akad. Nauk SSSR 146 (1962), 757 = Soviet Math. Dokl. 3 (1962), 1378.

[3] C. Chevalley, Sborn. per. Matematika 2 (1958), no. 1.

[4] L.E. Dickson, *Linear groups*, Dover, New York, 1958.

[5] I.M. Gel'fand and M. Neumark, Dokl. Akad. Nauk SSSR 54 (1946), 195.

[6] I.M. Gel' fand and M.A. Naımark, Mat. Sb. (N.S.) 21 (63) (1947), 405.

[7] R. Steinberg, Canad. J. Math. 3 (1951), 225.

[8] J.A. Green, Trans. Amer. Math. Soc. 80 (1955), 402.


Translated by:
J. N. Whitney

# 3.

## (with M. I. Graev)

# Representations of quaternion groups over locally compact and functional fields

Funkts. Anal. Prilozh. **2** (1) (1968) 20–35
[Funct. Anal. Appl. **2** (1968) 19–33]. MR **38**:4611. Zbl. **233**:20016

## INTRODUCTION

Irreducible unitary representations of groups of quaternions over some fields K are studied herein. Cases are examined when K is an arbitrary incoherent locally compact continuous field and when K is a quotient field of formal power series of one variable t with real coefficients.

Before giving the definition of these groups and formulating the fundamental results, let us recall the general definition of quaternion algebra.

A set of elements of the following kind:

$$x = x_0 + x_1 i + x_2 j + x_3 k, \tag{1}$$

where $\chi_\nu \in A$, $\nu = 0, 1, 2, 3$, is called [1] the algebra E of quaternions over the commutative ring A with unity. Addition of these elements and their multiplication with elements from A are carried out by components. Multiplication is in conformity with the following table:

$$i^2 = a,\ j^2 = b,\ k^2 = -ab,\ ij = -ji = k,\ jk = -kj = -bi,\ ki = -ik = -aj, \tag{2}$$

where a, b are fixed elements from A such that ab ≠ 0; the algebra E generally depends on the selection of these elements a, b.

Evidently, the quaternion algebra A is isomorphic to the algebra of all second-order matrices of the following kind:

$$\begin{pmatrix} \alpha & \beta \\ b\bar\beta & \bar\alpha \end{pmatrix}, \tag{3}$$

where $\alpha$, $\beta$ are elements of the commutative ring $A(\sqrt{a})$; $\bar\alpha$, $\bar\beta$ are elements from $A(\sqrt{a})$ which are conjugate to $\alpha$, $\beta$. The isomorphism is established as follows:

$$x_0 + x_1 i + x_2 j + x_3 k \mapsto \begin{pmatrix} x_0 + x_1\sqrt{a} & x_2 + x_3\sqrt{a} \\ b(x_2 - x_3\sqrt{a}) & x_0 - x_1\sqrt{a} \end{pmatrix}. \tag{4}$$

The quaternion $\bar{x} = x_0 - x_1 i - x_2 j - x_3 k$ is called conjugate to the quaternion (1); the product

$$N(x) = x\bar{x} = x_0^2 - ax_1^2 - bx_2^2 + abx_3^2 \tag{5}$$

is called the norm of the quaternion x. Let us note that N(x) equals the determinant of the matrix corresponding to x under the mapping (4). If A is a field, then for E to be an algebra with division it is necessary and sufficient that the following condition be satisfied

$$N(x) \neq 0, \quad \text{when } x \neq 0.$$

---

Lomonosov Moscow State University and Institute of Applied Mathematics of the Academy of Sciences of the USSR. Translated from Funktsional'nyi Analiz i Ego Prilozheniya, Vol. 2, No. 1, pp. 20–35, January–March, 1968. Original article submitted October 16, 1967.

Principally the case when A = K is a continuous incoherent locally compact field is studied herein. It is hence assumed that the characteristic of the finite field $F_q = O/P$, where O is the subring of integer elements in K and P is the maximum ideal in O, does not equal 2. It is known that in this case there exists, to the accuracy of isomorphism, just one field of quaternions over K.† We obtain this field by setting $a = \varepsilon$, $b = \rho$ in the multiplication table (2), where $\varepsilon$ is a fixed element from $O \setminus P$, which is not quadratic in K, and $\rho$ is a fixed generating element of the ideal P.

Let us connect the following groups with the field of quaternions E. Let E* be a multiplicative group of all quaternions $x \neq 0$, G a group of all quaternions x for which $\| N(x) \| = 1$ ($\| a \|$ denotes the norm of the element $a \in K$), SG the group of all quaternions x with norm $N(x) = 1$; furthermore, let PE*, PG, PSG denote factor groups of the corresponding groups by their centers. With the exception of E* all these groups are compact and have similar structure. Namely, as is easy to see, the following exact sequences of homomorphisms hold:

$$0 \to PG \to PE^* \to Z_2 \to 0, \quad 0 \to Z_2 \to SG \to PSG \to 0, \quad 0 \to PSG \to PG \to Z_2 \to 0.$$

One of these groups, the group PG, is considered in Section 1. A description of all its irreducible representations is obtained, and the characters of these representations are calculated. Let us note that an analogous description of the irreducible representations may be made for the other listed groups also.

A brief exposition of the results of this section has been given in [2].

The case when K is the quotient field of the ring $R\{t\}$ of formal power series in t with coefficients from R is considered in Section 2. In this case there exists several nonisomorphic algebras E of quaternions over K, including three nonisomorphic algebras with division. These algebras with division are obtained if we set $a = b = -1$; $a = -1$, $b = t$; $a = -1$, $b = t$; respectively, in the multiplication tables (2).

We shall discuss the first case when $a = b = -1$ (the construction for the other cases is analogous). In this case the group SE* of quaternions with norm $N(x) = 1$ is isomorphic to the group SU $(2, C\{t\})$ of matrices

$$\begin{pmatrix} \alpha & \beta \\ -\bar{\beta} & \bar{\alpha} \end{pmatrix}, \quad \alpha\bar{\alpha} + \beta\bar{\beta} = 1,$$

whose elements belong to the ring $C\{t\}$ of formal power series over C ($\bar{\alpha}$ denotes the series obtained from $\alpha$ by replacing all the coefficients by the complex conjugates). Representations of this group are studied in Section 2. We limit ourselves here just to the construction itself of the representations. The proof of their irreducibility, as well as other questions concerning these representations (the proof of the completeness, formulas for the characters, the Plansherel theorem), will be the topic of a separate article.

A new concept of derivative for representations is introduced in Section 3. Namely, a group G of matrices

$$u(t) = \begin{pmatrix} \alpha(t) & \beta(t) \\ -\overline{\beta(t)} & \overline{\alpha(t)} \end{pmatrix}, \quad |\alpha(t)|^2 + |\beta(t)|^2 = 1,$$

whose elements are complex valued infinitely differentiable functions, is considered. The irreducible unitary representation $T(u(t))$ of the group G is called the nth order derivative at the point $t_0$ if it depends only on $u(t_0), u'(t_0), \ldots, u^{(n)}(t_0)$, where the dependence on $u^{(n)}(t_0)$, is essential. It is easy to see that the description of such representations reduces to the description of representations of the group SU(2, $C\{t\}$, which are considered in Section 2. It will be shown in Section 3 that the first-order derivative may be obtained by a passage to the limit (exactly as the derivative of a function obtained from the function itself. Namely, the irreducible of group G is taken which depends only on $u(0)$ and $u(t_0)$: $T(u(t)) = T_l\,(u(0)) \otimes T_l\,(u(t_0))$, where $T_l$ is the irreducible (finite

†Moreover, to isomorphism accuracy there exists exactly one quaternion algebra over K which is not an algebra with division, namely, the algebra of all second-order matrices over K. Let us note that representations of the groups GL(2,K) and SL(2,K) of second-order matrices over K were described completely in [4].

dimensional) representation of a unitary group of weight $l$. It turns out that if $l \to \infty$ and simultaneously $t_0 \to 0$, where $\lim t_0 l = \alpha \neq 0$, then the representation $T_l(u(0)) \otimes T_l(u, (t_0))$ goes over into the first-order derivative in the limit. The exact meaning of this assertion is elucidated in Section 3.

In conclusion, the authors are grateful to Professor Harish Chandra, as well as S. I. Gel'fand, for discussions which stimulated their interest in these problems.

## 1. Representations of Groups of Quaternions over an Incoherent Locally Compact Field K

### 1. Classification of Representations.

Let E be the field of quaternions

$$x = x_0 + x_1 i + x_2 j + x_3 k \tag{1}$$

over an incoherent locally compact continuous field K. Multiplication of the quaternions is defined by the following table:

$$i^2 = \varepsilon,\ j^2 = p,\ k^2 = -\varepsilon p,\ ij = -ji = k,\ jk = -kj = -pi,\ ki = -ik = -\varepsilon j. \tag{2}$$

Here $\varepsilon$ is a fixed element of the field K with norm 1 which is not a square in K; p is a fixed generating element of the maximum ideal P in the ring O of integer elements from K. It is assumed throughout that the characteristic of the field O/P is not 2. Let the real number $|x| = \|N(x)\|^{1/2}$ be called the modulus of the quaternion x; here $N(x) = \overline{x}x$, and $\|u\|$ is the norm of the element $u \in K$. Let G be the group of all quaternions with modulus $|x| = 1$ (evidently the coordinates $x_\nu$ of these quaternions are integers); PG the factor group of the group G with its center Z. A description of irreducible unitary representations of the group PG is given in this section. Evidently, the description of irreducible representations of the group PG is equivalent to a description of irreducible representations $T(x)$ of the group G, such that $T(x) \equiv 1$ at the center Z of the group G.

Let us introduce the concept of rank of an irreducible representation. Let q be the number of elements of a finite field of residues O/P (we assume q to be odd). Then $|p| = q^{-1}$, $|j| = |k| = q^{-1/2}$, and hence $|x|$ can only take on the values 0 and $q^{n/2}$, $n = 0, \pm 1, \ldots$.

Let $G_\nu$, $\nu = 0, 1/2, 1, 3/2, \ldots$, be a subgroup of quaternions $x \in G$ of the form $x = a + x'$, where $a \in K$, $|x'| \le q^{-\nu}$. Evidently, $G_\nu$ are normal divisors in G and they generate a decreasing chain

$$G = G_0 \supset G_{1/2} \supset G_1 \supset G_{3/2} \supset \ldots,$$

whose intersection is the center Z of the group G. It hence follows directly: If $T(x)$ is an irreducible representation of the group, such that $T(x) \equiv 1$ on Z, then $T(x) \equiv 1$ on the subgroups $G_\nu$ for sufficiently large $\nu$.

We agree to call the least $\nu$ for which $T(x) \equiv 1$ on $G_\nu$ the rank of the representation $T(x)$. Therefore, irreducible representations of rank $\nu$ may be identified with irreducible representations of the finite factor group $G/G_\nu$, which are not identically unity in $G_{\nu-1/2}/G_\nu$.

Evidently, the unit representation is the single representation of rank $\nu = 0$. Furthermore, it is easy to see directly that $G/G_{1/2}$ is a commutative group of order q + 1. Hence, representations of rank $1/2$ are one-dimensional, and their number is q. Each of the representations of rank $1/2$ is given by the following mapping:

$$x_0 + x_1 i + x_2 j + x_3 k \mapsto \pi(x_0 + \sqrt{\varepsilon} x_1),$$

where $\pi$ is a multiplicative character in $K(\sqrt{\varepsilon})$ which satisfies the following conditions: 1) $\pi(x) \not\equiv 1$, 2) $\pi(x) \equiv 1$ in K, 3) $\pi(1 + x) \equiv 1$ for $|x| < 1$. (Evidently there are exactly q such characters.) A description of higher rank representations is given in the next paragraphs.

### 2. Description of Irreducible Representations of Rank $\nu$, where $\nu$ is an Integer.

a) Representations of the First Series. Let $U_p$ be a commutative subalgebra of quaternions of the form $\alpha = a + bj$, $a, b \in K$. Evidently, we represent any quaternion x uniquely as $x = \alpha + \beta i$, where $\alpha$, $\beta \in U_p$. Let us introduce the subgroup $A'_\nu \subset G$ of quaternions $x = \alpha + \beta i$ for which $|\beta| \le q^{-\nu/2}$. Let us give one-dimensional representations of the group $A'_\nu$.

Let $\pi(\alpha)$ be an arbitrary character in the multiplicative group $U_\rho^*$ of elements $\alpha \in U_p$ with modulus $|\alpha| = 1$ satisfying the following conditions:

1) $\pi(\alpha) \equiv 1$ for $\alpha \in K$;

2) the rank is $\pi = \nu$; this means that $\pi(1+\alpha) \equiv 1$ for $|\alpha| \le q^{-\nu}$, but $\pi(1+\alpha) \not\equiv 1$ for $|\alpha| = q^{-\nu+1/2}$.

It is verified directly that the mapping

$$\pi:\ \alpha + \beta i \mapsto \pi(\alpha) \tag{3}$$

is a homomorphism of the subgroup $A'_\nu$, i.e., yields its one-dimensional representation. Let $T'_\pi(x)$ be a representation of the group G induced by the representation (3) of the subgroup $A'_\nu$. This means that $T'_\pi(x)$ acts in the space of all functions $f(x)$ in G which satisfy the condition $f(yx) = \pi(y)f(x)$ for any $y \in A'_\nu$ ($\pi(y)$ is a function in $A'_\nu$ defined by the mapping (3)); the operators $T'_\pi(x)$ are given by the following formula:

$$(T'_\pi(x_0)f)(x) = f(xx_0).$$

Let us show that the rank of $T'_\pi$ is $\nu$. In fact, since $G_\nu \subset A'_\nu$ and $\pi(x_0) \equiv 1$ in $G_\nu$, we then have for any $x_0 \in G_\nu$

$$(T'_\pi(x_0)f)(x) = f(xx_0) = f(xx_0x^{-1}x) = \pi(xx_0x^{-1})f(x) = f(x).$$

On the other hand, it is not difficult to see that $T'_\pi(x_0) \not\equiv 1$ in $G_{\nu-1/2}$. We call the representations $T'_\pi$ the first series of representations of rank $\nu$.

b) Representations of the Second Series. Let us introduce the subalgebra $U_\rho$ of quaternions $U_{\varepsilon\rho}$ in place of $\alpha = a + bj'$, where $j' = x_2j + x_3k$ is a fixed quaternion satisfying the relation $j'^2 = \varepsilon\rho$ (i.e., $x_3^2 - \varepsilon x_3^2 = \varepsilon$). Evidently we represent any quaternion x uniquely as $x = \alpha + \beta i$, where $\alpha, \beta \in U_{\varepsilon\rho}$. Let us introduce the subgroup $A''_\nu \subset G$ of quaternions $x = \alpha + \beta i$, $\alpha, \beta \in U_{\varepsilon\rho}$ for which $|\beta| \le q^{-\nu/2}$. Replacing the algebra $U_p$ by $U_{\varepsilon p}$ and the subgroup $A'_\nu$ by $A''_\nu$ in the previous discussions, we obtain the second series of representations $T''_\pi$ of rank $\nu$.

THEOREM 1. 1) The representations $T'_\pi$, $T''_\pi$ are irreducible; 2) two representations $T_{\pi_1}$, $T_{\pi_2}$ are equivalent if and only if they belong to the very same series, where either $\pi_1 = \pi_2$, or $\pi_1 = \pi_2^{-1}$; 3) dim $T'$ = dim $T''_\pi = (q+1)q^{\nu-1}$; 4) the number of pairwise nonequivalent representations in each series equals $\frac{1}{2}(q-1)q^{\nu-1}$.

Proof of Assertions 1) and 2). Let $T_{\pi_1}, T_{\pi_2}$ be representations of the very same series, the first, say, and let $H_{\pi_1}$, $H_{\pi_2}$ be the spaces in which they act. Let us find all linear mappings $K: H_{\pi_1} \to H_{\pi_2}$ such that $KT_{\pi_1}(x) = T_{\pi_2}(x)K$ for any $x \in G$.

Each such mapping is given by the formula

$$(Kf)(x) = \sum_{x' \in G/A'_\nu} K(x, x')f(x'),$$

where $K(x, x')$ satisfies the following conditions:

$$K(yx, y'x') = \pi_2(y)\pi_1^{-1}(y')K(x, x')$$

for any $y, y' \in A'_\nu$;

$$K(xx_0, x'x_0) = K(x, x')$$

for any $x_0 \in G$. From these conditions it follows that $K(x, x') = F(xx'^{-1})$, where the function F satisfies the relation

$$F(y_2xy_1) = \pi_2(y_2)F(x)\pi_1(y_1) \tag{4}$$

for any $y_1, y_2, \in A'_\nu$. Thus, there remains to find all functions F satisfying condition (4).

LEMMA. If $x = \alpha + \beta i \notin A'_\nu \cup A'_\nu i$, then $F(x) = 0$.

Indeed, let $u \in U_\rho$ run through the set of elements for which $|u| \le q^{-\nu/2}$. Then the quaternions $y = 1 + ui$ and $x^{-1}yx$ belong to $A'_\nu$, where direct calculation yields

$$x^{-1}yx = \left(1 + \varepsilon \frac{\bar{\alpha}\bar{\beta}u - \alpha\beta\bar{u}}{\alpha\bar{\alpha} - \varepsilon\beta\bar{\beta}}\right) + \frac{\bar{\alpha}^2 u - \varepsilon\beta^2\bar{u}}{\alpha\bar{\alpha} - \varepsilon\beta\bar{\beta}} i.$$

From (4) it follows that $\pi_2(y)F(x) = F(x)\pi_1(x^{-1}yx)$, i.e.,

$$F(x) = F(x)\pi_1\left(1 + \varepsilon \frac{\bar{\alpha}\bar{\beta}u - \alpha\beta\bar{u}}{\alpha\bar{\alpha} - \varepsilon\beta\bar{\beta}}\right) \tag{5}$$

for any $u \in U_p$ with modulus $|u| \le q^{-\nu/2}$. Since $\alpha + \beta i \notin A'_\nu \cup A_\nu i$ by the assumption in the lemma, then $|\alpha\beta| > q^{-\nu/2}$. Hence, and from the assumption on the character $\pi_1$ also, it follows easily that $\pi_1\left(1 + \varepsilon \frac{\bar{\alpha}\bar{\beta}u - \alpha\beta\bar{u}}{\alpha\bar{\alpha} - \varepsilon\beta\bar{\beta}}\right) \ne 1$ when $u \in U_p$ runs through the set of elements with modulus $|u| \le q^{-\nu/2}$. Therefore, we have $F(x) = 0$ for any $x \notin A'_\nu \cup A'_\nu i$ by virtue of (5).

There remains to determine $F(x)$ on the sets $A'_\nu$ and $A'_\nu i$. By virtue of the relation (4), the function F is defined uniquely by its values $F(1)$ and $F(i)$, where the relations

$$F(1) = \pi_1(\alpha)\pi_2^{-1}(\alpha)F(1), \quad F(i) = \pi_1(\alpha)\pi_2(\alpha)F(i)$$

are satisfied for any $\alpha \in U^*_p$.

It follows from these relations that for $\pi_1 = \pi_2$ and for $\pi_1 = \pi_2^{-1}$ there exists, to the accuracy of a factor, one nonzero solution of the functional equation (4); no nonzero solution exists in the other cases. Assertions 1) and 2) are thereby proved for representations belonging to the same series. The nonequivalence of representations belonging to different series is established by analogous reasoning; let us note that it also follows from the formulas for the characters (see Section 6).

Proof of Assertion 3). It is evident that $T'_\pi = G : A'_\nu$. Let us note that $A'_1 = G_{1/2}$, and, hence, $G : A'_1 = q + 1$. Furthermore, it is easy to see that $A'_\nu : A'_{\nu+1} = q$. It hence follows that $G : A'_\nu = (q+1)q^{\nu-1}$.

Proof of Assertion 4). It is easy to see that the number of different characters $\pi(\alpha)$ of rank not greater than $\nu$ in $U^*_\rho$ is $(q-1)q^{2\nu-1}$, and the number of different characters of rank not greater than $\nu$, equal to the unit identically on K, is $(q-1)q^{2\nu-1}/(q-1)q^{\nu-1} = q^\nu$. It hence follows that the number of characters $\pi(\alpha)$ on $U_p$ which satisfy conditions 1), 2) before is $q^\nu - q^{\nu-1} = (q-1)q^{\nu-1}$. Since equivalent representa- the pairs $\pi$ and $\pi^{-1}$, and only to such pairs, then the number of different representations of the first series is $\frac{1}{2}(q-1)q^{\nu-1}$. The proof is literally the same for the second-series representations.

## 3. Description of Irreducible Representations of Rank $\nu = 2m + \frac{1}{2}$, $m = 0, 1, 2, \ldots$

Let $U_\varepsilon$ be a commutative subalgebra of quaternions of the form $\alpha = a + bi$, $a, b \in K$. It is evident that any quaternion x is uniquely representable as $x = \alpha + \beta j$, where $\alpha, \beta \in U_\varepsilon$.

Let us introduce the subgroup $A_\nu \subset G$ of quaternions $x = \alpha + \beta j$, $\alpha, \beta \in U_\varepsilon$ in which $|\beta| \le q^{-m}$. Let us give one-dimensional representations of the group $A_\nu$.

Let $\pi(\alpha)$ be an arbitrary character on the multiplicative group $U^*_\varepsilon$ of elements $\alpha \in U_\varepsilon$ with modulus $|\alpha| = 1$, which satisfies conditions 1), 2) of Section 2. It is easy to see that the mapping

$$\pi : x = \alpha + \beta j \mapsto \pi(\alpha) \tag{6}$$

yields a one-dimensional representation of the group $A_\nu$.

Let $T_\pi(x)$ be a representation of the group G induced by the representation (6) of the subgroup $A_\nu$. From the definition it follows that the rank of $T_\pi$ is $\nu$.

THEOREM 2. (1). The representations $T_\pi$ are irreducible and pairwise not equivalent; 2) $\dim T_\pi = q^{\nu-1/2}$; 3) the number of representations $T_\pi$ is $(q^2-1)q^{\nu-3/2}$.

The proof is carried out analogously to the proof of Theorem 1.

## 4. Description of Irreducible Representations of Rank $\nu = 2m + 3/2$, m = 0, 1, 2, ...

Let us introduce the subgroup $A_\nu \subset G_{1/2}$ of quaternions $x = \alpha + \beta j$, $\alpha, \beta \in U_\varepsilon$ for which $|\beta| \leq q^{-m}$, $|\alpha - \bar{\alpha}| < 1$, $|\beta - \bar{\beta}| < q^{-m}$. Let $\pi(\alpha)$ be an arbitrary character on $\mathring{U}_\varepsilon \cap A_\nu$, which satisfies conditions 1, 2) of Section 2. It is easy to see that the mapping (6) is a representation of the subgroup $A_\nu$.

Let us consider the representation $T_\pi(x)$ of the subgroup $G_{1/2}$ induced by this representation. By definition, $T_\pi(x)$ acts in the space $H_\pi$ of all functions $f(x)$ in $G_{1/2}$, which satisfy the condition: $f(yx) = \pi(y) f(x)$ for any $y \in A_\nu$; the operator $T_\nu(x_0)$ is given by the following formula: $(T_\pi(x_0) f)(x) = f(xx_0)$.

By the same reasoning as in the proof of Theorem 1, it can be seen that the representations $T_\pi(x)$ of the group $G_{1/2}$ are irreducible and not pairwise equivalent.

Now, let us continue each of the representations $T_\pi(x)$ of the subgroup $G_{1/2}$ to a representation of the whole group G. To do this we use the following simple remark. As has already been noted, the factor group $G/G_{1/2}$ is cyclic of order q + 1. Let us take a generating element therein, and let z be the prototype of this element in the group G. It is easy to see that z can be chosen so that $z \in U_\varepsilon^*$ and $z^{q+1} \in K$, i.e., $z^{q+1}$ belongs to the center of the group $G_{1/2}$. From the definition of z it follows that any element $x \in G$ may be represented, and uniquely besides, as $x = z^s x'$, s = 0, 1, ..., q; $x' \in G_{1/2}$.

Let us compare the linear operator $B_z$ in $H_\pi$, defined by the formula

$$(B_z f)(x') = \sum_{y \in A_\nu / G_\nu} \pi^{-1}(y) f(z^{-1} y x' z) \tag{7}$$

to the element z.

It follows directly from (7) that

1) the operator $B_z$ maps $H_\pi$ into $H_\pi$;

2) for any $x' \in G_{1/2}$ the equality

$$B_z T_\pi(x') = T_\pi(z x' z^{-1}) B_z \tag{8}$$

is valid.

Let us prove that $B_z \neq 0$; then from the irreducibility of $T_\pi$ there will follow that $B_z$ is a nondegenerate operator.

Let us take the following function $f(x') \in H_\pi$ : $f(x') = \pi(x')$, when $x' \in A_\nu$; $f(x') = 0$, when $x' \notin A_\nu$. We then have

$$(B_z f)(1) = \sum_{y \in A_\nu / G_\nu} \pi^{-1}(y) f(z^{-1} y z) = \sum \pi^{-1}(y) \pi(z^{-1} y z), \tag{9}$$

where the summation is taken only over those $y \in A_\nu / G_\nu$ for which $z^{-1} y z \in A_\nu$ (the set of such y is known not to be empty). Since $z \in U^*_\varepsilon$, then we have $\pi(y^{-1} y z) = \pi(y)$ for these elements y. Therefore, $(B_z f)(1) \neq 0$ by virtue of (9), and, hence, $B_z \neq 0$.

It follows from (8) that $B_z^{q+1} T_\pi(x') = T_\pi(z^{q+1} x' z^{-q-1}) B_z^{q+1} = T_\pi(x') B_z^{q+1}$. Therefore, since $T_\pi(x')$ is an irreducible representation, the operator $B_z^{q+1}$ is a multiple of the unit operator

$$B_z^{q+1} = \lambda E, \tag{10}$$

where $\lambda \neq 0$. We shall not evaluate the $\lambda$.

Let $\mu_1, \ldots, \mu_{q+1}$ be a set of roots of the equation $\lambda t^{q+1} = 1$, where $\lambda$ is the coefficient in (10). Let us compare the following representation $T_{\pi,k}(x)$ of the group G to each $k = 1, \ldots, q + 1$, where $T_{\pi,k}$ acts in the space $H_\pi$ of representations $T_\pi$ of the subgroup $G_{1/2}$. The operator $T_{\pi,k}$ corresponding to the element $x = z^s x'$, where $s = 0, 1, \ldots, q, x' \in G_{1/2}$ is defined by the formula

$$T_{\pi,k}(x) = (\mu_k B_z)^s T_\pi(x'). \tag{11}$$

An elementary verification shows that the operators $T_{\pi,k}(x)$ actually form a representation, and that the rank of this representation is $\nu$.

THEOREM 3.1) The representations $T_{\pi,k}$ are irreducible and pairwise not equivalent; 2) dim $T_{\pi,k} = q^{\nu-1/2}$; 3) the number of representations $T_{\pi,k}$ equals $(q^2-1)q^{\nu-3/2}$.

Proof. The representations $T_{\pi,k}$ are irreducible since their constraints $T_\pi$ on the subgroup $G_{1/2}$ are irreducible. Furthermore, let $T_{\pi_1,k_1}$ and $T_{\pi_2,k_2}$ be equivalent. Then their constraints $T_{\pi_1}$ and $T_{\pi_2}$ are equivalent on the subgroup $G_{1/2}$, and, consequently, $\pi_1 = \pi_2 = \pi$. Since $T_{\pi,k_1}(z) = \mu_{k_1} B_z$, $T_{\pi,k_2}(z) = \mu_{k_2} B_z$, it then follows from the equivalence of the representations that $\mu_{k_1} = \mu_{k_2}$, i.e., $k_1 = k_2$.

The proof of assertions 2) and 3) is carried out by computing the indices of the subgroups, exactly as has been done in the proof of Theorem 1.

## 5. Completeness Theorem.

THEOREM 4. All irreducible representations of the group G which equal one at its center are exhausted by the representations described above.

Proof. Let us note that the representations of rank not greater than $\nu$ form a complete system of irreducible representations of the factor group $G/G_\nu$. Therefore, by the Burnside theorem, the sum of the squares of the dimensionalities of these representations equals $G : G_\nu$. There hence results that the sum of the squares of the dimensionalities of all representations or rank $\nu$ equals the difference $G : G_\nu - G : G_{\nu-1/2}$. Let us evaluate this difference. We have $G : G_{1/2} = q + 1$; $G_{\nu-1/2} : G_\nu = q^2$, when $\nu$ is an integer; $G_{\nu-1/2} : G_\nu = G_\nu = q$, when $\nu$ is a half-integer. We therefore obtain

$$G:G_\nu - G:G_{\nu-1/2} = \begin{cases} (q+1)^2(q-1)\,q^{2(\nu-1)}, & \text{when } \nu \text{ is an integer;} \\ (q^2-1)\,q^{2\nu-3/2}, & \text{when } \nu \text{ is a half integer.} \end{cases}$$

To prove the theorem it is sufficient to show that the sum of the squares of the dimensionalities of the representations of rank $\nu$ we have found will equal $G : G_\nu - G : G_{\nu-1/2}$. But this is easily verified by a direct calculation since the number of representations of rank $\nu$ and their dimensionalities were already evaluated (see Theorems 1, 2, 3).

## 6. Characters of Irreducible Representations.

Without proof, let us note that in each class of conjugate elements of the group G there is an element belonging to one of the subgroups $U_\varepsilon^*$, $U_p^*$, $U_{\varepsilon p}^*$. Therefore, it is sufficient to indicate the values of the characters on only these subgroups.

THEOREM 5. Let T be an irreducible representation of rank $\nu$; let $\alpha$ run through the set $U_\varepsilon^* \cup U_p^* \cup U_{\varepsilon p}^*$. Then tr $T(\alpha)$ = dim T for $|\alpha - \bar{\alpha}| \le q^{-\nu}$; tr $T(\alpha)$ for $|\alpha - \bar{\alpha}| > q^{-\nu}$ is compact only in one of the subgroups $U_\varepsilon^*$, $U_p^*$, $U_{\varepsilon p}^*$, namely, for integer $\nu$ for the first series representations tr $T(\alpha)$ is compact on $U_p^*$, for the second series representations, on $U_{\varepsilon p}^*$, for half-integer $\nu$ the tr $T(\alpha)$ is compact on $U_\varepsilon^*$.

Explicit expressions for tr $T(\alpha)$ on the appropriate subgroups are presented below for $|\alpha - \bar{\alpha}| > q^{-\nu}$.

a) $\nu$ an integer, $T_\pi$ a first series representation, $\alpha \in U_p$.

$$\operatorname{tr} T_\pi(\alpha) = \operatorname{sign}_p(-1)^\nu \cdot c_\pi |\alpha - \bar{\alpha}|^{-1} \left[\operatorname{sign}_p\left(\frac{\alpha^2 - \bar{\alpha}^2}{j}\right)\pi(\alpha) + \operatorname{sign}_p\left(\frac{\bar{\alpha}^2 - \alpha^2}{j}\right)\pi(\bar{\alpha})\right] \text{ npu } |\alpha - \bar{\alpha}| > q^{-\nu+1/2}.$$

Here $\mathrm{sign}_\tau x = 1$ when $x \in K$ is representable as $x = a^2 - \tau b^2$, $a, b \in K$, $\mathrm{sign}_\tau x = 1$, otherwise, $c_\pi = q^{-1/2} \sum \pi(1 - 2\varepsilon \mathfrak{p}^{\nu-1} x^2 j)$, where the summation is over $x \in O/P$; it can be shown that $c_\pi = \pm 1$ if $-1$ is a square in K, and $c_\pi = \pm\sqrt{-1}$ otherwise.

$$\operatorname{tr} T_\pi(\alpha) = (q-1)^{-1} q^{\nu-1} \sum \pi\left(a + \mathfrak{p}^{\nu-1} b \frac{x^2 + \varepsilon y^2}{x^2 - \varepsilon y^2} j\right)$$

for $|\alpha - \bar{\alpha}| = q^{-\nu+1/2}$, $\alpha = a + \mathfrak{p}^{\nu-1} bj$ ; summation is over $x, y \in O/P$, $(x, y) \neq 0$.

b) $\nu$ an integer; $T_\pi$ a second series representation. Formulas for the characters are obtained from the preceding by replacing $\mathfrak{p}$ by $\varepsilon_1 \mathfrak{p}$ and $j$ by $j'$.

c) $\nu = 2m + 1/2$, $\alpha \in U_\varepsilon^\bullet$ $(|\alpha - \bar{\alpha}| > q^{-\nu})$.

$$\operatorname{tr} T_\pi(\alpha) = |\alpha - \bar{\alpha}|^{-1} \operatorname{sign}_\varepsilon\left(\frac{\alpha - \bar{\alpha}}{i}\right) \pi(\alpha).$$

d) $\nu = 2m + 3/2$, $\alpha \in U_\varepsilon^\bullet$ $(|\alpha - \bar{\alpha}| > q^{-\nu})$.

$$\operatorname{tr} T_{\pi,k} = -|\alpha - \bar{\alpha}|^{-1} \operatorname{sign}_\varepsilon\left(\frac{\alpha - \bar{\alpha}}{i}\right) \pi_k(\alpha).$$

Here $\pi_k(\alpha)$ is a character on $U_\varepsilon^*$ defined by the following conditions:

$$\pi_k(\alpha) = \pi(\alpha) \quad \text{on } U_\varepsilon^\bullet \cap A_\nu; \quad \pi_k(z) = \mu_k \mu^{-1},$$

where $\mu$ is some fixed root of the equation $\lambda t^{q+1} = 1$, which is independent of k (the notation is from Section 4).

Proof. The derivation of the formulas for the characters is based on the Frobenius formula for the characters of induced representations of finite groups. This derivation is simple but technically rather awkward. Hence, we present it just for one case, the representation of rank $\nu = 2m + 1/2$. The calculations for other representations are analogous.

Thus, let $T_\pi$ be a representation of rank $\nu = 2m + 1/2$, $m = 0, 1, 2, \ldots$. According to the Frobenius formula for the characters of induced representations of finite groups, we have

$$\operatorname{tr} T_\pi(\alpha) = \frac{1}{h} \sum \pi(y^{-1} \alpha y), \tag{12}$$

where $h = A_\nu : G_\nu = (q+1) q^{2\nu - 1}$; the summation is over the set of elements $y \in G/G_\nu$ for which $y^{-1} \alpha y \in A_\nu / G_\nu$.

If $|\alpha - \bar{\alpha}| \le q^{-\nu}$, then $\alpha \in G_\nu$, and it follows directly from (12) that $\operatorname{tr} T_\pi(\alpha) = \dim T_\pi$.

Now, let $\alpha \in U_p^*$, $|\alpha - \bar{\alpha}| > q^{-\nu}$, i.e., $\alpha = 1 + \mathfrak{p}^k aj$, where $k < 2m$, $a \in K$, $|a| = 1$. Let us show that $\operatorname{tr} T_\pi(\alpha) = 0$.

Setting $y = u + vj$, $u\, v \in U_\varepsilon$. we obtain

$$y^{-1} \alpha y = (u + vj)^{-1} (1 + \mathfrak{p}^k aj)(u + vj) = 1 + \mathfrak{p}^{k+1} \frac{\bar{u}\bar{v} - uv}{u\bar{u} - \mathfrak{p} v\bar{v}} a + \mathfrak{p}^k \frac{\bar{u}^2 - \mathfrak{p} v^2}{u\bar{u} - \mathfrak{p} v\bar{v}} aj.$$

Since $\left| \frac{\bar{u}^2 - \mathfrak{p} v^2}{u\bar{u} - \mathfrak{p} v\bar{v}} a \right| = 1$, then $y^{-1} \alpha y \in A_\nu$ if and only if $k \ge m$, i.e., $\alpha \in A_\nu$. Therefore, if $k < m$, then $\operatorname{tr} T_\pi(\alpha) = 0$; if $m < k < 2m$, then

$$\operatorname{tr} T_\pi(\alpha) = \frac{1}{h}\sum \pi\left(1 + \mathfrak{p}^{k+1}\frac{\bar{u}\bar{v} - uv}{u\bar{u} - \mathfrak{p}v\bar{v}}a\right),$$

where the summation is all over $u + vj \in G/G_\nu$. It is easy to see that this sum is zero.*

It has thus been proved that $\operatorname{tr} T_\pi(\alpha) = 0$ when $\alpha \in U^*_\rho$, $|\alpha - \bar{\alpha}| > q^{-\nu}$. Analogously we see that $\operatorname{tr} T_\pi(\alpha) = 0$ when $\alpha \in U^*_{\varepsilon p}$, $|\alpha - \bar{\alpha}| > q^{-\nu}$.

Finally, let us evaluate $\operatorname{tr} T_\pi(\alpha)$, when $\alpha \in U^*_\varepsilon$, $|\alpha - \bar{\alpha}| > q^{-\nu}$. Setting $y = u + vj$, $u, v \in U_\varepsilon$, we have

$$y^{-1}\alpha y = \alpha + \mathfrak{p}\frac{v\bar{v}(\alpha - \bar{\alpha})}{u\bar{u} - \mathfrak{p}v\bar{v}} + \frac{\bar{u}v(\alpha - \bar{\alpha})}{u\bar{u} - \mathfrak{p}v\bar{v}}j.$$

Since $|u| = 1$, then $y^{-1}\alpha y \in A_\nu$ if and only if $|v(\alpha - \bar{\alpha})| \leq q^{-m}$. Therefore,

$$\operatorname{tr} T_\pi(\alpha) = (q+1)^{-1}q^{-(2\nu-1)}\sum \pi\left(\alpha + \mathfrak{p}\frac{v\bar{v}(\alpha - \bar{\alpha})}{u\bar{u} - \mathfrak{p}v\bar{v}}\right);$$

the summation is over all $u + vj \in G/G_\nu$ satisfying the condition $|v(\alpha - \alpha)| \leq q^{-m}$. Inserting the new summation variable $t = vu^{-1}(1 - \rho(vu^{-1})(\overline{vu^{-1}}))^{-1/2}$ instead of v, we obtain (taking into account that u runs through a set of $(q+1)q^{\nu-1/2}$ elements).

$$\operatorname{tr} T_\pi(\alpha) = q^{-(\nu-1/2)}\sum \pi(\alpha + \mathfrak{p}(\alpha - \bar{\alpha})t\bar{t}) = \pi(\alpha)q^{-(\nu-1/2)}\sum \pi\left(1 + \mathfrak{p}\frac{\alpha - \bar{\alpha}}{\alpha}t\bar{t}\right);$$

the summation is over $t \in U_\varepsilon/\mathfrak{p}^{2m}U_\varepsilon$ for which $|t| \leq 1$, $|t(\alpha - \bar{\alpha})| \leq q^{-m}$.

Let us show that the sum $I = \sum \pi\left(1 + \mathfrak{p}\frac{\alpha - \bar{\alpha}}{\alpha}t\bar{t}\right)$ depends only on $|\alpha - \bar{\alpha}|$. In fact, let $|\alpha - \bar{\alpha}| = q^{-k}$, $\alpha = a + \mathfrak{p}^k bi$, where $|b| = 1$ (and $|a| = 1$ for $k > 0$). After suitable changes of the summation variable, we obtain

$$I = \sum \pi(1 + \mathfrak{p}^{k+1}\bar{\alpha}t\bar{t}i) = \sum \pi(1 + \mathfrak{p}^{k+1}at\bar{t}i - \varepsilon\mathfrak{p}^{2k+1}bt\bar{t}).$$

Since $|\mathfrak{p}^{2k+1}t\bar{t}| \leqslant q^{-(2m+1)}$, the last member in the parentheses may be discarded. Therefore,

$$I = \sum \pi(1 + \mathfrak{p}^{k+1}at\bar{t}i) = \sum \pi(1 + \mathfrak{p}^{k+1}t\bar{t}i),$$

where the summation is over elements t for which $|t| \leqslant \min(1, q^{k-m})$.

---

* Namely, by the change of variable $v = \bar{u}t$ this sum is reduced to $I = \sum \pi\left(1 + \mathfrak{p}^{k+1}\frac{\bar{t} - t}{1 - \mathfrak{p}t\bar{t}}a\right)$. Then if the change of variable $t \to t + \mathfrak{p}^{2m-k-1}xi$, $x \in K$, $|x| = 1$, is made, after simple manipulations we will obtain that $I = \pi(1 - 2\mathfrak{p}^{2m}axi)I$. Since $\pi(1 - 2\mathfrak{p}^{2m}axi) \neq 1$, then $I = 0$.

An elementary calculation, which we omit, yields $I = (-1)^k q^{\nu-1/2+k}$. Since $(-1)^k = \mathrm{sign}_\varepsilon p^k =$ $\mathrm{sign}_\varepsilon \frac{\alpha-\bar{\alpha}}{i}$, then $I = q^{\nu-1/2}|\alpha-\bar{\alpha}|^{-1}\mathrm{sign}_\varepsilon\left(\frac{\alpha-\bar{\alpha}}{i}\right)$. It has thereby been proved that

$$\mathrm{tr}\, T_\pi(\alpha) = |\alpha-\bar{\alpha}|^{-1}\,\mathrm{sign}_\varepsilon\left(\frac{\alpha-\bar{\alpha}}{i}\right)\pi(\alpha).$$

## 2. Representations of a Group of Unitary Matrices over a Ring of Formal Power Series

Let $C\{t\}$ be a ring of formal power series with the complex coefficients

$$\alpha = a_0 + a_1 t + \ldots + a_n t^n + \ldots, \tag{1}$$

provided with a natural topology.† Let us introduce the group $G = SU(2, C\{t\})$ of matrices of the form

$$x = \begin{pmatrix} \alpha & \beta \\ -\bar{\beta} & \bar{\alpha} \end{pmatrix}, \quad \alpha\bar{\alpha} + \beta\bar{\beta} = 1, \tag{2}$$

with coefficients from $C\{t\}$. (Here $\bar{\alpha}$ is a power series obtained from the series $\alpha$ by replacing all the coefficients by the complex conjugates.)

Let us give a decreasing chain of subgroups in G

$$G = G_0 \supset G_1 \supset \ldots \supset G_n \supset \ldots,$$

where $G_n$ is a subgroup of matrices from G of the form $x = e + t^n x'$, where e is the unit matrix, x' a matrix with elements from $C\{t\}$. Evidently, $G_n$ are normal divisors in G, and the group G is the projective limit of the sequence of factor groups

$$\ldots \to G/G_n \to \ldots \to G/G_1 \to G/G_0 = E.$$

Let us note that the groups $G/G_n$ are Lie groups, while the group G itself is not locally compact.‡

Let T(x) be an irreducible unitary representation of the group G. We say that T(x) is a representation of rank n if $T(x) \equiv 1$ on $G_n$, but $T(x) \not\equiv 1$ on $G_{n-1}$. Hence, T(x) is a representation of the factor group $G/G_n$ which is not identically unity on $G_{n-1}/G_n$.

Evidently, the single representation of rank 0 is the unit representation. Furthermore, since $G/G_1 \cong SU(2,C)$, the representation of rank 1 is the well known representation of the customary group of unitary second-order matrices.

By analogy with the group of quaternion matrices over an unconnected locally compact field K, representations of a group G of arbitrary rank will be constructed in this section.

Description of Representations of Even Rank n = 2m. Let us introduce the subgroup $A_{2m} \subset SU(2, C\{t\})$ of matrices of the form

† $C\{t\}$ is the projective limit of the sequence $\ldots \to C_n\{t\} \ldots \to C_1\{t\} \to C_0\{t\}$, where $C_n\{t\}$ is a ring of polynomials of degree not greater than n (multiplication in $C_n\{t\}$ is defined by the modulus of the polynomials divided by $t^{n+1}$).

‡ Evidently the group $G/G_n$ is isomorphic to the group of matrices of the form (2) with elements from the ring $C_{n-1}\{t\}$ of polynomials of degree not greater than $n-1$ (with multiplication by means of the modulus of the polynomials divided by $t^n$).

$$y = \begin{pmatrix} \alpha & t^m\beta \\ -t^m\bar{\beta} & \bar{\alpha} \end{pmatrix}, \quad \alpha, \beta \in C\{t\}.$$

Let us assign a one-dimensional representation to this group.

Let $C^*\{t\}$ be a multiplicative group of the ring $C\{t\}$ consisting of all power series of the form (1) for which $|a_0| = 1$. We call the continuous function $\pi(\alpha)$ in $C^*\{t\}$ a character of rank n if the following conditions are satisfied:

1) $\pi(\alpha_1\alpha_2) = \pi(\alpha_1)\pi(\alpha_2)$ for any $\alpha_1, \alpha_2 \in C^*\{t\}$ ;

2) $|\pi(\alpha)| \equiv 1$;

3) $\pi(\alpha)$, where $\alpha = a_0 + a_1 t + \ldots + a_n t^n + \ldots$, depends only on $a_0, \ldots, a_n$, where $\pi(1 + a_{n-1}t^{n-1}) \not\equiv 1$, when $a_{n-1}$ runs over the imaginary axis.

It is easy to see that if $\pi(\alpha)$ is a character of rank $n = 2m$, then the mapping

$$\pi: y = \begin{pmatrix} \alpha & t^m\beta \\ -t^m\bar{\beta} & \bar{\alpha} \end{pmatrix} \to \pi(\alpha) \tag{3}$$

is a one-dimensional representation of the subgroup $A_{2m}$.

Let us consider the representation $T_\pi(x)$ of the group G induced by the representation (3) of the subgroup $A_{2m}$. This representation acts in the space of functions $f(x)$ in G which satisfy the following conditions:

$$f(yx) = \pi(y) f(x) \text{ for any } y \in A_{2m}, x \in G; \qquad \int_{G/A_{2m}} |f(x)|^2 dx < \infty.$$

(Integration is over the invariant measure in $G/A_{2m}$, which exists, as is easy to see.) The representation operator is given by $(T_\pi(x_0)f)(x) = f(xx_0)$.

The representations $T_\pi(x)$ are irreducible. Two representations $T_{\pi_1}$ and $T_{\pi_2}$ are equivalent if and only if either $\pi_1 = \pi_2$, or $\pi_1 = \pi_2^{-1}$.

The proof is carried out along the same scheme as for the group of quaternions over an unconnected locally compact field.

Description of Representations of Odd Rank n = 2M + 1. Let us introduce the subgroup $A_{2m+1} \subset G_1$ of matrices of the form

$$y = \begin{pmatrix} 1 + t\alpha & t^m b + t^{m+1}\beta \\ -t^m b - t^{m+1}\bar{\beta} & 1 + t\bar{\alpha} \end{pmatrix}, \quad \alpha, \beta \in C\{t\}, b \in \mathbf{R}.$$

Let $\pi(\alpha)$ be a character of rank $n = 2m + 1$ on $C^*\{t\}$, which satisfies the following additional condition: $\pi(1 + at^{n-1}) \equiv 1$ for $a \in R$. Then the mapping

$$y \mapsto \pi(1 + t\alpha) \tag{4}$$

is a one-dimensional representation of the group $A_{2m+1}$.

Let us consider the representation $T_\pi(x)$ of the subgroup $G_1$ induced by the representation (4) of the subgroup $A_{2m+1}$. It can be proved that the representations $T_\pi(x)$ are irreducible and pairwise not equivalent.

Our problem is to start from the representations $T_\pi(x)$ of the subgroup $G_1$ and to construct irreducible unitary representations of the whole group G. This construction is based upon the following obvious fact: The group G is a semi-direct product $G = SU(2, C) \cdot G_1$ of the subgroup SU(2, C) of matrices with coefficients from C and the normal divisor $G_1$.

Let $D \subset SU(2, C)$ be a subgroup of all matrices $\delta$, such that the representation $T_\pi^\delta(x) = T_\pi(\delta^{-1}x\delta)$ is equivalent to the original representation $T_\pi(x)$. It can be shown that D consists of all diagonal matrices and, therefore, is isomorphic to the group of rotations of a circle.

From the irreducibility of $T_\pi(x)$ it follows that for each $\delta \in D$ there exists a unitary operator $B(\delta)$ in the space of representations $T_\pi$ such that $B(\delta)T_\pi(\delta^{-1}x\delta) = T_\pi(x)B(\delta)$ for any $x \in G_1$; the operator $B(\delta)$ is hence defined uniquely to the accuracy of a scalar factor. The mapping

$$\delta \to B(\delta) \tag{5}$$

is a continuous projective representation of the group D.† But any continuous projective representation for a group of rotations of a circle is equivalent to the usual representation. In other words, by multiplying the operator $B(\delta)$ with a suitable function $\lambda(\delta)$, it is possible to have (5) be the usual group representation.

Let us construct irreducible representations of the subgroup $D \cdot G_1$. To do this, let us assign an arbitrary character $\chi(\delta)$ to the subgroup D, and let us compare the pair $(\pi\chi)$ of the next representation $T_{\pi,\chi}$ of the group $D \cdot G_1$. This representation acts in the same space as does the representation $T_\pi$ of the subgroup $G_1$. The operator $T_{\pi,\chi}(\delta x)$, where $\delta \in D$, $x \in G_1$ is given by the following formula $T_{\pi,\chi}(\delta x) = \chi(\delta)B(\delta)T_\pi(x)$.

The fact that $T_{\pi,\chi}$ is a representation follows directly from the definition of the operators $B(\delta)$. This representation is irreducible since its constraint in $G_1$ is irreducible.

Finally, let us take the representation of the group G induced by the representation $T_{\pi,\chi}$ of the subgroup $D \cdot G_1$. We obtain the desired series of irreducible unitary representations of rank $n = 2m + 1$ of the group G.

Detailed proofs of the facts presented here and an explicit formula for the operators $B(\delta)$ will be given in another paper.

## 3. Local Representations of a Group of Unitary Matrices with Elements from a Ring of Infinitely Differentiable Functions

Let us consider the ring K of all functions of the form $c + f(t)$, where $f(t)$ are complex infinitely differentiable, finite (or rapidly decreasing) functions of the real variable t, c are constants, multiplication is defined in the customary way.

With the ring K let us associate a group SU (2, K) of unitary matrix-functions

$$u(t) = \begin{pmatrix} \alpha(t) & \beta(t) \\ -\overline{\beta(t)} & \overline{\alpha(t)} \end{pmatrix}, \quad |\alpha(t)|^2 + |\beta(t)|^2 = 1,$$

$\alpha(t), \beta(t) \in K$. We shall be interested here only in local representations of the group SU (2, K), i.e., such representations $T(u(t))$ as depend only on the values of the function u(t) itself and a finite number of its derivatives at a finite number of points $t_1, \ldots, t_n$.

If the irreducible unitary representation $T(u(t))$ depends only on $u(t_0), u'(t_0), \ldots, u^{(n)}(t_0)$, where the dependence on $u^{(n)}(t_0)$ is essential, then we call it the derivative of order n at the point $t_0$.

There is a simple connection between the local representations of the group SU (2,K) and the representations of the groups SU $(2, C\{t\})$ considered in Section 2. First of all, let us note that any local irreducible unitary representation $T(u(t))$ of the group SU (2,K) is a tensor product of a finite number of local representations each of which is compact at just one point.

In fact, let $T(u(t))$ depend only on $u(t_i), u'(t_i), \ldots, u^{(m_i)}(t_i), i = 1, \ldots, n$. Let us introduce the subgroup $G_{t_1,\ldots,t_n}^{m_1,\ldots,m_n}$ of elements u(t) satisfying the conditions

†That is, for any $\delta_1, \delta_2 \in D$ the relation $B(\delta_1\delta_2) = \sigma(\delta_1\delta_2)B(\delta_1)B(\delta_2)$ is satisfied, where $\sigma$ is some scalar function.

$$u(t_i) = e,\ u'(t_i) = \ldots = u^{(m_i)}(t_i) = 0,\ i = 1, \ldots, n,$$

here e is the unit matrix. Evidently, $G_{t_1,\ldots,t_n}^{m_1,\ldots,m_n}$ is a normal divisor in SU (2, K), and $T(u(t)) \equiv 1$ on $G_{t_1,\ldots,t_n}^{m_1,\ldots,m_n}$. herefore, T(u(t)) is an irreducible unitary representation of the factor group $\tilde{G}_{t_1,\ldots,t_n}^{m_1,\ldots,m_n} = SU(2, K)/G_{t_1,\ldots,t_n}^{m_1,\ldots,m_n}$. ut evidently, $G_{t_1,\ldots,t_n}^{m_1,\ldots,m_n}$ is isomorphic to the direct product:

$$\tilde{G}_{t_1,\ldots,t_n}^{m_1,\ldots,m_n} \cong \tilde{G}_{t_1}^{m_1} \times \ldots \times \tilde{G}_{t_n}^{m_n},$$

rom which our assertion follows directly.

Furthermore, the natural isomorphism $\tilde{G}_{t_i}^{m} \cong SU(2, C\{t\})/G_{m+1}$ holds, where $G_{m+1}$ is a subgroup of the roup SU (2, C{t}) defined in Section 2. Therefore, the description of the irreducible unitary representations f the group SU (2, K) as derivatives of order m reduces to a description of representations of rank m + 1 f the group SU (2, C{t}); this description has been given in Section 2.

It will be shown here that the representation of the group SU (2, K), which is the first-order derivaive, can be obtained by a passage to the limit exactly as the derivative of a function is obtained from the unction itself. Namely, let us consider a local, irreducible unitary representation of the group SU (2,K), hich depends only on u(0) and $u(t_0)$, where $t_0$ is close to 0. Any such representation has the form $T_{l_1,l_2}(u(t)) = T_{l_1}(u(0)) \otimes T_{l_2}(u(t_0))$, where $T_{l_1}(u)$, $T_{l_2}(u)$ are irreducible representations of the group SU(2,C) f weights $l_1$ and $l_2$, respectively (their dimensionalities are respectively $2l_1 + 1$ and $2l_2 + 1$). For simplicity, e will henceforth assume that $l_1, l_2$ are integers and $l_1 = l_2$. We show that if $t_0 \to 0$ and $l_1 \to \infty$ simultaneously, hen $T_{l_1}(u(0)) \otimes T_{l_1}(u(t_0))$ goes over, in the limit, into a representation, which is a first-order derivative. he exact meaning of this assertion will be clarified later.

Let $H_{l_1}$ be the space of representations $T_{l_1}(u)$ of the group SU (2, C), $\mathcal{H}_{l_1}$ the space of representations $T_{l_1}(u(0)) \otimes T_{l_1}(u(t_0))$ of the group SU (2, K). Since $\mathcal{H}_{l_1} = H'_{l_1} \otimes H''_{l_1}$, then (see [5]) $\mathcal{H}_{l_1}$ is decomposed into the irect sum: $\mathcal{H}_{l_1} = \sum_{l=0}^{2l_1} H_l$. As $H_l$ let us take the canonical basis $\{e_m\}$, consisting of eigenvectors relative the diagonal subgroup. The canonical basis $g_m^l$, $m = -l, -l+1, \ldots, l$; $l = 0, 1, \ldots, 2l_1$, where $g_m^l$ runs hrough vectors of the canonical basis in $H_l$ for fixed $l$, has thereby been defined in $\mathcal{H}_{l_1}$. The following relaons hold between $g_m^l$ and $e'_m$, $e''_m$ of the canonical bases in $H'_{l_1}$ and $H''_{l_1}$:

$$g_m^l = \sum_{m_1+m_2=m} C(l_1, l_1, l; m_1, m_2)\, e'_{m_1} e''_{m_2}; \quad e'_{m_1} e''_{m_2} = \sum_{l=|m_1+m_2|}^{2l_1} C(l_1, l_1, l; m_1, m_2)\, g^l_{m_1+m_2},$$

here $C(l_1, l_1, l;\ m_1, m_2)$ are Clebsch-Gordan coefficients.

For any pair $l_1$, $l_1'$, where $l_1' > l_1$, let us give the linear isometric mapping $\mathcal{H}_{l_1} \to \mathcal{H}_{l_1'}$, which compares ne vector $g_m^l \in \mathcal{H}_{l_1}$, to each basis vector $g_m^l \in \mathcal{H}_{l_1'}$. Therefore, a sequence of imbeddings holds

$$\mathcal{H}_0 \to \mathcal{H}_1 \to \ldots \to \mathcal{H}_n \to \ldots;$$

et $\mathcal{H}$, be the inductive limit of this sequence. We now construct a representation $T(u(0), u'(0))$ of the group U(2, K) in the space $\mathcal{H}$, which is the limit of the representation $T_{l_1}(u(0)) \otimes T_{l_1}(u(t_0))$ as $l_1 \to \infty$ and $t_0 \to 0$.

Let us first establish the following, which is of independent interest. Let $a_{m,m'}^{l,l'}(s; l_1, t_0)$ be matrix elenents in the basis $g_m^l$ of the operator $T_{l_1}(e) \otimes T_{l_1}(u_s(t_0))$, where $u_s(t_0) = \begin{pmatrix} e^{ist_0} & 0 \\ 0 & e^{-ist_0} \end{pmatrix}$, and e is the unit natrix. Then if $t_0 \to 0$ and $l_1 \to \infty$ simultaneously, such that $\lim l_1 t_0 = \alpha$, then $a_{m,m'}^{l,l'}(s; l_1, t_0)$ tends to some mit $a_{m,m'}^{l,l'}(s)$; this limit will be evaluated.

For the proof, let us note that the matrix elements $a^{l,l'}_{m,m'}(s; l_1, t_0)$ are given by the following formula:

$$a^{l,l'}_{m,m'}(s; l_1, t_0) = \delta_{m,m'} \sum_{m_1} C(l_1, l_1, l; m_1, m - m_1) \times C(l_1, l_1, l', m_1, m - m_1) \exp\{2i(m - m_1) st_0\}.$$

Let us substitute explicit expressions for the Clebsch-Gordan coefficients here:

$$C(l_1, l_1, l; m_1, m - m_1) = \left(\frac{(2l_1+1)(l+m)!\,(2l_1-l)!\,(l!)^2}{(2l_1+l+1)!\,(l_1+m_1)!\,(l_1+m-m_1)!\,(l_1-m_1)!\,(l_1-m+m_1)!}\right)^{1/2}$$
$$\times \sum_{k=\max(0,-m)}^{\min(l-m,l)} \frac{(-1)^{l-m-k}(l+l_1-m-k)!\,(l_1+m_1+k)!}{k!\,(l-m-k)!\,(l-k)!\,(m+k)!},$$

and let us replace $t_0$ by $\alpha/l_1$. It is easy to see that an expression is obtained as a result which is equivalent to an integral sum of the form $\sum_{m_1=-l_1}^{l_1} f\left(\frac{m_1}{l_1}\right)\frac{1}{l_1}$, where $f$ is some continuous function in $[-1, 1]$. Therefore, the limit of this expression as $l_1 \to \infty$ exists and equals $\int_{-1}^{1} f(x)\,dx$. Omitting computations, let us present the final expression for this limit:

$$a^{l,l'}_{m,m'}(s) = \lim a^{l,l'}_{m,m'}(s; l_1, t_0) = \delta_{m,m'} \int_{-1}^{1} P_l^m(x)\, P_{l'}^m(x)\, e^{2i\alpha sx}\,dx, \tag{1}$$

where $P_l^m(x)$ is the normalized associated Legendre function [5].

Let us now construct the representation $T(u(0), u'(0))$ of the group $G = SU(2, K)$ in the space $\mathcal{H}$. Let $\pi(\alpha)$ be a character of rank 2 on $C^*\{t\}$, which satisfies the additional condition $\pi(a) \equiv 1$ for $a \in C$. The construction presented in Section 2 compares the irreducible unitary representation $T(u(0), u'(0))$ of the group $G = SU(2, K)$ to the character $\pi$. This representation may be realized as follows. It acts in the space $\mathcal{H}'$ of functions $f(v)$ on $SU(2, C)$ which satisfy the following conditions: 1) $f(\delta v) = f(v)$ for any $\delta \in D$, where $D \subset SU(2, K)$ is a subgroup of diagonal matrices; 2) $\int |f(v)|^2 dv < \infty$. The representation operator is given by the following formula:

$$(T(u(0), u'(0))f)(v) = f(v \cdot u(0)) \cdot \exp[2\alpha(vu'(0)u^{-1}(0)v^{-1})_{11}] \tag{2}$$

where $(\;)_{11}$ is the matrix element of the appropriate matrix.

Let us note that the space $\mathcal{H}'$ of functions $f(v)$ is isomorphic to $\mathcal{H}$. Indeed, it is known that it decomposes into the direct sum $\mathcal{H}' = \sum_{l=0}^{\infty} H_l$; but $\mathcal{H}$ also decomposes into such a direct sum.

Let us find the matrix elements of the operators $T(u(0), u'(0))$ in the basis $g_m^l$. Let us note that these operators are expressed in terms of the operators $T(u(0), 0)$, and $T(e, u'_s(0))$, where e is the unit matrix, and $u_s(0) = \begin{pmatrix} is & 0 \\ 0 & -is \end{pmatrix}$. The operators $T(u(0), 0)$ have a simple structure: their constraint in each of the subspaces $\mathcal{H}_l$ coincides with $T_l(u(0)) \otimes T_l(u(0))$. We will hence be interested only in the operators $T(e, u'_s(0))$.

It turns out that matrix elements of the operator $T(e, u'_s(0))$ in the basis $g_m^l$ are expressed by means of (1), i.e., they are obtained by a passage to the limit from the matrix elements of the operator $T_{l_1}(e) \otimes T_{l_1}(u_s(t_0))$ as $l_1 \to \infty$, $t_0 \to 0$ if $\lim l_1 t_0 = \alpha$.

In order to see this, let us note that the space SU (2, C)/D, in which the functions $f(v) \in \mathcal{H}$ are assigned, s the sphere $S^2$, and the vectors $g_m^l$ are the spherical functions $g_m^l = Y_l^m(\varphi, \theta) = \frac{1}{\sqrt{2\pi}} e^{im\varphi} P_l^m(\cos\theta)$. There esults from (2) that $T(e, u_s'(0))$ is the operator of multiplication by $\exp\{2i\alpha s \cos\theta\}$. Therefore, matrix lements of this operator in the basis $g_m^l$ are expressed by means of (1).

## LITERATURE CITED


. N. Bourbaki, Algebra [Russian translation], Fizmatgiz, Moscow (1962).
. I. M. Gel'fand and M. I. Graev, "Representations of groups of quaternions over an unconnected locally compact continuous field," Dokl. Akad. Nauk SSSR, 177, No. 1 (1967).
. G. Mackey, "Infinite dimensional group representations," Matematika, 6, No. 6, 57-103 (1962).
. I. M. Gel'fand and M. I. Graev, "Representations of groups of second-order matrices with elements from a locally compact field," Usp. Mate. Nauk, 18, No. 4, 29-99 (1962). Also see
I. M. Gel'fand, M. I. Graev, and I. I. Pyatetskii-Shapiro, Representation Theory and Automorphic Functions [in Russian], Nauka, Moscow (1966).
. I. M. Gel'fand, R. A. Minlos, and Z. Ya. Shapiro, Representations of Rotation Groups and Lorentz Groups [in Russian], Fizmatgiz, Moscow (1958).

## 4.

(with I. N. Bernstein and S. I. Gelfand)

# A new model for representations of finite semisimple algebraic groups

Usp. Mat. Nauk **29** (3) (1974) 185–186. MR **53**:5760. Zbl. **354**:20031

1. An important problem in representation theory is to construct the so-called model, i.e. representations of the given group $G$ that contain almost every irreducible representation of $G$ exactly once. Up to now, only relatively few models are known (see [1] for $GL_2$, [3] for $\mathrm{Sp}_4$, [2,4] for an arbitrary algebraic group). Being interesting in itself, every model leads also to an important theory of gamma- and zeta-factors [1,3,4]. The construction of the model we describe below can be applied to an arbitrary split algebraic group over a local field; however, we formulate the result and give the proof only for groups over finite fields.

The note of Lusztig [7] was very useful to us during the writing of this note.

2. Let $G$ be a split semisimple algebraic group over a finite field $\mathbb{F}_q$, and let $r$ be the rank of $G$. For any extension $\mathbb{F}_q$, $q=p^a$, of the field $\mathbb{F}_p$ denote by $G=G(q)$ the group of $\mathbb{F}_q$ points of $G$. We are interested in properties of $G(q)$ for large $q$'s. In the following $f(q)=O(q^k)$ means that $f(q)<cq^k$ where $c$ depends only on the type of group $G$.

We introduce some notation. Let $H$ be a split Cartan subgroup of $G$, $\Delta$ be the root system of $G$ with respect to $H$, $\Sigma$ be the set of simple roots, and $\Delta_+$ be the set of positive roots (in some ordering). Also let $W$ be the Weyl group of $\Delta$ and $l(w)$ be the length of an element $w\in W$ with respect to the set of simple reflections. Denote by $s_0$ the unique element of maximum length in $W$. For any root $\gamma$ denote by $X_\gamma$ the unipotent one-parameter (isomorphic to $\mathbb{F}_q$) subgroup in $G$, corresponding to $\gamma$. Let $U=\prod X_\gamma(\gamma\in\Delta_+)$ and $B=HU$.

Define a Coxeter element $c\in W$. Let us choose some ordering $\alpha_1,\dots,\alpha_r$ of simple roots and set $c=\sigma_{\alpha_1}\dots\sigma_{\alpha_r}$. Such elements $c$ (for various orderings of $\Sigma$) are called Coxeter elements; the number of Coxeter elements is $2^{n-i}$ where $i$ is the number of simple components of $G$ (see [5]). For any Coxeter element $c$ consider the following subgroups $B_c$, $V_c$ in $B$: $B_c=B\cap cBc^{-1}$, $V_c=U\cap c(s_0Us_0^{-1})c^{-1}$. It is clear that $B_c\cap V_c=\{e\}$, $B_cV_c=B$. Next, one show that $V_c$ is a commutative group of order $q^r$. Let $\mathscr{E}(G)$ be the set of equivalence classes of complex irreducible representations of $G$. For any representation $T$ of the group $G$ and for any $\omega\in\mathscr{E}(G)$ denote by $i(\omega,T)$ the multiplicity of $\omega$ in $T$.

For any Coxeter element $c\in W$ denote by $T_c$ the representation of $G$ induced by the trivial representation of $B_c$.

**Theorem 1.** *Let $f(q)$ be the number of $\omega\in\mathscr{E}(G)$ satisfying $i(\omega,T_c)=1$. Then $f(q)=q^r+O(q^{r-1})$.*

As card $\mathscr{E}(G) = q^r + O(q^{r-1})$ [6] we see that $T_c$ contains almost all irreducible representations of $G$ with multiplicity 1.

3. To prove Theorem 1 we use the following result. Let $T_1$ and $T_2$ be two arbitrary representations of a finite group $G$. Write $R_{ij} = \mathrm{Hom}_G(T_i, T_j)$ and consider the algebra $R = \oplus R_{ij}$ $(i, j = 1, 2)$.

**Proposition 1.** *Let $\theta$ be an anti-automorphism of the algebra $R$ such that (i) $\theta^2 = 1$, (ii) $\theta(R_{ij}) = R_{\bar{j}\bar{i}}$ (where $\bar{1} = 2$, $\bar{2} = 1$). Let $R_{12}^- = \{x \in R_{12}, \theta(x) = -x\}$. Then*

$$\frac{1}{2} \sum_{\omega \in \mathscr{E}(G)} i(\omega, T_1)[i(\omega, T_1) - 1] \leqq \dim R_{11} - \dim R_{12} + 2 \dim R_{12}^-.$$

4. Now we use the existence of the particular anti-automorphism $g \to g'$ of the group $G$ (it is often used in similar situations; see [2, 4, 8]). This anti-automorphism has the following properties: (i) $(g')' = g$, (ii) $B' = B$, (iii) $x' = x$ for $x \in s_0 H$ (see [8]; for $g = PGL_n$ this anti-automorphism is the reflection of a matrix with respect to the second diagonal). Now let $c \in W$ be an arbitrary Coxeter element. Then $c''$ is again a Coxeter element and $B_{c'} = (B_c)'$, $V_{c'} = (V_c)'$. Let $T_1 = T_c$, $T_2 = T_{c'}$; identify elements of the space $R_{ij} = \mathrm{Hom}(T_i, T_j)$ with functions on $G$ that are left invariant with respect to $B_i$ and right invariant with respect to $B_j$ (with $B_1 = B_c$, $B_2 = B_{c'}$). For any function $f$ on $G$ define $\theta f$ by $\theta f(g) = f(g')$. Then $\theta$ is an anti-automorphism of the algebra $R = \oplus R_{ij}$ satisfying the conditions (i), (ii) of proposition 1.

Choose an element $v_1 \in V_c$ such that $h v_1 h^{-1} = v_1$ for all $h \in H$, $h \neq 1$, and write $v_2 = v_1'$. For any $x \in s_0 H$ denote by $K_{ij}(x) \in B_i \backslash G / B_j$ the two-sided coset $B_i v_i x v_j B_j$ $(i, j = 1, 2)$.

**Proposition 2.** (i) *$K_{ij}(x_1) \neq K_{ij}(x_2)$ for $x_1 \neq x_2$.* (ii) *The number of cosets in $B_i \backslash G / B_j$ that are different from all $K_{ij}(x)$ is $O(q^{r-1})$.*

Let us note now that denoting by $\varphi_x \in R_{12}$ the characteristic function of the coset $K_{12}(x)$, we have $\theta(\varphi_x) = \varphi_x$. Theorem 1 following easily from this remark and from Propositions 1, 2.

5. Let us now give a generalization of Theorem 1. Again let $\alpha_1, \ldots, \alpha_r$ be some ordering of $\Sigma$; write $w = \sigma_{\alpha_1} \ldots \sigma_{\alpha_k}$ where $0 \leqq k \leqq r$. Define subgroups $U_w \subset U$, $H_w \subset H$ by $U_w = U \cap w U w^{-1}$, $H_w = \{h \in H, \alpha_i(h) = 1 \text{ for } k+1 \leqq i \leqq r\}$. Let $\Theta$ be an one-dimensional representation of $U_w$ such that $\Theta|_{x_{\alpha_i}} \not\equiv 1$ for $k+1 \leqq i \leqq r$ and $\Theta|_{x_\gamma} \equiv 1$ for all other $X_\gamma \subset U_w$. Also Let $\Pi$ be an arbitrary one-dimensional representation of $H_w$. Then $\Psi(hu) = \Theta(u)\Pi(h)$ is a one-dimensional representation of the group

$$A_w = H_w U_w \subset B$$

**Theorem 2.** *The conclusion of Theorem 1 remains true for the representation*

$$T_w^\Psi = \mathrm{Ind}_{Aw}^G(\Psi)$$

*instead of $T_c$.*

For $k = r, \Pi \equiv 1$ the representation $T_w^\Psi$ becomes $T_c$ from Theorem 1, for $k = 0$ it becomes the representation studied in [2], for $G = GL(n, \mathbb{F}_q)$, $w = (12\ldots n) \in W, \Pi \equiv 1$ it becomes the Lusztig representation [7], for $G = \mathrm{Sp}_4(\mathbb{F}_q)$, $k = 1$, $w = \sigma_\alpha$ (where $\alpha$ is the short simple root) it becomes the representation from [3].

For any $\omega \in \mathscr{E}(G)$ denote by $m(\omega)$ the Schur index (over $\mathbb{Q}$) of the representation $\omega$. Theorem 1 implies

**Corollary.** *$m(\omega) = I$ for "almost all" $w \in \mathscr{E}(G)$.*

## References

1. Jacquet, H., Langlands, R.: Automorphic forms on $GL(2)$. Lecture Notes in Mathematics, vol. 114. Springer, Berlin Heidelberg New York, 1970.
2. Gelfand, I.M., Graev, M.I.: The construction of irreducible representation of simple algebraic groups over a finite field. Dokl. Akad. Nauk SSSR (3) **147** (1962) 529–532.
3. Novodvorskij, M.E., Piateski-Shapiro, I.I.: Generalized Bessel models for the rank 2 symplictic group. Mat. Sb. **90** (132); 2 (1973) 246–256.
4. Gelfand, I.M., Kazhdan, D.A.: Representations of the group $GL(n, K)$ where $K$ is a local field. Funkt. Anal. Prilozh. (4) **6** (1972) 73–74.
5. Bourbaki, N.: Groups et algebras de Lie, Chaps. IV–VI. Hermann, Paris 1968.
6. Borel, A. (ed.): Seminar on algebraic groups and related finite groups. Lecture Notes in Mathematics, vol 131. Springer, Berlin Heidelberg New York 1970.
7. Lusztig, G.: On the discrete series representations of the general linear group over finite field. Bull. Am. Math. Soc. **79** (1973) 550–554.
8. Steinberg, R.: Lectures on Chevalley groups. Yale Univ. 1967.
9. Curtis, C.W., Reiner, I.: Representation theory of finite groups and associative algebras. Interscience, New York London 1962.

Received by the Society January 10, 1974

# 5.

## (with M. I. Graev)

# Irreducible unitary representations of the group of unimodular second-order matrices with elements from a locally compact field

Dokl. Akad. Nauk SSSR **149** (1963) 499–502 [Sov. Math., Dokl. 4 (1963) 397–400].
Zbl. **119**:270

1. We consider the group $G = SL(2, K)$ of second-order matrices with determinant one, whose ele- ents are from an arbitrary locally compact nondiscrete field $K$. In this paper we shall construct ir- ducible unitary representations of the group $G$ and compute their traces.* As a preliminary we formu- te some known facts on locally compact nondiscrete fields.

A. Any connected locally compact field is isomorphic either to the field $R$ of the real, or to the eld $C$ of the complex numbers. Any nonconnected locally compact nondiscrete field of characteristic ro is isomorphic to a finite extension of the field $Q_p$ of the p-adic numbers (where $p$ is a prime mber). Any nonconnected locally compact nondiscrete field of characteristic $p \neq 0$ is isomorphic to finite extension of the field $F_p(t)$ of the formal power series $a_{-s}t^{-s} + \cdots + a_0 + a_1 t + \cdots$ with co- ficients from the finite field of order $p$ (cf. [3]).

B. Denote by $K^+$ the additive group of the field $K$; by $K^*$ its multiplicative group; by $dx$ the ar measure on $K^+$. On $K$ we define the function $|a|$ by $d(xa) = |a|\, dx$. Evidently the function $|a|$ s the following properties: 1) $|0| = 0$ and $|a| > 0$ if $a \neq 0$; 2) $|ab| = |a|\,|b|$ for any $a, b$ in $K$. e function $|a|$ is a norm in $K$ except when $K = C$. In the case $K = C$ it is found that $|a|$ is the uare of a norm. Clearly $d^*x = |x|^{-1}dx$ represents the Haar measure on $K$. Hence it is readily seen at by the linear fractional transformation $x' = xg = \dfrac{\alpha x + \gamma}{\beta x + \delta}$ the measure is transformed according to e formula $dx' = |\beta x + \delta|^{-2}\,|\det g|\, dx$.

C. Assume $K$ to be disconnected. Then its multiplicative group $K$ is isomorphic to the direct prod- t $Z \times Z_q \times A$, where $Z$ is the infinite cyclic group, $Z_q$ the finite cyclic group of the order $= p^n - 1$ ($p$ prime), and $A$ the group of the elements $x$ of $K$ for which $|x - 1| < 1$. It is easy to show at for $p \neq 2$ every element of $A$ can be represented as the square of some other element of $A$. Hence the field $K$ $(p \neq 2)$ there are exactly three quadratic extensions: $K(\sqrt{r_0})$, $K(\sqrt{r_q})$, and $K(\sqrt{r_0 r_q})$, ere $r_0$, $r_q$ are generating elements of $Z$, $Z_q$ respectively.

D. Every quadratic extension $K(\sqrt{r})$ of the field $K$ will be called a "complex" plane and its ele- nts $z = x + \sqrt{r}\,y$ are "complex" numbers. By $\bar{z}$ we shall denote the complex conjugate of $z$. The ele- nts of $K^*$ which can be represented in the form $z\bar{z} = x^2 - ry^2$, $x, y \in K$, will be called positive (with gard to the given extension) and the others negative. We introduce on $K^*$ the function $\operatorname{sign} x$: We de- e $\operatorname{sign} x = 1$ if $x$ is positive, $\operatorname{sign} x = -1$ if $x$ negative. It is easy to show that the set of the nega- e numbers is not empty and that the function $\operatorname{sign} x$ is a character on $K$, i.e., $\operatorname{sign}(xy) = \operatorname{sign} x \operatorname{sign} y$ any $x, y \in K^*$.

---

* The representations of the group of second-order matrices over a finite field have been studied in [2], se over the complex and the real field in [1] and [5].

E. The additive group $K^+$ of a locally compact nondiscrete field is self-dual. We shall denote b $\chi(x)$ the characters on $K^+$ ($\chi(x+y) = \chi(x)\chi(y)$, $|\chi(x)| = 1$). Let $\chi \not\equiv 1$ be a certain character on $K^+$. Then any character on $K^+$ is uniquely defined by $\chi_u(x) = \chi(ux)$ where $u \in K^+$.

2. Now we describe the fundamental continuous series of irreducible unitary representations of the group $G$. The construction of these representations is well known in the case of the field $K$ of th complex numbers [1]; the construction can be carried over to the case of an arbitrary locally compact nondiscrete field $K$. A representation of a continuous series is defined by a character $\pi(x)$ on the group $K$ (i.e., a continuous complex-valued function that satisfies the condition $\pi(xy) = \pi(x)\pi(y)$, $|\pi(x)| = 1$); it is formed in the space of the functions $f(x)$ on $K$ for which $\|f\|^2 = \int |f(x)|^2\, dx < \infty$. The operator of the representation $T_\pi(g)$ which corresponds to the matrix $g = \begin{bmatrix} \alpha & \beta \\ \gamma & \delta \end{bmatrix}$, is given by

$$T_\pi(g)\, f(x) = f\left[\frac{\beta + \delta x}{\alpha + \gamma x}\right] \pi(\alpha + \gamma x)\, |\alpha + \gamma x|^{-1}.$$

It can be shown that the representations $T_\pi(g)$ are irreducible (cf. [1], [5], [4]). *Two representations of a continuous series corresponding to the characters $\pi_1$, $\pi_2$ are equivalent if and only if either $\pi_1 = \pi_2$ or $\pi_1 = \bar{\pi}_2$*

Another realization of the representations of a continuous series will be obtained if from the func tion $f(x)$ we go over to its Fourier representation $\phi(u) = \int f(x)\chi(-ux)\, dx$ where $\chi \not\equiv 1$ is a certain character on $K^+$. In this realization the operator $T_\pi(g)$ is given by $T_\pi(g)\,\phi(u) = \int K(g;\, u, v)\,\phi(v)\, dv$ where the kernel $K(g;\, u, v)$ is a generalized function defined by

$$K(g;\, u, v) = |\gamma|^{-1} \chi\left[\frac{u\alpha + v\delta}{\gamma}\right] \int \chi\left[-\frac{1}{\gamma}(ux + vx^{-1})\right] \pi(x)\, d^*x \qquad (1)$$

if $\gamma \neq 0$, and

$$K(g;\, u, v) = |\alpha| \cdot \pi(\alpha)\,\chi(\alpha\beta u)\,\delta(v - \alpha^2 u)$$

if $\gamma = 0$ ($\delta(u)$ represents the delta function on $K^+$). The integral in (1) will be called a Bessel function of the first kind (cf. [2]).

3. In connection with every quadratic extension $K(\sqrt{\tau})$ of the field $K$ we shall have a discrete series of irreducible unitary representations of the group $G = SL(2, K)$. The formulas for the operator of this series will be obtained from (1) by "analytic continuation". Indeed we replace in (1) the character $\pi$ on $K^*$ by an arbitrary character on the "complex" plane $K^*(\sqrt{\tau})$, and we integrate in (1) not over the "real axis" $K$, but over the contour on $K(\sqrt{\tau}) \setminus K$ along which the expression $-\gamma^{-1}(uz + vz^{-1})$ has its values within the original field $K$. It is readily seen that this contour consists of those $z$ for which $z\bar{z} = vu^{-1}$. (We point out that for negative $vu^{-1}$ such a contour does not exist, and in this case we have to take $K(g;\, u, v) = 0$.)

*To every multiplicative character $\pi(z)$ on $K(\sqrt{\tau})$ corresponds a pair of irreducible unitary representations $T_\pi^+(g)$ and $T_\pi^-(g)$ of the group $G$. The representation $T_\pi^+(g)$ operates in the space of the functions $\phi(u)$, vanishing for $\operatorname{sign} u = -1$, for which $\|\phi\|^2 = \int |\phi(u)|^2\, du < \infty$. Similarly $T_\pi^-(g)$ operates in the space of the functions $\phi(u)$ vanishing for $\operatorname{sign} u = 1$. The operators $T_\pi^+(g)$, $T_\pi^-(g)$ are given by $T_\pi^+(g)\,\phi(u) = \int K(g;\, u, v)\,\phi(v)\, dv$, $T_\pi^-(g)\,\phi(u) = \int K(g;\, u, v)\,\phi(v)\, dv$, where*

$$K(g;\, u, v) = c_\chi \operatorname{sign} u \cdot \operatorname{sign}\gamma \cdot |\gamma|^{-1} \chi\left[\frac{u\alpha + v\delta}{\gamma}\right] \int_{z\bar{z} = vu^{-1}} \chi\left[-\frac{1}{\gamma}(z + z^{-1})\right] \pi(z)\, d^*z \qquad (2)$$

*if $\gamma \neq 0$, $\operatorname{sign}(vu^{-1}) = 1$;*

$$K(g;\, u,\, v) = \operatorname{sign}\alpha \cdot |\alpha| \cdot \pi(\alpha) \cdot \chi(\alpha\beta u)\, \delta(v - \alpha^2 u),$$

$\gamma = 0$; $K(g;\, u,\, v) = 0$, *if* $\operatorname{sign}(vu^{-1}) = -1$.

We put $c_\chi^{-1} = \int \chi(z\bar z)\, dz$ (the integral to be taken over the complex plane $K(\sqrt{r})$). It is not dif-
cult to show that either $c_\chi = \pm 1$ or $c_\chi = \pm i$. In (2) we integrate over the circle $z\bar z = vu^{-1}$ with the
easure $d^*z$ defined by the conditions $\int d^*z = 1$, $d^*(zz_0) = d^*z$ for every $z_0$ satisfying $z_0 \bar z_0 = 1$. If
e circle $\bar z z = vu^{-1}$ is compact, then the integral is convergent. It is natural to call
$\int_{=vu^{-1}} \chi\left[-\frac{1}{\gamma}(z + z^{-1})\right] \pi(z)\, d^*z$ a Bessel function of the second kind (in the case of the field of
e real numbers this integral equals $J_n(\sqrt{uv/\gamma})$ where $n$ is the number of the character $\pi$ and $J_n$ the
essel function of index $n$.

*The representations* $T^+_{\pi_1}(g)$ and $T^+_{\pi_2}(g)$ (*correspondingly* $T^-_{\pi_1}(g)$ *and* $T^-_{\pi_2}(g)$) *are equivalent if and
ly if* $\pi_1(z) = \pi_2(z)$ *or* $\pi_1(z) = \pi_2^{-1}(z)$ *on the circle* $z\bar z = 1$. Thus the set of mutually nonequivalent
presentations $T^\pm_\pi(g)$ is isomorphic to the set of the characters on the circle $z\bar z = 1$ where the char-
ters $\pi$ and $\pi^{-1}$ are considered as identified; consequently this set of representations is discrete. *Two
presentations* $T^+_{\pi_1}(g)$ *and* $T^-_{\pi_2}(g)$ *are not equivalent. Two representations from different series are
so never equivalent.*

*With the representations of the continuous and the discrete series all irreducible unitary repre-
ntations of the group* $G$ *are exhausted, as far as they are contained in the regular representation.*

4. Let $T(g)$ be any one of the representations of $G$ considered above. Then the trace $\operatorname{tr} T(g)$ of
e operator $T(g)$ is defined as a generalized function in the appropriate space of the fundamental
nctions on $G$. Below formulas will be given for $\operatorname{tr} T(g)$ on the set of the regular elements of the
oup $G$ (that is the set of all matrices $g$ with distinct eigenvalues).

Let $T_\pi(g)$ be a representation of the continuous series, corresponding to the character $\pi(x)$ on
If the eigenvalues $\lambda$, $\lambda^{-1}$ of the matrix $g$ lie in $K$, then

$$\operatorname{tr} T_\pi(g) = \frac{\pi(\lambda) + \pi(\lambda^{-1})}{|\lambda - \lambda^{-1}|};$$

the opposite case $\operatorname{tr} T_\pi(g) = 0$. Now let $T^+_\pi(g)$, $T^-_\pi(g)$ be representations of a discrete series as
scribed in section 3; if $\lambda$, $\lambda^{-1}$ are the eigenvalues of the matrix $g$, then

$$\operatorname{tr} T^+_\pi(g) - \operatorname{tr} T^-_\pi(g) = c_\chi \operatorname{sign}\gamma \cdot \frac{\pi(\lambda) + \pi(\lambda^{-1})}{|\lambda - \lambda^{-1}|},$$

ere $\lambda$ and $\lambda^{-1}$ are elements of $K(\sqrt{r})$ and not of $K$; in the other cases $\operatorname{tr} T^+_\pi(g) - \operatorname{tr} T^-_\pi(g) = 0$.
rther we have

$$\operatorname{tr} T^+_\pi(g) + \operatorname{tr} T^-_\pi(g) = 2 \int_{z\bar z = 1} \frac{\operatorname{sign}(\alpha + \delta - z - \bar z)}{|\alpha + \delta - z - \bar z|}\, \pi(z)\, d^*z.$$

is integral can easily be evaluated if $K$ is the field of the real numbers. In the case of a discon-
cted field $K$ it can be shown that if the eigenvalues of the matrix $g$ are elements either of $K$ or of a
adratic extension of $K$, different from $K(\sqrt{r})$, then $\operatorname{tr} T^+_\pi(g) = \operatorname{tr} T^-_\pi(g) = 0$ for all $\pi$, with the pos-
ble exception of a finite number of them (depending on $g$).

Received 12/DEC/62

**BIBLIOGRAPHY**

[1] I. M. Gel'fand and M. A. Naĭmark, Izv. Akad. Nauk SSSR Ser. Mat. 11 (1947), 411.

[2] I. M. Gel'fand and M. I. Graev, Dokl. Akad. Nauk SSSR 146 (1962), 757 = Soviet Math. Dokl. 3 (1962), 1378.

[3] L. S. Pontrjagin, *Continuous groups*, 2nd ed., GITTL, Moscow, 1954. (Russian)

[4] H. Boseck, Math. Nachr. 24 (1962), 229.

[5] V. Bargmann, Ann. of Math. (2) 48 (1947), 568.

Translated by:
Hans Schwerdtfeger

# 6.
## (with M. I. Graev)

# Plancherel's formula for the groups of the unimodular second-order matrices with elements in a locally compact field

Dokl. Akad. Nauk SSSR **151** (1963) 262–264 [Sov. Math., Dokl. **4** (1963) 397–400].
Zbl. **204**:141

1. We consider the group $G$ of unimodular matrices of the second order with elements in a con-nuous locally compact field **K**. In [1] a description has been given of the irreducible unitary repre-ntations of the group $G$. In particular, it has been established that for the group $G$ there are several eries of irreducible unitary representations. One of these series (the "continuous"series) is connect-d with the fundamental field **K**; each of the other (the "discrete") series is connected with one of the adratic extensions of the field **K**. Thus if **K** is the field of the complex numbers, there is only one eries (since the field of the complex numbers does not admit extension); if **K** is the field of real num-ers, there are two series of representations (since the field of the real numbers has exactly one quad-tic extension); if **K** is a disconnected field, there are four series of representations (because a dis-onnected field has three quadratic extensions).* Within each series a representation is given by a ertain multiplicative character. More accurately, a representation of the continuous series is given a multiplicative character $\pi$ on **K**; for this character $\pi$ and $\pi^{-1}$ corresponds to equivalent represen-tions. A representation of the discrete series connected with the quadratic extension $K(\sqrt{\tau})$ of the ld **K** is given by the character $\pi$ on the "unit circle" $t\bar{t} \equiv x^2 - \tau y^2 = 1$, $t = x + \sqrt{\tau}y$; for this aracter also $\pi$ and $\pi^{-1}$ correspond to equivalent representations.

In [1] we have also found the traces of the irreducible unitary representations.

We point out that in the case of a disconnected field **K** there is another special representation of e group $G$ which had not been indicated in [1]. This representation is realized in the space of e functions $\phi(x)$ on **K** for which

$$\int \varphi(x)\, dx = 0,$$

$$(\varphi, \varphi) = \int \ln|x_1 - x_2|\, \varphi(x_1)\overline{\varphi(x_2)}\, dx_1\, dx_2 < \infty.$$

The operator of the representation corresponding to the matrix $g = \begin{pmatrix} \alpha & \beta \\ \gamma & \delta \end{pmatrix}$, had the form**

$$T_0(g)\varphi(x) = \varphi\left(\frac{\delta x + \beta}{\gamma x + \alpha}\right)|\gamma x + \alpha|^{-2}$$

The trace $\operatorname{Tr} T_0(g)$ of the operator of the special representation can be expressed by the following rmula. If the eigenvalues $\lambda$, $\lambda^{-1}$ of the matrix $g$ are elements of the field **K**, then

---

* In the following we exclude the particular case where the finite field $O/P$ of the residues has the character-tic 2 (we denote by $O$ the ring of the integral elements of **K** and by $P$ a maximum ideal of $O$). In this particular se the field **K** has more than three quadratic extensions (even infinitely many if the characteristic of **K** is not ero).

** In the case of the field of the real numbers this representation belongs to the discrete series.

$$\operatorname{Tr} T_0(g) = \frac{|\lambda| + |\lambda^{-1}|}{|\lambda - \lambda^{-1}|} - 1;$$

if $\lambda$ and $\lambda^{-1}$ are not elements of **K**, then $\operatorname{Tr} T_0(g) = 1$. Here we denote by $|\lambda|$ the norm of the element $\lambda$ in **K**.

It is the purpose of the present article to expand the functions on $G$ in series of irreducible representations. An accurate setting of the problem will be given in §2.

2. Let $f(g)$ be a finite function on the group $G$. Together with each irreducible unitary representation $T_\pi(g)$ of the group $G$ we consider the operator

$$T_\pi(f) = \int f(g) T_\pi(g)\, dg. \tag{1}$$

The problem is to obtain an inversion of the formula (1), that is, to establish the function $f(g)$ if the operator $T_\pi(f)$ is known. We shall solve the problem for a disconnected locally compact field **K**. *

We shall assume that the residue field $O/P$ connected with **K** is not of characteristic 2; by $O$ we denote the ring of the integers in **K**, by $P$ a maximum ideal in $O$. In this case **K** has three quadratic extensions: $\mathbf{K}(\sqrt{\mathfrak{p}})$, $\mathbf{K}(\epsilon\mathfrak{p})$ and $\mathbf{K}(\sqrt{\epsilon})$, where $\mathfrak{p}$ is generating element of the ideal $P$ and $\epsilon$ an element of the finite order $q-1$ in **K**, where $q$ is the order of the field $O/P$.

Let us introduce the following notations. Let $\pi$ be the multiplicative character on **K**; let $\pi_\tau$ be a character on the circle $t\bar{t} = 1$ in $\mathbf{K}(\sqrt{\tau})$. Let $d^*t$ be the measure on **K**, invariant under multiplicatio and $d_\tau^* t$ the invariant measure on the circle $t\bar{t} = 1$ in $\mathbf{K}(\sqrt{\tau})$; by $d\pi$, $d\pi_\tau$ we denote the invariant measures on the corresponding character groups. We normalize these measures as follows:

$$\int_{|t|\leqslant 1} |t|\, d^*t = 1,\ \int d_\tau^* t = 1,\ \int f(t)\,\pi(t)\, d^*t\, d\pi = f(1),\ \int f(t)\,\pi_\tau(t)\, d_\tau^* t\, d\pi_\tau = f(1).$$

We denote by $T_\pi(g)$ the representation which in the continuous series corresponds to the character $\pi$; by $T_{\pi_\tau}(g)$ we denote the direct sum of the representations of the discrete series $T_{\pi_\tau}^+(g)$ and $T_{\pi_\tau}^-(g)$, corresponding to the character $\pi_\tau$ (cf. [1]). Finally, let $T_0(g)$ denote the operator of the special representation.

*The inversion formula is then given by*

$$cf(g) = \int \mu(\pi) \operatorname{Tr}(T_\pi(f) T_\pi^{-1}(g))\, d\pi$$
$$+ \sum_{\tau=\mathfrak{p},\, \epsilon\mathfrak{p},\, \epsilon} \int \mu(\pi_\tau) \operatorname{Tr}(T_{\pi_\tau}(f) T_{\pi_\tau}^{-1}(g))\, d\pi_\tau + 2 \operatorname{Tr}(T_0(f) T_0^{-1}(g)), \tag{2}$$

*where*

$$\mu(\pi) = -\int_K \pi(t)\, |t|\, |1-t|^{-2}\, d^*t\ ^{**}, \tag{3}$$

$$\mu(\pi_\epsilon) = -\int_{t\bar{t}=1} \pi_\epsilon(t)\, |1-t|^{-2}\, d_\epsilon^* t, \tag{4}$$

$$\mu(\pi_\tau) = -\int_{t\bar{t}=1,\ |1-t|<1} \pi_\tau(t)\, [\,|1-t|^{-2} + 1]\, d_\tau^* t \quad (\tau = \mathfrak{p},\ \epsilon\mathfrak{p}); \tag{5}$$

---

* For connected fields the solution has been obtained earlier: in [2] for the field of the complex numbers and in [3] for the field of the real numbers.

** For $\pi(x) = |x^{ip}|$ the function $\mu(\pi)$ has been determined in [4] for fields of characteristic zero.

$= q^{-1} (q-1)^{-1} (q+1)$; Tr $A$ *is the trace of the operator* $A$.

We point out that the integrals (3), (4), (5) are divergent and therefore they have to be understood n the sense of a regularized value. For instance, $\mu(\pi_\epsilon)$ is the value of the analytic function of the ariable $\nu$ which for $\nu = -2$ is given by $\phi(\nu) = \int \pi_\epsilon(t)\, |1-t|^\nu\, d^*_\epsilon t$.

From the formula (2) easily follows the Plancherel formula

$$c\int |f(g)|^2 dg = \int \mu(\pi)\, \mathrm{Tr}\,(T_\pi(f)\, T^*_\pi(f))\, d\pi + \sum_{\tau=p,\,\varepsilon p,\,\varepsilon} \int \mu(\pi_\tau)\, \mathrm{Tr}\,(T_{\pi_\tau}(f)\, T^*_{\pi_\tau}(f))\, d\pi_\tau + 2\,\mathrm{Tr}\,(T_0(f)\, T^*_0(f)), \tag{6}$$

here $\mu(\pi)$, $\mu(\pi_\tau)$ are given by the formulas (3)–(5). This formula is valid for any function $f(g)$ ıose square is integrable on $G$.

3. For connected fields the measure $\mu(\pi)$ occurring in Plancherel's formula has been computed in ], [3]. In the case of the field of the complex numbers it has the form $\mu(\pi) = c(\rho^2 + n^2)$, where $z) = |z|^{i\rho} e^{in \arg z}$. In the case of the field of the real numbers it has the following form. For a rep- sentation of the continuous series $\mu(\pi) = c\rho \operatorname{cth} \frac{\pi\rho}{2}$ if $\pi(x) = |x|^{i\rho}$; $\mu(\pi) = c\rho \operatorname{th} \frac{\pi\rho}{2}$ if $x) = |x|^{i\rho} \operatorname{sign} x$. For the representation of the discrete series $\mu(\pi) = c\,|n|$, where $\pi(z) = e^{in \arg z}$.

It is easy to show that *for the measure* $\mu(\pi)$ *in both cases, the field of the complex as well as the al numbers, one has the single formula*

$$\mu(\pi) = c\int \pi(t)\, |t|\, |1-t|^{-2} d^*t. \tag{7}$$

In the case of a representation of the continuous series the integral (7) refers to the fundamental eld **K**, but in the case of a representation of the discrete series, to the circle $t\bar{t} = 1$; $d^*_\epsilon t$ is the easure, invariant under multiplication.*

Thus Plancherel's measure is given by the same formula in both cases, for a connected and for a sconnected field.

4. The inversion formula (2) can be obtained by immediate calculation. First we point out that it equivalent to the following formula for generalized functions

$$\mathcal{J}(g) \equiv \int \mu(\pi)\, \mathrm{Tr}\, T_\pi(g)\, d\pi + \sum_{\tau=p,\,\varepsilon p,\,\varepsilon} \int \mu(\pi_\tau)\, \mathrm{Tr}\, T_{\pi_\tau}(g)\, d\pi_\tau + 2\,\mathrm{Tr}\, T_0(g) = c\delta(g), \tag{8}$$

here $\delta(g)$ is the delta-function on the group $G$.

We substitute on the left-hand side of the equation the expressions for $\mu(\pi)$ and $\mu(\pi_\tau)$ and the pressions for the traces (cf. [1]). For $g = e$, where $e$ is a unit of $G$, all the integrals in (8) can be mediately evaluated. Thereby it turns out that $\mathcal{J}(g) = 0$ if $g \neq e$. Thus it follows that the general- ed function is concentrated in the point $g = e$ and therefore $\mathcal{J}(g) = c\,\delta(g)$.

---

*For an arbitrary field **K** we determine the function $|x|$ by the formula $d(xx_0) = |x_0|\, dx$, where $dx$ is the dition-invariant measure on **K**. Thus in the case of the field of the complex numbers $|x|$ is the square of the dulus of the complex number $x$.

**The same argument can be used in the case of a connected field K.**

Received 9/ APR/ 63

BIBLIOGRAPHY

[1] I. M. Gel'fand and M. I. Graev, Dokl. Akad. Nauk SSSR 149 (1963), 499 = Soviet Math. Dokl. 4 (1963), 397.
[2] I. M. Gel'fand and M. A. Naĭmark, Izv. Akad. Nauk SSSR Ser. Mat. 11 (1947), 411.
[3] V. Bargmann, Ann. of Math. (2) 48 (1947), 568.
[4] F. I. Mautner, Amer. J. Math. 80 (1958), 441.

Translated by:
Hans Schwerdtfeger

# 7.
## (with D. A. Kazhdan)

# On the representation of the group $GL(n, K)$, where $K$ is a local field

Funkts. Anal. Prilozh. **6** (4) (1972) 73–74 [Funct. Anal. Appl. **6** (1972) 315–317].
Zbl. **288**:22024

In this paper we present a theorem on the representation of the group GL(n, K) of n-th order matrices over a local non-Archimedian field K and their application to the construction of a $\Gamma$-function representation. A conjecture is formulated concerning the relation between the representation of the group GL(n, K) and the Shafarevich-Weyl group of the field K. This conjecture is a sharpening of the Weyl-Langlands conjecture concerning the non-commutative law of reciprocity [1-3]. The work was announced with detailed proofs by the theory of representations (Budapest, 1971).

In what follows, we let $G_n$ denote the group GL(n, K), $Z_n$ denote the subgroup of the upper triangular matrices with 1 on the diagonal (i.e., $Z_n$ is the maximum unipotent subgroup in $G_n$), $P_n \subset G_n$ is the subgroup of matrices p of the form $p = \|p_{ik}\|$, where $p_{in} = 0$ for $i = 1, \ldots, n-1$, $p_{nn} = 1$; $M_n \subset P_n$ is the unipotent radical in $P_n$, i.e., the matrices m of the form $m = \|m_{ik}\|$, where $m_{ik} = \delta_{ik}$ for $k < n$, $m_{in} = m_i$ for $i < n$, $m_{nn} = 1$; by $G_{n-1} \subset G_n$ we mean the standard imbedding of GL(n − 1, K) in GL(n, K). We have $P_n = G_{n-1}M_n$.

Furthermore, $(\pi, V)$ denotes the nonreducible admissible representation (see [6]) of the group $G_n$ in the space V. By $(\hat{\pi}, V)$ we mean the representation of the group $G_n$, defined by the formula $\hat{\pi}(g) = \pi^{-1}(g)$, where g' is the matrix conjugate to g.

THEOREM 1. There exists in V a nonsingular bilinear form $\langle , \rangle$ such that $\langle \pi(g)\xi_1, \hat{\pi}(g)\xi_2\rangle = \langle \xi_1, \xi_2\rangle$ for all $g \in G_n$, $\xi_1, \xi_2 \in V$.

Let $\psi$ be an additive nontrivial character of K. Set $\theta(z) = \psi(\sum z_{i,i+1})$, where the summation is over i from 1 to n and $z = \|z_{ik}\| \in Z_n$.

Definition. The representation $(\pi, V)$ is said to be nondegenerate if there is a linear functional $l$ on V, $l \neq 0$, such that for all $z \in Z_n$, $\xi \in V$

$$l(\pi(z)\xi) = \theta(z)\, l(\xi). \qquad (*)$$

THEOREM 2. Each cuspidal nonreducible representation $(\pi, V)$ of the group $G_n$ is nondegenerate. (The definition of a cuspidal representation is found in [3].)

THEOREM 3. Let $(\pi, V)$ be any nonreducible admissible representation of $G_n$. Then any linear functionals $l_1$ and $l_2$, satisfying (*), are proportional.

Theorem 3 can be formulated in another manner. Denote by $C^{\infty}_{C,\theta}$ the space of functions $f(g)$ on $G_n$ such that : 1) $f(z^{-1}g) = \theta(z)f(g)(z \in Z_n)$, 2) for each $f$ there exists an open subgroup $N_f \subset G_n$ such that $f(gn) = f(g)$ for all $g \in G_n$, $n \in N_f$, and 3) there exists a compacta $M_f \subset G_n$ such that $f \subset Z_n M_f$. The space, which is dual to $C^{\infty}_{C,e}$, we denote by $C^{-\infty}_{\theta}(G_n)$. Denote by $\rho(g_0)$ the operator $\rho(g_0)$: $C^{\infty}_{C,\theta} \to C^{\infty}_{C,\theta}(G_n)$, defined by by the formula $\rho(g_0)f(g) = f(gg_0)$ $(g_0 \in G_n)$, and denote by $\rho'(g_0)$ the operator in the space $G^{-\infty}_{e}$, which is dual to $\rho(g_0)$.

THEOREM 4. Let $(\pi, V)$ be a nondegenerate admissible nonreducible representation of the group $G_n$. Then there is a unique subspace $W_n(\pi) \subset C^{-\infty}_{\theta}$ which is invariant with respect to $\rho(g)$ (for all $g \in G_n$) such that the representation $(\rho, W_n)$ is isomorphic to the representation $(\pi, V)$.

---

Moscow State University. Translated from Funktsional'nyi Analiz i Ego Prilozheniya, Vol. 6, No. 4, pp. 73-74, October-December, 1972. Original article submitted June 21, 1972.

For finite field analogies to Theorems 3 and 4 see [4].

We now consider the restriction of the nondegenerate representation $(\pi, V)$ to the subgroup $P_n \subset G_n$. We denote by $C^{\infty}_{C,\theta}(P_n)$ the space of functions $f(p)$ $(p \in P_n)$ such that 1) $f(z^{-1}p) = \theta(z) f(p) (z \in Z_n, p \in P_n)$, 2) for every $f$ there exists an open subgroup $N_f \subset P_n$ such that $f(pn) = f(p) (p \in P_n, n \in N_f)$ and 3) there exists a compacta $M_f \subset P_n$ such that $f \subset Z_n M_f$. The space, dual to $C^{\infty}_{C,\theta}$, we denote by $C^{-\infty}_{\theta}(P_n)$. Denote by $\rho(p_0)$: $C^{\infty}_{C,\theta}(P_n) \to C^{\infty}_{C,\theta}(P_n)$ the $p_0$ right shift operator. Let $(\pi, V)$ be a nondegenerate representation of $G_n$, and let $l$ be a functional on V such that $l(\pi(z)\xi) = \theta(z) l(\xi) (z \in Z_n, \xi \in V)$. Construct the mapping $\varphi(\pi): V \to C^{-\infty}_{\theta}(P_n)$ in the following way: for every $\xi \in V$ set the element $\varphi_{\pi}(\xi) \in C^{-\infty}_{\theta}(P_n)$ equal to $\varphi_{\pi}(\xi)(p) = l(\pi(p)\xi)$. From the definition it is clear that $\varphi_{\pi}(\pi(p_0)\xi) = p'(p_0)\varphi_{\pi}(\xi)$ for all $p_0 \in P_n$, $\xi \in V$.

THEOREM 5. If $(\pi, V)$ is a nonreducible cuspidal representation, then $\varphi_{\pi}$ is a $P_n$-isomorphism between the space V and the space $C^{\infty}_{C,\theta}(P_n)$.

In particular, it follows from the theorem that all cuspidal nonreducible representations of the group $G_n$ when restricted to $P_n$ are isomorphic.

Hypothesis 1. For any nondegenerate representation $(\pi, V)$ of the group $G_n$, the mapping $\varphi_{\pi}: V \to C^{-\infty}_{\theta}(P_n)$ does not have a kernel.

In proving Theorem 5 we use the following theorem which has independent interest.

THEOREM 6. Every nontrivial subspace in $C^{-\infty}_{\theta}(P_n)$ which is invariant with respect to right shifts contains $C^{\infty}_{C,\theta}(P_n)$.

Obviously, as a $G_{n-1}$-module $C^{\infty}_{C,\theta}(P_n)$ is canonically isomorphic to $C^{\infty}_{C,\theta}(G_{n-1})$ and therefore $C^{-\infty}_{\theta}(P_n)$ is canonically isomorphic to $C^{-\infty}_{\theta}(G_{n-1})$. This follows because by Theorem 4, for each nonreducible, admissible, nondegenerate representation $(\tau, U)$ of the group $G_{n-1}$ there exists a unique subspace $W_{n-1}(\tau) \subset C^{-\infty}_{\theta}(P_n)$ such that right shifts on the elements of $G_{n-1}$ induce in $W_{n-1}(\tau)$ a representation isomorphic to $(\tau, U)$.

We now turn to the structure of the Γ-function. Let $(\pi, V)$ be a nonreducible cuspidal representation of the group $G_n$. Set $s_n = \sum_{i=1}^{n} (-1)^{i-1} e_{i,n-i+1}$, where $e_{i,k}$ is the matrix with the unit in the $(i, k)$ place and zeros in the remaining places. According to Theorem 2 there exists a functional $l$ such that $l(\pi(z)\xi) = \theta(z) l(\xi)$ for $z \in Z_n$, $\xi \in V$.

We define the functional $\hat{l}$ on V by the formula $\hat{l}(\xi) = l(\pi(s_n)\xi)$ $(\xi \in V)$. Then $l(\pi(z)\xi) = \theta(z)\hat{l}(\xi)$ for any $z \in Z_n$, $\xi \in V$. Define $\varphi_{\pi}$ and $\varphi_{\hat{\pi}}$: $V \to C^{\infty}_{C,\theta}(P_n)$ by the formulas $\varphi_{\pi}(\xi)(p) = l(\pi(p)\xi)$ and $\varphi_{\hat{\pi}}(\xi)(p) = l(\hat{\pi}(p)\xi)$. By Theorem 5 $\varphi_{\pi}$ and $\varphi_{\hat{\pi}}$ are isomorphisms. Denote by $K(\pi)$ the operator $K(\pi) = \varphi_{\hat{\pi}} \circ \varphi_{\pi}^{-1}$: $C^{\infty}_{C,\theta}(P_n) \to C^{\infty}_{C,\theta}(P_n)$. Since by Theorem 3 the functional $l$ is defined uniquely to within a factor, then $K(\pi)$ is defined correctly. Since $\varphi_{\pi}$ and $\varphi_{\hat{\pi}}$ are $P_n$-invariants, then $\rho(g)K(\pi) = K(\pi)\rho(g'^{-1})$ for all $g \in G_{n-1} \subset P_n$, where $\rho(g)$ is the right shift operator in $C^{\infty}_{C,\theta}(P_n)$. It is possible to "correct" $K(\pi)$ so that it commutes with $\rho(g)$. For this we define the operator A: $C^{\infty}_{C,\theta}(P_n) \to C^{\infty}_{C,\theta}(P_n)$ in the following manner. Let $f(p) \subset C^{\infty}_{C,\theta}(P_n)$. Represent p in the form $p = mg_{n-1}(m \in M_n, g_{n-1} \in G_{n-1} \subset P_n)$. Assume that $Af(p) = f(p_1)$, where $p_1 = ms_{n-1}g^{-1}_{n-1}$. We introduce the operator $C(\pi) = A \circ K(\pi)$; $C(\pi)$: $C^{\infty}_{C,\theta}(P_n) \to C^{\infty}_{C,\theta}(P_n)$; in this connection $C(\pi) \circ \rho(g) = \rho(g) \circ C(\pi)$ for $g \in G_{n-1}$.

We denote by $C'(\pi)$ the dual operator which operates in the space $C^{-\infty}_{\theta}(P_n)$. Let $(\tau, U)$ be a nonreducible, admissible, nondegenerate representation of $G_{n-1}$. By Theorem 4 there is a unique subspace $W_{n-1}(\tau) \subset C_{\theta}(G_{n-1})$ such that in $W_{n-1}(\tau)$ the representation $(\tau)$ the representation $(\tau, U)$ is realized by right shifts. From the uniqueness of $W_{n-1}(\tau)$ it follows that the operator $C'(\pi)$ translates it into itself, and thus, because of the irreducibility of $W_{n-1}(\tau)$, the restriction of $C'(\pi)$ to $W_{n-1}(\tau)$ is the operator of multiplication by a number. We denote this number by $\Gamma(\pi, \tau)$ and call it the Γ-function of the pair of representations $\pi$ and $\tau$.

We now turn to a formulation of a more precise noncommutative law of reciprocity. Let $\sigma$ be a finite dimensional representation of the Shafarevich-Weil group $W_K$ of the field K. We define the function $\Gamma_{\psi}(\sigma) = L(1/2, \sigma)/L(1/2, \hat{\sigma})\varepsilon(\psi, \hat{\sigma})$, where $\hat{\sigma}$ is the representation; contragradient to $\sigma$. For the definition of the functions $L(s, \delta)$, $\varepsilon(\psi, \sigma)$ see [5].

Hypothesis 2. There exists an embedding $\varkappa_n$: $(\hat{W}_k)_n \to \Pi_n$ of the space $(\hat{W}_k)_n$ of n-dimensional representations of the group $W_k$ in the space $\Pi_n$ of nonreducible, admissible, nondegenerate representations of the group GL(n, K) such that $\Gamma_\psi(\pi, \hat{\tau}) = \Gamma_\psi(\varkappa_n, \pi, \varkappa_{n-1}\tau)$ for each $\pi \in (\hat{W}_k)_n$, $\tau \in (\hat{W}_k)_{n-1}$; the mapping $\varkappa_n$ is bijective.

This hypothesis is a more precise form of a conjecture of Langlands [2] for G = GL(n, K).

## LITERATURE CITED

1. A. Weil, Math. Ann. 168, 149-156 (1967).
2. R. P. Langlands, "Problems in the theory of automorphic forms," Preprint [Russian translation] Matematika, 15, No. 2, 57-83 (1971).
3. H. Jacuet and R. P. Langlands, Automorphic Forms on GL(Z), Lecture Notes in Math., 114, Springer-Verlag, 1970.
4. I. M. Gel'fand and M. I. Graev, Dokl. Akad. Nauk. SSSR, 147, No. 3, 529-532 (1962).
5. R. P. Langlands, "On Artin's L-functions," Rice University Studies, Complex Analysis, 56, No. 2, 23-28 (1970).
6. Harish-Chandra (Notes by van Dijk), Harmonic Analysis on Reductive p-adic Groups, Lecture notes in Math., 114, Springer Verlag (1970).

# 8.

## (with I. N. Bernstein and S. I. Gelfand)

# Models of representations of Lie groups*

Proc. Petrovskij Semin. **2** (1976) 3–21. [Sel. Math. Sov. **1** (2) (1981) 121–142].
Zbl. **499**:22004

Starting with the classical works of E. Cartan and H. Weyl, representations of compact Lie groups have been studied in sufficient detail. Yet although the characters of the irreducible representations have been described with exhaustive thoroughness and lucidity, the construction of the representations themselves is in a less satisfactory state. Just about the only general construction, inspired by the theory of infinite-dimensional representations of groups, is as follows. Let $U$ be a compact Lie group. Consider the so-called principal affine space $N\backslash G$, where $G$ is the complex Lie group corresponding to $U$ and $N$ is a maximal unipotent subgroup of $G$. Then one can realize the irreducible representations of the group $U$ in the space of homogeneous analytic functions on the principal affine space (in other words, in the space of analytic sections of some one-dimensional fiber space over the quotient space of the group $U$ by a maximal torus).

We are dealing here with a model of the representations; namely one can introduce a scalar product on the space of analytic functions on the principal affine space so that in the decomposition of the resulting unitary representation of $U$ into irreducible factors, all the irreducible representations of $U$ occur with multiplicity one (see [1, 4]). Granted the naturalness of this approach, just about the only shortcoming of this construction is that in this model we require the functions to be analytic.

Let us consider an example to clarify this. Let $U = SO_3$ be the group of rotations of 3-space. Then the model associated with the principal affine space is as follows. Consider the space of even analytic functions $f(z_1, z_2)$ in two complex variables with the scalar product

$$\langle f, g\rangle = \int f\bar{g}e^{-|z_1|^2-|z_2|^2}\,dz_1\,d\bar{z}_1\,dz_2\,d\bar{z}_2.$$

*Originally published in Proceedings of the I. G. Petrovsky Seminar on Differential Equations and Mathematical Problems of Physics, Moscow University, 2 (1976), 3–21. Translated by Mikhail Katz.

Decomposing the functions in this space into homogeneous functions, we get every irreducible representation of $SO_3$ exactly once. However, long before this construction one knew how to construct the representations of $SO_3$ in the space of all (not just analytic) square-integrable functions on the 2-sphere. This also gives a model of the representations of $SO_3$; i.e., every irreducible representation of $SO_3$ occurs once.

In this article a similar model of its representations is constructed for every semisimple compact Lie group $U$; i.e., a homogeneous space is described, such that every irreducible representation of $U$ is contained exactly once in the space of all (not just analytic) suitably chosen square-integrable vector functions on the homogeneous space. For the group $SO_3$ this realization coincides with the one described above.

In our realization the homogeneous space is the compact symmetric space $X$ corresponding to the group $U$ (see [5]). Let $K_{\mathbf{R}}$ be the stationary subgroup of some point of $x_0 \in X$. We will construct a representation $\tau$ of the group $K_{\mathbf{R}}$ such that the induced representation of $U$ contains every irreducible representation of $U$ exactly once. For a simply connected group $U$ we have $\dim \tau = 2^l$, where $l$ is the rank. (The space $L^2(X)$ contains one $2^l$th part of all the irreducible representations of $U$, the representations with even highest weight.)

In this article the necessary representations of the stationary subgroup $K_{\mathbf{R}}$ are constructed separately for each of the simple Lie groups. This method has its advantages, because the resulting spaces and representations of $K_{\mathbf{R}}$ are of independent interest.

As an example, let us show what our models look like for the classical Lie groups. The representations given here differ in appearance from those constructed in Section 4, because simply connected simple groups will be considered there. For this reason we have had to use the apparatus of Clifford algebras in Section 4. Recall that we are considering the representation of the group $U$ induced by the representation $\tau$ of the stationary subgroup $K_{\mathbf{R}}$:

1. $U = U_n$, $K_{\mathbf{R}} = O_n$, $\tau$ is the natural representation of $O_n$ in the space $\oplus_{i=0}^{n} \Lambda^i(\mathbf{R}^n)$.
2. $U = O_{2n+1}$, $K_{\mathbf{R}} = O_{n+1} \times O_n$, $\tau$ is the representation of $K_{\mathbf{R}}$ in the space $\oplus_{i=0}^{n+1} \Lambda^i(\mathbf{R}^{n+1})$, trivial on the second factor.
3. $U = USp_{2n}$, $K_{\mathbf{R}} = U_n$, $\tau$ is the natural representation of $K_{\mathbf{R}}$ in $\oplus_{i=0}^{n} \Lambda^i(\mathbf{C}^n)$.
4. $U = O_{2n}$, $K_{\mathbf{R}} = O_n \times O_n$, $\tau$ is the representation of $K_{\mathbf{R}}$ in the space $\oplus_{i=0}^{n} \Lambda^i(\mathbf{R}^n)$, trivial on the second factor.

This paper is an introduction to the study, from a unified viewpoint, of the representations of noncompact real Lie groups and of semisimple algebraic groups over various fields (see [2]).

## 1. Statement of the main results

*1.* Let $G$ be a connected algebraic reductive group over the field of complex numbers. Fix a Cartan subgroup $H$ of $G$. Denote by $R_H$ the lattice of weights of the group $H$, consisting of the algebraic homomorphisms $H \to \mathbf{C}^*$. Let $\Delta \subset R_H$ be the root system of the group $G$ relative to $H$; for every root $\gamma$, denote by $N_\gamma$ the one-parameter unipotent subgroup of $G$ which corresponds to the root $\gamma$. Fix a system of positive roots $\Delta_+ \subset \Delta$.

Let $\theta$ be a fixed Cartan involution of the group $G$, i.e., an algebraic antiautomorphism such that $\theta^2 = \mathrm{id}$ and $\theta(h) = h$ for all $h \in H$ ( it is easy to check that any two such involutions are conjugate by an inner automorphism corresponding to some element of $H$). Clearly, $\theta(N_\gamma) = N_{-\gamma}$ for all $\gamma \in \Delta$.

The group $K = \{g \in G | \theta(g) = g^{-1}\}$ will be called an *involutory subgroup* of $G$. For example, if $G = \mathrm{GL}_n(\mathbf{C})$, and $H$ is the subgroup of diagonal matrices, then $\theta$ can be taken to be the map $g \mapsto g^T$ (transposition); in this case $K = O_n(\mathbf{C})$.

Let $i$ be an antilinear automorphism of the group $G$ that maps $H$ to itself and preserves the lattice of weights $R_H$ (i.e., $\chi(i(h)) = \overline{\chi(h)}$ for all $\chi \in R_H$, $h \in H$). Then $i(N_\gamma) = N_\gamma$ for all $\gamma \in \Delta$, so that the subgroup $G_R = \{g \in G | i(g) = g\}$ is a split real form of the group $G$. Assume that $i$ commutes with $\theta$ or, equivalently, $\theta(G_R) = G_R$. Then the subgroup $U = \{g \in G | i(\theta(g)) = g^{-1}\}$ is compact; it is the compact form of the group $G$. The compact subgroup $K_R = U \cap G_R = \{u \in U | \theta(u) = u^{-1}\}$ will play an important role. Let us call it the *involutory subgroup* of the group $U$.

The group $S = K \cap H = \{h \in H | h^2 = 1\}$ is also essential here. Since every element $h \in H$ is determined by the numbers $\chi(h)$, $\chi \in R_H$, it follows that $i(s) = s$ for all $s \in S$, so that $S \subset K_{\mathbf{R}}$. From the definition it follows that $S$ is a finite commutative group and $\operatorname{card} S = 2^{\operatorname{rk} G}$, where $\operatorname{rk} G$ is the rank of $G$.

**Example.** $G = GL_n(\mathbf{C})$, $H$ is the diagonal subgroup, $\theta$ is the transposition of matrices, $i$ is the passage to the complex conjugate matrix. Then

$$K = O_n(\mathbf{C}); \qquad G_{\mathbf{R}} = GL_n(\mathbf{R}), \qquad K_{\mathbf{R}} = O_n,$$

and $S$ consists of diagonal matrices with $\pm 1$ on the diagonal.

*2.* Let us state the main results of this work. Let $\tau$ be a finite-dimensional representation of the group $K_{\mathbf{R}}$. For every irreducible representation $\pi$ of the group $U$, we are interested in the multiplicity with which the representation $\pi$ occurs in the representation $\operatorname{Ind}_{K_{\mathbf{R}}}^U(\tau)$, i.e., the number

$$\dim \operatorname{Hom}_U\big(\pi, \operatorname{Ind}_{K_{\mathbf{R}}}^U(\tau)\big).$$

**Proposition 1.** *Let $m \in R_H$ be the highest weight of the irreducible representation $\pi$ and $m|_S$ the restriction of $m$ to $S$, the corresponding one-dimensional* *representation of S. Then*

$$\dim \operatorname{Hom}_U\big(\pi, \operatorname{Ind}^U_{K_{\mathbf{R}}}(\tau)\big) \leqslant \dim \operatorname{Hom}_S(m|_S, \tau|_S). \tag{1}$$

We say that a representation of a compact group is multiplicity-free if it can be decomposed into the sum of pairwise inequivalent irreducible representations.

**Corollary 1.** *If* $\tau|_S$ *is multiplicity-free then so is* $\operatorname{Ind}^U_{K_{\mathbf{R}}}(\tau)$.

It turns out that for "almost all" irreducible representations of the group $U$, strict equality holds in formula (1).

More precisely, let $C \in R_H$ be the collection of highest weights of all irreducible representations of the group $U$ (i.e., the Weyl chamber in $R_H$ relative to the ordering given by the system $\Delta_+$).

**Proposition 2.** *Let* $\tau$ *be a fixed representation of the group* $K_{\mathbf{R}}$. *Then there exists a weight* $l \in C$ *such that for all weights* $m \in C + l$ *one has*

$$\dim \operatorname{Hom}_U\big(\pi, \operatorname{Ind}^U_{K_{\mathbf{R}}}(\tau)\big) = \dim \operatorname{Hom}_S(m_S, \tau|_S), \tag{2}$$

*where* $m_S = m|_S$ *and* $\pi$ *is the irreducible representation of* $U$ *with highest weight* $m$.

**Corollary 2.** *Let* $\tau$ *be a representation of the group* $K_{\mathbf{R}}$ *such that* $\tau|_S$ *is the regular representation of the group* $S$. *Then for any irreducible representation* $\pi$ *of the group* $U$, $\dim \operatorname{Hom}_U(\pi, \operatorname{Ind}^U_{K_{\mathbf{R}}}(\tau)) \leqslant 1$, *and there exists a weight* $l \in C$ *such that* $\dim \operatorname{Hom}_U(\pi, \operatorname{Ind}^U_K(\tau)) = 1$ *for all irreducible representations* $\pi$ *with highest weight* $m \in C + l$.

We want to find the representations $\tau$ such that equality (2) holds for all irreducible representations of the group $U$. It is easy to point out one such example.

**Proposition 3.** *If* $\tau = 1$ *is the trivial representation of the group* $K_{\mathbf{R}}$, *an irreducible representation* $\pi$ *of group* $U$ *occurs in* $\operatorname{Ind}^U_{K_{\mathbf{R}}}(1)$ *if and only if* $m|_S = 1$, *i.e.,* $m$ *is an even weight* ($m \in 2R_H$).

Propositions 1–3 are proved in Section 2.

**Definition.** A representation of a compact group is called a model if any irreducible representation occurs in it exactly once.

One could say that if $\tau$ is any representation of the group $K_{\mathbf{R}}$ such that $\tau|_S$ is a regular representation of $S$, then $\operatorname{Ind}^U_{K_{\mathbf{R}}}(\tau)$ is "almost a model." An important result of our work is the construction of a representation $\tau$ of the group $K_{\mathbf{R}}$ for every group $G$ such that $\operatorname{Ind}^U_{K_{\mathbf{R}}}(\tau)$ is a model for the group $U$. Namely, we have:

**Theorem 1** (on models). *There exists a representation* $\tau$ *of the group* $K_{\mathbf{R}}$ *such that the representation* $\operatorname{Ind}^U_{K_{\mathbf{R}}}(\tau)$ *is a model; i.e., every irreducible representation of* $U$ *occurs exactly once in the decomposition.*

This theorem is proved in Sections 3 and 4. In Section 3 we state a condition on the representation $\tau$ under which $\mathrm{Ind}_{K_{\mathbf{R}}}^{U}(\tau)$ is a model (Theorem 1), and reduce the problem to the case of simple groups. In Section 4 we construct the representation $\tau$ for each simple group separately.

## 2. Proof of Propositions 1–3

*1.* The proof of the propositions formulated above is based on an analysis of the dimensions

$$\dim \mathrm{Hom}_U\big(\pi, \mathrm{Ind}_{K_{\mathbf{R}}}^{U}(\tau)\big) = \dim \mathrm{Hom}_{K_{\mathbf{R}}}(\pi|_{K_{\mathbf{R}}}, \tau) \quad \text{and} \quad \dim \mathrm{Hom}_S(m|_S, \tau|_S).$$

First of all, by Frobenius' duality

$$\dim \mathrm{Hom}_U\big(\Pi, \mathrm{Ind}_{K_{\mathbf{R}}}^{U}(\tau)\big) = \dim \mathrm{Hom}_{K_{\mathbf{R}}}(\Pi|_{K_{\mathbf{R}}}, \tau).$$

Further, every finite-dimensional representation of the group $U$ can be extended to an algebraic representation of the group $G$; we thus get a one-to-one correspondence between the representations of the groups $U$ and $G$ (in the following we will consider only the algebraic representations of $G$). Since $K_{\mathbf{R}}$ is a compact form of the group $K$, a similar assertion also holds for the groups $K_{\mathbf{R}}$ and $K$. Thus our problem is reduced to comparing the numbers $\dim \mathrm{Hom}_K(\pi|_K, \tau)$ with the numbers $\dim \mathrm{Hom}_S(m|_S, \tau|_S)$ in a purely algebraic setting. In the following we will only be concerned with algebraic groups and representations.

*2.* We introduce some additional notation. Let $N = \prod_{\gamma \in \Delta_+} N_\gamma$ be the maximal unipotent subgroup, $N^- = \prod_{\gamma \in \Delta_+} N_{-\gamma}$ the opposite subgroup, $B = HN$ the Borel subgroup, and let $B^- = HN^-$. Clearly, $\theta(B) = B^-$, $\theta(N) = N^-$.

**Lemma 1.** *$B \cap K = S$, and $KB$ is an open dense subset of $G$.*

**Proof.** Clearly, $B \cap K = \{g \in B \,|\, \theta(g) = g^{-1}\}$. Since $\theta(B) = B^-$ and $B \cap B^- = H$, we have $B \cap K = H \cap K = S$. Considering the tangent spaces at the identity, it is easy to convince oneself that $KB$ is an open and (because $G$ is connected) dense subset of $G$.

Suppose we are given a representation $\tau$ of the group $K$ on the space $L$. We will say that the rational function $f$ on the group $G$ with values in $L$ is $\tau$-equivariant if $f(kg) = \tau(k)f(g)$ for all $k \in K$, $g \in G$.

Suppose $m$ is a weight of the group $H$. Extend it to a character of the group $B$, setting $m(N) = 1$; we say that the function $f$ on $G$ has weight $m$ if $f(gb) = m(b)f(g)$ for all $b \in B$, $g \in G$.

Since $KB$ is dense in $G$, every $\tau$-equivariant rational function $f$ of weight $m$ is regular on $KB$.

*3.* The following lemma reduces the study of the multiplicities we are interested in to the study of the dimensions of certain spaces of algebraic functions on $G$.

**Lemma 2.**

a. *Let $\tau$ be a representation of the group $K$ in the space $L$ and $m \in R_H$. Then the space* $\mathrm{Hom}_S(m|_S, \tau|_S)$ *is isomorphic to the space of rational $\tau$-equivariant functions of weight $m$.*
b. *Let $m$ be the highest weight of an irreducible representation $\pi$ of the group $G$. Then the space* $\mathrm{Hom}_K(\pi|_K, \tau)$ *is isomorphic to the space of regular $\tau$-equivariant functions of weight $m$.*

**Proof.**

a. Let $L$ be the space of the representation $\tau$, and $\varphi : \mathbf{C} \to L$ be a homomorphism from $\mathrm{Hom}_S(m|_S, \tau|_S)$. Define a regular function $\tilde{f}$ on $K \times B$ by the formula $\tilde{f}(k,b) = m(b) \cdot \tau(k)\varphi(1)$.
Clearly, $\tilde{f}(ks, b) = \tilde{f}(k, sb)$; i.e., $\tilde{f}$ depends only on the product $kb$; therefore, $\tilde{f}(k,b) = f(kb)$, where $f$ is a certain regular $\tau$-equivariant function of weight $m$ on $KB$. $f$ can be viewed as a rational function on $G$. Conversely, for every such function one constructs a homomorphism $\varphi \in \mathrm{Hom}_S(m|_S, \tau|_S)$ by the formula $\varphi(1) = f(1)$.
b. Let $V$ be the space of the representation $\pi$, and let $v^+$ be the highest weight vector in $V$. For every $\varphi \in \mathrm{Hom}_K(\pi|_K, \tau)$ one constructs a regular $\tau$-equivariant function of weight $m$ on $G$ by

$$f(g) = \varphi(\pi(g)v^+).$$

To construct the inverse mapping, consider the representation $\pi^*$ on the space $V^*$ that is dual to $\pi$. Define the mapping $\psi$ from the space $V^*$ to the space of complex regular functions of weight $m$ on $G$ by the formula $\psi(v^*)(g) = (v^*, \pi(g)v^+)$. The mapping $\psi$ is an isomorphism [7, 9].

Now let $f$ be a regular $\tau$-equivariant function of weight $m$. Then to every vector $l^* \in L^*$ there corresponds a regular function of weight $m$ $u_{l^*}(g) = (l^*, f(g))$ and therefore the element $\psi^{-1}(u_{l^*}) \in V^*$. In this way we have obtained a mapping $\varphi^* : L^* \to V^*$ and the dual mapping $\varphi : V \to L$. Clearly, $\varphi \in \mathrm{Hom}_K(\pi|_K, \tau)$. The mappings just constructed define an isomorphism between the space of regular $\tau$-equivariant functions of weight $m$ and the space $\mathrm{Hom}_K(\pi|_K, \tau)$. The lemma is proved.

Proposition 1 immediately follows from Lemma 2.

4. *Proof of Proposition 3.* It follows from Lemma 2 that if a representation $\pi$ with highest weight $m$ occurs in $\mathrm{Ind}_K^U(1)$, then $m|_S = 1$; i.e., $m$ is an even weight. Conversely, suppose that $m$ is an even weight; i.e., $m = 2l$ where $l \in C$. We have to prove that there exists a nonzero regular function $q_m(g)$ such that $q_m(kgb) = m(b)q_m(g)$ for all $k \in K$, $g \in G$, $b \in B$.

Let $\rho$ be an irreducible representation of the group $U$ with highest weight

$l$, $V$ the space of the representation $\rho$, $V^*$ the dual space, $v^+$ a vector of highest weight in $V$, $v^{*-}$ a vector of lowest weight in $V^*$. Set $a_l(g) = (v^{*-}, \rho(g)v^+)$. Then

$$a_l(\theta(b_1)gb_2) = l(b_1b_2)a(g).$$

Indeed,

$$\begin{aligned} a_l(\theta(b_1)gb_2) &= (v^{*-}, \rho(\theta(b_1))\rho(g)\rho(b_2)v^+) \\ &= (\rho^*(\theta(b_1))^{-1}v^{*-}, \rho(g)\rho(b_2)v^+) \\ &= l(b_1)l(b_2)(v^{*-}, \rho(g)v^+) = l(b_1b_2)a_l(g). \end{aligned}$$

Now set $q_{2l}(g) = a_l(\theta(g)g)$. Then $q_{2l}(kg) = q_{2l}(g)$ for all $k \in K$, $g \in G$ and

$$q_{2l}(gb) = a_l(\theta(b)gb) = l(b^2)q_{2l}(g) = m(b)q_{2l}(g),$$

as required.

5. *Proof of Proposition 2.* Let $\tau$ be a representation of the group $K$. By Lemma 2 it suffices to prove that there exists a weight $l_0 \in C$ such that every $\tau$-equivariant rational function of weight $m \in l_0 + C$ is regular. In Section 2.4 we constructed, for every even weight $m \in 2R_H$, a regular function $q_m$ on $G$ satisfying $q_m(kgb) = m(b)q_m(g)$. Clearly, the mapping $f \to q_m f$ establishes an isomorphism between the space of $\tau$-equivariant functions of weight $l$ and the space of $\tau$-equivariant functions of weight $l + m$.

Note that $R_H/2R_H$ is a finite set, and the space of $\tau$-equivariant rational functions of any given weight is finite-dimensional. Since any $\tau$-equivariant weight function $f$ is regular on $KB$, Proposition 2 follows from the following lemma.

**Lemma 3.** *There exists a weight* $m \in 2R_H$ *such that* $q_m(g) = 0$ *for all* $g \in G$, $g \notin KB$.

**Proof.** We first prove that $g \in KB$ if and only if $\theta(g)g \in B^-B = N^-HN$. Indeed, if $g = kb \in KB$, then $\theta(g)g = \theta(b)\theta(k)kb = \theta(b)b \in B^-B$. Conversely, let

$$g_1 = \theta(g)g, \quad \text{and} \quad g_1 = u^-hu, \quad u^- \in N^-, \quad h \in H, \quad u \in N. \tag{3}$$

Clearly, $\theta(g_1) = g_1$; i.e., $g_1 = \theta(u)\theta(h)\theta(u^-)$. Since $\theta(u) \in N^-$, $\theta(h) = h \in H$, $\theta(u^-) \in N$, and the decomposition (3) is unique, we have $\theta(u) = u^-$. Since the group $G$ is connected, there exists an element $h_1 \in H$ such that $h_1^2 = h$. Set $b = h_1u$ and $k = gb^{-1}$. Then it is clear that $\theta(b)b = g_1$ and $\theta(k)k = 1$; i.e., $k \in K$. Therefore, $g = kb \in KB$.

To complete the proof we use the formula $q_{2l}(g) = a_l(\theta(g)g)$, where $a_l$ is the function introduced in Section 2.4.

Let $l$ be an arbitrary regular highest weight, i.e., $wl \neq l$ for any element $w$ of the Weyl group $W$ of the group $G$. Then it is known (see, for example,

[7]) that $a_l(g) = 0$ when $g \notin N^- HN$. Therefore, $q_{2l}(g) = a_l(\theta(g)g) = 0$ when $g \notin KB$.

### 3. Shallow representations

This section and Section 4 are devoted to proving the theorem on models. Here we present a method that allows us to determine when a representation $\tau$ of the group $K_{\mathbf{R}}$ is a model (Theorem 1). In Section 4 we will turn to the construction of models using this method.

*1.* We first reduce the problem of constructing a model to the case of a simply connected group $G$. Let $p: G_1 \to G$ be the universal cover; $\theta$, $\theta_1$ and $i$, $i_1$ agree with $p$. Then $pU_1 = U$, $pK_{1\mathbf{R}} \subset K_{\mathbf{R}}$. Suppose $\tau_1$ defines a model for $K_{1\mathbf{R}}$ (i.e., $\operatorname{Ind}_{K_{1\mathbf{R}}}^{U_1}(\tau_1)$ is a model for $U_1$) and $\tau' \subset \tau_1$ is the largest subrepresentation which is trivial on $\operatorname{Ker} p \cap K_{1\mathbf{R}}$. Then $\tau'$ can be viewed as a representation of $pK_{1\mathbf{R}} \subset K_{\mathbf{R}}$. Now it is easy to show that $\tau = \operatorname{Ind}_{pK_{1\mathbf{R}}}^{K_{\mathbf{R}}}(\tau')$ defines a model for $K_{\mathbf{R}}$.

*2.* By Propositions 1 and 2 and Lemma 2 the representation $\tau$ of the group $K$ defines a model if and only if the following two properties hold:

1. $\tau|_S$ is a regular representation of $S$.
2. Suppose that $m \in C$ and $f$ is a $\tau$-equivariant rational function on $G$ of weight $m$, then $f$ is a regular function.

In this section we will find a condition on the representation $\tau$ that guarantees condition 2.

It is convenient for us to consider the simplest case first—$G = SL_2(\mathbf{C})$. Let $i$ be complex conjugation, $i(g) = \bar{g}$, and $\theta$ transposition. Then

$$U = SU_2, \quad K = SO_2(C), \quad K_{\mathbf{R}} = SU_2 \cap SL_2(\mathbf{R}) = SO_2, \quad S = \{\pm e\},$$

$$H = \left\{\begin{pmatrix} t & 0 \\ 0 & t^{-1} \end{pmatrix}\right\}, \qquad B = \left\{\begin{pmatrix} t & x \\ 0 & t^{-1} \end{pmatrix}\right\}.$$

The subgroups $H$ and $K$ are isomorphic to the multiplicative group $\mathbf{C}^*$. Hence their one-dimensional (algebraic) representations are given by an integer, the degree $n$ of the representation $t \to t^n$.

**Lemma 4.** *If $g = \theta(g)$, then $g = \theta(g_1)g_1$ for some $g_1 \in G = SL_2(\mathbf{C})$.*

The proof is by direct verification.

**Lemma 5.** *Let $\tau$ be a one-dimensional representation of $K$ of degree $j =$ 0, $+1$, or $-1$; and let $m$ be the highest weight of $B$, $m\begin{pmatrix} t & x \\ 0 & t^{-1} \end{pmatrix} = t^n$, $n \geqslant 0$. Let $f$ be a rational $\tau$-equivariant function on $G$. Then $f$ is a regular function.*

**Proof.** Since $KB$ is dense in $G$, $f(e) \neq 0$. Now, setting $b = k = -e$, $g = e$ in

$$f(kgb) = \tau(k)m(b)f(g), \tag{4}$$

we see that the weight $m$ must be even if $i = 0$ and odd if $i = \pm 1$.

In the space of all regular functions on $G$ that satisfy $f(gb) = m(b)f(g)$, left translations realize the $(n+1)$-dimensional representation of the group $G = SL_2(\mathbf{C})$ ($n$ is the degree of weight $m$). But the $(n+1)$-dimensional representation of $SL_2(\mathbf{C})$ contains all the weights of $H$ of degrees between $-n$ and $n$ whose parity coincides with that of $n$. Therefore, the existence of a nonzero regular function $f$ satisfying the condition follows from the fact that the groups $K$ and $H$ are conjugate in $SL_2(\mathbf{C})$. The lemma now follows from the fact that a rational function $f$ is determined uniquely up to a factor by (4).

**Corollary.** *Let $\tau$ be a representation of $K$ which is a direct sum of representations of degrees* 0, $+1$, *and* $-1$, *and $f$ a rational $\tau$-equivariant function on $G$ of weight $\geqslant 0$ with respect to $B$. Then $f$ is a regular function.*

*3.* Let us return to the general case. For every simple root $\alpha$ of the group $G$ consider the corresponding homomorphism $\varphi_\alpha : SL_2(\mathbf{C}) \to G$ which agrees with the choice of $i$ and $\theta$. Since the group $G$ is simply connected, $\varphi_\alpha$ is an embedding [7]. Set

$$G_\alpha = \varphi_\alpha(SL_2(\mathbf{C})) \quad \text{and} \quad K_\alpha = \varphi_\alpha(SO_2(\mathbf{C})) \subset K.$$

A representation $\tau$ of the group $K$ is called shallow if for any simple root $\alpha$ the restriction of $\tau$ to $K_\alpha \cong SO_2(\mathbf{C})$ contains only one-dimensional representations of $K_\alpha$ of degrees 0, $+1$, and $-1$.

**Theorem 2.** *Let $\tau$ be a shallow representation of the group $K$, $m \in C$, and $f$ a $\tau$-equivariant rational function on $G$ of weight $m$. Then $f$ is a regular function.*

**Proof.** Since $KB$ is dense in $G$, the function $f$ is regular on $KB$. Further, $G_\alpha \cap KB$ is dense in $G_\alpha$, and therefore the function $f$ can be restricted to $G_\alpha$. By the corollary to Lemma 5, $f|_{G_\alpha}$ is a regular function on $G_\alpha$. Therefore, the function $f$ is regular on the set $P = \cup_\alpha KG_\alpha B$. Therefore, Theorem 1 follows from the following lemma:

**Lemma 6.** $\dim(G\backslash P) \leqslant \dim G - 2$.

**Proof.** Let $G_\theta = \{g \in G \mid g = \theta(g)\}$. Consider the mapping $r: G \to G_\theta$ given by $r(g) = \theta(g)g$. The inverse image of any point $x \in G_\theta$ is either empty or homeomorphic to $K$. Since $\dim G = \dim K + \dim C_\theta$, to prove the lemma it suffices to show that $\dim(G_\theta \backslash r(P)) \leqslant \dim G_\theta - 2$. To this end we use the Bruhat decomposition of the group $G$. Let $WH \subset G$ be the normalizer of $H$ in $G$, and for each $w \in W$ pick a representative $x_w \in wH$. Define the subgroups

$$N_w^+ = N^+ \cap wN^+w^{-1}, \qquad \tilde{N}_w^+ = N^+ \cap wN^-w^{-1},$$
$$N_w^- = N^- \cap w^{-1}N^-w, \qquad \tilde{N}_w^- = N^- \cap w^{-1}N^+w,$$

so that $N^- = N_w^-\tilde{N}_w^-$, $N_w^- \cap \tilde{N}_w^- = \{e\}$. The Bruhat decomposition asserts that every element $g \in G$ can be uniquely represented as $g = u^-hx_wu^+$, where $w \in W$, $u^+ \in N_w^+$, $h \in H$, $u^- \in N^-$. Set $G^w = N^- H x_w N_w^+$; now consider $G^w \cap G_\theta$ for each $w$.

a. If $n \in wH$ for some $w$, then $\theta(n) \in w^{-1}H$. Since $\theta(N^-) = N^+$, $\theta(N^+) = N^-$, we have $\theta(G^w) = G^{w^{-1}}$. Therefore, $G^w \cap G_\theta \neq \emptyset$ only if $w \neq w^{-1}$.

b. Let $w = w^{-1}$. Then $\theta(N_w^+) = N_w^-$, $\theta(\tilde{N}_w^+) = \tilde{N}_w^-$.
Let

$$g \in G^w \cap G_\theta \quad \text{and} \quad g = u^- \tilde{u}^- n u^+$$

be its Bruhat decomposition. We have

$$\theta(g) = \theta(u^+)\theta(n)\theta(\tilde{u}^-)\theta(u^-),$$

and the equality $g = \theta(g)$ is equivalent to the following set of equalities:

$$\theta(n) = n; \qquad \theta(u^-) = u^+; \qquad \theta(\tilde{u}^-) = n^{-1}\tilde{u}^- n.$$

Therefore

$$\dim G^w \cap G_\theta = d_1^w + d_2^w,$$

where

$$d_2^w = \dim(G_\theta \cap wH) \leqslant \dim H, \qquad d_2^w = \dim\{\tilde{u}^- \in \tilde{N}_w^-, \theta(\tilde{u}^-) = n^{-1}\tilde{u}^- n\}$$

(one can prove that $d_1^w$ depends only on $w$, not on $n \in wH$). Let us estimate $d_1^w$ and $d_2^w$ separately.

c. Let $n_0$ be a fixed element of $G_\theta \cap wH$ and $n = hn_0 \in G_\theta \cap wH$. Then $hn_0 = n = \theta(n) = \theta(n_0)\theta(h) = whw^{-1} \cdot n_0$, i.e., $h = whw^{-1}$. Therefore, $d_1^w = \dim H$ {multiplicity of the eigenvalue $+1$ of the element $w$}. It is well known [7] that if $w \neq w_\gamma$ for some root $\gamma$, then $w$ is not a reflection in a hyperplane, and hence $d_1^w \leqslant \dim H - 2$. Therefore,

$$\dim(G^w \cap G_\theta) \leqslant \dim H + \dim N^- - 2 = \dim G_\theta - 2$$

where

$$w \neq w_\nu, w \neq e.$$

d. Let $w = w_\gamma$ for some positive root $\gamma$. Then $d_1^w = \dim H - 1$. Let us estimate $d_2^w$. Fix $h \in wH \cap G_\theta$. Then $d_2^w$ equals the dimension of the space $X_f$ of fixed points of the map $f(\tilde{u}) = n^{-1}\theta(\tilde{u})n$ of the subgroup $\tilde{N}_w^-$ into itself. Since for every root subgroup $N_\beta$ of $G$, $\theta(N_\beta) = N_{-\beta}$, we have $f(N_\beta) = N_{-w\beta}$. Among the negative roots $\beta$ there is exactly one for which $-w_\gamma\beta = \beta$, namely $\beta = -\gamma$, it is clear that $N_{-\gamma} \in \tilde{N}_w^-$. For all the other negative roots we have $-w_\gamma\beta \neq \beta$. Since whenever $-w_\gamma\beta \neq \beta$ the component of the element $\tilde{u} \in X_f$ in $N_{-w\beta}$ is determined by the component of $\tilde{u}$ in $N_\beta$, it follows that $\dim X_f = 1/2(\dim \tilde{N}_w^- + 1)$. If $w = w_\gamma$ and $\gamma$ is not a simple root, then $\dim N_w^- \geqslant 3$; therefore $d_1^w \leqslant \dim \tilde{N}_w^- - 1$. Hence in this case too, one has $\dim(G^w \cap G_\theta) \leqslant \dim G_\theta - 2$.

e. To prove the lemma it remains to show that $G^e \cap G_\theta \subset r(P)$ and that for any simple root $\alpha$, $G^{w_\alpha} \cap G_\theta \subset r(P)$.

First, suppose that $g \in G^e \cap G_\theta$. Then

$$g = u^- h u^+, \quad u^- \in N^-, \quad h \in H, \quad u^+ \in N^+ \quad \text{and} \quad \theta(u^+) = u^-.$$

Since $H$ is a connected torus, there exists an element $h_1 \in H$ such that $h = h_1^2 = h_1\theta(h_1)$. Then

$$g = u^- h_1^2 u^+ = r(hu^+) \in r(P),$$

because $hu^+ \in B$.

f. Now suppose that $w = w_\alpha$, and $\alpha$ is a simple root. Any element $x \in G^w$ can be represented as $x = u^- hg_\alpha u^+$, where $u^- \in U_w^-$, $u^+ \in U_w^+$, $h \in Z_H^0(HG_\alpha)$ (connected component), $g_\alpha \in G_\alpha$. Here $u^+$, $u^-$, and $hg_\alpha$ are uniquely defined. If $x = \theta(x)$, then $u_1 = \theta(u_2)$ and $\theta(hg_\alpha) = hg_\alpha$. But $\theta(hg_\alpha) = \theta(g_\alpha)h = h\theta(g_\alpha)$ and therefore $g_\alpha = \theta(g_\alpha)$. Since $G_\alpha \cong SL_2(\mathbf{C})$, $g_\alpha = \theta(g_1)g_1$ for some $g_1 \in G_2$ (Lemma 4). Further, $Z_H^0(HG_\alpha)$ is a connected torus; therefore, there exists an $h_1 \in Z_H^0(HG_\alpha)$ such that $h_1^2 = h$. Then

$$x = r(g_1 h_1 u_1^+) \in r(P),$$

because $g_1 h_1 u^+ \in G_\alpha B$.

**Corollary 3.** *Let $\tau$ be a shallow representation of the group $K$ that is regular when restricted to $S$. Then $\mathrm{Ind}_{K_{\mathbf{R}}}^{U}(\tau)$ is a model.*

## 4. Constructing models

*1.* Using Theorem 1 for every semisimple group $G$ we construct a model representation of the corresponding subgroup $K$. Clearly, it suffices to construct such a representation for a simple, simply connected group $G$. We cannot yet offer a general construction for such a representation, so we will have to consider each case separately.

We first list the simple groups $G$ and the corresponding subgroups $K \subset G$ (see [8]). (For the exceptional types, $G$ is taken to be the simple, simply connected group of the type given. See text for the precise description of the subgroup $\mathbf{Z}_2$):

$$A_n : G = SL_n(\mathbf{C}), K = SO_n(\mathbf{C});$$
$$B_n : G = \mathrm{Spin}_{2n+1}, K = (\mathrm{Spin}_{n+1} \times \mathrm{Spin}_n)/\mathbf{Z}_2;$$
$$C_n : G = \mathrm{Sp}_{2n}(\mathbf{C}), K = GL_n(\mathbf{C});$$
$$D_n : G = \mathrm{Spin}_{2n}, K = (\mathrm{Spin}_n \times \mathrm{Spin}_n)/\mathbf{Z}_2;$$
$$E_6 : K = \mathrm{Sp}_8(\mathbf{C})/\mathbf{Z}_2;\ E_7 : K = SL_8(\mathbf{C})/\mathbf{Z}_2;\ E_8 : K = \mathrm{Spin}_{16}/\mathbf{Z}_2;$$
$$F_4 : K = (SL_2(\mathbf{C}) \times \mathrm{Sp}_6(\mathbf{C}))/\mathbf{Z}_2;\ G_2 : K = (SL_2(\mathbf{C}) \times SL_2(\mathbf{C}))/\mathbf{Z}_2.$$

*2.* We will use Clifford algebras in constructing the representations that define a model. We recall the essential facts about these algebras [6].

Let $E$ be an $n$-dimensional complex space, $e_1, \ldots, e_n$ a basis for $E$.

Consider the form $\Sigma x_i^2$ on $E$, and let $O(E)$ and $SO(E)$ be the orthogonal and the special orthogonal groups with respect to this form. Denote by $C_n = C(E)$ the algebra with generators $e_1, \ldots, e_n$ and relations $e_i^2 = -1$, $e_ie_j = -e_je_i$, $i, j = 1, \ldots, n$; $i \neq j$. The group $O(E)$ acts on the algebra $C(E)$ by changing coordinates in the space generated by $e_1, \ldots, e_n$.

We list the basic properties of Clifford algebras.

a. The elements $e_{i_1} \cdots e_{i_k}$, where $1 \leqslant i_1 < i_2 < \cdots < i_k \leqslant n$ form a basis for $C_n$; in particular $\dim C_n = 2^n$.

b. If $n = 2k$, then $C_n$ is isomorphic to the matrix algebra of order $2^k$ and the center $C_n$ consists of scalars.

If $n = 2k + 1$, then $C_n$ is isomorphic to the direct sum of two matrix algebras of order $2^k$; its center consists of elements $x + ye$ where $e = e_1e_2 \cdots e_n$; $x, y \in \mathbf{C}$.

c. Let $C_n^+$ and $C_n^-$ be the subspaces of $C_n$ generated by the monomials with an even or an odd number of factors, respectively. Then

$$C^+ \cdot C^+ = C^- \cdot C^- = C^+, \qquad C^+ \cdot C^- = C^- \cdot C^+ = C^-.$$

d. Consider the mapping $\varphi_n : C_{n-1} \to C_n$ defined on the generators by $\varphi_n(e_i) = e_i \cdot e_n$ $(i = 1, \ldots, n-1)$. This mapping defines an isomorphism of $C_{n-1}$ onto $C_n^+$.

e. Let $* : C_n \to C_n$ be the antiautomorphism which is the identity on the generators $e_i^* = e_j$. Then

$$(e_{i_1} \cdots e_{i_k})^* = e_{i_1} \cdots e_{i_k} \cdot (-1)^{k(k-1)/2}.$$

If $x = \Sigma x_ie_i$, then $x \cdot x = -\Sigma x_i^2$; in particular, if $\Sigma x_i^2 = 1$, then $x$ is invertible in $C_n$. For each vector $x = \Sigma x_ie_i \in E$ such that $\Sigma x_i^2 = 1$, denote by $\varphi(x)$ the transformation of the space $E$ given by $\varphi(x)y = xyx^*$; this is the reflection in the plane orthogonal to $x$. In particular, $\varphi(x) \in O(E)$. Denote by $\mathrm{pin}_n$ the subgroup of the multiplicative group of the algebra $C_n$ generated by all the elements $x \in E$ for which $x \cdot x = -1$, and extend the mapping $\varphi$ to a homomorphism $\varphi : \mathrm{pin}_n \to O(E)$. Then set

$$\mathrm{Spin}_n = \{u \in \mathrm{pin}_n | uu^* = 1\}.$$

It is easy to check that

$$\mathrm{Spin}_n = \{u \in C_n^+ | uu^* = 1, uEu^* = E\}.$$

f. The homomorphism $\varphi = \varphi_n$ maps $\mathrm{Spin}_n$ to $SO(E) \approx SO_n(\mathbf{C})$. The kernel of the mapping $\varphi$ is $\{\pm 1\}$. The group $\mathrm{Spin}_n$ is connected, and for $n > 2$ it is also simply connected.

Every element $u \in \mathrm{Spin}_n$ defines an automorphism of the algebra $C_n(x \to uxu^{-1})$ which coincides with the action of the element $\varphi(u) \in SO_n$.

g. Denote by $\delta_{2k}$ the standard action of the algebra $C_{2k}$ on the space of dimension $2^k$.

If $n = 2k + 1$, then the mapping $\delta_{n-1}\varphi_n^{-1}$ defines a $2^k$-dimensional representation of the algebra $C_n^+$ and therefore a $2^k$-dimensional representation of the group $\mathrm{Spin}_n$, called the spinor representation; it is an irreducible representation.

If $n = 2k$, then $\delta_n$ defines a $2^k$-dimensional representation of the group $\mathrm{Spin}_n$, which is also called a spinor representation. This representation is reducible: it can be decomposed as the direct sum of two irreducible $2^{k-1}$-dimensional representations (which correspond to the different eigenvalues of the operator $\delta_n(e_1, \ldots, e_n)$. These representations are called the half-spinor representations.

*3. Type* $\mathbf{A}_n$. Let $G = SL_n(\mathbf{C})$; let $H$ be the diagonal subgroup of $G$; and $\theta$ transposition. Then $K = SO_n(\mathbf{C})$, and $S$ is the group of diagonal matrices (of determinant 1) with $\pm 1$ on the diagonal.

Let $E$ be an $n$-dimensional space on which the group $G$ acts; $e_1, \ldots, e_n$ a basis of it. Let $e = e_1 \cdots e_n \in C(E)$. Then $e^2 = (-1)^{n(n-1)/2} = \epsilon^2$.

We split up the space $C(E)$ into the direct sum of the subspaces $C_\epsilon$ and $C_{-\epsilon}$ annihilated by left multiplication by $e - \epsilon$ and $e + \epsilon$ respectively. Since $k(e) = e$ for all $k \in K = SO_n(E)$ and $k'(e) = -e$ for all $k' \in O(E) - SO(E)$, it follows that the subspaces $C_\epsilon$ and $C_{-\epsilon}$ are invariant under $SO(E)$ and are interchanged by the elements of $O(E) \setminus SO(E)$.

Let us take $\tau$ to be the representation of the group $SO(E)$ in the space $C_\epsilon$ and prove that it defines a model.

a. Let $S_0$ be the subgroup of all diagonal matrices with $\pm 1$ on the diagonal. Then $\mathrm{card}(S_0/S) = 2$, and all the elements of $S_0 - S$ interchange $C_\epsilon$ and $C_{-\epsilon}$. To prove that $\tau|_S$ is a regular representation it then suffices to check that the representation $\tau'$ of the entire group $S_0$ in $C(E) = C_\epsilon \oplus C_{-\epsilon}$ is regular. This is verified directly.

b. It suffices to prove that $\tau'$ is a shallow representation of $SO_n$. But $C(E)$ can be naturally identified with $\Lambda^*(E)$, and the action of $SO(E)$ on $C(E) \cong \Lambda^*(E)$ can be extended to an action of the whole group $G$. It remains to check that for all $\gamma \in \Delta$ the representation in the space $\Lambda^*(E)$ is shallow with respect to $H_\gamma$. This is also easy to verify directly.

*4. Type* $\mathbf{B}_n$. $G = \mathrm{Spin}_{2n+1}$. Let us realize the group $G$ as a subgroup of the multiplicative group of the algebra $C_{2n+1} = C(E)$; let $e_1, \ldots, e_{2n+1}$ be a basis for $E$.

For every pair of indices $i, i + n$ $(i = 1, \ldots, n)$ set

$$H_i = \{\alpha + \beta e_i e_{i+n}, \alpha^2 + \beta^2 = 1\};$$

$H_i$ is a commutative subgroup of $G$, and $H = \prod_{i=1}^n H_i$ is a Cartan subgroup of $G$.

Set $\hat{e} = e_1 \cdots e_n$, and define an antiautomorphism $\theta$ of the group $G$ by $\theta(g) = (\hat{e}g\hat{e}^{-1})^* = (\hat{e}g\hat{e}^{-1})^{-1}$; since $\theta|_H = \mathrm{id}$, $\theta$ is a Cartan involution. The subgroup $K$ corresponding to $\theta$ has the form

$$K = \{ g \in G \,|\, \hat{e}g\hat{e}^{-1} = g^{-1} \}.$$

In particular, $\varphi : G \to SO_n$ maps $K$ to the operators on $E$ that commute with $\varphi(e)$, i.e., preserve the subspaces $E'$ and $E''$ generated by the vectors $(e_1, \ldots, e_n)$ and $(e_{n+1}, \ldots, e_{2n+1})$, respectively. Since $\operatorname{Ker}\varphi$ lies in the center of $G$, it follows that $\varphi(K) = \{$connected component of the group $SO(E) \times SO(E')\}$. Therefore, $K$ is generated by the groups $\mathrm{Spin}_n$ and $\mathrm{Spin}_{n+1}$ corresponding to $E'$ and $E''$; i.e., $K = (K_1 \times K_2)/\{1, \rho\}$, where $K_1 \cong \mathrm{Spin}_n$, $K_2 \cong \mathrm{Spin}_{n+1}$, $\rho = (\rho_1, \rho_2)$, and $\rho_1 \in K_1$ and $\rho_2 \in K_2$ are the nontrivial elements of the kernel of the map $\mathrm{Spin} \to SO$.

The group $S$ consists of the elements of the form

$$\pm e_{i_1} e_{i_2} \cdots e_{i_k} \cdot e_{i_1+n} e_{i_2+n} \cdots e_{i_k} + n,$$

where $k$ is even; $\operatorname{ord} S = 2^n$.

Let us construct the representation $\tau$. Denote by $\tau_1'$ the $2^{n-1}$-dimensional representation of the group $SO_n$ on the space $C_{n,\epsilon}$; and by $\tau_1$ the representation of the group $K$ obtained from $\tau_1'$ by means of the projection $K \to SO_n$.

Let $\rho$ be the spinor representation of the group $\mathrm{Spin}_{2n+1}$ in the space $V$, $\dim V = 2^n$. Let $z = e_1 \cdots e_n$ for even $n$, and let $z = e_{n+1} \cdots e_{2n+1}$ for odd $n$. Then $z \in \mathrm{Spin}_{2n+1}$, $z$ centralizes $K$, and $z^2 = \epsilon^2$ is a scalar. Therefore $V$ can be written as the sum $V_\epsilon \oplus V_{-\epsilon}$ of $K$-invariant subspaces, where

$$V_\epsilon = \operatorname{Ker}(\rho(z) - \epsilon), \qquad V_{-\epsilon} = \operatorname{Ker}(\rho(z) + \epsilon).$$

Set $\tau_2 = \rho|_{v_\epsilon}$ and then $\tau = \tau_1 \oplus \tau_2$. We prove that $\tau$ generates a model.

a. Clearly, $S = \{i, \nu\} \cdot S'$, where $\nu = -1$ and $S'$ is the subgroup of $S$ consisting of the monomials $e_{i_1} \cdots e_{i_k} e_{i_1+n} \cdots e_{i_k+n}$, taken with the plus sign. We have $\tau_1(\nu) = 1$, $\tau_2(\nu) = -1$. Therefore, to prove that $\tau|_S$ is regular we must check that $\tau_1|_{S'}$ and $\tau_2|_{S'}$ are regular representations of $S'$.

For the representation $\tau_1$ this has already been proved in Section 4.3. Consider $\tau_2$. Let

$$S_0' = \{ e_{i_1} \cdots e_{i_k} e_{i_1+n} \cdots e_{i_k+n}, \text{ where } 0 < i_1 < i_2 < \cdots < i_k \leqslant n \}.$$

Then $S'$ is a subgroup of $S_0'$ of index 2, and for all elements $s \in S_0' \backslash S'$, we have $-sz = zs$. Therefore, all the elements of $S_0' \backslash S'$ interchange the spaces $V_\epsilon$ and $V_{-\epsilon}$, and it suffices to prove that all the characters of $S_0'$ occur in the restriction of $\rho$ to $S_0'$. This follows from the fact that elements of $S_0'$ generate in $\operatorname{End} V = C_{2n+1}^+$ a subspace of dimension $\operatorname{card} S_0' = 2^n$.

b. Since $\tau_2$ occurs in the representation $\rho|_{v_\epsilon}$, it suffices to check that the representation $\rho$ is shallow with respect to all the subgroups $H_\gamma \subset \mathrm{Spin}_{2n+1}$, but this follows immediately from the description of its highest weight [3].

Consider the representation $\tau_1$. We can consider a larger representation $\tau_1'$; this is a representation in the space of all forms in $n$-space.

Let $G' \cong \mathrm{Spin}_3$ be the subgroup of $G$ generated by the linear combinations of $e_1$, $e_{n+1}$ and $e_{2n+1}$. The short root $\alpha$ of the group $G$ is realized in $G'$. Since $G' \cap K \subset K_2$, it follows that $\tau_1|_{K_2} \equiv 1$.

Consider the space $\tilde{E}$ with basis $\tilde{e}_1$, $\tilde{e}_2$, $\tilde{e}_3$, $\tilde{e}_4$, and the corresponding group $\tilde{G} \cong \mathrm{Spin}_4$. Embed $\tilde{G}$ in $G$ by the map $\tilde{e}_1 \to e_1$, $\tilde{e}_2 \to e_2$, $\tilde{e}_3 \to e_{n+1}$, $\tilde{e}_4 \to e_{n+2}$. Then the group $\tilde{K} = K(\tilde{G})$ is isomorphic to $(\tilde{K}_1 \times \tilde{K}_2)/(1,\rho)$, where $\tilde{K}_1 \cong \tilde{K}_2 \cong \mathrm{Spin}_2$.

Since $\tilde{G}$ contains the long root $\alpha$, it suffices to prove that $\tau_1|_{\tilde{G}}$ is a shallow representation. But the restriction of $\tau_1$ to $\tilde{G}$ is proportional to the four-dimensional representation $\sigma$ obtained in the following way. Consider the standard representation of the group $SO_2$ on the two-dimensional space $\tilde{E}$, and consider the corresponding representation in $\Lambda^*\tilde{E}$ from which $\sigma$ is obtained by means of the projection $K_1 \times K_2 \to SO_2$ (where $K_1 \to SO_2$ is the natural projection and $K_2$ is mapped to 1); here we are using the fact that

$$\Lambda^*(E_1 \oplus E_2) = \Lambda^*(E_1) \otimes \Lambda^*(E_2)).$$

One has to prove that the weights of $\sigma$ restricted to $K_\alpha$ equal 0, $\pm 1$.

Let $\hat{K}_\alpha$ be a two-sheeted covering of the group $K_\alpha$. Then the embedding of $K_\alpha$ into $K$ can be lifted to a mapping from $\hat{K}_\alpha$ to $K_1 \times K_2$; let $k$ and $l$ be the degrees of the respective projections of $\hat{K}_\alpha$ to $K_1$ and to $K_2$. Note that neither $k$ nor $l$ is zero, since the dimension of the centralizer of $K_\alpha$ is 4 (because $\mathrm{Spin}_4 \cong SL_2(\mathbf{C}) \times SL_2(\mathbf{C})$) but the centralizers of $K_1$ and $K_2$ coincide with $K$.

Consider the spinor representation $\rho$ of the group $\tilde{G}$. It can be decomposed into the sum of two half-spinor representations $\rho_1$ and $\rho_2$. Under the isomorphism $\tilde{G} \cong SL_2 \times SL_2$, $\rho_1$ and $\rho_2$ become the standard representations of the two components of the group $\tilde{G}$. In particular, in the representation $\rho$ the weights of the group $K_\alpha$ coincide with the weights of the group $H_\alpha$ and equal 0, 0, $+1$, $-1$, and the weights of $\hat{K}_2$ are 0, 0, $\pm 2$.

On the other hand, the weights of groups $K_1$ and $K_2$ are $\{\pm 1, \pm 1\}$; i.e., the weights of $\hat{K}_\alpha$ are $\pm k \pm l$. Hence $|k| = |l| = 1$. Since $K_1$ has weights 0, 0, $\pm 2$ in the representation $\sigma$, it follows that $\hat{K}_\alpha$ has weights 0, 0, $\pm 2$, and $K_\alpha$ has weights 0, 0, $\pm 1$, as required.

5. *Type* $\mathbf{C}_n$. Let us realize the group $G = \mathrm{Spin}_{2n}$ as the group of transformations of a $2n$-dimensional space $E$, which preserve the form

$$I = \begin{vmatrix} 0 & 1 & & & & & \\ -1 & 0 & 0 & 1 & & & \\ & & -1 & 0 & & & \\ & & & & \ddots & & \\ & & & & & 0 & 1 \\ & & & & & -1 & 0 \end{vmatrix},$$

i.e., $gIg^* = I$. Pick $H$ to be the subgroup of diagonal elements in $G$:

$$\begin{Vmatrix} \lambda_1 & & & & & & \\ & \lambda_1^{-1} & & & & & \\ & & \lambda_2 & & & & \\ & & & \lambda_2^{-1} & & & \\ & & & & \ddots & & \\ & & & & & \lambda_n & \\ & & & & & & \lambda_n^{-1} \end{Vmatrix}$$

Set $\theta(g) = g^*$. In this case

$$K = \{ g \in GL_{2n} | g = g^{*-1}, gIg^{-1} = I \} = \{ g \in \mathrm{Sp}_{2n} | g = g^{*-1} \}.$$

The operator $I$ has eigenvalues $+i$ and $-i$; denote by $E_+$ and $E_-$ the corresponding $n$-dimensional spaces. Clearly, $E_+$ and $E_-$ are invariant under $K$. To every element $k$ of $K$ we associate the corresponding transformation $\tau'(k)$ on the space $E_+$ (this gives an isomorphism $K \to GL(E_+)$). Denote by $\tau$ the resulting representation of the group $K$ on the space $\Lambda^*(E)$. We prove that it defines a model.

a. Clearly, $S$ is the subset of matrices in $H$ whose weights $\lambda_i$ are $\pm 1$. Choose a basis $e_1^+, \ldots, e_n^+$ for $E_+$, where $e_j^+ = e_{2j-1} + ie_{2j}$; then, in this basis, $S$ is identified with the set of diagonal matrices with $\pm 1$s on the diagonal. Therefore $\tau|_S$ is a regular representation.

b. Let $\alpha$ be a long root. Then it is realized in the group $\mathrm{Sp}_2 \cong SL_2$, embedded in $\mathrm{Sp}_{2n}$. Since $E_+ = E'_+ \oplus E''_+$, where $E'_+$ is generated by $e_1^+$ and $E''_+$ by $e_2^+, \ldots, e_n^+$, it follows that $\Lambda^*(E_+) = \Lambda^*(E'_+) \otimes \Lambda(E''_+)$; i.e., it is a multiple of $\Lambda^*(E'_+)$. In the space generated by $e_1$ and $e_2$, $K_\alpha$ can be written as

$$\left\{ \begin{pmatrix} \alpha & \beta \\ -\beta & \alpha \end{pmatrix} \middle| \alpha^2 + \beta^2 = 1 \right\}.$$

Therefore $K_\alpha$ has weight 1 in the one-dimensional space $E'_+$ and weights 0 and 1 in $\Lambda^*(E'_+)$.

Now suppose that $\alpha$ is a short root. Then (just as before) everything can be reduced to the case $G = \mathrm{Sp}_4$.

Let $E^{\nu}$ be the space generated by $e_1$ and $e_3$, and $E^{\nu\nu}$ the space generated by $e_2$ and $e_4$. The form $I$ defines a pairing of $E^{\nu}$ and $E^{\nu\nu}$. If $g$ is any unimodular transformation of $E^{\nu}$ and $g^{*-1}$ is one of $E^{\nu\nu}$, then the resulting transformation of $E$ preserves the form $I$. We thus get an embedding

$\varphi_\alpha : SL_2 \to G;$

$$\begin{pmatrix} a & b \\ c & d \end{pmatrix} \to \begin{bmatrix} a & 0 & b & 0 \\ 0 & d & 0 & -c \\ c & 0 & d & 0 \\ 0 & -b & 0 & d \end{bmatrix}.$$

This embedding corresponds to the short root $\alpha$, so that $K_\alpha$ has the form

$$\begin{bmatrix} a & 0 & b & 0 \\ 0 & a & 0 & b \\ -b & 0 & a & 0 \\ 0 & -b & 0 & a \end{bmatrix}.$$

In the space $E_+$, $K_\alpha$ has weights $\pm 1$; i.e., in $\Lambda^*(E_+)$ it has weights 0, $\pm 1$, as required.

*6. Type* $\mathbf{D}_n$. Consider the natural embedding of the group $G = \mathrm{Spin}_{2n}$ into the group $\tilde{G} = \mathrm{Spin}_{2n+1}$ of type $\mathbf{B}_n$. This mapping defines an isomorphism of Cartan subgroups. Therefore, the restriction of the representation $\tau$ constructed for $\tilde{G}$ in Section 4.4 is the required representation on $G$.

*7. Type* $\mathbf{G}_2$. Let $G$ be a group of type $\mathbf{G}_2$. Then $K = (K_1 \times K_2)/\{1, \rho\}$, where $K_1$ and $K_2$ are isomorphic to $SL_2(\mathbf{C})$, $\rho_1$ and $\rho_2$ are the nontrivial elements in the centers of $K_1$ and $K_2$, and $\rho = (\rho_1, \rho_2)$ (see Tits [8]). Consider the root decomposition of the group $G$ with respect to a Cartan subgroup $H$ contained in $K$. Clearly, the roots corresponding to the unipotent subgroups of $K_1$ and $K_2$ are orthogonal. Assume that the roots $\pm\beta$ correspond to $K_1$; and the roots $\pm(\beta + 2\alpha)$, to $K_2$.

Consider the representation of the group $K_1 \times K_2$ of the form $\tau = 1 \otimes 1 \oplus 1 \otimes \tau_2$, where $\tau_2$ is the three-dimensional representation of the group $K_2$. Let us prove that $\tau$ defines a model.

a. Consider the root decomposition with respect to the group $H$, and construct the mapping $\varphi : \tilde{G} \to G$ (where $\tilde{G} = SL_3(\mathbf{C})$ is mapped to the subgroup generated by the long roots). It is easy to check that this is an embedding that defines an isomorphism of Cartan subgroups; assume that it agrees with the involution $\theta$.

Therefore, $\varphi$ defines an isomorphism $\tilde{S} \to S$, where $\tilde{S}$ is the subgroup of elements of order 2 in the Cartan subgroup of $G$.

Consider the involutory subgroup $\tilde{K}$ in $\tilde{G}$. $\tilde{K} \cong PSL_2(\mathbf{C})$, and $\varphi$ is an embedding of $\tilde{K}$ into $K$. Any such embedding is conjugate to the embedding induced by the diagonal map

$$SL_2(\mathbf{C}) \to SL_2(\mathbf{C}) \times SL_2(\mathbf{C}).$$

As verified in Section 4.3, $\tau|_{\tilde{S}}$ is a regular representation.

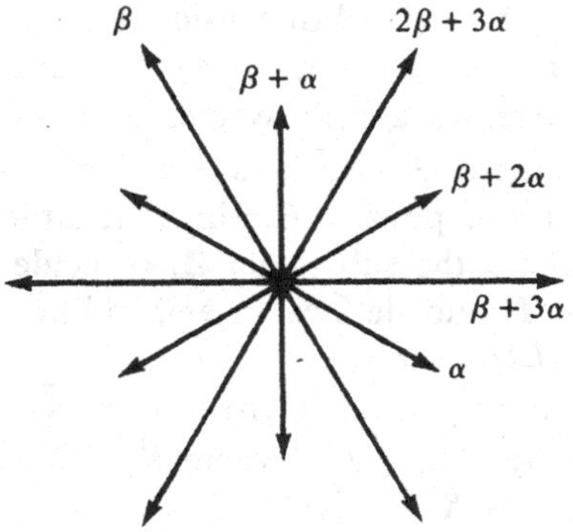


b. We check that $\tau$ is a shallow representation. The group $G_\beta \subset \varphi(\tilde{G})$, and it easily follows from Section 4.3 that all the weights in $\tau$ with respect to $K$ are 0, $\pm 1$.

Consider the group $K_\alpha$. Let $\tilde{K}_\alpha$ be a two-sheeted covering of $K_\alpha$, and let $\psi: \tilde{K}_\alpha \to K_1 \times K_2$ be the mapping which corresponds to the embedding

$$K_\alpha \to K = (K_1 \times K_2)/\{1, \rho\}.$$

If $H_1$ and $H_2$ are Cartan subgroups of $K_1$ and $K_2$, we can assume that $\psi(\tilde{K}_\alpha) \subset H_1 \times H_2$ and that the mapping $\psi$ is determined by the two integers $k$ and $l$, the degrees of the mappings of $\tilde{K}$ to $H_1$ and $H_2$. Note that $k$ and $l$ have the same parity, and they may be assumed to be nonnegative. Let us find these numbers.

For this purpose we consider the weights of the adjoint action of the group $\tilde{K}_\alpha$ on the Lie algebra $\mathfrak{g}$ of the group $G$. On the one hand, $K_\alpha$ is conjugate to $H_\alpha$, so that the weights of $K_\alpha$ in $\mathfrak{g}$ are 0, 0, 0, 0, $\pm 1$, $\pm 1$, $\pm 2$, $\pm 3$, $\pm 3$; on the other hand, those weights are $\pm k \pm l$, $\pm k \pm 3l$, $\pm 2k$, $\pm 2l$, 0, 0 (this can be seen from the computation of weights with respect to the groups $H_1$, $H_2 \subset \hat{H}$). If one of the numbers $k$ or $l$ were zero, then $K_\alpha$ would be contained in one of the subgroups $K_1$ or $K_2$; then the centralizer of $K_\alpha$ in $K$ would have dimension 4. However, the centralizers of $K_\alpha$ in $K$ and in $G$ have different dimensions, since the centralizer of $K_\alpha$ in $G$ contains the subgroup $G_{3\alpha+2\beta}$, so that $k, l > 0$. Hence $k = 3$, $l = 1$. Therefore, the weights of the group $\tilde{K}_\alpha$ in the representation $\tau$ are 0, $\pm 2$, so that the weights of the group $K_\alpha$ in this representation are 0, $\pm 1$.

8. *Type* $\mathbf{F}_4$. Let $G$ be a group of type $\mathbf{F}_4$. Then $K = (K_1 \times K_2)/\{1, \rho\}$, where $K_1 \cong SL_2(\mathbf{C})$, $K_2 = \mathrm{Sp}_6(\mathbf{C})$, $\rho_1, \rho_2$ are the nontrivial elements in the centers of $K_1$ and $K_2$, and $\rho = (\rho_1, \rho_2)$ (see Tits [8]). Set $\tau = \tau_1 \oplus \tau_2$. Here $\tau_1$ is trivial on $K_2$; on $K_1$ it is the sum of the trivial and the 3-dimensional representations, and $\tau_2 = \Phi_1 \otimes \Phi_2$, where $\Phi_1, \Phi_2$ are the standard representations of $K_1$ and $K_2$ on 2-dimensional and 6-dimensional spaces. We prove that $\tau$ generates a model.

The root system of $\mathbf{F}_4$ in a 4-dimensional space with basis $\epsilon_i$ ($i = 1, \ldots, 4$) has the form $\pm\epsilon_i$, $\pm\epsilon_i \pm \epsilon_j$, $\frac{1}{2}(\pm\epsilon_1 \pm \epsilon_2 \pm \epsilon_3 \pm \epsilon_4)$ (see [3]). The roots $\pm\epsilon_i$, $\pm\epsilon_i \pm \epsilon_j$ generate a root system of type $\mathbf{B}_4$. This defines a morphism $\varphi : \tilde{G} \to G$, where $G = \mathrm{Spin}_9$ is a group of type $\mathbf{B}_4$. Here $\tilde{H}$ (a Cartan subgroup of $\tilde{G}$) is mapped to $H$. Since the lattices of weights for the system of type $\mathbf{F}_4$ and for the subsystem $\mathbf{B}_4$ coincide [3], $\varphi$ is an isomorphism between $\tilde{H}$ and $H$ and defines an embedding $\tilde{G} \to G$. This induces an embedding of $\tilde{K} = K(\tilde{G})$ into $K$.

The group $\tilde{K}$ equals $(\tilde{K}_1 \times \tilde{K}_2)/\{1, \tilde{\rho}\}$, where $\tilde{K}_1 = \mathrm{Spin}_4$, $\tilde{K}_2 = \mathrm{Spin}_5$. But $\mathrm{Spin}_5 \cong \mathrm{Spin}_4$, $\mathrm{Spin}_4 = \tilde{K}_1' \times \tilde{K}_1''$, where $\tilde{K}_1'$ and $\tilde{K}_1''$ are isomorphic to $SL_2$. Therefore $\tilde{K} = (\tilde{K}_1' \times \tilde{K}_1'' \times \tilde{K}_2)/\{1, \tilde{\rho}\}$, where $\tilde{\rho} = (\tilde{\rho}_1', \tilde{\rho}_1'', \tilde{\rho}_2)$ and $\tilde{\rho}_1'$, $\tilde{\rho}_1''$, $\tilde{\rho}_2$ are the nontrivial elements of the centers of $\tilde{K}_1'$, $\tilde{K}_1''$, and $\tilde{K}_2$.

Consider the embedding $\tilde{K} \to K$ and lift it to a mapping

$$\psi : \tilde{K}_1' \times \tilde{K}_1'' \times \tilde{K}_2 \to K_1 \times K_2.$$

Let $\psi_1$, $\psi_2$ be the projections of $\psi$ to $K_1$ and $K_2$. Then $\psi_1(\tilde{K}_2) = 1$. Further, $\psi$ is nontrivial on one of the groups $\tilde{K}_1'$ and $\tilde{K}_1''$; for instance, $\tilde{K}_1'$. Then $\psi_1 : \tilde{K}_1' \to K$ is an isomorphism; since $\tilde{K}_1''$ commutes with $\tilde{K}_1'$, it follows that $\psi_1(K_1'') = 1$.

Therefore, $\mathrm{Ker}\, \psi_2|_{\tilde{K}_1'' \times \tilde{K}_2}$ has at most two elements. But we then have an embedding (since all the 6-dimensional representations of the group $SL_2 \times \mathrm{Sp}_4$ can easily be enumerated). Since $\psi_2(\tilde{K}_1')$ commutes with $\psi_2(\tilde{K}_1'' \times \tilde{K}_2)$, it follows that $\psi_2(\tilde{K}_1') = 1$.

Thus, $\psi$ maps $\tilde{K}_1'$ isomorphically to $K_1$, and embeds $\tilde{K}_1'' \times \tilde{K}_2 = \mathrm{Sp}_2 \times \mathrm{Sp}_4$ in a natural way into $K_2 = \mathrm{Sp}_6$. Recall the structure of the representation $\tilde{\tau}$ for the group $\tilde{K}$ (see Section 4.4). It had the form $\tilde{\tau}_1 \oplus \tilde{\tau}_2$, with $\tilde{\tau}_1$ and $\tilde{\tau}_2$ 8-dimensional representations, where $\tilde{\tau}_1$ is trivial on $\tilde{K}_2$ and $\tilde{\tau}_2(\tilde{\rho}_2) = -1$. This immediately implies that $\tilde{\tau}_2$ is the tensor product of the standard representations of $\tilde{K}_2 = \mathrm{Sp}_4$ and the standard representation of one of the groups $\tilde{K}_1'$ and $\tilde{K}_1''$. Since these groups are interchanged by an inner automorphism of the group $G$ corresponding to some element of $H$ (any element in $S_0 - S$ in the notation of Section 4.4), we may assume that $\tilde{\tau}_2 = \tilde{\Phi}_1' \oplus \tilde{\Phi}_2$ where $\tilde{\Phi}_2'$ and $\tilde{\Phi}_2$ are the standard representations of the groups $\tilde{K}_1'$ and $\tilde{K}_2$.

It is easy to check that $\tilde{\tau}_1 = \tilde{\tau}_1' \oplus \tilde{\tau}_2''$, where $\tilde{\tau}_1'$ is the sum of the trivial representations and the 3-dimensional representation $T$ of $\tilde{K}_1'$, and $\tilde{\tau}_2'' = \tilde{\Phi}_1' \otimes \tilde{\Phi}_2''$ where $\tilde{\Phi}_1'$ and $\tilde{\Phi}_2''$ are the standard representations of $\tilde{K}_1'$ and $\tilde{K}_2''$ ($\tilde{K}_1'$ and $\tilde{K}_1''$ may have to be interchanged here). In this way,

$$\tilde{\tau} = 1 \oplus (T \otimes 1 \otimes 1) \oplus (\tilde{\Phi}_1' \otimes (\tilde{\Phi}_1'' \otimes 1 \oplus 1 \otimes \tilde{\Phi}_2)),$$

i.e., $\tilde{\tau} = \tilde{\tau}|_G$, as required.

9. *Type* $\mathbf{E}_8$. Let $G$ be a group of type $\mathbf{E}_8$. Then $K$ is a group of type $\mathbf{D}_8$; the kernel of the mapping $\mathrm{Spin}_{16} \to K$ has two elements and *does not coincide* with the kernel of the spinor representation $\mathrm{Spin}_{16} \to SO_{16}$ (see [8]).

Consider the spinor representation $\rho$ of the group $\mathrm{Spin}_{16}$ on the space $V$, and denote by $\tau$ the resulting representation of $\mathrm{Spin}_{16}$ in End $V$. Since $\tau$ is trivial on the center, it can be viewed as a representation of the group $K$. We prove that it defines a model.

There exists an embedding of the root system $\mathbf{A}_8$ into $\mathbf{E}_8$, because the extended Dynkin diagram of $\mathbf{E}_8$, obtained by adjoining the lowest weight, contains a subdiagram of type $\mathbf{A}_8$. It induces a mapping $\varphi: \tilde{G} \to G$, where $\tilde{G} \cong SL_9$, which maps the Cartan subgroup $\tilde{H}$ of the group $\tilde{G}$ onto the Cartan subgroup $H$. Moreover, $\varphi$ defines a mapping of $\tilde{K} = K(\tilde{G})$ to $K$. It suffices to check that $\tau|_{\tilde{K}}$ defines a model for $\mathbf{A}_8$.

Let us analize in greater detail the mapping of $\tilde{K} = SO_9$ to $K$. Let $\psi': \mathrm{Spin}_9 \to \mathrm{Spin}_{16}$ be a lifting of it, and let $\psi'': \mathrm{Spin}_9 \to SO_{16}$ be the composition of $\psi'$ and the projection $\mathrm{Spin}_{16} \to SO_{16}$. The mapping $\psi''$ defines a 16-dimensional representation of the group $\mathrm{Spin}_9$. $\mathrm{Spin}_9$ has only three representations of dimension no greater than 16: the trivial one 1, the standard one $\Phi$ of dimension 9, and the spinor one $\rho$ of dimension 16. Clearly $\psi''$ is nontrivial; nor can it be the sum of $\Phi$ and trivial representations, because in such a case $\psi': \mathrm{Spin}_9 \to \mathrm{Spin}_{16}$ would be the natural mapping, and the mapping $\psi''': \mathrm{Spin}_9 \to K$ would be nontrivial on the kernel of the mapping $\mathrm{Spin}_9 \to \mathrm{Spin}_9$; i.e., it would not factor through the mapping $SO_9 \to K$.

Thus $\psi''$ is a spinor representation on the space $V$, and $\tau|_K$ is given by the natural action of $\mathrm{Spin}_9$ in End $V$. The space End $V$ can be identified with $C_9^+$ (by definition of the spinor representation). But $C_9^+$ is isomorphic to the representation of the group $SO_9$ constructed in Section 4.2, as required.

*10. Type* $\mathbf{E}_7$. Let $G$ be a simply connected group of type $\mathbf{E}_7$. Then $K$ is isomorphic to the group $SL_8/\{\pm 1\}$ (see [8]).

Consider the standard representation $\Phi$ of the group $SL_8$ in the space $E$, $\dim E = 8$. Then the representation $\tau = \Phi \otimes (\Phi \oplus \Phi^*)$ can be viewed as a representation of the group $K$. We prove that it generates a model.

The root system of $\mathbf{E}_7$ can be embedded in the root system of $\mathbf{E}_8$. Let $\varphi: G \to \tilde{G}$ be the corresponding mapping of groups which agrees with $\theta$, and $\varphi(H) \subset \tilde{H}$, $\varphi(K) \subset \tilde{K}$. The mapping $\varphi$ defines a homomorphism $\varphi: K \to \tilde{K}$ with discrete kernel, and its lifting to the universal covers, $\varphi': SL_8 \to SO_{16}$. Since all the representations of $SL_8$ except $\Phi$, $\Phi^*$, and the trivial representation have dimension greater than 16, and there is no invariant bilinear form on the representations $\Phi \oplus \Phi$ and $\Phi^* \oplus \Phi^*$, it follows that the 16-dimensional representation of the group $SL_8$ defined by the mapping $\varphi'$ has the form $\Phi \oplus \Phi^*$ and is defined on the space $E \oplus E^*$. In this way, the representation of the group $SL_8$ on the space $E \oplus E^*$ can be extended to a representation of the group $\mathrm{Spin}_{16}$. It follows, in particular, that $\varphi: K \to \tilde{K}$ is an embedding, so that the group $S$ can be viewed as a subgroup of index 2 in $\tilde{S} = \tilde{K} \cap \tilde{H}$. Let $S' \subset SL_8$ be the inverse image of the group $S$ under the projection $SL_8 \to K$, $\tilde{S}' \subset \mathrm{Spin}_{16}$ the inverse image of $\tilde{S}$ under the projection $\mathrm{Spin}_{16} \to \tilde{K}$.

Since $\mathrm{End}(E \oplus E^*) = (E \otimes E) \oplus (E \otimes E^*) \oplus (E^* \otimes E) \oplus (E^* \otimes E^*)$, and the representation of $\tilde{K}$ in End $(E \oplus E^*)$ is shallow, the representation $\tau$ is also shallow.

We prove that $\tau|_S$ is a regular representation. Indeed, the representation of the group $\tilde{S}'$ on $E \oplus E^*$ is irreducible (since the space of invariants of the group $\tilde{S}'$ in $\mathrm{End}(E \oplus E^*)$ is 1-dimensional by the preceding section). Therefore, if $\tilde{s} \in \tilde{S}' \backslash S'$, then $\tilde{S}(E)$ does not intersect $E$ (since $sS' \cup S' = \tilde{S}'$ and $\tilde{s}^2 \in S'$). Hence $\tilde{s}$ interchanges the spaces $E \otimes (E \oplus E^*)$ and $\tilde{s}E \otimes (E \oplus E^*)$, whose sum gives us

$$(E \oplus E^*) \otimes (E \oplus E^*) \cong \mathrm{End}(E \oplus E^*).$$

Since the representation of the group $\tilde{S}$ in $\mathrm{End}(E \oplus E^*)$ is regular, the representation of $S$ on $E \otimes (E \oplus E^*)$ is also regular, as required.

*11. Type* $\mathbf{E}_6$. Let $G$ be the simply connected group of type $\mathbf{E}_6$. Then $K \cong \mathrm{Sp}_8/\{\pm 1\}$ (see [8]). Let $\Phi$ be the standard 8-dimensional representation of the group $\mathrm{Sp}_8$ on the space $E$. The representation $\tau = \Phi \otimes \Phi^*$ in the space End $V$ can be viewed as a representation of the group $K$. Let us prove that it defines a model.

Embed the root system $\mathbf{E}_6$ into $\mathbf{E}_7$, and consider the corresponding mapping $\varphi: G \to \tilde{G}$. Here $\varphi(H) \subset \tilde{H}$ and $\varphi(K) \subset \tilde{K}$. The mapping $K \to \tilde{K}$ can be lifted to a mapping $\mathrm{Sp}_8 \to SL_8$ that obviously coincides with the standard representation. In particular, $\varphi: K \to \tilde{K}$ is an embedding. By means of this embedding we can identify $S$ with a subgroup of $\tilde{S}$ of index 2. Since the representation $\tau$ of $\tilde{K}$ in the space $(E^* \otimes E) \oplus (E^* \otimes E)$ is shallow, it follows that it is also shallow with respect to $K$.

We now prove that $\tau|_S$ is a regular representation. Indeed, since $S$ is a subgroup of $\tilde{S}$ of index 2 and the representation $\tilde{\tau}|_{\tilde{S}}$ is regular, $\tilde{\tau}|_S$ is the doubled regular representation. But $\Phi|_{S'} = \Phi^*|_{S'}$, where $S' \subset \mathrm{Sp}_8$ is the inverse image of the group $S$ under the projection $\mathrm{Sp}_8 \to K$. Therefore, $\tilde{\tau}|_S$ is isomorphic to $\tau|_S \oplus \tau|_S$. It follows that $\tau|_S$ is a regular representation, as required.

## References


1. I. N. Bernstein, I. M. Gelfand, S. I. Gelfand, *Differential operators on the base affine space and a study of* $\mathfrak{g}$*-modules*, in: Lie Groups and Their Representations, Akademiai Kiado, Budapest, 1975, and Adam Hilger, London, 1975, 21–64.
2. I. N. Bernstein, I. M. Gelfand, S. I. Gelfand, *Models for representations of compact Lie groups* (in Russian), Funktsional. Anal. i Prilozhen. 9 (1975), 61–62. See also I. N. Bernstein, I. M. Gelfand, S. I. Gelfand, *A new model for*

*the representations of finite semisimple algebraic groups* (in Russian), Uspekhi Mat. Nauk 29 (1974), 185–186.

3. N. Bourbaki, Groupes et algèbres de Lie. Paris, Hermann, 1960, 1968.
4. I. M. Gelfand, *The cohomology of infinite dimensional Lie algebras, some questions of integral geometry*, Actes Congrès Int. Math., 1, Nice, 1970.
5. S. Helgason, Differential Geometry and Symmetric Spaces, New York, Academic Press, 1962.
6. D. Husemoller, Fibre Bundles, New York, McGraw-Hill, 1966.
7. R. Steinberg, Lectures on Chevalley Groups, Yale Univ., 1967.
8. J. Tits, *Tabellen zu den einfachen Lie Gruppen und ihren Darstellungen*, Lect. Notes Math. 40, 1967.
9. D. P. Zhelobenko, Compact Lie Groups and their Representations (in Russian), Moscow, Nauka, 1970.

# 9.

## (with S. G. Gindikin)

# Complex manifolds whose skeletons are semisimple real Lie groups, and analytic discrete series of representations

Funkts. Anal. Prilozh. **11** (4) (1977) 20–28 [Funct. Anal. Appl. **11** (1977) 258–265].
MR **58**:11230. Zbl. **444**:22006

This paper consists of two parts. In the first part a set of complex manifolds is associated with an arbitrary real semisimple Lie group G. In the case when these manifolds contain Stein manifolds, certain series of unitary representations of G can be realized as holomorphic functions on them.

In the second part, this construction is carried out in detail for the example of G = SL(2, ℝ), and this part of the paper (beginning with Sec. 3) can be read independently of the first.

Let G be a real semisimple Lie group with finite center, $L^2(G)$ the Hilbert space of functions with norm

$$\|f\|^2 = \int_G |f(g)|^2\, dg, \tag{1}$$

where dg is an invariant measure. The problem of harmonic analysis on G consists in decomposing the regular representation $T_{g_1}f(g) = f(gg_1)$ in $L^2(G)$ into irreducible representations. As is well known, the irreducible unitary representations of G fall into a finite number of series, each of which is associated with a class of equivalent Cartan subgroups of G. It is therefore natural to break up the expansion of $L^2(G)$ into two stages. In the first stage, $L^2(G)$ is decomposed into a direct sum of subspaces

$$L^2(G) = L_1 \oplus \ldots \oplus L_k, \tag{2}$$

corresponding to the various series, and then each $L_j$ is decomposed in terms of the representations of a single series. It seems important to have natural methods for obtaining the decomposition (2). It can be shown that the decomposition problem for an individual $L_j$ is simpler since $f \in L_j$ can be recovered completely if we know the integrals of f over the classes of conjugate elements associated with the corresponding Cartan subgroup $H_j$ and over their translates, and the decomposition of $L_j$ thus reduces to the problem of harmonic analysis on the commutative group $H_j$.

As was already remarked, we construct a set of complex manifolds $G_1, \ldots, G_k$, whose skeleton (Shilov boundary) is the group G, and to which the action of G × G extends. As will be shown in subsequent publications, each $G_j$ is associated with a single series of unitary representations of G (with a single $L_j$). In this paper we consider only the case when some $G_j$ is a Stein manifold. It can be shown that this happens if and only if it is possible to introduce into D = G/U, where U is a maximal compact subgroup, the structure of a Hermitian symmetric space (symmetric domain in $\mathbb{C}^n$). In this case we can define an analog $\mathcal{H}^2(G_j)$ of the Hardy space with, moreover, the space of boundary values of functions in $\mathcal{H}^2(G_j)$ being one of the subspaces $L_j$ in $L^2(G)$ in (2). The irreducible representations into which the restriction of the regular representation to $L_j$ decomposes are realized in spaces of holomorphic vector-functions on D = G/U (see [2]).

The main construction of this paper generalizes to pseudo-Riemannian symmetric spaces (see p.

The authors are grateful to V. I. Arnol'd, I. N. Bernstein, and S. I. Gel'fand for useful discussions and advice.

---

Moscow State University. Translated from Funktsional'nyi Analiz i Ego Prilozheniya, Vol. 11, No. 4, pp. 19-27, October-December, 1977. Original article submitted June 6, 1977.

Complex Manifolds Having a Real Semisimple Group G as Skeleton

Let $G_C$ be the complexification of G. We construct complex manifolds as subdomains of :.

Let $\sigma$ be the involutive automorphism in $G_C$ which distinguishes G, i.e., $G = \{g \in G_C:$ $(g) = g\}$, and let $Y = \{g \in G_C: \sigma(g) = g^{-1}\}$. We choose representatives $H^1, \ldots, H^k$ from ich equivalence class of Cartan subgroups in G (relative to inner automorphisms). Let :, ..., $H_C^k$, $H_C^j \supset H^j = G \cap H_C^j$ be the Cartan subgroups in $G_C$. Of course, the $H_C^j$, $1 \leqslant j \leqslant k$, 'e equivalent in $G_C$. Let the $H_j = H_C^j \cap Y$ be subgroups in $H_C^j$ complementary to $H^j$.

The Weyl group W of $G_C$ acts on every subgroup $H_C^j$. We let $W_\sigma^j$ denote the subgroup of ıtomorphisms in W which commute with $\sigma$. Then the group $W_\sigma^j$ acts on both $H^j$ and $H_j$. Let $W^j$ ınote the real Weyl group (the group of automorphisms $h \mapsto g^{-1}hg$, $g \in G$, preserving $H^j$, $H_j$ ıctored by the group of inner automorphisms). We have: $W^j$ is a normal subgroup in $W_\sigma^j$. : say that elements of $H_j$ are regular elements if they are not left fixed by any $w \in W_\sigma^j \setminus$ .* Consider the Weyl chambers in $H_j$, i.e., the connected components of the set of regular .ements: $H_j^{(1)}, \ldots, H_j^{(s)}$. The group $W_\sigma^j/W^j$ acts simply transitively on the set $\{H_j^{(1)}, \ldots, H_j^{(s)}\}$. : extend $W^j$ by adjoining the operations of multiplication by elements of $H^j \cap H_j$ (i.e., $\in H^j$, $h^2 = e$). We denote the resulting extension of $W^j$ by $M^j$. The automorphisms in $M^j$ 'eserve each of the components $H_j^{(m)}$.

Proposition 1 (on the Cartan Decomposition). 1) Almost every element $g \in G_C$ can be 'itten in the form

$$g = g_1^{-1} h g_2, \text{ where } g_1, \; g_2 \in G, \; h \in H_1 \cup H_2 \cup \ldots \cup H_k. \tag{3}$$

2) For $g \in G_C$ of general type in (3), $h \in H_j^{(m)}$ for certain j, m. In this case h is ıfined uniquely up to the substitution

$$h \mapsto w(h), \quad w \in M^j, \quad h \in H_j^{(m)}. \tag{4}$$

3) Under these same conditions the pair $(g_1, g_2)$ in (3) is defined uniquely up to the ıbstitution

$$(g_1, g_2) \mapsto (g_1 h_1, \; g_2 h_1), \quad h_1 \in H^j.$$

Main Definition. Put

$$G_j^{(m)} = \{g \in G_C: g = g_1^{-1} h g_2, \; g_1, \; g_2 \in G, \; h \in H_j^{(m)}\}. \tag{5}$$

COROLLARY. The sets $G_j^{(m)}$ are domains in $G_C$; they are mutually disjoint and $u \overline{\bigcup_{j,m} G_j^{(m)}} =$ :. The group G × G acts on every domain $G_j^{(m)}$:

$$g \mapsto g_1^{-1} g g_2, \quad g_1, \; g_2 \in G, \quad g \in G_j^{(m)}. \tag{6}$$

.e group G lies on the boundary of $G_j^{(m)}$ and is the skeleton.

Proposition 2. The manifold $G_j^{(m)}$ is a Stein manifold if and only if all elements $g \in$ $^{n)}$ are regular in $G_C$. In this case the functions $\alpha(g) = \alpha(h)$ are holomorphic on $G_j^{(m)}$, .ere $\alpha$ is a root on $H_C$, a Cartan subgroup in $G_C$, $g = g_1^{-1} h g_1$, $g_1 \in G_C$, $h \in H_C$.

It can be shown that under the conditions of Proposition 2, the group G is the group automorphisms of a complex symmetric domain in $C^n$. Here the group $H^j$ is necessarily com- .ct, and $H_j$ has no compact subgroups.

COROLLARY. If $G_j^{(m)}$ is a Stein manifold, then there exists a series of unitary repre- ıntations whose characters are the boundary values in the sense of $\mathcal{D}'(G)$ of holomorphic .nctions on $G_j^{(m)}$.

---

*ne must not confuse the regularity on $H_j$ with respect to the group $W_\sigma^j/W^j$ defined here with ıe regular elements in the complex semisimple group $G_C$ (see Proposition 2 below), i.e., .th elements of the form $g^{-1}hg$, $g \in G_C$, $h \in H_C^j$, $w(h) \neq h$ for all $w \in W$, $w \neq e$.

## 2. Hardy Spaces $\mathcal{H}^2(G_j^{(m)})$ on the Manifolds $G_j^{(m)}$

Let $G_j^{(m)}$ be a Stein manifold and $\varphi(g)$ a holomorphic function on $G_j^{(m)}$. We say that $\varphi(g)\in\mathcal{H}^2(G_j^{(m)})$, if

$$\sup_{g_1\in G,\, h\in H_j^{(m)}}\int_G |\varphi(g_1^{-1}hg_1g)|^2\,dg<+\infty. \quad (\ldots)$$

THEOREM 3. Let $\varphi\in\mathcal{H}^2(G_j^{(m)})$. Then

$$\lim_{h\to e}\varphi(g_1^{-1}hg_1g)=\varphi_b(g),\quad g\in G, \quad (8)$$

exists in $L^2(G)$ and does not depend on $g_1$, where the limit is taken in the $L^2(G)$ norm for a fixed $g_1$.

Thus, the operator given by taking boundary values

$$b:\mathcal{H}^2(G_j^{(m)})\to L^2(G);\quad \varphi\mapsto\varphi_b \quad (9)$$

is defined. We let $L_j^{(m)}$ denote the image of $\mathcal{H}^2(G_j^{(m)})$ under this map.

COROLLARY. If $G_j^{(m)}$ is a Stein manifold, then $L^{(m)}$ is a nonzero subspace of $L^2(G)$ invariant under the regular representation of G.

Proposition 4. The restriction of the regular representation to $L_j^{(m)}$ decomposes into a discrete direct sum of irreducible unitary representations which can be realized in the space of holomorphic vector-functions on the symmetric domain $D = G/U$ in $\mathbf{C}^n$.

It turns out that one can give the explicit form of the projection $L^2(G)\to L_j^{(m)}$. In this paper we give it below for SL (2; R).

Remark. The construction given admits the following generalization. Let $E = G/V$ be a pseudo-Riemannian symmetric space, i.e., G is a real semisimple Lie group and V the subgroup determined by an involutive automorphism $\tau$. We may assume that $E=\{g\in G:\ \tau(g)=g^{-1}\}$. In the case when $G = G_1\times G_1$ and $V = G_1$ is the diagonal of the direct product $(\tau(g_1, g_2) = (g_2, g_1))$, we obtain the case considered above.

Consider the complexification of $E$: $E_{\mathbf{C}}=G_{\mathbf{C}}/V_{\mathbf{C}}$, $G=\{g\in G_{\mathbf{C}}:\sigma(g)=g\}$, $V_{\mathbf{C}}=\{g\in G_{\mathbf{C}}:\ \tau(g)=g\}$, $E_{\mathbf{C}}=\{g\in G_{\mathbf{C}}:\ \tau(g)=g^{-1}\}$, $V=V_{\mathbf{C}}\cap G$, $E=E_{\mathbf{C}}\cap G$. Let $F=\{g\in E_{\mathbf{C}}:\ \sigma(g)=g^{-1}\}$. There is a complex structure on $E_{\mathbf{C}}$. Using the same scheme as above, we can consider inequivalent Cartan subgroups in E: $H^1,\ldots,H^k$, the subgroups in F: $H_1,\ldots,H_k$ complementary to them, the Weyl chambers in $H_j$: $H_j^{(1)},\ldots,H_j^{(s)}$, and the domain

$$E_j^{(m)}=\{g\in E_{\mathbf{C}}:\ g=g_1^{-1}h\tau(g_1),\ g_1\in G,\ h\in H_j^{(m)}\}.$$

If there are Stein manifolds among the $E_j^{(m)}$, then as above, the Hardy spaces $\mathcal{H}^2(E_j^{(m)})$ of holomorphic spaces on $E_j^{(m)}$ having boundary values on E in the topology of $L^2(E)$ (with respect to an invariant measure) are defined. The spaces $L_j^{(m)}$ of boundary values will in this case be nonzero invariant subspaces in $L^2$ under the regular representation on $E$: $f(x)\mapsto f(g^{-1}x\tau(g))$, $x\in E$, $g\in G$.

We realize the construction given above for SL (2; R) (see [1, 3, 4]).

## 3. The Manifolds $G_j^{(m)}$ for G = SL (2; R)

We take the Cartan subgroups in SL(2; R) to be

$$H^1=\left\{\begin{pmatrix} c & d\\ -d & c\end{pmatrix},\ c^2+d^2=1\right\},\quad H^2=\left\{\begin{pmatrix} c & d\\ d & c\end{pmatrix},\ c^2-d^2=1\right\}. \quad (10)$$

The subgroup SL (2; R) is distinguished in SL(2; C) by the automorphism $\sigma(g) = \bar{g}$.

Put

$$Y=\{g\in SL(2;\mathbf{C});\ \bar{g}=g^{-1}\}=\left\{\begin{pmatrix} a & ib\\ ic & \bar{a}\end{pmatrix},\ a\in\mathbf{C};\ b,c\in\mathbf{R},\ |a|^2+bc=1\right\}.$$

The subgroups complementary to $H^1$, $H^2$ in Y are

$$H_1=\left\{\begin{pmatrix} a & ib \\ -ib & a\end{pmatrix}, a^2-b^2=1\right\}, \quad H_2=\left\{\begin{pmatrix} a & ib \\ ib & a\end{pmatrix}, a^2+b^2=1\right\}. \tag{11}$$

notice that $H^1$ and $H_2$ are compact.

The groups $W^1_\sigma$, $W^2_\sigma$ are generated by a single element

$$w(a, b)=(a, -b), \tag{12}$$

ich corresponds to conjugation by the matrix $g=\begin{pmatrix} i & 0 \\ 0 & -i\end{pmatrix}$, i.e., $W^1_a = W^2_a = \{e, w\}$. Moreover, $= W^2_\sigma$, and $W^1$ consists only of a single element ($w \in W^2_\sigma$, but not $w \in W^1_\sigma$, can be obined by conjugation by $g=\begin{pmatrix} 0 & 1 \\ -1 & 0\end{pmatrix}$). In accordance with this there is a unique Weyl chamr $H_2^{(1)} = H_2 \setminus \{e\}$ in $H_2$ and two Weyl chambers in $H_1$:

$$H_1^{(1)}=\left\{\begin{pmatrix} a & ib \\ -ib & a\end{pmatrix}, a^2-b^2=1, ab>0\right\},$$

$$H_1^{(2)}=\left\{\begin{pmatrix} a & ib \\ -ib & a\end{pmatrix}, a^2-b^2=1, ab<0\right\}. \tag{13}$$

have: $H_1 = H_1^{(1)} \cup H_1^{(2)} \cup \{e\}$.

Further, $H_1 \cap H^1 = H_2 \cap H^2 = \{e, -e\}$. As in Sec. 1 we let $M^1$, $M^2$ denote the group obined by extending $W^1$, $W^2$, respectively, by the automorphism $g \mapsto -g$. As a result, $M^2$ consts of four elements $\{(a, b) \mapsto (a, b), (a, b) \mapsto (a, -b), (a, b) \mapsto (-a, -b), (a, b) \mapsto (-a, b)\}$, and has the two elements $\{(a, b) \mapsto (a, b), (a, b) \mapsto (-a, -b)\}$.

Proposition 1'. 1) Almost every element $g \in SL(2; \mathbf{C})$ can be written in the form

$$g = g_1^{-1} h g_2, \quad g_1, g_2 \in SL(2; \mathbf{R}), \; h \in H_1 \text{ or } h \in H_2. \tag{3'}$$

2) If $h \in H_1$, $h \neq e$ in (3'), then h is determined up to the replacement $h \mapsto -h$ (siltaneous change of signs of $a$ and b); if $h \in H_2$, $h \neq e$, then h is determined up to a transrmation $h \mapsto w(h)$, $w \in M^2$ (which allows any change of the signs of $a$ and b).

3) If $h \in H_j$, $h \neq e$ in (3'), then the $(g_1, g_2)$ are determined up to a substitution $(g_1, \mapsto (g_1h_1, g_2h_1)$, $h_1 \in H^j$.

This proposition is a special case of Proposition 1 in Sec. 1 and can easily be proved rectly.

We further consider three domains on $SL(2; \mathbf{C})$: $G_+ = G_1^{(1)}$, $G_- = G_1^{(2)}$, $G_0 = G_2^{(1)}$, where the n) are defined in (4): $g \in G_j^{(m)}$, if g can be taken into $H_j^{(m)}$ by the transformation $g \mapsto gg_2$, $g_1, g_2 \in SL(2; \mathbf{R})$. Clearly, $SL(2; \mathbf{R})$ lies on the boundary of all three domains since $\in SL(2; \mathbf{R})$ corresponds in (3') to h = e.

Remark. We have $G_+ \to G_-$, $G_- \to G_+$, $G_0 \to G_0$ for each of the transformations $g \mapsto \bar{g}$; $g \mapsto g^{-1}$.

We break Proposition 2 for SL (2; R) into several assertions.

Proposition 5. The manifolds $G_+$, $G_-$, $G_0$ are, respectively, biholomorphically equivait to the tube domains in $\mathbf{C}^3$ given by:

$$\begin{aligned} D_+ &= \{(z_1, z_2, z_3): y_1y_2 - y_3^2 > 0, \; y_1 < 0\}, \\ D_- &= \{(z_1, z_2, z_3): y_1y_2 - y_3^2 > 0, \; y_1 > 0\}, \\ D_0 &= \{(z_1, z_2, z_3): y_1y_2 - y_3^2 < 0\}, \; y_j = \operatorname{Im} z_j. \end{aligned} \tag{14}$$

Sketch of Proof. It can be shown that $G_+$, $G_-$, $G_0$ consist of the complex matrices $\begin{pmatrix} \alpha & \beta \\ \gamma & \delta\end{pmatrix}$, $-\beta\gamma = 1$, for which

$$\begin{aligned} &G_+: \{\operatorname{Im}(\alpha\bar{\beta}) \operatorname{Im}(\delta\bar{\beta}) - (\operatorname{Im}\beta)^2 > 0, \operatorname{Im}(\bar{\alpha}, \beta) > 0\}, \\ &G_-: \{\operatorname{Im}(\alpha\beta) \operatorname{Im}(\delta\beta) - (\operatorname{Im}\beta)^2 > 0, \operatorname{Im}(\bar{\alpha}\beta) < 0\}, \\ &G_0: \{\operatorname{Im}(\alpha\bar{\beta}) \operatorname{Im}(\delta\bar{\beta}) - (\operatorname{Im}\beta)^2 < 0\}. \end{aligned}$$

ing the change of variables

$$z_1 = \alpha/\beta, \quad z_2 = \delta/\beta, \quad z_3 = -1/\beta, \tag{15}$$

we obtain the domains (4).

COROLLARY. The manifolds $G_+$ and $G_-$ are Stein manifolds while $G_0$ is not. The group $G = SL(2; \mathbf{R})$ is the skeleton of $G_+$, $G_-$, $G_0$.

Proof. The first part of the assertion follows from the fact that the tube domains $D_+$, $D_-$ have convex bases and are hence domains of holomorphy. On the other hand, the base of the domain $D_0$ is not convex. The second assertion follows from the fact that under transformation (15) the points $g = \begin{pmatrix} \alpha & \beta \\ \gamma & \delta \end{pmatrix} \in SL(2; \mathbf{R})$, $\beta \neq 0$, go precisely into points of the common skeleton of the domains $D_\pm$, $D_0$, and the closure of this image gives the entire skeleton.

Remark. We have proved that the manifolds $G_+$ and $G_-$ are biholomorphically equivalent to the domains $D_+$ and $D_-$ in $\mathbf{C}^3$. These are well-known symmetric domains (they are of course biholomorphically equivalent to one another); they are called Siegel half planes. In order to recover the usual notation, we write $(z_1, z_2, z_3)$ as elements of the symmetric matrix $z = \begin{pmatrix} z_1 & z_3 \\ z_3 & z_2 \end{pmatrix}$. Then $D_-$ consists of z such that $y = \operatorname{Im} z$ is a positive-definite symmetric matrix (written $y \gg 0$). The full group of analytic automorphisms of $D_-$ is $SP(4; \mathbf{R})$. Our constructions are related to the fact that $SL(2; \mathbf{R}) \times SL(2; \mathbf{R})$ can be imbedded as a subgroup in $Sp(4; \mathbf{R})$. Different realizations of the domain $D_-$ may be convenient in different problems. In particular, this is the case in the problem which interests us concerning the Hardy spaces. The point is that although the full automorphism group of $D_-$, i.e., $Sp(4; \mathbf{R})$, acts on the skeleton of $D_-$, there is no invariant measure with respect to this group. Distinct variants of the Hardy space are associated with various rather large subgroups $G_0$ in $Sp(4; \mathbf{R})$ having open trajectories on the skeleton with an invariant measure. It is technically convenient here to have a realization of the manifold which is such that $G_0$ acts transitively on the set of endpoints of the skeleton. We consider some examples.

1) The group $\{z \mapsto z + a\}$ acts transitively on the skeleton of the Siegel half plane $D_-$ $(\Omega(D_-) = \{z: \operatorname{Im} z = 0\})$, where the $a$ are real symmetric matrices. The measure $dx = dx_1 dx_2 d$ is invariant under these transformations, and we can consider the Hardy space $\mathcal{H}^2(D_-)$, consisting of holomorphic functions $\varphi(z)$ in $D_-$ for which

$$\sup_{y \gg 0} \int_{\mathbf{R}^3} |\varphi(x + iy)|^2 \, dx < +\infty.$$

Such $\varphi(z)$ have boundary values in $L^2(\mathbf{R}^3)$ (with respect to the measure dx). This Hardy space has frequently been considered in the literature (see [5, 6]).

2) It is well known that $D_-$ can be mapped biholomorphically onto the domain

$$C_- = \{z: (e - z\bar{z}) \gg 0\}$$

in $\mathbf{C}^3$ considered as a space of matrices $\begin{pmatrix} z_1 & z_3 \\ z_3 & z_2 \end{pmatrix}$. The skeleton of $C_-$ consists of the matrices: $\Omega(C_-) = \{z: z = z', z\bar{z} = e\}$. The group $U(2)$, which is a maximal compact subgroup of $Sp(4; \mathbf{R})$, acts transitively on this compact manifold. There is an invariant measure with respect to $U(2)$ on $\Omega(C_-) = U(2)/O(2)$. As $\mathcal{H}^2(C_-)$ we can take the space of holomorphic functions $\varphi(z)$, $z \in C_-$, for which

$$\sup_{h \in H} \int_{U(2)} \varphi(uhu')\, du < +\infty,$$

where H consists of $h = \begin{pmatrix} h_1 & 0 \\ 0 & h_2 \end{pmatrix}$, $0 < h_i < 1$, and du is an invariant measure of $U(2)$. As $h_1$, $h_2 \to 1$, the functions in $\mathcal{H}^2(C_-)$ have boundary values in $L^2(U(2)/O(2))$.

3) The realization of the symmetric domain $Sp(4; \mathbf{R})/U(2)$ in the form of domains $G_+ \subset SL(2; \mathbf{C})$ which we consider leads to the skeleton $\Omega(G_\pm) = (SL(2; \mathbf{R}) \times SL(2; \mathbf{R}))/SL(2; \mathbf{R})$ and Hardy spaces $\mathcal{H}^2(G_\pm)$, defined by Eq. (7), where $G = SL(2; \mathbf{R})$. Apparently, this realization and the Hardy spaces associated with it have not been considered previously.

In concluding our remarks, we emphasize that the three realizations above lead to different variants of the Hardy space over the same Hermitian symmetric space.

In order to get the second part of Proposition 2 in the case of SL(2; **R**), we prove the llowing lemma.

LEMMA 6. Every element $g \in G_\pm$ has an eigenvalue $\lambda(g)$ such that $|\lambda(g)| > 1$.

Proof. In other words, we must prove that $g \in G_\pm$ cannot have eigenvalues equal to 1 modulus, or, equivalently, we cannot simultaneously have tr g real and $|\mathrm{tr}\ g| < 2$.

Let $g = g_1^{-1} h g_2$, $h \in H_1$, $h \neq e$, $g_1, g_2 \in \mathrm{SL}(2; \mathbf{R})$. Since the element $hg_2g_1^{-1}$ has the same genvalues as g, we may restrict ourselves to the case $g = hg_2$. Let $h = \begin{pmatrix} a & ib \\ -ib & a \end{pmatrix}$, $a^2 - b^2 =$ $g_2 = \begin{pmatrix} \alpha & \beta \\ \gamma & \delta \end{pmatrix}$, $\alpha\delta - \beta\gamma = 1$. Then $\mathrm{tr}\ g = \mathrm{tr}\ hg_2 = a(\alpha + \delta) + ib(\gamma - \beta)$. If $\mathrm{Im\ tr}\ g = 0$ then $= \beta$, $\alpha\delta = 1 + \beta^2 \geqslant 1$, $\alpha + \delta \geqslant 2$. That is, $|\mathrm{tr}\ g| = |a|\,|\alpha + \delta| \geqslant 2|a|$. Since $h \neq e$, we have $| > 1$ and $|\mathrm{tr}\ g| > 2$.

COROLLARY. The function $\lambda(g)$ (from Lemma 6) is holomorphic in $G_\pm$.

Direct calculations show that $\lambda(g)$ has continuous boundary values on $G = \mathrm{SL}(2; \mathbf{R})$. We t $\lambda_\pm(g)$ denote the boundary values on SL(2; **R**) in the domains of $G_\pm$, respectively.

LEMMA 7. Let $g \in \mathrm{SL}(2; \mathbf{R})$. Then $\lambda_+(g) = \lambda_-(g) = \lambda$, if g has real eigenvalues $\{\lambda, \lambda^{-1}\}$, $|> 1$; $\lambda_+(g) = e^{i\varphi}$, and $\lambda_-(g) = e^{-i\varphi}$, if g has complex eigenvalues $\{\lambda, \lambda^{-1}\}$, $\lambda| = 1$, and reces to the form $\begin{pmatrix} \cos\varphi & \sin\varphi \\ -\sin\varphi & \cos\varphi \end{pmatrix}$.

We recall that elements of the first type are called hyperbolic, and those of the secd type, elliptic.

It is not hard to derive the next proposition from Lemma 7.

Proposition 8. The characters of the analytic (respectively, antianalytic) discrete ries $\pi_\pm^n$ are the boundary values (in the topology of $\mathcal{D}'(G)$) of holomorphic functions in (respectively, $G_-$)

$$\pi_\pm^n(g) = \frac{[\lambda(g)]^{-n}}{\lambda(g) - 1/\lambda(g)}, \quad g \in G_\pm, \quad n = 1, 2, \ldots \tag{16}$$

Thus, the characters of both series are given by the same analytic formulas (16), alough in the case of the analytic series the corresponding function must be taken in $G_+$, d for the antianalytic series, in $G_-$. By Lemma 7, this leads for the same value of n to entical boundary values on the hyperbolic elements and to different boundary values on the liptic elements.

Since the domains $G_+$ and $G_-$ are invariant under translation by elements of G, we have e next corollary.

COROLLARY. The functions $\pi_\pm^n(gg_1)$, $g \in G_\pm$, $g_1 \in G$ are holomorphic in $G_\pm$.

We remark that $\pi_\pm^n(g)$ can be holomorphically extended outside $G_\pm$, but since the characrs are important as the kernels of the projections onto the corresponding representation, is important to know how to extend all the $\pi_\pm^n(gg_1)$ in terms of g simultaneously for $\in G$. The domains $G_\pm$ are remarkable precisely because all the translates of the correonding characters extend analytically in them, and hence in $G_\pm$ all the functions in the bspaces of $L^2(G)$ in which the representations corresponding to $\pi_\pm^n$ are realized extend alytically.

We note the formula

$$\pi_\pm^n(g) = \int\limits_{|t|=1} \frac{t^{-n-1}dt}{(t + t^{-1} - \mathrm{sp}\ g)}, \quad g \in G_\pm \tag{17}$$

ee [4] for the characters of the continuous series).

In concluding this section we give another way of describing the domains $G_\pm$.

Proposition 9. An element $g = \begin{pmatrix} \alpha & \beta \\ \gamma & \delta \end{pmatrix} \in \mathrm{SL}(2; \mathbf{C})$ belongs to $G_+$ (respectively, $G_-$) if and ly if $\mathrm{Im}\frac{\alpha z + \gamma}{\beta z + \delta} > 0$ for all z in the half plane $\{\mathrm{Im}\ z > 0\}$ (respectively, $\mathrm{Im}\frac{\alpha z + \gamma}{\beta z + \delta} < 0$ r $\mathrm{Im}\ z < 0$).

Proof. Consider the transformation $z \mapsto g(z) = \frac{\alpha z + \gamma}{\beta z + \delta}$. Since for $g \in SL(2; \mathbf{R})$ this transformation preserves the half planes $\{\operatorname{Im} z > 0\}$, $\{\operatorname{Im} z < 0\}$, it suffices to study the transformation $z \mapsto g(z)$ for $g \in H_1$, $g \in H_2$. For $g \in H_1$, we study the image of the real axis: $\operatorname{Im} \frac{ax + ib}{-ibx + a} = \frac{ab(1 + x^2)}{a^2 + b^2x^2}$. For $ab > 0 \, (g \in H_1^{(1)})$, the points of the real axis go into points of the half plane $\{\operatorname{Im} z > 0\}$: for $ab < 0 \, (g \in H_1^{(2)})$, their images lie in the half plane $\{\operatorname{Im} z < 0\}$. For $g \in H_2$ we have: $\operatorname{Im} \frac{ax + ib}{ibx + a} = \frac{ab(1 - x^2)}{a^2 - b^2x^2}$, i.e., the image of the real axis intersects the real axis.

## 4. The Hardy Spaces $\mathcal{H}^2(G_\pm)$ for G = SL(2; R)

In accordance with Theorem 3 of Sec. 2, we construct the Hardy spaces $\mathcal{H}^2(G_+)$, $\mathcal{H}^2(G_-)$, and define the operators $\{b_\pm : \mathcal{H}^2(G_\pm) \to L^2(G)\}$ obtained by passing to the boundary values (9) and the images $L_+$, $L_-$ under this transformation. In this section we construct the projections $P_+ : L^2(G) \to L_+$, $P_- : L^2(G) \to L_-$.

THEOREM 10. The operators $P_+$, $P_-$ have the form

$$P_\pm \varphi(g) = b_\pm \circ \mathcal{K}_\pm, \quad \mathcal{K}_\pm : L^2(G) \to \mathcal{H}^2(G_\pm), \quad \mathcal{K}_\pm \varphi(g) = \int_G K_\pm(g, g_1)\varphi(g_1)dg_1,$$

$$\mathcal{K}_\pm(g, g_1) = \frac{(\lambda(gg_1^{-1}))^2}{(\lambda(gg_1^{-1}) - 1)^3(\lambda(gg_1^{-1}) + 1)}, \tag{18}$$

where $b_\pm$ is the operator obtained by taking boundary values (9) in $G_\pm$, the kernels $K_\pm(g, g_1)$ are holomorphic for $g \in G_\pm$ and $\lambda(g)$ is the eigenvalue of g of largest modulus.

Remarks. 1. The operators $\mathcal{K}_\pm$ commute with translations by $g \in G$, whence we have $K_\pm(g \cdot g_1)$ for the kernel $K_\pm(g, g_1) = K_\pm(gg_2, g_1g_2)$, $g_2 \in G$, $K_\pm(g, g_1) = K_\pm(gg_1^{-1}, e)$.

2. The kernels $K_\pm(g, g_1)$ can be interpreted as Cauchy–Szegö kernels for $\mathcal{H}^2(G_\pm)$. A certain difficulty in calculating them arises from the fact that although the complex manifolds $G_\pm$ are homogeneous, the subgroup of transformations preserving $\mathcal{H}^2(G_\pm)$ is not transitive. At the same time, the problem of calculating the Cauchy–Szegö kernels for $\mathcal{H}^2(C_-)$, $\mathcal{H}^2(D_-)$ is simpler (see [6]).

3. On the other hand, $K_\pm(g, e)$ can be considered as the analytic continuation into the domain $G_\pm$ of the character of the regular representation in $L^2(G)$ restricted to $L_\pm$. For this reason

$$K_\pm(g, e) = \sum_{n=1}^{\infty} n\pi_\pm^n(g), \quad g \in G_\pm, \tag{19}$$

where the characters of $\pi_\pm^n$ are defined in (16) and the series in (19) converges absolutely.

4. Let $K_{\pm, b}(g, e) \in \mathcal{D}'(G)$ be the boundary values of $K_\pm(g, e)$. We study them on the regular elements. By Lemma 7,

$$\begin{aligned} K_{+,b}(g, e) &= K_{-,b}(g, e), \quad \text{if } g \text{ is a hyperbolic element}, \\ K_{+,b}(g, e) &= -K_{-,b}(g, e), \quad \text{if } g \text{ is an elliptic element}. \end{aligned} \tag{20}$$

Let $L_0$ be the orthogonal complement to $L_+ \oplus L_-$. Then we have for the character of the regular representation on $L_0$:

$$K_0(g, e) = \delta(g) - K_{+,b}(g, e) - K_{-,b}(g, e). \tag{21}$$

It follows from (21) and (18) that $K_0(g, e) = 0$ on the elliptic elements; $K_0(g, e) = \frac{-2\lambda^2}{(\lambda - 1)^3(\lambda +}$ if g has real eigenvalues $(\lambda, \lambda^{-1})$, $|\lambda| > 1$.

We observe that $L_0$ decomposes with respect to the representations of the continuous series, and the last fact agrees with the result that the characters of the representations of the continuous series are equal to zero on the elliptic elements.

The projection onto $L_0$ is given by the closure of the operator of convolution with $K_0(g, e)$.

5. The functions $K_{\pm}(g, e)$, $g \in G_{\pm}$, can be rewritten in the form

$$K_{\pm}(g, e) = [(\mathrm{tr}\ (g) - 2)^3 (\mathrm{tr}\ (g) + 2)]^{-1/2}, \tag{22}$$

ere a unique branch of the square root is chosen from the condition: $K_{\pm}(g, e) > 0$ for $(\mathrm{tr}\, g) = 0$, $|\mathrm{tr}\, g| > 2\}$.

## LITERATURE CITED


V. Bargmann, "Irreducible unitary representations of the Lorentz group," Ann. Math., 48, 568-640 (1947).

Harish-Chandra, "Representations of semisimple Lie groups. VI," Am. J. Math., 78, 564-628 (1956).

S. Lang, $SL_2(\mathbb{R})$, Addison-Wesley (1975).

I. M. Gel'fand, M. I. Graev, and I. I. Pyatetskii-Shapiro, Representation Theory and Automorphic Functions, Saunders (1969).

S. Bochner, "Group invariance of Cauchy's formula in several variables," Ann. Math., 45, 686-707 (1944).

S. G. Gindikin, "Analysis in homogeneous spaces," Usp. Mat. Nauk, 19, No. 4, 3-92 (1964).

# 10.
## (with A.V. Zelevinskij)

# Models of representations of classical groups and their hidden symmetries

Funkts. Anal. Prilozh. **18** (3) (1984) 14–31 [Funct. Anal. Appl. **18** (1984) 183–198].
Zbl. **556**:22003

## 0. The Main Results

Three important constructions lie behind the motivation of this paper:

1. The classical realization of irreducible representations of the group $SO_3$ acting on functions on the two-dimensional sphere. We may say that the space of functions on the two-dimensional sphere is a model of representations of $SO_3$ (meaning that its decomposition into irreducible components contains all the irreducible representations of $SO_3$, each appearing with multiplicity one).

2. A recent construction of Biederharn and Flath [1]: they built a model (in the sense indicated above) of finite-dimensional irreducible representations of the Lie algebra $sl(3, \mathbf{C})$; they also found that the action of $sl(3, \mathbf{C})$ on this model extends to an action of the larger Lie algebra $so(8, \mathbf{C})$.

3. The starting point of the R. Penrose's twistor program (see [2]): the complexification of the Minkowski space $\mathbf{R}^4$ followed by compactification leads to the Grassman manifold of 2-planes in $\mathbf{C}^4$.

We show here that these constructions are different aspects of a unifying construction of models of representations which is carried out below for all classical groups. A fourth important aspect of this construction is a remarkable parallelism between exterior and symmetric algebras; one of its consequences is that Lie supergroups and superalgebras arise naturally in the "purely even" problem.

Let us give a systematic description of the content of this paper, beginning with the results concerning the first construction.

Let G be a reductive algebraic group over $\mathbf{C}$. A model of representations of the group G is defined as a representation of G which decomposes into the direct sum of all (finite-dimensional) irreducible algebraic representations in which each such representation appears with multiplicity one.* One of the most natural ways of realizing a model is to express it as an induced representation $\mathrm{Ind}_M^G \tau$. This realization is the most convenient when $\tau = 1$: in this case the model is realized in the space of regular functions on the homogeneous space G/M. A

*H. Weyl's "unitary trick" shows that constructing such a model is equivalent to constructing a model of representations of a compact form of G. In this paper we use the language of complex groups; henceforth, by group we shall always mean, without further mention, an algebaic group over $\mathbf{C}$.

Moscow State University. Translated from Funktional'nyi Analiz i Ego Prilozheniya, Vol. 18, No. 3, pp. 14-31, July-September, 1984. Original article submitted March 15, 1984.

assical example of such a model is the representation of $SO_3$ acting on functions on the two-nensional sphere.

The systematic investigation of models was initiated in [3]. A model of the form $\mathrm{Ind}_M^G \tau$ s constructed there for every connected reductive group G, where the homogeneous space M is the complexification of the maximal-rank symmetric space corresponding to G. In this per we deal with classical groups only, i.e., G will be one of the groups GL(n, C) ($n \geqslant 2$), n, C) ($n \geqslant 3$), or Sp(2n, C) ($n \geqslant 1$) (in what follows we omit the symbol C to simplify nota-ons). For classical groups the result of [3] can be formulated as follows:

THEOREM 1 ([3]). For each of the classical groups consider its subgroup M and the rep-sentation $\tau$ of M displayed in the following table (in which $\lambda(n)$ stands for the representa-on of GL(n) in the exterior algebra $\Lambda^*(C^n)$, as well as for its restriction to an arbitrary bgroup of GL(n):

| $G$ | $GL(n)$ | $O(2n+1)$ | $O(2n)$ | $Sp(2n)$ |
|---|---|---|---|---|
| $M$ | $O(n)$ | $O(n+1)\times O(n)$ | $O(n)\times O(n)$ | $GL(n)$ |
| $\tau$ | $\lambda(n)$ | $\lambda(n+1)\otimes 1$ | $\lambda(n)\otimes 1$ | $\lambda(n)$ |

Then in each of these cases $\mathrm{Ind}_M^G \tau$ is a model of representations of the group G.

The first main result of this paper is the construction of a new series of models which, a certain sense, are dual to the models provided by Theorem 1.

THEOREM 2. For each of the classical groups G, except for GL(2n + 1) and O(2n + 1), nsider the subgroup M and the representation $\tau$ of M displayed in the following table (in ich $\sigma(n)$ stands for the representation of GL(n) in the symmetric algebra $S^*(C^n)$, as well for its restriction to an arbitrary subgroup of GL(n)):

| $G$ | $GL(2n)$ | $O(2n)$ | $Sp(4n+2)$ | $Sp(4n)$ |
|---|---|---|---|---|
| $M$ | $Sp(2n)$ | $GL(n)$ | $Sp(2n+2)\times Sp(2n)$ | $Sp(2n)\times Sp(2n)$ |
| $\tau$ | $\sigma(2n)$ | $\sigma(n)$ | $\sigma(2n+2)\otimes 1$ | $\sigma(2n)\otimes 1$ |

Then in each of these cases $\mathrm{Ind}_M^G \tau$ is a model of representations of the group G.

In Theorem 2 spaces G/M are, as above, complexified symmetric spaces; the latter, how-er, are no longer of maximal rank.

We remark that between the models provided by Theorems 1 and 2 there exists a peculiar rrespondence. Thus, it is readily verified that each of the models of Theorem 1 turns into e of the models of Theorem 2 if we replace all groups, subgroups, and representations as llows: GL(n) is replaced by GL(2n), O(n) by Sp(2n), and Sp(2n) by O(4n), where $\lambda(n)$ is placed by $\sigma(2n)$. The geometric meaning of this "doubling" procedure is still obscure.

The models provided by Theorem 2 have the "drawback" that the induced representation infinite dimensional. To eliminate it, one may choose in each case a subgroup $M_0$ in M ch that $\tau = \mathrm{Ind}_{M_0}^M 1$; in this way the given model of representations of G is realized as $\mathrm{d}_{M_0}^G 1$. It turns out that such a realization exists also for classical groups which are not vered by Theorem 2, i.e., for G = GL(2n + 1), O(2n + 1), and hence it is natural to join ese and other groups in a single series. The subgroup $M_0$ is conveniently described using e following terminology. We let GL(n − 1/2) and Sp(2n − 1) denote the affine subgroup of (n) and Sp(2n), respectively, i.e., in each the case, the stabilizer of the null linear nctional on the space of the standard representation. Thus, the index n for GL(n) may now sume both integer and half-integer values, and for Sp(n), arbitrary positive integer values.

THEOREM 2'. For each classical group G we define the subgroup $M_0$ of G as in the follow-ng table:

| $G$ | $GL(n)$ | $O(n)$ | $Sp(2n)$ |
|---|---|---|---|
| $M_0$ | $Sp(n-1)$ | $GL\left(\frac{n-1}{2}\right)$ | $Sp(n-1)\times Sp(n)$ |

Then in each case $\mathrm{Ind}_{M_0}^{G} 1$ is a model of representations of G.

We remark that subgroup $M_0$ is reductive only for those groups which are not covered by Theorem 2: for $G = GL(2n+1)$, $M_0 = Sp(2n)$, whereas for $G = O(2n+1)$, $M_0 = GL(n)$. These two models were described by Krämer in [4]; therein he also observed that there are no other reductive subgroups $M_0$ with this property in simple groups G ([4] is written in the language of compact groups).

Theorems 2 and 2' are proved in Sec. 1.

In Sec. 2, inspired by the aforementioned construction of Biederharn and Flath, we examine a different approach to models which in some sense are dual to the preceding ones. Roughly speaking, we want to realize the model of representations of group G as the restriction of a certain remarkable representation of an "overgroup" L containing G. More precisel we produce such models in a somewhat weaker sense: we construct a representation of the Lie algebra $\mathfrak{l}$ of L (which, as a rule, cannot be integrated to a representation of L), whose restriction to the Lie algebra $\mathfrak{g}$ of the group G is the representation of $\mathfrak{g}$ corresponding to the model of representations of G.† These models will be referred to as Res-models, in contrast to the Ind-models discussed previously.

In this paper we shall use the following geometric construction of Res-models. Let L be a group containing G, $\Gamma = L/P$ a homogeneous space of L, $\pi$ a vector L-bundle over $\Gamma$, and $\Omega$ an open G-invariant subset of $\Gamma$. It is clear that in the space $\mathcal{T}(\Omega, \pi)$ of regular sections there are compatible actions of the group G and the Lie algebra $\mathfrak{l}$; it is in this space that we will realize our model. As in the case of Ind-models, this realization is the most convenient when $\pi$ is a one-dimensional trivial bundle.

The Res-models of the indicated type and the Ind-models of the group G are intimately related. Thus, suppose that in the Res-model described above G acts transitively on $\Omega$. Let M be the stationary subgroup of G at some point of $\Omega$, and let $\tau$ be the natural representation of M in the fiber of the bundle $\pi$ over this point. It is clear that the representation of G in the space $\mathcal{T}(\Omega, \pi)$ is naturally isomorphic to $\mathrm{Ind}_M^G \tau$. Therefore, given the Res-model one can construct the Ind-model. The second main result of this paper is the construction for al classical groups G of the Res-models corresponding to the Ind-models discussed above.

The "overgroup" L in our construction will always be a product of classical groups (some times it is convenient to replace O(n) by its component of identity SO(n)). Also, for the homogeneous spaces $\Gamma$ we shall take compact Hermitian symmetric spaces of special form corresponding to L: the so-called "fundamental Grassmanians." For each classical group L we define its fundamental Grassmanian $\Gamma(L)$ as follows: $\Gamma(GL(2n))$ is the Grassman manifold of n-dimensional subspaces in $\mathbf{C}^{2n}$, $\Gamma(GL(2n+1))$ is the Grassman manifold of $(n+1)$-dimensional subspaces of $\mathbf{C}^{2n+1}$, $\Gamma(O(2n))$ is the Grassman manifold of n-dimensional isotropic subspaces of $\mathbf{C}^{2n+1}$, $\Gamma(O(2n+1))$ is the Grassman manifold of n-dimensional isotropic subspaces of $\mathbf{C}^{2n+1}$, and, finally, $\Gamma(Sp(2n))$ is the Grassman manifold of n-dimensional Lagrangian subspaces of $\mathbf{C}^{2n}$ All spaces $\Gamma(L)$ are connected, except for $\Gamma(O(2n))$, which has two components. The latter are naturally isomorphic, and each is a homogeneous space of SO(2n); we denote them by $\Gamma(SO(2n))$. Then $\Gamma(SO(2n))$ and $\Gamma(O(2n-1))$ can be naturally identified (see Sec. 1 for details). Finall if $L = L_1 \times L_2$ is a product of classical groups $L_1$ and $L_2$, we set $\Gamma(L) = \Gamma(L_1) \times \Gamma(L_2)$.

Apparently, the fundamental Grassmanians (and their superanalogs, discussed below) are very interesting geometric objects whose importance goes far beyond the framework in which th are applied here.

We are now ready to formulate our definition of Res-models.

<u>THEOREM 3.</u> For each classical group G we define an "overgroup" $L_0 \supset G$ and a bundle $\pi$ ov the space $\Gamma_0 = \Gamma(L_0)$ by means of the following table (if $\Gamma_0$ is given as a subset in a Grassmanian of subspaces of a certain dimension in $\mathbf{C}^n$, then we denote by $\Lambda(n)$ the bundle over $\Gamma_0$ whose fiber over the point $V \in \Gamma_0$, i.e., "over" the subspace V of $\mathbf{C}^n$, is the exterior algebra $\Lambda^*(V)$):

†In the case where $G = O(n)$, i.e., for G not connected, a certain cautiousness should be observed: in this case the spectrum of the resulting representation of the Lie algebra $\mathfrak{g}$ is no of multiplicity one.

| | | | |
|---|---|---|---|
| $G$ | $GL(n)$ | $O(n)$ | $Sp(2n)$ |
| $L_0$ | $Sp(2n)$ | $GL(n)$ | $Sp(2n) \times Sp(2n)$ |
| $\pi$ | $\wedge(2n)$ | $\wedge(n)$ | $\wedge(2n) \otimes 1$ |

Then in each case the action of the group G on $\Gamma_0$ has a unique open (dense) orbit $\Omega_0$, the representation of G in the space of regular sections, $\mathcal{T}(\Omega_0, \pi)$, is a Res-model of resentations of G.†

THEOREM 4. For each classical group G we define the "overgroup" L by means of the lowing table (the embedding of G in L will be indicated in Sec. 2):

| | | | | |
|---|---|---|---|---|
| $G$ | $GL(n)$ | $O(2n+1)$ | $O(2n)$ | $Sp(2n)$ |
| $L$ | $SO(2n+2)$ | $O(2n+2) \times SO(2n+2)$ | $SO(2n+2) \times O(2n)$ | $GL(2n+1)$ |

Then in each case the action of G on $\Gamma = \Gamma(L)$ has a unique open (dense) orbit $\Omega$, and representation of G in the space of regular functions on $\Omega$ is a Res-model of representans of G.

We remark that the Res-models of Theorem 3 correspond in the sense indicated above to Ind-models of Theorem 1, whereas the Res-models of Theorem 4 correspond to the Indels of Theorem 2'. We see that in Theorem 4 (3) the models are realized in spaces of ctions on the Grassmanian $\Gamma(L)$ (respectively, in spaces of sections of vector bundles). restore the symmetry it is necessary to generalize the notion of Res-model, considering "overgroup" L a supergroup rather than a group. More precisely, in each of the cases ated by Theorem 3 the sections of the bundle may be regarded as (regular) functions on a ermanifold $\Gamma$ for which $\Gamma_0$ is the "underlying" manifold (this may be taken as the definition $\Gamma$; for the terminology on supermanifolds and Lie supergroups and superalgebras see [5, 6]).

Supermanifold $\Gamma$ is a Hermitian symmetric superspace in the sense of paper [7]; it can be lized naturally as a homogeneous space of the supergroup L with the group $L_0 \times \mathbf{C}^*$ as underng manifold. It is natural to consider $\Gamma$ as the fundamental Grassmanian of the supergroup nd denote it by $\Gamma = \Gamma(L)$.

By the foregoing discussion, we can reformulate Theorem 3 as follows:

THEOREM 3'. For each classical group we define the supergroup L (over C) by the followtable (the notations are explained in [6]):

| | | | |
|---|---|---|---|
| $G$ | $GL(n)$ | $O(n)$ | $Sp(2n)$ |
| $L$ | $OSp(2 \mid 2n)$ | $GL(1 \mid n)$ | $OSp(2 \mid 2n) \times Sp(2n)$ |

Let $\Gamma = \Gamma(L)$ be the fundamental Grassmanian of the supergroup L (thus, $\Gamma_0$ serves as underlying manifold of L, while the regular functions on $\Gamma$ are sections of the bundle escribed in Theorem 3). Then the action of G on $\Gamma$ has a unique open orbit $\Omega$, and the resentation of G in the space of regular functions on $\Omega$ is a Res-model of representations G.

COROLLARY. On the Res-model of representations of G constructed in Theorem 3 there is a ural action of the Lie superalgebra $\mathfrak{l}$ of supergroup L, which has the Lie algebra $\mathfrak{l}_0 \oplus \mathbf{C}$ as even part.

The proof of Theorem 3' will be published separately, together with a study of the ulting representations of the Lie (super)algebra $\mathfrak{l}$.

There is an alternative way of looking at the correspondence between Ind- and Res-models. E be the (abstract) model of representations of the group G, i.e., the sum of all irreible representations of G, each taken with multiplicity one. A priori, the space E is owed only with a structure of G-module. The realization of E as the Ind-models of orems 1, 2, or 2' yields a supplementary structure of commutative (super)algebra on which

we shall see soon it is natural to reserve the notations L, $\Gamma$, and $\Omega$ for the overobjects vering" $L_0$, $\Gamma_0$, and $\Omega_0$.

G and its Lie algebra $\mathfrak{g}$ act by automorphisms and respectively, derivations. The Res-model compatible with the Ind-model in the sense indicated above defines on E an action of the larger Lie (super)algebra $\mathfrak{l}$ by (super)derivations, which extends the action of $\mathfrak{g}$. We may say that the elements of $\mathfrak{l}$ define the hidden symmetries of the model of representations of G.

For the reader's convenience we list together the Ind- and Res-models in correspondence in the following tables:

Table 1 (Theorems 1, 3, and 3')

| | | | | |
|---|---|---|---|---|
| $L$ | $OSp(2\|2n)$ | $GL(1\|2n+1)$ | $GL(1\|2n)$ | $OSp(2\|2n)\times Sp(2n)$ |
| $L_0$ | $Sp(2n)$ | $GL(2n+1)$ | $GL(2n)$ | $Sp(2n)\times Sp(2n)$ |
| $G$ | $GL(n)$ | $O(2n+1)$ | $O(2n)$ | $Sp(2n)$ |
| $M$ | $O(n)$ | $O(n+1)\times O(n)$ | $O(n)\times O(n)$ | $GL(n)$ |
| $\tau$ | $\lambda(n)$ | $\lambda(n+1)\otimes 1$ | $\lambda(n)\otimes 1$ | $\lambda(n)$ |

Table 2 (Theorems 2, 2' and 4)

| | | | | |
|---|---|---|---|---|
| $L$ | $SO(4n+2)$ | $SO(4n+4)$ | $SO(2n+2)\times O(2n)$ | $O(2n+2)\times SO(2n+2)$ |
| $G$ | $GL(2n)$ | $GL(2n+1)$ | $O(2n)$ | $O(2n+1)$ |
| $M$ | $Sp(2n)$ | — | $GL(n)$ | — |
| $M_0$ | $Sp(2n-1)$ | $Sp(2n)$ | $GL\left(n-\frac{1}{2}\right)$ | $GL(n)$ |

| | | |
|---|---|---|
| $L$ | $GL(4n+1)$ | $GL(4n+3)$ |
| $G$ | $Sp(4n)$ | $Sp(4n+2)$ |
| $M$ | $Sp(2n)\times Sp(2n)$ | $Sp(2n+2)\times Sp(2n)$ |
| $M_0$ | $Sp(2n-1)\times Sp(2n)$ | $Sp(2n+1)\times Sp(2n)$ |

Now let us consider a more geometric approach to our results. To establish the relatio between the Res- and Ind-models we construct embeddings of the homogeneous spaces G/M as ope subsets of homogeneous spaces L/P of larger groups. Generally speaking, given such an embedding, as we have explained earlier, a larger Lie algebra $\mathfrak{l}$ acts in the space of the induc representation $\mathrm{Ind}_M^G 1$, along with the group G, as well as in other induced representations $\mathrm{Ind}_M^G \tau$ (independently on whether there are models or not). Therefore, these induced representations possess a hidden symmetry "encoded" in the given embedding $G/M \subset L/P$.

The aforementioned embeddings are of independent interest besides having applications i the theory of representations. The geometric construction on which Theorems 1 and 3 are bas amounts to complexification with subsequent compactification of all classical Riemannian max mal-rank symmetric spaces. It produces compact Hermitian symmetric spaces of a special form (the fundamental Grassmanians defined above). The last result of this paper is to provide a construction of this type for all classical symmetric spaces.

By classical compact group we shall always mean one of the groups $U(n, \mathbb{R})$, $U(n, \mathbb{C})$, or $U(n, \mathbb{H})$ (i.e., a compact orthogonal, unitary, or symplectic group, respectively) and also a product of several such groups; the complexifications of these groups are $O(n)$, $GL(n)$, and $Sp(2n)$, respectively. Any space of the form $X = U_0/U_0^\theta$, where $U_0$ is the component of identit of the classical compact group U and $U_0^\theta$ is the group of fixed points of an involutive automosphism $\theta$ of U, will be referred to as a classical compact symmetric space. The complexifi tion $X_{\mathbb{C}}$ of the space X can be realized as $G_0/G_0^\theta$, where $G_0$ and $G_0^\theta$ are the complexifications o the groups $U_0$ and $U_0^\theta$. Theorem 5 below contains a list of irreducible spaces X and $X_{\mathbb{C}}$; we remark that it differs from the usual list (see [8]) because our spaces X are not always simpl connected.

<u>THEOREM 5</u>. Let X be a classical compact symmetric space, and let $X_{\mathbb{C}} = G_0/G_0^\theta$ be its complexification. We define a group L containing $G_0$ and a homogeneous L-space $\Gamma$ by means of the following table (the fundamental Grassmanians $\Gamma(L)$ were defined above, and $\Gamma_k^n$ denotes th Grassmanian of k-dimensional subspaces of $\mathbb{C}^n$):

| No. | 1 | 2 | 3 | 4 |
|---|---|---|---|---|
| $X$ | $U(n, \mathbf{C})/U(n, \mathbf{R})$ | $U(n, \mathbf{H})/U(n, \mathbf{C})$ | $U_0(2n, \mathbf{R})/U(n, \mathbf{C})$ | $U(2n, \mathbf{C})/U(n, \mathbf{H})$ |
| $X_{\mathbf{C}}$ | $GL(n)/O(n)$ | $Sp(2n)/GL(n)$ | $SO(2n)/GL(n)$ | $GL(2n)/Sp(2n)$ |
| $L$ | $Sp(2n)$ | $Sp(2n) \times Sp(2n)$ | $SO(2n) \times SO(2n)$ | $SO(4n)$ |
| $\Gamma$ | $\Gamma(L)$ | $\Gamma(L)$ | $\Gamma(L)$ | $\Gamma(L)$ |

| No. | 5 | 6 | 7 |
|---|---|---|---|
| $X$ | $U_0(n, \mathbf{R})/(U(k, \mathbf{R}) \times U(n-k, \mathbf{R}))_0$ | $U(n, \mathbf{C})/U(k, \mathbf{C}) \times U(n-k, \mathbf{C})$ | $U(n, \mathbf{H})/U(k, \mathbf{H}) \times U(n-k, \mathbf{H})$ |
| $X_{\mathbf{C}}$ | $SO(n)/S(O(k) \times O(n-k))$ | $GL(n)/GL(k) \times GL(n-k)$ | $Sp(2n)/Sp(2k) \times Sp(2(n-k))$ |
| $L$ | $GL(n)$ | $GL(n) \times GL(n)$ | $GL(2n)$ |
| $\Gamma$ | $\Gamma^n_k$ | $\Gamma^n_k \times \Gamma^n_{n-k}$ | $\Gamma^{2n}_{2k}$ |

| No. | 8 | 9 | 10 |
|---|---|---|---|
| $X$ | $(U_0(n, \mathbf{R}) \times U_0(n, \mathbf{R}))/U_0(n, \mathbf{R})$ | $(U(n, \mathbf{C}) \times U(n, \mathbf{C}))/U(n, \mathbf{C})$ | $(U(n, \mathbf{H}) \times U(n, \mathbf{H}))/U(n, \mathbf{H})$ |
| $X_{\mathbf{C}}$ | $(SO(n) \times SO(n))/SO(n)$ | $(GL(n) \times GL(n))/GL(n)$ | $(Sp(2n) \times Sp(2n))/Sp(2n)$ |
| $L$ | $SO(2n)$ | $GL(2n)$ | $Sp(4n)$ |
| $\Gamma$ | $\Gamma(L)$ | $\Gamma(L)$ | $\Gamma(L)$ |

Then in each case the action of $G_0$ on $\Gamma$ has a (unique) open (dense) orbit; this orbit is morphic with $X_{\mathbf{C}}$ as a $G_0$-space.

Thus, the construction which yields $\Gamma$ given X is analogous to the R. Penrose's twistor struction.

Theorems 3, 4, and 5 are proved in Sec. 2.

This work, like [3], is part of a unifying program for investigating the representations reductive groups over various fields. We conclude the introduction by a brief account of er results relevant to this program.

1. Let B be a Borel subgroup of the connected complex reductive group G. A simple eval-ion of functional dimension shows that in order that $\mathrm{Ind}_M^G \tau$ be a model of representations G it is necessary that the subgroup M satisfy the condition $\dim M \geqslant \dim G/B$, with equality n $\dim \tau < \infty$. Another natural requirement is that M be transverse, in some sense, to B, example, that the action of M on G/B be locally transitive. This condition is met by all subgroups involved in Theorems 1, 2, and 2'. We remark that all these subgroups are re-tive or close to reductive. For simple groups G a complete list of all reductive subgroups ch act locally transitively on G/B was obtained in [12] (actually, paper [12] uses a differ- terminology; the geometrical interpretation to the result of [12] given here was obtained [10]). Interesting results were also obtained in another extreme situation, where the sub-up M is solvable. A classical model of representations is $\mathrm{Ind}_U^G 1$, where U is a maximal uni-ent subgroup of G, i.e., the model is realized in the space of regular functions on the damental affine space G/U (the assertion that it is a model is essentially equivalent to Cartan's theory of the highest weight). An interesting class of solvable subgroups M own as the Coxeter subgroups) was constructed in [13].

2. Let $\mathrm{Ind}_M^G \tau$ be the model of representations of group G constructed in [3] (for classi- groups G the pairs (M, $\tau$) were indicated in Theorem 1). Representation $\tau$ is, generally aking, not uniquely determined. In paper [3] an important class of such representations isolated: the so-called fine representations. In [3] these representations were structed separately for each case; a general construction (valid only in the case where the kin diagram of G has no multiple arrows) was proposed in [14]. An interesting application fine representations is given in [15], where it is shown that the nonuniqueness of their ice controls the decomposition of the unitary principal series of the real form of group G o irreducible representations.

3. We should also mention two works on models for the group $G = GL(n, F_q)$: paper [16], in which an analog of the construction of Theorem 1 is studied, and paper [17] in which a construction, closely related to our Theorem 2, is given. Klyachko's work [17] deserves special attention because it constructs a true model of all representations of G, using a "mixture" of a symplectic and maximal unipotent subgroup of G, whereas [16] provides only an "almost-model."

The authors are grateful to V. V. Serganova and D. A. Leites for a consultation on super manifolds and to A. B. Goncharov for useful discussion. In particular, Theorem 5 was obtained in cooperation with A. B. Goncharov.

## 1. Ind-Models

This section is devoted to the proofs of Theorems 2 and 2' formulated above.

We remind the reader that $GL(n - 1/2)$ is the affine subgroup of $GL(n)$: the semidirect product of $GL(n - 1)$ and the Abelian normal subgroup $\mathbf{C}^{n-1}$. Similarly, $Sp(2n - 1)$ is the affine subgroup of $Sp(2n)$: the semidirect product of $Sp(2n - 2)$ and the Heisenberg group corresponding to the $(2n - 2)$-dimensional symplectic space over $\mathbf{C}$ (see [9]).

The embedding of $Sp(2n)$ in $GL(2n + 1)$ is the composition of the standard embeddings $Sp(2n) \subset GL(2n) \subset GL(2n + 1)$. Similarly, the embedding of $GL(n)$ in $O(2n + 1)$ is the composition of the standard embeddings $GL(n) \subset SO(2n) \subset SO(2n + 1) \subset O(2n + 1)$. All the other embeddings appearing in Theorems 2 and 2' are standard.

First of all, we show that Theorem 2 is a consequence of Theorem 2'. To see this, it suffices to verify that in each case the representation $\tau$ of the group M, defined in Theorem 2, is isomorphic to $\mathrm{Ind}_{M_0}^{M} 1$; this will imply, by the transitivity property of the induction operation, that $\mathrm{Ind}_{M}^{G}\tau = \mathrm{Ind}_{M_0}^{G} 1$ is a model of representations of G.

Let us examine here the case $M = Sp(2n)$, $M_0 = Sp(2n - 1)$. We denote by V a 2n-dimensional symplectic space in which these two groups act. By definition, the representation $\mathrm{Ind}_{M_0}^{M} 1$ acts in the space of regular functions $\mathbf{C}[M/M_0]$; obviously, $M/M_0 = V^* \setminus \{0\}$ is the space of nonzero linear functions on V. But since $\dim V^* \geqslant 2$, every regular function on $V^* \setminus \{0\}$ admits an extension to a regular function on $V^*$. Therefore, the space $\mathbf{C}[M/M_0]$ can be naturally identified with the symmetric algebra $S(V)$, and hence representation $\mathrm{Ind}_{M_0}^{M} 1$ can be identified with representation $\tau = \sigma(2n)$ of Theorem 2. The other cases are treated similarly.

Let us now turn to the proof of Theorem 2' and examine first the case $G = GL(n)$, $M_0 = Sp(n - 1)$. The first step is to provide a convenient realization of the homogeneous space $G/M_0$. Let V be an n-dimensional vector space over $\mathbf{C}$ in which the group G acts. Set $\Phi = V^* \oplus \Lambda^2(V^*)$, i.e., the elements of $\Phi$ are pairs $(\xi, \varphi)$, where $\xi$ is a linear functional and $\varphi$ is a skew-symmetric bilinear form on V. G acts naturally on $\Phi$: $g\cdot(\xi, \varphi) = (g^{*-1}\xi, g^{*-1}\varphi g^{-1})$.

LEMMA 1. The action of G on $\Phi$ has a unique open (dense) orbit: $\Omega = \{(\xi, \varphi) \in \Phi\colon \xi \neq 0, \operatorname{Ker}\xi \cap \operatorname{Ker}\varphi = 0\}$. Subgroup $M_0$ is the stabilizer of a point of $\Omega$; hence, the space $G/M_0$ can be naturally identified with $\Omega$.

The straightforward proof is left to the reader. The pairs $(\xi, \varphi) \in \Omega$ will be referred as as nondegenerate.

To verify that representation $\mathrm{Ind}_{M_0}^{G} 1$ has multiplicity-one spectrum, we use the following test.

LEMMA 2. Let G be a connected reductive group over $\mathbf{C}$, B a Borel subgroup of G, and M an algebraic subgroup of G. In order that representation $\mathrm{Ind}_{M}^{G} 1$ have multiplicity-one spectrum it suffices that the following two conditions be satisfied:

(1) $M \cap B = \{1\}$,

(2) dim M + dim B = dim G.

In fact, (1) and (2) imply that the action of M on G/B is locally transitive. By Coroly 2 of [10], it follows that the spectrum of $\mathrm{Ind}_M^G 1$ has multiplicity one.

Now let us verify that conditions (1) and (2) are fulfilled for G = GL(n) and $M = M_0 =$ 2n − 1). In this case the Borel subgroup B is the stabilizer in G of some complete flag $= V_0 \subset V_1 \subset \ldots \subset V_{n-1} \subset V_n = V$. Applying Lemma 1 we see that

$$\dim G - \dim M_0 = \dim G/M_0 = \dim \Phi = n + \frac{n(n-1)}{2} = \frac{n(n+1)}{2} = \dim B,$$

ch proves (2). To prove (1), we must be more specific concerning the choice of subgroup or, equivalently, of that point of Ω with $M_0$ as stabilizer.

LEMMA 3. Suppose $(\xi, \varphi) \in \Omega$ has the property that the pair $(\xi \mid V_i, \varphi \mid V_i)$ is nondegenerate all i = 1, 2, ..., n. If $M_0$ is the stabilizer of $(\xi, \varphi)$, then $M_0 \cap B = \{1\}$.

Proof. Pick $g \in M_0 \cap B$. By induction on n we may assume that the restriction of g to $_1$ is trivial. Let $v_n \in V \setminus V_{n-1}$, and write $gv_n = cv_n + v'$ with $c \in \mathbf{C}$ and $v' \in V_{n-1}$. Set $(1-c)v_n - v'$. We can express the fact that the pair $(\xi, \varphi)$ is a fixed point of g in form:

$$(*)\ \xi(v) = 0, \quad \varphi(v, v'') = 0 \text{ . for all } \quad v'' \in V_{n-1}.$$

$c \neq 1$, we get $0 \neq v \in \operatorname{Ker} \xi \cap \operatorname{Ker} \varphi$, which contradicts the nondegeneracy of the pair $(\xi, \varphi)$. refore, c = 1 and $v = -v' \in V_{n-1}$. Applying (*) once more, we see that $v' \in \operatorname{Ker} \xi \mid V_{n-1} \cap$ $\varphi \mid V_{n-1}$. Since the pair $(\xi \mid V_{n-1}, \varphi \mid V_{n-1})$ is nondegenerate by hypothesis, v' = 0, i.e., $gv_n =$ and hence g = 1, as asserted.

Thus, we have shown that representation $\mathrm{Ind}_{M_0}^G$ has multiplicity-one spectrum. To verify t it contains all irreducible representations of G, we explicitly indicate the functions m $\mathbf{C}[\Omega]$ which are highest vectors for these representations.

Recall that the irreducible algebraic representations of G = GL(n) are parametrized by hest weights $\mathbf{m} = (m_1, \ldots, m_n)$, where $m_i$ are integers satisfying $m_1 \geqslant m_2 \geqslant \ldots \geqslant m_n$ e, e.g., [11]). Corresponding representation $\rho_{\mathbf{m}}$ has a unique (modulo a factor) highest tor $f_{\mathbf{m}}$, i.e., a common eigenvector of all transformations $b \in B$; moreover, if b acts on $V_{i-1}$ as multiplication by a scalar $b_i$, then

$$\rho_{\mathbf{m}}(b) f_{\mathbf{m}} = \Big(\prod_{i=1}^{n} b_i^{m_i}\Big) f_{\mathbf{m}}.$$

Let us correspond to each pair $(\xi, \varphi) \in \Phi$ a collection of forms $\omega_r(\xi, \varphi) \in \wedge^r (V^*)$ (r = 2, ..., n) by the rule: $\omega_{2k}(\xi, \varphi) = \varphi \wedge \varphi \wedge \cdots \wedge \varphi$ (k times), and $\omega_{2k+1}(\xi, \varphi) = \omega_{2k}(\xi, \varphi) \wedge \xi$. ect a basis $v_1, \ldots, v_n$ in V such that for each r = 1, 2, ..., n the vectors $v_1, \ldots, v_r$ m a basis in subspace $V_r$. Set

$$f_r(\xi, \varphi) = \omega_r(\xi, \varphi)\ (v_1 \wedge \cdots \wedge v_r).$$

is clear that all functions $f_1, \ldots, f_n$ are regular on Φ and $f_n$ does not vanish on Ω. It a direct consequence of the definitions that $f_r$ is a highest vector of weight (1, ..., 1, ..., 0) (r times 1). We conclude that for every highest weight $\mathbf{m} = (m_1, \ldots, m_n)$ the ction $f_{\mathbf{m}} = \prod_{r=1}^{n} f_r^{m_r - m_{r+1}}$ belongs to C[Ω] and is a highest vector of weight **m**.

This completes the proof of Theorem 2' in the case $G = GL(n)$. We record a corollary of the proof that will be important in the sequel.

LEMMA 4. The natural representation of $G = GL(n)$ in the space $\mathbf{C}[\Phi]$ of regular functions of $\Phi = V^* \oplus \wedge^2 V^*$ is the sum of all irreducible polynomial representations of G, each taken with multiplicity one (we recall that a representation $\rho_{\mathbf{m}}$ with highest weight $\mathbf{m} = (m_1, \ldots, m_n)$ is a polynomial when $m_n \geqslant 0$; weights with this property will be termed polynomial weights

In fact, we have the obvious embedding $\mathbf{C}[\Phi] \subset \mathbf{C}[\Omega]$; it is clear that the function $f_{\mathbf{m}}$ constructed above belongs to $C[\Phi]$ if and only if $m_n \geqslant 0$.†

Now let G be one of the groups $O(2n)$ $(n \geqslant 2)$, $O(2n + 1)$, or $Sp(2n)$. Let W be a vector space in which G is realized ($\dim W = 2n$ or $2n + 1$), and let $\langle,\rangle$ be a nondegenerate bilinear form on W preserved by G (this form is symmetric if G is an orthogonal group and skew-symmetric if G is symplectic). We denote by $G_0$ the component of the identity in G, i.e., $G_0$ is one of the groups $SO(2n)$, $SO(2n + 1)$, or $Sp(2n)$. We begin with a description of the irreducible representations of G and $G_0$ that suits the goals of this paper.

We fix an n-dimensional subspace $V \subset W$ such that the restriction of the form $\langle,\rangle$ to it equals 0. We define the parabolic subgroup P of $G_0$ as the stabilizer of the subspace V. We denote by U the unipotent radical of P: subgroup U consists of those transformations $g \in P$ which act trivially on V. Let G' denote the factor group P/U; we identify G' with the group of all invertible linear transformations of V, i.e., $G' = GL(n)$.

Let $\pi$ be a representation of $G_0$ in a space E. We denote by $E^U$ the subspace of U-invaria vectors in E. It is clear that G' acts naturally in $E^U$; we denote the representation of $G' = GL(n)$ by $\pi^U$. Obviously, the correspondence $\pi \to \pi^U$ between the representations of the groups $G_0$ and G' is functorial and takes direct sums into direct sums.

LEMMA 5. The correspondence $\pi \to \pi^U$ takes irreducible (finite-dimensional) algebraic representations of $G_0$ into irreducible algebraic representations of G'. Representation $\pi$ is uniquely specified up to equivalence by $\pi^U$. Therefore, the irreducible representations of $G_0$ are parametrized by certain highest weights $\mathbf{m} = (m_1, \ldots, m_n)$ (we call them admissible): to the highest weight $\mathbf{m}$ corresponds the representation $\pi_{\mathbf{m}}$ such that $\pi^U_{\mathbf{m}} = \rho_{\mathbf{m}}$. For $G_0 = SO(2n + 1$ or $Sp(2n)$ the admissible weights are precisely the polynomial $\mathbf{m}$ (i.e., the weights $\mathbf{m}$ with $m_n \geqslant 0$). For $G_0 = SO(2n)$ the admissible weights are those satisfying the (weaker) requiremen $m_{n-1} \geqslant |m_n|$.

This lemma is a straightforward corollary of the familiar description of the irreducible representations of the group $G_0$ in terms of highest weights ([11], Chapter 8). In fact, a Borel subgroup B of $G_0$ is the preimage of a Borel subgroup B' of G' under the natural projectio $P \to G'$; hence, f is the highest vector of representation $\pi$ of $G_0$ if and only if f is both U-invariant and a highest vector of representation $\pi^U$.

Lemma 5 allows us to give a simple description of the irreducible (algebraic) representa tions of G:

a) If $G = Sp(2n)$, then $G = G_0$ and hence the irreducible representations of G are exactl $\pi_{\mathbf{m}}$, where $\mathbf{m}$ runs over the polynomial highest weights of $GL(n)$.

b) $G = O(2n + 1)$ is the direct product of $G_0$ and the group with two elements $\{\pm I\}$, wher I is the identity transformation of the space W. We denote by $\pi^{\pm}_{\mathbf{m}}$ the representation of G specified by the conditions $\pi^{\pm}_{\mathbf{m}}|_{G_0} = \pi_{\mathbf{m}}$ and $\pi^{\pm}_{\mathbf{m}}(-I) = \pm 1$. Clearly, the irreducible representations of G are exactly $\pi^{\pm}_{\mathbf{m}}$, where $\mathbf{m}$ runs over the polynomial highest weights of $GL(n)$.

c) The case $G = O(2n)$ is slightly more complicated. The general theory of representations of groups with a normal subgroup of index 2 leads to the following result. Let $\mathbf{m} = (m_1, \ldots, m_n)$ be a polynomial highest weight. If $m_n > 0$, then representation $\mathrm{Ind}^G_{G_0} \pi_m$ is

†The assertion of Lemma 4 was known, in one form or another, to Schur; its version in the language of characters is given, for example, in [18], Example 1.5.4.

reducible, and we denote it by $\pi_m^0$. If, however, $m_n = 0$, then $\mathrm{Ind}_{G_0}^G \pi_m$ is the sum of two distinct irreducible representations $\pi_m^{\pm}$. The representations of type $\pi_m^0$ and $\pi_m^{\pm}$ are mutually nonequivalent and exhaust all irreducible representations of G = O(2n).

From this description we obtain immediately the following statement.

LEMMA 6. Let G be one of the classical groups O(2n), O(2n + 1) or Sp(2n), let $G_0$ be the component of the identity in G, and let τ be a representation of $G_0$ which decomposes into the direct sum of representations $\pi_m$, where m runs over all polynomial highest weights of GL(n). Then $\mathrm{Ind}_{G_0}^G \tau$ is a model of representations of the group G.

With these preparations we are now ready to start the proof of Theorem 2' for orthogonal and symplectic groups. First of all, we notice that in all cases subgroup $M_0$ is contained in the component of identity $G_0$ of G. By the transitivity of the induction operation, $\mathrm{Ind}_{M_0}^G 1 = \mathrm{Ind}_{G_0}^G \mathrm{Ind}_{M_0}^{G_0} 1$. Hence, in view of Lemma 6, it suffices to verify that $\mathrm{Ind}_{M_0}^{G_0} 1 = \Sigma\pi_m$, where the sum is taken over all polynomial highest weights. Recalling Lemmas 4 and 5 we see that we have reduced the proof to the proof of the following lemma.

LEMMA 7. In the above notation, $C[G_0/M_0]^U$ and $C[\Phi]$ are isomorphic as G'-modulus.†

We begin with the description of the spaces $G/M_0$.

a) G = O(2n), $G_0$ = SO(2n), $M_0$ = GL(n — 1/2). We realize the space $G/M_0$ as the manifold of triples (X, Y, η), where X and Y are complementary n-dimensional isotropic subspaces of W and η is a nonzero linear functional on X.

b) G = O(2n + 1), $G_0$ = SO(2n + 1), $M_0$ = GL(n). We realize $G/M_0$ as the manifold of triples (X, Y, z), where X and Y are nonintersecting n-dimensional isotropic subspaces of W and z is a unit-length vector in W orthogonal to both X and Y.

c) G = $G_0$ = Sp(2n), with even n = 2m, $M_0$ = Sp(2m — 1) × Sp(2m). We realize $G/M_0$ as the manifold of triples (Z, Z', η), where Z and Z' are mutually orthogonal n-dimensional subspaces of W such that the restriction of the form <,> to each of them is nondegenerate, and η is a nonzero linear functional on Z.

d) G = $G_0$ = Sp(2n), with odd n = 2m + 1, $M_0$ = Sp(2m + 1)× Sp(2m). We realize $G/M_0$ as the manifolds of triples (Z, Z', η), where Z and Z' are mutually orthogonal subspaces of W, such that the restriction of the form <,> to each of them is nondegenerate, dim Z = n + 1, dim Z' = n — 1, and η is a nonzero linear functional on Z.

To describe the spaces $G_0/M_0$ for orthogonal groups we need some simple facts concerning isotropic Grassmanians. Since these facts will be used again in the sequel, we formulate them as a lemma.

LEMMA 8. Let G = O(2n) be realized as the group of transformations in the 2n-dimensional vector space W preserving the form <,> . Let Γ = Γ(G) be the fundamental Grassmanian of G, i.e., the Grassman manifold of n-dimensional isotropic subspaces of W (see Sec. 0).

a) Γ has exactly two components, which are orbits of $G_0$ = SO(2n). Two isotropic subspaces, $V_1$ and $V_2$, lie in the same component of Γ if and only if $\dim(V_1 \cap V_2)$ has the same parity as n.

b) Let V and V' be complementary n-dimensional isotropic subspaces of W. Using the form <,> we identify V' with the dual V* of V. Then the subspaces X ∈ Γ which are transverse to V', are in a one-to-one correspondence with the skew-symmetric bilinear forms on V: the form $\varphi \in \wedge^2 V^*$ can be regarded as an operator V → V* = V', and the corresponding subspace X(φ) is the graph of this operator.

---

†For G = O(2n) the subgroup $M_0$ is uniquely determined modulo conjugation only in G, but not in $G_0$. For this reason, its choice must be made more precise, which is done in the proof.

c) Let $W^1$ be a nondegenerate codimension-one subspace of W, and let $z_0$ be a unit length vector in W which is orthogonal to $W_1$. Embed the group $G^1 = O(2n-1) = O(W^1, \langle,\rangle)$ in $G_0 = SO(2n)$, letting $g \in G^1$ act on the vector $z_0$ by the rule $g \cdot z_0 = \det g \cdot z_0$. Let $\Gamma^1 = \Gamma(G^1)$ be the fundamental Grassmanian of the group $G^1$, i.e., the set of $(n-1)$-dimensional isotropic subspaces of $W^1$. Then $\Gamma$ is isomorphic to the space of pairs $(X^1, z)$, where $X^1 \in \Gamma^1$ and z is a unit-length vector in $W^1$ which is orthogonal to $X^1$: to the pair $(X^1, z)$ corresponds to the subspace $X = X^1 \oplus \mathbf{C}(z + iz_0)$. Moreover, the projection $X \to X^1 = X \cap W^1$ defines a $G_1$-space isomorphism of each of the components of $\Gamma$ onto $\Gamma^1$; in particular, $\Gamma^1$ is connected.

The (easy) proof is left to the reader.

Let us return to the spaces $G_0/M_0$.

a) $G = O(2n)$, $G_0 = SO(2n)$. From Lemma 8 it follows easily that $G/M_0$ has exactly two components, which are orbits of the group $G_0$. Moreover, two triples $(X_1, Y_1, \eta_1)$ and $(X_2, Y_2, \eta_2)$ lie in the same component of $G/M_0$ if and only if $X_1$ and $X_2$ (or, equivalently, $Y_1$ and $Y_2$) lie in the same component of $\Gamma(G)$. We choose the subgroup $M_0$ such that $G_0/M_0$ consists of triples $(X, Y, \eta)$ with X lying in the same connected component of $\Gamma(G)$ as the subspace V fixed above.

b) $G = O(2n+1)$, $G_0 = SO(2n+1)$. In this case too $G/M_0$ consists of two components which are orbits of $G_0$. The projection $(X, Y, Z) \to (X, Y)$ gives an isomorphism (of homogeneous $G_0$-spaces) of either of these components onto the space $\{(X, Y) \in \Gamma(G) \times \Gamma(G) : X \cap Y = 0\}$. Hence, we can realize $G_0/M_0$ as any of these components.

To prove Theorem 7 we explicitly construct an isomorphism of $G'$-modules $\mathbf{C}[\Phi] \cong \mathbf{C}[G_0/M_0]^U$ (recall that the space $\Phi = V^* \oplus \wedge^2 V^*$ is based on the n-dimensional space V, which in each of the considered cases is embedded into the 2n- or (2n + 1)-dimensional space W in which G is realized). To this end, we exhibit a regular mapping $p: G/M_0 \to \Phi$, separately for each case.

a) $G = O(2n)$, $G/M_0 = \{(X, Y, \eta) : X, Y \in \Gamma(G),\ X \cap Y = 0,\ \eta \in X^* \setminus \{0\}\}$. We denote by $p_{Y,X}$ the projection of W on Y along X. If $x = (X, Y, \eta) \in G/M_0$, we define the pair $p(x) = (\xi(x), \varphi(x))$ by the rule

$$\xi(x)(v) = \eta(p_{X,Y}(v)),$$
$$\varphi(x)(v_1, v_2) = \langle p_{Y,X} v_1, v_2\rangle \qquad (v, v_1, v_2 \in V).$$

b) $G = O(2n+1)$, $G/M_0 = \{(X, Y, z) : X, Y \in \Gamma(G),\ X \cap Y = 0,\ z \in W,\ \langle z, z\rangle = 1,\ z \perp (X \oplus Y)\}$. We denote by $p_{Y,X}$ the projector of W on Y along $X \oplus \mathbf{C}z$. If $x = (X, Y, z) \in G/M_0$, we define the pair $p(x) = (\xi(x), \varphi(x))$ by the rule

$$\xi(x)(v) = \langle v, z\rangle, \quad \varphi(x)(v_1, v_2) = \langle p_{Y,X} v_1, v_2\rangle + \frac{\langle v_1, z\rangle \cdot \langle v_2, z\rangle}{2}.$$

c) $G = Sp(2m)$. $G/M_0$ consists of the triples $(Z, Z', \eta)$ described above. If $x = (Z, Z', \eta) \in G/M_0$, we define the pair $p(x) = (\xi(x), \varphi(x))$ by the rule

$$\xi(x)(v) = \eta(p_{Z,Z'}, v), \quad \varphi(x)(v_1, v_2) = \langle p_{Z',Z} v_1, p_{Z',Z} v_2\rangle.$$

It is readily verified that the formulas given above define indeed a mapping $p: G/M_0 \to \Phi$ and hence also $p: G_0/M_0 \to \Phi$ (i.e., the forms $\varphi(x)$ constructed on V are skew-symmetric in all cases). It is clear that this mapping is regular and hence induces a mapping $p^*: \mathbf{C}[\Phi] \to \mathbf{C}[G_0/M_0]$. It is verified directly that if $g \in P = \mathrm{Stab}_{G_0}(V)$ and $\bar{g}$ denotes the image of g under the natural projection $P \to G' = GL(V)$, then $p(gx) = \bar{g}p(x)$ for $x \in G/M_0$. This shows that $p^*$ takes $\mathbf{C}[\Phi]$ into $\mathbf{C}[G_0/M_0]^U$ and is a morphism of $G'$-modules. We clain that $p^*: \mathbf{C}[\Phi] \to \mathbf{C}[G_0/M_0]^U$ is an isomorphism in all the considered cases.

To prove this we exhibit an open subset of $G_0/M_0$ with the property that the restriction of p to this subset has a rather simple structure. In each case we define open (dense) subsets $\Gamma_0 \subset G_0/M_0$ and $\Phi_0 \subset \Phi$ as follows:

a) $G_0 = SO(2n)$. $\Gamma_0 = \{(X, Y, \eta) : V \cap Y = 0\}$, $\Phi_0 = \{(\xi, \varphi) : \xi \neq 0\}$;
b) $G_0 = SO(2n+1)$. $\Gamma_0 = \{(X, Y, z) : V \cap Y = 0\}$, $\Phi_0 = \Phi$;

c) $G_0 = Sp(2n)$, n even. $\Gamma_0 = \{(Z, Z', \eta)\colon V \cap Z' = 0\}$, $\Phi_0 = \Omega$ (Lemma 1).

d) $G_0 = Sp(2n)$, n odd. $\Gamma_0 = \{(Z, Z', \eta)\colon V \cap Z' = 0, \operatorname{Ker} \eta \cap V = 0\}$, $\Phi_0 = \Omega$ (Lemma 1).

LEMMA 9. In all cases $\Gamma_0$ is P-invariant and the restriction of the mapping p to $\Gamma_0$ is rincipal U-bundle over $\Phi_0$ which is isomorphic to the product $\Phi_0 \times U \to \Phi_0$.

Proof. The P-invariance of $\Gamma_0$ is obvious. Next, let us construct a section $s\colon \Phi_0 \to \Gamma_0$. this end we select, in all cases, an n-dimensional isotropic subspace V' transverse to V, then use the form <,> to identify V' with V*. This permits us to regard the components nd $\varphi$ of any pair $(\xi, \varphi) \in \Phi_0$ as an element of V' and a skew-symmetric operator $V \to V'$, pectively. Using this interpretation we define the point $s(\xi, \varphi)$ in each case separately.

a) $G_0 = SO(2n)$. We define $s(\xi, \varphi)$ as the triple $(X, Y, \eta)$, where $Y = V'$, $X = \{v -$ $)\colon v \in V\}$ is the graph of the operator $-\varphi\colon V \to V'$, and $\eta(x) = \langle \xi, x \rangle$.

b) $G_0 = SO(2n - 1)$. Fix a unit-length vector $z_0 \in W$ orthogonal to both V and V'. define $s(\xi, \varphi)$ as the triple $(X, Y, z)$, where $Y = V'$, $X = \{v - \varphi(v) - \frac{\langle \xi, v \rangle}{2} \cdot \xi - \langle \xi, v \rangle z_0\colon$ $\in V\}$, $z = z_0 + \xi$.

c) $G_0 = Sp(2n)$ with n even. We define $s(\xi, \varphi)$ as the pair $(Z, Z', \eta)$, where $Z =$ $- 2\varphi(v)\colon v \in V\}$, $Z' = \{v + 2\varphi(v)\colon v \in V\}$, $\eta(z) = 2\langle \xi, z \rangle$.

d) $G = SO(2p)$ with n odd. We define $s(\xi, \varphi)$ as the triple $(Z, Z', \eta)$, where $Z =$ $- 2\varphi(v) + c\xi\colon v \in V, c \in \mathbf{C}\}$, $Z' = \{v + 2\varphi(v)\colon v \in V, \langle \xi, v \rangle = 0\}$, $\eta(z) = \langle \xi, z \rangle$.

It is readily verified that the constructed triple $s(\xi, \varphi)$ belongs indeed to $\Gamma_0$ and mapping $s\colon \Phi_0 \to \Gamma_0$ is a section, i.e., $p \circ s = \mathrm{id}$ [for example, in case a) the fact that ce X is isotropic follows from Lemma 8, b); we also remark that in cases b) and c) the degeneracy of the form <,> on the subspaces Z and Z' follows from the nondegeneracy of pair $(\xi, \varphi)$].

Proving the assertion of Lemma 9 is equivalent to showing that for every point $x \in \Gamma_0$ re exists a (unique) transformation $u \in U$ which brings x to the form $s(\xi, \varphi)$ for some $\varphi) \in \Phi_0$. The latter can be easily verified.

We are now able to prove that the mapping $p^*\colon \mathbf{C}[\Phi] \to \mathbf{C}[G_0/M_0]^U$ is an isomorphism. The t that $p^*$ is monomorphic follows immediately from Lemma 9: if $p^*f = 0$ for some $f \in \mathbf{C}[\Phi]$, n the restriction of f to the open (dense) subset $\Phi_0$ equals 0 (it can be written as $s^*p^*f$), hence $f = 0$.

For $G_0 = SO(2n)$ or $SO(2n + 1)$, the fact that $p^*$ is epimorphic is an easy consequence of ma 9. Let $F \in \mathbf{C}[G_0/M_0]^U$. Then $s^*F \in \mathbf{C}[\Phi_0]$. Recall that for $G_0 = SO(2n + 1)$, $\Phi_0 = \Phi$, reas for $G_0 = SO(2n)$, $\Phi_0$ is specified by the condition $\xi \neq 0$. We see that in these cases $\Phi_0$ has codimension $\geqslant 2$ in $\Phi$ (for this reason we must exclude $G = O(2)$ from the list of ssical groups since the last assertion is not true for it). Therefore, function $s^*F$ ex- ds to a regular function $f \in \mathbf{C}[\Phi]$. Using the U-invariance of F, we conclude that F and agree on the open (dense) subset $\Gamma_0 \subset G_0/M_0$, and hence everywhere.

If $G = Sp(2n)$, then the above reasoning fails, and we are forced to make a small detour. already verified fact that $p^*$ is monomorphic, in conjunction with Lemmas 4 and 5, shows t representation $\operatorname{Ind}_{M_0}^{G} 1$ contains all irreducible representations of the group $G = Sp(2n)$. refore, in order to prove that $p^*$ is an epimorphism, it suffices to verify that the spec- m of representation $\operatorname{Ind}_{M_0}^{G} 1$ has multiplicity one. To this end we apply the test of Lemma We take as a Borel subgroup B in G the preimage of a Borel subgroup B' in $G' = GL(V)$ under the natural projection $P \to G'$. Pick a point $(\xi, \varphi) \in \Omega$ such that its stabilizer $M_0'$ in $G'$ intersects $B'$ only at the identity (Lemma 3) and take as the subgroup $M = M_0$ in $G$ the stabilizer of the point $s(\xi, \varphi) \in \Gamma_0$. In view of this choice, both conditions (1) and (2) of Lemma 2 follow immediately from Lemma 9.

This completes the proof of Lemma 7, and hence of Theorem 2'.

## 2. Res-Models

This section is devoted to the proof of Theorems 3, 4, and 5 formulated in Sec. 0. Recall that in Theorem 3 a group $L_0 \supset G$ and a bundle $\pi$ over the fundamental Grassmanian $\Gamma_0 = \Gamma(L_0)$ are associated with every classical group. We must show that the action of G on $\Gamma_0$ has a unique open orbit $\Omega_0$, the stabilizer of G at some point $x \in \Omega_0$ is the subgroup M appearing in Theorem 1, and the representation of M in the fiber of bundle $\pi$ over x is the representation appearing in Theorem 1. In Theorem 4 we are given the group $L \supset G$ and we must show that the action of G on $\Gamma = \Gamma(L)$ has a unique open orbit $\Omega$ and that $\Omega$ is isomorphic, as a G-space, with the space $G/M_0$ of Theorem 2'. Let us examine again all the possible cases.

Proof of Theorem 3. a) The case $G = GL(n)$, $L_0 = Sp(2n)$. Suppose $L_0$ is realized in the 2n-dimensional vector space W with bilinear form $\langle , \rangle$. The group G is embedded in $L_0$ as the stabilizer of a pair of complementary Lagrangian subspaces, V and V' and is identified with GL(V). The fundamental Grassmanian $\Gamma_0 = \Gamma(L_0)$ is the Grassmanian of Lagrangian subspaces in W. We denote by $\Omega_0$ the set of subspaces $X \in \Gamma_0$ which are transverse to both V and V'. Clearly, $\Omega_0$ is open in $\Gamma_0$ and G-invariant. It is readily verified that the elements $X \in \Omega_0$ are exactly the graphs of invertible symmetric operators $\psi: V \to V'$ (as usual, we identify V' and $V^*$ by means of the form $\langle , \rangle$); also, the element $g \in G$ takes the graph of $\psi$ into the graph of $g^{*-1}\psi g^{-1}$. That is to say, $\Omega_0$ can be identified (as a G-space) with the space of nondegenerate symmetric forms on V; in particular, $\Omega_0$ is indeed a G-orbit, and the stability subgroup of G at the point $x \in \Omega_0$ is $M = O(n)$.

b) The case $G = O(n)$, $L_0 = GL(n)$. Suppose $L_0$ is a transformation group in the n-dimensional vector space W and G is the subgroup of $L_0$ preserving a nondegenerate symmetric form $\langle , \rangle$ on W. The fundamental Grassmanian $\Gamma_0 = \Gamma(L_0)$ is the Grassmanian of all m-dimensional subspaces of W, where $m = n/2$ for n even and $m = (n+1)/2$ for n odd. We denote by $\Omega_0 \subset \Gamma_0$ the set of the m-dimensional subspaces on which the restriction of the form $\langle , \rangle$ is nondegenerate. It is clear that $\Omega_0$ is an open G-orbit in $\Gamma_0$ and that the stabilizer M of the point $x \in \Omega_0$ in G is $O(m) \times O(m)$ for $n = 2m$, and $O(m) \times O(m-1)$ for $n = 2m-1$.

c) The case $G = Sp(2n)$, $L_0 = Sp(2n) \times Sp(2n)$. We first realize G as a transformation group in a 2n-dimensional symplectic space $(W, \langle , \rangle)$, and regard $L_0$ simply as $G \times G$. The embedding $G \subset L_0$ is the diagonal embedding (we do not need here the realization of $L_0$ as a linear group). The Grassmanian $\Gamma_0 = \Gamma(L_0) = \Gamma(G) \times \Gamma(G)$ consists of pairs of Lagrangian subspaces of W (and $L_0$ acts on $\Gamma_0$ by the rule $(g_1, g_2)(V_1, V_2) = (g_1V_1, g_2V_2)$). Set $\Omega_0 = \{(V, V') \in \Gamma_0 : V \cap V' = 0\}$. The clearly $\Omega_0$ is an open G-orbit in $\Gamma_0$ and the stabilizer M of the point $x \in \Omega_0$ in G is the subgroup GL(n).

The fact that in all the cases the representation of $M = \mathrm{Stab}_G x$ in the fiber of the bundle $\pi$ over $x \in \Omega_0$ is the representation $\tau$ described in Theorem 1 is an immediate consequence of the definitions. Thus, Theorem 3 follows from Theorem 1.

Proof of Theorem 4. a) $G = GL(n)$, $L = SO(2n+2)$. Let W be a $(2n+2)$-dimensional vector space with bilinear form $\langle , \rangle$ in which we realize the group L. Pick two complementary isotropic subspaces in W, $\overline{V}$ and $\overline{V}'$, and fix a decomposition $\overline{V} = V \oplus \mathbf{C}v_0$ into the sum of an n-dimensional space and a one-dimensional space. Also, set $V' = \{v \in \overline{V}' : \langle v, v_0 \rangle = 0\}$ and let $v_0' \in \overline{V}'$ be a vector orthogonal to V satisfying the constraint $\langle v_0, v_0' \rangle = 1$. Then we can write $W = V \oplus \mathbf{C}v_0 \oplus V' \oplus \mathbf{C}v_0'$. Next, we embed G into L as the subgroup preserving the subspaces V, V' and the vectors $v_0$, $v_0'$, and identify G with GL(V).

Now we realize the fundamental Grassmanian $\Gamma = \Gamma(L)$ as the component of $\overline{V}$ in the space of $(n+1)$-dimensional isotropic subspaces of W (see Lemma 8, a)). Denote by $\Omega$ the set of spaces $X \in \Gamma$ which are transverse to all three subspaces $V$, $Cv_0$, and $\overline{V}'$. Then $\Omega \subset \Gamma$ is :arly open and G-invariant. It remains to verify that the G-space $\Omega$ is isomorphic to the pace $G/M_0$ described in Sec. 1, i.e., to space of nondegenerate pairs $(\xi, \varphi) \in V^* \oplus \wedge^2 V^*$ e Lemma 1, Sec. 1). In fact, by Lemma 8, b) every subspace $X \in \Gamma$ transverse to $\overline{V}'$ is the ph of a skew-symmetric operator $\overline{\varphi}: \overline{V} \to \overline{V}'$; the latter admits the matrix representation

$$\begin{array}{cc} & \begin{array}{cc} V & Cv_0 \end{array} \\ \begin{array}{c} V' \\ Cv_0' \end{array} & \begin{pmatrix} \varphi & \xi \\ -\xi & 0 \end{pmatrix} \end{array},$$

h $\xi \in V^*$ and $\varphi \in \wedge^2 V^*$. Then clearly $X$ is transverse to $V$ and $Cv_0$ if and only if the r $(\xi, \varphi)$ is nondegenerate. Thus, we have exhibited an isomorphism of $\Omega$ onto the space of degenerate pairs $(\xi, \varphi)$, and it is readily verified that this isomorphism commutes with action of G.

b) $G = O(2n)$, $L = SO(2n + 2) \times O(2n)$. Denote by $L_0$ the subgroup $O(2n + 1) \times O(2n)$ of the embedding of $O(2n + 1)$ in $SO(2n + 2)$ is described in Lemma 8, c)). By Lemma 8, c), fundamental Grassmanians $\Gamma(L)$ and $\Gamma(L_0)$ can be naturally identified as $L_0$-spaces; we pre- to work with $L_0$ and $\Gamma(L_0)$ rather than $L$ and $\Gamma(L)$ (although this is not obligatory). Let group $\overline{G} = O(2n + 1)$ be realized in the $(2n + 1)$-dimensional vector space $W$ with bilinear m $\langle , \rangle$. Fix a codimension-one nondegenerate subspace $W$ of $\overline{W}$ and a unit length vector $\in W$ orthogonal to $\overline{W}$, and then realize $G$ as the subgroup of $\overline{G}$ preserving $\overline{W}$ and $z_0$. The up $L_0$ is simply regarded as the product $\overline{G} \times G$ (we do not need its realization as a linear up), and $G$ is embedded in $L_0$ diagonally.

The space $\Gamma = \Gamma(L_0)$ consists of pairs $(\overline{X}, \overline{Y})$ with $\overline{X}$ an n-dimensional isotropic subspace $\overline{W}$, and $\overline{Y}$ an n-dimensional isotropic subspace of $W$ (and $L_0$ acts on the pairs $(\overline{X}, \overline{Y})$ coordi- ewise). Set $\Omega = \{(\overline{X}, \overline{Y}) \in \Gamma: \overline{X}$ is transverse to $\overline{Y}$ and $W\}$. It is clear that $\Omega$ is open, se, and G-invariant in $\Gamma$. It remains to show that the G-space $\Omega$ is isomorphic to the G- ce $G/M_0$ considered in Sec. 1, i.e., to the space of triples $(X, Y, \eta)$, where $X$ and $Y$ transverse n-dimensional isotropic subspaces of $W$ and $\eta$ is a nonzero linear functional on As in Sec. 1, we shall identify the functionals on $X$ with vectors from $Y$ using the form >. We associate with each triple $(X, Y, \eta)$ a pair $(\overline{X}, \overline{Y})$ by setting $\overline{Y} = Y$, $\overline{X} = \{x - \frac{\langle \eta, x\rangle}{2} \cdot$ $- \langle \eta, x\rangle \cdot z_0 : x \in X\}$. Using construction b) of Lemma 9, 1, it is readily verified that this pping is an isomorphism of G-spaces from $G/M$ to $\Omega$.

c) $G = O(2n + 1)$, $L = O(2n + 2) \times SO(2n + 2)$. As in case b), we replace $L$ by the sub- up $L_0 = O(2n + 2) \times O(2n + 1)$. By Lemma 8, c), $\Gamma(L)$ and $\Gamma(L_0)$ can be identified as $L_0$- ces. Suppose the group $\overline{G} = O(2n + 2)$ is realized in the $(2n + 2)$-dimensional space $\overline{W}$ with inear form $\langle , \rangle$. Fix a codimension-one nondegenerate subspace $W$ in $\overline{W}$ and a unit-length tor $z_0 \in \overline{W}$ orthogonal to $W$, and regard $G$ as the subgroup of $\overline{G}$ preserving $W$ and $z_0$. Embed n $L_0 = \overline{G} \times G$ diagonally.

The space $\Gamma = \Gamma(L_0)$ consists of all pairs $(\overline{X}, \overline{Y})$ with $\overline{X}$ an $(n + 1)$-dimensional isotropic space of $\overline{W}$, and $\overline{Y}$ an n-dimensional isotropic subspace of $W$ ($L_0$ acts on $\Gamma$ coordinate-wise). $\Omega = \{(\overline{X}, \overline{Y}) \in \Gamma: \overline{X} \cap \overline{Y} = 0\}$. The clearly $\Omega$ is an open, dense, and G-invariant subset of $\Gamma$. remains to show that $\Omega$ is isomorphic, as a G-space, to the space $G/M_0$ introduced in Sec. 1, ., the space of triples $(X, Y, z)$, where $X$ and $Y$ are n-dimensional isotropic transverse sub- ces of $W$ and $z$ is a unit-length vector of $W$ which is orthogonal to $X$ and $Y$. We associate h each triple $(X, Y, z)$ a pair $(\overline{X}, \overline{Y})$ setting $\overline{Y} = Y$, $\overline{X} = X \oplus C(z + iz_0)$. From Lemma 8, c) Sec. 1 it follows that this mapping is an isomorphism of G-spaces from $G/M_0$ to $\Omega$.

d) $G = Sp(2n)$, $L = GL(2n + 1)$. We realize L as a transformation group in the (2n + 2)-dimensional space $\overline{W}$. Pick a codimension-one subspace W of $\overline{W}$, a symplectic form <, > on W, and a vector $z_0$ in $\overline{W} \setminus W$. Embed G in L as the subgroup preserving W, <, >, and $z_0$. The space $\Gamma = \Gamma(L)$ is the Grassmanian of all (n + 1)-dimensional subspaces $\overline{Z}$ of $\overline{W}$.

d1) n even. Set $\Omega = \{\overline{Z} \in \Gamma$: $\overline{Z}$ is transverse to both W and $\mathbf{C}z_0$, and the restriction of <,> to $\overline{Z} \cap W$ is nondegenerate$\}$. It is clear that $\Omega$ is an open (dense) G-invariant subset of $\Gamma$. It remains to show that $\Omega$, as a G-space, is isomorphic to the space $G/M_0$ discussed in Sec 1, i.e., to the space of triples $(Z, Z', \eta)$, where Z and Z' are mutually orthogonal n-dimensional subspaces in W, and $\eta$ is a nonzero linear functional on Z. We associate with each trip $(Z, Z', \eta)$ the space $\overline{Z} = Z' \oplus \mathbf{C}(z + z_0)$, where z is the vector of Z associated with the functional $\eta$ via the form <, >. Then it is readily verified that this mapping is an isomorphism of the G-spaces $G/M_0$ and $\Omega$.

d2) n odd. Denote by $p_W$ the projection of $\overline{W}$ on W along $\mathbf{C}z_0$, and set $\Omega = \{\overline{Z} \in \Gamma$: $\overline{Z}$ is transverse to both W and $\mathbf{C}z_0$, and the restriction of the form <, > to $p_W(\overline{Z})$ is nondegenerate$\}$. Again, it is clear that $\Omega$ is an open (dense) G-invariant subset of $\Gamma$, and it remains to verif that it is isomorphic, as a G-space, to the space $G/M_0$ discussed in Sec. 1, i.e., to the spac of triples $(Z, Z', \eta)$, where Z and Z' are mutually orthogonal nondegenerate subspaces of W with $\dim Z = n + 1$, $\dim Z' = n - 1$, and $\eta$ is a nonzero linear function on Z. To this end we associate with each triple $(Z, Z', \eta)$ the subspace $\overline{Z} = \{z + \eta(z)z_0 : z \in Z\}$. It is readily verified that this mapping is an isomorphism of the G-spaces $G/M_0$ and $\Omega$.

The proof of Theorem 4 is now complete.

Proof of Theorem 5. The case where X is of maximal rank, i.e., points 1 and 2 of Theore 5, as well as the particular cases of point 5, where $k = n/2$ for n even and $k = (n + 1)/2$ for n odd, were already treated while proving Theorem 3. The remaining case can be settled similarly.

As we have already observed in Sec. 0, there is a natural action of the Lie algebra $\mathfrak{l}$ of the group L (respectively, of the Lie superalgebra $\mathfrak{l}$ containing $\mathfrak{l}_0 \oplus \mathbf{C}$ as its even part) on the model of representations of the group G from Theorem 4 (respectively, Theorem 3). We conclude with the explicit formulas of the action of the Lie algebra $\mathfrak{l} = so(2n + 2)$ on the model of representations of $G = GL(n)$.

Recall that the model of representations of G acts in the space $\mathbf{C}[\Omega]$ of regular function on the set $\Omega$ of nondegenerate pairs $(\xi, \varphi) \in V^* \oplus \wedge^2 V^*$. The embedding of $\Omega$ in the Grassmanian $\Gamma = \Gamma(SO(2n + 2))$ is described in step a) of the proof of Theorem 4. Using the above notations one can write every element of the Lie algebra so(2n + 2) in the matrix form:

$$\begin{array}{c} \\ V \\ \mathbf{C}v_0 \\ V' \\ \mathbf{C}v_0' \end{array}\begin{array}{c} \begin{array}{cccc} V & \mathbf{C}v_0 & V' & \mathbf{C}v_0' \end{array} \\ \begin{pmatrix} A & a & B & b \\ \alpha & c & -b & 0 \\ C & -\gamma & -A^* & -\alpha \\ \gamma & 0 & -a & -c \end{pmatrix} \end{array},$$

where $A \in \mathfrak{gl}(V)$, $B \in \mathrm{Hom}(V^*, V)$, $B^* = -B$, $C \in \mathrm{Hom}(V, V^*)$, $C^* = -C$, $a, b \in V$, $\alpha, \gamma \in V^*$, $c \in \mathbf{C}$ (we use the natural identifications $V' = V^*$, $\mathrm{Hom}(\mathbf{C}, V) = \mathrm{Hom}(V', \mathbf{C}) = V$, $\mathrm{Hom}(V, \mathbf{C}) = \mathrm{Hom}(\mathbf{C}, V') = V^*$). We write the vector fields on $\Omega$ in the form $\psi(\xi, \varphi)\,\partial_\varphi + \eta(\xi, \varphi)\,\partial_\xi$, where $\psi(\xi, \varphi) \in \wedge^2 V^*$, and $\eta(\xi, \varphi) \in V^*$. Then the representation of the various generators of the Lie the algebra so(2n + 2) as vector fields on $\Omega$ is given by

$$A \mapsto (A^*\varphi + \varphi A)\,\partial_\varphi + (A^*\xi)\,\partial_\xi,$$
$$B \mapsto (\varphi B \varphi)\,\partial_\varphi + (\varphi B \xi)\,\partial_\xi, \quad C \mapsto -C\partial_\varphi,$$
$$a \mapsto \varphi(a)\,\partial_\xi,$$

$$b \mapsto (\varphi(b) \wedge \xi)\partial_\varphi - \xi(b)\,\xi\partial_\xi,$$
$$c \mapsto c\xi\partial_\xi,$$
$$\alpha \mapsto (\alpha \wedge \xi)\,\partial_\varphi,$$
$$\gamma \mapsto \gamma\partial_\xi$$

e normalize the exterior product $\wedge$: $V^* \times V^* \to \wedge^2 V^*$, setting $(\alpha_1 \wedge \alpha_2)(v_1, v_2) = \alpha_1(v_1)\,\alpha_2(v_2)$ $\alpha_1(v_2)\,\alpha_2(v_1))$ . These formulas are direct consequences of the definitions.

## LITERATURE CITED


. L. C. Biederharn and D. Flath, "Beyond the enveloping algebra of $sl_3$," Preprint, Duke Univ. (1982).
. Twistors and Gauge Fields [in Russian], Collection of Papers, Mir, Moscow (1983).
. I. M. Gel'fand, I. N. Bernshtein, and S. I. Gel'fand, "Models of representations of compact Lie groups," Tr. Sem. Petrovskogo, No. 2, 3-21 (1976) [Selecta Math. Sov., 1, No. 2, 121-142 (1981)].
. M. Krämer, "Some remarks suggesting an interesting theory of harmonic functions on SU(2n + 1)/Sp(n) and SO(2n + 1)/U(n)," Arch. Math., 33, No. 1, 76-79 (1979).
. V. G. Kac, "Lie superalgebras," Adv. Math., 26, No. 1, 8-96 (1977).
. D. A. Leites, Theory of Supermanifolds [in Russian], Petrozavodsk (1983).
. V. V. Serganova, "Classification of simple real Lie superalgebras and symmetric superspaces," Funkts. Anal. Prilozhen., 17, No. 3, 46-54 (1983).
. S. Helgason, Differential Geometry and Symmetry Spaces, Academic Press, New York—London (1962).
. G. Lions and M. Vergne, The Weil Representation, Maslov Index, and Theta-Series, Progress in Math., Vol. 6, Birkhäuser, Boston (1980).
. É. B. Vinberg and B. N. Kimel'fel'd, "Homogeneous domains on flag manifolds and spherical subgroups of semisimple Lie groups," Funkts. Anal. Prilozhen., 12, No. 3, 12-19 (1978).
. A. O. Barut and R. Raczka, Theory of Group Representations and Its Applications, Second revised edition, Polish Scientific Publishers, Warsaw (1980).
. M. Krämer, "Sphärische Untergruppen in kompakten zusammenhängenden Liegruppen," Comp. Math., 39, 129-153 (1979).
. I. N. Bernshtein, I. M. Gel'fand, and S. I. Gel'fand, "A new model of representations of finite semisimple algebraic groups," Usp. Mat. Nauk, 29, No. 3, 185-186 (1974).
. T. G. Khovanova, "Models of representations and generalized Clifford algebras," Funkts. Anal. Prilozhen., 16, No. 4, 90-91 (1982).
. D. Vogan, Representations of Real Reductive Lie Groups, Birkhäuser, Boston (1981).
. V. V. Shikheeva, "Construction of models of representations of the general linear group over a finite field of odd characteristic," Usp. Mat. Nauk, 34, No. 5, 233-234 (1979).
. A. A. Klyachko, "Models for complex representations of GL(n, q)," Mat. Sb., 120, 371-386 (1983).
. I. G. Macdonald, Symmetric Functions and Hall Polynomials, Clarendon Press, Oxford (1979).

# Part V

# Verma modules; resolutions of finite-dimensional representations

# 1.

## (with I. N. Bernstein and S. I. Gelfand)

# Differential operators on the base affine space and a study of $\mathfrak{g}$-modules

Prepr. 77, IPM Akad. Nauk SSSR (1972) (English transl. in: Lie groups and their representations. Proc. Summer School in Group Representations. Bolyai Janos Math. Soc., Budapest 1971, pp. 21–64, New York: Halsted 1975). Zbl. **338**:58019

The present work consists of two parts. In the first part we study the ring of regular differential operators on the base affine space of a complex semisimple group. By the *base affine space* of a group we mean the quotient space $A = N_+ \setminus G$ of the group $G$ by a maximal unipotent subgroup $N_+$. Experience in representation theory suggests that for many problems in representation theory the solution results from a careful study of the base affine space. In particular, the structure of the ring of regular differential operators on $A$ seems to be closely connected with the representations of the real forms of the group $G$. In addition to the connections with representation theory, the study of this ring yields an instructive and rather advanced example for the study of the rings of regular differential operators on algebraic varieties, an area in which not much is known so far.

We approach the study of the differential operators on $A$ by establishing a connection between the regular functions on the group $G$ and the regular differential operators on the base affine space $A$. We would also like to draw the reader's attention to Conjecture II, where the notion of the generalized Segal—Bargmann space for a representation of a compact Lie group is introduced.

The second part of the work is formally independent of the first and is devoted to the algebraic study of modules over the Lie algebra $\mathfrak{g}$ of the group $G$. We restrict ourselves to a category of $\mathfrak{g}$-modules, which is closely connected with the theory of highest weight. We shall call this category of $\mathfrak{g}$-modules the category $O$. The category $O$ contains in a natural way every finite-dimensional representation of the Lieal gebra $\mathfrak{g}$. The fundamental result of this part lies in constructing a resolution for finite-dimensional $\mathfrak{g}$-modules. The simplest objects of the category $O$ are the modules $M_\chi$ and it seems important that the resolution consists of modules which are direct sums of these simplest modules. The description of the modules occurring in the composition series of the modules $M_\chi$, which is given in the Appendix, is also useful. Unfortunately, the complete structure of these composition series is not known to us yet.

We think that the methods developed in the second part of this paper may turn out to be useful in the further study of questions considered in the first part.

The fundamental content of this work is concentrated in Theorem 6.3 and Conjectures I and II in the first part and Theorems 8.12 and 10.1 in the second.

## § 1. Notations and preliminaries

$\mathfrak{g}$ is a semisimple Lie algebra of rank $r$ over $\mathbf{C}$, $\mathfrak{h}$ is a Cartan subalgebra of $\mathfrak{g}$. $\Delta$ denotes the root system of $\mathfrak{g}$ corresponding to $\mathfrak{h}$, with a fixed ordering, $\Delta_+$ and $\Delta_-$ the system of the positive and negative roots, respectively, $\Sigma$ the set of simple roots, and $\varrho=\frac{1}{2}\sum_{\gamma\in\Delta_+}\gamma$. $E_\gamma\in\mathfrak{g}$ is the root vector corresponding to the root $\gamma\in\Delta$. Here we have $\gamma([E_\gamma, E_{-\gamma}])=2$.

$\mathfrak{n}_+$ is the subalgebra of $\mathfrak{g}$ spanned by the vectors $E_\gamma$, $\gamma\in\Delta_+$, while $\mathfrak{n}_-$ is the subalgebra of $\mathfrak{g}$ spanned by $E_\gamma$, $\gamma\in\Delta_-$. $\mathfrak{b}=\mathfrak{h}\oplus\mathfrak{n}_+$. $U(\mathfrak{g})$, $U(\mathfrak{n}_+)$, $U(\mathfrak{n}_-)$ are the universal enveloping algebras of $\mathfrak{g}$, $\mathfrak{n}_+$, $\mathfrak{n}_-$, respectively; $Z(\mathfrak{g})$ is the centre of $U(\mathfrak{g})$.

$\mathfrak{h}^*$ is the dual space of $\mathfrak{h}$.

$G$ is a complex semisimple Lie group with Lie algebra $\mathfrak{g}$; $H$, $N_+$, $N_-$ and $B$ are the subgroups of $G$ corresponding to the subalgebras $\mathfrak{h}$, $\mathfrak{n}_+$, $\mathfrak{n}_-$ and $\mathfrak{b}$, respectively.

$A=N_+\backslash G$ is the base affine space of the group $G$.

Additional notations, used in Part 2.

$\mathbf{Z}_+$ is the set of non-negative integers.

$\mathfrak{h}^*_{\mathbf{R}}$ denotes the real linear subspace of $\mathfrak{h}^*$ spanned by all roots $\gamma\in\Delta$.

$\langle,\rangle$ is the scalar product in $\mathfrak{h}^*$ constructed with the help of the Killing form of the algebra $\mathfrak{g}$; $\|\cdot\|$ is the corresponding norm in $\mathfrak{h}^*_{\mathbf{R}}$.

$\mathfrak{h}^*_{\mathbf{Z}}$ is the lattice in $\mathfrak{h}^*_{\mathbf{R}}$ consisting of those $\chi\in\mathfrak{h}^*$ for which $2\langle\chi,\gamma\rangle/\langle\gamma,\gamma\rangle\in\mathbf{Z}$ for all $\gamma\in\Delta$.

$$K=\{\chi\in\mathfrak{h}^*\,|\,\chi=\sum_{\alpha\in\Sigma}n_\alpha\cdot\alpha,\ n_\alpha\in\mathbf{Z}_+\};\quad K\subset\mathfrak{h}^*_{\mathbf{Z}}.$$

$\chi_1\geqq\chi_2$ means that $\chi_1-\chi_2\in K$ $(\chi_1,\chi_2\in\mathfrak{h}^*)$.

$W$ is the Weyl group of the algebra $\mathfrak{g}$, $\sigma_\gamma\in W$ is the reflexion corresponding to the root $\gamma\in\Delta$, i.e. $\sigma_\gamma\chi=\chi-2\langle\chi,\gamma\rangle\langle\gamma,\gamma\rangle^{-1}\gamma$. We note that $\sigma_\alpha\varrho=\varrho-\alpha$ for $\alpha\in\Sigma$.

$\chi_1\sim\chi_2$ for $\chi_1,\chi_2\in\mathfrak{h}^*$ means that there exists an element $w\in W$ such that $\chi_1=w\chi_2$.

$l(w)$ is the length of the element $w\in W$, i.e. the smallest possible number of factors in a decomposition $w=\sigma_{\alpha_1}\cdot\ldots\cdot\sigma_{\alpha_k}$, $\alpha_i\in\Sigma$.

$$W^{(i)}=\{w\in W\,|\,l(w)=i\}.$$

$\Xi_\gamma=\{\chi\in\mathfrak{h}^*_{\mathbf{R}}\,|\,\langle\chi,\gamma\rangle=0\}$; the connected components of $\mathfrak{h}^*_{\mathbf{R}}\backslash(\bigcup_{\gamma\in\Delta}\Xi_\gamma)$ are called the *Weyl chambers;* $\bar{C}$ is the closure of the Weyl chamber $C$; $C^+$ is the Weyl chamber containing $\varrho$. The group $W$ acts on the set of Weyl chambers simply transitively. Two Weyl chambers $C_1$ and $C_2$ are called *neighbouring* if $\dim(\bar{C}_1\cap\bar{C}_2)=\dim\mathfrak{h}^*_{\mathbf{R}}-1$. In this case there exists a unique element $\gamma\in\Delta_+$ such that $\sigma_\gamma C_1=C_2$ and the hyperplane $\Xi_\gamma$ separates $C_1$ and $C_2$;

$$D=\mathfrak{h}^*_{\mathbf{Z}}\cap C^+.$$

An element $\chi \in \mathfrak{h}_{\mathbf{R}}^*$ is called *regular* if $\langle \chi, \gamma \rangle \neq 0$ for all $\gamma \in \Delta$.
Let $M$ be a $\mathfrak{h}$-module, $\chi \in \mathfrak{h}^*$. Put

$$M^{(\chi)} = \{f \in M \mid xf = \chi(x) \cdot f \text{ for all } x \in \mathfrak{h}\};$$
$$P(M) = \{\chi \in \mathfrak{h}^* \mid M^{(\chi)} \neq 0\}.$$

Let $M$ be a $\mathfrak{g}$-module and $0 = M_0 \subset M_1 \subset \ldots \subset M_k = M$ its Jordan—Hölder composition series, $L_i = M_i/M_{i-1}$ are simple $\mathfrak{g}$-modules. The collection of the modules $L_i$, with multiplicity, is called the Jordan—Hölder decomposition of $M$ and is denoted by $JH(M)$.

PART I

# DIFFERENTIAL OPERATORS

## § 2. Introduction

Let $G$ be a connected complex semisimple Lie group of rank $r$, $B$ a Borel subgroup, $N_+$ the unipotent radical of $B$, $H \subset B$ the Cartan subgroup of $B$. The quotient space $A = N_+ \diagdown G$ plays a fundamental rôle in representation theory; it is called the *base affine space* of the group $G$.

The aim of this part is to study the ring $\mathscr{D}(A)$ of regular differential operators on $A$.

First of all we study the possible connections between the space of regular differential operators on $A$ and the space $\mathscr{E}(G)$ of regular functions on $G$. More precisely, Conjecture I claims the possibility of embedding $\mathscr{E}(G)$ into $\mathscr{D}(A)$ (operation $f \to \hat{f}$); here $\mathscr{D}(A) = U(\mathfrak{h}) \underset{\mathbf{C}}{\otimes} \mathscr{E}(G)$, where $U(\mathfrak{h})$ is the universal enveloping algebra of the Lie algebra $\mathfrak{h}$ of the group $H$. In this part we prove a result (Theorem 6.6) weaker than Conjecture I, namely we construct an isomorphism between the $L$-modules $L \underset{\mathbf{C}}{\otimes} \mathscr{E}(G)$ and $L \underset{U(\mathfrak{h})}{\otimes} \mathscr{D}(A)$, where $L$ is the quotient field of the ring $U(\mathfrak{h})$.

Further on we construct a scalar product in the ring $\mathscr{E}(A)$ of regular functions on $A$ which is invariant under the action of the maximal compact subgroup $K \subset G$. The completion of the space $\mathscr{E}(A)$ by this scalar product consists of analytic functions on the complex manifold $A$. This space is a generalization of the Segal—Bargmann space. Conjecture II states that an operator adjoint to a differential operator is again a differential operator, hence the ring $\mathscr{D}(A)$ is selfadjoint with respect to the introduced scalar product. It has to be noted that the introduced involution in $\mathscr{D}(A)$ does not preserve the order of a differential operator. For instance, in the case of the group of matrices of order $h$, the adjoint to the simplest operator of order zero will be an operator of order $h-1$.

## § 3. Regular differential operators

In this section the rings of regular differential operators on $G$ and $A$ are introduced. We shall consider $G$ and $A$ as algebraic varieties over $\mathbf{C}$. The projection $\pi : G \to A$ is a morphism of algebraic varieties. The rings of regular functions on $G$ and $A$ will be denoted by $\mathscr{E}(G)$ and $\mathscr{E}(A)$, respectively. Let $\pi^* : \mathscr{E}(G) \to \mathscr{E}(A)$

be the embedding induced by the mapping $\pi$. The image $\pi^* \mathscr{E}(A)$ consists of exactly those functions which are constant on the left cosets of the subgroup $N$.

The variety $A$ is non-singular and quasi-affine. More precisely, let

$$\hat{A} = \text{Spec max } \mathscr{E}(A)$$

be the affine algebraic variety corresponding to $\mathscr{E}(A)$. Then there is a natural isomorphism between $A$ and a dense open subset of $\hat{A}$.

Let us define in the space $\mathscr{E}(G)$ the left and right representations $L^G$ and $R^G$ of the group $G$ by the usual formulas

$$(L^G_{g_0} f)(g) = f(g_0^{-1} g), \quad (R^G_{g_0} f)(g) = f(g g_0), \quad g, g_0 \in G.$$

It is a common property of both of these representations that every element $f \in \mathscr{E}(G)$ is contained in some finite-dimensional invariant subspace.

The group $G$ acts naturally on the space $A$ (right translations). In addition one can also define the left-hand action of the Cartan subgroup on $A$ by associating to each element $h \in H$ the transformation $x \to hx$. The element $hx \in A$ is well defined because $H$ normalizes $N$.

Let us define in the space $\mathscr{E}(A)$ the representations $L^A$ of the group $H$ and $R^A$ of the group $G$ by means of the formulas

$$(L^A_h f)(x) = f(h^{-1} x), \quad (R^A_g f)(x) = f(xg); \; x \in A, \; h \in H, \; g \in G.$$

Obviously, $L^G_h \pi^* = \pi^* L^A_h$ and $R^G_h \pi^* = \pi^* R^A_h$.

By differentiating $R^A$ we obtain for each $X \in \mathfrak{g}$ an operator $R^A_X: \mathscr{E}(A) \to \mathscr{E}(A)$. These operators determine a representation of $\mathfrak{g}$ that extends to a representation of $U(\mathfrak{g})$; here the operator corresponding to an element $X \in U(\mathfrak{g})$ will also be denoted by $R^A_X$. We define the similar representations $R^G_X$ and $L^G_X$ of the algebra $U(\mathfrak{g})$ in $\mathscr{E}(G)$ and the representation $L^A_Y$ of the algebra $U(\mathfrak{g})$ in $\mathscr{E}(A)$.

*Definition* 3.1. Let $X$ be a quasi-affine variety and $\mathscr{E}(X)$ be the ring of regular functions on $X$. A linear mapping $D: \mathscr{E}(X) \to \mathscr{E}(X)$ is called a *regular differential operator of order* $\leqq k$ $(k \geqq 0)$ on $X$ if it satisfies the condition

$$[f_{k+1}[f_k \ldots [f_1, D] \ldots]] = 0 \tag{3.1}$$

for any $f_1, f_2, \ldots, f_{k+1} \in \mathscr{E}(X)$. In (3.1) $f_i$ denotes the operator of multiplication by $f_i$.

The differential operators on $X$ form a ring which will be denoted by $\mathscr{D}(X)$.

*Remark.* The definition given here coincides with that of a differential operator on an arbitrary algebraic variety given in [1].

It is easy to see that a differential operator of order zero is an operator of multiplication by a function $f \in \mathscr{E}(X)$.

Any vector field over $X$ determines a differential operator of order $\leqq 1$ on $X$. It can be verified that on composing an operator of order $\leqq k$ with an operator of order $\leqq l$ we obtain a differential operator of order $\leqq k+l$, moreover the commutator of two such operators is of order $\leqq k+l-1$.

We shall use the following properties of differential operators (see [1]).

*Lemma* 3.2. 1. *If $Y \subset X$ is a dense open subset (in the Zariski topology) of a quasi-affine variety $X$, than every regular differential operator can be restricted to $Y$. More precisely, there exists a unique differential operator $D': \mathscr{E}(Y) \to \mathscr{E}(Y)$ whose restriction to $\mathscr{E}(X) \hookrightarrow \mathscr{E}(Y)$ coincides with $D$.*

2. *Let $X$ be a non-singular variety and $Z_1, \ldots, Z_n$ be a system of vector fields over $X$ which defines a basis in the tangent space at each point $x \in X$. Then every differential operator $D$ on $X$ has a unique representation in the form*

$$D = \sum a_{i_1 i_2 \ldots i_n}(x) Z_1^{i_1} Z_2^{i_2} \ldots Z_n^{i_n},$$

*where $i_j$ are non-negative integers, $a_{i_1 \ldots i_n}$ are regular functions only a finite number of which are different from zero.*

For any element $X \in U(\mathfrak{g})$ the operators $R_X^G$ and $L_X^G$ are differential operators on $G$, moreover $L_Y^A$, $Y \in U(\mathfrak{h})$ and $R_X^A$ are differential operators on $A$. The description of the ring of differential operators on $G$ yields the following proposition.

*Proposition* 3.3. The mapping

$$\vartheta: \mathscr{E}(G) \underset{\mathbf{C}}{\otimes} U(\mathfrak{g}) \to \mathscr{D}(G)$$

given by the formula

$$\vartheta\left(\sum f_i \otimes X_i\right) = \sum f_i L_{X_i}^G$$

is an isomorphism of left $\mathscr{E}(G)$-modules.

The proof of Proposition 3.3. follows easily from Lemma 3.2.

In what follows the element $\vartheta^{-1}(D) \in \mathscr{E}(G) \otimes U(\mathfrak{g})$ will be called the *standard form* of $D \in \mathscr{D}(G)$.

Now let us turn to the study of the ring $\mathscr{D}(A)$. In this ring we can define a representation of the group $G$. Indeed, we put

$$D^g = R_g^A D R_{g^{-1}}^A, \quad g \in G, \quad D \in \mathscr{D}(A).$$

Similarly, we put

$${}^h D = L_h^A D L_{h^{-1}}^A, \quad h \in H, \quad D \in \mathscr{D}(A).$$

We notice that the variety $A$ is smooth, but it is not an affine variety and there may exist differential operators on $A$ which cannot be expressed by operators of the first order. (See e.g. Example 2.)

We shall say that a differential operator $D'$ on $G$ is a *lifting* of the operator $D$ on $A$, if

$$D' \pi^* f = \pi^* D f, \quad f \in \mathscr{E}(A).$$

*Theorem* 3.4. [2]. *Every differential operator on $A$ can be lifted to $G$.*

In order to prove this theorem we need the following lemma.

*Lemma* 3.5. *Let $0=\pi(e)\in A$. There exists a mapping $\eta:\mathscr{D}(A)\to U(\mathfrak{g})/\mathfrak{n}_+ U(\mathfrak{g})$ such that for all $D\in\mathscr{D}(A)$ and all $f\in\mathscr{E}(A)$*

$$(R_X^A f)(0) = (Df)(0) \tag{3.2}$$

*for every element $X\in U(\mathfrak{g})$ belonging to the coset $\eta(D)$.*

*Proof.* It is easy to see that if $X\in\mathfrak{n}_+$ then $R_X^A f(0)=0$ for all $f\in\mathscr{E}(A)$. Therefore the validity of the equality (3.2) does not depend on the choice of the element $X$ in a $\mathfrak{n}_+ U(\mathfrak{g})$-coset.

Now let $X_1, \ldots, X_N$ be a basis in $\mathfrak{h}\oplus\mathfrak{n}_-$. Then the vector fields $R_{X_i}^A$ form a basis in the tangent space at each point of a certain affine neighbourhood $V$ of the point $0\in A$. According to Lemma 3.2, in this neighbourhood $V$ the operator $D$ can be expressed in the form

$$D = \sum a_{i_1 i_2 \ldots i_N} (R_{X_1}^A)^{i_1} \cdot \ldots \cdot (R_{X_N}^A)^{i_N}.$$

Put

$$X = \sum a_{i_1 i_2 \ldots i_N}(0) X_1^{i_1} \ldots X_N^{i_N} \in U(\mathfrak{g}).$$

It is easy to see that the image $\eta(D)$ of the element $X$ in $U(\mathfrak{g})/\mathfrak{n}_+ U(\mathfrak{g})$ satisfies the condition of the lemma.

Note that the element $\eta(D)$ is uniquely determined by the equality (3.2) (see [2]).

Let $\tau: U(\mathfrak{g})\to U(\mathfrak{g})$ be an anti-automorphism such that $\tau(X)=-X$ for $X\in\mathfrak{g}$. Obviously, $\tau(\mathfrak{n}_+ U(\mathfrak{g}))=U(\mathfrak{g})\mathfrak{n}_+$, so $\tau$ determines an isomorphism

$$\bar{\tau}:\ U(\mathfrak{g})/\mathfrak{n}_+ U(\mathfrak{g}) \to U(\mathfrak{g})/U(\mathfrak{g})\mathfrak{n}_+.$$

*Definition* 3.6. Let $D\in\mathscr{D}(A)$. Define a function $\sigma_D(g)$ on $G$ with values in $U(\mathfrak{g})/U(\mathfrak{g})\mathfrak{n}_+$ by the formula

$$\sigma_D(g) = \bar{\tau}\eta(D^g).$$

One can verify [2, § 8] that $\sigma_D(g)$ is a regular function on $G$, i.e.

$$\sigma_D(g)\in\mathscr{E}(G)\otimes U(\mathfrak{g})/U(\mathfrak{g})\mathfrak{n}_+.$$

Now we are ready to complete the proof of the theorem. Let $D\in\mathscr{D}(A)$ and consider an arbitrary element $\sigma'_D\in\mathscr{E}(G)\otimes U(\mathfrak{g})$ which is sent into $\sigma_D(g)$ by the natural projection

$$\mathscr{E}(G)\otimes U(\mathfrak{g})\to\mathscr{E}(G)\otimes U(\mathfrak{g})/U(\mathfrak{g})\mathfrak{n}_+.$$

Put

$$D' = \vartheta(\sigma'_D(g))\in\mathscr{D}(G).$$

We shall show that $D'$ is a lifting of $D$ to $G$, i.e. that

$$(D'\pi^* f)(g) = (\pi^* Df)(g), \quad f\in\mathscr{E}(A),\ g\in G \tag{3.3}$$

(3.3) can be rewritten in the form

$$(R_g^G D' R_{g^{-1}}^G \pi^* R_g^A f)(e) = (\pi^* D^g R_g^A f)(e).$$

It is easy to see that $R_g^G D' R_{g^{-1}}^G$ is a regular differential operator on $G$ and that $\vartheta^{-1}(R_g^G D' R_{g^{-1}}^G)$ is projected into $\sigma_{D^g}$ under the natural mapping

$$\mathscr{E}(G)\otimes U(\mathfrak{g}) \to \mathscr{E}(G)\otimes U(\mathfrak{g})/U(\mathfrak{g})\mathfrak{n}_+.$$

Therefore (replacing $D$ by $D^g$ and $f$ by $R_g^A f$) it suffices to prove that $(D'\pi^* f)(e) = (\pi^* Df)(e)$.

Let $X = \sigma'_D(e) \in U(\mathfrak{g})$. Clearly, then $(L_X^G f)(e) = (D'f)(e)$ for $f \in \mathscr{E}(G)$. Moreover, the image of $\tau^{-1}(X)$ in $U(\mathfrak{g})/\mathfrak{n}_+ U(\mathfrak{g})$ is equal to $\eta(D)$, hence

$$(\pi^* Df)(e) = (Df)(0) = (R_{\tau^{-1}X}^A f)(0) = (R_{\tau^{-1}X}^G \pi^* f)(e), \quad f \in \mathscr{E}(A).$$

Therefore, the required equality is implied by the following lemma.

*Lemma* 3.7. *If* $Y \in U(\mathfrak{g})$, $f \in \mathscr{E}(G)$ *then*

$$(L_{\tau Y}^G f)(e) = (R_Y^G f)(e).$$

*Proof.* If $Y \in \mathfrak{g}$ then the lemma follows from the definitions of $L^G$ and $R^G$. Assume now that the lemma is valid for $Y_1, Y_2 \in U(\mathfrak{g})$. Then we have

$$(R_{Y_1 Y_2}^G f)(e) = (R_{Y_1}^G R_{Y_2}^G f)(e) = (L_{\tau(Y_1)}^G R_{Y_2}^G f)(e) = (R_{Y_2}^G L_{\tau(Y_1)}^G f)(e) =$$
$$= (L_{\tau(Y_2)}^G L_{\tau(Y_1)}^G f)(e) = (L_{\tau(Y_1 Y_2)}^G f)(e).$$

(Here we use the fact that $L_Y^G$ and $R_{Y'}^G$ commute for any $Y, Y' \in U(\mathfrak{g})$.) Hence we have the lemma for $Y = Y_1 \cdot Y_2$, and the proof is complete.

*Proposition* 3.8. Let us denote by $I_+$ the left ideal in the ring $\mathscr{D}(G)$ spanned by the operators $L_X^G$, $X \in \mathfrak{n}_+$.

*1)* Let $D \in \mathscr{D}(G)$. Then $D$ gives rise to a differential operator on $A$ (i.e. $D(\mathscr{E}(A)) \subset \mathscr{E}(A)$) if and only if

$$[L_X^G, D] \in I_+ \quad \text{for} \quad X \in \mathfrak{n}_+.$$

*2)* $D \in \mathscr{D}(G)$ gives rise to the zero operator on $A$ if and only if $D \in I_+$.

The proof of this proposition is rather simple and is left to the reader.

We shall now describe how these conditions can be expressed in terms of the standard form of the operator $D$. We remark that

*1)* $D \in I_+$ if and only if

$$\vartheta^{-1}(D) \in \mathscr{E}(G)\otimes U(\mathfrak{g})\mathfrak{n}_+.$$

*2)* If $\vartheta^{-1}(D) = \sum f_i \otimes X_i \in \mathscr{E}(G)\otimes U(\mathfrak{g})$ and $X \in \mathfrak{g}$ then

$$\vartheta^{-1}([L_X^G, D]) = \sum L_X^G f_i \otimes X_i + \sum f_i \otimes [X, X_i].$$

*Corollary* 3.9. Let $D \in \mathcal{D}(A)$. Then it has a unique lifting $D'$ such that

$$\vartheta^{-1}(D') \in \mathcal{E}(G) \otimes U(\mathfrak{h} \oplus \mathfrak{n}_-).$$

*Definition* 3.10. *1)* The element $\vartheta^{-1}(D') \in \mathcal{E}(G) \otimes U(\mathfrak{h} \oplus \mathfrak{n}_-)$ is called the *standard form* of the operator $D \in \mathcal{D}(A)$ and will be denoted by $s(D)$.

*2)* The *lowest term* $s_0(D)$ of an operator $D \in \mathcal{D}(A)$ is defined as the element of $\mathcal{E}(G) \otimes U(\mathfrak{h})$ which is equal to the projection of $s(D)$ under the decomposition

$$U(\mathfrak{h} \oplus \mathfrak{n}_-) = U(\mathfrak{h}) \oplus \mathfrak{n}_- U(\mathfrak{h} \oplus \mathfrak{n}_-)$$

*Definition* 3.11. Let us denote by $Wu$ the subring of $\mathcal{D}(A)$ consisting of the operators

$$L_X^A, \ X \in U(\mathfrak{h}).$$

## § 4. Conjecture I; examples

*Conjecture* I. *There exists a mapping $\mathcal{E}(G) \to \mathcal{D}(A)$, $f \to \tilde{f}$ with the following properties:*

*1) $\tilde{f} = f$ for $f \in \mathcal{E}(A) \subset \mathcal{E}(G)$.*

*2) The mapping $f \to \tilde{f}$ commutes with the representations $R_g$ and $L_h$.*

*3) The mapping $\mathcal{E}(G) \otimes Wu \to \mathcal{D}(A)$ given by the formula*

$$\sum f_i \otimes Z_i \to \sum \tilde{f}_i Z_i, \ f_i \in \mathcal{E}(G), \ Z_i \in Wu$$

*is an isomorphism of Wu-modules.*

*4) Let $s_0(\tilde{f}) = \sum f_i \otimes X_i$. Then $f = \sum f_i \cdot X_i(-\varrho)$* (here $\varrho \in \mathfrak{h}^*$ is half-sum of the positive roots; $X_i \in U(\mathfrak{h})$ is considered as a polynomial function on $\mathfrak{h}^*$).

A weakened version of Conjecture I is

*Conjecture* I'. *$\mathcal{D}(A)$ is a free Wu-module.*

We remark that the mapping $f \to \tilde{f}$ (if it exists) is not uniquely determined by the properties *1)—4)*. We assume however that there exists a "natural" mapping $f \to \tilde{f}$. The following examples will perhaps illuminate to the reader what we have in mind.

We shall now present several examples illustrating the notions and facts expounded above.

*Example* 1. $G = SL(2, \mathbf{C})$ is the group of $2 \times 2$ matrices of determinant 1. We choose as $N_+ \subset G$ the subgroup of all matrices of the form

$$\begin{pmatrix} 1 & x \\ 0 & 1 \end{pmatrix}.$$

In this case $A=N_+\diagdown G$ can be identified with the punctured complex plane i.e. $A=\mathbf{C}^2\diagdown\{(0,0)\}$, while the mapping $\pi:G\to A$ is of the form

$$g=\begin{pmatrix}u_1 & u_2\\ z_1 & z_2\end{pmatrix}\to(z_1,z_2).$$

Let

$$E_-=\begin{pmatrix}0&0\\1&0\end{pmatrix},\quad H=\begin{pmatrix}1&0\\0&-1\end{pmatrix},\quad E_+=\begin{pmatrix}0&1\\0&0\end{pmatrix}$$

be a basis of $\mathfrak{g}$. Then

$$L^G_{E_-}=-u_1\frac{\partial}{\partial z_1}-u_2\frac{\partial}{\partial z_2},$$

$$L^G_{E_+}=-z_1\frac{\partial}{\partial u_1}-z_2\frac{\partial}{\partial u_2},$$

$$L^G_H=z_1\frac{\partial}{\partial z_1}+z_2\frac{\partial}{\partial z_2}-u_1\frac{\partial}{\partial u_1}-u_2\frac{\partial}{\partial u_2}.$$

The algebra of regular functions $\mathscr{E}(G)$ is $\mathscr{E}(G)=\mathbf{C}(u_1,u_2,z_1,z_2)/(u_1z_2-u_2z_1-1)$ and $\mathscr{E}(A)\subset\mathscr{E}(G)$ consists of the functions $f\in\mathscr{E}(G)$ which satisfy

$$-L^G_{E_+}f=z_1\frac{\partial f}{\partial u_1}+z_2\frac{\partial f}{\partial u_2}=0.$$

The ring $Wu\subset\mathscr{D}(A)$ coincides with the ring of polynomials of the single generator

$$L^A_H=z_1\frac{\partial}{\partial z_1}+z_2\frac{\partial}{\partial z_2}.$$

Now we show how to construct the mapping $f\to\bar{f}$. Let $T$ be an irreducible representation of $SL(2,\mathbf{C})$. We denote by $\mathscr{E}_T$ the largest subspace of $\mathscr{E}(G)$ such that the restriction of $R^G$ to this subspace is a multiple of $T$. Furthermore, for each $n\in\mathbf{Z}$ we denote by $\mathscr{E}^n_T$ the subspace of $\mathscr{E}_T$ consisting of all functions $f\in\mathscr{E}_T$ for which $L^G_Hf=nf$.

*Lemma* 4.1. *1)* $\mathscr{E}(G)=\bigoplus_{T,n}\mathscr{E}^n_T$.

*2) Let* $\dim T=l+1$. *Then* $\dim\mathscr{E}^n_T=l+1$ *for* $n=-l,-l+2,\dots,l-2,l$ *and* $\mathscr{E}^n_T=0$ *for the remaining values of n. Those n for which* $\mathscr{E}^n_T\neq0$ *are weights of T.*

*3)* $\mathscr{E}^n_T$ *is invariant under* $R^G$ *and the restriction of* $R^G$ *to* $\mathscr{E}^n_T$ *is equivalent to T.*

*4)* $\mathscr{E}^l_T$ *consists of vectors of highest weight with respect to* $L^G$ *(i.e.* $L^G_Xf=0$ *for all* $f\in\mathscr{E}^l_T$ *and* $X\in\mathfrak{n}_+$*).*

The proof of this lemma follows easily from simple properties of the representations of $SL(2,\mathbf{C})$.

It suffices to construct the mapping $f\to\bar{f}$ on each space $\mathscr{E}^n_T$ separately. We may assume that the restriction of $L^G$ to the smallest invariant subspace of $\mathscr{E}(f)$

containing $f$ is equivalent to $T$. Let us put $f_i=(L^G_{E_+})^i f$. Then obviously $f_i\in\mathscr{E}_T^{n+2i}$, and in particular $f_i=0$ for $i>\frac{1}{2}(l-n)$. In accordance with property 2 of the mapping $f\to\tilde{f}$ (see Conjecture I), we shall look for the operator $\tilde{f}$ in the form

$$\tilde{f}=\sum_{i=0}^{\frac{1}{2}(l-n)} f_i(L^G_{E_-})^i\alpha_i \tag{4.1}$$

where $\alpha_i\in Wu$.

Let us assume now that the right-hand side of (4.1) defines an operator on $A$. Applying the relations

$$[E_+, E_-^i]=E_-^{i-1}\cdot i(H-i+1)$$

we find easily that the elements $\alpha_i\in Wu$ satisfy the equations

$$\alpha_{i-1}+i(H-i+1)\alpha_i=0.$$

Therefore, putting $\frac{1}{2}(l-n)=p$ we obtain

$$\alpha_i=(-1)^{p-i}\frac{p!}{i!}(H-p+1)\cdot\ldots\cdot(H-i)\alpha_p.$$

Consequently, we have the operator $\tilde{f}$ if we determine $\alpha_p$. It is clear that property 2) requires $\alpha_p\in\mathbf{C}$. Moreover, $s_0(\tilde{f})=f\otimes\alpha_0$ and $H(\varrho)=1$. Therefore, to assure the validity of property 1) we have to take

$$\alpha_p=(p!)^{-2}.$$

Then all statements 1)—4) of Conjecture I will be valid.

*Example* 2. Let $G=SL(3,\mathbf{C})$ and $N$ be the subgroup of upper triangular matrices with units on the diagonal.

Let $T_i$ be the $i$-th fundamental representation of $G$, $i=1,2$. Both representations $T_i$ are three-dimensional and the spaces $\mathscr{E}_{T_i}$ (see Example 1) are of dimension 9. Here $\mathscr{E}_{T_1}$ consists of the linear combinations of the matrix elements $g_{ij}$, $1\leq i,j\leq 3$ of the matrix $g\in G$, and $\mathscr{E}_{T_2}$ consists of the linear combinations of the second order minors of $g$. Let us construct the mapping $f\to\tilde{f}$ for $f\in\mathscr{E}_{T_1}$. Put $f_i=\alpha_1 g_{i1}+\alpha_2 g_{i2}+\alpha_3 g_{i3}$, $i=1,2,3$. Then for arbitrary $\alpha_1$, $\alpha_2$, $\alpha_3$ the elements $f_i$ form a subspace of $\mathscr{E}_{T_1}$ such that the restriction of $R^G$ to this subspace is equivalent to $T_1$. Here $f_3$ is a vector of highest weight, that is $\tilde{f}_3=f_3$. Let us now give formulas for $\tilde{f}_1$ and $\tilde{f}_2$. Let $E_{ij}$ $(i\neq j)$, $E_{11}-E_{22}$ and $E_{22}-E_{33}$ be the basis in $\mathfrak{g}$. We put

$$L^G_{E_{ij}}=\hat{E}_{ij},\quad L^G_{E_{ii}-E_{jj}}=\hat{Z}_{ij}.$$

These operators act on the functions $f_i$ as follows

$$\hat{E}_{ij}f_k=-\delta_{jk}f_i,$$

$$\hat{Z}_{ij}f_k=(\delta_{jk}-\delta_{ik})f_k,$$

where $\delta_{ij}$ is the Kronecker symbol.

Let us consider in $\mathscr{D}(G)$ the following operators:

$$\tilde{f}_3 = f_3$$

$$\tilde{f}_2 = f_2 \hat{Z}_{32} - f_3 \hat{E}_{32}$$

$$\tilde{f}_1 = f_1 \hat{Z}_{12}(\hat{Z}_{13}+1) + f_2 \hat{E}_{21}(\hat{Z}_{13}+1) + f_3(\hat{E}_{32}\hat{E}_{21} + \hat{E}_{31}\hat{Z}_{12}).$$

It is easy to verify that the operators $\tilde{f}_1, \tilde{f}_2, \tilde{f}_3$ belong to $\mathscr{D}(A)$ (i.e. $\tilde{f}_i(\mathscr{E}(A)) \subset \mathscr{E}(A)$) and the mapping $f_i \to \tilde{f}_i$ satisfies all conditions of Conjecture I. Thus the mapping $f \to \tilde{f}$ has been constructed for $f \in \mathscr{E}_{T_1}$.

We also remark that the operator $\tilde{f}_1$ is an operator of the second order on $A$ which cannot be expressed in term of first order operators on $A$.

The mapping $f \to \tilde{f}$ for $f \in \mathscr{E}_{T_2}$ can be constructed in the same way.

In Examples 1 and 2 we were dealing with representations $T$ for which all weight subspaces were one-dimensional. It is a more difficult task to construct the operation $f \to \tilde{f}$ in the case when these subspaces are not one-dimensional. Now let us see a simple example of that kind.

*Example* 3. Let us put, as in Example 2, $G = SL(3, \mathbf{C})$ and $T$ be the adjoint representation of $G$ in its Lie algebra $\mathfrak{g}$. Let $f$ be a vector of highest weight with respect to $L^G$ in $\mathscr{E}_T$. We shall introduce the following notations

$$f_{13} = f, \quad f_{23} = \hat{E}_{21} f, \quad f_{12} = -\hat{E}_{32} f,$$

$$h_{12} = -\hat{E}_{21} f_{12}, \quad h_{23} = -\hat{E}_{32} f_{23},$$

$$f_{21} = -\hat{E}_{21} h_{23}, \quad f_{32} = -\hat{E}_{32} h_{12}, \quad f_{31} = \hat{E}_{32} f_{21}.$$

Then the restriction of $L^G$ to the subspace spanned by $f_{ij}$ and $h_{ij}$ is equivalent to $T$. Here $h_{12}$ and $h_{23}$ generate a two-dimensional weight subspace of $T$. We define the operation $f \to \tilde{f}$ in the following way

$$\tilde{h}_{12} = h_{12}\left(\hat{Z}_{12}\hat{Z}_{23} + \frac{2}{3}\hat{Z}_{12} - \frac{2}{3}\hat{Z}_{23}\right) + h_{23}\left(-\hat{Z}_{12}\hat{Z}_{23} + \frac{1}{3}\hat{Z}_{12} - \frac{4}{3}\hat{Z}_{23}\right) +$$

$$+ f_{12}\hat{E}_{21}(3\hat{Z}_{23}+1) - f_{23}\hat{E}_{32}(3\hat{Z}_{12}+2) + f_{13}(3\hat{E}_{21}\hat{E}_{32} + \hat{E}_{31}).$$

$$\tilde{h}_{23} = h_{12}\left(-\hat{Z}_{12}\hat{Z}_{23} - \frac{4}{3}\hat{Z}_{12} + \frac{1}{3}\hat{Z}_{23}\right) + h_{23}\left(\hat{Z}_{12}\hat{Z}_{23} - \frac{2}{3}\hat{Z}_{12} + \frac{2}{3}\hat{Z}_{23}\right) -$$

$$- f_{12}\hat{E}_{21}(3\hat{Z}_{23}+2) + f_{23}\hat{E}_{32}(3\hat{Z}_{12}+1) - f_{13}(3\hat{E}_{21}\hat{E}_{32} + 2\hat{E}_{31}).$$

We remark that in the subspace spanned by $\tilde{h}_{12}$ and $\tilde{h}_{23}$, there is, up to multiplication, only one operator of first order namely $\tilde{h}_{12} - \tilde{h}_{23}$; other operators in this space are of order 2.

## § 5. The generalized Segal—Bargmann spaces

In this section we shall consider a generalization of Bargmann's construction [3] of representations of the group $SU(2)$.

Let $K$ be a maximal compact subgroup of the group $G$, $\mathfrak{k}$ be its Lie algebra, and $\mathfrak{i}: \mathfrak{g} \to \mathfrak{g}$ be the corresponding Cartan involution. We assume that $K$ is chosen so that $\mathfrak{i}(\mathfrak{n}_+) = \mathfrak{n}_-$. Let $T_i$, $1 \leqq i \leqq r$ be the representations of $G$ corresponding to the fundamental highest weights of $\mathfrak{g}$. Let $f_i$ be a vector of highest weight in $T_i$. We define a function $H_i(g)$ on $G$ by the formula

$$H_i(g) = \|T(g^{-1})f_i\|_i^2,$$

where $\|\cdot\|_i$ is a $K$-invariant norm on $T_i$ such that $\|f_i\|_i = 1$. Clearly, $H_i(ng) = H_i(g)$ for $n \in N_+$, hence $H_i$ can also be considered as a function on $A$. Let $u(t)$ be a positive, rapidly decreasing function defined for $t > 0$. Let $\varrho_i$ $(1 \leqq i \leqq r)$ be positive numbers. We define the weight function $\varrho(x)$ on $A$ by the formula

$$\varrho(x) = u\left(\sum \varrho_i H_i(x)\right).$$

*Definition* 5.1. The *Segal—Bargmann space* of the group $G$ is defined as the completion of $\mathscr{E}(A)$ with respect to the scalar product

$$\{f, g\} = \int_A f(x)\bar{g}(x)\varrho(x)\omega,$$

where $\omega$ denotes the $G$-invariant measure on $A$.

It is obvious that this scalar product in $\mathscr{E}(A)$ is invariant under $R_k^A$, $k \in K$.

*Conjecture* II. *There exists a function $u(t)$ such that for any $D \in \mathscr{D}(A)$ we have $D^* \in \mathscr{D}(A)$; here $D^*$ denotes the adjoint operator of $D$ with respect to $\{,\}$.*

Let us consider the simplest case $G = SL(2, \mathbf{C})$, $K = SU(2)$. In this case $A$ is the plane $\mathbf{C}^2$ without the point $(0, 0)$, and $\mathscr{E}(A)$ is the space of polynomials of two variables $z_1$, $z_2$. The scalar product is introduced in $\mathscr{E}(A)$ by the formula

$$\{f, g\} = \int e^{-\varrho(|z_1|^2 + |z_2|^2)} f\bar{g}\, dz_1\, dz_2\, d\bar{z}_1\, d\bar{z}_2.$$

Now the ring $\mathscr{D}(A)$ is generated by the operators $z_i$ and $\dfrac{\partial}{\partial z_i}$, $i = 1, 2$. It is easy to verify that $\left(\dfrac{\partial}{\partial z_i}\right)^* = \varrho z_i$ and therefore

$$(z_i)^* = \varrho^{-1}\frac{\partial}{\partial z_i}.$$

This shows that Conjecture II is true in the present case.

The above construction for $SU(2)$ was suggested by Bargmann [3].

Using the examples given in § 4 we can show that Conjecture II is also valid for $G = SL(3, \mathbf{C})$. As $u(t)$ we take the decreasing positive solution of the equation

$$t\frac{d^2u}{dt^2} + 2\frac{du}{dt} + u = 0.$$

It seems that for $G=SL(n, \mathbf{C})$ the function $u(t)$ has to satisfy the equation

$$\frac{d^{n-1}}{dt^{n-1}}(ut^{n-2})+(-1)^n u = 0.$$

A more precise version of Conjecture II is the following

*Conjecture* II′. *Let $g \to g^*$ be the anti-automorphism of $G$ corresponding to the anti-involution $(-\mathfrak{i}):\mathfrak{g}\to\mathfrak{g}$. The numbers $\varrho_i$ can be chosen in such a way that for any function $f\in\mathscr{E}(G)$ the equality $(\tilde{f})^*=(\widetilde{f^*})$ is satisfied. Here $f^*(g)=f(g^*)$.*

## § 6. The mapping $\pi_*$

In this section we shall give a construction which yields a weaker version of Conjecture I. More precisely, for every function $f\in\mathscr{E}(G)$ a collection of regular differential operators on $A$ will be constructed. In addition to this, we shall show that all differential operators on $A$ can be obtained in this way. In the construction we apply an operation $\pi_*$ which maps functions on $G$ into functions on $A$ and which, we believe, is of independent interest. This operation is an algebraic analogue of the averaging operation over a subgroup (which is unipotent in the present case). It is remarkable that $\pi_*$ transfers the operation of multiplication by a function $f(g)$ into an "almost" differential operator on $A$. The exact formulation of these facts is given in Theorems 6.3 and 6.5.

*Lemma and Definition* 6.1. *There exists a unique mapping $\pi_*:\mathscr{E}(G)\to\mathscr{E}(A)$ such that*

*1) $\pi_* R_g^G = R_g^A \pi_*$ and $\pi_* L_h^G = L_h^A \pi_*$*

*for all $g\in G$, $h\in H$,*

*2) $\pi_*\pi^*\varphi=\varphi$ for all $\varphi\in\mathscr{E}(A)$.*

*Proof.* First we prove that $\pi_* f$ is uniquely determined by the conditions *1)* and *2)*. It is enough to consider the case when $f$ lies in a subspace $V$ irreducible and invariant with respect to $L^G$; further we can assume that $f$ is a weight function of weight $\chi$ with respect to the restriction of $L^G$ to $H$. If $\chi$ is a highest weight of the given irreducible representation, then $f\in\operatorname{Im}\pi^*$, i.e. $f=\pi^*\varphi$, hence in view of *2)*, $\pi_* f=\varphi$.

Assume now that $\chi$ is not a highest weight. Let us denote by $f_0$ a vector of highest weight in $V$ and by $\chi_0$ the corresponding highest weight. Then, under the action of $R^G$, $f$ and $f_0$ are transformed by the same irreducible representation of $G$. From *1)* and the fact that every irreducible representation of $G$ occurs in $\mathscr{E}(A)$ only once (see [2]) it follows that $\pi_* f$ and $\pi_* f_0$ belong to the same subspace, irreducible and invariant with respect to $R^A$. But then the weights of

$\pi_* f$ and $\pi_* f_0$ with respect to $L^A$ coincide (see [2]). From *1)* it follows that the weight of $\pi_* f$ is equal to $\chi$ and the weight of $\pi_* f_0$ is equal to $\chi_0 \neq \chi$. Since $\pi_* f_0 \neq 0$ we have $\pi_* f = 0$.

From this proof we immediately obtain a construction for $\pi_*$. Indeed, if $f$ is a weight vector, not of highest weight, lying in a subspace, irreducible and invariant under $L^G$, then we put $\pi_* f = 0$. On the other hand, if $f$ is a vector of highest weight then $f = \pi^* \varphi$ and we put $\pi_* f = \varphi$. The lemma is proved.

Let us denote for each $\gamma \in \Delta$ the operator $L^G_{E_\gamma}$ by $\hat{E}_\gamma$. It follows from the construction of $\pi_*$ that $\pi_*(\hat{E}_\gamma f) = 0$ for all $f \in \mathscr{E}(G)$ and $\gamma \in \Delta_-$.

By means of the mapping $\pi_*$ one can construct differential operators on $A$ in the following way.

*Definition* 6.2. Let $f \in \mathscr{E}(G)$. We define an operator $\hat{f}$ in the space $\mathscr{E}(A)$ by the formula $\hat{f}(\varphi) = \pi_*(f \cdot \pi^* \varphi)$, $\varphi \in \mathscr{E}(A)$.

*Theorem* 6.3. *There exists a non-zero element $Z \in Wu$ such that $Z\hat{f}$ is a regular differential operator on $A$.*

*Proof.* A differential operator $D$ on $G$ is called a *chain* if it can be expressed in the form $\hat{E}_{\gamma_1} \hat{E}_{\gamma_2} \dots \hat{E}_{\gamma_k}$, $\gamma_i \in \Delta_+$. The weight of this chain is defined to be $\gamma_1 + \gamma_2 + \dots + \gamma_k \in \mathfrak{h}^*$. Let us denote by $\Xi$ the set of all chains $D$ such that $Df \neq 0$ and by $\Xi_0 \subset \mathfrak{h}^*$ the set of all their weights. Obviously, $\Xi$ and $\Xi_0$ are finite sets. For any function $\varphi \in \mathscr{E}(A)$ and any chain $D$ we have $D(f\pi^* \varphi) = Df \cdot \pi^* \varphi$, since $\hat{E}_\gamma \pi^* \varphi = 0$ for $\gamma \in \Delta_+$. Therefore, if $D \notin \Xi$ then $D(f\pi^* \varphi) = 0$. Let us denote by $U$ the subspace of $\mathscr{E}(G)$ consisting of all functions $u$ such that $Du = 0$ for any chain $D \notin \Xi$.

*Lemma* 6.4. *There exist a regular differential operator $T$ on $G$ and an element $Z \in Wu$ such that $Tu = Z\pi^* \pi_* U$ for all $u \in U$.*

The theorem is an immediate consequence of this lemma, since for any function $\varphi \in \mathscr{E}(A)$ we have $f \cdot \pi^* \varphi \in U$, and so

$$T(f\pi^* \varphi) = Z\pi^* \pi_*(f \cdot \pi^* \varphi) = \pi^* Z\hat{f}(\varphi),$$

i.e. $T \circ f \circ \pi^* = \pi^* \circ Z \circ \hat{f}$. It follows from this equality that the differential operator $T \circ f$ preserves $\mathscr{E}(A) \subset \mathscr{E}(G)$, or in other words, that $Z\hat{f}$ is a differential operator on $A$.

*Proof* of the lemma. Let $H_1, \dots, H_r$ denote a basis in $\mathfrak{h}$. The elements of $Wu$ are polynomials of $H_1, \dots, H_r$ and, as above, we may consider them as polynomial functions on $\mathfrak{h}^*$. Let $\Delta$ be the Laplace operator of the second order on $G$ (constructed by means of the Killing form). Then there exists an element $P \in Wu$ such that for any vector $\varphi \in \mathscr{E}(A)$ the equality $\Delta\varphi = P\varphi$ is satisfied, or equivalently, $\Delta\varphi = P(\chi_0) \cdot \varphi$, where $\chi_0$ is the weight of the vector $\varphi$.

Let $B$ be the restriction of the Killing form of the algebra $\mathfrak{g}$ to $\mathfrak{h}$, and $Q(\chi)$ the dual quadratic form on $\mathfrak{h}^*$. It follows from results of Harish-Chandra [4] that $P(\chi)=Q(\chi+\varrho)-Q(\varrho)$, where $\varrho$ is half-sum of the positive roots.

For an arbitrary weight $\beta$ we denote by $P_\beta$ and $Z_\beta$ the elements of $Wu$ corresponding to the polynomial functions

$$P_\beta(\chi) = P(\chi+\beta)$$

and

$$Z_\beta(\chi) = 2\langle\beta, \chi+\varrho\rangle - \langle\beta,\beta\rangle$$

respectively ($\langle , \rangle$ denotes the scalar product in $\mathfrak{h}^*$ corresponding to the quadratic form $Q$). Let $T=\prod_\beta (P_\beta-\Delta)$, $Z=\prod_\beta Z_\beta$, where $\beta$ runs through $\Xi_0\setminus\{0\}$.

We shall show that for all $u\in U$ the equality $Tu=Z\pi^*\pi_* u$ is satisfied.

Clearly, it suffices to verify this equality when $u$ is a weight vector lying in a subspace $V$ which is irreducible and invariant under $L^G$. Let $u_0$ be a vector of highest weight in $V$, $\chi$ and $\chi_0$ be the weights of $u$ and $u_0$, respectively, with respect to $\mathfrak{h}$. It follows from the uniqueness of the vector of highest weight that $u_0=cDu$ where $D$ is an appropriate chain and $c\in\mathbf{C}$, and therefore $\chi_0-\chi\in\Xi_0$.

*Case* 1, $\chi\neq\chi_0$. The restriction of $\Delta$ to $V$ is multiplication by $P(\chi_0)$. Therefore

$$(P_\beta-\Delta)u = (P_\beta(\chi)-P(\chi_0))u = (P(\chi+\beta)-P(\chi_0))u = 0,$$

if $\beta=\chi_0-\chi\in\Xi_0\setminus\{0\}$. Thus $Tu=0$, and since $\pi_* u=0$ in this case, we have $Tu=$ $=Z\pi^*\pi_* u=0$.

*Case* 2, $\chi=\chi_0$. Then

$$(P_\beta-\Delta)u = (P(\chi_0+\beta)-P(\chi_0))u = (Q(\chi_0+\beta+\varrho)-Q(\chi_0+\varrho))u =$$
$$= (2\langle\chi_0+\varrho, \beta\rangle + \langle\beta,\beta\rangle)u = Z_\beta(\chi_0)u,$$

hence

$$Tu = (\pi(P_\beta-\Delta))u = \pi Z_\beta(\chi_0)u = Zu = Z\pi^*\pi_* u.$$

The proof of the lemma and of Theorem 6.3 is complete.

It can be shown that the order of the operator $Z\hat{f}$ is equal to card $\Xi_0-1$, i.e. the order of $Z'$.

*Theorem* 6.5. *Every regular differential operator $D$ on $A$ can be written in the form $D=\sum Z_j\hat{f}_j$, where $Z_j\in Wu$ and $f_j\in\mathscr{E}(G)$.*

*Proof.* Let $D'$ be a lifting of $D$ to $G$ (Theorem 3.4). By means of simple transformations $D'$ can be transformed to the form

$$D' = \sum Z_j f_j + \sum A_k \hat{E}_{\gamma_k} + \sum \hat{E}_{\delta_i} B_i,$$

where $A_k$ and $B_i$ are differential operators on $G$, $Z_j\in Wu$, $f_j\in\mathscr{E}(G)$, $\gamma_k\in\Delta_+$, $\delta_i\in\Delta_-$.

Since $\hat{E}_{\gamma_k}\pi^*\varphi = 0$ and $\pi_* E_{\delta_i} f = 0$, we have

$$D\varphi = \pi_*\pi^* D\varphi = \pi_* D'\pi^*\varphi = \sum Z_j \pi_*(f_j \pi^*\varphi) = \sum Z_j \hat{f}_j(\varphi),$$

which proves theorem 6.5.

Let $L$ be the quotient field of the ring $Wu$. Then the operation $f \to \hat{f}$ extends to a mapping $\Omega: L \underset{\mathbf{C}}{\otimes} \mathscr{E}(G) \to L \underset{Wu}{\otimes} \mathscr{D}(A)$. Here $\Omega$ commutes with the actions of $R_g$ and $L_g$.

*Theorem* 6.6. *$\Omega$ is an isomorphism of L-modules $L \otimes \mathscr{E}(G)$ and $L \underset{Wu}{\otimes} \mathscr{D}(A)$.*

*Proof.* It follows from Theorem 6.5 that $\Omega$ is a surjection.

In Theorem 6.3 we constructed for every function $f \in \mathscr{E}(G)$ an operator $Z_f \in Wu$ such that $Z_f \cdot \hat{f} \in \mathscr{D}(A)$.

*Lemma* 6.7.

$$s_0(Z_f \cdot \hat{f}) = f \otimes Z_f + \sum f_i \otimes Z_i,$$

*where* $\deg Z_i < \deg Z_f$.

*Proof.* The lifting of the operator $Z_f \cdot \hat{f}$ has the form $T \circ f$, where $T = \prod_\beta (P_\beta - \Delta)$ (see the proof of Theorem 6.3). It follows from [4] that

$$\Delta = P + \sum_{\gamma \in \Delta_+} c_\gamma \hat{E}_{-\gamma} \hat{E}_\gamma, \quad c_\gamma \in \mathbf{C}.$$

Consequently

$$P_\beta - \Delta = Z_\beta - \sum_{\gamma \in \Delta_+} c_\gamma \hat{E}_{-\gamma} \hat{E}_\gamma.$$

Let $K = \operatorname{card}(\Xi_0 \setminus \{0\})$. It is easy to show by induction on $K$ that

$$T = \prod_\beta Z_\beta + \sum_\lambda c_\lambda H_\lambda X_\lambda Y_\lambda,$$

where $c_\lambda \in \mathbf{C}$, and $H_\lambda$, $X_\lambda$ and $Y_\lambda$ are products of suitably chosen operators $H_i \in Wu$, $\hat{E}_{-\gamma}$, $\gamma \in \Delta_+$, and $\hat{E}_\gamma$, $\gamma \in \Delta_+$, respectively. We also have $\deg H_\lambda + \deg X_\lambda \leq K$ for all $\lambda$, and $\deg H_\lambda < K$ if $X_\lambda = 1$. Moreover, $Z_f = \prod_\beta Z_\beta$. From this it follows that the operator $T \circ f$ is of the form $f \cdot Z_f + \sum f_\mu H_\mu X_\mu Y_\mu$, where $f_\mu \in \mathscr{E}(G)$, $H_\mu$, $X_\mu$ and $Y_\mu$ satisfy the same conditions as $H_\lambda$, $X_\lambda$, $Y_\lambda$. Therefore, $s_0(Z_f \cdot \hat{f})$ (see Definition 3.10) is of the form $s_0(Z_f \cdot \hat{f}) = f \otimes Z_f + \sum f_i \otimes Z_i$, where $\deg Z_f = K$, $\deg Z_i < K$. The lemma is proved.

Let us now consider the element $D = \sum \tilde{Z}_i \otimes f_i \in \mathscr{E}(G)$, where $\tilde{Z}_i \in L$, $\tilde{Z}_i \neq 0$, and the $f_i$ are linearly independent. We shall show that $\Omega(D) \neq 0$. Let us multiply $D$ by an element $Z \in Wu$ such that $ZD = \sum Z_i \otimes f_i$, where $Z_i \in W_u$ and $Z_i$ is divisible by $Z_{f_i}$. Let $l = \max \deg Z_i$. Then $\Omega(ZD) \in \mathscr{D}(A)$ and by Lemma 6.7

$$s_0(\Omega(ZD)) = \sum f_i \otimes Z_i + \sum f_j' \otimes Z_j',$$

where $\deg Z_j' < l$. This implies that $s_0(\Omega(ZD)) \neq 0$, consequently $\Omega(D) \neq 0$.

We have now proved that $\Omega$ is an injection and therefore an isomorphism.

PART II

# THE RESOLUTION OF A FINITE-DIMENSIONAL $\mathfrak{g}$-MODULE

## § 7. Introduction

Let $\mathfrak{g}$ be, as above, a complex semisimple Lie algebra. Let $V$ be a finite-dimensional irreducible $\mathfrak{g}$-module. The cohomology groups $H^i(\mathfrak{n}_-, V)$ play an important rôle in the theory of representations (see [5] and [6]). They have the following properties expressed in Bott's theorem [5]. Let $W$ be the Weyl group of the algebra $\mathfrak{g}$. Then $\dim H^i(\mathfrak{n}_-, V)=\text{card}\, W^{(i)}$, where $W^{(i)}=\{w\in W: l(w)=i\}$ and $l(w)$ is the length of the element $w\in W$.

The fundamental result of this part is Theorem 10.1 which improves on Bott's theorem.

For any $\chi\in\mathfrak{h}^*$ we denote by $M_\chi$ the $U(\mathfrak{g})$-module generated by a vector of highest weight $\chi-\varrho$ (the exact definition is given below). In Theorem 10.1 we construct a resolution

$$0\leftarrow V\leftarrow C_0\leftarrow C_1\leftarrow\ldots\leftarrow C_s\leftarrow 0$$

of the $\mathfrak{g}$-module $V$ such that

$$C_i=\bigoplus_{w\in W^{(i)}} M_{w\chi},$$

where $\chi-\varrho$ is the highest weight of $V$. Bott's theorem follows from Theorem 10.1 since $M_\chi$ is a free $U(\mathfrak{n}_-)$-module with one generator.

In this part we shall make a systematic use of a certain category of $\mathfrak{g}$-modules which we call category $O$ (see [12]). § 8 is devoted to the exposition of the properties of category $O$.

In § 9 several results concerning the cohomologies of Lie algebras are presented. In particular, in this section a purely algebraic proof of Bott's theorem is given which, it seems to us, is simpler than the proofs presented in [5] and [6]. This proof has several points of contact with Kostant's proof [6], but it does not make any use of the Hermitian structure. The observant reader will notice that the resolution constructed in the proof of Bott's theorem is dual to a part of the de Rham resolution well-known from the theory of formal differential forms.

In the Appendix we describe the modules occurring in the Jordan—Hölder decomposition of the modules $M_\chi$. The study of the structure of the modules $M_\chi$ was initiated in Verma's work [7]. We remark that the works [7], [8] and [9] contain everything we know about the modules $M_\chi$. All these facts are also contained in Theorems 8.7, 8.8, 8.12 and 10.1 of the present work.

## § 8. The category $O$

*Definition* 8.1. Category $O$ is the full subcategory of the category of left $U(\mathfrak{g})$-modules consisting of all modules $M$ such that

*1)* $M$ is a finitely generated $U(\mathfrak{g})$-module,

*2)* $M$ can be made $\mathfrak{h}$-diagonal, i.e. there exists a basis in $M$ consisting of weight vectors,

*3)* $M$ is $U(\mathfrak{n}_+)$-finite, i.e. for any $f \in M$ the space $U(\mathfrak{n}_+)f$ is finite-dimensional.

*Definition* 8.2. Let $\chi \in \mathfrak{h}^*$. We denote by $J_\chi$ the left ideal in $U(\mathfrak{g})$ generated by the elements $E_\gamma$, $\gamma \in \Delta_+$ and $H - \chi(H) + \varrho(H)$, $H \in \mathfrak{h}$. Let us put $M_\chi = U(\mathfrak{g})/J_\chi$. We shall denote by $f_\chi$ the image of $1 \in U(\mathfrak{g})$ in $M_\chi$.

For the sake of convenience we formulate the elementary properties of the category $O$ and the modules $M_\chi$ in Propositions 8.3, 8.5, 8.6. These propositions are simple consequences of Harish–Chandra's theorem about Laplace operators (see [4]).

*Proposition* 8.3. *1)* The category $O$ is closed under taking submodules, factor modules and finite direct sums.

*2)* Let $M \in O$. Then all the spaces $M^{(\psi)}$ $\psi \in \mathfrak{h}^*$ are finite-dimensional and $M = \bigoplus_{\psi \in \mathfrak{h}^*} M^{(\psi)}$. In addition to this $P(M)$ is contained in a finite union of sets $\chi_i - K$, $\chi_i \in \mathfrak{h}^*$.

*3)* Every element $M \in O$ has a finite Jordan—Hölder composition series.

*4)* $M_\chi$ is a free $U(\mathfrak{n}_-)$-module with $f_\chi$ as a generator.

*5)* $M_\chi \in O$.

*6)* In $M_\chi$ there exists a maximal proper submodule. The corresponding irreducible factor module will be denoted by $L_\chi$.

*7)* Every irreducible module in the category $O$ is of the form $L_\chi$, $\chi \in \mathfrak{h}^*$.

We denote by $Z(\mathfrak{g})$ the centre of the algebra $U(\mathfrak{g})$ and by $\Theta$ the set of all homomorphisms $\vartheta : Z(\mathfrak{g}) \to \mathbf{C}$.

*Definition* 8.4. Let $M$ be an arbitrary $\mathfrak{g}$-module. To each element $f \in M$ which is an eigenvector with respect to all the operators $z \in Z(\mathfrak{g})$ we can assign a homomorphism $\vartheta_f \in \Theta$ such that $zf = \vartheta_f(z) \cdot f$ for all $z \in Z(\mathfrak{g})$. The set of all such homomorphisms $\vartheta$ will be denoted by $\Theta(M)$.

*Proposition* 8.5. *1)* $\Theta(M_\chi)$ consists of a single element which we shall denote by $\vartheta_\chi$.

*2)* $\vartheta_{\chi_1} = \vartheta_{\chi_2}$ if and only if $\chi_1 \sim \chi_2$.

*Proposition* 8.6. Let $M \in O$. Then

*1)* $\Theta(M)$ is finite.

*2)* For any $\vartheta \in \Theta$ we put $I_\vartheta = \operatorname{Ker} \vartheta \subset Z(\mathfrak{g})$. Let $M_\vartheta^{(n)} = \{f \in M \mid I_\vartheta^n f = 0\}$. Then $M_\vartheta^{(n)}$ stabilizes for large values of $n$. The obtained submodule of $M$ will be denoted by $M_\vartheta$.

*3)* $\Theta(M_\vartheta) = \{\vartheta\}$.

*4)* $M = \bigoplus_{\vartheta \in \Theta(M)} M_\vartheta$.

*5)* The mapping $M \to M_\vartheta$ is an exact functor in $O$.

Now we shall pass on to the study of the modules $M_\chi$. The following two theorems give a complete description of the homomorphisms between modules $M_\chi$.

*Theorem* 8.7 [7]. *Let* $\chi, \psi \in \mathfrak{h}^*$. *Then either*

*1)* $\mathrm{Hom}_{U(\mathfrak{g})}(M_\chi, M_\psi) = 0$.

*or*

*2)* $\mathrm{Hom}_{U(\mathfrak{g})}(M_\chi, M_\psi) = \mathbf{C}$, *and every non-trivial homomorphism* $M_\chi \to M_\psi$ *is an injection.*

*Theorem* 8.8. *Let* $\chi, \psi \in \mathfrak{h}^*$,

*then*

$$\mathrm{Hom}_{U(\mathfrak{g})}(M_\chi, M_\psi) = \mathbf{C}$$

*if and only if there exists a sequence of roots* $\gamma_1, \ldots, \gamma_k \in \Delta_+$ *satisfying the following condition (A) for the pair* $(\chi, \psi)$.

*Condition (A).*

*1)* $\chi = \sigma_{\gamma_k} \cdot \sigma_{\gamma_{k-1}} \cdot \ldots \cdot \sigma_{\gamma_1} \psi$.

*2) Put* $\chi_0 = \psi$, $\chi_i = \sigma_{\gamma_i} \cdot \ldots \cdot \sigma_{\gamma_1} \psi$. *Then* $\chi_{i-1} - \chi_i = n\gamma_i$, *where n is a non-negative integer.*

In particular $\mathrm{Hom}_{U(\mathfrak{g})}(M_\chi, M_\psi) \neq 0$ only if $\chi \sim \psi$ and $\chi \leqq \psi$.

Theorem 8.8 was formulated in [7] as a conjecture; a proof of the sufficiency of condition (*A*) was also given there. A complete proof of Theorem 8.8 was given in [9].

The structure of the submodules of the modules $M_\chi$ is most interesting when $\chi \in D$. We shall study this case in more detail. For this purpose we introduce the following partial ordering in the Weyl group $W$.

*Definition* 8.9. If $w_1, w_2 \in W$ and $\gamma \in \Delta_+$ then $w_1 \xrightarrow{\gamma} w_2$ means that $w_1 = \sigma_\gamma w_2$ and $l(w_1 = l(w_2) + 1$. (Sometimes we shall omit the symbol $\gamma$ above the arrow.) We put $w < w'$ if there exists a sequence $w_1, w_2, \ldots, w_k$ of elements of $W$ such that

$$w \to w_1 \to w_2 \to \ldots \to w_k \to w'.$$

*Theorem* 8.8′. *Let* $\chi \in D$, $w_1, w_2 \in W$. *Then* $\mathrm{Hom}_{U(\mathfrak{g})}(M_{w_1\chi}, M_{w_2\chi}) = \mathbf{C}$ *if and only if* $w_1 \leqq w_2$.

In what follows all modules $M_{w\chi}$, $w \in W$, $\chi \in D$ will be considered as submodules of $M_\chi$.

Theorem 8.8′ is an immediate consequence of Theorem 8.8 and the following lemmas.

*Lemma* 8.10. *Let* $\chi \in D$, $\gamma \in \Delta$, $w \in W$. *Then* $w\chi - \sigma_\gamma w\chi = n\gamma$, *where* $n \in \mathbf{Z}$, $n \neq 0$, *while* $n > 0$ *if and only if* $l(\sigma_\gamma w) > l(w)$.

*Lemma* 8.11. *Let* $w \in W$ *and* $\gamma \in \Delta$ *be such that* $l(\sigma_\gamma w) < l(w)$. *Then* $w < \sigma_\gamma w$.

The proofs of these lemmas will be presented in § 11.

The complete structure of the submodules of the modules $M_\chi$ is not yet known. Some information is contained in the following theorem.

*Theorem* 8.12. *Let* $\psi, \chi \in \mathfrak{h}^*$. *Then* $L_\chi \in JH(M_\psi)$ *if and only if there exists a sequence* $\gamma_1, \gamma_2, \ldots, \gamma_k \in \Delta_+$ *satisfying condition (A) for the pair* $(\chi, \psi)$ (see Theorem 8.8).

Since we shall need this theorem in § 10, we shall give the proof in the Appendix.

*Corollary.* Let $\chi \in D$. Then the Jordan—Hölder decomposition of the module $M_{w\chi}$ consists of the modules $L_{w'\chi}$, where $w' \leqq w$ (possibly counted with multiplicities). $L_{w\chi}$ occurs in this decomposition exactly once. The example in [9] shows that the modules $M_\psi$ may contain submodules $M$ which are not generated by submodules $M_\chi \subset M$.

In fact, even the following proposition can be proved.

*Proposition* 8.13. Let $\mathfrak{g} = sl(4, \mathbf{C})$. Consider the module $M_\psi$ corresponding to a weight $\psi \in D$. Then $M_\psi$ contains a submodule $M$ such that

$$M \neq \sum_{M_{w\psi} \subset M} M_{w\psi}.$$

This statement is equivalent to the fact that the number of elements in $JH(M_\psi)$ is greater than the number of elements of $W$.

## § 9. Cohomology of Lie algebras

In this section we shall recall a number of results concerning the cohomology of Lie algebras. Moreover, Bott's theorem will also be proved here.

Let $\mathfrak{a}$ be an arbitrary complex Lie algebra and $M$ an $\mathfrak{a}$-module. The exact sequence of $\mathfrak{a}$-modules

$$0 \longleftarrow M \longleftarrow C_0 \xleftarrow{d_1} C_1 \xleftarrow{d_2} \ldots, \tag{9.1}$$

where each module $C_i$ is free over $U(\mathfrak{a})$ is called a *free resolution* of $M$.

Let $N$ be another $\mathfrak{a}$-module. Consider the complex

$$0 \longrightarrow \mathrm{Hom}_{\mathfrak{a}}(C_0, N) \xrightarrow{d_1'} \mathrm{Hom}_{\mathfrak{a}}(C_1, N) \xrightarrow{d_2'} \ldots$$

and put

$$\mathrm{Ext}^i(M, N) = \mathrm{Ker}\, d_{i+1}' / \mathrm{Im}\, d_i'.$$

Let $\tau: U(\mathfrak{a}) \to U(\mathfrak{a})$ be the anti-automorphism defined by the formula $\tau(X) = -X$ for $X \in \mathfrak{a}$. We denote by $N^\tau$ the right $U(\mathfrak{a})$-module whose underlying space coincides with that of $N$ and on which the action of $U(\mathfrak{a})$ is defined by the formula $f \cdot X = \tau(X) f$ for $f \in N$, $X \in U(\mathfrak{a})$.

Now consider the complex

$$0 \longleftarrow N^\tau \underset{\mathfrak{a}}{\otimes} C_0 \xleftarrow{d_1''} N^\tau \underset{\mathfrak{a}}{\otimes} C_1 \xleftarrow{d_2''} \dots$$

and put

$$\operatorname{Tor}_i(N^\tau, M) = \operatorname{Ker} d_i'' / \operatorname{Im} d_{i+1}''.$$

The following standard facts hold (see [10]).

*1)* The groups $\operatorname{Tor}_i(N^\tau, M)$ and $\operatorname{Ext}^i(M, N)$ are independent of the choice of the resolution (9.1).

*2)* Let $\mathfrak{a}$ be a finite-dimensional Lie algebra and $M$, $N$ be finite-dimensional $\mathfrak{a}$-modules. Then

a) $[\operatorname{Ext}^i(M, N)]^* = \operatorname{Tor}_i(N^*, M)$,

where $N^* = \operatorname{Hom}_{\mathbf{C}}(N, \mathbf{C})$ is a right $\mathfrak{a}$-module ([10], Chapter XI, § 3, Proposition 3.3),

b) $\operatorname{Tor}_i(N^*, M) = \operatorname{Tor}_i(M^\tau, (N^\tau)^*)$.

([10], Chapter VI, § 1).

The cohomology group of $\mathfrak{a}$ with coefficients in $M$ is defined by the formula $H^i(\mathfrak{a}, M) = \operatorname{Ext}^i(\mathbf{C}, M)$, where $\mathbf{C}$ is the trivial one-dimensional $\mathfrak{a}$-module.

The computation of the cohomology groups is done by means of the standard resolution $V(\mathfrak{a})$ of the module $\mathbf{C}$, which is defined in the following way.

We put

$$C_k = U(\mathfrak{a}) \underset{\mathbf{C}}{\otimes} \Lambda^k(\mathfrak{a}), \quad k = 0, 1, \dots$$

Then we define a homomorphism $d_k: C_k \to C_{k-1}$ of $\mathfrak{a}$-modules by means of the formula

$$d_k(X \otimes X_1 \wedge \dots \wedge X_k) = \sum_{i=1}^{k} (-1)^{i+1} (XX_i \otimes X_1 \wedge \dots \wedge \hat{X}_i \wedge \dots \wedge X_k) +$$

$$+ \sum_{1 \le i < j \le k} (-1)^{i+j} (X \otimes [X_i, X_j] \wedge X_1 \wedge \dots \wedge \hat{X}_i \wedge \dots \wedge \hat{X}_j \wedge \dots \wedge X_k).$$

Here $X \in U(\mathfrak{a})$, $X_i \in \mathfrak{a}$ and the symbol ^ means that the corresponding element is to be omitted. Furthermore, we define $\varepsilon: C_0 \to \mathbf{C}$ by the formula $\varepsilon(X) =$ (the constant part of $X$), $X \in U(\mathfrak{a})$.

As was shown in [10], Chapter XIII, § 7, the sequence

$$0 \longleftarrow \mathbf{C} \xleftarrow{\varepsilon} C_0 \xleftarrow{d_1} C_0 \xleftarrow{d_2} \dots$$

is a resolution of the $\mathfrak{a}$-module $\mathbf{C}$.

Subsequently we shall need a generalization of the resolution $V(\mathfrak{a})$ for the case of relative cohomology.

Let $\mathfrak{a}$ be a complex Lie algebra and $\mathfrak{p}$ a subalgebra. The adjoint action of $\mathfrak{p}$ in $\mathfrak{a}$ yields a representation $\vartheta$ of the algebra $\mathfrak{p}$ in the linear space $\mathfrak{a}/\mathfrak{p}$. The corresponding representations of $\mathfrak{p}$ in the linear spaces $\Lambda^k(\mathfrak{a}/\mathfrak{p})$ will be denoted by the same symbol $\vartheta$.

Let us consider for each $k$, $k=0, 1, 2, \ldots$, the module

$$D_k = U(\mathfrak{a}) \underset{U(\mathfrak{p})}{\otimes} \Lambda^k(\mathfrak{a}/\mathfrak{p}).$$

We define the operators $d_k: D_k \to D_{k-1}$ in the following way. Let $X_1, \ldots, X_k$ be elements of $\mathfrak{a}/\mathfrak{p}$. Let $Y_1, Y_2, \ldots, Y_k \in \mathfrak{a}$ be arbitrary representatives for $X_1, \ldots, X_k$, respectively, and put

$$d_k(X \otimes X_1 \wedge \ldots \wedge X_k) = \sum_{i=1}^{k} (-1)^{i+1}(XY_i \otimes X_1 \wedge \ldots \wedge \hat{X}_i \wedge \ldots \wedge X_k) +$$

$$+ \sum_{1 \leq i < j \leq k} (-1)^{i-j}(X \otimes \overline{[Y_i, Y_j]} \wedge X_1 \wedge \ldots \wedge \hat{X}_i \wedge \ldots \wedge \hat{X}_j \wedge \ldots \wedge X_k).$$

Here $X \in U(\mathfrak{a})$, and $\bar{Y}$ is the image of the element $Y \in \mathfrak{a}$ in $\mathfrak{a}/\mathfrak{p}$. It is easy to verify that the operator $d_k$ is well defined, i.e. independent of the choice of the representatives $Y_i$.

In addition to this, we introduce the augmentation $\varepsilon: D_0 \to \mathbf{C}$ by putting $\varepsilon(X \otimes 1) =$ (the constant part of $X$). Thus we have constructed a sequence $V(\mathfrak{a}, \mathfrak{p})$ of $U(\mathfrak{a})$-modules

$$0 \longleftarrow \mathbf{C} \xleftarrow{\varepsilon} D_0 \xleftarrow{d_1} D_1 \longleftarrow \ldots.$$

Direct computation shows that this sequence is a complex, i.e. $d_{i-1}d_i = 0$, $\varepsilon d_0 = 0$. We shall call this complex $V(\mathfrak{a}, \mathfrak{p})$ the *relative chain complex* of the algebra $\mathfrak{a}$ with respect to the subalgebra $\mathfrak{p}$. Clearly, $V(\mathfrak{a}, 0) = V(\mathfrak{a})$.

*Theorem* 9.1. *The complex* $V(\mathfrak{a}, \mathfrak{p})$ *is exact.*

*Proof.* Our proof will be similar to the proof of exactness of the standard complex $V(\mathfrak{a})$, given in [10].

We define a filtration in $V(\mathfrak{a}, \mathfrak{p})$ by writing $A \in D_k^{(l)}$ if $A \in D_k$ can be written in the form

$$A = \sum c_i(X^{(i)} \otimes X_1^{(i)} \wedge \ldots \wedge X_k^{(i)}),$$

where

$$c_i \in \mathbf{C},\ X^{(i)} \in U(\mathfrak{a}),\ X_j^{(i)} \in \mathfrak{a}/\mathfrak{p} \quad \text{and} \quad \deg X \leq l-k.$$

It is clear that $d_k(D_k^{(l)}) \subset D_{k-1}^{(l)}$. Therefore, to prove the theorem, it is enough to show that for every $l$ the complex

$$0 \longleftarrow M^{(l)} \longleftarrow D_0^{(l)}/D_0^{(l-1)} \xleftarrow{d_1^{(l)}} D_1^{(l)}/D_1^{(l-1)} \xleftarrow{d_2^{(l)}} \ldots$$

is exact. Here $M^{(0)} = \mathbf{C}$ and $M^{(l)} = 0$ if $l > 0$. It follows from the Poincaré—Birkhoff—Witt theorem that $D_k^{(l)}/D_k^{(l-1)} = S_{l-k}(\mathfrak{a}/\mathfrak{p}) \otimes \Lambda^k(\mathfrak{a}/\mathfrak{p})$, where $S_{l-k}(\mathfrak{a}/\mathfrak{p})$ denotes the set of all homogeneous elements of degree $l-k$ in the symmetric algebra of the space $\mathfrak{a}/\mathfrak{p}$. The operator

$$d_k^{(l)}: D_k^{(l)}/D_k^{(l-1)} \to D_{k-1}^{(l)}/D_{k-1}^{(l-1)}$$

is given by the formula

$$d_k^{(l)}(X \otimes X_1 \wedge \ldots \wedge X_k) = \sum_{i=1}^{k} (-1)^{i+1} (XX_i \otimes X_1 \wedge \ldots \wedge \hat{X}_i \wedge \ldots \wedge X_k).$$

Therefore, the complex Gr $V(\mathfrak{a}, \mathfrak{p})$:

$$0 \leftarrow \mathbf{C} \leftarrow \bigoplus_l D_0^{(l)}/D_0^{(l-1)} \leftarrow \bigoplus_l D_1^{(l)}/D_1^{(l-1)} \leftarrow \ldots$$

coincides with the Koszul complex (see [10], Chapter VIII, § 4) of the space $\mathfrak{a}/\mathfrak{p}$. This means that the complex Gr $V(\mathfrak{a}/\mathfrak{p})$ and also the complex $V(\mathfrak{a}/\mathfrak{p})$ are exact. Theorem 9.1 is proved.

*Proposition* 9.2. Let $\mathfrak{p}$ and $\mathfrak{q}$ be subalgebras of $\mathfrak{a}$ such that $\mathfrak{a} = \mathfrak{p} \oplus \mathfrak{q}$ (as a linear space). Then $V(\mathfrak{a}, \mathfrak{p}) \approx V(\mathfrak{q})$ as complexes of $U(\mathfrak{q})$-modules.

*Proof.* We define a mapping of complexes $\varphi: V(\mathfrak{q}) \to V(\mathfrak{a}, \mathfrak{p})$ by the formula

$$\varphi_k(X \otimes X_1 \wedge \ldots \wedge X_k) = X \otimes \bar{X}_1 \wedge \ldots \wedge \bar{X}_k,$$

where $X \in U(\mathfrak{q})$, $X_i \in \mathfrak{q}$ and $\bar{X}_i$ is the image of $X_i$ in $\mathfrak{a}/\mathfrak{p}$. The theorem of Poincaré—Birkhoff—Witt implies that $\varphi$ is an isomorphism.

*Remark* 1. Proposition 9.2 owes its significance to the following fact. Assume that $\mathfrak{q}$ is a subalgebra of $\mathfrak{a}$ and that there exists in $\mathfrak{a}$ a subalgebra $\mathfrak{p}$ which is complementary to $\mathfrak{q}$. Then the action of the algebra $\mathfrak{q}$ on $V(\mathfrak{q})$ can be extended to the action of the whole algebra $\mathfrak{a}$. We remark that this extension depends essentially on the choice of $\mathfrak{p}$.

*Remark* 2. Let $A$ be a complex Lie group, $P$ a Lie subgroup of $A$, $\mathfrak{a}$ and $\mathfrak{p}$ the Lie algebras of $A$ and $P$ respectively. Let us consider the de Rham complex $\Omega = \{\Omega^k\}$ of formal analytic differential forms at the point $e$ on the space $A/P$. More precisely, let $z_1, \ldots, z_n$ be a system of coordinates on the complex manifold $A/P$ in a neighbourhood of the point $e$. Then $\Omega^k$ consists of the forms

$$\omega = \sum a_{i_1 i_2 \ldots i_k}(z)\, dz_{i_1} \wedge \ldots \wedge dz_{i_k},$$

where $a_{i_1 i_2 \ldots i_k}(z)$ are formal power series of the variables $z_1, \ldots, z_n$.

The group $A$ acts on the space $A/P$. Of course, it cannot act on the complex $\Omega$. However, the Lie algebra $\mathfrak{a}$ acts on $\Omega$. It is easy to verify that the complex $\Omega$ is dual to the complex $V(\mathfrak{a}, \mathfrak{p})$ constructed above. Therefore the exactness of $V(\mathfrak{a}, \mathfrak{p})$ is also a consequence of the exactness of de Rham's complex $\Omega$.

In what follows we shall be interested in the case when $\mathfrak{a} = \mathfrak{g}$ is a complex semisimple Lie algebra, $\mathfrak{p} = \mathfrak{b} = \mathfrak{h} \oplus \mathfrak{n}_+$ is a Borel subalgebra of $\mathfrak{g}$. We shall study the structure of the members $D_k$ of the complex $V(\mathfrak{g}, \mathfrak{b})$.

*Lemma* 9.3. *Let $V$ be a $\mathfrak{b}$-module and let $V^{\mathfrak{g}}=U(\mathfrak{g})\underset{U(\mathfrak{b})}{\otimes} V$. The mapping $V\to V^{\mathfrak{g}}$ generates an exact functor from the category of $\mathfrak{b}$-modules to the category of $\mathfrak{g}$-modules. If in addition $V$ is a one-dimensional module, $Hv=\chi(H)v$, $E_\gamma v=0$ for $H\in\mathfrak{h}$, $\gamma\in\Delta_+$, $v\in V$ then $V^{\mathfrak{g}}=M_{\chi+\varrho}$.*

The proof follows easily from the fact that $U(\mathfrak{g})$ is a free $U(\mathfrak{b})$-module. The second statement follows from the definition of $M_{\chi+\varrho}$.

Lemma 9.3 enables us to present an easy description of the modules $D_k$ occuring in the complex $V(\mathfrak{g}, \mathfrak{b})$.

*Definition* 9.4. Let $\Psi$ be a finite collection of weights (we allow that some of them coincide). We shall say that the module $M$ is of *type* $\Psi$ if there exists a filtration $0=M^{(0)}\subset M^{(1)}\subset\ldots\subset M^{(l)}=M$ such that $M^{(i)}/M^{(i-1)}=M_{\psi_i}$ and the collection of weights $\{\psi_i\}$ coincides with $\Psi$.

*Lemma* 9.5. *Let $N$ be a finite-dimensional $\mathfrak{h}$-diagonalizable $\mathfrak{b}$-module, $\Psi(N)=\{\varphi+\varrho\}$, where $\varphi$ runs through all weights of $N$ (with multiplicities). Then the module $U(\mathfrak{g})\underset{U(\mathfrak{b})}{\otimes} N$ is of type $\Psi(N)$.*

The proof is an immediate consequence of Lemma 9.3.

*Corollary*. The module $D_k$ in the complex $V(\mathfrak{g}, \mathfrak{b})$ is of type $\Psi(\Lambda^k(\mathfrak{g}/\mathfrak{b}))$.

Since the Lie algebra $\mathfrak{g}$ acts on the complex $V(\mathfrak{g}, \mathfrak{b})$, we can distinguish in it a subcomplex corresponding to the "zero" eigenvalues of the elements of $Z(\mathfrak{g})$. More precisely, let $\vartheta=\vartheta_\varrho\in\Theta$. We consider the subcomplex $V_\vartheta(\mathfrak{g}, \mathfrak{b})$ of $V(\mathfrak{g}, \mathfrak{b})$ consisting of the submodules $(D_k)_\vartheta\subset D_k$ and the module $\mathbf{C}_\vartheta=\mathbf{C}$. It follows from Proposition 8.6, §5 that the complex $V_\vartheta(\mathfrak{g}, \mathfrak{b})$ is exact.

*Proposition* 9.6. Let $\Psi_k=\{w\varrho|w\in W^{(k)}\}$. Then $(D_k)_\vartheta$ is of type $\Psi_k$.

First we prove the following lemma.

*Lemma* 9.7. *Let $M$ be a module of type $\Psi$ and $\vartheta\in\Theta$. Then the module $M_\vartheta$ is of type $\Psi_\vartheta$, where $\Psi_\vartheta$ is the collection of all weights $\psi\in\Psi$ such that $\vartheta_\psi=\vartheta$.*

*Proof.* Let $0=M^{(0)}\subset M^{(1)}\subset\ldots\subset M^{(k)}=M$ be a filtration of $M$ for which $M^{(i)}/M^{(i-1)}=M_{\psi_i}$, $\psi_i\in\Psi$. From the exactness of the functor $M\to M_\vartheta$ it follows that the modules $M^{(i)}_\vartheta$ form a filtration of $M_\vartheta$ and $M^{(i)}_\vartheta/M^{(i-1)}_\vartheta=(M_{\psi_i})_\vartheta$. It follows from Proposition 8.5 that $(M_{\psi_i})_\vartheta=M_{\psi_i}$ if $\vartheta_{\psi_i}=\vartheta$ and $(M_{\psi_i})_\vartheta=0$ if $\vartheta\neq\vartheta_{\psi_i}$. This implies Lemma 9.7.

It follows from Lemma 9.7 that $(D_k)_\vartheta$ is of type $[\Psi(\Lambda^k(\mathfrak{g}/\mathfrak{b}))]_\vartheta$. Now we shall study this set.

Let $\Phi$ be a subset of $\Delta$. Put $|\Phi|=\sum_{\gamma\in\Phi}\gamma$. Since the set of weights of $\mathfrak{g}/\mathfrak{b}$ coincides with $\Delta_-$, the collection of weights of $\Lambda^k(\mathfrak{g}, \mathfrak{b})$ (with multiplicities) coincides with the collection of weights of the form $-|\Phi|$ for all $\Phi\subset\Delta_+$ such that card $\Phi=k$. Therefore,

$$[\Psi(\Lambda^k(\mathfrak{g}/\mathfrak{b}))]_\vartheta=\{\varrho-|\Phi|\,|\,\Phi\subset\Delta_+,\ \text{card}\,\Phi=k,\ (\varrho-|\Phi|)\sim\varrho\}.$$

For any element $w\in W$ we put $\Phi_w=\{\gamma\in\Delta_+|w^{-1}\gamma\subset\Delta_-\}$. Then card $\Phi_w=k$, for $w\in W^{(k)}$ (Lemma 11.1). Thus Proposition 9.6 is a consequence of the following lemma.

*Lemma* 9.8. *Let* $w\in W$, $\Phi\subset\Delta_+$. *Then* $\varrho-w\varrho=|\Phi|$ *if and only if* $\Phi=\Phi_w$.

A proof of this lemma is presented in § 11, see also [6].

Now we can formulate the main theorem of this section.

*Theorem* 9.9. *Let* $V$ *be an irreducible finite-dimensional* $\mathfrak{g}$*-module with highest weight* $\lambda$. *Then there exists an exact sequence of* $U(\mathfrak{g})$*-modules*

$$0\leftarrow V\leftarrow B_0^V\leftarrow B_1^V\leftarrow\ldots\leftarrow B_s^V\leftarrow 0$$

*where* $s=\dim\mathfrak{n}_-$ *and* $B_k$ *is a module of type* $\Psi_k(\lambda)=\{w(\lambda+\varrho)|w\in W^{(k)}\}$.

*Proof.* In Proposition 9.6 the required exact sequence of the $\mathfrak{g}$-modules $B_k^{\mathbf{C}}$ was constructed for the case $V=\mathbf{C}$ (i.e. $\lambda=0$). In the general case we consider the exact sequence $B_*^{\mathbf{C}}\underset{\mathbf{C}}{\otimes} V$ and put

$$B_k^V=(B_k^{\mathbf{C}}\otimes V)_{\theta_{\lambda+\varrho}}.$$

Now we prove that the sequence

$$0\leftarrow V\leftarrow B_0^V\leftarrow B_1^V\leftarrow\ldots$$

satisfies the conditions of the theorem. Its exactness follows from the fact that $M\to(M\otimes V)_\theta$ is an exact functor. Now we show that $B_k^V$ is of type $\Psi_k(\lambda)$.

*Lemma* 9.10. *Let* $\chi\in\mathfrak{h}^*$, $V$ *be a finite-dimensional* $\mathfrak{g}$*-module. Let us denote by* $\Psi$ *the set* $\{\lambda+\chi\}$ *where* $\lambda$ *runs through all weights of* $V$ *with the corresponding multiplicities. Then* $M_\chi\otimes V$ *is of type* $\Psi$.

*Proof.* Let $e_1,\ldots,e_l$ be a basis in $V$ consisting of weight vectors and $\lambda_1,\lambda_2,\ldots,\lambda_l$ be the corresponding weights. We choose an enumeration of the vectors $e_i$ such that $\lambda_i<\lambda_j$ implies $i>j$. Let $a_i=f_\chi\otimes e_i\in M_\chi\otimes V$ and $M^{(k)}=U(\mathfrak{g})(a_1,\ldots,a_k)$. Then $0=M^{(0)}\subset M^{(1)}\subset\ldots\subset M^{(l)}$.

To prove Lemma 9.10 it will suffice to show that

$$M^{(i)}/M^{(i-1)}=M_{\lambda_i+\chi}\quad\text{and}\quad M^{(l)}=M_\chi\otimes V.$$

Let $\bar a_k$ denote the image of $a_k$ in $M^{(k)}/M^{(k-1)}$. It is obvious that $\bar a_k$ is a generator of $M^{(k)}/M^{(k-1)}$ of weight $\chi+\lambda_k-\varrho$ and $E_\gamma\bar a_k=0$ for $\gamma\in\Delta_+$. Therefore $M^{(k)}=U(\mathfrak{n}_-)(a_1,\ldots,a_k)$. We shall show that $M^{(k)}$ is a free $U(\mathfrak{n}_-)$-module with generators $a_1,\ldots,a_k$. Let $X_i\in U(\mathfrak{n}_-)$, $1\leqq i\leqq k$ and let $p$ be the largest among the degrees of $X_i$ (with respect to the natural filtration in $U(\mathfrak{n}_-)$). Then

$$\sum_{i=1}^k X_ia_i=\sum_{i=1}^k X_if_\chi\otimes e_i+\sum_{j=1}^l Y_jf_\chi\otimes e_j\neq 0,$$

since the degrees of the elements $Y_j$ are less than $p$ and $M_\chi$ is a free $U(\mathfrak{n}_-)$-module.

This means that $M^{(k)}/M^{(k-1)}$ is a free $U(\mathfrak{n}_-)$-module with the single generator $\bar{a}_k$, i.e. $M^{(k)}/M^{(k-1)} \approx M_{\chi+\lambda_k}$.

Similar considerations will show that $M^{(l)} = M_\chi \otimes V$. This completes the proof of Lemma 9.10.

It follows from this lemma that $B_k^{\mathbf{C}} \otimes V$ is of type $\Psi$, where $\Psi = \{\lambda_i + w\varrho \,|\, \lambda_i$ are the weights of $V$ (with multiplicities), $w \in W^{(k)}\}$. Therefore, Theorem 9.9 follows from Lemma 9.7 and the following lemma.

*Lemma* 9.11. *Let $V$ be a finite-dimensional irreducible* $\mathfrak{g}$*-module with highest weight $\lambda$. Then for every $w \in W$ there exists exactly one weight $\mu \in P(V)$ such that $\mu + w\varrho \sim \lambda + \varrho$. Moreover the weight $\mu$ has multiplicity one in $V$.*

*Proof.* Let $\mu \in P(V)$, $w_1, w_2 \in W$ be such that $w_1(\mu + w\varrho) = \lambda + \varrho$; since $w_1\mu$ is a weight of $V$, we have $w_1\mu \leqq \lambda$.

In addition, $w_1 w\varrho \leqq \varrho$. Therefore $w_1\mu = \lambda$ and $w_1 w\varrho = \varrho$. This implies $w_1 = w^{-1}$ and thus $\mu = w\lambda$. The last statement of the lemma follows from the fact that the multiplicity of the weight $\mu$ in $V$ is equal to the multiplicity of the highest weight $\lambda$.

*Corollary* (Bott's theorem [5], [6]). *Let $V$ be a finite-dimensional irreducible* $\mathfrak{g}$*-module. Then*

$$\dim H^i(\mathfrak{n}_-, V) = \operatorname{card} W^{(i)}.$$

*Proof.* We know that

$$H^i(\mathfrak{n}_-, V) = \operatorname{Ext}^i_{\mathfrak{n}_-}(\mathbf{C}, V) = \operatorname{Tor}_i^{\mathfrak{n}_-}(V^*, \mathbf{C})^* = \operatorname{Tor}_i^{\mathfrak{n}_-}(\mathbf{C}, (V^\tau)^*)^*.$$

Let us construct the resolution $\{B_i^{V_1}\}$ for the module $V_1 = (V^\tau)^*$. Then $\operatorname{Tor}_i^{\mathfrak{n}_-}(\mathbf{C}, V_1)$ will be the homologies of the complex

$$0 \longleftarrow \bar{B}_0^{V_1} \xleftarrow{\bar{d}_1} \dots \longleftarrow \bar{B}_s^{V_1} \longleftarrow 0.$$

The algebra $\mathfrak{h}$ acts on this complex in a natural way. In view of Theorem 9.9 we have here that $\bar{B}_i^{V_1} = B_i^{V_1}/\mathfrak{n}_- B_i^{V_1}$ is a finite-dimensional space whose weights with respect to $\mathfrak{h}$ are equal to $w(\lambda+\varrho)$, $w \in W^{(i)}$, and each of these weights occurs with multiplicity one. Therefore $\dim \bar{B}_i^{V_1} = \operatorname{card} W^{(i)}$, and every $\bar{d}_i$ is the null-mapping. The corollary is proved.

## § 10. Construction of the resolution of a finite-dimensional $\mathfrak{g}$-module

The present section is devoted to the proof of Theorem 10.1, which yields a sharpening of Theorem 9.9.

*Theorem* 10.1. *Let $V$ be an irreducible finite-dimensional* $\mathfrak{g}$*-module with highest weight $\lambda$. Then there exists an exact sequence of* $\mathfrak{g}$*-modules*

$$0 \longleftarrow V \xleftarrow{\varepsilon} C_0^V \xleftarrow{d_1} C_1 \longleftarrow \dots \longleftarrow C_s^V \longleftarrow 0$$

where

$$s = \dim \mathfrak{n}_-, \quad C_k = \bigoplus_{w \in W^{(k)}} M_{w(\mu+\varrho)}.$$

First we shall present an explicit construction of the mappings $d_i$ and $\varepsilon$. Let us put $\chi=\lambda+\varrho$. Then $\chi\in D$ and by Theorem 8.8′ every submodule of the module $M_{w\chi}$, $w\in W$ can be considered as a submodule of $M_\chi$. By Theorem 8.7, any mapping $M_{w_1\chi}\to M_{w_2\chi}$ is a multiple of the canonical imbedding for $w_1<w_2$, hence it can be determined by a complex number $c_{w_1w_2}$. Therefore, any mapping $C_i^V\to C_{i-1}^V$ can be represented by a complex matrix $(c_{w_1w_2})$, $w_1\in W^{(i)}$ and $w_2\in W^{(i-1)}$. Thus in order to construct the mappings $d_i$ it will suffice to define the corresponding matrices $(d^{(i)}_{w_1w_2})$.

*Definition* 10.2. Let us call a quadruple $(w_1, w_2, w_3, w_4)$ of elements of $W$ a *square* if

$$w_1\to w_2\to w_4 \quad\text{and}\quad w_1\to w_3\to w_4$$

(see Definition 8.9).

It will be convenient to consider the finite directed graph corresponding to $W$, whose vertices are the members of $W$ and in which an arc leads from $w_1$ to $w_2$ if $w_1\to w_2$.

*Lemma* 10.3.* *Let $w_1, w_2\in W$ and $l(w_1)-2=l(w_2)$. Then the number of elements $w'\in W$ such that $w_1\to w'\to w_2$ is equal to either zero or two.*

*Lemma* 10.4. *To each arrow $w_1\to w_2$ we can assign a number $s(w_1, w_2)=\pm 1$ in such a way that for every square $(w_1, w_2, w_3, w_4)$ the product of the numbers assigned to the four arrows occurring in it is equal to $-1$.*

The proofs of Lemmas 10.3 and 10.4 will be given in § 11.

Now we can improve on Theorem 10.1 in the following way.

*Theorem* 10.1′. *With the notation of Theorem* 10.1. *we define the mapping $d_i: C_i\to C_{i-1}$ by means of the matrix $(d^{(i)}_{w_1w_2})$, $w_1\in W^{(i)}$, $w_2\in W^{(i-1)}$, where $d^{(i)}_{w_1w_2}=s(w_1, w_2)$ if $w_1\to w_2$ and $d^{(i)}_{w_1w_2}=0$ otherwise. Let us denote by $\varepsilon: C_0\to V$ the natural surjection. Then the sequence*

$$0\longleftarrow V\xleftarrow{\ \varepsilon\ } C_0^V\xleftarrow{\ d_1\ } C_1^V\longleftarrow \dots\xleftarrow{\ d_s\ } C_s^V\longleftarrow 0 \tag{10.1}$$

*is exact.*

*Proof.* It follows immediately from Lemmas 10.3 and 10.4 that $d_i\circ d_{i+1}=0$ for $i=1, \dots, s-1$.

We remark that $W^{(1)}=\{\sigma_\alpha, \alpha\in\Sigma\}$. Therefore, Harish-Chandra's theorem on ideals [4] implies the exactness of the sequence (10.1) at its members $V$ and $C_0$.

Assume now that we have already shown the exactness of the sequence at the members $C_0, \dots, C_{i-1}$. We shall prove that it is also exact at the member $C_i$, i.e. that $\operatorname{Ker} d_i=\operatorname{Im} d_{i+1}$. Let us put $K=\operatorname{Ker} d_i$. The desired equality $d_{i+1}(C_{i+1})=K$ is obviously a consequence of the following three lemmas.

* Lemma 10.3 follows easily from certain unpublished results of D.-N. Verma concerning the Möbius function on the Weyl group.

(We recall that $C_{i+1} = \bigoplus_{w \in W^{(i+1)}} M_{w\chi}$ is a free $U(\mathfrak{n}_-)$-module.)

*Lemma* 10.5. *Let $C$ be a free $U(\mathfrak{n}_-)$-module with generators $f_1, \ldots, f_n$ and $\zeta: C \to K$ be a homomorphism of $U(\mathfrak{n}_-)$-modules such that $\zeta(f_i)$ is a weight vector in $K$ (with respect to $\mathfrak{h}$). Then $\zeta$ is a surjection if and only if the induced mapping $\bar{\zeta}: C/\mathfrak{n}_- C \to K/\mathfrak{n}_- K$ is surjective.*

*Lemma* 10.6. *The mapping*

$$\bar{d}_{i+1}: C_{i+1}/\mathfrak{n}_- C_{i+1} \to K/\mathfrak{n}_- K$$

*is an injection.*

*Lemma* 10.7.

$$\dim_{\mathbf{C}} C_{i+1}/\mathfrak{n}_- C_{i+1} = \dim_{\mathbf{C}} K/\mathfrak{n}_- K < \infty.$$

*Proof of Lemma* 10.5. Clearly if $\zeta$ is surjective then so is $\bar{\zeta}$. Conversely, assume that $\bar{\zeta}$ is surjective but $\zeta$ is not. Consider the weight vector $f$ in $K$ of weight $\psi$ having the following properties:

a) $f \notin \operatorname{Im} \zeta$,

b) any vector $f'$ of weight $\psi' > \psi$ belongs to $\operatorname{Im} \zeta$.

There always exists such a vector $f$ since $K \in O$, and therefore for any weight $\psi \in \mathfrak{h}^*$ there are finitely many weights $\psi' > \psi$ such that $K^{(\psi')} \neq \{0\}$.

Let $\bar{f}$ be the image of $f$ in $K/\mathfrak{n}_- K$. Then $\bar{f} = \sum c_i \bar{\zeta}(f_i)$. Since $\mathfrak{n}_- K$ is invariant under $\mathfrak{h}$, $\mathfrak{h}$ acts on $K/\mathfrak{n}_- K$. Therefore we can assume that $c_i \neq 0$ only for those indices $i$ for which the weight of $\zeta(f_i)$ is equal to $\psi$. Moreover, $g = f - \sum c_i \zeta(f_i)$ is a weight vector lying in $\mathfrak{n}_- K$, and thus $g = \sum_{\gamma \in \Delta_+} E_{-\gamma} g_\gamma$, where the weight of $g_\gamma$ is $\psi + \gamma > \psi$. According to the construction of $f$, $g_\gamma \in \operatorname{Im} \zeta$, and therefore $f \in \operatorname{Im} \zeta$. Lemma 10.5 is proved.

*Proof of Lemma* 10.6. The quotient space $C_{i+1}/\mathfrak{n}_- C_{i+1}$ is a linear space over $\mathbf{C}$ for which $\{\bar{f}_{w\chi} | w \in W^{(i+1)}\}$ forms a basis.* Since the homomorphism $\bar{d}_{i+1}$ commutes with the action of $\mathfrak{h}$ and all of the vectors $\bar{f}_{w\chi}$ have different weights it will suffice to prove that $\bar{d}_{i+1}(\bar{f}_{w\chi}) \neq 0$ for any $w \in W^{(i+1)}$.

The proof of this proposition is divided in a natural way into two steps.

*Lemma* 10.6a. *The irreducible modules occurring in the Jordan—Hölder decomposition of the module $K$ are of the form $L_{w\chi}$, $l(w) > i$.*

*Proof.* For the proof of Lemma 10.6a we shall make use of the exact sequence

$$0 \leftarrow V \leftarrow B_0^V \leftarrow B_1^V \leftarrow \ldots \leftarrow B_s^V \leftarrow 0 \tag{10.2}$$

constructed in § 9. It is clear that for all $j$

$$JH(B_j^V) = \bigcup_{w \in W^{(j)}} JH(M_{w\chi}) = JH(C_j^V).$$

* Let us recall, that $f_{w\chi}$ is the generator of $M_{w\chi}$ (see Def. 8.2).

Since both of the sequences (10.1) and (10.2) are exact at their members with numbers less than $i$, we have $JH(K)=JH(K_B)$, where $K_B$ denotes the kernel of the mapping $B_i^V \to B_{i-1}^V$. As the sequence (10.2) is exact at its member $B_i^V$, $K_B$ is equal to the image of $B_{i+1}^V$. Therefore

$$JH(K_B) \subset JH(B_{i+1}^V) = \bigcup_{w \in W^{(i+1)}} JH(M_{w\chi}),$$

and thus Lemma 10.6a is implied by the Corollary of Theorem 8.12.

*Lemma* 10.6b. *Let $w_0 \in W$ and the module $M \in O$ be given. We assume that $l(w) \geqq l(w_0)$ for any $L_{w\chi}$ occurring in $JH(M)$. Let $\tau: M_{w_0\chi} \to M$ be a homomorphism such that $\tau(f_{w_0\chi}) \neq 0$. Then the image of $\tau(f_{w_0\chi})$ in $M/\mathfrak{n}_- M$ is not* 0.

*Proof.* We shall use induction on the number of elements in $JH(M)$. Let $f \in M$ be an element of maximal weight $\psi$ and $N \subset M$ be the submodule generated by $f$. Then the module $N$ is isomorphic to a factor module of the module $M_\psi$. We shall distinguish two cases.

*Case* 1. $\tau(f_{w_0\chi}) \in N$. In this case

$$L_{w_0\chi} \subset JH(N) \subset JH(M_\psi).$$

Therefore (as it follows from Theorem 8.12) $\psi = w_1\chi$, where $w_1 \geqq w_0$. On the other hand

$$L_\psi \subset JH(N) \subset JH(M).$$

Thus, according to the condition of the lemma $w_1 = w_0$, i.e. $\psi = w_0\chi$. Since $\psi$ is the maximal weight of $M$,

$$\tau(f_{w_0\chi}) \notin \mathfrak{n}_- M.$$

*Case* 2. $\tau(f_{w_0\chi}) \notin N$. In this case the statement of the lemma for $M$ can be reduced to the similar statement for the module $M/N$. Since $JH(M/N) \subsetneqq JH(M)$, we can apply the induction hypothesis. Lemma 10.6b is proved.

To complete the proof of Lemma 10.6 it suffices to apply Lemma 10.6b to the module $M=K$.

*Proof of Lemma* 10.7. The module $K$ has only a finite number of generators (as a $U(\mathfrak{n}_-)$-module). Therefore $\bar{K} = K/\mathfrak{n}_- K$ is a finite-dimensional space over $\mathbf{C}$. Let us choose weight vectors $f_1, \ldots, f_n \in K$ whose images in $\bar{K}$ form a basis in $\bar{K}$. Let us consider the free $U(\mathfrak{n}_-)$-module $C$ with $n$ generators $g_1, \ldots, g_n$ and define a homomorphism of $U(\mathfrak{n}_-)$-modules $\vartheta: C \to K$ by the formula $\vartheta(g_i) = f_i$. By virtue of Lemma 10.5 $\vartheta$ is surjective.

Let us consider the exact sequence

$$0 \leftarrow V \leftarrow C_0^V \xleftarrow{d_1} \ldots \leftarrow C_i^V \xleftarrow{\vartheta} C$$

of $U(\mathfrak{n}_-)$-modules.

Since $C_j^V$ and $C$ are free $U(\mathfrak{n}_-)$-modules, this sequence can be augmented to a free resolution

$$0 \longleftarrow V \longleftarrow C_0^V \longleftarrow \ldots \longleftarrow C_i^V \xleftarrow{\vartheta} C \xleftarrow{\eta} D \longleftarrow \ldots$$

of the $U(\mathfrak{n}_-)$-module $V$.

Now, consider the sequence

$$\bar{D} \xrightarrow{\bar{\eta}} \bar{C} \xrightarrow{\bar{\vartheta}} \bar{C}_i^V \xrightarrow{\bar{d}_i} \bar{C}_{i-1}^V$$

where, for any $U(\mathfrak{n}_-)$-module $M$, $\bar{M}$ denotes $1 \underset{U(\mathfrak{n}_-)}{\otimes} M = M/\mathfrak{n}_- M$. By definition,

$$\mathrm{Tor}_i^{\mathfrak{n}_-}(\mathbf{C}, V) = \mathrm{Ker}\,\bar{\vartheta}/\mathrm{Im}\,\bar{\eta}.$$

We shall show that $\bar{\vartheta}$ and $\bar{\eta}$ are equal to 0.

From the exact sequence

$$D \xrightarrow{\eta} C \xrightarrow{\tau} K \longrightarrow 0$$

we obtain the exact sequence

$$\bar{D} \xrightarrow{\bar{\eta}} \bar{C} \xrightarrow{\bar{\tau}} \bar{K} \longrightarrow 0.$$

But $\bar{\tau}$ is an isomorphism, hence $\bar{\eta}=0$.

Furthermore, we have the exact sequence

$$C \xrightarrow{\vartheta} C_i^V \xrightarrow{d_i} K_{i-1} \longrightarrow 0$$

and thus the sequence

$$\bar{C} \xrightarrow{\bar{\vartheta}} \bar{C}_i^V \xrightarrow{\bar{d}_i} \bar{K}_{i-1} \longrightarrow 0,$$

where $K_{i-1} = \mathrm{Ker}\, d_{i-1} = \mathrm{Im}\, d_i$. Applying Lemma 10.6 to the mapping $d_i: C_i \to K_{i-1}$ we see that $\bar{d}_i$ is an isomorphism, hence $\bar{\vartheta}=0$.

Thus we have

$$\dim \mathrm{Tor}_i^{\mathfrak{n}_-}(\mathbf{C}, V) = \dim \bar{C} = \dim (K/\mathfrak{n}_- K).$$

On the other hand, the Corollary to Theorem 9.9 shows that

$$\dim \mathrm{Tor}_i^{\mathfrak{n}^-}(\mathbf{C}, V) = \mathrm{card}\, W^{(i)} = \dim (C_i/\mathfrak{n}_- C_i).$$

Lemma 10.7 is proved.

This also completes the proof of Theorem 10.1′ and Theorem 10.1.

## § 11. Proof of the lemmas

First we shall present several results clarifying the properties of the function $l(w)$ in more detail (see [11]).

Let us put $\Phi_w = \Delta_+ \cap w(\Delta_-)$.

*Lemma* 11.1.

*1)* $l(w) = \text{card}\, \Phi_w$.

*2) If* $\gamma_1, \gamma_2 \in \Phi_w$ *and* $\gamma_1 + \gamma_2 = n\gamma$ *for* $n \in \mathbf{Z}_+$, *then* $\gamma \in \Phi_w$.

*Proof.* Statement *1)* is proved in [11]. To verify statement *2)* it is enough to prove it separately for $\Delta_+$ and $w\Delta_-$, and it is obvious in both cases.

*Proof of Lemma* 8.10. Plainly,

$$w\chi - \sigma_\gamma w\chi = \frac{2\langle w\chi, \gamma\rangle}{\langle \gamma, \gamma\rangle}\gamma.$$

Since $\chi \in \mathfrak{h}^*_{\mathbf{Z}}$ and the element $\chi$ is regular,

$$n = \frac{2\langle w\chi, \gamma\rangle}{\langle \gamma, \gamma\rangle} \in \mathbf{Z} \setminus 0.$$

We assume that $n > 0$ and prove then that $l(\sigma_\gamma w) > l(w)$. By *1)* of Lemma 11.1, it suffices to prove that $\text{card}\, \Phi_w < \text{card}\, \Phi_{\sigma_\gamma w}$.

a) $\gamma \notin \Phi_w$. Indeed, $\langle \chi, w^{-1}\gamma\rangle = \langle w\chi, \gamma\rangle > 0$, and therefore $w^{-1}\gamma \notin \Delta_-$.

b) $\gamma \in \Phi_{\sigma_\gamma w}$. Indeed,

$$(\sigma_\gamma w)^{-1}\gamma = w^{-1}(-\gamma) = -w^{-1}\gamma \in \Delta_-.$$

c) Let $\delta \in \Phi_w$ and $\sigma_\gamma \delta \in \Delta_+$. Then $\sigma_\gamma \delta \in \Phi_{\sigma_\gamma w}$. In fact,

$$(\sigma_\gamma w)^{-1}(\sigma_\gamma \delta) = w^{-1}\sigma_\gamma^{-1}\sigma_\gamma \delta = w^{-1}\delta \in \Delta_-.$$

d) Let $\delta \in \Phi_w$, $\sigma_\gamma \delta \in \Delta_-$. We shall show that $\delta \in \Phi_{\sigma_\gamma w}$. Assume, on the contrary that $w^{-1}\sigma_\gamma \delta \in \Delta_+$. Then $w^{-1}(-\sigma_\gamma \delta) \in \Delta_-$, and thus $-\sigma_\gamma \delta \in \Phi_w$. But $\delta + (-\sigma_\gamma \delta) = \frac{2\langle \gamma, \delta\rangle}{\langle \delta, \delta\rangle}\gamma$, while $\frac{2\langle \gamma, \delta\rangle}{\langle \delta, \delta\rangle} > 0$. By *2)* of Lemma 11.1 this implies $\gamma \in \Phi_w$, and that contradicts a).

Propositions a)—d) imply that if $n > 0$ then $l(w) < l(\sigma_\gamma w)$.

By interchanging $w$ and $\sigma_\gamma w$ we obtain that if $n < 0$ then $l(w) > l(\sigma_\gamma w)$. Lemma 8.10 is proved.

*Lemma* 11.2.

*1) Let* $w \in W$, $\alpha \in \Sigma$. *Then* $\alpha \in \Phi_w$ *implies* $w \xrightarrow{\alpha} \sigma_\alpha w$, *and* $\alpha \notin \Phi_w$ *implies* $\sigma_\alpha w \xrightarrow{\alpha} w$.

*2) If* $\sigma_\alpha w \xrightarrow{\alpha} w$ *for all* $\alpha \in \Sigma$, *then* $w = 1$.

*3) There exists a unique element* $s \in W$ *such that* $s \to \sigma_\alpha s$ *for all* $\alpha \in \Sigma$.

This lemma follows immediately from the theorems proved in [11].

*Lemma* 11.3. *Let* $w_1, w_2 \in W$, $\gamma \in \Delta_+$ *and* $\alpha \in \Sigma$, $\alpha \neq \gamma$.

*We put*

$$\sigma_\alpha w_1 \xrightarrow{\alpha} w_1 \quad \text{and} \quad \sigma_\alpha w_1 \xrightarrow{\gamma} w_2. \tag{11.1}$$

*Then*

$$w_2 \xrightarrow{\alpha} \sigma_\alpha w_2 \quad \text{and} \quad w_1 \xrightarrow{\sigma_\alpha \gamma} \sigma_\alpha w_2. \tag{11.2}$$

*Conversely,* (11.2) *implies* (11.1)

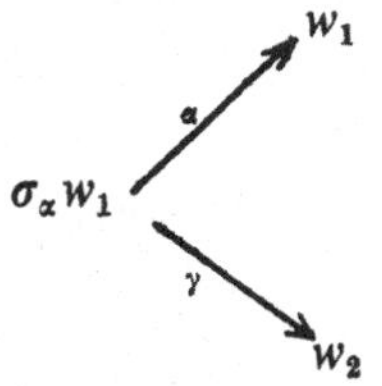


Diagram 1

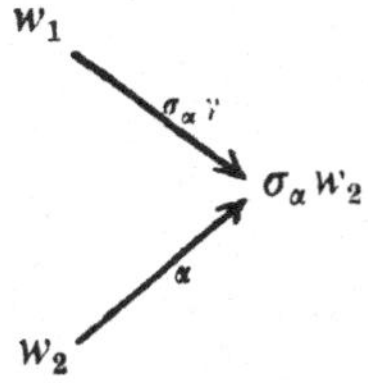


Diagram 2

*Proof.* We have to prove that Diagram 1 is equivalent to Diagram 2. Let $\gamma'=\sigma_\alpha\gamma$. Then $\gamma'\in\Delta_+$ and $\sigma_\alpha w_2=\sigma_\gamma w_1$. Since $l(w_1)=l(w_2)$, formula (11.2) is equivalent to $l(\sigma_\alpha w_2)=l(w_2)-1$. Let $\chi\in D$. Then, by Lemma 8.10,

$$w_2\chi-\sigma_\alpha w_1\chi=n\gamma,\quad n>0.$$

Applying $\sigma_\alpha$ to this equality we obtain

$$\sigma_\alpha w_2\chi-w_1\chi=n\gamma'.$$

Using Lemma 8.10 again we obtain that $l(\sigma_\alpha w_2)<l(w_1)$. Therefore, (11.1) implies (11.2).

We can prove similarly that (11.2) implies (11.1).

*Proof of Lemma* 8.11. The proof will be performed by induct on on $l(w)$. If $l(\sigma_\gamma w)=l(w)-1$, then by definition $\sigma_\gamma w>w$. Since $l(w)$ and $l(\sigma_\gamma w)$ are of different parities, it remains to consider the case in which $l(\sigma_\gamma w)\leqq l(w)-3$.

Let $\alpha\in\Sigma$ be a root such that $w\to\sigma_\alpha w$. Obviously, $\alpha\neq\gamma$ and $\sigma_\alpha\sigma_\gamma w=\sigma_{\gamma'}\sigma_\alpha w$, where $\gamma'=\sigma_\alpha\gamma\in\Delta_+$.

Here we have

$$l(\sigma_{\gamma'}\sigma_\alpha w)=l(\sigma_\alpha\sigma_\gamma w)\leqq l(\sigma_\gamma w)+1\leqq l(w)-2<l(\sigma_\alpha w).$$

By the induction hypothesis there exists a chain

$$\sigma_\alpha w\to w_1\to w_2\to\ldots\to w_2\to\sigma_\alpha\sigma_\gamma w$$

and thus also a chain

$$w_{-1}\xrightarrow{\gamma_0}w_0\xrightarrow{\gamma_1}w_1\longrightarrow\ldots\longrightarrow w_k\xrightarrow{\gamma_k+1}\sigma_\alpha\sigma_\gamma w \tag{11 3}$$

(where $w_{-1}=w$, $w_0=\sigma_\alpha w$ and $\gamma_0=\alpha$). There are two possibilities.

*Case* 1. $\sigma_\alpha\sigma_\gamma w\to\sigma_\gamma w$. Then there exists a chain

$$w=w_{-1}\to w_0\to w_1\to\ldots\to w_k\to\sigma_\alpha\sigma_\gamma w\to\sigma_\gamma w,$$

hence $w<\sigma_\gamma w$.

*Case* 2. $\sigma_\gamma w \to \sigma_\alpha \sigma_\gamma w$. Let $i$ be the largest number such that in the chain (11.3) $\gamma_i = \alpha$. Applying Lemma 11.3 several times we obtain that

$$\sigma_\alpha w_k \xrightarrow{\sigma_\alpha \gamma_{k+1}} \sigma_\gamma w,$$

$$\sigma_\alpha w_{k-1} \xrightarrow{\sigma_\alpha \gamma_k} \sigma_\alpha w_k, \ldots, \sigma_\alpha w_i \xrightarrow{\sigma_\alpha \gamma_{i+1}} \sigma_\alpha w_{i+1}.$$

But $\sigma_\alpha w_i = w_{i-1}$. Hence we obtain a chain

$$w = w_1 \xrightarrow{\gamma_0} w_0 \xrightarrow{\gamma_1} \ldots \longrightarrow w_{i-1} \xrightarrow{\sigma_\alpha \gamma_{i+1}} \sigma_\alpha w_{i+1} \longrightarrow \ldots \xrightarrow{\sigma_\alpha \gamma_{k+1}} \sigma_\gamma w.$$

Therefore, $w < \sigma_\gamma w$, and Lemma 8.11 is proved.

*Proof of Lemma* 9.8. We want to prove that if $w \in W$, $\Phi \subset \Delta_+$ and $\varrho - w\varrho = |\Phi|$, then $\Phi = \Phi_w = \Delta_+ \cap w\Delta_-$. If $w = e$, this is obvious. We shall perform the proof by induction on $l(w)$. Let $l(w) = i > 0$. We choose an $\alpha \in \Sigma$ such that $l(\sigma_\alpha w) = i - 1$, i.e. $\alpha \in \Phi_w$. Then

$$|\sigma_\alpha \Phi| = \sigma_\alpha \varrho - \sigma_\alpha w\varrho = \varrho - \sigma_\alpha w \varrho - \alpha.$$

$\Phi \subset \Delta_+$ implies $\alpha \notin \sigma_\alpha \Phi$. Therefore, putting $\sigma_\alpha w = w'$, we have

$$\varrho - w'\varrho = \varrho - \sigma_\alpha w\varrho = |\sigma_\alpha \Phi \cup \{\alpha\}|.$$

We shall show that $\alpha \in \Phi$. Indeed, assume that this is not true. Then $\sigma_\alpha \Phi \cup \{\alpha\} \subset \Delta_+$ and by the induction hypothesis $\Phi_{w'} = \sigma_\alpha \Phi \cup \{\alpha\}$, i.e. $\alpha \in \Phi_{w'}$. According to Lemma 11.2. 1 $\sigma_\alpha w \to w$, which contradicts the choice of $\alpha$.

Thus $\alpha \in \Phi$. Let us put $\Phi = \Phi - \{\alpha\}$. Then $\varrho - \sigma_\alpha w\varrho = |\sigma_\alpha \Phi'|$ and $\sigma_\alpha \Phi' \subset \Delta_+$. By the induction hypothesis $\Phi_{w'} = \sigma_\alpha \Phi'$, i.e. $\Phi = \sigma_\alpha \Phi_{w'} \cup \{\alpha\}$. It is easy to verify that $\sigma_\alpha \Phi_{w'} \cup \{\alpha\} = \Phi_w$. Lemma 9.8 is proved.

*Proof of Lemma* 10.3. We shall prove this by induction on $l(w_1)$. Since $l(w_1) \geqq 2$, we can choose an $\alpha \in \Sigma$ such that $w_1 \to \sigma_\alpha w_1$. There are two possibilities.

*Case* 1. $w_2 \to \sigma_\alpha w_2$. Let us assign to every chain

$$w_1 \xrightarrow{\gamma_1} w \xrightarrow{\gamma_2} w_2 \tag{11.4}$$

a chain

$$\sigma_\alpha w_1 \xrightarrow{\delta_1} w' \xrightarrow{\delta_2} \sigma_\alpha w_2 \tag{11.5}$$

in the following way.

a) If $\gamma_1 \neq \alpha$, then put $\delta_1 = \sigma_\alpha \gamma_1$, $\delta_2 = \sigma_\alpha \gamma_2$, $w' = \sigma_\alpha w$. It follows from Lemma 11.3 that

$$\sigma_\alpha w_1 \xrightarrow{\delta_1} w' \xrightarrow{\delta_2} \sigma_\alpha w_2.$$

b) If $\gamma_1 = \alpha$, then put $\delta_1 = \gamma_2$, $\delta_2 = \alpha$, $w' = w_2$. Using Lemma 11.3 again, it is easy to see that we have constructed a one-to-one correspondence between the chains of form (11.4) and the chains of form (11.5). As $l(\sigma_\alpha w_1) < l(w_1)$, we can apply the induction hypothesis.

*Case* 2. $\sigma_\alpha w_2 \to w_2$. If $w_1 \xrightarrow{\gamma_1} w \xrightarrow{\gamma_2} w_2$, then by Lemma 11.3, we have either $\gamma_1=\alpha$ or $\gamma_2=\alpha$. The same lemma shows us that $w_1 \to \sigma_\alpha w_2$ is equivalent to $\sigma_\alpha w_1 \to w_2$. Therefore, either there is no chain of form (11.5) or there exist exactly two chains $w_1 \to \sigma_\alpha w_1 \to w_2$ and $w_1 \to \sigma_\alpha w_2 \to w_2$. Lemma 10.3 is proved.

*Proof of Lemma* 10.4. For any element $w \in W$ we put $I(w)=\{w' \in W \mid w' \geqq w\}$. We shall prove by induction on $l(w)$ that $s(w_1, w_2)$ can be defined for $w_1, w_2 \in I(w)$ in such a way that the condition of the lemma be satisfied for any square $(w_1, w_2, w_3, w_4)$ with $w_1 \in I(w)$. This will imply the statement of Lemma 10.4 since $I(s)=W$ (the element $s \in W$ is defined in Lemma 11.2. 3).

Thus assume $w \in W$. We choose an $\alpha \in \Sigma$ such that $w \to \sigma_\alpha w$. By the induction hypothesis, the function $s(w_1, w_2)$ has been defined for $w_1, w_2 \in I(\sigma_\alpha w)$.

*1)* Let

$$w = w_0 \xrightarrow{\gamma_1} w_1 \longrightarrow \dots \xrightarrow{\gamma_k} w_k,$$

where $\gamma_i \neq \alpha$ for all $i$. Then by Lemma 11.3 $\sigma_\alpha w_0 \to \sigma_\alpha w_1 \to \dots \to \sigma_\alpha w_k$ and $w_1 \to \sigma_\alpha w_i$ for all $i$. In particular, $\sigma_\alpha w_i \in I(\sigma_\alpha w)$ for all $i$.

*2)* Let $w' \in I(w) \setminus I(\sigma_\alpha w)$ and let

$$w = w_0 \xrightarrow{\gamma_1} w_1 \longrightarrow \dots \xrightarrow{\gamma_k} w_k = w'$$

be an arbitrary chain leading from $w$ to $w'$. Then $\gamma_i \neq \alpha$ for all $i$, because otherwise $w' \in I(\sigma_\alpha w)$ would hold in view of *1)*. In particular, we have $w' \to \sigma_\alpha w'$ and $\sigma_\alpha w' \in I(\sigma_\alpha w)$.

*3)* Let us now define the function $s(w_1, w_2)$ for all arrows $w_1 \xrightarrow{\gamma} w_2$, $w_1 \in I(w)$. Here we assume that $s(w_1, w_2)$ has already been defined if $w_1 \in I(\sigma_\alpha w)$.

Let $w_1 \in I(w) \setminus I(\sigma_\alpha w)$. If $w_2 = \sigma_\alpha w_1$, then we put $s(w_1, w_2)=1$. Let $w_2 \neq \sigma_\alpha w_1$. Then Lemma 11.3.2 implies

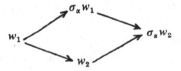

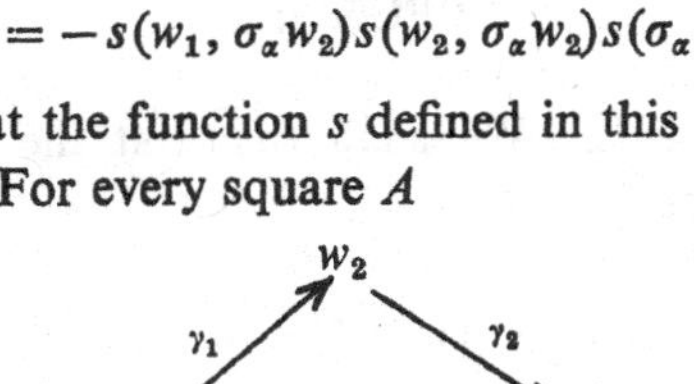


where $\sigma_\alpha w_1, \sigma_\alpha w_2 \in I(\sigma_\alpha w)$. In this case we put

$$s(w_1, w_2) = -s(w_1, \sigma_\alpha w_2) s(w_2, \sigma_\alpha w_2) s(\sigma_\alpha w_1, \sigma_\alpha w_2). \tag{11.6}$$

*4)* Now we prove that the function $s$ defined in this manner satisfies the conditions of Lemma 10.4. For every square $A$

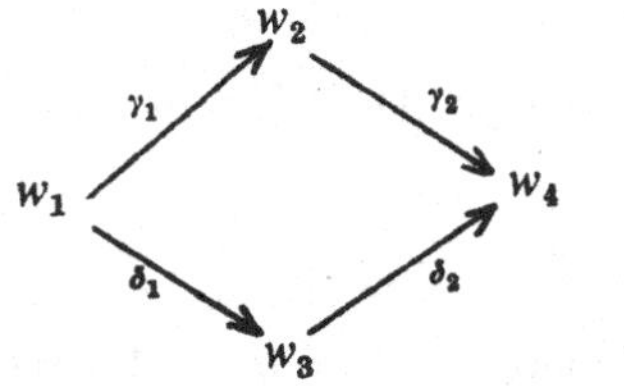

we put

$$p(A) = s(w_1, w_3)s(w_1, w_4)s(w_2, w_4)s(w_3, w_4).$$

We have to prove that $p(A)=-1$. There are the following possibilities.

a) $w_1 \in I(\sigma_\alpha w)$. In this case $p(A)=-1$ by the induction hypothesis.

b) $w_1 \notin I(\sigma_\alpha w)$ and $\gamma_1=\delta=\alpha$. Then $p(A)=-1$ in view of formula (11.6).

c) $w_1 \notin I(\sigma_\alpha w)$ and $\gamma_1 \neq \alpha$, $\delta_1 \neq \alpha$. Since $w_2 \neq w_3$, we have $\gamma_2 \neq \delta_2$, and therefore one of the elements $\gamma_2, \delta_2$ must be different from $\alpha$. Then by Lemma 11.3 $w_4 \to \sigma_\alpha w_4$, and thus $\gamma_2 \neq \alpha$ and $\delta_2 \neq \alpha$. According to Lemma 11.3 we have five squares

$$A_1 = (w_1, w_2, \sigma_\alpha w_1, \sigma_\alpha w_2),$$

$$A_2 = (w_1, w_3, \sigma_\alpha w_1, \sigma_\alpha w_3),$$

$$A_3 = (w_2, w_4, \sigma_\alpha w_2, \sigma_\alpha w_4),$$

$$A_4 = (w_3, w_4, \sigma_\alpha w_3, \sigma_\alpha w_4),$$

$$A_5 = (\sigma_\alpha w_1, \sigma_\alpha w_2, \sigma_\alpha w_3, \sigma_\alpha w_4).$$

Here $p(A_i)=-1$, $i=1, \dots, 5$, in view of a) and b). It is easy to verify that $p(A) \cdot \prod_{i=1}^{5} p(A_i)=1$, hence $p(A)=-1$, too.

d) $\gamma_1 = \alpha$, $\delta_2 \neq \alpha$.

By Lemma 11.3, $w_3 \to \sigma_\alpha w_3$, $w_4 \to \sigma_\alpha w_4$ and we have three squares

$$A_1 = (w_1, w_2, w_3, \sigma_\alpha w_3),$$

$$A_2 = (w_3, w_4, \sigma_\alpha w_3, \sigma_\alpha w_4),$$

$$A_3 = (w_2, w_4, \sigma_\alpha w_3, \sigma_\alpha w_4).$$

In view of a) and b) we have $p(A_i)=-1$, $i=1, 2, 3$. Since $p(A)p(A_2)=p(A_1)p(A_3)$, we also have $p(A)=-1$.

Thus we have considered all the possibilities for the roots $\gamma_1, \gamma_2, \delta_1, \delta_2$ (up to interchanging $\gamma_i$ with $\delta_i$ and $w_2$ with $w_3$). The proof of Lemma 10.4 is complete.

## APPENDIX

Let $\chi, \psi \in \mathfrak{h}^*$, $\gamma_1, \dots, \gamma_k \in \Delta_+$. We shall say that the sequence $\gamma_2, \dots, \gamma_k$ satisfies condition $(A)$ for the pair $(\chi, \psi)$ if

*1)* $\chi=\sigma_{\gamma_k} \dots \sigma_{\gamma_1} \psi$,

*2)* Let $\chi_0=\psi$ and $\chi_i=\sigma_{\gamma_i} \dots \sigma_{\gamma_1} \psi$. Then for all $i$ we have $\chi_{i-1}-\chi_i=n_i \gamma_i$, where

$$n_i = \frac{2\langle \chi_{i-1}, \gamma_i \rangle}{\langle \gamma_i, \gamma_i \rangle} \in \mathbf{Z}_+.$$

In the Appendix we are going to prove the following theorem.

*Theorem* A1. *Let* $L_\chi \in JH(M_\psi)$. *Then there exists a sequence* $\gamma_1, \dots, \gamma_k$ *which satisfies condition* (A) *for the pair* $(\chi, \psi)$.

In the proof of Theorem A1 we shall use the following lemma.

*Lemma* A2. *Let* $\chi \in \mathfrak{h}^*$. *Then*

$$JH(M_\chi) = \bigcup_{\substack{\varphi \sim \chi \\ \varphi \leqq \chi}} c_\varphi L_\varphi, \quad c_\varphi \in \mathbf{Z}_+,$$

*where* $c_\chi = 1$.

This lemma follows easily from Propositions 8.5 and 8.6.

*Proof of Theorem* A1. Let $\lambda \in \mathfrak{h}^*_{\mathbf{Z}}$. Let us denote by $F_\lambda$ an irreducible finite-dimensional $\mathfrak{g}$-module for which $\lambda$ is an extremal weight. We recall that if $\mu \in P(F_\lambda)$, then $\|\mu\| \leqq \|\lambda\|$ and $\|\mu\| = \|\lambda\|$ implies $\mu \sim \lambda$.

*Definition* A3. Let $\chi \in \mathfrak{h}^*$, $\lambda \in \mathfrak{h}^*_{\mathbf{Z}}$. The pair $(\chi, \lambda)$ is called *admissible* if there are no $w \in W$ and $\mu \in P(F_\lambda)$ such that

a) $w\chi < \chi$,

b) $w(\chi + \mu) \geqq \chi + \lambda$,

c) $w(\chi + \mu) \sim \chi + \lambda$.

The meaning of Definition A3 is illuminated by the following proposition, whose proof we postpone until later.

*Proposition* A4. *Suppose* $\chi, \psi \in \mathfrak{h}^*$, $\lambda \in \mathfrak{h}^*_{\mathbf{Z}}$ *and the pair* $(\chi, \lambda)$ *is admissible. Furthermore, let* $L_\chi \in JH(M_\psi)$. *Then there exists a weight* $\mu \in P(F_\lambda)$ *such that* $L_{\chi+\lambda} \in JH(M_{\psi+\mu})$.

Proposition A4 includes all the information about the modules $M_\chi$ and $L_\chi$ that we need to prove Theorem A1. The subsequent arguments will only concern the geometric structure of the space $\mathfrak{h}^*$.

*Lemma* A5. *1) There exists a constant* $c_1 > 0$ *with the following property: if* $\varphi > \psi$, $\varphi + \vartheta_1 < \psi + \vartheta_2$ $(\varphi, \psi, \vartheta_1, \vartheta_2 \in \mathfrak{h}^*)$, *then* $c_1(\|\vartheta_1\| + \|\vartheta_2\|) > \|\varphi - \psi\|$.

*2) There exists a constant* $c_2$ *with the following property: let* $\varphi \in \mathfrak{h}^*_{\mathbf{Z}}$; *then one can find a sequence of weights* $0 = \varphi_0, \varphi_1, \dots, \varphi_k = \varphi$, $\varphi_i \in \mathfrak{h}^*_{\mathbf{Z}}$, *such that*

$$\|\varphi_i - \varphi_{i-1}\| < c_2 \quad \text{and} \quad d(\varphi_i, [0, \varphi]) < c_2,$$

*where* $[0, \varphi]$ *is the segment in* $\mathfrak{h}^*$ *connecting* 0 *with* $\varphi$, *and* $d(\ ,\ )$ *denotes the distance in* $\mathfrak{h}^*_{\mathbf{R}}$.

*3) Let* $\varphi, \psi \in \bar{C}^+$, $w \in W$. *Then* $\langle w\varphi, \psi \rangle \leqq \langle \varphi, \psi \rangle$. *If in addition* $\varphi, \psi \in C^+$, *then the equality is satisfied for* $w = e$ *only.*

*4) If* $C$ *is an arbitrary Weyl chamber,* $\varphi, \psi \in C$ *and* $\varphi \sim \psi$, *then* $\varphi = \psi$.

*5) Let* $\sum n_i \gamma_i = c\gamma$, *where* $\gamma_i, \gamma \in \Delta$, $n_i \in \mathbf{Z}$. *Then* $c \in \mathbf{Z}$ *as well.*

*Proof.* *1)* It is well-known that $\langle\varrho, \alpha\rangle > 0$ for all $\alpha \in \Sigma$. Therefore there exists a $c > 0$ such that $\langle\varrho, \alpha\rangle > c\|\alpha\|$ for all $\alpha \in \Sigma$. This implies that for any $\varkappa \in K$ we have $\langle\varrho, \varkappa\rangle \geqq c\|\varkappa\|$. Moreover, $\varphi - \psi > 0$ and $\vartheta_1 - \vartheta_2 > \varphi - \psi$. Consequently,

$$\|\varrho\|(\|\vartheta_1\| + \|\vartheta_2\|) \geqq \langle\varrho, \vartheta_1 - \vartheta_2\rangle > \langle\varrho, \varphi - \psi\rangle \geqq c\|\varphi - \psi\|.$$

Therefore we can put $c_1 = \|\varrho\| c^{-1}$.

*2)* It suffices to put $c_2 = 2d$, where $d$ is the diameter of an arbitrary fundamental domain of the group $\mathfrak{h}_{\mathbf{Z}}^*$ in $\mathfrak{h}_{\mathbf{R}}^*$.

*3)* Let us write $w$ in the form $w = \sigma_{\alpha_k} \dots \sigma_{\alpha_1}$ with $k = l(w)$ and put $\varphi_0 = \varphi$, $\varphi_i = w_i \varphi$, where $w_i = \sigma_{\alpha_i} \dots \sigma_{\alpha_1}$. Then $\varphi_i - \varphi_{i+1} = c_i \alpha_i$ where

$$c_i = \frac{2\langle\varphi_i, \alpha_i\rangle}{\langle\alpha_i, \alpha_i\rangle} = \frac{2\langle\varphi, w_i^{-1}\alpha_i\rangle}{\langle\alpha_i, \alpha_i\rangle}.$$

It follows from Lemma 11.2.1 that $c_i \geqq 0$. It is also clear that $c_i \neq 0$ if $\varphi \in C$. Therefore $\langle\varphi, \psi\rangle - \langle w\varphi, \psi\rangle = \sum c_i \langle\alpha_i, \psi\rangle$. This completes the proof.

*4)* It can be shown that $C = C^*$. Now let $\psi = w\varphi$. According to 3, $\langle\varphi, \psi\rangle \geqq \geqq \langle\psi, \psi\rangle$. But $\|\varphi\| = \|\psi\|$, hence $\varphi = \psi$.

*5)* We can assume that $\gamma \in \Sigma$. In this case the lemma follows from the fact that the elements of $\Sigma$ form a basis in which every root has integer coordinates.

The proof of Lemma A5 is complete.

Let $\chi \in \mathfrak{h}^*$. We shall denote by $Y(\chi)$ the following proposition:

$Y(\chi)$. *For every* $\psi \in \mathfrak{h}^*$ *such that* $L_\chi \in JH(M_\psi)$ *there exists a sequence* $\gamma_1, \dots, \gamma_k \in \Delta_+$ *which satisfies condition* (A) *for the pair* $(\chi, \psi)$.

Theorem A1 says that $Y(\chi)$ is valid for all $\chi \in \mathfrak{h}^*$. We shall prove it in several steps. For any $\gamma \in \Delta_+$ we shall denote $\Xi_\gamma$ the hyperplane in $\mathfrak{h}_{\mathbf{R}}^*$ orthogonal to $\gamma$.

*Definition* A6. Let $c_3 = 3c_1c_2$. The element $\varphi \in \mathfrak{h}^*$ is called *strongly regular* if $d(\operatorname{Re}\varphi, \Xi_\gamma) > c_3$ for all $\gamma \in \Delta_+$.

*Step* 1. Let $\chi, \varphi \in \mathfrak{h}^*$ be strongly regular while $\chi - \varphi \in \mathfrak{h}_{\mathbf{Z}}^*$ and $\operatorname{Re}\chi$, $\operatorname{Re}\varphi$ belong to the same Weyl chamber. Then propositions $Y(\chi)$ and $Y(\varphi)$ are equivalent.

*Step* 2. Let $\varphi, \chi \in \mathfrak{h}^*$ be strongly regular elements with $\chi - \varphi \in \mathfrak{h}_{\mathbf{Z}}^*$. In addition to this, let $\operatorname{Re}\chi \in C$, $\operatorname{Re}\varphi \in \sigma_\gamma C$, where $C$ and $\sigma_\gamma C$ $(\gamma \in \Delta_+)$ are neighbouring Weyl chambers and $\langle\operatorname{Re}\chi, \gamma\rangle < 0$. Then $Y(\varphi)$ implies $Y(\chi)$.

It is easy to see that Steps 1 and 2 enable us to reduce the proof of $Y(\chi)$ for a strongly regular $\chi \in \mathfrak{h}^*$ to the case $\operatorname{Re}\chi \in C^+$. Now we are going to prove $Y(\chi)$ in this case. Let $\psi \in \mathfrak{h}^*$ and $L_\chi \in JH(M_\psi)$. It follows from Lemma A2 that $\chi \sim \psi$ and $\chi \leqq \psi$. This means that $\psi = \chi + \varkappa$, where $\varkappa \in K$. Then $\langle\varphi, \psi\rangle = \langle\chi, \chi\rangle + \langle\varkappa, \varkappa\rangle + 2\langle\chi, \varkappa\rangle$. Now

$$\langle\psi, \psi\rangle = \langle\chi, \chi\rangle, \quad \operatorname{Re}\langle\chi, \varkappa\rangle = \langle\operatorname{Re}\chi, \varkappa\rangle \geqq 0,$$

and $\langle\varkappa, \varkappa\rangle \geqq 0$. Consequently $\langle\varkappa, \varkappa\rangle = 0$, hence $\chi = \psi$.

Therefore Steps 1 and 2 prove $Y(\chi)$ for all the strongly regular weights $\chi \in \mathfrak{h}^*$. The general case is reduced to this one by means of the following Step 3.

*Step* 3. Let $\chi \in \mathfrak{h}^*$. Then there exists a strongly regular element $\varphi \in \mathfrak{h}^*$ such that $Y(\varphi)$ implies $Y(\chi)$.

Thus the proof of Theorem A1 reduces to the proofs of Steps 1, 2, 3 and Proposition A4.

In the proof of Steps 1, 2, 3 it will be convenient for us to make use of the following lemma.

*Lemma* A7. *Let* $C_1$ *and* $C_2$ *be two Weyl chambers,* $\gamma, \psi \in \mathfrak{h}^*$ *such that* $\operatorname{Re}\gamma \in \bar{C}_1$, $\operatorname{Re}\psi \in \bar{C}_2$, $\chi \sim \psi$ *and* $\lambda \in \mathfrak{h}^*$, $\mu \in \mathfrak{h}^*_{\mathbf{Z}}$ *be chosen so that* $\operatorname{Re}\chi+\lambda \in C_1$, $\operatorname{Re}\psi+\mu \in C_2$. *Suppose that the sequence of roots* $\gamma_1, \ldots, \gamma_k \in \Delta_+$ *satisfies condition* (A) *for the pair* $(\chi+\lambda, \psi+\mu)$. *Then it satisfies condition* (A) *for the pair* $(\chi, \psi)$ *as well.*

*Proof.* Item *2)* of condition (A) follows immediately from the fact that the sign of $\langle\chi, \gamma\rangle$ is constant in a Weyl chamber. In order to verify item *1)* we put $\chi_k = \sigma_{\gamma_k} \ldots \sigma_{\gamma_1}\psi$. Then $\chi \sim \psi \sim \chi_k$ and $\operatorname{Re}\chi$, $\operatorname{Re}\chi_k \in C_1$. By Lemma A5. 4, $\operatorname{Re}\chi = \operatorname{Re}\chi_k$. Moreover, $\chi - \chi_k \in \mathfrak{h}^*_{\mathbf{Z}}$ and consequently $\operatorname{Im}\chi = \operatorname{Im}\chi_k$, i.e. $\chi = \chi_k$.

*Proof of Step* 1. Lemma A5. 2 enables us to reduce the proof of Step 1 to the proof of the following proposition. Let $\chi \in \mathfrak{h}^*$, $\lambda \in \mathfrak{h}^*_{\mathbf{Z}}$ be such that $d(\operatorname{Re}\chi, \Xi) > 2c_1c_2$, where $\Xi = \bigcup_{\gamma \in \Delta_+} \Xi_\gamma$ and $\|\lambda\| < c_2$. Then $Y(\chi+\lambda)$ implies $Y(\chi)$.

First we show that such a pair $(\chi, \lambda)$ is admissible. Indeed, assume that there exist $w \in W$ and $\mu \in P(F_\gamma)$ for which $w\chi < \chi$ and $w(\chi+\mu) > \chi+\lambda$. Then by Lemma A5. 1,

$$2c_2 > 2\|\lambda\| \geqq \|\lambda\| + \|\mu\| > c_1^{-1}\|\chi - w\chi\|.$$

On the other hand

$$\|\chi - w\chi\| \geqq d(\operatorname{Re}\chi, \Xi) > 2c_1c_2.$$

Therefore, the pair $(\chi, \lambda)$ is indeed admissible.

Now let $L_\chi \in JH(M_\psi)$. Then according to Proposition A4, $L_{\chi+\lambda} \in JH(M_{\psi+\mu})$ for a certain $\mu \in P(F_\lambda)$. It follows from $Y(\chi+\lambda)$ that there exists a sequence of roots $\gamma_1, \ldots, \gamma_k \in \Delta_+$ which satisfies condition (A) for the pair $(\chi+\lambda, \psi+\mu)$. Since $\|\mu\| < c_2$, $\psi$ and $\psi+\mu$ lie in the same Weyl chamber. Lemma A7 implies then that the same sequence $\gamma_1, \ldots, \gamma_k$ satisfies condition (A) for the pair $(\chi, \psi)$ as well.

*Proof of Step* 2. Applying Step 1, we can reduce the proof to the case when

$$3c_1\|\varphi - \chi\| < d(\operatorname{Re}\varphi, \Xi_\delta) \tag{A.1}$$

for all

$$\delta \in \Delta_+ \setminus \gamma.$$

We put $\lambda = \varphi - \chi$ and prove that the pair $(\chi, \lambda)$ is admissible. Indeed, suppose we have $w \in W$ and $\mu \in P(F_\lambda)$ such that

$$w\chi < \chi, \quad w(\chi+\mu) > \chi+\lambda.$$

$w\chi < \chi$ implies $w \neq \sigma_\gamma$. Therefore,

$$\|w\chi - \chi\| > d(\operatorname{Re}\varphi, \bigcup_{\delta \in \gamma} \Xi_\delta) > 2c_1\|\lambda\|.$$

But then, by Lemma A5. 1,

$$2\|\lambda\| \geqq \|\lambda\| + \|\mu\| \geqq c_1^{-1}\|w\chi - \chi\| > 2\|\lambda\|.$$

Thus the pair $(\chi, \lambda)$ is indeed admissible.

Now assume $L_\chi \in JH(M_\psi)$, $\psi \in \mathfrak{h}^*$. Then $L_{\chi+\lambda} \in JH(M_{\psi+\mu})$ for some $\mu \in P(F_\lambda)$. $Y(\chi+\lambda)$ implies the existence of a sequence $\gamma_1, \ldots, \gamma_k \in \Delta_+$ satisfying condition (A) for the pair $(\chi+\lambda, \psi+\mu)$. To complete the proof of Step 2 we have to construct a sequence of roots which satisfies condition (A) for the pair $(\chi, \psi)$.

Let $w \in W$ be such that $\psi = w\chi$. Then $w^{-1}(\psi+\mu) = \chi + w^{-1}\mu$. Since $\|w^{-1}\mu\| \leqq \|\lambda\|$, we have either $\operatorname{Re} w^{-1}(\psi+\mu) \in C$ or $\operatorname{Re} w^{-1}(\psi+\mu) \in \sigma_\gamma C$. Let us consider these two cases separately.

*1)* $\operatorname{Re} w^{-1}(\psi+\mu) \in C$. We have

$$w^{-1}(\psi+\mu) \sim \chi+\lambda \sim \sigma_\gamma(\chi+\lambda)$$

and

$$\operatorname{Re}\sigma_\gamma(\chi+\lambda) \in C.$$

By Lemma A5. 4,

$$\sigma_\gamma(\chi+\lambda) = w^{-1}(\psi+\mu) = \chi + w^{-1}\mu.$$

Consequently

$$\sigma_\gamma\chi - \chi = w^{-1}\mu - \sigma_\gamma\lambda.$$

Since $w^{-1}\mu$ and $\sigma_\gamma\lambda$ are weights of $F_\lambda$, we have

$$\sigma_\gamma\chi - \chi = c\gamma = \sum_{\gamma_i \in \Delta} n_i\gamma_i,$$

where $n_i \in \mathbf{Z}$. By Lemma A5. 5 $c \in \mathbf{Z}$, and in addition

$$c = -\frac{2\langle \chi, \gamma\rangle}{\langle \gamma, \gamma\rangle} > 0.$$

By Lemma A7, the sequence $\gamma_1, \ldots, \gamma_k$ satisfies condition (A) for the pair $(\sigma_\gamma\chi, \psi)$, hence the sequence $\gamma_1, \ldots, \gamma_k, \gamma$ satisfies condition (A) for the pair $(\chi, \psi)$.

*2)* $\operatorname{Re} w^{-1}(\psi+\mu) \in \sigma_\gamma C$. Let us put

$$\chi_i = \sigma_{\gamma_i} \ldots \sigma_{\gamma_1}\psi,$$

$$\bar{\chi}_i = \sigma_{\gamma_i} \ldots \sigma_{\gamma_1}(\psi+\mu).$$

We remark that all $\chi_i$ and $\bar{\chi}_i$ are congruent modulo $\mathfrak{h}_{\mathbf{Z}}^*$. Thus for each $i$ we have either $\chi_i < \chi_{i-1}$ or $\chi_i > \chi_{i-1}$. If $\chi_i < \chi_{i-1}$ for all $i$, then the sequence $\gamma_1, \ldots, \gamma_k$ satisfies condition (A) for the pair $(\chi, \psi)$. In the opposite case we denote by $i_0$ the smallest index such that $\chi_{i_0} \geqq \chi_{i_0-1}$. We shall show that the sequence of roots

$$\gamma_1, \ldots, \gamma_{i_0-1}, \gamma_{i_0+1}, \ldots, \gamma_k, \gamma$$

satisfies condition (A) for the pair $(\chi, \psi)$.

First of all $\operatorname{Re}\bar{\chi}_{i_0}$ and $\operatorname{Re}\chi_{i_0-1}$ lie in the same Weyl chamber. Indeed,

$$\|\chi_{i_0-1}-\bar{\chi}_{i_0-1}\| < \|\lambda\|,$$
$$\|\chi_{i_0}-\bar{\chi}_{i_0}\| \leqq \|\lambda\|,$$

and by Lemma A2. 1,

$$\|\bar{\chi}_{i_0-1}-\bar{\chi}_{i_0}\| < 2c_1\|\lambda\|.$$

In addition to this, the real parts of the weights

$$\bar{\chi}_{i_0-1} \quad\text{and}\quad \bar{\chi}_{i_0} = \sigma_{\gamma_{i_0}}\bar{\chi}_{i_0-1}$$

as well as of the weights

$$\chi_{i_0-1} = \sigma_{\gamma_{i_0-1}}\dots\sigma_{\gamma_1}\psi \quad\text{and}\quad \bar{\chi}_{i_0-1} = \sigma_{\gamma_{i_0-1}}\dots\sigma_{\gamma_1}(\psi+\mu)$$

belong to different Weyl chambers. However, on account of condition (A1) the ball of radius $2c_1\|\lambda\|$ about the point $\operatorname{Re}\bar{\chi}_{i_0-1}$ intersects exactly two Wey chambers

$$\sigma_{\gamma_{i_0}}\dots\sigma_{\gamma_k}C \quad\text{and}\quad \sigma_{\gamma_{i_0}}\dots\sigma_{\gamma_k}\sigma_\gamma C.$$

Therefore, $\operatorname{Re}\bar{\chi}_{i_0}$ and $\operatorname{Re}\chi_{i_0-1}$ lie in the same Weyl chamber.

By Lemma A7 the sequence $\gamma_1, \dots, \gamma_{i_0-1}$ satisfies condition (A) for the pair $(\chi_{i_0-1}, \psi)$, and the sequence $\gamma_{i_0+1}, \dots, \gamma_k$ satisfies condition (A) for the pair $(\sigma_\gamma\chi, \chi_{i_0-1})$. In exactly the same way as in case 1) we can show that $\chi<\sigma_\gamma\chi$. Consequently the sequence of roots

$$\gamma_1, \gamma_2, \dots, \gamma_{i_0-1}, \gamma_{i_0+1}, \dots, \gamma_k, \gamma$$

satisfies condition (A) for the pair $(\chi, \psi)$.

*Proof of Step* 3. Let us first consider the case $\operatorname{Re}\chi=0$. Then the assumption $L_\chi\in JH(M_\psi)$ and Lemma A2 imply $\psi=\chi$.

Assume now that $\operatorname{Re}\chi\neq 0$. We can choose a weight $\lambda\in\mathfrak{h}_{\mathbf{Z}}^*$ in such a way that the following conditions are satisfied.

a) $\operatorname{Re}\chi$ and $\lambda$ belong to the closure $\bar{C}$ of the same Weyl chamber $C$.

b) The weight $\chi+\lambda$ is strongly regular.

c) Let $c_4=\min\{\|\varkappa\| \mid \varkappa\in K\setminus\{0\}\}$. Then

$$\|\lambda-n\operatorname{Re}\chi\| \leqq nc_4/2c_1 \tag{A.2}$$

for some $n\in\mathbf{Z}_+$.

*Lemma* A8. *Let* $\nu\in P(F_\lambda)$ *such that* $\chi+\nu\sim\chi+\lambda$. *Then and* $\nu$ Re $\chi$ *lie in the same Weyl chamber and*

$$\|\nu-n\operatorname{Re}\chi\| \leqq nc_4/2c_1. \tag{A.3}$$

*Proof.* Let $w_1, w_2\in W$ be chosen in such a way that $w_1C=C^+$ and $w_1(\chi+\lambda)=$ $=w_2(\chi+\nu)$. Then $w_1\lambda$ is the highest weight of the representation $F_\lambda$ and therefore $w_1\lambda-w_2\nu\in K$. Thus

$$\langle\varrho, w_1\lambda\rangle \geqq \langle\varrho, w_2\nu\rangle,$$

where the equality holds only if $w_2\nu=w_1\lambda$. On the other hand, by Lemma A2. 3,

$$\mathrm{Re}\langle\varrho, w_2\chi\rangle \leqq \mathrm{Re}\langle\varrho, w_1\chi\rangle.$$

Comparing these inequalities with the equality $w_1(\chi+\lambda)=w_2(\chi+\nu)$, we obtain $w_1\lambda=w_2\nu$ and $w_1\chi=w_2\chi$. Now applying the element $w_2^{-1}w_1$ to the inequality (A2) we obtain the statement of the lemma.

Now we prove that the pair $(\chi, \lambda)$ is admissible. Indeed, let $w\in W$ and $\mu\in P(F_\lambda)$ such that

$$w\chi<\chi,\ w(\chi+\mu)\sim\chi+\lambda,\ w(\chi+\mu)>\chi+\lambda.$$

Since $w\chi<\chi$, we have $\|w\chi-\chi\|\geqq c_4$. This implies

$$\|w(n+1)\chi-(n+1)\chi\| \geqq (n+1)c_4$$

and

$$w((n+1)\chi) < (n+1)\chi.$$

Applying Lemma A5. 1 we obtain that

$$c_1(\|\lambda-n\chi\|+\|\mu-n\chi\|) \geqq (n+1)c_4,$$

with contradicts (A2) and (A3). Consequently, the pair $(\chi, \lambda)$ is admissible.

Now let $L_\chi\in JH(M_\psi)$. Then in view of Proposition A4, $L_{\chi+\lambda}\in JH(M_{\psi+\mu})$ for some $\mu\in P(F_\lambda)$. Applying Lemma A8 again we find that $\psi$ and $\psi+\mu$ lie in the same Weyl chamber. Let $\gamma_1, \dots, \gamma_k\in\Delta_+$ be a sequence of roots which satisfies condition (A) for the pair $(\chi+\lambda, \psi+\mu)$. Then by Lemma A7 it also satisfies condition (A) for the pair $(\chi, \psi)$.

*Proof of Proposition* A4. Let $\lambda=\mu_1, \mu_2, \dots, \mu_l$ be the weights of the module $F_\lambda$ (with the corresponding multiplicities). Then, as follows from Lemma 9.10,

$$JH(M_\chi\otimes F_\lambda) = \bigcup_{i=1}^{l} JH(M_{\chi+\mu_i}).$$

*Lemma* A9. *Let $M_1, M_2\in O$, while $JH(M_1)\subset JH(M_2)$ and $F$ be a finite-dimensional* g*-module. Then*

$$JH(M_1\otimes F)\subset JH(M_2\otimes F).$$

The proof follows easily from the fact that tensor product by $F$ is an exact functor.

Now let the pair $(\chi, \lambda)$ be admissible and $L_\chi\subset JH(M_\psi)$. Then

$$JH(L_\chi\otimes F_\lambda)\subset JH(M_\psi\otimes F_\lambda) = \bigcup_{i=1}^{l} JH(M_{\psi+\mu_i}).$$

Therefore, in order to show $L_{\chi+\lambda}\subset JH(M_{\psi+\mu})$, for some weight $\mu\in P(F_\lambda)$ it suffices to prove that

$$L_{\chi+\lambda}\subset JH(L_\chi\otimes F_\lambda).$$

Let $M$ be a maximal proper submodule in $M_\chi$ such that $L_\chi=M_\chi/M$. Then

$$L_\chi\otimes F_\lambda = M_\chi\otimes F_\lambda/M\otimes F_\lambda.$$

Since $L_{\chi+\lambda}\in JH(M_\chi\otimes F_\lambda)$, it suffices to prove that $L_{\chi+\lambda}\notin JH(M\otimes F_\lambda)$.

By Lemma A2,

$$JH(M) \subset \bigcup_{\substack{\varphi \sim \chi \\ \varphi < \chi}} JH(M_\varphi),$$

consequently

$$JH(M \otimes F_\lambda) \subset \bigcup_{\substack{\varphi \sim \chi \\ \varphi < \chi}} c_\varphi JH(M_\varphi \otimes F_\lambda).$$

Thus it suffices to show that for any $\varphi \in \mathfrak{h}^*$ with $\varphi \sim \chi$, $\varphi < \chi$ and any $\mu \in P(F_\lambda)$, $L_{\chi+\lambda} \notin JH(M_{\varphi+\mu})$.

Assume, on the contrary, that $L_{\chi+\lambda} \in JH(M_{\varphi+\mu})$. Let $\varphi = w\chi$, $w \in W$. Then by lemma A2,

$$\chi + \lambda \sim \varphi + \mu = w(\varphi + w^{-1}\mu)$$

and

$$\chi + \lambda < w(\varphi + w^{-1}\mu).$$

But $w^{-1}\mu \in P(F_\lambda)$ which contradicts our assumption on admissibility.

This completes the proof of Proposition A4 and thus of Theorem A1.

*Theorem* A10. *Let* $\chi, \psi \in \mathfrak{h}^*$ *and* $\gamma_1, \ldots, \gamma_k \in \Delta_+$ *be a sequence of roots satisfying condition* (A) *for the pair* $(\chi, \psi)$. *Then* $\mathrm{Hom}\,(M_\chi, M_\varphi) = \mathbf{C}$.

A proof of Theorem A10 is contained in [8] and [9].

Theorems 8.8 and 8.10 follow immediately from Theorems A1 and A10.

## REFERENCES


[1] Grothendieck, A., Éléments de géométrie algébrique IV (Troisième partie), *Publ. Math. I.H.E.S.* **28** (1966), 1—248.

[2] Gelfand, I. M. and Kirillov, A. A. (Гельфанд, И. М., Кириллов, А. А.), Структура тела Ли, связанного с полупростой расщепимой алгеброй Ли (The structure of the Lie division ring associated with semisimple split Lie algebras), *Функц. анализ*, **3**:1 (1969), 7—26.

[3] Bargmann, V., On a Hilbert space of analytic functions and an associated integral transform, Part I, *Comm. Pure Appl. Math.*, **14** (1961), 187—214.

[4] Harish-Chandra, On some application of the universal enveloping algebra of a semisimple Lie algebra, *Trans. Amer. Math. Soc.*, **70** (1951), 28—96.

[5] Bott, R., Homogeneous vector bundles, *Ann. of Math.*, **66** (1957), 203—248.

[6] Kostant, B. Lie algebra cohomology and the generalized Borel—Weil Theorem, *Ann. of Math.*, **74** (1961), 329—387.

[7] Verma, D.-N., Structure of certain induced representations of complex semisimple Lie algebras, *Bull. Amer. Math. Soc.*, **74** (1968), 160—166.

[8] Verma, D.-N., *Structure of certain induced representations of complex semisimple Lie algebras, Dissertation*, Yale Univ. 1966.

[9] Bernstein, I. N., Gelfand, I. M. and Gelfand, S. I. (Бернштейн И. Н., Гельфанд И. М., Гельфанд С. И.), Структура представлений, порожденных векторами старшего веса, (The structure of representations generated by a vector of highest weight), *Функц. анализ*, **5**:1 (1971) 1—9.

[10] Cartan, H. and Eilenberg, S., *Homological Algebra*, Princeton, 1956. (Картан А., Эйленберг С., *Гомологическая алгебра*, И. Л., Москва, 1960.)

[11] Bourbaki, N., *Groupes et algèbres de Lie, ch. 4—6*, Hermann, Paris 1968.

[12] Gelfand, I. M., The cohomology of infinite-dimensional Lie algebras; some questions of integral geometry, in *Actes, Congrès intern. math., 1970*, Vol. 1, pp. 95—111, Hermann, Paris 1971.

# 2.

## (with I. N. Bernstein and S. I. Gelfand)

# Structure of representations generated by vectors of heighest weight

Funkts. Anal. Prilozh. **5** (1) (1971) 1-9 [Funct. Anal. Appl. **5** (1971) 1-8].
MR **45**:298. Zbl. **246**:17008

## §1. Introduction

Let $\mathfrak{g}$ be a complex semisimple Lie algebra. Despite the fact that finite representations of the algebra $\mathfrak{g}$ were one of the very first objects of study, there exist even simpler representations. Here a host of properties of finite representations are essentially consequences of analogous properties of these simpler modules. Categories of these modules, such as the so-called O-modules, were introduced by the authors, and their definition was given in [1]. In this work there will be studied elementary objects of that character, the modules $M_\chi$ (see [1]). This work can be read independently of [1]. Study of the $M_\chi$ module was begun in Verma's work* which obtained a series of deep results on $M_\chi$ modules.

The $M_\chi$ modules (for a precise definition, see below) are of interest since they are the simplest moduli generated by a single vector of highest weight $\chi-\rho$, where $\rho$ as usual denotes a half-sum of positive roots. All other modules generated by a single vector of highest weight, including also irreducible finite representations, are factor modules of the modules $M_\chi$. In the present work there is a complete description of the categories of the modules $M_\chi$. This is to say, on the basis of the Verma Theorem, Hom($M_{\chi_1}$, $M_{\chi_2}$) is either 0 or C. The question arises, for such $\chi_1$, $\chi_2$ pairs, whether there exists a nontrivial mapping of $M_{\chi_1}$ to $M_{\chi_2}$. The main result of our article is the establishment of necessary conditions for the existence of such a mapping (Theorem 2). The proof of Theorem 2 is fairly involved; the authors have been unable to come upon an easier proof. The form of the hypothesis in [2], for Theorem 2, was retained.†

The results obtained on $M_\chi$ modules permit one to understand from a single point of view the greater part of the classical results in complex semisimple Lie algebras, in particular the Kostant theorem or the equivalence to it of the Weil formula for characters, the Borel-Weil Theorem, etc.

## §2. Notation and Background‡

$\mathfrak{g}$ is a complex semisimple Lie algebra of rank $r$; $\mathfrak{h}$ is a Cartan subalgebra of the algebra $\mathfrak{g}$;

$\Delta$ is a system of roots of $\mathfrak{g}$ relative to $\mathfrak{h}$ with a fixed ordering; $\Sigma$ is the sum of the simple roots; $\Delta_+$ is the set of positive roots, $\rho=\frac{1}{2}\sum_{\gamma\in\Delta_+}\gamma$;

$E_\gamma$ is the root vector corresponding to the root $\gamma\in\Delta$, where $\gamma([E_\gamma, E_{-\gamma}])=2$. It is known that $[E_\alpha, E_{-\beta}]=0$ for $\alpha,\beta\in\Sigma$, $\alpha\neq\beta$;

---

*The authors wish to express their gratitude to Prof. J. Dixmier for bringing the exceptionally interesting work [2] to their notice.

†We note that as formulated in [2] the theorem asserting that every submodule M in $M_\chi$ has the form $M=\bigcup M_{\chi_i}$ is false. Counterexamples can be constructed, in particular for Lie algebra of the group SL (4, C); see our last page.

‡For precise definitions and proofs, see also [4] and [5].

---

Moscow State University. Translated from Funktsional'nyi Analiz i Ego Prilozheniya, Vol. 5, No. 1, pp. 1-9, January-March, 1971. Original article submitted October 5, 1970.

$\mathfrak{n}_-$ is the subalgebra of $\mathfrak{g}$, natural in $E_{-\gamma}$, $\gamma \in \Delta_+$;

$U(\mathfrak{g})$, $U(\mathfrak{n}_-)$ are enveloping algebras of $\mathfrak{g}$ and $\mathfrak{n}_-$, respectively; $Z(\mathfrak{g})$ is the center of $U(\mathfrak{g})$;

$U(\mathfrak{g})(x_1, \ldots, x_k)$ for elements $x_1, \ldots, x_k$ from the $U(\mathfrak{g})$-module M denotes the $U(\mathfrak{g})$-submodule in M generated by $x_1, \ldots, x_k$. The meaning of $U(\mathfrak{n}_-)(x_1, \ldots, x_k)$ is analogous;

$\mathfrak{h}^*$ is the space dual to $\mathfrak{h}$; $\mathfrak{h}^*_R$ is the real linear subspace in $\mathfrak{h}^*$, natural for all roots $\gamma \in \Delta$; $\mathfrak{h}^* = \mathfrak{h}^*_R + i\mathfrak{h}^*_R$;

$<\cdot, \cdot>$ is the scalar product in $\mathfrak{h}^*$ constructed according to the Killing form of the algebra $\mathfrak{g}$, $|\cdot|$ is the corresponding norm in $\mathfrak{h}^*_R$. We note that $\chi([E_\gamma, E_{-\gamma}]) = 2<\chi, \gamma>/<\gamma, \gamma>$, $\chi \in \mathfrak{h}^*$, $\gamma \in \Delta_+$;

$\mathfrak{h}^*_Z$ is the lattice in $\mathfrak{h}^*$ constructed from those $\chi$ for which $2<\chi, \gamma>/<\gamma, \gamma> \in Z$ for all $\gamma \in \Delta$;

$f^{(\alpha)}$ is the fundamental weight corresponding to the root $\alpha \in \Sigma$, i.e., $2<f^{(\alpha)}, \alpha>/<\alpha, \alpha> = 1$, $<f^{(\alpha)}, \beta> = 0$ for $\beta \in \Sigma\setminus\alpha$. The weights $f^{(\alpha)}$, $\alpha \in \Sigma$, generate $\mathfrak{h}^*_Z$;

$\chi_1 \gg \chi_2$ for $\chi_1, \chi_2 \in \mathfrak{h}^*$ signifies that $\chi_1 - \chi_2 = \sum_{\alpha \in \Sigma} n_\alpha \alpha$, $n_\alpha \in Z$, $n_\alpha \geqslant 0$;

W is the Weil group of the algebra $\mathfrak{g}$;

$\sigma_\gamma \in W$ is the reflection corresponding to the root $\gamma \in \Delta$, i.e., $\sigma_\gamma \chi = \chi - 2<\chi, \gamma><\gamma, \gamma>^{-1}\gamma$, $\chi \in \mathfrak{h}^*$. We recall that if $\alpha \in \Sigma$, $\gamma \in \Delta_+$, $\gamma \neq \alpha$, then $\sigma_\alpha \gamma \in \Delta_+$;

$l(w)$ is the length of the element $w \in W$, i.e., the least number of factors in a representation $w = \sigma_{\alpha_1}, \ldots, \sigma_{\alpha_k}$, $\alpha_i \in \Sigma$;

$\chi_1 \sim \chi_2$ for $\chi_1, \chi_2 \in \mathfrak{h}^*$ signifies the existence of an element $w \in W$ for which $\chi_1 = w\chi_2$;

$\Xi \subset \mathfrak{h}^*$ is the union of hyperplanes $<\operatorname{Re} \chi, \gamma> = 0$ for all $\gamma \in \Delta$; Weil chambers are connected components $\mathfrak{h}^*\setminus\Xi$; $\bar{C}$ is the closure of the Weil chamber C; $C^+$ is the Weil chamber containing $\rho$. The group W acts simply transitively on the set of Weil chambers. It can be shown that if $\chi \in \bar{C}$ and $w\chi \in \bar{C}$, then $\operatorname{Re} \chi = \operatorname{Re} w\chi$. Two Weil chambers $C_1$ and $C_2$ are called adjacent if $\dim_R(\bar{C}_1 \cap \bar{C}_2) = 2r-1$. There then exists an element $\gamma \in \Delta_+$, such that $\sigma_\gamma C_1 = C_2$;

$P(F) \in \mathfrak{h}^*_Z$ is the set of weights of the finite-dimensional module F.

## §3. The Modules $M_\chi$

Definition 1. Suppose $\chi \in \mathfrak{h}^*$. We denote by $J_\chi$ the left ideal (in $U(\mathfrak{g})$), generated by the elements $E_\gamma$, $\gamma \in \Delta_+$, and $H - \chi(H) + \rho(H)$, $H \in \mathfrak{h}$. We further put $M_\chi = U(\mathfrak{g})/J_\chi$.

The following readily verified lemma describes the simplest properties of the modules $M_\chi$.

LEMMA 1. 1) A $U(\mathfrak{g})$-module $M_\chi$ is generated by one generator $f_\chi$ of weight $\chi - \rho$ such that $E_\alpha f_\chi = 0$ for $\alpha \in \Sigma$; 2) $U(\mathfrak{n}_-)$ acts on $M_\chi$ without zero divisors; 3) $U(\mathfrak{n}_-)(f_\chi) = M_\chi$.

Besides, Properties 1) and 2) uniquely characterize the module $M_\chi$.

THEOREM 1. (Verma, [2]). Let $\chi, \psi \in \mathfrak{h}^*$. Two cases are possible: 1) $\operatorname{Hom}_{U(\mathfrak{g})}(M_\chi, M_\psi) = 0$, 2) $\operatorname{Hom}_{U(\mathfrak{g})}(M_\chi, M_\psi) = C$, and every nontrivial homomorphism $M_\chi \to M_\psi$ is an imbedding.

In the second case we can consider $M_\chi$ as a submodule of $M_\psi$.

We shall study in the present note for what $\chi, \psi \in \mathfrak{h}^*$ the module $M_\chi$ can be imbedded in $M_\psi$. This is obviously possible only in case $\chi \ll \psi$.

LEMMA 2. Let $\alpha \in \Sigma$, $\psi = \sigma_\alpha \chi$ and $\psi - \chi = n\alpha$, where $n \in Z$, $n \geq 0$. Then $M_\chi \subset M_\psi$.

Proof. It is easy to show that $E_\alpha(E^k_{-\alpha} f_\psi) = k(n-k)E^{k-1}_{-\alpha} f_\psi$. Put $f = E^n_{-\alpha} f_\psi$. Then $f$ has weight $\chi - \rho$, $E_\alpha f = 0$ and, for any $\beta \in \Sigma$, $\beta \neq \alpha$, $E_\beta f = E_\beta E^n_{-\alpha} f_\psi = E^n_{-\alpha} E_\beta f_\psi = 0$. By Lemma 1, the submodule in $M_\psi$, natural on $f$ is isomorphic to $M_\chi$, which demonstrates the lemma.

Suppose $z \in Z(\mathfrak{g})$. Then $z \cdot f_\chi$ is proportional to $f_\chi$, i.e., $z \cdot f_\chi = \theta_\chi(z) f_\chi$, $\theta_\chi(z) \in C$. Since $f_\chi$ is a generator in $M_\chi$, $z \cdot x = \theta_\chi(z)x$ for all $x \in M_\chi$. In this way we get the homomorphism $\theta_\chi : Z(\mathfrak{g}) \to C$.

LEMMA 3. $\theta\chi_1 = \theta\chi_2$ if and only if $\chi_1 \sim \chi_2$.

The proof of Lemma 3 follows readily from a theorem of Harish-Chandra on eigenvalues of Laplace operators (see [3] and [8]).

COROLLARY. If $M\chi \subset M\psi$ then $\chi \sim \psi$.

Central to this work is Theorem 2 giving, along with Theorem 3, necessary and sufficient conditions that $M\chi$ is imbedded in $M\psi$.

## §4. Necessity of Condition (A)

Suppose $\chi, \psi \in \mathfrak{h}^*$, $\gamma_1, \gamma_2, \ldots, \gamma_k \in \Delta_+$. We shall say that the sequence $\gamma_1, \ldots, \gamma_k$ satisfies Condition (A) for the pair $(\chi, \psi)$, if

1) $\chi = \sigma_{\gamma_k} \ldots \sigma_{\gamma_1}\psi$.

2) Putting $\chi_0 = \psi$, $\chi_i = \sigma_{\gamma_i} \ldots \sigma_{\gamma_1}\psi$, it is then true that $2\langle \chi_{i-1}, \gamma_i\rangle / \langle \gamma_i, \gamma_i\rangle \in Z$.

3) $\langle \chi_{i-1}, \gamma_i\rangle \geq 0$.

THEOREM 2. Let $\chi, \psi \in \mathfrak{h}^*$ be such that $M\chi \subset M\psi$. Then there exists a sequence $\gamma_1, \ldots, \gamma_k \in \Delta_+$, satisfying Condition (A) for the pair $(\chi, \psi)$.

We preface the proof of the theorem by a series of lemmas.

LEMMA 4. Let M be some $U(\mathfrak{g})$-module, and $0 = M_0 \subset M_1 \subset \ldots \subset M_n = M$ such that its filtration by submodules $M_i$ obeys $M_k/M_{k-1} \approx M\chi_k$. Let $\chi_s$ be maximal* in the set of those $\chi_k$, for which $\chi_s \sim \chi_k$. Then there is, in M, a submodule isomorphic to $M\chi_s$.

Proof. Let $J_i \subset Z(\mathfrak{g})$ be the kernel of the homomorphism $\theta\chi_i$ and $J = J_1 \ldots J_n$. Then J annihilates M, which is to say that there operates in M a commutative finite-dimensional algebra $Z(\mathfrak{g})/J$. (Finiteness of $Z(\mathfrak{g})/J$ follows from the Noether character of $Z(\mathfrak{g})$.) Since $Z(\mathfrak{g})/J$ decomposes into a direct sum of local rings (see [12]), $M = \oplus N^{(j)}$ where the submodules $N^{(j)}$ correspond to the distinct ones among the maximal ideals $J_i \in Z(\mathfrak{g})$. Suppose submodule $N^{(1)}$ corresponds to $J_s$; we put $L_k = N^{(1)} \cap M_k$. Then $L_k/L_{k-1} = M\chi_k$, if $J_k = J_s$ (i.e., $\chi_k \sim \chi_s$), and $L_k/L_{k-1} = 0$, if $J_k \neq J_s$. Hence there is in $N^{(1)}$ an element $f$ of weight $\chi_s - \rho$ and no elements of greater weight (since there are no such from the module $L_k/L_{k-1}$). By Lemma 1, the submodule in M generated by $f$ is isomorphic to $M\chi_s$.

LEMMA 5. Let F be a finite $U(\mathfrak{g})$-module and let $\chi \in \mathfrak{h}^*$. Then in $M\chi \otimes F$ there exists a filtration by submodules $0 = L_0 \subset L_1 \subset L_n = M\chi \otimes F$ such that $L_k/L_{k-1} \approx M\chi + \lambda_k$, where $\lambda_k \in P(F)$.

Proof. Let $e_1, \ldots, e_n$ be a basis of F consisting of weight elements with weights $\lambda_1, \ldots, \lambda_n$, and suppose the indexing chosen so that $i \leq j$, if $\lambda_i \gg \lambda_j$. Let $a_i = f\chi \otimes e_i \in M\chi \otimes F$ and $L_k = U(\mathfrak{g})\ (a_1, \ldots, a_k)$. We verify that the modules $L_k$ satisfy the hypotheses of the lemma. In fact the image $\bar{a}_k$ of the element $a_k$ in $L_k/L_{k-1}$ is a generator in $L_k/L_{k-1}$ of weight $\chi + \lambda_k - \rho$ and $E_\alpha \bar{a}_k = 0$ for $\alpha \in \Sigma$. Hence $L_k = U(\mathfrak{n}_-)\ (a_1, \ldots, a_k)$. We verify that $L_k$ is a free $U(\mathfrak{n}_-)$-module. Suppose $X_i \in U(\mathfrak{n}_-)$, $1 \leq i \leq k$, and $l$ is the largest of the degrees of elements $X_i$ (relative to the natural filtration in $U(\mathfrak{n}_-)$). Then $\sum_{i=n}^{k} X_i a_i = \sum_{i=0}^{k} X_i f_\chi \otimes e_i + \sum_{j=1}^{n} Y_j f_\chi \otimes e_j \neq 0$, since the degrees of the elements $Y_j \in U(\mathfrak{n}_-)$ are less than $l$. Applying Lemma 1, we conclude that $L_k/L_{k-1} \approx M\chi + \lambda_k$.

The following lemma provides the key to the proof of Theorem 2.

LEMMA 6. Suppose $\chi, \psi \in \mathfrak{h}^*$ are such that $M\chi \subset M\psi$, that F is a finite-dimensional $U(\mathfrak{g})$-module, and that $\lambda \in P(F)$ is a weight such that the weight $\chi + \lambda$ is maximal in the set $W(\chi + \lambda) \cap (\chi + P(F))$. Then there exists a weight $\mu \in P(F)$ such that $M\chi + \lambda \subset M\psi + \mu$.

Proof. It follows from Lemmas 4 and 5 that there is a submodule M in $M\chi \otimes F$ which is isomorphic to $M\chi + \lambda$. Clearly, $M\chi \otimes F$ is imbedded in $M\psi \otimes F$. Suppose that $0 = L_0 \subset L_1 \subset \ldots \subset L_n = M\psi \otimes F$ is the filtration entering in Lemma 5, and that $L_k$ is the least submodule containing the image of M. Then the image of M, isomorphic to $M\chi + \lambda$, maps nontrivially to $L_k/L_{k-1} \approx M\psi + \mu$, $\mu \in P(F)$, and, according to Theorem 1, this mapping is an imbedding.

*We say that weight $\chi$ is maximal in the set $D \subset \mathfrak{h}$ if, from the condition $\varphi \in D$, $\varphi \gg \chi$ there follows $\varphi = \chi$.

LEMMA 7. Let $C_1$ and $C_2$ be two Weil chambers, $\chi \in C_1$, $\chi' \in \overline{C}_1$, $\psi \in C_2$, $\psi' \in \overline{C}_2$, $\chi' - \chi \in \mathfrak{h}^*_Z$, $\psi' - \psi \in \mathfrak{h}^*_Z$, $\chi \sim \psi$, $\chi' \sim \psi'$ and let the roots $\gamma_1, \ldots, \gamma_k \in \Delta_+$ satisfy Condition (A) for the pair $(\chi, \psi)$. Then they satisfy Condition (A) for the pair $(\chi', \psi')$ too.

Proof. Point 2) of Condition (A) is obvious, Point 3) follows from the fact that for any $\gamma$ the sign of $<\mathrm{Re}\ \chi, \gamma>$ is constant in a Weil chamber, and Point 1) follows from the fact that $\chi' \sim \chi'_k = \sigma_{\gamma_k} \ldots \sigma_{\gamma_1}\psi'$, $\chi' \in \overline{C}_1$, $\chi'_k \in \overline{C}_1$, whence $\mathrm{Re}\ \chi' = \mathrm{Re}\ \chi'_k$. Since $\chi' - \chi'_k \in \mathfrak{h}^*_Z$, $\mathrm{Im}\ \chi' = \mathrm{Im}\ \chi'_k$, which is to say $\chi' = \chi'_k$.

Lemma 7 shows us how to extend Condition (A) for one pair $(\chi, \psi)$ to its confirmation for another pair $(\chi', \psi')$.

LEMMA 8. Let $C_1$ and $C_2$ be adjacent Weil chambers, and let $\varphi_1 \in C_1$, $\varphi_2 \in C_2$ and $\gamma \in \Delta_+$ be such that $<\mathrm{Re}\ \varphi_1, \gamma> < 0$, $<\mathrm{Re}\ \varphi_2, \gamma> > 0$. Then $\sigma_\gamma C_1 = C_2$.

Proof. Suppose the assertion of the lemma does not hold. Then we can choose $\varphi'_1 \in C_1 \cap \mathfrak{h}^*_R$ and $\varphi'_2 \in C_2 \cap \mathfrak{h}^*_R$ such that $|\varphi'_2 - \varphi'_1|$ is much smaller than $<\varphi'_2, \gamma>$. Since the sign of $<\mathrm{Re}\ \varphi, \gamma>$ is constant in a Weil chamber, the inequalities $<\varphi'_1, \gamma> < 0$, $<\varphi'_2, \gamma> > 0$, hold, whence $<\varphi'_2 - \varphi'_1, \gamma> > <\varphi'_2, \gamma>$, which leads to a contradiction.

Proof of the Theorem. Suppose $\chi \in \mathfrak{h}^*$. Denote by $Y(\chi)$ the following assertion:

$Y(\chi)$: for every $\psi \in \mathfrak{h}^*$ for which $M_\chi \subset M_\psi$, there exists a sequence $\gamma_1, \ldots, \gamma_k \in \Delta_+$, satisfying Condition (A) for the pair $(\chi, \psi)$.

We proceed to prove $Y(\chi)$ for all $\chi$. We first note that $Y(\chi)$ is true for all $\chi \in \overline{C}^+$, since in this case there follows from $\psi \gg \chi$ and $\psi \sim \chi$ the fact that $\psi = \chi$. The arbitrary case is handled by the following steps.

Step 1. Let C be a Weil chamber, let $\chi \in C$ and let $\gamma \in \Delta_+$ be a root such that $<\gamma, \mathrm{Re}\ \chi> < 0$; let $\sigma_\gamma C$ be a Weil chamber adjoint to C. Let F be a finite dimensional $U(\mathfrak{g})$-module for which $\chi + P(F) \subset \overline{C} \cup \sigma_\gamma \overline{C}$ and $\lambda \in P(F)$. Then, if $\chi + \lambda \in \sigma_\gamma C$, there follows from $Y(\chi + \lambda)$ the truth of $Y(\chi)$.

Proof. Suppose $M_\chi \subset M_\psi$ and $w \in W$ are such that $\psi = w\chi$. It follows from the hypothesis that there exist no more than two weights $\mu$ of the module F such that $\chi + \lambda \sim \psi + \mu$. Suppose $\mu_1$ is that one of these weights for which $\psi + \mu_1 \in w\sigma_\gamma C$ (it always exists) and $\mu_2$ is the one for which $\psi + \mu_2 \in wC$ (it may not exist). By Lemma 6, either $M_{\chi+\lambda} \subset M_{\psi+\mu_1}$, or $M_{\chi+\lambda} \subset M_{\psi+\mu_2}$. We give detailed consideration to both cases.

1. Suppose $M_{\chi+\lambda} \subset M_{\psi+\mu_1}$. It follows from $Y(\chi + \lambda)$ that there is a sequence $\gamma_1, \ldots, \gamma_k$ of elements of $\Delta_+$, satisfying Condition (A) for the pair $(\chi + \lambda, \psi + \mu_1) = (\overline{\chi}, \overline{\psi})$. We shall construct a sequence satisfying Condition (A) for the pair $(\chi, \psi)$. Put $\chi_i = \sigma_{\gamma_i}, \ldots, \sigma_{\gamma_1}\psi$, $\overline{\chi}_i = \sigma_{\gamma_i}, \ldots, \sigma_{\gamma_1}\overline{\psi}$. We remark that $\chi_i$ and $\overline{\chi}_i$ are always different on the element $\mathfrak{h}^*_Z$. Hence for any i, either $\chi_i \ll \chi_{i-1}$, or $\chi_i \gg \chi_{i-1}$. If $\chi_i \ll \chi_{i-1}$ for every i, then the sequence $\gamma_1, \ldots, \gamma_k$ satisfies Condition (A) for the pair $(\chi, \psi)$. In the contrary case we denote by $i_0$ the first number for which $\chi_{i_0} \gg \chi_{i_0-1}$. We show that the sequence $\gamma_1, \ldots, \gamma_{i_0-1}, \gamma_{i_0+1}, \ldots, \gamma_k, \gamma$ satisfies Condition (A) for the pair $(\chi, \psi)$. It follows from Lemma 8 that the elements $\chi_{i_0-1}$ and $\overline{\chi}_{i_0}$ lie in the same Weil chamber. This means that $\sigma_\gamma \sigma_{\gamma_k} \ldots \sigma_{\gamma_{i_0+1}} \cdot \sigma_{\gamma_{i_0-1}} \ldots \sigma_{\gamma_1} = w^{-1}$, and Point 1) of Condition (A) is fulfilled. Moreover, $\sigma_\gamma = \sigma_{\gamma_k} \ldots \sigma_{\gamma_{i_0+1}} \sigma_{\gamma_{i_0}} \sigma_{\gamma_{i_0+1}} \ldots \sigma_{\gamma_k}$, i.e., $\gamma = \pm\sigma_{\gamma_k} \ldots \sigma_{\gamma_{i_0+1}} \gamma_{i_0}$. Hence $2<\chi, \gamma>/<\gamma, \gamma> = \pm 2<\chi_{i_0-1}, \gamma_{i_0}>/<\gamma_{i_0}, \gamma_{i_0}> \in \mathbb{Z}$. Fulfillment of Point 2) of Condition (A) for the remaining elements of the sequence is obvious. Fulfillment of Point 3) of Condition (A) follows at once from Lemma 8 and Lemma 7.

2. Now suppose that $M_{\chi+\lambda} \subset M_{\psi+\mu_2}$ and that $\gamma_1, \ldots, \gamma_k$ is a sequence of roots satisfying Condition (A) for the pair $(\chi + \lambda, \psi + \mu_2)$. From Lemma 7 and from the condition $<\mathrm{Re}\ \chi, \gamma> < 0$ it at once follows that the sequence $\gamma_1, \ldots, \gamma_k, \gamma$ satisfies this condition for the pair $(\chi, \psi)$.

Step 2. Let C be a Weil chamber, let $\chi \in C$, and let F be a finite-dimensional module such that $\chi + P(F) \subset C$, and $\lambda \in P(F)$. Then $Y(\chi)$ follows from $Y(\chi + \lambda)$.

The proof of this assertion is analogous to the one conducted above, but simpler.

We note that applying Steps 1 and 2 in the required order, we can show $Y(\chi)$ for every $\chi \in \mathfrak{h}^*$, "sufficiently far" from the set $\Xi$. More precisely, d being distance in $\mathfrak{h}^*_R$, suppose that $\chi \in \mathfrak{h}^*$ is a weight such that $d(\mathrm{Re}\ \chi, \Xi) > 3|\rho|$. Then we can construct a sequence $\chi = \chi_0, \chi_1, \ldots, \chi_k$ of elements of $\mathfrak{h}^*$ such that $d(\mathrm{Re}\ \chi_i, \Xi) > 2|\rho|$, $\chi_i - \chi_{i+1} \in \mathfrak{h}^*_Z$, $\chi_k \in C^+$ and such that for every i one of the following two conditions is fulfilled:

1) $\chi_i$ and $\chi_{i+1}$ lie in two adjacent Weil chambers C and $\sigma_\gamma C$, $\gamma \in \Delta_+$, and moreover $\langle \mathrm{Re}\, \chi_i, \gamma \rangle < 0$ and $|\chi_{i+1} - \chi_i|$ is a much smaller distance from $\chi_i$ than all other Weil chambers;

2) $\chi_i$ and $\chi_{i+1}$ lie in the same Weil chamber and $|\chi_{i+1} - \chi_i| < 2|\rho|$.

Considering the finite-dimensional modules $F_i$ of least weights $\chi_{i+1} - \chi_i$ and applying respectively Steps 1 and 2, we extend $Y(\chi)$ to truth of the assertion $Y(\chi_k)$. Here the fact is used that the length of all weights $F_i$ does not surpass $|\chi_{i+1} - \chi_i|$ (see [5]).

The case of arbitrary $\chi$ is analyzed with the help of the following considerations.

Step 3. Let $\chi \in \mathfrak{h}^*$, $\mathrm{Re}\, \chi \neq 0$. For each $a > 0$ we denote by $D_a$ the cone in $\mathfrak{h}^*_R$ consisting of those nonzero $\varphi$ such that the angle between $\varphi$ and $\mathrm{Re}\, \chi$ is less than $a$. We can choose $a$ small enough that the Weil chamber C intersects $D_a$ if and only if $\chi \in \bar{C}$. We choose $\lambda \in \mathfrak{h}^*_Z$, such that 1) $d(\mathrm{Re}(\chi + \lambda), \Xi) > 3|\rho|$, 2) $\mathrm{Re}(\chi + \lambda)$ is maximal in the set $W(\mathrm{Re}(\chi + \lambda)) \cap D_a$.

We now show that $Y(\chi)$ follows from $Y(\chi + \lambda)$.

Suppose F is a finite-dimensional module with least weight $\lambda$. We show that for all weights $\chi + \lambda$ the premises of Lemma 6 are fulfilled. Suppose the weight $\mu \in P(F)$ is such that $\mathrm{Re}\,(\chi + \lambda) \sim \mathrm{Re}\,(\chi + \mu)$. Since $|\mathrm{Re}\,(\chi + \lambda)| = |\mathrm{Re}\,(\chi + \mu)|$ and $|\lambda| \geq |\mu|$, it follows that $\mathrm{Re}(\chi + \mu) \in D_a$, which means the inequality $\mathrm{Re}(\chi + \lambda) \ll \mathrm{Re}(\chi + \mu)$ cannot be fulfilled. Hence we can take Lemma 6 and the assertion that if $M_\chi \subset M_\psi$, then $M_{\chi+\lambda} \subset M_{\psi+\mu}$ for some $\mu \in P(F)$. Reasoning analogous to that adduced above shows that the angle between $\mathrm{Re}\,(\psi + \mu)$ and $\mathrm{Re}\, \psi$ is not greater than $a$ and hence $\psi + \mu$ and $\psi$ lie in the same Weil chamber. We may now apply Lemma 7 which completes the proof of Step 3 and of the whole of Theorem 2.

## §5. Sufficiency of Condition (A)

THEOREM 3 (see [11]). Let $\chi, \psi \in \mathfrak{h}^*$. If there exists a sequence of roots $\gamma_1, \ldots, \gamma_k \in \Delta_+$, satisfying Condition (A) for the pair $(\chi, \psi)$, then $M_\chi \subset M_\psi$.

Proof.* It suffices to consider the case where the sequence consists of a single root $\gamma \in \Delta_+$.

LEMMA 9. If $\chi, \psi \in \mathfrak{h}^*$ and $\alpha \in \Sigma$ are such that $M_\chi \subset M_\psi$ and $M_{\sigma_\alpha \chi} \subset M_\chi$, then $M_{\sigma_\alpha \chi} \subset M_{\sigma_\alpha \psi}$.

Proof. It follows immediately from the conditions of the lemma that $\psi - \sigma_\alpha \psi = n\alpha$, where $n \in Z$. By Lemma 2 it is enough to consider the case $n > 0$, where $M_{\sigma_\alpha \psi} \subset M_\psi$. Suppose $\bar{f}_\psi$ is the image of $f_\psi$ in $M_\psi / M_{\sigma_\alpha \psi}$. Then $E^n_{-\alpha} \bar{f}_\psi = 0$. The image of the element $f_\chi$ in $M_\psi / M_{\sigma_\alpha \psi}$ has the form $\bar{f}_\chi = X\bar{f}_\psi$, where $X \in U(\mathfrak{g})$. The element $E^k_{-\alpha} X$ can be written in the form $E^k_{-\alpha} X = X_1 E^{k_1}_{-\alpha}$, where $k_1$ increases unboundedly with k. Hence $E^k_{-\alpha} \bar{f}_\chi = 0$ for large k. Since $E_\alpha E^k_{-\alpha} \bar{f}_\chi = k(n' - k) E^{k-1}_{-\alpha} \bar{f}_\chi$ (where $n' = 2\langle \chi, \alpha \rangle / \langle \alpha, \alpha \rangle \in Z$), it follows that $E^{n'}_{-\alpha} \bar{f}_\chi = 0$, i.e., that $M_{\sigma_\alpha \chi} \subset M_{\sigma_\alpha \psi}$.

LEMMA 10. The assertion of the theorem is true for $\chi \in \mathfrak{h}^*_Z$, $\psi = \sigma_\gamma \chi$.

Proof. We can suppose that $\chi \neq \psi$. We choose elements $\alpha_1, \ldots, \alpha_k \in \Sigma$, such that in the sequence of weights $\chi_i = \sigma_{\alpha_i} \ldots \sigma_{\alpha_1} \chi$ the relations $\chi_{i+1} \gg \chi_i$, $\chi_k \in \bar{C}^+$ are fulfilled. Suppose $\psi_i = \sigma_{\alpha_i} \ldots \sigma_{\alpha_1} \psi$. Each $\psi_i$ is obtained from $\chi_i$ by some mapping which we denote by $\sigma_{\gamma_i}$, $\gamma_i \in \Delta_+$. By Lemma 9 it is enough to show that $M_{\chi_i} \subset M_{\psi_i}$ for some i.

We take as i the last index for which $\psi_i \gg \chi_i$ ($i < k$, since $\chi_k \gg \psi_k$ and $\chi_k \neq \psi_k$). Then $\psi_i - \chi_i = n\gamma_i$, where $n > 0$, and $\psi_{i+1} - \chi_{i+1} = \sigma_{\alpha_{i+1}}(\psi_i - \chi_i) = n\sigma_{\alpha_{i+1}} \gamma_i \ll 0$, which is only possible if $\alpha_{i+1} = \gamma_i$, but then $M_{\chi_i} \subset M_{\psi_i}$ by Lemma 2.

We now prove Theorem 3. Let $\gamma \in \Delta_+$, $\chi, \psi \in \mathfrak{h}^*$ be such that $\chi = \sigma_\gamma \psi$, $\psi - \chi = n\gamma$, where $n \in Z$. Denote by $\mathfrak{n}_{n\gamma}$ the finite dimensional subspace in $U(\mathfrak{n}_-)$, consisting of elements of weight $-n\gamma$.

The module $M_\psi$, considered as a $U(\mathfrak{n}_-)$-module, is a free module with image $f_\psi$. Hence for the proof of the theorem we need to find a nonnull element $X \in \mathfrak{n}_{n\gamma}$ such that $E_\alpha X f_\psi = 0$ for all $\alpha \in \Sigma$. The equations written give a system of linear homogeneous equations in the space $\mathfrak{n}_{n\gamma}$, whose coefficients depend linearly on $\psi$. Consider the hyperplane S in $\mathfrak{h}^*$ consisting of those weights $\varphi$ such that $2\langle \varphi, \gamma \rangle / \langle \gamma, \gamma \rangle = n$. By Lemma 10 the given system has a nontrivial solution for all $\varphi \in S \cap \mathfrak{h}^*_Z$. If $\gamma = w\alpha$, $w \in W$, $\alpha \in \Sigma$, then

*This proof differs from that in [11]. In particular it does not use enumerations of simple Lie algebras.

$S \cap \mathfrak{h}^*_Z = \{w(nf^{(\alpha)} + \sum_{\beta \in \Sigma \setminus \alpha} n_\beta f^{(\beta)}),\ n_\beta \in Z\}$, i.e., it constitutes an (n−1)-dimensional lattice in S. Hence our system has a nontrivial solution for all $\varphi \in S$ and in particular for $\varphi = \Psi$. This completely proves Theorem 3.

Theorems 2 and 3 permit introduction of the following partial ordering into the group W.

Definition 2. Let $\chi_0 \in C^+ \cap \mathfrak{h}^*_Z$. We put $w_1 < w_2$ for elements $w_1, w_2 \in W$, if $M_{w_1\chi_0} \subset M_{w_2\chi_0}$.

THEOREM 4. 1) The ordering introduced in Theorem 4 does not depend on the choice of $\chi_0 \in C^+ \cap \mathfrak{h}^*_Z$.

2) Suppose $w_1, w_2 \in W$. The inequality $w_1 < w_2$ is fulfilled if and only if there exists a sequence of images $\sigma_{\gamma_1}, \ldots, \sigma_{\gamma_k} \in W$, such that $w_1 = \sigma_{\gamma_k} \ldots \sigma_{\gamma_1} w_2$ and $l(\sigma_{\gamma_{i+1}} \ldots \sigma_{\gamma_1} w_2) > l(\sigma_{\gamma_i} \ldots \sigma_{\gamma_1} w_2)$ for every i. Here $l(w)$ is the length of the element $w \in W$.

The proof of the first part of the theorem follows immediately from Theorems 2 and 3 and Lemma 7. For proving the second part we need the following result on constructing the system of roots (see [9]). Suppose $w \in W$. Then $l(w)$ coincides with the number of those roots $\gamma \in \Delta_+$, for which $w\gamma \in -\Delta_+$.

As remarked in [2], the adduced ordering can be given the following geometrical meaning. Let G be a simply-connected Lie group with Lie algebra $\mathfrak{g}$, let $B \subset G$ be the Borel subgroup corresponding to the subalgebra generated by $\mathfrak{h}$ and $E_\gamma$, $\gamma \in \Delta_+$. Suppose moreover that $P = G/B$ is the fundamental projective space of the group G. Each element $w \in W$ we put in correspondence with the subspace $P_w = BwB \subset P$. These subspaces are the key to the splitting of P (see [10]). Here $w_1 < w_2$ if and only if $P_{w_1} \supset P_{w_2}$. We also note the connection of this ordering with the Paley-Wiener Theorem for complex semisimple Lie groups G ([6]).

## §6. Appendix. Multiplicity of the Weight of a Finite-Dimensional Representation

We introduce here a simple algebraic proof of the Kostant formula for the multiplicity of a weight of a finite-dimensional representation (see [7]). To do this we shall use only the definition of the module $M_\chi$ (Def. 1) and Lemmas 1 and 3. To make it possible to use Lemma 3 we note that in [8] there was introduced a proof of the Harish-Chandra Theorem on the eigenvalues of a Laplace operator without using the formula of H. Weil for characters.

Let $\tilde{\mathscr{E}}$ be the space of all functions in $\mathfrak{h}^*$. For $u \in \tilde{\mathscr{E}}$ we place $\operatorname{supp} u = \{\chi \in \mathfrak{h}^* | u(\chi) = 0\}$. The group W acts in $\tilde{\mathscr{E}}$ by the formula $(wu)(\chi) = u(w^{-1}\chi)$. Let $\mathscr{E} \subset \tilde{\mathscr{E}}$ be a subspace consisting of all functions u for which supp u is contained in the union of a finite number of sets of the form $\{\chi - \sum_{\gamma \in \Delta_+} n_\gamma \gamma | n_\gamma \in Z,\ n_\gamma > 0\}$.

The space $\mathscr{E}$ is a commutative algebra with respect to the convolution operation $u_1 * u_2(\chi) = \sum_{\psi \in \mathfrak{h}^*} u_1(\chi - \psi) u_2(\psi)$ (in this sum only a finite number of terms are different from zero). Let $\delta_\chi \in \mathscr{E}$ be a function such that $\delta_\chi(\chi) = 1$, $\delta_\chi(\psi) = 0$ for $\psi \neq \chi$. Then $\delta_0$ is the unit of the ring $\mathscr{E}$.

Definition 3. 1) $Q(\chi)$ is the number of families $\{n_\gamma\}_{\gamma \in \Delta_+}$, $n_\gamma \in Z$, $n_\gamma \geq 0$, such that $\sum_{\gamma \in \Delta_+} n_\gamma \gamma = -\chi$.

2) $L = \prod_{\gamma \in \Delta_+} \left(\delta_{\frac{\gamma}{2}} - \delta_{-\frac{\gamma}{2}}\right)$.

We shall call the function $Q(\chi)$ the Kostant function.*

LEMMA 11. $L * Q * \delta_{-\rho} = \delta_0$.

Proof. Let $a_\gamma = \delta_0 + \delta_{-\gamma} + \delta_{-2\gamma} + \ldots \in \mathscr{E}$. Then clearly $Q = \prod_{\gamma \in \Delta_+} a_\gamma$ и $(\delta_0 - \delta_\gamma) * a_\gamma = \delta_0$. The lemma at once follows from the fact that $L = \prod_{\gamma \in \Delta_+} (\delta_0 - \delta_{-\gamma}) * \delta_\rho$.

LEMMA 12. $wL = (-1)^{l(w)} L$ for $w \in W$.

*Our definition of $Q(\chi)$ differs from that provided in [7] in replacement of $\chi$ by $-\chi$.

Proof. It suffices to verify that $\sigma_\alpha L = -L$ for $\alpha \in \Sigma$. Since $\sigma_\alpha$ transposes elements of the set $\Delta_+ \setminus \alpha$ and carries $\alpha$ to $-\alpha$, $\sigma_\alpha L = \left(\delta_{-\frac{\alpha}{2}} - \delta_{\frac{\alpha}{2}}\right) * \prod_{\gamma \in \Delta_+ \setminus \alpha} \left(\delta_{\frac{\gamma}{2}} - \delta_{-\frac{\gamma}{2}}\right) = -L.$

Definition 4. Suppose the $U(\mathfrak{g})$-module M is a direct sum of weights of spaces $V_\chi$ of weight $\chi$ and $\dim V_\chi < \infty$ for all $\chi \in \mathfrak{h}^*$. We call the function $\pi_M \in \mathfrak{F}$, given by the formula $\pi_M(\chi) = \dim V_\chi$ by the name character of M.

LEMMA 13. $\pi_{M_\chi}(\psi) = Q(\psi - \chi + \rho)$.

Proof. Suppose $\gamma_1, \ldots, \gamma_s$ is an arbitrary indexing of the roots $\gamma \in \Delta_+$. From the theorem of Poincaré-Birkhoff-Witt (see [4], Ch. I) it follows from Lemma 1 that the elements $E_{-\gamma_1}^{n_1} \ldots E_{-\gamma_s}^{n_s} f_\chi$ ($n_i \in Z$, $n_i \geq 0$) form a basis in $M_\chi$, and Lemma 13 follows from the definition of Q.

COROLLARY. $L * \pi_{M_\chi} = \delta_\chi$.

LEMMA 14. Let $U(\mathfrak{g})$-module M and element $\chi_0 \in \bar{C}^+$ be such that 1) there is a basis in M of weight vectors; 2) $z \cdot x = \theta_{\chi_0}(z)x$ for every $x \in M$, $z \in Z(\mathfrak{g})$; 3) $\pi_M$ exists and $\pi_M \in \mathscr{E}$.

Let $D_M = \{\chi \in \mathfrak{h}^* | \chi \sim \chi_0, \chi \ll \psi + \rho$ for some $\psi \in \operatorname{supp} \pi_M\}$.

Then $\pi_M = \sum_{\chi \in D_M} c_\chi \pi_{M_\chi}, c_\chi \in Z$.

Proof. In view of Condition 3) there exists in $\operatorname{supp} \pi_M$ at least one maximal element $\chi$. Then $E_\alpha x = 0$ for any $x \in V_\chi$, $\alpha \in \Sigma$. Hence for any $z \in Z(\mathfrak{g})$, $v \in V_\chi$ we have $z \cdot x = \theta_{\chi+\rho}(z)x$, and hence, $\chi + \rho \sim \chi_0$, i.e., $\chi + \rho \in D_M$. Let $k = \dim V_\chi$. By definition of $M_\chi$ it follows that we can construct a mapping $\varkappa$: $(M_{\chi+\rho})^k \to M$, translating the generators $f_{\chi+\rho}$ to k linearly independent elements $V_\chi$. Suppose L and N are the kernel and co-kernel of $\varkappa$. It follows from the exact sequence $0 \to L \to (M_{\chi+\rho})^k \xrightarrow{\varkappa} M \to N \to 0$ that $\pi_M = \pi_L - \pi_N - k\pi_{M_{\chi+\rho}}$. The moduli L and M satisfy the hypotheses of Lemma 14 and in addition $D_L$ and $D_N$ lie in $D_M \setminus (\chi + \rho)$. Induction on the number of elements of $D_M$ concludes the proof of Lemma 14.

Remark. The foregoing reasoning establishes that the moduli $M_\chi$ generate the group of Grothendieck category of the moduli M satisfying the conditions of Lemma 14.

COROLLARY. Under the hypotheses of Lemma 14, $L * \pi_M \subset D_M$.

Suppose now that F is a continuous finite-dimensional $U(\mathfrak{g})$-module with highest weight $\lambda$. Then $\pi_F$ is invariant with respect to W (see [5]). By Lemma 12, $w(L * \pi_F) = (-1)^{l(w)} (L * \pi_F)$. By the Corollary to Lemma 14, $L * \pi_F = \sum c_w \delta_{w(\lambda+\rho)}$. Since $\dim V_\lambda = 1$ in the module F, it follows that $c_e = 1$ (see the proof of Lemma 14), hence $L * \pi_F = \sum_{w \in W} (-1)^{l(w)} \delta_{w(\lambda+\rho)}$. Applying Lemma 11, we get $\pi_F = Q * \delta_{-\rho} \sum_{w \in W} (-1)^{l(w)} \delta_{w(\lambda+\rho)}$. In conclusion, we get the following theorem.

THEOREM 5 (Kostant formula). Let F be a finite-dimensional continuous representation of $\mathfrak{g}$ with highest weight $\lambda$. Then

$$\pi_F(\mu) = \sum_{w \in W} (-1)^{l(w)} Q(\mu + \rho - w(\lambda + \rho)).$$

Note. In conclusion we append an example of a submodule M in $M_\chi$ which does not have the form $\cup M_{\chi_i}$. Let $\mathfrak{g}$ be the Lie algebra of the group $SL(4, C)$, $E_{ik}$ ($i \neq k$), and let $E_{ii} - E_{jj}$ be a basis of $\mathfrak{g}$ with $E_{ik}(i < k)$ corresponding to roots from $\Delta_+$. We consider the module $M_\chi$, wherein all weights $\chi$ are given by the equations $\chi(E_{11} - E_{12}) = \chi(E_{33} - E_{44}) = 0$, $\chi(E_{22} - E_{33}) = 1$. Suppose $f_1 = E_{32} f_\chi$ and $\tilde{M} = U(\mathfrak{g})(f_1)$. It follows from Theorem 2 that if $E_{ik} a = 0$ for $i < k$, then $a = a_1 + c f_\chi$, where $a_1 \in \tilde{M}$, $c \in C$. Put $x = E_{42}E_{21}f_\chi + E_{43}E_{31}f_\chi$. Then $x \notin \tilde{M}$ and $E_{ik}x \in \tilde{M}$ for $i < k$. Hence the submodule generated by x and $\tilde{M}$ is a proper submodule of $M_\chi$ not having the form $\cup M_\chi$.

## LITERATURE CITED


1. I. M. Gel'fand, "Cohomologies of infinite-dimensional Lie algebras; some questions of integral geometry" [in Russian], Report to the International Mathematical Congress (1970).

2. D.-N. Verma, "Structure of certain induced representations of complex semisimple Lie algebras," Bull. Amer. Math. Soc., 74, 160-166 (1968).
3. Harish-Chandra, "On some applications of universal enveloping algebra of a semisimple Lie algebra," Trans. Amer. Math. Soc., 70, 28-96 (1951).
4. Lie Topological Groups. Theory of Lie Algebras [Russian translation], IL, Moscow (1962).
5. J.-P. Serre, Lie Algebras and Lie Groups [Russian translation], Mir, Moscow (1969).
6. I. M. Gel'fand and M. I. Graev, "Fourier transforms of rapidly decreasing functions on complex semisimple groups," Dokl. Akad. Nauk SSSR, 131, No. 3, 496-499 (1960).
7. B. Kostant, "A formula for the multiplicity of a weight," Trans. Amer. Math. Soc., 93, 53-73 (1959).
8. K. R. Parthasarathy, Rao R. Ranga, and V. S. Varadarajan, "Representations of complex semisimple Lie groups and Lie algebras," Ann. Math., 85, 383-429 (1967).
9. N. Bourbaki, Groupes et Algebres de Lie, Hermann (1968), Chaps. 4-6.
10. B. Kostant, "Lie algebra cohomology and generalized Schubert cells," Ann. Math., 77, 72-144 (1963).
11. D.-N. Verma, "Structure of certain induced representations of complex semisimple Lie algebras," Dissertation, Yale Univ. (1966).
12. O. Zariski and P. Samuel, Commutative Algebra, Vol. 1, Van Nostrand (1958).

# 3.

(with I. N. Bernstein and S. I. Gelfand)

## Differential operators on a cubic cone

Usp. Mat. Nauk **27** (1) (1972) 185–190 [Russ. Math. Surv. **27** (1) (1972) 169–174].
Zbl. **257**:58010

Consider in the space $\mathbf{C}^3$ with the coordinates $x_1, x_2, x_3$ the surface $X$ defined by the equation $x_1^3 + x_2^3 + x_3^3 = 0$. We prove the following theorem:

THEOREM 1. *Let $D(X)$ be the ring of regular differential operators on $X$, and $D_a$ the ring of germs at the point 0 of analytic operators on $X$. Then*

1°. *the rings $D(X)$ and $D_a$ are not Noetherian;*

2°. *for any natural number $k$ the rings $D(X)$ and $D_a$ are not generated by the subspaces $D_k$ ($D_{ak}$, respectively) of operators of order not exceeding $k$. In particular, the rings $D(X)$ and $D_a$ are not finitely generated.*

Theorem 1 answers questions raised in Malgrange's survey article [1]. The ring $D(X)$ has an interesting structure (see Proposition 1).

We denote by $E(X)$ the ring of regular functions on $X (E(X) = \mathbf{C}[x_1, x_2, x_3]/[x_1^3 + x_2^3 + x_3^3])$ and by $D(X)$ the ring of regular differential operators on $X$. By $D_k$ we denote the space of operators of order not exceeding $k$. Setting $a_\lambda(f)(x) = f(\lambda x)$ and $b_\lambda(\mathscr{D})(f) = a_\lambda(Da_{\lambda-1}(f))$ for $\lambda \in \mathbf{C}^*$ we define an action of the group $\mathbf{C}^*$ in the spaces $E(X)$ and $D(X)$. It is clear that $E(X) = \bigoplus_{i=0}^{\infty} E^i(X)$, where $E^i(X)$ is the finite-dimensional space of homogenous functions of degree $i$ on $X$. We call an operator $\mathscr{D} \in D(X)$ homogenous of degree $i$ ($i \in \mathbf{Z}$) if $b_\lambda(\mathscr{D}) = \lambda^i \mathscr{D}$ for all $\lambda \in \mathbf{C}^*$ (equivalent definition: $\mathscr{D}(E^n(X)) \subset E^{n+i}(X)$ for all $n$). We denote by $D^i$ the space of all such operators and set $D_k^i = D^i \cap D_k$.

LEMMA 1. a) $D_k = \bigoplus_{i=-\infty}^{\infty} D_k^i$; b) $D(X) = \bigoplus_{i=-\infty}^{\infty} D^i$.

PROOF. a) Let $\mathscr{D} \in D_k$. We define the operator $\mathscr{D}^{(i)}$ in $E(X)$ as follows: if $f = E^n(X)$, then $\mathscr{D}^{(i)}f = (\mathscr{D}f)^{(n+i)}$ (that is, the component of degree of homogeneity $n + i$ of the function $\mathscr{D}f$ ).

It is easy to verify that $\mathscr{D}^{(i)} \in D_k$ and therefore $\mathscr{D}^{(i)} \in D_k^i$. Since any operator of order not exceeding $k$ is defined by its values on the space $\bigoplus_{j=0}^{k} E^j(X)$, it follows that $\mathscr{D}^{(i)}$ is not equal to 0 for finite $i$. Clearly $\mathscr{D} = \Sigma \mathscr{D}^{(i)}$.

b) This statement follows directly from a):

We set $I = x_1 \frac{\partial}{\partial x_1} + x_2 \frac{\partial}{\partial x_2} + x_3 \frac{\partial}{\partial x_3}$.

PROPOSITION 1.

1°. $D^i = 0$ *for* $i < 0$.

2°. $D^0$ *is generated by the elements* $1, I, I^2, \ldots$

3°. $D^1_k/(D^1_{k-1} + ID^1_{k-1}) = \mathbf{C}^3$ $(k = 0, 1, 2, \ldots)$.

We derive Theorem 1 from Proposition 1.

It is easy to verify that $D_k \cdot D_l \subset D_{k+1}$ and $D^i \cdot D^j \subset D^{i+j}$. Moreover, if $\mathscr{D} \in D^i$, then $[I, \mathscr{D}] = I\mathscr{D} - \mathscr{D}I = i\mathscr{D}$.

For any natural number $k$ we set $J_k = \sum_{n \geq 0} I^n D^1_k + \sum_{i \geq 2} D^i$ and $A_k = D^0 + J_k$. It follows from Proposition 1 and the formulae above that $J_k$ is a two-sided ideal in the ring $D(X)$, and that $A_k$ is a subring of $D(X)$.

If $l > k$, then $J_l \underset{\neq}{\supset} J_k$. From this it follows that the ring $D(X)$ is not Noetherian. Since $A_l \underset{\neq}{\supset} A_k \supset D_k$ for $l > k$, the ring $D(X)$ is not generated by the subspace $D_k$.

Consider the ring $D_a$ of germs at 0 of analytic differential operators on $X$. The group $\mathbf{C}^*$ acts in this ring. Every element can be expanded in a convergent series $\mathscr{D} = \sum_{i=-\infty}^{\infty} \mathscr{D}^{(i)}$, where $\mathscr{D}^{(i)} \in D_a$ is a homogenous operator of degree $i$, and the order of $\mathscr{D}^{(i)}$ does not exceed that of $\mathscr{D}$ (specifically, $\mathscr{D}^{(i)} = \frac{1}{2\pi i} \int b_\lambda(\mathscr{D})\lambda^{-i-1}\, d\lambda$, where the integral is taken over the unit circle in in the $\lambda$-plane).

If $f \in E^n(X)$, then $\mathscr{D}^{(i)} f$ is a homogenous analytic function of degree of homogeneity $n + i$; hence $\mathscr{D}^{(i)} f \in E^{n+i}(X)$. Therefore we may assume that $\mathscr{D}^{(i)} \in D^i \subset D(X)$ (it is clear that if $\mathscr{D}^{(i)} f = 0$ for all $f \in E(X)$, then $\mathscr{D}^{(i)} = 0$). It follows from Proposition 1 that every operator $\mathscr{D} \in D_a$ can be expanded in a series $\mathscr{D} = \sum_{i=0}^{\infty} \mathscr{D}^{(i)}$, where $\mathscr{D}^{(i)} \in D^i$.

Let $J_{ak} = \{\mathscr{D} \in D_a \mid \mathscr{D}^{(i)} \in J_k$ for all $i\}$, and let $A_{ak} = D^0 + J_{ak}$. Then $J_{ak}$ is a two-sided ideal in the ring $D_a$, and $A_{ak}$ is a subring of $D_a$. Since $J_{al} \underset{\neq}{\supset} J_{ak}$ and $A_{al} \underset{\neq}{\supset} A_{ak} \supset D_{ak}$ for $l > k$, it follows that $D_a$ is not Noetherian and is not generated by a subspace $D_{ak}$, where $k$ is any natural number. Theorem 1 is now proved.

PROOF OF PROPOSITION 1. Consider the non-singular algebraic manifold $X_0 = X \setminus 0$.

LEMMA 2. a) *The embedding of $X_0$ in $X$ induces an isomorphism $E(X) \to E(X_0)$ of rings of regular functions on $X$ and $X_0$.*

b) *The embedding of $X_0$ in $X$ induces an isomorphism $D(X) \to D(X_0)$ of regular differential operator rings.*

PROOF. a) follows from the fact that $X$ is a normal manifold and that codim $\{0\}$ in $X$ is greater than 1. Since $X$ is an affine manifold, a) implies b).

Denote by $\mathcal{D}_k^i$ the sheaf of germs of the differential operators $\mathcal{D}$ on $X_0$ of order not exceeding $k$ that satisfy the condition $[I, \mathcal{D}] = i\mathcal{D}$. Then $D_k^i = \Gamma(X_0, \mathcal{D}_k^i)$.

Consider the projective manifold $\overline{X} = X_0/\mathbf{C}^*$. It is known that $\overline{X}$ is an elliptic curve. By $\pi$ we denote the natural projection $\pi\colon X_0 \to \overline{X}$.

Consider the sheaves $\Delta_k^i = \pi_*(\mathcal{D}_k^i)$ on the manifold $\overline{X}$. It is clear that $\Gamma(\overline{X}, \Delta_k^i) = \Gamma(X_0, \mathcal{D}_k^i) = D_k^i$.

By $\widetilde{\mathcal{L}}$ we denote the sheaf of functions on $X_0$ that are homogenous of degree 1, and we set $\mathcal{L} = \pi_*(\widetilde{\mathcal{L}})$; $\mathcal{L}$ is a sheaf on $\overline{X}$.

The following facts are easy to verify:

1°. $\mathcal{L}$ and $\Delta_k^i$ are sheaves of modules over the sheaf of rings $O_{\overline{X}}$.

2°. $\mathcal{L}$ is an invertible sheaf; $\Gamma(\overline{X}, \mathcal{L}) = \mathbf{C}^3$.

3° $\Delta_k^i = \Delta_k^0 \otimes \mathcal{L}^i$, this isomorphism being consistent with the natural embeddings $\Delta_k^i \to \Delta_l^i$ for $l > k$.

4°. Set $\sigma_k = \Delta_k^0/\Delta_{k-1}^0$. Then $\sigma_k = S^k(\sigma_1)$ (where $S^k$ is the $k$-th symmetric power of the sheaf).

5°. By $\widetilde{N}$ we denote a subsheaf in $\mathcal{D}_1^0$, whose sections on every neighbourhood are defined as $\{f(x)I\}$, where $f(x)$ is a function of degree of homogeneity 0. Then the sheaf $N = \pi_*(\widetilde{N})$ on $\overline{X}$ is a subsheaf of $\Delta_1^0$. We regard $N$ as a subsheaf of $\sigma_1 = \Delta_1^0/\Delta_0^0$.

6°. The map $1 \mapsto I$ defines the isomorphism of sheaves $O_{\overline{X}} \xrightarrow{\sim} N$.

7°. Set $\mathcal{K} = \sigma_1/N$. Then $\mathcal{K}$ is an invertible sheaf naturally isomorphic to the tangent sheaf to $\overline{X}$.

The tangent sheaf on an elliptic curve $\overline{X}$ is known to be isomorphic to $O_{\overline{X}}$. We fix a certain non-zero global section $k$ of $\mathcal{K}$.

LEMMA 3. *For every $n > 0$ there exists an exact sequence $V_n$ of sheaves on $\overline{X}$*

$$0 \to N^n \xrightarrow{\varphi} \sigma_n \xrightarrow{\psi} \mathcal{K} \otimes \sigma_{n-1} \to 0.$$

*Here the diagram*

$$\begin{array}{ccccccccc} 0 \to & N^n & \to & \sigma_n & \to & \mathcal{K} \otimes \sigma_{n-1} & \to 0 \\ & \downarrow & & \downarrow & & \downarrow & \\ 0 \to & N^{n+1} & \to & \sigma_{n+1} & \to & \mathcal{K} \otimes \sigma_n & \to 0, \end{array} \tag{1}$$

*where every vertical homomorphism is obtained by multiplying $I \in \Gamma(\overline{X}, N)$ by the section, commutes.*

PROOF. Construct the maps $\varphi$ and $\psi$ in a neighbourhood $U \subset \overline{X}$. Let $I$ be a global generating element of the sheaf $N$, and let $\widetilde{k}$ be a local section over $U$ of the sheaf $\sigma_1$ which becomes $k \in \Gamma(\overline{X}, \mathscr{K})$ under the map $\sigma_1 \to \mathscr{K}$.

Since the sequence $0 \to N \to \sigma_1 \to \mathscr{K} \to 0$ is exact, it follows that the restriction of $\sigma_1$ to $U$ is a free sheaf over $O_{\overline{X}}$ with the generators $I$ and $\widetilde{k}$. Then $\sigma_n = S^n(\sigma_1)$ is a free sheaf on $U$ with the generators $\widetilde{k}^i I^{n-i}$ $(i = 0, 1, \dots, n)$. We define the maps $\varphi$ and $\psi$ by the formulae: $\varphi(I^n) = I^n$, $\psi(\widetilde{k}^i I^{n-i}) = ik \otimes (\widetilde{k}^{i-1} I^{n-i})$.

It is easy to verify that the sequence $(V_n)$ is exact. It is also clear that $\varphi$ does not depend on the choice of $\widetilde{k}$. Let us prove that $\psi$ does not depend on the choice of $\widetilde{k}$. In fact, let $\hat{k}$ be another section of the sheaf $\sigma_2$ on $U$. Then $\hat{k} = \widetilde{k} + fI$, where $f \in \Gamma(U, O_X)$,

$$\psi(\hat{k}^i I^{n-i}) = \psi\Big(\Big(\sum_{j=0}^{i} C_i^j \widetilde{k}^{i-j} f^j\Big) I^{n-i+j}\Big) = k \otimes \sum_j C_i^j (i-j)\, \widetilde{k}^{i-j-1} I^{n-i+j} f^j =$$
$$= i\,(k \otimes (\hat{k} + fI)^{i-1} I^{n-i}).$$

So the exact sequence $(V_n)$ is defined globally.

It follows from the construction of the homomorphisms $\varphi$ and $\psi$ that the diagram (1) is commutative. This proves Lemma 3.

We are interested in the spaces $H^0(\overline{X}, \sigma_k) = \Gamma(\widetilde{X}, \sigma_k)$ and $H^1(\overline{X}, \sigma_k)$.

LEMMA 4. $\dim H^0(\overline{X}, \sigma_1) = 1$.

PROOF. Every first-order operator $\mathscr{D}$ can be split uniquely into a sum $\mathscr{D} = f + \mathscr{D}'$, where $f$ is the operator of multiplication by the function $f = \mathscr{D}(1)$ and $\mathscr{D}' = \mathscr{D} - f$ is differentiation in the ring of functions. From this it follows that $\Delta_1^0 = \sigma_1 \oplus \Delta_0^0$. Since $\Gamma(\overline{X}, \Delta_0^0) = \mathbf{C}$, we need only show that $\Gamma(\overline{X}, \Delta_1^0)$ is two-dimensional.

Let $\mathscr{D} \in D_1^0$. Then $\mathscr{D} = f + \mathscr{D}'$, where $f = \mathscr{D}(1)$, and where $f$ and $\mathscr{D}$ have the degree of homogeneity 0; in particular, $f \in \mathbf{C}$.

Let us prove that $\mathscr{D}' = cI$ where $c \in \mathbf{C}$. Set $f_1' = \mathscr{D}'x_1$, $f_2' = \mathscr{D}'x_2$, $f_3' = \mathscr{D}'x_3$. Then $f_i' \in E^1(X)$ and we can extend them to linear functions $f_i$ on $\mathbf{C}^3$. The operator $\mathscr{D}'$ coincides with $\widetilde{\mathscr{D}} = f_1 \frac{\partial}{\partial x_1} + f_2 \frac{\partial}{\partial x_2} + f_3 \frac{\partial}{\partial x_3}$.

Therefore $\widetilde{\mathscr{D}}(x_1^3 + x_2^3 + x_3^3) = 3(f_2 x_1^2 + f_2 x_2^2 + f_3 x_3^2) = c(x_1^3 + x_2^3 + x_3^3)$, where $c \in \mathbf{C}[x_1, x_2, x_3]$ (here equality is considered in the ring $\mathbf{C}[x_1, x_2, x_3]$). Since $f_1, f_2, f_3$ are linear functions, we have $c \in \mathbf{C}$ and $f_i = cx_i$, that is, $\mathscr{D}' = cI$.

So we have shown that every operator $\mathscr{D} \in D_1^0$ is of the form $\mathscr{D} = c_1 + cI$, where $c_1, c \in \mathbf{C}$. The lemma is now proved.

We shall use the following well-known facts on the cohomology of coherent sheaves on an elliptic curve (see [2]).

1°. If $\mathscr{F}$ is a coherent sheaf on $\overline{X}$, then $H^i(\overline{X}, \mathscr{F})$ for $i \geqslant 2$.

2°. $\dim H^0(\overline{X}, O_X) = \dim H^1(\overline{X}, O_X) = 1$.

3°. If $\mathscr{L}$ is an invertible sheaf on $\overline{X}$ and $\dim H^0(\overline{X}, \mathscr{L}) > 1$, then $H^1(\overline{X}, \mathscr{L}) = 0$ and $H^0(\overline{X}, \mathscr{L}^i) = 0$ for $i < 0$.

LEMMA 5. *Consider the exact sequence of sheaves*

$$0 \to N^n \to \sigma_n \to \mathscr{K} \otimes \sigma_{n-1} \to 0 \qquad (V_n).$$

*Then* $\dim H^0(\overline{X}, \sigma_n) = 1$, *and the boundary homomorphism* $\delta_n\colon H^0(\overline{X}, \mathscr{K} \otimes \sigma_{n-1}) \to H^1(\overline{X}, N^n)$ *is an isomorphism.*

PROOF. We prove the lemma by induction on $n$.

Let $n = 1$. We write out the exact cohomology sequence

$$0 \to H^0(\overline{X}, N) \xrightarrow{\widetilde{\varphi}_1} H^0(\overline{X}, \sigma_1) \xrightarrow{\widetilde{\psi}_1} H^0(\overline{X}, \mathscr{K}) \xrightarrow{\delta_1}$$
$$\to H^1(\overline{X}, N) \xrightarrow{\varphi_1'} H^1(\overline{X}, \sigma_1) \xrightarrow{\psi_1'} H^1(\overline{X}, \mathscr{K}) \to 0.$$

Recall that $N$ and $\mathscr{K}$ are isomorphic to $O_{\overline{X}}$. Hence $\widetilde{\varphi}_1$ is an isomorphism (because $\dim H^0(X, \sigma_2) = \dim H^0(\overline{X}, N) = 1$). This means that $\widetilde{\psi}_1 = 0$. Therefore $\delta_1$ is an isomorphism. Hence $\varphi_1' = 0$, and $\psi_1'$ is an isomorphism.

Suppose that the lemma has been proved for the sequence $V_n$; let us prove it for $V_{n+1}$. We write out the exact cohomology sequences that correspond to the sequences $V_n$ and $V_{n+1}$, and connect them according to diagram (1) (see Lemma 3)

$$\begin{array}{ccccccccccc}
0 \to H^0(N^n) & \longrightarrow & H^0(\sigma_n) & \to & H0(\mathscr{K} \otimes \sigma_{n-1}) & \xrightarrow{\delta_n} & H^1(N^n) & \to & H^1(\sigma_n) & \to & H^1(\mathscr{K} \otimes \sigma_{n-1}) \to 0 \\
\downarrow & & \downarrow & & \downarrow \tau & & \downarrow \eta & & \downarrow & & \downarrow \\
0 \to H^0(N^{n+1}) & \xrightarrow{\widetilde{\varphi}_{n+1}} & H^0(\sigma_{n+1}) & \to & H^0(\mathscr{K} \otimes \sigma_n) & \xrightarrow{\delta_{n+1}} & H^1(N^{n+1}) & \to & H^1(\sigma_{n+1}) & \to & H^1(\mathscr{K} \otimes \sigma_n) \to 0.
\end{array}$$

It is clear that $\eta$ is an isomorphism. By the inductive hypothesis $\delta_n$ is an isomorphism. Since $\eta\delta_n = \delta_{n+1}\tau \neq 0$, it follows that $\delta_{n+1} \neq 0$. Here $\mathscr{K} \otimes \sigma_n \approx \sigma_n$, and by the inductive hypothesis $\dim H^0(\overline{X}, \mathscr{K} \otimes \sigma_n) = 1$. Therefore $\delta_{n+1}$ is an isomorphism. It is now clear that $\widetilde{\varphi}_{n+1}$ is an isomorphism and that $\dim H^0(\overline{X}, \sigma_{n+1}) = 1$. Lemma 5 is now proved.

Statement 2° of Proposition 1 is a direct consequence of this lemma. For it follows at once from the exact sequence $0 \to \Delta_{n-1}^0 \to \Delta_n^0 \to \sigma_n \to 0$ that $\dim D_n^0 \leqslant \dim D_{n-1}^0 + 1$ and therefore $\dim D_n^0 \leqslant n + 1$. Hence $D_n^0$ is generated by the elements $1, I, I^2, \ldots, I^n$.

LEMMA 6. 1°. *Let* $i < 0$. *Then* a) $H^0(\overline{X}, \sigma_n \otimes \mathcal{L}^i) = 0$; b) $H^0(X, \Delta_n^i) = 0$.

Bearing in mind that $H^0(\overline{X}, \mathcal{L}^i) = 0$, it is easy to prove the lemma by induction over $n$.

Lemma 6 implies Statement 1° of Proposition 1.

LEMMA 7. *For any* $n$ ($n = 0, 1, \ldots$) we have

1°. $H^1(\overline{X}, \sigma_n \otimes \mathcal{L}) = 0$.

2°. *We consider the natural map* $\theta$: $\sigma_{n-1} \otimes \mathcal{L} \to \sigma_n \otimes \mathcal{L}$ *(multiplication by* $I$*) and denote by* $\theta'$ *the corresponding cohomology map* $\theta'$: $H^0(\overline{X}, \sigma_{n-1} \otimes \mathcal{L}) \to H^0(\overline{X}, \sigma_n \otimes \mathcal{L})$. *Then* $\theta'$ *is an embedding, and* $H^0(\overline{X}, \sigma_n \otimes \mathcal{L})/\mathrm{Im}\,\theta'$ *is three-dimensional.*

3°. $H^1(\overline{X}, \Delta_n^1) = 0$.

4°. $\dim (D_n^1/(D_{n-1}^1 + ID_{n-1}^1)) = 3$.

PROOF. From the exact sequence of sheaves

$$0 \to \sigma_{n-1} \otimes \mathcal{L} \xrightarrow{\theta} \sigma_n \otimes \mathcal{L} \to \mathcal{K}^n \otimes \mathcal{L} \to 0$$

we obtain the exact cohomology sequence

$$0 \to H^0(\overline{X}, \sigma_{n-1} \otimes \mathcal{L}) \xrightarrow{\theta'} H^0(\overline{X}, \sigma_n \otimes \mathcal{L}) \to H^0(\overline{X}, \mathcal{K}^n \otimes \mathcal{L}) \to$$
$$\to H^1(\overline{X}, \sigma_{n-1} \otimes \mathcal{L}) \to H^1(\overline{X}, \sigma_n \otimes \mathcal{L}) \to H^1(\overline{X}, \mathcal{K}^n \otimes \mathcal{L}) \to 0.$$

Since $H^1(\overline{X}, \mathcal{K}^n \otimes \mathcal{L}) = H^1(\overline{X}, \mathcal{L}) = 0$, we find by induction on $n$ that $H^1(\overline{X}, \sigma_n \otimes \mathcal{L}) = 0$ ($n = 0, 1, \ldots$). Here $\theta'$ is an embedding and $H^0(\overline{X}, \sigma_n \otimes \mathcal{L})/\mathrm{Im}\,\theta' = H^0(\overline{X}, \mathcal{K}^n \otimes \mathcal{L}) = H^0(\overline{X}, \mathcal{L}) = \mathbf{C}^3$.

From the exact cohomology sequence corresponding to the exact sequence of sheaves

$$0 \to \Delta_{n-1}^1 \to \Delta_n^1 \to \sigma_n \otimes \mathcal{L} \to 0$$

we find by induction on $n$ that $H^1(\overline{X}, \Delta_n^1) = 0$ and $D_n^1/D_{n-1}^1 = H^0(\overline{X}, \sigma_n \otimes \mathcal{L})$ ($n = 0, 1, \ldots$). Therefore $D_n^1/(D_{n-1}^1 + ID_{n-1}^1) = H^0(\overline{X}, \sigma_n \otimes \mathcal{L})/\mathrm{Im}\,\theta' = \mathbf{C}^3$. The lemma is now proved.

This lemma contains Statement 3° of Proposition 1.

NOTE. By the same method as in Lemma 7 we can show that $D_k^i = x_1 D_k^{i-1} + x_2 D_k^{i-1} + x_3 D_k^{i-1}$ for $i > 1$ and for any $k$.

## References


[1] B. Malgrange, Analytic spaces.
= Uspekhi Mat. Nauk **27** : 1 (1972), 147–184.

[2] J. Serre, Groupes algébriques et corps de classes, Hermann & Cie, Paris 1959.

# 4.

(with I. N. Bernstein and S. I. Gelfand)

# Schubert cells and cohomology of the spaces *G/P*

Usp. Mat. Nauk **28** (3) (1973) 3–26 [Russ. Math. Surv. **28** (3) (1973) 1–26].
Zbl. **289**:57024

We study the homological properties of the factor space $G/P$, where $G$ is a complex semisimple Lie group and $P$ a parabolic subgroup of $G$. To this end we compare two descriptions of the cohomology of such spaces. One of these makes use of the partition of $G/P$ into cells (Schubert cells), while the other consists in identifying the cohomology of $G/P$ with certain polynomials on the Lie algebra of the Cartan subgroup $H$ of $G$. The results obtained are used to describe the algebraic action of the Weyl group $W$ of $G$ on the cohomology of $G/P$.

## Contents



## Introduction

Let $G$ be a linear semisimple algebraic group over the field $\mathbf{C}$ of complex numbers and assume that $G$ is connected and simply-connected. Let $B$ be a Borel subgroup of $G$ and $X = G/B$ the fundamental projective space of $G$.

The study of the topology of $X$ occurs, explicitly or otherwise, in a large number of different situations. Among these are the representation theory of semisimple complex and real groups, integral geometry and a number of problems in algebraic topology and algebraic geometry, in which analogous spaces figure as important and useful examples. The study of the homological properties of $G/P$ can be carried out by two well-known methods. The first of these methods is due to A. Borel [1] and involves the identification of the cohomology ring of $X$ with the quotient ring of the ring of polynomials on the Lie algebra $\mathfrak{h}$ of the Cartan subgroup

$H \subset G$ by the ideal generated by the $W$-invariant polynomials (where $W$ is the Weyl group of $G$). An account of the second method, which goes back to the classical work of Schubert, is in Borel's note [2] (see also [3]); it is based on the calculation of the homology with the aid of the partition of $X$ into cells (the so-called Schubert cells). Sometimes one of these approaches turns out to be more convenient and sometimes the other, so naturally we try to establish a connection between them. Namely, we must know how to compute the correspondence between the polynomials figuring in Borel's model of the cohomology and the Schubert cells. Furthermore, it is an interesting problem to find in the quotient ring of the polynomial ring a symmetrical basis dual to the Schubert cells. These problems are solved in this article. The techniques developed for this purpose are applied to two other problems. The first of these is the calculation of the action of the Weyl group on the homology of $X$ in a basis of Schuberts cells, which turns out to be very useful in the study of the representations of the Chevalley groups.

We also study the action of $W$ on $X$. This action is not algebraic (it depends on the choice of a compact subgroup of $G$). The corresponding action of $W$ on the homology of $X$ can, however, be specified in algebraic terms. For this purpose we use the trajectories of $G$ in $X \times X$, and we construct explicitly the correspondences on $X$ (that is, cycles in $X \times X$) that specify the action of $W$ on $H_*(X, \mathbf{Z})$. The study of such correspondences forms the basis of many problems in integral geometry.

At the end of the article, we generalize our results to the case when $B$ is replaced by an arbitrary parabolic subgroup $P \subset G$. When $G = GL(n)$ and $G/P$ is the Grassmann variety, analogous results are to be found in [4].

B. Kostant has previously found other formulae for a basis of $H^*(X, \mathbf{Z})$, $X = G/B$, dual to the Schubert cells. We would like to express our deep appreciation to him for drawing our attention to this series of problems and for making his own results known to us.

The main results of this article have already been announced in [13].

We give a brief account of the structure of this article. At the beginning of § 1 we introduce our notation and state the known results on the homology of $X = G/B$ that are used repeatedly in the paper. The rest of § 1 is devoted to a statement of our main results.

In § 2 we introduce an ordering on the Weyl group $W$ of $G$ that arises naturally in connection with the geometry of $X$, and we investigate its properties.

§ 3 is concerned with the ring $R$ of polynomials on the Lie algebra $\mathfrak{h}$ of the Cartan subgroup $H \subset G$. In this section we introduce the functionals $D_w$ on $R$ and the elements $P_w$ in $R$ and discuss their properties.

In § 4 we prove that the elements $D_w$ introduced in § 3 correspond to the Schubert cells of $X$.

§5 contains generalizations and applications of the results obtained, in particular, to the case of manifolds $X(P) = G/P$, where $P$ is an arbitrary parabolic subgroup of $G$. We also study in §5 the correspondences on $X$ and in particular, we describe explicitly those correspondences that specify the action of the Weyl group $W$ on the cohomology of $X$. Finally, in this section some of our results are put in the form in which they were earlier obtained by B. Kostant, and we also interpret some of them in terms of differential forms on $X$.

## §1. Notation, preliminaries, and statement of the main results

We introduce the notation that is used throughout the article.

$G$ is a complex semisimple Lie group, which is assumed to be connected and simply-connected;

$B$ is a fixed Borel subgroup of $G$;

$X = G/B$ is a fundamental projective space of $G$;

$N$ is the unipotent radical of $B$;

$H$ is a fixed maximal torus of $G$, $H \subset B$;

$\mathfrak{G}$ is the Lie algebra of $G$; $\mathfrak{h}$ and $\mathfrak{N}$ are the subalgebras of $\mathfrak{G}$ corresponding to $H$ and $N$;

$\mathfrak{h}^*$ is the space dual to $\mathfrak{h}$;

$\Delta \subset \mathfrak{h}^*$ is the root system of $\mathfrak{h}$ in $\mathfrak{G}$;

$\Delta_+$ is the set of positive roots, that is, the set of roots of $\mathfrak{h}$ in $\mathfrak{N}$, $\Delta_- = -\Delta_+$, $\Sigma \subset \Delta_+$ is the system of simple roots;

$W$ is the Weyl group of $G$; if $\gamma \in \Delta$, then $\sigma_\gamma : \mathfrak{h}^* \to \mathfrak{h}^*$ is an element of $W$, a reflection in the hyperplane orthogonal to $\gamma$. For each element[1] $w \in W = \mathrm{Norm}(H)/H$, the same letter is used to denote a representative of $w$ in $\mathrm{Norm}\,(H) \subset G$.

$l(w)$ is the length of an element $w \in W$ relative to the set of generators $\{\sigma_\alpha, \alpha \in \Sigma\}$ of $W$, that is, the least number of factors in the decomposition

$$w = \sigma_{\alpha_1}\sigma_{\alpha_2} \dots \sigma_{\alpha_l}, \quad \alpha_i \in \Sigma. \tag{1}$$

The decomposition (1), with $l = l(w)$, is called reduced; $s \in W$ is the unique element of maximal length, $r = l(s)$;

$N_- = sNs^{-1}$ is the subgroup of $G$ "opposite" to $N$.

For any $w \in W$ we put $N_w = w\,N_-w^{-1} \cap N$.

HOMOLOGY AND COHOMOLOGY OF THE SPACE $X$. We give at this point two descriptions of the homological structure of $X$. The first of these (Proposition 1.2) makes use of the decomposition of $X$ into cells, while the second (Proposition 1.3) involves the realization of two-dimensional cohomology classes as the Chern classes of one-dimensional bundles.

We recall (see [5]) that $N_w = w\,N_-w^{-1} \cap N$ is a unipotent subgroup of

[1] Norm $H$ is the normalizer of $H$ in $G$.

$G$ of (complex) dimension $l(w)$.

1.1. PROPOSITION (see [5]), *Let $o \in X$ be the image of $B$ in $X$. The open and closed subvarieties $X_w = Nwo \subset X$, $w \in W$, yield a decomposition of $X$ into $N$-orbits. The natural mapping $N_w \to X_w$ ($n \mapsto nwo$) is an isomorphism of algebraic varieties.*

Let $\bar{X}_w$ be the closure[1] of $X_w$ in $X$, $[\bar{X}_w] \in H_{2l(w)}(\bar{X}_w, \mathbf{Z})$ the fundamental cycle of the complex algebraic variety $\bar{X}_w$ and $s_w \in H_{2l(w)}(X,\mathbf{Z})$ the image of $[\bar{X}_w]$ under the mapping induced by the embedding $\bar{X}_w \hookrightarrow X$.

1.2. PROPOSITION (see [2]). *The elements $s_w$ form a free basis of $H_*(X,\mathbf{Z})$.*

We now turn to the other approach to the description of the cohomology of $X$. For this purpose we introduce in $\mathfrak{h}$ the root system $\{H_\gamma, \gamma \in \Delta\}$ dual to $\Delta$. (This means that $\sigma_\gamma \chi = \chi - \chi(H_\gamma)\gamma$ for all $\chi \in \mathfrak{h}^*$, $\gamma \in \Delta$). We denote by $\mathfrak{h}_\mathbf{Q} \subset \mathfrak{h}$ the vector space over $\mathbf{Q}$ spanned by the $H_\gamma$. We also set $\mathfrak{h}^*_\mathbf{Z} = \{\chi \in \mathfrak{h}^* \mid \chi(H_\gamma) \in \mathbf{Z} \text{ for all } \gamma \in \Delta\}$ and $\mathfrak{h}^*_\mathbf{Q} = \mathfrak{h}^*_\mathbf{Z} \otimes_\mathbf{Z} \mathbf{Q}$.

Let $R = S(\mathfrak{h}^*_\mathbf{Q})$ be the algebra of polynomial functions on $\mathfrak{h}_\mathbf{Q}$ with rational coefficients. We extend the natural action of $W$ on $\mathfrak{h}^*$ to $R$. We denote by $I$ the subring of $W$-invariant elements in $R$ and set $I_+ = \{f \in I \mid f(0) = 0\}$, $J = I_+R$.

We construct a homomorphism $\alpha$: $R \to H^*(X, \mathbf{Q})$ in the following way. First let $\chi \in \mathfrak{h}^*_\mathbf{Z}$. Since $G$ is simply-connected, there is a character $\theta \in \operatorname{Mor}(H, \mathbf{C}^*)$ such that $\theta(\exp h) = \exp \chi(h)$, $h \in \mathfrak{h}$. We extend $\theta$ to a character of $B$ by setting $\theta(n) = 1$ for $n \in N$. Since $G \to X$ is a principal fibre space with structure group $B$, this $\theta$ defines a one-dimensional vector bundle $E_\chi$ on $X$. We set $\alpha_1(\chi) = c_\chi$, where $c_\chi \in H^2(X, \mathbf{Z})$ is the first Chern class of $E_\chi$. Then $\alpha_1$ is a homomorphism of $\mathfrak{h}^*_\mathbf{Z}$ into $H^2(X, \mathbf{Z})$, which extends naturally to a homomorphism of rings $\alpha$: $R \to H^*(X, \mathbf{Q})$.

Note that $W$ acts on the homology and cohomology of $X$. Namely, let $K \subset G$ be a maximal compact subgroup such that $T = K \cap H$ is a maximal torus in $K$. Then the natural mapping $K/T \to X$ is a homeomorphism (see [1]). Now $W$ acts on the homology and cohomology of $X$ in the same way as on $K/T$.

1.3. PROPOSITION ([1], [8]). (i) *The homomorphism $\alpha$ commutes with the action of $W$ on $R$ and $H^*(X, \mathbf{Q})$.*

(ii) Ker $\alpha = J$, *and the natural mapping $\bar{\alpha}$: $R/J \to H^*(X, \mathbf{Q})$ is an isomorphism.*

In the remainder of this section we state the main results of this article.

The integration formula. We have given two methods of describing the

---

[1] As $X_w$ is an open and closed variety, its closure in the Zariski topology is the same as in the ordinary topology.

cohomological structure of $X$. One of the basic aims of this article is to establish a connection between these two approaches. By this we understand the following. Each Schubert cell $s_w \in H_*(X, \mathbf{Z})$ gives rise to a linear functional $\hat{D}_w$ on $R$ according to the formula

$$\hat{D}_w(f) = \langle s_w, \alpha(f)\rangle$$

(where $\langle\ ,\ \rangle$ is the natural pairing of homology and cohomology). We indicate an explicit form for $\hat{D}_w$.

For each root $\gamma \in \Delta$, we define an operator $A_\gamma : R \to R$ by the formula

$$A_\gamma f = \frac{f - \sigma_\gamma f}{\gamma}$$

(that is, $A_\gamma f(h) = [f(h) - f(\sigma_\gamma h)]/\gamma(h)$ for all $h \in \mathfrak{h}$). Then we have the following proposition.

PROPOSITION. *Let* $w = \sigma_{\alpha_1} \ldots \sigma_{\alpha_l}$, $\alpha_i \in \Sigma$. If $l(w) < l$, *then* $A_{\alpha_1} \ldots A_{\alpha_l} = 0$. *If* $l(w) = l$, *then the operator* $A_{\alpha_1} \ldots A_{\alpha_l}$ *depends only on* $w$ *and not on the representation of* $w$ *in the form* $w = \sigma_{\alpha_1} \ldots \sigma_{\alpha_l}$; *we put* $A_w = A_{\alpha_1} \ldots A_{\alpha_l}$.

This proposition is proved in § 3 (Theorem 3.4).

The functional $\hat{D}_w$ is easily described in terms of the $A_w$: we define for each $w \in W$ another functional $D_w$ on $R$ by the formula $D_w f = A_w f(0)$. The following theorem is proved in § 4 (Theorem 4.1).

THEOREM. $D_w = \hat{D}_w$ *for all* $w \in W$.

We can give another more explicit description of $D_w$ (and thus of $\hat{D}_w$). To do this, we write $w_1 \xrightarrow{\gamma} w_2$, $w_1, w_2 \in W$, $\gamma \in \Delta_+$, to express the fact that $w_1 = \sigma_\gamma w_2$ and $l(w_2) = l(w_1) = l + 1$.

THEOREM. *Let* $w \in W$, $l(w) = l$.

(i) *If* $f \in R$ *is a homogeneous polynomial of degree* $k \neq l$, *then* $\hat{D}_w(f) = 0$.

(ii) If $\chi_1, \ldots, \chi_l \in \mathfrak{h}_{\mathbf{Q}}^*$, then $\hat{D}_w(\chi_1 \ldots \chi_l) = \sum \chi_1(H_{\gamma_1}) \ldots \chi_l(H_{\gamma_l})$, where the sum is taken over all chains of the form

$$e = w_0 \xrightarrow{\gamma_1} w_1 \xrightarrow{\gamma_2} \ldots \xrightarrow{\gamma_l} w_l = w^{-1}$$

(see Theorem 3.12 (i), (v)).

The next theorem describes the basis of $H^*(X, \mathbf{Q})$ dual to the basis $\{s_w \mid w \in W\}$ of $H^*(X, \mathbf{Z})$. We identify the ring $\bar{R} = R/J$ with $H^*(X, \mathbf{Q})$ by means of the isomorphism $\bar{\alpha}$ of Proposition 1.3. Let $\{P_w \mid w \in W\}$ be the basis of $\bar{R}$ dual to the basis $\{s_w \mid w \in W\}$ of $H_*(X, \mathbf{Z})$. To specify $P_w$, we note that the operators $A_w : R \to R$ preserve the ideal $J \subset R$ (lemma 3.3 (v)), and so the operators $\bar{A}_w : \bar{R} \to \bar{R}$ are well-defined.

THEOREM. (i) *Let* $s \in W$ *be the element of maximal length*, $r = l(s)$ *Then* $P_s = \rho^r/r! \pmod J = |W|^{-1} \prod_{\gamma\in\Delta_+} \gamma \pmod J$, (*where* $\rho \in \mathfrak{h}_{\mathbf{Q}}^*$ *is half* *the sum of the positive roots and* $|W|$ *is the order of* $W$)

(ii) *If* $w \in W$, *then* $P_w = \bar{A}_{w^{-1}s}P_s$ (see Theorem 3.15, Corollary 3.16, Theorem 3.14(i)).

Another expression for the $P_w$ has been obtained earlier by B. Kostant (see Theorem 5.9).

The following theorem gives a couple of important properties of the $P_w$.

THEOREM (i). *Let* $\chi \in \mathfrak{h}_{\mathbf{Q}}^*$, $w \in W$. *Then* $\chi \cdot P_w = \sum_{w \xrightarrow{\gamma} w'} w\chi(H_\gamma)\, P_{w'}$

(see Theorem 3.14 (ii)).

(ii) *Let* $\mathcal{P} : H_*(X, \mathbf{Q}) \to H^*(X, \mathbf{Q})$ *be the Poincaré duality. Then* $\mathcal{P}(s_w) = \bar{\alpha}(P_{ws})$ (see Corollary 3.19).

THE ACTION OF THE WEYL GROUP. The action of $W$ on $H^*(X, \mathbf{Q})$ can easily be described using the isomorphism $\bar{\alpha}: R/J \to H^*(X, \mathbf{Q})$, but we are interested in the problem of describing the action of $W$ on the basis $\{s_w\}$ of $H_*(X, \mathbf{Q})$.

THEOREM. *Let* $\alpha \in \Sigma$, $w \in W$. *Then* $\sigma_\alpha s_w = -s_w$ *if* $l(w\sigma_\alpha) = l(w) - 1$ *and* $\sigma_\alpha s_w = -s_w + \sum_{w' \xrightarrow{\gamma} w\sigma_\alpha} w'\alpha(H_\gamma)\, s_{w'}$, *if* $l(w\sigma_\alpha) = l(w) + 1$ (see Theorem 3.12 (iv)).

In § 5 we consider some applications of the results obtained. To avoid overburdening the presentation, we do not make precise statements at this point. We merely mention that Theorem 5.5 appears important to us, in which a number of results is generalized to the case of the varieties $X(P) = G/P$ ($P$ being an arbitrary parabolic subgroup of $G$), and also Theorem 5.7, in which we investigate certain correspondences on $X$.

## §2. The ordering on the Weyl group and the mutual disposition of the Schubert cells

2.1 DEFINITION (i) *Let* $w_1, w_2 \in W$, $\gamma \in \Delta_+$. *Then* $w_1 \xrightarrow{\gamma} w_2$ *indicates the fact that* $\sigma_\gamma w_1 = w_2$ *and* $l(w_2) = l(w_1) + 1$.

(ii) *We put* $w < w'$ *if there is a chain*

$$w = w_1 \to w_2 \to \ldots \to w_k = w'.$$

It is helpful to picture $W$ in the form of a directed graph with edges drawn in accordance with Definition 2.1 (i).

Here are some properties of this ordering.

2.2 LEMMA. *Let* $w = \sigma_{\alpha_1} \ldots \sigma_{\alpha_l}$ *be the reduced decomposition of an element* $w \in W$. *We put* $\gamma_i = \sigma_{\alpha_1} \ldots \sigma_{\alpha_{i-1}}(\alpha_i)$. *Then the roots* $\gamma_1, \ldots, \gamma_l$ *are distinct and the set* $\{\gamma_1, \ldots, \gamma_l\}$ *coincides with* $\Delta_+ \cap w\Delta_-$.

This lemma is proved in [6].

2.3 COROLLARY. (i) *Let* $w = \sigma_{\alpha_1} \ldots \sigma_{\alpha_l}$ *be the reduced decomposition and let* $\gamma \in \Delta_+$ *be a root such that* $w^{-1}\gamma \in \Delta_-$. *Then for some* i

(2) $$\sigma_\gamma\sigma_{\alpha_1} \ldots \sigma_{\alpha_i} = \sigma_{\alpha_1} \ldots \sigma_{\alpha_{i-1}}.$$

(ii) *Let* $w \in W$, $\gamma \in \Delta_+$. *Then* $l(w) < l(\sigma_\gamma w)$, *if and only if* $w^{-1}\gamma \in \Delta_+$.

PROOF (i) From Lemma 2.2 we deduce that $\gamma = \sigma_{\alpha_1} \ldots \sigma_{\alpha_{i-1}} (\alpha_i)$ *for some* $i$, and (2) follows.

(ii) If $w^{-1}\gamma \in \Delta_-$, then by (2) $\sigma_\gamma w = \sigma_{\alpha_1} \ldots a_{\alpha_{i-1}} \sigma_{\alpha_{i+1}} \ldots \sigma_{\alpha_l}$, that is $l(\sigma_\gamma w) < l(w)$. Interchanging $w$ and $\sigma_\gamma w$, we see that if $w^{-1}\gamma \in \Delta_+$, then $l(w) < l(\sigma_\gamma w)$.

2.4 LEMMA. *Let* $w_1$, $w_2 \in W$, $\alpha \in \Sigma$, $\gamma \in \Delta_+$, *and* $\gamma \neq \alpha$. *Let* $\gamma' = \sigma_\alpha \gamma$. If

(3) 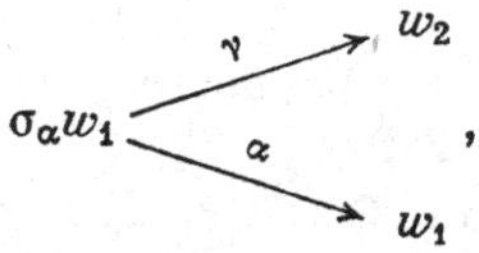


*then*

(4) 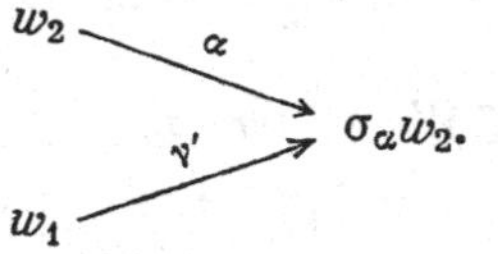


*Conversely*, (3) *follows from* (4).

PROOF. Since $\alpha \in \Sigma$ and $\gamma \neq \alpha$, we have $\gamma' = \sigma_\alpha \gamma \in \Delta_+$. It is therefore sufficient to show that $l(\sigma_\alpha w_2) > l(w_2) = l(w_1)$. This follows from Corollary 2.3, because $\sigma_\alpha w_2 = \sigma_{\gamma'} w_1$ and $(\sigma_\alpha w_2)^{-1}\gamma' = w_2^{-1}\sigma_\alpha\gamma' = w_2^{-1}\gamma \in \Delta_-$ by (3). The second assertion of the lemma is proved similarly.

2.5 LEMMA. *Let* $w$, $w' \in W$, $\alpha \in \Sigma$ *and assume that* $w < w'$. *Then*

a) *either* $\sigma_\alpha w \leqslant w'$ *or* $\sigma_\alpha w < \sigma_\alpha w'$,

b) *either* $w \leqslant \sigma_\alpha w'$ *or* $\sigma_\alpha w < \sigma_\alpha w'$.

PROOF a) Let

$$w = w_1 \to w_2 \to \quad \ldots \quad \to w_k = w'.$$

We proceed by induction on $k$. If $\sigma_\alpha w < w$ or $\sigma_\alpha w = w_2$, the assertion is obvious. Let $w < \sigma_\alpha w$, $\sigma_\alpha w \neq w_2$. Then $\sigma_\alpha w < \sigma_\alpha w_2$ by Lemma 2.4. We obtain a) by applying the inductive hypothesis to the pair $(w_2, w')$.

b) is proved in a similar fashion.

2.6. COROLLARY. *Let* $\alpha \in \Sigma$, $w_1 \xrightarrow{\alpha} w_1'$, $w_2 \xrightarrow{\alpha} w_2'$. *If one of the elements* $w_1$, $w_1'$ *is smaller (in the sense of the above ordering) than one of* $w_2$, $w_2'$, *then* $w_1 \leqslant w_2 < w_2'$ *and* $w_1 < w_1' \leqslant w_2'$.

The property in Lemma 2.5 characterizes the ordering $<$. More precisely, we have the following proposition:

2.7 PROPOSITION. *Suppose that we are given a partial ordering* $w \dashv w'$ *on* $W$ *with the following properties:*

a) *If* $\alpha \in \Sigma$, $w \in W$ *with* $l(\sigma_\alpha w) = l(w) + 1$, *then* $w \dashv \sigma_\alpha w$.

b) *If* $w \dashv w'$, $\alpha \in \Sigma$, *then either* $\sigma_\alpha w \dashv w'$ *or* $\sigma_\alpha w \dashv \sigma_\alpha w'$.

*Then* $w \dashv w'$ *if and only if* $w \leqslant w'$.

PROOF. Let $s$ be the element of maximal length in $W$. It follows from a) that $e \dashv w \dashv s$ for all $w \in W$.

I. We prove that $w \leqslant w'$ implies that $w \dashv w'$. We proceed by induction on $l(w')$. If $l(w') = 0$, then $w' = e$, $w = e$ and so $w \dashv w'$. Let $l(w') > 0$ and let $\alpha \in \Sigma$ be a root such that $l(\sigma_\alpha w') = l(w') - 1$. Then by Lemma 2.5 a), either $\alpha_\alpha w \leqslant \sigma_\alpha w'$ or $w \leqslant \sigma_\alpha w'$.

(i) $w \leqslant \sigma_\alpha w' \Rightarrow w \dashv \sigma_\alpha w'$ (by the inductive hypothesis), $\Rightarrow w \dashv w'$. (using a)).

(ii) $\sigma_\alpha w \leqslant \sigma_\alpha w' \Rightarrow \sigma_\alpha w \dashv \sigma_\alpha w'$ (by the inductive hypothesis), $\Rightarrow$ either $w \dashv \sigma_\alpha w'$ or $w \dashv w'$ (applying b) to the pair $(\sigma_\alpha w, \sigma_\alpha w')$), $\Rightarrow w \dashv w'$.

II. We now show that $w \dashv w'$ implies that $w \leqslant w'$. We proceed by backward induction on $l(w)$. If $l(w) = r = l(s)$, then $w = s$, $w' = s$, and so $w \leqslant w'$. Let $l(w) < r$ and let $\alpha$ be an element of $\Sigma$ such that $l(\sigma_2 w) = l(w) + 1$. By b) either $\sigma_\alpha w \dashv w'$ or $\sigma_\alpha w \dashv \sigma_\alpha w'$.

(i) $\sigma_\alpha w \dashv w' \Rightarrow \sigma_\alpha w \leqslant w'$ (by the inductive hypothesis) $\Rightarrow w \leqslant w'$.

(ii) $\sigma_\alpha w \dashv \sigma_\alpha w' \Rightarrow \sigma_\alpha w \leqslant \sigma_\alpha w' \Rightarrow w \leqslant w'$ (by Corollary 2.6).

Proposition 2.7 is now proved.

2.8 PROPOSITION. *Let $w \in W$ and let $w = \sigma_{\alpha_1} \ldots \sigma_{\alpha_l}$ be the reduced decomposition of $w$.*

a) *If $1 \leqslant i_1 < i_2 < \ldots < i_k \leqslant l$ and*

$$w' = \sigma_{\alpha_{i_1}} \ldots \sigma_{\alpha_{i_k}}, \tag{5}$$

*then $w' \leqslant w$.*

b) *If $w' < w$, then $w'$ can be represented in the form* (5) *for some indexing set $\{i_j\}$.*

c) *If $w' \to w$, then there is a unique index $i$, $1 \leqslant i \leqslant l$, such that*

$$w' = \sigma_{\alpha_1} \ldots \sigma_{\alpha_{i-1}} \sigma_{\alpha_{i+1}} \ldots \sigma_{\alpha_l}. \tag{6}$$

PROOF. Let us prove c). Let $w' \to w$. Then by Lemma 2.2 there is at least one index $i$ for which (6) holds. Now suppose that (6) holds for two indices $i$, $j$, $i < j$. Then $\sigma_{\alpha_{i+1}} \ldots \sigma_{\alpha_j} = \sigma_{\alpha_i} \ldots \sigma_{\alpha_{j-1}}$. Thus, $\sigma_{\alpha_i} \ldots \sigma_{\alpha_j} = \sigma_{\alpha_{i+1}} \ldots \sigma_{\alpha_{j-1}}$, which contradicts the assumption that the decomposition $w = \sigma_{\alpha_1} \ldots \sigma_{\alpha_l}$ is reduced.

b) follows at once from c) if we take into account the fact that the decomposition (6) is reduced. We now prove a) by induction on $l$. We treat two cases separately.

(i) $i_1 > 1$. Then by the inductive hypothesis $w' \leqslant \sigma_{\alpha_2} \ldots \sigma_{\alpha_l}$, that is, $w' \leqslant \sigma_{\alpha_1} w < w$.

(ii) $i_1 = 1$. Then, by the inductive hypothesis, $\sigma_{\alpha_1} w' = \sigma_{\alpha_{i_2}} \ldots \sigma_{\alpha_{i_k}} \leqslant \sigma_{\alpha_1} w = \sigma_{\alpha_2} \ldots \sigma_{\alpha_l}$. By Corollary 2.6, $w' \leqslant w$.

Proposition 2.8 yields an alternative definition of the ordering on $W$ (see [7]). The geometrical interpretation of this ordering is very interesting and useful in what follows.

2.9 THEOREM. *Let $V$ be a finite-dimensional representation of a Lie algebra $\mathfrak{G}$ with dominant weight $\lambda$. Assume that all the weights $w\lambda$ $w \in W$, are distinct and select for each $w$ a non-zero vector $f_w \in V$ of weight $w\lambda$. Then*

$$w' \leqslant w \Leftrightarrow f_{w'} \in U(\mathfrak{N}) f_w$$

*(where $U(\mathfrak{N})$ is the enveloping algebra of the Lie algebra $\mathfrak{N}$).*

PROOF. For each root $\gamma \in \Delta$ we fix a root vector $E_\gamma \in \mathfrak{G}$ in such a way that $[E_\gamma, E_{-\gamma}] = H_\gamma$. Denote by $\mathfrak{A}_\gamma$ the subalgebra of $\mathfrak{G}$, generated by $E_\gamma$, $E_{-\gamma}$, and $H_\gamma$. $\mathfrak{A}_\gamma$ is isomorphic to the Lie algebra $sl_2(\mathbf{C})$. Let $w' \xrightarrow{\gamma} w$ and let $\widetilde{V}$ be the smallest $\mathfrak{A}_\gamma$-invariant subspace of $V$ containing $f_{w'}$.

2.10 LEMMA. *Let $n = w'\lambda(H_\gamma) \in \mathbf{Z}$, $n > 0$. The elements $\{E^i_{-\gamma} f_{w'} \mid i = 0, 1, \ldots, n\}$ form a basis of $\widetilde{V}$. Put $\tilde{f} = E^n_{-\gamma} f_{w'}$. Then $E_{-\gamma}\tilde{f} = 0$, $E^n_\gamma \tilde{f} = c' f_{w'}$ $(c' \neq 0)$ and $f_w = c\tilde{f}$ $(c \neq 0)$.*

PROOF. By Lemma 2.2, $w'^{-1}\gamma \in \Delta_+$, hence $E_\gamma f_{w'} = = cE_\gamma w' f_e = cw' E_{w'^{-1}\gamma} f_e = 0$, that is, $f_{w'}$ is a vector of dominant weight relative to $\mathfrak{A}_\gamma$. All the assertions of the lemma, except the last, follow from standard facts about the representations of the algebra $\mathfrak{A}_\gamma \cong sl_2(\mathbf{C})$. Furthermore, $\tilde{f}$ and $f_w$ are two non-zero vectors of weight $w\lambda$ in $V$, and since the multiplicity of $w\lambda$ in $V$ is equal to 1, these vectors are proportional. The lemma is now proved.

To prove Theorem 2.9 we introduce a partial ordering on $W$ by putting $w' \dashv w$ if $f_{w'} \in U(\mathfrak{N}) f_w$. Since all the weights $w\lambda$ are distinct, the relation $\dashv$ is indeed an ordering; we show that it satisfies conditions a) and b) of Proposition 2.7.

a) Let $\alpha \in \Sigma$ and $l(\sigma_\alpha w) = l(w) + 1$. Then $w \xrightarrow{\alpha} \sigma_\alpha w$, and by Lemma 2.10, $f_w \in U(\mathfrak{N}) f_{\sigma_\alpha w}$, that is, $w \dashv \sigma_\alpha w$.

b) Let $w \dashv w'$. We choose an $\alpha \in \Sigma$ such that $w \xrightarrow{\alpha} \sigma_\alpha w$. Replacing $w'$ by $\sigma_\alpha w'$, if necessary, we may assume that $\sigma_\alpha w' \to w'$. We prove that $\sigma_\alpha w \dashv w'$, that is, $f_{\sigma_\alpha w} \in U(\mathfrak{N}) f_{w'}$. It follows from Lemma 2.10 that $E_{-\alpha} f_{w'} = 0$ and $f_{\sigma_\alpha w} = cE^n_{-\alpha} f_w$. Let $\mathfrak{P}_\alpha$ be the subalgebra of $\mathfrak{G}$ generated by $\mathfrak{N}$, $\mathfrak{h}$ and $\mathfrak{A}_\alpha$. Since $w \dashv w'$, $f_w \in U(\mathfrak{N}) f_{w'}$ and so $f_{\sigma_\alpha w} = cE^n_{-\alpha} f_w = X f_{w'}$, where $X \in U(\mathfrak{P}_\alpha)$. Any element $X$ of $U(\mathfrak{P}_\alpha)$ can be represented in the form $X = \sum_{i=1}^{k} Y_i Y'_i + \widetilde{Y} E_{-\alpha}$, where $Y_i \in U(\mathfrak{N})$, $Y'_i \in U(\mathfrak{h})$, $\widetilde{Y} \in U(\mathfrak{P}_\alpha)$. Therefore, $f_{\sigma_\alpha w} = \sum Y_i Y'_i f_{w'} = \sum c_i Y_i f_{w'} \in U(\mathfrak{N}) f_{w'}$ and Theorem 2.9 is proved.

We use Theorem 2.9 to describe the mutual disposition of the Schubert cells.

2.11. THEOREM (Steinberg [7]). *Let $w \in W$, $X_w \subset X$ a Schubert cell, and $\overline{X}_w$ its closure. Then $X_{w'} \subset \overline{X}_w$ if and only if $w' \leqslant w$.*

To prove this theorem, we give a geometric description of the variety $X_w$.

Let $V$ be a finite-dimensional representation of $G$ with regular dominant weight $\lambda$ (that is, all the weights $w\lambda$ distinct). As above, we choose for each $w \in W$ a non-zero vector $f_w \in V$ of weight $w\lambda$. We consider the space $P(V)$ of lines in $V$; if $f \in V$, $f \neq 0$, then we denote by $[f] \in P(V)$ a line passing through $f$. Since $\lambda$ is regular, the stabilizer of the point $[f_e] \in P(V)$ under the natural action of $G$ on $P(V)$ is $B$. The $G$-orbit of $[f_e]$ in $P(V)$ is therefore naturally isomorphic to $X = G/B$. In what follows, we regard $X$ as a subvariety of $P(V)$.

For each $w \in W$ we denote by $\phi_w$ the linear function on $V$ given by $\phi_w(f_w) = 1$, $\phi_w(f) = 0$ if $f \in V$ is a vector of weight distinct from $w\lambda$.

2.12 LEMMA. *Let $f \in V$ and $[f] \in X$. Then*

$$[f] \in X_w \Leftrightarrow f \in U(\mathfrak{N}) f_w, \ \varphi(f) \neq 0.$$

PROOF. We may assume that $f = gf_e$ for some $g \in G$.

Let $[f] \in X_w$, that is, $g \in NwB$. Then $f = c_1 \exp(Y)wf_e$ for some $Y \in \mathfrak{N}$, hence $f \in U(\mathfrak{N})f_w$, and $\phi_w(f) \neq 0$.

On the other hand, it is clear that for each $f \in V$ there is at most one $w \in W$ such that $f \in U(\mathfrak{N}) f_w$ and $\phi_w(f) \neq 0$. The Lemma now follows from the fact that $X = \bigcup_{w \in W} X_w$.

We now prove Theorem 2.11

a) Let $X_{w'} \subset \overline{X}_w$. Then $[f_{w'}] \in \overline{X}_w$, and by Lemma 2.12, $f_{w'} \in U(\mathfrak{N})f_w$. So $w' \leqslant w$, by Theorem 2.9.

b) To prove the converse it is sufficient to consider the case $w' \xrightarrow{\gamma} w$. Let $n = w\lambda(H_\gamma) \in \mathbf{Z}$. Just as in the proof of Theorem 2.9, a) we can show that $n > 0$, $E_\gamma^n f_w = cf_{w'}$ and $E_\gamma^{n+1} f_w = 0$.

Therefore $\lim_{t \to \infty} t^{-n} \exp(tE_\gamma) f_w = \frac{c}{n!} f_{w'}$, that is, $[f_{w'}] \in X_w$. Hence, $X_{w'} \subset \overline{X}_w$.

## §3. Discussion of the ring of polynomials on $\mathfrak{h}$

In this section we study the rings $R$ and $\overline{R}$. For each $w \in W$ we define an element $P_w \in \overline{R}$ and a functional $D_w$ on $R$ and investigate their properties. In the next section we shall show that the $D_w$ correspond to Schubert cells, and that the $P_w$ yield a basis, dual to the Schubert cell basis, for the cohomology of X.

3.1 DEFINITION. (i) *$R = \oplus R_i$ is the graded ring of polynomial functions on $\mathfrak{h}_\mathbf{Q}$ with rational coefficients. $W$ acts on $R$ according to the rule $wf(h) = f(w^{-1}h)$.*

(ii) *$I$ is the subring of $W$-invariant elements in $R$,*

$$I_+ = \{f \in I \mid f(0) = 0\}.$$

(iii) *$J$ is the ideal of $R$ generated by $I_+$.*
(iv) $R = R/J$.

3.2 DEFINITION. *Let $\gamma \in \Delta$. We specify an operator $A_\gamma$ on $R$ by the rule*

$$A_\gamma f = \frac{f - \sigma_\gamma f}{\gamma}.$$

*$A_\gamma f$ lies in $R$, since $f - \sigma_\gamma f = 0$ on the hyperplane $\gamma = 0$ in $\mathfrak{h}_Q$.*

The simplest properties of the $A_\gamma$ are described in the following lemma.

3.3 LEMMA.(i) $A_{-\gamma} = -A_\gamma$, $A_\gamma^2 = 0$.
(ii) $wA_\gamma w^{-1} = A_{w\gamma}$.
(iii) $\sigma_\gamma A_\gamma = -A_\gamma\sigma_\gamma = A_\gamma$, $\sigma_\gamma = -\gamma A_\gamma + 1 = A_\gamma\gamma - 1$.
(iv) $A_\gamma f = 0 \Leftrightarrow \sigma_\gamma f = f$.
(v) $A_\gamma J \subset J$.
(vi) *Let $\chi \in \mathfrak{h}_Q^*$. Then the commutator of $A_\gamma$ with the operator of multiplication by $\chi$ has the form $[A_\gamma, \chi] = \chi(H_\gamma)\sigma_\gamma$.*

PROOF. (i) – (iv) are clear. To prove (v), let $f = f_1 f_2$, where $f_1 \in I_+$, $f_2 \in R$. It is then clear that $A_\gamma f = f_1 . A_\gamma f_2 \in J$. As to (vi), since $\sigma_\gamma\chi = \chi - \chi(H_\gamma)\gamma$, we have

$$[A_\gamma, \chi] f = A_\gamma (\chi f) - \chi A_\gamma (f) = \frac{1}{\gamma} (\chi f - \sigma_\gamma\chi\cdot\sigma_\gamma f - \chi f + \chi\sigma_\gamma f) =$$

$$= \frac{\chi - \sigma_\gamma\chi}{\gamma}\cdot\sigma_\gamma f = \chi (H_\gamma)\cdot\sigma_\gamma f.$$

The following property of the $A_\gamma$ is fundamental in what follows.

3.4 THEOREM. *Let $\alpha_1, \ldots, \alpha_l \in \Sigma$, and put $w = \sigma_{\alpha_1} \ldots \sigma_{\alpha_l}$; $A_{(\alpha_1, \ldots, \alpha_l)} = A_{\alpha_1} \ldots A_{\alpha_l}$.*

a) *If $l(w) < l$, then $A_{(\alpha_1, \ldots, \alpha_l)} = 0$.*

b) *If $l(w) = l$, then $A_{(\alpha_1, \ldots, \alpha_l)}$ depends only on $w$ and not on the set $\alpha_1, \ldots, \alpha_l$. In this case we put $A_w = A_{(\alpha_1, \ldots, \alpha_l)}$.*

The proof is by induction on $l$, the result being obvious when $l = 1$.

For the proof of a), we may assume by the inductive hypothesis that $l(\sigma_{\alpha_1} \ldots \sigma_{\alpha_{l-1}}) = l - 1$, consequently $l(\sigma_{\alpha_1} \ldots \sigma_{\alpha_{l-1}} \sigma_{\alpha_l}) = l - 2$. Then $\sigma_{\alpha_i} \sigma_{\alpha_{i+1}} \ldots \sigma_{\alpha_{l-1}} = \sigma_{\alpha_{i+1}} \ldots \sigma_{\alpha_{l-1}} \sigma_{\alpha_l}$ for some $i$ ( we have applied Corollary 2.3 to the case $w = \sigma_{\alpha_{l-1}} \ldots \sigma_{\alpha_1}$, $\gamma = \alpha_l$). We show that $A_{(\alpha_i, \ldots, \alpha_l)} = 0$.

Since $l - i < l$, the inductive hypothesis shows that $A_{\alpha_i} A_{\alpha_{i+1}} \ldots A_{\alpha_{l-1}} = A_{\alpha_{i+1}} \ldots A_{\alpha_{l-1}} A_{\alpha_l}$, and so by lemma 3.3 (i) $A_{\alpha_i} \ldots A_{\alpha_l} = A_{\alpha_{i+1}} \ldots A_{\alpha_l} A_{\alpha_l} = 0$.

To prove b), we introduce auxiliary operators $B_{(\alpha_1, \ldots, \alpha_l)}$, by setting

$$B_{(\alpha_1, \ldots, \alpha_l)} = \sigma_{\alpha_l} \ldots \sigma_{\alpha_1} A_{(\alpha_1, \ldots, \alpha_l)}.$$

We put $w_i = \sigma_{\alpha_l} \ldots \sigma_{\alpha_i}$. Then in view of Lemma 3.3 (ii, iii) we have

(7) $$B_{(\alpha_1,\ldots,\alpha_l)} = A_{\alpha_1}^{w_2} A_{\alpha_2}^{w_3} \ldots A_{\alpha_{l-1}}^{w_l} A_{\alpha_l}$$

(where $A_\gamma^w$ stands for $wA_\gamma w^{-1}$).

3.5 LEMMA. *Let $\chi \in \mathfrak{h}_Q^*$. The commutator of $B_{(\alpha_1,\ldots,\alpha_l)}$ with the operator of multiplication by $\chi$ is given by the following formula:*[1]

(8) $$[B_{(\alpha_1,\ldots,\alpha_l)}, \chi] = \sum_{i=1}^{l} \chi(w_{i+1}H_{\alpha_i})\, w_{i+1}w_i^{-1} B_{(\alpha_1,\ldots,\hat{\alpha}_i,\ldots,\alpha_l)}.$$

PROOF. We have

$$[B_{(\alpha_1,\ldots,\alpha_l)}, \chi] = [A_{\alpha_1}^{w_2} A_{\alpha_2}^{w_3} \ldots A_{\alpha_l}, \chi] =$$
$$= \sum_{i=1}^{l} A_{\alpha_1}^{w_2} A_{\alpha_2}^{w_3} \ldots [A_{\alpha_i}^{w_{i+1}}, \chi] \ldots A_{\alpha_l} = \sum_{i=1}^{l} T_i.$$

By Lemma 3.3 (ii, vi), $[A_{\alpha_i}^{w_{i+1}}, \chi] = \chi(w_{i+1}H_{\alpha_i})\, \sigma_{w_{i+1}\alpha_i}$.

Since $\sigma_{w_{i+1}\alpha_i} = w_{i+1}w_i^{-1}$, we have

$$T_i = \chi(w_{i+1}H_{\alpha_i})\, A_{\alpha_1}^{w_2} \ldots A_{\alpha_{i-1}}^{w_i} w_{i+1}w_i^{-1} A_{\alpha_{i+1}}^{w_{i+2}} \ldots A_{\alpha_l}.$$

We want to move the term $w_{i+1}w_i^{-1}$ to the left. To do this we note that for $j < i$

$$A_{\alpha_j}^{w_{j+1}} w_{i+1}w_i^{-1} = w_{i+1}w_i^{-1} (A_{\alpha_j}^{w_{j+1}})^{w_i w_{i+1}^{-1}} = w_{i+1}w_i^{-1} A_{\alpha_j}^{w_i w_{i+1}^{-1} w_{j+1}} =$$
$$= w_{i+1}w_i^{-1} A_{\alpha_j}^{\sigma_{\alpha_l}\ldots\hat{\sigma}_{\alpha_i}\ldots\sigma_{\alpha_{j+1}}}.$$

Therefore,

$$T_i = \chi(w_{i+1}H_{\alpha_i})\, w_{i+1}w_i^{-1} A_{\alpha_1}^{\sigma_{\alpha_l}\ldots\hat{\sigma}_{\alpha_i}\ldots\sigma_{\alpha_2}} \ldots A_{\alpha_{i-1}}^{\sigma_{\alpha_l}\ldots\sigma_{\alpha_{i+1}}} A_{\alpha_{i+1}}^{\sigma_{\alpha_l}\ldots\sigma_{\alpha_{i+2}}} \ldots A_{\alpha_l}.$$

By (7), applied to the sequence or roots $(\alpha_1, \ldots, \hat{\alpha}_i, \ldots, \alpha_l)$, we have

$$T_i = \chi(w_{i+1}H_{\alpha_i})\, w_{i+1}w_i^{-1} B_{(\alpha_1,\ldots,\hat{\alpha}_i,\ldots,\alpha_l)},$$

and Lemma 3.5 is proved.

If $l(\sigma_{\alpha_1} \ldots \hat{\sigma}_{\alpha_i} \ldots \sigma_{\alpha_l}) < l - 1$, then $T_i = 0$ by the inductive hypothesis. If $l(\sigma_{\alpha_1} \ldots \hat{\sigma}_{\alpha_i} \ldots \sigma_{\alpha_l}) = l - 1$, then, putting $w' = \sigma_{\alpha_1} \ldots \hat{\sigma}_{\alpha_i} \ldots \sigma_{\alpha_l}$ and $\gamma = \sigma_{\alpha_1} \ldots \sigma_{\alpha_{i-1}}(\alpha_i)$, we see from Lemma 2.2 that $w' \xrightarrow{\gamma} w$, and also

$$\chi(w_{i+1}H_{\alpha_i}) = w'\chi(w'w_{i+1}H_{\alpha_i}) = w'\chi(\sigma_{\alpha_1}\ldots\sigma_{\alpha_{i-1}}H_{\alpha_i}) = w'\chi(H_\gamma)$$

and

$$w_{i+1}w_i^{-1}B_{(\alpha_1,\ldots,\hat{\alpha}_i,\ldots,\alpha_l)} = w_{i+1}w_i^{-1}w'^{-1}A_{(\alpha_1,\ldots,\hat{\alpha}_i,\ldots,\alpha_l)} = w^{-1}A_{(\alpha_1,\ldots,\hat{\alpha}_i,\ldots,\alpha_l)}.$$

[1] ^ indicates that the corresponding term must be omitted.

Using Proposition 2.8 c) and the inductive hypothesis, (8) can be rewritten in the following form:

$$[B_{(\alpha_1, \ldots, \alpha_l)}, \chi] = \sum_{w' \xrightarrow{\gamma} w} w'\chi(H_\gamma) w^{-1} A_{w'}.$$

The right-hand side of this formula does not depend on the representation of $w$ in the form of a product $\sigma_{\alpha_1} \ldots \sigma_{\alpha_l}$. The proof of theorem 3.4 is thus completed by the following obvious lemma.

3.6. LEMMA. *Let $B$ be an operator in $R$ such that $B(1) = 0$ and $[B, \chi] = 0$ for all $\chi \in \mathfrak{h}_Q^*$. Then $B = 0$.*

3.7. COROLLARY. *The operators $A_w$ satisfy the following commutator relation:*

$$[w^{-1}A_w, \chi] = \sum_{w' \xrightarrow{\gamma} w} w'\chi(H_\gamma) w^{-1} A_{w'}.$$

We put $S_i = R_i^*$ (where $R_i \subset R$ is the space of homogeneous polynomials of degree $i$) and $S = \oplus S_i$. We denote by ( , ) the natural pairing $S \times R \to Q$. Then $W$ acts naturally on $S$.

3.8 DEFINITION. (i) *For any $\chi \in \mathfrak{h}_Q^*$ we let $\chi^*$ denote the transformation of $S$ adjoint to the operator of multiplication by $\chi$ in $R$.*

(ii) *We denote by $F_\gamma$: $S \to S$ the linear transformation adjoint to $A_\gamma$: $R \to R$.*

The next lemma gives an explicit description of the $F_\gamma$.

3.9 LEMMA. *Let $\gamma \in \Delta$. For any $D \in S$ there is a $\tilde{D} \in S$ such that $\gamma^* (\tilde{D}) = D$. If $\tilde{D}$ is any such operator, then $\tilde{D} - \sigma_\gamma \tilde{D} = F_\gamma(D)$, (in particular, the left-hand side of this equation does not depend on the choice of $\tilde{D}$).*

PROOF. The existence of $\tilde{D}$ follows from the fact that multiplication by $\gamma$ is a monomorphism of $R$. Furthermore, for any $f \in R$ we have

$$(\tilde{D} - \sigma_\gamma \tilde{D}, f) = (\tilde{D}, f - \sigma_\gamma f) = (\tilde{D}, A_\gamma f \cdot \gamma) = (\gamma^* (\tilde{D}), A_\gamma f) = (D, A_\gamma f),$$

*hence* $\tilde{D} - \sigma_\gamma \tilde{D} = F_\gamma$.

REMARK. It is often convenient to interpret $S$ as a ring of differential operators on $\mathfrak{h}$ with constant rational coefficients. Then the pairing ( , ) is given by the formula $(D, f) = (Df)(0)$, $D \in S$, $f \in R$. Also, it is easy to check that $\chi^*(D) = [D, \chi]$, where $\chi \in \mathfrak{h}_Q^*$ and $D \in S$ are regarded as operators on $R$.

Theorem 3.4 and Corollary 3.7 can be restated in terms of the operators $F_\gamma$.

3.10 THEOREM. *Let $\alpha_1, \ldots, \alpha_l \in \Sigma$, $w = \sigma_{\alpha_1} \ldots \sigma_{\alpha_l}$.*

(i) *If $l(w) < l$, then $F_{\alpha_l} \ldots F_{\alpha_1} = 0$.*

(ii) *If $l(w) = l$, then $F_{\alpha_l} \ldots F_{\alpha_1}$ depends only on $w$ and not on $\alpha_1, \ldots, \alpha_l$. In this case the transformation $F_{\alpha_l} \ldots F_{\alpha_1}$ is denoted by $F_w$. (Note that $F_w = A_w^*$).*

(iii) $$[\chi^*, F_w w] = \sum_{w' \xrightarrow{\gamma} w} w'\chi(H_\gamma) F_{w'} w.$$

3.11. DEFINITION. *We set* $D_w = F_w$ (1).

As we shall show in § 4, the functionals $D_w$ correspond to the Schubert cells in $H_*$ $(X, \mathbf{Q})$ in the sense that $(D_w, f) = \langle s_w, \alpha(f)\rangle$ *for all* $f \in R$.

The properties of the $D_w$ are listed in the following theorem.

3.12. THEOREM. (i) $D_w \in S_{l(w)}$.

(ii) *Let* $w \in W$, $\alpha \in \Sigma$. *Then*

$$F_\alpha D_w = \begin{cases} 0 & \text{if } l(w\sigma_\alpha) = l(w) - 1, \\ D_{w\sigma_\alpha} & \text{if } l(w\sigma_\alpha) = l(w) + 1. \end{cases}$$

(iii) *Let* $\chi \in \mathfrak{h}^*_{\mathbf{Q}}$. *Then*

$$\chi^*(D_w) = \sum_{w' \xrightarrow{\gamma} w} w'\chi(H_\gamma) D_{w'}.$$

(iv) *Let* $\alpha \in \Sigma$. *Then*

$$\sigma_\alpha D_w = \begin{cases} -D_w, & \text{if } l(w\sigma_\alpha) = l(w) - 1, \\ -D_w + \sum_{w' \xrightarrow{\gamma} w\sigma_\alpha} w'\alpha(H_\gamma) D_{w'} & \text{if } l(w\sigma_\alpha) = l(w) + 1. \end{cases}$$

(v) *Let* $w \in W$, $l(w) = l$, $\chi_1, \ldots, \chi_l \in \mathfrak{h}^*_{\mathbf{Q}}$. *Then* $(D_w, \chi_1, \ldots, \chi_l) = \sum \chi_1(H_{\gamma_1}) \ldots \chi_l(H_l)$, *where the summation extends over all chains*

$$e \xrightarrow{\gamma_1} w_1 \xrightarrow{\gamma_2} w_2 \rightarrow \ldots \xrightarrow{\gamma_l} w_l = w^{-1}.$$

PROOF. (i) and (ii) follow from the definition of $D_w$ and Theorem 3.10 (i).

(iii) $\chi^*(D_w) = \chi^* F_w w(1) = [\chi^*, F_w w]$ (1) (since $\chi^*(1) = 0$), and (iii) follows from Theorem 3.10 (iii).

It follows from Lemma 3.3 (iii) that $\sigma_\alpha = \alpha^* F_\alpha - 1$. Thus, (iv) follows from (ii) and (iii).

(v) We put $\widetilde{D}_w = D_{w^{-1}}$. Then the $\widetilde{D}_w$ satisfy the relation

(9) $$\chi^*(\widetilde{D}_w) = \sum_{w' \xrightarrow{\gamma} w} \chi(H_\gamma) \widetilde{D}_{w'}.$$

Since $(D, \chi f) = (\chi^*(D), f)$, (v) is a consequence of (9) by induction on $l$.

Let $\mathscr{H}$ be the subspace of $S$ orthogonal to the ideal $J \subset R$. It follows from Lemma 3.3 (vi) that $\mathscr{H}$ is invariant with respect to all the $F_\gamma$. It is also clear that $1 \in \mathscr{H}$. Thus, $D_w \in \mathscr{H}$ for all $w \in H$.

3.13. THEOREM. *The functionals* $D_w$, $w \in W$, *form a basis for* $\mathscr{H}$.

PROOF. a) We first prove that the $D_w$ are linearly independent. Let $s \in W$ be the element of maximal length and $r = l(s)$. Then, by Theorem 3.12 (v), $D_s(\rho^r) > 0$ and so $D_s \neq 0$. Now let $\sum c_w D_w = 0$ and let $\widetilde{w}$ be one of the elements of maximal length for which $c_w \neq 0$. Put $l = l(\widetilde{w})$.

There is a sequence $\alpha_1, \ldots, \alpha_{r-l}$ for which $\tilde{w}\sigma_{\alpha_1} \ldots \sigma_{\alpha_{r-l}} = s$. Let $F = F_{\alpha_{r-l}} \ldots F_{\alpha_1}$. It follows from Theorem 3.10 that $FD_{\tilde{w}} = D_s$ and $FD_w = 0$ if $l(w) \geqslant l$, $w \neq \tilde{w}$. Therefore $F(\sum c_w D_w) = c_{\tilde{w}} D_s \neq 0$.

b) We now show that the $D_w$ span $\mathscr{H}$. It is sufficient to prove that if $f \in R$ and $(D_w, f) = 0$ for all $w \in W$, then $f \in J$. We may assume that $f$ is a homogeneous element of degree $k$. For $k = 0$ the assertion is clear.

Now let $k > 0$ and assume that the result is true for all polynomials $f$ of degree less than $k$. Then for all $\alpha \in \Sigma$ and $w \in W$, $(D_w, A_\alpha f) = (F_\alpha D_w, f) = 0$, by Theorem 3.10 (i) and (ii). By the inductive hypothesis, $A_\alpha f \in J$, that is, $f - \sigma_\alpha f = \alpha A_\alpha f \in J$. Hence for all $w \in W$, $f \equiv wf \pmod J$. Thus, $|W|^{-1} \sum_{w \in W} wf \equiv f \pmod J$. Since the left-hand side belongs to $I_+$, we see that $f \in J$. Theorem 3.13 is now proved.

The form ( , ) gives rise to a non-degenerate pairing between $\overline{R} = R/J$ and $\mathscr{H}$. Let $\{P_w\}$ be the basis of $\overline{R}$ dual to $\{D_w\}$. The following properties of the $P_w$ are immediate consequences of Theorem 3.12.

3.14 THEOREM. (i) *Let* $w \in W$, $\alpha \in \Sigma$. *Then*

$$A_\alpha P_w = \begin{cases} 0 & \text{if } l(w\sigma_\alpha) = l(w) + 1, \\ P_{w\sigma_\alpha} & \text{if } l(w\sigma_\alpha) = l(w) - 1. \end{cases}$$

(ii) $\chi P_w = \sum\limits_{w \xrightarrow{\gamma} w'} w\chi(H_\gamma) P_{w'}$ *for* $\chi \in \mathfrak{h}_{\mathbf{Q}}^*$.

(iii) *Let* $\alpha \in \Sigma$. *Then*

$$\sigma_\alpha P_w = \begin{cases} P_w & \text{if } l(w\sigma_\alpha) = l(w) + 1, \\ P_w - \sum\limits_{w\sigma_\alpha \xrightarrow{\gamma} w'} w\alpha(H_\gamma) P_{w'} & \text{if } l(w\sigma_\alpha) = l(w) - 1. \end{cases}$$

From (i) it is clear that all the $P_w$ can be expressed in terms of the $P_s$. More precisely, let $w = \sigma_{\alpha_1} \ldots \sigma_{\alpha_l}$, $l(w) = r - l$. Then

$$P_w = A_{\alpha_l} \ldots A_{\alpha_1} P_s.$$

To find an explicit form for the $P_w$ it therefore suffices to compute the $P_s \in \overline{R}$.

3.15 THEOREM. $P_s = |W|^{-1} \prod\limits_{\gamma \in \Delta_+} \gamma \pmod J$.

PROOF. We divide the proof into a number of steps. We fix an element $h \in \mathfrak{h}$ such that all the $wh$, $w \in W$, are distinct.

1. We first prove that there is a polynomial $Q \in R$ of degree $r$ such that

$$(10) \qquad Q(sh) = 1, \; Q(wh) = 0 \text{ for } w \neq s.$$

For each $w \in W$ we choose in $R$ a homogeneous polynomial $\widetilde{P}_w$ of degree

$l(w)$ whose image in $\bar{R} = R/J$ is $P_w$. Since $\{P_w\}$ is a basis of $R$, any polynomial $f \in R$ can be written in the form $f = \sum \tilde{P}_w f_w$, where $f_w \in I$ (this is easily proved by induction on the degree of $f$). Now let $Q' \in R$ be an arbitrary polynomial satisfying (10) and let $Q' = \sum \tilde{P}_w g_w$, $g_w \in I$. It is clear that $Q = \sum_w g_w(h) \tilde{P}_w$ meets our requirements.

2. Let $\bar{Q}$ be the image of $Q$ in $\bar{R}$, and let $\bar{Q} = \sum c_w P_w$ be the representation of $\bar{Q}$ in terms of the basis $\{P_w\}$ of $R$. We now prove that $c_s = (-1)^r \prod_{\gamma \in \Delta_+} (\gamma(h))^{-1}$.

To prove this we consider $A_s \bar{Q}$. On the one hand $A_s \bar{Q} = c_s$, by Theorem 3.13 (i); on the other hand, $A_s Q$ is a constant, since $Q$ is a polynomial of degree $r$. Hence, $A_s Q = c_s$.

We now calculate $A_s Q$, Let $s = \sigma_{\alpha_1} \ldots \sigma_{\alpha_r}$ be the reduced decomposition. We put $w_i = \sigma_{\alpha_i} \ldots \sigma_{\alpha_1}$ (in particular, $w_0 = e$), $\gamma_i = w_{i-1}^{-1} \alpha_1$, $Q_i = A_{\alpha_{i+1}} \ldots A_{\alpha_r} Q$.

LEMMA. *$Q_i$ is a polynomial of degree $i$,*

$$Q_i(w_i h) = (-1)^{r-i} \cdot \prod_{r \geqslant j > i} (\gamma_j(h))^{-1}$$

*and $Q_i(wh) = 0$ if $w \not\geqslant w_i$.*

PROOF. We prove the lemma by backward induction on $i$. For $i = r$ we have $w_r = s$, $Q_r = Q$, and the assertion of the lemma follows from the definition of $Q$.

We now assume the lemma proved for $Q_i$, $i > 0$. In the first place, it is clear that $Q_{i-1} = A_{\alpha_i} Q_i$ is a polynomial of degree $i - 1$.

Furthermore,

$$Q_{i-1}(wh) = A_{\alpha_i} Q_i(wh) = \frac{Q_i(wh) - Q_i(\sigma_{\alpha_i} wh)}{\alpha_i(wh)}.$$

If $w = w_{i-1}$, then $w < w_i$, $\sigma_{\alpha_i} w = w_i$ and $\alpha_i(w_{i-1} h) = (w_{i-1}^{-1} \alpha_i)(h) = -(w_i^{-1} \alpha_i)(h) = -\gamma_i(h)$. Therefore, using the inductive hypothesis, we have

$$Q_{i-1}(w_{i-1} h) = -\frac{Q_i(w_i h)}{\alpha_i(w_{i-1} h)} = (-1)^{r-i+1} \prod_{r \geqslant j > i-1} (\gamma_j(h))^{-1}.$$

But if $w \not\geqslant w_{i-1}$, Corollary 2.6 implies that $w \not\geqslant w_i$ and $\sigma_{\alpha_i} w \not\geqslant w_i$. So $Q_{i-1}(wh) = 0$, and the lemma is proved.

Note that by Lemma 2.2, as $i$ goes from 1 to $r$, $\gamma_i$ ranges over all the positive roots exactly once. Therefore

$$c_s = A_s Q = Q_0 = (-1)^r \prod_{\gamma \in \Delta_+} (\gamma(h))^{-1}.$$

3. Consider the polynomial $\mathrm{Alt}(Q) = \sum (-1)^{l(w)} wQ$; $\mathrm{Alt}(Q)$ is skew-symmetric, that is, $\sigma_\alpha \mathrm{Alt}(Q) = -\mathrm{Alt}(Q)$ for all $\gamma \in \Delta$. Therefore $\mathrm{Alt}(Q)$ is divisible

(in $R$) by $\prod_{\gamma\in\Delta_+}\gamma$. Since the degrees of $\mathrm{Alt}(Q)$ and $\prod_{\gamma\in\Delta_+}\gamma$ are equal (to $r$), $\mathrm{Alt}(Q) = \lambda \prod_{\gamma\in\Delta_+}\gamma$. Furthermore, $\mathrm{Alt}(Q)\,(h) = (-1)^r$, so that

$$\text{(11)}\qquad \mathrm{Alt}\,(Q) = (-1)^r \prod_{\gamma\in\Delta_+} (\gamma(h))^{-1} \prod_{\gamma\in\Delta_+}\gamma.$$

4. We put $\mathrm{Alt}\,(\overline{Q}) = \sum (-1)^{l(w)}\, w\overline{Q}$. By Theorem 3.14 (iii), $\mathrm{Alt}(P_s) = \sum(-1)^{l(w)}\, wP_s = |W|\, P_s$. Therefore $\mathrm{Alt}(\overline{Q}) = c_s\, |W|\, P_s$ + terms of smaller degree. Since $\mathrm{Alt}(Q)$ is a homogeneous polynomial of degree $r$, we have

$$\text{(12)}\qquad \mathrm{Alt}\,(\overline{Q}) = c_s\,|W|\,P_s.$$

By comparing (11) and (12) we find that

$$P_s = |W|^{-1} \prod_{\gamma\in\Delta_+}\gamma \ (\mathrm{mod}\ J).$$

The theorem is now proved.

3.16 COROLLARY. *Let $\rho$ be half the sum of the positive roots. Then $P_s = \rho^r/r!$ (mod $J$).*

PROOF. For each $\chi \in \mathfrak{h}^*$ we consider the formal power series $\exp\chi$ on $\mathfrak{h}$ given by

$$\exp\chi = \sum_{n=0}^{\infty} \chi^n/n!.$$

Then we have (see [9])

$$\sum_{w\in W} (-1)^{l(w)} \exp\,(w\rho) = \prod_{\gamma\in\Delta_+} \left[\exp\frac{\gamma}{2} - \exp\left(-\frac{\gamma}{2}\right)\right].$$

Comparing the terms of degree $r$ we see that

$$\frac{1}{r!}\sum (-1)^{l(w)}\,(w\rho)^r = \prod_{\gamma\in\Delta_+}\gamma.$$

If $\rho^r(\mathrm{mod}\ J) = \lambda P_s$, $\lambda \in \mathbf{C}$, then $(w\rho^r)\ (\mathrm{mod}\ J) = \lambda wP_s = \lambda(-1)^{l(w)}P_s$. Thus, $\frac{1}{|W|}\sum (-1)^{l(w)}\,(w\rho)^r = \lambda P_s\,(\mathrm{mod}\ J)$. The result now follows from Theorem 3.15.

To conclude this section we prove some results on products of the $P_w$ in $\overline{R}$.

3.17. THEOREM. (i) *Let $\alpha \in \Sigma$, $w \in W$. Then*

$$P_{\sigma_\alpha} P_w = \sum_{w \xrightarrow{\gamma} w'} \chi_\alpha\,(H_{w^{-1}\gamma})\, P_{w'},$$

*where $\chi_\alpha \in \mathfrak{h}^*_{\mathbf{Z}}$ is the fundamental dominant weight corresponding to the root $\alpha$ (that is, $\chi_\alpha\,(H_\beta) = 0$ for $\alpha \neq \beta \in \Sigma$, $\chi_\alpha\,(H_\alpha) = 1$).*

(ii) *Let $w_1, w_2 \in W$, $l(w_1) + l(w_2) = r$. Then $P_{w_1}\ P_{w_2} = 0$ for*

$w_2 \neq w_1 s$, $P_{w_1} P_{w_1 s} = P_s$.

(iii) *Let* $w \in W$, $f \in \overline{R}$. *Then* $f P_w = \sum_{w' \geqslant w} c_{w'} P_{w'}$.

(iv) *If* $w_1 \nleqslant w_2 s$, *then* $P_{w_1} P_{w_2} = 0$.

PROOF. (i) By Theorem 3.12 (v), $P_{\sigma_\alpha} = \chi_\alpha \pmod J$. Therefore (i) follows from Theorem 3.14 (ii).

(ii) The proof goes by backward induction on $l(w_2)$. If $l(w_2) = r$, then $w_2 = s$, $w_1 = e$ and $P_{w_1} = 1$.

To deal with the general case we find the following simple lemma useful, which is an easy consequence of the definition of the $A_\gamma$.

3.18 LEMMA. *Let* $\gamma \in \Delta$, $f, g \in R$. *Then* $A_\gamma(A_\gamma f \cdot g) = A_\gamma f \cdot A_\gamma g$.

Thus, let $w_2 \in W$, $l(w_2) = l < r$, and choose $\alpha \in \Sigma$ so that $w_2 \xrightarrow{\alpha} \sigma_\alpha w_2$. We consider two cases separately.

A) $w_1 \overset{\alpha}{\nrightarrow} \sigma_\alpha w_1$. We observe that the following equation holds for any $w \in W$

$$l(ws) = r - l(w). \tag{13}$$

Since in our case $l(\sigma_\alpha w_2) = l + 1$ and $l(\sigma_\alpha w_1) = r - l + 1$, we see that $\sigma_\alpha w_1 s \neq \sigma_\alpha w_2$, and so $w_1 s \neq w_2$. On the other hand, $P_{w_2} = A_\alpha P_{\sigma_\alpha w_2}$ and $P_{w_1} = A_\alpha P_{\sigma_\alpha w_1}$ by Theorem 3.14 (i). Therefore, an application of Lemma 3.18 shows that

$$P_{w_1} P_{w_2} = A_\alpha P_{\sigma_\alpha w_1} \cdot A_\alpha P_{\sigma_\alpha w_2} = A_\alpha (P_{\sigma_\alpha w_1} \cdot A_\alpha P_{\sigma_\alpha w_2}) = A_\alpha (P_{\sigma_\alpha w_1} \cdot P_{w_2}).$$

Since $l(\sigma_\alpha w_1) + l(w_2) = r - l + 1 + l > r$, we have $P_{\sigma_\alpha w_1} P_{w_2} = 0$. Hence $P_{w_1} P_{w_2} = 0$ as well.

B) $\sigma_\alpha w_1 \overset{\alpha}{\nrightarrow} w_1$. In this case, $P_{\sigma_\alpha w_1} = A_\alpha P_{w_1}$ and $P_{w_2} = A_\alpha P_{\sigma_\alpha w_2}$, by Theorem 3.14 (i). Again applying Lemma 3.18, we have

$$A_\alpha (P_{w_1} P_{w_2}) \doteq A_\alpha (P_{w_1} \cdot A_\alpha P_{\sigma_\alpha w_2}) = A_\alpha P_{w_1} \cdot A_\alpha P_{\sigma_\alpha w_2} = \tag{14}$$

$$= A_\alpha (A_\alpha P_{w_1} \cdot P_{\sigma_\alpha w_2}) = A_\alpha (P_{\sigma_\alpha w_1} \cdot P_{\sigma_\alpha w_2}).$$

Since the $P_w$ form a basis of $\overline{R}$, any element $f$ of degree $r$ in $\overline{R}$ has the form $f = \lambda P_s$, $\lambda \in \mathbf{C}$. Furthermore, $A_\alpha P_s = P_{\sigma_\alpha s} \neq 0$. But $\deg P_{w_1} P_{w_2} = \deg P_{\sigma_\alpha w_1} \cdot P_{\sigma_\alpha w_2} = r$. Therefore (14) is equivalent to

$$P_{w_1} P_{w_2} = P_{\sigma_\alpha w_1} P_{\sigma_\alpha w_2}.$$

Applying the inductive hypothesis to the pair $(\sigma_\alpha w_1, \sigma_\alpha w_2)$, we obtain part (ii) of the theorem.

(iii) is an immediate consequence of Theorem 3.14 (ii).

(iv) follows from (ii) and (iii).

We define the operator $\mathcal{P}: \overline{R} \to \mathcal{H}$ of Poincaré duality by the formula $(\mathcal{P}f)(g) = D_s(fg)$, $f, g \in \overline{R}$, $\mathcal{P}f \in \mathcal{H}$.

3.19. COROLLARY. $\mathcal{P} P_w = D_{ws}$.

## §4. Schubert cells

We prove in this section that the functionals $D_w$, $w \in W$ introduced in §3 correspond to Schubert cells $s_w$, $w \in W$.

Let $s_w \in H_*(X, \mathbf{Q})$ be a Schubert cell. It gives rise to a linear functional on $H^*(X, \mathbf{Q})$, which, by means of the homomorphism $\alpha: R \to H^*(X, \mathbf{Q})$ (see Theorem 1.3), can be regarded as a linear functional on $R$. This functional takes the value 0 on all homogeneous components $P_k$ with $k \neq l(w)$, and thus determines an element $\hat{D}_w \in S_{l(w)}$.

4.1. THEOREM. $\hat{D}_w = D_w$ (cf. Definition 3.11).

This theorem is a natural consequence of the next two propositions.

PROPOSITION 1. $\hat{D}_e = 1$, *and for any* $\chi \in \mathfrak{h}_{\mathbf{Z}}^*$

$$\chi^* (\hat{D}_w) = \sum_{w' \xrightarrow{\gamma} w} w'\chi (H\gamma) \hat{D}_{w'}. \tag{15}$$

PROPOSITION 2. *Suppose that for each* $w \in W$ *we are given an element* $\hat{D}_w \in S_{l(w)}$, *with* $\hat{D}_e = 1$, *for which* (15) *holds for any* $\chi \in \mathfrak{h}_{\mathbf{Z}}^*$. *Then* $\hat{D}_w = D_w$.

Proposition 2 follows at once from Theorem 3.12 (iii) by induction on $l(w)$.

We turn now to the proof of Proposition 1.

We recall (see [10]) that for any topological space $Y$ there is a bilinear mapping

$$H^i (Y, \mathbf{Q}) \times H_j (Y, \mathbf{Q}) \xrightarrow{\frown} H_{j-i} (Y, \mathbf{Q})$$

(the cap-product). It satisfies the condition:

$$1.\ \langle c \frown y, z \rangle = \langle y, c \cdot z \rangle \tag{16}$$

for all $y \in H_j (Y, \mathbf{Q})$, $z \in H^{j-i} (Y, \mathbf{Q})$, $c \in H^i (Y, \mathbf{Q})$.

2. Let $f: Y_1 \to Y_2$ be a continuous mapping. Then

$$f_* (f^*c \frown y) = c \frown f_*y \tag{17}$$

for all $y \in H_j (Y_1, \mathbf{Q})$, $c \in H^i (Y_2, \mathbf{Q})$.

By virtue of (17) we have for any $\chi \in \mathfrak{h}_{\mathbf{Z}}^*$, $f \in R$

$$(\chi^* (\hat{D}_w), f) = (\hat{D}_w, \chi f) = \langle s_w, \alpha_1 (\chi) \alpha (f) \rangle = \langle s_w \frown \alpha_1 (\chi), \alpha (f) \rangle.$$

Therefore (15) is equivalent to the following geometrical fact.

PROPOSITION 3. For all $\chi \in \mathfrak{h}_{\mathbf{Z}}^*$

$$s_w \frown \alpha_1 (\chi) = \sum_{w' \xrightarrow{\gamma} w} w'\chi (H_\gamma) s_{w'}. \tag{18}$$

We restrict the fibering $E_\chi$ to $\bar{X}_w \subset X$ and let $c_\chi \in H^2 (\bar{X}_w, \mathbf{Q})$ be the first Chern class of $E_\chi$. By (17) and the definition of the homomorphism $\alpha_1: \mathfrak{h}_{\mathbf{Z}}^* \to H^2 (X, \mathbf{Q})$, it is sufficient to prove that

(19) $$s_w \cap c_\chi = \sum_{w' \xrightarrow{\gamma} w} w'\chi(H_\gamma)\, s_{w'}.$$

in $H_{2l(w)-2}(\overline{X}_w, \mathbf{Q})$.

To prove (19), we use the following simple lemma, which can be verified by standard arguments involving relative Poincaré duality.

4.2 LEMMA. *Let $Y$ be a compact complex analytic space of dimension $n$, such that the codimension of the space of singularities of $Y$ is greater than 1. Let $E$ be an analytic linear fibering on $Y$, and $c \in H^2(Y, \mathbf{Q})$ the first Chern class of $E$. Let $\mu$ be a non-zero analytic section of $E$ and $\sum m_i Y_i = \operatorname{div} \mu$ the divisor of $\mu$. Then $[Y] \cap c = \sum m_i [Y_i] \in H_{2n-2}(Y, \mathbf{Q})$, where $[Y]$ and $[Y_i]$ are the fundamental classes of $Y$ and $Y_i$.*

Let $w \in W$, and let $X_w \subset X$ be the corresponding Schubert cell. From Lemma 4.2 and Theorem 2.11 it is clear that to prove Proposition 3 it is sufficient to verify the following facts.

4.3. PROPOSITION. *Let $w' \xrightarrow{\gamma} w$. Then $\overline{X}_w$ is non-singular at points $x \in X_{w'}$.*

4.4. PROPOSITION. *There is a section $\mu$ of the fibering $E_\chi$ over $\overline{X}_w$ such that*

$$\operatorname{div} \mu = \sum_{w' \xrightarrow{\gamma} w} w'\chi(H_\gamma)\, \overline{X}_{w'}.$$

To verify these facts we use the geometrical description of Schubert cells given in 2.9. We consider a finite-dimensional representation of $G$ on a space $V$ with regular dominant weight $\lambda$, and we realize $X$ as a subvariety of $P(V)$. For each $w \in W$ we fix a vector $f_w \in V$ of weight $w\lambda$.

PROOF OF PROPOSITION 4.3. For a root $\gamma \in \Delta_+$ we construct a three-dimensional subalgebra $\mathfrak{A}_\gamma \subset \mathfrak{G}$ (as in the proof of Theorem 2.9). Let $i$: $SL_2(\mathbf{C}) \to G$ be the homomorphism corresponding to the embedding $\mathfrak{A}_\gamma \to \mathfrak{G}$. Consider in $SL_2(\mathbf{C})$ the subgroups $B' = \left\{\begin{pmatrix} a & b \\ 0 & a^{-1}\end{pmatrix}\right\}$, $H' = \left\{\begin{pmatrix} a & 0 \\ 0 & a^{-1}\end{pmatrix}\right\}$ and $N'_- = \left\{\begin{pmatrix} 1 & 0 \\ x & 1\end{pmatrix}\right\}$ and the element $\sigma = \begin{pmatrix} 0 & 1 \\ -1 & 0\end{pmatrix}$. We may assume that $i(H') \subset H$, $i(B') \subset B$.

Let $\widetilde{V}$ be the smallest $\mathfrak{A}_\gamma$-invariant subspace of $V$ containing $f_{w'}$. It is clear that $\widetilde{V}$ is invariant under $i(SL_2(\mathbf{C}))$, and that the stabilizer of the line $[f_{w'}]$ is $B'$. This determines a mapping $\delta$: $SL_2(\mathbf{C})/B' \to X$. The space $SL_2(\mathbf{C})/B'$ is naturally identified with the projective line $\mathbf{P}^1$. Let $o, \infty \in \mathbf{P}^1$ be the images of $e, \sigma \in SL_2(\mathbf{C})$.

We define a mapping $\xi$: $N_{w'} \times \mathbf{P}^1 \to X$ by the rule

$$(x, z) \longmapsto x \cdot \delta(z).$$

4.5. LEMMA. *The mapping $\xi$ has the following properties:*

(i) $\xi(N_{w'} \times \{o\}) = X_{w'}$, $\xi(N_{w'} \times (\mathbf{P}^1 \setminus o)) \subset X_w$.

(ii) *The restriction of* $\xi$ *to* $(N_{w'} \times \mathbf{P}^1 \setminus \infty))$ *is an isomorphism onto a certain open subset of* $\bar{X}_w$.

Proposition 4.3 clearly follows from this lemma.

PROOF OF LEMMA 4.5. The first assertion of (i) follows at once from the definition of $X_{w'}$. Since the cell $X_w$ is invariant under $N$, the proof of the second assertion of (i) is reduced to showing that $\delta(z)\cdot \in X_w$ for $z \in \mathbf{P}^1 \setminus o$. Let $h \in SL_2(\mathbf{C})$ be an inverse image of $z$. Then $h$ can be written in the form $h = b_1 \sigma b_2$, where $b_1, b_2 \in B'$. It is clear that $i(b_2)f_{w'} = c_1 f_{w'}$ and $i(\sigma)f_{w'} = c_2 f_w$, where $c_1$, $c_2$ are constants. Therefore $i(h)f_{w'} = c_1 c_2 i(b_1)f_w$, that is, $\delta(z) \in X_w$.

To prove (ii), we consider the mapping

$$w'^{-1}\circ\xi\colon N_{w'} \times (\mathbf{P}^1 \setminus \infty) \to X.$$

The space $\mathbf{P}^1\setminus\infty$ is naturally isomorphic to the one-parameter subgroup $N'_- \subset SL_2(\mathbf{C})$.

The mapping $\xi\colon N_{w'} \times N'_- \to X$ is given by the rule

$$\xi(n, n_1) = ni(n_1)\,[f_{w'}],\ n \in N_{w'},\ n_1 \in N'.$$

Thus,

$$w'^{-1}\circ\xi(n, n_1) = (w'^{-1}nw')(w'^{-1}i(n_1)\,w')\,[f_e].$$

We now observe that $w'^{-1}N_{w'}w' \subset N_-$ (by definition of $N_{w'}$), and $w'^{-1}i(N')w' \in N_-$ (since $w'^{-1}\ \gamma \in \Delta_+$). Furthermore, the intersection of the tangent spaces to these subgroups consists only of 0, because $N_{w'} \subset N$, $i(N'_-) \subset N_-$. The mapping $N_- \to X$ ($n \mapsto n[f_e]$) is an isomorphism onto an open subset of $X$. Therefore (ii) follows from the next simple lemma, which is proved in [5], for example.

4.6. LEMMA. *Let* $N_1$ *and* $N_2$ *be two closed algebraic subgroups of a unipotent group* $N$ *whose tangent spaces at the unit element intersect only in* 0. *Then the product mapping* $N_1 \times N_2 \to N$ *gives an isomorphism of* $N_1 \times N_2$ *with a closed subvariety of* $N$.

This completes the proof of Proposition 4.3.

PROOF OF PROPOSITION 4.4. Any element of $\mathfrak{h}^*_{\mathbf{Z}}$ has the form $\chi = \lambda - \lambda'$, where $\lambda$, $\lambda'$ are regular dominant weights. In this case, $E_\chi = E_\lambda \otimes E_{\lambda'}^{-1}$, and it is therefore sufficient to find a section $\mu$ with the required properties in the case $\chi = \lambda$.

We consider the space $P(V)$, where $V$ is a representation of $G$ with dominant weight $\lambda$. Let $\eta_V$ be the linear fibering on $P(V)$ consisting of pairs $(P, \phi)$, where $\phi$ is a linear functional on the line $P \subset V$. Then $E_\lambda = i^*(\eta_V)$, where $i\colon X \to P(V)$ is the embedding described in § 2.

The linear functional $\phi_w$ on $V$ (see the proof of Theorem 2.11) yields a section of the bundle $\eta$. We shall prove that the restriction of $\mu$ to this section on $\bar{X}_w$ is a section of the fibering $E_\lambda$ having the requisite properties.

By Lemma 2.12, $\mu(x) \neq 0$ for all $x \in X_w$. The support of the divisor div $\mu$ is therefore contained in $\bar{X}_w \setminus X_w = \bigcup_{w' \xrightarrow{-\gamma} w} \bar{X}_{w'}$.

Since $\overline{X}_{w'}$ is an irreducible variety, we see that $\operatorname{div} \mu = \sum_{w' \xrightarrow{\gamma} w} a_\gamma \overline{X}_{w'}$, where $a_\gamma \in \mathbf{Z}$, $a_\gamma \geqslant 0$. It remains to show that $a_\gamma = w'\chi(H_\gamma)$.

In view of Lemma 4.5 (i) and (ii), the coefficient $a_\gamma$ is equal to the multiplicity of zero of the section $\delta^*(\mu)$ of the fibering $\delta^*(E_\lambda)$ on $\mathbf{P}^1$ at the point $o$, that is, the multiplicity of zero of the function $\psi(t) = \phi_{w'}((\exp tE_{-\gamma})f_{w'})$ for $t = 0$. It follows from Lemma 2.10 that $\psi(t) = ct^n$, hence $a_\gamma = n = w'\chi(H_\gamma)$. This completes the proof of Proposition 4.4 and with it of Theorem 4.1.

## §5. Generalizations and supplements

1. **Degenerate flag varieties.** We extend the results of the previous sections to spaces $X(P) = G/P$, where $P$ is an arbitrary parabolic subgroup of $G$. For this purpose we recall some facts about the structure of parabolic subgroups $P \subset G$ (see [7]).

Let $\Theta$ be some subset of $\Sigma$, and $\Delta_\Theta$ the subset of $\Delta_+$ consisting of linear combinations of elements of $\Theta$. Let $G_\Theta$ be the subgroup of $G$ generated by $H$ together with the subgroups $N_\gamma = \{\exp tE_\gamma \mid t \in \mathbf{C}\}$ for $\gamma \in \Delta_\Theta \cup -\Delta_\Theta$, and let $N_\Theta$ be the subgroup of $N$ generated by the $N_\gamma$ for $\gamma \in \Delta_+ \backslash \Delta_\Theta$. Then $G_\Theta$ is a reductive group normalizing $N_\Theta$, and $P_\Theta = G_\Theta N_\Theta$ is a parabolic subgroup of $G$ containing $B$.

It is well known (see [7], for example) that every parabolic subgroup $P \subset G$ is conjugate in $G$ to one of the subgroups $P_\Theta$. We assume in what follows that $P = P_\Theta$, where $\Theta$ is a fixed subset of $\Sigma$. Let $W_\Theta$ be the Weyl group of $G_\Theta$. It is the subgroup of $W$ generated by the reflections $\sigma_\alpha$, $\alpha \in \Theta$.

We describe the decomposition of $X(P)$ into orbits under the action of $B$.

5.1. PROPOSITION. (i) *$X(P) = \bigcup_{w \in W} Bwo$, where $o \in X(P)$ is the image of $P$ in $G/P$.*

(ii) *The orbits $Bw_1o$ and $Bw_2o$ are identical if $w_1w_2^{-1} \in W_\Theta$ and otherwise are disjoint.*

(iii) *Let $W_\Theta^1$ be the set of $w \in W$ such that $w\Theta \subset \Delta_+$. Then each coset of $W/W_\Theta$ contains exactly one element of $W_\Theta^1$. Furthermore, the element $w \in W_\Theta^1$ is characterized by the fact that its length is less than that of any other element in the coset $wW_\Theta$.*

(iv) *If $w \in W_\Theta^1$, then the mapping $N_w \to X(P)$ $(n \to nwo)$ is an isomorphism of $N_w$ with the subvariety $Bwo \subset X(P)$.*

PROOF. (i)–(ii) follow easily from the Bruhat decomposition for $G$ and $G_\Theta$. The proof of (iii) can be found in [7], for example, and (iv) follows at once from (iii) and Proposition 1.1.

Let $w \in W_\Theta^1$, $X_w(P) = Bwo$, let $\overline{X}_w(P)$ be the closure of $X_w(P)$ and $[\overline{X}_w(P)] \in H_{2l(w)}(\overline{X}_w(P), \mathbf{Z})$ its fundamental class. Let $s_w(P) \in H_{2l(w)}(X(P), \mathbf{Z})$ be the image of $[\overline{X}_w(P)]$ under the mapping induced by the embedding $\overline{X}_w(P) \hookrightarrow X(P)$. The next proposition is an analogue of Proposition 1.2.

5.2. PROPOSITION ([2]). *The elements $s_w(P)$, $w \in W^1_\Theta$, form a free basis in $H_*(X(P), \mathbf{Z})$.*

5.3. COROLLARY. *Let $\alpha_P$: $X \to X(P)$ be the natural mapping. Then $(\alpha_P)_* s_w = 0$ if $w \notin W^1_\Theta$, $(\alpha_p)_* s_w = s_w(P)$ if $w \in W^1_\Theta$.*

5.4. COROLLARY. $(\alpha_P)_*$: $H_*(X, \mathbf{Z}) \to H_*(X(P), \mathbf{Z})$ *is an epimorphism, and* $(\alpha_P)^*$: $H^*(X(P), \mathbf{Z}) \to H^*(X, \mathbf{Z})$ *is a monomorphism.*

5.5. THEOREM. (i) $\mathrm{Im}(\alpha_P)^* \subset H^*(X, \mathbf{Z}) = \overline{R}$ *coincides with the set of $W_\Theta$-invariant elements of $\overline{R}$.*

(ii) $P_w \in \mathrm{Im}(\alpha_P)^*$ *for $w \in W^1_\Theta$ and $\{(\alpha_P)^{*-1}P_w\}_{w\in W^1_\Theta}$ is the basis in $H^*(X(P), \mathbf{Z})$ dual to the basis $\{s_w(P)\}_{w\in W^1_\Theta}$ in $H_*(X(P), \mathbf{Z})$.*

PROOF. Let $w \in W^1_\Theta$. Since $\langle P_w, s_{w_1}\rangle = 0$ for $w_1 \notin W^1_\Theta$, $P_w$ is orthogonal to $\mathrm{Ker}(\alpha_P)_*$, that is, $P_w \in \mathrm{Im}(\alpha_P)^*$. Now (ii) follows from the fact that $\langle(\alpha_P)^*P_w, s_{w'}(P)\rangle = \langle P_w, s_{w'}\rangle$ for $w, w' \in W^1_\Theta$. To prove (i), it is sufficient to verify that the $P_w$, $w \in W^1_\Theta$, form a basis for the space of $W_\Theta$-invariant elements of $\overline{R}$. We observe that an element $f \in \overline{R}$ is $W_\Theta$-invariant if and only if $A_\alpha f = 0$ for all $\alpha \in \Theta$. Since $w \in W^1_\Theta$ if and only if $l(w\sigma_\alpha) = l(w) + 1$ for all $\alpha \in \Theta$, (i) follows from Theorem 3.14(i).

2. CORRESPONDENCES. Let $Y$ be a non-singular oriented manifold. An arbitrary element $z \in H_*(Y \times Y, \mathbf{Z})$ is called a correspondence on $Y$. Any such element $z$ gives rise to an operator $z_*$: $H_*(Y, \mathbf{Z}) \to H_*(Y, \mathbf{Z})$, according to

$$z_*(c) = (\pi_2)_*((\pi_1)^*(\mathcal{P}c) \cap z), \quad c \in H_*(Y, \mathbf{Z}),$$

where $\pi_1, \pi_2$: $Y \times Y \to Y$ are the projections onto the first and second components, and $\mathcal{P}$ is the Poincaré duality operator. We also define an operator $z^*$: $H^*(Y, \mathbf{Z}) \to H^*(Y, \mathbf{Z})$ by setting $z^*(\xi) = \mathcal{P}[(\pi_1)_*((\pi_2)^*(\xi) \cap z)]$, $\xi \in H^*(Y, \mathbf{Z})$. It is clear that $z_*$ and $z^*$ are adjoint operators.

Let $z$ be assigned to a (possibly singular) submanifold $Z \subset Y \times Y$, in such a way that $z$ is the image of the fundamental cycle $[Z] \in H_*(Z, \mathbf{Z})$ under the mapping induced by the embedding $Z \hookrightarrow Y \times Y$. Then

$$z_*(c) = (\rho_2)_* ([Z] \cap (\rho_1)^*\mathcal{P}c),$$

where $\rho_1, \rho_2$: $Z \to Y$ are the restrictions of $\pi_1, \pi_2$ to $Z$.

If, in this situation, $\rho_1$: $Z \to Y$ is a fibering and $c$ is given by a submanifold $C \subset Y$, then the cycle

$$[Z] \cap (\rho_1)^*\mathcal{P}c$$

is given by the submanifold $\rho_1^{-1}(C) \subset Z$.

We want to study correspondences in the case $Y = X = G/B$.

5.6. DEFINITION. *Let $w \in W$. We put $Z_w = \{(gwo, go)\} \subset X \times X$ and denote by $z_w$ the correspondence $z_w = [\overline{Z}_w] \subset H_*(X \times X, \mathbf{Z})$.*

5.7. THEOREM. $(z_w)_* = F_w$.

PROOF. We calculate $(z_w)_* (s_{w'})$.

Since the variety $\overline{Z}_w$ is $G$-invariant and $G$ acts transitively on $X$, the mapping $\rho_1 : \overline{Z}_w \to X$ is a fibering. Thus,

$$(z_w)_* (s_{w'}) = (\rho_2)_* [\rho_1^{-1} (\overline{X}_{w'})].$$

It is easily verified that $\rho_1^{-1} (\overline{X}_{w'}) = \overline{\pi_1^{-1} (X_{w'}) \cap Z_w}$. We put $Y = \pi_1^{-1} (X_{w'}) \cap Z_w \subset X \times X$. Then

(20) $$Y = \{(nw'o,\ nw'bwo) \mid n \in N,\ b \in B\}.$$

Since the dimension of the fibre of $\rho_1 : Z_w \to X$ is equal to $2l(w)$, we see that $\dim Y = 2l(w) + 2l(w')$. It is clear from (20) that

$$\rho_2(Y) = \{nw'bwo \mid n \in N,\ b \in B\} = Bw'Bwo.$$

It is well known (see [6], Ch. IV, § 2.1 Lemma 1) that

$$Bw'Bwo = Bw'wo \cup \Big( \bigcup_{l(w_1)<l(w)+l(w')} Bw_1o \Big).$$

Thus, two cases can arise.

a) $l(w'w) < l(w') + l(w)$. In this case, $\dim \rho_2 (Y) < 2l(w') + 2l(w)$, and so $(z_w)_*(s_{w'}) = (\rho_2)_*[Y] = 0$.

b) $l(w'w) = l(w') + l(w)$. In this case, $\rho_2 (Y) = X_{w'w} + X'$, where $\dim X' < \dim X_{w'w} = 2l(w') + 2l(w)$. Thus, $(\rho_2)_*[Y] = [\overline{X}_{w'w}]$, that is, $(z_w)_*(s_{w'}) = s_{w'w}$. Comparing the formulae obtained with 3.12 (ii), we see that $(z_w)_* = F_w$.

5.8. COROLLARY. $z_w = \sum s_{w's} \otimes s_{w'w}$, *where the summation extends over those* $w' \in W$ *for which* $l(w'w) = l(w) + l(w')$.

In § 1 we have defined an action of $W$ on $H_*(X, \mathbf{Z})$. This definition depended on the choice of a compact subgroup $K$. Using Theorem 5.7 we can find explicitly the correspondences giving this action.

In fact, it follows from Lemma 3.3 (iii) that $\sigma_\alpha = \alpha^* F_\alpha - 1$ for any $\alpha \in \Sigma$. The transformation $F_\alpha$ is given by the correspondence $Z_{\sigma_\alpha}$. The operator $\alpha^*$ can also be given by a correspondence: if $U_\alpha = \Sigma c_i U_i$ is a divisor in $X$ giving the cycle $\mathscr{P}(\alpha) \in H_{2r-2}(X, \mathbf{Z})$ (for example, $U_\alpha = \sum_{\beta \in \Sigma} \alpha(H_\beta) X_{\sigma_\beta}$), then the cycle $\widetilde{U}_\alpha = \Sigma c_i \widetilde{U}_i$, where $\widetilde{U}_i = \{(x, x) \mid x \in U_i\} \subset X \times X$, determines the correspondence that gives the operator $\alpha^*$. The operator $\sigma_\alpha$ in $H_*(X, \mathbf{Z})$ is therefore given by the correspondence $\widetilde{U}_\alpha * Z_{\sigma_\alpha} - 1$ (where $*$ denotes the product of correspondences, as in [11]). Using the geometrical realization of the product of correspondences (see [11]), we can explicitly determine the correspondence $S_\alpha$ that gives the transformation $1 + \sigma_\alpha$ in $H_*(X, \mathbf{Z})$, namely, $S_\alpha = \Sigma c_i \hat{U}_i$ where $\hat{U}_i = \{(x, y) \in X \times X \mid x \in U_i,\ \tilde{x}^{-1}\tilde{y} \in P_{\{\alpha\}}\}$. In this expression, $\tilde{x}, \tilde{y} \in G$ are arbitrary representatives of $x$, $y$, and $P_{\{\alpha\}}$ is the parabolic subgroup corresponding to the root $\alpha$.

3. B. Kostant has described the $P_w$ in another way. We state his result.

Let $h \in \mathfrak{h}_{Q}^{*}$ be an element such that $\alpha(h) > 0$ for all $\alpha \in \Sigma$. Let $J_h = \{f \in R \mid f(wh) = 0$ for all $w \in W\}$ be an ideal of $R$.

5.9. THEOREM. (i) *Let $w \in W$, $l(w) = l$. There is a polynomial $Q_w \in R$ of degree $l$ such that*

(21) $\qquad Q_w(wh) = 1, \; Q_w(w'h) = 0 \quad if \quad l(w') \leqslant l(w), \; w' \neq w.$

*The $Q_w$ are uniquely determined by* (21) *to within elements of $J_h$.* (ii) *Let $Q_w^0$ be the form of highest degree in the polynomial $Q_w$. The image of $Q_w^0$ in $\bar{R}$ is equal to* $\prod_{\gamma \in \Delta_- \cap w^{-1}\Delta_+} (\gamma(h))^{-1} \cdot P_w$.

The proof is analogous to that of Theorem 3.15.

4. We choose a maximal compact subgroup $K \subset G$ such that $K \cap B \subset H$ (see § 1). The cohomology of $X$ can be described by means of the $K$-invariant closed differential forms on $X$. For let $\chi \in \mathfrak{h}_{Z}^{*}$, and let $E_\chi$ be the corresponding one-dimensional complex $G$-fibering on $X$. Let $\tilde{\omega}_\chi$ be the 2-form on $X$ which is the curvature form of connectedness associated with the $K$-invariant metric on $E_\chi$ (see [12]). Then the class of the form $\omega_\chi \frac{1}{2\pi i} \tilde{\omega}_\chi$ is $c_\chi \in H^2(X, \mathbf{Z})$. The mapping $\chi \to \omega_\chi$ extends to a mapping $\theta: R \to \Omega_{ev}^{*}(X)$, where $\Omega_{ev}^{*}$ is the space of differential forms of even degree on $X$. One can prove the following theorem, which is a refinement of Proposition 1.3 (ii) and Theorem 3.17.

5.10. THEOREM (i) Ker $\theta = J$, *that is, $\theta$ induces a homomorphism of rings $\bar{\theta}: \bar{R} \to \Omega_{ev}^{*}(X)$.* (ii) *Let $w_1, w_2 \in W$, $w_1 \nleqslant w_2 s$. Then the restriction of the form $\bar{\theta}(P_{w_1})$ to $X_{w_2}$ is equal to* 0. (iii) *Let $w_1, w_2 \in W$, $w_1 \nleqslant w_2 s$. Then $\bar{\theta}(P_{w_1})\, \bar{\theta}(P_{w_2}) = 0$.*

## References

[1] A. Borel, Sur la cohomologie des espaces fibrés principaux et des espaces homogènes des groupes de Lie compacts, Ann. of Math. (2) **57** (1953), 115–207; MR 14 #490. Translation: in '*Rassloennye prostranstva*', Inost. lit., Moscow 1958

[2] A. Borel, Kählerian coset spaces of semisimple Lie groups, Proc. Nat. Acad, Sci. U.S.A. **40** (1954), 1147–1151; MR 17 #1108.

[3] B. Kostant, Lie algebra cohomology and generalized Schubert cells, Ann. of Math. (2) 77 (1963), 72–144; MR 26 #266.

[4] G. Horrocks, On the relations of S-functions to Schubert varieties, Proc. London Math. Soc. (3) 7 (1957), 265–280; MR 19 #459.

[5] A. Borel, Linear algebraic groups, Benjamin, New York, 1969; MR #4273. Translation: *Lineinye algebraicheskie gruppy*, 'Mir', Moscow 1972.

[6] N. Bourbaki, Groupes et algebras de Lie, Ch. 1–6, Éléments de mathématique, 26, 34, 36, Hermann & Cie, Paris, 1960–72. Translation: *Gruppyi algebry Li*, 'Mir', Moscow 1972.

[7] R. Steinberg, Lectures on Chevalley groups, Yale University Press, New Haven, Conn. 1967.

[8] M. F. Atiyah and F. Hirzebruch, Vector bundles and homogeneous spaces, Proc.

Sympos. Pure Math.; Vol. III, 7–38, Amer. Math. Soc., Providence, R. I., 1961; MR 25 # 2617.
= Matematika **6**: 2 (1962), 3–39.

[9] P. Cartier, On H. Weyl's character formula, Bull. Amer. Math. Soc. **67** (1961), 228–230; MR 26 # 3828.
= Matematika **6**: 5 (1962), 139–141.

[10] E. H. Spanier, Algebraic topology, McGraw-Hill, New York 1966; MR 35 # 1007
Translation: *Algebraicheskaya topologiya,* 'Mir', Moscow 1971.

[11] Yu.I. Manin, Correspondences, motives and monoidal transformations, Matem. Sb. **77** (1968), 475–507; MR 41 # 3482
= Math. USSR–Sb. **6** (1968), 439–470.

[12] S. S. Chern, Complex manifolds, Instituto de Fisca e Matemática, Recife 1959; MR 22 # 1920.
Translation: *Kompleksnye mnogoobraziya*, Inost. Lit., Moscow 1961.

[13] I. N. Bernstein, I. M. Gel'fand and S. I. Gel'fand, Schubert cells and the cohomology of flag spaces, Funkts. analiz **7**: 1 (1973), 64–65.

Received by the Editors 13 March 1973

Translated by D. Johnson

# 5.

# (with I. N. Bernstein and S. I. Gelfand)

# Category of $\mathfrak{g}$-modules

Funkts. Anal. Prilozh. **10** (2) (1976) 1–8. [Funct. Anal. Appl. **10** (1976) 87–92].
MR **53**:10880. Zbl. **353**:18013

## §1. INTRODUCTION

Let A be a finite-dimensional associative algebra with identity over a field K, and let $\mathcal{A}$ be the category of finite-dimensional A modules. An important invariant of such an algebra is its Cartan matrix, which is defined as follows. Let $L_1, \ldots, L_k$ be a complete collection of irreducible A modules. For each $L_i$ there exists a unique (up to isomorphism) indecomposable projective A-module $P_i$ that covers $L_i$; i.e., Hom $(P_i, L_i) \neq 0$. Let $c_{ij} = (P_i : L_j)$ be the number of occurrences of $L_j$ in the Jordan–Hölder series of $P_i$. The integral matrix $C = \|c_{ij}\|$, $i, j = 1, \ldots, k$, is called the Cartan matrix of A (or $\mathcal{A}$).

In certain cases C is a symmetric, positive-definite matrix and, moreover, can be represented in the form $C = D^t \cdot D$, where D is some other integral matrix (not necessarily square).

This fact is ordinarily a reflection of some duality principle; to wit, the equality $C = D^t \cdot D$ means that there exists a class of modules $M_1, \ldots, M_l$, such that each $P_i$ has a composition series with factors isomorphic to $M_j$, and for any i, j the number of occurrences of $M_j$ in the series for $P_i$ is equal to the number of occurrences of $L_i$ in the Jordan–Hölder series for $M_j$. Thus, it can be said that the modules $M_j$ occupy an intermediate position between the projective modules $P_i$ and the simple modules $L_i$; they are their "mean geometric."

The elucidation of the reason why $C = D^t \cdot D$ and the intrinsic (in terms of $\mathcal{A}$) characterization of the modules $M_j$ are highly interesting problems, approaches to which are absolutely unclear at present.

At present two classes of categories $\mathcal{A}$ are known for which the Cartan matrix C has this property.

Case 1. Let char $K = p > 0$, and let $A = KG$ be the group algebra of some finite group G, so that $\mathcal{A}$ is the category of finite-dimensional G modules over K. Let $V_1, \ldots, V_l$ be a complete collection of irreducible representations of G over the field C of complex numbers, and let $M_1, \ldots, M_l$ be the A modules obtained by their reduction to characteristic p (see [1]).

If $L_1, \ldots, L_k$ are a complete collection of irreducible A modules (i.e., irreducible representations of G over K) and D is the matrix with entries $d_{ij} = (M_i : L_j)$, then $C = D^t \cdot D$ (for more details see [1, §§82, 83]).

Case 2. Let $\mathfrak{g}$ be a semisimple complex Lie algebra, $\bar{\mathfrak{g}}$ the Lie algebra over a closed field of characteristic $p > 0$ obtained by the reduction of $\mathfrak{g}$, and $A = U^0(\bar{\mathfrak{g}})$ a bounded universal enveloping algebra of $\bar{\mathfrak{g}}$. Finite-dimensional A modules were studied by Humphreys [2], who constructed a collection of A modules $M_1, \ldots, M_l$ that occupy an "intermediate position" between projective and simple modules as described above and proved that the Cartan matrix of the category of A modules can be represented in the form $C = D^t \cdot D$.

The purpose of this article is to construct a category of $\mathfrak{g}$ modules having the same property for each semisimple Lie algebra $\mathfrak{g}$ over C. Simple objects of this category can be indexed by the elements of the Weyl group W of $\mathfrak{g}$ (to $w \in W$ corresponds a simple module $L_w$ and a projective module $P_w$). In addition, to each $w \in W$ corresponds some $\mathfrak{g}$ module $M_w$ (the so-called Verma module; see [3, 4, 5, 8]). If we now set $C = \|c_{w,w'}\|$ where $c_{w,w'} = (P_w : L_{w'})$, and $D = \|d_{w,w'}\|$, where $d_{w,w'} = (M_w : L_{w'})$, then $C = D^t \cdot D$. In addition, $\mathcal{A}$ has other good properties (D is unipotent and all objects of $\mathcal{A}$ have a finite cohomological dimension).

Note also that a detailed study of the relatively simple "finite-dimensional" category $\mathcal{A}$ makes it possible to obtain certain information about the structure of submodules of Verma modules, which is a difficult and interesting problem.

Moscow State University. Translated from Funktsional'nyi Analiz i Ego Prilozheniya, Vol. 10, No. 2, pp. 1-8, April-June, 1976. Original article submitted December 22, 1975.

We would like to dedicate our article to J. Dixmier, who accomplished a great deal in the study of the tructure of enveloping algebras for Lie algebras. His papers and book [5] were most responsible for crys-allizing this branch of mathematics into a separate and important direction.

We are also greatly indebted to J. Humphreys [2], who significantly instigated our paper..

## §2. SOME NOTATION

$\mathfrak{g}$ is a complex semisimple Lie algebra, and $\mathfrak{h}$ is a Cartan subalgebra of $\mathfrak{g}$.

$\Delta$ is the root system of $\mathfrak{g}$ with respect to $\mathfrak{h}$; $\Sigma$ is the set of simple roots; $\Delta_+$ is the set of positive roots; nd $\rho = \frac{1}{2} \sum_{\gamma \in \Delta_+} \gamma$.

$\mathfrak{n}_+$, and $\mathfrak{n}_-$ are the subalgebras of $\mathfrak{g}$, spanned by the root vectors corresponding to the roots in $\Delta_+$ (resp. $\Delta_+$).

$U(\mathfrak{g})$, and $U(\mathfrak{n}_+)$ are enveloping algebras of $\mathfrak{g}$ and $\mathfrak{n}_+$, respectively, and $Z(\mathfrak{g})$ is the center of $U(\mathfrak{g})$.

$\mathfrak{h}^*$ is the space dual to $\mathfrak{h}$; $\mathfrak{h}_Z^{\bullet}$ is the integral lattice in $\mathfrak{h}^*$, consisting of the weights of the finite-dimen-ional representations of $\mathfrak{g}$; and $C^\circ \subset \mathfrak{h}^*$ is the positive Weyl chamber.

W is the Weyl group of $\mathfrak{g}$ with respect to $\mathfrak{h}$; $\sigma_\gamma$ is the reflection corresponding to $\gamma \in \Delta_+$; and $l(w)$ is the ngth of $w \in W$, i.e., the smallest number of factors in the representation $w = \sigma_{\alpha_1} \ldots \sigma_{\alpha_l}$, $\alpha_i \in \Sigma$.

$K(\chi)$ for $\chi \in \mathfrak{h}^*$ is Kostant's function, i.e., the number of representations of $\chi$ in the form $\chi = \sum_{\alpha \in \Delta_+} n_\alpha \alpha$; $\in Z$, $n_\alpha \geqslant 0$; $\Gamma \subset \mathfrak{h}_Z^{\bullet}$ is the set of elements $\chi$ such that $K(\chi) \neq 0$; for $\chi, \psi \in \mathfrak{h}^*$, $\chi < \psi$ means that $-\chi \in \Gamma$.

JH(M) denotes the collection of simple modules (with multiplicities) occurring in the Jordan–Hölder eries of a $\mathfrak{g}$ module M.

## §3. ELEMENTARY PROFERTIES OF THE CATEGORY $\mathcal{O}$

Definition 1 (the Category $\mathcal{O}$). The objects of the category $\mathcal{O}$ are left $\mathfrak{g}$ modules M having the following roperties:

1) M is a finitely generated $\mathfrak{g}$ module;

2) M is $\mathfrak{h}$ diagonalizable;

3) M is $\mathfrak{n}_+$ finite (i.e., $\dim_C U(\mathfrak{n}_+)f < \infty$ for any $f \in M$).

The morphisms of $\mathcal{O}$ are arbitrary $\mathfrak{g}$ module morphisms.

Important objects of $\mathcal{O}$ are the so-called Verma modules (the modules $M_\chi$; see [3, 4]).

Let $\chi \in \mathfrak{h}^*$. We denote by $J_\chi$ the left ideal in $U(\mathfrak{g})$ generated by $\mathfrak{n}_+$ and $\{H - \chi(H) + \rho(H),\ H \in \mathfrak{h}\}$, and e set $M_\chi = U(\mathfrak{g}) J_\chi$. It is clear that $M_\chi \in \mathcal{O}$.

Let us list the basic properties of $\mathcal{O}$ and of Verma modules.

1) $\mathcal{O}$ is an Abelian category with finite direct sums. The space Hom(M, M') is finite-dimensional for any , $M' \in \mathcal{O}$.

2) Let $M \in \mathcal{O}$, $\psi \in \mathfrak{h}^*$. We denote by $M^{(\psi)} \subset M$ the subspace of vectors of weight $\psi$ in M. Then $\dim_C$ $^{(\psi)} < \infty$. Let $P(M) = \{\psi \in \mathfrak{h}^* | M^{(\psi)} \neq 0\}$. Then there exists a finite number of weights $\psi_1, \ldots, \psi_k$ such that $(M) \subset \cup (\psi_j + (-\Gamma))$, where $\Gamma$ is the semigroup generated by positive roots (for example, $P(M_\chi) = \chi - \rho - \Gamma$). particular, if $M \neq 0$, then P(M) contains at least one maximal weight $\chi$, i.e., such that $P(M) \cap \{\chi + \Gamma\} = \chi$.

3) Let us describe all of the simple objects of $\mathcal{O}$. Let $\chi \in \mathfrak{h}^*$. For a Verma module $M_\chi$ there is a u-que simple factor module (see [5]), which we denote by $L_\chi$. The modules $L_\chi$ are pairwise nonisomorphic and re exhausted by all of the simple objects in $\mathcal{O}$.

4) Let $Z(\mathfrak{g})$ be the center of $U(\mathfrak{g})$, and let $\Theta$ be the set of characters of $Z(\mathfrak{g})$ (i.e., homomorphisms $(\mathfrak{g}) \to C$). For each $\theta \in \Theta$ consider the complete subcategory $\mathcal{O}_\theta$ of $\mathcal{O}$, consisting of modules M satisfying the llowing condition: For each $z \in Z(\mathfrak{g})$ the module M is annihilated by some power of $[z - \theta(z)]$. Then $\mathcal{O} = \bigoplus_{\theta \in \Theta}$

$\mathcal{O}_\theta$, i.e., each $M \in \mathcal{O}$ decomposes into a finite sum $M = \bigoplus_{\theta \in \Theta} M(\theta)$, $M(\theta) \in \mathcal{O}_\theta$, and for $\theta_1 \neq \theta_2$ Hom$(M_1, M_2) = 0$ for any $M_1 \in \mathcal{O}_{\theta_1}$, $M_2 \in \mathcal{O}_{\theta_2}$.

5) For each $\chi \in \mathfrak{h}$ there exists a character $\theta_\chi$ such that $zf = \theta_\chi(z)f$ for all $f \in M_\chi$, $z \in Z(\mathfrak{g})$. Therefore, $M_\chi \in \mathcal{O}_{\theta_\chi}$. Each $\theta \in \Theta$ has the form $\theta_\chi$ for some $\chi \in \mathfrak{h}^*$; $\theta_{\chi_1} = \theta_{\chi_2} \Leftrightarrow \chi_1 = w\chi_2$ for some $w \in W$. Let $\Lambda(\theta) = \{\chi \in \mathfrak{h}^*, \theta = \theta_\chi\}$.

6) Using 1), 4), and 5), one can prove that each $M \in \mathcal{O}$ has finite length (see [5]). Therefore, each $M \in \mathcal{O}$ decomposes into the direct sum of indecomposable objects, where the summands are uniquely defined up to isomorphism and rearrangement (the Krull–Schmidt theorem [1]).

7) Let $\theta = \theta_\chi$ and $M \in \mathcal{O}_\theta$. Then $P(M) \subset \bigcup_{w \in W} \{w(\chi - \rho) - \Gamma\}$. It suffices to verify this assertion for simple objects in $\mathcal{O}_\theta$, for which it follows from 1), 3), and 5).

## §4. PROJECTIVE OBJECTS OF $\mathcal{O}$

THEOREM 1. Each object in $\mathcal{O}$ is a factor object of a projective object.

We shall prove the more precise Theorem 2.

THEOREM 2. Let $\chi \in \mathfrak{h}^*$, $\theta \in \Theta$. There exist a module $Q = Q(\theta, \chi) \in \mathcal{O}_\theta$ and a vector $q \in Q^{(\chi)}$ such that for any $M \in \mathcal{O}$ the mapping Hom$(Q, M) \to (M(\theta))^{(\chi)}$, defined by $\varphi \to \varphi(q)$, is an isomorphism.

It follows from Theorem 2 that the functor $M \to$ Hom$(Q, M)$ is isomorphic to the functor $M \to (M(\theta))^{(\chi)}$ and hence, is faithful; i.e., $Q$ is projective (in $\mathcal{O}$). On the other hand, each $M \in \mathcal{O}$ is generated by a finite number of vectors $f_i \in (M(\theta_i))^{(\chi_i)}$, and by Theorem 2, M is a factor module of the projective module $\bigoplus_i Q(\theta_i, \chi_i)$.

Therefore, it suffices to prove Theorem 2.

Proof. Let $I \subset U(\mathfrak{g})$ be the left ideal generated by $H - \chi(H)$, $H \in \mathfrak{h}$, and $(\mathfrak{n}_+)^N$ for sufficiently large N (N will be chosen later). Let $\hat{Q} = U(\mathfrak{g})/I$ and let $\hat{q} \in \hat{Q}$ be the image of $1 \in U(\mathfrak{g})$ in $\hat{Q}$. It is clear that $\hat{Q} \in \mathcal{O}$.

Let us prove that if N is sufficiently large, then for any $M \in \mathcal{O}_\theta$ the mapping $(\hat{Q}, M) \to M^{(\chi)}$ $(\varphi \mapsto \varphi(\hat{q}))$ is an isomorphism. Since $\hat{q}$ is a generator of $\hat{Q}$, this mapping is an imbedding. Conversely, let $f \in M^{(\chi)}$. Consider $\alpha : U(\mathfrak{g}) \to M$, $\alpha(X) = Xf$. It is clear that $\alpha(H - \chi(H)) = 0$, $H \in \mathfrak{h}$. On the other hand, let $\theta = \theta_\psi$ for some $\psi \in \mathfrak{h}^*$. Then by virtue of 7), $P(M) \subset \bigcup_{w \in W} \{w(\psi - \rho) - \Gamma\}$, and hence, for sufficiently large N (not depending on M or f) $(\mathfrak{n}_+)^N f = 0$. Therefore, $\alpha((\mathfrak{n}_+)^N) = 0$; i.e., $\alpha$ defines a mapping $\hat{\alpha} : \hat{Q} \to M$, such that $\hat{\alpha}(\hat{q}) = f$.

It is now clear that the module $Q = \hat{Q}(\theta)$, and the projection q of $\hat{q}$ onto Q satisfy the condition of Theorem 2.

COROLLARY 1. If $P \in \mathcal{O}$ is an indecomposable projective module, then P has a unique maximal submodule P', so that the simple module L = P/P' corresponds to it. We obtain a one-to-one correspondence $P \longleftrightarrow P/P'$ between indecomposable projective objects and simple objects of $\mathcal{O}$.

We denote the projective module corresponding to the simple module $L_\chi$ by $P_\chi$. It is called the projective covering of $L_\chi$.

This corollary follows from Theorem 1 and Property 6) (see [1], Theorem 54.11).

Proposition 1. Let $\chi \in \mathfrak{h}^*$, $M \in \mathcal{O}$. Then dim Hom$(P_\chi, M) = (M : L_\chi)$.

Proof. Let $0 \to M_1 \to M \to M_2 \to 0$ be an exact sequence in $\mathcal{O}$. Since $P_\chi$ is a projective object, the validity of the proposition for M follows from its validity for $M_1$ and $M_2$. Therefore, we can assume that $M = L_\psi$ is a simple object. It now follows from Corollary 1 that if $\chi \neq \psi$, then both sides of the equality equal zero. But if $\chi = \psi$, then Hom$(P_\chi, L_\chi) =$ Hom$(L_\chi, L_\chi) = \mathbb{C}$.

Remark. The category $\mathcal{O}$ is self-dual, i.e., equivalent to the dual category $\mathcal{O}^0$. This equivalence is defined by the following functor F. Let $i : \mathfrak{g} \to \mathfrak{g}$ be an anti-involution such that $i(H) = H$, $H \in \mathfrak{h}$. Then $i(\mathfrak{n}_+) = \mathfrak{n}_-$. Let $M \in \mathcal{O}$ and let $M^*$ be the space of linear functionals on M. We define the action of $\mathfrak{g}$ on $M^*$ by $(X\xi)(f) = \xi(i(X)f)$, $\xi \in M^*$, $f \in M$, $X \in \mathfrak{g}$. As F(M) we take the submodule of $M^*$ generated by the vectors characteristic with respect to $\mathfrak{h}$. It can be proved that $F(M) \in \mathcal{O}$, and that Hom$(M_1, M_2) =$ Hom$(F(M_2), F(M_1))$. In addition, F(F(M)) is naturally isomorphic to M. It is clear from the construction that $\dim(F(M))^{(\chi)} = \dim M^{(\chi)}$ for any $\chi \in \mathfrak{h}^*$. In particular, $F(L_\chi) = L_\chi$, and hence $F(\mathcal{O}_\theta) = \mathcal{O}_\theta$ for any $\theta \in \Theta$.

It is clear that the modules $I_\chi = F(P_\chi)$ are indecomposable injective objects in $\mathcal{O}$, and that each indecomposable injective object in $\mathcal{O}$ is isomorphic to one of the $I_\chi$.

## §5. THE CATEGORY $\mathcal{O}_\theta$

It follows from Property 4) in §3 that many properties of $\mathcal{O}$ can be studied "locally," i.e., inside the categories $\mathcal{O}_\theta$ for various $\theta \in \Theta$. Each of the $\mathcal{O}_\theta$ is a "finite-dimensional" category in the following sense:

THEOREM 3. Let $\theta = \theta_\chi$ for some $\chi \in \mathfrak{h}^*$, and let $P = \bigoplus_{w \in W} n_w P_{w\chi}$ be a projective object in $\mathcal{O}_\theta$, such that $n_w > 0$ for all $w \in W$. Consider the finite-dimensional algebra $A = \mathrm{Hom}(P, P)$. The functor $M \to \mathrm{Hom}(P, M)$ defines an equivalence of $\mathcal{O}_\theta$ with the category of finite-dimensional right A modules.

The proof follows immediately from [6] (Theorem II.1.3) and the fact that the $P_{w\chi}$, $w \in W$, exhaust all of the indecomposable projective modules in $\mathcal{O}_\theta$.

Remark. It can be deduced from the explicit construction of $F : \mathcal{O}_\theta \to \mathcal{O}_\theta^0$ (see the remark in §4) that there exists an antiautomorphism $i: A \to A$ such that $i^2 = 1$.

Let $\theta = \theta_\chi$. If $\chi$ is a character such that $w\chi - \chi \notin \mathfrak{h}_Z^*$ for all $w \in W$ (i.e., $\chi$ is a character in general position), then any simple object in $\mathcal{O}_\theta$ is projective, and $\mathcal{O}_\theta$ is arranged very simply.

When $\chi$ is a character not in general position, the linkages appear among the various simple objects in $\mathcal{O}_\theta$. The most complicated and interesting case is when $\chi$ is a regular integral weight; i.e., $\chi \in \mathfrak{h}_Z^*$ and $w\chi \neq \chi$ for $w \in W$, $w \neq e$.

THEOREM 4. Let $\chi$, $\chi'$ be two regular integral weights, and let $\theta = \theta_\chi$, $\theta' = \theta_{\chi'}$. Then $\mathcal{O}_\theta$ and $\mathcal{O}_{\theta'}$ are equivalent.

This theorem is proved by the same methods as Theorem 2 in [4] and Theorem E1 in [7]. We shall not prove it.

## §6. THE CARTAN MATRIX AND THE DUALITY THEOREM

Our purpose in this section is to investigate the number of occurrences of the simple modules $L_\chi$ in the Jordan–Hölder series of an indecomposable projective module $P_\psi$. We shall fix $\theta \in \Theta$ and operate inside one category $\mathcal{O}_\theta$; i.e., we shall assume that $\chi, \psi \in \Lambda(\theta)$.

Definition 2. 1) The Cartan matrix $C = \| c_{\chi\psi} \|$, $\chi, \psi \in \Lambda(\theta)$, is defined by $c_{\chi\psi} = (P_\chi : L_\psi)$.

2) The decomposition matrix $D = \| d_{\chi\psi} \|$, $\chi, \psi \in \Lambda(\theta)$, is defined by $d_{\chi\psi} = (M_\chi : L_\psi)$.

THEOREM 5. 1) Let $\chi_1, \ldots, \chi_s$ be an ordering of the weights $\chi \in \Lambda(\theta)$, such that $\chi_i < \chi_j \Rightarrow i > j$. Then $D$ is an upper triangular matrix with ones on the diagonal.

2) $C = D^t \cdot D$. In particular, C is a symmetric matrix.

The proof of 1) is obvious. To prove 2) we introduce the concept of a p filtration.

Definition 3. Let $M \in \mathcal{O}$. A filtration $0 = M_0 \subset M_1 \subset M_2 \subset \ldots \subset M_k = M$ is called a p-filtration if $M_i/M_{i-1} \simeq M_{\chi_i}$ for some $\chi_i \in \mathfrak{h}^*$. In this case we denote by $(M : M_\chi)$ the number of i such that $\chi_i = \chi$. (It follows from the first part of Theorem 5 that the numbers $(M : M_\chi)$ do not depend on the choice of p filtration.)

Proposition 2. 1) Each positive module $P \in \mathcal{O}$ admits a p filtration.

2) The duality $(P_\chi : M_\psi) = d_{\psi\chi} = (M_\psi : L_\chi)$ holds.

It is clear that the second part of Theorem 5 follows from this proposition.

Proof of Proposition 2. a) LEMMA 1. Suppose that M admits a p filtration, $\chi$ is a maximal weight in $\Pi(M)$, $f \in M^{(\chi)}$, and $M' = U(\mathfrak{g}) f$. Then $M' \simeq M_{\chi-\rho}$ and $M/M'$ admits a p filtration.

Using induction on the length of a p filtration, we can assume that $M = M_k$, $f \notin M_{k-1}$. Then we obtain a nontrivial mapping of M' into $M_k/M_{k-1} \simeq M_{\chi_k}$. Since $\chi + \rho \nless \chi_k$, this mapping is an isomorphism, and $\chi_k = \chi + \rho$. Therefore, $M' = M_{\chi+\rho}$ and $M/M' \simeq M_{k-1}$ admits a p filtration.

b) If $M = M_1 \oplus M_2$ admits a p filtration, then each of $M_1$, $M_2$ admits a p filtration.

Again let $\chi$ be maximal among the weights in P(M). We can assume that $M_1^{(\chi)} \neq 0$. Choose $0 \neq f \in M_1^{(\chi)}$ and set $M' = U(\mathfrak{g}) f$. Then $M' \simeq M_{\chi,0}$ and $M/M' \simeq M_1/M' \oplus M_2$ admits a p filtration. Induction on the length of M completes the proof.

c) For any $\chi \in \mathfrak{h}^*$ the module $\hat{Q}$, constructed in the proof of Theorem 2, admits a p filtration.

We choose in $U(\mathfrak{n}^+)$ a collection of weight elements $x_1, \ldots, x_s$ with weights $\lambda_1, \ldots, \lambda_s$ so that the images of $x_i$ in $U(\mathfrak{n}_+)/U(\mathfrak{n}_+)(\mathfrak{n}_+)^N$ define a basis there and $\lambda_i < \lambda_j \Rightarrow i > j$. Let $\hat{Q}_j \subset \hat{Q}$ be the submodule generated by $(x_1\hat{q}, \ldots, x_j\hat{q})$. It follows from the Poincaré–Birkhoff–Witt theorem that the $\hat{Q}_j$ form a p filtration in $\hat{Q}$. In particular, $(\hat{Q} : M_\psi) = K(\psi - \chi - \rho)$, where K is Kostant's function (see §2).

d) Since the $Q(\chi, \theta)$ (see §4) are direct summands of $\hat{Q}$, they admit a p filtration. Since any projective module $P \in \mathcal{O}_\theta$ is a direct summand of some $\bigoplus_i Q(\chi_i, \theta)$, the first part of Proposition 2 is proved.

e) By virtue of Proposition 1, to prove the second part of Proposition 2 it suffices to show that

$$(P : M_\psi) = \dim \operatorname{Hom}(P, M_\psi) \tag{1}$$

for any projective object $P \in \mathcal{O}_\theta$ and any $\psi \in \Lambda(\theta)$.

Let $Q(\chi, \theta) = \sum_{\psi \in \Lambda(\theta)} n_\psi(\chi) P_\psi$. Then $n_\psi = \dim \operatorname{Hom}(Q(\chi, \theta), L_\psi) = (L_\psi)^{(\chi)}$. In particular, $n_\psi(\chi) = 0$, if $\chi + \rho \neq \psi$, and $n_{\chi+\rho}(\chi) = 1$. Therefore, since (1) is linear in P, it suffices to verify it for $Q = Q(\chi, \theta)$.

Both sides of the equality are unchanged when Q is replaced by $\hat{Q}$. Here $(\hat{Q} : M_\psi) = K(\psi - \chi - \rho)$ and $\dim \operatorname{Hom}(\hat{Q}, M_\psi) = \dim (M_\psi)^{(\chi)} = K(\psi - \chi - \rho)$; i.e., (1) and Proposition 2 are thereby proved.

## §7. COHOMOLOGICAL DIMENSION OF $\mathcal{O}$

In this section we prove that $\mathcal{O}$ has finite cohomological dimension and indicate what it equals. Let us recall the definition of cohomological dimension.

Let $\mathscr{C}$ be an Abelian category and M an object of $\mathscr{C}$. By the cohomological dimension dh(M), we mean the smallest number $l$, such that there exists a projective resolution of M of length $l$, i.e., an exact sequence

$$0 \leftarrow M \leftarrow P_0 \leftarrow P_1 \leftarrow \ldots \leftarrow P_l \leftarrow 0,$$

where the $P_i$ are projective.

THEOREM 6. For any $M \in \mathcal{O}$ $\operatorname{dh}(M) \leqslant 2S$, where S is the maximum length of an element in the Weyl group W.

LEMMA 2. Let $0 \to M_1 \to M \to M_2 \to 0$ be an exact sequence. Then:

a) $\operatorname{dh}(M) \leq \max(\operatorname{dh}(M_1), \operatorname{dh}(M_2))$,

b) $\operatorname{dh}(M_2) \leq \max(\operatorname{dh}(M_1) + 1, \operatorname{dh}(M))$.

The lemma is proved by standard homological arguments (see, e.g., [6]).

We shall prove Theorem 6 in three stages.

1. Let $\chi \in C^0$. Then $\operatorname{dh}(M_{w\chi}) \leqslant l(w)$.

Proof. Consider a p filtration of $P_{w\chi}$. Its factors have the form $M_{w'\chi}$, $w' \in W$. By virtue of Theorem 5, only the $M_{w'\chi}$, such that $L_{w\chi} \in JH(M_{w'\chi})$, occur in this filtration. As follows from [7], in this case either $w' = w$, or $l(w') < l(w)$, and $w'\chi > w\chi$. Using Lemma 1, we obtain an exact sequence

$$0 \to M \to P_{w\chi} \to M_{w\chi} \to 0,$$

where M has a p filtration with factors $M_{w'\chi}$, $l(w') < l(w)$. By virtue of the induction assumption and Lemma 2a), $\operatorname{dh}(M) \leq l(w) - 1$. By Lemma 2b), $\operatorname{dh}(M_{w\chi}) \leq l(w)$.

2. $\operatorname{dh}(L_{w\chi}) \leq 2S - l(w)$.

Proof. We shall use induction up to $l(w)$. For $l(w) = S$ we have $M_{w\chi} = L_{w\chi}$, and hence $\operatorname{dh}(L_{w\chi}) \leq S$. For arbitrary w we have an exact sequence

$$0 \to M' \to M_{w\chi} \to L_{w\chi} \to 0,$$

here, by virtue of [7] (see [7], Appendix), JH(M') consists of the modules $L_{w\chi}$ with $l(w') > l(w)$. Just as be-re, by induction we obtain $dh(L_{w\chi}) \leq 2S - l(w)$.

3. Since each object has finite length, the theorem follows from Lemma 2a) and the inequality $dh(L) \leq$ S, which is valid for each simple object $L \in \mathcal{O}$.

Remark. It can be shown that if $\chi$ is a regular integral highest weight, i.e., a weight such that $L_\chi$ is nite-dimensional, then $dh(L_\chi) = 2S$. More precisely, it can be shown that $\mathrm{Ext}^{\bullet}_{\mathcal{O}}(L_\chi, L_\chi)$ is isomorphic to the ohomology algebra $H^*(X, C)$, where $X = G/B$ is the base projective space of a group G with Lie algebra $\mathfrak{g}$, is the Borel subgroup of G (see [7]). In particular, $\mathrm{Ext}^{2S}_{\mathcal{O}}(L_\chi, L_\chi) = C$.

## LITERATURE CITED


. C. W. Curtis and I. Reiner, Representation Theory of Finite Groups and Associative Algebras, Wiley, New York (1962).
. J. E. Humphreys, "Modular representations of classical Lie algebras and semisimple groups," J. Algebra, 19, 51-79 (1971).
. D.-N. Verma, "Structure of certain induced representations of complex semisimple Lie algebras," Bull. Amer. Math. Soc., 74, 160-166 (1968).
. I. N. Bernshtein, I. M. Gel'fand, and S. I. Gel'fand, "The structure of representations generated by vectors of highest weight," Funktsional'. Analiz i Ego Prilozhen., 5, No. 1, 1-9 (1971).
. J. Dixmier, Algebras Enveloppantes, Gauthier-Villars, Paris (1974).
. H. Bass, Algebraic K-Theory, W. A. Benjamin, New York (1968).
. I. N. Bernshtein (Bernstein), I. M. Gel'fand (Gelfand), and S. I. Gel'fand (Gelfand), "Differential operators on the base affine space and a study of $\mathfrak{g}$-modules," Proc. of the Summer School on Group Representations, Bolyai Janos Math. Soc. (Budapest, 1971), Akademiai Kiado, Budapest (1975), pp. 21-64.
. I. M. Gel'fand (Gelfand), "The cohomology of infinite-dimensional Lie algebras; some questions of integral geometry," Intern. Congress. Math., Reports, Vol. 1, Nice (1970).

# 6.

## (avec I. N. Bernstein et S. I. Gelfand)

# Structure locale de la catégorie des modules de Harish-Chandra I

C. R. Acad. Sci., Paris, Ser. **A 286** (1978) 435–437. MR **58**:16966. Zbl. **416**:22018

La catégorie de modules de Harish-Chandra d'un groupe de Lie réel semi-simple G a algèbre de Lie $\mathfrak{g}$ est étudiée. On démontre que cette catégorie est localement équivalente à la catégorie des modules de dimension finie sur une certaine algèbre Q, finitement engendrée par le centre Z ($\mathfrak{g}$) de l'algèbre enveloppante U ($\mathfrak{g}$).

*The category of Harish-Chandra modules over a real semisimple Lie group* G *with Lie algebra* $\mathfrak{g}$ *is considered. It is shown that this category is locally equivalent to the category of finite dimensional modules over a* C*-algebra* Q *finitely generated over the center* Z($\mathfrak{g}$) *of the enveloping algebra* U($\mathfrak{g}$).

1. La théorie des modules de Harish-Chandra, introduite dans ([1]), est une des variantes de la théorie algébrique des représentations des groupes de Lie semi-simples. En dehors de l'intérêt évident que présente l'étude des modules de Harish-Chandra pour cette partie de la théorie des représentations, il nous semble que les structures qui apparaissent dans cette étude peuvent servir de modèles pour une théorie plus générale, qui pourraît être appelée « géométrie algébrique non commutative ». Il est important de noter que, même dans une théorie aussi non commutative que la théorie des représentations, la non commutativité s'avère de dimension finie. On peut donc espérer qu'une géométrie algébrique non commutative suffisamment profonde peut apparaître, même dans le cadre d'une non commutativité de dimension finie.

Fixons une fois pour toutes un groupe de Lie linéaire connexe semi-simple G et un sous-groupe compact maximal $K \subset G$; soient $\mathfrak{g}$ et $\mathfrak{k}$ les algèbres de Lie de G et K.

2. Modules de Harish-Chandra. — Désignons par $\mathscr{E}$ (G) l'espace des distributions sur G; pour chaque sous-groupe $H \subset G$ posons $\mathscr{E}_H(G) = \{\varphi \in \mathscr{E}(G) \mid \text{supp } \varphi \subset H\}$. L'inclusion $H \subsetneq G$ détermine l'inclusion $\mathscr{E}(H) \subsetneq \mathscr{E}(G)$. On a aussi $\mathscr{E}(H) \subset \mathscr{E}_H(G)$. L'opération de convolution munit $\mathscr{E}(K)$ et $\mathscr{E}_K(G)$ d'une structure de **C**-algèbre. Soit $\{e\}$ le sous-groupe trivial de G. Nous identifierons l'algèbre $\mathscr{E}_{\{e\}}(G)$ avec l'algèbre enveloppante U($\mathfrak{g}$). Remarquons que $\mathscr{E}_K(G) \supset \mathscr{E}_{\{e\}}(G) = U(\mathfrak{g})$. On peut démontrer l'égalité $\mathscr{E}_K(G) = U(\mathfrak{g})\, \mathscr{E}(K)$.

L'élément $\varphi \in \mathscr{E}_K(G)$ sera dit K-fini, si $\dim_{\mathbf{C}} \mathscr{E}(K)\, \varphi < \infty$. L'ensemble de tous les éléments K-finis est un idéal dilatère $\mathscr{H} = \mathscr{H}(G, K)$ (l'analogue de l'algèbre de Hecke). On voit immédiatement que $\mathscr{H}^2 = \mathscr{H}$. Le $\mathscr{E}_K(G)$-module V sera dit algébrique quand $\mathscr{H}\, V = V$.

D'après une représentation donnée du groupe G dans l'espace de Banach V, on construit canoniquement un $\mathscr{E}_K(G)$-module algébrique déterminé par l'action naturelle de $\mathscr{E}_K(G)$ dans le sous-espace $V^{alg} \subset V$ des vecteurs différentiables K-finis.

Nous appelerons module de Harish-Chandra tout $\mathscr{E}_K(G)$-module algébrique V à un nombre fini de générateurs, tel que $\dim_{\mathbf{C}} hV < \infty$ pour chaque $h \in \mathscr{H}$. Cette notion a été introduite dans ([1]) [*voir* aussi ([2])] sous une autre forme. La catégorie de tous les modules de Harish-Chandra sera désignée par C.

Soit Z ($\mathfrak{g}$) le centre de U ($\mathfrak{g}$). Notons $\Theta = \text{Spec max } Z(\mathfrak{g})$ l'ensemble des caractères (homomorphismes) $\theta : Z(\mathfrak{g}) \to \mathbf{C}$. C'est une variété algébrique isomorphe à $\mathbf{C}^r$, $r = \text{rang } \mathfrak{g}$

[(3), 7 . 3 . 8]. Nous aurons besoin de la notion d'élément régulier $\theta \in \Theta$. Soit $\mathfrak{f}$ une sous-algèbre de Cartan de $\mathfrak{g}$, $\mathfrak{f}_{\mathbf{C}}^* = \mathrm{Hom}_{\mathbf{R}}(\mathfrak{f}, \mathbf{C})$. L'homomorphisme de Harish-Chandra $Z(\mathfrak{g}) \to S(\mathfrak{f})$ induit l'application $\mathfrak{f}_{\mathbf{C}}^* \to \Theta$ [*voir* (3), 7 . 4 . 7]. Si l'image inverse de l'élément $\theta \in \Theta$ par cette application contient un nombre maximal de points, $\theta$ sera dit régulier.

Le support [(4), II, 4 . 4] de tout $Z(\mathfrak{g})$-module M sera désigné par supp $M \subset \Theta$.

Chaque module de Harish-Chandra est un $Z(\mathfrak{g})$-module pour lequel supp V est un ensemble fini.

Pour tout sous-ensemble $D \subset \Theta$, nous notons $C_D$ la sous-catégorie pleine de C constituée de tous les modules V tels que supp $V \subset D$.

3. Une $Z(\mathfrak{g})$-algèbre Q sera dite $Z(\mathfrak{g})$-finie si elle possède un nombre fini de générateurs en tant que $Z(\mathfrak{g})$-module. Notons Mod Q la catégorie des Q-modules M tels que $\dim_{\mathbf{C}} M < \infty$. Si $D \subset \Theta$, $\mathrm{Mod}_D\, Q$ désignera la sous-catégorie complète de Mod Q constituée des modules M pour lesquels supp $M \subset D$.

Théorème 1. — *Soit* $D \subset \Theta$ *un ouvert connexe borné. Il existe alors une telle algèbre* $Z(\mathfrak{g})$-*finie* Q, *que les catégories* $C_D$ *et* $\mathrm{Mod}_D\, Q$ *sont équivalentes en tant que* $Z(\mathfrak{g})$-*catégories.*

4. Dans cette section nous donnons une esquisse de la démonstration du théorème 1.

Soit $\delta$ une représentation de dimension finie du sous-groupe compact maximal $K \subset G$ dans l'espace L. Construisons d'après $\delta$ la $Z(\mathfrak{g})$-algèbre $Q_\delta$ de la manière suivante. Soit $(\mathrm{End}_{\mathbf{C}}\, L)^0$ l'algèbre duale à l'algèbre des endomorphismes de l'espace L. Posons

$$U_\delta = U(\mathfrak{g}) \otimes_{\mathbf{C}} (\mathrm{End}_{\mathbf{C}}\, L)^0.$$

Désignons par $J_\delta$ l'idéal à gauche de $U_\delta$ engendré par les éléments de la forme $X \otimes 1 - 1 \otimes \delta(X)$, $X \in \mathfrak{k}$. Soit $N_\delta = \{u \in U_\delta \mid J_\delta\, u \subset J_\delta\}$ le normalisateur de $J_\delta$. Posons $Q_\delta = N_\delta / J_\delta$.

Pour un $E_K(G)$-module algébrique V, soit $r_\delta(V) = \mathrm{Hom}_K(L, V)$. Définissons l'action de $Q_\delta$ dans l'espace $r_\delta(V)$ de la manière suivante. Soient $\varphi \in r_\delta(V)$, $q \in Q_\delta$ et soit $\varphi = \sum X_i \otimes Y_i$ un représentant de $q$ dans $N_\delta$. Définissons l'élément $q(\varphi) \in r_\delta(V) = \mathrm{Hom}_K(L, V)$ par l'égalité $q(\varphi)\,\xi = \sum X_i\, \varphi(\delta(Y_i)\xi)$, $\xi \in L$.

Proposition I. — (i) $Q_\delta$ *est une algèbre* $Z(\mathfrak{g})$-*finie. L'application* $V \mapsto r_\delta(V)$ *définit un foncteur* $r_\delta : C \to \mathrm{Mod}\, Q_\delta$;

(ii) *soit* D *un ouvert connexe borné de* $\Theta$. *Il existe alors une telle représentation* $\delta$ *du groupe* K *que* $r_\delta$ *détermine une équivalence des catégories* C *et* $\mathrm{Mod}_D\, Q_\delta$.

La première partie de cette proposition se démontre d'une manière analogue à (3), 9 . 4. Pour démontrer la seconde, il suffit de vérifier qu'en prenant $\delta$ somme directe d'un nombre suffisamment grand de représentation deux à deux non équivalentes du groupe K on obtient $r_\delta(V) \neq \{0\}$ pour tout module non nul $V \in C_D$.

Le théorème 1 découle de la proposition 1.

5. Nous voulons choisir l'algèbre Q du théorème 1 aussi simple que possible. Dans cette section, nous montrerons, en se servant des représentations de la série principale, que Q peut être réalisée sous forme d'algèbre de fonctions polynômiales à valeurs non commutatives.

Définissons les représentations de la série principale pour le groupe G de la manière suivante. Soit P un sous-groupe parabolique minimal de G, $P = MAU$ la décomposition de Langlands, où U est le radical unipotent de P, $A \cong \mathbf{R}_+^l$ — la partie vectorielle connexe du sous-groupe de Cartan, et $M = P \cap K$ centralise A. Soit $\mathfrak{a}$ l'algèbre de Lie du groupe A, $\mathfrak{a}^*$ duale à $\mathfrak{a}$, W le groupe de Weyl de $\mathfrak{a}$ dans $\mathfrak{g}$. Soit $(\rho, L)$ une représentation irréductible de dimension finie du groupe MA, $\rho'$ une représentation de P dans l'espace L, telle que $\rho'(u)=1$ pour $u \in U$ et $\rho'(ma)=\rho(ma).\,\mu(a)$, où $\mu(a)=d(aua^{-1})/du$. Soit $\mathscr{E}_M(P)$ l'algèbre des distributions sur P concentrées sur M. Cette algèbre agit dans l'espace L à l'aide de la représentation $\rho'$ et dans l'algèbre $\mathscr{H}$ par convolution.

DÉFINITION. — On appelle représentation de la série principale le $\mathscr{E}_K(G)$-module $T^\rho=\mathscr{H}\otimes_{\mathscr{E}_{M(P)}} L$.

PROPOSITION 2. — (i) $T^\rho$ *est un module de Harish-Chandra;* supp $T^\rho$ *se réduit à un seul point* $\theta_\rho \in \Theta$;

(ii) *le module* $T^\rho$ *est irréductible pour presque toutes les représentations* $\rho$;

(iii) *pour presque toutes les représentations* $\rho_1$, *le module* $T^{\rho_1}$ *est équivalent à* $T^{\rho_2}$ *si et seulement si* $\rho_1$ *et* $\rho_2$ *sont conjuguées relativement à* W;

(iv) *chaque module de Harish-Chandra irréductible est un sous-quotient d'un des modules* $T^\rho$ [([3]), 9.4].

Chaque représentation irréductible $\rho$ du groupe MA est de la forme $\rho(ma)=\sigma(m)\, e^{\lambda(\log a)}$, ou $\sigma$ est une représentation irréductible de M et $\lambda \in \mathfrak{a}^*$. L'algèbre $\mathscr{E}(M)$ des distributions sur M agit naturellement dans l'espace L de la représentation $\rho$. Par ailleurs, notons $T^\rho|_K$ le $\mathscr{E}$ (K)-module obtenu de $T^\rho$ par l'inclusion $\mathscr{E}(K) \subset \mathscr{E}_K(G)$. On vérifie facilement que le $\mathscr{E}$ (K)-module $T^\rho|_K$, $\rho=(\sigma, \lambda)$, est canoniquement isomorphe au $\mathscr{E}$(K)-module $(\mathscr{H}\cap\mathscr{E}(K))\otimes_{\mathscr{E}(M)} L$.

Par conséquent, on peut identifier tous les $Q_\delta$-modules $r(T^\rho)$, $\rho=(\sigma, \lambda)$, pour $\sigma$ fixe, avec un même espace linéaire (de dimension finie) que nous noterons $L_\delta^\sigma$.

LEMME. — *L'action des éléments* $q \in Q_\delta$ *dans* $L_\delta^\sigma$ *est polynômiale relativement à* $\lambda$.

Nous avons donc construit un homomorphisme d'algèbres :

$$\psi_\delta^\sigma : \quad Q_\delta \to (\text{End } L_\delta^\sigma)\, [\mathfrak{a}^*].$$

Pour une représentation fixé $\delta$, l'espace $L_\delta^\sigma$ est non nul seulement pour un nombre fini de représentations $\sigma$ du groupe M. L'algèbre $B_\delta=\oplus_\sigma$ End $L_\delta^\sigma$ est donc de dimension finie.

THÉORÈME 2. — *L'homomorphisme* $\psi_\delta=\oplus_\sigma \psi_\delta^\sigma : Q_\delta \to B_\delta[\mathfrak{a}^*]$ *est injectif.*

La démonstration s'obtient facilement de la proposition 2 , (iv).

Dans un article suivant, nous étudions en détail l'image de l'inclusion $\psi_\delta$.

(*) Séance du 9 janvier 1978.

([1]) I. M. GELFAND, V. A. PONOMAREV, *Uspekni Mat. Nauk*, 23, 1968, n° 2, p. 3-60.
([2]) J. LEPOWSKY, *Trans. Amer. Math. Soc.*, 176, 1973, p. 1-44.
([3]) J. DIXMIER, *Algèbres enveloppantes*. Gautier-Villars, Paris, 1974.
([4]) N. BOURBAKI, *Algèbre commutative*, Hermann, Paris, 1965.

*Laboratoire de Méthodes Mathématiques en Biologie, Université d'État de Moscou, Moscou 117.234, U.R.S.S.*

# 7.

## (avec I. N. Bernstein et S. I. Gelfand)

# Structure locale de la catégorie des modules de Harish-Chandra II

C. R. Acad. Sci., Paris, Ser. A **286** (1978) 495–497. MR **81** e:22026. Zbl. **431**:22013

Cette Note est la suite de (1). On y introduit la notion d'algèbre agréable de fonctions analytiques dans un domaine $D \subset \mathbf{C}^l$ à valeurs dans une **C**-algèbre B de dimension finie. On démontre que la catégorie des modules de Harish-Chandra d'un groupe de Lie semi-simple réel G est localement équivalente à la catégorie des modules de dimension finie sur une certaine algèbre agréable R. Les exemples $G = SL_2(\mathbf{R})$, $SL_3(\mathbf{R})$, $SL_2(\mathbf{C})$ sont considérés.

*This Note is the continuation of* (1). *The notion of agreeable algebra of analytic functions in a domain* $D \subset \mathbf{C}^l$ *taking values in a finite dimensional* **C**-*algebra is introduced. It is shown that the category of Harish-Chandra modules over a real semisimple Lie group* G *is locally equivalent to the category of finite dimensional modules over an agreeable algebra* R. *The examples* $G = SL_2(\mathbf{R})$, $SL_3(\mathbf{R})$, $SL_2(\mathbf{C})$ *are considered.*

1. Le présent article est la suite de (1), dont les résultats et les notations sont systématiquement employés ici.

Soit D un ouvert connexe de l'espace $\Theta = \text{Spec max } Z(\mathfrak{g})$. Notons $\mathcal{O}(D)$ l'anneau des fonctions analytiques sur D. Pour chaque algèbreQ $Z(\mathfrak{g})$-finie, appelons localisation de Q à D l'algèbre $Q_D = Q \underset{Z(\mathfrak{g})}{\otimes} \mathcal{O}(D)$. Il est clair que la catégorie $\text{Mod}_D\, Q$ est équivalente à la catégorie des $Q_D$-modules de dimension finie.

Nous voulons construire une algèbre Q qui vérifie les hypothèses du théorème 1 de (1), dont la structure locale serait aussi simple que possible.

2. Algèbres agréables. — Soit $\mathbf{C}^l$ l'espace complexe aux coordonnées $z_1, \ldots, z_l$, E un voisinage du point $(0, \ldots, 0) \in \mathbf{C}^l$ et B une **C**-algèbre unitaire semi-simple. Notons $B(E) = B \underset{\mathbf{C}}{\otimes} \mathcal{O}(E)$ l'algèbre des fonctions analytiques $f : E \to B$.

Définition. — Une sous-algèbre $R \subset B(E)$ est dite agréable, si

(i) $R \supset \mathcal{O}(E)\,1$;

(ii) pour presque tous les points $z \in E$ (i.e. pour les points situés en-dehors d'une certaine sous-variété $E_1 \subset E$ de dimension $l-1$), l'ensemble des valeurs $f(z)$, $f \in R$, coïncide avec B.

Théorème 1. — *Soit* $\theta \in \Theta$ *un élément régulier. On peut alors choisir un voisinage* $D \subset \Theta$ *de l'élément* $\theta$ *et une algèbre* Q *vérifiant les hypothèses du théorème* 1 *de* (1), *de sorte que l'algèbre* $Q_D$ *soit isomorphe a une certaine algèbre agréable* R.

*Remarques.* — (*a*) L'algèbre agréable R du théorème 1 peut être réalisée sous forme d'algèbre de fonctions sur un ouvert connexe E de dimensions $l = \text{rang}_{\mathbf{R}} G$. Dans le cas où $l = \text{rang}\, \mathfrak{g}$ (i.e. le groupe G est décomposable sur **R**), on peut prendre $E = D$.

(*b*) Il découle du théorème 1 et de la remarque précédente que les supports de tous les modules $V \in C_D$ remplissent la réunion d'un nombre fini de sous-variétés de dimension L d'un connexe D de dimension *r*.

(*c*) Dans la démonstration du théorème 1, l'isomorphisme de l'algèbre R sur l'algèbre $Q_D$ est choisi de sorte que les fonctions de coordonnées $z_1, \ldots, z_l$ sur E correspondent à des éléments $z_1^*, \ldots, z_l^* \in Z(\mathfrak{g})$ tels que $\theta(z_i^*) = 0$.

3. Nous dirons qu'une sous-algèbre $R \subset B(E)$ est *fort agréable*, si R est un $\mathcal{O}(E)$-module libre. Les algèbres fort agréables sont utiles parce qu'elles sont entièrement déterminées par des conditions de codimension 1. Plus précisément, soient R, R′ deux sous-algèbres fort agréables de B(E). Supposons que R et R′ coïncident hors de codimension 2, i. e. il existe un sous-ensemble $X \subset E$, codim $X \geqq 2$, tel que $R_{E-X} = R'_{E-X}$. On démontre alors facilement que $R = R'$.

Hypothèse. – L'algèbre R du théorème 1 peut être choisie fort agréable.

4. Montrons comment on déduit le théorème 1 du théorème 2 de [1]. Fixons une représentation irréductible $\sigma$ du groupe M. L'application $\xi_\sigma : \lambda \to \theta_\rho$, $\rho = (\sigma, \lambda)$ de la proposition 2 (i) de [1] est une application polynômiale $\mathfrak{a}^*$ dans $\Theta$. Soit $\Xi \subset \Theta$ la réunion des images de tous les $\xi_\sigma$. C'est une sous-variété localement algébrique de dimension 1 dans $\Theta$. D'après le théorème de Harish-Chandra [[1], prop. 2, (iv)], $\bigcup_{V \in C} \operatorname{supp} V = \Xi$.

Soit $\theta \in \Theta$ un caractère régulier. Désignons par $\tilde{\Phi}$ l'ensemble de toutes les représentations $\rho = (\sigma, \lambda)$ telles que $\theta_\rho = \theta$. Le groupe de Weyl W agit sur $\tilde{\Phi}$, et librement (en vertu de la régularité de $\theta$). Fixons l'ensemble $\Phi = (\rho_1, \ldots, \rho_t)$ de représentants des orbites de W dans $\tilde{\Phi}$. Pour chaque $\rho_i = (\sigma_i, \lambda_i) \in \Phi$ l'application $\xi_{\sigma_i}$ sera une immersion dans un voisinage du point $\lambda_i \in \mathfrak{a}^*$. Il existe donc $l$ éléments $z_1, \ldots, z_l$, $\theta(z_i) = 0$, qui déterminent un système de coordonnées dans des voisinages de chacun des points $\lambda_i$. Soit E un petit voisinage de l'origine de l'espace $\mathbf{C}^l$ avec les coordonnées $(z_1, \ldots, z_l)$, $E_i$ le voisinage correspondant des points $\lambda_i$, et $D \subset \Theta$ un tel voisinage de $\theta$ que $D \cap \Xi = \bigcup_{i=1}^{t} \xi_{\sigma_i}(E_i)$.

Soit $B = \bigoplus_{i=1}^{t} \operatorname{End} L_\delta^{\sigma_i}$. L'homomorphisme $\psi_\delta$ du théorème 2 de [1] se prolonge à un plongement $\tilde{\Psi}_\delta : (Q_\delta)_D \to B(E)$. Il découle de la proposition 2 (ii), (iii) de [1] que $R = \tilde{\Psi}_\delta((Q_\delta)_D)$ est une sous-algèbre agréable de B(E). Le théorème 1 découle maintenant de la proposition 1 de [1].

5. Exemple :

$$G = PSL_2(\mathbf{R}), \qquad K = SO_2/(\pm 1), \qquad M = \{1\},$$
$$A = \left\{ \begin{pmatrix} a & 0 \\ 0 & a^{-1} \end{pmatrix}, a > 0 \right\}, \qquad N = \left\{ \begin{pmatrix} 1 & n \\ 0 & 1 \end{pmatrix} \right\}.$$

Soit $\theta$ le caractère correspondant à la représentation unité I du groupe G. Les représentations de la série principale sont indexées par les homomorphismes

$$\lambda_s : \begin{pmatrix} a & 0 \\ 0 & a^{-1} \end{pmatrix} \to a^s, \qquad s \in \mathbf{C}.$$

On a alors $\tilde{\Phi} = \{\lambda_1, \lambda_{-1}\}$, tandis que W applique $\lambda_1$ dans $\lambda_{-1}$, de sorte que l'on peut poser $\Phi = \{\lambda_1\}$. Le module $T = T^{\lambda_1}$ contient le sous-module unité I, et $T/I = T^+ \oplus T^-$, ou $T^+$, $T^-$ sont des modules irréductibles non équivalents. En outre, tout sous-module non trivial $V \subset T$ contient I. Les modules $T^{\lambda_s}$ sont irréductibles et deux à deux non équivalents lorsque $s$ est proche de 1, mais $s \neq 1$.

Les représentations irréductibles du groupe K sont unidimentionnelles et se déterminent par un entier unique – leur degré. Désignons la représentation de degré $i$ par $\delta_i$. On sait que

$$T^{\lambda_s}|_K = \bigoplus_{i=-\gamma}^{\gamma} \delta_i, \qquad T^+|_K = \bigoplus_{i>0} \delta_i, \quad T^-|_K = \bigoplus_{i<0} \delta_i.$$

Posons $\delta=\delta_{-1}\oplus\delta_0\oplus\delta_1$. Soit $D\subset\Theta$ un voisinage suffisamment petit du point $\theta$. La proposition 2 (iv) de ([1]) implique $\mathrm{Hom}_K(V,\delta)\neq 0$ pour tout module $V\in C_D$. La représentation $\delta=\delta_{-1}\oplus\delta_0\oplus\delta_1$ satisfait à la proposition 1 (ii) de ([1]). Il est clair que l'espace L sera de dimension trois, avec la base naturelle $e_{-1}$, $e_0$, $e_1$. Il existe donc un isomorphisme naturel de l'algèbre $B=\mathrm{End}\,L_\delta$ sur l'algèbre $M_3$ des matrices d'ordre trois.

Choisissons un élément $z^*\in Z(\mathfrak{g})$ qui définit la coordonnée $z$ dans un voisinage du point $s=1$ (par exemple, $z=s^2-1$). Décrivons l'algèbre $R=\tilde{\Psi}((Q_\delta)_D)\subset M_3$ (*voir* le n° 4). Soient $E_{ij}\in M_3$, $-1\leqq i,j\leqq 1$ les matrices élémentaires. Il est clair que $E_{ii}\in R$ pour $i=-1,0,1$. Donc $R=\bigoplus_{i,j}R_{ij}$, où $R_{ij}=\{r\in R, r=f(z)E_{ij}\}$. Puisque $\mathcal{O}(D)\,1\subset R$, on a $R_{ij}=J_{ij}E_{ij}$, où $J_{ij}$ est un idéal de $\mathcal{O}(D)$. La fibre de R coïncide en tout point $z\neq 0$ avec $M_3$, de sorte que $J_{ij}$ est engendré par l'élément $z^{k_{ij}}$. Il ne reste qu'à trouver $k_{ij}$.

Évidemment, $k_{ii}=0$, $i=-1,0,1$. Puisque chaque sous-module de T contient I, on a $k_{0i}=0$ pour tous les $i$, et $k_{ij}>0$ pour $i\neq j$, $i\neq 0$. En calculant les coefficients matriciaux $f_{-1,0}$ et $f_{1,0}$ des représentations de la série principale $T^\lambda$, comme fonctions de $s$, on peut montrer qu'elles ont un zéro d'ordre un pour $s=1$. Donc $k_{-1,0}=k_{1,0}=1$. D'où $k_{-1,1}=k_{1,-1}=1$.

Ainsi

$$R=\{a_{ij}(z),\ z\in D\,|\,a_{ij}(0)=0 \text{ pour } (i,j)=(-1,0),(1,0),(-1,1),(1,-1)\}.$$

La catégorie $C_{\{0\}}$ est alors équivalents à la catégorie des R-modules de dimension finie dans lesquels l'opérateur $z-1$ est nilpotent. On vérifie facilement que ceci coïncide avec la description de $C_{\{0\}}$ dans ([2]).

6. Autres exemples. — Dans tous les exemples, nous prendrons en guise de $\theta$ le caractère de $Z(\mathfrak{g})$ correspondant à la représentation unité de G.

(*a*) $G=PGL_2(\mathbf{R})$, $\dim E=1$, $B=M_2\oplus M_2$; R consiste des couples de fonctions matricielles $\{a_{ij}(z), b_{ij}(z)\}\in B(E)$ tels que $a_{12}=b_{12}=0$, $a_{11}=b_{11}$ pour $z=0$.

(*b*) $G=SL_2(\mathbf{C})$, $\dim E=1$, $B=M_2\oplus M_1$; R consiste des couples $\{a_{ij}(z), b(z)\}\in B(E)$ tels que $a_{12}=0$, $a_{11}=b$ pour $z=0$. La description correspondante de la catégorie $C_{\{\theta\}}$ coïncide avec celle donnée dans ([3]).

(*c*) $G=SL_3(\mathbf{R})$, E est un domaine de $\mathbf{C}^2$ à coordonnées $z_1$, $z_2$, $B=M_4\oplus M_3\oplus M_3\oplus M_1$; R consiste de quadruplets de fonctions $\{a_{ij}, b_{ij}, c_{ij}, d\}\in B(E)$ tels que

(i) pour $z_1=0$: $a_{13}=a_{14}=a_{23}=a_{24}=b_{12}=b_{13}=0$, $a_{33}=b_{22}$, $a_{34}=b_{23}$, $a_{43}=b_{32}$, $a_{44}=b_{33}$;

(ii) pour $z_2=0$: $a_{12}=a_{14}=a_{32}=a_{34}=c_{12}=c_{13}=0$, $a_{22}=c_{22}$, $a_{24}=c_{23}$, $a_{42}=c_{32}$, $a_{44}=c_{33}$;

(iii) pour $z_1+z_2=0$: $b_{13}=b_{23}=c_{13}=c_{23}=0$, $b_{33}=c_{33}$.

(*) Séance du 9 janvier 1978.

([1]) I. N. Bernstein, I. M. Gelfand, S. I. Gelfand, *Comptes rendus*, 286, série A, 1978, p. 435.

([2]) I. M. Gelfand, *Actes du Congrès international des mathématiciens*, Nice, 1970, I, 95-111, Gauthier-Villars, Paris.

([3]) I. M. Gelfand, V. A. Ponomarev, *Uspeki Nat. Nauk*, 23, 1968, n° 2, p. 3-60.

*Laboratoire de Méthodes Mathématiques en Biologie, Université d'État de Moscou, Moscou 117.234, U.R.S.S.*

# 8.

## (with I. N. Bernstein and S. I. Gelfand)

# Algebraic bundles over $P^n$ and problems of linear algebra

Funkts. Anal. Prilozh. **12** (3) (1978) 66–68 [Funct. Anal. Appl. **12** (1978) 212–214].
MR **80** c: 14010 a, Zbl. **402**:14005

1. The description of the algebraic vector bundles over projective space $P^n$ has attracted the attention of many specialists in algebraic geometry (see [1-3]). Recently, interest in this problem has increased even more in connection with the remarkable papers of Atiyah and Ward [4] and Belavin and Zakharov [5], in which the connection of bundles over $CP^3$ with gauge fields on the four-dimensional sphere is described. In the present note it is shown how the classification of bundles over $P^n$ reduces to a problem of linear algebra, viz., to the classification of finite-dimensional graded representations of the exterior (Grassman) algebra on (n + 1) variables. There are special cases of such a reduction in Barth [2] and Drinfel'd and Manin [3]. Independently obtained, Beilinson [6] is close to our result. We want to express profound gratitude to Yu. I. Manin, whose report on [3] stimulated our interest in these questions.

2. Let $\Xi$ be an (n + 1)-dimensional linear space over an algebraically closed field k, $\Lambda$ be the exterior algebra on the space $\Xi$. We introduce a grading on $\Lambda$, by setting deg $\xi = -1$ for $\xi \in \Xi$. By a $\Lambda$-module we shall mean a finitely generated graded $\Lambda$-module; notation $V = \bigoplus_j V_j$. Let $\mathcal{P}$ be the class of free $\Lambda$-modules; we shall call $\Lambda$-modules V, V' $\mathcal{P}$-equivalent, if $V \oplus P = V' \oplus P'$ for some $P, P' \in \mathcal{P}$.

3. Let P be the projective space corresponding to $\Xi$. We shall construct, for each $\Lambda$-module V, a complex L(V) of vector bundles over P. Namely, we set $L_j = V_{-j} \otimes \mathcal{O}(j)$, where $\mathcal{O}(j)$ is the j-th power of the Hopf bundle; by definition, a section of the bundle $L_j$ is a homo-

Moscow State University. Translated from Funktsional'nyi Analiz i Ego Prilozheniya, Vol. 12, No. 3, pp. 66-67, July-September, 1978. Original article submitted January 10, 197

eous function $f(\xi)$ of degree of homogeneity with values in $V_{-j}$. We define the differential $L_j \to L_{j+1}$, by setting $df(\xi) = \xi(f(\xi))$.

If $\xi \in \Xi$, $\xi \neq 0$, then the fiber $L_\xi(V)$ of complex $L(V)$, corresponding to the point $\bar{\xi} \in P$, ncides with the complex vector spaces $L_{\bar{\xi}}(V) = (\ldots \to V_1 \xrightarrow{\xi} V_0 \xrightarrow{\xi} V_{-1} \to \ldots)$. We call the $\Lambda$-module aithful if $H^i(L_\xi(V)) = 0$ for $i \neq 0$ for all $0 \neq \xi \in \Xi$. In this case $H^0(L(V))$ is a vector dle over P; its fiber at the point $\bar{\xi}$ coincides with $H^0(L_\xi(V))$. We denote this bundle by ).

THEOREM 1. Any algebraic vector bundle over P has the form $\Phi(V)$ for some faithful $\Lambda$-ule V. Here $\Phi(V) \approx \Phi(V')$ if and only if V and V' are $\mathcal{P}$-equivalent.

Remarks. 1) The map $V \mapsto \Phi(V)$ (for exact $\Lambda$-modules V) commutes with tensor products, ing symmetric and exterior powers, and passage to the dual module.

2) Let $0 \to V \to P \to V' \to 0$ be an exact sequence of $\Lambda$-modules, where $P \in \mathcal{P}$, V is a faith- module. Let W be the $\Lambda$-module obtained from V' by the grading shift: $W_j = V'_{j+1}$. Then s a faithful $\Lambda$-module and $\Phi(W) = \Phi(V) \otimes \mathcal{O}(1)$.

3) Let $\xi_0, \ldots, \xi_n$ be a basis in $\Xi$, $\omega = \xi_0 \ldots \xi_n \in \Lambda$. It is easy to verify that each $\Lambda$-ule V can be represented in the form $V = V^0 \oplus P$, where $P \in \mathcal{P}$, $\omega V^0 = 0$, to $\mathcal{P}$-equivalent ules V correspond isomorphic modules $V^0$. Hence vector bundles over P are classified by thful modules over the algebra $\Lambda/(\omega)$.

3. To formulate a more precise result we need the machinery of derived categories (see ). Let Coh be the category of coherent sheaves on P, $C^b(\mathrm{Coh})$ be the category of bounded plexes of objects of Coh and $D^b(\mathrm{Coh})$ be the derived category.

Let $\mathcal{M}(\Lambda)$ be the category of $\Lambda$-modules. Considering, for each $V \in \mathcal{M}(\Lambda)$, L(V) as a com- x of sheaves on P, we get a functor $L: \mathcal{M}(\Lambda) \to C^b(\mathrm{Coh})$. By $L_D$ we denote the composite func- $\mathcal{M}(\Lambda) \to C^b(\mathrm{Coh}) \to D^b(\mathrm{Coh})$. It is easy to verify that for $V \in \mathcal{P}$ the complex L(V) is acyclic, that $L_D(V) \approx 0$. Hence the functor $L_D$ factors through some functor $L'_D: \mathcal{M}(\Lambda)/\mathcal{P} \to D^b(\mathrm{Coh})$, re $\mathcal{M}(\Lambda)/\mathcal{P}$ is the quotient category of $\mathcal{M}(\Lambda)$ by the family of morphisms, factoring through ects $P \in \mathcal{P}$ (see [7, 8]).

THEOREM 2. The functor $L'_D: \mathcal{M}(\Lambda)/\mathcal{P} \to D^b(\mathrm{Coh})$ is an equivalence of categories.

Remarks. 1) Let $\mathcal{N}$ be the complete subcategory of $\mathcal{M}(\Lambda)/\mathcal{P}$, consisting of these modules such that $H^i(L(V)) = 0$ for $i \neq 0$. Then it follows from Theorem 2 that the functor $V \mapsto$ L(V)) defines an equivalence of the category $\mathcal{N}$ with the category Coh. Whence it is easy derive Theorem 1.

2) The equivalence $L'_D$ defines on $\mathcal{M}(\Lambda)/\mathcal{P}$ a structure of triangulated **category.** This ucture is characterized by the condition that for any exact sequence $0 \to V' \to V \to V'' \to 0$ $\mathcal{M}(\Lambda)$ the morphisms $V' \to V \to V''$ are included in a triangle in $\mathcal{M}(\Lambda)/\mathcal{P}$, while in this way gets all pairs of morphisms contained in triangles. In particular, if $V \in \mathcal{P}$, then V'' = ), where T is the translation functor.

3) Let k be the trivial $\Lambda$-module of degree 0, V be a faithful $\Lambda$-module. Then $H^i(P, \Phi(V)) =$ $\mathrm{m}_{\mathcal{M}(\Lambda)/\mathcal{P}}(k, T^iV)$. For $i \neq 0$ this group is equal to $\mathrm{Ext}^i_{\mathcal{M}(\Lambda)}(k, V)$.

4. We shall explain the scheme of the proof of Theorem 2. Let $X = \Xi^*$, $S = S(X)$ be the metric algebra on the space X with its ordinary grading $S = \bigoplus_{j \geq 0} S_j$, $\mathcal{M}(S)$ be the category of ded finitely generated S-modules. We denote by $C^b(S)$ and $C^b(\Lambda)$ the categories of bounded plexes of objects from $\mathcal{M}(S)$ and $\mathcal{M}(\Lambda)$, while in the case $\mathcal{M}(\Lambda)$ it will be assumed that differential $\partial$ in the complex satisfies the condition $\partial\xi = -\xi\partial$ for $\xi \in \Xi$.

We construct a function F: $C^b(\Lambda) \to C^b(S)$. A complex $(V, \partial) \in C^b(\Lambda)$ will be considered a bigraded space $V = \oplus V^i_j$, where i is the number of the module in the complex, j is the ading in $\mathcal{M}(\Lambda)$; analogously for complexes $(W, d) \in C^b(S)$. The differentials $\partial$ and d have bi- gree (1, 0). We set $F(V) = W = S \otimes V$ (tensor product over k). We define the differential in W by the formula $d(s \otimes v) = \Sigma x_i s \otimes \xi_i v + s \otimes \partial v$, where $\{x_i\}$, $\{\xi_i\}$ are dual bases in X and $\Xi$; define the bidegree in W as follows: if $s \in S_k$, $v \in V^i_j$, then $s \otimes v \in W^{i-j}_{j+k}$.

Let $D^b(\Lambda)$ and $D^b(S)$ be the derived categories corresponding to $C^b(\Lambda)$ and $C^b(S)$.

THEOREM 3. The functor $F: C^b(\Lambda) \to C^b(S)$ extends to a functor $F_D: D^b(\Lambda) \to D^b(S)$; the func- r $F_D$ is an equivalence of triangulated categories.

To prove Theorem 3 it is necessary to consider the adjoint functor $G: C(S) \to C(\Lambda)$. It i defined as follows: $G(W) = V = \mathrm{Hom}_k(\Lambda, W)$; $\partial(v)\lambda = -\Sigma x_i v(\xi_i \lambda) + d(v(\lambda))$; $V^i_j(\Lambda_k) \subset W^{i-j-k}_{j+k}$. Although th image $G(C^b(S))$ does not lie in $C^b(\Lambda)$, G allows one to define a functor $G_D: D^b(S) \to D^b(\Lambda)$. Usin the Koszul complex, it is easy to verify that the functor $G_D$ is inverse to the function $F_D$.

5. Let $\mathcal{F}, \mathcal{I}$ be the full subcategories in $D^b(S)$ and $D^b(\Lambda)$, generated by the complexes, consisting of finite-dimensional (respectively free) modules. It is easy to verify that $F_D^{-1}(\mathcal{F}) = \mathcal{I}$, so that $F_D$ defines an equivalence of categories $D^b(\Lambda)/\mathcal{I} \to D^b(S)/\mathcal{F}$ (the quotient categories in the sense of Verdier [7]).

Using Serre's theorem, describing the category Coh in terms of $\mathcal{M}(S)$ (see [9]), it is easy to get that the category $D^b(\mathrm{Coh})$ is equivalent with $D^b(S)/\mathcal{F}$. Thus, from Theorem 3 fol lows

THEOREM 4. The categories $D^b(\mathrm{Coh})$ and $D^b(\Lambda)/\mathcal{I}$ are equivalent.

6. Proposition. The natural imbedding $\mathcal{M}(\Lambda) \to D^b(\Lambda)$ defines an equivalence of categor- ies $\mathcal{M}(\Lambda)/\mathcal{P} \to D^b(\Lambda)/\mathcal{I}$.

The proposition follows from the fact that free $\Lambda$-modules are projective and injective Theorem 2 follows from this proposition and Theorem 4.

7. Theorems 1-4 are true for any field k; Theorems 3 and 4 are true if k is replaced by an arbitrary basis Z, $\Xi$ by a locally free sheaf of $\mathcal{O}_Z$-modules, P by a projective spectru of sheaves of algebras $S = S(X)$, where $X = \Xi^*$.

## LITERATURE CITED


1. G. Horrocks, Proc. London Math. Soc., 14, 689-713 (1964).
2. W. Barth, Invent. Math., 42, 63-92 (1977).
3. V. G. Drinfel'd and Yu. I. Manin, Usp. Mat. Nauk, 33, No. 3, 165-166 (1978).
4. M. Atiyah and R. Ward, Commun. Math. Phys., 55, 117-124 (1977).
5. A. A. Belavin and V. I. Zakharov, Preprint IF, Chernogolovka (1977).
6. A. A. Beilinson, Funkts. Anal. Prilozhen., 12, No. 3, 68-69 (1978).
7. J.-L. Verdier, Lecture Notes Math., 569, 262-311 (1977).
8. M. Auslander and I. Reiten, Lecture Notes Math., 488, 1-8 (1975).
9. J. P. Serre, in: Fiber Spaces [in Russian], IL, Moscow (1957), pp. 372-453.

# Part VI

# Enveloping algebras and their quotient skew-fields

# 1.
## (with A. A. Kirillov)
# Fields associated with enveloping algebras of Lie algebras

Dokl. Akad. Nauk SSSR **167** (1966) 503–505 [Sov. Math., Dokl. 7 (1966) 407–409].
Zbl. **149**:29

Noncommutative rings are naturally encountered in many problems of analysis. As examples one can mention ıgs of operators in Hilbert space, rings of differential operators on a smooth manifold, rings of opera- rs in quantum field theory, group rings of a group and enveloping algebras of Lie algebras. One of e most common and interesting examples is the ring of all differential operators on a smooth manifold that are invariant with respect to a given group $G$ of diffeomorphisms of $X$. A study of the algebraic ructure of this ring is essential for analysis.

In the present note we confine ourselves to the special case of those rings which are enveloping gebras of Lie algebras. As in algebraic geometry, the situation here is simplified if we adopt a rational point of view and in place of rings consider quotient fields. Let us proceed to the exact finitions.

Suppose $G$ is a Lie algebra defined over a commutative field $L$ of characteristic 0 and $U(G)$ is e enveloping algebra of the algebra $G$. We can show that $U(G)$ is a (two-sided) Ore domain, i. e. ring without divisors of zero in which any two elements have a common (right and left) multiple. It therefore possible to define a quotient field of the algebra $U(G)$, each element of which can be writ- n in the forms $ab^{-1}$ and $c^{-1}d$, where $a, b, c, d \in U(G)$. We will denote this field by $D(G)$ and call it e Lie field of the algebra $G$. The center of $D(G)$ will be denoted by $C(G)$.

Now suppose $K$ is an arbitrary field over a commutative field $L$. We define the dimension of $K$ er $L$ by the following formula:

$$\mathrm{Dim}_L K = \sup_{\alpha} \inf_{b \neq 0} \overline{\lim_{N \to \infty}} \frac{\ln d(\alpha b, N)}{\ln N}, \tag{1}$$

here $\alpha = (a_1, a_2, \ldots, a_s)$ is any finite collection of elements of $K$, $\alpha b = (a_1 b, a_2 b, \cdots, a_s b)$, $d(\alpha, N)$ the dimension of the subspace in $K$ consisting of all elements which can be written in the form of a lynomial of degree $\leq N$ (with coefficients in $L$) in elements of the collection $\alpha$. For the case when e field $K$ is commutative the concept of dimension introduced here coincides, as is easy to verify, ith the degree of transcendence.

Theorem 1. *Let $A$ be an algebra over a commutative field $L$ of characteristic 0. Suppose that ere exists in $A$ a filtration*

$$L = A_0 \subset A_1 \subset \cdots \subset A_k \subset \cdots; \quad A_k A_l \subset A_{k+l},$$

*ch that the associated graded algebra*

$$\mathrm{gr}\, A = \sum A_k / A_{k-1}$$

*isomorphic to the algebra of all polynomials in $n$ variables (the grading in which need not coincide* *with the standard grading).*

*Then $A$ is an Ore domain and the quotient field $K$ of the algebra $A$ has dimension $n$.*

The proof of this theorem makes use of the following property of the field $K$. Let $p_i$ be the natural projection of $A_i$ onto $A_i/A_{i-1} \in \operatorname{gr} A$. For each nonzero element $a \in A$ there exists a unique integer $i$ for which $p_i(a) \neq 0$. We will call $p_i(a)$ the *leading term of the element* $a$ and denote it by $[a]$.

Each element $b \in K$ can be written in the form $a_1 a_1^{-1}$, $a_1, a_2 \in A$. It turns out that the rational function $[a_1]/[a_2]$ depends only on the element $b$ and not on its notation in the form $a_1 a_2^{-1}$. We denote this function by $[b]$.

Lemma 1. *The mapping $b \to [b]$ is a homomorphism of the multiplicative group of the field $K$ into the group of rational functions of the form $PQ^{-1}$, where $P$ and $Q$ are homogeneous polynomials (in general, of different degree) in $n$ variables.*

This lemma is also useful in studying the automorphisms of the field $K$.

Now let $G$ be an algebraic Lie algebra. Then we have

Theorem 2. *The numbers $n$ and $k$ defined by the equations*

$$2n + k = \operatorname{Dim}_L D(G), \quad k = \operatorname{Dim}_L C(G), \tag{2}$$

*are nonnegative integers. In addition, the equations*

$$2n + k = \dim G, \quad k = \operatorname{codim} O, \tag{3}$$

*are valid, where $O$ is an orbit of general position in the conjugate of the adjoint representation of the algebraic group corresponding to the Lie algebra $G$.*

Our basic conjecture is that for algebraic Lie algebras the numbers $n$ and $k$ define the field $D(G)$ to within isomorphism.

We denote by $D_{n,k}(L)$ the field generated over the commutative field $L$ by the elements $x_1, \cdots, x_n, \partial/\partial x_1, \cdots, \partial/\partial x_n, y_1, \cdots, y_k$ with the natural relations

$$x_i y_j - y_j x_i = 0, \quad y_i \frac{\partial}{\partial x_j} - \frac{\partial}{\partial x_j} y_i = 0, \quad \frac{\partial}{\partial x_j} x_i - x_i \frac{\partial}{\partial x_j} = \delta_{ij}.$$

The field $D_{n,k}(L)$ plays the role of a standard model. The center of $D_{n,k}(L)$, as is easily verified, coincides with the subfield $D_{0,k}(L) \subset D_{n,k}(L)$. It follows from Theorem 1 that

$$\operatorname{Dim}_L D_{n,k}(L) = 2n + k.$$

The basic conjecture discussed above can be stated more precisely in the following way.

(T). The field $D(G)$ is isomorphic to $D_{n,k}(L)$, where the numbers $n$ and $k$ are defined by equations (2) or (3).

Theorem 3. *The conjecture* (T) *is valid in the following cases:*

1) *$G$ is an arbitrary nilpotent Lie algebra,*

2) *$G$ is a full matrix Lie algebra or a Lie algebra of matrices with zero trace,*

3) *$G$ is a semisimple Lie algebra of rank 2 over an algebraically closed commutative field.*

The proof of Theorem 3 is carried out in a completely different manner in each of these cases. We will indicate the basic steps of the proofs in each case.

1) The conjecture (T) is replaced by the following assertion. In the algebra $U(G)$ there exists elements $x_1, \cdots, x_n, y_1, \cdots, y_n; z_1, \cdots, z_k$ which generate the field $D(G)$ and satisfy the relations

$$x_i x_j - x_j x_i = y_i y_j - y_j y_i = z_i z_j - z_j z_i = x_i z_j - z_j x_i = y_i z_j - z_i y_j = 0; \quad x_i y_j - y_j x_i = \delta_{ij} C,$$

ere $C$ is a nonzero element of the center of $U(G)$.

This assertion is proved by induction on the dimension of the algebra $G$. Here we use certain re-lts of Dixmier [1] and one of the present authors [2] concerning the connection between $U(G)$ and $(G_0)$, where $G_0$ is an ideal of codimension 1 in $G$.

2) We consider the auxiliary Lie algebra $G_n$, isomorphic to the algebra of matrices of order $n+l$ hose last row consists of zeros. The validity of the conjecture (T) for matrix algebras $G$ follows om the results of one of the present authors on the structure of the center of the algebra $U(G)$ [3] ad from the validity of the conjecture for $G_n$. The latter assertion is proved by induction on $n$ on e basis of the following lemma.

Lemma 2. *Let* $e_{ik}$, $1 \leq i \leq n$, $1 \leq k \leq n+1$, *be the natural basis in* $G_n$. *Let* $D'$ *denote the sub-eld of* $D(G_n)$ *consisting of all elements that commute with* $e_{ii}$ *and* $e_{i,\,n+1}$, $1 \leq i \leq n$. *Then the eld* $D'$ *is isomorphic to* $D(G_{n-1})$.

From this it follows that $D(G_n) \simeq D_{n(n+1)/2,0}(L)$.

3) The proof is obtained by a direct construction of a basis in $D(G)$ having the necessary prop-es. The construction is facilitated by the fact that the field $D(G)$ is generated by the subfield anned by the maximal solvable subalgebra $M$ in $G$, and by the subfield consisting of the elements at commute with the elements of $M$.

Received 8/DEC/65

BIBLIOGRAPHY


1] J. Dixmier, Bull. Soc. Math. France 85 (1957), 325. MR 20 #1928.

2] A. A. Kirillov, Uspehi Mat. Nauk 17 (1962), no. 4(106), 57. MR 25 #5396.

3] I. M. Gel'fand, Mat. Sb. 26(68) (1950), 193. MR 11, 498.


Translated by:
S. Smith

# 2.

## (avec A. A. Kirillov)

# Sur les corps liés aux algèbres enveloppantes des algèbres de Lie

Publ. Math., Inst. Hautes Etud. Sci. **31** (1966) 509–523. MR **33**:7731. Zbl. **144**:21

Il est bien connu que des anneaux non-commutatifs interviennent de façon assez naturelle dans plusieurs questions d'analyse. Les opérateurs de la théorie des champs quantiques, les opérateurs différentiels, les algèbres (group algebras) de groupes de Lie, les algèbres enveloppantes des algèbres de Lie, sont autant d'exemples de tels anneaux, en plus de l'exemple classique de l'anneau de tous les opérateurs dans l'espace de Hilbert qui vient immédiatement à l'esprit. Du point de vue algébrique l'exemple de l'anneau des opérateurs différentiels sur une variété indéfiniment différentiable invariants par rapport à un groupe de difféomorphismes donné est très intéressant et général. La méthode algébrique dans ce cas consiste à étudier la structure de cet anneau.

L'emploi des transformations

$$x \to i\frac{d}{dx}, \qquad \frac{d}{dx} \to ix \qquad \text{(transformation de Fourier)}$$

ou bien

$$x + \frac{d}{dx} = y, \qquad x - \frac{d}{dx} = -2\frac{d}{dy}$$

devient tout à fait indispensable lorsqu'on aborde la question de cette façon. Les transformations ci-dessus doivent être considérées au même niveau que les changements de coordonnées ordinaires.

Dans cet article nous ne considérons qu'un seul type d'anneaux : les algèbres enveloppantes $\mathscr{U}(G)$ des algèbres de Lie G, autrement dit, l'anneau des opérateurs différentiels sur un groupe de Lie invariants par rapport aux translations à gauche. Il existe une multitude d'anneaux de tel type qui ne sont pas isomorphes entre eux. Comme le montre l'exemple de la géométrie algébrique, la situation est simplifiée si l'on passe à la classification birationnelle, c'est-à-dire si l'on identifie les anneaux dont les corps des quotients coïncident. Les anneaux non-commutatifs $\mathscr{U}(G)$ que l'on étudie ici, admettent également un corps des quotients $\mathscr{D}(G)$.

Soit L un corps (commutatif) de caractéristique nulle et K un corps quelconque non commutatif sur L, dont le centre sera désigné par Z(K). Nous définissons la notion de dimension du corps K sur le corps L (voir § 4), cette dimension est notée $\mathrm{Dim}_L K$. Nous faisons correspondre à chaque corps deux nombres $n$ et $k$, donnés par les formules suivantes

$$k = \mathrm{Dim}_L Z(K)$$
$$2n + k = \mathrm{Dim}_L K$$

Dans le cas $K = \mathscr{D}(G)$ où G est une algèbre de Lie algébrique les nombres $n$ et $k$ sont des entiers finis non-négatifs. Le nombre $2n+k$ est égal à la dimension de l'algèbre G, le nombre $k$ est égal à la codimension de l'orbite générique dans la représentation duale de la représentation adjointe du groupe de Lie dont l'algèbre de Lie est G.

Nous avançons l'hypothèse suivante :

Pour une algèbre de Lie algébrique G les nombres $n$ et $k$ déterminent le corps $\mathscr{D}(G)$ à un isomorphisme près.

Pour chaque couple d'entiers $n$ et $k$ non-négatifs nous construisons un corps canonique $\mathscr{D}_{n,k}(L)$, dont les invariants sont exactement $n$ et $k$.

Il en résulte en particulier que deux corps $\mathscr{D}_{n,k}(L)$ et $\mathscr{D}_{n',k'}(L)$ ne sont isomorphes que lorsque $n = n'$, $k = k'$. Une affirmation un peu plus faible (portant sur les anneaux) donne la réponse à une question posée par Dixmier. Le corps $\mathscr{D}_{n,k}(L)$ est engendré sur L par les éléments $x_1, \ldots, x_n, \frac{\partial}{\partial x_1}, \ldots, \frac{\partial}{\partial x_n}, y_1, \ldots, y_k$ avec les relations de commutation ordinaires pour $x$ et $\frac{\partial}{\partial x}$. L'hypothèse que nous avons mentionnée ci-dessus peut être précisée de la façon suivante :

Si G est une algèbre de Lie algébrique sur le corps L, on a $\mathscr{D}(G) \simeq \mathscr{D}_{n,k}(L)$.

Cette affirmation est démontrée ici, dans le cas où G est une algèbre de matrices ainsi que dans le cas d'une algèbre nilpotente quelconque. D'ailleurs pour tous les exemples examinés par les auteurs de cet article, cette hypothèse s'est révélée vraie.

Les nombres $n$ et $k$ jouent un rôle très important dans la théorie des représentations de dimension infinie. Ils sont bien connus des spécialistes. Ces nombres caractérisent les représentations « génériques ». Notamment la représentation générique est réalisée de façon naturelle dans l'espace des fonctions de $n$ variables et elle dépend de $k$ paramètres.

A la fin de l'article nous donnons deux exemples d'algèbres de Lie non-algébriques : pour la première notre hypothèse est satisfaite, pour la seconde elle ne l'est pas.

## 1. Préliminaires.

Nous allons rappeler quelques faits bien connus de la théorie des anneaux. Pour rendre l'exposé plus complet et en faciliter la lecture, nous allons donner leur démonstration.

Soit A un anneau à filtration croissante :

$$A_0 \subset A_1 \subset \ldots \subset A_n \subset \ldots, \quad \bigcup_{i=0}^{\infty} A_i = A,$$

cette filtration étant liée avec la loi de multiplication par la relation

$$A_i A_j \subset A_{i+j} \tag{1}$$

Posons

$$\mathrm{gr}^i A = A_i / A_{i-1}, \qquad \mathrm{gr}\, A = \sum_{i=0}^{\infty} \mathrm{gr}^i A$$

Désignons par $\pi_i$ la projection naturelle $A_i \to \mathrm{gr}^i A$. Si $x \in A_i$, $y \in A_j$ l'élément $\pi_{i+j}(xy) \in \mathrm{gr}^{i+j} A$ ne dépend que de $\pi_i(x) \in \mathrm{gr}^i A$ et $\pi_j(y) \in \mathrm{gr}^j A$ en vertu de (1). Pour $i$ et $j$ arbitraires nous obtenons ainsi une application $\mathrm{gr}^i A \times \mathrm{gr}^j A \to \mathrm{gr}^{i+j} A$. Il est évident que ces applications définissent sur gr A une structure d'anneau gradué.

*Lemme 1. — Si l'anneau* gr A *est un anneau noethérien sans diviseurs de zéro, alors l'anneau considéré* A *a les mêmes propriétés.*

*Démonstration.* — Soit $\mathfrak{I}$ un idéal (à gauche, à droite ou bilatère) de l'anneau A. Posons $\mathrm{gr}\, \mathfrak{I} = \sum_i (\mathfrak{I}_i / \mathfrak{I}_{i-1})$, $\mathfrak{I}_i$ étant l'intersection de $\mathfrak{I}$ et $A_i$. Il est clair que gr $\mathfrak{I}$ est un idéal (resp. à gauche, à droite ou bilatère) de l'anneau gr A. Soit

$$\mathfrak{I}^{(1)} \subset \mathfrak{I}^{(2)} \subset \ldots \subset \mathfrak{I}^{(n)} \subset \ldots$$

une suite croissante d'idéaux de A. Alors

$$\mathrm{gr}\, \mathfrak{I}^{(1)} \subset \mathrm{gr}\, \mathfrak{I}^{(2)} \subset \ldots \subset \mathrm{gr}\, \mathfrak{I}^{(n)} \subset \ldots$$

est une suite croissante d'idéaux de gr A. L'anneau gr A étant noethérien, on a, à partir d'une valeur assez grande de $n$ les égalités suivantes

$$\mathrm{gr}\, \mathfrak{I}^{(n)} = \mathrm{gr}\, \mathfrak{I}^{(n+1)} = \ldots$$

Nous allons montrer que dans ces conditions

$$\mathfrak{I}^{(n)} = \mathfrak{I}^{(n+1)} = \ldots$$

En effet, soit $x$ un élément de $\mathfrak{I}^{(n+1)}$. Désignons par $k$ le plus petit entier pour lequel $x \in \mathfrak{I}_k^{(n+1)} = \mathfrak{I}^{(n+1)} \cap A_k$. Comme $\mathrm{gr}^k \mathfrak{I}^{(n+1)} = \mathrm{gr}^k \mathfrak{I}^{(n)}$, il existe un élément $y \in \mathfrak{I}^{(n)}$ tel que $\pi_k(y) = \pi_k(x)$. Par suite $x - y \in \mathfrak{I}_{k-1}^{(n+1)}$. De même on peut trouver un élément $y' \in \mathfrak{I}^{(n)}$ tel que la différence $x - y - y' \in \mathfrak{I}_{k-2}^{(n+1)}$. En continuant ce procédé nous arrivons à la conclusion que $x \in \mathfrak{I}^{(n)}$.

Maintenant nous allons montrer que l'anneau A n'a pas de diviseurs de zéro. Soient $x_1$, $x_2$ deux éléments différents de zéro de l'anneau A, et soient $k_1$, $k_2$ les plus petits entiers pour lesquels $x_i \in A_{k_i}$, $i = 1, 2$. Alors $\pi_{k_i}(x_i) \neq 0$. Par suite

$$\pi_{k_1+k_2}(x_1 x_2) = \pi_{k_1}(x_1) \pi_{k_2}(x_2) \neq 0,$$

d'où l'on déduit $x_1 x_2 \neq 0$.

*Lemme 2. — Un anneau noethérien (à gauche) quelconque qui n'a pas de diviseurs de zéro est un anneau d'Ore.* (C'est-à-dire que deux éléments non nuls quelconques de l'anneau ont un multiple commun à gauche).

*Démonstration.* — Soient $a \neq 0$, $b \neq 0$ deux éléments de l'anneau A. Désignons par $\mathfrak{I}_n$ l'idéal à gauche de A, engendré par les éléments $a, ab, \ldots, ab^n$. Comme l'anneau A est noethérien il existe une valeur de $n$ pour laquelle $\mathfrak{I}_n = \mathfrak{I}_{n+1}$. Il en résulte que

$$ab^{n+1} = x_0 a + x_1 ab + \ldots + x_n ab^n \tag{2}$$

Soit $k$ le plus petit entier pour lequel $x_k \neq 0$. Alors, divisant à droite par $b^k$ les deux membres de la formule (2), nous obtenons

$$ab^{n+1-k} = x_k a + x_{k+1} ab + \ldots + x_n ab^{n-k}$$

ou encore

$$x_k a = (-x_{k+1} a - \ldots - x_n ab^{n-k-1} + ab^{n-k}) b$$

d'où l'on déduit que les éléments $a$ et $b$ ont le multiple à gauche commun $x_k a$.

Pour chaque anneau A de Ore à gauche et à droite, on peut définir un corps des quotients de la façon suivante. Considérons l'ensemble des expressions de type $ba^{-1}$ ou de type $a^{-1}b$, $a$ et $b$ étant des éléments de l'anneau A, $a \neq 0$. Nous identifions les expressions $a^{-1}b$ et $cd^{-1}$ si $ac = bd$.

L'anneau A étant un anneau de Ore, il en résulte qu'une « fraction gauche » quelconque $a^{-1}b$ peut être écrite sous forme de « fraction droite » $cd^{-1}$, que chaque couple de fractions $a^{-1}b$, $c^{-1}d$ peut être réduit à un « dénominateur commun » c'est-à-dire que l'on peut trouver des éléments $x, y_1, y_2$ tels que $a^{-1}b = x^{-1}y_1$, $c^{-1}d = x^{-1}y_2$.

Pour les fractions gauches ayant un dominateur commun on peut définir de façon naturelle les opérations d'addition, de soustraction et de division :

$$x^{-1}y_1 \pm x^{-1}y_2 = x^{-1}(y_1 \pm y_2)$$
$$(x^{-1}y_1)^{-1}(x^{-1}y_2) = y_1^{-1}y_2$$

Ensuite on peut définir la multiplication par un élément $a^{-1}b$ en considérant cette opération comme la division par l'élément inverse $b^{-1}a$.

On vérifie sans aucune difficulté que l'ensemble des éléments ainsi obtenus muni des opérations définies ci-dessus est en effet un corps.

## 2. Définition de l'anneau $R_n(A)$ et du corps $\mathcal{D}_n(A)$.

Soit maintenant A un anneau noethérien quelconque sans diviseurs de zéro. Nous désignons par $R_n(A)$ l'algèbre $R_n \otimes_{\mathbf{Z}} A$, où $R_n$ est l'algèbre sur $\mathbf{Z}$ à $2n$ générateurs $p_1, \ldots, p_n, q_1, \ldots, q_n$ soumis aux relations

$$p_i p_j - p_j p_i = 0, \qquad q_i q_j - q_j q_i = 0, \qquad p_i q_j - q_j p_i = \delta_{ij} 1 \tag{3}$$

Dans l'anneau $R_n(A)$ on peut définir la filtration croissante

$$A = (R_n(A))_0 \subset (R_n(A))_1 \subset \ldots$$

où $(R_n(A))_i$ est l'ensemble de tous les éléments de $R_n(A)$ qui peuvent s'écrire sous forme de polynômes (non commutatifs) en les $p_1, \ldots, p_n, q_1, \ldots, q_n$, à coefficients dans A et de degré $\leqslant i$.

*Lemma 3.* — *L'algèbre* $R_n(A)$ *est un* A-*module libre dont une base est composée de tous les monômes de type*

$$p^k q^l = p_1^{k_1} \dots p_n^{k_n} q_1^{l_1} \dots q_n^{l_n} \tag{4}$$

La démonstration de ce lemme se simplifie essentiellement si nous supposons que l'anneau A est une algèbre sur un corps (commutatif) de caractéristique o. C'est précisément ce cas qui sera important pour nous dans ce qui va suivre; nous nous bornerons donc à ce cas dans la démonstration.

Montrons d'abord que les monômes de type (4) engendrent le A-module $R_n(A)$. Puisque $R_n(A)$ est la réunion des sous-modules $(R_n(A))_k$, nous pouvons raisonner par récurrence sur $k$. Supposons que pour $k<k_0$ notre affirmation soit démontrée. Des relations (3) on déduit immédiatement qu'un monôme quelconque de degré $k$ est égal à un monôme de type (4) modulo $(R_n(A))_{k-2}$. Alors notre affirmation est vraie pour $k=k_0$. Mais pour $k=0$ elle est évidemment vraie parce que $(R_n(A))_0 = A$.

Il reste à montrer que les monômes de type (4) sont linéairement indépendants sur A. Soit

$$x = \sum_{k,l} a_{k,l} p^k q^l = 0.$$

Ordonnons suivant l'ordre lexicographique les indices $(k, l)$ et soit $(k^0, l^0)$ le plus grand de ces indices, tel que $a_{k,l} \neq 0$. Désignons par ad $y$ l'opérateur dans $R_n(A)$, déterminé par la formule

$$\mathrm{ad}\, y . z = yz - zy$$

Un calcul simple montre que

$$\prod_{i=1}^{n} (\mathrm{ad}\, q_i)^{k_i^0} \prod_{i=1}^{n} (\mathrm{ad}\, p_i)^{l_i^0} x = \prod_{i=1}^{n} (-1)^{k_i^0} k_i^0 ! \, l_i^0 ! \, a_{k^0, l^0} \neq 0$$

d'où contradiction avec l'égalité $x=0$.

Il résulte immédiatement du lemme 3 que l'anneau gradué gr $R_n(A)$ est isomorphe à l'anneau polynomial [1]

$$A[p_1, \dots, p_n, q_1, \dots, q_n].$$

Il s'ensuit que gr $R_n(A)$ est un anneau noethérien sans diviseurs de zéro. Vu les résultats du § 1, l'anneau $R_n(A)$ possède les mêmes propriétés, donc il peut se plonger dans un corps des quotients. Ce corps sera noté $\mathscr{D}_n(A)$.

Il est clair que si deux anneaux $A_1$ et $A_2$ ont le même corps des quotients K, alors

$$\mathscr{D}_n(A_1) = \mathscr{D}_n(A_2) = \mathscr{D}_n(K).$$

Remarquons aussi les identités facilement vérifiables :

$$R_n(R_m(A)) = R_{m+n}(A) = R_n(A) \otimes_A R_m(A)$$
$$\mathscr{D}_n(\mathscr{D}_m(A)) = \mathscr{D}_{m+n}(A).$$

---

[1] Par « anneau polynomial » A[$x$] où A est un anneau non-commutatif nous entendons ici l'algèbre $A \otimes_{\mathbf{Z}} \mathbf{Z}[x]$. Le théorème fondamental de Hilbert reste encore valable dans ce cas : si l'anneau A est noethérien (à gauche. à droite ou bilatère), alors l'anneau A[$x$] possède la même propriété.

**3. L'anneau $R_{n,k}(L)$ et le corps $\mathscr{D}_{n,k}(L)$.**

Utilisons la construction décrite au § 2 dans le cas où l'anneau A est l'algèbre des polynômes en $k$ variables sur un corps commutatif L de caractéristique nulle.

Nous noterons :

$$R_{n,k}(L) = R_n(L[x_1, \ldots, x_k])$$
$$\mathscr{D}_{n,k}(L) = \mathscr{D}_n(L[x_1, \ldots, x_k]) = \mathscr{D}_n(L(x_1, \ldots, x_k)).$$

Rappelons que dans l'anneau $R_n(A)$, et, par conséquent, dans l'anneau $R_{n,k}(L)$, nous avons défini une filtration croissante telle que l'anneau gradué associé $\operatorname{gr} R_{n,k}(L)$ est isomorphe à l'anneau des polynômes dépendant des variables $p_1, \ldots, p_n, q_1, \ldots, q_n$, dont les coefficients sont choisis dans l'anneau $L[x_1, \ldots, x_k]$. Pour chaque élément non nul $a \in R_{n,k}(L)$ il existe un entier $i$ (et un seul) tel que $\pi_i(a) \in \operatorname{gr}^i R_{n,k}(L)$ est bien défini et non nul. Nous appellerons cet élément partie principale de $a$; il sera noté $[a]$. Il découle de cette définition, que $[a]$ est un polynôme homogène dépendant des variables $p_1, \ldots, p_n, q_1, \ldots, q_n$, à coefficients dans $L[x_1, \ldots, x_k]$.

Il est évident que quels que soient $a_1, a_2 \in R_{n,k}(L)$ on a :

$$[a_1 a_2] = [a_1][a_2] \tag{5}$$

c'est-à-dire que la partie principale du produit est égale au produit des parties principales des facteurs de ce produit.

Soit $b$ un élément quelconque du corps $\mathscr{D}_{n,k}(L)$. Cet élément est de la forme $a_1^{-1}a_2$, où $a_i \in R_{n,k}(L)$.

*Lemme 4. — La fonction rationnelle* $\dfrac{[a_2]}{[a_1]}$ *ne dépend que de l'élément* $b \in \mathscr{D}_{n,k}(L)$ *(et ne dépend pas de la façon dont il a été représenté sous forme de quotient d'éléments de* $R_{n,k}(L)$*). Notons cette fonction par* $[b]$*. Pour* $b_1, b_2 \in \mathscr{D}_{n,k}(L)$ *quelconques on a l'égalité :*

$$[b_1 b_2] = [b_1][b_2]$$

*Démonstration.* — Supposons que $b$ soit représenté par la fraction droite $a_1 a_2^{-1}$ et la fraction gauche $a_3^{-1}a_4$. Montrons que dans ce cas $\dfrac{[a_1]}{[a_2]} = \dfrac{[a_4]}{[a_3]}$. En effet, d'après l'égalité $a_1 a_2^{-1} = a_3^{-1} a_4$ on a par définition $a_3 a_1 = a_4 a_2$. D'où $[a_3][a_1] = [a_4][a_2]$ ce qui démontre la première partie du lemme. Il est clair qu'il suffit de démontrer la seconde partie dans le cas où $b_1 \in R_{n,k}(L)$. Mais dans ce cas nous avons :

$$b_1 b_2 = b_1(a_1 a_2^{-1}) = (b_1 a_1) a_2^{-1}$$

Donc

$$[b_1 b_2] = \frac{[b_1 a_1]}{[a_2]} = [b_1]\frac{[a_1]}{[a_2]} = [b_1][b_2]$$

*Corollaire. — La fonction* $[b]$ *est invariante relativement aux automorphismes intérieurs du corps.*

En effet $[aba^{-1}] = [a][b][a]^{-1} = [b]$.

Remarquons maintenant que l'anneau $R_{n,k}(L)$ admet une autre filtration naturelle. A savoir, on prend pour $(R_{n,k}(L))_i$ l'ensemble de tous les éléments qui peuvent être représentés sous forme de polynôme (non commutatif) à coefficients dans L, de degré $\leqslant i$, en les générateurs $p_1, \ldots, p_n, q_1, \ldots, q_n, x_1, \ldots, x_k$. L'anneau gradué associé à cette filtration est évidemment isomorphe à l'anneau des polynômes à $2n+k$ variables. Comme nous l'avons fait plus haut, nous pouvons déterminer une application $b \to [b]$ qui est un homomorphisme du groupe multiplicatif du corps $\mathcal{D}_{n,k}(L)$ dans le groupe des fonctions rationnelles de forme $PQ^{-1}$, où P, Q sont des polynômes homogènes (en général de degrés différents) de $2n+k$ variables. Cette application sera utilisée au paragraphe suivant pour démontrer le théorème de non-isomorphisme.

## 4. Théorème de non-isomorphisme.

Nous montrerons que les anneaux $R_{n,k}(L)$ et les corps $\mathcal{D}_{n,k}(L)$ que nous avons construits ne sont pas isomorphes (en tant qu'algèbres sur L) pour différentes valeurs de $n$ et $k$. La démonstration est basée sur le concept de dimension de corps et d'anneau.

Commençons par le cas plus simple de l'anneau. Soit A une algèbre sur le corps commutatif L, $\alpha = (a_1, \ldots, a_s)$ un ensemble fini quelconque d'éléments de A. Considérons l'ensemble de tous les éléments de notre algèbre qui peuvent être représentés sous forme de polynôme non commutatif à coefficients dans L, de degré $\leqslant N$, à variables choisies dans $\alpha$. Cet ensemble est évidemment un espace vectoriel sur L, dont la dimension sera notée $d(\alpha, N)$.

L'expression

$$\sup_{\alpha} \overline{\lim_{N \to \infty}} \frac{\log d(\alpha, N)}{\log N} \tag{6}$$

sera appelée *dimension* de l'algèbre A et notée $\mathrm{Dim}_L A$.

*Lemme 5.* — $\mathrm{Dim}_L R_{n,k}(L) = 2n+k$.

*Démonstration.* — Prenons pour $\alpha$ l'ensemble

$$\alpha_0 = (p_1, \ldots, p_n, q_1, \ldots, q_n, x_1, \ldots, x_k)$$

On calcule sans difficulté $d(\alpha_0, N)$ qui est égal au coefficient binomial $C_{N+2n+k}^{2n+k}$. (Rappelons que la dimension de l'espace des polynômes de degré $\leqslant N$ à $m$ variables est $C_{N+m}^{m}$).

Donc :

$$\mathrm{Dim}_L R_{n,k}(L) \geqslant \lim_{N \to \infty} \frac{\log C_{N+2n+k}^{2n+k}}{\log N} = 2n+k.$$

Soit maintenant $\alpha = (a_1, \ldots, a_s)$ un ensemble fini quelconque d'éléments de $R_{n,k}(L)$. Chaque $a_i$ peut être représenté sous forme de polynôme à variables dans $\alpha_0$. Soit $m$ le degré maximal de ces polynômes. Chaque polynôme de degré $\leqslant N$ relativement aux variables $a_i$ est un polynôme de degré $\leqslant mN$ relativement aux générateurs.

Donc :

$$d(\alpha, \mathrm{N}) \leqslant d(\alpha_0, m\mathrm{N}).$$

Mais

$$\lim_{\mathrm{N}\to\infty} \frac{\log d(\alpha_0, m\mathrm{N})}{\log \mathrm{N}} = \lim_{\mathrm{N}\to\infty} \frac{\log \mathrm{C}^{2n+k}_{m\mathrm{N}+2n+k}}{\log \mathrm{N}} = 2n+k,$$

d'où le lemme.

Considérons maintenant un corps $\mathscr{D}$ quelconque sur le corps commutatif L. Soit $\alpha=(a_1, \ldots, a_s)$ un ensemble fini quelconque d'éléments de ce corps. L'ensemble $(a_1 b, \ldots, a_s b)$, où $b \in \mathscr{D}$, sera noté $\alpha b$. Définissons la dimension du corps $\mathscr{D}$ sur L de la manière suivante :

$$\mathrm{Dim}_{\mathrm{L}} \mathscr{D} = \sup_{\alpha} \inf_{b\neq 0} \overline{\lim_{\mathrm{N}\to\infty}} \frac{\log d(\alpha b, \mathrm{N})}{\log \mathrm{N}} \tag{7}$$

*Remarque 1.* — La formule (7) définissant la dimension d'un corps est également applicable à tout anneau sans diviseur de zéro. On vérifie aisément que pour l'anneau $\mathrm{R}_{n,k}(\mathrm{L})$ les formules (6) et (7) donnent la même valeur numérique pour la dimension.

*Remarque 2.* — Si le corps $\mathscr{D}$ est commutatif, on vérifie aisément que la grandeur $\dim_{\mathrm{L}} \mathscr{D}$ est égale au degré de transcendance de $\mathscr{D}$ sur L.

Calculons maintenant la dimension du corps $\mathscr{D}_{n,k}(\mathrm{L})$.

*Lemme 6.* — $\mathrm{Dim}_{\mathrm{L}} \mathscr{D}_{n,k}(\mathrm{L}) = 2n+k$.

*Démonstration.* — Soit $\alpha=(a_1, \ldots, a_s)$ un ensemble fini quelconque d'éléments de $\mathscr{D}_{n,k}(\mathrm{L})$. Il existe un élément $b \neq 0$ tel que tous les éléments $a_i b$ appartiennent à $\mathrm{R}_{n,k}(\mathrm{L})$. En raisonnant comme dans la démonstration du lemme 5 on obtient l'inégalité

$$\mathrm{Dim}_{\mathrm{L}} \mathscr{D}_{n,k}(\mathrm{L}) \leqslant 2n+k.$$

Pour obtenir l'inégalité de sens contraire, il est naturel de considérer l'ensemble

$$\alpha_0 = (p_1, \ldots, p_n, q_1, \ldots, q_n, x_1, \ldots, x_k)$$

et de démontrer que pour un élément $b \neq 0$ quelconque les monômes

$$\mathrm{P}_{k,l,m} = (p_1 b)^{k_1} \ldots (p_n b)^{k_n} (q_1 b)^{l_1} \ldots (q_n b)^{l_n} (x_1 b)^{m_1} \ldots (x_k b)^{m_k}$$

sont linéairement indépendants. Malheureusement ils ne le sont pas, par exemple pour $b = x_1^{-1}$. On peut néanmoins tourner cette difficulté de la manière suivante. Nous montrerons que les monômes $\mathrm{P}_{k,l,m}$ peuvent être dépendants seulement lorsque la fonction homogène $[b]$ (cf. la fin du § 3) a un degré d'homogénéité égal à $-1$. Il en résulte que les monômes

$$\widetilde{\mathrm{P}}_{k,l,m} = (p_1 ab)^{k_1} \ldots (p_n ab)^{k_n} (q_1 ab)^{l_1} \ldots (q_n ab)^{l_n} (x_1 ab)^{m_1} \ldots (x_k ab)^{m_k}$$

où $a \neq 0$ est un élément quelconque de $\mathscr{D}_{n,k}(\mathrm{L})$ ne peuvent être linéairement dépendants que si $[ab]$ est une fonction homogène de degré $-1$. Soit $a$ un élément du corps $\mathscr{D}_{n,k}(\mathrm{L})$ tel que le degré de $[a]$ soit non nul. Il suit de ce que nous venons de dire que, soit les $\mathrm{P}_{k,l,m}$

soit les $\widetilde{P}_{k,l,m}$ forment un système linéairement indépendant. Par conséquent, en choisissant $\alpha$ égal à la réunion des ensembles $\alpha_0$ et $\alpha_0 a$ on obtient

$$d(\alpha, N) \geqslant C_{N+2n+k}^{2n+k}$$

d'où l'on déduit l'inégalité cherchée

$$\mathrm{Dim}_L \mathscr{D}_{n,k}(L) \geqslant 2n+k.$$

Voyons maintenant dans quels cas les monômes $P_{k,l,m}$ peuvent être linéairement dépendants. Pour un ensemble fini quelconque de ces monômes, il existe un élément $Q \in R_{n,k}(L)$ tel que toutes les expressions $P_{k,l,m}Q$ appartiennent à $R_{n,k}(L)$. Considérons la partie principale du produit $P_{k,l,m}Q$. D'après le § 3 nous avons

$$[P_{k,l,m}Q] = p_1^{k_1} \dots p_n^{k_n} q_1^{l_1} \dots q_n^{l_n} x_1^{m_1} \dots x_k^{m_k} [b]^d [Q] \tag{8}$$

où $d$ désigne le degré du monôme $P_{k,l,m}$ :

$$d = k_1 + \dots + k_n + l_1 + \dots + l_n + m_1 + \dots + m_k.$$

Donc :

$$\deg[P_{k,l,m}Q] = (1 + \deg[b])d + \deg[Q]$$

Si $\deg[b] \neq -1$ les polynômes $[P_{k,l,m}Q]$ ont des degrés différents pour des valeurs différentes de $d$.

Supposons que ces polynômes soient linéairement dépendants :

$$\Sigma c_{k,l,m} P_{k,l,m} = 0$$

Supposons que $\deg[b] > -1$ (respectivement $\deg[b] < -1$) et soit $d$ le degré maximum (respectivement minimum) des monômes dotés d'un coefficient $c_{k,l,m}$ non nul.

On a alors l'égalité

$$\sum_{\deg P_{k,l,m} = d} c_{k,l,m} [P_{k,l,m}Q] = 0 \tag{9}$$

obtenue en considérant les parties principales dans l'expression

$$\Sigma c_{k,l,m} P_{k,l,m} Q = 0$$

En se rappelant l'expression (8) pour $[P_{k,l,m}Q]$, nous voyons que (9) est impossible pour des coefficients $c_{k,l,m}$ non nuls. La démonstration du lemme est terminée.

*Remarque 3.* — En démontrant la formule $\mathrm{Dim}_L \mathscr{D}_{n,k}(L) = 2n+k$ nous ne nous sommes servis que des propriétés suivantes de ce corps :

1) $\mathscr{D}_{n,k}(L)$ est le corps des quotients d'une certaine algèbre A (à savoir $R_{n,k}(L)$);

2) L'algèbre A admet une filtration telle que l'algèbre graduée associée gr A soit isomorphe à l'algèbre polynomiale à $2n+k$ variables.

Par conséquent nous avons en fait démontré la proposition plus générale suivante :

*Théorème 1. — Soit* A *une algèbre sur le corps commutatif* L, *munie d'une filtration telle que l'algèbre graduée associée* gr A *soit isomorphe à l'algèbre polynomiale à m variables. Notons* $\mathscr{D}$(A) *le corps des quotients de l'algèbre* A. *On a alors* $\mathrm{Dim}_L A = m$.

Le résultat obtenu permet de résoudre entièrement la question de l'isomorphisme des anneaux $R_{n,k}(L)$ et des corps $\mathscr{D}_{n,k}(L)$ pour différentes valeurs de $n$ et $k$.

*Théorème 2. — Les isomorphismes des* L-*algèbres*

$$R_{n,k}(L) \simeq R_{n',k'}(L) \quad \textit{et} \quad \mathscr{D}_{n,k}(L) \simeq \mathscr{D}_{n',k'}(L)$$

*ont lieu seulement si* $n = n'$, $k = k'$.

*Démonstration.* — Le centre de l'anneau $R_{n,k}(L)$ est $R_{0,k}(L)$, le centre de l'anneau $\mathscr{D}_{n,k}(L)$ est $\mathscr{D}_{0,k}(L)$. Des corps et anneaux isomorphes ont les mêmes dimensions. Par conséquent l'isomorphisme est possible seulement si $2n+k = 2n'+k'$, $k = k'$.

## 5. Définition du corps $\mathscr{D}(G)$.

Soit G une algèbre de Lie sur un corps commutatif L de caractéristique nulle. Notons $\mathscr{U}(G)$ l'algèbre enveloppante de l'algèbre G, c'est-à-dire l'algèbre obtenue en factorisant l'algèbre associative libre engendrée par les éléments de l'algèbre G par l'idéal engendré par les éléments de la forme

$$xy - yx - [x, y]$$

où $x, y \in G$.

Chaque élément de $\mathscr{U}(G)$ peut être représenté sous forme de polynôme à variables dans G et à coefficients dans L. On obtient ainsi une filtration croissante dans $\mathscr{U}(G)$

$$L = (\mathscr{U}(G))_0 \subset (\mathscr{U}(G))_1 \subset \ldots \subset (\mathscr{U}(G))_n \subset \ldots$$

en posant que $(\mathscr{U}(G))_i$ est l'ensemble de tous les éléments de $\mathscr{U}(G)$ qui peuvent être représentés sous forme de polynômes de degré $\leqslant i$ à variables dans G.

Il découle du théorème de Poincaré-Birkhoff-Witt (cf. par exemple [1]) que gr $\mathscr{U}(G)$ est l'algèbre polynomiale à $m$ variables, où $m = \dim G$. Comme nous l'avons vu dans le § 1, on peut en déduire que $\mathscr{U}(G)$ est un anneau d'Ore sans diviseurs de zéro. Le corps des fractions de l'algèbre $\mathscr{U}(G)$ sera noté $\mathscr{D}(G)$ et appelé corps de Lie de l'algèbre G.

Des résultats du paragraphe précédent (cf. théorème 1) on déduit l'égalité

$$\mathrm{Dim}_L \mathscr{D}(G) = \dim G.$$

*Hypothèse fondamentale. — Le corps* $\mathscr{D}(G)$ *est isomorphe à un des corps* $\mathscr{D}_{n,k}(L)$ *lorsque* G *est une algèbre de Lie algébrique.*

## 6. Démonstration de l'hypothèse fondamentale pour les algèbres de matrices.

Nous montrerons que l'hypothèse fondamentale est satisfaite lorsque G est l'algèbre de toutes les matrices à trace nulle ou bien l'algèbre de toutes les matrices. La démonstration est basée sur la proposition suivante.

*Lemme 7. — Soit* $G_n^0$ *l'algèbre de toutes les matrices d'ordre n dont la dernière ligne est entièrement composée de zéros. On a alors* $\mathscr{D}(G_n^0) \simeq \mathscr{D}_{\frac{n(n-1)}{2},0}$.

Montrons d'abord que ce lemme implique l'hypothèse fondamentale pour les algèbres de matrices. Il est bien connu que le centre de l'algèbre enveloppante de

518

l'algèbre $G_n$ de toutes les matrices d'ordre $n$ (respectivement de l'algèbre $\widetilde{G}_n$ des matrices à trace nulle) est engendrée par les éléments $\Delta_1, \ldots, \Delta_n$ (respectivement $\Delta_2, \ldots, \Delta_n$), où $\Delta_i(g)$ est la trace de la $i$-ème puissance extérieure de la matrice $g$. (Nous employons la réalisation de l'algèbre $\mathcal{U}(G_n)$ sous forme de polynômes sur $G_n$, cf. [2]). Les éléments de la ligne inférieure interviennent au premier degré dans les polynômes $\Delta_i$. Il est clair que $\mathscr{D}(G_n)$ (respectivement $\mathscr{D}(\widetilde{G}_n)$ est engendrée par $\mathscr{D}(G_n^0)$ et par les éléments $\Delta_1, \ldots, \Delta_n$ (respectivement $\Delta_2, \ldots, \Delta_n$). Par conséquent

$$\mathscr{D}(G_n) = \mathscr{D}_{\frac{n(n-1)}{2}, n}, \qquad \mathscr{D}(\widetilde{G}_n) = \mathscr{D}_{\frac{n(n-1)}{2}, n-1}.$$

Démontrons maintenant le lemme 7 par récurrence sur $n$. Considérons dans $G_{n+1}^0$ une base naturelle composée d'éléments $e_{ik}$, $1 \leqslant i \leqslant n$, $1 \leqslant k \leqslant n+1$, $(e_{ik})$ étant la matrice dont tous les éléments sont nuls sauf l'élément à l'intersection de la $i$-ème ligne et de la $k$-ème colonne qui est égal à 1).

Posons

$$q_i = e_{i,n+1}, \qquad p_i = e_{ii} q_i^{-1}, \qquad \widetilde{e}_{ik} = e_{ik} q_i^{-1} q_k.$$

Un calcul direct montre que

$$[q_i, q_j] = 0, \qquad [p_i, p_j] = 0, \qquad [p_i, q_j] = \delta_{ij} \cdot 1,$$
$$[\widetilde{e}_{ik}, q_j] = \delta_{kj} q_j, \qquad [\widetilde{e}_{ik}, p_j] = -\delta_{kj} p_j.$$

Par conséquent si les coefficients $c_{ik}$ $(1 \leqslant i, k \leqslant n)$ satisfont aux conditions

$$\sum_i c_{ik} = 0, \qquad k = 1, 2, \ldots, n \tag{10}$$

l'élément $\sum_{i,k} c_{ik} \widetilde{e}_{ik}$ est permutable à $q_i$ et $p_i$, $i = 1, 2, \ldots, n$.

Considérons l'ensemble G de toutes les matrices $C$ d'ordre $n$ formées d'éléments $c_{ik}$ pour lesquels (10) est vérifié. Faisons correspondre à toute matrice de ce type l'élément $\alpha(C) = \sum_{i,k} c_{ik} \widetilde{e}_{ik} \in \mathscr{D}(G_{n+1}^0)$. Il se trouve que l'application $\alpha$ est telle que l'on a

$$\alpha([C_1, C_2]) = [\alpha(C_1), \alpha(C_2)]$$

Nous laissons au lecteur la vérification longue mais facile de cette égalité. Il est clair que le corps $\mathscr{D}(G_{n+1}^0)$ est engendré par les éléments de la forme $\alpha(C)$, $C \in G$ et les éléments $p_1, \ldots, p_n, q_1, \ldots, q_n$. Pour démontrer le lemme il suffit de vérifier que le corps engendré par les éléments $\alpha(G)$ est isomorphe à $\mathscr{D}_{\frac{n(n-1)}{2}, 0}(L)$. Mais l'ensemble G est une algèbre de Lie isomorphe à $G_n^0$, et le fait qui reste à vérifier est vrai par l'hypothèse de récurrence. Le lemme est démontré [1].

[1] Comme l'a remarqué le référent, pour que cette affirmation soit vraie, il faut savoir que $\alpha$, prolongé comme homomorphisme de $\mathcal{U}(G)$ dans $\mathscr{D}(G_{n+1}^0)$, est injectif. Ce fait peut être établi de façon suivante. Choisissons une base de l'algèbre G de telle façon que les images par l'application $\alpha$ des éléments de cette base soient $\approx{e}_{ik} = \widetilde{e}_{ik} - e_{nk}$, $1 \leqslant i \leqslant n-1$, $1 \leqslant k \leqslant n$. Ordonnons l'ensemble de ces éléments suivant l'ordre lexicographique et considérons les monômes formés à partir de ces éléments, les éléments correspondants pour chaque monôme étant rangés dans l'ordre croissant. Il faut montrer que ces monômes soit linéairement indépendants. Cela se fait de la même manière que celle utilisée pour la démonstration de l'indépendance des monômes $P_{k,l,m}$ aux pages 12-13.

## 7. Démonstration de l'hypothèse fondamentale pour les algèbres de Lie nilpotentes.

Nous aurons besoin de certains résultats sur la structure de l'algèbre $\mathscr{U}(G)$ et du corps $\mathscr{D}(G)$, où G est une algèbre de Lie nilpotente sur un corps commutatif de caractéristique nulle.

*Lemme 8 (Dixmier). — Le centre du corps* $\mathscr{D}(G)$ *est le corps des quotients du centre* $Z(G)$ *de l'algèbre* $\mathscr{U}(G)$; *il est isomorphe au corps des fonctions rationnelles à k variables. L'entier k est pair et impair en même temps que* dim G. *Si* $G_0$ *est un idéal de codimension* 1 *dans* G, *nous avons soit* $Z(G_0)\subset Z(G)$, *soit* $Z(G)\subset Z(G_0)$.

La démonstration des faits énumérés dans le lemme peut être obtenue à partir des résultats de [3], [4]; cf. également [5].

*Lemme 9. — Il existe dans l'algèbre* $\mathscr{U}(G)$ *des éléments* $x_1, \ldots, x_n, y_1, \ldots, y_n, z_1, \ldots, z_k$ *(les entiers n et k dépendant de l'algèbre* G) *avec les propriétés suivantes :*

(*) *Le corps de Lie* $\mathscr{D}(G)$ *est engendré par les éléments* $x_i, y_i, z_j$, $1\leqslant i\leqslant n$, $1\leqslant j\leqslant k$;

(**) *On a les relations de commutation*

$$[x_i, x_j]=[y_i, y_j]=[x_i, z_j]=[y_i, z_j]=[z_i, z_j]=0, \qquad [x_i, y_j]=\delta_{ij}.c,$$

*où c est un élément non nul de* $Z(G)$.

Il est clair que le lemme implique l'hypothèse fondamentale pour l'algèbre G : il suffit de poser $p_i=x_i c^{-1}$, $q_i=y_i$.

Démontrons le lemme 9 par récurrence sur la dimension de G. Soit $G_0$ un idéal de codimension 1 dans G, $x$ un élément de G n'appartenant pas à $G_0$.

*1^er^ cas :* $Z(G_0)\subset Z(G)$. Dans ce cas il existe un élément de $Z(G)$ qui n'appartient pas à $Z(G_0)$. Cet élément peut être représenté (d'une manière unique) sous la forme :

$$y=a_0x^n+a_1x^{n-1}+\ldots+a_n, \qquad a_i\in\mathscr{U}(G_0).$$

Puisque $[y, b]=0$, pour un $b$ quelconque nous avons

$$[a_0, b]x^n+(na_0[x, b]+[a_1, b])x^{n-1}+\ldots=0$$

d'où

$$[a_0, b]=0, \qquad [na_0x+a_1, b]=0$$

c'est-à-dire

$$a_0\in Z(G_0), \qquad na_0x+a_1\in Z(G).$$

Par l'hypothèse de récurrence il existe dans $\mathscr{U}(G_0)$ des éléments $x_1, \ldots, x_n$, $y_1, \ldots, y_n, z_1, \ldots, z_k$ qui satisfont aux conditions du lemme. Posons $z_{k+1}=na_0x+a_1$.

Il est évident que le système d'éléments $x_1, \ldots, x_n, y_1, \ldots, y_n, z_1, \ldots, z_{k+1}$ jouit des propriétés (*) et (**).

*2^e^ cas :* $Z(G)\subset Z(G_0)$. Dans ce cas il existe un élément $y\in Z(G_0)$ qui n'appartient pas à $Z(G)$. On a alors $[x,y]\neq 0$. Soit $k$ le nombre maximum pour lequel $(\mathrm{ad}\, x)^k y\neq 0$. Remplaçant, s'il le faut, $y$ par $(\mathrm{ad}\, x)^{k-1}y$, on peut supposer que $k=1$. Ainsi, il existe

*520*

dans $Z(G_0)$ un élément $y$ tel que $[x,y]=z\neq 0$ mais $[x,z]=0$ c'est-à-dire $z\in Z(G)$. Notons $\widetilde{\mathscr{D}(G_0)}$ le sous-corps de $\mathscr{D}(G_0)$ formé par les éléments qui commutent avec $x$. Nous montrerons que l'application

$$\varphi=\exp\left(-\frac{y}{z}\operatorname{ad} x\right)=\sum_k \frac{(-1)^k}{k!}\left(\frac{y}{z}\right)^k (\operatorname{ad} x)^k$$

est un homomorphisme de l'algèbre $\mathscr{U}(G_0)$ dans $\widetilde{\mathscr{D}(G_0)}$. En effet :

$$\varphi(ab)=\sum_k \frac{(-1)^k}{k!}\left(\frac{y}{z}\right)^k (\operatorname{ad} x)^k ab=\sum_k \sum_{i+j=k} \frac{(-1)^k}{k!}\left(\frac{y}{z}\right)^k \frac{k!}{i!\,j!}(\operatorname{ad} x)^i a(\operatorname{ad} x)^j b=$$

$$\sum_{i,j}\frac{(-1)^{i+j}}{i!\,j!}\left(\frac{y}{z}\right)^{i+j}(\operatorname{ad} x)^i a(\operatorname{ad} x)^j b=\varphi(a)\varphi(b);$$

$$(\operatorname{ad} x)\varphi(a)=\sum_k \frac{(-1)^k}{k!}\operatorname{ad} x\left[\left(\frac{y}{z}\right)^k(\operatorname{ad} x)^k a\right]=$$

$$\sum_k \frac{(-1)^k}{(k-1)!}\left(\frac{y}{z}\right)^{k-1}(\operatorname{ad} x)^k a+\sum_k \frac{(-1)^k}{k!}\left(\frac{y}{z}\right)^k(\operatorname{ad} x)^{k+1}a=0.$$

Soit maintenant $x_1, \ldots, x_n, y_1, \ldots, y_n, z_1, \ldots, z_k$ un système de générateurs de $\mathscr{D}(G_0)$ jouissant de la propriété (**). Posons

$$\tilde{x}_i=z^{N}\varphi(x_i),\qquad \tilde{y}_i=z^{N}\varphi(y_i),\qquad 1\leqslant i\leqslant n,$$
$$\tilde{x}_{n+1}=xz^{2N-1}\varphi(c),\qquad \tilde{y}_{n+1}=y$$

et choisissons pour $\tilde{z}_1, \ldots, \tilde{z}_{k-1}$ un ensemble d'éléments de $Z(G)$ qui engendrent le centre du corps $\mathscr{D}(G)$. Nous montrerons qu'on peut choisir l'entier N et l'élément $y$ de sorte que le système $\tilde{x}_i$, $\tilde{y}_i$, $\tilde{z}_j$ jouit des propriétés (*) et (**).

Démontrons d'abord qu'un choix approprié de $y$ implique que $\varphi(c)\neq 0$. En effet, $y$ avait été choisi de sorte que $y\in Z(G_0)$, $[x,y]=z\neq 0$, $z\in Z(G)$. Par conséquent le rôle de $y$ peut être joué par $y_\tau=y+\tau z$ pour tout $\tau\in L$. Considérons

$$\varphi_\tau(c)=\sum_k \frac{(-1)^k}{k!}\left(\frac{y_\tau}{z}\right)^k(\operatorname{ad} x)^k c \tag{11}$$

et supposons que, pour tout $\tau$, cette expression est égale à 0.

Il est clair que l'expression (11) est un polynôme en $\tau$. Le coefficient de $\tau^j$ dans ce polynôme se calcule aisément, il est égal à

$$\frac{(-1)^j}{j!}\varphi((\operatorname{ad} x)^j c).$$

Soit $j$ l'entier maximum pour lequel $(\operatorname{ad} x)^j c$ est non nul. On a alors $(\operatorname{ad} x)^j c\in Z(G)$. Or l'application $\varphi$ est l'identité sur $Z(G)$. Par conséquent $\varphi((\operatorname{ad} x)^j c)=(\operatorname{ad} x)^j c\neq 0$, ce qui est en contradiction avec le fait que notre polynôme est identiquement nul.

Désormais nous supposons que $y$ a été choisi de sorte que $\varphi(c) \neq 0$. La condition (**) découle directement du fait que $\varphi$ est un homomorphisme de $\mathscr{U}(G_0)$ dans $\widetilde{\mathscr{D}(G_0)}$. Il reste à démontrer que le système $\tilde{x}_i$, $\tilde{y}_i$, $\tilde{z}_j$ engendre le corps $\mathscr{D}(G)$. Soit $g_1, \ldots, g_m$ une base de G telle que l'opérateur ad $x$ soit de forme

$$(\mathrm{ad}\, x)g_i = \sum_{j>i} a_{ij} g_i$$

On a alors :

$$\varphi(g_i) = g_i + \sum_{j>i} b_{ij}(y, z) g_j.$$

On voit donc que $\mathscr{D}(G_0)$ est engendré par $\varphi(\mathscr{U}(G_0))$ et les éléments $y$ et $z$. Puisque les éléments $\varphi(z_i)$ et $z$ appartiennent à Z(G), ils s'expriment en fonction de $\tilde{z}_1, \ldots, \tilde{z}_{k-1}$.

On voit donc que le corps $\mathscr{D}(G)$ est engendré par les éléments $x, y, \varphi(x_i), \varphi(y_i)$, $\tilde{z}_i$, $1 \leqslant i \leqslant n$, $1 \leqslant j \leqslant k-1$. Il suffit maintenant de remarquer que pour N suffisamment grand les éléments $\tilde{x}_i$, $\tilde{y}_i$ appartiennent à $\mathscr{U}(G)$. Le lemme est démontré.

## 8. Exemples.

Dans ce paragraphe, L est soit le corps **R** des réels, soit le corps **C** des nombres complexes. Soit G l'algèbre de Lie sur L de dimension 3 engendrée par $x, y, z$ avec les relations :

$$[x, y] = y, \qquad [x, z] = \alpha z, \qquad [y, z] = 0$$

où $\alpha$ est un nombre irrationnel. Le groupe linéaire de Lie associé a pour éléments les matrices de la forme

$$\begin{pmatrix} \exp a & 0 & b \\ 0 & \exp \alpha a & c \\ 0 & 0 & 1 \end{pmatrix}$$

où $a, b, c \in L$. Ce groupe n'est évidemment pas algébrique. Dans ce cas le corps de Lie $\mathscr{D}(G)$ n'est isomorphe à aucun des corps $\mathscr{D}_{n,k}(L)$. En effet, le centre du corps $\mathscr{D}(G)$ coïncide avec le corps L (ceci peut être démontré sans difficulté, par exemple, en se servant des résultats obtenus dans [6] sur la structure du centre de $\mathscr{D}(G)$ pour les algèbres de Lie résolubles). Par conséquent, si le corps $\mathscr{D}(G)$ était isomorphe à $\mathscr{D}_{n,k}(L)$ on aurait les égalités :

$$2n + k = \mathrm{Dim}_L \mathscr{D}(G) = \dim G = 3$$
$$k = \mathrm{Dim}_L L = 0$$

ce qui est impossible pour $n$ entier.

Considérons maintenant l'extension $\tilde{G}$ de l'algèbre G par une algèbre T de dimension 1, à élément de base $t$, et avec les relations

$$[y, z] = t, \qquad [x, t] = (1 + \alpha)t, \qquad [y, t] = (z, t] = 0$$

Puisque $\widetilde{G}/T = G$ l'algèbre $\widetilde{G}$ n'est également pas algébrique. Néanmoins on peut vérifier que les éléments

$$p_1 = yt^{-1}, \qquad q_1 z, \qquad p_2 = (1+\alpha)^{-1}t, \qquad q_2 = yzt^{-2}xt^{-1}$$

engendrent $\mathscr{D}(G)$ et jouissent de la propriété suivante

$$[p_i, p_j] = [q_i, q_j] = 0, \qquad [p_i, q_j] = \delta_{ij}.1$$

On a ainsi que $\mathscr{D}(\widetilde{G})$ est isomorphe à $\mathscr{D}_{2,0}(L)$.

## BIBLIOGRAPHIE


[1] *Séminaire Sophus Lie*, Paris, 1955.

[2] I. M. Gelfand, Центр инфинитезимального группового кольца, *Mat. Sbornik*, 1950, t. 26, 103-112.

[3] J. Dixmier, Sur les représentations unitaires des groupes de Lie nilpotents, II, *Bull. Soc. Math. France*, 85 (1957), 325-388.

[4] J. Dixmier, Sur l'algèbre enveloppante d'une algèbre de Lie nilpotente, *Arch. Math.*, 1959, vol. 10, 321-326.

[5] A. A. Kirillov, Унитарные представления нильпотентных групп Ли, *Uspekhi Matem. Nauk*, 1962, t. 17, 57-101.

[6] P. Bernat, Sur le corps enveloppant d'une algèbre de Lie résoluble, *C. r. Acad. Sci.*, t. 258, 2713-2715.


*Manuscrit reçu le 19 janvier 1966.*

# 3.

## (with A. A. Kirillov)

# On the structure of the field of quotients of the enveloping algebra of a semisimple Lie algebra

Dokl. Akad. Nauk SSSR **180** (1968) 775–777 [Sov. Math., Dokl. **9** (1968) 669–671].
MR **37**:5260. Zbl. **244**:17006

1. Let $\mathfrak{g}$ be a semisimple Lie algebra, $U(\mathfrak{g})$ be its enveloping algebra, $D(\mathfrak{g})$ be the field of quo-ents of the algebra $U(\mathfrak{g})$. In the paper [1] the authors introduced a conjecture on the field $D(\mathfrak{g})$ for arbitrary algebraic Lie algebra. This conjecture was proven in [1] for nilpotent Lie algebras, for a mplete matrix algebra and for the algebra of all matrices with zero trace. In this paper we shall in-stigate the case of an arbitrary complex semisimple Lie algebra. *

Our approach to the study of the structure of $D(\mathfrak{g})$ suggests the theory of infinite-dimensional presentations of complex semisimple Lie groups. It is based on the realization of the algebra $U(\mathfrak{g})$ a ring of differential operators on a basis for affine space [2] and the extension of this algebra with e help of the introduction of virtual Laplace operators. ** The basic result consists of the descrip-on of the extension of the enveloping algebra $\widetilde{U}(\mathfrak{g})$ (Theorem 1) and its field of quotients $\widetilde{D}(\mathfrak{g})$ heorem 2).

2. Let us introduce notation: $\mathfrak{h}$ is the Cartan subalgebra in $\mathfrak{g}$; $\Delta(\Delta_+, \Delta_-)$ is the union of the ots (positive roots, negative roots) of the algebra $\mathfrak{g}$; $X_\alpha$ is the root vector corresponding to the root $\in \Delta$; $\mathfrak{n}_+ (\mathfrak{n}_-)$ is the subalgebra generated by all $X_\alpha$, $\alpha \in \Delta_+(\Delta_-)$. Let $G$ be a simply connected e group for which $\mathfrak{g}$ is the Lie algebra. We will denote by $H, N_+, N_-$, the subgroups corresponding the subalgebras $\mathfrak{h}, \mathfrak{n}_+, \mathfrak{n}_-$. The uniform space $A = N_-\backslash G$ of right adjacent classes in $G$ by $N_-$ e will call a basic affine space. Since the subgroup $H$ is normal in $N_-$, it operates on $A$ as a left anslation: an element $h \in H$ transfers the adjacent class $N_-g$ to $N_-hg$. As is well known [3], the bset $N_-HN_+$ is open and compact in $G$. Therefore, any open compact subset of $A$ is naturally iden-ied with a subgroup $HN_+$. In this subset we define coordinates $r_i$, $i = 1, 2, \cdots, k = \dim H$; $t_j$, $j =$ 2, $\cdots$, $n = \dim N_+$, satisfying the condition:

$$\tau_i(hn_+) = \tau_i(h) = e^{\langle \sigma_i, \ln h\rangle};$$

$(hn_+) = t_j(n_+) =$ coefficient by $X_{\alpha_j}$ in the expansion of $\ln n_+$. Here ln denotes some preimage of the ement of the Lie group in the corresponding Lie algebra with respect to the canonical mapping; $, \cdots, \sigma_k$ are the highest weights of the elements of the representations of the group $G$; $\alpha_1, \cdots, \alpha_n$ e the positive roots of the algebra $\mathfrak{g}$.

The manifold $A$ is naturally provided the structure of an affine algebraic manifold. Namely, the ng $F$ of regular functions on $A$ by definition consists of the functions $f$ having the following condi-on: the linear span of the functions which are obtained from $f$ by all mappings from the group $G$ has nite dimension.

---

* In fact we will employ our method to any decomposable reductive algebra over a field of characteristic zero.

** The use of these operators is similar to the introduction of generators for the study of the ring of cohomol-ies classifying the spaces $BG$. Also, as in topology, the introduction of virtual Laplace operators simplifies any formulas in the theory of representations.

**Lemma 1.** *The ring $F$ is contained in $C[r_1, \cdots, r_k, t_1, \cdots, t_n]$, and its field of quotients coincides with $(r_1, \cdots, r_k, t_1, \cdots, t_n)$.*

2. The basic unit of our discussion will be the ring $R$ of differential operators on $A$ preserving the ring $F$. We introduce the following additional structure in the ring $R$:

1) A filtration, $F = R_0 \subset R_1 \subset \cdots \subset R_m \subset \cdots \subset R$, is associated with the order of a differential operator.

2) A gradation, $R = \Sigma_\lambda {}^\lambda R$, is associated by means of left translation. To each integral $k$-dimensional vector $\lambda = (\lambda_1, \cdots, \lambda_n)$ corresponds the subspace ${}^\lambda R$ of those operators which by left translation on $h \in H$ are multiplied on $r^\lambda(h) = \prod_{i=1}^k r_i^{\lambda_i}(h)$.

3) The decomposition in the direct sum $R = \Sigma R^\mu$ is associated by means of right translations. To each $k$-dimensional vector $\mu = (\mu_1, \cdots, \mu_k)$ with integral nonnegative coordinates corresponds the subspace $R^\mu$ in which the representation of $G$, a multiple of the irreducible representation $T_\mu$ with highest weight $\mu$, is realized (we identify the integral vector $\mu$ with the linear functional $\mu_1\sigma_1 + \cdots + \mu_k\sigma_k$ on $\mathfrak{h}$).

By the letter $R$ with several indices we denote the intersection of the subsets of $R$ corresponding to each of these indices, thus, ${}^\lambda R_m^\mu$ denotes $R_m \cap {}^\lambda R \cap R^\mu$.

In the following the ring ${}^0R^0$, which we will call the ring of virtual Laplace operators and will denote by $\widetilde{Z}(\mathfrak{g})$, plays a critical role. To each element $X \in \mathfrak{h}$ there corresponds an infinitesimal operator as a left translation $L_X$:

$$L_X f(N_- g) = \frac{d}{dt} f(N_- \exp(tX) g)|_{t=0},$$

obviously belonging to ${}^0R_1^0$. We will denote by $\psi$ the homomorphism of the algebra $U(\mathfrak{h})$ to $\widetilde{Z}(\mathfrak{g})$, sending $X \in \mathfrak{h}$ to $L_X + \langle \rho, X \rangle$, where $\rho = \frac{1}{2} \Sigma_{\alpha \in \Delta^+} \alpha = \Sigma_{i=1}^k \sigma_i$. In particular, if $\alpha_1^*, \cdots, \alpha_k^*$ is a basis of $\mathfrak{h}$ dual to the basis $\sigma_1, \cdots, \sigma_k$, then $\psi(\alpha_i^*) = r_i \partial/\partial r_i + 1$.

Let $\phi: U(\mathfrak{g}) \longrightarrow R$ be the natural representation of the enveloping algebra in the ring of differential operators on $A$ corresponding to the operation of the group $G$ on $A$ by right translation. It is clear that the image of the algebra $U(\mathfrak{g})$ is contained in ${}^0R$ and the image of its center $Z(\mathfrak{g})$ is contained in $\widetilde{Z}(\mathfrak{g})$.

The subring in ${}^0R$ generated by $\phi(U(\mathfrak{g}))$ and $\widetilde{Z}(\mathfrak{g})$ will be called the extended enveloping algebra and will be denoted by $\widetilde{U}(\mathfrak{g})$.* In the algebra $\widetilde{U}(\mathfrak{g})$ it is possible to define the operation of the Weyl group $W$ in the following way. We recall ([4], Chapter III, §5) that to each element $s \in W$ corresponds a Weyl operator $B_s$ operating in some space of smooth rapidly decreasing functions on $A$ by the formula

$$B_s f(N_- g) = \int_{N_- \cap s^{-1} N_+ s} f(N_- s^{-1} n g)\, dn.$$

It is proved that for any operator $L \in \widetilde{U}(\mathfrak{g})$ the operator $L^s = B_s L B_s^{-1}$ also belongs to $\widetilde{U}(\mathfrak{g})$. We will denote by $\widetilde{U}(\mathfrak{g})^W$ the union of the operators $L \in \widetilde{U}(\mathfrak{g})$ satisfying the condition: $L^s = L$ for all $s \in W$.

**Theorem 1.** *The mapping $\phi$ determines an isomorphism of $U(\mathfrak{g})$ and $\widetilde{U}(\mathfrak{g})^W$. The algebra $\widetilde{U}(\mathfrak{g})$ is isomorphic in a natural way to the tensor product $U(\mathfrak{g}) \otimes \widetilde{Z}(\mathfrak{g})$. The mapping $\psi$ is an isomorphism of the $W$-modules $U(\mathfrak{h})$ and $\widetilde{Z}(\mathfrak{g})$.*

We now pass to the field of quotients $\widetilde{D}(\mathfrak{g})$ of the algebra $\widetilde{U}(\mathfrak{g})$. Since the algebra $\widetilde{U}(\mathfrak{g})$ is

* Apparently $\widetilde{U}(\mathfrak{g})$ in fact coincides with ${}^0R$, however we are not ready to prove this.

nbedded in $R$, the field $\widetilde{D}(\mathfrak{g})$ is embedded in the quotient field $D$ of the algebra $R$ which, by Lemma , is isomorphic to the standard field $D_{n+k,0}$, generated by the operators $r_i$, $\partial/\partial r_i$, $1 \le i \le k$, $t_j$, $\partial/\partial t_j$, $\le j \le n$. Let ${}^0D$ be the union of the elements of $D$ permutable with left translations. It is clear that $D$ is generated by the operators $r_i$, $\partial/\partial r_i$, $1 \le t \le k$, $t_j$, $\partial/\partial t_j$, $1 \le j \le n$ and, hence, is isomorphic to e standard field $D_{n,k}$.

Theorem 2. *The field* $\widetilde{D}(\mathfrak{g})$ *coincides with* ${}^0D$.

The proof of Theorem 2 relies on the study of the structure of the space ${}^\lambda R^\mu$ as a $\widetilde{Z}(\mathfrak{g})$-module. his is done with the help of the duality of Frobenius-Cartan for induced representations.

It is possible to formulate the result we obtain like this.

Theorem 3. *Let $K$ be the field of quotients of the ring $\widetilde{Z}(\mathfrak{g})$. The span of ${}^\lambda R^\mu \times_{\widetilde{Z}(\mathfrak{g})} K$ has imension $d_\mu \cdot m_\mu(\lambda)$, where $d_\mu$ is the dimension of the irreducible representation $T_\mu$, and $m_\mu(\lambda)$ is e multiplicity of the weight of $\lambda$ in the space of transformations $T_\mu$.*

We shall indicate briefly the idea of the derivation of Theorem 2 from Theorem 3. First it is posible to verify that each element $x \in {}^0D$ may be written in the form $x = ab^{-1}$ where $a$ and $b$ belong ${}^\lambda R$ for some $\lambda$. Replacing $a$ and $b$ with $ac$ and $bc$ where $c \in {}^{-\lambda}R$, we obtain that ${}^0D$ coincides ith the field of quotients of the ring ${}^0R$. We note that here we are using a nonempty ${}^{-\lambda}R$, following om Theorem 3. It remains to verify that the rings ${}^0R$ and $\widetilde{U}(\mathfrak{g})$ have one and the same field of quoents. For this it is sufficient to prove that they coincide with the vector spaces ${}^0R \otimes_{\widetilde{Z}(\mathfrak{g})} K$ and $(\mathfrak{g}) \otimes_{\widetilde{Z}(\mathfrak{g})} K$. Since each space is contained in the first, it is necessary only to compare the dimenions of the intersection of these infinite-dimensional spaces with each of the finite-dimensional paces ${}^0R^\mu \otimes_{Z(\mathfrak{g})} K$. The dimension of the other is calculated similarly and is found to be the same.*

Moscow State University Received 11/MAR/68

## BIBLIOGRAPHY

.] I. M. Gel'fand and A. A. Kirillov, Inst. Hautes Études Sci. Publ. Math. No. 31 (1966), 5. MR 34 #7731.

2] I. M. Gel'fand and M. I. Graev, Dokl. Akad. Nauk SSSR 131 (1960), 496 = Soviet Math. Dokl. 1 (1960), 276. MR 22 #9876.

3] Harish-Chandra, J. Math. Pures Appl. (9) 35 (1956), 203. MR 18, 137.

4] I. M. Gel'fand, M. I. Graev and I. I. Pjateckiĭ-Šapiro, *Generalized functions*, Vol. 6, "Nauka", (1966). (Russian)

] B. Kostant, Amer. J. Math. 85 (1963), 327. MR 28 #1252.

Translated by:
R. L. Johnson

*In fact, that this dimension is equal to $d_\mu \cdot m_\mu(0)$ also follows from the paper of Kostant [5].

# 4.

## (with A. A. Kirillov)

# The structure of the Lie field connected with a split semisimple Lie algebra

Funkts. Anal. Prilozh. 3 (1) (1969) 7–26 [Funct. Anal. Appl. 3 (1969) 6–21].
MR 39:2827. Zbl. 244:17007

## 1. INTRODUCTION

Let $\mathfrak{g}$ be a split semisimple Lie algebra over a field K of characteristic 0. We denote its enveloping algebra by $U(\mathfrak{g})$ and the quotient field of $U(\mathfrak{g})$ by $D(\mathfrak{g})$. In [1] we conjectured that $D(\mathfrak{g})$ depends comparatively weakly on the original algebra $\mathfrak{g}$ and that it is isomorphic to one of the standard fields $D_{n,k}(K)$ (for the definition of these fields see section 13].*

In the present article we replace the enveloping algebra $U(\mathfrak{g})$ by its finite extension $\tilde{U}(\mathfrak{g})$. This extension is obtained from $U(\mathfrak{g})$ by adding the "virtual Laplace operators" that belong to the algebraic extension of the center $Z(\mathfrak{g})$ of $U(\mathfrak{g})$. In many problems in representation theory it is natural and convenient to replace $U(\mathfrak{g})$ by $\tilde{U}(\mathfrak{g})$.

We show that our conjecture is true in the case of the extended algebra $\tilde{U}(\mathfrak{g})$, that is, that the corresponding field $\tilde{D}(\mathfrak{g})$ is isomorphic to $D_{n,k}(K)$. The method we employ is suggested by the theory of infinite-dimensional representations. It is based on a detailed study of the ring R of differential operators on a "basic affine space" A (see [2]). The extended algebra $\tilde{U}(\mathfrak{g})$ is embedded in the subring $R^0 \subset R$ consisting of the differential operators that commute with left translations on A. We conjecture that $\tilde{U}(\mathfrak{g})$ in fact coincides with $R^0$; however, in the present article we only prove the weaker statement that the localizations of $\tilde{U}(\mathfrak{g})$ and $R^0$ with respect to their common center $\tilde{Z}(\mathfrak{g})$ coincide. In passing we obtain information about the structure of the ring R as a G-$\tilde{Z}(\mathfrak{g})$-bimodule. These results generalize the well-known work of B. Konstant [3] concerning the structure of $U(\mathfrak{g})$ as a G-$Z(\mathfrak{g})$-bimodule.

The results of the present article were briefly reported in [4].

The authors thank M. I. Graev and P. Delin' for valuable discussions.

## 2. NOTATION

$\mathfrak{g}$ denotes a semisimple Lie algebra over the field K of characteristic 0;

$\mathfrak{h}$ denotes a Cartan subalgebra of $\mathfrak{g}$;

$\Sigma$ denotes the system of roots of $\mathfrak{g}$ with respect to $\mathfrak{h}$ with a fixed order relation;

$X_\alpha$ denotes a characteristic vector corresponding to the root $\alpha$;

$\mathfrak{n}_+ (\mathfrak{n}_-)$ denotes the nilpotent subalgebra generated by all the $X_\alpha$, $\alpha > 0$ ($\alpha < 0$);

$T_\mu$ denotes the irreducible finite-dimensional representation of $\mathfrak{g}$ with the highest weight $\mu$; we use the same letter to denote the corresponding representations of $U(\mathfrak{g})$ and G;

$V_\mu$ denotes the representation space of $T_\mu$;

$m_\mu(\lambda)$ denotes the multiplicity of the weight $\lambda$ in the representation $T_\mu$;

$\alpha_1, \ldots, \alpha_k$ denote the simple roots of $\mathfrak{g}$;

$\sigma_1, \ldots, \sigma_k$ denote the highest weights of the fundamental representations of $\mathfrak{g}$;

*In [1] we conjectured that $D(\mathfrak{g})$ is isomorphic to $D_{n,k}(K)$ for any algebraic Lie algebra $\mathfrak{g}$; in this connection we had in mind that K is an algebraically closed field of characteristic 0.

Moscow State University. Translated from Funktsional'nyi Analiz i Ego Prilozheniya, Vol. 3, No. 1, pp. 7-26, January-March, 1969. Original article submitted September 30, 1968.

P denotes the semigroup of the highest weights of the irreducible representations of $\mathfrak{g}$;

P' denotes the semigroup generated by the positive roots;

G denotes a simply connected algebraic group corresponding to $\mathfrak{g}$ (for the details see section 3);

H, $N_+$, $N_-$ denote the subgroups of G corresponding to the subalgebras $\mathfrak{h}$, $\mathfrak{n}_+$, $\mathfrak{n}_-$;

$A = N_-\backslash G$ denotes the "basic affine space";

F denotes the ring of regular functions on A (see section 4);

R denotes the ring of regular differential operators on A (see section 6);

$R^\lambda \subset R$ denotes the proper subspace corresponding to the weight $\lambda$; here R is considered as an H-module relative to left translations (see section 6).†

An asterisk denotes the transition to the dual space (that is, to the space of K-linear mappings into K ) or the transition to the adjoint operator.

## 3. RESULTS ON SEMISIMPLE LIE ALGEBRAS AND LIE GROUPS

We quote some well known facts from the theory of semisimple Lie algebras and Lie groups (see [5-7]).

Let T be an irreducible finite-dimensional representation of the Lie algebra $\mathfrak{g}$. The space V, in which T acts, is the direct sum of the subspaces $V^\lambda$ which are eigen-subspaces for the operators T(X), $X \in \mathfrak{h}$:

$$T(X)\xi = \langle\lambda, X\rangle\xi \quad \text{for} \quad \xi \in V^\lambda.$$

The space $V^\lambda$ is non-zero for only a finite subset in $\mathfrak{h}^*$; this subset is called the system of weights of the representation T. The following holds

$$T(X_\alpha)V^\lambda \subset V^{\lambda+\alpha}, \tag{3.1}$$

where the left-hand side vanishes only when $V^{\lambda+\alpha} = 0$, that is, when $\lambda + \alpha$ is not a weight.

There is exactly one of the subspaces $V^\lambda$ which is invariant relative to $T(\mathfrak{n}_+)$; this space is one-dimensional and the corresponding weight $\mu$ is called the highest weight. All the remaining weights of the representation T have the form $\mu - \sum_{\alpha>0} k_\alpha \cdot \alpha$, where the coefficients $k_\alpha$ are non-negative integers. The representation T is defined with accuracy up to an equivalence by its highest weight. The highest weights of the irreducible representations form a free semigroup P with the generators $\sigma_1, \ldots, \sigma_k$, where k is the rank of $\mathfrak{g}$ (that is, the dimension of $\mathfrak{h}$).

For each semisimple algebra $\mathfrak{g}$ we canonically construct a linear algebraic group G (when K is the field of complex numbers, G is the simply connected Lie group corresponding to the Lie algebra $\mathfrak{g}$). The algebraic subgroups $N_+$, $N_-$, H of G correspond to the subalgebras $\mathfrak{n}_+$, $\mathfrak{n}_-$, $\mathfrak{h}$ of $\mathfrak{g}$. Below we present those of their properties that will be used in what is to follow. The groups $N_-$ and $N_+$ are nilpotent and, as algebraic manifolds, are isomorphic to $K^n$ where n is the number of positive roots of $\mathfrak{g}$; as an algebraic group H is isomorphic to the direct product of k copies of the multiplicative group $K^*$ of K.

We let $n_+(t)$, $n_-(s)$, $h(\tau)$ denote the elements of $N_+$, $N_-$, H with the coordinates $t = \{t_\alpha\}_{\alpha>0}$, $s = \{s_\alpha\}_{\alpha<0}$, $\tau = \{\tau_i\}_{1\leq i\leq k}$. To each finite-dimensional representation T of $\mathfrak{g}$ corresponds a representation of G that acts in the same representation space V and will be denoted by the same letter. These representations are connected by the equations

$$T(n_+(t)) = \exp\Big(\sum_{\alpha>0} t_\alpha T(X_\alpha)\Big), \quad T(n_-(s)) = \exp\Big(\sum_{\alpha<0} s_\alpha T(X_\alpha)\Big),$$
$$T(h(\tau))\xi = \tau^\lambda\xi \quad \text{for } \xi \in V^\lambda, \tag{3.2}$$

where $\tau^\lambda = \prod_{i=1}^{k} \tau_i^{\lambda_i}$, and the numbers $\lambda_i$ are determined from the expansion $\lambda = \sum_{i=1}^{k} \lambda_i\sigma_i$.

†In [4] this space was denoted by ${}^\lambda R$.

The groups $N_+$, $N_-$ and H generate G. The group H normalizes $N_+$ and $N_-$, moreover

$$h(\tau)n_+(t)h(\tau)^{-1} = n_+(t'), \text{ where } t'_\alpha = \tau^\alpha t_\alpha,$$
$$h(\tau)n_-(s)h(\tau)^{-1} = n_-(s'), \text{ where } s'_\alpha = \tau^\alpha s_\alpha. \quad (3.3)$$

Let N(H) be the normalizer of H in G. The factor-group W = N(H)/H is called the Weyl group; it acts naturally on the Lie algebra $\mathfrak{h}$ of the group H and on the dual space $\mathfrak{h}^*$, preserving the lattice of weights and roots. The actions of W on H and on $\mathfrak{h}^*$ are connected by the relation $(\tau^w)^\lambda = \tau^{w(\lambda)}$, where $\tau^w$ is determined from the equation† $wh(\tau)w^{-1} = h(\tau^w)$.

The ring of functions regular on G coincides with the set of G-finite functions‡ and consists of linear combinations of the matrix elements of the irreducible finite-dimensional representations.

The homogeneous space $A = N_-\backslash G$ is a quasiaffine algebraic manifold (that is, it can be realized as a locally closed subset in affine space). The ring F of functions regular on A coincides with the set of G-finite functions.

The manifold A, relative to the action of the Borel subgroup $B = HN_+$, splits into a finite number of orbits; each orbit is an affine manifold and contains exactly one point of the form $N_-w$, $w \in W$. In particular, the orbit $A_0$ of maximal dimension corresponds to a unique element of W; this orbit is open in A and B is simply transitive on it.

## 4. THE STRUCTURE OF THE RING F

Let V be the finite-dimensional subspace of functions from F in which the group G acts according to the irreducible representation $T_\mu$. Then, to each element $\xi$ from the representation space $V_\mu$ of $T_\mu$ corresponds a function $f_\xi \in V$ so that

$$f_\xi(xg) = f_{T_\mu(g)\xi}(x). \quad (4.1)$$

Let $x_0$ be the initial point of the homogeneous space $A = N_-\backslash G$ corresponding to the coset $\{N_-\}$. The correspondence $\xi \to f_\xi(x_0)$ is a linear form on $V_\mu$ and, consequently, is of the form $\langle\xi, \eta\rangle$ where $\eta \in V_\mu^*$. Since $x_0 n_- = x_0$ for all $n_- \in N_-$, the vector $\eta$ must be $N_-$-invariant. But there is only one such vector (with accuracy up to proportionality) in the indecomposable space $V_\mu^*$ and this is the vector of lowest weight. Conversely, if $\eta$ is the vector of lowest weight in $V_\mu^*$, then the functions

$$f_\xi(g) = \langle T_\mu(g)\xi, \eta\rangle, \quad \xi \in V_\mu, \quad (4.2)$$

depend only on the coset $N_-g$ and, consequently, can be regarded as functions on A. The set of these functions is a finite-dimensional subspace $V \subset F$ in which G acts according to the representation $T_\mu$. Thus, we have proved

<u>LEMMA 4.1.</u> The ring F is the direct sum of the subspaces $F^\mu$ in which G acts according to the irreducible representation $T_\mu$:

$$F = \sum_{\mu \in P} F^\mu.$$

Let us note that we can define the action of the group H on the space A with the help of "left translation."

In fact, since H normalizes the stationary subgroup $N_-$, the left translation $g \to hg$ in the group G, by the element $h \in H$, determines the mapping $N_-g \to N_-hg$ of the space A; this mapping will also be called a left translation.

Thus, we can define a representation S of the group H in the space F by

$$[S(h)f](x) = f(hx). \quad (4.3)$$

†Here and in what follows we use the single letter w to denote an element of the Weyl group and an element of N(H) which is a representative of the appropriate class.

‡A function $f$ on the G-space X is said to be <u>G-finite</u> if its translations by means of transformations from G generate a finite-dimensional space.

LEMMA 4.2. For $f \in F^{\mu}$ the following holds

$$S(h(\tau))f = \tau^{\mu} f. \tag{4.4}$$

Proof. Each $f \in F^{\mu}$ has the form (4.2); hence, if $x = N_{-}g$, then

$$f(h(\tau)x) = \langle T_{\mu}(h(\tau)g)\xi, \eta\rangle = \langle T_{\mu}(g)\xi, T^{*}_{\mu}(h(\tau))\eta\rangle = \tau^{\mu}\langle T_{\mu}(g)\xi, \eta\rangle.$$

Here we have used the fact that the weights of the representation $g \to T^{*}_{\mu}(g)^{-1}$, contragredient to $T_{\mu}$, are inverse to the weights of $T_{\mu}$.

In particular, $-\mu$ is the lowest weight of this representation so that $T^{*}_{\mu}(h(\tau))\eta = T^{*}_{\mu}(h(\tau^{-1}))^{-1}\eta = (\tau^{-1})^{-\mu}\eta = \tau^{\mu}\eta$.

COROLLARY 4.1. The decomposition $F = \sum_{\mu \in P} F^{\mu}$ is a polygradation of the ring F: $F^{\lambda} \cdot F^{\mu} = F^{\lambda+\mu}$.

COROLLARY 4.2. The space $\sum_{i=1}^{k} F^{\sigma_i}$ generates the ring F.

## 5. THE COORDINATES $\tau$, t

As we remarked in section 3 the manifold A contains an open subset $A_0$ on which the subgroup $B = HN_{+}$ is simply transitive. This allows us to introduce coordinates into $A_0$, assigning the coordinates $(\tau, t)$ to the point $x = x_0 h(\tau) n_{+}(t)$. Let us recall that $\tau = \{\tau_i\} \in (K^{*})^{k}$, $t = \{t_{\alpha}\} \in K^{n}$. We are going to explain how the functions belonging to F are expressed in these coordinates.

LEMMA 5.1. The ring F is contained in $K[\tau, t]$, and the quotient field of F coincides with $K(\tau, t)$.

Proof. Let $f \in F^{\mu}$. Then

$$f(\tau, t) = \langle T_{\mu}(h(\tau)n_{+}(t))\xi, \eta\rangle = \langle T_{\mu}(n_{+}(t))\xi, T^{*}_{\mu}(h(\tau))\eta\rangle = \Big\langle \exp\Big(\sum_{\alpha>0} t_{\alpha}T_{\mu}(X_{\alpha})\Big)\xi, \tau^{\mu}\eta\Big\rangle = P_{\xi}(t_{\alpha})\tau^{\mu},$$

where $P_{\xi}$ is a polynomial. The first assertion of the lemma is thus proved.

To prove the second statement we take special forms of the vectors $\xi$. First, if $\xi_0$ is the vector of highest weight then $T(X_{\alpha})\xi_0 = 0$ for all $\alpha > 0$ and so $P_{\xi_0} = \text{const}$. Hence $F^{\mu}$ contains the monomial $\tau^{\mu}$. Next let $\xi = T(X_{-\alpha})\xi_0$. Then it follows from (3.1), (3.2) and from the commutation relations in $\mathfrak{g}$ that $P_{\xi} = (\mu, \alpha)\, t_{\alpha} + P_{\xi}'$, where $P_{\xi}'$ is a polynomial in the variables $t_{\beta}$, $\beta < \alpha$. The second assertion in the lemma is deduced directly from this last result.

Let us note that in fact we have proved the following more precise statement. The space $F^{\mu}$ consists of polynomials of the form $\tau^{\mu} \cdot P(t)$, where $P(t)$ belongs to some subspace $L_{\mu} \subset K[t]$. The spaces $L_{\mu}$ form a (poly)filtration of the ring $K[t]$: $L_{\mu} \subset L_{\nu}$, if $\mu \ll \nu$ (that is, if $\nu - \mu \in P$), $L_{\mu}L_{\nu} \subset L_{\mu+\nu}$, $K[t] = \bigcup_{\mu \in P} L_{\mu}$.

Next let us note that, at the initial point $x_0$, the coordinates $(\tau, t)$ have the values $\tau_i = 1$, $1 \le i \le k$, $t_{\alpha} = 0$, $\alpha > 0$. Hence the local ring $F_{x_0}$ consists of all the rational functions of $\tau_i$, $t_{\alpha}$ that are finite at the point $\tau_i = 1$, $t_{\alpha} = 0$, and the completion of this ring coincides with the ring of power series $K[[\tau - 1, t]]$. In the language of algebraic geometry the results we have obtained can be expressed as follows:

COROLLARY 5.1. The manifold A is rational and all of its points are regular points.

## 6. THE DIFFERENTIAL OPERATORS ON A

By a differential operator of order not greater than k on A we mean a linear mapping D of F into itself satisfying the identity

$$[f_{k+1} \ldots [f_2, [f_1, D]] \ldots ] = 0 \quad \text{for any} \quad f_1, \ldots, f_{k+1} \in F. \tag{6.1}$$

In expanded form this identity is:

$$D(f_1 \ldots f_{k+1}) - \sum_i f_i D(f_1 \ldots \hat{f}_i \ldots f_{k+1}) + \sum_{i,j} f_i f_j \times D(f_1 \ldots \hat{f}_i \ldots \hat{f}_j \ldots f_{k+1}) - \ldots - (-1)^k f_1 \ldots f_{k+1} D(1) = 0. \tag{6.2}$$

In particular, the operator of zero order is the multiplication by a function, and the operator of the first order splits into the sum of a differentiation and the operator of multiplication by a function. We can verify (see [8]) that the composition of a differential operator, of order not greater than k, and of an operator, of order not greater than $l$, is a differential operator of order not greater than $k + l$.

Let us also note (see [8]) that if $F_1$ is the localization of F with respect to any multiplicative system, then a differential operator D, of order not greater than k, can be uniquely extended to a mapping $F_1 \to F_1$ that satisfies (6.2) for any $f_1, \ldots, f_{k+1} \in F_1$. This extension will be denoted by the same letter D.

LEMMA 6.1. The set $R_k$ of operators of order not greater than k is a finite F-module.

Proof. By Corollary 4.2 the ring F has a finite number of generators. It follows from (6.2) that each operator $D \in R_k$ is completely determined by its values on the elements of the form $f_1 \ldots f_k$, where the $f_i$ are taken from the system of generators. The set of the values of D on these elements is a submodule of a finite F-module. The assertion of the lemma now follows from the result that F is a Noetherian ring (as is any K-algebra with a finite number of generators).

A differential operator D is said to be homogeneous of weight $\lambda$ if $DF^{\mu} \subset F^{\mu+\lambda}$ for all $\mu$. The set of homogeneous operators of weight $\lambda$ will be denoted by $R^{\lambda}$. It is obvious that $R^{\lambda} \cdot R^{\mu} \subset R^{\lambda+\mu}$.

LEMMA 6.2. The ring R is the direct sum of the subspaces $R^{\lambda}$. The subspaces $R_k^{\lambda} = R^{\lambda} \cap R_k$ are finite-dimensional.

Proof. For any $D \in R$ we define $D^{\lambda}$ as the mapping of F into itself that maps the function $f \in F^{\mu}$ into the component of $Df$ in the space $F^{\lambda+\mu}$. It is easy to verify that $[f, D^{\lambda}] = [f, D]^{\lambda+\mu}$ for all $f \in F^{\mu}$. This shows that $D^{\lambda}$ is a differential operator whose order is not greater than the order of D. The first part of the lemma is thus proved. To prove the second part we use the previous lemma which asserts that $R_k$ is a finite F-module. This means that each operator $D \in R_k$ has the form $\sum_{i=1}^{N} f_i D_i$, where, without loss of generality, we can take the $D_i$ to be homogeneous: $D_i \in R^{\lambda_i}$. Then each operator from $R_k^{\lambda}$ has the form $\sum_{i=1}^{N} f_i D_i$, where $f_i \in F^{\lambda-\lambda_i}$. Because all the spaces $F^{\lambda}$ are finite-dimensional the lemma is proved.

LEMMA 6.3. The gradation $R = \sum R^{\lambda}$ is connected with the action of the left translations by the relation

$$S(h(\tau))\, D\, S(h(\tau)^{-1}) = \tau^{\lambda} D \quad \text{for} \quad D \in R^{\lambda}. \tag{6.3}$$

Proof. This follows at once from Lemma 4.2 and from the definition of $R^{\lambda}$.

With each $X \in \mathfrak{g}$ we associate the mapping $i(X)$ of the ring F into itself which, in the notation of section 4, has the form

$$i(X) f_{\xi} = f_{T_{\mu}(X)\xi} \quad \text{for} \quad f_{\xi} \in F^{\mu}. \tag{6.4}$$

It is easy to check that $X \to i(X)$ is a representation of the Lie algebra $\mathfrak{g}$ in the algebra of differentiations of F (this representation is associated in the sense of section 3, with the representation T of the group G in the space F). Therefore, it can be extended to a homomorphism of the enveloping algebra $U(\mathfrak{g})$ into the ring R. We shall see below (section 8) that this homomorphism is an embedding.

Next, to each $X \in \mathfrak{h}$ corresponds the differentiation $j(X)$ of the ring F given by

$$j(X) f = \langle \mu, X \rangle f \quad \text{for} \quad f \in F^{\mu}. \tag{6.5}$$

The correspondence $X \to j(X)$ is a representation of the algebra $\mathfrak{h}$ (associated with the representation S of the group H with the help of left translations) and can be extended to a homomorphism j of the algebra $U(\mathfrak{h})$ into R which is obviously an embedding.

Let us note that all the operators we have constructed preserve the gradation in F, that is, they are homogeneous differential operators of zero weight.

To conclude this section we give a table of the commutation relations between the operators we have introduced and the operators of the representations T and S:

$$T(g)i(X)T(g)^{-1}=i(\mathrm{Ad}\,g\,X),\quad S(h)i(X)=i(X)S(h),\quad i(X)j(X')=j(X')i(X),$$
$$T(g)\cdot j(X')=j(X')T(g),\quad S(h)j(X')=j(X')S(h), \tag{6.6}$$

where $g\in G$, $h\in H$, $x\in U(\mathfrak{g})$, $X'\in U(\mathfrak{h})$.

## 7. THE SPACE Y

We are going to introduce the analog of generalized functions on A with support at a point.

Let $F_x$ be the local ring of the point $x\in A$ and let $J_x$ be a maximal ideal of this ring. A linear form on the space $F_x/J_x^{k+1}$ will be called a generalized function of order k with support at the point x.

In particular, for $k=0$ there is a unique (with accuracy up to a multiplier) generalized function $\delta_x$ defined by $\langle\delta_x, f\rangle=f(x)$.

We define the action of the differential operators on generalized functions by putting

$$\langle D^*\sigma, f\rangle=\langle\sigma, Df\rangle \tag{7.1}$$

for any operator D and any generalized function $\sigma$.

LEMMA 7.1. If $\sigma$ is a generalized function of order k with support at the point x and if D is a differential operator of order not greater than $l$, than $D^*\sigma$ is a generalized function of order not greater than $k+l$ with support at x.

Proof. This follows from the inclusion $DJ_x^m\subset J_x^{m-l}$ which, in turn, follows from (6.2).

It is obvious that the correspondence $D\to D^*$ is an antirepresentation of the ring R in the space of generalized functions.

The set of generalized functions of order not greater than m with support at the initial point $x_0=\{N_-\}$ will be denoted by $Y_m$ and we set $Y=\bigcup_{m\geqslant 0}Y_m$. The stationary subgroup of the point $x_0$ acts on $F_{x_0}$ and sends $J_{x_0}$ into itself. Hence, it acts (with the help of a contragredient representation) also on Y and preserves each of the subspaces $Y_m$. Thus, Y is an $N_-$ module with a filtration. The structure of this module plays an essential role in what follows.

LEMMA 7.2. Let $\alpha$ be the mapping of $U(\mathfrak{g})$ into Y defined by

$$\alpha(X)=i(X)^*\delta_{x_0}. \tag{7.2}$$

Then we have the following exact sequence of filtered $N_-$ modules:

$$0\to\mathfrak{n}_-U(\mathfrak{g})\to U(\mathfrak{g})\xrightarrow{\alpha}Y\to 0, \tag{7.3}$$

where $U(\mathfrak{g})$ is endowed with the natural filtration† and with the natural action of the group $N_-$ which arises from the adjoint representation Ad of the group G in $U(\mathfrak{g})$.

Proof. We first verify that $\alpha$ is a homomorphism of the filtered $N_-$ modules. If $X\in\mathfrak{g}$, then $i(X)$ is a differentiation in F, that is, an operator of the first order. This means that, if $X\in U(\mathfrak{g})_k$, then $i(X)$ is an operator of order not greater than k and, by Lemma 7.1, $\alpha(X)\in Y_k$. Next, the representation $\rho$ of the group $N_-$ in $Y_k$ is contragredient to the representation T acting in $F_{x_0}/J_{x_0}^{k+1}$. Hence

$$\langle\rho(n_-)\alpha(X), f\rangle=\langle\alpha(X), T(n_-)^{-1}f\rangle=[i(X)T(n_-)^{-1}f](x_0)=[T(n_-)i(X)T(n_-)^{-1}f](x_0)=[i(\mathrm{Ad}\,n_-X)f](x_0)=\langle\alpha(\mathrm{Ad}\,n_-X), f\rangle.$$

Thus, $\rho(n_-)\circ\alpha=\alpha\circ\mathrm{Ad}\,n_-$, that is, $\alpha$ is a homomorphism of the $N_-$ modules.

We now show that $\alpha$ is an epimorphism. By section 5 we can identify the space $F_{x_0}/J_{x_0}^{k+1}$ with the space of polynomials, of degree not greater than k, in the variables $\tau_i$, $1\leq i\leq k$, $t_\alpha$, $\alpha>0$.

The operators $i(X)$, $X\in\mathfrak{g}$, define differentiations of the ring of rational functions of $\tau$, t; consequently they have the form

---

† $U(\mathfrak{g})_m$ is the set of elements that can be written as a polynomial of degree not greater than m, in the elements of $\mathfrak{g}$.

$$\sum_{i=1}^{k} P_i(\tau, t)\frac{\partial}{\partial \tau_i} + \sum_{\alpha>0}' Q_\alpha(\tau, t)\frac{\partial}{\partial t_\alpha},$$

where P and Q are rational functions of the variables $\tau$, t.

To complete the proof of Lemma 7.2 we need a result concerning the functions $P_i$ and $Q_\alpha$; we present this as a separate lemma.

LEMMA 7.3. Let $I \subset R$ be the set of differential operators that send $F_{x_0}$ into $J_{x_0}$. Then

$$\begin{aligned} i(X_\alpha) &\equiv \frac{\partial}{\partial t_\alpha} \bmod I && \text{for } \alpha > 0, \\ i(X) &\equiv \sum \langle \sigma_i, X\rangle \frac{\partial}{\partial \tau_i} \bmod I && \text{for } X \in \mathfrak{h}, \\ i(X_\alpha) &\equiv 0 \bmod I && \text{for } \alpha < 0. \end{aligned} \tag{7.4}$$

Proof. Because operators of the first order map $J^2_{x_0}$ into $J_{x_0}$, for two such operators to coincide (modulo I) it is sufficient to verify that they coincide on the subspace complementary to $J^2_{x_0}$ in $F_{x_0}$. For this subspace we can take the linear hull of the functions 1, $\tau_i$, $1 \leq i \leq k$, $t_\alpha$, $\alpha > 0$. The calculations required are analogous to those made in section 5 and so we omit them. Lemma 7.3 is proved.

COROLLARY 7.1. The restriction of the mapping $\alpha$ to $U(\mathfrak{h} + \mathfrak{n}_+)$ is an isomorphism.

In fact it follows at once from (7.4) and the structure of the local ring $F_{x_0}$ described in section 5, that for each k the mapping $\alpha$ defines an isomorphism of the factor-spaces $U(\mathfrak{h} + \mathfrak{n}_+)_{k+1}/U(\mathfrak{h} + \mathfrak{n}_+)_k$ and $Y_{k+1}/Y_k$.

Let us return to the proof of Lemma 7.2. It remains to verify that the kernel of the mapping $\alpha$ coincides with $\mathfrak{n}_-U(\mathfrak{g})$. It is clear that this kernel is a right ideal. From the last of the relations at (7.4) we see that $\mathfrak{n}_-$ lies in the kernel of $\alpha$.

Finally, it follows from the Poincaré-Birkhoff-Witt Theorem (see [7]) that $U(\mathfrak{g})$ is the direct sum of the subspaces $U(\mathfrak{h} + \mathfrak{n}_+)$ and $\mathfrak{n}_-U(\mathfrak{g})$. Lemma 7.2 now follows from Corollary 7.1.

## 8. THE LOCAL EXPRESSION FOR A DIFFERENTIAL OPERATOR

At each point $x \in A$ the differential operator $D \in R$ defines the generalized function $D^*\delta_x$ and is itself uniquely defined by the set of these generalized functions. Since the manifold A is homogeneous, instead of the set of generalized functions $D^*\delta_x$ at various points, we can consider another set consisting of generalized functions with support at the single point $x_0$. For each $D \in R$ and $g \in G$ we put

$$\varphi_D(g) = [T(g)DT(g)^{-1}]^*\delta_{x_0}. \tag{8.1}$$

Thus, with each differential operator D we have associated a function $\varphi_D$ on the group G with values in Y. D itself is expressed in terms of $\varphi_D(g)$ by

$$(Df)(x_0 g) = \langle \varphi_D(g), T(g)f\rangle, \tag{8.2}$$

which we obtain by applying both sides of (8.1) to the function $T(g)f$.

LEMMA 8.1. For a function $\varphi: G \to Y$ to correspond to a differential operator $D \in R$ it is necessary and sufficient that the following should hold:

a) $\varphi$ is a regular vector function on G (that is, $\varphi \in O \otimes_K Y$, where O denotes the ring of functions regular on G).

b) for any $n_- \in N_-$ we have $\varphi(n_-g) = \rho(n_-)\varphi(g)$.

Proof. Because the ring O consists of all G-finite functions (see section 3) condition a) is equivalent to the G-finiteness of $\varphi$. In view of the equation

$$\varphi_D(g_1 g) = \varphi_{T(g)DT(g)^{-1}}(g_1) \tag{8.3}$$

this is equivalent to the G-finiteness of the operator D as an element of the G-module R. By Lemma 6.2 each operator D is the sum of homogeneous operators, and each homogeneous operator belongs to one of the finite-dimensional G-invariant spaces $R^\lambda_k$. We have thus proved the necessity of a).

Next, the points $x_0$ and $x_0n_-$ coincide for all $n_- \in N_-$. Hence $Df(x_0g) = Df(x_0n_-g)$. It follows from this result and from (8.2) that $\langle\varphi_D(g), T(g)f\rangle = \langle\varphi_D(n_-g), T(n_-)T(g)f\rangle$. By replacing $T(g)f$ by $f_1$ and by recalling the definition of the representation $\rho$, we can write $\langle\varphi_D(g), f_1\rangle = \langle\rho(n_-)^{-1}\varphi_D(n_-g), f_1\rangle$, the necessity of b) follows from this last result.

Finally, if $\varphi$ satisfies both of the conditions of the lemma then we can define a mapping D: F → F by

$$Df(x_0g) = \langle\varphi(g), T(g)f\rangle. \tag{8.4}$$

In fact, condition b) shows that $Df$ is properly defined on A and condition a) ensures that this function is G-finite.

Let us note next that condition a) implies the existence of a number k such that $\varphi(g) \in Y_k$ for all $g \in G$. We are going to show that in this case D is a differential operator of order not greater than k. To do this we must check that (6.2) holds. With the help of (8.4) we can reduce the problem of verifying (6.2) to that of verifying that the identity

$$\langle\sigma, f_1 \ldots f_{k+1}\rangle - \sum_i f_i(x_0)\langle\sigma, f_1 \ldots \hat{f}_i \ldots f_{k+1}\rangle$$
$$+ \sum_{i,j} f_i(x_0) f_j(x_0)\langle\sigma, f_1 \ldots \hat{f}_i \ldots \hat{f}_j \ldots f_{k+1}\rangle - \ldots + (-1)^{m+1} f_1(x_0) \ldots f_{k+1}(x_0)\langle\sigma, 1\rangle \tag{8.5}$$

holds for any $\sigma \in Y_k$. It is obvious that (8.5) is valid for those $\sigma$ of the form $D_1^*\delta_{x_0}$, where $D_1$ is a differential operator of order not greater than k. But, by Lemma 7.2, each element $\sigma \in Y_k$ has the form $\alpha(X) = i(X)^*\delta_{x_0}$ where $X \in U(\mathfrak{g})_k$.

It remains to verify that $\varphi = \varphi_D$. This follows at once from a comparison of (8.1) and (8.4) and the lemma is proved.

We next clarify the additional properties that $\varphi_D$ has when D is homogeneous.

LEMMA 8.2. For the function $\varphi$: G → Y to correspond to a homogeneous differential operator of weight $\lambda$ it is necessary and sufficient that, in addition to a) and b) of Lemma 8.1, the following condition should hold:

c) $$\varphi_D(h(\tau)g) = \tau^{\lambda}\rho(h(\tau))\varphi_D(g) \quad \text{for all} \quad h(\tau) \in H,$$

where $\rho$ is the representation of the group H in the space $Y \approx U(\mathfrak{g})/\mathfrak{n}_-U(\mathfrak{g})$ that arises from the adjoint representation Ad of the group G in $U(\mathfrak{g})$.

Proof. By Lemma 6.3 the following relation is a characteristic property of homogeneous operators of weight $\lambda$:

$$S(h(\tau))DS(h(\tau))^{-1} = \tau^{\lambda}D.$$

Hence, the assertion of the lemma follows from the equation

$$\varphi_{S(h)DS(h)^{-1}}(g) = \rho(h)^{-1}\varphi_D(hg) \quad \text{for all} \quad h \in H,\ D \in R, \tag{8.6}$$

which we are now going to prove. We use the result that, by Lemma 7.2, each $\sigma \in Y$ has the form $\sigma = \alpha(X) = i(X)^*\delta_{x_0}$, $X \in U(\mathfrak{g})$. Suppose, for example, that

$$\varphi_{S(h)DS(h)^{-1}}(g) = \alpha(X_1), \qquad \varphi_D(hg) = \alpha(X_2). \tag{8.7}$$

Then $[S(h)DS(h)^{-1}f](x_0g) = [i(X_1)T(g)f](x_0)$. On the other hand, $[S(h)DS(h)^{-1}f](x_0g) = [DS(h)^{-1}f](x_0hg) = [i(X_2)T(hg)S(h)^{-1}f](x_0)$.

It follows from the commutation relations (6.6) that

$$i(X_2)T(hg)S(h)^{-1} = S(h)^{-1}T(h)\,i(\mathrm{Ad}(h^{-1})X_2)T(g).$$

Since we have $[S(h)^{-1}T(h)f](x_0) = f(x_0)$ for any $f \in F$, we obtain that

$$[i(X_1)T(g)f](x_0) = [i(\mathrm{Ad}(h)^{-1}X_2)T(g)f](x_0)$$

for any $f \in F$. By the definitions of the representation $\rho$ and of the mapping $\alpha$ we can rewrite our result as

$$\langle\alpha(X_1), f_1\rangle = \langle\rho(h^{-1})\alpha(X_2), f_1\rangle,$$

where $f_1 = T(g)f$. Since $f_1$ is arbitrary, it follows from here that $\alpha(X_1) = \rho(h^{-1})\alpha(X_2)$, and this equation is seen to be the same as (8.6) when we recall the notation (8.7); thus the lemma is proved.

Next we are interested in determining which functions $\varphi$ correspond to operators from $i(U(\mathfrak{g}))$ and $j(U(\mathfrak{h}))$ (see section 6). An answer is given by the following lemma.

LEMMA 8.3. For any $X \in U(\mathfrak{g})$, $X' \in U(\mathfrak{h})$ we have

$$\varphi_{i(X)}(g) = \alpha(\mathrm{Ad}\, g\, X), \qquad \varphi_{j(X')}(g) = \alpha(X').$$

Proof. The first equation follows from the definitions of the mappings i and $\alpha$ [see (6.4) and (7.2)] and from the commutation relations (6.6).

To prove the second equation it is necessary, in addition, to note that $j(X')\delta_{x_0} = i(X')\delta_{x_0}$; this result can be verified directly.

COROLLARY 8.1. The mapping $i\colon U(\mathfrak{g}) \to R$ is an embedding.

In fact, the kernel I of i consists of those $X \in U(\mathfrak{g})$ for which $\mathrm{Ad}\, gX \in \mathfrak{n}_- U(\mathfrak{g})$ for all $g \in G$.

We are going to show that, under any finite-dimensional representation, the elements of I map into the zero. It is sufficient to verify this for all the irreducible representations $T_\mu$. We consider the subspace $W_\mu \subset V_\mu$ generated by vectors of the form $T(X)\xi$, where $X \in I$, $\xi \in V_\mu$. Let $\eta$ be the vector of lowest weight in $V_\mu^*$. Then for any $X \in \mathfrak{n}_- U(\mathfrak{g})$ and for any $\xi \in V_\mu$ we have $\langle T(X)\xi, \eta\rangle = 0$. Hence, the vectors of $W_\mu$ are orthogonal to $\eta$ and, consequently, $W_\mu \neq V_\mu$. On the other hand, $W_\mu$ is invariant with respect to the operators of the representation because I is invariant with respect to Ad G. Hence, $W_\mu = 0$, as we required to show. It remains to note that $U(\mathfrak{g})$ has a complete set of finite-dimensional representations (see, for example, [9], Chapter 6, Exercise 7). Hence $I = 0$ and the mapping i is an embedding.

## 9. THE EXTENDED CENTER $\tilde{Z}(\mathfrak{g})$

The ring $\tilde{Z}(\mathfrak{g})$ of all differential operators on A that commute with the action of the group G is said to be the extended center of the algebra $U(\mathfrak{g})$.

Equation (8.3) shows that the operator D belongs to $\tilde{Z}(\mathfrak{g})$ if and only if its local expression $\varphi_D(g)$ is independent of g.

We can regard the usual center $Z(\mathfrak{g})$ as a subring of $\tilde{Z}(\mathfrak{g})$. In fact if $X \in Z(\mathfrak{g})$ then $\mathrm{Ad}\, gX = X$ for all $g \in G$. It follows from what has been said and from Lemma 8.3 that $i(X)$ lies in $\tilde{Z}(\mathfrak{g})$.

The following lemma describes the algebraic structure of $\tilde{Z}(\mathfrak{g})$.

LEMMA 9.1. The mapping j is an isomorphism of the algebras $U(\mathfrak{h})$ and $\tilde{Z}(\mathfrak{g})$.

Proof. We have noted already in section 6 that j is an embedding of $U(\mathfrak{h})$ in R. It follows from Lemma 8.3 that $j(U(\mathfrak{h}))$ lies in $\tilde{Z}(\mathfrak{g})$. Let us prove the reverse inclusion. Let $D \in \tilde{Z}(\mathfrak{g})$. Since D commutes with the action of the group G and since the representations of G in the spaces $F^\mu$ are irreducible and pairwise inequivalent, D leaves the subspace $F^\mu$ invariant, that is, it is a homogeneous operator of zero weight. By Lemma 8.2 it follows from this last result that $\varphi_D(hg) = \rho(h)\varphi_D(g)$. On the other hand we know that, for $D \in \tilde{Z}(\mathfrak{g})$, $\varphi_D$ is constant. Hence, the value of $\varphi_D$ is an H-invariant element of Y. But, as an H-module, Y is isomorphic to $U(\mathfrak{h}+\mathfrak{n}_+)$ (see Corollary 7.1 and the definition of the action of H given in Lemma 8.2). It is obvious that the set of H-invariant elements in $U(\mathfrak{h}+\mathfrak{n}_+)$ coincides with $U(\mathfrak{h})$. Thus, there exists an $X \in U(\mathfrak{h})$ such that $\varphi_D(g) \equiv \alpha(X)$; from here we find that $D = j(X)$ and so the lemma is proved.

Let us note the following fact which will be required later on.

COROLLARY 9.1. The set of $N_-$-invariant elements in Y coincides with $\alpha(U(\mathfrak{h}))$.

In fact, it follows from Lemma 8.1 that the constant function $\varphi(g) \equiv \sigma$ corresponds to a differential operator $D \in R$ if and only if $\sigma$ is an $N_-$-invariant element in Y.

The correspondence between D and $\varphi_D$ and the operation of multiplying differential operators are connected in a comparatively complex manner. We show, however, that the situation can be substantially simplified if one of the factors belongs to $\tilde{Z}(\mathfrak{g})$. In this case the local expression of the product is the same as the product of the local expressions of the factors.

A precise statement of this fact is given in

LEMMA 9.2. We transfer the structure of a ring from $U(\mathfrak{h}+\mathfrak{n}_+)$ to Y with the help of the isomorphism $\alpha$ (see Corollary 7.1). Then for any $D_1 \in R$, $D_2 \in \tilde{Z}(\mathfrak{g})$ and $g \in G$ we have

$$\varphi_{D_1 D_2}(g) = \varphi_{D_1}(g)\,\varphi_{D_2}(g).$$

Proof. We easily deduce from the definition of $\varphi_D$ that

$$\varphi_{D_1 D_2}(g) = [T(g)\,D_2 T(g)^{-1}]^*\,[\varphi_{D_1}(g)] \text{ . for any } \quad D_1,\ D_2 \in R. \tag{9.1}$$

When $D_2 \in \tilde{Z}(\mathfrak{g})$ this equation takes the form

$$\varphi_{D_1 D_2}(g) = D_2^*[\varphi_{D_1}(g)]. \tag{9.2}$$

Next let g be fixed. We put

$$\varphi_{D_1}(g) = \alpha(X_1), \quad \varphi_{D_2}(g) = \alpha(X_2), \quad \varphi_{D_1 D_2}(g) = \alpha(X_3),$$

where $X_1, X_2, X_2 \in U(\mathfrak{h}+\mathfrak{n}_+)$. We must prove that $X_3 = X_1X_2$. By the definition of the mapping $\alpha$ we can rewrite (9.2) as:

$$i(X_3)^*\delta_{x_0} = D_2^* i(X_1)^*\delta_{x_0}.$$

It follows from Lemmas 8.3 and 9.1 that $X_2 \in U(\mathfrak{h})$ and that $D_2 = j(X_2)$. It remains to note that $j(X_2)$ and $i(X_1)$ commute [see (6.6)] and that $j(X_2)^*\delta_{X_0} = i(X_2)^*\delta_{X_0}$. [This follows from the definitions of $i(X_2)^*$ and $j(X_2)^*$ and from (4.2).] The lemma is proved.

Next we are going to explain the connection between $Z(\mathfrak{g})$ and $\tilde{Z}(\mathfrak{g})$ in greater detail.

LEMMA 9.3. The algebra $\tilde{Z}(\mathfrak{g})$ is a free module of rank $|W|$ over $Z(\mathfrak{g})$. The quotient field of $\tilde{Z}(\mathfrak{g})$ is the normal algebraic extension of the quotient field of $Z(\mathfrak{g})$. The Galois group of this extension coincides with the Weyl group of the algebra $\mathfrak{g}$.

Proof. We consider the mapping $j^{-1}i$: $Z(\mathfrak{g}) \to U(\mathfrak{h})$. This mapping was (by a somewhat different method) defined and studied by Harish-Chandra [10]. We can state Harish-Chandra's result as follows. Let $\delta$ be the automorphism of $U(\mathfrak{h})$ which is defined on the generators $X \in \mathfrak{h}$ by

$$\delta(X) = X + \langle\rho, X\rangle,$$

where $\rho = \frac{1}{2}\sum_{\alpha>0}\alpha$ (an equivalent definition: $\rho = \sum_{i=1}^{k}\sigma_i$ ). Let $I(\mathfrak{h})$ denote the subalgebra of $U(\mathfrak{h})$ consisting of elements which are invariant with respect to the Weyl group. Then

$$j^{-1}i(Z(\mathfrak{g})) = \delta(I(\mathfrak{h})). \tag{9.3}$$

Another proof of (9.3) based on the concept of an orispherical transformation was given by the present authors in [4].

Let us return to the proof of the lemma. In view of (9.3) it is sufficient to verify that $U(\mathfrak{h})$ is a free module of rank $|W|$ over $I(\mathfrak{h})$, that the quotient field of $U(\mathfrak{h})$ is a normal extension of the quotient field of $I(\mathfrak{h})$ and that the Galois group of this extension is W. All these statements have been proved by Chevalley in [11].

## 10. THE EXTENDED ALGEBRA $\tilde{U}(\mathfrak{g})$

We define $\tilde{U}(\mathfrak{g})$ as the subring of R that is generated by operators from $i(U(\mathfrak{g}))$ and $j(U(\mathfrak{h}))$. Another (abstract) definition is given in Remark 10.2 below.

We are going to consider the enveloping algebra $U(\mathfrak{g})$ as a subalgebra of $\tilde{U}(\mathfrak{g})$ by identifying $U(\mathfrak{g})$ with $i(U(\mathfrak{g}))$. Thus, $\tilde{U}(\mathfrak{g})$ is generated by its subalgebras $U(\mathfrak{g})$ and $\tilde{Z}(\mathfrak{g})$. A more precise result is given by

THEOREM 1. The extended algebra $\tilde{U}(\mathfrak{g})$ is isomorphic to the tensor product $U(\mathfrak{g}) \underset{Z(\mathfrak{g})}{\otimes} \tilde{Z}(\mathfrak{g})$. We can define the action of the Weyl group W in $\tilde{U}(\mathfrak{g})$ so that $U(\mathfrak{g})$ coincides with the set $\tilde{U}(\mathfrak{g})^W$ consisting of elements of $\tilde{U}(\mathfrak{g})$ that are invariant with respect to W.

Before proving this theorem we are going to make several remarks. By definition the algebra $\widetilde{U}(\mathfrak{g})$ is an algebra of operators in the space F. Because $\widetilde{U}(\mathfrak{g})$ lies in $R^0$, the subspaces $F^{\mu}$ are invariant with respect to $\widetilde{U}(\mathfrak{g})$. For each $\mu \in P$ we obtain a representation of the extended algebra $\widetilde{U}(\mathfrak{g})$ in the space $F^{\mu}$; we denote this representation by $T_{\mu}^{(e)}$. It is obvious that the restriction of this representation to $U(\mathfrak{g})$ coincides with $T_{\mu}$. By using the action of the group W on $\widetilde{U}(\mathfrak{g})$ referred to in Theorem 1, we obtain $|W|$ different* representations $T_{\mu}^{(w)}$, $w \in W$, each of which is an extension of $T_{\mu}$. On the other hand, any irreducible representation of $\widetilde{U}(\mathfrak{g})$ is scalar on $\widetilde{Z}(\mathfrak{g})$ and, consequently, is still irreducible when restricted to $U(\mathfrak{g})$. From this follows

COROLLARY 10.1. Each irreducible finite-dimensional representation of the algebra $\widetilde{U}(\mathfrak{g})$ can be extended to a representation of the extended algebra $\widetilde{U}(\mathfrak{g})$ in exactly $|W|$ ways. All the irreducible representations of $\widetilde{U}(\mathfrak{g})$ are obtained in this way.

Thus, the set of irreducible representations of $\widetilde{U}(\mathfrak{g})$ is an "unramified covering" of the set of irreducible representations of $U(\mathfrak{g})$.

REMARK 10.1. We have only referred to finite-dimensional representations of the algebras $U(\mathfrak{g})$ and $\widetilde{U}(\mathfrak{g})$. It is interesting to enquire how the infinite-dimensional representations of these algebras are connected. In particular we hope that the description of the Harish-Chandra modules (see [12]) over $\widetilde{U}(\mathfrak{g})$ is simpler than the analogous problem for $U(\mathfrak{g})$.

REMARK 10.2. The results of Lemma 9.3 and of Theorem 1 allow us to give an "abstract" definition of the algebra $\widetilde{U}(\mathfrak{g})$ without using its realization as a ring of differential operators. We can define $\widetilde{U}(\mathfrak{g})$ as $U(\mathfrak{g}) \otimes_{Z(\mathfrak{g})} U(\mathfrak{h})$, where $Z(\mathfrak{g})$ is identified with the subalgebra $\delta(I(\mathfrak{h}))$ of $U(\mathfrak{h})$ with the help of the mapping $j^{-1}i$.

Let us turn now to the proof of Theorem 1. We use a theorem of Konstant [3] that asserts that $U(\mathfrak{g})$ is a free $Z(\mathfrak{g})$-module. Let $X_i$, $i = 1, 2, \ldots$, be a basis for this module. Since $\widetilde{U}(\mathfrak{g})$ is generated by the subalgebras $U(\mathfrak{g})$ and $\widetilde{Z}(\mathfrak{g})$, each $X \in \widetilde{U}(\mathfrak{g})$ can be written as

$$X = \sum \tilde{z}_i X_i, \text{ where } \tilde{z}_i \in \widetilde{Z}(\mathfrak{g}). \tag{10.1}$$

Let us verify that the expression at (10.1) is unique. Suppose that for some set of $\tilde{z}_i \in \widetilde{Z}(\mathfrak{g})$ we have

$$\sum_i \tilde{z}_i X_i = 0. \tag{10.2}$$

Let $\{\tilde{z}_w\}$, $w \in W$, denote a basis in $\widetilde{Z}(\mathfrak{g})$ considered as a $Z(\mathfrak{g})$-module (as we remarked in section 9 this module is a free module by Chevalley's Theorem). Then $\tilde{z}_i = \sum_w z_{iw} \tilde{z}_w$, where $z_{iw} \in Z(\mathfrak{g})$, and from (10.2) we obtain that

$$\sum_w \tilde{z}_w X_w = 0, \text{ where } X_w = \sum_i z_{iw} X_i. \tag{10.3}$$

As is well known (see [7], Chapter 14), the space $U(\mathfrak{g})$ is the direct sum of $Z(\mathfrak{g})$ and of the subspace $U_0(\mathfrak{g})$ generated by the elements of the form $XY - YX$, where $X, Y \in U(\mathfrak{g})$. Let p denote the projection of $U(\mathfrak{g})$ onto $Z(\mathfrak{g})$ that is parallel to $U_0(\mathfrak{g})$. We show that $\sum_w \tilde{z}_w p(X_w) = 0$. Because this expression lies in $\widetilde{Z}(\mathfrak{g})$ it is sufficient to check that, under the representations $T_{\mu}$, $\mu \in P$, it is mapped onto the zero; then, for any $\tilde{z} \in \widetilde{Z}(\mathfrak{g})$, the operator $T_{\mu}(\tilde{z})$ is scalar. Hence, it is sufficient to verify that $\operatorname{tr} \sum_w T_{\mu}(\tilde{z}_w p(X_w)) = 0$. But, if $\tilde{z} \in \widetilde{Z}(\mathfrak{g})$ and $X \in U_0(\mathfrak{g})$, then $\operatorname{tr} T_{\mu}(\tilde{z}X) = 0$. From here we have that $\operatorname{tr} \sum_w T(\tilde{z}_w p(X_w)) = \operatorname{tr} \sum_w T(\tilde{z}_w X_w) = 0$ in view of (10.3).

Next let us note that, by the definition of the $\tilde{z}_w$, $\sum_w \tilde{z}_w p(X_w) = 0$ only when $p(X_w) = 0$ for all $w \in W$. Thus, we have proved that (10.3) implies that

---

*The fact that these representations are different follows from the equation $T_{\mu}^{(w)}(j\,\delta(X)) = \langle w(\mu + \rho), X \rangle E$ for $X \in \mathfrak{h}$, $w \in W$, which in turn follows from the definitions of the representation $T_{\mu}^{(w)}$ and of the automorphism $\delta$.

$$p(X_w) = 0 \text{ for all } w \in W. \tag{10.4}$$

By multiplying (10.3) on the right by any $X \in U(\mathfrak{g})$ we obtain from (10.4) that

$$p(X_w X) = 0 \text{ for all } X \in U(\mathfrak{g}). \tag{10.5}$$

As is well known, the last equation is possible only when $X_w = 0$.* This means that (10.3) implies that $X_w = 0$ for all $w \in W$. By recalling the definition of $X_w$, we obtain that $z_{iw} = 0$ for all $i$ and for all $w$. From here we see that $\tilde{z}_i = 0$, thus proving the uniqueness of the expansion (10.1). The first assertion of the theorem is proved.

The second assertion follows at once from the first and from Lemma 9.3 and the theorem is proved.

REMARK 10.3. It would be interesting to define the action of the Weyl group in $\tilde{U}(\mathfrak{g})$ directly, without relying on the first part of Theorem 1. For the case when the ground field K is R or C, the present authors proposed such a definition in [4]; it is based on the concept of an orispherical transformation in the space of functions on A.† Unfortunately, the operators of orispherical transformations cannot be applied to functions from F; hence, the method suggested in [4] does not directly carry over to the case of an arbitrary field.

REMARK 10.4. We conjecture that the extended algebra $\tilde{U}(\mathfrak{g})$ coincides with the ring $R^0$. Theorem 2, which is proved below, and its Corollaries 12.1 and 12.2 give some support to this conjecture; in addition, it is true in the simplest example when $\mathfrak{g} = sl(2, K)$.

## 11. THE PAIRING $U(\mathfrak{g}) \times U(\mathfrak{g}) \to U(\mathfrak{h})$

In this section we are going to prove some statements about the structure of the algebra $U(\mathfrak{g})$ which it is convenient to formulate in terms of the pairing (that is, the bilinear mapping) $U(\mathfrak{g}) \times U(\mathfrak{g})$ into $U(\mathfrak{h})$.

Let $X_1, \ldots, X_n$; $Y_1, \ldots, Y_n$; $Z_1, \ldots, Z_n$ be bases in the spaces $\mathfrak{n}_-$, $\mathfrak{h}$, $\mathfrak{n}_+$ respectively.

By the Poincaré-Birkhoff-Witt Theorem monomials of the form

$$P_{i,l,m}(X, Y, Z) = X_1^{i_1} \ldots X_n^{i_n} Y_1^{l_1} \ldots Y_n^{l_k} Z_1^{m_1} \ldots Z_n^{m_n}$$

form a basis in $U(\mathfrak{g})$. We define the mapping $\varepsilon$: $U(\mathfrak{g}) \to U(\mathfrak{h})$ by

$$\varepsilon(P_{i,l,m}) = \begin{cases} P_{i,l,m}, & \text{if } i = 0,\ m = 0, \text{ т. е. } P_{i,l,m} \in U(\mathfrak{h}), \\ 0 & \text{otherwise.} \end{cases} \tag{11.1}$$

We introduce a (poly)gradation in $U(\mathfrak{g})$ by assigning zero weight to the elements of $\mathfrak{h}$ and weight $\alpha$ to the element $X_\alpha$. Then $U(\mathfrak{g}) = \sum U(\mathfrak{g})^\lambda$. It is clear from (11.1) that the mapping $\varepsilon$ is zero on all the $U(\mathfrak{g})^\lambda$ except when $\lambda = 0$.

Next we define the pairing $U(\mathfrak{g}) \times U(\mathfrak{g}) \to U(\mathfrak{h})$ by putting, for $A, B \in U(\mathfrak{g})$,

$$\{A, B\} = \varepsilon(AB). \tag{11.2}$$

We state some obvious properties of this pairing:

a) $\{AC, B\} = \{A, CB\}$ for any $C \in U(\mathfrak{g})$;

b) $\{CA, B\} = \{A, BC\} = C\{A, B\}$ for $C \in U(\mathfrak{h})$; (11.3)

c) if $A \in U(\mathfrak{g})^\lambda$, $B \in U(\mathfrak{g})^\mu$ and $\lambda + \mu \neq 0$, then $\{A, B\} = 0$;

d) if $A \in \mathfrak{n}_- U(\mathfrak{g})$ or $B \in U(\mathfrak{g})\mathfrak{n}_+$, then $\{A, B\} = 0$.

It turns out that the last property has a converse.

---

* For completeness of presentation we give a simple proof of this fact. For any irreducible representation $T_\mu$ the operator $T_\mu(X_w)$ has the property

$$\operatorname{tr}[T_\mu(X_w) T_\mu(X)] = 0 \text{ for all } X \in U(\mathfrak{g}).$$

Hence $T_\mu(X_w) = 0$ and $X_w = 0$.

†The general concept of an orispherical transformation has been suggested by I. M. Gel'fand and M. I. Graev in [13]. The concept of an orispherical transformation used in [4] is given in [14] (see [14], Chapter III, §5).

LEMMA 11.1. If $\{A, B\} = 0$ for all $B \subset U(\mathfrak{g})$ [respectively for all $A \in U(\mathfrak{g})$], then $A \in \mathfrak{n}_-U(\mathfrak{g})$ [respectively $B \in U(\mathfrak{g})\mathfrak{n}_+$].

Proof. Let $U(\mathfrak{g})^\perp$ be the set of those A in $U(\mathfrak{g})$ for which $\{A, B\} = 0$ for all B.

Because $U(\mathfrak{g})^\perp$ contains $\mathfrak{n}_-U(\mathfrak{g})$ we can define the factor-space $Y_0 = U(\mathfrak{g})^\perp/\mathfrak{n}_-U(\mathfrak{g})$ in the space $Y = U(\mathfrak{g})/\mathfrak{n}_-U(\mathfrak{g})$. Let us recall that Y can be identified with $U(\mathfrak{h} + \mathfrak{n}_+)$.

By the same token we can introduce the gradation $Y = \sum_{\lambda \in P} Y^\lambda$ in Y; this gradation is connected with the representation $\rho$ of the group H in Y by the relation

$$\rho(h(\tau))\sigma = \tau^\lambda \sigma \text{ for } \sigma \in Y^\lambda.$$

Because the subspace $Y_0$ is invariant with respect to the action of H we can write $Y_0 = \sum Y_0^\lambda$.

Let $\lambda_0$ be the smallest weight for which $Y_0^\lambda$ is different from zero. We are going to show that the elements of $Y_0^{\lambda_0}$ are invariant with respect to the group $N_-$. Let us recall that the action of $N_-$ in Y arises from the adjoint representation by going over to a factor-space. The corresponding representation of the Lie algebra $\mathfrak{n}_-$ acts according to the formula

$$\rho(X_{-\alpha})X \equiv [X_{-\alpha}, X] \bmod \mathfrak{n}_-U(\mathfrak{g}) \equiv -XX_{-\alpha} \bmod \mathfrak{n}_-U(\mathfrak{g}) \tag{11.4}$$

and, obviously, has the property $\rho(X_{-\alpha})Y^\lambda \subset Y^{\lambda-\alpha}$. It follows from here that $\rho(X_{-\alpha})Y_0^{\lambda_0} = 0$, which is equivalent to $N_-$-invariance.

By Corollary 9.1 all the $N_-$-invariant elements of Y lie in $Y^0 = U(\mathfrak{h})$. Hence, the weight $\lambda_0$ must be zero. But $Y_0^0 = 0$ because, by the definition of the pairing, $\{A, 1\} = A$ when $A \in U(\mathfrak{h})$. Thus, we have proved the first assertion of the lemma. We obtain the second assertion from the first if we note that there is an automorphism of $\mathfrak{g}$ which sends the subalgebras $\mathfrak{n}_-$, $\mathfrak{h}$, $\mathfrak{n}_+$ into $\mathfrak{n}_+$, $\mathfrak{h}$, $\mathfrak{n}_-$, respectively.

COROLLARY 11.1. The pairing we have introduced is non-degenerate when we consider it as the mapping

$$U(\mathfrak{h} + \mathfrak{n}_+)^\lambda \times U(\mathfrak{h} + \mathfrak{n}_-)^{-\lambda} \to U(\mathfrak{h}).$$

In fact, because $U(\mathfrak{g}) = \mathfrak{n}_-U(\mathfrak{g}) + U(\mathfrak{h} + \mathfrak{n}_+) = U(\mathfrak{g})\mathfrak{n}_+ + U(\mathfrak{h} + \mathfrak{n}_-)$, by the lemma we have just proved our pairing is non-degenerate as a mapping of $U(\mathfrak{h} + \mathfrak{n}_+) \times U(\mathfrak{h} + \mathfrak{n}_-)$ into $U(\mathfrak{h})$. The required result follows from property (11.3), c).

## 12. THE STRUCTURE OF $\mathrm{Hom}_G(V_\mu, R^\lambda)$

We are next going to study the structure of $R^\lambda$ as a G-module. To do this we consider the space $\mathrm{Hom}_G(V_\mu, R^\lambda)$ that consists of mappings $A: V_\mu \to R^\lambda$ which commute with the action of the group G. Since $R^\lambda$ is a $\tilde{Z}(\mathfrak{g})$-module (with respect to multiplication on the right) and since the elements of $\tilde{Z}(\mathfrak{g})$ are invariant with respect to the action of G, we can give the space being considered the structure of a $\tilde{Z}(\mathfrak{g})$-module.

The following theorem describes the structure of this module.

THEOREM 2. $\mathrm{Hom}_G(V_\mu, R^\lambda)$ is a finite $\tilde{Z}(\mathfrak{g})$-module of rank $m_\mu(\lambda)$ (= the multiplicity of the weight $\lambda$ in the representation $T_\mu$).

Proof. We begin by establishing the natural isomorphism of $\tilde{Z}(\mathfrak{g})$-modules:

$$\mathrm{Hom}_G(V_\mu, R^\lambda) \simeq \mathrm{Hom}_{HN_-}(V_\mu, Y \otimes E_\lambda), \tag{12.1}$$

where $E_\lambda$ is the one-dimensional $HN_-$-module corresponding to the homomorphism of $HN_-$ into $K^*$ defined by $h(\tau)n_- \to \tau^\lambda$. The required isomorphism is determined as follows.

Let $A \in \mathrm{Hom}_G(V_\mu, R^\lambda)$. To each vector $\xi \in V_\mu$ corresponds an operator $A(\xi) \in R^\lambda$. We consider the local expression $\varphi_{A(\xi)}(g)$ of $A(\xi)$ and set

$$B(\xi) = \varphi_{A(\xi)}(e) \otimes 1 \in Y \otimes E_\lambda. \tag{12.2}$$

Thus we have constructed a mapping $B: V_\mu \to Y \otimes E_\lambda$. Let us show that this mapping commutes with the action of the group $HN_-$. This follows from the following equations which are based on results in section 8:

$$B(T(n_-)\xi) = \varphi_{A(T(n_-)\xi)}(e)\otimes 1 = \varphi_{T(n_-)A(\xi)T(n_-)^{-1}}(e)\otimes 1 = \varphi_{A(\xi)}(n_-)\otimes 1 = \rho(n_-)\varphi_{A(\xi)}(e)\otimes 1 = \beta(n_-)B(\xi);$$

$$B(T(h(\tau))\xi)) = \varphi_{A(T(h(\tau))\xi)}(e)\otimes 1 = \varphi_{T(h(\tau))A(\xi)T(h(\tau))^{-1}}(e)\otimes 1$$

$$= \varphi_{A(\xi)}(h(\tau))\otimes 1 = \tau^{\lambda}\rho(h(\tau))\varphi_{A(\xi)}(e)\otimes 1 = \rho(h(\tau))\varphi_{A(\xi)}(e)\otimes\tau^{\lambda} = \beta(h(\tau))B(\xi).$$

Here $\beta$ denotes the action of $HN_-$ in $Y\otimes E_\lambda$.

Hence $B\in \mathrm{Hom}_{HN_-}(V_\mu, Y\otimes E_\lambda)$. Conversely, with each such B we can associate an $A\in \mathrm{Hom}_G(V_\mu, R^\lambda)$ by the equation

$$\varphi_{A(\xi)}(g)\otimes 1 = B(T(g)\xi). \tag{12.3}$$

Lemmas 8.1 and 8.2 show that the function $\varphi_{A(\xi)}(g)$, defined by (12.3) is, in fact, the local expression for an operator $A(\xi)\in R^\lambda$. It follows from (8.3) that the mapping $\xi\to A(\xi)$ is a homomorphism of G-modules.

To prove that (12.1) is an isomorphism it remains to note that the transition from A to B, given by (12.2), obviously commutes with the action of $\tilde{Z}(\mathfrak{g})$.

Later we identify the algebra $\tilde{Z}(\mathfrak{g})$ with $U(\mathfrak{h})$ by means of the embedding j defined in section 6; we also identify $Y\otimes E_\lambda$ with $U(\mathfrak{h}+\mathfrak{n}_+)$ with the help of the mapping $\alpha\otimes 1$ which sends $X\in U(\mathfrak{h}+\mathfrak{n}_+)$ into $\alpha(x)\otimes 1\in Y\otimes E_\lambda$. By Lemma 9.2 the structure of the $U(\mathfrak{h})$-module appearing on $U(\mathfrak{h}+\mathfrak{n}_+)$ coincides with the natural structure of the right $U(\mathfrak{h})$-module on $U(\mathfrak{h}+\mathfrak{n}_+)$: the element $X\in U(\mathfrak{h})$ defines the transformation $X'\to X'X$ of $U(\mathfrak{h}+\mathfrak{n}_+)$.

The next step is to replace $\mathrm{Hom}_{HN_-}(V_\mu, Y\otimes E_\lambda)$ by a subspace $W_\mu^\lambda\subset U(\mathfrak{h}+\mathfrak{n}_+)$. To achieve this we assign to each $B\in \mathrm{Hom}_{HN_-}(V_\mu, Y\otimes E_\lambda)$ the element $B(\xi_0)\in U(\mathfrak{h}+\mathfrak{n}_+)$, where $\xi_0$ is the vector of highest weight in $V_\mu$. Because $\xi_0$ is cyclic with respect to the subgroup $HN_-$, the passage from B to $B(\xi_0)$ is one-to-one. We next clarify the structure of the space $W_\mu^\lambda$ which is the image of $\mathrm{Hom}_{HN_-}(V_\mu, Y\otimes E_\lambda)$ under the mapping $B\to B(\xi_0)$. First of all, since B commutes with the action of H, it follows that $W_\mu^\lambda\in U(\mathfrak{h}+\mathfrak{n}_+)^{\mu-\lambda}$.

Let $X\to\check{X}$ be the antiautomorphism of the algebra $U(\mathfrak{n}_-)$ defined by: $\check{X}=-X$ for $X\in\mathfrak{n}_-$.

It follows from the fact that B commutes with the action of $N_-$, from (11.4) and from the first of the equations at (3.2) that

$$B(T_\mu(X)\xi_0) = B(\xi_0)\check{X} \text{ for } X\in U(\mathfrak{n}_-). \tag{12.4}$$

Equation (12.4) allows us to recover B, knowing $B(\xi_0)$, because the vector $\xi_0\in V_\mu$ is cyclic with respect to $U(\mathfrak{n}_-)$. However, to recover B correctly it is necessary (and obviously sufficient) that the element $X=B(\xi_0)$ should satisfy the condition:

$$\alpha(X\check{X}_1)=0, \quad \text{if} \quad T_\mu(X_1)\xi_0=0. \tag{12.5}$$

Let $I_\mu\subset U(\mathfrak{n}_-)$ be the left ideal consisting of those $X_1$ for which $T_\mu(X_1)\xi_0=0$. We are going to show that, in terms of the pairing introduced in section 11, (12.5) takes the form

$$\{X,\check{X}_1\}=0 \text{ for all } X_1\in I_\mu. \tag{12.6}$$

In fact, if X satisfies (12.5) and $X_1\in I_\mu$, then $\alpha(X\check{X}_1)=0$ and so $\{X,\check{X}_1\}=0$.

Conversely, if (12.6) holds, then for any $X_2\in U(\mathfrak{n}_-)$ we have

$$\{X\check{X}_1, X_2\} = \{X,\check{X}_1X_2\} = \{X,(\check{X}_2X_1)^\vee\} = 0,$$

because $I_\mu$ is a left ideal. By Lemma 11.1 it follows from here that $X\check{X}_1\in\mathfrak{n}_-U(\mathfrak{g})$ and so $\alpha(X\check{X}_1)=0$.

We have proved that $W_\mu^\lambda$ consists of all the elements of $U(\mathfrak{h}+\mathfrak{n}_+)^{\mu-\lambda}$ that are orthogonal to $(I_\mu)^{\lambda-\mu}$. Next let $X_1,\ldots,X_m$ be a basis in $U(\mathfrak{n}_+)^{\mu-\lambda}$, $X_1',\ldots,X_l'$ be a basis in $(I_\mu)^{\lambda-\mu}$ and let $X=\sum_{i=1}^m C_iX_i$, where $C_i\in U(\mathfrak{h})$. Then the condition for X to be orthogonal to the space $(I_\mu)^{\lambda-\mu}$ becomes

$$\sum_{i=1}^m C_ia_{ij}=0, \quad 1\leqslant j\leqslant l, \text{ where } a_{ij}=\{X_i,X_j'\}. \tag{12.7}$$

It is clear from here that $W_\mu^\lambda$ is a $U(\mathfrak{h})$-module of rank $m-l$. But we easily deduce from the definition of $I_\mu$ that $m-l = m_\lambda(\mu)$, and Theorem 2 is proved.

COROLLARY 12.1. Let L be the quotient field of the ring $\widetilde{Z}(\mathfrak{g})$. Then

$$\widetilde{U}(\mathfrak{g}) \underset{\widetilde{Z}(\mathfrak{g})}{\otimes} L = R^0 \underset{\widetilde{Z}(\mathfrak{g})}{\otimes} L.$$

In fact, both sides of this equation are vector spaces over L and the group G acts in these spaces. Moreover, the second space contains the first. Hence, it is sufficient to prove that, for any $\mu \in P$, the representation $T_\mu$ occurs in the decomposition of both spaces with the same multiplicity, that is, that

$$\dim_L \mathrm{Hom}_G(V_\mu \underset{K}{\otimes} L, \widetilde{U}(\mathfrak{g}) \underset{\widetilde{Z}(\mathfrak{g})}{\otimes} L) = \dim_L \mathrm{Hom}_G(V_\mu \underset{K}{\otimes} L, R^0 \underset{\widetilde{Z}(\mathfrak{g})}{\otimes} L).$$

By Theorem 2 the right-hand side is equal to the multiplicity of the zero weight in $T_\mu$. The left-hand side has the same magnitude in view of the results of Konstant [3].

COROLLARY 12.2. The quotient field $D(\mathfrak{g})$ of the algebra $U(\mathfrak{g})$ coincides with the quotient field of the ring $R^0$.

Remark 12.1. It would be interesting to explain, in greater detail, the structure of $W_\mu^\lambda$ as a $U(\mathfrak{h})$-module (for example, to determine the number of generators and their degrees). It is obvious that an answer depends on the structure of the matrix $\{a_{ij}\}$ in (12.7).

## 13. THE ISOMORPHISM OF $\widetilde{D}(\mathfrak{g})$ AND $D_{n,k}(K)$

Let D denote the quotient field of the ring R.

LEMMA 13.1. The field D is isomorphic to the standard field $D_{n+k,0}(K)$.

Proof. Let us recall that by the standard field $D_{n,k}(K)$ we mean [1] the field generated, over (K) by the generators $p_1, \ldots, p_n, q_1, \ldots, q_n, z_1, \ldots, z_k$, subject to the restriction $p_iq_j - q_jp_i = \delta_{ij}$ and that all simpler commutators are zero.

We show that D coincides with the quotient field of the ring M of all differential operators with polynomial coefficients in the n + k variables $(\tau, t)$. It is obvious that M is isomorphic to the standard ring $R_{n+k,0}(K)$ which is generated, over K, by the same generators and relations as the field $D_{n+k,0}(K)$. (In fact, for the standard generators we can take the operators $\partial/\partial\tau_i$, $\partial/\partial t_\alpha$, $\tau_i$, $t_\alpha$, where $1 \le i \le k$, $\alpha > 0$.)

By Lemma 5.1 the field D contains the field $\Pi$ of rational functions of the variables $(\tau, t)$. Since each operator from R can be extended to a differential operator over $\Pi$, R is contained in $\Pi \underset{F}{\otimes} M$. On the other hand, we can show that M is contained in $\Pi \underset{F}{\otimes} R$.

In fact, each operator A from M sends a generator of the ring F into polynomials in $(\tau, t)$, that is, into elements of $\Pi$. Because $\Pi$ is the quotient field of the ring F, there is a function $f \in F$ such that $fA$ sends a generator of F again into elements of F. Then $fA \in R$ and so $A \in \Pi \underset{F}{\otimes} R$.

It obviously follows from the inclusions $R \subset \Pi \underset{F}{\otimes} M$, $M \subset \Pi \underset{F}{\otimes} R$ that R and M have the same quotient field and so the lemma is proved.

Next let us consider the action of the group H in D that is induced by the actions of left translations in R. Let $D^0$ be the subfield consisting of all H-invariant elements.

LEMMA 13.2. The field $D^0$ is isomorphic to the standard field $D_{n,k}(K)$.

Proof. We show that $D^0$ is generated by the operators $\partial/\partial t_\alpha$, $t_\alpha$, $\tau_i\,\partial/\partial\tau_i$ which, obviously, we can take as the standard generators of $D_{n,k}(K)$.

It follows directly from the definition of a left translation that the element $h(\tau_1)$ sends the point with the coordinates $(\tau, t)$ into the point with the coordinates $(\tau_1\tau, t)$. Hence, the operators $\partial/\partial t_\alpha$, $t_\alpha$ and $\tau_i\,\partial/\partial\tau_i$ lie in $D^0$. Let $D'$ denote the field generated by them. Let A be an arbitrary element of $D^0$, we can write it in the form $PQ^{-1}$, where P and Q are polynomials in $\tau_i$ with coefficients from $D'$. Thus, we obtain that $P = AQ$ or, in more detail,

$$\sum_\lambda a_\lambda \tau^\lambda = A \sum_\lambda b_\lambda \tau^\lambda,$$

where $a_\lambda$ and $b_\lambda$ lie in D'. By operating on this equation with $h(\tau_1) \in H$ we obtain that

$$\sum_\lambda a_\lambda \tau_1^\lambda \tau^\lambda = A \sum_\lambda b_\lambda \tau_1^\lambda \tau^\lambda.$$

Because this equation holds for all $\tau_1$, it follows from here that $a_\lambda = Ab_\lambda$ for each $\lambda$. At least one of the coefficients $b_\lambda$ is nonzero. Hence $A = a_\lambda b_\lambda^{-1} \in D'$. Consequently, $D^0 = D'$ and the lemma is proved.

LEMMA 13.3. The field $D^0$ coincides with the quotient field of the ring $R^0$.

Proof. We can show by precisely the same arguments as were used in proof of the preceding lemma that each element of $D^0$ can be expressed as $PQ^{-1}$, where P and Q belong to the same space $R^\lambda$. Let S be a nonzero element of the space $R^{-\lambda}$; the existence of such an element is guaranteed by Theorem 2 because $-\lambda$ is the weight of some irreducible representation $T_\mu$ (for example, for $\mu$ we can take the weight $w(-\lambda)$, where w is the element of the Weyl group that sends $-\lambda$ into the dominant Weyl chamber). We put $P_1 = PS$, $Q_1 = QS$. Then $P_1$ and $Q_1$ belong to $R^0$ and $PQ^{-1} = P_1Q_1^{-1}$. Thus, the lemma is proved.

Collecting the results of Lemmas 13.2, 13.3, and of Corollary 12.2 we obtain the following

THEOREM 3. The quotient field $\tilde{D}(\mathfrak{g})$ is generated by the operators $\partial/\partial t_\alpha$, $t_\alpha$, $\tau_i\, \partial/\partial\tau_i$, $1 \le i \le k$, $\alpha > 0$ and, consequently, is isomorphic to the standard field $D_{n,k}(K)$.

## LITERATURE CITED


1. I. M. Gel'fand and A. A. Kirillov, "On fields connected with enveloping Lie algebras," Dokl. Akad. Nauk SSSR, 167, No. 3, 503-506 (1966).
2. I. M. Gel'fand and M. I. Graev, "The Fourier transformation of rapidly decreasing functions on complex semisimple groups," Dokl. Akad. Nauk SSSR, 131, No. 3, 496-499 (1960).
3. B. Konstant, "Lie group representations on polynomial rings," Amer. J. Math., 85, No. 3, 327-404 (1963).
4. I. M. Gel'fand and A. A. Kirillov, "On the structure of the quotient field of the enveloping algebra of a semisimple Lie algebra," Dokl. Akad. Nauk SSSR, 180, No. 4, 775-777 (1961).
5. A. Borel and J. Tits, "Groupes reductifs," Publs. Math. Inst. des Hautes-Etudes Sci., 27, 55-151 (1965).
6. J. Tits, "Classification of algebraic semisimple groups," Symposium, Boulder, Colorado (1965).
7. The Theory of Lie Algebras. The Topology of Lie Groups [Russian translation], IL, Moscow (1962).
8. A. Grothendieck, "Eléments de géometrie algébrique. IV (Troisiéme partie)," Publs. Math. Inst. des Hautes-Etudes Sci., 28, 1-248 (1966).
9. N. Jacobson, Lie Algebras, Interscience, New York (1962).
10. Harish-Chandra, "On some applications of the universal enveloping algebra of a semisimple Lie algebra," Trans. Amer. Math. Soc., 70, 28-96 (1951).
11. C. Chevalley, "Invariants of finite groups generated by reflections," Amer. J. Math., 77, No. 4, 778-782 (1955).
12. I. M. Gel'fand and V. A. Ponomarev, "The category of Harish-Chandra modules over the Lie algebra of the Lorentz group," Dokl. Akad. Nauk SSSR, 176, No. 2, 243-246 (1967). [For more detail see: Uspekhi Mat. Nauk, 23, No. 2, 3-60 (1968)].
13. I. M. Gel'fand and M. I. Graev, "The geometry of homogeneous spaces, the representation of groups in homogeneous spaces, and related problems of integral geometry," Trudy Mosk. Matem. Obshch., 8, 321-390 (1959).
14. I. M. Gel'fand, M. I. Graev, and I. I. Pyatetski-Shapiro, Generalized Functions. Vol. 6. The Theory of Representations and Automorphic Functions [in Russian], "Nauka," Moscow (1966).

# Part VII

# Finite-dimensional representations

# 1.

(with M. L. Tsetlin)

# Finite-dimensional representations of the group of unimodular matrices

Dokl. Akad. Nauk SSSR **71** (1950) 825–828. Zbl. **37**:153
(Submitted by Academician A.N. Kolmogorov, February 16, 1950)

1. Let $K_n$ be a Lie algebra of all matrices of order $n$, i.e. the set of all $n \times n$ matrices $e, f, \dots$ for which the operations of addition and multiplication by scalars are defined in the usual way and the commutator is defined by the formula $[f, g] = fg - gf$.

We say that a finite-dimensional representation of the Lie algebra $K_n$ is given if for each $f \in K_n$ there is a matrix $F$ of order $N$ defined in such a way that $f \to F$ and $g \to G$ implies that $\lambda f + \mu g \to \lambda F + \mu G$ and $[f, g] \to [F, G]$. The aim of this paper is to find all irreducible representations of $K_n$. A system of invariants which uniquely define a representation has been given by Cartan. He has also shown that it is feasible to construct for each such system of invariants (so-called highest weight of the representation) the representation which corresponds to that system (i.e., to the given highest weight). However, one could not say that the problem of describing each representation effectively had been solved[1].

In this paper we give explicit formulas which effectively define the representation.

Denote by $e_{ik}$ the matrix of order $n$ which has 1 at the intersection of the $i$-th row and $k$-th column and zeros in all other places. Denote by $E_{ik}$ the matrix of order $N$ which, under our representation, corresponds to the element $e_{ik} \in K_n$. Since each matrix in $K_n$ is a linear combination of $e_{ik}$, the set of $E_{ik}$ uniquely defines the representation. Thus the task of finding all irreducible representations of $K_n$ can be reformulated in a very simple way: *find $n^2$ matrices $E_{ik}$ $(i, k = 1, \dots n)$ of order $N$ satisfying the following conditions:*

$$[E_{ik}, E_{kl}] = E_{il} \quad (i \neq l), \qquad [E_{ik}, E_{ki}] = E_{ii} - E_{kk},$$
$$[E_{i_1 k_1}, E_{i_2 k_2}] = 0 \quad \text{for } k_1 \neq i_2 \text{ and } i_1 \neq k_2.$$

The system of matrices $E_{ik}$ is required to be irreducible (which means that there is no subspace invariant for all $E_{ik}$).

The effective enumeration of all representations is now achieved by actual explicit description of linear transformations (matrices) $E_{ik}$.

2. Let us now proceed to the enumeration of all representations. For the sake of clarity we first consider the formulas for $n = 2$ and $n = 3$.

---

[1] To see that the existing construction of irreducible representations is far from perfect, it is sufficient to say that the dimensions of irreducible representations have not been computed from their constructions but have been obtained much later from completely different considerations.

The case $n=2$ is trivial and has been well known for a long time. The answer here is: each representation is defined by two integers $m_1$ and $m_2$ $(m_1 \geqq m_2)$. The operators $E_{ik}$ $(i,k=1,2)$ act in the space $R$ and one can choose such a basis consisting of vectors $\xi_q$ (where $q$ is an integer such that $m_1 \geqq q \geqq m_2$) in $R$ in which the representation is given by the formulas

$$\begin{aligned} &E_{11}\xi_q = q\xi_q, \quad E_{22}\xi_q = (m_1+m_2-q)\xi_q, \\ &E_{12}\xi_q = \sqrt{(m_1-q)(q-m_2+1)}\,\xi_{q+1}, \\ &E_{21}\xi_q = \sqrt{(m_1-q+1)(q-m_2)}\,\xi_{q-1}. \end{aligned} \tag{1}$$

Let us now write the formulas for $n=3$. It turns out that in that case one can also write formulas similar to formulas (1) in the case $n=2$. Each irreducible representation for $n=3$ is defined by three integers $m_1, m_2, m_3$ $(m_1 \geqq m_2 \geqq m_3)$. Vectors of the basis in the representation space $R$ (where the operators $E_{ik}$ act) are now conveniently numbered by triples $p_1, p_2, q$ (and not by the sole integer $q$ as in (1)). We shall denote vectors of that basis in the following way: $\begin{pmatrix} p_1 & & p_2 \\ & q & \end{pmatrix}$. Here $p_1, p_2, q$ are any integers which satisfy the inequalities

$$m_3 \leqq p_2 \leqq m_2 \leqq p_1 \leqq m_1$$

and

$$p_2 \leqq q \leqq p_1.$$

The representation is now defined by the following formulas:

$$E_{11}\begin{pmatrix} p_1 & & p_2 \\ & q & \end{pmatrix} = q\begin{pmatrix} p_1 & & p_2 \\ & q & \end{pmatrix},$$

$$E_{22}\begin{pmatrix} p_1 & & p_2 \\ & q & \end{pmatrix} = (p_1+p_2-q)\begin{pmatrix} p_1 & & p_2 \\ & q & \end{pmatrix},$$

$$E_{12}\begin{pmatrix} p_1 & & p_2 \\ & q & \end{pmatrix} = \sqrt{(p_1-q)(q-p_2+1)}\begin{pmatrix} p_1 & & p_2 \\ & q+1 & \end{pmatrix},$$

$$E_{21}\begin{pmatrix} p_1 & & p_2 \\ & q & \end{pmatrix} = \sqrt{(p_1-q+1)(q-p_2)}\begin{pmatrix} p_1 & & p_2 \\ & q-1 & \end{pmatrix}, \tag{2}$$

$$\begin{aligned} E_{23}\begin{pmatrix} p_1 & & p_2 \\ & q & \end{pmatrix} = &\sqrt{\frac{(m_1-p_1)(m_2-p_1-1)(m_3-p_1-2)(p_1-q+1)}{(p_1-p_2+2)(p_1-p_2+1)}}\begin{pmatrix} p_1+1 & & p_2 \\ & q & \end{pmatrix} \\ &+\sqrt{\frac{(m_1-p_2+1)(m_2-p_2)(m_3-p_2-1)(p_2-q)}{(p_1-p_2+1)(p_1-p_2)}}\begin{pmatrix} p_1 & & p_2+1 \\ & q & \end{pmatrix}, \end{aligned}$$

$$\begin{aligned} E_{32}\begin{pmatrix} p_1 & & p_2 \\ & q & \end{pmatrix} = &\sqrt{\frac{(m_1-p_1+1)(m_2-p_1)(m_3-p_1-1)(p_1-q)}{(p_1-p_2+1)(p_1-p_2)}}\begin{pmatrix} p_1-1 & & p_2 \\ & q & \end{pmatrix} \\ &+\sqrt{\frac{(m_1-p_2+2)(m_2-p_2+1)(m_3-p_2)(p_2-q)}{(p_1-p_2+2)(p_1-p_2+1)}}\begin{pmatrix} p_1 & & p_2-1 \\ & q & \end{pmatrix}, \end{aligned}$$

$$E_{33}\begin{pmatrix} p_1 & & p_2 \\ & q & \end{pmatrix} = (m_1+m_2+m_3-p_1-p_2)\begin{pmatrix} p_1 & & p_2 \\ & q & \end{pmatrix}.$$

We see that the formulas for $E_{11}, E_{22}, E_{21}, E_{12}$ coincide with the formulas (1) for $n=2$. Due to lack of space, we do not write formulas for $E_{13}$ and $E_{31}$. They can be obtained from the relations $E_{13}=[E_{12}, E_{23}]$, $E_{31}=[E_{32}, E_{21}]$.

3. Now we write explicit formulas for irreducible representations for arbitrary $n$. We have, therefore, to define linear transformations $E_{ik}$ $(i, k=1,\dots,n)$ which act in some linear space $R$. Each representation is defined by $n$ integers $m_1, m_2, \dots, m_n$ $(m_1 \geqq m_2 \geqq \dots \geqq m_n)$. These numbers coincide with the highest weight as defined by Cartan.

Each vector of the basis in $R$ is defined by the set $\alpha$ of numbers $m_{pq}$, $p \leqq q$, $q=1,\dots,n-1$ arranged in the following pattern:

$$\alpha=\begin{pmatrix} m_{1,n-1} & m_{2,n-1} & \dots & m_{n-1,n-1} \\ \multicolumn{4}{c}{\dots\dots\dots\dots\dots\dots} \\ & m_{13} \quad m_{23} \quad m_{33} & & \\ & m_{12} \quad m_{22} & & \\ & m_{11} & & \end{pmatrix}. \tag{3}$$

The numbers $m_{pq}$ are arbitrary integers which satisfy the inequalities $m_{p,q+1} \geqq m_{pq} \geqq m_{p+1,q+1}$. The numbers $m_1, \dots, m_i, \dots, m_n$ which define the representation are denoted here by $m_{in}$. The vector of the basis defined by the pattern $\alpha$ shall be written simply as $(\alpha)$.

Let $m_{pq}-p=l_{pq}$ and

$$a^j_{k-1,k}=\left[(-1)^{k-1}\frac{\prod\limits_{i=1}^{k}(l_{i,k}-l_{j,k-1})\prod\limits_{i=1}^{k-2}(l_{i,k-2}-l_{j,k-1}-1)}{\prod\limits_{i\neq j}(l_{i,k-1}-l_{j,k-1})(l_{i,k-1}-l_{j,k-1}-1)}\right]^{1/2} \tag{4}$$

Denote now by $\alpha^i_{k-1,k}$ the pattern obtained from the pattern $\alpha$ by replacing $m_{i,k-1}$ with $m_{i,k-1}+1$. Then the operator $E_{k-1,k}$ acts on the vector $(\alpha)$ in the following way

$$\begin{aligned} E_{k-1,k}(\alpha) &= \sum_j a^j_{k-1,k}(\alpha^j_{k-1,k}), \\ E_{kk}(\alpha) &= \left(\sum_{i=1}^{k} m_{ik}-\sum_{i=1}^{k-1} m_{i,k-1}\right)(\alpha), \\ E_{k,k-1}(\alpha) &= \sum_j b^j_{k,k-1}(\bar{\alpha}^j_{k,k-1}). \end{aligned} \tag{5}$$

Here $\bar{\alpha}^j_{k,k-1}$ denotes the pattern obtained from $\alpha$ by replacing $m_{j,k-1}$ with $m_{j,k-1}-1$ and $b^j_{k,k-1}$ is defined by the formula

$$b^j_{k,k-1}=\left[(-1)^{k-1}\frac{\prod\limits_{i=1}^{k}(l_{i,k}-l_{j,k-1}+1)\prod\limits_{i=1}^{k-2}(l_{i,k-2}-l_{j,k-1})}{\prod\limits_{i\neq j}(l_{i,k-1}-l_{j,k-1}+1)(l_{i,k-1}-l_{j,k-1})}\right]^{1/2}. \tag{4'}$$

Formulas (5) completely define the representation since any operator $E_{pq}$ can be obtained by commuting operators of the form (5).

4. We shall, nevertheless, write out the formulas for $E_{pq}$. Let $p < q$. Denote by $\alpha_{pq}^{i_p \dots i_{q-1}}$ the pattern obtained from $\alpha$ by increasing each of the indices $m_{i_p p}, \dots, m_{i_{q-1} q-1}$ by 1 while leaving all other indices unchanged.

Then

$$E_{pq}(\alpha) = \sum_{(i_p, \dots, i_{q-1})} a_{pq}^{i_p, \dots, i_{q-1}} (\alpha_{pq}^{i_p, \dots, i_{q-1}}),$$

where

$$a_{pq}^{i_p, \dots, i_{q-1}} = \pm \frac{\prod_{k=p}^{q-1} a_{k,k+1}^{i_k}}{\prod_{k=p}^{q-2} [(l_{i_k k} - l_{i_{k+1} k+1})(l_{i_k k} - l_{i_{k+1} k+1} - 1)]^{1/2}}.$$

The sign is determined by the number of inversions in the sequence $i_p, \dots, i_{q-1}$. The action of $E_{pq}$ for $p > q$ is defined in a similar way. In applications of representation theory it is usually required only to know those $(\alpha)$ which can be obtained from that given under the action of $E_{ik}$. The coefficients $a_{pq}^{j \dots}$ are needed much more rarely.

Note that if the basis $(\alpha)$ is assumed to be orthonormal, then $(E_{pq})^* = E_{qp}$.

Received February 13, 1950

## 2.

(with M. L. Tsetlin)

# Finite-dimensional representations of the group of orthogonal matrices

Dokl. Akad. Naūk SSSR **71** (1950) 1017–1020. Zbl. **37**:153
(Submitted by Academician I.G. Petrovsky, February 23, 1950)

1. The aim of this paper is to give explicit formulas which effectively define irreducible representations of the group of orthogonal matrices of order $n$ similar to those which have been given in [1] for the group of unimodular matrices. The representations will be described infinitesimally.

Consider the Lie algebra $K_n$ of skew-symmetric matrices of order $n$, i.e. the set of skew-symmetric matrices of order $n$ with operations of addition and multiplication by scalars and the commutator $[a, b] = ab - ba$. To find a representation of $K_n$ means to specify for each $a \in K_n$ another matrix $A$ of order $N$ in such a way that $a \to A$ and $b \to B$ implies $\alpha a + \beta b \to \alpha A + \beta B$ and $[a, b] \to [A, B]$. In this paper we give effective formulas for irreducible representations[1].

Denote by $a_{ik}$ the skew-symmetric matrix which has 1 in the $(i, k)$-th position, $-1$ in the $(k, i)$-th position and zeros in all other places. Under our representation the matrix $a_{ik}$ goes into some matrix $I_{ik}$ of order $N$. Each skew-symmetric matrix is a linear combination of $a_{ik}$ and, therefore, to define a representation it is sufficient to know the matrices $I_{ik}$.

2. For $n = 3$ (i.e. for the group of orthogonal matrices of order 3) such formulas have been well known for a long time. In that case each representation is defined by a single integer or half-integer $p$. The matrices $I_{ik}$ are of order $(2p + 1)$. We shall denote the vectors of the basis in the representation space by $\xi_q$ $(q = -p, -p + 1, \ldots, p)$. The matrices $I_{ik}$ in that basis are defined by the formulas

$$\left.\begin{aligned} &I_{21}\xi_q = iq\xi_q, \\ &I_{32}\xi_q = \sqrt{(p-q)(p+q+1)}\,\xi_{q+1} - \sqrt{(p-q+1)(p+q)}\,\xi_{q-1}, \\ &I_{31}\xi_q = -i\sqrt{(p-q)(p+q+1)}\,\xi_{q+1} - i\sqrt{(p-q+1)(p+q)}\,\xi_{q-1}. \end{aligned}\right\} \qquad (1)$$

We shall now give similar formulas for arbitrary $n$.

Each representation of a group of orthogonal matrices of even order $(n = 2k + 2)$ is defined by $k + 1$ numbers $m_1, \ldots, m_{k+1}$ $(m_1 \geqq m_2 \geqq \ldots \geqq m_{k+1})$, and representations of a group of odd order $(n = 2k + 1)$ are defined by $k$ numbers $m_1, \ldots, m_k$ $(m_1 \geqq m_2 \geqq \ldots \geqq m_k)$. The numbers $m_i$ are either all integers or all half-integers.

---

1 A representation is called irreducible if there is no subspace invariant for $A$ corresponding to all $a \in K_n$.

We shall denote by $\alpha$ the following pattern for $n = 2k + 2$:

$$\alpha = \begin{pmatrix} m_{2k,1} & \cdots\cdots\cdots & m_{2k,k} \\ m_{2k-1,1} & \cdots\cdots\cdots & m_{2k-1,k} \\ & \cdots\cdots\cdots & \\ & \cdots\cdots\cdots & \\ & m_{41} \quad m_{42} & \\ & m_{31} \quad m_{32} & \\ & m_{21} & \\ & m_{11} & \end{pmatrix} \tag{2}$$

and for $n = 2k + 1$ the pattern

$$\alpha = \begin{pmatrix} m_{2k-1,1} & \cdots\cdots\cdots & m_{2k-1,k} \\ & \cdots\cdots\cdots & \\ & m_{41} \quad m_{42} & \\ & m_{31} \quad m_{32} & \\ & m_{21} & \\ & m_{11} & \end{pmatrix}. \tag{2'}$$

Here the numbers $m_{ij}$ are either all integers or all half-integers satisfying the following inequalities:

$$\left.\begin{aligned} m_{2p+1,i+1} \leqq m_{2p,i} \leqq m_{2p+1,i} \quad & (i = 1, \ldots, p), \\ m_{2p,i+1} \leqq m_{2p-1,i} \leqq m_{2p,i} \quad & (i = 1, \ldots, p-1), \\ -m_{2p,p} \leqq m_{2p-1,p} \leqq m_{2p,p}. & \end{aligned}\right\} \tag{3}$$

The numbers which define the representation, that is $m_1, \ldots, m_{k+1}$ for $n = 2k + 2$ and $m_1, \ldots, m_k$ for $n = 2k + 1$, are here denoted by $m_{2k+1,1}, \ldots, m_{2k+1,k+1}$ and $m_{2k,1}, \ldots, m_{2k,k}$ respectively.

To each pattern $\alpha$ allowed by the above conditions we associate a vector of the basis in the representation space. That vector will be denoted $\xi(\alpha)$. Denoted now by $\xi^+(\alpha_r^j)$ the vector which corresponds to the pattern obtained from $\alpha$ by replacing $m_{rj}$ with $m_{rj} + 1$ and by $\xi^-(\alpha_r^j)$ the vector which correspond to the pattern obtained from $\alpha$ by replacing $m_{rj}$ with $m_{rj} - 1$. Then the action of the operators $I_{2p+1,2p}$ and $I_{2p+2,2p+1}$ on the vector $\xi(\alpha)$ is defined by the formulas[2]

$$\begin{aligned} I_{2p+1,2p}\xi(\alpha) &= \sum_{j=1}^{p} A(m_{2p-1,j})\xi^+(\alpha_{2p-1}^j) - \sum_{j=1}^{p} A(m_{2p-1,j} - 1)\xi^-(\alpha_{2p-1}^j), \\ I_{2p+2,2p+1}\xi(\alpha) &= \sum_{j=1}^{p} B(m_{2p,j})\xi^+(\alpha_{2p}^j) - \sum_{j=1}^{p} B(m_{2p,j} - 1)\xi^-(\alpha_{2p}^j) + iC_{2p}\xi(\alpha). \end{aligned} \tag{4}$$

To define the coefficients $A, B$ and $C$ appearing in the above formulas it is more convenient to introduce the numbers $l_{ik}$ in the following way:

[2] The coefficients $A(m_{2p-1,j})$, $B(m_{2p,j})$, $A(m_{2p-1,j} - 1)$ should more appropriately be written as $A_{2p-1,j}(m_{2p-1,j})$, $B_{2p,j}(m_{2p,j})$, $A_{2p-1,j}(m_{2p-1,j} - 1)$.

$$\left.\begin{array}{ll} m_{2p-1,p} = l_{2p-1,p}, & m_{2p,p}+1 = l_{2p,p}, \\ m_{2p-1,p-1}+1 = l_{2p-1,p-1}, & m_{2p,p-1}+2 = l_{2p,p-1}, \\ \dots\dots\dots\dots & \dots\dots\dots\dots \\ m_{2p-1,1}+p-1 = l_{2p-1,1}, & m_{2p,1}+p = l_{2p,1}. \end{array}\right\} \quad (5)$$

Then the coefficients $A, B$ and $C$ are computed as follows:

$$\left.\begin{array}{l} A(l_{2p-1,j}) = \left| \dfrac{\prod\limits_{r=1}^{p-1}(l_{2p-1,r}-l_{2p-1,j}-1)(l_{2p-2,r}+l_{2p-1,j}) \prod\limits_{r=1}^{p}(l_{2p,r}-l_{2p-1,j}-1)(l_{2p,r}+l_{2p-1,j})}{\prod\limits_{r\neq j}(l_{2p-1,r}^2-l_{2p-1,j}^2)[l_{2p-1,r}^2-(l_{2p-1,j}+1)^2]} \right|^{1/2}, \\ B(l_{2p,j}) = \left| \dfrac{\prod\limits_{r=1}^{p}(l_{2p-1,r}^2-l_{2p,j}^2)\prod\limits_{r=1}^{p+1}(l_{2p+1,r}^2-l_{2p,j}^2)}{l_{2p,j}^2(4l_{2p,j}^2-1)\prod\limits_{r\neq j}(l_{2p,r}^2-l_{2p,j}^2)[(l_{2p,r}-1)^2-l_{2p,j}^2]} \right|^{1/2}, \end{array}\right\} \quad (6)$$

$$C_{2p} = \frac{\prod\limits_{r=1}^{p} l_{2p-1,r}\prod\limits_{r=1}^{p+1} l_{2p+1,r}}{\prod\limits_{r=1}^{p} l_{2p,r}(l_{2p,r}-1)}.$$

The formulas (4) completely define the representation since any other operator $I_{pq}$ can be obtained by commuting operators of the form (4).

We also note that if the basis $\xi(\alpha)$ is assumed to be orthonormal then $(I_{pq})^* = -I_{pq}$.

3. We now write out separately formulas for the orthogonal groups of orders 4 and 5.

For the orthogonal group of order 4 it is sufficient to specify the action of the operators $I_{21}, I_{32}, I_{43}$:

$$I_{21}\xi\binom{p}{q} = iq\xi\binom{p}{q},$$

$$\left.\begin{array}{l} I_{32}\xi\binom{p}{q} = \sqrt{(p-q)(p+q+1)}\,\xi\binom{p}{q+1} - \sqrt{(p-q+1)(p+q)}\,\xi\binom{p}{q-1}, \\ I_{43}\xi\binom{p}{q} = \sqrt{\dfrac{(p+q+1)(p-q+1)(m_1-p)(m_1+p+2)(p-m_2+1)(p+m_2+1)}{(2p+1)(2p+3)(p+1)^2}} \\ \qquad \cdot\xi\binom{p+1}{q} + iq\dfrac{(m_1+1)m_2}{p(p+1)}\xi\binom{p}{q} \\ \qquad - \sqrt{\dfrac{(p+q)(p-q)(m_1-p+1)(m_1+p+1)(p-m_2)(p+m_2)}{(2p+1)(2p-1)p^2}}\,\xi\binom{p-1}{q}. \end{array}\right\}$$

(7)

Here $m_1 \geqq p \geqq m_2, p \geqq q \geqq -p$; $m_1, m_2, p, q$ are either all integers or all half-integers and $m_1, m_2$ are the numbers which define the representation.

Representations of the orthogonal group of order 5 are completely defined by the action of the operators $I_{21}, I_{32}, I_{43}, I_{54}$.

These operators act on the basis vector $\xi\begin{pmatrix} m_1 & & m_2 \\ & p & \\ & q & \end{pmatrix}$ is the following way:

$$I_{21}\xi\begin{pmatrix} m_1 & & m_2 \\ & p & \\ & q & \end{pmatrix} = iq\xi\begin{pmatrix} m_1 & & m_2 \\ & p & \\ & q & \end{pmatrix},$$

$$I_{32}\xi\begin{pmatrix} m_1 & & m_2 \\ & p & \\ & q & \end{pmatrix} = \sqrt{(p-q)(p+q+1)}\xi\begin{pmatrix} m_1 & & m_2 \\ & p & \\ & q+1 & \end{pmatrix} - \sqrt{(p-q+1)(p+q)}\xi\begin{pmatrix} m_1 & & m_2 \\ & p & \\ & q-1 & \end{pmatrix},$$

$$I_{43}\xi\begin{pmatrix} m_1 & & m_2 \\ & p & \\ & q & \end{pmatrix}$$

$$= \sqrt{\frac{(p+q+1)(p-q+1)(m_1-p)(m_1+p+2)(p-m_2+1)(p+m_2+1)}{(2p+1)(p+1)^2(2p+3)}}\xi\begin{pmatrix} m_1 & & m_2 \\ & p+1 & \\ & q & \end{pmatrix}$$

$$+ iq\frac{(m_1+1)m_2}{p(p+1)}\xi\begin{pmatrix} m_1 & & m_2 \\ & p & \\ & q & \end{pmatrix}$$

$$- \sqrt{\frac{(p+q)(p-q)(m_1-p+1)(m_1+p+1)(p-m_2)(p+m_2)}{p^2(2p-1)(2p+1)}}\xi\begin{pmatrix} m_1 & & m_2 \\ & p-1 & \\ & q & \end{pmatrix}, \tag{8}$$

$$I_{54}\xi\begin{pmatrix} m_1 & & m_2 \\ & p & \\ & q & \end{pmatrix}$$

$$= \sqrt{\frac{(m_1-p+1)(m_1+p+2)(n_1-m_1)(n_1+m_1+3)(m_1-n_2+1)(m_1+n_2+2)}{(m_1+m_2+1)(m_1+m_2+2)(m_1-m_2+1)(m_1-m_2+2)}}$$

$$\cdot\xi\begin{pmatrix} m_1+1 & & m_2 \\ & p & \\ & q & \end{pmatrix}$$

$$+ \sqrt{\frac{(p-m_2)(m_2+p+1)(n_2-m_2)(n_2+m_2+1)(n_1-m_2+1)(n_1+m_2+2)}{(m_1+m_2+1)(m_1+m_2+2)(m_1-m_2)(m_1-m_2+1)}}$$

$$\cdot\xi\begin{pmatrix} m_1 & & m_2+1) \\ & p & \\ & q & \end{pmatrix}$$

$$-\sqrt{\frac{(m_1+p+1)(m_1-p)(n_1-m_1+1)(n_1+m_1+2)(m_1-n_2)(m_1+n_2+1)}{(m_1+m_2)(m_1+m_2+1)(m_1-m_2)(m_1-m_2+1)}}$$

$$\cdot\xi\begin{pmatrix} m_1-1 & m_2 \\ & p \\ & q \end{pmatrix}$$

$$-\sqrt{\frac{(p-m_2+1)(m_2+p)(n_2-m_2+1)(n_2+m_2)(n_1-m_2+2)(m_2+n_1+1)}{(m_1+m_2)(m_1+m_2+1)(m_1-m_2+2)(m_1-m_2+1)}}$$

$$\cdot\xi\begin{pmatrix} m_1 & m_2-1 \\ & p \\ & q \end{pmatrix}$$

where $n_1 \geqq m_1 \geqq n_2$, $n_2 \geqq m_2 \geqq -n_2$, $m_1 \geqq p \geqq m_2$. $p \geqq q \geqq -p$; $n_1, n_2, m_1, m_2, p, q$ are either all integers or half-integers and $n_1, n_2$ are the numbers which defines the representation.

Received February 20, 1959

**References**

1. Gelfand, I.M., Tsetlin, M.L.: Dokl. Akad. Naūk (5) **77** (1950).

# 3.
## (with M. I. Graev)

# Finite-dimensional irreducible representations of the unitary and the full linear groups, and related special functions

Izv. Akad. Nauk SSSR, Ser. Mat. **29** (1965) 1329–1356
[Transl., II. Ser., Am. Math. Soc. **64** (1965) 116–146]. MR **34**:1450. Zbl. **139**:307

In this paper an effective description of the operators of irreducible finite-dimensional representations of the group $G_n$ of nonsingular matrices of order $n$ is obtained. Special functions which are related to these representations and which, in a sense, are generalizations of the ordinary beta-functions are introduced and studied. Discrete series of infinite-dimensional irreducible representations of the real forms of the group $G_n$ are also considered.

## §1. Introduction

This paper is devoted to the effective description of operators of finite-dimensional irreducible representations of the group $G_n = GL(n)$ of nonsingular complex matrices of the $n$th order. The first essential step in this direction was taken in [1] where effective formulas for the infinitesimal operators of the irreducible representations were obtained. One can regard the present paper, in a sense, as a continuation of [1].

When speaking of irreducible representations of the group $G_n$, we have in mind only the complex-analytic representations of this group. It is well known that any finite-dimensional irreducible representation of the group $G_n$ appears as the tensor product $T_1 \otimes \bar{T}_2$ where $T_1$ and $T_2$ are complex-analytic irreducible representations and $\bar{T}_2$ is the representation conjugate to $T_2$.

It is known that the finite-dimensional representations of the group $U_n$ of unitary matrices (as also of any other real form of the group $G_n$) are given by the same formulas as the representations of the group $G_n$. However, the study of the representations of the group $G_n$ has one significant advantage: in the group $G_n$ one can find one-parameter subgroups, generating $G_n$, for which the representation operators have the most simple form (see Theorem 1). In order to obtain formulas for the representation operators of the group $U_n$, it is sufficient to factor the elements of the group $U_n$ into a product of elements from these one-parameter subgroups.

Let us recall the description of the irreducible complex-analytic representations for the simplest case, namely that of the group $G_2$ of matrices of the second order (this description is well known). Every finite-dimensional irreducible representation of the group $G_2$ is constructed in the following manner. We consider the space $H_{m_1,m_2}$ of polynomials $P(z)$ of degree not exceeding $m_1 - m_2$; here $m_1 \geq m_2$ are a pair of integers specifying the representation. The dimension of this space is evidently $m_1 - m_2 + 1$.

The representation is obtained by associating with every matrix $g = \begin{bmatrix} \alpha & \beta \\ \gamma & \delta \end{bmatrix}$ the following operator $T(g)$ in the space $H_{m_1,m_2}$:

$$T(g)P(z) = P\left(\frac{\alpha z + \gamma}{\beta z + \delta}\right)(\beta z + \delta)^{m_1 - m_2}\Delta^{m_2}, \tag{1}$$

where $\Delta = \det g$. It is easy to see that this operator $T(g)$ takes every polynomial $P(z)$ from $H_{m_1,m_2}$ into a polynomial in $H_{m_1,m_2}$.

We now choose in the space $H_{m_1,m_2}$ of polynomials, a basis consisting of the monomials

$$f_m = z^{m-m_2}, \quad m_2 \leqslant m \leqslant m_1, \tag{2}$$

and consider the matrix elements of the operators $T(g)$ in this basis. Because of (1) we have

$$T(g) f_m = (\alpha z + \gamma)^{m-m_2}(\beta z + \delta)^{m_1 - m}\Delta^{m_2}. \tag{3}$$

Expanding the right-hand side in powers of $z$, we get

$$T(g) f_m = \sum_{m'} (m', m \mid g) f_{m'}. \tag{4}$$

By (3) the coefficients of expansion (4) may be presented in the form of either of the following formulas:

$$(m', m \mid g) = \frac{\Delta^{m_2}}{(m' - m_2)!}\frac{d^{m'-m_2}}{dz^{m'-m_2}}[(\alpha z + \gamma)^{m-m_2}(\beta z + \delta)^{m_1 - m}]\big|_{z=0}; \tag{5}$$

$$(m', m \mid g) = \sum_{\substack{s+t=m'-m_2,\\ 0\leqslant s\leqslant m-m_2\\ 0\leqslant t\leqslant m_1-m}} \frac{(m-m_2)!\,(m_1-m)!}{s!\,t!\,(m-m_2-s)!\,(m_1-m-t)!}\Delta^{m_2}\alpha^s\beta^t\gamma^{m-m_2-s}\delta^{m_1-m-t}. \tag{6}$$

We note that formula (6) is considerably simplified if one of the elements of the matrix $g$ is equal to zero; in this case all the terms in the sum (6) become zero with the possible exception of one term.

It follows from formula (5) that the matrix elements $(m', m|g)$ of the operator $T(g)$ can be expressed in terms of the Jacobi polynomials $G_n(p, q; x)$ i.e. ultimately in terms of the Gauss hypergeometric function. We recall that the Jacobi polynomials are defined by the formula

$$G_n(p, q; x) = \frac{x^{1-q}(1-x)^{q-p}}{q(q+1)\dots(q+n-1)}\frac{d^n}{dx^n}[x^{q+n-1}(1-x)^{p+n-q}]. \tag{7}$$

In order to obtain the required expression, we introduce in (5) the new variable $x = \alpha(\beta z + \delta)/\Delta$ where $\Delta = \det g$. Then

$$\alpha z+\gamma=-\frac{\Delta}{\beta}(1-x), \quad \frac{d}{dz}=\frac{\alpha\beta}{\Delta}\frac{d}{dx}.$$

Consequently, (5) takes the form

$$(m', m\mid g)=\frac{(-1)^{m-m_2}}{(m'-m_2)!}\alpha^{m+m'-m_1-m_2}\beta^{m'-m}\Delta^{m_1-m'}\frac{d^{m'-m_2}}{dx^{m'-m_2}}[x(1-x)]\Big|_{x=\frac{\alpha\delta}{\Delta}}. \quad (8)$$

Comparing (7) and (8), we get

$$(m', m\mid g)=(-1)^{m'-m_2}\frac{(m_1-m)(m_1-m-1)\ldots[(m_1-m)-(m'-m_2-1)]}{(m'-m_2)!}$$
$$\times\gamma^{m-m'}\delta^{m_1+m_2-m-m'}\Delta^{m'-m_2}G_{m'-m_2}\left(m_1+m_2-2m'+1, m_1+m_2-m-m'; \frac{\alpha\delta}{\Delta}\right), (9)$$

We now obtain the formula for matrix elements $(m', m|g)$ in a different way. It is well known that any matrix $g=\begin{bmatrix}\alpha & \beta\\ \gamma & \delta\end{bmatrix}$, where $\alpha\neq 0$, can be uniquely factored into the product

$$g=z\delta\zeta, \quad (10)$$

where $z$, $\delta$, $\zeta$ are matrices of the form

$$z=\begin{pmatrix}1 & 0\\ z & 1\end{pmatrix}, \quad \zeta=\begin{pmatrix}1 & \zeta\\ 0 & 1\end{pmatrix}, \quad \delta=\begin{pmatrix}\lambda_1 & 0\\ 0 & \lambda_2\end{pmatrix}$$

(it is convenient to denote matrices $z$, $\zeta$ as well as their elements by the same letter *). Thus we have

$$T(g)=T(z)\,T(\delta)\,T(\zeta),$$

i.e.

$$(m^{(1)}, m^{(2)}\mid g)=\sum_{m,m'}(m^{(1)}, m'\mid z)\,(m, m\mid\delta)\,(m, m^{(2)}\mid\zeta). \quad (11)$$

*Therefore, to describe the matrix element $(m', m|g)$, it is sufficient to know the matrix elements of the operators $T(\delta)$, $T(z)$, $T(\zeta)$.* We write these matrix elements.

By (6), we obtain the elementary formulas:

$$(m', m\mid\delta)=\lambda_1^{m}\lambda_2^{m_1+m_2-m}\delta_{m,m'}, \quad (12)$$

* It is easy to see that the elements of the matrices $z$, $\zeta$ and $\delta$ are expressed in terms of the elements of the matrix $g$ as follows: $z=\gamma/\alpha$, $\zeta=\beta/\alpha$, $\lambda_1=\alpha$, $\lambda_2=\Delta/\alpha$, where $\Delta=\det g$.

where $\delta_{m,m'}$ is the Kronecker symbol;

$$(m', m \mid z) = \begin{cases} \dfrac{(m-m_2)!}{(m'-m_2)!\,(m-m')!}\, z^{m-m'}, & \text{if } m' \leqslant m, \\ 0, & \text{if } m' > m; \end{cases} \tag{13}$$

$$(m', m \mid \zeta) = \begin{cases} \dfrac{(m_1-m)!}{(m_1-m')!\,(m'-m)!}\, \zeta^{m'-m}, & \text{if } m' \geqslant m, \\ 0, & \text{if } m' < m. \end{cases} \tag{14}$$

Consequently, the formula for matrix elements becomes

$$(m', m \mid g) = \sum_{s \geqslant m,\, m'} \frac{(s-m_2)!\,(m_1-m)!}{(m'-m_2)!\,(s-m')!\,(m_1-s)!\,(s-m)!}\, \gamma^{s-m'} \beta^{s-m} \Delta^{m_1+m_2-s}. \tag{15}$$

One uses similar arguments for the group of matrices of the $n$th order. Namely, to describe the representation operators, one can confine oneself to operators corresponding to triangular matrices. In fact, not all triangular matrices are needed in this connection; only the most simple ones which generate subgroups of triangular matrices. The principal result of this paper is Theorem 1 in which explicit formulas for representation operators corresponding to these matrices are obtained. On the basis of this result, formulas for arbitrary representation operators can be obtained.

## §2. Canonical basis in the space of irreducible representation and formulas for the matrix elements of representations

**1. Canonical basis in the space of irreducible representation.** We explain the construction of the canonical basis in the space of irreducible representation (cf. [1]). This construction is based on the following two facts in the theory of finite-dimensional representations.

A. *A finite-dimensional irreducible representation of the group $G_n$ is uniquely determined by the n-dimensional vector $m_n = (m_{1n}, \cdots, m_{nn})$ with integral coordinates $m_{in}$, where $m_{1n} \geq m_{2n} \geq \cdots \geq m_{nn}$ (the highest weight of the representation).*

This vector $m_n$ will be referred to as the label (number) of this irreducible representation. The space in which the irreducible representation with the label $m_n$ operates, will be denoted by $H_{m_n}$, and the operators of the representation will be denoted by $T_{m_n}(g)$.

Let us agree to regard the group $G_{n-1}$ as a subgroup of the group $G_n$, imbedding it in $G_n$ in the following manner:

$$g_{n-1} \to \begin{pmatrix} g_{n-1} & 0 \\ 0 & 1 \end{pmatrix}.$$

B. *In the decomposition of the finite-dimensional irreducible representation of the group* $G_n$ *into irreducible representations of the subgroup* $G_{n-1}$, *the latter enter with multiplicity not greater than one. In the decomposition for* $H_{m_n}$, $m_n = (m_{1n}, \ldots, m_{nn})$ *those and only those subspaces* $H_{m_{n-1}}$, $m_{n-1} = (m_{1,n-1}, \ldots, m_{n-1,n-1})$ *enter here for which*

$$m_{i,n} \geqslant m_{i,n-1} \geqslant m_{i+1,n}, \quad i = 1, \ldots, n-1.$$

The construction of the canonical basis in the space $H_{m_n}$ is carried out as follows:

We consider the decreasing sequence of groups

$$G_n \supset G_{n-1} \supset \ldots \supset G_1.$$

We decompose the space $H_{m_n}$ into subspaces, invariant and irreducible with respect to the subgroup $G_{n-1}$, decompose each of these subspaces into subspaces, invariant and irreducible with respect to the subgroup $G_{n-2}$ and so forth, arriving finally at $G_1$. Consequently we obtain a decomposition of $H_{m_n}$ into one-dimensional subspaces (so long as the last of the subgroups, i.e. $G_1$, is commutative). By virtue of statement B, this decomposition is uniquely determined. Each of the one-dimensional subspaces obtained is given by the decreasing sequence of subspaces

$$H_{m_n} \supset H_{m_{n-1}} \supset \ldots \supset H_{m_1},$$

where $H_{m_k}$ is the subspace which is irreducible with respect to the subgroup $G_k$.

Thus each of the one-dimensional subspaces can be characterized by means of the triangular array

$$m = \begin{bmatrix} m_{1,n} \quad m_{2,n} \ldots m_{n-1,n} \quad m_{n,n} \\ m_{1,n-1} \quad \ldots \quad m_{n-1,n-1} \\ \ldots\ldots\ldots\ldots\ldots \\ m_{12} \quad m_{22} \\ m_{11} \end{bmatrix}. \tag{1}$$

In this array the first row is the label of the initial representation. In the other rows there can be arbitrary integers, satisfying the inequalities

$$m_{i,j} \geqslant m_{i,j-1} \geqslant m_{i+1,j}. \tag{2}$$

We choose vectors in each of the one-dimensional subspaces constructed and take these vectors as the canonical basis in the space $H_{m_n}$. In what follows, we will identify each of the basis vectors by the array $m$.

*Thus, the basis vectors in the space of irreducible representation appears as all possible admissible arrays of the form* (1) *of which the upper row* $m_n = (m_{1n}, \ldots, m_{nn})$ *is fixed; this upper row gives the label of the representation.*

The problem is how to evaluate matrix elements of the representation operators $T_{m_n}(g)$ in this basis.

2. **Formulas for infinitesimal operators.** Let us introduce a class of one-parameter subgroups of the group $G_n$.

Let $e_{pq}$ denote the matrix of $n$th order having the entry 1 at the intersection of the $p$th row and $q$th column and zero elsewhere. Further, let $I$ denote the unit matrix.

We consider the matrices $I + te_{pq}$, $p \neq q$ and the corresponding representation operators $T(g) = T_{m_n}(g)$:

$$T_{pq}(t) = T(I + te_{pq}), \qquad p \neq q. \tag{3}$$

We also put

$$T_{pp}(t) = T(I + (t-1)e_{pp}). \tag{4}$$

It is well known that every matrix $g \in G_n$ can be represented as a product of the diagonal matrices $I + (t-1)e_{pp}$, $p = 1, \cdots, n$ and the matrices $I + te_{p,p-1}$, $I + te_{p-1,p}$, $p = 2, \cdots, n$. Therefore, in order to specify the irreducible representation $T(g)$, it is sufficient to specify the operators $T_{pp}(t)$, $p = 1, \cdots, n$ and the operators $T_{p,p-1}(t)$, $T_{p-1,p}(t)$, $p = 2, \cdots, n$.

Let us introduce the infinitesimal operators

$$E_{pq} = \left.\frac{dT_{pq}(t)}{dt}\right|_{t=0}, \qquad p \neq q; \qquad E_{pp} = \left.\frac{dT_{pp}(t)}{dt}\right|_{t=1}. \tag{5}$$

It is easy to verify that the operators $T_{pq}(t)$, $p \neq q$ are expressed in terms of the operators $E_{pq}$ by the formula

$$T_{pq}(t) = \exp(tE_{pq}). \tag{6}$$

We now list explicit formulas for the operators $E_{pq}$. On the basis of these formulas, in §3 we obtain formulas for the operators $T_{pp}(t)$, $T_{p,p-1}(t)$ and $T_{p-1,p}(t)$.

*The infinitesimal operators* $E_{pq}$ *are given by the formulas*

1) $E_{pp}m = (k_p - k_{p-1})m,$ (7)

*where*

$$k_p = m_{1p} + \ldots + m_{pp} \qquad (p = 1, \ldots, n), \qquad k_0 = 0. \tag{8}$$

2) *For* $p > q$

$$E_{pq}m = \sum_{i_q, \ldots, i_{p-1}} \left\{ \frac{a_{p-1}^{i_{p-1}} b_q^{i_q} \prod_{k=q}^{p-2} c_{k,k+1}^{i_k, i_{k+1}}}{\prod_{k=q}^{p-1} d_k^{i_k}} \right\} m_{q,q+1,\ldots,p-1}^{i_q, i_{q+1}, \ldots, i_{p-1}}, \tag{9}$$

*where*

$$m_{q,\, q+1,\, \ldots,\, p-1}^{i_q,\, i_{q+1},\, \ldots,\, i_{p-1}}$$

*is the array which is obtained from the array* $m$ *if each of the indices*

$$m_{i_{q-1},\, q-1}, \ldots, m_{i_p,\, p}$$

*is decreased by* 1 *while all the others are unchanged, and where*

$$a_{p-1}^{i_{p-1}} = \prod_{j > i_{p-1}} (l_{i_{p-1},\, p-1} - l_{j,\, p} - 1); \tag{10}$$

$$b_q^{i_q} = \prod_{j \geqslant i_q} (l_{i_q,\, q} - l_{j,\, q-1}) \text{ npu } q > 1; \; b_1^1 = 1; \tag{11}$$

$$c_{k,\, k+1}^{i_k,\, i_{k+1}} = \operatorname{sign}(m_{i_{k+1},\, k+1} - m_{i_k,\, k}) \prod_{\substack{j \geqslant i_{k+1} \\ j \neq i_k}} (l_{i_{k+1} k+1} - l_{j,\, k}) \prod_{\substack{j > i_k \\ j \neq i_{k+1}}} (l_{i_k,\, k} - l_{j,\, k+1} - 1); \tag{12}$$

$$d_k^{i_k} = \prod_{j < i_k} (l_{jk} - l_{i_k,\, k}) \prod_{j \geqslant i_k} (l_{i_k,\, k} - l_{jk}); \tag{13}$$

$$l_{ij} = m_{ij} - 1. \tag{14}$$

**In the sequel, of all the operators $E_{pq}$ only $E_{p,p-1}$ and $E_{p-1,p}$ will play a fundamental role. We have**

$$E_{p,\, p-1} m = \sum_{i=1}^{p-1} \frac{a_{p-1}^i b_{p-1}^i}{d_{p-1}^i} m_{p-1}^i. \tag{15}$$

3) *For* $p < q$

$$E_{pq}\, m = \sum_{i_p,\, i_{p+1},\, \ldots,\, i_{q-1}} \left\{ \frac{\hat{a}_{q-1}^{i_{q-1}} \hat{b}_p^{i_p} \prod_{k=p}^{q-2} \hat{c}_{k,\, k+1}^{i_k,\, i_{k+1}}}{\prod_{k=p}^{q-1} d_k^{i_k}} \right\} \hat{m}_{p,\, p+1,\, \ldots,\, q-1}^{i_p,\, i_{p+1},\, \ldots,\, i_{q-1}}, \tag{16}$$

*where*

$$\hat{m}_{p,\, p+1,\, \ldots,\, q-1}^{i_p,\, i_{p+1},\, \ldots,\, i_{q-1}}$$

*is the array which is obtained from the array* $m$ *if each of the indices*

$$m_{i_{q-1},\, q-1}, \ldots, m_{i_p,\, p}$$

*is increased by* 1 *while all the others are unchanged, and where*

$$\hat{a}_{q-1}^{i_{q-1}} = \prod_{j \leqslant i_{q-1}} (l_{j,\, q} - l_{i_{q-1},\, q-1}); \tag{17}$$

$$\hat{b}_p^{i_p} = \prod_{j < i_p} (l_{j,\, p-1} - l_{i_p,\, p} - 1) \text{ npu } p > 1; \; \hat{b}_1^1 = 1; \tag{18}$$

$$\hat{c}_{k,\,k+1}^{i_k,\,i_{k+1}} = \operatorname{sign}(m_{i_{k+1},\,k+1} - m_{i_k,\,k})$$

$$\times \prod_{\substack{j<i_{k+1} \\ j\neq i_k}} (l_{j,k} - l_{i_{k+1},\,k+1} - 1) \prod_{\substack{j\leqslant i_k \\ j\neq i_{k+1}}} (l_{j,\,k+1} - l_{i_k,\,k}); \tag{19}$$

*the coefficients* $d_k^{i\,k}$ *are given by the formula* (13).

In particular, we have

$$E_{p-1,\,p}\, m = \sum_{i=1}^{p-1} \frac{\hat{a}_{p-1}^{i}\, \hat{b}_{p-1}^{i}}{d_{p-1}^{i}}\, \hat{m}_{p-1}^{i}. \tag{20}$$

One can show that the operators $E_{pq}$, defined by formulas (7), (9) and (16), are infinitesimal operators of our representation by directly verifying that the operators satisfy the commutation relations

$$[E_{pq}, E_{qr}] = E_{pr} \quad (r \neq p), \quad [E_{pq}, E_{qp}] = E_{pp} - E_{qq},$$
$$[E_{p_1 q_1}, E_{p_2 q_2}] = 0, \qquad \text{if } p_1 \neq q_2,\ q_1 \neq p_2. \tag{21}$$

Therefore it follows that they define the representation. Further, it is easy to verify that in the representation space there exists only one vector of highest weight and that this highest weight is equal to $(m_{1,n}, \ldots, m_{nn})$. Hence this representation is irreducible and equivalent to the initial one.

We note that the operators $E_{pq}$, determine the infinitesimal representation operators of the group $U_n$ of unitary matrices. Namely, these infinitesimal operators are $E_{pp}$, $E_{pq} - E_{qp}$ and $i(E_{pq} + E_{qp})$, $q < p$.

Formulas for the operators $E_{pq}$ (in a rather different form) were first obtained in [1]. * As the basis there were taken vectors $\widetilde{m}$, differing from the vectors $m$ by

* In [1] formulas for representations were published without an indication of the proof. In this connection, to show, as in the present paper, also, that the operators $E_{pq}$ give the representation, that it is irreducible, and that in this way all the irreducible representations are obtained, it is sufficient to do the following:

1) To verify the commutation relations (21); this is done by direct calculation. Similarly, it is proved that the operators give the representation.

2) To prove that there exists only one vector of highest weight; it is easy to verify this directly too. This proves the irreducibility of the representations constructed.

3) To show that by such a construction one can obtain the irreducible representation corresponding to any given vector of the highest weight; this follows immediately from the formulas mentioned in [1]. Similarly it is proved that all the representations are constructed.

We note the interesting paper [4] where another derivation of these formulas is given and, in the following part, there are a number of interesting applications, in particular to the problem of decomposition of the Kronecker product of two irreducible representations.

a scalar multiplier $m = \lambda(m)\widetilde{m}$, where

$$\lambda(m) = \prod_{k=2}^{n}\left[\frac{\prod\limits_{i>j}(l_{j,k-1}-l_{ik}-1)!\prod\limits_{i\leqslant j}(l_{i,k}-l_{j,k-1})!}{\prod\limits_{i<j}(l_{i,k-1}-l_{j,k-1})}\right]^{\frac{1}{2}}. \tag{22}$$

The basis $\widetilde{m}$ is more convenient for investigating the representation of the group $U_n$ of unitary matrices, since in this basis the representation operators of the group $U_n$ are expressed by unitary matrices.

3. **Integral formulas.** The following fundamental theorem will be proved here.

**Theorem 1.** *The operators of irreducible representation*

$$T_{pp}(t) = T(I+(t-1)e_{pp}), \quad T_{p,p-1}(t) = T(I+te_{p,p-1})$$

*and*

$$T_{p-1,p}(t) = T(I+te_{p-1,p})$$

*are expressed by the formulas:*

$$T_{pp}(t)\,m = t^{k_p-k_{p-1}}m, \tag{23}$$

*where*

$$k_p = m_{1,p}+\ldots+m_{p,p} \quad (p=1,\ldots,n), \quad k_0 = 0; \tag{24}$$

$$T_{p,p-1}(t)\,m = \sum_{m'} B^{(1)}_{p-1}(m',m)\,t^{k_{p-1}-k'_{p-1}}m', \tag{25}$$

*where* $B^{(1)}_{p-1}(m',m)$ *is a numerical coefficient, not depending on* $t$, *defined by the formula*

$$B^{(1)}_{p-1}(m',m) = \prod_{i<j}\frac{(l_{i,p-1}-l_{j,p}-1)!}{(l'_{i,p-1}-l_{j,p}-1)!}\prod_{i\leqslant j}\frac{(l_{i,p-1}-l_{j,p-2})!}{(l'_{i,p-1}-l_{j,p-2})!}$$
$$\times\frac{\prod\limits_{i<j}(l'_{i,p-1}-l_{j,p-1}-1)!}{\prod\limits_{i\leqslant j}(l_{i,p-1}-l'_{j,p-1})!}\prod_{i<j}(l'_{i,p-1}-l'_{j,p-1}); \tag{26}$$

$l_{ij} = m_{ij} - i$; *the summation is carried out over all the arrays* $m'$ *in which* $m'_{ij} = m_{ij}$ *for* $j \neq p-1$, $m'_{i,p-1} \leq m_{i,p-1}$, $i = 1, \cdots, p-1$;

$$T_{p-1,p}(t)\,m = \sum_{m'} B^{(2)}_{p-1}(m',m)\,t^{k'_{p-1}-k_{p-1}}m', \tag{27}$$

*where* $B^{(2)}_{p-1}(m',m)$ *is a numerical coefficient, not depending on* $t$, *defined by the formula*

$$\mathrm{B}^{(2)}_{p-1}(m', m) = \prod_{i\leqslant j} \frac{(l_{i,p}-l_{j,p-1})!}{(l_{i,p}-l'_{j,p-1})!} \prod_{i<j} \frac{(l_{i,p-2}-l_{j,p-1}-1)!}{(l_{i,p-2}-l'_{j,p-1}-1)!}$$

$$\times \frac{\prod\limits_{i<j}(l_{i,p-1}-l'_{j,p-1}-1)!}{\prod\limits_{i\leqslant j}(l'_{i,p-1}-l_{j,p-1})!} \prod_{i<j}(l'_{i,p-1}-l'_{j,p-1}); \tag{28}$$

*the summation is carried out over all the arrays* $m'$ *in which* $m'_{ij} = m_{ij}$ *for* $j \neq p-1$, $m'_{i,p-1} \geq m_{i,p-1}$, $i = 1, \cdots, p-1$.

**Proof.** The formula for operator $T_{pp}(t)$ immediately follows from formula (7) for the infinitesimal operator $E_{pp}$. Therefore we only have to prove the formulas for the operators $T_{p,p-1}(t)$ and $T_{p-1,p}(t)$. We shall give the derivation of formula (25) (the derivation of formula (27) is carried out similarly).

Thus it is necessary to prove that

$$T_{p,p-1}(t) = \exp(tE_{p,p-1}), \tag{29}$$

where $E_{p,p-1}$ is the operator defined by formula (15). Evidently, relation (29) is equivalent to the two relations

$$T_{p,p-1}(0) = E, \tag{30}$$

where $E$ is the identity operator, and

$$\frac{dT_{p,p-1}(t)}{dt} = E_{p,p-1}T_{p,p-1}(t). \tag{31}$$

Relation (30) follows immediately from formulas (25) and (26) for the operator $T_{p,p-1}(t)$; it is therefore sufficient to prove relation (31).

We renormalize the basis vectors $m$, replacing $m$ by

$$\Big[\prod_{i<j}(l_{i,p-1}-l_{j,p}-1)! \prod_{i\leqslant j}(l_{i,p-1}-l_{j,p-2})!\Big]^{-1} m.$$

Then the formulas for $E_{p,p-1}$ and $T_{p,p-1}(t)$ take the simpler form

$$E_{p,p-1}m = \sum_{i=1}^{p-1} \frac{1}{\prod\limits_{j=1}^{i-1}(l_{j,p-1}-l_{i,p-1}) \prod\limits_{j=i+1}^{p-1}(l_{i,p-1}-l_{j,p-1})} m^{i}_{p-1}; \tag{32}$$

$$T_{p,p-1}(t)m = \sum_{m'} \frac{\prod\limits_{i<j}(l'_{i,p-1}-l_{j,p-1}-1)!}{\prod\limits_{i\leqslant j}(l_{i,p-1}-l'_{j,p-1})!} \prod_{i<j}(l'_{i,p-1}-l'_{j,p-1})\, t^{k_{p-1}-k'_{p-1}} m'. \tag{33}$$

We commute the operator $E_{p,p-1}T_{p,p-1}$. Since the arrays $m$ and $m'$ in (33) differ only in the elements of the $(p-1)$th row from the bottom, we shall henceforth indicate only this row and write $(m_{1,p-1}, \cdots, m_{p-1,p-1})$ in place of $m$.

Then, because of (32) and (33), the formula for $E_{p,p-1}T_{p,p-1}$ can be written as

$$E_{p,\,p-1}T_{p,\,p-1}\cdot(m_{1,\,p-1},\,\ldots,\,m_{p-1,\,p-1}) = \sum_{i=1}^{p-1}\sum_{m'}\frac{\prod\limits_{k<j}(l'_{k,\,p-1}-l_{j,\,p-1}-1)!}{\prod\limits_{k\leqslant j}(l_{k,\,p-1}-l'_{j,\,p-1})!}\prod_{k<j}(l'_{k,\,p-1}-l'_{j,\,p-1})$$

$$\times\frac{1}{\prod\limits_{j=1}^{i-1}(l'_{j,\,p-1}-l'_{i,\,p-1})\prod\limits_{j=i+1}^{p-1}(l'_{i,\,p-1}-l'_{j,\,p-1})}t^{k_{p-1}-k'_{p-1}}$$

$$\times(m'_{1,\,p-1},\,\ldots,\,m'_{i,\,p-1}-1,\,\ldots,\,m'_{p-1,\,p-1}),$$

i.e.

$$E_{p,\,p-1}T_{p,\,p-1}\cdot(m_{1,\,p-1},\,\ldots,\,m_{p-1,\,p-1}) = \sum_{i=1}^{p-1}\sum_{m'}\frac{\prod\limits_{k<j}(l'_{k,\,p-1}-l_{j,\,p-1}-1)!}{\prod\limits_{k\leqslant j}(l_{k,\,p-1}-l'_{j,\,p-1})!}\prod_{\substack{k<j\\ k\neq i;\ j\neq i}}(l'_{k,\,p-1}-l'_{j,\,p-1})\,t^{k_{p-1}-k'_{p-1}}$$

$$\times(m'_{1,\,p-1},\,\ldots,\,m'_{i,\,p-1}-1,\,\ldots,\,m'_{p-1,\,p-1}). \tag{34}$$

Under the summation sign, we replace $m'_{i,p-1}-1$ by $m'_{i,p-1}$. Consequently, (34) takes the form

$$E_{p,\,p-1}T_{p,\,p-1}\cdot(m_{1,\,p-1},\,\ldots,\,m_{p-1,\,p-1}) = \sum_{m'}\left\{\frac{\prod\limits_{k<j}(l'_{k,\,p-1}-l_{j,\,p-1}-1)!}{\prod\limits_{k\leqslant j}(l_{k,\,p-1}-l'_{j,\,p-1})!}\prod_{\substack{k<j\\ k\neq i;\ j\neq i}}(l'_{k,\,p-1}-l'_{j,\,p-1})\right.$$

$$\left.\times\sum_{i=1}^{p-1}\left[\prod_{j=1}^{i}(l_{j,\,p-1}-l'_{i,\,p-1})\prod_{j=i+1}^{p-1}(l'_{i,\,p-1}-l_{j,\,p-1})\right]\right\}t^{k_{p-1}-k'_{p-1}-1}$$

$$\times(m'_{1,\,p-1},\,\ldots,\,m'_{p-1,\,p-1}). \tag{35}$$

We compare this formula with the following formula for the operator $dT_{p,p-1}/dt$ which is an immediate consequence of (33):

$$\frac{dT_{p,\,p-1}}{dt}\cdot(m_{1,\,p-1},\,\ldots,\,m_{p-1,\,p-1}) = \sum_{m'}\frac{\prod\limits_{k<j}(l'_{k,\,p-1}-l_{j,\,p-1}-1)!}{\prod\limits_{k\leqslant j}(l_{k,\,p-1}-l'_{j,\,p-1})!}$$

$$\times\prod_{k<j}(l'_{k,\,p-1}-l'_{j,\,p-1})(k_{p-1}-k'_{p-1})t^{k_{p-1}-k'_{p-1}-1}(m'_{1,\,p-1},\,\ldots,\,m'_{p-1,\,p-1}). \tag{36}$$

We see that the relation

$$\frac{dT_{p,p-1}}{dt} = E_{p,p-1} T_{p,p-1}$$

is equivalent to

$$\sum_{i=1}^{p-1} \frac{\prod_{j=1}^{i} (l_{j,p-1} - l'_{i,p-1}) \prod_{j=i+1}^{p-1} (l'_{i,p-1} - l_{j,p-1})}{\prod_{j=1}^{i-1} (l'_{j,p-1} - l'_{i,p-1}) \prod_{j=i+1}^{p-1} (l'_{i,p-1} - l'_{j,p-1})} = \sum_{i=1}^{p-1} (l_{i,p-1} - l'_{i,p-1}). \tag{37}$$

Thus it is sufficient to prove relation (37).

We rewrite the left side of this relation in the form

$$\mathcal{I} = \sum_{i=1}^{p-1} \frac{\prod_{j=1}^{p-1} (l_{j,p-1} - l'_{i,p-1})}{\prod_{\substack{j=1 \\ j\neq i}}^{p-1} (l'_{j,p-1} - l'_{i,p-1})}. \tag{38}$$

It is evident that this expression is a symmetric function of the variables $l_{i,p-1}$ and the variables $l'_{i,p-1}$. Reducing the sum (38) to a common denominator, we obtain

$$\mathcal{I} = \frac{P(l, l')}{\prod_{i<j} (l'_{i,p-1} - l'_{j,p-1})}, \tag{39}$$

where $P(l, l')$ is some polynomial in $l_{i,p-1}$, $l'_{i,p-1}$ of degree one higher than the degree of the polynomial in the denominator. This polynomial $P$ must be divisible by the polynomial in the denominator since, as is easily seen, for $l'_{i_1,p-1} = l'_{i_2,p-1}$, $i_1 \neq i_2$, the expression $\mathcal{I}$ has no singularity. This shows that $\mathcal{I}$ is a polynomial of the first degree, symmetric relative to the variables $l'_{i,p-1}$ and to the variables $l_{i,p-1}$. Consequently,

$$\mathcal{I} = \alpha \sum_{i=1}^{p-1} l_{i,p-1} - \beta \sum_{i=1}^{p-1} l'_{i,p-1}. \tag{40}$$

We shall determine the coefficients $\alpha$ and $\beta$. Assuming

$$l'_{i,p-1} = l_{i,p-1}, \quad i = 2, \ldots, p-1,$$

we obtain from formula (38),

$$\mathcal{I} = l_{1,p-1} - l'_{1,p-1},$$

while from formula (40) we have

$$\mathcal{J} = \alpha l_{1,\,p-1} - \beta l'_{1,\,p-1} + (\alpha - \beta) \sum_{i=2}^{p-1} l_{i,\,p-1}.$$

Equating these two expressions, we obtain $\alpha = \beta = 1$.

Thus we have proved that

$$\mathcal{J} = \sum_{i=1}^{p-1} (l_{i,\,p-1} - l'_{i,\,p-1}).$$

This also proves relation (25).

4. **Matrix elements for the case $n = 3$.** In conclusion, we will cite formulas for the matrix elements of representations of the group $G_3$.

From Theorem 1, the matrix elements of the operator $T(I + te_{32})$ have the form

$$\left[\begin{matrix} m_{13} & m_{23} & m_{33} & m_{13} & m_{23} & m_{33} \\ & m'_{12} \quad m'_{22} & & & m_{12} \quad m_{22} & \\ & m'_{11} & & & m_{11} & \end{matrix}\ \middle|\ I + te_{32}\right]$$

$$= \frac{(l_{12} - l_{23} - 1)!\,(l_{12} - l_{33} - 1)!\,(l_{22} - l_{33} - 1)!\,(l_{12} - l_{11})!}{(l'_{12} - l_{23} - 1)!\,(l'_{12} - l_{33} - 1)!\,(l'_{22} - l_{33} - 1)!\,(l'_{12} - l_{11})!}$$

$$\times \frac{(l'_{12} - l_{22} - 1)!}{(l_{12} - l'_{12})!\,(l_{12} - l'_{22})!\,(l_{22} - l'_{22})!}\,(l'_{12} - l'_{22})\,\delta(m'_{11}, m_{11})\, t^{(m_{12}+m_{22})-(m'_{12}+m'_{22})}, \tag{41}$$

where $\delta(m'_{11}, m_{11})$ is the usual Kronecker delta. The matrix elements of the operator $T(I + te_{23})$ have the form

$$\left[\begin{matrix} m_{13} & m_{23} & m_{33} & m_{13} & m_{23} & m_{33} \\ & m'_{12} \quad m'_{22} & & & m_{12} \quad m_{22} & \\ & m'_{11} & & & m_{11} & \end{matrix}\ \middle|\ I + te_{23}\right]$$

$$= \frac{(l_{13} - l_{12})!\,(l_{13} - l_{22})!\,(l_{23} - l_{22})!\,(l_{11} - l_{22} - 1)!}{(l_{13} - l'_{12})!\,(l_{13} - l'_{22})!\,(l_{23} - l'_{22})!\,(l_{11} - l'_{22} - 1)!}$$

$$\times \frac{(l_{13} - l_{22}' - 1)!}{(l'_{12} - l_{12})!\,(l'_{12} - l_{22})!\,(l'_{22} - l_{22})!}\,(l'_{12} - l'_{22})\,\delta(m'_{11}, m_{11})\, t^{(m_{12}'+m_{22}')-(m_{12}+m_{22})}. \tag{42}$$

We now write formulas for the matrix elements of $T(I + te_{31})$ and $T(I + te_{13})$. For this we use the relations

$$\begin{aligned} T(I + te_{31}) &= T^{-1}(s)\, T(I + te_{32})\, T(s), \\ T(I + te_{13}) &= T^{-1}(s)\, T(I + te_{23})\, T(s), \end{aligned} \tag{43}$$

where $s = \begin{pmatrix} 0 & 1 \\ 1 & 0 \end{pmatrix}$.

The operator $T(s)$ is given by the formula

$$T(s)\begin{bmatrix} m_{13} & & m_{23} & & m_{33} \\ & m_{12} & & m_{22} & \\ & & m_{11} & & \end{bmatrix} = (-1)^{m_{22}}\begin{bmatrix} m_{13} & & m_{23} & & m_{33} \\ & m_{12} & & m_{22} & \\ & & m_{12}+m_{22}-m_{11} & & \end{bmatrix}, \tag{44}$$

which is easily obtained from the formulas for the representations of the group $G_2$, mentioned in §1.

Thus, by (43) and (44), we obtain the following formulas for the matrix elements of $T(I + te_{31})$ and $T(I + te_{13})$:

$$\left[\begin{matrix} m_{13} & m_{23} & m_{33} \\ m'_{12} & m'_{22} & \\ & m'_{11} & \end{matrix}\;\begin{matrix} m_{13} & m_{23} & m_{33} \\ m_{12} & m_{22} & \\ & m_{11} & \end{matrix}\;\middle|\; I + te_{31}\right]$$

$$= (-1)^{m_{22}-m'_{22}}\left[\begin{matrix} m_{13} & m_{23} & m_{33} \\ m'_{12} & m'_{22} & \\ & m'_{12}+m'_{22}-m'_{11} & \end{matrix}\;\begin{matrix} m_{13} & m_{23} & m_{33} \\ m_{12} & m_{22} & \\ & m_{12}+m_{22}-m_{11} & \end{matrix}\;\middle|\; I + te_{32}\right]; \tag{45}$$

$$\left[\begin{matrix} m_{13} & m_{23} & m_{33} \\ m'_{12} & m_{22} & \\ & m'_{11} & \end{matrix}\;\begin{matrix} m_{13} & m_{23} & m_{33} \\ m_{12} & m_{22} & \\ & m_{11} & \end{matrix}\;\middle|\; I + te_{13}\right]$$

$$= (-1)^{m_{22}-m'_{22}}\left[\begin{matrix} m_{13} & m_{23} & m_{33} \\ m'_{12} & m'_{22} & \\ & m'_{12}+m'_{22}+m'_{11} & \end{matrix}\;\begin{matrix} m_{13} & m_{23} & m_{33} \\ m_{12} & m_{22} & \\ & m_{12}+m_{22}-m_{11} & \end{matrix}\;\middle|\; I + te_{23}\right]. \tag{46}$$

## §3. General beta-functions related to irreducible finite-dimensional representations

On the basis of the formulas of §2, it is possible to write the matrix elements of an arbitrary representation operator of the group $G_n$. But these matrix elements are expressed by fairly complicated sums. This section is devoted to additional considerations which reveal the deeper structure of the operator $T(g)$ for an arbitrary element $g$.

1. The subgroups $Z^+$ and $Z^-$; canonical factorization of elements of the group $G_n$. We will write matrices $g_n \in G_n$ in block form:

$$g_n = \begin{pmatrix} \alpha & \beta \\ \gamma & \delta \end{pmatrix},$$

where $\alpha$ is a square matrix of order $n - 1$, $\beta$ is a column consisting of $n - 1$ elements, etc.

We denote by $Z^-$ the subgroup of matrices of the form

$$z = \begin{pmatrix} I & 0 \\ z_n & 1 \end{pmatrix},$$

where $I$ is the unit matrix, $z_n = (z_{n1}, \cdots, z_{n,n-1})$ and we denote by $Z^+$ the subgroup of matrices of the form

$$\zeta = \begin{pmatrix} I & \zeta_n \\ 0 & 1 \end{pmatrix},$$

where $\zeta_n = \begin{pmatrix} \zeta_{1n} \\ \vdots \\ \zeta_{n-1,n} \end{pmatrix}$. We also recall that the group $G_{n-1}$ is assumed to be imbedded in the group $G_n$ as follows:

$$g_{n-1} \to \begin{pmatrix} g_{n-1} & 0 \\ 0 & 1 \end{pmatrix}.$$

The following assertion is well known.

Any element $g_n \in G_n$, with the exception of a submanifold of elements having a lower dimension, can be uniquely factored into a product of the form

$$g_n = \lambda z g_{n-1} \zeta, \tag{1}$$

where $g_{n-1} \in G_{n-1}$, $z \in Z^-$, $\zeta \in Z^+$ and $\lambda$ is a scalar factor.

On the basis of (1) for operators of the irreducible representation we obtain the factorization

$$T(g_n) = T(\lambda I) T(z) T(g_{n-1}) T(\zeta). \tag{2}$$

We observe that the matrix elements of the operator $T(\lambda I)$ are known (see formula (23) of §2). Thus if we know the operators $T(z)$ and $T(\zeta)$, then the description of the matrix elements for the group of $n$th order is reduced, by formula (2), to the description of matrix elements for the group of $(n-1)$th order.

This subsection deals with the description of the operators $T(z)$ and $T(\zeta)$.

The operators $T(z)$ and $T(\zeta)$ play an important role in the theory of representations. In them is essentially contained the whole general theory of special functions associated with complex semisimple Lie groups.

2. Properties of $T(z)$ and $T(\zeta)$; general beta-functions $F^{(1)}(z)$ and $F^{(2)}(\zeta)$.

Let $z$ be a vector in a $(n-1)$-dimensional space $R$

$$z = (z_1, \ldots, z_{n-1}),$$

and let $\zeta$ be a vector in the dual space $R'$

$$\zeta = \begin{pmatrix} \zeta_1 \\ \vdots \\ \zeta_{n-1} \end{pmatrix}$$

With these vectors we construct the $n$th order matrices

$$\begin{pmatrix} I & 0 \\ z & 1 \end{pmatrix}, \quad \begin{pmatrix} I & \zeta \\ 0 & 1 \end{pmatrix},$$

which, for brevity, we also denote respectively by $z$ and $\zeta$.

We introduce the notation

$$\begin{aligned} T(z) &= F^{(1)}(z) = F^{(1)}(z_1, \ldots, z_{n-1}), \\ T(\zeta) &= F^{(2)}(\zeta) = F^{(2)}(\zeta_1, \ldots, \zeta_{n-1}). \end{aligned} \tag{3}$$

Thus, $F^{(1)}$ and $F^{(2)}$ are functions in an $(n-1)$-dimensional space, whose values are operators in the space of irreducible representations of the group $G_n$. If we know these functions, then the problem of finding the matrix elements of representations of the group of $n$th order can be reduced to the study of representations of the group of $(n-1)$th order.

Let us determine which property of the functions $F^{(1)}$ and $F^{(2)}$ is equivalent to the fact that the operators $T(g)$ constructed for them give the representation of the group $G_n$.

So let us suppose we already know all the finite-dimensional irreducible representations of the group $G_{n-1}$. Every such representation is specified by the set of $n-1$ integers

$$m_n = (m_{1,n-1}, \ldots, m_{n-1,n-1}),$$

where $m_{1,n-1} \geq m_{2,n-1} \geq \cdots \geq m_{n-1,n-1}$.

Let us introduce the notion of an "interval" of representations. For this we prescribe $n$ fixed numbers

$$m_n = (m_{1,n}, \ldots, m_{n,n}), \quad m_{1,n} \geqslant m_{2,n} \geqslant \ldots \geqslant m_{n,n},$$

playing the role of the "boundaries" of the interval and we consider all irreducible representations of the group $G_{n-1}$ whose number $m_{n-1}$ satisfies the inequalities

$$m_{1,n} \geqslant m_{1,n-1} \geqslant m_{2,n} \geqslant m_{2,n-1} \geqslant m_{3,n} \geqslant \ldots \geqslant m_{n-1,n} \geqslant m_{n-1,n-1} \geqslant m_{n,n}.$$

We call the direct sum of these representations the interval of representations of the group $G_{n-1}$. This is the module

$$H_{m_n} = \sum_{m_{n-1}} H_{m_{n-1}},$$

on which the operators $T(g_{n-1})$ of representation of the group $G_{n-1}$ act.

We consider the auxiliary $(n-1)$-dimensional space $R$ and its dual space $R'$. Let $z \in R$ and $\zeta \in R'$. In these spaces the group $G_{n-1}$ acts as follows:

$$z = (z_1, \ldots, z_{n-1}) \rightarrow zg_{n-1} = (z_1', \ldots, z_{n-1}'),$$

where

$$z_i' = z_1 g_{1i} + \ldots + z_{n-1} g_{n-1,i}, \quad i = 1, \ldots, n-1,$$

$$\zeta = \begin{pmatrix} \zeta_1 \\ \cdots \\ \zeta_{n-1} \end{pmatrix} \rightarrow g_{n-1}^{-1}\zeta = \begin{pmatrix} \zeta_1' \\ \cdots \\ \zeta_{n-1}' \end{pmatrix},$$

where

$$\zeta_i' = \hat{g}_{i1}\zeta_1 + \ldots + \hat{g}_{i,n-1}\zeta_{n-1}, \quad i = 1, \ldots, n-1,$$

$\|\hat{g}_{ij}\|$ is the inverse of matrix $g_{n-1} = \|g_{ij}\|$.

We consider the continuous function $F^{(1)}(z)$, $z \in R$ and $F^{(2)}(\zeta)$, $\zeta \in R'$ whose values are operators in the space $H_{m_n}$, the interval of representations of the group $G_{n-1}$.

We call these functions *generalized beta-functions of rank* $n-1$ if they satisfy the conditions

1) $F^{(1)}(z' + z'') = F^{(1)}(z')F^{(1)}(z'')$, (4′)

$F^{(2)}(\zeta' + \zeta'') = F^{(2)}(\zeta')F^{(2)}(\zeta'')$ (4″)

for arbitrary $z', z'' \in R$, $\zeta', \zeta'' \in R'$.

2) $T^{-1}(g_{n-1})F^{(1)}(z)T(g_{n-1}) = F^{(1)}(zg_{n-1})$, (5′)

$T^{-1}(g_{n-1})F^{(2)}(\zeta)T(g_{n-1}) = F^{(2)}(g_{n-1}^{-1}\zeta)$ (5″)

for arbitrary $g_{n-1} \in G_{n-1}$.

3) The operators $F^{(1)}(0)$ and $F^{(2)}(0)$ are nonsingular.

We observe that since any nonzero vector is the image by a nonsingular linear transformation of the vector $(0, \cdots, 0, 1)$, it follows that by property 2), the values of the functions $F^{(1)}(z)$ and $F^{(2)}(\zeta)$ are completely determined by their values at $(0, \cdots, 0, 1)$.

Later, we will show, on the basis of property 2), that the matrix elements of the function $F^{(1)}(z)$ are, in the canonical basis, monomials in the components of the vector $z$:

$$C z_1^{k_1} \dots z_{n-1}^{k_{n-1}},$$

where the exponents $k_1, \cdots, k_{n-1}$ can be precisely determined. The expression for the coefficient $C$ turns out to be very important. The same also holds for the function $F^{(2)}(\zeta)$.

A pair of generalized beta-functions $F^{(1)}(z)$ and $F^{(2)}(\zeta)$ is called an *associated pair* if in addition to conditions 1), 2) and 3), they also satisfy

$$\text{4)} \quad F^{(2)}(\zeta) F^{(1)}(z) = (1+z\zeta)^{-(m_{1n}+\dots+m_{nn})} T(I+\zeta z) F^{(1)}(z) F^{(2)}(\zeta) T((1+z\zeta) I). \quad (6)$$

Here $I$ is the unit matrix of order $n-1$ while the products $z\zeta$ and $\zeta z$ are to be regarded as matrix products.

Condition 4) gives us the permutation rule for the cofactors $F^{(2)}$ and $F^{(1)}$.

We note that, because of condition 2), it is sufficient that 4) hold for $z = (0, \cdots, 0, 1)$.

We call two associated pairs of beta-functions $(F^{(1)}, F^{(2)})$ and $\hat{F}^{(1)}, \hat{F}^{(2)})$ equivalent if they take values on the same interval of representations $H_{m_n}$, connected by the relations

$$\hat{F}^{(1)}(z) = A^{-1}F^{(1)}(z) A, \qquad \hat{F}^{(2)}(\zeta) = A^{-1}F^{(2)}(\zeta) A,$$

where $A$ is some nonsingular operator in the space $H_{m_n}$.

Theorem 2. *Corresponding to every interval $H_{m_n}$ of representations of the group $G_{n-1}$ there is, up to equivalence, a unique associated pair of general beta-functions $F^{(1)}, F^{(2)}$.*

Proof. First of all, let us observe that the representation $T(g_{n-1})$ of the group $G_{n-1}$ in the space $H_{m_n}$ can be extended to the irreducible representation $T(g_n)$ of the group $G_n$, which is defined uniquely up to equivalence. This follows immediately from assertions A and B of §2.1.

Now let $z$ be a vector in the $(n-1)$-dimensional space $R$ and $\zeta$ be a vector in a dual space $R'$. With these vectors we construct the following matrices of the $n$th order which, for brevity, we denote by the same letters:

$$z = \begin{pmatrix} I & 0 \\ z & 1 \end{pmatrix}, \quad \zeta = \begin{pmatrix} I & \zeta \\ 0 & 1 \end{pmatrix}.$$

Let

$$F^{(1)}(z) = T(z), \qquad F^{(2)}(\zeta) = T(\zeta).$$

By the above, this construction determines the functions $F^{(1)}(z)$ and $F^{(2)}(\zeta)$ uniquely up to equivalence.

We shall show that $F^{(1)}(z)$ and $F^{(2)}(\zeta)$ form an associated pair of general beta-functions, i.e. they satisfy conditions 1)–4).

Condition 1) follows immediately from the matrix relations

$$\begin{pmatrix} I & 0 \\ z' & 1 \end{pmatrix}\begin{pmatrix} I & 0 \\ z'' & 1 \end{pmatrix}=\begin{pmatrix} I & 0 \\ z'+z'' & 1 \end{pmatrix},$$
$$\begin{pmatrix} I & \zeta' \\ 0 & 1 \end{pmatrix}\begin{pmatrix} I & \zeta'' \\ 0 & 1 \end{pmatrix}=\begin{pmatrix} I & \zeta'+\zeta'' \\ 0 & 1 \end{pmatrix},$$

while condition 2) follows from

$$\begin{pmatrix} g_{n-1} & 0 \\ 0 & 1 \end{pmatrix}^{-1}\begin{pmatrix} I & 0 \\ z & 1 \end{pmatrix}\begin{pmatrix} g_{n-1} & 0 \\ 0 & 1 \end{pmatrix}=\begin{pmatrix} I & 0 \\ zg_{n-1} & 1 \end{pmatrix},$$
$$\begin{pmatrix} g_{n-1} & 0 \\ 0 & 1 \end{pmatrix}^{-1}\begin{pmatrix} I & \zeta \\ 0 & 1 \end{pmatrix}\begin{pmatrix} g_{n-1} & 0 \\ 0 & 1 \end{pmatrix}=\begin{pmatrix} 1 & g_{n-1}^{-1}\zeta \\ 0 & 1 \end{pmatrix}.$$

Condition 3) is evident.

Finally, condition 4) follows immediately from the simple matrix relation

$$\begin{pmatrix} I & \zeta \\ 0 & 1 \end{pmatrix}\begin{pmatrix} I & 0 \\ z & 1 \end{pmatrix}$$
$$=(1+z\zeta)^{-1}\begin{pmatrix} I+\zeta z & 0 \\ 0 & 1 \end{pmatrix}\begin{pmatrix} I & 0 \\ z & 1 \end{pmatrix}\begin{pmatrix} I & \zeta \\ 0 & 1 \end{pmatrix}\begin{pmatrix} (1+z\zeta)I & 0 \\ 0 & 1 \end{pmatrix}.$$

This proves the theorem.

Conversely, given the interval of representations $H_{m_n}$ and an associated pair of beta-functions $F^{(1)}(z)$, $F^{(2)}(\zeta)$, one can construct from them an irreducible representation of the group $G_n$ in the following manner.

With the $(n-1)$-dimensional vectors $z \in R$ and $\zeta \in R'$ we construct matrices of the $n$th order $z = \begin{bmatrix} I & 0 \\ z & 1 \end{bmatrix}$ and $\zeta = \begin{bmatrix} I & \zeta \\ 0 & 1 \end{bmatrix}$ and put

$$\begin{aligned} \hat{T}(z) &= F^{(1)}(z), \quad \hat{T}(\zeta) = F^{(2)}(\zeta), \\ \hat{T}(g_{n-1}) &= T(g_{n-1}), \quad \hat{T}(\lambda I_n) = \lambda^{m_{1n}+\dots+m_{nn}}, \end{aligned} \tag{7}$$

where $I_n$ is the unit matrix of order $n$.

If $g$ is an arbitrary element of $G_n$, then we write it as

$$g = \lambda z g_{n-1}\zeta, \tag{8}$$

where $z \in Z^-$, $\zeta \in Z^+$, $g_{n-1} \in G_{n-1}$ and $\lambda$ is a scalar factor * and put

* We know from subsection 1 that all the elements $g$, with the exception of a submanifold of lower dimension, are expressible in this form and that such a factorization is unique.

$$\hat{T}(g) = \hat{T}(\lambda I_n)\hat{T}(z)\hat{T}(g_{n-1})\hat{T}(\zeta). \tag{9}$$

Thus, $\hat{T}(g)$ is an operator in the space $H_{m_n}$.

Theorem 3. *If $F^{(1)}(z)$, $F^{(2)}(\zeta)$ is an associated pair of beta-functions, then formula* (9) *gives irreducible representation of the group $G_n$. In addition, any irreducible representation can be obtained in this way.*

Proof. We will show that the operators $\hat{T}(g)$ form the representation. In other words, we will show that $\hat{T}(I_n)$ is the identity operator and that

$$\hat{T}(g'g'') = \hat{T}(g')\hat{T}(g'') \tag{10}$$

for any two matrices $g'$ and $g''$ of $G_n$. The assertion that $\hat{T}(I_n)$ is the identity operator is evident. Further, it is easily seen that relation (10) is equivalent to the following simpler relations:

a) $\hat{T}(g'_{n-1}g''_{n-1}) = \hat{T}(g'_{n-1})\hat{T}(g''_{n-1})$ for arbitrary $g'_{n-1}, g''_{n-1} \in G_{n-1}$;

b) $\hat{T}(\lambda g) = \hat{T}(\lambda I_n)\hat{T}(g) = \hat{T}(g)\hat{T}(\lambda I_n)$;

c) $\hat{T}(z'z'') = \hat{T}(z')\hat{T}(z'')$, $z', z'' \in Z^-$;

$$\hat{T}(\zeta'\zeta'') = \hat{T}(\zeta')T(\zeta''), \quad \zeta', \zeta'' \in Z^+;$$

d) $\hat{T}(z)\hat{T}(g_{n-1}) = \hat{T}(g_{n-1})\hat{T}(g_{n-1}^{-1}zg_{n-1})$;

$$\hat{T}(\zeta)\hat{T}(g_{n-1}) = \hat{T}(g_{n-1})\hat{T}(g_{n-1}^{-1}\zeta g_{n-1});$$

e) from the relation $\zeta z = \lambda' z' g_{n-1}\zeta'$, where $\zeta, \zeta' \in Z^+$, $z, z' \in Z^-$, $g_{n-1} \in G_{n-1}$, and $\lambda'$ is a scalar, it follows that

$$\hat{T}(\zeta)\hat{T}(z) = \hat{T}(\lambda I_n)\hat{T}(z')\hat{T}(g_{n-1})\hat{T}(\zeta').$$

So we have to verify relations a)–e). First of all, we observe that relations a) and b) follow immediately from the definition of the operator $\hat{T}(g)$. Further, relation c) immediately follows from condition 1) and relation d) from condition 2) on the functions $F^{(1)}$ and $F^{(2)}$.

Finally, it is easy to see that relation e) follows immediately from conditions 4) and 2).

Thus it is proved that the operators $\hat{T}(g)$ form a representation of the group $G_n$ in the space $H_{m_n}$. By virtue of its construction, this representation is an extension of the representation $T(g_{n-1})$ of the subgroup $G_{n-1}$. Hence (as was already remarked at the beginning of the proof of Theorem 2) the representation $\hat{T}(g)$ is irreducible and, moreover, its "number" is $m_n = (m_{1n}, \cdots, m_{nn})$.

Since $m_n$ was chosen arbitrarily, therefore it is clear that all the irreducible representations of the group $G_n$ can be obtained by means of the construction described. This proves the theorem.

3. **Matrix beta-functions.** Suppose $F^{(1)}(z)$, $F^{(2)}(\zeta)$ are an associated pair of beta-functions. We denote the matrix elements of the operators $F^{(1)}(z)$ and $F^{(2)}(\zeta)$ in the canonical basis by $\mathrm{B}^{(1)}(m', m|z)$ and $\mathrm{B}^{(2)}(m', m|\zeta)$ respectively, and we call them matrix beta-functions.

Thus, the operators $F^{(1)}(z)$ and $F^{(2)}(\zeta)$ act on the basis vectors $m$ as follows:

$$F^{(1)}(z)\,m = \sum_{m'} \mathrm{B}^{(1)}(m', m \mid z)\, m', \tag{11'}$$

$$F^{(2)}(\zeta)\,m = \sum_{m'} \mathrm{B}^{(2)}(m', m \mid \zeta)\, m'. \tag{11''}$$

*We shall show the functions* $\mathrm{B}^{(1)}(m', m|z)$ *and* $\mathrm{B}^{(2)}(m', m|\zeta)$ *are monomials in the coordinates of the vectors* $z$ *and* $\zeta$. *Namely, the following formulas hold:*

$$\mathrm{B}^{(1)}(m', m \mid z) = \mathrm{B}^{(1)}(m', m) \prod_{i=1}^{n-1} z_i^{r_i - r_i'}, \tag{12'}$$

$$\mathrm{B}^{(2)}(m', m \mid \zeta) = \mathrm{B}^{(2)}(m', m) \prod_{i=1}^{n-1} \zeta_i^{r_i' - r_i}, \tag{12''}$$

*where*

$$r_i = k_i - k_{i-1}, \quad k_i = m_{1i} + \ldots + m_{ii} \quad (i = 1, \ldots, n), \quad k_0 = 0, \tag{13}$$

*while the coefficients* $\mathrm{B}^{(1)}(m', m)$ *and* $\mathrm{B}^{(2)}(m', m)$ *are defined by*

$$\mathrm{B}^{(1)}(m', m) = \mathrm{B}^{(1)}(m', m \mid z_0), \quad z_0 = (1, \ldots, 1), \tag{14'}$$

$$\mathrm{B}^{(2)}(m', m) = \mathrm{B}^{(2)}(m', m \mid \zeta_0), \quad \zeta_0 = (1, \ldots, 1). \tag{14''}$$

**Proof.** Suppose first that all the numbers $z_1, \cdots, z_{n-1}$ are different from zero. Let us consider the diagonal matrix $\delta \in G_{n-1}$:

$$\delta = \begin{pmatrix} z_1 & 0 & \ldots & 0 \\ 0 & z_2 & \ldots & 0 \\ \cdot & \cdot & \cdot & \cdot \\ 0 & 0 & \ldots & z_{n-1} \end{pmatrix}.$$

By condition 2) we have the relation

$$F^{(1)}(z_1, \ldots, z_{n-1}) = T^{-1}(\delta)\, F^{(1)}(1, \ldots, 1)\, T(\delta). \tag{15}$$

The operator $T(\delta)$, according to Theorem 1, is given by

$$T(\delta) m = \left( \prod_{i=1}^{n} z_i^{r_i} \right) m.$$

Hence, passing, in equation (15), from the operators their matrix elements, we obtain formula (12').

Formula (12′) is proved under the assumption that $z_1, \cdots, z_{n-1}$ are non-zero. However, since the matrix elements are continuous functions of $z_1, \cdots$ $\cdots, z_{n-1}$, the formula remains valid also for arbitrary $z_1, \cdots, z_{n-1}$.

The derivation of formula (12″) is similar.

Henceforth by the term "matrix beta-functions" we will mean the functions $B^{(1)}(m', m|z)$ and $B^{(2)}(m', m|\zeta)$ themselves as well as the coefficients $B^{(1)}(m', m)$, $B^{(2)}(m', m)$.

We observe that since the right-hand side of equation (12′) is a continuous function of $z_1, \cdots, z_{n-1}$, we have

$$B^{(1)}(m', m) = 0,$$

when $r_i - r_i' < 0$ for at least one $i$. Similarly

$$B^{(2)}(m', m) = 0,$$

when $r_i' - r_i < 0$ for at least one $i$.

It is possible to show (this follows immediately, for example, from the recurrence relations which will be obtained in subsection 4) that the matrices $B^{(1)}(m', m)$ and $B^{(2)}(m', m)$ have triangular form. Namely, $B^{(1)}(m', m) \neq 0$ only in the case when $m_{ij} \geq m_{ij}'$ for all $i, j$; $B^{(2)}(m', m) \neq 0$ only in the case when $m_{ij} \leq m_{ij}'$ for all $i, j$.

Remark. The functions $B^{(1)}(m', m)$ and $B^{(2)}(m', m)$ are defined only for such pairs of arrays $m'$, $m$ for which the top rows coincide. It is convenient, however, to consider them as defined for arbitrary pairs $m'$, $m$, taking

$$B^{(1)}(m', m) = B^{(2)}(m', m) = 0,$$

if the top rows of the arrays $m'$ and $m$ are different. With this definition, the functions $B^{(1)}(m', m)$ and $B^{(2)}(m', m)$ can be regarded as matrices of operators, acting on the *universal* module $H = \Sigma_{m_n} H_{m_n}$, where the summation is carried out over all the intervals $m_n$.

Let us now rewrite the conditions 1), 2) and 4) satisfied by the functions $F^{(1)}(z)$ and $F^{(2)}(\zeta)$ in the form of conditions on their matrix elements. We obtain the following relations.

Condition 1

$$\sum_m B^{(1)}(m', m)\, B^{(1)}(m, m'') \prod_{i=1}^{n-1} \alpha_i^{r_i - r_i'} \beta_i^{r_i'' - r_i} = B^{(1)}(m', m'') \prod_{i=1}^{n-1} (\alpha_i + \beta_i)^{r_i'' - r_i'}, \tag{16'}$$

$$\sum_m B^{(2)}(m', m) B^{(2)}(m, m'') \prod_{i=1}^{n-1} \alpha_i^{r_i'-r_i} \beta_i^{r-r_i''} = B^{(2)}(m', m'') \prod_{i=1}^{n-1} (\alpha_i+\beta_i)^{r_i'-r_i''}, \tag{16''}$$

where

$$r_i = k_i - k_{i-1}, \quad k_i = m_{1i} + \ldots + m_{ii} \quad (i = 1, \ldots, n), \quad k_0 = 0;$$

and $\alpha_i$, $\beta_i$ are arbitrary numbers.

Relations (16′) and (16″) can be given in another form. For this we remove the parentheses in the right-hand sides of the equations and equate coefficients of equal powers of $\alpha_i$ and $\beta_i$. As a result we obtain

$$\sum_m B^{(1)}(m', m) B^{(1)}(m, m'') = B^{(1)}(m', m'') \prod_{i=1}^{n-1} \frac{(r_i'' - r_i')!}{(r_i'' - r_i)!\,(r_i - r_i')!}, \tag{17'}$$

$$\sum_m B^{(2)}(m', m) B^{(2)}(m, m'') = B^{(2)}(m', m'') \prod_{i=1}^{n-1} \frac{(r_i' - r_i'')!}{(r_i' - r_i)!\,(r_i - r_i'')!}. \tag{17''}$$

The sums extended over all the arrays for which the sum $m_{1i} + \cdots + m_{ii} = k_i$ is a given fixed number.

Condition 2

$$\sum_{m', m''} (m^{(1)}, m' \mid g_{n-1}^{-1}) B^{(1)}(m', m'') (m'', m^{(2)} \mid g_{n-1})$$

$$\times \prod_{i=1}^{n-1} \alpha_i^{r_i''-r_i} = B^{(1)}(m^{(1)}, m^{(2)}) \prod_{i=1}^{n-1} (\alpha_1 g_{1i} + \ldots + \alpha_{n-1} g_{n-1,i})^{r_i^{(2)}-r_i^{(1)}}, \tag{18'}$$

$$\sum_{m', m''} (m^{(1)}, m' \mid g_{n-1}^{-1}) B^{(2)}(m', m'') (m'', m^{(2)} \mid g_{n-1}) \prod_{i=1}^{n-1} \alpha^{r_i'-r_i''}$$

$$= B^{(2)}(m^{(1)}, m^{(2)}) \prod_{i=1}^{n-1} (\hat{g}_{i1}\alpha_1 + \ldots + \hat{g}_{i,n-1}\alpha_{n-1})^{r_i^{(1)}-r_i^{(2)}}, \tag{18''}$$

where $\|\hat{g}_{ij}\|$ is the inverse of $g_{n-1} = \|g_{ij}\|$ and $(m', m'' | g_{n-1})$ are the matrix elements of the operator $T(g_{n-1})$, $g_{n-1} \in G_{n-1}$.

Condition 4

$$\sum_{m} B^{(2)}(m^{(1)}, m)\, B^{(1)}(m, m^{(2)}) \prod_{i=1}^{n-1} \alpha_i^{r_i^{(1)}-r_i} \beta_i^{r_i^{(2)}-r_i}$$

$$= \sum_{m, m'} (m^{(1)}, m \mid g_{n-1})\, B^{(1)}(m, m')\, B^{(2)}(m', m^{(2)})$$

$$\times (1 + \alpha_1\beta_1 + \ldots + \alpha_{n-1}\beta_{n-1})^{-r_n^{(2)}} \prod_{i=1}^{n-1} \alpha_i^{r_i'-r_i^{(2)}} \beta_i^{r_i'-r_i}, \qquad (19)$$

where

$$g_{n-1} = \begin{pmatrix} 1+\alpha_1\beta_1 & \alpha_1\beta_2 & \ldots \alpha_1\beta_{n-1} \\ \alpha_2\beta_1 & 1+\alpha_2\beta_2 & \ldots \alpha_2\beta_{n-1} \\ \cdots & \cdots & \cdots \\ \alpha_{n-1}\beta_1 & \alpha_{n-1}\beta_2 & \ldots \alpha_{n-1}\beta_{n-1} \end{pmatrix}$$

Let us indicate explicit formulas for beta-functions for special values of $m'$ and $m$. For this, we note that

$$F^{(1)}(0, \ldots, 0, 1) = T(I + e_{n, n-1}),$$

$$F^{(2)}(0, \ldots, 0, 1) = T(I + e_{n-1, n}).$$

Consequently, by Theorem 1, we conclude that if the arrays $m'$, $m$ differ only in the elements of the second row from the top, then $B^{(1)}(m'; m)$ and $B^{(2)}(m', m)$ are expressed, respectively, by formulas (26) and (28) of §2 where $p = n$.

**4. Recurrence relations for beta-functions.** In this subsection we will obtain recurrence relations expressing beta-functions of rank $n-1$ by means of beta-functions of rank $n-2$.

As a preliminary, we make an observation concerning beta-functions of the lower rank. Let $F_{n-2}^{(1)}$, $F_{n-2}^{(2)}$ of beta-functions of rank $n-2$. By definition, the values of these functions are operators acting in the interval $H_{m_{n-1}}$ of representations of the group $G_{n-2}$. We now consider the arbitrary interval

$$H_{m_n} = \sum_{m_{n-1}} H_{m_{n-1}}$$

of representations of the group $G_{n-1}$. Since the operators $F_{n-2}^{(1)}$, $F_{n-2}^{(2)}$ are given in each of the subspaces $H_{m_{n-1}}$, they can be extended to operators on the whole space $H_{m_n}$. The operators on $H_{m_n}$ thus obtained, will, as before, be denoted by $F_{n-2}^{(1)}$ and $F_{n-2}^{(2)}$.

Repeating this argument, we can regard every pair of beta-functions $F_k^{(1)}(z)$, $F_k^{(2)}(\zeta)$ of rank $k < n-1$ as operators in the space $H_{m_n}$.

We pass on to the derivation of the recurrence relation. From the definition, we have

$$F^{(1)}_{n-2}(1, \ldots, 1) = T(z_{n-1}),$$

where $z_{n-1}$ is a matrix of $G_{n-1}$ in which we have unity in the lower row and in the diagonal and zero at the remaining places. Evidently, this matrix transforms the $(n-1)$-dimensional vector $(0, \cdots 0, 1)$ into the vector $(1, \cdots, 1)$. Consequently, by virtue of relation (5′) for beta-functions, we get

$$F^{(1)}_{n-1}(1, \ldots, 1) = F^{(1)^{-1}}_{n-2}(1, \ldots, 1)\, F^{(1)}_{n-1}(0, \ldots, 0, 1)\, F^{(1)}_{n-2}(1, \ldots, 1). \quad (20')$$

We observe that the operator $F^{(1)}_{n-1}(0, \cdots, 0, 1)$ is given by an explicit formula. Thus, formula (20′) expresses the beta-function of rank $n-1$ by means of the beta-function of rank $n-2$.

Similarly, we have

$$F^{(2)}_{n-1}(1, \ldots, 1) = F^{(2)}_{n-2}(1, \ldots, 1)\, F^{(2)}_{n-1}(0, \ldots, 0, 1)\; F^{(2)^{-1}}_{n-2}(1, \ldots, 1). \quad (20'')$$

Let us write the relations obtained in matrix form. For this we note that the matrix elements $B^{(1)}_{n-2}(m', m)$; $B^{(2)}_{n-2}(m', m)$ of the operators $F^{(1)}_{n-2}(1, \cdots, 1)$ and $F^{(2)}_{n-2}(1, \cdots, 1)$ are expressed by the following formulas.

Let $\overline{m}'$, $\overline{m}$ be the arrays obtained from the arrays $m'$, $m$ by omitting the top row. Then

$$B^{(1)}_{n-2}(m', m) = B^{(1)}_{n-2}(\overline{m}', \overline{m}), \quad (21')$$

$$B^{(2)}_{n-2}(m', m) = B^{(2)}_{n-2}(\overline{m}', \overline{m}), \quad (21'')$$

where in the right-hand sides we have beta-functions of rank $n-2$.

Further, we note that the operator $F^{(1)^{-1}}_{n-2}(1, \cdots, 1)$ is related to the operator $F^{(1)}_{n-2}(1, \cdots, 1)$ as follows:

$$F^{(1)^{-1}}_{n-2}(1, \ldots, 1) = T^{-1}(\delta) F^{(1)}_{n-2}(1, \ldots, 1)\, T(\delta),$$

where $\delta \in G_{n-1}$ is a diagonal matrix in which $\delta_{ii} = 1$ for $i < n-1$, $\delta_{-1,n-1} = 1$. Therefore, on the basis of formula (23) §2, we find that the matrix elements of the operator $F^{(1)^{-1}}_{n-2}(1, \cdots, 1)$ have the form

$$(-1)^{k'_{n-2} - k_{n-2}} B^{(1)}_{n-2}(m', m).$$

Thus, we have proved

**Theorem 4.** *We have the following recurrence relation expressing a matrix beta-function of rank $n-1$ by means of the beta-function of rank $n-2$*

$$B^{(1)}_{n-1}(m^{(1)}, m^{(2)}) = \sum_{m', m} \prod_{i<j} \frac{(l_{i,n-1} - l_{jn} - 1)!}{(l'_{i,n-1} - l_{jn} - 1)!} \prod_{i \leqslant j} \frac{(l_{i,n-1} - l_{j,n-2})!}{(l'_{i,n-1} - l_{j,n-2})!}$$
$$\times \frac{\prod\limits_{i<j} (l'_{i,n-1} - l_{j,n-1} - 1)!}{\prod\limits_{i \leqslant j} (l_{i,n-1} - l'_{j,n-1})!} \prod_{i<j} (l'_{i,n-1} - l'_{j,n-1}) B^{(1)}_{n-2}(m^{(1)}, m') B^{(1)}_{n-2}(m, m^{(2)}). \quad (22')$$

*Similarly*

$$B^{(2)}_{n-1}(m^{(1)}, m^{(2)}) = \sum_{m', m} \prod_{i \leqslant j} \frac{(l_{i,n} - l_{j,n-1})!}{(l_{i,n} - l'_{j,n-1})!} \prod_{i<j} \frac{(l_{i,n-2} - l_{j,n-1} - 1)!}{(l_{i,n-2} - l'_{j,n-1} - 1)!}$$
$$\times \frac{\prod\limits_{i<j} (l_{i,n-1} - l'_{i,n-1} - 1)!}{\prod\limits_{i \leqslant j} (l'_{i,n-1} - l_{j,n-1})!} \prod_{i<j} (l'_{i,n-1} - l'_{j,n-1}) B^{(2)}_{n-2}(m^{(1)}, m') B^{(2)}_{n-2}(m, m^{(2)}). \quad (22'')$$

It is easily seen that the summation is actually carried out over arrays $m'$ and $m$ for which the two top rows are fixed

$$m'_{i,n-1} = m^{(1)}_{i,n-1}, \quad m_{i,n-1} = m^{(2)}_{i,n-1}, \quad i = 1, \ldots, n-1,$$

and the remaining rows coincide:

$$m'_{ij} = m_{ij} \quad \text{for } j < n-1.$$

## Supplement

**Discrete series of irreducible representations of real forms of the group $G_n$**

1. We now consider, instead of the group $G_n$ of nonsingular complex matrices of the $n$th order, its Lie algebra $L$. We will show here that with each finite-dimensional irreducible representation of the algebra $L$ is associated a finite set of infinite-dimensional irreducible representations which are, in a sense, analytic continuations of the finite-dimensional representation. The operators of these representations will be described by explicit formulas for their matrix elements in the Gel'fand-Cetlin basis.

On the basis of this result, we will construct discrete series of irreducible "unitary" representations of real forms of the algebra $L$.

Namely, let $G^{p,q}$ be the group of complex matrices of the $n$th order preserving the Hermitian form

$$|x_1|^2 + \ldots + |x_p|^2 - |x_{p+1}|^2 - \ldots - |x_{p+q}|^2, \quad p + q = n,$$

and let $L^{p,q}$ be the real Lie algebra of the group $G^{p,q}$. It is known that $G^{p,q}$ is one of the real forms of the group $G$ and, therefore, the algebra $L^{p,q}$ is one of the real forms of the algebra $L$. We call the representation of the algebra $L^{p,q}$ unitary if the operators of the representation are skew-hermitian (for the reason that the operators of the corresponding representation of the Lie group, if it exists, are unitary).

We will show that among the infinite-dimensional representations, associated with a given finite-dimensional representation, there are unitary representations of the algebra $L^{p,q}$. By the same token, "discrete series" of irreducible unitary representations of the algebra $L^{p,q}$, related to the finite-dimensional representations, will be constructed. Each of these representations is specified by a set of integers, the highest weight of the corresponding finite-dimensional representation and the index, distinguishing this representation among all the representations associated with a given finite-dimensional representation.

2. We shall say that a representation of the algebra $L$ in the space $H$ is given if one has a mapping $f \to F$, associating, with every element $f \in L$, the linear operator $F$ in $H$ and satisfying the following conditions:

1) from $f_1 \to F_1$, $f_2 \to F_2$, it follows that

$$\lambda_1 f_1 + \lambda_2 f_2 \to \lambda_1 F_1 + \lambda_2 F_2,$$

where $\lambda_1 \lambda_2$ are arbitrary complex numbers, and

$$[f_1, f_2] \to [F_1, F_2] \equiv F_1 F_2 - F_2 F_1;$$

2) the operators $F$ are defined on the set of all vectors of the form

$$\lambda_1 e_1 + \ldots + \lambda_n e_n,$$

where $e_1, \cdots, e_n, \cdots$ is some orthogonal basis in $H$, and transform this set into itself.

We know that the algebra $L$ is the algebra of all complex matrices of the $n$th order. Let us take as a basis in $L$ the matrices $e_{kl}$, $k, l = 1, \cdots, n$ defined in §2. The algebra $L$ is completely determined by the commutation relations among the basis elements $e_{kl}$, which have the form

$$[e_{ik}, e_{kl}] = e_{il} \quad \text{for} \quad i \neq l,$$

$$[e_{ik}, e_{ki}] = e_{ii} - e_{kk},$$

$$[e_{i_1 k_1}, e_{i_2 k_2}] = 0 \quad \text{for} \quad k_1 \neq i_2, \quad i_1 \neq k_2.$$

Thus, giving the representation of the algebra $L$ is equivalent to giving the operator $E_{kl}$, $k, l = 1, \cdots, n$ in the Hilbert space $H$, satisfying the commutation relations

$$\begin{gathered} [E_{ik}, E_{kl}] = E_{il} \quad \text{for} \quad i \neq l; \quad [E_{ik}, E_{ki}] = E_{ii} - E_{kk}; \\ [E_{i_1 k_1}, E_{i_2 k_2}] \quad \text{for} \quad k_1 \neq i_2, \quad i_1 \neq k_2. \end{gathered} \tag{1}$$

We note that, in fact, it is sufficient to specify only the system of operators $E_{kk}$, $E_{k,k-1}$, $E_{k-1,k}$ satisfying the following commutation relation (K):

$$[E_{ii}, E_{kk}] = 0;$$
$$[E_{ii}, E_{k,k-1}] = [E_{ii}, E_{k-1,k}] = 0 \quad \text{for } i < k-1;$$
$$[E_{ii}, E_{i,i-1}] = [E_{i,i-1}, E_{i-1,i-1}] = E_{i,i-1};$$
$$[E_{i-1,i-1}, E_{i-1,i}] = [E_{i-1,i} E_{ii}] = E_{i-1,i};$$
$$[E_{i,i-1}, E_{i-1,i}] = E_{ii} - E_{i-1,i-1}; \quad [E_{i,i-1}, E_{k-1,k}] = 0 \quad \text{for } k \neq i;$$
$$[E_{i,i-1}, E_{k,k-1}] = [E_{i-1,i}, E_{k-1,k}] = 0 \quad \text{for } k \neq i \pm 1.$$

If these operators are given, then the operators $E_{kl}$ are determined, by induction on $|k-l|$, from the formulas

$$E_{k,k-p} = [E_{k,k-1}, E_{k-1,k-p}], \quad E_{k-p,k} = [E_{k-p,k-1}, E_{k-1,k}].$$

It is easily seen that the operators $E_{kl}$ constructed satisfy the commutation relations (1).

Henceforth, in describing the representations, we will restrict ourselves to formulas for the operators $E_{k,k}$, $E_{k,k-1}$ and $E_{k-1,k}$ only.

3. To construct the infinite-dimensional representations of the algebra $L$, we begin with the formulas for finite-dimensional representations of the algebra $L$, obtained in [1].

Every finite-dimensional irreducible representation of the algebra $L$ is specified by a set of $n$ integers $m_{1n} \geq \cdots \geq m_{nn}$. It acts in the finite-dimensional vector space in which, as orthonormal basis, we have all possible triangular arrays

$$m = \begin{bmatrix} m_{1n} & m_{2n} \ldots & m_{nn} \\ & m_{1,n-1} \ldots m_{n-1,n-1} & \\ & m_{11} & \end{bmatrix}, \tag{2}$$

where $m_{ij}$ are integers satisfying the conditions

$$m_{ij} \geqslant m_{i,j-1} \geqslant m_{i+1,j}.$$

In these arrays, the top row is fixed; it specifies the representation itself. The elements of the remaining rows can assume arbitrary admissible values.

The representation operators $E_{kk}$, $E_{k,k-1}$, $E_{k-1,k}$ are given by the formulas

$$E_{kk}\, m = (r_k - r_{k-1})\, m, \tag{3}$$

where $r_k = m_{1k} + \cdots + m_{kk}$ for $k = 1, \cdots, n$; $r_0 = 0$;

$$E_{k,k-1} m = a^1_{k-1} m^1_{k-1} + \ldots + a^{k-1}_{k-1} m^{k-1}_{k-1}, \tag{4}$$

where $m^j_{k-1}$ is the array obtained from $m$ on replacing $m_{j,k-1}$ by $m_{j,k-1}-1$;

$$a^j_{k-1}=\left[-\frac{\prod_{i=1}^{k}(m_{ik}-m_{j,k-1}-i+j+1)\prod_{i=1}^{k-2}(m_{i,k-2}-m_{j,k-1}-i+j)}{\prod_{i\neq j}(m_{i,k-1}-m_{j,k-1}-i+j+1)(m_{i,k-1}-m_{j,k-1}-i+j)}\right]^{\frac{1}{2}}; \quad (5)$$

$$E_{k-1,k}\,m=b^1_{k-1}\hat{m}^1_{k-1}+\ldots+b^{k-1}_{k-1}\hat{m}^{k-1}_{k-1}, \quad (6)$$

where $\hat{m}^j_{k-1}$ is the array obtained from $m$ on replacing $m_{j,k-1}$ by $m_{j,k-1}+1$;

$$b^j_{k-1}=\left[-\frac{\prod_{i=1}^{k}(m_{ik}-m_{j,k-1}-i+j)\prod_{i=1}^{k-2}(m_{i,k-2}-m_{j,k-1}-i+j-1)}{\prod_{i\neq j}(m_{i,k-1}-m_{j,k-1}-i+j)(m_{i,k-1}-m_{j,k-1}-i+j-1)}\right]^{\frac{1}{2}} \quad (7)$$

**4. Description of infinite-dimensional representations of the algebra $L$.** Let us introduce arrays of a more general form than (2). For every $k=1,\cdots,n-1$, we assign a pair of indices $i_k$, $i'_k\,(i_k<i'_k)$ which can assume the values $i_k=0, 1,\cdots,k$; $i'_k=1,\cdots,k+1$. With every such set of indices $i_k$, $i'_k$ we associate the set of schemes of the form (2), whose elements satisfy the inequalities

1) $m_{j,k}\geqslant m_{j+1,k}$;
2) $m_{j-1,k+1}+1\geqslant m_{jk}\geqslant m_{j,k+1}+1$ for $j\leqslant i_k$, where $m_{0,k+1}=+\infty$;
3) $m_{j,k+1}\geqslant m_{j,k}\geqslant m_{j+1,k+1}$ for $i_k<j<i'_k$;
4) $m_{j+1,k+1}-1\geqslant m_{j,k}\geqslant m_{j+2,k+1}-1$ for $j\geqslant i'_k$, where $m_{k+2,k+1}=-\infty$.

We note that the original array corresponds to the case $i_k=0$, $i'_k=k+1$; these arrays will be called normal arrays.

Let us consider all possible arrays $m$ in which the top row is fixed and the elements of the remaining rows satisfy conditions 1)–4).

We introduce the Hilbert space $H\{i_k, i'_k\}$ in which these arrays form an orthonormal basis. The operators $E_{kk}$, $E_{k-1,k}$, $E_{k,k-1}$ on $H\{i_k, i'_k\}$ are given by the same formulas (3)–(7) as in the case of a finite-dimensional representation. It is easily seen that these operators transform every basis vector of $H$ again into a linear combination of basis vectors of $H$.

We observe that formulas (5)–(7) determine the coefficients $a^j_{k-1}$, $b^j_{k-1}$ up to sign.

We indicate the principle for choosing the sign. Suppose that $N^j_k$, $\hat{N}^j_k$ are the number of factors in the expressions for $a^j_{k-1}$, $b^j_{k-1}$ respectively, whose signs are changed in passing from a normal array to the given array $m$. It is evident that $N^j_k=\hat{N}^j_k$ and moreover, this number is the same for all arrays of the space

$H\{i_k, i'_k\}$.

Let us suppose that

$$\arg a^j_{k-1} = \arg b^j_{k-1} = \frac{\pi}{2} N^j_k. \tag{8}$$

It is easily seen that for this determination of the coefficients $a^j_{k-1}, b^j_{k-1}$, the operators $E_{kk}, E_{k,k-1}, E_{k-1,k}$ satisfy the commutation relation (K); consequently, they give the representation of the Lie algebra $L$ in the Hilbert space $H\{i_k, i'_k\}$. This representation is irreducible.

*Thus with every finite-dimensional irreducible representation of the algebra $L$, we have associated a set of infinite-dimensional representations of $L$.* As a simple calculation shows, the total number of these representations (including the original finite-dimensional representation) is equal to $2^{-n} n! (n+1)!$.

5. **Discrete series of unitary representations** of real forms of the algebra $L$. To every partition $n = p + q$, $p \geq q$ of the number $n$ as the sum of integral nonnegative numbers corresponds a real form $L^{p,q}$ of the algebra $L$. The space of the algebras $L^{p,q}$ is a real linear subspace in $L$, spanned by the vectors $ie_{kk}$, $k = 1, \cdots, n$ and the vectors $h^+_{kl}, h^-_{kl}$ $(k > l)$ defined by the formulas

$$h^+_{kl} = e_{kl} - e_{lk}, \quad h^-_{kl} = i(e_{kl} + e_{lk})$$

for $k, l \leq p$ and for $k, l > p$;

$$h^+_{kl} = e_{kl} + e_{lk}, \quad h^-_{kl} = i(e_{kl} - e_{lk})$$

for $k > p$, $l \leq p$.

A representation of the algebra $L$ in the space $H\{i_k, i'_k\}$ induces a representation of the algebra $L^{p,q}$; it is irreducible if the original representation of $L$ is irreducible. Let us determine those values of the indices $i_k, i'_k$ for which this representation of the algebra $L^{p,q}$ is unitary.

We observe that the condition for the representation of $L^{p,q}$ to be unitary is equivalent to the following relations for the operators $E_{kl}$:

1) $E^*_{kk} = E_{kk}$; $k = 1, \dots, n$;

2) $E^*_{k,k+1} = E_{k+1,k}$ for $k \neq p$; $E^*_{p,p+1} = -E_{p+1,p}$.

The following assertion is easily verified.

*The operators $E_{kk}, E_{k,k-1}$ and $E_{k-1,k}$, defined by formulas (3)–(7), satisfy relations 1) and 2) for those and only those spaces $H\{i_k, i'_k\}$ for which the indices $i_k, i'_k$ are determined by the conditions*

$$i_k = 0, \quad i'_k = k + 1 \quad \textit{for } k < p;$$

$$i'_p = i_p + 1, \quad i_p = 0, 1, \dots, p; \quad i_k = i_p, \quad i'_k = i'_p + k - p \quad \textit{for } k > p.$$

In this way, the spaces $H\{i_k, i'_k\}$ are determined completely by the index $i_p$, which may assume the values $0, 1, \cdots, p$.

*As a result, we have obtained* $p+1$ *series of irreducible unitary representations of the algebra* $L^{p,q}$. *The number of the series is given by the index* $i_p$. *For each series, the representation is specified by a set of n integers—the highest weight of the corresponding finite-dimensional representation.*

One can easily see that all *the representations constructed are mutually inequivalent.*

6. In the case of the algebra $L^{n-1,1}$, it is possible to obtain a similar description of the representations of the continuous series. We consider triangular arrays $m$ of the form (2), characterizable by the conditions

1) $m_{1n} = -(n-1)/2 + \sigma$, $m_{nn} = (n-1)/2 + \bar{\sigma}$ where $\sigma$ is a complex number; the other elements of the array $m$ are integers;

2) for all elements $m_{ij}$, with the exception of $m_{1,n-1}$ and $m_{n-1,n-1}$, we have the inequalities $m_{i,j+1} \geq m_{ij} \geq m_{i+1,j+1}$.

Let us take the set of those arrays for which the top row is fixed; we take these arrays as an orthonormal basis in the Hilbert space $H$. In this space, we define operators $E_{kk}$, $E_{k,k-1}$ and $E_{k-1,k}$ by the formulas (3)–(7) (here the coefficients $a^j_{k-1}$, $b^j_{k-1}$ are, for $k < n-1$, real positive numbers while $a^j_{n-1}$, $b^j_{n-1}$ are purely imaginary numbers; for definiteness, let $\arg a^j_{n-1} = \arg b^j_{n-1} = \pi/2$). It is easy to verify that *the operators* $E_{kk}$, $E_{k,k-1}$, $E_{k-1,k}$ *give, in the space* $H$, *an irreducible unitary representation of the algebra* $L^{n-1,1}$. This representation is determined by specifying the top row of the array $m$ i.e. by the complex number $\sigma$ and the set of $n-2$ integers $m_{2,n} \geq \cdots \geq m_{n-1,n}$.

## BIBLIOGRAPHY


[1] I. M. Gel'fand and M. L. Cetlin, *Finite-dimensional representations of the group of unimodular matrices*, Dokl. Akad. Nauk SSSR 71 (1950), 825. (Russian) MR 12, 9.

[2] I. M. Gel'fand and M. A. Naĭmark, *Unitary representations of the classical groups*, Trudy Mat. Inst. Steklov., Vol. 36, Izdat. Akad. Nauk SSSR, Moscow, 1950. (Russian); German transl., Akademie Verlag, Berlin, 1957. MR 13, 722; MR 19, 13.

[3] I. M. Gel'fand, R. A. Minlos and Z. Ja. Šapiro, *Representations of the rotation group and of the Lorentz group*, Fizmatgiz, Moscow, 1958. (Russian) MR 22 #5694.

[4] G. E. Baird and L. C. Biedenharn, *On the representations of the semi-simple Lie groups*. II, J. Math. Phys. 4 (1963), 1449. MR 28 #165.


Translated by V. N. Singh

# Part VIII

## Indecomposable representations of semisimple Lie groups and of finite-dimensional algebras; problems of linear algebra

# 1.
## (with V. A. Ponomarev)
# Indecomposable representations of the Lorentz group

Usp. Mat. Nauk **23** (2) 3-60 (1968) [Russ. Math. Surv. **23** (2) (1968) 1-58].
MR **38**:5325. Zbl. **236**:22012

Let $L$ be the Lie algebra of the Lorentz group or, what is the same, of the group $SL(2,C)$. We denote by $L_k$ the Lie algebra of its maximal compact subgroup, that is, of $SU(2)$. Let $M_i$ be the finite-dimensional irreducible $L_k$-modules (the finite-dimensional representations of $L_k$). Consider an $L$-module $M$. The authors call $M$ a *Harish-Chandra module* if, regarded as $L_k$-module, it can be written as a sum

$$M = \bigoplus_i M_i$$

of finite-dimensional irreducible $L_k$-modules $M_i$. Here, for each $M_{i_0}$, only finitely many $L_k$-submodules equivalent to $M_{i_0}$ are supposed to occur in the decomposition of $M$.

A Harish-Chandra module is called indecomposable if it cannot be decomposed into the direct sum of $L$-submodules. In this paper the indecomposable Harish-Chandra modules over $L$ are completely described. We find that there are two types of indecomposable Harish-Chandra modules. The modules of the first type are the non-singular Harish-Chandra modules and are defined by the following invariants: an integer $2l_0 (l_0 \geqslant 0)$, a complex number $l_1$, and an integer $n$. The first two of these invariants are already known as invariants of the irreducible representations of the Lorentz group (see [2]). The case of non-singular modules has been investigated earlier by Zhelobenko [3] from a somewhat different approach.

The case of singular Harish-Chandra modules is of the greatest interest. The solution of this problem reduces to a non-trivial problem of linear algebra, which is investigated in detail in Chapter 2. The invariants of singular indecomposable modules are, as before, numbers $l_0$, $l_1$, $l_0 \geqslant 0$, $2l_0$ integral and $2l_0 - |l_1|$ integral.

However, instead of the one additional invariant $n$, there are now more invariants. Two types of singular modules are possible: those of the first and those of the second kind.

Singular modules of the first kind are characterized, in addition to the invariants $l_0$ and $l_1$, by a sequence of integers of arbitrary length. Singular indecomposable modules of the second kind are characterized by the following collection of invariants: the numbers $l_0$, $l_1$ given above, a set of integers $j_1, j_2, \ldots, j_k$, an integer $q$, and a further arbitrary complex parameter $\mu$. The presence of this parameter is particularly interesting, because it indicates the possibility of deforming an indecomposable module with $l_0$ and $l_1$ fixed.

The problems of linear algebra that are used in establishing the facts set out above are of independent interest in that the authors develop and use the apparatus of MacLane's theory of linear relations [4].

## Contents



## Introduction

1. In this paper we investigate a certain fairly general class of representations of the Lorentz group. While *irreducible* representations of the Lorentz group have long been thoroughly investigated, the study of other representations of this group involves considerable difficulties. Thus, for example, the indecomposable representations of the Lorentz group have not been classified as yet.

However, this situation should not cause surprise. Already for the simplest non-compact group – the additive group $R$ of real numbers – the study of the finite-dimensional representations is more difficult than the study of the irreducible ones. The irreducible representations of this group are all one-dimensional and can, of course, easily be enumerated. But the study of the finite-dimensional representations of $R$ reduces to the transformation of matrices into Jordan form, and this, naturally, is more difficult.

Finite-dimensional representations are known to decompose into direct sums of indecomposables. Each indecomposable representation of the group $R$ is characterized by a corresponding irreducible and one further invariant – the dimension of the Jordan block.

As for infinite-dimensional representations of $R$, this question is equivalent to the problem of canonical forms of operators in infinite-dimensional spaces. And here it is not even clear whether the question of a canonical form is a reasonable one.

2. We consider representations of the Lorentz group that are, so to speak, analogous to finite-dimensional representations of $R$. These

representations correspond to the modules we have called Harish-Chandra modules. Certainly we cannot demand that such a representation acts in a finite-dimensional space. For even the irreducible representations of the Lorentz group are, as a rule, infinite-dimensional. The representations we shall study (Harish-Chandra modules) are obtained by glueing together a finite number of irreducible representations.

The indecomposable representations of the Lorentz group may be divided into two kinds – "singular" and "non-singular".[1] For "non-singular" representations the situation coincides almost exactly with the Jordan normal form. A non-singular indecomposable representation is obtained by "glueing together" a few equivalent irreducible ones. Such a representation is completely characterized by the irreducible representations contained in it, and a number $n$ – the size of the block. The investigation of this case was completely carried out by Zhelobenko [3].

Considerably more difficult and interesting is the so-called singular case. Whilst non-singular representations are obtained by "glueing together" a few equivalent irreducibles, in the singular case continguous irreducible representations are glued together. Two types of indecomposable representation of the proper Lorentz group are possible. One type is called an "*open chain*", and the other, a "*closed chain*".

Whereas a non-singular indecomposable representation has only one additional invariant, an integer $n$, in the singular case the number of invariants is considerably larger. Thus, in the case of an open chain, the set of invariants is a sequence of integers of arbitrary length. As an invariant of a closed chain, in addition to a sequence of integers, there also appears a complex number $\mu$.[2]

The investigation of the indecomposable representations of the Lorentz group leads to a certain problem of linear algebra. It seems to us that this problem is of independent interest. Its solution (Chapter II) may be read independently.

We think that the application of the machinery of the theory of relations introduced by MacLane to the solution of this problem merits attention. We also feel that the machinery developed in Chapter II will be useful for a number of other questions.

This reduction of a problem of representation theory to one of linear algebra is also interesting because in our linear algebra problem we can replace finite-dimensional spaces by infinite-dimensional ones: normed, barrelled, and so on, and thus proceed to a much larger class of infinite-dimensional representations.

---

[1] The general definition of "singular" and "non-singular" representations was introduced in essence in [1]. The idea of "contiguous" irreducible representations was also introduced there. These are representations that cannot be separated by neighbourhoods in the representation space. In this terminology, the non-singular representations are those having no contiguous representations, and the special representations are those having contiguous representations. However, we shall not use this in this paper.

[2] Thus, these structures, because of the presence of the number $\mu$, admit deformations, a fact that was difficult to predict beforehand.

We restrict our investigation of the representations of the proper Lorentz group. Furthermore, instead of representations of the proper Lorentz group, we discuss the problem of representations of the Lie algebra of this group.

We are very grateful to A.P. Lavut for his help in editing this paper

## Chapter I

1. We recall that a Lie algebra is a finite-dimensional space over a field $K$, in which an operation of commutation is given satisfying the following requirements.

1) The commutator $[x, y]$ of the elements $x, y$ depends bilinearly on $x$ and $y$, that is,

$$[x_1+x_2, y]=[x_1, y]+[x_2, y], \quad [x, y_1+y_2]=[x, y_1]+[x, y_2],$$
$$\alpha[x, y]=[\alpha x, y]=[x, \alpha y] \quad (\alpha \in K);$$

2) for any elements $x, y, z \in L$,

$$[[x, y], z]+[[y, z], x]+[[z, x], y]=0 \quad \text{(Jacobi identity)}$$

3) $[x, y]+[y, x]=0.$

In particular, the Lie algebra of the proper Lorentz group is a six-dimensional space over $R$. In this space, a basis of six elements $a_1, a_2, a_3, b_1, b_2, b_3$ can be chosen so that the commutators of these elements have the following form:

$$\left.\begin{array}{l}
[a_1, a_2]=a_3, \quad [a_2, a_3]=a_1, \quad [a_3, a_1]=a_2, \\
{[b_1, b_2]=-a_3, \quad [b_2, b_3]=-a_1, \quad [b_3, b_1]=-a_2,} \\
{[a_1, b_1]=[a_2, b_2]=[a_3, b_3]=0, \quad [a_1, b_2]=b_3, \quad [a_1, b_3]=-b_2,} \\
{[a_2, b_3]=b_1, \quad [a_2, b_1]=-b_3,} \\
{[a_3, b_1]=b_2, \quad [a_3, b_2]=-b_1.}
\end{array}\right\} \tag{1}$$

The Lorentz group contains a subgroup isomorphic to the group of rotations of a three-dimensional space. In accordance with this, the Lie algebra $L$ contains a subalgebra $L_k$ with basis $a_1, a_2, a_3$, isomorphic to the Lie algebra of the group of rotations of a three-dimensional space.

It is convenient to consider, instead of the Lie algebra $L$ over $R$, the complexification of this algebra, which we also denote by $L$. In what follows the symbol $L$ only has this meaning.

The Lie algebra $L$ is an algebra of dimension six over the complex numbers C. It is given by the commutation relations (1) between the basis elements $a_1, a_2, a_3, b_1, b_2, b_3$.

In the Lie algebra over C it is convenient to take, instead of the basis elements $a_i, b_i$, a basis consisting of the following elements:

$$\left.\begin{array}{lll}
h_+=ia_1-a_2, & h_-=ia_1+a_2, & h_3=ia_3, \\
f_+=ib_1-b_2, & f_-=ib_1+b_2, & f_3=ib_3.
\end{array}\right\} \tag{2}$$

It is easy to obtain the commutation relations between these basis elements:

$$\left.\begin{array}{llll} [h_+, h_3] = -h_+, & [h_-, h_3] = h_-, & [h_+, h_-] = 2h_3, & \\ [h_+, f_+] = [h_-, f_-] = [h_3, f_3] = 0, & & [h_+, f_3] = -f_+, & [h_-, f_3] = f_-, \\ [h_+, f_-] = 2f_3, & [h_3, f_-] = -f_-, & [h_-, f_+] = -2f_3, & [h_3, f_+] = f_+, \\ [f_+, f_3] = h_+, & [f_-, f_3] = -h_-, & [f_+, f_-] = -2h_3. & \end{array}\right\} \quad (3)$$

We denote by $L_k$ the subalgebra of $L$ with basis $h_+$, $h_-$, $h_3$. This subalgebra is determined by the rotation subgroup of the Lorentz group.

2. A representation of the Lie algebra $L$, or a module over $L$, is a linear space $M$ in which, to each element $a \in L$ there corresponds a linear mapping $A$ of $M$, such that if $[a, b] = c$, then $AB - BA = C$. In accordance with the algebraic setting of this paper, we do not give $M$ a topology. We do not assume that $M$ is finite-dimensional. Indeed, in all the most interesting cases, $M$ is infinite-dimensional. However, an investigation of all representations $M$, without restriction, is probably not very reasonable. We restrict ourselves to the modules we shall call Harish-Chandra modules, which are defined below.

First let us consider modules (representations) corresponding to the rotation subgroup, that is, to the subalgebra $L_k$ with generators $h_+$, $h_-$, $h_3$. If such a module is finite-dimensional, then it is the direct sum of irreducible modules.

We recall the form of the finite-dimensional irreducible $L_k$-modules. Each such module is given by a number $l$, an integer or half an integer. An irreducible $L_k$-module is a finite-dimensional linear space of dimension $2l + 1$ over C in which linear mappings $H_+$, $H_-$, $H_3$ operate corresponding to the elements $h_+$, $h_-$, $h_3$ and satisfying the commutation relations:

$$[H_+, H_3] = -H_+, \quad [H_-, H_3] = H_-, \quad [H_+, H_-] = 2H_3.$$

In this space we chose a basis $\{e_m\}$ $(m = -l, -l+1, \ldots, l-1, l)$ of eigenvectors of $H_3$.

$$H_3 e_m = m e_m. \quad (4)$$

The operators $H_+$ and $H_-$ are given in this basis by the formulae:

$$H_+ e_m = \sqrt{(l+m+1)(l-m)}\, e_{m+1}, \qquad H_- e_m = \sqrt{(l+m)(l-m+1)}\, e_{m-1}. \quad (5)$$

By $R_l$ we mean a finite-dimensional $L$-module which is the direct sum of irreducible $L_k$-modules with the same integer $l$.

3. As we have already said, a module for $L$ is a vector space over C in which operators $H_+$, $H_-$, $H_3$, $F_+$, $F_-$, $F_3$ act satisfying the commutation relations obtained from (3) by the substitution $h \to H$, $f \to F$.

An $L$-module can be regarded as an $L_k$-module. We discuss only modules $M$ that are an algebraic[1] direct sum $M = \bigoplus_i M_i$ of finite-dimensional

---

1 The term "algebraic" means that each vector $\xi$ in $M$ is the sum $\xi = \sum_i \xi_i$ of a finite number of elements $\xi_i \in M_i$.

irreducible $L_k$-submodules, where for each $M_{i_0}$ the decomposition $M=\bigoplus_i M_i$ only contains a finite number of submodules equivalent to $M_{i_0}$. We call such modules for $L$, Harish-Chandra modules.

Thus, a module $M$ for the Lie algebra $L$ of the Lorentz group is a Harish-Chandra module if as $L_k$-module it is a direct sum of submodules $R_l$.

This definition may be carried over to any semisimple Lie algebra $L$. Let $G$ be a semisimple Lie group and $G_k$ its maximal compact subgroup. Let $L$ denote the Lie algebra of $G$, and let $L_k$ be the subalgebra corresponding to $G_k$. We call a module $M$ for $L$ a Harish-Chandra module if it is the algebraic direct sum of $L_k$-submodules $M_i$. Here each $M_i$ is an irreducible $L_k$-module, and for each $M_{i_0}$ the decomposition $M=\bigoplus_i M_i$ contains only a finite number of submodules equivalent to $M_{i_0}$.

In this paper we are concerned only with representations of the Lie algebra of the Lorentz group. This means that we must find a module $M$ such that:

a) operators $H_+$, $H_-$, $H_3$, $F_+$, $F_-$, $F_3$ act on $M$ and satisfy the commutation relations

$$\left.\begin{array}{llll}[H_+,\ H_3]=-H_+, & [H_-H_3]=H_-, & [H_+,\ H_-]=2H_3, & \\ [H_+,\ F_+]=[H_-,\ F_-]=[H_3,\ F_3]=0, & & [H_+,\ F_3]=-F_+, & [H_-,\ F_3]=F_-, \\ [H_+,\ F_-]=2F_3, & [H_3,\ F_-]=-F_-, & [H_-;\ F_+]=-2F_3, & [H_3,F_+]=F_+, \\ [F_+,\ F_3]=H_+, & [F_-,\ F_3]=-H_-, & [F_+,\ F_-]=-2H_3. & \end{array}\right\} \tag{6}$$

b) $M$ is a Harish-Chandra module.

## §1. Formulae for the operators $H_+$, $H_-$, $H_3$, $F_+$, $F_-$, $F_3$ and irreducible Harish-Chandra modules.

In this section we give expressions for $H_i$ and $F_i$ in terms of auxiliary mappings $E_+$, $E_-$, $D_+$, $D_-$, $D_0$. This enables us to determine the irreducible Harish-Chandra modules.

**1. The subspace $R_{l,m}$.** Let us describe how the operators $H_+$, $H_-$, $H_3$, which correspond to the basis of $L_k$, operate on $M$. By definition, $M$ is the direct sum of finite-dimensional subspaces $R_l$ such that on each subspace $R_l$, the representation of $L_k$ is multiply irreducible. In accordance with the formulae (4) and (5) for irreducible representations, $R_l$ can be written as the direct sum of subspaces $R_{l,m}$, where $R_{l,m}$ is an eigenspace for $H_3$.

$$H_3\xi=m\xi\,(m=-l,\ -l+1,\ \dots,\ l_0)\ \text{for each}\ \xi\in R_{l,\ m}. \tag{1.1}$$

The operators $H_+$ and $H_-$ map the $R_{l,m}$ into each other:

$$\left.\begin{array}{llll} H_+: & R_{l,m}\to R_{l,m+1}\quad(-l\leqslant m<l), & H_+: & R_{l,l}\to 0, \\ H_-: & R_{l,m}\to R_{l,m-1}\quad(-l<m\leqslant l), & H_-: & R_{l,-l}\to 0. \end{array}\right\} \tag{1.2}$$

For each $\xi\in R_{l,m}$ we have the relations:

$$H_-H_+\xi=(l+m+1)(l-m)\xi,\qquad H_+H_-\xi=(l+m)(l-m+1)\xi, \tag{1.3}$$

which are consequences of (5).

We use the decomposition $M = \bigoplus_{l,m} R_{l,m}$ throughout the article.

2. The auxiliary mappings $E_+$ and $E_-$. We require mappings giving isomorphisms between the subspaces $R_{l,m}(m = -l, -l+1, \ldots, l-1, l)$ of $R_l$. We define these mappings $E_+$ and $E_-$ by the following formulae:

$$\left.\begin{aligned} H_+\xi &= \sqrt{(l+m+1)(l-m)}\, E_+\xi, && \text{when} \quad \xi \in R_{l,m}, \quad m \neq l, \\ E_+\xi &= 0, && \text{when} \quad \xi \in R_{l,l}; \end{aligned}\right\} \tag{1.4}$$

$$\left.\begin{aligned} H_-\xi &= \sqrt{(l+m)(l-m+1)}\, E_-\xi, && \text{when} \quad \xi \in R_{l,m} \quad m \neq -l, \\ E_-\xi &= 0, && \text{when} \quad \xi \in R_{l,-l}. \end{aligned}\right\} \tag{1.5}$$

Thus, $E_+$ maps $R_{l,m}$ into $R_{l,m+1}$ and $E_-$ maps $R_{l,m}$ into $R_{l,m-1}$. It is easy to verify the following relations:

$$E_-E_+\xi = \xi \quad (\xi \in R_{l,m};\ m \neq l), \quad E_+E_-\xi = \xi \quad (\xi \in R_{l,m};\ m \neq -l).$$

Consequently, $E_+$: $R_{l,m} \to R_{l,m+1}$ is an isomorphism for $m \neq l$ and $E_-$: $R_{l,m+1} \to R_{l,m}$ is its inverse.

3. Formulae for the operators $F_+$, $F_-$, $F_3$. The auxiliary mappings $D_0$, $D_+$, $D_-$. The operators $F_+$, $F_-$, $F_3$ correspond to the basis elements $f_+$, $f_-$, $f_3$ of $L$. In order to determine how these operators act on $M$, we must find all solutions of the following commutation relations:

$$\left.\begin{aligned} &[H_+, F_+] = [H_-, F_-] = [H_3, F_3] = 0, \\ &[H_+, F_3] = -F_+, \quad [H_-, F_3] = F_-, \\ &[H_+, F_-] = 2F_3, \quad [H_3, F_-] = -F_-, \\ &[H_-, F_+] = -2F_3, \quad [H_3, F_+] = F_+, \end{aligned}\right\} \tag{1.6}$$

$$[F_+, F_3] = H_+, \quad [F_-; F_3] = -H_-, \quad [F_+, F_-] = -2H_3. \tag{1.7}$$

And so the situation is the following. We are given a countable number of finite-dimensional spaces $R_{l,m}$ and, in the direct sum $\bigoplus_{l,m} R_{l,m}$ isomorphisms

$$\begin{aligned} E_+:&\quad R_{l,m} \to R_{l,m+1} \quad (m = -l, -l+1, \ldots, l-1), \\ E_-:&\quad R_{l,m} \to R_{l,m-1} \quad (m = -l+1, \ldots, l), \\ E_+:&\quad R_{l,l} \to 0, \quad E_-: R_{l,-l} \to 0 \end{aligned}$$

such that $E_+E_- = I$ for $R_{l,m}$, $m \neq -l$, $E_-E_+ = I$ for $R_{l,m}$, $m \neq l$. Then operators $H_+$, $H_-$, $H_3$ are given by the formulae (1.1), (1.4), (1.5) and we have to find operators $F_3$, $F_+$, $F_-$ satisfying the relations (1.6), (1.7). We obtain the solution from certain auxiliary operators $D_+$, $D_-$ and $D_0$ defined[1] in $\bigoplus R_{l,m}$ and satisfying the following relations:

[1] More precisely, $D_+$ and $D_0$ are defined on $\bigoplus R_{l,m}$, but $D_-$ is defined on the analogous sum, where the summands $R_{l,l}$ and $R_{l,-l}$ are excluded.

1° $D_0 R_{l,m} \subset R_{l,m}$, $\quad D_+ R_{l,m} \subset R_{l+1,m}$, $\quad D_- R_{l,m} \subset R_{l-1,m}$.

2° The diagrams

$$\left.\begin{array}{l}
\begin{array}{ccc} R_{l-1,m+1} & \xleftarrow{D_-} & R_{l,m+1} \\ {\scriptstyle E_+}\uparrow & & \uparrow{\scriptstyle E_+} \\ R_{l-1,m} & \xleftarrow[D_-]{} & R_{l,m} \end{array} \quad (-l+1 \leqslant m < l-1), \\[2ex]
\begin{array}{ccc} R_{l,m+1} & \xleftarrow{D_0} & R_{l,m+1} \\ {\scriptstyle E_+}\uparrow & & \uparrow{\scriptstyle E_+} \\ R_{l,m} & \xleftarrow{D_0} & R_{l,m} \end{array} \quad (-l \leqslant m < l), \\[2ex]
\begin{array}{ccc} R_{l,m+1} & \xrightarrow{D_+} & R_{l+1,m+1} \\ {\scriptstyle E_+}\uparrow & & \uparrow{\scriptstyle E_+} \\ R_{l,m} & \xrightarrow{D_+} & R_{l+1,m} \end{array} \quad (-l \leqslant m < l)
\end{array}\right\} \tag{1.8}$$

are commutative.[1]

The operators $F_3$, $F_+$ and $F_-$ may be expressed in terms of these operators in the following way: For $\xi \in R_{l,m}$,

$$\left.\begin{aligned}
F_3\xi &= \sqrt{l^2-m^2}\, D_-\xi - mD_0\xi - \sqrt{(l+1)^2-m^2}\, D_+\xi, \\
F_+\xi &= \sqrt{(l-m)(l-m-1)}\, D_-E_+\xi - \sqrt{(l-m)(l+m+1)}\, D_0E_+\xi + \\
&\quad + \sqrt{(l+m+1)(l+m+2)}\, E_+D_+\xi, \\
F_-\xi &= -\sqrt{(l+m)(l+m-1)}\, D_-E_-\xi - \sqrt{(l+m)(l-m+1)}\, D_0E_-\xi - \\
&\quad - \sqrt{(l-m+1)(l-m+2)}\, E_-D_+\xi.
\end{aligned}\right\} \tag{1.9}$$

REMARK. The operator $D_-$ is not defined on the subspaces $R_{l,l}$ and $R_{l,-l}$. However, this causes no ambiguity in (1.9), because if an expression containing $D_-$ does not make sense, then it appears with the coefficient zero.

The formulae (1.9) for the operators $F_+$, $F_-$, $F_3$ together with the formulae (1.1), (1.4), (1.5) for $H_+$, $H_-$, $H_3$ define a representation of $L$ (in other words, they satisfy the commutation relations (6)), if and only if $D_0$, $D_+$, $D_-$ satisfy the following relations:

$$\left.\begin{array}{l}
3^\circ\ lD_+D_0\xi = (l+2)\, D_0D_+\xi, \\
4^\circ\ (l+1)\, D_-D_0\xi = (l-1)\, D_0D_-\xi \quad (\text{where } \xi \in R_{l,m}), \\
5^\circ\ (2l-1)\, D_+D_-\xi - (2l+3)\, D_-D_+\xi - D_0^2\xi = \xi.
\end{array}\right\} \tag{1.10}$$

**4. Irreducible Harish-Chandra modules for $L$.** As an example, let us find those Harish-Chandra modules in which each subspace $R_{l,m}$ in the decomposition $M = \bigoplus_{l,m} R_{l,m}$ has dimension at most one. We shall deduce

[1] Since the mapping $E_-$: $R_{l,m} \to R_{l,m-1} (m \neq -l)$ is inverse to $E_+$: $R_{l,m-1} \to R_{l,m}$, the analogous diagrams, with $E_-$ instead of $E_+$, are commutative.

from later results that, in fact, every irreducible Harish-Chandra module for $L$ satisfies this condition.

Thus, let $M$ be irreducible and suppose that each non-trivial subspace $R_{l,m}$ in $M = \bigoplus_{l,m} R_{l,m}$ is one-dimensional. In this case each $L_k$-module $R_l$ is irreducible.

Let us denote by $l_0$ the minimal index $l$ occurring in the decomposition[1] $M = \bigoplus_l R_l$. From the formula (1.9) for the operators $F_+$, $F_-$, $F_3$ it is easy to see that the subspace $M' = \bigoplus_{l'} R_{l'}$, where the index $l'$ ranges over the values $l' = l_0$, $l_0 + 1$, $l_0 + 2$, ..., is invariant under $E_+$, $E_-$, $D_0$, $D_+$, $D_-$. Consequently, $M'$ is a submodule of $M$. By assumption $M$ is irreducible, therefore $M = M'$. Thus, the index $l$ in the decomposition $M = \bigoplus_l R_l$ of an irreducible module $M$ takes either only integral values, or only half-integral values.

Let us show that the kernel of $D_-$ in the irreducible module $M$ is $R_{l_0}$. Assume the contrary and suppose that $D_- R_{l',m_0} = 0$ for some index $l' > l_0$. Then $D_-$ also maps the other subspaces $R_{l',m}$ with the same index $l'$ onto zero. This follows from the fact that the diagrams

$$\begin{array}{ccc} R_{l-1,m} & \stackrel{D_-}{\longleftarrow} & R_{l,m} \\ E_+\uparrow & & \uparrow E_+ \\ R_{l-1,m-1} & \stackrel{D_-}{\longleftarrow} & R_{l,m-1} \end{array}$$

are commutative and that $E_+ \colon R_{l,m} \to R_{l,m+1}$ is an isomorphism. But then the subspace $M' = R_{l'} \oplus R_{l'+1} \oplus R_{l'+2} \oplus \dots$ is invariant under $D_+$ and $D_-$, that is, $M'$ is a submodule of $M$, which contradicts the irreducibility of $M$.

In the same manner it can be shown that if the irreducible module $M$ is infinite-dimensional, then $D_+$ has no kernel. For if the kernel of $D_+$ is some subspace $R_{l'}$, then $M$ is finite-dimensional and $l'$ is the maximal index in the decomposition $M = \bigoplus_l R_l$.

CASE 1. *An infinite-dimensional irreducible module* $M$. We choose in $M$ a basis $\{\xi_{l,m}\}$ of the following kind. We demand that $\xi_{l,m}$ belongs to $R_{l,m}$, and that the following relations hold:

$$E_+\xi_{l,m} = \xi_{l,m+1} \qquad (-l \leqslant m < l),$$
$$D_+\xi_{l,m} = \xi_{l+1,m} \qquad (l = l_0,\ l_0+1,\ l_0+2,\ \dots).$$

Such a basis can be chosen, because the diagram (1.8) commutes for the operators $D_+$, $E_+$.

[1] We recall that $R_l = \bigoplus_{m=-l}^{m=l} R_{l,m}$.

The mappings $E_-$, $D_0$, $D_-$ are expressed in the basis $\{\xi_{l,m}\}$ as follows:

$$E_-\xi_{l,m} = \xi_{l,m-1},$$

$$D_0\xi_{l,m} = d_l^0\xi_{l,m}, \qquad D_-\xi_{l,m} = d_l^-\xi_{l-1,m}, \qquad D_-\xi_{l_0,m} = 0.$$

Formulae of this form, together with the fact that the numbers $d_l^0$ and $d_l^-$ do not depend on $m$, easily follow form the commutativity of the diagrams (1.8) for $D_0$, $D_+$ and $D_-$.

The numbers $d_l^-$, $d_l^0$ can be found from the relations (1.10) among $D_-$, $D_0$, $D_+$. These relations, in the basis $\xi_{l,m}$, take the form

$$\left.\begin{array}{c} l d_l^0 = (l+2)\, d_{l+1}^0, \qquad (l+1)\, d_l^- d_l^0 = (l-1)\, d_l^- d_{l-1}^0, \\ (2l-1)\, d_l^- - (2l+3)\, d_{l+1}^- - (d_l^0)^2 = 1 \qquad (l = l_0,\ l_0+1,\ l_0+2,\ \ldots), \\ d_l^- = 0. \end{array}\right\} \quad (1.11)$$

It is not difficult to verify that the solution of these recurrence relations is the following:

$$d_l^0 = \frac{c}{l(l+1)}, \qquad \text{when} \quad l_0 \neq 0,$$

$$d_0^0 = c_1;\ d_l^0 = 0, \quad \text{when} \quad l_0 = 0, \quad l > 0,$$

where $c$ is some constant.

To unify these expressions, and to give a symmetric form to later formulae we set $c = il_0l_1$ and $c_1 = il_1$,

$$d_l^0 = \frac{il_0l_1}{l(l+1)} \qquad (l = l_0,\ l_0+1,\ l_0+2,\ \ldots).$$

Substituting this expression in (1.11) we obtain

$$(2l-1)\, d_l^- - (2l+3)\, d_{l+1}^- = 1 - \frac{l_0^2 l_1^2}{l^2(l+1)^2},$$

and, multiplying both sides of this equation by $(2l+1)$ we have

$$(4l^2-1)\, d_l^- - (4(l+1)^2-1)\, d_{l+1}^- = (2l+1) - l_0^2 l_1^2\left(\frac{1}{l^2} - \frac{1}{(l+1)^2}\right),$$

from which finally:

$$d_l^- = -\frac{(l^2-l_0^2)(l^2-l_1^2)}{l^2(4l^2-1)} \qquad (l > l_0).$$

We have shown earlier that the kernel of $D_-$ is $R_{l_0}$. Consequently, the coefficient $d_l^-$ must be different from zero for all $l > l_0$. Since $l = l_0,\ l_0+1, \ldots$, this means that $|l_1| - l_0$ cannot be a positive integer.

Let us consider the formulae for the operators $H_+$, $H_-$, $H_3$, $F_+$, $F_-$, $F_3$ in our infinite-dimensional irreducible module $M = \bigoplus_l R_l$ $(l = l_0,\ l_0+1, \ldots)$. The operators $H_+$, $H_-$, $H_3$ on $R_l$ give an irreducible representation of $L_k$. In the basis $\xi_{l,m}$ these operators are given by the formulae

$$H_3\xi_{l,m} = m\xi_{l,m}, \qquad H_+\xi_{l,m} = \sqrt{(l+m+1)(l-m)}\,\xi_{l,m+1},$$

$$H_-\xi_{l,m} = \sqrt{(l+m)(l-m+1)}\,\xi_{l,m-1}.$$

The operators $F_+$, $F_-$, $F_3$ act on the basis $\xi_{l,m}$ according to the formulae

$$F_3\xi_{l,m}=\sqrt{l^2-m^2}\,d_l^-\xi_{l-1,m}-m\,d_l^0\xi_{l,m}-\sqrt{(l+1)^2-m^2}\,d_l^+\xi_{l+1,m},$$

$$F_+\xi_{l,m}=\sqrt{(l-m)(l-m-1)}\,d_l^-\xi_{l-1,m+1}-\sqrt{(l-m)(l+m+1)}\,d_l^0\xi_{l,m+1}+$$
$$+\sqrt{(l+m+1)(l+m+2)}\,d_l^+\xi_{l+1,m+1},$$

$$F_-\xi_{l,m}=-\sqrt{(l+m)(l+m-1)}\,d_l^-\xi_{l-1,m-1}-\sqrt{(l+m)(l-m+1)}\,d_l^0\xi_{l,m-1}-$$
$$-\sqrt{(l-m+1)(l-m+2)}\,d_l^+\xi_{l+1,m-1},$$

where $d_l^0=i\,\frac{l_0 l_1}{l(l+1)}$, $d_l^-=-\frac{(l^2-l_0^2)(l^2-l_1^2)}{l^2(4l^2-1)}$, $d_l^+=1$, and the number $l_1$ is such that $|l_1|-l_0$ is not a positive integer.

CASE 2. *A finite-dimensional irreducible module M.* Here $M$ is the direct sum $\bigoplus_{l,m} R_{l,m}$ of one-dimensional subspaces $R_{l,m}$, where $l_0 \leqslant l < |l_1|$, with $l_1$ real and such that $|l_1| \geqslant l_0$ and $|l_1|-l_0$ is integral. As in case 1, a basis $\xi_{l,m}$ can be chosen in $M$ on which $F_+$, $F_-$, $F_3$ act according to the formulae obtained for case 1. The difference lies in the fact that the indices $l$ now take values from $l_0$ to $|l_1|-1$. Here $d_l^+ \equiv 1$ if $l_0 \leqslant l \leqslant |l_1|-1 < |l_1| = 0$.

## §2. Laplace operators: the decomposition of modules into indecomposables.

To decompose an arbitrary module $M$ into the direct sum of indecomposable modules we use the so-called Laplace operators.

**1. Laplace operators and their properties.** With a Lie algebra $L$ there is connected an associative algebra, the so-called enveloping algebra of the Lie algebra. To define it, we consider arbitrary polynomials in the elements $h_+$, $h_-$, $h_3$ and $f_+$, $f_-$, $f_3$ and introduce the multiplication rules $f_+f_- - f_-f_+ = [f_+, f_-]$ etc.

So we have obtained an associative algebra $U(L)$. It is clear that our module for $L$ is a module for the associative algebra $U(L)$. It can be shown that the centre of $U(L)$ is isomorphic to the ring of polynomials with the two generators

$$\Delta_1=\frac{1}{2}(h_-f_++f_-h_+)+h_3f_3,\qquad \Delta_2=h_-h_+-f_-f_++h_3^2-f_3^2+2h_3.$$

A direct verification easily shows that these elements actually commute with $h_i$ and $f_i$. Only this fact will be needed. The corresponding operators on $M$ will again be denoted by $\Delta_1$ and $\Delta_2$. They are expressed in terms of the operators $H_i$ and $F_i$ by the formulae:

$$\Delta_1=\frac{1}{2}(H_-F_++F_-H_+)+H_3F_3,\qquad \Delta_2=H_-H_+-F_-F_++H_3^2-F_3^2+2H_3. \quad (2.1)$$

If the expressions (1.1), (1.4), (1.5) and (1.9) for the operators $H_i$ and $F_i$ are substituted in these formulae, then after a certain amount of calculation we obtain

$$\Delta_1\xi = -l(l+1)D_0\xi \qquad (\xi \in R_l), \tag{2.2}$$
$$\Delta_2\xi = (l^2-1)\xi - (l+1)^2 D_0^2\xi + (4l^2-1)D_+D_-\xi \qquad (\xi \in R_l). \tag{2.3}$$

In addition, we sometimes need another formula for $\Delta_2$:

$$\Delta_2\xi = ((l+1)^2-1)\xi - l^2D_0^2\xi + (4(l+1)^2-1)D_-D_+\xi \qquad (\xi \in R_l), \tag{2.3'}$$

which can easily be obtained from the relation 5° (p.8) by applying the preceding. From the expressions (2.2) and (2.3) it is easy to see that we have

PROPOSITION 2.1. *Each subspace $R_{l,m}$ is invariant under the Laplace operators $\Delta_1$ and $\Delta_2$.*

The following proposition will also be useful.

PROPOSITION 2.2. *The auxiliary operators $D_+$, $D_-$, $D_0$, $E_+$, $E_-$ commute with the Laplace operators $\Delta_1$ and $\Delta_2$*[1]*.*

PROOF. We denote by $(\Delta_i)_{l,m}$ the restriction of $\Delta_i$ to $R_{l,m}$. From Proposition 2.1 it follows that $\Delta_i$ is the sum $\bigoplus\limits_{l,m}(\Delta_i)_{l,m}$ of its restrictions. The operators $(\Delta_i)_{l,m}$ are expressible in terms of $D_0$, $D_-$, and $D_+$ (formulae (2.2) and (2.3)). Since $D_0$, $D_+$ and $D_-$ commute with $E_+$ and $E_-$, the Laplace operators $\Delta_1$ and $\Delta_2$ commute with $E_+$ and $E_-$.

Multiplying the relations 3° (p.8) by $(l+1)$, we have

$$l(l+1)D_+D_0\xi = (l+1)(l+2)D_0D_+\xi \qquad (\xi \in R_{l,m}),$$

since $\Delta_1\xi = -l(l+1)D_0\xi$ for $\xi \in R_{l,m}$. It follows from this that $D_+\Delta_1\xi = \Delta_1D_+\xi$, that is, $\Delta_1$ commutes with $D_+$. (We remark that if $\xi \in R_{l,m}$, then $D_+\xi \in R_{l+1,m}$). Similarly, from relation 4° (p.8) we find that $\Delta_1$ also commutes with $D_-$.

Now let us show that $D_+$ and $D_-$ also commute with the Laplace operator $\Delta_2$. When we replace in the expression (2.3) for $\Delta_2$ the vector $\xi$ by $\xi' = D_+\xi$, we obtain
$$\Delta_2D_+\xi = ((l+1)^2-1)D_+\xi - (l+2)^2D_0^2D_+\xi + (4(l+1)^2-1)D_+D_-D_+\xi \quad (\xi \in R_{l,m}).$$
Replacing $D_0D_+$ by $D_+D_0$ from 3° (p.  ), we have

$$\Delta_2D_+\xi = ((l+1)^2-1)D_+\xi - l^2D_+D_0^2\xi + (4(l+1)^2-1)D_+D_-D_+\xi =$$
$$= D_+[((l+1)^2-1)\xi - l^2D_0^2\xi + (4(l+1)^2-1)D_-D_+\xi] \qquad (\xi \in R_{l,m}).$$

Combining this with the expression (2.3'), for $\Delta_2$ we obtain

$$\Delta_2D_+\xi = D_+\Delta_2\xi.$$

The assertion for $D_-$ is proved similarly.

**2. The decomposition of a Harish-Chandra module into indecomposables. The properties of Laplace operators in indecomposable modules.**

THEOREM 2.1. *A Harish-Chandra module for the Lie algebra L of the proper Lorentz group is decomposable into the direct sum of a countable number of indecomposable modules. On each indecomposable module, the Laplace operators $\Delta_1$ and $\Delta_2$ have each one eigenvalue, $\lambda_1$ and $\lambda_2$, respectively.*

[1] The commutativity of $D_-$ with $\Delta_1$ and $\Delta_2$ means that if $\xi \in \operatorname{Def} D_-$, then $\Delta_i\xi \in \operatorname{Def} D_-$ and $\Delta_iD_-\xi = D_-\Delta_i\xi$. (Here Def $D_-$ is the domain of definition of $D_-$.)

PROOF. We have already said that each of the subspaces $R_{l,m}$ is invariant under the Laplace operators $\Delta_1$ and $\Delta_2$. Since these operators commute with each other $R_{l,m}$ can be written as the direct sum of subspaces $R_{l,m}(\lambda_1^i, \lambda_2^i)$ on each of which each of the operators $\Delta_1$ and $\Delta_2$ has one eigenvalue $\lambda_1^i$ and $\lambda_2^i$.

Consider fixed numbers $\lambda_1$ and $\lambda_2$ and the set of those $l,m$ for which there exist subspaces $R_{l,m}(\lambda_1^i, \lambda_2^i)$ such that $\lambda_1^i = \lambda_1$, $\lambda_2^i = \lambda_2$. Let usus denote by $M(\lambda_1, \lambda_2)$ the submodule of $M$ for which $M(\lambda_1, \lambda_2) = \bigoplus R_{l,m}(\lambda_1, \lambda_2)$, where the sum ranges over all those $l,m$ for which $\lambda_1^i = \lambda_1$, $\lambda_2^i = \lambda_2$. Thus, in $M(\lambda_1, \lambda_2)$ each of $\Delta_1$ and $\Delta_2$ has one eigenvalue, $\lambda_1$ and $\lambda_2$, respectively. Let us show that the subspace $M(\lambda_1, \lambda_2)$ is a submodule of $M$. For this purpose it is sufficient to show that $M(\lambda_1, \lambda_2)$ is invariant under each of the auxiliary mappings $E_+$, $E_-$, $D_+$, $D_-$, $D_0$. But this is a trivial consequence of Proposition (2.2). The theorem is now proved.

THEOREM 2.2. *Let $M$ be a Harish-Chandra module in which each of the Laplace operators $\Delta_1$ and $\Delta_2$ has one eigenvalue. Then there exists an integral or half-integral number $l_0 \geqslant 0$ and a complex number $l_1$ such that the eigenvalues $\lambda_1$ and $\lambda_2$ have the form*

$$\lambda_1 = -il_0 l_1, \qquad \lambda_2 = l_0^2 + l_1^2 - 1. \tag{2.4}$$

PROOF. Let us denote by $l_0$ the minimal index appearing in the decomposition $M = \bigoplus_l R_l$ of $M$ into $L_k$-submodules $R_l$. We recall that the representation of $L_k$ on $R_l$ is multiply irreducible. It is clear that the operator $D_-$ maps $R_{l_0}$ to zero (see p.8). Hence for any vector $\xi \in R_{l_0}$, we can write, corresponding to (2.2) and (2.3):

$$\left.\begin{aligned} \Delta_2\xi &= (l_0^2 - 1)\,\xi - (l_0+1)^2 D_0^2\xi \qquad (\xi \in R_{l_0}), \\ \Delta_1\xi &= -l_0(l_0+1) D_0\xi. \end{aligned}\right\} \tag{2.5}$$

The subspace $R_{l_0}$ is invariant under $D_0$. This means that we can find in it an eigenvector for $D_0$: $D_0\xi_0 = \mu\xi_0$. Substituting in (2.5) $\xi = \xi_0$, we obtain the eigenvalues $\lambda_1$ and $\lambda_2$ of the Laplace operators, expressed in terms of $l_0$ and $\mu$ as follows:

$$\lambda_1 = -l_0(l_0+1)\mu, \qquad \lambda_2 = (l_0^2 - 1) - (l_0+1)^2\mu^2.$$

Putting $(\lambda_0 + 1)\mu = l_1$, we have

$$\lambda_1 = -il_0 l_1, \qquad \lambda_2 = l_0^2 + l_1^2 - 1.$$

By hypothesis, each of $\Delta_1$ and $\Delta_2$ has one eigenvalue on $M$. Consequently, these eigenvalues are expressed in terms of $l_0$ and $l_1$ by the formulae (2.4), and the theorem is proved.

We remark that all such eigenvalues $\lambda_1$ and $\lambda_2$ for the Laplace operators $\Delta_1$ and $\Delta_2$ are possible, since in the case of irreducible representations (see §1. 4) $\Delta_1$ and $\Delta_2$ are multiples of $I$ and have the form

$$\Delta_1 = -il_0 l_1 I, \qquad \Delta_2 = (l_0^2 + l_1^2 - 1I.)$$

In what follows we also have to know the eigenvalues of the operators $D_+D_-$, $D_-D_+$ and $D_0$. Let us calculate them.

THEOREM 2.3. *Let $M$ be a Harish-Chandra module in which the Laplace operators $\Delta_1$ and $\Delta_2$ have only one eigenvalue $\lambda_1$ and $\lambda_2$. Then on each subspace $R_l$ the operators $D_+D_-$, $D_-D_+$ and $D_0$ have only one eigenvalue $d_l^-$, $d_l^+$ and $d_l^0$, respectively. Here the numbers $d_l^-$, $d_l^+$ and $d_l^0$ are expressed in terms of $l_0$ and $l_1$ in the following way:*

$$d_{l_0}^- = 0, \quad d_l^- = \frac{(l^2-l_0^2)(l_1^2-l^2)}{(4l^2-1)\,l^2}, \quad d_l^+ = \frac{((l+1)^2-l_0^2)(l_1^2-(l+1)^2)}{(4(l+1)^2-1)(l+1)^2}, \tag{2.6}$$

$$d_l^0 = \frac{il_0l_1}{l(l+1)} \text{ and } d_0^0 = il_1.$$

PROOF. We give the proof for the eigenvalue $d_l^-$ of $D_+D_-$ on $R_l$. We write out (2.2) (2.3) in a slightly modified form:

$$\left.\begin{aligned} (4l^2-1)\,D_+D_-\xi &= \Delta_2\xi-(l^2-1)\,\xi+(l+1)^2 D_0^2\xi \quad (\xi\in R_l), \\ (l+1)\,D_0\xi &= -\left(\frac{\Delta_1}{l}\right)\xi \quad (l\neq 0). \end{aligned}\right\}$$

From this

$$(4l^2-1)\,D_+D_-\xi = \Delta_2\xi + \frac{\Delta_1^2}{l^2}\,\xi-(l^2-1)\,\xi \qquad (\xi\in R_l), \qquad l\neq 0.$$

Thus, the operator $D_+D_-$ on the subspace $R_{l,m}$ is a linear combination of the commuting operators $\Delta_1$ and $\Delta_2$. Since $\Delta_1$ and $\Delta_2$ have only one eigenvalue, $D_+D_-$ has only one eigenvalue $d_l^-$ on $R_{l,m}$:

$$d_l^- = \frac{1}{(4l^2-1)}\left(\lambda_2+\frac{\lambda_1^2}{l^2}-(l^2-1)\right).$$

When we substitute in this the expression for the eigenvalues $\lambda_1$ and $\lambda_2$ in terms of $l_0$ and $l_1$ (see Theorem 2.2), we obtain

$$d_l^- = \frac{(l^2-l_0^2)(l_1^2-l^2)}{(4l^2-1)\,l^2} \qquad \left(l\neq 0;\quad l\neq\frac{1}{2}\right).$$

We remark that $d_0^-$ and $d_{\frac{1}{2}}^-$ are zero. This follows from the fact that $D_-$ maps the subspaces $R_0$ and $R_{1/2}$ to zero (provided, of course, that they occur in the decomposition).

The formula for the eigenvalues $d_l^+$ and $d_l^0$ is proved similarly.

3. Let $C_s(\lambda_1, \lambda_2)$ ($s = 1$ or $\frac{1}{2}$) denote the set of all Harish-Chandra modules for $L$ in which the Laplace operators have the eigenvalues $\lambda_1$ and $\lambda_2$, and where, in case $s=1$, for every $M\in C_1(\lambda_1, \lambda_2)$, every index $l$ in the decomposition $M=\bigoplus_l R_l$ is integral, and in case $s=\frac{1}{2}$, every index $l$ is half-integral.

THEOREM 2.4. *Let* $M\in C_s(\lambda_1, \lambda_2)$, $M'\in C_{s'}(\lambda_1', \lambda_2')$, *where* $(s, \lambda_1, \lambda_2)\neq(s', \lambda_1', \lambda_2')$. *Then* $\mathrm{Hom}(M, M')=0$.

PROOF. If $\gamma\in\mathrm{Hom}_L(M, M')$, then $\gamma R_{l,m}\subset R_{l,m}'$ and $\Delta_i'\gamma = \gamma\Delta_i$ ($i = 1, 2$). But this contradicts the assumption that

$(s, \lambda_1, \lambda_2) \neq (s', \lambda_1', \lambda_2')$. Thus, it is sufficient to study modules $M \in C_s(\lambda_1, \lambda_2)$.

REMARK. In all places in the sequel, when a precise knowledge of the index $s$ is not important, we write $C(\lambda_1, \lambda_2)$ instead of $C_s(\lambda_1, \lambda_2)$.

DEFINITION. The category of modules $C(\lambda_1, \lambda_2)$ is called singular if the numbers $l_0$ and $l_1$ constructed from $\lambda_1$ and $\lambda_2$ (see (2.4)) are such that $l_1 - l_0$ is an integer. Otherwise it is called non-singular.

We see that the study of the non-singular categories $C(\lambda_1, \lambda_2)$ is much simpler than that of the singular. All the interesting peculiarities of Harish-Chandra modules are hidden in the singular categories $C(\lambda_1, \lambda_2)$.

## §3. The non-singular category $C(\lambda_1, \lambda_2)$

1. Let us consider an $L$-module $M \in C(\lambda_1, \lambda_2)$, where $(\lambda_1, \lambda_2)$ is a non-singular pair (that is, $l_1 - l_0$ is not an integer). In this section we show that this module is completely determined by a certain finite-dimensional space and a nilpotent mapping $a$ in it, where isomorphism of the modules is equivalent to similarity of the linear mappings $a$.

Let $M$ be a module in $C(\lambda_1, \lambda_2)$. We consider a subspace $R_{l_0, m_0}$, where $m_0$ is any one of the numbers $-l_0, -l_0+1, \dots, l_0$, and define in this finite-dimensional subspace $R_{l_0, m_0}$ a linear mapping $a: R_{l_0, m_0} \to R_{l_0, m_0}$ by the formula

$$a\xi = D_0\xi - \frac{il_1}{l_0+1}\xi, \qquad \xi \in R_{l_0, m_0}. \tag{3.1}$$

The mapping $a$ is nilpotent, because by Theorem 2.3 $D_0$ has the eigenvalue $d_{l_0}^0 = \frac{il_1}{l_0+1}$ on $R_{l_0}$.

We show further that the module $M \in C(\lambda_1, \lambda_2)$ is completely determined by the finite-dimensional space $R_{l_0, m_0}$ and the nilpotent mapping $a$ in it, if the category $C(\lambda_1, \lambda_2)$ is non-singular.

However, first we require some auxiliary results.

2. Auxiliary propositions. LEMMA 3.1. *In the module $M$ from the non-singular category $C(\lambda_1, \lambda_2)$, the mappings*

$$\begin{aligned} D_+ &: R_{l,m} \longrightarrow R_{l+1,m}, \\ D_- &: R_{l+1,m} \longrightarrow R_{l,m} \end{aligned} \qquad (l \geqslant l_0)$$

*are isomorphisms.*

PROOF. We write out again the formulae for the eigenvalues $d_l^-$ and $d_l^+$ of $D_+D_-$ and $D_-D_+$ on $R_l$:

$$d_l^- = \frac{(l^2-l_0^2)(l_1^2-l^2)}{(4l^2-1)\,l^2}, \qquad d_l^+ = \frac{((l+1)^2-l_0^2)(l_1^2-(l+1)^2)}{(4(l+1)^2-1)(l+1)^2}.$$

By hypothesis, the module $M$ is non-singular, consequently the difference $l_1^2 - l_0^2$ is non-zero for all $l$. Thus, the eigenvalues $d_l^+$ and $d_l^-$ are different from zero. This means that the mappings $D_+D_-: R_{l,m} \to R_{l,m} (l \neq l_0)$ and $D_-D_+: R_{l,m} \to R_{l,m}$ are non-singular. Consequently, both the mappings $D_-: R_{l+1,m} \to R_{l,m}$ and

$D_+: R_{l,m} \to R_{l+1,m}$ $(l \geqslant l_0)$ are monomorphisms. This is possible only when the dimensions of all the spaces $R_{l,m}$ ($l = l_0, l_0 + 1, l_0 + 2, \ldots$) are equal. From this it follows that all the mappings $D_-: R_{l+1,m} \to R_{l,m}$ and $D_+: R_{l,m} \to R_{l+1,m}$ are isomorphisms, and the Lemma is proved.

LEMMA 3.2. *In a non-singular module $M \in C(\lambda_1, \lambda_2)$ the Laplace operators $\Delta_1$ and $\Delta_2$ are such that each operator $(\Delta_i)_{l,m}$ is similar to $(\Delta_i)_{l_0,m_0}$.*

PROOF. We choose a subspace $R_{l_0,m_0}$ and denote by $(\Delta_1)_{l_0,m_0}$ and $(\Delta_2)_{l_0,m_0}$ the restrictions of $\Delta_1$ and $\Delta_2$ to this subspace. We recall that the mappings

$$E_+: R_{l_0,m} \to R_{l_0,m+1} \qquad (-l_0 \leqslant m < l_0),$$
$$E_-: R_{l_0,m} \to R_{l_0,m-1} \qquad (-l_0 < m \leqslant l_0)$$

are isomorphisms. The Laplace operators commute with these mappings. Consequently, each restriction $(\Delta_i)_{l_0,m}$ of $\Delta_i$ onto $R_{l_0,m}$ is similar to $(\Delta_i)_{l_0,m_0}$.

By Lemma 3.1 the mappings $D_+: R_{l,m} \to R_{l+1,m}$ are also isomorphisms. Moreover, $D_+$ commutes with the Laplace operators $\Delta_1$ and $\Delta_2$. Consequently, the mapping $(\Delta_i)_{l_0+1,m}: R_{l_0+1,m} \to R_{l_0+1,m}$ is similar to the restriction $(\Delta_i)_{l_0,m}$ and is expressed in terms of it by means of the isomorphism $D_+$:

$$(\Delta_i)_{l_0+1,m}\xi = D_+(\Delta_i)_{l_0,m}(D_+)^{-1}\xi,$$

where $\xi$ is a vector from $R_{l_0+1,m}$.

Continuing these arguments we have no difficulty in establishing that each restriction $(\Delta_i)_{l,m}$ of $\Delta_i$ onto $R_{l,m}$ is similar to $(\Delta_i)_{l_0,m_0}$.

LEMMA 3.3. *If $M \in C(\lambda_1, \lambda_2)$ is a non-singular module, then the operators $\Delta_1$ and $\Delta_2$ are connected on the whole of $M$ by the relation*

$$\Delta_1^2 + l_0^2\Delta_2 - l_0^2(l_0^2 - 1)I = 0. \tag{3.2}$$

PROOF. Suppose that $l_0 \neq 0$. Then from the identities (2.5) it follows that for $\xi \in R_{l_0,m_0}$

$$\left(\Delta_2 + \frac{\Delta_1^2}{l_0^2} - (l_0^2 - 1)I\right)\xi = 0.$$

By (3.2), $(\Delta_i)_{l,m}$ is similar to $(\Delta_i)_{l_0,m_0}$. Consequently, the relation (3.2) is valid for any vector $\xi \in R_{l,m}$, and hence in the whole of $M$.

Suppose that $l_0 = 0$. Then from (2.5) it follows that $\Delta_1 = 0$ on $R_{0,0}$. Consequently $\Delta_1 = 0$ on any subspace $R_{l,m}$, that is, $\Delta_1 \equiv 0$. The lemma is now proved.

WARNING. Lemmas 3.1, 3.2 and 3.3 are not true in the singular case!

3. We have seen that to each non-singular module $M \in C(\lambda_1, \lambda_2)$ there corresponds a nilpotent linear mapping $a$ of a certain finite-dimensional space $P = R_{l_0,m_0}$

$$a\xi=\left(D_0-\frac{il_1}{l_0+1}I\right)\xi \quad \text{for} \quad \xi\in R_{l_0,m_0}.$$

Let us denote by the symbol $\tilde{A}$ the pair $(P, a)$ consisting of a finite-dimensional space $P$ and a nilpotent mapping $a$.

THEOREM 3.1. *To each pair $\tilde{A}$ and non-singular pair $(\lambda_1, \lambda_2)$ of numbers there corresponds a module $M \in C(\lambda_1, \lambda_2)$ over $L$ such that $P = R_{l_0,m_0}$ and $a$ is related to $D_0$ by (3.1).*

Let us change the name of $P$ into $R_{l_0,m_0}$ and consider the transformation $D_0$ of it, defined by the formula

$$D_0\xi=a\xi+\frac{il_1}{(l_0+1)}\xi, \text{ where } \xi\in R_{l_0,m_0}.$$

Consider the space $M=\bigoplus_{l,m} R_{l,m}$ $(l=l_0, l_0+1, l_0+2, \ldots; m=-l, -l+1, \ldots, l-1, l)$, which is the direct sum of subspaces whose dimensions are all equal to dim $P$.

Take arbitrary isomorphisms $E_+ : R_{l,m}\to R_{l,m+1} (m\neq l)$ and put $E_+: R_{l,l}\to 0$. Define an isomorphism $E_-: R_{l,m+1}\to R_{l,m}$ such that it is inverse to $E_+: R_{l,m}\to R_{l,m+1}$, and put $E_-: R_{l,-l}\to 0$. Take arbitrary isomorphisms $D_+: R_{l,m_0}\longrightarrow R_{l+1,m_0} (l=l_0, l_0+1, \ldots;$ the index $m_0$ fixed). On all the remaining subspaces $R_{l,m}$, define mappings $D_+: R_{l,m}\to R_{l+1,m}$, so that the following diagrams are commutative:

$$\begin{array}{ccc} R_{l,m+1} & \xrightarrow{D_+} & R_{l+1,m+1} \\ {\scriptstyle E_+}\uparrow & & \uparrow{\scriptstyle E_+} \\ R_{l,m} & \xrightarrow{D_+} & R_{l+1,m} \end{array} \qquad (l>m\geqslant -l).$$

It now remains to construct the mappings $D_0$ and $D_-$ of $M$.

To do this we define on $M$ the mappings $\Delta_1$ and $\Delta_2$. First, on the fixed subspace $R_{l_0,m_0}$ set

$$\Delta_1\xi=-l_0(l_0+1)a\xi-il_1l_0\xi,$$
$$\Delta_2\xi=(l_0^2+l_1^2-1)\xi-(l_0+1)^2\left(a^2\xi+2\frac{il_1}{l_0+1}a\xi\right), \qquad \xi\in R_{l_0,m_0}.$$

Between an arbitrary subspace $R_{l,m}$ and $R_{l_0,m_0}$ there is an isomorphism $J_{l,m}: R_{l_0,m_0}\longrightarrow R_{l,m}$, given by the formula $J_{l,m}=(E_+)^{m-m_0}(D_+)^{l-l_0}$. Let us define the mappings $\Delta_1$ and $\Delta_2$ on the subspaces $R_{l,m}$ by the following formulae:

$$(\Delta_i)_{l,m}\xi=J_{l,m}(\Delta_i)_{l_0,m_0}J^{-1}\xi \qquad (\xi\in R_{l,m}, i=1, 2). \tag{3.3}$$

Here $(\Delta_i)_{l_0,m_0}$ are the operators $\Delta_1$ and $\Delta_2$ on $R_{l_0,m_0}$. So we have defined $\Delta_1$ and $\Delta_2$ on the whole of $M$.

Now define the mapping $D_0: R_{l,m}\longrightarrow R_{l,m}$ by the formula

$$D_0\xi = -\frac{1}{l(l+1)}\Delta_1\xi \qquad (\xi \in R_{l,m}),$$

and $D_+D_-\colon R_{l,m} \to R_{l,m}$ by the formula

$$D_+D_-\xi = \frac{1}{4l^2-1}(\Delta_2\xi - (l^2-1)\xi + (l+1)^2 D_0^2\xi) =$$

$$= \frac{1}{4l^2-1}\left(\Delta_2\xi - (l^2-1)\xi + \frac{\Delta_1^2}{l^2}\xi\right) \qquad (\xi \in R_{l,m};\ l \neq l_0).$$

After this, the mapping $D_-\colon R_{l,m} \to R_{l-1,m}$ is put equal to $(D_+)^{-1}(D_+D_-)$ for $l \neq l_0$, and equal to zero for $l = l_0$.

The mappings $E_+$, $E_-$, $D_0$, $D_+$, $D_-$ so constructed satisfy the relations $1^\circ$– $5^\circ$ on p.8. Then the operators $F_+$, $F_-$, $F_3$ and $H_3$, $H_+$, $H_-$ constructed from these mappings by means of (1.9) give a module $M$ for $L$. From the very method of construction it is clear that $M$ is uniquely determined by the specification of $a$. The module obtained is non-singular, because we assumed at the beginning $\lambda_1$ and $\lambda_2$ to be such that the numbers $l_0$ and $l_1$ related to them did not lead to an integral difference $l_1 - l_0$.

COROLLARY 3.1. *For the modules $M$ and $M'$ from the non-singular category $C(\lambda_1, \lambda_2)$ to be equivalent it is necessary and sufficient that the subspaces $R_{l_0,m_0}$ and $R'_{l_0,m_0}$ in these modules have the same dimension, and that the mappings $D_0$: $R_{l_0,m_0} \to R_{l_0,m_0}$ and $D_0'$: $R'_{l_0,m_0} \to R'_{l_0,m_0}$ are similar.*

4. We have established a correspondence between modules $M \in C(\lambda_1, \lambda_2)$ in the non-singular category $C(\lambda_1, \lambda_2)$ and pairs $\tilde{A}$ consisting of a finite-dimensional space $P$ and a nilpotent operator $a$ on it. We show now that this correspondence is functorial. Let us formulate this precisely. We introduce a category $S$.

The objects of $S$ are pairs $A = (P, a)$ where $P$ is a space over the complex numbers and $a$ is a nilpotent linear transformation $a\colon P \to P$. The morphisms $\gamma\colon A \to A'$ are those linear mappings $\gamma: P \to P'$, for which the diagram

$$\begin{array}{ccc} P & \xrightarrow{a} & P \\ \gamma\downarrow & & \downarrow\gamma \\ P' & \xrightarrow{a'} & P' \end{array}$$

is commutative.

THEOREM 3.2. *The non-singular category $C(\lambda_1, \lambda_2)$ is equivalent to the category $S$.*

PROOF. A correspondence between the objects $M \in C(\lambda_1, \lambda_2)$ and the objects $A \in S$ was established in Theorem 3.1. It remains for us to establish a correspondence between the morphisms. Let $\Gamma\colon M \to M'$ be a morphism of the module $M \in C(\lambda_1, \lambda_2)$ into $M' \in C(\lambda_1, \lambda_2)$. Then from the fact that $\Gamma$ commutes with $H_3$, $H_+$, $H_-$ it follows that $\Gamma R_{l,m} \subset R'_{l,m}$ (here $M = \bigoplus_{l,m} R_{l,m}$, $M' = \bigoplus_{l,m} R'_{l,m}$). Thus, $\Gamma$ is defined by a set of morphisms $\gamma_{l,m}\colon R_{l,m} \to R'_{l,m}$. If we choose indices $l_0$, $m_0$ and put $\gamma_{l_0,m_0} = \gamma$, then we obtain a morphism of the pair $\tilde{A} = (D_0, R_{l_0,m_0})$ into $(D_0', R'_{l_0,m_0})$, that is, $\gamma$ is a morphism $\gamma: \tilde{A} \to \tilde{A}'$.

Now suppose, conversely, that we are given a mapping $\gamma: \tilde{A} \to \tilde{A}'$, that is, $\gamma: P \to P'$ is such that the diagram

$$\begin{array}{ccc} P & \xrightarrow{\gamma} & P' \\ {\scriptstyle a}\downarrow & & \downarrow{\scriptstyle a'} \\ P & \xrightarrow{\gamma} & P' \end{array}$$

is commutative. Identifying first of all $P$ with $R_{l_0, m_0}$ and $P'$ with $R'_{l_0, m_0}$, where the modules $M$ and $M'$ are constructed corresponding to $\tilde{A}$ and $\tilde{A}'$, we obtain a morphism $\gamma: R_{l_0, m_0} \to R'_{l_0, m_0}$ such that $\gamma D_0 = D'_0\gamma$. Let us now construct morphisms $\gamma_{l,m}: R_{l,m} \to R'_{l,m}$. The mapping $J_{l,m} = E_+^{m-m_0}D_+^{l-l_0}$ is an isomorphism $J_{l,m}: R_{l_0, m_0} \to R_{l,m}$. Therefore we put

$$\gamma_{l,m} = J'_{l,m}\gamma J_{l,m}^{-1}, \tag{3.4}$$

where $J'_{l,m}$ is the analogous mapping: $J'_{l,m}: R'_{l,m} \to R'_{l,m}$. Let us show that $\gamma_{l,m}: R_{l,m} \to R'_{l,m}$ gives a morphism of the module $M = \bigoplus R_{l,m}$ into $M' = \bigoplus R'_{l,m}$.

For this purpose we recall that the Laplace operators $(\Delta_i)_{l,m}$ $(i = 1, 2)$ in $R_{l,m}$ are similar to the operators $(\Delta_i)_{l_0, m_0}$, and that $(\Delta_i)_{l,m} = J_{l,m}(\Delta_i)_{l_0, m_0}J_{l,m}^{-1}$. From this it follows immediately that $(\Delta'_i)_{l,m}\gamma_{l,m} = \gamma_{l,m}(\Delta_i)_{l,m}$. Since

$$\left.\begin{aligned} D_0\xi &= -\frac{1}{l(l+1)}\Delta_1\xi, \\ D_+D_- &= \frac{1}{4l^2-1}\left(\Delta_2\xi - (l^2-1)\xi + \frac{\Delta_1^2}{l^2}\xi\right) \end{aligned}\right\} \tag{3.5}$$

for $\xi \in R_{l,m}$, we have

$$(D'_0)_{l,m}\gamma_{l,m} = \gamma_{l,m}(D_0)_{l,m}$$

and

$$(D'_+D'_-)_{l,m}\gamma_{l,m} = \gamma_{l,m}(D_+D_-)_{l,m}.$$

By substitution in (3.4) it follows directly that $E_+$ and $D_+$ commute with the morphisms $\gamma_{l,m}$. From this we see that all the necessary maps commute. Since the operators $F$ can be expressed in terms of the mappings $D$ and $E$ we have indicated, the theorem is proved.

COROLLARY 3.2. *To an indecomposable module $M$ in the non-singular category $\mathcal{C}(\lambda_1, \lambda_2)$, there corresponds an indecomposable object $A$ in the category $S$.*

The form of indecomposable objects $A \in S$ is well known. They are finite-dimensional spaces $P$ and nilpotent transformations $a: P \to P$ whose matrices in a suitable basis have the form of a single Jordan block.

Thus, *an indecomposable non-singular Harish-Chandra module has, in comparison with an irreducible module, one further complementary invariant, an integer $n$: the dimension of the Jordan block.*

5. Let us write out the explicit form of $E_+$, $E_-$, $D_+$, $D_-$, $D_0$ in a non-singular indecomposable module $M \in C(\lambda_1, \lambda_2)$. We denote by $[E_+]_{l,m}$ the matrix of the mapping $E_+: R_{l,m} \to R_{l,m+1}$, by $[E_-]_{l,m}$ the matrix of $E_-: R_{l,m} \to R_{l,m-1}$ and so on.

THEOREM 3.3. *Let $M$ be a non-singular indecomposable Harish-Chandra module for the Lie algebra L. Then all the subspaces $R_{l,m}(l = l_0, l_0 + 1, l_0 + 2, \ldots)$ have the same dimension. Bases can be chosen in them so that $[E_+]_{l,m}$ $(m \neq l)$, $[E_-]_{l,m}$ $(m \neq -l)$, $[D_+]_{l,m}$ are unit matrices.*[1] *The matrices $[D_0]_{l,m}$ and $[D_-]_{l,m}$ can be expressed in terms of a matrix $[a_0]$ of the form*

$$[a_0] = \begin{pmatrix} 0 & 1 & 0 & \ldots & 0 \\ 0 & 0 & 1 & \ldots & 0 \\ \cdot & \cdot & \cdot & \cdot & \cdot \\ 0 & 0 & 0 & \ldots & 1 \\ 0 & 0 & 0 & \ldots & 0 \end{pmatrix}.$$

*The expressions for $[D_0]_{l,m}$, $[D_-]_{l,m}$ are the following:*

$$[D_0]_{l,m} = i\frac{l_0 l_1}{l(l+1)}[I] + \frac{l_0(l_0+1)}{l(l+1)}[a_0],$$

$$[D_-]_{l,m} = \frac{(l_0^2 - l^2)}{l^2(4l^2-1)}((l^2 - l_1^2)[I] + 2il_1(l_0+1)[a_0] + (l_0+1)^2[a_0]^2);$$

*where $[I]$ is the identity matrix of the same dimension as $[a_0]$.*

The proof of this theorem is a repetition of that of Theorem 3.1. That the matrices of $E_+$, $E_-$, $D_+$ can be reduced to the identity matrix by a choice of bases in the $R_{l,m}$ follows from the fact that these mappings are isomorphisms and the corresponding diagrams commute.

In such a basis, the matrices $[\Delta_1]_{l,m}$ and $[\Delta_2]_{l,m}$ of the Laplace operators on the subspaces $R_{l,m}$ do not depend on the indices $l$ and $m$, and may be expressed in terms of $[a_0]$ in accordance with the formulae (2.5) and (2.6).

$$[\Delta_1]_{l,m} = -l_0(l_0+1)[a_0] - il_1 l_0 [I],$$

$$[\Delta_2]_{l,m} = (l_0^2 + l_1^2 - 1)[I] - 2il_1(l_0+1)[a_0] - (l_0+1)^2[a_0]^2.$$

After this, the matrices of $D_0$ and $D_+D_-$ are found from (3.5).

$$[D_0]_{l,m} = \frac{l_0(l_0+1)}{l(l+1)}[a_0] + \frac{il_0 l_1}{l(l+1)}[I],$$

$$[D_+D_-]_{l,m} = \frac{(l_0^2 - l^2)}{(4l^2-1)l^2}((l^2 - l_1^2)[I] + 2il_1(l_0+1)[a_0] + (l_0+1)^2[a_0]^2).$$

---

[1] We denote the matrices of maps $a$, $E$, $D$, etc, by $[a]$, $[D]$, $[E]$, etc.

Since we have arranged $[D_+]_{l,m}$ to be the identity matrix, $[D_-]_{l,m}$ is equal to $[D_+D_-]_{l,m}$, and the theorem is proved.

## §4. The singular category $C(\lambda_1, \lambda_2)$

1. The aim of this section is to study the singular category $C(\lambda_1, \lambda_2)$, that is, (representation) modules $M$ for $L$, corresponding to the case when the pair $(\lambda_1, \lambda_2)$ is singular. We recall that the pair $\lambda_1, \lambda_2$ is called singular if the numbers $l_0, l_1$ related to $\lambda_1, \lambda_2$, where $\lambda_1 = -il_0l_1, \lambda_2 = l_0^2 - l_1^2 - 1$, are such that $l_1 - l_0$ is an integer. (We recall that $l_0 \geqslant 0$ and $2l_0$ is always an integer.)

It turns out that if $M$ is a singular indecomposable module, then it has, in comparison with an irreducible, quite a non-trivial set of invariants. This set consists of a sequence of integers, which may be arbitrarily long (but finite), and an arbitrary complex number $\mu$ (in a number of cases $\mu$ may be absent).

In contrast to the non-singular case, where $M \in C(\lambda_1, \lambda_2)$ is given completely by a pair consisting of a space $P$ and a nilpotent transformation $a$ on it, in the singular case the description of the objects $M \in C(\lambda_1, \lambda_2)$ is more complicated. We define presently a category $S_0$ whose objects serve to classify the objects of $C(\lambda_1, \lambda_2)$. The objects $\widetilde{A}$ of $S_0$ are pairs of finite-dimensional spaces $P_1$ and $P_2$ with four linear maps:

$$d_+:\ P_1 \to P_2, \quad d_-:\ P_2 \to P_1, \quad \delta_1:\ P_1 \to 0, \quad \delta_2:\ P_2 \to P_2. \tag{4.1}$$

These maps must satisfy the conditions

$$1^\circ\ d_-\delta_2 = \delta_2 d_+ = 0, \tag{4.2}$$

$$2^\circ\ \delta_2 \text{ and } d_+d_- \text{ are nilpotent.} \tag{4.3}$$

A morphism $\gamma: \widetilde{A} \to \widetilde{A}'$ is a pair of linear maps $\gamma = (\gamma_1, \gamma_2)$, $\gamma_1:\ P_1 \to P_1'$, $\gamma_2:\ P_2 \to P_2'$, such that the diagram

$$\begin{array}{ccccccc} P_1 & \xrightarrow{d_+} & P_2 & \xrightarrow{\delta_2} & P_2 & \xrightarrow{d_-} & P_1 \\ {\scriptstyle\gamma_1}\downarrow & & {\scriptstyle\gamma_2}\downarrow & & {\scriptstyle\gamma_2}\downarrow & & {\scriptstyle\gamma_1}\downarrow \\ P_1' & \xrightarrow{d_+'} & P_2' & \xrightarrow{\delta_2'} & P_2' & \xrightarrow{d_-'} & P_1' \end{array}$$

commutes.

The main result in this section is Theorem 4.1, which states that the singular category $C(\lambda_1, \lambda_2)$ is equivalent to the category $S_0$.

2. Before proving this theorem we prove some lemmas.

LEMMA 4.1. *In the module $M$ of the singular category $C(\lambda_1, \lambda_2)$, all the subspaces $R_{l,m}$ for $l_0 \leqslant l \leqslant |l_1| - 1$ and all admissible $m$ have the same dimension. The subspaces $R_{l,m}(l \geqslant |l_1|$, any admissible $m)$ also have the same dimension.*[1] *Furthermore, the mappings*

[1] In particular, it may happen that $\dim R_{l_0,m} = 0$ and $\dim R_{|l_1|,m} \neq 0$. These are the modules $M$ to which there corresponds a pair of numbers $(|l_1|, (\mathrm{sign}\, l_1)\cdot l_0)$.

$$\left.\begin{array}{lll} D_+: & R_{l,m} \to R_{l+1,m} & (l \neq |l_1|-1), \\ D_-: & R_{l,m} \to R_{l-1,m} & (l \neq l_0;\ l \neq |l_1|) \end{array}\right\} \qquad (4.4)$$

*are isomorphisms.*

PROOF. Previously (in Theorem 2.3, (2.6)), we have found formulae for the eigenvalues $d_l^-$ and $d_l^+$ of $D_+D_-$ and $D_-D_+$ on $R_{l,m}$: Here are these formulae:

$$d_l^+ = \frac{((l+1)^2 - l_0^2)(l_1^2 - (l+1)^2)}{(4(l+1)^2 - 1)(l+1)^2}, \qquad d_l^- = \frac{(l^2 - l_0^2)(l_1^2 - l^2)}{(4l^2 - 1)l^2}.$$

The module $M$ is singular, therefore $l_1$ is real and the difference $(|l_1| - l_0)$ is a positive integer. Consequently, $d_l^+$ is equal to zero only for $l = |l_1| - 1$, and $d_l^-$ is equal to zero only for $l = l_0$ and $l = |l_1|$.

Thus, the following maps are non-singular:

$$\begin{array}{lll} (D_-D_+): & R_{l,m} \to R_{l,m}, & l \neq |l_1|-1, \\ (D_+D_-): & R_{l,m} \to R_{l,m}, & l \neq l_0, \quad l \neq |l_1|, \text{ and also } |m| \neq l. \end{array}$$

This is possible only if the maps

$$\left.\begin{array}{lll} D_+: & R_{l,m} \to R_{l+1,m} & (l \neq |l_1|-1), \\ D_-: & R_{l,m} \to R_{l-1,m} & (l \neq l_0;\ l \neq |l_1|, \text{ and also } |m| \neq l) \end{array}\right\} \qquad (4.5)$$

are monomorphisms. To fulfil this condition, the subspaces $R_{l,m}$ for $l = l_0, l_0 + 1, \ldots, |l_1| - 1$ must all have the same dimension, and similarly, the dimensions of the subspaces $R_{l,m}$ for $l \geqslant |l_1|$ must all be equal. This means that the maps (4.5) are isomorphisms.

REMARK. From the formulae (2.6) it is clear that in a singular module $M_1$ the maps

$$\begin{array}{ll} (D_-D_+): & R_{|l_1|-1,m} \to R_{|l_1|-1,m}, \\ (D_+D_-): & R_{|l_1|,m} \to R_{|l_1|,m}, \quad |m| \neq l_1, \end{array}$$

are nilpotent.

We have shown earlier (Proposition 2.2 §2) that the maps $D_+$, $D_-$, $E_+$, $E_-$ commute with the Laplace operators $\Delta_1$ and $\Delta_2$. Therefore from Lemma 4.1 we deduce directly:

LEMMA 4.2. *In a singular module $M$ all the operators $(\Delta_1)_{l,m}$ with $l_0 \leqslant l \leqslant |l_1| - 1$ are similar to $(\Delta_1)_{l_0,m_0}$, and the operators $(\Delta_1)_{l,m}$ with $l \geqslant |l_1|$ are similar to $(\Delta_1)_{|l_1|,m_0}$. Corresponding assertions hold for the operators $(\Delta_2)_{l,m}$.*

Now we prove another important lemma.

LEMMA 4.3. *Define in the singular module $M$ an operator $\delta$ by the formula*

$$\delta = \frac{1}{l_0^2 - l_1^2}\left(\Delta_2 + \frac{\Delta_1^2}{l_0^2} - (l_0^2 - 1)I\right) \qquad (l_0 \neq 0). \qquad (4.6)$$

*Then on the subspaces $R_{l,m}$ for which $l_0 \leqslant l \leqslant |l_1| - 1$ the operator $\delta$ is zero, and on the remaining subspaces $R_{l,m}$, $\delta$ is nilpotent.*

PROOF. Since on each of the subspaces $R_{l,m}$ the operators $\Delta_1$ and $\Delta_2$ have only one eigenvalue $\lambda_1 = -il_0l_1$, $\lambda_2 = l_0^2 + l_1^2 - 1$, respectively, it is clear that on each of them $\delta$ is nilpotent. Let us now prove that on the subspaces $R_{l,m}$, $l_0 \leqslant l < |l_1|$, $\delta = 0$.

On $R_{l_0,m_0}$, the operator $D_-$ is zero. Thus, for any vector $\xi \in R_{l_0,m_0}$, in accordance with (2.5) we can write $\Delta_1\xi = -l_0(l_0+1)D_0\xi$;
$(\Delta_2 + (l_0+1)^2 D_0^2 - (l_0^2-1)I)\xi = 0$. By hypothesis, $l_0 \neq 0$. Therefore

$$\frac{1}{l_0^2 - l_1^2}\left(\Delta_2 + \frac{(\Delta_1)^2}{l_0^2} - (l_0^2-1)I\right)\xi = \delta\xi = 0, \qquad \xi \in R_{l_0,m_0}.$$

The operator $\delta$ is a linear combination of $\Delta_2$, $\Delta_1^2$, and $I$. Therefore, by Lemma 4.2, $(\delta)_{l,m}$ for $l_0 \leqslant l \leqslant |l_1| - 1$ is similar to $\delta_{l_0,m_0}$. Thus, $\delta = 0$ for $l_0 \leqslant l \leqslant |l_1| - 1$.

REMARK. As was proved in §3, $\delta \equiv 0$ in the non-singular case. This is the principal difference between the singular and non-singular cases.

We now state a property of the operator $\delta$ that is important to us.

LEMMA 4.4.

$$\delta D_+\xi = 0 \quad \text{for} \quad \xi \in R_{|l_1|-1,m}, \tag{4.7}$$

$$D_-\delta\xi = 0 \quad \text{for} \quad \xi \in R_{|l_1|,m}. \tag{4.8}$$

PROOF. The operator $\delta$, as a linear combination of $\Delta_2$, $\Delta_1^2$ and $I$, commutes with $D_+$ and $D_-$. Therefore $D_+\delta\xi = \delta D_+\xi$ if $\xi \in R_{|l_1|-1,m}$, and $D_-\delta\xi = \delta D_-\xi$ if $\xi \in R_{|l_1|,m}$.

The assertion follows therefore from the fact that $\delta = 0$ on $R_{|l_1|-1,m}$.

3. We now transfer our attention to the connection between modules $M \in C(\lambda_1, \lambda_2)$ of the singular category, and objects of $S_0$. First we construct for each singular module $M$ such an object. For this purpose, if $M$ is a singular module, we select the subspaces $R_{|l_1|-1,m_0}$ and $R_{|l_1|,m_0}$, where $m_0$ is an admissible value of $m$. Let us consider the maps

$$D_+:\quad R_{|l_1|-1,m_0} \longrightarrow R_{|l_1|,m_0}, \qquad D_-:\quad R_{|l_1|,m_0} \longrightarrow R_{|l_1|-1,m_0}$$

and

$$\delta:\quad R_{|l_1|,m_0} \longrightarrow R_{|l_1|,m_0}, \qquad \delta:\quad R_{|l_1|-1,m_0} \longrightarrow 0. \tag{4.9}$$

We denote these spaces by $P_1 = R_{|l_1|-1,m_0}$, $P_2 = R_{|l_1|,m_0}$, and the maps by $d_+$, $d_-$, $\delta_1$, $\delta_2$, respectively. The sextuple $(P_1, P_2, d_+, d_-, \delta_1, \delta_2)$ is an element $\tilde{A}$ of the category $S_0$. In fact, the conditions $d_-\delta_2 = \delta_2 d_+ = 0$, $\delta_2$ and $d_+d_-$ nilpotent, come respectively from Lemmas 4.4, 4.3, and the remark preceding Lemma 4.2. Thus, the conditions (4.2) and (4.3) are satisfied. It is also easy to see that to each morphism $\Gamma: M \longrightarrow M'$, $M, M' \in C(\lambda_1, \lambda_2)$, there corresponds a morphism $\gamma = (\gamma_1, \gamma_2)$ of $\tilde{A}$ into $\tilde{A}'$.

We now show that to each sextuple $\tilde{A} = (P_1, P_2, d_+, d_-, \delta_1, \delta_2) \in S_0$ (that is, satisfying the conditions (4.1), (4.2), (4.3)), there corresponds a module $M \in C(\lambda_1, \lambda_2)$, $\lambda_1 = -il_0l_1$, $\lambda_2 = l_0^2 + l_1^2 - 1$. Let us construct the module $M$. First we choose a number $m_0$ compatible with $|l_1| - 1$, and put

$P_1 = R_{|l_1|-1, m_0}$, $P_2 = R_{|l_1|, m_0}$. Then, on $R_{|l_1|-1, m_0}$, and $R_{|l_1|, m_0}$ we put

$$D_+ \equiv d_+ : \ R_{|l_1|-1, m_0} \to R_{|l_1|, m_0}, \qquad D_- \equiv d_- : \ R_{|l_1|, m_0} \to R_{|l_1|-1, m_0}.$$

Consider the space $M = \bigoplus_{l, m} R_{l, m}$, where $l = l_0, l_0+1, \ldots, |l_1|-1, |l_1|, \ldots,$ and $m$ takes, for each $l$, all admissible values, $\dim R_{l,m} = \dim P_1$ for all $l \leqslant |l_1| - 1$, and $\dim R_{l,m} = \dim P_2$ for $l \geqslant |l_1|$. Let us now construct $E_+, E_-, D_+, D_-, D_0$. We take $E_+ : R_{l, m} \to R_{l, m+1}$ $(m \neq l)$ as any isomorphism between these spaces, $E_+ :$ $R_{l, l} \to 0$, $E_- : R_{l, m} \to R_{l, m-1}$ as the isomorphism inverse to $E_+$, $E_- :$ $R_{l, -l} \to 0$. We will take $D_+ : R_{l, m} \to R_{l+1, m}$ for $l \neq |l_1| - 1$ to be an arbitrary isomorphism. For $l = |l_1| - 1$, $D_+$ is defined for $m = m_0$ by the formula (4.4), and this definition carries over to the other values of $m$ by means of the isomorphism $E_+$. We must now define $D_0$, $D_-$. For this purpose let us first construct $\Delta_2$ and $\Delta_1$. In §2 (formulae (2.2), (2.3)), we had the following expressions for the Laplace operators:

$$\Delta_1 \xi = -l(l+1) D_0 \xi \qquad (\xi \in R_{l, m}), \tag{4.10}$$

$$\Delta_2 \xi = (l^2 - 1)\xi - (l+1)^2 D_0^2 \xi + (4l^2 - 1) D_+ D_- \xi \qquad (\xi \in R_{l, m}). \tag{4.11}$$

Since for $l = |l_1|, m = m_0$ $D_+, D_-$ and $\delta = \frac{1}{l_0^2 - l_1^2}\left(\Delta_2 + \frac{\Delta_1^2}{l_0^2} - (l_0^2 - 1) I\right)$ are given, we define

$$\left.\begin{aligned} \Delta_2 \xi &= (l_1^2 + l_0^2 - 1)\xi + l_1^2 \frac{4l_1^2 - 1}{l_1^2 - l_0^2} D_+ D_- \xi + l_0^2 \delta \xi, \\ \Delta_1^2 \xi &= -l_0^2 l_1^2 \left(\xi + \frac{4l_1^2 - 1}{l_1^2 - l_0^2} D_+ D_- \xi + \delta \xi\right), \qquad \xi \in R_{|l_1|, m_0}; \end{aligned}\right\} \tag{4.12}$$

By hypothesis, the operators $D_+$, $D_-$ and $\delta$ are nilpotent and $l \neq 0$, and $\Delta_1$ has only one eigenvalue, which is non-zero: thus, $\Delta_1$ is uniquely determined.

Consider now the map $J_{k_1, k_2} = (E_+)^{k_1} (D_+)^{k_2}$ where for $k_1 < 0$,

$$(E_+)^{k_1} = (E_-)^{|k_1|}.$$

For $l \geqslant |l_1|$ we now put

$$(\Delta_i)_{l, m} = J_{m - m_0, l - |l|_1} (\Delta_i)_{|l_1|, m_0} J^{-1}_{m - m_0, l - |l_1|} \qquad (i = 1, 2). \tag{4.13}$$

Next, we construct the operators $D_0$, $D_-$. To satisfy the condition $\Delta_1 = -l(l+1)(D_0)_{l, m}$ we put

$$(D_0)_{l, m} = -\frac{1}{l(l+1)} (\Delta_1)_{l, m}. \tag{4.14}$$

Similarly, for $\xi \in R_{l,m}$, we define

$$D_+D_-\xi = \frac{1}{4l^2-1}\left(\Delta_2\xi + \frac{\Delta_1^2\xi}{l^2} - (l^2-1)\,\xi\right). \tag{4.15}$$

As $D_+$ is given, we can find $D_-$ from it.

In a completely similar manner, $D_0$ and $D_-$ can be found for $l_0 \leqslant l < |l_1|$. We only have to use, instead of (4.11), the formula

$$\Delta_2\xi = ((l+1)^2-1)\,\xi - l^2D_0^2 + (4\,(l+1)^2-1)\,D_-D_+\xi \qquad (\xi \in R_{l,\,m}) \tag{4.16}$$

and to put $\delta = 0$.

It can be verified that the operators $F_+$, $F_-$, $F_3$ and $H_+$, $H_-$, $H_3$, constructed from the above $D_+$, $D_-$, $D_0$, $E_+$, $E_-$, give a representation. We shall not verify that there is a correspondence between the morphisms (this may be done just as in §3). Thus, Theorem 4.1 holds. The singular category $\mathfrak{C}(\lambda_1, \lambda_2)$ is equivalent to $S_0$.

4. In this section we write down explicit formulae for all the operators.

Let $M$ be a singular Harish-Chandra module to which there corresponds the pair of numbers $l_0$, $l_1$. Then, if matrices $[d_+]_{|l_1|-1,\,m_0}$, $[d_-]_{|l_1|,\,m_0}$, $[\delta]_{|l_1|,\,m_0}$ and $[\delta]_{|l_1|-1,\,m_0}=0$ are given, in the subspaces $R_{l,m}$ a basis can be chosen in which the matrices $[D_+]_{l,\,m}$, $[D_-]_{l,\,m}$, $[D_0]_{l,\,m}$ are defined by the formulae below. We write

$$[a_0] = \frac{4l_1^2-1}{l_1^2-l_0^2}\,[d_-d_+]_{|l_1|-1,\,m_0},$$

$$[a_1] = \frac{4l_1^2-1}{l_1^2-l_0^2}\,[d_+d_-]_{|l_1|,\,m_0},$$

$$[\delta] = [\delta]_{|l_1|,\,m_0},$$

$[I_0]$ – the identity matrix of dimension $n_0 \times n_0$, where $n_0 = \dim R_{|l_1|-1,\,m_0}$,

$[I_1]$ – the identity matrix of dimension $n_1 \times n_1$, where $n_1 = \dim R_{|l_1|,\,m_0}$.

Then

$$[D_0]_{l,\,m} = \frac{il_0l_1}{l\,(l+1)}\sqrt{[a_0]+[I_0]}, \qquad \text{when} \quad l_0 \leqslant l \leqslant |l_1|-1,$$

$$[D_0]_{l,\,m} = i\,\frac{l_0l_1}{l\,(l+1)}\sqrt{[a_1]+[I_1]+[\delta]}, \qquad \text{when} \quad l \geqslant |l_1|,$$

$$[D_+]_{l,\,m} = [I_0], \qquad \text{when} \quad l_0 \leqslant l < |l_1|-1,$$

$$[D_+]_{l,\,m} = [I_1], \qquad \text{when} \quad l \geqslant |l_1|,$$

$$[D_+]_{|l_1|-1,\,m} = [d_+],$$

$$[D_-]_{l,\,m} = \frac{(l^2-l_0^2)\,(l_1^2-l^2)}{(4l^2-1)\,l^2}\left([I_0] + \frac{l_1^2}{l_1^2-l^2}\,[a_0]\right), \qquad \text{when} \quad l_0 \leqslant l \leqslant |l_1|-1,$$

$$[D_-]_{l,\,m} = \frac{(l^2-l_0^2)\,(l_1^2-l^2)}{(4l^2-1)\,l^2}\left([I_0] + \frac{l_1^2}{l_1^2-l^2}\,[a_1] - \frac{l_0^2}{l^2-l_0^2}\,[\delta]\right), \qquad \text{when} \quad l > |l_1|,$$

$$[D_-]_{|l_1|,\,m} = [d_-].$$

5. For a complete solution of the question of a canonical basis we must solve the following problem: given are two linear spaces $P_1$ and $P_2$, and maps $d_+: P_1 \to P_2$, $d_-: P_2 \to P_1$, $\delta_2: P_2 \to P_2$ such that $d_-\delta_2 = \delta_2 d_+ = 0$, and $\delta_2$ and $d_+d_-$ are nilpotent: we have to find bases of $P_1$ and $P_2$ in which the matrices of these maps have the canonical form, and we have to find a complete system of invariants.

## Chapter II

In this chapter we discuss the problem of linear algebra to which we were led in Chapter I in trying to classify the singular modules. At the Mathematical Congress of 1967 Szekeres communicated to us a similar solution. However, we do not know his method, and therefore we are not aware whether the more general problem we need may be solved in the same way. We use here MacLane's notion of relations, which is a generalization of the notion of a linear map. It is remarkable that to work with relations is easier than with linear maps.

And so the problem is the following. We consider two linear spaces $P_1$ and $P_2$ and three linear maps:

$$d_+:\ P_1 \to P_2, \quad d_-:\ P_2 \to P_1, \quad \delta:\ P_2 \to P_2, \quad \delta:\ P_1 \to 0$$

such that $\delta d_+ = 0$, $d_-\delta = 0$, $\delta$ and $d_-d_+$ are nilpotent. We have to bring this system to a canonical form.

If we introduce the space $P = P_1 \oplus P_2$ and consider in it the maps $a$ and $b$ given by the matrices

$$a = \begin{pmatrix} 0 & d_+ \\ d_- & 0 \end{pmatrix}, \quad b = \begin{pmatrix} 0 & 0 \\ 0 & \delta \end{pmatrix},$$

then $ab = ba = 0$, and $a$ and $b$ are nilpotent.

More generally, suppose that we have $r$ linear spaces $P_s$ $(s = 1, \ldots, r)$, and in $P = \bigoplus_{s=1}^{r} P_s$ we consider two linear maps:

1° $bP_s \subset P_s$, $aP_s = P_{s+1}$, $aP_r = P_1$,
2° $ab = ba = 0$,
3° $a$, $b$ nilpotent.

We now introduce the important idea of a homogeneous subspace of $P = \bigoplus_{s=1}^{r} P_s$. We say that a subspace $\varphi \subseteq P$ is homogeneous if $\bigoplus_{s=1}^{r} \varphi \cap P_s = \varphi$.

Our task is to find a canonical form for $a$ and $b$ in $P$ with the grading of $P$ given above (that is, the decomposition $P = \bigoplus_{s=1}^{r} P_s$). In what follows, we write $C^n$ instead of $P$, and $C_s$ instead of $P_s$.

The plan of this chapter is the following.

In §1 the apparatus of relations is developed in the form in which we need it for the solution of our problem, and also the so-called stabilized sequences and elementary intervals are introduced and studied.

If the reader is only interested in the problem of a pair of nilpotent operators $a$ and $b$ for which $ab = ba = 0$, then he may §2, although for the solution of the problem that arose in Chapter I (the same problem in graded spaces) we need §2.

In §3 an account is given of certain fairly simple techniques which we use repeatedly in §§4 and 5. The reader may omit §3 on a first reading and refer to it only when necessary.

Thus, the basic material is contained in §§1, 4 and 5, while §§2 and 3 play an auxiliary role.

## §1 The apparatus of relations and the construction of stabilized sequences.

**1.** Definition: Let $C^n$ be a $n$-dimensional complex vector space. Then we mean by a relation any subspace of $C^n \oplus C^n$.

To a linear map $a$: $C^n \longrightarrow C^n$ there corresponds the subspace of pairs $(x, ax)$ in $C^n \oplus C^n$, where $x$ ranges over $C^n$. Thus, a relation is a generalization of a linear map. In particular, the zero map $\theta$: $C^n \to 0$ corresponds to the relation $\theta$, consisting of the pairs $(x, 0)$, where $x$ ranges over $C^n$.

The identity map $I$: $C^n \to C^n$ corresponds to the relation I, consisting of the pairs $(x, x)$, as $x$ ranges over $C^n$.

We denote the relation corresponding to a linear map $a$ by the same symbol $a$.

If $A$ is a relation consisting of some set of pairs $(x, y)$, then the set of all pairs $(y, x)$ defines a new subspace of $C^n \oplus C^n$, called the inverse relation $A^{\#}$. It is easy to verify that if $a$ is an invertible linear map, then $a^{\#} = a^{-1}$.

We define the dimension of a relation $A$ as the dimension of the subspace $A$ of $C^n \oplus C^n$. It is denoted by $\dim A = m(A)$.

Let $A$ and $B$ be two relations. Their product $AB$ is defined in the following manner:[1] $(x, z) \in AB$ if and only if there exists a $y$ such that $(x, y) \in B$ and $(y, z) \in A$.

It is easy to verify that if the relations are defined by linear maps, then this definition coincides with the generally accepted definition of multiplication of maps.

We say that $A \subseteqq B$ if $A$ is contained in $B$, and $A \subset B$ if $A$ is contained in $B$, but is not equal to it.

Any relation $\alpha$ such that $\alpha \subseteqq \theta$ determines uniquely a subspace of $C^n \oplus \{0\}$. We denote these by small Greek letters and call them, for brevity, subspaces of $C^n$.

[1] We ask the reader to excuse this unnatural order. However, in fact, it agrees with the usually accepted way of multiplying linear maps.

The multiplication of relations is associative: $A(BC) = (AB)C$. Furthermore, from $A \subseteq B$ follows $AC \subseteq BC$ and $CA \subseteq CB$.

Note that an inequality $A \subset B$ may not carry over under multiplication.

We introduce the trivial relation $\Omega$, consisting of the single pair $(0, 0)$. Let $A$ be any relation. Then we call[1] $\Omega A$ the kernel of $A$, and $\theta A$ the domain of definition of $A$. If $a$ is a linear map, then $\theta a = \theta$, $\Omega a$ is the kernel of $a$ in the usual sense.

From $\Omega \subset \theta$ it follows that $\Omega A \subseteq \theta A$. If $\Omega A = \theta A$, that is, if the kernel coincides with the domain of definition, then $A$ is called a null relation. Thus, a relation $A$ is non-null if

$$\Omega A \subset \theta A.$$

A pair of subspaces $\alpha$ and $\beta$ of $\mathbf{C}^n$, $\alpha \subseteq \theta$, $\beta \subseteq \theta$ is denoted by $(\alpha, \beta)$ and is called an interval if $\alpha \subseteq \beta$. In case $\alpha = \beta$, the interval is called trivial. In particular, every relation $A$ defines the interval $(\Omega A, \theta A)$.

We call the number $\rho(\alpha, \beta) = \dim \beta - \dim \alpha$ the length of the interval. Somewhat later (see p. 37) we introduce the notion of the image and pre-image of a subspace $\varphi$ with respect to a relation $A$. Here is some further terminology. Let $A$ be a relation. Then

$\Omega A$ is called the kernel of $A$,

$\theta A$ is called the domain of definition of $A$,

$A\Omega$ is called the indeterminacy of $A$,

$A\theta^{\#}$ is called the domain of values of $A$.

(Note that $\Omega^{\#} = \Omega$.) We suggest that the reader should examine what these notions mean when $A = a$ or $A = a^{\#}$, where $a$ is a linear map.

2. In this subsection we state some propositions about relations which we use throughout the chapter.

We have already said that each relation $D$ defines a pair of subspaces (an interval) $(\Omega D, \theta D)$, where $(\Omega D, \theta D)$ is non-trivial if $D$ is non-null.

PROPOSITION 1.1. *The interval* $(\Omega CD, \theta CD)$ *is contained in* $(\Omega D, \theta D)$.

PROOF. Multiplying the inequalities $\Omega \subset \Omega C$ and $\theta C \subseteq \theta$ on the right by $D$, we obtain

$$\Omega D \subseteq \Omega CD, \qquad \theta CD \subseteq \theta D, \tag{1.1}$$

which is Proposition 1.1.

Our subsequent study of nilpotent maps $a$ and $b$ for which $ab = ba = 0$ is based on introducing monomials $D$ of the form

$$D = a^{k_1} (b^{\#})^{k_2} \dots a^{k_{l-1}} (b^{\#})^{k_l}, \tag{1.2}$$

where $k_1$ and $k_l \geqslant 0$, and the remaining $k_i$ are positive.

This technique of dealing with monomials in $a$ and $b^{\#}$ is very convenient. For example, the reader would certainly find it rather lengthy to define in words the subspace $\Omega a b^{\#} a$.

We define the degree of a monomial $D$ as the number $s(D) = \sum_i k_i$. If

[1] Thus, the kernel $\Omega A$ consists of the pairs $(x, 0) \in A$. The domain of definition of $A$ consists of the pairs $(x, 0)$ for which there exists a $y$, such that $(x, y) \in A$.

$s(D)=0$, this means that $D=1$. Of course, we could also consider monomials containing $a^{\#}$ and $b$.

PROPOSITION 2.1. *The relation $ab=0$ is equivalent to $\theta b^{\#}\subseteqq \Omega a$.*

PROOF. $b^{\#}$ consists of the pairs $(bx, x)$, $\theta b^{\#}$ consists of all pairs of the form $(bx, 0)$. $\Omega a$ consists of all pairs $(y, 0)$ such that $ay=0$. Since $ab=0$, $\theta b^{\#}\subseteqq \Omega a$. We often use this inequality. Let us write it out in the following form:

$$\Omega=\Omega b^{\#}\subseteqq \theta b^{\#}\subseteqq \Omega a\subseteqq \theta a=\theta. \tag{1.3}$$

Multiplying these inequalities on the right by $D$ we obtain

$$\Omega D=\Omega b^{\#}D\subseteqq \theta b^{\#}D\subseteqq \Omega aD\subseteqq \theta aD=\theta D. \tag{1.4}$$

Similarly we can prove

PROPOSITION 1.3. $ba=0$ is equivalent to $a\theta^{\#}\subseteqq b^{\#}\Omega$.

3. Let us consider the set of all subspaces of the forms $\Omega D$ and $\theta D$, where $D$ ranges over the set $\Sigma$ of all monomials of the form (1.2).

THEOREM 1.1. *The number of distinct subspaces of the form $\Omega D$ and $\theta D$ $(D\in\Sigma)$ is finite. They form an ordered sequence*

$$\Omega\subset\beta_1\subset\beta_2\subset\ldots\subset\beta_{s-1}\subset\beta_s=\theta. \tag{1.5}$$

This sequence is called *stabilized.*

Before proving the theorem we prove a lemma.

LEMMA 1.1. *Let $D_1$ and $D_2$ be two monomials. Then two types of relations are possible between the intervals $(\Omega D_1, \theta D_1)$ and $(\Omega D_2, \theta D_2)$.*

*1° The intervals follow one after the other, that is, $\Omega D_1\subseteqq\theta D_1\subseteqq\Omega D_2\subseteqq\theta D_2$ (or the same thing with $D_1$ and $D_2$ interchanged).*

*2° One interval contains the other, that is, $\Omega D_1\subseteqq\Omega D_2\subseteqq\theta D_2\subseteqq\theta D_1$.*

PROOF. Let us read off from right to left the factors $a$ and $b^{\#}$, occurring in $D_1$ and $D_2$. We denote by $D_0$ the largest common part of these monomials (it is, of course, possible that $D_0=1$). Then the following cases can occur: 1) $D_1=D_1'b^{\#}D_0$, $D_2=D_2'aD_0$, 2) $D_1=D_0$, $D_2=D_2'D_0$ or cases 3) and 4), differing from 1) and 2) by the interchange of $D_1$ and $D_2$.

Consider case 1). As $\Omega\subseteqq\Omega D_1'\subseteqq\theta D_1'\subseteqq\theta$, multiplying on the right by $b^{\#}$ we have $\Omega D_1'b^{\#}\subseteqq\theta D_1'b^{\#}\subseteqq\theta b^{\#}$.

Similarly $\Omega a\subseteqq\Omega D_2'a\subseteqq\theta D_2'a$. Since $\theta b^{\#}\subseteqq\Omega a$, by combining we have $\Omega D_1'b^{\#}\subseteqq\theta D_1'b^{\#}\subseteqq\Omega D_2'a\subseteqq\theta D_2'a$. Multiplying on the right by $D_0$ we obtain $\Omega D_1\subseteqq\theta D_1\subseteqq\Omega D_2\subseteqq\theta D_2$, that is, $(\Omega D_1, \theta D_1)$ lies to the left of the interval $(\Omega D_2, \theta D_2)$.

Thus, to find out which interval precedes the other we must go from right to left along $D_1$ and $D_2$ until the first difference is reached. The interval for which $D$ contains $b^{\#}$ at this point lies to the left.

We now make a remark we shall find useful later. Consider the intervals I $(\Omega b^{\#}D_0, \theta b^{\#}D_0)$ and II $(\Omega aD_0, \theta aD_0)$.

We know that one follows after the other (see (1.4)). If we replace $b^{\#}D_0$ by $D_1'b^{\#}D_0$ and $aD_0$ by $D_2'aD_0$, then the intervals can only move away from each other, because $(\Omega D_1' b^{\#} D_0,\ \theta D_1' b^{\#} D_0)$ lies inside I, and $(\Omega D_2' a D_0,\ \theta D_2' a D_0)$ inside II.

Now we go over to case 2), when $D_2$ is divisible on the right by $D_1$. As $D_2 = D_2' D_1$, the interval $(\Omega D_2' D_1,\ \theta D_2' D_1)$ lies inside $(\Omega D_1,\ \theta D_1)$, that is, $\Omega D_1 \subseteqq \Omega D_2 \subseteqq \theta D_2 \subseteqq \theta D_1$.

PROOF OF THEOREM 1.1. Consider the set $\Sigma$ of all monomials and the set of all subspaces $\Omega D$ and $\theta D$, as $D$ ranges over $\Sigma$. Since the whole space $\theta$ is finite-dimensional, this set contains only finitely many distinct subspaces, which we denote by $\Omega, \beta_1, \ldots, \beta_s = \theta$. So we obtain a stabilized sequence $\Omega \subset \beta_1 \subset \beta_2 \subset \ldots \subset \beta_{s-1} \subset \beta_s = \theta$.

PROPOSITION 1.4. *If $D_1$ and $D_2$ are non-null monomials and the intervals $(\Omega D_1, \theta D_1)$ and $(\Omega D_2, \theta D_2)$ coincide, then one of the monomials is a right divisor of the other.*

PROOF. If this were not so, then we would have case 1) of Lemma 1.1. But this is impossible, since in case 1), $\Omega D_1 \subseteqq \theta D_1 \subseteqq \Omega D_2 \subseteqq \theta D_2$, hence the intervals can coincide only if they are trivial.

PROPOSITION 1.5. *If*

$$\Omega D \subset \Omega D_1 \subset \theta D,$$

*then $D$ divides $D_1$ on the right, that is, $D_1 = D'D$.*

PROOF. From this inequality, by Lemma 1.1 it follows that

$$\Omega D \subset \Omega D_1 \subseteqq \theta D_1 \subseteqq \theta D.$$

But (by the same lemma) this is possible only if $D_1$ is divisible on the right by $D$.

Similarly, if $\Omega D \subset \theta D_1 \subset \theta D$, then $D_1$ is divisible on the right by $D$.

4. Elementary intervals of the first and second kind. With each non-null monomial $D$ there is associated an interval $(\Omega D, \theta D)$. It turns out that this interval can be divided into three. For we have the inequality (1.4).

$$\Omega D \subseteqq \theta b^{\#} D \subseteqq \Omega a D \subseteqq \theta D.$$

If the middle one of these intervals, that is, $(\theta b^{\#} D, \Omega a D)$, is non-trivial, then we call $D$ a monomial of the first kind. We denote by $\Sigma_1$ the set of such monomials. Thus, $\Sigma_1$ consists of the monomials $D$ for which

$$\theta b^{\#} D \subset \Omega a D. \tag{1.6}$$

From (1.6), it is clear that any monomial $D \in \Sigma_1$ is non-null. The monomials $D$ in $\Sigma_1$ have the following remarkable property.

PROPOSITION 1.6. *For every monomial $D$ in $\Sigma_1$ the interval $(\theta b^{\#} D, \Omega a D)$ is elementary. In other words, $\theta b^{\#} D = \beta_{i-1}$, $\Omega a D = \beta_i$, where $\beta_{i-1}$ and $\beta_i$ are neighbouring subspaces in the stabilized sequence (1.5).*

PROOF. Suppose the contrary. Let $\beta_{i_0}$ be a subspace in the stabilized sequence that is larger than $\theta b^{\#} D$ and smaller than $\Omega a D$:

$$\theta b^{\#} D \subset \beta_{i_0} \subset \Omega a D.$$

Since the interval $(\theta b^{\#}D, \Omega aD)$ is contained inside $(\Omega D, \theta D)$, we have a fortiori that $\Omega D \subset \beta_{i_0} \subset \theta D$.

The subspace $\beta_{i_0}$ is either the kernel or the domain of definition of some monomial. For the sake of definiteness let us assume that $\beta_{i_0}$ has the form $\Omega D_1$. This means that we have $\Omega D \subset \Omega D_1 \subset \theta D$.

By Proposition 1.5 this implies that $D_1$ is divisible on the right by $D$, that is, $D_1 = D_1' D$, $D_1' \neq 1$ Here we must distinguish two cases. In the first case, $D_1$ has the form $D_1' = D_1'' b^{\#}$; and in the second case, $D_1' = D_1'' a$.

CASE 1. Suppose that $D_1 = D_1'' b^{\#} D$. Since $D_1$ is divisible on the right by $b^{\#}D$, the second variant of Lemma 1.1 holds, and we have $\Omega b^{\#}D \subseteqq \Omega D_1 \subseteqq \theta D_1 \subseteqq \theta b^{\#}D$.

We recall that $\Omega D_1 = \beta_{i_0}$, consequently $\beta_{i_0} \subseteqq \theta b^{\#}D$. However, this is in contradiction to our basic assumption that $\theta b^{\#}D \subset \beta_{i_0} \subset \Omega aD$. Thus, in case 1 the proposition is proved.

CASE 2. Suppose that $D_1 = D_1'' aD$. In this case the monomial $D_1$ is divisible on the right by $aD$, and we can write:

$$\Omega aD \subseteqq \Omega D_1 \subseteqq \theta D_1 \subseteqq \theta aD$$

or

$$\Omega aD \subseteqq \beta_{i_0},$$

which contradicts our hypothesis $\theta b^{\#}D \subset \beta_{i_0} \subset \Omega aD$.

Thus, Proposition 1.6 is proved under the assumption that $\beta_{i_0} = \Omega D_1$. Under the assumption that $\beta_{i_0} = \theta D_2$ the proof proceeds similarly.

So we have shown that to each monomial $D \in \Sigma_1$ there corresponds a certain elementary interval $(\beta_{i-1}, \beta_i) = (\theta b^{\#}D, \Omega aD)$.

PROPOSITION 1.7. *To distinct monomials $D_1$ and $D_2$ of $\Sigma_1$ there correspond distinct intervals.*

PROOF. We show that for distinct monomials $D_1$ and $D_2$ only two types of inequality are possible: either $\theta b^{\#}D_1 \subset \Omega aD_1 \subseteqq \theta b^{\#}D_2 \subset \Omega aD_2$, or the same inequality with $D_1$ and $D_2$ interchanged. This means that the elementary intervals $(\theta b^{\#}D_1, \Omega aD_1)$ and $(\theta b^{\#}D_2, \Omega aD_2)$ are distinct.

As in Lemma 1.1 we consider two cases. In the first case the monomials $D_1$ and $D_2$ may be written in the form $D_1 = D_1' b^{\#} D_0$ and $D_2 = D_2' a D_0$ or similarly with $D_1$ and $D_2$ interchanged. In the second case one of these monomials divides the other on the right.

CASE 1. $D_1 = D_1' b^{\#} D_0$, $D_2 = D_2' a D_0$. By Lemma 1.1 we have $\Omega D_1 \subset \theta D_1 \subseteqq \Omega D_2 \subset \theta D_2$. Since the interval $(\theta b^{\#}D_i, \Omega aD_i)$ lies inside $(\Omega D_i, \theta D_i)$, the proposition is proved in this case.

CASE 2. $D_1$ divides $D_2$ on the right. We must now consider the two possibilities: a) $D_2 = D_2' b^{\#} D_1$; b) $D_2 = D_2' a D_1$.

a) Suppose that $D_2 = D_2' b^{\#} D_1$. Then $D_2$ is divisible on the right by $b^{\#}D_1$, and therefore, by Lemma 1.5, we may conclude that $\Omega b^{\#}D_1 \subseteqq \Omega D_2 \subset \theta D_2 \subseteqq \theta b^{\#}D_1$.

We know (1.4) that the interval $(\theta b^{\#}D_2, \Omega a D_2)$ lies inside $(\Omega D_2, \theta D_2)$. Therefore we obtain from the last inequality: $\theta b^{\#}D_2 \subset \Omega a D_2 \subseteq \theta b^{\#}D_1$ and so $\theta b^{\#}D_2 \subset \Omega a D_2 \subseteq \theta b^{\#}D_1 \subset \Omega a D_1$, that is, under the hypothesis that $D_2 = D_2' b^{\#}D_1$, the proposition is proved.

b) Suppose that $D_2 = D_2' a D_1$. In this case we find in a similar manner that $\theta b^{\#}D_1 \subset \Omega a D_1 \subseteq \theta b^{\#}D_2 \subset \Omega a D_2$.

The proposition is now completely proved. More than that, we have shown how to decide, knowing the form of the monomials $D_1$ and $D_2$, which of the intervals $(\theta b^{\#}D_1, \Omega a D_1)$ and $(\theta b^{\#}D_2, \Omega a D_2)$ lies to the left of the other.

An elementary interval $(\beta_{i-1}, \beta_i)$ of the form $(\theta b^{\#}D, \Omega a D)$ for some $D \in \Sigma_1$ is said to be of the first kind. All the remaining elementary intervals are said to be of the second kind.

So we have set up a one-to-one correspondence between the set $\Sigma_1$ and the set of elementary intervals of the first kind. From this it is clear that the set $\Sigma_1$ is finite. We denote by $k_1$ the maximal degree of a monomial $D$ in $\Sigma_1$.

5. Now we study the elementary intervals of the second kind. We shall see that there exists a number $m$ such that for every $k \geqslant m$ and for every interval $(\beta_{i-1}, \beta_i)$ of the second kind we may select a uniquely-determined monomial $D$, $s(D) = k$, such that $(\beta_{i-1}, \beta_i) = (\Omega D, \theta D)$.

We shall also see that to each non-null monomial $D$ of degree $k \geqslant m$ there corresponds an elementary interval $(\Omega D, \theta D)$ and that this elementary interval is of the second kind.

For the study of elementary intervals of the second kind we introduce an important operation, which we call the operation of completion. According to (1.4), we have

$$\Omega D = \Omega b^{\#}D \subseteq \theta b^{\#}D \subseteq \Omega a D \subseteq \theta a D = \theta D.$$

The operation of completion consists in replacing the interval $(\Omega D, \theta D)$ by the three intervals $(\Omega b^{\#}D, \theta b^{\#}D)$, $(\theta b^{\#}D, \Omega a D)$, $(\Omega a D, \theta a D)$.

If the middle one $(\theta b^{\#}D, \Omega a D)$ of the newly obtained intervals is non-trivial, then it is an elementary interval of the first kind. The two other intervals have the form $(\Omega D', \theta D')$, where $D'$ is a monomial whose degree exceeds that of $D$ by 1. In one of these intervals $D'$ is equal to $b^{\#}D$, and in the other to $aD$.

REMARK. *It is easy to see that the completion of a trivial interval leads to trivial intervals.*

We shall show that, by means of the operation of completion, we obtain a stabilized sequence from the interval $(\Omega, \theta)$. The monomials of degree 1 are $a$ and $b^{\#}$. They give the following decomposition of $(\Omega, \theta)$ into three intervals:

$$\Omega = \Omega b^{\#} \subseteq \theta b^{\#} \subseteq \Omega a \subseteq \theta a = \theta.$$

The interval $(\theta b^{\#}, \Omega a)$, if it is non-trivial, is already elementary (see

Proposition 1.6). The two other intervals $(\Omega b^{\#}, \theta b^{\#})$ and $(\Omega a, \theta a)$ have the form $(\Omega D, \theta D)$, where $D$ is a monomial of degree 1. These intervals can be completed further by monomials of degree two, and so on. By induction, we see that at the $k$-th step any interval has one of the two forms:

$(\Omega D, \theta D)$, $s(D) = k$, or $(\theta b^{\#} D, \Omega a D)$, $s(D) \leqslant k-1$.

Suppose that for $k = k_0$ we obtain the stabilized sequence. Then every elementary interval in the stabilized sequence has one of two forms.

Either it may be written in the form $(\theta b^{\#} D, \Omega a D)$, when it is of the first kind, or it has the form $(\Omega D, \theta D)$, $s(D) = k_0$.

Although at the point $k = k_0$ we have already obtained the stabilized sequence, we may carry the process of completion further, that is, we may replace a non-trivial interval $(\Omega D, \theta D)$ by the three intervals

$$(\Omega b^{\#} D, \theta b^{\#} D), \quad (\theta b^{\#} D, \Omega a D), \quad (\Omega a D, \theta a D). \tag{1.7}$$

(We recall that $\Omega D = \Omega b^{\#} D$, $\theta D = \theta a D$.) If $s(D) \geqslant k_0$, then the interval $(\Omega D, \theta D)$ is elementary, and therefore, *of the intervals* (1.7) *one and only one is non-trivial*.

Let us now recall that there are only finitely many monomials of the first kind, and that $k_1 = \sup_{D \in \Sigma_1} s(D)$ is the largest of their degrees. We put $m = \sup(k_0, k_1)$, and call $m$ the index of terminal stability. For $s(D) = k > m$, we may say[1] that the elementary interval $(\theta b^{\#} D, \Omega a D)$ is trival, and therefore, for $k > m$ the elementary interval $(\Omega D, \theta D)$ is identical with one and only one of the intervals $(\Omega b^{\#} D, \theta b^{\#} D)$ and $(\Omega a D, \theta a D)$, which have the same form as the original.

Thus, with each elementary interval of the second kind we can associate an infinite sequence of monomials $D_m, D_{m+1}, \ldots$, such that $D_j = d_j \cdot D_{j-1}$ ($d_j = a$ or $b$) and such that the interval has the form

$$(\Omega D_m, \theta D_m) = (\Omega D_{m+1}, \theta D_{m+1}) = \ldots \tag{1.8}$$

Let $\Sigma_2$ denote the set of infinite sequences $\{D_j\}$ ($j = m, m+1, \ldots$,) for which $D_{j+1} = d_{j+1} D_j$, corresponding to an elementary interval of the second kind. This set is finite, since the elements of $\Sigma_2$ are in one-to-one correspondence with the elementary intervals of the second kind.

The arguments above also lead to the following remarks:

1° If $s(D) > m$, then the interval $(\Omega D, \theta D)$ is either trivial or elementary.

2° If the sequence $D_j$, where $D_{j+1} = d_{j+1} D_j$, $s(D_j) = j$ $(j = m+1, m+2, \ldots)$, does not belong to $\Sigma_2$, then beginning with some value of $j$, the interval $(\Omega D_j, \theta D_j)$ is trivial.

It is convenient to state our results about intervals of the second kind by introducing the concept of an infinite word

---

[1] For $k \leqslant k_1$ the interval $(\theta b^{\#} D, \Omega a D)$ may be non-trivial, that is, an elementary interval may be " transformed " from the form $(\Omega D, \theta D)$ into an interval of the first kind $(\theta b^{\#} D, \Omega a D)$.

$$\tilde{D} = \ldots d_p d_{p-1} \ldots d_1,\ d_i = a \text{ or } b^{\#}. \tag{1.9}$$

Denoting by $D_j$ the "segment" $D_j = d_j d_{j-1} \ldots d_1$ of the word, we obtain a nested sequence of intervals $(\Omega D_i, \theta D_i)$

As the preceding discussion shows, these intervals stop decreasing at some point $i$. We denote the interval so obtained by $(\Omega\tilde{D}, \theta\tilde{D})$. We know that $(\Omega\tilde{D}, \theta\tilde{D})$ is either trivial, or elementary (of the second kind). If $(\Omega\tilde{D}, \theta\tilde{D})$ is non-trivial, then from $(\Omega\tilde{D}, \theta\tilde{D}) = (\Omega\tilde{D}_1, \theta\tilde{D}_1)$ it follows[1] that $D = D_1$.

If the interval $(\Omega\tilde{D}, \theta\tilde{D})$ is trivial, then we naturally call the word null. Thus, there are only finitely many non-null words of infinite length.

6. **The main results of §1.** We have shown that in the subspace $\theta$ consisting of all pairs of the form $(x, 0)$ there is a sequence of subspaces $\Omega \subset \beta_1 \subset \beta_2 \subset \ldots \subset \beta_{s-1} \subset \theta$ such that the kernel $\Omega D$ and the domain of definition $\theta D$ of any monomial coincides with some subspace $\beta_i$ in the sequence above. This sequence is called the stabilized sequence.

The pair of subspaces $(\beta_{i-1}, \beta_i)$ is called an elementary interval.

It turns out that an elementary interval may have one of two forms. The interval $(\beta_{i-1}, \beta_i)$ is said to be of the first kind if there is a monomial $D$ such that $\beta_{i-1} = \theta b^{\#} D$, $\beta_i = \Omega a D$.

The set of monomials having the above property is denoted by $\Sigma_1$. There is a one-to-one correspondence between monomials $D \in \Sigma_1$ and elementary intervals of the first kind.

All the remaining intervals $(\beta_{i-1}, \beta_i)$ are said to be of the second kind. To each interval of the second kind we can associate uniquely an infinite word $\tilde{D}$, such that the interval has the form $(\Omega\tilde{D}, \theta\tilde{D})$. (The definition of $(\Omega\tilde{D}, \theta\tilde{D})$ for an infinite word is given at the end of the preceding subsection.)

We denote by $\Sigma_2$ the set of infinite words corresponding to elementary intervals of the second kind. $\Sigma_2$ is finite, and there is a one-to-one correspondence between the $\tilde{D} \in \Sigma_2$ and the elementary intervals of the second kind. We shall show later that the words $\tilde{D} \in \Sigma_2$ are periodic.

## §2. Relations in modules with a grading.

Since the situation considered in the preceding section carries over to graded spaces, the aim of this section is to give an account of some propositions which simplify our task in graded spaces.

1. Consider a decomposition of $\mathbb{C}^n$ into a direct sum $\mathbb{C}^n = \bigoplus_t C_t$ a so-called grading. Note that

[1] We recall that it follows from Lemma 1.5 that if $s(D) = s(D_1)$ and $(\Omega D, \theta D) = (\Omega D_1, \theta D_1)$ is a non-trivial interval, then $D = D_1$.

$$\mathbf{C}^n \oplus \mathbf{C}^n = \bigoplus_t (C_t \oplus C_t). \tag{2.1}$$

We say that a relation $A$, that is, a subspace of $\mathbf{C}^n \oplus \mathbf{C}^n$, is homogeneous if

$$A = \bigoplus A \cap (C_t \oplus C_t).$$

$A \cap (C_t \oplus C_t) = A_t$ is then a relation on $C_t$, and we say that $A$ is decomposed into the direct sum of the relations $A_t$: $A = \bigoplus A_t$. If we denote by $e_t$ the identity relation on $C_t$, that is, the set of all pairs of the form $(x_t, x_t) \in \mathbf{C}^n \oplus \mathbf{C}^n$, where $x_t \in C_t$, then we may write $A_t = e_t A e_t$ (more generally, for any relation $A$, $A \cap (C_s \oplus C_t) = e_t A e_s$).

The homogeneity of $A$ means that

$$A = \bigoplus e_t A e_t. \tag{2.2}$$

It follows easily from (2.2) that if $A$ is homogeneous, then $A^{\#}$ is also homogeneous. The relation $\Omega$ is homogeneous, and $\Omega_t = \Omega$. $\theta$ is also homogeneous, and $\theta = \bigoplus \theta_t$, where $\theta_t = C_t \oplus 0$.

PROPOSITION 2.1. *If $A$ and $B$ are two homogeneous relations, then $AB$ is also homogeneous, and*

$$AB = \bigoplus A_t B_t. \tag{2.3}$$

PROOF. Let $(x, z) \in AB$. There exists a $y$ such that $(x, y) \in B$, $(y, z) \in A$ Then $(x, y) = \Sigma (x_t, y_t)$, $(x_t, y_t) \in B_t$, $(y, z) = \Sigma (y_t, z_t)$, $(y_t, z_t) \in A_t$. Then $(x, z) = \Sigma (x_t, z_t)$, that is, $AB \subseteq \bigoplus (AB)_t$. Since always $\bigoplus (AB)_t \subseteq AB$, we have $AB = \Sigma (AB)_t$, that is, $AB$ is homogeneous. From the argument it is also clear that $(AB)_t = A_t B_t$.

LEMMA 2.1. *If the subspaces $C_t$ in the decomposition $\mathbf{C}^n = \bigoplus C_t$ are invariant under linear maps $a$ and $b$, then the relation $D = a^{k_1} (b^{\#})^{k_2} a^{k_3} \ldots$ is homogeneous with respect to the grading $\mathbf{C}^n \oplus \mathbf{C}^n = \bigoplus (C_t \oplus C_t)$, that is,*

$$D = \bigoplus D_t, \text{ where } D_t = e_t D e_t. \tag{2.4}$$

PROOF. $a$ and $b$ are homogeneous, consequently $b^{\#}$ is also homogeneous. This means, by Proposition 2.1, that the monomial $D$ is homogeneous.

THEOREM 2.1. *Under the hypotheses of the previous lemma all the subspaces $\beta_i$ in the stabilized sequence are homogeneous, that is, $\beta_i = \bigoplus (\beta_i)_t$.*

PROOF. $\Omega D$ and $\theta D$, as products of homogeneous relations, are homogeneous. Thus, the subspaces $\beta_i$ of the stabilized sequence are also homogeneous. We can construct from the monomials $D$ the monomials $D_t$ (2.4) on $C_t \oplus C_t$. It follows from 2.1 that

$$D_t = (a_t)^{k_1} (b_t^{\#})^{k_2} \ldots \tag{2.5}$$

Thus, if $\beta_i$ is the stabilized sequence for $a$ and $b$, then the $(\beta_i)_t$ form the stabilized sequence for $a_t$ and $b_t$, apart from the fact that some of the $(\beta_i)_t$ may coincide. We clearly have

$$\rho(\beta_{i-1}, \beta_i) = \sum_t \rho((\beta_{i-1})_t, (\beta_i)_t). \tag{2.6}$$

It is also clear that if $(\beta_{i-1}, \beta_i)$ is an interval of the first kind, generated by the monomial $D$ (that is, having the form $(\theta b^{\#}D, \Omega aD)$), and if the interval $((\beta_{i-1})_t, (\beta_i)_t)$ is non-trivial, then it is also of the first kind and is generated by the monomial $D_t$.

A module is said to be of the first kind if all its elementary intervals are of the first kind. Similarly, if they are all of the second kind, the module is said to be of the second kind.

We use the simple arguments above in showing that any module may be decomposed into a direct sum of modules of the first and second kinds.

2. The situation given at the beginning of this chapter was the following. We have a space $\mathbf{C}^n = \bigoplus C_s\ (s = 1, \ldots, k)$ and two maps $a$ and $b$. Further, $bC_s \subset C_s$, and there exists a permutation $\pi(s)$ of the indices $(1, \ldots, k)$ such that $aC_s \subseteqq C_{\pi(s)}$. Also $ab = ba = 0$, and $a$ and $b$ are nilpotent.

The aim of this subsection is to prove the following theorem.

THEOREM 2.2. *If $a$ and $b$ satisfy the conditions above, then the $\beta_i$ of the stabilized sequence are homogeneous.*

We precede the proof of this theorem by a few remarks of independent interest.

Suppose that we are given a decomposition $\mathbf{C}^n = \bigoplus_{s=1}^{k} C_s$ and that $A$ is a relation, that is, a subspace of $\mathbf{C}^n \oplus \mathbf{C}^n$.

If $\pi$ is a permutation of the indices 1, 2, ..., $k$, then $A$ is called $\pi$-homogeneous if

$$A = \bigoplus_{s=1}^{k} e_{\pi(s)} A e_s, \tag{2.7}$$

that is, if any pair $(x, y) \in A$ may be written uniquely in the form

$$(x, y) = \sum_{s=1}^{k} (x_s, y_{\pi(s)}),\quad x_i,\ y_i \in C_i. \tag{2.8}$$

From this point of view, the homogeneous relations introduced in the preceding subsection are 1-homogeneous (where 1 is the identity permutation).

PROPOSITION 2.2. *If $A$ is $\pi$-homogeneous, then $A^{\#}$ is $\pi^{-1}$-homogeneous.*

The proof is clear.

The proof of the following proposition is similar to that of the analogous lemma in 2.1.

PROPOSITION 2.2. *If $A_1$ is $\pi_1$-homogeneous and $A_2$ is $\pi_2$-homogeneous, then $A_1A_2$ is $\pi_1\pi_2$-homogeneous.*

PROPOSITION 2.4. *If $\varphi \subseteq \theta$ and $\varphi$ is $\pi$-homogeneous, then $\varphi$ is 1-homogeneous.*

PROOF. For every element of $\varphi$ has the form $(x, 0)$. The proof comes directly from (2.8).

THEOREM 2.3 *Each subspace $\beta_i$ in the stabilized sequence is 1-homogeneous (that is, homogeneous).*

PROOF. Each such subspace has the form $\Omega D$ or $\theta D$, where $D = a^{k_1}(b^{\#})^{k_2}a^{k_3} \dots D$ is $\pi_1$-homogeneous, with $\pi_1 = \pi^{k_1+k_3+k_5+\cdots}$, and $\pi$ is the characteristic of homogeneity of $a$. Thus, $\theta D$ is also $\pi$-homogeneous. As $\theta D \subset \theta$, hence $\theta D$ is 1-homogeneous, that is, simply homogeneous. A similar statement holds for $\Omega D$, and the theorem is proved.

## §3. Further results about relations.

1. We have already pointed out that four characteristic subspaces are associated with each relation $A$: the kernel $\Omega A$, the domain of definition $\theta A$, the indeterminacy $A\Omega$ and the domain of values $A\theta^{\#}$. The dimensions of these spaces have undoubted interest. We have already introduced a number $m(A) = \dim A$, the so-called dimension of the relation. We now introduce another important number $\rho = \dim \theta A - \dim \Omega A$, which we call the rank of the relation and denote by $\rho(A)$. It turns out that the dimensions of the subspaces $\Omega A$, $\theta A$, $A\Omega$, $A\theta^{\#}$ are connected with the numbers $m$ and $\rho$ in the following way:

$$\left.\begin{aligned} \dim \theta A + \dim A\Omega &= m, \\ \dim A\theta^{\#} + \dim \Omega A &= m, \\ \dim \theta A - \dim \Omega A &= \rho, \\ \dim A\theta^{\#} - \dim A\Omega &= \rho. \end{aligned}\right\} \tag{3.1}$$

Thus, the dimensions of the spaces $\Omega A$, $\theta A$, $A\theta^{\#}$, $A\Omega$ are completely determined if the dimension $m(A)$ and the rank $\rho(A)$ are given.

We call a relation $A$ regular if $\theta A \supset \theta A^{\#}$. In other words, the regular relations are those for which the image is contained in the domain of definition. If $A$ is regular, then by replacing $\mathbf{C}^n$ by $\theta A$ we may regard $A$ as a relation on $\theta A \oplus \theta A$, that is, we may restrict $A$ to the subspace.

A relation $A$ is called non-singular (an isomorphism) if its kernel and domain of indeterminacy are zero, that is,

$$\Omega A = \Omega \text{ and } A\Omega = \Omega. \tag{3.2}$$

Such a relation determines an isomorphic map of the domain of definition $\theta A$ onto the image $A\theta^{\#}$. If $A$ is non-singular, then it is clear from (3.1) that $A^{\#}$ is also non-singular. From (3.1) it is clear that $A$ is non-singular if and only if $m(A) = \rho(A)$.

2. Let $\varphi \subseteq \theta$ be a subspace, and $C$ a relation. We introduce the subspaces $\varphi C^{\#}$ and $\varphi C$, which are called, respectively, the image and the pre-image of $\varphi$ under $C$.

When $C$ is a linear operator, this definition coincides with the usual one. Now let $(\varphi_1, \varphi_2)$ be an interval, that is, $\varphi_1 \subset \varphi_2 \subseteqq \theta$. Then $(\varphi_1 C, \varphi_2 C)$ is also an interval.

Next we prove a lemma connecting these two intervals.

LEMMA 3.1. *Let $\gamma \subseteqq \theta$ be such that $\varphi_1 C \oplus \gamma \subseteqq \varphi_2 C$. Then there exists a non-singular relation (isomorphism) $C_1$, $C_1 \subseteqq C$, and a subspace $\gamma' \subset \theta$ such that*

$$\gamma' = \theta C_1^{\#}, \qquad \gamma = \theta C_1, \tag{3.3}$$

$$\varphi_1 \oplus \gamma' \subseteqq \varphi_2. \tag{3.4}$$

(It goes without saying that $\gamma$ and $\gamma'$ are connected by $\gamma = \gamma' C_1$, and that $C_1$ is an isomorphic map of $\gamma'$ onto $\gamma$, so that we may say that $C_1$ is chosen so that $\varphi_1 C \oplus \gamma' C_1 \subseteqq \varphi_2 C$ implies that $\varphi_1 \oplus \gamma' \subseteqq \varphi_2$.)

PROOF. Let $\{(e_i, 0)\}$ be a basis of $\gamma \subset \theta$. Since $\gamma \subset \varphi_2 C$, there exists for each pair $(e_i, 0)$ a vector $f_i$ such that $(f_i, 0) \in \varphi_2$, $(e_i, f_i) \in C$. Let $C_1$ denote the relation generated by the pairs $(e_i, f_i)$. It is clear that $C_1 \subseteqq C$.

We show that if $(e, 0) \in C_1$, that is $(e, 0) \in \Omega C_1$, then $e = 0$. If $(e, 0) \in C_1$, then by definition of $C_1$, $(e, 0) \in \gamma$. If $e \neq 0$, then $\gamma \cap \Omega C_1 \neq \Omega$. Hence $\gamma \cap \varphi C_1 \supset \gamma \cap \Omega C_1 \neq \Omega$. This contradicts the fact that, by hypothesis, $\gamma \cap \varphi C_1 = \Omega$. Thus, $e = 0$. This proves that the vectors $f_i$ are also linearly independent, as otherwise we could find a linear combination of the vectors $(e_i, f_i)$ of the form $(e, 0)$ with $e \neq 0$.

Thus, the relation $C_1$ is a non-singular linear map, given by the formula[1] $C_1 e_i = f_i$. If we denote by $\gamma'$ the subspace spanned by the $f_i$, then (3.3) is proved. Let us prove (3.4). Firstly by construction $(f_i, 0) \in \varphi_2$, and so $\gamma' = \theta C_1^{\#} \subseteqq \varphi_2$. We now show that $\gamma' \cap \varphi_1 = \Omega$. Let $(f, 0) \in \varphi_1$ and $(f, 0) \in \gamma'$. Then by definition of $C_1$ it follows from $(f, 0) \in \gamma'$ that there exists $e \neq 0$ such that $(e, f) \in C_1$. From $(e, f) \in C_1$ and $(f, 0) \in \varphi_1$ it follows that $(e, 0) \in \varphi_1 C_1$. But since $\varphi_1 C_1 \cap \gamma = \Omega$ by hypothesis, this gives $e = 0$. (3.4) is thus proved, and the proof of the lemma is complete.

PROPOSITION 3.1.

$$\rho(\varphi_1, \varphi_2) \geqslant \rho(\varphi_1 C, \varphi_2 C). \tag{3.5}$$

PROOF. In Lemma 3.1, choose $\gamma$ so that $\varphi_1 C \oplus \gamma = \varphi_2 C$. Then $\dim \gamma = \rho(\varphi_1 C, \varphi_2 C)$. Since also $\varphi_1 \oplus \gamma' \subseteqq \varphi_2$, and $\dim \gamma' = \dim \gamma$, we have $\rho(\varphi_1, \varphi_2) \geqslant \rho(\varphi_1 C, \varphi_2 C)$.

3. In the following sections we need a more precise form of Lemma 3.1 for an interval $(\varphi_1, \varphi_2)$ in $\theta$.

LEMMA 3.2. *Let $a$ and $b$ be maps satisfying the condition $ba = 0$, that is, $\theta a^{\#} \subseteqq \Omega b$. Consider the intervals $(\varphi_1 a, \varphi_2 a)$ and $(\varphi_1 b^{\#}, \varphi_2 b^{\#})$. Let $\gamma_1$ and $\gamma_2$ be any subspaces satisfying the conditions*

[1] We denote a relation and the linear map generated by it by one and the same symbol.

$$\varphi_1 a \oplus \gamma_1 = \varphi_2 a, \quad \varphi_1 b^{\#} \oplus \gamma_2 = \varphi_2 b^{\#}. \tag{3.6}$$

*Then there exist non-singular relations (isomorphisms)* $a_1 \subseteq a$, $b_1^{\#} \subseteq b^{\#}$ *and subspaces* $\gamma_1'$ *and* $\gamma_2'$, *such that*

$$\theta a_1 = \gamma_1, \quad \theta a_1^{\#} = \gamma_1', \quad \theta b_1^{\#} = \gamma_2, \quad \theta b_1 = \gamma_2'. \tag{3.7}$$

*In addition we have the relations*

$$\varphi_1 \oplus \gamma_1' \oplus \gamma_2' \subseteq \varphi_2. \tag{3.8}$$

*Also* $\gamma_1' a_1 = \gamma_1$, $\gamma_2' b_1^{\#} = \gamma_2$.

**PROOF.** As in Lemma 3.1, we construct non-singular relations (isomorphisms) $a_1$ and $b_1^{\#}$ such that $\theta a_1 = \gamma_1$, $\theta b_1^{\#} = \gamma_2$. Putting $\gamma_1' = \theta a_1^{\#}$, $\gamma_2' = \theta b_1$ (then $\gamma_1 = \gamma_1' a$ and $\gamma_2 = \gamma_2' b_1^{\#}$, as was proved in Lemma 3.1), we have $\varphi_1 \oplus \gamma_1' \subseteq \varphi_2$, $\varphi_1 \oplus \gamma_2' \subseteq \varphi_2$.

To show that (3.8) holds: $\varphi_1 \oplus \gamma_1' \oplus \gamma_2' \subseteq \varphi_2$, we use the following little "trick". We have

$$\gamma_1' b^{\#} = \gamma_1 a^{\#} b^{\#} = \gamma_1 (ba)^{\#} = \gamma_1 0^{\#} = \Omega.$$

Hence, $(\varphi_1 \oplus \gamma_1') b^{\#} = \varphi_1 b^{\#}$. This means that we may think of $\gamma_2'$ as constructed from the interval $((\varphi_1 \oplus \gamma_1') b^{\#}, \varphi_2 b^{\#})$, since if we replace $\varphi_1$ by $\varphi_1 \oplus \gamma_1$ in the second equation of (3.6), nothing is changed. But in this case $(\varphi_1 \oplus \gamma_1') \oplus \gamma_2' \subseteq \varphi_2$. The lemma is now proved.

**REMARK.** There exists a subspace $\bar{\gamma} \subset \Omega b$ such that

$$\varphi_1 \oplus \gamma_1' \oplus \gamma_2 \oplus \bar{\gamma} = \varphi_2. \tag{3.9}$$

For the proof we remark that we may replace the equation $(\varphi_1 \oplus \gamma_1') b^{\#} = \varphi_1 b^{\#}$ by $(\varphi_1 + \Omega b \cap \varphi_2) b^{\#} = \varphi_1 b^{\#}$. For we have $\gamma_1' \subset \Omega b$, that is, $\gamma_1' \subset \Omega b \cap \varphi_2$.

Let us now denote by $\bar{\gamma}$ a complement to $\gamma_1'$ in $\Omega b \cap \varphi_2$. We then have $\varphi_1 + \Omega b \cap \varphi_2 = \varphi_1 \oplus \gamma_1' \oplus \bar{\gamma}$. We show that $(\varphi_1 + \Omega b \cap \varphi_2) \oplus \gamma_2' = \varphi_2$. Assume the contrary: that $(\varphi_1 + \Omega b \cap \varphi_2) \oplus \gamma_2' \subset \varphi_2$. Choose $\bar{\gamma}_2 \supset \gamma_2'$ such that $(\varphi_1 + \Omega b \cap \varphi_2) \oplus \bar{\gamma}_2 = \varphi_2$. Since $\bar{\gamma}_2 \subset \varphi_2$, $\varphi' = \varphi_1 + \Omega b \cap \varphi_2 \supset \Omega b \cap \varphi_2$, the images of $\bar{\gamma}_2$ and $\varphi'$ do not intersect, that is, $\bar{\gamma}_2 b^{\#} \cap \varphi' b^{\#} = \Omega$. But $\dim \bar{\gamma}_2 b^{\#} = \dim \bar{\gamma}_2$ and $\dim \bar{\gamma}_2 b^{\#} \leqslant \rho(\varphi_1 b^{\#}, \varphi_2 b^{\#})$. Hence $\rho(\varphi_1, \varphi_2) \leqslant \rho(\varphi_1 b^{\#}, \varphi_2 b^{\#})$. But $\rho(\varphi_1 b^{\#}, \varphi_2 b^{\#}) = \dim \gamma_2 = \dim \gamma_2' < \dim \bar{\gamma}_2$. So we have obtained a contradiction.

**PROPOSITION 3.2.**

$$\rho(\varphi_1, \varphi_2) \geqslant \rho(\varphi_1 a, \varphi_2 a) + \rho(\varphi_1 b^{\#}, \varphi_2 b^{\#}). \tag{3.10}$$

**PROOF.** $\dim \gamma_1' = \dim \gamma_1 = \rho(\varphi_1 a, \varphi_2 a)$, $\dim \gamma_2' = \dim \gamma_2 = \rho(\varphi_1 b^{\#}, \varphi_2 b^{\#})$. Our equation then follows directly from (3.8).

4. We now show that our basic Lemma 3.1 is true also when $C^n$ has a grading $C^n = \bigoplus_s C_s$.

LEMMA 3.3. *Let $(\varphi_1, \varphi_2)$ be a homogeneous interval in $\theta$ and $C$ be a homogeneous relation. Consider the interval $(\varphi_1 C, \varphi_2 C)$ and a homogeneous subspace $\gamma$ that satisfies the relation $\varphi_1 C \oplus \gamma \subseteqq \varphi_2 C$.*

*Then there is a non-singular $\pi$-homogeneous relation $C_1$ with the following properties:*

$$\left.\begin{array}{l} \text{a) } \theta C_1 = \gamma, \\ \text{b) } \varphi_1 \oplus \gamma_1' \subseteqq \varphi_2, \text{ where } \gamma_1' = \theta C_1^{\#}. \end{array}\right\} \qquad (3.11)$$

PROOF. It follows from Propositions 2.3 and 2.4 that the subspaces $\varphi_1 C$ and $\varphi_2 C$ are homogeneous. The subspaces $\varphi_1 C$ and $\varphi_2 C$ may be represented as sums of homogeneous components:[1]

$$\varphi_1 C = \sum_s (\varphi_1)_{s'} C_{s', s} \quad (\text{where } s' = \pi s);$$

similarly $\varphi_2 C = \sum_s (\varphi_2)_{s'} C_{s', s}$.

Therefore we may consider the intervals $(\varphi_1, \varphi_2)_{s'}$ and $((\varphi_1)_{s'} C_{s', s}, (\varphi_2)_{s'} C_{s', s})$ and a component $\gamma_s$, such that $(\varphi_1)_{s'} C_{s's} \oplus \gamma_s \subseteqq (\varphi_2)_{s'} C_{s', s}$.

Applying Lemma 3.1 to these intervals and the subspace $\gamma_s$, we find that there exists a non-singular relation $(C_1)_{s', s}$ with the following properties:

a) $\theta_{s'} (C_1)_{s', s} = \gamma_s$.

b) $(\varphi_1)_{s'} \oplus \theta_s (C_1)^{\#}_{s, s'} \subseteqq (\varphi_2)_{s'}$.

It is clear that the subspace $C_1$ required is $\sum_s (C_1)_{s', s}$.

5. A theorem on the decomposition of an arbitrary relation. In this subsection we state without proof a theorem we shall use in §5. However, first we mention some simple facts about relations.

Let $A$ be an arbitrary relation. For $\Omega A^i$ and $\Omega A^{i+1}$, the kernels of the powers of this relation, it is clear that $\Omega A^i \subseteqq \Omega A^{i+1}$. Furthermore, it follows from Proposition 3.1 that $\rho(\Omega A^i, \Omega A^{i+1}) \geqslant \rho(\Omega A^{i+1}, \Omega A^{i+2})$. Therefore the kernels of the powers of a relation form a sequence of subspaces of the following form:

$$\Omega \subset \Omega A \subset \Omega A^2 \subset \ldots \subset \Omega A^{k-1} \subset \Omega A^k = \Omega A^{k+1} = \ldots = \Omega A^{k+j} = \ldots$$

The kernel $\Omega A^k$ at which the kernels of the powers of the relation stop increasing, is called the stable kernel of $A$ and is denoted by $\Omega A^\infty$.

Similarly, the domains of definition of the powers of a relation form a sequence of subspaces of the following form:

$$\ldots = \theta A^{l+1} = \theta A^l \subset \theta A^{l-1} \subset \ldots \subset \theta A^2 \subset \theta A \subset \theta.$$

We call $\theta A^l$ the stable domain of definition of $A$ and denote it by $\theta A^\infty$. It is easy to see that $\Omega A^\infty \subseteqq \theta A^\infty$.

[1] In the formulae that are used here, the subspaces $\varphi_i$ and $\gamma$ have only one index, $s$ or $s'$. This is because each pair that belongs to these subspaces has the form $(x, 0) = \sum_s (x_s, 0)$. Therefore we need not write the second index.

It is known that any linear operator in a finite-dimensional space can be written as a direct sum of an invertible and a nilpotent operator. For relations the decomposition is, in a certain sense, similar. In fact, any relation can be written as the direct sum of a regular invertible relation and a certain non-regular part.

We say that the relation $A$ is reducible into a pair of subspaces $C_1$ and $C_2$, $C_1 \oplus C_2 = \mathbf{C}^n$, if the relation is homogeneous with respect to the decomposition $\mathbf{C}^n = C_1 \oplus C_2$. By (2.2), in this case $A = e_1 A e_1 \oplus e_2 A e_2$, where $e_i$ are the diagonals in the pair $(C_i \oplus C_i)$.

THEOREM 3.1. *Let $A$ be any relation on $\mathbf{C}^n \oplus \mathbf{C}^n$. Then there exist two subspaces $C_1$ and $C_2$ with $C_1 \oplus C_2 = \mathbf{C}^n$, with respect to which the relation $A$ is reducible. These subspaces $C_1$ and $C_2$ may be chosen so that the relation $e_1 A e_1$ is regular and invertible. Then the subspace $\overline{C}_1 = (C_1 \oplus 0)$ complements the stable kernel $\Omega A^\infty$ in the stable domain of definition $\Theta A^\infty$, that is,*

$$\Omega A^\infty \oplus \overline{C}_1 = \Theta A^\infty. \tag{3.12}$$

REMARK. The subspaces $C_1$ and $C_2$ are not necessarily uniquely determined. But if we choose two other subspaces $C_1'$ and $C_2'$ satisfying the requirements of Theorem 3.1, then the relations $e_1 A e_1$ and $e_1' A e_1$ are similar,[1] and also the relations $e_2 A e_2$ and $e_2' A e_2$ are similar.

## §4 Separation and examination of submodules of the first kind

In this section we consider the set of intervals of the first kind in the stabilized sequence of subspaces. We recall that to each such interval $(\beta_{i-1}, \beta_i)$, there corresponds uniquely a certain monomial $D \in \Sigma_1$ such that $\beta_{i-1} = \Theta b^\# D$, $\beta_i = \Omega a D$.

Let $(\beta_{i-1}, \beta_i)$ be an elementary interval. Let $\gamma_i$ be a subspace[2] complementing $\beta_{i-1}$ in $\beta_i$ If this interval is of the first kind and $D$ is the corresponding monomial in $\Sigma_1$, then we also denote $\gamma_i$ by $\gamma(D)$. In other words, for $D \in \Sigma_1$, $\gamma(D)$ is a subspace satisfying the conditions

$$\gamma(D) \cap \Theta b^\# D = \Omega, \qquad \gamma(D) \oplus \Theta b^\# D = \Omega a D. \tag{4.1}$$

It is easy to see that in any sum $\sum_D \gamma(D)$, where $D$ ranges over a subset of $\Sigma_1$, $\gamma(D)$ occurs as a direct summand.

In this section we begin by describing a certain canonical form for modules of the first kind. We give this canonical form by bases of "open sequence" type. We then show that the subspaces $\gamma(D)$, where $D$ ranges

---

1 Two relations $A$ and $A'$ on the spaces $\mathbf{C}^n \oplus \mathbf{C}^n$ and $(\mathbf{C}^n \oplus \mathbf{C}^n)'$ are similar if there is an isomorphism $\mathbf{C}^n \to \mathbf{C}^n$, that induces an isomorphism of $A$ onto $A'$.

2 We introduce such $\gamma_i$ for intervals of the first and second kinds.

over $\Sigma_1$, may be chosen so that the subspace $\Sigma\gamma(D)$ is invariant under $a$ and $b$ and decomposes into a direct sum of submodules defined by open sequences. Further, in this and the following sections we construct a decomposition of $\theta$ into the direct sum of canonical submodules of the first and second kinds. [1]

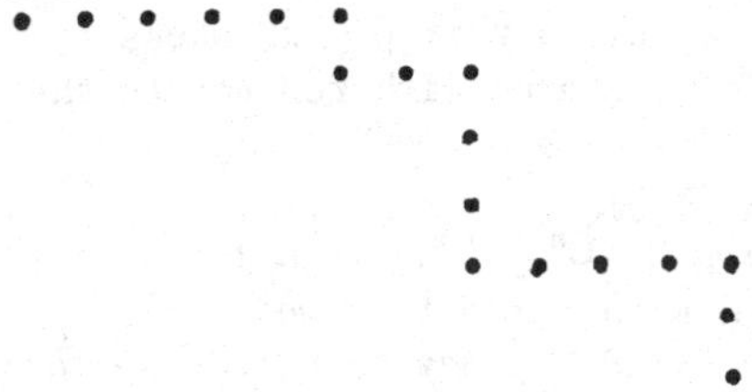

Fig. 1

**1. Definition of "open sequences."** Consider a set $M$ consisting of $n$ integral points in the plane, arranged as[2] "stairs" (Fig. 1). We associate with each point $(i, j) \in M$ a basis vector $e(i, j) \in \mathbf{C}^n$ and define operators $a$ and $b$ in $\mathbf{C}^n$ by putting

$$\left.\begin{aligned} ae(i, j) &= \begin{cases} e(i+1, j) & \text{if} \quad (i+1, j) \in M, \\ 0 & \text{if} \quad (i+1, j) \notin M, \end{cases} \\ be(i, j) &= \begin{cases} e(i, j+1) & \text{if} \quad (i, j+1) \in M, \\ 0 & \text{if} \quad (i, j+1) \notin M. \end{cases} \end{aligned}\right\} \tag{4.2}$$

DEFINITION 4.1. The space $\mathbf{C}^n$ with the ring of operators generated by $a$ and $b$ is called a *canonical module of the first kind,* defined (with the help of (4.2)) by the "open sequence" $M$. We denote it by $K(M)$.

It is easy to see that the operators $a$ and $b$ on $K(M)$ are nilpotent and that $ab = ba = 0$.

Presently we describe, without proof, the stabilized sequence and the set $\Sigma_1$ that correspond to an "open chain".

Any elementary interval $(\beta_{i-1}, \beta_i)$ in the stabilized sequence of a canonical module of the first kind is of the first kind, and its length $\rho(\beta_{i-1}, \beta_i)$ is 1. The stabilized sequence has the form $\Omega \subset \beta_1 \subset \beta_2 \subset \ldots \subset \beta_n = \theta$, where the dimension of $\beta_i$ is $i$, and $n$ is the dimension of the module.

In this case the set $\Sigma_1$ is linearly ordered by the degrees of the monomials. Let $D_0$ be the monomial of maximal degree $m = n - 1$ in $\Sigma_1$. Then any monomial $D \in \Sigma_1$ is a left divisor of $D_0$, that is, $D_0 = D \cdot D'$. We say that $D_0$ generates the whole open sequence, and the module $\theta$ of the first kind which it defines.

---

1 We recall that a module of the first (second) kind is a module whose stabilized sequence contains only elementary intervals of the first (second) kind.

2 The statement that the set $M$ of integral points in the plane is arranged as "stairs" means that for $(i, j) \in M$ there are three mutually exclusive possibilities:

$$1)\ (i+1, j) \in M, \qquad 2)\ (i, j-1) \in M, \qquad 3)\ (i+1, j) \notin M \text{ and } (i, j-1) \notin M,$$

where the third possibility only holds for one point (the lower end of the "stairs".)

EXAMPLE 4.1. The open sequence shown in Fig. 1 is generated by the monomial $D_0 = (b^{\#})^2 a^4 (b^{\#})^3 a^2 b^{\#} a^5$ of degree 17. $\Sigma_1$ consists of 18 monomials, $1, b^{\#}, (b^{\#})^2, (b^{\#})^2 a, \ldots$ the left divisors of $D_0$. We shall not write out in full the stabilized sequence $\{\beta_i\}$ $(i = 0, 1, \ldots, 18)$ and the correspondence between elementary intervals $(\beta_{i-1}, \beta_i)$ and monomials in $\Sigma_1$; we only remark that this correspondence is not simple. Thus, the monomial 1 corresponds to the elementary interval $(\beta_6, \beta_7)$ and $b^{\#}$ corresponds to $(\beta_3, \beta_4)$. Here $\beta_6 = \theta b^{\#}$ is the 6-dimensional space spanned by the vectors $(e_6, e_9, e_{10}, e_{11}, e_{16}, e_{17})$, $\beta_7 = \Omega a$ is spanned by $(e_6, e_9, e_{10}, e_{11}, e_{16}, e_{17}, e_{18})$, $\beta_2 = \theta (b^{\#})^2$ is spanned by $(e_9, e_{10}, e_{16})$ and $\beta_3 = \Omega b^{\#} a$ by $(e_q, e_{10}, e_{16}, e_{17})$ (the basis vectors are numbered by going along the stairs in Fig. 1 from top to bottom).

EXERCISE. Find the elementary intervals corresponding to the following monomials in $\Sigma_1$: $(b^{\#})^2, (b^{\#})^2 a, \ldots$

We now describe the form of the open sequence when the space $\mathbf{C}^n$ is given a grading $\mathbf{C}^n = \sum_{s=1}^{r} C_s$, where each component $C_s = \mathbf{C}^{n_s}$ occurs in the sum as a direct summand. Here $bC_s \subset C_s$ and $aC_s \subseteqq C_{s+1}$, that is, the map $a$ corresponds to a cyclic permutation $\pi$, $\pi(s) = s + 1 \pmod r$. Exactly as in the case without grading, an open sequence is given by a set $M$, consisting of $n$ integer points in the plane, arranged as "stairs"; first of all, with each point $(i, j) \in M$ we associate a basis vector $e(i, j) \in \mathbf{C}^n$, and we define operators $a$ and $b$ by the formulae (4.2). But now, so that the decomposition of $\mathbf{C}^n$ into the $C_s$ agrees with the action of the operators $a$ and $b$, we postulate the following conditions: a) if $e(i, j) \in C_s$ and $(i + 1, j) \in M$, then $e(i + 1, j) \in C_{s+1}$; b) if $e(i, j) \in C_s$ and $(i, j - 1) \in M$, then also $e(i, j - 1) \in C_s$.

2. **A partial ordering in $\Sigma_1$.** Consider the set $\Sigma$ of all monomials. Let us define on $\Sigma$ a relation of order according to which a monomial $D_1$ precedes a monomial $D_2$ if $D_1$ is a left divisor[1] of $D_2$ : $D_2 = D_1 \cdot D'$. We call $D_1$ an immediate predecessor of $D_2$ if $D_2 = D_1 \cdot d$ and $s(d) = 1$ (that is, $d$ is either $a$ or $b^{\#}$). It is clear that for each monomial $D \neq 1$ there is only one monomial that immediately precedes it.

Let us now construct the directed graph of the partially ordered set $\Sigma_1$, taking each monomial $D \in \Sigma_1$ as a vertex of the graph, and connecting two monomials $D_1$, $D_2$ by an interval (an edge of the graph) if $D_1$ immediately precedes $D_2$.

THEOREM 4.1. *Suppose that $\Sigma_1$ is not empty. Then the graph of the partially ordered set $\Sigma_1$ is a tree (that is, a connected graph containing no cycles) in which not more than two edges leave from any vertex. The lowest vertex of the tree is the monomial 1 (the identity relation).*

---

[1] In contrast to §1, where *right divisors* were considered, the partial ordering introduced in this section is defined by *left divisors*.

PROOF. Let $D \in \Sigma_1$, that is, $\theta b^{\#}D \subset \Omega aD$. We show then that any monomial $D_1$ that is a left divisor of $D$ ($D = D_1 \cdot D'$) also belongs to $\Sigma_1$. For otherwise we would have $\theta b^{\#}D_1 = \Omega aD_1$. But then, multiplying both sides by $D'$, we obtain $\theta b^{\#}D = \Omega aD$, which contradicts the hypothesis $D \in \Sigma_1$.

We have shown that, together with each $D \in \Sigma_1$, the graph $\Sigma_1$ contains all its predecessors (in $\Sigma$!), and among them 1. This shows that the graph is connected. The absence of cycles follows from the uniqueness of the monomial that immediately precedes a given monomial $D \neq 1$. Since not more than two monomials follow each $D \in \Sigma_1$, not more than two edges leave from each vertex.[1]

EXAMPLE 4.2. Suppose that the set $\Sigma_1$ consists of the monomials $1,\ a,\ b^{\#},\ (b^{\#})^2,\ b^{\#}a,\ b^{\#}a^2,\ b^{\#}ab^{\#}$. Then the tree $\Sigma_1$ has the form shown in Fig. 2.

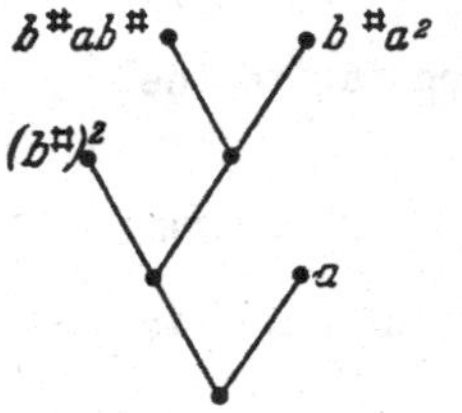

Fig. 2.

A monomial $D \in \Sigma_1$ is called final if both $Da \notin \Sigma_1$ and $Db^{\#} \notin \Sigma_1$. It is clear that $\Sigma_1$ is uniquely determined if all its final monomials are given. Thus, in example 4.2 (Fig. 2), $\Sigma_1$ is determined by its final monomials $a,\ (b^{\#})^2,\ b^{\#}a^2,\ b^{\#}ab^{\#}$.

We denote the length of the interval $(\theta b^{\#}D, \Omega aD)$ by $\rho_1(D)$

$$\rho_1(D) = \dim \Omega aD - \dim \theta b^{\#}D. \tag{4.3}$$

If $D \in \Sigma_1$, then $\rho_1(D)$ is the length of the corresponding elementary interval of the first kind. For $D \notin \Sigma_1$ we have $\rho_1(D) = 0$. Later we show that the values of the function $\rho_1(D)$ for all monomials $D$ form a complete set of invariants of the maximal submodule of the first kind contained in $\theta$.

**3. A compatible choice of the subspaces $\gamma(D)$.** The separation of a submodule of the first kind is based on the following theorem.

THEOREM 4.2. *The set of subspaces $\{\gamma(D);\ D \in \Sigma_1\}$ can be chosen compatibly, that is, so that for each monomial $D$ the subspaces $\gamma(D),\ \gamma(Da),\ \gamma(Db^{\#})$ satisfy the following conditions:*

$$\gamma(Da)\,a^{\#} \subseteqq \gamma(D), \tag{4.4}$$

$$\gamma(D)\,b^{\#} = \gamma(Db^{\#}). \tag{4.5}$$

The proof of the theorem consists in constructing the set $\{\gamma(D)\}$, beginning with the final vertices of the tree $\Sigma_1$. The descent from the final vertices to the remaining ones can be made (on the basis of Lemma 3.2), by making $\gamma(D)$ agree with $\gamma(Da)$ and $\gamma(Db^{\#})$ at each vertex $D$.

Let us make some preliminary remarks.

1) We note that the compatibility conditions (4.4), (4.5) mean that for three neighbouring vertices of the tree $\Sigma$

[1] For $D \in \Sigma_1$, $Da$ and $Db^{\#}$ need not belong to $\Sigma_1$.

$$\begin{array}{ccc} Db^{\#} & & Da \\ \cdot & & \cdot \\ & \diagdown\ \diagup & \\ & \dot{D} & \end{array} \tag{4.6}$$

the corresponding subspaces $\gamma$ are related by the diagram

$$\begin{array}{ccc} \gamma(Db^{\#}) & & \gamma(Da) \\ & \nwarrow^{b} \quad {}^{a}\swarrow & \\ & \gamma(D) & \end{array} \tag{4.7}$$

2) Although we only need to construct $\gamma(D)$ for $D \in \Sigma_1$, it will be more convenient for the exposition if we do not exclude monomials (or vertices of the tree $\Sigma$) not belonging to $\Sigma_1$, since for $D \in \Sigma_1$ it may happen that $Da \notin \Sigma_1$ or $Db^{\#} \notin \Sigma_1$. It is useful to remember here that $\gamma(D) = \Omega$ for $D \notin \Sigma_1$.

3) For the subspaces forming the elementary intervals of the first kind (or null intervals) corresponding to the monomials $Db^{\#}$, $D$, $Da$, we introduce the notation:

$$\left.\begin{array}{lll} (\text{for } Db^{\#}) & \beta_{i-1} = \theta b^{\#} Db^{\#}, & \beta_i = \Omega a Db^{\#}, \\ (\text{for } D) & \beta_{j-1} = \theta b^{\#} D, & \beta_j = \Omega a D, \\ (\text{for } Da) & \beta_{k-1} = \theta b^{\#} Da, & \beta_k = \Omega a Da. \end{array}\right\} \tag{4.8}$$

In this notation

$$\left.\begin{array}{l} (\beta_{i-1}, \beta_i) = (\beta_{j-1} b^{\#}, \beta_j b^{\#}), \\ (\beta_{k-1}, \beta_k) = (\beta_{j-1} a, \beta_j a), \end{array}\right\} \tag{4.9}$$

that is, the interval $i$ is the image of $j$ under $b$, and the interval $k$ is the pre-image of $j$ under $a$.

The following lemma demonstrates the possibility of making $\gamma(D)$ agree with a previous choice of the subspaces $\gamma(Da)$ and $\gamma(Db^{\#})$ (the descent "along" the arrow $a$ and "against" the arrow $b$ in diagram (4.7)) and makes the properties of the maps in (4.7) more precise.

LEMMA 4.1. *Let $\gamma(Db^{\#})$ and $\gamma(Da)$ be subspaces corresponding to the intervals $(\beta_{i-1}, \beta_i)$, $(\beta_{k-1}, \beta_k)$ (see (4.8)), that is, such that*

$$\beta_{i-1} \oplus \gamma(Db^{\#}) = \beta_i, \quad \beta_{k-1} \oplus \gamma(Da) = \beta_k.$$

*Then the subspace $\gamma(D)$ corresponding to $(\beta_{j-1}, \beta_j)$ can be chosen compatibly with $\gamma(Db^{\#})$ and $\gamma(Da)$, that is, satisfying the conditions (4.4), (4.5), and so that in addition the map $\gamma(D) \overset{a}{\longleftarrow} \gamma(Da)$ is a monomorphism and the map $\gamma(Db^{\#}) \overset{b}{\longleftarrow} \gamma(D)$ is an epimorphism.*

PROOF OF THE LEMMA. The equations (4.9) allow us to apply Lemma 3.2 to the intervals $(\beta_{i-1}, \beta_i)$, $(\beta_{j-1}, \beta_j)$, $(\beta_{k-1}, \beta_k)$ By this lemma, there exist non-singular relations $a_1 \subset a$ and $b_1^{\#} \subset b^{\#}$ with the properties

$$\left.\begin{aligned} &\theta a_1 = \gamma(Da), \quad \theta b_1^{\#} = \gamma(Db^{\#}), \\ &\beta_{j-1} \oplus \theta a_1^{\#} \oplus \theta b_1 \subseteqq \beta_j. \end{aligned}\right\} \tag{4.10}$$

Now to extend the left-hand side of the last inequality to $\beta_j$ it is convenient to use the subspace $\varphi'$ introduced in the proof of Lemma 3.2, which in our case has the form $\varphi' = \beta_{j-1} + (\beta_j \cap \Omega b)$. It was shown that $\varphi' \oplus \theta b_1 = \beta_j$, and also that $\beta_{j-1} \oplus \theta a_1^{\#} \subseteqq \varphi'$. Let us now choose a subspace $\bar{\gamma} \subset \beta_j \cap \Omega b$ so that $\beta_{j-1} \oplus \theta a_1^{\#} \oplus \bar{\gamma} = \varphi'$. It is clear that such a choice is possible, from the definition of $\varphi'$ and the inequality $\theta a_1^{\#} \subset \theta a^{\#} \subset \Omega b$. Then from the relations just written down we obtain $\beta_{j-1} \oplus \theta a_1^{\#} \oplus \theta b_1 \oplus \gamma = \beta_j$.

Let us now show that the subspace so constructed

$$\gamma(D) = \theta a_1^{\#} \oplus \theta b_1 \oplus \bar{\gamma}$$

satisfies (4.4) and (4.5); this will prove the lemma.

Since $\gamma(Da) = \theta a_1$ and $a_1 \subset a$, the image of $\gamma(Da)$ under $a$ coincides with the image of $\gamma(Da)$ under $a_1$: $\gamma(Da)\,a^{\#} = \gamma(Da)\,a_1^{\#} = \theta a_1 a_1^{\#}$. But $a_1$ is a non-singular linear operator, and so this subspace coincides with the domain of definition of $a_1^{\#}$, that is,

$$\gamma(Da)\,a^{\#} = \theta a_1^{\#} \subseteqq \gamma(D).$$

Thus, condition (4.4) for $\gamma(D)$, $\gamma(Da)$ is fulfilled. Furthermore, since $a_1$ is a non-singular operator, the map $a_1\colon \gamma(Da) \longrightarrow \theta a_1^{\#}$ is an isomorphism, that is, the map $a\colon \gamma(Da) \to \gamma(D)$ in diagram (4.7) is a monomorphism.

It is also easy to verify that the second compatibility condition (4.5) is satisfied. From the non-singularity of $b_1 \subset b$ it follows that $\theta b_1 b^{\#} = \theta b_1 b_1^{\#} = \theta b_1^{\#}$. But this subspace is $\gamma(Db^{\#})$ (4.10). We recall that $\theta a_1^{\#} \subset \Omega b$ and $\bar{\gamma} \subset \Omega b$, whence $(\theta a_1^{\#} \oplus \bar{\gamma})\,b^{\#} = \Omega$. Consequently, $\gamma(D)\,b^{\#} = \gamma(Db^{\#})$. The non-singularity of the operator $b_1$ shows that the map $b_1\colon \theta b_1 \longrightarrow \gamma(Db^{\#}) = \theta b_1^{\#}$ is an isomorphism, and then the map $b\colon \gamma(D) \longrightarrow \gamma(Db^{\#})$ in diagram (4.7) is an epimorphism. The lemma is now proved.

PROOF OF THEOREM 4.2. We begin the construction of the subspaces $\gamma(D)$ with the final monomials $D \in \Sigma_1$, that is, those for which $Da \notin \Sigma_1$, $Db^{\#} \notin \Sigma_1$. In this case the subspaces $\gamma(Da) = \gamma(Db^{\#}) = \Omega$ are already known. By Lemma 4.1, $\gamma(D)$ can then be chosen satisfying the compatibility conditions.

We recall that the monomials in $\Sigma_1$ have degrees not exceeding a certain number $m$. All monomials $D \in \Sigma_1$ of degree $m$ are final, and for these $\gamma(D)$ is already chosen. Suppose that $\gamma(D)$ has been constructed for all $D$ with $s(D) = k$. Let us now go over to monomials of degree $k - 1$. Each of these precedes two monomials of degree $k$, and again by Lemma 4.1, $\gamma(D)$ can be chosen compatibly with $\gamma(Da)$, $\gamma(Db^{\#})$. In this way we construct the whole set $\{\gamma(D);\ D \in \Sigma_1\}$ compatibly. This completes the proof of the theorem.

Using a diagram of the form (4.7) for each vertex $D$ of the tree $\Sigma_1$, we may put the assertions of Theorem 4.2 and Lemma 4.1 into the following form.

COROLLARY 4.1. *The subspaces $\gamma(D)$, $D \in \Sigma_1$ can be chosen so that they are connected by maps $a$ and $b$ in a diagram $S$, obtained from the tree $\Sigma_1$ by replacing each vertex $D$ by the subspace $\gamma(D)$, each edge going from $D$ to $Da$ by an arrow $a$, and each edge going from $D$ to $Db^{\#}$, by an arrow $b$. Here each map $D \overset{a}{\longleftarrow} Da$ is a monomorphism, and $Db^{\#} \overset{b}{\longleftarrow} D$ is an epimorphism.*

As we did not wish to complicate the proof, we have formulated Theorem 4.2 about a compatible choice of the $\gamma(D)$ without relating this choice to the grading in the space $\Theta$. However, the generalization to the case when a grading $\Theta = \bigoplus \Theta_s$, is given, presents no difficulty. In fact, Lemma 3.2, on which the proof of the theorem was based, is true in a stronger form when carried over to the graded case, with homogeneous operators $a$, $b$ and homogeneous subspaces $\varphi_1$, $\varphi_2$. In addition, we know (§2) that the subspaces in the stabilized sequence are homogeneous. Therefore, the following stronger form of Theorem 4.2 is true.

THEOREM 4.3. *Suppose that a grading $\Theta = \bigoplus \Theta_s$ is given with respect to which the operator $b$ is homogeneous $(\Theta_s b^{\#} \subseteqq \Theta_s)$ and that $a$ is a weakly homogeneous operator $(\Theta_s a^{\#} \subseteqq \Theta_{s+1})$. Then the subspaces $\gamma D$, $D \in \Sigma_1$, can be chosen to be homogeneous and compatible, that is, so that the conditions*

$$\gamma_s(Da)\, a^{\#} \subseteq \gamma_{s+1}(D), \tag{4.11}$$

$$\gamma_s(D)\, b^{\#} = \gamma_s(Db^{\#}), \tag{4.12}$$

*are satisfied, where $\gamma_s(D) = \gamma(D) \cap \Theta_s$.*

Let us consider the subspaces $\Theta_1 = \bigoplus \gamma(D)$. The following theorem shows the significance of a compatible choice of the $\gamma(D)$ for this subspace.

THEOREM 4.4. *If the subspaces $\gamma(D)$, $D \in \Sigma_1$ are homogeneous and compatible, then the subspace $\Theta_1 = \bigoplus_{D \in \Sigma_1} \gamma(D)$ is homogeneous and invariant under the operators $a$ and $b$.*

PROOF. Let us show that for each $D \in \Sigma_1$ the image of $\gamma(D)$ under $a$ or $b$ is contained in one of the subspaces of the set $\{\gamma(D), D \in \Sigma_1\}$. From this, taking into account the fact that $\gamma(D)$ is homogeneous, it follows immediately that $\Theta_1$ is homogeneous and invariant.

For the operator $b$ the above assertion is already contained in the compatibility condition (4.5) $\gamma(D)\, b^{\#} = \gamma(Db^{\#})$ (or (4.12) for the graded case). For the operator $a$, if also $D = D'a$, the assertion is contained in condition (4.4) (or (4.11)), applied to $D'$: $\gamma(D)\, a^{\#} = \gamma(D'a)\, a^{\#} \subseteqq \gamma(D')$. It remains to consider the image $\gamma(D)\, a^{\#}$, of $\gamma(D)$ under $a$, when $D = 1$ or $D = D'b^{\#}$. But in these two cases, $\gamma(D) \subset \Omega a$, and since $\Omega a$ is the kernel of $a$, $\gamma(D)a^{\#}=\Omega$. In fact, for $D=1$ $\gamma(D)\subseteqq\beta_j=\Omega a$ and for $D=D'b^{\#}$ $\gamma(D)\subseteqq\beta_j=\Omega aD'b^{\#}\subseteqq\theta b^{\#}\subset\Omega a$.

The theorem is now proved.

It will be shown later that the invariant subspace $\theta_1\subset\theta$ also has an invariant direct complement $\theta_2$. From this it follows that $\theta_1$ is a submodule of $\theta$ (§2). From the results of §2 it also follows that $\theta_1$ is a submodule of the first kind.

**4. Decomposition of the submodule $\theta_1$ into the direct sum of canonical submodules of the first kind.** In this subsection we show that the submodule $\theta_1$ can be decomposed into a direct sum of indecomposable submodules each of which is a canonical module of the first kind, defined by an open sequence; we also give the construction of these open sequences in terms of the subspaces $\gamma(D)$ already known.

Fig. 3

Let us first isolate a particular case. Assume that the set $\Sigma_1$ is linearly ordered and that the dimension of each subspace $\gamma(D)$, $D\in\Sigma_1$ is 1.

Since $\Sigma_1$ is linearly ordered, it only contains one final monomial, and any other monomial is a left divisor of it. Thus, if the degree of the final monomial is $m$, then $\Sigma_1$ consists of $m+1$ monomials $D(j)$ $(j=0,1,2,\ldots,m)$, $s(D_j)=j$; here $D(0)=1$, $D(j)=D(j-1)d_j$, where $d_j$ is either $a$ or $b^{\#}$, and the final monomial $D_m=d_1\cdot d_2\ldots d_m$.

Let us construct the tree $\Sigma_1$ and, as described in Corollary (4.1), convert it into a diagram $S$, assuming that the $\gamma(D)$ have been chosen compatibly.

EXAMPLE 4.3. $\Sigma_1$ is generated by the single final monomial $D(7)=b^{\#}a^4(b^{\#})^2$. The diagram $S$ is shown in Fig. 3.

If the diagram $S$ is suitably arranged on the integral lattice in the plane,[1] then it can be thought of as defining a certain open sequence.

The submodule $\theta_1$, which is the direct sum of the one-dimensional subspaces $\gamma(D)$, is isomorphic to the canonical module of the first kind defined by this open sequence.

EXERCISE. Prove this assertion.

We now treat the general case and sketch a method by which in the module $\theta_1=\Sigma\gamma(D)$ a basis can be constructed that breaks up into bases of open sequence type, in such a way that the module is the direct sum of the corresponding canonical modules of the first kind. If a basis $E(D)=\{\xi_1(D),\xi_2(D),\ldots,\xi_{\rho(D)}(D)\}$ is given in each subspace $\gamma(D)$, then

$$E_1=\bigcup_{D\in\Sigma_1}E(D)$$

is a basis of $\theta_1$. In order that the basis $E_1$ should break up into open sequences, the bases $E(D)$ must agree in a specified way with the action of the operators $a$ and $b$ on $\gamma(D)$.

[1] The $i$-axis must be directed along the arrow $a$, and the $j$-axis along the arrow $b$.

For the construction of the basis $E(D)$, $D \in \Sigma_1$, we use again the method of descent from the final vertices of the tree $\Sigma_1$ (or of the diagram $S$) that was used in constructing the subspaces $\gamma(D)$. Here the properties of the mappings in the diagram $S$ are used in an essential way.

For all the final monomials $D \in \Sigma_1$ we choose the bases $E(D)$ arbitrarily. Let $m = \sup_{D \in \Sigma_1} s(D)$. All the monomials of degree $m$ are final, and for these the bases have been chosen. Let us now go over to monomials $D$ of degree $m-1$. If $Da \notin \Sigma_1$, $Db^{\#} \notin \Sigma_1$, then $D$ is a final monomial and we already have $E(D)$. Consider the case when $D$ is not a final monomial. In the diagram $Db^{\#} \xleftarrow{b} D \xleftarrow{a} Da$, $a$ is a monomorphism and $b$ is an epimorphism. Furthermore, by Lemma 4.3, $\gamma(D) = \theta a_1^{\#} \oplus \theta b_1 \oplus \overline{\gamma}$ (4.12), and $a\colon \gamma(Da) \longrightarrow \theta a_1^{\#}$ $b\colon \theta b_1 \longrightarrow \gamma(Db^{\#})$ are isomorphisms. Therefore we can carry the basis $E(Da)$ into $\theta a_1^{\#}$, by means of $a$, carry the basis $E(Db^{\#})$ into $\theta b_1$ by the inverse of $b_1$, and extend these bases to a basis of $\gamma(D)$ by an arbitrary basis of $\overline{\gamma}$.

If we continue, by induction, to choose bases $E(D)$ in this way for all monomials $D$ of lower degree, we construct a basis $E = \bigcup_{D \in \Sigma_1} E(D)$ in the space $\theta$.

Let us now show how, having such a basis, we can " draw out open sequences ". We first turn our attention to sequences generated by final monomials. Let $\gamma(D)$ correspond to a final monomial $D$. Take any vector $\xi$ in $E(D)$ and denote it by $\xi(0, 0)$.

The monomial $D$ can be written in the form $D = D'd_1$, where $d$ is either $a$ or $b^{\#}$. By construction, the basis $E(D')$ of $\gamma(D')$ contains a vector $\xi'$, such that the mapping $d_1$ takes $\xi(0, 0)$ to $\xi'$. We denote the vector $\xi'$ by $\xi(1, 0)$ if $d_1 = a$, and by $\xi(0, -1)$ if $d_1 = b^{\#}$. The monomial $D'$ in its turn may be written in the form $D' = D''d_2$. In $E(D'')$ we can find a vector $\xi''$ such that $d_2\colon \xi' \to \xi''$, and so on.

Continuing this process we obtain a finite sequence of vectors $\xi^{(k)} = \xi(i, j)$ such that $d_k\colon \xi^{(k-1)} \to \xi^{(k)}$. The indices $i$ and $j$ here are constructed by the rule: if $d_k = a$, then we increase $i$ by 1, leaving $j$ fixed; if $d_k = b^{\#}$, then we decrease $j$ by 1, leaving $i$ fixed. It is not difficult to see that the vectors $\xi^{(k)}$ form an open sequence, and that the subspace which these vectors span is isomorphic to the canonical module of the first kind $K(M)$, where $M$ is the "staircase " consisting of all $(i, j)$ occurring in the construction.

However, open sequences are not only generated by final monomials $D$. Let $D$ be a monomial in $\Sigma_1$ such that $\dim \gamma(D) > \dim \gamma(Da) + \dim \gamma(Db^{\#})$.

Then, by construction, the basis $E(D)$ contains vectors that are not basis vectors carried over from $\gamma(Da)$ and $\gamma(Db^{\#})$. These are basis vectors from the subspace $\overline{\gamma}$. It is clear that each such vector provides the start of a new open sequence. We say that such a sequence is generated by the monomial $D$.

This process of constructing a basis in the subspace $\theta_1 = \bigoplus_{D \in \Sigma_1} \gamma(D)$ may be carried out just as successfully when we are given a grading of $C^n$, and thus of $\theta = \bigoplus \theta_s$. Here each subspace $\gamma(D)$ in the homogeneous and compatible set is homogeneous, and the bases $E(D)$ in $\gamma(D)$ can also be chosen homogeneously ($E = \bigcup E_s$, $E_s$ a basis in $\gamma_s$). As a result we obtain a decomposition of $\theta$ into a direct sum of graded subspaces, isomorphic to canonical modules of the first kind.

THEOREM 4.5. *The submodule*

$$\theta_1 = \bigoplus_{D \in \Sigma_1} \gamma(D),$$

*where the subspaces $\gamma(D)$ are homogeneous and compatible, is decomposable into a direct sum of homogeneous canonical modules of the first kind.*

## §5. Separation and examination of submodules of the second kind.

In this section we are concerned with the elementary intervals of the second kind. We recall that to each such interval $(\beta_{i-1}, \beta_i)$ there corresponds uniquely an infinite sequence of monomials $\{D_k\}$ $(k = s(D_k))$, such that

$$\beta_{i-1} = \Omega D_k, \quad \beta_i = \theta D_k \qquad (k = m, m+1, m+2, \ldots), \tag{5.1}$$

Here $D_{k+1} = d_{k+1} D_k$, where $d_{k+1} = a$ or $b^{\#}$.

As in §4, we choose in each interval a subspace $\gamma_i$ such that $\beta_{i-1} \oplus \gamma_i = \beta_i$. We shall show that, under a certain compatibility assumption, the subspace $\theta_2 = \bigoplus \gamma_i$, where the sum ranges over all intervals of the second kind, is invariant under $a$ and $b$ and, as a module, may be decomposed into a direct sum of canonical modules of the second kind, defined by "closed sequences".

In the next subsection we describe the construction of these modules.

1. The construction of "closed sequences". Consider a set $M$ consisting of $t$ integral points $(i, j)$ and having the form of a "staircase", as described in §4.1 in the construction of "open sequences". Let $(0, 0)$ be the first point and $(k, l)$ the last point of $M$. In addition we assume that the staircase finishes with a vertical part, that is, $(k, l-1) \in M$.

We obtain a "closed sequence" of the simplest form from the open sequence $\{e(i, j), (i, j) \in M\}$ by changing only the action of the operator $a$ on $e(k, l)$. Put[1] $ae(k, l) = \mu e(0, 0)$, $\mu \neq 0$. This equation closes the sequence.

In our simple example of a "closed sequence", we note one point that will help us to understand the more general construction.

Along the staircase we label the points $(0, 0), \ldots, (k, l)$ by the numbers $\nu = 0, \ldots, t-1$. Next, we consider the one-dimensional subspaces $L_\nu$ spanned by the vectors

$e_\nu = e(i, j)$, $(i, j) \in M$ $(\nu = 0, \ldots, t-1)$; $e_1 = e(0, 0)$, $\ldots$, $e_{t-1} = e(k, l)$.

The $L_\nu$ are connected by the diagram $L_0 \to L_1 \to \ldots \to L_{t-1}$, in which each

[1] For $\mu = 0$ the sequence remains open.

map is an isomorphism $e_\nu \to e_{\nu+1}$ defined either by $a$ or by $b^{-1}$ (having in mind the restriction of $b$ to the corresponding $L_{\nu+1}$). Having completed the diagram by the closure condition $L_{t-1} \to L_0$ ($e_{t-1} \overset{a}{\to} \mu e_0$), we define a map $T: L_0 \to L_0$ given by the formula $Te_0 = \mu e_0$. But the subspace $L_0$ is not singled out from the set $\{L_\nu\}$ by this property. Beginning with any $L_{\nu_0}$ and going along the cycle of maps in the completed diagram, we obtain an analogous map $L_{\nu_0} \to L_{\nu_0}$ ($e_{\nu_0} \to \mu e_{\nu_0}$).

In the general case, to which we now go over, the subspaces $L_\nu = L(i, j)$ have arbitrary dimension $q$ (but the same for all $(i, j)$), and the map $L_0 \to L_0$ is given by a matrix consisting of one Jordan block with eigenvalue $\mu$. Let us now describe this construction more accurately.

Suppose that we are given a "staircase" $M$ (described at the beginning of this subsection), a complex number $\mu$, and a natural number $q$. With each point $(i, j) \in M$ we associate a $q$-dimensional space $L(i, j)$ and we define operators $a$ and $b$ on $\mathbf{C}^n = \bigoplus_{(i,j)\in M} L(i, j)$ by the following relations.

For each $(i, j) \in M$:

$a$: $L(i, j) \to L(i+1, j)$ is an isomorphism if $(i+1, j) \in M$.
$a$: $L(i, j) \to 0$ if $(i+1, j) \notin M$ and $(i, j) \neq (k, l)$.
$a$: $L(k, l) \to L(0, 0)$ is an isomorphism.
$b$: $L(i, j) \to L(i, j+1)$ is an isomorphism if $(i, j+1) \in M$.
$b$: $L(i, j) \to 0$ if $(i, j+1) \notin M$.

In addition, we require that the map $L(0, 0) \to L(0, 0)$, which is the product of the isomorphisms in the cyclic diagram

$$L(0, 0) \longrightarrow \ldots \longrightarrow L(k, l) \xrightarrow{a} L(0, 0), \tag{5.2}$$

is represented in some basis by a single Jordan block with eigenvalue $\mu$. Here each isomorphism (except the last) in the cyclic diagram is either $a$: $L(i, j) \to L(i+1, j)$, or $b^{-1}$: $L(i, j+1) \to L(i, j)$.

If the subspaces $L(i, j)$ in the "staircase" are labelled in order, beginning with $L(0, 0)$, by an index $\nu$ ($\nu = 0, \ldots, t-1$), then the subspaces $L_\nu$ are connected by isomorphisms $d_\nu$: $L_{\nu-1} \to L_\nu$ in the cyclic diagram

$$L_0 \xrightarrow{d_1} L_1 \xrightarrow{d_2} L_2 \longrightarrow \ldots \longrightarrow L_{t-1} \xrightarrow{d_t} L_0. \tag{5.3}$$

Here each isomorphism $d_\nu$: $L_{\nu-1} \to L_\nu$ is either $a$ or $b^{-1}$ (more precisely, in the second case, $d_\nu^{-1}$: $L_\nu \to L_{\nu-1}$ is the map $b$). With each such diagram we can associate the word (monomial) $T = d_t d_{t-1} \ldots d_2 d_1$, where $d_i$ is either $a$ or $b^{\#}$. We may assume that the monomial $T$ defines the map $T(0)$: $L_0 \to L_0$.

We say that the monomial $T$ generates the closed sequence.

The direct sum of the subspaces $L(i, j)$ with the ring of operators generated by $a$ and $b$ on the space $\bigoplus_{(i,j)} L(i, j)$, is called a canonical module of the second kind if three conditions are fulfilled: 1) the monomial corresponding to it is non-periodic, 2) $T(0)$: $L_0 \to L_0$ has one eigenvalue $\mu \neq 0$, and 3) the matrix of this map in a suitable basis is a single Jordan block.

Thus, a canonical module of the second kind is uniquely determined by specifying the monomial $T$, the integer $q = \dim L_\nu$, and the arbitrarily chosen complex number $\mu$. We denote such a module by $K(T, q, \mu)$. In this section we show that the module $K(T, q, \mu)$ is irreducible and, conversely, that each irreducible module of the second kind is of the type $K(T, q, \mu)$.

We remark that, in the cyclic diagram (5.3), each map $T(\nu)$: $L_\nu \to L_\nu$, equal to $d_\nu \dots d_1 d_t \dots d_{\nu+2} d_{\nu+1}$, is similar to $T(0)$: $L_0 \to L_0$. Thus, two canonical modules $K(T, q, \mu)$ and $K(T', q', \mu')$ are equivalent if $(q, \mu) = (q', \mu')$ and the monomial $T'$ is a cyclic permutation of $T$.

Let us describe (without proof) the form of the stabilized sequence in the subspace $\theta$ of $\mathbf{C}^n \oplus \mathbf{C}^n$ $(n = q \cdot s(T))$, that corresponds to the module $K(T, q, \mu)$. It can be shown that: a) the stabilized sequence of subspaces consists of $t = s(T)$ elementary intervals of the second kind; b) each interval $(\beta_{i-1}, \beta_i)$ has length $q$; c) to each elementary interval $(\beta_{i-1}, \beta_i)$ there corresponds a monomial $T(\nu)$ such that $(\beta_{i-1}, \beta_i) = (\Omega T^\infty(\nu), \theta T^\infty(\nu))$; d) the number of such monomials $T(\nu)$ is $t$, and each of them is a cyclic permutation of $T(0)$.

We now define the graded canonical module of the second kind with $r$ components $K_s$, the homogeneous operator $b$, and the $\pi$-homogeneous operator $a$. We recall that $aK_s \subset K_{s+1}$, $aK_r \subset K_1$. For this purpose we change only the requirement 1) for $T$. We allow $T$ to be periodic; however, we require:

1a) The sum of the exponents of the powers of $a$ in $T$ is divisible by $r$.

1b) If $T$ can be written in the form $T = (T')^q$ $(q \neq 1)$, then the sum of the powers of $a$ in $T'$ is not divisible by $r$.

**3. The cycles of elementary intervals of the second kind.** We already know (§1.5) that with each elementary interval of the second kind there is associated uniquely an infinite monomial $\tilde{D} = \dots d_p \dots d_3 d_2 d_1$, where $d_j = a$ or $b^{\#}$, so that

$$(\beta_{k-1}, \beta_k) = (\Omega\tilde{D}, \theta\tilde{D}).$$

Here, if $D_j = d_j \dots d_2 d_1$ denotes a finite "segment" of the word, then $(\beta_{k-1}, \beta_k) = (\Omega D_j, \theta D_j)$ for all $j > m$.

The set of infinite monomials corresponding to elementary intervals of the second kind is denoted by $\Sigma_2$. It is finite, and there is a one-to-one correspondence between members $\tilde{D}$ of $\Sigma_2$ and elementary intervals of the second kind.

In this subsection we show that every $\tilde{D} \in \Sigma_2$ is periodic.

Let $\tilde{D} \in \Sigma_2$ ; we denote by $\tilde{D}(j)$ the infinite word for which

$$\tilde{D} = \tilde{D}(j) D_j, \text{ where } s(D_j) = j. \tag{5.4}$$

From the definition of an infinite word it follows that

$$\Omega\tilde{D} = (\Omega\tilde{D}(j)) D_j \text{ and } \theta\tilde{D} = (\theta\tilde{D}(j)) D_j. \tag{5.5}$$

THEOREM 5.1 *If $\tilde{D} \in \Sigma_2$, then each $\tilde{D}(j)$ belongs to $\Sigma_2$ $(j = 1, 2, \dots)$.*

PROOF. Assume the contrary. Then for some $j$ we have $\Omega\tilde{D}(j) = \theta\tilde{D}(j)$. Multiplying this equation on the right by $D(j)$ we obtain (see (5.5))

$\Omega\tilde{D} = \theta\tilde{D}$. But this contradicts the fact that $\tilde{D} \in \Sigma_2$.

The sequence $\{(\Omega\tilde{D}(j),\ \theta\tilde{D}(j))\}$ of elementary intervals of the second kind can contain only finitely many distinct intervals. But an elementary interval defines uniquely the infinite momomial that corresponds to it. Therefore, the sequence $\{\tilde{D}(j)\}$ contains only finitely many distinct monomials. Hence the monomial $\tilde{D}$, which is the beginning of this sequence, may be written in the form $\tilde{D} = \ldots TTTD_k$, where $s(D_k) \geqslant 0$ and $T$ is a finite non-periodic monomial.

Somewhat later we shall show that $s(D_k) = 0$, that is, that $\tilde{D}$ is periodic. If $\tilde{D}$ is a periodic monomial, then the monomials $\tilde{D}(j)$ (see (5.4)) are also periodic. The sequence of monomials $\{\tilde{D}(j)\}$ $(j = 0, 1, 2, \ldots)$ that is obtained from the periodic monomial $\tilde{D}$, is called "cyclic". The corresponding sequence of elementary intervals is also called cyclic. The number $t = s(T)$ is called the period of the cycle.

LEMMA 5.1. *The lengths of the elementary intervals in a cyclic sequence are equal.*

PROOF. Applying the inequality of Proposition 3.2 we obtain (using (5.4))

$$\rho(\Omega\tilde{D}(j+1),\ \theta\tilde{D}(j+1)) \geqslant \rho(\Omega\tilde{D}(j+1)\,d_j,\ \theta\tilde{D}(j+1)\,d_j) = \rho(\Omega\tilde{D}(j),\ \theta\tilde{D}(j)).$$

Putting $j = 1, 2, \ldots, t$, where $t$ is the period of the cycle, we find

$$\rho(\Omega\tilde{D}(t),\ \theta\tilde{D}(t)) \geqslant \ldots \geqslant \rho\,\Omega\tilde{D}(1),\ \theta\tilde{D}(1)) \geqslant \rho(\Omega\tilde{D},\ \theta\tilde{D})).$$

By the periodicity, the outside terms are equal, and so all the intermediate ones are equal, and the lemma is proved.

THEOREM 5.2. *All the monomials $\tilde{D} \in \Sigma_2$ are periodic.*

PROOF. We have shown that each $\tilde{D} \in \Sigma_2$ has the form $\tilde{D} = T^{\infty}D_k$. Let $T = d_t \ldots d_1$ and $D_k = d'D_{k-1}$. Assume that $D_k$ violates the periodicity of $\tilde{D}$, that is, $s(D_k) \neq 0$ and $d' \neq d_t$.

Consider the three monomials $T^{\infty}$, $T^{\infty}d_t$ and $T^{\infty}d'$. They can all be obtained from $\tilde{D}$ by division on the right, and so, by Theorem 5.1, they also belong to $\Sigma_2$. By hypothesis, $d_t$ and $d'$ are different; this means that one of them is equal to $a$, and the other to $b^{\#}$. Therefore, by Proposition 3.2, we have:

$$\rho(\Omega T^{\infty},\ \theta T^{\infty}) \geqslant \rho(\Omega T^{\infty}d_t,\ \theta T^{\infty}d_t) + \rho(\Omega T^{\infty}d',\ \theta T^{\infty}d').$$

It follows from Lemma 5.1 that $\rho(\Omega T^{\infty},\ \theta T^{\infty}) = \rho(\Omega T^{\infty}d_t,\ \theta T^{\infty}d_t)$. Thus, $\rho(\Omega T^{\infty}d',\ \theta T^{\infty}d') = 0$; in other words, $T^{\infty}d' \notin \Sigma_2$. We have obtained a contradiction, and the theorem is proved.

REMARK. *If $\tilde{D} = \ldots TTT$ is an infinite periodic monomial in $\Sigma_2$, then $s(T) > 1$.*

For suppose that $T = a$. The operator $a$ is nilpotent, therefore there exists a number $N$ such that $\Omega a^N = \theta$. In addition, $\theta a^j = \theta$. Thus, the interval $(\Omega a^{\infty},\ \theta a^{\infty}) = (\theta,\ \theta)$ is trival. Similarly, if $T = b^{\#}$, then the interval $(\Omega(b^{\#})^{\infty},\ \theta(b^{\#})^{\infty}) = (\Omega, \Omega)$ is also trivial. This means that the infinite monomials $\tilde{D} = \ldots aaa$ and $\tilde{D} = \ldots b^{\#}b^{\#}b^{\#}$ do not belong to $\Sigma_2$. So, we have shown that each infinite monomial $\tilde{D} \in \Sigma_2$ has the form $\tilde{D} = \ldots TTT = T^{\infty}$, where $T$ is a non-periodic monomial. We say that the monomial $T$ generates the infinite monomial $\widetilde{D}$. As all the infinite monomials $\widetilde{D} \in \Sigma_2$ are distinct, their generating monomials $T$ are also distinct. Here the elementary interval $(\beta_{k-1}, \beta_k)$ corresponding to the monomial $\widetilde{D} \in \Sigma_2$ may be written in the form

$$(\beta_{k-1}, \beta_k) = (\Omega T^\infty, \theta T^\infty). \tag{5.6}$$

In accordance with the definition of §3, $\beta_{k-1}$ is the stable kernel of the relation $T$, and $\beta_k$ is the stable domain of definition of this relation. We denote by $T(j)$ the following cyclic permutation of the monomial $T(0) = d_t \ldots d_2 d_1$ :

$$T(j) = d_j \ldots d_1 d_t \ldots d_{j+2} d_{j+1}. \tag{5.7}$$

It is not difficult to verify that if $\{\widetilde{D}(j)\}$ $(j = 0, 1, \ldots, t-1)\}$ is a cyclic sequence and $\widetilde{D}(0)$ is generated by the monomial $T(0)$, then $\widetilde{D}(j)$ is generated by $T(j)$ (see (5.7)).

4. **A compatible choice of subspaces.** We have established that if $(\beta_{k-1}, \beta_k)$ is an elementary interval of the second kind, then $\beta_{k-1}$ is the stable kernel $\Omega T^\infty$ of a generating monomial (relation) $T$, and $\beta_k$ is the stable domain of definition of this relation. By Theorem 3.1 there exists a subspace $\delta$ such that 1) $\Omega T^\infty \oplus \delta = \theta T^\infty$ and 2) $\delta$ is a domain of regularity of the relation $T$. This means that for each non-zero vector $(x, 0) \in \delta$ there exists a unique non-zero vector $(y, 0) \in \delta$ such that $(x, y) \in T$. If a subspace is a domain of regularity for a relation $T$, then we denote it by $\delta(T)$.

THEOREM 5.3. *Let* $\{(\Omega\widetilde{D}(j), \theta\widetilde{D}(j))\} = \{(\Omega T^\infty(j), \theta T^\infty(j))\}$ $(j = 0, 1, 2, \ldots, t-1)$ *be a cyclic sequence of elementary intervals of the second kind. Then the subspaces* $\delta_j = \delta(T(j)), (\Omega T^\infty(j) \oplus \delta_j = \theta T^\infty(j))$ *can be chosen compatibly. Namely, if* $\widetilde{D}(j-1) = \widetilde{D}(j)a$, *then* $\delta_{j-1}a^\# = \delta_j$, *and if* $\widetilde{D}(j-1) = \widetilde{D}(j)b^\#$, *then* $\delta_j b^\# = \delta_{j-1}$.

PROOF. Choose a subspace $\delta_0$, $\Omega T^\infty \oplus \delta_0 = \theta T^\infty$, that is a domain of regularity for the relation $T$.

We now show how to construct from $\delta_0$ all the remaining subspaces $\delta_j$. We take in $\delta_0$ a basis $\{(e_{0,k}, 0)\}$ $(k = 1, 2, \ldots, p)$. The subspace $\delta_0$ is a domain of regularity for $T$. Therefore, for each $(e_{0,k}, 0)$, there exists a unique $(f_{0,k}, 0) \in \delta_0$ such that $(e_{0,k}, f_{0,k}) \in T$. The vectors $\{(f_{0,k}, 0)\}$ form a new basis for $\delta_0$.

The relation $T$ can be written as a product $T = d_t d_{t-1} \ldots d_1$ $s(d_j) = 1$. Hence for each pair $(e_{0,k}, f_{0,k}) \in T$ we can select vectors $e_{1,k}, e_{2,k}, \ldots, e_{t-1,k}$ such that $(e_{j-1,k}, e_{j,k}) \in \delta_j$. The pairs $\{(e_{j,k}, 0)\}$ $(k = 1, 2, \ldots, p)$ span a subspace, which we denote by $\delta_j$.

Let us show that the $\delta_j$ so constructed are the required subspaces $\delta(T(j))$. We denote the product $d_j \ldots d_2 d_1$ by $G$, and $d_t \ldots d_{j+2} d_{j+1}$ by $H$. Then $T$ can be written in the form $T = HG$, and $T(j) = d_j \ldots d_1 d_t \ldots d_{j+1}$ can be written in the form $T(j) = GH$. In accordance with this,

$$T^{\infty} = T^{\infty}(j)\,G, \qquad T^{\infty}(j) = T^{\infty}H. \tag{5.8}$$

Let us show that $\theta_j \in \theta T^{\infty}(j)$. By construction, we have $(e_{j,k}, f_{0,k}) \in H$ and $(f_{0,k}, 0) \in \delta_0 \subset \theta T^{\infty}$. Consequently, $(e_{j,k}, 0) \in \theta T^{\infty} H$, and by (5.8), $(e_{j,k}, 0) \in \theta T^{\infty}(j)$. Thus, $\delta_j$, which is spanned by the pairs $(e_{j,k}, 0)$, is contained in $\theta T^{\infty}(j)$.

To establish the remaining properties of $\delta_j$ we use the following lemma, which is a variant of Lemma 3.1.

LEMMA 5.2. *Let $(\varphi_1, \varphi_2)$ and $(\varphi_1 G, \varphi_2 G)$ be two intervals in $\theta$, where $G$ is some relation, and let $\delta$ be a subspace satisfying the condition $\delta \oplus \varphi_1 G = \varphi_2 G$.*

*If a relation $G_1 \subset G$ is chosen so that $\theta G_1 = \delta$, $\theta G_1^{\#} \subset \varphi_2$ and $\dim G_1 = \dim \delta$, then $G_1$ is non-singular, and $\theta G_1^{\#} \cap \varphi_1 = \Omega$.*

The proof is similar to that of Lemma 3.1.

Let us denote the interval $(\Omega T^{\infty}(j), \theta T^{\infty}(j))$ by $(\varphi_1, \varphi_2)$. Then by the formula (5.8) we obtain $(\Omega T^{\infty}, \theta T^{\infty}) = (\Omega T^{\infty}(j)\,G, \theta T^{\infty}(j)\,G) = (\varphi_1 G, \varphi_2 G)$. Let $G_1$ denote the relation spanned by the pairs $(e_{0,k}, e_{j,k})$ $(k = 1, 2, \dots, p)$. It is clear that $G_1 \subset G$, $\theta G_1 = \delta_0$ and $\theta G_1^{\#} = \delta_j$, $\dim G_1 = \dim \delta_0$. So we have shown that $\delta_j = \theta G_1^{\#} \subset \varphi_2$. Thus, the situation is completely analogous to that of Lemma 5.2, and therefore the relation $G_1$ is non-singular and $\theta G_1^{\#} \cap \varphi_1 = \Omega$ or, in the previous notation, $\delta_j \cap \Omega T^{\infty}(j) = \Omega$. The relation $G_1$ is non-singular, and so the dimension of its domain of values is equal to $\dim G_1$. In other words, $\dim \delta_j = \dim \delta_0$, and hence the vectors $\{(e_{j,k}, 0)\}$ $(k = 1, 2, \dots, p)$ are linearly independent.

So we have shown that the subspace $\delta_j$ has these properties: $\delta_j \subset \theta T^{\infty}(j)$, $\delta_j \cap \Omega T^{\infty}(j) = \Omega$, $\dim \delta_j = \dim \delta_0$. But since the length of the interval $(\Omega T^{\infty}(j), \theta T^{\infty}(j))$ is equal to $\rho(\Omega T^{\infty}, \theta T^{\infty})$,

$$\delta_j \oplus \Omega T^{\infty}(j) = \theta T^{\infty}(j). \tag{5.9}$$

It remains to show that the subspace $\delta_j$ is a domain of regularity for $T(j)$ and that the subspaces $\delta_j$ $(j = 0, 1, 2, \dots, t-1)$ have been chosen compatibly.

We have shown that the vectors $\{(e_{j,k}, 0)\}$ $(k = 1, 2, \dots, p)$ are linearly independent. From this it follows that the relation generated by the pairs $\{(e_{j-1,k}, e_{j,k})\}$ $(k = 1, 2, \dots, p)$ is non-singular. Let $\bar{d}_j$ denote this relation. It is clear that $\bar{d}_j \subseteq d_j$, where $d_j$ is either $a$ or $b^{\#}$. Thus, the relation $T_1(j) = \bar{d}_j \dots \bar{d}_1 \bar{d}_t \dots \bar{d}_{j+1}$, which is contained in $T(j)$, is also non-singular. It is easy to see that $\theta T_1(j) = \delta_j$ and $\theta T_1^{\#}(j) = \delta_j$.

In conjunction with (5.9) this shows that $\delta_j$ is a domain of regularity for $T(j)$. If $d_j = a$, then $(e_{j-1,k}, e_{j,k}) \in a$, that is, $e_{j,k} = a e_{j-1,k}$. This means that $\delta_j = \delta_{j-1} a^{\#}$. Similarly, in the case $d_j = b^{\#}$ we obtain $\delta_{j-1} = \delta_j b^{\#}$. Consequently, the subspaces $\delta_j$ have been chosen compatibly. The theorem is now proved.

If $T$ is some monomial generating a cycle, then we denote by $\theta_2(T)$

the sum $\bigoplus_{j=0}^{t-1}\delta_j$ of all the subspaces $\delta_j$ in the cycle. We show that $\theta_2(T)$ is invariant under the mappings $a$ and $b$. In the course of proving Theorem 5.3 we constructed isomorphisms $d_j$: $\delta_{j-1}\to\delta_j$, where $d_j$ is either $a$ or $b^{\#}$. But this still does not determine how the operators $a$ and $b$ act on the whole space $\theta_2(T)$.

Suppose that $d_j = a$. Then the isomorphism $d_j$: $\delta_{j-1}\to\delta_j$ shows that $\delta_j=\delta_{j-1}a^{\#}$. From the inequality $\delta_{j-1}a^{\#}\subseteqq\theta a^{\#}\subseteqq\Omega b$ it follows that in this case $\delta_j$ lies in the kernel $\Omega b$ of $b$. Similarly, if $d_j=b^{\#}$, then $\delta_{j-1}=\delta_j b^{\#}$. In this case, from the inequality $\delta_j b^{\#}\subseteqq\theta b^{\#}\subseteqq\Omega a$ it follows that $\delta_{j-1}$ is contained in the kernel $\Omega a$ of $a$.

This shows that for an arbitrary subspace $\delta_j$ there are the possibilities: either $\delta_j a^{\#}=\delta_{j+1}$, or $\delta_j a^{\#}=\Omega$. Similarly, for $b$, either $\delta_j b^{\#}=\delta_{j-1}$, or $\delta_j b^{\#}=\Omega$. From this it is clear that $\theta_2(T)$ is invariant under $a$ and $b$. Hence we have

THEOREM 5.4. *Let* $\{(\Omega\tilde{D}(j),\ \theta\tilde{D}(j))\}$ $(j = 0, 1, 2, \ldots, t-1)$ *be elementary intervals of the second kind, forming a cycle. If the subspaces* $\delta_i$ $(\Omega\tilde{D}(j)\oplus\delta_j=\theta\tilde{D}(j))$ *are compatibly chosen, then the subspace* $\theta_2(T)=\sum_{j=0}^{t-1}\delta_j$ *is invariant under* $a$ *and* $b$.

**5. The decomposition of $\theta_2(T)$ into a direct sum of canonical modules of the second kind.** Let $\delta_0$ be a subspace such that $\Omega T^\infty(0)\oplus\delta_0=\theta T^\infty(0)$. By construction, $\delta_0$ is a domain of regularity for the relation $T(0)$. This means that, for any pair $(x, 0)\in\delta_0$, there exists one and only one pair $(y, 0)\in\delta_0$ such that $(x, y)\in T(0)$. Thus, a non-singular linear map is defined on $\delta_0$, which we also denote by $T(0)$. It follows from Theorem 3.2 that if we choose another subspace $\delta_0'$ as domain of regularity, then the map $T'(0)$: $\delta_0'\to\delta_0'$ is similar to $T(0)$.

We remark that the map $T(0)$: $\delta_0\to\delta_0$ is the product $d_t\ldots d_2d_1$ of isomorphisms $d_j$: $d_{j-1}\to\delta_j$. From this it follows that each map $T(j)$: $\delta_j\to\delta_j$ is similar to $T(0)$: $\delta_0\to\delta_0$.

The subspace $\delta_0$ can be decomposed into a direct sum of invariant subspaces $\delta_0(\lambda_i)$ $(i = 1, 2, \ldots, l)$ on each of which the matrix of $T(0)$, in a suitable basis, is a single Jordan block. By means of the isomorphism $d_1$: $\delta_0\to\delta_1$, the decomposition $\delta_0=\bigoplus_i\delta_0(\lambda_i)$ may be carried over into the subspace $\delta_1$. So we find that $\delta_1=\bigoplus_i\delta_1(\lambda_i)$. It is clear that the decomposition can be carried over to all the subspaces $\delta_j$ in the cycle. Here the subspace $\delta_j(\lambda_i)$ is invariant under $T(j)$.

We denote the direct sum $\bigoplus_{j=0}^{t-1}\delta_j(\lambda_i)$ by $\theta_2(\lambda_i)$. Each such subspace is invariant under $a$ and $b$. This follows from the fact that $d_j$ is equal either to $a$ or to $b^{\#}$ and maps $\delta_{j-1}(\lambda_i)$ isomorphically onto $\delta_j(\lambda_i)$. Thus, $\theta_2(\lambda_i)$ is a submodule. Furthermore, by construction, $T(j)$: $\delta_j(\lambda_i)\to\delta_j(\lambda_i)$ has one eigenvalue $\lambda_i$, and in a suitable basis the matrix of this map is a single

Jordan block. Thus, $\theta_2(\lambda_i)$ is a canonical module of the second kind, and we have proved

THEOREM 5.5. *The submodule $\theta_2(T)$ can be decomposed into a direct sum of canonical submodules of the second kind.*

In order not to complicate the text, we have not given here the proofs of Theorems 5.2 - 5.5 for the case when $C^n$ has a grading. The statements of these theorems remain unchanged. The difference lies in the fact that a canonical module of the second kind with a grading is defined not by the monomial $T$, but by some power $T^q$ of it.

6. Summarizing the results of §§4 and 5 we may say that the module $C^n$ can be represented as a direct sum of canonical modules of the first and second kinds. We recall that a module of the first kind is one in which every interval $(\beta_{i-1}, \beta_i)$ in the stabilized sequence $\{\beta_i\}$ is of the first kind. Modules of the second kind are defined analogously. The form of canonical modules of the first and second kinds was given in §4.1 and §5.2. Similar assertions hold for graded modules. Using the results of §3, it is not difficult to show that the canonical modules are indecomposable.

CONCLUSION. 1°. In this chapter we have solved the problem of carrying to canonical form two linear maps $a$ and $b$ such that $ab = ba = 0$. We have shown that the indecomposable modules of this form may be of two kinds.

An indecomposable module of the first kind is completely determined when we are given the finite monomial $D = (b^{\#})^{i_k} \dots a^{i_3} (b^{\#})^{i_2} a^{i_1}$, that generates it. The sequence of numbers $(i_1, i_2, \dots, i_k)$; $i_1 \geqslant 0$, $i_k \geqslant 0$, $i_j > 0$, corresponding to such a monomial $D$ is a complete set of invariants for an indecomposable module of the first kind. Two indecomposable modules for which these sequences are equal are equivalent. The canonical form of such a module was described on p. 52.

An indecomposable module of the second kind is uniquely determined when we are given a finite monomial $T = (b^{\#})^{j_l} \dots a^{j_3} (b^{\#})^{j_2} a^{j_1}$ and two further numbers: a positive integer $q$ and an arbitrary complex number $\mu$. Here the monomial $T$ may only be given to within a cyclic permutation.

The sequence $(j_1, j_2, \dots, j_l, q, \mu)$ is a complete set of invariants for an indecomposable module of the second kind. The canonical form of such modules was described on p. 52.

2°. In addition, the analogous problem in the so-called graded case has been solved. Namely, suppose that we are given $r$ spaces $K_1, K_2, \dots, K_r$ and linear maps $a$ and $b$ defined on the space $C^n = \bigoplus_{s=1}^{r} K_s$ by the following formulae: $b\colon K_s \to K_s$; $a\colon K_{s-1} \to K_s$; $a\colon K_r \to K_1$ (assuming here that the maps $a\colon C^n \to C^n$ and $b\colon C^n \to C^n$ are nilpotent and such that $ab = ba = 0$). The graded modules can also be of two kinds. An indecomposable module of the first kind is uniquely determined by specifying a monomial $D = (b^{\#})^{i_u} \dots (b^{\#})^{i_2} a^{i_1}$ and a positive integer $s$ $(1 < s \leqslant r)$. The canonical form of such a module was described in §4.1.

An indecomposable module of the second kind is uniquely determined by a monomial $T = (b^{\#})^{j_l} \dots a^{j_3} (b^{\#})^{j_2} a^{j_1}$, a positive integer $s$ $(1 < s \leqslant r)$, and by two further numbers: a positive integer $q$ and an arbitrary complex number $\mu$. Here the monomial $T$ is defined to within an arbitrary cyclic permutation (see p. 52).

3°. In the first chapter the problem on representations of the Lorentz group was reduced to the following problem of linear algebra.[1]

We are given two spaces $K_1$ and $K_2$ and linear maps $a_+: K_1 \to K_2$; $a_-: K_2 \to K_1$; $b_1: K_1 \to 0$ and $b_2: K_2 \to K_2$ such that the maps $b_2$ and $a_- a_+$ are nilpotent and $b_2 a_+ = a_- b_2 = 0$. This corresponds to the graded case, which was described in the preceding section. Here the number $r$ of subspaces $K_s$ is 2, and we have a further restriction on $b$, namely that $b$ is identically zero on $K_1$. The canonical form of such indecomposable modules can easily be obtained from the description of canonical graded modules.

## References

[1] I.M. Gel'fand, Some questions of analysis and differential equations, Uspekhi Mat. Nauk 14 : 3 (1959), 3–19.
= Amer Math. Soc. Transl. (2) 26 (1963), 201–219.

[2] I.M. Gel'fand and R.A. Minlos and Z. Ya Shapiro, *Predstavleniya gruppy vrashchenii i gruppy Lorentsa, ikh. primeneniya,* Gosudarstv. Izdat. Fiz.-Mat. Lit., Moscow 1958.
Translation: Representations of the rotation and Lorentz groups and their applications, Pergamon Press, Oxford 1963.

[3] D.P. Zhelobenko, Linear representations of the Lorentz group, Dokl. Akad. Nauk SSSR 126 (1959), 935–938.

[4] S. MacLane, An algebra of additive relations, Proc. Nat. Acad. Sci. USA. 47 (1961), 1043–1051.
= Sb. perevodov. Matem. 7: 6 (1963), 1–12.

Received by the Editors December 18, 1967.

Translated by B. Hartley.

[1] In the statement we give here, the subspaces and maps are denoted by other symbols than in Chapter II.

# 2.

## (with V. A. Ponomarev)

# Remarks on the classification of a pair of commuting linear transformations in a finite-dimensional space

Funkts. Anal. Prilozh. **3** (4) (1969) 81–82 [Funct. Anal. Appl. 3 (1969) 325–329].
MR **40**:7279. Zbl. **204**:453

There is extensive literature devoted to the problem of the canonical form of a pair of commuting linear transformations a and b in a finite-dimensional space relative to the transformations $a' = cac^{-1}$ and $b' = cbc^{-1}$, where c is a non-singular linear transformation. We will show that attempts of direct determination of the canonical form do not make sense. Namely, the solution of this problem would imply the solution of the problem of classifying any finite number of arbitrary non-commuting linear transformations. By using any finite set of matrices $A_1, \ldots, A_k$, we will construct a pair of commuting matrices a and b, $a = a(A_1, \ldots, A_k)$ and $b = b(B_1, \ldots, B_k)$, such that the similarity of the pairs a, b and a', b' is equivalent to the similarity of the systems $A_1, \ldots, A_k$ and $A_1', \ldots, A_k'$.

A somewhat more general statement is: we will consider the category G of all Artin modules over the ring K[X, Y] of polynomials of two variables (i.e., of modules which are finite-dimensional spaces over the field K). Then the category G contains a complete subcategory which is isomorphic to the category $L_K$ of Artin modules over the free algebra of k variables.

Now we turn to the construction. For simplicity we consider the case k = 3. We will consider the space E over some field K to be equal to the direct sum $E = \bigoplus_{i=1}^{7} E_i$ of seven subspaces $E_i$ of the same dimensionality dim $E_i = n$. We will define linear transformations a and b in the space E by using the following commutative diagram D:

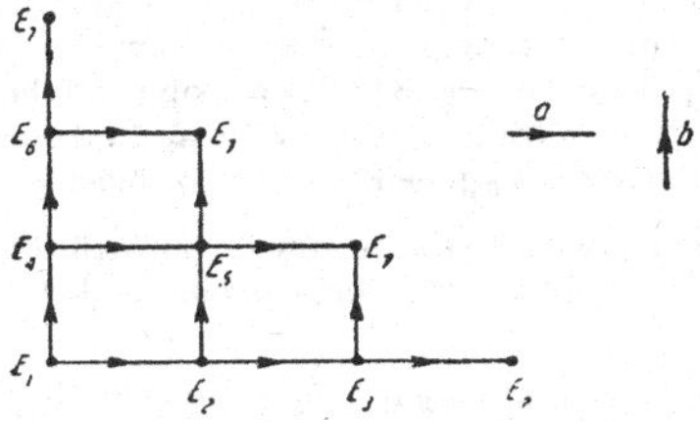

The points $E_i (i = 1, 2, \ldots, 7)$ in the diagram D represent the subspaces $E_i$. If two points, for example $E_4$ and $E_5$, are connected by a horizontal arrow, this means the transformation a defines an isomorphism between the subspaces $E_4$ and $E_5$: $aE_4 = E_5$. If two points, for example $E_2$ and $E_5$, are connected by a vertical arrow, this means the transformation b defines an isomorphism between $E_2$ and $E_5$: $bE_2 = E_5$. Moreover, we will require that the following condition be satisfied:

$$\text{if} \quad x \in E_7, \text{ then } ax = bx = 0. \tag{1}$$

It follows from the commutativity of the diagram D that the linear transformations a and b commute, i.e., ab = ba.

---

Moscow State University. Institute of Biophysics, Academy of Sciences of the USSR. Translated from Funktsional'nyi Analiz i Ego Prilozheniya, Vol. 3, No. 4, pp. 81–82, October–December, 1969. Original article submitted June 23, 1969.

We write these linear transformations in matrix form. In order to do this we will define a basis $e_{i,j}$ ($j = 1, 2, \ldots, n$; $i = 1, 2, \ldots, 7$) in each of the subspaces $E_i$. Then the previously mentioned description can be replaced by the following system of equations:

$$
\begin{aligned}
&ae_{1,j} = e_{2,j}, \quad be_{1,j} = e_{4,j}, && ae_{5,j} = \sum_{k=1}^{n} \alpha_{j,k} e_{7,k}, \quad be_{5,j} = \sum_{k=1}^{n} \beta_{j,k} e_{7,k}, \\
&ae_{2,j} = e_{3,j}, \quad be_{2,j} = e_{5,j}, && \\
&ae_{3,j} = e_{7,j}, \quad be_{3,j} = \sum_{k=1}^{n} \alpha_{j,k} e_{7,k}, && ae_{6,j} = \sum_{k=1}^{k} \beta_{j,k} e_{7,k}, \quad be_{6,j} = \sum_{k=1}^{n} \gamma_{j,k} e_{7,k}, \\
&ae_{4,j} = e_{5,j}, \quad be_{4,j} = e_{6,j}, && ae_{7,j} = 0, \quad be_{7,j} = 0,
\end{aligned}
$$

where $A = (\alpha_{j,k})$, $B = (\beta_{j,k})$, and $C = (\gamma_{j,k})$ are arbitrary $n \times n$ matrices with elements from the field K.

Thus, by means of the matrices A, B, and C in the n-dimensional space F, it is possible to construct a commuting pair of matrices a and b in a space of n-dimensions of E. We note that the three linear transformations A, B, and C in the space F can be arbitrary and, in particular, do not necessarily have to commute with one another.

It can be proved that if (E, a, b) and (E', a', b') are equivalent, then their corresponding systems (F, A, B, C) and (F', A', B', C') (where F and F' are linear spaces; A, B, and C are linear transformation are also equivalent. In other words, the existence of $x$: $E \to E'$ such that

$$a' = xax^{-1}, \quad b' = xbx^{-1} \tag{3}$$

implies the existence of $y$: $F \to F'$ such that

$$A' = yAy^{-1}, \quad B' = yBy^{-1}, \quad C' = yCy^{-1}. \tag{4}$$

The converse is also true: if (F, A, B, C) and (F', A', B', C') are equivalent, then their correspondin systems (E, a, b) and (E', a', b') are also equivalent.

Thus, we see that the problem of classifying a pair of commuting linear transformations contains the problem of classifying any three linear transformations as a special case.

Remark. A system (E, a, b) can be constructed which does not necessarily depend on three, but on any finite number k of arbitrary linear transformations A, B, C, D, . . . in the space F ($\dim F < \dim E$).

We state the result just obtained in a different form. It is known that a system (E, a, b), consisting of a finite-dimensional vector space E over the field K and two commutating linear transformations a and b, can be considered as a module $E_{a,b}$ over the ring K[X, Y] of polynomials of two variables over the field K. Let G denote the category of all K[X, Y]-modules. Thus, objects in the category G are all K[X, Y]-modules and morphisms are all homomorphisms $f$: $E_{a,b} \to E'_{a',b'}$ of one module into another.

Let $\tilde{G}$ denote a complete subcategory of the category G, in which the objects are all K[X, Y]-modules, determined by the diagram D and the conditions (1), and morphisms between two modules are all morphisms between objects.

The system (F, A, B, C) of the space F over the field K and of three linear transformations A, B, and C in this space can be considered as a module F over the free non-commutative algebra $A_3$ over the field K where $A_3$ is an algebra with three generators. Let $L_3$ denote the category of $A_3$-modules.

THEOREM. The complete subcategory $\tilde{G}$ of K[X, Y]-modules, defined by the diagram D and the condition (1), is isomorphic to the category $L_3$ of modules over the free algebra $A_3$ with three generators.

COROLLARY 1. The module $E_{a,b,D}$ is indecomposable if and only if the $A_3$-module corresponding to it is indecomposable.

COROLLARY 2. Let the $A_3$-module F be decomposable into a direct sum of submodules $F_s$ ($s = 1, 2, \ldots, p$). Then the module $E_{a,b,D}$ corresponding to it is also decomposable. This decomposition can be chosen ($E_{a,b,D} = \bigoplus_s E_{a,b,D}(s)$) such that the module $E_{a,b,D}(s)$ corresponds to the module $F_s$.

# 3.

## (with M. I. Graev and V. A. Ponomarev)

# The classification of the linear representations of the group $SL(2, C)$

Dokl. Akad. Nauk SSSR **194** (1970) 1002–1005
[Sov. Math., Dokl. **11** (1970) 1319–1323]. MR **43**:2162. Zbl. **229**:22024

0. The irreducible representations of the group $SL(2, C)$ are well known. However, in contrast to mpact groups, the study of any representation of $SL(2, C)$ cannot be reduced to a study of an educible representation. In the present work, all the representations of the group $SL(2, C)$, "com-sed" in the natural sense from a finite number of irreducible representations, are classified. We all call these representations Harish-Chandra modules (see the precise definitions below). We give complete description of such representations, namely, each such representation is decomposed into a ect sum of a finite number of nondecomposable representations. In the article, all the nondecom-sable representations are defined up to equivalence. These representations may be singular or non-ngular, where the singular case is the more interesting (see the formulation of the result in §5°). We so present a method of forming nondecomposable representations with given invariants from some ementary blocks which are formed simply from irreducible representations. The nonsingular case has en investigated earlier by Želobenko [6], who also gives some additional results. The articles in ich the infinitesimal classification of Harish-Chandra modules is obtained [1–3] form the basis of e present work.

1°. The definition of the Harish-Chandra $G$-modules. The linear topological space $H$, in which e continuous representation of the connected semisimple Lie group $G$ acts, will be called a $G$-dule. Suppose $U$ is a compact group. By an algebraic $U$-module we shall mean the linear space $H_0$ ithout a topology) in which the representation of group $U$ acts and for which the following condition fulfilled: the representation in $H_0$ may be decomposed into a direct sum of irreducible (finite-mensional) representations, where each of these representations enters into the decomposition with finite multiplicity. The $G$-module $H$ will be called a *Harish-Chandra module*, if $H$ contains the gebraic $U$-module $H_0$ as an everywhere dense subspace, where $U$ is the maximal compact subgroup the group $G$. Note that $H_0$ is the smallest linear subspace in $H$ containing all the irreducible $U$-bmodules in $H$. Thus, $H_0$ is unambiguously defined in $H$.

Suppose $\mathfrak{g}$ is a Lie algebra of the group $G$. It may be shown that the operators on $H$ correspond-g to elements of $\mathfrak{g}$ are defined on all of $H_0 \subset H$ and that $H_0$ is invariant with respect to these erators. Thus, the space $H_0$ is given the structure of a $\mathfrak{g}$-module. The Harish-Chandra $G$-module $H$ ill be called *nondecomposable* if the $\mathfrak{g}$-module $H_0$ corresponding to it is nondecomposable, and $H$ ill be called *irreducible* if the $\mathfrak{g}$-module $H_0$ is irreducible. Two Harish-Chandra $G$-modules $H'$ and $H''$ will be called *equivalent* if the corresponding $\mathfrak{g}$-modules $H_0'$ and $H_0''$ are isomorphic. Note that e definition of a Harish-Chandra $G$-module, as also the definitions of nondecomposability, irreducibil-y and equivalence, do not depend on the choice of the maximal compact subgroup $U \subset G$.

2°. Elementary $G$-modules. We shall construct the simplest class of nondecomposable Harish-andra $G$-modules, where $G = SL(2, C)$. Let $\pi = (n_1, n_2)$ be an arbitrary pair of complex numbers whose difference is an integer. We shall say that the function $f(z_1, z_2)$, $z_1, z_2 \in \mathbb{C}$, ($(z_1, z_2) \neq (0, 0)$), is *homogeneous of degree* $\pi$, if $f(\lambda z_1, \lambda z_2) = \lambda^{n_1-1}\bar{\lambda}^{n_2-1} f(z_1, z_2)$ for any $\lambda \neq 0$. Homogeneous functions will also be called adjoint homogeneous functions of order zero. We shall call the function $f(z_1, z_2)$ *an adjoint homogeneous function of order* $m$ *and of degree* $\pi$ ($m = 1, 2, \cdots$) if for any $\lambda \neq 0$ the difference $f(\lambda z_1, \lambda z_2) - \lambda^{n_1+1}\bar{\lambda}^{n_2-1} f(z_1, z_2)$ is an adjoint homogeneous function of order $m-1$ and of degree $\pi$.

Let $D_\pi^m$ be the space of all infinitely differentiable adjoint homogeneous functions of order $m$ and of degree $\pi$. In $D^m$, we shall fix the representation $T(g)$ of the group $G = SL(2, \mathbb{C})$ as follows:

$$(T(g)f)(z_1, z_2) = f(\alpha z_1 + \gamma z_2, \beta z_1 + \delta z_2),\ g = \begin{pmatrix} \alpha & \beta \\ \gamma & \delta \end{pmatrix}.$$

The constructed $G$-modules $D_\pi^m$ will be called *elementary*.* It can be shown that all of them are nondecomposable. We formulate an assertion concerning the construction of their composition series.

Clearly, $D_\pi^0 \subset D_\pi^1 \subset \cdots \subset D_\pi^m \subset \cdots$. Here $D_\pi^m / D_\pi^{m-1} \cong D_\pi^0$ for any $m$. We shall call the pair $\pi = (n_1, n_2)$ *a singular point* if $n_1, n_2$ are nonzero integers and have the same sign. If $\pi$ is a nonsingular point, then the $G$-module $D_\pi^0$ is irreducible and, therefore, the series $0 \subset D_\pi^0 \subset D_\pi^1 \subset \cdots \subset D_\pi^m$ is a composition series for $D_\pi^m$. If $\pi = (n_1, n_2)$ is a singular point, then for any $m$, there exists a unique $G$-module $F_\pi^m$ (different from $D_\pi^{m-1}$ and $D_\pi^m$) such that $D_\pi^{m-1} \subset F_\pi^m \subset D_\pi^m$.** Thus in the singular case the composition series for $D_\pi^m$ is of the form

$$0 \subset F_\pi^0 \subset D_\pi^0 \subset F_\pi^1 \subset D_\pi^1 \subset \ldots \subset F_\pi^m \subset D_\pi^m.^{***} \tag{1}$$

The factors of series (1) are the $G$-modules $F_\pi^0$ and $F_{\pi^{-1}}^0$, where $\pi^{-1} = (-n_1, -n_2)$ (i.e., $F_\pi^k / D_\pi^{k-1} \cong F_\pi^0$, $D_\pi^k / F_\pi^k \cong F_{\pi^{-1}}^0$). One of these factors if finite-dimensional and the other is infinite-dimensional (for example, in the case in which $n_1 > 0$, $n_2 > 0$, $F_\pi^0$ is the (finite-dimensional) module of all homogeneous polynomials of degree $\pi$ and $F_{\pi^{-1}}^0 \cong D_{\pi'}^0$, where $\pi' = (n_1, -n_2)$).

In the singular case, by elementary modules we shall mean both the $G$-modules $D_\pi^m$ and the $G$-modules $F_\pi^m$ obtained from them by "shortening". Note that the $G$-modules $D_{\pi^{-1}}^{m-1}$ and $F_{\pi^{-1}}^m$ may also be obtained by "shortening" $D_\pi^m$, i.e. $F_{\pi^{-1}}^m \cong D_\pi^m / F_\pi^0$, $D_{\pi^{-1}}^{m-1} \cong F_\pi^m / F_\pi^0$.

We now give the definition of an elementary $G$-module for the case of an arbitrary reducible Lie group $G$. Let $N$ be the maximal unipotent subgroup of the group $G$ and let $B$ be its normalizer in $G$.

---

* For the $G$-module $D_\pi^0$, the construction of the irreducible representations of Gel'fand-Naĭmark [5] is well known.

** If $n_1 > 0$, $n_2 > 0$, then the space $F_\pi^m$ may be defined as follows. Fix the function $\phi_m \in D_{\pi_0}^m$, $\pi_0 = (1, 1)$ ($\phi_m \neq 0$) such that for any $g \in G$, $T(g)\phi_m = \phi_m$ for the module $D_{\pi_0}^{m-1}$ (for example, one can take $\phi_m(z) = \ln^m(|z_1|^2 + |z_2|^2)$). We shall call the function $f \in D_\pi^m$ an *m-quasi-polynomial*, if $f = P\phi_m$ for the module $D_\pi^{m-1}$, where $P$ is a polynomial (it is not difficult to verify that the definition does not depend on the choice of $\phi_m$). The space of all *m*-quasi-polynomials is our $F_\pi^m$. Note that the space $F_{\pi^{-1}}^m$ may be defined from considerations of duality.

*** In the singular, as in the nonsingular case, $D_\pi^m$ has only one composition series.

›nsider any finite-dimensional nondecomposable representation $r$ of the factor group $B/N$ acting in e space $E$ and lift it in the trivial way onto the group $B$. Let $H_r$ be the space of all infinitely dif-rentiable functions $f(g)$ on $G$ with values in $E$ satisfying the following condition: $f(bg) = r(b)f(g)$ r any $b \in B$. Assign to $H_r$ the representation $T(g)$ of the group $G$ by the formula $(g_0)f)(g) = f(gg_0)$.

The constructed $G$-modules $H_r$ will be called *elementary*. It is not difficult to show that in the ase of the group $SL(2, \mathbf{C})$, they coincide with the $G$-modules $D^m_\pi$ defined above. Hypothesis: The -modules $H_r$ are nondecomposable.

3°. The classification of $G$-modules in the nonsingular case. We shall call a Harish-Chandra $G$-odule nonsingular, if all the factors of its composition series are infinite-dimensional modules. learly, all the $G$-modules $D^m_\pi$, where $\pi$ is a nonsingular point, are nonsingular. It can be shown 3], and also [6]) that these exhaust, up to equivalence, the nonsingular nondecomposable Harish-handra $G$-modules. Note that the two nonsingular $G$-modules $D^{m_1}_{\pi_1}$ and $D^{m_2}_{\pi_2}$ are equivalent if and only $m_1 = m_2$ and either $\pi_2 = \pi_1$ or $\pi_2 = \pi_1^{-1}$.

4°. The elementary operations over $G$-modules. In the singular case, the nondecomposable arish-Chandra $G$-modules will be constructed from the elementary ones by means of the following three lementary operations.

a) Gluing. Let $H_1$, $H_2$ be two $G$-modules, let $H_1' \subset H_1$ and $H_2' \subset H_2$ be isomorphic submodules, nd let $a: H_1 \to H_2$ be the given isomorphism. The $G$-module $H(a) = (H_1 \oplus H_2)/H$, where $H$ is the ubmodule consisting of the pairs $(x, ax)$, $x \in H_1$, is called the gluing of $H_1$ and $H_2$ over the sub-odules $H_1'$ and $H_2'$. We shall refer to this operation as *operation A*.

b) The dual operation. Let $H_1$ and $H_2$ be two $G$-modules, let $H_1' \subset H_1$, $H_2' \subset H_2$ be submodules uch that between the factor modules we have the isomorphism $a: H_1/H_1' \to H_2/H_2'$. The submodule $(a) \subset H_1 \oplus H_2$ consisting of all pairs $(h_1, h_2)$, $h_1 \in H_1$, $h_2 \in H_2$ such that $a\bar{h}_1 = \bar{h}_2$, where $\in H_i/H_i'$ is the image of the element $h_i \in H_i$ under the natural homomorphism, will be called the *luing of* $H_1$ and $H_2$ over the factor modules $H_1/H_1'$ and $H_2/H_2'$. We shall refer to this operation as *peration B*.

Remark. Both operations are invariant with respect to the replacement of $a$ by $\lambda a$, where $\lambda \neq 0$ is n arbitrary number. From this it follows that the gluing over irreducible submodules (or, correspond-ngly, over irreducible factor modules) does not depend on the choice of the isomorphism $a$.

c) Polymerization. Suppose that $H$ is a $G$-module and that $H_1 \neq H_2$ are two isomorphic sub-odules of $H$. Fix the isomorphism $a: H_1 \to H_2$. We shall call the $G$-module $H^{(m)}(\lambda, a) =$ $\overset{m}{\oplus} H)/H^\lambda$, where $H^\lambda$ is a submodule consisting of the elements

$$(ah_1 - \lambda h_1, ah_2 - \lambda h_2 - h_1, \ldots, ah_m - \lambda h_m - h_{m-1}), \tag{2}$$

here $h_1, \ldots, h_m \in H_1$ and $\lambda \neq 0$ is an arbitrary fixed complex number, a polymerization of $m$ copies f the $G$-module $H$ $(m = 1, 2, \ldots)$. It is easy to verify that $H^{(m)}(\lambda_0\lambda, \lambda_0 a) = H^{(m)}(\lambda, a)$ for any $_0 \neq 0$.

5°. The classification of nondecomposable $G$-modules in the singular case. Consider the fixed singular point $\pi = (n_1, n_2)$, where $n_1 > 0$, $n_2 > 0$. We shall enumerate the singular nondecomposable larish-Chandra $G$-modules which have as factors of their composition series $F^0_\pi$ and $F^0_{\pi^{-1}}$.

1) The elementary modules $D_{\pi}^{m}$, $F_{\pi}^{m}$, $D_{\pi-1}^{m}$, $F_{\pi-1}^{m+1}$.*

2) The $G$-modules $D_{\pi}^{m_1,\dots,m_k}$, where $m_0 \geq 0$, $m_i > 0$ $(1 \leq i \leq k)$, $k > 1$. They are constructed from the $k$ elementary modules

$$F_{\pi-1}^{m_1},\ D_{\pi'}^{m_2},\ F_{\pi-1}^{m_3},\ D_{\pi'}^{m_4},\ \dots, \tag{3}$$

where in the odd-numbered places are modules of the $F_{\pi-1}^{m}$ type and in the even-numbered places, modules of the $D_{\pi'}^{m}$ type $(\pi' = (n_1, -n_2))$. This construction is carried out thus. In each module of chain (3), an irreducible submodule and an irreducible factor module (both isomorphic to $F_{\pi-1}^{0}$) are uniquely determined. The first two members of chain (3) are glued together by operation $B$ (see §4, b)) over the irreducible factor modules. We obtain module $D_{\pi}^{m_1,m_2}$. With the irreducible submodule from $D_{\pi'}^{m_2}$ we associate in the natural way the irreducible submodule $H_2' \subset D_{\pi}^{m_1,m_2}$.** We glue together $D_{\pi}^{m_1,m_2}$ and $F_{\pi-1}^{m_3}$ by operation $A$ (see §4, a)) over the irreducible submodules $H_2' \subset D_{\pi}^{m_1,m_2}$ and $F_{\pi-1}^{0} \subset F_{\pi-1}^{m_3}$. We obtain the module $D_{\pi}^{m_1,m_2,m_3}$. We associate the irreducible factor module $D_{\pi}^{m_1;m_2,m_3}/H_3'$ in a natural way with the irreducible factor module $F_{\pi-1}^{m_3}/D_{\pi-1}^{m_3-1}$.*** We glue together $D_{\pi}^{m_1,m_2,m_3}$ and $D_{\pi'}^{m_4}$ by operation $B$ over the irreducible factor modules $D_{\pi}^{m_1,m_2,m_3}/H_3'$ and $D_{\pi'}^{m_4}/D_{\pi'}^{m_4-1}$. We obtain the module $D_{\pi}^{m_1,m_2,m_3,m_4}$, etc.

3) The $G$-modules $D_{\pi,-+}^{m_1,\dots,m_k}$, $D_{\pi,+-}^{m_1,\dots,m_{2p+1}}$, $D_{\pi,--}^{m_1,\dots,m_{2p+1}}$, which essentially do not differ from $D_{\pi}^{m_1,\dots,m_k}$. They are obtained by gluing together chain (3) with one or both of the extreme numbers eliminated, namely, in constructing $D_{\pi,-+}^{m_1,\dots,m_k}$, the first member of the chain, $F_{\pi-1}^{m_1}$, is replaced by the member $F_{\pi-1}^{m_1}/F_{\pi-1}^{0} \cong D_{\pi}^{m_1-1}$, in constructing $D_{\pi,+-}^{m_1,\dots,m_{2p+1}}$, the last member, $F_{\pi-1}^{m_{2p+1}}$ is replaced by the member $D_{\pi-1}^{m_{2p+1}-1} \subset F_{\pi-1}^{m_{2p+1}}$, and, finally, in constructing $D_{\pi,--}^{m_1,\dots,m_{2p+1}}$, both ends of chain (3) are shortened.

4) The $G$-modules $D_{\pi}^{m_1,\dots,m_{2s};m,\lambda}$ (which are given by the collection of $2s$ $(s > 0)$ integers $m_i > 0$,**** another integer $m > 0$, and a complex number $\lambda \neq 0$). These are obtained by a polymerization of $m$ copies of the modules $D_{\pi}^{m_1,\dots,m_{2s}}$, namely, in the $G$-modules $F_{\pi-1}^{m_1}$ and $D_{\pi'}^{m_{2s}}$, which are the extreme members of chain (3), irreducible submodules are fixed (isomorphic to $F_{\pi-1}^{0}$). Let $H_1$, $H_2$ be the images of these submodules in $D_{\pi}^{m_1,\dots,m_{2s}}$ after the gluing of chain (3). According to [3], the isomorphism $a: H_1 \to H_2$ may be defined in the canonical form. We assume that $D_{\pi}^{m_1,\dots,m_{2s};m,\lambda} = (\overset{m}{\oplus} D_{\pi}^{m_1,\dots,m_{2s}})/H^{\lambda}$, where $H^{\lambda}$ is a submodule consisting of elements of the form (2).

On the sequences $(m_1, \dots, m_{2s})$ considered in this construction, an additional condition of nonperiodicity is placed: there does not exist a divisor $r$ of the number $s$ $(r < s)$ such that $m_p = m_q$ for any $p$ and $q$ which are congruent modulo $2r$.

---

* The module $F_{\pi-1}^{0} \cong D_{\pi'}^{0}$ is excluded, since according to our classification this module is nonsingular. For the same reason, in the second group the modules $D_{\pi}^{0,m_2} \cong D_{\pi'}^{m_2}$ are excluded.

** $H_2'$ consists of all the pairs $(0, h) \in D_{\pi}^{m_1,m_2}$, where $h$ runs through the irreducible submodule of $D_{\pi'}^{m_2}$.

*** $H_3'$ is the image of $D_{\pi}^{m_1,m_2} \oplus D_{\pi-1}^{m_3-1}$ $(F_{\pi-1}^{m_3}/D_{\pi-1}^{m_3-1}$ is an irreducible factor module) under the natural mapping $D_{\pi}^{m_1,m_2} \oplus F^{m_3} \to D_{\pi}^{m_1,m_2,m_3}$.

**** For the case in which $m_1 = 0$, the gluing of $F_{\pi-1}^{0} \cong D_{\pi'}^{0}$ with $D_{\pi'}^{m_2}$ gives $D_{\pi'}^{m_2}$. Therefore, for $m_1 = 0$, we can assume that chain (3) starts at once with $D_{\pi'}^{m_2}$.

**Theorem.** 1) *The modules constructed above are nondecomposable.* 2) *They are mutually non-uivalent, with the exception of the case in which both modules belong to the fourth class. Two odules of the fourth class* $D_{\pi_1}^{m_1,\cdots,m_{2s};m,\lambda}$ *and* $D_{\pi_2}^{m'_1,\cdots,m'_{2s'};m',\lambda'}$ *are equivalent if and only if* $= \pi_2$, $\lambda = \lambda'$, $m = m'$, $s = s'$, *and* $(m'_1, \cdots, m'_{2s})$ *is obtained from* $(m_1, \cdots, m_{2s})$ *by a cyclic rmutation on an even number of numbers.* 3) *The constructed G-modules exhaust, up to equivalence, l the singular nondecomposable Harish-Chandra G-modules.*

Institute of Applied Mathematics

Academy of Sciences of the USSR

Received 1/JUNE/70

## BIBLIOGRAPHY

[1] I. M. Gel'fand and V. A. Ponomarev, *The category of Harish-Chandra modules over the Lie algebra of the Lorentz group*, Dokl. Akad. Nauk SSSR 176 (1967), 243–246 = Soviet Math. Dokl. 8 (1967), 1065–1068. MR 36 #6552.

[2] ———, *Classification of indecomposable infinitesimal representations of the Lorentz group*, Dokl. Akad. Nauk SSSR 176 (1967), 502–505 = Soviet Math. Dokl. 8 (1967), 1114–1117. MR 36 #2739.

[3] ———, *Indecomposable representations of the Lorentz group*, Uspehi Mat. Nauk 23 (1968), no. 2 (140), 3–60. (Russian) MR 37 #5325.

[4] ———, Funkcional. Anal. i Priložen. 3 (1969), no. 4, 81. (Russian)

[5] I. M. Gel'fand, M. I. Graev and N. Ja. Vilenkin, *Generalized functions*. Vol. 5: *Integral geometry and representation theory*, Fizmatgiz, Moscow, 1962; English transl., Academic Press, New York, 1966. MR 28 #3324; 34 #7726.

[6] D. P. Želobenko, *Linear representations of the Lorentz group*, Dokl. Akad. Nauk SSSR 126 (1959), 935–938. (Russian) MR 22 #906.

Translated by:
Zdanna K. Skalsky

# 4.

## (with V. A. Ponomarev)

# Quadruples of subspaces of a finite-dimensional vector space

Dokl. Akad. Nauk SSSR **197** (1971) 762–765 [Sov. Math., Dokl. **12** (1971) 535–539].
MR **44**:2762. Zbl. **294**:15001

The present article deals with the problem of classifying quadruples of subspaces of a finite-dimensional vector space. This problem naturally includes as special cases many problems of linear algebra, among which are the determination of the canonical form of a linear transformation, the determination of the canonical form of an additive relation, the Kronecker indices and several others.

1. Let $P$ be a finite-dimensional vector space over an algebraically closed field $K$ of characteristic 0. Further, let $\{E_t\}$, $t = 1, 2, \cdots, r$, be some set of subspaces of $P$. Such a system is denoted by $S = \{P, E_t\}$, $t = 1, 2, \cdots, r$. By a *morphism* from $S$ to $S'$, we shall mean a linear transformation $a$: $P \to P'$ whose restriction $a_t$ to $E_t$ maps $E_t$ into $E'_t$: $a_t = a|E_t: E_t \to E'_t$. We will say that $S$ is isomorphic to $S'$, $S \sim S'$, if there exists such a transformation $a$ with $a$ and $a_t$ invertible.

The problem of classifying the systems $\{P, E_1, \cdots, E_r\}$ is trivial in the case $r = 2$ and is easily solved for $r = 3$. The object of this article is the classification of the systems $\{P, E_1, \cdots, E_4\}$, which we will call quadruples. The classification of quadruples is apparently the most important.

We mention several examples of quadruples.

1) Let $A$: $E \to E$ be a given linear transformation of a finite-dimensional vector space $E$. We associate with the pair $E$, $A$ a system $S = \{P, E_1, \cdots, E_4\}$ by setting $P = E \oplus E$, $E_1 = E \oplus \{0\}$ (that is, the subspace $E_1$ consists of all pairs $(x, 0)$, $x \in E$), $E_2 = \{0\} \oplus E$, $E_3$ is the graph of the transformation $A$, that is, $E_3$ is the set of pairs $(x, Ax)$: $E_3 = \{(x, Ax) \mid x \in E\}$ and $E_4$ is the diagonal subspace of $E \oplus E$, that is, $E_4$ is the set of all pairs $(x, x)$, $x \in E$.

2) To a given pair of spaces $E$ and $F$ and two linear transformations $A$: $E \to F$ and $B$: $F \to E$ (see [3]) there corresponds a system $S$:

$$P = E \oplus F; \quad E_1 = E \oplus \{0\}; \quad E_2 = \{0\} \oplus F;$$
$$E_3 = \{(x, Ax) \mid x \in E\}; \quad E_4 = \{(By, y) \mid y \in F\}.$$

3) To a given pair of spaces $E$ and $F$ and two linear transformations $A_1$: $E \to F$ and $A_2$: $E \to F$ [1] there corresponds a system $S$: $P = E \oplus F$; $E_1 = E \oplus \{0\}$; $E_2 = \{0\} \oplus F$; $E_3$ is the graph of $A_1$; $E_4$ is the graph of $A_2$.

4) To a given additive relation* $A$ on the space $E$, that is, any subspace $A$ of

* The concept of an additive relation is introduced by MacLane in [2].

the direct sum $E \oplus E$, there corresponds a system $S$:

$$P = E \oplus E; \quad E_1 = E \oplus \{0\}; \quad E_2 = \{0\} \oplus E; \quad E_3 = A,$$
$$E_4 \text{ is the diagonal subspace of } E \oplus E.$$

5) To a given pair of spaces $E$ and $F$ with two additive relations $A_1 \subset E \oplus F$ and $A_2 \subset E \oplus F$ there corresponds a quadruple $S$:

$$P = E \oplus F; \quad E_1 = E \oplus \{0\}; \quad E_2 = \{0\} \oplus F; \quad E_3 = A_1; \quad E_4 = A_2.$$

Example 5) contains all the previous examples as special cases. It will be shown that among the systems $S = \{P, E_1, \cdots, E_4\}$ there exist very interesting quadruples which do not reduce to case 5). Thus, this paper is a generalization of the problem of classifying pairs of additive relations $A_1 \subset E \oplus F$ and $A_2 \subset E \oplus F$, which had already been solved by V. A. Ponomarev.

It is easy to see that any quadruple $S$ decomposes into a direct sum of indecomposables. We will give below a canonical description of all indecomposable quadruples. By the same token, the classification problem for quadruples will be solved.

2. **Description of the indecomposable quadruples.** With any quadruple $S = \{P, E_1, \cdots, E_4\}$ we can associate two numbers $n(S) = \dim P$ and $\rho(S) = \sum_{i=1}^{4} \dim E_i - 2 \dim P$. $\rho(S)$ will be called the defect of the quadruple. It turns out that if $S$ is an indecomposable quadruple, then its defect is equal to one of the five numbers: $-2, -1, 0, 1, 2$. It will be seen from our description that quadruples of defect 0 correspond precisely to those of examples 2) and 1), while quadruples of defect $-1, 0, 1$ correspond precisely to those of example 5). Indecomposable quadruples of defect $-2$ or 2 cannot be described in terms of relations.

We will first give a canonical form for indecomposable quadruples for which $n(S) = \dim P$ is an even number: $n(S) = 2k$. We choose a basis for $P$ which de denote by $e_1, \cdots, e_k, f_1, \cdots, f_k$.

We now describe the systems $S_i(2k, \rho)$, $S_{i,j}(2k, 0)$, $S(2k, 0; \lambda)$, where $i, j = 1, 2, 3, 4$; $\rho = -1, 1$; $\lambda \in K$. We first define $S_3(2k, 1)$, $S_{1,3}(2k, 0)$ and $S(2k, 0; \lambda)$, recalling that $P = \{e_1, e_2, \cdots, e_k, f_1, \cdots, f_k\}$:

$S_3(2k, -1)$:

$$E_1 = \{e_1, \ldots, e_k\},$$
$$E_2 = \{f_1, \ldots, f_k\},$$
$$E_3 = \{(e_2 + f_1), (e_3 + f_2), \ldots, (e_k + f_{k-1})\},$$
$$E_4 = \{(e_1 + f_1), (e_2 + f_2), \ldots, (e_k + f_k)\}.$$

$S_3(2k, 1)$:

$$E_1 = \{e_1, \ldots, e_k\},$$
$$E_2 = \{f_1, \ldots, f_k\},$$
$$E_3 = \{e_1, (e_2 + f_1), \ldots, (e_k + f_{k-1}), f_k\},$$
$$E_4 = \{(e_1 + f_1), (e_2 + f_2), \ldots, (e_k + f_k)\}.$$

$S_{1,3}(2k, 0)$:

$$E_1 = \{e_1, \ldots, e_k\},$$
$$E_2 = \{f_1, \ldots, f_k\},$$
$$E_3 = \{e_1, (e_2 + f_1), \ldots, (e_k + f_{k-1})\},$$
$$E_4 = \{(e_1 + f_1), (e_2 + f_2), \ldots, (e_k + f_k)\}.$$

$S(2k, 0; \lambda)$:

$$E_1 = \{e_1, \ldots, e_k\},$$
$$E_2 = \{f_1, \ldots, f_k\},$$
$$E_3 = \{(e_1 + \lambda f_1), (e_2 + f_1 + \lambda f_2), \ldots, (e_k + f_{k-1} + \lambda f_k)\},$$
$$E_4 = \{(e_1 + f_1), (e_2 + f_2), \ldots, (e_k + f_k)\}.$$

In the expression $S(n, \rho)$, $n$ stands for the dimension of $P$ and $\rho$ for the defect of $S$; $\{\xi_1, \xi_2, \cdots, \xi_k\}$ denotes the subspace spanned by the vectors $\xi_1, \cdots, \xi_k$.

All the remaining systems $S_i$ and $S_{i,j}$ can be obtained from those just described by permuting the subscripts of the subspaces $E_t$. Let $\sigma$ be some permutation of the subscripts 1, 2, 3, 4. We will denote by $S' = \sigma S$ the system $S = \{P, E_1', \cdots, E_4'\}$ which is obtained from $S = \{P, E_1, \cdots, E_4\}$ by changing the subscripts of the subspaces $E_t$: $E_t' = E_{\sigma^{-1}(t)}$.

We put $S_i(2k, \rho) = \sigma_{3,i}(2k, \rho)$, $\rho = -1, 1$ and further, $S_{i,j}(2k, 0) = \sigma_{1,i}\sigma_{3,j}S_{1,3}(2k, 0)$ where $i, j = 1, 2, 3, 4$. Here, $\sigma_{i,j}$ is the permutation which transposes $i$ and $j$ and leaves the other subscripts fixed.

It can be shown that

1°. $\sigma S_i(2k, \rho) \sim S_{\sigma(i)}(2k, \rho), \quad \rho = -1, 1,$

2°. $\sigma S_{i,j}(2k, 0) \sim S_{\sigma(i), \sigma(j)}(2k, 0),$

3°. $S_{ij}(2k, 0) \sim S_{j,i}(2k, 0),$

4°. $S(2k, 0; 0) \sim S_{1,3}(2k, 0),$

5°. $S(2k, 0; 1) \sim S_{3,4}(2k, 0),$

6°. $\sigma S(2k, 0; \lambda) \sim S(2k, 0; \sigma(\lambda)),$

where $\sigma(\lambda) \in \{\lambda, 1-\lambda, 1/(1-\lambda), (\lambda-1)/\lambda, 1/\lambda\}$ (the number $\lambda$ varies under the action of $\sigma$ in accordance with the same rule as the cross-ratio of four points on the projective line under the permutation $\sigma$ of the points).

Proposition 1. *If the system* $S = \{P, E_1, \cdots, E_4\}$ *is indecomposable and* $\dim P = 2k$, *then* $S$ *is isomorphic to one of the systems*:

$$S_i(2k, -1), \quad S_i(2k, 1), \quad S_{i,j}(2k, 0) \qquad (i < j,\ i, j = 1, 2, 3, 4),$$
$$S(2k, 0; \lambda), \quad \lambda \neq 0, \quad \lambda \neq 1.$$

We now give a canonical description of the indecomposable quadruples for which $n(S) = \dim P$ is an odd number, $n(S) = 2k-1$. We describe systems $S_i(2k-1, \rho)$, $\rho = -1, 1$. We choose a basis $e_1, \cdots, e_k, f_1, \cdots, f_{k-1}$ for $P$, and describe first the systems $S_1(2k-1, -1)$, $S_2(2k-1, 1)$, $S_{1,3}(2k-1, 0)$ $S(2k-1, -2)$, $S(2k-1, 2)$.

$S_1(2k-1, -1)$:

$$E_1 = \{e_1, \ldots, e_k\},$$
$$E_2 = \{f_1, \ldots, f_{k-1}\},$$
$$E_3 = \{(e_2 + f_1), (e_3 + f_2), \ldots, (e_k + f_{k-1})\},$$
$$E_4 = \{(e_1 + f_1), \ldots, (e_{k-1} + f_{k-1})\}.$$

$S_2(2k-1, 1)$:

$$E_1 = \{e_1, \ldots, e_k\},$$
$$E_2 = \{f_1, \ldots, f_{k-1}\},$$
$$E_3 = \{e_1, (e_2 + f_1), (e_3 + f_2), \ldots, (e_k + f_{k-1})\},$$
$$E_4 = \{(e_1 + f_1), (e_2 + f_2), \ldots, (e_{k-1} + f_{k-1}), e_k\}.$$

$S_{1,3}(2k-1, 0)$:

$$E_1 = \{e_1, \ldots, e_k\},$$
$$E_2 = \{f_1, \ldots, f_{k-1}\},$$
$$E_3 = \{e_1, (e_2 + f_1), (e_3 + f_2), \ldots, (e_k + f_{k-1})\},$$
$$E_4 = \{(e_1 + f_1), (e_2 + f_2), \ldots, (e_{k-1} + f_{k-1})\}.$$

$S(2k-1,-2)$:

$$E_1=\{e_1,\ldots,e_{k-1}\},\quad E_2=\{f_1,\ldots,f_{k-1}\},\quad E_3=\{(e_2+f_1),(e_3+f_2),\ldots,(e_k+f_{k-1})\},$$
$$E_4=\{(e_1+f_2),(e_2+f_3),\ldots,(e_{k-2}+f_{k-1}),(e_{k-1}+e_k)\}.$$

$S(2k-1,2)$:

$$E_1=\{e_1,\ldots,e_k\},\quad E_2=\{f_1,\ldots,f_{k-1},e_k\},\quad E_3=\{e_1,(e_2+f_1),\ldots,(e_k+f_{k-1})\},$$
$$E_4=\{f_1,(e_1+f_2),(e_2+f_3),\ldots,(e_{k-2}+f_{k-1}),(e_{k-1}+e_k)\}.$$

We put $S_i(2k-1,-1)=\sigma_{1,i}S_1(2k-1,-1)$; $S_i(2k-1,1)=\sigma_{2,i}S_2(2k-1,1)$; $S_{i,j}(2k-1,0)=\sigma_{1,i}\sigma_{3,j}S_{1,3}(2k-1,0)$; $i,j=1,2,3,4$. It can be shown that

1°. $\sigma S_i(2k-1,\rho)\sim S_{\sigma(i)}(2k-1,\rho)$, $\rho=-1,1$;

2°. $\sigma S_{i,j}(2k-1,0)\sim S_{\sigma(i),\sigma(j)}(2k-1,0)$;

3°. $S_{i,j}(2k-1,0)\sim S_{j,i}(2k-1,0)$;

4°. $\sigma S(2k-1,\rho)\sim S(2k-1,\rho)$, $\rho=-2,2$.

Proposition 2. *If the system $S=\{P,E_1,\cdots,E_4\}$ is indecomposable and $\dim P=2k-1$, then $S$ is isomorphic to one of the systems:*

$$S_i(2k-1,-1),\quad S_i(2k-1,1),\quad S_{ij}(2k-1,0)\quad (i<j;\ i,j=1,2,3,4),$$
$$S(2k-1,-2),\quad S(2k-1,2).$$

We note that the systems $S_i(k,\rho)$, $S_{i,j}(k,0)$, $S(2k,0;\lambda)$, $S(2k-1,-2)$, $S(2k-1,2)$ $(\rho=\pm1,\ \lambda=0,\ \lambda\neq1,\ \lambda\in K,\ k=1,2,3,\cdots)$ are pairwise nonisomorphic.

3. Theorem 1. *Let $S=\{P;E_1,E_2,E_3,E_4\}$ be a system consisting of a finite-dimensional vector space $P$ over an algebraically closed field $K$ and of a quadruple of subspaces $E_i\subset P$. Then $S$ decomposes into a direct sum of subsystems of the form:*

$$S(2k-1,-2),\quad S(2k-1,2),\quad S_i(k,-1),\quad S_i(k,1),\quad S_{ij}(k,0)\quad (i<j)$$
$$S(2k,0;\lambda),\quad \lambda\in K\ (\lambda\neq0,\lambda\neq1,i,j=1,2,3,4;\ k=1,2,3,\ldots).$$

4. We describe the decomposition of the systems in examples 1)–5) of § 1. The enumeration of the subscripts of the subspaces $E_t$ is chosen in the same way as in the corresponding examples of §1.

1) Suppose the quadruple $S$ is constructed from the pair $E$, $A$, where $A$ is a linear transformation $A\colon E\to E$. Then this quadruple decomposes into a direct sum of indecomposable quadruples of the types

$$S_{1,3}(2k,0),\quad S_{3,4}(2k,0),\ S(2k,0;\lambda),\quad \lambda\in K,\quad \lambda\neq0,\quad \lambda\neq1.$$

2) Let the quadruple $S$ be constructed from the two spaces $E$ and $F$ and the linear transformations $A\colon E\to F$ and $B\colon F\to E$ (see 2), § 1). Then $S$ decomposes into a direct sum of quadruples of the types

$$S_{1,3}(k,0),\quad S_{2,4}(k,0),\quad S_{3,4}(2k,0),\quad S(2k,0;\lambda),\quad \lambda\neq0,\quad \lambda\neq1.$$

3) Suppose the quadruple $S$ is constructed as in 3), § 1, from two spaces $E$ and $F$ and two transformations $A_1\colon E\to F$ and $A_2\colon E\to F$. Then $S$ decomposes into a direct sum of quadruples of the types

$$S_{1,3}(2k,0),\quad S_{1,4}(2k,0),\quad S_{3,4}(2k,0),\quad S(2k,0;\lambda),\quad \lambda\neq 0,\quad \lambda\neq 1,$$
$$S_2(2k-1,-1),\quad S_2(2k-1,1).$$

4) Suppose the quadruple $S$ is constructed from the space $E$ and the additive relation $A\subset E\oplus E$. Then $S$ decomposes into a direct sum of quadruples of the types

$$S_{1,3}(2k,0),\quad S_{2,3}(2k,0),\quad S_{3,4}(2k,0),\quad S(2k,0;\lambda),\quad \lambda\neq 0,\quad \lambda\neq 1,$$
$$S_3(2k,-1),\quad S_3(2k,1).$$

5) Let the quadruple $S$ be constructed from two spaces $E$ and $F$ and two relations $A_1\subset E\oplus F$ and $A_2\subset E\oplus F$. Then $S$ decomposes into a direct sum of quadruples of the types.

$$S_{1,3}(k,0),\quad S_{1,4}(k,0),\quad S_{2,3}(k,0),\quad S_{2,4}(k,0),\quad S_{3,4}(k,0),$$
$$S(2k,0;\lambda),\quad \lambda\neq 0,\quad \lambda\neq 1,\quad \lambda\in K,\quad S_3(2k,\rho),\quad S_4(2k,\rho),$$
$$S_1(2k-1,\rho),\quad S_2(2k-1,\rho),\quad \rho=\pm 1.$$

We see that the systems $S(2k-1,-2)$ and $S(2k-1,2)$ of defects $\pm 2$ do not occur among these examples. These quadruples are of great interest.

Institute of Applied Mathematics
Academy of Sciences of the USSR

Received 21/OCT/70

## BIBLIOGRAPHY

[1] L. Kronecker, S.-B. Preuss Akad. Berlin 1890, 763.

[2] S. MacLane, Proc. Nat. Acad. Sci. U.S.A. 47 (1961), 1043; Russian transl., Matematika 7 (1963), no. 6, 3. MR 23 #A3773.

[3] N. M. Dobrovol'skaja and V. A. Ponomarev, Uspehi Mat. Nauk 20 (1965), no. 6,(126), 81. MR 36 #2631.

Translated by:
D. L. Johnson

# 5.

## (with I. N. Bernstein and V. A. Ponomarev)

## Coxeter functors and Gabriel's theorem

Usp. Mat. Nauk **28** (2) (1973) 19–33
[Russ. Math. Surv. **28** (2) (1973) 17–32]. Zbl. **269**:08001

It has recently become clear that a whole range of problems of linear algebra can be formulated in a uniform way, and in this common formulation there arise general effective methods of investigating such problems. It is interesting that these methods turn out to be connected with such ideas as the Coxeter–Weyl group and the Dynkin diagrams.

We explain these connections by means of a very simple problem. We assume no preliminary knowledge. We do not touch on the connections between these questions and the theory of group representations or the theory of infinite-dimensional Lie algebras. For this see [3]–[5].

Let $\Gamma$ be a finite connected graph; we denote the set of its vertices by $\Gamma_0$ and the set of its edges by $\Gamma_1$ (we do not exclude the cases where two vertices are joined by several edges or there are loops joining a vertex to itself). We fix a certain orientation $\Lambda$ of the graph $\Gamma$; this means that for each edge $l \in \Gamma_1$ we distinguish a starting-point $\alpha(l) \in \Gamma_0$ and an end-point $\beta(l) \in \Gamma_0$.

With each vertex $\alpha \in \Gamma_0$ we associate a finite-dimensional linear space $V_\alpha$ over a fixed field $K$. Furthermore, with each edge $l \in \Gamma_1$ we associate a linear mapping $f_l$: $V_{\alpha(l)} \to V_{\beta(l)}$ ($\alpha(l)$ and $\beta(l)$ are the starting-point and end-point of the edge $l$). We impose no relations on the linear mappings $f_l$. We denote the collection of spaces $V_\alpha$ and mappings $f_l$ by $(V, f)$.

DEFINITION 1. Let $(\Gamma, \Lambda)$ be an oriented graph. We define a *category* $\mathcal{L}(\Gamma, \Lambda)$ in the following way. An *object* of $\mathcal{L}(\Gamma, \Lambda)$ is any collection $(V, f)$ of spaces $V_\alpha$ $(\alpha \in \Gamma_0)$ and mappings $f_l$ $(l \in \Gamma_1)$. A morphism $\varphi$: $(V, f) \to (W, g)$ is a collection of linear mappings $\varphi_\alpha$: $V_\alpha \to W_\alpha$ $(\alpha \in \Gamma_0)$ such that for any edge $l \in \Gamma_1$ the following diagram

$$\begin{array}{ccc} V_{\alpha(l)} & \xrightarrow{f_l} & V_{\beta(l)} \\ \downarrow{\scriptstyle \varphi_{\alpha(l)}} & & \downarrow{\scriptstyle \varphi_{\beta(l)}} \\ W_{\alpha(l)} & \xrightarrow[g_l]{} & W_{\beta(l)} \end{array}$$

is commutative, that is, $\varphi_{\beta(l)} f_l = g_l \varphi_{\alpha(l)}$.

Many problems of linear algebra can be formulated in these terms. For example, the question of the canonical form of a linear transformation $f\colon V \to V$ is connected with the diagram

The classification of a pair of linear mappings $f_1 : V_1 \to V_2$ and $f_2 : V_1 \to V_2$ leads to the graph

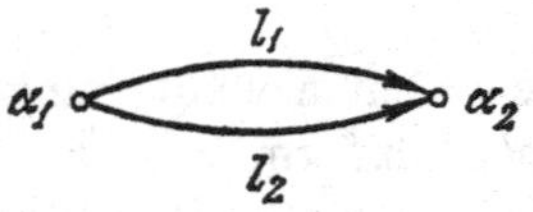


A very interesting problem is that of the classification of quadruples of subspaces in a linear space, which corresponds to the graph

This last problem contains several problems of linear algebra.[1]

Let $(\Gamma, \Lambda)$ be an oriented graph. The direct sum of the objects $(V, f)$ and $(U, g)$ in $\mathscr{L}(\Gamma, \Lambda)$ is the object $(W, h)$, where $W_\alpha = V_\alpha \oplus U_\alpha$, $h_l = f_l \oplus g_l$ $(\alpha \in \Gamma_0, l \in \Gamma_1)$.

We call a non-zero object $(V, f) \in \mathscr{L}(\Gamma, \Lambda)$ indecomposable if it cannot be represented as the direct sum of two non-zero objects. The simplest indecomposable objects are the irreducible objects $L_\alpha$ $(\alpha \in \Gamma_0)$, whose structure is as follows: $(L_\alpha)_\gamma = 0$ for $\gamma \neq \alpha$, $(L_\alpha)_\alpha = K$, $f_l = 0$ for all $l \in \Gamma_1$.

It is clear that each object $(V, f)$ of $\mathscr{L}(\Gamma, \Lambda)$ is isomorphic to the direct sum of finitely many indecomposable objects.[2]

In many cases indecomposable objects can be classified.[3]

In his article [1] Gabriel raised and solved the following problem: to find all graphs $(\Gamma, \Lambda)$ for which there exist only finitely many non-isomorphic indecomposable objects $(V, f) \in \mathscr{L}(\Gamma, \Lambda)$. He made the following

---

[1] Let us explain how the problem of the canonical form of a linear operator $f\colon V \to V$ reduces to that of a quadruple of subspaces. Consider the space $W = V \oplus V$ and in it the graph of $f$, that is, the subspace $E_4$ of pairs $(\xi, f\xi)$, where $\xi \in V$. The mapping $f$ is described by a quadruple of subspaces in $W$, namely $E_1 = V \oplus 0$, $E_2 = 0 \oplus V$, $E_3 = \{(\xi, \xi) \mid \xi \in V\}$ ($E_3$ is the diagonal) and $E_4 = \{(\xi, f\xi) \mid \xi \in V\}$ – the graph of $f$. Two mappings $f$ and $f'$ are equivalent if and only if the quadruples corresponding to them are isomorphic. In fact, $E_1$ and $E_2$ define "coordinate planes" in $W$, $E_3$ establishes an identification between them, and then $E_4$ gives the mapping.

[2] It can be shown that such a decomposition is unique to within isomorphism (see [6], Chap. II, 14, the Krull–Schmidt theorem).

[3] We believe that a study of cases in which an explicit classification is impossible is by no means without interest. However, we should find it difficult to formulate precisely what is meant in this case by a "study" of objects to within isomorphism. Suggestions that are natural at first sight (to consider the subdivision of the space of objects into trajectories, to investigate versal families, to distinguish "stable" objects, and so on) are not, in our view, at all definitive.

surprising observation. For the existence of finitely many indecomposable objects in $\mathcal{L}(\Gamma, \Lambda)$ it is necessary and sufficient that $\Gamma$ should be one of the following graphs:

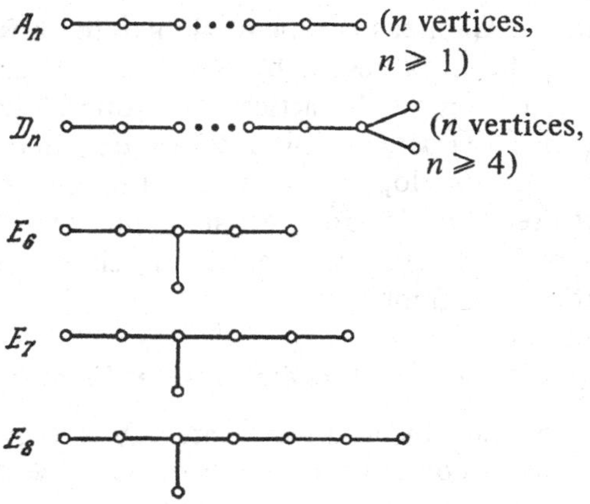


(this fact does not depend on the orientation $\Lambda$).The surprising fact here is that these graphs coincide exactly with the Dynkin diagrams for the simple Lie groups.[1]

However, this is not all. As Gabriel established, the indecomposable objects of $\mathcal{L}(\Gamma, \Lambda)$ correspond naturally to the positive roots, constructed according to the Dynkin diagram of $\Gamma$.

In this paper we try to remove to some extent the "mystique" of this correspondence. Whereas in Gabriel's article the connection with the Dynkin diagrams and the roots is established a posteriori, we give a proof of Gabriel's theorem based on exploiting the technique of roots and the Weyl group. We do not assume the reader to be familiar with these ideas, and we give a complete account of the necessary facts.

An essential role is played in our proof by the functors defined below, which we call Coxeter functors (the name arises from the connection of these functors with the Coxeter transformations in the Weyl group). For the particular case of a quadruple of subspaces these functors were introduced in [2] (where they were denoted by $\Phi^+$ and $\Phi^-$). Essentially, our paper is a synthesis of Gabriel's idea on the connection between the categories of diagrams $\mathcal{L}(\Gamma, \Lambda)$ with the Dynkin diagrams and the ideas of the first part of [2], where with the help of the functors $\Phi^+$ and $\Phi^-$ the "simple" indecomposable objects are separated from the more "complicated" ones.

---

[1] More precisely, Dynkin diagrams with single arrows.

We hope that this technique is useful not only for the solution of Gabriel's problem and the classification of quadruples of subspaces, but also for the solution of many other problems (possibly, not only problems of linear algebra).

Some arguments on Gabriel's problem, similar to those used in this article, have recently been expressed by Roiter. We should also like to draw the reader's attention to the articles of Roiter, Nazarova, Kleiner, Drozd and others (see [3] and the literature cited there), in which very effective algorithms are developed for the solution of problems in linear algebra. In [3], Roiter and Nazarova consider the problem of classifying representations of ordered sets; their results are similar to those of Gabriel on the representations of graphs.

## § 1. Image functors and Coxeter functors

To study indecomposable objects in the category $\mathscr{L}(\Gamma, \Lambda)$ we consider "image functors", which construct for each object $V \in \mathscr{L}(\Gamma, \Lambda)$ some new object (in another category); here an indecomposable object goes either into an indecomposable object or into the zero object. We construct such a functor for each vertex $\alpha$ at which all the edges have the same direction (that is, they all go in or all go out). Furthermore, we construct the "Coxeter functors" $\Phi^+$ and $\Phi^-$, which take the category $\mathscr{L}(\Gamma, \Lambda)$ into itself.

For each vertex $\alpha \in \Gamma_0$ we denote by $\Gamma^\alpha$ the set of edges containing $\alpha$. If $\Lambda$ is some orientation of the graph $\Gamma$, we denote by $\sigma_\alpha \Lambda$ the orientation obtained from $\Lambda$ by changing the directions of all edges $l \in \Gamma^\alpha$.

We say that a vertex $\alpha$ is $(-)$-accessible (with respect to the orientation $\Lambda$) if $\beta(l) \neq \alpha$ for all $l \in \Gamma_1$ (this means that all the edges containing $\alpha$ start there and that there are no loops in $\Gamma$ with vertex at $\alpha$). Similarly we say that the vertex $\beta$ is $(+)$-accessible if $\alpha(l) \neq \beta$, for all $l \in \Gamma_1$.

DEFINITION 1.1 1) Suppose that the vertex $\beta$ of the graph $\Gamma$ is $(+)$-accessible with respect to the orientation $\Lambda$. From an object $(V, f)$ in $\mathscr{L}(\Gamma, \Lambda)$ we construct a new object $(W, g)$ in $\mathscr{L}(\Gamma, \sigma_\beta \Lambda)$.

Namely, we put $W_\gamma = V_\gamma$ for $\gamma \neq \beta$.

Next we consider all the edges $l_1, l_2, \ldots, l_k$ that end at $\beta$ (that is, all edges of $\Gamma^\beta$). We denote by $W_\beta$ the subspace in the direct sum $\bigoplus_{i=1}^{k} V_{\alpha(l_i)}$ consisting of the vectors $v = (v_1, \ldots, v_k)$ (here $v_i \in V_{\alpha(l_i)}$) for which $f_{l_i}(v_1) + \ldots + f_{l_k}(v_k) = 0$. In other words, if we denote by $h$ the mapping $h\colon \bigoplus_{i=1}^{k} V_{\alpha(l_i)} \to V_\beta$ defined by the formula

$h(v_1, v_2, \ldots, v_k) = f_{l_1}(v_1) + \ldots + f_{l_k}(v_k)$, then $W_\beta = \operatorname{Ker} h$.

We now define the mappings $g_l$. For $l \notin \Gamma^\beta$ we put $g_l = f_l$. If $l = l_j \in \Gamma^\beta$, then $g_l$ is defined as the composition of the natural embedding of $W_\beta$ in $\bigoplus V_{\alpha(l_i)}$ and the projection of this sum onto the term $V_{\alpha(l_j)} = W_{\alpha(l_j)}$. We note that on all edges $l \in \Gamma^\beta$ the orientation has been changed, that is, the resulting object $(W, g)$ belongs to $\mathcal{L}(\Gamma, \sigma_\beta\Lambda)$. We denote the object $(W, g)$ so constructed by $F_\beta^+(V, f)$.

2) Suppose that the vertex $\alpha \in \Gamma_0$ is $(-)$-accessible with respect to the orientation $\Lambda$. From the object $(V, f) \in \mathcal{L}(\Gamma, \Lambda)$ we construct a new object $F_\alpha^-(V, f) = (W, g) \in \mathcal{L}(\Gamma, \sigma_\alpha\Lambda)$. Namely, we put

$$W_\gamma = V_\gamma \text{ for } \gamma \neq \alpha$$
$$g_l = f_l \text{ for } l \notin \Gamma^\alpha$$

$W_\alpha = \bigoplus_{i=1}^{k} V_{\beta(l_i)}/\operatorname{Im} \widetilde{h}$, where $\{l_1, \ldots, l_k\} = \Gamma^\alpha$, and the mapping

$\widetilde{h}: V_\alpha \to \bigoplus_{i=1}^{k} V_{\beta(l_i)}$ is defined by the formula $\widetilde{h}(v) = (f_{l_1}(v), \ldots, f_{l_k}(v))$.

If $l \in \Gamma^\alpha$, then the mapping $g_l\colon W_{\beta(l)} \to W_\alpha$ is defined as the composition of the natural embedding of $W_{\beta(l)} = V_{\beta(l)}$ in $\bigoplus_{i=1}^{k} V_{\beta(l_i)}$ and the projection of this direct sum onto $W_\alpha$.

It is easy to verify that $F_\beta^+$ (and similarly $F_\alpha^-$) is a functor from $\mathcal{L}(\Gamma, \Lambda)$ into $\mathcal{L}(\Gamma, \sigma_\beta\Lambda)$ (or $\mathcal{L}(\Gamma, \sigma_\alpha\Lambda)$, respectively). The following property of these functors is basic for us.

THEOREM 1.1 1) *Let* $(\Gamma, \Lambda)$ *be an oriented graph and let* $\beta \in \Gamma_0$ *be a vertex that is* (+)*-accessible with respect to* $\Lambda$. *Let* $V \in \mathcal{L}(\Gamma, \Lambda)$ *be an indecomposable object. Then two cases are possible:*

a) $V \approx L_\beta$ *and* $F_\beta^+ V = 0$ (*we recall that* $L_\beta$ *is an irreducible object, defined by the condition* $(L_\beta)_\gamma = 0$ *for* $\gamma \neq \beta$, $(L_\beta)_\beta = K$, $f_l = 0$ *for all* $l \in \Gamma_1$).

b) $F_\beta^+(V)$ *is an indecomposable object,* $F_\beta^- F_\beta^+(V) = V$, *and the dimensions of the spaces* $F_\beta^+(V)_\gamma$ *can be calculated by the formula*

$$\dim F_\beta^+(V)_\gamma = \dim V_\gamma \text{ for } \gamma \neq \beta, \qquad (1.1.1)$$
$$\dim F_\beta^+(V)_\beta = -\dim V_\beta + \sum_{l \in \Gamma^\beta} \dim V_{\alpha(l)}.$$

2) *If the vertex* $\alpha$ *is* $(-)$*-accessible with respect to* $\Lambda$ *and if* $V \in \mathcal{L}(\Gamma, \Lambda)$ *is an indecomposable object, then two cases are possible:*

a) $V \approx L_\alpha$, $F_\alpha^-(V) = 0$.

b) $F_\alpha^-(V)$ *is an indecomposable object,* $F_\alpha^+ F_\alpha^-(V) = V$,

(1.1.2) $$\dim F_\alpha^-(V)_\gamma = \dim V_\gamma \text{ for } \gamma \neq \alpha,$$

$$\dim F_\alpha^-(V)_\alpha = -\dim V_\alpha + \sum_{l \in \Gamma^\alpha} \dim V_{\beta(l)}.$$

PROOF. If the vertex $\beta$ is (+)-accessible with respect to $\Lambda$, then it is (−)-accessible with re  't to $\sigma_\beta \Lambda$, and so the functor $F_\beta^- F_\beta^+$: $\mathscr{L}(\Gamma, \Lambda) \to \mathscr{L}(\Gamma, \Lambda)$ is uefined. For each object $V \in \mathscr{L}(\Gamma, \Lambda)$ we construct a morphism $i_V^\beta$: $F_\beta^- F_\beta^+(V) \to V$ in the following way.

If $\gamma \neq \beta$, then $F_\beta^- F_\beta^+(V)_\gamma = V_\gamma$, and we put $(i_V^\beta)_\gamma = \mathrm{Id}$, the identity mapping.

For the definition of $(i_V^\beta)_\beta$ we note that in the sequence of mappings

$F_\beta^+(V)_\beta \xrightarrow{\tilde{h}} \bigoplus_{l \in \Gamma^\beta} V_{\alpha(l)} \xrightarrow{h} V_\beta$ (see definition 1.1) $\operatorname{Ker} h = \operatorname{Im} \tilde{h}$; we take for $(i_V^\beta)_\beta$ the natural mapping

$$F_\beta^- F_\beta^+(V)_\beta = \bigoplus_{l \in \Gamma^\beta} V_{\alpha(l)}/\operatorname{Im} \tilde{h} = \bigoplus_{l \in \Gamma^\beta} V_{\alpha(l)}/\operatorname{Ker} h \to V_\beta.$$

It is easy to verify that $i_V^\beta$ is a morphism. Similarly, for each (−)-accessible vertex $\alpha$ we construct a morphism $p_V^\alpha$: $V \to F_\alpha^+ F_\alpha^-(V)$. Now we state the basic properties of the functors $F_\alpha^-$, $F_\beta^+$ and the morphisms $p_V^\alpha$, $i_V^\beta$.

LEMMA 1.1. 1) $F_\alpha^\pm(V_1 \oplus V_2) = F_\alpha^\pm(V_1) \oplus F_\alpha^\pm(V_2)$. 2) $p_V^\alpha$ *is an epimorphism and* $i_V^\beta$ *is a monomorphism.* 3) *If* $i_V^\beta$ *is an isomorphism, then the dimensions of the spaces* $F_\beta^+(V)_\gamma$ *can be calculated from* (1.1.1). *If* $p_V^\alpha$ *is an isomorphism, then the dimensions of the spaces* $F_\alpha^-(V)_\gamma$ *can be calculated from* (1.1.2). 4) *The object* Ker $p_V^\alpha$ *is concentrated at* $\alpha$ *(that is,* $(\operatorname{Ker} p_V^\alpha)_\gamma = 0$ *for* $\gamma \neq \alpha$*). The object* $V/\operatorname{Im} i_V^\beta$ *is concentrated at* $\beta$. 5) *If the object* $V$ *has the form* $F_\alpha^+ W$ *(*$F_\beta^- W$*, respectively), then* $p_V^\alpha$ *(*$i_V^\beta$*) is an isomorphism.* 6) *The object* $V$ *is isomorphic to the direct sum of the objects* $F_\beta^- F_\beta^+(V)$ *and* $V/\operatorname{Im} i_V^\beta$ *(similarly,* $V \approx F_\alpha^+ F_\alpha^-(V) \oplus \ker p_V^\alpha$*).*

PROOF. 1), 2), 3), 4) and 5) can be verified immediately. Let us prove 6).

We have to show that $V \approx F_\beta^- F_\beta^+(V) \oplus \tilde{V}$, where $\tilde{V} = V/\operatorname{Im} i_V^\beta$. The natural projection $\varphi'_\beta$: $V_\beta \to \tilde{V}_\beta$ has a section $\varphi_\beta$: $\tilde{V}_\beta \to V_\beta$ ($\varphi'_\beta \cdot \varphi_\beta = \mathrm{Id}$). If we put $\varphi_\gamma = 0$ for $\gamma \neq \beta$, we obtain a morphism $\varphi$: $\tilde{V} \to V$. It is clear that the morphisms $\varphi$: $\tilde{V} \to V$ and $i_V^\beta$: $F_\beta^- F_\beta^+(V) \to V$ give a decomposition of $V$ into a direct sum. We can prove similarly that $V \approx F_\alpha^+ F_\alpha^-(V) \oplus \operatorname{Ker} p_V^\alpha$.

We now prove Theorem 1.1. Let $V$ be an indecomposable object of the category $\mathscr{L}(\Gamma, \Lambda)$, and $\beta$ a (+)-accessible vertex with respect to $\Lambda$. Since

$V \approx F_\beta^- F_\beta^+ (V) \oplus V/\mathrm{Im}\, i_V^\beta$ and $V$ is indecomposable, $V$ coincides with one of the terms.

CASE I). $V = V/\mathrm{Im}\, i_V^\beta$. Then $V_\gamma = 0$ for $\gamma \neq \beta$ and, because $V$ is indecomposable, $V \approx L_\beta$.

CASE II). $V = F_\beta^- F_\beta^+(V)$, that is, $i_V^\beta$ is an isomorphism. Then (1.1.1) is satisfied by Lemma 1.1. We show that the object $W = F_\beta^+(V)$ is indecomposable. For suppose that $W = W_1 \oplus W_2$. Then $V = F_\beta^- (W_1) \oplus F_\beta^- (W_2)$ and so one of the terms (for example, $F_\beta^-(W_2)$) is 0. By 5) of Lemma 1.1, the morphism $p_V^\beta : W \to F_\beta^+ F_\beta^-(W)$ is an isomorphism, but $p_V^\beta(W_2) \subset F_\beta^+ F_\beta^-(W_2) = 0$, that is, $W_2 = 0$.

So we have shown that the object $F_\beta^+(V)$ is indecomposable. We can similarly prove 2) of Theorem 1.1.

We say that a sequence of vertices $\alpha_1, \alpha_2, \ldots, \alpha_k$ is (+)-accessible with respect to $\Lambda$ if $\alpha_1$ is (+)-accessible with respect to $\Lambda$, $\alpha_2$ is (+)-accessible with respect to $\sigma_{\alpha_1}\Lambda$, $\alpha_3$ is (+)-accessible with respect to $\sigma_{\alpha_2}\sigma_{\alpha_1}\Lambda$, and so on. We define a (−)-accessible sequence similarly.

COROLLARY 1.1. *Let* $(\Gamma, \Lambda)$ *be an oriented graph and* $\alpha_1, \alpha_2, \ldots, \alpha_k$ *a (+)-accessible sequence.*

1) *For any* $i$ $(1 \leqslant i \leqslant k)$, $F_{\alpha_1}^- \cdot \ldots \cdot F_{\alpha_{i-1}}^- (L_{\alpha_i})$ *is either* 0 *or an indecomposable object in* $\mathscr{L}(\Gamma, \Lambda)$ *(here* $L_{\alpha_i} \in \mathscr{L}(\Gamma, \sigma_{\alpha_{i-1}}\sigma_{\alpha_{i-2}} \ldots \sigma_{\alpha_1}\Lambda)$*).*[1]

2) *Let* $V \in \mathscr{L}(\Gamma, \Lambda)$ *be an indecomposable object, and*

$$F_{\alpha_k}^+ F_{\alpha_{k-1}}^+ \cdot \ldots \cdot F_{\alpha_1}^+ (V) = 0.$$

*Then for some* $i$

$$V \approx F_{\alpha_1}^- F_{\alpha_2}^- \cdot \ldots \cdot F_{\alpha_{i-1}}^- (L_{\alpha_i}).$$

We illustrate the application of the functors $F_\beta^+$ and $F_\alpha^-$ by the following theorem.

THEOREM 1.2. *Let* $\Gamma$ *be a graph without cycles (in particular, without loops), and* $\Lambda$, $\Lambda'$ *two orientations of it.*

1) *There exists a sequence of vertices* $\alpha_1, \ldots, \alpha_k$, *(+)-accessible with respect to* $\Lambda$, *such that* $\sigma_{\alpha_k}\sigma_{\alpha_{k-1}} \cdot \ldots \cdot \sigma_{\alpha_1}\Lambda = \Lambda'$.

2) *Let* $\mathscr{M}$, $\mathscr{M}'$ *be the sets of classes (to within isomorphism) of indecomposable objects in* $\mathscr{L}(\Gamma, \Lambda)$ *and* $\mathscr{L}(\Gamma, \Lambda')$, $\widetilde{\mathscr{M}} \subset \mathscr{M}$ — *the set of classes of objects* $F_{\alpha_1}^- F_{\alpha_2}^- \cdot \ldots \cdot F_{\alpha_{i-1}}^- (L_{\alpha_i})$ $(1 \leqslant i \leqslant k)$, *and* $\widetilde{\mathscr{M}}' \subset \mathscr{M}'$ *the set of classes of objects* $F_{\alpha_k}^+ \cdot \ldots \cdot F_{\alpha_{i+1}}^+ (L_{\alpha_i})$ $(1 \leqslant i \leqslant k)$. *Then the functor* $F_{\alpha_k}^+ \cdot \ldots \cdot F_{\alpha_1}^+$ *sets up a one-to-one correspondence between* $\mathscr{M} \setminus \widetilde{\mathscr{M}}$ *and* $\mathscr{M}' \setminus \widetilde{\mathscr{M}}'$.

---

[1] Where it cannot lead to misunderstanding, we denote by the same symbol $L_\alpha$ irreducible objects in all categories $\mathscr{L}(\Gamma, \Lambda)$, omitting the indication of the orientation $\Lambda$.

This theorem shows that, knowing the classification of indecomposable objects for $\Lambda$, we can easily carry it over to $\Lambda'$; in other words, problems that can be obtained from one another by reversing some of the arrows are equivalent in a certain sense.

Examples show that the same is true for graphs with cycles, but we are unable to prove it.

PROOF OF THEOREM 1.2. It is clear that 2) follows at once from 1) and Corollary 1.1. Let us prove 1).

It is sufficient to consider the case when the orientations $\Lambda$ and $\Lambda'$ differ in only one edge $l$. The graph $\Gamma \setminus l$ splits into two connected components. Let $\Gamma'$ be the one that contains the vertex $\beta(l)$ ($\beta(l)$ is taken with the orientation of $\Lambda$). Let $\alpha_1, \ldots, \alpha_k$ be a numbering of the vertices of $\Gamma'$ such that for any edge $l' \in \Gamma'_1$ the index of the vertex $\alpha(l')$ is greater than that of $\beta(l')$. (Such a numbering exists because $\Gamma'$ is a graph without cycles.) It is easy to see that the sequence of vertices $\alpha_1, \ldots, \alpha_k$ is the one required (that is, it is (+)-accessible and $\sigma_{\alpha_k} \cdot \ldots \cdot \sigma_{\alpha_1} \Lambda = \Lambda'$). This proves Theorem 1.2.

It is often convenient to use a certain combination of functors $F_\alpha^\pm$ that takes the category $\mathcal{L}(\Gamma, \Lambda)$ into itself.

DEFINITION 1.2. Let $(\Gamma, \Lambda)$ be an oriented graph without oriented cycles. We choose a numbering $\alpha_1, \ldots, \alpha_n$ of the vertices of $\Gamma$ such that for any edge $l \in \Gamma_1$ the index of the vertex $\alpha(l)$ is greater than that of $\beta(l)$. We put $\Phi^+ = F_{\alpha_n}^+ \cdot \ldots \cdot F_{\alpha_2}^+ F_{\alpha_1}^+$, $\Phi^- = F_{\alpha_1}^- \cdot F_{\alpha_2}^- \cdot \ldots \cdot F_{\alpha_n}^-$. We call $\Phi^+$ and $\Phi^-$ *Coxeter functors.*

LEMMA 1.2. 1) *The sequence $\alpha_1, \ldots, \alpha_n$ is (+)-accessible and $\alpha_n, \ldots, \alpha_1$ is (−)-accessible.* 2) *The functors $\Phi^+$ and $\Phi^-$ take the category $\mathcal{L}(\Gamma, \Lambda)$ into itself.* 3) *$\Phi^+$ and $\Phi^-$ do not depend on the freedom of choice in numbering the vertices.*

The proof of 1) and 2) is obvious. We prove 3) for $\Phi^+$. We note firstly that if two different vertices $\gamma_1, \gamma_2 \in \Gamma_0$ are not joined by an edge and are (+)-accessible with respect to some orientation, then the functors $F_{\gamma_1}^+$ and $F_{\gamma_2}^+$ commute (that is, $F_{\gamma_2}^+ F_{\gamma_1}^+ = F_{\gamma_1}^+ F_{\gamma_2}^+$).

Let $\alpha_1, \ldots, \alpha_n$ and $\alpha'_1, \ldots \alpha'_n$ be two suitable numberings and let $\alpha_1 = \alpha'_m$. Then the vertices $\alpha'_1, \alpha'_2, \ldots, \alpha'_{m-1}$ are not joined to $\alpha_1$ by an edge (if $\alpha_1$ and $\alpha'_i$ $(i < m)$ are joined by an edge $l$, then $\alpha(l) = \alpha'_m = \alpha_1$ by virtue of the choice of the numbering of $\alpha'_1, \ldots, \alpha'_n$, but this contradicts the choice of the numbering of $\alpha_1, \ldots, \alpha_n$). Therefore $F_{\alpha'_m}^+ \cdot \ldots \cdot F_{\alpha'_1}^+ = F_{\alpha'_{m-1}}^+ \cdot \ldots \cdot F_{\alpha'_1}^+ F_{\alpha_1}^+$. Carrying out a similar argument with $\alpha_2$, then with $\alpha_3$, and so on, we prove that $F_{\alpha'_n}^+ \cdot \ldots \cdot F_{\alpha'_1}^+$ $F_{\alpha'_n}^+ \cdot \ldots \cdot F_{\alpha'_1}^+ = F_{\alpha_n}^+ \cdot \ldots \cdot F_{\alpha_1}^+$.

The proof is similar for the functor $\Phi^-$.

Following [2] we can introduce the following definition.

DEFINITION 1.3. Let $(\Gamma, \Lambda)$ be an oriented graph without oriented cycles. We say that an object $V \in \mathcal{L}(\Gamma, \Lambda)$ is (+)-(respectively, (−)-) irregular if $(\Phi^+)^k V = 0$ ($(\Phi^-)^k V = 0$) for some $k$. We say that an object $V$ is regular if $V \approx (\Phi^-)^k (\Phi^+)^k V \approx (\Phi^+)^k (\Phi^-)^k V$ for all $k$.

NOTE 1. Using the morphisms $p_V^\alpha$ and $i_V^\beta$ introduced in the proof of Theorem 1.1, we can construct a canonical epimorphism $p_V^k\colon V \to (\Phi^+)^k (\Phi^-)^k V$ and monomorphism $i_V^k\colon (\Phi^-)^k (\Phi^+)^k V \to V$. The object $V$ is regular if and only if for all $k$ these morphisms are isomorphisms.

NOTE 2. If an object $V$ is annihilated by the functor $F^+_{\alpha_s} \cdot \ldots \cdot F^+_{\alpha_1}$ ($\alpha_1, \ldots, \alpha_s$ is some (+)-accessible sequence), then this object is (+)-irregular. Moreover, the sequence $\alpha_1, \ldots, \alpha_s$ can be extended to $\alpha_1, \ldots, \alpha_s, \alpha_{s+1}, \ldots, \alpha_m$ so that $F^+_{\alpha_m} \cdot \ldots \cdot F^+_{\alpha_{s+1}} \cdot F^+_{\alpha_s} \cdot \ldots \cdot F^+_{\alpha_1} = (\Phi^+)^s$.

THEOREM 1.3. *Let $(\Gamma, \Lambda)$ be an oriented graph without oriented cycles.* 1) *Each indecomposable object $V \in \mathcal{L}(\Gamma, \Lambda)$ is either regular or irregular.* 2) *Let $\alpha_1, \ldots, \alpha_n$ be a numbering of the vertices of $\Gamma$ such that for any $l \in \Gamma_1$ the index of $\alpha(l)$ is greater than that of $\beta(l)$. Put* $V_i = F^-_{\alpha_1} F^-_{\alpha_2} \cdot \ldots \cdot F^-_{\alpha_{i-1}} (L_{\alpha_i}) \in \mathcal{L}(\Gamma, \Lambda)$, $\hat{V}_i = F^+_{\alpha_n} \cdot \ldots \cdot F^+_{\alpha_{i+1}} (L_{\alpha_i}) \in \mathcal{L}(\Gamma, \Lambda)$ *(here $1 \leq i \leq n$). Then $\Phi^+(V_i) = 0$ and any indecomposable object $V \in \mathcal{L}(\Gamma, \Lambda)$ for which $\Phi^+(V) = 0$ is isomorphic to one of the objects $V_i$. Similarly, $\Phi^-(\hat{V}_i) = 0$, and if $V$ is indecomposable and $\Phi^-(V) = 0$, then $V \approx \hat{V}_i$ for some $i$.* 3) *Each* (+)-*(respectively,* (−)-*) irregular indecomposable object $V$ has the form $(\Phi^-)^k V_i$ (respectively, $(\Phi^+)^k \hat{V}_i$) for some $i$, $k$.*

Theorem 1.3 follows immediately from Corollary 1.1.

With the help of this theorem it is possible, as was done in [2] for the classification of quadruples of subspaces, to distinguish "simple" (irregular) objects from more "complicated" (regular) objects; other methods are necessary for the investigation of regular objects.

## § 2. Graphs, Weyl groups and Coxeter transformations

In this section we define Weyl groups, roots, and Coxeter transformations, and we prove results that are needed subsequently. We mention two differences between our account and the conventional one.

a) We have only Dynkin diagrams with single arrows.

b) In the case of graphs with multiple edges we obtain a wider class of groups than, for example, in [7].

DEFINITION 2.1. Let $\Gamma$ be a graph without loops.

1) We denote by $\mathscr{E}_\Gamma$ the linear space over $\mathbf{Q}$ consisting of sets $x = (x_\alpha)$ of rational numbers $x_\alpha$ ($\alpha \in \Gamma_0$).

For each $\beta \in \Gamma_0$ we denote by $\bar{\beta}$ the vector in $\mathscr{E}_\Gamma$ such that $(\bar{\beta})_\alpha = 0$ for $\alpha \neq \beta$ and $(\bar{\beta})_\beta = 1$.

We call a vector $x = (x_\alpha)$ *integral* if $x_\alpha \in \mathbf{Z}$ for all $\alpha \in \Gamma_0$.

We call a vector $x = (x_\alpha)$ *positive* (written $x > 0$) if $x \neq 0$ and

$x_\alpha \geqslant 0$ for all $\alpha \in \Gamma_0$.

2) We denote by $B$ the quadratic form on the space $\mathscr{E}_\Gamma$ defined by the formula $B(x) = \sum_{\alpha \in \Gamma_0} x_\alpha^2 - \sum_{l \in \Gamma_1} x_{\gamma_1(l)} \cdot x_{\gamma_2(l)}$, where $x = (x_\alpha)$, and $\gamma_1(l)$ and $\gamma_2(l)$ are the ends of the edge $l$. We denote by $\langle\, ,\, \rangle$ the corresponding symmetric bilinear form.

3) For each $\beta \in \Gamma_0$ we denote by $\sigma_\beta$ the linear transformation in $\mathscr{E}_\Gamma$ defined by the formula $(\sigma_\beta x)_\gamma = x_\gamma$ for $\gamma \neq \beta$, $(\sigma_\beta x)_\beta = -x_\beta + \sum_{l \in \Gamma^\beta} x_{\gamma(l)}$, where $\gamma(l)$ is the end-point of the edge $l$ other than $\beta$.

We denote by $W$ the semigroup of transformations of $\mathscr{E}_\Gamma$ generated by the $\sigma_\beta$ $(\beta \in \Gamma_0)$.

LEMMA 2.1. 1) *If $\alpha, \beta \in \Gamma_0$, $\alpha \neq \beta$, then $\langle \bar{\alpha}\, \bar{\alpha} \rangle = 1$ and $2\langle \bar{\alpha}\, \bar{\beta} \rangle$ is the negative of the number of edges joining $\alpha$ and $\beta$.* 2) *Let $\beta \in \Gamma_0$. Then $\sigma_\beta(x) = x - 2\langle \bar{\beta}, x \rangle \bar{\beta}$, $\sigma_\beta^2 = 1$. In particular, $W$ is a group.* 3) *The group $W$ preserves the integral lattice in $\mathscr{E}_\Gamma$ and preserves the quadratic form $B$.* 4) *If the form $B$ is positive definite (that is, $B(x) > 0$ for $x \neq 0$), then the group $W$ is finite.*

PROOF. 1), 2) and 3) are verified immediately; 4) follows from 3).

For the proof of Gabriel's theorem the case where $B$ is positive definite is interesting.

PROPOSITION 2.1. *The form $B$ is positive definite for the graphs $A_n$, $D_n$, $E_6$, $E_7$, $E_8$ and only for them* (see [7], Chap. VI).

We give an outline of the proof of this proposition.

1. If $\Gamma$ contains a subgraph of the form

(*)

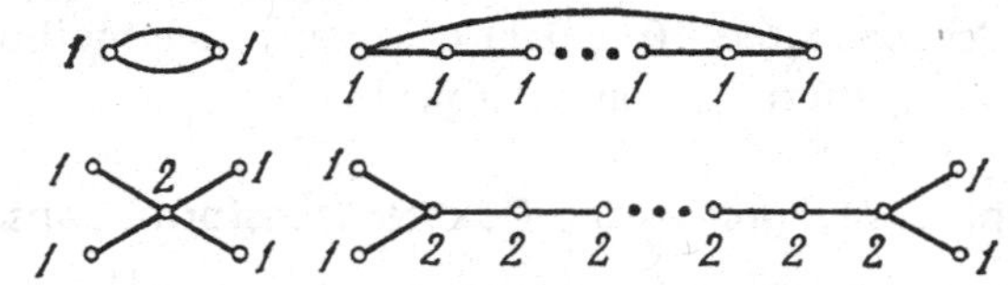

then the form $B$ is not positive definite, because when we complete the numbers at the vertices in Fig. (*) by zeros, we obtain a vector $x \in \mathscr{E}_\Gamma$ for which $B(x) \leqslant 0$. Hence, if $B$ is positive definite, then $\Gamma$ has the form

(**)

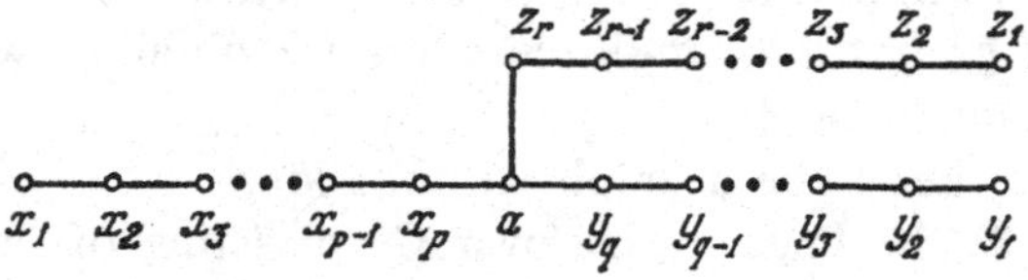


where $p$, $q$, $r$ are non-negative integers.

2 For each non-negative integer $p$ we consider the quadratic form in $(p+1)$ variables $x_1, \ldots, x_{p+1}$

$$C_p(x_1, \ldots, x_{p+1}) = -x_1x_2 - x_2x_3 - \ldots - x_px_{p+1} + x_1^2 + \ldots + x_p^2 + \frac{p}{2(p+1)} x_{p+1}^2.$$

This form is non-negative definite, and the dimension of its null space is 1. Moreover, any vector $x \neq 0$ for which $C_p(x) = 0$ has all its coordinates non-zero.

To prove these facts it is sufficient to rewrite $C_p(x)$ in the form

$$C_p(x) = \sum_{i=1}^{p} \frac{i}{2(i+1)} \left(x_{i+1} - \frac{i+1}{i} x_i\right)^2.$$

3. We place the numbers $x_1, \ldots, x_p, y_1, \ldots, y_q, z_1, \ldots, z_r\ a$ at the vertices of $\Gamma$ in accordance with Fig. (**). Then

$$B(x_i, y_i, z_i, a) = C_p(x_1, \ldots, x_p, a) + C_q(y_1, \ldots, y_q, a) + \\ + C_r(z_1, \ldots, z_r, a) + \left(1 - \frac{p}{2(p+1)} - \frac{q}{2(q+1)} - \frac{r}{2(r+1)}\right) a^2.$$

Hence it is clear that $B$ is positive definite if and only if

$\frac{p}{2(p+1)} + \frac{q}{2(q+1)} + \frac{r}{2(r+1)} < 1$, that is, $\frac{1}{p+1} + \frac{1}{q+1} + \frac{1}{r+1} > 1$.

4. We may suppose that $p \leqslant q \leqslant r$. We examine possible cases.

a) $p = 0$, $q$ and $r$ arbitrary. $A = \frac{1}{p+1} + \frac{1}{q+1} + \frac{1}{r+1} > 1$, that is, $B$ is positive definite (series $A_n$).

b) $p = 1$, $q = 1$, $r$ arbitrary. $A > 1$ (series $D_n$),

c) $p = 1$, $q = 2$, $r = 2, 3, 4$. $A > 1$ ($E_6$, $E_7$, $E_8$),

d) $p = 1$, $q = 2$, $r \geqslant 5$. $A \leqslant 1$,

$p = 1$, $q = 3$, $r \geqslant 3$. $A \leqslant 1$,

$p \geqslant 2, q \geqslant 2, r \geqslant 2$. $A \leqslant 1$.

Thus $B$ is positive definite for the graphs $A_n$, $D_n$, $E_6$, $E_7$, $E_8$ and only for them.

DEFINITION 2.2 A vector $x \in \mathscr{E}_\Gamma$ is called a *root* if for some $\beta \in \Gamma_0$, $w \in W$ we have $x = w\bar{\beta}$. The vectors $\bar{\beta}$ ($\beta \in \Gamma_0$) are called *simple roots*. A root $x$ is called positive if $x > 0$ ( see Definition 2.1).

LEMMA 2.2 1) *If $x$ is a root, then $x$ is an integral vector and $B(x) = 1$. 2) If $x$ is a root, then $(-x)$ is a root. 3) If $x$ is a root, then either $x > 0$ or $(-x) > 0$.*

PROOF. 1) follows from Lemma 2.1; 2) follows from the fact that $\sigma_\alpha(\bar{\alpha}) = -\bar{\alpha}$ for all $\alpha \in \Gamma_0$.

3) is needed only when $B$ is positive definite and we prove it only in this case.

We can write the root $x$ in the form $\sigma_{\alpha_1} \sigma_{\alpha_2} \cdot \ldots \cdot \sigma_{\alpha_k} \bar{\beta}$, where $\alpha_1, \ldots, \alpha_k, \beta \in \Gamma_0$. It is therefore sufficient to show that if $y > 0$ and $\alpha \in \Gamma_0$, then either $\sigma_\alpha y > 0$ or $y = \bar{\alpha}$ (and $-\sigma_\alpha y = +\bar{\alpha} > 0$).

Since $\|y\| = \|\bar{\alpha}\| = 1$, we have $|\langle\bar{\alpha}, y\rangle| \leqslant 1$. Moreover, $2\langle\bar{\alpha}, y\rangle \in \mathbf{Z}$. Hence $2\langle\bar{\alpha}, y\rangle$ takes one of the five values 2, 1, 0, −1, −2.

a) $2\langle\bar{\alpha}, y\rangle = 2$. Then $\langle\bar{\alpha}, y\rangle = 1$, that is, $y = \bar{\alpha}$.

b) $2\langle\bar{\alpha}, y\rangle \leqslant 0$. Then $\sigma_\alpha(y) = y - 2\langle\bar{\alpha}, y\rangle\,\bar{\alpha} > 0$.

c) $2\langle\bar{\alpha}, y\rangle = 1$. Since $2\langle\bar{\alpha}, y\rangle = 2y_\alpha - \sum_{l\in\Gamma^\alpha} y_{\gamma(l)}$ ($\gamma(l)$ is the other end-point of the edge $l$), we have $y_\alpha > 0$, that is, $y_\alpha \geqslant 1$. Hence $\sigma_\alpha y = y - \bar{\alpha} > 0$.

This proves Lemma 2.2.

DEFINITION 2.3. Let $\Gamma$ be a graph without loops, and let $\alpha_1, \ldots, \alpha_n$ be a numbering of its vertices. An element $c = \sigma_{\alpha_n} \cdot \ldots \cdot \sigma_{\alpha_1}$ ($c$ depends on the choice of numbering) of the group $W$ is called a *Coxeter transformation*.

LEMMA 2.3. *Suppose that the form $B$ for the graph $\Gamma$ is positive definite:*

1) *the transformation $c$ in $\mathscr{E}_\Gamma$ has non non-zero invariant vectors;*

2) *if $x \in \mathscr{E}_\Gamma$, $x \neq 0$, then for some $i$ the vector $c^i x$ is not positive.*

PROOF. 1) Suppose that $y \in \mathscr{E}_\Gamma$, $y \neq 0$ and $cy = y$. Since the transformations $\sigma_{\alpha_n}, \sigma_{\alpha_{n-1}}, \ldots, \sigma_{\alpha_2}$ do not change the coordinate corresponding to $\alpha_1$ (that is, for any $z \in \mathscr{E}_\Gamma$ $(\sigma_{\alpha_i} z)_{\alpha_1} = z_{\alpha_1}$ for $i \neq 1$), we have $(\sigma_{\alpha_1} y)_{\alpha_1} = (cy)_{\alpha_1} = y_{\alpha_1}$. Hence $\sigma_{\alpha_1} y = y$ Similarly we can prove that $\sigma_{\alpha_2} y = y$, then $\sigma_{\alpha_3} y = y$, and so on.

For all $\alpha \in \Gamma_0$, $\sigma_\alpha y = y - 2\langle\bar{\alpha}, y\rangle\bar{\alpha} = y$, that is $\langle\bar{\alpha}, y\rangle = 0$. Since the vectors $\bar{\alpha}(\alpha \in \Gamma_0)$ form a basis of $\mathscr{E}_\Gamma$ and $B$ is non-degenerate, $y = 0$.

2) Since $W$ is a finite group, for some $h$ we have $c^h = 1$. If all the vectors $x, cx, \ldots, c^{h-1}x$ are positive, then $y = x + cx + \ldots + c^{h-1}x$ is non-zero. Hence $cy = y$, which contradicts 1).

## §3. Gabriel's theorem

Let $(\Gamma, \Lambda)$ be an oriented graph. For each object $V \in \mathscr{L}(\Gamma, \Lambda)$ we regard the set of dimensions $\dim V_\alpha$ as a vector in $\mathscr{E}_\Gamma$ and denote it by $\dim V$.

THEOREM 3.1 (Gabriel [1]). 1) *If in $\mathscr{L}(\Gamma, \Lambda)$ there are only finitely many non-isomorphic indecomposable objects, then $\Gamma$ coincides with one of the graphs $A_n$, $D_n$, $E_6$, $E_7$, $E_8$.*

2) *Let $\Gamma$ be a graph of one of the types $A_n$, $D_n$, $E_6$, $E_7$, $E_8$, and $\Lambda$ some orientation of it. Then in $\mathscr{L}(\Gamma, \Lambda)$ there are only finitely many non-isomprphic indecomposable objects. In addition, the mapping $V \mapsto \dim V$ sets up a one-to-one correspondence between classes of isomorphic indecomposable objects and positive roots in $\mathscr{E}_\Gamma$.*

We start with a proof due to Tits of the first part of the theorem.

TITS'S PROOF. Consider the objects $(V, f) \in \mathscr{L}(\Gamma, \Lambda)$ with a fixed dimension $\dim V = m = (m_\alpha)$.

If we fix a basis in each of the spaces $V_\alpha$, then the object $(V, f)$ is completely defined by the set of matrices $A_l$ $(l \in \Gamma_1)$, where $A_l$ is the matrix of the mapping $f_l$: $V_{\alpha(l)} \to V_{\beta(l)}$. In each space $V_\alpha$ we change the basis by means of a non-singular $(m_\alpha \times m_\alpha)$ matrix $g_\alpha$. Then the matrices $A_l$ are replaced by the matrices

(*) $$A_l' = g_{\beta(l)}^{-1} A_l \, g_{\alpha(l)}.$$

Let $A$ be the manifold of all sets of matrices $A_l$ ($l \in \Gamma_1$) and $G$ the group of all sets of non-singular matrices $g_\alpha$ ($\alpha \in \Gamma_0$). Then $G$ acts on $A$ according to (*); clearly, two objects of $\mathcal{L}(\Gamma, \Lambda)$ with given dimension $m$ are isomorphic if and only if the sets of matrices $\{A_l\}$ corresponding to them lie in one orbit of $G$.

If in $\mathcal{L}(\Gamma, \Lambda)$ there are only finitely many indecomposable objects, then there are only finitely many non-isomorphic objects of dimension $m$. Therefore the manifold $A$ splits into a finite number of orbits of $G$. It follows[1] that $\dim A \leqslant \dim G - 1$ (the $-1$ is explained by the fact that $G$ has a 1-dimensional subgroup $G_0 = \{g(\lambda) \mid \lambda \in K^*\}$, $g(\lambda)_\alpha = \lambda \cdot 1_{V_\alpha}$, which acts on $A$ identically). Clearly, $\dim G = \sum_{\alpha \in \Gamma_0} m_\alpha^2$, $\dim A = \sum_{l \in \Gamma_1} m_{\alpha(l)} m_{\beta(l)}$.

Therefore the condition $\dim A \leqslant \dim G - 1$ can be rewritten in the form[2] $B(m) > 0$ (if $m \neq 0$). In addition, it is easy to verify that $B((x_\alpha)) \geqslant B((|x_\alpha|))$ for all $x = (x_\alpha) \in \mathcal{E}_\Gamma$.

So we have shown that if in $\mathcal{L}(\Gamma, \Lambda)$ there are finitely many indecomposable objects, then the form $B$ in $\mathcal{E}_\Gamma$ is positive definite.

As we have shown in Proposition 2.1, this holds only for the graphs $A_n$, $D_n$, $E_6$, $E_7$, $E_8$.

We now prove the second part of Gabriel's theorem.

LEMMA 3.1. *Suppose that* $(\Gamma, \Lambda)$ *is an oriented graph,* $\beta \in \Gamma_0$ *a (+)-accessible vertex with respect to* $\Lambda$, *and* $V \in \mathcal{L}(\Gamma, \Lambda)$ *an indecomposable object. Then either* $F_\beta^+(V)$ *is an indecomposable object and* $\dim F_\beta^+(V) = \sigma_\beta(\dim V)$, *or* $V = L_\beta$, $F_\beta^+(V) = 0$, $\dim F_\beta^+(V) \neq \sigma_\beta(\dim V) < 0$. *A similar statement holds for a (−)-accessible vertex* $\alpha$ *and the functor* $F_\alpha^-$.

This lemma is a reformulation of Theorem 1.1.

COROLLARY 3.1. *Suppose that the sequence of vertices* $\alpha_1, \ldots, \alpha_k$ *is (+)-accessible with respect to* $\Lambda$ *and that* $V \in \mathcal{L}(\Gamma, \Lambda)$ *is an indecomposable object. Put* $V_j = F_{\alpha_j}^+ F_{\alpha_{j-1}}^+ \cdot \ldots \cdot F_{\alpha_1}^+ V$, $m_j = \sigma_{\alpha_j} \sigma_{\alpha_{j-1}} \cdot \ldots \cdot \sigma_{\alpha_1}(\dim V)$ $(0 \leqslant j \leqslant k)$. *Let* $i$ *be the last index such that* $m_j > 0$ *for* $j \leqslant i$. *Then the* $V_j$ *are indecomposable objects for* $j \leqslant i$, *and* $V = F_{\alpha_1}^- \cdot \ldots \cdot F_{\alpha_j}^- V_j$. *If* $i < k$, *then* $V_{i+1} = V_{i+2} = \ldots = V_k = 0$, $V_i = L_{\alpha_{i+1}}$, $V = F_{\alpha_1}^- \cdot \ldots \cdot F_{\alpha_i}^- (L_{\alpha_{i+1}})$. *Similar statements are true when (+) is replaced by (−).*

We now show that in the case of a graph $\Gamma$ of type $A_n$, $D_n$, $E_6$, $E_7$ or $E_8$ (that is, $B$ is positive definite), indecomposable objects correspond to positive roots.

a) Let $V \in \mathcal{L}(\Gamma, \Lambda)$ be an indecomposable object.

---

[1] This argument is suitable only for an infinite field $K$. If $K = \mathbf{F}_q$ is a finite field, we must use the fact that the number of non-isomorphic objects of dimension $m$ increases no faster than a polynomial in $m$, and the number of orbits of $G$ on the manifold $A$ is not less than $C \cdot q^{\dim A - (\dim G - 1)}$.

[2] We can clearly restrict ourselves to graphs without loops.

We choose a numbering $\alpha_1, \alpha_2, \ldots, \alpha_n$ of the vertices of $\Gamma$ such that for any edge $l \in \Gamma_1$ the vertex $\alpha(l)$ has an index greater than that of $\beta(l)$. Let $c = \sigma_{\alpha_n} \cdot \ldots \cdot \sigma_{\alpha_1}$ be the corresponding Coxeter transformation.

By Lemma 2.3, for some $k$ the vector $c^k(\dim V) \in \mathscr{E}_\Gamma$ is not positive.

If we consider the (+)-accessible sequence $\beta_1, \beta_2, \ldots, \beta_{nk} = (\alpha_1, \ldots, \alpha_n, \alpha_1, \ldots, \alpha_n, \ldots, \alpha_1, \ldots, \alpha_n)$ ($k$ times), then we have $\sigma_{\beta_{nk}} \cdot \ldots \cdot \sigma_{\beta_1}(\dim V) = c^k(\dim V) \not> 0$. From Corollary 3.1 it follows that there is an index $i < kn$ (depending only on $\dim V$) such that $V = F^-_{\beta_1} \cdot F^-_{\beta_2} \cdot \ldots \cdot F^-_{\beta_i}(L_{\beta_{i+1}})$, $\dim V = \sigma_{\beta_1} \cdot \ldots \cdot \sigma_{\beta_i}(\bar{\beta}_{i+1})$. It follows that $\dim V$ is a positive root and $V$ is determined by the vector $\dim V$.

b) Let $x$ be a positive root.

By Lemma 2.3, $c^k x \not> 0$ for some $k$. Consider the (+)-accessible sequence $\beta_1, \beta_2, \ldots, \beta_{nk} = (\alpha_1, \ldots, \alpha_n, \ldots, \alpha_1, \ldots, \alpha_n)$ ($k$ times). Then $\sigma_{\beta_{nk}} \cdot \ldots \cdot \sigma_{\beta_1}(x) = c^k(x) \not> 0$. Let $i$ be the last index for which $\sigma_{\beta_i} \sigma_{\beta_{i-1}} \cdot \ldots \cdot \sigma_{\beta_1}(x) > 0$. It is obvious from the proof of 3) in Lemma 2.2 that $\sigma_{\beta_i} \cdot \ldots \cdot \sigma_{\beta_1}(x) = \bar{\beta}_{i+1}$.

It follows that Corollary 3.1 that $V = F^-_{\beta_1} F^-_{\beta_2} \cdot \ldots \cdot F^-_{\beta_i}(L_{\beta_{i+1}}) \in \mathscr{L}(\Gamma, \Lambda)$ is an an indecomposable object and $\dim V = \sigma_{\beta_1} \cdot \ldots \cdot \sigma_{\beta_i}(\bar{\beta}_{i+1}) = x$.

This concludes the proof of Gabriel's theorem.

NOTE 1. When $B$ is positive definite, the set of roots coincides with the set of integral vectors $x \in \mathscr{E}_\Gamma$ for which $B(x) = 1$ (this is easy to see from Lemma 2.3 and the proof of Lemma 2.2).

NOTE 2. It is interesting to consider categories $\mathscr{L}(\Gamma, \Lambda)$, for which the canonical form of an object of dimension $m$ depends on fewer than $C \cdot |m|^2$ parameters (here $|m| = \Sigma\ |m_\alpha|$, $\alpha \in \Gamma_0$). From the proof it is obvious that for this it is necessary that $B$ should be non-negative definite.

As in Proposition 2.1 we can show that $B$ is non-negative definite for the graphs $A_n$, $D_n$, $E_6$, $E_7$, $E_8$ and $\hat{A}_0$, $\hat{A}_n$, $\hat{D}_n$, $\hat{E}_6$, $\hat{E}_7$, $\hat{E}_8$, where

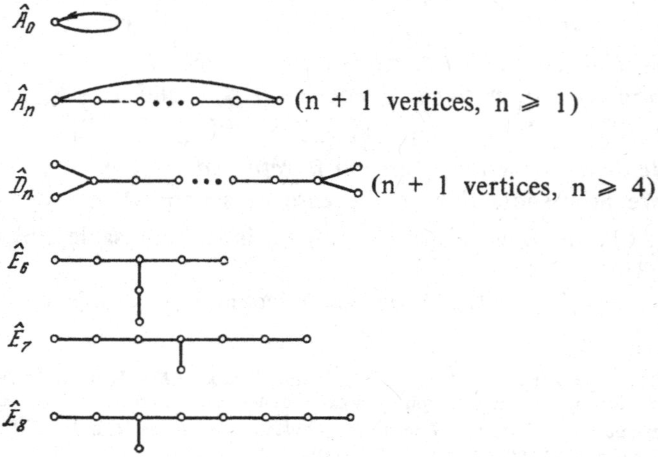

(the graphs $\hat{A}_n$, $\hat{D}_n$, $\hat{E}_6$, $\hat{E}_7$, $\hat{E}_8$ are extensions of the Dynkin diagrams (see [7])).

In a recent article Nazarova has given a classification of indecomposable objects for these graphs. In addition, she has shown there that such a classification for the remaining graphs would contain a classification of pairs of non-commuting operators (that is, in a certain sense it is impossible to give such a classification).

## § 4. Some open questions

Let $\Gamma$ be a finite connected graph without loops and $\Lambda$ an orientation of it.

CONJECTURES. 1) Suppose that $x \in \mathscr{E}_\Gamma$ is an integral vector, $x > 0$, $B(x) > 0$ and $x$ is not a root. Then any object $V \in \mathscr{L}(\Gamma, \Lambda)$ for which $\dim V = x$ is decomposable.

2) If $x$ is a positive root, then there is exactly one (to within isomorphism) indecomposable object $V \in \mathscr{L}(\Gamma, \Lambda)$, for which $\dim V = x$.

3) If $V$ is an indecomposable object in $\mathscr{L}(\Gamma, \Lambda)$ and $B(\dim V) \leqslant 0$, then there are infinitely many non-isomorphic indecomposable objects $V' \in \mathscr{L}(\Gamma, \Lambda)$ with $\dim V' = \dim V$ (we suppose that $K$ is an infinite field).

4) If $\Lambda$ and $\Lambda'$ are two orientations of $\Gamma$ and $V \in \mathscr{L}(\Gamma, \Lambda')$ is an indecomposable object, then there is an indecomposable object $V' \in \mathscr{L}(\Gamma, \Lambda')$ such that $\dim V' = \dim V$.

We illustrate this conjecture by the example of the graph $(\Gamma, \Lambda)$

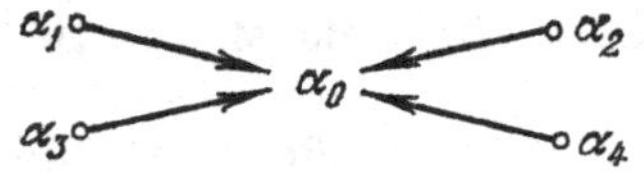


(quadruple of subspaces).

For each $x \in \mathscr{E}_\Gamma$ we put $\rho(x) = -2\langle\bar{\alpha}_0, x\rangle$ (if $x = (x_0, x_1, x_2, x_3, x_4)$, then $\rho(x) = x_1 + x_2 + x_3 + x_4 - 2x_0$).

In [2] all the indecomposable objects in the category $\mathscr{L}(\Gamma, \Lambda)$ are described. They are of the following types.

1. Irregular indecomposable objects (see the end of § 1). Such objects are in one-to-one correspondence with positive roots $x$ for which $\rho(x) \neq 0$.

2. Regular indecomposable objects $V$ for which $B(\dim V) \neq 0$. These objects are in one-to-one correspondence with positive roots $x$ for which $\rho(x) = 0$.

3) Regular objects $V$ for which $B(\dim V) = 0$. In this case $\dim V$ has the form $\dim V = (2n, n, n, n, n)$, $\rho(\dim V) = 0$. Indecomposable objects with fixed dimension $m = (2n, n, n, n, n)$ depend on one parameter. If $m \in \mathscr{E}_\Gamma$ is an integral vector such that $m > 0$ and $B(m) = 0$, then it has the form $m = (2n, n, n, n, n)$ $(n > 0)$ and there are indecomposable objects $V$ for which $\dim V = m$.

If $f$ is a linear transformation in $n$-dimensional space consisting of one

Jordan block then the quadruple of subspaces corresponding to it (see the Introduction) is a quadruple of the third type.

## References

[1] P. Gabriel, Unzerlegbare Darstellungen I, Manuscripta Math. **6** (1972), 71–103.
[2] I. M. Gelfand and V. A. Ponomarev, Problems of linear algebra and classification of quadruples of subspaces in a finite-dimensional vector space, Colloquia Mathematica Societatis Ianos Bolyai, 5, Hilbert space operators, Tihany (Hungary), 1970, 163–237 (in English). (For a brief account, see Dokl. Akad. Nauk SSSR **197** (1971), 762–765 = Soviet Math. Doklady **12** (1971), 535–539.)
[3] L. A. Nazarova and A. V. Roiter, Representations of partially ordered sets, in the collection "Investigations in the theory of representations", Izdat. Nauka, Leningrad 1972, 5–31.
[4] I. M. Gelfand, The cohomology of infinite-dimensional Lie algebras. Actes Congrès Internat. Math. Nice 1970, vol. 1. (1970), 95–111 (in English).
[5] I. M. Gelfand and V. A. Ponomarev, Indecomposable representations of the Lorentz group, Uspekhi Mat. Nauk **23**: 2 (1968), 3–60, MR **37** # 5325. = Russian Math. Surveys **23**: 2 (1968), 1–58.
[6] C. W. Curtis and I. Reiner, Representation theory of finite groups and associative algebras, Interscience, New York–London 1962, MR **26** # 2519. Translation: *Teoriya predstavlenii konechnykh grupp i assotsiativnykh algebr,* Izdat. Nauka, Moscow 1969.
[7] N. Bourbaki, Éléments de mathématique, XXVI, Groupes et algèbres de Lie, Hermann & Co., Paris 1960, MR **24** # A2641. Translation: *Gruppy i algebry Li,* Izdat. Mir, Moscow 1972.

Received by the Editors, 18 December 1972.

Translated by E. J. F. Primrose.

# 6.
## (with V. A. Ponomarev)
# Model algebras and representations of graphs

Funkts. Anal. Prilozh. **13** (3) (1979) 1–12. [Funct. Anal. Appl. **13** (1980) 157–166].
Zbl. **437**:16020

## . Definitions and Statement of Results

1.1. Graphs and Their Representations. A combinatorial graph Γ is a pair (S, A), ere S is the set of vertices of the group Γ and the set A consists of two-element subsets S, i.e., $(l \in A) \Leftrightarrow (l = \{i, j\} \mid i, j \in S)$. An unordered pair {i, j} is called an edge joining e vertices i and j. In this case, we also say that the vertices i and j are joined by e edge $l = \{i, j\}$.

In what follows, we assume that S is a finite set.

An orientation Λ on the graph Γ is a function $\Lambda : A \to S \times S$ such that $\Lambda\{i, j\}$ is equal either (i, j) or (j, i). A graph Γ equipped with an orientation Λ will be denoted by . Following Gabriel, a representation ρ of an oriented graph $\Gamma_\Lambda$ over a field K is a set linear spaces $\{V_i\}_{i\in S}$ over K and linear maps $\varphi_{ij} : V_j \to V_i$, where $(i, j) \in \Lambda A$. If $\rho = \{V_i; \varphi_{ij}\}$ d $\rho' = \{V_i'; \varphi_{ij}'\}$ are two representations of the same graph $\Gamma_\Lambda$, a morphism $f : \rho \to \rho'$ is a set $\{f_i\}_{i\in S}$ of linear maps $f_i : V_i \to V_i'$ such that for every pair $(i, j) \in \Lambda A$ the diagram

$$\begin{array}{ccc} V_j & \xrightarrow{\varphi_{ij}} & V_i \\ {\scriptstyle f_j}\downarrow & & \downarrow{\scriptstyle f_i} \\ V_j' & \xrightarrow{\varphi_{ij}'} & V_i' \end{array}$$

commutative. It is easy to see that this defines a category $\mathcal{L}_K(\Gamma_\Lambda)$ of representations a graph $\Gamma_\Lambda$ over a field K.

In [1] the Coxeter functors $\Phi^+$ and $\Phi^-$ ($\Phi^{(\cdot)} : \mathcal{L}_K(\Gamma_\Lambda) \to \mathcal{L}_K(\Gamma_\Lambda)$) were defined, having the llowing properties: 1) $\Phi^+\rho = 0$ if and only if the object ρ is projective, and correspondingly, $\Phi^-\rho = 0$ if ρ is injective; 2) if a representation ρ is indecomposable and is not proective, then the representation $\Phi^+\rho$ is also indecomposable and not equal to zero, and $\Phi^-$ $\Phi^+\rho) \cong \rho$; analogously, if ρ is indecomposable and not injective, then $\Phi^-\rho \neq 0$ and $\Phi^-\rho$ is decomposable; moreover, $\Phi^+(\Phi^-\rho) \cong \rho$. The definition of the functors $\Phi^+$ and $\Phi^-$ will be given low in Sec. 5.

We will prove below that for a tree $\Gamma_\Lambda = (S, \Lambda A)$, the indecomposable projective reprentations can be listed in a natural way by means of the vertices $i \in S$ of $\Gamma_\Lambda$. These reprentations will be denoted by $\rho_i^{(1)}$ $(i \in S)$.

We put inductively $\rho_i^{(n+1)} = \Phi^-\rho_i^{(n)}, n=1,2,\ldots$. By the properties of the functors $\Phi^-$, the presentations $\rho_i^{(n)}$, where $i \in S$, $n \in \mathbb{N}$, are indecomposable and such that $\Phi^+\rho_i^{(1)} = 0$ and $\Phi^+\rho_i^{(n+1)} \cong$ $^{n)}$, $n = 1, 2, \ldots$. Our main goal will be to construct in terms of the graph Γ = (A, S) a cerin algebra $B_K(\Gamma)$ over the field K and study the properties of $B_K(\Gamma)$. As we shall show, $(\Gamma)$ is a direct sum $\bigoplus_{n\in\mathbb{N}} \bigoplus_{i\in S} \rho_i^{(n)}$ of the representations $\rho_i^{(n)}$. It therefore gives, in a certain nse, a concrete realization of the representations $\rho_i^{(n)}$. We believe that $B_K(\Gamma)$ is of imrtance in studying other properties of the representations $\rho_i^{(n)}$.

1.2. The Model Algebra $B_K(\Gamma)$ and Representations $\rho_{B,\Lambda}$ and $\rho_{i,\Lambda}^{(n)}$. Let K be a field. We nstruct in terms of an unoriented graph Γ = (A, S) an algebra $B_K(\Gamma)$ with identity e as llows. $B_K(\Gamma)$ is the algebra over K with generators $b_i$ (indexed by the vertices $i \in S$ of e graph Γ) and relations: 1) $b_i^2 = 0$ for all $i \in S$, 2) $b_i b_j = 0$ if the vertices i and j are t joined in Γ; 3) $b_i\left(\sum_{j\in S(i)} b_j\right) b_i = 0$ for every $i \in S$ (where S(i) denotes the star of the verx i, i.e., $j \in S(i)$ means that $\{i, j\} \in A$).

---

Moscow State University. Translated from Funktsional'nyi Analiz i Ego Prilozheniya, l. 13, No. 3, pp. 1–12, July–September, 1979. Original article submitted March 23, 1979.

We will sometimes denote the model algebra $B_K(\Gamma)$ by B.

It is easy to see that the right ideals $b_iB$ of B considered as vector spaces are linearly independent and $B \cong Ke \oplus (\bigoplus_{i\in S} b_iB)$. For any orientation $\Lambda$ of the graph $\Gamma$, it is therefore possible to define a representation $\rho_{B,\Lambda}$ of $\Gamma_\Lambda$ by putting $V_i = b_iB$ and defining linear maps $\varphi_{ij}: V_j \to V_i$ by the formula $\varphi_{ij}(b_jx) = b_ib_jx$.

A monomial $b_{i_1}\dots b_{i_m}$ in the algebra $B_K(\Gamma)$ will be said to be reduced if the sequence $(i_1, \dots, i_m)$ is a path in $\Gamma$, i.e., $\{i_s, i_{s+1}\} \in A$, $s = 1, \dots, m-1$. The number m is called the degree of the reduced monomial.

We denote by $V_{i,t,m}$, where $i, t \in S$, $m = 1, 2, \dots$, the subspace of $B_K(\Gamma)$ generated by the reduced monomials y of degree m having the form $y = b_ixb_t$, i.e., by monomials of length m starting with $b_i$ and ending with $b_t$. In particular, if $\{i, t\} \in A$, $V_{i,t,2} = Kb_ib_t$. We define in addition $V_{i,i,1} = Kb_i$ and $V_{i,t,1} = 0$, if $i \neq t$. We introduce another $V_{i,t,0}$, by putting $V_{i,t,0} = 0$. We prove below that $B = Ke \oplus \bigoplus_{i,t\in S, m\in N} V_{i,t,m}$.

We now construct representations $\rho^{(n)}_{t,\Lambda}$ of an oriented graph $\Gamma_\Lambda$. In order to do this, we first construct for each t and n an integer-valued function $\chi^{(n)}_t : S \to \mathbf{Z}$. on S. This function (in the case when $\Gamma$ is a tree) is uniquely defined by the following conditions: a) if $(i, j) \in$ then $\chi^{(n)}_t(j) = \chi^{(n)}_t(i) - 1$; b) $\chi^{(n)}_t(t) = 2n - 1$. The representation $\rho^{(n)}_{t,\Lambda}$ is now defined as follows. We put $V_i = V_{i,t,m_i}$, where $m_i = \chi^{(n)}_t(i)$, and define $\varphi_{ij}: V_{j,t,m_j} \to V_{i,t,m_i}$ by $\varphi_{ij}(b_jxb_t) = b_ib_jxb_t$, where x is such that $b_jxb_t \in V_{j,t,m_j}$.

Our main goal is the proof of the following two theorems.

THEOREM 1. Let $\Lambda$ be an arbitrary orientation of the tree $\Gamma$. Then $\rho_{B,\Lambda} \cong \bigoplus_{n\in N} (\bigoplus_{t\in S} \rho^{(n)}_{t,\Lambda})$.

THEOREM 2. For every oriented tree $\Gamma_\Lambda$ and all $t \in S$ and $n \geqslant 1$, we have $\rho^{(n+1)}_{t,\Lambda} \cong \Phi^+\rho^{(n)}_{t,\Lambda}$, $\Phi^+\rho^{(n+1)}_{t,\Lambda} \cong \rho^{(n)}_{t,\Lambda}$, $\Phi^+\rho^{(1)}_{t,\Lambda} \cong 0$.

## 2. Algebra $P_K(\Gamma)$. Compatible Functions

2.1. We will prove here that the sum $B_K(\Gamma) = Ke + \sum_{m\in N} \sum_{i,t\in S} V_{i,t,m}$ is direct.

We define an auxiliary algebra $P_K(\Gamma)$ by taking $P_K(\Gamma)$ to be the K-algebra with identity e and generators $c_i$, where $i \in S$, with the following relations: 1) $c_i^2 = 0$; 2) $c_ic_j = 0$, if the vertices i and j are not joined in the graph $\Gamma$, i.e., $\{i, j\} \notin A$.

A reduced monomial is a product $c_{i_1}c_{i_2}\dots c_{i_m}$ of generators $c_j$ such that $\{i_\alpha, i_{\alpha+1}\} \in A$ for all $\alpha = 1, \dots, m-1$. The number m is called the degree of the reduced monomial. It is clear that an arbitrary monomial $c_{j_1}c_{j_2}\dots c_{j_m}$ is nonzero if and only if it is reduced. Thus, the set of reduced monomials in the algebra $P_K(\Gamma)$ is naturally isomorphic to the set of paths in the graph $\Gamma$. (A path joining the vertices a and b is a sequence $(i_1, i_2, \dots, i_n)$ of vertices $i_\alpha \in S$, such that $a = i_1$, $b = i_n$ and any two vertices $i_\alpha$, $i_{\alpha+1}$ are joined in $\Gamma$. The number $l = n - 1 \geqslant 0$ is usually called the length of the path.) The algebra $P_K(\Gamma)$ will be called the path algebra of $\Gamma$.

We will denote by $W_{i,t,m}$ $(i, t \in S, m \in N)$ the space in $P_K(\Gamma)$ generated by the reduced monomials y of the form $y = b_ixb_t$ of degree m. We also put $W_{i,t,2} = Kc_ic_t$ if $(i, t) \in A$ and $W_{i,t,1} \stackrel{\text{def}}{=} Kc_i$, and take $W_{i,j,2} = 0$ if $i \neq j$.

It is easy to see that $P_K(\Gamma)$ splits as a vector space into a direct sum:

$$P_K(\Gamma) = Ke \oplus (\bigoplus_{m\in N} (\bigoplus_{i,t\in S} W_{i,t,m})). \tag{1}$$

It follows from the definitions that the previously defined algebra $B_K(\Gamma)$ is isomorphic to the quotient $P_K(\Gamma)/J$, where J is the two-sided ideal generated by the elements $c_i \sum_{j\in S(i)} c_jc_i \in W_{i,i,3}$, where S(i) is the star of i. It is obvious that J is homogeneous with respect to the grading (1), i.e., $J = \sum_{i,t,m} (J \cap W_{i,t,m})$. It clearly follows that the sum $B_K(\Gamma) = Ke \oplus \sum_{i,t,m} V_{i,t,m}$ is direct.

2.2. Graphs and Compatible Functions. An integer-valued function $\chi: S \to \mathbf{Z}$ is said to compatible with the orientation $\Lambda$ on a graph $\Gamma_\Lambda = (S, \Lambda A)$ if $(i, j) \in \Lambda A$ implies that $\chi(j) = \chi(i) - 1$. We note that functions with such properties may not exist for an arbitrary graph . However, if $\Gamma_\Lambda$ is a tree, then as we shall prove below, compatible functions on $\Gamma_\Lambda$ always exist.

Assertion 1. Let $\Gamma_\Lambda = (S, \Lambda A)$ be a connected oriented graph and $\chi$ and $\chi'$ be two functions compatible with $\Lambda$. Then there exists an integer k such that for all $i \in S$, $\chi'(i) = \chi(i) + k$.

The proof is obvious.

Assertion 2. For every oriented tree $\Gamma_\Lambda = (S, \Lambda A)$ there exists a uniquely determined compatible function $h_\Lambda: S \to \mathbf{Z}$ with the following properties: a) $h_\Lambda(i) \geqslant 0$ for all i, and b) there exists a $t \in S$ for which $h_\Lambda(t) = 0$.

The proof is based on the following lemma.

LEMMA 3. Let $\Gamma = (S, A)$ be a tree, S' be some connected subset of S and let $j \notin S'$ be a vertex such that there exists $i \in S'$ with $\{i, j\} \in A$. Then the point i is uniquely determined by j and S'.

We first show that at least one compatible function $\chi: S \to \mathbf{Z}$. can be constructed for an oriented tree $\Gamma_\Lambda$. Let $p \in S$ be an arbitrary vertex. We put $\chi(p) = 0$. The construction will be carried out by induction on the number of elements for which the function has already been constructed. Assume that $\chi$ has been defined on a connected subset S' containing the vertex p. Since $\Gamma_\Lambda$ is connected, there exists a $j \notin S'$ such that there is an $i \in S$ with $\{i, j\} \in A$. By Lemma 3, the vertex i is unique. By the induction hypothesis, we assume that $\chi(i)$ has already been defined. We set $\chi(j) = \chi(i) - 1$ if $(i, j) \in \Lambda A$, and $\chi(j) = \chi(i) + 1$ if $(j, i) \in \Lambda A$. Continuing this process, we define the function $\chi$ on the entire set S.

It is easy to see that the function $\chi$ thus constructed is compatible with the orientation $\Lambda$. It follows from Assertion 1 that the function $\chi$ is uniquely determined by its value at the single point p (where we have put $\chi(p) = 0$).

Let $k = \min_{j \in S}(\chi(j))$. We define a function $h_\Lambda$ by putting $h_\Lambda(i) = \chi(i) - k$. This is the required construction of $h_\Lambda$.

The function $h_\Lambda$ will be called the height on the graph $\Gamma_\Lambda$.

COROLLARY 4. Let $\chi_\Lambda: S \to \mathbf{Z}$ be a compatible function on an oriented tree $\Gamma_\Lambda$ and t some fixed vertex $(t \in S)$. Then $\chi_\Lambda = h_\Lambda + (\chi_\Lambda(t) - h_\Lambda(t))$.

A function $\chi: S \to \mathbf{Z}$ on an unoriented graph $\Gamma = \{S, A\}$ will be said to be compatible if $\{i, j\} \in A$ implies that $|\chi(i) - \chi(j)| = 1$.

Assertion 5. Let $\chi: S \to \mathbf{Z}$ be a compatible function on an unoriented graph $\Gamma = \{S, A\}$. Then there is a uniquely determined orientation $\Lambda$ of $\Gamma$ such that $\chi$ is compatible with $\Lambda$.

Proof. Let $\{i, j\} \in A$ and $\chi(j) - \chi(i) = 1$. We set in this case $\Lambda\{i, j\} = (j, i)$. If $\chi(j) - \chi(i) = -1$, we put $\Lambda\{i, j\} = (i, j)$.

Proof of Theorem 1

We prove that $\rho_{B,\Lambda} \cong \bigoplus_{\substack{t \in S \\ n \in N}} \rho_{t,\Lambda}^{(n)}$ for any orientation $\Lambda$ of the graph $\Gamma$.

We denote by $R_t^{(n)}$ the subspace of the algebra B given by the direct sum

$$R_t^{(n)} = \bigoplus_i V_{i, t; \chi_t^{(n)}(i)}. \tag{1}$$

In order to prove the theorem, it suffices to verify that

$$Bb_t \cong \bigoplus_{n=1}^{\infty} R_t^{(n)}, \tag{2}$$

where $Bb_t$ is the left ideal generated by the element $b_t$. We denote the set of triples $\alpha = (i, t, \chi_t^{(n)}(i))$, where $t \in S$ and n are fixed and i varies over all of S, by $r_t^{(n)}$. (It is clear that $r_t^{(n)} \subset S \times S \times \mathbf{Z}$.)

Since $B = Ke \oplus (\bigoplus_{\alpha \in S\times S\times \mathbf{Z}} V_\alpha)$, in order to prove Eq. (2) it suffices to check that the various subsets $r_t^{(n)}$ do not intersect and that $\sum_{n=1}^{\infty} R_t^{(n)} = Bb_t$. We divide the proof into three steps

1) We prove that the different subsets $r_t^{(n)}$ do not intersect. By construction, $r_t^{(n)}$ is isomorphic to the graph of a compatible function $\chi_t^{(n)}$. It is seen from Corollary 4 that for all $i \in S$,

$$\chi_t^{(n)}(i) = h_\Lambda(i) + 2n - 1 - h_\Lambda(t). \tag{3}$$

Consequently, different subsets $r_t^{(n)}$ do not intersect.

2) We prove that $\sum_{n\in\mathbf{Z}} R_t^{(n)} = Bb_t$. In order to do this, we prove that each subspace $V_{i,t,m} = V_\alpha$ is such that $\alpha \in r_t^{(n)}$ for some n. By definition, $V_\alpha \cong W_\alpha / W_\alpha \cap J$, where $W_\alpha$ is the space of paths. Let $(x_1, \ldots, x_m)$ be a path joining the vertices $x_1 = t$ and $x_m = i$. We find the numbers $h_\Lambda(i) - h_\Lambda(t) = h_\Lambda(x_m) - h_\Lambda(x_1) = \sum_{j=1}^{m-1} (h_\Lambda(x_{j+1}) - h_\Lambda(x_j))$. By the properties of compatible functions, $|h_\Lambda(x_{j+1}) - h_\Lambda(x_j)| = 1$ for adjacent vertices $j+1$ and $j$. Consequently,

$$h_\Lambda(i) - h_\Lambda(t) = \sum_{j=1}^{m-1} (h_\Lambda(x_{j+1}) - h_\Lambda(x_j)) = m - 1 \pmod 2. \tag{4}$$

This means that there exists an integer n such that $m - 1 = 2n + h_\Lambda(i) - h_\Lambda(t)$. By Eq. (3) this means that $m = \chi_t^{(n+1)}(i)$. In other words, $\alpha = (i, t, m) \in r_t^{(n+1)}$ or $V_\alpha \subset R_t^{(n+1)}$.

3) We prove that the subspace $V_\alpha = V_{i,t,m}$ is equal to zero if $\alpha \in r_t^{(n)}$ and $n \leqslant 0$. It follows immediately that if $n \leqslant 0$, then $R_t^{(n)} = \bigoplus_{\alpha\in r_t^{(n)}} V_\alpha = 0$, i.e., only $R_t^{(n)}$ with $n > 0$ appear in the decomposition.

Let t, i be vertices in the graph $\Gamma = (S, A)$ and $(x_1, \ldots, x_l)$ a minimal path joining the vertices $x = t$ and $x_l = i$. The number $l - 1$ is called the distance between t and i.

<u>LEMMA 6.</u> Assume that the distance between the vertices t and i in the graph $\Gamma = (S, A$ is equal to $l - 1$. Then: 1) $\dim V_{i,t,l} = 1$; 2) if $m < l$, then $V_{i,t,m} = 0$.

<u>Proof of the Lemma.</u> By definition, $V_\alpha = W_\alpha/(W_\alpha \cap J)$, where $W_\alpha = W_{i,t,m}$ is the subspace of paths of length $m - 1$ joining the vertices i, t. Since $\Gamma$ is a tree, there exists precisely one minimal path $(x_1, \ldots, x_l)$, jointing $x_1 = t$ and $x_l = i$. Hence, $\dim W_{i,t,l} = 1$ and if $m < l$ then $W_{i,t,m} = 0$. By definition, the ideal J is generated by the elements $q_j = c_j(\sum_{i\in S(j)} c_i) c_j$ where S(j) is the star of j. Since a reduced monomial corresponds to a minimal path, we therefore have $W_{i,t,l} \cap J = 0$. Thus, $\dim(V_{i,t,l}) = \dim W_{i,t,l} = 1$. The lemma is proved.

Since $\chi_t^{(n)}(i) = 2n - 1 + h_\Lambda(i) - h_\Lambda(t)$ and by hypothesis $n \leqslant 0$, we have $m = \chi_t^{(n)}(i) \leqslant -1 + h_\Lambda(i) - h_\Lambda(t)$.

Let $(x_1, \ldots, x_l)$ be the minimal path joining the vertices $x_1 = t$ and $x_l = i$. Then $h_\Lambda(i) - h_\Lambda(t) = h_\Lambda(x_l) - h_\Lambda(x_1) = \sum_{j=1}^{l-1} (h_\Lambda(x_{j+1}) - h_\Lambda(x_j)) \leqslant \sum_{j=1}^{l-1} |h_\Lambda(x_{j+1}) - h_\Lambda(x_j)| = l - 1$. From this we obtain $m = \chi_t^{(n)}(i) \leqslant -1 + l - 1 = l - 2 < l$. Thus, by Lemma 6 the subspace $V_\alpha = V_{i,t,m}$ is zero.

## 4. Objects $\rho_{t,\Lambda}^{(k)}$

In this section we prove that the indecomposable projective objects of the category $\mathcal{L}_K(\Gamma_\Lambda)$ are precisely the representations $\rho_{t,\Lambda}^{(1)}$.

<u>4.1. Category $\mathcal{L}_K(\Gamma_\Lambda)$.</u> We give an equivalent description of the category $\mathcal{L}_K(\Gamma_\Lambda)$ as a category of modules over a certain ring.

We define a partial ordering $S \leq$ on the set S of vertices of an oriented graph $\Gamma_\Lambda = (S, \Lambda A)$. We will put 1) $i = i$ for all $i \in S$; 2) if $(i, j) \in \Lambda A$, then $i < j$; 3) if $i \leqslant j$ and $j \leqslant k$, then $i \leqslant k$. Thus, the relation $i \leqslant k$ means that there exists a path $(x_1, \ldots, x_l)$, joining the points $i = x_1$ and $k = x_l$, such that $(x_j, x_{j+1}) \in \Lambda A$ for all $1 \leqslant j < l$.

The set consisting of pairs (i, j) such that $i \leqslant j$ will be denoted by $\overline{\Lambda A}$. It is easily shown that $\overline{\Lambda A}$ is the smallest of all sets $X \subset S \times S$, having the following properties: 1) $\Lambda A \subset X$; 2) $(i, i) \in X$ for all $i \in S$; 3) if $(i, j) \in X$ and $(j, k) \in X$, then $(i, k) \in X$.

We define the K-algebra $L_\Lambda$ to be the algebra with generators $a_{ij}$ such that $i \leqslant j$, and relations 1) $a_{ij}a_{kl} = 0$ for $j \neq k$; 2) $a_{ij}a_{jl} = a_{il}$.

We remark that it follows from relation 2), in particular, that $a_{ii}^2 = a_{ii}$, i.e., the $a_{ii}$ are idempotents in $L_\Lambda$. We set $e = \Sigma a_{ii}$. It is easily seen that e is the identity element of $L_\Lambda$.

A left module M over the algebra $L_\Lambda$ may be viewed as a vector space V over the field K on which the operators $a_{ij}$, $(i, j) \in \overline{\Lambda A}$ act. The category of left modules over the algebra $L_\Lambda$ will be denoted by $C(L_\Lambda)$.

Assertion 7. The categories $\mathcal{L}_K(\Gamma_\Lambda)$ and $C(L_\Lambda)$ are equivalent.

Proof. We construct in terms of the left module V over the algebra $L_\Lambda$ a representation $\rho$ of $\Gamma_\Lambda$. We set $V_i = a_{ii}V$. Since $a_{ii}^2 = a_{ii}$, $\Sigma a_{ii} = e$ and $a_{ii}a_{jj} = 0$ for $i \neq j$, we have $V \cong \bigoplus_{i \in S} V_i$. Since $a_{ij}a_{tt} = \delta_{jt}a_{it}$, it is easily seen that $a_{ij}V_j \subseteq V_i$ and $a_{ij}V_t = 0$ for $j \neq t$.

We now define a representation $\rho$ of $\Gamma_\Lambda$ by setting $\rho = \{V_i, \varphi_{ij}\}$, where $\varphi_{ij}$, $i \neq j$, $(i, j) \in \Lambda A$, is the restriction of $a_{ij}$ to $V_j$, i.e., $a_{ij}: V_j \to V_i$.

It is easy to verify that this construction is functorial, i.e., we have constructed a functor $T: C(L_\Lambda) \to \mathcal{L}_K(\Gamma_\Lambda)$.

We define the functor $Q: \mathcal{L}_K(\Gamma_\Lambda) \to C(L_\Lambda)$ as follows. Let $\rho = \{V_i, \varphi_{ij}\} \in \mathcal{L}_K(\Gamma_\Lambda)$. We put $V = \bigoplus_{i \in S} V_i$ and let $x \in V = \bigoplus_{i \in S} V_i$. We will assume that $V_i$ is canonically imbedded in V; then x can be written in the form $x = \sum_{i \in S} x_i$, $x_i \in V_i$. We define operators $a_{ii}$ by $a_{ii}x = x_i$. For $(i, j) \in \Lambda A$ we set $a_{ij}x = \varphi_{ij}x_j$, where, as before, $x = \sum_{i \in S} x_i$, $x_i \in V_i$. Now let $(j, k) \in \overline{\Lambda A}$. By definition, this means that there exists a path $(x_1, \ldots, x_l)$ such that $j = x_1$, $k = x_l$ and $(x_s, x_{s+1}) \in \Lambda A$ for $1 \leqslant s < l$. We put $a_{jk} = a_{x_1x_2}a_{x_2x_3} \cdots a_{x_{l-1}x_l}$. It is easy to verify that we have thereby constructed a functor $Q: \mathcal{L}_K(\Gamma_\Lambda) \to L_\Lambda$ and that $TQ \cong 1_{\mathcal{L}}$. Thus, the categories $\mathcal{L}_K(\Gamma_\Lambda)$ and $C(L_\Lambda)$ are equivalent.

4.2. It follows from Assertion 7 that the projective objects of the category $\mathcal{L}_K(\Gamma_\Lambda)$ correspond to projective modules over the algebra $L_\Lambda$.

Assertion 8. The indecomposable modules over the K-algebra $L_\Lambda$ are indexed in a natural way by the elements of S (the vertices of the tree $\Gamma_\Lambda$).

Proof. Each indecomposable projective module $M \in C(L_\Lambda)$ is an indecomposable direct summand of the algebra $L_\Lambda$ considered as a left module over itself. By definition, the module $L_\Lambda$ is a vector space with basis $a_{ij}$, $(i, j) \in \overline{\Lambda A}$. We denote by $P_t$ the subspace of $L_\Lambda$ with basis $\{a_{it}\}$, where the index t is fixed and $(i, t) \in \overline{\Lambda A}$. It follows from the formulas $a_{ij}a_{jt} = a_{it}$ and $a_{ik}a_{jt}|_{k \neq j} = 0$ that the subspace $P_t$ is a left submodule of $L_\Lambda$. Moreover, it follows from the equality $a_{tt}(a_{tt}) = a_{tt}$, which holds for every basis element $a_{it} \in P_t$, that the submodule $P_t$ is indecomposable.

Thus, the free left module $L_\Lambda$ splits into a direct sum of projective modules $P_t$, $L_\Lambda = \bigoplus_{t \in S} P_t$. The submodule $P_t$ as a vector space has the basis $\{a_{it}\}$, $i \in S$, $(i, t) \in \overline{\Lambda A}$, consisting of generators $a_{it}$ of the algebra $L_\Lambda$ with fixed second index t.

Assertion 9. The representation $\rho_{t,\Lambda}^{(1)} = \{V_{i,t,m_i}^{(1)}; \varphi_{ij}\}$, corresponding to a vertex $t \in S$, is such that

a) $\dim V_{i,t,m_i}^{(1)} = 1$ if and only if $i \leqslant t$,

b) $V_{i,t,m_i}^{(1)} = 0$ in all other cases, i.e., when either $t < i$, or i and t are not comparable under the partial ordering $\leq$ generated by the orientation $\Lambda$.

Proof. The set of triples $(i, t, m_i) \in S \times S \times \mathbf{Z}$ corresponding to the representation $\rho_t^{(1)}$, was denoted by us as $r_t^{(1)}$. It is isomorphic to the graph of the compatible function $\chi_{t,\Lambda}^{(1)}(i)$ such that $m_i = \chi_{i,\Lambda}^{(1)}(i)$ and $\chi_{t,\Lambda}^{(1)}(t) = 1$. Let $(x_1, \ldots, x_l)$ be the minimal path joining the points $x_1 = i$ and $x_l = t$. The value $\chi = \chi_{t,\Lambda}^{(1)}$ at the vertex $i = x_1$ can be found from the equation

$$\chi(t) = \chi(i) - \chi(t) + \chi(t) = (\chi(x_1) - \chi(x_l)) + 1 = 1 + \sum_{s=1}^{l-1} (\chi(x_s) - \chi(x_{s+1})). \tag{5}$$

Since $\chi$ is compatible and the vertices $x_s$ and $x_{s+1}$ $(1 \leqslant s < l)$ are joined in the graph $\Gamma$, we have $|\chi(x_s) - \chi(x_{s+1})| = 1$. Hence it is clear from Eq. (5) that $\chi(i) \leqslant l$. Moreover, $\chi(i) = l$ if and only if $\chi(x_s) - \chi(x_{s+1}) = 1$ for all s. By definition, the function $\chi = \chi^{(1)}_{t,\Lambda}$ is compatible with the orientation $\Lambda$ and, therefore, $\chi(x_{s+1}) = \chi(x_s) - 1 \Leftrightarrow (x_s, x_{s+1}) \in \Lambda A$, i.e., $x_s < x_{s+1}$. Thus, the equality $\chi^{(1)}_{t,\Lambda}(i) = l$ holds only when $i \leqslant t$. In all the remaining cases, $\chi^{(1)}_t(i) < l$. Applying Lemma 6, we see that the assertion has been proved.

The nonzero spaces $V^{(1)}_{i,t,m_i}$ can also be described as follows:

a) $V^{(1)}_{t,t,m_t} = V^{(1)}_{t,t,1} = Kb_t$,

b) for all vertices $i \in S$, $(i, t) \in \Lambda A$, we have $V^{(1)}_{i,t,m_i} = V^{(1)}_{i,t,2} = Kb_ib_t$,

c) for all vertices $j \in S$ such that $(j, t) \notin \Lambda A$, but $j < t$, $V^{(1)}_{j,t,m_j} = V^{(1)}_{j,t,l} = Kb_jb_{x_2} \ldots b_{x_{l-1}}b_t$, where $(j, x_2, \ldots, x_{l-1}, t)$ is the minimal path joining the vertices j and t.

These equalities have been proved, essentially, in Lemma 6.

THEOREM 3. In the category $\mathscr{L}_K(\Gamma_\Lambda)$, the number of distinct indecomposable projective representations is equal (up to equivalence) to the number of vertices in the graph $\Gamma$, and these representations are given by the $\rho^{(1)}_{t,\Lambda}$, $t \in S$.

Proof. We have determined $R^{(1)}_t = \sum_{i \in S} V^{(1)}_{i,t,m_i}$. In accordance with Assertion 9, we can choose a basis $\xi_{it}$ in the space $R^{(1)}_t$ (where $i \leqslant t$ and t is fixed) as follows: a) $\xi_{tt} = b_t$; b) if $(i, t) \in \Lambda A$, then $\xi_{it} = b_ib_t$; c) if $j < t$ but $(j, t) \notin \Lambda A$, then $\xi_{jt} = b_jb_{x_2} \ldots b_{x_{l-1}}b_t$, where $(j, x_2, \ldots, x_{l-1}, t)$ is the minimal path joining the vertices j and t.

We defined previously the projective module $P_t$ over the algebra $L_\Lambda$ as the submodule of $L_\Lambda$ generated by the elements $a_{it}$ $(i \leqslant t)$ of the algebra $L_\Lambda$. It is seen from the above description of the space $R^{(1)}_t$ that there exists a natural isomorphism $\varphi : R^{(1)}_t \to P_t$, defined by $\varphi\xi_{it} = a_{it}$ for all $i \leqslant t$.

We define an action of the operators $a_{ij} \in L_\Lambda$, where $i \leqslant j$, in $R^{(1)}_t$ by setting $a_{kl}\xi_{jt}|_{l \neq j} =$ and $a_{kj}\xi_{jt} = \xi_{kt}$. It is easy to see that this turns $R^{(1)}_t$ into an $L_\Lambda$-module isomorphic to the module $P_t$.

It is also easy to see that the construction just given, which given a representation $\rho^{(1)}_t$ constructs a module over the algebra $L_\Lambda$, coincides with the functor Q described above. We have thus shown that $Q\rho^{(1)}_{t,\Lambda} \simeq P_t$. The theorem is proved.

## 5. Representations $\rho^{(n)}_{t,\Lambda}$ and Coxeter Functors

The goal of this section is to prove Theorem 2, which asserts that $\rho^{(n+1)}_{t,\Lambda} \simeq \Phi^-\rho^{(n)}_{t,\Lambda}$.

5.1. We first give the definition of the functors $\Phi^+$, $\Phi^-$ in [1]. These functors are defined in terms of the reflection functors $F^+_i$, $F^-_i$ $(i \in S)$

A vertex $p \in S$ of an oriented graph $\Gamma_\Lambda$ is said to be (−)-accessible relative to the orientation $\Lambda$ if all arrows joining p to neighboring vertices $j \in S(p)$ emanate from p, i.e., $(j, p) \in \Lambda A$ for all $j \in S(p)$. Analogously, a vertex $q \in S$ is said to be (+)-accessible if for all $i \in S(q)$ $(q, i) \in \Lambda A$ (i.e., all arrows joining q to $i \in S(q)$ are directed toward q).

We denote by $\sigma_i\Lambda$ the orientation obtained from $\Lambda$ by reversing the direction of all the arrows joining the point i to the points $j \in S(i)$.

Let $\rho \in \mathscr{L}(\Gamma_\Lambda)$ be a representation, $\rho = \{V_i, \varphi_{ij}\}$, and let p be some (−)-accessible vertex of the graph $\Gamma_\Lambda$. We construct using the representation $\rho$ a new representation $F^-_p\rho$ as follows. Since the vertex p is (−)-accessible, there are m linear maps $\varphi_{lp} : V_p \to V_l$, defined in the representation $\rho$, where $l \in \{l_1, \ldots, l_m\} = S(p)$. We have thereby defined a linear map h: $V_p \to \bigoplus_{l \in S(p)} V_l$, given by the formula $h(x_p) = (\varphi_{l_1p}(x_p), \varphi_{l_2p}(x_p), \ldots, \varphi_{l_mp}(x_p))$. We set $\bar{V}_p = \operatorname{Coker} h = (\bigoplus_{l \in S(p)} V_l)/\operatorname{Im} h$ and denote by $\psi$ the natural map $\psi : \bigoplus V_l \to \bar{V}_p$. We have thus constructed an exact sequence $V_p \xrightarrow{h} \bigoplus_{l \in S(p)} V_l \xrightarrow{\psi} \bar{V}_p \to 0$.

We denote by $\widetilde{\varphi}_{pl}$ the composition $\psi i_l$ of the natural inclusion $i_l : V_l \to \bigoplus_{l \in S(p)} V_l$ with the imorphism $\psi$.

The spaces obtained can be depicted in the diagram

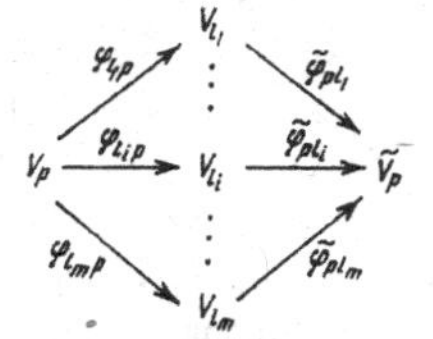


The new representation $F_p^- \rho = \{V_i^-, \varphi_{ij}^-\}$ is defined as follows: $V_p^- = \widetilde{V}_p$; $V_i^- = V_i$, if $i \neq p$; $_j = \widetilde{\varphi}_{pj}$ for all $(p, j) \in \sigma_p\Lambda$ and $\varphi_{ij}^- = \varphi_{ij}$ for all $(i, j) \in \Lambda A$ such that $j \neq p$. It is easy to rify that we have thereby defined a functor $F_p^- : \mathcal{L}_K(\Gamma_\Lambda) \to \mathcal{L}_K(\Gamma_{\sigma_p\Lambda})$, called the reflection nctor.

The reflection functor $F_q^+$ is constructed by duality. Let the vertex $q \in S$ be (+)-acssible in the graph $\Gamma\Lambda$. We construct in terms of the representation $\rho = \{V_i, \varphi_{ij}\} \in \mathcal{L}(\Gamma_\Lambda)$ new representation $F_q^+\rho = \{V_i^+, \varphi_{ij}^+\}$ as follows. For $j \neq q$ we put $V_j^+ = V_j$; and for pairs (i, j) $\Lambda A$ such that $i \neq q$, we put $\varphi_{ij}^+ = \varphi_{ij}$. The new space $V_q^+$ and maps $\varphi_{iq}^+ : V_q^+ \to V_i$, $i \in S(q)$, e defined as follows. Let $h: \bigoplus_{i \in S(q)} V_i \to V_q$ be the map given by $h(v_1, \ldots, v_m) = \sum_{i \in S(q)} \varphi_{qi} v_i$ (where $\in V_i$ and $\varphi_{qi} v_i \in V_q$). We set $V_q^+ = \operatorname{Ker} h$. We have constructed the following exact sequence:

$$0 \to V_q^+ \xrightarrow{\mu} \bigoplus_{i \in S(q)} V_i \xrightarrow{h} V_q.$$

denote by $\varphi_{iq}^+$ the composition $\pi_i \mu$ of the inclusion $\mu$: $V_q^+ \to \bigoplus_{i \in S(p)} V_i$ and the projection $\pi_i$ : $\bigoplus_{S(q)} V_j \to V_i$. It is easy to verify that we have thereby defined a functor $F_q^+ : \mathcal{L}_k(\Gamma_\Lambda) \to \mathcal{L}_k$ $_{\sigma_q\Lambda})$.

A sequence $(x_1, x_2, \ldots, x_m)$ of vertices of the graph $\Gamma\Lambda$ will be called (+)-accessle relative to the orientation $\Lambda$ if the vertex $x_1$ is (+)-accessible relative to $\Lambda$ and for ery $1 < j \leqslant m$ the vertex $x_j$ is (+)-accessible relative to the orientation $\Lambda' = \sigma_{x_{j-1}} \cdots \sigma_{x_1}\Lambda$. quences which are (−)-accessible are defined analogously. Let $\alpha = \{x_1, \ldots, x_r\}$ be a (−)-cessible sequence of vertices $x_j$ in the graph $\Gamma\Lambda$ (where r is the number of vertices in S). e functor $\Phi_\alpha^- = F_{x_r}^- F_{x_{r-1}}^- \cdots F_{x_1}^-$ is called the Coxeter functor. It is easy to check that if e graph $\Gamma\Lambda$ contains different (−)-accessible sequences $\alpha$ and $\beta$, then the functors $\Phi_\alpha^-$ and are isomorphic. We will therefore denote this functor by $\Phi^-$. Clearly, $\Phi^- : \mathcal{L}(\Gamma_\Lambda) \to \mathcal{L}(\Gamma_\Lambda)$.

5.2. The proof of Theorem 2 is obtained in elementary fashion from the following propition.

Proposition 10. Assume the vertex p is (−)-accessible in the graph $\Gamma\Lambda$. Then $F_p^- \rho_{t,\Lambda}^{(n)} \cong$ $^{n')}_{\sigma_p\Lambda}$, where $n' = n + \delta_{pt}$ and $\delta_{pt}$ is the Kronecker symbol.

The representation $\rho_{t,\Lambda}^{(n)}$ corresponds to a compatible function $\chi : S \to \mathbf{Z}$, $\chi = \chi_{t,\Lambda}^{n}$, so that $_{t,m_i} \in \rho_{t,\Lambda}^{(n)}$ only if $m_i = \chi(i)$.

LEMMA 11. The compatible functions $\chi$ and $\chi'$ corresponding to the representations $\rho_{t,\Lambda}^{(n)}$ $^{n')}_{,\Lambda'}$ (where $\Lambda' = \sigma_p\Lambda$ and $n' = n + \delta_{pt}$) are such that: 1) for all i not equal to p, $\chi'(i) = \chi(i)$; $\chi'(p) = \chi(p) + 2$.

Proof. We define a function $\bar{\chi}: S \to Z$ by $\bar{\chi} = \chi + 2\delta_p$, where the function $\delta_p : S \to \mathbf{Z}$ is fined by $\delta_p(i) = 0$ if $i \neq p$, and $\delta_p(p) = 1$. We show that $\bar{\chi}$ is a function compatible with e orientation $\Lambda' = \sigma_p\Lambda$ and corresponds to the representation $\rho_{t,\Lambda'}^{(n')}$, i.e., $\bar{\chi} = \chi_{t,\Lambda'}^{(n')} = \chi_{t,\sigma_p\Lambda}^{n+\delta_{pt}}$. hypothesis, the point p is (−)-accessible in the graph $\Gamma\Lambda$, i.e., $(j, p) \in \Lambda A$ for all $\{p,$ $\in A$. In terms of the function $\chi = \chi_{t,\Lambda}^{(n)}$, this means that for all $j \in S(p)$

$$\chi(p) = \chi(j) - 1. \tag{6}$$

The new function $\bar{\chi} = \chi + 2\delta_p$ is such that if $j \neq p$, then $\bar{\chi}(j) = \chi(j)$ and $\bar{\chi}(p) = \chi(p) + 2$. Using these properties and Eq. (6), we obtain that $\bar{\chi}(p) = \chi(p) + 2 = \chi(j) + 1 = \bar{\chi}(j) + 1$. This equality is true for all points $j \in S(p)$. Thus, $\bar{\chi}$ is compatible with the orientation $\sigma_p\Lambda$.

Using Corollary 4, we can thus write

$$\bar{\chi}(j) = 2n' - 1 + h_{t,\Lambda'}(j) - h_{t,\Lambda'}(t) = \chi^n_{t,\Lambda}(j) + 2\delta_p(j) = 2n - 1 + h_{t,\Lambda}(j) - h_{t,\Lambda}(t) + 2\delta_p(j).$$

Putting j = t in this equality, we obtain $2n' - 1 = 2n - 1 + 2\delta_p(t)$, i.e., $n' = n + \delta_p(t) = n + \delta_p$ The lemma is proved.

It follows from Lemma 11 that the sets of spaces $\{V_{i,t,m_i}\}$ and $\{V_{i,t,m'_i}\}$, corresponding to the representations $\rho^{(n)}_{t,\Lambda}$ and $\rho^{(n')}_{t,\Lambda'}$, differ only in one space. Namely, $\rho^{(n)}_{t,\Lambda}$ contains $V_{p,t,m_p}$, and $\rho^{(n')}_{t,\Lambda'}$ contains $V_{p,t,m'_p} = V_{p,t,m_p+2}$. All the remaining spaces $V_{i,t,m_i}$ and $V_{i,t,m'_i}$ $(i \neq p)$ are the same.

We consider the sequence of spaces

$$V_{p,t,m_p} \xrightarrow{\varphi_\Sigma} \sum_{j\in S(p)} V_{j,t,m_j+1} \xrightarrow{b_p} V_{p,t,m_p+2}, \tag{7}$$

where $\varphi_\Sigma$ and $b_p$ are the maps in the algebra $B_K(\Gamma)$ corresponding to left multiplication by the elements $(\sum_{j\in S(p)} b_j)$ and $b_p$, respectively (i.e., $\varphi_\Sigma(x) = \sum_{j\in S(p)} b_j x$ for $x \in V_{p,t,m_p}$).

The preceding arguments show that the subspaces corresponding to the left-hand and middle terms of this sequence occur in the representation $\rho^{(n)}_{t,\Lambda}$, while the subspaces corresponding to the middle and right-hand terms of the sequence appear in $\rho^{(n')}_{t,\Lambda'}$.

It therefore follows from the definition of the functor $F^-_p$ that in order to prove Proposition 10, it suffices to prove that the sequence (7) is exact at the middle term and that $V_{p,t,m_p+2} \simeq \operatorname{Coker} \varphi_\Sigma$. This in turn is a special case of the following proposition.

Proposition 12. Let $V_{i,t,m}$ be the subspace in the algebra $B_K(\Gamma)$ generated by the reduced monomials of the form $b_i y b_t$ of degree m. Then

$$V_{i,t,m} \xrightarrow{\varphi_\Sigma} \sum_{j\in S(i)} V_{j,t,m+1} \xrightarrow{b_i} V_{i,t,m+2} \to 0 \tag{8}$$

is an exact sequence, where $\varphi_\Sigma$ and $b_i$ are the linear maps in $B_K(\Gamma)$ given by left multiplication by $\sum_{j\in S(i)} b_j$ and $b_i$, respectively.

Proof. a) Let $x \in V_{i,t,m}$. This means that $x = b_i y$ (where $y \in B_K(\Gamma)$ has degree $m-1$). By definition, $\varphi_\Sigma = \sum_{j\in S(i)} b_j$. Hence, $(b_i\varphi_\Sigma)x = b_i(\sum_{j\in S(i)} b_j)b_i y = 0$, since $b_i(\sum_{j\in S(i)} b_j)b_i = 0$. We have proved that $\operatorname{Im}\varphi_\Sigma \leqslant \operatorname{Ker} b_i$.

b) We prove that the map $b_i$ in this sequence is an epimorphism. By definition, the space $V_{i,t,m+2}$ is generated by the reduced monomials $z = b_i v b_t$ of degree m + 2. Let $m \geqslant 1$. Then the reduced monomial $b_i v b_t = b_i b_j w b_t$, where $\{i, j\} \in A$, i.e., $j \in S(i)$. It is clear that $b_j w b_t \in V_{j,t,m+1}$, in other words, the vector $z = b_i v b_t \in \operatorname{Im} b_i$. Since the space $V_{i,t,m+2}$ is generated by reduced monomials z of the form $b_i v b_t$, every vector in $V_{i,t,m+2}$ has an inverse image in $\sum_{j\in S(i)} V_{j,t,m+1}$ under the map $b_i$. Thus, the map $b_i$ in diagram (8) is an epimorphism. The proof of this fact in the case when m = 0 is elementary.

c) It remains to prove that sequence (8) is exact at the middle term. Each space $V_\alpha$ ($\alpha$ = (i, t, m)) is a quotient space $W_\alpha / W_\alpha \cap J$, where $W_\alpha = W_{i,t,m}$ is the space of homogeneous monomials $c_i x c_t$ of degree m in the algebra $P_K(\Gamma)$.

Consider the diagram

$$\begin{array}{ccccc} W_{i,t,m} & \xrightarrow{\psi_\Sigma} & \sum_{j\in S(i)} W_{j,t,m+1} & \xrightarrow{c_i} & W_{i,t,m+2} \\ \downarrow{\scriptstyle \pi_\alpha} & & \downarrow{\scriptstyle \pi_\Sigma} & & \downarrow{\scriptstyle \pi_{\alpha'}} \\ V_{i,t,m} & \xrightarrow{\varphi_\Sigma} & \sum_{j\in S(i)} V_{j,t,m+1} & \xrightarrow{b_i} & V_{i,t,m+2} \to 0, \end{array} \tag{9}$$

ere $\varphi_\Sigma$ and $c_i$ denote the maps in the algebra $P_K(\Gamma)$ corresponding to left multiplication $\sum_{j\in S(i)} c_j$ and $c_i$, respectively, and where $\pi_\alpha: W_\alpha \to V_\alpha$ is the natural projection with kernel $_\alpha = W_\alpha \cap J$, with $\pi_\Sigma$ and $\pi_{\alpha'}$ defined analogously.

We remark that the map $c_i$ in (9) is an isomorphism. Indeed, we can choose the set of duced monomials $\sum_{j\in S(i)} W_{j,t,m+1}$ of degree m + 1 to be a basis in the set $\{c_j y c_t\}$. The map $c_i$ kes each such monomial into a reduced monomial $c_i c_j y c_t$ of degree m + 2. These monomials turn form a basis in $W_{i,t,m+2}$.

For brevity, we will denote the spaces $W_{i,t,m}$, $\sum_{j\in S(i)} W_{j,t,m+1}$, and $W_{i,t,m+2}$ by $W_\alpha$, $W_\Sigma$, and $W_{\alpha'}$, spectively.

We denote the kernels of the epimorphisms $\pi_\alpha$, $\pi_\Sigma$, and $\pi_{\alpha'}$ by $K_\alpha$, $K_\Sigma$, and $K_{\alpha'}$. As we have marked above, $K_\alpha = W_\alpha \cap J$ and $K_\Sigma = W_\Sigma \cap J$, where J is the two-sided ideal in $P_K(\Gamma)$ generated by the elements $c_i(\sum_{j\in S(i)} c_j)c_i$, $i \in S$.

We depict these spaces in the following commutative diagram,

$$\begin{array}{ccccc}
0 & & 0 & & 0 \\
\downarrow & & \downarrow & & \downarrow \\
K_\alpha & & K_\Sigma & & K_{\alpha'} \\
\downarrow & & \downarrow & & \downarrow \\
W_\alpha & \xrightarrow{\psi_\Sigma} & W_\Sigma & \xrightarrow{c_i} & W_{\alpha'} \\
\downarrow & & \downarrow & & \downarrow \\
V_\alpha & \xrightarrow{\varphi_\Sigma} & V_\Sigma & \xrightarrow{b_i} & V_{\alpha'} \to 0 \\
\downarrow & & \downarrow & & \downarrow \\
0 & & 0 & & 0
\end{array}$$

In order to prove the equality $\operatorname{Im}\varphi_\Sigma = \operatorname{Ker} b_i$, it suffices to prove that the spaces $\pi_\Sigma^{-1}(\operatorname{Im}\varphi_\Sigma)$ and $\pi_\Sigma^{-1}(\operatorname{Ker} b_i)$ are equal in the space $W_\Sigma$ (where $\pi_\Sigma^{-1}(X)$ denotes the complete inverse image of the subspace X under the map $\pi_\Sigma$: $W_\Sigma \to V_\Sigma$). It is easy to verify that $\pi_\Sigma^{-1}(\operatorname{Im}\varphi_\Sigma) = K_\Sigma + \psi_\Sigma W_\alpha$, $\pi_\Sigma^{-1}(\operatorname{Ker} b_i) = (c_i)^{-1}(K_{\alpha'})$. Therefore, in order to prove that $\operatorname{Im}\varphi_\Sigma = \operatorname{Ker} b_i$ we ve to prove that $K_\Sigma + \omega_\Sigma W_\alpha = c_i^{-1}(K_{\alpha'})$. Applying the isomorphism $c_i$ to both sides of this uality, we obtain the equivalent equality

$$c_i K_\Sigma + c_i \psi_\Sigma W_\alpha = K_{\alpha'}. \tag{10}$$

In order to prove (10), we remark that the ideal J in $P_K(\Gamma)$ is generated by the elements $d_i = c_i(\sum_{j\in S(i)} c_j)c_i$. An arbitrary element $w \in J$ can therefore be written as a sum $w = \sum_\beta \sum_i$ $x_{\beta i} d_i y_\beta$, where $a_{\beta i} \in K$, $i \in S$, $x_\beta$ and $y_\beta$ are certain monomials in the algebra $P_K(\Gamma)$. Elements the form $x d_i y$, where x and y are monomials, will be called reduced elements of J.

The space $K_{\alpha'} = W_{\alpha'} \cap J$, where $\alpha' = (i, t, m + 2)$, is spanned by all reduced elements degree m + 2 of the form $x_\beta d_j y_\beta$.

We show that every such reduced element either belongs to the subspace $c_i K_\Sigma$, or to the bspace $c_i \psi_\Sigma W_\alpha$. We consider two cases.

1) m = 1, i.e., $W_{\alpha'} = W_{i,t,3}$. In this case, the subspace $K_{\alpha'} = W_{\alpha'} \cap J$ is either onemensional (when i = t) or else 0 (if i ≠ t). In the case i = t we have $K_{\alpha'} = Kd_t = K(c_t \sum_{j\in S(t)}$ $c_t)$. Here the equality $c_t\psi_\Sigma W_\alpha = K_{\alpha'}$ is obvious, since $W_\alpha = W_{t,t,1} = K(c_t)$ and $c_t\psi_\Sigma(c_t) = c_t(\sum_{j\in S(t)}$ $c_t = d_t$.

2) Assume that m > 1. Then each reduced monomial $x_\beta d_j y_\beta \in K_{\alpha'}$ has degree m + 2 > 3, so at the product $x_\beta y_\beta$ of monomials has degree m − 1 > 0. We must now consider two variants: $x_\beta \neq 1$ and $x_\beta = 1$.

2a) Let $x_\beta \neq 1$, i.e., assume the degree of the monomial $x_\beta$ is greater than or equal to . Since the monomial $x_\beta d_j y_\beta \in W_{\alpha'} = W_{i,t,m+2}$, it is clear that $x_\beta = c_i z_\beta$, i.e., $x_\beta d_j y_\beta = c_i z_\beta d_j y_\beta$. is now easy to show that $c_i z_\beta d_j y_\beta \in c_i(K_\Sigma)$.

2b) Assume that $x_\beta = 1$, i.e., we consider a reduced element of the form $d_j y_\beta \in K_{\alpha'} \subset W_{i,t,m+2}$. Since $d_j y_\beta \in W_{i,t,m+2}$ and each element in $W_{i,t,m+2}$ has the form $c_i x$, this means that $j = i$, i.e., we consider a reduced element $d_i y_\beta = c_i(\sum_{j \in S(i)} c_j) y_\beta$. It follows from the definition of $c_i$ and $\psi_\Sigma$ that $d_i y_\beta = c_i \psi_\Sigma (c_i y_\beta)$, i.e., $d_i y_\beta \in c_i \psi_\Sigma W_\alpha$.

We have thus proved that each reduced element $x_\beta d_j y_\beta$ in the subspace $K_{\alpha'} = W_{\alpha'} \cap J$ either belongs to the subspace $c_i K_\Sigma$, or else to $c_i \psi_\Sigma W_\alpha$. Since the reduced elements of the form $x_\beta d_j y_\beta$ generate $K_{\alpha'}$, Eq. (10) and also Proposition 12 are proved.

## LITERATURE CITED

1. I. N. Bernshtein, I. M. Gel'fand, and V. A. Ponomarev, "Coxeter functors and a theorem of Gabriel," Usp. Mat. Nauk, 28, No. 2, 19-33 (1973).
2. I. M. Gel'fand and V. A. Ponomarev, "Structures, representations, and their associated algebras. I, II," Usp. Mat. Nauk, 31, No. 5, 71-88 (1976); Usp. Mat. Nauk, 32, No. 1, 85-106 (1977).

# Part IX

# Representations of infinite-dimensional groups

# 1.

(with M. I. Graev and A. M. Vershik)

# Representations of the group $SL(2, \mathbf{R})$, where R is a ring of functions

Usp. Mat. Nauk **28** (5) (1973) 82–128 [Russ. Math. Surv. **28** (5) (1973) 87–132].
Zbl. **297**:22003

We obtain a construction of the irreducible unitary representations of the group of continuous transformations $X \to G$, where $X$ is a compact space with a measure $m$ and $G = PSL(2, \mathbf{R})$, that commute with transformations in $X$ preserving $m$.

This construction is the starting point for a non-commutative theory of generalized functions (distributions). On the other hand, this approach makes it possible to treat the representations of the group of currents investigated by Streater, Araki, Parthasarathy, and Schmidt from a single point of view.

## Contents



## Introduction

One stimulus to the present work was the desire to extend the theory of generalized functions to the non-commutative case. Let us explain what we have in mind.

Let $\mathbf{R}$ be the real line, $X$ a compact manifold, and $f(x)$ an infinitely differentiable function on $X$ with values in $\mathbf{R}$, that is, a mapping $X \to \mathbf{R}$. A group structure arises naturally on the set of functions $f(x)$, which we denote by $\mathbf{R}^X$. Irreducible unitary representations of this group are defined by the formula $f(x) \mapsto e^{il(f)}$, where $l$ is a linear functional in the space of "test" functions $f(x)$. Thus, to each generalized function (distribution) there corresponds an *irreducible* representation of $\mathbf{R}^X$. If we replace $\mathbf{R}$ by any other Lie group $G$, then it is natural to ask for the construction of *irreducible* unitary representations of the group $G^X$, regarded as a natural non-commutative analogue to the theory of distributions. Such an attempt was made in [1], §3.

However, our progress was only partial. We succeeded in defining distributions with support at a single point or at a finite number of points (for the group $SU(2)$) – analogues to the delta function and its derivatives; we were also able to introduce the concept of a derivative and show that $\delta'$ is the derivative of $\delta$. The work came to a halt because we did not succeed in introducing the concept of an integral, without which the theory of generalized functions cannot go on.

The problem of constructing an integral for $G^X$ can be stated as follows. Suppose that an $X$ measure $m$ is given. We have to find *irreducible* unitary representations of $G^X$ that go over into equivalent ones under transformations of $X$ preserving $m$. Reducible representations of this kind can be constructed without special difficulty. However, even the case $G = \mathbf{R}$ indicates that for our purposes reducible representations are unsuitable.

For a long time it was not clear to the authors whether such irreducible representations exist for semisimple groups $G$. Finally we succeeded in constructing such representations for a number of semisimple groups, namely, groups in which the identity representation is not isolated in the set of all irreducible unitary representations.

In this paper we analyze in detail only the case of the group $SL(2, \mathbf{R})$. The fact is, as experience with representation theory shows, that an understanding of any new situation is impossible without a preliminary study of the group $SL_2$ from all points of view.

We have performed the construction of the integral several times, each time from a somewhat different standpoint. The order in which we have written down the various constructions corresponds more or less to the order in which we thought them out. The first construction proceeds from a very simple idea: to obtain the multiplicative integral as the limit of a tensor product of representations, each member of the product being a closer approximation to the identity representation than the last, more precisely, to the point of the representation space to which the identity representation is attached.

From the last few sections it is clear that this representation can also be interpreted in terms of the cocycles of Streater, Araki, Parthasarathy and Schmidt (more precisely, it is not the 1-cocycles that play the fundamental role, but rather the reducible representations from $\mathrm{Ext}^1$). The proof of the irreducibility of these representations is a new feature in our constructions.

At the end of §6 we construct two other projective unitary representations of the group $(PSL(2, \mathbf{R}))^X$.

The construction of the integral for all other groups $G^X$ for which $G$ satisfies the condition that the identity representation is not isolated in the set of all irreducible representations will be presented elsewhere.

The integral constructed in this paper provides us with a constructive representation of the group $G^X$, which in the terminology of mathematical physics is the group of currents of $G$.

Thus, this paper can also be regarded as a survey, from a somewhat different standpoint, of work on the representations of the group of currents.

Representations of the group of currents have been widely studied by a number of authors (Streater, Araki, Parthasarathy and Schmidt). See [4], [7]–[11], [13], [14], [15], and the further literature cited in the survey papers [6] and [12].

## §1. Some information on the representations of the group of real 2 X 2 matrices

1. **Representations of the supplementary series.** We consider here the group $G = PSL(2, \mathbf{R})$ of real matrices $g = \begin{pmatrix} \alpha & \beta \\ \gamma & \delta \end{pmatrix}$ with determinant 1 in which $g$ and $-g$ are identified. This group is known to be isomorphic to the group of complex matrices of the form $\begin{pmatrix} \alpha & \beta \\ \bar{\beta} & \bar{\alpha} \end{pmatrix}$, where $|\alpha|^2 - |\beta|^2 = 1$ and $g$ and $-g$ are again identified. In what follows we use either the first or the second definition of $G$, as convenient.

Let $G$ be given in the second form.

We introduce the space $K$ of continuous infinitely differentiable functions on the unit circle $|\zeta| = 1$ in the complex plane. With each real number $\lambda$ in the interval $0 < \lambda < 1$ we associate a representation $T_\lambda$ of $G$ in $K$.

$$\left(T_\lambda \begin{pmatrix} \alpha & \beta \\ \bar{\beta} & \bar{\alpha} \end{pmatrix} f\right)(\zeta) = f\left(\frac{\alpha\zeta + \bar{\beta}}{\beta\zeta + \bar{\alpha}}\right) |\beta\zeta + \bar{\alpha}|^{\lambda - 2}. \tag{1}$$

In $K$ there is a positive definite Hermitian form $(f_1, f_2)_\lambda$ that is invariant under the operators $T_\lambda$:[1)]

$$(f_1, f_2)_\lambda = \frac{\Gamma\left(1 - \frac{\lambda}{2}\right)}{4\sqrt{\pi}\,\Gamma\left(\frac{1-\lambda}{2}\right)} \int_0^{2\pi}\int_0^{2\pi} \left|\sin\frac{\varphi_1 - \varphi_2}{2}\right|^{-\lambda} f_1(e^{i\varphi_1})\overline{f_2(e^{i\varphi_2})}\, d\varphi_1\, d\varphi_2 \tag{2}$$

We denote by $H_\lambda$ the completion of $K$ in the norm $\|f\|_\lambda^2 = (f, f)_\lambda$.

---

1) The numerical factor is chosen so that $(1, 1)_\lambda = 1$, where 1 is the function identically equal to unity.

It is evident that the $T_\lambda$ can be extended to unitary operators in $H_\lambda$.

DEFINITION. A unitary representation in the Hilbert space $H_\lambda$, as defined by (1), is called a representation $T_\lambda$ of the supplementary series of $G$.

It is known that all representations $T_\lambda$, $0 < \lambda < 1$, of the supplementary series are irreducible and pairwise inequivalent.[1)]

It is sometimes convenient to specify the representations of the supplementary series in another manner. Let $G$ be given in the first form. For each $\lambda$, $0 < \lambda < 1$, we introduce the space $\mathcal{D}_\lambda$ of continuous real functions such that $f(x) = O(|x|^{\lambda-2})$ as $x \to \pm\infty$. In $\mathcal{D}_\lambda$ we introduce the positive definite Hermitian form $(f_1, f_2)_\lambda$:

$$(f_1, f_2)_\lambda = \int_{-\infty}^{+\infty}\int_{-\infty}^{+\infty} |x_1 - x_2|^{-\lambda} f_1(x_1)\overline{f_2(x_2)}\,dx_1\,dx_2. \tag{3}$$

A representation of the supplementary series acts in the Hilbert space obtained by completing $\mathcal{D}_\lambda$ in the norm $\|f\|_\lambda^2 = (f, f)_\lambda$. The representation operators have the following form:

$$\left(T_\lambda\begin{pmatrix}\alpha & \beta\\ \gamma & \delta\end{pmatrix} f\right)(x) = f\left(\frac{\alpha x+\gamma}{\beta x+\delta}\right)|\beta x+\delta|^{\lambda-2}. \tag{4}$$

**2. Canonical representations of $G$.** Some unitary representations, which we call canonical, of the group $G$ of $2 \times 2$ real matrices play an important role in our work. These very pretty representations of the matrix group are interesting for their own sake. We present two methods of specifying the canonical representations.

2a. THE FIRST METHOD. We specify $G$ in the second form. Further, we let $K \subset G$ be the maximal compact subgroup in $G$ consisting of the matrices of the form $\begin{pmatrix}e^{it} & 0\\ 0 & e^{-it}\end{pmatrix}$.

Of fundamental importance for the first specification of the canonical representations is the function

$$\psi(g) = \left(\frac{4}{\operatorname{Sp} gg^* + 2}\right)^{1/2}.$$

THEOREM 1.1. *For any $\lambda > 0$ the function $\psi^\lambda(g)$ is positive definite on $G$ and constant on the double cosets of $K$.*

PROOF. The fact that $\psi^\lambda(g)$ is constant on the double cosets of $K$ is obvious. That it is positive definite is a consequence of the following two lemmas.

---

1) (1) defines a unitary representation $T_\lambda$ also in the interval $1 < \lambda < 2$. The scalar product $(f_1, f_2)_\lambda$ in the space of this representation is defined as the analytic continuation of the function of $\lambda$ defined in the domain Re $\lambda < 1$ by the convergent integral (2).

The representations $T_\lambda$ and $T_{2-\lambda}$, $0 < \lambda < 2$, are known to be equivalent; hence we can always restrict our attention to the interval $0 < \lambda < 1$.

We denote by $\phi_\lambda(g)$ the zonal spherical function of the representation $T_\lambda$ of the supplementary series with respect to $K$, normalized so that $\phi_\lambda(e) = 1$.

LEMMA 1.1. *The function $\phi_\lambda(g)$ is continuous and differentiable in $\lambda$ at $\lambda = 0$. Also $\lim_{\lambda\to 0} \phi_\lambda(g) = 1$ and*

$$\lim_{\lambda\to 0} \frac{d\varphi_\lambda(g)}{d\lambda} = \ln \psi(g).$$

The proof follows easily from the explicit form of $\phi_\lambda(g)$:

$$\varphi_\lambda(g) = P_{-\lambda/2}\left(\frac{1}{2}\operatorname{Sp}(gg^*)\right),$$

where

$$P_{-\lambda/2}(x) = \frac{1}{\pi}\int_0^\pi (x + \sqrt{(x^2 - 1 \cos \phi)})^{-\lambda/2} d\phi$$

is the Legendre function.

LEMMA 1.2. *The function*

$$\psi^\mu(g) = \exp\left(\mu \frac{d\varphi_\lambda(g)}{d\lambda}\Big|_{\lambda=0}\right)$$

*is positive definite for $\mu > 0$.*

PROOF. By Lemma 1.1, $\frac{d\varphi_\lambda(g)}{d\lambda}\Big|_{\lambda=0} = \lim_{\lambda\to 0}\frac{\varphi_\lambda(g)-1}{\lambda}$. It is evident that $\frac{\varphi_\lambda(g)-1}{\lambda}$ is conditionally positive definite, that is, $\sum_{i,j} \frac{\varphi_\lambda(g_i g_j^{-1})-1}{\lambda}\xi_i\bar{\xi}_j \geqslant 0$ under the condition that $\sum \xi_i = 0$. It then follows that $\exp\left(\mu\frac{\varphi_\lambda(g)-1}{\lambda}\right)$ is positive definite for $\mu > 0$. Since the positive definite functions form a weakly closed set, the limit $\exp\left(\mu\frac{d(\varphi_\lambda(g)-1)}{d\lambda}\Big|_{\lambda=0}\right) = \lim_{\lambda\to 0}\exp\left(\mu\frac{\varphi_\lambda(g)-1}{\lambda}\right)$ is positive definite.

DEFINITION. The unitary representation of $G$ defined by the positive definite function $\psi^\lambda(g)$, $\lambda > 0$, is called *canonical.* A cyclic vector $\xi_\lambda$ in the space of a canonical representation for which $(T(g)\xi_\lambda, \xi_\lambda) = \psi^\lambda(g)$ is called *canonical.*

Let us construct a canonical representation. We consider the space $Y = K\backslash G$, which is a Lobachevskii plane. Let $y_0$ be the point of $Y$ that corresponds to the coset of the identity element. We define the kernel $\Psi^\lambda(y_1, y_2)$, where $\lambda > 0$, on the Lobachevskii plane by the formula

$$\Psi^\lambda(y_1, y_2) = \psi^\lambda(g_1 g_2^{-1}),$$

where $y_1 = y_0 g_1$, $y_2 = y_0 g_2$. By Theorem 1.1 this kernel is positive definite.

We consider the space of all finite continuous functions on $Y$. We denote by $L_\lambda$ the completion of this space in the norm:

$$\|f\|^2 = \int \Psi^\lambda(y_1, y_2) f(y_1)\overline{f(y_2)}\, dy_1\, dy_2,$$

where $dy$ is an invariant measure on $Y$.

A canonical representation of $G$ is defined by operators in $L_\lambda$ of the form

$$(T(g)f)(y) = f(yg).$$

(That the operators $T(g)$ are unitary and form a representation of $G$ is obvious.)

THEOREM 1.2. *If $\lambda > 1$, then a canonical representation in $L_\lambda$ splits into a direct integral over the representations of the principal continuous series of G. If $0 < \lambda < 1$, then*

$$L_\lambda = H_\lambda \oplus L_\lambda^0,$$

*where $H_\lambda$ is the space of the representation $T_\lambda$ of the supplementary series, and $L_\lambda^0$ splits into representations of the principal continuous series only.*

PROOF. It suffices to verify that $\psi^\lambda(g)$ can be expanded in zonal spherical functions of the corresponding irreducible representations. We may limit our attention to the matrices $g = \begin{pmatrix} \cosh t & \sinh t \\ \sinh t & \cosh t \end{pmatrix}$. For these matrices we have $\psi^\lambda(g) = \left(\frac{2}{\cosh 2t + 1}\right)^{\lambda/2}$. Furthermore, we know that the zonal spherical functions of the representations of the principal continuous series have the following form:

$$\varphi_{1+i\rho}(g) = P_{\frac{-1-i\rho}{2}}(\cosh 2t),$$

where $P_{\frac{-1-i\rho}{2}}$ is the Legendre function.

Let $\lambda > 1$; then $\left(\frac{2}{x+1}\right)^{\lambda/2}$ is square integrable on $[1, \infty)$ and can therefore be expanded in an integral of functions $P_{\frac{-1-i\rho}{2}}(x)$ (the Fock-Mehler transform). Thus, we have

$$\psi^\lambda(g) = \int_0^\infty a_\lambda(\rho)\, \varphi_{1+i\rho}(g)\, d\rho.$$

The coefficients $a_\lambda(\rho)$ in this expression can be calculated by the inversion formula for the Fock-Mehler transform; we obtain

$$a_\lambda(\rho) = \frac{1}{2\pi^2} \frac{\Gamma\left(\lambda - \frac{1}{2} - i\rho\right) \Gamma\left(\lambda - \frac{1}{2} + i\rho\right)}{\Gamma^2(\lambda)} \rho \tanh \pi\rho$$

Now let $0 < \lambda < 1$. It is known that the zonal spherical function $\phi_\lambda(g) = P_{-\lambda/2}(\cosh 2t)$ of the representation $T_\lambda$ of the supplementary series has the following asymptotic form:

$$P_{-\lambda/2}(x) = \frac{2^{-\lambda/2}\Gamma\left(\frac{1-\lambda}{2}\right)}{\sqrt{\pi}\,\Gamma\left(1 - \frac{\lambda}{2}\right)} x^{-\lambda/2} + O\left(\frac{1}{x}\right) \quad \text{as } x \to \infty.$$

It follows that the function

$$\left(\frac{2}{x+1}\right)^{\lambda/2}-\frac{\sqrt{\pi}\,2^{\lambda}\Gamma\left(1-\frac{\lambda}{2}\right)}{\Gamma\left(\frac{1-\lambda}{2}\right)}P_{-\lambda/2}(x)$$

is square integrable on $[1, \infty)$ and can therefore be expanded in an integral of functions $P_{\frac{-1-i\rho}{2}}(x)$.

In what follows we say that the canonical representation $L_\lambda$ for $0 < \lambda < 1$ is congruent to the representation $H_\lambda$ of the supplementary series modulo representations of the principal series.

The explicit separation of the component of the supplementary series in $L_\lambda$ will be carried out a little later (see Theorem 1.3).

In conclusion we give another two expressions for the kernel $\Psi^\lambda(y_1, y_2) = \psi^\lambda(g_1 g_2^{-1})$, where $y_1 = y_0 g_1$, $y_2 = y_0 g_2$.

From the definition of $\psi(g)$ it follows easily that

$$\Psi^\lambda(y_1, y_2) = \cosh^{-\lambda}\rho(y_1, y_2),$$

where $\rho(y_1, y_2)$ is the invariant metric on the Lobachevskii plane.

We suppose further that the Lobachevskii plane $Y$ is realized as the interior of the unit disk $|z| < 1$ in the complex plane, where $G$ acts by fractional linear transformations: $z \to \frac{\alpha z+\bar{\beta}}{\beta z+\bar{\alpha}}$. It is easy to verify that in this realization $\Psi^\lambda$ has the following form:

$$\Psi^\lambda(z_1, z_2) = \left[\frac{(1-|z_1|^2)(1-|z_2|^2)}{|1-z_1\bar{z}_2|^2}\right]^{\lambda/2}.$$

We observe that the invariant measure on the unit disk is $(1-|z|^2)^{-2}dz\,d\bar{z}$. Thus, in the realization on the unit disk the norm in the space of the canonical representation has the following form:

$$(5)\quad \|f\|^2 = \\ = \int\left[\frac{(1-|z_1|^2)(1-|z_2|^2)}{|1-z_1\bar{z}_2|^2}\right]^{\lambda/2}(1-|z_1|^2)^{-2}(1-|z_2|^2)^{-2}f(z_1)\overline{f(z_2)}\,dz_1\,d\bar{z}_1\,dz_2\,d\bar{z}_2.$$

2b. A SECOND METHOD OF SPECIFYING A CANONICAL REPRESENTATION. Suppose that the Lobachevskii plane is realized as the interior of the unit disk $|z| < 1$. Then the norm in the space $L_\lambda$ of a canonical representation is given by (5). If we now go over from the functions $f(z)$ to the functions $(1-|z|^2)^{\frac{1}{2}\lambda-2}f(z)$, then we obtain a new and very convenient realization of a canonical representation.

In this realization the space $L_\lambda$ of a canonical representation is the completion of the space of finite (that is, vanishing close to the boundary) continuous functions in the unit disk $|z| < 1$ with respect to the norm

(6) $$\|f\|^2 = \int\limits_{|z_1|<1,\ |z_2|<1} |1 - z_1\bar{z}_2|^{-\lambda} f(z_1)\overline{f(z_2)}\, dz_1\, d\bar{z}_1\, dz_2\, d\bar{z}_2.$$

The representation operators act according to the formula

(7) $$\left(T\begin{pmatrix}\alpha & \beta\\ \bar{\beta} & \bar{\alpha}\end{pmatrix} f\right)(z) = f\left(\frac{\alpha z + \bar{\beta}}{\beta z + \bar{\alpha}}\right) |\beta z + \bar{\alpha}|^{\lambda - 4}.$$

We note that for finite continuous functions $f(z)$ in the disk $|z| < 1$ the norm $\|f\|^2$ can be written in the following convenient form:

(8) $$\|f\|^2 = \sum_{m,n=0}^{\infty} c_{mn}(\lambda) \left| \int\limits_{|z|<1} f(z) z^m \bar{z}^n\, dz\, d\bar{z} \right|^2,$$

where

$$c_{mn}(\lambda) = \frac{\Gamma\left(\frac{\lambda}{2} + m\right)\Gamma\left(\frac{\lambda}{2} + n\right)}{m!\, n! \left(\Gamma\left(\frac{\lambda}{2}\right)\right)^2}.$$

In particular, if $f$ depends only on the modulus $r$, then the norm $\|f\|^2$ takes the following simple form:

$$\|f\|^2 = 2\pi \sum_{n=0}^{\infty} c_{nn}(\lambda) \left| \int_0^1 f(r) r^{2n+1}\, dr \right|^2.$$

To derive (8) it is sufficient to represent the kernel $|1 - z_1\bar{z}_2|^{-\lambda}$ as the product $(1 - z_1\bar{z}_2)^{-\lambda/2}(1 - \bar{z}_1 z_2)^{-\lambda/2}$, and then to expand each factor in a binomial series.

This space is very interesting. We shall see now that it contains a large store of generalized functions.

We consider the space $K$ of test functions $\phi(z)$ that are continuous and infinitely differentiable in the closed disk $|z| \leqslant 1$. Let $l(\phi)$ be a generalized function, that is, a linear functional on $K$. From $l$ we construct a new functional in the space $K_0$ of functions $f(z)$ that are infinitely differentiable and finite in the disk $|z| < 1$ (that is, vanish near the boundary):

$$(l, f) = \sum_{m,n=0}^{\infty} c_{mn}(\lambda)\, l(z^m\bar{z}^n) \int\limits_{|z|<1} f(z) z^m \bar{z}^n\, dz\, d\bar{z}.$$

If this series converges absolutely and $|(l, f)| \leqslant c\, \|f\|$, then the functional $(l, f)$ can be extended to a continuous linear functional in $L_\lambda$ and therefore specifies an element $l$ in $L_\lambda$.

We claim that the delta function $\xi_\lambda = \delta(z)$ concentrated at the point $z = 0$ belongs to $L_\lambda$. In fact, since $\xi_\lambda(1) = 1$ and $\xi_\lambda(z^m\bar{z}^n) = 0$ for $m + n > 0$, we have

(9) $$(\xi_\lambda, f) = (\delta, f) = \int f(z) dz\, d\bar{z}.$$

Consequently, by the definition of the norm in $L_\lambda$, $|(\xi_\lambda, f)| \leqslant \|f\|$ and hence $\delta(z) \in L_\lambda$.

It can be shown that $\xi_\lambda$ is a canonical vector in $L_\lambda$, that is,

$$(T(g)\xi_\lambda, \xi_\lambda) = \psi^\lambda(g).$$

This is easily derived from (9) if we replace $f$ by $T(g)f_n$, where $f_n$ is a sequence of test functions converging to $\xi_\lambda$, and then proceed to the limit.[1]

It is not difficult to verify that $L_\lambda$ contains not only $\delta(z)$, but also all its derivatives, $\delta^{(m,n)}(z) = \frac{\partial^{m+n}\delta(z)}{\partial z^m \partial \bar{z}^n}$. In particular, all derivatives $\delta^{(2n+1)}(r) = \frac{d^{2n+1}\delta(r)}{dr^{2n+1}}$ lie in the subspace of $L_\lambda$ consisting of functions[2] that depend only on the modulus $r$.

We now look at the generalized functions $l = a(z)\delta(1 - |z|^2)$, where $a(z)$ is a continuous function on the circle $|z| = 1$ (that is, $l(\varphi) = \frac{1}{2}\int_0^{2\pi} a(e^{it})\varphi(e^{it})\,dt$)

---

1) We can obtain this result formally by substituting in (9)

$$f(z) = T(g)\xi_\lambda = \delta\left(\frac{\alpha z + \bar{\beta}}{\beta z + \bar{\alpha}}\right)|\beta z + \bar{\alpha}|^{\lambda-4} = \delta\left(z + \frac{\bar{\beta}}{\alpha}\right)|\alpha|^{-\lambda}.$$

2) The functions $\delta^{(m,n)}(z)$ form an orthogonal basis in $L_\lambda$. Thus, each element $l$ of this space can be written in the form

$$l = \sum_{m,n=0}^{\infty} b_{mn} \frac{\partial^{m+n}\delta(z)}{\partial z^m \partial \bar{z}^n},$$

with

$$\|l\|^2 = \sum_{m,n=0}^{\infty} (m!\,n!)^2 c_{mn}(\lambda)\,|b_{mn}|^2.$$

In particular, in the subspace of functions depending only on the modulus $r$ there is an orthogonal basis consisting of the functions $\delta^{(2n+1)}(r) = \frac{d^{2n+1}\delta(r)}{dr^{2n+1}}$. Thus, any element of this subspace can be written in the form $l = \sum_{n=0}^{\infty} b_n \frac{d^{2n+1}\delta(r)}{dr^{2n+1}}$, with

$$\|l\|^2 = \sum_{n=0}^{\infty} [(2n+1)!]^2 c_{mn}(\lambda)\,|b|^2.$$

LEMMA 1.3. *The functions $l = a(z)\delta(1 - |z|^2)$ belong to the space $L_\lambda$ for $0 < \lambda < 1$.*

PROOF. Since

$$l(z^m \bar{z}^n) = \frac{1}{2}\int_0^{2\pi} a(e^{it})\, e^{i(m-n)t}\, dt = \alpha_{m-n},$$

we have

$$|(l, f)| \leqslant \sum_{m,n=0}^{\infty} c_{mn}(\lambda)\, |\alpha_{m-n}| \left| \int_{|z|<1} f(z) z^m \bar{z}^n\, dz\, d\bar{z} \right|.$$

Hence, by the Cauchy inequality

$$|(l, f)| \leqslant \Big( \sum_{m,n=0} c_{mn}(\lambda)\, |\alpha_{m-n}|^2 \Big)^{1/2} \|f\|.$$

Thus, it remains to prove the convergence of the series

$$\sum_{m,n=0}^{\infty} c_{mn}(\lambda)\, |\alpha_{m-n}|^2 = \sum_{k=-\infty}^{+\infty} |\alpha_k|^2 \Big( \sum_{m-n=k} c_{mn}(\lambda) \Big). \tag{10}$$

To do so we use the following estimate for the coefficients $c_{mn}(\lambda) = \dfrac{\Gamma\left(\frac{\lambda}{2}+m\right)\Gamma\left(\frac{\lambda}{2}+n\right)}{m!\, n!\, (\Gamma(\lambda/2))^2}$:

$$c_{mn}(\lambda) \leqslant C m^{\frac{\lambda}{2}-1} n^{\frac{\lambda}{2}-1}.$$

If follows from this estimate that for $0 < \lambda < 1$ the series $\sum_{n=0}^{\infty} c_{nn}(\lambda)$ converges. On the other hand, it is not hard to see that $\sum_{m-n=k} c_{mn}(\lambda) \leqslant \sum_{n=0}^{\infty} c_{nn}(\lambda)$ and, therefore, $\sum_{m-n=k} c_{mn}(\lambda) \leqslant C_1$, where $C_1$ does not depend on $k$. Since $\sum_{k=0}^{\infty} |\alpha_k|^2$ converges, the convergence of the series (10) follows.[1)]

---

[1)] We note that the function $\delta(z)$ can also be obtained by a limit passage. Let us consider the sequence of functions $f_n \in L_\lambda$ of the form

$$f_n(z) = \begin{cases} c_n & \text{for} \quad |z| < \frac{1}{n}, \\ 0 & \text{for} \quad \frac{1}{n} < |z| < 1, \end{cases}$$

where $c_n > 0$ is defined by the condition $\|f_n\| = 1$. It is easy to verify that this is a fundamental sequence in $L_\lambda$ and that its limit is equal to $\delta(z)$. Similarly, we can obtain $\delta(1 - |z|^2)$ as the limit of a fundamental sequence of functions of the form

$$f_n(z) = \begin{cases} c_n, & \text{for} \quad \frac{n-1}{n} < |z| < \frac{n}{n+1}, \\ 0 & \text{if} \quad |z| < \frac{n-1}{n} \quad \text{or} \quad \frac{n}{n+1} < |z| < 1 \end{cases}$$

where $c_n$ is determined from the relation $\int f_n(z)\, dz\, d\bar{z} = \pi$.

We state without proof two simple propositions on the functions $\xi_\lambda = \delta(z)$ and $l = f(z)\,\delta(1 - |z|^2)$.

LEMMA 1.4. *Let* $l_1 = f_1(z)\,\delta(1 - |z|^2)$, $l_2 = f_2(z)\,\delta(1 - |z|^2)$, *then*

$$(11) \qquad 1)\ (l_1, l_2)_\lambda = 2^{-\lambda-2} \int_0^{2\pi}\int_0^{2\pi} \left|\sin \frac{\varphi_1-\varphi_2}{2}\right|^{-\lambda} f_1(e^{i\varphi_1})\,\overline{f_2(e^{i\varphi_2})}\,d\varphi_1\,d\varphi_2.$$

$$(12) \qquad 2)\ (\delta(z),\ \delta(1-|z|^2))_\lambda = \pi.$$

From (11), it follows, in particular, that

$$(13) \qquad \|\delta(1-|z|^2)\|^2 = 2^{-\lambda}\pi^{3/2}\,\frac{\Gamma\left(\frac{1-\lambda}{2}\right)}{\Gamma\left(1-\frac{\lambda}{2}\right)}.$$

LEMMA 1.5. *The representation operator $T(g)$, where* $g = \begin{pmatrix}\alpha & \beta\\ \bar\beta & \bar\alpha\end{pmatrix}$, *takes the function* $f(z)\delta(1 - |z|^2)$ *into* $f\left(\frac{\alpha z+\bar\beta}{\beta z+\bar\alpha}\right)|\beta z + \bar\alpha|^{\lambda-2}\delta(1 - |z|^2)$.

The next result follows immediately from Lemmas 1.4 and 1.5.

THEOREM 1.3. *For $0 < \lambda < 1$ the subspace $H_\lambda \subset L_\lambda$ generated by the functions $f(z)\,\delta(1 - |z|^2)$ is invariant and the representation of $G$ acting in it is the representation $T_\lambda$ of the supplementary series.*

In what follows we find it useful to know the projection of the canonical vector $\xi_\lambda$ onto $H_\lambda$. This projection is obviously invariant under the maximal compact subgroup and is consequently proportional to $\delta(1 - |z|^2)$. Let

$$\eta_\lambda = c(\lambda)\delta(1 - |z|^2), \qquad \|\eta_\lambda\| = 1,$$

where $c(\lambda) = \|\delta(1 - |z|^2)\|^{-1}$. Then, according to Lemma 1.4, 2) we have $(\xi_\lambda, \eta_\lambda) = \pi c(\lambda)$; consequently the projection of the canonical vector $\xi_\lambda$ onto the subspace $H_\lambda$ is equal to $\pi c(\lambda)\eta_\lambda$, where $c^2(\lambda) = \|\delta\,(1 - |z|^2)\|^{-2} = 2^\lambda\pi^{-3/2}\ \Gamma(1 - \lambda/2)/\Gamma(\frac{1}{2}(1 - \lambda))$.

The next result follows easily from standard estimates for $\Gamma(\lambda)$.

LEMMA 1.6. $(\xi_\lambda, \eta_\lambda) = 1 + O(\lambda^2)$ *as* $\lambda \to 0$.

COROLLARY. *As $\lambda \to 0$, the distance from the canonical vector $\xi_\lambda$ in $L_\lambda$ to $H_\lambda$ is $O(\lambda^2)$.*

**3. Theorems on tensor products of representations of** $G$. THEOREM 1.4. *Let $T_{\lambda_1}$ and $T_{\lambda_2}$ be two representations of the supplementary series. If $\lambda_1 + \lambda_2 < 1$, then $T_{\lambda_1} \otimes T_{\lambda_2} = T_{\lambda_1+\lambda_2} \oplus T$, where $T$ splits into representations of the principal continuous and the discrete series only.*

*If $\lambda_1 + \lambda_2 \geqslant 1$, then $T_{\lambda_1} \otimes T_{\lambda_2}$ splits into representations of the principal continuous and the discrete series only.*

*The tensor product of two irreducible unitary representations of which at least one belongs to the principal continuous or the discrete series splits into representations of the principal continuous and the discrete series only.*

For the proof see [12].

We now let $G_n = \underbrace{G \times \ldots \times G}_{n}$. Since $G$ is of type 1, it is standard knowledge that any irreducible unitary representation $T$ of $G_n$ can be obtained as follows. We are given irreducible unitary representations $T^{(1)}, \ldots, T^{(n)}$ of $G$, acting in Hilbert spaces $H_1, \ldots, H_n$, respectively. A representation $T$ of $G_n$ acts in the tensor product $H_1 \otimes \ldots \otimes H_n$ according to the following formula:

(14) $$T(g_1, \ldots, g_n)(\xi_1 \otimes \ldots \otimes \xi_n) = (T^{(1)}(g_1)\xi_1) \otimes \ldots \otimes (T^{(n)}(g_n)\xi_n).$$

Here two representations $T'$ and $T''$ of $G_n$ are equivalent if and only if all the corresponding representations $T'^{(i)}$ and $T''^{(i)}$ of $G$ are equivalent, $i = 1, \ldots, n$.

We say that a representation $T$ of $G_n$ of the form (14) is *purely of the supplementary series* if all the $T^{(i)}$ are representations of $G$ of the supplementary series. In this case, if $T^{(i)} = T_{\lambda_i}$, $i = 1, \ldots, n$, then the corresponding representations of the group $G_n$ are denoted by $T_{\lambda_1, \ldots, \lambda_n}$, and the representation space by $H_{\lambda_1, \ldots, \lambda_n}$.

The next theorem follows from the one stated above.

THEOREM 1.5. *Let $T_{\lambda'_1, \ldots, \lambda'_n}$ and $T_{\lambda''_1, \ldots, \lambda''_n}$ be two representations of $G_n$ purely of the supplementary series: and let $\lambda'_i + \lambda''_i < 1$, $i = 1, \ldots, n$. Then $T_{\lambda'_1, \ldots, \lambda'_n} \otimes T_{\lambda''_1, \ldots, \lambda''_n} = T_{\lambda'_1+\lambda''_1, \ldots, \lambda'_n+\lambda''_n} \oplus T$, where in the decomposition of $T$ into irreducible representations there are no representations purely of the supplementary series.*

In what follows we find it useful to specify explicitly an embedding of $H_{\lambda_1+\lambda_2}$ $(\lambda_1 + \lambda_2 < 1)$ in the tensor product $H_{\lambda_1} \otimes H_{\lambda_2}$.

Let $G$ be defined in the first form. Then $H_{\lambda_1+\lambda_2}$ is the completion of the space of finite continuous real functions $f(x)$ with the norm

$$\| f \|^2 = \int_{-\infty}^{+\infty}\int_{-\infty}^{+\infty} | x - x' |^{-\lambda_1-\lambda_2} f(x)\overline{f(x')}\, dx\, dx'.$$

Now $H_{\lambda_1} \otimes H_{\lambda_2}$ is the completion of the space of finite continuous functions $F(x_1, x_2)$ of two variables with the norm

$$\| F \|^2 = \int | x_1 - x_1' |^{-\lambda_1} | x_2 - x_2' |^{-\lambda_2} F(x_1, x_2)\overline{F(x_1', x_2')}\, dx_1\, dx_2\, dx_1'\, dx_2'.$$

In $H_{\lambda_1} \otimes H_{\lambda_2}$ there are many generalized functions. The precise meaning of this statement is the following.

Let $l(\tilde{F})$ be a linear functional on the space of infinitely differentiable functions $\tilde{F}(x_1, x_2)$ such that $|\tilde{F}(x_1, x_2)| < C(1 + x_1^2)^{-\lambda_1/2}(1 + x_2^2)^{-\lambda_2/2}$. By means of $l$ we construct the following linear functional $\tilde{l}$ on the space of finite infinitely differentiable functions $F(x_1, x_2)$: $(\tilde{l}, F) = l(\tilde{F})$, where

$$\tilde{F}(x_1, x_2) = \int_{-\infty}^{+\infty}\int_{-\infty}^{+\infty} | x_1 - x_1' |^{-\lambda_1} | x_2 - x_2' |^{-\lambda_2} F(x_1', x_2')\, dx_1'\, dx_2'.$$

If the functional $(\tilde{l}, F)$ is defined and continuous in the norm of $H_{\lambda_1} \otimes H_{\lambda_2}$, then we identify the generalized function $l$ with the vector $\tilde{l} \in H_{\lambda_1} \otimes H_{\lambda_2}$.

LEMMA 1.7. *If $\lambda_1 > 0$, $\lambda_2 > 0$, $\lambda_1 + \lambda_2 < 1$ and $l$ is a generalized function having the form $l = f(x_1)\,\delta(x_1 - x_2)$ ($f(x)$ finite), that is,*

$$l(F) = \int_{-\infty}^{+\infty} F(x, x) f(x)\, dx,$$

*then the functional $(\tilde{l}, F)$ is continuous in the norm of $H_{\lambda_1} \otimes H_{\lambda_2}$.*

This lemma is proved by standard calculations involving the Fourier transform, and we omit the proof. Next we can establish the following lemma.

LEMMA 1.8. *Let $\tilde{l}_1$ and $\tilde{l}_2$ be defined by the generalized functions $l_1 = f_1(x_1)\,\delta(x_1 - x_2)$ and $l_2 = f_2(x_1)\,\delta(x_1 - x_2)$. Then*

$$(l_1, l_2) = \int_{-\infty}^{+\infty}\int_{-\infty}^{+\infty} |x_1 - x_2|^{-\lambda_1 - \lambda_2} f_1(x_1) f_2(x_2)\, dx_1\, dx_2.$$

THEOREM 1.6. *If $\lambda_1 > 0$, $\lambda_2 > 0$, $\lambda_1 + \lambda_2 < 1$, then the mapping defines an isometric embedding of $H_{\lambda_1 + \lambda_2}$ in $H_{\lambda_1} \otimes H_{\lambda_2}$, consistent with the action of $G$ on these spaces.*

This theorem follows immediately from the lemmas stated above.

We need a somewhat more general theorem, which can be proved similarly.

THEOREM 1.7. *Let* $\lambda_1, \lambda_2, \ldots, \lambda_k > 0$ and $\lambda_1 + \ldots + \lambda_k < 1$. *Then the mapping*

$$f(x) \mapsto \delta(x_1 - x_k, \ldots, x_{k-1} - x_k) f(x_k)$$

*defines an isometric embedding*

$$H_{\lambda_1 + \cdots + \lambda_k} \to H_{\lambda_1} \otimes \ldots \otimes H_{\lambda_k},$$

*consistent with the action of $G$.*

The meaning of the concepts and mappings introduced is the same as that explained earlier for $k = 2$.

The mappings indicated above are consistent; namely, if

$$\lambda = \sum_i \lambda_i, \quad \lambda_i = \sum_j \lambda_{ij}, \quad \lambda_{ij} > 0, \quad \lambda < 1,$$

then the mapping

$$H_\lambda \to \bigotimes_{i,j} H_{\lambda_{ij}}$$

is the composition of the mappings

$$H_\lambda \to \bigotimes_i H_{\lambda_i} \text{ and } H_{\lambda_i} \to \bigotimes_j H_{\lambda_{ij}}.$$

## §2. Construction of the multiplicative integral of representations of $G = PSL(2, \mathbf{R})$.

Let $G = PSL(2, \mathbf{R})$ and let $X$ be a compact topological space with a given measure $m$. We define a group operation on the set of functions

$g$: $X \to G$ as pointwise multiplication: $(g_1g_2)(x) = g_1(x)g_2(x)$. We define $G^X$ as the group of all continuous functions $g$: $X \to G$ with the topology of uniform convergence.

We give here a construction of an irreducible unitary representation of the group $G^X$, which we call the *multiplicative integral of representations of G.*

Following this definition of the integral as closely as possible, we replace $G^X$ by the group of step functions and define an integral on it as a tensor product of representations. As we decrease the length of the intervals of subdivision and simultaneously allow the parameter on which the representations in the tensor product depend to approach a certain limit, we obtain a representation of $G^X$, which we call the integral of representations. It is remarkable that the integral of representations is an irreducible representation.

We now proceed to precise definitions.

**1. Definition of the group $G^0$.** For every Borel subset $X' \subset X$ we denote by $G_{X'}$ the group of functions $g$: $X \to G$ that are constant on $X'$ and equal to 1 on the complement of $X'$. It is obvious that there exists a natural isomorphism $G_{X'} \cong G$.

A partition $\nu$: $X = \cup X_i$ of $X$ into finitely many disjoint Borel subsets is called *admissible.*

On the set of admissible partitions we define an ordering, setting $\nu_1 < \nu_2$ if $\nu_2$ is a refinement of $\nu_1$. It is obvious that the set of admissible partitions is directed (that is, for any $\nu_1$ and $\nu_2$ there exists a $\nu$ such that $\nu_1 \subset \nu$ and $\nu_2 < \nu$).

For any admissible partition $\nu$: $X = \bigcup_{i=1}^{k} X_i$ we denote by $G_\nu$ the group of functions $g$: $X \to G$ that are constant on each of the subsets $X_i$. It is obvious that

$$G_\nu = G_{X_1} \times \dots \times G_{X_k}.$$

Observe that for $\nu_1 < \nu_2$ there is a natural embedding: $G_{\nu_1} \to G_{\nu_2}$. We define the group of step functions $G^0$ as the inductive limit of the $G_\nu$:

$$G^0 = \lim_{\longrightarrow} G_\nu.$$

In this section we construct a representation of $G^0$. We make the transition from this representation to a representation of $G^X$ in §3.

**2. Construction of a representation of $G^0$.** Let $m$ be a positive finite measure on $X$, defined on all Borel subsets of $X$. We always assume that $m$ is countably additive.

Let us consider the Hilbert spaces in which the representations $T_\lambda$ of the supplementary series act. In §1 we denoted these spaces with the action of $G$ defined on them by $H_\lambda$, $0 < \lambda < 1$. Next, we denote by $H_0$ the one-dimensional space in which the identity representation of $G$ acts.

Let $\nu$: $X = \bigcup_{i=1}^{k} X_i$ be an arbitrary admissible partition such that

$\lambda_i = m(X_i) < 1$. We set

$$\mathcal{H}_\nu = H_{\lambda_1} \otimes \ldots \otimes H_{\lambda_k}.$$

In $\mathcal{H}_\nu$ we define a representation of the group

$$G_\nu = G_{X_1} \times \ldots \times G_{X_k},$$

supposing that $G_{X_i} \cong G$ ($i = 1, \ldots, k$) acts in $H_{\lambda_i}$.

The representation of $G_\nu$ in $\mathcal{H}_\nu$ is irreducible (see § 1.3); we have agreed to call such representations purely of the supplementary series.

Note that since $G_{\nu_1} \subset G_{\nu_2}$ for $\nu_1 < \nu_2$, a representation of each of the groups $G_{\nu'}$, $\nu' < \nu$, is also defined in $\mathcal{H}_\nu$.

LEMMA 2.1. *If $\nu_1 < \nu_2$, then $\mathcal{H}_{\nu_2}$ splits into the direct sum of subspaces invariant under $G_{\nu_1}$: $\mathcal{H}_{\nu_2} = \mathcal{H}_{\nu_1} \oplus \mathcal{H}'$ where $\mathcal{H}'$ does not contain invariant subspaces in which a representation of $G_{\nu_1}$ purely of the supplementary series acts.*

PROOF. Let $\nu_2 > \nu_1$, that is, $\nu_1$: $X = \bigcup_{i=1}^{k} X_i$, $\nu_2$: $X = \bigcup_{i,j} X_{ij}$, where $X_i = \bigcup_j X_{ij}$. We set $\lambda_i = m(X_i)$, $\lambda_{ij} = m\,(X_{ij})$; thus, $\mathcal{H}_{\nu_1} = \bigotimes_i H_{\lambda_i}$, $\mathcal{H}_{\nu_2} = \bigotimes_{i,j} H_{\lambda_{ij}}$. We also set $\mathcal{H}^i_{\nu_2} = \bigotimes_j H_{\lambda_{ij}}$; then $\mathcal{H}_{\nu_2} = \bigotimes_i \mathcal{H}^i_{\nu_2}$. It is evident that $G_{X_i}$ acts diagonally in $\mathcal{H}^i_{\nu_2} = \bigotimes_j H_{\lambda_{ij}}$ (that is, acts simultaneously on each factor $H_{\lambda_{ij}}$). Thus, the representation of $G_{X_i} \cong G$ in $\mathcal{H}^i_{\nu_2}$ is a tensor product of representations $T_{\lambda_{ij}}$ of the supplementary series.

From this it follows that $\mathcal{H}^i_{\nu_2} = H_{\lambda_i} \oplus H^i_{\nu_2}$, $\lambda_i = \sum_j \lambda_{ij}$, where $H_{\lambda_i}$ is the space in which the representation $T_{\lambda_i}$ of the supplementary series acts, and $H^i_{\lambda_2}$ splits only into representations of the principal, the continuous, and the supplementary series (see § 1.3). Forming the tensor product of the spaces $\mathcal{H}^i_{\nu_2}$ and bearing in mind $\bigotimes_i H_{\lambda_i} = \mathcal{H}_{\nu_1}$ and $G_{X_1} \times \ldots \times G_{X_k} = G_{\nu_1}$, we obtain: $\mathcal{H}_{\nu_2} = \mathcal{H}_{\nu_1} \oplus \mathcal{H}'$, where $\mathcal{H}'$ does not contain representations purely of the supplementary series.

THEOREM 2.1. *There exist morphisms of Hilbert spaces*

$$j_{\nu_2\nu_1}: \mathcal{H}_{\nu_1} \to \mathcal{H}_{\nu_2},$$

*defined for each pair $\nu_1 < \nu_2$ of admissible partitions of $X$ satisfying the following conditions:*

1) $j_{\nu_2\nu_1}$ *commutes with the action of $G_{\nu_1}$ in $\mathcal{H}_{\nu_1}$ and $\mathcal{H}_{\nu_2}$;*

2) $j_{\nu_3\nu_2} \circ j_{\nu_2\nu_1} = j_{\nu_3\nu_1}$ *for any $\nu_1 < \nu_2 < \nu_3$.*

*These morphisms are determined uniquely to within factors $c_{\nu_2\nu_1}$ ($|c_{\nu_2\nu_1}| = 1$).*

PROOF. From Lemma 2.1 it follows that for each pair $\nu_1 < \nu_2$ there exists a morphism $j_{\nu_2\nu_1}$: $\mathcal{H}_{\nu_1} \to \mathcal{H}_{\nu_2}$ that commutes with the action of $G_{\nu_1}$, and that this morphism is uniquely determined to within a factor. We claim that the morphism $j_{\nu_2\nu_1}$ can be chosen so that condition 2) is satisfied.

Let $\nu_2 > \nu_1$, that is, $\nu_1$: $X = \bigcup_{i=1}^{k} X_i$, $\nu_2$: $X = \bigcup_{i,j} X_{ij}$, where $X_i = \bigcup_j X_{ij}$. We set $\lambda_i = m(X_i)$, $\lambda_{ij} = m(X_{ij})$; thus, $\mathcal{H}_{\nu_1} = \bigotimes_{i=1}^{k} H_{\lambda_i}$, $\mathcal{H}_{\nu_2} = \bigotimes_{i,j} H_{\lambda_{ij}}$.

For each $i = 1, 2, \ldots, k$ we define a mapping $H_{\lambda_i} \to \bigotimes_{i,j} H_{\lambda_{ij}}$, $\lambda_i = \sum_j \lambda_{ij}$, compatible with the action of $G_{X_i} \cong G$, just as this was done in §1.3. These mappings induce the mapping

$$j_{\nu_2\nu_1} : \mathcal{H}_{\nu_1} \to \mathcal{H}_{\nu_2},$$

which is compatible with the action of $G_\nu$ in these spaces.

From the definition of the mappings $H_{\lambda_i} \to \bigotimes_j H_{\lambda_{ij}}$ it follows easily that the mappings $j_{\nu_2\nu_1}$ so defined satisfy the compatibility requirement 2) of the Theorem.

REMARK. Another method of specifying the compatibility of the system of morphisms $j_{\nu_2\nu_1}$ will be given in §3.

DEFINITION. We assign to all possible pairs $\nu_1 < \nu_2$ of admissible partitions of $X$ the morphisms $j_{\nu_2\nu_1} : \mathcal{H}_{\nu_1} \to \mathcal{H}_{\nu_2}$, which commute with the action of $G_{\nu_1}$ in $\mathcal{H}_{\nu_1}$ and $\mathcal{H}_{\nu_2}$ and satisfy the compatibility condition $j_{\nu_3\nu_2} \circ j_{\nu_2\nu_1} = j_{\nu_3\nu_1}$ for $\nu_1 < \nu_2 < \nu_3$. We introduce the space

$$\mathcal{H}^0 = \varinjlim \mathcal{H}_\nu,$$

and let $\mathcal{H}$ be the completion of $\mathcal{H}^0$ in the norm defined in $\mathcal{H}^0$.

Then there is a natural way of defining a unitary representation $U$ of $G^0$ in $\mathcal{H}$.

(Specifically, we have the natural embeddings $G_\nu \hookrightarrow G^0$, $\mathcal{H}_\nu \hookrightarrow \mathcal{H}^0$. Let $\tilde{g} \in G^0$ and $\xi \in \mathcal{H}^0$. Then there exists an admissible partition $\nu$ such that $\tilde{g} \in G_\nu$, $\xi \in \mathcal{H}_\nu$, and we set $U_{\tilde{g}}\xi = T(\tilde{g})\xi$, where $T$ is a representation operator of $G_\nu$ in $\mathcal{H}_\nu$. It is evident that this definition does not depend on the choice of $\nu$. The unitary operator $U_{\tilde{g}}$ on $\mathcal{H}^0$ so constructed can be extended by continuity from $\mathcal{H}^0$ to its completion $\mathcal{H}$.)

LEMMA 2.2. *The representation $U_{\tilde{g}}$ does not depend on the choice of the morphisms $j_{\nu_2\nu_1}$.*

PROOF. Let $j'_{\nu_2\nu_1}: \mathcal{H}_{\nu_1} \to \mathcal{H}_{\nu_2}$ be another system of morphisms satisfying the conditions of the definition, and let $\mathcal{H}'^0 = \varinjlim \mathcal{H}_\nu$ be the inductive limit constructed with respect to this system of morphisms. We claim that the representations of $G^0$ in $\mathcal{H}^0$ and $\mathcal{H}'^0$ are equivalent.

By Theorem 2.1. we have $j'_{\nu_2\nu_1} = c_{\nu_2\nu_1} j_{\nu_2\nu_1}$, where $c_{\nu_2\nu_1}$ are numerical factors satisfying $c_{\nu_3\nu_2} c_{\nu_2\nu_1} = c_{\nu_3\nu_1}$ for $\nu_1 < \nu_2 < \nu_3$. It follows from this condition that $c_{\nu_2\nu_1} = c_{\nu_2\nu_0} c^{-1}_{\nu_1\nu_0}$, where $\nu_0$ is a fixed admissible partition ($\nu_0 < \nu_1 < \nu_2$). For $\nu > \nu_0$ we specify isomorphisms of the spaces $\mathcal{H}_\nu \to \mathcal{H}_\nu$ in the following manner: $\xi \mapsto c_{\nu\nu_0}\xi$. Since $j'_{\nu_2\nu_1} = c_{\nu_2\nu_0} c^{-1}_{\nu_1\nu_0} j_{\nu_2\nu_1}$, they take

$j_{\nu_2\nu_1}$ into $j'_{\nu_2\nu_1}$ and consequently induce an isomorphism of the spaces $\mathscr{H}^0$ and $\mathscr{H}'^0$ compatible with the action of $G^0$.

THEOREM 2.2. *The representation U of $G^0$ in $\mathscr{H}$ is irreducible.*

PROOF. Let $\nu$: $X = \bigcup_{i=1}^{k} X_i$ be an admissible partition such that $\lambda_i = m(X_i) < 1$. We restrict the representation of $G^0$ in $\mathscr{H}$ to $G_\nu$.

An irreducible representation of $G_\nu$ acts in $\mathscr{H}_\nu = \bigotimes_{i=1}^{k} H_{\lambda_i}$. By Lemma 2.1 $\mathscr{H}_\nu$ occurs with multiplicity 1 in each $\mathscr{H}_{\nu'}$ for $\nu' > \nu$. Since $\mathscr{H}_\nu$ is irreducible, it follows that it occurs with multiplicity[1] in the whole space $\mathscr{H}$.

We now suppose that $\mathscr{H}$ splits into the direct sum of invariant subspaces, $\mathscr{H} = \mathscr{H}' \oplus \mathscr{H}''$. Then $\mathscr{H}_\nu$ is contained in one of the summands, for example, in $\mathscr{H}'$. Now let $\nu' > \nu$. Since $\mathscr{H}_{\nu'} \supset \mathscr{H}_\nu$ and an irreducible representation of $G_{\nu'}$ acts in $\mathscr{H}_{\nu'}$, we have $\mathscr{H}_{\nu'} \subset \mathscr{H}'$. Consequently, $\mathscr{H}$ contains all the subspaces $\mathscr{H}_{\nu'}$, $\nu' > \nu$, and therefore coincides with $\mathscr{H}$. This completes the proof.

THEOREM 2.3. *Let $m_1$ and $m_2$ be two positive measures on X, $U^{(1)}$ and $U^{(2)}$ representations of $G^0$ defined on these measures. If $m_1 \neq m_2$, then the representations $U^{(1)}$ and $U^{(2)}$ are inequivalent.*

PROOF. We denote by $\mathscr{H}^{(1)}$ and $\mathscr{H}^{(2)}$ the representation spaces of $U^{(1)}$ and $U^{(2)}$. Since $m_1 \neq m_2$, there exists an admissible partition $\nu$: $X = \bigcup_{i=1}^{k} X_i$ such that $\lambda_i^{(1)} = m_1(X_i) < 1$, $\lambda_i^{(2)} = m_2(X_i) < 1$ for $i = 1, \ldots, k$ and $m_1(X_i) \neq m_2(X_i)$ at least for one $i$.

We claim that the representations of $G_\nu \subset G^0$ in $\mathscr{H}^{(1)}$ and $\mathscr{H}^{(2)}$ are inequivalent. It then follows that the representations of the whole group $G^0$ in these spaces a fortiori are inequivalent.

Let us first consider the spaces

$$\mathscr{H}_\nu^{(1)} = \bigotimes_{i=1}^{k} H_{\lambda_i^{(1)}} \text{ and } \mathscr{H}_\nu^{(2)} = \bigotimes_{i=1}^{k} H_{\lambda_i^{(2)}},$$

in which the irreducible representations of $G_\nu$ act. Since $\lambda_i^1 \neq \lambda_i^2$ for at least one $i$, the representations of $G_\nu$ in $\mathscr{H}_\nu^{(1)}$ and $\mathscr{H}_\nu^{(2)}$ are inequivalent.

Furthermore, $\mathscr{H}_{\nu'}^{(2)} = \mathscr{H}_\nu^{(2)} \oplus \mathscr{H}'_\nu$, for every $\nu' > \nu$, and $\mathscr{H}'_\nu$ does not contain representations purely of the supplementary series of $G_\nu$, therefore does not contain representations equivalent to $\mathscr{H}_\nu^{(1)}$ (see Lemma 2.1). Consequently the whole space $\mathscr{H}_{\nu'}^{(2)}$ does not contain representations equivalent to $\mathscr{H}_\nu^{(1)}$. But then $\mathscr{H}^{(2)}$ also does not contain representations of $G_\nu$ equivalent to $\mathscr{H}_\nu^{(1)}$. Since obviously $\mathscr{H}_\nu^{(1)} \subset \mathscr{H}^{(1)}$, the representations of $G_\nu$ in $\mathscr{H}^{(1)}$ and $\mathscr{H}^{(2)}$ are inequivalent.

## §3. Another construction of the multiplicative integral of representations of $G = PSL(2, \mathbf{R})$.

The concept of the multiplicative integral of representations of $G = PSL(2, \mathbf{R})$ introduced in §2 can also be obtained starting out from the canonical representations of $G$. Here we explain this second method. It is surprising that although the representations in the product are significantly more "massive", their product turns out to be the same as before.

1. **Construction of the representation.** As before, let $X$ be a compact topological space on which a positive finite measure $m$ is given, defined on all Borel subsets and countably additive.

We consider the canonical representations of $G$ in the Hilbert spaces $L_\lambda$ introduced in §1.2. Next we denote by $L_0$ the one-dimensional space in which the identity representation $T_0$ of $G$ acts.

We recall that in each space $L_\lambda$ we have fixed a cyclic vector $\xi_\lambda$ which we have called canonical. For this vector

$$(T(g)\xi_\lambda, \xi_\lambda) = \psi^\lambda(g),$$

where $\psi(g)$ is the function defined in §1.2.

With each admissible partition $\nu$: $X = \bigcup_{i=1}^{k} X_i$, we associate a Hilbert space

$$\mathcal{L}_\nu = L_{\lambda_1} \otimes \ldots \otimes L_{\lambda_k},$$

where $\lambda_i = m(X_i)$. We define in $\mathcal{L}_\nu$ a unitary representation of

$$G_\nu = G_{X_1} \otimes \ldots \otimes G_{X_k},$$

assuming that each group $G_{X_i} \cong G$ acts in $L_{\lambda_i}$ in accordance with the corresponding canonical representation, and trivially on the remaining factors $L_{\lambda_j}$, $j \neq i$. We observe that since $G_{\nu_1} \subset G_{\nu_2}$ for $\nu_1 < \nu_2$, an action of each of the groups $G_{\nu'}$, $\nu' < \nu$, is also defined in $\mathcal{L}_\nu$.

For each admissible partition $\nu$: $X = \bigcup_{i=1}^{k} X_i$ we specify a vector $\xi_\nu \in \mathcal{L}_\nu$:

$$\xi_\nu = \xi_{\lambda_1} \otimes \ldots \otimes \xi_{\lambda_k},$$

where $\xi_{\lambda_i}$ is the canonical vector in $L_{\lambda_i}$. It is obvious that $\xi_\nu$ is a cyclic vector in $\mathcal{L}_\nu$.

LEMMA 3.1. *For any pair of partitions $\nu_1 < \nu_2$ the mapping $\xi_{\nu_1} \mapsto \xi_{\nu_2}$ can be extended to a morphism*

$$j_{\nu_2\nu_1} : \mathcal{L}_{\nu_1} \to \mathcal{L}_{\nu_2},$$

*which commutes with the action of $G_{\nu_1}$.*

PROOF. It is sufficient to verify that

$$(T(g_{\nu_1})\xi_{\nu_1}, \xi_{\nu_1})_{\mathcal{L}_{\nu_1}} = (T(g_{\nu_1})\xi_{\nu_2}, \xi_{\nu_2})_{\mathcal{L}_{\nu_2}}$$

for any $g_{\nu_1} \in G_{\nu_1}$, where the parentheses denote the scalar product in the corresponding space.

According to hypothesis we have $\nu_1$: $X = \bigcup_{i=1}^{k} X_i$, $\nu_2$: $X = \bigcup_{i,j} X_{ij}$, where $X_i = \bigcup_j X_{ij}$. Let $g_{\nu_1} \in G_{\nu_2}$, that is, $g_{\nu_1} = g_1 \ldots g_n$, where $g_i \in G_{X_i} \cong G$. Then

$$(1) \qquad (T(g_{\nu_1})\xi_{\nu_1}, \xi_{\nu_1})_{\mathscr{L}_{\nu_1}} = \prod_{i=1}^{k} (T(g_i)\xi_{\lambda_i}, \xi_{\lambda_i})_{L_i} = \prod_{i=1}^{k} \psi^{\lambda_i}(g_i),$$

where $\lambda_i = m(X_i)$. Similarly, let $g_{\nu_2} \in G_{\nu_2}$, that is, $g_{\nu_2} = \prod_{i,j} g_{ij}$, where $g_{ij} \in G_{X_{ij}} \cong G$. Then

$$(2) \qquad (T(g_{\nu_2})\xi_{\nu_2}, \xi_{\nu_2})_{\mathscr{L}_{\nu_2}} = \prod_{i,j} \psi^{\lambda_{ij}}(g_{ij}),$$

where $\lambda_{ij} = m(X_{ij})$.

If now $g_{\nu_2} = g_{\nu_1}$, this means that $g_{ij} = g_i$ for every $i = 1, \ldots, k$. In addition, since $\lambda_i = \sum_j \lambda_{ij}$, for any $i = 1, \ldots, k$ we have $\prod_j \psi^{\lambda_{ij}}(g_{ij}) = \psi^{\lambda_i}(g_i)$. Consequently the expressions (1) and (2) are the same and the lemma is proved.

It is obvious that the morphisms $j_{\nu_2 \nu_1}$ satisfy the compatibility condition $j_{\nu_3\nu_2} \circ j_{\nu_2\nu_1} = j_{\nu_3\nu_1}$ for $\nu_1 < \nu_2 < \nu_3$.

DEFINITION. We denote by $\mathscr{L}^0$ the inductive limit of the spaces $\mathscr{L}_\nu$, $\mathscr{L}^0 = \varinjlim \mathscr{L}_\nu$ and by $\mathscr{L}$, the completion of $\mathscr{L}^0$ in the norm defined in $\mathscr{L}^0$.

There is a natural way of defining in $\mathscr{L}$ a unitary representation of the group $G^0 = \varinjlim G_\nu$, which we denote by $U_{\tilde{g}}$.

2. **Definition of a vacuum vactor in $\mathscr{L}$.** Let $K$ be a maximal compact subgroup of $G$, and let $K^0 \subset G^0$ be the subgroup of step functions on $X$ with values in $K$. Vectors in $\mathscr{L}$ of unit norm that are invariant under $K^0$ are called *vacuum* vectors.

We claim that a vacuum vector exists in $\mathscr{L}$.

Let $j_\nu$ be the natural mapping $\mathscr{L}_\nu \to \mathscr{L}$. It follows from the definition of $\xi_\nu$ that their images $j_\nu \xi_\nu$ in $\mathscr{L}$ coincide, that is, there exists a vector $\xi_0 \in \mathscr{L}$ such that $\xi_0 = j_\nu \xi_\nu$ for any admissible partition $\nu$.

Since each vector $\xi_\nu \in \mathscr{L}_\nu$ is invariant under $K_\nu \subset G_\nu$, it follows that $\xi_0 = \varinjlim \xi_\nu$ is invariant under $K^0 = \varinjlim K_\nu$, so that it is a vacuum vector in $\mathscr{L}$.

In addition, since each $\xi_\nu \in \mathscr{L}_\nu$ is a cyclic vector in $\mathscr{L}_\nu$ under $G_\nu$, $\xi_0 = \varinjlim \xi_\nu$ is a cyclic vector in $\mathscr{L}$ under $G^0 = \varinjlim G_\nu$.

In §3.3 we shall prove the uniqueness of the vacuum vector in $\mathscr{L}$ (Theorem 3.2).

Let us calculate the spherical function of the representation $U_{\tilde{g}}$:

$$\Psi(\tilde{g}) = (U_{\tilde{g}}\xi_0, \xi_0)_{\mathscr{L}},$$

where $\tilde{g} = g(\cdot) \in G^0$.

LEMMA 3.2.

$$\Psi(\tilde{g}) = \exp\left( \int_X \ln \psi(g(x))\, dm(x) \right),$$

*where $\psi(g)$ is the function on $G$ introduced in* §1.2.

PROOF. For any $\tilde{g} = g(\cdot) \in G^0$ we can find an admissible partition $\nu$: $X = \bigcup_{i=1}^{k} X_i$ of $X$ such that $\tilde{g} \in G_\nu$, that is, $\tilde{g} = g_1 \ldots g_k$, where $g_i \in G_{X_i} \cong G$. But then, setting $\lambda_i = m(X_i)$, we have

$$(U_{\tilde{g}}\xi_0, \xi_0)_{\mathscr{L}} = (T(\tilde{g})\xi_\nu, \xi_\nu)_{\mathscr{L}_\nu} = \prod_{i=1}^{k} (T(g_i)\xi_{\lambda_i}, \xi_{\lambda_i})_{L_{\lambda_i}} =$$

$$= \prod_{i=1}^{k} \psi^{\lambda_i}(g_i) = \exp\left( \sum_{i=1}^{k} \ln(\psi(g_i))\, m(X_i) \right) =$$

$$= \exp\left( \int \ln(\psi(g(x))\, dm(x) \right).$$

**3. Equivalence of the two constructions of the representations of $G^0$.** We claim that the unitary representation of $G^0$ in $\mathscr{L}$ constructed here is equivalent to the representation in $\mathscr{H}$ constructed in §2.

LEMMA 3.3. *There is an embedding morphism $\mathscr{H} \to \mathscr{L}$ that is compatible with the action of $G^0$.*

PROOF. In §1.2 we have shown that for $0 < \lambda < 1$ the canonical representation $L_\lambda$ of $G$ is congruent to the representation $H_\lambda$ of the supplementary series modulo representations of the principal continuous series; in other words, $L_\lambda = H_\lambda \oplus L'_\lambda$, where $L'_\lambda$ can be expanded in an integral over representations of the principal continuous series.

Hence it follows that the representation space $\mathscr{L}_\nu = \bigotimes_{i=1}^{k} L_{\lambda_i}$ of $G_\nu = G_{X_1} \times \ldots \times G_{X_k}$, where $\nu$: $X = \bigcup_{i=1}^{k} X_i$, $\lambda_i = m(X_i) < 1$, splits into a direct sum of invariant subspaces:

$$\mathscr{L}_\nu = \mathscr{H}_\nu \oplus \mathscr{L}'_\nu,$$

where $\mathscr{H}_\nu = \bigotimes_{i=1}^{k} H_{\lambda_i}$, and $\mathscr{L}'_\nu$ for any $\nu' < \nu$ does not contain representations purely of the supplementary series of $G_{\nu'}$.

This also shows that under the morphism $j_{\nu_2\nu_1}: \mathscr{L}_{\nu_1} \to \mathscr{L}_{\nu_2}$, $\nu_1 < \nu_2$, the subspace $\mathscr{H}_{\nu_1} \subset \mathscr{L}_{\nu_1}$ is mapped into $\mathscr{H}_{\nu_2} \subset \mathscr{L}_{\nu_2}$. Consequently, $\mathscr{L}^0 = \varinjlim \mathscr{L}_\nu$ contains $\mathscr{H}^0 = \varinjlim \mathscr{H}_\nu$. Going over to the completions, we see that $\mathscr{L}$ contains $\mathscr{H}$.

REMARK. We have incidentally constructed a compatible system of morphisms $j_{\nu_2\nu_1}: \mathscr{H}_{\nu_1} \to \mathscr{H}_{\nu_2}$, $\nu_1 < \nu_2$, that commute with the action of $G_{\nu_1}$.

In §2 we have given another method of specifying such a compatible system of morphisms. However, the method given there has the advantage that it does not make use of the concept of a vacuum vector and does not depend on the choice of a maximal compact subgroup.

THEOREM 3.1. *The morphism $\mathcal{H} \to \mathcal{L}$ defined above is an isomorphism. Thus, the representations of $G^0$ in $\mathcal{H}$ and $\mathcal{L}$ are equivalent.*

PROOF. Since the vacuum vector $\xi_0 \in \mathcal{L}$ is cyclic in $\mathcal{L}$, it is sufficient for us to verify that $\xi$ belongs to $\mathcal{H}$. To do this we find the projection of $\xi_0$ onto each $\mathcal{H}_\nu$.

Let $\nu$: $X = \bigcup_{i=1}^{k} X_i$ be an arbitrary admissible partition such that $\lambda_i = m(X_i) < 1$. Then we represent $\xi_0$ as an element of $\mathcal{L}_\nu = \bigotimes_{i=1}^{k} L_{\lambda_i}$ in the form $\xi_0 = \bigotimes_{i=1}^{k} \xi_{\lambda_i}$, where $\xi_{\lambda_i}$ is the canonical vector in $L_{\lambda_i}$.

According to §1.2b the projection of $\xi_\lambda \in L_\lambda$ onto $H_\lambda \subset L_\lambda$ is equal to $c_\lambda \eta_\lambda$, where $\eta_\lambda$ is the fixed unit vector of $H_\lambda$ explicitly constructed there; here $1 - c_\lambda = O(\lambda^2)$ as $\lambda \to 0$.

Hence it follows that the projection of $\xi_0 = \bigotimes_{i=1}^{k} \xi_{\lambda_i}$ onto $\mathcal{H}_\nu = \bigotimes_{i=1}^{k} H_{\lambda_i}$ is equal to

$$\eta'_\nu = \Big(\prod_{i=1}^{k} c_{\lambda_i}\Big)\eta_\nu,$$

where $\eta_\nu = \bigotimes_{i=1}^{k} \eta_{\lambda_i}$ is a vector of unit norm.

From the estimate $c_\lambda = 1 + O(\lambda^2)$ it follows that for an indefinite refinement of $\nu$, as $\max \lambda_i$ tends to zero, the norm of $\eta'_\nu$ tends to 1, hence that $\eta'_\nu$ itself tends to $\xi_0$, and the theorem is proved.

THEOREM 3.2. *The vacuum vector in $\mathcal{L}$ is uniquely determined to within a factor.*

PROOF. Let $\xi'_0$ and $\xi''_0$ be two vacuum vectors in $\mathcal{L}$, and let $\eta'_\nu$ and $\eta''_\nu$ be their projections onto $\mathcal{H}_\nu = \bigotimes_{i=1}^{k} H_{\lambda_i}$, where $\nu$: $X = \bigcup_{i=1}^{k} X_i$, $\lambda_i = m(X_i) < 1$. Since $\xi'_0$ and $\xi''_0$ are invariant under $K_\nu = K_{X_1} \times \ldots \times K_{X_k} \subset G_\nu$, their projections $\eta'_\nu$ and $\eta''_\nu$ onto $\mathcal{H}_\nu$ have the same property. But in $\mathcal{H}_\nu$ there is, to within a factor, only one vector that is invariant under $K_\nu$; consequently, $\eta'_\nu$ and $\eta''_\nu$ are proportional.

On the other hand, since $\mathcal{H}^0 = \varinjlim \mathcal{H}_\nu$ is everywhere dense in $\mathcal{L}$ (by Theorem 3.1), the vectors $\eta'_\nu$ and $\eta''_\nu$ converge, respectively, to $\xi'_0$ and $\xi''_0$ when $\nu$ is refined indefinitely provided that $\max \lambda_i \to 0$. Consequently, since $\eta'_\nu$ and $\eta''_\nu$ are proportional, so are the limit vectors $\xi'_0$ and $\xi''_0$. This completes the proof.

REMARK. We emphasize that the vacuum vector $\xi_0$ is contained in each subspace $\mathcal{L}_\nu$ whereas it is not contained in any of the subspaces $\mathcal{H}_\nu$.

**4. A representation of $G^X$.** So far we have constructed a representation $U_{\tilde{g}}$ of the group $G^0$ of step functions $X \to G$. We now show how a representation of the group $G^X$ of continuous functions on $X \to G$ can be defined in terms of this representation. Namely, we claim that the representation $U_{\tilde{g}}$ of $G^0$ (second construction) can be extended to a representation of a complete metric group containing both $G^0$ and $G^X$ as everywhere dense subgroups. This then defines an irreducible unitary representation of $G^X$.

For simplicity we assume further that the support of $m$ is the whole space $X$.

We first construct a certain metric on $G^0$. Let $\rho(y_1, y_2)$ be the invariant metric on the Lobachevskii plane $Y = K\backslash G$. We define on $G$ a metric $d(g_1, g_2)$ invariant under right translations and such that

$$d(g_1, g_2) \geqslant \rho(y_0 g_1, y_0 g_2)$$

for any $g_1, g_2 \in G$, where $y_0$ is the point on the Lobachevskii plane $Y$ that corresponds to the unit coset. (Such a metric exists; for example, we may set $d(g_1, g_2) = \rho(y_0 g_1, y_0 g_2) + \rho(y_1 g_1, y_1 g_2)$, where $y_1 \neq y_0$.) We now introduce a metric $\delta$ on the group $G^0$ of step functions, setting

$$\delta(g_1(\cdot), g_2(\cdot)) = \int d(g_1(x), g_2(x))dm(x).$$

Completing $G^0$ in this metric we obtain a complete metric group $\bar{G}^X$, consisting of all $m$-measurable functions $g(\cdot)$ for which

$$\int d(g(x), e)\, dm(x) < \infty.$$

Observe that the completion of $G^0$ in the metric $\delta$ contains, in particular, the group $G^X$ of continuous functions; $G^X$ is everywhere dense in this completion.

We claim that the representation $U_{\tilde{g}}$ of $G^0$ constructed above can be extended to a representation of $\bar{G}^X$.

For this purpose we consider the functional $\Psi$ on $G^0$ introduced earlier:

$$\Psi(\tilde{g}) = (U_{\tilde{g}}\xi_0, \xi_0)_{\mathscr{L}},$$

where $\xi_0 \in \mathscr{L}$ is a vacuum vector. It was shown above that

$$\Psi(\tilde{g}) = \exp\left(\int \ln \psi(g(x))\, dm(x)\right). \tag{3}$$

LEMMA 3.4. *The functional $\Psi(\tilde{g})$ can be extended from $G^0$ to a continuous functional on the whole group $\bar{G}^X$.*

PROOF. It is sufficient to verify that $\Psi(\tilde{g})$ in (3) is defined and continuous on the whole group $\bar{G}^X$.

We use the following expression for $\psi(g)$ introduced in §1.2a: $\psi(g) = \cosh^{-1}\rho(y_0 g, g)$.

From this expression it follows that $|\ln \psi(g)| \leqslant \rho(y_0 g, y_0) \leqslant d(g, e)$, therefore the integral $\int \ln \psi(g(x)) dm(x)$ converges absolutely.

The continuity of $\Psi(\tilde{g})$ follows immediately from the following estimate.

$$\left| \int \ln \psi(g_1(x)) dm(x) - \int \ln \psi(g_2(x)) dm(x) \right| \leqslant \delta(g_1(\cdot), g_2(\cdot)).$$

We shall prove this inequality. We set $\tau_i(x) = \rho(y_0 g_i(x), y_0)$ and use the bound $\left| \ln \frac{\cosh \tau_1}{\cosh \tau_2} \right| \leqslant \left| \tau_1 - \tau_2 \right|$. We have

$$\left| \int \ln \psi(g_1(x)) dm(x) - \int \ln \psi(g_2(x)) dm(x) \right| \leqslant \int | \ln \psi(g_1(x)) - \ln \psi(g_2(x)) | dm(x) =$$

$$= \int \left| \ln \frac{\cosh \tau_1(x)}{\cosh \tau_2(x)} \right| dm(x) \leqslant \int |\tau_1(x) - \tau_2(x)| dm(x) \leqslant \int \rho(y_0 g_1(x), y_0 g_2(x)) dm(x) \leqslant$$

$$\leqslant \int d(g_1(x), g_2(x)) dm(x) = \delta(g_1(\cdot), g_2(\cdot)).$$

THEOREM 3.3. *The representation $U_{\tilde{g}}$ of $G^0$ can be extended by continuity to a unitary representation of $\bar{G}^x$.*

This follows immediately from the preceding lemma and the following proposition.

LEMMA 3.5. *Let $G$ be a topological group satisfying the first axiom of countability, $G^0 \subset G$ a subgroup everywhere dense in $G$, and $T$ a continuous unitary representation of $G^0$ in a Hilbert space $H$ with a cyclic vector $\xi$. Further, let $\Phi(g) = (T(g)\xi, \xi)$, $g \in G^0$. If $\Phi(\cdot)$ can be extended to a continuous function on $G$, then the representation $T$ of $G^0$ can be extended by continuity to a unitary representation of $G$.*

PROOF OF THE LEMMA. We first show that for any $\eta_1, \eta_2 \in H$ the function $\Phi_{\eta_1 \eta_2}(g) = (T(g)\eta_1, \eta_2)$ can be extended by continuity from $G^0$ to $G$.

Let $H_0$ be the space consisting of finite linear combinations of vectors $T(g)\xi$, $g \in G^0$. Since $\xi$ is a cyclic vector, $H_0$ is everywhere dense in $H$.

It is evident that if $\eta_1, \eta_2 \in H_0$, then $\Phi_{\eta_1 \eta_2}(\cdot)$ can be extended by continuity from $G^0$ to $G$. For if $\eta_1 = \Sigma a_i T(g_i')\xi$, $\eta_2 = \Sigma b_j T(g_j'')\xi$, then

$$\Phi_{\eta_1 \eta_2}(g) = \sum a_i \bar{b}_j \Phi(g_j''^{-1} g g_i').$$

We now let $\eta_1, \eta_2$ be arbitrary; without loss of generality we may suppose that $\|\eta_1\| = \|\eta_2\| = 1$. For any $\eta_1', \eta_2' \in H$, $\|\eta_1'\| = \|\eta_2'\| = 1$, we have

$$|\Phi_{\eta_1 \eta_2}(g) - \Phi_{\eta_1' \eta_2'}(g)| \leqslant \|\eta_1 - \eta_1'\| + \|\eta_2 - \eta_2'\|.$$

Hence the family $\Phi_{\eta_1 \eta_2}(g)$ is equicontinuous in $\eta_1, \eta_2$, and since it can be extended to all $g \in G$ for an everywhere dense set of vectors $\eta_1, \eta_2$, this proves that $\Phi_{\eta_1 \eta_2}(\cdot)$ can be extended to $G$ for all $\eta_1, \eta_2$.

We claim that the operators $T(g)$, $g \in G$, so obtained are unitary, that is,

$(T(g)\xi, T(g)\xi) = (\xi, \xi)$ for any $\xi \in H$. Let $\{g_n\}$ be a sequence of elements of $G^0$ that converges to $g$.

For any $\varepsilon > 0$ there exists an $N$ such that for $m, n > N$

$$(4) \qquad |(T(g_m)\xi, T(g_n)\xi) - (\xi, \xi)| < \varepsilon$$

(since $(T(g_m)\xi, T(g_n)\xi) = (T(g_n g_m)\xi, \xi)$, hence converges to $(\xi, \xi)$).

On the other hand, we can find $m$ and $n$, greater than $N$, such that

$$(5) \qquad |(T(g_m)\xi, T(g_n)\xi) - (T(g)\xi, T(g_n)\xi)| + \\ + |(T(g)\xi, T(g_n)\xi) - (T(g)\xi, T(g)\xi)| < \varepsilon.$$

Comparing (4) and (5), we see that

$$|(T(g)\xi, T(g)\xi) - (\xi, \xi)| < 2\varepsilon,$$

which proves that $T(g)$ is unitary.

From the fact that the $T(g)$ are unitary and weakly convergent it follows that $\|T(g_n)\xi - T(g)\xi\| \to 0$, as $g_n \to g$, $g_n \in G^0$. From this it follows automatically that $T(g_1)T(g_2)\xi = T(g_1 g_2)\xi$ for all $g_1, g_2 \in G$.

Thus, we have constructed an irreducible unitary representation of the complete metric group $G^X$ in $\mathcal{L}$. Restricting it to the everywhere dense subgroup $G^X$ of $\bar{G}^X$, we obtain the required irreducible unitary representation of $G^X$ in $\mathcal{L}$.

Since $G^X$ is dense in $\bar{G}^X$, the vacuum vector $\xi_0$ is also cyclic relative to $G^X$, and for every $\tilde{g} \in G^X$ we have

$$(6) \qquad (U_{\tilde{g}}\xi_0, \xi_0) = \exp\left(\int \ln \psi(g(x))\, dm(x)\right).$$

We see that the metric $\delta$ does not figure in (6). Hence our representation of $G^X$ does not depend on the choice of the metric $\delta$ in the construction.

**5. The representation $U_{\tilde{g}}$ commutes with the transformations of $X$ that preserve the measure $m$.** We consider continuous transformations $\sigma: x \mapsto x^\sigma$ of $X$. They induce automorphisms of $G^X$:

$$\tilde{g} = g(\cdot) \mapsto \tilde{g}^\sigma = g^\sigma(\cdot), \quad \text{where } g^\sigma(x) = g(x^{\sigma^{-1}}).$$

If $U_{\tilde{g}}$ is the representation of $G^X$ constructed in this section and $\sigma$ is an arbitrary continuous transformation of $X$, then we can define a new representation $U_{\tilde{g}}^\sigma$ of $G^X$ by setting $U_{\tilde{g}}^\sigma = U_{\tilde{g}^\sigma}$.

THEOREM 3.4. *The representation $U_{\tilde{g}}^\sigma$ is equivalent to the representation of $G^X$ defined in terms of the measure $m^\sigma$ on $X$, where $m^\sigma(X') = m(X'^\sigma)$ for any measurable subset $X' \subset X$. In particular, if $\sigma$ preserves $m$, then $U_{\tilde{g}}^\sigma$ and $U_{\tilde{g}}$ are equivalent.*

PROOF. Let $U'_{\tilde{g}}$ be the representation of $G^X$ defined in terms of the measure $m^\sigma$ on $X$. We compare the spherical functions $(U'_{\tilde{g}}\xi_0, \xi_0)$ and $(U_{\tilde{g}}^\sigma\xi_0, \xi_0)$, where $\xi_0$ is the vacuum vector. On the one hand,

$$(U'_{\tilde{g}}\xi_0, \xi_0) = \exp\left(\int \ln\cosh^{-1}\psi(g(x))dm^\sigma(x)\right).$$

On the other hand,

$$(U^\sigma_{\tilde{g}}\xi_0, \xi_0) = (U_{\tilde{g}^\sigma}\xi_0, \xi_0) = \exp\left(\int \ln\cosh^{-1}\psi(g(x^{\sigma^{-1}}))dm(x)\right) =$$

$$= \exp\left(\int \ln\cosh^{-1}\psi(g(x))dm^\sigma(x)\right).$$

Thus, $(U'_{\tilde{g}}\xi_0, \xi_0) = (U^\sigma_{\tilde{g}}\xi_0, \xi_0)$ for any $\tilde{g} \in G^X$, hence $U'_{\tilde{g}}$ and $U^\sigma_{\tilde{g}}$ are equivalent.

**6. Invariant definition of a canonical representation.** In §1 a canonical representation of $G$ was defined constructively. We wish to demonstrate that its connection with the representation of $G^X$ in $\mathcal{L}$ we have constructed is not accidental.

There is a natural embedding of $G$ in $G^X$. When we restrict our representation of $G^X$ in $\mathcal{L}$ to $G$, we obtain a certain representation of $G$. We look at the vacuum vector $\xi_0$ in $\mathcal{L}$, that is, the vector $\xi_0$, $\|\xi_0\| = 1$, that is invariant under $K^X$. We consider the minimal $G$-invariant subspace that contains $\xi_0$ and denote it by $L_\lambda$. It follows from the construction performed in §3.1 that *$L_\lambda$ is a canonical representation of* $G^X$ with $\lambda = m(X)$. We draw attention to the following interesting fact.

If in $\mathcal{L}$ we consider the restriction of the representation of $G^X$ to $G$, naturally embedded in $G^X$, then for $\lambda = m(X) < 1$ there is precisely one representation of the supplementary series $H_\lambda$, with $\lambda = m(X)$, that occurs as a discrete component in the decomposition. The orthogonal complement to this space splits into representations only of the principal continuous and the discrete series. For $m(X) > 1$ this representation of the supplementary series is absent. This follows from the construction of the representation of $G^X$ carried out in §2.

## §4. A representation of $G^X$ associated with a Lobachevskii plane.

Here we give an explicit form of the multiplicative integral of representations of $G = PSL(2, \mathbf{R})$.

**1. Construction of a representation of $G^X$.** Let $X$ be a compact topological space, $m$ a positive finite measure on $X$ defined on all Borel subsets and countably additive. For simplicity we assume that the support of $m$ is the whole space $X$.

Let $Y$ be a Lobachevskiĭ plane on which the group of motions $G$ acts transitively. We consider the set $Y^X$ of all continuous mappings $\tilde{y} = y(\cdot)$: $X \to Y$. We introduce the linear space $\mathcal{H}^0$, whose elements are formal finite linear combinations of such mappings:

$$\sum \lambda_i \circ \tilde{y}_i, \quad \lambda_i \in \mathbf{C}, \quad \tilde{y}_i \in Y^X.$$

In other words, $\mathcal{H}^0$ is a free linear space over $\mathbf{C}$ with $Y^X$ as a set of generators.

We introduce a scalar product in $\mathcal{H}^0$. Let $\rho(y_1, y_2)$ be the invariant metric on the Lobachevskiĭ plane. For any pair of mappings in $Y^X$, $\tilde{y}_1 = y_1(\cdot)$ and $\tilde{y}_2 = y_2(\cdot)$ we set

$$(\tilde{y}_1, \tilde{y}_2) = \exp\left(\int \ln \cosh^{-1} \rho(y_1(x), y_2(x)) dm(x)\right)$$

and then extend this scalar product by linearity to the whole space $\mathcal{H}^0$. The Hermitian form so defined on $\mathcal{H}^0$ is positive definite (for a proof see the end of §4.2 below). Let $\mathcal{H}$ be the completion of $\mathcal{H}^0$ in the norm $\|\xi\|^2 = (\xi, \xi)$.

We define a unitary representation $U_{\tilde{g}}$ of $G^X$ in $\mathcal{H}$. For this purpose we observe first that an action of $G^X$ on the set $Y^X$ of continuous mappings $X \to Y$ is naturally defined. Namely, an element $\tilde{g} = g(\cdot) \in G^X$ takes $\tilde{y} = y(\cdot)$ into $\tilde{y}\tilde{g} = y_1(\cdot)$, where $y_1(x) = y(x)g(x)$.

We assign to each $\tilde{g} \in G^X$ the following operator $U_{\tilde{g}}$ in $\mathcal{H}^0$:

$$U_{\tilde{g}}\left(\sum \lambda_i \circ \tilde{y}_i\right) = \sum \lambda_i \circ (\tilde{y}_i \tilde{g}^{-1}).$$

THEOREM 4.1. *The operators $U_{\tilde{g}}$ are unitary on $\mathcal{H}^0$ and form a representation of $G^X$.*

PROOF. The fact that the operators $U_{\tilde{g}}$ form a representation is obvious. That they are unitary follows immediately from the invariance of $\rho(y_1, y_2)$ on $Y$.

Since the operators $U_{\tilde{g}}$ are unitary on $\mathcal{H}^0$, they can be extended to unitary operators in the whole space $\mathcal{H}$. So we have constructed a unitary representation of $G^X$ in $\mathcal{H}$.

2. **Realization in the unit disk.** We provide explicit expressions for the scalar product in $\mathcal{H}^0$ and for the operator $U_{\tilde{g}}$ when $Y$ is realized as the interior of the unit disk $|z| < 1$.

Let $G$ be given as the group of matrices $g = \begin{pmatrix} \alpha & \beta \\ \bar{\beta} & \bar{\alpha} \end{pmatrix}$, let $Y$ be the interior of the unit disk $|z| < 1$, and let $G$ act in the unit disk in the following manner: $z \to zg^{-1} = \dfrac{\alpha z + \beta}{\bar{\beta} z + \bar{\alpha}}$.

Then $\mathcal{H}^0$ is the space of finite formal linear combinations

$$\sum \lambda_i \circ z_i(\cdot),$$

where $z(\cdot)$ are continuous mappings of $X$ into the unit disk $|z| < 1$. For a pair of mappings $z_1(\cdot)$ and $z_2(\cdot)$ the scalar product in $\mathcal{H}^0$ has the following form:

$$(z_1(\cdot), z_2(\cdot)) = \exp \int \ln \left( \frac{(1-|z_1(x)|^2)(1-|z_2(x)|^2)}{|1-z_1(x)\overline{z_2(x)}|^2} \right)^{1/2} dm(x). \tag{1}$$

The representation operator $U_{\tilde{g}}$, $\tilde{g} = \begin{pmatrix} \alpha(\cdot) & \beta(\cdot) \\ \overline{\beta(\cdot)} & \overline{\alpha(\cdot)} \end{pmatrix}$, takes $z(\cdot)$ into $z(\cdot)\tilde{g}^{-1} = z_1(\cdot)$, where

$$z_1(x) = \frac{\alpha(x) z(x) + \beta(x)}{\overline{\beta(x)} z(x) + \overline{\alpha(x)}}. \tag{2}$$

We indicate another convenient realization of the representation (1). (It can be obtained from the first by the transformation $z(\cdot) \to \lambda(z(\cdot)) \circ z(\cdot)$, where $\lambda(z(\cdot)) = \exp\int \ln(1-|z(x)|^2)^{-1}\,dm(x).)$

In this realization, as before, the elements of $\mathcal{H}^0$ are formal finite linear combinations of continuous transformations of $X$ into the unit disc $|z|<1$:

$$\sum \lambda_i \circ z_i(\cdot);$$

but the scalar product has the simpler form:

$$(3)\qquad (z_1(\cdot),\ z_2(\cdot)) = \exp\int \ln|1-z_1(x)\overline{z_2(x)}|^{-1}\,dm(x).$$

The representation operator $U_{\tilde{g}}$ is given by the formula:

$$U_{\tilde{g}}z(\cdot) = \exp\left(\int \ln|\overline{\beta(x)}z(x)+\overline{\alpha(x)}|^{-1}\,dm(x)\right)\circ z_1(\cdot),$$

where $z_1(x)$ is defined by (2).

In conclusion we verify that the Hermitian form introduced in $\mathcal{H}^0$ is positive definite. It is simplest to confirm this for the Hermitian form given by (3).

We introduce the following notation:

$$f_i(x,n) = \begin{cases} z_i^n(x) & \text{for } n>0,\\ \overline{z_i^{|n|}(x)} & \text{for } n<0,\end{cases}$$

$$F_i(x_1,\dots,x_k;\ n_1,\dots,n_k) = \prod_{s=1}^{k} f_i(x_s,n_s), \text{ where } i=1,2.$$

It is not hard to check that the scalar product (3) can be represented in the following form:

$$(4)\quad (z_1(\cdot),\ z_2(\cdot)) = \sum_{k=0}^{\infty}\ \sum_{n_1,\dots,n_k\ (n_j\neq 0)} \frac{2^{-k}}{k!}\frac{1}{|n_1\dots n_k|}\times$$

$$\times\int F_1(x_1,\dots,x_k;n_1,\dots,n_k)\overline{F_2(x_1,\dots,x_k;\ n_1,\dots,n_k)}\,dm(x_1)\ \dots\ dm(x_k).$$

To obtain this expression from (3) we have to expand first the function $\ln|1-z_1(x)\overline{z_2(x)}|^{-1}$ in a series:

$$\ln|1-z_1(x)\overline{z_2(x)}|^{-1} = \frac{1}{2}\sum_{n\neq 0}\frac{1}{|n|}\int f_1(x,n)\overline{f_2(x,n)}\,dm(x).$$

Then we expand $\exp u$ in a power series, where

$$u = \frac{1}{2}\sum_{n\neq 0}\frac{1}{|n|}\int f_1(x,n)\overline{f_2(x,n)}\,dm(x),$$

and obtain the required expression (4).

It is evident that each term in (4) gives a positive definite Hermitian form on $\mathcal{H}^0$; consequently, the Hermitian form given by (4) for any pair of mappings $z_1(\cdot)$ and $z_2(\cdot)$ is positive definite.

**3. Equivalence of the representation constructed here with the preceding ones.**

THEOREM 4.2. *The representation $U_{\tilde{g}}$ of $G^X$ in $\mathcal{H}$ is equivalent* *to the representation constructed in* §3. *Hence it follows, in particular, that* $U_{\tilde{g}}$ *is irreducible.*

Let $y_0$ be the point of the Lobachevskii plane $Y = K\backslash G$ (where $K$ is a fixed maximal compact subgroup) corresponding to the unit coset. We denote by $\tilde{y}_0 = y_0(\cdot)$ the mapping that takes $X$ into $y_0$. The vector $\tilde{y}_0$ belongs to $\mathcal{H}$, and it is clear that

$$U_{\tilde{g}}\tilde{y}_0 = \tilde{y}_0$$

for every $\tilde{g} \in K^X$. Thus $\tilde{y}_0$ is a vacuum vector in $\mathcal{H}$.

LEMMA 4.1. *The vector* $\tilde{y}_0$ *is cyclic in* $\mathcal{H}$.

PROOF. It is sufficient to verify that as $\tilde{g}$ ranges over $G^X$, $U_{\tilde{g}}\tilde{y}_0$ ranges over the whole of $Y^X$.

It is known that the natural fibration $G \to Y = K\backslash G$ is trivial, hence there exists a continuous cross section $s$: $Y \to G$. Now $s$ induces the mapping $Y^X \to G^X$ under which $\tilde{y} = y(\cdot) \in Y^X$ goes into $\tilde{g} = g(\cdot) \in G^X$, where $g(x) = s[y(x)]$. It is also clear that $y(\cdot) = \tilde{y}_0 g(\cdot)$. This completes the proof.

Let us find the spherical function $(U_{\tilde{g}}\tilde{y}_0, \tilde{y}_0)$, where $\tilde{y}_0$ is a vacuum vector. Since $U_{\tilde{g}}\tilde{y}_0 = \tilde{y}_0\tilde{g}^{-1}$, we obtain by the formula for the scalar product in $\mathcal{H}$

$$(U_{\tilde{g}}\tilde{y}_0, \tilde{y}_0) = \exp \int \ln \cosh^{-1} \rho(y_0 g^{-1}(x), y_0)\, dm(x) =$$

$$= \exp \int \ln \cosh^{-1} \rho(y_0 g(x), y_0)\, dm(x) = \exp \int \ln \psi(g(x))\, dm(x).$$

We proceed now to the proof of Theorem 4.2. In §3 the representation $U_{\tilde{g}}$ of $G^X$ was defined in the Hilbert space $\mathcal{L}$ with the cyclic vacuum vector $\xi_0$. It was also established that $(U_{\tilde{g}}\xi_0, \xi_0) = \exp \int \ln \psi(g(x))dm(x)$.

So we see that $(U_{\tilde{g}}\xi_0, \xi_0)_{\mathcal{L}} = (U_{\tilde{g}}\tilde{y}_0, \tilde{y}_0)_{\mathcal{H}}$. Since the vectors $\xi_0$ and $\tilde{y}_0$ are cyclic in their respective spaces, it follows that the mapping $\xi_0 \mapsto \tilde{y}_0$ can be extended to an isomorphism $\mathcal{L} \to \mathcal{H}$ that commutes with the action of $G^X$ in $\mathcal{L}$ and $\mathcal{H}$. This proves Theorem 4.2.

## §5. A representation of $G^X$ associated with a maximal compact group $K \subset G$

**1. Construction of a representation of** $G$. We take $G$ to be the group of matrices $\begin{pmatrix} \alpha & \beta \\ \bar{\beta} & \bar{\alpha} \end{pmatrix}$, $|\alpha|^2 - |\beta|^2 = 1$. As before, let $X$ be a compact topological space with positive finite measure $m$. For simplicity we assume that the support of $m$ is the whole space $X$ and that $m(X) = 1$. Henceforth we write $dx$ instead of $dm(x)$.

Although the method of construction that we use here for the representation of $G^X$ is cumbersome, it has the advantage that all the formulae can be written out explicitly and completely and are to some extent analogous to the expression of representations of the rotation group by means of spherical functions.

We suggest that on first reading the reader should omit the simple but tedious proof in the second half of § 5.1 of the fact that the formulae gives a unitary representation.

Formulae for representations of the Lie algebra of $G^X$ are given in two forms at the end of this section.

We introduce the Hilbert space $\mathcal{H}$ whose elements are all the sequences

$$F = (f_0, f_1, \ldots, f_k, \ldots),$$

where $f_0 \in \mathbf{C}$, and $f_k$ for $k > 0$ are functions $\underbrace{X \times \ldots \times X}_{k} \times \underbrace{\mathbf{Z} \times \ldots \times \mathbf{Z}}_{k} \to \mathbf{C}$, satisfying the following conditions:[1]

1) $f_k(x_1, \ldots, x_k; n_1, \ldots, n_k)$ is symmetric with respect to permutations of the pairs $(x_i, n_i)$, $(x_j, n_j)$;

2) $f_k(x_1, \ldots, x_k; n_1, \ldots, n_k)\,|_{n_i=0} =$

$f_{k-1}(x_1, \ldots, \hat{x}_i, \ldots, x_k; n_1, \ldots, \hat{n}_i, \ldots, n_k)$ $(i = 1, \ldots, k)$;

(1) 3) $\| F \|^2 = | f_0 |^2 +$

$$+\sum_{k=1}^{\infty} \sum_{\substack{n_1, \ldots, n_k \\ (n_i \neq 0)}} \frac{1}{k!} \frac{1}{| n_1 \ldots n_k |} \int | f_k(x_1, \ldots, x_k; n_1, \ldots, n_k) |^2 \, dx_1 \ldots dx_k < \infty.$$

REMARK. Nothing would be changed in the definition of $\mathcal{H}$ if we were to assume that all the integral indices $n_i$ are non-zero. Then, of course, condition 2) is unnecessary, and the norm, as before, is given by (1).

We construct a unitary representation of $G^X$ in $\mathcal{H}$. First we introduce on $G$ functions $P_{mn}(g)$ and $p_n(g)$. We define $P_{mn}(g)$ for $n \geqslant 0$ as the coefficient of $z^m$ in the power series expansion of $\left(\frac{\alpha z+\beta}{\bar{\beta} z+\bar{\alpha}}\right)^n$, where $g = \begin{pmatrix} \alpha & \beta \\ \bar{\beta} & \bar{\alpha} \end{pmatrix}$. Thus.

$$\left(\frac{\alpha z+\beta}{\bar{\beta} z+\bar{\alpha}}\right)^n = \sum_m P_{mn}(g) z^m, \qquad n \geqslant 0$$

For $n \leqslant 0$ we define $P_{mn}(g)$ by:

$$\left(\frac{\bar{\alpha}\bar{z}+\bar{\beta}}{\beta\bar{z}+\alpha}\right)^{|n|} = \sum_m P_{-m,\, n}(g)\, \bar{z}^m, \qquad n \leqslant 0.$$

---

[1] $\hat{x}_i$, $\hat{n}_i$ indicate that the corresponding variables are omitted.

From this definition it follows that 1) $P_{-m,-n}(g) = P_{mn}(g)$; 2) $P_{mn}(g) = 0$ if $mn < 0$; 3) $P_{m0}(g) = 1$ for $m = 0$ and $P_{m0}(g) = 0$ for $m \neq 0$; 4) $P_{0n}(g) = \left(\frac{\beta}{\bar{\alpha}}\right)^n$ for $n > 0$ and $P_{0n}(g) = \left(\frac{\bar{\beta}}{\alpha}\right)^{|n|}$ for $n < 0$.

Next we set $p_n(g) = \left(\frac{\bar{\beta}}{\bar{\alpha}}\right)^n$ for $n > 0$, $p_n(g) = \left(\frac{\beta}{\alpha}\right)^{|n|}$ for $n < 0$, $p_0(g) = 1$.

Let $\tilde{g} = \begin{pmatrix} \alpha(\cdot) & \beta(\cdot) \\ \bar{\beta}(\cdot) & \bar{\alpha}(\cdot) \end{pmatrix}$ be an arbitrary element of $G^X$. We associate with it the operator $U_{\tilde{g}}$ in $\mathscr{H}$ that is given by the formula

$$U_{\tilde{g}}\{f_k\} = \{\varphi_k\},$$

where

$$\varphi_k(x_1, \ldots, x_k; n_1, \ldots, n_k) = \Psi(\breve{g}) \times \tag{2}$$

$$\times \sum_{m_1, \ldots, m_k} \left\{ \prod_{i=1}^{k} P_{m_i n_i}(g(x_i)) \times \sum_{s=0}^{\infty} \sum_{\substack{l_1, \ldots, l_s \\ (l_i \neq 0)}} \frac{(-1)^{l_1+\ldots+l_s}}{s! \, |l_1 \ldots l_s|} \times \right.$$

$$\left. \times \int \prod_{j=1}^{s} p_{l_j}(g(t_j)) f_{k+s}(x_1, \ldots, x_k, t_1, \ldots, t_s; m_1, \ldots, m_k, l_1, \ldots, l_s) \, dt_1 \ldots dt_s \right\},$$

and

$$\Psi(\breve{g}) = \exp \int \ln \psi^2(g(x)) \, dx. \tag{3}$$

Here $\psi(g)$ denotes $|\alpha|^{-1}$.

THEOREM 5.1. *The operators $U_{\tilde{g}}$ are unitary and form a representation of $G^X$, that is,*

$$(U_{\tilde{g}} F_1, U_{\tilde{g}} F_2) = (F_1, F_2)$$

*for any $\tilde{g} \in G^X$ and $F_1, F_2 \in \mathscr{H}$ and*

$$U_{\tilde{g}_1} U_{\tilde{g}_2} F = U_{\tilde{g}_1 \tilde{g}_2} F$$

*for any $\tilde{g}_1, \tilde{g}_2 \in G^X$ and $F \in \mathscr{H}$.*

We verify these relations for the vectors $F$ of a certain space $\mathscr{H}^0$ everywhere dense in $\mathscr{H}$, which we now introduce.

We denote by $M$ the set of sequences of the form

$$F = (1, f_1, f_2, \ldots, f_k, \ldots),$$

where

$$f_k(x_1, \ldots, x_k; n_1, \ldots, n_k) = u(x_1, n_1) \ldots u(x_k, n_k),$$

and $u(x, n)$ is a function continuous in $x$ for any fixed $n$ such that $u(x, 0) = 1$ and

(4) $$\sum_{n\neq 0}\frac{1}{|n|}\int |u(x, n)|^2\,dx<\infty.$$

We denote by $\mathscr{H}^0$ the space of all finite linear combinations of elements of $M$.

We must verify that $\mathscr{H}^0 \subset \mathscr{H}$. Now it is evident that the vectors $F \in M$ satisfy conditions 1) and 2) in the definition of $\mathscr{H}$. Furthermore, if

(5) $$F = (1, u(x, n), \ldots, u(x_1, n), \ldots, u(x_k, n_k), \ldots)$$

is a vector of $M$, then its norm $\|F\|$ can be represented in the form:

(6) $$\|F\|^2=\exp\Big(\sum_{n\neq 0}\frac{1}{|n|}\int |u(x, n)|^2\,dx\Big).$$

Consequently, by (4), $F$ also satisfies condition 3), and hence $F \in \mathscr{H}$. We observe that if $F \in M$, that is, if it has the form (4), then the expression for $U_{\tilde{g}}F$ reduces to the following simple form:

(7) $$U_{\tilde{g}}F = \lambda(\tilde{g}, u)(1, v(x, n), \ldots, v(x_1, n_1), \ldots, v(x_k, n_k), \ldots),$$

where

(8) $$\lambda(\tilde{g}, u)=\Psi(\tilde{g})\exp\Big(\sum_{l\neq 0}\frac{(-1)^l}{|l|}\int p_l(g(x))\,u(x, l)\,dx\Big),$$

(9) $$v(x, n)=\sum_m P_{mn}(g(x))\,u(x, m).$$

LEMMA 5.1. *The space $\mathscr{H}^0$ is everywhere dense in $\mathscr{H}$.*

PROOF. We assume that all the indices $n_i$ are non-zero (see the Remark on p. 115). Suppose that $\mathscr{H}^0$ is not dense in $\mathscr{H}$, hence that there exists a non-zero vector $F = \{f_k^0\}$, orthogonal to $\mathscr{H}^0$. We consider in $\mathscr{H}$ the vectors of the form $F_\lambda = \{\lambda^k f_k\}$, where $f_0 = 1$, $f_k(x_1, \ldots, x_k, n_1, \ldots, n_k) = u(x_1, n_1) \ldots u(x_k, n_k)$ for $k > 0$, $u(x, n)$ is a continuous function, and $\lambda$ is an arbitrary number.

Since $F_\lambda \in \mathscr{H}^0$, we have $(F_\lambda, F) = 0$, that is,

(10) $$\sum_{k=0}^{\infty}\frac{\lambda^k}{k!}\Big(\sum_{\substack{n_1,\ldots,n_k\\(n_i\neq 0)}}|n_1\ldots n_k|^{-1}\times$$
$$\times\int f_k(x_1,\ldots,x_k;\,n_1,\ldots,n_k)\,\overline{f_k^0(x_1,\ldots,x_k;\,n_1,\ldots,n_k)}\,dx_1\ldots dx_k\Big)=0$$

for any $\lambda$. Hence it follows that for every $k = 0, 1, \ldots$ we have the relation

$$\sum_{\substack{n_1,\ldots,n_k\\(n_i\neq 0)}}|n_1\ldots n_k|^{-1}\times$$
$$\times\int f_k(x_1,\ldots,x_k;\,n_1,\ldots,n_k)\,\overline{f_k^0(x_1,\ldots,x_k;\,n_1,\ldots,n_k)}\,dx_1\ldots dx_k=0,$$

or

$$(11) \qquad \sum_{\substack{n_1, \ldots, n_k \\ (n_i = 0)}} |n_1 \ldots n_k|^{-1} \times$$

$$\times \int u(x_1, n_1) \ldots u(x_k, n_k) f_k^0(x_1, \ldots, x_k;\ n_1, \ldots, n_k)\, dx_1 \ldots dx_k = 0$$

for any continuous function $u(x, n)$ such that

$$\sum_{n \neq 0} \frac{1}{|n|} \int |u(x, n)|^2\, dx < \infty.$$

Since $f_k^0(x_1, \ldots, x_k;\ n_1, \ldots, n_k)$ is symmetric under permutations of the pairs $(x_i, n_i)$ and $(x_j, n_j)$, it follows from (11) that $f_k^0 \equiv 0$ $(k = 0, 1, \ldots)$. Thus, $F = 0$, in contradiction to the hypothesis.

To prove Theorem 5.1 we need certain relations for the functions $P_{mn}(g)$ and $p_l(g)$:

a) $\dfrac{(-1)^n}{|n|} P_{mn}(g) = \dfrac{(-1)^m}{|m|} P_{nm}(g)$ for $m \neq 0$, $n \neq 0$;

b) for any compact subset $V \subset G$ there exist constants $C > 0$ and $r$, $0 < r < 1$, such that $|P_{mn}(g)| < Cr^{|m|+|n|}$;

c) $\sum_{m'} P_{m'n}(g_1) P_{mm'}(g_2) = P_{mn}(g_1 g_2)$;

d)

$$\sum_{l \neq 0} \frac{(-1)^l}{|l|} p_l(g_1) P_{ml}(g_2) = \begin{cases} \dfrac{(-1)^m}{|m|} (p_m(g_1 g_2) - p_m(g_2)) & \text{for } m \neq 0, \\ -2 \ln \dfrac{\psi(g_1)\psi(g_2)}{\psi(g_1 g_2)} & \text{for } m = 0; \end{cases}$$

e)

$$\sum_{n \neq 0} \frac{1}{|n|} P_{mn}(g) \overline{P_{m'n}(g)} = \begin{cases} |m|^{-1} \delta_{mm'} & \text{for } m \neq 0,\ m' \neq 0, \\ -(-1)^m |m|^{-1} p_m(g) & \text{for } m \neq 0,\ m' = 0, \\ -(-1)^{m'} |m'|^{-1} \overline{p_{m'}(g)} & \text{for } m = 0,\ m' \neq 0, \\ -4 \ln \psi(g) & \text{for } m = m' = 0 \end{cases}$$

($\delta_{mm'}$ is the Kroneker delta).[1])

LEMMA 5.2. *The function $\lambda(\tilde{g}, u)$ defined by* (8) *satisfies the following functional relation*:

$$(12) \qquad \lambda(\tilde{g}_1, v)\lambda(\tilde{g}_1, u) = \lambda(\tilde{g}_1 \tilde{g}_2, u),$$

*where*

$$v(x, n) = \sum_m P_{mn}(g_2(x)) u(x, m).$$

---

[1]) We can derive a) from the relation $P_{mn}(g) = \dfrac{1}{2\pi i} \displaystyle\int_{|z|=1} \left(\frac{\alpha z + \beta}{\bar{\beta} z + \bar{\alpha}}\right)^n z^{-m-1}\, dz$, $n > 0$; the bound b) for $P_{mn}(g)$ follows from the fact that the radius of convergence of the series $\sum_m P_{mn}(g) z^m$ is greater than 1; the relation c) follows immediately from the definition of the functions $P_{mn}$; d) follows easily from the definition of $P_{mn}$ and $p_l$ and a).

PROOF. It follows from the definition of $\lambda$ that

$$\lambda(\tilde{g}_1, v)=\Psi(\tilde{g}_1)\exp\Big(\sum_{l\neq 0}\frac{(-1)^l}{|l|}\int p_l(g_1(x))\,v(x, l)\,dx\Big)=$$
$$=\Psi(\tilde{g}_1)\exp\Big(\sum_{l\neq 0}\sum_{m}\frac{(-1)^l}{|l|}\int p_l(g_1(x))\,P_{ml}(g_2(x))\,u(x, m)\,dx\Big).$$

We sum over $l$, apply d),[1] and obtain

$$\lambda(\tilde{g}_1, v)=\Psi(\tilde{g}_1)\exp\Big\{\sum_{m\neq 0}\frac{(-1)^m}{|m|}\int p_m((g_1g_2)(x))\,u(x, m)\,dx-$$
$$-\sum_{m\neq 0}\frac{(-1)^m}{|m|}\int p_m(g_2(x))\,u(x, m)\,dx\Big\}\frac{\Psi(\tilde{g}_1\tilde{g}_2)}{\Psi(\tilde{g}_1)\Psi(\tilde{g}_2)}=\frac{\lambda(\tilde{g}_1\tilde{g}_2, u)}{\lambda(\tilde{g}_2, u)}.$$

LEMMA 5.3. $U_{\tilde{g}_1}U_{\tilde{g}_2} = U_{\tilde{g}_1\tilde{g}_2}F$ *for any* $F\in M$ *and any* $\tilde{g}_1, \tilde{g}_2 \in G^X$.

PROOF. Let

$$F = (1, u(x, n), \ldots, u(x_1, n_1), \ldots, u(x_k, n_k), \ldots).$$

Then

$$U_{\tilde{g}_2}F = \lambda(\tilde{g}_2, u)(1, v(x, n), \ldots, v(x_1, n_1) \ldots v(x_k, n_k), \ldots),$$

where

$$v(x, n) = \sum_m P_{mn}(g_2(x))u(x, m);$$

$$U_{\tilde{g}_1}U_{\tilde{g}_2}F = \lambda(\tilde{g}_1, v)\lambda(\tilde{g}_2, u)(1, w(x, n), \ldots, w(x_1, n_1), \ldots, w(x_k, n_k), \ldots),$$

where

$$w(x, n)=\sum_{m'}P_{m'n}(g_1(x))\,v(x, m')=\sum_{m', m}P_{m'n}(g_1(x))\,P_{mm'}(g_2(x))\,u(x, m).$$

It follows from Lemma 5.2 that $\lambda(\tilde{g}_1, v)\lambda(\tilde{g}_2, u) = \lambda(\tilde{g}_1\tilde{g}_2, u)$. On the other hand, by c) for $P_{mn}(g)$, we have $w(x, n) = \sum_m P_{mn}((g_1g_2)(x))u(x, m)$. Consequently, $U_{\tilde{g}_1}U_{\tilde{g}_2}F = U_{\tilde{g}_1\tilde{g}_2}F$.

COROLLARY. *The operators* $U_{\tilde{g}}$ *form a representation of* $G^X$ *in* $\mathcal{H}^0$.

LEMMA. 5.4. $(U_{\tilde{g}}F_1, U_{\tilde{g}}F_2) = (F_1, F_2)$ *for any* $F_1, F_2 \in M$ and $\tilde{g}\in G^X$.

PROOF. Let

$$F_1 = (1, u_1(x, n), \ldots, u_1(x_1, n_1), \ldots, u_1(x_k, n_k), \ldots),$$
$$F_2 = (1, u_2(x, n), \ldots, u_2(x_1, n_1), \ldots, u_2(x_k, n_k), \ldots).$$

Then

$$(F_1, F_2)=\exp\Big(\sum_{n\neq 0}\frac{1}{|n|}\int u_1(x, n)\,\overline{u_2(x, n)}\,dx\Big).$$

On the other hand, using the expression (7) for $U_{\tilde{g}}F$ we have

$$(U_{\tilde{g}}F_1, U_{\tilde{g}}F_2)=\exp\Big\{4\int \ln\psi(g(x))\,dx+$$
$$+\sum_{l\neq 0}\frac{(-1)^l}{|l|}\int p_l(g(x))\,u_1(x, l)\,dx+\sum_{l\neq 0}\frac{(-1)^l}{|l|}\int \overline{p_l(g(x))\,u_2(x, l)}\,dx+$$
$$+\sum_{n\neq 0}\frac{1}{|n|}\sum_{m, m'}\int P_{mn}(g(x))\,\overline{P_{m'n}(g(x))}\,u_1(x, m)\,\overline{u_2(x, m')}\,dx\Big\}.$$

---

[1] Reversal of the order of the summations is permissible in view of b).

In the last expression we sum over $n$ under the exponential sign, then use e) and $u_i(x, 0) = 1$, and obtain $(U_{\tilde{g}}F_1, U_{\tilde{g}}F_2) = (F_1, F_2)$ after some elementary simplifications.

COROLLARY. *The $U_{\tilde{g}}$ are unitary operators in $\mathcal{H}^0$.*

**2. Irreducibility of the representation $U_{\tilde{g}}$.** The representation operators $U_{\tilde{g}}$ assume a specially simple form when restricted to the subgroup of matrices

$$\tilde{k} = \begin{pmatrix} e^{i\varphi(\cdot)/2} & 0 \\ 0 & e^{-i\varphi(\cdot)/2} \end{pmatrix}.$$

Namely,

$$U_{\tilde{k}} \{f_k\} = \{f_k'\}, \tag{13}$$

where $f_k'(x_1, \ldots, x_k; n_1, \ldots, n_k) = \exp\left(i \sum_{s=1}^{k} n_s \varphi(x_s)\right) f_k(x_1, \ldots, x_k; n_1, \ldots, n_k)$.

From this expression it is clear that *the family of commuting operators $U_{\tilde{k}}$ has a simple spectrum in $\mathcal{H}$. The vacuum vector in $\mathcal{H}$ is*

$$\xi_0 = (1, f_1(x, n), \ldots, f_k(x_1, \ldots, x_k; n_1, \ldots, n_k), \ldots),$$

*where $f_k(x_1, \ldots, x_k; n_1, \ldots, n_k) = 0$ if $|n_1| + \ldots + |n_k| > 0$.*

THEOREM 5.2. *The representation $U_{\tilde{g}}$ of $G^X$ in $\mathcal{H}$ is irreducible.*

PROOF. Let $A$ be a bounded operator in $\mathcal{H}$ that commutes with all the operators $U_{\tilde{g}}$, in particular, with the $U_{\tilde{k}}$ of the form (13). Since the family of operators $U_{\tilde{k}}$ has a simple spectrum, an operator A that commutes with them has the form

$$A\{f_k\} = \{a_k f_k\}, \tag{14}$$

where $a_k(x_1, \ldots, x_k; n_1, \ldots, n_1)$ are measurable functions (satisfying the same relations as the $f_k$). Let $\xi_0$ be the vacuum vector in $\mathcal{H}$ defined above. It follows from (14) that $A\xi_0 = a_0\xi_0$.

We apply to $\xi_0$, the operator $U_{\tilde{g}}$, $\tilde{g} \in G$, where

$$\tilde{g} = \begin{pmatrix} \cosh\tau & \sinh\tau \\ \sinh\tau & \cosh\tau \end{pmatrix}, \quad \tau \neq 0,$$

and $\tau$ does not depend on $x$. We obtain

$$U_{\tilde{g}}\xi_0 = \{f_k\},$$

where $f_k(x_1, \ldots, x_k; n_1, \ldots, n_k) = \cosh^{-1}\tau \prod_{i=1}^{k} \tanh^{|n_i|}\tau$.

From $U_{\tilde{g}}A\xi_0 = AU_{\tilde{g}}\xi_0$ it follows that $a_0\{f_k\} = \{a_k f_k\}$, hence $a_0 f_k = a_k f_k$. Consequently, since the $f_k$ do not vanish, $a_k = a_0$ for every $k$, that is, $A$ is a multiple of the unit operator.

**3. Equivalence of the representation $U_{\tilde{g}}$ to representations constructed earlier.** THEOREM 5.3. *The representation $U_{\tilde{g}}$ of $G^X$ in $\mathcal{H}$ is equivalent to* *the representations constructed in the preceding sections.*

PROOF. It follows easily from the definition of $U_{\tilde{g}}$ and the scalar product in $\mathscr{H}$ that the spherical function corresponding to the vacuum vector $\xi_0$ has the form

$$(U_{\tilde{g}}\xi_0, \xi_0) = \exp\left(2\int \ln \psi(g(x))dx\right).$$

So we see that this spherical function coincides (with suitable agreement of measures on $X$) with the spherical functions of the representations of $G^X$ constructed in the preceding sections. The theorem follows immediately from this fact and the irreducibility of $U_{\tilde{g}}$.

**4. Infinitesimal formulae for the representation.** We give formulae for the representation operators of the Lie algebra of $G^X$ in $\mathscr{H}$ (that is, of the algebra $\mathfrak{G}^X$ of continuous mappings $X \to \mathfrak{G}$ of $X$ to the Lie algebra $\mathfrak{G}$ of $G$ with the natural commutation relations).

We take the following matrices as generators of $\mathfrak{G}$:

$$a_{\tilde{\varphi}}^0 = \begin{pmatrix} i\frac{\varphi(\cdot)}{2} & 0 \\ 0 & -i\frac{\varphi(\cdot)}{2} \end{pmatrix}, \quad a_{\tilde{\tau}} = \begin{pmatrix} 0 & \tau(\cdot) \\ \tau(\cdot) & 0 \end{pmatrix}, \quad a_{i\tilde{\tau}} = \begin{pmatrix} 0 & i\tau(\cdot) \\ -i\tau(\cdot) & 0 \end{pmatrix},$$

where $\varphi(\cdot)$, $\tau(\cdot)$ are continuous mappings $X \to \mathbf{R}$. We denote the Lie operators in $\mathscr{H}$ corresponding to these elements by $A_{\tilde{\varphi}}^0$, $A_{\tilde{\tau}}$ and $A_{i\tilde{\tau}}$

Expressions for $A_{\tilde{\varphi}}^0$, $A_{\tilde{\tau}}$, and $A_{i\tilde{\tau}}$ are easily obtained from the formula (1) for the representation operators $U_{\tilde{g}}$ of $G^X$. Namely,

$$(A_{\tilde{\varphi}}^0 f)_k(x_1, \dots, x_k; n_1, \dots, n_k) = i\left(\sum_{s=1} n_s\varphi(x_s)\right) f_k(x_1, \dots, x_k; n_1, \dots, n_k);$$

$$\begin{aligned}(A_{\tilde{\tau}} f)_k(x_1, \dots, x_k; n_1, \dots, n_k) = \\ = -\sum_{s=1}^{k} n_s\tau(x_s)(f_k(x_1, \dots, x_k; n_1, \dots, n_s+1, \dots, n_k) - \\ - f_k(x_1, \dots, x_k; n_1, \dots, n_s-1, \dots, n_k)) - \\ - \int \tau(t)(f_{k+1}(x_1, \dots, x_k, t; n_1, \dots, n_k, 1) + \\ + f_{k+1}(x_1, \dots, x_k, t; n_1, \dots, n_k, -1))\,dt,\end{aligned}$$

$$\begin{aligned}(A_{i\tilde{\tau}} f)_k(x_1, \dots, x_k; n_1, \dots, n_k) = \\ = i\sum_{s=0}^{k} n_s\tau(x_s)(f_k(x_1, \dots, x_k, n_1, \dots, n_s+1, \dots, n_k) + \\ + f_k(x_1, \dots, x_k, n_1, \dots, n_s-1, \dots, n_k)) + \\ + i\int \tau(t)(f_{k+1}(x_1, \dots, x_k, t; n_1, \dots, n_k, 1) - \\ - f_{k+1}(x_1, \dots, x_k, t; n_1, \dots, n_k, -1))\,dt.\end{aligned}$$

It is convenient to go from $A_{\tilde{\tau}}$ and $A_{i\tilde{\tau}}$ to

$$A_{\tilde{\tau}}^+ = \frac{1}{2}(A_{\tilde{\tau}} + iA_{i\tilde{\tau}}), \quad A_{\tilde{\tau}}^- = \frac{1}{2}(A_{\tilde{\tau}} - iA_{i\tilde{\tau}}).$$

The operators $A_{\tilde{\tau}}^{+}$ and $A_{\tilde{\tau}}^{-}$ act in the following manner:

$$(A_{\tilde{\tau}}^{+}f)_k(x_1, \ldots, x_k;\ n_1, \ldots, n_k) =$$
$$= -\sum_{s=1}^{k} n_s\tau(x_s) f_k(x_1, \ldots, x_k;\ n_1, \ldots, n_s+1, \ldots, n_k) -$$
$$- \int \tau(t) f_{k+1}(x_1, \ldots, x_k, t;\ n_1, \ldots, n_k, 1)\, dt,$$

$$(A_{\tilde{\tau}}^{-}f)_k(x_1, \ldots, x_k;\ n_1, \ldots, n_k) =$$
$$= \sum_{s=1}^{k} n_s\tau(x_s) f_k(x_1, \ldots, x_k;\ n_1, \ldots, n_s -1, \ldots, n_k) -$$
$$- \int \tau(t) f_{k+1}(x_1, \ldots, x_k, t;\ n_1, \ldots, n_k, -1)\, dt.$$

Since $A_{\tilde{\phi}}^{0}$, $A_{\tilde{\tau}}^{+}$, and $A_{\tilde{\tau}}^{-}$ form a representation of the algebra $\mathfrak{G}^X$, the same commutation relations hold for them as for the corresponding elements of the Lie algebra $\mathfrak{G}^X$, namely,

$$[A_{\tilde{\tau}}^{+}, A_{\tilde{\varphi}}^{0}] = -iA_{\tilde{\tau}\tilde{\varphi}}^{+}, \quad [A_{\tilde{\tau}}^{-}, A_{\tilde{\varphi}}^{0}] = iA_{\tilde{\tau}\tilde{\varphi}}^{-}, \quad [A_{\tilde{\tau}_1}^{+}, A_{\tilde{\tau}_2}^{-}] = -2iA_{\tilde{\tau}_1\tilde{\tau}_2}^{0}.$$

**5. Another method of realizing the operators $A_{\tilde{\phi}}^{0}$, $A_{\tilde{\tau}}^{+}$ and $A_{\tilde{\tau}}^{-}$.** The method of realization proposed here seems very interesting to us. We specify the elements of $\mathscr{H}$ not as sequences of functions of $x_1, \ldots, x_k$, $n_1, \ldots, n_k$, but as sequences of functions of the $x$-parameters alone.

Let us consider, for example, the function $f_3(x_1, x_2, x_3;\ 2, 1, -4)$. We assign to it the function $f_{3,4}(x_1, x_1, x_2;\ y_1, y_1, y_1, y_1)$. More generally, if a function $f_k(x_1, \ldots, x_k;\ n_1, \ldots, n_k)$ is given, we first discard the zeros among the numbers $n_1, \ldots, n_k$. We then pick out the positive numbers among the $n_i$ and denote their sum by $m$ and the sum of the absolute values of the negative $n_i$ by $n$. $x_1$ is then repeated $|n_1|$ times, $x_2$ is repeated $|n_2|$ times, etc. Because of the symmetry of $f$ in the pairs $(x_i, n_i)$ we can write down first all the arguments $x_i$ with positive $n_i$, then all those with negative $n_i$; we denote the resulting function by

$$f_{mn}(\underbrace{x_1, \ldots, x_1}_{|n_1|}, \ldots, \underbrace{x_k, \ldots, x_k}_{|n_k|}).$$

Now we give a precise definition that does not depend on these arguments.

We introduce the space $\mathscr{H}^0$, whose elements are infinite sequences $f = \{f_{mn}\}_{m,n=0,1,\ldots}$, *where* $f_0 \in \mathbf{C}$, $f_{mn}\colon \underbrace{X \times \ldots \times X}_{m+n} \to \mathbf{C}$ *for* $m + n > 0$,

satisfying the following conditions:

1) the functions $f_{mn}(x_1, \ldots, x_m;\ y_1, \ldots, y_n)$ are continuous;

2) the functions $f_{mn}(x_1, \ldots, x_m; y_1, \ldots, y_n)$ are symmetric in the first $m$ arguments and in the last $n$ arguments:

3) $$\|f\|^2 = \sum_{m,n=0}^{\infty} \sum \frac{1}{(k+l)!\,|m|\,|n|} \int |f_{mn}(\underbrace{x_1, \ldots, x_1}_{m_1}, \ldots, \underbrace{x_k, \ldots, x_k}_{m_k}; \underbrace{y_1, \ldots, y_1}_{n_1}, \ldots, \underbrace{y_l, \ldots, y_l}_{n_l})|^2\, dx_1 \ldots dx_k\, dy_1 \ldots dy_l < \infty,$$

where the inner sum is taken over all partitions $m = m_1 + \ldots + m_k$, $n = n_1 + \ldots + n_k$, and $|m| = m_1, \ldots, m_k$, $|n| = n_1, \ldots, n_k$. (For $m = 0$ we take $|m| = 1$.)

An isometric correspondence between $\mathscr{H}^0$ and the previous space is given by

$$\{f_{mn}\} \longmapsto \{f_k\},$$

where

$$f_k(x_1, \ldots, x_p, y_1, \ldots, y_q; m_1, \ldots, m_p, n_1, \ldots, n_q) = \\ = f_{mn}(\underbrace{x_1, \ldots, x_1}_{m_1}, \ldots, \underbrace{x_p, \ldots, x_p}_{m_p}; \underbrace{y_1, \ldots, y_1}_{|n_1|}, \ldots, \underbrace{y_q, \ldots, y_q}_{|n_q|})$$

$$(p+q=k;\ m_i>0;\ n_i<0,\ m=m_1+\ldots+m_p,\ n=|n_1|+\ldots+|n_q|).$$

Now $A^0_{\varphi}$, $A^+_{\tau}$, and $A^-_{\tau}$ act in $\mathscr{H}^0$ according to the formulae:

$$(A^0_{\varphi}f)_{mn}(x_1, \ldots, x_m; y_1, \ldots, y_n) = \\ = i\Big(\sum_{s=1}^{m} \varphi(y_s) - \sum_{s=1}^{n} \varphi(y_s)\Big) f_{mn}(x_1, \ldots, x_m; y_1, \ldots, y_n),$$

$$(A^+_{\tau}f)_{mn}(x_1, \ldots, x_m; y_1, \ldots, y_n) = \\ = -\sum_{s=1}^{m} \tau(x_s) f_{m+1,n}(x_1, \ldots, x_m, x_s; y_1, \ldots, y_n) + \\ + \sum_{s=1}^{n} \tau(y_s) f_{m,n-1}(x_1, \ldots, x_m; y_1, \ldots, \hat{y}_s, \ldots, y_n) - \\ - \int \tau(t) f_{m+1,n}(x_1, \ldots, x_m, t; y_1, \ldots, y_n)\, dt,$$

$$(A^-_{\tau}f)_{mn}(x_1, \ldots, x_m; y_1, \ldots, y_n) = \\ = \sum_{s=1}^{m} \tau(x_s) f_{m-1,n}(x_1, \ldots, \hat{x}_s, \ldots, x_m; y_1, \ldots, y_n) - \\ - \sum_{s=1}^{n} \tau(y_s) f_{m,n+1}(x_1, \ldots, x_m; y_1, \ldots, y_n, y_s) - \\ - \int \tau(t) f_{m,n+1}(x_1, \ldots, x_m; y_1, \ldots, y_n, t)\, dt.$$

## §6. Another method of constructing a representation of $G^X$

**1. The general construction.** a) *Construction of the representation space.* Let $G$ be an arbitrary topological group and $H$ a linear topological space in which a representation $T(g)$ of $G$ is defined. Further, Let $X$ be a compact topological space with a positive finite measure $m$. As before, we assume that $m$ is a countably additive, non-negative Borel measure whose support is $X$.

We suppose that in $H$ there is a linear functional $l$ ($l \neq 0$) invariant under $T(g)$. Then we construct a representation $U_{\tilde{g}}$ of the group $G^X$ of continuous mappings $X \to G$ from the representation $T(g)$ of $G$ and the functional $l$.

We denote by $H^X$ the set of all continuous mappings $\tilde{f} = f(\cdot)\colon X \to H$ such that $l(f(x))$ does not depend on $x$ and $l(f(x)) = 0$.

We introduce a new linear space $\mathscr{H}^0$ whose elements are formal finite sums of elements of $H^X$:

$$\tilde{f}_1 \dotplus \ldots \dotplus \tilde{f}_n. \tag{1}$$

Here we set $\lambda_1 \tilde{f} \dotplus \lambda_2 \tilde{f} = (\lambda_1 + \lambda_2)\tilde{f}$ if $\lambda_1 + \lambda_2 \neq 0$, and $\tilde{f} \dotplus (-\tilde{f}) = 0$. We emphasize that if $f_1(x)$ and $f_2(x)$ are not proportional, then $\tilde{f} = \tilde{f}_1 + \tilde{f}_2$ and $\tilde{f}_1 \dotplus \tilde{f}_2$ are regarded as distinct elements.

In $\mathscr{H}^0$ operations of addition and multiplication by a factor $\lambda \in \mathbf{C}$ are defined in the natural way. Namely, the product of $\tilde{f} = f(\cdot)$ by $\lambda \in \mathbf{C}$ is defined as $\lambda \circ f(x) = \lambda f(x)$ if $\lambda \neq 0$, and $0 \circ f(x) = 0$. As a result, $\mathscr{H}$ becomes a linear space.

b) *Action of the operators in* $\mathscr{H}^0$. We define a representation $U_{\tilde{g}}$ of $G^X$ in $\mathscr{H}^0$ by

$$(U_{\tilde{g}} f)(x) = \lambda T(g(x)) f(x), \tag{2}$$

where $\lambda(\tilde{g}, \tilde{f})$ is a function of $\tilde{g}$ and $\tilde{f}$ such that $\lambda(\tilde{g}, c\tilde{f}) = \lambda(\tilde{g}, \tilde{f})$ for any $c \neq 0$, and we extend $U_{\tilde{g}}$ by additivity to all elements (1).

LEMMA 6.1. *The operators $U_{\tilde{g}}$ form a representation of $G^X$ if and only if the function $\lambda(\tilde{g}, \tilde{f})$ satisfies the following additional condition for any $\tilde{g}_1, \tilde{g}_2 \in G^X$ and $\tilde{f} \in H^X$: $\lambda(\tilde{g}_1, \tilde{f}_1)\lambda(\tilde{g}_2, \tilde{f}) = \lambda(\tilde{g}_1\tilde{g}_2, \tilde{f})$ where $f_1(x) = T(g_2(x))f(x)$.*

*If the weaker relation*

$$\lambda(\tilde{g}_1, \tilde{f}_1)\lambda(\tilde{g}_2, \tilde{f}) = c(\tilde{g}_1, \tilde{g}_2)\lambda(\tilde{g}_1\tilde{g}_2, \tilde{f})$$

*holds, then the $U_{\tilde{g}}$ form a projective representation.*

The proof is obvious.

c) *Construction of a unitary representation.* Let $H_0 \subset H$ be the set of elements $\xi$ such that $l(\xi) = 0$. By the invariance of $l$, $H_0$ is an invariant subspace of $H$. Suppose that an invariant positive definite scalar product $(\xi_1, \xi_2)$ is defined in $H$.

We construct a scalar product in the space $\mathcal{H}^0$ from the scalar product $(\xi_1, \xi_2)$ in $H$. To do this we first fix a vector $\xi_0 \in H$ such that $l(\xi_0) = 1$. For any pair of elements $\tilde{f}_1 = f_1(\cdot)$ and $\tilde{f}_2 = f_2(\cdot)$ of $H^X$ we define our scalar product as follows:

$$\langle \tilde{f}_1, \tilde{f}_2 \rangle = l(\tilde{f}_1)\overline{l(\tilde{f}_2)} \exp\left(\frac{1}{l(\tilde{f}_1)\overline{l(\tilde{f}_2)}} \int (f_1'(x), f_2'(x))\, dm(x)\right), \tag{3}$$

where $f_i'(x) = f_i(x) - l(\tilde{f}_i)\xi_0$ are elements of $H_0$ for any $x \in X$. (We recall that $l(f(x))$ is independent of $x$. Instead of $l(f(x))$ we write $l(\tilde{f})$, where $\tilde{f} = f(\cdot)$.)

We extend this scalar product to the whole space $\mathcal{H}^0$ by linearity.

LEMMA 6.2. *The Hermitian form on $\mathcal{H}^0$ defined by* (3) *is positive definite.*

PROOF. We choose arbitrary elements $\tilde{f}_1, \ldots, \tilde{f}_n$ of $H^X$ and prove that the matrix $\langle \tilde{f}_i, \tilde{f}_j \rangle$ is positive definite. We have

$$\frac{\langle \tilde{f}_i, \tilde{f}_j \rangle}{l(\tilde{f}_i)\overline{l(\tilde{f}_j)}} = \sum_{n=0}^{\infty} \frac{1}{n!} \left(\frac{1}{l(\tilde{f}_i)\overline{l(\tilde{f}_j)}} \int (f_i'(x), f_j'(x))\, dm(x)\right)^n. \tag{4}$$

Since $a_{ij} = \frac{1}{l(\tilde{f}_i)\overline{l(\tilde{f}_j)}} \int (f_i'(x), f_j'(x))dm(x)$ is positive definite, by Schur's lemma each term of (4) is positive definite, and the lemma is proved.

Thus, $U_{\tilde{g}}$ acts in a pre-Hilbert space. Let us see how the multiplier $\lambda(\tilde{g}, \tilde{f})$ can be chosen so that the representation is unitary.

For this purpose we first construct from $\xi_0$ a function $\beta(g)$ with values in $H$: $\beta(g) = T(g)\xi_0 - \xi_0$. Since $l$ is invariant we have $l(\beta(g)) = l(T(g)\xi_0) - l(\xi_0) = 0$, that is, $\beta(g) \in H_0$ for any $g \in G$. It is not hard to see that $\beta(g)$ is a cocycle with values in $H_0$, in other words, it satisfies the relation $\beta(g_1) + T(g_1)\beta(g_2) = \beta(g_1 g_2)$ for any $g_1, g_2 \in G$.

We observe that $\beta(g)$ depends on the way we have fixed the vector $\xi_0 \in H$.

LEMMA 6.3. *If we set*

$$\lambda(\tilde{g}, \tilde{f}) = c(\tilde{g}) \exp\left(-\frac{1}{l(\tilde{f})} \int \left[(T(g(x)) f'(x), \beta(g(x))) + \frac{1}{2}\|\beta(g(x))\|^2\right] dm(x)\right), \tag{5}$$

*where $f'(x) = f(x) - l(\tilde{f})\xi_0$, $|c(\tilde{g})| = 1$, then the operators $U_{\tilde{g}}$ defined by* (2) *are unitary and form a projective representation. Specifically, $U_{\tilde{g}_1} U_{\tilde{g}_2} = c(\tilde{g}_1, \tilde{g}_2) U_{\tilde{g}_1 \tilde{g}_2}$, where*

$$c(\tilde{g}_1, \tilde{g}_2) = \frac{c(\tilde{g}_1)c(\tilde{g}_2)}{c(\tilde{g}_1\tilde{g}_2)} \times \exp\left(i \int \operatorname{Im}(T(g(x))\beta(g_2(x)), \beta(g_1(x)))\, dm(x)\right). \tag{6}$$

The proof comes from a direct verification.

REMARK. This condition on $\lambda(\tilde{g}, \tilde{f})$ is also necessary.

We now state our final result. A linear topological space $H$ is given and also a representation $T(g)$ of $G$ in $H$. A linear functional $l$ is given in $H$ that is invariant under the action of $G$, that is, $l(T(g)\xi) = l(\xi)$ for any $g \in G$ and $\xi \in H$. We define a scalar product $(\xi_1, \xi_2)$ in the subspace $H_0$ of elements $\xi$ such that $l(\xi) = 0$.

A representation of $G^X$ is constructed as follows. We consider continuous mappings $\tilde{f} = f(\cdot)\colon X \to H$ such that $l(f(x)) = \text{const} \neq 0$. We introduce the space $\mathscr{H}^0$ whose elements are the formal sums $\tilde{f}_1 \dotplus \ldots \dotplus \tilde{f}_n$ with the relations $\lambda_1\tilde{f} + \lambda_2\tilde{f} = (\lambda_1 + \lambda_2)\tilde{f}$ if $\lambda_1 + \lambda_2 \neq 0$, $\tilde{f} + (-\tilde{f}) = 0$. We construct the scalar product:

$$\langle \tilde{f}_1, \tilde{f}_2\rangle = l(\tilde{f}_1)\overline{l(\tilde{f}_2)} \exp\left(\frac{1}{l(\tilde{f}_1)\overline{l(\tilde{f}_2)}} \int (f_1'(x), f_2'(x))\, dm(x)\right),$$

where $f_i'(x) = f_i(x) - l(\tilde{f}_i)\xi_0$, and $\xi_0$ is a fixed vector in $H$ such that $l(\xi_0) = 1$. This scalar product is then extended to the whole space $\mathscr{H}^0$. We denote by $\mathscr{H}$ the completion of $\mathscr{H}^0$ in this scalar product

The operators $U_{\tilde{g}}$ are defined by the formula $(U_{\tilde{g}}f)(x) = \lambda(\tilde{g}, \tilde{f})T(g(x))f(x)$, where

$$\lambda(\tilde{g}, \tilde{f}) = \\ = c(\tilde{g}) \exp\left(-\frac{1}{l(\tilde{f})} \int \left[(T(g(x))f'(x), \beta(g(x))) + \frac{1}{2}\|\beta(g(x))\|^2\right] dm(x)\right),$$

$$\beta(g) = T(g)\xi_0 - \xi_0, \quad |c(\tilde{g})| = 1,$$

and are extended by additivity to sums of the form (1) and then to the completion. These operators are unitary and form a projective representation of $G^X$, namely, $U_{\tilde{g}_1} U_{\tilde{g}_2} = c(\tilde{g}_1\tilde{g}_2)U_{\tilde{g}_1\tilde{g}_2}$, where $c(\tilde{g}_1, \tilde{g}_2)$ is defined by (6).

2. **Construction of a representation of $G^X$, where $G = PSL(2, \mathbf{R})$.** We now apply the general construction described above to the case of the group $G = PSL(2, \mathbf{R})$, given in the second form.

We define $H$ as the space of all continuous functions on the circle $|\xi| = 1$ in which the representation acts according to the following formula:

(7) $$(T(g)f)(\zeta) = f\left(\frac{\alpha\zeta + \bar{\beta}}{\beta\zeta + \bar{\alpha}}\right)|\beta\zeta + \bar{\alpha}|^{-2},$$

and the invariant linear functional $l$ is

$$l(f) = \frac{1}{2\pi}\int_0^{2\pi} f(e^{it})\, dt.$$

In the subspace $H_0$ of functions $f(\zeta)$ for which $l(f) = 0$ we specify a scalar product as follows:

(8) $$\|f\|^2 = \sum_{n\neq 0} \frac{1}{|n|}|a_n|^2, \quad \text{where} \quad f(e^{it}) = \sum_{n\neq 0} a_n e^{int},$$

or, in integral form,

(9) $$\|f\|^2 = c\int_0^{2\pi}\int_0^{2\pi} \ln\left|\sin\frac{t_1 - t_2}{2}\right| f(e^{it_1})\overline{f(e^{it_2})}\, dt_1\, dt_2.$$

It is clear from (8) that this scalar product is positive definite, and from

(9) that it is invariant under the operators $T(g)$ of the form (7). (We recall that $\int_0^{2\pi} f(e^{it})\,dt = 0$.)

We now construct $\mathcal{H}^0$. We fix in $H$ the function $\xi_0 = f_0(\zeta) \equiv 1$. Then $\beta(g) = T(g)f_0 - f_0$; hence, $\beta(g, \zeta) = |\beta\zeta + \bar{\alpha}|^{-2} - 1$, where $g = \begin{pmatrix} \alpha & \beta \\ \bar{\beta} & \bar{\alpha} \end{pmatrix}$.

Note that $\beta(g, \zeta)$ takes only real values. We examine the set $H^X$ of continuous functions $f(x, \zeta)$ satisfying the following condition:

$$\frac{1}{2\pi}\int_0^{2\pi} f(x, e^{it})\,dt = 1 \quad \text{for all} \quad x \in X.$$

The elements of $\mathcal{H}^0$ are all possible finite formal linear combinations of such functions: $\sum \lambda_i \circ f_i(x, \zeta)$, $\lambda_i \in \mathbf{C}$. A scalar product is defined for any pair of functions $\tilde{f}_1 = f_1(x, \zeta)$ and $\tilde{f}_2 = f_2(x, \zeta)$ in $H^X$ by the formula

$$\langle \tilde{f}_1, \tilde{f}_2\rangle = \exp\left(c \int \ln\left|\sin\frac{t_1 - t_2}{2}\right| f_1'(x, e^{it_1})\overline{f_2'(x, e^{it_2})}\,dt_1\,dt_2\,dm(x)\right),$$

where $f_i' = f_i - 1$, and is then extended by linearity to the whole space $\mathcal{H}^0$.

The representation operator $U_{\tilde{g}}$ is defined by the formula

$$(U_{\tilde{g}} f)(x, \zeta) = \lambda(\tilde{g}, \tilde{f}) \circ f\left(x, \frac{\alpha(x)\zeta + \overline{\beta(x)}}{\beta(x)\zeta + \overline{\alpha(x)}}\right) |\beta(x)\zeta + \overline{\alpha(x)}|^{-2},$$

$f \in H^X$, where

$$\lambda(\tilde{g}, \tilde{f}) = \exp\Big(-c\int \ln\left|\sin\frac{t_1 - t_2}{2}\right| \times$$
$$\times \Big(T(g(x))f'(x, e^{it_1}) + \frac{1}{2}\beta(g(x), e^{it_1}), \beta(g(x), e^{it_2})\Big)\,dt_1\,dt_2\,dm(x)\Big),$$

$f' = f - 1$, and is then extended by linearity first to the whole space $\mathcal{H}^0$, and then to its completion $\mathcal{H}$.

We note that in the case considered here the scalar product of the vectors $T(g_1(x))\beta(g_2(x))$ and $\beta(g_1(x))$ is real. Therefore, by Lemma 6.3 (see the expression for $c(\tilde{g}_1, \tilde{g}_2)$), the operators $U_{\tilde{g}}$ form a representation of $G^X$.

We make here an essential remark. We can take for $H$ a subspace of functions on the unit circle $|\zeta| = 1$ that are close to being analytic (or anti-analytic) in the interior of the unit disc. Then we obtain other representations of $G^X$, which are projective.

**3. Another construction of a representation of $G^X$, where $G = PSL(2, \mathbf{R})$.** It is sometimes convenient to define representations of $G$ not on functions on the circle but on functions on the line. Let $G$ be given in the first form.

We consider the space $H$ of all real continuous functions that satisfy the following condition: $f(t) = O(t^{-2})$ as $t \to \pm\infty$; and we define a representation of $G$ in $H$ by the following formula:

$$(T(g)f)(t)=f\left(\frac{\alpha t+\gamma}{\beta t+\delta}\right)|\beta t+\delta|^{-2}.$$

We further define in $H$ a $G$-invariant linear functional $l(f)$:

$$l(f)=\int_{-\infty}^{+\infty} f(t)\,dt.$$

Let $H_0 \subset H$ be the subspace of functions on which $l(f) = 0$. In $H_0$ there is an invariant positive definite scalar product

$$(f_1, f_2)=-\int_{-\infty}^{+\infty}\int_{-\infty}^{+\infty} \ln|t_1-t_2|\, f_1(t_1)\overline{f_2(t_2)}\,dt_1\,dt_2.$$

We now fix the function $f_0(t) = \frac{\pi^{-1}}{1+t^2}$ in $H$, for which $l(f_0) = 1$. We also set $\beta(g, t) = (T(g)f_0)(t) - f_0(t)$.

We proceed to the construction of $\mathcal{H}^0$. We consider the set $H^X$ of continuous functions $f(x, t)$, $x \in X$, $t \in \mathbf{R}$, satisfying the following conditions:

1) $f(x, t) = O(t^{-2})$ as $t \to \pm\infty$;

2) $\int_{-\infty}^{+\infty} f(x, t)dt = 1$ for any $x \in X$.

The elements of $\mathcal{H}^0$ are all possible formal linear combinations of such functions: $\sum \lambda_i \circ f_i(x, t)$, $\lambda_i \in \mathbf{C}$.

A scalar product is defined for any pair $f_1(x, t)$, $f_2(x, t)$ by the following formula:

$$\langle f_1(x, t), f_2(x, t)\rangle=\exp\left(-\int \ln|t_1-t_2|\, f_1'(x, t_1)\overline{f_2'(x, t_2)}\,dt_1 dt_2 dm(x)\right),$$

where $f_i'(x, t)=f_i(x, t)-\frac{\pi^{-1}}{1+x^2}$.

The representation operator acts as follows:

$$U_{\tilde{g}} f(x, t)=\lambda(\tilde{g}, \tilde{f})\circ f\left(x, \frac{\alpha(x)t+\gamma(x)}{\beta(x)t+\delta(x)}\right)|\beta(x)t+\delta(x)|^{-2},$$

where

$$\lambda(\tilde{g}, \tilde{f})=\exp\Bigl(\int \ln|t_1-t_2| \times$$
$$\times\left(T(g(x))f'(x, t_1)+\frac{1}{2}\beta(g(x), t), \beta(g(x), t_2)\right)dt_1\,dt_2\,dm(x)\Bigr).$$

REMARK. If we use instead of $H$ only the subspace $H^+$ (or $H^-$) of functions close to analytic in the upper (or lower) half-plane, then we obtain other (projective) representations of $G^X$.

**4. Equivalence of the representation $U_{\tilde{g}}$ to the representation constructed in §5.** THEOREM 6.2 *The representation $U_{\tilde{g}}$ of $G^X$, $G = PSL(2, \mathbf{R})$ constructed here is equivalent to the representation constructed in §5. Hence* *it follows that $U_{\tilde{g}}$ is irreducible.*

PROOF. For the proof it is sufficient to construct an isometric mapping of $\mathcal{H}^0$ into the representation space of §5 that commutes with the action of $G^X$ in these spaces.

Let us examine the construction of the representation in §6.2. In it $\mathcal{H}^0$ consists of formal linear combinations of functions $f(x, \zeta)$, $|\zeta| = 1$, such that $\frac{1}{2\pi}\int_0^{2\pi} f(x, e^{it})dt = 1$ for any $x \in X$. We expand $f(x, e^{it})$ in a Fourier series in $t$: $f(x, e^{it}) = 1 + \sum_{n\neq 0} a_n(x)e^{int}$. We associate with $f(x, t) \in \mathcal{H}^0$ an element of the space $\mathcal{H}_V$ constructed in §5: $F = (1, u(x, n), \ldots, \ldots, u(x_1, n_1), \ldots, u(x_k, n_k), \ldots)$, where $u(x, n) = (-1)^n a_n(x)$. We extend this to a linear mapping of the whole space $\mathcal{H}^0$ onto $\mathcal{H}_V$. From the definition of the norm in these spaces it follows easily that the mapping so constructed is an isometry. Furthermore, it can be shown that it commutes with the action of $G^X$ in these spaces.

## §7. Construction with a Gaussian measure

1. To explain the construction of this section we find it convenient to make some modifications in the general constructions in §6.1.

Let $G$ be a topological group, $E$ a real Hilbert space, and suppose that an orthogonal representation $T(g)$ of $G$ in $E$ and a cocycle with values in $E$ are given, that is, a function $\beta$: $G \to E$ satisfying the relation $\beta(g_1) + T(g_1)\beta(g_2) = \beta(g_1g_2)$.

From the part $(T, \beta)$ we construct a new unitary representation of $G$. In what follows we change the notation for the group and write $\Gamma$ in place of $G$, because in the examples $\Gamma$ can be both $G$ and $G^X$.

First we construct a new (complex) space $\mathcal{H}^0$ whose elements are formal finite linear combinations of elements $\xi_i \in E$:

$$\lambda_1 \circ \xi_1 + \ldots + \lambda_n \circ \xi_n, \qquad \lambda_i \in \mathbf{C}. \tag{1}$$

In contrast to the preceding section, $\lambda \circ \xi$ and $\lambda\xi$ are now regarded as distinct.

Thus, the original space $E$ has a natural embedding in $\mathcal{H}^0$ as a subset (not as a subspace!).

We now define a scalar product in $\mathcal{H}^0$. Namely, for elements $\xi_1, \xi_2 \in H$ we define a scalar product by the formula $\langle \xi_1, \xi_2\rangle = \exp(\xi_1, \xi_2)$, where the round parentheses denote the scalar product in $E$, and then we extend this scalar product by linearity to all formal linear combinations like (1). It is easy to verify that the Hermitian form so introduced is positive definite (see the proof of Lemma 6.2).

Let $\mathcal{H}$ be the completion of $\mathcal{H}^0$ in the scalar product just introduced.

We define a representation $U_\gamma$ of $\Gamma$ in $\mathcal{H}^0$. The action of operators $U_\gamma$,

$\gamma \in \Gamma$, on elements $\xi \in E$ is given as follows:

$$U_{\gamma^{-1}}\xi = \exp\left(-\frac{1}{2}\|\beta(\gamma)\|^2 - (T_\gamma\xi, \beta(\gamma)\right) \circ (T(g)\xi + \beta(g)),$$

and then we extend these operators by linearity to the whole space $\mathcal{H}^0$.

It can easily be established that the $U_\gamma$ are unitary operators, hence can be extended to the whole space $\mathcal{H}$, and that these operators form a representation of $\Gamma$. Later we shall see how this construction is related to that of Araki and Streater.

2. Let us consider two examples.

a) Let $\Gamma = PSL(2, \mathbf{R})$. Let $T(g)$ be the representation of $PSL(2, \mathbf{R})$ constructed in §6.3 (or §6.2), and $\beta(g)$ the cocycle defined there. We denote by $E$ the real subspace of $H_0$ also given there. From it we construct $\mathcal{H}$ and the representation $U_g$.

THEOREM 7.1. *The representation so constructed coincides with the canonical representation introduced in* §1.

b) Let $\Gamma = (PSL(2, \mathbf{R}))^X$. We consider the space of all mappings $\tilde{f}$: $X \to E$ that are measurable on $X$ and satisfy the condition

$$\|\tilde{f}\|^2 = \int \|f(x)\|^2\, dm(x) < \infty.$$

This is a real Hilbert space. We define in it a representation $\tilde{T}$ of $\Gamma$ by the formula $\tilde{T}(\tilde{g})\tilde{f} = T(g(x))f(x)$, where $T$ is the representation of example a). Next, we introduce a cocycle $\tilde{\beta}$, setting $\tilde{\beta}(\tilde{g}) = \beta(g(x))$, where $\beta(g)$ is the cocycle of example a), and we construct from the pair $(\tilde{T}, \tilde{\beta})$ a representation $U_{\tilde{g}}$ by the procedure indicated above.

THEOREM 7.2. *The representation so constructed coincides with the representations* $U_{\tilde{g}}$ *constructed in* §§2–6.

3. We explain briefly another method of constructing a representation of the group, which can be specialized to yield the construction explained at the beginning of this section.

Let $K$ be a real Hilbert space and $K'$ its dual space. We consider the functional $\chi(\xi') = e^{-\|\xi'\|^2}$. It is normed, continuous in the norm, and positive definite on $K'$, as on an additive group. Therefore it is the Fourier transform of a weak distribution in $K$: $\chi(\xi') = \int\limits_K e^{i\langle \xi', \xi\rangle}\, d\nu(k)$. (A weak distribution is a finitely additive, normalized, non-negative measure defined on the algebra of cylinder sets in $K$, that is, sets of the level of Borel functions of finitely many linear functionals.)

It is known that $\nu$ can be extended to a countably additive measure in an arbitrary nuclear extension $\tilde{K}$ of $K$. We call this the standard Gaussian measure, and we quote two properties of this measure that we shall need.

1) The standard Gaussian measure in $\tilde{K}$ is equivalent (that is, mutually absolutely continuous) to its translations by elements of $K$.

2) Every orthogonal transformation of $K$ can be uniquely extended to a linear and measurable transformation in $\tilde{K}$ that is defined almost everywhere and preserves the Gaussian measure.

We now consider the space $E$, introduce the pair $(T, \beta)$ (See §7.1), and choose the standard Gaussian measure $\mu$ in some nuclear extension $\tilde{E}$ of $E$. We examine the space $L^2(\tilde{E}, \mu)$ of all square integrable complex-valued functionals on $\tilde{E}$ (more precisely, of classes of functionals that coincide almost everywhere). From the pair $(T, \beta)$, we construct in $L^2(\tilde{E}, \mu)$ a representation $U_g$:

$$(U_{g^{-1}}F)(\varphi) = e^{-\frac{1}{2}\|\beta(g)\|^2 - \langle T(g)\beta(g), \varphi\rangle} F(T(g)\varphi + \beta(g)),$$

where $F \in L^2(\tilde{E}, \mu)$, $\varphi \in \tilde{E}$.

The following theorem can be proved:

THEOREM 7.3. *The correspondence $g \to U_g$ is a unitary representation of $G$ in $L^2(\tilde{E}, \mu)$.*

REMARK. If there is another cocycle $\beta'$ in $E' \cong E$, then we can construct the more general representation:

$$(U_{g^{-1}}^{(\beta', \beta)}F)(\varphi) = e^{i\langle\beta', \varphi\rangle} (U_{g^{-1}}F)(\varphi).$$

4. We indicate briefly the connection between the representation constructed here and that constructed in §7.1. Let $\xi \in E$. To $\xi$ we assign the following function in $L^2(\tilde{E}, \mu)$: $\xi \longmapsto F(\varphi) = ce^{(\xi, \varphi)}$, where the number $c$ is determined by the condition $c^2 \int e^{2(\xi, \varphi)}\, d\mu = 1, \quad c > 0.$

It is easy to verify that $c = c_0 e^{-\|\varphi\|^2}$, where $c_0$ is an absolute constant.

Let $F_1$ and $F_2$ correspond to the elements $\xi_1$ and $\xi_2$ of $E$. We compute

$$(F_1, F_2) = \int F_1(\varphi)F_2(\varphi)d\mu(\varphi).$$

It is easy to see that $(F_1, F_2) = e^{(\xi_1, \xi_2)}$, and thus is the isometric mapping of $\mathcal{H}$ into $L^2(\tilde{E}, \mu)$ given in §7.1. It can be verified that this mapping is an isomorphism. An elementary calculation shows that for a given $T(g)$ the representations in $\mathcal{H}$ and $L^2(\tilde{E}, \mu)$ are equivalent.

It is well known that the space $L^2(\tilde{E}, \mu)$, where $\mu$ is the standard Gaussian measure, can be represented in the form

$$L^2(\tilde{E}, \mu) = \exp H \cong \sum_{n=0}^{\infty} \oplus \frac{1}{n!} \underbrace{H \otimes \ldots \otimes H}_{n},$$

where $H$ is the complexification of $E'$ and $\underbrace{H \otimes \ldots \otimes H}_{n}$ is the subspace of generalized Hermite polynomials of degree $n$. Hence the preceding investigations show that our representation of $G^X$ is realized in $\mathcal{H} = \exp H$ by means of the cocycle $\beta$ (See §7.1). This realization coincides with the general scheme of Streater and Araki, which they have examined, however, only for certain soluble groups. In the terms used here the problem of the construction of a representation reduces to that of the discovery of the cocycle $\beta$ and to the proof of the irreducibility of the representation of $G^X$. We have done this in the present paper for $G = PSL(2, \mathbf{R})$.

## References

[1] I. M. Gel'fand and M. I. Graev, Representations of the quaternion groups over locally compact fields and function fields, Funktsional. Anal. i Prilozhen. 2 (1968) no. 1, 20–35. MR **38** #4611.

[2] I. M. Gel'fand and I. Ya. Vilenkin, *Nekotorye primeneniya garmonicheskogo analiza. Osnashchennye gilbertory prostranstva*, Gos. Izdat. Fiz.-Mat. lit. Moscow 1961. MR **26** #4173.
Translation: Generalized Functions, Vol. 4: Applications of harmonic analysis, Academic Press, New York and London 1964. MR **30** #4152.

[3] I. M. Gel'fand, M. I. Graev, and I. Ya. Vilenkin, *Integral'naya geometriya i svyazannye s nei voprosy teorii predstavlenii* Gos. Izdat. Fiz.-Mat. lit. Moscow 1962. MR **28** # 3324.
Translation: Generalized Functions, Volume 5: Integral geometry and representation theory, Academic Press, New York and London 1966. MR **34** #7726.

[4] H. Araki, Factorizable representations of current algebra, Publ. Res. Inst. Math. Sci. **5** (1969/70), 361–422. MR **41** #7931.

[5] I. Dixmier, Les C*-algèbres et leurs representations, second ed., Gauthier-Villars, Paris 1969. MR **30** #1404, **39** #7442.

[6] A. Guichardet, Symmetric Hilbert spaces and related topics, Lecture Notes in Mathematics, Springer-Verlag, Berlin–Heidelberg–New York 1972.

[7] D. Mathon, Infinitely divisible projective representations of the Lie Algebras, Proc. Cambridge Philos. Soc. **72** (1972) 357–368.

[8] K. R. Parthasarathy, Infinitely divisible representations and positive functions on a compact group, Comm. Math. Phys. **16**, 148–156 (1970).

[9] K. R. Parthasarathy and K. Schmidt, Infinitely divisible projective representations, cocycles, and Levy-Khinchine-Araki formula on locally compact groups, Research Report 17, Manchester–Sheffield School of Probability and Statistics, 1970.

[10] K. R. Parthasarathy and K. Schmidt, Factorizable representations of current groups and the Araki-Woods embedding theorem, Acta Math. **128**, 53–71 (1972).

[11] K. R. Parthasarathy and K. Schmidt, Positive definite kernels, continuous tensor products, and central limit theorems of probability theory, Lecture Notes in Mathematics, Springer-Verlag, Berlin–Heidelberg–New York 1972.

[12] L. Pukanszky, On the Kroneker products of irreducible representations of the 2 x 2 real unimodular group, Part I, Trans. Amer. Math. Soc. **100** (1961) 116–152.

[13] R. F. Streater, Current commutation relations, continuous tensor products, and infinitely divisible group representations, Rend. Sci. Ist. Fis. E. Fermi, **11** (1969), 247–263.

[14] R. F. Streater, Continuous tensor products and current commutation relations, Nuovo Cimento **A 53** (1968), 487.

[15] R. F. Streater, Infinitely divisible representations of Lie algebras, Z. Wahrscheinlichkeitstheorie und Verw. Gebiete **19**, 1971, 67–80.

Received by the Editors, 15 June 1973

Translated by W. J. Holman

# 2.

(with M. I. Graev and A. M. Vershik)

# Representations of the group of smooth mappings of a manifold $X$ into a compact Lie group

Compos. Math. **35** (1977) 299–334. MR **58**:28257. Zbl. **368**:53034


## Abstract

Some important nonlocal representations of the group $G^X$ consisting of $C^\infty$-mappings of a Riemannian manifold $X$ to a compact semisimple Lie group $G$ are constructed. The irreducibility, as well as non-equivalence of the introduced representations corresponding to different Riemannian metrics are proved. The ring of representations is calculated.


## Introduction

In this paper the unitary representations of the group $G^X$ of smooth functions on a manifold $X$ taking values in a compact Lie group $G$ are being built.

According to the idea of a paper [20], one can construct irreducible nonlocal representations of the group of measurable $G$-valued functions on $X$, $G$ being a Lie group, in the case when the unity representation is not an isolated point in the space of irreducible unitary representations. Such a construction for groups $G = SU(n, 1)$ and $G = SO(n, 1)$ is realized in [20] and [21]. The other examples of groups of this kind represent the groups of isometries of a Eucledean space, as well as solvable and nilpotent groups (cf. Araki [2], Streater [17], Guichardet [8], [9], Delorme [5], Parthasarathy-Schmidt [13] et al.).

In the case of compact Lie groups $G$ the unity representation is an isolated point, and it is impossible to follow the scheme of [20]. If, however, one takes the group $G^X$ of continuously differentiable mappings instead group of measurable mappings then it is possible to construct for this group a series of irreducible nonlocal representations. The idea of this construction is to consider at first the group $\theta^1(X;G)$ of smooth sections of the 1-jet fibre bundle $j^1(X;G)\to X$ (cf. §I); the initial group $G^X$ is naturally imbedded in the $\theta^1(X;G)$ as a subgroup. As $\theta^1(X;G)$ may be regarded as a group of functions taking values in the skew-product $G\cdot(\underbrace{\mathfrak{G}\times\cdots\times\mathfrak{G}}_{\dim X})$, $\mathfrak{G}$ being the Lie algebra of the group $G$ (cf. §I), and as the unity representation of this skew-product is not isolated, so it is possible, following the pattern of [20], to build nonlocal irreducible representation for $\theta^1(X;G)$ and to restrict it on the subgroup $G^X$. Here in this paper we study the representations of the group $G^X$ obtained in this way.

The constructions of this kind of the representations of the group $G^X$ have been found, after the appearance of [20], by the authors of the present paper and, independently, by Parthasarathy and Schmidt [14], R.S. Ismagilov [11], Albeverio and Hoeg-Krohn [1]. The irreducibility of these representations for $G=SU(2)$ in the case $\dim X\geq 5$ had been proved by R.S. Ismagilov [11].

As communicated to the authors A. Guichardet, P. Delorme had been studied the representations of the group $G^X$, where $G$ is a compact Lie group. *It is being proved in the present paper the irreducibility of the representations for the group $G^X$, where $G$ is any compact semisimple Lie group and* $\dim X\geq 2$.

The case $\dim X=1$ remains open at the moment, the difficulty being connected with the more complicated, than for the spaces $X$ of greater dimension, character of the restriction of the representation of the group $G^X$ on the subgroup $A^X$, where $A\subset G$ is the Cartan subgroup.

We use in this paper, in a systematic way, the technique of Gaussian measures, which has been for the first time used in the study of nonlocal irreducible representations of current groups in [20], [21] and which has proved very useful for verifying the irreducibility and non-equivalence of functional group representations (see also [22]).

Here is the brief contents of the paper.

We consider, in §1, jet fibrations over $X$ and the groups connected with them and we explain the main idea of the construction of the representation. We also introduce there the Maurer–Cartan cocycle, which plays a significant role in this construction. The §2 is of subsidiary character. There we give an account of what is connected with the construction $\mathrm{EXP}_\beta T$. This construction is more or less explicitly described in [2], [8], [13], but the usage of Gaussian measures, which began in [20], [21], allows to develop a systematic theory. It will be set forth in detail somewhere else. In §3 we construct the representations of the group $G^X$ and formulate the principal results of the paper. The proofs of the main theorems are given in §§4 and 5.

In some papers of physical nature (see, for instance, [18], [3]) it was considered the so-called Sugawara algebra. The authors have noticed that the corresponding group is the central $\mathbb{R}^1$-extension of the group $\theta^1(X; G)$ (see above). (If we take $G = \mathbb{R}^1$, then we get the generalized functional Heisenberg group). We construct, in §6, nonlocal irreducible unitary representations for this group as well.

## §1. The jets of smooth functions with values in a Lie group and Maurer–Cartan cocycle

Here we introduce the principal definitions concerning the group $C^\infty(X; G)$: the fibre bundle of $k$-jets, the group of sections of this fibre bundle, $k$-jet imbeddings etc. The description of the most important classes of representations of the group $C^\infty(X; G)$, both local and nonlocal, becomes more transparent if one passes to the group of sections of the $k$-jet fibre bundle (cf. section 3 of this paragraph). The most important for our purposes is the definition of the Maurer–Cartan cocycle, given in section 2.

1. *The group $G^X = C^\infty(X; G)$ and its jet extentions.* Let $G$ be an arbitrary real Lie group, $X$ – a real connected $C^\infty$-manifold. Consider the set $C^\infty(X; G)$ of $C^\infty$-mappings $\tilde{g}: X \to G$ such that $g(x) = 1$ outside of some compact set (depending of $\tilde{g}$). Let us supply the set $C^\infty(X; G)$ with the natural topology. The group operation in $C^\infty(X; G)$ is defined pointwise: $(g_1g_2)(x) = g_1(x)g_2(x)$. We shall denote the topological group $C^\infty(X; G)$ by $G^X$.

Let us define now the $k$-jet imbedding of the group $G^X$ ($k = 0, 1, \ldots$). Recall that the $k$-jet of a mapping $X \to G$ in a point $x_0 \in X$ is, by definition, the class of smooth mappings $X \to G$, all of them taking the same value at $x_0$, and such that all corresponding partial derivatives of these mappings up to the $k$-th order, taken at $x_0$, coincide. One can define the $k$-jet space at a point $x_0 \in X$ as follows. Consider a subgroup $G^X_{x_0,k}$ of $G^X$, consisting of such functions $\tilde{g}: X \to G$ that $g(x_0) = 1$ and all partial derivatives of $\tilde{g}$ up to the $k$-th order are equal to zero in a point $x_0$. It is easy to verify that $G^X_{x_0,k}$ is a normal subgroup of the group $G^X$; the space of $k$-jets in a point $x_0$ is naturally identified with the factor group $G^X/G^X_{x_0,k}$.

It is clear that the factor-groups $G^X/G^X_{x,k}$ corresponding to different $x \in X$ are isomorphic. The group $G^X/G^X_{x,k}$ is, moreover, uniquely defined, up to isomorphism, by the group $G$, number $k$ and dimension $m$ of the manifold $X$. We shall denote this group by $G^k_m$ and call it *the Leibnitz group of order k and degree m of the group G.*

In what follows we consider for the most part the case $k = 1$. Then $G^1_m \cong G \cdot (\underbrace{\mathfrak{G} \times \cdots \times \mathfrak{G}}_{m})$, with $\mathfrak{G}$ – the Lie algebra of $G$, i.e. $G^1_m$ is a skew-product of the direct sum of $m$ copies of Lie algebra $\mathfrak{G}$ and the group $G$; representation of the group $G$ on $\mathfrak{G} \times \cdots \times \mathfrak{G}$ is adjoint representation. The group $G^k_1$ can easily be described as well (see [14]). In the general case the group $G^k_m$ is a skew-product of $G$ and a nilpotent group $U^k_m$, which space is a sum of several copies of the space $\mathfrak{G}$; the formula of the group rule in $G^k_m$ for arbitrary $k$ and $m$ is actually rather complicated.

Let us define a $k$-jet fibre bundle $j^k(X; G) \to X$, which is a fibration over $X$ with fibre corresponding to $x \in X$ consisting of all the $k$-jets in a point $x$. The structure of a fibre bundle is introduced in $j^k(X; G)$ in a natural way.[1]

The fibre bundle $j^k(X; G)$ gives reason to the following definition generalizing the definition of a vector bundle:

DEFINITION: Let $H$ be a connected Lie group. A smooth fibre bundle $\xi$ with the standard fibre $H$ and the structure group Aut $H$ (= the group of all continuous group automorphisms of $H$) will be called a group bundle with a group $H$.

A group bundle with an additive group $\mathbb{R}^n$ (or $\mathbb{C}^n$) is a vector bundle in usual sense, that is why our notion presents a 'noncommutative' analogue of a vector bundle. The sections of a group bundle form a group, since each fibre has a group structure. A trivial group bundle with a group $H$ and base $X$ is a fibering $H \times X \to X$, and the group of all its smooth sections is the group $H^X$.

The jet fibre bundle $j^k(X; G) \to X$ defined above is a group bundle in the sense of our definition with the group $G^k_m$, $m = \dim X$. This bundle is not trivial unless the tangent bundle be so. Every fibre of this bundle over $x \in X$ is canonically provided with a structure of the group $G^X/G^X_{x,k} \cong G^k_m$.

Let us denote by $\theta^k(X; G)$ a space of all differentiable sections,

[1] It is a more traditional approach to consider $j^k(X; G)$ to be a fibre bundle over $X \times G$. The definition in the text is however, more suitable for us.

with compact support, of $k$-jet bundle $j^k(X;G)\to X$ supplied with usual topology. The space $\theta^k(X;G)$ is a group with regard to pointwise multiplication in the fibres.

Notice that there are defined natural group epimorphisms $\theta^{k+1}(X;G)\to\theta^k(X;G)$, $k=0,1,\ldots$; in particular, epimorphism $\theta^1(X;G)\to\theta^0(X;G)=G^X$.

Definition: We call a $k$-jet imbedding a map

$$\mathfrak{J}^k: G^X\to\theta^k(X;G)$$

defined by: $(\mathfrak{J}^k\tilde{g})(x)$ is the $k$-jet of a function $\tilde{g}$ in a point $x\in X$.

The following is fairly evident: $\mathfrak{J}^k$ *is a group monomorphism.*

Remark: All proposition of this section remains true if one substitutes the group $G^X$ by the group of differentiable sections of any group bundle over $X$ with the group $G$, $j^k(X;G)$ by the corresponding $k$-jet fibre bundle etc.

2. *The group $\theta^1(X;G)$ and Maurer–Cartan cocycle.* As we shall deal with the case $k=1$ let us examine the group $\theta^1(X;G)$ in greater detail. For every $x\in X$ the elements of a factor group $G^X/G^X_{x,1}$ may be considered as the pairs $(g(x), a(x))$, where $g(x)\in G$ and $a(x)$ is a linear mapping of tangential spaces $T_xX\to T_{g(x)}G$, i.e. $a(x)\in \operatorname{Hom}(T_x, T_{g(x)})$. Consequently, one can put into correspondence to each element of the group $\theta^1(X;G)$ a map $TX\to TG$ of the tangential sheaves which is linear on the fibres. It is easy to see that this correspondence is an isomorphism of the group $\theta^1(X;G)$ and the group $(TG)^{TX}$ of all differentiable mappings $TX\to TG$ with compact support, which are linear on the fibres.

Further, the right trivialization of the tangential shief $TG$ permits us to identify each tangential space $T_gG$ with the space $T_eG\cong\mathfrak{G}$ and thus defines an isomorphism $TG\cong G\cdot\mathfrak{G}$ of the group $TG$ and a semidirect product $G\cdot\mathfrak{G}$. Therefore the group $(TG)^{TX}$ is isomorphic to a skew-product of the group $G^X$ and an additive group $\Omega^1(X;\mathfrak{G})$ of all differentiable mappings $TX\to\mathfrak{G}$ with compact support and linear on the fibres, i.e. a group of $\mathfrak{G}$-valued 1-forms on $X$.

Consequently, there are determined the canonical isomorphisms of the groups:

$$\theta^1(X;G)\cong(TG)^{TX}\cong G^X\cdot\Omega^1(X;\mathfrak{G}),$$

where $\Omega^1(X;\mathfrak{G})$ is the space of all differentiable $\mathfrak{G}$-valued 1-forms with compact support on $X$, and the action $V$ of $G^X$ on $\Omega^1(X;\mathfrak{G})$ is an adjoint action in every fibre:

$$(V(\tilde{g})\omega)(x) = \mathrm{Ad}\, g(x) \circ \omega(x).$$

REMARK: We have two different imbedding of $G^X$ into $\theta^1(X;G)$: $\mathfrak{J}^1 : G^X \to \theta^1(X;G)$ and $G^X \to G^X \cdot 0 \subset G^X \cdot \Omega^1(X;\mathfrak{G})$.

Since $G^X$ is acting on $\Omega^1(X;\mathfrak{G})$ and there is a natural epimorphism $\theta^1(X;G) \to G^X$, then it is defined in $\Omega^1(X;\mathfrak{G})$ also an action of the group $\theta^1(X;G)$.

Let us introduce the Maurer–Cartan 1-cocycle. Making use of the isomorphism $\theta^1(X;G) \to G^X \cdot \Omega^1(X;\mathfrak{G})$, define a 1-cocycle $\alpha$ of $\theta^1(X;G)$ taking values in $\Omega^1(X;\mathfrak{G})$ by the following formula. Given $\xi \in \theta^1(X;G)$, i.e. $\xi = (\tilde{g}, \omega)$, where $\tilde{g} \in G^X$, $\omega \in \Omega^1(X;\mathfrak{G})$, let

$$\alpha(\xi) = \omega.$$

DEFINITION: A restriction of the 1-cocycle $\alpha$ on the subgroup $\mathfrak{J}^1 G^X \cong G^X$ will be called *Maurer–Cartan cocycle* and denoted by $\beta$. Therefore

$$\beta(\tilde{g}) = \alpha(\mathfrak{J}^1 \tilde{g}).$$

For $\beta$ to be a 1-cocycle means that for every $\tilde{g}_1, \tilde{g}_2 \in G^X$ it satisfies

$$\beta(\tilde{g}_1\tilde{g}_2) = \beta\tilde{g}_1 + V(\tilde{g}_1)\beta\tilde{g}_2.$$

It is not hard to prove that the conception of the Maurer–Cartan cocycle is natural with respect to the diffeomorphisms of manifolds $X \to Y$.

Owing to the extreme importance of the Maurer–Cartan cocycle we shall give a direct definition of it. Denote by $d\tilde{g}$ a differential of a map $\tilde{g} : X \to G$,

$$d\tilde{g} : TX \to TG.$$

Let $R : TG \to \mathfrak{G}$ be the right trivialization. Then we have

$$\beta\tilde{g} = R \circ d\tilde{g}. \tag{1}$$

REMARK 1: Let $\sigma: TX \to TX$ be a map, preserving each fibre and linear on it. Then the map

$$\tilde{g} \to (\beta\tilde{g}) \circ \sigma \in \Omega^1(X; \mathfrak{G}), \quad \tilde{g} \in G^X \tag{2}$$

is a 1-cocycle to the group $G^X$ again. The set of such cocycles forms a linear space. It is interesting to find out whether every 1-cocycle of $G^X$ with values in $\Omega^1(X; \mathfrak{G})$ is cohomological to a cocycle of the form (2) with $\beta$ the Maurer–Cartan cocycle.

REMARK 2: The group $G^1_m \cong G \cdot (\underbrace{\mathfrak{G} \times \cdots \times \mathfrak{G}}_{m})$ is represented with invariant scalar-product in the space $\underbrace{\mathfrak{G} \times \cdots \times \mathfrak{G}}_{m}$ and possesses a 1-cocycle with values in this space, defined by

$$\alpha_0(g, a) = a, \quad g \in G, \quad a \in \underbrace{\mathfrak{G} \times \cdots \times \mathfrak{G}}_{m}.$$

The 1-cocycle $\alpha$ of $\theta^1(X; G)$ introduced above can be defined locally in the following way. Assume that we have some local coordinates in $X$, and so the isomorphism $\mathrm{Hom}\,(T_x, \mathfrak{G}) \cong \underbrace{\mathfrak{G} \times \cdots \times \mathfrak{G}}_{m}$, $m = \dim X$ is defined. Thus the formula for the cocycle $\alpha$ may be written as follows:

$$\alpha(\tilde{g}, \omega)(x) = \alpha_0(g(x), \omega(x)).$$

Thus the existence of the Maurer–Cartan cocycle for a group $G^X$ is connected with the existence of the 1-cocycle $\alpha_0$ for $G^1_m$.

3. *Representations of the group* $G^X$. By using $k$-jet imbeddings $\mathfrak{J}^k: G^X \to \theta^k(X; G)$ one can construct various representations of $G^X$.

Consider at first arbitrary unitary representation $\pi$ of the group $G^k_m$. Fix a point $x_0 \in X$ and define an isomorphism $G^X/G^X_{x_0,k} \cong G^k_m$. A representation of the group $G^X$ can be defined by

$$T_\pi(\tilde{g}) = \pi((\mathfrak{J}^k\tilde{g})(x_0)). \tag{3}$$

The representation (3) is an analogue of the partial derivative in $x_0$. For instance, let $G = \mathbf{R}^1$, then $G^1_m = \mathbf{R}^1 \oplus \mathbf{R}^m$; if $\pi$ is a one-dimensional representation of $G^1_m: (\xi_0, \xi_1, \ldots, \xi_m) \mapsto \exp\,(i\sum_{s=0}^m \alpha_s\xi_s)$, then $T_\pi(\tilde{g})$ is a one-dimensional representation $\tilde{g} \mapsto \exp\,[i(\alpha_0 g(x) + \sum_{s=1}^m \alpha_s(\partial g/\partial x^s))_{x=x_0}]$. For $G = SU(2)$, $k = 1$ and $m = 1$ such representations were described for the first time in [6]; they were constructed there directly by passing to a limit, like that used in the definition of a derivative.

The representations (3) are local ones since they depend on $k$-jets in a point $x_0 \in X$ only. Similarly, one can define representation of $G^X$ depending on $k$-jets in a finite number of points.

More interesting representations of the group $G^X$ arise due to the existence of nonlocal irreducible representations of the group $\theta^k(X; G)$, the restrictions of which to $\mathfrak{J}^k G^X$ remain irreducible.

In [20] there was given a construction of nonlocal representations for a current group, i.e. a group of functions on $X$ taking values in some Lie group. For that construction to be applicable is necessary that the unity representation of the coefficient group be not isolated in the space of all its unitary representations, or, in more general terms,[1] that the first cohomology group with values in the space of some unitary representation of the coefficient group be non-trivial.

Since the local structure of the group $\theta^k(X; G)$ is similar to that of the group $C^\infty(X: G^k_m)$, $m = \dim X$, the construction of [20], [21] can be applied to $\theta^k(X; G)$ if the unity representation of $G^k_m$ is not isolated. That is the case when $G$ is a compact semisimple Lie group and $k = 1$, for $G^1_m$ then possesses an orthogonal representation in the space $\underbrace{\mathfrak{G} \times \cdots \times \mathfrak{G}}_{m}$ and, as we have seen, there exists a non-trivial 1-cocycle of the group $G^1_m$ with values in that space.

For the group $G^X$, with $G$ a compact semisimple Lie group, there arises, consequently, a nonlocal unitary representation. We emphasize that though this representation is not local, thus resembling the representations constructed in [20], it does not admit any extention even to the group of continuous mappings $X \to G$, because its construction makes use of the 1-jet imbedding.

## §2. Some subsidiaries preliminary facts

In the present paragraph we set forth some general definitions and statements, which will be used for the construction of representations for $G^X$, as well as in the proofs of the main theorems.

1. *The space* EXP $\bar{H}$. Let $H$ be a real nuclear countably Hilbert-

[1] In all examples known to the authors the existence of a unitary representation of a Lie group with non-trivial first cohomology group accompanies the non-isolation of the unitary representation in the space of unitary representations. Moreover, such a representation admits a deformation which is at the same time a deformation of the unity representation. However, the necessity of such coincidence is not yet established.

norm space, and let $\langle , \rangle$ be some inner product in it. Denote by $\bar{H}$ a completion of $H$ in the norm $\|h\| = \langle h, h\rangle^{1/2}$ and by $H'$ a dual to $H$. Then there are natural embeddings $H \subset \bar{H} \subset H'$.

Let us define in $H'$ a Gaussian measure $\mu$ with the zero mean and a correlation functional $B(h_1, h_2) = \langle h_1, h_2\rangle$, by its Fourier transform

$$\int_{H'} e^{i\langle F,h\rangle} d\mu(F) = e^{-1/2\langle h,h\rangle} \quad (h \in H).$$

This measure $\mu$ will be called a *standard Gaussian measure* in $H'$.

Let us introduce a complex Hilbert space $L^2_\mu(H')$ of all square-integrable with respect to $\mu$ functionals on $H'$. The functional $\Omega \in L^2_\mu(H')$, identically equal to unity on $H'$, will be called *the vacuum vector.*

We shall determine a natural isomorphism of the space $L^2_\mu(H')$ and another space which we call an exponential of $H'$ and denote EXP $\bar{H}$.

Let $\bar{H}_{\mathrm{C}}$ be a complexification of $\bar{H}$ and $S^n\bar{H}_{\mathrm{C}}$ $(n = 1, 2, \ldots)$ the symmetrized tensor product of $n$ copies of $\bar{H}_{\mathrm{C}}$. Let also $S^0\bar{H}_{\mathrm{C}} = \mathbb{C}$. Call an *exponential* EXP $\bar{H}$ *of a space* $\bar{H}$, *or Fock space corresponding to* $\bar{H}$, a complex Hilbert space

$$\mathrm{EXP}\,\bar{H} = \bigoplus_{n=0}^{\infty} S^n\bar{H}_{\mathrm{C}}.$$

To establish an isomorphism $\mathrm{EXP}\,\bar{H} \cong L^2_\mu(H')$ consider in the space EXP $\bar{H}$ the set of vectors

$$\exp h = 1 \oplus h \oplus \frac{1}{\sqrt{2!}} h \otimes h \oplus \frac{1}{\sqrt{3!}} h \otimes h \otimes h \oplus \cdots \quad (h \in H).$$

The set is known to be total in EXP $\bar{H}$ [8], i.e. its linear span is dense in EXP $\bar{H}$. Consider a mapping of the set $\{\exp h\}$ into the space $L^2_\mu(H')$:

$$\exp h \mapsto e^{\|h\|^2/2} e^{i\langle\cdot,h\rangle} \in L^2_\mu(H'). \tag{1}$$

One proves that this mapping conserves the inner product and so, by virtue of the totality of the sets $\{\exp h\}$ and $\{e^{i\langle\cdot,h\rangle}\}$ in EXP $\bar{H}$ and $L^2_\mu(H')$ correspondingly, the mapping (1) can be uniquely prolonged up to an isomorphism of the spaces:

$$\bigoplus_{n=0}^{\infty} S^n \bar{H}_C \cong L^2_\mu(H').$$

We call thus defined isomorphism *the canonical isomorphism* between the space $L^2_\mu(H')$ and its Fock model $\mathrm{EXP}\,\bar{H} = \bigoplus_{n=0}^{\infty} S^n \bar{H}_C$. Notice that the canonical isomorphism puts in correspondence the vacuum vector $\Omega \in L^2_\mu(H')$ and the vector $1 = \exp 0$, as well as the subspace of deneralized Hermitian polynomials of degree $n$ and the subspace $S^n \bar{H}_C$ in $\mathrm{EXP}\,\bar{H}$ (see, for instance, [23], [12]).

REMARK: If there are two triples of spaces $H_1 \subset \bar{H}_1 \subset H'_1$, $H_2 \subset \bar{H}_2 \subset H'_2$ with $\bar{H}_1 = \bar{H}_2$, and $\mu_1$, $\mu_2$ are standard Gaussian measures in $H'_1$ and $H'_2$ correspondingly, then there exists a canonical isomorphism $L^2_{\mu_1}(H'_1) \cong L^2_{\mu_2}(H'_2)$. It arises from canonical isomorphisms between this space and the Fock space.

2. *The representation* $\mathrm{EXP}_\beta\,V$. Consider a topological group $G$ and an orthogonal representation $V$ of $G$ in the space $H$. Let us extend the representation $V$ to the space $H' \supset H$ dual to $H$, by

$$\langle V(g)F, h\rangle = \langle F, V^{-1}(g)h\rangle$$

for any $F \in H'$, $h \in H$.

Consider a 1-cocycle $\beta$ of the group $G$ with values in $H$, i.e. a continuous mapping $\beta : G \to H$ which for every $g_1, g_2 \in G$ satisfies

$$\beta(g_1 g_2) = \beta g_1 + V(g_1)\beta g_2.$$

Given a representation $V$ and a 1-cocycle $\beta$, we shall construct a new representation, $U$, of the group $G$ in the space $L^2_\mu(H') \cong \mathrm{EXP}\,\bar{H}$, by

$$(U(g)\Phi)(F) = e^{i\langle F, \beta g\rangle}\Phi(V^{-1}(g)F). \tag{2}$$

Call the representation $U$ an *exponential* of the initial representation $V$ of $G$ (with respect to the 1-cocycle $\beta$) and denote it $\mathrm{EXP}_\beta\,V$. Now let us point out some simple properties of the representations $\mathrm{EXP}_\beta\,V$.

(1) Let the 1-cocycle $\beta, \beta' : G \to H$ be cohomological, i.e. there exists such a vector $h_0 \in H$ that $\beta' g - \beta g = V(g)h_0 - h_0$ for each $g \in G$. In this case the representations $U = \mathrm{EXP}_\beta\,V$ and $U' = \mathrm{EXP}_{\beta'}\,V$ are equivalent.

We can observe, indeed, that for every $g \in G$, $U'(g) = A_{h_0} U(g) A_{h_0}^{-1}$, where $A_{h_0}$ is defined by $(A_{h_0}\Phi)(F) = e^{i\langle F, h_0\rangle}\Phi(F)$.

(2) $\mathrm{EXP}_{\beta_1} V_1 \otimes \mathrm{EXP}_{\beta_2} V_2 = \mathrm{EXP}_{\beta_1 \oplus \beta_2} (V_1 \oplus V_2)$.

(3) LEMMA 1. *Let $V_C$ be the complexification of a representation $V$, $S^n V_C$ $(n = 1, 2, \ldots)$ a symmetrized tensor product of $n$ copies of the representation $V_C$, $S^0 V_C$ a unity representation. If $\beta = 0$, then*

$$\mathrm{EXP}_\beta V \cong \bigoplus_{n=0}^{\infty} S^n V_C.$$

To prove the lemma it suffices to pass to the Fock model of the representation space.

Notice now that given a 1-cocycle $\beta: G \to H$ and an arbitrary bounded linear operator $A$ in $H$ which commutes with the operators of the representation $V$, the function

$$(A\beta)g = A(\beta g)$$

is a 1-cocycle of $G$ taking values in $H$, too. Thus, with every 1-cocycle $\beta$ one can connect a family of unitary representations of $G$ in $L^2_\mu(H')$:

$$U_A = \mathrm{EXP}_{A\beta} V,$$

$A$ being any bounded linear operator in $H$, commuting with the operators of the representation $V$.

LEMMA 2 (of a tensor product): *Let $\alpha = \begin{pmatrix} \alpha_{11} & \alpha_{12} \\ \alpha_{21} & \alpha_{22} \end{pmatrix}$ be an arbitrary matrix of bounded linear operators $\alpha_{ij}: H \to H$, commuting with the operators of a representation $V$ and satisfying the condition: $\alpha^*_{1i}\alpha_{1j} + \alpha^*_{2i}\alpha_{2j} = \delta_{ij} E$ $(i, j = 1, 2)$, $E$ being the unity operator (i.e. $\alpha^*\alpha = 1$). Then*

$$U_{A_1} \otimes U_{A_2} \cong U_{\alpha_{11}A_1 + \alpha_{12}A_2} \otimes U_{\alpha_{21}A_1 + \alpha_{22}A_2}.$$

PROOF (cf. [22]): Define an operator in $L^2_\mu(H') \otimes L^2_\mu(H')$:

$$(R\Phi)(F_1, F_2) = \Phi(\alpha^*_{11}F_1 + \alpha^*_{21}F_2, \alpha^*_{12}F_1 + \alpha^*_{22}F_2).$$

The operator $R$ is unitary since, as it is easy to see,

$$\int e^{i[\langle \alpha^*_{11}F_1 + \alpha^*_{21}F_2, f_1\rangle + \langle \alpha^*_{12}F_1 + \alpha^*_{22}F_2, f_2\rangle]} d\mu(F_1) d\mu(F_2)$$

$$= \int e^{i[\langle F_1, f_1\rangle + \langle F_2, f_2\rangle]} d\mu(F_1) d\mu(F_2).$$

For every $g \in G$,

$$R(U_{A_1}(g) \otimes U_{A_2}(g))R^{-1} = U_{\alpha_{11}A_1+\alpha_{12}A_2}(g) \otimes U_{\alpha_{21}A_1+\alpha_{22}A_2}(g),$$

whereof the statement of the lemma follows.

COROLLARY: *If $A_1$, $A_2$ are invertible operators in $H$, then*

$$U_{A_1} \otimes U_{A_2} \cong U_A \otimes U_0,$$

*where $A = (A_1^* A_1 + A_2^* A_2)^{1/2}$, $U_0$-the representation, which corresponds to zero cocycle.*

Indeed, put, for short, $B_1 = A_1^* A_1$, $B_2 = A_2^* A_2$ and notice that $B_1 - B_1 A^{-2} B_1 = B_2 - B_2 A^{-2} B_2$, this operator being self-adjoint and positively definite. Consider operators $\alpha_{11} = A^{-1} A_1^*$, $\alpha_{12} = A^{-1} A_2^*$,

$$\alpha_{21} = (B_1 - B_1 A^{-2} B_1)^{1/2} A_1^{-1}, \quad \alpha_{22} = -(B_2 - B_2 A^{-2} B_2)^{1/2} A_2^{-1}.$$

One easily proves that these operators satisfy the conditions of Lemma 2. On the other hand, $\alpha_{11}A_1 + \alpha_{12}A_2 = A$, $\alpha_{21}A_1 + \alpha_{22}A_2 = 0$, q.e.d.

3. *The definition of the representation for a group G with respect to a pair of* 1*-cocycles.* Let $U = \mathrm{EXP}_\beta V$ be the representation of $G$ in the space $L^2_\mu(H')$ defined in the section 2. It is not difficult to verify that this representation is equivalent to the following representation $U_\lambda$ in the space $L^2_\mu(H')$:

$$(U_\lambda(g)\Phi)(F) = e^{i\lambda[\langle F, \beta g\rangle + \mathrm{Im}\,\lambda \|\beta g\|^2]} \Phi(V^{-1}(g)F - 2\,\mathrm{Im}\,\lambda.\ \beta(g^{-1})),$$

$\lambda$ being any complex number, $|\lambda| = 1$. Namely, $U_\lambda(g) = A_\lambda U(g) A_\lambda^{-1}$, where $A_\lambda$ is the operator, uniquely defined by its action on the functionals $e^{i\langle \cdot, h\rangle}$:

$$A_\lambda : e^{i\langle \cdot, h\rangle} \mapsto e^{(\lambda^2 - 1)\|h\|^2/2} e^{i\lambda\langle \cdot, h\rangle}.$$

Given another 1-cocycle $\beta'$ of the group $G$ with values in $H$ we define the operators $\tilde{U}_\lambda(g)$ in $L^2_\mu(H')$ by

$$(\tilde{U}_\lambda(g)\Phi)(F) = e^{i\langle F, \beta' g\rangle} (U_\lambda(g)\Phi)(F), \quad g \in G.$$

THEOREM: *The operators* $\tilde{U}_\lambda(g)$ are unitary and constitute a projective representation of the group G, namely

$$\tilde{U}_\lambda(g_1 g_2) = e^{2\,\mathrm{Im}\,\lambda\cdot\alpha(g_1,g_2)}\tilde{U}_\lambda(g_1)\tilde{U}_\lambda(g_2)$$

*for every* $g_1, g_2 \in G$, *where*

$$\alpha(g_1, g_2) = \langle \beta g_1^{-1}, \beta' g_2 \rangle.$$

The proof is immediate.

We want to define now an extension of the additive group $\mathbb{R}^+$ by the group $G$. For this purpose observe that the function $\alpha$ for every $g_1, g_2, g_3 \in G$ satisfies

$$\alpha(g_1, g_2) + \alpha(g_1 g_2, g_3) = \alpha(g_1, g_2 g_3) + \alpha(g_2, g_3).$$

Consequently, $\alpha$ is a 2-cocycle of the group $G$ with values in $\mathbb{R}$ and therefore defines an extension $\tilde{G}$ of $\mathbb{R}^+$ by $G$. The elements of $\tilde{G}$ are the pairs $(g, c)$, $g \in G$, $c \in \mathbb{R}^+$ with the following multiplication rule:

$$(g_1, c_1)(g_2, c_2) = (g_1 g_2, c_1 + c_2 - 2\,\mathrm{Im}\,\lambda\alpha(g_1, g_2)).$$

It corresponds to the projective representation of $G$ in the space $L^2_\mu(H')$ defined above an (affine) unitary representation of the group $\tilde{G}$, given by

$$(\tilde{U}_\lambda(g, c)\Phi)(F) = e^{i[\lambda(\langle F,\beta g\rangle + \mathrm{Tm}\,\lambda\|\beta g\|^2) + (c + \langle F, \beta' g\rangle)]}\Phi(V^{-1}(g)F - 2\,\mathrm{Tm}\,\lambda\cdot\beta(g^{-1})).$$

4. *The singularity conditions for two measures.* Recall that two measures, $\mu$ and $\nu$, in the space $X$ are said to be equivalent if for every measurable set $A \subset X$ the conditions $\mu(A) = 0$ and $\nu(A) = 0$ hold simultaneously. The measures $\mu$ and $\nu$ are called mutually singular if there is a measurable set $A \subset X$ such that $\mu(A) = 0$, $\nu(X - A) = 0$.

The lemma which follows is well known in the theory of the Gaussian measure spaces.

LEMMA 3: *Let $\mu$ be the standard Gaussian measure in the space $H'$. The measures $\mu$ and $\mu(\cdot + x)$ are mutually singular if and only if $x \notin \bar{H}$.*

PROOF: We may assume that $\bar{H} = l^2$ and $\mu$ is a product measure, $\mu = m_1 \times \cdots \times m_n \times \cdots$ where $dm_i(t) = (2\pi)^{-1/2} e^{-t^2/2} dt$; $x = (x_1, \ldots, x_n, \ldots)$. The measure $\mu(\cdot + x)$ is a product measure either and therefore, by virtue of the zero-one law, the measures $\mu$ and $\mu(\cdot + x)$

are either mutually singular or equivalent. By virtue of Kakutani theorem the measures $\mu$ and $\mu(\cdot + x)$ are equivalent if and only if $\Pi_k \int \sqrt{dm_k \cdot dm_k(\cdot + x)} > 0$. The easy verification shows that this happens if and only if $\Sigma_k x_k^2 < \infty$, i.e. $x \in \bar{H}$.

LEMMA 4: *Let $\mu$ be the standard Gaussian measure in $H'$ and $\nu$ a measure in $H'$ satisfying $\nu(\bar{H}) = 0$. Then the measure $\mu$ is mutually singular with the convolution $\mu * \nu$ of the measures $\mu$ and $\nu$.*

(A convolution of the measures $\mu$ and $\nu$ is defined by $(\mu * \nu)(\cdot) = \int \mu(\cdot - x) d\nu(x)$.)

PROOF: Assume that the measures $\mu$ and $\mu_1 = \mu * \nu$ are not mutually singular. Then $(d\mu_1/d\mu) = p > 0$ on a set of positive $\mu$-measure. Since $\mu_1(\cdot) = \int \mu(\cdot - x) d\nu(x)$ then, by the Fubini theorem, $p(y) = \int [d\mu(y - x)/d\mu(y)] d\nu(x) > 0$ for $y$ in the set mentioned above. Hence, there exists a set $B$, $\nu(B) > 0$, such that $[d\mu(\cdot - x)/d\mu] > 0$ for each $x \in B$. But then $\mu$ and $\mu(\cdot - x)$ are not mutually singular, and consequently, by Lemma 3, $x \in \bar{H}$, in which case $\nu(\bar{H}) > 0$. We came to a contradiction with the assumption.

5. *The spectral measures.* Let $G$ be an abelian (not necessarily locally compact) topological group possessing a countable base of open sets, $\hat{G}$ the group of its continuous characters, $U$ a unitary representation of $G$ in the complex Hilbert space $\mathcal{H}$. By applying the spectral theorem (see, for example, [4]) to the $C^*$-algebra, generated by the operators of the representation $U$, we get the following decomposition: *There exists an isomorphism of the space $\mathcal{H}$ onto the direct integral of Hilbert spaces,*

$$T: \mathcal{H} \to \int_{\hat{G}}^{\oplus} \mathcal{H}_\chi d\mu(\chi),$$

*with $\mu$ a Borel measure on $\hat{G}$, which transfers the operators $U(g)$, $g \in G$ into the operators*

$$(TU(g)T^{-1}f)(\chi) = \chi(g) f(\chi). \tag{3}$$

The measure $\mu$ on $\hat{G}$ is defined by $U$ uniquely up to equivalence and is called the *spectral measure* of the representation (3). The representation (3) of $G$ in the space $\int_{\hat{G}}^{\oplus} \mathcal{H}_\chi d\mu(\chi)$ is called the *spectral decomposition* of the initial representation $U$. If $\dim \mathcal{H}_\chi = 1$ for a.e. $\chi$ it is said that the representation $U$ has a simple spectre.

We give here some statements of the spectral measures.

(1) Two representations of $G$ are disjoint if and only if their spectral measures are mutually singular.

(2) The spectral measure of the sum of two representations of $G$ is equivalent to the sum of their spectral measures.

(3) The spectral measure of the tensor product of two representations of $G$ is equivalent to the convolution of their spectral measures.

(4) The weakly closed algebra generated by the operators (3) in the space $\int_{\hat{G}}^{\oplus} \mathcal{H}_\chi d\mu(\chi)$ coincides with the algebra of the operators of multiplication by an arbitrary $\mu$-measurable function $a(\chi)$: $f(\chi) \mapsto a(\chi)f(\chi)$ (i.e. this algebra is isomorphic to $L_\mu^\infty(\hat{G})$, see, for example, [4]).

In the §4 we shall make use of the following generalization of the statement 4):

Assume that a unitary representation of the group $G$ can be decomposed into a direct integral of representations

$$U = \int_{\Xi}^{\oplus} U_\xi d\nu(\xi)$$

($\Xi$ is a measure space with the measure $\nu$).

It means that $U$ is equivalent to the representation in the direct integral of Hilbert spaces $\mathcal{H} = \int_{\Xi}^{\oplus} \mathcal{H}_\xi d\nu(\xi)$ given by

$$(U(g)f)(\xi) = U_\xi(g)f(\xi),$$

$U_\xi$ being a representation of $G$ in $\mathcal{H}_\xi$.

Lemma 5: *If the representations $U_{\xi_1}$, $U_{\xi_2}$, are disjoint for almost every (with respect to $\nu$) $\xi_1 \neq \xi_2$, then the weakly closed operator algebra generated by the operators $U(g)$, $g \in G$ contains the operators of multiplication by every bounded $\nu$-measurable function $a(\xi): f(\xi) \mapsto a(\xi)f(\xi)$.*

Corollary: *Let a representation $U$ of $G$ be decomposed in a tensor product of two representations, $U = U' \otimes U''$, with the corresponding spectral measures $\mu'$ and $\mu''$. If the measures $\mu'(\cdot + \chi_1)$ and $\mu'(\cdot + \chi_2)$ are mutually singular for almost every $\chi_1 \neq \chi_2$ (with respect to $\mu''$), then the weakly closed operator algebra generated by the operators $U(g)$, $g \in G$, contains all operators $E \otimes U''(g)$, $g \in G$ (and, therefore, all operators $U'(g) \otimes E$).*

Indeed, let $U'' = \int_{\hat{G}}^{\oplus} U''_\chi d\mu''(\chi)$ be the spectral decomposition of the representation $U''$. Then $U = \int_{\hat{G}}^{\oplus} (U' \otimes U''_\chi) d\mu''(\chi)$. Since the spectral measure of the representation $U' \otimes U''_\chi$ is $\mu'(\cdot + \chi)$, then, by the conjecture made above, the representations $U' \otimes U''_{\chi_1}$ and $U' \otimes U''_{\chi_2}$ are disjoint for almost all (with respect to $\mu''$) $\chi_1 \neq \chi_2$. It follows from the Lemma 5 that the weakly closed operator algebra generated by the operators $U(g)$, $g \in G$, contains the operators of multiplication by the functions $a_g(\chi) = \chi(g)$, i.e. the operators $E \otimes U''(g)$.

## §3. Nonlocal representations of the Group $G^X$. The ring of representations

1. *Construction of the representations of the group $G^X$.* Let us begin to study of the representations of $G^X$ which are connected with the Maurer–Cartan cocycle (see §1). *From now on we shall consider only those Lie groups G for which their Lie algebra possesses an inner product invariant under the adjoint action of the group G.* In particular, all compact and all abelian Lie groups satisfy this condition.

In order to construct a representation of $G^X$ we assume that $X$ has a structure of a Riemannian manifold. This structure induces an orthogonal structure in the tangent bundle $TX$, as well as a strictly positive smooth measure $dx$ on $X$.

Let us introduce an inner product in the space $\Omega^1(X)$ of $\mathbb{R}$-valued 1-forms of the $C^\infty$ class, with compact support by the formula

$$\langle \omega_1, \omega_2 \rangle = \int_X \langle \omega_1(x), \omega_2(x) \rangle_x dx,$$

where $\langle , \rangle_x$ is an inner product in the conjugate tangential space $T_x^* X$. Let us also fix an inner product in the Lie algebra $\mathfrak{G}$ of $G$ which is invariant under the adjoint action of the group $G$.

Consider now the space $\Omega^1(X; \mathfrak{G}) = \Omega^1(X) \otimes \mathfrak{G}$ of differentiable $\mathfrak{G}$-valued 1-forms on $X$ with compact support. The orthogonal structures in the spaces $\Omega^1(X)$ and $\mathfrak{G}$ introduced above induce the orthogonal structure in their tensor product $\Omega^1(X; \mathfrak{G})$. It is clear that the latter inner product in $\Omega^1(X; \mathfrak{G})$ is $G^X$-invariant.

Denote by $H$ the pre-Hilbert space $\Omega^1(X; \mathfrak{G})$, by $\bar{H}$ its completion in the norm introduced in $H$, by $\mathcal{F}$ a space conjugate to $H$ and by $\mu$ the standard Gaussian measure in $\mathcal{F}$.

Let us define, according to the general definition given in the section 2 of §2, the new unitary representation $U = \mathrm{EXP}_\beta V$ of the group $G^X$. For that purpose we shall extend the representation $V$ of $G^X$ from $H$ to the space $\mathcal{F} \supset H$ by the formula

$$\langle V(\tilde{g})F, \omega\rangle = \langle F, V^{-1}(\tilde{g})\omega\rangle, \quad \omega \in H.$$

Let $\beta$ be a Maurer–Cartan 1-cocycle, that is $\beta\tilde{g} = R \circ d\tilde{g}$ (see §1). Define a unitary representation $U$ of the group $G^X$ in the space $L^2_\mu(\mathcal{F})$ by

$$(U(\tilde{g})\Phi)(F) = e^{i\langle F, \beta\tilde{g}\rangle}\Phi(V^{-1}(\tilde{g})F).$$

Note that the representation $U$ depends on the Riemann space structure in $X$.

We formulate now the main results of the paper.

THEOREM 1: *If $G$ is a compact semisimple Lie group and* dim $X \geq 2$, *then the representation $U = \mathrm{EXP}_\beta V$ of the group $G^X$ is irreducible.*

THEOREM 2: *Let $G$ be a compact semisimple Lie group. Then the representations $U$ of $G^X$ corresponding to different Riemannian metrics on $X$ are not equivalent.*

The proof of the theorems 1, 2 will be given in §5.[1] It rests upon the results of §4 where the restriction of the representation $U$ of $G^X$ to an abelian subgroup is studied. Note that the main results of §4 are valid for the case dim $X \geq 2$ only. The problem of irreducibility for the representation $U$ in the case dim $X = 1$ still remains open.

REMARK 1: The representation $U$ of $G^X$ is a restriction to $\mathfrak{Z}^1G^X \cong G^X$ of the representation $\tilde{U}$ of the group $\theta^1(X; G) \cong G^X \cdot \Omega^1(X; \mathfrak{G})$ in the space $L^2_\mu(\mathcal{F})$ defined in the following way (see §1).

Let $\alpha \in \theta^1(X; G)$, i.e. $\alpha = (\tilde{g}, \omega)$, $\tilde{g} \in G^X$, $\omega \in \Omega^1(X; \mathfrak{G})$. Then

$$(\tilde{U}(\alpha)\phi)(F) = e^{i\langle F, \omega\rangle}\Phi(V^{-1}(\tilde{g})F).$$

The representation $\tilde{U}$ is an "integral of representations" in the sense of [20, 21]. Its irreducibility can be easily derived from the theorems of [20, 21]. Thus theorem 1 is asserting that the representation $\tilde{U}$ remains irreducible when restricted to the image of $G^X$.

[1] The notion of a spherical function of the representation $U$ given in §5 allows one to prove Theorem 2 in a different way (independently of §4).

REMARK 2: The construction of the representation $U$ of $G^X$ presented above can be transferred in a natural way to the group $\theta(\xi)$ of all differentiable sections with compact support of an arbitrary fibre bundle $\xi$ over $X$ with a fibre $G$. One can easily see that the theorems 1 and 2 are true for groups $\theta(\xi)$ as well.

Indeed, there is in $X$ an open dense submanifold $X_0 \subset X$ such that the fibration $\xi|_{X_0}$ is trivial and consequently $\theta(\xi|_{X_0}) \cong G^{X^0}$. When $X$ is replaced by $X_0$, the space of the representation remains the same, and so the statements concerning the irreducibility and non-equivalence of the representations of the group $\theta(\xi)$ become analogous to those concerning representations of $G^X$, namely to the Theorems 1 and 2.

REMARK 3: Denote $W_2^1(X; G)$ the completion of $G^X$ in the metric

$$d(\tilde{g}_1, \tilde{g}_2) = \langle \beta(\tilde{g}_1^{-1}\tilde{g}_2), \beta(\tilde{g}_1^{-1}\tilde{g}_2)\rangle^{1/2} + \langle \beta(\tilde{g}_1\tilde{g}_2^{-1}), \beta(\tilde{g}_1\tilde{g}_2^{-1})\rangle^{1/2}.$$

$W_2^1(X; G)$ is an analogue of Sobolev space $\mathring{W}_2^1(X)$.[1]

(For example, if $G = SU(n)$ and $\dim X = 1$, then $W_2^1(X; G) = \{g(\cdot) \mid \int_X (\sum_{i,k} |g'_{ik}(x)|^2) dx < \infty\}$.)

One sees from the formulae defining the representation $U$ of $G^X$ that this representation can be extended to a representation of the group $W_2^1(X; G)$.

REMARK 4: Representations $U = \mathrm{EXP}_\beta V$ of $G^X$ induce the Hermitian representations of its Lie algebra $\mathfrak{G}^X$. The latter representations can be extended to the representations of the complexification $(\mathfrak{G}_C)^X$ of $\mathfrak{G}^X$. The explicit formulae for the operators of these (non-Hermitian) representations of the algebra $(\mathfrak{G}_C)^X$ can be easily put down as finite sums, if one uses the Fock model of the representation space. If, for example, $G = SO(n)$, then $\mathfrak{G}_C = s\ell(n, \mathbb{C})$. Therefore, our representations give rise to (non-Hermitian) nonlocal representations of the current algebra $s\ell(n, \mathbb{C})$, which depend on 1-jets.

2. *Representations of the group $G^X$ connected with subbundles of the tangent bundle TX.* Let us define now a more wide class of representations of $G^X$. Let $E$ be an arbitrary differentiable subbundles of the tangent bundle $TX$. Consider the restrictions of 1-forms $\omega \in \Omega^1(X; \mathfrak{G})$

[1] This analogy is noticed also in [1], where the representation under consideration (defined independently and in a different way) is called the energy representation. There is in [1] useful realization in the space of functions on the trajectories of Group -Wiener process (for $\dim X = 1$).

to the subbundle $E$. They form a linear space which we denote by $H_E$. In this space, as in the initial one, there is a naturally defined representation, $V_E$, of the group $G^X$. Let $\beta$ be the Maurer–Cartan cocycle. Then for any $\tilde{g} \in G^X$ define $\beta_E\tilde{g}$ as a restriction of the mapping $\beta\tilde{g}: TX \to \mathfrak{G}$ to the subbundle $E$. It is clear that $\beta_E$ is a 1-cocycle of $G^X$ taking values in $H_E$.

Let there be a Riemannian manifold structure $\tau$ on $X$. This structure induces an inner product in the space $H_E$ which is invariant under the representation $V_E$ of $G^X$. Denote by $U_{E,\tau}$ an exponential of the representation $V_E$, which is connected with the cocycle $\beta_E$:

$$U_{E,\tau} = \mathrm{EXP}_{\beta_E} V_E.$$

The next theorem is a consequence the theorem 1.

THEOREM 3: *If $G$ is a compact semisimple Lie group and* $\dim X \geq 2$ *then the representation $U_{E,\tau}$ of $G^X$ is irreducible.*

To prove it let us decompose the fibre bundle $TX$ into an orthogonal sum $TX = E \oplus E_\perp$ of the fibration $E$ and its orthogonal complement $E_\perp$. Evidently,

$$\mathrm{EXP}_{\beta_E} V_E \otimes \mathrm{EXP}_{\beta_{E\perp}} V_{E_\perp} \cong \mathrm{EXP}_\beta V.$$

Therefore, the irreducibility of $U_{E,\tau} = \mathrm{EXP}_{\beta_E} V_E$ is an immediate consequence of that of the representation $\mathrm{EXP}_\beta V$.

THEOREM 4: *Let $G$ be a compact semisimple Lie group. Then the representations $U_{E_1,\tau_1}$ and $U_{E_2,\tau_2}$ of $G^X$ are equivalent if and only if $E_1 = E_2$ and the inner products in the space $H_{E_1} = H_{E_2}$ induced by the Riemannian structures $\tau_1, \tau_2$ on $X$ coincide.*

The proof of this theorem is similar to that of the theorem 2 (cf. §5).

3. *Decomposition of the tensor product of representations $U = \mathrm{EXP}_\beta V$.* The representations $U = \mathrm{EXP}_\beta V$ of $G^X$ defined in the first section depend on the Riemannian space structure $\tau$ on $X$. To emphasize this circumstance we shall denote them by $U_\tau = \mathrm{EXP}_\beta (V, \tau)$. We may, on the contrary, consider the Riemannian metric as fixed, the parameter of a representation being the 1-cocycle of the group $G^X$ originated from the Maurer–Cartan cocycle as it was explained in the Remark 2, section 2 of §1.

More exactly, let $\tau_0$ be a fixed Riemannian metric on $X$, $(,)_x$ an inner product it induces in $T_x$, $x \in X$. Consider an arbitrary Riemannian metric $\tau$ on $X$ and let $(,)_{\tau,x}$ stand for the inner product in $T_x$, $x \in X$ induced by $\tau$. The latter inner product can be represented in the form

$$(\xi_1, \xi_2)_{\tau,x} = (\sigma_\tau(x)\xi_1, \sigma_\tau(x)\xi_2)_x,$$

$\sigma_\tau(x): T_x \to T_x$ being a self-adjoint positive linear operator. Observe that the function $x \mapsto \sigma_\tau(x)$ defines the Riemannian metric $\tau$ on $X$ in a unique way.

Define the operator $A_\tau$ in the space $\Omega^1(X; \mathfrak{G})$ by

$$(A_\tau \omega)(x) = |\sigma_\tau(x)|^{1/2} \omega(x) \circ \sigma_\tau^{-1}(x),$$

where $|\sigma_\tau(x)| = \det \sigma_\tau(x)$. Evidently, $A_\tau$ commutes with the operators of the representation $V$. It follows that if $\beta$ is the Maurer–Cartan cocycle, then the function

$$(A_\tau \beta)\tilde{g} = A_\tau(\beta \tilde{g}), \quad \tilde{g} \in G^X$$

is a 1-cocycle of $G^X$, too.

LEMMA 1: *There is an equivalence of the representations of* $G^X$:

$$\mathrm{EXP}_\beta (V, \tau) \cong \mathrm{EXP}_{A_\tau \beta} (V, \tau_0).$$

PROOF: Let $\mu_{\tau_0}$, $\mu_\tau$ be the standard Gaussian measures in $\mathcal{F} = (\Omega^1(X; \mathfrak{G}))'$ which are induced by the Riemannian metrics $\tau_0$, $\tau$ on $X$. It easily follows from the definition of the operator $A_\tau$ that the correlation functionals $B_{\tau_0}$, $B_\tau$ of these measures are connected by

$$B_\tau(\omega_1, \omega_2) = B_{\tau_0}(A_\tau \omega_1, A_\tau \omega_2).$$

Evidently, the mapping $e^{i\langle \cdot, \omega \rangle} \mapsto e^{i\langle \cdot, A_\tau \omega \rangle}$ extends to the isomorphism of Hilbert spaces $L^2_{\mu_\tau}(\mathcal{F}) \to L^2_{\mu_{\tau_0}}(\mathcal{F})$ which transfers operators of the representation $\mathrm{EXP}_\beta (V, \tau)$ into operators of the representation $\mathrm{EXP}_{A_\tau \beta} (V, \tau_0)$.

In what follows we shall consider representations $\mathrm{EXP}_{A\beta} (V, \tau_0)$ with $A$ an arbitrary self-adjoint positively definite linear operator in the space $\Omega^1(X; \mathfrak{G})$ commuting with the operators of the representation $V$ of $G^X$. Using Lemma 2, §2, of tensor products, we obtain.

LEMMA 2:

$$\mathrm{EXP}_{A_1\beta}(V,\tau_0)\otimes\mathrm{EXP}_{A_2\beta}(V,\tau_0)\cong\mathrm{EXP}_{A\beta}(V,\tau_0)\otimes\mathrm{EXP}_{0}\cdot V,$$

*where* $A=(A_1^2+A_2^2)^{1/2}$, $\mathrm{EXP}_0\,V$ *is the representation corresponding the zero cocycle (it does not depend on the Riemannian space structure on* $X$*).*

The following theorem presents a decomposition of the tensor product of the representations which are examined and enables us to calculate the additive generators of the ring of representations.

THEOREM 5: *The representation* $\mathrm{EXP}_{A_1\beta}(V,\tau_0)\otimes\mathrm{EXP}_{A_2\beta}(V,\tau_0)$ *of the group* $G^X$ *can be decomposed into a continual direct sum of the representations of the form*

$$\mathrm{EXP}_{A\beta}(V,\tau_0)\otimes V^{x_1}\otimes\cdots\otimes V^{x_n}\quad(n=0,1,\ldots),$$

*where* $A=(A_1^2+A_2^2)^{1/2}$ *and* $V^{x_0}$ $(x_0\in X)$ *is a representation of* $G^X$ *in the space* $\mathfrak{G}_{\mathrm{C}}$, *given by*

$$V^{x_0}(\tilde g)=\mathrm{Ad}\,g(x_0).$$

PROOF OF THEOREM 5: It is a consequence of Lemma 1 of §2 that $\mathrm{EXP}_0\,V=\bigoplus_{n=0}^{\infty}S^nV_{\mathrm{C}}$ where $V_{\mathrm{C}}$ is the complexification of the representation $V$, $S^nV_{\mathrm{C}}$ is a simmetrized tensor product of $n$ copies of the representation $V_{\mathrm{C}}$. The representation $S^nV_{\mathrm{C}}$ may, in its turn, be decomposed into a continual direct sum of the representations $V^{x_1}\otimes\cdots\otimes V^{x_n}$. Namely, $S^nV_{\mathrm{C}}$ is equivalent to finite multiple of a continual direct sum of the representations

$$\int_{\tilde X^n}^{\oplus}V^{x_1}\otimes\cdots\otimes V^{x_n}dx_1\ldots dx_n,$$

where the integral is taken over a domain $\tilde X^n\subset X^n$ which is fundamental with respect to the permutation group of $x_1,\ldots,x_n$. To complete the proof of the theorem one has to make use of Lemma 2.

COROLLARY: *Let* $A$ *be a symmetric positively-definite linear operator in the space* $\Omega^1(X;\mathfrak{G})$, *commuting with the action of the group* $G^X$, *and* $W$ *a local finitely-dimensional representation of* $G^X$. *Then the representations of the form* $\mathrm{EXP}_{A\beta}(V,\tau_0)\otimes W$ *are the additive generators in the ring of representations they generate.*

We shall point out that Theorem 5 is proved for a manifold $X$ of any dimension and an arbitrary Lie group $G$ satisfying the conditions given in the beginning of the Section 1.

REMARK 1: If $G$ is a compact semisimple Lie group, $\dim X \geq 2$ and $x_1, \dots, x_n$ are different points of $X$, then the representation $\mathrm{EXP}_{A_\beta}(V, \tau_0) \otimes V^{x_1} \otimes \cdots \otimes V^{x_n}$ of $G^X$ is irreducible. The proof of this statement can be given along the same lines as for Theorem 1.

REMARK 2: Lemma 2 and Theorem 5 can be easily formulated in terms of Riemannian metrics on $X$.

LEMMA 2′: *If* $\dim X \neq 2$, *then*

$$\mathrm{EXP}_\beta(V, \tau_1) \otimes \mathrm{EXP}_\beta(V, \tau_2) \cong \mathrm{EXP}_\beta(V, \tau) \otimes \mathrm{EXP}_0 V,$$

*where* $\tau$ *is a Riemannian metric uniquely defined by the equation*

$$(1) \qquad |\sigma_\tau(x)|\sigma_\tau^{-2}(x) = |\sigma_{\tau_1}(x)|\sigma_{\tau_1}^{-2}(x) + |\sigma_{\tau_2}(x)|\sigma_{\tau_2}^{-2}(x);$$

for the definition of $\sigma_\tau(x)$ see p. 20.[1]

THEOREM 5′: *If* $\dim X \neq 2$, *then the representation* $\mathrm{EXP}_\beta(V, \tau_1) \otimes \mathrm{EXP}_\beta(V, \tau_2)$ *can be decomposed into a continual direct sum of the representations of the form* $\mathrm{EXP}_\beta(V, \tau) \otimes V^{x_1} \otimes \cdots \otimes V^{x_n}$ *(*$n = 0, 1, \dots$*) the Riemannian metric* $\tau$ *being defined as in the Lemma* 2.

The similar statements take place for a tensor product $U_{E,\tau_1} \otimes U_{E,\tau_2}$ of the representations described in section 2, where $E$ is an arbitrary subbundle of the tangent bundle $TX$.

Let us formulate now some statements concerning tensor products $U_{E_1,\tau_1} \otimes U_{E_2,\tau_2}$ with $E_1 \neq E_2$.

If $E_1 \cap E_2 = 0$, then there exists such a Riemannian space structure $\tau$ on $X$ that $U_{E_1,\tau_1} \otimes U_{E_2,\tau_2} \cong U_{E_1+E_2,\tau}$. This Riemannian structure is defined by the conditions: (a) $E_1$ and $E_2$ are mutually orthogonal in the metric on $TX$ induced by $\tau$, (b) the Riemannian structures $\tau$, $\tau_i$ induce the same inner product in the space $H_{E_i}$ ($i = 1, 2$) (see section 2).

[1] The equation (1) can fail to have a solution $\sigma_\tau(x)$ in the case $\dim X = 2$.

If $E_1 \cap E_2 = E \neq 0$ and $E$ is also a subbundle of the tangent bundle, then let $E_i^\perp \subset E_i$ designate the orthogonal complementation to $E$ in $E_i$ with respect to the metric induced by $\tau_i$ $(i = 1, 2)$. Then we have an isomorphism $U_{E_i,\tau_i} \cong U_{E_i^\perp,\tau_i} \otimes U_{E,\tau_i}$.

Consequently,

$$U_{E_1,\tau_1} \otimes U_{E_2,\tau_2} \cong (U_{E_1^\perp,\tau_i} \otimes U_{E_2^\perp,\tau_2}) \otimes (U_{E,\tau_1} \otimes U_{E,\tau_2}).$$

This reduces the problem of decomposition of the tensor product $U_{E_1,\tau_1} \otimes U_{E_2,\tau_2}$ to the cases considered above.

## §4. Restriction of the representation $U$ of the group $G^X$ to an abelian subgroup

Let $G$ be a compact semisimple Lie group, $X$ a connected open manifold of the class $C^\infty$, $U = \mathrm{EXP}_\beta V$ the representation of the group $G^X$ in the space $L^2_\mu(\mathcal{F})$ constructed in section 1 of §2.

Let $\mathfrak{A}$ be an arbitrary Cartan subalgebra of the algebra $\mathfrak{G}$, $A \subset G$ – the Cartan subgroup corresponding to $\mathfrak{A}$. Let us put $\mathfrak{A}^X$ for the additive group of differentiable $C^\infty$ mappings $a: X \to \mathfrak{A}$ with a compact support and exp for the exponential mapping $\mathfrak{A}^X \to A^X$.

Define the representation $W$ of the group $\mathfrak{A}^X$ in the space $L^2_\mu(\mathcal{F})$ by

$$W(a) = U(\exp a), \quad a \in \mathfrak{A}^X.$$

Since $\beta(\exp a) = da$, the operators $W(a)$ have the following form:

$$(W(a)\Phi)(F) = e^{i\langle F, da\rangle}\Phi(V^{-1}(\exp a)F). \tag{1}$$

We proceed now to the calculation of the spectral measure of the representation $W$. Notice that the character group $(\mathfrak{A}^X)^\wedge$ of $\mathfrak{A}^X$ is isomorphic to $(\mathfrak{A}^X)' = (C^\infty(X))' \otimes \mathfrak{G}$ (the isomorphism is given by the correspondence $F \mapsto \chi_F(\cdot) = e^{i\langle F, \cdot\rangle}$). The spectral measure can be therefore considered as a measure in the space $(\mathfrak{A}^X)'$ conjugate to $\mathfrak{A}^X$.

Let $\mathfrak{m}$ denote the orthogonal complementation in $\mathfrak{G}$ to $\mathfrak{A}: \mathfrak{G} = \mathfrak{A} \oplus \mathfrak{m}$. Let $\mathcal{F}_\mathfrak{A} \subset \mathcal{F}$, $\mathcal{F}_\mathfrak{m} \subset \mathcal{F}$ be the subspaces of, correspondingly, $\mathfrak{A}$-valued and $\mathfrak{m}$-valued generalized 1-forms on $X$, and $\mu_\mathfrak{A}$, $\mu_\mathfrak{m}$ be the standard Gaussian measures in $\mathcal{F}_\mathfrak{A}$ and $\mathcal{F}_\mathfrak{m}$. It is clear that $\mathcal{F} = \mathcal{F}_\mathfrak{A} \oplus \mathcal{F}_\mathfrak{m}$, $\mu = \mu_\mathfrak{A} \times \mu_\mathfrak{m}$ and

$$L^2_\mu(\mathcal{F}) = L^2_{\mu_\mathfrak{A}}(\mathcal{F}_\mathfrak{A}) \otimes L^2_{\mu_\mathfrak{m}}(\mathcal{F}_\mathfrak{m}).$$

LEMMA 1: *The representation $W$ of the group $\mathfrak{A}^X$ can be decomposed into the tensor product $W = W_{\mathfrak{A}} \otimes W_{\mathfrak{m}}$ of the representations in the spaces $L^2_{\mu\mathfrak{A}}(\mathcal{F}_{\mathfrak{A}})$ and $L^2_{\mu_{\mathfrak{m}}}(\mathcal{F}_{\mathfrak{m}})$ which are defined by*

$$(W_{\mathfrak{A}}(a)\Phi(F) = e^{i\langle F, da\rangle}\Phi(F), \tag{2}$$

$$(W_{\mathfrak{m}}(a)\Phi)(F) = \Phi(V^{-1}(\exp a)F). \tag{3}$$

This lemma is a straight consequence of (1), if one notes that $V(\exp a)$ is acting trivially on $\mathcal{F}_{\mathfrak{A}}$.

COROLLARY: *The spectral measure of the representation $W$ is equivalent to the convolution of the spectral measures of $W_{\mathfrak{A}}$ and $W_{\mathfrak{m}}$.*

LEMMA 2: *The spectral measure of the representation $W_{\mathfrak{A}}$ is equivalent to thie Gaussian measure on $(\mathfrak{A}^X)'$ with the zero mean and the correlation functional*

$$B(a_1, a_2) = \langle da_1, da_2\rangle, \quad a_1, a_2 \in \mathfrak{A}^X \tag{4}$$

(the angular brackets denote the inner product in $\Omega^1(X; \mathfrak{A}) \subset \Omega^1(X; \mathfrak{G})$).

PROOF: Consider the differentiation operator

$$d: \mathfrak{A}^X \to \Omega^1(X; \mathfrak{A}).$$

Evidently, its kernel is zero.[1] Let

$$d^*: (\Omega^1(X; \mathfrak{A}))' = \mathcal{F}_{\mathfrak{A}} \to (\mathfrak{A}^X)'$$

be the mapping conjugate to $d$. It follows from the formula (2) for operators $W_{\mathfrak{A}}(a)$ that the image $d^*\mu_{\mathfrak{A}}$ of $\mu_{\mathfrak{A}}$ is the spectral measure of $W_{\mathfrak{A}}$. It is well known that a linear transformation transfers a Gaussian measure into a Gaussian one, the correlation functionals of these measures being connected by the formula

$$B_{d^*\mu_{\mathfrak{A}}}(a_1, a_2) = B_{\mu_{\mathfrak{A}}}(da_1, da_2).$$

Consequently, $B_{d^*\mu_{\mathfrak{A}}}(a_1, a_2) = \langle da_1, da_2\rangle$. The lemma is proved.

[1] Since $X$ is open, $\mathfrak{A}^X$ does not contain constants.

Our aim now is to find the spectral measure of the representation $W_{\mathfrak{m}}$. Let $H_{\mathrm{C}}$ be the complexification of the space $\Omega^1(X, \mathfrak{m})$, $V_{\mathfrak{m}}$ – the representation of $\mathfrak{A}^X$ in $H_{\mathrm{C}}$ given by

$$(V_{\mathfrak{m}}(a)\omega)(x) = \mathrm{Ad}\,(\exp a(x)) \circ \omega(x).$$

Denote by $S^n V_{\mathfrak{m}}$ the symmetrized tensor product of $n$ copies of the representation $V_{\mathfrak{m}}$, $n = 1, 2, \ldots$, and by $S^0 V_{\mathfrak{m}}$ the unity representation.

Lemma 3: $W_{\mathfrak{m}} \cong \bigoplus_{n=0}^{\infty} S^n V_{\mathfrak{m}}$.

This lemma follows immediately from Lemma 1, §2.

Corollary: $W \cong W_{\mathfrak{A}} \oplus (\bigoplus_{n=1}^{\infty} (W_{\mathfrak{A}} \otimes S^n V_{\mathfrak{m}}))$.

Define for any root $\alpha$ of the algebra $\mathfrak{G}$ (with respect to $\mathfrak{A}$) and any $x_0 \in X$ a distribution $\varphi_{x_0}^{\alpha} \in (\mathfrak{A}^X)'$ by

$$\langle \varphi_{x_0}^{\alpha}, a \rangle = \alpha(a(x_0)). \tag{5}$$

Lemma 4: *The spectral measure $\nu_n$ of the representation $S^n V_{\mathfrak{m}}$ is concentrated on the subset of distributions of the form $\varphi_{x_1}^{\alpha_1} + \cdots + \varphi_{x_n}^{\alpha_n}$. Moreover, on each subset $\{\varphi_{x_1}^{\alpha_1} + \cdots + \varphi_{x_n}^{\alpha_n} \mid x_1, \ldots, x_n \in X\}$ where $\alpha_1, \ldots, \alpha_n$ are fixed, the measure $\nu_n$ is equivalent to the measure $dx_1 \ldots dx_n$.*

Proof: It suffices to check the statement for the case $n = 1$. Let $\mathfrak{m}_{\mathrm{C}}$ denote the complexification of $\mathfrak{m}$ and $\mathfrak{G}^{\alpha} \subset \mathfrak{m}_{\mathrm{C}}$ the rooted subspace corresponding to the root $\alpha$. Consider the subspaces $H_{\mathrm{C}}^{\alpha} = \Omega^1(X) \otimes \mathfrak{G}^{\alpha}$ of the space $H_{\mathrm{C}} = \Omega^1(X) \otimes \mathfrak{m}_{\mathrm{C}}$. They are orthogonal for different $\alpha$, $\mathfrak{A}^X$-invariant and

$$H_{\mathrm{C}} = \bigoplus_{\alpha \in \Delta} H_{\mathrm{C}}^{\alpha}$$

($\Delta$ is the set of all roots).

The representation operators $V_{\mathfrak{m}}$ are given on each subspace $H_{\mathrm{C}}^{\alpha}$ by

$$(V_{\mathfrak{m}}(a)\omega)(x) = e^{i\alpha(a(x))}\omega(x).$$

It is clear that the spectral measure of the representation $V_{\mathfrak{m}}$ in the subspace $H_{\mathrm{C}}^{\alpha}$ is concentrated on the subset of the distributions $\varphi_x^{\alpha}$,

$x \in X$ defined by (5), this measure being equivalent to $dx$ under the identification $\varphi_x^\alpha \mapsto x \in X$. It follows that the spectral measure of the representation $V_m$ in the whole space $H_C$ is concentrated on the set $\{\varphi_x^\alpha | \alpha \in \Delta, x \in X\}$ and equivalent (under the identification $\varphi_x^\alpha \mapsto (x, \alpha) \in X \times \Delta$) to the product of the measure $dx$ on $X$ and a uniform measure on $\Delta$. Lemma is proved.

COROLLARY: *The Riemannian structure on the manifold X changed, the spectral measure of the representation $W_m$ becomes equivalent to the former measure.*

LEMMA 5: *Let $\varphi \in (\mathfrak{A}^X)'$ be a distribution,*

(6) $$\varphi = \lambda_1 \varphi_{x_1}^{\alpha_1} + \cdots + \lambda_n \varphi_{x_n}^{\alpha_n}, \ \lambda_k \in \mathbb{R}, \ \varphi \neq 0,$$

*where $\varphi_{x_k}^{\alpha_k}$ are given by* (5). *If* dim $X \geq 2$, *then $\varphi$ is not an element of the completion $\overline{\mathfrak{A}^X}$ of the space $\mathfrak{A}^X \subset (\mathfrak{A}^X)'$ in the norm $\|a\| = \langle da, da \rangle^{1/2}$.*

PROOF: As the distributions $\varphi_x^\alpha$ are local with respect to $x$, it suffices to prove that for every $\alpha \in \Delta$, $x_0 \in X$ and a neighbourhood $X_0$ of $x_0$ the distribution $\varphi = \varphi_x^\alpha$ is not an element of the completion of the space $\mathfrak{A}^{X_0} = C^\infty(X_0) \otimes \mathfrak{A}$.

Let $X_0$ be a sufficiently small neighbourhood of $x_0 \in X$; $x^1, \ldots, x^n$ local coordinates in $X_0$; $e_1, \ldots, e_\tau$ and orthonormal basis in $\mathfrak{A}$. We do not lose in generality if assume that $X_0$ is the open unit ball with centre $x_0$. Let us represent the elements $a \in \mathfrak{A}^{X_0}$ in the form

$$a = \sum_{k=1}^{\tau} a_k e_k, \quad a_k \in C^\infty(X_0).$$

In the chosen coordinates $\|a\|$ and $\langle \varphi_{x_0}^\alpha, a \rangle$ are expressed as follows:

(7) $$\|a\|^2 = \sum_{k=1}^{\tau} \int_{X_0} \left( \sum_{i,j=1}^{m} \tau_{ij}(x) \frac{\partial a_k}{\partial x^i} \frac{\partial a_k}{\partial x^j} \right) dx,$$

where $\tau_{ij}$ is the metric tensor in $X_0$;

(8) $$\langle \varphi_{x_0}^\alpha, a \rangle = \sum_{k=1}^{\tau} \alpha(e_k) a_k(0).$$

We observe that $\alpha(e_k) \neq 0$ for at least one $k$.

Suppose now that $\varphi_{x_0}^{a}$ is an element of the completion $\mathfrak{A}^{X_0}$ in the norm $\|.\|$. Then, by virtue of (7) and (8), there exist such functions $\omega_1, \ldots, \omega_m$ on $X_0$, that

$$\int_{X_0} \left( \sum_{i,j=1}^{m} \tau_{ij}(x)\omega_i(x)\omega_j(x) \right) dx < \infty \tag{9}$$

and at the same time

$$\int_{X_0} \left( \sum_{i,j=1}^{m} \tau_{ij}(x) \frac{\partial f}{\partial x^i} \omega_j \right) dx = f(0) \tag{10}$$

for any $f \in C^\infty(X_0)$. (In other words, delta function $\delta(x)$ is an element of the completion of the space $C^\infty(X_0)$ in the 'energy norm' $\|f\|^2 = \int_{X_0} [\Sigma_{i,j} \tau_{ij}(\partial f/\partial x^i)(\partial f/\partial x^j)]dx$. It is known (see, for example, [16]) that it is impossible, if dim $X \geq 2$.[1] q.e.d.

Corollary 1: *Let $\mu$ be the spectral measure of the representation $W_{\mathfrak{A}}$ of $\mathfrak{A}^X$. If* dim $X \geq 2$ *then for every distributions $\varphi_1 \neq \varphi_2$ of the form* (6) *the measures $\mu(\cdot + \varphi_1)$ and $\mu(\cdot + \varphi_2)$ are mutually singular.*

It follows from Lemma 2, indeed, that $\mu$ is a Gaussian measure in $(\mathfrak{A}^X)'$ with the correlation functional $B(a_1, a_2) = \langle da_1, da_2 \rangle$. As we have seen, $\varphi_1 - \varphi_2$ is not an element of the completion $\overline{\mathfrak{A}}^X$ of $\mathfrak{A}^X$ in the norm $\|a\| = \langle da, da \rangle^{1/2}$. Therefore, the statement is an immediate consequence of Lemma 3, §2.

Corollary 2: *The spectral measure $\mu$ of the representation $W_{\mathfrak{A}}$ and the spectral measure $\mu * \nu_n$ of the representation $W_{\mathfrak{A}} \otimes S^n V_m$ $(n = 1, 2, \ldots)$ are mutually singular.*

Indeed, in view of Lemma 4 and 5, $\nu_n(\overline{\mathfrak{A}}^X) = 0$, *and the statement is an immediate consequence of Lemma* 4, §2.

Lemma 6: *Let $\mathfrak{U}_{\mathfrak{A}}$ be a weakly closed operator algebra generated by operators of the representation $W$ of the group $\mathfrak{A}^X : (W(a)\Phi)(F) = e^{i\langle F, da \rangle}\Phi(V^{-1}(\exp a)F)$. If* dim $X \geq 2$, *then the algebra $\mathfrak{U}_{\mathfrak{A}}$ contains every shift operator $\Phi(F) \mapsto \Phi(V^{-1}(\exp a)F)$, $a \in \mathfrak{A}^X$, and consequently, every operator $\Phi(F) \mapsto e^{i\langle F, da \rangle}\Phi(F)$, $a \in \mathfrak{A}^X$.*

Proof: Let $\mu$, $\nu$ be the spectral measures of the representations

[1] If dim $X \geq 2$, $W_2^1(X) \not\subset C(X)$.

$W_{\mathfrak{A}}$, $W_{\mathfrak{m}}$ of $\mathfrak{A}^X$ composing the tensor product $W = W_{\mathfrak{A}} \otimes W_{\mathfrak{m}}$ introduced in Lemma 1. It follows from Lemma 4 and Lemma 5 (Corollary 1) that the measures $\mu(\cdot + \varphi_1)$ and $\mu(\cdot + \varphi_2)$ are mutually singular for almost every (with respect to measure $\nu$) pair of functions $\varphi_1 \neq \varphi_2$ from $(\mathfrak{A}^X)'$. Hence, by Lemma 5, §2 (Corollary), the weakly closed operator algebra generated by the operators $W(a)$, $a \in \mathfrak{A}^X$ contains operators $E \otimes W_{\mathfrak{m}}(a)$, $a \in \mathfrak{A}^X$ ($E$ stands for unity operator). Finally, observe that the operator $E \otimes W_{\mathfrak{m}}(a)$ is the shift operator $\Phi(F) \mapsto \Phi(V^{-1})(\exp a)F)$.

LEMMA 7: *Let $\tau_1$, $\tau_2$ be two different Riemannian metrics on $X$ and $W^1$, $W^2$ the corresponding representations given by* (1) *of the group $\mathfrak{A}^X$. Then the spectral measures of $W^1$, $W^2$ are mutually singular.*

We shall reduce now Lemma 7 to a simpler proposition. Let us start with the remark that if $Y \subset X$ is an arbitrary neighbourhood where the metrics $\tau_1$ and $\tau_2$ do not coincide, it suffices to prove lemma for the restrictions of the representations $W^1$, $W^2$ on a subgroup $\mathfrak{A}^Y \subset \mathfrak{A}^X$. Therefore, one can without loss of generality assume that $X$ is a unit ball.

We decompose now the representation $W^i$ $(i = 1, 2)$ into a tensor product: $W^i = W^i_{\mathfrak{A}} \otimes W^i_{\mathfrak{m}}$ (Lemma 1), and let $\mu^i$, $\nu^i$ be the spectral measures of the representations $W^i_{\mathfrak{A}}$ and $W^i_{\mathfrak{m}}$ correspondingly. Recall that, by Lemma 2, $\mu^i$ is a Gaussian measure in $(\mathfrak{A}^X)'$ with the zero mean and the correlation functional $B^i(a, a) = \langle da, da \rangle_i$, where $\langle , \rangle_i$ is an inner product in $\Omega^1(X; \mathfrak{A})$ induced in $X$ by $\tau_i$ $(i = 1, 2)$.

It is true that if the measures $\mu^1$, $\mu^2$ are mutually singular, the same holds for the spectral measures $\mu^1 * \nu^1$ and $\mu^2 * \nu^2$ of the representations $W^1$, $W^2$. It is known, on the one hand, that if two Gaussian measures $\mu^1$, $\mu^2$ are mutually singular, then so are any shifts $\mu^1(\cdot + \varphi_1)$, $\mu^2(\cdot + \varphi_2)$ of these measures (see, for example, [15], pp. 117–118). On the other hand, in view of Lemma 4 (Corollary), the measures $\nu^1$ and $\nu^2$ are equivalent. Hence, the measures $\mu^1 * \nu^1$ and $\mu^2 * \nu^2$ are mutually singular. Therefore, to prove the Lemma we need to show the mutual singularity of $\mu^1$ and $\mu^2$.

At the end, let us remark that $\mathfrak{A}^X = \mathbf{R}^X_1 \oplus \cdots \oplus \mathbf{R}^X_r$, $\mathbf{R}_i = \mathbf{R}$ ($r = \dim \mathfrak{A}$), and the representations $W^i_{\mathfrak{A}}$ of the group $\mathfrak{A}^X$ are tensor products of the representations of the groups $\mathbf{R}^X_i$. Thus, it suffices to prove the singularity for spectral measures of the representations of each of the latter subgroup $\mathbf{R}^X_i$.

Consequently, making use of the explicit expressions (7) for the correlation functionals $\langle da, da \rangle_i$ we have reduced the proof of Lemma 7 to the proof of the following statement.

PROPOSITION: *Let $X$ be the open unity ball in $\mathbb{R}^m$ and $\tau^1_{kl}(x)$, $\tau^2_{kl}(x)$ $(k, l = 1, \ldots, m)$ be differentiable functions on $X$ with the matrices $\tau^1(x) = \|\tau^1_{kl}(x)\|$, $\tau^2(x) = \|\tau^2_{kl}(x)\|$ positively definite in every $x \in X$. Let*

$$B^i(f, f) = \int_X \left( \sum_{k,l=1}^m \tau^i_{kl}(x) \frac{\partial f}{\partial x^l} \frac{\partial f}{\partial x^l} \right) dx, \quad f \in C^\infty(X) \quad (i = 1, 2).$$

*Suppose $\tau^1(x) \neq \tau^2(x)$. Then the Gaussian measures in $(C^\infty(X))'$ with the zero mean and the correlation functionals $B^1$ and $B^2$ are mutually singular.*

PROOF: Consider the operators

$$B^i = -\sum_{k,l=1}^m \frac{\partial}{\partial x^l} \left( \tau^i_{kl} \frac{\partial}{\partial x^k} \right) \quad (i = 1, 2). \tag{11}$$

in the $a$ space $C^\infty(X)$ of differentiable functions on $X$ with compact support. Note that the inner products $B^1(f, g) = \langle B^1 f, g \rangle$ and $B^2(f, g) = \langle B^2 f, g \rangle$ are mutually equivalent in the space $C^\infty(X)$ and also equivalent to any inner product defined by an elliptic operator of the form (11), and by the operator $\Delta = \sum_{k=1}^m \partial^2/(\partial x^k)^2$, in particular. The completion of $C^\infty(X)$ in the inner product defined by $\Delta$ is the Sobolev space $\mathring{W}^1_2$.

By theorem of Feldman, the Gaussian measures $\mu^1$, $\mu^2$ are equivalent if and only if the operator $B^1 - B^2$ is of Hilbert–Schmidt class with respect to the inner product determined by each of the forms $B^1(f, g)$, $B^2(f, g)$.[1] It is equivalent thing to say that $B^1 - B^2$ is a Hilbert–Schmidt operator in $\mathring{W}^1_2$. But since $B^1 - B^2$ is a differentiation operator of the form (11) again, it is possible in the case $B^1 = B^2$ only. The Proposition is proved.

REMARK: All results of this paragraph can also be formulated without considerable change for the representations $U_{E,\tau}$ of the group $G^X$ determined by arbitrary subbundles $E$ of the tangent bundle $TX$.

[1] Formulated in a different way (see [15], p. 130, Theorem 4), the Feldman's theorem says that the measures $\mu^1$ and $\mu^2$ are equivalent if and only if $B^1 - B^2 = (B^2)^{1/2} \Gamma (B^2)^{1/2}$ with $\Gamma$ a Hilbert–Schmidt operator.

The analogue of Lemma 7 for these representations is the following lemma. Let

$$W_{E_i,\tau_i}(a) = U_{E_i,\tau_i}(\exp a), a \in \mathfrak{A}^X \quad (i = 1, 2).$$

LEMMA 8: *The spectral measures of the representations $W_{E_1,\tau_1}$ and $W_{E_2,\tau_2}$ of the group $\mathfrak{A}^X$ are equivalent if and only if $E_1 = E_2$ and the inner products in the space $H_{E_1} = H_{E_2}$ induced by the Riemannian metrics $\tau_1$, $\tau_2$, coincide.*

(The definitions of the space $H_E$ and the representation $U_{E,\tau}$ see in section 2 of §3).

## §5. Proof of the main theorems

1. To begin with, notice that it suffices to prove theorems 1 and 2 under the assumption that $X$ is an open manifold diffeomorphic to $\mathbb{R}^m$. Indeed, every smooth connected manifold $X$, dim $X = m$, contains an open submanifold $Y$, which is everywhere dense in $X$ and diffeomorphic to $\mathbb{R}^m$. It is evident that the set $\mathcal{F}_Y = (\Omega^1(Y; \mathfrak{G}))'$, is a subset of full Gaussian measure $\mu$ in $\mathcal{F} = (\Omega^1(X; \mathfrak{G}))'$. Hence, we have the coincidence $L^2_\mu(\mathcal{F}_Y) = L^2_\mu(\mathcal{F})$ and thus the assertions of Theorems 1 and 2 about representations of $G^X$ reduce to those of representations of $G^Y$.

Let, therefore, $G$ be a compact semisimple Lie group, $X$ an open connected manifold, and let $U = \mathrm{EXP}_\beta V$ be the unitary representation of $G^X$ in the space $L^2_\mu(\mathcal{F})$ defined in section 1 of §3:

$$(U(\tilde{g})\Phi)(F) = e^{i\langle F, \beta\tilde{g}\rangle}\Phi(V^{-1}(\tilde{g})F).$$

LEMMA 1: *If* dim $X \geq 2$, *then the cyclic subspace $H \subset L^2_\mu(\mathcal{F})$ of the group $G^X$ generated by the vacuum vector $\Omega \in L^2_\mu(\mathcal{F})$ is irreducible.*

PROOF: Let $\mathfrak{A} \subset \mathfrak{G}$ be an arbitrary Cartan subalgebra, $\mathfrak{m}$ its orthogonal complementation in $\mathfrak{G}$. Let us decompose the space $L^2_\mu(\mathcal{F})$ into the tensor product $L^2_\mu(\mathcal{F}) = L^2_{\mu_\mathfrak{A}}(\mathcal{F}_\mathfrak{A}) \otimes L^2_{\mu_\mathfrak{m}}(\mathcal{F}_\mathfrak{m})$ according to Lemma 1, §4. Let $H_\mathfrak{A}$ denote the space of all functionals $\Phi \in L^2_\mu(\mathcal{F})$ such that $\Phi(\cdot + F_\mathfrak{m}) = \Phi(\cdot)$ for every $F_\mathfrak{m} \in \mathcal{F}_\mathfrak{m}$. It is also evident that $H_\mathfrak{A}$ is invariant under the action of operators $W(a) = U(\exp a)$, $a \in \mathfrak{A}^X$, and the restriction of the representation $W$ of $\mathfrak{A}^X$ to the subspace $H_\mathfrak{A}$ is equivalent to the representation $W_\mathfrak{A}$ (see Lemma 1, §4).

It follows from Lemma 3, §4 (Corollary) and Lemma 5, §4 (Corollary 2) that the restrictions of the representation $W$ of the group $\mathfrak{A}^X$ to the subspace $H_{\mathfrak{A}}$ and its orthogonal complement are disjoint. Consequently, every bounded linear operator $C$ in $L^2_\mu(\mathcal{F})$ which commutes with the operators $U(\tilde{g})$, $\tilde{g} \in G^X$ and, therefore, with the operators $W(a)$, $a \in \mathfrak{A}^X$ leaves the subspace $H_{\mathfrak{A}}$ invariant.

Let now $\mathfrak{A}_1$ be another Cartan subalgebra in $\mathfrak{G}$, such that $\mathfrak{A} \cap \mathfrak{A}_1 = 0$ (it certainly exists), let $\mathfrak{m}_1$ be the orthogonal complementation to $\mathfrak{A}_1$ in $\mathfrak{G}$, $H_{\mathfrak{A}_1} \subset L^2_\mu(\mathcal{F})$ a subspace corresponding to $\mathfrak{A}_1$. We shall verify that $H_{\mathfrak{A}} \cap H_{\mathfrak{A}_1} = \{c\Omega\}$. Indeed, if $\mathfrak{A} \cap \mathfrak{A}_1 = 0$, then $\mathfrak{m} + \mathfrak{m}_1 = \mathfrak{G}$ and $\mathcal{F}_{\mathfrak{m}} + \mathcal{F}_{\mathfrak{m}_1} = \mathcal{F}$. Hence, if $\Phi \in H_{\mathfrak{A}} \cap H_{\mathfrak{A}_1}$, that is $\Phi(\cdot + F_{\mathfrak{m}}) = \Phi(\cdot + F_{\mathfrak{m}_1}) = \Phi(\cdot)$ for every $F_{\mathfrak{m}} \in \mathcal{F}_{\mathfrak{m}}$, $F_{\mathfrak{m}_1} \in \mathcal{F}_{\mathfrak{m}_1}$, then $\Phi(\cdot + F) = \Phi(\cdot)$ for every $F \in \mathcal{F}$ and, consequently, $\Phi = \text{const}$.

Let $C$ be an arbitrary bounded linear operator in $L^2_\mu(\mathcal{F})$ which commutes with the operators $U(\tilde{g})$, $\tilde{g} \in G^X$. Since $C$, as was demonstrated above, leaves both subspaces $H_{\mathfrak{A}}, H_{\mathfrak{A}_1}$ invariant and since $H_{\mathfrak{A}} \cap H_{\mathfrak{A}_1} = \{c\Omega\}$, then $C\Omega = c\Omega$. The statement of the Lemma follows.

Lemma 2. *If* $\dim X \geq 2$, *then the weakly closed operator algebra* $\mathfrak{U}$ *generated by the operators* $U(\tilde{g})$, $\tilde{g} \in G^X$ *in the space* $L^2_\mu(\mathcal{F})$, *contains the operators of multiplication by the functionals of the form*

$$F \mapsto e^{i\langle F, \Sigma_{k=1}^n V(\tilde{g}_k) du_k\rangle}, \; \tilde{g}_k \in G^X, \; u_k \in \mathfrak{A}^X, \; n = 1, 2, \ldots.$$

Proof: In view of Lemma 6, §4, $\mathfrak{U}$ contains the operators of multiplication by $e^{i\langle\cdot, da\rangle}$, $a \in \mathfrak{A}^X$, where $\mathfrak{A}$ is an arbitrary Cartan subalgebra in $\mathfrak{A}$. Consequently, $\mathfrak{U}$ also contains every operator $A_u$ of multiplication by $e^{i\langle\cdot, du\rangle}$, $u \in \mathfrak{G}^X$. It is evident that the operator of multiplication by $e^{i\langle F, \Sigma_{k=1}^n V(\tilde{g}_k) du_k\rangle}$ equals to the product of the operators $U(\tilde{g}_k) A_{u_k} U^{-1}(\tilde{g}_k)$ and, therefore, is also an element of $\mathfrak{U}$.

Lemma 3. *The set* $M = \{\Sigma_{k=1}^n V(\tilde{g}_k) du_k | \tilde{g}_k \in G^X, \; u_k \in \mathfrak{G}^X, \; n = 1, 2, \ldots\}$ *is dense in the space* $\Omega^1(X; \mathfrak{G})$ *of* 1*-forms in the norm introduced there.*

Proof: The representation $V$ of $G^X$ in the space $\Omega^1(X; \mathfrak{G})$ is a continual direct sum $V = \int_X^{\oplus} \tilde{V}^x dx$, $\tilde{V}^x$ being a representation in the space $H^x$ of 1-forms in $x \in X$ taking their values in $\mathfrak{G}$.

It is evident that these representations $\tilde{V}^x$ are pairwise disjoint. In the meantime one proves that the component of the set $M$ in every space $H^x \cong \oplus^m \mathfrak{G}$ ($m = \dim X$) coincides with $H^x$. It follows that $M$ is dense in $\Omega^1(X;\mathfrak{G})$.

REMARK: It is a consequence of Lemma 3 that the set $\{\beta\tilde{g}|\tilde{g} \in G^X\}$ is a total one in $\Omega^1(X;\mathfrak{G})$ with respect to the norm introduced there. This fact can be, however, established directly.

PROOF OF THEOREM I. Let $\mathcal{H} \subset L^2_\mu(\mathcal{F})$ be the cyclic subspace for the group $G^X$ which is generated by the vacuum vector $\Omega$. In view of Lemma I, $\mathcal{H}$ is irreducible. By Lemma 2, $\mathcal{H}$ contains all the functionals $e^{i\langle\cdot,\omega\rangle}$, $\omega \in M$, where $M = \{\Sigma_{k=1}^n V(\tilde{g}_k)du_k | \tilde{g}_k \in G^X,\ u_k \in \mathfrak{G}^X,\ n = 1, 2, \ldots\}$. Using Lemma 3, $M$ is dense in $\Omega^1(X;\mathfrak{G})$ and, therefore, the functionals $e^{i\langle\cdot,\omega\rangle}$, $\omega \in M$ form the set total in $L^2_\mu(\mathcal{F})$. Consequently, $\mathcal{H} = L^2_\mu(\mathcal{F})$ and Theorem I is proved.

PROOF OF THEOREM 2. Let $\tau_1, \tau_2$ be different Riemannian metrics on $X$, $U^1$ and $U^2$ the corresponding representations of the group $G^X$. Consider the representations of the group $\mathfrak{A}^X$, where $\mathfrak{A}$ a Cartan subalgebra of $\mathfrak{G}$: $W^i(a) = U^i(\exp a)$, $i = 1, 2$. In view of Lemma 7, §4, the representations $W^1$ and $W^2$ are disjoint and, consequently, the same is true for the representations $U^1$ and $U^2$ of $G^X$. The Theorem is proved.

Along these lines Theorem 4 of the equivalence conditions for the representations $U_{E_1,\tau_1}$, $U_{E_2,\tau_2}$ of $G^X$ can be deduced from Lemma 8, §4.

2. *Spherical function of the representation* $U = \mathrm{EXP}_\beta V$. We show here that *if $G$ is a compact semisimple Lie group and* $\dim X \geq 2$, *then the vacuum vector $\Omega$ is invariantly defined in the space of representation* $U = \mathrm{EXP}_\beta V$ *of* $G^X$.

Let us, indeed, consider the subspace $H_\mathfrak{A} \subset L^2_\mu(\mathcal{F})$ introduced in the proof of Lemma I. We have established above that (a) $H_\mathfrak{A}$ is invariant with respect to the representation $W(a)$ of the group $\mathfrak{A}^X$; (b) the spectral measure of the representation $W(a)$ in the subspace $H_\mathfrak{A}$ is a Gaussian measure in $(\mathfrak{A}^X)'$ with the zero mean; (c) the restrictions $W(a)$ to the subspace $H_\mathfrak{A}$ and its orthogonal complementation are mutually disjoint. On the contrary, it easily follows from the results of §4 that the subspace $H_\mathfrak{A} \subset L^2_\mu(\mathcal{F})$ is determined in invariant way by the conditions (a), (b) and (c).

Since the intersection of the subspaces $H_{\mathfrak{A}_i}$, where $\mathfrak{A}_i$ runs over different Cartan subalgebras in $\mathfrak{G}$, contains the multiples of $\Omega$ only, then the vector $\Omega$ is also invariantly determined, up to a multiplier, in the space of representation.

We shall call *the spherical function of the representation* $U = \mathrm{EXP}_\beta V$ the following function on $G^X$:

$$\psi(\tilde{g}) = \langle U(\tilde{g})\Omega, \Omega\rangle.$$

It follows from the definition of operators $U(\tilde{g})$ that

$$\psi(\tilde{g}) = \exp\left(-\tfrac{1}{2}\|\beta\tilde{g}\|^2\right).$$

The representation $U$ is uniquely determined by its sperical function and it is a consequence of the invariant definition of the vacuum vector that equivalent representations $U$ determine equal spherical functions $\psi$. One easily deduces from this fact the statement of Theorem 2 in the case $\dim X \geq 2$.

## §6. Extension of the group $\theta^1(X; G)$ of section of 1-jet fibre bundle and its representations

In the papers of physical character [18], [3] and others the so-called Sugawara algebra, accompanying a Lie algebra, is considered. It turns out that Sugawara algebra is the Lie algebra of a certain infinite-dimensional group which can be described in terms of the present paper. Let us give the precise definitions.

Let $X$ be a Riemannian manifold, $G$ a real Lie group such that its Lie algebra $\mathfrak{G}$ possesses an inner product invariant under the adjoint action of the group $G$. Let us consider the group

$$\theta^1(X; G) = G^X \cdot \Omega^1(X; \mathfrak{G})$$

of all differentiable sections with compact support of the 1-jet fibre bundle $j^1(X; G) \to X$ (see §I). The group $\theta^1(X; G)$ is acting in the space $\Omega^1(X; \mathfrak{G})$ of differentiable $\mathfrak{G}$-valued 1-forms on $X$ with compact support.

Let us introduce two cocycles of the group $\theta^1(X; G)$ taking values in $\Omega^1(X; \mathfrak{G})$:

$$\alpha(\tilde{g}, \tilde{a}) = \tilde{a},$$
$$\tilde{\beta}(\tilde{g}, \tilde{a}) = \beta\tilde{g} \quad (\tilde{g} \in G^X, \tilde{a} \in \Omega^1(X; \mathfrak{G})),$$

where $\beta$ is the Maurer–Cartan cocycle.

Define for any elements $f_1=(\tilde{g}_1, \tilde{a}_1)$ and $f_2=(\tilde{g}_2, \tilde{a}_2)$ of $\theta^1(X; G)$:

$$\gamma(f_1, f_2)(x) = \langle(\tilde{\beta} f_1^{-1})(x), (\alpha f_2)(x)\rangle_x,$$

$$\gamma_0(f_1, f_2) = \langle\tilde{\beta} f_1^{-1}, \alpha f_2\rangle = \int_X \gamma(f_1, f_2)(x)dx.$$

It is clear that for every $f_1, f_2, f_3 \in \theta^1(X; G)$ the following is true:

$$\gamma(f_1, f_2) + \gamma(f_1 f_2, f_3) = \gamma(f_1, f_2 f_3) + \gamma(f_2, f_3).$$

Therefore, $\gamma$ is a 2-cocycle of the group $\theta^1(X; G)$ with values in $\mathbb{R}^X$. Similarly, $\gamma_0$ is a 2-cocycle of $\theta^1(X; G)$ with values in $\mathbb{R}$.

Let us verify the non-triviality of 2-cocycles $\gamma$ and $\gamma_0$. Suppose, for example, that the 2-cocycle $\gamma$ is trivial. It means that there exists a mapping $c: \theta^1(X; G) \to \mathbb{R}^X$ such that

(1) $$\gamma(f_1, f_2) = c(f_1 f_2) - c(f_1) - c(f_2)$$

for every $f_1, f_2 \in \theta^1(X; G)$. Let $\mathfrak{A} \subset \mathfrak{G}$ be an arbitrary abelian subalgebra, $A \subset G$ a corresponding abelian subgroup. Consider the abelian subgroup $\theta^1(X; A) \subset \theta^1(X; G)$. It follows from (1) that $\gamma(f_1, f_2) = \gamma(f_2, f_1)$ for every $f_1, f_2 \in \theta^1(X; A)$. In the meantime it is clear that $\gamma(f_1, f_2) \neq \gamma(f_2, f_1)$ on $\theta^1(X; A)$. Therefore, $\gamma \not\sim 0$. One proves the non-triviality of the cocycle $\gamma_0$ likewise.

The cocycles $\gamma$ and $\gamma_0$ define nontrivial extensions of the additive groups $\mathbb{R}^X$ and $\mathbb{R}$ by means of the group $\theta^1(X; G)$. Let $S(X; G)$ and $S^0(X; G)$, correspondingly, denote these extensions. Thus, the elements of the group $S(X; G)$ are the pairs $(f, c)$, $f \in \theta^1(X; G)$, $c \in \mathbb{R}^X$ with the multiplication rule

$$(f_1, c_1)(f_2, c_2) = (f_1 f_2, c_1 + c_2 + \gamma(f_1, f_2));$$

the elements of the group $S^0(X; G)$ are the pairs $(f, c)$, $f \in \theta^1(X; G)$, $c \in \mathbb{R}$ with the multiplication rule

$$(f_1, c_1)(f_2, c_2) = (f_1 f_2, c_1 + c_2 + \gamma_0(f_1, f_2)).$$

Note that the group $S^0(X; G)$ is isomorphic to the factor group of $S(X; G)$ with the subgroup of elements type $(1, c)$, $c \in \mathbb{R}^X$ with $\int_X c(x)dx = 0$.

To indicate the connection of the groups constructed above with Sugawara algebra, let us suppose $G$ a compact Lie group. Then Lie algebra of the group $S^0(\mathbb{R}^4; G\times G)$ is Sugawara àlgebra, its generators being described in [3].

Let us construct now representations of the groups $S(X;G)$ and $S^0(X;G)$. Let $\mathcal{F}=(\Omega^1(X;\mathfrak{G}))'$, and $\mu$ be the standard Gaussian measure on $\mathcal{F}$. According to Sect. 3, §2 one defines a unitary representation of the group $S^0(X;G)$ in the space $L^2_\mu(\mathcal{F})$ for a pair of cocycles $\alpha,\tilde\beta$ of the group $\theta^1(X;G)$ by the following:

$$(U(f,c)\Phi)(F)=e^{is(c+\langle F,\alpha f\rangle)+\frac{1}{2}\langle F,\tilde\beta f\rangle-\frac{1}{4}\|\tilde\beta f\|^2}\Phi(V^{-1}(\tilde g)F+\tilde\beta f^{-1}), \tag{2}$$

where $f=(\tilde g,\tilde a)\in\theta^1(X;G)$ and $s\neq 0$ is a real parameter. If $G$ is a compact semisimple Lie group and $\dim X\geq 2$, then, making use of Theorem I, one proves that the representation (2) is irreducible.

As the group $S^0(X;G)$ is isomorphic to the factor group of $S(X;G)$, the representation (2) of the group trivially extends to a representation of the group $S(X;G)$, the result being the representation of $S(X;G)$ in the space $L^2_\mu(\mathcal{F})$ given by

$$\begin{aligned}(\tilde U(f,c)\Phi)(F)&=e^{is(\int_X c(x)dx+\langle F,\alpha f\rangle)+\frac{1}{2}\langle F,\tilde\beta f\rangle-\frac{1}{4}\|\tilde\beta f\|^2}\\ &\quad\times\Phi(V^{-1}(\tilde g)F+\tilde\beta f^{-1}).\end{aligned} \tag{3}$$

EXAMPLE: $X=\mathbb{R}$, $G=\mathbb{R}^+$. In this case the elements of the group $S(X;G)$ are the triples $(g(x), a(x), c(x))$, $x\in\mathbb{R}$ with the multiplication rule:

$$(g_1,a_1,c_1)(g_2,a_2,a_2)=\left(g_1+g_2,a_1+a_2,c_1+c_2+\frac{dg_1}{dx}a_2\right).$$

The factor group of $S(X;G)$ with subgroup $G_0$ of elements $(g,0,0)$ with $g=\text{const.}$, is isomorphic to the functional Heisenberg group with $\delta'$-commutation law. The representation (3) of $S(X;G)$ is trivial on $G_0$ and therefore it defines the representation of this Heisenberg group.

Authors are grateful to A. Lodkin for his English translation of the paper.

## REFERENCES

[1] S. ALBEVERIO, R. HOEGH-KROHN: *The Energy Representation of Sobolev-Lie Groups.* Universität Bielefeld, 1976.

[2] H. ARAKI: Factorisable representations of current algebra. *Publ. of RIMS., Kyoto Univ., Ser. A., 5, no. 3* (1970) 361.

[3] S. COLEMAN, D. GROSS, R. JACKIW: Fermion avatars of the Sugawara Model. *Phys. Rev., 180, no. 5* (1969) 1359.

[4] N. DUNFORD and J.T. SCHWARTZ: *Linear Operators, Part 1: General Theory.* Interscience Publ., N.-Y; 1958.

[5] P. DELORME: *1-Cohomologie des Groupes localement compacts et Produits tensoriels continus de Representations.* These de 3° cicle, Paris, 1975.

[6] I.M. GELFAND, M.I. GRAEV: Representations of the group of quaternions over local compact and functional fields. *Funct. Anal. and its Appl., 2, no. 1* (1968) 20.

[7] I.M. GELFAND, N.Ya. VILENKIN: Generalized functions, Vol. 4: *Applications of Harmonic Analysis.* Academic Press, New York, 1964.

[8] A. GUICHARDET: Symmetric Hilbert spaces and related topics. *Lect. Notes in Math., no. 261,* Springer, Berlin, 1972.

[9] A. GUICHARDET: Cogomologie des groupes localement compacts et produits tensoriels continus de representations. Paris, 1975 (preprint).

[10] A. GUICHARDET: *Representations de $G^X$ selon Gelfand et Delorme.* Séminare Bourbaki, 28-e année 1975/76, no. 486 (1976).

[11] R.S. ISMAGILOV: On unitary representations of the group $C_0^\infty(X, G)$, $G = SU_2$: *Mat. Sbornik, 100, no. 1* (1976) 117.

[12] K. ITO: Multiple Wiener integral. *J. Math. Soc. Japan, 3* (1951) 157.

[13] K.R. PARTHASARATHY and K. SCHMIDT: Positive definite kernels, continuous tensor products and central limit theorems of probability theory. *Lect. Notes in Math., no. 272,* Springer, Berlin, 1972.

[14] K.R. PARTHASARATHY and K. SCHMIDT: A new method for constructing factorisable representations for current groups and current algebras. *Commun. Math. Phys; 51, no. 1* (1976) 1.

[15] A.V. SKOROHOD: *Integration in a Hilbert Spaces.* Moscow, "Nauka", 1975 (Russian).

[16] S.L. SOBOLEV: *Some Applications of the Functional Analysis in Mathematical Physics.* Leningrad, 1950 (Russian).

[17] R.F. STREATER: Current commutation relations, continuous tensor products and infinitely divisible group representations. *Rendiconti di Sc. Int. di Fisica E. Fermi, 11* (1969) 247.

[18] H. SUGAWARA: A field theory of currents. *Phys. Rev. 170, no. 5* (1968) 1659.

[19] A.M. VERŠIK: On the theory of normal dynamic systems. *Doklady Acad. Nauk USSR, 144, no. 1* (1962) 9.

[20] A.M. VERŠIK, I.M. GELFAND and M.I. GRAEV: Representations of the group SL (2, R), where R is a ring of functions. *Russ. Math. Surv. 28, no. 5* (1973) 83.

[21] A.M. VERŠIK, I.M. GELFAND and M.I. GRAEV: Irreducible representations of the group $G^X$ and cogomology. *Funct. Anal. and its Appl. 8, no. 2* (1974) 67.

[22] A.M. VERŠIK, I.M. GELFAND and M.I. GRAEV: Representations of the group of diffeomorphisms. *Uspehi Matemat. Nauk, 30, no. 6* (1975) 3.

[23] N. WIENER: *Nonlinear Problems in Random Theory.* Wiley, N.-Y, 1958.

(Oblatum 14–XII–1976)

Institute of Applied Mathematics
Academy of Sciences USSR
Moscow, 125047, Miusskaya pl., 4
Leningrad State University

# 3.

## (with M. I. Graev and A. M. Vershik)

# Irreducible representations of the group $G^X$ and cohomologies

Funkts. Anal. Prilozh. **8** (2) (1974) 67–69 [Funct. Anal. Appl. **8** (1974) 151–153].
MR **50**:530. Zbl. **299**:22004.

## 1. Representation of $EXP_\beta T$

Let $\Gamma$ be a topological group and let T be a representation of $\Gamma$ in the real Hilbert space H. We shall assume that all the representations considered are orthogonal, i.e., that the scalar product is conserved. Let $\beta:\Gamma\to H$ be a 1-cocycle, i.e., a continuous mapping satisfying the relation $\beta(\gamma_1\gamma_2)=\beta(\gamma_1)\beta(\gamma_2)$. In [1, 5] a new representation U of $\Gamma$ is constructed for the pair (T, $\beta$). We shall cite this construction.

The representation U acts in the Hilbert space $\mathrm{EXP}\,H=\mathbf{R}\oplus H\oplus S^2H\oplus\ldots\oplus S^nH\oplus\ldots$, where $S^nH$ is the n-th symmetrized tensor order of H (i.e., the subspace of tensor order $\overset{n}{\otimes}H$ generated by the vectors $v\otimes\ldots\otimes v$). In order to define the operators of the representation explicitly, we introduce the vectors $\mathrm{EXP}\,v=1\oplus v\oplus(2!)^{-1/2}v\otimes v\oplus(3!)^{-1/2}v\otimes v\otimes v\oplus\ldots$, where $v\in H$. It has been proved [3] that these vectors belong to EXP H, are linearly dependent, and that linear combinations of them are everywhere dense in EXP H. For every $\gamma\in\Gamma$ and $v\in H$ we put

$$U(\gamma)\,\mathrm{EXP}\,v=\exp\left(-\frac{1}{2}\|\beta(\gamma)\|^2-(T(\gamma)v,\beta(\gamma))\right)\mathrm{EXP}\,(T(\gamma)v+\beta(\gamma)). \tag{1}$$

It has been proved that the operators $U(\gamma)$ may be extended to linear operators on the whole space EXP H and that the operators so constructed on EXP H form a representation of group $\Gamma$.

We denote the representation so constructed by $EXP_\beta T$ and call it the canonical representation of $\Gamma$ corresponding to the pair (T, $\beta$). It is not difficult to verify that the homolog 1-cocycles $\beta_1:\Gamma\to H$ and $\beta_2:\Gamma\to H$ correspond to equivalent representations $EXP_{\beta_1}T$ and $EXP_{\beta_2}T$.

We shall use this construction here to construct irreducible representations of the group $G^X$. It leads to interesting and important representations, however, not only for infinite groups but also in the case of classical groups – for representations of semisimple Lie groups [for example $\Gamma=SL(2,\mathbf{R})$] (see §7 in [5]).

## 2. The Group $G^X$

Let G be a topological group, and let X be a compact topological space without isolated points. We introduce the topological group $G^X$, the elements of which are all possible continuous mappings $f:X\to G$, the group operation is defined as pointwise multiplication: $(f_1f_2)(x)=f_1(x)f_2(x)$, and the topology as the topology of uniform convergence. We shall assume that for any finite subset $\{x_1,\ldots,x_k\}\subset X$ ($x_i\neq x_j$ for $i\neq j$) and any $g_1,\ldots,g_k\in G$ there exists $f\in G^X$ such that $f(x_i)=g_i$, $i=1,\ldots,k$.

## 3. Construction of Representations of $G^X$

Let the representation T of the group G in real Hilbert space H and the 1-cocycle $\beta:G\to H$ be given. Further, let the continuous denumerably additive positive finite Borel measure m be given on X. For each $x_0\in X$ we define the representation $T^{x_0}$ of $G^X$ in the space $H: T^{x_0}(f)=T(f(x_0))$. Now let $T^X=\int\oplus T^x\,dm(x)$.

---

Leningrad State University. Institute of Applied Mathematics, Academy of Sciences of the USSR.
Translated from Funktsional'nyi Analiz i Ego Prilozheniya, Vol. 8, No. 2, pp. 67–69, April–June, 1974. Original article submitted December 10, 1973.

This means that the Hilbert space $H^X$ of mappings $F: X \to H$ for which $\int \|F(x)\|^2 dm(x) < \infty$ is given; the representation $T^X$ acts in $H^X$ according to the expression $(T^X F)(x) = T(f(x)) F(x)$. We associate with the 1-cocycle $\beta: G \to H$ the 1-cocycle $\tilde{\beta}: G^X \to H^X$ defined as follows: $[\tilde{\beta}(f)](x) = \beta(f(x))$. We introduce the following canonical representation of the group $G^X$: $U^{m, T, \beta} = \mathrm{EXP}_{\tilde{\beta}} T^X$.

THEOREM 1. Let $K \subset G$ be a fixed compact subgroup and let the cocycle $\beta: G \to H$ vanish identically on K. We assume that: 1) the vectors $\beta(g)$, $g \in G$ generate H, and 2) H contains no nonzero vectors invariant with respect to K. Then the representation $U^{m,T,\beta}$ of $G^X$ is irreducible.

THEOREM 2. The representations $U^{m_1, T_1, \beta_1}$ and $U^{m_2, T_2, \beta_2}$ are equivalent if and only if there exists $c \neq 0$ such that: 1) $m_2 = c^2 m_1$, and 2) there exists an isomorphism $\nu: H_1 \to H_2$ of the spaces of representations $T_1$, $T_2$ such that $\nu \circ T_1(g) = T_2(g) \circ \nu$ and $\nu \beta_1(g) = c\beta_2(g)$ for any $g \in G$.

Remark 1. The representation $U^{m,T,\beta}$ may be extended to the representation of the semidirect product $G^X \cdot \mathscr{X}$, where $\mathscr{X}$ is a group of homeomorphisms $\psi: X \to X$ conserving the measure m.

Remark 2. Theorems 1 and 2 remain valid if the real space EXP $H^X$ is replaced by the complex space $\mathbf{C} \otimes_{\mathbf{R}} \mathrm{EXP}\, H^X$.

In Sec. 5 (Lemma) other irreducible representations of $G^X$ will be defined which together with the representations $U^{m,T,\beta}$ form an additive basis of some ring of representations of $G^X$.

We note that the construction of the canonical representation U of $G^X$ is carried out in two stages: first of all, using the representation T of G and the cocycle $\beta$, the representation $T^X = \int \oplus T^x dm(x)$ of $G^X$ and the cocycle $\tilde{\beta}$ are constructed; then the canonical representation $U = \mathrm{EXP}_{\tilde{\beta}} T^X$ is constructed for the pair $(T^X, \tilde{\beta})$. The representation U may also be obtained by another method: first of all the representation T of G and the cocycle $\beta$ are used to construct the canonical representation $\mathrm{EXP}_\beta T$ of G, and then the continuous tensor product $\bigotimes_X (\mathrm{EXP}_\beta T)^x dm(x)$ with respect to measure m of the representations $(\mathrm{EXP}_\beta T)^x$ is considered (for the definition of the continuous tensor product of representations see, e.g., [3]). It can be proved that these constructions coincide, i.e.,

$$\mathrm{EXP}_{\tilde{\beta}}\left(\int \oplus T^x \, dm(x)\right) \cong \bigotimes_X (\mathrm{EXP}_\beta T)^x \, dm(x).$$

We note that by the definition of $\mathrm{EXP}_\beta T$ in the space of each representation $(\mathrm{EXP}_\beta T)^x$ the vector EXP0 is fixed.

## 4. Examples

a) $G = SO(n, 1)$ (the group of linear transformations in $R^{n+1}$ with determinant 1 conserving the quadratic form $x_1^2 + \ldots + x_n^2 - x_{n+1}^2$); $K = O(n) \subset G$. Let $H^0$ be the space of real continuous functions $\varphi(\omega)$ on the sphere $|\omega|^2 = \omega_1^2 + \ldots + \omega_n^2 = 1$ satisfying the condition $\int \varphi(\omega)\, d(\omega) = 0$, where $d\omega$ is an invariant measure on the sphere. We define H as the completion of $H^0$ with respect to the norm $\|\varphi\|^2 = \int \ln(1 - (\omega, \omega'))\, \varphi(\omega) \cdot \varphi(\omega')\, d\omega\, d\omega'$. In H there acts the representation T of group G: $(T(g)\varphi)(\omega) = \varphi(\omega') |\omega_1 g_{1,n+1} + \ldots + \omega_n g_{n,n+1} + g_{n+1,n+1}|^{1-n}$, where $\omega_i' = (\omega_1 g_{1i} + \ldots + \omega_n g_{ni} + g_{n+1,i})(\omega_1 g_{1,n+1} + \ldots + \omega_n g_{n,n+1} + g_{n+1,n+1})^{-1}$. We put $\beta(g, \omega) = |\omega_1 g_{1,n+1} + \ldots + \omega_n g_{n,n+1} + g_{n+1,n+1}|^{1-n} - 1$. Then $\beta$ is the 1-cocycle $G \to H$ and the pair $(T, \beta)$ satisfies the conditions of Theorem 1.

b) $G = SU(n, 1)$ (the group of linear transformations in $C^{n+1}$ with determinant 1 conserving the Hermitian form $z_1\bar{z}_1 + \ldots + z_n\bar{z}_n - z_{n+1}\bar{z}_{n+1}$); $K = U(n) \subset G$. We define $H^0$ as the space of real continuous functions $\varphi(\omega)$ on the sphere $|\omega|^2 = |\omega_1|^2 + \ldots + |\omega_n|^2 = 1$ satisfying the condition $\int \varphi(\omega) d\omega = 0$; let H be the completion of $H^0$ with respect to the norm $\|\varphi\|^2 = \int \ln|1 - (\omega, \omega')|\, \varphi(\omega)\, \varphi(\omega')\, d\omega\, d\omega'$, where $(\omega, \omega') = \omega_1\bar{\omega}_1' + \ldots + \omega_n\bar{\omega}_n'$. In H there acts the representation T of the group G: $(T(g)\varphi)(\omega) = \varphi(\omega') |\omega_1 g_{1,n+1} + \ldots + \omega_n g_{n,n+1} + g_{n+1,n+1}|^{2-2n}$, where $\omega_i'$ is defined just as for $G = SO(n, 1)$. We put $\beta(g, \omega) = |\omega_1 g_{1,n+1} + \ldots + \omega_n g_{n,n+1} + g_{n+1,n+1}|^{2-2n} - 1$. Then $\beta$ is the 1-cocycle $G \to H$ and the pair $(T, \beta)$ satisfies the conditions of Theorem 1.

The special case $G = SO(2, 1) \cong PSU(1, 1)$ was worked out earlier in [5].

## 5. Expansion of Tensor Products of Representations into Irreducibles

Here we shall assume that the pair X, G satisfies the following condition. For every $g_1, g_2 \in G$ there will be found a compact subset $F \subset G$ such that for any pair of closed subsets $X_1, X_2 \subset X$, $X_1 \cap X_2 = \phi$ there exists a function $f \in G^X$ such that $f(X_1) = g_1$, $f(X_2) = g_2$, and $f(X) \subset F$.

LEMMA. Let the pair $(T, \beta)$ satisfy the conditions of Theorem 1 and let $T_1, \ldots, T_n$ be arbitrary irreducible representations of the group G. We choose fixed pairwise distinct points $x_1, \ldots, x_n \in X$. Then the representation

$$U^{m,T,\beta} \otimes T_1^{x_1} \otimes \ldots \otimes T_n^{x_n} \tag{2}$$

is irreducible.

THEOREM 3. Assume that the measures $m_1, m_2$ are equivalent and let the representation T of G be expanded into a discrete direct sum of irreducible representations: $T = \bigoplus_i T_i$. Then the tensor product $U^{m_1,T,\beta} \otimes U^{m_2,T,\beta}$ is decomposed into irreducible representations as follows:

$$U^{m_1,T,\beta} \otimes U^{m_2,T,\beta} \cong U^{m,T,\beta} \oplus \sum_{k=1}^{\infty} \sum_{i_1,\ldots,i_k} \int_{\tilde{X}^k} \oplus (U^{m,T,\beta} \otimes T_{i_1}^{x_1} \otimes \ldots \otimes T_{i_k}^{x_k})\, dm(x_1) \ldots dm(x_k),$$

where $m = m_1 + m_2$; the integration is effected over an arbitrary fundamental domain $\tilde{X}^k \subset X^k$ with respect to the permutation group of the coordinates $x_1, \ldots, x_k$.

The expansion of the tensor products of two representations of form (2) is defined analogously.

THEOREM 4. Let us assume that G is in a group of type 1 and the pair $(T, \beta)$ satisfies the conditions of Theorem 1. We consider a ring of representations of the group $G^X$ generated by representations of form (2) where m runs through the set of nonnegative Borel continuous measures on X, and $T_i$ is any irreducible representation of the group G. Then the representations of form (2) form an additive basis in this ring.

The results of this note may be transferred without essential changes to representations T in complex Hilbert space. The unitary representations $U^{m,T,\beta}$ are, generally speaking, projective representations.

## LITERATURE CITED

1. R. F. Streater, Nuovo Cimento, 53A, 487 (1968).
2. H. Araki, Publ. of RIMS Kyoto Univ., 5, No. 3, 361–422 (1970).
3. A. Guichardet, Lecture Notes in Math., Springer (1972).
4. K. R. Parthasarathy and K. Schmidt, Lecture Notes in Math., Springer (1972).
5. A. M. Vershik, I. M. Gel'fand, and M. I. Graev, Uspekhi Mat. Nauk, 28, No. 5, 83-128 (1973).

# 4.

## (with M. I. Graev and A. M. Vershik)

# Representations of the group of diffeomorphisms

Usp. Mat. Nauk **30** (6) (1975) 3–50 [Russ. Math. Surv. **30** (6) (1975 1–50].
MR **53**:3188. Zbl. **317**:58009

**This article contains a survey of results on representations of the diffeomorphism group of a non-compact manifold $X$ associated with the space $\Gamma_X$ of configurations (that is, of locally finite subsets) in $X$. These representations are constructed from a quasi-invariant measure $\mu$ on $\Gamma_X$. In particular, necessary and sufficient conditions are established for the representations to be irreducible. In the case of the Poisson measure $\mu$ a description is given of the corresponding representation ring.**

Contents

## Introduction

This article is a survey of results on the representations of the group Diff $X$ of finite diffeomorphisms of a smooth non-compact manifold $X$. As for many infinite groups, it is rather difficult to see what the complete stock of irreducible unitary representations of this group might be. Therefore, it is of some interest to single out certain natural classes of representations.

We consider the space $\Gamma_X$ of infinite configurations (that is, locally finite subsets) in $X$ on which the group Diff $X$ acts in a natural way. If $\mu$ is a quasi-invariant measure in $\Gamma_X$ and $\rho$ is a representation of the symmetric group $S_n (n = 1, 2, \ldots)$, then we construct a unitary representation of Diff $X$ from $\mu$ and $\rho$, which we call elementary. There is, therefore, a close connection between the theory of elementary representations of Diff $X$ and the theories of quasi-invariant measures on $\Gamma_X$ and representations of the symmetric groups. We note that quasi-invariant measures on $\Gamma_X$ are studied in statistical physics (Gibbs measures and the simplest of them – the Poisson measure) (see, for example, [12]); and in the theory of point processes (see, for example, [17] and elsewhere). The space of infinite configurations $\Gamma_X$ is, in its own right, a very important example of an infinite-dimensional manifold, and its study is one of the interesting problems of topology, analysis, and statistical physics.

Representations of Diff $X$ that are of finite functional dimension, that is, representations associated with the space of finite configurations, were considered in [8] and [9]. In §1 we incidentally prove by a new method that the representations of a wide class are irreducible. However, we are basically interested in representations of infinite functional dimension associated with $\Gamma_X$; they can be regarded as limits of "partially finite" representations.

In this paper necessary and sufficient conditions are obtained for elementary representations to be irreducible. In the case when $\mu$ is the Poisson measure it is proved that the set of elementary representations is multiplicatively closed, that is, the tensor product of two elementary representations splits into the sum of elementary representations, and the structure of the corresponding representation ring is described.

An important property of Diff $X$, which distinguishes it from locally compact groups and which will become apparent in the situations we discuss, is that to a single orbit of Diff $X$ in $\Gamma_X$ there is no corresponding representation; however, one can construct a representation from a measure on $\Gamma_X$ that is ergodic with respect to the action of Diff $X$. More interesting and more widely studied is the class of representations associated with the Poisson measure on $\Gamma_X$ (see §4). The representation of Diff $X$ in the space $L^2_\mu(\Gamma_X)$, where $\mu$ is the Poisson measure, arose (as an $N/V$ limit) in [15]; however, the role of the Poisson measure was not noted here. It is remarkable that this same representation can be realized in a Fock space as $\mathrm{EXP}_\beta T$, where $T$ is a representation of Diff $X$ in $L^2_m(X)$ and $\beta$ is a certain cocycle (see §4). This circumstance links the theory that we discuss here with [1] and [2];

Representations of Diff $X$ associated with the Poisson measure on $\Gamma_X$ are studied by another method in [22]. As far as we know, up to now, no measures, and in particular no Gibbs measures apart from the Poisson measures, have been discussed in connection with representations of Diff $X$.

Representations of the cross product of the additive group $C^\infty(X)$ and the group Diff $X$ are investigated in a number of very interesting physics papers (see [15] for a list of references; see also [19] and [20]). All the representations of Diff $X$ discussed in this article extend to representations of the cross product $C^\infty(X) \cdot \mathrm{Diff}\, X$; the mathematical part of the results of [15] is contained in this paper.

## §0. Basic definitions and some preliminary information

**1. The group Diff** $X$. Everywhere, $X$ denotes a connected manifold of class $C^\infty$. Diff $X$ denotes the group of all diffeomorphisms $\psi: X \to X$ that are the identity outside a compact set (depending on $\psi$). The group Diff $X$ is assumed to be furnished with the natural topology: a sequence $\psi_n$ is regarded as tending to $\psi$ if $\psi$ and every $\psi_n$ is the identity outside a certain compact set $K$ and if $\psi_n$, together with all its derivatives, tends to $\psi$ uniformly on $K$. If $Y \subset X$ is an arbitrary open subset, then Diff $Y$ denotes the subgroup of diffeomorphisms $\psi \in \mathrm{Diff}\, X$ that are the identity on $X \setminus Y$.

**2. The groups** $S^\infty$ **and** $S_\infty$. We denote by $S^\infty$ the group of all permutations of the sequence of natural numbers, by $S_\infty \subset S^\infty$ the subgroup of all finite permutations, and by $S^n$ the group of all permutations of the numbers $1, \ldots, n$ $(n = 1, 2, \ldots)$. We regard the $S_n$ as subgroups of $S_\infty$; thus, $S_1 \subset \ldots \subset S_n \subset \ldots$ and $S_\infty = \varinjlim S_n$. In what follows, $S_0$ is understood to mean the trivial group.

**3. The configuration spaces** $\Gamma_X$ **and** $\mathrm{B}_X$. Any locally finite subset of $X$ is called a configuration[1] in $X$, that is a subset $\gamma \subset X$ such that $\gamma \cap K$ is finite for any compact set $K \subset X$. By this definition, any configuration is either a finite or a countable subset of $X$; if $X$ is compact, then all configurations in $X$ are finite.

Let us denote by $\Gamma_X$ the space of all infinite and by $\mathrm{B}_X$ the space of all finite configurations in $X$. The group of diffeomorphisms Diff $X$ acts naturally on $\Gamma_X$ and $\mathrm{B}_X$. The space of finite configurations $\mathrm{B}_X$ decomposes

[1] Sometimes a configuration is defined differently, allowing points $x \in X$ to be included in $\gamma$ with repetitions; with such a definition a configuration is not a subset of $X$.

into a countable union of subsets that are transitive under Diff $X$: $B_X = \bigsqcup_{n\geq 0} B_X^{(n)}$, where $B_X^{(n)}$ is the collection of all $n$-point subsets in $X$. We note that $B_X^{(0)}$ consists of a single element – the empty set $\phi$.

For any subset $Y \subset X$ with compact closure the space $\Gamma_X$ splits into the product $\Gamma_X = B_Y \times \Gamma_{X\setminus Y}$. Consequently, since $B_Y = \bigsqcup_{n\geq 0} B_Y^{(n)}$, we have $\Gamma_X = \bigsqcup_{n\geq 0} B_Y^{(n)} \times \Gamma_{X\setminus Y}$, and all the subsets in this decomposition are invariant under Diff $Y$.

**4. The space $\widetilde{X}^\infty$ and the topology in $\Gamma_X$.** Let us consider the infinite product $X^\infty = \prod_{i=1}^{\infty} X_i$, $X_i = X$, furnished with the weak topology. The group $S^\infty$ acts naturally on $X^\infty$. We define the subset $\widetilde{X}^\infty \subset X^\infty$ as the set of all sequences $(x_1, \ldots, x_n, \ldots) \in X^\infty$ such that: 1) $x_i \neq x_j$ when $i \neq j$ and 2) the sequence $x_1, \ldots, x_n, \ldots$ has no accumulation points in $X$. The space $\widetilde{X}^\infty$ is invariant under the action of Diff $X$ and $S^\infty$, and the $S^\infty$-orbit of any point of $\widetilde{X}^\infty$ is closed.

There is a natural bijection $\widetilde{X}^\infty/S^\infty \to \Gamma_X$. We introduce the corresponding quotient topology in $\Gamma_X$; this topology is Hausdorff and metrizable. Similarly, the bijections $\widetilde{X}^n/S_n \to B_X^{(n)}$, where

$$\widetilde{X}^n = \{(x_1, \ldots, x_n) \in X^n;\ x_i \neq x_j \text{ when } i \neq j\},$$

and $S_n$ acts on $X^n$ as the permutation group of the coordinates, give the topology on $B_X^{(n)} (n = 1, 2, \ldots)$ and hence on $B_X = \bigsqcup_{n\geq 0} B_X^{(n)}$.

It is easy to see that $\Gamma_X$ is, as a topological space, the projective limit of the spaces $B_K$. Namely, $\Gamma_X = \varprojlim (B_K, \pi_{KK'})$, where $K$ runs through the open submanifolds in $X$ with compact closures, and $\pi_{KK'}$: $B_K \to B_{K'} (K' \subset K)$ is the restriction of the configuration $\gamma \in B_K$ to $K'$, that is, $\pi_{KK'}\, \gamma = \gamma \cap K'$.

**5. Quasi-invariant and ergodic measures.** Let $G$ be a group acting on a space $Y$. A measure $\mu$ given on some $G$-invariant $\sigma$-algebra in $Y$ is said to be quasi-invariant under $G$ if the inverse image of any measurable set of positive measure, under any transformation $g$: $Y \to Y$ with $g \in G$, has positive measure. If $\mu$ is quasi-invariant, then the measures $\mu$ and $g\mu$ (where $g\mu$ is defined as the image of $\mu$, that is, $g\mu(C) = \mu(g^{-1}C)$) are equivalent; the density of $g\mu$ with respect to $\mu$ at a point $y \in Y$ is denoted by $\frac{d\mu(g^{-1}y)}{d\mu(y)}$. The class of $\sigma$-finite measures equivalent to $\mu$ is called the type of $\mu$.

A quasi-invariant measure $\mu$ in $Y$ is said to be ergodic with respect to the action of $G$ if every measurable set $A \subset Y$ such that $\mu(gA \,\Delta\, A) = 0$ for any $g \in G$ is either a null set or a set of full measure.

We discuss measures on $\Gamma_X$, and other spaces connected with $\Gamma_X$, that are quasi-invariant under the action of Diff $X$, and we construct from these measures unitary representations of Diff $X$.

**6. Measures in the configuration space $\Gamma_X$.** We define,[1] as usual, the $\sigma$-algebra $\mathfrak{A}(\Gamma_X)$ of Borel sets on $\Gamma_X$. Henceforth, when we talk of measures on $\Gamma_X$, we mean[2] complete, non-negative, Borel, normalized, countably-additive measures $\mu$. Since the structure of a complete metric space can be introduced in $\Gamma_X$, for any Borel measure $\mu$ the space $(\Gamma_X, \mu)$ is (after taking the completion of the $\sigma$-algebra $\mathfrak{A}(\Gamma_X)$ with respect to $\mu$) a Lebesgue space [11], and the technique of conditional decomposition (conditional measures, and so on) can be applied. The same applies to other spaces and fibre bundles over $\Gamma_X$ that occur in this paper.

Many measures on $\Gamma_X$ arising for various reasons in statistical physics and probability theory turn out to be quasi-invariant and ergodic under Diff $X$. The following example is classical.

POISSON MEASURE. Given any positive[3] smooth measure $m$ on a manifold $X$, we consider the union $\Delta_X = B_X \cup \Gamma_X$ of all configurations on $X$. We define the measure of each subset $\{\gamma \in \Delta_X; |\gamma \cap U| = n\}$ by

$$\mu\{\gamma \in \Delta_X;\ |\gamma \cap U| = n\} = \frac{[\lambda m(U)]^n}{n!} e^{-\lambda m(U)},$$

where $\lambda > 0$ is fixed. By Kolmogorov's theorem there exists a unique measure on $\mathfrak{A}(\Gamma_X)$ defined by these conditions. It is called the Poisson measure with parameter $\lambda$ (associated with the measure $m$ on $X$).

Let us note the following important properties of the Poisson measure $\mu$, which follow immediately from its definition.

1) When $m(X) < \infty$, the measure $\mu$ is concentrated on the set $B_X$ of finite configurations, and when $m(X) = \infty$, it is concentrated on $\Gamma_X$.

2) Suppose that the manifold $X = X_1 \cup \ldots \cup X_n$ is split arbitrarily into finitely many disjoint measurable subsets, that $\Delta_X = \Delta_{X_1} \times \ldots \times \Delta_{X_n}$ is the corresponding decomposition of $\Delta_X$ into a direct product, and that $\mu_i$ is the projection of the Poisson measure $\mu$ onto $\Delta_{X_i}$ $(i = 1, \ldots, n)$. Then $\mu = \mu_1 \times \ldots \times \mu_n$. This property of the Poisson measure is called *infinite decomposability*.

3) The Poisson measure is quasi-invariant under Diff $X$ and invariant under the subgroup Diff$(X, m) \subset$ Diff $X$ of diffeomorphisms preserving $m$. Here,

---

1 Note that $\mathfrak{A}(\Gamma_X)$ is $\sigma$-generated by sets of the form $C_{U,n} = \{\gamma \in \Gamma_X; |\gamma \cap U| = n\}$, where $U$ runs over the compact sets in $X$ $(n = 0, 1, \ldots)$.

2 In statistical physics a measure $\mu$ on $\Gamma_X$ is usually called a state (see, for example, [12]) and in probability theory and the theory of mass observation it is usually called a point random process (see, for example, [17]).

3 By a positive smooth measure we mean a measure with positive density at all points $x \in X$.

$$(1) \qquad \frac{d\mu(\psi^{-1}\gamma)}{d\mu(\gamma)} = \prod_{x\in\gamma} \frac{dm(\psi^{-1}x)}{dm(x)}$$

(the product makes sense, because by the finiteness of $\psi$, almost all the factors are equal to 1).

4) If $m(X) = \infty$, then the Poisson measure $\mu$ is ergodic with respect to Diff $X$. Furthermore (see §4), if dim $X > 1$, then the Poisson measure is ergodic with respect to Diff $(X, m)$.

Any measure in $B_X$ that is quasi-invariant under Diff $X$ is equivalent to a sum of smooth positive measures on $B_X^{(n)}$. In particular, any two quasi-invariant measures on $B_X^{(n)}$ are equivalent. Let us note that for any $Y \subset X$, where $\overline{Y}$ is compact, the projection of any quasi-invariant measure in $\Gamma_X$ onto $B_Y = \bigsqcup_{n\geq 0} B_Y^{(n)}$ is non-zero for all $n$.

## §1. The ring of representations of Diff $X$ associated with the space of finite configurations

We discuss here the simplest class of representations of Diff $X$. These representations have finite functional dimension; from the point of view of orbit theory they have been discussed in detail by Kirillov [9].

1. **The representations** $V^\rho$. We associate with each pair $(n, \rho)$, where $\rho$ is a unitary representation of the symmetric group $S_n$ in a space $W (n = 0, 1, \dots)$, a unitary representation $V^\rho$ of Diff $X$. The construction of $V^\rho$ is similar to Weyl's construction of the irreducible finite-dimensional representations of the general linear group.

Given a positive smooth measure $m$ on $X$, we define $m_n$ in $X^n$ to be the product measure: $m_n = m \times \dots \times m$. We consider the space $L^2_{m_n}(X^n, W)$ of functions $F$ on $X^n$ with values in the representation space $W$ of $\rho$ such that

$$\|F\|^2 = \int \|F(x_1, \dots, x_n)\|_W^2 \, dm(x_1) \dots dm(x_n) < \infty.$$

A unitary representation $U_n$ of Diff $X$ is given on $L^2_{m_n}(X^n, W)$ by the formula

$$(1) \qquad (U_n(\psi)F)(x_1, \dots, x_n) = \prod_{k=1}^{n} J_\psi^{1/2}(x_k)\, F(\psi^{-1}x_1, \dots, \psi^{-1}x_n),$$

where $J_\psi(x) = \dfrac{dm(\psi^{-1}x)}{dm(x)}$. Let us denote by $H_{n,\rho}$ the subspace of functions $F \in L^2_{m_n}(X^n, W)$ such that $F(x_{\sigma(1)}, \dots, x_{\sigma(n)}) = \rho^{-1}(\sigma)F(x_1, \dots, x_n)$ for any $\sigma \in S_n$. It is obvious that $H_{n,\rho}$ is invariant under Diff $X$.

We define the representation $V^\rho$ of Diff $X$ as the restriction of $U_n$ from $L^2_{m_n}(X^n, W)$ to $H_{n,\rho}$.

In the particular case when $\rho$ is the unit representation of $S_n$, then $V^\rho$ acts by (1) on the space of scalar functions $F(x_1, \ldots, x_n)$ that are symmetric in all the arguments.

It is obvious that if $m$ is replaced on $X$ by any other smooth positive measure, each $V^\rho$ is replaced by an equivalent representation.

Let us construct another realization of $V^\rho$, which will be useful later on. Let $\widetilde{X}^n \subset X^n$ be the submanifold of points $(x_1, \ldots, x_n) \in X^n$ with pairwise distinct coordinates. We consider the fibration $p$ of $\widetilde{X}^n$ by the orbits of $S_n$, $p: \widetilde{X}^n \to \mathrm{B}_X^{(n)}$. Note that $p \circ \psi = \psi \circ p$ for any $\psi \in \mathrm{Diff}\, X$. Suppose that we are given any measurable cross section $s: \mathrm{B}_X^{(n)} \to \widetilde{X}^n$. Obviously, for any $\psi \in \mathrm{Diff}\, X$ and $\gamma \in \mathrm{B}_X^{(n)}$ the elements $s(\psi^{-1}\gamma)$ and $\psi^{-1}(s\gamma)$ lie in the same fibre of $p$, and we define a function $\sigma$ on $\mathrm{Diff}\, X \times \mathrm{B}_X^{(n)}$ with values in $S_n$ by the formula $s(\psi^{-1}\gamma) = [\psi^{-1}(s\gamma)]\,\sigma(\psi, \gamma)$, where[1] $(x_1, \ldots, x_n)\sigma = (x_{\sigma(1)}, \ldots, x_{\sigma(n)})$.

Let $\mu = pm_n$ be the projection onto $\mathrm{B}_X^{(n)}$ of the measure $m_n = m \times \ldots \times m$ on $X^n$. We denote by $L^2_\mu(\mathrm{B}_X^{(n)}, W)$ the space of functions $F$ on $\mathrm{B}_X^{(n)}$ with values in $W$ such that

$$\| F \|^2 = \int \| F(\gamma) \|^2_W \, d\mu(\gamma) < \infty.$$

We define the representation $V^\rho$ of $\mathrm{Diff}\, X$ in $L^2_\mu(\mathrm{B}_X^{(n)}, W)$ by

$$(V^\rho(\psi)F)(\gamma) = \left(\frac{d\mu(\psi^{-1}\gamma)}{d\mu(\gamma)}\right)^{1/2} \rho(\sigma(\psi, \gamma))\, F(\psi^{-1}\gamma). \tag{2}$$

It is not difficult to check that this representation is equivalent to the one constructed earlier. To see this it is sufficient to consider the map $s^*: H_{n,\sigma} \to L^2_\mu(\mathrm{B}_X^{(n)}, W)$ induced by the cross section $s$, $((s^*F)(\gamma) = F(s\gamma))$. It is easy to verify that $s^*$ is an isomorphism and that the operators $V^\rho(\psi)$ in $H_{n,\rho}$ go over under $s^*$ to operators of the form (2).

In the particular case when $\rho$ is the unit representation of $S_n$, then $V^\rho$ acts on $L^2_\mu(\mathrm{B}_X^{(n)})$ according to the formula

$$(V^\rho(\psi)F)(\gamma) = \left(\frac{d\mu(\psi^{-1}\gamma)}{d\mu(\gamma)}\right)^{1/2} F(\psi^{-1}\gamma).$$

**2. Properties of the representations $V^\rho$.** From the definition of $V^\rho$ we obtain immediately the following result.

PROPOSITION 1. *For any representations $\rho_1$ and $\rho_2$ of $S_n$ $(n = 0, 1, \ldots)$ there is an equivalence* $V^{\rho_1 \oplus \rho_2} \cong V^{\rho_1} \oplus V^{\rho_2}$.

DEFINITION (see [18]). The *exterior product* $\rho_1 \circ \rho_2$ of representations $\rho_1$ of $S_{n_1}$ and $\rho_2$ of $S_{n_2}$ is the representation of $S_{n_1+n_2}$ induced by the

---

[1] $\sigma$ is a 1-cocycle of Diff $X$ with values in the set of measurable maps $\mathrm{B}_X^{(n)} \to S_n$ (see Appendix 2).

representation $\rho_1 \times \rho_2$ of $S_{n_1} \times S_{n_2}$: $\rho_1 \circ \rho_2 = \mathrm{Ind}_{S_{n_1} \times S_{n_2}}^{S_{n_1+n_2}} (\rho_1 \times \rho_2)$. We are assuming that $S_{n_1}$ and $S_{n_2}$ are embedded in $S_{n_1+n_2}$ as the subgroups of permutations of $1, \ldots, n_1$ and of $n_1 + 1, \ldots, n_n + n_2$, respectively. Note (see [18]) that exterior multiplication is commutative and associative. The following fact parallels standard results about representations of the classical groups in the Weyl realization.

PROPOSITION 2. *For any* $n_1, n_2 = 0, 1, 2, \ldots$ *and any representations* $\rho_1$ *and* $\rho_2$ *of* $S_{n_1}$ *and* $S_{n_2}$, *respectively, there is an equivalence* $V^{\rho_1 \circ \rho_2} \cong V^{\rho_1} \otimes V^{\rho_2}$.

COROLLARY. *The set of representations* $V^\rho$ *is closed under the operation of tensor multiplication.*

THEOREM 1. 1) *If* $\rho$ *is an irreducible representation of* $S_n$, *then the representation* $V^\rho$ *of* Diff $X$ *is irreducible.* 2) *Two representations* $V^{\rho_1}$ *and* $V^{\rho_2}$, *where* $\rho_1$ *and* $\rho_2$ *are irreducible representations of* $S_{n_1}$ *and* $S_{n_2}$, *respectively, are equivalent if and only if* $n_1 = n_2$ *and* $\rho_1 \sim \rho_2$.

PROOF. We consider $V^{\rho_n}$, where $\rho_n$ is the regular representation of $S_n$ $(n = 0, 1, 2, \ldots)$. It is easy to see that $V^{\rho_n}$ is equivalent to the representation in $\overset{n}{\otimes} L^2_m(X)$ given by

$$(V^{\rho_n}(\psi) F)(x_1, \ldots, x_n) = \prod_{k=1}^{n} J_\psi^{1/2}(x_k) F(\psi^{-1}x_1, \ldots, \psi^{-1}x_n).$$

Results of Kirillov ([9], Theorem 4) imply that the $V^{\rho_n}$ are pairwise disjoint and that the number of interlacings of $V^{\rho_n}$ is $n!$, that is, equal to the number of interlacings of $\rho_n$. Hence and from Proposition 1 the assertion of the theorem follows immediately.

When dim $X > 1$, a stronger assertion is true, which we prove independently of the results in [9]. Namely, let $m$ be an arbitrary smooth positive measure on $X$ such that $m(X) = \infty$. We denote by Diff$(X, m)$ the subgroup of diffeomorphisms $\psi \in$ Diff $X$ that leave $m$ invariant.

THEOREM 2. *If* dim $X > 1$, *then the assertion of Theorem* 1 *is true for the restrictions of the* $V^\rho$ *to* Diff$(X, m)$.

The proof will depend on the following two assertions.

LEMMA 1. *For any natural number n and any set of distinct points* $x_1, \ldots, x_n$ *in X there exist neighbourhoods* $O_1, \ldots, O_n$, *corresponding to* $x_1, \ldots, x_n$, *with the following properties*:

1) *the closure* $\bar{O}_1$ *of* $O_i$ *is* $C^\infty$*-diffeomorphic to a disc,* $\bar{O}_i \cap \bar{O}_j = \phi$ *when* $i \neq j$ *and* $m(O_1) = \ldots = m(O_n)$;

2) *for any permutation* $(k_1, \ldots, k_n)$ *of* $1, \ldots, n$ *there is a diffeomorphism* $\psi \in$ Diff$(X, m)$ *such that* $\psi(\bar{O}_i) = \bar{O}_{k_i}$ $(i = 1, \ldots, n)$.

PROOF. It is sufficient to consider the case when $X$ is an open ball and $m$ is the Lebesgue measure in $X$. In this case it is easy to check that for any $x_i$ and $x_j$, $i \neq j$, there is a diffeomorphism $\psi_{ij} \in$ Diff$(X, m)$ with the following properties:

1) for any sufficiently small $\varepsilon > 0$ we have
$\psi_{ij} D^{\varepsilon}_{x_i} = D^{\varepsilon}_{x_j}$, $\psi_{ij} D^{\varepsilon}_{x_j} = D^{\varepsilon}_{x_i}$, where $D^{\varepsilon}_{x}$ is a disc of radius $\varepsilon$ with centre at $x \in X$;

2) the diffeomorphism $\psi_{ij}$ is the identity in neighbourhoods of $x_k$ for which $k \neq i, j$.

Hence the assertion of the lemma follows immediately.

LEMMA 2. *For any open connected submanifold $Y \subset X$ compact closure, the subspace $\widetilde{L}^2_m(Y) \subset L^2_m(Y)$ of functions $f$ on $Y$ such that $\int_Y f(y)dm(y) = 0$ is irreducible under the operators of the representation of* $\mathrm{Diff}(Y, m)$: $(U(\psi)f(y) = f(\psi^{-1}y)$.

PROOF. First we claim that for any non-trivial invariant subspace $\mathcal{L} \subset \widetilde{L}^2_m(Y)$ and any neighbourhood $O \subset Y$, where $O$ is $C^\infty$-diffeomorphic to a disc, there is a vector $f \in \mathcal{L}$, $f \neq 0$, such that $\mathrm{supp}\, f \subset O$. For let us take an arbitrary vector $f^{(1)} \in \mathcal{L}$, $f^{(1)} \neq 0$. Since $f^{(1)} =$ const on $Y$, there is a $y_0 \in Y$ such that $f^{(1)} \neq$ const in any neighbourhood $O'$ of $y_0$. Consequently, there exists a diffeomorphism $\psi \in \mathrm{Diff}(Y, m)$ such that $\mathrm{supp}\, \psi \subset O'$ and $f^{(1)}(\psi y) \not\equiv f^{(1)}(y)$. We put $f^{(2)}y = f^{(1)}(\psi y) - f^{(1)}(y)$. Then $f^{(2)} \in \mathcal{L}$, $f^{(2)} \neq 0$ and $\mathrm{supp}\, f^{(2)} \subset O'$. If the neighbourhood $O'$ is sufficiently small, then, by Lemma 1, there is a diffeomorphism $\psi_1 \in \mathrm{Diff}(Y, m)$ with $\psi_1 O' \subset O$ that carries $f^{(2)}$ into a vector $f$ with $\mathrm{supp}\, f \subset O$.

Let us suppose that $\widetilde{L}^2_m(Y) = \mathcal{L}_1 \oplus \mathcal{L}_2$, where $\mathcal{L}_1$ and $\mathcal{L}_2$ are non-zero invariant subspaces. We fix neighbourhoods $O$ and $O'$ in $Y$ such that $\bar{O}$ and $\bar{O}'$ are $C^\infty$-diffeomorphic to discs, $\bar{O} \cap \bar{O}' = \phi$, and $m(O) = m(O')$. From what has been proved, there are $f_i \in \mathcal{L}_i$, $f_i \neq 0$, such that $\mathrm{supp}\, f_i \subset O$ $(i = 1, 2)$.

It is obvious that we can find a neighbourhood $O_1 \subset O$, where $\bar{O}_1$ is $C^\infty$-diffeomorphic to a disc, and a diffeomorphism $\psi \in \mathrm{Diff}(O, m)$ such that $\int_{O_1} f_1(\psi y)\overline{f_2(y)}dm(y) \neq 0$; without loss of generality we may assume that $\psi = 1$. For any $\varepsilon > 0$ we can write $O = O_1 \cup O^\varepsilon \cup (O \setminus (O_1 \cup O^\varepsilon))$, where $\bar{O}^\varepsilon$ is $C^\infty$-diffeomorphic to a disc, $\bar{O}_1 \cap \bar{O}^\varepsilon = \phi$, and $m(O \setminus (O_1 \cup O^\varepsilon)) < \varepsilon$. It is not difficult to prove that there is a diffeomorphism $\psi_\varepsilon \in \mathrm{Diff}(Y, m)$ that is the identity on $O_1$ and such that $m(\psi_\varepsilon O^\varepsilon \setminus O') < \varepsilon$ (see, for example, [3], Lemma 1.1). Since $O \cap O' = \phi$, we have

$$\int_O f_1(\psi_\varepsilon y)\overline{f_2(y)}\,dm(y) = \int_{O_1} f_1(y)\overline{f_2(y)}\,dm(y) + \int_{O^\varepsilon \setminus \psi_\varepsilon^{-1}O'} f_1(\psi_\varepsilon y)\overline{f_2(y)}\,dm(y) + \int_{O \setminus (O_1 \cup O^\varepsilon)} f_1(\psi_\varepsilon y)\overline{f_2(y)}\,dm(y).$$

Consequently, because $\mathscr{L}_1$ and $\mathscr{L}_2$ are orthogonal,

$$\int_{O_1} f_1(y)\overline{f_2(y)}\,dm(y) + \int_{O^\varepsilon \setminus \psi_\varepsilon^{-1}O'} f_1(\psi_\varepsilon y)\overline{f_2(y)}\,dm(y) + \int_{O\setminus(O_1\cup O^\varepsilon)} f_1(\psi_\varepsilon y)\overline{f_2(y)}\,dm(y) = 0.$$

Since the second and third terms in this equation can be made arbitrarily small, we have $\int_{O_1} f_1(y)\overline{f_2(y)}dm(y) = 0$, which is a contradiction.

PROOF OF THEOREM 2. Let us realize the representation $V^\rho = V^{n,\rho}$ of Diff $X$ as acting on the subspace $H_{n,\rho} \subset L^2_{m_n}(X^n, W)$, where $W$ is the space of the representation $\rho$ of $S_n$ ( for the definition of $H_{n,\rho}$, see §1.1). In this realization the operators of the representation of Diff$(X, m)$ have the following form:

$$(V^{n,\rho}(\psi)F)(x_1, \ldots, x_n) = F(\psi^{-1}x_1, \ldots, \psi^{-1}x_n), \quad \psi \in \text{Diff}\,(X, m).$$

Let $O_1, \ldots, O_n$ be arbitrary disjoint neighbourhoods in $X$ satisfying conditions 1 and 2 of Lemma 1. We denote by $H^{n,\rho}_{O_1,\ldots,O_n}$ the subspace of functions of $H_{n,\rho}$ that are concentrated on $\bigcup_{(k_1,\ldots,k_n)} (O_{k_1} \times \ldots \times O_{k_n}) \subset X^n$ where $(k_1, \ldots, k_n)$ runs over all permutations of $(1, \ldots, n)$; obviously there is a natural isomorphism

$$H^{n,\rho}_{O_1,\ldots,O_n} \cong L^2_m(O_1) \otimes \ldots \otimes L^2_m(O_n) \otimes W.$$

We consider the subspace

$$\widetilde{H}^{n,\rho}_{O_1,\ldots,O_n} \cong \widetilde{L}^2_m(O_1) \otimes \ldots \otimes \widetilde{L}^2_m(O_n) \otimes W,$$

where $\widetilde{L}^2_m(O_i) \subset L^2_m(O_i)$ is the orthogonal complement to the subspace of contsants. From the definition it follows that $\widetilde{H}^{n,\rho}_{O_1,\ldots,O_n}$ is invariant under under the subgroup $G_{O_1,\ldots,O_n}$ of diffeomorphisms $\psi \in$ Diff$(X, m)$ such that $\psi(O_1 \cup \ldots \cup O_n) = O_1 \cup \ldots \cup O_n$. We denote by $V^{n,\rho}_{O_1,\ldots,O_n}$ the restriction of the representation $V^{n,\rho}$ of $G_{O_1,\ldots,O_n}$ to $H^{n,\rho}_{O_1,\ldots,O_n}$.

Note that the subgroup $G^0_{O_1,\ldots,O_n} \subset G_{O_1,\ldots,O_n}$ of diffeomorphisms that are the identity on $O_1, \ldots, O_n$ acts trivially on $H^{n,\rho}_{O_1,\ldots,O_n}$ and that by Lemma 1 the factor group $G_{O_1,\ldots,O_n}/G^0_{O_1,\ldots,O_n}$ is isomorphic to the cross product of Diff$(O_1, m) \times \ldots \times$ Diff$(O_n, m)$ with $S_n$. The assertion how follows easily from this and from Lemma 2.

*The representations $V^{n,\rho}_{O_1,\ldots,O_n}$ of $G_{O_1,\ldots,O_n}$, where $\rho$ runs over the inequivalent irreducible representations of $S_n$, are irreducible and pairwise* *inequivalent.*

We now claim that *the representation* $V^{n,\rho}_{O_1,\ldots,O_n}$ *of* $G_{O_1,\ldots,O_n}$ *occurs in* $V^{n,\rho}$ *with multiplicity* 1 *and not at all in representations* $V^{n,\rho'}$, *where* $\rho' \neq \rho$, *nor in representations* $V^{n',\rho'}$, *where* $n' < n$.

For let $H'$ be the orthogonal complement to $\widetilde{H}^{n,\rho}_{O_1,\ldots,O_n}$ in $H_{n,\rho}$. We split $H'$ into the sum of subspaces that are primary with respect to $\mathrm{Diff}(O_1, m)\times \ldots \times \mathrm{Diff}(O_n, m)$. It is not difficult to see that in each of these subspaces at least one of the subgroups $\mathrm{Diff}(O_i, m)$ $(i = 1, \ldots, n)$ acts trivially. But the representation of each subgroup $\mathrm{Diff}(O_i, m)$ in $\widetilde{H}^{n,\rho}_{O_1,\ldots,O_n}$ is a multiple of a non-trivial irreducible representation. Consequently, the representations $V^{n,\rho'}_{O_1,\ldots,O_n}$ are not contained in $H'$, nor for the same reason in $H_{n',\rho}$, $n' < n$.

From the properties of $V^{n,\rho}_{O_1,\ldots,O_n}$ we have just established it follows immediately that the representations $V^{\rho}$ of $\mathrm{Diff}(X, m)$ are pairwise inequivalent. We claim that they are irreducible.

Let $\mathcal{L} \subset H_{n,\rho}$ be a subspace invariant under $\mathrm{Diff}(X, m)$, $\mathcal{L} \neq 0$. Then for any collection $O_1, \ldots, O_n$ of disjoint neighbourhoods satisfying conditions 1 and 2 of Lemma 1 either $\widetilde{H}^{n,\rho}_{O_1,\ldots,O_n} \subset \mathcal{L}$, or $\widetilde{H}^{n,\rho}_{O_1,\ldots,O_n} \cap \mathcal{L} = 0$. It is not difficult to see that the spaces $\widetilde{H}^{n,\rho}_{O_1,\ldots,O_n}$ generate $H_{n,\rho}$, therefore, $\widetilde{H}^{n,\rho}_{O_1,\ldots,O_n} \subset \mathcal{L}$ for some collection $O_1, \ldots, O_n$. But then, by Lemma 1, $\mathcal{L}$ contains the whole of $\widetilde{H}^{n,\rho}_{O_1,\ldots,O_n}$ and hence coincides with $H_{n,\rho}$. The theorem is now proved.

REMARK. Let us denote by $\mathfrak{A}$ the group of all (classes of coinciding mod 0) invertible measurable transformations of $X$ that preserve the measure $m$ (the dimension of $X$ is arbitrary); we furnish $\mathfrak{A}$ with the weak topology. The representation $V^{\rho}$ of $\mathrm{Diff}(X, m) \subset \mathfrak{A}$ extends naturally to a representation of $\mathfrak{A}$ and the resulting representation $V^{\rho}$ of $\mathfrak{A}$ is continuous in the weak topology. It is easy to show that in the weak topology $\mathrm{Diff}(X, m)$, for $\dim X > 1$, is everywhere dense in $\mathfrak{A}$. This makes it possible to prove Theorem 2 anew, reducing its proof to those of the analogous assertions for $\mathfrak{A}$, which are easily verified. On the other hand, this path enables us to establish Theorem 1 for any weakly dense subgroup of $\mathfrak{A}$, that is, to prove the following proposition.

THEOREM 3. *The assertions of Theorem* 1 *are true for the restrictions of the representations* $V^{\rho}$ *of* $\mathfrak{A}$ *to any subgroup* $G \subset \mathfrak{A}$ *that is weakly dense in* $\mathfrak{A}$.

3. **The representation ring** $V^{\rho}$. We consider the free module $\mathcal{R}$ over $\mathbf{Z}$ on the set of all pairwise inequivalent irreducible representations $V^{\rho}$ of Diff $X$ as basis. By the propositions in § 1.2, the tensor product $V^{\rho_1} \otimes V^{\rho_2}$ of irreducible representations $V^{\rho_1}$ and $V^{\rho_2}$ decomposes into a sum of irreducible representations $V^{\rho}$ and therefore is an element of $\mathcal{R}$.

In this way a ring structure is defined in $\mathcal{R}$, where multiplication is the tensor product.

Let us introduce another ring $R(S)$ associated with the representations of the symmetric groups $S_n$ (see [18]). We denote by $R(S_n)$ the free module over $\mathbf{Z}$ on the set of pairwise inequivalent irreducible representations of $S_n$ ($n = 0, 1, 2, \ldots$) as basis (where $R(S_0) = \mathbf{Z}$). We consider the $\mathbf{Z}$-module $R(S) = \bigoplus_{n=0}^{\infty} R(S_n)$ and give a ring structure to $R(S)$ by defining multiplication as the exterior product. From the propositions in §1.2 we obtain immediately the following result.

THEOREM 4. *The ring $\mathcal{R}$ generated by the representations $V^\rho$ of* Diff $X$ *is isomorphic to* $R(S)$.

For the map $\rho \to V^\rho$, where $\rho$ runs over the representations of $S_n$ ($n = 0, 1, \ldots$) extends to a ring isomorphism $R(S) \to \mathcal{R}$.

REMARK. There exists a natural ring isomorphism

$$\theta \colon R(S) \to \mathbf{Z}[a_1, a_2, \ldots],$$

where $a_n$ is the $n$-th elementary symmetric function in an infinite number of unknowns, $n = 1, 2, \ldots$; for the definition of $\theta$ see, for example, [18]. By the theorem we have proved, there is a ring isomorphism $\mathcal{R} \to \mathbf{Z}[a_1, a_2, \ldots]$, where to each representation $V^\rho$ there corresponds the symmetric function $\theta(\rho)$.

These symmetric functions in an infinite number of unknowns have the usual properties of characters: each representation $V^\rho$ is uniquely determined by its symmetric function, on adding two representations their corresponding symmetric functions are added, and on taking the tensor product they are multiplied.

## §2. Quasi-invariant measures in the space of infinite configurations

Before turning to the discussion of representations of Diff $X$ associated with the space of infinite configurations $\Gamma_X$, we ought first of all to study in detail measures in $\Gamma_X$, and in fibrations over it, that are quasi-invariant under Diff $X$. We have already recalled that there are many such measures with various properties (see §0.6); these measures arise (in another connection) in statistical physics, probability theory, and elsewhere.

The ergodic theory for infinite dimensional groups differs in many ways from the theory for locally compact groups (see, for example, [6]). In particular, the action of Diff $X$ in $\Gamma_X$ is such that in $\Gamma_X$ there is no quasi-invariant measure that is concentrated on a single orbit.[1] In addition, care is needed because an infinite-dimensional group can act transitively, but not ergodically, on an infinite-dimensional space [13]. This explains the

[1] For locally compact groups such a measure exists and is equivalent to the transform of the Haar measure on the group.

somewhat lengthy proof of the lemma in §2.1, which at first glance would appear obvious.

1. **Lemma on quasi-invariant measures on $B_Y^{(n)} \times \Gamma_{X-Y}$.** LEMMA 1. *Let $Y \subset X$ be a connected open submanifold with compact closure, let $\mu_n$ be a measure on $B_Y^{(n)} \times \Gamma_{X-Y}$ that is quasi-invariant under the subgroup* Diff $X$, *and let $\mu_n'$ and $\mu_n''$ be the projections of $\mu_n$ onto $B_Y^{(n)}$ and $\Gamma_{X-Y}$, respectively. Then $\mu_n$ is equivalent to $\mu_n' \times \mu_n''$ $(n = 0, 1, 2, \dots)$.*

REMARK. If $\mu_n$ is the restriction to $B_Y^{(n)} \times \Gamma_{X-Y}$ of a fixed quasi-invariant measure $\mu$ on $\Gamma_X$, then the measures $\mu_n''$ on $\Gamma_{X-Y}$ are, generally speaking, not equivalent. It is easy to show that the equivalence of the measures $\mu_n''$ on $\Gamma_{X-Y}$ $(n = 0, 1, 2, \dots)$ corresponds precisely to the equivalence of the measures $\mu$ and $\mu' \times \mu''$ on $\Gamma_X$, where $\mu'$ and $\mu''$ are the projections of $\mu$ onto $B_Y$ and $\Gamma_{X-Y}$.[1]

First we prove the following geometrically obvious proposition.

PROPOSITION 1. *In* Diff $Y$ *there is a countable set of one-parameter subgroups $G_l$ such that the group $G \subset$* Diff $Y$ *generated by them acts transitively in $B_Y^{(n)}$ $(n = 1, 2, \dots)$.*

PROOF. We suppose first that dim $Y = 1$. We specify in $Y$ a countable basis of neighbourhoods $U_r$, $\bar{U}_r \subset Y$, that are diffeomorphic to $\mathbf{R}^1$. We fix for each $r$ a diffeomorphism $\varphi_r$: $\mathbf{R}^1 \to U_r$. Under $\varphi_r$ the group of translations on $\mathbf{R}^1$ goes over into a one-parameter group of diffeomorphisms $x \to f_t(x)$ on $U_r$ $(-\infty < t < \infty)$, which acts transitively on $U_r$. The map $\varphi_r$ can always be chosen so that the diffeomorphisms $x \to f_t(x)$ on $U_r$ extend trivially to a diffeomorphism on the whole of $Y$. It is not difficult to check that the sequence of groups $\{G_l\}$ constructed in this way satisfies the required condition.

Now let dim $Y = p$, where $p > 1$. We specify in $\mathbf{R}^p$ a countable set of one-parameter subgroups $H_l \subset$ Diff $\mathbf{R}^p$ such that the group generated by them acts transitively in $\mathbf{R}^p$; the construction of such a family presents no difficulty.

Now we take a countable basis of neighbourhoods $U_r$ in $Y$, diffeomorphic to $\mathbf{R}^p$, and fix diffeomorphisms $\varphi_r$: $\mathbf{R}^p \to U_r$. Let us denote by $G_{lr}$ the image of $H_l$ under $\varphi_r$. The elements of $G_{lr}$ can be extended trivially to diffeomorphisms over the whole of $Y$, and so $G_{lr}$ can be regarded as a one-parameter subgroup of Diff $Y$.

For a fixed $r$, the subgroups $G_{lr}$ generate a group which acts transitively in $U_r$ and leaves the points of $Y \setminus U_r$ fixed. Hence it is obvious that the group $G \subset$ Diff $Y$ generated by all the $G_{lr}$ acts $n$-transitively in $Y$ $(n = 1, 2, \dots)$, and Proposition 1 is proved.

The following proposition is concerned with the theory of measurable currents of a quasi-invariant measure.

---

[1] When $\mu$ is the Poisson measure, the equivalence $\mu \sim \mu' \times \mu''$ is a direct consequence of the property of being infinitely decomposable. The assertion of the lemma in this case is trivial.

PROPOSITION 2. *Suppose that* $\mathbf{R}^1$ *acts measurably*[1] *on the Lebesgue space* $(X, \mu)$ *with quasi-invariant measure* $\mu$ *and that* $\zeta$ *is a measurable partitioning of* $(X, \mu)$ *that is fixed* mod 0 *under* $\mathbf{R}^1$. *Then for almost all* $C \in \zeta$ *the conditional measures* $\mu^C$ *on* $C$ *are quasi-invariant under* $\mathbf{R}^1$.

PROOF. For any $t \in \mathbf{R}$ and $C \in \zeta$ we put

$$q(t, C) = \inf_A \mu^C (T^t A),$$

where the inf is taken over all subsets $A \subset C$, with $\mu^C(A) = 1$. We also use the notation $q_0(t, C) = q(T, C)q(-t, C)$. Obviously, the condition $\mu^C \sim T^t\mu^C$ is equivalent to $q_0(t, C) = 1$.

Since the action of $\mathbf{R}^1$ on $(X, \mu)$ is measurable and $\zeta$ is a measurable partitioning, $q(t, C)$, and hence also $q_0(t, C)$, are measurable as functions on $\mathbf{R}^1 \times X_\zeta$ $(X_\zeta = X/\zeta)$.

Since $\mu$ is quasi-invariant, for any fixed $t \in \mathbf{R}^1$ we have $\mu^C \sim T^t\mu^C$ for almost all $C \in \zeta$ with respect to the measure $\mu_\zeta$ on $X_\zeta$ ($\mu_\zeta$ is the projection of $\mu$); hence $q_0(t, C) = 1$ almost everywhere with respect to $\mu_\zeta$ on $X_\zeta$. Hence, by Fubini's theorem for $(\mathbf{R}^1 \times X_\zeta, m \times \mu_\zeta)$, where $m$ is the Lebesgue measure on $\mathbf{R}^1$, for almost all $C \in \zeta$ with respect to $\mu_\zeta$ we have: $m\{ t \in \mathbf{R}^1; q_0(t, C) \neq 1 \} = 0$.

On the other hand, the set of $t \in \mathbf{R}^1$ for which $\mu^C \sim T^t\mu^C$ (for a fixed $C$) forms a group. Thus, for almost all $C \in \zeta$ the set $\{ t \in \mathbf{R}^1; q_0(t, C) = 1\}$ is a subgroup of $\mathbf{R}^1$ of full Lebesgue measure. But every subgroup of a locally compact group with full Haar measure coincides with the whole group [4]. Consequently, for almost all $C \in \zeta$ $\{ t \in \mathbf{R}^1; q_0(t, C) = 1 \} = \mathbf{R}^1$, that is, for almost all $C \in \zeta$ the measure $\mu^C$ is quasi-invariant under the action of $\mathbf{R}^1$, and Proposition 2 is proved.

PROOF OF LEMMA 1. Let $\{G_l\}$ be a countable set of one-parameter subgroups of Diff $Y$ such that the group $G$ generated by them acts transitively in $\mathrm{B}_Y^{(n)}$; such a set exists by Proposition 1.

Since $G_l \cong \mathbf{R}^1$, it follows from Proposition 2 that for almost all, (in the sense of $\mu_n''$), configurations $\gamma \in \Gamma_{X-Y}$ the conditional measure $\mu_n^\gamma$ on $\mathrm{B}_Y^{(n)}$ is quasi-invariant under each $G_l$ $(l = 1, 2, \ldots)$, hence also under the whole group $G$ generated by them.

On the other hand, the measures on $\mathrm{B}_Y^{(n)}$ that are quasi-invariant under $G$ are all equivalent to each other, consequently, to $\mu_n'$. For on $\mathrm{B}_Y^{(n)}$, as on every smooth manifold, there is, up to equivalence, a unique measure that is quasi-invariant under a group of diffeomorphisms acting transitively, namely, the smooth measure with everywhere positive density.

---

[1] That is the map $\mathbf{R}^1 \times X \to X$ $((g, x) \to gx)$ is measurable as a map between spaces with measures $m \times \mu$ and $\mu$ respectively, where $m$ is the Lebesgue measure in $\mathbf{R}^1$.

Thus, for almost all $\gamma \in \Gamma_{X-Y}$ (in the sense of $\mu''_n$) the conditional measure $\mu^\gamma_n$ on $B^{(n)}_Y$ is equivalent to $\mu'_n$, and the lemma is proved.

**2. Measurable indexings in $\Gamma_X$.** We say that $i$ is an indexing in $\Gamma_X$ if for each configuration $\gamma \in \Gamma_X$ there is a bijective map $i(\gamma, \cdot)\colon \gamma \to N$, $N = \{1, 2, \ldots\}$.

We denote by $\Gamma_{X,1}$ the subset of elements $(\gamma, x) \in \Gamma_X \times X$ such that $x \in \gamma$, and we associate with each indexing $i$ a bijective map $\Gamma_{X,1} \to \Gamma_X \times N$, defined by $(\gamma, x) \to (\gamma, i(\gamma, x))$. If this map is measurable in both directions (with respect to Borel $\sigma$-algebras on $\Gamma_{X,1}$ and $\Gamma_X \times N$), then the indexing $i$ is called *measurable*.[1]

Let $i$ be a measurable indexing. We introduce a sequence of measurable maps $a_k\colon \Gamma_X \to X$ $(k = 1, 2, \ldots)$ defined by the conditions: $a_k(\gamma) \in \gamma$, $i(\gamma, a_k(\gamma)) = k$ (that is, $a_k(\gamma)$ is the $k$-th element of the configuration $\gamma$). We associate with $i$ a cross section $s\colon \Gamma_X \to \widetilde{X}^\infty$, defined by $s(\gamma) = (a_1(\gamma), \ldots, a_n(\gamma), \ldots)$. It is not difficult to verify that the set $s\,\Gamma_X$ is measurable and that the bijective map $\Gamma_X \to s\Gamma_X$ is measurable in both directions.

For any $\psi \in \operatorname{Diff} X$ and $\gamma \in \Gamma_X$, the elements $s(\psi^{-1}\gamma) \in \widetilde{X}^\infty$ and $\psi^{-1}(s\gamma) \in \widetilde{X}^\infty$ belong to the same $S^\infty$-orbit in $\widetilde{X}^\infty$. We define a map $\sigma\colon \operatorname{Diff} X \times \Gamma_X \to S^\infty$ by $s(\psi^{-1}\gamma) = [\psi^{-1}(s\gamma)]\,\sigma(\psi, \gamma)$; the notation here means $(x_1, \ldots, x_n, \ldots)\,\sigma = (x_{\sigma(1)}, \ldots, x_{\sigma(n)}, \ldots)$.

Let us now introduce the idea of an *admissible indexing.* We are given an increasing sequence $X_1 \subset \ldots \subset X_k \subset \ldots$ of connected open subsets with compact closures such that $X = \bigcup_k X_k$.

DEFINITION. We say that a measurable indexing $i$ is admissible (with respect to the given sequence $X_1 \subset \ldots \subset X_n \subset \ldots$) if the map $\sigma\colon \operatorname{Diff} X \times \Gamma_X \to S^\infty$ defined by it satisfies the following condition: if $\operatorname{supp} \psi \subset X_k$ and $|\gamma \cap X_k| = n$, then $\sigma(\psi, \gamma) \in S_n$ $(k = 1, 2, \ldots;$ $n = 0, 1, \ldots)$.

In particular, $\sigma(\psi, \gamma) \in S$ for any $\psi \in \operatorname{Diff} X$ and $\gamma \in \Gamma_X$.

It is not difficult to construct examples of admissible indexings. For example, the following indexing, which was proved to be measurable in [17], is admissible.

Let a continuous metric be given on $X$. With each positive integer $k$ we associate a covering $(X_{kl})_{l=1,2,\ldots}$ of $X$ by disjoint measurable subsets with diameters not exceeding $1/k$, satisfying the following two conditions.

1) the partitioning $X = X_{k1} \cup \ldots \cup X_{kl} \cup \ldots$ is a refinement of $X = X_1 \cup (X_2 \setminus X_1) \cup \ldots \cup (X_n \setminus X_{n-1}) \cup \ldots$;

2) Each set $X_n$ is covered by finitely many of the sets $X_{kl}$.

---

[1] If a measure $\mu$ is given in $\Gamma_X$, then indexings need be given only on subsets of full measure in $\Gamma_X$, and we make no distinction between indexings that coincide mod 0.

It is obvious that such a covering exists. We number its elements so that if $X_{ki} \subset X_n$ and $X_{kj} \subset X \setminus X_n$, then $i < j$ ($n = 1, 2, \ldots$). For any $x \in X$ and $k \in N$ we put $f_k(x) = l$, if $x \in X_{kl}$. The correspondence $x \to (f_1(x), \ldots, f_k(x), \ldots)$ is a morphism from $X$ to the set of all sequences of positive integers. We define an ordering in $X$ by putting $x' \prec x''$ if $(f_1(x'), \ldots f_k(x'), \ldots) < (f_1(x''), \ldots, f_k(x''), \ldots)$ in the lexicographic ordering.

For any $\gamma \in \Gamma_X$ and $x \in \gamma$, the set $\{x' \in \gamma; x' \prec x\}$ is finite, because it is contained in the compact set $\bar{X}_{11} \cup \ldots \cup \bar{X}_{1,f_1(x)}$. Consequently, for any $\gamma \in \Gamma_X$, the set of elements $x \in \gamma$ is a sequence with respect to the ordering introduced in $X$; we denote by $i(\gamma, x)$ the number of elements $x \in \gamma$ in this sequence. The map $(\gamma, x) \to i(\gamma, x)$ so constructed is a measurable indexing (see [17]). It is not difficult to prove that it is also admissible.

**3. Convolution of measures.** DEFINITION. The convolution $\mu_1 * \mu_2$ (see, for example, [17]) of two measures $\mu_1$ and $\mu_2$ on the space of all configurations $\Delta_X = \Gamma_X \cup B_X$ is defined as the image of the product measure $\mu_1 \times \mu_2$ on $\Delta_X \times \Delta_X$ under the map $(\gamma_1, \gamma_2) \to \gamma_1 \cup \gamma_2$.

REMARK. This definition agrees with the usual definition for the convolution of two measures in the space $\mathscr{F}(X)$ of generalized functions on $X$($\Delta_X$ is embedded in $\mathscr{F}(X)$ by $\gamma \to \sum_{x \in \gamma} \delta_x$), because the union of (disjoint) configurations corresponds to the sum of their images in $\mathscr{F}(X)$.

It is obvious that $\psi(\mu_1 * \mu_2) = \psi\mu_1 * \psi\mu_2$ for any $\psi \in \operatorname{Diff} X$, where $\psi\mu$ is the image of $\mu$ under the diffeomorphism $\psi$. Hence *the convolution of quasi-invariant* (*under* Diff $X$) *measures is itself quasi-invariant.*

Note that *if $\mu_1$ and $\mu_2$ are Poisson measures with parameters $\lambda_1$ and $\lambda_2$, respectively, then their convolution $\mu_1 * \mu_2$ is the Poisson measure with parameter $\lambda_1 + \lambda_2$*. (This fact follows easily from the definition of the Poisson measure).

Later on we shall be interested in the case when one of the factors is a quasi-invariant measure concentrated on $\Gamma_X$, and the second is a smooth positive measure $m_n$ concentrated on $B_X^{(n)}$ ($n = 1, 2, \ldots$). Since all smooth positive measures $m_n$ on $B_X^{(n)}$ are equivalent, the type of the measure $\mu * m_n$ depends only on the type of $\mu$ and on $n$.

Let us agree to call the type of $\mu * m_n$ on $\Gamma_X$ the *n-point augmentation* of $\mu$ and to denote it by $n \circ \mu$. Thus, with each measure $\mu$ on $\Gamma_X$ there is associated a sequence of measures $0 \circ \mu \sim \mu, 1 \circ \mu, \ldots, n \circ \mu, \ldots$ defined up to equivalence. Note that $n_1 \circ (n_2 \circ \mu) \sim (n_1 + n_2) \circ \mu$ for any $n_1$ and $n_2$.

Here we establish the following properties of the operation $\circ$.

1) *For any quasi-invariant measure $\mu$ on $\Gamma_X$ there exists a quasi-invariant measure $\mu'$ such that $1 \circ \mu' \sim \mu$.*

2) *If a measure $\mu$ on $\Gamma_X$ is ergodic, then $1 \circ \mu$ is also ergodic.*

To prove this we give an admissible indexing $i$ on $\Gamma_X$ and let $s\colon \Gamma_X \to \widetilde{X}^\infty$ be the cross section defined by this indexing (see §2.2). We denote by $Y_s$ the image of $\Gamma_X$ under $s$ and by $\Delta_s$ the minimal $S_\infty$-invariant subset of $\widetilde{X}^\infty$ containing $Y_s$; obviously, $\Delta_s$ is the disjoint union $\Delta_s = \bigcup_{\sigma \in S_\infty} Y_s\sigma$. Since $Y_s \subset \widetilde{X}^\infty$ is measurable and $S_\infty$ countable, $\Delta_s$ is a measurable subset of $\widetilde{X}^\infty$. Since the indexing i is admissible, it follows that $\Delta_s$ is invariant under Diff $X$.

Let $\mu$ be a measure on $\Gamma_X$ that is quasi-invariant under Diff $X$. Let $c$ be an arbitrary positive function on $S_\infty$ such that $\sum_{\sigma \in S_\infty} c(\sigma) = 1$; we introduce a measure $\widetilde{\mu}$ on $\widetilde{X}^\infty$ by the formula:

$$\widetilde{\mu} = \sum_{\sigma \in S_\infty} c(\sigma)(s\mu)\sigma,$$

where $(s\mu)\sigma$ is the image of $\mu$ under the map $\gamma \to (s\gamma)\sigma$. In other words, for any measurable subset $A \subset \widetilde{X}^\infty$

$$\widetilde{\mu}(A) = \sum_{\sigma \in S_\infty} c(\sigma)\, \mu\,[p(A \cap Y_s\sigma)] \tag{1}$$

where $p$ is the projection $\widetilde{X}^\infty \to \Gamma_X$.

Obviously, $\widetilde{\mu}(\Delta_s) = 1$. Note that the choice of the positive function $c$ on $S_\infty$ does not play a role in defining $\widetilde{\mu}$, because the measures on $\widetilde{X}^\infty$ constructed from two such functions are equivalent. From the definition of $\widetilde{\mu}$ it follows easily that:

a) $p\widetilde{\mu} = \mu$ where $p\widetilde{\mu}$ is the projection of $\widetilde{\mu}$ onto $\Gamma_X$.

b) the measure $\widetilde{\mu}$ on $\widetilde{X}^\infty$ is quasi-invariant under both Diff $X$ and $S_\infty$.

We cite without proof two further simple assertions.

PROPOSITION 3. *If a normalized measure* $\mu_1$ *on* $\widetilde{X}^\infty$ *is quasi-invariant under* Diff $X$ *and if* $p\mu_1 = \mu$ *and* $\mu_1(\Delta_s) = 1$, *then* $\mu_1 \sim \widetilde{\mu}$.

PROPOSITION 4. *If a measure* $\mu$ *in* $\Gamma_X$ *is ergodic, then* $\widetilde{\mu}$ *is also ergodic with respect to* Diff $X$.

Let us decompose the space $X^\infty = \prod_{i=1}^{\infty} X_i$, where $X_i = X$, into the direct product $X^\infty = X \times \prod_{i=2}^{\infty} X_i$ and consider the induced map $h = X \times \widetilde{X}^\infty \to \widetilde{X}^\infty$ (that is, $h(x; \{x_k\}_{k=1}^{\infty}) = (\{x'_k\}_{k=1}^{\infty}; x'_1 = x,\ x'_k = x_{k-1}$ when $k > 1$)).

PROPOSITION 5. $h(m \times \mu) \sim m * \mu$, *where* $m$ *is an arbitrary smooth positive measure on* $X$.

PROOF. Consider the diagram

$$\begin{array}{ccc} X \times \widetilde{X}^\infty & \xrightarrow{\;h\;} & \widetilde{X}^\infty \\ \Big\downarrow{\scriptstyle p_1} & & \Big\downarrow{\scriptstyle p} \\ X \times \Gamma_X & \xrightarrow{\;\widetilde{h}\;} & \Gamma_X \end{array},$$

where $p_1 = \mathrm{Id} \times p$, $\widetilde{h}(x, \gamma) = \gamma \cup \{x\}$. Obviously, this is commutative, and $(p \circ h)(m \times \widetilde{\mu}) = (\widetilde{h} \circ p_1)(m \times \widetilde{\mu}) = m * \mu$. Further, the measure $h(m \times \widetilde{\mu})$ is quasi-invariant and concentrated on $\Delta_s$ (since the indexing $i$ is admissible, the point $h(x, s\gamma)$ belongs to the same $S_\infty$-orbit as $s(\gamma \cup \{x\})$). Consequently, by Proposition 3, $h(m \times \widetilde{\mu}) \sim m * \mu$.

The proof of the following assertion is similar to that of Lemma 1 in §2.1.

PROPOSITION 6. *Every measure $\mu$ in $\widetilde{X}^\infty$ that is quasi-invariant under* Diff $X$ *is equivalent to the product $m_n \times \mu_n$ of its projections in the factorization $\widetilde{X}^\infty = \widetilde{X}^n \times \widetilde{X}^\infty_{n+1}$; moreover, $m_n$ is equivalent to a positive smooth measure on $X^n$, and $\mu_n$ is quasi-invariant under* Diff $X$.

COROLLARY. *A quasi-invariant measure $\widetilde{\mu}$ in $\widetilde{X}^\infty$ is ergodic if and only if it is regular (that is, satisfies the* $0 - 1$ *law).*

PROOF OF PROPERTY 1). Let $\mu$ be a quasi-invariant measure on $\Gamma_X$. By Proposition 6, $\mu \sim h(m \times \mu_1)$, where $m$ is a smooth positive measure on $X$, $\mu_1$ is a quasi-invariant measure on $\widetilde{X}^\infty$, and $h$: $X \times \widetilde{X}^\infty \to \widetilde{X}^\infty$ is the map induced by the direct product (see above). Since $i$ is admissible, $\mu_1$, like $\mu$, is concentrated on $\Delta_s$; consequently, by Proposition 3, $\mu_1 \sim \mu'$ is a quasi-invariant measure in $\Gamma_X$. By Proposition 5, $\widetilde{\mu} \sim h(m \times \widetilde{\mu}') \sim m * \mu'$; consequently, $\mu \sim m * \mu'$, as required.

PROOF OF PROPERTY 2). If the measure $\mu$ in $\Gamma_X$ is ergodic, then by Proposition 4, the measure $\widetilde{\mu}$ in $\widetilde{X}^\infty$ is ergodic; consequently, by the corollary to Proposition 6, $\widetilde{\mu}$ is regular. Obviously, $m \times \widetilde{\mu}$ is then also regular and therefore ergodic. Consequently, the measure $m * \mu \sim h(m \times \widetilde{\mu})$ is also ergodic and hence, so is its projection $m * \mu$.

DEFINITION. We say that a quasi-invariant measure $\mu$ is *saturated* if $1 \circ \mu \sim \mu$ (and consequently, $n \circ \mu \sim \mu$ for any $n$).

It is not difficult to verify that *the Poisson measure is saturated* (this follows from the property of being infinitely decomposable).

We now give a criterion for a measure $\mu$ to be saturated. The map $T$: $X^\infty \to X^\infty$, defined by $(Tx)_i = x_{i+1}$ $(i = 1, 2, \ldots)$ is called *left translation* in $X^\infty$. Obviously, the subset $\widetilde{X}^\infty$ is $T$-invariant.

PROPOSITION 7. *For a quasi-invariant measure on $\Gamma_X$ to be saturated it is necessary and sufficient that the measure $\widetilde{\mu}$ on $\widetilde{X}^\infty$ corresponding to it (defined by means of a fixed admissible indexing) is quasi-invariant under the left translation $T$.*

PROOF. From the definition of the left translation $T$ it follows that $\widetilde{\mu} \sim h(m \times T\widetilde{\mu})$. On the other hand, by Proposition 3, $1 \circ \mu \sim h(m \times \widetilde{\mu})$. Hence it is obvious that the condition $1 \circ \mu \sim \mu$ is equivalent to $\widetilde{\mu} \sim T\widetilde{\mu}$.

An example of a non-saturated measure $\mu$ will be given in Appendix 1.

**4. The space $\Gamma_{X,n}$ and Campbell's measure on $\Gamma_{X,n}$.** We consider the Cartesian product $\Gamma_X \times X^n$ $(n = 1, 2, \ldots)$ and denote by $\Gamma_{X,n}$ the set of elements $(\gamma; x_1, \ldots, x_n) \in \Gamma_X \times X^n$, where $\gamma \in \Gamma_X$, $x_i \in X$, such that $x_i \in \gamma$ $(i = 1, \ldots, n)$ and $x_i \neq x_j$ when $i \neq j$. Further, we put $\Gamma_{X,0} = \Gamma_X$.

Obviously, $\Gamma_{X,n}$ is closed in $\Gamma_X \times X^n$.

Now $\Gamma_{X,n}$ can be regarded as a fibre space, $\pi\colon \Gamma_{X,n} \to \Gamma_X$, whose fibre over a point $\gamma \in \Gamma_X$ is the collection of all ordered $n$-point subsets in $\gamma$.

Let us denote by $\mathfrak{A}_n$ the $\sigma$-algebra of all Borel sets in $\Gamma_{X,n}$. We associate with each subset $C \in \mathfrak{A}_n$ a function on $\Gamma_X$: $\nu_C(\gamma) = \{$the number of points $(x_1, \dots, x_n) \in X^n$ such that $(\gamma\colon x_1, \dots, x_n) \in C\}$.

From the continuity of $\pi$ it follows that $\nu_C$ is a Borel function.

DEFINITION. Let $\mu$ be a measure on $\Gamma_X$. The *Campbell measure* on $\Gamma_{X,n}$ associated with $\mu$ is the measure $\widetilde{\mu}$ on $\mathfrak{A}_n$ defined by

$$\widetilde{\mu}(C) = \int\limits_{\Gamma_X} \nu_C(\gamma)\, d\mu(\gamma), \qquad C \in \mathfrak{A}_n.$$

A Campbell measure $\widetilde{\mu}$ induces on the fibres of the fibration $\pi\colon \Gamma_{X,n} \to \Gamma_X$ a uniform measure, which is 1 at each point of the fibre.

We define in $\Gamma_{X,n}$ the actions of the groups Diff $X$ and $S_n$:

$$\psi\colon (\gamma;\, x_1, \dots, x_n) \mapsto (\psi^{-1}\gamma;\, \psi^{-1}x_1, \dots, \psi^{-1}x_n),$$
$$\sigma\colon (\gamma;\, x_1, \dots, x_n) \mapsto (\gamma;\, x_{\sigma(1)}, \dots, x_{\sigma(n)}).$$

Obviously, $\psi$ and $\sigma$ are continuous and $\psi \circ \sigma = \sigma \circ \psi$ for any $\psi \in \text{Diff}\, X$ and $\sigma \in S_n$. The next result is easy to establish.

LEMMA 2. *The Campbell measure $\widetilde{\mu}$ on $\Gamma_{X,n}$ corresponding to a measure $\mu$ on $\Gamma_X$ is invariant under $S_n$. If the measure $\mu$ on $\Gamma_X$ is invariant under* Diff $X$, *then the Campbell measure $\widetilde{\mu}$ is also quasi-invariant under* Diff $X$, *and*

$$\left.\frac{d\widetilde{\mu}(\psi^{-1}c)}{d\widetilde{\mu}(c)}\right|_{c=(\gamma;\, x_1, \dots, x_n)} = \frac{d\mu(\psi^{-1}\gamma)}{d\mu(\gamma)}.$$

Now let $i$ be a measurable indexing in $\Gamma_X$. We denote by $\widetilde{N}^n$ the set of all $n$-tuples of natural numbers $(i_1, \dots, i_n)$, where $i_p \neq i_q$ when $p \neq q$ $(p, q = 1, \dots, n)$. We define a map

$$\Gamma_{X,n} \to \Gamma_X \times \widetilde{N}^n \tag{2}$$

by

$$(\gamma;\, x_1, \dots, x_n) \mapsto (\gamma;\, i(\gamma, x_1), \dots, i(\gamma, x_n)).$$

This map is bijective, measurable in both directions, and carries the Campbell measure $\widetilde{\mu}$ on $\Gamma_{X,n}$ into the measure $\mu \times \nu$ on $\Gamma_X \times \widetilde{N}^n$, where $\nu$ is the measure on $\widetilde{N}^n$, that is equal to 1 at each point on $\widetilde{N}^n$. Thus, the space $(\Gamma_{X,n}, \widetilde{\mu})$ can be identified with $(\Gamma_X \times N^n, \mu \times \nu)$.

Under this identification the actions of $S_n$ and Diff $X$ go over from $\Gamma_{X,n}$ to $\Gamma_X \times \widetilde{N}^n$. It is not difficult to verify that the action of these groups on $\Gamma_X \times \widetilde{N}^n$ are given by:

$$\sigma\colon (\gamma, a) \mapsto (\gamma, a\sigma),$$
$$\psi\colon (\gamma, a) \mapsto (\psi^{-1}\gamma, \sigma(\psi, \gamma)a),$$

where $a = (i_1, \ldots, i_n)$; $a\sigma = (i_{\sigma(1)}, \ldots, i_{\sigma(n)})$, $\sigma \in S_n$;
$\sigma a = (\sigma(i_1), \ldots, \sigma(i_n))$ $\sigma \in S^\infty$; and $\sigma(\psi, \gamma)$ is the function on
Diff $X \times \Gamma_X$ with values in $S^\infty$ defined by $i$ (see §2.2).

**5. The map** $\Gamma_X \times X^n \to \Gamma_{X,n}$. Let us consider the spaces $\Gamma_X \times X^n$ and
$\Gamma_{X,n}$ together with their $\sigma$-algebras of Borel subsets (see §2.4). Let $\mu$ be a
quasi-invariant measure on $\Gamma_X$, and $m_n$ a smooth, positive measure on $X^n$.
In $\Gamma_X \times X^n$ we specify the measure $\mu \times m_n$ and in $\Gamma_{X,n}$ the Campbell
measure $\widetilde{n \circ \mu}$ corresponding to $n \circ \mu \sim \mu * m_n$ on $\Gamma_X$.

A map $\alpha\colon \Gamma_X \times X^n \to \Gamma_{X,n}$ is given by the following formula:

$$\alpha(\gamma; x_1, \ldots, x_n) = (\gamma \cup \{\dot{x}_1, \ldots, x_n\}; x_1, \ldots, x_n);$$

$\alpha$ is taken to be defined on the subset of elements
$(\gamma\colon x_1, \ldots, x_n) \in \Gamma_X \times X^n$ for which $\gamma \cap \{x_1, \ldots, x_n\} = \phi$; it is not
difficult to verify that this subset and its image in $\Gamma_{X,n}$ are sets of full
measure. Nor is it difficult to check that $\alpha$ is measurable in both directions
and commutes with the action of Diff $X$ on both $\Gamma_X \times X^n$ and $\Gamma_{X,n}$.

THEOREM. *The image* $\alpha(\mu \times m_n)$ *of the measure* $\mu \times m_n$ *on* $\Gamma_X \times X^n$
*under* $\alpha$ *is equivalent to the Campbell measure* $\widetilde{n \circ \mu}$.

PROOF. We carry out the proof for the case $n = 1$; the arguments for
arbitrary $n$ are similar.

We define maps $\alpha_1\colon \Gamma_X \times X \to \Gamma_X$ and $\alpha_2\colon \Gamma_{X,1} \to \Gamma_X$ by
$\alpha_1(\gamma, x) = \gamma \cup \{x\}$, $\alpha_2(\gamma, x) = \gamma$. It is obvious that the following diagram
commutes:

$$\begin{array}{ccc} \Gamma_X \times X & \xrightarrow{\ \alpha\ } & \Gamma_{X,1} \\ {\scriptstyle\alpha_1}\searrow & & \swarrow{\scriptstyle\alpha_2} \\ & \Gamma_X & \end{array}$$

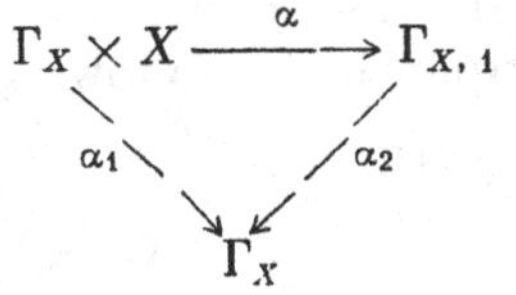

The image of the measure $\mu \times m$ on $\Gamma_X \times X$ under $\alpha_1 = \alpha_2\alpha$ is $1 \circ \mu$.
Hence it follows that $\alpha(\mu \times m) \prec \widetilde{1 \circ \mu}$. It remains to prove that
$\alpha(\mu \times m) \succ \widetilde{1 \circ \mu}$.

First we construct a measurable indexing in $\Gamma_X$ in the following way. We
fix a continuous metric $\rho$ in $X$ and a point $\dot{x}_0 \in X$. We consider the subset of $\Gamma_X$

(3)      $\{\gamma \in \Gamma_X\colon \rho(x_0, x) \neq \rho(x_0, x')$ for any $x \neq x'$ in $\gamma\}$

and the preimage of (3) under $\alpha_2$ in $\Gamma_{X,1}$. As is easy to see, these subsets
are of full measure in $\Gamma_X$ and $\Gamma_{X,1}$, respectively, and it is to be understood in what follows that it is these subsets which are meant by $\Gamma_X$ and
$\Gamma_{X,1}$.

We prescribe an ordering on each configuration $\gamma \in \Gamma_X$, putting $x \prec x'$
for any $x, x' \in \gamma$, if $\rho(x_0, x) < \rho(x_0, x')$. For any $(\gamma, x) \in \Gamma_{X,1}$ we
denote by $i(\gamma, x)$ the number of the element $x \in \gamma$, as given by the

ordering on $\gamma$. It is not difficult to see that $i$ is a measurable indexing. We now introduce a sequence of measurable maps $a_k : \Gamma_X \to X$ ($i = 1, 2, \dots$), where $a_k(\gamma)$ is the $k$-th element in the configuration $\gamma$.

Now let $C \subset \Gamma_{X,1}$ be an arbitrary measurable set of positive Campbell measure: $\widetilde{1 \circ \mu}(C) > 0$; we have to prove that then $(\mu \times m)(\alpha^{-1} C) > 0$. Since $\Gamma_{X,1}$ splits into the countable union of subsets $\{(\gamma, a_k(\gamma)); \gamma \in \Gamma_X\}$ ($k = 1, 2, \dots$), we may assume without loss of generality that $C \subset \{(\gamma, a_k(\gamma))\}$ for some $k$. We introduce the notation $\widetilde{C}_n = \{(\gamma', x) \in \Gamma_X \times X; \gamma' \cup \{x\} \in \alpha_2 C, a_n(\gamma \cup \{x\})) = x\}$ ($n = 1, 2, \dots$).

Note that the condition $\widetilde{1 \circ \mu}(C) > 0$ is equivalent to

$$(4) \qquad (\mu \times m)\{(\gamma', x) \in \Gamma_X \times X;\ \gamma' \cup \{x\} \in \alpha_2 C\} > 0.$$

In its turn (4) is equivalent to the existence of a natural number $l$, for which $(\mu \times m)$ $\{(\gamma', x) \in \Gamma_X \times X,\ \gamma' \cup \{x\} \in \alpha_2 C,\ a_l(\gamma' \cup \{x\}) = x\} > 0$, that is, $(\mu \times m)$ $(\widetilde{C}_l) > 0$.

On the other hand, since $C \subset \{(\gamma, a_k(\gamma))\}$, it follows that

$$\alpha^{-1} C = \{(\gamma', x) \in \Gamma_X \times X;\ \gamma' \cup \{x\} \in \alpha_2 C,\ a_k(\gamma' \cup \{x\}) = x\},$$

that is, $\alpha^{-1} C = \widetilde{C}_k$. Thus, the proof of the lemma reduces to proving the following assertion: *for any natural numbers k and l the conditions* $(\mu \times m)$ $(\widetilde{C}_k) > 0$ *and* $(\mu \times m)$ $(\widetilde{C}_l) > 0$ *are equivalent.*

Let us prove this assertion. We write $X_r = \{x \in X; \rho(x_0, x) < r\}$, where $r > 0$ is an arbitrary rational number ($X_0 = \phi$). We fix a positive integer $n \geqslant \max(k, l)$ and introduce the following subsets in $\Gamma_X \times X$:

$$U^p_{r_1, \dots, r_n} = \{(\gamma', x) \in \Gamma_X \times X;\ |\gamma' \cap (X_{r_i} \setminus X_{r_{i-1}})| = 1 \text{ when } i \neq p,$$
$$|\gamma' \cap (X_{r_p} \setminus X_{r_{p-1}})| = 0,\ x \in X_{r_p} \setminus X_{r_{p-1}}\},$$

where $r_1, \dots, r_n$ are rational numbers such that $0 = r_0 < r_1 < \dots < r_n$ ($p = 1, \dots, n$). It is obvious that the sets $U^p_{r_1, \dots, r_n}$ cover $\widetilde{C}_p$; consequently, the condition $(\mu \times m)$ $(\widetilde{C}_l) > 0$ amounts to the existence of some $U^l_{r_1, \dots, r_n}$ such that

$$(5) \quad (\mu \times m)\{(\gamma', x) \in U^l_{r_1, \dots, r_n};\ \gamma' \cup \{x\} \in \alpha_2 C,\ a_l(\gamma' \cup \{x\}) = x\} > 0$$

We note that $U^l_{r_1, \dots, r_n} \subset (\mathrm{B}^{(n-1)}_{X_{r_n}} \times \Gamma_{X \setminus X_{r_n}}) \times X$ and make use of the fact that by Lemma 1 of §2.1 the restriction $\mu_{n-1}$ of $\mu$ to $\mathrm{B}^{(n-1)}_{X_{r_n}} \times \Gamma_{X \setminus X_{r_n}}$ is equivalent to the product $\underbrace{m \times \dots \times m}_{n-1} \times \mu''_{n-1}$, where $\mu''_{n-1}$ is the projection of $\mu_{n-1}$ onto $\Gamma_{X \setminus X_{r_n}}$. So we obtain that (5) is equivalent to the following condition:

$$(6) \qquad (m \times \ldots \times m \times \mu''_{n-1})\{(x_1, \ldots, x_n; \gamma') \in X^n \times \Gamma_{X \setminus X_{r_n}}; \\ x_i \in X_{r_i} \setminus X_{r_{i-1}}, \ i = 1, \ldots, n; \ \gamma' \cup \{x_1, \ldots, x_n\} \in \alpha_2 C\} > 0.$$

Thus, $(\mu \times m)(\widetilde{C}_l) > 0$ amounts to the condition that (6) is satisfied for some collection of rational numbers $0 = r_0 < r_1 < \ldots < r_n$. But from the same arguments it follows that the condition $(\mu \times m)(\widetilde{C}_k) > 0$ also is equivalent to (6). Consequently, the conditions $(\mu \times m)(\widetilde{C}_l) > 0$ and $(\mu \times m)(\widetilde{C}_k) > 0$ are equivalent, as required.

## §3. Representations of Diff $X$ defined by quasi-invariant measures in the space of infinite configurations (elementary representations)

**1. Definition of elementary representations.** Let $\mu$ be a quasi-invariant measure in the space of infinite configurations $\Gamma_X$. We introduce a series of unitary representations of Diff $X$ associated with $\mu$. First we consider the space $L^2_\mu(\Gamma_X)$. In it a unitary representation $U_\mu$ of Diff $X$ is defined by[1]

$$(1) \qquad (U_\mu(\psi) f)(\gamma) = \left(\frac{d\mu(\psi^{-1}\gamma)}{d\mu(\gamma)}\right)^{1/2} f(\psi^{-1}\gamma)$$

We do not study the properties of $U_\mu$ separately, but examine straightaway a wider class – the elementary representations. For the Poisson measure $\mu$ these representations are additive generators in the representation ring determined by $U_\mu$ (see §4).

Although the proof of the irreducibility and other properties of $U_\mu$ are simpler than in the general case, we prefer to study all the elementary representations simultaneously.

DEFINITION. A representation of Diff $X$ is called elementary if it is of the form $U_\mu \otimes V^\rho$, where $U_\mu$ is the representation in $L^2_\mu(\Gamma_X)$ given by (1), and $V^\rho$ is the representation defined in §1.

Thus, each elementary representation is given by a quasi-invariant measure $\mu$ on $\Gamma_X$ and a representation $\rho$ of the symmetric group $S_n$ $(n = 0, 1, 2, \ldots)$.

THEOREM 1. *If $\mu$ is an ergodic measure on $\Gamma_X$ and $\rho$ is an irreducible representation of $S_n$, then the elementary representation $U_\mu \otimes V^\rho$ of Diff $X$ is irreducible.*

REMARK 1. The converse assertion is obvious.

REMARK 2. Another convenient formulation of Theorem 1 is: When $\mu$ is ergodic, then $U_\mu$ is *absolutely irreducible*, that is, remains irreducible after taking the tensor product with any irreducible representation $V^\rho$.

Essentially, the whole of §3 is devoted to a proof of Theorem 1. But first we construct some other useful realizations of elementary representations.

---

[1] If $\mu$ is concentrated not on $\Gamma_X$, but on $B_X^{(n)}$, then (1) gives the representation $V^{\rho_n^0}$ (see §1), where $\rho_n^0$ is the unit representation of $S_n$.

**2. The representations $U_\mu^\rho$.** Let $\rho$ be a unitary representation of $S_n$ in a space $W$ ($n = 0, 1, \ldots$). We consider the space $L^2_{\tilde\mu}(\Gamma_{X,n}, W)$ of functions $F$ on $\Gamma_{X,n}$ with values in $W$ such that

$$\| F \| = \int_{\Gamma_{X,n}} \| F(c) \|^2_W \, d\tilde\mu(c) < \infty;$$

$\tilde\mu$ is the Campbell measure on $\Gamma_{X,n}$ corresponding to the measure $\mu$ on $\Gamma_X$ (see §2.4). A unitary representation $U$ of Diff $X$ is given in $L^2_{\tilde\mu}(\Gamma_{X,n}, W)$ by

$$(U(\psi)F)(\gamma; x_1, \ldots, x_n) = \left(\frac{d\mu(\psi^{-1}\gamma)}{d\mu(\gamma)}\right)^{1/2} F(\psi^{-1}\gamma; \psi^{-1}x_1, \ldots, \psi^{-1}x_n).$$

We denote by $H_{\mu, n, \rho}$ the subspace of functions $F \in L^2_{\tilde\mu}(\Gamma_{X,n}, W)$ such that $F(\gamma, x_{\sigma(1)}, \ldots, x_{\sigma(n)}) = \rho^{-1}(\sigma)F(\gamma, x_1, \ldots, x_n)$ for any $\sigma \in S_n$. Obviously, $H_{\mu, n, \rho}$ is invariant under Diff $X$.

DEFINITION. The restriction of the representation $U$ of Diff $X$ from $L^2_\mu(\Gamma_{X,n}, W)$ to $H_{\mu, n, \rho}$ is denoted by $U_\mu^\rho$.

In the particular case when $\rho$ is the unit representation of $S_0$, $U_\mu^\rho \cong U_\mu$, where $U_\mu$ is the representation in $L^2_\mu(\Gamma_X)$ defined by (1).

REMARK. If in this construction of $U_\mu^\rho$ the space $\Gamma_X$ of infinite configurations is replaced by the space $B_X^{(k)}$ of $k$-point configurations ($k \geqslant n$), then we obtain instead of $U_\mu^\rho$ the representation $V^{\rho \circ \overset{\circ}{\rho}_{k-n}}$ defined in §1, where $\rho^0_{k-n}$ is the unit representation of $S_{k-n}$.

THEOREM 2. $U_\mu \otimes V^\rho \cong U^\rho_{n \circ \mu}$, where $\rho$ is a representation of $S_n$. (For the definition of $n \circ \mu$, see §2.3).

PROOF. In §2.5 we have established an isomorphism between spaces with measures

$$(2) \qquad (\Gamma_X \times X^n, \mu \times m_n) \to (\Gamma_{X,n}, \widetilde{n \circ \mu})$$

($\widetilde{n \circ \mu}$ is the Campbell measure on $\Gamma_{X,n}$ corresponding to $n \circ \mu$), which commutes with the action of Diff $X$. Let us consider the isomorphism of Hilbert spaces $L^2_{\widetilde{n\circ\mu}}(\Gamma_{X,n}, W) \to L^2_\mu(\Gamma_X) \otimes L^2_{m_n}(X^n, W)$ induced by (2), where $W$ is the space of the representation $\rho$ of $S_n$. It is easy to verify that the image of $H_{\mu, n, \rho} \subset L^2_{\widetilde{n\circ\mu}}(\Gamma_{X,n}, W)$ is $L^2_\mu(\Gamma_X) \otimes H_{n,\rho}$, where $H_{n,\rho} \subset L^2_{m_n}(X^n, W)$ is the subspace of the representation $V^\rho$ (see §1) and that the operators $U_{n\circ\mu}(\psi)$ in $H_{\mu,n,\rho}$ go over to $U_\mu(\psi) \otimes V^\rho(\psi)$.

COROLLARY 1. *The class of the representations $U_\mu^\rho$ is the same as that of the elementary representations $U_\mu \otimes V^\rho$.*

For on the one hand, $U_\mu \otimes V^\rho \cong U^\rho_{n\circ\mu}$; and on the other hand, for any quasi-invariant measure $\mu$ on $\Gamma_X$ there is another quasi-invariant measure $\mu'$ such that $\mu \sim n \circ \mu'$ (see §2.3) and, consequently,

$U^\rho_\mu \cong U_{\mu'} \otimes V^\rho$.

COROLLARY 2. *If $\mu$ is a saturated measure (that is, $1 \circ \mu \sim \mu$) and, in particular, if $\mu$ is the Poisson measure, then $U^\rho_\mu \cong U_\mu \otimes V^\rho$.*

THEOREM 3. *If $\mu$ is an ergodic measure on $\Gamma_X$ and $\rho$ an irreducible representation of $S_n$, then the representation $U^\rho_\mu$ of* Diff $X$ *is irreducible.*

We note that Theorem 1 follows immediately from Theorems 2 and 3. For let $\rho$ be an irreducible representation of $S_n$ and $\mu$ an ergodic measure on $\Gamma_X$. Since $n \circ \mu$ is also ergodic (see § 2.3), $U_\mu \otimes V^\rho \cong U^\rho_{n\circ\mu}$ is irreducible by Theorem 3.

**3. Another realization of $U^\rho_\mu$.** Let $\rho$ be a unitary representation of $S_n$ in $W$. We consider the set $\widetilde{N}^n$ of all $n$-tuples $a = (i_1, \ldots, i_n)$ of natural numbers, where $i_p \neq i_q$, when $p \neq q$. We define an action of $S^\infty$ on $\widetilde{N}^n$ by $a \to \sigma a = (\sigma(i_1), \ldots, \sigma(i_n))$, $\sigma \in S^\infty$. We denote by $l^2(\widetilde{N}^n, W)$ the space of functions $\varphi$ on $\widetilde{N}^n$ with values in $W$ such that

$$\|\varphi\|^2 = \sum_{a\in\widetilde{N}^n} \|\varphi(a)\|^2_W < \infty.$$

We consider in $l^2(\widetilde{N}^n, W)$ the subspace $H^\rho$ of all functions $\varphi \in l^2(\widetilde{N}^n, W)$ such that $\varphi(i_{\sigma(1)}, \ldots, i_{\sigma(n)}) = \rho^{-1}(\sigma)\varphi(i_1, \ldots, i_n)$ for any $\sigma \in S_n$. Obviously, $H^\rho$ is invariant under the action of $S^\infty$. Now let $i$ be an arbitrary measurable indexing in $\Gamma_X$ and $\sigma$ the map Diff $X \times \Gamma_X \to S^\infty$ defined by it.

LEMMA 1. *The representation $U^\rho_\mu$ of* Diff $X$ *is equivalent to the representation in $L^2_\mu(\Gamma_X) \otimes H^\rho$, defined by*

$$(3) \qquad (U^\rho_\mu(\psi) f)(\gamma, a) = \left(\frac{d\mu(\psi^{-1}\gamma)}{d\mu(\gamma)}\right)^{1/2} f(\psi^{-1}\gamma, \sigma(\psi, \gamma) a).$$

PROOF. In § 2.4 an isomorphism was established between spaces with measures:

$$(4) \qquad (\Gamma_X \times \widetilde{N}^n, \mu \times \nu) \to (\Gamma_{X,n}, \tilde{\mu})$$

($\nu$ is the measure on $\widetilde{N}^n$ that is 1 at each point). We consider the isomorphism of Hilbert spaces $L^2_{\tilde{\mu}}(\Gamma_{X,n}, W) \to L^2_\mu(\Gamma_X) \otimes l^2(\widetilde{N}^n, W)$ induced by (4). It is easy to verify that the image of the subspace $H_{\mu,n,\rho} \subset L^2_{\tilde{\mu}}(\Gamma_{X,n}, W)$ on which $U^\rho_\mu$ acts is $L^2_\mu(\Gamma_X) \otimes H^\rho$ and that the operators $U^\rho_\mu(\psi)$ in $H_{\mu,n,\rho}$ go over to operators of the form (3).

**4. The decomposition of the space of the representation $U^\rho_\mu$ of Diff $X$ into a sum of subspaces that are primary with respect to the subgroup Diff $X_k$.** Let

$$(5) \qquad X_1 \subset \ldots \subset X_k \subset \ldots$$

be an increasing sequence of open connected subsets with compact closures such that $X = \bigcup_k X_k$.

We fix an admissible indexing $i$ (with respect to (5)); let $\sigma(\psi, \gamma)$ be the map Diff $X \times \Gamma_X \to S_\infty$ defined by it. By Lemma 1, the representation $U^\rho_\mu$ may be realized in $L^2_\mu(\Gamma_X) \otimes H$; here $H = H^\rho$ is the space of functions $\varphi$ on $\widetilde{N}^n$ with values in the space $W$ of the representation $\rho$ of $S_n$ such that $\| \varphi \|^2 = \sum_{a \in \widetilde{N}^n} \| \varphi(a) \|^2_W < \infty$ and $\varphi(a\sigma) = \rho^{-1}(\sigma)\varphi(a)$ for any $\sigma \in S_n$. The representation operators are given by (3).

We decompose $L^2_\mu(\Gamma_X) \otimes H$ into a direct sum of subspaces that are primary with respect to the subgroups Diff $X_k \subset$ Diff $X$ (that is, those that are the identity on $X \setminus X_k$) ($k = 1, 2, \ldots$).

First we decompose $\Gamma_X$ into a countable union of spaces that are invariant under Diff $X_k$:

$$\Gamma_X = \bigsqcup_{r=0}^{\infty} B^{(r)}_{X_k} \times \Gamma_{X \setminus X_k}, \tag{6}$$

where $B^{(r)}_{X_k}$ is the space of $r$-point subsets in $X_k$. It follows from (6) that

$$L^2_\mu(\Gamma_X) \otimes H = \bigoplus_{r=0}^{\infty} (L^2_{\mu_r}(B^{(r)}_{X_k} \times \Gamma_{X \setminus X_k}) \otimes H),$$

where $\mu_r$ is the restriction of $\mu$ to the subset $B^{(r)}_{X_k} \times \Gamma_{X \setminus X_k} \subset \Gamma_X$. It remains to decompose each term in this sum into a direct sum of invariant subspaces that are primary with respect to Diff $X_k$.

Next we split $H$ into the direct sum of subspaces that are primary with respect to the symmetric group $S_r \subset S_\infty$. This decomposition can be presented in the following way: $H = \bigoplus_i (W^i_r \otimes C^i_r)$, where $W^i_r$ are the spaces in which the irreducible and pairwise inequivalent representations $\rho^i_r$ of $S_r$ act; $C^i_r$ is the space on which $S_r$ acts trivially.

As a result, we obtain a decomposition into the direct sum:

$$L^2_{\mu_r}(B^{(r)}_{X_k} \times \Gamma_{X \setminus X_k}) \otimes H = \bigoplus_i (L^2_{\mu_r}(B^{(r)}_{X_k} \times \Gamma_{X \setminus X_k}) \otimes W^i_r \otimes C^i_r).$$

All the terms of this decomposition are invariant under Diff $X$. For since the indexing is admissible, it follows from $\psi \in$ Diff $X_k$ and $\gamma \in B^{(r)}_{X_k} \times \Gamma_{X \setminus X_k}$ that $\sigma(\psi, \gamma) \in S_r$. We claim that these subspaces are primary and disjoint.

We denote by $\mu'_r$ and $\mu''_r$ the projections of $\mu_r$ onto $B^{(r)}_{X_k}$ and $\Gamma_{X \setminus X_k}$, respectively. By Lemma 1 of §2, the measure $\mu_r$ on $B^{(r)}_{X_k} \times \Gamma_{X \setminus X_k}$ is equivalent to the product $\mu'_r \times \mu''_r$ of $\mu'_r$ and $\mu''_r$. Consequently, there is an isomorphism

$$\tau_r \colon L^2_{\mu_r}(B^{(r)}_{X_k} \times \Gamma_{X \setminus X_k}) \to L^2_{\mu'_r}(B^{(r)}_{X_k}) \otimes L^2_{\mu''_r}(\Gamma_{X \setminus X_k}),$$

defined by

$$\tau_r F = \left( \frac{d\mu_r}{d\mu'_r \, d\mu''_r} \right)^{1/2} F.$$

We denote by the same letter the trivial extension of $\tau_r$ to an isomorphism

$$\tau_r\colon L^2_{\mu_r}(B^{(r)}_{X_k}\times\Gamma_{X\setminus X_k})\otimes W^i_r\otimes C^i_r\to(L^2_{\mu'_r}(B^{(r)}_{X_k})\otimes W^i_r)\otimes(L^2_{\mu''_r}(\Gamma_{X\setminus X_k})\otimes C^i_r).$$

We denote the elements of $B^{(r)}_{X_k}$ by $\gamma^{(r)}$ and define a map $\sigma_r$: Diff $X_k \times B^{(r)}_{X_k} \to S_r$ by $\sigma_r(\psi, \gamma^{(r)}) = \sigma(\psi, \gamma)$, where $\gamma \in B^{(r)}_{X_k} \times \Gamma_{X\setminus X_k}$, $\gamma \cap X_k = \gamma^{(r)}$. This is well defined, because if $\gamma \cap X_k = \gamma' \cap X_k$, then $\sigma(\psi, \gamma) = \sigma(\psi, \gamma')$. Immediately from the definition of $\tau_r$ we derive the next result.

LEMMA 2. *Under the isomorphism $\tau_r$ the operators $U(\psi) = U^\rho_\mu(\psi)$, $\psi \in$ Diff $X_k$ go over to operators $\tau_r U(\psi)\tau_r^{-1}$ of the following form: $\tau_r U(\psi)\tau_r^{-1} = U^i_r(\psi) \otimes I$, where $I$ is the unit operator in $L^2_{\mu''_r}(\Gamma_{X\setminus X_k}) \otimes C^i_r$ and $U^i_r(\psi)$ is an operator in $L^2_{\mu'_r}(B^{(r)}_{X_k}) \otimes W^i_r$, that is, in the space of functions on $B^{(r)}_{X_k}$ with values in $W^i_r$ defined by*

$$(U^i_r(\psi)F)(\gamma^{(r)}) = \left(\frac{d\mu'_r(\psi^{-1}\gamma^{(r)})}{d\mu'_r(\gamma^{(r)})}\right)^{1/2}\rho^i_r(\sigma_r(\psi,\gamma^{(r)}))\,F(\psi^{-1}\gamma^{(r)}).$$

The representation $U^i_r$ of Diff $X_k$ is equivalent to $V^{\rho^i_r}$ defined in § 1.1. By Proposition 3 of § 1.2, all the representations $V^{\rho^i_r}$ of Diff $X_k$ are irreducible and mutually pairwise inequivalent. Therefore Lemma 2 has the following corollaries.

COROLLARY 1. *The representations of* Diff $X_k$ *in the subspaces $L^2_{\mu_r}(B^{(r)}_{X_k} \times \Gamma_{X\setminus X_k}) \otimes W^i_r \otimes C^i_r$ $(r = 0, 1, 2, \ldots;\ i = 1, 2, \ldots)$ are primary and disjoint.*

COROLLARY 2. *Any invariant subspace under* Diff $X_k$

$$\tilde{\mathcal{L}}^i_{k,r}\subset(L^2_{\mu'_r}(B^{(r)}_{X_k})\otimes W^i_r)\otimes(L^2_{\mu''_r}(\Gamma_{X\setminus X_k})\otimes C^i_r)$$

*is of the form $\tilde{\mathcal{L}}^i_{k,r} = (L^2_{\mu'_r}(B^{(r)}_{X_k}) \otimes W^i_r) \otimes D^i_r$, where $D^i_r \subset L^2_{\mu''_r}(\Gamma_{X\setminus X_k}) \otimes C^i_r$.*

COROLLARY 3. *Any subspace $\mathcal{L} \subset L^2_\mu(\Gamma_X) \otimes H$ that is invariant under* Diff $X_k$ *splits into the direct sum $\mathcal{L} = \bigoplus_{r,i} \mathcal{L}^i_{k,r}$, where*

$$\mathcal{L}^i_{k,r}=\mathcal{L}\cap(L^2_{\mu_r}(B^{(r)}_{X_k}\times\Gamma_{X\setminus X_k})\otimes W^i_r\otimes C^i_r). \qquad (7)$$

**5. Proof of Theorem** 3. We use the notation and results of the preceding subsection. We denote by $L^\infty_\mu(\Gamma_X)$ the space of essentially bounded functions on $\Gamma_X$ with respect to $\mu$. Now $L^\infty_\mu(\Gamma_X)$ is a ring with the usual multiplication. Further, if $f \in L^2_\mu(\Gamma_X) \otimes H$ and $\varphi \in L^\infty_\mu(\Gamma_X)$, then $\varphi f \in L^2_\mu(\Gamma_X) \otimes H$.

LEMMA 3. *If $\mathcal{L} \subset L^2_\mu(\Gamma_X) \otimes H$ is invariant under* Diff $X$, *then $\mathcal{L}$ is invariant under multiplication by elements of $L^\infty_\mu(\Gamma_X)$.*

PROOF. We denote by $\mu_{X_k}$ the projection of $\mu$ onto $B_{X_k}$, and by $L^\infty_{\mu_{X_k}}(B_{X_k})$ the space of essentially bounded functions of $B_{X_k}$ with respect to the measure $\mu_{X_k}$ $(k = 1, 2, \ldots)$. We identify each space $L^\infty_{\mu_{X_k}}(B_{X_k})$ with its image under the natural map $L^\infty_{\mu_{X_k}}(B_{X_k}) \to L^\infty_\mu(\Gamma_X)$ (that is, the space of essentially bounded functions on $\Gamma_X$ that are constant on the fibres of the fibration $\Gamma_X \to B_{X_k}$). We consider the union $\bigcup_k L^\infty_{\mu_{X_k}}(B_{X_k})$.

It is obvious that for any $f \in L^2_\mu(\Gamma_X) \otimes H$ and $\varphi \in L^\infty_\mu(\Gamma_X)$ the product $\varphi f \in L^2_\mu(\Gamma_X) \otimes H$ is approximated in $L^2_\mu(\Gamma_X) \otimes H$ by elements $\varphi' f$, where $\varphi' \in \bigcup_k L^\infty_{\mu_{X_k}}(B_{X_k})$. Therefore, to prove the lemma it is sufficient to check that $\mathscr{L}$ is invariant under multiplication by elements of $L^\infty_{\mu_{X_k}}(B_{X_k})$ $(k = 1, 2, \ldots)$.

We fix $k$ and denote by $L^\infty_{\mu'_n}(B^{(n)}_{X_k})$ the subspace of functions in $L^\infty_{\mu_{X_k}}(B_{X_k})$ that are concentrated on $B^{(n)}_{X_k} \times \Gamma_{X-X_k}$, (here $B^{(n)}_{X_k}$ is the subspace of $n$-point subsets in $X_k$) $(n = 0, 1, \ldots)$. Obviously, for any $f \in L^2_\mu(\Gamma_X) \otimes H$ and $\varphi \in L^\infty_{\mu_{X_k}}(B_{X_k})$ the product $\varphi f$ is approximated in $L^2_\mu(\Gamma_X) \otimes H$ by finite sums of elements $\varphi_n f$, where $\varphi_n \in L^\infty_{\mu'_n}(B^{(n)}_{X_k})$. Thus, the proof of the lemma reduces to the following assertion:

*If* $\mathscr{L} \subset L^2_\mu(\Gamma_X) \otimes H$ *is invariant under* Diff $X_k$, *then* $\mathscr{L}$ *is invariant under multiplication by elements of* $L^\infty_{\mu'_n}(B^{(n)}_{X_k})$ $(n = 0, 1, 2, \ldots)$.

Let us prove this assertion. We use the decomposition $\mathscr{L} = \bigoplus_{r, i} \mathscr{L}^i_{k, r}$, where

$$\mathscr{L}^i_{k, r} = \mathscr{L} \cap (L^2_{\mu_r}(B^{(r)}_{X_k} \times \Gamma_{X \setminus X_k}) \otimes W^i_r \otimes C^i_r). \tag{8}$$

It is sufficient to prove the assertion for each subspace $\mathscr{L}^i_{k, r}$ separately. Note that when $r \neq n$, the supports of the functions in $L^r_{\mu_r}(B^{(r)}_{X_k} \times \Gamma_{X \setminus X_k})$ and in $L^\infty_{\mu'_n}(B^{(n)}_{X_k})$ do not intersect. Therefore, it is only necessary to consider the case $r = n$.

Let $\widetilde{\mathscr{L}}^i_{k, n}$ be the image of $\mathscr{L}^i_{k, n}$ under $\tau_n$. By Corollary 2 to Lemma 2, $\widetilde{\mathscr{L}}^i_{k, n}$ has the form $\widetilde{\mathscr{L}}^i_{k, n} = L^2_{\mu'_n}(B^{(n)}_{X_k}) \otimes W^i_r \otimes D^i_r$.

Hence it is clear that $\widetilde{\mathscr{L}}^i_{k, n}$ is invariant under multiplication by functions in $L_{\mu'_n}(B^{(n)}_{X_k})$. Since the corresponding elements in $\mathscr{L}^i_{k, n}$ and $\widetilde{\mathscr{L}}^i_{k, n}$ differ only by the factor $\left(\dfrac{d\mu'_n\, d\mu''_n}{d\mu_n}\right)^{1/2}$, the space $\mathscr{L}^i_{k, n}$ is also invariant under multiplication by elements of $L^\infty_{\mu'_n}(B^{(n)}_{X_k})$, and the lemma is proved.

We consider the space $H = H^\rho$, a factor in the tensor product $L^2_\mu(\Gamma_X) \otimes H$. A representation $R$ of $S_\infty$ is defined in $H$ by

$(R(\sigma)\varphi)(a) = \varphi(\sigma^{-1}a)$.

Note that $R \cong \mathrm{Ind}_{S_n \times S_\infty^n}^{S_\infty} (\rho \times I)$, where $S_\infty^n$ is the subgroup of finite permutations leaving $1, \ldots, n$ fixed and $I$ is the unit representation of $S^n$. Hence the next result follows easily.

LEMMA 4. *The representation $R$ of $S_\infty$ is irreducible.*

LEMMA 5. *Every subspace $\mathcal{L} \subset L^2_\mu(\Gamma_X) \otimes H$ that is invariant under* Diff $X$ *is also invariant under the operators $I \otimes R(\sigma)$, $\sigma \in S_\infty$, where $I$ is the unit operator in $L^2_\mu(\Gamma_X)$.*

PROOF. Since $S_\infty = \varinjlim S_p$, it is sufficient to prove that $\mathcal{L}$ is invariant under the operators $I \otimes R(\sigma)$, $\sigma \in S_p$ $(p = 1, 2, \ldots)$.

We use the notation and results of §3.4. Let $p$ be any fixed positive integer. We consider the subspace

$$\mathcal{L}_{k,p} = \sum_{r=p}^{\infty} \sum_i \oplus \mathcal{L}^i_{k,r} \quad (k = 1, 2, \ldots),$$

where $\mathcal{L}^i_{k,r}$ is defined by (8).

Clearly the union $\bigcup_{k=1}^{\infty} \mathcal{L}_{k,p}$ is everywhere dense in $\mathcal{L}$. Therefore, it is sufficient to prove that each space $\mathcal{L}^i_{k,r}$, $r \geqslant p$, is invariant under $I \otimes R(\sigma)$.

We consider the image $\widetilde{\mathcal{L}}^i_{k,r}$ of $\mathcal{L}^i_{k,r}$ under $\tau_r$. By Corollary 2 to Lemma 2 $\widetilde{\mathcal{L}}^i_{k,r}$ and hence its preimage $\mathcal{L}^i_{k,r}$ is invariant under $I \otimes R(\sigma)$, $\sigma \in S_r$. Since $p \leqslant r$, we have $S_p \subset S_r$, and so $\mathcal{L}^i_{k,r}$ is also invariant under the operators $I \otimes R(\sigma)$, $\sigma \in S_p$. The lemma is now proved.

LEMMA 6. *Every subspace $\mathcal{L} \subset L^2_\mu(\Gamma_X) \otimes H$ invariant under* Diff $X$ *is of the form $\mathcal{L} = L^2_\mu(A) \otimes H$, where $A \subset \Gamma_X$ is a measurable subset.*

PROOF. It follows from Lemmas 4 and 5 that

$$\mathcal{L} = E \otimes H,$$

where $E \subset L^2_\mu(\Gamma_X)$, and from Lemma 3 that $E$ is invariant under multiplication by functions from $L^\infty_\mu(\Gamma_X)$. Consequently, $E = L^2_\mu(A)$, where $A$ is a measurable subset in $\Gamma_X$, and the lemma is proved.

The assertion of Theorem 3 follows immediately from Lemma 6. For let $\mathcal{L} \subset L^2_\mu(\Gamma_X) \otimes H$ be an invariant subspace with respect to Diff $X$. Then from Lemma 6 we have $\mathcal{L} = L^2_\mu(A) \otimes H$, where $A \subset \Gamma_X$ is a measurable subset. Consequently, since $\mu$ is ergodic, either $\mathcal{L} = 0$ or $\mathcal{L} = L^2_\mu(\Gamma_X) \otimes H$. This proves Theorem 3, and with it Theorem 1.

REMARK 1. The assertion of the theorem remains true for any subgroup $G \subset$ Diff $X$ satisfying the following requirements:

1) $G$ acts ergodically on $\Gamma_X$;

2) for any open connected submanifold $Y \subset X$ with compact closure, Theorem 1 of §1 about the representations $V^\rho$ of Diff $Y$ remains true for

the restriction of the $V^p$ to $G \cap \text{Diff } Y$.

REMARK 2. The proof of Theorem 3 can be simplified considerably in the case $n = 0$, that is, for the representation $U_\mu$ in $L^2_\mu(\Gamma_X)$. In this case it reduces to proving Lemma 1 of §2.1 and establishing a functional version of the $0 - 1$ law. In the simplest case when $\mu \sim \mu'_{X_k}$ $(k = 1, 2, \dots)$ $(\mu'_{X_k}, \mu''_{X_k}$ are the projections of $\mu$ onto $B_{X_k}$ and $\Gamma_{X-X_k}$, respectively), this law consists of the following. Let $\mathcal{L}$ be a subspace of $L^2_\mu(\Gamma_X)$. If in terms of the decomposition $L^2_\mu(\Gamma_X) \cong L^2_{\mu'_{X_k}}(B_{X_k}) \otimes L^2_{\mu''_{X_k}}(\Gamma_{X-X_k})$ the subspace $\mathcal{L} \subset L^2_\mu(\Gamma_X)$ for any $k$ has the form

$$\mathcal{L} \cong L^2_{\mu'_{X_k}}(B_{X_k}) \otimes C_k, \quad C_k \subset L^2_{\mu''_{X_k}}(\Gamma_{X-X_k}),$$

then[1] either $\mathcal{L} = 0$ or $\mathcal{L} = L^2_\mu(\Gamma_X)$. An extra difficulty comes from the fact that there is no equivalence $\mu \sim \mu'_{X_k} \times \mu''_{X_k}$, generally speaking, and only the weaker relation $\mu \sim \sum_{r=0}^{\infty} \mu'_r \times \mu''_r$ is true (for the definition of $\mu'_r$ and $\mu''_r$ see p. 25).

## §4. Representations of Diff $X$ generated by the Poisson measure

**1. Properties of the Poisson measure.** Let $X$ be a non-compact manifold with a smooth positive measure $m$, $m(X) = \infty$, and let $\mu = \mu_\lambda$ be the Poisson measure on $\Gamma_X$ with parameter $\lambda$ corresponding to the measure $m$ on $X$ (for the definition of the Poisson measure, see §0.2).

Some basic properties of Poisson measure were stated in §0.6.

LEMMA 1. *If* $\dim X > 1$, *then for any two $\mu$-measurable sets* $A_1, A_2 \subset \Gamma_X$ *with positive measure there exists a diffeomorphism* $\psi \in \text{Diff}(X, m)$ *such that*[2] $\mu(A_1 \cap \psi A_2) > \frac{1}{2}\mu(A_1)\mu(A_2)$. ($\text{Diff}(X, m)$ *is the subgroup of diffeomorphisms preserving* $m$.)

PROOF. First we recall some definitions and facts. By a cylindrical set in $\Gamma_X$ we mean a set $A \subset \Gamma_X$, of positive measure, of the form $A = \pi_Y^{-1} A'$, where $Y$ is a compact set in $X$, $A' \subset B_Y$ is a measurable subset and $\pi_Y$ is the natural map $\Gamma_X \to B_Y$ $(\pi_Y \gamma = \gamma \cap Y)$; $Y$ is called the *carrier*[3] of $A$. Since the Poisson measure $\mu$ is infinitely decomposable, it follows that if the carriers of two cylindrical sets $A_1$ and $A_2$ intersect in a set of measure 0, then $\mu(A_1 \cap A_2) = \mu(A_1)\mu(A_2)$.

We recall that any $\mu$-measurable set $C \subset \Gamma_X$ can be approximated by cylindrical sets (that is, for any $\varepsilon > 0$ there is a cylindrical set $A$ such that $\mu(C \Delta A) < \varepsilon$).

---

[1] We recall that in these terms the usual $0 - 1$ law would be formulated as follows. Let $f \in L^2_\mu(\Gamma_X)$. If $f = 1_k \otimes f_k$ for any $k$, where $1_k$ is the constant in $L^2_{\mu'_{X_k}}(B_{X_k})$ and $f_k \in L^2_{\mu''_{X_k}}(\Gamma_{X-X_k})$, then $f = \text{const}$.

[2] For $X = \mathbf{R}^1$ the lemma is false, because in this case $\text{Diff}(X, m)$ is trivial.

[3] Of course, the carrier of a cylindrical set is not uniquely defined.

Hence it is clear that it is sufficient to prove the assertion of the lemma for cylindrical sets $A_1$ and $A_2$. Without loss of generality we may suppose further that $X = \mathbf{R}^n$, $n > 1$.

Let us establish the following property of Diff$(X, m)$: if $Y_1$ and $Y_2$ are two compact sets in $X$, then there is a diffeomorphism $\psi \in$ Diff$(X, m)$ such that $Y_1 \cap \psi Y_2 = \phi$.

For if $Y_1$, $Y_2$ are compact in $X = \mathbf{R}^n$ and $m$ is the Lebesgue measure in $\mathbf{R}^n$, then there exists a disc containing $Y_1$ and $Y_2$ and a rotation $\psi$ of the disc (this preserves $m$) such that $Y_1 \cap \psi Y_2 = \phi$; this rotation $\psi$ can be extended beyond the boundary of the disc to a finite diffeomorphism of $\mathbf{R}^n$ preserving $m$. If now $m$ is an arbitrary smooth positive measure in $\mathbf{R}^n$, then it is sufficient to use a lemma (see [21]), which states that any open ball in $\mathbf{R}^n$, $n > 1$, with a smooth measure $m$ can be mapped diffeomorphically onto itself so that $m$ goes over to the Lebesgue measure.

Let $Y_1$ and $Y_2$ be the carriers of $A_1$ and $A_2$. By what has just been proved, we can find a diffeomorphism $\psi \in$ Diff$(X, m)$ such that $Y_1 \cap \psi Y_2 = \phi$. But then $\mu(A_1 \cap \psi A_2) = \mu(A_1)\mu(\psi A_2) = \mu(A_1)\mu(A_2)$, and hence $\mu(A_1 \cap \psi A_2) > \frac{1}{2}\mu(A_1)\mu(A_2)$. The lemma is now proved.

THEOREM 1. *If* $\dim X > 1$, *then the Poisson measure* $\mu$ *in* $\Gamma_X$ *is ergodic with respect to* Diff$(X, m)$.

PROOF. Let $A \subset \Gamma_X$, $\mu(A) > 0$, be a subset that is invariant mod 0 under Diff$(X, m)$; we must prove that $\mu(A) = 1$. Suppose the contrary: that $\mu(\Gamma_X \setminus A) > 0$. Then by Lemma 1 there is a $\psi \in$ Diff$(X, m)$ such that $\mu(\Gamma_X \setminus A) \cap \psi A) > 0$; hence, since $A$ is invariant, $\mu((\Gamma_X \setminus A) \cap A) > 0$, which is false. This proves the theorem.

**2. The representation of Diff $X$ generated by the Poisson measure.** Let $\mu = \mu_\lambda$ be the Poisson measure on $\Gamma_X$ with parameter $\lambda > 0$. In §3 we have associated with each quasi-invariant measure $\mu$ on $\Gamma_X$ a unitary representation $U_\mu$ of Diff $X$ in $L^2_\mu(\Gamma_X)$ defined by

$$(U_\mu(\psi) f)(\gamma) = \left(\frac{d\mu(\psi^{-1}\gamma)}{d\mu(\gamma)}\right)^{1/2} f(\psi^{-1}\gamma),$$

and also a set of elementary representations $U^\rho_\mu$. For the Poisson measure $\mu_\lambda$ the theory of such representations can be advanced considerably further than in the general case. In particular, it is possible to describe the corresponding representation ring. Furthermore, for $\mu_\lambda$ the representations $U_{\mu_\lambda}$ can be realized in the form $\mathrm{EXP}_\beta\, T$ (see [1] and [2]).

*In what follows we write* $U_\lambda$ *(instead of* $U_{\mu_\lambda}$*) for representation generated by the Poisson measure* $\mu_\lambda$.

Since $\mu_\lambda$ is ergodic, by Theorem 1 of §3, $U_\lambda$ is irreducible.

**3. The spherical function of the representation $U_\lambda$.** Let us assume that $\dim X > 1$. We consider the subgroup Diff$(X, m) \subset$ Diff $X$ of diffeomorphisms preserving the measure $m$; for us this subgroup will play a role similar to that of maximal compact subgroups in the theory of representations of semisimple Lie groups.

Since $\mu$ is invariant under Diff($X$, $m$) and $U_\lambda$ is infinitely decomposable, the restriction of $U_\lambda$ to Diff($X$, $m$) is given by

$$(1)\qquad (U_\lambda(\psi)f)(\gamma) = f(\psi^{-1}\gamma), \quad \psi \in \text{Diff}\,(X, m).$$

In view of Theorem 1 there is in $L^2_{\mu_\lambda}(\Gamma_X)$ one, and up to a multiplicative factor, only one vector that is invariant under Diff($X$, $m$), namely, $f_0 \equiv 1$.

DEFINITION. The following function on Diff $X$ is called the *spherical function* of $U_\lambda$:

$$u_\lambda(\psi) = \langle U_\lambda(\psi)f_0, f_0\rangle,$$

where the brackets denote the inner product in $L^2_{\mu_\lambda}(\Gamma_X)$.

Let us find an explicit form for the spherical function. Let supp $\psi \subset Y$, $m(Y) < \infty$. We denote by $\tilde{\mu}_\lambda$ the projection of $\mu_\lambda$ onto $B_Y$ and by $\tilde{\mu}^n_\lambda$ the restriction of $\tilde{\mu}_\lambda$ to $B_Y^{(n)}$. Then

$$u_\lambda(\psi) = \int_{\Gamma_X} \Big(\prod_{x\in\gamma} \mathcal{J}_\psi^{1/2}(x)\Big)\, d\mu_\lambda(\gamma) = \int_{B_Y} \Big(\prod_{x\in\gamma} \mathcal{J}_\psi^{1/2}(x)\Big)\, d\tilde{\mu}_\lambda(\gamma) =$$

$$= \sum_{n=0}^{\infty} \int_{B_Y^{(n)}} \Big(\prod_{x\in\gamma} \mathcal{J}_\psi^{1/2}(x)\Big)\, d\tilde{\mu}^n_\lambda(\gamma) =$$

$$= \sum_{n=0}^{\infty} \frac{e^{-\lambda m(Y)}[\lambda m(Y)]^n}{n!} \Big(\frac{1}{m(Y)} \int_Y \mathcal{J}_\psi^{1/2}(x)\, dm(x)\Big)^n =$$

$$= e^{-\lambda m(Y)} e^{\lambda \int_Y \mathcal{J}_\psi^{1/2}(x)\,dm(x)} = e^{\lambda \int_Y (\mathcal{J}_\psi^{1/2}(x) - 1)\, dm(x)}$$

So we obtain

$$(2)\qquad u_\lambda(\psi) = \exp\Big(\lambda \int_X (\mathcal{J}_\psi^{1/2}(x) - 1)\, dm(x)\Big),$$

where $\mathcal{J}_\psi(x) = \dfrac{dm(\psi^{-1}x)}{dm(x)}$. Since $f_0$ is defined invariantly in the representation space $L^2_{\mu_\lambda}(\Gamma_X)$ and since, by (2), $u_{\lambda_1} \neq u_{\lambda_2}$ when $\lambda_1 \neq \lambda_2$, we obtain the following theorem.

THEOREM 2. *The representations* $U_{\lambda_1}$ *and* $U_{\lambda_2}$ *of* Diff $X$ (dim $X > 1$) *are not equivalent when* $\lambda_1 \neq \lambda_2$.

**4. The Gaussian form of the representation $U_\lambda$ of Diff $X$.** Let us consider the *real* Hilbert space $H = L^2_m(X)$, where $m$ is a smooth positive measure on $X$. A unitary representation $T$ of Diff $X$ is given in $H$ by $(T(\psi)f)(x) = \mathcal{J}_\psi^{1/2}(x)f(\psi^{-1}x)$, where $\mathcal{J}_\psi(X) = \dfrac{dm(\psi^{-1}x)}{dm(x)}$. A 1-cocycle

$\beta$: Diff $X \to H$ is given by $[\beta(\psi)](x) = \mathcal{J}_\psi^{1/2}(x) - 1$.

In accordance with [1] and [2] this is a way of constructing a new representation $\widetilde{U} = \mathrm{EXP}_\beta T$ of Diff $X$.

We denote by $\widetilde{\mu}$ a measure in the space $\mathcal{F}(X)$ of generalized functions on $X$ given by its characteristic functional:

(3) $$\int_{\mathcal{F}(X)} e^{i\langle F, f\rangle} d\widetilde{\mu}(F) = e^{-\|f\|^2/2},$$

where $\|\cdot\|$ is the norm in $L^2_m(X)$. We call $\widetilde{\mu}$ the *standard Gaussian measure* in $\mathcal{F}(X)$.

The representation $\widetilde{U} = \mathrm{EXP}_\beta T$ is given in $L^2_{\widetilde{\mu}}(\mathcal{F}(X))$ by

$$(\widetilde{U}(\psi)\Phi)(F) = e^{i\langle F, \beta(\psi)\rangle}\Phi(T^*(\psi)F),$$

where the operator $T^*(\psi)$ is defined by $\langle T^*(\psi)F, f\rangle = \langle F, T(\psi)f\rangle$.

LEMMA 2. *The vector $\Phi_0 \equiv 1$ in $L^2_{\widetilde{\mu}}(\mathcal{F}(X))$ is cyclic with respect to* Diff $X$.

PROOF. Since $(\widetilde{U}(\psi)\Phi_0)(F) = e^{i\langle F, \beta(\psi)\rangle}$, the assertion of the lemma is just that the set of functionals $e^{i\langle F, \beta(\psi)\rangle}$, $\psi \in$ Diff $X$, is total in $L^2_{\widetilde{\mu}}(\mathcal{F}(X))$, that is, the minimal linear subspace $\mathcal{L} \subset L^2_{\widetilde{\mu}}(\mathcal{F}(X))$ containing them is $L^2_{\widetilde{\mu}}(\mathcal{F}(X))$ itself. It is known [1] that the functionals of the form

(4) $$\Phi(F) = \prod_{i=1}^{n} \langle F, f_i\rangle,$$

where $f_1, \ldots, f_n$ are smooth finite functions on $X$ ($n = 0, 1, 2, \ldots$) form a total set in $L^2_{\widetilde{\mu}}(\mathcal{F}(X))$; therefore, it is sufficient to prove that $\mathcal{L}$ contains all functionals of the form (4).

Let $f$ be any smooth finite function on $X$ satisfying $\int f(x)dm(x) = 0$, let $\tau$ be any real number such that $1 - \tau f(x) > 0$ for all $x \in X$. A measure $m_\tau$ in $X$ is given by $dm_\tau(x) = (1 - \tau f(x))dm(x)$. The measures $m$ and $m_\tau$ coincide outside a compact set $Y \supset \mathrm{supp}\, f$, and $m(Y) = m_\tau(Y)$. Therefore, by a theorem of Moser [21], there exists a diffeomorphism $\psi \in$ Diff $X$ carrying $m$ to $m_\tau$, that is, $\mathcal{J}_\psi(x) = 1 - \tau f(x)$. But then $\beta(\psi) = \sqrt{(1 - \tau f(x))} - 1$, and hence the functional $e^{i\langle F, \sqrt{(1-\tau f(x))} - 1\rangle}$ belongs to $\mathcal{L}$ for any sufficiently small $\tau$. Hence all the terms in the expansion of $e^{i\langle F, \sqrt{(1-\tau f(x))} - 1\rangle}$ as a power series in $\tau$ also belong to $L$.

The coefficient of $\tau^n$ in this expansion is $c_n\langle F, f\rangle^n$, $c_n \neq 0$, apart from terms of the form $\prod_{i=1}^{k} \langle F, f_i\rangle$, $k < n$. Therefore, by induction on $n$, we can verify that $\mathcal{L}$ contains all functionals of the form $\langle F, f\rangle^n$, where $\int f(x)dm(x) = 0$, and hence those of the form $\langle F, f\rangle^n$, where $f$ is an arbitrary smooth finite function. Since the functionals (4) can be presented as linear combinations of the $\langle F, f\rangle^n$, they also belong to $\mathcal{L}$, and the lemma is proved.

If in the definition of $\widetilde{U}$ we replace the 1-cocycle $\beta$ by $s\beta$, where $s$ is any real number, we obtain a one-parameter family of unitary representations of Diff $X$: $\widetilde{U}_s = \mathrm{EXP}_{s\beta} T$.

It is obvious that the assertion of lemma 2 remains true for all representations $\widetilde{U}_s$, $s \neq 0$.

THEOREM 3. *If* $s \neq 0$, *then* $\widetilde{U}_s \cong \widetilde{U}_{s^2}$, *where* $\widetilde{U}_{s^2}$ *is the representation of* Diff $X$ *generated by the Poisson measure with parameter* $s^2$ (see ≠ 4.2).

PROOF. We compute the matrix element $\langle U_s(\psi)\Phi_0, \Phi_0\rangle$, where $\Phi_0 \equiv 1$. From formula (3) for the characteristic functional of $\widetilde{\mu}$ we immediately obtain

$$\langle U_s(\psi)\Phi_0, \Phi_c\rangle = \exp\left(s^2 \int\limits_X (\mathcal{J}_\psi^{1/2}(x) - 1)\, dm(x)\right) = u_{s^2}(\psi),$$

where $u_{s^2}$ is the spherical function of $U_{s^2}$. The assertion of the theorem follows from this and the fact that $\Phi_0$ is cyclic.

Let us now consider the special case $s = 0$. Then (see [1]) $\widetilde{U}_0$ splits into the direct sum: $\widetilde{U}_0 = T^0 \oplus T^1 \oplus \ldots \oplus T^n \oplus \ldots$, where $T^0$ is the unit representation and $T^n$ ($n \geqslant 1$) is the $n$-th symmetrized power of $T$ introduced at the beginning of §3.4. In the notation of §1 we have $T^n = V^{\rho_n^0}$, where $\rho_n^0$ is the unit representation of $S_n$. Thus, $\widetilde{U}_0 = \bigoplus_{n=0}^{\infty} V^{\rho_n^0}$.

**5. Elementary representations of Diff $X$ associated with the Poisson measure.** According to §3, the tensor product $U_\lambda^\rho = U_\lambda \otimes V^\rho$ of representations $U_\lambda$ and $V^\rho$ of Diff $X$ is called an *elementary representation*. By Theorem 1 of §3, an *elementary representation* $U_\lambda^\rho$ *is irreducible if and only if* $\rho$ *is an irreducible representation of* $S_n$ $(n = 0, 1, 2, \ldots)$.

THEOREM 4. *Two irreducible elementary representations* $U_{\lambda_1}^{\rho_1}$ *and* $U_{\lambda_2}^{\rho_2}$ *of* Diff $X$, $\dim X > 1$, *are equivalent if and only if* $\lambda_1 = \lambda_2$ *and* $\rho_1$ *and* $\rho_2$ *are equivalent representations of* $S_n$.

PROOF. We restrict $U_\lambda^\rho$ to Diff$(X, m)$. From §4.4 it follows that $\overline{U}_\lambda \cong \bigoplus_{n=0}^{\infty} V^{\rho_n}$, where $\rho_n^0$ is the unit representation of $S_n$ (the bar denotes the restriction to Diff$(X, m)$). Consequently,

$$\overline{U}_\lambda^\rho \cong \overline{V}^\rho \oplus \sum_{n=1}^{\infty} \overline{V}^{\rho\circ\rho_n^0}. \tag{5}$$

From the decomposition (5) it follows easily on the basis of Theorem 2 in §2 that $\overline{U}_{\lambda_1}^{\rho_1} \not\sim \overline{U}_{\lambda_2}^{\rho_2}$ when $\rho_1 \not\sim \rho_2$. Hence, a fortiori, $U_{\lambda_1}^{\rho_1} \not\sim U_{\lambda_2}^{\rho_2}$ when $\rho_1 \not\sim \rho_2$. It only remains to discuss the representations $U_{\lambda_1}^{\rho}$ and $U_{\lambda_2}^{\rho}$.

Let $U_{\lambda_1}^{\rho} \not\sim U_{\lambda_2}^{\rho}$. We denote by $f_{\lambda_i}$ the vacuum vector in the representation space of $U_{\lambda_i}$ $(i = 1, 2)$ and by $F$ an arbitrary vector in the representation space of $V^\rho$. From (5) it follows easily that under an isomorphism of the representation spaces of $U^\rho_{\lambda_1}$ and $U^\rho_{\lambda_2}$ the vector $f_{\lambda_1} \otimes F$ goes over into the vector $f_{\lambda_2} \otimes F$; therefore,

$$\langle U^\rho_{\lambda_1}(\psi) f_{\lambda_1} \otimes F,\ f_{\lambda_1} \otimes F \rangle = \langle U^\rho_{\lambda_2}(\psi) f_{\lambda_2} \otimes F,\ f_{\lambda_2} \otimes F \rangle$$

for any $\psi \in \mathrm{Diff}\, X$. Hence $u_{\lambda_1} = u_{\lambda_2}$, where $u_\lambda$ is the spherical function for $U_\lambda$ (see §4.3), and therefore, $\lambda_1 = \lambda_2$.

## §5. The ring of elementary representations generated by the Poisson measure

**1. The decomposition of the tensor product $U_{\lambda_1} \otimes U_{\lambda_2}$ of representations of Diff $X$ into irreducible representations.** First we prove a general theorem about representations $\mathrm{EXP}_\beta T$. Let $G$ be an arbitrary group and $T$ a unitary representation of $G$ in a real space $H$; let $\beta$: $G \to H$ be a 1-cocycle. Then a new unitary representation $U_s = \mathrm{EXP}_{s\beta} T$ can be defined as in [1] and [2] where $s$ is an arbitrary real number. Let $H'$ be the dual space to $H$ and $\widetilde{\mu}$ a measure in any nuclear completion $\widetilde{H}$ of $H'$, defined by the characteristic functional:

$$\int e^{i\langle F,\ f\rangle}\, d\mu(F) = e^{-\|f\|^2/2}.$$

The operators of the representation $\widetilde{U}_s = \mathrm{EXP}_{s\beta} T$ act in the complex Hilbert space $L^2_{\widetilde{\mu}}(\widetilde{H})$ according to the formula:

$$(\widetilde{U}_s(g)\Phi)(F) = e^{is\langle F,\ \beta(g)\rangle}\Phi(T^*(g)F).$$

THEOREM 1. *If $s_1^2 + s_2^2 = s_1'^2 + s_2'^2$, then $\widetilde{U}_{s_1} \otimes \widetilde{U}_{s_2} \cong \widetilde{U}_{s_1'} \otimes \widetilde{U}_{s_2'}$.*

PROOF. We define operators $A_t$, $t \in \mathbf{R}$, in $L^2_{\widetilde{\mu}}(\widetilde{H}) \otimes L^2_{\widetilde{\mu}}(\widetilde{H})$ by the formula:

$$(A_t\Phi)(F_1,\ F_2) = \Phi(\cos t\, F_1 + \sin t\, F_2,\ \ -\sin t\, F_1 + \cos t\, F_2).$$

From the definition of the Gaussian measure $\widetilde{\mu}$ it follows that $A_t$ for any $t \in \mathbf{R}$ is a unitary operator. Further, from the definition of $\widetilde{U}_s$ it follows easily that

$$A_t^{-1}(\widetilde{U}_{s_1}(g) \otimes \widetilde{U}_{s_2}(g))\, A_t = \widetilde{U}_{s_1 \cos t + s_2 \sin t}(g) \otimes \widetilde{U}_{-s_1 \sin t + s_2 \cos t}(g).$$

Consequently, $\widetilde{U}_{s_1} \otimes \widetilde{U}_{s_2} \cong \widetilde{U}_{s_1 \cos t + s_2 \sin t} \otimes \widetilde{U}_{-s_1 \sin t + s_2 \cos t}$ for any $t \in \mathbf{R}$; hence the assertion of the theorem follows immediately.

COROLLARY. $\widetilde{U}_{s_1} \otimes \widetilde{U}_{s_2} \cong \widetilde{U}_{\sqrt{(s_1^2 + s_2^2)}} \otimes \widetilde{U}_0$.

Now let $U_\lambda$ be the representation of Diff $X$ generated by the Poisson measure with parameter $\lambda$ (see §4), let $V^\rho$ be the representation of Diff $X$ defined in §1 ($\rho$ runs over the representations of $S_n$).

THEOREM 2.

$$U_{\lambda_1} \otimes U_{\lambda_2} \cong \bigoplus_{n=0}^{\infty} (U_\lambda \otimes V^{\rho_n^0}), \tag{1}$$

*where* $\lambda = \lambda_1 + \lambda_2$, *and* $\rho_n^0$ *is the unit representation of* $S_n$. *Every term in* (1) *is an irreducible representation of* Diff $X$.

PROOF. Let $T$ be a representation of Diff $X$ in $L_m^2(X)$ defined by $(T(\psi)f)(x) = \mathcal{J}_\psi^{1/2}(x)f(\psi^{-1}x)$. A 1-cocycle $\beta$: Diff $X \to L_m^2(X)$ is given by $[\beta(\psi)](x) = \mathcal{J}_\psi^{1/2}(x) - 1$. We consider the representation $\widetilde{U}_s = \mathrm{EXP}_{s\beta} T$ of Diff $X$. By Theorem 1 we have, for any $\lambda_1 > 0$, $\lambda_2 > 0$

$$\widetilde{U}_{\sqrt{\lambda_1}} \otimes \widetilde{U}_{\sqrt{\lambda_2}} \cong \widetilde{U}_{\sqrt{\lambda_1+\lambda_2}} \otimes \widetilde{U}_0. \tag{2}$$

On the other hand, it was proved in §4.4 that $\widetilde{U}_{\sqrt{\lambda}} \cong U_\lambda$ for any $\lambda > 0$, and $\widetilde{U}_0 = \bigoplus_{n=0}^{\infty} V^{\rho_n^0}$. Consequently, (2) implies that

$U_{\lambda_1} \otimes U_{\lambda_2} \cong \bigoplus_{n=0}^{\infty} (U_\lambda \otimes V^{\rho_n^0})$, where $\lambda = \lambda_1 + \lambda_2$.

The irreducibility of $U_\lambda \otimes V^{\rho_n^0}$ follows from the main theorem of §3.

2. **The decomposition of the tensor product of two elementary representations of Diff $X$ associated with the Poisson measure.**

THEOREM 3. $U_{\lambda_1}^{\rho_1} \otimes U_{\lambda_2}^{\rho_2} \cong \bigoplus_{n=0}^{\infty} U_{\lambda_1+\lambda_2}^{\rho_1 \circ \rho_2 \circ \rho_n^0}$.

(For the definition of the operation $\rho_1 \circ \rho_2$, see §1.)

PROOF. By definition, $U_{\lambda_1}^{\rho_1} = U_{\lambda_1} \otimes V^{\rho_1}$, $U_{\lambda_2}^{\rho_2} = U_{\lambda_2} \otimes V^{\rho_2}$. Consequently, $U_{\lambda_1}^{\rho_1} \otimes U_{\lambda_2}^{\rho_2} \cong (U_{\lambda_1} \otimes U_{\lambda_2}) \otimes (V^{\rho_1} \otimes V^{\rho_2})$. By Theorem 2,

$U_{\lambda_1} \otimes U_{\lambda_2} = \bigoplus_{n=0}^{\infty} (U_{\lambda_1+\lambda_2} \otimes V^{\rho_n^0})$. Further, $V^{\rho_1} \otimes V^{\rho_2} \cong V^{\rho_1 \circ \rho_2}$ for any $\rho_1, \rho_2$ (see §1). Hence the required result is obtained straightaway.

COROLLARY. *The set of representations of* Diff $X$ *that split into the direct sum of irreducible representations of the form* $U_\lambda \otimes V^\rho$ *is closed under the operation of taking the tensor product.*

## §6. Representations of Diff $X$ associated with infinitely divisible measures

The group Diff $X$ acts naturally in the space $\mathcal{F}(X)$ of generalized functions on $X$. Therefore, representations of Diff $X$ can be constructed for any quasi-invariant measure $\widetilde{\mu}$ on $\mathcal{F}(X)$. We have already noted earlier that the configuration space $\Gamma_X$ has a natural embedding in $\mathcal{F}(X)$, and, therefore, the representations considered earlier are part of a considerably wider class of representations. Here we consider a special class of measures in $\mathcal{F}(X)$ (infinitely decomposable measures), which generate the same stock of representations as the representations $U_\lambda \otimes V^\rho$ studied in §§4 and 5.

**1. The measure $\widetilde{\mu}_\tau$ in the space of generalized functions on $X$.** Let $\mathcal{F}(X)$ be the space of generalized functions on $X$; we define an action of Diff $X$ on $\mathcal{F}(X)$ by $\langle \psi^* F, f\rangle = \langle F, f \circ \psi\rangle$.

Let us consider a positive definite function $\lambda_\tau$ on $\mathbf{R}$ of the form

$$\chi_\tau(t) = \exp\left(\int_{-\infty}^{+\infty} (e^{i\alpha t} - 1)\, d\tau(\alpha)\right),$$

where $\tau$ is a non-negative finite measure on $\mathbf{R}$ (not necessarily normalized). ($\chi_\tau(t)$ is the Fourier transform of a certain infinitely divisible measure on $\mathbf{R}$.) Let $m$ be a fixed smooth positive measure on $X$, $m(X) = \infty$. A measure $\widetilde{\mu} = \widetilde{\mu}_\tau$ is given on $\mathcal{F}(X)$ by the characteristic functional

$$(1) \quad L_\tau(f) = \exp\left(\int_X \ln \chi_\tau(f(x))\, dm(x)\right) = \exp\left(\int_{\mathbf{R}}\int_X (e^{i\alpha f(x)} - 1) dm(x) d\tau(\alpha)\right).$$

We list some basic properties of $\widetilde{\mu}$, which follow easily from this definition.

1) If $X = X_1 \cup \cdots \cup X_n$ is a finite partitioning of $X$ and $\mathcal{F}(X) = \mathcal{F}(X_1) \oplus \ldots \oplus \mathcal{F}(X_n)$ the corresponding decomposition of $\mathcal{F}(X)$ into a direct sum, then $\widetilde{\mu} = \widetilde{\mu}_1 \times \ldots \times \widetilde{\mu}_n$, where $\widetilde{\mu}_i$ is the projection of $\mu$ onto the subspace $\mathcal{F}(X_i)$, $i = 1, \ldots, n$ (infinite decomposability).

2) The measure $\widetilde{\mu} = \widetilde{\mu}_\tau$ is concentrated on the set $\mathcal{F}^0(X)$ of generalized functions of the form $\sum_{k=1}^{\infty} \alpha_k \delta_{x_k}$, $\alpha_k \neq 0$, where $\alpha_k \in \operatorname{supp} \tau$, and $\{x_k\}$ is a set without accumulation points in $X$ (that is, a configuration in $X$); $\delta_x$ denotes the delta-function on $X$ concentrated at $x \in X$.[1]

3) The measure $\mu$ is quasi-invariant under Diff $X$, and

$$\left.\frac{d\widetilde{\mu}(\psi^* F)}{d\widetilde{\mu}(F)}\right|_{F = \sum_{k=1}^{\infty} \alpha_k \delta_{x_k}} = \prod_{k=1}^{\infty} \mathcal{J}_\psi(x_k),$$

where $\mathcal{J}_\psi(x) = \dfrac{dm(\psi^{-1}x)}{dm(x)}$. (Since $\psi$ is finite, only finitely many factors $\mathcal{J}_\psi(x_k)$ are distinct from 1.)

Let us note the particular case when $\tau$ is concentrated at one point, $\alpha = 1$. Then $\mathcal{F}^0(X) = \{\sum \delta_{x_i};\ \{x_i\} \in \Gamma_X\}$, where $\Gamma_X$ is the configuration $\mathcal{F}^0(X) \to \Gamma_X$ in which each generalized function $\Sigma\delta_{x_i}$ is associated with a configuration $\{x_i\} \in \Gamma_X$. It is not difficult to verify that the image of $\mu$ under this map is the Poisson measure on $\Gamma_X$ with parameter $\lambda = \tau(\mathbf{R})$.

[1] The converse is also true: any infinitely decomposable measure in $\mathcal{F}(X)$ concentrated on a set of the type indicated is a measure $\mu_\tau$ for a certain $\tau$; when $X = \mathbf{R}^1$, this fact is very well known (see, for example, J.L. Doob, Stochastic processes, Wiley & Sons, New York 1953.

**2. The representation of Diff $X$ associated with $\tilde{\mu}_\tau$.** A unitary representation of Diff $X$ is given in $L^2_{\tilde{\mu}}(\mathcal{F}(X))$ by the formula

$$(V_\tau(\psi)\Phi)(F) = \left(\frac{d\tilde{\mu}_\tau(\psi^* F)}{d\tilde{\mu}_\tau(F)}\right)^{1/2}\Phi(\psi^* F). \tag{2}$$

In the particular case when the measure $\tau$ on $\mathbf{R}$ is concentrated at a single point and is equal to $\lambda$ at this point, the representation $V_\tau$ so constructed is equivalent to the representation $U_\lambda$ corresponding to the Poisson measure with parameter $\lambda$ (see the remark above). Here we shall obtain the decomposition of $V_\tau$ into elementary representations.

We denote by $\mathcal{F}_0(X)$ the set of all generalized functions of the form $\Sigma\, \alpha_k \delta_{x_k}$, $\alpha_k \neq 0$, where $\{x_k\} \in \Gamma_X$; it was mentioned above that $\mathcal{F}_0(X)$ is a subset of full measure in $\mathcal{F}(X)$. We introduce the space $\mathbf{R}^\infty = \prod_{i=1}^{\infty} \mathbf{R}_i$, $\mathbf{R}_i = \mathbf{R}$ with measure $\nu = \tau_0 \times \ldots \times \tau_0 \times \ldots$, where $\tau_0$ is the normalized measure on $\mathbf{R}$: $\tau_0 = \frac{1}{\lambda}$, $\lambda = \tau(\mathbf{R})$.

Next, let $i(\gamma, x)$ be an admissible indexing (for the definition see §2) in $X$ and consider the sequence of maps $a_k\colon \Gamma_X \to X$ $(k = 1, 2, \ldots)$ defined by $a_k(\gamma) \in \gamma$, $i(\gamma, a_k(\gamma)) = k$. We define a map

$$\pi\colon \mathcal{F}_0(X) \to \Gamma_X \times \mathbf{R}^\infty$$

by

$$\pi\left(\sum_{x\in\gamma} \alpha_x \delta_x\right) = (\gamma;\ \alpha_{a_1(\gamma)}, \ldots, \alpha_{a_k(\gamma)}, \ldots).$$

Standard arguments establish the following result.

LEMMA 1. *The map $\pi$ is measurable in both directions; the image of $\mathcal{F}_0(X)$ is a subset of full measure in $(\Gamma_X \times \mathbf{R}^\infty; \mu \times \nu)$; the image of $\tilde{\mu}_\tau$ under $\pi$ is the product measure $\mu \times \nu$, where $\mu$ is the Poisson measure on $\Gamma_X$ with parameter $\lambda = \tau(\mathbf{R})$.*

By means of $\pi$ the action of Diff $X$ can be carried over from $\mathcal{F}_0(X)$ to $\Gamma_X \times \mathbf{R}^\infty$. It is not difficult to see that the action of Diff $X$ on $\Gamma_X \times \mathbf{R}^\infty$ is given by

$$\psi\colon (\gamma, \alpha) \mapsto (\psi^{-1}\gamma,\ \alpha\sigma(\psi, \gamma)), \tag{3}$$

where $\alpha = (\alpha_1, \ldots, \alpha_n, \ldots)$, and $\alpha\sigma = (\alpha_{\sigma(1)}, \ldots, \alpha_{\sigma(n)}, \ldots)$; here $\sigma(\psi, \gamma)$ is the map Diff $X \times \Gamma_X \to S_\infty$ defined by $i$ (see §2).

Lemma 1 and (3) imply the next result.

LEMMA 2. *The representation $V_\tau$ of* Diff $X$ *is equivalent to the representation acting on $L^2_\mu(\Gamma_X) \otimes L^2_\nu(\mathbf{R}^\infty)$ according to the formula*

$$(V_\tau(\psi) f)(\gamma, \alpha) = \left(\frac{d\mu(\psi^{-1}\gamma)}{d\mu(\gamma)}\right)^{1/2} f(\psi^{-1}\gamma,\ \alpha\sigma(\psi, \gamma)). \tag{4}$$

Here $\mu$ is the Poisson measure with parameter $\lambda = \tau(\mathbf{R})$, and $\nu = \tau_0 \times \ldots \times \tau_0 \times \ldots$

Let us now consider the space $L^2_\nu(\mathbf{R}^\infty) = L^2_{\tau_0}(\mathbf{R}) \otimes \ldots \otimes L^2_{\tau_0}(\mathbf{R}) \otimes \ldots$ with a given unitary representation $T$ of $S_\infty$: $(T(\sigma)\varphi)(\alpha) = \varphi(\alpha\sigma)$. We split $L^2_\nu(\mathbf{R}^\infty)$ into invariant subspaces.

We fix an orthonormal basis $e_0, e_1, \ldots$ in $L^2_{\tau_0}(\mathbf{R})$, where $e_0 \equiv 1$ (that is, $e_0$ is the function on $\mathbf{R}$ everywhere equal to 1). It is known that the vectors $e_{i_1} \otimes \ldots \otimes e_{i_k} \otimes \ldots$, where $\Sigma i_k < \infty$ (so that all the indices $i_k$ apart from finitely many are zero) form a basis in $L^2_\nu(\mathbf{R})$. Now $S_\infty$ permutes these vectors. We divide the set $A$ of basis vectors into orbits under $S_\infty$.

Let us consider all possible collections of natural numbers of the form

$$(n_1, \ldots, n_k; i_1, \ldots, i_k), \tag{5}$$

where $n_1 \geqslant \ldots \geqslant n_k$, $i_p \neq i_q$ when $p \neq q$, and $i_p > i_{p+q}$ if $n_p = n_{p+1}$; $k = 0, 1, \ldots$ ($k = 0$ corresponds to the empty set). With each collection (5) we associate a basis vector in $A$:

$$e_{i_1} \underbrace{\otimes \ldots \otimes e_{i_1}}_{n_1} \otimes \ldots \otimes e_{i_k} \underbrace{\otimes \ldots \otimes e_{i_k}}_{n_k} \otimes e_0 \otimes e_0 \otimes \ldots . \tag{6}$$

We denote by $A^{i_1, \ldots, i_k}_{n_1, \ldots, n_k}$ the orbit of $S_\infty$ in $A$ generated by (6) and by $H^{i_1, \ldots, i_k}_{n_1, \ldots, n_k}$ the subspace of $L^2_\nu(\mathbf{R}^\infty)$ spanned by the vectors of $A^{i_1, \ldots, i_k}_{n_1, \ldots, n_k}$.

It is easy to establish that the orbits $A^{i_1, \ldots, i_k}_{n_1, \ldots, n_k}$ are pairwise distinct and that their union is the whole of $A$. Hence it follows that $L^2_\nu(\mathbf{R}^\infty)$ splits into a direct sum of invariant subspaces

$$L^2_\nu(\mathbf{R}^\infty) = \oplus H^{i_1, \ldots, i_k}_{n_1, \ldots, n_k}; \tag{7}$$

the sum is taken over all collections $(n_1, \ldots, n_k; i_1, \ldots, i_k)$ of the type (5). Note that representations of $S_\infty$ in $H^{i_1, \ldots, i_k}_{n_1, \ldots, n_k}$ and in $H^{i'_1, \ldots, i'_k}_{n_1, \ldots, n_k}$ are equivalent.

The decomposition (7) we have just obtained leads to the following result.

LEMMA 3. *The space $L^2_\mu(\Gamma_X) \times L^2_\nu(\mathbf{R}^\infty)$ decomposes into the direct sum of $V_\tau$-invariant subspaces*:

$$L^2_\mu(\Gamma_X) \otimes L^2_\nu(\mathbf{R}^\infty) = \sum \oplus L^2_\mu(\Gamma_X) \otimes H^{i_1, \ldots, i_k}_{n_1, \ldots, n_k}. \tag{8}$$

*The representations of* Diff $X$ *in* $L^2_\mu(\Gamma_X) \otimes H^{i_1, \ldots, i_k}_{n_1, \ldots, n_k}$ and in $L^2_\mu(\Gamma_X) \otimes H^{i'_1, \ldots, i'_k}_{n_1, \ldots, n_k}$ *are equivalent.*

We denote by $V^{n_1, \ldots, n_k}_\tau$ the restriction of $V_\tau$ to $L^2_\mu(\Gamma_X) \otimes H^{i_1, \ldots, i_k}_{n_1, \ldots, n_k}$.

LEMMA 4. $V^{n_1, \ldots, n_k}_\tau \cong U^{\rho^0_{n_1} \circ \ldots \circ \rho^0_{n_k}}_\lambda$ *where* $\lambda = \tau(\mathbf{R})$, *and* $\rho^0_n$ *is the unit representation of* $S_n$.

The assertion of the lemma is easily established if we use the realization of elementary representations introduced in §3.3.

Lemmas 3 and 4 give the next result.

THEOREM 1. *The representation* $V_\tau$ *of* Diff $X$ *in* $L^2_{\mu_\tau}(\mathcal{F}(X))$ *defined by* (2) *splits into a discrete direct sum of irreducible elementary representations of the form* $U^\rho_\lambda$, *where* $\lambda = \tau(\mathbf{R})$.

REMARK 1. If $\tau$ is concentrated at two points on **R**, and if the measures of these points are $\lambda_1$ and $\lambda_2$, respectively, then, as is easy to show, $V_{\tilde{\tau}} \cong U_{\lambda_1} \otimes U_{\lambda_2}$. In this way we obtain from Lemmas 3 and 4, in particular, the decomposition of the tensor product $U_{\lambda_1} \otimes U_{\lambda_2}$ into irreducible representations. This was obtained by another method in §5.

REMARK 2. The representation $V_\tau$ can be treated as a continual tensor product of Poisson representations $U_\lambda$; Theorem 1 then gives a decomposition of the continual tensor product of $U_\lambda$ into irreducible representations.

REMARK 3. The representation so constructed is a cyclic subrepresentation in $\mathrm{EXP}_\beta T$, where $T$ is the representation of Diff $X$ in the real space $L^2(X \times \mathbf{R}, m \times \tau)$ and $\beta$ is the 1-cocycle: $[\beta(\psi)](x, \alpha) = \mathcal{J}^{1/2}_\psi(x) - 1$, see §7. If $\tau = \delta_{x_0}$, $x_0 \neq 0$, then we obtain the Gaussian form for the Poisson representation $U_\lambda$ (see §4.4).

**3. Criteria for representations $V_\tau$ of Diff $X$ to be equivalent.** By Lemma 3, the multiplicity with which $V_\tau^{n_1, \ldots, n_k}$ occurs in $V_\tau$ depends only on the numbers $n_1, \ldots, n_k$ and on the dimension of $L^2_{\tau_0}(\mathbf{R})$. Therefore, Lemmas 3 and 4 also imply the following result.

THEOREM 2. *Let* $\tau'$ *and* $\tau''$ *be two non-negative finite measures on* **R** *such that*

1) $\tau'(\mathbf{R}) = \tau''(\mathbf{R})$;

2) *the supports of* $\tau'$ *and* $\tau''$ *are either both infinite or consistent of the same finite number of points.*

*Then the representations* $V_{\tau'}$ *and* $V_{\tau''}$ *of* Diff $X$ *are equivalent.*

By Theorem 2, each representation $V_\tau$ is given, up to equivalence, by a pair of numbers: the parameter $\lambda = \tau(\mathbf{R})$ of the Poisson measure ($0 < \lambda < \infty$) and the index $h$, which is equal to $n$ if $\tau$ is concentrated on $n$ points, and is $\infty$ if supp $\tau$ is infinite. It is convenient, therefore, to denote these representations by $V_{\lambda,h}$ (instead of the previous notation $V_\tau$).

**4. The tensor product of representations $V_\tau = V_{\lambda,h}$.** THEOREM 3. $V_{\lambda_1,h_1} \otimes V_{\lambda_2,h_2} \cong V_{\lambda_1+\lambda_2,\, h_1+h_2}$; *thus, the set of representations* $V_\tau = V_{\lambda,h}$ *is closed under the operation of tensor multiplication.*

PROOF. We have $V_{\lambda_i,h_i} \cong V_{\tau_i}$, where $\tau_i$ is any non-negative finite measure on **R** scuh that $\tau_i(\mathbf{R}) = \lambda_i$ and $|\mathrm{supp}\,\tau_i| = h_i$ if $h_i < \infty$, and supp $\tau_i$ is any infinite set if $h_i = \infty$ ($i = 1, 2$). The measures $\tau_1$ and $\tau_2$ can always be chosen so that supp $\tau_1 \cap$ supp $\tau_2 = \phi$. Let us consider $\tau = \tau_1 + \tau_2$ on **R**. Obviously, $V_\tau = V_{\lambda_1+\lambda_2,\, h_1+h_2}$. Therefore, it is sufficient for us to check that $V_\tau \cong V_{\tau_1} \otimes V_{\tau_2}$.

We denote by $\tilde{\mu}_1$, $\tilde{\mu}_2$, and $\tilde{\mu}$ the measures on $\mathcal{F}(X)$ corresponding to $\tau_1$, $\tau_2$, and $\tau$ on **R**, respectively, and by $L_{\tau_1}(f)$, $L_{\tau_2}(f)$, and $L_\tau(f)$ their characteristic functionals. Since $\operatorname{supp} \tau_1 \cap \operatorname{supp} \tau_2 = \phi$, we have $L_\tau(f) = L_{\tau_1}(f)\, L_{\tau_2}(f)$. Hence

$$L^2_{\tilde{\mu}}(\mathscr{F}(X)) \cong L^2_{\tilde{\mu}_1}(\mathscr{F}(X)) \otimes L^2_{\tilde{\mu}_2}(\mathscr{F}(X)),$$

and so $V_\tau \cong V_{\tau_1} \otimes V_{\tau_2}$, as required.

## §7. Representations of the cross product $\mathscr{G} = C^\infty(X)\cdot\mathrm{Diff}\, X$

**1. Definition of $\mathscr{G}$ and the construction of representations.** Let us introduce the (additive) group $C^\infty(X)$ of all real finite functions $f$ on $X$ of class $C^\infty$. Now Diff $X$ acts on $C^\infty(X)$ as a group of automorphisms $f \to f \circ \psi^{-1}$. In this way we can define the cross product $\mathscr{G} = C^\infty(X)\cdot\mathrm{Diff}\, X$ of $C^\infty(X)$ with the multiplication: $(f_1, \psi_1)(f_2, \psi_2) = (f_1 + f_2 \circ \psi^{-1}, \psi_1\psi_2)$.

Let $\tilde{\mu}$ be an arbitrary quasi-invariant (under Diff $X$) measure in the space $\mathscr{F}(X)$ of generalized functions on $X$. We consider $L^2_{\tilde{\mu}}(\mathscr{F}(X))$ and associate with each element $(f, \psi)\ \mathscr{G}$ the following operator $V(f, \psi)$ in $L^2_{\tilde{\mu}}(\mathscr{F}(X))$:

$$(1) \qquad (V(f, \psi)\Phi)(F) = e^{i\langle F, f\rangle}\left(\frac{d\mu(\psi^*F)}{d\mu(F)}\right)^{1/2}\Phi(\psi^*F).$$

It is easy to check that the $V(f, \psi)$ are unitary and form a representation of $\mathscr{G}$. This representation of $\mathscr{G}$ is cyclic with respect to $C^\infty(X)$ (the constant is a cyclic vector). It is irreducible if and only if the measure $\tilde{\mu}$ on $\mathscr{F}(X)$ is ergodic with respect to Diff $X$.

**2. Representations associated with infinitely decomposable measures.** From now on we restrict ourselves to the measure $\tilde{\mu}$ on $\mathscr{F}(X)$ introduced in §6, that is, measures with characteristic functionals of the form

$$L_\tau(f) = \exp\left(\int\limits_{\mathbf{R}}\int\limits_X (e^{i\alpha f(x)} - 1)\, dm(x)\, d\tau(\alpha)\right),$$

where $m$ is a smooth positive measure on $X$, $m(X) = \infty$, and $\tau$ is a non-negative finite measure on $\mathbf{R}$. The representation of $\mathscr{G}$ corresponding to this measure is now denoted by $V_\tau$ (the measure $m$ on $X$ is assumed to be fixed).

It is not difficult to prove that these measures are ergodic with respect to Diff $X$; consequently, *the representations $V_\tau$ of $\mathscr{G} = C^\infty(X)\cdot$Diff $X$ are irreducible.*

LEMMA 1. *If* dim $X > 1$, *then $\tilde{\mu}$ is ergodic with respect to the subgroup* Diff $(X, m) \subset$ Diff $X$ *of diffeomorphisms preserving $m$.*

The proof goes as for Poisson measures (see §4).

COROLLARY 1. *The restriction of $V_\tau$ to the subgroup $C^\infty(X)\cdot$Diff$(X, m)$ is irreducible.*

COROLLARY 2. *The only vectors in* $L^2_{\tilde{\mu}}(\mathcal{F}(X))$ *that are invariant under* Diff $(X, m)$ *are the constants.*

Let $\Phi_0$ be the function in $L^2_{\tilde{\mu}}(\mathcal{F}(X))$ that is identically equal to 1. The following function on $\mathcal{G}$ is called the *spherical function* of $V_\tau$:

$$u_\tau(f, \psi) = \langle V_\tau(f, \psi)\Phi_0, \Phi_0\rangle,$$

where the brackets denote the scalar product in $L^2_{\tilde{\mu}}(\mathcal{F}(X))$. Since $\Phi_0$ is a cyclic vector in $L^2_{\tilde{\mu}}(\mathcal{F}(X))$, the representation $V_\tau$ is uniquely determined by $u_\tau(f, \psi)$.

By a simple calculation we obtain

$$u_\tau(f, \psi) = \exp\left(\int_{\mathbf{R}}\int_{X} (\mathcal{J}_\psi^{1/2}(x)\, e^{i\alpha f(x)} - 1)\, dm(x)\, d\tau(\alpha)\right), \tag{2}$$

where $\mathcal{J}_\psi(x) = \dfrac{dm(\psi^{-1}x)}{dm(x)}$.

Obviously, if $\tau_1 = \tau_2$, then $u_{\tau_1} \neq u_{\tau_2}$. Since $\Phi_0$ is defined invariantly in $L^2_{\tilde{\mu}}(\mathcal{F}(X))$, we have the following result.

LEMMA 2. *If* $\tau_1 \neq \tau_2$, *the representations* $V_{\tau_1}$ *and* $V_{\tau_2}$ *of* $\mathcal{G} = C^\infty(X)\cdot\text{Diff}$ $(\dim X > 1)$ *are inequivalent.*

**3. The Gaussian form of the representations $V_\tau$.** Let us consider the complex Hilbert space $H$ of functions on $X \times \mathbf{R}$ with the norm

$$\|\varphi\|^2 = \int\int |\varphi(x, \alpha)|^2\, dm(x)\, d\tau(\alpha).$$

A unitary representation $T$ of $\mathcal{G}$ is given in $H$ by

$$(T(f, \psi)\varphi)(x, \alpha) = e^{i\alpha f(x)}\mathcal{J}_\psi^{1/2}(x)\,\varphi(\psi^{-1}x, \alpha).$$

We define map $\beta$: $\mathcal{G} \to H$ by

$$[\beta(f, \psi)](x, \alpha) = e^{i\alpha f(x)}\mathcal{J}_\psi^{1/2}(x) - 1.$$

It is easy to verify that for any $g_1, g_2 \in \mathcal{G}$ we have: $\beta(g_1g_2) = \beta(g_1) + T(g_1)\beta(g_2)$, so that $\beta$ is a 1-cocycle.

Let us construct from $T$ and the 1-cocycle $\beta$ a new representation $\widetilde{V}_\tau = \text{EXP}_\beta T$ of $\mathcal{G}$ (see [1], [2]). We denote the dual space of $H$ by $H'$. The standard Gaussian measure on the completion $\widetilde{H}'$ of $H'$ is the measure $\mu$ with the characteristic functional

$$\int e^{i\,\mathrm{Re}\,\langle F, f\rangle}\, d\mu(F) = e^{-\|f\|^2/2}, \qquad f \in H.$$

The representation $\widetilde{V}_\tau$ of $\mathcal{G}$ is given on $L^2_\mu(\widetilde{H}')$ by

$$(\widetilde{V}_\tau(g)\Phi)(F) = e^{i\,\mathrm{Re}\,\langle F, \beta(g)\rangle}\Phi(T^*(g)F),$$

where the operator $T^*(g)$ is given by $\langle T^*(g)F, f\rangle = \langle F, T(g)f\rangle$.

THEOREM. *Let $\tau$ be a measure on* $\mathbf{R}$ *such that $\int e^{i\alpha t}d\tau(\alpha)$ is a real function of $t$. Then the restriction of the representation $\widetilde{V}_\tau$ of $\mathcal{G} = C^\infty(X)\cdot$ Diff $X$ to the cyclic subspace $\mathcal{H} \subset L^2_\mu(H')$ generated by the vector $\Omega \equiv 1$ is equivalent to the representation of $V_\tau$ defined in* §7.2.

PROOF. From the definition of $\widetilde{V}_\tau$ it follows that

$$\langle \widetilde{V}_\tau(f, \psi)\Omega, \Omega\rangle = \exp\left(\int\int (\cos(\alpha f(x))\, \mathcal{J}_\psi^{1/2}(x) - 1)\, dm(x)\, d\tau(\alpha)\right)$$

(here the diamond brackets denote the inner product in $L^2_\mu(H')$). Consequently, from the hypothesis of the theorem,

$$\langle V_\tau(f, \psi)\Omega, \Omega\rangle = \exp\left(\int\int (e^{i\alpha f(x)}\mathcal{J}_\psi^{1/2}(x) - 1)\, dm(x)\, d\tau(\alpha)\right) = u_\tau(f, \psi),$$

where $u_\tau$ is the spherical function of $V_\tau$ (see §7.2). Hence the assertion of the theorem follows immediately.

REMARK 1. A representation of $C^\infty(X)\cdot$ Diff $X$ was constructed in [15] by means of the $N/V$ limit. This representation coincides with that constructed here for the Poisson measure (the connection with the Poisson measure was apparently not noticed), and the transition to a Fock model in [15] is equivalent to the realization of this representation as $\mathrm{EXP}_\beta T$ (see above).

We emphasize that a representation of the cross product can be constructed for any measure in $\mathcal{F}(X)$ that is quasi-invariant under Diff $X$. However, only those that are constructed from an infinitely decomposable measure have the structure $\mathrm{EXP}_\beta T$, because it is only in this case that there is a vacuum vector.

REMARK 2. Instead of $C^\infty(X)$ we can consider an arbitrary group of smooth functions $C^\infty(X, G) = G^X$ on $X$ with values in a Lie group $G$ and the cross product $C^\infty(X, G)\cdot$ Diff $X$.

If a unitary representation $\pi$ of $G$ is given on a space $H$, then the representation $T$ of this cross product acts naturally in the space $\mathcal{H} = \int \oplus H_x dm(x)$, $H_x = H$. This is irreducible if $\pi$ is irreducible. If $\beta$: $C^\infty(X, G)\cdot$ Diff $X \to \mathcal{H}$ is a non-trivial cocycle (see [1]), then in EXP $\mathcal{H}$ we get a representation $\mathrm{EXP}_\beta T$ of $C^\infty(X, G)\cdot$ Diff $X$.

## APPENDIX 1

### On the methods of defining measures on the configuration space $\Gamma_X$

1. Let $X_1 \subset \ldots \subset X_n \subset \ldots$ be a sequence of open submanifolds in $X$ with compact closures such that $X = \bigcup_n X_n$. The projections $p_k$: $\Gamma_X \to \mathrm{B}_{X_k}$

are given by putting $p_k \gamma = \gamma \cap X_k$. Let $\mu$ be a measure on $\Gamma_X$ and $\mu_k = p_k \mu$ its projection on $B_{X_k} (k = 1, 2, \dots)$. Then the measures $\mu_k$ are mutually compatible, that is, for any $k > l$ we have $p_{lk} \mu_k = \mu_l$, where $p_{lk}: B_{X_k} \to B_{X_l}$ is the natural projection. A well known theorem of Kolmogorov about the extension of measures enables us to establish the converse: if $\{\mu_k\}$ is any compatible sequence of measures on $\{B_{X_k}\}$, then there is a unique measure on $\Gamma_X$ such that $p_k \mu = \mu_k (k = 1, 2, \dots)$.

We can abandon the compatibility conditions and consider sequences of measures $\mu_k$ on $B_{X_k}$ for which $\lim_{k \to \infty} p_{lk} \mu_k = \mu^{(l)}$ exists (in the weak sense) for all $l$. In this case the measures $\mu^{(l)}$ on $B_{X_l}$ are compatible and define a measure $\mu$ on $\Gamma_X$.

We also recall that $\Gamma_X$ is naturally embedded in the space of generalized functions ($\gamma \to \sum_{x \in \gamma} \delta_x$), therefore, the methods for defining measures in linear spaces are applicable here (by means of the characteristic functional and so on); see, for example, [5].

2. A fundamentally different method describing measure on $\Gamma_X$ has received attention in statistical physics [7]. It generalizes the method of specifying Markov measures (by transition probabilities). It consists in giving conditional measures on $B_Y$ (or their densities with respect to Poisson measure) as functions on $\Gamma_{X \setminus Y}$ for all compact domains $Y \subset X$ by means of a single function (the potential) on $B_X$. The question of existence and uniqueness of the measure on $\Gamma_X$ with a given system of conditional measures is, as a rule, very difficult. Curiously enough, in this case the condition for a measure $\mu$ on $\Gamma_X$ to be quasi-invariant under Diff $X$ can be formulated very simply: all the conditional measures on $B_Y$, where $Y$ is any open set with compact closure, must be equivalent to a quasi-invariant (under Diff $Y$) measure on $B_Y$. By now there are many such measures known in statistical physics that are not equivalent to the Poisson measure (Gibbs measures).

A measure on $\Gamma_X$ can also be given with the help of so-called correlation functions on $B_X$; a correlation function defines uniquely an initial measure on $\Gamma_X$ (see, for example, [12]).

3. Let us introduce yet another method of defining measures on $\Gamma_X$. We say that a normalized Borel measure $\mu$ on $X^\infty$ (see §0.3) is admissible if:

1) $\mu(\widetilde{X}^\infty) = 1$, that is, $\widetilde{X}^\infty$ is a subset of full measure in $X^\infty$;

2) $\mu$ is quasi-invariant under Diff $X$.

If $\mu$ is an admissible measure on $X^\infty$, then its projection $\widetilde{\mu} = p\mu$ on $\Gamma_X$ (that is, $\widetilde{\mu}(C) = \mu(p^{-1}C)$ for any measurable set $C \subset \Gamma_X$) is a quasi-invariant measure on $\Gamma_X$. This method of defining a measure on $\Gamma_X$ is of limited interest, however, it is convenient for constructing various examples.

It is easy to show that an admissible measure $\mu$ on $X^\infty$ is ergodic if and only if it is regular (regularity means that it satisfies the $0 - 1$ law). In particular, any admissible product measure $\mu = m_1 \times \dots \times m_n \times \dots$ is ergodic. The following lemma is analogous to the Borel-Cantelli lemma.

LEMMA. *The product measure* $\mu = m_1 \times \ldots \times m_n \times \ldots$ *on* $X^\infty$ *is admissible if and only if* $\sum_{i=1}^{\infty} m_i(Y) < \infty$ *for any compact set* $Y \subset X$.

EXAMPLE OF AN ADMISSIBLE MEASURE. Let $X = \mathbf{R}^n$. We denote by $m_a$ the Gaussian measure with centre at $a \in \mathbf{R}$ and with unit correlation matrix: $dm_a(x) = (2\pi)^{-n/2} e^{-\|x-a\|^2/2}\, dx$, where $dx$ is the Lebesgue measure on $\mathbf{R}^n$. By a direct computation it is not difficult to check that $\mu = m_{a_1} \times \ldots \times m_{a_n} \times \ldots$ is admissible if, for example, $\|a_n\| \geqslant c \log n$, $n = 1, 2, \ldots$.

If $\|a_n\| \to \infty$ sufficiently quickly, then it is easy to verify that $\mu$ is concentrated on the set $\Delta_s = \prod_{\sigma \in S_\infty} (s\Gamma_X)\sigma$, where $s\colon \Gamma_X \to \widetilde{X}^\infty$ is the cross-section corresponding to a certain admissible indexing $i$ in $\Gamma_X$. However, $\mu$ is not invariant under left translations in $\widetilde{X}^\infty$: $(Tx)_1 = x_{i+1}$. Hence its projection $\widetilde{\mu} = p\mu$ is a non-saturated measure in $\Gamma_X$ (see Proposition 7 in §2.3).

Another example refers to the group Diff $X$, where $X$ is a compact manifold. In this case, let $\widetilde{X}^\infty$ denote the set of all sequences in $X$ that converge in $X$. It is easy to verify that $\widetilde{X}^\infty/S^\infty \equiv \Gamma_X$ is the union of all countable subsets of $X$ with a unique limit point (one for each subset). Let $x_0 \in X$, let $\rho$ be a continuous metric in $X$, and let $\{m_n\}$ be a sequence of smooth measures in $X$ such that $\lim_{n\to\infty} \int \sigma(x, x_0)dm_n(x) = 0$. It is clear that the product measure $m^{x_0} = \prod_n m_n$ is concentrated on $\widetilde{X}^\infty$. We introduce the measure $\widetilde{m} = \int m^{x_0}\, dx_0$ – the mixing of the measures $m^{x_0}$ in $\widetilde{X}^\infty$. This $\widetilde{m}$ projects onto a measure $m$ on $\Gamma_X$ that is quasi-invariant and ergodic under Diff $X$. Another more complicated example of a measure on the countable subsets of a compact manifold that is quasi-invariant and ergodic under Diff $X$ was constructed earlier in [8].

## APPENDIX 2

### $S_\infty$-cocyles and Fermi representations

According to the standard definition, a 1-cocycle on Diff $X$ with values in the group $S_\infty(\Gamma_X)$ of measurable maps $\Gamma_X \to S_\infty$ is a map $\sigma$: Diff $X \to S_\infty(\Gamma_X)$ satisfying the following condition:

$$\sigma(\psi_1, \gamma)\sigma(\psi_2, \psi_1^{-1}\gamma) = \sigma(\psi_1\psi_2, \gamma). \tag{1}$$

Two 1-cocycles $\sigma_1$ and $\sigma_2$ are said to be cohomologous if there exists a measurable map $\sigma_0\colon \Gamma_X \to S_\infty$ such that

(2) $$\sigma_2(\psi, \gamma) = \sigma_0(\gamma)\sigma_1(\psi, \gamma)\sigma_0^{-1}(\psi^{-1}\gamma).$$

Let $i$ be a measurable indexing in $\Gamma_X$, $a_k = \Gamma_X \to X$ $(k = 1, 2, \ldots)$ the sequence of measurable maps defined by $i$ (see § 2.2); and $s\colon \Gamma_X \to \widetilde{X}^\infty$ the cross-section of the fibration[1] $\widetilde{X}^\infty \to \Gamma_X$ defined by $i$:

$$s(\gamma) = (a_1(\gamma), \ldots, a_k(\gamma), \ldots).$$

Further, let $\Delta_s = \prod_{\sigma \in S_\infty} (s\Gamma_X)\sigma$.

We say that a measurable indexing $i$ is *correct* if for any $\gamma, \gamma' \in \Gamma_X$ the conditions $|\gamma \cap K| = |\gamma' \cap K|$ and $\gamma \cap (X \setminus K) = \gamma' \cap (X \setminus K)$ for a certain compact set $K \subset X$ imply that $a_k(\gamma) = a_k(\gamma')$ for all indices $k$ except finitely many.[2] A cross-section $s\colon \Gamma_X \to \widetilde{X}^\infty$ defined for a correct indexing $i$ is also called *correct.* If $i$ is correct, then the set $\Delta_s$ is invariant under Diff $X$.

To each correct cross-section $s\colon \Gamma_X \to \widetilde{X}^\infty$ there corresponds a 1-cocycle $\sigma_s$ defined by the following relation (see § 2.2):

$$s(\psi^{-1}\gamma) = [\psi^{-1}(s\gamma)]\sigma_s(\psi, \gamma).$$

REMARK. There are examples of cocycles that are not generated by correct cross-sections.

Cocycles $\sigma_s$ generated by correct cross-sections $s$ are also called *correct.*

We give, without proof, some properties of correct cocycles $\sigma_s$.

1) The cross-section $s$ is uniquely determined by the correct cocycle $\sigma_s$ corresponding to it.

2) Any two correct cocycles $\sigma_{s_1}$ and $\sigma_{s_2}$ are cohomologous as cocycles with values in $S^\infty(\Gamma_X)$, that is,

(3) $$\sigma_{s_2}(\psi, \gamma) = \sigma_0(\gamma)\sigma_{s_1}(\psi, \gamma)\sigma_0^{-1}(\psi^{-1}\gamma),$$

where $\sigma_0$ is the measurable map $\Gamma_X \to S^\infty$.

3) No correct 1-cocycle is cohomologous to the trivial cocycle.

4) Let $\sigma_s$ be a correct cocycle, $\sigma_0\colon \Gamma_X \to S^\infty$ a measurable map, and $\sigma$ a 1-cocycle defined by

$$\sigma(\psi, \gamma) = \sigma_0(\gamma)\sigma_s(\psi, \gamma)\sigma_0^{1-}(\psi^{-1}\gamma).$$

For $\sigma$ to be correct it is necessary and sufficient that $\sigma_0(\gamma)\sigma_0^{-1}(\psi^{-1}\gamma) \in S^\infty$ for any $\psi \in$ Diff $X$ and $\gamma \in \Gamma_X$.

5) Two correct cocycles $\sigma_{s_1}$ and $\sigma_{s_2}$ are cohomologous as cocycles with values in $S_\infty(\Gamma_X)$ if and only if the cross-sections $s_1$ and $s_2$ are cofinal, that is, $s_2 = \sigma_0 \circ s_1$ where $\sigma_0 \in S_\infty(\Gamma_X)$.

---

[1] Note that, generally speaking, the fibration $\widetilde{X}^\infty \to \Gamma_X$ has no continuous cross-sections. It can be shown that this fibration has no continuous quotient fibrations with fibre $\mathbf{Z}_2$.

[2] The condition of correctness is weaker than the condition of admissibility introduced in § 2.2

For each 1-cocycle $\sigma$: Diff $X \to S_\infty(\Gamma_X)$ we define a corresponding 1-cocycle $\alpha_0$: Diff $X \to \mathbf{Z}_2(\Gamma_X)$ by

$$\alpha_\sigma(\psi, \gamma) = \text{sign}\, \sigma(\psi, \gamma) \tag{4}$$

(sign $\sigma$ is the parity of $\sigma \in S_\infty$).

It is obvious that when $\sigma_1$ and $\sigma_2$ are cohomologous, then the correspond- cocycles $\alpha_{\sigma_1}$ and $\alpha_{\sigma_2}$ are cohomologous. However, there exist cocycles $\sigma_1$ and $\sigma_2$ (even correct ones) that are not cohomologous, although the corresponding cocycles $\alpha_{\sigma_1}$ and $\alpha_{\sigma_2}$ are.

EXAMPLE. The correct cocycles $\sigma_1(\psi, \gamma)$ and $\sigma_2(\psi, \gamma) = \sigma_0\sigma(\psi, \gamma)\sigma_0^{-1}$, where $\sigma_0 \notin S_\infty$. Since sign $(\sigma_0\sigma\sigma_0^{-1})$ = sign $\sigma$, we have $\alpha_{\sigma_1} = \alpha_{\sigma_2}$. However, the cocycles $\sigma_1$ and $\sigma_2$ are themselves not cohomologous (see property 4).

A SUFFICIENT CONDITION FOR COCYCLES TO BE COHOMOLOGOUS. Let $\sigma_2(\psi, \gamma) = \sigma_0(\gamma)\sigma_1(\psi, \gamma)\sigma_0^{-1}(\psi^{-1}\gamma)$. If $\sigma_0(\gamma) = \sigma_0\tilde{\sigma}_0(\gamma)$, where $\tilde{\sigma}_0 \in S_\infty(\Gamma_X)$ and $\sigma_0$ is an arbitrary element of $S^\infty$, then $\alpha_{\sigma_1} \sim \alpha_{\sigma_2}$.

Note that each $\mathbf{Z}_2$-cocycle $\alpha$ defines a $\mathbf{Z}_2$-covering of $\Gamma_X$ with a given action of Diff $X$ on it. The elements of this covering are pairs $(\varepsilon, \gamma)$, $\gamma \in \Gamma_X$, $\varepsilon = \pm 1$; and Diff $X$ acts by $\psi(\varepsilon, \gamma) = (\varepsilon\,\alpha(\psi, \gamma), \psi^{-1}\gamma)$.

$\mathbf{Z}_2$-cocycles of the form (4), where $\sigma$ is a correct cocycle, are called *correct* $\mathbf{Z}_2$*-cocycles,* and the $\mathbf{Z}_2$-coverings of $\Gamma_X$ defined by them are also called correct.

LEMMA. *Correct* $\mathbf{Z}_2$*-cocycles are non-trivial.*

Let $i$ be a correct indexing in $\Gamma_X$, $s$: $\Gamma_X \to \widetilde{X}^\infty$ the cross-section defined by $i$, $\mu$ a quasi-invariant measure in $\Gamma_X$, and $T$ a unitary representation of $S_\infty$ acting on a Hilbert space $H$. We associate with the triple $(i, \mu, T)$ a unitary representation $V$ of Diff $X$ in the space $\mathscr{H} \subset L^2_\mu(\Delta_s, H)$ of functions $f$: $\Delta_s \to H$ such that

$$\|f\|^2 = \int_{\Gamma_X} \|f(s\gamma)\|^2_H\, d\mu(\gamma) < \infty; \qquad f(x\sigma) = T^{-1}(\sigma) f(x)$$

for every $\sigma \in S_\infty$; the representation operators are defined by

$$(V(\psi) f)(x) = \left(\frac{d\mu(\psi^{-1}px)}{d\mu(px)}\right)^{1/2} f(\psi^{-1}x), \tag{5}$$

where $p$ is the projection $\Delta_s \to \Gamma_X$.

ALTERNATIVE DEFINITION: $V$ is given in the space $L^2_\mu(\Gamma_X, H)$ of functions $f$: $\Gamma_X \to H$ for which $\|f\|^2 = \int \|f(\gamma)\|^2_H d\mu(\gamma) < \infty$. the operators $V(\psi)$ have the form

$$(V(\psi) f)(\gamma) = \left(\frac{d\mu(\psi^{-1}\gamma)}{d\mu(\gamma)}\right)^{1/2} T(\sigma_s(\psi, \gamma)) f(\psi^{-1}\gamma), \tag{6}$$

where $\sigma_s$ is the 1-cocycle generated by a correct cross-section $s$: $\Gamma_X \to \widetilde{X}^\infty$.

It is obvious that the representations of Diff $X$ so defined are equivalent.

Note that $V = \text{Ind}_{G_\gamma}^{\text{Diff}\, X}(T \circ \pi)$, where $G_\gamma$ is the subgroup of all diffeomorphisms $\psi \in \text{Diff}\, X$ for which $\psi\gamma = \gamma$, and $\pi$ is the projection $G_\gamma \to G_\gamma \setminus G_\gamma^0 \cong S_\infty$. ($G_\gamma^0$ is the subgroup leaving every point $x \in \gamma$ fixed).

Two 1-cocycles $\sigma$ and $\sigma'$ are said to be equivalent with respect to a measure $\mu$ if $\sigma(\psi, \gamma) = \sigma'(\psi, \gamma)$ mod 0 with respect to $\mu$ for any $\psi \in \text{Diff}\, X$. Note that the right-hand side of (6) does not change if the 1-cocycle is replaced by any equivalent cocycle. Therefore, the 1-cocycles $\sigma$ need only be defined up to an equivalence.

The properties 1) – 5) of 1-cocycles can be reformulated without difficulty for equivalence classes of 1-cocycles. Moreover, 2) can be made more precise as follows: the map $\sigma_0$ in (3) is uniquely determined mod 0 if $\mu$ is ergodic.

Let us consider the particular case when $T = \text{Ind}_{S_n \times S_\infty^n}^{S_\infty} (\rho \times I)$, where $\rho$ is a unitary representation of $S_n$ and $I$ is the unit representation of $S_\infty^n$ ($S_\infty^n$ is the group of finite permutations of $n + 1, n + 2, \ldots$). Comparing the definition of $V$ with that of the elementary representations $U_\mu^\rho$ (see §3) it is easy to check that $V \cong U_\mu^\rho$. Hence, in particular, in this example all the correct cocycles lead to equivalent representations.[1]

Quite a different example is the Fermi representation. Let us consider a $\mathbf{Z}_2$-cocycle $\alpha(\psi, \gamma)$ (see above) and define a representation of Diff $X$ in $L_\mu^2(\Gamma_X)$ by

$$(V(\psi) f)(\gamma) = \alpha(\psi, \gamma) \left( \frac{d\mu(\psi^{-1}\gamma)}{d\mu(\gamma)} \right)^{1/2} f(\psi^{-1}\gamma).$$

This is called a *Fermi representation* of Diff $X$.

When the $\mathbf{Z}_2$-cocycle is generated by a correct cross-section $s\colon \Gamma_X \to \widetilde{X}^\infty$, that is, when $\alpha(\psi, \gamma) = \text{sign}\, \sigma_s(\psi, \gamma)$, where $\sigma_s\colon$ Diff $X \to S_\infty(\Gamma_X)$ is a correct cocycle, then there is another convenient realization of $V$: the representation is given in $\mathcal{H}$ by a function $f(x)$ on $\Delta_s \subset X^\infty$ such that

1) $f(x\sigma) = \text{sign}\, \sigma f(x)$ for any $\sigma \in S_\infty$ (an "odd" function).

2) $\|f\|^2 = \int_{\Gamma_X} |f(s\gamma)|^2\, d\mu(\gamma) < \infty$.

The representation operators are defined by (5). In this case it can be shown by the same arguments as in §3 that *if $\mu$ is ergodic and the cross-section $s$: $\Gamma_X \to \widetilde{X}^\infty$ is generated by an admissible indexing $i$* (in the sense of §2.2), *then the Fermi representation $V(\psi)$ is irreducible.*

REMARK 1. Apparently, there exist non-equivalent Fermi representations of Diff $X$ constructed from the same measure $\mu$ on $\Gamma_X$ (for the construction of a Fermi representation by means of an $N/V$ limit, see [19] and [20]).

REMARK 2. The group Diff $X$ has factor representations of type II – it is sufficient to take a representation of $S_\infty$ of type II in $H$ (for example,

[1] There is a more general fact: if a representation $T$ of $S_\infty$ can be extended to a representation of $S^\infty$, then the representation $V$ of Diff $X$ corresponding to $T$ is uniquely determined, up to equivalence, by $T$ and a measure $\mu$ on $\Gamma_X$.

the regular representation) and to construct a representation of Diff $X$ in $L^2_\mu(\Gamma_X, H)$ from it.

# APPENDIX 3

## Representations of Diff $X$ associated with measures in the tangent bundle of the space of infinite configurations

The representations of Diff $X$ discussed above are of "zero order", that is, they do not depend on the differentials of the diffeomorphisms. In this context let us note that these representations can be extended to representations of the group of measurable finite transformations of a manifold $X$ with a quasi-invariant measure.

However, one can construct representations of Diff $X$ of positive order (by a representation of order $k$ we mean one depending essentially on the $k$-jet of the diffeomorphisms). A number of representations of Diff $X$ in spaces of finite functional dimension were defined in [9]; in the terminology of this paper these representations are connected with the space $B_X$ of finite configurations. But there are also representations of positive order connected with the space $\Gamma_X$ of infinite configurations. Let us take as an example a representation of order 1 connected with this space.

Let $\mu$ be a measure in $\Gamma_X$ that is quasi-invariant under the action of Diff $X$. We consider the "tangent bundle" $T\Gamma_X$ over $\Gamma_X$, that is a fibre bundle over $\Gamma_X$, where the fibre over $\gamma = \{x_i\}$ is the direct product $\prod_{i=1}^{\infty} T_{x_i} X$ of the tangent spaces at the points $x_i \in \gamma$.

The space $T\Gamma_X$ can be regarded as the factor space $\widetilde{TX^\infty}/S^\infty$, where $\widetilde{TX^\infty}$ is the subset of $(TX)^\infty = \prod_{i=1}^{\infty} TX_i$, $X_i = X$, consisting of points $\{(x_i, v_i);\ v_i \in T_{x_i}X_i,\ \{x_i\} \in \widetilde{X}^\infty\}$. The topology in $T\Gamma_X$ and the $\sigma$-algebra of Borel sets are induced from $\widetilde{TX^\infty}$.

Let $\lambda_i^\gamma$ be a normalized measure in $T_{x_i}X$ that is equivalent to the Lebesgue measure, and let $\lambda^\gamma = \prod_{i=1}^{\infty} \lambda_i^\gamma$ be the product measure in $\prod_{i=1}^{\infty} T_{x_i}X$. In this way a measure $\lambda$ is introduced in $T\Gamma_X$ such that its projection onto $\Gamma_X$ is $\mu$ and that the conditional measures in $T^\gamma\Gamma_X = \prod_{i=1}^{\infty} T_{x_i}X$ are $\lambda^\gamma$. The action of Diff $X$ on $(T\Gamma_X, \lambda)$ is defined by $\psi(\gamma, a) = (\psi\gamma, d\psi a)$, where $a = \prod_{x\in\gamma} a_x$, $a_x \in T_xX$ and $d\psi$ is the natural action on $T\Gamma_X$. The measure $\lambda$ is quasi-invariant under this action. This leads to a unitary representation of Diff $X$ in $L^2_\lambda(T\Gamma_X)$. A proof that this is irreducible for ergodic measures $\mu$ can be modelled on §3. The parameters of the representations just constructed are the measure $\mu$ in $\Gamma_X$ and the measures

$\lambda^\gamma$ in $T^\gamma \Gamma_X$.

It is easy to see how to construct representations of order 1 analogous to the elementary representations. We do not say much about representations of higher order, because difficulties in describing them arise even in the case of a finite number of particles (see [9]).

REMARK. The representations of Diff $X$ listed in this appendix can be extended to representations of the cross-product $C^\infty(X) \cdot$ Diff $X$.

## References

[1] A. M. Vershik, I. M. Gel'fand and M. I. Graev, Representations of the group $SL(2, R)$, where $R$ is a ring of functions, Uspekhi Mat. Nauk **28**:5 (1973), 83–128. = Russian Math. Surveys **28**:5 (1973), 87–132.

[2] A. M. Vershik, I. M. Gel'fand and M. I. Graev, Irreducible representations of the group $G^X$ and cohomology, Functsional. Anal. i Prilozhen. 8:2 (1974), 67–69. MR **50** # 530. = Functional Anal. Appl. **8** (1974), 151–153.

[3] D. B. Anosov and A. B. Katok, New examples in smooth ergodic theory. Ergodic diffeomorphisms, Trudy Moskov. Mat. Obshch. **23** (1970), 3–36.

[4] A. Weil, L'intégration dans les groupes topologiques et ses applications, Actual. Sci. Ind. 869, Hermann & Cie, Paris 1940. MR **3** # 198.
Translation: *Integrirovanie v topologicheskikh gruppakh i ego prilozheniya,* Izdat. Inost. Lit., Moscow 1950.

[5] I. M. Gel'fand and N.Ya. Vilenkin, *Obobshchennye funktsii, vyp.4. Nekotorye primeneniya garmonicheskogo analiza. Oskashchennye gil'bertovy prostranstva* Gos. Izdat. Fiz.-Mat. Lit., Moscow 1961. MR **26** #4173.
Translation: Generalized functions, vol.4, Some applications of harmonic analysis, Equipped Hilbert spaces, Academic Press, New York-London 1964.

[6] A. M. Vershik, Description of invariant measures for the actions of some infinite-dimensional groups, Dokl. Akad. Nauk SSSR **218** (1974), 749–752. = Soviet Math. Dokl. **15** (1974), 1396–1400.

[7] R. L. Dobrushin, R. A. Minlos and Yu. M. Sukhov, *Prilozhenie k knige Ryuelya: Statisticheskaya mekhanika* (Supplement to Ruelle's book *Statistical mechanics*). Mir, Moscow 1971.

[8] R. S. Ismagilov, Unitary representations of the group of diffeomorphisms of the circle, Funktsional Anal. i Prilozhen. 5:3 (1971), 45–53. = Functional. Anal. Appl. **5** (1971), 209–216.

[9] A. A. Kirillov, Unitary representations of the group of diffeomorphisms and some of its subgroups, Preprint IPM, No.82 (1974).

[10] A. A. Kirillov, Dynamical systems, factors, and group representations, Uspekhi Mat. Nauk **22**:5 (1967), 67–80. = Russian Math. Surveys **22**:5 (1967), 63–75.

[11] V. A. Rokhlin, On the fundamental ideal of measure theory, Mat. Sb. **25** (1949), 107–150. MR **11** # 18

[12] D. Ruelle, Statistical mechanics. Rigorous results, W. A. Benjamin Inc., Amsterdam 1969.
Translation: *Statisticheskaya mekhanika. Strogie rezul'taty,* Mir, Moscow 1971.

[13] S. V. Fomin, On measures invariant under a certain group of transformations, Izv. Akad. Nauk SSSR Ser. Mat. **14** (1950), 261–274. MR **12** # 33.
[14] G. Goldin, Non-relativistic current algebras as unitary representations of groups, J. Mathematical Phys. **12** (1971), 462–488. MR **44** # 1330.
[15] G. Goldin, K. J. Grodnik, R. Powers and D. Sharp, Non-relativistic current algebra in the $N/V$ limit, J. Mathematical Phys. **15** (1974), 88–100.
[16] A. Guichardet, Symmetric Hilbert spaces and related topics, Lecture Notes in Math. **261**, Springer-Verlag, Berlin-Heidelberg-New York 1972.
[17] J. Kerstan, K. Mattes and J. Mecke, Unbegrenzt teilbare Punktprozesse, Berlin 1974.
[18] D. Knutson, λ-rings and the representation theory of the symmetric group, Lecture Notes in Math. **308**, Springer-Verlag, Berlin-Heidelberg-New York, 1973.
[19] R. Menikoff, The hamiltonian and generating functional for a non-relativistic local current algebra, J. Mathematical Phys. **15** (1974), 1138–1152. MR **49** # 10285.
[20] R. Menikoff, Generating functionals determining representations of a non-relativistic local current algebra in the $N/V$ limit, J. Mathematical Phys. **15** (1974), 1394–1408.
[21] J. Moser, On the volume elements on a manifold, Trans. Amer. Math. Soc. **120** (1965), 286–294. MR **32** # 409
[22] R. S. Ismagilov, Unitary representations of the group of diffeomorphisms of the space $\mathbf{R}^n$, $n \geqslant 2$, Funktsional. Anal. i Prilozhen. **9**:2 (1975), 71–72. = Functional Anal. Appl. **9** (1975), 144–145.

Received by the Editors, 15 May 1975

Translated by A. West

# 5.

(with M. I. Graev and A. M. Vershik)

# Representations of the group of functions taking values in a compact Lie group

Compos. Math. **42**, 217–243 (1981). MR **83** g:22002. Zbl. **449**:22019

## Introduction

In our paper [3] a series of important unitary representations of the group $G^X$ of all smooth functions on Riemannian manifold $X$ taking values in a compact semisimple Lie group $G$ has been constructed and considered. Those representations can be obtained in the following way.

Let $\Omega(X; \mathfrak{g})$ be the space of smooth 1-forms $\omega(x)$ on $X$ taking values in the Lie algebra $\mathfrak{g}$ of the group $G$, that is $\omega(x)$ is a linear operator from the tangent space $T_xX$ into $\mathfrak{g}$. Let us introduce the norm in $\Omega(X; \mathfrak{g})$ by the formula

$$\|\omega\|^2 = \int_X \mathrm{Sp}(\omega(x)\omega^*(x))dx,$$

where $\omega^*(x): \mathfrak{g} \to T_xX$ is the operator conjugate to $\omega(x)$ (it is defined since $T_xX$ and $\mathfrak{g}$ have natural structures of Euclidean spaces), $dx$ is the Riemannian measure on $X$.

Define the unitary representation $V(\tilde{g})$ of the group $G^X$ in the space $\Omega(X; \mathfrak{g})$ by

$$(V(\tilde{g})\omega)(x) = \mathrm{Ad}g(x) \circ \omega(x).$$

Define the Maurer–Cartan cocycle $\beta\tilde{g}$ on $G^X$ taking values in $\Omega(X; \mathfrak{g})$ by

$$(\beta\tilde{g})(x) = (dg(x))g^{-1}(x)$$

(here $dg(x): T_xX \to T_{g(x)}G$ and $(\beta\tilde{g})(x): T_xX \to T_eG = \mathfrak{g}$).

The representation $U$ of the group $G^X$ is $\mathrm{EXP}_\beta\, V$ in the sense of [1], [2]. It means that $U$ acts in the Fock space $\mathrm{EXP}\, H = \bigoplus_{n=0}^{\infty} S^n H_C$ where $H = \overline{\Omega(X;\mathfrak{g})}$ is the completion of $\Omega(X;\mathfrak{g})$, $H_C$ is the complexification of $H$, $S^n H_C$ is the symmetrized tensor product of $n$ copies of $H_C$. The action of the operator $U(\tilde{g})$ on the vectors $\mathrm{EXP}\,\omega = \bigoplus_{n=0}^{\infty} \frac{1}{\sqrt{n!}} \underbrace{\omega \otimes \cdots \otimes \omega}_{n}$, $\omega \in \Omega(X;\mathfrak{g})$ (which form a total set in $\mathrm{EXP}\, H$) is defined by

$$U(\tilde{g})\,\mathrm{EXP}\,\omega = e^{(-\|\beta\tilde{g}\|^2/2) - \langle V(\tilde{g})\omega, \beta\tilde{g}\rangle}\,\mathrm{EXP}(V(\tilde{g})\omega + \beta\tilde{g}).$$

(A more convenient realization of this representation using the Gaussian measure is given in §5).

The representation $U$ of the group $G^X$ draws a great interest. After the paper [1], it has been almost simultaneously discovered by several authors: [17], [7], [3], [15]. There are a lot of variants and modifications of this construction in [3]: constructing of representations by a vector field, a fibre-bundle and so on.

In this paper we continue to consider these representations and correct a mistake which took place in the proof of its irreducibility in [3] (the statement of lemma 4 §2 of [3] is false). This correction requires a new development of the treatment of measures on an infinite-dimensional space of functions on $X$. It is interesting that some properties of the measures depend on the dimension of $X$. Let us mention one of the results (lemma 10): if $\dim X \geq 4$, $\mu$ is the Gaussian measure on the space of distributions on $\mathring{X}$ with Fourier transform $\exp(-\frac{1}{2}\|\ \|^2_{\mathring{W}^1_2(X)})$, $\nu_1$, $\nu_2$ are two singular measures concentrated on the set of generalized functions of the form $\Sigma\, \lambda_i \delta_{x_i}$, then the measures $\mu * \nu_i$ and $\mu * \nu_2$ are singular. A proof of this result is based on the following property of the Sobolev space $\mathring{W}^1_2(X)$ (lemma 3): if $X$ is a compact Riemannian manifold and $\dim X \geq 4$ then there exists a Hilbert–Schmidt extension of $\mathring{W}^1_2(X)$ not containing the generalized functions of the form $\Sigma\, \lambda_i \delta_{x_i}$.

The main result of this work is the proof of the irreducibility of the representations $U(\tilde{g})$ for $\dim X \geq 4$. Thus not only the question of irreducibility for $\dim X = 1$ is open, as it has been stated in [3], but also that for the cases $\dim X = 2$ and $\dim X = 3$. Irreducibility for $\dim X \geq 5$, $G = SU(2)$, $X$ being an open set in $\mathbb{R}^m$ has been proved before in [7]. Besides, for $\dim X \geq 4$ we give a new proof of the non-equivalence of the representations corresponding to different Riemannian metrics on $X$.

In the main, the plane of the proof of the irreducibility is a repetition of that of [3]. The analysis of the spectral function of the Laplace–Beltrami operator is the crucial point (§1), dimension 4 being critical (see footnote to §6). This paper can be read independently of [3], though its contents doesn't include that of [3].

Let us give briefly the plan of the paper. In §1, we prove the main lemmas about Hilbert–Schmidt extensions of the space $\mathring{W}_2^1(X)$, $\dim X \geq 4$. There are auxiliary facts about representations of abelian groups in §2. In §3, we propose an example of non-singularity of the Gaussian measure with its convolution and a criterion of singularity of measures. §4 contains the main results about convolutions of the Gaussian measures generated by the Laplace–Beltrami operators on $X$ whose measures are concentrated on the delta-functions. In §5, the spectrum of the restriction of the representation of $G^X$ to a commutative subgroup is considered. At last in §6, we prove the theorems of irreducibility and non-equivalence of the representations of $G^X$ for $\dim X \geq 4$ and consider the difficulties appearing in dimensions 1, 2, 3. In an appendix the formulas for the representations of the Lie algebra of $G^X$ are given.

The authors are grateful to R. Hoegh-Krohn for pointing out the flaw in lemma 4 of §2 in [3] and M. Solomyak for a useful consultation.

## §1. The extensions of the space $\mathring{W}_2^1(X)$

Let $X$ be a compact Riemannian manifold with a boundary $\partial X \neq \phi$, $\mathring{X} = X \setminus \partial X$, $dx$ a Riemannian measure on $X$, $\Delta$ the Laplace–Beltrami operator. We consider the real Sobolev space $\mathring{W}_2^1(X)$ (i.e. the completion of the space of compactly supported functions on $\mathring{X}$ with respect to the norm $\|f\| = |\int_X \Delta f \cdot f dx|^{1/2}$).

Let $\{u_k\}_{k=1}^{\infty}$ be the orthonormal basis in $\mathring{W}_2^1(X)$ of the eigenfunctions of the Dirichlet problem for the Laplace–Beltrami operator $\Delta$, ordered accordingly non-decrease of the eigenvalues.

Lemma 1: *If* $\dim X \geq 4$ *then there exists* $c = \{c_k\} \in l^2$, $c_k > 0$ *such that*

$$\sum_{k=1}^{\infty} c_k u_k^2(x) = \infty$$

*everywhere on* $\mathring{X}$.

PROOF: Let us take $c_k = \dfrac{1}{\sqrt{k+1}\,\ln(k+1)}$. Then $\Sigma_{k=1}^{\infty} c_k^2 < \infty$. Let $m = \dim X$, $\lambda_k$ be the $k$th eigenvalue of $\Delta$. We will prove that the following estimate is true in $\mathring{X}$:

$$\sum_{\lambda_k \le \lambda} \frac{1}{\sqrt{k+1}\,\ln(k+1)} u_k^2(x) \sim C p(x) \varphi_m(\lambda), \quad \varphi_m(\lambda) = \int^{\lambda} \frac{d\lambda}{\lambda^{2-m/4} \ln \lambda}; \qquad (1)$$

with $C > 0$ and $p(x) > 0$ on $\mathring{X}$. Since $\varphi_m(\lambda) = \ln \ln \lambda$ for $m = 4$ and $\varphi_m(\lambda) > \lambda^{1/4-\epsilon}$ for $m > 4$, this estimate implies the lemma.

Denote by $v_k(x)$ the eigenfunctions of the Laplace–Beltrami operator, normalized in $L^2(X, dx)$, then $u_k(x) = \dfrac{v_k(x)}{\sqrt{\lambda_k}}$. Since $\lambda_k \sim C_1 k^{2/m}$ we have $k + 1 \sim C_2 \lambda_k^{m/2}$. Therefore

$$\sum_{\lambda_k \le \lambda} \frac{1}{\sqrt{k+1}\,\ln(k+1)} u_k^2(x) = \sum_{\lambda_k \le \lambda} \frac{v_k^2(x)}{\lambda_k \sqrt{k+1}\,\ln(k+1)}$$
$$\sim C_3 \sum_{a < \lambda_k \le \lambda} \frac{v_k^2(x)}{\lambda_k^{1+m/4} \ln \lambda_k}. \quad (a > 1)$$

Let us express the right-hand sum in terms of the spectral function $E(x, x; \lambda)$ of $\Delta$. Since $E(x, x; \lambda) = \Sigma_{\lambda_k \le \lambda} v_k^2(x)$ we have

$$\sum_{a < \lambda_k \le \lambda} \frac{v_k^2(x)}{\lambda_k^{1+m/4} \ln \lambda_k} = \int_a^{\lambda} \frac{dE(x, x; \lambda)}{\lambda^{1+m/4} \ln \lambda}$$
$$= \frac{E(x, x; \lambda)}{\lambda^{1+m/4} \ln \lambda} - \int_a^{\lambda} E(x, x; \lambda) d\left(\frac{1}{\lambda^{1+m/4} \ln \lambda}\right).$$

Now we use the classical asymptotic formula of Carleman for the spectral function $E(x, x; \lambda)$ [14]:

$$E(x, x; \lambda) \sim C_4 p(x) \lambda^{m/2}, \quad p(x) > 0.$$

We get

$$\int_a^{\lambda} \frac{dE(x, x; \lambda)}{\lambda^{1+m/4} \ln \lambda} \sim C_5 p(x) \int_a^{\lambda} \frac{d\lambda}{\lambda^{2-m/4} \ln \lambda}$$

and therefore

$$\sum_{\lambda_k \le \lambda} c_k u_k^2(x) \sim C_6 p(x) \int_a^{\lambda} \frac{d\lambda}{\lambda^{2-m/4} \ln \lambda}.$$

COROLLARY. *If* $\dim X \geq 4$ *then* $\sum_k u_k^4(x) = \infty$.

PROOF: By the Cauchy inequality

$$\infty = \Big(\sum_k c_k u_k^2(x)\Big)^2 \leq \sum_k c_k^2 \cdot \sum_k u_k^4(x),$$

and the statement follows from $\sum_k c_k^2 < \infty$.

REMARK: It can be proved that the series $\sum_k u_k^4(x)$ diverges as a power one for $\dim X > 4$ and as a logarithmic one for $\dim X = 4$.

Now we prove a statement generalizing lemma 1.

LEMMA 2: *Let* $n \in \mathbb{Z}_+$ *and* $x_1, \ldots, x_n$ *be mutually disjoint points in* $\mathring{X}$. *Then for any* $\lambda_1, \ldots, \lambda_n \in \mathbb{R}$, $\sum \lambda_i^2 \neq 0$ *we have*

$$\sum_{k=1}^{\infty} \frac{1}{\sqrt{k+1}\,\ln(k+1)} (\lambda_1 u_k(x_1) + \cdots + \lambda_n u_k(x_n))^2 = \infty.$$

PROOF: Owing to (1) it is sufficient to prove that

$$\sum_{\lambda_k \leq \lambda} \frac{1}{\sqrt{k+1}\,\ln(k+1)} u_k(x) u_k(y) = o\left(\int_a^\lambda \frac{d\lambda}{\lambda^{2-m/4} \ln \lambda}\right) \tag{2}$$

on any compact subset of the complement of the diagonal in $\mathring{X} \times \mathring{X}$.

Repeating the same reasonings as in the proof of lemma 1 we get

$$\sum_{\lambda_l \leq \lambda} \frac{1}{\sqrt{k+1}\,\ln(k+1)} u_k(x) u_k(y) \sim C \int_a^\lambda \frac{dE(x, y; \lambda)}{\lambda^{1+m/4} \ln \lambda}$$

$$= C\left(\frac{E(x, y; \lambda)}{\lambda^{1+m/4} \ln \lambda} - \int_a^\lambda E(x, y; \lambda)\, d\left(\frac{1}{\lambda^{1+m/4} \ln \lambda}\right)\right). \tag{3}$$

Let us use the following estimate for the spectral function which is true on any compact subset of $\mathring{X} \times \mathring{X} \backslash \text{diag}$:

$$|E(x, y; \lambda)| < C(1+\lambda)^{m/2} \tag{4}$$

(see [12]). It is clear that (3) and (4) imply (2).

Let $C^\infty(\mathring{X})$ the space of all compactly supported $C^\infty$-functions on $\mathring{X}$ and $\mathscr{F}(X) = [C^\infty(\mathring{X})]'$ the space of Schwartz's distributions. Every nonzero element $f \in \mathring{W}_2^1(X)$ defines the nonzero functional on $C^\infty(\mathring{X})$ by the formula:

$$\langle f, \varphi \rangle = \int_X f(x)\varphi(x)dx.$$

Then we can imbed $\mathring{W}_2^1(X)$ in $\mathscr{F}(X)$.

LEMMA 3: *Let $X$ be a compact Riemannian manifold,* $\dim X \geq 4$. *There exists a Hilbert–Schmidt extension $\tilde{H}$ of $\mathring{W}_2^1(X)$, $\tilde{H} \subset \mathscr{F}(X)$ which doesn't contain the generalized functions of the form $\Sigma\, \lambda_i \delta_{x_i}$, $x_i$ being mutually disjoint points in $\mathring{X}$ and $\Sigma\, \lambda_i^2 \neq 0$.*

PROOF: Let us consider the (generalized) decomposition of the space of distributions on $\mathring{X}$ with respect to the system $\{u_k\}_{k=1}^\infty$, i.e. the mapping

$$\mathscr{F}(X) \xrightarrow{T} \mathbb{R}^\infty : f \mapsto \{\langle f, u_k \rangle\}_{k=1}^\infty. \tag{5}$$

Its natural domain of definition in $\mathscr{F}(X)$ contains any Hilbert-Schmidt extension of $\mathring{W}_2^1(X)$. Indeed, the image of $T$ contains any space being a Hilbert–Schmidt extension of $l^2$.

$T$ maps the space $\mathring{W}_2^1(X)$ into $l^2$, a delta-function $\delta_x$ into the sequence $\{u_k(x)\}_{k=1}^\infty$ and any linear combination $\Sigma_i\, \lambda_i \delta_{x_i}$ into the sequence $\{\Sigma_i\, \lambda_i u_k(x_i)\}_{k=1}^\infty$. Let us take $c_k$ from lemma 2 and consider the extension $H'$ of the space $l^2$ with respect to the operator $\Gamma : \Gamma e_k = c_k e_k$. It is a Hilbert–Schmidt extension and by lemma 2 the space $H' \subset \mathbb{R}^\infty$ doesn't contain the sequence $\{\Sigma_i\, \lambda_i u_k(x_i)\}_{k=1}^\infty$, $x_i$ being mutually disjoint points in $\mathring{X}$ and $\Sigma_i\, \lambda_i^2 \neq 0$. The space $\tilde{H} = T^{-1}H'$ is the required Hilbert–Schmidt extension of $\mathring{W}_2^1(X)$.

REMARK: As a matter of fact the statement of lemma 3 is true for any Riemann manifolds $X$, $\dim X \geq 4$. This result can be got from lemma 3 by partitioning $X$ into compact manifold. However we need not obtain the result in this paper.

Now we make clear the difference between the cases $\dim X = 4$ and $\dim X > 4$. For $\dim X > 4$ the space $\tilde{H}$ from lemma 3 can be described explicitly. Namely, it is known from the Sobolev space theory [11] that if and only if $l \leq \frac{m}{2}$ then $\mathring{W}_2^l(X)$ as subset of $\mathcal{F}(X)$ doesn't contain delta-functions $\delta_x$ and their linear combinations. On the other hand, if and only if $l > 1 + \frac{m}{4}$ the implication $\mathring{W}_2^l(X) \subset \mathring{W}_2^l(X)$ is a Hilbert–Schmidt extension. Consequently if $\frac{m}{2} \geq l > 1 + \frac{m}{4}$ then $\mathring{W}_2^l(X)$ is a space required in lemma 3. This inequality can be solved for $m \geq 5$ $\left(l = \frac{m}{2}\right)$.

This reasoning shows that for $m = 4$ $\tilde{H}$ can not be described in terms of $\mathring{W}_2^l(X)$ (and any "power" terms, but only "logarithmic" terms). It is clear that a choice of an extension $\left(c_k = \frac{1}{\sqrt{k+1}\ln(k+1)}\right)$ suitable for $m = 4$ suits for $m > 4$ as well.

## §2. Disjointness of spectral measures

Let $X$ be a Borel space and $\{\mu_\alpha\}$ be a measurable family of measures on $X$ where $\alpha$ runs through a space $A$ with a measure $\nu$. (Measurability of $\{\mu_\alpha\}$ means that the mapping $(\alpha, Y) \mapsto \mu_\alpha(Y)$, where $\alpha \in A$, $Y \in \mathfrak{A}$, $\mathfrak{A}$ is the algebra of the measurable sets in $X$, is measurable on $A \times \mathfrak{A}$.)

DEFINITION: The family $\{\mu_\alpha\}$ is called $\nu$-singular mod 0 if for almost all (with respect to $\nu \times \nu$) pairs $(\alpha', \alpha'')$ the measures $\mu_{\alpha'}$ and $\mu_{\alpha''}$ are singular. The family $\{\mu_\alpha\}$ is called $\nu$-disjoint (or disjoint if it is clear what measure $\nu$ is considered) if for any measurable subsets $A_1$ and $A_2$ in $A$ of positive $\nu$-measure, such that $\nu(A_1 \cap A_2) = 0$, the measures $\int_{A_1} \mu_\alpha d\nu(\alpha)$ and $\int_{A_2} \mu_\alpha d\nu(\alpha)$ are singular.

REMARK: As easy examples show, generally speaking singularity of $\{\mu_\alpha\}$ doesn't imply disjointness.

Let $G$ be an abelian topological group possessing a sufficient set of continuous characters $\chi : G \to S^1$. We assume that for any continuous unitary representation $U$ of $G$ in a complex Hilbert space $\mathcal{H}$ there exists an isomorphism $T$ of $\mathcal{H}$ onto a direct integral of Hilbert spaces,

$$T : \mathcal{H} \to \int_{\hat{G}}^{\oplus} \mathcal{H}_\chi d\mu(\chi)$$

with $\mu$ a Borel measure on $\hat{G}$ ($\hat{G}$ is the space of the measurable characters), which transfers $U(g)$, $g \in G$ into the operators

$$(TU(g)T^{-1}f)(\chi) = \chi(g)f(\chi).$$

This assumption is true of course for any locally compact group and for some others. In particular it is true for the group $\mathfrak{a}^X$ of §5.[1] The rest of §2 relates to the groups with the assumption being true.

The measure $\mu$ on $\hat{G}$ is defined by $U$ uniquely up to equivalence and is called the spectral measure of $U$. The realization of the representation of $G$ in the space $\int_{\hat{G}}^{\oplus} \mathcal{H}_\chi d\mu(\chi)$ is called the spectral decomposition of the initial representation $U$.

We give here two statements about the spectral measures.

(1) Two unitary representations of $G$ are disjoint (that is they contain no equivalent subrepresentations) if and only if their spectral measures are singular.

(2) The spectral measures of the direct sum and the tensor product of two representations of $G$ are equivalent to correspondingly the sum and the convolution of their spectral measures.

DEFINITION: A measurable family of unitary representations $U_\alpha$ of $G$ where $\alpha$ runs through $(A, \nu)$ is called disjoint if the family of the spectral measures $\mu_\alpha$ of $U_\alpha$ is disjoint (with respect to the given Borel measure $\nu$ on $A$).

In other words the set of representations $U_\alpha$ is called disjoint if for any subsets $A_1, A_2$ of $A$ with positive $\nu$-measure and $\nu(A_1 \cap A_2) = 0$, the representations $\int_{A_1}^{\oplus} U_\alpha d\nu(\alpha)$ and $\int_{A_2}^{\oplus} U_\alpha d\nu(\alpha)$ are disjoint.

Let us assume that a representation $U$ of $G$ is decomposed into a direct integral of representations

$$U = \int_A^{\oplus} U_\alpha d\nu(\alpha).$$

It means that $U$ is equivalent to the representation in a direct integral of Hilbert spaces

$$H = \int_A^{\oplus} H_\alpha d\nu(\alpha)$$

[1] It is false for example for the multiplicative group of classes of mod 0 measurable mappings $S^1 \to S^1$.

given by

$$(U(g)f)(\alpha) = U_\alpha(g)f(\alpha),$$

$U_\alpha$ being a representation of $G$ in $H_\alpha$.

The next lemma is a simple measure theoretic variant of Schur's lemma.

LEMMA 4: *If a family of representations of G is disjoint then the commutant of U (that is the ring of operators in H commuting with U(g)) consists of*

$$B = \int_A B_\alpha d\nu(\alpha)$$

*where* $B_\alpha$ *is a measurable operator function taking values in the commutant of* $U_\alpha$.

PROOF: Let an operator $B$ commute with the operators of the representation. Then for any $A_1 \subset A$ with $\nu(A_1) > 0$ the space $\int_{A_1}^{\oplus} H_\alpha d\nu(\alpha)$ is invariant with respect to $B$ (owing to disjointness with its direct complement). Consequently $B$ is decomposed into a direct integral.

COROLLARY: *Under the assumption of the lemma the W*-algebra generated by the operators U(g) contains the operators of multiplication by every bounded ν-measurable function* $a(\alpha): f(\alpha) \mapsto a(\alpha)f(\alpha)$.[2]

Indeed, every such operator commutes with the operators $B$ i.e. it belongs to the bicommutant of $U$ which by von-Neumann's theorem is the weak closure of the algebra generated by operators $U(g)$.

LEMMA 5: *Let a representation U of G be decomposed in a tensor product* $U = U' \otimes U''$ *of representations* $U'$ *and* $U''$ *with the corresponding spectral measures* $\mu'$ *and* $\mu''$. *If the family of measures* $\{\mu'_\chi\}$, $\mu'_\chi = \mu'(\cdot - \chi)$ *is* $\mu''$*-disjoint then the weakly closed operator*

[2] In [3] this statement (§2 lemma 5) contained the condition of singularity (instead of disjointness) of the $U_\alpha$. Notice that without the assumption of singularity of the $U_\alpha$ the statement is false.

*algebra generated by the operators $U(g), g \in G$ contains all operators $E \otimes U''(g)$ (and therefore all operators $U'(g) \otimes E$).*

PROOF: Let $U'' = \int_{\hat{G}}^{\oplus} U''_x d\mu''(x)$ be the spectral decomposition of the representation $U''$. Then

$$U = \int_{\hat{G}}^{\oplus} (U' \otimes U''_\chi) d\mu''(\chi).$$

Since the spectral measure of $U' \otimes U''_\chi$ is $\mu'_\chi = \mu'(\cdot - \chi)$ it follows that by the condition of the lemma the family of representations $U' \otimes U''_\chi$ is disjoint. It follows from corollary of lemma 4 that the $W^*$-algebra generated by $U(g)$ contains the operators of multiplication by functions $a_g(\chi) = \langle \chi, g \rangle$, i.e. the operators $E \otimes U''(g)$.

We emphasize that lemma 5 is true for any multiplicities of the spectra of $U'$ and $U''$.

## §3. A condition of singularity of the Gaussian measure together with its convolution

Let $\mu$ be the standard Gaussian measure in $\mathbb{R}^\infty$, i.e. the Gaussian measure with zero mean and Fourier transform

$$\int_{\mathbb{R}^\infty} e^{i(f,x)} d\mu(x) = \exp(-\tfrac{1}{2}\|f\|^2_{l^2}).$$

The following proposition is widely known. The measure $\mu$ and its translation $\mu_y = \mu(\cdot - y)$ are equivalent (that is mutually absolute continuous) if and only if $y \in l^2$. It is also well known that if two Gaussian measures are not equivalent, then they are singular (see for example [13]).

Let now $\nu$ be a Borel measure in $\mathbb{R}^\infty$. Let us consider the convolution $\mu * \nu$ that is the measure which is defined on cylindrical sets $A$ by

$$(\mu * \nu)(A) = \int_{\mathbb{R}^\infty} \mu_y(A) d\nu(y) = \int_{\mathbb{R}^\infty} \mu(A - y) d\nu(y).$$

It is clear that singularity of $\mu$ and $\mu * \nu$ implies that $\nu(l^2) = 0$. Taking in mind the statement given above one could assume that the opposite implication is true as well, that is $\nu(l^2) = 0$ implies singularity of $\mu$ and $\mu * \nu$. (It is just that what was affirmed in lemma 4 of §2 in [3]. The mistake was the unjustified passage to the limit in the expression for the density.) However this is wrong. A counterexample can be given even when $\nu$ is a Gaussian measure.

EXAMPLE: Let $\nu$ be a Gaussian measure with zero mean whose correlation operator $\Gamma$ is Hilbert-Schmidt but not nuclear. The convolution $\mu * \nu$ is a Gaussian measure as well the correlation operator $C = E + \Gamma$ ($E$ is the identity operator). Owing to Feldman's theorem [10] the measures $\mu$ and $\mu * \nu$ are equivalent. On the other hand, since $\Gamma$ is not nuclear, $\nu(l^2) = 0$ by the Minlos-Sazonov theorem (see [10]).

A criterion of singularity of the standard Gaussian measure $\mu$ together with its convolution can be gotten from the following lemma generalizing Ismagilow's lemma [7].

LEMMA 6: *Let $\Gamma$ be a strictly positive Hilbert-Schmidt operator in $\ell^2$, $\tilde{H} \subset \mathbb{R}^\infty$ be the completion of $l^2$ with respect to the norm $\|y\|_\Gamma = \langle \Gamma y, y\rangle^{1/2}$. Then there exists a subset $A \subset \mathbb{R}^\infty$ such that $\mu(A) = 1$ and $\mu(A - z) = 0$ for any $z \notin \tilde{H}$.*

PROOF: Let us introduce the expression

$$\Phi_\Gamma(x) \equiv :\langle \Gamma x, x\rangle: = \lim_n (\langle \Gamma_n x, x\rangle - \mathrm{Sp}\, \Gamma_n),$$

where $\Gamma_n = P_n \Gamma P_n$, $P_n$ is the projection onto the space spanned by the first $n$ eigenvectors of $\Gamma$ (about Wick regularisation – the sign : :– see for example [9]). It is known (see [9]) that if $\Gamma$ is a Hilbert-Schmidt operator then $\Phi_\Gamma(x)$ is defined and finite almost everywhere with respect to $\mu$. (Notice that if $\mathrm{Sp}\, \Gamma = \infty$, the sequences $\langle \Gamma_n x, x\rangle$ and $\mathrm{Sp}\, \Gamma_n$ don't converge.) Let us put

$$A = \{x : \Phi_\Gamma(x) \text{ is finite}\}.$$

According to what has been said above, $\mu(A) = 1$. We prove that $\mu(A - z) = 0$ for any $z \notin \tilde{H}$.

For that purpose we use the following simple fact about the standard Gaussian measure $\mu$ in $\mathbb{R}^\infty$ (see for example [6]). The series

$\Sigma_{n=1}^{\infty} d_n y_n$ either converges almost everywhere with respect to $\mu$ (if $d \in l^2$) or has neither a finite nor infinite sum almost everywhere (if $d \notin l^2$). Thus $\mu\{y : \Sigma_{n=1}^{\infty} d_n y_n = \infty\} = 0$ for any $d \in \mathbb{R}^{\infty}$.

Since $\mu$ doesn't change by orthogonal transformations, the operator $\Gamma$ in lemma can be assumed to be diagonal: $\Gamma e_n = c_n e_n$, $c_n > 0$, where $e_n = (0, \ldots, 0, 1, 0, \ldots)$, the unit is on the $n$th place. According to the remark above

$$\mu\{x : \lim_n \langle \Gamma_n x, z\rangle = \infty\} = \mu\left\{x : \sum_{n=1}^{\infty} c_n z_n x_n = \infty\right\} = 0$$

for any $z \in \mathbb{R}^{\infty}$. Let $z \notin \tilde{H}$. Then since $\lim_n \langle \Gamma_n z, z\rangle = \langle \Gamma z, z\rangle = \infty$, the following implication is true:

$$\{x : \lim_n (2\langle \Gamma_n x, z\rangle + \langle \Gamma_n z, z\rangle) \text{ is finite}\} \subset \{x : \lim_n \langle \Gamma_n x, z\rangle = \infty\}.$$

Consequently

$$\mu\{x : \lim_n (2\langle \Gamma_n x, z\rangle + \langle \Gamma_n z, z\rangle) \text{ is finite}\} = 0.$$

Since $\mu\{x : \lim_n (\langle \Gamma_n x, x\rangle - \operatorname{Sp} \Gamma_n) \text{ is finite}\} = 1$ it follows that $\mu\{x : \Phi_\Gamma(x+z) \text{ is finite}\} = \mu\{x : \lim_n (\langle \Gamma_n x, x\rangle - \operatorname{Sp} \Gamma_n + 2\langle \Gamma_n x, z\rangle + \langle \Gamma_n z, z\rangle) \text{ is finite}\} = 0$ i.e. $\mu(A - z) = 0$.

Corollary 1: *With the notations of lemma 6, if $\nu$ is a Borel measure on $\mathbb{R}^{\infty}$ such that $\nu(\tilde{H}) = 0$ then $\mu$ and $\mu * \nu$ are singular.*

Indeed $(\mu * \nu)(A) = \int_{\mathbb{R}^{\infty} \setminus H} \mu(A - z) d\nu(z) = 0$ because, by lemma 6, $\mu(A - z) = 0$ if $z \notin \tilde{H}$. Since $\mu(A) = 1$ the measures $\mu * \nu$ and $\mu$ are singular.

A convenient criterion for the singularity of the measures is given by the next corollary.

Corollary 2: *Let $\mu$ be the standard Gaussian measure in $\mathbb{R}^{\infty}$. If there exists an element $c = \{c_k\} \in l^2$, $c_k > 0$ such that $\nu\{x : \Sigma_{k=1}^{\infty} c_k x_k^2 < \infty\} = 0$ then $\mu$ and $\mu * \nu$ are singular.*

Indeed one can apply corollary 1 to the case when $\Gamma e_n = c_n e_n$, $n = 1, 2, \ldots$.

## §4. Convolutions with a Gaussian measure in the space of distributions on $X$

Let $X$ be a Riemannian manifold with a boundary $\partial X$ ($\partial X$ may be empty), $\mathring{X} = X \setminus \partial X$, $\mu$ be the Gaussian measure in the space $\mathcal{F}(X)$ of distributions on $\mathring{X}$ associated with the Laplace–Beltrami operator, i.e. the measure with Fourier transform $\exp(-\frac{1}{2}\| \quad \|^2_{\mathring{W}^1_2})$. We introduce for convenience the following notations:

$$\Phi = \left\{ \sum_{i=1}^{n} \lambda_i \delta_{x_i} : x_i \in \mathring{X}, \lambda_i \in \mathbb{Z}, n \in \mathbb{Z}_+ \right\},$$

$$\Phi_{\mathbb{R}} = \left\{ \sum_{i=1}^{n} \lambda_i \delta_{x_i} : x_i \in \mathring{X}, \lambda_i \in \mathbb{R}, n \in \mathbb{Z}_+ \right\}.$$

LEMMA 7: *If $X$ is a compact Riemannian manifold,* $\dim X \geq 4$ *then there exists a subset $A \subset \mathcal{F}(X)$ such that $\mu(A) = 1$ and $\mu(A - \varphi) = 0$ for any $\varphi \in \Phi_{\mathbb{R}}$, $\varphi \neq 0$.*

The statement is a straight consequence of lemmas 3 and 6.

COROLLARY: *If $X$ is a compact Riemannian manifold,* $\dim X \geq 4$, *and $\nu$ is a measure on the space $\mathcal{F}(X)$ such that $\nu(\Phi_{\mathbb{R}} \setminus \{0\}) = 1$ then the measures $\mu$ and $\mu * \nu$ are singular.*

Now we consider a Riemannian manifold $X$. Let $Y \subset X$ be an open subset with compact closure $\bar{Y}$. Let us consider the functions with supports in $Y$ and restrict to them distributions from $\mathcal{F}(X)$. We get the projection $\pi : \mathcal{F}(X) \to \mathcal{F}(Y)$ and $\pi\mu_X = \mu_Y$ where $\mu_X, \mu_Y$ are the Gaussian measures associated with the corresponding Laplace–Beltrami operators on $X$ and $Y$.

LEMMA 8: *Let $Y \subset X$ be the same as above,* $\dim X \geq 4$. *Then there exists a subset $A \subset \mathcal{F}(X)$ such that $\mu_X(A) = 1$, $\mu_Y(\pi A - \pi\varphi) = 0$ for any $\varphi \in \Phi$ provided $\pi\varphi \neq 0$.*

The statement follows from lemma 7.

COROLLARY: *Let $X$ be a Riemannian manifold,* $\dim X \geq 4$, $\{Y_n\}_{n=1}^{\infty}$ *be a countable base of open sets in $X$ with compact closure, $\pi_n$ be the projection $\mathcal{F}(X) \to \mathcal{F}(Y)$. There exists a subset $A \subset \mathcal{F}(X)$ such that $\mu_X(A) = 1$ and $\mu_{Y_n}(\pi_n A - \pi_n\varphi) = 0$ for any $n$ and $\varphi \in \Phi$ provided $\pi_n\varphi \neq 0$.*

Let us put for any $n$ $A_n = \pi_n^{-1}(\pi_n A)$ and introduce for any $\varphi \in \Phi$ the set

$$A^\varphi = A \backslash \Big( \bigcup_{n,\psi} (A_n - \psi) \Big),$$

where $\psi \in \Phi$, $\operatorname{supp} \psi \subset \operatorname{supp} \varphi$ and $\pi_n \psi \neq 0$ (similar sets were considered in [7]).

LEMMA 9: *Let $A$ be a set as in the corollary of lemma* 8 ($\dim X \geq 4$). *Then $\varphi_1, \varphi_2 \in \Phi$ and $\varphi_1 \neq \varphi_2$ imply $(A^{\varphi_1} + \varphi_1) \cap (A^{\varphi_2} + \varphi_2) = \phi$.*

PROOF: Let us assume on the contrary that there exist $\varphi_1 \neq \varphi_2$ such that $(A^{\varphi_1} + \varphi_1) \cap (A^{\varphi_2} + \varphi_2) \neq \phi$, i.e. there exist $a_i \in A^{\varphi_i}, i = 1, 2$, such that $a_1 + \varphi_1 = a_2 + \varphi_2$. Since $\varphi_1 \neq \varphi_2$ there exists a point $x_1$ belonging to the support of $\varphi_1$ with a coefficient $\lambda$ and to that of $\varphi_2$ with a coefficient $\lambda' \neq \lambda$. Let for definiteness $\lambda \neq 0$, $\lambda'$ can be equal to 0. Let $Y_n$ be a basis of neighbourhood of $x_1$ and let $Y_n$ contain no other points of the supports of $\varphi_1$ and $\varphi_2$. Then from the equality $\pi_n(a_1 + \varphi_1) = \pi_n(a_2 + \varphi_2)$ it follows that $\pi_n a_1 = \pi_n a_2 + k\delta_{x_1}$, where $k \neq 0$. But this is impossible since on the one hand $\pi_n a_1 \in \pi_n A \backslash (\pi_n A + k\delta_{x_1})$ and on the other $\pi_n a_2 + k\delta_{x_1} \in \pi_n A + k\delta_{x_1}$.

LEMMA 10: *Let $X$ be a Riemannian manifold,* $\dim X \geq 4$, *$\mu$ be the Gaussian measure in the space of distributions on $\mathring{X}$ with Fourier transform* $\exp(-\frac{1}{2}\| \quad \|^2_{W_2^1})$, *$\nu_1$ and $\nu_2$ be singular measures on $\mathcal{F}(X)$,* $\nu_1(\Phi) = \nu_2(\Phi) = 1$. *Then the measures $\mu * \nu_1$ and $\mu * \nu_2$ are singular. In particular if* $\nu_1(\{0\}) = 0$ *then the measures $\mu * \nu_1$ and $\mu$ are singular.*

PROOF: We take $A$ in the same way as above and put $B^\varphi = A^\varphi + \varphi (\varphi \in \Phi)$. The mapping $\varphi \mapsto B^\varphi$ is measurable as a mapping of $\Phi$ into the family $\mathcal{B}(\mathcal{F}(X))$ of the Borel subsets of $\mathcal{F}(X)$ (both spaces are provided with the natural Borel structure). Therefore if $Q \subset B$ is a Borel set then $B^Q = \bigcup_{\varphi \in Q} B^\varphi$ is measurable with respect to every Borel measure. Let $Q_1 \cap Q_2 = \phi$ and $\nu_1(Q_1) = \nu_2(Q_2) = 1$. By lemma 9 $B^{Q_1} \cap B^{Q_2} = \bigcup_{\substack{\varphi_1 \in Q_1 \\ \varphi_2 \in Q_2}} (B^{\varphi_1} \cap B^{\varphi_2}) = \phi$. On the other hand

$$(\mu * \nu_1)(B^{Q_1}) = \int_{Q_1} \mu \Big( \bigcup_{\varphi \in Q_1} B^{\varphi_1} - \psi \Big) d\nu_1(\psi) =$$

$$= \int_{Q_1} \mu \Big( \bigcup_{\varphi \in Q_1} (A^\varphi + \varphi - \psi) \Big) d\nu_1(\psi) \geq \int_{Q_1} \mu(A^\psi) d\nu_1(\psi) = 1$$

because $\mu(A^{\psi}) = \mu(A) = 1$. By analogy $(\mu * \nu_2)(B^{Q_2}) = 1$ and the lemma is proved.

Using the disjointness of a measure family (see §2) one can re-formulate the statement of lemma 10 in the following way.

LEMMA 10′: *Let $X$ and $\mu$ be the same as in lemma 10 and let $\nu$ be a measure in the space of distributions on $X$ such that $\nu(\Phi) = 1$. Then the family of measures $\mu_\varphi = \mu(\cdot - \varphi)$ is $\nu$-disjoint.*

REMARK: All lemmas of this section will remain true if the set $\mathbb{Z}$ in the definition of $\Phi$ is substituted by any countable subset of $\mathbb{R}$.

LEMMA 11: *Let two Riemannian structures $\tau_1, \tau_2$ be given on a compact manifold $X$, $\dim X \geq 4$, and $\mu_1, \mu_2$ be Gaussian measures generated by these structures in the space of distributions $\mathcal{F}(X)$. Then there exists a subset $A \subset \mathcal{F}(X)$ which satisfies one of the following conditions:*

(i) $\mu_2(A) = 1$ *and* $\mu_1(A - \varphi) = 0$ *for any* $\varphi \in \Phi_{\mathbb{R}} \setminus \{0\}$,
(ii) $\mu_1(A) = 1$ *and* $\mu_2(A - \varphi) = 0$ *for any* $\varphi \in \Phi_{\mathbb{R}} \setminus \{0\}$.

PROOF: The Riemannian structure $\tau_i$ generates an inner product in the function space on $\mathring{X}$:

$$(f, g)_i = -\int_X \Delta_i f \cdot g d_i x$$

$\Delta_i$ being the Laplace-Beltrami operator on the Riemannian manifold $(X, \tau_i)$, $i = 1, 2$. It is clear that these inner products are equivalent, i.e. $(f, g)_2 = (Bf, g)_1$ where $B$ is a positive bounded invertible operator. Let, as above, $\mathring{W}_2^1(X)$ be a completion of the space of compactly supported functions on $\mathring{X}$, $\{u_k\}_{k=1}^{\infty}$ be the orthonormal basis in $\mathring{W}_2^1(X)$ of the eigenfunctions of the Dirichlet problem for $\Delta_1$. Let us consider the mapping (5) from the space of distributions on $X$ into $\mathbb{R}^\infty$. It maps $\mathring{W}_2^1(X)$ into $l^2$, functions $\varphi = \Sigma_i \lambda_i \delta_{x_i}$ into sequences $\{\Sigma_i \lambda_i u_k(x_i)\}_{k=1}^{\infty}$, the measures $\mu_1, \mu_2$ correspondingly into the standard Gaussian measure $\mu_1'$ and the Gaussian measure $\mu_2'$ with the correlation operator $B = \|b_{ij}\|$.

Let us consider the following sets in $\mathbb{R}^\infty$:

$$A_1' = \left\{x : \sum_k c_k(x_k^2 - 1) < \infty\right\}, \qquad A_2' = \left\{x : \sum_k c_k(x_k^2 - b_{kk}) < \infty\right\},$$

where $c_k = \dfrac{1}{\sqrt{k+1}\ln(k+1)}$. Since $\Gamma = \|c_k\delta_{ik}\|$ is a Hilbert–Schmidt operator with respect to the inner products $(,)_1$ and $(,)_2$, the expressions $\Sigma_k c_k(x_k^2-1)$ and $\Sigma_k c_k(x_k^2-b_{kk})$ give almost everywhere (correspondingly with respect to $\mu_1'$ and $\mu_2'$) finite quadratic functionals with zero mean. Therefore $\mu_i'(A_i')=1, i=1,2$.

We assume now that $\Sigma_k c_k(1-b_{kk}) \neq -\infty$ and show that it implies $\mu_1'(A_2'-a)=0$ for any $a=\{a_k\}$, $a_k = \Sigma_i \lambda_i u_k(x_i)$. Indeed

$$A_2'-a = \left\{x : \sum_k c_k[(x_k+a_k)^2-b_{kk}] < \infty\right\} =$$

$$= \left\{x : \sum_k c_k[(x_k^2-1)+2a_kx_k+(1-b_{kk})+a_k^2] < \infty\right\}.$$

Since $\Sigma_k c_k a_k^2 = \infty$ by lemma 2 and since $\Sigma_k c_k(1-b_{kk}) \neq -\infty$, the series $\Sigma_k c_k(1-b_{kk}+a_k^2)$ diverges. On the other hand $\Sigma_k c_k(x_k^2-1)<\infty$ almost everywhere with respect to $\mu_1'$ and since $\Sigma_k c_k^2a_k^2<\infty$ it follows that $\Sigma_k c_ka_kx_k<\infty$ almost everywhere with respect to $\mu_1'$. Consequently $\mu_1'(A_2'-a)=0$.

Let $A_2$ be the inverse image of $A_2'$ in $\mathcal{F}(X)$. Then $\mu_2(A_2)=1$ and the statement proved above implies that $\mu_1(A_2-\varphi)=0$ for any $\varphi = \Sigma_i \lambda_i\delta_{x_i}$ $(\varphi \neq 0)$.

If $\Sigma_k c_k(1-b_{kk}) = -\infty$ then similar reasonings prove that the inverse image $A_1$ of $A_1'$ satisfies $\mu_2(A_1-\varphi)=0$ for any $\varphi = \Sigma_i \lambda_i\delta_{x_i}$ $(\varphi \neq 0)$.

COROLLARY 1: *Under the conditions of lemma* 11, *if $\nu$ is a measure on $\mathcal{F}(X)$ concentrated on $\Phi_R\backslash\{0\}$ then either $\mu_1 \perp \mu_2 * \nu$ or $\mu_2 \perp \mu_1 * \nu$.*

COROLLARY 2: *Under the same conditions, if $\mu_1 \perp \mu_2$ and $\nu$ is a measure on $\mathcal{F}(X)$ concentrated on $\Phi_R$ and satisfying $\nu(\{0\})>0$ then $\mu_1 * \nu$ and $\mu_2 * \nu$ are not equivalent.*

Indeed, by the corollary 1 and the singularity of the measures $\mu_1, \mu_2$ either $\mu_1 \perp \mu_2 * \nu$ or $\mu_2 \perp \mu_1 * \nu$. On the other hand $\nu(\{0\})>0$ implies that $\mu_1 < \mu_1 * \nu$, $\mu_2 < \mu_2 * \nu$. Consequently $\mu_1 * \nu \not\sim \mu_2 * \nu$.

REMARK: Notice that convolutions of two singular Gaussian measures $\mu_1, \mu_2$ with a measure $\nu$ can turn out equivalent.

EXAMPLE: Let $\mu_1$ be the standard Gaussian measure on $\mathbb{R}^\infty$, $\mu_2$ and $\nu$ be the Gaussian measures with correlation functionals $\Sigma_n (1-n^{-1/2})x_n^2$ and $\Sigma_n nx^2$. Then $\mu_1 \perp \mu_2$ and at the same time $\mu_1 * \nu \sim \mu_2 * \nu$.

## §5. On the spectrum of a representation of the abelian group $\mathfrak{a}^X$

We consider the group $G^X$ of smooth mappings of a Riemannian manifold $X$ into a compact semisimple Lie group $G$. Let $U$ be the unitary representation of $G^X$ being defined in the Introduction. We will use here another model of this representation (see [3]).

Let $\Omega(X;\mathfrak{g})$ be the pre-Hilbert space of $\mathfrak{g}$-values 1-forms on $X$ with the unitary representation $V$ of $G^X$ acting there and $\beta: G^X \to \Omega(X;\mathfrak{g})$ be the Maurer–Cartan cocycle (see Introduction). Let further $\mathscr{F}$ be the conjugate space to $\Omega(X;\mathfrak{g})$ and $\mu$ be the Gaussian measure on $\mathscr{F}$ with Fourier transform $\chi(\omega) = \exp(-\frac{1}{2}\|\omega\|^2)$. $U$ acts in the Hilbert space $L^2_\mu(\mathscr{F})$ by the following formula

$$(U(\tilde{g})\Phi)(F) = e^{i\langle F,\beta\tilde{g}\rangle}\Phi(V^{-1}(\tilde{g})F).$$

(see [3] about equivalence of the representations).

Let now $\mathfrak{a} \subset \mathfrak{g}$ be a Cartan subalgebra of the Lie algebra $\mathfrak{g}$ of $G$, $\mathfrak{a}^X$ be the abelian group of compactly supported $C^\infty$-mappings $X \to \mathfrak{a}$. $\mathfrak{a}^X$ is said in §2 to satisfy the condition formulated there and therefore for any unitary representation of $\mathfrak{a}^X$ the spectral measure is defined. Let us define the unitary representation of $\mathfrak{a}^X$ in $L^2_\mu(\mathscr{F})$ by the formula

$$W(a) = U(\exp a), \quad a \in \mathfrak{a}^X.$$

Here we give a summary of properties of $W$ and its spectral measure (see [3] §4 lemmas 1–4).

(1) Let $\mathfrak{m}$ denote the orthogonal complement in $\mathfrak{g}$ to $\mathfrak{a}$, $\mathscr{F}_\mathfrak{a}$, $\mathscr{F}_\mathfrak{m}$ be the subspaces of, correspondingly, $\mathfrak{a}$-valued and $\mathfrak{m}$-valued generalized 1-forms on $X$, $\mu_\mathfrak{a}$, $\mu_\mathfrak{m}$ be the standard Gaussian measures on $\mathscr{F}_\mathfrak{a}$ and $\mathscr{F}_\mathfrak{m}$. Then

$$W = W_\mathfrak{a} \otimes W_\mathfrak{m},$$

where $W_\mathfrak{a}$, $W_\mathfrak{m}$ are the representations in the corresponding spaces

$L^2_{\mu_{\mathfrak{a}}}(\mathcal{F}_{\mathfrak{a}})$ and $L^2_{\mu_{\mathfrak{m}}}(\mathcal{F}_{\mathfrak{m}})$ which are defined by

$$(W_{\mathfrak{a}}(a)\Phi)(F) = e^{i\langle F, da\rangle}\Phi(F),$$
$$(W_{\mathfrak{m}}(a)\Phi)(F) = \Phi(\mathrm{Ad}^{-1}(\exp a)F).$$

Thus the spectral measure of $W$ is equal to the convolution of the spectral measures of $W_{\mathfrak{a}}$ and $W_{\mathfrak{m}}$.

(2) $$W_{\mathfrak{m}} \cong \bigoplus_{n=0}^{\infty} S^n V_{\mathfrak{m}},$$

where $V_{\mathfrak{m}}$ is the representation in the space $\overline{\Omega(X;\mathfrak{m})_C}$ given by

$$(V_{\mathfrak{m}}(a)\omega)(x) = \mathrm{Ad}(\exp a(x)) \cdot \omega(x),$$

$S^n V_{\mathfrak{m}}$ $(n > 0)$ is the symmetrized tensor product of $n$ copies of $V_{\mathfrak{m}}$; $S^0 V_{\mathfrak{m}}$ is the unity representation.

(3) The spectral measure $\mu$ of $W_{\mathfrak{a}}$ is equivalent to the Gaussian measure on $(\mathfrak{a}^X)'$ with zero mean and Fourier transform $\chi(a) = \exp(-\frac{1}{2}\langle da, da\rangle)$, $a \in \mathfrak{a}^X$.

(4) By the decomposition $W_{\mathfrak{m}} \cong \bigoplus_{n=0}^{\infty} S^n V_{\mathfrak{m}}$, the spectral measure $\nu$ of $W_{\mathfrak{m}}$ is equivalent to the sum of the spectral measure $\nu_0$ of the unity representation (i.e. the measure concentrated on the point 0) and the spectral measures $\nu_n$ of $S^n V_{\mathfrak{m}}$ $(n = 1, 2, \ldots)$.

Let $\Sigma$ denote the set of the roots of $\mathfrak{g}$ with respect to $\mathfrak{a}$. Let $\nu_n^{\alpha_1, \ldots, \alpha_n}$, $\alpha_i \in \Sigma$, $n = 1, 2, \ldots$, denote the measure in $(\mathfrak{a}^X)'$ concentrated on the set of distributions of the form $\varphi_{x_1}^{\alpha_1} + \cdots + \varphi_{x_n}^{\alpha_n}$ where $\langle \varphi_x^{\alpha}, a\rangle = \alpha(a(x))$ and equivalent on this set to the measure $dx_1 \ldots dx_n$. Then $\nu_n$ is equivalent to the sum of $\nu_n^{\alpha_1, \ldots, \alpha_n}$ $(n = 1, 2, \ldots)$.

We denote now by $\Psi$ the set of distributions of the form $\varphi_{x_1}^{\alpha_1} + \cdots + \varphi_{x_n}^{\alpha_n}$ where $\alpha_i \in \Sigma$, $x_i \in \mathring{X}$, $n = 0, 1, \ldots$ and note that $\nu(\Psi) = 1$.

Lemma 12: *If* $\dim X \geq 4$ *then the family of measures* $\mu_\varphi = \mu(\cdot - \varphi)$, $\varphi \in \Psi$ *is* $\nu$*-disjoint (see* §2).

Proof: Let us fix a unit vector $e \in \mathfrak{a}$ such that the numbers $\alpha(e)$ are mutually distinct as $\alpha$ runs through the non-zero elements of $\Sigma$. Let $\mathbb{R}_1 \subset \mathfrak{a}$ be the one-dimensional space spanned by $e$, $\mathbb{R}_1^X \subset \mathfrak{a}^X$ be the space of compactly supported $C^\infty$-smooth real functions on $X$, let $(\mathbb{R}_1^X)'$ be the space conjugate to $\mathbb{R}_1^X$. We consider the projection

$$\tau : (\mathfrak{a}^X)' \to (\mathbb{R}_1^X)'$$

and denote by $\mu'$, $\nu'$ the projections of $\mu$ and $\nu$ on $(\mathbb{R}_1^X)'$. Owing to the properties (3), (4) of the measures $\mu$ and $\nu$, $\mu'$ is the Gaussian measure on $(\mathbb{R}_1^X)'$ with zero mean and Fourier transform $\chi(f) = \exp(-\frac{1}{2}\|f\|_{W_1^2}^2)$, $\nu'$ is concentrated on the set $\tau\Psi = \{\alpha_1(e)\delta_{x_1} + \cdots + \alpha_n(e)\delta_{x_n} : x_i \in \mathring{X}, \alpha_i \in \Sigma, n = 0, 1, 2, \ldots\}$.

Consequently, by lemma 10′ the family of measures $\mu'_{\varphi'} = \mu'(\cdot - \varphi')$, $\varphi' \in \tau\Psi$ is $\nu'$-disjoint. Then the family of measures $\mu''_\varphi = \mu'_{\tau\varphi}$ is $\nu$-disjoint since, by the choice of $e$, the mapping $\Psi \to \tau\Psi$ is a bijection and $\tau\nu = \nu'$.

It follows now from this disjointness that the family of measures $\mu_\varphi = \mu(\cdot - \varphi)$, $\varphi \in \Psi$ is $\nu$-disjoint on $\mathfrak{a}^X$, i.e. for any $A_1, A_2 \subset \Psi$ of positive $\nu$-measures with $\nu(A_1 \cap A_2) = 0$ the measures $\int_{A_i} \mu_\varphi d\nu(\varphi)$, $i = 1, 2$ are mutually singular. Indeed, since the family of measures $\mu''_\varphi$ is $\nu$-disjoint on $\Psi$ then there exists $B_1, B_2 \subset (\mathbb{R}_1^X)'$ such that $B_1 \cap B_2 = \phi$ and $\int_{A_i} \mu''_\varphi(B_i) d\nu(\varphi) = \nu(A_i)$. Let $\bar{B}_i$ denote the inverse image of $B_i$ in $(\mathfrak{a}^X)'$. Then for any $\varphi \in \Psi$ the set $\bar{B}_i + \varphi$ is the inverse image of $B_i + \tau\varphi$ and therefore $\mu_\varphi(\bar{B}_i) = \mu''_\varphi(B_i)$. Consequently $\int_{A_i} \mu_\varphi(\bar{B}_i) d\nu(\varphi) = \nu(A_i)$, $i = 1, 2$ and since $\bar{B}_1 \cap \bar{B}_2 = \phi$ the measures $\int_{A_i} \mu_\varphi d\nu(\varphi)$, $i = 1, 2$ are mutually singular.

COROLLARY 1: *The spectral measure of* $W_\mathfrak{a}$ *and those of* $W_\mathfrak{a} \otimes S^n V_\mathfrak{m}$, $n = 1, 2, \ldots$ *are mutually singular.*

COROLLARY 2: *The weakly closed operator algebra generated by the operators of* $W$ *contains all operators of multiplication by* $e^{i\langle \cdot, da \rangle}$.

That follows from lemmas 12 and 5.

LEMMA 13: *Let* $X$ *be a smooth manifold,* $\dim X \geq 4$, $\tau_1, \tau_2$ *be two Riemannian structures on* $X$, $W^1, W^2$ *be the representations of* $\mathfrak{a}^X$ *corresponding to these structures. If* $\tau_1 \neq \tau_2$ *then* $W^1$ *and* $W^2$ *are not equivalent.*

PROOF: Without a loss of generality we can assume $X$ to be a compact manifold. Indeed, in the opposite case one can take an arbitrary neighbourhood $Y \subset X$ with a compact closure on which $\tau_1 \neq \tau_2$, and consider instead of $W^i$ their restrictions to $\mathfrak{a}^Y$.

We note that $\mathfrak{a}^X = \mathbb{R}_1^X \oplus \cdots \oplus \mathbb{R}_r^X$, $\mathbb{R}_k = \mathbb{R}$ $(r = \dim \mathfrak{a})$. Therefore it is sufficient to prove that the restrictions $\overline{W^i}$ of $W^i$ to $\mathbb{R}_1^X$ are not equivalent. For this purpose we will find the spectral measures of $\overline{W^i}$.

We have $W^i = W^i_{\mathfrak{a}} \otimes W^i_{\mathfrak{m}}$ and $W^i_{\mathfrak{a}}$ is a tensor product of representations of the groups $\mathbb{R}^X_k$. Let us denote by $\overline{W}^i_{\mathfrak{a}}$ and $\overline{W}^i_{\mathfrak{m}}$ the restrictions of $W^i_{\mathfrak{a}}$ and $W^i_{\mathfrak{m}}$ to $\mathbb{R}^X_1$. Then $\overline{W}^i = \overline{W}^i_{\mathfrak{a}} \otimes \overline{W}^i_{\mathfrak{m}}$.

By definition, the spectral measure of $\overline{W}^i_{\mathfrak{a}}$ is the standard Gaussian measure $\bar{\mu}_i$ in the space of distributions $\mathscr{F}(X)$ generated by the Riemannian structure $\tau_i$. The inequality $\tau_1 \neq \tau_2$ implies $\bar{\mu}_1 \perp \bar{\mu}_2$ (for a proof, see [3], p. 326). The spectral measures $\bar{\nu}_i$ of $\overline{W}^i_{\mathfrak{m}}$ are equivalent to each other ($\bar{\nu}_1 \sim \bar{\nu}_2 \sim \bar{\nu}$) and are concentrated on the set $\Phi = \left\{\sum_{i=1}^{n} \lambda_i \delta_{x_i} : x_i \in \mathring{X},\ \lambda_i \in Z,\ n \in Z_+\right\}$ with $\bar{\nu}(\{0\}) > 0$. Therefore by corollary 2 of lemma 11 $\bar{\mu}_1 * \bar{\nu}_1$ and $\bar{\mu}_2 * \bar{\nu}_2$ are not equivalent. Those are the spectral measures of $\overline{W}^i$ and consequently the representations $\overline{W}^1$ and $\overline{W}^2$ are not equivalent.

## §6. Irreducibility and non-equivalence of representations of $G^X$

THEOREM 1: *If* $\dim X \geq 4$ *then the representation* $U$ *of* $G^X$ *is irreducible.*

A proof reduces to a check-up of the following two statements:

(1) the cyclic subspace generated by the vacuum vector $\Phi \equiv 1$ is irreducible,

(2) the vacuum vector is cyclic in the space of $U$.

PROOF OF STATEMENT 1: Let $\mathfrak{a} \subset \mathfrak{g}$ be an arbitrary Cartan subalgebra. We denote by $H_{\mathfrak{a}}$ the space of functionals $\Phi \in L^2_\mu(\mathscr{F})$ such that $\Phi(\cdot + F) = \Phi(\cdot)$ for any $F \in \mathscr{F}_{\mathfrak{m}}$ (see §5). By definition $W$, $H_{\mathfrak{a}}$ is invariant under the action of $W$ and the representations of $\mathfrak{a}^X$ in $H_{\mathfrak{a}}$ and in its orthogonal complement are equivalent, correspondingly, to $W_{\mathfrak{a}}$ and $\bigoplus_{n=1}^{\infty}(W_{\mathfrak{a}} \otimes S^n V_{\mathfrak{m}})$. It follows from corollary 1 of lemma 12 that the representations of $\mathfrak{a}^X$ in $H_{\mathfrak{a}}$ and in its orthogonal complement are disjoint.

Let us verify now that if $\mathfrak{a}_1, \mathfrak{a}_2$ are two Cartan subalgebra with $\mathfrak{a}_1 \cap \mathfrak{a}_2 = 0$ then $H_{\mathfrak{a}_1} \cap H_{\mathfrak{a}_2} = \{c\Omega\}$, $\Omega$ being the vacuum vector. Indeed, let $\mathfrak{m}_1, \mathfrak{m}_2$ be the orthogonal complements in $\mathfrak{g}$, correspondingly, of $\mathfrak{a}_1$ and $\mathfrak{a}_2$. Since $\mathfrak{a}_1 \cap \mathfrak{a}_2 = 0$ we have $\mathfrak{m}_1 + \mathfrak{m}_2 = \mathfrak{g}$ and therefore $\mathscr{F}_{\mathfrak{m}_1} + \mathscr{F}_{\mathfrak{m}_2} = \mathscr{F}$. Hence if $\Phi \in H_{\mathfrak{a}_1} \cap H_{\mathfrak{a}_2}$ that is $\Phi(\cdot + F_1) = \Phi(\cdot + F_2) = \Phi(\cdot)$ for any $F_1 \in \mathscr{F}_{\mathfrak{m}_1}$ and $F_2 \in \mathscr{F}_{\mathfrak{m}_2}$, then $\Phi(\cdot + F) = \Phi(\cdot)$ for any $F \in \mathscr{F}$ that is $\Phi(\cdot) = c\Omega$.

Let $A$ be an arbitrary operator in $L^2_\mu(\mathcal{F})$ which commutes with operators of $U$. Then $A$ commutes with operators of $W$. Since the representations of $\mathfrak{a}^X$ in $H_\mathfrak{a}$ and in its orthogonal complements are disjoint, the operator $A$ leaves every subspace $H_\mathfrak{a}$ invariant and therefore $A\Omega = c\Omega$. Statement 1 follows.[3]

PROOF OF STATEMENT 2: Follows from corollary 2 of lemma 12. Namely, the corollary implies that the weak closed operator algebra generated by operators $U(\tilde{g})$, $\tilde{g} \in G^X$ in $L^2_\mu(\mathcal{F})$ contains the operators of multiplication by functionals of the form

$$(6) \quad \exp\left(i\left\langle F, \sum_{k=1}^{n} V(\tilde{g}_k)du_k\right\rangle\right), \quad \tilde{g}_k \in G^X, \quad u_k \in \mathfrak{g}^X, \quad n = 1, 2, \ldots.$$

Then one can prove that the set of vectors of the form $\Sigma_{k=1}^n V(\tilde{g}_k)du_k$ is dense in $\Omega(X; \mathfrak{g})$ which implies that the functionals (6) form a total set in $L^2_\mu(\mathcal{F})$ and statement 2 follows (see [3] §5 lemmas 2, 3).

THEOREM 2: *Let $G$ be a compact semisimple Lie group, $X$ be a manifold,* $\dim X \geq 4$. *If $\tau_1, \tau_2$ are two different Riemannian structures on $X$ then the corresponding representations $U$ of $G^X$ are not equivalent.*

Indeed, by lemma 13 even the restrictions of these representations to an abelian subgroup $\mathfrak{a}^X$ are not equivalent.

REMARK: It can be proved by quite other reasonings that theorem 2 stands for $\dim X = 1$. In this case the spectral measure of $W = W_\mathfrak{a} \otimes W_\mathfrak{m}$ is equivalent to that of $W_\mathfrak{a}$. At the same time the spectral measures of the representations $W_\mathfrak{a}$ corresponding to different Riemannian structures on $X$ are singular. Therefore the spectral measures of the corresponding representations $W^\cdot$ are singular as well. (The spectral measures of $W$ and $W_\mathfrak{a}$ are likely to be equivalent for $\dim X = 2$ as well, see the remark below.)

Another proof of nonequivalence of the representations of $G^X$ for

[3] As a matter of fact we repeated here the proof of lemma 1 of §5 in [3] with the difference that the reference to corollary 2 of lemma 5 of §4 in [3] was changed by the reference to corollary 2 of lemma 12 of this paper.

$G = SU(2)$ and $X$ a manifold of an arbitrary dimension can be found in [7].

In conclusion we give some remarks on the cases $\dim X \leq 3$.

1. Let $X$ be a compact Riemannian manifold, $\mu$ be standard Gaussian measure in the space of distributions on $X$, $\nu$ be a measure on the set of delta-functions $\delta_x$ equivalent to $dx$ on this set.

Our proof of irreducibility is based on the statement that for $\dim X \geq 4$ the measures $\mu$ and $\mu * \nu^n$ $(n = 1, 2, \ldots)$ are mutually singular. Let us show that this statement fails for $\dim X \leq 2$ and therefore the methods of this work can not be used for the dimensions 1 and 2.

PROPOSITION: *If* $\dim X = 1$ *then* $\mu$ *and* $\mu * \nu^n$ $(n = 1, 2, \ldots)$ *are equivalent.*

PROOF: If $\dim X = 1$ then $\mathring{W}_2^1(X) \subset C$ (see [11]). Therefore $\delta_x \in (\mathring{W}_2^1(X))'$ for any $x \in \mathring{X}$ and consequently the measure $\nu^n$ is concentrated on $(\mathring{W}_2^1(X))'$. It is known that the Gaussian measure $\mu$ is quasi-invariant under the translations by vectors from $(\mathring{W}_2^1(X))'$. Hence the convolution of every measure, concentrated on $(\mathring{W}_2^1(X))'$, with $\mu$ is equivalent to $\mu$. In particular $\mu * \nu^n \sim \mu$.

The case $\dim X = 2$ is more subtle: since in this case $\delta_x \notin (\mathring{W}_2^1(X))'$ for all $x \in X$, any translation of $\mu$ by a linear combination of $\delta$-functions is singular with $\mu$. However it has been shown by Hoegh–Krohn that for $X$ being an Euclidean disk the measures $\mu * \nu$ and $\mu$ are equivalent[4] and $\dfrac{d(\mu * \nu)}{d\mu} \in L^2_\mu$. We give here a sketch of a proof.

Let $\{u_k\}_{k=1}^{\infty}$ be the orthonormal basis in $\mathring{W}_2^1(X)$ of eigenfunctions of Dirichlet problem for the Laplace operator which are ordered accordingly non-decrease of the eigenvalues, $\mu_n$, $\nu_n$ be the projections of $\mu$ and $\nu$ on the finite-dimensional space generated by $u_1, \ldots, u_n$.

We put $p = \dfrac{d(\mu * \nu)}{d\mu}$, $p_n = \dfrac{d(\mu_n * \nu_n)}{d\mu_n}$ and calculate the density $p$. It is known (see [6], the theorem about convergence of martingales) that $p$ does exist almost everywhere with respect to $\mu$ and $p = \lim p_n$. Generally speaking $\int p d\mu \leq 1$ and $\int p d\mu = 1$ if and only if $\mu * \nu$ and $\mu$ are equivalent.

[4] A more simple example of the appearance is given in §3 where $\nu$ is a Gaussian measure.

To prove the equivalence of $\mu$ and $\mu * \nu$ it is sufficient to prove that $p_n$ converges in $L^2_\mu$. Indeed if $p_n \to \tilde{p}$ in $L^2_\mu$ then $\tilde{p} = p$ and therefore $\int p d\mu = \int \tilde{p} d\mu = \lim_n \int p_n d\mu = 1$ (since $\int p_n d\mu = 1$ for all $n$). Let us calculate $\|p_n - p_m\|_{L^2_\mu}$.

Recall that if $\mu_n$ is the standard Gaussian measure in $\mathbb{R}^n$ and $\mu_n(\cdot + u)$ is its translation by a vector $u$, then $\frac{d\mu_n(\cdot + u)}{d\mu_n}(y) = e^{-\langle y,u\rangle - (1/2)\|u\|^2_n}$. Therefore, since $\delta_x \sim \Sigma_k u_k(x)$ we get

$$p_n(y) = \frac{d(\mu_n * \nu_n)}{d\mu_n}(y) = \int_X \exp\left(-\sum_1^n y_k u_k(x) - \tfrac{1}{2}\sum_1^n u_k^2(x)\right)dx.$$

It implies

$$\langle p_n, p_m\rangle = \frac{1}{(2\pi)^{m/2}} \int_{\mathbb{R}^m} \int_X \int_X \exp\left(-\sum_1^n y_k u_k(x) - \tfrac{1}{2}\sum_1^n u_k^2(x)\right.$$
$$\left. - \sum_1^m y_k u_k(x') - \tfrac{1}{2}\sum_1^m u_k^2(x') - \tfrac{1}{2}\sum_1^m y_k^2\right) dx dx' dy. \quad (m \geq n).$$

Integrating by $y$ we get

$$\langle p_n, p_m\rangle = \int_X \int_X \exp\left(\sum_1^n u_k(x) u_k(x')\right) dx dx' \quad (m \geq n).$$

In particular $\langle p_n, p_m\rangle = \langle p_n, p_n\rangle$ for $m \geq n$. Hence

$$\begin{aligned} \|p_n - p_m\|^2 &= \langle p_m, p_m\rangle - \langle p_n, p_m\rangle \\ &= \int_X \int_X \exp\left(\sum_1^n u_k(x) u_k(x')\right) \\ &\quad \times \left[\exp\left(\sum_{n+1}^m u_k(x) u_k(x')\right) - 1\right] dx dx' \quad (m \geq n). \end{aligned}$$

To prove that $p_n$ converges in $L^2_\mu$, i.e. $\|p_n - p_m\| \to 0$ for $m, n \to \infty$, it is sufficient to verify that $\lim_n \int_X \int_X \exp(\Sigma_1^n u_k(x) u_k(x')) dx dx'$ does exist. For that it is sufficient now to verify that

$$\int_X \int_X \exp G(x, x') dx dx' < \infty,$$

where $G(x, x') = \lim_n \Sigma_1^n u_k(x) u_k(x')$ is the Green function for the Laplace operator. It is known that when $X$ is an Euclidean disk the

Green function has a logarithmic singularity on the diagonal:

$$G(x, x') \sim \frac{1}{2\pi} \ln \frac{1}{\rho(x, x')},$$

$\rho$ being a Euclidean metric on the plane (see for example [8]). Consequently

$$\int_X \int_X \exp G(x, x') dx dx' \sim \int_{|x| \leq 1} \int_{|x'| \leq 1} [\rho(x, x')]^{-1/2\pi} dx dx',$$

and it is clear that the last integral does converge.

2. For $\dim X = 3$ equivalence of $\mu$ and $\mu * \nu$ can not be proved by the same means as for $\dim X = 2$ since $\int_X \int_X \exp G(x, x') dx dx'$ diverges. On the other hand nonequivalence of those can not be deduced from lemma 6 since for $\dim X = 3$ $\Sigma_k c_k u_k^2(x)$ converges almost everywhere for any $\{c_k\} \in l^2$.

REMARK: In [3] a projective representation of the group $G^X \cdot \Omega(X; \mathfrak{g})$ (Sugawara's group) was constructed. This representation is irreducible for $\dim X \geq 4$ since its restriction to $G^X$ is irreducible due to theorem 1. Yet this representation is irreducible for a manifold $X$ of an arbitrary dimension including the case $\dim X = 1$. This statement easily follows from an analysis of the decomposition of the representation with respect to the abelian normal subgroup $\Omega(X; \mathfrak{g})$ and the lemmas about the spectrum of the representation of $\mathfrak{a}^X \subset G^X$.

## Appendix. The Lie algebra $\mathfrak{g}^X$ and its representation

Let as above $X$ be a Riemannian manifold, $\mathfrak{g}$ be a Lie algebra of a compact semisimple Lie group $G$. We consider the space $C^\infty(X; \mathfrak{g})$ of all compactly supported $C^\infty$-mapping $X \to \mathfrak{g}$ provided with the usual topology. Let us define in $C^\infty(X; \mathfrak{g})$ the Lie algebra structure by $[a_1, a_2](x) = [a_1(x), a_2(x)]$. This Lie algebra is called the Lie algebra of $G^X$ and denoted by $\mathfrak{g}^X$.

Using the exponential mapping

$$\exp: \mathfrak{g}^X \to G^X$$

we may assign to any unitary representation $U$ of $G^X$ the representation $L$ of its Lie algebra in the same space, defined by

$$L(a)\xi = \lim_{t\to 0}\left(\frac{1}{t}(U(\exp ta) - E)\xi\right).$$

(The operators $L(a)$ are defined on the set of such vectors in the space of $U$ that this limit does exist.)

Let $U = \mathrm{EXP}_\beta V$ be the representation of $G^X$ which is considered in the paper. Let us give the formulas for the operators $L$ of the corresponding representation of $\mathfrak{g}^X$.

Let $U$ be realized in the Fock space

$$\mathrm{EXP}\, H = \bigoplus_{n=0}^{\infty} S^n H_{\mathrm{C}},$$

where $H = \overline{\Omega(X;\mathfrak{g})}$, $H_{\mathrm{C}}$ is the complexification of $H$, $S^n H_{\mathrm{C}}$ is the symmetrized tensor product of $n$ copies of $H_{\mathrm{C}}$. The formulas for operators of $U$ in this realization is given in the Introduction. From the formulas one can easily get an expression for operators of the representation of $\mathfrak{g}^X$.

Let us put

$$s^n(\omega_1 \otimes \cdots \otimes \omega_n) = \frac{1}{n!}\sum \omega_{i_1} \otimes \cdots \otimes \omega_{i_n},$$

where an ordered set $(i_1, \ldots, i_n)$ in the sum runs through the family of all permutations of $\{1, \ldots, n\}$. If $\omega_i \in H_{\mathrm{C}}$ then $s^n(\omega_1 \otimes \cdots \otimes \omega_n) \in S^n H_{\mathrm{C}}$.

PROPOSITION: *The operators $L(a)$ of the representation of $\mathfrak{g}^X$ induced by $U$ are defined on all vectors $s^n(\omega_1 \otimes \cdots \otimes \omega_n)$, $\omega_i \in H_{\mathrm{C}}$ and are given by the formula*

$$\begin{aligned} L(a)&(s^n(\omega_1 \otimes \cdots \otimes \omega_n)) \\ &= -\sqrt{n}\sum_{i=1}^{n} \langle \omega_i, da\rangle s^{n-1}(\omega_1 \otimes \cdots \otimes \omega_{i-1} \otimes \omega_{i+1} \otimes \cdots \otimes \omega_n) \\ &\quad + n\sum_{i=1}^{n} s^n(\omega_1 \otimes \cdots \otimes \omega_{i-1} \otimes [a, \omega_i] \otimes \omega_{i+1} \otimes \cdots \otimes \omega_n) \\ &\quad + \sqrt{n+1}\sum_{i=1}^{n+1} s^{n+1}(da \otimes \omega_1 \otimes \cdots \otimes \omega_n). \end{aligned} \tag{7}$$

In particular

$$L(a)(\underbrace{\omega \otimes \cdots \otimes \omega}_{n}) = -\sqrt{n}\langle \omega, da\rangle \underbrace{\omega \otimes \cdots \otimes \omega}_{n-1} + \sum_{i=1}^{n} \underbrace{\omega \otimes \cdots \otimes \omega}_{i-1} \otimes [a, \omega] \otimes \underbrace{\omega \otimes \cdots \otimes \omega}_{n-i} + \frac{1}{\sqrt{n+1}} \sum_{i=0}^{n} \underbrace{\omega \otimes \cdots \otimes \omega}_{i} \otimes da \otimes \underbrace{\omega \otimes \cdots \otimes \omega}_{n-i}.$$

REMARK: The formulas (7) give not only the representation of $\mathfrak{g}^X$ but also the representation (which is not Hermitian now) of its complexification $\mathfrak{g}_{\mathbb{C}}^X$ and therefore $\mathfrak{g}_\tau^X$ where $\mathfrak{g}_\tau$ is an arbitrary semisimple Lie algebra.

**Added in proof**

At the last time S. Albeverio, R. Hoegh-Krohn and D. Testard proved irreducibility of $U(\tilde{g})$ in the case $\dim X \geq 3$ and reducibility in the case $\dim X = 1$ (preprint 1980, Bochum).

REFERENCES


[1] A.M. Veršik, I.M. Gelfand and M.I. Graev: Representations of the group $SL(2, R)$ where $R$ is a ring of functions. *Russ. Math. Surv. 28, no. 5* (1973) 83.

[2] A.M. Veršik, I.M. Gelfand and M.I. Graev: Irreducible representations of the group $G^X$ and cohomology. *Funct. Anal. and its Appl. 8, no 2* (1974) 67.

[3] I.M. Gelfand, M.I. Graev, A.M. Veršik: Representations of the group of smooth mappings of a manifold $X$ into a compact Lie group. *Compositio Math. 35, Fasc. 3* (1977) 299.

[4] I.M. Gelfand, N.Ya. Vilenkin: *Generalized functions, Vol. 4: Applications of Harmonic Analysis.* Academic Press, New York, 1964.

[5] K. Maurin: *Metody przestreni Hilberta.* Warsawa, 1959.

[6] J.L. Doob: *Stochastic processes.* New York, 1953.

[7] R.S. Ismagilov: On unitary representations of the group $C_0^\infty(X, G)$, $G = SU_2$. *Mat. Sbornik, 100, no. 1* (1976) 117.

[8] R. Courant: *Partial differential equations.* New York, 1962.

[9] B. Simon: *The $P(\varphi_2)$ Euclidean (quantum) field theory.* Princeton Univ. Press, 1974.

[10] A.V. Skorohod: *Integration in Hilbert spaces.* Moscow, "Nauka", 1975.

[11] S.L. Sobolev: *Some Applications of the Functional Analysis in Mathematical Physics.* Leningrad, 1950.

[12] L. Hormander: The spectral function of an elliptic operator. *Acta Math. 121, no. 3–4* (1968) 193.

[13] G.E. Shilov, Fan Dik-Tin: *Integral, Measure, and Derivative on linear spaces.* Moscow, "Nauka", 1967.

[14] M.A. Shubin: Pseudodifferential Operators and the Spectral Theory. Moscow, "Nauka", 1978.

[15] S. ALBEVERIO, R. HOEGH-KROHN: The Energy representation of Sobolev–Lie groups. Preprint *Universitat Bielefeld*, 1976.

[16] A. GUICHARDET: *Symmetric Hilbert spaces and related topics. Lect. Notes in Math. N 261*, Springer, Berlin, 1972.

[17] K.R. PARTHASARATHY and K. SCHMIDT: A new method for constructing factorisable representation for current groups and current algebras, *Commun. Math. Phys.*, *51, no. 1* (1976) 1.

(Oblatum 18-VII-1979)

Institute of Applied Mathematics
Academy of Sciences USSR
Moscow, 125047, Miusskaya pl., 4.
Leningrad State University

# 6.

## (with M. I. Graev and A. M. Vershik)

# A commutative model of representation of the group of flows $SL(2, R)^X$ that is connected with a unipotent subgroup

Funkts. Anal. Prilozh. **17** (2) (1983) 70–72 [Funct. Anal. Appl. **17** (1983) 137–139].
Zbl. **536**:22008

Let G be an arbitrary metrizable topological group, $G_0$ be a commutative subgroup of it, and $\pi$ be a unitary representation of G. Let us suppose that the restriction of $\pi$ to the subgroup $G_0$ is cyclic. Then by an isometric operator we can reduce the operators $\{\pi(a), a \in G_0\}$ to diagonal form, i.e., we can realize the space of the representation $\pi$ as an $L^2$-space with respect to the spectral measure of the representation $\pi|_{G_0}$. After this, we will try to describe all representations in this realization. We call such a realization a *commutative model* of the representation $\pi$ with respect to $G_0$. The commutative models of irreducible representations of the group SL(2, R) with respect to its commutative subgroups — orthogonal, unipotent, and diagonal, are well known (see, e.g., [3]).

In the present note, we construct a commutative model of the fundamental representation of the group of flows $SL(2, R)^X$ with respect to its unipotent subgroup $\left\{\begin{pmatrix}1 & 0\\ \gamma(\cdot) & 1\end{pmatrix}\right\}$. Let us recall that $SL(2, R)^X$ is the group of bounded measurable functions on a manifold X with a smooth measure m with values in SL(2, R). Here the irreducible unitary representation of the group $SL(2, R)^X$, constructed in [1], is said to be fundamental. This representation is noteworthy since it is nonlocal: The representation operator $U(\tilde{g})$ "depends on the function $\tilde{g}$: $X \to SL(2, R)$ as a whole" and not on its values at separate points. In [1] the authors have constructed several models, including the commutative model with respect to the subgroup $O(2)^X$; it has been reduced to the Fock (Boson) realization of the fundamental representation. The commutative model with respect to a unipotent subgroup is more contensive. This model reduces, in particular, to an astonishing measure, generated by an infinitely divisible discrete random process (i.e., to a measure, concentrated on linear combinations of $\delta$-functions): After the introduction of suitable density, the corresponding $\sigma$-finite measure is invariant, up to a scalar, with respect to multiplication by positive principal functions (see Proposition 3)!

We start from the spherical function of the fundamental representation $U: \Psi(\tilde{g}) = \langle U(\tilde{g})\xi_0, \xi_0\rangle$, where $\xi_0$ a cyclic (vacuum) vector in the representation space. By [1], this function is expressed by the equation

$$\Psi(\tilde{g}) = \exp\left(-\frac{1}{2}\int_X \ln\left[\frac{1}{4}(\mathrm{Sp}(g(x)g^*(x)) + 2)\right] dm(x)\right). \tag{1}$$

Its restriction to the unipotent subgroup $\left\{\begin{pmatrix}1 & 0\\ \gamma(\cdot) & 1\end{pmatrix}\right\}$ has the form

$$L(\gamma) = \exp\left(\int_X \ln l(\gamma(x))\, dm(x)\right),$$

where $l(y) = (1 + y^2/4)^{-1/2}$. Since L is a positive-definite functional on the space $\mathcal{F}$ of the principal function on X, it is the Fourier transform of a finite normalized measure $\mu$ on the set $\Phi$ of generalized functions (see [2]).

Let us establish some properties of the measure $\mu$. To this end, we introduce auxiliary positive-definite functionals of $\mathcal{F}$:

---

Applied Mathematics Institute, Academy of Sciences of the USSR. Leningrad State University. Translated from Funktsional'nyi Analiz i Ego Prilozheniya, Vol. 17, No. 2, pp. 70–72, April–June, 1983. Original article submitted December 1, 1982.

$$L_{\pm}(\gamma) = \exp\left(\int_X \ln l_{\pm}(\gamma(x))\, dm(x)\right),$$

where $l_+(y) = (1 - iy/2)^{-1/2}$ and $l_-(y) = (1 + iy/2)^{-1/2}$; let $\mu_\pm$ denote the finite normalized measures on $\Phi$ that correspond to these functionals. Since $l = l_+ \cdot l_-$, it follows that $L = L_+ \cdot L_-$, and therefore $\mu$ is the convolution of the measures $\mu_+$ and $\mu_-$.

Proposition 1. The supports of the measures $\mu_+$, $\mu_-$, and $\mu$ are concentrated, respectively, on the sets $\Phi_0^{(\pm)} \{\sum c_k \delta_{x_k}, c_k \gtrless 0, \sum |c_k| < \infty\}$ and $\Phi_0 = \Phi_0^+ + \Phi_0^-$, where $\delta_x$ is the delta-function on X, concentrated at the point x.

To prove this proposition, it is necessary to use the Levy–Khinchin representation of the functions $l_+$, $l_-$, and $l$ and to use a well-known theorem on discrete fields (see [4, p. 94, Theorem 3]; this theorem is valid for an arbitrary measure space (X, m)).

Proposition 2. The following equations are valid for each function $z(x) = \gamma(x) - i\alpha(x)$ where $\gamma, \alpha \in \mathcal{F}$ and $\alpha < 2$:

$$\int e^{i\langle \xi, z\rangle} d\mu_+(\xi) = \exp\left(\int_X \ln(1 - iz(x))^{-1/2} dm(x)\right),$$
$$\int e^{i\langle \xi, \bar z\rangle} d\mu_-(\xi) = \exp\left(\int_X \ln(1 + i\overline{z(x)})^{-1/2} dm(x)\right). \qquad (2$$

Proof. For $\alpha \equiv 0$ the equations follow from the definition of the measures $\mu_+$ and $\mu_-$. For arbitrary $\alpha$, these equations can be obtained by analytic continuation.

COROLLARY.

$$\int \exp\left(\int_X \alpha(x)\,|\xi(x)|\, dm(x) + i\langle \xi, \gamma\rangle\right) d\mu(\xi) = \exp\left(\int_X \ln\left(\frac{1}{4}((2-\alpha(x))^2 + \gamma^2(x))\right)^{-1/2} dm(x)\right). \qquad (3$$

The integral $\int \alpha(x)\,|\xi(x)|\, dm(x)$, where $\xi = \Sigma c_k \delta_{x_k}$, must be understood here as $\Sigma \alpha(x_k)|c_k|$.

Definition. The functional v on $\Phi_0$ is defined as follows: $v(\sum c_k \delta_{x_k}) = \sum |c_k|$, and the $\sigma$-finite (not finite) measure $\nu$ on $\Phi_0$ is defined by the equality $(d\nu/d\mu)(\xi) = \exp(2v(\xi))$.

It follows from (3) that for arbitrary $\alpha, \gamma \in \mathcal{F}, \alpha > 0$, we have

$$\int \exp\left(-\int_X \alpha(x)\,|\xi(x)|\, dm(x) + i\langle \xi, \gamma\rangle\right) d\nu(\xi) = \exp\left(\int_X \ln\left(\frac{1}{4}(\alpha^2(x) + \gamma^2(x))\right)^{-1/2} dm(x)\right). \qquad (4$$

Proposition 3. The measure $\nu$ is projective-invariant with respect to the transformations $\xi \mapsto \varepsilon^2\xi$, where $\varepsilon$ is an arbitrary principal function ("relatively invariant" in the terminology of Bourbaki), i.e., it is quasiinvariant with respect to these transformations and, in addition, $d\nu(\xi)$ and $d\nu(\varepsilon^2\xi)$ differ only by a constant factor for each fixed function $\varepsilon$. In addition,

$$\frac{d\nu(\varepsilon^2\xi)}{d\nu(\xi)} = \exp\left(\int_X \ln \varepsilon^2(x)\, dm(x)\right). \qquad (5)$$

Construction of the Representation. We introduce the space $L^2_\nu(\Phi)$ of the complex-valued functions on $\nu$ that are square-integrable with respect to the measure $\Phi$. Let us associate the following operator in $L^2_\nu(\Phi)$ with each triangular matrix $\tilde g = \begin{pmatrix} \varepsilon^{-1}(\cdot) & 0 \\ \gamma(\cdot) & \varepsilon(\cdot)\end{pmatrix}$: $(U(\tilde g) f)(\xi) = \exp \times \left(\int_X \ln|\varepsilon(x)|\, dm(x) + i\langle \xi, \varepsilon\gamma\rangle\right) \cdot f(\varepsilon^2\xi)$. In particular, $\left(U\begin{pmatrix} 1 & 0 \\ \gamma(\cdot) & 1\end{pmatrix} f\right)(\xi) = e^{i\langle \xi, \gamma\rangle} f(\xi)$. It follows easily from (5) that these operators are unitary and, obviously, they form a representation of the triangular subgroup.

In order to extend this representation to a representation of the whole group $SL(2, \mathbf{R})^X$, we define the set of the functionals $f_z$ in $L^2_\nu(\Phi)$, where $z(\cdot) = p(\cdot) + iq(\cdot)$, $p, q \in \mathcal{F}$, $q > 0$ (i.e. z is a function on X with values in the Lobachevskii plane):

$$f_z(\xi) = \exp(-v(q\xi) + i\langle \xi, p\rangle).$$

ıey form a total set in $L_\nu^2(\Phi)$. It follows from (4) that

$$\langle f_{z_1}, f_{z_2}\rangle = \exp\left(-\int_X \ln\left(\frac{1}{2}|z_1(x) - \overline{z_2(x)}|\right) dm(x)\right). \tag{6}$$

ǝt us define the operators $U(\tilde{g})$, $\tilde{g} = \begin{pmatrix} \alpha(\cdot) & \beta(\cdot) \\ \gamma(\cdot) & \delta(\cdot) \end{pmatrix} \in SL(2, R)^X$ on the set of the functionals $f_z$ by ıe equation

$$U(\tilde{g}) f_z = \exp\left(-\int_X \ln|\beta(x) z(x) + \alpha(x)| dm(x)\right) \cdot f_{(\beta z+\alpha)^{-1}(\delta z+\gamma)} \tag{7}$$

it follows from (7) that the system $\{f_z\}$ is coherent with respect to the operators $U(\tilde{g})$]. t follows from (6) that the operators $U(\tilde{g})$ preserve the scalar products $\langle f_{z_1}, f_{z_2}\rangle$, and ıerefore they can be uniquely extended to unitary operators on the whole space $L_\nu^2(\Phi)$. It s obvious that these operators form a unitary representation of the group $SL(2, R)^X$. It is asily verified that the restriction of this representation to the triangular subgroup coinides with the above-defined representation.

THEOREM. The constructed representation of the group $SL(2, R)^X$ is equivalent to the ındamental representation and is, by the same token, its commutative model with respect to unipotent subgroup.

Indeed, it follows from (6) and (7) that the spherical function $\Psi(\tilde{g}) = \langle U(g) f_{\sqrt{-1}}, f_{\sqrt{-1}}\rangle$ f this representation is expressed by Eq. (1), i.e., it coincides with the spherical funcion of the fundamental representation. The desired statement follows from the cylicity of $\sqrt{-1}$.

## LITERATURE CITED


. A. M. Vershik, I. M. Gel'fand, and M. I. Graev, "Representations of the group SL(2, R), where R is a ring of functions," Usp. Mat. Nauk, 28, No. 5, 83-128 (1973).
. I. M. Gel'fand and N. Ya. Vilenkin, Applications of Harmonic Analysis, Academic Press (1964).
. I. M. Gel'fand, M. I. Graev, and I. I. Pyatetskii-Shapiro, Representation Theory and Automorphic Functions [in Russian], Nauka, Moscow (1966).
. A. V. Skorokhod, Random Processes with Independent Increments [in Russian], Nauka, Moscow (1964).

# Appendix

# 1.
## Graeme Segal

# Two papers on representation theory

Representation Theory. Lecture Note Series, vol. 69, London Math. Soc., Cambridge University Press, 1982, pp. 1–13

These two papers are devoted to the representation theory of two infinite dimensional Lie groups, the group $SL_2(\mathbf{R})^X$ of continuous maps from a space $X$ into $SL_2(\mathbf{R})$, and the group Diff($X$) of diffeomorphisms (with compact support) of a smooth manifold $X$. Almost nothing of a systematic kind is known about the representations of infinite dimensional groups, and the mathematical interest of studying these very natural examples hardly needs pointing out.

Nevertheless the stimulus to the work came from physics, and I shall try to indicate briefly how the representations arise there. Physicists encountered not the groups but their Lie algebras, the algebra $\mathfrak{g}^X$ of maps from $X$ to the Lie algebra $\mathfrak{g}$ of $SL_2(\mathbf{R})$, and the algebra Vect($X$) of vector fields on $X$. The space $X$ is physical space $\mathbf{R}^3$. Choosing a basis in $\mathfrak{g}$, to represent $\mathfrak{g}^X$ is to associate linearly to each real-valued function $f$ on $\mathbf{R}^3$ three operators $J_i(f)$ ($i = 1, 2, 3$), such that

$$[J_i(f), J_j(g)] = \sum_k c_{ijk} J_k(fg),$$

where $c_{ijk}$ are the structural constants of $\mathfrak{g}$. In quantum field theory one writes $J_i(f)$ as $\int_{\mathbf{R}^3} f(x) j_i(x)\,dx$, where $j_i$ is an operator-valued distribution. Then the relations to be satisfied are

$$[j_i(x), j_j(y)] = \sum_k c_{ijk}\, \delta(x-y) j_k(y) \qquad (*)$$

where $\delta$ is the Dirac delta-function.

Similarly, to represent Vect($\mathbf{R}^3$) is to associate operators $P(f)$ to vector-valued functions $f$: $\mathbf{R}^3 \to \mathbf{R}^3$ so that $[P(f), P(g)] = P(h)$, where

$$h = \sum_i \left( f_i \frac{\partial g}{\partial x_i} - g_i \frac{\partial f}{\partial x_i} \right).$$

Writing $P(f) = \sum_i \int f_i(x) p_i(x) dx$ this becomes

$$[p_i(x), p_j(y)] = \delta_i(x-y)p_j(y) - \delta_j(x-y)p_i(x), \qquad (**)$$

where $\delta_k = \partial\delta/\partial x_k$.

Operators with the properties of $j_i(x)$ and $p_i(x)$ arise commonly in quantum field theory in the guise of "current algebras". For example, if one has a complex scalar field given by operators $\psi(x)$ (for $x \in \mathbf{R}^3$) which satisfy either commutation or anticommutation relations of the form $[\psi^*(x), \psi(y)]_\pm = \delta(x-y)$, then the "current-like" operators $p_i(x)$ defined by

$$p_i(x) = \frac{1}{2}\left\{ \psi^*(x) \frac{\partial\psi(x)}{\partial x_i} - \frac{\partial\psi^*(x)}{\partial x_i} \psi(x) \right\}$$

satisfy (**). Similarly if one has an $N$-component field $\psi$ satisfying $[\psi_\alpha^*(x), \psi_\beta(y)]_\pm = \delta_{\alpha\beta}\delta(x-y)$, and $\sigma_1, \ldots, \sigma_n$ are $N \times N$ matrices representing the generators of a Lie algebra $\mathfrak{g}$ then the operators $j_i(x) = \psi^*(x)\sigma_i\psi(x)$ satisfy (*). (These examples are taken from [3].)

In connection with the quantization of gauge fields it is also worth mentioning that, as we shall see below, the most natural representation of the group of all smooth automorphisms of a fibre bundle is its action on $L^2(E)$, where $E$ is the space of connections ("gauge fields") in the bundle, endowed with a Gaussian measure.

## Representations of the group $SL(2, \mathbf{R})^X$.

This paper is concerned with the construction of a single irreducible unitary representation of the group $G^X$ of continuous maps from a space $X$ equipped with a measure into the group $G = SL_2(\mathbf{R})$. (In this introduction I shall always think of $G$ as $SU_{1,1}$, i.e. as the complex matrices $\begin{pmatrix} a & b \\ \bar{b} & \bar{a} \end{pmatrix}$ such that $|a|^2 - |b|^2 = 1$.)

An obvious way of obtaining an irreducible representation of $G^X$ is to choose some point $x$ of $X$ and some irreducible representation of $G$ by operators $\{U_g\}_{g \in G}$ on a Hilbert space $H$, and to make $G^X$ act on $H$ through the evaluation-map at $x$, i.e. to make $f \in G^X$ act on $H$ by $U_{f(x)}$. This representation can be regarded as analogous to a "delta-function" at $x$. More generally, for any finite set of points $x_1, \ldots, x_n$ in $X$ and corresponding irreducible representations $g \mapsto U_g^{(i)}$ of $G$ on Hilbert spaces $H_1, \ldots, H_n$ one can make $G$ act irreducibly on the tensor product $H_1 \otimes \cdots \otimes H_n$ by assigning to $f \in G^X$ the operator $U_{f(x_1)}^{(1)} \otimes \cdots \otimes U_{f(x_n)}^{(n)}$. The object of the paper is to generalize this construction and produce a representation on a "continuous tensor product" of a family of Hilbert spaces $\{H_x\}$ indexed by the points of $X$ (and weighted by the measure on $X$). There is a simple criterion for deciding whether a representation is an acceptable solution of the problem, in view of the following remark. For any representation $U$ of $G^X$ and any continuous map $\phi: X \to X$ there is a twisted representation $\phi^*U$ given by $(\phi^*U)_f = U_{f\phi}$. The representation to be constructed ought to have the property that $\phi^*U$ is equivalent to $U$ whenever $\phi$ is a measure-preserving homeomorphism of $X$, i.e. for each such $\phi$ there should be a unitary operator $T$ such that $U_{f\phi} = T_\phi U_f T_\phi^{-1}$.

The paper describes six different constructions of the representation, but only three are essentially different. Of these, one, described in §4 of the paper, is extremely simple, but not very illuminating because it is a construction a posteriori. I shall deal with it first. For any group $\Gamma$ and any cyclic unitary representation of $\Gamma$ on a Hilbert space $H$ with cyclic vector $\xi \in H$ ("cyclic" means that the vectors $U_\gamma \xi$, for all $\gamma \in \Gamma$, span a dense subspace of $H$) one can reconstruct the Hilbert space and the representation from the complex-valued function $\gamma \mapsto \Psi(\gamma) = \langle \xi, U_\gamma \xi \rangle$ on $\Gamma$. To see this, consider the abstract vector space $H_0$ whose basis is a collection of formal symbols $U_\gamma \xi$ indexed by $\gamma \in \Gamma$. An inner product can be introduced in $H_0$ by prescribing it on the basis elements:

$$\langle U_{\gamma_1} \xi, U_{\gamma_2} \xi \rangle = \Psi(\gamma_1^{-1} \gamma_2).$$

The group $\Gamma$ has an obvious natural action on $H_0$, preserving the inner product. Then $H$ is simply the Hilbert space completion of $H_0$. The function $\Psi$ is called the *spherical function* of the representation corresponding to $\xi \in H$.

In our case the group $\Gamma = G^X$ has an abelian subgroup $K^X$, where $K = SO_2$ is the maximal compact subgroup of $G$, and it turns out that the desired representation $H$ contains (up to a scalar multiple) a unique unit vector $\xi$ invariant under $K^X$. The corresponding spherical function is easy to describe. The orbit of $\xi$ can be identified with $G^X/K^X$, i.e. with the maps of $X$ into $G/K$, which is the Lobachevskii plane. (I shall always think of $G/K$ as the open unit disk in **C** with the Poincaré metric.) Given two maps $f_1, f_2: X \to G/K$ the corresponding inner product is

$$\exp \int_X \log \operatorname{sech} \rho(f(x_1), f(x_2))\, dx,$$

where $\rho$ is the $G$-invariant Lobachevskii or Poincaré metric on $G/K$. This means that the spherical function $\Psi$ is given by

$$\Psi(f) = \exp \int_X \log \psi(f(x))\, dx,$$

where, if $g = \begin{pmatrix} a & b \\ \bar{b} & \bar{a} \end{pmatrix} \in G$, then $\psi(g) = |a|^{-1}$. To see that this construction does define a representation of $G^X$ the only thing needing to be checked is that the inner product is positive. That is done in §4.2. But of course it is not clear from this point of view that the representation is irreducible.

A more illuminating construction of the representation is to realize the continuous tensor product as a limit of finite tensor products. To do this we actually represent the group of $L^1$ maps from $X$ to $G$, i.e. the group obtained by completing the group of continuous maps in the $L^1$ metric (cf. §3.4). The $L^1$ maps contain as a dense subgroup the group of step-functions $X \to G$, and it is on the subgroup of step-functions that the representation is concretely defined.

If one is to form a limit from the tensor products of increasing numbers of vector spaces then the vector spaces must in some sense get "smaller". It happens that the group $SL_2(\mathbf{R})$ has the comparatively unusual property (cf. below) of possessing a family (called the "supplementary series") of irreducible representations $H_\lambda$ (where $0 < \lambda < 1$) which do in a certain sense "tend to" the trivial one-dimensional representation as $\lambda \to 0$. Furthermore there is an isometric embedding $H_{\lambda+\mu} \to H_\lambda \otimes H_\mu$ whenever $\lambda + \mu < 1$. Now for any partition $\nu$ of $X$ into parts $X_1, \ldots, X_n$ of measures $\lambda_1, \ldots, \lambda_n$ one can consider the group $G_\nu$ of those step-functions $X \to G$ which are constant on the steps $X_i$. The group $G_\nu$ acts on $\mathscr{H}_\nu = H_{\lambda_1} \otimes \cdots \otimes H_{\lambda_n}$; and when a partition $\nu'$ is a refinement of $\nu$ then $\mathscr{H}_\nu$ is naturally contained in $\mathscr{H}_{\nu'}$. Accordingly, the group $\bigcup_\nu G_\nu$ of all step-functions acts on $\bigcup_\nu \mathscr{H}_\nu$, and the desired representation is the completion of this.

The construction just outlined is carried out in §2 of the paper. A variant is described in §3, where the representations $\{H_\lambda\}$ of the supplementary series are replaced by another family $\{L_\lambda\}$ with analogous properties – the so-called "canonical" representations. These are cyclic but not irreducible, and $L_\lambda$ contains $H_\lambda$ as a summand. In terms of their spherical functions $L_\lambda$ tends to $H_\lambda$ as $\lambda \to 0$. The spherical function $\psi_\lambda$ of $L_\lambda$ is very simple, given by $\psi_\lambda(g) = |a|^{-\lambda}$ when $g = \begin{pmatrix} a & b \\ \bar{b} & \bar{a} \end{pmatrix}$. In other words, $L_\lambda$ is spanned by vectors $\xi_u$ indexed by $u$ in the unit disk $G/K$, and $\langle \xi_u, \xi_{u'} \rangle = \operatorname{sech}^\lambda \rho(u, u')$. Notice that the size of the generating $G$-orbit $\{\xi_u\}$ in $L_\lambda$ tends to 0 as $\lambda \to 0$.

The remaining constructions exploit a quite different idea, which is useful in other situations too, as we shall see. I shall explain it in general terms.

*Gaussian measures on affine spaces*

Suppose that a group $\Gamma$ has an affine action on a real vector space $E$ with an inner product; i.e. to each $\gamma \in \Gamma$ there corresponds a transformation of $E$ of the form $v \mapsto T(\gamma)v + \beta(\gamma)$, where $T(\gamma): E \to E$ is linear and orthogonal, and $\beta(\gamma) \in E$. Then there is an induced unitary action of $\Gamma$ on the space $L^2(E)$ of functions on $E$ which are square-summable with respect to the standard Gaussian measure $e^{-\|v\|^2} dv$. Because this measure is not translation-invariant we have to define $U_\gamma: L^2(E) \to L^2(E)$ bv

$$(U_{\gamma^{-1}} f)(v) = \Phi_\gamma(v) f(T(\gamma)v + \beta(\gamma)),$$

where the factor

$$\Phi_\gamma(v) = e^{\frac{1}{2}\|v\|^2 - \frac{1}{2}\|T(\gamma)v + \beta(\gamma)\|^2} = e^{-\langle T(\gamma)v, \beta(\gamma)\rangle - \frac{1}{2}\|\beta(\gamma)\|^2}$$

is to achieve unitarity.

The importance of this construction is that the representation of $\Gamma$ on $L^2(E)$ may be irreducible even when the underlying linear action on $E$ by $\gamma \mapsto T(\gamma)$ is highly reducible. If the linear action $T$ is given then the affine action is evidently described by the map $\beta: \Gamma \to E$. This is a "cocycle", i.e. $\beta(\gamma\gamma') = \beta(\gamma) + T(\gamma)\beta(\gamma')$, and it is easy to see that the affine space is precisely described up to isomorphism by the cohomology class of $\beta$ in $H^1(\Gamma; E)$. One sometimes speaks of "twisting" the action of $\Gamma$ on $L^2(E)$ by means of $\beta$.

Apart from the description just given there are two other useful ways of looking at $L^2(E)$. The first of these is as a "Fock space". For the Gaussian measure on $E$ the polynomial functions are square-summable, and are dense in $L^2(E)$. So $L^2(E)$ can be identified with the Hilbert space completion of the symmetric algebra $S(E)$ of $E$. (A little care is necessary here: to make the natural inner product in $S(E)$ correspond to the Gaussian inner product in $L^2(E)$ one must identify $S^n(E)$ not with the homogeneous polynomials on $E$ of degree $n$, but with the "generalized Hermite polynomials" of degree $n$.)

The other way of approaching $L^2(E)$ is to observe that it contains (and is spanned by) elements $e^v$ for each $v \in E$, with the property that

$$\langle e^v, e^{v'} \rangle = e^{\langle v, v' \rangle}. \qquad \ldots (\dagger)$$

This means that $L^2(E)$ can be obtained from the abstract free vector space whose basis is a set of symbols $\{e^v\}_{v \in E}$ by completing it using the inner product defined by (†). Better still, one can start with symbols $\epsilon_v$ and define

$$\langle \epsilon_v, \epsilon_{v'} \rangle = e^{-\frac{1}{2}\|v - v'\|^2}:$$

this makes it plain that the construction uses only the *affine* structure of $E$. (Of course $\epsilon_v = e^{-\frac{1}{2}\|v\|^2} e^v$.)

The group $SU_{1,1}$ acts on the circle $S^1$, and has a very natural affine action on the space $E_\lambda$ of smooth measures on $S^1$ with integral $\lambda$. $E_\lambda$ is a coset of the vector space $E_0$ of smooth measures with integral 0. The invariant norm in $E_0$ is given by

$$\|\alpha\|^2 = \sum_{n>0} \frac{1}{n} |a_n|^2 \text{ when } \alpha = \sum_{n \neq 0} a_n e^{in\theta} d\theta.$$

Then $L^2(E_\lambda)$ is the "canonical representation" $L_{\lambda^2}$ mentioned above. (This is stated, not quite precisely, as Theorem (7.1) of the paper.) For if $\alpha_u$ is the Dirichlet measure on $S^1$ corresponding to $u$ in the unit disk $G/K$ (i.e. $\alpha_u$ is the transform of $d\theta$ by any element of $G$ which takes 0 to $u$) then $\lambda\alpha_u \in E_\lambda$ and

$$\frac{1}{2}\| \lambda\alpha_u - \lambda\alpha_{u'}\|^2 = \lambda^2 \log\cosh \rho(u, u').$$

Returning to the group we are studying, $G^X$ has an affine action (pointwise) on $E_1^X$, the space of maps $X \to E_1$. (Notice that the linear action of $G^X$ on $E_0^X$ is highly reducible.) The space $L^2(E_1^X)$ has the appropriate multiplicative property with respect to $X$: for if $X$ is a disjoint union $X = X_1 \cup X_2$ then $E_1^{X_1 \cup X_2} = E_1^{X_1} \times E_1^{X_2}$ and $L^2(E_1^{X_1 \cup X_2}) = L^2(E_1^{X_1}) \otimes L^2(E_1^{X_2})$. In the paper the equivalence of the representations on $L_2(E_1^X)$ and on the continuous tensor product of § §2 and 3 is proved by calculating the spherical functions, but it is quite easy to give an explicit embedding of the continuous tensor product in $L^2(E_1^X)$. For if $Y$ is a part of $X$ with measure $\lambda$ then the map $E_1^Y \to E_{\sqrt{\lambda}}$ given by $f \mapsto \lambda^{-\frac{1}{2}} \int_Y f$ is compatible with the Gaussian measures, so that $L^2(E_{\sqrt{\lambda}})$ is a subspace of $L^2(E_1^Y)$, and for any partition $X = X_1 \cup \ldots \cup X_n$ with $m(X_i) = \lambda_i$ we have

$$\begin{aligned} L_{\lambda_1} \otimes \cdots \otimes L_{\lambda_n} &= L^2(E_{\sqrt{\lambda_1}}) \otimes \cdots \otimes L^2(E_{\sqrt{\lambda_n}}) \\ &\subset L^2(E_1^{X_1}) \otimes \cdots \otimes L^2(E_1^{X_n}) \\ &= L^2(E_1^X). \end{aligned}$$

Indeed it is pointed out in [10] that for any affine space $E$ the space $L^2(E^X)$ can always be interpreted as a continuous tensor product of copies of $L^2(E)$ indexed by the points of $X$.

The irreducibility of the representation can be seen very easily in the Fock version. For the cocycle $\beta$ vanishes on the abelian subgroup $K^X$, and so under $K^X$ the representation breaks up into its components $S^n(E_0^X)$, on which $K^X$ acts just by multiplication operators. The characters of $K^X$ which arise are all distinct, so the irreducibility of the representation follows from the fact that the vacuum vector is cyclic, which is easily proved (cf. §5.2).

The three approaches to $L^2(E_1^X)$ are described in § §5, 6 and 7 of the paper. In connection with §6 notice that to give an affine action of $\Gamma$ on a vector space $E$ is the same thing as to give a linear action on a vector space $H$ together with an invariant linear form $l: H \to \mathbf{R}$ such that $l^{-1}(0) = E$: the affine space is then $l^{-1}(1)$. (There is no point, in §6, in considering functions $f: X \to H$ other than those satisfying $l(f(x)) = 1$, and the formulae become less cumbersome under that assumption.)

§5 describes the Fock space version, but not quite in the standard form. The space $E_1$ of measures on the circle can be identified (by Fourier series) with a space of maps $\mathbf{Z} \to \mathbf{C}$. Accordingly $E_1^X$ is a space of maps $X \times \mathbf{Z} \to \mathbf{C}$, and the symmetric power $S^k(E_1^X)$ is a space of maps

$$\underbrace{X \times \ldots \times X}_{\leftarrow k \rightarrow} \times \underbrace{\mathbf{Z} \times \ldots \times \mathbf{Z}}_{\leftarrow k \rightarrow} \to \mathbf{C}$$

which are symmetric in the obvious sense. The effect of this point of view is to identify $L^2(E_1^X)$ with a space of functions on the free abelian group generated by the space $X$, i.e. on the space whose points are "virtual finite subsets" $\Sigma n_i x_i$ of $X$, with $n_i \in \mathbf{Z}$. This is intriguing, but whether it is more than a curiosity it is hard to say.

That concludes my account of the contents of the paper itself; but I shall mention some related matters. The most obvious question to ask is what class of groups $G$ the method applies to. As it stands it evidently does not work for groups for which the trivial representation is isolated in the space of all irreducible representations. This excludes all compact groups, as for them the irreducible representations form a discrete set. The isolatedness of the trivial representation has been cleverly investigated by Kazhdan [5], who proved in particular that among semisimple groups the trivial representation is isolated if the group contains $SL_3(\mathbf{R})$ as a subgroup. The only simple groups not excluded by Kazhdan's criteria are $SO_{n,1}$ and $SU_{n,1}$ – notice that $PSL_2(\mathbf{R}) \cong SO_{2,1}$ and $PSL_2(\mathbf{C}) \cong SO_{3,1}$. For these the method works just as for $SL_2(\mathbf{R})$. (For example $SO_{n,1}$ is the group of all conformal transformations of $S^{n-1}$, and the affine space $E_\lambda$ used above can be replaced by the space of measures on $S^{n-1}$ with integral $\lambda$.)

A class of groups for which the trivial representation is not isolated consists of the semidirect products $G \tilde{\times} V$, where $G$ is a compact group with an orthogonal action on a real vector space $V$. Indeed if $\Omega$ is a $G$-orbit in $V$ an element $g \in G$ acts naturally on $L^2(\Omega)$, and $v \in V$ can be made to act by multiplication by the function $e^{i\langle v, \omega \rangle}$. When the compact orbit $\Omega$ is close to the origin in $V$ the representation $L^2(\Omega)$ is close to the trivial representation (in the sense of its spherical function). Furthermore $G \tilde{\times} V$ has an obvious affine action on $V$: the induced action on $L^2(V)$ is the direct integral of the irreducible representations $L^2(\Omega)$ for all orbits $\Omega \subset V$.

Thus the methods of the paper apply to all groups of the form $G \tilde{\times} V$. The importance of this is that it provides a way of constructing a representation of the group $(G^X)_{sm}$ of *smooth* maps from a manifold $X$ to a compact group $G$. For a smooth map $f: X \to G$ induces a map of tangent bundles $Tf: TX \to TG$, and this can be regarded as a map which to each point $x \in X$ assigns a "1-jet" $j(x) \in J_x G = G \tilde{\times} (T_x^* X \otimes \mathfrak{g})$ where $\mathfrak{g}$ is the Lie algebra of $G$. As the groups $J_x G$ is of the form $G \tilde{\times} V$ the method of the paper provides a representation of the group $\Gamma$ of bundle maps $TX \to TG$. (The fact that $J_x G$ depends on $x$, giving rise to a bundle of groups on $X$, is not important.) $\Gamma$ contains $(G^X)_{sm}$ as a subgroup, and it turns out that the representation constructed remains irreducible when restricted to $(G^X)_{sm}$, at least when $\dim(X) \geqslant 4$. That is proved in the papers [1] and [2].

It is interesting to notice that the group of bundle maps $\Gamma$ is just the semidirect product $(G^X)_{sm} \tilde{\times} \Omega^1(X; \mathfrak{g})$, where $\Omega^1(X; \mathfrak{g})$ is the space of 1-forms on $X$ with values in $\mathfrak{g}$ ; and the associated affine space $E$ is the space of connections in the trivial $G$-bundle on $X$. The fact that the space of the representation of $(G^X)_{sm}$ is $L^2(E)$ is, of course, suggestive from the point of view of gauge theories in physics.

## Representations of the group of diffeomorphisms

This paper is devoted to the representation theory of the group Diff($X$) of diffeomorphisms with compact support of a smooth manifold $X$. (A diffeomorphism has compact support if it is the identity outside a compact region.)

The most obvious unitary representation of Diff($X$) is its natural action on $H = L^2(X)$, the space of square-summable $\frac{1}{2}$-densities on $X$. (By choosing a smooth measure $m$ on $X$ one can identify $L^2(X)$ with the usual space of functions $f$ on $X$ which are square-summable with respect to $m$. Then the action of a diffeomorphism $\psi$ on $f$ will be $f \mapsto \tilde{f}$, where

$$\tilde{f}(x) = J_\psi(x)^{\frac{1}{2}} f(\psi^{-1} x)$$

and $J_\psi(x) = dm(\psi^{-1} x)/dm(x)$. But it is worth noticing that $L^2(X)$ is canonically associated to $X$, and does not involve $m$.)

From $H$ a whole class of irreducible representations of Diff($X$) can be obtained by the well-known method introduced by Weyl to construct the representations of the general linear groups. For any integer $n$ the symmetric group $S_n$ acts on the $n$-fold tensor product $H^{\otimes n} = H \otimes \cdots \otimes H$ by permuting the factors, and the action commutes with that of Diff($X$). It turns out that under Diff($X$) $\times$ $S_n$ the tensor product decomposes

$$H^{\otimes n} = \bigotimes_\rho V^\rho \otimes W_\rho ,$$

where $\{W_\rho\}$ is the family of all irreducible representations of $S_n$, and $V^\rho$ is a certain irreducible representation of Diff($X$). More explicitly, $V^\rho$ is the space of $L^2$ functions $\underbrace{X \times \ldots \times X}_{\leftarrow n \rightarrow} \to W_\rho$ which are equivariant with respect to $S_n$: thus it makes sense even when $\rho$ is not irreducible, and $V^{\rho \oplus \rho'} \cong V^\rho \oplus V^{\rho'}$. The class of representations $\{V^\rho\}$, which were first studied by Kirillov, is closed under the tensor product: if $\rho$ and $\sigma$ are representations of $S_n$ and $S_m$ then $V^\rho \otimes V^\sigma \cong V^{\rho \cdot \sigma}$, where $\rho \cdot \sigma$ is the representation of $S_{n+m}$ induced from $\rho \otimes \sigma$. All of this is explained in § 1 of the paper.

It is then natural to ask, especially when $X$ is not compact, whether new representations of Diff($X$) can be constructed by forming some kind of infinite tensor product $H^{\otimes \infty}$ and decomposing it under the infinite symmetric group $S^\infty$ of all permutations of $\{1, 2, 3, \ldots\}$. This question is the main subject of the paper, and it is considered in the following way.

The $L^2$ functions $X^n \to W_\rho$ are the same as those $\tilde{X}^n \to W_\rho$ where

$\widetilde{X}^n \subset X^n$ is the space of $n$-triples of *distinct* points. The symmetric group $S_n$ acts on $\widetilde{X}^n$, and the quotient space is $B_X^{(n)}$, the space of $n$-point subsets of $X$. Diff($X$) acts transitively on $B_X^{(n)}$, and there is a unique class of quasi-invariant measures on it. The representation $V^\rho$ can be regarded as the space of sections of a vector bundle on $B_X^{(n)}$ whose fibre is $W_\rho$. An appropriate infinite analogue of $B_X^{(n)}$ is the space $\Gamma_X$ of infinite "configurations" in $X$, i.e. the space of countable subsets $\gamma$ of $X$ such that $\gamma \cap K$ is finite for every compact subset $K$ of $X$. This space, and the probability measures on it, play an important role in both statistical mechanics and probability theory. One can imagine the points of a configuration as molecules of a gas filling $X$, or as faulty telephones.

Diff($X$) does not act transitively on $\Gamma_X$: two configurations are in the same orbit only if they coincide outside a compact region. Nevertheless one can define (in many ways) measures on $\Gamma_X$ which are quasi-invariant and ergodic under Diff($X$). For each such measure $\mu$ there is an irreducible representation $U_\mu$ of Diff($X$) on $L^2(\Gamma_X; \mu)$. More generally, for each representation $\rho$ of a finite symmetric group $S_n$ there is an irreducible representation $U_\mu^\rho$: it is the space of sections of the infinite dimensional vector bundle on $\Gamma_X$ whose fibre is the representation $H^\rho$ of $S^\infty$ induced from the representation $\rho \otimes 1$ of $S_n \times S_n^\infty$. ($S_n^\infty$ denotes the subgroup of permutations in $S^\infty$ which leave $1, 2, \ldots, n$ fixed.) More explicitly, one can consider a covering space $\Gamma_{X,n}$ of $\Gamma_X$ defined by

$$\Gamma_{X,n} = \{(\gamma; x_1, \ldots, x_n) \in \Gamma_X \times X^n : x_i \in \gamma \text{ for } i = 1, \ldots, n\}.$$

$\Gamma_{X,n}$ is locally homeomorphic to $\Gamma_X$, and therefore a measure $\mu$ on $\Gamma_X$ defines a measure $\widetilde{\mu}$, the "Campbell measure", on $\Gamma_{X,n}$. The space of the representation $U_\mu^\rho$ is the space of maps $\Gamma_{X,n} \to W_\rho$ which are $S_n$-equivariant and square-summable for $\widetilde{\mu}$.

The simplest and most important measures on $\Gamma_X$ are the Poisson measures $\mu_\lambda$ (parametrized by $\lambda > 0$), for which the measure of the set $\{\gamma \in \Gamma_X : \text{card}(\gamma \cap K) = n\}$ is $\left(\frac{\lambda m}{n!}\right)^n e^{-\lambda m}$, where $m$ is the measure of $K$. More can be said about the representations $U_\lambda^\rho = U_{\mu_\lambda}^\rho$ in the Poisson case:

(i) They form a closed family under the tensor product, and have the following simple behaviour

(a) $U_\lambda^\rho \cong U_\lambda \otimes V^\rho$, and

(b) $U_\lambda \otimes U_{\lambda'} = U_{\lambda + \lambda'}$.

(ii) $U_\lambda$ is what is called in statistical mechanics an "N/V limit". In other words, if $X$ is the union of an expanding sequence $X_1 \subset X_2 \subset X_3 \subset \ldots$ of open relatively compact submanifolds such that $X_N$ has volume $\lambda^{-1}N$ then $L^2(\Gamma_X; \mu_\lambda)$ is the limit as $N \to \infty$ of the spaces $L^2_{\text{sym}}((X_N)^N)$ of symmetric $L^2$ functions of $N$ points in $X_N$. (This is explained in [4], [7], [8].)

(iii) $U_\lambda$ has a more concrete realization as $L^2(E_\lambda)$, where $E_\lambda$ is an affine space with a Gaussian measure (and an affine action of Diff($X$)). $E_\lambda$ is the space of $\frac{1}{2}$-densities $f$ on $X$ which are close to the standard Lebesgue $\frac{1}{2}$-density $f_\lambda = (\lambda\, dx)^{\frac{1}{2}}$ as $x \to \infty$, in the sense that $f - f_\lambda$ belongs to $H = L^2(X)$. This is an affine space associated to the vector space $H$ and the cocycle $\beta$: $\mathrm{Diff}(X) \to H$ given by

$$\beta(\psi) = \lambda^{\frac{1}{2}}(J_\psi^{\frac{1}{2}} - 1),$$

where $J_\psi(x) = dm(\psi^{-1}x)/dm(x)$ as before. As we have seen when discussing the representations of $G^X$, $L^2(E_\lambda)$ can also be regarded as a Fock space $S(H) = \bigotimes_{n \geq 0} L^2_{\mathrm{sym}}(X^n)$, but with the natural action of $\mathrm{Diff}(X)$ twisted by the cocycle $\beta$. Because $\beta$ vanishes on the subgroup $\mathrm{Diff}(X, m)$ of measure-preserving diffeomorphisms we see that in the Poisson case the representations associated to infinite configurations break up and give us nothing new when restricted to $\mathrm{Diff}(X, m)$.

In the paper the equivalence of $L^2(E_\lambda)$ and $L^2(\Gamma_X)$ is proved by considering the spherical functions, but it can also be described explicitly as a sequence of maps $L^2_{\mathrm{sym}}(X^n) \to L(\Gamma_X)$. In fact $L^2(X) \to L(\Gamma_X)$ takes $\lambda^{\frac{1}{2}} f$ to the function

$$\gamma \mapsto \sum_{X \in \gamma} f(x) - \lambda \int_X f(x)dx,$$

while $L^2_{\mathrm{sym}}(X \times X) \to L^2(\Gamma_X)$ takes $\lambda f$ to

$$\gamma \mapsto \sum_{x, y \in \gamma} f(x, y) - 2\lambda \sum_{x \in \gamma} \int_X f(x, y)dy + \lambda^2 \int\int_{X \times X} f(x, y)dxdy,$$

and so on.

The fact that there is a Gaussian realization of the representation is closely connected with the property of the Poisson measure $\mu_\lambda$ called "infinite divisibility". The latter means that if $X$ is the disjoint union of two pieces $X_1$ and $X_2$, so that $\Gamma_X = \Gamma_{X_1} \times \Gamma_{X_2}$ up to sets of measure zero, then $\mu_\lambda = \mu_\lambda^{(1)} \times \mu_\lambda^{(2)}$, where $\mu_\lambda^{(i)}$ is the projection of $\mu_\lambda$ on $\Gamma_{X_i}$. This implies that when the representation $U_\lambda$ of $\mathrm{Diff}(X)$ is restricted to the subgroup $\mathrm{Diff}(X_1) \times \mathrm{Diff}(X_2)$ it becomes $U_\lambda^{(X_1)} \otimes U_\lambda^{(X_2)}$, a property which must certainly be possessed by a construction of the type of $L^2(E_\lambda)$.

The reader may at first be confused by the fact that the affine action on $L^2(E_\lambda)$ used in this paper is the Fourier transform of the natural one used in the paper on $G^X$. Perhaps it is worth pointing out explicitly that if a group $G$ acts orthogonally on a real vector space $H$ with an inner product, and $\beta$: $G \to H$ is a cocycle, and $L^2(H)$ is formed using the standard Gaussian measure, then the following two unitary actions of $G$ on $L^2(H)$ are unitarily equivalent

(a) $g \mapsto A_g$, where $(A_g\phi)(h) = e^{\frac{1}{4}\|h\|^2 - \frac{1}{4}\|h - \beta(g)\|^2}\, \phi(g^{-1}(h - \beta(g)))$,

(b) $g \mapsto B_g$, where $(B_g\phi)(h) = e^{i\langle\beta(g),h\rangle}\,\phi(g^{-1}h)$.

The automorphism of $L^2(H)$ relating them is characterized by

$$e^{\langle a,h\rangle - \frac{1}{2}\|a\|^2} \longleftrightarrow e^{i\langle a,h\rangle}$$

for all $a \in H$. The important thing to notice about it is that it takes polynomials to polynomials.

I shall conclude this account by drawing attention to the matters treated rather sketchily in Appendix 2, as I think they are interesting and deserve to be investigated further. The representations we obtained from $\Gamma_X$ were constructed from a particularly simple family $\{H^\rho\}$ of irreducible unitary representations of the uncountable discrete group $S^\infty$. But the group which seems more obviously relevant – because a diffeomorphism with compact support can move only finitely many points of a configuration – is the countable group $S_\infty$ of the permutations of the natural numbers which leave almost all fixed. The representations $H^\rho$ restrict to irreducible representations of $S_\infty$; but most representations of $S_\infty$, notably the one-dimensional sign representation, do not extend to $S^\infty$. (There is a natural compact convex set of primary representations of $S_\infty$ which has been elegantly described by Thoma [9]. It is the family of all primary representations which admit a finite trace. It contains the trivial representation, the sign representation, and the regular representation. All members are of type II except for the two one-dimensional representations.) Menikoff [8] has constructed a representation of Diff($X$) corresponding to the sign representation of $S_\infty$ as an $N/V$ limit of the fermionic space $L^2_{\text{skew}}((S_N)^N)$ of antisymmetric functions of $N$ particles in $X_N$.

Can one construct a representation of Diff($X$) corresponding to any unitary representation $H$ of $S_\infty$? A possible method is described in Appendix 2. Let us choose an arbitrary rule for ordering the points of each configuration $\gamma \in \Gamma_X$. This gives us a map $s\colon \Gamma_X \to \widetilde{X}^\infty$ (where $\widetilde{X}^\infty$ is the space of ordered configurations), which clearly cannot be continuous. We require of the ordering only "correctness": if $\gamma$ and $\gamma'$ differ only in a compact region then the sequences $s(\gamma)$ and $s(\gamma')$ are required to coincide after finitely many terms. Consider the subspace $\Delta_s = S_\infty \cdot s(\Gamma_X)$ of $\widetilde{X}^\infty$. It is invariant under Diff($X$) $\times$ $S_\infty$, and $\Delta_s/S_\infty = \Gamma_X$. It was proved in §2.3 of the paper that for any quasi-invariant ergodic measure $\mu$ on $\Gamma_X$ there is a quasi-invariant ergodic measure $\widetilde{\mu}$ on $\Delta_s$. Then the space of $S_\infty$-invariant maps $\Delta_s \to H$ which are square-summable with respect to $\widetilde{\mu}$ affords a unitary representation of Diff($X$) associated to ($\mu$, $H$, $s$). The extent of its dependence on the arbitrary and inexplicit choice of $s$ is rather unclear, as is its relation to the $N/V$ limit of the physicists. But the method does seem to produce, at least, a large supply of type II representations of Diff($X$).

Vershik and Kerov [11] have proved that Thoma's family of representations can be obtained as limits of finite-dimensional representations of the finite symmetric groups. (One associates representations of $S_{n+1}$ to representations of $S_n$ by induction.) I imagine that this description should permit one both to construct the corresponding representations of Diff($X$) as $N/V$ limits, and to describe them in terms of a Gaussian measure.

**Note.** The definition of the topology of $\Gamma_X$ given in the paper does not seem quite correct. One method of obtaining it is as follows.

The topology on the space $B_X$ of finite configuration in $X$ is obvious and uncontroversial. For a connected open manifold $X$ the connected components of $B_X$ are the $B_X^{(n)}$ for $n = 0, 1, 2, \ldots$ If $Y$ is an open relatively compact submanifold of $X$ let us topologize $\Gamma_{\bar{Y}}$ as a quotient space of $B_X$. Then if $Y$ is connected so is $\Gamma_{\bar{Y}}$. Now define $\Gamma_X$ as $\varprojlim \Gamma_{\bar{Y}}$, where $Y$ runs through all open relatively compact submanifolds of $X$. This means that a configuration moves continuously precisely when it appears to move continuously to every observer with a bounded field of vision.

An alternative definition is: $\Gamma_X$ has the coarsest topology such that $\tilde{f}: \Gamma_X \to \mathbf{R}$ is continuous for every continuous function $f: X \to \mathbf{R}$ with compact support, where $\tilde{f}(\gamma) = \sum_{x \in \gamma} f(x)$.

It is easy to see (cf. [6]) that $\Gamma_X$ is metrizable, separable, and complete. On the other hand if $\tilde{X}^\infty$ is given the product topology then the map $\tilde{X}^\infty \to \Gamma_X$ is not continuous, and I do not see how to obtain $\Gamma_X$ as a quotient space from a sensible topology on $\tilde{X}^\infty$. (It does not seem, however, that the topology on $\tilde{X}^\infty$ plays a significant role in the paper.)

In conclusion, notice that the fundamental group of $\Gamma_X$ is $S^\infty$; but of course $\Gamma_X$ is not locally simply connected. The map $\Gamma_{X,n} \to \Gamma_X$, for example, is a local homeomorphism, but not a locally trivial fibration.

## References

[1] I. M. Gelfand, M. I. Graev and A. M. Veršik, Representations of the group of smooth mappings of a manifold $X$ into a compact Lie group. Compositio Math., **35** (1977), 299–334.

[2] I. M. Gelfand, M. I. Graev and A. M. Veršik, Representations of the group of functions taking values in a compact Lie group. Compositio Math., **42** (1981), 217–243.

[3] G. Goldin, Non-relativistic current algebras as unitary representations of groups. J. Math. Phys., **12** (1971), 462–488.

[4] G. Goldin, K. J. Grodnik, R. Powers and D. Sharp, Non-relativistic current algebras in the $N/V$ limit. J. Math. Phys., **15** (1974), 88–100.

[5] D. A. Kazhdan, The connection of the dual space of a group with the structure of its closed subgroups. Functsional. Anal. i Prilozhen **1** (1967), 71–74.
= Functional Anal. Appl. **1** (1967), 63–66.

[6] K. Matthes, J. Kerstan and J. Mecke, Infinitely divisible point processes. John Wiley, 1978.

[7] R. Menikoff, The hamiltonian and generating functional for a non-relativistic local current algebra. J. Math. Phys., **15** (1974), 1138–1152.

[8] R. Menikoff, Generating functionals determining representations of a non-relativistic local current algebra in the $N/V$ limit. J. Math. Phys., **15** (1974), 1394–1408.

[9] E. Thoma, Die unzerlegboren, positiv-definiten Klassenfunktionen der abzählbar unendlichen symmetrischen Gruppe. Math. Z., **85** (1964), 40–61.

[10] A. M. Vershik, I. M. Gelfand and M. I. Graev, Irreducible representations of the group $G^X$ and cohomology. Functsional. Anal. i Prilozhen., **8** (1974), 67–69. = Functional Anal. Appl., **8** (1974), 151–153.

[11] A. M. Vershik and S. V. Kerov, Characters and factor-representations of the infinite symmetric group. Doklady AN SSSR **257** (1981), 1037–1040.

# 2.

## Claus Michael Ringel

# Four papers on problems in a linear algebra

Representation Theory. Lecture Note Series, vol. 69, London Math. Soc., Cambridge University Press, 1982, pp. 141–156

This volume contains four papers on problems in linear algebra. They form part of a general investigation which was started with the famous paper [Q] on the four subspace problem. The $r$ subspace problem asks for the determination of the possible positions of $r$ subspaces in a vector space, or, equivalently, of the indecomposable representations of the following oriented graph

(*)

with $r + 1$ vertices. For $r \geqslant 5$, this problem seems to be rather hard to attack, however one may try to obtain at least partial results dealing with special kinds of representations. Also, the $r$ subspace problem can be used as a test problem for more elaborate problems in linear algebra. This seems to be the case for some of the investigations published in this volume, they have been generalized recently to the case of arbitrary oriented graphs [M, S].

Three of the four papers deal with the $r$ subspace problem. (We should remark that there is a rather large overlap of [F] and [I, II]. However, the main argument of [F], the proof given in section 7, is not repeated in [I, II], whereas [I, II] give the details for the complete irreducibility of the representations $\rho_{t,l}$ which only was announced in [F]. We also recommend the survey given by Dlab [8].) Given $r$ subspaces $E_1, \ldots, E_r$ of a finite-dimensional vector space $V$, we obtain a lattice homomorphism $\rho$ from the free modular lattice $D^r$ with $r$ generators $e_1, \ldots, e_r$ into the lattice $L(V)$ of all subspaces of $V$ given by $\rho(e_i) = E_i$. Such a lattice homomorphism is called a representation of $D^r$. In [F], Gelfand and Ponomarev introduce a set of indecomposable representations $\rho_{t,l}$ with $0 \leqslant t \leqslant r$ and $l \in \mathbf{N}$, which we will call the preprojective representations (in [F], the representations $\rho_{t,l}$ with $1 \leqslant t \leqslant r$ are called representations of the first kind, those of the form $\rho_{o,l}$ representations of the second kind; in [I, II] there may arise some confusion: $\rho_{t,l}$ is denoted by $\rho^+_{t,l}$, whereas the symbol $\rho_{t,l}$ used in [I, II] stands for the same type of representation but with a shift of the indices, see Proposition 8.2 in [II]). For the construction of the preprojective representations, we refer to section 1.4 of

[F]: one first defines a finite set $A_t(r, l)$ (which later we will identify with a set of paths in some oriented graph), considers the vector space with basis the set $A_t(r, l)$, and also a subspace $Z_t(r, l)$ generated by certain sums of the canonical base elements of $A_t(r, l)$. The residue classes of the canonical base elements of $A_t(r, l)$ in $V_{t,l} = A_t(r, l)/Z_t(r, l)$ will be denoted by $\xi_\alpha$ (with $\alpha \in A_t(r, l)$). Now, the representation $\rho_{t,l}$ is given by the vector space $V_{t,l}$ together with a certain $r$ tuple of subspaces of $V_{t,l}$, all being generated by some of the generators $\xi_\alpha$. Note that this implies that $\rho_{t,l}$ is defined over the prime field $k_0$ of $k$. (Gelfand and Ponomarev usually assume that the characteristic of $k$ is zero, thus $k_0 = \mathbb{Q}$. However, all results and proofs remain valid in general.)

The main result concerning these representations $\rho_{t,l}$ asserts that in case $\dim V_{t,l} > 2$, the representation $\rho_{t,l}$ is completely irreducible. This means that the image of $D^r$ under the lattice homomorphism $\rho_{t,l}\colon D^r \to L(V_{t,l})$ is the set of all subspaces of $V_{t,l}$ defined over the prime field $k_0$, thus $\rho_{t,l}(D^r)$ is a projective geometry over $k_0$. The first essential step in the proof of this result is to show that the subspaces $k\xi_\alpha$ are of the form $\rho(e_\alpha)$ for some $e_\alpha \in D^r$. (In [F], this is only announced, but it is an immediate consequence of theorem 8.1 in [II].)

The second step is to show that any subspace of $V_{t,l}$ which is defined over the prime field, lies in the lattice of subspaces generated by the $k\xi_\alpha$ provided $\dim V_{t,l} > 2$. Combining both assertions, we conclude that $\rho_{t,l}$ is completely irreducible unless $\dim V_{t,l} \leq 2$. The proof of the second step occupies section 9 of [II]. Here, one considers the following situation: there is given a set $R = \{\xi_\alpha \mid \alpha\}$ of non-zero vectors of a vector space $V$ (= $V_{t,l}$), with the following properties:

(1) $R$ generates $V$

(2) $R$ is indecomposable (there is no proper direct decomposition $V = V' \oplus V''$ with $R = (R \cap V') \cup (R \cap V'')$), and

(3) $R$ is defined over the prime field (there exists a basis of $V$ such that any $\xi_\alpha \in R$ is a linear combination of the base vectors with coefficients in the prime field $k_0$).

Then it is shown that the lattice of subspaces of $V$ generated by the one-dimensional subspaces $k\xi_\alpha$, is isomorphic to the lattice of subspaces of $k_0^n$, with $n = \dim V$.

Perhaps we should add that the representations $\rho\colon D^r \to L(V)$ with $V$ being generated by the one-dimensional subspaces of the form $\rho(a)$, $a \in D^r$, seem to be of special interest. In this case, the one-dimensional subspaces of the form $\rho(a)$, $a \in D^r$ determine completely $\rho(D^r)$. (Namely, let $b \in D^r$, and $U$ the subspace generated by all one-dimensional subspaces of the form $\rho(x)$, $x \in D^r$, satisfying $\rho(x) \subseteq \rho(b)$, and choose $x_1, \ldots, x_s$ such that

$$\rho(b) \subseteq U \oplus \rho(x_1) \oplus \ldots \oplus \rho(x_s) = U \oplus \rho(\sum_{i=1}^{s} x_i). \text{ Thus, } \rho(b) = U \oplus (\rho(\sum_{i=1}^{s} x_i) \cap \rho(b)).$$

Assume, $U$ is a proper subspace of $\rho(b)$. Then there exists $t \leq s$ with

$\rho(\sum_{i=1}^{t-1} x_i) \cap \rho(b) = 0$, whereas $\rho(\sum_{i=1}^{t} x_i) \cap \rho(b)$ is non-zero, and therefore one-dimensional. This however implies that $\rho(\sum_{i=1}^{t} x_i) \cap \rho(b) = \rho(b \sum_{i=1}^{t} x_i)$ is contained in $U$, a contradiction. Thus $\rho(b) = U$.). For $r \geqslant 4$, there always are indecomposable representations which do not have this property.

In the case $r = 4$, we may give the complete list of all lattices of the form $\rho(D^4)$, where $\rho$ is an indecomposable representation. Besides the projective geometries over any prime field, and of arbitrary finite dimension $\neq 1$, and the lattice 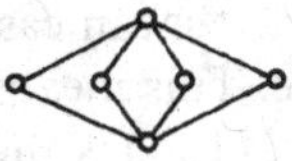, we obtain all the lattices $S(n, 4)$ introduced by Day, Herrmann and Wille in [6]. Let us just copy $S(14, 4)$ and note that any interval $[c_n, c_m]$ is again of the form $S(n - m, 4)$.

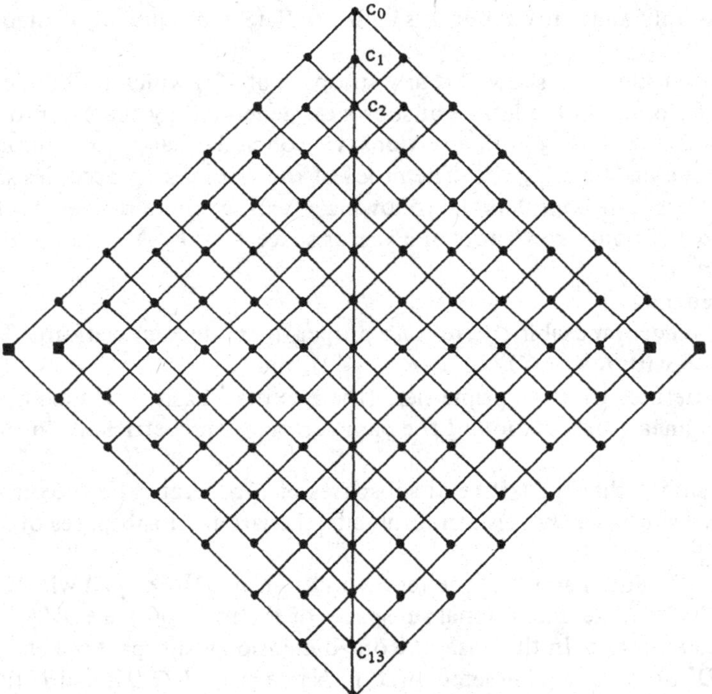


(In fact, in case either $\rho: D^4 \to L(V)$ or its dual is preprojective and dim $V > 2$, we have seen above that $\rho(D^4)$ is the full projective geometry over the prime field. If neither $\rho$ nor its dual is preprojective, $\rho$ is said to be regular. If $\rho$ is regular and non-homogeneous, say of regular length $n$ (see [9]), then $\rho(D^4) \approx S(n, 4)$, whereas for $\rho$ homogeneous, we have

$\rho(D^4) \approx$ 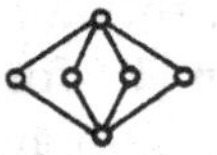 .)

Gelfand and Ponomarev use the representations $\rho_{t,l}$ of $D^r$ in order to get some insight into the structure of $D^r$. The existence of a free modular lattice with a given set of generators is easily established, however the mere existence result does not say anything about the internal structure of $D^r$. In fact, it has been shown by Freese [14] that for $r \geqslant 5$, the word problem in $D^r$ is unsolvable. The free modular lattice $D^3$ in 3 generators $e_1, e_2, e_3$ was first described by Dedekind [7], it looks as follows:

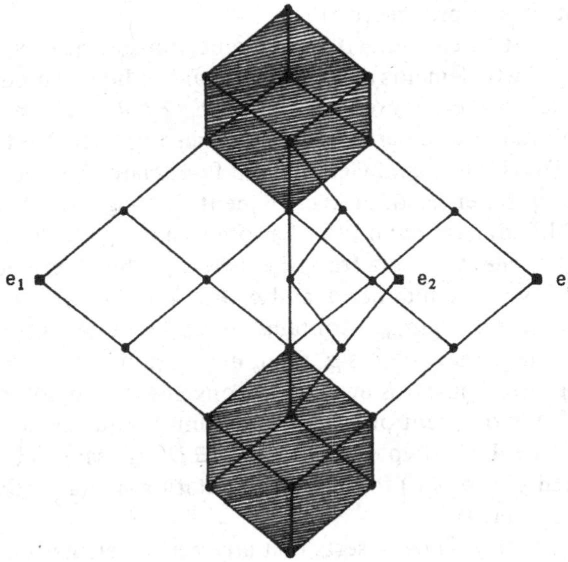


We have shaded two parts of $D^3$, both being Boolean lattices with $2^3$ elements. For $r \geqslant 4$, Gelfand and Ponomarev have constructed two countable families of Boolean sublattices $B^+(l)$ and $B^-(l)$ with $2^r$ elements, where $l \in \mathbf{N}$, and such that

$$B^-(1) < B^-(2) < \ldots < B^-(l) < B^-(l+1) < \ldots$$

and

$$\ldots B^+(l+1) < B^+(l) < \ldots < B^+(2) < B^+(1),$$

called the lower and the upper cubicles, respectively. Let $B^- = \bigcup_{l \in \mathbf{N}} B^-(l)$, and

$B^+ = \bigcup_{l \in \mathbf{N}} B^+(l).$

The elements of these cubicles have an important property: they are perfect. This notion has been introduced by Gelfand and Ponomarev in [F] for the following property: $a$ is said to be perfect if $\rho(a)$ is either $O$ or $V$ for any indecomposable representation $\rho: D^r \to L(V)$. This means that for any representation, the image of $a$ is a direct summand. For any perfect element $a$, let $N_k(a)$ be the set of all indecomposable representations $\rho: D^r \to L(V)$, with $V$ a finite dimension vector space over the field $k$ and which satisfy $\rho(a) = 0$. It is shown in [F] that for $a \in B^+$, the set $N_k(a)$ is finite and contains only preprojective representations. Dually, for $a \in B^-$, the set $N_k(a)$ contains all but a finite number of indecomposable representations, and all indecomposable representations not in $N_k(a)$ are preinjective (the representations dual to preprojective ones are called preinjective).

In dealing with perfect elements it seems to be convenient to work modulo linear equivalence. Two elements $a, a' \in D^r$ are said to be linear equivalent provided $\rho(a) = \rho(a')$ for any representation $\rho: D^r \to L(V)$. Of course, any element linear equivalent to a perfect element is also perfect. Up to linear equivalence, one has $B^- < B^+$ and Gelfand and Ponomarev have conjectured that, up to linear equivalence, all perfect elements belong to $B^- \cup B^+$. However, this has to be modified. Herrmann [19] has pointed out that there are additional perfect elements arising from the different characteristics of fields. For example, for any prime number $p$, and $m \geqslant 2$, there is some perfect element $d_{pm} \in D^r$ such that $N_k(d_{pm})$ contains all representations $\rho_{t,l}$ with $l < m$, and, in case the characteristic of $k$ is $p$, then, in addition, the representation $\rho_{o,m}$, and nothing else. Thus, it is even more convenient to work in the free $p$-linear lattice $D_p^r$, the quotient of $D^r$ modulo $p$-linear equivalence where $p$ is either zero or a prime. Here, two elements, $a, a' \in D^r$ are said to be $p$-linear equivalent provided $\rho(a) = \rho(a')$ for any representation $\rho$ in a vector space over a field of characteristic $p$.

The modified conjecture now asserts that any perfect element is $p$-linearly equivalent to an element in $B^- \cup B^+$. This indeed is true, as we want to show. Thus, assume there exists a perfect element $a \in D^r$ which is not $p$-linear equivalent to an element of $B^- \cup B^+$. Gelfand and Ponomarev have shown that then $N(a) = N_k(a)$ contains all preprojective representations and no preinjective representation. In a joint paper [10] with Dlab, we have shown that for $r \geqslant 5$, the set $N(a)$ either contains only the preprojective representations or else all but the preinjective representations. The elements $x \in D^r$ are given by lattice polynomials in the variables $e_1, \ldots, e_r$. Of course, there will be many different lattice polynomials which define the same element $x$. A lattice polynomial with minimal number of occurrences of variables defining $x$ will be called a reduced expression of $x$ and this number of variables in a reduced expression will be called the complexity $c(x)$ of $x$. Now, let $\rho: D^r \to L(V)$ be a representation, $U$ a one-dimensional subspace of $V$, and

$\rho'\colon D^r \to L(V/U)$ the induced representation, with $\rho'(e_i) = (\rho(e_i) + U)/U$ for the generators $e_i$, $1 \leqslant i \leqslant r$. We claim that for $x \in D^r$, we have

$$\dim \rho'(x) \leqslant c(x) - 1 + \dim \rho(x).$$

[For the proof, we consider instead of $\rho'$ the representation $\rho''\colon D^r \to L(V)$ with $\rho''(x)$ the full inverse image of $\rho'(x)$ under the projection $V \to V/U$, thus $\dim \rho''(x) = 1 + \dim \rho'(x)$, for $x \in D^r$. Also note that $\rho(x) \subseteq \rho''(x)$ for all $x$. By induction on $c(x)$, we show the formula

$$\dim \rho''(x) - \dim \rho(x) \leqslant c(x).$$

Since $\dim U = 1$, this clearly is true for $x = e_i$, with $\rho''(e_i) = \rho(e_i) + U$. Now assume the formula being valid both for $x_1$ and $x_2$. For $x = x_1 + x_2$ with $c(x) = c(x_1) + c(x_2)$, we have

$$\begin{aligned}\dim \rho''(x) &= \dim \rho''(x_1 + x_2) \leqslant \dim \rho(x_1 + x_2) + c(x_1) + c(x_2)\\ &= \dim \rho(x) + c(x).\end{aligned}$$

Similarly, for $x = x_1 x_2$ with $c(x) = c(x_1) + c(x_2)$, we have

$$\begin{aligned}\dim \rho''(x) &= \dim \rho''(x_1 x_2) = \dim \rho''(x_1) + \dim \rho''(x_2) - \dim \rho''(x_1 + x_2)\\ &\leqslant \dim \rho(x_1) + c(x_1) + \dim \rho(x_2) + c(x_2) - \dim \rho(x_1 + x_2)\\ &= \dim \rho(x_1 x_2) + c(x_1) + c(x_2) = \dim \rho(x) + c(x).\end{aligned}$$

This finishes the proof.]

It is now sufficient to find a preprojective representation $\rho\colon D^r \to L(V)$ with $\dim V > c(a)$ and a one-dimensional subspace $U$ of $V$ such that the induced representation $\rho'$ in $V/U$ has no preprojective direct summand. Namely, our considerations above imply that $\dim \rho'(a) \leqslant c(a) - 1 < \dim V/U$, due to the fact that $\rho(a) = 0$, and therefore there exists at least one indecomposable representation $\sigma$ in $N(a)$ which is not preprojective. As a consequence, in case $r \geqslant 5$, we know that $N(a)$ contains all but the preinjective representations. By duality, we similarly show that $N(a)$ contains only the preprojective representations, thus we obtain a contradiction. So, let us construct a suitable preprojective representation with the properties mentioned above. In fact, instead of considering representations of $D^r$, we will work inside the abelian category of representations of the oriented graph (*). We denote by $P_{t,l} = (V_{t,l}; \rho_{t,l}(e_1), \ldots, \rho_{t,l}(e_r))$ the graph representation corresponding to $\rho_{t,l}$. Take any homomorphism $\varphi\colon P_{0,1} \to P_{0,2}$ such that $R = \operatorname{Cok} \varphi$ is regular (that is, has no non-zero preprojective or preinjective direct summand. For example, there always exists such a $\varphi$ with $R$ being the direct sum of two indecomposable representations of dimension types $(1; 1, 1, 0, \ldots, 0)$ and $(r-3; 0, 0, 1, 1, \ldots, 1)$.) Now apply $\Phi^{-i}$ for $i \in \mathbf{N}$. We obtain exact sequences

$$0 \to P_{0,i+1} \xrightarrow{\Phi^{-i}(\varphi)} P_{0,i+2} \to \Phi^{-i} R \to 0,$$

thus, the inclusion

$$\varphi_i = \Phi^{-i}(\varphi) \circ \ldots \circ \Phi^{-1}(\varphi) \circ \varphi : P_{0,1} \to P_{0,i+2}$$

has regular cokernel (extensions of regular representations being regular, again). We now only have to choose $i$ such that $\dim V_{0,i+2} > c(a)$. This finishes the proof in case $r \geqslant 5$. (For $r = 4$, we again take $\varphi: P_{0,1} \to P_{0,2}$ with Cok $\varphi$ being the direct sum of two representations of dimension types $(1; 1, 1, 0, 0)$ and $(1, 0, 0, 1, 1)$, and form $\varphi_i$. The indecomposable summands of Cok $\varphi_i$ all belong to one component $C$ of the Auslander–Reiten quiver, thus we conclude as above that $C \subseteq N(a)$. By duality, one similarly shows that there are representations in $C$ which do not belong to $N(a)$, so again we obtain a contradiction. Note that in case $r = 4$, the conjecture has been solved before by Herrmann [19].)

We consider now the general problem of representations of an oriented graph $(\Gamma, \Lambda)$. We do not recall the definition of the category $L(\Gamma, \Lambda)$ of representations of $(\Gamma, \Lambda)$ over some fixed field $k$, nor the typical examples, but just refer to the first two pages of [BGP]. We only note that $L(\Gamma, \Lambda)$ can also be considered as the category of modules over the path algebra $k(\Gamma, \Lambda)$, see [17], and $k(\Gamma, \Lambda)$ is a finite-dimensional $k$-algebra if and only if $(\Gamma, \Lambda)$ does not have oriented cycles. In [15], Gabriel had shown that $(\Gamma, \Lambda)$ has only finitely many indecomposable representations if and only if $\Gamma$ is the disjoint union of graphs of the form $A_n$, $D_n$, $E_6$, $E_7$ and $E_8$ (they are depicted on the third page of [BGP]). It turned out that in case $\Gamma$ is of the form $A_n$, $D_n$, $E_6$, $E_7$ or $E_8$, the indecomposable representations of $(\Gamma, \Lambda)$, with $\Lambda$ an arbitrary orientation, are in one-to-one correspondence to the positive roots of $\Gamma$. It is the aim of the paper [BGP] to give a direct proof of this fact. It introduces appropriate functors which produce all indecomposable representations from the simple ones in the same way as the canonical generators of the Weyl group produce all positive roots from the simple ones. We later will come back to these functors and their various generalizations.

Given a finite graph $\Gamma$, let $E_\Gamma$ be the $\mathbf{Q}$-vector space of functions $\Gamma_0 \to \mathbf{Q}$, an element of $E_\Gamma$ being written as a tuple $x = (x_\alpha)$ indexed by the elements $\alpha \in \Gamma_0$. For $\beta \in \Gamma_0$, we denote its characteristic function by $\overline{\beta}$ (thus $\overline{\beta}_\alpha = 0$ for $\alpha \neq \beta$, and $\overline{\beta}_\beta = 1$). Any representation $V$ of $(\Gamma, \Lambda)$ gives rise to an element $\dim V$ in $E_\Gamma$, its dimension type. For any orientation $\Lambda$ of $\Gamma$, and any $\beta \in \Gamma_0$, there is a unique simple representation $L_\beta$ of dimension type $\dim L_\beta = \overline{\beta}$. In case there are no oriented cycles in $(\Gamma, \Lambda)$, we obtain in this way all simple representations of $(\Gamma, \Lambda)$, thus, in this case, $E_\Gamma$ may be identified with the rational Grothendieck group $G_0(\Gamma, \Lambda) \underset{\mathbf{Z}}{\otimes} \mathbf{Q}$ (here, $G_0(\Gamma, \Lambda)$ is the factor group of the free abelian group with basis the set of all representations of $(\Gamma, \Lambda)$

modulo all exact sequences) with dim being the canonical map (sending a representation to the corresponding residue class). On $E_\Gamma$, there is defined a quadratic form $B$. In fact, for any orientation $\Lambda$ of $\Gamma$, we may consider the (non-symmetric) bilinear form $B_\Lambda$ on $E_\Gamma$ given by

$$B_\Lambda(x, y) = \sum_{l \in \Gamma_0} x_\alpha y_\alpha - \sum_{l \in \Gamma_1} x_{\alpha(l)} y_{\beta(l)}$$

and $B$ is the corresponding quadratic form $B(x) = B_\Lambda(x, x)$. Note that $B$ is positive definite if and only if $\Gamma$ is the disjoint union of graphs of the form $A_n, D_n, E_6, E_7$ and $E_8$, and in these cases, the root system for $\Gamma$ is by definition just the set of solutions of the equation $B(x) = 1$.

For $k$ algebraically closed and $B$ being positive definite we will outline a direct proof that dim: $L(\Gamma, \Lambda) \to E_\Gamma$ induces a bijection between the indecomposable representations of $(\Gamma, \Lambda)$ and the positive roots. There is the following algebraic-geometric interpretation of $B$ due to Tits [15]: The representations of $(\Gamma, \Lambda)$ of dimension type $x$ may be considered as the algebraic variety

$$m^x(\Gamma, \Lambda) = \prod_{l \in \Gamma_1} \operatorname{Hom}(k^{\alpha(l)}, k^{\beta(l)}),$$

and there is an obvious action on it by the algebraic group

$$G^x = \prod_{\alpha \in \Gamma_0} \operatorname{GL}(\alpha, k)/\Delta$$

with $\Delta$ being the multiplicative group of $k$ diagonally embedded as group of scalars. Clearly

$$B(x) = \dim G^x + 1 - \dim m^x(\Gamma, \Lambda).$$

Using this interpretation, Gabriel has shown in [16] that it only remains to prove that the endomorphism ring of any indecomposable representation is $k$. So assume $V$ is indecomposable, and that there are non-zero nilpotent endomorphisms. Then $V$ contains a subrepresentation $U$ with $\operatorname{End}(U) = k$ and $\operatorname{Ext}^1(U, U) \neq 0$. [Namely, let $0 \neq \varphi$ be an endomorphism with image $S$ of smallest possible length, thus $\varphi^2 = 0$, and let $W = \bigoplus_{i=1}^{r} W_i$ be the kernel of $\varphi$, with all $W_i$ indecomposable. Now $S \subseteq W$, thus the projection of $S$ into some $W_i$ must be non-zero. Since $S$ was an image of a non-zero endomorphism of smallest length, we see that $S$ embeds into this $W_i$. We may assume $i = 1$. Thus there is an inclusion $\iota: S \to W_1$. If $W_1$ has non-zero nilpotent endomorphisms, we use induction. Otherwise $\operatorname{End}(W_1) = k$. Also, $\operatorname{Ext}^1(W_1, W_1) \neq 0$, since on the one hand $\operatorname{Ext}^1(S, W_1) \neq 0$ due to the exact sequence

$$0 \to \bigoplus_{i=1}^{r} W_i \to V \to S \to 0$$

and, on the other hand, the inclusion $\iota$ gives rise to a surjection $\mathrm{Ext}^1(\iota, W)$. Here we use that $L(\Gamma, \Lambda)$ is a hereditary category]. The bilinear form $B_\Lambda$ has the following homological interpretation [25]:

$$B_\Lambda(\dim V, \dim V') = \dim_k \mathrm{Hom}(V, V') - \dim_k \mathrm{Ext}^1(V, V'),$$

for all representations $V, V'$. Consequently, the existence of a representation $U$ satisfying $\mathrm{End}(U) = k$, $\mathrm{Ext}^1(U, U) \neq 0$ would imply that

$$B(\dim U) = B_\Lambda(\dim U, \dim U) \leqslant 0,$$

contrary to the assumption that $B$ is positive definite. This finishes the proof.

For any finite connected graph $\Gamma$ without loops, Kac [21, 22] gave a purely combinatorial definition of its root system $\Lambda$. Note that $\Lambda$ is a subset of $E_\Gamma$ containing the canonical base vectors $\overline{\beta}$, for $\beta \in \Gamma_0$, and being stable under the Weyl group $W$, the group generated by the reflections $\sigma_\beta$ along $\overline{\beta}$ with respect to $B$. The set $\Delta$ can also be interpreted in terms of root spaces of certain (usually infinite dimensional) Lie algebras [21]. Denote by $\Delta_+$ the set of roots with only non-negative coordinates with respect to the canonical basis. Then $\Delta$ is the union of $\Delta_+$ and $\Delta_- = -\Delta_+$. In case $\Gamma$ is of type $A_n$, $D_n$, $E_6$, $E_7$ or $E_8$, the root system is finite and coincides with the set of solutions of $B(x) = 1$. Otherwise the root system is infinite and will contain besides certain solutions of $B(x) = 1$ also some solutions of $B(x) \leqslant 0$. The elements $x$ of the root system which satisfy $B(x) = 1$ are called real roots, they are precisely the elements of the $W$-orbits of the canonical base elements. The remaining elements of the root system are called imaginary roots, and Kac has determined a fundamental domain for this set, the fundamental chamber.

Now, one has the following results (at least if $k$ is either finite or algebraically closed): For any finite graph $\Gamma$ without loops, and any orientation $\Lambda$, the set of dimension types of indecomposable modules is precisely the set $\Delta_+$ of positive roots. For any positive real root $x$, there exists precisely one indecomposable representation $V$ of $(\Gamma, \Lambda)$ with $\dim V = x$. For any positive imaginary root $x$, the maximal dimension $\mu_x$ of an irreducible component in the set of isomorphism classes of indecomposable representations of dimension $x$ is precisely $1 - B(x, x)$. (Note that the subset of indecomposable representations in $m^x(\Gamma, \Lambda)$ is constructible, and $G^x$-invariant, thus we can decompose it as a finite disjoint union of $G^x$-invariant subsets each of which admits a geometric quotient. By definition, $\mu_x$ is the maximum of the dimensions of these quotients.) In particular, we see that the number of indecomposable representations (or of the maximal dimension of families of indecomposable representations) of $(\Gamma, \Lambda)$ does not depend on the orientation $\Lambda$. For $\Gamma$ of the form $A_n$, $D_n$, $E_6$, $E_7$ or $E_8$, this is Gabriel's theorem (of course, there are no imaginary roots). For $\Gamma$ of the form $\widetilde{A}_n$, $\widetilde{D}_n$, $\widetilde{E}_6$, $\widetilde{E}_7$, or $\widetilde{E}_8$, the so called tame cases, these results have been shown by Donovan–Freislich [13] and Nazarova [23], see also [9]; in fact, in these cases one obtains a full classification of all indecomposable representations; also, it is possible in these cases to describe completely the rational invariants of the action of $G^x$ on $m^x(\Gamma, \Lambda)$, for any dimension type $x$, see [27]. Of course, the oriented graphs of finite or tame representation type are rather special ones. It has been known since some time that the remaining $(\Gamma, \Lambda)$ are wild: there always is a full exact subcategory of $L(\Gamma, \Lambda)$ which is equivalent to the category $\mathcal{M}_{k\langle X, Y\rangle}$ of $k\langle X, Y\rangle$-modules ($k\langle X, Y\rangle$ being the polynomial ring in two non-commuting indeterminates). In this situation, the results above are due to Kac [21, 22]. Note that this solves all the conjectures of Bernstein–Gelfand–Ponomarev formulated in [BGP]. However, there remain many open questions concerning wild graphs $(\Gamma, \Lambda)$. One does not expect to obtain a complete classification of the indecomposable representations of such a graph, but one would like to have some more knowledge about certain classes of representations. For example, there does not yet exist a combinatorial description of the set of those roots which are dimension types of representations $V$ with $\mathrm{End}(V) = k$.

We have mentioned above that the root system $\Delta$ of $\Gamma$ is stable under the Weyl group $W$ and that any $W$-orbit of $\Delta$ contains either one of the base vectors $\bar{\beta}$ (with $\beta \in \Gamma_0$) or an element of the fundamental chamber. One therefore tries to find operations which associate to an indecomposable representation $V$ of $(\Gamma, \Lambda)$ with $\Lambda$ an orientation, and a Weyl group element $w \in W$ a new indecomposable representation of $(\Gamma, \Lambda')$, where $\Lambda'$ is a possibly different orientation of $\Gamma$. By now, several such operations are known (see [BGP, 21, 28]), the first one being the reflection functors $F_\beta^-$, $F_\beta^+$ introduced by Bernstein, Gelfand and Ponomarev in [BGP]. Here, for the definition of $F_\beta^+$, the vertex $\beta$ is supposed to be a sink, thus the simple representation $L_\beta$ with dimension vector $\dim L_\beta = \bar{\beta}$ is projective. This concept has been generalized by Auslander, Platzeck and Reiten [1] dealing with any finite dimensional algebra $A$ (or even an artin algebra) with a simple projective module $L$. For this, we need the Auslander–Reiten translates $\tau, \tau^{-1}$. Recall that $\tau X_A$ is defined for any $A$-module $X_A$: let $P_1 \xrightarrow{p} P_0 \to X_A \to 0$ be a minimal projective resolution of $X_A$, then $\mathrm{Tr}\, X_A$ is by definition the cokernel of the map $\mathrm{Hom}(p, A_A)$ and $\tau X = D\,\mathrm{Tr}\,X$, $\tau^{-1} X = \mathrm{Tr}\,D\,X$, with $D$ the usual duality with respect to the base field $k$. So assume $L$ is a simple projective $A$-module, let $P$ be the direct sum of one copy of each of the indecomposable projective modules different from $L$, and $B = \mathrm{End}(P \oplus \tau^{-1} L)$. The functor considered by Auslander, Platzeck and Reiten is $F = \mathrm{Hom}_A(P \oplus \tau^{-1} L, -)$ from the category $\mathcal{M}_A$ of $A$-modules to $\mathcal{M}_B$. The functor induces an equivalence of the full subcategory $T$ of $\mathcal{M}_A$ of all modules which do not have $L$ as a direct summand and a certain full subcategory of $\mathcal{M}_B$. Note that $P \oplus \tau^{-1} L$ is a tilting module in the sense of [18], except in the trivial case of $L$ being, in addition, injective. (A tilting module $T_A$ is defined by the following three properties:

(1) proj. dim. $T_A \leqslant 1$, (2) there exists an exact sequence $0 \to A_A \to T' \to T'' \to 0$, with $T'$, $T''$ being direct sums of direct summands of $T_A$, and (3) $\mathrm{Ext}^1(T_A, T_A) = 0$. Now, if $L_A$ is simple projective and not injective, the middle term $Y$ of the Auslander–Reiten sequence

$$0 \to L \to Y \to \tau^{-1}L \to 0$$

starting with $L$ is projective. This sequence shows, on the one hand, that proj. dim. $\tau^{-1}L = 1$. On the other hand, it also gives an exact sequence of the form needed in (2). Finally, $\mathrm{Ext}^1_A(P \oplus \tau^{-1}L, P \oplus \tau^{-1}L) \approx \approx D\,\mathrm{Hom}(P \oplus \tau^{-1}L, L) = 0$, since any non-zero homomorphism from a module to $L$ is a split epimorphism.)

A certain composition of the reflection functors $F^+_\beta$ (or $F^-_\beta$, respectively) is of particular interest, the Coxeter functor $\Phi^+$ (or $\Phi^-$). An explicit calculation for the $r$-subspace situation is given in [F], in the special case of the 4-subspace problem it had been defined before in [Q]. The Coxeter functors are endofunctors of $L(\Gamma, \Lambda)$, they are only defined in case $(\Gamma, \Lambda)$ does not have oriented cycles (non-oriented cycles are allowed, see [9]). Note that the assignment of an orientation $\Lambda$ without oriented cycles is equivalent to the choice of a partial ordering of $\Gamma_0$ (let $\alpha \leqslant \beta$ iff there exists an oriented path $\alpha = \alpha_0 \leftarrow \alpha_1 \leftarrow \ldots \ldots \leftarrow \alpha_m = \beta$), and also to the choice of a Coxeter transformation: this is a Weyl group element of the form $c = \sigma_{\alpha_n} \ldots \sigma_{\alpha_1}$ with $\alpha_1, \ldots, \alpha_n$ being the elements of $\Gamma_0$ in some fixed ordering (take an ordering of $\Gamma$ which refines the given partial ordering). So assume from now on that $(\Gamma, \Lambda)$ is a connected oriented graph without oriented cycles, and let $c$ be the corresponding Coxeter element. The Coxeter functors $\Phi^+$ and $\Phi^-$ defined in [BGP] have the following properties: if $V$ is an indecomposable representation of $(\Gamma, \Lambda)$, then either $V$ is projective and then $\Phi^+(V) = 0$, or else $V$ is not projective, and then $\Phi^+(V)$ again is indecomposable, $\Phi^-\Phi^+(V) \approx V$ and $\dim \Phi^+(V) = c \dim V$. Thus the Coxeter functor $\Phi^+$ realizes the action of the Coxeter transformation on the set of all representations without non-zero projective direct summands. The usefulness of the Coxeter functors seems to have its origin in their relation to the Auslander–Reiten translation $\tau$. Namely, Gabriel ([17], Prop. 5.3, see also [1,5]) has shown that $\tau$ can be identified with $C^+ \circ T$, where $T$ is the functor which maps the representation $(V, f)$ to $(V, -f)$. In particular, for $\Gamma$ being a tree, we can identify $\tau$ with $C^+$ itself.

In order to explain the value of the Auslander–Reiten translation $\tau$ (and therefore of the Coxeter functors), we have to recall the definition of the Auslander–Reiten quiver of a finite dimensional algebra $A$. Its vertices are the isomorphism classes $[X]$ of the indecomposable $A$-modules $X$, and, if $X$, $Y$ are indecomposable modules, then there is an arrow $[X] \to [Y]$ iff there exists an irreducible map $X \to Y$ (a map $f$ is said to be irreducible provided it is neither a split monomorphism nor a split epimorphism, and for any factorization $f = f'' \circ f'$, we have that $f'$ is a split monomorphism or $f''$ is a split epimorphism [2]). Now, the Auslander–Reiten quiver is a translation quiver with respect to $\tau$: if $X$ is indecomposable and not projective, then there exists an irreducible map $Y \to X$ iff there exists an irreducible map $\tau X \to Y$.

For the finite dimensional hereditary algebras $A$, the structure of the Auslander–Reiten quiver is known. We will recall this result in the special case of $A = k(\Gamma, \Lambda)$. First, we need some notation. Define $\mathbf{Z}(\Gamma, \Lambda)$ as follows: its vertices are the elements of $\Gamma_0 \times \mathbf{Z}$, and for any arrow ${}^{i}\circ \xleftarrow{\alpha} \circ^{j}$, there are arrows $(i, z) \xrightarrow{(\alpha, z)} (j, z)$ and $(j, z) \xrightarrow{(\alpha^*, z)} (i, z+1)$, for all $z \in \mathbf{Z}$, see [24] and also [17, 29]. Note that in case $\Gamma$ is a tree, $\mathbf{Z}(\Gamma, \Lambda)$ does not depend on the orientation $\Lambda$ and just may be denoted by $\mathbf{Z}\Gamma$. If $I \subseteq \mathbf{Z}$, let $I(\Gamma, \Lambda)$ be the full subgraph of all vertices $(i, z)$ with $i \in I$. In particular, we will have to consider $\mathbf{N}(\Gamma, \Lambda)$ and $\mathbf{N}^-(\Gamma, \Lambda)$, where $\mathbf{N} = \{1, 2, 3, \ldots\}$ and $\mathbf{N}^- = \{-1, -2, -3, \ldots\}$. Also, denote by $A_\infty$ the following infinite graph

$$\circ \!-\!\!-\! \circ \!-\!\!-\! \circ \ldots \circ \!-\!\!-\! \circ \ldots \; .$$

The result is as follows: in case $\Gamma$ is of the form $A_n$, $D_n$, $E_6$, $E_7$ or $E_8$, the Auslander–Reiten quiver of $k(\Gamma, \Lambda)$ is a finite full connected subquiver of $\mathbf{Z}\Gamma$. (In case $D_n$ with $n \equiv 0(2)$, the Auslander–Reiten quiver of $k(\Gamma, \Lambda)$ is $[1, n-1]\ (\Gamma, \Lambda)$, in case of $E_7$ or $E_8$, it is $[1,9]\ (\Gamma, \Lambda)$ or $[1,15]\ (\Gamma, \Lambda)$, respectively; in the remaining cases, it is slightly more difficult to describe, see [17, 29]). In all other cases, the Auslander–Reiten quiver of $k(\Gamma, \Lambda)$ has infinitely many components, all but two being quotients of $\mathbf{Z}A_\infty$ (see [26]), the remaining two being of the form $\mathbf{N}(\Gamma, \Lambda)$ and $\mathbf{N}^-(\Gamma, \Lambda)$. The component of the form $\mathbf{N}(\Gamma, \Lambda)$ contains the indecomposable projective modules: in fact, the indecomposable projective module $P_i$ corresponding to the vertex $i \in \Gamma_0$ appears as indexed by $(i, 1)$, and the module indexed by $(i, z)$, $z \in \mathbf{N}$, is just $\Phi^{-z+1}(P_i)$, this component is called the preprojective component. Similarly, the component of the form $\mathbf{N}^-(\Gamma, \Lambda)$ is called the preinjective component, it contains the indecomposable injective module $J_i$ corresponding to $i \in \Gamma_0$ as indexed by $(i, -1)$, and the module indexed by $(i, -z)$, $z \in \mathbf{N}$, is just $\Phi^{+z-1}(J_i)$.

Let us consider in more detail a preprojective component $\mathscr{P}$, and the modules belonging to $\mathscr{P}$; they will be called preprojective modules. In case $\Gamma$ is of type $A_n$, $D_n$, $E_6$, $E_7$, or $E_8$, we let $\mathscr{P}$ denote the full Auslander–Reiten quiver; in any case, we note that an indecomposable representation of $(\Gamma, \Lambda)$ is said to be preprojective iff it is of the form $\Phi^{-z}P$, with $P$ indecomposable projective and $z \geqslant 0$. (A general theory of preprojective modules has been developed by Auslander and Smalø, see [3]). For an indecomposable preprojective representation $X$, there are only finitely many indecomposable modules $Y$ such that $\mathrm{Hom}(Y, X) \neq 0$, all of them are preprojective again, and any non-invertible homomorphism $Y \to X$ is a sum of compositions of irreducible maps. In particular, if $X$, $Y$ are indecomposable and preprojective and

Hom$(X, Y) \neq 0$, then there is an oriented path $[X] \to \dots \to [Y]$ in $\mathscr{P}$. In fact, the complete categorical structure of the full subcategory of preprojective modules can be read off from the combinatorial description of $\mathscr{P}$ as a translation quiver: the category of all preprojective modules is equivalent to the quotient category $\langle\rangle\mathscr{P}$ of the path category of $\mathscr{P}$ modulo the so called mesh relations (see [4, 24, 17]). Note that the category $\langle\rangle\mathscr{P}$ allows to reconstruct all the modules in $\mathscr{P}$. Namely, any module $X_A$ is isomorphic to Hom$({}_AA_A, X_A)$, thus, if $A_A = \bigoplus_{i \in \Gamma_0} P_i^{n_i}$, then $X_A$ can be identified with $\bigoplus_{i \in \Gamma_0}$ Hom$(P_i, X)^{n_i}$, and Hom$(P_i, X)$ can be calculated inside $\langle\rangle\mathscr{P}$, since both $P_i$, $X$ are preprojective.

Starting from the preprojective component $\mathscr{P}$ of $k(\Gamma, \Lambda)$, one may define a (usually infinite-dimensional) algebra $\Pi$ as follows: Take the direct sum of all homomorphism spaces Hom$((j, 1), (t, l))$ in $\langle\rangle\mathscr{P}$ and define the product of two residue classes $\bar{w}$, $\bar{w}'$ of paths $w: (j, 1) \to \dots \to (t, l)$ and $w': (j', 1) \to \dots \to (t', l')$ as follows: in case $t = j'$, let $\bar{w}\,\bar{w}'$ be the residue class of the composed path $\tau^{-l+1}(w') \circ w: (j, 1) \to \dots \to (t', l + l' - 1)$, and 0 otherwise. There is a purely combinatorial description of $\Pi$ in terms of $(\Gamma, \Lambda)$ due to Gelfand and Ponomarev, see [R]. Let $\hat{\Gamma}$ be obtained from $(\Gamma, \Lambda)$ by adding to each arrow $\alpha: i \to j$ an additional arrow $\alpha^*: j \to i$. We clearly can identify $\Pi$ with the factor algebra of the path algebra $k\hat{\Gamma}$ modulo the ideal generated by the element $\sum_{\alpha \in \Gamma_1} \alpha \cdot \alpha^* + \sum_{\alpha \in \Gamma_1} \alpha^* \cdot \alpha$. Note that this description is independent of the choice of the orientation $\Lambda$. Also, we see from both descriptions that $\Pi$ contains as a subalgebra $k(\Gamma, \Lambda)$, thus we may consider $\Pi$ as a right $k(\Gamma, \Lambda)$-module, and the first description now shows that the $k(\Gamma, \Lambda)$-module $\Pi_{k(\Gamma,\Lambda)}$ decomposes as the direct sum of all preprojective representations of $(\Gamma, \Lambda)$ each occurring with multiplicity one, and therefore is called the preprojective algebra of $\Gamma$. (For the proper generalisation to the case of a species, we refer to [11]. We also should note the slight deviation of the preprojective algebra from the model algebra defined in [M], which reduces to the algebra $A^r$ given in [I, II] in the case of the $r$-subspace situation. Namely, here the constant paths have square zero, whereas they are idempotents in $\Pi$. Now, in $\Pi$ the sum of the constant paths is the identity element. In order also to have an identity element, Gelfand and Ponomarev add to the direct sum of all preprojective modules an additional one-dimensional space $k\epsilon$. There is a change of definition proposed in [S], using the constant paths as idempotents as in $\Pi$, but adding again an additional identity element.) Since $\Pi$ is the direct sum of the preprojective representations of $(\Gamma, \Lambda)$, it follows that $\Pi$ is finite dimensional if and only if $\Gamma$ is of the form $A_n$, $D_n$, $E_6$, $E_7$, or $E_8$. In [12], the tame cases $\widetilde{A}_n$, $\widetilde{D}_n$, $\widetilde{E}_6$, $\widetilde{E}_7$ and $\widetilde{E}_8$ have been characterized by the fact that the Gelfand–Kirillov dimension of $\Pi$ is 1, whereas it is $\infty$ for the wild cases.

Let us return to the special case of the $r$ subspace graph (*), with $r \geqslant 4$. The description above gives that the preprojective component $\mathscr{P}$ is of the form

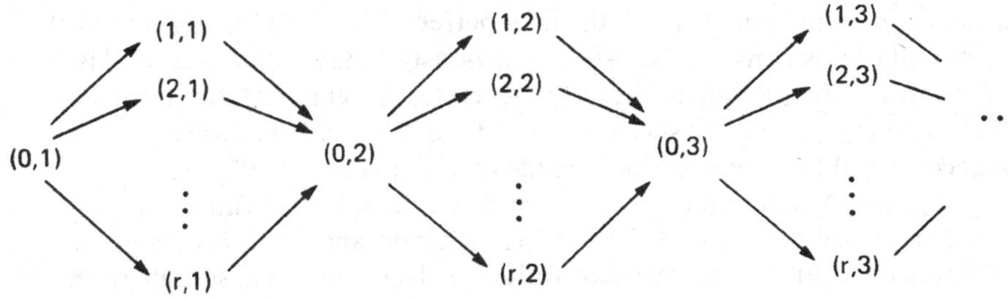


If we denote the arrows in the following way:

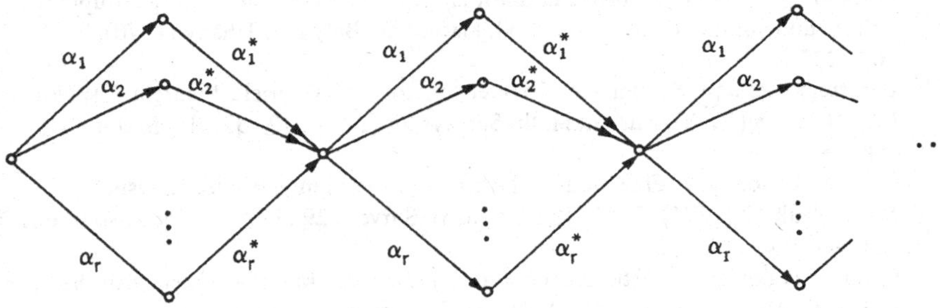

then the mesh relations are as follows: $\alpha_i \, \alpha_i^* = 0$ for all $i$, and $\sum_i \alpha_i^* \, \alpha_i = 0$.

Thus, if we want to determine the total space of the representation labelled $(t, l)$, we have to calculate $\mathrm{Hom}((0, 1), (t, l))$ inside the category $\langle\rangle\mathscr{P}$, and this amounts to the calculation of all possible paths from $(0, 1)$ to $(t, l)$, taking this as the basis of a vector space and factoring out the mesh relations. However, taking from the beginning into account the relations $\alpha_i \, \alpha_i^* = 0$, we just as well may work with the vector space generated by the set $A_t(r, l)$ and factoring out the remaining mesh relations. This shows that we obtain as total space the vector space $V_{t,l}$. Similarly, the $r$ different subspaces of the representation labelled $(t, l)$ are given by the various $\mathrm{Hom}((j, 1), (t, l))$, $1 \leqslant j \leqslant r$, again calculated in $\langle\rangle\mathscr{P}$, and therefore coincide with the subspaces $\rho_{t,l}(e_j)$. In this way, we obtain directly the description of the preprojective representations of $D^r$ given by Gelfand and Ponomarev (and a direct proof of Proposition 8.2 in [F]).

Finally, let us note in which way the preprojective component of $D^r$ determines the lattice $B^+$ of perfect elements belonging to the upper cubicles. For any perfect element $a$, we have denoted by $N(a)$ the set of indecomposable representation $\rho$ satisfying $\rho(a) = 0$. We claim that for $a \in B^+$, the set $N(a)$ is a finite, predecessor closed subset of $\mathscr{P}$ (an element $x$ is said to be a predecessor of $y$ in case there is an oriented path $x \to \ldots \to y$. For the proof, we first note that clearly $N(a) \cap \mathscr{P}$ is predecessor closed, since for indecomposable representations $\rho, \rho'$ with $\mathrm{Hom}(\rho, \rho') \neq 0$, and $a$ perfect, $\rho' \in N(a)$ implies $\rho \in N(a)$. Since not all of $\mathscr{P}$ is contained in $N(a)$, it obviously follows that $N(a) \cap \mathscr{P}$ is finite. However, any complete slice of $\mathscr{P}$ generates all representations outside of $\mathscr{P}$, thus taking a complete slice of $\mathscr{P}$ outside of $N(a) \cap \mathscr{P}$, we easily see that no indecomposable representation outside of $\mathscr{P}$ can belong to $N(a)$, thus $N(a) \subseteq \mathscr{P}$.) Thus $N$ determines a map from $B^+$ to the set of all finite, predecessor closed subsets of $\mathscr{P}$. This map is bijective and order-reversing, thus $B^+$ is anti-isomorphic to the lattice of finite, predecessor closed subsets of $\mathscr{P}$.

## References


[Q] Gelfand, Ponomarev: Problems in linear algebra and classification of quadruples in a finite dimensional vector space. Coll. Math. Soc. Bolyai 5, Tihany (1970), 163–237.

[BGP] Bernstein, Gelfand, Ponomarev: Coxeter functors and Gabriel's theorem. Uspekhi Mat. Nauk **28** (1973), Russian Math. Surveys **28** (1973), 17–32, also in this volume.

[F] Gelfand, Ponomarev: Free modular lattices and their representations. Uspekhi Math. Nauk **29** (1974), 3–58. Russian Math. Surveys **29** (1974), 1–56, also in this volume.

[I] Gelfand, Ponomarev: Lattices, representations and algebras connected with them. I. Uspekhi Math. Nauk **31** (1976), 71–88. Russian Math. Surveys **31** (1976), 67–85, also in this volume.

[II] Gelfand, Ponomarev: Lattices, representations and algebras connected with them. II. Uspechi Math. Nauk **32** (1977), 85–106. Russian Math. Surveys **32** (1977), 91–114, also in this volume.

[M] Gelfand, Ponomarev: Model algebras and representations of graphs. Funkc. Anal. i Pril. **13**.3 (1979), 1–12. Funct. Anal. Appl. **13** (1979), 157–166.

[R] Rojter: Gelfand–Ponomarev algebra of a quiver. Abstract, 2nd ICRA (Ottawa 1979).

[S] Gelfand, Ponomarev: Representations of graphs. Perfect subrepresentations. Funkc. Anal. i Pril. **14**.3 (1980), 14–31. Funct. Anal. Appl. **14** (1980), 177–190.

[1] Auslander, Platzek, Reiten: Coxeter functors without diagrams. Trans. Amer. Math. Soc. **250** (1979), 1–46.

[2] Auslander, Reiten: Representation theory of artin algebras III, IV, V. Comm. Algebra **3** (1975), 239–294; **5** (1977), 443–518; **5** (1977), 519–554.

[3] Auslander, Smalø: Preprojective modules over artin algebras. J. Algebra (to appear).

[4] Bautista: Irreducible maps and the radical of a category. Preprint.

[5] Brenner, Butler: The equivalence of certain functors occurring in the representation theory of artin algebras and species. J. London Math. Soc. **14** (1976), 183–187.

[6] Day, Herrmann, Wille: On modular lattices with four generators. Algebra Universalis **3** (1972), 317–323.

[7] Dedeking: Über die von drei Moduln erzeugte Dualgruppe. Math. Ann. **53** (1900), 371–403.

[8] Dlab: Structure des treillis linéaires libres. Seminaire Dubreil. Springer LNM **795** (1980), 10–34.

[9] Dlab, Ringel: Indecomposable representations of graphs and algebras. Mem. Amer. Math. Soc. **173** (1976).

[10] Dlab, Ringel: Perfect elements in the free modular lattices. Math. Ann. **247** (1980), 95–100.

[11] Dlab, Ringel: The preprojective algebra of a modulated graph. Springer LNM **832** (1980), 216–131.

[12] Dlab, Ringel: Eigenvalues of Coxeter transformations and the Gelfand–Kirillov dimension of the preprojective algebra. Proc. Amer. Math. Soc. (to appear).

[13] Donovan, Freislich: The representation theory of finite graphs and associated algebras. Carleton Lecture Notes **5** (1973).

[14] Freese: Free modular lattices. Trans. AMS **261** (1980), 81–91.

[15] Gabriel: Unzerlegbare Darstellungen I. Manuscripta Math. **6** (1972), 71–103.

[16] Gabriel: Indecomposable representations II. Symposia Math. Inst. Naz. Alta Mat. **11** (1973), 81–104.

[17] Gabriel: Auslander–Reiten sequences and representation finite algebras. Springer LNM **831** (1980), 1–71.

[18] Happel, Ringel: Tilted algebras. Trans. Amer. Math. Soc. (to appear).

[19] Herrmann: Rahmen und erzeugende Quadrupeln in modularen Verbänden. To appear in Algebra Universalis.

[20] Hutchinson: Embedding and unsolvability theorems for modular lattices. Algebra Universalis 7 (1977), 47–84.

[21] Kac: Infinite root systems, representations of graphs and invariant theory. Inv. Math. **56** (1980), 57–92. part II: preprint.

[22] Kac: Some remarks on representations of quivers and infinite root systems. Springer LNM **832** (1980), 311–327.

[23] Nazarova: Representations of quivers of infinite type. Izv. Akad. Nauk. SSSR. Ser. Mat. **37** (1973), 752–791.

[24] Riedtmann: Algebren, Darstellungsköcher, Überlagerungen und Zurück. Comment. Math. Helv. **55** (1980), 199–224.

[25] Ringel: Representations of $K$-species and bimodules. J. Algebra **41** (1976), 269–302.

[26] Ringel: Finite dimensional algebras of wild representation type. Math. Z. **161** (1978), 235–255.

[27] Ringel: The rational invariants of tame quivers. Inv. Math. **58** (1980), 217–239.

[28] Ringel: Reflection functors for hereditary algebras. J. London Math. Soc. (2) **21** (1980), 465–479.

[29] Ringel: Tame algebras. Springer LNM **831** (1980), 137–287.

# Table of contents for volumes I and III

## Volume I

### Part I. Survey lectures and articles of general content



### Part II. Banach spaces and normed rings

**Part III. Differential equations and mathematical physics**



**Essays in honour of Izrail M. Gelfand**

# Volume III

## Part I. Integral geometry



## Part II. Cohomology and characteristic classes

**Part III. Functional integration; probability; information theory**



**Part IV. Mathematics of computation; cybernetics; biology**

## Part V. General theory of hypergeometric functions

# Bibliography

The bibliography in these *Collected Papers* is a revised and updated version of the "List of I.M. Gelfand's Publications" published in the birthday addresses which can be found at the end of volume I. Hence the numbering of this Bibliography does not match that of these original 'Lists'.

The articles and monographs are listed in chronological order. The numbers in the right-hand column indicate where an article can be found in these Collected Papers, for example I.II.1 means volume I, part II, article 1.

From 1933–1947 the *Doklady* were published both in Russian and in a foreign language edition entitled *Comptes Rendus (Doklady) de l'Académie de l'URSS*.

Articles marked with an asterisk (*) were originally published in Russian and translated especially for this publication.

References to the reviews published in Mathematical Reviews (MR) and Zentralblatt für Mathematik (Zbl.) have been given as far as could be ascertained.


**1936**

1. Sur un lemme de la théorie des espaces linéaires. Izv. Nauchno-Issled. Inst. Mat. Khar'kov Univ., Ser. 4, **13** (1936) 35–40. Zbl. **14**:162 I.II.1

**1937**

2. Zur Theorie abstrakter Funktionen. Dokl. Akad. Nauk SSSR **17** (1937) 243–245. Zbl. **18**:71
3. Operatoren und abstrakte Funktionen. Dokl. Akad. Nauk SSSR **17** (1937) 245–248. Zbl. **18**:72

**1938**

4. Abstrakte Funktionen und lineare Operatoren. Mat. Sb., Nov. Ser. **4** (46) (1938) 235–284. Zbl. **20**:367 I.II.2

**1939**

5. (with A.N. Kolmogorov) On rings of continuous functions. Dokl. Akad. Nauk SSSR **22** (1939) 11–15. Zbl. **21**:411
6. On normed rings. Dokl. Akad. Nauk SSSR **23** (1939) 430–432. Zbl. **21**:294 I.II.4

7. To the theory of normed rings. II. On absolutely convergent trigonometrical series and integrals. Dokl. Akad. Nauk SSSR **25** (1939) 570–572. Zbl. **22**:357 I.II.5

8. To the theory of normed rings. III. On the ring of almost periodic functions. Dokl. Akad. Nauk SSSR **25** (1939) 573–574. Zbl. **22**:357 I.II.6

9. On one-parametrical groups of operators in a normed space. Dokl. Akad. Nauk SSSR **25** (1939) 713–718. Zbl. **22**:358 I.II.3

**1940**

10. (with D.A. Rajkov) On the theory of characters of commutative topological groups. Dokl. Akad. Nauk SSSR **28** (1940) 195–198. Zbl. **24**:120 I.II.7

**1941**

11. Normierte Ringe. Mat. Sb., Nov. Ser. **9** (51) (1941) 3–23. Zbl. **24**:320 I.II.8

12. (with G.E. Shilov) Über verschiedene Methoden der Einführung der Topologie in die Menge der maximalen Ideale eines normierten Ringes. Mat. Sb., Nov. Ser. **9** (51) (1941) 25–38. Zbl. **24**:321 I.II.9

13. Ideale und primäre Ideale in normierten Ringen. Mat. Sb., Nov. Ser. **9** (51) (1941) 41–48. Zbl. **24**:322 I.II.10

14. Zur Theorie der Charaktere der Abelschen topologischen Gruppen. Mat. Sb., Nov. Ser. **9** (51) (1941) 49–50. Zbl. **24**:323 I.II.11

15. Über absolut konvergente trigonometrische Reihen und Integrale. Mat. Sb., Nov. Ser. **9** (51) (1941) 51–66. Zbl. **24**:323 I.II.12

**1942**

16. (with M.A. Najmark) On the embedding of normed rings into the ring of operators in Hilbert space. Mat. Sb., Nov. Ser. **12** (54) (1942) 197–213. Zbl. **60**:270 I.II.13

17. (with D.A. Rajkov) Irreducible unitary representations of locally bicompact groups. Mat. Sb. 13 (55), 301–316 (1942) [Transl., II. Ser., Am. Math. Soc. **36** (1964) 1–15]. Zbl. **166**:401 II.I.1

**1943**

18. (with D.A. Rajkov) Irreducible unitary representations of locally compact groups. Dokl. Akad. Nauk SSSR **42** (1943) 199–201. Zbl. **61**:253

**1946**

19. (with M.A. Najmark) Unitary representations of the Lorentz group. J. Phys., Acad. Sci. USSR **10** (1946) 93–94. Zbl. **61**:253 II.II.1

20. (with M.A. Najmark) On unitary representations of the complex unimodular group. Dokl. Akad. Nauk SSSR **54** (1946) 195–198. Zbl. **29**:5 II.II.3

21. (with D.A. Rajkov and G.E. Shilov) Commutative normed rings. Usp. Mat. Nauk **1**, 2 (1946) 48–146 [Transl., II. Ser., Am. Math. Soc. 5 (1957) 115–220]. Zbl. **201**:457 I.II.14

22. (with M.A. Najmark) Unitary representations of semisimple Lie groups I. Mat. Sb., Nov. Ser. **21** (63) (1946) 405–434 [Transl., II. Ser., Am. Math. Soc. (reprinted) **9** (1962) 1–14]. Zbl. **38**:17

*23. (with M.A. Najmark) Complementary and degenerate series of representations of the complex unimodular group. Dokl. Akad. Nauk SSSR **58** (1946) 1577–1580. Zbl. **37**:304 II.II.5

**1947**

24. (with M.A. Najmark) Unitary representations of the group of linear transformations of the straight line. Dokl. Akad. Nauk SSSR **55** (1947) 567–570. Zbl. **29**:5 II.I.2

25. (with M.A. Najmark) The principal series of irreducible representations of the complex unimodular group. Dokl. Akad. Nauk SSSR **56** (1947) 3–4. Zbl. **29**:5 II.II.4

*26. (with M.A. Najmark) Unitary representations of the Lorentz group. Izv. Akad. Nauk SSSR, Ser. Mat. **11** (1947) 411–504. Zbl. **37**:153 II.II.2

**1948**

27. Lectures on linear algebra. Moscow-Leningrad: Gostekhizdat 1948 (in Russian). Zbl. **38**:156

28. (with M.A. Najmark) Normed rings with an involution and their representations. Izv. Akad. Nauk SSSR, Ser. Mat. **12**, 445–480 (1948) (English translation in: Commutative normed rings, I.M. Gelfand, D.A. Rajkov and G.E. Shilov, pp. 240–274. Chelsea 1964). Zbl. **31**:34 I.II.15

29. (with A.M. Yaglom) General relativistically invariant equations and infinite-dimensional representations of the Lorentz group. Dokl. Akad. Nauk SSSR **59** (1948) 655–659 (in Russian). Zbl. **37**:127

30. Integral equations. Article in the Great Soviet Encyclopedia (1948) (in Russian)

*31. (with A.M. Yaglom) General relativistically invariant equations and infinite-dimensional representations of the Lorentz group. Zh. Ehksp. Teor. Fiz. **18** (1948) 703–733 II.II.9

*32. (with M.A. Najmark) The trace in principal and complementary series representations of the complex unimodular group. Dokl. Akad. Nauk SSSR **61** (1948) 9–11. Zbl. **35**:299 II.II.6

33. (with A.M. Yaglom) Relativistically invariant equations corresponding to a definite charge and a definite energy. Dokl. Akad. Nauk SSSR **63** (1948) 371–374 (in Russian). Zbl. **31**:95

34. (with A.M. Yaglom) Pauli's theorem for general relativistically invariant equations. Zh. Ehksp. Teor. Fiz. **18** (1948) 1096–1104 (in Russian)

35. (with A.M. Yaglom) Charge conjugacy for general relativistically invariant equations. Zh. Ehksp. Teor. Fiz. **18** (1948) 1105–1111 (in Russian)

*36. (with M.A. Najmark) On the connection between representations of complex semisimple Lie groups and its maximal compact subgroup. Dokl. Akad. Nauk SSSR **63** (1948) 225–228. Zbl. **35**:15 II.II.7

*37. (with M.A. Najmark) An analogue of the Plancherel formula for the complex unimodular group. Dokl. Akad. Nauk SSSR **63** (1948) 609–612. Zbl. **38**:18 II.II.8

**1950**

*38. Center of the infinitesimal group ring. Mat. Sb., Nov. Ser. **26** (68) (1950) 103–112. Zbl. **35**:300 II.I.3

39. (with M.A. Najmark) The connexion between the unitary representations of the complex unimodular group and those of its unitary subgroup. Izv. Akad. Nauk SSSR, Ser. Mat. **14** (1950) 239–260 (in Russian). Zbl. **37**:15

40. Spherical functions on symmetric Riemannian spaces. Dokl. Akad. Nauk SSSR 70, (1950) 5–8 [Transl., II. Ser., Am. Math. Soc. **37** (1964) 39–43]. Zbl. **38**:274 II.I.4

*41. (with M.L. Tsetlin) Finite-dimensional representations of the group of unimodular matrices. Dokl. Akad. Nauk SSSR **71** (1950) 825–828. Zbl. **37**:153 II.VII.1

*42. (with M.L. Tsetlin) Finite-dimensional representations of the group of orthogonal matrices. Dokl. Akad. Nauk SSSR **71** (1950) 1017–1020. Zbl. **37**:153 II.VII.2

*43. Eigenfunction expansions for equations with periodic coefficients. Dokl. Akad. Nauk SSSR **73** (1950) 1117–1120. Zbl. **37**:345 I.III.1

44. (with M.A. Najmark) Unitary representations of classical groups. Tr. Mat. Inst. Steklova **36** (1950) 1–288 (in Russian). Zbl. **41**:362

(English transl. of the Introduction, Chap. 9 'Spherical functions' and Chap 18 'Transitivity classes for the set of pairs. Another way of describing representations of the complementary series') II.II.11

**1951**

45. (with S.V. Fomin) Unitary representations of Lie groups and geodesic flows on surfaces of constant negative curvature. Dokl. Akad. Nauk SSSR **76** (1951) 771–774 (in Russian). Zbl. **45**:388

46. Lectures on linear algebra, 2nd ed. Moscow-Leningrad: Gostekhizdat 1951 (English transl.: New York: Interscience 1961). Zbl. **98**:11

47. (with B.M. Levitan) On the determination of a differential equation from its spectral function. Izv. Akad. Nauk SSSR, Ser. Mat. **15** (1951) 309–361 [Transl., II. Ser., Am. Math. Soc. **1** (1955) 253–304]. Zbl. **44**:93 I.III.2

48. Remark on N.K. Bari's paper 'Biorthogonal system and bases in a Hilbert space'. Uch. Zap. Mosk. Gos. Univ., Ser. Mat. **140** (4) (1951) 224–225 (in Russian)

**1952**

49. (with Z.Ya. Shapiro) Representations of the group of rotations of 3-dimensional space and their applications. Usp. Mat. Nauk **7** (1) (1952) 3–117 (in Russian). Zbl. **49**:157

50. (with S.V. Fomin) Geodesic flows on manifolds of constant negative curvature. Usp. Mat. Nauk **7** (1) (1952) 118–137 [Transl., II. Ser., Am. Math. Soc. **1** (1955) 49–65]. Zbl. **66**:361 II.III.2

51. (with M.A. Najmark) Unitary representations of the unimodular group that contain an identity representation of the unitary subgroup. Tr. Mosk. Mat. O.-va **1** (1952) 423–473 (in Russian). Zbl. **49**:358

*52. (with M.I. Graev) Unitary representations of the real simple Lie groups. Dokl. Akad. Nauk SSSR **86** (1952) 461–463. Zbl. **49**:358 II.II.12

53. On the spectrum of non-selfadjoint differential operators. Usp. Mat. Nauk **7** (6) (1952) 183–184 (in Russian). Zbl. **48**:96

**1953**

*54. (with B.M. Levitan) On a simple identity for eigenvalues of second order differential operator. Dokl. Akad. Nauk SSSR **88** (1953) 593–596. Zbl. **53**:60 I.III.3

*55. (with M.I. Graev) Unitary representations of real unimodular groups (Principal non-degenerate series). Izv. Akad. Nauk SSSR, Ser. Mat. **17** (1952) 189–249. Zbl. **52**:341 II.II.13

*56. (with M.I. Graev) On a general method for decomposition of the regular representation of a Lie group into irreducible representations. Dokl. Akad. Nauk SSSR **92** (1952) 221–224. Zbl. **53**:15 II.II.14

57. (with M.I. Graev) The analogue of Plancherel's theorem for real unimodular groups. Dokl. Akad. Nauk SSSR **92** (1952) 461–464 (in Russian). Zbl. **53**:15

58. (with G.E. Shilov) Fourier transforms of rapidly increasing functions and questions of the uniqueness of the solution of Cauchy's problem. Usp. Mat. Nauk **8** (6) (1952) 3–54 [Transl., II. Ser., Am. Math. Soc. **5** (1957) 221–274]. Zbl. **52**:116

**1954**

*59. (with R.A. Minlos) Solution of quantum field equations. Dokl. Akad. Nauk SSSR **97** (1954) 209–212. Zbl. **58**:232 I.III.4

**1955**

60. (with V.B. Lidskij) On the structure of the regions of stability of linear canonical systems of differential equations with periodic coefficients. Usp. Mat. Nauk **10** (1) (1955) 3–40 [Transl., II. Ser., Am. Math. Soc. **8** (1958) 143–181]. Zbl. **64**:89 I.III.5

*61. Generalized random processes. Dokl. Akad. Nauk SSSR **100** (1955) 853–856. Zbl. **68**:112 III.III.1

62. (with M.I. Graev) The traces of unitary representations of the real unimodular group. Dokl. Akad. Nauk SSSR **100** (1955) 1037–1040 (in Russian). Zbl. **64**:111

63. (with M.I. Graev) The analogue of Plancherel's formula for the classical groups. Tr. Mosk. Mat. O.-va **4** (1955) 375–404 [Transl., II. Ser., Am. Math. Soc. **9** (1958) 123–154. Zbl. **66**:20

64. (with Z.Ya. Shapiro) Homogeneous functions and their applications. Usp. Mat. Nauk **10** (3) (1955) 3–70 [Transl., II. Ser., Am. Math. Soc. **8** (1958) 21–85]. Zbl. **65**:101

65. (with G.E. Shilov) On a new method in theorems concerning the uniqueness of the solution of Cauchy's problem for systems of linear partial differential equations. Dokl. Akad. Nauk SSSR **102** (1955) 1065–1068 (in Russian). Zbl. **67**:72

*66. (with A.G. Kostyuchenko) Eigenfunction expansions for differential and other operators. Dokl. Akad. Nauk SSSR **103** (1955) 349–352. Zbl. **65**:104 I.III.6

**1956**

67. (with A.M. Yaglom) Integration in functional spaces and its applications in quantum physics. Usp. Mat. Nauk **11** (1) (1956) 77–114 [J. Math. Phys. **1** (1960) 48–69]. Zbl. **92**:451 III.III.2

*68. On identities for eigenvalues of a second order differential operator. Usp. Mat. Nauk **11** (1) (1956) 191–198. Zbl. **70**:83 I.III.7

69. (with F.A. Berezin) Some remarks on the theory of spherical functions on symmetric Riemannian manifolds. Tr. Mosk. Mat. O.-va **5** (1956) 311–351 [Transl., II. Ser., Am. Math. Soc. **21** (1962) 193–238]. Zbl. **72**:17 II.III.1

70. (with G.E. Shilov) Quelques applications de la théorie des fonctions généralisées. J. Math. Pures et Appl., IX. Ser. **35** (1956) 383–413. Zbl. **75**:285

*71. (with A.N. Kolmogorov and A.M. Yaglom) To the general definition of the amount of information. Dokl. Akad. Nauk SSSR **111** (1956) 745–748 (German transl. in: Arbeiten zur Informationstheorie II, Mathematische Forschungsberichte.

Berlin: VEB Deutscher Verlag der Wissenschaften 1958). Zbl. **71**:345 III.III.3

72. On some problems of functional analysis. Usp. Mat. Nauk **11** (6) (1956) 3–12 [Transl., II. Ser., Am. Math. Soc. **16** (1960) 315–324]. Zbl. **100**:321 I.I.2
73. (with F.A. Berezin, M.I. Graev, and M.A. Najmark) Group representations. Usp. Mat. Nauk **11** (6) (1956) 13–40 [Transl., II. Ser., Am.Math. Soc. **16** (1960) 325–353]. Zbl. **74**:103
74. (with N.N. Chentsov) On the numerical evaluation of continuous integrals. Zh. Ehksp. Teor. Fiz. **31** (1956) 1106–1107 (in Russian)
75. (with M.L. Tsetlin) On quantities with anomalous parity. Zh. Ehksp. Teor. Fiz. **31** (1956) 1107–1109 (in Russian)

**1957**

76. (with A.M. Yaglom) Calculation of the amount of information about a random function contained in another such function. Usp. Mat. Nauk **12** (1) (1957) 3–52. [Transl., II. Ser., Am. Math. Soc. **12** (1959) 199–246] III.III.4
77. On the subrings of the ring of continuous functions. Usp. Mat. Nauk **12** (1) (1957) 247–251 [Transl., II. Ser., Am. Math. Soc. **16** (1957) 477–479]. Zbl. **100**:322
78. Some aspects of functional analysis and algebra. Proc. Int. Congr. Math. 1954, Amsterdam **1** (1957) 253–276. Zbl. **79**:326 I.I.1

**1958**

79. (with G.E. Shilov) Generalized functions 1. Properties and operations. Moscow: Fizmatgiz 1958 (English transl.: New York: Academic Press 1964; German transl.: Berlin: VEB Deutscher Verlag der Wissenschaften 1960; French transl.: Paris: Dunod 1962). Zbl. **91**:111
80. (with G.E. Shilov) Generalized functions 2. Spaces of fundamental and generalized functions. Moscow: Fizmatgiz 1958 (English transl.: New York: Academic Press 1968; German transl.: Berlin: VEB Deutscher Verlag der Wissenschaften 1960; French transl.: Paris: Dunod 1964). Zbl. **91**:111
81. (with G.E. Shilov) Generalized functions 3. Theory of differential equations. Moscow: Fizmatgiz 1958 (English transl.: New York: Academic Press 1967; German transl.: Berlin: VEB Deutscher Verlag der Wissenschaften 1964; French transl.: Paris: Dunod 1964). Zbl. **91**:111
82. (with K.I. Babenko) Some observations on hyperbolic systems. Nauchn. Dokl. Vyssh. Shk. **1** (1958) 12–18 (in Russian). Zbl. **144**:139

83. (with R.A. Minlos and Z.Ya. Shapiro) Representations of the rotation and Lorentz groups. Moscow: Fizmatgiz 1958 (English transl.: London: Pergamon Press 1963). Zbl. **108**:220
84. (with R.A. Minlos and A.M. Yaglom) Path integrals. Proceedings of the Third All-Union Mathematics Congress **3** (1958) 521–531 (in Russian)
85. (with F.A. Berezin, M.I. Graev and M.A. Najmark) Representations of Lie groups. Proceedings of the Third All-Union Mathematics Congress **3** (1958) 246–254 (in Russian). Zbl. **97**:109
86. (with A.N. Kolmogorov and A.M. Yaglom) The amount of information and entropy for continuous distributions. Proceedings of the Third All-Union Mathematics Congress **3** (1958) 300–320 (in Russian). Zbl. **92**:340
87. (with I.G. Petrovskij and G.E. Shilov) The theory of systems of partial differential equations. Proceedings of the Third All-Union Mathematics Congress **3** (1958) 65–72 (in Russian). Zbl. **107**:74
88. (with S.I. Braginskij and R.P. Fedorenko) The theory of the compression and pulsation of a plasma column under a powerful pulse discharge. Fiz. Plazmy Probl. Upr. Termoyad. Reakts. **4** (1958) 201–222 (in Russian)
89. (with N.N. Chentsov and A.S. Frolov) The computation of continuous integrals by the Monte Carlo method. Izv. Vyssh. Uchebn. Zaved., Mat. **5** (6) (1958) 32–45 (in Russian). Zbl. **139**:323

**1959**

90. Some problems in the theory of quasilinear equations. Usp. Mat. Nauk **14** (2) (1959) 87–158 [Transl., II. Ser., Am. Math. Soc. **29** (1963) 295–381]. Zbl. **96**:66 I.III.8
91. (with I.I. Piatetski-Shapiro) Theory of representations and theory of automorphic functions. Usp. Mat. Nauk **14** (2) (1959) 171–194 [Transl., II. Ser., Am. Math. Soc. **26** (1963) 173–200]. Zbl. **121**:306
92. Some questions of analysis and differential equations. Usp. Mat. Nauk **14** (3) (1959) 3–19 [Transl., II. Ser., Am. Math. Soc. **26** (1963) 201–219]. Zbl. **91**:88 I.I.3
93. (with M.I. Graev) Geometry of homogeneous spaces, representations of groups in homogeneous spaces and related questions of integral geometry. Tr. Mosk. Mat. O.-va **88** (1959) 321–390 [Transl., II. Ser., Am. Math. Soc. **37** (1964) 351–429]. Zbl. **136**:434 II.III.3
94. (with M.I. Graev) The decomposition into irreducible components of representations of the Lorentz group in the spaces

of functions defined on symmetric spaces. Dokl. Akad. Nauk SSSR **127** (1959) 250–253 (in Russian). Zbl. **99**:321

*95. (with I.I. Piatetski-Shapiro) On a theorem of Poincaré. Dokl. Akad. Nauk SSSR **127** (3) (1959) 490–493. Zbl. **107**:171 I.III.9

*96. (with M.I. Graev) On the structure of the ring of rapidly decreasing functions on a Lie group. Dokl. Akad. Nauk SSSR **124** (1959) 19–21. Zbl. **103**:336 II.II.10

**1960**

97. (with N.N. Chentsov, S.M. Fejnberg, and A.S. Frolov) On the application of the method of random tests (the Monte Carlo method) for the solution of the kinetic equation. Proceedings of the Second International Conference on the Peaceful Use of Atomic Energy. (Geneva, 1958), **2** (1960) 628–683 III.IV.1

98. (with Sya Do-Shin) On positive definite distributions. Usp. Mat. Nauk **15** (1) (1960) 185–190 (in Russian). Zbl. **97**:314

99. Integral geometry and its relation to the theory of group representations. Usp. Mat. Nauk **15** (2) (1960) 155–164 [Russ. Math. Surv. **15** (2) (1960) 143–151]. Zbl. **119**:177 I.I.4

100. (with M.I. Graev) Fourier transforms of rapidly decreasing functions on complex semisimple groups. Dokl. Akad. Nauk SSSR **131** (1960) 496–499 (in Russian). Zbl. **103**:337

101. (with M.L. Tsetlin) On continuous models of control systems. Dokl. Akad. Nauk SSSR **131** (1960) 1242–1245 (in Russian)

102. On elliptic equations. Usp. Mat. Nauk **15** (3) (1960) 121–132 [Russ. Math. Surv. **15** (1960) 113–123]. Zbl. **95**:78 I.I.5

103. On a paper by K. Hoffmann and I.M. Singer. Usp. Mat. Nauk **15** (3) (1960) 239–240 (in Russian). Zbl. **154**:386

104. (with D.A. Rajkov and G.E. Shilov) Commutative normed rings. Moscow: Fizmatgiz 1960 (English transl.: New York: Chelsea 1964; German transl.: Berlin: VEB Deutscher Verlag der Wissenschaften 1964; French transl.: Paris: Gauthier-Villars 1964). Zbl. **134**:321

105. (with M.I. Graev) Integrals over hyperplanes of fundamental and generalized functions. Dokl. Akad. Nauk SSSR **135** (1960) 1307–1310 [Sov. Math., Dokl. **1** (1960) 1369–1372]. Zbl. **108**:296

**1961**

106. (with N.Ya. Vilenkin) Generalized functions 4. Applications of harmonic analysis. Moscow: Fizmatgiz 1961 (English transl.: New York: Academic Press 1964; German transl.: Berlin: VEB Deutscher Verlag der Wissenschaften 1960; French transl.: Paris: Dunod 1967). Zbl. **136**:112

107. (with S.V. Fomin) Calculus of Variations. Moscow: Fizmatgiz (1961) (English transl.: Englewood Cliffs, N.J.: Prentice Hall 1963). Zbl. **127**:54

108. (with V.A. Borovikov, A.F. Grashin, and I.Ya. Pomeranchuk) Phase-shift analysis of *pp*-scattering at 95 *Mev*. Zh. Ehksp. Teor. Fiz. **40** (1961) 1106–111 [Soviet Physics **13** (4) (1961) 780–784]

109. (with A.F. Grashin and L.N. Ivanova) Phase-shift analysis of *pp*-scattering at an energy of 150 Mev. Zh. Ehksp. Teor. Fiz. **40** (5) (1961) 1338–1342 (in Russian)

110. (with V.S. Gurfinkel and M.L. Tsetlin) Some considerations on the tactics of making movements. Dokl. Akad. Nauk SSSR **139** (1961) 1250–1253 (in Russian)

111. (with M.I. Graev et al.) Magnetic surfaces of the 3-path helical magnetic field excited by a crimped field. Zh. Tekh. Fiz. **31** (1961) 1164–1169 (in Russian)

112. (with M.I. Graev) Integral transformations connected with straight line complexes in a complex affine space. Dokl. Akad. Nauk SSSR **138** (1961) 1266–1269 [Sov. Math., Dokl. **2** (1961) 809–812]. Zbl. **109**:151 III.I.1

113. (with M.L. Tsetlin) The principle of non-local search in automatic optimization systems. Dokl. Akad. Nauk SSSR **137** (1961) 295–298 [Sov. Phys., Dokl. **6** (1961) 192–194]

**1962**

114. (with M.L. Tsetlin) Some methods of control for complex systems. Usp. Mat. Nauk **17** (1) (1962) 3–25 [Russ. Math. Surv. **17** (1) (1962) 95–117]. Zbl. **107**:299 III.IV.3

115. (with A.I. Morozov, N.M. Zueva et al.) An example of the theoretical determination of a magnetic field that does not have magnetic surfaces. Dokl. Akad. Nauk SSSR **143** (1962) 81–83 (in Russian)

116. (with M.I. Graev) An application of the horysphere method to the spectral analysis of functions in real and imaginary Lobachevskii space. Tr. Mosk. Mat. O.-va **11** (1962) 243–308 (in Russian). Zbl. **176**:443

117. (with M.I. Graev and N.Ya. Vilenkin) Generalized functions 5. Integral geometry and representation theory. Moscow: Fizmatgiz 1962 (English transl.: New York: Academic Press 1966; French transl.: Paris: Dunod 1970). Zbl. **115**:167

118. (with M.I. Graev) Categories of group representations and the problem of classifying irreducible representations. Dokl. Akad. Nauk SSSR **146** (1962) 757–760 [Sov. Math., Dokl. **3** (1962) 1378–1381] II.IV.1

119. (with I.I. Piatetski-Shapiro) Unitary representations in homogeneous spaces with discrete stationary groups. Dokl. Akad. Nauk SSSR **147** (1962) 17–20 [Sov. Math., Dokl. **3** (1962) 1528–1531]. Zbl. **119**:270 II.III.4

120. (with I.I. Patetski-Shapiro) Unitary representations in a space $G/\Gamma$, where $G$ is a group of $n$-by-$n$ real matrices and $\Gamma$ is a subgroup of integer matrices. Dokl. Akad. Nauk SSSR **147** (1962) 275–278 [Sov. Math., Dokl. **3** (1962) 1574–1577]. Zbl. **119**:271 II.III.5

121. (with M.I. Graev) Construction of irreducible representations of simple algebraic groups over a finite field. Dokl. Akad. Nauk SSSR **147** (1962) 529–532 [Sov. Math., Dokl. **3** (1962) 1646–1649]. Zbl. **119**:269 II.IV.2

122. (with V.S. Gurfinkel and M.L. Tsetlin) On the techniques of control of complex systems and their relation to physiology. Symposium 'Biological aspects of cybernetics', pp. 66–73. Moscow: Publ. Akad. Nauk SSSR (1962) [Transl., II. Ser., Am. Math. Soc. **111** (1978) 213–219]

123. (with O.V. Lokytsievskij) On difference schemes for the solution of the equation of thermal conductivity. The 'double sweep' method for the solution of difference equations. Appendices I and II to 'Theory of difference schemes. An introduction.' by S.K. Godunov and V.S. Ryaben'kij. Moscow: Fizmatgiz 1962. Zbl. **106**:319. (Amsterdam: North-Holland 1964, pp. 232–262) III.IV.2

**1963**

124. (with L.M. Chailakhyan and S.A. Kovalev) Intracellular irritation of the various compartments of a frog's heart. Dokl. Akad. Nauk SSSR **148** (1963) 973–976 (in Russian)

125. (with M.I. Graev, A.I. Morozov et al.) On the structure of a toroidal magnetic field that does not have magnetic surfaces. Dokl. Akad. Nauk SSSR **148** (1963) 1286–1289 (in Russian)

126. (with M.I. Graev) Irreducible unitary representations of the group of unimodular second-order matrices with elements from a locally compact field. Dokl. Akad. Nauk SSSR **149** (1963) 499–502 [Sov. Math., Dokl. **4** (1963) 397–400]. Zbl. **119**:270 II.IV.5

127. (with I.I. Piatetski-Shapiro) Automorphic functions and the theory of representations. Tr. Mosk. Mat. O.-va **12** (1963) 389–412 [Trans. Mosc. Math. Soc. **12** (1965)]. Zbl. **136**:73

128. Automorphic functions and the theory of representations. Proc. Int. Congr. Math. Stockholm (1962) 74–85. Zbl. **138**:71 I.I.6

129. (with M.I. Graev) Plancherel's formula for the groups of the unimodular second-order matrices with elements in a locally compact field. Dokl. Akad. Nauk SSSR **151** (1963) 262–264 [Sov. Math., Dokl. **4** (1963) 397–400]. Zbl. **204**:141 II.IV.6

130. (with M.I. Graev) Representations of a group of second order matrices with elements from a locally compact field and special functions on locally compact fields. Usp. Mat. Nauk **18** (4) (1963) 29–99 [Russ. Math. Surv. **18** (4) (1963) 29–100]. Zbl. **166**:402

131. (with A.Ya. Fridenstein) On the possible mechanism of change of immunological tolerance. Usp. Sov. Biol. **55** (1963) 428–429 (in Russian)

132. (with V.S. Gurfinkel, Ya.M. Kots, M.L. Shik, and M.L. Tsetlin) On the synchronization of motor units and some related ideas. Biofizika **8** (1963) 475–488 (in Russian)

133. (with Yu.G. Fedorov and I.I. Piatetski-Shapiro) Determination of crystal structure by the method of nonlocal search. Dokl. Akad. Nauk SSSR **152** (1963) 1045–1048 [Sov. Math. Dokl. **4** (1963) 1487–1490] III.IV.4

*134. (with Yu.G. Fedorov, R.A. Kayushina, and B.K. Vainstein) Determination of crystal structures by the method of the *R*-factor minimization. Dokl. Akad. Nauk SSSR **153** (1963) 93–96 III.IV.5

135. (with I.I. Piatetski-Shapiro and M.L. Tsetlin) On certain classes of games and automata games. Dokl. Akad. Nauk SSSR **152** (1963) 845–848. [Transl., II. Ser., Am. Math. Soc. **87** (1970) 275–280]. MR **28**:1068. Zbl. **137**:143 III.IV.6

136. (with M.I. Graev) The structure of the ring of finite functions on the group of second-order unimodular matrices with elements from a disconnected locally compact field. Dokl. Akad. Nauk SSSR **153** (1963) 512–515 [Sov. Math. Dokl. **4** (1963) 1679–1700]. MR **33**:4183. Zbl. **199**:200

137. (with G.E. Shilov) Categories of finite-dimensional spaces. Vestn. Mosk. Univ., Ser. I. **4** (1963) 27–48 (in Russian). MR **28**:1223. Zbl. **161**:27

**1964**

138. (with V.I. Guelstein, A.G. Malenkov, and Yu.M. Vasil'ev) Characteristics of cell complexes of ascitic mouse hepatoma 22. Dokl. Akad. Nauk SSSR **156** (1964) 168–170 (in Russian)

139. (with M.I. Graev and I.I. Piatetski-Shapiro) Representations of adèle groups. Dokl. Akad. Nauk SSSR **156** (1964) 487–490. [Sov. Math., Dokl. **5** (1964) 657–661]. MR **29**:2237. Zbl. **133**:294

140. (with V.S. Gurfinkel) Investigation of recognition activity. Biofizika **9** (1964) 710–717 (in Russian)

141. (with V.I. Bryzgalov, V.S. Gurfinkel, and M.L. Tsetlin) Homogeneous automata games and their simulation on digital

computers. Avtom. Telemekh. **25** (1964) 1572–1580 (in Russian). MR **30**:1897. Zbl. **141**:339

142. (with E.G. Glagoleva and A.A. Kirillov) The coordinate method. Moscow: Nauka 1964 (English transl.: New York: Gordon & Breach 1967; German transl.: Leipzig: Teubner 1968; Czech. Transl.: Bratislava: ALFA 1976)

**1965**

143. (with M.I. Graev) Finite-dimensional irreducible representations of the unitary and the full linear groups, and related special functions. Izv. Akad. Nauk SSSR, Ser. Mat. **29** (1965) 1329–1356 [Transl., II. Ser., Am. Math. Soc. **64** (1965) 116–146]. MR **34**:1450. Zbl. **139**:307 II.VII.3
144. (with V.I. Guelstein, A.G. Malenkov, and Yu.M. Vasil'ev) Cell complexes in ascitic hepatomata of mice and rats, in the collection "Cell differentiation and induction mechanisms". Moscow: Nauka (1965) 220–232 (in Russian)
145. (with E.G. Glagoleva and E.E. Schnol) Functions and their graphs. Moscow: Nauka 1965 (English transl.: New York: Gordon & Breach 1967; German transl.: Leipzig: Teubner 1971). Zbl. **129**:267

**1966**

146. (with M.I. Graev and I.I. Piatetski-Shapiro) Theory of representations and automorphic functions. Moskau: Nauka 1966. (English transl.: Philadelphia London Toronto: Saunders 1969). MR **36**:3725. Zbl. **138**:72
147. (with A.A. Kirillov) Fields associated with enveloping algebras of Lie algebras. Dokl. Akad. Nauk SSSR **167** (1966) 503–505 [Sov. Math., Dokl. **7** (1966) 407–409]. Zbl. **149**:29 II.VI.1
148. (avec A.A. Kirillov) Sur les corps liés aux algèbres enveloppantes des algèbres de Lie., Publ. Math., Inst. Hautes Etud. Sci. **31** (1966) 509–523. MR **33**:7731. Zbl. **144**:21 II.VI.2
149. (with V.I. Guelstein, A.G. Malenkov, and Yu.M. Vasil'ev) Local interactions of cells in cell complexes of ascitic hepatoma 22. Dokl. Akad. Nauk SSSR **167** (1966) 437–439 (in Russian)
150. (with V.I. Guelstein, A.G. Malenkov, and Yu.M. Vasil'ev) Interrelationships of contacting cells in the cell complexes of mouse ascites hepatoma. Int. J. Cancer **1** (1966) 451–462 III.IV.12
151. (with M.I. Graev and E.Ya. Shapiro) Integral geometry on a manifold of $k$-dimensional planes. Dokl. Akad. Nauk SSSR **168** (1966) 1236–1238 [Sov. Math., Dokl. **7** (1966) 801–804]. Zbl. **168**:201
152. (with L.V. Erofeeva, and Yu.M. Vasil'ev) The behaviour of fibroblasts of a cell culture on removal of part of the mono-

layer. Dokl. Akad. Nauk SSSR **171** (1966) 721–724 (in Russian)

153. (with M.L. Tsetlin) Mathematical simulation of mechanisms of the central nervous system. In the collection: Models of structure-functional organisation of certain biological systems, pp. 9–26. Moscow: Nauka 1966 (in Russian)
154. (with V.S. Gurfinkel, M.L. Tsetlin, and M.L. Shik) Some problems in the analysis of movement. In the collection: Models of structure-functional organisation of certain biological systems, pp. 264–276, Moscow: Nauka 1966 (In: Models of the Structural-Functional Organization of Certain Biological Systems, pp. 329–345, Cambridge, Massachusetts, London: MIT Press 1971) III.IV.7
155. (with Yu.G. Fedorov, S.L. Ginzburg, and E.B. Vul) The ravine method in problems of X-ray structure analysis, pp. 1–77, Moscow: Nauka 1966 (in Russian)
156. Lectures on linear algebra, pp. 1–280. Moscow: Nauka 1966 (in Russian). MR **34**:4274. Zbl. **158**:297

**1967**

157. (with M.I. Graev and Z.Ya. Shapiro) Integral geometry on *K*-dimensional planes. Funkts. Anal. Prilozh. **1** (1) (1967) 15–31 [Funct. Anal. Appl. **1** (1967) 14–27]. MR **35**:3620. Zbl. **164**:231 III.I.2
158. (with V.A. Ponomarev) Categories of Harish-Chandra models over the Lie algebra of the Lorentz group. Dokl. Akad. Nauk SSSR **176** (1967) 243–246 [Sov. Math., Dokl. **8** (1967) 1065–1068]. MR **36**:6552. Zbl. **241**:22025
159. (with V.A. Ponomarev) Classification of indecomposable infinitesimal representations of the Lorentz group. Dokl. Akad. Nauk SSSR **176** (1967) 502–505 [Sov. Math., Dokl. **8** (1967) 114–1117]. MR **36**:2739. Zbl. **246**:22013
160. (with D.B. Fuks) Cohomology of Lie groups with real coefficients. Dokl. Akad. Nauk SSSR **176** (1967) 24–27 [Sov. Math., Dokl. **8** (1967) 1031–1034]. MR **37**:2252a. Zbl. **169**:547
161. (with D.B. Fuks) Topology of noncompact Lie groups. Funkts. Anal. Prilozh. **1** (4) (1967) 33–45 [Funct. Anal. Appl. **1** (1967) 285–295]. MR **37**:2253. Zbl. **169**:547 III.II.2
162. (with M.I. Graev) Representations of the quaternion group over a disconnected locally compact continuous field. Dokl. Akad. Nauk SSSR **177** (1967) 17–20 [Sov. Math., Dokl. **8** (1967) 1346–1349]. MR **36**:2742. Zbl. **225**:22009
163. (with Yu.I. Arshavskij, M.B. Berkinblit, and V.S. Yakobson) Functional organization of afferent connections of Purkinje cells of the paramedian lobe of the cerebellum. Dokl. Akad. Nauk SSSR **177** (1967) 732–753 (in Russian)
164. (with D.B. Fuks) Topological invariants of non-compact Lie

groups connected with infinite-dimensional representations. Dokl. Akad. Nauk SSSR **177** (1967) 763–766 [Sov. Math., Dokl. **8** (1967) 1483–1486]. MR **37**:2252b. Zbl. **169**:548

165. (with V.S. Imshennik, L.G. Khazin, O.V. Lokytsievskij, V.S. Ryaben'kij, and N.M. Zueva) The theory of non-linear oscillation of electron plasma. Zh. Vychisl. Mat. Mat. Fiz. **7** (1967) 322–347 (in Russian). Zbl. **181**:575

166. (with M.I. Graev) Irreducible representations of the Lie algebras of the group $U(p, q)$. In the collection: High energy physics and the theory of elementary particles, pp. 216–226. Kiev: Naukova Dumka 1967 (in Russian). MR **37**:3814

**1968**

167. (with M.I. Graev) Complexes of $k$-dimensional planes in the space $\mathbf{C}^n$ and Plancherel's formula for the group $GL(n, \mathbf{C})$. Dokl. Akad. Nauk SSSR **179** (1968) 522–525 [Sov. Math., Dokl. **9** (1968) 394–398]. MR **37**:4764. Zbl. **198**:271 III.I.3

168. (with M.I. Graev) Complexes of straight lines in the space $\mathbf{C}^n$. Funkts. Anal. Prilozh. **2**(3) (1968) 39–52 [Funct. Anal. Appl. **2** (1968) 219–229]. MR **38**:6522. Zbl. **179**:509 III.I.4

169. (with V.A. Ponomarev) Indecomposable representations of the Lorentz group. Usp. Mat. Nauk **23** (2) 3–60 (1968) [Russ. Math. Surv. **23** (2) (1968) 1–58]. MR **38**:5325. Zbl. **236**:22012 II.VIII.1

170. (with Yu.M. Vasil'ev) Surface changes disturbing intracellular homeostasis as a factor inducing cell growth and division. Curr. Mod. Biol. **2** (1968) 43–55

171. (with A.A. Kirillov) On the structure of the field of quotients of the enveloping algebra of a semisimple Lie algebra. Dokl. Akad. Nauk SSSR **180** (1968) 775–777 [Sov. Math., Dokl. **9** (1968) 669–671]. MR **37**:5260. Zbl. **244**:17006 II.VI.3

172. (with D.B. Fuks) On classifying spaces for principal fiberings with Hausdorff bases. Dokl. Akad. Nauk SSSR **181** (1968) 515–518 [Sov. Math., Dokl. **9** (1968) 851–854]. MR **38**:716. Zbl. **181**:266 III.II.1

173. (with D.B. Fuks) The cohomologies of the Lie algebra of the vector fields in a circle. Funkts. Anal. Prilozh. **2** (4) (1968) 92–93 [Funct. Anal. Appl. **2** (1968) 342–343]. MR **39**:6348a. Zbl. **176**:115 III.II.3

174. (with Yu.M. Vasil'ev) Change of cellular surface – the basis of biological singularities of a tumor cell. Vestn. Akad. Med. Nauk SSSR **3** (1968) 45–49 (in Russian)

175. (with M.I. Graev) Representations of quaternion groups over locally compact and functional fields. Funkts. Anal. Prilozh. **2** (1) (1968) 20–35 [Funct. Anal. Appl. **2** (1968) 19–33]. MR **38**:4611. Zbl. **233**:20016 II.IV.3

176. (with L.V. Domnina, R.I. Rapoport, and Yu.E.M. Vasil'ev) Wound healing in cell cultures. Exp. Cell Res. **54** (1968) 83–93

177. (with A.B. Fel'dman, V.S. Gurfinkel, G.N. Orlovskij, F.V. Severin, and M.L. Shik) The control of certain types of movement. In the collection: Material of the international symposium IFAK on technical and biological problems of control (1968) (in Russian)

178. (with L.V. Erofeeva, O.Yu. Ivanova, I.L. Slavnaya, Yu.M. Vasil'ev, and A.A. Yaskovets) Factors controlling the proliferation of normal and tumour cells. In: Connective tissue in normal and pathological conditions, pp. 212–215. Novosibirsk: Nauka 1969 (in Russian)

**1969**

179. (with A.A. Kirillov) The structure of the Lie field connected with a split semisimple Lie algebra. Funkts. Anal. Prilozh. **3** (1) (1969) 7–26 [Funct. Anal. Appl. **3** (1969) 6–21]. MR **39**:2827. Zbl. **244**:17007 II.VI.4

180. (with M.I. Graev and Z.Ya. Shapiro) Differential forms and integral geometry. Funkts. Anal. Prilozh. **3** (2) (1969) 24–40 [Funct. Anal. Appl. **3** (1969) 101–114]. MR **39**:6232. Zbl. **191**:528 III.I.5

181. (with D.B. Fuks) Cohomology of the Lie algebra of vector fields on a manifold. Funkts. Anal. Prilozh. **3** (2) (1969) 87 [Funct. Anal. Appl. **3** (1969) 155]. MR **39**:6348 b. Zbl. **194**:246

182. (with D.B. Fuks) Cohomologies of the Lie algebra of tangential vector fields of a smooth manifold. Funkts. Anal. Prilozh. **3** (3) (1969) 32–52 [Funct. Anal. Appl. **3** (1969) 194–210]. MR **41**:1067. Zbl. **216**:203 III.II.4

183. (with A.S. Mishchenko) Quadratic forms over commutative group rings and the K-theory. Funkts. Anal. Prilozh. **3** (4) (1969) 28–33 [Funct. Anal. Appl. **3** (1969) 277–281]. MR **41**:9243. Zbl. **239**:55004 III.II.8

184. (with V.A. Ponomarev) Remarks on the classification of a pair of commuting linear transformations in a finite-dimensional space. Funkts. Anal. Prilozh. **3** (4) (1969) 81–82 [Funct. Anal. Appl. **3** (1969) 325–329]. MR **40**:7279. Zbl. **204**:453 II.VIII.2

185. (with Yu.I. Arshavskij, M.B. Berkinblit, and V.S. Yakobson) Two types of granular cell in the cortex of the cerebellum. Nejrofiziologiya **1** (1969) 167–176 (in Russian)

186. (with E.K. Fetisova, V.I. Guelstein, and Yu.M. Vasil'ev) Stimulation of synthesis of DNA in mouse embryo fibroblastlike cells in vitro by factors of different character. Dokl. Akad. Nauk SSSR **187** (1969) 913–915 (in Russian)

187. (with Sh.A. Guberman, M.L. Shik, and Yu.M. Vasil'ev) Interaction in biological systems. Priroda **6**, 13–21; **7**, 24–33 (1969) (in Russian)

188. (with N.M. Chebotareva, Sh.A. Guberman, M.L. Izvekova, E.I. Kandel, T.V. Lebedeva, D.K. Luhev, and I.F. Nikolaeva) Prognostic matematic alevelutiei ictusorilor hemorogice in scepul preciazavii indicatiiler tratementului chirurgical. K. Accidouteler vasculare cerebrale, pp. 44–45, Bucuresti (1969)

189. (with N.M. Chebotareva, Sh.A. Guberman, M.L. Izvekova, E.I. Kandel, T.V. Lebedeva, D.K. Lunev, and I.F. Nikolaeva) Computer prognosis of spontaneous intracerebral haemorrhage for the purpose of its surgical treatment. IV. Int. Congr. Neurosurg., Excerpta Med. **32** (1969)

**1970**

190. (with M.I. Graev and Z.Ya. Shapiro) Integral geometry in projective space. Funkts. Anal. Prilozh. **4** (1) (1970) 14–32 [Funct. Anal. Appl. **4** (1970) 12–28]. MR **43**:6856. Zbl. **199**: 255 III.I.6

191. (with M.I. Graev and Z.Ya. Shapiro) A problem of integral geometry connected with a pair of Grassmann manifolds. Dokl. Akad. Nauk SSSR **193** (1970) 259–262 [Sov. Math., Dokl. **11** (1970) 892–896]. MR **42**:3728. Zbl. **209**:267 III.I.7

192. (with M.I. Graev and V.A. Ponomarev) The classification of the linear representations of the group SL (2, **C**). Dokl. Akad. Nauk SSSR **194** (1970) 1002–1005 [Sov. Math., Dokl. **11** (1970) 1319–1323]. MR **43**:2162. Zbl. **229**:22024 II.VIII.3

193. (with D.B. Fuks) Cohomology of the Lie algebra of formal vector fields. Dokl. Akad. Nauk SSSR **190** (1970) 1267–1270 [Sov. Math., Dokl. **11** (1970) 268–271]. MR **44**:2247. Zbl. **264**:17005

194. (with D.B. Fuks) Cohomology of the Lie algebra of formal vector fields. Izv. Akad. Nauk SSSR, Ser. Mat. **34** (1970) 322–337 [Math. USSR, Izv. **34** (1970) 327–342]. MR **44**:1103. Zbl. **216**:203 III.II.5

195. (with D.B. Fuks) Cohomologies of Lie algebra of tangential vector fields II. Funkts. Anal. Prilozh. **4** (4) (1970) 23–31 [Funct. Anal. Appl. **4** (1970) 110–116]. MR **44**:2248. Zbl. **208**:514 III.II.6

196. (with D.B. Fuks) Cohomologies of Lie algebra of vector fields with nontrivial coefficients. Funkts. Anal. Prilozh. **4** (3) (1970) 10–25 [Funct. Anal. Appl. **4** (1970) 181–192]. MR **44**:7752. Zbl. **222**:58001 III.II.7

197. The cohomology of infinite dimensional Lie algebras; some questions of integral geometry. Int. Congr. Math., Nice 1970, **1** (1970) 95–111. Zbl. **239**:58004 I.I.7

198. (with I.N. Bernstein and S.I. Gelfand) Differential operators on the base affine space. Dokl. Akad. Nauk SSSR **195** (1970) 1255–1258 [Sov. Math., Dokl. **11** (1970) 1646–1649]. MR **43**:3402. Zbl. **217**:369

199. (with E.K. Fetisova, V.I. Guelstein, and J.M. Vasil'ev) Stimulation of DNA synthesis in cultures of mouse embryo fibroblastlike cells. J. Cell Physiol. **75** (1970) 305–313

200. (with Yu.I. Arshavskij, B.M. Berkinblit, and V.S. Yakobson) Organization of afferent connections of intercalary neurons in the paramedian lobe of the cerebellum of a cat. Dokl. Akad. Nauk SSSR **193** (1970) 250–253 (in Russian)

201. (with D.B. Fuks) Cycles representing cohomology classes of the Lie algebra of formal vector fields. Usp. Mat. Nauk **25** (5) (1970) 239–240. MR **45**:2737 (in Russian). Zbl. **216**:204

202. (with D.B. Fuks) Upper bounds for cohomology of infinite-dimensional Lie algebras. Funkts. Anal. Prilozh. **4** (4) (1970) 70–71 [Funct. Anal. Appl. **4** (1970) 323–324]. MR **44**:4792. Zbl. **224**:18013 III.II.9

203. (with N.M. Chebotareva, Sh.A. Guberman, M.L. Izvekova, E.I. Kandel', N.V. Lebedeva, D.K. Lunev, and I.F. Nikolaeva) Mathematical prognosis of the outcome of haemorrhages with the aim of determining evidence for surgical treatment. Zh. Nevropatol. Psikhiatr. **2** (1970) 177–181 (in Russian)

204. (with Yu.I. Arshavskij, M.B. Berkenblit, O.I. Fukson, and V.S. Yakobson) Features of the influence of the lateral reticular nucleus of medulla oblongata on the cortex of the cerebellum. Nejrofiziologiya **2** (1970) 581–586 (in Russian)

205. (with L.V. Domnina, O.Yu. Ivanova, S.G. Komm, L.V. Ol'shevskaya, and Yu.M. Vasil'ev) Effect of colcemid on the locomotory behaviour of fibroblasts. J. Embryol. Exp. Morphol. **24** (1970) 625–640

206. (with V.A. Ponomarev) Problems of linear algebra and classification of quadruples of subspaces in a finite-dimensional vector space. Colloq. Math. Soc. Janos Bolyai **5**. Hilbert space operators. Tihany, Hungary 1970 (1972). Zbl. **294**:15002

207. (with S.L.Ginzburg, G.V. Gurskaya, G.M. Lobanova, M.G. Nejgauz, and L.A. Novakovskaya) Crystal structure of paroxyacetophenon. Dokl. Akad. Nauk SSSR **195** (1970) 341–344 [Sov. Phys., Dokl. **15** (1970) 999–1002]

**1971**

208. (with V.A. Ponomarev) Quadruples of subspaces of a finite-dimensional vector space. Dokl. Akad. Nauk SSSR **197** (1971)

762–765 [Sov. Math., Dokl. **12** (1971) 535–539]. MR **44**:2762. Zbl. **294**:15001 II.VIII.4

209. (with I.N. Bernstein and S.I. Gelfand) Structure of representations generated by vectors of heighest weight. Funkts. Anal. Prilozh. **5** (1) (1971) 1–9 [Funct. Anal. Appl. **5** (1971) 1–8]. MR **45**:298. Zbl. **246**:17008 II.V.2

210. (with V.I. Guelstein and Yu.M. Vasil'ev) Initiation of DNA synthesis in cultures of mouse fibroblastlike cells under the action of substances that disturb the formation of microtubes. Dokl. Akad. Nauk SSSR **197** (1971) 1425–1428 (in Russian)

211. (with Yu.I. Arshavskij, M.B. Berkinblit, and O.I. Fukson) Organization of projections of somatic nerves in different regions of the cortex of the cerebellum of a cat. Nejrofiziologiya **3** (2) (1971) (in Russian)

212. (with Yu.I. Arshavskij, M.B. Berkinblit, I.A. Keder-Stepanova, E.M. Smelyanskaya, and V.S. Yakobson) Background activity of Purkinje cells in intact and deafferentized frontal lobes of the cortex of the cerebellum of a cat. Biofizika **16** (1971) 684–691 (in Russian)

213. (with Yu.I. Arshavskij, M.B. Berkinblit, O.I. Fukson, and V.S. Yakobson) Afferent connections and interaction of neurons of the cortex of the cerebellum. In the collection: Structural and functional organization of the cerebellum, pp. 40–47. Moscow: Nauka (1971) (in Russian)

214. (with Yu.I. Arshavskij, M.B. Berkinblit, O.I. Fukson, and V.S. Yakobson) The reticular afferent system of the cerebellum and its functional significance. Izv. Akad. Nauk SSSR, Ser. Biol. **3** (1971) 375–383 (in Russian)

215. (with L.B. Margolis, V.I. Samojlov, and Yu.M. Vasil'ev) A quantitative estimate of the form and orientation of cell nuclei in a culture. Ontogenez **2** (1971) 138–144 (in Russian)

216. (with T.A. Fajn, Sh.A. Guberman, G.G. Guelstein, and I.M. Rotvajn) An estimate of the pressure in the pulmonary artery from electro- and phonocardiographical data under a defect of the intraventricular partition. Kardiologiya **5** (1971) 84–87 (in Russian)

217. (with Yu.I. Arshavskij, M.B. Berkinblit, O.I. Fukson, and V.S. Yakobson) Functional role of the reticular afferent system of the cerebellum. Prepr. IPM Akad. Nauk SSSR (1971) (in Russian)

218. (with V.I. Guelstein and Yu.M. Vasil'ev) Initiation of DNA synthesis in cell cultures by colcemid. Proc. Natl. Acad. Sci. USA **68** (1971) 977–979 III.IV.13

219. (with V.Ya. Brodskij, L.V. Domnina, V.I. Guelstein, L.B. Klempner, T.L. Marshak, and Yu.M. Vasil'ev) The kinetics

of proliferation in cultures of mouse embryo fibroblastlike cells. Tsitologiya **13** (1971) 1362–1377 (in Russian)

220. (with D.A. Kazhdan) Representations of the group $GL(n, K)$, where $K$ is a local field. Prepr. **71**, IPM Akad. Nauk SSSR (1971) (English transl. in: Lie groups and their representations. Proc. Summer School in Group Representations. Bolyai Janos Math. Soc., Budapest 1971, pp. 95–118. New York: Halsted 1975). Zbl. **348**:22011

221. (with D.A. Kazhdan) Some questions of differential geometry and the computation of the cohomology of Lie algebras of vector fields. Dokl. Akad. Nauk SSSR **200** (1971) 269–272 [Sov. Math., Dokl. **12** (1971) 1367–1370]. MR **44**:4770. Zbl. **238**:58001

222. (with T.L. Fajn, Sh.A. Guberman, G.G. Guelstein, I.M. Rotvajn, V.A. Silin, and V.K. Sukhov) Recognition of the degree of pulmonary hypertonia under a defect of the intraventricular partition with the aid of the EVM. Krovoobrashchenie **6** (1971) (in Russian)

**1972**

223. (with L.V. Domnina, O.Yu. Ivanova, and Yu.M. Vasil'ev) The action of metaphase inhibitors on the form and movement of interphase fibroblasts in a culture. Tsitologiya **14** (1972) 80–88 (in Russian)

224. (with Yu.I. Arshavskij, M.B. Berkinblit, O.I. Fukson, and V.S. Yakobson) Suppression of reactions of Purkinje cells under preceding activation of the reticulo-cerebellar path. Fiziol. Zh. SSSR Im. I.M. Sechenova **58** (1972) 208–214 (in Russian)

225. (with I.N. Bernstein and S.I. Gelfand) Differential operators on a cubic cone. Usp. Mat. Nauk **27** (1) (1972) 185–190 [Russ. Math. Surv. **27** (1) (1972) 169–174]. Zbl. **257**:58010 II.V.3

226. (with D.B. Fuks and D.A. Kazhdan) The actions of infinite-dimensional Lie algebras. Funkts. Anal. Prilozh. **6** (1) (1972) 10–15 [Funct. Anal. Appl. **6** (1972) 9–13]. MR **46**:922. Zbl. **267**:18023 III.II.10

227. (with I.N. Bernstein and S.I. Gelfand) Differential operators on the base affine space and a study of $\mathfrak{g}$-modules. Prepr. **77**, IPM Akad. Nauk SSSR (1972) (English transl. in: Lie groups and their representations. Proc. Summer School in Group Representations. Bolyai Janos Math. Soc., Budapest 1971, pp. 21–64. New York: Halsted 1975). Zbl. **338**:58019 II.V.1

228. (with L.V. Domnina, O.Yu. Ivanova, L.B. Margolis, L.V. Ol'shevskaya, Yu.A. Rovenskij, and Yu.M. Vasil'ev) Defective formation of the lamellar cytoplasm in neoplastic fibroblasts. Proc. Natl. Acad. Sci. USA **69** (1972) 248–252

229. (with Sh.A. Guberman, M.L. Izvekova, V.J. Kejlis-Borok, and E.Ya. Rantsman) Criteria of high seismicity, determined by pattern recognition. Proc. Final Symp. Upper Mantle Project **13** (1972)

230. (with Yu.I. Arshavskij, M.B. Berkinblit, O.I. Fukson, and G.N. Orlovskij) Activity of the neurons of the dorsal spinocerebellar tract under locomotion. Biofizika **17** (1972) 487–494 (in Russian)

231. (with Yu.A. Arshavskij, M.B. Berkinblit, O.I. Fukson, and G.N. Orlovskij) Activity of the neurons of the ventral spinocerebellar tract under locomotion. Biofizika **17** (1972) 883–896 (in Russian)

232. (with Yu.I. Arshavskij, M.B. Berkenblit, O.I. Fukson, and G.N. Orlovskij) Activity of the neurons of the ventral spinocerebellar tract under locomotion of cats with deafferentized hind legs. Biofizika **17** (1972) 1113–1119 (in Russian)

233. (with Sh.A. Guberman, M.L. Izvekova, V.I. Kejlis-Borok, E.Ya. Rantsman) Criteria of high seismicity. Dokl. Akad. Nauk SSSR **202** (1972) 1317–1320 (in Russian)

234. (with Yu.I. Arshavskij, M.B. Berkenblit, O.I. Fukson, and G.N. Orlovskij) Recordings of neurones of the dorsal spinocerebellar tract during evoked locomotion. Brain Res. **43** (1972) 272–275 III.IV.8

235. (with Yu.I. Arshavskij, M.B. Berkenblit, O.I. Fukson, and G.N. Orlovskij) Origin of modulation in neurones of the ventral spinocerebellar tract during locomotion. Brain Res. **43** (1972) 276–279 III.IV.9

236. (with D.B. Fuks and D.I. Kalinin) Cohomology of the Lie algebra of Hamiltonian formal vector fields. Funkts. Anal. Prilozh. **6** (63) (1972) 25–29 [Funct. Anal. Appl. **6** (1972) 193–196]. MR **47**:1088. Zbl. **259**:57023 III.II.11

237. (with D.A. Kazhdan) On the representation of the group $GL(n, K)$, where $K$ is a local field. Funkts. Anal. Prilozh. **6** (4) (1972) 73–74 [Funct. Anal. Appl. **6** (1972) 315–317]. Zbl. **288**:22024 II.IV.7

238. (with Sh.A. Guberman, M.S. Kaletskaya, V.I. Kejlis-Borok, E.Ya. Rantsman, and M.P. Zhidkov) An attempt to carry over criteria of heigh seismicity from Central asia to Anatolia and adjoining regions. Dokl. Akad. Nauk SSSR **210** (1972) 327–330 (in Russian)

239. (with L.V. Domnina, V.I. Guelstein, and Yu.M. Vasil'ev) Regulation of the behaviour of connective tissue cells in multicell systems. In: Histophysiology of connective tissue, vol. **1**, pp. 31–36. Novosibirsk 1972

240. (with L.V. Domnina, O.Yu. Ivanova, S.G. Komm, L.V. Ol'shevskaya, and Yu.M. Vasil'ev) The action of metaphase inhibitors on the form and movement of fibroblasts in a culture. Tsitologiya **14** (1972) 80–88 (in Russian)

**1973**

241. (with Sh.A. Guberman, M.L. Izvekova, V.I. Kejlis-Borok, and E.Ya. Rantsman) Recognition of places of possible origin of powerful earthquakes (in Eastern Central Asia). Vychisl. Seismol. **6** (1973) (in Russian)
242. (with Yu.I. Arshavskij, M.V. Berkenblit, O.I. Fukson, and G.N. Orlovskij) Activity of neurons of the cuneo-cerebellar tract under locomotion. Biofizika **18** (1973) 126–131 (in Russian)
243. (with Yu.M. Vasil'ev) Interactions of normal and neoplastic fibroblasts with the substratum. Ciba Foundation Symposium on Cell Locomotion (1973) 312–331
244. (with I.N. Bernstein and V.A. Ponomarev) Coxeter functors and Gabriel's theorem. Usp. Mat. Nauk **28** (2) (1973) 19–33 [Russ. Math. Surv. **28** (2) (1973) 17–32]. Zbl. **269**:08001 II.VIII.5
245. (with I.N. Bernstein and S.I. Gelfand) Schubert cells and cohomology of flag spaces. Funkts. Anal. Prilozh. **7** (1) (1973) 64–65 [Funct. Anal. Appl. **7** (1973) 53–55]. MR **47**:6713. Zbl. **282**:20035
246. (with L.V. Domnina, O.Yu. Ivanova, L.B. Margolis, and Yu.M. Vasil'ev) Intracellular interaction in cultures of transformed fibroblasts of strain L and normal mouse fibroblasts.Tsitologiya **15** (1973) 1024–1028 (in Russian)
247. (with I.N. Bernstein and S.I. Gelfand) Schubert cells and cohomology of the spaces $G/P$. Usp. Mat. Nauk **28** (3) (1973) 3–26 [Russ. Math. Surv. **28** (3) (1973) 1–26]. Zbl. **289**:57024 II.V.4
248. (with M.I. Graev and A.M. Vershik) Representations of the group $SL(2, \mathbf{R})$, where $\mathbf{R}$ is a ring of functions. Usp. Mat. Nauk. **28** (5) (1973) 82–128 [Russ. Math. Surv. **28** (5) (1973) 87–132]. Zbl. **297**:22003 II.IX.1
249. (with Yu.M. Vasil'ev) Disturbance of morphogenetic reactions of cells under tumorous transformation. Vestn. Akad. Med. Nauk SSSR **4** (1973) 61–69 (in Russian)
250. (with V.I. Guelstein, O.Yu. Ivanova, L.B. Margolis, and Yu.M. Vasil'ev) Contact inhibition of movement in the cultures of transformed cells. Proc. Natl. Acad. Sci. USA **70** (1973) 2011–2014
251. (with L.V. Domnina, E.E. Krivitska, L.V. Ol'shevskaya, Yu.A. Rovenskij, and Yu.M. Vasil'ev) The structure of the lamellar cytoplasm of normal and tumorous fibroblasts.

Papers from a Soviet-French symposium. In: Ultrastructure of cancerous cells, pp. 49–71. Moscow: Nauka 1973 (in Russian)

252. (with L.V. Domnina, E.K. Fetisova, O.Yu. Pletyushkina, and Yu.M. Vasil'ev) Comparative study of density dependent inhibition of growth in the cultures of normal and neoplastic fibroblast-like cells. Abstracts 6th meeting of the European study group for cell proliferation, p. 15. Moscow: Nauka 1973

253. (with Yu.M. Vasil'ev) Factors inducing DNA synthesis and mitosis in normal and neoplastic cell culture. Abstracts 6th meeting of the European study group for cell proliferation, p. 61. Moscow: Nauka 1973

254. (with Sh.A. Guberman, V.I. Kejlis-Borok, E.Ya. Rantsman, I.M. Rotvajn, and M.I. Zhidkov) Determination of criteria of high seismism by means of recognition algorithms. Vestn. Mosk. Gos. Univ. **5** (1973) 78–83 (in Russian)

255. (with A.D. Bershadskij, L.V. Domnina, V.I. Guelstein, O.Yu. Ivanova, S.G. Komm, L.B. Margolis, and Yu.M. Vasil'ev) Interactions of normal and neoplastic cells with various surfaces. Neoplasma **20** (1973) 583–585

256. (with D.B. Fuks) PL Foliations. Funkts. Anal. Prilozh. **7** (4) (1973) 29–37 [Funct. Anal. Appl. **7** (1973) 278–284]. MR **49**:3958. Zbl. **294**:57016. III.II.12

257. (with Yu.I. Arshavskij, M.B. Berkinblit, O.I. Fukson, G.N. Orlovskij, and B.S. Yakobson) Some peculiarities of the organization of afferent links of the cerebellum. In: 4th International Biophysical Congress, Pushchino Symp. **3** (1973) 327–346 (in Russian)

258. (with Yu.I. Arshavskij, O.I. Fukson, and G.N. Orlovskij) Activity of the neurons of the cuneo-cerebellar tract for locomotion. Biofizika **18** (1973) 126–131 (in Russian)

**1974**

259. (with M.I. Graev and A.M. Vershik) Irreducible representations of the group $G^X$ and cohomologies. Funkts. Anal. Prilozh. **8** (2) (1974) 67–69 [Funct. Anal. Appl. **8** (1974) 151–153]. MR **50**:530. Zbl. **299**:22004. II.IX.3

260. (with B.L. Feigin and D.B. Fuks) Cohomologies of the Lie algebra of formal vector fields with coefficients in its adjoint space and variations of characteristic classes of foliations. Funkts. Anal. Prilozh. **8** (2) (1974) 13–29 [Funct. Anal. Appl. **8** (1974) 99–112]. MR **50**:8553. Zbl. **298**:57011. III.II.13

*261. (with I.N. Bernstein and S.I. Gelfand) A new model for representations of finite semisimple algebraic groups. Usp. Mat. Nauk **29** (3) (1974) 185–186. MR 53:5760. Zbl. **354**:20031 II.IV.4

262. (with Yu.N. Arshavskij, M.B. Berkinblit, A.M. Smelyanskij, and V.S. Yakobson) Background activity of Pourkynje cells of the paramedian part of the cortex of the cerebellum of a cat. Biofizika **19** (1974) 903–907 (in Russian)

263. (with Yu.I. Arshavskij, M.B. Berkinblit, O.I. Fukson, and G.N. Orlovskij) Differences in the working of spino-cerebral tracts in artificial irritation and locomotion. In: Mechanisms of the union of neurons in the nerve centre, pp. 99–105. Leningrad: Nauka 1974 (in Russian)

264. (with Sh.A. Guberman, M.S. Kaletska, V.I. Kejlis-Borok, E.Ya. Rantsman, I.M. Rotvajn, and M.P. Zhidkov) Recognition of places where strong earthquakes are possible. II. Four regions of Asia Minor and South-East Europe. Vychisl. Seismol. **7** (1974) 3–39 (in Russian)

265. (with Sh.A. Guberman, V.I. Kejlis-Borok, E.Ya. Rantsman, I.M. Rotvajn, and M.P. Zhidkov) Recognition of places where strong earthquakes are possible. III. The case when the boundaries of disjunctive nodes are not known. Vychisl. Seismol. **7** (1974) 41–62 (in Russian)

266. (with V.I. Guelstein, O.Yu. Ivanova, S.G. Komm, and L.B. Margolis, and Yu.M. Vasil'ev) The results of intercellular impacts in cultures of normal and transformed fibroblasts. Tsitologiya **16** (1974) 752–756 (in Russian)

267. (with D.B. Fuks) PL Foliations. II. Funkts. Anal. Prilozh. **8** (3) (1974) 7–11 [Funct. Anal. Appl. **8** (1974) 197–200]. MR **54**:6159. Zbl. **316**:57010 III.II.15

268. (with V.A. Ponomarev) Free modular lattices and their representations. Usp. Mat. Nauk **29** (6) (1974) 3–58 [Russ. Math. Surv. **29** (6) (1974) 1–56]. MR **53**:5393. Zbl. **314**:15003

269. (with O.Yu. Ivanova, L.B. Margolis, and Yu.M. Vasil'ev) Orientation of mitosis of fibroblasts is determined in the interphase. Proc. Natl. Acad. Sci. USA **71** (1974) 2032

270. (with Yu.I. Arshavskij, M.B. Berkinblit, O.I. Fukson, and G.N. Orlovskij) Peculiarities of information entering the cortex of the cerebellum via different afferent paths, structural and functional organization of the cerebellum, pp. 34–41. Kiev: Naukova Dumka 1974 (in Russian)

**1975**

271. (with L.V. Domnina, A.V. Lyubimov, Yu.M. Vasil'ev, and O.S. Zakharova) Contact inhibition of phagocytosis in epithelial sheets: alterations of cell surface properties induced by cell-cell contacts. Proc. Natl. Acad. Sci. USA **72** (1975) 719–722

272. (with A.P. Chern and Yu.M. Vasil'ev) Spreading of normal and transformed fibroblasts in dense cultures. Exp. Cell Res. **90** (1975) 317–327

273. (with A.M. Gabrielov and M.V. Losik) The combinatorial computation of characteristic classes. Funkts. Anal. Prilozh. **9** (1975) 54–55 [Funct. Anal. Appl. **9** (1975) 48–49]. MR **51**:1839. Zbl. **312**:57016

274. Quantitative evaluation of cell orientation in culture. J. Cell. Sci. **17** (1975) 1–10

275. (with M.I. Graev and A.M. Vershik) Representations of the group of diffeomorphisms connected with infinite configurations. Prepr. Inst. Appl. Math. **46** (1975) 1–62 (in Russian)

276. (with M.I. Graev and A.M. Vershik) The square roots of quasiregular representations of the group $SL(2, k)$. Funkts. Anal. Prilozh. **9** (2) (1975) 64–66 [Funct. Anal. Appl. **9** (1975) 146–148]. MR **51**:8338. Zbl. **398**:22010

277. (with I.N. Bernstein and S.I. Gelfand) Models of representations of compact Lie groups. Funkts. Anal. Prilozh. **9** (4) (1975) 61–62 [Funct. Anal. Appl. **9** (1975) 322–324]. MR **54**:2884. Zbl. **339**:22009

278. (with I.N. Bernstein and S.I. Gelfand) Models of representations of Lie groups. Proc. Petrovskij Semin. **2** (1976) 3–21. [Sel. Math. Sov. **1** (2) (1981) 121–142] Zbl. **499**:22004 II.IV.8

279. (with A.M. Gabrielov and M.V. Losik) Combinatorial computation of characteristic classes. Funkts. Anal. Prilozh. **9** (2) (1975) 12–28 [Funct. Anal. Appl. **9** (1975) 103–115]. MR **53**:14504a. Zbl. **312**:57016 III.II.16

280. (with A.M. Gabrielov and M.V. Losik) Combinatorial computation of characteristic classes. Funkts. Anal. Prilozh. **9** (3) (1975) 5–26 [Funct. Anal. Appl. **9** (1975) 186–202]. MR **53**:14504a. Zbl. **341**:57017 III.II.17

281. (with L.A. Dikij) Asymptotic behaviour of the resolvent of Sturm-Liouville equations and the algebra of the Korteweg-de Vries equations. Usp. Mat. Nauk **30** (5) (1975) 67–100 [Russ. Math. Surv. **30** (5) (1975) 77–113]. MR **58**:22746. Zbl. **461**:35072. I.III.12

282. (with I.S. Tint and Yu.M. Vasil'ev) Processes determining the changes of shape of a cell after its separation from the epigastrium. Tsitologiya **5** (1975) 633–638 (in Russian)

283. (with O.Yu. Pletyushkina and Yu.M. Vasil'ev) Neoplastic fibroblasts sensitive to growth inhibition by parent normal cells. Br. J. Cancer **31** (1975) 535–543

284. (with L.B. Margolis, V.I. Samojlov, and Yu.M. Vasil'ev) Methods of measuring the orientation of cells. Ontogenez **6** (1) (1975) 105–110 (in Russian)

285. (with E.K. Fetisoba, O.Yu. Pletyushkina, and Yu.M. Vasil'ev) Insensibility of dense cultures of transformed mice fibroblasts to the action of agents, stimulating the synthesis of DNA in cultures of normal cells. Tsitologiya **17** (1975) 442–446 (in Russian)

286. (with Yu.I. Arshavskij, G.N. Orlovskij, and G.A. Pavlova) The activity of neurons of the ventral spino-cerebral tract in "fictitious scratching". Biofizika **20** (1975) 748–749 (in Russian)

287. (with Yu.I. Arshavskij, G.N. Orlovskij, and G.A. Pavlova) Origin of the modulation of the activity of vestibular-spinal neurons in scratching. Biofizika **20** (1975) 946–947 (in Russian)

288. (with Yu.I. Arshavskij, M.B. Berkinblit, T.G. Delyagina, A.G. Fel'dman, O.I. Fukson, G.N. Orlovskij, and G.A. Pavlova) On the role of the cerebellum in regulating some rhythmic movements (locomotion, scratching). Summaries 12th meeting of the All-Union Physiological Society, pp. 15–16. Tbilisi 1975 (in Russian)

289. (with T.G. Delyagina, A.G. Fel'dman, and G.N. Orlovskij) On the role of the central program and afferent inflow in generation of scratching movements in the cat. Brain Res. **100** (1975) 297–313

290. (with M.I. Graev and A.M. Vershik) Representations of the group of diffeomorphisms. Usp. Mat. Nauk **30** (6) (1975) 3–50 [Russ. Math. Surv. **30** (6) (1975) 1–50]. MR **53**:3188. Zbl. **317**:58009 II.IX.4

291. (with Sh.A. Guberman, M.S. Kaletska, V.I. Kejlis-Borok, E.Ya. Rantsman, I.M. Rotvajn, and L.P. Zhidkov) Prognosis of a place where strong earthquakes occur, as a problem of recognition. In: Modelling of training and behaviour, pp. 18–25. Moscow: Nauka 1975 (in Russian)

**1976**

292. (with D.B. Fuks and A.M. Gabrielov) The Gauss-Bonnet theorem and the Atiyah-Patodi-Singer functionals for the characteristic classes of foliations. Topology **15** (1976) 165–188. MR **55**:4201. Zbl. **347**:57009 III.II.14

293. (with A.M. Gabrielov and M.V. Losik) A local combinatorial formula for the first class of Pontryagin. Funkts. Anal. Prilozh. **10** (1) (1976) 14–17 [Funct. Anal. Appl. **10** (1976) 12–15]. MR **53**:14504b. Zbl. **328**:57006 III.II.18

294. (with L.A. Dikij) A Lie algebra structure in a formal variational calculation. Funkts. Anal. Prilozh. **10** (1) (1976) 1–8

[Funct. Anal. Appl. **10** (1976) 16–22]. MR **57**:7670. Zbl. **347**:49023

295. (with I.N. Bernstein and S.I. Gelfand) Category of g-modules. Funkts. Anal. Prilozh. **10** (2) (1976) 1–8 [Funct. Anal. Appl. **10** (1976) 87–92]. MR **53**:10880. Zbl. **353**:18013 II.V.5

296. (with A.M. Gabrielov and M.V. Losik) Atiyah-Patodi-Singer functionals for characteristic functionals for tangent bundles. Funkts. Anal. Prilozh. **10** (2) (1976) 13–28 [Funct. Anal. Appl. **10** (1976) 95–107]. MR **54**:1245. Zbl. **344**:57008 III.II.19

297. (with L.A. Dikij) Fractional powers of operators and Hamiltonian systems. Funkts. Anal. Prilozh. **10** (4) (1976) 13–29 [Funct. Anal. Appl. **10** (1976) 259–273]. MR **55**:6484. Zbl. **346**:35085 I.III.10

298. (with Yu.I. Manin and M.A. Shubin) Poisson brackets and the kernel of the variational derivative in the formal calculus of variations. Funkts. Anal. Prilozh. **10** (4) (1976) 30–34 [Funct. Anal. Appl. **10** (1976) 274–278]. MR **55**:13486. Zbl. **395**:58005

299. (with V.A. Ponomarev) Lattices, representations, and algebras connected with them. I. Usp. Mat. Nauk **31** (5) (1976) 71–88 [Russ. Math. Surv. **31** (5) (1976) 67–85]. MR **58**:16779a. Zbl. **358**:06020

*300. (with M.V. Losik) Computing characteristic classes of combinatorial vector bundles. Prepr. Inst. Appl. Mat. **99** (1976) III.II.20

301. (with Sh.A. Guberman, V.I. Kejlis-Borok, L. Knopov, E. Press, E.Ya. Rantsman, I.M. Rotvajn, and A.M. Sadovskij) Conditions for the occurence of strong earthquakes (California and some other areas). Vychisl. Seismol. **9** (1976) 3–92 (in Russian)

302. (with Sh.A. Guberman, V.I. Kejlis-Borok, L. Knopov, F. Press, E.Ya. Rantsman, I.M. Rotvajn, and A.M. Sadovskij) Pattern recognition applied to earthquake epicenters in California. Phys. Earth Planet. Inter. **11** (1976) 277–283

303. (with M.A. Alekseevskaya, I.V. Martynov, and V.M. Sablin) First results of the prognostication of the effect of transmural (large focal) myocardial infarcts. Aktual'nye voprosy kardiologii, Otdelennye rezul'taty lecheniya elokachestvennykh opukholej, 19–24. Moscow: Nauka 1976 (in Russian)

304. (with E.E. Bragina and Yu.M. Vasil'ev) Formation of bundles of microfilaments during spreading of fibroblasts on the substratum. Exp. Cell Res. **97** (1976) 241–248

305. (with A.D. Bershadskij, V.I. Gelfand, V.I. Guelstein, and Yu.M. Vasil'ev) Serum dependence of expression of the transformed phenotype experiments with subline of mouse *L* fibro-

blasts adapted to growth in serum-free medium. Int. J. Cancer (1976) 84–92

306. (with O.Yu. Ivanova, L.B. Margolis, and Yu.M. Vasil'ev) Effect of colcemid on the spreading of fibroblasts in culture. Exp. Cell Res. **101** (1976) 207–219
307. (with L.V. Domnina, N.A. Dorfman, O.Yu. Pletyushkina, and Yu.M. Vasil'ev) Active cell edge and movements of concanavalin *A* receptors on the surface of epithelial and fibroblastic cells. Proc. Natl. Acad. Sci. USA **73** (1976) 4085–4089
308. (with N.M. Chebotareva, Sh.A. Guberman, M.L. Izvekova, E.I. Kandel, N.V. Lebedeva, D.K. Lunev, I.F. Nikolaeva, and E.V. Shmidt) A computer study of prognosis of cerebral haemorrhage for choosing optimal treatment. Eur. Congr. Neurosurg., pp. 71–72. Edinburgh (1976)
309. (with N.M. Chebotareva, Sh.A. Guberman, M.L. Izvekova, E.I. Kandel, and E.V. Shmidt) Prognostication of the results of surgical treatment of haemorrhaging lesions by means of a computer. Vopr. Nejrokhir. **3** (1976) 20–23 (in Russian)
310. (with O.Yu. Ivanova, L.B. Margolis, and Yu.M. Vasil'ev) Effect of colcemid on spreading of fibroblast in culture. Exp. Cell Res. **101** (1976) 207–219
311. (with Yu.M. Vasil'ev) Effects of colcemid on morphogenetic processes and locomotion of fibroblasts. Cell Motility **3** (1976) 279–304

**1977**

312. (with M.A. Alekseevskaya, Sh.A. Guberman, I.V. Martynov, I.M. Rotvajn, and V.M. Sablin) Prognostication of the result of a large focal myocardial infarct by means of learning program. Kardiologiya **17** (1977) 26–31 (in Russian)
313. (with M.A. Alekseevskaya, L.D. Golovnya, Sh.A. Gubermann, M.L. Izvekova, and A.L. Syrkin) Prognostication of the result of a myocardial infarct by means of the program "Cortex-3". Kardiologiya **17** (6) (1977) 13–23 (in Russian)
314. (with M.Yu. Melikova, S.G. Gindikin, and M.L. Izvekova) Prognostication of the healing of duodenal ulcers. Aktual. Vopr. Gastroenterol. **10** (1977) 42–51 (in Russian)
315. (with M.A. Alekseevskaya, L.D. Golovnya, M.L. Izvekova, I.V. Martynov, V.M. Sablin, and A.L. Syrkin) A general guide-line or a general method for creating one (On ways of applying mathematical methods in medicine). Summaries of lectures at the All-Union Conf. on the theory and practice of automatic electrocardiological and clinical investigations, pp. 3–5. Kaunas (1977) (in Russian)

316. (with M.A. Alekseevskaya, E.S. Klyushin, A.V. Nedostup and A.L. Syrkin) On the methodology of creating a formalized description of the patient (using the example of prognostication of remote results of electro-impulsive treatment of a constant form of flickering arrhythmy). Summaries of lectures at the All-Union Conf. on the theory and practice of automatic electrocardiological and clinical investigations, pp. 5–8. Kaunas (1977) (in Russian)

317. (with L.V. Domnina, O.Yu. Pletyushkina, and Yu.M. Vasil'ev) Effects of antitubilins on redistribution of cross-linked receptors on the surface of fibroblasts and epithelial cells. Proc. Natl. Acad. Sci. USA **74** (1977) 2865–2868

318. (with O.Yu. Ivanova, S.G. Komm, L.B. Margolis, and Yu.M. Vasil'ev) The influence of colcemid on the polarization of cells on narrow strips of the adhesive substratum. Tsitologiya **19** (1977) 357–360 (in Russian)

319. (with A.D. Bershadskij, V.I. Gelfand, and Yu.M. Vasil'ev) The influence of serum on the development of cell transformation. Vestn. Akad. Mech. Nauk **3** (1977) 55–59 (in Russian)

320. (with A.D. Bershadskij, A.D. Lyubimov, Yu.A. Rovenskij, and Yu.M. Vasil'ev) Contact interaction of cell surfaces. Lectures at the Soviet-Italian symposium "Tissue proteinases in normal and pathological state", Moscow 22–27 September 1977 (in Russian)

321. (with Yu.M. Vasil'ev) Mechanisms of morphogenesis in cell cultures. Int. Rev. Cytol. **50** (1977) 159–274 III.IV.14

322. (with Yu.I. Arshavskij, M.B. Berkinblit, and V.S. Yakobson) A formula for the analysis of histograms of intercellular intervals of Pourkine cells. Biofizika **22** (1977) (in Russian)

323. (with M.A. Alekseevskaya, A.M.Gabrielov, A.D. Gvishiani, and E.Ya. Rantsman) Morphological division of mountainous countries by formalized criteria. Vychisl. Seismol. **10** (1977) 33–79 (in Russian)

324. (with M.A. Alekseevskaya, A.M. Gabrielov, A.D. Gvishiani, and E.Ya. Rantsman) Formal morphostructural zoning at mountain territories. J. Geophys. **43** (1977) 227–235

325. (with L.A. Dikij) The Resolvent and Hamiltonian systems. Funkts. Anal. Prilozh. **11** (2) (1977) 11–27 [Funct. Anal. Appl. **11** (1977) 93–105]. MR **56**:1359. Zbl. **357**:58005. I.III.13

326. (with S.G. Gindikin) Nonlocal inversion formulas in real integral geometry. Funkts. Anal. Prilozh. **11** (3) (1977) 12–19 [Funct. Anal. Appl. **11** (1977) 173–179]. MR **56**:16265. Zbl. **385**:53056 III.I.8

327. (with V.A. Ponomarev) Representation lattices and the algebras connected with them. Usp. Mat. Nauk **32** (1) (1977)

85–107 [Russ. Math. Surv. **32** (1) (1977) 91–114]. Zbl. **358**:06021

328. (with S.G. Gindikin) Complex manifolds whose skeletons are semisimple real Lie groups, and analytic discrete series of representations. Funkts. Anal. Prilozh. **11** (4) (1977) 20–28 [Funct. Anal. Appl. **11** (1977) 258–265]. MR **58**:11230. Zbl. **444**:22006. II.IV.9
329. (with M.I. Graev and A.M. Vershik) Representations of the group of smooth mappings from a manifold *X* into a compact Lie group. Dokl. Akad. Nauk SSSR **323** (1977) 745–748 [Sov. Math., Dokl. **18** (1977) 118–121]. MR **55**:10602. Zbl. **393**:22012
330. (with M.I. Graev and A.M. Vershik) Representations of the group of smooth mappings of a manifold *X* into a compact Lie group. Compos. Math. **35** (1977) 299–334. MR **58**:28257. Zbl. **368**:53034 II.IX.2
331. (with Yu.I. Arshavskij, G.N. Orlovskij, and G.A. Pavlova) Activity of neurons of the lateral reticular nucleus in scratching. Biofizika **22** (1) (1977) (in Russian)

**1978**

332. (with L.V. Domnina, O.Yu. Pletyushkina, and Yu.M. Vasil'ev) Influence of agents, destroying microtubules, on the distribution of receptors of the surface of cultured cells. Tsitologiya **20** (1978) 796–801 (in Russian)
333. (with Yu.I. Arshavskij, G.N. Orlovskij, and G.A. Pavlova) Messages conveyed by spino-cerebellar pathways during scratching in the cat. 1. Activity of neurons of lateral reticular nucleus. Brain Res. **151** (1978) 479–491
334. (with Yu.I. Arshavskij, G.N. Orlovskij, and G.A. Pavlova) Messages conveyed by spino-cerebellar pathways during scratching in the cat. 2. Activity of neurons of the ventral spino-cerebellar tract. Brain Res. **151** (1978) 493–506
335. (with M.B. Berkinblit, T.G. Delyagina, A.G. Fel'dman, and G.N. Orlovskij) Generation of scratching. I. Activity of spinal interneurons during scratching. J. Neurophysiol. **41** (1978) 1040–1057 III.IV.10
336. (with M.B. Berkinblit, T.G. Delyagina, A.G. Fel'dman, and G.N. Orlovskij) Generation of scratching. 2. Non-regular regimes of generation. J. Neurophysiol. **41** (1978) 1058–1069
337. (avec I.N. Bernstein et S.I. Gelfand) Structure locale de la catégorie des modules de Harish-Chandra I. C.R. Acad. Sci., Paris, Ser. A **286** (1978) 435–437. MR **58**:16966, Zbl. **416**:22018 II.V.6

338. (avec I.N. Bernstein et S.I. Gelfand) Structure locale de la catégorie des modules de Harish-Chandra II. C.R. Acad. Sci., Paris, Ser. A **286** (1978) 495–497. MR **81e**:22026. Zbl. **431**:22013 II.V.7

339. (with Yu.M. Vasil'ev) Mechanisms of non-adhesiveness of endothelial and epithelial surfaces. Nature **275** (1978) 710–711

340. (with Yu.M. Vasil'ev, A.D. Bershadskij, V.A. Rozenblat, and I.S. Tint) Microtubular system in cultured mouse epithelial cells. Cell Biol. Int. Rep. **2** (1978) 345–351

341. (with L.A. Dikij) Variational calculus and the Korteweg-de Vries equations. Partial differential equations. Proc. All-Union Conf., Moscow 1976, dedic. I.G. Petrovskij pp. 81–83 (1978) (in Russian). Zbl. **498**:35074

342. (with L.A. Dikij) Calculus of jets and non-linear Hamiltonian systems. Funkts. Anal. Prilozh. **12** (2) (1978) 8–23 [Funct. Anal. Appl. **12** (1978) 81–94]. MR **58**:18561, Zbl. **388**:58009.

*343. (with L.A. Dikij) A family of Hamiltonian structures related to nonlinear integrable differential equations. Prepr. Inst. Appl. Mat. **136** (1978). MR **81**:58027 I.III.11

344. (with I.N. Bernstein and S.I. Gelfand) Algebraic bundles over $P^n$ and problems of linear algebra. Funkts. Anal. Prilozh. **12** (3) (1978) 66–68 [Funct. Anal. Appl. **12** (1978) 212–214]. MR **80c**:14010a, Zbl. **402**:14005. II.V.8

345. (with B.L. Feigin and D.B. Fuks) Cohomology of infinite-dimensional Lie algebras and Laplace operators. Funkts. Anal. Prilozh. **12** (4) (1978) 1–5 [Funct. Anal. Appl. **12** (1978) 243–247]. MR **80i**:58050, Zbl. **396**:17008. III.II.21

346. (with S.G. Gindikin, M.L. Izvekova, and M.Yu. Melikova) On one approach to formalization of the diagnostic attitude of a doctor (using the prognosis of the healing of duodenal ulcers). Summaries of lectures at the All-Union Conf. biological and medical cybernetics, vol. 2, pp. 27–31 (1978) (in Russian)

347. (with S.G. Gindikin, M.L. Izvekova, and M.Yu. Melikova) On some questions of mathematical diagnostics: examples of problems from gastroenterology. Summaries of lectures at the Second All-Union Cong. on Gastroenterology, vol. 2, pp. 57–58, Moscow-Leningrad: Nauka (1978) (in Russian)

348. (with Yu.I. Manin) Dualità, Enciclopedia Einaudi, Vol. 5, pp. 126–178. Einaudi, Torino (1978)

349. (with Yu.I. Arshavskij, G.N. Orlovskij, and G.A. Pavlova) Messages conveyed by descending tract during scratching in the cat. I. Activity of vestibulospinal neurons. Brain Res. **159** (1978) 88–110

350. (with M.A. Alekseevskaya, E.S. Klyushin, A.V. Nedostup, and A.L. Syrkin) A new approach to the problem of the choice of information and formalization of the description of the patient for the solution of medical problems on a computer. Prepr. Inst. Appl. Math. **144** (1978) (in Russian)

**1979**

351. (with M.A. Alekseevskaya, E.S. Klyushin, and A.V. Nedostup) The gathering of medical information for processing on a computer (manual). Inst. Appl. Math. (1979) (in Russian)
352. (with S.G. Gindikin, M.L. Izvekova, and M.Yu. Melikova) One method of formalizing the diagnostic attitude of a doctor (examples of prognosis of the healing of a duodenal ulcer). Prepr. Acad. Sci. USSR Sci. Committee on the complex problem "Cybernetics" (1979) (in Russian)
353. (with L.B. Margolis, E.J. Vasil'eva, and Yu.M. Vasil'ev) Upper surfaces of epithelial sheets and of fluid lipid films are non-adhesive for platelets. Proc. Natl. Acad. Sci. USA **76** (1979) 2303–2305
354. (with A.D. Bershadskij, V.I. Gelfand, V.A. Rozenblat, I.S. Tint, and Yu.M. Vasil'ev) Morphology of microtubular systems in epithelial cells in the kidney of a mouse. Ontogenez **10** (1979) 231–235 (in Russian)
355. (with L.A. Dikij) Integrable nonlinear equations and the Liouville theorem. Funkts. Anal. Prilozh. **13** (1) (1979) 8–20 [Funct. Anal. Appl. **13** (1979) 6–15]. MR **80i**:58027. Zbl. **423**:34003. I.III.14
356. (with S.G. Gindikin and Z.Ya. Shapiro) A local problem of integral geometry in a space of curves. Funkts. Anal. Prilozh. **13** (2) (1979) 11–31 [Funct. Anal. Appl. **13** (1980) 87–102]. MR **80k**:53100. Zbl. **415**:53046. III.I.9
357. (with V.A. Ponomarev) Model algebras and representations of graphs. Funkts. Anal. Prilozh. **13** (3) (1979) 1–12. [Funct. Anal. Appl. **13** (1980) 157–166]. Zbl. **437**:16020 II.VIII.6
358. (with I.Ya. Dorfman) Hamiltonian operators and algebraic structures related to them. Funkts. Anal. Prilozh. **13** (4) (1979) 13–31 [Funct. Anal. Appl. **13** (1980) 248–262]. MR **81c**:58035. Zbl. **428**:58009 I.III.15
359. (with S.G. Gindikin and M.I. Graev) A problem of integral geometry in $RP^n$, connected with the integration of differential forms. Funkts. Anal. Prilozh. **13** (4) (1979) 64–67 [Funct. Anal. Appl. **13** (1980) 288–290]. MR **83a**:43006. Zbl. **423**:58001 III.I.10
360. (with Yu.I. Arshavskij) The role of the brain stem and cerebellum in the regulation of rhythmic movements. Proc. 13 Congr. Pavlov Physiol. Soc., vol. 1, pp. 474–475. Leningrad: Nauka 1979 (in Russian)

361. (with Yu.I. Arshavskij, M.B. Berkinblit, and G.N. Orlovskij) The significance of signals passing along the various spinocerebral pathways for the work of locomotive centres of the brain stem in scratching. In: Nejronnye mekhanizmy integrativnoj deyatel'nosti mozzhechka, pp. 88–91. Erevan 1979 (in Russian)

362. (with Yu.I. Arshavskij, M.B. Berkinblit, G.N. Orlovskij) Signalling mechanisms of the scratching reflex and their interaction with the cerebellum. In: Nejronnye mekhanizmy integrativnoj deyatel'nosti mozzhechka, pp. 92–96. Erevan 1979 (in Russian)

**1980**

363. (with M.Ya. Ratner, B.I. Rozenfeld, and V.V. Serov) The problem of classifying glomerule kidneys. Prepr. Acad. Sci. USSR Sci. Committee on the complex problem "Cybernetics" (1980) (in Russian)

364. (with V.A. Ponomarev) Representations of graphs. Perfect sub-representations. Funkts. Anal. Prilozh. **14** (3) (1980) 14–31 [Funct. Anal. Appl. **14** (1980) 177–190]. MR **83c**:05113. Zbl. **453**:05027

365. (with I.Ya. Dorfman) The Schouten bracket and Hamiltonian operators. Funkts. Anal. Prilozh. **14** (3) (1980) 71–74 [Funct. Anal. Appl. **14** (1980) 223–226]. MR **82e**:58039. Zbl. **444**:58010. I.III.16

366. (with M.I. Graev) Admissible $n$-dimensional complexes of curves in $\mathbf{R}^n$. Funkts. Anal. Prilozh. **14** (4) (1980) 36–44 [Funct. Anal. Appl. **14** (1980) 274–281]. MR **82**:53013. Zbl. **454**:53042.

367. (with S.G. Gindikin and M.I. Graev) Integral geometry for one-dimensional fibrations of general form over $\mathbf{RP}^n$. Prepr. Inst. Appl. Math. **60** (1980) 1–24 (in Russian). MR **82g**:53081

368. (with S.G. Gindikin and M.I. Graev) Integral geometry in affine and projective spaces. Itogi Nauki Tekh., Ser. Sovrem. Probl. Mat. 16, 53–226, Moscow: VINITI (1980) [J. Sov. Math. **18** (1980) 39–167]. MR **82m**:43017. Zbl. **465**:52005. III.I.11

369. (with I.V. Cherednik and S.A. Chernyakevich) A formalized differentiated description of the motor of the stomach and the duodenal intestine. Prepr. Acad. Sci. USSR Sci. Committee on the complex problem "Cybernetics", pp. 1–34 (1980) (in Russian)

370. (with G.G. Guelstein, I.P. Lukashevich, and M.A. Shifrin) Study of the correlation between electrocardiograph and coronary data. Prepr. Acad. Sci. USSR Sci. Committee on the complex problem "Cybernetics", pp. 1–28 (1980) (in Russian)

371. (with Zh.L. Bliokh, L.V. Domnina, O.Yu. Ivanova, O.Yu. Pletyushkina, T.S. Svitkina, V.V. Smolyaninov, and Yu.M. Vasil'ev) Spreading of fibroblasts in a medium containing cytochalasin *B*: Formation of lamellar cytoplasm as a combination of several functionally different processes. Proc. Natl. Acad. Sci. USA **77** (1980) 5919–5922

**1981**

372. (with E.V. Pomerantsev, V.M. Sablin, M.N.Starkova, V.A. Sulimova, A.L. Syrkin, and V.L. Vakhlyaev) Prognostication of complications and classification of patients with severe myocardial infarcation. Summaries of lectures at the second All-Union Conf. "Theory and practice of automation of electrocardiological and clinical studies", pp. 274–276. Kaunas 1981 (in Russian)
373. (with G.G. Guelstein, I.P. Lukashhevich, M.A. Shifrin, and L.S. Zingerman) Expressibility of electrocardiograph changes in severe disease of the coronary artery in patients with chronic ischemic heart disease. Summaries of lectures at the second All-Union Conf. "Theory and practice of automation of electron-cardiological and clinical studies", pp. 304–307. Kaunas 1981 (in Russian)
374. (with S.M. Khoroshkin, E.V. Pomerantsev, B.I. Rozenfel'd, V.A. Sulimov, A.L. Syrkin, and V.L. Vaklyaev) Choice of information for the classification of patients with myocardial infarcation and choice of medical tactics. Summaries of lectures at the second All-Union Conf. "Theory and practice of automation of electrocardiological and clinical studies", pp. 267–278. Kaunas 1981 (in Russian)
375. (with Yu.M. Vasil'ev) Neoplastic and normal cells in culture. London-Sydney: Cambridge University Press 1981
376. (with S.G. Gindikin, M.L. Izvekova, and M.Yu. Melikova) The immediate prognosis for healing of duodenal ulcers (control of classification). Trans. Second Moscow Med. Inst. Ser. "Surgery" **32** (1981) 73–80 (in Russian)
377. (with Yu.M. Vasil'ev) Interaction of normal and tumorous cells with the medium. Moscow: Nauka 1981 (in Russian)
378. (with V.M. Alekseev, M.A. Alekseevskaya, E.E. Gogin, L.D. Golovnya, M.L. Izvekova, E.S. Klyushin, I.V. Martynov, V.A. Ponomarev, I.V. Sablin, A.L. Syrkin, and R.M. Zaslavska) Multi-purpose chart of a patient with myocardial infarction (for setting up a data bank in a computer). Prepr. Acad. Sci. USSR Sci. Committee on the complex problem "Cybernetics" (1981) (in Russian)
379. (with V.A. Ponomarev) Gabriel's theorem is also true for representations of graphs endowed with relations. Funkts. Anal.

Prilozh. **15** (2) (1981) 71–22 [Funct. Anal. Appl. **15** (1981) 132–133]. Zbl. **479**:18003

380. (with I.Ya. Dorfman) Hamiltonian operators and infinite-dimensional Lie algebras. Funkts. Anal. Prilozh. **15** (3) (1981) 23–40 [Funct. Anal. Appl. **15** (1982) 173–187]. MR **82j**:58045. Zbl. **478**:58013
381. (with Yu.I. Manin) Simmetria, Enciclopedia Einaudi, Vol. 12, pp. 916–943. Einaudi, Torino 1981
382. (with A.D. Bershadskij, Zh.L. Bliokh, L.V. Domnina, O.Yu. Ivanova, V.V. Smolyahinov, T.M. Svitkina, I.S. Tint, and Yu.M. Vasil'ev) Mechanisms of morphological reactions determining the shape and movement of normal and transformed cells in culture. In: Nemyshechnie sistemy, pp. 65–75. Moscow: Nauka 1981 (in Russian)
383. (with O.Yu. Ivanova, S.G., Komm, and Yu.M. Vasil'ev) Stabilization independent of micropipelets of the cell surface of normal and transformed connective tissue cells. Tsitologiya **23** (1981) 62–65 (in Russian)
384. (with M.I. Graev and A.M. Vershik) Representations of the group of functions taking values in a compact Lie group. Compos. Math. **42**, 217–243 (1981). MR **83g**:22002. Zbl. **449**:22019. II.IX.5

**1982**

385. (with M.I. Graev) Integral transformations connected with two remarkable complexes in projective space. Prepr. Inst. Appl. Math. **93** (1982) (in Russian)
386. (with M.I. Graev and R. Rosu) Non-local inversion formulae in a problem of integral geometry connected with $p$-dimensional planes in real projective space. Funkts. Anal. Prilozh. **16** (3) (1982) 49–51 [Funct. Anal. Appl. **16** (1982) 196–198]. Zbl. **511**:53072
387. (with M.N. Starkova and A.L. Syrkin) Classification of patients and prognosis of healing in myocardial infarction. Prepr. Acad. Sci. USSR Sci. Committee on the complex problem "Cybernetics" (1982) (in Russian)
388. (with M.L. Izvekova, M.N. Starkova, and A.L. Syrkin) The methodology of comparing material from two hospitals and the construction of a single guide-line for the prognosis of the effect of a strong focal myocardial infarction. Prepr. Acad. Sci. USSR Sci. Committee on the complex problem "Cybernetics" (1982) (in Russian)
389. (with M.A. Brodskij, M.Ya. Ratner, B.I. Rozenfel'd, V.V. Serov, I.I. Stenina, and V.A. Varshavskij) Determination of a morphological picture of glomerule kidney from clinical-

functional data (by means of a formal scheme modelling the diagnosis of kidney consultants). Prepr. Acad. Sci. USSR Sci. Committee on the complex problem "Cybernetics" (1982) (in Russian)

390. (with S.M. Khorshkin, B.I. Rozenfeld, V.A. Sulimov, A.L. Syrkin, and V.D. Vakhlyaev) Selection of information for the classification of patients with myocardial infarcation and choice of medical tactics. Prepr. Acad. Sci. USSR Sci. Committee on the complex problem "Cybernetics" (1982) (in Russian)

391. (with I.Ya. Dorfman) Hamiltonian operators and the classical Yang-Baxter equation. Funkts. Anal. Prilozh. **16**(4) (1982) 1–9 [Funct. Anal. Appl. **16** (1982) 241–248]. Zbl. **527**:58018 I.III.17

392. (with Yu.I. Arshavskij and G.N. Orlovskij) The cerebellum and control of rhythmical movements. Trends Neurosci. **6** (10) (1983) 417–422 III.IV.11

393. (with Yu.I. Arshavskij, I.N. Beloozerova, G.N. Orlovskij, and Yu.V. Panchin) Neural mechanisms in the generation of nutritional rhythmics in molluscs. Lecture at the First All-Union Biophysical Congress. Moscow 1982 (in Russian)

394. (with Yu.M. Vasil'ev) Possible common mechanism of morphological and growth-related alterations accompanying neoplastic transformation. Proc. Natl. Acad. Sci. USA **79** (1982) 2594–2597 III.IV.15

395. (with M.I. Graev and A.M. Vershik) A commutative model of the basic representation of the group $SL(2, R)X$ connected with a unipotent subgroup. Prepr. Inst. Appl. Math. **169** (1982) (in Russian)

396. (with B.I. Rozenfeld and M.A. Shifrin) Structural organisation of data in problems of medical diagnosis and prognosis. Prepr. Acad. Sci. USSR Sci. Committee on the complex problem "Cybernetics" (1982) (in Russian)

397. (with R.G. Ajrapetyan, M.I. Graev, and G.R. Oganesyan) The Plancherel formula for the integral transformation connected with a complex of lines intersecting an algebraic straight line in $\mathbf{C}^3$ and $\mathbf{CF}^3$. Dokl., Akad. Nauk Arm. SSR **75** (1) (1982) 9–15 (in Russian). Zbl. **504**:43009

398. (with L.V. Domnina, V.I. Gelfand, O.Yu. Ivanova, O.Yu. Pletyushkina, and Yu.M. Vasil'ev) Effects of small doses of cytochalasins on fibroblasts: preferential changes of active edges and focal contacts. Proc. Natl. Acad. Sci. USA **79** (1982) 7754–7757

399. (with R.D. MacPherson) Geometry in Grassmannians and a generalization of the dilogarithm. Adv. Math. **44** (1982) 279–312. Zbl. **504**:57021 III.II.22

**1983**

400. (with I.V. Cherednik) An abstract Hamiltonian formalism for the classical Yang-Baxter bundles. Usp. Mat. Nauk **38** (3) (1983) 3–21 [Russ. Math. Surv. **38** (3) (1983) 1–22]. Zbl. **536**:58006

401. (with R.G. Ajrapetyan, M.I. Graev, and G.R. Oganesyan) Plancherel theorem for the integral transformation connected with a complex of $p$-dimensional planes in $\mathbf{CF}^n$. Dokl. Akad. Nauk SSSR **268**, 265–268 (1983) [Sov. Math., Dokl. **27** (1983) 47–50]. Zbl. **527**:53045

402. (with G.S. Shmelev) Geometric structures of double bundles and their relation to certain problems in integral geometry. Funkts. Anal. Prilozh. **17** (2) (1983) 7–22 [Funct. Anal. Appl. **17** (1983) 84–96]. Zbl. **519**:53058 III.I.12

403. (with M.I. Graev and A.M. Vershik) A commutative model of representation of the group of flows $SL(2,\mathbf{R})^X$ that is connected with a unipotent subgroup. Funkts. Anal. Prilozh. **17** (2) (1983) 70–72 [Funct. Anal. Appl. **17** (1983) 137–139]. Zbl. **536**:22008 II.IX.6

404. (with A.Yu. Lyuiko, M.N. Starkova, and A.L. Syrkin) Retrospective estimate of non-stable cardiac angina in various forms of myocardial infarction. Klin. Med. **3** (1983) 28–31 (in Russian)

405. (with A.A. Grinberg, M.L. Izvekova, and V.P. Lakhtina) Prognosis of recidive haemorrhaging in patients with ulcerous disease of the stomach and duodenal intestine. Vestn. Khir., Grekov **130** (4) (1983) 21–24 (in Russian)

**1984**

406. (with A.V. Zelevinskij) Models of representations of classical groups and their hidden symmetries. Funkts. Anal. Prilozh. **18** (3) (1984) 14–31 [Funct. Anal. Appl. **18** (1984) 183–198]. Zbl. **556**:22003 II.IV.10

407. (Yu.L. Daletskij) Some formal differential structures related to Lie superalgebras. Prepr. Inst. Math. **85** (1984) (in Russian)

408. (with R.G. Ajrapetyan, M.I. Graev, and G.R. Oganesyan) The Plancherel theorem for the integral transformation connected to a pair of Grassmannians. Izv. Akad. Nauk Arm. SSR, Mat. **19** (6) (1984) 467–483 [Sov. J. Contemp. Math. Anal., Arm. Acad. Sci. **18** (4) (1983) 21–32]. MR **86c**:53046. Zbl. **577**:44002

409. (with M.I. Graev and R. Rosu). The problem of integral geometry and intertwining operators for a pair of real Grassmannian manifolds. J. Oper. Theory **12** (2) (1984) 359–383. MR **86c**:22016. Zbl. **551**:53034 III.I.13

410. (with M.A. Brodskij, M.Ya. Ratner, B.I. Rozenfeld, I.I. Stenina, and V.A. Varshavskij) Morphological-clinical variants of chronic glomerulonephritis and their role in evaluation of serenity of disease. Arkh. Patol. **11** (1984) 46–52 (in Russian)

411. Functions of the cerebellum for the control of rhythmic movements. Today's views on the function of the cerebellum, pp. 181–188. Erevan 1984 (in Russian)

412. (with Yu.I. Arshavskij and G.N. Orlovskij) The cerebellum and the control of rhythmic movements. Moscow: Nauka 1984 (in Russian)

413. (with Yu.I. Arshavskij, G.N. Orlovskij, G.A. Pavlova, and L.B. Popova) Origin of signals convate by the ventral spinocerebellar tract and spino-reticulo-cerebellar pathway. Exp. Brain Res. **54** (3) (1984) 426–431

414. (with L.V. Domnina, O.Yu. Ivanova, O.Yu. Pletyushkina, T.M. Svitkina, and Yu.M. Vasil'ev) Formation of processes in the spreading of fibroblasts in a medium with Cytochalasin B in vitro. Ontogenez **15** (3) (1984) 275–282 (in Russian)

415. (with Yu.M. Vasil'ev) Membrane-cytoskeleton interactions during cell spreading on non-cellular surfaces. 16th meeting of the Federation of European Biochemical Societies, Abstracts, p. 34 (1984) (in Russian)

416. (with A.D. Bershadskij, V.I. Gelfand, L.A. Lyass, A.S. Serpinskaya, and Yu.M. Vasil'ev) Multinucleation induced improvements of the spreading of the transformed cells on the substratum. Proc. Natl. Acad. Sci. USA **81** (1984) 3098–3102

**1985**

417. (with M.I. Graev) On some families of irreducible unitary representations of the group $U(\infty)$. Prepr. Inst. Appl. Math. **51** (1985) (in Russian)

418. (with A.V. Zelevinskij) Polyhedra in the scheme space and the canonical basis for irreducible representations of $gl_3$. Funkts. Anal. Prilozh. **19**(2) (1985) 72–75 [Funct. Anal. Appl. **19** (1985) 141–144]. Zbl. **606**:17006

419. (with A.V. Zelevinskij) The canonical basis in irreducible representations of $gl_3$ and applications. In: Proc. III. Int. Semin. on Group-Theoretical Methods in Physics, Yurmala, 1985 (in Russian)

420. (with A.V. Zelevinskij) Multiplicities and good bases for $gl_n$. In: Proc. III. Int. Semin. on Group-Theoretical Methods in Physics, Yurmala, 1985 (in Russian)

421. (with B.I. Rozenfel'd and M.A. Shifrin) Structural organization of data in medical diagnostics and prognosis. In: Problems of medical diagnostics and prognosis from the point of

view of a mathematician, I.M. Gelfand (ed.). Vopr. Kibern., Mosk. 112 (1985) 5–64 (in Russian)

422. (with N.M. Chebotareva, S.G. Gindikin, Sh.A. Guberman, M.L. Izvekova, E.I. Kandel, M.Yu. Melikova, B.I. Rozenfel'd, M.N. Starkova, and A.L. Syrkin) Some problems of classification and prognosis from various area of the medicin. In: Problems of medical diagnostics and prognosis from the point of view of a mathematician, I.M. Gelfand (ed.). Vopr. Kibern., Mosk. **112** (1985) 65–127 (in Russian)

423. (with G.I. Dzuba, Sh.A. Guberman, and L.V. Kuznetsov) Applications of the global approach to the discrimination of objects in the automatized analysis of chest x-rays. In: Problems of medical diagnostic and prognosis from the point of view of mathematician. I.M. Gelfand (ed.). Vopr. Kibern., Mosk. **112** (1985) 148–171 (in Russian)

424. (with B.I. Rozenfeld, and M.A. Shifrin) "Diagnostic games" in medical diagnostics and prognosis. Psikhol. Zh. **5** (1985) (in Russian)

425. (with Yu.I. Arshavskij, G.N. Orlovskij, Yu.V. Panchin, G.A. Pavlova, and L.B. Popova) Regeneration of neurons in pedal ganglia of pteropodial mollusc Clione limacina. Nejrofiziologiya **17** (4) (1985) 449–455 (in Russian)

426. (with A.V. Zelevinskij) Representation models for classical groups and their higher symmetries. In: Elie Cartan et les mathématiques d'aujourdhui. The mathematical heritage of Elie Cartan, Sémin. Lyon 1984, Astérisque, No. Hors Sér., 117–128 (1985). Zbl. **594**:22007

**1986**

427. (with G.N. Orlovskij and M.L. Shik) Locomotion and stratching in tetrapods. In: Neural control of rhythmic movements, J. Wiley, N.Y., 1986

428. General theory of hypergeometric functions. Dokl. Akad. Nauk SSSR **288** (1) (1986) 14–18 [Sov. Math., Dokl. **33** (1986) 573–577] III.V.1

429. (with S.I. Gelfand) Generalized hypergeometric equations. Dokl. Akad. Nauk SSSR **288** (2) (1986) 279–283 [Sov. Math., Dokl. **33** (1986) 643–646] III.V.2

430. (with M.I. Graev) A duality theorem for general hypergeometric functions. Dokl. Akad. Nauk SSSR **289** (1) (1986) 19–23 [Sov. Math., Dokl. **34** (1987) 9–13]. Zbl. **619**:33006 III.V.3

431. (with A.B. Goncharov) On a characterization of Grassmann manifolds. Dokl. Akad. Nauk SSSR **289** (5) (1986) 1047–1052 [Sov. Math., Dokl. **34** (1987) 189–193] III.I. 14

432. (with A.V. Zelevinskij) Algebraic and combinatorial aspects of the general theory of hypergeometric functions. Funkts. Anal. Prilozh. **20** (3) (1986) 17–34 [Funct. Anal. Appl. **20** (1986) 183–197]. Zbl. **619**:33004 III.V.4
433. (with V.A. Vasil'ev and A.V. Zelevinskij) Behaviour of general hypergeometric functions in complex domain. Dokl. Akad. Nauk SSSR **290** (2) (1986) 277–281 [Sov. Math., Dokl. **34** (1987) 268–272]. Zbl. **619**:33005
434. (with A.B. Goncharov) Reconstruction of a function with compact support by its integrals over lines intersecting the given set of points in the space. Dokl. Akad. Nauk SSSR **290** (5) (1986) 1037–1040 [Sov. Math., Dokl. **34** (1987) 373–376]. Zbl. **621**:53052
435. (with V.V. Minakhin and V.N. Shander) Integration on supermanifolds and Radon supertransformations. Funkts. Anal. Prilozh. **20** (4) (1986) 67–69 [Funct. Anal. Appl. **20** (1986) 310–312]
436. (with A.V. Zelevinskij) Canonical basis in irreducible representations of $gl_3$ and its applications. In: Group theoretic methods in physics, vol. 2, pp. 31–45. Moscow: Nauka 1986 (in Russian)
437. (with A.V. Zelevinskij) Multiplicities and regular bases for $gl_n$. In: Group theoretic methods in physics. Moscow: Nauka, vol. 2 (1986) 22–31 (in Russian)
438. (with G.I. Dzuba, Sh.A. Guberman, and L.V. Kuznetsov) The experience in the development of the system for the automatized chest radiograph processing. Proc. 1st All-Union Seminar "Algorithms and Software for the Data Analysis in Medical and Biological Studies", Pushchino (1986) (in Russian)
439. (with B.I. Rozenfeld, A.L. Syrkin, and M.A. Shifrin) Clinical course types of the myocardium infarction and their prognostic value. Kardiologiya **9** (1986) 9–12 (in Russian)
440. (with A.V. Alekseevskij, M.A. Shifrin, and M.A. Rainer) Algorithms for differential diagnosis of purulent meningitis of various etiologies in children of the first year of life. Preprint of the Scientific Council on Cybernetics, Moscow (1986) 49 pp. (in Russian)
441. (with V.B. Dugina and Yu. M. Vasil'ev) Reversible reorganization of cultured cell cytoskeleton induce byphorbol ether. Dokl. Akad. Nauk SSSR **291** (1) (1986) 985–988 (in Russian)
442. (with M.B. Berkenblit and A.G. Fel'dman) A model for the control of polyartrial extremity movements. Biofizika **31** (1) (1986) 128–138 (in Russian)
443. (with M.B. Berkenblit and A.G. Fel'dman) A model for the aiming phase of the wiping reflex. In: Neurobiology of verte-

brate locomotion. Ed. S. Grillner. Macmillan Press (1986) 217–230

444. (with Yu.I. Arshavskij, T.G. Delyagina, G.N. Orlovskij, G.A. Pavlova, Yu.V. Panchin, and L.B. Popova) The influence of the locomotion system of pedal ganglia of the pteropodial mollusc upon isolated neurons. Nejrofiziologiya **18** (6) (1986) 756–763 (in Russian)

**1987**

445. (with M. Goresky, R.D. MacPherson, V.V. Serganova) Combinatorial geometries, convex polyhedra, and Schubert cells. Adv. Math. **63** (3) (1987) 301–316. Zbl. **622**:57014 III.V.5

446. (with V.V. Serganova) On the general definition of a matroid and a greedoid. Dokl. Akad. Nauk SSSR **292** (1) (1987) 15–19 [Sov. Math., Dokl. **35** (1987) 6–10]

447. (with V.V. Serganova) Strata of a maximal torus in a compact homogeneous space. Dokl. Akad. Nauk SSSR **292** (3) (1987) 524–528 [Sov. Math., Dokl. **35** (1987) 63–66] III.V.6

*448. (with V.V. Serganova) Combinatorial geometries and torus strata on homogeneous compact manifolds. Usp. Mat. Nauk **42** (2) (1987) 107–133 III.V.7

*449. (with V.A. Vasil'ev, A.V. Zelevinskij) General hypergeometric functions on complex Grassmannians. Funkts. Anal. Prilozh. **21** (1) (1987) 23–38. Zbl. **614**:33008 III.V. 8

450. (with M.I. Graev) General hypergeometric functions on the Grassmannian $G_{3,6}$. Preprint IFM **123** (1987) 27pp. (in Russian)

451. (with M.I. Graev) Strata in $G_{3,6}$ and corresponding hypergeometric functions. Preprint IFM **127** (1987) 25pp. (in Russian)

*452. (with M.I. Graev and A.V. Zelevinskij) Holonomic systems of equations and series of hypergeometric type. Dokl. Akad. Nauk SSSR **295** (1) (1987) 14–19 III.V.9

*453. (with A.N. Varchenko) On heaviside functions of configuration of hyperplanes. Funkts. Anal. Prilozh. **21** (4) (1987) 1–18 III.V.10

*454. (with T.V. Alekseevskaya and A.V. Zelevinskij) Arrangements of real hyperplanes and the associated partition function. Dokl. Akad. Nauk SSSR **297** (6) (1987) 1289–1293 III.V.11

455. (with B.I. Rozenfeld and M.A. Shifrin) Data structuring in medical problems. Preprint of the Scientific Council on Cybernetics, Moscow (1987) 45 pp. (in Russian)

456. (with V.B. Dugina, T.M. Svitkina, and Yu.M. Vasil'ev) Special type of morphological reorganization induced by phorbol ester: reversible partition of cell into mobile and stable domains. Proc. Natl. Acad. Sci. USA **84** (1987) 4122–4125 III.IV.16

457. (with M.I. Graev) Hypergeometric functions related to the grassmannian $G_{3,6}$. Dokl. Akad. Nauk SSSR **293** (2) (1987) 288–292 [Sov. Math., Dokl. **35** (1987) 298–303]

458. (with B.V. Lidskij and V.A. Ponomarev) Preprojective reduction of the free modular lattice $D^r$. Dokl. Akad. Nauk SSSR **293** (3) (1987) 524–528 [Sov. Math., Dokl. **35** (1987) 334–338]

459. (with Yu.M. Vasil'ev) Membrane and cytoskeleton interrelations for cells binding to non-cellular surfaces. In: Proc. 16th FEBS Conf., Moscow, vol. 1 (1987) 164–166 (in Russian)

460. (with Yu.I. Arshavskij, T.G. Delyagina, E.S. Meiserov, G.N. Orlovskij, G.A. Pavlova, Yu.V. Panchin, and L.B. Popova) Growth of neuritis and development of connections in the neuron culture of pteropodial mollusc. Nejrofiziologiya **19** (1) (1987) 81–86 (in Russian)

461. (with Yu.I. Arshavskij, T.G. Delyagina, E.S. Meiserov, G.N. Orlovskij, G.A. Pavlova, Yu.V. Panchin, and L.B. Popova) Neuronal mechanisms controlling locomotion in the pteropodial mollusc Clione Limacina. Zh. Ob. Biologii **48** (3) (1987) 325–339 (in Russian)

**1988**

462. (with V.S. Retakh and V.V. Serganova) Generalized Airy functions, Schubert cells and Jordan groups. Dokl. Akad. Nauk SSSR **298** (1) (1988) 17–21 III.V.12

# Acknowledgements

We would like to thank the original publishers of I. M. Gelfand's papers for granting permission to reprint them here.

The numbers following each source correspond to the numbering of the article in this volume.

Reprinted from *Compos. Math.* © Martinus Nijhoff: IX.2, IX.5

Reprinted from *C. R. Acad. Sci.* © Acad. of Sc.: V.6, V.7

Reprinted from *Dokl. Akad. Nauk SSSR:* I.2, III.3, III.4

Reprinted from *Funct. Anal. Appl.* © Consultants Bureau: IV.3, IV.7, IV.9, V.2, V.5, V.8, VI.4, VIII.2, VIII.6, IX.3, IX.6

Reprinted from *J. Phys.* ©: II.1

Reprinted from *Lie Groups and Their Representations* © Akademia Kiado: V.1

Reprinted from *Publ. Math.* © IHES: VI.2

Reprinted from *Representation Theory* © Cambridge University Press: Appendix

Reprinted from *Russ. Math. Surv.* © British Library: V.4, VIII.1, VIII.5, IX.1, IX.4

Reprinted from *Selecta Math. Sov.* © Birkhäuser: IV.8

Reprinted from *Sov. Math., Dokl.* © American Mathematical Society: III.4, III.5, IV.1, IV.2, IV.5, IV.6, VI.1, VI.3, VIII.3, VIII.4

Reprinted from *Transl., II. Ser., Am. Math. Soc.* © American Mathematical Society: I.1, I.4, III.1, III.2, III.3, VII.3